AN

ARABIC-ENGLISH LEXICON

AN

ARABIC-ENGLISH LEXICON

BY

EDWARD WILLIAM LANE

IN EIGHT PARTS

PART 2 ج - خ

مَدُّ القَامُوسِ

AN

ARABIC-ENGLISH LEXICON,

DERIVED FROM THE BEST AND THE MOST COPIOUS EASTERN SOURCES;

COMPRISING A VERY LARGE COLLECTION
OF WORDS AND SIGNIFICATIONS OMITTED IN THE ĶÁMOOS,
WITH SUPPLEMENTS TO ITS ABRIDGED AND DEFECTIVE EXPLANATIONS,
AMPLE GRAMMATICAL AND CRITICAL COMMENTS,
AND EXAMPLES IN PROSE AND VERSE:

COMPOSED BY MEANS OF THE MUNIFICENCE OF THE MOST NOBLE
ALGERNON,
DUKE OF NORTHUMBERLAND, K.G.,
ETC. ETC. ETC.,
AND THE BOUNTY OF
THE BRITISH GOVERNMENT:

BY EDWARD WILLIAM LANE,
CORRESPONDENT OF THE INSTITUTE OF FRANCE, ETC.

IN TWO BOOKS:

THE FIRST CONTAINING ALL THE CLASSICAL WORDS AND SIGNIFICATIONS COMMONLY KNOWN
TO THE LEARNED AMONG THE ARABS:
THE SECOND, THOSE THAT ARE OF RARE OCCURRENCE AND NOT COMMONLY KNOWN.

BOOK I.—PART 2.

ج — خ

KALPAZ

ISBN: 978-93-5128-814-5(Set)
ISBN: 978-93-5128-824-4(PB)

First Published in 1865
Indian Reprint in 2017

Published by

Kalpaz Publications
C-30, Satyawati Nagar,
Delhi – 110052
E-mail: kalpaz@hotmail.com
Website: kalpazpublications.com
Ph.: +91-9212142040

Cataloging in Publication Data—DK

Courtesy: D.K. Agencies (P) Ltd. <docinfo@dkagencies.com>

Lane, Edward William, 1801-1876, **author.**

Madd al-qa˜mu˜s = An Arabic-English lexicon : derived from the best and the most copious eastern sources, comprising a very large collection of words and significations omitted in the Kamoos, with supplements to its abridged and defective explanations, ample grammatical and critical comments, and examples in prose and verse : composed by means of the munificence of the most noble algernon, Duke of Northumberland, K. G., etc. etc. etc., and the bounty of the British Government / by Edward William Lane.

volumes cm

English and Arabic.

Parts 6-8 edited with a memoir by Stanley Lane Poole.

"First published in 1863"—Title page verso.

"In two books: the first containing all the words and significations commonly known to the learned among the Arabs : the second, those that are of rare occurrence and not communly known."
ISBN 9789351288145 (set :

1. Arabic language—Dictionaries—English. 2. English language—Dictionaries—Arabic. I. Lane-Poole, Stanley, 1854-1931, editor. II. Title. III. Title: Arabic-English lexicon.

PJ6640.L36 2017 DDC 492.7321 23

ج

The fifth letter of the alphabet: called جِيمْ, which is one of the names of letters of the fem. gender, but which it is allowable to make masc.: it is one of the letters termed مَجْهُورَة [or vocal, i. e. pronounced with the voice, and not with the breath only]: and of the number of the letters termed مَحْقُورَة, and حُرُوفُ القَلْقَلَة, because it cannot be uttered in a case of pause without a strong compression, and a strong sound: and it is also one of those termed شَجْرِيَّة, from الشَّجْر, which is the place of opening of the mouth. (TA.) — It is sometimes substituted for ى, when the latter letter is doubled, (Ḳ,) or is so substituted by some of the Arabs; (AA, Ṣ;) as in فُقَيْمِجّ, for فُقَيْمِىّ; (AA, Ṣ, Ḳ;) and مُرِّجّ, for مُرِّىّ. (AA, Ṣ.) An Arab of the desert recited to Khalaf El-Aḥmar,

• خَالِي عُوَيْفٌ وَأَبُو عَلِجِّ • اَلْمُطْعِمَانِ اللَّحْمَ بِالْعَشِجِّ •

[*My maternal uncle is 'Oweyf, and Aboo-'Alijj, who feed with flesh-meat at nightfall*]; meaning عَلِىّ and عَشِىّ. (Ṣ.) It is also sometimes substituted for a single ى. (Ṣ, Ḳ.) AZ gives the following ex.:

• يَا رَبِّ إِنْ كُنْتَ قَبِلْتَ حَجَّتِجْ •
• فَلَا يَزَالُ شَاحِجٌ يَأْتِيكَ بِجْ •

[*O my Lord, if Thou accept my plea, a brayer* (or *mule*) *shall not cease to bring me to Thee* (i. e. to thy temple)]; (Ṣ;) meaning حَجَّتِى (Ḳ) [and بِى]. أَمْسَجَتْ and أَمْسَجَا are also mentioned as occurring in a verse, for أَمْسَتْ and أَمْسَى [because originally أَمْسَيَتْ and أَمْسَيَ]. (Ṣ.) But all these substitutions are abominable, (Ṣ, Ibn-'Oṣfoor,) and only allowable in cases of poetical necessity. (Ibn-'Oṣfoor.) It is further said that some of the Arabs, among whom were the tribe of Ḳuḍá'ah, changed ى, when occurring immediately after ع, into ج; and said, for رَاعٍ, [originally رَاعِىٌ,] رَاعِجٌ: this is what is termed عَجْعَجَة: Fr attributes the substitution of ج for ى to the tribe of Ṭeiyi, and some of the tribe of Asad. (TA.) — Some of the Arabs also changed it into ى; saying شَيَرَة for شَجَرَة, and جَيْبَات for جَثْجَاث, and يَصّ for جَصّ. (AZ, Ṣ in art. بص.) ═ [As a numeral, ج denotes *Three;* and, as such, is generally written without the dot, but thus ح, or thus ـــ, to distinguish it from ح, which denotes eight.]

Bk. I.

جأ

جُؤْجُؤٌ The *breast* (صَدْر, Zj in his "Khalḳ el-Insán," Ṣ, Ḳ, TA) of a human being, (TA,) and of a bird, and ‡of a ship: (Ṣ, TA:) or the *sternum,* or *breast-bone:* or the *middle of the breast:* or the *part where the heads of the bones of the breast come together;* as in the Nh and M: (TA:) pl. جَآجِئُ. (Ṣ, Ḳ.) An Arab is related to have said, مَا أَطْيَبَ جُوذَابَ الأَرُزِّ بِجَآجِئِ الإِوَزِّ [*How delicious is* جوذاب *of rice* (i. e. *rice prepared with sugar and flesh-meat*) *with the breasts of geese!*]. (TA.) And you say, شَقَّتِ السَّفِينَةُ بِجُؤْجُؤِهَا ‡ [*The ship clave the water with her breast*]. (TA.)

جأب

1. جَأَبَ, aor. ـَ, (Ṣ, Ḳ,) inf. n. جَأْبٌ, (Ṣ,) *He gained, earned,* or *acquired,* (Ṣ, Ḳ,) *wealth,* or *property:* (Ḳ:) but [SM says,] I have not seen that any of the leading lexicologists has mentioned this addition of wealth, or property. (TA. [See, however, جَأْبٌ, below.]) The rájiz (Ru-beh Ibn-El-'Ajjáj, TA) says,

• وَاللّٰهُ رَاعٍ عَمَلِى وَجَأْبِى •

[*And God is mindful of my work and my earning*]. (Ṣ, TA.) ═ Also *He sold* جَأْب, i. e. مَغْرَة [*red ochre*]; (IAṭr, Ḳ;) and so جَبَأَ. (IAṭr, TA.)

جَأْبٌ *Thick, gross, big,* or *bulky:* (Ṣ, Ḳ:) or *strong:* (A:) applied to an ass, (A, Ḳ,) or to a wild ass: (Ṣ, Ḳ:) as also جَابٌ, without ء: (Ṣ:) pl. جُؤُوبٌ. (TA.) Accord. to the Ḳ [and the A], it signifies also Whatever is *rude,* or *coarse; thick, gross, big,* or *bulky:* (كُلُّ جَافٍ غَلِيظٍ:) but in the L, we find كَاهِلٌ جَأْبٌ غَلِيظٌ [meaning that جَأْب applied to the part of the back termed كاهل signifies *thick,* or *big*]: and خَلْقٌ جَأْبٌ as meaning *a thick, gross, big,* or *bulky, make.* (TA.) — [Hence,] الجَأْبُ *The lion.* (A, Ṣgh, Ḳ.) — And جَأْبَةُ المِدْرَى, (Ṣ, A, Ḳ,) or, accord. to AO (Ṣ) and the Mj (TA) and Sh, (TA in art. جوب, q. v.,) without ء, (Ṣ, TA,) A doe-gazelle *having her horn just come forth;* because the horn when it first comes forth is thick, and afterwards becomes slender; (Ṣ, Ḳ;) thus showing her to be young: (Ṣ:) or a doe-gazelle, and a cow, *strong in the horn.* (A.) [See also art. جوب.] — You say also, فُلَانٌ شَخْتُ الآلِ جَأْبُ الصَّبْرِ † *Such a one is slender in body, or person,* [but] *great in patience.* (Ṣ.) ═ The *navel.* (Ḳ.) ═ *Red ochre;* syn. مَغْرَة [read by Golius مَعزة]; (Mj, Ḳ;) with and without ء. (Mj, TA.)

جَأْبَةُ البَطْنِ, (Ibn-Buzurj, Ḳ,) as also جَبْأَةٌ, مَأْنَةُ البَطْنِ, (Ibn-Buzurj, TA,) *i. q.* البَطْنِ مَأْنَةُ, (Ḳ,) i. e. *The part of the belly that is between the navel and the pubes.* (TA.)

جُؤْبَةٌ: see what next follows.

جُؤُوبَةٌ (Ḳ) and † جُؤْبَةٌ (Ḳ accord. to some copies, but not in others nor in the TA) *A grinning, and frowning,* or *contracting, of the face;* or *looking sternly, austerely,* or *morosely.* (Ḳ.)

جَأْآبٌ *A gainer, an earner,* or *an acquirer, of wealth,* or *property.* (TA voce جَوَّابٌ.)

جاثليق

جَاثَلِيقٌ [an arabicized word, from the Greek καθολικὸς, The *catholicos;* i. e.] the *primate of the Christians in the country of El-Islám,* [*residing*] *in the* [*chief*] *city of El-Islám: under him is the* بِطْرِيق [or بَطْرِك, or بَطْرَك, i. e. *patriarch*] *of Antioch: then, under him, is the* مَطْرَان [or *metropolitan*]; *under whom is the* أُسْقُف [or *bishop*], *in every province: then, the* قِسِّيس [or *priest*]: *then, the* شَمَّاس [or *deacon*]: (Ḳ:) accord. to Ṣgh, *a judge,* or *ruler:* in the Tekmileh, *a wise man,* or *sage.* (TA.) ج and ق do not occur in any one word, unless it is arabicized or a word imitative of a sound: (Ṣ and Ḳ* at the beginning of the section in which this word is mentioned:) accord. to El-Jawáleeḳee, they do not occur in any Arabic word unless separated, as in جَلَوْبَق and جَرَنْدَق: accord. to Lth, they occur in many words, most of which are arabicized. (TA ib.)

جار

1. جَأَرَ, aor. ـَ, inf. n. جُؤَارٌ (Ṣ, A, Ḳ) and جَأْرٌ, (Ḳ,) *He,* (a bull, Ṣ and Ḳ, or a calf, A,) and *she,* (a cow, Ḳ,) *lowed.* (Ṣ, A, Ḳ.) جُؤَارٌ is like خُوَارٌ; and is substituted for the latter in a reading of the Ḳur vii. 146 and xx. 90. (Akh, Ṣ.) — Also, (Ṣ, A, Ḳ,) inf. ns. as above, (Ḳ,) *He* (a man praying, A and TA) *raised his voice in prayer,* or *supplication:* (Th, Ḳ:) *he cried out:* (Es-Suddee, TA:) *he cried out, calling for aid,* or *succour; humbled,* or *abased, himself, and raised his voice:* (A:) *he humbled,* or *abased, himself, with earnest supplication;* (Ṣ, Ḳ;) إِلَى اللّٰهِ

to God; (Ṣ;) *and cried out*, or *called, for aid*, or *succour*. (Ḳ.) [Accord. to Ḳatádeh, يَجْأَرُونَ in the Ḳur xxiii. 66 signifies يَجْزَعُونَ, as written in the TA; but this is app. a mistranscription for يَجْزَعُونَ, *They manifest grief and agitation*; &c.] — Also ‡ *It* (a plant) *grew tall*; (A, Ḳ;) like as one says, صَاحَتِ الشَّجَرَةُ. (A.) And جَأَرَتِ الأَرْضُ ‡ *The plants*, or *herbage, of the land grew tall*. (A, Ḳ.)

جَأْرٌ, applied to herbage, (Az, A, Ḳ,) ‡ *Fresh, juicy*, or *sappy*: (Ḳ,* TA:) or *tall, and full-grown*: (Az, TA:) and *abundant*. (A, Ḳ.) — غَيْثٌ جَأْرٌ, (Ḳ,) and ▼ جُؤْرٌ, (Aṣ, Ṣ, A, Ḳ,) and, accord. to Aṣ, ▼ جُؤَارٌ, (TA voce جور,) as also جُورٌ, (Ḳ,) ‡ *A copious rain*; (Aṣ, Ṣ, A, Ḳ;) *that makes the plants*, or *herbage, to grow tall*: (A:) or the second, *a rain that makes a sound*, or *noise*. (TA.)

جُؤْرٌ: } see جَأْرٌ.
جُؤَارٌ: }

هُوَ جَأَّرٌ بِاللَّيْلِ [*He is one who cries out, calling for aid, or succour; who humbles, or abases, himself, and raises his voice; much*, or *often, by night*]. (A, TA.)

جاش

1. جَاشَتْ نَفْسُهُ, aor. ـَ, *His soul rose*, or *heaved, by reason of grief* or *fear*; (Aṣ, Ḳ;) a dial. var. of جَاشَتْ, aor. تَجِيشُ. (TA.) — Also, both of these, *His* (a coward's) *soul purposed flight*: or *was frightened*. (TA in art. جيش.) — جَأَشَ إِلَيْهِ, aor. ـَ, *He came, came forward*, or *advanced, towards him*. (Ḳ.)

جَأْشٌ The *return to its place*, (رُوَاعٌ, Lth, Ṣ, Ḳ,) or the *fright*, (رَوْعٌ, as in a copy of the A, [which latter is agreeable with explanations of the verb given above,]) *of the heart, when in a state of commotion on an occasion of fright*. (Lth, Ṣ, A, Ḳ.) ═ The *soul* (نَفْس) of a man: (IDrd, A, Ḳ:) or his *heart*: or *firmness*, or *constancy, of heart*: or *strength of heart when one hears a thing and knows not what it is*: (TA:) sometimes, [جَاشٌ,] without ء: (Ḳ:) pl. جُؤُوشٌ (Ḳ) and جِئَاشٌ. (TA.) You say, فُلَانٌ رَابِطُ الجَأْشِ *Such a one restrains his soul*, or *himself, from flight, by reason of his courage*; (Ṣ, A;*) *is strong in heart*; as also رَبِيطُ الجَأْشِ: (Ṣ in art. ربط:) or both signify *is courageous*. (Ḳ in art. ربط, q. v.) And وَاهِي الجَأْشِ [*Infirm*, or *weak, in soul*, or *heart*]. (A, TA.) And رَبَطَ لِذٰلِكَ الأَمْرِ جَأْشًا, (A,) or جَاشًا only, [without ء,] (ISk, TA,) see art. ربط. And رَبَطَ جَأْشُهُ *His heart became strong*. (Ḳ in art. ربط, q. v.) — See also جُؤْشُوشٌ.

جُؤْشٌ: see what follows.

جُؤْشُوشٌ The *breast*, or *chest*; (Ṣ, A, Ḳ;) as also ▼ جَأْشٌ and ▼ جُؤْشٌ: (A:) or *its* حَيْزُوم, q. v. (Ibn-'Abbád, Ḳ.) — The *forepart* (صَدْر) of the night; accord. to which explanation it is tropical: or *what is between the beginning and the third* thereof: or *a while* thereof: (TA:) or *a portion* thereof; (Lḥ, Ḳ;) and of people. (Ḳ.) ═ Also A *thick*, or *gross*, or *coarse*, man. (Ibn-'Abbád, Ḳ.)

جأل

ضَبُعٌ جَيْأَلُ *The* ضبع [or *female hyena*]; (Ṣ, Ḳ;) a name thereof, of the measure فَيْعَلُ, determinate without ال, (Ṣ,) imperfectly decl.; (Ḳ;) as also ▼ جَيْأَلَةُ, (Ṣ, Ḳ,) accord. to Ks; (Ṣ;) and جَيَلُ, without ء, (Ṣ, Ḳ,) the ى not being changed into ا as in نَابٌ and the like because the ء, though literally suppressed, is considered as though meant to be retained, and because the ى is considered as though meant to be quiescent; (Aboo-'Alee the grammarian, Ṣ, TA;) and الجَيْأَلُ, (Ḳ,) like the first, but with ال. (TA.) — Also, الجَيْأَلُ, accord. to Ibn-Es-Seed, *The wolf*: but MF deems this strange. (TA.)

جَيْأَلَةُ: see above.

جأم

جَأْمٌ: see art. جوم.

جاموس

جَامُوسٌ: see art. جمس.

جأن

جُؤْنَةٌ *A receptacle of the kind termed* سَفَط, *covered with skin, for the perfumes of the seller of perfumes*; as also جُونَةٌ: originally with ء: pl. like صُرَدٌ [i. e. جُوَنٌ: thus in the TA, without ء]. (Ḳ.) See also art. جون.

جاه

جَاهٌ: see art. جوه.

جاورس

جَاوَرْسٌ: see art. جرس.

Quasi جاى

3. جَاآنِي: see 3 in art. جيأ.

جب

1. جَبَّهُ, aor. ـُ, (Mṣb, TA,) inf. n. جَبٌّ (Ṣ, A, Mgh, Mṣb, Ḳ) and جِبَابٌ, (A, Ḳ, MF,) *He cut it*; or *cut it off*; (Ṣ, A, Mgh, Mṣb, Ḳ;) as also ▼ اِجْتَبَّهُ. (Ḳ,* TA.) — جَبَّ خُصَاهُ, inf. n. جَبٌّ and جِبَابٌ, *He cut off entirely*, or *extirpated, his testicles*; (TA;) [as also ▼ اِجْتَبَّهَا; for] جَبٌّ (A, Ḳ) and جِبَابٌ and اِجْتِبَابٌ (TA) signify the *cutting off entirely*, or *extirpating*, (A, Ḳ, TA,) *of the testicle*, (Ḳ, TA,) or *of the genitals*: (A:) [or] جِبَابٌ signifies [or signifies also, as inf. n. of جُبَّ,] the *having the testicles*, (Ṣ, TA,) or *genitals*, (Mṣb,) *entirely cut off*. (Ṣ,* Mṣb, TA.) You say also, جَبَبْتُهُ, meaning *I cut off entirely*, or *extirpated, his genitals*; (Mṣb;) [or *his testicles*; or *his penis*; as is implied in the TA:] and جُبَّ, inf. n. جَبٌّ, (Mgh, TA,) [or جِبَابٌ,] *he had his penis and his testicles* [or *either of these*] *cut off entirely*, or *extirpated*. (Mgh, TA.*) — جَبَّ السَّنَامَ, aor. ـُ, inf. n. جَبٌّ; and ▼ اِجْتَبَّهُ; *He cut off the hump* of the camel: accord. to Lth, جَبٌّ signifies the *cutting off entirely*, or *extirpating, of the hump*. (TA.) ═ جَبَّ النَّخْلَ, (Aṣ, Ṣ, Mṣb, TA,) [aor. ـُ,] inf. n. جَبٌّ, (A, Ḳ,) or جِبَابٌ, (Ṣ, TA,) or جَبَابٌ, (A,) or both the second and last, (Mṣb, [the first is disallowed by MF,]) *He fecundated the palm-trees* [*with the pollen of the male tree*]. (Aṣ, Ṣ, A, Mṣb, TA.) You say, جَآءَ زَمَنُ الجِبَابِ, (Ṣ,) or الجَبَابِ, with fet-ḥ, (A,) or both, (Mṣb,) [*The time of the fecundating of the palm-trees came*]. ═ جَبَّ القَوْمَ, (Ṣ,) aor. ـُ, (TA,) inf. n. جَبٌّ, (Ḳ,) *He surpassed*, or *overcame, the people*, or *company of men*; (Ṣ, Ḳ,* TA;) accord. to some, in grounds of pretension to respect or honour, or in beauty, and in any or every manner. (TA.) And جَبَّتِ النِّسَآءَ *She surpassed the* [other] *women in her beauty*. (TA.) The saying

* جَبَّتْ نِسَآءَ العَالَمِينَ بِالسَّبَبْ *

[*She overcame the women of the whole world by means of the string*] relates to a woman who measured round her hinder parts with a string, and then threw it to the women of the tribe, that they might do with it the like; but they found it to be much exceeding their measures. (TA.) See 3, in three places. ═ See also 2.

2. تَجْبِيبٌ The *reaching of the* [*whiteness termed*] تَحْجِيل, in a horse, *to the knee and the hock*: (Ṣ:) or the *rising of the whiteness to* [*the extent of*] *what is termed* الجُبَبُ. (Ḳ.) You say of a horse, فِيهِ تَجْبِيبٌ [*In him is a rising of the* تحجيل *to the knee and the hock*]: and in this case, the horse is said to be مُجَبَّبٌ: and the subst. is ▼ جُبَبٌ [meaning *a whiteness of the legs rising to the knee and the hock*]. (Ṣ.) [See مُجَبَّبٌ.] ═ The act of *shrinking* [from a thing]; or the *being averse* [from it]; or the act of *withdrawing*; (Ṣ, Ḳ, TA;) *outwardly* or *inwardly*. (TA.) You say of a man, جَبَّبَ فَذَهَبَ [*He shrank*, or *was averse*, or *withdrew, and went away*]. (Ṣ.) And جَبَّبَ النَّاسُ عَنْ طَاعَةِ اللّٰهِ *The people forsook*, or *relinquished, the obeying of God*. (TA from a trad.) — The act of *fleeing*. (Ḳ.) You say of a man, جَبَّبَ *He fled*. (TA.) El-Ḥoṭei-ah says,

* وَنَحْنُ إِذَا جَبَّبْتُمْ عَنْ نِسَائِكُمْ *
* كَمَا جَبَّبَتْ مِنْ عِنْدِ أَوْلَادِهَا الحُمُرْ *

[*And we, when ye flee from your women, like as the wild asses have fled from the presence of their young ones*]. (TA.) And ▼ جَبَّ, said of a man, [if not a mistranscription for جَبَّبَ,] signifies *He went quickly, fleeing from a thing*. (TA.) ═ The act of *satisfying with water* (Ḳ, TA) the earth, (الجَبُوب, TA,) or cattle. (Ḳ, TA.)

3. جَابَّ The act of *vying*, or *contending for superiority, in goodliness*, or *beauty, &c.*, (Ḳ,) as, for instance, in grounds of pretension to respect or honour, and in lineage: (TA:) and مُجَابَّة the *vying*, or *contending for superiority, in goodliness*, or *beauty*, (Ḳ,) *&c.*, (TA,) and *in food:* (Ḳ:) but [SM says,] I know not whence this addition, respecting food, is derived. (TA. [See, however, what follows, from the A.]) You say, جَابَّنِى فَجَبَبْتُهُ ♦ *He vied with me*, or *contended with me for superiority, and I overcame him.* (TA.) And جَابَّتِ الْمَرْأَةُ صَاحِبَتَهَا فَجَبَّتْهَا ♦ حُسْنًا *The woman vied*, or *contended for superiority, with her fellow, and surpassed her in beauty.* (TA.) And جَابَّهُ فِى القِرَى فَجَبَّهُ ♦ *He vied with him*, or *contended with him for superiority, in the entertainment of guests, and he overcame him* therein. (A.)

4. اجبّ *It* (camels' milk) *had*, or *produced, what is termed* جُبَاب [q. v.]. (Ḳ.)

5. تجبّب *He clad himself with a* جُبَّة [q. v.]. (MA.) [And so, app., ♦ اجتبّ, explained by Golius, on the authority of Ibn-Maạroof, as signifying *He put on a vest*, or *tunic.*]

8: see 1, in three places: — and see also 5.

R. Q. 1. جَبْجَبَ *He dealt*, or *trafficked, in* جَبَاجِب [pl. of جُبْجُبَة, q. v.]. (TA.)

R. Q. 2. تَجَبْجَبَ *i. q.* اِتَّشَقَ; (Ṣ, TA;) i. e. *He prepared what is called* جُبْجُبَة: (TA:) or *he put what is called* خَلْع *into a* جُبْجُبَة [q. v.]. (AZ, TA.)

جُبٌّ *A well:* (A, Ḳ:) or *a well not cased with stone or the like:* (Ṣ, A, Mṣb, Ḳ:) or *a well containing much water:* or *a deep well:* (A, Ḳ:) or *of some other description:* (A:) or *a well in a good situation with respect to pasture:* or *one that people have found; not one that they have dug:* (Ḳ:) or *a well that is not deep:* (Lth, TA:) or *a well that is wide*, or *ample:* (El-Kilábeeyeh, TA:) or *a well that is cut through rock*, or *smooth rock*, or *stones*, or *smooth stones*, or *hard and smooth and large stones:* (Aboo-Ḥabeeb, TA:) of the masc. gender; (Mṣb, TA;) [not fem. like بِئْر;] or masc. and fem.: (Fr, Mṣb:) pl. [of pauc.] أَجْبَابٌ (Mṣb, Ḳ) and [of mult.] جِبَابٌ and جِبَبَةٌ. (Ṣ, Mṣb, Ḳ.) — *A well that is dug wherein a grape-vine is planted; like as one is dug for the shoot of a palm-tree:* pl. جِبَابٌ. (ISh, TA.) — The *inside of a well, from its bottom to its top, whether cased with stone or the like or not.* (Sh, TA.) — The جُرْن of a well [app. meaning *A hollowed stone*, or *stone basin, for water, placed at the mouth* of a well: or, perhaps, *a hollowed stone placed over the mouth;* for many a well has such a stone, forming a kind of parapet]. (Zeyd Ibn-Kuthweh, TA.) — [*A kind of leathern bag;*] a مَزَادَة *of which one part is sewed to another*, (Ḳ, TA,) *wherein they used to prepare the beverage termed* نَبِيذ, *until, by use, it acquired strength for that purpose;* mentioned in a trad., forbidding the use of it; and also called ♦ مَجْبُوبَة. (TA.) — The *spathe*, or *envelope*, of the spadix, or flowers, of the palm-tree; also called جُفّ: the former word was unknown to A'Obeyd: both occur, accord. to different readings, in a trad., where it is said that a charm contrived to bewitch Mohammad was put into the جُبّ, or جُفّ, of a طَلْعَة: accord. to Sh, (TA,) it means the *inside* of a طَلْعَة [which latter here app. signifies, as it does in some other instances, the spathe, not the spadix, of a palm-tree]; (Ḳ, TA;) in like manner as the inside of a well, from its bottom to its top, is called جُبّ: the pl. is جِبَابٌ. (TA.) Hence the well-known prov., جِبَابٌ فَلَا تَعَنَّ أَبْرًا [They are merely *envelopes of the flowers of palm-trees; therefore weary not thyself to effect fecundation*]; applied to a man in whom is little or no good; meaning he is like the spathes of the palm-tree in which are no flowers; therefore weary not thyself by attempting to make him good; لَا تَعَنَّ being for لَا تَتَعَنَّ. (MF.)

جُبَّةٌ *A well-known garment* [or *coat*], (Mṣb, Ḳ, TA,) *of the kind of those called* مُقَطَّعَات: (TA:) accord. to 'Iyáḍ, *a garment cut out and sewed:* accord. to Ibn-Ḥajar and others, *a double garment quilted with cotton;* or, sometimes, if *of wool, a single garment, not quilted with anything:* (MF:) [most probably not so much resembling the modern garment more generally known by the same name (for a description and representation of which see my "Modern Egyptians," ch. i.,) as a kind of جُبَّة still worn in Northern Africa, described in this Lexicon voce مِدْرَعَةٌ: accord. to Golius, "*tunica ex panno gossipino*, cui pallium seu toga imponitur, *cum subductitio panno et intercedente gossipio punctim consuta:* Italis consona voce *giuppa:* si ita cum gossipio consuta non sit, دُرَّاعَةٌ tunica illa gossipina dicitur:"] pl. جُبَبٌ (Mṣb, Ḳ) and جِبَابٌ. (Ṣ, Ḳ.) — *I. q.* دِرْعٌ [*A coat of mail;* or *any coat of defence*]: (Ḳ:) pl. جُبَبٌ. (TA.) Er-Rá'ee says,

لَنَا جُبَبٌ وَأَرْمَاحٌ طِوَالٌ
بِهِنَّ نُمَارِسُ الحَرْبَ الشَّطُونَا

[*We have coats of mail*, or *of defence, and long spears: with them we ply distant war*]. (TA.) — The *part* of a spear-head *into which the shaft enters:* (Ṣ, Ḳ:) and the ثَعْلَب is the part of the spear-shaft that enters into the head. (TA.) — [In the TA, جُبَّةُ الرُّمْحِ is also explained as meaning ما دخل من السنان فيه *The part of the spear-head that enters into the shaft:* but it seems that من has been inserted here by a mistake of the copyist; and that the true meaning intended is *the part of the spear-shaft into which the head enters;* though in general the shaft enters into the head.] — The *part in which is the* مُشَاشَة [q. v.] of a horn. (Zeyd Ibn-Kuthweh, TA.) — The حِجَاج [or *bone that surrounds the cavity* (see art. حج)] *of the eye.* (Ḳ.) — The *contents* (حَشْو) *of the solid hoof:* or the *horny box* (قَرْن) *of the solid hoof:* or the *joint between the* ساق [which seems to mean here, as it does in many other instances, the *hind shank*,] *and the thigh:* (Ḳ:) or the *shank-joint* of a horse or the like (مَوْصِلُ الوَظِيفِ) [commonly applied, as in the Ṣ and Ḳ voce رُسْغ, to the *upper extremity of the pastern*, i. e. the *fetlock-joint*, which seems to be the meaning intended in this instance,]) *in the* ذِرَاع [which here app. means the *fore leg*, not the arm]: or, accord. to Aṣ, the *part where the* وظيف [or *shank*] *is set into the hoof:* (Ṣ:) or the *part of the* رسغ [or *pastern*], *of a horse, where the* وظيف [or *shank*] *joins upon the* حَوْشَب [which seems here to mean the *upper pastern-bone*]: or, as AO says, the *part where a horse's* وظيف *joins to the upper part of the* حوشب: or, as he says in another place, the *place where each tibia and hind shank, of a horse, meet;* [the *hock-joint;*] expl. by ملتقى ساقيه ووظيفى رجليه: and the *place of junction of any two bones, except in the back-bone.* (TA.) — Accord. to Lth, *Whiteness of the* بطانية [a word which I have not found anywhere but in this instance] *of a horse or similar beast, extending to the hairs that surround the hoof.* (TA.)

جَبٌّ *A cutting off of the hump of a camel:* (Ḳ:) or *a cutting in the hump of a camel:* (TA:) [or the *state of having the hump cut off;* as seems to be indicated in the Ṣ:] or *an erosion of the hump of a camel, by the saddle, so that it does not grow large.* (Ḳ, TA.) — See also 2.

جُبَابٌ *Butter*, or *what is produced by churning, of camels' milk;* like as زُبْد is what is produced by churning of cows' or sheep's or goats' milk: (Mṣb in art. زبد:) *what rises upon the surface*, (T, Ṣ,) or *what has collected together* [or *coagulated*], (Ḳ,) *of the milk of camels, resembling* زُبْد, (T, Ṣ, Ḳ,) which camels' milk has not: (Ṣ, Ḳ:) when a camel shakes about a skin of camels' milk, suspended to him, what is termed جباب collects at the mouth of the skin. (T.)

جَبُوبٌ The *earth*, (Lḥ, Ḳ,) *in general;* (Lḥ;) sometimes written جَبُوبُ, as a proper name, without the article, and imperfectly decl., like شَعُوبُ: (TA:) so called because it is cut, i. e. dug; or because it cuts, i. e. dissunders, the bodies of those buried in it: (Sub, TA:) and hence ♦ جَبَّانٌ and ♦ جَبَّانَةٌ, signifying *a burial-ground;* from الجَبّ and الجَبُوب; accord. to Kh; but others derive these two words from جبن: (TA:) or *rugged land:* (Aṣ, Ṣ, Ḳ:) or *hard* or *rugged land, composed of rock, not of soil:* (IAạr, TA:) or *earth*, or *dust:* (Lḥ, Ḳ:) or the *surface of the earth;* (ISh, Ṣ, Ḳ;) *whether plain or rugged or mountainous:* (ISh:) a word without a pl.: (Ṣ:) also *coarse*, or *big, lumps* or *clods of clay* or *mud;* or *of dry*, or *tough*, or *cohesive, clay* or *mud; plucked from the surface of the ground:* (TA:) or *crumbled clods of clay* or *mud;* or *of dry*, or *tough*, or *cohesive, clay* or *mud:* (IAạr, TA:) and with ة, *a lump*, or *clod, of clay* or *mud;* or *of dry*, or *tough*, or *cohesive, clay* or *mud.* (Ḳ.)

جُبَابِىّ: see what next follows.

جُبِّىٌّ [app. a contraction of جُبَبِىّ], or ♦ جُبَابِىّ, *A seller of* جِبَاب [pl. of جُبَّة, q. v.]. (Ḳ.)

جَبَّانٌ : } see جَبُوبٌ ; and see art. جبن.
جَبَّانَةٌ : }

جُبْجُبَةٌ : see what next follows, in two places.

جُبْجُبَةٌ, (Ṣ,) or ♦ جَبْجَبَةٌ, (A,) or both, (Ḳ,) and جَبَاجِبُ [which is the pl.], (L, TA,) *The stomach of a ruminant animal*, (Ṣ, A, Ḳ, TA) *in which* خَلْع [q. v.] *is put*, (Ṣ, TA,) i. e., (TA,) *in which is put flesh-meat cut in pieces*; (Ḳ, TA;) or *in which is put flesh-meat to be used as provision in travelling*; (TA;) or *in which melted grease* (Ṣ, Ḳ) *is collected* (Ṣ) or *put*: (Ḳ:) or the *skin of the side of a camel, cut out in a round form, in which is prepared flesh-meat*, (Ḳ, TA,) *such as is called* وَشِيقَةٌ, (TA,) *which is flesh-meat that is boiled once, and then cut into strips, and dried, or salted and sun-dried*; the most lasting of all provision [of the kind]: (Ṣ, TA:) or the first and second both signify *tripe*; in Persian, إِشْكَنْبه or شكنبه. (MA.) A coward is likened to a جبجبة in which خلع is put; because of his turgidness and his little profitableness. (TA.) — Also, the first, *A vessel, or receptacle, made of skin, in which water is given to camels, and in which one macerates* هَبِيد [i. e. *colocynths*, or *the pulp thereof*, or *the seeds thereof*]. (TA.) — And *A basket*, (Ṣ, Ḳ, TA,) *of small size*, (TA,) *made of skins*, (Ṣ, Ḳ, TA,) *in which dust*, or *earth, is removed*: (Ṣ, TA:) or, accord. to Ḳt, it is [♦ جَبْجَبَةٌ,] with fet-ḥ: (TA:) pl. جَبَاجِبُ. (Ṣ.) — And *A drum*: pl. جَبَاجِبُ [which is explained in the Ḳ as meaning "a drum" instead of "drums"]: as in the saying, ضُرِبَتْ عَلَى بَابِهِ الجَبَاجِبُ [*The drums were beaten at his door*]. (A.)

جُبْجُبِىٌّ *A tripe-seller*. (Golius from Meyd. [See جُبْجُبَةٌ.])

جُبْجُبِيَّةٌ *Food made with tripe*; in Persian, شِكَنْبَه وَا; (Golius from Meyd;) in Turkish, شْكَنْبَه شُورْبَاسِى. (MA.)

أَجَبُّ A camel *having his hump cut off*: (Ṣ, Ḳ:) or *having his hump eroded by the saddle, so that it does not grow large*: (Ḳ:) or *having no hump*: (A, TA:) fem. جَبَّاءُ. (A, Ḳ.) — And [hence,] the fem., ‡ A woman *not having* [*prominent*] *buttocks*: (Ḳ:) or *i. q.* رَسْحَاءُ [i. e. *having small buttocks sticking together*; or *having little flesh in her posteriors and thighs*]: (ISh, TA:) or *whose bosom and breasts have not become large*: (Ḳ:) or *whose breast has not become large*: (Sh, TA:) or *small in the breast*; from the same epithet applied to a she-camel; (A;) for a woman having small breasts is like the camel that has no hump: (TA:) or *having no thighs*; (Ḳ;) i. e. *having lean thighs; as though having no thighs*. (TA.) Also, the masc., †A pubes *having little flesh*. (TA.) — [Hence, also,] الأَجَبُّ *i. q.* الفَرْجُ [as meaning *The pudendum muliebre*]; (Ḳ;) from the same word as applied to a camel [having no hump]. (TA.)

مَجَبَّةٌ The *middle*, or *main part*, (جَادَّةٌ,) *of a road*. (Ṣ.)

مُجَبَّبٌ A horse *in which the* [*whiteness termed*] تَحْجِيل *reaches to the knee and the hock*; (Ṣ;) [i. e.] *in which the whiteness* [*of the lower part of the leg*] *reaches to the knee and the hock* or *the knees and the hocks*: (TA:) or *in which the* تحجيل *reaches to his knees*: (Lth, TA:) or *in which the whiteness rises to* [*the extent of*] *what is termed* الجُبَبُ; (Ḳ, TA;) or *more than this*, [perhaps a mistake of a copyist for *less than this*,] *so as not to reach to the knees*: or *in which the whiteness reaches to the hairs that surround his hoof*. (TA.) — بِئْرٌ مُجَبَّبَةُ الجَوْفِ *A well having in the middle a part wider than the rest, hollowed out like a cupola*. (Fr, TA.)

مَجْبُوبٌ *Having the genitals*, (Mṣb,) or *the testicles* (Ṣ,* Mgh, TA) *and the penis*, (Mgh,) *cut off entirely*, or *extirpated*: (Ṣ,* Mgh, Mṣb, TA:) or *having the penis cut off*. (TA.)

مَجْبُوبَةٌ : see جُبٌّ.

جبأ

1. جَبَأَ and جَبُوءَ, aor. ـَ, *He restrained*, or *withheld, himself*; *refrained, forbore*, or *abstained*; or *turned back*, or *reverted*. (Ḳ, TA.) You say, جَبَأَ عَنْهُ, and جَبُوءَ, meaning *He restrained*, or *withheld, himself*, &c., *from him*, or *it*; and *regarded him*, or *it, with reverence, veneration, dread, awe*, or *fear*: (TA:) [or,] accord. to AZ, جَبَأْتُ عَنِ الرَّجُلِ, inf. n. جَبْءٌ and جُبُوءٌ, [to which Golius adds جُبُوءٌ and جَبَأَةٌ, but, I suspect, from incorrect MSS.,] means *I drew*, or *held*, or *hung, back from the man*; or *remained behind him*; or *shrank from him*; or *shrank from him and hid myself*: and he cites (from Nuṣeyb Ibn-Miḥjen, TA)

* فَهَلْ أَنَ إِلَّا مِثْلُ سَيِّقَةِ العِدَى
* إِنِ ٱسْتَقْدَمَتْ نَحْرٌ وَإِنْ جَبَأَتْ عَقْرُ

[*And am I otherwise than like the beasts driven away by the enemy? If they go before, slaughter* befalls them; *and if they remain behind, hocking*]. (Ṣ, TA.) You say also, مَا جَبَأَ عَنْ شَتْمِى *He did not draw back from reviling me; did not desist*, or *abstain, therefrom*. (TA.) — *It* (a sword) *recoiled*, or *reverted, without penetrating*, or *without effect*: (Ḳ:) or so the former verb [only]. (TA.) — *It* (the sight, or the eye,) *recoiled*, or *reverted*: (Ḳ:) or so the former verb [only]; *and disliked*, or *disapproved*, or *hated, the thing* [*that was before it*]. (TA.) You say, جَبَأَتْ عَيْنِى عَنِ الشَّىْءِ *My eye recoiled*, or *reverted, from the thing*. (Ṣ.) And of a woman of displeasing aspect you say, إِنَّ العَيْنَ لَتَجْبَأُ عَنْهَا [*Verily the eye recoils from her with dislike*]. (Aṣ, TA.) — *He disliked, disapproved*, or *hated*: (Ḳ:) or so the former verb [only]. (TA.) You say, جَبَأَ الشَّىْءَ *He disliked*, &c., *the thing*. (TA.) — *He inclined* his neck: (Ḳ:) or so the former verb [only]. (TA.) — *He hid himself*; (Ḳ, TA;) [app. *from fear*;] as, for instance, a ضَبّ [q. v.] in its hole. (TA.) — *He*, or *it, came*, or *went, forth*, or *out*: (Ḳ:) [or so the former verb only.] You say of a serpent, جَبَأَ عَلَيْهِ *It came forth upon him from its hole* (Ṣ, TA) *so as to frighten him*; and in like manner one says of a hyena, and a ضَبّ, and a jerboa. (TA.) And جَبَأَ عَلَى القَوْمِ *He came forth unexpectedly upon the people*, or *company of men*. (TA.) And جَبَأَ الجَرَادُ *The locusts invaded*, or *came suddenly upon, the country*. (TA.)

4. أَجْبَأَتْ said of a land, (Ṣ,) or اجبأ said of a place, (Ḳ,) *It abounded with* [*the kind of truffles called*] كَمْأَة, (Ṣ,) or كَمْء, (so in some copies of the Ḳ,) or [rather] جِبَأَة [a pl. or quasi-pl. n. of جَبْء]. (So in other copies of the Ḳ.) = اجبأ *He hid* a thing. (Ḳ.) And hence, *He hid* his camels *from the collector of the poor-rate*. (IAạr, TA.) — *He sold* seed-produce *before it showed itself to be in a good state*, (Ṣ, Ḳ, TA,) or *before it came to maturity*. (TA.) Hence, in a trad., مَنْ أَجْبَى فَقَدْ أَرْبَى [*He who sells* seed-produce *before it shows itself to be in a good state*, or *before it has come to maturity, practices* the like of *usury*]: (Ṣ, TA:) originally with ء, (Ṣ,) which is suppressed for the purpose of assimilation [to اربى]. (TA. [See 4 in art. جبو and جبى.]) = اجبأ عَلَى القَوْمِ *He overlooked the people*, or *company of men*; or *commanded*, or *had, a view of them*; or *came in sight of them*; syn. أَشْرَفَ. (Ḳ.)

جَبْءٌ sing. of جِبَأَةٌ, like as فَقْعٌ is of فِقَعَةٌ, and فَرْدٌ of فِرَدَةٌ: (Ṣ:) or *i. q.* كَمْأَةٌ: (Ḳ:) or n. un. of ♦ جَبْأَةٌ, which is a coll. gen. n., like كَمْأَةٌ: (MF and TA, voce قَعْبٌ:) [J says,] جِبَأَةٌ signifies *Red* كَمْأَة [or *truffles*]: or, accord. to El-Aḥmar, *those* [*truffles*] *that incline to redness*; كَمْأَةٌ signifying those that incline to dust-colour and blackness; and فِقَعَةٌ, the white; and بَنَاتُ أَوْبَرَ, the small: (Ṣ:) accord. to AḤn, ♦ جَبْأَةٌ signifies *a white thing resembling a* كَمْء, *of which no use is made*: but accord. to IAạr, the *black* كَمْأَة; which, he says, are the *best of* كمأة: (TA:) the pl. of جَبْءٌ is أَجْبُؤٌ, (Ṣ, Ḳ,) a pl. of pauc., (Ṣ,) and جِبَأَةٌ, [as mentioned above,] or, accord. to Sb, this is a quasi-pl. n., (TA,) and ♦ جَبَأٌ, (Ḳ,) or this also is a quasi-pl. n. (TA.) — *I. q.* أَكَمَةٌ [q. v., i. e. *A hill*, or *mound*, &c.]: pls. as above. (Ḳ.) — *A hollow*, or *cavity*, (T, Ḳ,) *in a mountain*, (TA,) *in which the water* (T, Ḳ) *of the rain* (TA) *stagnates*, (T,) or *collects*: (Ḳ:) pl. as above. (Ḳ.)

جَبَأٌ : see the next preceding paragraph.

جَبْأَةٌ : see جَبْءٌ, in two places. = Also *A shoemaker's board*, (Ṣ, Ḳ,) *on which he cuts his leather*; also called قُرْزُومٌ. (Ṣ.) = And The *place where the false ribs of the camel end, and thence as far as the navel and udder*. (Ḳ.) — And The *part* of the belly *called the* مَأْنَة thereof; as also جَبَابَةٌ; (Ibn-Buzurj, TA;) i. e. the *part between the navel and the pubes*. (TA in art. جأب.)

جُبَّأٌ (Ṣ, Ḳ) and ♦ جُبَّاءٌ (Sb, Ḳ) *Fearful*, or *cowardly*: (Ṣ, Ḳ:) fem. with ة: and therefore the pl. is formed by the addition of و and ن

(Sb, TA.) Mafroog Ibn-'Amr Esh-Sheybánee says,

* فَمَا أَنَا مِنْ رَيْبِ المَنُونِ بِجُبَّأٍ *
* وَلَا أَنَا مِنْ سَيْبِ الإِلٰهِ بِآيِسِ *

[*But I am not fearful of the vicissitudes of fortune, nor despairing of the favour of God*]. (Ṣ, TA.)

جُبَّأٌ : see what next precedes.

جَابِئٌ The *locust*, or *locusts*: (Ṣ, Ḳ:) so called because of the coming forth thereof [suddenly or unexpectedly: see 1, last two sentences]: (Ṣ, TA:) as also جَابٍ [q. v.]. (TA.)

أَرْضٌ مَجْبَأَةٌ *A land abounding with* [*the truffles called*] جَبْأَة. (Ṣ.)

جبت

الجِبْتُ, not a pure Arabic word, because it comprises the letters ج and ت without any of the letters of the kind called ذَوْلَقِى [which are ر and ل and ن]; (Ṣ;) *The idol*: (Ṣ, Ḳ:) or *idols*: (Ksh in iv. 54:) or the name of *a certain idol*, (Bḍ and Jel on that verse,) *belonging to Ḳureysh*; as also الطَّاغُوتُ: (Jel:) and *that which is worshipped instead*, or *to the exclusion, of God; whatever it be*: (Ksh, Bḍ, Ḳ:) said to be originally الجِبْسُ, i. e., (Bḍ,) *he*, or *that, wherein is no good*: (Bḍ, Ḳ:) and *the diviner*: (Ṣ, Ḳ:) and *the enchanter*: (Ṣ, Ḳ, Kull:) and *the like thereof*: (Ṣ:) or *the Devil; Satan*: (Kull:) and *enchantment*. (Esh-Shaabee, Ḳ.) Accord. to Esh-Shaabee, يُؤْمِنُونَ بِالجِبْتِ وَالطَّاغُوتِ, in the Ḳur [iv. 54], means *They believe in enchantment and the Devil*: or, accord. to I'Ab, by الجبت is meant Ḥoyei Ibn-Akhtab; and by الطاغوت, Kaab Ibn-El-Ashraf: (TA:) or the words relate to these two men, Jews, who, in order to induce Ḳureysh to join with them in a league against Mohammad, prostrated themselves to the gods of Ḳureysh: (Ksh, Bḍ:) or to certain Jews, who said that the worship of idols is more pleasing to God than that to which Mohammad invited. (Bḍ.) It is said in a trad. that what are termed الطِّيَرَةُ and العِيَافَةُ and الطَّرْقُ are مِنَ الجِبْتِ [app. meaning *of things wherein is no good*: or *kinds of divination*: or *from the Devil*]. (Ṣ.)

جبذ

1. جَبَذَهُ, (Ṣ, A, Mgh,* L, Mṣb, Ḳ,*) aor. ِ, (Mgh, Mṣb, Ḳ,) inf. n. جَبْذٌ, (T, Mgh, Mṣb, Ḳ, &c.,) *i. q.* جَذَبَهُ (T,* Ṣ, A, Mgh,* L, Mṣb, Ḳ,* &c.) i. e. *He drew it*; &c.: (T, TA:) formed by transposition from the latter; (A'Obeyd, Ṣ, A;) accord. to some: (Mṣb:) or it is a dial. var. of the latter; (M, L, Ḳ, &c.;) of the dial. of Temeem; (T, Mṣb;) not formed from the latter by transposition, (Ibn-Es-Sarráj, IJ, M, L, Ḳ,) for both are equally conjugated: (Ibn-Es-Sarráj, IJ, L, Mṣb:) and اِجْتِبَاذٌ [inf. n. of اجتبذ] signifies the same as جَبْذٌ. (Ḳ.) جَبَذَنِى رَجُلٌ مِنْ خَلْفِى, meaning *A man pulled me from behind me*, occurs in a trad. (L.) — You say also, جَبَذَتْهُ and جَذَبَتْهُ, meaning, ‡ *She repelled him*, or *rejected him*; namely, a man who sought her in marriage. (T and TA in art. جذب.)

7. اِنْجِبَاذٌ [inf. n. of انجبذ] *i. q.* اِنْجِذَابٌ, (Ḳ,) meaning *Quick going* or *journeying* or *travelling*. (TA.)

8: see 1.

جبر

1. جَبَرَ, (Ṣ, Mṣb, Ḳ, &c.,) aor. ُ, (Mṣb,) inf. n. جَبْرٌ (Ṣ, A, Mṣb, Ḳ, &c.) and جُبُورٌ, (M, Ḳ,) which latter, accord. to MF, is an inf. n. of the intrans. verb only, but it has been heard as an inf. n. of the trans. verb also, (TA,) and جِبَارَةٌ, (Lḥ, Ḳ,) *He set* a bone; *reduced* it *from a fractured state*; (Ṣ, A, Mṣb, Ḳ, &c.;) as also ▼جَبَّرَ, (A, IAmb, Ḳ,) inf. n. تَجْبِيرٌ; (TA;) and ▼اجبر, (Ibn-Ṭalḥah, MF, TA,) but this is extremely strange, and not found in the lexicons of celebrity, (MF,) and not heard by AO; (TA;) [and ▼اجتبر.] One says also, ▼جَبَّرَ يَدَهُ, (A, IAmb,) or جَبَرَهَا, (Mṣb,) *He* (a bone-setter) *set his arm*, or *reduced it from a fractured state*: (A:) or *put upon it the* جَبِيرَة [or *splints*]. (Mṣb.) — Hence, (TA,) جَبَرَ, (AAF, M, Ḳ, &c.,) inf. n. جَبْرٌ (Ṣ, A, Ḳ) and جُبُورٌ [but respecting this latter see above] and جِبَارَةٌ; (Ḳ;) and ▼جَبَّرَ, (Ḳ,) inf. n. تَجْبِيرٌ; (TA;) and ▼اجبر; (Ibn-Ṭalḥah, MF, TA; [but respecting this form see above;]) and ▼اجتبر; (Ḳ;) ‡ *He restored* a man *from a state of poverty to wealth*, or *competence*, or *sufficiency*: (AAF, Ṣ, A, Ḳ, &c.:) or *he benefited* a poor man; *conferred a benefit*, or *benefits, upon* him: (M, Ḳ:) but the former is the more appropriate explanation: (AAF, TA:) and this signification is tropical; (IDrst, MF, TA;) the poor man being likened to one who has a broken bone, and his restoration to wealth, or competence, being likened to the setting of the bone; wherefore he is called فَقِيرٌ, as though the vertebræ of his back were broken: (IDrst, TA:) in the A it is mentioned as proper, not tropical; but the author of the A afterwards mentions جَبَرْتُ فُلَانًا as tropical in the sense of نَعَشْتُهُ ‡ [*I recovered such a one from his embarrassment*, &c.; *repaired his broken fortune*, or *his condition*]. (TA.) One says also, جَبَرْتُ فَاقَةَ الرَّجُلِ ‡ [*I repaired the broken fortune of the man*;] *I restored the man to wealth*, or *competence*, or *sufficiency*. (AHeyth, TA.) And جَبَرْتُ اليَتِيمَ † [*I put the affairs of the orphan into a right*, or *good, state*: or] *I gave to the orphan*. (Mṣb.) And جَبَرَ ‡ *He restored* anything *to a sound, right*, or *good, state*. (IDrst, TA.) And جَبَرَهُ اللّٰهُ † [*May God render him sound, and strong*]: said in relation to a child. (Ṣ and Ḳ in art. زرع.) And جَبَرْتُ نِصَابَ الزَّكَاةِ بِكَذَا † *I made the amount of the property equal to that which renders it incumbent on the possessor to pay the poor-rate, by* [adding] *such a thing*: the name of that thing is جبران [app. ▼جُبْرَان]: and the person who does this is termed ▼جَابِرٌ. (Mṣb.) — جَبَرَ also signifies *He compelled*, or *constrained*, another. (B.) You say, جَبَرَهُ عَلَى الأَمْرِ, (Lḥ, Az, Mṣb, Ḳ,) aor. ُ, inf. n. جَبْرٌ and جُبُورٌ, (Mṣb,) a chaste form of the verb, of the dial. of El-Ḥijáz, (Az, TA,) or of the Benoo-Temeem and of many of the people of El-Ḥijáz, (Mṣb,) or of Temeem alone; (Lḥ, TA;) [but said in the Mgh to be of weak authority;] and ▼اجبرهُ; (Th, Ṣ, Mṣb, Ḳ, &c.;) both these forms of the verb mentioned by AZ, Fr, A'Obeyd, and others, (Mṣb,) but the latter is the form used by the generality of the Arabs, (Lḥ, TA,) and by the grammarians [in general]; (TA;) *He compelled him, against his will, to do the thing*: (Lḥ, Th, Az, Ṣ, Mṣb, Ḳ:) ▼إِجْبَارٌ originally signifying the *inciting, urging*, or *inducing*, another *to restore a thing to a sound, right*, or *good, state*. (B.) And ▼اجبره عَلَى الحُكْمِ *He* (a judge) *compelled him to submit to*, or *to perform, the sentence*. (L.) — Also جَبَرَ, [aor. ُ,] inf. n. جُبُورٌ (Ṣ, Mṣb, Ḳ) and جَبْرٌ, (Mṣb, Ḳ,) which latter, accord. to MF [and the Mgh], is an inf. n. of the trans. verb only, but it has been heard as an inf. n. of the intrans. verb also; (TA;) and ▼انجبر, (T, Ṣ, Ḳ,) and ▼اجتبر, (T, Ṣ,) and ▼تجبّر; (Ḳ;) *It* (a bone) *became set*, or *reduced from a fractured state*. (T, Ṣ, Mṣb, Ḳ.) — And [hence,] the first of these verbs, with the same inf. ns.; (Ḳ;) and ▼اجتبر, (Ṣ,* Ḳ,) and ▼انجبر, and ▼تجبّر, and ▼استجبر; (Ḳ;) ‡ *He* (a poor man, Ḳ, and an orphan, TA) *became restored from a state of poverty to wealth*, or *competence*, or *sufficiency*: (Ṣ,* Ḳ:) or *received a benefit*, or *benefits*: (Ḳ:) ▼اجتبر is syn. with انتعش ‡ [*he recovered*, or *became recovered, from his embarrassment*, &c.]. (A.) [And † *It* (anything) *became restored to a sound, right*, or *good, state*.] El-'Ajjáj has used جَبَرَ transitively and intransitively in the same sentence, saying,

* قَدْ جَبَرَ الدِّينَ الإِلٰهُ فَجَبَرْ *

[† *God hath restored the religion to a sound, right*, or *good, state, and it hath become restored thereto*]: (Ṣ:) or, accord. to some, the second verb is corroborative of the first; the meaning being, *God hath desired*, or *purposed, to restore the religion*, &c., *and hath completed its restoration*. (B.)

2: see 1, in three places.

4: see 1, in five places. — اجبرهُ also signifies *He imputed to him* [*the tenet of*] الجَبْر; (Ṣ,* L, Ḳ;*) *he called him a* جَبْرِىّ: (L:) like as اكفره signifies "he imputed to him infidelity." (Ṣ.)

5. تجبّر: see 1, latter part, in two places. Also † *What had gone from him* (a man) *returned to him*: (Ḳ:) or *some of his property that had gone from him returned to him*. (T, TA.) † *He* (a sick man) *became in a good state*. (Ḳ.) † *It* (a plant, TA, and a tree, Ḳ) *became green, and put forth leaves* (Ḳ, TA) *and fresh green twigs, when dry*: *produced fresh shoots in its dry parts*: (TA:) *it* (herbage) *became somewhat restored to a good state after having been eaten*: (Ḳ,* TA:) or *grew after having been eaten*. (Ṣ.) — *He* (a man, Ṣ) *magnified himself; behaved proudly, haughtily*, or *insolently*. (Ṣ, A, Ḳ.) — † *He* (a man) *obtained wealth*, or *property*: (Ḳ:) but

Lḥ explains it as meaning, intransitively, *he obtained wealth*, or *property*. (TA.)

7: see 1, latter part, in two places.

8: see 1, in five places. You say also, أَصَابَتْهُ مُصِيبَةٌ لَا يَجْتَبِرُهَا [† *A calamity befell him from which he will not recover*]; i. e. لَا مُجْتَبَرَ ♦ مِنْهَا [† *there is no recovering from it*]. (TA.)

10: see 1, latter part. — استجبره ‡ *He exerted himself much*, or *exceedingly*, or *to the utmost, in paying frequent attention to him, taking care of him*, or *putting his affairs into a right*, or *proper, state.* (A.)

جَبْرٌ, in computation, † The *addition of something for the purpose of reparation.* (TA.) [Hence, الجَبْرُ † *Algebra*; more commonly called الجَبْرُ وَالمُقَابَلَةُ *perfective addition and compensative subtraction*; or *restoration and compensation*; because of the frequency of these operations in the reduction of equations.] — The *contr. of* قَدَرٌ: (Ṣ, Mṣb, Ḳ:) it is the *assertion that God compels his servants*, or *mankind, to commit acts of disobedience*; (Mṣb;) the *virtual denial that actions proceed from man, and attributing them to God*; the sect that hold the tenet thus termed asserting that man, with respect to his actions, is like the feather suspended in the air; whereas قَدَرٌ signifies the "virtual attributing of optional, or voluntary, actions to man; asserting that man creates his own optional, or voluntary, actions:" (IbrD:) A 'Obeyd says that it is a post-classical term. (Ṣ.) — *A king*; (AA, T, M, Ḳ;) of uncertain derivation: (M:) and *a slave*, or *servant*: (A 'Obeyd, Kr, Ḳ, &c.:) thus bearing two contr. significations: (Ḳ:) and *a man*: (AA, A 'Obeyd, Ḳ, &c.:) [see جَبْرَئِيلُ:] and *a young man*: and [a] *courageous* [man]. (Ḳ.) — [Also, app., *Aloes-wood*:] الجَبْرُ is explained in the Ḳ as signifying العُودُ, which means *wood* in general, as well as aloes-wood in particular; and to this is added in the TA, الذى يُجْبَرُ بِهِ, as though the meaning were the *wood with which one sets bones*; but I think that يُجْبَرُ is a mistranscription for يُجْمَرُ; and that the meaning is *aloes-wood with which one fumigates.*]

جَبْرِىٌّ or جَبَرِىٌّ: see جَبْرِيَّةٌ.

جَبْرُوَةٌ and جَبَرُوَةٌ and جَبَرُوتٌ &c.: see what next follows.

جَبْرِيَّةٌ (Ṣ, Ḳ) and جَبَرِيَّةٌ and جِبِرِيَّةٌ and ♦ جِبْرِيَآءُ (Ḳ) and ♦ جَبَرِيَّآءُ (Aboo-Naṣr, TA) and ♦ جَبْرُوَةٌ (Ṣ, Ḳ) and ♦ جَبَرُوَةٌ (Ḳ) and ♦ جَبَرُوتٌ (Ṣ, Mṣb, Ḳ, one of the forms most known, of the measure فَعَلُوتٌ, like مَلَكُوتٌ and رَهَبُوتٌ and رَغَبُوتٌ and رَحَمُوتٌ, said to be the only other words of this measure, though, as MF says, this requires consideration, TA) and ♦ جَبْرُوتٌ (Ḳ) and ♦ جَبَرُّوتٌ (Et-Tedmuree, TA) and ♦ جَبَرُوتَى (Ḳ, like رَحَمُوتَى [&c.], TA) and ♦ جَبُّورَةٌ (Ṣ, Ḳ) and ♦ جَبُّورٌ (Lḥ, Kr) and ♦ جُبُّورٌ (Lḥ, TA) and ♦ جُبُورَةٌ and ♦ تَجْبَارٌ, (Ḳ,) all inf. ns., (TA,) [or simple substs.,] meaning The *quality denoted by the epithet* جَبَّارٌ; (Ḳ;) i. e. *self-magnification, pride, haughtiness*, or *insolence*; or *proud, haughty*, or *insolent, behaviour*; (Ṣ, Mṣb, Ḳ;) &c. (Ḳ, TA.) Hence, مَا كَانَتْ نُبُوَّةٌ إِلَّا تَنَاسَخَتْهَا مُلْكٌ جَبْرِيَّةٌ [*There has been no prophetic office but a kingly office has succeeded in its place through* some one's *self-magnification, pride, haughtiness*, or *insolence*]; i. e., but kings have magnified themselves, or behaved proudly or haughtily or insolently, after it. (A, TA.) — الجَبَرِيَّةُ (Ṣ, Ḳ) and الجَبْرِيَّةُ, (Th, Mṣb,) or the latter is a mispronunciation, or is the correct form, (Ḳ,) and the former is so pronounced in order to assimilate it to القَدَرِيَّةُ; (Mṣb, Ḳ;*) the latter is the pronunciation of the scholastic theologians of the persuasion of Esh-Sháfi'ee (El-Ḥáfidh in the "Tabṣeer," B) in old times, but the term used in the conventional language of the modern scholastic theologians is ♦ المُجْبِرَةُ; (B;) and الجبرية, also, is a post-classical term; (TA;) *The contr. of* القَدَرِيَّةُ; (Ṣ, Ḳ;) *the sect who hold the tenet termed* جَبْرٌ [q. v.]; (Mṣb;) *a sect of those who follow their own natural desires, whose founder was El-Ḥoseyn Ibn-Moḥammad En-Nejjár El-Baṣree, who assert that man has no power; that* [what are termed] *voluntary motions are of the same predicament as a tremour; though this does not oblige them to deny the imposition of duties*; (Lb, TA;) *a sect who assert that God compels his servants*, or *mankind, to commit sins*: (AHeyth, TA:) n. un. ♦ جَبْرِىٌّ or جَبَرِىٌّ. (Mṣb.)

جِبْرَالُ and جَبْرَالُ: see جَبْرَئِيلُ.

جُبْرَانٌ: see 1.

جَبْرِيلُ and جِبْرِيلُ &c.: } see جَبْرَئِيلُ.
جَبْرِينُ and جِبْرِينُ: }

جَبَرِيَّةٌ and جِبْرِيَآءُ: see جَبْرِيَّةٌ.

جَبْرَئِيلُ, (Ṣ, Mṣb, Ḳ, &c.,) imperfectly decl., because having the quality of a proper name and that of a foreign word, or being a compound regarded as forming a single word, as some say, (TA,) originally Syriac, or Hebrew, [גַּבְרִיאֵל,] (Esh-Shiháb [El-Khafájee],) *A proper name of an angel*; (TA;) [*Gabriel*: and also, *of a man*:] signifying *the servant of God*: (A 'Obeyd, Ṣ, Mṣb, Ḳ, TA:) or (rather, TA) *the man of God*: (A 'Obeyd, TA:) being said to be composed of جَبْرٌ, (Ṣ, Mṣb, TA,) signifying "servant," or "slave," (Mṣb, TA,) or rather "man," (TA,) and إِيلُ, (Ṣ, Mṣb, TA,) signifying "God:" (Mṣb, TA:) or both together signify *the servant of the Compassionate*: or *the servant of the Mighty*, or *Glorious*: (TA:) this form of the word is of the dialects of Ḳeys and Temeem: (TA:) and there are other dial. vars.; namely, ♦ جَبْرَئِلُ, without ى, and ♦ جَبْرَئِلُّ, (Ṣ, Ḳ,) and ♦ جَبْرَئِلُ, and ♦ جَبْرَائِيلُ, (Ḳ,) and ♦ جَبْرَايِيلُ, (Es-Suyooṭee, TA,) and ♦ جَبْرَائِلُ, (Ḳ,) and ♦ جَبْرَايِلُ, (Es-Suyooṭee, TA,) and ♦ جِبْرِيلُ, (Ṣ, Mṣb, Ḳ, which is the form most known and most chaste, and is of the dial. of El-Ḥijáz, TA,) and ♦ جَبْرِيلُ, (Mṣb, Ḳ, reckoned of weak authority by Fr, because the measure فَعْلِيل [or فَعْلِيل] does not exist in the language, for as to سَمْوِيل, mentioned by Esh-Shiháb as against the objection of Fr, it is of the measure فَعْوِيل, MF, TA,) and ♦ جَبْرِئِلُ, and ♦ جَبْرَالُ, and ♦ جِبْرَالُ, (Ḳ,) and ♦ جِبْرِينُ, and ♦ جَبْرِينُ, (Ṣ, Ḳ,) and ♦ جَبْرَائِينُ. (Es-Suyooṭee, MF.)

جَبْرَيِيلُ: }
جَبْرَايِلُ and جَبْرَائِلُ: } see جَبْرَئِيلُ.
جَبْرَايِيلُ and جَبْرَائِيلُ: }
جَبْرَائِينُ: }

جُبَارٌ A thing *of which no account*, or *for which no revenge* or *retaliation* or *mulct, is taken.* (Ṣ, A, Mṣb, Ḳ, TA.) You say, ذَهَبَ دَمُهُ جُبَارًا *His blood went unrevenged, unretaliated*, or *unexpiated by a mulct.* (Ṣ, A.) And جُرْحٌ جُبَارٌ *A wound for which is no retaliation, nor any expiatory mulct.* (A, TA.) And حَرْبٌ جُبَارٌ *A war in which is no retaliation*, (Ḳ, TA,) *nor any expiatory mulct.* (TA.) And المَعْدِنُ جُبَارٌ [*The mine is a thing for which no mulct is exacted*]: i. e., if the mine fall in upon him who is working in it, and he perish, his hirer is not to be punished for it. (Ṣ and Mṣb from a trad.) And البِئْرُ جُبَارٌ [*The well is a thing for which no mulct is exacted*]: i. e., if a man fall into an ancient well, and perish, his blood is not to be expiated by a mulct: (TA:) or, as some say, it relates to a hired man's descending into a well to cleanse it, or to take forth something from it, if he fall into it and die. (TA in art. بأر.) And جُرْحُ العَجْمَآءِ جُبَارٌ *The wound of the speechless beast*, if it get loose and wound a man or other thing while loose, *is a thing for which no retaliation or expiatory mulct is exacted.* (T, A,* Mṣb.*) — *Clear*, or *quit*, of a thing: so in the saying, أَنَا مِنْهُ خَلَاوَةٌ وَجُبَارٌ [*I am clear, or quit, of it*]. (Ḳ. [See also فَالِجٌ.]) — *A torrent.* (Ḳ.) — *Anything that corrupts*, or *mars, and destroys*; (so accord. to some copies of the Ḳ, and the TA;) as the torrent, &c.: (TA:) or *anything that is corrupted*, or *marred, and destroyed.* (So accord. to other copies of the Ḳ.) — *Tuesday*; (Ṣ, Ḳ;) an ancient name thereof, (Ṣ,) used in the Time of Ignorance; (TA;) as also ♦ جِبَارٌ. (Ḳ.)

جِبَارٌ: see what next precedes.

جِبَارَةٌ and ♦ جَبِيرَةٌ *Splints; pieces of wood with which bones are set*, or *reduced from a fractured state*: (Ṣ, Ḳ:) or *bones which are put upon a diseased part of the person, to reduce it to a sound state*: pl. جَبَائِرُ. (Mṣb.) — Also, both words, *A wide bracelet*; syn. يَارَقٌ: (Ṣ, Ḳ:) *a bracelet* (سِوَارٌ) *of gold* or *silver*: pl. جَبَائِرَةُ [or جَبَائِرُ, as above?]. (A 'Obeyd, TA.)

جَبُّورَةٌ: see جَبْرِيَّةٌ.

جَبِيرَةٌ: see جِبَارَةٌ.

جَبَّارٌ One *who magnifies himself*, or *behaves*

proudly or *haughtily* or *insolently, and does not hold any one to have any claim upon him, or to deserve anything of him:* (Ḳ:) one *who slays when in anger:* (Ṣ, A:) one *who slays unjustly:* (Ḳ:) *imperious,* or *domineering, by absolute force and power; overbearing; tyrannical; a tyrant:* (TA:) any one *who exalts himself,* or *is insolent and audacious, in pride and in acts of rebellion* or *disobedience; who is bold,* or *audacious, and immoderate, inordinate,* or *exorbitant;* or *excessively, immoderately,* or *inordinately, proud,* or *corrupt,* or *unbelieving,* or *disobedient,* or *rebellious; or who exalts himself and is inordinate in infidelity;* or *who is extravagant in acts of disobedience and in wrongdoing;* or *who is refractory,* or *averse from obedience;* (Ḳ,* TA;) as also ▼ جِبِّيرٌ: (Ḳ:) or this latter signifies one *who magnifies himself much,* or *behaves very proudly* or *haughtily* or *insolently:* (Ṣ:) and the former, one *who proudly, haughtily,* or *insolently, disdains the service of God:* (Lḥ, TA:) fem. with ة: pl. masc. جَبَّارُونَ and جَبَابِرَةٌ. (A, TA.) — الجَبَّارُ [*A name of*] *God;* so called because of his magnifying Himself [above every other being], (Ḳ,) and his highness: (TA:) meaning *the Compeller of his creatures to do whatsoever He willeth:* (Bḍ and Jel in lix. 23:) or *the Compeller of his creatures to obey the commands and prohibitions which He pleaseth to impose upon them:* (Mṣb, TA:) accord. to Fr, from أَجْبَرَ, and the only instance known to him of an epithet of the measure فَعَّالٌ from a verb of the measure أَفْعَلَ except دَرَّاكٌ [q. v.] from أَدْرَكَ: (Az, TA:) or, accord. to Fr, from جَبَرَ as syn. with أَجْبَرَ: (Mṣb:) it is also explained as meaning *the Supreme; the High above his creatures:* (Az, TA:) or *the Unattainable;* and hence applied to the palm-tree [of which the branches cannot be reached by the hand]: (IAmb, TA:) or it may signify ‡ *the Restorer of the poor to wealth* or *competence* or *sufficiency.* (Az, TA.) [God is also called] جَبَّارُ القُلُوبِ عَلَى فِطَرَاتِهَا † *The Establisher of hearts according to their natural constitutions which He hath given them in the mothers' wombs, disposing them to know Him and to confess Him, both the unfortunate of them and the fortunate.* (TA from a trad. of 'Alee.) — Also ‡ *A name of* الجَوْزَآءُ [*the constellation Orion*]; (A, Ḳ;) because it is [represented] in the form of a crowned king upon a throne. (A.) — ذِرَاعُ الجَبَّارِ † *The cubit of the king:* (A, TA:) or *the long cubit:* or, as Ḳt thinks, by الجبّار is here meant a certain foreign king whose fore arm was of full length. (TA.) — قَلْبٌ جَبَّارٌ ‡ *A heart that receives not admonition:* (A:) or *that admits not compassion.* (Ḳ.) — جَبَّارٌ, (Seer, Ḳ,) without ة, (Seer, TA,) applied to a palm-tree (نَخْلَةٌ), signifies ‡ *Tall and young;* (Seer, Ḳ, TA;) as also ▼ جُبَّارٌ: (Ḳ:) or is applied to palm-trees collectively (نَخْلٌ), and signifies *tall, and above the reach of the hand;* (T, Ṣ;) and the epithet applied to a single palm-tree is with ة; (Ṣ, A;) in this sense; meaning *less than* سَحُوقٌ: (A:) or, with ة, it signifies a *young palm-tree, that has attained its utmost height and has borne fruit:* (M:) or *that has been ascended* [*for the purpose of cutting off its fruit*], *and retains its excellence, surpassing therein other palm-trees.* (AḤn, TA.) — Also, hence, as Az thinks, (TA,) ‡ *Huge, tall, and strong; a giant.* (T, A,* Ḳ.) — And, with ة, (Ṣ, A,) and also without ة, (A,) applied to a she-camel, ‡ *Great* (Ṣ, A) *and fat.* (Ṣ.)

جُبَّارٌ: see جَبَّارٌ.

جُبُورٌ and جَبُّورٌ: see جَبْرِيَّةٌ.

جِبِّيرٌ: see جَبَّارٌ.

جَبُّورَةٌ: see جَبْرِيَّةٌ.

جَابِرٌ, (Ṣ,) and جَابِرُ بْنُ حَبَّةَ, (Ṣ, A, Ḳ,) names of ‡ *Bread;* (Ṣ, A, Ḳ;) and أَبُو جَابِرٍ is a surname thereof; (Ṣ, Ḳ;) and so أُمُّ جَابِرٍ: which last also signifies *the ear of corn:* (T in art. ام:) and i. q. الهَرِيسَةُ [*grain,* or *wheat, bruised,* or *brayed, and then cooked*]. (Ḥar p. 227.) — فُلَانٌ جَابِرٌ لِى i. q. ▼ مُسْتَجْبِرٌ ‡ [*Such a one exerts himself much,* or *exceedingly,* or *to the utmost, in paying frequent attention to me, taking care of me,* or *putting my affairs into a right,* or *proper, state*]. (A.) — See also 1.

تَجْبَارٌ: see جَبْرِيَّةٌ.

مَجْبَرٌ [an inf. n. of 1]: see 8.

المُجْبَرَةُ: see جَبْرِيَّةٌ.

مُجَبِّرٌ One *who sets bones,* or *reduces them from a fractured state; a bone-setter.* (Ṣ, A, Ḳ.)

مَجْبُورَةٌ A woman *possessed by a jinnee,* or *genie;* syn. مَجْنُونَةٌ; but this is held to be of weak authority. (Mgh.)

المُتَجَبِّرُ *The lion.* (Ḳ.)

مُسْتَجْبِرٌ: see جَابِرٌ.

جبرئيل

جَبْرَئِيلُ and its vars.: see art. جبر.

جبس

جِبْسٌ [*Gypsum;*] i. q. جِصٌّ, or جَصٌّ, (Ḳ,) *with which one builds.* (Kr, TA.*)

جَبَّاسَةٌ The *place of* جِبْس; [i. e., *in which gypsum is found,* or *prepared;*] as also مجبّسة [i. e. ▼ مُجَبَّسَةٌ, or, more probably, it is a mistranscription, for ▼ مَجْبَسَةٌ, like مَجْصَاةٌ (originally مَجْصَصَةٌ) and مَبْقَلَةٌ &c.]. (TA.)

مُجَبَّسَةٌ or مَجْبَسَةٌ: see جَبَّاسَةٌ.

جبل

1. جَبَلَهُ, (Ṣ, Mṣb, Ḳ,) aor. ِ (Mṣb, Ḳ) and ُ, (Ḳ,) inf. n. جَبْلٌ, (KL,) *He* (God) *created him.* (Ṣ, Mṣb, Ḳ, KL.) So in the phrase, جَبَلَهُ عَلَى كَذَا, (Mṣb,) or على الشَّىْءِ, (Ḳ,) *He* (God) *created him with an adaptation,* or *a disposition, to such a thing,* or *to the thing; adapted him,* or *disposed him, by nature thereto.* (Mṣb.) It is said in a trad., جُبِلَتِ القُلُوبُ عَلَى حُبِّ مَنْ أَحْسَنَ إِلَيْهَا وَبُغْضِ مَنْ أَسَآءَ إِلَيْهَا [*Hearts are created with a disposition to the love of him who does good to them, and the hatred of him who does evil to them*]. (TA.) — Also, (Ḳ,) inf. n. as above, (TA,) i. q. جَبَرَهُ [evidently as meaning *He compelled him, against his will,* عَلَى الأَمْرِ *to do the thing;* for he who is created with a disposition to do a thing is as though he were compelled to do it]; and so ▼ اجبلهُ, (Ḳ, TA,) inf. n. إِجْبَالٌ. (TA.) ═ جَبِلَ † *He* (a man) *became like a mountain* (جَبَل) *in bigness, thickness, coarseness,* or *roughness.* (TA.) — جَبِلَ حَدِيدُهُمْ, (Ḳ, TA; in the CḲ, جَبَلَ; and in a MS. copy of the Ḳ, without any vowels;) † *Their iron was,* or *became, blunt, such as would not penetrate.* (Ḳ,* TA.)

3. جابل *He* (a man) *alighted,* or *descended and abode,* or *sojourned,* or *settled, in a mountain.* (AA, TA.)

4. اجبل *He came,* or *went,* or *betook himself, to the mountain.* (ISk, Ṣ, Ḳ.) — ‡ *He* (a digger) *reached a hard place,* (Ṣ, Ḳ,) or *stone,* (Mgh,) *in his digging.* (TA. [الحَافِر, meaning "the digger," Golius seems to have misunderstood as meaning "the hoof" of a horse.]) — [Hence,] ‡ *He* (a poet) *experienced difficulty in diction,* (Ḳ, TA,) *so that he said nothing original, nor anything in the way of repetition.* (TA.) — And طَلَبَ حَاجَةً فَأَجْبَلَ † *He sought a thing that he wanted, and failed of attaining it.* (TA.) — And سَأَلْنَاهُمْ فَأَجْبَلُوا ‡ *We asked them, and they refused, and did not give.* (Ibn-'Abbád, Z, TA.) — And أَجْبَلُوا ‡ *Their iron became blunt, so that it would not penetrate.* (Ḳ,* TA.) ═ اجبلهُ ‡ *He found him to be a* جَبْل, i. e. *a niggard:* (Ḳ, TA:) it is considered as implying fixedness. (TA.) — See also 1.

5. تجبّلوا *They entered a mountain:* (Ḳ:) or, accord. to the O, you say, تَجَبَّلَ القَوْمُ الجِبَالَ, meaning, *the people,* or *company of men, entered the mountains.* (TA.)

جَبْلٌ † *Big, thick, coarse,* or *rough;* (TA;) as also ▼ جَبِلٌ, applied to a thing (Ṣ, O, Ḳ) of any kind: (Ḳ:) or this latter is applied to an arrow, signifying † *coarsely, roughly,* or *rudely, pared.* (Ḳ.) You say رَجُلٌ جَبْلُ الرَّأْسِ, (Ḳ, TA, [in the CḲ, erroneously, جَبَلُ الرأس,]) and الوَجْهِ, (TA,) ‡ *A man having a big, thick, coarse,* or *rough, head, and face;* (TA;) *having little sweetness.* (Ḳ, TA.) [See also جَبِيل.] And ▼ رَجُلٌ مِجْبَالٌ † *A big, thick, coarse,* or *rough, and heavy, man.* (Ḥam p. 818.) And اِمْرَأَةٌ جَبْلَةٌ (Ḳ [in one place in the CḲ جَبَلَةٌ and جَبِلَةٌ, but only جَبْلَةٌ accord. to the TA,]) and ▼ مِجْبَالٌ (Ṣ, Ḳ) ‡ *A woman big, thick, coarse,* or *rough,* (Ṣ, Ḳ, TA,) *in make;* (Ṣ;) *large in make.* (TA.) And خِلْقَةٌ جَبْلَةٌ † *A big, thick, coarse,* or *rough, make.* (Ḥam p. 821.) And نَاقَةٌ جَبْلَةُ السَّنَامِ ‡ *A she-camel having an increasing hump.* (TA.) And سَيْفٌ جَبْلٌ and ▼ مِجْبَالٌ † *A sword not made thin.* (TA.) ═ Also (Ḳ, TA, [in the CḲ, جَبَلٌ,]) *A court* [of a house]; syn. سَاحَةٌ. (Ḳ.)

جَبْل : see جَبَل : — and جِبِلّ — Also *Dry trees.* (K.)

جِبْل *Much;* or *numerous;* (S, K;) as also ▾ جُبْل. (K.) So in the phrases مَالٌ جِبْلٌ [*Much property;* or *numerous cattle*]; and حَيٌّ جِبْلٌ *A numerous tribe.* (S.) — See also جِبِلّ, in two places.

جَبَل [*A mountain:* or] *any of the mountains* أَوْتَاد [lit. "pegs," or "stakes," a term applied to the mountains because they are supposed to make the earth firm, or fast,]) *of the earth, that is great and long;* (Mgh, K;) or, as some say, only *such as is long;* (Mṣb;) such as is isolated being called أَكَمَة, or قُنَّة: (K:) [and also applied to *a rocky tract; any rocky elevation, however little elevated:*] and sometimes it means *stone;* [or *rock;*] such, for instance, as is reached by the digger: and hence it is applied to Eṣ-Ṣafà and El-Marweh: (Mgh:) pl. [of mult.] جِبَالٌ (S, Mṣb, K) and (of pauc., Mṣb) أَجْبُلٌ (Mṣb, K) and أَجْبَالٌ. (K.) — [Hence,] † *A man who does not remove from his place:* you say of such a one, هُوَ جَبَلٌ. (TA.) — ‡ *A niggard.* (K, TA.) [See 4.] — ‡ The *lord,* or *chief, of a people,* or *company of men:* and *their learned man.* (Fr, K, TA.) — اِبْنَةُ الجَبَلِ † *The serpent:* (K:) because it keeps to the جَبَل. (TA.) — † *Calamity,* or *misfortune.* (K.) — † *The bow that is made from the tree called* نَبْع; (K, TA;) because this is one of the trees of the جَبَل. (TA.) — † *The echo.* (Ḥar p. 472.)

جُبْل : see جِبْل.

جَبِل : see جَبَل. — Also, applied to the iron head, or blade, of an arrow, or of a spear, or of a sword, &c., ‡ *Blunt; that will not penetrate into a thing:* (Ibn-'Abbád, K,* TA:) and so, with ة, applied to a فَأْس. (TA.)

جُبُل : see جِبِلّ.

جَبْلَة (K, TA, [in the CK جَبَلَة,]) and ▾ جُبْلَة The *face:* or the بَشَرَة [or *external skin*] *thereof:* or the *part thereof that is turned towards one.* (K.) — Also, (K,) or the former, (TA,) *A vice, fault, defect,* or *blemish.* (K.) — And *Strength.* (K.) — And *Hardness of the earth,* or *ground.* (Lth, K.) — See also جُبْلَة: — and see جِبِلَّة.

جُبْلَة *A camel's hump;* (S, K;) as also ▾ جَبْلَة. (K.) — See also جِبِلّ: — and see جِبِلَّة, in two places.

جِبْلَة : see جِبِلَّة. — Also The *origin,* or *stock,* (K, TA,) of any created thing; (TA;) and so ▾ جُبُلَّة. (K, TA.) — The *fundamental nature,* or *composition,* of a mountain. (TA.) — ثَوْبٌ جَيِّدُ الجِبْلَةِ ‡ *A garment,* or *piece of cloth, good in respect of the thread* (K, TA) *and the weaving.* (TA.) — رَجُلٌ ذُو جِبْلَةٍ † *A big, thick, coarse,* or *rough, man.* (S, K.) — See also جِبِلّ, in two places: — and see جَبْلَة.

جَبَلَة : see جِبِلَّة.

جُبُلّ : see what next follows.

جِبِلّ and ▾ جُبُلّ and ▾ جُبْل [accord. to the CK like عَقْل, but correctly like عُقُل,] and ▾ جِبْل and ▾ جُبُل, (S, K,) accord. to different readings of the instance occurring in the Kur xxxvi. 62, the first being the reading of the people of El-Medeeneh, (S,) [and the most common,] *A great company of men;* as also ▾ جِبِلَّة and ▾ جَبِيل: (K:) or [simply] *a company of men;* (S;) as also ▾ جُبُل, accord. to Kh; (Sgh, TA;) and so ▾ جِبْلَة and ▾ جُبْلَة and ▾ جِبِلَّة: which last three signify also the *same as* أُمَّة [*a nation,* or *people,* &c.]: (K:) it is said [by some] that جِبِلّ is pl. [or coll. gen. n.] of ▾ جِبِلَّة meaning *a numerous company:* (TA:) جِبَلَة is pl. of ▾ جِبْل: one says, قَبَّحَ ٱللّٰهُ جِبِلَّتَكُمْ [*May God remove far from prosperity,* or *success,*] *your companies:* (Fr, TA:) and جِبَل is pl. of ▾ جِبْلَة. (Bḍ in xxxvi. 62.)

جِبِلَّة *Much,* or *an abundance,* or *a large quantity* or *number,* of anything; as also ▾ جِبِلَّة. (K.) — See also جِبِلّ: — and see what next follows, in two places.

جِبِلَّة (S, Mṣb, K) and ▾ جُبُلَّة and ▾ جَبِيلَة (Sgh, MF) and ▾ جِبْلَة (AA, S, K) and ▾ جُبْلَة and ▾ جِبِلَة and ▾ جَبَلَة, (K,) but this last, accord. to MF, is unknown, (TA,) *Nature;* or *natural, native, innate,* or *original, constitution, disposition, temper,* or other *quality* or *property; idiosyncrasy;* syn. خَلِيقَة (AA, S, Sgh, Mṣb, K) and طَبِيعَة (Mṣb, K) and غَرِيزَة; all these signifying the same: (Mṣb:) pl. of the first جِبِلَّاتٌ. (S.) Hence, in the Kur [xxvi. 184], وَٱلْجِبِلَّةَ ٱلْأَوَّلِينَ, (S,) meaning الخَلِيقَة, (Jel,) or ذَوِي الجِبِلَّة, i. e. *And the preceding created beings:* (Bḍ:) El-Ḥasan read with ḍamm [i. e. ▾ الجُبُلَّة or ▾ الجُبْلَة]. (S.) — See also جِبِلّ, in three places: — and see جِبْلَة.

جَبَلِيّ *Of,* or *relating to, a mountain* or *mountains; contr. of* سُهْلِيّ. (The Lexicons &c. passim.)

جِبِلِّيّ *Natural;* i. e. *of,* or *relating to, the natural, native, innate,* or *original, constitution, disposition, temper,* or other *quality* or *property;* like طَبِيعِيّ; i. e. *essential; resulting from the Creator's ordering of the natural disposition in the body.* (Mṣb.)

جِبَال ‡ The *body, with,* or *without, the members;* syn. جَسَد and بَدَن; (K, TA;) as being likened to a mountain in bigness [?]. (TA.) One says, أَحْسَنَ ٱللّٰهُ جِبَالَهُ, meaning, ‡ [*May God render beautiful*] *his body* (جَسَدَهُ): and [*render good*] *his created* خُلُق [or *mind, with its qualities and attributes:* but I rather think that خُلُق is here a mistranscription for خَلْق, meaning *make*]. (Ibn-'Abbád, TA.)

جَبِيل : see جِبِلّ. — جَبِيلُ الوَجْهِ ‡ A man *having a bad,* or *an ugly, face.* (K, TA.) [See also جَبْلَة.]

جَبِيلَة : see جِبِلَّة.

مِجْبَال : see جَبَل, in three places.

مَجْبُول, applied to a man, † *Great, large,* or *big,* (K, TA,) *in make;* as though he were a mountain. (TA.)

جبن

1. جَبُنَ, (S, Mṣb, K,) aor. ـُ, inf. n. جُبْنٌ (Mṣb, K) and جُبُنٌ (K) and جَبَانَةٌ; (Mṣb, K;) and جَبَنَ, (S, ISd, Mṣb,) aor. ـُ; (Mṣb, TA;) *He* (a man) *was,* or *became, such as is termed* جَبَان (S, Mṣb, K) and جَبِين; (S, K;) i. e. *cowardly,* (K,) or *weak hearted.* (Mṣb.) And جَبُنَ عَنْهُ *He held back,* or *refrained, from him,* or *it, through cowardice.* (TA in art. عرس.)

2. جَبَّنَهُ, inf. n. تَجْبِينٌ, *He attributed to him cowardice* (جُبْن). (S.) And هُوَ يُجَبَّنُ, inf. n. as above, *He is accused of cowardice.* (K.)

4. اجبنهُ *He found him to be such as is termed* جَبَان; (S, Mṣb, K;) i. e. *a coward,* or *cowardly,* (K,) or *weak-hearted:* (Mṣb:) or *he reckoned him a coward;* (M, K;) as also ▾ اجتبنهُ. (K.)

5. تجبّن *It* (milk) *became like* جُبْن [i. e. *cheese*]. (K.) — And hence, perhaps, (TA,) † *He* (a man) *became thick, gross, coarse,* or *big.* (S, TA.)

8. اجتبنهُ *He made cheese of it;* i. e. of milk. (T, K.) — See also 4.

جُبْن and ▾ جُبُن and ▾ جُبُنّ, (S, Mṣb, K,) the first of which is the most approved, and the last the most rare, and said by some to be used only in a case of necessity in poetry, (Lth, Mṣb,) [*Cheese;*] *a certain thing that is eaten,* (S, Mṣb) *well known:* (K:) n. un. جُبْنَة, (TA,) a word having a more particular signification than جُبْن, (S,) meaning *a* قُرْص [or *round, flattened, loaf*] *thereof,* (Mgh,) [or *a cheese,* or *piece of cheese,*] as also جُبُنَّة (TA) and جُبُنَة. (S, TA.) — Also جُبْن and ▾ جُبُن, [inf. ns. of جَبُنَ, used as simple substs.,] *Cowardice; weak-heartedness;*] the *quality denoted by* جَبَان. (S.)

جُبُن : see جُبْن, in two places.

جُبُنّ : see جُبْن.

جُبْنِيّ *A seller of* جُبْن [i. e. *cheese*]. (TA.) — And a rel. n. from سُوقُ الجُبْنِ [*The cheese-market*] in Damascus. (K.)

جَبَان, (S, Mṣb, K,) an epithet from جَبُنَ, (S,) applied to a man and to a woman, (S, Mṣb, K,) in the latter case like حَصَان and رَزَان, (Ibn-Es-Sarráj, S,) and with ة also applied to a woman; (M, Mṣb, K;) and ▾ جَبِين, (S, K,) from جَبُنَ, (S,) applied to a man and to a woman; and ▾ جُبَّان; (K;) *A coward;* or *cowardly;* i. e. *wont to dread things, so as not to venture upon them boldly,* (K, TA,) *by night or by day;* (TA;) *weak-hearted:* (Mṣb:) جَبَان is *contr. of* شُجَاع: (Mṣb in art. شجع:) pl. masc. جُبَنَاء, (Mṣb, K,) [properly of جَبِينٌ,] and fem. جَبَانَاتٌ. (Lth, Mṣb, TA.) هُوَ جَبَانُ الكَلْبِ [*He is one whose dog is cowardly,*] means ‡ *he is extremely generous:* (Ḳ, TA:) because, by reason of guests' coming to him, his dog does not growl. (TA.) And you say, فُلَانٌ شُجَاعُ القَلْبِ جَبَانُ الوَجْهِ ‡ [app. meaning *Such a one is courageous in heart, mild in face*]. (TA. [Expl. by اجنى الوجه, which seems to be a mistranscription.])

جَبِينٌ: see جَبَانٌ. — Also *The part above the temple, on the right of the forehead,* and *on the left thereof;* the two being called جَبِينَانِ: (Ṣ:) *the side of the forehead,* [so Bḍ in xxxvii. 103,] *from the part over against the place where the hair falls off, to the temple, on the right of the forehead,* and *on the left thereof:* so say Az and IF and others: the forehead (الجَبْهَةُ) is between the جَبِينَانِ: (Mṣb:) or the جَبِينَانِ are *the two borders of the forehead, on either side thereof, in the part between the two eyebrows* (فِيمَا بَيْنَ الحَاجِبَيْنِ [so in the copies of the Ḳ, a mistake for فيما يَلِى الحاجبين *in the part next to the two eyebrows*]), *rising to the place where the growth of the hair terminates:* (Ḳ:) or *between the place where the growth of the hair terminates and the eyebrows:* (TA:) or the جبين is the *borders* (in the T, the *border,* TA) *of the forehead, between the two temples, uniting with the* نَاصِيَة [or *place where the hair grows in the fore part of the head,* or the *hair of that part*]: (Ḳ, TA:) and it sometimes occurs as meaning the *forehead:* (MF, TA:) [see an ex. voce قَرِبَ, where it is used in this last sense, and is fem., perhaps because syn. with جَبْهَة, for] Lḥ says that it is always masc.: (TA:) pl. [of mult.] جُبُنٌ and [of pauc.] أَجْبِنَةٌ (Mṣb, Ḳ) and أَجْبُنٌ. (Ḳ.)

جَبَّانٌ: see جَبَانٌ. — Also *One who keeps,* or *guards, the produce of land in the desert.* (TA.) — See also what next follows.

جَبَّانَةٌ (Ṣ, Mgh, Mṣb, Ḳ) and † جَبَّانٌ, (Ṣ, Mṣb, Ḳ,) the former of which is the more common, (Mṣb,) *A place of prayer,* (Mṣb,) or *common place of prayer,* (Mgh,) *in a* صَحْرَآء [or *desert tract*]. (Mgh, Mṣb.) — *A burial-ground:* (Ḳ:) this is sometimes called جبّانة because the place of prayer is generally in the burial-ground: (Mṣb:) accord. to Kh, these two words, in this sense, are from الجَبُّ and الجَبُوبُ; but others derive them from جبن. (TA in art. جب.) — *A* [*desert tract such as is termed*] صَحْرَآء. (Ṣ, Ḳ.) — *A place that produces much herbage:* and *level, elevated land:* (AḤn, Ḳ:) or the latter, *level, elevated land, that produces much herbage:* (Aboo-Kheyreh, TA:) accord. to ISh, *it is smooth, without trees; but it may have in it hills, and a tract abounding with trees: and sometimes the* جبّانة *is level, without hills and without any tract abounding with trees; but it is not in sand nor in mountains, though it may be in* [*high grounds such as are termed*] قِفَاف [pl. of قُفّ] *and in* [*what are termed*] شَقَائِق [pl. of شَقِيقَةٌ]. (TA.)

جَبَّانِىٌّ *A dweller in the* جَبَّان, meaning صَحْرَآء. (TA.)

أَجْبَنُ [*More,* and *most, cowardly,* or *weak-hearted*]. (TA.) You say أَجْبَنُ مِنْ صَافِرٍ, i. e. [*More cowardly*] *than a whistling bird:* (Ṣ in art. صفر:) or, as some say, *than a thief.* (TA in that art. [See also Freytag's Arab. Prov., i. 326.])

مَجْبَنَةٌ [*A cause of cowardice,* or *weak-heartedness*]. One says, الوَلَدُ مَجْبَنَةٌ مَبْخَلَةٌ [*Children are a cause of cowardice* and *a cause of niggardliness*]; because one loves continuance of life, and property, on account of them. (Ṣ, TA.)

جبه

1. جَبَهَهُ, (Ṣ, Mṣb, Ḳ,) aor. َ, (Mṣb, Ḳ,) inf. n. جَبْهٌ, (Ḥam p. 355,) *He slapped,* (Ṣ,) or *struck,* (Ḳ,) or *hit,* (Mṣb,) or *struck him on,* (Ḥam ubi suprà,) *his* جَبْهَة [or *forehead*]. (Ṣ, Mṣb, Ḳ, and Ḥam ubi suprà.) — And [hence], (Ḳ,) aor. and inf. n. as above, (TA,) ‡ *He turned him back* (Ḳ, TA) *from the thing that he wanted:* (TA:) or ‡ *he met him, or encountered him, with,* or *he said to him,* or *did to him, a thing disliked,* or *hated:* (Ḳ, TA, TḲ:) or ‡ *he encountered him,* or *confronted him, with speech in which was roughness:* (JK, M, TA:) or ‡ *he encountered him with that which he disliked,* or *hated;* or *he accused him thereof to his face;* (M, TA, and Ḥam ubi suprà;) or so جَبَهَهُ بِالمَكْرُوهِ. (Ṣ, TA.) — جَبَهَ المَآءَ, (JK, Ṣ, Ḳ,) inf. n. as above, (JK, Ṣ,) ‡ *He came to the water,* (JK, Ṣ, Ḳ,) *when there was at it no apparatus for drawing,* (Ṣ,) i. e. *no pulley &c.,* (JK, TA,) or *having no means of drawing, so that he only looked at the surface of the water.* (Z, Ḳ.) [See also جَبِيهَةٌ.] — جَبَهَ القَوْمَ ‡ *It* (winter) *came upon the people unprepared for it.* (A, Ḳ.)

2. تَجْبِيهٌ The *mounting a fornicator,* or *an adulterer, upon an ass, and turning his face towards the tail;* whence the trad. of the Jews respecting the adulterer, يُحَمَّمُ وَيُجَبَّهُ وَيُجْلَدُ [*He shall have his face blackened with charcoal, and be mounted on an ass with his face towards the tail, and be flogged*]: (Mgh:) or the *mounting two fornicators,* or *adulterers, upon an ass, and turning the backs of their heads towards each other,* (JK, Mgh,) *and parading them round about;* thus accord. to the Tekmileh: (Mgh:) or the *blackening* [*with charcoal*] *the faces of two fornicators,* or *adulterers,* (أَنْ يُحَمَّرَ in the Ḳ being a mistake for أَنْ يُحَمَّمَ, i. e. يُسَوَّدَ, TA,) *and mounting them upon a camel or an ass, and turning their faces in contrary directions:* accord. to analogy, it should mean *turning their faces towards each other;* for it is from الجَبْهَةُ: تَجْبِيهٌ also signifies the *bending, inclining, lowering,* or *hanging down, the head towards the ground:* and it may be hence; (i. e. it may be hence that it is applied to him who is mounted on a beast in the manner described above; TA;) because he to whom this is done bends down his head in shame and confusion: or it may be from جَبَهَهُ meaning "he did to him a thing disliked, or hated." (Ḳ.)

8. اِجْتِبَاهٌ The *deeming* [one] *rude, coarse, unkind, hard,* or *churlish.* (JK.) — And The *fearing* [a person or thing]. (JK.) — And اجتبهه *He disliked,* or *hated, and feared, it,* or *him;* namely, a country, or town, and a man. (JK.) — *He disapproved it,* or *disliked it, and did not find it wholesome;* namely, water, (En-Nawádir, Ḳ,) &c. (Ḳ.)

جَبَهٌ *Largeness,* (Ṣ,) or *width* (JK, M, Ḳ) *and beauty,* (M, Ḳ,) *of the* جَبْهَة [or *forehead*]: (JK, Ṣ, M, Ḳ:) or *protuberance,* or *prominence, thereof.* (M, Ḳ.)

جَبْهَةٌ [The *forehead;*] the *part of the face which is the place of prostration:* (Aṣ, Mṣb, Ḳ:) or the *even part that is between the eyebrows and the* نَاصِيَة [or *place where the hair grows in the fore part of the head*]: (Kh, JK, Mṣb, Ḳ:) so of a man, (Ṣ, Mṣb,) and of others: (Ṣ:) [or,] of a horse; the *part that is below the ears and above the eyes:* (TA:) pl. جِبَاهٌ. (Mṣb, TA.) — [Hence,] الجَبْهَةُ † *A certain Mansion of the Moon;* [*the Tenth Mansion;*] (Ṣ, Ḳ, Ḳzw;) *consisting of four stars;* (Ṣ;) *the four stars* [ζ, γ, η, and α,] *in the neck and heart of Leo;* [regarded by the Arabs as *the fore-part,* or *forehead, of Leo;*] (Ḳzw in his descr. of Leo;) *between each of which and the next to it is the space of a whip's length; the northernmost of them called by astrologers* الأَسَد. (Ḳzw in his descr. of the Mansions of the Moon.) — Also † *The moon* (Ḳ, TA) *itself:* but [it rather seems to mean the *upper part* of the disc of the moon; for] it is said in the M that a certain unknown poet has metaphorically assigned a جبهة to the moon. (TA.) — Also جَبْهَةٌ, ‡ The *chief* of a people, or company of men; (JK, Mgh, Ḳ, TA;) like as one says the وَجْه thereof. (Mgh, TA.) — ‡ The *generous and manly,* or *manly and noble, persons* of a people, or company of men: or *men exerting themselves in the case of a bloodwit or a debt or other obligation,* (Ḳ, TA,) *or in repairing the condition of a poor man,* (TA,) *and who come not to any one but he is ashamed to turn them back,* (Ḳ, TA,) or *who are seldom or never turned back by anyone:* so, accord. to Aboo-Sa'eed, in a trad. in which it is said that there shall be no poor-rate in the case of the جبهة. (TA.) — † *A company,* or *collected number,* of men, (JK, Ṣ, Mṣb,) and of horses: (JK, Mṣb:) or, of horses, the *best:* (TA:) and [simply] *horses;* (Lth, Ṣ, M, Mgh, Ḳ;) a word having no sing., or n. un.: (M, Ḳ:) accord. to Lth, (TA,) having this last meaning in the trad. above mentioned; (Ṣ, Mgh, TA;) because horses are the best of beasts. (Mgh.) — ‡ *Abjectness,* or *ignominy;* (JK, M, Ḳ, TA;) and *a state of annoyance,* or *molestation:* (Z, TA:) thought by ISd to be from جَبَهَهُ meaning "he encountered him with that which he disliked or hated," or "he accused him thereof to his face;" because the doing this causes one to experience abjectness, or ignominy. (TA.) It is said to have this meaning in a trad., in which it is said, فَإِنَّ ٱللّٰهَ قَدْ أَرَاحَكُمْ مِنَ الجَبْهَةِ وَالسَّجَّةِ وَالبَجَّةِ, i. e. *For God hath relieved you from abjectness,* or *ignominy,* &c. *and milk diluted with water, and blood drawn from a vein* [*of a camel*], which the Arabs used to eat: or in this trad., (TA,) الجِبْهَةُ is the name of *a certain idol* (ISd, Ḳ, TA) *that was worshipped in the Time of Ignorance:* (TA:) and السَّجَّةُ and البَجَّةُ were two idols. (Ṣ and Ḳ in art. سج.)

وَرَدْنَا مَآءً لَهُ جَبْهَةٌ *We came to a water that was salt, so that the drinking thereof did not take away the thirst of our cattle:* (ISk, JK, Ṣ:) or *that was altered for the worse in taste and colour, from some such cause as long standing, though still drinkable;* or *covered with the green substance called* طُحْلُب *and with leaves:* or *that was deep in the bottom, difficult to give to drink.* (ISk, Ṣ.) [See also 1, last meaning but one.]

جُبْهَةٌ i. q. جُبَّأٌ, (Ḳ,) i. e. A *fearful*, or *cowardly*, man. (TA.)

جَابِهٌ, applied to a gazelle, (JK,) or to a bird or a wild animal, (Ḳ,) *That meets one with its face or forehead;* and such is *of evil omen:* (Ḳ:) contr. of قَعِيدٌ. (JK.) — *Coming to water:* so in a saying cited and explained in art. اذن, conj. 2. (IAạr, TA.)

أَجْبَهُ, applied to a man, *Large,* (Ṣ,) or *wide* (JK, M, Ḳ) *and beautiful,* (M, Ḳ,) *in the* جَبْهَة [or *forehead*]: (JK, Ṣ, M, Ḳ:) or *protuberant,* or *prominent, therein:* (M, Ḳ:) applied to a horse, *having a protuberant,* or *prominent, forehead, rising beyond the bone of the nose:* (TA:) fem. جَبْهَآءُ; (Ṣ, Ḳ;) of which the dim. is جُبَيْهَآءُ. (Ṣ.) — الأَجْبَهُ *The lion;* (Ḳ;) because of the width of his جَبْهَة. (TA.)

جبى and جبو

[The words belonging to the former of these two arts. cannot well be classed by themselves, being intimately connected with those of the latter, which are the more numerous and common, and from which they are generally easily distinguishable.]

1. جَبَى, (Ḳ,) first pers. جَبَيْتُ, (Ks, Ṣ, Er-Rághib,) aor. ـِ; (Ḳ;) and جَبَى, aor. ـَ, (Ḳ,) said by MF to be unknown, and also, because neither the second nor the third radical is a faucial letter, unreasonable, but mentioned by Sb, though held by him to be of weak authority, and mentioned also by IAạr, as extr., like أَبَى, aor. يَأْبَى; (TA;) and جَبَا, (TA,) first pers. جَبَوْتُ, (Ks, Ṣ, TA,) aor. ـُ; (TA;) inf. n. [of the first and second] جَبْىٌ (Sh, Ḳ) and [probably of the last only] جَبًا and جِبًا and [of the last, but whether in the first or the second of the senses here following is not shown,] جِبْوَةٌ; (Ḳ; [or this last has a different application, explained below;]) *He collected* water in a trough or tank for beasts &c. [Ks, Ṣ, Ḳ, Er-Rághib.] — And (hence, metaphorically, Er-Rághib, TA) the first of these verbs, (Ṣ, Mgh, Mṣb, Ḳ, Er-Rághib,) and the second, (Ḳ,) and the third; (Ṣ, Mṣb, TA;) inf. n. (of the first, Ṣ, Mṣb) جِبَايَةٌ (Ṣ, Mgh, Mṣb, Ḳ, Er-Rághib) and (of the last, Ṣ, Mṣb) جِبَاوَةٌ (Ṣ, Mṣb, Ḳ) and جُبُوٌّ (TA) [and probably جِبْوَةٌ also, which see above]; ‡ *He collected* the [tax called] خَرَاج, (Ṣ, Mgh, Mṣb, Ḳ, Er-Rághib,) and [other] property. (Mṣb, TA.) The last of these verbs is said in the Ṣ to be originally with ء, though pronounced without ء; but IB says that this is not the case, and that it has not been heard with ء. (TA.) You say also, جَبَاهُ القَوْمَ † [*He collected it from the people,* or *company of men*]; (M, Ḳ,* TA;) and جَبَى مِنْهُمْ † [*He collected from them*]. (M, Ḳ, TA.) — Also جَبَا [or جَبَى], (TA,) first pers. جَبَيْتُ, (Zj, TA,) † *He appropriated* a thing *purely to himself, exclusively of any partner; chose it,* or *took it in preference, for himself.* (Zj, TA.) And hence, (Zj, TA,) ↓ اجتباهُ † *He chose it,* or *selected it,* (Zj, Ṣ, Ḳ,) لِنَفْسِهِ *for himself.* (TA.) — And جَبَى and جَبَا, [originally جَبَأَ,] *He returned, receded, retreated,* or *went back.* (TA.)

2. جَبَّى, inf. n. تَجْبِيَةٌ, *He placed his hands upon his knees,* (Ḳ, TA,) in prayer; (TA;) or *upon the ground:* or *he fell prostrate;* or *fell upon his face:* (Ḳ:) or *he lowered his body and his hands, and raised his buttocks:* (Ḥam p. 801:) [or] تجبية signifies a man's *standing* [*with the hands upon the knees*] *in the manner of the* رَاكِع: (Ṣ, Ḳ:*) accord. to A'Obeyd, what is thus termed is of two kinds: one is the *placing the hands upon the knees, while standing:* the other, the *prostrating oneself,* or *falling upon the face, lying down;* which is سُجُود: (Ṣ:) or the *bending down, and placing the hands upon the knees;* because it is a bringing-together of the limbs. (Mgh.) — جَبَّاهُ, inf. n. as above, meaning *He gave to him,* is vulgar. (TA.)

4. اجبى in the trad. مَنْ أَجْبَى فَقَدْ أَرْبَى is originally أَجْبَأَ [q. v.]: (Ṣ:) accord. to IAth, it is a corruption of the relater, or the ء is suppressed to assimilate the verb to اربى. (TA.) The inf. n., إِجْبَآءٌ, is variously explained, as follows: (TA:) The *selling seed-produce before it shows itself to be in a good state:* (A'Obeyd, Th, Ṣ, Ḳ:) and a man's *hiding his camels from the collector of the poor-rate:* (IAạr, A'Obeyd, Ḳ:) and *i. q.* عِينَةٌ; i. e. the *selling to a man a commodity for a certain price to be paid at a certain period, then buying it of him with ready money for a less price than that for which it was sold.* (TA.)

8. اِجْتِبَآءٌ † The *drawing forth* property *from the places in which it is known,* or *presumed,* or *accustomed, to be.* (TA.) — † The *collecting in the way of choice,* or *selection.* (Er-Rághib, TA.) — See also 1, last sentence but one. — Also † God's *particular,* or *peculiar, distinguishing* of men *by abundant bounty, from which various blessings result to them without their labour;* as happens to prophets and some others. (Er-Rághib, TA.) — Also اجتباهُ, † *He forged it:* and *he extemporized it.* (TA.) Hence, in the Ḳur [vii. 202], قَالُوا لَوْلَا ٱجْتَبَيْتَهَا † *They say, Wherefore hast thou not forged it,* (Fr, TA,) or *produced it,* (Th, TA,) or *invented it,* (Jel,) or *put it together by forgery,* (Bḍ,) *of thyself?* (Fr, Th, Bḍ, Jel, TA:) or *wherefore hast thou not sought it,* or *demanded it,* of God? (Bḍ.)

جَبًا, or جَبَى, (as in different copies of the Ḳ,) written with ا and with ى, (TA,) The camel-waterer's *going in advance of the camels a day before their coming to the water, and collecting for them water in the drinking-trough, and then bringing them to it* (IAạr, Ḳ, TA) *on the morrow.* (IAạr, TA.) [App. an inf. n., of which, in this sense, the verb is not mentioned.] So in the verse,

بِالرَّيْثِ مَا أَرْوَيْتُهَا لَا بِالعَجَلْ
وَبِالجَبَا أَرْوَيْتُهَا لَا بِالقَبَلْ

[*Slowly I satisfied their thirst; not hastily: and by going in advance of them a day, and collecting for them water in the trough, and bringing them to it on the morrow, I satisfied their thirst; not by pouring the water into the trough while they were drinking, without having prepared any for them beforehand*]. (IAạr, TA.) — Also the former, *Water collected* [*in a trough*]; and so ↓ جُبْوَةٌ, with ḍamm. (TA. [See جِبًا.]) — *A watering-trough* (Ḳ, TA) *in which water is collected:* (TA:) or the *station of the drawer of water, upon the* [*upper part of the*] *casing.* (Ḳ.) — The *place where a well is dug:* (Ḳ:) in this sense, and in the next, also written جَبَى. (TA.) — The *brink,* or *margin,* of a well. (Aboo-Leylà, Ḳ.) The *earth that is around a well, that is seen from afar;* (Ṣ;) originally جَبَأٌ: (TA:) *what is around a well:* (Ḳ:) and *what is around a watering-trough:* (TA:) pl. أَجْبَآءٌ. (Ḳ.) — جَبًا meaning *A gift without compensation* is a vulgar word. (TA.)

جِبًا *Water collected* (T, Ṣ, Ḳ) *in a trough,* (T, Ḳ,) *being drawn from a well,* (T, TA,) *for camels;* (Ṣ;) [like جَبًا;] as also ↓ جِبْوَةٌ, (Ḳ, TA, and so in a copy of the Ṣ, but omitted in the CḲ,) or ↓ جُبْوَةٌ, (so in two copies of the Ṣ,) and ↓ جَبْوَةٌ also, (so in a copy of the Ṣ, [see جَبًا,]) and ↓ جِبَاوَةٌ, (Ḳ, TA, and so in a copy of the Ṣ.) and ↓ جَبَاوَةٌ, (Ḳ, TA, and so in another copy of the Ṣ, but omitted in the CḲ,) and ↓ جِبَاءَةٌ: (TA as from the Ḳ, but not in the CḲ: [perhaps a mistranscription for ↓ جِبَايَةٌ:]) but accord. to IAmb, جِبًا is pl. of ↓ جِبْيَةٌ. (TA.)

جِبْوَةٌ: see جِبًا.

جُبْوَةٌ: see جَبًا and جِبًا.

جَبْوَةٌ: see جِبًا. — Also † *A mode,* or *manner, of collecting the* [*tax called*] خَرَاج; and so ↓ جِبْيَةٌ, which Lḥ calls an inf. n. (TA.)

جِبْيَةٌ: see what next precedes: — and see جِبًا.

جِبَاءَةٌ: see جِبًا.

جِبَاوَةٌ and جَبَاوَةٌ: see جِبًا.

جِبَايَةٌ: see جِبًا. — † The *tax called* إِتَاوَة [or خَرَاج]. (TA in art. اتو.) [Originally an inf. n.]

جَبَايَا [a pl. of which the sing. is not mentioned] *Wells which are dug, and in which the shoots of grape-vines are set.* (AḤn, Ḳ.)

جَابٍ A collector of water for camels: belonging to arts. جبو and جبى. (TA.) — † The locust (Ḳ) that collects everything by eating it; as also جَابِئٌ [q. v.]. (TA.) The Arabs say, إِذَا جَآءَتِ السَّنَةُ جَآءَ مَعَهَا الجَابِى وَالجَانِى, i. e. † [When the year of drought comes,] the locust and the wolf [come with it]. (IAạr, TA.)

جَابِيَةٌ A watering-trough, (Ṣ,) or large watering-trough, (Ḳ,) in which water is collected (Ṣ) for camels: (TA:) or a watering-trough that collects water: (Er-Rághib, TA:) pl. جَوَابٍ. (Ṣ.) Hence, in the Ḳur [xxxiv. 12], وَجِفَانٍ كَالْجَوَابِى [And bowls like watering-troughs, or great watering-troughs]. (Ṣ.) — † A company of men. (Ḳ,* TA.)

لُؤْلُؤَةٌ مُجَبَّاةٌ A hollowed pearl: (Ibn-Wahb, TA:) the latter word thought by El-Khaṭṭábee to be formed by transposition from مُجَوَّبَةٌ. (TA.)

جث

1. جَثَّهُ, (Ṣ, Mṣb, TA,) aor. ـُ, (Mṣb, TA,) inf. n. جَثٌّ, (A, Ḳ,) He pulled it up, or out; as also ▼اجتثّهُ; (Ṣ, Mṣb;) or the latter denotes a quicker action than the former; and properly signifies he took its whole جُثَّة [or body]: (TA:) or the former, he cut it; or cut it off: (A, L, Ḳ:) or he cut it off from its root: (L:) or he pulled it up, or out, by the root; namely, a tree: (A, Ḳ:) he uprooted it, or eradicated it. (A.) ═ جَثَّ said of a collector of honey, He took the honey with its جَث and its مَحَارِين, i. e., the bees that had died in it. (IAạr, TA.)

7. انْجَثَّ (M, L, TA) and ▼اُجْتُثَّ (M, A, L, TA) It was, or became, pulled up, or out: properly, its whole جُثَّة [or body] was taken; said of a tree: (TA in explanation of the latter:) it was, or became, cut, or cut off; (A, L;) or cut off from its root; (L;) pulled up, or out, by the root; uprooted, or eradicated: said of a tree. (A.)

8: see 1 and 7.

جَثٌّ, so in the Ṣ [and L] and other lexicons, but in the Ḳ it is implied that it is ▼جُثٌّ, (TA,) Bees' wax: or any particles, of the wings of the bees, (Ṣ, Ḳ,) and of their bodies, (Ṣ,) intermixed with the honey: (Ṣ, Ḳ:) [or] the خُرْشَاء of honey; (Ḳ;) i. e. the young bees, or the wings, that are upon honey; as in the M and L &c.: (TA:) or the bees that have died in the honey. (IAạr, TA.) — Also Dead locusts. (IAạr, Ḳ.)

جُثٌّ: see جَثٌّ. ═ Also Elevated ground (Ṣ, TA, but not in all the copies of the former) such as has a form visible from a distance: (TA:) or ground that is elevated so as to be like a small [hill of the kind called] أَكَمَة. (Ḳ.) ═ The envelope of a fruit; (Ḳ;) [or of the spadix of a palm-tree;] like جُفّ; the ث being a substitute for ف. (TA.)

جُثَّةٌ The body, or corporeal form or figure, (شَخْص,) of a man, (Ṣ, A, Mṣb, Ḳ,) [absolutely, or] sitting, (Ṣ, A, Mṣb,) or sleeping, [by which is meant, as in many other instances, lying down,] (Ṣ, Mṣb,) or reclining, or lying on the side: (TA:) that of a man standing erect being termed طَلَل (Mṣb) or قَامَة; (TA;) and شَخْص applying in common to what is termed جُثَّة and what is termed طلل, in relation to a man: (Mṣb:) or جُثَّة is used only in relation to a man upon a horse's or camel's saddle, wearing a turban: so says IDrd on the authority of Abu-l-Khaṭṭáb El-Akhfash; but he adds that this has not been heard from any other: (TA:) pl. [of mult.] جُثَث (A, TA) and [of pauc.] أَجْثَاث; the latter as though formed from جُثّ, without regard to the augmentative letter [ة]; or it may be pl. of جُثَث, and thus a pl. pl. (TA.) — Also A body; [a corpse;] syn. جَسَد; as in the saying, in a trad., اَللّٰهُمَّ جَافِ الأَرْضَ عَنْ جُثَّتِهِ [O God, remove the earth from his body, or corpse: i. e., let it not press against his sides in the grave]. (TA.) — [And The body of a tree: see 7.]

جَثِيثٌ [a coll. gen. n.] Young palm-trees, or shoots of palm-trees, that are cut off from the mother-trees, or plucked forth from the ground, and planted: n. un. with ة: it is thus called until it yields fruit; when it is called نَخْلَة: (Ṣ:) or what are planted, of the shoots of palm-trees; (AHn, Ḳ;) not what are set, of the stones: (AHn, TA:) or shoots of palm-trees when they are first pulled off from the mother-trees: (Aṣ, TA: [as also قَثِيثٌ:]) or, with ة, it signifies a palm-tree produced from a date-stone, for which a hole is dug, and which is transplanted with the earth adhering to its root: (AA, TA:) or what falls in succession from [app. a mistake for at] the roots, or lower parts, of palm-trees. (Abu-l-Khaṭṭáb, TA.) — And Grapes that fall at the roots, or lower parts, of the vine. (ISd, TA.)

مِجَثَّةٌ and ▼مِجْثَاثٌ A thing with which جَثِيث [q. v.] are uprooted: (M, Ḳ:) an iron implement with which young palm-trees, or shoots of palm-trees, are pulled up or off. (Ṣ.)

مِجْثَاثٌ: see what next precedes.

شَجَرَةٌ مُجْتَثَّةٌ A tree [pulled up, or out, by the root: or] that has no root in the ground. (A.) — بَحْرُ المُجْتَثِّ The fourteenth metre of verse; as though it were cut off from the خَفِيف; (TA;) the metre consisting of

* مُسْتَفْعِلُنْ فَاعِلَاتُنْ فَاعِلَاتُنْ *

(Ḳ. [So originally; but in usage, the last of the three feet is cut off. Accord. to the TA, the first foot is properly written مُسْتَفْعِ لُنْ, as in some copies of the Ḳ.])

جثل

1. جَثُلَ, aor. ـُ; (Lth, Mṣb, Ḳ;) and جَثِلَ, aor. ـَ; (Ḳ;) inf. n. جُثُولَة and جَثَالَة; (Mṣb, Ḳ;) both of the former verb; (Mṣb, TA;) [but both mentioned in the Ṣ as simple substs.;] It was, or became, such as is termed جَثْل, explained below. (Mṣb, Ḳ.)

Q. Q. 4. اِجْثَأَلَّ He (a bird) ruffled his feathers (Ṣ, Ḳ) by reason of the cold. (TA.) — † He (a man) became angry, and prepared himself for fighting. (Ṣ, Ḳ.) — ‡ It (a plant, or herbage,) became tall, and tangled, or luxuriant, or abundant and dense: (Z, Ḳ:) or became tall (اهتزّ, q. v.), and such as might be grasped with the hand. (AZ, Ṣ, Ḳ.) — It (the plumage of a bird) became ruffled. (Ḳ.)

جَثْلٌ (AZ, Ṣ, Mṣb, Ḳ) and ▼جَثِيلٌ (Ḳ) A thing, (Mṣb,) or hair, (AZ, Ṣ, Ḳ,) and ‡ trees (شَجَر), (Ḳ, TA,) much, or abundant, (AZ, Ṣ, Mṣb, Ḳ,) and thick, or coarse, or rough, (Mṣb,) or tangled, or luxuriant and dense, (Ḳ,) and soft: (TA:) or thick, or coarse, or rough, and short: or dense and black: (Ḳ:) or the blackest hair: or the thickest, or coarsest, or roughest: (Lth, TA:) or anything large, big, or bulky, and dense, and tangled, or luxuriant. (Ḳ.) You say لِحْيَةٌ جَثْلَةٌ An abundant, thick, or coarse, or rough, beard. (Mṣb:) or a thick, or dense, beard. (TA.) And نَاصِيَةٌ جَثْلَةٌ A horse's forelock moderate in respect of quantity and length: such is approved. (Ṣ.) And شَجَرَةٌ جَثْلَةٌ ‡ A many-leaved, big, tree. (Ṣ, Ḳ, TA.) ═ جَثْلٌ A species of ant, large and black; as also جُفْلٌ: (IDrd, TA:) or جَثْلَة signifies a black ant: (Ṣ:) or a large ant: and جَثْل is its pl. [or coll. gen. n.]. (Ḳ.)

جَثِيلٌ: see جَثْلٌ.

جُثَالَةٌ Leaves that have become gradually scattered, or strewn. (Ḳ.)

مُجْثَئِلٌّ Broad, or wide. (Ḳ.) — Standing erect. (Ṣ, Ḳ.)

جثم

1. جَثَمَ, aor. ـُ, (Ṣ, Mgh, Mṣb, Ḳ) and ـِ, (Ṣ, Ḳ,) inf. n. جُثُومٌ (Ṣ, Mgh, Mṣb, Ḳ) and جَثْمٌ, (Ḳ,) said of a bird, (Ṣ, Mgh, Mṣb, Ḳ,) and of a hare, and sometimes of a gazelle, (Mṣb,) or of a [young gazelle such as is termed] خِشْف, (Ḳ,) and of a camel, (Mṣb, Ḳ,) and a jerboa, (Ḳ,) and a man, (Ṣ, Ḳ,) He clave to the ground: (Ṣ, Ḳ:) or kept to his place, not quitting it: (Ḳ:) or fell upon his breast: (Mṣb,* Ḳ:) جُثُومٌ in the case of a bird and a hare is like بُرُوك in the case of a camel: (Mṣb:) or in the case of a bird it is like جُلُوس in the case of a man [so that the verb means he sat]. (Mgh.) — Also, (Ḳ,) aor. ـُ, (TA,) said of seed-produce, It rose a little from the ground. (Ḳ,* TA.) — Also, (AHn, Ḳ,) aor. ـِ, (AHn, TA,) inf. n. جُثُومٌ, said of a raceme of a palm-tree, Its unripe, or ripening, dates became somewhat large: (AHn, Ḳ:*) or it became large, and kept its place. (T, TA.) — Also, inf. n. جُثُومٌ, said of the night, ‡ It became half spent. (Th, Ḳ, TA.) ═ جَثَمَ also signifies He collected clay, or mould, and earth, or dust, and ashes. (Ḳ.)

2. جثّم, (Ṣ, Mgh,) inf. n. تَجْثِيمٌ, (KL,) [He made a bird, and a hare, and the like, to cleave to the ground, then to be shot at, or cast at, and so killed: see مُجَثَّمَة:] he kept, or held, a bird confined, that it might [be shot at, or cast at, and] die: (KL:) he turned an animal on his side to be slaughtered. (Golius, as from the KL, but not in my copy of that work.)

5. تجثّم He (a bird) mounted his female for the purpose of copulation. (TA.)

جُثْمُرٌ Seed-produce *rising a little from the ground;* as also ▼ جُثْمُرٌ. (Ḳ,* TA.) —— A raceme of a palm-tree *having its unripe,* or *ripening, dates becoming somewhat large.* (Ḳ,* TA.)

جُثْمُرٌ: see the next preceding paragraph.

جُثْمُرٌ: see جَثَّامَةٌ: —— and جُثَامٌ.

جُثْمَةٌ: see جُثُومٌ.

جُثْمَةٌ *Clay,* or *mould,* and *earth,* or *dust,* and *ashes, collected.* (Ḳ.)

جُثَمَةٌ: see جُثُومٌ.

جُثَمَةٌ: see جَثَّامَةٌ: —— and جُثَامٌ.

جُثْمَانٌ The *body, with the limbs* or *members;* syn. جِسْمٌ: and *i. q.* شَخْصٌ [app. as meaning *a person;* not, as J seems to have held, *a corporeal,* or *material, form* or *thing* or *substance, such as is seen from a distance;* see جِسْمٌ]: (Ḳ:) or, accord. to Aṣ, it has the latter meaning, *i. q.* شَخْصٌ; and جُسْمَانٌ has the former meaning, that of جِسْمٌ (Ṣ, Mṣb) and جَسَدٌ: (Mṣb:) or, accord. to AZ, جُثْمَانٌ is *syn. with* جُسْمَانٌ, (Ṣ, Mṣb,) i. e. جَسَدٌ. (Ṣ.) One says, مَا أَحْسَنَ جُثْمَانَ الرَّجُلِ and جُسْمَانَهُ, meaning [*How goodly,* or *beautiful, is*] *the body,* or *person, of the man!* (AZ, Ṣ.) J cites, as an ex. of this word in the sense of شخص, from a verse of Bishr,

* سَنَامٌ كَجُثْمَانِ البَنِيَّةِ أَتْلَعَا *

observing that by البنيّة is meant the Kaabeh: but IB says that the right reading, as found in his poetry, is البَلِيَّةِ, and أَتْلَعَ; and that the meaning is, *A* [*long*] *hump like the* جثمان [or *body*] *of the she-camel that is placed* [*and confined without food or water until she dies*] *at the grave of a dead man.* (TA.) One says also, جَآءَنَا بِثَرِيدٍ كَجُثْمَانِ القَطَاةِ [*He brought us crumbled bread moistened with broth and piled up, like the body of the bird of the kind called* قطا]. (Ṣ.)

جُثْمَانِيَّةُ المَآءِ, as used in the saying of El-Farajeeyeh, (Ḳ,) so in the copies of the Ḳ, [or El-Faraheeyeh, accord. to the CḲ,] but correctly of El-Farezdaḳ, (TA,)

* وَبَاتَتْ بِجُثْمَانِيَّةِ المَآءِ نِيبُهَا *

* إِلَى ذَاتِ رَحْلٍ كَالْمَآتِمِ حُسَّرَا *

means *The water itself:* or *the middle thereof:* or *the place where it collected.* (Ḳ, TA.) [The poet says, *And her aged she-camels passed the night in the water,* &c., ... *like the companies of mourning women having the head,* or *the face,* &c., *uncovered:* but what is meant by الى ذات رحل, unless it be *with one having a saddle upon her,* (إِلَى) being sometimes used in the sense of مَعَ,) I am unable to conjecture. In the CḲ, نِيبُهَا and كَالْمَآتِمِ are erroneously put for نَيْبُهَا and كَالْمَآتِمِ.]

جُثَامٌ *Incubus,* or *nightmare;* (T, Ḳ;) *what comes upon a man when he is sleeping;* (T, TA;) *what comes upon a man in the night, preventing him from speaking;* *i. q.* نَيْدُلَانٌ; (IAạr, TA;) as also ▼ جَاثُومٌ (T, Ḳ) and ▼ جَثَّامَةٌ and ▼ جُثْمُرٌ and ▼ جُثَمَةٌ. (T, TA.)

جُثُومٌ: see جَاثِمٌ. —— Hence, † *A hare.* (TA.)

جُثُومٌ pl. of جَاثِمٌ [q. v.]. (TA.) —— Also *A hill such as is called* أَكَمَةٌ; and so ▼ جُثْمَةٌ (Ḳ) and ▼ جُثَمَةٌ. (TA.)

جُثَامٌ: see جَاثِمٌ.

جَثَّامَةٌ: see جَاثِمٌ. —— Hence, ‡ A man *who keeps to the region of cities, towns, villages,* or *cultivated land, and does not travel:* (Mṣb:) a man *who sleeps much, and does not travel;* as also ▼ جُثَمَةٌ (Ṣ, Ḳ) and ▼ جُثْمُرٌ and ▼ جَاثُومٌ: (Ḳ:) [see also جَاثِمَةٌ:] *stupid, dull, wanting in intelligence;* or *not penetrating, sharp, vigorous,* or *effective, in the performing of affairs:* and *a forbearing,* or *clement, personage, chief,* or *man of rank* or *quality.* (Ḳ.) —— See also جُثَامٌ.

جَاثِمٌ A bird, (Mṣb, Ḳ,) and a hare, and sometimes a gazelle, (Mṣb,) or a [young gazelle such as is termed] خِشْف, (Ḳ,) and a camel, (Mṣb, Ḳ,) and a jerboa, and a man, (Ḳ,) *cleaving to the ground:* or *keeping to his place, not quitting it:* (Ḳ:) or *falling upon his breast:* (Mṣb,* Ḳ:) as also ▼ جُثُومٌ: (Ḳ:) [or the latter] and ▼ جُثَامٌ *doing so much,* or *often:* and ▼ جَثَّامَةٌ *doing so very much,* or *very often:* (Mṣb:) and the first, also, *sitting upon his legs like a bird:* pl. جُثُومٌ (TA) [and جُثَّمٌ, accord. to Freytag]. فَأَصْبَحُوا فِي دَارِهِمْ جَاثِمِينَ, in the Ḳur [vii. 76, &c.], means [*And they became, in their abode,*] *bodies cast upon the ground:* (TA:) or *extinct,* or *motionless;* and *dead.* (Bḍ.) —— الجُثُومُ ‡ *The stars composing the constellation of the Scorpion;* also called البُرُوكُ: see بَرَكَ. (L and TA in art. برك.)

جَاثِمَةٌ One *who does not quit his house,* or *tent.* (Lth, TA.) [See also جَثَّامَةٌ.]

جَاثُومٌ: see جَثَّامَةٌ: —— and جُثَامٌ.

مَجْثَمٌ [and مَجْثِمٌ *A place where a bird,* &c., *cleaves to the ground:* or *to which it keeps:* or *where it falls upon its breast.* And particularly,] The *seat,* or *form,* of a hare: (TA:) [pl. مَجَاثِمُ.]

مُجَثَّمَةٌ *A bird,* and *a hare,* and *the like, that is confined,* or *set up, to be killed;* (A 'Obeyd, Ṣ, Mgh;) *that is made to cleave to the ground* (تُجَثَّمُ), *and then shot at,* or *cast at, until it is killed;* (Ṣ, Mgh;) which manner of killing is forbidden: (Ṣ:) or *any animal that is set up and shot at,* or *cast at, and* [*so*] *killed:* (A'Obeyd, TA:) or *a sheep,* or *goat, that is shot at with arrows:* ('Ikrimeh, Mgh:) or *a sheep,* or *goat, that is stoned* (Sh, Mgh, TA) *until it dies, and is then eaten.* (TA.)

جثى and جثو

1. جَثَا and جَثَى (Ṣ, Mṣb, Ḳ) عَلَى رُكْبَتَيْهِ, (Ṣ, Mṣb,) aor. ـُ and ـِ, inf. n. جُثُوٌّ and جُثِيٌّ, (Ṣ, Mṣb, Ḳ,) *He sat upon his knees;* (Ḳ, TA;) for the purpose of contention or disputation, or the like: (TA:) or جثا [and جثى] *he kneeled; put himself in a kneeling posture;* which is the mode of sitting of him who is contending or disputing: (AZ, Ḥar p. 512:) [or *he put down his knees upon the ground and raised his buttocks;* i. e. *he kneeled with his body and thighs erect,* or *nearly so:* see جَاثٍ:] and جَثَا لِرُكْبَتَيْهِ *he fell* [*upon his knee*]; and جَثَوْا لِلرُّكَبِ [*they fell upon the knees, and sank backwards so as to rest the body upon the heels* or *upon the left foot bent sideways beneath;* for] جُثُوٌّ is the *manner of sitting of the* مُتَشَهِّد [*in prayer*]: (Ḥam p. 287:) or جَثَا, (Ḳ, TA,) inf. n. جَثْوٌ and جُثُوٌّ, (TA,) *he stood upon the extremities of his toes;* (Ḳ;) like جَذَا; from which AO reckons it to be formed by substitution [of ث for ذ]; but IJ says that they are two dial. vars. (TA.) Aboo-Thumámeh says,

* أُخَاصِمُهُمْ مَرَّةً قَائِمًا * وَأَجْثُو إِذَا مَا جَثَوْا لِلرُّكَبْ *

[*I contending,* or *disputing, with them one time standing, and falling upon my knees when they fell upon their knees*]. (Ḥam p. 287.) —— جَثَوْتُ, (Ṣgh, Ḳ,) inf. n. جَثْوٌ; (TA;) and جَثَيْتُ, (Ṣgh, Ḳ,) inf. n. جَثْيٌ; (TA;) *I collected* camels, and sheep or goats. (Ṣgh, Ḳ.)

3. جَاثَيْتُ رُكْبَتِي إِلَى رُكْبَتِهِ, (Ḳ, and so in some copies of the Ṣ,) or جَاثَيْتُهُ رُكْبَتِي إِلَى رُكْبَتِهِ; (so in other copies of the Ṣ,) [*I sat,* or *sat with him, with my knee to his knee, each of us sitting upon his knees,* in contending or disputing: see 1]: and جَاثَيْتُهُ alone, (Ṣ voce حَاضَرْتُهُ,) [signifies the same,] inf. n. مُجَاثَاةٌ (Ḳ and TA voce مُحَاضَرَةٌ) [and جِثَآءٌ: see also 6].

4. اجثاه (Ṣ, Ḳ) *He made him to sit upon his knees:* [see 1:] or *he made him to stand upon the extremities of his toes.* (Ḳ.)

6. تَجَاثَوْا عَلَى الرُّكَبِ [*They sat together upon their knees*], (Ṣ, Ḳ,) in contending or disputing; inf. n. مُجَاثَاةٌ and جِثَآءٌ, which are [properly inf. ns. of 3, but are] thus used as inf. ns. of a verb to which they do not conform. (TA.) —— التَّجَاثِى فِي إِشَالَةِ الحَجَرِ is like التَّجَاذِى [*The vying, one with another, in lifting the stone, for trial of strength*]. (TA.)

جُثًا, or جُثًى, [pl. of جُثْوَةٌ, q. v. —— Also] *A company,* or *congregated body,* of men; (TA;) or so ▼ جُثْوَةٌ (Bḍ in xlv. 27) [or ▼ جِثْوَةٌ]: and *companies,* or *congregated bodies,* thereof. (TA.) It has the former meaning in a trad., where it is said, يَصِيرُونَ يَوْمَ القِيَامَةِ جُثًا كُلُّ أُمَّةٍ تَتْبَعُ نَبِيَّهَا [*They shall become, on the day of resurrection, a company,* or *congregated body, each people following its prophet:* or here the pl. meaning is more reasonable]: and the latter in the trad., فُلَانٌ مِنْ جُثَى جَهَنَّمَ [*Such a one is of the companies,* or *congregated bodies, of Hell,* or *Hell-fire*], accord. to one recital: otherwise, ▼ مِنْ جُثِيِّ جَهَنَّمَ *of those that sit upon the knees therein.* (TA.) —— الجُثَا is also said to have been *A certain idol, to which sacrifices were performed.* (TA.)

جُثْوَةٌ: see what next follows, in two places: —— and see جُثًا.

جُثْوَةٌ and ↓ جِثْوَةٌ and ↓ جَثْوَةٌ *Stones collected together:* (Ṣ, Ḳ:) or the *stones of earth collected together like the* [*mound over a*] *grave:* and the first, *a hillock:* or *a heap of earth:* (TA:) or *collected earth:* (Ḥam p. 399:) or *a quantity collected* of earth &c.: (Ḥam p. 381:) and (hence, Ḥam p. 381) *a grave:* (TA, Ḥam pp. 381 and 399:) pl. جُثًا, (TA, Ḥam p. 399,) or جُثِىٌّ. (Ḥam ib.) It is said in a trad., رَأَيْتُ قُبُورَ الشُّهَدَآءِ جُثًا *I saw the tombs of the martyrs* [to be] *collections of earth.* (TA.) And جُثَى الحَرَمِ (pl. of جُثْوَةٌ, TA) and جِثَى الحَرَمِ (pl. of ↓ جِثْوَةٌ, TA) signify *What are collected, in the sacred territory, of the stones of the* جِمَار [or *pebbles cast at Minè*]: (Ṣ:) or this is a mistake; (Ḳ;) pointed out by Ṣgh in the TṢ: (TA:) the meaning is, *what are collected together of the stones that are set* [*in heaps*] *at the limits of the sacred territory:* or *the* أَنْصَاب [or *stones set up around the Kaabeh*] *upon which victims were slain in sacrifice.* (Ḳ, TA.) — Also *i. q.* جَذْوَةٌ [*A live coal;* or *piece of fire;* &c.]: (Ḳ:) or so ↓ جِثْوَةٌ and ↓ جَثْوَةٌ: (TA:) or جُثْوَةٌ مِنْ نَارٍ: (Fr, TA:) asserted by Yaạḳoob to be formed by substitution [of ث for ذ]. (TA.) — And The *middle* [of a thing]. (IAạr, Ḳ, TA: but omitted in the CḲ and in a MS. copy of the Ḳ.) — And The *body, with the limbs* or *members;* syn. جَسَدٌ: (Ḳ:) or so جُثْوَةٌ: pl. جُثًى. (Sh, TA.) — And جُثْوَةٌ, *A great,* or *large, man.* (ISh, TA.) — See also جُثًا.

جِثْوَةٌ: see جُثْوَةٌ, in three places.

جُثَآءٌ *i. q.* شَخْصٌ [app. as meaning *A person;* or the *body* of a man, like جُثْوَةٌ and جُثَّةٌ]; as also ↓ جُثَآءٌ. (Sgh, Ḳ.) — [And hence, perhaps,] *Incubus,* or *nightmare.* (TA. [But in this sense it is written in the TA جثا, without ء, and without any syll. sign.]) — Also *i. q.* جَزَآءٌ [*Requital,* or *compensation*]. (Ḳ.) — And *Quantity, measure, size, bulk,* or *extent;* and *amount, sum,* or *number,* (Ḳ, TA,) as, for instance, of a people, or company of men. (TA.)

جُثَآءٌ: see the next preceding paragraph.

جَاثٍ *Sitting upon his knees:* or *standing upon the extremities of his toes:* (Ḳ:) and [simply] *sitting:* or [*kneeling with his body and thighs erect,* or *nearly so;* i. e.] *putting down his knees* [*upon the ground*] *and raising his buttocks:* (TA:) [see also 1, of which it is the part. n.:] pl. جُثِىٌّ and جِثِىٌّ; (Ḳ;) or these may be pls., like بُكِىٌّ and بِكِىٌّ, pls. of بَاكٍ; or inf. ns. used as epithets [as is indicated in the Ṣ]. (Er-Rághib, TA.) You say قَوْمٌ جُثِىٌّ [*A company of men sitting upon their knees*]; (Ṣ, Mṣb;) like as you say قَوْمٌ جُلُوسٌ and جَلَسَ جُلُوسًا. (Ṣ.) And hence, in the Ḳur [xix. 73], وَنَذَرُ الظَّالِمِينَ فِيهَا جُثِيًّا, and also, with kesr to the ج because of the kesr of the letter following it, جِثِيًّا, [*And we will leave the wrongdoers therein, sitting upon their knees.*] (Ṣ.) And فُلَانٌ مِنْ جُثَى جَهَنَّمَ: see جُثًا. (TA.) And, in the Ḳur [xlv. 27], وَتَرَى كُلَّ أُمَّةٍ جَاثِيَةً (TA) *And thou shalt see every people sitting upon the knees,* (Bḍ, Jel,) *in an upright posture, not at ease:* (Bḍ:) or *congregated;* (Bḍ, Jel;) from جُثْوَةٌ signifying "a company," or "congregated body." (Bḍ.) Whence, (TA,) سُورَةُ الجَاثِيَةِ *The* [*forty-fifth*] *chapter, of the Ḳur-án, next after that called* الدُّخَان. (Ṣ, TA.) — [الجَاثِى عَلَى رُكْبَتَيْهِ, or الجَاثِى, † *The constellation Hercules.*]

[مَجْثًى *A place of sitting upon the knees.*]

جح

R. Q. 1. جَحْجَحَ *He mentioned a* جَحْجَاح [or *chief,* &c.] *of his people.* (TA.) — جَحْجَحَتْ *She gave birth to a* جحجاح. (TA.) And جحجحت بِوَلَدِهَا *She brought forth her child a* جحجاح. (A.)

جَحْجَحٌ: see the following paragraph. — Also *A low,* an *ignoble,* a *vile,* a *mean,* or a *sordid, man; possessing no manly qualities.* (AA, T, Ḳ.) [Thus it bears two contr. significations.]

جَحْجَاحٌ (Ṣ, A, Ḳ) and ↓ جَحْجَحٌ (Ḳ) *A chief, lord, master,* or *man of rank* or *quality* or *distinction:* (Ṣ, A, Ḳ:) or one *who is liberal, bountiful,* or *munificent;* or one *who is noble,* or *generous:* (TA:) and one *who hastens* [*to render aid*] *in cases of evil:* (A:) pl. (of the former, Ṣ, A, TA, [or of the latter accord. to analogy,]) جَحَاجِحُ (Ṣ, A, Ḳ) and (of the same, A, TA) جَحَاجِحَةٌ (A, Ḳ) and جَحَاجِيحُ: (Ḳ:) it is said in the Ṣ that these two are pls. of the first pl.; the ة in the former of them being substituted for the ى in the latter of them, which is rejected; for one of these two letters must be retained, but both cannot be together: this assertion in the Ṣ, however, is well refuted by MF. (TA.)

جحد

1. جَحَدَ, aor. ـَ, inf. n. جَحْدٌ and جُحُودٌ, *He denied* a thing; *disacknowledged it;* (L, MF;) in an absolute sense, whether knowing it to be otherwise than as he represented it to be or not. (MF.) [It is used by grammarians, and often by others, as relating to something past, or supposed or asserted to be past; and thus, in a more restricted sense than نَفَى.] You say, جَحَدَهُ حَقَّهُ, and بِحَقِّهِ, inf. ns. as above; [and ↓ جاحدهُ; (see 3 in art. كبر, where جَاحَدَهُ is used in explaining كَابَرَهُ; and see what follows;)] *He denied,* or *disacknowledged, his right,* or *due, knowing it to be such,* (Ṣ, A,* Mṣb, Ḳ, MF,) and also, *not knowing it;* (MF;) the doing of which is also termed مُكَابَرَةٌ: (TA:) but accord. to some, it is made trans. by means of ب only by its being made to imply the meaning of كَفَرَ. (MF.) — Also جَحَدَهُ, *He found him to be niggardly,* or *avaricious:* (Ḳ:) or *he found him to possess little good;* i. e., *to be* either *niggardly* or *poor.* (TA.) — جَحِدَ, (Ṣ, Ḳ,) aor. ـَ, (Ḳ,) inf. n. جَحَدٌ, (Ṣ,) *He* (a man) *was,* or *became, niggardly,* or *avaricious;* (Ṣ;) *possessed little good;* (Ṣ, Ḳ;) as also ↓ اجحد: (Ṣ:) or *his property became dissipated* or *dispersed, and passed away;* and so ↓ the latter verb. (AA, TA.) — *It* (anything, TA) *was,* or *became, little in quantity,* or *scanty.* (Ḳ, TA.) — *It* (a person's life, TA) *was,* or *became, strait, and difficult.* (Ḳ,* TA.) — *It* (a plant) *was,* or *became, scanty;* (Ṣ;) *did not grow tall.* (Ṣ, Ḳ.) — جَحِدَتِ الأَرْضُ *The land became dry, and of no good.* (L.) — جَحِدَ عَامُنَا [*Our year was,* or *became, one of little rain:* see جَحِدٌ]. (A.)

3: see 1.

4: see 1, in two places.

جَحْدٌ and ↓ جُحْدٌ and ↓ جَحَدٌ *Paucity,* or *scantiness, of good;* (Ṣ, Ḳ;) which means both *niggardliness* and *poverty:* (A:) *straitness of the means of subsistence;* as also ↓ جُحُودٌ. (TA.) One says, ↓ نَكَدًا لَهُ وَجَحَدًا (Ṣ) and نُكْدًا لَهُ ↓ وَجُحْدًا (L in art. نكد) [May God decree *straitness,* or *difficulty, to him, and poverty*]: a form of imprecation. (TA.) — جَحْدٌ as an epithet, fem. with ة: see جَحِدٌ, in three places.

جُحْدٌ: } see جَحْدٌ, in four places.
جَحَدٌ: }

جَحِدٌ (Ṣ, Ḳ) and ↓ جَحْدٌ and ↓ أَجْحَدُ (Ḳ) A man *niggardly,* or *avaricious;* (Ṣ;) *possessing little good.* (Ṣ, Ḳ.) [Hence,] ↓ أَرْضٌ جَحْدَةٌ *Dry land, in which is no good.* (L.) And عَامٌ جَحِدٌ, (Ṣ,) or ↓ جَحْدٌ, (A,) *A year in which is little rain.* (Ṣ.) — Also جَحِدٌ, *A thick and short horse:* fem. with ة: pl. جِحَادٌ. (Ḳ.)

جُحُودٌ: see جَحْدٌ.

جُحَادٌ (applied to a man, TA) *Slow in emitting his seminal fluid;* syn. بَطِىْءُ الإِنْزَالِ. (Ḳ.)

أَجْحَدُ: see جَحِدٌ.

جحر

1. جَحَرَ, (A, Ḳ,) aor. ـَ; (Ḳ;) and ↓ انجحر, (Ṣ, Mṣb, Ḳ,) and ↓ تجحّر, (Ḳ,) and ↓ استجحر, (A,) said of a [lizard of the kind called] ضَبّ, (A, Mṣb, Ḳ,) and of a jerboa, and of a serpent, (Mṣb,) [&c., (see جُحْرٌ,)] *It entered its burrow,* or *hole;* (Ṣ, A, Ḳ;) *betook itself to it for refuge;* or *resorted to it.* (Mṣb.) — [Hence,] جَحَرَتْ عَيْنُهُ [and ↓ تجحّرت or ↓ انجحرت (see جُحْرَآءُ)] ‡ *His eye sank,* or *became depressed, in his head.* (Ṣ, A, Ḳ.) — جَحَرَتِ الشَّمْسُ لِلْغُيُوبِ † [*The sun set,* or *became near to setting*]. (TA.) — And جَحَرَتِ الشَّمْسُ † *The sun rose high,* (Ḳ,) *so that the shade receded and contracted.* (TA.) — جَحَرَ said of a man, † *He retreated,* or *retired; remained behind;* or *held back.* (Ṣ, TA.) — جَحَرَ الرَّبِيعُ ‡ *The* [rain called] ربيع *withheld itself:* (A:) [or] *the* [season called] ربيع *did not give us rain.* (Ḳ.) — And جَحَرَ عَنَّا الخَيْرُ † *Good,* or *prosperity, kept back from us,* (Ḳ,* TA,) *and did not betide us.* (TA.) — See also 4.

4. اجحرهُ *He made it* (a [lizard of the kind called] ضَبّ [&c.]) *to enter its burrow,* or *hole;*

(Ṣ,* Ḳ;) as also ▼ جَحْرَهُ : (Ḳ:) and *it* (rain) *constrained it* (a ضَبّ [&c.]) *to enter its burrow,* or *hole.* (A,* TA.) — [Hence,] أَجْحَرَهُمُ الفَزَعُ ‡ [*Fright drove them into their dwellings*]. (A.) — اجحرت السَّنَةُ النَّاسَ ‡ *Drought,* or *dearth, brought the people into strait,* or *narrow, circumstances.* (A.) — And اجحرهُ إِلَى كَذَا † *He constrained him,* or *compelled him, to have recourse to,* or *to betake himself to,* or *to do, such a thing.* (Ḳ,* TA.) = اجحر القَوْمُ ‡ *The people,* or *company of men, entered upon a time of drought,* (Ḳ,) *and difficulty.* (TA.) — اجحرت النُّجُومُ ‡ *The stars* (i. e. the stars of winter, TA) *occasioned no rain.* (Ḳ, TA.)

5: } see 1; each in two places.
7: }

8. اجتحر لِنَفْسِهِ جُحْرًا *It* [a ضَبّ &c.] *made for itself a burrow,* or *hole.* (Ṣ, Ḳ.)

10: see 1.

جَحْرٌ *A deep-bottomed cavern.* (Ḳ.)

جُحْرٌ (Ṣ, A, Mgh, Mṣb, Ḳ) and ▼ جُحْرَانٌ (Ṣ, Ḳ) The *burrow,* or *hole,* (M, Ḳ,) of a [lizard of the kind called] ضَبّ, (A, Mgh, Mṣb, MF,) and ‡ of a jerboa, and ‡ of a serpent, (Mgh, Mṣb, MF,*) and † of any venomous reptile or the like, and wild beast, (M, Ḳ,) or of any creature that is not of a large size; (TA;) and [the *den*] of a hyena: (Ḳ in art. وجر:) pl. [of mult.] of the former, جِحَرَةٌ (Ṣ, Mṣb, Ḳ [in the CḲ جَحَرَةٌ]) and [of pauc.] أَجْحَارٌ. (Ṣ, Ḳ.) — And [hence,] the former, (A,) or ▼ the latter, or both, (IAth, TA,) ‡ The *vulva* of a woman; the *pudendum muliebre:* and ‡ the *anus.* (IAth, TA.) — You say, حَصِّنِى جُحْرَكِ ‡ [*Protect thou* (O woman) *thy pudendum*]. (A.) And it is said in a trad., (by 'Áïsheh, A,) ▼ إِذَا حَاضَتِ المَرْأَةُ حَرُمَ الجُحْرَانُ (Ṣ, TA) ‡ *When a woman has the menstrual discharge, the vulva is forbidden:* (TA:) or, (accord. to one reading, TA,) حَرُمَ الجُحْرَانِ, i. e. *both* (A) *the vulva and the anus* (TA) *are forbidden;* (A, TA;) one having been forbidden before. (TA.) — [Hence likewise,] the former signifies also † *A hole,* or *aperture,* (ثُعْلُب,) *whence rainwater flows.* (Ḳ and TA in art. ثعلب.)

جَحْرَةٌ (Ṣ, Ḳ) and جَحَرَةٌ (Ḳ) † *A severe, hard,* or *distressful, year;* (Ṣ, Ḳ;) *one of drought, dearth,* or *unfruitfulness,* (Ḳ,) and *of little rain;* because it drives the people into the tents, or houses. (TA.)

عَيْنٌ جَحْرَاءُ i. q. ▼ مُتَجَحِّرَةٌ or ▼ مُنْجَحِرَةٌ, (accord. to different copies of the Ḳ,) i. e. ‡ *An eye deep,* or *depressed, in its socket.* (TA.) It occurs in a trad., in a description of Ed-Dejjál; but Az says that [in this instance] it is correctly جَخْرَاءُ, with خ. (TA.)

جُحْرَانٌ: see جُحْرٌ, in three places.

جَاحِرٌ [*Entering a burrow,* or *hole:* and also] † *remaining behind, not having come up to others;* (Ḳ, TA;) applied to a horse or the like, &c. (TA.) جَوَاحِرُ [is its pl., signifying] *Entering into burrows,* or *holes,* (Ṣ, Ḳ,) and *hiding-places:* (Ṣ:) *entering secretly into* [*their*] *habitations:* (KL:) and also † *remaining behind;* applied to wild animals &c. (TA.)

مَجْحَرٌ, (Ḳ,) pl. مَجَاحِرُ, (Ṣ, A,) ‡ *A hiding-place;* (Ṣ, A, Ḳ;) *a place of refuge.* (Ḳ.)

عَيْنٌ مُتَجَحِّرَةٌ or مُنْجَحِرَةٌ: see جَحْرَاءُ.

جحش

1. جَحَشَهُ, (Ks, Ṣ, Mgh, Ḳ,*) aor. ـَ, (Mgh, Ḳ,) inf. n. جَحْشٌ, (Ṣ, Ḳ,) *It scratched it,* or *the like,* (namely the skin, Ṣ, Mgh, Ḳ, or a man's side, Ks,) *so as to abrade the surface,* (Mgh, Ḳ,) or *so as to abrade the skin;* (Ks;) syn. سَحَجَهُ, (Ks, Ḳ,) and قَشَرَهُ: (Mgh, Ḳ:) or i. q. خَدَشَهُ: or it signifies *more than this last:* (Ks, Ḳ:) or *less than this last:* (Lth, Ḳ:) and *it* (an arrow) *made a mark upon it;* [or *grazed it;*] namely, a wall. (Mgh.) You say, أَصَابَهُ شَىْءٌ فَجَحَشَ وَجْهَهُ [*A thing struck him, and abraded the surface of the skin of his face*]: and بِهِ جَحْشٌ [*in him, or it, is an abrasion of the skin*]: (Ṣ, TA:) or جَحْشٌ is not in the face, nor [anywhere] in the body [except in the side]. (L, TA.) It is said in a trad., respecting Moḥammad, سَقَطَ مِنْ فَرَسٍ فَجُحِشَ شِقُّهُ *He fell from a horse, and the skin of his side was scratched,* or *lacerated,* or *abraded.* (Mgh,* TA.) [See also مَجْحُوشٌ.]

جَحْشٌ *A young ass;* (Ṣ, Mṣb, Ḳ;) *domestic* and *wild:* or *before it becomes big:* (TA:) or *from the time when it is brought forth until it becomes big from sucking:* when it has completed the year, it is called تَوْلَبٌ: (Aṣ:) [or the latter is applied to a wild ass of that age:] pl. [of pauc.] أَجْحَاشٌ (so in a copy of the Ṣ) and [of mult.] جِحَاشٌ and جِحْشَانٌ (Ṣ, Mṣb, Ḳ) and جُحُوشٌ (Mṣb) and جِحَشَةٌ: (Aṣ, TA:) [dim. جُحَيْشٌ:] and fem. جَحْشَةٌ. (Ṣ, Ḳ.) It is said in a prov., الجَحْشَ لَمَّا بَذَّكَ الأَعْيَارُ, (A, TA,) i. e., Seek thou, or pursue thou, *the young ass when the full-grown asses outstrip thee:* applied to him who seeks much, and it escapes him; so one says to him, Seek thou less than that. (TA.) [Meyd gives فَاتَكَ in the place of بَذَّكَ.] — Also ‡ *A mare's colt;* (A, Ḳ;) as being likened to a young ass. (TA.) — And ‡ *A gazelle;* (Ibn-'Abbád, Ḳ;) in the dial. of Hudheyl: (TA:) or *a young gazelle;* (A, TA;) in that dial.; occurring in a poem of Aboo-Dhu-eyb; but accord. to one relation, the word there is خِشْف. (TA.)

جَحِيشٌ The *side,* (Ḳ, TA,) of a man: (TA:) and *a lateral,* or *an adjacent, part,* or *place,* or *tract.* (Sh, Ḳ.) You say, أُصِيبَ جَحِيشُهُ *His side was hit,* or *hurt.* (TA.) And نَزَلَ فُلَانٌ الجَحِيشَ *Such a one alighted in the adjacent part* or *tract.* (TA.) = A man *who retires to a distance, apart from others:* (Ṣ:) *who alights apart from others, and does not mix with them:* (IDrd, Ḳ:) *who lives alone, with none to incommode him in his house.* (AḤn.) You say, نَزَلَ فُلَانٌ جَحِيشًا *Such a one alighted alone; apart from others.* (TA.)

جُحَيْشٌ [dim. of جَحْشٌ]. You say, هُوَ جُحَيْشُ وَحْدِهِ ‡ *He is one who follows his own opinion only,* (Ṣ, A, Ḳ,) *who has his gain to himself exclusively,* (TA,) *and does not consult others, nor mix with them;* (Ḳ, TA;) as also عُيَيْرُ وَحْدِهِ; [q. v.;] meaning dispraise; (Ṣ, A, TA;) the man being thereby likened to a [little] young ass. (TA.)

بَيْتٌ جَاحِشٌ [*A tent*] *apart from the tribe.* (TA.)

مَجْحُوشٌ One *whose side* (جَحِيشُهُ, TA) *is hit,* or *hurt.* (Ḳ, TA.)

جحظ

1. جَحَظَتْ عَيْنُهُ, aor. ـَ, (Ṣ, Ḳ,) inf. n. جُحُوظٌ (T, Ṣ, TA) and جِحَاظٌ, (M, TA,) *His eyeball, the globe of his eye, was prominent* (T, M, Ḳ, TA) *and apparent:* (TA:) or *was large* (Ṣ, Ḳ, TA) *and prominent;* (Ṣ, TA;) *as though a large pearl came forth from the eyelids.* (Jm, TA.) = جَحَظَ إِلَيْهِ عَمَلَهُ ‡ *He looked into,* or *examined, his deed, and saw the evil that he had done:* (Ḳ:) and it may mean *he looked into his face, and reminded him of the evil of his deed.* (Az, TA.) The Arabs also say, لَأُجَحِّظَنَّ إِلَيْكَ أَثَرَ يَدِكَ, meaning † *I will assuredly show thee the evil of the effect of thy hand.* (Az, TA.)

2. جحّظ, inf. n. تَجْحِيظٌ, *He looked sharply,* or *intently.* (Ḳ.)

جَاحِظَتَانِ: see جَحْظَتَانِ.

جَحْظَمٌ: see جَاحِظٌ.

جِحَاظٌ The *part* [*which is next below,* or *around, the eye, and*] *which is called the* مَحْجِر *of the eye.* (IDrd, Az, L, Ḳ.) — And, (Az, Ḳ,) in one copy [of the work of IDrd, i. e. the Jm,] (Az,) The *edge of the gland of the penis.* (Az, Ḳ.) — جِحَاظَانِ: see جَاحِظَتَانِ.

جَاحِظَتَانِ: see جِحَاظَتَانِ.

جَاحِظٌ A man *having the eyeball,* or *globe of the eye, prominent and apparent;* (TA;) or *large and prominent;* (Ṣ, TA;) as also ▼ جَحْظَمٌ, in which the م is augmentative. (Ṣ, TA.) And جَاحِظُ العَيْنَيْنِ A man *whose blacks of his eyes are prominent.* (TA.) You say also, فُلَانٌ جَاحِظٌ إِلَىَّ بِعَيْنِهِ, and ▼ مُجَحِّظٌ, meaning *Such a one is looking at me intently.* (T, TA in art. زنر.) And جُحَّظٌ and جُحُظٌ, [which are pls. of جَاحِظٌ,] applied to men, signify *Raising the eyes, and looking fixedly;* or *stretching and raising the sight;* or *opening the eyes and not moving the eyelids.* (L, TA.)

جَاحِظَتَانِ, (so in copies of the Ṣ, and in the L,) or ▼ جَحْظَتَانِ, (so in a copy of the Ṣ, and so accord. to a copy of the KL, in which the sing. is written جَحَاظَةٌ, though Golius, on the authority of that work, writes it جِحَاظَةٌ,) or ▼ جَحْظَتَانِ, accord. to Lth, (TA,) or ▼ جِحَاظَانِ, (as written in one copy of the Ṣ,) The *two blacks of the eye* [or rather *of the two eyes*]. (Lth, Ṣ, L, TA.)

مُجَحَّظٌ : see جَاحِظٌ.

جحف

1. جَحَفَهُ, aor. ـَ, (Ḳ,) inf. n. جَحْفٌ, (TA,) *He, or it, stripped off, scraped off,* or *otherwise removed, its superficial part*; (Ḳ;) [and so ▼اجتحفه; for you say,] اجتحف السَّيْلُ الوَادِى *The torrent stripped off* [or *swept away*] *the superficial parts of the valley.* (TA.) — *He, or it, took away, carried away,* or *removed, the whole of it,* or *the greater part of it,* or *much of it*; or *swept it away*: (Ḳ, TA:) or, as some say, *vehemently.* (TA.) And مُجَاحَفَةٌ [an inf. n. of ▼جاحف] signifies The *taking* a thing, *taking* it *away, removing* it, or *sweeping* it *away.* (TA.) — *He, or it, destroyed,* and *extirpated, him,* or *it*; as also ▼اجتحفه and ▼اجحف به. (Mgh.) — *It* (a bucket) *took it and bore it away*; namely, water. (Ṣ, Ḳ.) — *He collected it,* لنفسه *for himself.* (Ḳ.) — *He laded it out with his hand* or *with a ladle,* namely, food, (IAạr, Ḳ,) and beverage, (TA,) لَهُ *for him.* (IAạr, Ḳ.) — *He threw it* (a thing, IDrd) *by kicking it* with his foot. (IDrd, Ḳ.) — جَحَفَ الكُرَةَ (Ḳ) and ▼اجتحفها (TA) *He snatched away the ball* (Ḳ, TA) from the ground. (TA in explanation of the former. See also جُحْفَةٌ.) — [The inf. n.] جَحْفٌ also signifies The act of *eating* (AA, Ṣ, TA) what is called ثَرِيد, (Ṣ,* TA,) or butter with dates, or with dried dates. (AA, Ṣ.) — And The act of *striking,* or *smiting,* with the sword. (AA, Ṣ, TA.) — And The act of *ejecting,* or *expelling.* (KL.) — And The *doing damage,* or *an injury.* (KL. [See also 4.]) — And جَحَفَ مَعَهُ *He inclined with him,* (Ḳ, TA,) عَلَى غَيْرِهِ *against another*: and in like manner, جَحَفَ لَهُ [*he inclined to him*]. (TA.) ═ جُحِفَ *He* (a man) *was affected with the flux of the belly termed* جُحَاف. (TA.)

3. جاحف, inf. n. مُجَاحَفَةٌ : see 1. — جاحفهُ, (Ṣ, Ḳ,) inf. n. as above, (TA,) also signifies *He pushed,* or *pressed, against,* or *upon, him,* or *it*; (Ṣ, Ḳ, TA;) and so جاحف به : (TA:) *and clave to him,* or *it*: (Ḥam p. 62:) and جِحَافٌ [which is also an inf. n. of the same verb] signifies the *pushing,* or *pressing, one against another,* or *one upon another, in war*: and the *striving, struggling, contending,* or *conflicting, in an affair.* (AA, TA.) Hence the saying of El-Aḥnaf, إِنَّمَا أَنَا بَيْنَ تَمِيمٍ كَعُلْبَةِ الرَّاعِى يُجَاحِفُونَ بِهَا يَوْمَ الوِرْدِ, i. e. [*I am, among Temeem, only like the milking-vessel of the pastor,*] *upon which they press,* or *crowd, together* [*on the day of coming to water*]. (TA.) — *He was,* or *became, near to him,* or *it.* (Ṣ, IF, Ḳ.) So in the phrase جاحف الذَّنْبَ [*He was,* or *became, near to committing the crime,* or *sin,* or *act of disobedience*]. (IF, TA.) [See also 4.] — جاحف عَنْهُ *He repelled from him.* (TA.) — [The inf. n.] جِحَافٌ also signifies The act of *fighting,* or *combating*: (Ḳ:) and *slaying.* (TA.) — And A bucket's *striking against the mouth of the well, so that its water pours out, and sometimes it becomes rent.* (Ṣ, Ḳ.)

4. اجحف به, (inf. n. إِجْحَافٌ, Mṣb,) *He,* or *it, took away, carried away,* or *removed, him,* or *it*; (Ṣ, Mṣb, Ḳ;) said, in this sense, of a torrent: (Mṣb:) and *extirpated him,* or *it*; (Ṣ, Mgh, Mṣb;) said of a torrent, (Ṣ, Mṣb,) and of time, or fortune, and of a calamity. (TA.) See also 1. — *He,* or *it, did damage,* or *an injury, to him.* (KL,* MA.) [See also 1.] It is said by one of the sages, مَنْ آثَرَ الدُّنْيَا أَجْحَفَ بِآخِرَتِهِ [*He who prefers the enjoyments of the present life mars his enjoyments of the life to come*]. (TA.) And you say, أَجْحَفَتْ بِهِ الفَاقَةُ *Want reduced him to poverty,* (Ḳ, TA,) *and caused his property to pass away.* (TA.) And أَجْحَفَتِ السَّنَةُ *The year was,* or *became, one of drought, and dearth,* or *sterility.* (Mṣb.) [Accord. to Fei,] إِجْحَافٌ is met. used as meaning ‡The *making to suffer excessive loss* or *detriment.* (Mṣb.) [It is also used as a simple subst., meaning *Damage, harm,* or *injury*: pl. إِجْحَافَاتٌ.] — *He imposed upon him,* or *tasked him with,* (namely, his slave, Mṣb,) *that which he was unable to do.* (Mṣb, TA.) — [Hence, perhaps, †*He strained it,* or *wrested it*; namely, a word, or an expression.] — *He approached him,* or *it*; *was,* or *became, near to him,* or *it.* (Ṣ, Ḳ.) [See also 3.] — *He approached it,* or *drew near to it,* (namely, a road,) *but did not enter it.* (TA.) And اجحف بِهِمْ, said of an enemy, and of a torrent, or rain, *He,* or *it, approached them,* or *drew near to them, but missed them.* (TA.) — *He was near to falling short of accomplishing it,* namely, an affair, or *of doing what was requisite therein*; or *was near to being remiss therein.* (TA.)

6. تجاحفوا *They reached,* or *hit, one another with staves,* (Ḳ, TA,) in the O, *with bows,* (TA,) *and swords,* (Ḳ, TA,) in fighting. (TA.) Hence the phrase, in a trad., إِذَا تَجَاحَفَتْ قُرَيْشٌ المُلْكَ بَيْنَهُمْ, i. e., *When Ḳureysh shall contend together in fight for dominion.* (TA.) — تجاحفوا الكُرَةَ بَيْنَهُمْ *They contended together in snatching away the ball* (تَخَاطَفُوهَا) *with the goff-sticks,* (Ḳ,* TA,) *after rolling it along.* (TA.)

8. اجتحفهُ : see 1, in three places. — Also *He seized it, took it,* or *carried it off, by force.* (Ḳ, TA.) — *He took it up,* namely, the food called ثَرِيد, *with the three fingers.* (Ṣgh, Ḳ.) — *He exhausted it,* namely, the water of a well, (Ḳ, TA,) *with the hand or with a vessel.* (TA.)

جَحْفٌ : see the next paragraph.

جُحْفَةٌ *A portion of clarified butter.* (Ṣgh, Ḳ.) — See also the next paragraph. ═ *An affection resembling* مَغْص [or *pain and griping*] *in the belly,* (Ḳ, TA,) *arising from indigestion.* (TA.) [See also جُحَاف.] ═ The *playing with the ball*; as also ▼جَحْفٌ [inf. n. of جَحَفَ, q. v.]. (Ḳ.)

جُحْفَةٌ *A portion of water remaining in the sides of a watering-trough* or *tank*; as also ▼جُحْفَةٌ. (Kr, Ḳ.) — The *water that is exhausted from a well*: or, *that remains in the well after the exhausting* [*of the rest*]. (Ḳ.) — *A small quantity of the food called* ثَرِيد, *in a vessel, not filling it.* (Ḳ.) — *The quantity that is laded out at once, of food*: or *a handful*: (IAạr, Ḳ:) pl. جُحَفٌ. (TA.) — *A portion of scattered herbage in the* قوز (a mistake for قُرْن, meaning the *most elevated part,* TA) *of a desert,* (Ḳ, TA,) *resembling waters on all its sides, such as that the seeker of water knows not which of the waters is the nearest to the extremity thereof.* (TA.)

جُحَافٌ *That carries away everything*; applied to a torrent; (Ṣ, Ḳ;) as also ▼جَاحِفٌ; (TA;) and to death. (Ṣ, Ḳ.) — *Death* [itself]. (Ṣ, Ḳ.) — *A flux of the belly, arising from indigestion*: (Ṣ, Ḳ:) or *a pain that attacks in consequence of eating flesh-meat without bread.* (TA.) [See also جُحْفَةٌ.]

جَحُوفٌ A bucket (دَلْوٌ) *that takes and bears away water.* (Ṣ, Ḳ.) — Food of the kind called ثَرِيد *remaining in the middle of a bowl.* (IAạr, Ḳ.)

جُحَافٌ : see جَاحِفٌ.

مَرَّ الشَّىْءُ مُضِرًّا وَمُجْحِفًا *The thing passed by approaching,* or *coming near.* (Ṣ.) — سَنَةٌ مُجْحِفَةٌ *A year that renders the cattle lean*: or *a year that destroys people,* or *impoverishes them,* or *injures them,* (تُجْحِفُ بِهِمْ,) *by slaughter,* or *by marring,* or *destroying, the cattle.* (TA.) — And مُجْحِفَةٌ, alone, *A calamity*; (Ḳ;) because it extirminates people. (TA.)

مَجْحُوفٌ A man *affected with the flux of the belly termed* جُحَاف. (Ṣ, Ḳ.)

جحفل

Q. 1. جَحْفَلَهُ, (Ṣ, Ḳ,) inf. n. جَحْفَلَةٌ, (TA,) *He prostrated him on the ground; threw him down*: (Ṣ, Ḳ:) and sometimes they said, جَعْفَلَهُ. (Ṣ.) ═ *He reproved, chid,* or *reproached, him for his deed*; or *did so severely.* (Ṣgh, Ḳ.)

Q. 2. تَجَحْفَلُوا *They congregated; collected themselves together.* (Ṣ, Ḳ.)

جَحْفَلٌ *An army*: (Ṣ:) or *a numerous army.* (Ḳ.) MF holds it to be formed, with an augmentative ل, from الجَحْفُ, meaning "the taking," or "carrying," a thing "away." (TA.) ═ *A great* man: (Ḳ:) or a man *of great estimation* or *dignity.* (Ṣ.) — *A generous, noble,* or *high-born, chief* or *lord.* (Ḳ.) — *Great in the sides.* (IAạr, Ḳ.)

جَحْفَلَةٌ The *lip* (Ṣ, Ḳ) of a solid-hoofed animal, (Ṣ,) [i. e.,] of a horse, a mule, and an ass: (Ḳ:) and metaphorically applied to that of a man, which is properly termed شَفَةٌ: (TA:) not, as some assert, peculiarly the upper lip: (MF:) pl. جَحَافِلُ. (TA.) — Also, (Ḳ,) جَحْفَلَتَانِ, (TA,) *Two callosities* (رَقْمَتَانِ) *in the two arms of the horse,* (Ḳ,) *resembling two marks made with a hot iron, facing each other, in the inner side of each arm.* (TA.)

جَحَنْفَلٌ (with an augmentative ن, Ṣ) *Thick-lipped.* (Ṣ, Ḳ.)

جحم

1. جَحَمَتِ النَّارُ, aor. ـَ, inf. n. جُحُومٌ; and جَحِمَت, aor. ـَ, inf. n. جَحَمٌ and جُحُومٌ [accord. to the CK جَحْمٌ] and جُحُومٌ; *The fire burned, burned up, burned brightly* or *fiercely, blazed*, or *flamed*; (K, TA;) and *had many live coals, and much flame*: (TA in explanation of the latter verb; and so the former or the latter is explained in a copy of the Ṣ, in which it is imperfectly written:) or the former signifies *it became great*: (TA:) and ▼ اجحمت *it became vehement*; said of fire, and also of war. (Ḥam p. 810.) ═ جَحَمَهُ, aor. ـَ, *He kindled* fire; *made* it *to burn, burn up, burn brightly* or *fiercely, blaze*, or *flame*. (K.)

4. اجحم عَنْهُ, (Ṣ, K,) inf. n. إِجْحَامٌ, (TA,) *He refrained, forbore, abstained*, or *desisted, from it*; (Ṣ, K;) namely, a thing; like احجم: (Ṣ:) but the former is a rare dial. var. (Ḥar p. 95.) Both these verbs bear contr. significations; being used as meaning *He advanced*, or *went forward*: and also *he receded*, or *drew back*. (MF.) ═ اجحم فُلَانًا *He*, or *it, was near to destroying*, or *killing, such a one*. (K.) ═ See also 1.

5. تجحّم *He burned with vehemence of desire*, or *covetousness, and niggardliness*; (K;) as also ▼ تجاحم: from جَاحِمُ الحَرْبِ. (TA.) — Hence, also, (TA,) *i. q.* تَضَايَقَ [app. meaning *He became straitened in disposition*]. (K.) You say also, هُوَ يَتَجَاحَمُ ▼ عَلَيْنَا, i. e., يَتَضَايَقُ [app., *He becomes straitened in disposition against us*]: a phrase mentioned by El-Mundhiree on the authority of Aboo-Ṭálib. (TA.)

6: see 5, in two places.

جَحْمَةٌ The *burning, burning brightly* or *fiercely, blazing*, or *flaming*, of fire; (Ḥam p. 77;) as also ▼ جَاحِمٌ: (TA:) or *vehemence of burning* or *blazing* or *flaming*: (Bḍ in xxxvii. 95:) or it is an epithet applied to fire because of its redness [or as meaning *red*]. (Ḥam ubi suprà.) — See also جَحِيمٌ.

جُحْمَةٌ: see جَحِيمٌ.

جَحِيمٌ *A fire burning*, or *blazing*, or *flaming, vehemently*; (K;) as also ▼ جَاحِمٌ: (Ḥam p. 810:) and *any fire having one part above another*; as also ▼ جَحْمَةٌ and ▼ جُحْمَةٌ; (K;) of which last the pl. is جُحَمٌ: (TA:) or *having many live coals, and flaming much*: (so in a copy of the Ṣ:) and *any great fire in a pit* or *the like*; (Ṣ, K;) from the saying in the Kur [xxxvii. 95], قَالُوا ابْنُوا لَهُ بُنْيَانًا فَأَلْقُوهُ فِى الجَحِيمِ [*They said, Build ye for him a building, and cast him into the great fire therein*]. (Ṣ.) And الجَحِيمُ is one of the names of *The fire* [*of Hell*]; (Ṣ, TA;) from which may God preserve us. (TA.) See also جَاحِمٌ.

جَحَّامٌ *Niggardly, tenacious, stingy, penurious*, or *avaricious*: (K:) from جَاحِمُ الحَرْبِ, meaning "the straitness, and vehemence, of war." (TA.)

جَاحِمٌ: see جَحِيمٌ. Also *Live coals* (جَمْرٌ) *vehemently burning* or *blazing* or *flaming*. (K.) And *a place vehemently hot*; (Ṣ, K;) as also ▼ جَحِيمٌ. (K.) El-Aạshà says, المَوْتُ جَاحِمٌ [app. meaning †*Death is like a burning*, or *fiercely-burning, fire*]. (Ṣ.) See also جَحْمَةٌ. — جَاحِمُ الحَرْبِ *The main part* [or *the thick*] *of the war* or *battle*: (K:) or *the straitness thereof*: (TA:) and *the vehemence of the fight* or *slaughter, in the scene thereof*. (K.) You say, اصْطَلَى بِجَاحِمِ الحَرْبِ ‡[*He warmed himself with the heat*, or *vehemence, of the battle*]. (TA.) — الجَاحِمَةُ *Fire*: (TA:) or, [as an epithet,] fire *burning, burning brightly* or *fiercely, blazing*, or *flaming*. (Ḥam p. 77.)

جد

1. جَدَّهُ, aor. ـُ, (Ṣ, Mṣb,) inf. n. جَدٌّ, (Ṣ, Mgh, Mṣb, K,) *He cut it*, or *cut it off*. (Ṣ, Mgh, Mṣb, K.) This is the primary signification. (Mgh.) You say of a weaver, جَدَّ ثَوْبًا *He cut off a piece of cloth* [sufficient for a garment or the like, from the web]. (Ṣ, K.) And جَدَّ النَّخْلَ, (Ṣ, Mgh, L,) aor. ـُ, (Ṣ, L,) inf. n. جَدٌّ (Ṣ, L, K) and جَدَادٌ (Lḥ, Mgh, L) and جِدَادٌ; (Lḥ, L; [in the L, the last two forms are mentioned as inf. ns., and the former of them is mentioned as inf. n. in the Mgh; but in the K, they are only mentioned as syn. with جَدٌّ; and in the Ṣ, it seems to be implied that they are simple substs., or quasi-inf. ns.;]) *i. q.* صَرَمَهُ; (Lḥ, Ṣ, Mgh, K;*) [like جَزَّهُ and جَدَّهُ;] i. e., *He cut off the fruit of the palm-trees*. (Mgh, L. [See also جَدَادٌ.]) And جُدَّتْ أَخْلَافُ النَّاقَةِ *The she-camel's teats were cut off by some accident that befell her*: (Aṣ, TA:) or, *in consequence of injury occasioned to her by the* صِرَار [q. v.]. (Ṣ.) And تَجْدِيدٌ [inf. n. of ▼ جدّد] signifies The *cutting off* the teat of a camel. (KL.) You say also, جَدَّ ثَدْيَا أُمِّكَ *May thy mother's breasts be cut off*: a form of imprecation against a man; and implying a wish for his separation. (Aṣ, L, from a trad.) — See also 5. ═ جَدَّ, aor. ـِ, inf. n. جِدَّةٌ, *It* (a garment, TA, or a thing, Ṣ, Mṣb, TA) *was new*; (Ṣ, L, Mṣb, K;) [as though *newly cut off from the web*;] from جَدَّ as signifying "he cut," or "cut off." (L.) [See also 5.] ═ جَدَّ, like تَعِبَ, (Mṣb,) sec. pers. جَدِدْتَ, [like its syn. حَظِظْتَ,] (L, Mṣb,) aor. يَجَدُّ; (Mṣb;) or جُدَّ, with ḍamm, (Mgh,) sec. pers. جُدِدْتَ, (Ṣ,) [aor. يُجَدُّ;] inf. n. جَدٌّ; (Ṣ,* Mgh, L, Mṣb;) *He was*, or *became, fortunate*, or *possessed of good fortune*, (Ṣ, Mgh, L, Mṣb,) or *of good worldly fortune*; (TA;) *he advanced in the world*, or *in worldly circumstances*; (Mgh;) بِالأَمْرِ *by the affair*, or *event*, whether good or evil; (L;) or بِالشَّىْءِ *by the thing*. (Mṣb.) And هُمْ يُجَدُّونَ بِهِمْ, as also يُحَظُّونَ بِهِمْ, *They become possessed of good fortune, and riches*, or *competence*, or *sufficiency*. (Ibn-Buzurj, L.) [You say also, جَدَّ جَدُّهُ ‡: so in a copy of the A: probably a mistranscription for جَدَّ جِدُّهُ, which see below: if not, meaning *His fortune became good*; or *his good fortune increased in goodness*: or, perhaps, *his dignity became great*; from what next follows]. ═ جَدَّ, فِى عَيْنِى, (Ṣ, A,) or فِى عُيُونِ النَّاسِ, and صُدُورِهِمْ, (Mgh,) aor. ـِ, inf. n. جَدٌّ, (Ṣ,) *He was*, or *became, great*, or *of great dignity* or *estimation, in my eye*, or *in the eyes of men*, and *their minds*. (Ṣ, A, Mgh.) It is said in a trad. of Anas, كَانَ الرَّجُلُ مِنَّا إِذَا قَرَأَ البَقَرَةَ وَآلَ عِمْرَانَ جَدَّ فِينَا, i. e., [*A man of us, when he recited* the chapter of *the Cow and* that of *the Family of 'Imrán* (the second and third chapters of the Kur-án),] *used to be great in our eyes*. (Ṣ.) ═ جَدَّ فِى الأَمْرِ, (Ṣ, A, K,) or فِى أَمْرِهِ, (L,) or فِى كَلَامِهِ, (Mṣb,) aor. ـِ (Ṣ, L, Mṣb, K) and ـُ, (L, K,) inf. n جِدٌّ, (Ṣ, K,) or جَدٌّ, (L, Mṣb,) جِدٌّ being a simple subst.; (Mṣb;) and ▼ اجدّ فيه; (L, K;) *He was serious*, or *in earnest*, (Ṣ, A, L, Mṣb, K,) *in the affair*, (Ṣ, A, K,) or *in his affair*, (L,) or *in his speech*; (Mṣb;) syn. حَقَّقَ; (L;) *contr. of* هَزَلَ. (L, Mṣb. [In the Ṣ and A and K, the inf. n. is said to signify the *contr. of* هَزْلٌ; and in the K, it is also said to be *syn. with* تَحْقِيقٌ.]) — And جَدَّ فِى الأَمْرِ, (Aṣ, Ṣ, L, Mṣb, K,) aor. ـِ and ـُ, (Ṣ, Mṣb, K,) inf. n. جِدٌّ, (Ṣ,* K,* TA,) or this is a simple subst., and the inf. n. is جَدٌّ; (Mṣb;) and ▼ اجدّ فيه; (As, Ṣ, L, K;) signify also *He strove, laboured*, or *toiled*; *exerted himself* or *his power* or *efforts* or *endeavours* or *ability*; *employed himself vigorously, strenuously, laboriously, diligently, studiously, sedulously, earnestly*, or *with energy*; *was diligent*, or *studious*; *took pains*, or *extraordinary pains*; *in the affair*. (Aṣ, Ṣ, L, Mṣb, K.) And جَدَّ فِى السَّيْرِ *He strove, laboured, toiled*, or *exerted himself, in going*, or *journeying*, or *in his course*, or *pace*; ‡*he hastened therein*: and in like manner, ▼ اجدّ السَّيْرَ † *he hastened his course*, or *pace*. (L.) And جَدَّ جِدُّهُ, [meaning *His labour*, or *exertion*, or *energy, was*, or *became, great*, or *extraordinary*: or] meaning ازْدَادَ جِدُّهُ جِدًّا [*his labour, &c., increased in labour, &c.*]: or it may mean *what was not* [*his*] جِدّ *became* جِدّ; wherefore, i. e. because it would be so eventually, it is here so called. (Ḥam p. 33. [See also جَدَّ جِدُّهُ, above.]) — جَدَّ بِهِ الأَمْرُ (A, L) ‡ *The affair*, or *event, distressed*, or *afflicted, him*. (L.) So in the saying of Aboo-Sahm,

* أَخَالِدُ لَا يَرْضَى عَنِ العَبْدِ رَبُّهُ *
* إِذَا جَدَّ بِالشَّيْخِ العُقُوقُ المُصَمِّمُ *

[*O Khálid, his Lord will not approve of the servant*, or *man*, (meaning the son,) *when cutting*, or *biting, disobedience to a parent distresses the old man*]. (L.) ═ جَدَّ, aor. ـِ, inf. n. جَدٌّ and جُدُودٌ, *It* (a house, or tent, بَيْت) *dripped*, or *let fall drops*. (K.)

2. جدّد, inf. n. تَجْدِيدٌ: see 1. ═ See also 4, in three places. ═ تجديد also signifies The *making* [or *weaving*] *stripes of different colours in a garment*. (KL.)

3. جَادَّهُ فِى الأَمْرِ, (Ṣ, L, Ḳ,*) inf. n. مُجَادَّةٌ, (L,) *i. q.* حَاقَّهُ (Ṣ, L) or حَاقَقَهُ (Ḳ) [*He contended with him respecting a thing, each of them asserting his right therein:* so accord. to explanations of حاقّه in the lexicons: but I think that the meaning intended here is, *he acted seriously,* or *in earnest, with him in the affair;* and this is confirmed by its being immediately added in the TA, after حاققه, "and أَجَدَّ" signifies "حَقَّقَ, as above mentioned:" see جَدَّ فِى الأَمْرِ expl. by حَقَّقَ as contr. of هَزَلَ]. Also *He exerted his full effort,* or *endeavour,* or *energy, with him in the affair.* (So accord. to an explanation of the inf. n., مجادّة, in the KL.)

4. اجدّ النَّخْلُ *The palm-trees attained to the time for the cutting off of the fruit.* (Ṣ, A,* L, Mṣb, Ḳ.) — [Hence, perhaps,] أَجْدَدْتُ قَرُونِى مِنْهُ *I (myself,* TA) *relinquished,* or *forsook, him,* or *it.* (Ḳ.) ═ اجدّهُ, and ▼ استجدّهُ, (Ṣ, A, L, Ḳ,) and ▼ جدّدهُ, (Ṣ, L, Ḳ,) *He made it new;* (Ṣ, A, L, Ḳ;) namely, a thing, (Ṣ,) or a garment: (A, TA:) or *he put it on,* or *wore it, new;* namely, a garment. (TA.) One says to him who puts on a new garment, أَبْلِ وَأَجِدَّ وَٱحْمَدِ الكَاسِى [*Wear out, and make,* or *put on, new, and praise the Clother,* meaning God]. (Ṣ.) And you say, بَلِىَ بَيْتُ فُلَانٍ فَأَجَدَّ بَيْتًا مِنْ شَعَرٍ [*The tent of such a one was,* or *became, rent,* or *pierced with holes; therefore he made a new tent of hair-cloth*]. (Ṣ.) And جدّد ▼ الأَمْرَ, and اجدّهُ, and ▼ استجدّهُ, *He originated,* or *innovated, the thing,* or *affair;* or *did it newly,* or *for the first time.* (Mṣb.) And جدّد ▼ الوُضُوْءَ ‡ [*He renewed the ablution termed* وضوء], and العَهْدَ ‡ [*the compact,* or *contract,* or *covenant,* &c.]. (TA.) — اجدّ فُلَانٌ أَمْرَهُ بِذٰلِكَ *Such a one established,* or *settled, firmly his affair,* or *case, thereby,* or *therein:* so says Aṣ, and he cites the following verse:

* أَجَدَّ بِهَا أَمْرًا وَأَيْقَنَ أَنَّهُ *
* لَهَا أَوْ لِأُخْرَى كَالطَّحِينِ تُرَابُهَا *

[*He established,* or *settled, firmly his case thereby,* or *therein, and knew certainly that he was for it,* (app. meaning a war, or battle, حَرْب, which is fem.,) *or for another whereof the dust would be like flour*]: Aboo-Naṣr says, It has been related to me that he said, اجدّ بها امرا means اجدّ أَمْرَهُ بِهَا; [and so this phrase is explained in the Ḳ;] but the former explanation I heard from himself: (L:) or this phrase means أَجَدَّ أَمْرُهُ بِهَا [so in two copies of the Ṣ, app., † *his affair,* or *case, became easy,* or *practicable, thereby,* like ground termed جَدَد, which is easy to walk, or travel, upon; see the next sentence]; امر being put in the accus. case as a specificative, like عَيْنًا in the phrase قَرِرْتُ بِهِ عَيْنًا, meaning قَرَّتْ بِهِ عَيْنِى. (Ṣ.) ═ اجدّ also signifies *It* (a road) *was,* or *became, what is termed* جَدَد [i. e. *hard,* or *level,* &c.]. (Ṣ, Ḳ.) And اجدّت لَكَ الأَرْضُ *The ground hath become to thee free from soft places, and clear to thy view.* (TA.) — Also *He walked along,* or *traversed, what is termed* جَدَد. (Ḳ.) And اجدّ القَوْمُ *The people,* or *company of men, came to what is so termed:* (Ṣ:) and *ascended upon the surface* (جَدِيد) *of the ground:* or *went upon sand such as is termed* جَدَد. (TA.) ═ See also 1, in three places.

5. تجدّد [originally *It became cut,* or *cut off.* — And hence,] *It* (an udder) *lost,* or *became devoid of, its milk:* (Ṣ, Ḳ:) and [in like manner] ▼ جُدَّ, aor. يُجَدُّ, inf. n. جَدَدٌ, *it,* (a breast, and an udder,) *became dry.* (AHeyth, TA.) — Hence also, [*It was newly made;* as though *newly cut off from the web;*] said of a garment: (TA:) and *it* (a thing, Ṣ, A) *became new:* (Ṣ, A, Ḳ:) and *it* (a thing, or an affair,) *originated; was originated,* or *innovated;* or *was done newly,* or *for the first time:* and sometimes ▼ استجدّ is used intransitively [in the same senses]. (Mṣb.) [Also † *It* (an action, as, for instance, ablution, and a compact, or the like,) *was renewed.* See جَدَّدَ as syn. with أَجَدَّ.]

10: see 4, in two places: ═ and see also 5.

جَدٌّ *Fortune,* or particularly *good fortune,* syn. حَظٌّ, (Ṣ, A, Mgh, L, Ḳ,) and بَخْتٌ, (Ṣ, A, L, Ḳ,) *in the world,* or *in worldly circumstances;* (TA;) *advance in the world,* or *in worldly circumstances:* (Mgh:) pl. [of mult.] جُدُودٌ (Ṣ) and [of pauc.] أَجْدَادٌ and أَجُدٌّ. (TA.) You say, فُلَانٌ ذُو جَدٍّ فِى كَذَا *Such a one is possessed of good fortune in such a thing.* (L.) And it is said in a trad. respecting the day of resurrection, وَإِذَا أَصْحَابُ الجَدِّ مَحْبُوسُونَ *And lo, the people who were possessed of good fortune and riches in the world were imprisoned.* (L.) And in a prayer, (L,) لَا يَنْفَعُ ذَا الجَدِّ مِنْكَ الجَدُّ *The good worldly fortune of him who is possessed of such fortune will not profit him,* (Mgh, L,) in the world to come, (L,) *in lieu of Thee;* (Mgh, L;*) i. e., *of obedience to Thee:* (Mgh, and Mughnee in art. مِنْ:) or *in lieu of the good fortune that cometh from Thee:* or, as some say, *will not defend him from Thee.* (Mughnee ubi suprà. [See also another explanation below.]) Hence, أَجِدَّكَ لَا تَفْعَلْ [or أَجَدَّكَ]; and, accord. to some, وَجَدِّكَ: see جِدٌّ. — *One's lot in life;* and *the means of subsistence that one receives from the bounty of God.* (L, Ḳ.) One says, لِفُلَانٍ فِى هٰذَا الأَمْرِ جَدٌّ *Such a one has in this thing,* or *state of affairs, means of subsistence.* (A 'Obeyd, L.) — *Richness; competence,* or *sufficiency;* or *the state of being in no need,* or *of having no wants,* or *of having few wants.* (Ṣ, L, Mṣb.) لا ينفع ذا الجدّ منك الجدّ, [explained above, is said to mean] *Riches,* &c., *will not profit the possessor thereof with Thee;* for nothing will profit him but acting in obedience to Thee: منك here signifies عِنْدَكَ. (Ṣ, Mṣb.) — *Greatness,* or *majesty;* (Mujáhid, Ṣ, Mgh, Mṣb, Ḳ;) accord. to some, specially of God: (TA:) so in the Ḳur lxxii. 3: (Ṣ, TA:) or his *freedom from all wants* or *the like;* syn. غِنًى. (Ṣ.) Hence, تَعَالَى جَدُّكَ, (Mgh, TA,) in a trad. respecting prayer, (TA,) *Exalted be thy greatness,* or *majesty.* (Mgh,* TA.) — See also أَجِدَّكَ, as an interrogative phrase, voce جِدٌّ. ═ Also, (Ṣ, Ḳ,) and ▼ مَجْدُودٌ, (Ṣ, A, Mgh, Ḳ,) and ▼ جَدِيدٌ, (Ṣ, Mṣb, Ḳ,) and ▼ جَدِّىٌّ, (Ṣ,) and ▼ جُدٌّ, and ▼ جُدِّىٌّ, the last two with ḍamm, (Ḳ,) applied to a man, *Fortunate;* or *possessed of good fortune;* (Ṣ, A, Mgh, Mṣb;) or *possessed of good worldly fortune:* (TA:) or *possessing great fortune,* or *great good fortune:* (Ḳ:) [the words here given from the Ṣ are there coupled with synonyms of the same form, thus; ▼ جَدِيدٌ, جَدٌّ حَظٌّ, ▼ مَجْدُودٌ مَحْظُوظٌ, حَظِيظٌ, and ▼ جَدِّىٌّ حَظِّىٌّ; on the authority of ISk:] ▼ جُدٌّ, with ḍamm, as an epithet applied to a man, is said by Sb to be syn. with مَجْدُودٌ; and its pl. is جُدُّونَ only. (L.) ═ Also جَدٌّ, *A grandfather; the father's father,* and *the mother's father:* (Ṣ, Mṣb, Ḳ:) and † *a higher ascendant; an ancestor:* (Mṣb:) and ▼ جَدَّةٌ *a grandmother; the father's mother,* and *the mother's mother:* (Ḳ:) [and † *a female ancestor:*] pl. of the former, أَجْدَادٌ [a pl. of pauc.] and جُدُودٌ and جُدُودَةٌ: (Ḳ:) and of the latter, جَدَّاتٌ. (TA.) Hence, accord. to some, وَجَدِّكَ لَا تَفْعَلْ: see جِدٌّ. ═ See also جُدَّةٌ: — and see جَدِيدٌ.

جُدٌّ: see جَدٌّ, in two places. ═ See also جُدَّةٌ. — Also *The side* (جَانِب) *of anything.* (Ḳ.) ═ And *A well in a place where is much herbage,* or *pasture:* (Ṣ, Mṣb, Ḳ:) *a well abounding with water;* (Ḳ;) [and] so ▼ جُدْجُدٌ; (KL;) but A 'Obeyd says that this is not known: (L:) and, contr., *a well containing little water: a scanty water,* or *water little in quantity: a water at the extremity of a* [*desert such as is called*] فَلَاة: (Ḳ:) *an old water:* (Th, Ḳ:) *an old well:* (KL:) pl. (in all these senses, TA) أَجْدَادٌ. (Mṣb, TA.)

جِدٌّ [accord. to some an inf. n., but accord. to others a simple subst., (see جَدَّ,)] *Seriousness,* or *earnestness, contr. of* هَزْلٌ, (Ṣ, A, Mṣb, Ḳ,) in speech. (Mṣb.) Hence, ثَلَاثٌ جِدُّهُنَّ جِدٌّ وَهَزْلُهُنَّ جِدٌّ [There are *three things in relation to which what is serious is serious and what is jesting is serious*]: a saying of Mohammad, whereby he forbade a man's divorcing and emancipating and marrying and then retracting, saying "I was jesting;" as was customary in the time of paganism. (Mṣb.) أَجِدَّكَ and ▼ أَجَدَّكَ signify the same; (Ṣ;) but the former is the more chaste; (TA;) جِدّ and جَدّ being thus used only as prefixed nouns: (Ṣ, Ḳ:) Aṣ says that the meaning is, أَبِجِدٍّ مِنْكَ هٰذَا [*Does this proceed from thee in seriousness,* or *in earnest?*]; and that جدّ is put in the accus. case because of the rejection of the [prep.] ب: AA says that the meaning is, مَا لَكَ أَجِدًّا مِنْكَ [*What aileth thee? Doth it proceed from thee in seriousness,* or *in earnest?*]; and that جدّ is put in the accus. case as an inf. n.: Th says that the phrase as it occurs in poetry is أَجِدَّكَ, with kesr: (Ṣ:) but when it occurs with وَ [in the place of أَ, or with أَ in the sense of وَ, as a particle denoting an oath,] it is

▼ وَجَدَّكَ [or أَجِدَّكَ], with fet-ḥ: (Ṣ, Ḳ:) you say, وَجَدِّكَ لَا تَفْعَلْ, (Ḳ, in the CḲ وَجَدَّكَ,) meaning, *By thy grandfather, do not* [such a thing]: or *by thy fortune*, or *good fortune, do not*: (TA:) also, when you say, أَجِدَّكَ لَا تَفْعَلْ, [or أَجِدَّكَ, for أَ (q. v.) is substituted for a particle of swearing, as in أَللّٰهِ لَأَفْعَلَنَّ,] the meaning is, I adjure thee *by thy truth*, (Lth, Ḳ,) and *by thy seriousness*, or *earnestness*, (Lth, TA,) *do not*: and when you say, أَجَدَّكَ ▼ لَا تَفْعَلْ, [or أَجِدَّكَ,] the meaning is, I adjure thee *by thy fortune*, or *good fortune, do not*: (Lth, Ḳ:) Aboo-'Alee Esh-Shalowbeenee asserts that it implies the signification of an oath. (MF.) In the phrase اجدك لا تفعل, AAF says, we may consider لا تفعل as put in the place of a denotative of state; or the phrase may be originally اجدك أَنْ لَا تَفْعَلَ, ان being suppressed, and its government annulled: [therefore it may be rendered, in the former case, *Is it with seriousness on thy part, thou doing* such a thing? and in the latter case, *Is it with seriousness on thy part* that *thou wilt not do* such a thing? i. e. *dost thou mean seriously* that *thou wilt not do* it? or in this case, اجدك may be used as a form of adjuration in one of the senses explained above, and لَا تَفْعَلُ may mean, that *thou do not* such a thing; or اجدك may mean وَجَدِّكَ, (explained above, and so in the three exs. below,) and لَا تَفْعَلُ, *thou wilt not do* it:] and, as AḤei says, there is here a nice point, which is this; that the noun [meaning the pronoun] to which جدّ is prefixed should agree in person with the verb which follows it; so that one should say, اجدّكَ لَا تَفْعَلُ, and اجدّى لَا أُكْرِمُكَ, and اجدّهُ لَا يَزُورُنَا; because جدّ is an inf. n. corroborating the proposition that follows it. (MF.) — Also, [and in this case, likewise, accord. to some an inf. n., but accord. to others a simple subst., (see, again, جُدٌّ,)] *A striving, labour*, or *toil; exertion of one's self*, or *of one's power* or *efforts* or *endeavours* or *ability; vigorousness, strenuousness, laboriousness, diligence, studiousness, sedulousness, earnestness*, or *energy; painstaking*, or *extraordinary painstaking*; (Ṣ, L, Mṣb, Ḳ;) in affairs, (Ṣ,) or in an affair. (Mṣb, Ḳ.) Hence, جِدًّا [meaning *In a great*, or *an extraordinary, degree; greatly, much, exceedingly*, or *extraordinarily; very; very greatly*, or *very much; extremely*]; as in the phrase, (Mṣb,) فُلَانٌ مُحْسِنٌ جِدًّا [*Such a one is beneficent in a great*, or *an extraordinary, degree; very, exceedingly*, or *extremely, beneficent*]: you should not say جَدًّا. (Ṣ, Mṣb.* [In my copy of the Mṣb, it is محسن جدّا بالفتح: but the context shows that there is an omission here, and that, after جدًّا, we should read, as in the Ṣ, وَلَا تَقُلْ جَدًّا.]) جِدًّا [in a phrase of this kind] is put in the accus. case as an inf. n. [of which the verb is understood; so that, in the ex. given above, the proper meaning is, يَجِدُّ فِى الإِحْسَانِ جِدًّا *striving in beneficence with a* great *striving*]; because it is not from the same root as the preceding word, nor is it identical with it [in meaning]. (L.) You say also, فِى هٰذَا خَطَرٌ جِدُّ عَظِيمٍ, meaning عَظِيمٌ جِدًّا [† *In this is a very*, or *an extremely, great danger*, or *risk*]. (Ṣ.) And هٰذَا العَالِمُ جِدُّ العَالِمِ *This is the learned* man, *the extremely* [or *the very*] *learned* man. (L.) And هٰذَا عَالِمٌ جِدُّ عَالِمٍ *This is a learned* man, *an extremely* [or *a very*] *learned* man. (L,* Ḳ.) — Also ‡ *Haste*. (Ṣ, L, Ḳ, TA.) So in the phrase فُلَانٌ عَلَى جِدِّ أَمْرٍ ‡ *Such a one is in haste in an affair*. (Ṣ, L, TA.) = Also *Executed seriously*, or *in earnest*, [*in which there is no jesting*,] *and excessive*; syn. مُحَقَّقٌ مُبَالَغٌ فِيهِ; [see جِدٌّ فِى أَمْرِهِ; meaning مُحَقَّقٌ فِيهِ وَمُبَالَغٌ فِيهِ; جِدٌّ thus used as an epithet having an intensive signification because it is originally an inf. n., or as some say, a simple subst.]: (L, Ḳ:) applied in this sense to a punishment: (L:) and also applied to a pace. (Ḳ in art. نص.) = See also جُدَّةٌ: — and see جَدِيدٌ.

جَدَّةٌ: see جَدٌّ, near the end of the paragraph.

جُدَّةٌ *The bank*, or *side*, of a river; as also ▼ جِدَّةٌ and ▼ جُدٌّ (IAth, L, Ḳ) and ▼ جُدٌّ (IAth, Mgh, L) and ▼ جُدُّ, (Mgh, L, Ḳ,) accord. to some, but correctly جُدٌّ; so called because cut off from the river, or because cut by the water, in like manner as it is called سَاحِلٌ because it is abraded by the water: (Mgh:) or the *part* of a river *that is near the land*; as also ▼ جُدَّةٌ: (L:) and the *shore* of the sea: (MF:) accord. to Aṣ, جُدَّةٌ is an arabicized word from the Nabathean كُدّ. (L.) — The *stripe*, or *streak, that is on the back of the ass, differing from his general colour*. (Ṣ, A,* Ḳ.) And ‡ *A streak* (Fr, Ṣ, Ḳ, TA) in anything, (TA,) as in a mountain, (Fr, Ṣ,) *differing in colour from the rest* of the mountain, (Ṣ,) white and black and red; (Fr, TA;) as also in the sky: (A, TA:) pl. جُدَدٌ, (Fr, Ṣ,) occurring in the Ḳur xxxv. 25; (Ṣ;) where some read جُدُدٌ, pl. of ▼ جديدة [app. جَدِيدَةٌ], which is syn. with جُدَّةٌ; and some, جَدَدٌ [q. v.]. (Bḍ.) — *A sign*, or *mark*, syn. عَلَامَةٌ, (Th, Ḳ,) of, or in, anything. (Th, TA.) — *A beaten way, marked with lines* [*cut by the feet of the men and beasts that have travelled along it*]: (Az, L:) or *a road*, or *way*: pl. جُدَدٌ: (Mṣb:) and جُدُودٌ, also, [app. another pl. of جُدَّةٌ,] signifies *paths*, or *tracks, forming lines upon the ground*. (Az, L.) See also جَادَّةٌ. — [Hence, app., but accord. to the Ṣ from the same word as signifying "a streak,"] رَكِبَ جُدَّةً مِنَ الأَمْرِ, (Ṣ, A, TA,) or جُدَّةَ الأَمْرِ, (Ḳ,) ‡ *He set upon a way*, or *manner, of performing the affair*: (A:) or *he formed an opinion respecting the affair*, or *case*. (Zj, Ṣ, A, Ḳ.) — See also جِدَّةٌ.

جِدَّةٌ: see جُدَّةٌ, in two places: — and see جَدِيدٌ. — Also *A rag; or piece torn off from a garment*; and so ▼ جُدَّةٌ: thus in the saying, مَا عَلَيْهِ جِدَّةٌ and جُدَّةٌ [*There is not upon him a rag*]. (Ḳ.) — *A collar upon the neck of a dog*: (Th, L, Ḳ:) pl. جِدَدٌ [like لِحًى pl. of لِحْيَةٌ, or perhaps a mistake for جُدَدٌ]. (L.)

جَدَدٌ *Hard ground*: (Ṣ:) or *hard level ground*: (Ḥar p. 522:) [see also جَدْجَدٌ:] or *rough level ground*: (Ḳ:) or *rough ground*: or *level ground*: (TA:) or *a level and spacious tract of land; a tract such as is called* صَحْرَآء, *and such as is called* فَضَآء, *containing no soft place in which the feet sink, nor any mountain, nor any* [*hill such as is called*] أَكَمَة; *sometimes wide, and sometimes of little width*: (ISh:) [and] *a conspicuous road*: (Bḍ in xxxv. 25:) pl. أَجْدَادٌ. (ISh.) It is said in a prov., مَنْ سَلَكَ الجَدَدَ أَمِنَ العِثَارَ [*He who walks along hard*, or *hard and level, ground is secure from stumbling*]; (Ṣ, TA;) meaning, he who pursues the course marked out by common consent is secure from stumbling. (TA.) And مَكَانٌ جَدَدٌ occurs in a trad., meaning *Level ground*. (TA.) — See also جَدِيدٌ. — Also Sand that is *thin*, or *fine*, (Ḳ, TA,) *and sloping down*. (TA.) — And *A thing resembling a* سِلْعَة [or *ganglion*] *in the neck of a camel*. (Ḳ.)

جَدَادٌ and ▼ جِدَادٌ *The cutting off of the fruit of palm-trees*. (Ṣ,* A,* L, Mṣb,* Ḳ.*) You say, هٰذَا زَمَنُ الجَدَادِ and الجِدَادِ [*This is the time*, or *season, of the cutting off of the fruit of the palm-trees*]. (Ṣ, A, Mṣb.*) Some say that جداد signifies particularly [as above,] the cutting off [of the fruit] of palm-trees; and جذاذ, the cutting off of all fruits, in a general sense: others say that they signify the same. (TA.) — Also *The time*, or *season, of the cutting off of the fruit of palm-trees*. (Ṣ,* L.) You say جَدَادٌ and جِدَادٌ, like صَرَامٌ and صِرَامٌ, and قَطَافٌ and قِطَافٌ; (Ks, Ṣ;) whence it seems as though the measures فَعَالٌ and فِعَالٌ were uniformly applicable to every noun signifying the time of the action; such nouns being likened to أَوَانٌ and إِوَانٌ. (Ṣ.)

جِدَادٌ: see جَدَادٌ, and جَادٌّ.

جَدُودٌ, (ISk, Ṣ, A, Ḳ,) or جَدُودَةٌ, (L,) *Having little milk*, (ISk, Ṣ, L, Ḳ,) *not in consequence of any injury*, (ISk, Ṣ,) or *not from any imperfection*; (L;) applied to a ewe, (ISk, Ṣ, Ḳ.) but not to a she-goat; the epithet مَصُورٌ being used in the latter case: (ISk, Ṣ:) or a ewe or she-goat *having no milk*; as also ▼ جَدَّآءُ: (A:) pl. جَدَائِدُ (Ṣ, L) and جِدَادٌ. (L.) — Also *A fat she-ass*: pl. جِدَادٌ. (AZ, Ḳ.)

جَدِيدٌ, of the measure فَعِيلٌ in the sense of the measure مَفْعُولٌ, [i. e. ▼ مَجْدُودٌ,] *Cut*, or *cut off*. (Ṣ, Mṣb.) A poet says,

* أَبَى حُبِّى سُلَيْمَى أَنْ يَبِيدَا *
* وَأَمْسَى حَبْلُهَا خَلَقًا جَدِيدَا *

[*My love of Suleymá hath refused to perish; but her cord* (i. e. her tie of affection to me) *hath become worn out and cut*]: (Ṣ:) [as جديد signifies "new" more commonly than "cut,"] this verse appears as though it involved a contradiction. (MF.) — Applied to a garment, or a piece of cloth [sufficient for a garment or the like], *Newly cut off* [*from the web*] *by the weaver:* (Ṣ, Ḳ:) and so (without ة, Ṣ) applied to a مِلْحَفَة; (Ṣ, A;) thus applied to a fem. n. because syn. with ▼ مَجْدُودَة; (Ṣ, ISd;) or, accord. to Sb, because by ملحفة in this case is meant إِزَار, and for a like reason in like cases; (Ḥam p. 555;) but one also says جَدِيدَةٌ; (ISd;) and accord. to some, جديد is of the measure فَعِيل in the sense of the measure فَاعِل, and therefore the ة is regularly affixed to it: (Ḥam ubi suprà:) the pl. is جُدُدٌ (Mbr, Th, Ṣ, A, Ḳ) and جُدَدٌ; (AZ, A'Obeyd, Mbr;) but the former is the more common. (TA.) — And hence, (L,) applied to a garment, (L, TA,) or a thing, (Ṣ, Mṣb,) *New; contr. of* قَدِيمٌ, (Mṣb,) or *contr. of* خَلَقٌ; (Ṣ, L;) from جِدَّةٌ as contr. of بِلًى; (Ḳ:) pl. [of pauc.] أَجِدَّةٌ and [of mult.] جُدُدٌ and جُدَدٌ. (L.) You say, أَصْبَحَتْ خُلْقَانُهُمْ جُدُدًا, a phrase mentioned by Lḥ, meaning خُلْقَانُهُمْ جُدُدًا [i. e. *Their old worn-out garments became replaced by new*]: or جُدُدًا may be here put for جَدِيدًا. (L.) — And hence, (TA,) الجَدِيدَانِ and ▼ الأَجَدَّانِ *The night and the day;* (Ṣ, Mṣb, Ḳ;) because they never become impaired by time. (TA.) You say, لَا أَفْعَلُهُ مَا ٱخْتَلَفَ الجَدِيدَانِ and ▼ الأَجَدَّانِ [*I will not do it while the day and the night succeed each other*]: (Ṣ:) or مَاكَرَّ الجَدِيدَانِ and الأَجَدَّانِ [*while the day and the night return time after time:* i. e., *ever*]. (A.) — Hence likewise, جَدِيدٌ also signifies A thing *of which one has had no knowledge.* (L.) — And hence, (L,) الجَدِيدُ signifies *Death:* (Ḳ:) or is applied as an epithet to death, in the dial. of Hudheyl. (L.) Accord. to Akh and El-Mugháfiṣ El-Báhilee, جَدِيدُ المَوْتِ means *The commencement of death.* (L.) ═ Also The *face,* or *surface,* of the earth, or ground; [as though it were cut;] (Ṣ, Ḳ, TA;) and so ▼ جَدَدٌ, and ▼ جُدَّةٌ, and ▼ جَدٌّ, (Ḳ,) and ▼ جُدٌّ. (TA.) ═ See also جَدٌّ, in two places.

جُدَادَةٌ *What is cut off from the roots,* or *eradicated,* of, or from, palm-trees &c. (Lḥ, TA.)

جَدِيدَةٌ The *kind of pad,* or *stuffed thing,* (رِفَادَة,) and the *felt, stuck,* or *attached, beneath the two boards* of a horse's saddle: there are two such things, called جَدِيدَتَانِ: (Ṣ:) or the جديدتان consist of the *felt that is stuck,* or *attached, in the inner side* of a horse's and of a camel's saddle: (L:) but جديدة thus applied is a post-classical word: the [classical] Arabs say جَدْيَةٌ, (Ṣ,) or, as in J's own handwriting, جَدِيَّةٌ. (So in the margin of a copy of the Ṣ.) ═ See also جُدَّةٌ.

جُدِّىٌّ: see جَدٌّ, in two places.

جِدِّىٌّ: see جِدٌّ.

جَدْجَدٌ *Hard level ground:* (Ṣ, Ḳ:) [see also جَدَدٌ:] *smooth ground:* and *rough ground:* (TA:) *a smooth tract such as is called* قُفّ. (AA, TA.)

جُدْجُدٌ [The *cricket;*] *i. q.* صَرَّارُ اللَّيْلِ, (Ṣ, M,) *a small flying thing,* (Ḳ,) *that leaps,* or *springs,* or *bounds, much,* (Ṣ, M,) *and creaks by night,* (TA,) *and bears a resemblance to the* جَرَاد [or *locust*]: (Ṣ, M, Ḳ:) and *a certain insect like the* جُنْدَب, (M, L, Ḳ,) *except that it is generally blackish, and short, but in some instances inclining to white; also called* صُرْصُر: (M, L:) or *i. q.* جُنْدَبٌ and صَدًى: (El-'Adebbes:) pl. جَدَاجِدُ. (Ṣ.) Accord. to IAạr, *A certain insect that clings to a skin,* or *hide, and eats it.* (TA.) ═ See also جُدٌّ.

جَادٌّ act. part. n. of جَدَّ; (Mgh, L;) *Cutting,* or *cutting off.* (Mgh.) ═ أَجَادٌّ أَنْتَ أَمْ هَازِلٌ *Art thou serious or jesting?* (A.) It is said in a trad., لَا يَأْخُذَنَّ أَحَدُكُمْ مَتَاعَ أَخِيهِ لَاعِبًا جَادًّا [*By no means shall any one of you take the property of his brother in play* and *in earnest*]; by which is meant taking a thing without meaning to steal it, but meaning to vex and anger the owner, so that the taker is in play with respect to theft, but in earnest in annoying. (TA in art. لعب.) — فُلَانٌ جَادٌّ *Such a one is striving, labouring,* or *toiling; exerting himself* or *his power* or *efforts* or *endeavours* or *ability;* &c. (TA.) And فُلَانٌ جَادٌّ ▼ مُجِدٌّ, thus with the two similar words together, (Aṣ, Ṣ, L,) signifies the same [in an intensive degree]. (L, TA.) ═ جَادٌّ مِائَةِ وَسْقٍ Land, or palm-trees, *of which the produce, cut therefrom, is a hundred camel-loads:* جَادٌّ being here used in the sense of ▼ مَجْدُودٌ. (L.) It is said in a trad. of Aboo-Bekr, نَحَلَ عَائِشَةَ جِدَادَ ▼ عِشْرِينَ وَسْقًا, meaning *He gave to 'Áïsheh palm-trees of which the quantity of the dates cut therefrom was a hundred camel-loads;* but the phrase heard from the Arabs is هٰذِهِ جَادُّ عِشْرِينَ: the former is like the saying الدَّرَاهِمُ ضَرْبُ الأَمِيرِ; and the latter, like عِيشَةٌ رَاضِيَةٌ. (Mgh.)

جَادَّةٌ The *main part of a road;* (Ṣ, Mgh, Mṣb, Ḳ;) *its middle:* (Mgh, Mṣb, and M voce جَرَجَة:) or *its even part:* or the *beaten track,* or *part along which one walks,* or *travels;* the *conspicuous part thereof:* or *a main road that comprises other roads,* or *tracks, and upon which one must pass:* (TA:) or *a road,* or *way,* absolutely; as also ▼ جُدَّةٌ: (Zj, MF:) or *a road leading to water:* (AHn, TA:) it is so called because it is marked with tracks, forming lines: (T, TA:) pl. جَوَادُّ, (Ṣ, A, Mgh, Mṣb, Ḳ,) occurring in poetry without teshdeed, but disapproved by Aṣ. (L.) فُلَانٌ عَلَى الجَادَّةِ means † *Such a one is following the right course of action* or *the like.* (Mgh.) You say also, هُوَ عَلَى جَادَّةِ الحَقِّ † [*He is on the road,* or *main road, of truth*]: not, however, عَلَى جَادَّةِ البَاطِلِ, but على مَزَلَّةِ البَاطِلِ, and مَزَلَّتِهِ, and مَهْلَكَتِهِ. (MF.)

أَجَدُّ [*Having some part,* or *parts, cut,* or *cut off:* fem. جَدَّآءُ]. — [Hence,] جَدَّآءُ A ewe, or she-goat, or she-camel, (TA,) *having her ear cut off.* (Ḳ, TA.) — A ewe, or she-goat, *having her teats cut off;* (Sh, TA;) as also ▼ مُجَدَّدَةٌ [q. v.], applied to a she-camel: (Aṣ, TA:) or *having her udder cut off.* (Khálid, TA.) — [And hence,] †A milch animal (TA [in the Ṣ app. restricted to a ewe]) *whose milk has passed away,* (ISk, Ṣ, Ḳ,) *by reason of some fault,* or *imperfection:* (ISk, Ṣ:) see also جَدُودٌ: or a ewe, or she-camel, or she-ass, *having little milk; having a dry udder:* or *having dry teats, being hurt by the* صِرَار [q. v.]: (L:) and أَجَدُّ † a breast *that has become dry.* (AHeyth.) — †A woman *small in the breast:* (Ṣ, Ḳ:) or *having short breasts.* (TA from a trad.) — †A desert, (فَلَاةٌ, Ṣ, Ḳ,) or land, (أَرْضٌ, A,) *in which is no water:* (Ṣ, A, Ḳ:) a desert (مَفَازَة) that is *dry.* (TA.) — سَنَةٌ جَدَّآءُ and عَامٌ أَجَدُّ †*A year of drought, and of dryness o the earth.* (TA.) ═ الأَجَدَّانِ: see جَدِيدٌ, in two places. ═ أَجَدُّ also signifies *More* [and *most*] *easy to walk or ride upon,* and *more* [and *most*] *plain* or *level;* applied to a road. (TA.) ═ And *More* [and *most*] *fortunate;* applied to a man. (ISd, A, L.)

مُجَدَّدَةُ الأَخْلَافِ A she-camel *having her teats cut off in consequence of injury occasioned to her by the* صِرَار [q. v.]. (Ṣ.) See also أَجَدُّ. ═ And كِسَآءٌ مُجَدَّدٌ A [garment of the kind called] *having stripes of different colours.* (Ṣ.)

مُجِدٌّ: see جَادٌّ. — إِنَّهَا لَمُجِدَّةٌ بِالرَّجُلِ, a phrase mentioned by Aṣ, said of a she-camel, meaning, *Verily she is quick in her pace with the man:* but Az says, I know not whether he said ▼ مَجَدَّةٌ or مُجِدَّةٌ: the former would be from جَدَّ; and the latter, from أَجَدَّ. (L.)

مَجَدَّةٌ: see what next precedes.

مَجْدُودٌ: see جَدِيدٌ, in two places; and جَادٌّ. ═ See also جَدٌّ, in two places.

جدب

1. جَدُبَ, (A, Mṣb, Ḳ,) aor. ُ, (Ḳ,) inf. n. جُدُوبَةٌ, (Ṣ, A, Mṣb, Ḳ,) *It* (a place, Ṣ, A, Ḳ, or a country, or region, Mṣb,) *was,* or *became, affected with drought, barrenness,* or *dearth;* or *with drought, and dryness of the earth;* (Ṣ, A, Mṣb, Ḳ;) as also جَدَبَ, (A,) inf. n. جَدْبٌ; (KL;) or جَدِبَ; (Ḳ;) and ▼ اجدب; (A, Ḳ;) or جَدِبَتْ, aor. َ; and ▼ أَجْدَبَتْ; both said of the earth or land (الأَرْضُ): (Mṣb:) and ▼ أَجْدَبَتِ البِلَادُ *the countries,* or *regions, were affected with drought, and the prices became high* [*therein*]. (TA.) ═ جَدَبَهُ, (Ṣ, M, A, Mṣb, Ḳ,) aor. ُ, (M, Mṣb, Ḳ) and ِ, (Ḳ,) inf. n. جَدْبٌ, (Mṣb,) *He found fault with it; dispraised it; expressed disapprobation of it.* (Ṣ, M, A, Mṣb, Ḳ.) So in the saying (Ṣ, A) relating to 'Omar, (A, TA,) in a trad., (Ṣ,) جَدَبَ السَّمَرَ بَعْدَ العِشَآءِ (Ṣ) or بَعْدَ العَتَمَةِ, (A) [*He expressed disapprobation of night-discourse after nightfall,* or *after the first third of the night reckoned from the disappearance of the redness of the twilight*].

3. جَادَبَتِ الإِبِلُ العَامَ, (ISk, Ṣ, A, TA,) inf. n. مُجَادَبَةٌ, (TA,) *The camels experienced,* or *have experienced, drought, and barrenness,* or *dryness*

of the earth, this year, and have become in such a state as not to eat anything but dry and black herbage, dry ثُمَام [or *panic grass*]: (ISk, Ṣ, TA:) or *have not met with, or found, anything but what was bad, by reason of drought, and barrenness, or dryness of the earth, this year.* (A.)

4: see 1, in three places. — أَجْدَبَتِ السَّنَةُ *The year became one of drought, barrenness, or dearth; or drought, and dryness of the earth.* (A,* TA.) — اجدب القَوْمُ *The people, or company of men, experienced drought, barrenness, or dearth; or drought, and dryness of the earth.* (Ṣ, A, Mṣb, Ḳ.) — [Hence,] نَزَلْنَا بِبَنِى فُلَانٍ فَأَجْدَبْنَا ‡ *We alighted as guests at the abode of the sons of such a one, and found not entertainment with them,* though they were in the enjoyment of plenty: (A:) [or] نَزَلْنَا فُلَانًا فَأَجْدَبْنَاهُ ‡ *we alighted as guests at the abode of such a one, and [found that] he did not entertain us.* (TA.) [The latter, if correct, is from what next follows.] = اجدب الأَرْضَ *He found the land to be affected with drought, barrenness, or dearth; or with drought, and dryness of the earth.* (Ṣ, A, Ḳ.)

5. مَا أَتَجَدَّبُ أَنْ أُصَاحِبَكَ ‡ *I do not deem it disagreeable, or unsuitable, to accompany thee;* syn. مَا أَسْتَوْخِمُ. (Ḳ.)

جَدْبٌ *Drought, barrenness,* or *dearth; contr. of* خِصْبٌ; (Ṣ;) i. q. مَحْلٌ, (A, Mṣb, Ḳ,) i. e. *drought,* or *suspension of rain, and dryness of the earth;* (Mṣb;) *dryness and barrenness of the earth:* (Ḥar p. 576:) and ▼ جَدَبٌ is a name, or subst., for الجَدْبُ, (Ḳ, TA,) meaning المَحْلُ; as in the saying of the rájiz, cited by Sb,

* لَقَدْ خَشِيتُ أَنْ أَرَى جَدَبَّا *
* فِى عَامِنَا ذَا بَعْدَ مَا أَخْصَبَّا *

[*Verily I feared to see drought,* or *barrenness,* &c., *in this our year, after it had been abundant in herbage*]; جَدَبًّا being used for الجَدْبَا; or, accord. to one reading, it is ▼ جَدْبَبًّا, with a doubled ب added; the change being made for the sake of the metre. (M, TA. [Respecting أَخْصَبَّا, see 4 in art. خصب.]) — Also *A place,* (Ṣ, A, Ḳ,) or *a country,* or *region,* (Mṣb,) *affected with drought, barrenness,* or *dearth; or with drought, and dryness of the earth;* and so ▼ جَدِيبٌ (Ṣ, A, Mṣb, Ḳ) and ▼ جَدُوبٌ and ▼ مَجْدُوبٌ, (Ḳ,) the last derived from جَدُبَ though this verb has not been used, (TA,) and ▼ مُجْدِبٌ, (M, A,) of which the pl. is مَجَادِيبُ. (A.) You say also أَرْضٌ جَدْبٌ [in which جدب is an inf. n. (though app. obsolete as such) and therefore applicable to a fem. subst.] (ISd, TA) and ▼ جَدْبَةٌ (Ṣ, A, Mṣb, Ḳ) and ▼ جَدِيبَةٌ (A, Mṣb) and ▼ جَدِيبٌ (Mṣb) and ▼ جَدُوبٌ (Lḥ, M, Mṣb) and ▼ مُجْدِبَةٌ, of which last the pl. is مَجَادِيبُ, (Mṣb,) *A land affected with drought,* &c.: (Ṣ, M, A, &c.:) and أَرَضُونَ جُدُوبٌ, (Ṣ, Ḳ,) as though to each part were applied the term جَدْبٌ [used as a subst.] from which is formed the pl. جُدُوبٌ, (TA,) and جَدْبٌ, (Ḳ,) which is here an inf. n. used as an epithet [and therefore applicable to a pl. subst.], (TA,) *lands affected with drought,* &c. (Ṣ, Ḳ.) And ▼ فَلَاةٌ جَدْبَآءُ [fem. of أَجْدَبُ] (M, Ḳ) *A desert affected with drought,* &c.; (Ḳ;) *in which is neither little nor much, neither pasture nor herbage.* (M, TA.) And ▼ فُلَانٌ جَدِيبُ الجَنَابِ *Such a one is environed by a tract affected with drought,* &c. (Ṣ. [But this phrase is generally used tropically, as meaning † *Such a one is ungenerous* or *illiberal* or *inhospitable.* See art. جنب.]) And سَنَةٌ جَدْبَةٌ (Ḳ in art. جوز) and ▼ عَامٌ جَدُوبٌ (M, TA) [*A year of drought,* &c.]. See also أَجَادِبُ, in two places. = Also i. q. عَيْبٌ [*A vice, fault, defect,* &c.]; (Ṣ, A, Ḳ;) a signification which may be either proper or tropical. (Er-Rághib, MF.)

جَدْبٌ: أَرْضٌ جَدْبَةٌ: see جَدْبٌ.

أَخَذَ فِى وَادِى جَدَبَاتٍ: see جَدَبَاتٌ, in art. جدب.

جَدَبٌ and جَدْبَبٌ: see جَدْبٌ.

جَدُوبٌ: see جَدْبٌ, in three places.

جَدِيبٌ: see جَدْبٌ, in three places.

جَادِبٌ *Finding fault, dispraising, expressing disapprobation:* whence the saying of Dhu-r-Rummeh,

* فَيَا لَكَ مِنْ خَدٍّ أَسِيلٍ وَمَنْطِقٍ *
* رَخِيمٍ وَمِنْ خَلْقٍ تَعَلَّلَ جَادِبُهْ *

meaning [*O thou smooth and even cheek, and gentle speech, and make*] *whereof he who dispraises it occupies himself vainly,* finding no defect in it. (Ṣ, TA.) — It is also said [as in the Ḳ &c.] to signify *Lying;* and the author of the 'Eyn says that it has no verb belonging to it [in this sense]; but this is a mistranscription [for خَادِبٌ]: AZ says that جَادِبٌ, with ج, has the signification here first given. (M, TA.)

جُنْدَبٌ and جُنْدُبٌ (Ṣ, Ḳ, &c.) and جِنْدَبٌ, like دِرْهَمٍ, (Sb, M, Ḳ,) the last of which is of weakest authority, because of a rare measure, whereof it has been said that there are only four examples: (TA:) in all of them the ن is said by some to be radical; but others, with more reason, hold it to be augmentative: (MF:) Sb says that it is augmentative: (Ṣ:) *A species of locust,* (Ṣ, Ḳ,) *well known:* (Ḳ:) or the *male locust:* or *small locust:* or, accord. to Seer, i. q. صَدًى [*a kind of cricket*], *that creaks by night, and hops and flies:* [but see صَدًى:] or, accord. to the M, it is *smaller than the* صدى, *and is found in the deserts:* pl. جَنَادِبُ. (TA.) صَرَّ الجندب [i. e. صَرَّ الجُنْدَبُ *The* جندب *creaked*] is a saying of the Arabs, used as a proverb; alluding to a difficult affair by which a person is troubled in mind; originating from the fact that the جندب, when its feet are scorched by the heated ground, does not keep them steadily upon it, and a creaking sound is consequently heard, produced by its legs. (TA.) — أُمُّ جُنْدُبٍ *The sand;* because the locust [or جندب] deposits its eggs therein: and the walker therein falls into evil [or encounters difficulty]. (TA.) — [Hence it signifies also] *Misfortune:* (Ṣ, M, Ḳ:) and *perfidy,* or *faithlessness,* or *treachery:* (M, Ḳ:) and *wrong,* or *injury:* (Ṣ, M, Ḳ:) and *evil conduct,* or *ill treatment.* (Ṣ.) You say, وَقَعَ فُلَانٌ فِى أُمِّ جُنْدُبٍ *Such a one fell into misfortune:* or *into perfidy.* (TA.) And وَقَعُوا فِى أُمِّ جُنْدُبٍ *They suffered wrong,* or *injury.* (AZ, Ṣ, Ḳ.) And وَقَعَ القَوْمُ بِأُمِّ جُنْدُبٍ *The people,* or *company of men, committed wrong,* or *injury, and slew him who was not a slayer:* (TA:) [as though they came with violence upon sand in which eggs of the جندب were deposited, and so destroyed the eggs, which had occasioned them no harm.] And رَكِبَ أُمَّ جُنْدُبٍ *He committed wrong,* or *injury.* (TA.)

[أَجْدَبُ i. q. جَدْبٌ as syn. with جَدِيبٌ: fem. جَدْبَآءُ. Hence,] فَلَاةٌ جَدْبَآءُ: see جَدْبٌ. — [Hence also,] سَنَةٌ جَدْبَآءُ *A year of much snow.* (L in art. شهب.) — أَجْدَبُ is [also] said in the M to be [used as] a subst. applied to what is termed مُجْدِبٌ [i. e. as syn. with the latter word used as an epithet in which the quality of a subst. is predominant; app. meaning *A place,* or *the like, affected with drought,* &c.]. (TA.) — [Also, as a comparative and superlative epithet, meaning *More,* and *most, affected with drought,* &c.; *contr.* of أَخْصَبُ.]

أَجَادِبُ, in a trad., where it is said, وَكَانَتْ فِيهِ أَجَادِبُ أَمْسَكَتِ المَآءَ, (Ḳ,* TA,) or وكانت فِيهَا, (TA,) [*And there were in it* أجادب *that retained the water*], is said to be pl. of أَجْدُبٌ, which is pl. of ▼ جَدْبٌ, (Ḳ, TA,) like as أَكَالِبُ is pl. of أَكْلُبٌ, which is pl. of كَلْبٌ; (TA;) and signifies *hard parts of the ground, that retain water, and do not imbibe it quickly;* or, as some say, *land having no plants* or *herbage,* from ▼ جَدْبٌ meaning "drought" &c.: the word is thus written in the two Ṣaheehs, of El-Bukháree and Muslim: (IAth, TA:) but some say that it is an anomalous pl. of جَدْبٌ, like as مَحَاسِنُ is of حُسْنٌ: and there are other readings; namely, أَجَادِبُ and أَحَادِبُ and أَخَارِبُ and أَجَارِدُ, pl. of أَجْرُدٌ, and إِخَاذَاتٌ, pl. of إِخَاذَةٌ. (MF, TA.)

مُجْدِبٌ, and its fem., with ة: see جَدْبٌ.

مِجْدَابٌ Land *scarcely ever,* or *never, abundant in herbage,* or *in the goods, conveniences,* or *comforts, of life; scarcely ever,* or *never, fruitful,* or *plentiful.* (Ḳ.)

مَجْدُوبٌ: see جَدْبٌ.

جدث

8. اجتدث *He made,* or *prepared, a* جَدَثٌ, i. e., *a grave,* or *sepulchre; or did so for himself.* (Ṣ, Ḳ, TA.)

جَدَثٌ *A grave; a sepulchre;* pl. أَجْدَاثٌ (Ṣ, Mṣb, Ḳ) and أَجْدُثٌ; (Ṣ, Ḳ;) of which latter, J cites an ex., but in this instance it is the proper name of a place. (TA.) It is of the dial. of Tihámeh: the people of Nejd say جَدَفٌ: (Mṣb:) or [as some say] the ف in the latter is a substitute for the ث in the former; for اجداث is used as a pl. by common consent, and اجداف is not used: (TA:) but Suh affirms, in the R, that the latter pl. is used by Ru-beh. (TA in art. جدف.)

جدح

1. جَدَحَ, aor. ـَ, inf. n. جَدْحٌ, *He mixed* anything. (L.) جَدَحَ السَّوِيقَ, (Ṣ, A, Mgh, L, Ḳ,) وَنَحْوَهُ, aor. and inf. n. as above; and ↓ جدّحهُ, inf. n. تَجْدِيحٌ; (L;) and ↓ اجتدحهُ, (Ṣ, L, Ḳ,) and ↓ اجدحهُ; (Ḳ;) *He stirred about the* سويق [or *meal made of parched barley* or *wheat*], *and the like, with water,* [or *milk,* (see what follows,) or *clarified butter,* or *fat of a sheep's tail, &c.,* (see لَتَّ,)] *until the whole became of a uniform consistence:* (L:) or *he stirred it about with a* مِجْدَح: (A, L:) or *he stirred about the* سويق *in milk, and the like, with a* مجدح, *until it became mixed:* (Lth, TA:) or *he beat and mixed the* سويق *with a* مجدح: (Mgh:) i. q. لَتَّهُ: (Ṣ, Ḳ:) and ↓ جدّحهُ, inf. n. تَجْدِيحٌ, *he mixed it;* in the Ḳ, لَطَّخَهُ; but the right reading is خَلَطَهُ, as in the L and other lexicons: (TA:) and ↓ اجتدحهُ *he drank it* (شربه [but this is perhaps a mistranscription for ضَرَبَهُ *he beat it*]) *with the* مجدح. (L, TA.)

2: see 1, in two places.

4: see 1. — احدح الإِبِلَ *He branded the camels on their thighs with the mark called* مِجْدَح. (Ḳ.)

8: see 1, in two places.

المُجْدَحُ: see the next paragraph.

مِجْدَحٌ The *instrument with which* سَوِيق *is stirred about with water &c.;* (Ṣ, A, Ḳ, &c.;) which is *a piece of wood the end whereof has several sides;* (Ṣ, L;) or *a piece of wood at the head of which are two cross pieces of wood;* (A, Mgh, L;) and *sometimes having three prongs:* (IAth, TA:) pl. مَجَادِحُ. (L.) — It is sometimes used tropically, as relating to evil, or mischief. (L.) [Thus it means ‡ *A stirrer-up* of evil or mischief; or *a thing that stirs up,* or *whereby one stirs up,* evil or mischief.] — Also † *Any one of the* مَجَادِيحُ السَّمَآءِ [or *stirrers-up of the sky,* or *of rain*]; (L;) these being *the* أَنْوَآء [or *stars,* or *asterisms, which, by their auroral settings or risings, were believed by the Pagan Arabs to bring rain &c.*]; (Ṣ, L, Ḳ;) *of those* انواء *that seldom or never failed* [*to bring rain*], *accord. to the Arabs:* (Mgh:) the ى in the pl. is added to give fulness to the sound of the kesreh; for the regular pl. is مَجَادِحُ, and the sing. of مجاديح should by rule be مِجْدَاحٌ. (A, IAth, Mgh.) One says, أَرْسَلَتِ السَّمَآءُ مَجَادِيحَهَا (L) or مَجَادِيحُ الغَيْثِ (A) † [*Its stirrers-up,* or *the stirrers-up of rain,* or *the stars* or *asterisms which were the bringers of it, sent forth rain*]. It is related of 'Omar, that he ascended the pulpit to pray for rain, and, having only offered a prayer for forgiveness, descended; whereupon it was said to him, "Thou hast not prayed for rain;" and he replied, لَقَدِ ٱسْتَسْقَيْتُ بِمَجَادِيحِ السَّمَآءِ † [*I have indeed prayed for rain by* words which are *the stirrers-up of rain*]; making the prayer for forgiveness to be a prayer for rain, in allusion to a passage in the Ḳur, lxxi. 9 and 10; and meaning thereby to deny the efficacy of the انواء. (A,* Mgh,* L.) المِجْدَحُ, also pronounced ↓ المُجْدَحُ, (Ṣ, Ḳ,) thus pronounced by El-Umawee, (Ṣ,) is moreover the name of † *A particular star* or *asterism, one of those which the Pagan Arabs asserted to be bringers of rain:* (L:) said to be الدَّبَرَانُ [*the Hyades;* or *the five chief stars thereof;* or *the brightest star thereof,* α *of Taurus*]; (Ṣ, A, L, Ḳ;) [which is called by this name of الدبران] because it rises latterly [with respect to the Pleiades], (Ṣ,) or because it follows يَدْبُرُ, i. e. يَتْبَعُ, the Pleiades: (T in art. دبر:) [whence] it is also called حَادِي النُّجُومِ ["the urger of the stars," properly, "with singing"], (Ṣ,) or حَادِي النَّجْمِ ["the urger of the asterism," meaning, "of the Pleiades"], and تَالِي النَّجْمِ ["the follower of the asterism," or, "of the Pleiades"], (Ḳzw,) and التَّالِي and التَّابِعُ ["the follower"]: (Sh:) or it is *a small star* or *asterism, between* الدبران *and* الثُّرَيَّا [or *the Pleiades*]: (IAṣr, Ḳ:) [perhaps meaning *the four stars that are the chief stars of the Hyades exclusively of a Tauri:*] or *three stars,* (Mgh, TA,) *like the three stones upon which a cooking-pot rests,* (TA,) *likened to a three-pronged* مِجْدَح; (Mgh, TA;) *on the* [*auroral*] *rising of which, heat is expected:* (TA:) the Arabs regarded it as one of the انواء which [by their auroral setting] foretokened rain. (IAth.) المِجْدَحَانِ is a name by which some of the Arabs called † *The two wings of* الجَوْزَآء [or *Orion*]. (Sh, TA.) — مِجْدَحٌ also signifies † *A certain mark made with a hot iron upon the thighs of camels.* (Ḳ.)

مُجَدَّحٌ Beverage, or wine, (شَرَاب,) *stirred about:* (Ṣ, Ḳ:) and in like manner, blood, when it is stirred about in the body of a gored animal by the goring horn. (L.)

مُجْدُوحٌ *Blood drawn from a vein, used in times of dearth,* or *drought,* (Ṣ, Ḳ,) *in the Time of Ignorance:* (Ṣ:) or *blood which was mixed with something else, and eaten in times of dearth:* (TA:) or *a kind of food of the Pagan Arabs, being blood obtained by opening a vein of a she-camel, which blood was received in a vessel, and drunk.* (T, TA.)

جدر

1. جَدَرَ, (Ḳ,) aor. ـُ, (TA,) *He made a* جِدَار [app. here meaning a *wall of enclosure*]; syn. حَوَّطَ: (Ḳ:) or *he built a* جدار: and *he founded* it. (Ḥam p. 818.) — *He concealed himself by means of a* جدار [or *wall*]. (Th, Ḳ.) — جُدِرَ, (A, Ḳ,) inf. n. جَدْرٌ; (TA;) and جَدَرَ, (Lḥ, Ḳ,) aor. ـُ, inf. n. جُدُورٌ; (Lḥ, TA;) and ↓ جُدِّرَ, (Ṣ, A, Ḳ,) which last some disallow, because this form denotes repetition, and the verb signifies the having a disease that befalls but once in a man's life; (MF;) *He* (a man, Ṣ, or a child, A) *had,* or *became attacked by,* جُدَرِىّ [or *small-pox*]. (Ṣ, A, Ḳ.) [And جَدَرَ الجُدَرِىُّ *The small-pox came forth,* or *broke out;* as in the TḲ: for its inf. n.] جُدُورٌ signifies the *coming forth,* or *breaking out, of the* جُدَرِىّ. (Ḳ.) — جَدُرَ, aor. ـُ, inf. n. جَدَارَةٌ, *He,* or *it, was,* or *became, adapted, disposed, apt, meet, suited, suitable, fitted, fit, competent, proper,* or *worthy.* (Ḳ.) You say, جَدُرَ بِهِ [and لَهُ] *He was,* or *became, adapted, disposed, apt, &c., for it.* (A.) [And جَدُرَ أَنْ يَفْعَلَ كَذَا *He was,* or *became, adapted, disposed, apt, &c., for doing such a thing.* See جَدِيرٌ.] — جَدَّرَهُ *He made,* or *called,* (جَعَلَ,) *him,* or *it, adapted, disposed, apt, meet, suited, suitable, fitted, fit, competent, proper,* or *worthy.* (Ḳ.)

2. جَدَّرَ بِنَآءَهُ: see 8. — جُدِّرَ: see 1.

4. مَا أَجْدَرَهُ بِالخَيْرِ [and لِلْخَيْرِ] *How well adapted* or *disposed,* or *how apt, meet, suited, suitable, fitted, fit, competent,* or *proper, is he for what is good!* or *how worthy is he of what is good!* (A.) And مَا أَجْدَرَهُ أَنْ يَفْعَلَ ذٰلِكَ, and أَجْدِرْ بِهِ, *How well adapted* or *disposed,* or *how apt, meet, &c., is he for doing that!* or *how worthy is he to do that!* (TA.) The usage of جَدُرَ, signifying "he was, or became, adapted, &c.," refutes the assertion of certain grammarians that these two forms of the verb deviate from general rule. (MF.)

8. اجتدر بِنَآءَهُ; and ↓ جدّرهُ, inf. n. تَجْدِيرٌ; (Ḳ) and مُجَدَّرٌ; (TA;) *He raised his building high;* or *constructed it firmly and strongly, and raised it high;* syn. شَيَّدَهُ. (Ḳ, TA.) [In the CḲ, we read اجْتَدَرَ بِنَاءَهُ, as though the pronoun ه referred to the word جِدار, which precedes, and thus the verb signified "he built a wall;" but it is shown in the TA that the right reading is that given above.]

Q. Q. 1. جَنْدَرَ الكِتَابَ *He passed the pen over what had become obliterated, of the writing,* (Ṣ, Ḳ,) *in order that it might become distinct.* (Ṣ.) And جندر الثَّوْبَ *He renewed the variegated,* or *figured, work of the garment,* or *piece of cloth, after it had gone.* (Ṣ, Ḳ.) [J says,] I think it to be an arabicized word. (Ṣ.)

جَدْرٌ *A wall;* or *a wall of enclosure;* syn. حَائِطٌ; as also ↓ جِدَارٌ [which is the more common]: (Ṣ, A, Mṣb, Ḳ:) pl. of the former, جُدُرٌ, (Ṣ, Mṣb, Ḳ,) sometimes used as a pl. of pauc., (Sb, TA,) and جُدْرٌ; (Ḳ;) and of the latter, جُدْرَانٌ. (Ṣ, Mṣb, Ḳ.) — The *basis,* or *foundation, of a wall:* (Ḳ:) and the *side of a wall:* (Lḥ, Ḳ:) pl., in both these senses, جُدُورٌ. (TA.) الجَدْرُ is applied to *The* [*wall called the*] حَطِيم (A, Ḳ) *of the Kaabeh;* (Ḳ;) because in it is a part of the [original] foundations of the house: (TA:) and it is also called الحِجْرُ. (A.) — † *A fence,* or *dam, raised of branches, to retain water;* likened to a wall: (Az, Mṣb:) or *a fence*, or *dam*, *to confine water*: pl. جُدُورٌ: (Suh, Mṣb:) and جُدْرٌ, [which is also a pl.,] signifies *fences*, or *dams*, *between houses*, *which retain water*. (TA.) — [The pl.] جُدُورٌ also signifies *Gardens*, or *walled gardens*, (حَوَائِط) of grapes. (TA.)

جَدْرَةٌ: see جَدِيرَةٌ.

جُدَرِىٌّ and جَدَرِىٌّ (Ṣ, Mṣb, Ḳ) [*Small-pox*;] *certain pustules* (Mṣb, Ḳ) *in the body*, (Ḳ,) *which break forth* (Mṣb, Ḳ) *from the skin, full of water, and afterwards opening*, (Mṣb,) *and generating thick purulent matter*; (Ḳ;) *a well-known disease, that attacks people once during life*. (TA.) — جُدَرِىُّ الأَرْضِ ‡ an appellation applied to *Truffles* (كَمْأَةٌ), denoting disapprobation. (TA from a trad.)

جِدَارٌ: see جَدْرٌ: and see also جَدِيرَةٌ.

جَدِيرٌ A place *having a wall built around it*; a *walled* place. (Ṣ, Ḳ.) — See also مُجَدَّرٌ. — Also *Adapted, disposed, apt, meet, suited, suitable, fitted, fit, proper, competent*, or *worthy*; syn. خَلِيقٌ (Ṣ, A, Mṣb, Ḳ) and حَقِيقٌ: (Mṣb:) fem. with ة: (TA:) pl. masc. جَدِيرُونَ and جُدَرَآءُ: (Ṣ, Ḳ:) pl. fem. جَدِيرَاتٌ and جَدَائِرُ. (TA.) You say, هُوَ جَدِيرٌ بِكَذَا (Ṣ, A, Mṣb) and لِكَذَا (TA) *He is adapted, disposed, apt, meet, &c., for such a thing*; (Ṣ, A, Mṣb;) and [*naturally*] *drawn to it*. (Ḥam p. 707.) And أَنْتَ جَدِيرٌ أَنْ تَفْعَلَ كَذَا *Thou art adapted, disposed, apt, meet, &c., for doing such a thing*; or *worthy to do it*. (Ṣ.) And إِنَّهُ لَمَجْدَرَةٌ ▼ أَنْ يَفْعَلَ, (Ḳ,) and in like manner you say of two persons, and of more, (TA,) and ▼مَجْدُورٌ, (Ḳ.) *Verily he is one who is adapted, disposed, apt, meet, &c., for doing* [such a thing]; or *worthy to do* [it]; syn. مَخْلَقَةٌ. (Ḳ.) [▼مَجْدَرَةٌ properly signifies *A place*, and hence *a thing, an affair*, and *a person, adapted, disposed, apt, meet, &c.*; like مَخْلَقَةٌ and مَحْرَاةٌ: and ▼مَجْدُورٌ, *Made*, or *called, adapted or disposed &c.*, though said by Aboo-Jaafar Er-Ruásee to be a pass. part. n. having no verb.] Also إِنَّهَا لَمَجْدَرَةٌ ▼ بِذٰلِكَ *Verily she is one who is adapted, disposed, apt, &c., for that*: and بِأَنْ تَفْعَلَ ذٰلِكَ *for doing that*: and in like manner you say of two persons, and of more. (TA.) And هٰذَا الأَمْرُ مَجْدَرَةٌ ▼ لِذٰلِكَ *This affair*, or *thing, is one that is adapted, apt, meet, &c., for that*; syn. مَحْرَاةٌ. (Ṣ.) And هٰذَا الأَمْرُ مَجْدَرَةٌ ▼ مِنْهُ *This affair*, or *thing, is one that is adapted, apt, meet, &c., for him to do*; i. e. he is adapted, apt, meet, &c., for doing it. (TA.)

جَدِيرَةٌ *An enclosure for camels*, (AZ, Ṣ, Ḳ,) *and for lambs and kids and calves &c.*, (TA,) *made of masses of stone*; (AZ, Ṣ;) as also ▼جَدْرَةٌ: (TA:) if *of mud*, or *clay*, it is called ▼جِدَارٌ: (AZ, TA:) or *an enclosure* (زَرْب) *for sheep or goats*. (TA.) — *Nature*; or *natural*, or *native, disposition, temper*, or other *property*. (Ḳ.)

هُوَ أَجْدَرُ بِهِ *He, or it, is more, or most, adapted, disposed, apt, meet, suited, suitable, fitted, fit, competent*, or *proper, for it*, or *him*; or *he is more*, or *most, worthy of it*. (A.) [See an ex. in a verse cited voce خُطَّةٌ.]

أَرْضٌ مَجْدَرَةٌ *A land in which is* جُدَرِىٌّ [or *small-pox*]: (Lḥ, Ṣ:) or *a land in which is much thereof*. (Ḳ.) — See also جَدِيرٌ, in five places.

مُجَدَّرٌ (Ṣ, Mgh, Mṣb, Ḳ) and ▼مَجْدُورٌ (Mgh, Ḳ) and ▼جَدِيرٌ (Mṣb, TA) *Having the* جُدَرِىّ [or *small-pox*]. (Ṣ, Mgh, Mṣb, Ḳ.) And ▼مَجْدُورُ الوَجْهِ [*Having the face marked with the small-pox*]. (A.)

مَجْدُورٌ: see مُجَدَّرٌ: — and see also جَدِيرٌ, in two places.

جدع

1. جَدَعَ الأَنْفَ, (Ṣ,* Mṣb, Ḳ,*) aor. َ, (Mṣb,) inf. n. جَدْعٌ, (Ṣ, Mṣb, Ḳ,) *He cut off the nose*; and in like manner, the ear; and the hand, or arm; and the lip; (Ṣ, Mṣb, Ḳ;) and a similar part: (TA:) and ▼أَجْدَعْتُ أَنْفَهُ signifies the same as جَدَعْتُ [*I cut off his nose*]: or جَدْعٌ signifies [absolutely] the *cutting off*; or *cutting so as to separate*. (TA.) In the following saying of a poet, the verb is used metaphorically,

* وَأَصْبَحَ الدَّهْرُ ذُو العِرْنِينِ قَدْ جُدِعَا *

[lit. *And nosed fortune became mutilated in the nose*; meaning, †*became marred*]. (TA.) And in the following phrase, occurring in a verse, كَأَنَّ اللّٰهَ يَجْدَعُ أَنْفَهُ وَعَيْنَيْهِ, the poet means, [*As though God cut off his nose*] *and put out his eyes*: see a similar saying in art. قلد, voce تَقَلَّدَ. (TA.) — جَدَعَهُ, (Ṣ, Ḳ,) aor. and inf. n. as above, (TA,) *He mutilated him*, or *maimed him, by cutting off his nose*, or *his ear*, or *his hand* or *arm*, or *his lip*, (Ṣ, Ḳ, TA,) or *the like*; (TA;) as also ▼جدّعهُ. (Ṣ, TA.) [Hence the phrase,] جَدْعًا ▼ لَهُ (Ṣ, Ḳ) [(*May God decree*) *to him mutilation*, or *maiming, by the cutting off of his nose*, or *the like*; or *cause it to befall him*: or] meaning أَلْزَمَهُ اللّٰهُ الجَدْعَ [†*may God make injury*, or *diminution of what is good, to cleave to him*]: (Ḳ:) said in imprecating a curse upon a man: similar to عَقْرًا لَهُ, q. v.: the first word being governed in the accus. case by a verb understood. (TA.) One says also, اِجْدَعْهُمْ بِالأَمْرِ حَتَّى يَذِلُّوا, a phrase mentioned by IAạr, but not explained by him; thought by ISd to mean, ‡ *Act thou, in commanding, as though thou mutilatedst them by cutting off their noses* [*until they become submissive*]. (TA.) In the phrase ▼صَوْتُ الحِمَارِ اليُجَدَّعُ [*The voice of the ass that has his ear, or ears, cut off*, (see مُجَدَّعٌ, below,)], occurring in a verse of Dhu-l-Khiraḳ Eṭ-Ṭuhawee, (Ṣ,) accord. to J, but not found by Ṣgh in the verses of that poet, and said to be in the Book [of Sb], though IB denies this, asserting it to be in the Nawádir of AZ, (TA,) Akh says, the poet means الَّذِى يُجَدَّعُ, like as you say, هُوَ اليَضْرِبُكَ, meaning الَّذِى يَضْرِبُكَ: Aboo-Bekr Ibn-es-Sarráj says, the poet, requiring refa for the rhyme, has changed the noun into a verb; and this is one of the worst of poetic licences. (Ṣ.) — [Hence,] السَّنَةُ تَجْدَعُ النَّبَاتَ ‡ [*The year of drought cuts off*, or *destroys, the herbage*]: (A, TA:) and تَجْدَعُ بِالمَالِ *destroys the camels* or *the like*. (Ṣ, O, Ḳ.) And جَدَّعَ ▼ القَحْطُ النَّبَاتَ ‡ *The drought prevented the growth*, or *increase, of the herbage*. (Ḳ, TA.) — [Hence also,] جَدَعَ عِيَالَهُ, inf. n. جَدْعٌ, †*He withheld good things from his family*, or *household*. (TA.) And جَدَعَتْهُ أُمُّهُ, aor. َ, inf. n. as above, ‡ *His mother fed him with bad food*; (Zj, Ḳ;) as also ▼أَجْدَعَتْهُ, (Ṣ, Ḳ,) inf. n. إِجْدَاعٌ; (TA;) and ▼جَدَّعَتْهُ, (Ḳ,) inf. n. تَجْدِيعٌ: (TA:) and ▼جدّعهُ and ▼اجدعهُ ‡ *He* (a pastor) *confined him* [a beast] *to bad pasture*. (TA.) جَدَعْتُهُ, (Ṣ,) inf. n. as above, (Ḳ,) also signifies †*I confined him, restricted him*, or *the like*; syn. حَبَسْتُهُ: and †*I imprisoned him*: (Ṣ, Ḳ:*) and so with ذ: (Ṣ:) or جَدْعٌ and جَذْعٌ both signify †the *confining*, or *restricting*, a person *with evil management, and with contemptuous treatment, and want of good care*. (AHeyth.) — جَدِعَ, aor. َ, (Mṣb,) inf. n. جَدَعٌ, (Ṣ, Mṣb, Ḳ,) *He* (a man) *was*, or *became, mutilated*, or *maimed, by the cutting off of his nose*, or *his ear*, (Ṣ,* Mṣb, Ḳ,*) or *his hand* or *arm*, or *his lip*, (Ṣ, Ḳ,) or *the like*: (TA:) or, accord. to some, you do not say جَدِعَ, but جُدِعَ: (TA:) and جَدِعَتِ الشَّاةُ *The sheep, or goat, was, or became, mutilated by having its ears entirely cut off*. (Mṣb.) — [Hence,] also, (Ṣ, Ḳ,) aor. as above, (Ḳ,) and so the inf. n., (Ṣ,) [as though meaning †*He was*, or *became, injured*;] ‡ *he* (a child) *had bad food*: (Ṣ, Ḳ, TA:) and *he* (a young weaned camel) *had bad food*: or *was ridden while* [*too*] *young, and in consequence became weak*. (TA.)

2. جدّعهُ, inf. n. تَجْدِيعٌ: see 1, in five places. — جَدَّعَهُ وَشَرَّاهُ ‡ *He made him to experience evil treatment, and derided him*; *as when one cuts off the ear of his slave, and sells him*. (TA.) — Also *He said to him* جَدْعًا لَكَ [explained above; see 1]. (Ṣ, Ḳ.*) [See also عَقَّرَهُ.]

3. جادع, inf. n. مُجَادَعَةٌ (Ṣ, Ḳ) and جِدَاعٌ, (Ḳ,) ‡ *He reviled, being reviled by another*, (Ḳ,* TA,) *saying* جَدْعًا لَكَ; *as though each of them cut off the nose of the other*: (TA:) and, (Ḳ,) or accord. to some, (TA,) ‡ *he contended in an altercation*; as also ▼تجادع; (Ṣ, Ḳ, TA;) [but the latter is said of a number of persons &c.] You say, تَرَكْتُ البِلَادَ تَجَادَعُ ▼ أَفَاعِيهَا, (Th, Ṣ,) and ▼تَجَدَّعُ also, (Th,) ‡ *I left the countries with their vipers eating one another*; (Th, Ṣ;) not meaning eating in reality, but *rending in pieces*, or *mangling, one another*: (Th:) and عَامٌ تَجَدَّعُ ▼ أَفَاعِيهَا, and ▼تَجَادَعُ, ‡ *A year in which the vipers eat one another*, by reason of its severity. (Th.)

4: see 1, in three places.

5: see 3, in two places.

6: see 3, in three places.

جَدْعٌ *What is cut off of the anterior parts of the nose, to its furthest*, or *uttermost, part*: (Aṣ,

TA:) an inf. n. used as a [proper] subst. (TA.) — جَدْعًا لَهُ : see 1. ═ ‡ *Unwholesomeness* in herbage. (Ḳ.)

جَدِعٌ ‡ A child *having bad food; or fed on bad food:* (Ṣ, Ḳ, TA:) pronounced by El-Mufaḍḍal with ذ; but Aṣ repudiated to him this pronunciation; (Ṣ, TA;) and his objection was confirmed by a young man of the Benoo-Asad called in as an umpire. (TA.)

جَدَعَةٌ *What remains, of the nose, ear, hand or arm, or lip, after the cutting off* [*of the rest*]: (Ṣ, Ḳ:) the *place of the cutting off thereof;* like عَرَجَةٌ from الأَعْرَجُ, and قَطَعَةٌ from الأَقْطَعُ. (TA.)

جَدَاعِ (Ṣ, A, Ḳ, &c.) and جُدَاعٌ (Ḳ, TA) ‡ *A year of drought;* because it cuts off, or destroys, (تَجْدَعُ,) the herbage, and abases men: (A, TA:) or *a severe,* or *calamitous, year, that destroys the camels or the like;* (Ṣ, O, Ḳ;) or *that destroys everything; as though it cut off its nose or the like.* (L.) — See also جُدَاعٌ.

جُدَاعٌ ‡ *Withered* herbage: (Ṣ:) or herbage *that is unwholesome to the feeder upon it:* (Ḳ:) or *tall, unwholesome, and withered.* (TA.) — And hence, الجُدَاعُ signifies ‡ *Death:* (Ḳ, TA:) written by some ▼الجَدَاعُ. (TA.)

أَجْدَعُ *Mutilated,* or *maimed, by having his nose cut off,* or *his ear,* (Ṣ, Mṣb, Ḳ,) or *his hand* or *arm,* or *his lip,* (Ṣ, Ḳ,) or *the like:* (TA:) fem. جَدْعَاءُ: (Ṣ, Mṣb:) and the latter, applied to a she-camel, *having the sixth part of her ear,* or *the fourth part of it, or more than that, to the half, cut off;* and to a she-goat, *having a third part,* or *more, of her ear cut off;* or, accord. to IAmb, any ewe or she-goat *having the ear lopped;* (TA;) or a ewe or she-goat *having her ear entirely cut off:* (Mgh, Mṣb:) and ▼مُجَدَّعٌ an ass *having the ear cut off,* (Ṣ,) or *having the ears cut off.* (Ḳ.) It is said in a prov., أَنْفُكَ مِنْكَ وَإِنْ كَانَ أَجْدَعَ [*Thy nose is a part of thee though it be cut off*]: applied with reference to him whose good and evil attaches to thee though he be not firmly connected with thee by relationship. (TA.) — الأَجْدَعُ one of the appellations applied to *The devil.* (Fr, Ḳ.*)

مُجَدَّعٌ: see أَجْدَعُ. — Also †A plant, or herbage, *of which the upper part has been eaten:* (Ṣ:) or *of which the upper part and the sides have been partly cut off* or *eaten.* (AḤn.)

جدف

1. جَدَفَهُ, aor. ِ, (IDrd, Ḳ,) inf. n. جَدْفٌ, (TA,) *He cut it; or cut it off:* (IDrd, Ḳ:) and so جَذَفَهُ. (TA.) ═ جَدَفَ, (Ks, Ṣ, Ḳ,) aor. ِ, (Ks, IDrd, Ṣ,) inf. n. جُدُوفٌ, (Ks, Ṣ, Ḳ,) or جَدْفٌ, (L as on the authority of Ks,) *He* (a bird) *flew* [*with his wings*] *clipped, appearing as though he turned his wings backward:* (Ks, Ṣ, Ḳ:) or *contracted his wing somewhat, in order to descend in his flight, and then inclined,* or *declined, in fear of the hawk:* (TA:) and *he* (a bird) *went quickly,* (Ḳ in art. جذف,) *with his wings;* generally *when one of the wings had been shortened;* (TA;) as also ▼اجدف and ▼انجدف: and so, all, with ذ. (Ḳ ib.) — [Hence,] جَدَفَ المَلَّاحُ بِالمِجْدَافِ [*The sailor rowed,* or *paddled, with the oar,* or *paddle*]. (AA, TA.) And جَدَفَ بِالسَّفِينَةِ, (TA,) or جَدَفَ السَّفِينَةَ, aor. ِ, inf. n. جَدْفٌ, (Mgh,) [*He rowed,* or *paddled, the ship,* or *boat;*] *he put the ship,* or *boat, in motion with the* مِجْدَفٌ [or مِجْدَافٌ]. (Mgh.) — Also جَدَفَ *He* (a man) *swung the arms;* (Ḳ, expl. by ضَرَبَ بِاليَدَيْنِ; in the O, بِاليَدِ, as is said in the TA;) as a man does in walking, moving them about: and the meaning seems to be, *he walked quickly:* (TA:) you say, جَدَفَ فِى مِشْيَتِهِ *he* (a man) *was quick in his manner of walking;* (AAF, TA;) and so with ذ: (Ṣ in art. جذف:) or جَدْفٌ signifies *a repeated interrupting of the voice* (تَقْطِيعُ الصَّوْتِ) *in singing to camels to urge or excite them.* (Ḳ,* TA.) — Also, (Ḳ,) inf. n. جَدْفٌ, (TA,) *He* (a gazelle) *went,* or *walked, with short steps.* (Ḳ,* TA.) And جَدَفَتْ *She* (a woman) *walked like those that are short:* and *she* (a gazelle, and a woman, TA) *went with short steps;* as also ▼اجدفت: and so, both, *with* ذ. (Ḳ in art. جذف.) — جَدَفَتِ السَّمَآءُ بِالثَّلْجِ *The sky cast down snow:* (Ḳ:) and so with ذ. (TA.)

2. جدّف, (Ṣ,) inf. n. تَجْدِيفٌ, (Ṣ, Ḳ,) *He denied,* or *disacknowledged, favours,* or *benefits;* or *was ungrateful,* or *unthankful, for them:* (Aṣ, Ṣ, Ḳ:) or *he deemed the gifts of God small:* (El-Umawee, Ṣ, Ḳ:) or *he said that he was in an evil state when he was in a good state:* (TA:) or *he said,* لَيْسَ لِى وَلَيْسَ عِنْدِى [app. meaning *There is nothing due to me nor by me*]; (Ḳ;) thus explained by Mohammad on his saying that the worst of deeds is التَّجْدِيفُ: (TA:) [accord. to Golius, *he blasphemed;* and identified by him, in this sense, with the Hebr. גִּדֵּף.] It is said in a trad., لَا تُجَدِّفُوا بِنِعْمَةِ ٱللّٰهِ (Ṣ, TA) *Deny not ye,* or *disacknowledge not,* or *be not ungrateful* or *unthankful for, the bounty of God, and deem it not small.* (TA.)

4: see 1, in two places. ═ اجدفوا *They raised cries, shouts, noises, a clamour,* or *confused cries* or *shouts* or *noises.* (Ḳ, TA.)

7: see 1.

جَدَفٌ *A grave; a sepulchre;* (Ṣ, Mṣb, Ḳ;) like جَدَثٌ; for the Arabs made ف and ث interchangeable: (Fr, Ṣ:) the former is of the dial. of Nejd; and the latter, of the dial. of Tihámeh: (Mṣb in art. جدث:) [accord. to some,] the former is formed from the latter by substitution [of ف for ث]: (Ṣ:) IJ argues that this is the case because the former has not أَجْدَافٌ for pl.: (TA:) but it has this pl., (Fr, Ṣ, R, TA,) used by Ru-beh. (R, TA.) ═ Also, said in a trad. to be the beverage of the jinn, or genii, (Ṣ, TA,) *Beverage that has not been covered* [*at night* according to a precept of the Prophet]: (Ḳatádeh, Ṣ, Ḳ:) or *of which the mouth of the skin containing it has not been tied* [*at night*]: (Ḳ:) or *a certain plant of El-Yemen, the eater of which needs not to drink after it:* (Ṣ, Ḳ:) or *a certain plant of El-Yemen, eaten by camels, which thereby become in no need of water:* (M, TA:) or the *froth,* or *floating particles, cast up by beverage;* (El-'Otbee, Hr, Ḳ;) as though it were cut off from the beverage. (El-'Otbee, Hr, TA.)

جَدَفَةٌ *Cries, shouts, noises, clamour,* or *a confusion of cries* or *shouts* or *noises:* and the *sound made in running.* (Ṣgh, Ḳ.)

جَوَادِفُ [pl. of جَادِفَةٌ,] Gazelles *going with short steps.* (Ṣgh, Ḳ.)

أَجْدَفُ *Short:* (Lth, Ḳ:) applied to a man. (TA.) — And [the fem.] جَدْفَآءُ A ewe, or she-goat, *having somewhat cut off from her ear.* (Ḳ.)

مِجْدَافٌ: see مِجْدَفٌ.

مُجَدَّفٌ *Straitened:* so in the saying, إِنَّهُ لَمُجَدَّفٌ عَلَيْهِ العَيْشُ [*Verily the means of living are rendered strait to him*]: (Ḳ:) but in the L, ▼لَمَجْدُوفٌ. (TA.)

مِجْدَافٌ The *wing* of a bird: (Ṣ, Mṣb, Ḳ:) sometimes with ذ. (Mṣb.) — And hence, (Ḳ,) [*An oar; a paddle;*] *a certain appertenance of a ship or boat;* (Aṣ, Ṣ, Mṣb, Ḳ;) *a piece of wood at the head of which is a broad board, with which one propels a ship* or *boat;* (M, TA;) and ▼مِجْدَفٌ [signifies the same;] *a certain thing with which a ship,* or *boat, is put in motion:* (Mgh:) pl. مَجَادِيفُ: (Mṣb:) from جَدَفَ said of a bird: (Aṣ, Ṣ, M:) also called مِجْذَافٌ (IDrd, Ṣ, Mṣb) and مِقْذَفٌ and مِقْذَافٌ. (TA.) — And hence, as being likened thereto, ‡ *A whip:* and so with ذ. (TA in this art. and in art. جذف.) — And for a similar reason, ‡ The *neck.* (TA.)

مَجْدُوفٌ A [skin of the kind called] زِقّ *having the legs cut off:* and so with ذ. (Ḳ,* TA.) And مَجْدُوفُ اليَدَيْنِ A man *having the arms,* or *hands, cut off.* (TA.) — And [hence,] the latter, †A *niggardly* man. (TA.) — And مَجْدُوفُ الكُمَّيْنِ, (Ḳ, TA,) and اليَدِ, and القَمِيصِ, and الإِزَارِ, (TA,) †*Short in respect of the sleeves,* (Ḳ, TA,) and *of the arm,* and *of the shirt,* and *of the waist-wrapper.* (TA.) — See also مُجَدَّفٌ.

جدل

1. جَدَلَهُ, aor. ُ (Ṣ, Ḳ) and ِ, (Ḳ,) inf. n. جَدْلٌ, (Ṣ,) *He twisted it firmly;* (Ṣ, Ḳ;) namely, a rope. (Ṣ.) — *He made it firm, strong,* or *compact.* (TA.) — [Hence,] جَارِيَةٌ حَسَنَةُ الجَدْلِ †[*A girl of beautiful compacture; of beautiful, compact make*]. (Ṣ.) — [Hence also,] شَاكِلَتِهِ الَّتِى جُبِلَ عَلَيْهَا جُبِلَ عَلَى † [*He did according to his own particular way, course, mode,* or *manner, of acting,* or *conduct, to which he was strongly disposed by nature*]. (TA.) ═ See also 2. ═ جَدَلَ, inf. n. جُدُولٌ, *It* (a thing) *was,* or *became, hard, and strong.* (Ḳ,* TA.) — جَدَلَ الحَبُّ فِى السُّنْبُلِ *The grain became strong in the ears:* (Ṣ, O, TA:) or, accord. to the Ḳ, it means وَقَعَ [i. e., *came* into the ears]. (TA.) — جَدَلَ said of a

young gazelle, &c., *He became strong, and followed his mother.* (Ḳ.) [See also جَادِلٌ.] —— جَدِلَ, aor. ـَ, inf. n. جَدَلٌ, [said in the Ṣ to be a subst. from 3, q. v.,] *He contended in an altercation, disputed,* or *litigated, vehemently,* or *violently.* (Mṣb.)

2. جَدَّلَهُ, (Ṣ, Mṣb, Ḳ,) inf. n. تَجْدِيلٌ, (Mṣb,) *He threw him down* (Ṣ, Mṣb, Ḳ) *upon the* جَدَالَة, (Mṣb, Ḳ,) i. e., (TA,) *upon the ground;* (Ṣ, TA;) as also ▼ جَدَلَهُ, (Ḳ,) inf. n. جَدْلٌ: (TA:) or the former signifies he did so *much,* or *often.* (TA.) You say, طَعَنَهُ فَجَدَّلَهُ [*He thrust him,* or *pierced him, with a spear* or *the like, and threw him down* &c.]. (Ṣ, Mṣb.) [See also 3.]

3. جادلهُ, inf. n. مُجَادَلَةٌ and جِدَالٌ, (Ṣ, Mgh, Ḳ,) *He contended in an altercation,* or *disputed,* or *litigated, with him:* (Ṣ, TA:) or did so *vehemently,* or *violently,* (Mgh, Ḳ,) and *ably,* or *powerfully:* (Ḳ:) [or he did so *obstinately,* or *merely for the purpose of convincing him;* for] مجادلة signifies the *disputing respecting a question of science for the purpose of convincing the opponent, whether what he says be wrong in itself or not:* (Kull p. 342:) [*he wrangled with him:*] or جادل, inf. n. مجادلة and جدال, as above, signifies originally *he contended in an altercation,* or *disputed,* or *litigated, by advancing what might divert the mind from the appearance of the truth and of what was right:* and accord. to a later usage, of the lawyers, *he compared evidences* [*in a discussion with another person,* or *other persons,*] *in order that it might appear which of those evidences was preponderant:* and the doing this is commendable if for the purpose of ascertaining the truth; but otherwise it is blameable: (Mṣb:) accord. to Er-Rághib, جدال signifies the *competing in disputation* or *contention, and in striving to overcome* [*thereby*]; from جَدَلْتُ الحَبْلَ, meaning, "I twisted the rope firmly;" as though each of the two parties twisted the other from his opinion: or, as some say, it originally means the act of *wrestling, and throwing down another upon the* جَدَالَة [or *ground*]: accord. to Ibn-El-Kemál, *a disputing that has for its object the manifesting and establishing of tenets* or *opinions.* (TA.) [See also جَدِلَ.]

4. اجدلت *She* (a gazelle) *had her young one* [*sufficiently grown to be*] *walking with her.* (Zj, Ḳ.)

5: see 7.

6. تجادلوا *They contended in an altercation, disputed,* or *litigated,* [or did so *vehemently,* or *violently,* &c., (see 3,)] *one with another.* (KL, MA, &c.,)

7. انجدل *He fell down upon the ground:* (Ṣ:) *he became thrown down upon the* جَدَالَة, i. e., the *ground;* and in like manner ▼ تجدّل, *he became thrown down,* &c., *much,* or *often.* (TA.)

8. اِجْتِدَالٌ The act of *building,* or *constructing.* (TA.) El-Kumeyt says,

مَجَادِلَ شَدَّ الرَّاصِفُونَ اجْتِدَالَهَا

(Ṣ, TA) i. e. [*Pavilions of which the masons have made strong*] *the building,* or *construction.* (TA.)

[Q. Q. 1. جَدْوَلَ *He ruled* a book *with lines;* such as are ruled round a page, &c. See جَدْوَلٌ.]

جَدْلٌ *Hard, and strong;* as also ▼ جَدِلٌ. (Ḳ,* TA.) —— Also, and ▼ جِدْلٌ, *A strong, firm,* or *compact, penis.* (Ḳ,* TA.) —— Also, (Ḳ,) or the former, (Ṣ, TA,) *Any member,* or *limb:* (Ṣ, Ḳ:) pl. جُدُولٌ. (Ṣ, TA.) —— Also, (Ḳ,) or the former, (TA,) *Any complete bone,* [app. *with its flesh,*] *not broken, nor mixed with aught beside:* pl. [of pauc.] أَجْدَالٌ and [of mult.] جُدُولٌ. (Ḳ, TA.) —— Also, (Ḳ,) or [the pl.] جُدُولٌ, (Lth, TA,) The *bones of the arms and legs* (Lth, Ḳ, TA) of a man: (Lth, TA:) and *of the fore and hind legs* of the victim termed عَقِيقَة. (TA from a trad.)

جِدْلٌ: see جَدْلٌ.

جَدَلٌ *Vehemence,* or *violence, in altercation* or *disputation* or *litigation;* (Ṣ, Ḳ;) and *ability,* or *power, to practise it:* (Ḳ:) [or simply *contention in an altercation; disputation;* or *litigation:*] a subst. from جَادَلَهُ: (Ṣ:) or inf. n. of جَدِلَ [q. v.]. (Mṣb.) —— Hence, as a term of logic, *A syllogism composed of things well known, or conceded; the object of which is to convince the opponent, and to make him to understand who fails to apprehend the premises of the demonstration.* (TA.)

جَدِلٌ: see جَدْلٌ. —— Also One *who contends in an altercation, disputes,* or *litigates, vehemently,* or *violently,* (Mṣb, Ḳ,) and *ably,* or *powerfully;* and so ▼ مِجْدَلٌ and ▼ مِجْدَالٌ. (Ḳ.)

جَدْلَآءُ fem. of أَجْدَلُ. —— Also syn., in two senses, with جَدِيلَةٌ, which see, in two places.

جَدْوَلٌ *A rivulet; a streamlet;* (Ṣ, Mṣb, Ḳ;) [whether *natural,* or *formed artificially for irrigation;* being often applied to *a streamlet for irrigation, in the form of a trench,* or *gutter;*] it is *less than a* سَاقِيَة; and this is less than a نَهْر: (Mgh in art. سقى:) as also جِدْوَلٌ: (Ḳ:) pl. جَدَاوِلُ. (Mṣb.) —— Hence, اِسْتَقَامَ جَدْوَلُهُمْ ‡ *Their affair,* or *case, was,* or *became, in a right, a regular,* or *an orderly, state; like the* جدول *when its flow is uniform and uninterrupted.* (TA.) And اِسْتَقَامَ جَدْوَلُ الحَاجِّ † *The caravan of the pilgrims formed an uninterrupted line.* (TA.) —— [Hence also جَدْوَلٌ as meaning † *A kind of small vein.* (Golius from Ibn-Seenà.)] —— Hence also جَدْوَلُ كِتَابٍ † [*A ruled line,* (such as is ruled round a page, &c.,) and *a column,* and *a table, of a book*]. (TA.)

جَدِيلٌ applied to a rope, *Firmly twisted;* as also ▼ مَجْدُولٌ. (TA.) —— *A camel's nose-rein* (Ṣ, Ḳ) *of hide,* or *leather,* (Ṣ,) *firmly twisted:* (Ṣ, Ḳ:) and *a cord of hide,* or *leather,* or *of* [*goats'*] *hair,* [*that is put*] *upon the neck of the camel:* (Ḳ:) and the [*kind of women's ornament termed*] وِشَاح (Ṣ, Ḳ) is sometimes thus called: (Ṣ:) pl. جُدُلٌ. (Ḳ.)

جَدَالَةٌ The *ground:* (Ṣ, Mṣb, Ḳ:) or *hard ground:* (TA:) or *ground having fine sand.* (Ḳ.)

جَدِيلَةٌ A رَهْط, [q. v.,] i. e., (TA,) *a thing like an* إِتْب, *of hide,* or *leather, which boys, and menstruous women, wear round the waist in the manner of an* إِزَار. (Ḳ, TA.) —— A [*tribe, such as is termed*] قَبِيلَة: and a *region, quarter,* or *tract;* syn. نَاحِيَة: (Ṣ, Ḳ:) and so ▼ جَدْلَآءُ, in both these senses, as used in the phrase, هٰذَا عَلَى جَدْلَائِهِ [*This is according to the way of his region,* and *of his tribe*]. (TA.) You say also, ذَهَبَ عَلَى جَدْلَائِهِ ▼, in the Ḳ, erroneously, جَدْلَائِهِ, (TA,) i. e., على وَجْهِهِ [*He went his own way*], (Ḳ, TA,) and نَاحِيَتِهِ [*towards his region,* or *quarter,* or *tract*]. (Ḳ.) —— *A state,* or *condition.* (Ḳ.) —— ‡ *A particular way, course, mode,* or *manner, of acting,* or *conduct;* syn. شَاكِلَة, (Ṣ, Ḳ,) and طَرِيقَة. (Ḳ.) You say, عَمِلَ عَلَى جَدِيلَتِهِ, i. e. [*He did according to his own particular way,* &c.; or] عَمِلَ عَلَى شَاكِلَتِهِ الَّتِى جُبِلَ عَلَيْهَا [explained above: see 1]. (TA.) —— ‡ *A determination of the mind.* (TA.) —— † The *management,* or *ordering, of* a people's *affairs;* the *exercise of the office of* عَرِيف. (AA, TA.)

جَادِلٌ A boy *becoming,* or *become, strong, vigorous,* or *robust.* (Ṣ.) —— A she-camel's young one *above such as is termed* رَاشِح, *which is such as has become strong, and walks with his mother.* (Aṣ, Ṣ.) [See also جَدَلَ.]

جَنْدَلٌ: and جُنَدِلٌ: &c.: see art. جندل.

أَجْدَلُ; fem. جَدْلَآءُ: see مَجْدُولٌ, in three places. —— Also, [accord. to most of the grammarians أَجْدَلُ, but accord. to some أَجْدَلٌ,] The *hawk;* syn. صَقْرٌ; (Ṣ, Ḳ;) as also ▼ أَجْدَلِىٌّ: (Ḳ:) or an epithet applied to the hawk [and therefore without tenween]: (TA:) pl. أَجَادِلُ. (Ḳ.)

أَجْدَلِىٌّ: see what next precedes.

مِجْدَلٌ A قَصْر [or *palace,* or *pavilion,* &c.,] (Ṣ, Ḳ, TA [in the CḲ القَصِيرُ is erroneously put for القَصْرُ]) *strongly constructed:* (TA:) pl. مَجَادِلُ. (Ṣ, Ḳ.) —— See also جَدِلٌ.

مِجْدَالٌ *A piece of rock or stone:* [*an oblong roofing-stone, of those which, placed side by side, form the roof of a subterranean passage,* &c.:] pl. مَجَادِيلُ. (TA.) —— See also جَدِلٌ.

مَجْدُولٌ: see جَدِيلٌ. [Hence,] دِرْعٌ مَجْدُولَةٌ ‡ *A compact coat of mail;* (Ṣ, TA;) as also ▼ جَدْلَآءُ: (Ṣ, Ḳ:) pl. [of the latter] جُدْلٌ. (Ḳ.) —— ‡ A man (Ḳ, TA) *of slender make,* (TA,) *slender in the* [*bones called*] قَصَب, *of firm,* or *compact, make* (مُحْكَمُ الفَتْلِ [as though firmly twisted]): (Ḳ, TA:) or *slender, slim, thin, spare, lean,* or *light of flesh; not from emaciation:* (Ṣ:) and مَجْدُولُ الخَلْقِ, as some say, *of firm,* or *compact, make.* (TA.) And مَجْدُولَةٌ † A woman *small in the belly, and compact in flesh:*

(A in art. فيض:) or مَجْدُولَةُ الخَلْقِ a girl *of beautiful compacture; of beautiful, compact make;* syn. حَسَنَةُ الجَدْلِ. (Ṣ.) Also سَاعِدٌ ↑أَجْدَلُ † [*A fore arm, or an upper arm,*] *of firm, or compact, make.* (Ḳ,* TA.) And سَاقٌ مَجْدُولَةٌ and ↑جَدْلَآءُ ‡ [*A shank of beautiful compacture;*] *well rounded; well turned;* syn. حَسَنَةُ الطَّىِّ. (Ḳ, TA.)

جدو

1. جَدَا عَلَيْهِ, (Mṣb, Ḳ,) and جَدَاهُ, first pers. جَدَوْتُ, (IB, TA,) aor. ـُ, (Ḳ,) inf. n. جَدْوٌ (Mṣb, TA) and جَدًا; (Mṣb;) and ↑اجدى عَلَيْهِ, (Mṣb, Ḳ,) and ↑اجداهُ, (Ṣ,) the prep. in the former of these two being suppressed in the latter; (TA;) and ↑اجتداهُ; (TA;) *He gave him a gift.* (Ṣ, IB, Mṣb, Ḳ,* TA.) — [Hence,] جَدَا عَلَيْهِ شُؤْمَهُ † *He drew his evil fortune, or ill luck, upon him:* an ironical expression; [for it literally means *he gave him, or bestowed upon him, his evil fortune.*] (TA.) — Hence also, اجدى ↑ عَلَيْكَ ‡ *It* (a thing) *sufficed thee.* (Mṣb.) مَا أَجْدَى ↑ فِعْلُهُ شَيْئًا ‡ *His deed, or act, did not profit him,* or *avail him, aught.* (Mṣb.) And مَا يُجْدِى ↑ عَنْكَ هٰذَا † *This does not stand thee in any stead; does not profit thee,* or *avail thee.* (Ṣ.) ═ جَدَوْتُهُ, (Ṣ, IB, Mṣb, Ḳ,*) [aor. ـُ,] inf. n. جَدْوٌ; (Ḳ;) and جَدَيْتُهُ; (Ḳ in art. جدى;) and ↑اِجْتَدَيْتُهُ, (Ṣ, Mṣb, Ḳ,*) and ↑اِسْتَجْدَيْتُهُ; *I sought,* or *demanded,* (Ṣ,) or *asked,* (IB, Mṣb, Ḳ,) *of him* (Ṣ, IB, Mṣb, Ḳ) *a gift,* (Ṣ,) or *a thing wanted.* (Ḳ.) [See an ex. of the last of these verbs in a verse cited in art. لا.] Hence, مُجَادَاةٌ [inf. n. of ↑جادى]: whence, in a trad., وَقَدْ عَرَفُوا أَنَّهُ لَيْسَ عِنْدَ مَرْوَانَ مَالٌ يُجَادُونَهُ عَلَيْهِ, meaning يَسْأَلُونَهُ عَلَيْهِ [i. e. *And they knew that there was not, in the possession of Marwán, property for which they should ask as owed by him*]. (TA.)

3: see 1.

4: see 1, in five places. ═ Also اجدى, *He obtained a gift.* (Ṣ, Mṣb.)

8: see 1, in two places.

10: see 1.

جَدًا *i. q.* جَدْوَى, q. v. — Hence, (Ḥar p. 32,) جَدًا, (Ḳ,) also written جَدَى, (ISk, TA,) or مَطَرٌ جَدًا, (Ṣ,) and, accord. to the Ḳ, ↑جَدْوَى, but this latter is not known except as signifying "a gift," (TA,) *A common,* or *general, rain;* (Ṣ, Ḳ, TA;) *of wide extent:* (TA:) or *of which the uttermost is not known.* (Ḳ.) One says also سَمَآءٌ جَدًا, meaning *A rain having a rain following it;* making the latter word masc. because it has the force of an inf. n. (TA.) And اَللّٰهُمَّ اسْقِنَا غَيْثًا غَدَقًا وَجَدًا طَبَقًا [*O God, water us with a copious rain, and a rain that shall cover the land*]: (Ṣ, TA:) occurring in a trad. respecting prayer for rain. (TA.) — And خَيْرٌ جَدًا *Ample good;* (Ḳ;) *of wide extent to men.* (TA.) ═ لَا آتِيكَ جَدَا الدَّهْرِ (Ṣ,* Ḳ,* TA) i. e. [*I will not come to thee*] *ever,* like يَدَ الدَّهْرِ; (S, TA;) or *to the end of time.* (Ḳ, TA.)

جَدْوَى *A gift;* (Ṣ, Mṣb, Ḳ;) as also ↑جَدًا: (Ṣ, Ḳ:) dual (of the former, TA) جَدْوَانِ and جَدْيَانِ; (Lḥ, M, Ḳ;) the former, regular; (M, TA;) the latter, anomalous, (M, Ḳ, TA,) formed by commutation. (M, TA.) You say, مَا أَصَبْتُ مِنْ فُلَانٍ جَدْوَى قَطُّ [*I have not obtained from such a one a gift ever*]. (TA.) And hence the prov., شَغَلَتْ شِعَابِى جَدْوَاىَ: see art. شعب. (Ṣ in that art.) — See also جَدًا.

جَدَآءٌ *Profit, utility,* or *avail.* (Ṣ, TA.) So in the saying, فُلَانٌ قَلِيلُ الجَدَآءِ عَنْكَ [*Such a one is of little profit, utility, or avail, to thee; will stand thee in little stead*]. (Ṣ.)

جَدِىٌّ [originally جَدِيوٌ] *Munificent,* or *bountiful.* (TA.)

جَادٍ *Asking, seeking,* or *demanding,* (Ṣ, Ḳ,) *a bounty,* or *benefit,* (Ṣ,) or *gift:* (Ḳ:) pl. جُدَاةٌ. (TA.)

أَجْدَى [*More, and most, profitable, useful,* or *availing*]. It is said in a prov., أَجْدَى مِنَ الغَيْثِ فِى أَوَانِهِ [*More profitable than rain in its season*]. (Meyd.)

جدى

1. جَدَيْتُهُ: see 1 in art. جدو. — One says of the locust, يَجْدِى كُلَّ شَىْءٍ, meaning *It eats everything.* (TA.) ═ لَمْ أَجْدِ مِنْ ذٰلِكَ بُدًّا *I found no means of avoiding,* or *escaping, that,* is sometimes said for لَمْ أَجِدْ الخ. (Kz, TA in art. وجد.)

2. جَدَّى الرَّحْلَ, inf. n. تَجْدِيَةٌ, *He made,* or *put, to the camel's saddle a* [جَدْيَة or] جَدِيَّة. (TA.)

4. اجدى *It* (a wound) *flowed* [*with blood:* see جَدِيَّةٌ]. (Ḳ.)

جَدْىٌ *A kid:* (Ṣ:) or *a male kid;* (IAmb, Mṣb, Ḳ;) the female being called عَنَاقٌ: (IAmb, Mṣb:) or *a kid in his first year;* (Mgh, Mṣb;) *not yet a year old:* (TA:) one should not say جِدْىٌ; (Ṣ;) this being a bad dial. var.: (Mṣb:) pl. (of pauc., TA, applied to three, Ṣ) أَجْدٍ (Ṣ, Mṣb, Ḳ) and (of mult., TA, applied to more than three, Ṣ) جِدَآءٌ (Ṣ, Mgh, Mṣb, Ḳ) and جِدْيَانٌ: (Ḳ:) جَدَايَا [as pl. of جَدْى] is not allowable. (Ṣ.) — Hence, as being likened thereto, (M, TA,) الجَدْىُ † *A certain star,* (Ṣ, Mṣb, Ḳ,) [*the star α of Ursa Minor,* commonly called *the pole-star,*] *that revolves with* بَنَاتُ نَعْشٍ, (Ḳ,) *by the side of the* [*north*] *pole, by which the ḳibleh is known,* (Ṣ,) or *according to which the ḳibleh is turned;* (Mṣb;) *the bright star at the extremity of the tail of the Lesser Bear;* (Ḳzw;) *the star of the ḳibleh;* (Mgh;) *also called* جَدْىُ الفَرْقَدِ; (Mgh, Mṣb;) *and called by the astronomers* ↑الجُدَىُّ, in the dim. form, *to distinguish it from what next follows.* (Mgh, MF.) [See also القُطْبُ.] — Hence also, (M, TA,) *A certain sign of the Zodiac;* (Ṣ, Ḳ;) [namely, *Capricornus;*] *the tenth of the signs of the Zodiac;* (Mgh;) *that next to the* دَلْو; *unknown to the Arabs* [*of the classical times*]. (Ḳ.) This and the former together are called [the] جَدْيَانِ. (TA.) ═ جَدْيَانِ is also an anomalous dual of جَدْوَى, q. v. (Lḥ, M, Ḳ.) ═ See also what next follows.

جَدْيَةٌ and ↑جَدِيَّةٌ, (Ṣ, Ḳ,) but not جَدِيدَةٌ, which is used by the vulgar, (Ṣ,) [*A kind of pad,* or] *a stuffed thing,* (Ṣ,) or *a stuffed piece* (Ḳ, TA) *of a* كِسَآء, (TA,) *that is put beneath a horse's saddle,* (Ḳ,) or *beneath the two boards* (الدَّفَّتَانِ) *of a horse's and of a camel's saddle;* [*one on either side;* for] there are two of such stuffed things: (Ṣ:) the pl. of the former is جَدَيَاتٌ, (Sb, Ṣ,) which may be used as a pl. of mult., (TA,) or جَدْيَاتٌ, so in [some of] the copies of the Ḳ, [but omitted in the CḲ and in my MS. copy of the Ḳ,] following the TṢ, as on the authority of A 'Obeyd and AA and En-Naḍr, (TA,) and ↑جَدْىٌ; (Ṣ, IB, [in some copies of the Ṣ جَدًا, but the former (which I find in two copies of the Ṣ) is said by IB to be the right; or rather this is a coll. gen. n.,]) like as شَرْىٌ is of شَرْيَةٌ: (IB, TA:) the pl. of ↑جَدِيَّةٌ is جَدَايَا. (Ṣ.) [See also جَدِيدَةٌ, and وِفَادَةٌ.]

جِدَآءٌ, (Ḳ in this art.,) or جِذَآءٌ, (A in art. جذر, and Ḳ in art. بوج,) [the latter is the term commonly known, *An arithmetical square;*] the *product of multiplication* [*of a number by itself*]; *as when you say, the* جداء [or جذاء] *of three* [in some copies of the Ḳ, *of three multiplied by three,*] *is nine;* (Ḳ,* TA;) *also called* مَالٌ. (Mṣb in art. جذر.) [See جَذْرٌ.]

الجُدَىُّ: see جَدْىٌ.

جَدَايَةٌ and جِدَايَةٌ *A young gazelle;* syn. غَزَالٌ: (Ḳ, and so in a copy of the Ṣ:) or *a young doe-gazelle;* syn. غَزَالَةٌ: (so in another copy of the Ṣ:) said by Aṣ to be *like the* عَنَاق *of goats:* (Ṣ:) or the *male,* and the *female, of the young of gazelles, when it has attained the age of six months,* or *seven, and has run, and become strong:* or, as some say, the *male thereof:* pl. جَدَايَا. (M, TA.)

جَدِيَّةٌ: see جَدْيَةٌ, in two places. ═ Also *Flowing blood;* (Lḥ, Ḳ;) blood not flowing being termed بَصِيرَةٌ: (Lḥ, TA:) or the former, *blood adhering to the body;* and the latter, blood upon the ground: (AZ, Ṣ:) or the former, *a streak of blood:* (Ṣ:) or the *first quantity that flows at once, of blood:* (TA:) pl. جَدَايَا. (Ṣ.) — *A piece of musk.* (Ḳ.) — The *colour* of the face. (Ḳ, TA.) You say, اِصْفَرَّتْ جَدِيَّةُ وَجْهِهِ [*The colour of his face became yellow*]. (TA.) ═ *I. q.* نَاحِيَةٌ [*A side; a lateral,* or *an outward* or *adjacent, part* or *portion, region, quarter,* or *tract;* &c.]. (Ḳ.) So in the saying, هُوَ عَلَى جَدِيَّتِهِ [app. meaning *He is keeping to his own side: he is following his own course;* like the phrase هُوَ عَلَى طَرِيقَتِهِ: or *he is by himself;* like هُوَ عَلَى وَحْدِهِ and جِدَتِهِ]. (TA.)

جَادٍ The *locust;* because it eats (يَجْدِى, i. e. يَأْكُلُ) everything: but the appellation [more] commonly known is جَابٍ. (TA.)

جَادِىٌّ [erroneously written in the CK, in this art. and in art. جود, without the sheddeh to the ى] *Saffron;* (Az, IF, K;) as also ↓ جَادِيَآءُ: (Sgh, K:) the former a rel. n. from الجَادِيَةُ, or جَادِيَةُ, a town of البَلْقَآء, in Syria, said to produce saffron: it is mentioned by Az and IF in this art., being held by them to be of the measure فَاعُولٌ [originally جَادُوىٌ]: by J, in art. جود, as being of the measure فَعْلِىٌّ. (TA.) —— Also ‡ *Wine;* (K, TA;) as resembling saffron in colour. (TA.)

جَادِيَآءُ: see what next precedes.

جذ

1. جَذَّهُ, (S, A, L, Msb,) aor. ُ, (L, Msb,) inf. n. جَذٌّ, (L, Msb, K,) *He cut it,* or *cut it off;* (S, A, L, Msb;) namely, a thing; (S, Msb;) as, for instance, a rope: (L:) or *he cut it off utterly;* (L, K;) and جَذْجَذَةٌ [inf. n. of ↓ جَذْجَذَ] signifies the same as جَذٌّ, i. e. the *cutting off utterly:* (K:) and اِجْتِذَاذٌ [inf. n. of ↓ اجتذّ], the act of *cutting:* (KL:) or جَذَّهُ signifies *he cut it off utterly and quickly;* and in like manner ↓ جذّذهُ [applied to many objects, or as meaning *he cut it,* &c., *repeatedly,* or *many times,* or *in many pieces,* or *much*]. (L.) [Hence,] جَذَّ النَّخْلَ, aor. ُ, inf. n. جَذٌّ and جَذَاذٌ and جِذَاذٌ, *He cut off the fruit of the palm-trees;* syn. صَرَمَ: (Lh, L:) or, accord. to some, جذاذ signifies the *cutting off* of all fruits, and جداد relates particularly to palm-trees. (TA in art. جد.) [Hence also,] جُذُّوهُمْ جَذًّا, in a trad. relating to the battle of Honeyn, †*Cut ye them off utterly; exterminate them by slaughter.* (L.) —— Also, (S, A, L, Msb,) aor. ُ, (L,) inf. n. جَذٌّ, (L, K,) *He broke it;* (S, A, L, Msb, K;) namely, a thing, (S, Msb,) or a hard thing: (M, L:) and اِجْتِذَاذٌ [inf. n. of ↓ اجتذّ], also, signifies the act of *breaking.* (KL.) —— Also, (L,) inf. n. جَذٌّ, (L, K,) *He hastened it;* or *hastened to it.* (L, K.*) It is said in a prov., respecting one who boldly ventures upon taking a false oath, جَذَّهَا جَذَّ العَيْرِ الصِّلِّيَانَةَ *He hastened to it as the ass hastens to the plant called* صلّيانة. (L.)

2: see 1.

5: see 7.

7. انجذّ *It became cut,* or *cut off:* (S, L, Msb, K:) or *cut off utterly:* or *cut off utterly and quickly:* and in like manner, ↓ تجذّذ [said of a number of things, or used in a frequentative or an intensive sense; being quasi-pass. of 2]. (L.)

8: see 1, in two places.

R. Q. 1. جَذْجَذَ, inf. n. جَذْجَذَةٌ: see 1.

جِذٌّ *A piece broken off; a fragment:* pl. أَجْذَاذٌ: so in the phrase, كَسَرْتُهُ أَجْذَاذًا [*I broke it in pieces,* or *fragments*]; occurring in a trad., relating to an idol. (L.) [See also جُذَاذٌ.]

مَا عَلَيْهِ جُذَّةٌ *There is not upon him a piece of rag;* i. e., *any garment to cover him:* (L:) or *anything* (S, L, K) *of clothing:* (S, L:) pl. جُذَذٌ. (Bd in xxi. 59.)

جُذَذٌ: see جُذَاذٌ.

جَذَاذٌ, with fet-h, The *act of separating a thing from another thing* (فَصْلُ شَىْءٍ عَنْ شَىْءٍ), accord. to some copies of the K and the TA; [see 1;] accord. to other copies of the K, [probably by mistranscription,] the *superiority of a thing over another thing,* (فَضْلُ شَىْءٍ عَلَى شَىْءٍ); as also ↓ جَذَاذَةٌ. (K.) ═ See also what next follows.

جُذَاذٌ and ↓ جِذَاذٌ, (S, L, K,) the former of which is the more chaste, (S, L,) and ↓ جَذَاذٌ, (L, K,) substs. from جَذٌّ the act of "breaking;" (K;) *What is broken, of,* or *from, a thing:* (S, L:) or *what is broken,* or *cut, in pieces:* or *broken pieces:* so in the Kur xxi. 59, in which the word is read in these three different forms: (L:) some also read جُذُذًا, which is pl. of ↓ جَذِيذٌ; and some, جُذَذًا, pl. of جُذَّةٌ: (Bd:) or the first signifies *fragments of a thing much broken;* [as a coll. gen. n.;] and the n. un. is ↓ جُذَاذَةٌ: (Lth, L:) [or] it is an extr. pl. of ↓ جَذِيذٌ; (L;) or *i. q.* مَجْذُوذٌ; and ↓ جِذَاذٌ is a dial. var.; or is pl. of ↓ جَذِيذٌ. (Bd.) [See also جِذٌّ.] —— Also the first, *Small pieces,* or *particles, of silver:* and ↓ جُذَاذَاتٌ, *pieces of* silver: (L:) or the latter, *cuttings,* or *clippings,* (S, L, K,) *of silver.* (M and L in art. قذ.) —— Also *Stones containing gold;* (Ks, S, L, K;) so called because they are broken: (Ks, S, L:) or *stones containing gold* &c. *which are broken;* as also ↓ جِذَاذٌ. (Msb.) —— And جذاذ [app. جُذَاذٌ or ↓ جِذَاذٌ], (L,) or ↓ جِذَذٌ, (TA,) *i. q.* فِرَقٌ [*Distinct parties,* or *portions,* &c., of men or things]. (L, TA.)

جِذَاذٌ: see جُذَاذٌ, in four places.

جَذِيذٌ and ↓ مَجْذُوذٌ *Cut;* or *cut off:* or *cut off utterly:* or *cut off utterly and quickly:* and also *broken:* (L:) pl. of the former, جُذُذٌ and جِذَاذٌ (Bd in xxi. 59) and جُذَاذٌ, which last is extr. (L. See the paragraph headed جُذَاذٌ, in three places.) —— Also the former, (A, K,) and شَرَابٌ جَذِيذٌ, (A,) and ↓ جَذِيذَةٌ, (S, L, K,) *Meal of parched barley* or *wheat;* syn. سَوِيقٌ; (S, A, L, K;) *so prepared* [*by being moistened with water* or *with clarified butter* &c.] *as to be drunk:* (A:) and ↓ جذيذة is [*a mess of the kind called*] جَشِيشَة *made of coarse* سويق; so called because it is broken and made into coarse particles: and *a quantity of* سويق, or *the like, such as a man eats or drinks at one time.* (L.)

جَذَاذَةٌ: see جَذَاذٌ.

جُذَاذَةٌ and its pl. جُذَاذَاتٌ: see جُذَاذٌ, in two places.

جَذِيذَةٌ: see جَذِيذٌ, in two places.

يَدٌ جَذَّآءُ [fem. of أَجَذُّ] ‡ *An amputated arm* or *hand:* used figuratively: see أَحَذُّ, in art. حذ. (L.) —— رَحِمٌ جَذَّآءُ † *A bond of relationship* [*cut,* or *severed,* or] *not made close by affection;* expl. by إِذَا لَمْ تُوصَلْ [a well-known phrase: erroneously explained by Golius as meaning "uterus infœcundus vel interruptæ conceptionis"]: (Fr, S, L, K:) as also حَذَّآءُ. (Fr, S, L.) —— يَمِينٌ جَذَّآءُ: see حَذَّآءُ. —— سِنٌّ جَذَّآءُ *A tooth much broken;* or *broken in pieces.* (K, TA.)

جَذَّانٌ (As, L, and so in some copies of the K: in other copies of the K, and in the TA, جُذَّانٌ:) *Soft stones:* (As, L, K:) as also كَذَّانٌ [q. v.]: (As, L:) n. un. (of both, As, L) with ة. (As, L, K.)

مَجْذُوذٌ: see جَذِيذٌ. ═ عَطَآءٌ غَيْرُ مَجْذُوذٍ [*A gift,* or *stipend,*] *not cut short,* or *not interrupted;* syn. غَيْرُ مَقْطُوعٍ. (A 'Obeyd, S, L.)

جذب

1. جَذَبَهُ, (S, A, Msb, K,) aor. ِ, (Msb, K,) inf. n. جَذْبٌ; (S, Msb;) as also جَبَذَهُ, (S, TA,) inf. n. جَبْذٌ, of the dial. of Temeem, (TA,) or formed by transposition; (S, TA; [but see art. جبذ;]) and ↓ اجتذبهُ; (S, A, K;) *He drew it; dragged it; pulled it; tugged it; strained it; extended it by drawing,* or *pulling* or *tugging; stretched it; extended, lengthened,* or *protracted, it;* (S, A, K;) namely, a thing; and sometimes relating to an ideal object. (TA.) —— جَذَبَ مِنَ الإِنَآءِ, (S,) or جَذَبَ مِنَ المَآءِ, (K,) or جَذَبَ المَآءَ, (A, Msb,) نَفَسًا, (S, A, Msb, K,) or نَفَسَيْنِ, (S, A, Msb,) ‡ *He drank* (S, K) *from the vessel,* (S,) or *of the water,* (K,) *by putting his mouth into it,* [*a draught,* or *two draughts:*] (S, K:) or *he conveyed* [or *drew up*] *into the innermost parts of his nose* [*a draught,* or *two draughts, of the water*]. (Msb.) And جَذَبَتْ لَبَنَهَا ‡ *She* (a camel) *drank her milk* when she was milked. (A. [But see what next follows: and see also 5.]) —— جَذَبَتْ, said of a she-camel, (S, K,) and of a she-ass also, aor. ِ, inf. n. جِذَابٌ, (TA,) † *She became scant of milk;* (S, K;) *she drew her milk* (جَذَبَتْ لَبَنَهَا) *from her udder so that it went away upwards.* (TA.) —— جَذَبَهُ, (K,) or جَذَبَهُ عَنْ أُمِّهِ, (S, A,) aor. ِ, inf. n. جَذْبٌ, (TA,) ‡ *He weaned him;* namely, a colt, (S, A, K,) and a young camel, and a lamb. (TA.) And one says of a mother, جَذَبَتْ وَلَدَهَا ‡ *She weaned her young one:* so accord. to Lh, who does not specify the kind. (ISd, TA.) And accord. to the T, جَذَبَ is said of a child, or of a lamb or kid, meaning ‡ *He was weaned.* (TA.) —— جَذَبَهُ and ↓ جَاذَبَهُ *He transferred,* or *removed, it* (a thing) *from its place.* (K.) —— جَذَبَتْهُ ‡ *She repelled him,* or *rejected him;* namely, a man who sought her in marriage; (T, A, TA;) as though from the saying جَاذَبَتْهُ فَجَذَبَتْهُ [which see below]; (T, TA;) [i. e.] as though she contended with him and overcame him, and thus he became separated from her; (T, A, TA;) as also جَبَذَتْهُ. (T, TA.) [Accord. to the TA, ↓ جَاذَبَتْهُ has the same meaning; but I think that this is a mistake of a copyist.] —— جَذَبَ فُلَانٌ حَبْلَ وِصَالِهِ † *Such a one severed the bond of his union.* (M, TA.) And جَذَبَ فُلَانٌ الحَبْلَ بَيْنَنَا ‡ *Such a one severed the bond of union between us.* (A, TA.) —— جَذَبَ النَّخْلَةَ, aor. ِ,

(AḤn, Ḳ,) inf. n. جَذْبٌ, (AḤn, TA,) *He cut off the* جَذَب [q. v.] *of the palm-tree,* (AḤn, Ḳ,) to eat it. (AḤn, TA.) — جَذَبَ العَدْوَ † *He ran quickly.* (L in art. معد.) See also 7. — جَذَبَ الشَّهْرُ, (Ṣ, A, Ḳ,) aor. ـِ, inf. n. جَذْبٌ, (TA,) ‡ *The greater part of the month* (عَامَّتُهُ, Ṣ, A, Ḳ, i. e. أَكْثَرُهُ, TA) *passed.* (Ṣ, A, Ḳ.) — جَذَبَهُ, aor. ـِ, *He overcame him in* المُجَاذَبَة (Ḳ) [lit. *drawing, dragging, pulling, &c.*; (see 3;) but also] used figuratively [as meaning ‡ *he overcame him in contention*]. (TA.) You say, ▼ جَاذَبْتُهُ فَجَذَبْتُهُ [*I contended with him in drawing, dragging, &c., and I overcame him therein:* and also] ‡ *I contended with him and I overcame him.* (T, A,* TA.)

3. جاذبهُ *He contended with him in drawing, dragging, pulling or tugging, straining, or stretching, &c.* (L in art. مد.) And جاذبهُ الحَبْلَ [*He contended with him in pulling the rope*]. (Mgh in art. نزع.) And جاذبهُ الثَّوْبَ *He contended with him in pulling the garment,* or *piece of cloth.* (A.) And جاذبوا الشَّىْءَ, inf. n. مُجَاذَبَةٌ, *They pulled the thing, every one of them to himself.* (Mṣb.) And جَاذَبْتُهُ فَجَذَبْتُهُ : see 1, last sentence: [a phrase having two meanings: for] you say, جَاذَبَا, (Ḳ, TA,) inf. n. مُجَاذَبَةٌ (TA) and جِذَابٌ, (Ḥar p. 636,) meaning † *They two contended* [in any manner], *each with the other:* (Ḳ, TA:) and [in like manner,] ▼ تجاذبا, (Ḳ,) inf. n. تَجَاذُبٌ, (Ṣ,) † *They two contended together.* (Ṣ,* Ḳ.) You say also, كَانَتْ بَيْنَهُمْ مُجَاذَبَاتٌ ثُمَّ ٱتَّفَقُوا ‡ [*There were contentions between them: then they agreed*]. (A, TA.) And جَاذَبْتُهُ الشَّىْءَ † *I contended with him for the thing.* (Ṣ.) — See also 1, in two places, beside the instance in the last sentence.

5. تجذّبهُ ‡ *He drank it;* (A, Ḳ;) namely, milk: said of a pastor. (A.)

6. تجاذبوا الثَّوْبَ *They contended together in pulling the garment,* or *piece of cloth.* (A.) [Hence,] تجاذبوا أَطْرَافَ الكَلَامِ ‡ [*They contended together in discourse, talk, or conversation*]. (A.) See also 3. — And see 7.

7. انجذب *It* (a thing) *was,* or *became, drawn, dragged, pulled, tugged, strained, extended by drawing* or *pulling* or *tugging,* or *stretched, &c.;* *it dragged,* or *trailed along;* syn. انجرّ. (Ṣ and Ḳ in art. جر.) — *It was,* or *became, transferred,* or *removed, from its place;* and so ▼ تجاذب. (Ḳ.) — اِنْجِذَابٌ also signifies † *Quick going* or *journeying* or *travelling.* (Ṣ.) [You say, انجذب, and ▼ جَذَبَ السَّيْرَ, (the latter occurring in the TA in art. خلج, &c., like جَذَبَ العَدْوَ, mentioned above, see 1,) † *He went,* or *journeyed,* or *travelled, quickly.*] And انجذبوا فِى السَّيْرِ and انجذب بِهِمِ السَّيْرُ ‡ *They brought,* or *purveyed, wheat,* or *corn,* or *provisions, from afar.* (A, TA.)

8. اجتذبهُ : see 1. — Also *He seized it,* or *took it,* or *carried it off, by force.* (Ḳ, TA.) — And † *He called, summoned,* or *invited, him.* (Ḥam p. 645.)

جَذْبٌ † *Quick* journeying or travelling. (ISd, Ḳ.) — And The *stopping,* or *a stoppage, of the flow of saliva* (اِنْقِطَاعُ الرِّيقِ). (Ṣ.)

جَذَبٌ The *pith that is at the head of the palm-tree, from which the* [*fibres called*] لِيف *are pulled off, and which is then eaten;* as though so called because pulled off [or cut] from the tree; (TA;) the *heart, pith,* or *cerebrum,* (جُمَّار,) *of the palm-tree;* (AḤn, Ṣ, TA;) so in some copies of the Ḳ; (TA;) i. e. the شَحْم *of the palm-tree:* (Ṣ:) or, as in some copies of the Ḳ, and in the M and L, only *such as is coarse:* (TA:) as also ▼ جِذَابٌ : (Ḳ:) n. un. جَذَبَةٌ. (Ṣ, Ḳ.) [See also جَذَمَةٌ.]

جَذْبَةٌ مِنْ غَزْلٍ *A portion that is drawn by a single pull of spun thread,* or *yarn.* (Ṣ.) [Hence,] مَا أَعْطَاهُ جَذْبَةَ غَزْلٍ ‡ *He gave him not aught.* (A, TA.) — بَيْنِى وَبَيْنَ المَنْزِلِ جَذْبَةٌ † *Between me and the place of alighting is a piece* [*of land* or *country,* or *a tract*], meaning *a distance:* (Ṣ:) or *a far-extending piece* [*of land* or *country*]. (Ḳ.) And بَيْنَنَا وَبَيْنَ بَنِى فُلَانٍ جَذْبَةٌ, and نَبْذَةٌ, † *Between us and the sons of such a one is a small space,* or *short distance;* i. e., they are near to us. (ISh, TA.)

أَخَذَ فِى وَادِى جَذَبَاتٍ [*He took his way into the valley of Jedhebát*]: (Ḳ:) or, as given by Meyd, [and in the A,] وَقَعُوا [*they fell* into the valley of Jedhebát]: a celebrated prov.: (TA:) applied to a man who has missed the object of his aim or pursuit; (Ḳ,* TA;) جذبات being said to be derived from جَذَبَ الصَّبِىَّ "he weaned the boy;" because, in weaning, a child sometimes dies: or from اِنْجَذَبَ بِهِمِ, or اِنْجَذَبُوا فِى السَّيْرِ, السَّيْرُ, explained above: or, accord. to some, the right reading is جَدَبَاتٍ : or, as Az says, on the authority of Aṣ, the most correct reading is خَدَبَاتٍ, from خَدَبَتْهُ الحَيَّةُ "the serpent bit him;" and the prov. is applied to him who falls into perdition, and to him who wanders in perplexity from the object of his aim or desire. (TA.) [See also another reading in art. خرب.]

جِذِبَّانٌ The *sandal-thong that is between the great and second toes.* (Ḳ, TA.) You say, مَا أَغْنَى عَنِّى جِذِبَّانًا *He did not stand me in stead of,* or *avail me as much as, a sandal-thong that is between the great and second toes.* (AA, TA.)

جَذَابِ, indecl., [as a proper name, changed in form from الجَاذِبُ,] (TA,) *Death:* (ISd, Ḳ:) so called because it draws away the soul. (ISd, TA.)

جِذَابٌ : see جَذَبٌ.

جَاذِبٌ : see جَذُوبٌ.

جَذَّابَةٌ *Hairs,* (TA,) or *coarse hairs,* or *a coarse hair,* (Ḳ,) *tied, and made into a snare,* (TA,) *for catching larks.* (Ḳ, TA.)

جَاذِبٌ ‡ A she-camel *that has exceeded the usual time of pregnancy, and passed beyond the time* [*of the year*] *when she had been covered:* (Lḥ, TA:) or a she-camel *that has extended,* or *protracted, the period of her pregnancy to eleven months.* (A, TA.) † A she-camel, (Ṣ, Ḳ,) and a she-ass, (TA,) *scant of milk;* (Ṣ, Ḳ, TA;) as also جَاذِبَةٌ and ▼ جَذُوبٌ : (Ḳ:) pl. [of the first and second] جَوَاذِبُ and جِذَابٌ. (Ṣ, Ḳ.)

جُوذَابٌ (M, Ḳ) and ذُوبَاجٌ, the latter formed by transposition, (L and TA in art. ذبج,) *A kind of food, prepared with sugar and rice and flesh-meat:* (M, Ḳ:) [from the Persian گوذاب, as observed by Golius:] it might be hastily imagined to be arabicized from جُوزَهْ آب ; but this is not the case: (TA:) [n. un. with ة: or] جُوذَابَةٌ is *a cake of bread* (خُبْزَةٌ) *put into the oven* (تَنُّور), *and having suspended over it a bird or some flesh-meat, the gravy of which flows upon it as long as it is cooking;* also called أُمُّ الفَرَجِ, because it removes one's anxiety for seasoning, or condiment. (Ḥar p. 227.)

جذر

1. جَذَرَ, (A, TA,) aor. ـُ, (TḲ,) inf. n. جَذْرٌ, (A, Ḳ,) *He cut,* or *cut off,* or *severed,* (Ḳ, TA,) a thing: (TA:) and (Ḳ) *he extirpated,* or *cut off entirely,* (A, Ḳ,) a thing; (A;) as also ▼ جذّر; (Ṣ;) and ▼ اجذر, inf. n. إِجْذَارٌ. (AZ, Ḳ.)

2: } see 1.
4: }

7. انجذر *It became cut,* or *cut off,* or *severed.* (Ḳ, TA.)

جَذْرٌ (Aṣ, IAạr, Ṣ, A, Mṣb, Ḳ) and ▼ جِذْرٌ (AA, Ṣ, Ḳ) The *root,* or *lower part,* (Aṣ, Ṣ, A, Mṣb, Ḳ,) of anything: (Aṣ, Ṣ, A:) or (so in the Ḳ, but in other lexicons "and") particularly, of the tongue: (Sh, A, Mṣb, Ḳ:) and of the penis: (Sh, Ḳ:) and of a horn (Ṣ,* A) of a cow (Ṣ) or of a bull; (A;) or the *horn* [itself] of a cow: (TA:) and the latter word, the *root,* or *foot,* or *lowest part,* of a tree: (TA:) and the former word, the *base* of the neck: (El-Hejeree, Ḳ:) pl. جُذُورٌ. (Ḳ.) Hence, نَزَلَتِ المَحَبَّةُ فِى جَذْرِ قَلْبِهِ *Love took up its abode in the bottom* (أَصْل) *of his heart.* (A.) And [hence] it is said in a trad., إِنَّ الأَمَانَةَ نَزَلَتْ فِى جَذْرِ قُلُوبِ الرِّجَالِ [app. meaning, *Verily reason,* or *intellect,* or rather *conscience,* each of which is a trust committed by God to man, and a faculty which renders him responsible for his faith and works, (see, in art. امن, an explanation of أَمَانَةٌ as used in the Ḳur xxxiii. 72,) *hath taken up its abode in the bottom of the hearts of men*]. (Ṣ.) — Also, both words, The *origin,* or *stock, from which one springs.* (TA.) — And the former, (Ṣ, A, Mgh, Mṣb, Ḳ,) and the latter, or the latter only, (Ḳ,) or the former only, (IAạr, TA,) *A root* of a number; (A;) *an arithmetical root;* (Mgh, Ḳ;) [*a square root;*] *a number that is multiplied by itself;* (Mṣb;) *as when you say that ten multiplied by ten is a hundred;* (Mgh, Mṣb;) and *three multiplied by three is nine;* (A;) *in the former of which cases, ten is the* جذر, (Mgh, Mṣb,) i. e., *the* جذر *of a hundred;* (Mgh;) *and in the latter, three;* (A;) *and in each case, the* [*square* or] *product of the multiplication is called the* مَال, (Mṣb,) or the جَذْرَآءُ, (A,) or the ▼مَجْذُور: (Mgh:) [pl. of pauc. أَجْذَار, and of mult. جُذُور.] It is of two kinds, نَاطِق [i. e. *rational*], and أَصَمّ [i. e. *surd*, or *irrational*]: the latter known only to God, accord. to a saying of 'Áisheh. (Mgh.)

جَذْر: see جِذْر.

جُوذَرٌ and جُؤْذَرٌ (Ṣ, Ḳ) and جَوْذَرٌ and جَوْذَرٌ and جُوذَرٌ and ▼جِيذَرٌ, (Ḳ, TA,) the last of which is written in some copies of the Ḳ [and in the CḲ] جَيْذَرٌ, (TA,) The *young one of a wild cow*: (Ṣ, Ḳ:) pl. of the first and second, جَآذِرُ. (Ṣ.) ISd thinks that جُوذَرٌ and جِيذَرٌ are Arabic, and that جُؤْذَرٌ and جُوْذَرٌ are Persian. (TA.) See also مُجْذِرٌ.

جِيذَرٌ or جَيْذَرٌ: see what next precedes.

مُجْذِرٌ A wild cow *having a young one*. (ISd, Ḳ.) Hence we decide that the ء in ▼جؤذر is augmentative; and because it often occurs as an augmentative in the second place. (ISd, TA.) [In the Ṣ it is regarded as a radical.]

مَجْذُورٌ: see جَذْرٌ.

جذع

4. اجذع, (Ṣ, Mgh, Mṣb, Ḳ,) inf. n. إِجْذَاعٌ, (Mgh, Mṣb,) *He* (a beast) *became such as is denoted by the term* جَذَعٌ; (TA;) said of the offspring of the sheep or goat, *he became in his second year*; of that of the cow, and of a solid-hoofed beast, *he became in his third year*; and of that of the camel, *he became in his fifth year*: (Ṣ, Mṣb, Ḳ:) but sometimes, when said of the offspring of the ewe, it means *he became six months old*, or *nine months old*; and such is allowable as a victim for sacrifice: (Ṣ:) IAạr says, it denotes a time, not a tooth (Mgh, Mṣb) growing or falling out: (Mṣb:) and said of a she-goat, اجذعت means *she became a year old*, and sometimes, *less than a year*, by reason of plenty of food; and of a sheep, اجذع means, when from young parents, *he became from six months old to seven*; and when from very old parents, *from eight months old to ten*. (Mgh, Mṣb.) [See جَذَعٌ, below.]

6. تجاذع ‡ *He* (a man) *pretended to be a* جَذَع [or *youth*]. (TA.)

جِذْعٌ The *trunk* of a palm-tree: (Ṣ,* Mṣb, Ḳ:) or, accord. to some, only *after it has become dry*: or, accord. to some, only *after it has been cut*: (TA:) or the *trunk* of a tree *when the head has gone*: (Ḥam p. 656:) in the Ḳur, xix. 23, it is applied to the trunk of a palm-tree which had become dry and was without a head; (Bḍ;) therefore this does not indicate any restriction nor the contrary: (TA:) pl. [of pauc.] أَجْذَاعٌ (Mṣb) and [of mult.] جُذُوعٌ. (Ṣ, Mṣb.) — The *beam of a roof*. (Mṣb, TA.)

جَذَعٌ A beast (Lth, Mgh) *before the* ثَنِىّ [q. v.], (Lth, Ṣ, Mgh, Mṣb, Ḳ,) *by one year*; *when it may for the first time be ridden and used*: (Lth:) fem. with ة: (Ṣ, Mgh, Mṣb, Ḳ:) pl. masc. [of pauc.] أَجْذَاعٌ (Yoo, O) and [of mult.] جُذْعَانٌ (Yoo, Ṣ, Mgh, Mṣb, Ḳ) and جِذْعَانٌ (L, Mṣb) and جِذَاعٌ (Ṣ, Mgh, Mṣb, Ḳ) and جُذَاعٌ; (Yoo, O;) and pl. fem. جَذَعَاتٌ: (Ṣ, Mṣb:) it is a name applied to the beast in a particular time, not denoting a tooth growing or falling out: (Ṣ, Ḳ:) but it differs in its application to different kinds of beasts: (Az:) applied to a sheep or goat, it means *a year old*; (IAạr;) *in his second year*: (Mgh:) or, applied to a sheep, *a year old*; and sometimes *less than a year*, by reason of plenty of food; (IAạr;) or *eight months old*, (Az, Mgh, TA,) or *nine*; (TA;) or, when from young parents, *from six months old to seven*; and when from very old parents *from eight months old to ten*; (IAạr, Mgh;) and the sheep thus called is a satisfactory victim for sacrifice: (Mgh, TA:) and applied to a goat, *a year old*; (Az, Mgh;) or *in its second year*; (AZ;) but the goat thus called is not a satisfactory victim for sacrifice: (Mgh:) applied to a bull, it means in like manner *in his second year*; (Mgh;) or *in his third year*; and the bull thus called is not a satisfactory victim for sacrifice: (TA:) applied to a horse, it means *in his third year*; (IAạr;) or *in his fourth year*: (Mgh:) [but see قَارِح:] and applied to a camel, *in his fifth year*; (Az, Mgh;) fem. with ة; and this (a جذعة) is what must be given for the poor-rate when the camels are more than sixty. (Az, TA.) [See also شَصْر.] — *A youth*, or *young man*. (Ḳ.) — † *One who is light-witted*, or *weak and stupid, like a youth*: opposed in this sense to بَازِل as meaning "old:" (IAạr, TA:) or *one whose teeth have fallen out, here and there*, [as though likened to a beast thus termed that has shed some of his first teeth,] because he has drawn near to his appointed term of life. (TA: [but it is not quite clear whether this explanation relate to جذع or to بازل.]) — فُلَانٌ ‡ [*A novice*, or *recent beginner*.] You say, فُلَانٌ فِى هٰذَا الأَمْرِ جَذَعٌ ‡ [*Such a one, in this affair, is a novice*, or *recent beginner*,] when he has begun it recently. (Ṣ, Z.) — الدَّهْرُ جَذَعٌ أَبَدًا ‡ *Time*, or *fortune, is ever new, like a youth*. (Ḳ,* TA.) — Hence, (TA,) الأَزْلَمُ الجَذَعُ ‡ *Time*, or *fortune*; (Ṣ, Ḳ;) as in the saying, أَهْلَكَهُمُ الأَزْلَمُ الجَذَعُ ‡ *Time*, or *fortune, destroyed them*; and لَا آتِيكَ الأَزْلَمَ الجَذَعَ ‡ *I will not come to thee ever*. (TA.) [See also art. زلم.] And accord. to some, (Ṣ,) *The lion*: (Ṣ, Ḳ:) but this is a mistake. (IB, L.) — And hence, (TA,) أُمُّ الجَذَعِ ‡ *Calamity*, or *misfortune*. (Ḳ, TA.) — أَعَدْتُ الأَمْرَ جَذَعًا ‡ *I renewed the thing*, or *affair, as it was at the first*: as, for instance, a war which had been extinguished. (TA.) And فُرَّ الأَمْرَ جَذَعًا [signifies, in like manner, ‡ *He recommenced the thing*: or] *he commenced the thing*. (TA.) And فُرَّ الأَمْرُ جَذَعًا ‡ *The thing was commenced*: (TA:) or *the thing returned to its first state; it recommenced*. (Ḳ in art. فر.) — جُذْعَانُ الجِبَالِ † *Small mountains*. (Ḳ.)

جُذُوعَةٌ [The *state of being what is denoted by the term* جَذَعٌ;] a subst. from إِجْذَاعٌ [inf. n. of اجذع]. (TA.)

جَذَعْمَةٌ *Young*; (Ṣ, Ḳ,* TA;) *not arrived at puberty*: (TA:) originally جَذَعَةٌ; (Ṣ, Ḳ;) the م being augmentative: (Ṣ:) the ة is either to give intensiveness to the meaning, or to denote the fem. gender; the word being considered as implying the meaning of نَفْس or جُثَّة. (TA.)

خَرُوفٌ مُتَجَاذِعٌ [*A lamb approaching the age in which the term* جَذَع *is applied to him*: expl. in some copies of the Ḳ by دَانٍ: in others, by وَانٍ: in the copies of the O, expl. by وَانٍ مِنَ الإِجْذَاعِ: in the TṢ and in the A, by دَانٍ, which is probably the right reading. (TA.)

(Quasi جذعم)

جَذَعْمَةٌ: see art. جذع.

جذف

1. جَذَفَهُ, (AA, Ṣ, Ḳ,) aor. ـِ, (Ḳ,) inf. n. جَذْفٌ, (AA, Ṣ,) *He cut it*; or *cut it off*: (AA, Ṣ, Ḳ:) and so with د. (TA.) — جَذَفَ said of a bird, a dial. var. of جَدَفَ: (Ṣ:) both signify *He* (a bird) *went quickly* (Ḳ, TA) *with his wings*; generally, *when one of the wings had been shortened*; (TA;) as also ▼اجذف and ▼انجذف: and so, both of these, with د. (Ḳ.) — [Hence, جَذَفَ بِالسَّفِينَةِ, or السَّفِينَةَ; and جَذَفَ بِالمِجْذَافِ; i. q. جَدَفَ, q. v.] — جَذَفَ فِى مِشْيَتِهِ *He* (a man) *was quick in his manner of walking*: (AO, Ṣ:) and so with د: (AAF, TA in art. جدف:) as also ▼تجذّف. (TA.) And جَذَفَتْ *She* (a woman) *walked like those that are short*: and *she* (a gazelle, and a woman, TA) *went with short steps*; as also ▼ اجذفت: and so, both, with د. (Ḳ.) — جَذَفَتِ السَّمَآءُ بِالثَّلْجِ *The sky cast down snow*: and so with د. (TA.) — جَذَفَ الشَّىْءَ i. q. جَذَبَهُ. (TA.)

4: see 1, in two places.

5: see 1.

7: see 1.

مِجْذَافٌ, (Ṣ, TA,) in the Ḳ ▼مِجْذَافَةٌ, but the former is the more proper, (TA,) i. q. مِجْدَافٌ; (Ṣ;) The *wing* of a bird: and so with د. (Mṣb in art. جدف.) — And [hence, *An oar*; *a paddle*;] *a certain thing with which a ship, or boat, is propelled*. (Ṣ,* TA.) — And hence, as being likened thereto, *A whip*: (Abu-l-Ghowth, Ṣ:) and so with د. (TA in art. جدف.)

زِقٌّ مَجْذُوفٌ A [skin of the kind called] زِقّ *having the legs cut off*: and so with د. (Ḳ* and TA in this art. and in art. جدف.)

مِجْذَافَةٌ: see مِجْذَافٌ.

جذل

1. جَذَلَ, (aor. ـُ, TḲ,) inf. n. جُذُولٌ, *It stood erect, and was firm*, (Ḳ, TA,) *like the* جِذْل *of a tree*. (TA.) And *He set himself up as an antagonist* to others, *in fight*. (TA in art. جحل.)

And said of a chameleon, *It became erect*; as also ▼استجذل. (TA.) You say also, بَاتَ يَسْتَجْذِلُ ▼ عَلَى ظَهْرِ دَابَّتِهِ ‡ *He slept* [*during the night*] *erect, without commotion, upon the back of his beast.* (TA.) ═ جَذِلَ, (Ṣ, Ḳ,) aor. ـَ, (Ḳ,) inf. n. جَذَلٌ, (Ṣ,) *He was*, or *became, joyful, glad*, or *happy*; as also ▼اجتذل. (Ṣ, Ḳ.)

4. اجذلهُ *He made him joyful, glad*, or *happy*. (Ṣ, Ḳ.)

8: see 1.

10: see 1, in two places.

جَذْلٌ: see what next follows, in two places.

جِذْلٌ (Ṣ, Ḳ) and ▼جَذْلٌ (Ḳ) The *trunk, stem, stump*, or *lower part*, (أَصْل,) *of a tree &c., after the branch or the like has gone*; pl. [of pauc.] أَجْذَالٌ and [of mult.] جِذَالٌ and جُذُولٌ and جُذُولَةٌ, (Ḳ,) which last is pl. of جَذْلٌ: (TA:) or *a large trunk*, or *lower portion, of a tree*; (Ṣ,* Ḳ;) pl. أَجْذَالٌ: (Ṣ:) and *a branch*, or *piece of wood, like the fruit-stalk of the raceme of a palm-tree.* (Ḳ.) [Hence,] عَادَ إِلَى جَذْلِهِ ▼ [or جِذْلِهِ] † *He returned to his original state*, or *condition.* (TA.) — Also, the former, *A post*, or *piece of wood, that is set up* (Ṣ, Ḳ) *in the place where camels lie down, at their watering-place*, (Ṣ,) *for the mangy camels to rub themselves against it.* (Ṣ, Ḳ.*) Hence, (Ṣ, Ḳ,) the saying of El-Ḥobáb Ibn-El-Mundhir, (Ṣ,) أَنَا جُذَيْلُهَا ▼ المُحَكَّكُ وَعُذَيْقُهَا المُرَجَّبُ † [*I am their much-rubbed little rubbing-post, and their propped little palm-tree loaded with fruit*, or *their honoured little palm-tree* &c.: see art. رجب]: (Ṣ,* Ḳ,* TA:) i. e., I am of those by means of whose counsel, or advice, people seek relief, like as the mangy camels seek relief from their mange by rubbing themselves against the post above mentioned, (TA,) and one having a family that will aid and defend me: (TA in art. رجب:) the dim. is here used for the purpose of aggrandizement. (Ḳ. [See also art. حك.]) And hence, (TA,) هُوَ جِذْلُ مَالٍ ‡ *He is a gentle manager of cattle*: (Ṣ,* Ḳ,* TA:) likened to the جذل that is set up. (TA.) One says also, إِنَّهُ جِذْلُ رِهَانٍ, i. e., صَاحِبُهُ [app. meaning, † *Verily he is one who is constantly engaged in contending for stakes*, or *wagers*; رهان being here an inf. n. of رَاهَنَ; not a pl. of رَهْنٌ; for if it were the latter, the explanation would be صَاحِبُهَا]. (Ḳ.) — *A small quantum* of property, or *a small number* of cattle; (Ḳ;) as though it were the original stock thereof. (TA.) — The *summit*, or *head, of a mountain*; and *a prominent portion thereof*: pl. أَجْذَالٌ. (Ḳ.) — The *side of a sandal.* (Ḳ.)

جَذِلٌ: see what next follows.

جَذْلَانُ [in copies of the Ḳ with tenween, but correctly without tenween, for the fem. is جَذْلَى,] *Joyful, glad*, or *happy*; (Ṣ, Ḳ;) as also ▼جَذِلٌ; (Ḳ;) and ▼جَاذِلٌ occurs in poetry: (IDrd, Ḳ:) pl. جُذْلَانٌ. (Ḳ.) You say, نَفْسُهُ جَذْلَى بِذٰلِكَ *His soul is joyful, glad*, or *happy, by means of that.* (TA.)

جُذَيْلٌ dim. of جِذْلٌ, q. v.

جَاذِلٌ [part. n. of جَذَلَ] † *Erect, in his place, not moving therefrom*; likened to the جذل that is set up in the place where camels lie down, at their watering-place, for the mangy camels to rub themselves against it. (Ṣ.) You say, بَاتَ جَاذِلًا عَلَى ظَهْرِ دَابَّتِهِ ‡ *He slept* [*during the night*] *erect, without commotion, upon the back of his beast.* (TA.) ═ See also جَذْلَانُ.

جذم

1. جَذَمَهُ, (Ṣ, Mṣb, Ḳ,) aor. ـِ (Mṣb, Ḳ, TA) and ـُ also, (accord. to some copies of the Ḳ,) inf. n. جَذْمٌ, (Ṣ, Mṣb, Ḳ,) *He cut it off*; (Ṣ, Mṣb, Ḳ;) and so ▼جذّمهُ: (Ḳ:) or جذّم signifies *he cut off* many things; or *cut off much*, or *frequently*: and جَذْمٌ signifies also the *cutting off quickly.* (TA.) [It is like حَذَمَهُ.] You say, جَذَمَ يَدَهُ, (Mṣb, Ḳ,) aor. ـِ, inf. n. as above, (Mṣb,) *He cut off*, or *amputated, his arm*, or *hand*; (Mṣb, Ḳ;) as also ▼اجذمها, (Ḳ,) inf. n. إِجْذَامٌ. (TA.) — [Hence,] جَذَمَ فُلَانٌ حَبْلَ وِصَالِهِ ‡ *Such a one severed the bond of his union*; as also جَذَبَهُ. (TA.) ═ جَذِمَ, (Ṣ, Mṣb,) aor. ـَ, (Mṣb,) inf. n. جَذَمٌ, *He* (a man) *had his arm*, or *hand, cut off*, or *amputated*; *was maimed of it.* (Ṣ, Mṣb.) You say, مَا الَّذِى أَجْذَمَهُ ▼ حَتَّى جَذِمَ [*What is it that has maimed him of his arm*, or *hand, so that he has become maimed of it?*]. (TA.) — And جَذِمَتِ اليَدُ, aor. ـَ, (Mṣb, Ḳ,) inf. n. جَذَمٌ, (Mṣb,) *The arm*, or *hand, was cut off*, or *amputated.* (Mṣb, Ḳ.) ═ جُذِمَ *He* (a man, Ṣ, Mṣb) *was*, or *became, affected*, or *smitten, with the disease termed* جُذَام. (Ṣ, Mgh, Mṣb, Ḳ.)

2: see 1.

4: see 1, in two places. ═ إِجْذَامٌ also signifies The *being quick* in pace, or going. (Lth, TA.) You say, اجذم فِى سَيْرِهِ, (Ṣ,) or اجذم السَّيْرَ, (Ḳ,) *He* (a camel, Ṣ) *hastened*, or *was quick, in his pace*, or *going.* (Ṣ, Ḳ.) And اجذم said of a horse, (Lḥ, Ḳ,) and the like, of such as run, (Lḥ, TA,) *He ran vehemently.* (Lḥ, Ḳ.) — اجذم عَنِ الشَّىْءِ *He abstained*, or *desisted, from the thing.* (Ṣ,* Ḳ.) — اجذم عَلَيْهِ *He decided, determined*, or *resolved, upon it.* (Ḳ.)

5: see 7.

7. انجذم *It was*, or *became, cut off*; (Ṣ, Ḳ;) as also ▼تجذّم: (Ḳ:) [or the latter is said of a number of things; or implies muchness, or frequency:] the two verbs are syn. [respectively] with اِنْقَطَعَ and تَقَطَّعَ. (TA.) — [Hence] you say, انجذم عَنِ الرَّكْبِ † *He was*, or *became, cut off from the company of riders upon camels.* (TA.) And En-Nábighah says,

* صَدَّتْ سُلَيْمَى وَأَمْسَى حَبْلُهَا انْجَذَمَا *

[*Suleymà has turned away, and the bond of her union* with me *has become severed*]. (Ṣ.)

جَذْمٌ: see the next paragraph. ═ Also *A cessation of the supply of corn or other provision.* (TA.) ═ A rope *cut off*, or *severed.* (TA.) — A man *whose extremities have fallen off in pieces, piece after piece, in consequence of the disease termed* جُذَام. (TA; but in this last sense, the word is there written without any syll. signs.)

جِذْمٌ The *root, source, origin*, or *original*, or the *fundamental* or *essential* or *principal part*, syn. أَصْلٌ, (Ṣ, Mṣb, Ḳ,) of a thing, (Ṣ, Mṣb, TA,) whatever that thing be; (TA;) as also ▼جَذْمٌ: (Ṣ, Ḳ:) pl. [of pauc.] أَجْذَامٌ and [of mult.] جُذُومٌ. (Ḳ.) — The *family* of a people; their *kinsfolk*: whence the saying, in a trad., لَمْ يَكُنْ رَجُلٌ مِنْ قُرَيْشٍ إِلَّا لَهُ جِذْمٌ بِمَكَّةَ [*There was not a man of Kureysh but he had kinsfolk in Mekkeh*]. (TA.) [And app. The *main stock from which tribes are derived*: for,] accord. to some, it ranks before شَعْبٌ. (TA voce بَطْنٌ.) — The *places* [or *place*] *of growth* of the teeth. (TA.) A poet says, (Ṣ,) namely, El-Ḥárith Ibn-Waaleh, (TA,)

* أَلْآنَ لَمَّا ٱبْيَضَّ مَسْرُبَتِى *

* وَعَضِضْتُ مِنْ نَابِى عَلَى جِذْمِ *

(Ṣ, TA,) [*Now, when the hair in the middle of my bosom, extending downwards to my navel, has become white, and I have bitten upon the place of growth of my canine tooth*]: i. e., I have become old, and eaten upon the جذم of my ناب. (TA.) — The *lower*, or *lowest, part*, or the *foundation*, of a wall: (Mgh from a trad.:) or the *remains* thereof: or *a portion* thereof. (TA.) — See also جِذْمَةٌ.

جِذْمٌ *Quick*; *swift.* (Ḳ.)

جَذْمَةٌ The *place of the arm*, or *hand, where it is cut off*, or *amputated*; as also ▼جَذَمَةٌ. (Ḳ.)

جُذْمَةٌ The *defect*, or *deficiency, of him who has had his arm*, or *hand, amputated*, or *who has lost the end-joints of his fingers*: so accord. to the copies of the Ḳ: but in the L, the *defect*, or *deficiency* [*resulting*] *from the amputation of the arm* or *hand* (مِنَ الإِجْذَامِ). (TA.) ═ مَا سَمِعْتُ لَهُ جُذْمَةً, with damm, meaning [*I heard him not utter*] *a word*, is not of established authority. (ISd, TA.)

جِذْمَةٌ *A piece cut off* (Ṣ, Ḳ) of a rope &c., (Ṣ,) or of a thing of which the extremity has been cut off, the lower, or principal, part remaining; (Ḳ;) as also ▼جِذْمٌ. (TA.) — *A whip*: (Ṣ, Ḳ:) because it becomes cut by that which is beaten with it. (TA.) — The *part of a whip of which the slender extremity has become much cut* [*by use*], *the lower*, or *principal, part remaining*; pl. جِذَمٌ: (L, TA:) or the *remaining part* of a whip; its *lower*, or *principal, portion.* (Aṣ, TA.) — *A thick piece of wood, having fire at the end of it or not*; [i. e. *a brand*, or *fire-brand*;] like جِذْوَةٌ. (AO, Ṣ and TA in art. جذو.) — † *A company of men* [as though cut off from others]. (TA.)

جَذَمَةٌ: see جَذْمَةٌ. — Also The *uppermost pith of the palm-tree*; *which is the best*; (Ḳ;) like جَذَبَةٌ. (TA.) — And *Dates that come forth upon one base.* (TA.)

جُذَامٌ [*Elephantiasis; a species of leprosy; the leprosy that pervaded Europe in the latter part of the Middle Ages;*] *a certain disease,* (Ṣ, Ḳ,) *arising from the spreading of the black bile throughout the whole person, so that it corrupts the temperament of the members, and the external condition thereof; and sometimes ending in the dissundering,* or *corrosion,* (so accord. to different copies of the Ḳ, TA,) *of the members, and their falling off, in consequence of ulceration;* (Ḳ, TA;) so called because *it dissunders the flesh, and causes it to fall off;* (Mṣb;) or because *the fingers,* or *toes, become cut off:* (TA:) it is *a cracking of the skin, and a dissundering, and gradual falling off, of the flesh.* (Mgh.)

نَوًى جَذُومٌ *A tract towards which one journeys separating lovers* or *objects of love.* (TA.)

جَذِيمٌ *Cut off; amputated.* (Ṣ, Mṣb, TA.) [See also أَجْذَمُ.]

جُذَامَةٌ *What remains,* of seed-produce, *after the reaping.* (Ṣ.) — [See also جُرَامَةٌ, in two places.]

جَذَّامٌ [an epithet] of the measure فَعَّالٌ from الجَذْمُ meaning القَطْعُ: so in the phrase جَذَّامُ حَبْلِ الهَوَى [*Wont to sever the bond of love*], in a verse of Ows Ibn-Thaạlabeh. (Ḥam p. 334.)

أَجْذَمُ *Having his arm,* or *hand, cut off,* or *amputated:* (Ṣ, Mṣb, Ḳ:) or *having lost the end-joints of his fingers:* (Ḳ:) fem. جَذْمَآءُ: (Mṣb:) pl. جَذْمَى. (Ṣ.) It is said in a trad., مَنْ تَعَلَّمَ القُرْآنَ ثُمَّ نَسِيَهُ لَقِىَ ٱللّٰهَ يَوْمَ القِيَامَةِ وَهُوَ أَجْذَمُ (A'Obeyd, Ṣ,*) i. e., [*He who learns the Ḳur-án and then forgets it shall meet God on the day of resurrection*] *having his arm,* or *hand, cut off:* (A'Obeyd, TA:) or *having lost all his limbs,* or *members:* (Ḳt, TA:) or † *having his plea cut off; having no tongue with which to speak, nor any plea in his hand:* (IAth, TA:) or † *having his means of access cut off:* (TA:) or † *with his hand devoid of good and of recompense.* (IAạr, El-Khaṭṭábee, TA.) And in another trad., كُلُّ خُطْبَةٍ لَيْسَ فِيهَا شَهَادَةٌ كَٱلْيَدِ الجَذْمَآءِ [*Every oration from the pulpit in which there is not an acknowledgment of the unity of God and of the mission of Mohammad is like the arm of which the hand is amputated*]. (TA.) — الكَفُّ الجَذْمَآءُ [*The amputated hand;*] † a name of *the star α of Cetus;* (so in the Egyptian Almanacs;) [i. e.] *the star that is in the head of Cetus:* so called because it is less extended [from the Pleiades] than that called الكَفُّ الخَضِيبُ. (Ḳzw. [See الكَفُّ الخَضِيبُ in art. خضب.]) — نَعْلٌ جَذْمَآءُ † *A sandal of which the* [thong *called*] قِبَال [*which is between two of the toes*] *is cut,* or *cut off,* or *severed.* (TA.) ═ See also مَجْذُومٌ.

مُجَذَّمٌ: see مَجْذُومٌ. ═ Also *A man tried,* or *proved, and strengthened by experience in affairs.* (TA.)

رَجُلٌ مِجْذَامٌ and ▼ مِجْذَامَةٌ, [the latter of a very rare measure, (see مِعْزَابَةٌ,)] † *A man who decides affairs.* (Ḳ.) ‡ *A man who, loving and being loved, when he is sensible of evil treatment quickly cuts the tie of affection:* (A, TA:) or the latter, ‡ *a man who quickly cuts the tie of love,* or *affection.* (Ṣ, Ḳ, TA.) رَجُلٌ مِجْذَامُ الرَّكْضِ فِى الحَرْبِ † *A man quick in running,* or *fleeing, in war.* (TA.) And رَجُلٌ مِجْذَامَةٌ ▼ لِلْحَرْبِ, and لِلسَّيْرِ, and لِلْهَوَى, † *A man who desists from, and relinquishes, war,* and *journeying,* and *love,* or *natural desire.* (Lḥ, TA.)

مَجْذُومٌ A man (Ṣ, Mṣb) *affected,* or *smitten, with the disease termed* جُذَام; (Ṣ, Mgh, Mṣb, Ḳ;) as also ▼ مُجَذَّمٌ (Ḳ) and ▼ أَجْذَمُ, (Kr, Ḳ,) which J erroneously disallows: (Ḳ:) J says, one does not say أَجْذَمُ: (TA:) [and Fei,] they say that أَجْذَمُ, of the measure of أَحْمَرُ, is not said in this sense. (Mṣb.)

مِجْذَامَةٌ: see مِجْذَامٌ, in two places.

جذمر

جِذْمَارٌ or جُذْمَارٌ: see what follows, in two places.

جُذْمُورٌ The *root,* or *lowest part,* of a thing: or the *first* thereof; (Ḳ;) the *beginning,* or *commencement, and fresh state,* thereof; its *first and fresh state.* (TA.) — *A piece,* or *portion,* (Ṣ, Ḳ,) *of the lowest part* (Ṣ) *of a palm-branch,* (Ṣ, Ḳ,) [i. e. *a stump thereof,*] *remaining upon the trunk when the* [*rest of the*] *branch has been cut off;* (Ṣ, Ḳ;) as also ▼ جُذْمَارٌ, (Ṣ,) or جِذْمَارٌ. (Ḳ.) [Accord. to the Ṣ, the ر is an augmentative letter.] — *The stump of a* [*tree of the kind called*] نَبْعَة, *remaining when the tree has been cut down.* (TA.) — *The stump of a hand of which the greater part has been cut off;* (TA;) *what remains of a hand that has been amputated, at the extremity of the two bones of the fore arm.* (T, TA.) One says, ضَرَبَهُ بِجُذْمُورِهِ and بِقُطْعَتِهِ [*He struck him with the stump of his amputated hand*]. (TA.) — *What remains of anything that has been cut off.* (IAạr, TA.) — [Pl. جَذَامِيرُ.] — You say, أَخَذَهُ بِجُذْمُورِهِ † *He took it altogether;* (Ḳ;) as also اخذه بجذاميره: (Ks, Ṣ, Ḳ:) or *he took it in its first and fresh state:* and Fr also mentions the phrases اخذه بِجِذْمِيرِهِ ▼ and بِجِذْمَارِهِ ▼. (TA.)

أَخَذَهُ بِجِذْمِيرِهِ: see what next precedes.

جذو

1. جَذَا, (Ṣ, Ḳ,) aor. ـُ, (TA,) inf. n. جَذْوٌ and جُذُوٌّ, (Ḳ,) *It* (a thing, TA) *stood firmly;* as also ▼ اجذى. (Ṣ, Ḳ.) You say also, جَذَا عَلَى شَىْءٍ *He* [or *it,* for instance, a stone of those (three in number) upon which a cooking-pot is placed, as is implied in the Ṣ, (see جَاذٍ,)] *remained firm upon a thing.* (Ṣ.) And جَذَا القُرَادُ فِى جَنْبِ البَعِيرِ *The ticks stuck,* and *clave, to the side of the camel:* (ISd, Ḳ, TA:) and جذا بالجمل *clung to the camel.* (TA.) — *I. q.* جَثَا [*He sat upon his knees; &c.*]; (AA, Fr, Ṣ, Ḳ;) as also ▼ اِجْذَوَى, [in form] like اِرْعَوَى; (TA;) except that جذا is more indicative of keeping to a place: (Kh, TA:) or *he stood upon the extremities of his toes:* (Aṣ, Ḳ:) accord. to Th, جُذُوٌّ is [the *standing*] *upon the extremities of the toes;* and جُثُوٌّ is [the sitting] upon the knees: (TA:) and accord. to IAạr, ▼ جَاذٍ means [a man *standing*] *upon his feet;* and جَاثٍ, [one sitting] upon his knees. (Ṣ, TA.) Accord. to El-Hasan Ibn-'Abd-Allah El-Kátib El-Iṣbahánee, جَذَا, inf. n. جُذُوٌّ, said of a bird, means *He stood upon the extremities of his toes, and warbled, and went round in his warbling;* which he does only when seeking the female: and said of a horse, it means *he stood upon his toes;* and in like manner when said of a man, whether for dancing or for some other purpose. (TA.) A poet says, (Ṣ,) namely, En-Noạmán Ibn-Nadleh, (TA,)

* إِذَا شِئْتُ غَنَّتْنِى دَهَاقِينُ قَرْيَةٍ *
* وَصَنَّاجَةٌ تَجْذُو عَلَى حَرْفِ مَنْسِمِ *

[*When I will, the husbandmen of a village sing to me, and a female player with the cymbals, standing upon the extremity of a toe*]. (Ṣ, TA: but in the latter, عَلَى كُلِّ مَنْسِمِ [*upon every toe*].) — Also, inf. n. جُذُوٌّ, *He,* or *it, was,* or *became, erect, and straight;* (TA;) and so ▼ اِجْذَوْذَى, inf. n. اِجْذِيذَآءٌ. (Az, TA.) See جَوَاذٍ, as applied to she-camels, in two places, voce جَاذٍ. — *It* (a camel's hump) *bore fat* [*so that it became elevated*]. (Ḳ.) — جَذَا مَنْخِرَاهُ *His nostrils were,* or *became, raised and extended.* (TA.) — Er-Rá'ee, describing a strong, or sturdy, she-camel, says,

* لَمْ يَجْذُ مِرْفَقُهَا فِى الدَّفِّ مِنْ زَوَرٍ *

meaning *Her elbow did not stand out far from the side by reason of* [*the distortion termed*] زَوَر. (TA.)

4. اجذى: see 1, first sentence. — Also *He* (a young camel) *bore fat in his hump.* (Ks, Ṣ, Ḳ.) El-Khansà says,

* يُجْذِينَ نِيًّا وَلَا يُجْذِينَ قِرْدَانَا *

They bear fat [*in their humps*], *and have not ticks clinging to them.* (IB, TA.) ═ *He lifted* a stone, (AA, Ṣ, TA,) *in order that he might know thereby his strength.* (TA.) One says, هُمْ يُجْذُونَ حَجَرًا وَيَتَجَاذَوْنَهُ ▼ [*They lift a stone, in order to prove their strength, and vie, one with another, in lifting it, for that purpose*]. (TA.) ▼ تَجَاذٍ, in the lifting of a stone, is *like* تَجَاثٍ: (Ṣ, TA:) تَجَاذَوْا حَجَرًا meaning تَرَابَعُوهُ لِيَرْفَعُوهُ [*They vied, one with another, to lift the stone, for trial of strength*]. (TA.) — اجذى طَرْفَهُ [in the CḲ, erroneously, طَرَفَهُ,] *He raised his eye,* or *sight, and cast it before him.* (Ḳ, TA.)

6: see 4, in two places.

9. اِجْذَوَى: see 1.

12. اِجْذَوْذَى: see 1.

جِذْوَةٌ and جُذْوَةٌ and جَذْوَةٌ (Ṣ, Mṣb, Ḳ) [*A brand,* or *fire-brand;*] *i. q.* جِذْمَةٌ, (AO, Ṣ, Ḳ,)

i. e. *a thick piece of wood, having fire at the end of it or not*: (AO, Ṣ, in explanation of the first:) or *a thick piece of fire-wood, not flaming*: (AO, TA, in explanation of جِذْوَةٌ مِنَ النَّارِ:) or *a thick stick, one end of which is a live coal*: (Aboo-Sa'eed, TA:) or *a thick stick upon which fire is taken* [*by kindling one end*]: (ISk, TA:) and *a piece* (قَبَسَةٌ, Ḳ, or قِطْعَةٌ, Ḥar p. 471) *of fire*: (Ḳ, Ḥar:) or *a portion of fire-wood that remains after flaming*: (Er-Rághib, TA:) and *a live coal*: (Ṣ, Ḳ:) or *a flaming live coal*: (Mṣb:) or جِذْوَةٌ مِنَ النَّارِ means *a piece of live coal*; so in the language of all the Arabs: (Mujáhid, Ṣ, TA:) pl. [of the first] جِذًى and [of the second] جُذًى (Ṣ, Mṣb, Ḳ) and (of the third, TA) جِذِىٌّ (Mṣb, TA, and so in a copy of the Ṣ) and جِذَآءٌ, (AAF, Ḳ,) which is held by ISd to be a pl. of the third. (TA.)

جِذَآءٌ: see جُذَآءٌ, in art. جذى.

جَاذٍ [part. n. of 1; fem. جَاذِيَةٌ: pl. masc. جُذَآءٌ; pl. fem. جَاذِيَاتٌ and جَوَاذٍ]. A rájiz says, (Ṣ,) namely, 'Amr Ibn-Jebel-El-Asadee, (TA,)

* لَمْ يُبْقِ مِنْهَا سَبَلُ الرَّذَاذِ *

* غَيْرَ أَثَافِى مِرْجَلٍ جَوَاذِى *

i. e. [*The continual fine rain left not thereof save the three stones that were the supports of a cooking-pot*] *remaining firm*. (Ṣ, TA.) — *I. q.* جَاثٍ [*Sitting upon his knees*; &c.]: (Fr, TA:) or *sitting upon his heels, with his feet upright*, [*resting*] *upon the extremities of his toes*: (Ṣ:) or *standing upon the extremities of the toes*: (AA, Ṣ:) see also 1: pl. [masc.] جُذَآءٌ. (Ṣ.) Aboo-Duwád describes mares as جَاذِيَاتٌ عَلَى السَّنَابِكِ, i. e. *Standing upon the toes*. (AA, Ṣ,* TA.) — جَوَاذٍ, (Ḳ,) applied to she-camels, (TA,) means *That bear themselves erect* (تَجْذُو ↓) *in their course, or pace, as though they lifted their feet clear from the ground*; (Ḳ, TA;) on the authority of Aboo-Leylà: (TA:) [the last words of the explanation in the Ḳ are كَأَنَّهَا تَقْلَعُ: in the TA, كانها تقلع الشُّهْرَ: I suppose that الشُّهْرَ is for فِى السَّيْرِ; and that the pret. of the aor. here used is قَلِعَ; for قَلِعٌ, which is of the regular form of a part. n. of such a verb as قَلِعَ, means "raising the feet clear from the ground in walking &c.:" but in one copy of the Ḳ, I find تُقْلِعُ: and another reading in some work seems to be تُقَلِّعُ; for] ISd says, I know not جَذَا with the meaning of أَسْرَعَ nor of أَقْلَعَ: and Aṣ says that جَوَاذٍ means *quick*, or *swift*, camels, *that do not stretch themselves forth in their course, or pace, but bear themselves erect* (يَجْذُونَ ↓ وَيَنْتَصِبْنَ). (TA.)

مُجْذٍ [act. part. n. of 4]. مَثَلُ الأَرْزَةِ المُجْذِيَةِ عَلَى الأَرْضِ, (Ṣ,) or كَالأَرْزَةِ المجذية على وجه الارض, (TA,) occurring in a trad., (Ṣ, TA,) describing the unbeliever, (TA,) means [*Like the pine-tree*] *that is firm* (Ṣ, TA) *and erect* [*upon the ground*]. (TA.)

مُجْذَوْذٍ [in the CḲ, erroneously, المُجْذُوذِىُّ,] *Keeping constantly to the dwelling*, or *to the camel's saddle* and *the dwelling*, (الرَّحْل and المَنْزِل, AA, Ṣ, Ḳ,) *not quitting it*: (AA, Ṣ:) like مُجْذَوْذٌ عَلَى الرَّحْلِ. (AA, Ṣ.) — And *A man who lowers, or abases, himself*; (El-Hejeree, ISd, TA;) as though he clave to the ground by reason of his abjectness; from جَذَا القُرَادُ فِى جَنْبِ البَعِيرِ [q. v.]. (ISd, TA.)

جر

1. جَرَّ, aor. ـُ, (Ṣ, A, Mṣb,) inf. n. جَرٌّ; (Ṣ, Ḳ;) and جرّر ↓, inf. n. تَجْرِيرٌ (Ṣ, Ḳ) [and app. تَجِرَّةٌ, said in the TA to be of the measure تَفْعِلَةٌ from الجَرّ], with teshdeed to denote repetition or frequency of the action, or its relation to many objects, or intensiveness; (Ṣ;) and اجترّ ↓, inf. n. اِجْتِرَارٌ; (Ṣ, L, Ḳ;) and اجدرّ ↓, inf. n. اِجْدِرَارٌ; (L, Ḳ;) in which the ت is changed into د, though you do not say اِجْدَرَأَ for اِجْتَرَأَ, nor اِجْدَرَحَ for اِجْتَرَحَ; (L;) and استجرّ ↓; (Ḳ;) *He dragged, drew, pulled, tugged, strained, extended by drawing* or *pulling or tugging*, or *stretched*, (A, L, Mṣb, Ḳ,) a thing, (A,) or a rope, (Ṣ, Mṣb,) and the like. (Mṣb.) You say, جَرُّوا أَذْيَالَهُمْ *They dragged along their hinder skirts*. (A.) And اجترّ ↓ الرُّمْحَ *He dragged, or drew along, the spear*. (TA.) And فُلَانٌ يَسْتَجِرُّ ↓ الحَدِيثَ مِنْ أَبَاعِدِ أَطْرَافِهِ †[*Such a one draws forth talk, or discourse, or news, or the like, from its most remote sources*]. (A in art. بعد.) And مَا الَّذِى جَرَّكَ إِلَى هٰذَا الأَمْرِ †[*What drew thee, led thee, induced thee, or caused thee, to do this thing*]. (TA in art. دعو.) — Also جَرَّ, aor. ـُ, (TA,) inf. n. جَرٌّ, (Ḳ,) ‡ *He drove* (camels and sheep or goats, TA) *gently*, (Ḳ, TA,) *letting them pasture as they went along*. (TA.) And جَرَّ الإِبِلَ عَلَى أَفْوَاهِهَا ‡ *He drove the camels gently, they eating the while*. (A.) — [Hence,] هَلُمَّ جَرًّا ↓ ‡ *At thine ease*. (TA.) El-Mundhiree explains هَلُمَّ جُرُّوا as meaning ‡ *Come ye at your ease*; from الجَرّ in driving camels and sheep or goats, as rendered above. (TA.) You say also, كَانَ ذٰلِكَ عَامَ كَذَا وَهَلُمَّ جَرًّا إِلَى اليَوْمِ (Ṣ, A, Mṣb,* TA) ‡ *That was in such a year, and has continued to this day*: (Mṣb, TA:) from الجَرّ meaning the act of "dragging," &c.: (TA:) or from أَجْرَرْتُهُ الدَّيْنَ, or from أَجْرَرْتُهُ الرُّمْحَ. (Mṣb.) جَرًّا is here in the accus. case as an inf. n., or as a denotative of state: but it is disputed whether this expression be classical or post-classical. (TA.) [See also art. هلم.] — جَرَّ الأَثَرَ, said of a numerous army, means †[*It made a continuous track, so that*] *it left no distinct foot-prints, or intervening* [*untrodden*] *spaces*. (TA.) جَرَّتِ الخَيْلُ الأَرْضَ بِسَنَابِكِهَا ‡ *The horses furrowed the ground with their hoofs*. (Aṣ, A, TA.) — جَرَّ جَرِيرَةً, (Ṣ, A, Mṣb, Ḳ,) aor. ـُ and ـِ, (Ḳ,) but the latter form is disallowed by MF as not authorised by usage nor by analogy, (TA,) inf. n. جَرٌّ, (Ḳ,) *He committed a crime, or an offence for which he should be punished*, or *an injurious action*, (Ṣ, Mṣb, Ḳ,*) against (عَلَى [and إِلَى, as in the Ḳ voce جَنَى,]) another or others, (Ṣ, Ḳ,) or himself; (A, Ḳ;) [as though *he drew it* upon the object thereof;] syn. جَنَى جِنَايَةً. (Ṣ, TA.) It is said in a trad., بَايَعَهُ عَلَى أَنْ لَا يَجُرَّ عَلَيْهِ إِلَّا نَفْسَهُ [*He promised*, or *swore, allegiance to him on the condition that he should not inflict an injury*, meaning *a punishment, upon him but* for an offence committed by *himself*;] i. e., that he should not be punished for the crime of another, of his children or parent or family. (TA.) — جَرَّ الفَصِيلَ: see 4, in two places. — جَرَّ الحَرْفَ فِى الإِعْرَابِ, aor. ـُ, inf. n. جَرٌّ, †*He made the* final *letter to have kesreh, in inflection*; *i. q.* خَفَضَ, q. v.: [الجَرّ is used in the conventional language of the Baṣrees; and الخَفْض, in that of the Koofees. (Kull p. 145.)] — جَرَّتْ, (Ṣ, A,) inf. n. جَرٌّ, (Ḳ,) ‡ *She exceeded the* [*usual*] *time of pregnancy*. (A.) ‡ *She* (a camel) *arrived at the time* [*of the year*] *in which she had been covered, and then went beyond it some days without bringing forth*: (Ṣ, TA:) or *withheld her fœtus in her womb after the completion of the year, a month, or two months, or forty days only*: (Ḳ,* TA:) Th says that she sometimes withholds her fœtus [beyond the usual time] a month. (TA. [See also جَرُورٌ.]) ‡ *She* (a mare) *exceeded eleven months and did not foal*: (Ḳ, TA:) the more she exceeds the usual term, the stronger is her foal; and the longest time of excess after eleven months is fifteen nights: accord. to AO, the time of a mare's gestation, after she has ceased to be covered, to the time of her foaling, is eleven months; and if she exceed that time at all, they say of her, جَرَّتْ. (TA.) ‡ *She* (a woman) *went beyond nine months without bringing forth*, (Ḳ, TA,) *exceeding that term by four days, or three*. (TA.) — †*It* (the night, اللَّيْلَةُ,) *was, or became, long*. (L in art. كبد.) — جَرَّ, aor. ـُ, (TA,) inf. n. جَرٌّ; (Ḳ;) and انجرّ ↓; (Ḳ;) †*He* (a camel) *pastured as he went along*: (IAạr, Ḳ: [if so, the aor. is contr. to analogy:]) or *he rode a she-camel and let her pasture* [*while going along*]. (Ḳ.) — جَرَّ النَّوْءُ بِالمَكَانِ †*The* نوء [or *auroral setting or rising of a star or asterism supposed to occasion rain*] *caused lasting rain in the place*. (TA.)

2: see 1, first sentence.

3. جارّهُ, (Ṣ, Ḳ,) inf. n. مُجَارَّةٌ, (TA,) or مُجَارَّةٌ, (TḲ,) *He delayed, or deferred, with him, or put him off, by promising him payment time after time*; syn. طَاوَلَهُ, (Ṣ,) or مَاطَلَهُ: (Ḳ:) or *he put off giving him his due, and drew him from his place to another*: (TA:) or *i. q.* جَانَاهُ, (so in copies of the Ḳ,) meaning, *he committed a crime against him*: (TḲ:) or حَابَاهُ. (TA, as from the Ḳ. [But this seems to be a mistranscription.]) It is said in a trad., لَا تُجَارِّ أَخَاكَ وَلَا تُشَارِّهِ, i. e. *Delay not*, or *defer not, with thy brother*, &c.: [*and do not act towards him in an evil*, or *inimical, manner*; or *do not evil to him, obliging him to do the like in return*; or *do not contend*, or *dispute, with him*:] or *bring not an injury upon him*: but accord. to one reading, it is

لَا تُجَارِهِ, without teshdeed, from الجَرْىُ, and meaning, contend not with him for superiority. (TA.)

4. اجرّهُ *He pierced him with the spear and left it in him so that he dragged it along:* (Ṣ, Ḳ:) or so اجرّهُ الرُّمْحَ: (A, Mṣb:) as though [meaning] *he made him to drag along the spear.* (TA.) —— *He put the* جَرِير, i. e. the *rope, upon his neck.* (Ḥar p. 308.) —— اجرّهُ جَرِيرَهُ [lit. *He made him to drag along his rope;* meaning,] ‡ *he left him to pasture by himself, where he pleased:* a prov. (L.) And اجرّهُ رَسَنَهُ [lit. *He made him to drag along his halter;* meaning,] ‡ *he left him to do as he would:* (Ṣ, Ḳ, TA:) *he left him to his affair.* (A, TA.) —— اجرّهُ الدَّيْنَ ‡ *He deferred for him the payment of the debt:* (Ṣ, A, Ḳ:) *he left the debt to remain owed by him.* (Mṣb.) —— اجرّهُ أَغَانِىَّ ‡ *He sang songs to him consecutively, successively,* or *uninterruptedly;* syn. تَابَعَهَا: (Ṣ, Ḳ, TA:) or ‡ *he sang to him a song and then followed it up with consecutive songs.* (A, TA.) —— اجرّ لِسَانَ الفَصِيلِ, (Ṣ,) or اجرّ الفَصِيلَ, (Aṣ, Ḳ,*) inf. n. إِجْرَارٌ; (Ḳ;) and ↓ جَرَّ الفَصِيلَ, (Aṣ, Ḳ,*) inf. n. جَرٌّ; (Ḳ;) ‡ *He slit the tongue of the young weaned camel, that it might not suck the teat:* (Ṣ, Ḳ, TA:) or إِجْرَارُ الفَصِيلِ signifies ‡ *the slitting the tongue of the young weaned camel, and tying upon it a piece of stick, that it may not suck the teat;* because it drags along the piece of stick with its tongue: or الإِجْرَارُ is like التَّفْلِيكُ, signifying † a pastor's *making, of coarse hair, a thing like the whirl,* or *hemispherical head, of a spindle, and then boring the tongue of the* [young] *camel, and inserting it therein, that it may not suck the teat:* so say some: (ISk, TA:) the animal upon which the operation has been performed is said to be ↓ مَجْرُورٌ and ↓ مُجَرٌّ. (TA.) [But sometimes ↓ جَرَّ signifies merely *He drew away* a young camel from its mother: see خَلِيَّةٌ voce خَلِىَ, in three places.] —— Hence, اجرّ لِسَانَهُ ‡ *He prevented him from speaking.* (A.) 'Amr Ibn-Maadee-Kerib Ez-Zubeydee says,

فَلَوْ أَنَّ قَوْمِى أَنْطَقَتْنِى رِمَاحُهُمْ
نَطَقْتُ وَلٰكِنَّ الرِّمَاحَ أَجَرَّتِ

[*And if the spears of my people had made me to speak, I had spoken; but the spears have prevented speech*]: i. e., had they fought, and shown their valour, I had mentioned that, and gloried in it, (Ṣ,) or in them; (TA;) but their spears have prevented my tongue from speaking, by their flight. (Ṣ,* TA.) ═ اجرّ as an intrans. verb: see 8. —— اجرّت البِئْرُ ‡ *The well was,* or *became, such as is termed* جَرُور. (Ibn-Buzurj, TA.)

7. انجرّ *It* (a thing, Ṣ) *was,* or *became, dragged, drawn, pulled, tugged, strained, extended by drawing* or *pulling* or *tugging,* or *stretched; it dragged,* or *trailed along;* syn. اِنْجَذَبَ. (Ṣ, Ḳ.) —— See also 1, last sentence but one.

8. اجترّ and اجدرّ: see 1, in three places. ═ اجترّ said of a camel, (Ṣ, Mṣb, Ḳ,) and any other animal having a كَرِش, (Ṣ, TA,) [i. e.] any cloven-hoofed animal, (Mṣb,) *He ejected the cud from his stomach and ate it again; ruminated; chewed the cud;* (Ṣ,* Mṣb,* Ḳ,* TA;) as also ↓ اجرّ. (Lḥ, Ḳ.)

10: see 1, in two places. ═ اِسْتَجْرَرْتُ لَهُ ‡ *I made him to have authority and power over me,* (Ḳ, TA,) *and submitted myself,* or *became submissive* or *tractable, to him;* (A, Ḳ, TA;) *as though I became to him one that was dragged,* or *drawn along.* (TA.) —— استجرّ عَنِ الرَّضَاعِ † *He* (a young camel) *refrained from sucking in consequence of a purulent pustule,* or *an ulcer, in his mouth or some other part.* (TA.)

R. Q. 1. جَرْجَرَ, (Ṣ, Mgh, Mṣb,) inf. n. جَرْجَرَةٌ, (Ṣ,* Ḳ,* TA,) *He* (a stallion-camel) *reiterated his voice,* or *cry,* (Ṣ,* Mgh, Mṣb, Ḳ,*) or *his braying,* (TA,) *in his windpipe.* (Ṣ,* Mgh, Mṣb, Ḳ.*) —— *He,* or *it, made,* or *uttered, a noise, sound, cry,* or *cries; he cried out; vociferated; raised a cry,* or *clamour.* (TA.) *It* (beverage, or wine,) *sounded,* or *made a sound* or *sounds,* (Ḳ, TA,) in the fauces. (TA.) And جَرْجَرَتِ النَّارُ † *The fire sounded,* or *made a sound* or *sounds.* (Mṣb.) ═ Also, (A, Mṣb,) inf. n. as above, (Ḳ,) *He poured water down his throat;* as also ↓ تَجَرْجَرَ: (Ḳ:) or *he swallowed it in consecutive gulps, so that it sounded,* or *made a sound* or *sounds;* (A, Mṣb, TA;) as also ↓ the latter verb. (Ḳ,* TA.) It is said in a trad., (of him who drinks from a vessel of gold or silver, Mgh, TA,) يُجَرْجِرُ فِى بَطْنِهِ نَارَ جَهَنَّمَ *He shall drink down into his belly the fire of Hell* (Az, A, Mgh, Mṣb) *in consecutive gulps, so that it shall make a sound or sounds:* (A:) or *he shall make the fire of Hell to gurgle reiteratedly in his belly;* from جرجر said of a stallion-camel. (Mgh.) Most read النَّارَ, as above; but accord. to one reading, it is النَّارُ, (Z, Mṣb,) and the meaning is, ‡ *The fire of Hell shall produce sounds in his belly like those which a camel makes in his windpipe:* the verb is here tropically used; and is masc., with ى, because of the separation between it and النار: (Z, TA:) but this reading and explanation are not right. (Mgh.) —— You say also, جَرْجَرَهُ المَآءَ *He poured water down his throat so that it made a sound or sounds.* (Ḳ,* TA.)

R. Q. 2: see R. Q. 1, in two places.

لَا ذَا جَرَ and لَا جَرَ, for لَا جَرَمَ and لَا ذَا جَرَمَ: see art. جرم.

جُرٌّ ‡ The *foot, bottom, base,* or *lowest part,* of a mountain; (Ṣ, A, Ḳ;) like ذَيْلٌ: (A, TA:) or the *place where it rises from the plain to the rugged part:* (IDrd, TA:) or الجُرُّ أَصْلُ الجَبَلِ is a mistranscription of Fr, and is correctly الجُرَاصِلُ الجَبَلُ [i. e. جُرَاصِلٌ signifies "a mountain"]: (Ḳ:) but جُرَاصِلٌ is not mentioned [elsewhere] in the Ḳ, nor by any one of the writers on strange words; and [SM says,] there is evidently no mistranscription: جُرُّ الجَبَلِ occurs in a trad., meaning *the foot, &c., of the mountain:* and its pl. is جِرَارٌ. (TA.) —— هَلُمَّ جَرًّا: see 1. ═ See also جَرَّةٌ. ═ لَا جُرَّ i. q. لَا جُرْمَ: see art. جرم. (TA.)

جَرَّةٌ [*A jar;*] *a well-known vessel;* (Mṣb;) *an earthen vessel; a vessel made of potters' clay:* (T, IDrd,* Ṣ,* Ḳ:*) or *anything made of clay:* (Mgh:) dim. جُرَيْرَةٌ: (TA:) pl. جِرَارٌ, (T, Ṣ, Mgh, Mṣb, Ḳ) and جَرَّاتٌ (Mṣb) and ↓ جَرٌّ, (T, Ṣ, Mṣb, Ḳ,) [or this last is rather a coll. gen. n., signifying *pottery,* or *jars,* &c.,] like تَمْرٌ in relation to تَمْرَةٌ; or, accord. to some, this is a dial. var. of جَرَّةٌ. (Mṣb.) Beverage of the kind called نَبِيذ made in such a vessel is forbidden in a trad.: (Mgh, TA:) but accord. to IAth, the trad. means a vessel of this kind glazed within, because the beverage acquires strength, and ferments, more quickly in a glazed earthen vessel. (TA.) ═ See also جُرَّةٌ: ═ and see what here next follows.

جُرَّةٌ (Ṣ, Ḳ) and ↓ جَرَّةٌ (Ḳ) *A small piece of wood,* (Ḳ,) or *a piece of wood about a cubit long,* (Ṣ,) *having a snare at the head,* (Ṣ, Ḳ,) *and a cord at the middle,* (Ṣ,) *with which gazelles are caught:* (Ṣ, Ḳ:) when the gazelle is caught in it, he strives with it awhile, and struggles in it, and labours at it, to escape; and when it has overcome him, and he is wearied by it, he becomes still, and remains in it; and this is what is termed [in a prov. mentioned below] his becoming at peace with it: (Ṣ,* TA:) or it is *a stuff,* or *stick, tied to a snare, which is hidden in the earth, for catching the gazelle; having cords of sinew;* when his fore leg enters the snare, the cords of sinew become tied in knots upon that leg; and when he leaps to escape, and stretches out his fore leg, he strikes with that stuff, or stick, his other fore leg and his hind leg, and breaks them. (AHeyth, TA.) نَاوَصَ الجُرَّةَ ثُمَّ سَالَمَهَا *He struggled with the* جُرَّة *and then became at peace with it* [see above] is a prov. applied to him who opposes the counsel, or opinion, of a people, and then is obliged to agree: (Ṣ,* TA:) or to him who falls into a case, and struggles in it, and then becomes still. (TA.) And it is said in another prov., هُوَ كَالبَاحِثِ عَنِ الجُرَّةِ [*He is like him who searches in the earth for the* جُرَّة]. (AHeyth, TA.) In the phrase إِذَا أَفْلَتَتْ مِنْ جُرَّتَيْهَا, in a saying of Ibn-Lisán-el-Ḥummarah, referring to sheep, [app. meaning *When they escape from their two states of danger,*] by جرّتيها he means their place of pasture (المَجَرّ) in a severe season [when they are liable to perish], and when they are scattered, or dispersed, by night, and [liable to be] attacked, or destroyed, by the beasts of prey: so says ISk: Az says that he calls their مجر two snares, into which they might fall, and perish. (TA.)

جِرَّةٌ *A mode,* or *manner, of dragging, drawing, pulling, tugging, straining,* or *stretching.* (Ḳ.) ═ The *stomach* of the camel, and of a cloven-hoofed animal: this is the primary signification: by extension of its meaning, it has the signification next following. (Mṣb.) —— The *cud which a camel* [or *cloven-hoofed animal*] *ejects from its stomach,* (Az, Ṣ,* IAth, Mgh, Mṣb, Ḳ,*) *and eats again,* (Ḳ,) or *chews,* or *ruminates,* (Az, IAth, Mṣb,) or *to chew,* or *ruminate;* (Ṣ;) as also

† جِرَّةٌ: (Ḳ:) it is said to belong to the same predicament as بَعْر. (Mgh.) Hence the saying, لَا أَفْعَلُ ذٰلِكَ مَا ٱخْتَلَفَتِ ٱلدِّرَّةُ وَٱلجِرَّةُ *I will not do that as long as the flow of milk and the cud go* [the former] *downwards and* [the latter] *upwards.* (Ṣ, A.*) [See also دِرَّةٌ.] And ٱجْتُلِبَتِ ٱلدِّرَّةُ بِالجِرَّةِ [*The flow of milk was procured by the cud*]: alluding to the beasts' becoming full of food, and then lying down and not ceasing to ruminate until the time of milking. (IAạr, TA.) And لَا يَحْنَقُ عَلَى جِرَّتِهِ † *He will not bear rancour, or malice, against his subjects:* or, as some say, † *he will not conceal a secret:* (TA:) and مَا يَحْنَقُ عَلَى جِرَّةٍ and مَا يَكْظِمُ على جرّة † *he does not speak when affected with rancour, or malice:* (TA in art. حنق:) [or the last has the contr. signification: for] لَا يَكْظِمُ عَلَى جِرَّتِهِ means ‡ *he will not be silent respecting that which is in his bosom, but will speak of it.* (TA in art. كظم.) —— Also The *mouthful with which the camel diverts and occupies himself until the time when his fodder is brought to him.* (Ḳ.)

جَرُورٌ ‡ A female *that exceeds the* [*usual*] *time of pregnancy.* (A.) ‡ A she-camel *that withholds her fœtus in her womb, after the completion of the year, a month, or two months, or forty days only;* (Ḳ,* TA;) or, *three months after the year:* they are the most generous of camels that do so: none do so but those that usually bring forth in the season called الرَّبِيع (المَرَابِيع); not those that usually bring forth in the season called الصَّيْف (المَصَايِيف): and only those do so that are red [or brown], and such as are of a white hue intermixed with red (الصُّهْب), and such as are ash-coloured: never, or scarcely ever, such as are of a dark gray colour without any admixture of white, because of the thickness of their skins, and the narrowness of their insides, and the hardness of their flesh. (IAạr, TA. [See also 1: and see خَضُوف.]) —— Also † A she-camel *that is made to incline to, and to suckle, a young one not her own; her own being about to die, they bound its fore legs to its neck, and put upon it a piece of rag, in order that she might know this piece of rag, which they then put upon another young one; after which they stopped up her nostrils, and did not unclose them until the latter young one had sucked her, and she perceived from it the odour of her milk.* (L.) —— Also, applied to a horse, (Ṣ, A, Ḳ,) and a camel, (Ḳ,) ‡ *That refuses to be led; refractory:* (Ṣ, A, Ḳ:) of the measure فَعُولٌ in the sense of the measure مَفْعُولٌ; or it may be in the sense of the measure فَاعِلٌ: (Az, TA:) or a *slow* horse, *either from fatigue or from shortness of step:* (A 'Obeyd, TA:) pl. جُرُرٌ. (TA.) —— And † A woman *crippled; or affected by a disease that deprives her of the power of walking:* (Sh, Ḳ:) because she is dragged upon the ground. (Sh, TA.) —— بِئْرٌ جَرُورٌ ‡ *A deep well;* (Sh, Ṣ, Ḳ;) *from which the water is drawn by means of the* سَانِيَة [q. v.], (Ṣ, A,) *and by means of the pulley and the hands;* like مَتُوحٌ and نَزُوعٌ: (A:) or *a well from which the water is drawn* [*by a man*] *upon a camel* [*to the saddle of which one end of the well-rope is attached*]; so called because its bucket is drawn upon the edge of the mouth thereof, by reason of its depth. (Aṣ, L.)

جَرِيرٌ *A rope:* pl. أَجِرَّةٌ. (Sh, TA.) *A rope for a camel, corresponding to the* عِذَار *of a horse,* (Ṣ, Ḳ,) *different from the* زِمَام. (Ṣ.) Also The *nose-rein of a camel;* syn. زِمَامٌ: (Ḳ:) or *a cord of leather, that is put upon the neck of a she-camel:* (Mṣb:) or *a cord of leather, like a* زِمَام: and applied also to *one of other kinds of plaited cords:* or, accord. to El-Hawázinee, [*a string*] *of softened leather, folded over the nose of an excellent camel or a horse.* (TA.) [See also خِطَامٌ.]

جِرَارَةٌ The *art of pottery:* the *art of making jars, or earthen vessels.* (TA. [See جَرَّةٌ.])

جَرِيرَةٌ *A crime; a sin; an offence which a man commits, and for which he should be punished; an injurious action:* (Ṣ,* Mṣb,* Ḳ,* TA:) syn. ذَنْبٌ, (Mṣb, Ḳ,) and جِنَايَةٌ: (Ṣ:) of the measure فَعِيلَةٌ in the sense of the measure مَفْعُولَةٌ: (Mṣb:) pl. جَرَائِرُ. (A.) See also what next follows.

فَعَلْتُ كَذَا مِنْ جَرَّاكَ, (Ṣ, A,* Ḳ,*) and من جَرَائِكَ, (Ḳ,) and من جَرَاكَ, and مِن جَرَّائِكَ, (Ṣ, Ḳ,) and † مِنْ جَرِيرَتِكَ, (Ḳ,) means مِنْ أَجْلِكَ, (Ṣ, A, Ḳ) i. e., [originally, *I did so*] *in consequence of thy committing it,* namely, a crime: and then, by extension of its application, [*because of thee,* or *of thine act &c.; on thine account; for thy sake;*] indicating any causation. (Bḍ in v. 35, in explanation of من جَرَّاكَ and من أَجْلِكَ.) One should not say مِجْرَاكَ, (Ṣ,) or بِجْرَاكَ. (A.)

جِرِّيٌّ (written in the Towsheeḥ with fet-ḥ to the ج also, TA,) [The *eel;*] *a kind of fish,* (Ṣ, Ḳ,) *long and smooth,* (Ḳ,) *resembling the serpent, and called in Persian* مَارْ مَاهِى; said to be a dial. var. of جِرِّيثٌ; (TA;) *not eaten by the Jews,* (Ḳ,) *and forbidden to be eaten by 'Alee;* (TA;) *having no scales:* (Ḳ:) or *any fish having no scales.* (Towsheeḥ, TA.)

جِرِّيَّةٌ The *stomach,* or *triple stomach,* or the *crop,* or *craw,* of a bird; syn. حَوْصَلَةٌ; (Ṣ, Ḳ;) as also جِرِّئَةٌ [q. v.] (Ḳ) and قِرِّيَّةٌ. (AZ, TA.) You say, أَلْقَاهُ فِى جِرِّيَّتِهِ, meaning, ‡ *He ate it.* (A, TA.) See also art. جرى.

جَرَّارٌ † *A man who leads a thousand.* (T, end of art. حفز.) —— جَيْشٌ جَرَّارٌ, (Ṣ, A,) and كَتِيبَةٌ جَرَّارَةٌ, (Ṣ, Ḳ,) ‡ *An army,* and *a troop of horse or the like, that marches heavily, by reason of its numbers:* (Aṣ, Ṣ, Ḳ:) or *dragging along the apparatus of war:* (A:) or *numerous.* (TA.) ═══ *A potter; a maker of jars, or earthen vessels.* (TA. [See جَرَّةٌ.])

جَرَّارَةٌ *A small,* (Ṣ, A, Ḳ, TA,) *yellow,* (A, TA,) *female* (TA) *scorpion,* (Ṣ, A, Ḳ, TA,) *like a piece of straw,* (TA, [thus I render على شكل التبنة, but I think that there must be here some mistranscription, as the words seem to be descriptive of form,]) *that drags its tail;* (Ṣ, Ḳ;) for which reason it is thus called; *one of the most deadly of scorpions to him whom it stings:* (TA:) pl. جَرَّارَاتٌ. (A, TA.)

جُرَّانٌ: see جَارٌّ, last sentence.

جَرْجَرٌ The *thing* [or *machine*] *of iron with which the reaped corn collected together is thrashed.* (Ḳ.) [See نَوْرَجٌ and مِدْوَسٌ.] ═══ See also جِرْجِرٌ.

جِرْجِرٌ: see جَرْجَارٌ. ═══ Also The *bean;* or *beans;* syn. فُولٌ; (Ṣ, Ḳ;) and so جُرْجُرٌ: (Ḳ:) of the dial. of the people of El-'Irák. (TA.) —— See also جِرْجِيرٌ.

جَرْجَرَةٌ, an onomatopœia: (Mṣb:) *A sound which a camel reiterates in his windpipe:* (Ṣ, Ḳ:) the *sound made by a camel when disquieted, or vexed:* (TA:) the *sound of pouring water into the throat:* (TA:) or the *sound of the descent of water into the belly:* (IAth, TA:) or the *sound of water in the throat when drunk in consecutive gulps.* (Mṣb.) [See R. Q. 1.]

جَرْجَارٌ A camel *that reiterates sounds in his windpipe:* (Ṣ:) or a camel *that makes much noise* [*or braying*]; as also † جِرْجِرٌ and † جُرَاجِرٌ. (Ḳ.) —— The *sound of thunder.* (Ḳ.) ═══ *A certain plant,* (Ṣ, Ḳ,) *of sweet odour;* (Ṣ;) *a certain herb having a yellow flower.* (AḤn, TA.)

جُرْجُورٌ *A large,* or *bulky,* camel: (Ḳ:) pl. جَرَاجِرُ, (Kr, Ḳ,) without ى [before the final letter], though by rule it should be with ى, except in a case of poetic necessity. (TA.) And, as a pl., *Large,* or *bulky,* camels; as also [its pl.] جَرَاجِرُ: (Ṣ:) or *large-bellied* camels: (TA:) and *generous,* or *excellent,* camels: (Ḳ, TA:) and *a herd,* or *collected number,* (Ḳ, TA,) of camels: (TA:) and مِائَةٌ جُرْجُورٌ *a complete hundred* (Ḳ, TA) of camels. (TA.)

جِرْجِيرٌ (Ṣ, Ḳ) and † جِرْجِرٌ (Ḳ) [The herb *eruca,* or *rocket;*] *a certain leguminous plant,* (Ṣ, Ḳ,) *well known:* (Ḳ;) *a plant of which there are two kinds; namely,* بَرِّىّ [i. e. *eruca sylvestris,* or *wild rocket*], and بُسْتَانِىّ [i. e. *eruca sativa,* or *garden-rocket*]; *whereof the latter is the better: its water, or juice, removes scars, and causes milk to flow, and digests food:* (TA:) AḤn says that the جرجير is the بَاقِلَّى [q. v.]; and that the جِرْجِير مِصْرِىّ is the تُرْمُس: [but see this last word.] (TA in art. ترمس.)

جَرْجَارَةٌ *A mill,* or *mill-stone;* syn. رَحًى: (Ḳ:) because of its sound. (TA.)

جُرَاجِرٌ: see جَرْجَارٌ. —— Also *That drinks much;* (Ḳ; [in the CḲ misplaced;]) applied to a camel: you say إِبِلٌ جُرَاجِرَةٌ. (IAạr, TA.) —— And hence, (TA,) Water *that makes a noise.* (Ḳ.)

جَارٌّ [act. part. n. of 1; *Dragging, drawing,* &c.]. —— جَارُّ الضَّبُعِ ‡ *Rain that draws the hyena from its hole by its violence:* or *the most violent rain;* as though it left nothing without dragging it along: (TA:) or *rain that leaves nothing without making it to flow, and dragging it along:*

(IAạr, TA:) or *the torrent that draws forth the hyena from its hole:* (A:) and in like manner, مَجْرُ ↓ الضَّبُعِ *the torrent that has torn up the ground;* as though the hyena were dragged along in it. (IAạr, Sh, TA.) You say also مَطَرٌ جَارُ الضَّبُعِ, and مَطْرَةٌ جَارَةُ الضَّبُعِ. (A.) — إِبِلٌ جَارَّةٌ ‡ *Working camels;* because they drag along burdens; (A, Mgh;) or tropically so called because they are dragged along by their nose-reins: (Mgh:) or *camels that are dragged along by their nose-reins:* (Ṣ, Ḳ, TA: [but in the copies of the Ṣ, and in those of the Ḳ, in my possession, تَجُرُّ is put for تُجَرُّ, though the latter is evidently meant, as is shown by what here follows:]) جَارَّةٌ is of the measure فَاعِلَةٌ in the sense of the measure مَفْعُولَةٌ: it is like as when you say عِيشَةٌ رَاضِيَةٌ in the sense of مَرْضِيَّةٌ, and مَآءٌ دَافِقٌ in the sense of مَدْفُوقٌ: (Ṣ:) or it means *such as carry goods,* or *furniture and utensils, and wheat,* or *food.* (AZ, TA voce حَانٌ, q. v.) It is said in a trad. that there is no poor-rate (صَدَقَة) in the case of such camels, (Ṣ, Mgh,) because they are the riding-camels of the people; for the poor-rate is in the case of pasturing camels, exclusively of the working. (Ṣ.) — لَا جَارَّ لِى فِى هٰذَا ‡ *There is no profit for me in this to attract me to it.* (A, TA.) — حَارٌّ جَارٌّ is an expression in which the latter word is an imitative sequent to the former; (Ṣ, Ḳ;) but accord. to A 'Obeyd, it was more common to say حَارٌّ يَارٌّ, with ى: (Ṣ:) and one says also ↓ حَرَّانُ يَرَّانُ جَرَّانُ. (TA in art. حر.)

جُورٌ is mentioned by Az in this art., meaning Rain *that draws along everything:* and rain *that occasions the herbage to grow tall:* and *a large and heavy* [bucket of the kind called] غَرْب; explained in this sense by AO: and *a bulky* camel; and, with ة, in like manner applied to a ewe: Fr says that the و in this word may be considered as augmentative or as radical. (TA.) [See also art. جور.]

جَارَّةٌ [fem. of جَارٌّ, q. v.: and, as a subst.,] *A road to water.* (Ḳ.)

جَارُورٌ *A river,* or *rivulet, of which the bed is formed by a torrent.* (Ṣ,* Ḳ,* TA.)

الأَجَرَّانِ *The jinn, or genii, and mankind.* (IAạr, Ḳ.)

مَجَرٌّ [The *place,* or *track, along which a thing is,* or *has been, dragged,* or *drawn*]. You say, رَأَيْتُ مَجَرَّ ذَيْلِهِ [*I saw the track along which his hinder skirt had been dragged*]. (A.) See also المَجَرَّةُ: and جَارٌّ. — *A place of pasture.* (TA.) — The جَائِز [or *beam*] *upon which are placed the extremities of the* عَوَارِض [or *rafters*]. (Ḳ)

مُجَرٌّ: see 4, in the latter portion of the paragraph.

المَجَرَّةُ ‡ [*The Milky Way in the sky;*] *the* شَرَج *of the sky;* (Ḳ;) *the whiteness that lies across in the sky, by the two sides of which are the* نَسْرَانِ [or *two constellations called* النَّسْرُ الطَّائِرُ and النَّسْرُ الوَاقِعُ]: or [*the tract called*] الطَّرِيقُ المَحْسُوسَةُ [which is probably the same; or *the tract*], *in the sky, along which* (مِنْهَا) *the* [*wandering*] *stars* [or *planets*] *take their ways:* (TA:) or *the gate of Heaven:* (Ḳ:) so called because it is like the trace of the مَجَرّ [or place along which a thing has been dragged, or drawn]. (Ṣ.) Hence the prov., سِطِى مَجَرْ ↓ تُرْطِبْ هَجَرْ ‡ *Reach the middle of the sky, O milky way,* (مجر being for مجرّة,) and the palm-trees of *Hejer will have ripe dates.* (A,* TA.)

مَجْرُورٌ [pass. part. n. of 1]: see 4, latter portion.

جرأ

1. جَرُؤَ, aor. ـُ, inf. n. جَرَاءَةٌ (Ṣ, Mṣb, Ḳ) and جَرَائِيَةٌ and جَرَايَةٌ, with ى [in the place of ء], which is extr., (Ḳ,) and جُرْأَةٌ and جُرَةٌ, (Ṣ, Ḳ,) thus sometimes, without ء, like as one says مَرْأَةٌ and مَرَةٌ, (Ṣ,) [all mentioned as inf. ns. in the TḲ, and app. as such in the Ḳ, but only the first is explicitly mentioned as an inf. n. in the Ṣ and Mṣb, and ↓ جُرْأَةٌ is said in the Mṣb to be a simple subst.,] *He was,* or *became, bold, daring, brave,* or *courageous;* (Ṣ, Mṣb,* Ḳ, TA;) *so as to attempt,* or *venture upon, a thing without consideration or hesitation:* (TA:) [said of a brute and the like, as well as of a man:] and ↓ استجرأ is syn. therewith. (IJ, W p. 146.)

2. جَرَّأَهُ عَلَيْهِ, (inf. n. تَجْرِئَةٌ, Ḳ,) *I emboldened him,* or *encouraged him, against him.* (Ṣ, Mṣb,* Ḳ, TA.)

5: see 8.

8. اجترأ عَلَيْهِ, (Ṣ, Ḳ,) or ↓ تجرّأ, (Mṣb,) *He became emboldened* or *encouraged,* or *he emboldened* or *encouraged himself, against him.* (Ṣ, Mṣb,* Ḳ, TA.) — اجترأ عَلَى القَوْلِ *He ventured upon the saying hastily and unhesitatingly.* (Mṣb.)

10: see 1.

جُرْأَةٌ *Boldness, daringness, bravery,* or *courage;* as also جُرَةٌ: (Ṣ: see 1:) the *quality of venturing* upon a saying [&c.] *hastily and unhesitatingly.* (Mṣb.)

جَرِىْءٌ *Bold, daring, brave,* or *courageous:* (Ṣ, Mṣb,* Ḳ, TA:) pl. أَجْرَآءُ, accord. to a MS. copy of the Ḳ; [and so in the CḲ;] but in the M, أَجْرِئَآءُ, with two hemzehs, on the authority of Lḥ; and so in some copies of the Ḳ; and sometimes جُرَآءُ, like حُلَمَآءُ, occurring in a trad., as some relate it; but the reading commonly known is حُرَآءُ, with the unpointed ح. (TA.) — جَرِىْءُ المُقْدَمِ *Bold, daring, brave,* or *courageous, in venturing* [against an adversary, or upon an undertaking]. (Ṣ.) — الجَرِىْءُ *The lion;* as also ↓ المُجْتَرِئُ. (O, Ḳ.)

جُرْئَةٌ *A chamber* (Ḳ, TA) *constructed of stones, with a stone placed over its entrance,* (TA,) *for the purpose of entrapping wild beasts:* (Ḳ, TA:) the piece of flesh-meat for the wild beast is put in the hinder part of the chamber; and when he enters to take the piece of meat, the stone falls upon the entrance, and closes it: (TA:) pl. جَرَائِى, (accord. to some copies of the Ḳ,) or جَرَائِئُ, (accord. to others,) mentioned by AZ as one of the forms of pl. repudiated by the Arabic grammarians except in some anomalous instances. (TA.)

الجِرِّيْئَةُ *The* قَانِصَة [here app. meaning the *stomach,* or *triple stomach,* or the *crop,* or *craw,* of a bird], and *the* حُلْقُوم [here app. meaning the *gullet* of a bird]; like الجِرِّيَّةُ; (Ḳ;) i. e. *the* حَوْصَلَة [meaning the *stomach,* or the *crop,* of a bird]: it is said in the T, on the authority of AZ, that الجِرِّيْئَةُ and القِرِّيَّةُ and النَّوْطَةُ signify *the* حَوْصَلَة *of a bird.* (TA.)

المُجْتَرِئُ: see جَرِىْءٌ.

جرب

1. جَرِبَ, (Ṣ, A, Mgh, Mṣb, Ḳ,) aor. ـَ, (Mgh, Mṣb, Ḳ,) inf. n. جَرَبٌ, (Mṣb, TA,) *He* (a camel, Ṣ, A, Mṣb, Ḳ, and a man, Ṣ, or other animal, Mṣb,) *was,* or *became, affected with what is termed* جَرَب [i. e. *the mange,* or *scab*]. (Ṣ, Mṣb, Ḳ.) مَا لَهُ جَرِبَ وَحَرِبَ is a form of imprecation against a man [meaning *What aileth him? may he have the scab, and be despoiled of all his wealth,* or *property:* or *may he have his camels affected with the mange,* or *scab, and be despoiled* &c.: or *may his camels be affected with the mange,* or *scab,* &c.]: it may express a wish that he may be affected with جَرَب: or جَرِبَ may be put for أَجْرَبَ, to assimilate it to حَرِبَ: or it may be for جَرِبَتْ إِبِلُهُ. (L.) — See 4. — Also ‡ *i. q.* هَلَكَتْ أَرْضُهُ [meaning *His land had its herbage dried up by drought;* or *became such as is termed* جَرْبَآء, fem. of أَجْرَب, q. v.]. (Ḳ.)

2. جرّبهُ, (A, Mṣb, Ḳ,) inf. n. تَجْرِبَةٌ, (M, A, Ḳ,) or تَجْرِيبٌ, the former, which see also below, being a simple subst., (Mṣb,) or both, but the former is irreg., are inf. ns., (TA,) *He tried, made trial of, made experiment of, tested, proved, assayed, proved by trial* or *experiment* or *experience, him,* or *it:* (A, Ḳ:) or *he tried it, made trial of it,* &c., namely, a thing, *time after time.* (Mṣb.) [You say also جرّب, for جرّب الأُمُورَ, meaning *He tried affairs:* and hence, *i. q.*] جَرَّبَ فِى الأُمُورِ [*He became experienced,* or *expert, in affairs*]. (T, TA.) And جَرَّبَتْهُ الأُمُورُ [*Affairs,* or *events, tried him,* &c.: and thus, *rendered him experienced,* or *expert*]. (Ṣ, TA.) And مَا جُرِّبَتْ عَلَيْهِ فَعْلَةٌ قَبِيحَةٌ قَطُّ [*A foul action was never found to be chargeable upon him*]. (Ṣ voce نَغْبَة.)

4. اجرب *He had his camels* [or *found them to be*] *affected with what is termed* جَرَب [i. e. *the mange,* or *scab*]; (Ṣ, A, L, Ḳ;) as also ↓ جَرِبَ, (L, Ḳ,) which may be for جَرِبَتْ إِبِلُهُ; or used for أَجْرَبَ, to assimilate it to حَرِبَ in a saying mentioned above; see 1. (L.)

Q. Q. 1. جَوْرَبَهُ *He put on him* [i. e., *on his*

(another's) *foot* or *feet,*] جَوْرَبٌ [i. e. *a sock* or *stocking*, or *a pair of socks* or *stockings*]. (Ṣ, Ḳ.)

Q. Q. 2. تَجَوْرَبَ *He put on* [i. e., *on his own foot* or *feet,*] جَوْرَب [i. e. *a sock* or *stocking*, or *a pair of socks* or *stockings*]. (Ṣ, Ḳ.) And in like manner, تجورب جَوْرَبَيْنِ [*He put on a pair of socks or stockings*]. (TA.)

جَرْبٌ : see جَرْبَةٌ.

جَرَبٌ [The *mange*, or *scab*;] *a certain disease*, (A,) *well known*; (Ṣ, A, Ḳ;) accord. to the medical books, (Mṣb,) *a gross humour, arising beneath the skin, from the mixture of the salt phlegm*, (Mṣb, MF,) or *the phlegm of the flesh*, (so in a copy of the Mṣb,) *with the blood, accompanied with pustules, and sometimes with emaciation, in consequence of its abundance*; (Mṣb, MF;) or [*an eruption consisting of*] *pustules upon the bodies of men and camels*. (M, TA.) You say, أَعْدَى مِنَ الجَرَبِ عِنْدَ العَرَبِ [*More transitive*, or *catching, than the mange*, or *scab, among the Arabs*]: (A, TA:) a proverb. (TA.) — † *Rust upon a sword*. (Ḳ.) — ‡ *A resemblance of rust upon the inner side of the* جَفْن [or *eyelid*], (M, Ḳ,) *sometimes covering the whole of it, and sometimes part of it*. (M.) You say, بِأَجْفَانِهِ جَرَبٌ ‡ [*In his eyelids is*] *a resemblance of rust upon their inner sides*. (A.) — † *A vice, a fault, a defect, an imperfection*, or *a blemish*. (IAạr, Ḳ.)

جَرِبٌ : see أَجْرَبُ.

جَرْبَةٌ *A place of seed-produce*; (Ṣ, Ḳ;) as also ▼ جَرِيبٌ : (Ḳ:) and *a tract of land such as is termed* قَرَاح [i. e. *a field*, or *land, sown* or *for sowing, without any building or trees in it*; or *land cleared for sowing and planting*; or *a separate piece of land in which palm-trees &c. grow*; &c.]: (Ḳ:) metaphorically applied by Imra-el-Keys to [*a grove of*] palm-trees, where he says

كَجَرْبَةِ نَخْلٍ أَوْ كَجَنَّةِ يَثْرِبِ

[*Like a grove of palm-trees, or like the plantation of Yethrib*]: (AḤn, TA:) or *land prepared for sowing* or *planting*: (AḤn, Ḳ:) or *a piece of land differing in condition from the land adjoining it*, [i. e. *a patch of land*,] *producing good plants* or *herbage*: (Lth, TA:) the pl. [or rather coll. gen. n.] is ▼ جَرْبٌ, (Lth, AḤn,) like تِبْنٌ as is of تِبْنَةٌ, and سِدْرٌ of سِدْرَةٌ: (AḤn:) or جَرْب signifies a قَرَاح; and its pl. is جِرَبَةٌ. (IAạr, TA.) — *A skin, or a mat, which is placed upon the brink of a well, lest the water should be scattered into the well* [app. in falling from the bucket into the channel of the tank or cistern &c.]: or (*a skin*, TA,) *that is placed in a rivulet* or *streamlet* (جَدْوَل [which is applied in the present day to an artificial streamlet for irrigation, in the form of a trench or gutter,]) *that the water may flow down over it* [app. from the well to the tank or cistern &c.]. (M, Ḳ.)

جَرِبَةٌ : see أَجْرَبُ, last sentence but one.

جَرْبَانٌ or جُرُبَّانٌ : see أَجْرَبُ : — and for the latter, see جُرُبَّانٌ.

جُرْبَانٌ and جُرُبَانٌ : see جُرُبَّانٌ, in five places.

جُرُبَّاءُ and جِرِبَّاءُ : see what next follows.

جُرُبَّانٌ (Ṣ, MF, TA) and جِرِبَّانٌ, (Mj, MF, TA,) which are the two forms commonly known, (MF, TA,) or, accord. to the Ḳ, ▼ جُرُبَانٌ and ▼ جُرْبَانٌ, or, accord. to the L, ▼ جُرْبَانٌ, and sometimes ▼ جُرُبَانٌ, or, accord. to some copies of the Ḳ, [and so in the CḲ,] ▼ جِرِبَّاءُ and ▼ جُرُبَّاءُ, which are evident mistranscriptions, or, accord. to the 'Ináyeh of El-Khafájee, جَرْبَانٌ, which is more strange, (MF,) but this last accords [most nearly] with its original, (TA,) [for it is] a Persian word arabicized, (Ṣ, TA,) originally كَرِيبَان; (TA;) The جَيْب [or *opening at the neck and bosom*] of a shirt: (Ḳ, TA:) or the *part around the neck, upon which are sewed the buttons*: (IB and TA in art. بنق:) or the [*part called*] لِبْنَة [q. v.] of a shirt. (Ṣ, TA.) — جُرُبَّانُ سَيْفٍ (Fr, Ṣ, Ḳ) and ▼ جُرْبَانُهُ, (Ḳ, TA,) or جِرِبَّانه, (CḲ,) *The edge* (حَدّ) *of a sword*: (Ḳ:) or *a thing* [i. e. *a case*] (Ḳ, TA) *of sewed leather* (TA) *in which are put a sword and its scabbard with the cords or belts by which it is suspended*; (Ḳ, TA;) *i. q.* قِرَابٌ: (Ṣ: [see also جِرَابٌ:]) or *a large sword-case in which are a man's sword and his whip and what else he requires*: (Fr, TA: [also called جُلُبَّان and جِلِبَّان and جُلْبَان:]) in the L, the first is [also] said to signify *the scabbard of a sword*. (TA.)

جِرْبِيَاءُ [a word of a very rare form, (see كِبْرِيَآءُ,)] The *north-west wind*; *a wind of the kind termed* نَكْبَآء, *that blows in a direction between that of the* [*north wind*, or *northerly wind, called*] شَمَال *and that of the* [*west wind*, or *westerly wind, called*] دَبُور, *and that dispels the clouds*: (Ṣ, TA:) it is a cold wind, and is sometimes attended by a little rain: (TA in art. نكب, q. v.:) or the [*north wind*, or *northerly wind, called*] شمال: or the *cold* of that wind: (Ḳ, TA:) or, (Ḳ,) as also أَزْيَبُ, (TA,) the *south-east wind*; the *wind that blows in a direction between that of the* [*south wind*, or *southerly wind, called*] جَنُوب *and that of the* [*east wind*, or *easterly wind, called*] صَبَا. (Ḳ, TA.) — Also, with the article ال, a name of *The seventh earth*: corresponding to العَرْبِيَآءُ, a name of "the seventh heaven." (TA.) — Also A *weak* man. (Ḳ.)

جِرَابٌ, (Ṣ, Mṣb, Ḳ, &c.,) not جَرَابٌ, (ISk, Mṣb, Ḳ,) or this latter is of weak authority, (Ḳ, TA,) or peculiar to the vulgar, (Ṣ, L,) *A provision-bag for travellers*: (Ḳ, Ḥar p. 174:) or *a bag*, or *receptacle, for travelling-provisions and for goods or utensils &c.*; syn. وِعَآء: (Ḳ, TA:) or *such a receptacle made of sheep-skin, in which nothing is kept but what is dry*: (TA:) pl. [of mult.] جُرُبٌ (Ṣ, Mṣb, Ḳ) and جُرْبٌ, (Ṣ, Ḳ,) the latter a contraction of the former, (TA,) and [of pauc.] أَجْرِبَةٌ. (Ṣ, Mṣb, Ḳ.) — ‡ *A sword-case*; or *a case*, or *receptacle, in which a sword is put with its scabbard and its suspensory belt* or *cord*; syn. قِرَابُ سَيْفٍ. (TA. [See also جُرُبَّانٌ.]) — † The *scrotum*. (Ḳ.) — جِرَابُ القَلْبِ † [*The pericardium*, or *heart-purse*]. (Ḳ in art. ثهت, &c.) — جِرَابُ البِئْرِ † *The cavity of the well*; (M, Ḳ;) or ‡ *its interior*, (Lth, Ṣ, M, A,) *from top to bottom*. (Lth, Ṣ, M.) You say, اطْوِ جِرَابَهَا بِالحِجَارَةِ *Case thou its interior with stones*. (A.)

جَرِيبٌ *A certain measure*, (M, A, Mgh, Ḳ,) or *quantity*, of wheat, (Ṣ, Mṣb,) *consisting of four* أَقْفِزَة [pl. of قَفِيز]: (M, A, Mṣb, Ḳ:) or *ten* أَعْشِرَآء; each قفيز thereof consisting of ten اقفزة [pl. of عَشِير]; so that the عشير is the hundredth part of the whole: (TA:) or, as some say, *a measure differing in different countries*; as is the case of the رِطْل and مُدّ and ذِرَاع &c. (MF, TA.) For the pl., see what follows. — Hence, (Mgh,) † *A certain quantity of land*; (Ṣ, Mgh, Mṣb;) *as much as is sown with the measure of seed so called*; (A, Mgh;) like as mules and the space that they travel are termed بَرِيد: (A, Mgh:*) it is *sixty cubits by sixty cubits*; accord. to Ḳudámeh, the *extent termed* أَشْل *multiplied by itself*; the اشل being sixty cubits; the cubit being six قَبَضَات; and the قَبْضَة, four أَصَابِع: the tenth part of the جريب is called قفيز, and the tenth of the قفيز is called عشير; so that the قفيز is ten اعشراء: (Mgh:) it is *a distinct portion of land, differing according to the different conventional usages of the people of different provinces*: it is said that the width of six moderate-sized barley-corns is called إِصْبَع; the قبضة is four اصابع; the ذِرَاع is six قبضات; ten أَذْرُع are called قَصَبَة; ten قَصَبَات are called اشل; and the جريب is the *extent termed* اشل *multiplied by itself*: the اشل multiplied by the قصبة is called قفيز; and the اشل multiplied by the ذراع is called عشير: so the جريب is *ten thousand cubits*: or, accord. to Ḳudámeh the Scribe, it is *three thousand and six hundred cubits*: (Mṣb:) pl. [of pauc.] أَجْرِبَةٌ and [of mult.] جُرْبَانٌ (Ṣ, Mṣb, Ḳ) and جُرُوبٌ. (R, TA.) See also جَرْبَةٌ. — Also *A valley*; (Lth, Mṣb, Ḳ; [accord. to the second of which, this is the primary signification;]) i. e., in an absolute sense; and, with the article ال, the name of a particular valley in the territory of Ḳeys: (TA:) pl. أَجْرِبَةٌ. (Lth, TA.)

جَوْرَبٌ [*A sock* or *stocking*, or *a pair of socks* or *stockings*;] the *wrapper of the foot* or *leg*: (Ḳ:) or *a pair of woollen envelopes for the feet, used for warmth*: (TA:) an arabicized word, (Ṣ, Mṣb,) from the Persian كُورَب, originally كُورْ پَا, i. e. "tomb of the foot:" (TA:) pl. جَوَارِبَةٌ and جَوَارِبُ; (Ṣ, A, Mṣb, Ḳ;) in the former of which, the ة is added because it is originally a foreign word. (Ṣ, TA.) You say, هُوَ أَنْتَنُ مِنْ رِيحِ الجَوْرَبِ [*He, or it, is more stinking than the smell of socks, or stockings*]. (A, TA.)

جَوَارِبِىٌّ *A maker of* جَوَارِب [i. e. *socks or stockings*]. (TA.)

أَجْرَبُ (Ṣ, A, Mgh, Mṣb, Ḳ) and ↑ جَرِبٌ (A, Mgh, Ḳ) and ↑ جَرْبَانُ or جَرْبَانٌ (Ḳ accord. to different copies) [*Mangy, or scabby;*] *affected with what is termed* جَرَب: (Ṣ, A, Mṣb, Ḳ:) applied to a camel, (A, Mṣb,) and to a man: (Ṣ, A:) fem. (of the first, Mṣb) جَرْبَآءُ (A, Mṣb) and [of the second] جَرِبَةٌ: (A:) pl. (of the first, Ṣ, Mṣb) جُرْبٌ (Ṣ, A, Mṣb, Ḳ) and (of the first, Ṣ, Mgh, TA, or of the second, Mgh, or of the third agreeably with analogy, TA) جَرْبَى (Ṣ, Mgh, Ḳ) and [of the first] أَجَارِبُ, which is like certain pls. of substantives, as أَجَادِلُ and أَنَامِلُ, (TA,) and (of the first contrary to rule, like عِجَافٌ and بِطَاحٌ and عِصَالٌ which are pls. of أَعْجَفُ and أَبْطَحُ and أَعْصَلُ, Mṣb, or of the second, IB, Ḳ, or of جُرْبٌ, which is pl. of the first, Ṣ) جِرَابٌ: (Ṣ, IB, Mṣb, Ḳ:) this last occurs in the following verse [of 'Amr, or 'Omeyr, Ibn-El-Ḥobáb, or El-Khabbáb; these variations being in different copies of the Ḳ; but in the TA art. نشر, and in a copy of the Ṣ in that art. and in the present one, 'Omeyr Ibn-El-Khabbáb]:

* وَفِينَا وَإِنْ قِيلَ ٱصْطَلَحْنَا تَضَاغُنٌ *
* كَمَا طَرَّ أَوْبَارُ الجِرَابِ عَلَى النَّشْرِ *

(Ṣ, Ḳ*) *Within us, though it be said that we have made peace, one with another,* and we are on good terms outwardly, *is mutual rancour: as the soft wool of the mangy camels* (while disease lurks beneath, within them, TA) *grows by reason of* [eating] *the* نشر [or *herbage*] *that becomes green at the end of summer* (*in consequence of rain falling upon it,* TA) *and is injurious to animals that pasture upon it:* (Ḳ, TA:) and it is said by IB, and in the Ḳ, that جراب, here, is pl. of جرب, not, as J says, of جُرْبٌ: but MF observes that فِعَالٌ is the pl. measure of several words of the measure فُعْلٌ, as رُمْحٌ and دُهْنٌ, and is even said by IHsh and Ibn-Málik and AḤei to be regularly applicable to sings. of this latter measure; whereas no grammarian nor Arabic scholar asserts that a word of the measure فُعْلٌ assumes فِعَالٌ as the measure of its pl. (TA.) — [Hence,] سَيْفٌ أَجْرَبُ ‡ *A sword reddened by much rust, which cannot be removed from it unless with a file.* (A.) — And أَرْضٌ جَرْبَآءُ ‡ *Land affected with drought:* (Ṣ, A, Mṣb, Ḳ:*) or *salt land, affected with drought, and containing nothing.* (ISd, TA.) — And الجَرْبَآءُ ‡ *The sky;* (Ṣ, M, A, Ḳ;) so called because of the stars (Ṣ, TA) and the milky way, (TA,) as though it were scabbed with stars; (Ṣ, IF, ISd;) its stars being likened to the marks of جَرَب; (A;) like as the sea is called أَجْرَدُ, and like as the sky is also called رَقِيعٌ because [as it were] patched with stars: (AAF, ISd:) or *that tract of the sky in which the sun and moon revolve:* (M, Ḳ:) or *the lowest heaven:* (AHeyth, TA:) and accord. to the M, جربة [so in the TA, app. ↑ جَرِبَةُ,] is applied as a determinate [proper] name to *the sky.* (TA.) — And جَرْبَآءُ †A *beautiful* girl; (IAạr, Ḳ;) so called because the women separate themselves from her, seeing that their goodly qualities are rendered foul by comparison with hers. (IAạr, TA.)

تَجْرِبَةٌ is a subst. from جَرَّبَ: (Mṣb:) or it is an inf. n. of that verb, (M, A, Ḳ,) and is one of the inf. ns. from which pls. are formed: (M, TA:) its pl. is تَجَارِبُ (M, Mṣb, TA) and تَجَارِيبُ. (M, TA.) En-Nábighah says,

* إِلَى الْيَوْمِ قَدْ جُرِّبْنَ كُلَّ التَّجَارِبِ *

[*To this day, they* (referring to females) *have been tried with every kind of tryings*]: and El-Aạshà says,

* كَمْ جَرَّبُوهُ فَمَا زَادَتْ تَجَارِبُهُمْ *
* أَبَا قُدَامَةَ إِلَّا المَجْدَ وَالفَنَعَا *

[*How often have they tried him, and their tryings of Aboo-Kudámeh have not increased* aught *save* his *glory and contentment!*]; تجارب being here a pluralized inf. n. made to govern an objective complement; which is a strange fact. (M, TA.) [But in this latter instance, we may consider ابا قدامة as a first objective complement of زادت, and شَيْئًا, understood before الّا, as a second objective complement of the same verb.]

مُجْرِبٌ A man *who has his camels affected with what is termed* جَرَب [i. e. *the mange,* or *scab*]: whence the prov., لَا إِلَاهَ لِمُجْرِبٍ [*There is no god to one who has his camels affected with the mange*]; as though he renounced his god by frequently swearing falsely by him that he had no pitch when it was demanded of him [for the purpose of curing other camels]: (A:) or لَا أَلِيَّةَ لِمُجْرِبٍ [*There is no oath to one who has his camels affected with the mange;* for the reason above mentioned, or because he is likely to deny that he has mangy camels lest his camels should be prevented from coming to water: and hence also,] أَكْذَبُ مِنْ مُجْرِبٍ [*More lying than one who has his camels affected with the mange*]; another prov. (Meyd. [See Freytag's Arab. Prov., ii. 382.])

مُجَرَّبٌ One *who has been tried,* or *proved, in affairs, and whose qualities have become known:* (T, TA:) or one *who has been tried,* or *proved, and strengthened by experience in affairs:* (Ṣ:) [*experienced,* or *expert, in affairs:*] or one *whose qualities have been tried,* or *proved.* (Ḳ, TA.) And ↑ مُجَرِّبٌ One *having experience in affairs.* (Ḳ, TA.) In general, but not always, (MF,) the Arabs used the former of these two epithets [which are virtually synonymous]. (Ṣ, MF.) — دَرَاهِمُ مُجَرَّبَةٌ *Weighed money.* (Kr, Ḳ.) — المُجَرِّبُ *The lion.* (Ṣgh, Ḳ.) ═ [It is also employed as an inf. n. of 2, in accordance with a usage of which there are many other instances; as in the saying,] أَنْتَ عَلَى المُجَرَّبِ [*Thou art about to have the proof,* or *experience*]: a prov., mentioned by AZ: said to him who asks respecting a thing which he is about to know of himself: originally said by a woman to a man who asked her an indecent question which he was himself about to resolve. (TA.)

مُجَرِّبٌ: see مُجَرَّبٌ.

جرث

جِرِّيثٌ *A kind of fish,* (Ṣ, Mgh, Ḳ,) *well known,* (TA,) *also called* جِرِّىٌّ and صِلَّوْرٌ, (Mgh, TA,) *resembling a serpent, called in Persian* مَارْ مَاهِى [*snake-fish,* or *eel*]; *forbidden to be eaten by the Jews:* whether it be lawful to the Muslims is disputed: (TA:) قِرِّيثٌ is a dial. var. (Ṣ in art. قرث.)

جرثم

Q. 2. تَجَرْثَمَ: see 3, in two places. ═ تجرثم الشَّىْءَ *He took the greater, main,* or *chief, part of the thing.* (Ḳ.)

Q. 3. اِجْرَنْثَمَ *He,* or *it, drew himself,* or *itself, together; contracted;* or *shrank:* (TA, Ḥar p. 207:) from جُرْثُومَةٌ signifying the "earth collected around a tree." (Ḥar ib.) Also, and ↑ تجرثم, *It* (a thing, Ṣ) *collected itself together,* or *became collected,* (Ṣ, Ḳ,) *and kept to a place.* (Ḳ.) Hence, in a trad., ↑ وَعَادَ لَنَا النِّقَادُ مُجْرَنْثِمًا *And the lambs, by reason of it,* namely, the vehement drought, *became collected,* or *drawn, together* [*in one place, and kept to it*]. (TA.) — *He* (a man, TA) *fell from a high,* or *higher, to a low,* or *lower, place;* as also ↑ تجرثم. (Ḳ, TA.)

جُرْثُمَةٌ: see what next follows.

جُرْثُومَةٌ [and accord. to Golius, as on the authority of the Ṣ, جُرْثُومٌ, but this I have not found in any Lexicon but his and that of Freytag,] The *root, lowest part, base, stock,* or *source,* syn. أَصْلٌ, of a thing; (Ṣ, Mgh, Ḳ;) whatever the thing be; (Mgh;) as also ↑ جُرْثُمَةٌ: (TA:) and the *place of collection* thereof: (Mgh, TA:) or the *earth that is collected at the roots,* or *lower parts, of trees:* (Ḳ:) or the *earth collected around a tree:* (Ḥar, p. 297:) or the *root of a tree to which the earth is collected:* (Lth, TA:) pl. جَرَاثِيمُ. (Mgh.) One is related to have said, الأَسْدُ جُرْثُومَةُ العَرَبِ فَمَنْ أَضَلَّ نَسَبَهُ فَلْيَأْتِهِمْ [*El-Asd are those, of the Arabs, to whom most others congregate; therefore whoever loses his genealogy, let him come to them*]: meaning الأَزْد. (TA.) And جَرَاثِيمُ جَهَنَّمَ means *The greater, main,* or *chief, degrees of the punishment of Hell.* (Mgh.) — The *base* (أَصْل) *of a sandhill overlooking what surrounds it.* (Ḥar p. 90.) And the pl. (جراثيم), *Places elevated above the ground, composed of clay and earth collected together.* (TA.) — The *earth collected by ants;* (TA;) *an ant-hill:* (Ḳ:) or جُرْثُومَةُ النَّمْلِ signifies *the ant-hill.* (Ṣ.) — The *earth,* or *dust, that the wind raises,* or *sweeps up and scatters.* (Ḳ.) — The غَلْصَمَة [or *epiglottis*]. (Ḳ.)

مُجْرَنْثِمٌ: see 3. — رَكَبٌ مُجْرَنْثِمٌ [in the CḲ, erroneously, رَكْبٌ مُجْرَنْثَمٌ,] i. q. مُسْتَهْدِفٌ, (Ḳ, TA, [in the CḲ, erroneously, مُسْتَهْدَفٌ,]) i. e., *A wide, elevated pubes.* (TA in art. هدف.)

جرح

1. جَرَحَهُ, (Ṣ, A, Ḳ, &c.,) aor. ـَ, (Ḳ, Mṣb, &c.,) inf. n. جَرْحٌ, (Ṣ, Mṣb, &c.,) *He wounded him;*

produced an effect, or *made an impression, upon him with a weapon:* (L:) *he cut him:* (A, MF:) or *clave,* or *rent, some part of his body:* (MF:) syn. كَلَمَهُ: (Ḳ:) and ↓ جرّحه, (Ṣ, Ḳ,) inf. n. تَجْرِيحٌ, (TA,) signifies the same (Ḳ) in an intensive sense, or as applying to several objects; (Ṣ;) or *he wounded him much.* (L.) — Also, (Ḳ,) or جَرَحَهُ بِلِسَانِهِ, (A, Mṣb,) inf. n. as above, (Mṣb,) [lit. *He wounded him with his tongue;* meaning] ‡ *he reviled him,* or *vilified him;* (A, Ḳ;) *he imputed to him a vice,* or *fault,* or *the like;* or *spoke against him.* (Mṣb.) And جَرَحُوهُ بِأَنْيَابٍ وَأَضْرَاسٍ [lit. *They wounded him with dog-teeth and grinders;* meaning] ‡ *they reviled him,* or *vilified him, and imputed to him vices* or *the like.* (A.) And hence, (Mṣb,) جَرَحَ الشَّاهِدَ (A, L, Mṣb, Ḳ) [and ↓ جرّحه, as in many of the law-books,] said of a judge, (A, L,) or other person, (L,) ‡ *He annulled the witness's claim to be legally credible,* (L, Ḳ,) *by happening to discover in him a falsehood &c.;* (L;) *he evinced in the witness something that caused his testimony to be rejected:* (Mṣb:) *he censured the witness, and rejected what he said.* (L.) And جَرَحَ الرَّجُلَ ‡ *He invalidated the man's testimony.* (L.) And جَرَحَ الشَّهَادَةَ ‡ [*He,* or *it, invalidated the testimony;* or *annulled its claim to be legally credible*]. (A, TA.) — Also جَرَحَ, and ↓ اجترح, ‡ *He gained, acquired,* or *earned;* (Ṣ, Mgh, Ḳ, TA;) or *applied himself with art and diligence to get, obtain, gain, acquire,* or *earn;* (Ṣ, Ḳ, TA;) a thing: (TA:) *he worked,* or *wrought, with his hand, and gained, acquired,* or *earned;* &c.: (Mṣb:) from جَارِحَةٌ. (Mgh.) You say, فُلَانٌ يَجْرَحُ لِعِيَالِهِ, and ↓ يَجْتَرِحُ, ‡ *Such a one* [*works, and earns sustenance,* or] *gains, acquires,* or *earns, and collects, for his family,* or *household.* (TA.) And بِئْسَ مَا جَرَحَتْ يَدَاكَ, and ↓ اجْتَرَحَتْ, ‡ *Very evil is that which thy hands have done,* or *wrought,* or *effected:* a metaphor taken from the signification of "cutting," or "wounding;" (A, TA;) accord. to El-Khafájee, a metaphorical meaning conventionally regarded as proper. (TA.) ↓ اِجْتَرَحُوا السَّيِّآتِ, in the Ḳur [xlv. 20], means ‡ *Have committed crimes, sins,* or *evil actions.* (TA.) ═ جَرِحَ, aor. as above, *He* (a man, TA) *received a wound.* (Ḳ, TA.) — And ‡ *He had his testimony rejected as not legally entitled to credit:* (Ḳ,* TA:) and so his relation. (TA.)

2: see 1, in two places.

8: see 1, in four places.

10. استجرح ‡ *He deserved that his claim to be legally credible should be annulled.* (A, TA.) And ‡ *It* (a tradition, or narrative, A, or a thing, Mṣb) *deserved to be rejected* [*as unworthy of credit* or *regard*]. (A, Mṣb.) اِسْتَجْرَحَتْ هٰذِهِ الأَحَادِيثُ means ‡ *These traditions deserved to be rejected* on account of their great number and the fewness of such as were true: (A:) or, by reason of their great number, *obliged those who were acquainted with them to annul the claim of some one or other of their relaters to be credited, and to reject his relation:* (L:) or *were corrupt:* (T, Ṣ,* TA:) [for] اِسْتِجْرَاحٌ signifies [also] the *being faulty, defective,* and *corrupt.* (Ṣ, Ḳ.) One says, قَدْ وَعَظْتُكُمْ فَلَمْ تَزْدَادُوا إِلَّا ٱسْتِجْرَاحًا: (Ṣ, A:) these words are from a خُطْبَة of 'Abd-El-Melik; and the meaning is, [*I have admonished you and ye have not increased save*] *in corrupt conduct:* or *in what gaineth for you censure.* (TA.)

جُرْحٌ: see the next paragraph.

جَرْحٌ a subst. from جَرَحَ; (Ṣ, L, Ḳ;) *A wound;* (L;) and so ↓ جُرْحٌ, in its original acceptation; but some of those skilled in the science of lexicology say that the former is employed to denote the effect produced upon bodies by iron instruments and the like; and the latter, that produced upon objects of the mind by the tongue: (MF:) the pl. of the former is جُرُوحٌ and أَجْرَاحٌ [which is a pl. of pauc.] (Ṣ, L, Ḳ) and جِرَاحٌ; (T, A, L;) but the second of these is of rare occurrence, (Ḳ,) only used in poetry: (Ṣ, L:) [respecting the third, see what follows:] ↓ جِرَاحَةٌ also signifies the same as جُرْحٌ; (Mṣb;) and its pl. is جِرَاحٌ (Ṣ, Mṣb, Ḳ) and جِرَاحَاتٌ (A, Mṣb) and جَرَائِحُ; (A;) or جِرَاحٌ is a coll. gen. n., of which جِرَاحَةٌ is the n. un.; or, accord. to Az, this last has not a sing. sense, as Lth asserts it to have, but is a pl. of جُرْحٌ, like as حِجَارَةٌ is of حَجَرٌ, and جِمَالَةٌ of جَمَلٌ, and حِبَالَةٌ of حَبْلٌ. (L.)

جُرْحَةٌ ‡ *A thing whereby testimony is invalidated, or its claim to be legally credible annulled:* as in the saying, هَلْ لَكَ جُرْحَةٌ ‡ [*Hast thou anything to adduce whereby to invalidate the testimony?*]. (A, TA.) أَقْصَصْتُكَ الجُرْحَةَ فَإِنْ كَانَ عِنْدَكَ مَا تَجْرَحُ بِهِ الحُجَّةَ فَهَلُمَّهَا, said by the judge of El-Medeeneh to one of the parties in a lawsuit, when about to give judgment against him, means ‡ *I authorize thee to adduce anything whereby to invalidate the testimony;* [*therefore, if thou have anything whereby thou mayest invalidate the allegation, adduce it.*] (A,* TA.)

جَرِيحٌ; pl. جَرْحَى; (Ṣ, A, Mṣb, Ḳ;) each of which is masc. and fem.; (Ṣ, Ḳ;) *Wounded.* (Mṣb.) The pl. is not formed by the addition of و and ن because the fem. is not formed by the addition of ة. (TA.)

جِرَاحَةٌ: see جُرْحٌ.

جُرَائِحِيٌّ: see what next follows.

جَرَّاحٌ *A surgeon that dresses wounds.* (Golius on the authority of Ibn-Maaroof; and so in the present day; as also ↓ جُرَائِحِيٌّ.)

جَارِحَةٌ sing. of جَوَارِحُ. (Mgh, L, Mṣb, TA.) — The latter signifies ‡ *Beasts, and birds, of prey;* or *that catch game:* (Ṣ, A,* Mgh, L, Mṣb, Ḳ:) thus the falcon is a جارحة, and so is the dog trained for hunting, because it gains for its owner: (L:) and this appellation is applied alike to the male and the female, like رَاحِلَةٌ and رَاوِيَةٌ. (Mṣb.) — And ‡ The *members,* or *limbs,* of a man, *with which things are gained* or *earned;* (Ṣ, Ḳ, TA;) or *with which one works;* (A;) as the hands or arms, and the feet or legs: (Ṣ, A, Ḳ, TA:) because they gain, or earn, or do, good and evil. (TA.) — [And ‡ The *organs* of the body: thus, for instance, جارحة is applied (in the Mṣb, art. بصر,) to the eye, which is termed (in the TA in that art.) the seeing جارحة (الجَارِحَةُ النَّاظِرَةُ).] — Also † *Mares:* [and *the like:*] because they bring gain to their owners by their breeding. (AA, T.) You say, مَا لَهُ جَارِحَةٌ † *He possesses not a female beast that bears young: he possesses not that which makes gain.* (TA.) And هٰذِهِ النَّاقَةُ مِنْ جَوَارِحِ المَالِ, and هٰذِهِ الأَتَانُ, (Ḳ,) and هٰذِهِ الفَرَسُ, (TA,) † *This she-camel,* and *this she-ass,* and *this mare, is young, unimpaired by age* (مُقْبِلَةٌ [i. q. مُقْتَبِلَةٌ]) *in the womb,* (Ḳ, TA,) *and in youthful vigour, and one of which the offspring is wished for.* (TA.)

جرد

1. جَرَدَ, aor. ـُ, inf. n. جَرْدٌ: see 2, in nine places. — جَرَدَ الجَرَادُ الأَرْضَ, (A, L, Mṣb,) aor. and inf. n. as above, (L,) ‡ *The locusts stripped the land of all its herbage;* (A,* L;) *ate what was upon the land.* (Mṣb.) — جَرَدَهُمُ الجَارُودُ ‡ [*The year of drought destroyed them*]. (A.) ═ جُرِدَتِ الأَرْضُ † *The land had its herbage eaten by locusts;* (Ṣ;) *was smitten by locusts.* (Mṣb.) — جُرِدَ said of seed-produce, † *It was smitten* [or *eaten*] *by locusts.* (Ḳ.) — And said of a man, (Ṣ,) † *He had a complaint of his belly from having eaten locusts.* (Ṣ, Ḳ.) ═ جَرِدَ, aor. ـَ, (Ḳ,) inf. n. جَرَدٌ, (TA,) ‡ *It* (a place) *was,* or *became, destitute of herbage.* (Ḳ, TA.) — † *He* (a man) *had no hair upon him* [i. e. *upon his body,* or, *except in certain parts:* see أَجْرَدُ]. (Ṣ: but only the inf. n. is there mentioned.) — ‡ *He* (a horse, Ḳ, TA, or similar beast, TA) *had short hair:* (TA:) or *had short and fine hair:* as also ↓ انجرد. (Ḳ, TA.) [See أَجْرَدُ.] — See also 7. — Also, (Ṣ, Ḳ,) inf. n. as above, (Ṣ,) † *He* (a man, Ṣ) *became affected with the cutaneous eruption termed* شَرًى, *from having eaten locusts.* (Ṣ, Ḳ.)

2. جرّد, (A, L,) inf. n. تَجْرِيدٌ, (Ṣ, A, L,) *He stripped, divested, bared,* or *denuded,* of garments, or clothes. (Ṣ, A, L.) You say, جرّده مِنْ ثِيَابِهِ, (A,) or مِنْ ثَوْبِهِ, (Th, L, Ḳ,) as also ↓ جَرَدَهُ, (Ḳ,) and جَرَدَهُ ثَوْبَهُ, (Th, L,) *He stripped, divested,* or *denuded, him of his garments,* or *of his garment:* (Th, A, L, Ḳ:) [this is the only signification of the verb given in the A as proper; its other significations given in that lexicon being there said to be tropical:] or جَرَّدْتُهُ مِنْ ثِيَابِهِ signifies *I pulled off from him his garments:* and ↓ جَرَدْتُ الشَّىْءَ, aor. ـُ, inf. n. جَرْدٌ, † *I removed from the thing that which was upon it.* (Mṣb.) — † *He peeled,* or *pared,* a thing; *divested* it *of its peel, bark, coat, covering,* or *the like;* as also ↓ جَرَدَهُ, (L, Ḳ,) aor. and inf. n. as above: (L:) and ↓ the latter, † *he peeled off* anything,

عَنْ شَيْءٍ from a thing. (Ṣ, L.) —— † He stripped skin of its hair; as also ▼ جَرَّدَ. (L, Ḳ.) —— ‡ It (drought) rendered the earth, or land, bare of herbage: so in the L and other lexicons: in the Ḳ, ▼ جَرَّدَ: but the former is the right. (TA.) —— † I. q. شَذَّبَ [generally signifying He pruned a tree or plant]. (Ṣ, TA.) —— ‡ [He bared a sword;] he drew forth a sword (Ṣ, A, Ḳ) from its scabbard; (A;) as also ▼ جَرَّدَ, (TA, and so in some copies of the Ḳ in the place of the former verb,) aor. as above. (TA.) —— [† He detached a company from an army: see جَرِيدَةٌ.] —— [† He divested a thing of every accessory, adjunct, appendage, or adventitious thing; rendered it bare, shere, or mere.] —— † He made the writing, or book, (L, Ḳ,) and the copy of the Ḳur-án, (L,) free from syllabical signs, (L, Ḳ,) and from additions and prefaces: (L:) he divested the Ḳur-án of the diacritical points, and of the vowel-signs of desinential syntax, and the like: (Ibráheem [En-Nakha'ee]:) or he wrote it, or read it, or recited it, without connecting with it any of the stories, or traditions, related by the Jews or Christians. (Ibn 'Oyeyneh, accord. to the L; or A 'Obeyd, accord. to the TA.) —— جَرَدَ القُطْنَ, and ▼ جَرَّدَهُ, † He separated the cotton from its seeds, with a مِحْلَاج: or separated and loosened it by means of a bow and a kind of wooden mallet, by striking the string of the bow with the mallet: syn. حَلَجَهُ. (Ḳ.) —— جَرَّدَ الحَجَّ, (ISh, Ḳ,) and ▼ تَجَرَّدَ بِالحَجِّ, (TA,) which latter alone is mentioned by Z and Ibn-El-Jowzee, (MF,) † He performed the rites and ceremonies of the pilgrimage (الحَجّ) separately from those of العُمْرَة [q. v.]: (ISh, Z, Ibn-El-Jowzee, Ḳ:) or the former signifies he made the performance of the pilgrimage to be free from the vitiations of worldly desires and objects. (Ḥar p. 392.) [See also 5.] —— جَرَّدَ لِلْقِيَامِ بِكَذَا: see 5. —— جَرَدَ القَوْمَ; (Ḳ;) and ▼ جَرَّدَهُمْ, (L, Ḳ,) aor. and inf. n. as above; (L;) † He asked, or begged, of the people, or company of men, and they refused him, or gave him against their will. (L, Ḳ.) ═ Also, (Ḳ,) inf. n. as above, (TA,) † He wore, or put on, جُرُود, i. e., old and worn-out garments. (Ḳ.)

5. تَجَرَّدَ He was, or became, stripped, divested, bared, or denuded, (Ṣ, A, L, Mṣb, Ḳ,) [and he stripped, divested, bared, or denuded, himself,] مِنْ ثِيَابِهِ of his clothes or garments, (A,* Mṣb,) or مِنْ ثَوْبِهِ of his garment; (L, Ḳ;*) as also ▼ انجرد, (A, L, Ḳ,) which latter, accord. to Sb, is not a quasi-pass. verb, (L,) [but it seems that he did not know جَرَدَ, in a sense explained above, (see 2, second sentence,) of which it is the quasi-pass., like as تَجَرَّدَ is of جَرَّدَ.] —— ‡ It (an ear of corn, A, Ḳ, and a flower, TA) came forth from its envelope, or calyx. (A, Ḳ, TA.) —— † It (expressed juice) ceased to boil, or estuate, (Ḳ,) [and so became divested of its froth, or foam.] —— † He (a man) was, or became, alone, by himself, apart from others; as though detached from the rest of men. (Ḥar p. 430.) —— ‡ He (a horse) outstripped the other horses in a race; as also ▼ انجرد, and انجرد عَنِ الخَيْلِ; like نَفَا الخَيْلَ; as though he threw off the others from himself as a man throws off his garment. (TA.) And † He (an ass) went forward from among the she-asses. (L.) —— تَجَرَّدَ لِلْأَمْرِ ‡ [He devoted himself to the affair, as though throwing aside all other things; he applied himself exclusively and diligently to it;] he strove or laboured, exerted himself or his power or efforts or endeavours or ability, employed himself vigorously or diligently or with energy, or took pains or extraordinary pains, in the affair, (Ṣ, A, Ḳ, and Ḥar p. 430,) not diverted therefrom by any other thing. (Ḥar ib.) And تَجَرَّدَ لِلْعِبَادَةِ ‡ [He devoted himself to, applied himself exclusively and diligently to, or strove &c. in, religious service, or worship]. (A.) And ▼ جَرَّدَ لِلْقِيَامِ بِكَذَا ‡ [He devoted himself to, applied himself exclusively and diligently to, or strove &c. in, the performance of such a thing]. (A.) And تَجَرَّدَ فِي السَّيْرِ, and ▼ انجرد, ‡ He strove or laboured, exerted himself or his power or efforts or endeavours or ability, in pace, or going; he hastened therein; like شَمَّرَ فِي سَيْرِهِ. (L, TA.) —— تَجَرَّدَ بِالحَجِّ: see 2. Accord. to Aḥmad, as related by Is-ḥáḳ Ibn-Manṣoor, (TA,) † He affected to be like, or he imitated, the pilgrim of Mekkeh, or the man performing the pilgrimage of Mekkeh. (Ḳ, TA.)

7. انجرد: see 5, first sentence. [Hence,] انجردت الإِبِلُ مِنْ أَوْبَارِهَا † The camels cast, or let fall, their fur, or soft hair. (L.) —— See also 1. —— † It (a garment, or piece of cloth,) became threadbare, or napless, (Ṣ, L, Ḳ,) and smooth; (Ṣ, L;) as also ▼ جَرِدَ. (L.) —— Said of a horse in a race: see 5. —— انجرد فِي السَّيْرِ: see 5. —— انجرد بِنَا السَّيْرُ, (Ṣ, A, L,) in the Ḳ, erroneously, انجرد بِهِ السَّيْلُ, (TA,) ‡ The journey, or march, (Ṣ, A, L,) became extended, (Ṣ, A, L, Ḳ,) and of long duration, [with us,] (Ṣ, L, Ḳ,) without our pausing or waiting for anything. (A.)

8. اجترد † The attacking one another with [drawn] swords. (KL.) [You say, اجتردوا † They so attacked one another; like as you say, اضطربوا.]

جَرْدٌ ‡ A garment old and worn out, (L, Ḳ, TA,) of which the nap has fallen off: or one between that which is new and that which is old and worn out: pl. جُرُودٌ. (L, TA.) You say بُرْدَةٌ جَرْدٌ, (A,) and ▼ جَرْدَةٌ [alone], (Ṣ, L, TA,) ‡ A [garment of the kind called] بردة worn so that it has become smooth. (Ṣ, A, L, TA.*) And [the pl.] جُرُودٌ, (Ḳ, TA, in the CḲ جَرُود,) as a subst., (TA,) † Old and worn-out garments. (Ḳ.) It is said in a trad. of Aboo-Bekr, لَيْسَ عِنْدَنَا مِنْ مَالِ المُسْلِمِينَ إِلَّا جَرْدُ هٰذِهِ القَطِيفَةِ, meaning † There is not in our possession, of the property of the Muslims, save this threadbare and worn-out قطيفة. (TA.) ═ † The pudendum, or pudenda; [app. because usually shaven, or depilated;] syn. فَرْجٌ, (Ḳ,) i. e. عَوْرَةٌ. (TA.) —— And † The penis. (Ḳ.) ═ † A shield. (Ḳ.) ═ † A remnant of property, or of cattle. (Ḳ.) ═ See also جَرِيدَةٌ.

جُرْدٌ: see جَرِيدَةٌ.

جَرَدٌ ‡ A wide, or spacious, tract of land in which is no herbage: (Ṣ, A, Ḳ:) an inf. n. used as an appellative subst. (A.) —— رُمِيَ عَلَى جَرَدِهِ and ▼ أَجْرَدِهِ † He (a man, TA) was shot, or struck with a missile, on his back. (Ḳ.) ═ See also what next follows.

أَجْرَدُ, (Ḳ,) fem. with ة; (Ṣ, Ḳ;) and ▼ جَرِدٌ, (Ṣ, A, Ḳ,) fem. جَرْدَآءُ; (A, Ḳ;) and ▼ جَرَدٌ, (TA, as from the Ḳ,) which last is an inf. n. used as an epithet; (TA;) ‡ A place (A, Ḳ) destitute of herbage: (Ṣ, A, Ḳ:) you say أَرْضٌ جَرِدَةٌ (Ṣ, Ḳ) and ▼ جَرْدَآءُ (A, Ḳ) and ▼ جَرِدِيَّةٌ, (TA,) and فَضَآءٌ ▼ أَجْرَدُ: of which last the pl. is [جُرْدٌ and] أَجَارِدُ. (Ṣ.) —— Also, the first, † A man affected with the cutaneous eruption termed شَرًى, from having eaten locusts. (TA.)

جَرْدَةٌ: see جَرْدٌ. —— Also † An old worn piece of rag: dim. ▼ جُرَيْدَةٌ. (TA from a trad.)

جُرْدَةٌ [The denuded, or unclad, part, or parts, of the body]. You say اِمْرَأَةٌ بَضَّةُ الجُرْدَةِ (A,* Ḳ) and ▼ المُتَجَرَّدِ (A, Ḳ) and ▼ المُجَرَّدِ, (T, A, Ḳ,) [A woman thin-skinned, or fine-skinned, and plump, in respect of the denuded, or unclad, part, or parts of the body: or] when divested of clothing: (T, A,* Ḳ:) the last of these words is here an inf. n.: if you say ▼ المُتَجَرِّدِ, with kesr, you mean, [in] the [denuded] body: (Ḳ:) [and so when you say الجُرْدَةِ, and المُجَرَّدِ; or this last may be regarded as an inf. n.:] المتجرد is more common than المتجرّد. (TA.) [In like manner,] you say, فُلَانٌ حَسَنُ الجُرْدَةِ and ▼ المُجَرَّدِ and ▼ المُتَجَرَّدِ; like as you say, حَسَنُ العُرْيَةِ and المُعَرَّى, which signify the same. (Ṣ.) It is said of Moḥammad, كَانَ أَنْوَرَ ▼ المُتَجَرَّدِ, i. e. He was bright in respect of what was unclad of his body, or person. (TA.) —— Also † Plain, or level, and bare, land. (Ṣ.)

الجُرْدَانُ (Ṣ, Ḳ) and ▼ المُجَرَّدُ and ▼ الأَجْرَدُ (Ḳ) † The yard of a horse &c.: (Ṣ:) or of a solid-hoofed animal: or it is of general application: (Ḳ:) or originally of a man; and metaphorically of any other animal: (TA:) pl. (of the first, TA) جَرَادِينُ. (Ḳ.)

جَرِدٌ: see جَرِدِيَّةٌ.

جَرَادٌ [a coll. gen. n., ‡ Locusts; the locust; a kind of insect] well known: (Ṣ, Mṣb, Ḳ:) so called from stripping the ground, (A, Mṣb,) i. e., eating what is upon it: (Mṣb:) n. un. with ة: (Ṣ, Mṣb:) applied alike to the male and the female: (Ṣ, Mṣb, Ḳ:) جراد is not the masc. of جرادة, but is a [coll.] gen. n.; these two words being like بَقَرٌ and بَقَرَةٌ, and تَمْرٌ and تَمْرَةٌ, and حَمَامٌ and حَمَامَةٌ, &c.: it is therefore necessary that the masc. should be [in my copies of the Ṣ, "should not be," but this is corrected in the margin of one of those copies,] of the same form as the fem., lest it should be confounded with the pl. [or rather the collective form]: (Ṣ:) but some say that جراد is the masc.; and جرادة, the fem.; and the saying رَأَيْتُ جَرَادًا عَلَى جَرَادَةٍ [as meaning I saw a male locust upon a female locust], like رَأَيْتُ نَعَامًا عَلَى نَعَامَةٍ, is cited: (TA:) it is first called سِرْوَةٌ; then, دَبًى; then, غَوْغَآءُ; then,

خَيْفَانٌ; then, كُتْفَانٌ; and then, جراد: (A 'Obeyd, TA:) Aṣ says that when the males become yellow and the females become black, they cease to have any name but جراد. (AḤn, TA.) [Hence,] اِبْنُ الجَرَادِ, (T in art. بنى,) or ابن الجَرَادَةِ, (TA in that art.,) † *The egg of the locust.* (T and TA ubi suprà.) — مَا أَدْرِى أَىُّ جَرَادٍ عَارَهُ, (Ṣ, Ḳ,) or أَىُّ الجَرَادِ, (A, L,) ‡ *I know not what man,* (Ṣ, Ḳ,) or *what thing,* (A,) *took him,* or *it, away.* (Ṣ, A, Ḳ.)

جَرِيدٌ [a coll. gen. n.], n. un. ▼ جَرِيدَةٌ: (Ṣ, Mṣb:) the latter is of the measure فَعِيلَةٌ in the sense of the measure مَفْعُولَةٌ; (Mṣb;) signifying ‡ *A palm-branch stripped of its leaves;* (Ṣ, A, Mṣb, Ḳ;) as long as it has the leaves on it, it is not called thus, but is called سَعَفَةٌ: (Ṣ:) or *a palm-branch in whatever state it be;* in the dial. of El-Ḥijáz: (TA:) or *a dry palm-branch:* (AAF, Ḳ:) or *a long fresh palm-branch:* (Ḳ:) pl. جَرَائِدُ. (TA.) — [Also, ▼ جَرِيدَةٌ, †*A tally, by which to keep accounts;* because a palm-stick is used for this purpose; notches being cut in it. — And hence, ▼ جَرِيدَةُ حِسَابٍ † *An account-book:* and ▼ جَرِيدَةُ الخَرَاجِ †*The register of the taxes,* or *of the land-tax.*] = إِبِلٌ جَرِيدَةٌ ‡ *Choice,* or *excellent,* (A, L,) *and strong,* (L,) *camels.* (A, L.) — See also أَجْرَدُ, in two places.

جُرَادَةٌ †*Anything that is peeled off,* or *pared, from another thing.* (Ṣ.)

جَرِيدَةٌ n. un. of جَرِيدٌ as a coll. gen. n.: see the latter in four places. — Also fem. of the latter as an epithet. — Also ‡ *A detachment of horsemen; a company* of horsemen *detached* (جُرِّدَتْ, Ṣ, A) *from the rest of the force,* (Ṣ,) or *from the main body of the horsemen,* (A,) *in some direction,* or *for some object:* (Ṣ, A:) or *a company* of horsemen *among whom are no foot-soldiers, nor any of the baser sort,* or *of those of whom no account is made:* (A:) or *horsemen among whom are no foot-soldiers;* (Ḳ;) as also ▼ جُرْدٌ [as though pl. of أَجْرَدُ], (Ḳ, TA,) with ḍamm, (TA,) or ▼ جَرْدٌ. (So in the CḲ.) [See an ex. under the word بَيْتٌ, last sentence.]

جُرَيْدَةٌ dim. of جَرْدَةٌ, q. v.

جُرَيْدَآءُ dim. of جَرْدَآءُ [fem. of أَجْرَدُ]: so in the phrase ▼ جُرَيْدَآءُ المَتْنِ † *The middle of the back of the neck, which is free from flesh.* (L.)

جَرَّادٌ † *One who polishes brazen vessels.* (Ḳ.)

جَارُودٌ ‡ *An unlucky* man; (Ṣ, Ḳ;) one *who strips off prosperity by his ill luck;* (A;) or as though he stripped off prosperity by his ill luck. (TA.) — Also, and ▼ جَارُودَةٌ, (A,) or سَنَةٌ جَارُودٌ, (Ṣ, Ḳ,) ‡ *A year of drought:* (A, Ḳ:) or *a year of severe drought and dryness of the earth;* (Ṣ;) as though it destroyed men. (TA.)

جَارُودَةٌ: see what next precedes.

الجَارُودِيَّةُ *A sect of the Zeydeeyeh,* (*of the Shee'ah,* TA,) *so called in relation to Abu-l-Járood Ziyád the son of Aboo-Ziyád:* (Ṣ, Ḳ:) Abu-l-Járood being he who was named by the Imám El-Báḳir "Surḥoob," explained by him as a devil inhabiting the sea: they held that Moḥammad appointed 'Alee and his descendants to the office of Imám, describing them, though not naming them; and that the Companions were guilty of infidelity in not following the example of 'Alee, after the Prophet: also that the appointment to the office of Imám, after El-Ḥasan and El-Ḥoseyn, was to be determined by a council of their descendants; and that he among them who proved himself learned and courageous [above others] was Imám. (MF.)

أَجْرَدُ ‡ A man *having no hair upon him;* (Ṣ, A, L, Ḳ;) i. e., *upon his body;* or *except in certain parts, as the line along the middle of the bosom and downwards to the belly, and the arms from the elbows downwards, and the legs from the knees downwards; contr. of* أَشْعَرُ, which signifies "having hair upon the whole of the body:" (IAth, L:) [fem. جَرْدَآءُ: and] pl. جُرْدٌ. (A, TA.) The people of Paradise are said (in a trad., TA) to be جُرْدٌ مُرْدٌ ‡ [*Having no hair upon their bodies, and beardless*]. (A, TA.) — Also applied to a horse, (Ṣ, A, Ḳ,) and any similar beast, (TA,) meaning ‡ *Having short hair:* (TA:) or *having short and fine hair.* (Ṣ, Ḳ.) This is approved, (Ṣ,) and is one of the signs of an excellent and a generous origin. (TA.) Pl. as above. (A.) In like manner, أَجْرَدُ القَوَائِمِ means ‡ *Having short,* or *short and fine, hair upon the legs.* (TA.) — Also ‡ A cheek *upon which no hair has grown.* (TA.) And †A sandal *upon which is no hair.* (L from a trad.) — Applied also to a place; and the fem., جَرْدَآءُ, to land: see جَرْدٌ, in three places. — Also ‡ Milk *free from froth.* (A.) And the fem., † Wine that is *clear,* (AḤn, Ḳ,) *free from dregs.* (AḤn, TA.) And † A sky *free from clouds.* (L.) — †*Smooth.* (Ḥam p. 413.) — †A heart *free from concealed hatred, and from deceit, dishonesty,* or *dissimulation.* (L.) — ‡ *Complete;* (A, Ḳ;) *free from deficiency;* (A, TA;) as also ▼ جَرِيدٌ; (Ṣ, A, Ḳ;) applied to a year (عَامٌ), (Ṣ, A,) and to a month, (Th, TA,) and to a day: (Ḳ:) fem. as above, applied to a year (سَنَةٌ). (A.) Accord. to Ks, (Ṣ,) you say, مَا رَأَيْتُهُ مُذْ أَجْرَدَانِ and ▼ مذ جَرِيدَانِ, meaning ‡ [*I have not seen him, or it, for,* or *during,*] *two days,* (Ṣ, A, Ḳ,) or *two months,* (Ṣ, Ḳ,) [or *two years,*] *complete.* (A, TA.) — ‡ A horse *wont to outstrip others;* (Ḳ;) *that outstrips others, and becomes separate from them by his swiftness.* (IJ, TA.) — And the fem., ‡ *A voracious* she-camel. (A.) = It is also used as a subst.: see جَرْدٌ: — and see الجُرْدَانُ. — Also † The *sea.* (AAF, M in art. جرب.) — And the fem., †*A smooth rock.* (Ṣ, TA.)

إِجْرِدٌّ, and sometimes without teshdeed, إِجْرِدٌ, *A certain plant which indicates the places where truffles* (كَمْأَة) *are to be found: a certain herb,* or *leguminous plant, said to have grains like pepper.* (En-Naḍr, TA.)

مُجْرَدٌ † A man *ejected from his property.* (IAạr, TA.)

مُجَرَّدٌ: see جَرْدَةٌ, in two places. — ‡ A *bare,* or *naked,* [or *drawn,*] sword. (A.) — [†*Divested of every accessory, adjunct, appendage,* or *adventitious thing; rendered bare, shere,* or *mere; abstract.* — In philosophy, *Bodiless; incorporeal; as though divested of body.*] = See also الجُرْدَانُ.

مَجْرُودٌ †*Peeled,* or *pared; divested of its peel, bark, coat, covering,* or *the like.* (Ṣ, L.) — أَرْضٌ مَجْرُودَةٌ †*Land of which the herbage has been eaten by locusts:* (Ṣ:) or *land smitten by locusts:* (Mṣb:) or *land abounding with locusts;* (A 'Obeyd, ISd, Ḳ;) a phrase similar to أَرْضٌ مَوْحُوشَةٌ; the epithet having the form of a pass. part. n. without a verb unless it be one that is imaginary. (ISd, TA.) — رَجُلٌ مَجْرُودٌ †*A man having a complaint of his belly from having eaten locusts.* (Ṣ.)

مُتَجَرِّدٌ and مُتَجَرَّدٌ: see جُرْدَةٌ, in four places: — and see what follows.

مُنْجَرِدٌ †A horse *having short, and little, hair:* (EM pp. 39 and 40:) or *sharp,* or *vigorous, in pace,* [*and*] *having little hair.* (Ḥar p. 455.) — مَا أَنْتَ بِمُنْجَرِدِ السِّلْكِ, (AZ, A, TA,) or ▼ بِمُتَجَرِّدِ السِّلْكِ, (so in a copy of the A,) said to one who is shy, or bashful, [meaning †*Thou art*] *not free from shyness in appearing* [*before others*]: (AZ, TA:) or ‡ *thou art not celebrated,* or *well-known.* (A, TA.)

جردب

Q. 1. جَرْدَبَ [from جَرْدَبَانٌ] *He put his hand upon the food* (Ḳ, TA) *that was before him on the table,* (TA,) *in order that no other person might take it:* (Ḳ, TA:) or *he ate with his right hand, and prevented* [*others from eating*] *with his left hand:* (IAạr, Ḳ:) also, (Ḳ,) or جردب عَلَى الطَّعَامِ, (TA,) *he ate greedily, gluttonously,* or *voraciously:* (Ḳ, TA:) or جردب فِى الطَّعَامِ *he put his left hand upon food that was before him on the table, in order that no other person might take it;* as also جَرْدَمَ: (Yaḳoob, Ṣ:) or جردب مَا فِى الإِنَآءِ *he ate, and made an end of, devoured,* or *consumed, what was in the vessel;* as also جردم [q. v.]. (Sh, TA.)

جُرْدُبِىٌّ: see what next follows, in two places.

جَرْدَبَانٌ, an arabicized word, (Ṣ, Ḳ,) from the Persian, (Ṣ,) originally گِرْدَهْ بَانْ "guardian of the cake of bread," (Ṣ, Ḳ,) and ▼ جُرْدُبِىٌّ and ▼ جُرْدُبَانٌ and ▼ مُجَرْدِبٌ, (Ḳ,) One *who puts his hand,* (Ḳ,) or *who puts his left hand,* (Ṣ,) *upon food,* (Ḳ,) or *upon a thing that is before him on the table,* (Ṣ,) *in order that no other person may take it:* (Ṣ, Ḳ:) or *who eats with his right hand, and prevents* [*others from eating*] *with his left hand:* and one *who eats greedily, gluttonously,* or *voraciously:* or the first and ▼ third signify, (Ḳ, TA,) or signify also, ‡ *a spunger;* (Ḳ, TA;) because of his greediness, gluttony, or voraciousness, and his boldness. (TA.) A poet says, (namely, El-Ghanawee, TA voce جَرْدَبِيلٌ,)

* إِذَا مَا كُنْتَ فِى قَوْمٍ شَهَاوَى *
* فَلَا تَجْعَلْ شِمَالَكَ جَرْدَبَانَا *

[*When thou art among a greedy company of men, put not thy left hand* upon the food *as a* جردبان]: (Fr, Ṣ:) or the last word is جَرْدَبِيلًا, meaning, (accord. to Sh, as also جَرْدَبَان, TA voce جرديبل,) one *that takes a fragment [of food] with his left hand, and eats with his right hand, and, when what the party have is consumed, eats what is in his left hand.* (TA.)

جَرْدَبَانٌ: see مُجَرْدِبٌ.

جرديبل

جَرْدَبِيلٌ: *i. q.* جَرْدَبَانٌ, explained in art. جردب.

جردق and جردق

جَرْدَقٌ and جَرْدَقٌ, (T, JK, TA,) or جَرْدَقَةٌ (Ṣ, Ḳ) and جَرْدَقَةٌ, (IAạr, Ḳ,) the last asserted by IAạr to have been heard by him from a man of chaste speech, (TA,) *A cake of bread:* (JK, Ṣ, Ḳ:) arabicized words, (T, JK, Ṣ, Ḳ,) from گِرْدَه, (Ḳ, TA,) which is Persian, meaning "round:" (TA:) pl. جَرَادِقُ and جَرَادِيقُ. (JK.)

جردم

Q. 1. جَرْدَمَةٌ, [inf. n. of جَرْدَمَ,] (Ṣ, Ḳ,) in relation to food, (Ṣ,) *i. q.* جَرْدَبَةٌ; (Ṣ, Ḳ;) i. e., The *covering the food that is before one with the left hand, [while eating with the right hand,] in order that no other person may take it:* accord. to Yaạkoob, the م is a substitute for the ب. (TA.) And جَرْدَمَ signifies *He ate, made an end of, devoured,* or *consumed,* what was in the bowl or vessel: (IAạr, Sh, Ḳ, TA:) and *he ate entirely* the bread. (Ḳ.) — Also جردم, *He talked much.* (Ṣ, Ḳ.) — And *He hastened, made haste, sped,* or *was quick;* (Kr, Ḳ;) as also جرذم. (Ḳ.) — جردم السِّتِّينَ *He passed the [age of] sixty.* (IAạr, Ḳ.)

جَرْدَمٌ [probably *Voracious:* see above: and hence,—] *Black locusts, having green heads.* (Ḳ.) — *Loquacious;* or *a great talker.* (Ḳ.)

جرذ

1. جَرِذَ, inf. n. جَرَذٌ, ‡ *He* (a horse [or similar beast]) *became affected with the kind of swelling termed* جَرَذٌ [q. v. infrà]. (A.) — جَرِذَتِ القَرْحَةُ † *The wound,* or *ulcer, formed itself into a knot,* or *lump,* (تَعَقَّدَتْ,) *like what is termed* جُرَذٌ, or جَرَذٌ. (Ḳ, accord. to different copies. [The former reading is app. the right.])

2. جرّذ ‡ *He trimmed* a tree, *as though by removing its* جُرَذ, meaning its *faulty parts,* or *knots,* which are likened to جِرْذَان [pl. of جُرَذٌ]: whence رَجُلٌ مُجَرَّذٌ ▼. (A.)—[And hence,] جرّذه الدَّهْرُ ‡ *Time,* or *fortune, tried and strengthened him by means of experience in affairs.* (T, L, TA.)

جَرَذٌ ‡ *Any swelling,* (A 'Obeyd, Ṣ, Mgh, L, Ḳ,) *and inflation of the sinews,* (A 'Obeyd, Ṣ, Mgh, L,) *in the hock* (A 'Obeyd, Ṣ, Mgh, L, Ḳ) *of a horse* (A 'Obeyd, Mgh, L) or *similar beast;* (Ṣ, Ḳ;) *and in the side of the hock-joint, externally and internally;* (A 'Obeyd, Mgh, L;) derived from جُرَذٌ, because resembling in form the rat (فأر) thus called: (Mgh:) or *an inflation of the sinews of a horse's legs, occasioning swellings which are likened to [the rats called]* جِرْذَان: (A:) or *a swelling in the side of a horse's hoof, and in his stifle-joint* (ثُفْنَة), or *in the hinder part of his hock, which grows so large as to prevent his walking and working;* also written جَرَزٌ; (ISh, L in arts. جرذ and جرز;) and *likewise affecting the camel:* (ISh, L in art. جرذ:) the original word is with ذ. (TA.) — Also ‡ The *faulty parts,* or *knots,* of a tree, *which are pared off;* likened to جِرْذَان. (A.)

أَرْضٌ جَرِذَةٌ *Land containing,* (Ṣ, L,) or *abounding with,* (Ḳ,) *[the large field-rats called]* جِرْذَان; (Ṣ, L, Ḳ;) like ارض فَئِرَةٌ. (A.) — دَابَّةٌ جَرِذٌ, (M, L,) or فَرَسٌ جَرِذٌ, (Mgh,) ‡ *A beast,* or *horse, affected with the kind of swelling termed* جَرَذٌ. (M, Mgh, L.) And رَجُلٌ جَرِذُ الرِّجْلَيْنِ ‡ [*A man whose legs are affected with similar swellings*]. (M, L, TA.)

جُرَذٌ [The *large field-rat;* so in the present day;] *a species of* فَأْر [or *rat*]: (Ṣ, A, Mgh, L, Ḳ:) or the *male* فأر: (T, M, IAmb, L, Mṣb:) or the *large male* فأر; *said to be larger than the jerboa, of a dusky colour, with a blackness in his tail:* (L:) or the *large* فأر *that is in the deserts,* or *uncultivated plains, and that does not frequent,* or *keep to, houses:* (Mṣb:) pl. جِرْذَانٌ, (Ṣ, A, Mgh, L, Mṣb, Ḳ,) or جُرْذَانٌ. (TA.) أَكْثَرَ اللّٰهُ جِرْذَانَ بَيْتِكَ [lit. *May God multiply the large rats of thy house,* or *tent,*] means ‡ *may God fill thy house,* or *tent, with wheat,* or *food.* (A.) And تَفَرَّقَتْ جِرْذَانُ بَيْتِهِ [lit. *The large rats of his house,* or *tent, became dispersed,*] has a contr. meaning. (Ḥar p. 274.) — أُمُّ جِرْذَانٍ *A sort of dates,* (L, Mṣb, Ḳ,) *of a large size:* before the fruit is cut [from the tree], rats collect beneath: so called *when fresh and ripe:* when dry, كُبَيْس: called in El-Koofeh مُوشَانٌ: (L:) and *a sort of palm-tree, the last in the time of the ripening of its fruit in El-Ḥijáz:* (Aṣ, AḤn, L:) or [simply] *the palm-tree.* (T in art. أمّ.) Hence the saying, إِذَا طَلَعَتِ الخَرَاتَانُ أُكِلَتْ أُمُّ جِرْذَانٍ [*When El-Kharátán* (the Eleventh Mansion of the Moon) *rises* aurorally (see مَنَازِلُ القَمَرِ in art. نزل), *the dates called* امّ جرذان *are eaten*]: for El-Kharátán rises [aurorally] in the last part of the hot season, after the [auroral] rising of سُهَيْل [or Canopus], and before the season called الصَّفَرِىّ. (AḤn, L.)

جِرْذَانَةٌ (in the CḲ جُرْذَانَةٌ) *A sort of date:* pl. جِرَاذِينُ. (Ḳ.)

مُجَرَّذٌ ‡ A man *freed from his faults by experience in affairs:* see 2: (A:) or an *intelligent,* or *a sagacious,* man: (M, L:) one *who has been tried and strengthened by experience in affairs.* (T, Ṣ, M, L, Ḳ.)

جرذق

جَرْذَقٌ and جَرْذَقَةٌ: see جردق.

جرذم

Q. 1. جَرْذَمَ, (Ḳ,) inf. n. جَرْذَمَةٌ, (L,) *He hastened, made haste, sped,* or *was quick,* (L, Ḳ,) in walking, or going, and in work; (L;) like جردم. (Ḳ.)

جرز

1. جَرَزَهُ, (Ṣ, Ḳ,* TA,) aor. ُ, inf. n. جَرْزٌ, (Ṣ, Mgh, TA,) *He cut it;* or *cut it off.* (Ṣ, Ḳ, Mgh, TA.) You say also, جُرِزَتِ الأَرْضُ, meaning, *The land had its herbage cut,* or *cut off:* (A:) or *became without herbage; its herbage having been consumed by the locusts or the sheep or goats or the camels or the like:* (Fr, TA:) and جَرِزَتِ الأَرْضُ, aor. َ, signifies [*the same,* or] *the land became what is termed* جُرُزٌ; as also أَجْرَزَتْ ▼. (TA.) — *It* (time, or fortune, الزَّمَانُ,) *destroyed, exterminated,* or *extirpated, him* or *it.* (A, TA.) — Also, (Ḳ,) aor. and inf. n. as above, (TA,) *He slew him.* (Ḳ,* TA.) = Also جَرَزَ, (Ḳ,) aor. and inf. n. as above, (TA,) *He ate quickly.* (Ḳ,* TA.) — And جَرُزَ, (Ḳ,) inf. n. جَرَازَةٌ, (TA,) *He ate much; was a great eater;* (Ḳ, TA;) *not leaving anything upon the table:* (TA:) or (Ḳ and TA, but in the CḲ "and") *he ate quickly; was a quick eater.* (Ḳ, TA.)

4. اجرز القَوْمُ, from أَرْضٌ جُرُزٌ, (Ṣ,) is a phrase similar to أَيْبَسُوا, (Ṣ, TA,) and signifies *The people journeyed in a land such as is termed* جُرُزٌ: (TA in art. يبس:) also (TA) *they were,* or *became, afflicted with dearth, scarcity, drought, sterility,* or *barrenness.* (Ḳ, TA.) — [And hence, app.,] اجرزت النَّاقَةُ *The she-camel became lean,* or *emaciated.* (Ḳ.)

جَرْزٌ:
جُرْزٌ: } see جُرُزٌ; the former, in two places.

جَرَزٌ: see جُرُزٌ. — Also سَنَةٌ جَرَزٌ *A year of dearth, scarcity, drought,* or *sterility:* (Ṣ,* Ḳ,* TA:) pl. أَجْرَازٌ. (Ṣ, TA.)

أَرْضٌ جُرُزٌ, (Ṣ, Mgh, Mṣb, Ḳ,) and جُرْزٌ ▼, and جَرْزٌ ▼, and جَرَزٌ ▼, (Ṣ, Ḳ,) and مَجْرُوزَةٌ ▼, (Ḳ,) *Land in which is no herbage;* (Fr, Ṣ, Mgh;) as though the herbage or the rain were cut off from it: (Ṣ:) or *from which the water is cut off, so that it is dried up and without herbage:* (Mṣb, in explanation of the first:) or *dried up, producing no herbage:* (Jel in xviii. 7, in explanation of the first:) or *of which the herbage has been cut:* (Bḍ ibid., in explanation of the first; and A in explanation of the last:) or *that produces no herbage:* (Ḳ:) or *of which the herbage has been eaten:* (Ḳ: or this is the signification of the last [only] according to the Ṣ:) or *not rained upon:* (Ḳ:) or, accord. to some of the expositions of the Ḳur, by the first is meant the *land of El-Yemen:* and by the second, accord. to El-'Otbee, [*land*] *that takes,* or *receives, much water, and does not dry it up:* and as to جُرْزٌ ▼, it may be an inf. n. used as an epithet, as though meaning ذَاتُ جُرُزٍ, i. e., *of which the herbage has been eaten*: also ▼ أَرْضٌ مَجْرُوزَةٌ is explained as signifying *land of which the herbage has been consumed by the locusts or the sheep or goats or the camels or the like*: (TA:) and ▼ مَفَازَةٌ مِجْرَازٌ *a barren desert*: (A,* Ḳ:) [see also جَارِزٌ:] the pl. of جُرُزٌ is جِرَزَةٌ, like as جِحَرَةٌ is pl. of جُحْرٌ; and that of جَرْزٌ is أَجْرَازٌ, like as أَسْبَابٌ is pl. of سَبَبٌ: (Ṣ:) and one says also أَرْضٌ أَجْرَازٌ, (Ḳ, TA,) as well as أَرَضُونَ أَجْرَازٌ. (TA.)

جَرْزَةٌ, (Ṣ, A,) or ▼ جَرَزَةٌ, with fet-ḥ to the ر, (Ḳ, TA,) *Extirpation*: (Ṣ, A:) or *destruction*. (Ḳ, TA.) Hence the saying, لَنْ تَرْضَى شَانِئَةٌ إِلَّا بِجَرْزَةٍ *A female that hates*, by reason of the vehemence of her hatred, *will not be content save with extirpation* [of the object of her hatred]: (Ṣ:) or لَمْ يَرْضَ شَانِئُهُ إِلَّا بِجَرْزَةٍ, (A,) or ▼ بِجَرَزَةٍ, (TA,) *His hater was not content save with extirpation* of him whom he hated. (A, TA.) You say also, ▼ رَمَاهُ ٱللّٰهُ بِشَرْزَةٍ وَجَرْزَةٍ, [or بِشَرَزَةٍ وَجَرَزَةٍ, thus, only, the former word is written in the Ḳ and TA in art. شرز, and I think that the latter is correctly written in like manner, as in the Ṣ and A,] *God smote him with destruction* [*and extirpation*]. (TA.)

جُرْزَةٌ *A bundle* of [the kind of trefoil called] قَتّ (A, Mgh, Mṣb, Ḳ) or the like: (Mgh, Mṣb, Ḳ:) or *a handful* thereof: (Mgh, Mṣb:) pl. جُرَزٌ. (A,* Mgh, Mṣb.)

جَرَزَةٌ: see جَرْزَةٌ, in three places.

جُرَازٌ, applied to a sword, (Ṣ, Ḳ,) *Cutting*, or *sharp*: (Ḳ:) or *cutting much*, or *very sharp*: (Ṣ:) or *penetrating*: or *that cuts off utterly*, or *entirely*. (TA.) — Hence you say of a she-camel, likening her to such a sword, إِنَّهَا لَجُرَازٌ لِلشَّجَرِ † *Verily she is one that eats and breaks the trees*. (TA.) See also جَرُوزٌ.

جَرُوزٌ ‡ *That eats much*; (A, Ḳ;) [when applied to a man,] *that leaves nothing upon the table when he eats*: (Ṣ, A:) or *that eats quickly*: (Ḳ:) applied to a man, and to a camel; (TA;) and to a woman; and to a she-camel, as also ▼ جُرَازٌ; (Ṣ;) or [any] female: (TA:) or, applied to a she-camel, *that eats everything*. (Aṣ, TA.)

جَارِزٌ, applied [to land, and hence,] to ‡ a woman, *Barren*: (Ṣ, A, Ḳ:) the woman is likened to land that produces no herbage. (TA.) [See also جُرُزٌ.] — أَرْضٌ جَارِزَةٌ *Dry and rugged land encompassed by sands or by a* [*level tract such as is called*] قَاع: (Ṣ, Ḳ:) pl. جَوَارِزُ: (Ṣ, TA:) mostly applied to islands of the sea. (TA.)

مُجْرِزَةٌ, applied to a she-camel, *Lean*, or *emaciated*. (Ḳ.)

مِجْرَازٌ: } see جُرُزٌ; the latter, in two places.
مَجْرُوزَةٌ: }

جرس

1. جَرَسَ, [aor., app., ـُ and ـِ, as seems to be implied in the Ḳ, inf. n. جَرْسٌ, which see below,] *He*, or *it*, *made a sound*; (TA;) as also ▼ اجرس: (Mgh, TA:) [or both signify *he*, or *it*, *made a low*, *gentle*, *slight*, or *soft*, *sound*; as appears from what follows.] You say, جَرَسَ بِٱلْكَلَامِ, (A,) or جَرَسَ ٱلْكَلَامَ, (Mṣb,) *He spoke in a low*, *gentle*, or *soft*, *voice or tone*; or *with modulation*, or *melody*; syn. نَغَمَ فِيهِ, (A,) or نَغَمَ بِهِ. (Mṣb.) And جَرَسَ, (Ṣ,) inf. n. جَرْسٌ; (Ḳ;) and ▼ تَجَرَّسَ; (Ṣ, Ḳ;) *He spoke*: (Ḳ:) or *he said a thing*, *and spoke in a low*, *gentle*, or *soft*, *voice* or *tone*, or *with modulation* or *melody*; expl. by تَكَلَّمَ بِشَيْءٍ وَتَنَغَّمَ. (Lth, Ṣ.) And ▼ اجرس is also used in the senses here following. *It* (a bird) *caused the sound of its passing to be heard*: (Ṣ, A,* Ḳ:) and in like manner it is said of a man. (Ḳ, accord. to the TA; but not found by me in any copy of the Ḳ.) ▼ And ‡ *It* (an ornament, حَلْيٌ,) *made a sound* (Ṣ, A,* Ḳ) *like that of a* جرس [or *bell*]; (TA;) as also ▼ انجرس. (A, TA.) ▼ And *It* (a tribe, حَيّ,) *made its sound* (جَرْس) *to be heard*: or, accord. to the T, *made the sound of the* جَرْس *of a thing to be heard*. (TA.) ▼ And *He* (a man) *raised his voice*. (TA.) ▼ And *He* (a camel-driver) *sang to camels for the purpose of urging* or *exciting*: (Ṣ, Ḳ:) or *raised his voice in doing so*. (A.) — [Hence, app.,] جَرَسَ, aor. ـُ (Lth, AO, Ṣ, Ḳ) and ـِ, (Ḳ,) inf. n. جَرْسٌ, (Lth, AO, Ḳ,) *He ate* [a thing: because a slight sound is made in doing so]: (AO, TA:) or *he licked* [a thing] *with his tongue*. (Ḳ.) You say, جَرَسَتِ ٱلنَّحْلُ ٱلْعُرْفُطَ *The bees ate the* [*trees called*] عرفط: (Ṣ:) and جَرَسَتِ ٱلنَّوْرَ, (Lth, A,) and ٱلْعَسَلَ [put tropically for ٱلنَّوْر because honey is made from flowers or blossoms], (Lth, TA,) *the bees ate the flowers*, or *blossoms*, *making a sound in so doing*: (A:) or *licked the flowers*, or *blossoms*, *and thence made honey*. (Lth, TA.) And جَرَسَتِ ٱلْمَاشِيَةُ ٱلشَّجَرَ, and ٱلْعُشْبَ, *The beasts licked the trees*, and *the herbage*. (TA.) And جَرَسَتِ ٱلْبَقَرَةُ وَلَدَهَا *The cow licked her young one*. (TA.)

2. جَرَّسَ بِٱلْقَوْمِ, inf. n. تَجْرِيسٌ, *He rendered the persons notorious*, or *infamous*; [as, for instance, *by parading them*, *and making public proclamation before them*; accord. to the usage of the verb in the present day;] syn. سَمَّعَ بِهِمْ, (Ḳ,) and نَدَّدَ, (Ibn-'Abbád, TA,) and صَوَّتَ. (A.) — جَرَّسَتْهُ ٱلْأُمُورُ, (Ṣ,) and ٱلدُّهُورُ, (TA,) inf. n. as above, (Ḳ,) ‡ [*Events*, and *misfortunes*,] *rendered him experienced*, or *expert*, *and sound*, or *firm*, *in judgment* &c. (Ṣ, Ḳ,* TA)

4. اجرس: see 1, in six places. — اجرس بِٱلْحَلْيِ ‡ [*He made a sound with the ornament*]: said of the owner [or wearer] of the ornament. (A.) — اجرس ٱلْجَرَسَ *He struck* [or *sounded*] *the bell*. (TA.) — أَجْرَسَنِى ٱلسَّبُعُ *The animal of prey heard my sound* (جَرْسِى): (ISk, Ṣ, A, Ḳ:) or *heard it from afar*. (TA.)

5: see 1.

7: see 1.

جَرْسٌ (Ṣ, A, Mṣb, Ḳ) and ▼ جِرْسٌ (Ṣ, A, Ḳ) and ▼ جَرَسٌ (Kr, ISd) *A sound*: (ISk, A, Ḳ:) or *a low*, *faint*, *gentle*, *slight*, or *soft*, *sound*: (IDrd, Ṣ, A, Ḳ:) such, for instance, as the sound of the beaks of birds, (Ṣ, A, Mṣb,) pecking, (A,) upon a thing which they are eating: (Ṣ:) and that of bees eating flowers or blossoms: (A:) and of a tribe [or crowd of men, more particularly as heard from some distance; i. e., *a hum*]: (TA:) and of a camel-driver singing to his beasts to urge or excite them: (A:) and the slight sound of a letter of the alphabet: (TA:) and *low*, *gentle*, or *soft*, *speech*: (Mṣb:) or when the word is used alone, [i. e., not coupled with another noun as it is in the second of the two examples here following,] it is with fet-ḥ: thus one says, مَا سَمِعْتُ لَهُ جَرْسًا; (A, Ḳ;) i. e., *I heard not any sound of him*, or *it*: (TA:) but you say, ▼ مَا سَمِعْتُ لَهُ حِسًّا وَلَا جِرْسًا, with kesr; (A, Ḳ;) i. e., *I heard not any motion*, *nor any sound*, *of him*, or *it*: (TA in art. حس:) pl. [app. of the third] أَجْرَاسٌ. (Ḥam p. 200.) [See also جَرْش.]

جِرْسٌ: see what next precedes, in two places.

جَرَسٌ [*A bell*;] *a thing well known*; (Mṣb;) the *thing that is hung to the neck of the camel* (Ṣ, Mgh, Ḳ) &c., *and that makes a sound*: (Mgh:) or, accord. to some, the [*little round bell called*] جُلْجُل: (TA:) and also *that which is struck* [*to make it sound*]: (Lth, Ṣ, Ḳ:) the *thing that is struck by the Christians at the times of prayers*: (Ḥar p. 616:) pl. أَجْرَاسٌ. (Mgh, Mṣb.) It is said in a trad., لَا تَصْحَبُ ٱلْمَلَائِكَةُ رُفْقَةً فِيهَا جَرَسٌ [*The angels will not accompany an assemblage of persons journeying together among whom is a bell*]: (Ṣ, TA:) the reason is said to be, because it guides others to them; for Mohammad liked not to let the enemy know of his approach until he came upon them suddenly. (TA.) — See also جَرْسٌ.

جُرْسَةٌ The *act of rendering* [a person or persons] *notorious*, or *infamous*. (TA.) [See 2.]

جَارُوسٌ *Voracious*. (IAạr, Ḳ.)

جَوَارِسُ [as though pl. of جَارِسَةٌ] *Bees*: (Ṣ:) or *bees eating flowers*, or *blossoms*, *and making a sound in doing so*: (A:) or جَوَارِسُ ٱلنَّحْلِ signifies *the males of bees*. (TA.)

جَاوَرْسٌ [*A species of millet*;] *a kind of grain*, (Mṣb, Ḳ,) *well known*, (Ḳ,) *resembling* ذُرَة, *but smaller*: (Mṣb:) or, accord. to some, *a species of* دُخْن: (Mṣb:) or i. q. دُخْن: (Ṣ in art. دخن, and TA in art. دخل:) or *a well-known grain*, *which is eaten*, *like* دُخْن, *of which there are three species*, *the best whereof is the yellow* [الاصفر, or the word may be الاصغر, *the smallest*,] *and weighty*: *it is likened to rice in its power*, or *virtue*, *is more astringent than* دُخْن, *promotes the flow of urine*, *and constringes*: the word is arabicized, from [the Persian] گَاوَرْس. (TA.)

مُجَرِّسٌ (Ṣ) and مُجَرَّسٌ (TA) ‡ A man (TA) *experienced*, or *expert*, *in affairs*, (Ṣ, TA,) *and rendered sound*, or *firm*, *in judgment* &c. (TA.) And with ة, ‡ A she-camel *tried and proved by use*, or *practice*, *in pace and riding*. (TA.)

مَجْرُوسٌ [*Uttered with a sound:* or *with a low, gentle, slight,* or *soft, sound*]. Every letter of the alphabet is مَجْرُوسَةٌ, except the soft letters, (A, TA,) namely, ا, و, and ى. (TA.)

جرش

1. جَرَشَهُ, (Ṣ, A, Ḳ,) aor. ـِ (MṢ, Ḳ) and ـُ, (Ḳ,) inf. n. جَرْشٌ, (A, TA,) *He bruised, brayed,* or *pounded, it,* (Ṣ, A, Ḳ,) and *he ground it,* namely, salt, and grain, (A,) *coarsely, not finely.* (Ṣ, A, Ḳ.) — *He stripped off, scraped off, rubbed off, abraded,* or *otherwise removed, its superficial part;* syn. قَشَرَهُ. (Ḳ.) — *He scratched, scraped, rubbed, grated, chafed,* or *fretted, it;* syn. حَكَّهُ; (Ḳ, TA;) like as the viper does its fangs; when its folds rub, or grate, together, causing a sound to be heard. (TA.) — *He scratched it* (حَكَّهُ, namely, his head,) with a comb, (Ṣ, A, Ḳ,) *so as to raise its scurf;* (Ṣ, Ḳ;) as also ↓ جَرَّشَهُ. (TA.) — *He rubbed and pressed it* (namely, the skin,) *with the hand, in order that it might become smooth* (Ḳ, TA) *and soft.* (TA.)

2: see 1, last signification but one.

جَرْشٌ The *sound of a viper's coming forth from the skin* [or *slough*] *when the former rubs,* or *grates, one part against another.* (Ḳ.) — And The *sound of a viper's fangs, when they rub,* or *grate* [*together*]. (TA.) — And The *sound arising from eating a rough thing:* or this is with س. (TA.)

جَرِيشٌ A thing, (Ṣ, Ḳ,) such as salt, (A,) *bruised, brayed,* or *pounded,* (Ṣ, A, Ḳ,) and *ground,* (A,) *coarsely, not finely:* (Ṣ, A, Ḳ:) or, applied to salt, it signifies مَا لَمْ يُطَيَّبْ [app. meaning *such as has not been purified*], (Ṣ, Ḳ, TA,) *that crumbles;* as though one part thereof were rubbed against another. (TA.) — Also *Coarse flour, such as is fit for* [*making the kind of food called*] خَبِيص مُرَمَّل. (TA.)

جُرَاشَةُ شَىْءٍ *What falls, of,* or *from, a thing coarsely bruised* or *brayed* or *pounded, when what is bruised* &c. *thereof is taken.* (Ṣ.) — جُرَاشَةٌ also signifies *What falls from the head when it is combed:* (A, TA:) and *what falls and becomes scattered from wood:* (A:) or *cuttings, chips, parings, and the like.* (TA.)

جَوَارِشُ [from the Persian گُوارِش, *A digestive stomachic;*] *a thing that causes food to digest;* as also هَاضُومٌ. (Ṣ in art. هضم.)

مَجْرُوشٌ A thing *having its superficial part stripped off, scraped off, rubbed off, abraded,* or *otherwise removed.* (TA.) — Skin *rubbed and pressed with the hand in order that it may become smooth and soft.* (TA.)

جرض

1. جَرِضَ بِرِيقِهِ, aor. ـَ, (IDrd, A, Ḳ,) inf. n. جَرَضٌ; (IDrd, Ḳ;) or جَرَضَ بِرِيقِهِ, aor. ـِ; (Ṣ;) but IḲṭṭ says that the former is the right; (IB;) *His throat,* or *fauces, became choked by his spittle; he was choked with his spittle:* (IDrd, A:) or, accord. to some, *he suffered the chokings* (غُصَص) *of death:* (TA:) or the verb is sometimes used in a general manner, in the former of the senses explained above, and in the place of غَصَّ and in that of شَرِقَ and in that of شَجِىَ: (MF in art. غص, q. v.:) and the above-mentioned phrase also signifies (A) *he swallowed his spittle with difficulty,* or *trouble,* or *labour, contending against anxiety,* or *grief.* (Ṣ, A, Ḳ.) You say also, هُوَ يَجْرَضُ بِنَفْسِهِ, (Ṣ,) or يَجْرُضُ بنفسه, (A,* TA,) inf. n. as above, (TA,) *He is near to dying;* accord. to Aṣ: (Ṣ, TA:) or *his soul reaches his fauces:* (TA:) or *he is at the point of death, his soul having reached his fauces.* (A.) And يَجْرِضُ عَلَى نَفْسِهِ *He dies.* (TA.) And جَرِضَتِ النَّاقَةُ بِجِرَّتِهَا [app. meaning *The she-camel was choked with her cud:* or *swallowed her cud with difficulty*]. (TA.) And جَرَضَ رِيقَهُ *He swallowed his spittle.* (A, TA.) And فُلَانٌ يَجْرِضُ عَلَيْكَ رِيقَهُ [app. meaning *Such a one swallows his choler against thee*]. (A.) — جَرَضَهُ *He strangled him.* (Ḳ.) You say, أُفْلِتَ مِنْهُمْ وَقَدْ جَرَضُوهُ *He escaped from them when they had strangled him* [*so as almost to kill him*]. (TA.)

4. اجرضهُ بريقه *He caused his throat,* or *fauces, to be choked by his spittle.* (Ṣ, Ḳ.)

جَرَضٌ *Spittle:* (A, Ḳ:) or *spittle by which one is choked.* (Ṣ.) — [The *choking,* or *strangulation,* or *rattles, immediately preceding death.* See 1: and see also جَرِيضٌ.] — *Difficulty, trouble,* or *labour;* syn. جَهْدٌ. (TA.)

جَرِيضٌ [A man *having his throat,* or *fauces, choked by his spittle.* — And hence,] *Having his soul reaching his fauces:* or *near to dying:* (TA:) or *at the point of death, his soul having reached his fauces, so that he is choked by it:* (A, TA:) or *dying:* (TA:) or *made,* or *suffered, to escape, after evil:* (Lth:) or *oppressed by grief* or *sorrow;* (Ṣ, Ḳ;) as also ↓ جِرْيَاضٌ and ↓ جِرْوَاضٌ: (ADk, Ḳ:) or *affected by intense anxiety* or *grief:* (TA:) [see also جُرَاضٌ:] pl. جَرْضَى; (A, Ḳ;) like as مَرْضَى is pl. of مَرِيضٌ. (A, TA.) You say, أُفْلِتَ, or أَفْلَتَ فُلَانٌ جَرِيضًا, *Such a one escaped,* or *was made to escape, being near to death;* (TA;) or *being at the point of death, his soul having reached his fauces, so that he was choked by it.* (A, TA.) And مَاتَ فُلَانٌ جَرِيضًا *Such a one died oppressed by grief* or *sorrow.* (Ṣ.) = In the following prov., حَالَ الجَرِيضُ دُونَ القَرِيضِ [*The* جريض *interposed as an obstacle in the way of the* قريض], الجريض signifies *the thing choking the throat* or *fauces;* (Ṣ, A;) and القريض signifies the cud: the meaning being, the thing choking the throat or fauces hindered from chewing the cud: (A:) or the former signifies *the choking,* or *having the throat,* or *fauces, obstructed;* and the latter, the poetry: (TA:) or the former, *the swallowing of spittle in dying;* and the latter, the sound, or voice, of a man in dying: (Er-Riyáshee:) or the former, *spittle swallowed:* (Ḥar p. 150, q. v.:) and also, *the chokings* (غُصَص) *of death:* [see also جَرَضٌ:] and *the moving to and fro of the two jaws at death:* (TA:) the prov. relates to an affair which is hindered by some obstacle: (A, O, Ḳ:) or it is said on the occasion of any affair which was possible and which has been hindered by the intervention of some obstacle: and the first who said it was 'Obeyd Ibn-El-Abraṣ, when El-Mundhir [on one of the days when it was his custom to slay whomsoever he met] desired him to recite some of his verses: (Zeyd Ibn-Kuthweh:) or the first who said it was Jowshan [in some copies of the Ḳ, Showshan, which, as is said in the TA, is a mistake,] El-Kilábee, when his father, having forbidden him to poetize, and seeing him sick of grief thereat, and at the point of death, gave him permission to do so: (Ḳ, TA:) whereupon, after saying these words, he recited some verses, and died. (TA.)

جُرَّاضٌ *A strangler.* (TA.) = *Suffering intense,* or *violent, grief.* (TA.) [See also جَرِيضٌ.]

جِرْوَاضٌ: } جِرْيَاضٌ: } see جَرِيضٌ.

جرع

1. جَرِعَ المَآءَ, aor. ـَ, (Ṣ, Mṣb, Ḳ,) inf. n. جَرْعٌ, (Ṣ, Ḳ,*) or جَرَعٌ; (Mṣb;) and جَرَعَهُ, aor. ـَ, inf. n. جَرْعٌ; (Ṣ, Mṣb, Ḳ;) but the latter is disallowed by Aṣ; (Ṣ;) *He swallowed the water;* (Mṣb, Ḳ;) as also ↓ اجترعهُ: (Mṣb:) or the latter signifies *he swallowed it at once.* (Ṣgh, Ḳ.) — See also 5.

2. [جَرَّعَهُ المَآءَ *He made him to swallow the water.*] تَجْرِيعٌ is The *pouring beverage into the throat against one's will:* but sometimes it is used of that which is not against one's will. (Ḥar p. 115.) And جرّعهُ signifies *He gave him to drink gulp after gulp,* or *sup after sup,* or *sip after sip.* (Ḥar p. 350.) — [And hence,] جرّعهُ الغُصَصَ, (Ḳ,) or جرّعهُ غُصَصَ الغَيْظِ, (Ṣ,) inf. n. تَجْرِيعٌ, (Ḳ,) ‡ *He made him to repress,* or *restrain, choking wrath,* or *rage.* (Ṣ, TA.)

4. اجرعهُ *He made it* (a rope or a bow-string) *to have one or more of its strands thick* [or rather *thicker than the others*]. (TA.)

5. تجرّع *He swallowed in consecutive portions, one time after another, like him who acts against his own will:* or, as IAth says, *he drank in haste:* or, accord. to some, *he drank by little and little.* (TA.) — [And hence,] تجرّع الغُصَصَ (Ṣ, Mṣb, Ḳ*) ‡ *He repressed,* or *restrained, choking wrath,* or *rage;* (Ṣ;) as though he swallowed it: (Mṣb:) and [in like manner] you say also, ↓ جَرِعَ الغَيْظَ ‡ *he repressed,* or *restrained, wrath,* or *rage.* (TA.)

8: see 1.

جَرَعٌ *A twisting in one of the strands of a rope,* (Ṣ, Ḳ,) or *of a bow-string,* (Ḳ,) *so that it appears above the other strands.* (Ṣ, Ḳ.) [It is app. an inf. n., of which the verb, if it have one, is جَرِعَ.] = See also جَرَعَةٌ.

جَرِعٌ A rope, (Ḳ,) or a bow-string, (TA,) *having the twisting termed* جَرَعٌ *in one of its strands;* as also ▼ مُجَرَّعٌ: (Ḳ:) or, accord. to IAṣr, a bow-string that is *even, except that there is a prominence in one part of it, wherefore it is rubbed and pulled with a piece of a* كِسَآء [q. v.] *until that prominence disappears:* and ▼ the latter, accord. to ISh, a bow-string *not uniformly nor well twisted, having in it prominences, so that one of its strands appears above the others,* or *some appear above others.* (TA.)

جَرْعَةٌ: see what next follows, in three places: — and see جَرَعَةٌ, in two places.

جُرْعَةٌ *A gulp,* or *as much as is swallowed at once,* of water; a جُرْعَة of water being like a لُقْمَة of food: (Mṣb:) or a *sup,* or *sip;* or *as much as is supped,* or *sipped, at once;* or *a mouthful of what is supped,* or *sipped;* (syn. حُسْوَةٌ;) of water; (Ṣ, Ḳ;) as also ▼ جَرْعَةٌ and ▼ جِرْعَةٌ: or جَرْعَةٌ and ▼ جُرْعَةٌ are substs. [signifying the *act of swallowing* water] from جَرِعَ المَآءَ "he swallowed the water:" (Ḳ:) or ▼ جَرْعَةٌ signifies *a single act of swallowing* water: (IAth, L:) and جُرْعَةٌ, *what one swallows:* (L, Ḳ:) or *a mouthful which one swallows:* (TA:) or *a small draught:* (IAth:) and its pl. is جُرَعٌ. (Mṣb, TA.) The dim. is ▼ جُرَيْعَةٌ. (Ṣ, Ḳ.) And hence the prov., أَفْلَتَ فُلَانٌ جُرَيْعَةَ الذَّقَنِ, (Ṣgh, Ḳ,) the verb being intrans., and جريعة being in the accus. case as a denotative of state, as though the speaker said, أَفْلَتَ قَادِرًا جُرَيْعَةَ الذَّقَنِ; (Ṣgh;) or بِجُرَيْعَةِ الذَّقَنِ; (Ṣ, Ḳ;) or ▼ بِجُرَيْعَائِهَا; (Ḳ;) *Such a one escaped* [from destruction] *when his spirit,* or *the remains thereof, had become in his mouth;* (L, Ḳ;) or *near thereto,* (Ḳ,) *as a sup* [or *little sup*] *of water to the chin* [of a person drinking]; (TA;) or *when death was as near to him as a little sup of water to the chin;* (L;) or *when at his last gasp:* (Fr, Ṣ:) applied to one who has been at the point of destruction, and then escaped: (Ṣ:) or, accord. to AZ, it is thus; أَفْلَتَنِى جُرَيْعَةَ الذَّقَنِ, which may mean *he made me to escape* &c., or *he escaped from me* &c.; in the latter case, افلتنى being for أَفْلَتَ مِنِّى; and [it is said that] جريعة is prefixed to الذقن because the motion of the chin indicates the nearness of the departure of the soul: or the meaning of the words related by AZ may be, he made me, i. e. the remains of my soul, to escape; the last two words being a substitute for the pronoun affixed to the verb. (Ṣgh.) One says also, أَفْلَتَنِى جُرَيْعَةَ الرِّيقِ, meaning *He outwent me,* [or *escaped me,*] *and I swallowed my spittle in wrath,* or *rage, against him.* (TA.) And مَا مِنْ جُرْعَةٍ أَحْمَدَ عُقْبَانًا مِنْ جُرْعَةِ غَيْظٍ تَكْظِمُهَا ‡ [*There is nothing that is swallowed more praiseworthy in its result than what is swallowed of wrath,* or *rage, which we repress,* or *restrain*]. (TA.)

جِرْعَةٌ: see جَرْعَةٌ.

جَرَعَةٌ (Ṣ, Ḳ) and ▼ جَرْعَةٌ (Ḳ) and ▼ جَرْعَآءُ (Ṣ, Ḳ) and ▼ أَجْرَعُ (Ḳ) *An even piece,* (Ṣ,) or *a round piece,* or *hill,* or *hillock,* (Ḳ,) *of sand, that produces no plants,* or *herbage;* (Ṣ, Ḳ;) and, as some add, *that retains no water:* (TA:) or *a piece,* or *tract, of sand, good for producing plants,* or *herbage, in which is no softness,* or *looseness:* (Ṣgh, L, Ḳ:) or *land in which is ruggedness, resembling sand:* (L, Ḳ:) or *a hill of which one side consists of sand, and one side of stones:* (Ḳ:) or what is termed ▼ جرعاء and ▼ اجرع is *larger than what is termed* جَرَعَة: ▼ جرعاء is also explained as signifying *sand of which the middle is elevated, and of which the sides are thin:* and, accord. to IAth, ▼ اجرع signifies *a wide place, in which is ruggedness:* (TA:) or this last, *a plain,* or *soft, place, intermixed with sand:* (Ḥam p. 574:) جَرَعَةٌ is sing., or n. un., of ▼ جَرَعٌ: (Ṣ, Ḳ:*) or, accord. to some, this last word is a sing., like اجرع; and its pl. [of pauc.] is أَجْرَاعٌ and [of mult.] جِرَاعٌ: the pl. of جَرَعَةٌ is جَرَعَانٌ: and the pl. of ▼ جَرْعَةٌ is جِرَاعٌ: and the pl. of ▼ جرعاء is جَرْعَاوَاتٌ: and the pl. of ▼ اجرع is أَجَارِعُ. (TA.)

جَرْعَآءُ: see جَرَعَةٌ, in four places.

جُرَيْعَةٌ dim. of جُرْعَةٌ, q. v.

جُرَيْعَآءُ: see جُرْعَةٌ.

أَجْرَعُ: see جَرَعَةٌ, in four places.

مُجْرِعٌ A she-camel *in which is not as much* [*milk*] *as will satisfy thirst, but only some sups:* (Ḳ:) pl. مَجَارِيعُ (L, Ḳ) and مَجَارِعُ: (L:) J explains the former pl. as signifying she-camels *having little milk; as though there were not in their udders more than some sups;* and the sing. he does not mention. (TA.)

مُجَرَّعٌ: see جَرِعٌ, in two places.

جرف

1. جَرَفَهُ, (Ṣ, Mṣb, Ḳ,) aor. ـُ, (Ṣ, Mṣb,) inf. n. جَرْفٌ (Ṣ, Mṣb, Ḳ) and جُرْفَةٌ, (Lḥ, Ḳ,) *He took away, carried away,* or *removed, the whole of it,* (Ṣ, Mṣb, Ḳ,) or *the greater part of it,* (Ṣ,) or *much of it:* (Ṣ, Ḳ:) and [in like manner ▼ جرّفهُ; for its inf. n.] تَجْرِيفٌ signifies the act of *carrying away wholly:* (KL:) and ▼ اجترفهُ *he took the whole of it.* (TA in art. جفت.) — Also, (inf. n. جَرْفٌ, TA,) *He swept it away,* namely, mud, (Ṣ, Ḳ,) from the surface of the earth; (TA;) and so ▼ جرّفهُ, (Ḳ,) inf. n. تَجْرِيفٌ; (TA;) and ▼ تجرّفهُ: (Ḳ:) or ▼ تَجْرِيفٌ signifies the act of *clearing away* mud or the like *well;* in Persian, نيك رنديدن: (KL: [Golius, app. misled by a mistranscription, has explained the verb, جرّف, as on the authority of the KL, by "bene effudit:"]) and ▼ اجترف الشَّىْءَ *he swept away the thing* (جَرَفَهُ) from the surface of the earth. (TA.) You say also, جَرَفَتْهُ السُّيُولُ, (Mṣb,) or ▼ جرّفتهُ, inf. n. تَجْرِيفٌ; (Ṣ;) and ▼ تجرّفتهُ; (Ṣ, Ḳ;) *The torrents swept it away;* (TA;) [or *swept it partially away;* or *wore it away;*] namely, a portion of land. (Ṣ, Mṣb, Ḳ. See جُرُفٌ.) And, of a death commonly prevailing, جَرَفَ النَّاسَ كَجَرْفِ السَّيْلِ ‡ [*It swept away,* or *destroyed, men, like the sweeping away of the torrent*]: (TA:) and ▼ يَجْتَرِفُ مَالَ القَوْمِ ‡ [*It sweeps away,* or *destroys, the cattle of the people*]. (Ṣ, TA.) — [*He shovelled it,* or *scooped it, away,* or *up,* or *out.*] You say, جَرَفَهُ بِكِلْتَا يَدَيْهِ [*He scooped it up,* or *out, with both his hands*]; i. e. something dry, as flour, and sand, and the like. (Ṣ in art. حفن.) — جُرِفَ *It* (herbage) *was eaten up utterly.* (TA.)

2. جرّفهُ, inf. n. تَجْرِيفٌ: see 1, in four places. — جَرَّفَهُ الدَّهْرُ † *Time,* or *fortune,* or *misfortune, destroyed,* or *exterminated, his property,* or *cattle, and reduced him to poverty.* (TA.) A poet (of the Benoo-Ṭeiyi, TA) says,

* فَإِنْ تَكُنِ الحَوَادِثُ جَرَّفَتْنِى *
* فَلَمْ أَرَ هَالِكًا كَٱبْنَىْ زِيَادِ *

† [*And if misfortunes have destroyed my property,* or *cattle, and reduced me to poverty, I have not seen any one in a state of perdition like the two sons of Ziyád*]. (Ṣ, TA.)

4. اجرف *It* (a place) *was invaded by a torrent such as is termed* جُرَافٌ. (Ḳ.)

5: see 1, in two places; and see جُرُفٌ.

8: see 1, in three places.

جَرْفٌ: see جُرُفٌ. — Also *A smooth side of a mountain.* (Aboo-Kheyreh, Ḳ.)

جَرَفٌ: see the next paragraph.

جُرُفٌ and ▼ جُرْفٌ, (Ṣ, Mṣb, Ḳ, &c.,) the latter a contraction of the former, (Mṣb,) [*An abrupt, water-worn, bank* or *ridge;*] *a bank* (جَانِبٌ Ksh and Jel in ix. 110) of a valley, *the lower part of which is excavated by the water, and hollowed out by the torrents, so that it remains uncompact, unsound,* or *weak;* (Ksh ib.;) *a bank,* or *an acclivity, of a water-course* of a valley and the like, *when the water has carried away from its lower part, and undermined it, so that it has become like what is termed a* دَحْل, *with its upper part overhanging;* (L;) *a portion of land* (or *sand,* Ṣ in art. تهر) *which the torrents have partially swept away,* or *worn away,* (▼ تَجَرَّفَتْهُ, Ṣ, Ḳ, or ▼ جَرَّفَتْهُ, Mṣb,) *and eaten;* (Ṣ, Mṣb, Ḳ;) *a portion of the lower part of the side of a valley,* and *of a river, eaten by the torrent;* (M, TA;) the *side of the bank of a river, that has been eaten by the water, so that some part of it every little while falls:* (Ḥar p. 47:) and the latter, [or each,] *a place which the torrent does not take away;* as also ▼ جَرْفٌ; (Ḳ;) [i. e. *a bank,* or *ridge, that remains rising abruptly by the bed of a torrent* or *stream:*] pl. [of pauc.] (of جُرُفٌ, TA) أَجْرَافٌ, (Ḳ,) like أَطْنَابٌ pl. of طُنُبٌ, (TA,) and [of mult.] (of جُرُفٌ, though it is implied in the Ḳ that it is of جُرْفٌ, TA) جِرَفَةٌ, like جِحَرَةٌ (Ṣ, Ḳ) pl. of جُحْرٌ, (Ṣ,) and جُرُوفٌ. (ISd, TA.)

جُرَافٌ A torrent *that carries away everything;* (Ṣ, Mṣb;) *i. q.* جُحَافٌ applied to a torrent; as also ▼ جُرُوفٌ; (Ḳ;) and ▼ جَارُوفٌ a torrent *that sweeps away that by which it passes, by reason of its copiousness, carrying away everything,* and so ▼ جَارِفٌ applied to rain. (TA.) — ‡ *A very voracious man:* (Ḳ, TA:) a man *who devours all the food:* (Ṣ:) one *who eats vehemently, leaving nothing remaining.* (M, TA.) —— ‡ *A man who marries much,* or *often, and is brisk, lively, sprightly,* or *active;* as also ▼ جَارُوفٌ. (K, TA.) —— ‡ *A sword that sweeps away everything.* (TA.) —— † *A sort of measure of capacity;* as also ▼ جِرَافٌ: (S, K:) *a certain large measure of capacity.* (ISk, TA.)

جِرَافٌ: see what next precedes.

نَيْطَلٌ جَرُوفٌ [*A capacious bucket:* see 3 in art. نهز]. (S in art. نهز.)

مِجْرَفَةٌ: see جُرَافَةٌ.

جَارِفٌ: see جُرَافٌ. —— Also ‡ *A death commonly,* or *generally, prevailing,* (S, K, TA,) *that sweeps away,* or *destroys,* (يَجْتَرِفُ,) *the cattle of the people.* (S, TA.) And † *Plague,* or *pestilence.* (K.) الجَارِفُ means † *A plague,* or *pestilence, that happened in the time of Ibn-Ez-Zubeyr;* (S;) or, as Lth says, الطَّاعُونُ الجَارِفُ means *the plague,* or *pestilence, that befel the people of El-'Irák* [*in the year of the Flight* 69], *spreading wide, and sweeping away the people like the sweeping away of the torrent.* (TA.) And ‡ *Evil fortune,* or *an affliction, that sweeps away,* or *destroys,* (Lth, K, TA,) *a people,* (K,) or *the cattle of a people.* (Lth, TA.)

جُورَفٌ: see جُرَافٌ. —— Hence, as being likened to the torrent thus termed, (TA,) ‡ *A quick,* or *swift,* بِرْذَوْن [or *hack,* &c.]. (K.) And † *An ass;* [app. meaning *a wild ass,* because of his swiftness.] (Sgh, K.) And, accord. to some, *A male ostrich:* (as in the K:) but this is a mistranscription for جُورَقٌ, with ق. (Abu-l-'Abbás, T, Sgh, L, TA.)

جَارُوفٌ: see جُرَافٌ, in two places. —— Also ‡ *Greedy; having an inordinate desire,* or *appetite, for food.* (K, TA.) —— And ‡ An *unfortunate* man. (K,* TA.)

مِجْرَفٌ: see مِجْرَفَةٌ. —— [Hence,] بَنَانٌ مِجْرَفٌ [*Fingers,* or *fingers' ends,*] *that take much food.* (IAar, TA.)

مِجْرَفَةٌ *A broom,* or *besom;* (K;) *a thing with which mud is swept away from the surface of the ground:* (S,* TA:) [applied in the present day to *a shovel:* and *a hoe:* and *a rake:*] as also ▼ مِجْرَفٌ: vulgarly, ▼ جُرَافَةٌ; [now applied by many to *a drag for dragging rivers &c.;*] of which the pl. is جَرَاوِيفُ. (TA.)

مُجْرَفٌ † A man *who has had his property,* or or *cattle, destroyed,* or *exterminated, and who has been reduced to poverty, by time,* or *fortune,* or *misfortune.* (TA.)

مُجْرَفٌ † *Lean,* or *emaciated.* (M, TA.) [See what next follows.]

مُتَجَرِّفٌ † A ram *whose general fatness has gone;* (Ibn-'Abbád, K;) and so a camel. (TA.) —— *Lean,* or *emaciated;* as also مُتَجَلِّفٌ. (TA in art. جلف.) You say, جَآءَ مُتَجَرِّفًا † *He* (a man, Ibn-'Abbád, TA) *came in a lean and lax state* (هَزِيلًا مُضْطَرِبًا). (Ibn-'Abbád, K.)

جرل

جِرْيَالٌ *A certain red dye.* (As, S, K.) —— The *redness* of gold. (S, K.) —— *Pure;* applied to red and other colours. (K.) —— Also, (S, K,) and ▼ جِرْيَالَةٌ, (K,) *Wine;* (S, K;) *inferior in goodness to such as is termed* سُلَافٌ: (S:) or the *colour* of wine. (S, K.) The phrase سَلَبْتُهَا جِرْيَالَهَا, used by El-Aashà, [lit. *I deprived it,* namely, wine, *of its colour,*] means *I drank it red, and discharged it in urine white.* (S.)

جِرْيَالَةٌ: see above.

جرم

1. جَرَمَهُ, aor. ـِ, (K,) inf. n. جَرْمٌ, (S,) [like جَزَمَهُ,] *He cut it,* or *cut it off.* (S, K.) —— جَرَمَ صُوفَ الشَّاةِ, (K,) or جَرَمَ الشَّاةَ, (S,) inf. n. as above, (TA,) *He shore,* or *sheared,* or *cut off the wool of, the sheep.* (S, K,* TA.) And جَرَمْتُ مِنْهُ *I took* [or *clipped somewhat*] *from it;* [namely, the wool;] like جَلَمْتُ. (S.) —— جَرَمَ النَّخْلَ, (S, Msb, K,) aor. as above, (TA,) inf. n. جَرْمٌ (K) and جَرَامٌ and جِرَامٌ, (S,* K,) *He cut the palm-trees;* (Msb;) [meaning] *he cut off the fruit of the palm-trees;* (S, K;) as also ▼ اجترمهُ: (S:) and in like manner, جَرَمَ التَّمْرَ *he cut off the dates.* (TA.) You say, هٰذَا زَمَنُ الجَرَامِ and الجِرَامِ, (S,) i. e., [*This is*] *the time of the cutting off of the fruit of the palm-trees.* (TA.) —— And جَرَمَ النَّخْلَ, inf. n. جَرْمٌ, *He computed by conjecture the quantity of fruit upon the palm-trees;* (K;) and so ▼ اجترمهُ: (Lh, K:) [like جَزَمَهُ and اجتزمهُ.] —— جَرَمَ, (S, K,) aor. ـِ, (S,) inf. n. جَرْمٌ, (TK,) also signifies *He gained, acquired,* or *earned,* [wealth, &c.,] (S, K,) لِأَهْلِهِ *for his family;* and so ▼ اجترم. (K.) And you say, يَجْرِمُ أَهْلَهُ and خَرَجَ يَجْرِمُ لِأَهْلِهِ, meaning *He went forth seeking* [sustenance], *and practising skill,* or *artifice, for his family.* (TA.) —— وَلَا يَجْرِمَنَّكُمْ شَنَآنُ قَوْمٍ, in the Kur [v. 3 and 11], is explained by some as meaning *And let not a people's hatred by any means occasion you,* or *cause you:* or it means *let not a people's hatred by any means induce you,* or *incite you.* (S, TA.) Some read ▼ لَا يُجْرِمَنَّكُمْ, with damm to the ى; and Zj says that جَرَمْتُ and أَجْرَمْتُ signify the same: but some say that the meaning is, *let it not by any means lead you into crime,* or *sin;* أَجْرَمْتُهُ being like آثَمْتُهُ, *I led him into sin,* &c. (TA.) —— Fr says that the asserting جَرَمْتُ to mean حَقَقْتُ [or rather حُقِقْتُ, for this is evidently, I think, the right reading, though I find حَقَقْتُ in the TA as well as in a copy of the S, in another copy of which I find جَرَمْتُ and حَقَقْتُ, suggesting that the right reading may perhaps be حُقَّتْ and جَرَمْتُ,] is nought: they who so explain it having been confused in their judgment by the saying of the poet Aboo-Asmà, (S, TA,) or, as some say, El-Howfazán, (TA,) or, accord. to some, 'Ateeyeh Ibn-'Ofeyf, (IB, TA,)

* وَلَقَدْ طَعَنْتَ أَبَا عُيَيْنَةَ طَعْنَةً *
* جَرَمَتْ فَزَارَةَ بَعْدَهَا أَنْ يَغْضَبُوا *

in which they made فزارة to be in the nom. case, as though the meaning were حُقَّ لَهَا الغَضَبُ [*it was right,* or *fit,* or *proper, for it,* (the tribe of Fezárah,) *to be angry;* nearly agreeing with an explanation of جَرَمَ given by Golius as on the authority of Ibn-Maaroof, namely, "meritus, dignus fuit"]: but, he says, فزارة is in the accus. case; the meaning being, جَرَمَتْهُمُ الطَّعْنَةُ أَنْ يَغْضَبُوا [which will be found explained, on the authority of IB, in what follows]: AO says that the meaning is, أَحَقَّتِ الطَّعْنَةُ عَلَيْهِمُ الغَضَبَ, i. e., فَزَارَةَ أَنْ يَغْضَبُوا, and حَقَّتْ also, [both having the same signification, i. e., *the thrust required Fezárah to be angry,*] from لَا جَرَمَ لَأَفْعَلَنَّ كَذَا meaning حَقًّا [*Verily I will do thus*]: (S, TA:) accord. to Fr, the meaning is, كَسَبَتْ فَزَارَةَ الغَضَبَ عَلَيْكَ, the right reading being, وَلَقَدْ طَعَنْتَ, with fet-h to the ت; [so that the verse means *And verily thou didst thrust Aboo-'Oyeyneh with a thrust* of thy spear *that occasioned,* or *caused, Fezárah, after it, to be angry* against thee:] for he is addressing Kurz El-'Okeylee, bewailing his death; and Kurz had thrust Aboo-'Oyeyneh, who was Hisn Ibn-Hudheyfeh Ibn-Bedr El-Fezáree. (IB, TA.) —— And جَرَمَ, (S, Msb, K,) aor. ـِ, inf. n. جَرْمٌ, (Msb,) *He committed a sin, a crime, a fault, an offence,* or *an act of disobedience;* (S, Msb, K;) syn. أَذْنَبَ, (Msb, K,) and اِكْتَسَبَ الإِثْمَ; (Msb;) [perhaps because he who does so brings upon himself the consequence thereof; as though originally جَرَمَ نَفْسَهُ او لِنَفْسِهِ أَثَرَ جُرْمٍ *he drew upon himself the effect of a sin,* &c.; (compare كَسَبَ and اِكْتَسَبَ;)] as also ▼ اجرم, (S, Msb, K,) inf. n. إِجْرَامٌ; (Msb;) and ▼ اجترم; (S, K:) and ▼ تجرّم. (El-'Okbaree, Har p. 207.) You say, جَرَمَ عَلَيْهِمْ جَرِيمَةً, and إِلَيْهِمْ, (K,) and بِهِمْ, used by a poet for عليهم or اليهم, (IAar, TA,) *He committed against them a crime,* or *an offence for which he should be punished;* as also ▼ اجرم. (K.) They said also, ▼ اجرم الذَّنْبَ [*He committed the sin,* or *crime,* &c.]; making the verb trans. (TA.) And a poet says,

* وَتَرَى اللَّبِيبَ مُحَسَّدًا لَمْ يَجْتَرِمْ *
* عِرْضَ الرِّجَالِ وَعِرْضُهُ مَشْتُومُ *

[*And thou seest the intelligent envied,* or *much envied: he has not injured the honour of men, while his honour is reviled*]. (Th, TA.) —— جَرِمَ, aor. ـَ, (K,) inf. n. جَرَمٌ, (TK,) *He* (a man, TA) *betook himself to eating the* جُرَامَة [in the CK, erroneously, جِرَامَة,] *of the palm-trees,* (AA, K,) [i. e., *the dates which had fallen in the cutting, and*] *which were among the branches.* (AA, TA.) —— جَرُمَ, said of a man, also signifies عَظُمَ جُرْمُهُ [*His sin,* or *crime,* &c., *was,* or *became, great*]; and so جَرُمَ, like كَرُمَ: [both are thus explained, in different places in this art., by the author of the TA; and the explanation in the latter case is followed by اى اذنب, i. e., *he committed a sin,* &c.; probably added by him to show that the reading found by him was جُرْمُهُ, not جِرْمُهُ: but I think that the right reading is عَظُمَ جِرْمُهُ *his body became great;* and this is confirmed by what here follows:] ↑اجرم explained in the copies of the Ḳ by عَظُمَ [in the TḲ عظم يعنى جرمه وجسده] should be جَرُمَ, a triliteral; and the meaning is عَظُمَ جُرْمُهُ: and in like manner, the three significations here following, assigned in the Ḳ to ↑اجرم, belong to جَرُمَ. (TA.) ═ *It* (his colour) *was*, or *became, clear.* (Ḳ,* TA.) — *He* (a man, TA) *was*, or *became, clear in his voice.* (Ḳ,* TA.) ═ جَرِمَ بِهِ *It* (blood) *stuck to him*, or *it*: (Ḳ,* TA, and so in a marginal note in a copy of the Ṣ:) and in like manner, tar to a camel. (The same marginal note.)

2. [جرّم *He cut off vehemently*, or *much*. (Golius, on the authority of a gloss in the KL.)] — جَرَّمْنَاهُمْ, inf. n. تَجْرِيمٌ, *We went forth from them.* (Lth, Ḳ.) — جَرَّمْنَا الشِّتَآءَ *We completed the winter.* (TA.) [See also 5.]

4. اجرم التَّمْرُ *The dates attained to the time for their being cut off.* (TA.) ═ See also 1, in six places.

5. تجرّم [*It became cut off.* — And hence,] ‡*It* (a year, حَوْلٌ) *became completed;* (Az, Ḳ, TA;) as though it became cut off from the preceding year: (Az, TA:) *it ended;* (Ṣ;) and so the winter: (TA:) and *it* (a night) *passed away,* (Ṣ, Ḳ,) *and became completed;* (Ḳ;) *it ended.* (TA.) ═ تَجَرَّمَ ثَمَانِيًا, a phrase used by Sá'ideh Ibn-Ju-eiyeh, means *He passed eight* nights. (TA.) [See also 2.] ═ تَجَرَّمَ عَلَيْهِ *He accused him of a sin, a crime, a fault, an offence,* or *an act of disobedience,* (Abu-l-'Abbás, Ṣ, Ḳ,) *which he had not committed,* (Abu-l-'Abbás, Ṣ,) or *though he had not committed any* (Ḳ.) — And تجرّم *He guarded against the commission of sin,* or *crime,* &c.; like تَأَثَّمَ. (Ḥar p. 207.) — See also 1. ═ Also *He called, cried out, shouted,* or *vociferated;* from جَرْمٌ meaning صَوْتٌ. (Ḥar p. 207. [But see جِرْمٌ.])

8: see 1, in five places.

جَرْمٌ *Hot;* syn. حَرٌّ, (Ṣ,) or [rather] حَارٌّ; (Ḳ;) *contr. of* صَرْدٌ; (Lth, TA;) a Persian word, (Ṣ,) arabicized; (Ṣ, Ḳ;) originally گَرْم. (TA.) You say أَرْضٌ جَرْمٌ *A warm land:* (AḤn, TA:) or *a hot land:* (IDrd, TA:) or *a vehemently hot land:* (Ḳ:) pl. جُرُومٌ, (AḤn, TA,) which, applied to countries, or regions, means the *contr. of* صُرُودٌ. (Ṣ.) ═ *A boat* (زَوْرَقٌ) *of El-Yemen;* (Ḳ;) *also called* نَقِيرَةٌ: (TA:) pl. as above. (Ḳ.) [In the dial. of Egypt, The *largest kind of Egyptian boat used on the Nile for the conveyance of grain and merchandise in general, but used only when the river is high, and also in the coasting-trade, and generally carrying from 5,000 to 15,000 bushels of grain.*]

جُرْمٌ *A sin, a crime, a fault, an offence,* or *an act of disobedience,* syn. ذَنْبٌ, (Ṣ, Mṣb,* Ḳ,) *whether intentional or committed through inadvertence;* (Kull voce إِثْمٌ;) as also ↑جَرِيمَةٌ; (Ṣ, Mṣb, Ḳ;) and ↑جُرْمَةٌ: (Ḳ:) *transgression:* (TA:) pl. [of pauc.] أَجْرَامٌ and [of mult.] جُرُومٌ, (Ḳ,) both of جُرْمٌ: the pl. of جَرِيمَةٌ is جَرَائِمُ. (TA.) ═ See also جِرَامٌ. ═ لَا جُرْمَ: see لَا جَرَمَ.

جِرْمٌ The *body;* syn. جَسَدٌ; (Ṣ, Mṣb, Ḳ;) or بَدَنٌ; (Th, TA;) as also ↑جِرْمَانٌ: (Ḳ:) or the أَلْوَاح [pl. of لَوْحٌ q. v.] and جُثْمَان [q. v.] of the جَسَد: (T, TA:) pl. (of pauc., TA) أَجْرَامٌ, (Mṣb, Ḳ,) which is also used as a sing., (TA,) and (of mult., TA) جُرُومٌ and جُرُمٌ. (Ḳ.) أَلْقَى عَلَيْهِ أَجْرَامَهُ is a phrase mentioned, but not explained, by Lḥ: ISd thinks that it means *He threw upon him the weight of his body;* as though the term جِرْمٌ applied to each separate part of his body. (TA.) — [Hence,] الأَجْرَامُ الفَلَكِيَّةُ *The* [*heavenly*] *bodies that are above the* عَنَاصِر, *of the orbs and stars.* (KT.) ═ The *throat,* or *fauces;* syn. حَلْقٌ. (Ḳ.) The phrase يَضِيقُ بِهِ الجِرْمُ, used by the poet Maan Ibn-'Ows, means †*It is a great,* or *formidable, thing,* or *matter:* [properly,] *the throat* (الحَلْقُ) *will not easily swallow it.* (TA.) — The *voice;* (Ṣ, Ḳ;) mentioned by ISk and others; (Ṣ;) and so explained as used in the phrase إِنَّ فُلَانًا لَحَسَنُ الجِرْمِ [*Verily such a one is good in respect of voice*]: (TA:) or *highness,* or *loudness,* of the voice: (Ḳ, TA:) you say, مَا عَرَفْتُهُ إِلَّا بِجِرْمِهِ [*I knew him not save by his voice,* or *his highness,* or *loudness,* of voice]: but some disapprove this: (TA:) AḤát says that the vulgar are addicted to saying, فُلَانٌ صَافِى الجِرْمِ *Such a one is clear in voice,* or *in throat:* but it is a mistake. (Ṣ, TA.) ═ *Colour.* (IAạr, Ṣ, Mṣb, Ḳ.) One may say, of نَجَاسَة [or filth], لَا جِرْمَ لَهَا, meaning *It has no colour.* (Mṣb.) ═ الأَجْرَامُ (app. as pl. of جِرْمٌ, TA) *The utensils,* or *apparatus, of the pastor.* (Ḳ.)

لَا جَرَمَ (Ṣ, Mṣb, Ḳ, &c.) and لَا ذَا جَرَمَ, (IAạr, Ḳ,) ذا being here a redundant connective as in several other instances, (IAạr, TA,) and لَا أَنْ ذَا جَرَمَ and لَا عَنْ ذَا جَرَمَ (Ḳ) and لَا جَرَ, (Ks, Ḳ, [in the CḲ لا جُرْمَ,]) in which the م is elided in consequence of frequency of usage, as the ى is in أَيْش for أَىُّ شَىْءٍ, and the ى and ء in حَاشَ لله for حَاشَى لله, (Ks, TA,) and لَا ذَا جَرَ (IAạr, TA) and ↑لَا جُرْمَ and ↑لَا جُرْمَ, (Ḳ,) originally *i. q.* لَا بُدَّ and لَا مَحَالَةَ [*There is no avoiding it; it is absolutely necessary;* &c.]: then, by reason of frequency of usage, employed in the manner of an oath, as meaning حَقًّا [*verily,* or *truly*]; wherefore, as in the case of an oath, ل is prefixed to its complement, (Fr, Ṣ, Mṣb, Ḳ,*) so that they say, لَا جَرَمَ لَآتِيَنَّكَ [*Verily I will come to thee*], (Fr, Ṣ, Ḳ,) and لَا جَرَمَ لَأَفْعَلَنَّ كَذَا [*Verily I will do thus*], (Ṣ, Mṣb,*) and لَا جَرَمَ لَقَدْ كَانَ كَذَا and لَا ذَا جَرَ and لَا ذَا جَرَمَ [*Verily it was thus,* or *verily such a thing happened*]: (IAạr, TA:) ISd says, Kh asserts that جَرَمَ [or لا جرم] is only a reply to something said before it; as when a man says, "They did such a thing," and you say, لَا جَرَمَ أَنَّهُمْ سَيَنْدَمُونَ, or أَنَّهُ سَيَكُونُ كَذَا وَكَذَا; and Az says that لا in لَا جَرَمَ is said to be a [mere] connective; and the meaning [of the former of the last two phrases] is كَسَبَ لَهُمْ عَمَلُهُمُ النَّدَمَ [*It* (their deed) *will earn for them,* or *occasion them, repentance;* and that of the latter, *it will occasion that such and such things shall happen*]: and some say that جَرَمَ means حَقَّ, and وَجَبَ, and that لا is a contradiction to the words preceding it, and that a new proposition then begins; as in the Ḳur [xvi. 64] where it is said, لَا جَرَمَ أَنَّ لَهُمُ النَّارَ, i. e., [*Nay,* or] *the case is not as they have said: the fire* [of Hell] *is their due.* (TA.)

لَا جُرْمَ: see the paragraph next preceding.

جِرْمَةٌ People *cutting off the fruit of palm-trees.* (Ṣ, Ḳ, TA.) [In this sense it is app. a pl. of pauc., or a quasi-pl. n., of جَارِمٌ, q. v.] — Also *Ripening dates cut off from the trees:* and this sense, not the former as is implied in the Ṣ, is meant by Imra-el-Ḳeys, where he says,

* عَلَوْنَ بِأَنْطَاكِيَّةٍ فَوْقَ عِقْمَةٍ *
* كَجِرْمَةِ نَخْلٍ أَوْ كَجَنَّةِ يَثْرِبِ *

[*They mounted, at Antioch, upon a variegated cloth, like the ripening dates cut off from palm-trees, or like the garden of Yethrib*]: he likens the variegated cloth and wool upon the هَوْدَج to red and yellow ripening dates, or to the garden of Yethrib because it abounded with palm-trees. (TA.)

جُرْمَةٌ: see جُرْمٌ.

جِرْمَانٌ: see جِرْمٌ.

جِرَامٌ (erroneously said in the Ḳ to be [جُرَامٌ] like غُرَابٌ, TA) and ↑جَرِيمٌ *Dry dates:* (AA, Ṣ, M, Ḳ:) mentioned by ISk among [syn.] words of the measures فَعَالٌ and فَعِيلٌ, like شَحَاحٌ and شَحِيحٌ, and بَجَالٌ and بَجِيلٌ, &c. (Ṣ.) — Also, both these words, (AA, Ṣ, Ḳ,*) but the former not heard in this sense by ISd, (TA,) *Date-stones;* (AA, Ṣ, Ḳ;) and so ↑جُرْمٌ: (mentioned in one copy of the Ṣ, but not in the TA, [probably an interpolation in the copy of the Ṣ above mentioned:]) and ↑جَرِيمَةٌ *a date-stone;* as in the saying of 'Ows Ibn-Ḥáritheh, لَا وَالَّذِى أَخْرَجَ العَذْقَ مِنَ الجَرِيمَةِ وَالنَّارَ مِنَ الوَثِيمَةِ [*No, by Him who has produced the palm-tree with its fruit from the date-stone, and fire from broken stones*]. (TA.)

جَرِيمٌ Dates (تَمْرٌ) *cut off from the tree;* (Ṣ, TA;) as also ↑مَجْرُومٌ. (TA.) And شَجَرَةٌ جَرِيمَةٌ *A cut tree.* (TA.) — See also جِرَامٌ, with which it is syn. in two senses: in the latter sense having جَرِيمَةٌ for its n. un. — Also *A thing with which date-stones are brayed,* or *crushed.* (TA.) ═ See also مُجْرِمٌ. ═ Also *Large-bodied;* (Ṣ,* Ḳ;) and so ↑مَجْرُومٌ: (Ḳ:) pl. (of the former, Ṣ) جِرَامٌ. (Ṣ, Ḳ.) The fem. of the former is with ة: (Ḳ:) [but] one says also جِلَّةٌ جَرِيمٌ, meaning *Large-bodied camels advanced in age.* (Ṣ.) ═ In El-Ḥijáz, The [*measure commonly termed*] مُدّ is thus called; accord. to Z, the مُدّ *of the Prophet.* (TA.)

جُرَامَةٌ *i. q.* جُذَامَةٌ; (Ḳ;) i. e., (TA,) The *dates that have fallen when they are cut off from the tree:* (Ṣ, TA:) so says Aṣ: (TA:) [but see the latter word as explained on the authority of the Ṣ in art. جذم:] and, (Ḳ,) or as some say, (TA,) *dates cut off from the tree:* or *what are gotten* (يُجْرَمُ) *thereof, after their being cut off, being picked up from the lower ends of the branches.* (Ḳ, TA. [See جَرِمٌ.]) And The قَصَل *of wheat and barley;* i. e., the *extremities thereof, which are bruised, and then cleared,* or *picked:* (Ḳ, TA:) but the term more known is جُذَامَةٌ, with ذال. (TA.)

جَرِيمَةٌ The *last of one's offspring:* (Ḳ:) as though there were a cutting off after it. (TA.) — See also جَارِمٌ: — and see جَرْمٌ: — and جُرَامٌ.

جَارِمٌ *Cutting off,* or *one who cuts off, the fruit of the palm-tree:* pl. جُرَّمٌ and جُرَّامٌ. (Ṣ.) [See also جَرِيمَةٌ and جُرْمَةٌ.] — جَارِمُ أَهْلِهِ (TA) and ▾ جَرِيمَةُ أَهْلِهِ (Ṣ, Ḳ*) *The gainer, acquirer,* or *earner,* [*of the sustenance*] *of his family.* (Ṣ, Ḳ, TA.) — See also مُجْرِمٌ, in two places.

مُجْرِمٌ *A sinner; a criminal; committing,* or *a committer of, a sin, a crime, a fault, an offence,* or *an act of disobedience;* as also ▾ جَرِيمٌ (Ḳ) and ▾ جَارِمٌ: (TA:) and المُجْرِمُونَ particularly signifies *the unbelievers:* (Zj, Ḳ:) so in the Ḳur vii. 38. (Zj, TA.) You say, هُوَ جَارِمٌ ▾ عَلَى نَفْسِهِ وَقَوْمِهِ, [as also مُجْرِمٌ,] *He is committing a crime,* or *an offence for which he should be punished, against himself and his people* or *party.* (TA.)

مُجْرَمٌ, (fem. with ة, Ṣ,) A *complete* year (Ṣ, Ḳ) and month; (Ibn-Hání, TA;) a year *past, completed.* (AZ, TA.)

مَجْرُومٌ: see جَرِيمٌ, in two places.

جرموق

جُرْمُوقٌ [*A kind of galoche;*] *a thing that is worn over the* [*kind of boot called*] خُفّ; (Ṣ, Mgh, Mṣb, Ḳ;) called in Persian خَرْكَشْ: (Mgh:) or *a small* خُفّ (JK, TA) *which is worn over the* [*ordinary*] خُفّ: (TA:) an arabicized word; (Ṣ;) [probably from the Persian چَرْم "leather," and مُوق, which is said by some to be arabicized, but by ISd to be a genuine Arabic word, meaning "a kind of خُفّ:" or it may be from سَرْمُوزَه, given as its Persian equivalent in the PṢ:] pl. جَرَامِيقُ. (Mṣb.)

جرن

1. جَرَنَ, (Ṣ, Ḳ,) aor. ـُ, (Ṣ,) inf. n. جُرُونٌ, (Ṣ, Ḳ,) said of a man, and of a beast, (ISk, Ṣ,) *He became accustomed, habituated,* or *inured, to a thing,* or *an affair.* (ISk, Ṣ, Ḳ.) And جَرَنَتْ يَدَاهُ عَلَى العَمَلِ, inf. n. as above, *His hands became accustomed,* or *inured, to the work.* (M, TA.) — Also, said of a garment, or piece of cloth, (Ṣ, Ḳ,) and of a coat of mail, (Ḳ,) and of a skin for water or milk, (Ṣ,) *It became threadbare,* or *worn, and soft,* or *smooth:* (Ṣ, Ḳ:) or, said of a skin, and of a book, or writing, *it became old and worn out.* (M, TA.) — جَرَنَ, (Ḳ,) inf. n. جَرْنٌ, (TA,) *He ground* grain (Ḳ, TA) *vehemently:* (TA:) of the dial. of Hudheyl. (TA.)

4. اجرن *He collected* dates *in the* جَرِين. (ISd, Ḳ.)

8. اجترن *He made,* or *prepared, a* جَرِين. (Ḳ.)

جُرْنٌ *A hollowed stone,* [or *stone basin,*] *from which the* [*ablution termed*] وُضُوْء *is performed;* (Ḳ;) *water being poured into it;* called by the people of El-Medeeneh مهرس [app. مِهْرَس, perhaps a dial. var. of مِهْرَاس, or a mistranscription for this]: so in the M: in the Jm, the مِهْرَاس *with which the* وضوء *is performed.* (TA.) — See also جُبٌّ. — [In the present day, applied also to *A stone mortar in which things are pounded.*] — See also جَرِينٌ, in two places.

جِرْنٌ The *body, with the limbs* or *members;* syn. جِسْمٌ; said to be a dial. var. of جِرْمٌ; or the ن may be a substitute for the م of جرم; but the former is the more probable, as the word has a pl., namely, أَجْرَانٌ, and this is scarcely ever the case when a word is formed by substitution. (TA.) Hence the saying, أَلْقَى عَلَيْهِ أَجْرَانَهُ, *i. q.* شَرَاشِرَهُ and القى عليه أَجْرَامَهُ [*He threw upon him,* or *it, the weight of his body*]: (Lḥ, TA: [see also شُرْشُرَةٌ, under which other explanations are given:]) or *he threw his weights* [meaning *his whole weight*] *upon him,* or *it;* and so القى عليه جِرَانَهُ ▾: or, accord. to the A, *he disposed,* or *subjected, his mind to it;* or *persuaded himself to do it;* namely, an affair. (TA.)

جِرَانٌ The *anterior* [or *under*] *part of the neck* of a camel, *from his* مَذْبَح [or the *part a little below the under jaw*] *to the place where he is stabbed:* (Ṣ, Mṣb, Ḳ:) and in like manner, of a horse; (Ṣ, TA;) the *inner* [or *under*] *part of the neck, from the pit of the uppermost part of the breast to the extremity of the neck at the head:* and, metaphorically, of a man: (TA:) pl. [of mult.] جُرُنٌ (Ṣ, Mṣb, Ḳ) and [of pauc.] أَجْرِنَةٌ; (Mṣb, TA;) which last is used by Ṭarafeh as a sing. (TA,* and EM p. 68.) You say, of a camel, أَلْقَى جِرَانَهُ بِالأَرْضِ [*He threw the under part of his neck upon the ground*]; meaning that he lay down, and stretched out his neck upon the ground. (Mṣb, TA.) See another ex. voce جِرْنٌ. [And see a verse cited in the first paragraph of art. حنو.] You say also, ضَرَبَ الحَقُّ بِجِرَانِهِ, meaning † *The truth,* or *right,* or *just claim, became established,* or *settled.* (T, TA.) — Also The *inner* [or *under*] *part* of the penis: pl. جُرُنٌ and أَجْرِنَةٌ, as above. (TA.)

جَرِينٌ *What one has ground* [of grain]: (Ḳ, TA:) of the dial. of Hudheyl. (TA.) — See also جَارِنٌ. — Also, and ▾ جُرْنٌ (T, Ṣ, M, Ḳ) and ▾ مِجْرَنٌ, (Ḳ,) or ▾ مُجْرَنٌ, (so in a copy of the Ṣ, but in other copies not mentioned,) The *place in which dates are dried:* (Ṣ:) or a بَيْدَر: (Ḳ:) or the جَرِين is *for grain;* and the بيدر, *for dates:* (Towsheeḥ, TA:) or the *place where dates are collected* [*and dried*] *when they are cut from the tree:* or, accord. to Lth, the *place of the* بيدر in the dial. of the people of El-Yemen, the generality of whom pronounce the word [جِرِين,] with kesr to the ج: (T, TA:) or the مِرْبَد; i. e. the *place in which fresh ripe dates are thrown to dry:* (Mgh:) or the بيدر *in which wheat is trodden out;* and also the *place in which fruits are dried:* (Mṣb:) the *place of wheat;* and sometimes [the *place*] *for* [*drying*] *dates and grapes:* (M, TA:) its pl. [of mult.] is جُرُنٌ, (Mgh, Mṣb, TA,) not جَرَائِنُ, (Mgh,) and [of pauc.] أَجْرِنَةٌ and أَجْرَانٌ: (TA:) A 'Obeyd says that مِرْبَدٌ and جَرِينٌ are of the dial. of El-Ḥijáz; and أَنْدَرٌ, of that of Syria; and بَيْدَرٌ, of El-'Irák: (TA in art. ربد:) ▾ جُرْنٌ is of the dial. of the people of Egypt, who use it as meaning the بيدر *of seed-produce, which is* [*sometimes*] *walled round;* and its pl. is أَجْرَانٌ. (TA.) [See also مِرْبَدٌ.]

جِرْيَانٌ a dial. var. of جِرْيَالٌ, (Ṣ, Ḳ,*) meaning *A certain red dye.* (ISd, TA.)

جَارِنٌ, applied to a garment, or piece of cloth, (T, Ṣ, Ḳ,) and to a skin for water or milk, &c., (T, TA,) *Old, and worn out:* (T, TA:) or *threadbare,* or *worn, and soft,* or *smooth:* and in like manner applied to a coat of mail: (Ṣ, Ḳ:) as also ▾ جَرِينٌ: (TA:) or, applied to a coat of mail (دِرْع), in which case it is with ة, *that has become smooth from much use:* (Ham p. 656:) pl. جَوَارِنُ: (Ṣ, TA:) and, applied to a commodity, or utensil, or an article of furniture, *used, and worn out:* and to a skin for water or milk, *dried up, and rough,* or *coarse, from use:* (TA:) and to a road, *worn,* or *effaced.* (Abu-l-Jarráḥ, Ṣ, Ḳ.) — Also The *young one of a serpent:* (Ṣ, Ḳ:) or *of a viper,* (Lth, M, TA,) *such as is smooth.* (Lth, TA.)

مُجْرَنٌ: see جَرِينٌ.

مِجْرَنٌ: see جَرِينٌ. — Also *Very voracious:* (Ḳ:) of the dial. of Hudheyl. (TA.)

مُجَرَّنٌ A whip *of which the thong has become soft,* or *smooth.* (Ḳ.) Az says, I have seen them make their whips from the جُرُن [pl. of جِرَانٌ q. v.] of camels such as are termed بُزْل [i. e. in the ninth year, or nine years old], because of the thickness thereof. (TA.)

جرو

4. أَجْرَتْ [in its primary sense app. signifies *She* (a bitch, and any female beast of prey,) *whelped;* or *had a whelp,* or *whelps:* see مُجْرٍ. (Accord. to Golius, as on the authority of J, *Cum fœtu abiit* vel *asportavit eum fera:* but I have not found it in any copy of the Ṣ, nor in any other lexicon.) — And hence,] said of a tree (شَجَرَة), *It had upon it* [*fruits such as are termed*] جِرَاء [pl. of جِرْوٌ]; (Aṣ, TA;) said [for instance] of a مَوْزَة [or banana-tree]: (AḤn, TA in art. موز:) and [in like manner] said of a herb,

or leguminous plant, (بَقْلَةٌ,) it had جِرَآء; (K in art. جرى; [in the CK, erroneously, جَرَاء;]) belonging to the present art., not to art. جرى. (TA.)

جِرْوٌ and جُرْوٌ and جَرْوٌ (of which the first is the most chaste, Msb) The *whelp*, or *young one*, of the dog, (S, Msb, K,) [and so, app., ▼جِرْوَةٌ, q. v.,] and of the lion, (K,) and of any beast of prey: (S, Msb:) pl. [of pauc.] أَجْرٍ, (S, K,) originally أَجْرُوٌ, (S,) and أَجْرِيَةٌ (Lh, K) and أَجْرَآءٌ (K) and [of mult.] جِرَآءٌ; (S, K;) أَجْرِيَةٌ, as pl. of جرو, being anomalous; (TA;) or it is pl. of جِرَآءٌ [and therefore not anomalous]. (S, TA.) — And the same, (K,) or the first of these, (S, El-Bári', Msb, TA,) only, (TA,) ‡ The *small* of anything, (El-Bári', Msb, K, TA,) as also ▼جِرْوَةٌ; (TA;) even, (K,) of the colocynth, (As, S, K,) and of the melon, and the like; (K;) as, for instance, (TA,) of the pomegranate, (S, TA,) and of the poppy, (AHn, TA in art. عتر,) and of the بَاذَنْجَان, (TA,) and of the cucumber, (S, TA,) as also ▼جِرْوَةٌ, (S, Msb,) likened to the whelps of dogs, because of their softness and smoothness: (Msb:) or *what is round* of the fruits of trees; as the colocynth and the like: (TA:) pl. [of pauc.] أَجْرٍ (Msb, K, TA [in the CK, erroneously, أَجْرُوْ]) and [of mult.] جِرَآءٌ. (As, Msb, K.) — † *Fruit when it first grows forth*, (AHn, K,) *in its fresh, juicy, state.* (AHn, TA.) — †The *seeds*, (M, TA,) or *envelope*, or *receptacle, of the seeds*, (K,) *of the* كَعَابِير [app. meaning the *round and compact pericarps* (in some of the copies of the K, erroneously, as is observed in the TA, عكابير,)] *that are at the heads of branches.* (M, K.) — ‡ *A tumour in a camel's hump*; and *in the withers*; so called by way of comparison [to a whelp]: (TA:) and *in the fauces.* (K.) — See also what next follows, last sentence.

جِرْوَةٌ: see جِرْوٌ, in three places. A huntsman beat, or disciplined, or trained, his dog for the purpose of the chase (ضَرَبَهُ عَلَى الصَّيْدِ); and thereupon it was said, ضَرَبَ جِرْوَتَهُ [app. meaning *He beat*, or *disciplined*, or *trained, his whelp*]: and hence this phrase became proverbially used, in instances here following. (Z, TA.) — *I. q.* نَفْسٌ [as meaning *Self*]. (TA.) You say, ضَرَبَ عَلَيْهِ جِرْوَتَهُ, meaning †*He disposed and subjected himself to it*; (S, TA;) namely, a thing, an affair, or a case: and *he endured it with patience*: and in like manner, ضَرَبَ جِرْوَةَ نَفْسِهِ: (TA:) and أَلْقَى فُلَانٌ جِرْوَتَهُ *he endured the thing with patience.* (S, TA.) Also ضَرَبْتُ جِرْوَتِى عَنْهُ †*I endured with patience the want*, or *loss, of it*: (TA:) or *my mind became at ease* [*respecting it*]. (AA, IB, TA.) ═ Also (K, TA, [in the CK جِرْوٌ,]) ‡ *A short she-camel.* (K, TA.)

مُجْرٍ and مُجْرِيَةٌ applied to a bitch, (S, K,) and a female beast of prey, (TA,) *Having a whelp*, or *young one*: (K:) or *having with her her whelps*, or *young ones.* (S.) El-Aalam says,

* وَتَجُرُّ مُجْرِيَةٌ لَهَا * لَحْمِى إِلَى أَجْرٍ حَوَاشِبْ *

And a hyena having young ones shall drag for her my flesh to young ones with swollen bellies, and short. (Skr p. 57.)

جرى

1. جَرَى, said of water (S, Mgh, Msb) &c., (S,) or of water and the like, (K,) more properly thus, as in the K, aor. يَجْرِى, (TA,) inf. n. جَرْىٌ (S, Mgh, Msb, K) and جَرَيَانٌ (S, K) and جِرْيَةٌ, (S,* Msb, K,) [which last see below,] *It ran*, or *passed along quickly*; originally said of water: (Er-Rághib, TA:) or *it flowed*; syn. سَالَ; *contr. of* سَكَنَ *and* وَقَفَ. (Msb.) — Said also of farina, in the phrase جَرَى الدَّقِيقُ فِى السُّنْبُلِ [*The farina pervaded the ears of wheat*]. (L in art. قمح.) — And of a horse (Mgh, Msb, K) and the like, (Msb, K,) aor. as above, (TA,) inf. n. جَرْىٌ (Msb, K) and جَرَيَانٌ (Msb) and جِرَآءٌ (Lth, K) and مَجْرًى, (S,) [*He ran;*] from the same verb said of water. (Mgh.) — And of a ship: you say, جَرَتِ السَّفِينَةُ, (S, TA,) inf. n. جَرْىٌ (TA) and مَجْرًى, (S, K,) [*The ship ran.*] — And of the sun, and a star: you say, جَرَتِ الشَّمْسُ, aor. يَجْرِى, inf. n. جَرْىٌ, [*The sun pursued its course:*] and جَرَتِ النُّجُومُ *The stars travelled*, or *passed along, from east to west.* (TA.) — جَرَى إِلَى كَذَا, (Msb, and Har p. 152,) inf. n. جَرْىٌ and جِرَآءٌ; (Msb;) and ▼اجرى إِلَيْهِ, inf. n. إِجْرَآءٌ; (Ham p. 224, and Har p. 152;) *He betook*, or *directed, himself to such a thing; made it his object; aimed at it; intended*, or *purposed, it*: (Msb, and Har ubi suprà:) and *he hastened to it*: (Msb:) but in the latter phrase, an objective complement is understood; and it is used in relation to something disapproved, or disliked; (Ham and Har;) properly, اجرى فِعْلَهُ إِلَيْهِ, (Ham,) or اجرى فِعْلَهُ بِالقَصْدِ إِلَيْهِ. (Har.) — Hence, perhaps, the saying, جَرَى الخِلَافُ فِى كَذَا [frequently used as meaning *A controversy ran*, or *ran on, respecting such a thing* between such and such persons]. (Msb.) — جَرَى لَهُ الشَّىْءُ, (Sh, TA,) and جَرَى عَلَيْهِ, (TA,) †*The thing was permanent*, or *continued, to him.* (Sh, TA.) [And, more commonly, †*The thing happened*, or *occurred, to him.* Whence, ▼مَاجَرَيَاتٌ, as pl. of مَاجَرَى, used as a single word, by late writers, meaning †*Events*, or *occurrences.*] — هُوَ يَجْرِى مَجْرَاهُ †*It is like it*, or *similar to it, in state, condition, case, or predicament.* (TA.) [*It* (a word or phrase) *follows the same rule* or *rules*, or *occupies the same grammatical place, as it* (another word or phrase). And similar to this is the saying,] الدَّيْنُ وَالرَّهْنُ يَتَجَارَيَانِ▼ مُجَارَاةَ المَبِيعِ وَالثَّمَنِ †[*The debt and the pledge are subject to the same laws as the thing sold and the price*]. (Mgh.) — [Also †*It acts as*, or *in a similar manner to, it*: and †*he acts in his stead*: see جَرِىٌّ. Hence the phrase, جَرَى مِنْهُ مَجْرَى كَذَا †*It acted upon him*, or *affected him, like*, or *in a similar manner to, such a thing*: as in the prov.,] جَرَى مِنْهُ مَجْرَى اللَّدُودِ †[*It acted upon him*, or *affected him, like*, or *similarly to, the medicine*, or *draught, called* لدود: منه here having the meaning of فيه]. (ISk, S in art. لد.) — [One says, also, of an inf. n., and of a part. n., that is regularly formed, يَجْرِى عَلَى الفِعْلِ, meaning †*It is conformable to the verb.*]

2. جرّى *He sent a deputy*, or *commissioned agent*; as also ▼اجرى. (K.) And جَرَّى جَرِيًّا *He made*, or *appointed, a deputy*, or *commissioned agent*; (ISk, S,* TA;) as also ▼استجراه. (S,* TA.) Hence the trad., (TA,) لَا يَسْتَجْرِيَنَّكُمُ▼ الشَّيْطَانُ (S, TA) *By no means let the Devil make you his followers and his commissioned agents.* (TA.) You say also, اجراه▼ فِى حَاجَتِهِ [*He sent him to accomplish his needful affair*]. (TA.)

3. جاراه, inf. n. مُجَارَاةٌ (S, Mgh, Msb, K) and جِرَآءٌ, (S, K,) *He ran with him.* (S, Mgh, Msb, K.) You say, جَارَيْتُهُ حَتَّى فُتُّهُ *I ran with him until I passed beyond him*, or *outwent him.* (TA in art. فوت.) — [*He vied, contended*, or *competed, with him in running*: and hence, †*in any affair*; like سَايَرَهُ.] You say, جاراهُ فِى كَذَا وَفَعَلَ مِثْلَ فعله †[*He vied, contended*, or *competed, with him in such an affair, and did like as he did*]. (Mgh in art. فوض.) And جاراهُ فِى الحَدِيثِ †[*He vied, contended*, or *competed, with him in discourse*]. (S.) And جَارَوْا فِى الحَدِيثِ (TA) and ▼تَجَارَوْا فيه (S, TA) †[*They vied, contended*, or *competed, one with another, in discourse*]. And it is said in a trad., مَنْ طَلَبَ العِلْمَ لِيُجَارِىَ بِهِ العُلَمَآءَ †*He who seeks knowledge in order that he may run* [i. e. *vie*] *with the learned in discussion and disputation, to show his knowledge to others, to be seen and heard.* (TA.) And in another trad., لَا تُجَارِ أَخَاكَ وَلَا تُشَارِهِ وَلَا تُمَارِهِ †[*Contend not for superiority with thy brother*, (so explained in the TA, voce جَارٍ, in art. جر,) *nor dispute with him, nor wrangle with him*]: (El-Jámi'-es-Sagheer:) or, as some relate it, لَا تُجَارِّ أَخَاكَ وَلَا تُشَارِّهِ. (TA in art. جر, q. v.)

4. اجراه *He made it to run*; (S, K,* TA;) said of water &c., (S,) or of water and the like. (K,* TA.) [Hence, اجرى دُمُوعًا, or دَمْعًا, *He shed tears.*] — Also *He made him to run*; namely, a horse (Mgh, Msb, K*) and the like: (Msb, K:*) in which sense مُجْرًى [as well as إِجْرَآءٌ] is used as an inf. n. (S.) — اجرى السَّفِينَةَ [*He made the ship to run*]: (S:) in this sense, also, مُجْرًى [as well as إِجْرَآءٌ] is used as an inf. n. (S, K.) — اجراهُ فِى حَاجَتِهِ: see 2. — اجرى إِلَيْهِ: see 1. — أَجْرَيْتُ عَلَيْهِ [and لَهُ] †*I made a thing permanent*, or *continual, to him.* (IAar, TA.) [And hence, both of these phrases, in the present day, †*I made him*, or *appointed him, a permanent*, or *regular, allowance of bread &c.; I provided for him*, or *maintained him.*] — [اجراه مُجْرَى كَذَا †*He made it to be like*, or *similar to, such a thing in state, condition, case, or predicament.* †*He made it* (a word or phrase) *to follow the same rule* or *rules*, or *to occupy the same grammatical place, as such another.* †*He made it to act as*, or *in a*

similar manner to, such a thing.] — [Hence,] لَا يَنْصَرِفُ i. q. اِسْمٌ لَا يُجْرَى † [A noun that is imperfectly declinable]. (TA in art. صمت, &c.) ═ أَجْرَتْ said of a herb, or leguminous plant, (بَقْلَة,) mentioned in this art. in the Ḳ: see art. جرو.

6. تَجَارَوْا فِى الحَدِيثِ: see 3. Hence, in a trad., تَتَجَارَى بِهِمُ الأَهْوَآءُ † [*Natural desires, or blamable inclinations, or erroneous opinions, contend with them for the mastery:* or] *they vie, or compete, one with another, in natural desires,* &c. (TA.) ═ See also 1.

10. اِسْتَجْرَاهُ *He demanded, or desired, that he should run.* (TA.) — See also 2, in two places.

لَا جَرَ and لَا ذَا جَرَ, for لَا جَرَمَ and لَا ذَا جَرَمَ: see art. جرم.

جُرَّةٌ and ↓ جِرَايَةٌ: see 1 in art. جرأ.

جَرَى: see جَرَآءٌ. ═ فَعَلْتُهُ مِنْ جَرَاكَ, and من ↓ جَرَائِكَ, *I did it because of thee,* or *of thine act; on thine account;* or *for thy sake; i. q.* من أَجْلِكَ; like من جَرَّاكَ [which see in art. جر]. (Ṣ, Ḳ.)

جَرْيَةٌ *i. q.* جَرْىٌ as inf. n. of جَرَى said of water (Mṣb, Ḳ) and the like: (Ḳ:) and also *A mode, or manner, of running* [thereof]. (TA.) You say, مَا أَشَدَّ جَرْيَةَ هٰذَا المَآءِ [*How vehement is the running, or manner of running, of this water!*]. (Ṣ.)

جَرَآءٌ and ↓ جِرَآءٌ (Ṣ, Ḳ) and ↓ جَرًى (Ḳ) and ↓ جَرَايَةٌ (Ṣ, Ḳ) and ↓ جَرَائِيَةٌ (IAạr, Ḳ, TA, [in the CḲ جِرَائِيَةٌ]) *Girlhood;* the *state of a* جَارِيَة. (Ṣ, Ḳ.) One says, كَانَ ذٰلِكَ فِى أَيَّامِ جَرَائِهَا *That was in the days of her girlhood.* (Ṣ.) ═ فَعَلْتُهُ مِنْ جَرَائِكَ: see جَرَى.

جِرَآءٌ: see the next preceding paragraph.

جَرِىٌّ *A commissioned agent; a factor; a deputy:* (Ṣ, Mgh, Ḳ:) because he runs in the affairs of him who appoints him, (Mgh,) or acts in his stead (يَجْرِى مَجْرَاهُ): (Ṣ, Mgh:) [in this and other senses following] used alike as sing. and pl., and also as [masc. and] fem.: (Ḳ:) but sometimes, though rarely, جَرِيَّةٌ is used for the fem., accord. to AḤát; and accord. to J [in the Ṣ, and Mṭr in the Mgh], it has أَجْرِيَآءُ for its pl. (TA.) And *A messenger,* or *person sent,* (Ṣ, Ḳ,) *that runs in an affair.* (TA.) But accord. to Er-Rághib, it is weaker [in signification, or in point of chasteness,] than رَسُولٌ and وَكِيلٌ [which are given as its syns. in the Ṣ and Ḳ]. (TA.) — *A servant.* (TA.) — *A hired man; a hireling.* (Kr, Ḳ.) — *A surety; a guarantee; one who is responsible, accountable,* or *answerable, for another.* (IAạr, Ḳ.) ═ The word signifying "bold," or "daring," is جَرِىْءٌ, with ء. (Ṣ.)

جِرَايَةٌ: see جَرَايَةٌ: ═ and جَرَآءٌ: ═ and جُرَّةٌ.

جَرَايَةٌ *The office of a* جَرِىّ, i. e. *a commissioned agent, factor, or deputy;* (Ṣ, Ḳ;) and *of a messenger:* (Ṣ:) as also ↓ جِرَايَةٌ. (TA.) ═ *A running* [or *permanent*] *daily allowance of food or the like.* (Ṣ, TA.) [Hence, in the present day, خُبْزُ جِرَايَة *Bread made of inferior flour, for servants and other dependants.*]

جَرَائِيَةٌ: see جَرَآءٌ.

جِرِيَّا: see إِجْرِيَّا.

جِرِّىٌّ [The eel;] *a certain fish, well known.* (Ḳ: mentioned also in art. جر, q. v.)

جِرِّيَّةٌ, like قِرِّيَّةٌ, (Ṣ,) The *stomach,* or *triple stomach,* or the *crop,* or *craw,* of a bird; syn. حَوْصَلَةٌ: (Ṣ, Ḳ: mentioned also in art. جر, q. v.:) so called because the food at the last runs into it, or because it is the channel through which the food runs: (Er-Rághib, TA:) thus pronounced by Fr, and by Th on the authority of Ibn-Nejdeh, without ء: by Ibn-Hání, [جِرِّيْئَةٌ,] with ء, on the authority of AZ. (TA.)

جَارٍ applied to water [and the like], [*Running,* or *flowing,* or] *pressing forward, in a downward* and *in a level course.* (Mṣb.) — Also, [as meaning *Running,*] applied to a horse and the like. (Mṣb.) — صَدَقَةٌ جَارِيَةٌ † *A permanent, continuous, charitable donation; such as the unalienable legacies provided for various benevolent purposes.* (TA.)

جَارِيَةٌ *A ship;* (Ṣ, Mṣb, Ḳ;) because of its running upon the sea: (Mṣb:) an epithet in which the quality of a subst. predominates: pl. جَوَارٍ. (TA.) — The *sun;* (Ḳ;) because of its running from region to region: (TA:) or the *sun's disk in the sky.* (T, TA.) And الجَوَارِى الكُنَّسُ *The stars.* (TA. [But see art. كنس.]) — The *wind:* pl. as above. (TA.) — *A girl,* or *young woman;* (Ṣ,* Mgh, Mṣb,* Ḳ;) *a female of which the male is termed* غُلَامٌ; so called because of her activity and running; opposed to عَجُوزٌ: (Mgh:) and ‡ *a female slave;* (Mgh voce غُلَامٌ;) [in this sense] applied even to *one who is an old woman, unable to work,* or *to employ herself actively;* alluding to what she was: (Mṣb:) pl. as above. (Mṣb, Ḳ.) — † The *eye* of any animal. (TA.) — † *A benefit, favour, boon,* or *blessing, bestowed by God* (Ḳ, TA) *upon his servants.* (TA.)

إِجْرِىٌّ *A kind of running:* pl. أَجَارِىُّ. (TA.) You say فَرَسٌ ذُو أَجَارِىَّ *A horse that has several kinds of running.* (TA.) — See also إِجْرِيَّا.

إِجْرِيَّةٌ: } see what next follows.
إِجْرِيَآءُ: }

إِجْرِيَّا The *act of running:* (Ṣ, and so in some copies of the Ḳ: [in this sense, erroneously said in the TA to be بتخفيف اليآء:]) or ↓ إِجْرِىٌّ. (So in this sense in some copies of the Ḳ.) — Also, (Ṣ, Ḳ,) and ↓ إِجْرِيَآءُ, (Ḳ,) † *A custom,* or *habit,* (Ṣ,) or *manner,* (Ḳ,) *that one adopts* (Ṣ, Ḳ) *and follows;* (Ḳ;) [like هِجْرِيَّا &c.;] and so ↓ إِجْرِيَةٌ without teshdeed: (TA:) and † *nature, constitution,* or *natural disposition;* [in the CḲ, الخَلْقُ is erroneously put for الخُلُقُ;] as also ↓ جِرِيَةٌ and ↓ إِجْرِيَّةٌ. (Ḳ.) One says, الكَرَمُ مِنْ إِجْرِيَّاهُ and ↓ من إِجْرِيَائِهِ † *Generosity is* [a quality] *of his nature,* &c. (Lḥ, TA.)

إِجْرِيَةٌ: see what next precedes, in two places.

مَجْرًى [*A place, and a time, of running,* &c.]. The *channel* of a river [and of a torrent &c.: *a conduit; a duct; any passage through which a fluid runs:* pl. مَجَارٍ]. (TA.) — Also an inf. n. of 1 [q. v.]. (Ṣ, Ḳ, &c.)

مُجْرٍ [*Making to run*]. It is said in a prov., كُلُّ مُجْرٍ فِى الخَلَآءِ يُسَرُّ [*Every one who makes* his horse *to run in the solitary place rejoices,* because no one can contradict his account of his horse's fleetness]. (Mgh.) [See Freytag's Arab. Prov., ii. 315 and 316, where two other readings are added: كلّ مجر بخلاءٍ مُجِيدٌ, i. e., *is possessor of a fleet horse;* and كلّ مجر بخلاءٍ سَابِقٌ, i. e., *is one who outstrips.*]

مَاجَرَيَاتٌ: see 1.

جز

1. جَزَّ, (Ṣ, A, Mṣb, Ḳ,) aor. ـُ, (Ṣ, Mṣb,) inf. n. جَزٌّ, (Ṣ, Mgh, Mṣb, Ḳ,) and ↓ جَزَّةٌ, (Ḳ,) [but the latter seems to be an inf. n. of un.,] *He cut* (Mgh, Mṣb, Ḳ) wool, (IDrd, Ṣ, Mgh, Mṣb,) [see حَلَقَ,] and, as some say, other things, (Mṣb,) or a dense thing, (Mgh,) or hair, (A, Ḳ,) and dry herbage, (Ḳ,) and seed-produce, (A,) and wheat, (Ṣ,) and palm-trees, (Ṣ, ISd, A, Mgh,) meaning their fruit; (Mgh;) as also ↓ اِجْتَزَّ. (Ḳ.) You say, جَزَزْتُ الشِّيحَ وَغَيْرَهُ, and ↓ اِجْتَزَزْتُهُ, in the sense of جَزَزْتُهُ [*I cut the sheeḥ,* a species of wormwood, *&c.*]. (Ṣ.) You say also, جَزَزْتُ الكَبْشَ وَالنَّعْجَةَ [*I shore,* or *sheared, the ram and the ewe*]; but of the she-goat and he-goat you say, حَلَقْتُهُمَا. (TA.) And جَزَّ النَّخْلَةَ, (Lḥ, A, Mgh,) aor. ـِ, inf. n. جَزٌّ and جَزَازٌ and جِزَازٌ, (Lḥ, TA,) [like جَدَّ and جَذَّ,] *He cut off the fruit of the palm-tree.* (Mgh, TA.) ═ See also 4, in four places.

2. جَزَّزَ, inf. n. تَجْزِيزٌ, *He dried* dates. (Mṣb.)

4. اجزّ *It attained to the proper time for being cut;* (Ṣ, Mṣb, TA;) said [app. of wool, and] of hair, and of herbage, (A,) and wheat, (AZ, Ṣ, Mṣb,) and barley; (AZ, Mṣb;) as also ↓ اِسْتَجَزَّ, (Ṣ, Mṣb, Ḳ,) said of wool, (Mṣb,) and of wheat; (Ṣ, Ḳ;) and ↓ جَزَّ, [aor., app., ـِ, as below,] said of wheat. (TA.) — اجزّ الشِّيحُ *The sheeḥ* [a species of wormwood] *attained to the proper time for being cut:* (L, TA:) or اجزّ الشَّيْخُ † *The old man attained to the proper time for dying.* (Ḳ.) [SM says,] الشَّيْخُ seems to be a mistranscription, for الشِّيحُ: if not, it is a tropical expression. (TA.) [But see 4 in art. جزر, and 8 in art. خضر.] — اجزّ الغَنَمُ *The sheep attained to the proper time for being s'orn;* (Ṣ, TA;) as also ↓ جَزَّ. (TA.) — اجزّ النَّخْلُ *The palm-trees attained to the proper time for having their fruit cut off;*

(Ṣ, Ḳ, TA;) as also ↑جَزَّ. (Ḳ.) —— [Hence, app.,] اجزّ التَّمْرُ The dates dried; (Ṣ, Ḳ;) as also ↑جَزَّ, aor. ـُ, (Ṣ, Mṣb, Ḳ,) inf. n. جُزُوزٌ, (Ṣ, Ḳ,) or جَزَّ: (Mṣb:) you say تَمْرٌ فِيهِ جُزُوزٌ *Dates in which is dryness.* (Ṣ, TA.) —— اجزّ القَوْمُ *The people attained to the proper time for the shearing of their sheep:* (Ḳ:) or *had their sheep shorn:* or *had their seed-produce cut.* (Ṣ, L.) ══ اجزّ الرَّجُلَ *He assigned to the man the* جِزَّة [or *wool, &c.,*] *of a sheep.* (Ḳ.)

8. اجتزّ and اجدزّ: see 1, in three places.

10: see 4.

جِزٌ i. q. جُزَّةٌ. (Bḍ in xv. 44.)

جَزَّةٌ [*A single cutting, or shearing, or the like.*]. (Ḳ.) See 1.

جِزَّةٌ [*A mode, or manner, of cutting, or shearing, or the like*]. You say, جَزَّهُ جِزَّةً حَسَنَةً [*He cut it, or sheared it, &c., in a good manner*]. (Ḳ, TA.) [In the CḲ, وَجِزَّةٌ is omitted after جِزَّةٌ حَسَنَةٌ; so that the reading there is جِزَّةٌ حَسَنَةٌ وَجِزَّةٌ.] ══ Also, (Ṣ, Ḳ,) and ↑جَزَزٌ and ↑جُزَازٌ and ↑جُزَازَةٌ, (Ḳ,) *What is cut,* of dates: (Ḳ:) or [*a fleece;* i. e.,] the *wool of a sheep* [*shorn*] *in one year:* (Ṣ, Ḳ:) [and the *hair of a goat, and of a camel,* (see تِمَّةٌ,) *when shorn; a shorn crop of wool,* and *of goat's hair,* and *of camel's hair:*] or the *wool of a ewe,* (Ḳ, TA,) or *of a ram,* (TA,) *when shorn, and not mixed with any other:* (Ḳ, TA:) so accord. to AHát: (TA:) or *wool not used after being shorn;* (Ḳ;) in which last sense you say also ↑صُوفٌ جَزَزٌ: (TA:) pl. جِزَزٌ and جَزَائِزُ: (Lḥ, Ḳ:) the latter like ضَرَائِرُ as pl. of ضَرَّةٌ, without regard to the difference of the vowels [in the sing.]. (TA.) One says to another, أَقْرِضْنِى جِزَّةً أَوْ جِزَّتَيْنِ [*Lend thou me the wool of a sheep, or of two sheep*]; and the latter gives him the wool of a sheep, or of two sheep. (Ṣ.) And one says of a man with a large beard, كَأَنَّهُ عَاضٌّ عَلَى جِزَّةٍ [*As though he were biting*] *the wool of a shorn sheep.* (Ḳ, TA: in the CḲ, عَلَى جَزَّةٍ.)

جَزَزٌ: see جِزَّةٌ, in two places.

جَزَازٌ and ↑جِزَازٌ The *act of cutting,* or *shearing, wool,* (Mgh,* Mṣb,) and *hair;* (Mgh;) and *of reaping;* (Fr, Ṣ, Ḳ;) and *of cutting seed-produce* (Mgh, Ḳ) *before it has attained to maturity;* (Ḳ;) and *of cutting off the fruit of palm-trees:* (Fr, Ṣ, Mgh:) and the *time,* or *season, of shearing sheep* [and *the like*]. (TA.) You say, هٰذَا زَمَنُ الجَزَازِ, and الجِزَازِ, *This is the time of cutting,* or *shearing, wool:* (Mṣb:) and *of reaping:* and *of cutting off the fruit of palm-trees.* (Ṣ.)

جُزَازٌ *What one cuts,* or *cuts off,* of anything; (Ḳ;) the *cuttings* of wool or any other thing: n. un. with ة: (TA:) [as, for instance,] *what is redundant of a skin or hide when it is cut:* (Ḳ:) or ↑جُزَازَةٌ signifies *what falls from a skin or hide* (Ṣ, A) *or other thing* (Ṣ) *when it is cut.* (Ṣ, A.) See also جِزَّةٌ.

جِزَازٌ: see جَزَازٌ.

جَزُوزٌ *What is cut,* or *shorn;* a masc. n.; and, as also ↑جَزُوزَةٌ, fem.: (Ḳ:) or the latter signifies *sheep of which the wool is shorn;* (Ṣ, A;) and is similar to رَكُوبَةٌ and حَلُوبَةٌ and عَلُوفَةٌ: (Ṣ:) Th says that a subst. of this class is only with ة, like the three words just mentioned: Lḥ says that it may be with ة and without ة; and that the pl. in both cases is of the measures فُعُلٌ and فَعَائِلُ: [in this instance, جُزُزٌ and جَزَائِزُ:] but ISd says, I hold that فُعُلٌ is the measure of the pl. of a subst. of this class without ة, such as رَكُوبٌ, of which the pl. is رُكُبٌ; and فَعَائِلُ, of that which is with ة, such as رَكُوبَةٌ, of which the pl. is رَكَائِبُ. (TA.)

جَزِيزٌ and ↑مَجْزُوزٌ *Cut,* or *shorn.* (Ḳ.) ══ See also جَزِيزَةٌ.

جُزَازَةٌ: see جِزَّةٌ: —— and جُزَازٌ.

جَزُوزَةٌ: see جَزُوزٌ.

جَزِيزَةٌ *A flock,* or *tuft, of wool;* as also ↑جُزْجُزَةٌ; (Ṣ, Ḳ;) which [latter] is *a tuft of wool,* or *of wool dyed of various colours,* (عِهْنَةٌ,) *that is hung upon a woman's camel-vehicle* (هَوْدَج): (Ṣ:) or the latter signifies *a tuft of wool tied with threads or strings, with which the woman's camel-vehicle* (هودج) *is ornamented:* and جَزَائِزُ [pl. of the former] and جَزَاجِزُ [pl. of the latter] signify *tufts of dyed wool which are hung upon the camel-vehicles* (هَوَادِج) *of women on the day of going forth on a journey;* also called ثُكَن: or جَزَائِزُ, (TA,) or ↑جَزِيزٌ, (L,) signifies *a kind of beads* (خَرَز) *with which the girls,* or *slave-girls,* (جَوَارٍ,) *of the Arabs of the desert are adorned, resembling* جَزْع: or *tufts of wool,* or *of wool dyed of various colours,* (عِهْن,) *which were used in the place of anklets.* (TA.)

جُزْجُزَةٌ: see جَزِيزَةٌ.

مِجَزٌّ *An instrument for cutting or shearing.* (Ṣ.)

مَجْزُوزٌ: see جَزِيزٌ.

جزأ

1. جَزَأَهُ, (Ṣ, Mṣb, Ḳ,) aor. ـَ, (Mṣb, Ḳ,) inf. n. جَزْءٌ, (Ṣ,) *He divided it* (a thing, Ṣ) *into parts,* or *portions;* (Ṣ, Ḳ;) *made it to consist of parts,* or *portions;* (Ṣ, Mṣb;) as also ↑جزّأه, (Ṣ,* Mṣb, Ḳ,) inf. n. تَجْزِئَةٌ, (Ṣ,) or تَجْزِىءٌ: (Mṣb:) when that which is divided is property, as, for instance, slaves, only this latter form of the verb, with teshdeed, is used. (TA.) —— Also, aor. and inf. n. as above, *He took a part,* or *portion, of it;* namely, a thing. (Ḥam p. 117.) And جَزَأَ الشِّعْرَ, inf. n. as above; and ↑جزّأه; *He curtailed the poetry of two feet in each verse:* or *he made the poetry to consist of two feet in each verse.* (TA. [See مَجْزُوءٌ.]) ══ Also *He made it firm, fast,* or *strong;* or *he bound it firmly, fast,* or *strongly;* (شَدَّهُ;) namely, a thing. (Ḳ.) ══ جَزَأَ بِهِ, (Ṣ, Ḳ,) aor. ـَ, (TA,) inf. n. جَزْءٌ, (Ṣ,) [and app. جُزْءٌ also,] *He was,* or *became, satisfied,* or *content, with it;* namely, a thing; (Ṣ, Ḳ;) as also جَزِىَ, a dial. var. mentioned by IAạr; (TA;) and ↑اجتزأ به, (Ṣ, Mṣb, Ḳ,) and ↑تجزّأ به. (Ṣ, Ḳ.) A poet says,

وَإِنَّ المَرْءَ يَجْزَأُ بِالكُرَاعِ *

[*And verily the man is satisfied,* or *content, with the shank of the sheep* or goat &c.]. (TA.) And you say طَعَامٌ لَا جُزْءَ لَهُ *Food whereof one is not satisfied with a little.* (TA.) And لَهُ فِى هٰذَا غَنَآءٌ وَجَزْءٌ [*He has, in this, competence and*] *sufficiency.* (Mgh.) And جَزَأَتِ الإِبِلُ بِالرُّطْبِ عَنِ المَآءِ, (Ṣ, Mgh, Ḳ,) or [simply] جَزَأَتِ الإِبِلُ, (Ḥar p. 475,) inf. n. جُزْءٌ, with ḍamm, (Ṣ, TA,) and جُزُوءٌ; (TA;) and جَزِئَتْ, (IAạr, Ḳ,) and ↑اجتزأت; (Mgh, and Ḥar ubi suprà;) *The camels were satisfied,* or *content, with green,* or *fresh, pasture* or *herbage* [*so as to be in no need of water*]. (Ṣ, Mgh, Ḳ, TA.) And ↑اجتزأ عَنِ امْرَأَتِهِ [*He was content to abstain from,* or *be without, conjugal intercourse with his wife*]. (M in art. ابل.)

2: see 1, in two places: ══ and see also 4.

4. اجزأهُ *It* (a thing) *satisfied, sufficed,* or *contented, him.* (Ṣ, Mgh, Ḳ.) [Hence,] اجزأ مُجْزَى غَيْرِهِ [or مُجْزَأَ غَيْرِهِ] *It* (a thing) *satisfied, sufficed,* or *contented, in lieu of another thing* or *other things; stood,* or *served, in stead thereof.* (Mṣb.) And أَجْزَأْتُ عَنْكَ مُجْزَأَ فُلَانٍ (Ṣ, Mgh, Ḳ) and مَجْزَأَ فلان and مَجْزَأَةَ فلان and مُجْزَأَةَ فلان, (Ṣ, Ḳ,) as also مُجْزَى فلان and مُجْزَاةَ فلان without ء and with ḍamm, and مَجْزَى فلان and مَجْزَاةَ فلان, (Ḳ in art. جزى,) *I satisfied, sufficed,* or *contented, thee as such a one; I stood thee,* or *served thee, in stead of such a one.* (Ṣ, Mgh, Ḳ.) And اجزأ الإِبِلَ بِالرُّطْبِ عَنِ المَآءِ, (Ṣ, Ḳ,) inf. n. إِجْزَآءٌ; (TA;) and ↑جزّأها, (Ṣ, Ḳ,) inf. n. تَجْزِئَةٌ, (Ṣ,) or تَجْزِىءٌ; (TA;) *He satisfied,* or *contented, the camels with green,* or *fresh, pasture* or *herbage* [*so that they were in no need of water*]. (Ṣ, Ḳ.) —— اجزأ is also *syn. with* جَزَى; the former being of the dial. of Temeem, and the latter of the dial. of El-Ḥijáz; (Akh, Mṣb;) and one may suppress the ء, and say أَجْزَى: (Mgh, Mṣb:) this last is used by some of the lawyers in the sense of [جَزَى, i. e.] قَضَى. (Az, Mgh, Mṣb.) One says, أَجْزَأَتْ عَنْكَ شَاةٌ *A sheep,* or *goat, made satisfaction for thee* (Ṣ, Mṣb,* Ḳ, TA) as a sacrifice; (TA;) syn. قَضَتْ; (Ṣ, Mṣb, Ḳ;) the verb being here a dial. var. of جَزَتْ. (Ṣ, Ḳ.) And البَدَنَةُ تُجْزِئُ عَنْ سَبْعَةٍ *The camel,* or *cow, makes satisfaction for seven:* or *serves in stead of seven.* (Mgh.) And هٰذَا يُجْزِئُ عَنْ هٰذَا [*This will make satisfaction, for this:* or *this will serve in stead of this*]: and, accord. to 'Alee Ibn-'Eesà, يُجْزِى also, suppressing the ء. (Mgh.) —— Also, said of pasture, or herbage, (Ḳ, TA,) and of a meadow, (TA,) † *It was,* or *became, luxuriant:* (Ḳ, TA:) because satisfying the beasts that feed upon it. (TA.) —— And, said of a company of men, *They had their camels satisfied with green,* or *fresh, pasture* or *herbage* [*so that they were in no need of water*]. (TA.) ══ أَجْزَأَتْ *She* (a woman) *brought forth*

females. (Ḳ. [But see جُزْءٌ, from which it is derived.]) ═ اجزأ *He furnished* an awl (مِخْصَف, Ṣ, Ḳ, or إِشْفَى, Ṣ), (Ṣ, Ḳ,) or a knife, (Mṣb,) *with a* جُزْأَة, i. e. *handle;* (Ṣ, Mṣb, Ḳ;) as also اجزى. (Mṣb.) — اجزأ الخَاتَمَ فِى إِصْبَعِهِ *He put the ring upon his finger.* (Ḳ.)

5. تجزّأ *It became divided into parts,* or *portions.* (Mṣb, KL.) ═ See also 1.

8: see 1, in three places.

جُزْءٌ: see جُزْءٌ ═ It is said by El-Khaṭṭábee to be a name for رُطْب [app. meaning رُطْبٌ, i. e. *Green,* or *fresh, pasture* or *herbage,* (see 1 and 4,)], with the people of El-Medeeneh; and occurs in a trad.; but the reading commonly known is جرو. (TA.)

جُزْءٌ *A part,* or *portion,* (Mṣb, Ḳ, TA,) or *division,* (TA,) of a thing; (Mṣb, TA;) properly and conventionally; (TA;) as also ▼ جُزْءٌ; (Ḳ;) *a constituent part* of a thing, as of a ship, and of a house or tent, and of a sum in reckoning; (B, TA;) [*an ingredient* of any compound or mixture;] *a share,* or *lot:* (TA:) pl. أَجْزَآءٌ: (Ṣ, Mṣb, Ḳ, &c.:) it has no other pl. (Sb, TA.) — [*A volume* of a book.] — *A foot* of a verse. (TA.) — In the Ḳur [xliii. 14], where it is said, وَجَعَلُوا لَهُ مِنْ عِبَادِهِ جُزْءًا, (Ḳ, TA,) or, as some read, جُزُؤًا, (Bḍ,) it means *Females;* (Ḳ, TA;) i. e., they asserted the angels to be the daughters of God: so says Th: and Aboo-Is-ḥáḳ says that it means, they asserted God's share of offspring to be the females; but that he had not found this in old poetry, nor had persons worthy of confidence related it on the authority of the Arabs [of the classical times]: Z disallows it, asserting it to be a lie against the Arabs; and Bḍ follows him: El-Khafájee says that the word may be used figuratively; for, as Eve was created of a part (جُزْء) of Adam, the word جزء may be applied to denote the female. (MF, TA.)

جُزْأَةٌ The *handle* of the [kind of awl called] مِخْصَف, (Ṣ, Ḳ,) and of the إِشْفَى: (Ṣ:) AZ says that it is not [the handle, or hilt,] of the sword, nor of the dagger; but is the *handle* of the مِيْسَمَة with which camels' feet are branded. (TA.) [See also ضَبَّةٌ.] — *A vine-prop;* (Ḳ, TA;) *a piece of wood with which a vine is raised from the ground.* (TA.) — In the dial. of the tribe of Sheybán, The *hinder,* or *hindermost,* شُقَّة [or *oblong piece of cloth*] *of a tent.* (TA.)

[جُزْئِىٌّ *Relating to a part* or *portion* or *division; partial; particular; contr. of* كُلِّىٌّ. — And, as a subst., *A particular:* pl. جُزْئِيَّاتٌ.]

[جُزْئِيَّةٌ The *quality of relating to a part* or *portion* or *division; relation to a part* &c.; *particularity.*]

جَزِىٌّ *Satisfying* food; as also ▼ مُجْزِئٌ; (Fr, Ḳ;) like شَبِيعٌ and مُشْبِعٌ. (Fr, TA.)

جَازِئٌ [act. part. n. of 1]. — هٰذَا رَجُلٌ جَازِئُكَ مِنْ رَجُلٍ *This is a man sufficing thee as a man.* (Ḳ,* TA.) — ظَبْيَةٌ جَازِئَةٌ *A doe-gazelle that is satisfied with green,* or *fresh, pasture* or *herbage* [*so as to be in no need of water*]: pl. جَوَازِئُ. (Ṣ.) The pl. is explained by IḲṭ as meaning *Gazelles:* (TA:) [or] it signifies [or signifies also] *Wild bulls* or *cows;* (Ḳ, TA;) because they are satisfied with green, or fresh, pasture or herbage so as to be in no need of water. (TA.) Also, the pl., *Palm-trees;* as not needing irrigation. (TA.)

أَجْزَأُ *More* [and *most*] *satisfying* or *sufficing* or *satisfactory:* hence, الفَارِسُ أَجْزَأُ مِنَ الرَّاجِلِ [*The horseman is more satisfactory than the footman*]. (Mgh.)

مَجْزَأٌ and مُجْزَأٌ are used as inf. ns. of 4 [q. v.]. (TA.)

مُجْزِئٌ: see جَزِىٌّ. — Also A *strong, fat,* camel; because sufficing for the wants of the rider and carrier. (TA.) ═ Also, and مُجْزِئَةٌ, A woman *who brings forth females.* (TA. [But see جُزْءٌ, from which the verb is derived.])

مَجْزَأَةٌ and مُجْزَأَةٌ are used as inf. ns. of 4 [q. v.]. (TA.)

مُجَزَّأٌ *Divided into parts,* or *portions.* (TA.) — [*Having a part,* or *portion, taken from it:* see 1.] — A verse *curtailed of two* [*of the original*] *feet:* [like the هَزَج and مُضَارِع &c., which were originally of six feet each, but of which every known example is of four only:] or a verse *consisting of two feet only:* [as a kind of the رَجَز, and two kinds of the مُنْسَرِح: to each of which, or, accord. to some, to the former of which only, when thus consisting of only two feet, the term مَنْهُوك is also applied:] the former is said to be عَلَى السَّلْبِ; and the latter, عَلَى الوُجُوبِ. (TA.)

جزر

1. جَزَرَ, aor. ـِ and sometimes ـُ, (Ḳ,) inf. n. جَزْرٌ, (Mgh, Ḳ,) *He cut,* or *cut off,* (Mgh, Ḳ,) a thing. (TA.) — جَزَرَ, (Ṣ, Mgh, Mṣb,) aor. ـُ, (Ṣ, Mṣb,) [inf. n. as above,] *He slaughtered* a camel (Ṣ, Mgh, Mṣb) or other animal, (Mṣb,) *and skinned* it; (Ṣ;) as also ▼ اجتزر. (Ṣ, TA.) You say also, جَزَرَ لَهُمْ, meaning *He slaughtered for them a camel.* (A.) And ▼ اجتزر القَوْمَ جَزُورًا *He slaughtered and skinned for the people a camel.* (TA.) — جَزَرَ النَّخْلَ, aor. ـُ, (Ṣ, Ḳ) and ـِ, (Ḳ,) inf. n. جَزْرٌ (Ṣ, Ḳ) and جَزَارٌ and جِزَارٌ, (Lḥ, Ḳ,) *He cut off the fruit of the palm-trees:* (Lḥ, Ṣ, Ḳ:) or, as some say, *he spoiled the palm-trees in fecundating them.* (TA.) — And جَزَرَ, (TA,) inf. n. جَزْرٌ, (Ḳ,) *He gathered* honey from the hive. (Ḳ, TA.) ═ جَزَرَ, aor. ـُ and ـِ, inf. n. جَزْرٌ, (Ṣ, Mṣb, Ḳ, &c.,) ‡ *It* (water) *sank, and disappeared; became low;* or *became remote;* (Ṣ, Ḳ;) *decreased; went away;* (TA;) *flowed away,* or *retired,* (A, Mgh,* Mṣb,) from the earth, or land: (A, Mgh:) *it* (the sea, and a river, Lth, ISd) *ebbed; contr. of* مَدَّ; (Ṣ, ISd, Ḳ; [but in this last sense, only ـِ is authorized by the Ḳ, and app. by ISd also, as the aor.;]) i. e., *retreated,* or *went back;* (Ṣ, Mṣb;*) as also ▼ انجزر; (ISd, TA;) or *ceased to increase.* (Lth, Mgh.)

4. اجزرهُ, (Ḳ,) or اجزر القَوْمَ, (ISk, Ṣ,) *He gave to him,* or *to the people, a sheep* or *a goat, for him,* or *for them, to slaughter;* (ISk, Ṣ, Ḳ;) meaning *a ewe* or *a ram* or *a she-goat;* (ISk, Ṣ;) or *a sheep,* or *goat, fit for slaughter:* (TA:) and أَجْزَرْتُهُ شَاةً *I gave to him a ewe* or *a ram* or *a she-goat, and he slaughtered it:* (ISk, TA:) and أَجْزَرْتُكَ بَعِيرًا, or شَاةً, *I gave to thee a camel,* or *a sheep* or *goat, that thou mightest slaughter it:* (A:) [but] accord. to ISk, one does not say أَجْزَرْتُهُمْ نَاقَةً, because a she-camel is fit for other purposes than that of slaughter: (Ṣ:) and accord. to some, one should not say اجزرهُ جَزُورًا, but اجزرهُ جَزَرَةً. (TA.) ═ اجزر *He* (a camel) *attained to the fit time for his being slaughtered.* (Ṣ, Ḳ.) — اجزر النَّخْلُ *The palm-trees attained to the fit time for the cutting off of the fruit.* (Ṣ, Ḳ.) — [And hence,] اجزر الشَّيْخُ ‡ *The old man attained to the fit time for his dying;* (Ḳ, TA;) being aged, and near to his perishing; like as the palm-tree attains to the fit time for having its fruit cut off. (TA.) Youths used to say to an old man أَجْزَرْتَ يَا شَيْخُ, meaning, *Thou hast attained to the fit time for thy dying, O old man:* and he would say, أَىْ بَنِىَّ وَتُخْتَضَرُونَ, i. e., "[O my sons, and] ye shall die youths:" but accord. to one way of relation, it is أَجْزَزْتَ; from أَجَزَّ البُرُّ "the wheat attained to the proper time for being cut." (Ṣ.) — اجزر القَوْمُ *The people attained to the fit time for the cutting off of the fruit of the palm-trees.* (Yz, TA.)

5: see 8.

6. تَجَازَرُوا ‡ *They reviled each other* (Ḳ, TA) *vehemently,* or *excessively.* (TA.)

7: see 1.

8. اجتزر: see 1, in two places. — اجتزروا فِى القِتَالِ and ▼ تجزّروا (Ḳ, TA) *They fought one another* [app. *so that they cut one another in pieces*]. (TA.) [In the Ḳ, this is immediately followed, as though for the purpose of explanation, by the words تَرَكُوهُمْ جَزَرًا لِلسِّبَاعِ أَىْ قِطَعًا: but there is evidently an omission in this place, at least of the conjunction وَ.] — And اجتزروا *They had a camel slaughtered for them.* (A.)

جَزْرٌ inf. n. of 1. (Ṣ, Ḳ, &c.) ═ And also ‡ The *sea* (Ḳ, TA) itself. (TA.)

جَزَرٌ, (not ▼ جَزْرٌ, Fr, Ṣ, [but see what follows,]) [a coll. gen. n.,] *Fat sheep or goats:* (Ṣ, Ḳ, TA:) n. un. جَزَرَةٌ: (Ṣ, Ḳ:) or *sheep,* or *goats, that are slaughtered;* (M;) as also ▼ جَزُورٌ: (Ḳ:) n. un. as above: (M: in the Ḳ جَزْرَةٌ:) or جَزَرَةٌ signifies *a sheep,* or *goat, fit for slaughter:* or *a sheep,* or *goat, to which the owners betake themselves and which they slaughter:* and *anything that is lawful to be slaughtered;* n. un. of جَزَرٌ, which is sometimes [written ▼ جَزْرٌ,] with fet-ḥ to the ز. (TA.) — جَزَرُ السِّبَاعِ *The flesh which beasts or birds of prey eat.* (Ṣ, Mgh.) One says, تَرَكُوهُمْ جَزَرًا (Ṣ, Ḳ) *They slew them:* (Ṣ:) or *they left them cut in pieces* لِلسِّبَاعِ [*for the beasts or birds of prey*]. (Ḳ.) And صَارُوا جَزَرًا لِلْعَدُوِّ [*They became a prey to the enemy, cut in pieces*]. (Mgh.) ═ See also جَزِيرَةٌ. ═ Also, and ▼جِزَرٌ, (Fr, Ṣ, Mṣb, Ḳ,) the latter with kesr to the ج, (Mṣb, Ḳ,) arabicized, (Ḳ,) from the Persian [گَزَر], (AḤn,) [coll. gen. ns., meaning *Carrots*, or the *carrot*;] *a certain root*, (أُرُومَة,) *which is eaten*, (Ṣ, Ḳ,) *well known*: (TA:) n. un. with ة; (Ḳ;) or جَزَرَةٌ: (Aṣ, Ṣ, Mṣb:) *the best kind is the red and sweet, which grows in winter: it is hot in the extreme of the second degree; moist in the first degree*; (TA;) *diuretic*; (Ḳ, TA;) *lenitive; emollient*; (TA;) *strengthening to the venereal faculty; emmenagogue: the putting of its pounded leaves upon festering ulcers is advantageous*: (Ḳ, TA:) *it is difficult of digestion; and engenders bad blood; but is made wholesome with vinegar and mustard.* (TA.) — [See also جِنْزَابٌ, in art. جزب.]

جَزَرٌ: see جَزَرٌ, in three places.

جِزَارٌ The *time of the cutting off of the fruit of palm-trees.* (Yz, TA.) [See also 1.]

جَزُورٌ *A camel* [*that is slaughtered*, or *to be slaughtered*]; (Ḳ;) applied to the *male* and the *female*: (Ṣ, Mṣb:) or (as some say, Ṣgh, Mṣb) properly *a she-camel that is slaughtered*: (Ṣgh, Mṣb, Ḳ:) but the former is the correct assertion; (TA;) though the word is fem., (IAmb, Ṣ, Mṣb, TA,) on the authority of hearsay; (TA;) therefore you say, رَعَتِ الجَزُورُ [*the camel for slaughter pastured*]: (IAmb, Mṣb:) or when used alone, it is fem., because what are slaughtered are mostly she-camels: (TA:) and when used as a common term, it implies the like of predominance [of the fem. gender]: (Ḥáshiyeh of Esh-Shiháb, TA:) [the shares into which the جزور is divided in the game called المَيْسِر are described voce بَدْءٌ:] pl. جُزُرٌ (Ṣ, Mṣb, Ḳ) and جَزَائِرُ and جُزُرَاتٌ, (Mṣb, Ḳ,) the last of which is pl. of جُزُرٌ, like as طُرُقَاتٌ is of طُرُقٌ. (TA.) — See also جَزَرٌ.

جُزَارَةٌ, of a camel, The *extremities*; (Ṣ, A;) namely, (Ṣ,) the *fore and hind legs*, (اليَدَانِ وَالرِّجْلَانِ, Ṣ, Ḳ,) *and the head*, (Ṣ,) or *neck*: (A, Ḳ:) because the slaughterer receives them; (Ṣ;) they being his hire, (Ṣ, Ḳ,) or right, (A,) not being included among the shares in the game called المَيْسِر. (TA.) But when a horse is said to be عَبْلُ الجُزَارَةِ, (Ṣ,) or ضَخْمُ الجُزَارَةِ, (M,) what is meant is thickness of the fore and hind legs, and abundance of sinews; and the head is not included, because largeness of the head, in a horse, is a fault. (Ṣ, M.)

جِزَارَةٌ The *trade of him who slaughters camels* (Mgh, Mṣb, Ḳ,* TA) *and other animals.* (Mṣb.)

جَزِيرَةٌ ‡ *An island; land in the sea* [or *in a river*], *from which the water has flowed away, so that it appears*; (Az, Mgh;) and in like manner, *land which a torrent does not overflow, but which it surrounds*; (Az, TA;) *land from which the tide retires*; as also ▼جَزَرٌ: (Ḳ:) so called because cut off from the main land: (Ṣ:) or because of the retiring of the water from it: (Mṣb:) pl. جَزَائِرُ: (Ṣ, Mgh:) [also, *a peninsula*:] and *a piece of ground* or *land*. (Kr, TA.)

جَزَّارٌ (Ṣ, A, Mgh, Mṣb, Ḳ) and ▼جَزِيرٌ (Ḳ) and ▼جَازِرٌ (A) *One who slaughters camels* (A, Mgh, Mṣb, Ḳ) *and other animals.* (Mṣb.)

جَزِيرٌ : } see what next precedes.
جَازِرٌ : }

مَجْزَرٌ, (Mṣb, Ḳ,) or مَجْزِرٌ, with kesr to the ز, (Ṣ, Ibn-Málik,) contr. to rule, as the aor. of the verb is with ḍamm, (Ibn-Málik, TA,) and sometimes ▼مَجْزَرَةٌ [or مَجْزِرَةٌ], (Mṣb,) *A place where camels are slaughtered*, (Ṣ, Mṣb, Ḳ,) *and other animals*, (Mṣb,) *namely, bulls and cows and sheep and goats, and where their flesh is sold*: pl. مَجَازِرُ. (TA.) In a trad. of 'Omar, persons are enjoined to avoid مَجَازِر, (Ṣ, TA,) meaning as above; because of their uncleanness; (TA;) or because the witnessing of the slaughter of animals hardens the heart and dispels mercy: (IAth, TA:) or the meaning is, *places of assembly*; because a camel is slaughtered only where people are collected together: (Ṣ, TA:) the ▼مَجْزَرَة is one of the places in which it is forbidden to perform the usual prayers. (Mgh.)

مَجْزَرَةٌ or مَجْزِرَةٌ: see مَجْزَرٌ, in two places.

جزع

1. جَزْعٌ [inf. n. of جَزَعَ] signifies The act of *cutting*; or *cutting off*. (TA.) [See also 8.] — [Hence,] جَزَعَ لَهُ جُزْعَةً مِنَ المَالِ *He cut off for him a portion of the property.* (Ṣ.) — And جَزَعَ الوَادِىَ, (Ṣ, Mṣb, Ḳ,) aor. َ, (Mṣb, Ḳ,) inf. n. جَزْعٌ, (Ṣ, Mṣb, Ḳ,) *He passed the valley to the other side*: (Mṣb:) or *he passed the valley* [*in any manner*]: (Ḳ:) or *he passed across it*; i. e., *crossed it*: (Ṣ, Ḳ:) and in like manner, الأَرْضَ *the land*: (Ḳ:) and المَفَازَةَ *the desert*: and المَوْضِعَ *the place.* (TA.) ═ جَزِعَ, (Ṣ, Mṣb, Ḳ,) aor. َ, (Mṣb, Ḳ,) inf. n. جَزَعٌ (Ṣ, Mṣb, Ḳ) and جُزُوعٌ, (Ḳ,) *He was*, or *became, impatient*, (Ṣ, Ḳ,) مِنَ الشَّىْءِ [*of the thing*]; (Ṣ;) and عَلَى فُلَانٍ [*on account of such a one*]; (Ṣ and Ḳ in art. اله, &c.;) جَزَع being the *contr. of* صَبْر: (Ṣ, Ḳ:) or *he had not sufficient strength to bear what befell him*, (O, Mṣb,) *and found not patience*: (Mṣb:) or *he manifested grief and agitation*: (TḲ:) or *he was*, or *became, affected with grief*: or *he was*, or *became, affected with most violent grief, such as prevented him and turned him from that to which he was directing himself*, or *from his object, and cut him off therefrom*: this meaning of cutting off being said by 'Abd-el-Ḳádir El-Baghdádee to be the primary signification. (TA.)

2. جَزَّعَ, inf. n. تَجْزِيعٌ, *It* (a full-grown unripe date) *became ripe to the extent of two thirds of it*: (Ṣ:) or *to the extent of half of it*; (Ḳ, TA;) *from the bottom*: (TA:) or *became partly ripe*: and in like manner one says of a grape. (TA.) — *It* (a watering-trough, or tank,) *had but little remaining in it.* (Ḳ,* TA.) — *He put a little water* into a skin. (TA.) ═ جَزَّعَ فُلَانًا, (Ḳ,) inf. n. as above, (TA,) *He caused the impatience* (جَزَع) *of such a one to cease*: (Ḳ:) *he said to him that which comforted him*, or *consoled him, and which caused his grief and fear to cease.* (IAth.)

4. اجزع جِزْعَةً, and جُزْعَةً, *He left*, or *caused to remain, a remainder*: (O, Ḳ:) or *less than half.* (TA.) ═ اجزعهُ *He caused him to be impatient*: (Ṣ, Ḳ:) or *he caused him to want sufficient strength to bear what befell him, and to be impatient.* (Mṣb.)

5: see 7, in two places. ═ تَجَزَّعُوا الغَنِيمَةَ *They divided among themselves the spoil.* (TA.)

7. انجزع *It* (a rope) *broke*, (Ḳ, TA,) *in any manner*: (TA:) or *broke in halves*; (Ḳ, TA;) but if it have broken at its extremity, one does not say انجزع. (TA.) And انجزعت العَصَا, and ▼تَجَزَّعَت, *The staff*, or *stick, broke* (Ḳ, TA) *in halves.* (TA.) ▼تَجَزَّعَ is also said of a spear, and of an arrow, &c., meaning *It broke in pieces.* (TA.)

8. اجتزعهُ *He broke it*, and *cut it off*: (Ḳ:) or *he broke it off*, and *cut it off, for himself*; namely, a branch, rod, or piece of wood, from a tree. (Ṣ.)

جَزْعٌ (Ṣ, Mṣb, Ḳ) and ▼جِزْعٌ, (Kr, Ḳ,) but IDrd ascribes the latter to the vulgar, (TA,) [The *onyx*; so called in the present day;] *certain beads*, or *gems*, (خَرَز,) (Mṣb,) the *beads*, or *gems*, (خَرَز, [here rendered by Golius "Murœna seu concha Veneris," though he also gives what I regard as the only correct signification, namely "onyx,"]) *of El-Yemen* (Ṣ, Ḳ) [and] *of China*, (Ḳ,) *in which are whiteness and blackness*, (Ṣ, Mṣb, Ḳ,) *and to which eyes are likened*, (Ṣ, Ḳ,) *and in particular*, by Imra-el-Ḳeys, *the eyes of wild animals*, because their eyes, while they are alive, are black, but when they die, their whiteness appears; (TA;) *a kind of stone having many colours, brought from El-Yemen and China*; (Ḳzw;) so called because *interrupted by various colours*; *its blackness being interrupted by its whiteness and its yellowness*: (IB:) 'Áisheh's necklace [which she lost on the occasion that subjected her to the accusation of adultery] was of جَزْع of Ḍhafári: (TA:) the wearing it in a signet induces anxiety, or disquietude of mind, and grief, and terrifying dreams, and altercation with men; and if the hair of one who experiences difficulty in bringing forth be wound upon it, she brings forth at once: (Ḳ: [and Ḳzw says the like, and more of a similar kind:]) n. un. جَزْعَةٌ (Mṣb, Ḳ,* TA) and جِزْعَةٌ. (Ḳ,* TA.) ═ See also what next follows.

جِزْعٌ, (Ṣ, O, L, Mṣb, Ḳ,) but AO says that it should be with fet-ḥ, [▼جَزْعٌ,] (Ḳ,) The *place of bending*, or *turning*, (مُنْعَطَف, Ṣ, Mṣb, Ḳ, or مُنْحَنًى, Aṣ, Ḳ,) *of a valley*: (Aṣ, Ṣ, Mṣb, Ḳ:) or the *middle thereof*: or the *place where it ends*: (IDrd, Ḳ:) or *its side*: (Mṣb:) or the *place of passing*, or *crossing*, of a valley: or *a widening part, of the narrow places*, thereof, *whether it produce plants &c. or do not produce them*: (TA:) or it is not so called unless [it be *a part*] *having width, and producing trees &c.*: (Mṣb, Ḳ:) or it may be *without plants*, or *herbage*, or *the like*: (TA:) or *a place, in a valley, in which are no trees*: (IAạr, Ḳ:) or *a place*, of a valley, *taking a round and wide form*: (TA:) pl. أَجْزَاعٌ. (Mṣb, Ḳ.) — *A place of alighting*, or *abiding, of a people*. (Ḳ.) — *Elevated land*, or *ground, by the side of which is a low*, or *depressed, part*. (Ḳ.) — *A bee-hive*: pl. as above. (Ibn-'Abbád, Ḳ.) — See also جَزْعٌ.

جَزِعٌ: see what next follows.

جَزِعٌ (Mṣb, Ḳ) and ↓ جَازِعٌ and ↓ جَزُعٌ (Ḳ) and ↓ جَزُوعٌ (Mṣb, Ḳ) and ↓ جُزَاعٌ (Ḳ) part. ns. of جَزِعَ; [*Impatient*; &c.;] (Mṣb, Ḳ;) but the last two have an intensive signification [*very impatient*, or *having much impatience*; &c.]. (IAạr.)

جُزْعَةٌ: see what next follows, in two places.

جِزْعَةٌ *A little*, or *small quantity*, of property, or wealth; and of water, (Ṣ, Ḳ,) remaining in a skin, (Lḥ, IDrd,) and in a leathern bottle, or other vessel, (IDrd,) and in a pool left by a torrent, but not in a well, (TA,) as also ↓ جُزْعَةٌ (IDrd, Ḳ) and [the dim.] ↓ جُزَيْعَةٌ, (IDrd,) and of milk, in a skin; (Lḥ;) or *a third part*, or *nearly that quantity*, of water, in a trough, or tank; (ISh;) or *a quantity* of water, and of milk, *less than the half of the skin* or *other vessel*, and *of the trough*; (TA;) and, as also ↓ جُزْعَةٌ, *somewhat remaining*; (O, Ḳ;) or the latter, particularly, of milk; (IAạr;) or both, accord. to some, [*a remainder consisting of*] *less than half*; (TA;) and the former, *a portion* [*not defined*] of property, or wealth; (Ṣ;) and particularly *a portion of a flock of sheep* or *goats*; (Aboo-Leylà, Ḳ;) as also ↓ جُزَيْعَةٌ; (Ṣ;) thus in the handwriting of Aboo-Sahl El-Harawee; but in the Mj of IF, ↓ جَزِيعَةٌ, of the measure فَعِيلَةٌ in the sense of the measure مَفْعُولَةٌ: (TA:) the pl. of جِزْعَةٌ is جِزَعٌ. (ISh.) — And [hence,] ‡ *A part*, or *portion*, of the night, (Ṣ, O, Ḳ,) *past* or *to come*, (TA,) *less than half*, (O, Ḳ,) *of the former part thereof* or *of the latter part*. (Ḳ.) — *A place in which is a collection of trees* (Ḳ, TA) *among which the camels or other beasts are made to rest at night from the cold, and are confined when they are hungry, or returning from water*, or *under rain*. (TA.) — Also n. un. of جَزْعٌ as syn. with جَزْعٌ. (TA.)

جُزَاعٌ: }
جَزُوعٌ: } see جَزِعٌ.

جَزِيعَةٌ and جُزَيْعَةٌ: see جِزْعَةٌ, in three places.

جَازِعٌ: see جَزِعٌ. — Also The *piece of wood which is placed in the trellis of a grape-vine, crosswise, upon which are laid the branches of the vine*; (Ṣ, Ḳ;) not known to Aboo-Sa'eed; (Ṣ;) *it is thus placed for the purpose of raising the branches from the ground*; and this piece of wood is also called خَشَبَةٌ جَازِعَةٌ; the latter word being thus used as an epithet. (TA.) Also *Any piece of wood that is put crosswise between two things for a thing to be borne upon it* (Ḳ, TA) is called its جَازِع. (TA.)

مُجَزَّعٌ *Interrupted by various colours* [*like the* جَزْع or *onyx*]: (IB:) or anything *in which are blackness and whiteness*; as also ↓ مُجْزَعٌ: (Ḳ:) and flesh-meat *in which are whiteness and redness*. (TA.) [Hence,] نَوًى مُجَزَّعٌ and ↓ مُجْزَعٌ *Date-stones of which some*, or *some parts, have been scraped*, or *abraded, so as to have become white, the rest being left of their* [*original*] *colour*: (Ḳ:) likened to the جَزْع. (TA.) And ↓ بُسْرٌ مُجْزِعٌ (Ṣ, Ḳ) and مُجَزَّعٌ; (Ḳ;) the former, says Sh, accord. to El-Ma'arree, but he adds that he himself held the latter to be the right; Az says that he heard the former from the people of Hejer, and it has the authority of A 'Obeyd; (TA;) *Full-grown dates that have ripened to the half*; (Ḳ, TA;) *from the bottom*: (TA:) or *to the extent of two thirds*: (Ṣ:) or *that have become partly ripe*: (TA:) fem. with ة: (Ṣ, Ḳ:) and in like manner you say ↓ تَمْرٌ مُتَجَزِّعٌ *dates that have ripened to the half*. (TA.)

مُجْزِعٌ: see مُجَزَّعٌ, in three places. — حَوْضٌ مُجْزِعٌ *A watering-trough*, or *tank, having but little water remaining in it*. (Ḳ.)

مُتَجَزِّعٌ: see مُجَزَّعٌ.

جزف

1. جَزَفَ [inf. n. of جَزَفَ] signifies The *taking* a thing [*in the manner termed*] جِزَافًا and مُجَازَفَةٌ [i. e. *by conjecture, not knowing the measure nor the weight*]: (Ṣ, TA:) or the *taking largely*, or *copiously*: (IF, Mṣb, TA:) and it is [from] a Persian word. (Mṣb. [See جُزَافٌ, below.]) And you say, جَزَفَ فِى الكَيْلِ, inf. n. جَزْفٌ, *He gave large measure*: (IKṭṭ, Mṣb:) and جَزَفَ لَهُ فِى الكَيْلِ *He gave him large measure*. (Jm, TA.)

3. جازف, (Mṣb, TA,) inf. n. مُجَازَفَةٌ (Ṣ, Mṣb, Ḳ, TA) and جِزَافٌ, (Ṣ, TA,) *He sold*, or *bought, a thing not knowing its measure nor its weight*: (Mṣb:) or *he conjectured in selling and buying*. (Ḳ.) — *He acted in an easy*, or *a facile, manner*, (Mṣb, TA,) in selling or buying. (Mṣb.) — And hence, جازف بنفسه ‡ *He perilled, endangered, jeoparded, hazarded*, or *risked, himself*; as though he acted in an easy manner with himself. (TA.)

5. تجزّف فيه *He picked out*, or *selected, the good in it*; syn. تَنَقَّدَ [in the CḲ, erroneously, تَنَقَّذَ]. (Ṣgh, Ḳ.)

8. اجتزف *He bought* a thing *by conjecture, not knowing the measure nor the weight*. (AA, Ḳ.)

جَزْفٌ *Of unknown quantity, whether measured or weighed*. (Nh, TA.)

جَزْفَةٌ *A portion* of a number of cattle: (Ḳ:) and of hair. (TA)

جَزَافٌ: see what next follows, in two places.

جُزَافٌ, accord. to the 'Eyn, in selling and buying, is [The *selling*, and *buying*,] *by conjecture, without measuring and without weighing*; and by rule should be ↓ جِزَافٌ, with kesr; i. e., if formed in accordance to the verb [which is جَازَفَ]: (Mgh:) or the *selling* or *buying a thing not knowing its measure nor its weight*: (Mṣb:) or *conjecture in selling and buying*; as also ↓ جِزَافٌ and ↓ جَزَافٌ and ↓ جِزَافَةٌ and ↓ جُزَافَةٌ and ↓ جَزَافَةٌ: (Ḳ, TA:) arabicized, from كُزَافْ, (Mṣb, Ḳ, TA,) which is Persian: (Mṣb, TA:) they say لَافْ وكُزَافْ, meaning "excess in speech, by conjecture:" accord. to the Jm, its primary signification is *muchness*, or *copiousness*: (TA:) some say that the most chaste form of جزاف is ↓ جِزَافٌ, with kesr; [because this is a regular inf. n. of جَازَفَ;] (MF, TA;) and some, that the triple vocalization of the ج in جزاف is a kind of جزاف [i. e. conjecture], since all assert it to be a Persian word arabicized, and it cannot be so and be also an inf. n., conformable to the verb and to rule: it seems that, when they arabicized it, its original was gradually forgotten, so they formed from it a verb, and derived from it, and made it analogous. (TA.) — بَيْعٌ جُزَافٌ and ↓ جِزَافٌ and ↓ جَزَافٌ and ↓ جَزِيفٌ (Ḳ) and ↓ مُجْتَزَفٌ (TA) *A thing sold*, or *bought, of unknown quantity, whether measured or weighed*. (TA.) [See also جَزْفٌ.]

جِزَافٌ: see جُزَافٌ, in four places.

جَزُوفٌ A pregnant female *exceeding the term of her bringing forth*. (Ḳ.)

جَزِيفٌ: see جُزَافٌ.

جِزَافَةٌ and جُزَافَةٌ and جَزَافَةٌ: see جُزَافٌ.

جَزَّافٌ *A fisherman*. (El-'Azeezee, Ḳ.)

جَازِفٌ فِى كَلَامِهِ † One *who pours forth his speech without rule*. (Mṣb.)

مِجْزَفَةٌ *A fishing-net*. (El-'Azeezee, Ḳ.)

مُجْتَزَفٌ: see بَيْعٌ جُزَافٌ.

جزل

1. جَزَلَهُ, (Ḳ,) or جَزَلَهُ جِزْلَتَيْنِ, (Ṣ,) aor. ـِ, (Ḳ,) inf. n. جَزْلٌ, (Ṣ,) *He cut it* (a thing, Ṣ) *in two pieces*, (Ṣ, Ḳ,) with a sword. (Ḳ.) — جَزَلُوا نَاقَةً *They cut the base of the neck of a she-camel*, that was slaughtered and dead, *in the part between the two shoulder-joints, in order that the neck might become relaxed; not cutting the whole of it*; previously to skinning. (Ḥam p. 689.) — جَزَلَهُ القَتَبُ, (Ḳ,* TA,) aor. and inf. n. as above; and ↓ اجزلهُ; *The saddle cut it*; namely, the withers of a camel. (Ḳ,* TA.) — جَزِلَ, aor. ـَ, (Ḳ,) inf. n. جَزَلٌ, (Ṣ, Ḳ,) *He* (a camel) *had a gall*, or *sore, in the withers, in consequence of which a bone came forth from it, and the place thereof became depressed*: (Ṣ, Ḳ:) or *he had his withers cut by the saddle*: (Ḳ:) or *he had a gall*, or *sore, in the withers, penetrating into the interior, and killing him*. (TA.) The epithet applied to a camel in this case is ↓ أَجْزَلُ; (Ṣ, Ḳ;) fem. جَزْلَآءُ: pl. جُزْلٌ. (Ḳ.) See also أَجْزَلُ. — جَزَلَ,

aor. ِ , (Mṣb, Ḳ,) inf. n. جَزَالَةٌ, (Mṣb,) said of firewood, (Mṣb, TA,) &c., (TA,) *It was thick and large.* (Mṣb, Ḳ.) —— Also, (Ḳ,) inf. n. as above, (Ṣ,) ‡ *He possessed good,* (Ṣ,* Ḳ, TA,) *strong, firm,* (TA,) *judgment,* (Ṣ,* Ḳ, TA,) [and *natural disposition,* and *intelligence;* for] جَزَالَةٌ is used in relation to judgment and natural disposition and intelligence: (Ḥam p. 770:) and جَزَالَةُ الرَّأْىِ signifies † *firmness of judgment:* (TA:) and اَلْجَزَالَةُ فِى المَنْطِقِ, † *chasteness,* or *clearness,* or *eloquence,* and *firmness, in speech.* (Ḥar p. 8.)

4: see 1. ═ اجزل عَطِيَّتَهُ ‡ *He made his gift large.* (TA.) And اجزل لَهُ مِنَ العَطَآءِ, (Ṣ,) or فِى العَطَآءِ, (Mṣb, TA,) ‡ *He gave to him largely.* (Ṣ, Mṣb, TA.)

10. استجزل رَأْيَهُ فِى هٰذَا † *He esteemed his judgment, or opinion, good* [*and strong and firm* (see جَزْلٌ)] *in this* [matter]; syn. اِسْتَجْوَدَهُ. (TA.)

جَزْلٌ *Large and dry firewood:* (Ṣ:) or *dry firewood:* (Ḳ:) or *thick, large firewood.* (Mṣb, Ḳ.) —— ‡ *Much* of a thing; as also ▼ جَزِيلٌ: (Ḳ:) or the latter, *great,* or *large;* [and so the former:] you say عَطَآءٌ جَزْلٌ and ▼ جَزِيلٌ ‡ [*a great,* or *large, gift*]: (Ṣ, TA:*) and ▼ ثَوَابٌ جَزِيلٌ ‡ [*a great,* or *large, recompense*]: (TA:) pl. جِزَالٌ; (Ṣ, Ḳ;) either of the former or of the latter. (TA.) And [the fem.] جَزْلَةٌ ‡ [A woman] *large in the posteriors.* (Ḳ, TA.) —— ‡ *Generous; munificent.* (Ḳ, TA.) —— ‡ *Intelligent; firm,* or *sound, in judgment.* (Ḳ, TA.) You say, فُلَانٌ جَزْلُ الرَّأْىِ ‡ [*Such a one is firm,* or *sound, in judgment*]. (Ṣ, Mṣb.) And when this is said to you, and you desire to deny it, say, بَلْ جَزِلُ ▼ الرَّأْىِ ‡ *Nay, unsound in judgment;* from جَزَلٌ, [inf. n. of جَزِلَ,] relating to a gall, or sore, in the withers [of a camel]. (A, TA.) You say also اِمْرَأَةٌ جَزْلَةٌ † *A woman possessing judgment:* (Ṣ:) or *intelligent; firm,* or *sound, in judgment;* as also ▼ جَزْلَآءُ: (Ḳ:) [but] IDrd says that جزلاء, [app. a mistranscription for جَزْلَآءُ,] as syn. with جَزْلَةٌ, is not of established authority. (TA.) —— Applied to a word, or an expression, (Ṣ, Ḳ, &c.,) ‡ *Strong,* (PṢ,) *sound, correct;* (PṢ, TḲ;) *contr. of* رَكِيكٌ. (Ṣ, Ḳ.) And applied to language, † *Chaste, clear,* or *eloquent, and comprehensive.* (TA.)

جِزْلٌ: see جِزْلَةٌ.

جَزِلٌ: see جَزْلُ الرَّأْىِ.

جِزْلَةٌ *A piece,* or *portion cut off.* (Ṣ, Ḳ.*) —— *A large portion of dates;* (Ṣ, Ḳ;) as also ▼ جِزْلٌ. (Ḳ.)

زَمَنُ الجِزَالِ (Ṣ, Ḳ) and الجَزَالِ (Ḳ) *The time of the cutting off of the fruit of the palm-trees.* (Ṣ, Ḳ.)

جَزِيلٌ: see جَزْلٌ, in three places.

أَجْزَلُ; fem. جَزْلَآءُ; pl. جُزْلٌ: see جَزِلَ: ═ and for the fem., see also جَزْلٌ.

جزم

1. جَزَمَهُ, (Ṣ, Mṣb, Ḳ,) aor. ِ , (Mṣb, Ḳ,) inf. n. جَزْمٌ, (Mṣb,) *He cut it,* or *cut it off;* (Ṣ, Mṣb, Ḳ;) namely, a thing: (Mṣb:) [like جَرَمَهُ &c.] —— جَزَمَ النَّخْلَ *He cut off the fruit of the palm-trees:* (Mṣb:) [like جَرَمَ النخل: but see another explanation, below.] And جَزَمَ مِنَ النَّخْلَةِ جَزْمًا [*He cut off a portion* of the fruit *from the palm-tree*]. (TA.) —— جَزَمَ الحَرْفَ, (Ṣ, ISd, Mṣb, Ḳ,) aor. as above, (TA,) and so the inf. n., (Ṣ,) *He made the letter quiescent;* (Ṣ, ISd, Mṣb, Ḳ;) i. e., the final letter of a declinable word; (Ṣ, ISd, Mṣb;) *he cut it off from motion:* (Mṣb:) or as though he cut off from it declinability: (Mbr, TA:) from جَزَمَ in the first of the senses explained above: جَزْمٌ in a declinable word being like سُكُونٌ in an indeclinable word. (Ṣ.) It is said in a trad. of En-Nakha'ee, التَّكْبِيرُ جَزْمٌ وَالتَّسْلِيمُ جَزْمٌ, meaning that neither should be prolonged in utterance, and that the last letter in each should be without a case-ending, i. e., be quiescent; so that one should not say [in prayer] اَللّٰهُ أَكْبَرُ [nor اَلسَّلَامُ عَلَيْكُمْ وَرَحْمَةُ اللّٰهِ, but أَكْبَرْ in the former instance, and اللّٰهْ in the latter]: or, accord. to Z, that one should not exceed the due bounds in the pronunciation of the hemzeh and the medd: (TA:) or that one should abstain from giving fulness and depth to the sound of the vowel, and should elide it entirely in the places of pausing, and avoid excess in the pronunciation of the hemzeh and the medd. (Mgh.) —— جَزَمَ عَلَى فُلَانٍ كَذَا وَكَذَا *He made such and such things to be binding,* or *obligatory, on such a one.* (Ḳ.) And جَزَمَ اليَمِينَ, (Ḳ,) inf. n. as above, (TA,) i. q. أَمْضَاهَا; (Ḳ;) i. e., *He made the oath to be unconditional, without exception, absolutely* or *decidedly* or *irreversibly binding;* (TḲ;) أَمْضَاهَا. (TA.) One says also, حَلَفَ يَمِينًا حَتْمًا جَزْمًا [*He swore an oath in an absolute, a decided,* or *an irreversible, manner*]. (TA.) And جَزَمَ الأَمْرَ, (Ḳ,) inf. n. as above, (TA,) *He decided the affair irreversibly.* (Ḳ.) And جَزَمْتُ مَا بَيْنِى وَبَيْنَهُ *I decided the matter between me and him.* (TA.) And أَفْعَلُ ذٰلِكَ جَزْمًا *I will do that decidedly; without any indulgence therein.* (Mṣb.) And جَزَمَ بِهِ [*He asserted it decisively*]. (TA passim.) And جَزَمَ عَلَى الأَمْرِ *He decided,* or *determined, upon,* or *upon doing, the thing,* or *affair.* (TA.) —— Also, this last phrase, *He was silent respecting the affair;* and so ▼ جزّم. (Ḳ,* TA.) —— And جَزَمَ عَنْهُ *He held back,* or *refrained, from it through cowardice; and was unable to do it;* and so ▼ جزّم: (Ḳ:) or ▼ جزّم القَوْمُ *the people lacked power or ability.* (Ṣ.) —— جَزَمَ البَعِيرُ فَمَا يَبْرَحُ [app., *The camel stopped, and would not quit his place*]. (TA: but the verb جزم is there without any syll. sign.) ═ جَزَمَ النَّخْلَ, (A'Obeyd, Ṣ, Ḳ,) inf. n. as above; (TA;) and ▼ اجتزمهُ; (Ḳ; and the act. part. n. of the latter is also mentioned in the Ṣ;) like جَرَمَهُ (Ṣ) [and اجترمهُ]; *He computed by conjecture the quantity of fruit upon the palm-trees.* (A'Obeyd, Ṣ, Ḳ.) —— And جَزْمٌ also signifies The *selling,* or *buying, fruit* [*by conjecture, while yet in a rudimental state,*] *in its calyxes, for money.* (IAạr, TA.) ═ Also جَزَمَ, (Ṣ, Ḳ,) inf. n. as above, (TA,) *He filled* a skin; (Ṣ, Ḳ;) and so ▼ جزّم, (Ṣ,* Ḳ,) inf. n. تَجْزِيمٌ. (Ṣ.) —— جَزَمَتِ الإِبِلُ, (Fr, Ḳ,) inf. n. as above, (Fr, TA,) *The camels satisfied their thirst* [as though they *filled themselves*] *with water.* (Fr, Ḳ.) —— And جَزَمَ, (IAạr, Ḳ,) aor. and inf. n. as above, (IAạr, TA,) *He ate one meal and was filled thereby:* (IAạr, Ḳ:) or *he ate one meal in every day and night.* (Th, Ḳ.) ═ جَزَمَ القِرَآءَةَ, (Lth, Ḳ,) inf. n. as above, (Lth, TA,) *He performed the reading,* or *recitation, so as to put the letters in their proper places, in a distinct,* or *perspicuous, and leisurely, manner.* (Lth, Ḳ.) —— And جَزْمٌ in writing means The *making the letters even.* (Ḳ.) ═ جَزَمَ بِسَلْحِهِ *He voided part of his excrement, part thereof remaining:* or *he cast forth his excrement.* (Ḳ.)

2: see 1, in four places.

4. اجزم نَخْلَهُ *He sold his palm-trees.* (TA.)

5. تجزّمت العَصَا *The staff became split* or *cracked.* (Ḳ.)

7. انجزم [*It became cut,* or *cut off.* —— And hence,] *It* (the final letter of a declinable word) *became,* or *was made, quiescent.* (Ṣ, TA.) —— *It* (a bone) *broke,* or *became broken.* (Ḳ.)

8: see 1. —— اجتزم جِزْمَةً مِنَ المَالِ *He took a portion of the cattle,* or *property, and left a portion.* (Ḳ.) —— اجتزم حَظِيرَتَهُ *He bought his* حظيرة [or *enclosure for camels* &c.]: (AḤn, Ḳ:) of the dial. of El-Yemámeh. (AḤn, TA.) —— اجتزم النَّخْلَةَ *He bought the fruit, only, of the palm-tree:* and اجتزم نَخْلَ فُلَانٍ *he bought the palm-trees of such a one.* (TA.)

جَزْمٌ [an inf. n. (see 1) used as an epithet]. You say حُكْمٌ جَزْمٌ *An indissoluble and irreversible decree* or *ordinance* or *sentence;* like قَضَآءٌ حَتْمٌ. (Mṣb.) —— A reed-pen (قَلَمٌ) *having the nib evenly, not obliquely, cut.* (Ṣ,* Ḳ,* TA.) —— The *modern Arabic character,* (Ṣ, Ḳ,) *composed of the letters of the alphabet:* (Ḳ:) accord. to AḤát, (TA,) so called because it was cut off from the character of Ḥimyer, (Ḳ, TA,) i. e., the مُسْنَد, which they have still in El-Yemen. (TA.) ═ *A thing that is stuffed into a she-camel's vulva,* (El-Umawee, Ṣ, Ḳ,) *that she may think it to be her young one,* [*when it is taken forth,*] *and incline to it,* [*and therefore yield her milk;*] *like the* دُرْجَة [q. v.]. (El-Umawee, Ṣ.) ═ A thing, or an event, *that comes before its time,* or *season:* (Ḳ:) that which comes in its time, or season, is termed وَزْمٌ. (TA.)

جِزْمٌ *A portion, share,* or *lot,* (Ḳ,) of palm-trees (TA) [and app. of the fruit of a palm-tree, &c.: see 1, third sentence].

جَزْمَةٌ [The *sign that is written over the final letter of a declinable word when it is quiescent*]. ═ *A single act of eating.* (Ṣ.)

جِزْمَةٌ *A hundred* [*head*] of cattle, and upwards: or *from ten to forty:* (Ḳ:) or it is peculiarly of camels; like صِرْمَةٌ: (TA:) or *such a portion as is termed* صِرْمَة of camels; and *such as is termed* فِرْقَة of sheep. (Ṣ, Ḳ.) [See also 8.]

جَازِمٌ *A full* water-skin or milk-skin; as also ▼ مُجْزَمٌ: (Ḳ,* TA:) and [the pl.] جَوَازِمُ *filled* milk-skins. (Ḳ.) —— Also, applied to a camel, and جَوَازِمُ applied to camels, *Satisfied with water.* (Ḳ.)

مُجْزَمٌ: see جَازِمٌ.

مَجْزُومٌ [*Cut,* or *cut off.* —— And hence,] applied to the final letter of a declinable word, *Made quiescent.* (TA.)

جزى

1. جَزَى, aor. ـِ, (Mṣb, Ḳ,) inf. n. جَزَآءٌ, (Mṣb,) *It* (a thing) *paid; gave,* or *rendered, as a satisfaction;* or *made, gave,* or *rendered, satisfaction:* (Mṣb:) or *satisfied; sufficed;* or *contented.* (Ḳ.) And جَزَيْتُ الدَّيْنَ *I paid the debt.* (Mṣb.) And جَزَيْتُ فُلَانًا حَقَّهُ *I paid such a one his right,* or *due.* (TA.) And مَا يَجْزِينِى هٰذَا الثَّوْبُ *This garment does not suffice me.* (TA.) —— And hence, (TA,) جَزَى عَنْهُ, (Ṣ, Mgh, Mṣb, Ḳ,) [aor. and] inf. n. as above, (Mgh,) *It* (a thing) *paid for him; gave,* or *rendered,* [a thing] *as a satisfaction for him; made, gave,* or *rendered, satisfaction for him;* (Ṣ, Mgh, Mṣb,* Ḳ;) and some of the lawyers use ▼ اجزى in this sense, like اجزأ: (Az, Mgh, Mṣb:) جَزَى is of the dial. of El-Ḥijáz, and اجزأ of the dial. of Temeem. (Akh, Mṣb.) Hence, in the Ḳur [ii. 45], لَا تَجْزِى نَفْسٌ عَنْ نَفْسٍ شَيْئًا [*A soul shall not give anything as a satisfaction for a soul,* i. e. *for another soul:* or *a soul shall not make satisfaction for a soul at all;* accord. to the latter rendering, شيئا being put in the accus. case after the manner of an inf. n.]. (Ṣ, Mṣb.) You say also, جَزَتْ عَنْكَ شَاةٌ *A sheep,* or *goat, made satisfaction for thee* [as a sacrifice]; (Ṣ, TA;) as also ▼ أَجْزَتْ: (TA:) Benoo-Temeem say أَجْزَأَتْ, with ء: (Ṣ, TA:) this last, thus explained, is a dial. var. mentioned by IḲṭṭ. (Mṣb.) And ▼ اجزى هٰذَا عَنْ هٰذَا *Such a thing stood,* or *served, in lieu, in the place,* or *in stead, of such a thing, without sufficing.* (Zj, Ḳ.) And ▼ يُجْزِى قَلِيلٌ مِنْ كَثِيرٍ; and هٰذَا مِنْ هٰذَا *A little stands,* or *serves, in lieu of much;* and *this, of this.* (IAạr, TA.) And ▼ اجزى عَنْهُ مُجْزَى فُلَانٍ and مُجْزَاةَ فُلَانٍ and (as though the augmentative letter [ا in اجزى] were imagined to be rejected, TA) مَجْزَى فلان and مَجْزَاةَ فلان *He satisfied, sufficed,* or *contented, him as such a one; he stood,* or *served, him in stead of such a one;* a dial. var. of اجزأ. (Ḳ.) And ▼ اجزى مُجْزَى غَيْرِهِ *It* (a thing) *satisfied, sufficed,* or *contented, as another thing; it stood,* or *served, in stead of another thing.* (Mṣb.) And ▼ اجزاك, with the [second] objective complement suppressed, *It was sufficient for thee.* (Mgh.) —— جَزَاهُ هٰذَا, (Mṣb,* TA,) and جَزَاهُ بِهِ, (Ḳ,) or بِمَا صَنَعَ, (Ṣ,) and عَلَيْهِ, (Ḳ,) [aor. and] inf. n. as above, (Ṣ, Ḳ,) *He repaid, requited, compensated,* or *recompensed, him* (Mṣb, Ḳ, TA) [*for such a thing, for it,* or *for what he had done*]; as also ▼ جَازَاهُ, (Ṣ, Ḳ,) inf. n. مُجَازَاةٌ and جِزَآءٌ: (Ḳ:) or, accord. to Fr, [contr. to many instances in the Ḳur,] the former verb relates only to good; and the latter, to good and to evil: but accord. to others, the former may relate to good and to evil; and the latter, to evil. (TA.) [See also جَزَآءٌ, below.] One says, in praying for another, جَزَاهُ اللّٰهُ خَيْرًا *May God repay him good:* and *requite,* or *recompense, him for good* [that he has done]. (Mṣb.) And ▼ جَازَيْتُهُ بِذَنْبِهِ *I punished him for his crime,* or *sin,* or *act of disobedience.* (Mṣb.) And جَزَى عَنْهُ فُلَانًا *He requited, compensated,* or *recompensed, for him, such a one.* (TA.) —— جَازَيْتُهُ فَجَزَيْتُهُ: see 3.

3: see 1, latter part, in two places. —— [جازاهُ خَيْرًا *He prayed for a reward for him from God:* or *said to him, May God reward thee.* (Golius, on the authority of Z.) —— جازى بِحَرْفٍ and بِظَرْفٍ, *He employed a particle,* and *an adverbial noun, as conditional; to denote that, with what follows, it expresses a condition with its complement.* For instance, in the Ṣ, voce حَيْثُ, it is said, حَيْثُ مِنَ الظُّرُوفِ الَّتِى لَا يُجَازَى بِهَا إِلَّا مَعَ مَا, i. e. حيث *is one of the adverbial nouns that are not employed conditionally,* or *to denote that, with what follows, they express a condition with its complement, unless with* ما, affixed thereto. See جَزَآءٌ, below.] —— ▼ جَازَيْتُهُ فَجَزَيْتُهُ [*I vied,* or *contended, with him in repaying, requiting, compensating,* or *recompensing, and*] *I overcame him* [therein]. (Ṣ.)

4. اجزى: see 1, in seven places. ═ Also *He furnished* a knife *with a handle;* a dial. var. of اجزأ: (Mṣb, Ḳ:) but ISd doubts its being so. (TA.)

6. تجازى دَيْنَهُ, and بِدَيْنِهِ, *He demanded payment of his debt.* (Ḳ.) You say, تَجَازَيْتُ دَيْنِى عَلَى فُلَانٍ *I demanded payment of my debt* [owed *by such a one*]. (Ṣ.) —— تَجَازَيَا [*They two repaid, requited, compensated,* or *recompensed, each other*]. (TA in art. قرض.)

8. اجتزاهُ *He sought,* or *demanded, of him repayment, requital, compensation,* or *recompense.* (Ḳ.)

جِزًى [a coll. gen. n., of which the n. un. is with ة]: see what next follows.

جِزْيَةٌ *The tax that is taken from the free non-Muslim subjects of a Muslim government;* (Ṣ, IAth, Mgh,* Mṣb, Ḳ;) *whereby they ratify the compact that ensures them protection:* (IAth:) [from جَزَى;] as though it were a compensation for their not being slain: (IAth, Mgh:) [or from the Persian گزِيت:] and also, (metaphorically, Mgh,) ‡ *a land-tax;* (Mgh, Ḳ;) *a tax that is paid by the owner of land:* (TA:) pl. جِزًى, (Ṣ,) or ▼ جِزْىٌ, (Mṣb,) or both, (Ḳ,) [but the latter is, properly speaking, a coll. gen. n.,] and جِزَآءٌ, (Ḳ, [in the CḲ, erroneously, جَزاءٌ,]) like كِتَابٌ. (TA.)

جَزَآءٌ *Repayment, requital, compensation,* or *recompense,* for a thing; as also ▼ جَازِيَةٌ; (Ḳ;) *a satisfaction, good for good,* and *evil for evil;* (Er-Rághib, TA;) sometimes *a reward,* and sometimes *a punishment:* (AHeyth, TA:) [the former word is an inf. n.; see 1;] the latter, a quasi-inf. n.: جَوَازٍ is pl. of the latter, or of the former, or of ▼ جَازٍ, accord. to different writers explaining the saying of El-Ḥoṭeí-ah,

مَنْ يَفْعَلِ الخَيْرَ لَا يَعْدَمْ جَوَازِيَهُ

[*Whoso doth good, he will not want his rewards,* or *his rewarders*]. (TA.) —— [Hence,] جَزَآءَ العُطَاسِ *In the time between the complimentary prayer addressed to a sneezer* (called التَّشْمِيت) *and the sneeze;* [or *as soon as one can compliment a sneezer by repeating the usual prayer of* يَرْحَمُكَ اللّٰهُ (*God have mercy on thee*).] (TA voce مُعَاقِبٌ.) —— [And جَزَآءُ شَرْطٍ *An apodosis;* the *complement,* or *correlative, of a condition;* also called جَوَابٌ, q. v. in art. جوب. —— And حَرْفُ جَزَآءٍ *A particle denoting compensation,* or *the complement of a condition.* And *A conditional particle;* as إِنْ; also termed حَرْفٌ لِلْجَزَآءِ, and جَزَآءٌ alone, and حَرْفُ شَرْطٍ.] —— هُوَ ذُو جَزَآءٍ *He is possessed of sufficiency,* or *competence,* or *wealth.* (TA.)

جَازٍ [act. part. n. of 1, q. v.]: see جَزَآءٌ. —— هٰذَا رَجُلٌ جَازِيكَ مِنْ رَجُلٍ *This is a man sufficient for thee as a man.* (Ṣ.)

جَازِيَةٌ: see جَزَآءٌ. ═ Also *Wild bulls,* or *cows.* (TA.) [See جَازِئٌ.]

مَجْزًى and مُجْزًى are used as inf. ns. of 4. [See 1.] (Ḳ.)

مُجْزٍ [act. part. n. of 4. It is said in the TA that مجزى, applied to a camel, signifies *Sufficing for a load* or *burden;* and its pl. is مجازى. And that مجزى لِأَمْرِهِ, applied to a man, signifies *Sufficing for his affair.* But مجزى in these instances is evidently a mistranscription, for مُجْزٍ; and مجازى, for مَجَازٍ. Mistranscriptions of this kind are of frequent occurrence in Lexicons.]

مَجْزَاةٌ and مُجْزَاةٌ are used as inf. ns. of 4. [See 1.] (Ḳ.)

جس

1. جَسَّهُ, (A, Mgh, Ḳ,*) or جَسَّهُ بِيَدِهِ, (Ṣ, Mṣb,) aor. ـُ, (Mgh, Mṣb,) inf. n. جَسٌّ, (A, Mgh, Mṣb, Ḳ,) *He felt it with his hand* (Ṣ, A, Mgh, Ḳ) *for the purpose of testing it, that he might form a judgment of it;* (Mgh, Mṣb;) as also ▼ اجتسّهُ. (Ṣ, Mṣb, Ḳ.) You say, جَسَّهُ الطَّبِيبُ, (Mgh,) and جَسَّ يَدَهُ, (A,) *The physician felt him,* (Mgh,) and *felt his arm,* or *hand,* (A,) *to know if he were hot or cold.* (Mgh.) And جَسَّ الشَّاةَ *He felt the sheep,* or *goat, to know if it were fat or lean.* (A, Mgh.) —— جَسٌّ is also, sometimes, with the eye. (IDrd, Ṣ, Mṣb.*) You say, جَسَّهُ بِعَيْنِهِ (IDrd, Ṣ, A, Ḳ) ‡ *He looked sharply,* or *intently,* or *attentively, at him,* or *it, for the purpose of investigation and clear perception.* (Ḳ, TA.) IDrd cites as an ex. a verse (of 'Obeyd, or 'Abeed, [for I find it written without any syll. signs,] the son of Eiyoob El-'Amberee, TA,) in which occurs

the expression جُسُّوهُ بِأَعْيُنِهِمْ : (Ṣ:) but Ṣgh says that the right reading is حُسُّوهُ. (TA.) — جَسَّ الأَرْضَ † *He traversed the land.* (Aboo-Sa'eed El-Yeshkuree, TA.) — جَسَّ الأَخْبَارَ, (Ṣ, Mṣb, Ḳ,) and ▼ تَجَسَّسَهَا, (Ṣ, A, Mṣb, Ḳ,) ‡ *He searched,* or *sought, for,* or *after, news* or *tidings; inquired,* or *sought for information, respecting news* or *tidings; searched* or *inquired* or *spied into, investigated, scrutinized,* or *examined, news* or *tidings:* (Ṣ, Ḳ:) *he searched,* or *sought, repeatedly,* or *leisurely and by degrees, for news* or *tidings.* (Mṣb.) You say also, تجسّس ▼ فُلَانًا, and مِنْ فُلَانٍ, ‡ *He inquired,* or *sought for information, respecting such a one;* as also تحسّس, which latter verb occurs in an extr. reading of verse 87 of ch. xii. of the Ḳur: or the former signifies *he sought after him for another;* and the latter, "he sought after him for himself:" or ▼ تجسّس signifies [*he acted as a spy;*] *he inquired respecting,* or *searched* or *inquired* or *spied into, things which others veiled or concealed by reason of disdainful pride or of shame or pudency;* and تحسّس, "he listened:" (TA:) in the last of the senses here assigned to it, the former verb is used in the Ḳur, xlix. 12, where it is said, ▼ وَلَا تَجَسَّسُوا, for تَتَجَسَّسُوا; (Mujáhid, Bḍ, Ḳ;) ▼ or the meaning here is, *and do not ye inquire respecting,* or *search* or *inquire* or *spy into, private circumstances:* or *take ye what appeareth, and leave ye what God hath concealed.* (Mujáhid, Ḳ.) You also say, اجتسّ ▼ لَهُمْ أَمْرَ القَوْمِ ‡ *He examined* or *spied into, and sought out, for them, the circumstances of the people.* (Mgh.)

5: see 1, in five places.

8: see 1, first and last sentences. — اجتسّت الإِبِلُ البَارِضَ, (A,) or الكَلَأَ, (Ḳ,) ‡ *The camels sought out the first sprouts of the herbage with their mouths:* (A:) or *cropped the herbage with their* مَجَاسّ, (Ḳ,) i. e., their *mouths.* (TA.)

جَسِيس : see جَاسُوس.

جَسَّاس : see جَاسُوس. — ‡ The *lion that marks his prey with his claws:* (Ḳ, TA:) or *one that traverses a land.* (Aboo-Sa'eed El-Yeshkuree, TA.) — الجَسَّاسَةُ *A certain beast, that will be in the islands, that will search after news,* or *tidings, and bring them to Ed-Dejjál* [or *Antichrist*]. (Lth, L, Ḳ.)

جَاسَّةٌ a dial. var. of حَاسَّةٌ, (Mṣb,) and sing. of جَوَاسّ, (Mṣb, TA,) which is syn. with حَوَاسّ, (Kh, Ṣ, A, Ḳ,) signifying The five *senses.* (TA.) See حَاسَّةٌ.

جَاسُوس ‡ *A spy, who searches for, and brings, information, news,* or *tidings:* (Ṣ,* Mṣb,* TA:) or *one who is acquainted with private affairs of an evil nature:* as also ▼ جَسِيس (Ḳ) [and ▼ جَسَّاس]: and نَامُوس signifies one who is acquainted with private affairs of a good nature: (TA:) pl. جَوَاسِيس. (A.)

مَجَسّ (A, TA) and ▼ مَجَسَّةٌ (Ṣ, A, Mgh, Mṣb, Ḳ) The *place which one feels with his hand,* (A,* Mgh, Ḳ, TA,) *for the purpose of testing it, that he may form a judgment of it:* (Mgh:) and ▼ the latter, the *place which the physician feels* [*to know if a patient be hot or cold*]: (Ṣ, Mṣb:) pl. مَجَاسّ. (Ṣ, A, Mṣb, Ḳ.) You say, ▼ مَجَسَّتُهُ حَارَّةٌ [*The place in which one feels him is hot*]. (A, TA.) And كَيْفَ تَرَى مَجَسَّهَا [*How dost thou find the place in which one feels her?*]; referring to a sheep or goat: to which one answers, "Indicative of fatness." (A.) — [Hence, † *Anything external which indicates the internal condition.*] It is said in a prov., (Ṣ, A, Mṣb, Ḳ,) relating to camels, (Mṣb,) أَفْوَاهُهَا مَجَاسُّهَا, (Ṣ, A, Ḳ,) or مَجَاسُّهَا أَفْوَاهُهَا, (Mṣb,) or أَحْنَاكُهَا مَجَاسُّهَا, (A, Ḳ,) † [*Their mouths,* or *their palates, are the things which indicate their internal condition:*] for if they eat well, he who looks at them sufficiently knows their fatness, (Ṣ, Mṣb, Ḳ,) without feeling them: (Ṣ, Ḳ:) if one see them eat well, it is as though he felt them: (A, TA:) or, accord. to AZ, they feel the herbage, to test it, with their heads [or mouths] and their palates: so that, accord. to his explanation, the term مجاسّ is tropically applied to these parts. (TA.) The prov. relates to external evidences of things explaining their internal qualities. (Ḳ.) [And hence,] رَعَتِ الإِبِلُ الكَلَأَ بِمَجَاسِّهَا (Ḳ) ‡ *The camels cropped the herbage with their mouths.* (TA.) — You say also, فُلَانٌ وَاسِعُ المَجَسِّ, like as you say رَحْبُ الذِّرَاعِ ‡ [app. meaning *Such a one is liberal, munificent,* or *generous*]: and in the contrary case, ضَيِّقُ المَجَسِّ, (A,) or ▼ المَجَسَّةِ, (Ḳ,) or both, (TA,) signifying غَيْرُ رَحْبِ الصَّدْرِ ‡ [app., *not liberal*]; (Ḳ, TA;) and *not* وَاسِعُ السَّرْبِ [which is explained as meaning *of ample bosom,* and *judgment,* and *love;* and *of ample way,* or *course of proceeding:* but I rather incline to think that the right reading is وَاسِعُ السِّرْبِ, and the meaning, *of ample,* or *large, mind,* or *heart*]. (TA.) You also say, إِنَّ فِى مَجَسَّتِكَ ▼ لَضِيقًا (A) or مَجَسِّكَ (TA) ‡ [app., *Verily in thy bosom,* or *mind,* or *heart, is narrowness;* or *in thee is illiberality.*]

مَجَسَّةٌ : see مَجَسّ, throughout. — It may also be used as an inf. n. of جَسَّهُ in the first of the senses assigned to it above. (Mgh.)

جسأ

1. جَسَأَ, aor. ـَ, inf. n. جُسُوْءٌ and جَسْأَةٌ, (Ḳ,) thus in the corrected copies of the Ḳ, but in some copies the latter inf. n. is written جَسَاءَةٌ, (TA,) *It* (a thing, TA) *was,* or *became, hard, tough, rigid,* or *stiff;* (Ḳ, TA;) as also جَسَا, aor. ـُ. (Mṣb and Ḳ in art. جسو.) — جَسَأَتْ يَدُهُ, aor. ـَ, inf. n. جَسْءٌ, *His hand became hard, tough, callous,* (Ṣ, TA,) *coarse,* or *rough,* (TA,) from work. (Ṣ, TA.) — جَسَأَتْ مَفَاصِلُهُ *His joints became rigid,* or *stiff.* (TA.) — جُسِئَتِ الأَرْضُ *The ground became hard,* or *hard and level, and rough:* from جَسْءٌ, explained below. (Ks, Ḳ.)

جَسْءٌ *Hard,* or *hard and level, and rough, ground,* (جَلَدٌ خَشِنٌ, Ks, Ḳ, TA,) *resembling small pebbles.* (TA.) [Accord. to some copies of the Ḳ, *Rough skin* (جِلْدٌ خَشِنٌ).] — *Frozen water,* (Ḳ, TA,) *resembling hail* [*strewed on the ground*]. (TA.)

جُسْأَةٌ *Hardness, toughness, callousness,* [*coarseness,* or *roughness,*] of the hand, from work. (Ṣ.) — *Rigidity,* or *stiffness, of the neck,* (Ṣ, Ḳ, TA,) in a horse or the like. (Ṣ, TA.)

يَدٌ جَسْأَةٌ *A hard, tough, callous, coarse,* or *rough, hand.* (Ḳ, TA.)

جَاسِئٌ [*Hard, tough, rigid,* or *stiff:* see 1]. You say جَبَلٌ جَاسِئٌ [*A hard,* or *hard and rugged, mountain*]. (TA.) And صُخُورٌ جَاسِئَةٌ [*Hard,* or *hard and rugged, rocks*]. (TA.) And نَبْتٌ جَاسِئٌ *A rigid,* or *tough, plant.* (TA.) And أَرْضٌ جَاسِئَةٌ *Hard,* or *hard and level, and rough, ground;* (TA;) as also ▼ مَجْسُوْءَةٌ. (Ks, Ḳ.) And مَكَانٌ جَاسِئٌ *A rugged place;* as also شَاسِئٌ. (TA.) And دَابَّةٌ جَاسِئَةُ القَوَائِمِ *A beast,* or *horse* or *the like, having rigid,* or *stiff, legs, that will hardly bend.* (TA.)

جَاسِئَاءُ [in the CḲ جاسِياء] *Hardness, toughness, rigidness,* or *stiffness;* and *ruggedness,* or *roughness.* (Ḳ,* TA.)

مَجْسُوْءَةٌ : see جَاسِئٌ.

جسد

1. جَسِدَ, aor. ـَ, (Ṣ, Ḳ,) inf. n. جَسَدٌ, (Ṣ, KL,) *It* (blood) *stuck,* or *adhered,* (Ṣ, Ḳ, KL,) بِهِ [*to him,* or *it*]: (Ṣ, Ḳ:) and *it* (blood) *became dry.* (KL.)

4. أُجْسِدَ *It* (a garment) *was made to stick,* or *adhere, to the* جَسَد [or *body*]. (Fr, Ṣ.) — Also, inf. n. إِجْسَادٌ, *It* (a garment) *was dyed so that it stood up by reason of* [*the thickness of*] *the dye.* (ISk, Ṣ.) [See مُجْسَد.]

5. تجسّد from الجَسَدُ is like تجسّم from الجِسْمُ: (Ṣ:) [apparently signifying *He became,* or *assumed, a* جَسَد, or *body;* or *became corporeal,* or *incarnate;* and thus it is used by late writers, and in the present day: but تجسّم generally signifies *he became corpulent:* accord. to the TA, تجسّد, said of a man, is like تجسّم: accord. to the KL, the former signifies *he became possessed of a body:* accord. to the PṢ, *he became corporeal,* or *corpulent.*]

جَسَدٌ The *body, with the limbs* or *members,* [or *whole person,*] of a human being, and of a jinnee (or genie), and of an angel: (El-Bárí', L, Mṣb, Ḳ:) it is thus applied only to the body of a rational animal; (El-Bárí', Mṣb;) to no body that is nourished with food except that of a human being; but to the body of every rational creature that neither eats nor drinks, such as the genii and the angels: (L:) [the genii, however, are commonly believed to eat and drink:] or, accord. to ISd, it seems to be applied to a body other than that of a human being tropically: (TA:) the [golden] calf of the Children of Israel cried, but did not eat nor drink; [wherefore it is termed

جسد ;] and in the Ḳur [vii. 146 and xx. 90], جَسَدًا is a substitute for عِجْلًا, or it may be meant for ذَا جَسَدٍ: (L:) [but see another explanation of جسدا, as there used, below:] or جَسَدٌ is *syn.* with بَدَنٌ [which is generally held to signify the *body without the head and arms and legs*]: (Ṣ, A:) the pl. is أَجْسَادٌ. (Lḥ, Mṣb, TA.) Lḥ mentions the saying, إِنَّهَا لَحَسَنَةُ الأَجْسَادِ [*Verily she is beautiful in respect of body*]; as though the sing. of which the pl. is thus used were applied to every separate part. (TA.) ═ *Saffron*; (Ṣ, Mṣb, Ḳ;) as also ▼جِسَادٌ: (Lth, IAạr, ISk, Ṣ, A, Mṣb, Ḳ:) or *a similar dye*: (Ṣ:) and ▼the latter signifies also *a similar dye*, (Lth, Mṣb,) *red*, or *intensely yellow*: (Lth, TA:) or *bastard saffron*; syn. عُصْفُرٌ; (Mṣb;) and so the former word. (L.) — Also, (Ṣ, L, Mṣb, Ḳ,) and ▼جَسِدٌ (L, Ḳ) and ▼جَاسِدٌ (L, Mṣb, Ḳ) and ▼جَسِيدٌ (L, Ḳ) and ▼جِسَادٌ, (R,) †*Blood*; (Ṣ;) as also ▼مُجْسَدٌ, as being likened to saffron: (Ḥam p. 215:) or *dry blood*: (R, L, Mṣb, Ḳ:) pl. of the first as above. (Ḥam p. 127.) — Some say that عِجْلًا جَسَدًا, in the Ḳur, [in two verses referred to above,] means *A red golden calf*. (Ṣ.)

جَسِدٌ : see جَسَدٌ : ═ and see also جَاسِدٌ.

جِسَادٌ : see جَسَدٌ, in three places.

جَسِيدٌ : see جَسَدٌ.

جَاسِدٌ Blood *sticking* to (ب) a person; as also ▼جَسِدٌ. (Ṣ.) ═ See also جَسَدٌ.

جَلْسَدٌ (L, Ḳ in art. جلسد) and الجَلْسَدُ (Ṣ, L, Ḳ, TA, in the CḲ الجُلْسُدُ) the name of *A certain idol*, (Ṣ, L, Ḳ,) *which was worshipped in the time of paganism*: (L:) the ل is an augmentative letter accord. to J and most of the leading lexicologists; but F disapproves of the mention of the word in this art. (TA.)

مُجْسَدٌ *Red.* (Ṣ.) — Also, (Ṣ, Mṣb, Ḳ,) and sometimes ▼مُجْسِدٌ, (IF, Mṣb,) and ▼مُجَسَّدٌ, (Ḳ,) A garment *dyed with saffron*: (Ḳ:) or *with bastard saffron*: (IAth, TA:) *dyed with* جِسَاد: (IF, Mṣb:) or *saturated with dye*: (Ṣ:) or one says, عَلَى فُلَانٍ ثَوْبٌ مُشْبَعٌ مِنَ الصِّبْغِ [Upon such a one is a garment saturated with dye]; and عَلَيْهِ ثَوْبٌ مُفْدَمٌ [Upon him is a garment saturated with red dye]; and when it is *standing up by reason of* [*the thickness of*] *the dye*, it is termed مُجْسَدٌ: (ISk, Ṣ:) the pl. of this word is مَجَاسِدُ. (Ṣ.) — See also what next follows: — and see جَسَدٌ.

مِجْسَدٌ, (Ṣ, A, Ḳ,) or ▼مُجْسَدٌ, (A,) the latter is the original form, because it is from أُجْسِدَ, meaning, "it was made to stick, or adhere, to the body," (Fr, Ṣ,) like مُطْرَفٌ and مِطْرَفٌ, and مُصْحَفٌ and مِصْحَفٌ, (Fr, TA,) *A garment worn next the body* (IAạr, Ṣ, A, Ḳ) *by a woman, so that she sweats in it*: (TA:) and *a garment worn next the body dyed with saffron*: (A:) pl. مَجَاسِدُ. (IAạr, A.) [Hence,] لَا يَخْرُجْنَ إِلَى المَسَاجِدِ فِى المَجَاسِدِ (IAạr, A) *They* (women) *shall by no means go forth to the mosques in the shirts that are next the body.* (IAạr.) — See also مُجْسَدٌ.

مُجَسَّدٌ : see مُجْسَدٌ.

جسر

1. جَسَرَ, (Ḳ,) aor. ـُ, (TA,) inf. n. جَسَارَةٌ (A, Ḳ, KL) and جُسُورٌ, said of a man, (Ḳ,) *He was daring, courageous*, or *bold*: (A, KL:) *he acted with penetrating energy*, or *sharpness, vigorousness, and effectiveness*; syn. مَضَى and نَفَذَ. (Ḳ.) You say, جَسَرَ عَلَى كَذَا, aor. ـُ, inf. n. جَسَارَةٌ, *He ventured upon such a thing daringly, courageously*, or *boldly*; (Ṣ, A;) as also ▼تجاسر: (Ṣ:) and ▼تجاسر عَلَيْهِ *he emboldened himself against it*, or *him*. (A, Ḳ.) And جَسَرَ عَلَى عَدُوِّهِ, (A, Mṣb,) aor. ـُ, inf. n. جُسُورٌ and جَسَارَةٌ, (Mṣb,) *He acted daringly, courageously*, or *boldly, against his enemy*. (A.) And لَا يَجْسُرُ أَنْ يَفْعَلَ كَذَا *He dares not to do such a thing*. (A.) ═ جَسَرَ, (Ḳ,) inf. n. جَسْرٌ, (TA,) *He* (a man) *arched*, or *vaulted*, *a* جسر [or *bridge*]. (Ḳ.) — It is said of [the giant] 'Ooj (عُوج), in a trad., وَقَعَ عَلَى نِيلِ مِصْرَ فَجَسَرَهُمْ سَنَةً, i. e., ‡[*He fell down upon the Nile of Egypt, and*] *became a bridge to them* [*for the space of a year*]. (A.) — And one says, جَسَرَتِ الرِّكَابُ المَفَازَةَ, and ▼اِجْتَسَرَتْهَا, ‡*The travelling-camels crossed*, or *passed over, the desert*, (A, Ḳ,) *as by a bridge*: (A:) and اجتسرت ▼السَّفِينَةُ البَحْرَ ‡*The ship crossed*, or *passed over, the sea*: (A:) or *rode upon, and passed through, the sea*. (Ḳ.)

2. جَسَّرَهُ, (A, Ḳ,) inf. n. تَجْسِيرٌ; (Ḳ;) [and ▼اجسره; (see أَغْرَهُ;)] *He encouraged him; emboldened him.* (A, Ḳ.)

4: see 2.

6. تجاسر : see 1, in two places. — Also *He stretched himself up, and raised his head.* (Ḳ.) — تجاسر لَهُ بِالعَصَا *He put himself in motion to him* (En-Nawádir, Ḳ) *with the staff*, or *stick*. (Ḳ.) — تجاسروا *They acted with mutual daring* or *courage* or *boldness*. (KL.) — *They journeyed* [app. *with boldness*, or *emulating one another in boldness*]. (TA.) — الخَيْلُ تَجَاسَرُ بِالكُمَاةِ [for تَتَجَاسَرُ] ‡*The horses convey the brave armed men away*, or *along*, or *across*. (A.)

8: see 1, last sentence, in two places.

جَسْرٌ : see جَسُورٌ, in two places. — Also, applied to a he-camel, *Sharp, spirited*, or *vigorous*; syn. مَاضٍ [as *contr. of* بَلِيدٌ]; as also ▼جَسْرَةٌ and ▼مُتَجَاسِرَةٌ applied to a she-camel: or (so in the Ḳ accord. to the TA; but in the CḲ, "and") *tall*: (Ḳ:) or *tall and bulky*: and with ة, applied to a she-camel, it has this last signification; (TA;) or signifies *strong, and bold to endure travel*: (A, TA:) the masc. epithet applied to a he-camel is rare. (Lth, TA.) — Also *Large*, or *bulky*; applied to a camel, (Ṣ, Ḳ,) &c., (Ṣ,) or to anything, (Ḳ,) or to any limb, or member: (TA:) fem. with ة. (Ṣ, Ḳ.) — جَسْرَةُ السَّوَاعِدِ, and المُخَدَّمِ, *Full* or *plump* [*in the fore arms*, and *in the place of the anklet*]; applied to a girl, or young woman. (A, TA.) ═ See also what next follows.

جِسْرٌ and ▼جَسْرٌ [*A bridge*; and *a dyke*, or *causeway*:] *that on which one crosses over a river or the like*; (Ṣ, Mgh, Mṣb, Ḳ;) *as a* قَنْطَرَةٌ *and the like*; (TA;) *whether built or not built*: (Mgh, Mṣb:) and *a bridge of boats; boats bound together, and tied to stakes in the bank, being over a river*; see قَنْطَرَةٌ: (TA:) pl. (of pauc., TA) أَجْسُرٌ (Ḳ) and (of mult., TA) جُسُورٌ. (Ṣ, Mṣb, Ḳ.) [Hence,] المَوْتُ جِسْرٌ يُوصِلُ الحَبِيبَ إِلَى الحَبِيبِ ‡[*Death is a bridge that conveys the friend to the friend*]. (TA.) And جَعَلَ طَاعَتَهُ جِسْرًا إِلَى نَجَاتِهِ ‡[*He made his obedience a bridge to his safety*]. (A, TA.)

جَسُورٌ *Daring, courageous*, or *bold*: (Ṣ, A:) or *courageous and tall*; as also ▼جَسْرٌ: (Ḳ:) or *courageous*; and also *tall and bulky*; applied to a man; and so ▼جَسْرٌ: (TA:) fem. of the former without, and sometimes with, ة: (Mṣb:) and of the latter with ة: (TA:) pl. of the former جُسُرٌ and جُسْرٌ. (Ḳ, TA.) It is not applied to a he-camel; but with ة is applied to a she-camel, meaning *Bold to traverse rugged*, or *difficult, tracts*. (Mṣb.)

جَسَّارٌ *Very daring* or *courageous* or *bold*. (TA.)

مُتَجَاسِرَةٌ : see جَسْرٌ.

جسم

1. جَسُمَ, (Ṣ, Mṣb, Ḳ, &c.,) aor. ـُ, (Ḳ,) inf. n. جَسَامَةٌ, (Mṣb, TA,) *He, or it*, (a thing, Ṣ,) *was*, or *became, great*, or *large*: (Ṣ, Ḳ:) or so جَسِمَ, aor. ـَ, inf. n. جَسَمٌ: and the former, *it* (a thing) *was*, or *became, great, big*, or *bulky*: (Mṣb:) or *he*, or *it, was*, or *became, great*, or *large, in body*: (KL:) or *he*, or *it, was*, or *became, corpulent*; or *corporeal*, or *bodied*; as also ▼تجسّم. (MA, PṢ.)

2. جسّم, inf. n. تَجْسِيمٌ, *He*, or *it, made*, or *rendered, corporeal*; or *great, large, big*, or *bulky*. (KL.) — [*He made to be solid*, or *to have length and breadth and thickness*.]

5: see 1. — [Hence,] تجسّم فِى عَيْنِى كَذَا ‡ *Such a thing assumed a form*, or *shape*, [or *an embodiment*,] *in my eye*. (TA.) And تجسّم فُلَانٌ مِنَ الكَرَمِ ‡[app. *Such a one was*, or *became, an embodiment of generosity*]. (TA.) And كَأَنَّهُ كَرَمٌ قَدْ تَجَسَّمَ ‡[app. *As though he were generosity embodied*]. (TA.) ═ تجسّم فُلَانًا ‡ *He chose such a one* (Ṣ, Ḳ, TA) مِنْ بَيْنِ القَوْمِ [*from among the people*, or *party*], (Ṣ,) or مِنَ العَشِيرَةِ فَأَرْسَلَهُ [*from the kinsfolk*, or *tribe*, &c., *and sent him*]: (TA:) as though he directed his course, or aim, to, or towards, his جِسْم [or body]; like as you say, تَأَيَّيْتُهُ, meaning "I directed my course, or aim, to, or towards, his آيَة, and his شَخْص." (Ṣ.) [See also 5 in art. جثم.] One says also, تَجَسَّمْهَا نَاقَةً مِنَ الإِبِلِ فَانْحَرْهَا [*Choose thou her, a she-camel from among the camels, and stick her*]. (TA.) — تجسّم الأَرْضَ † *He betook himself towards the land*, or *country*, (Ṣ, Ḳ,) *desiring to go thither.* (Ṣ.) — تجسّم الرَّمْلَ, (Ṣ, Ḳ,) and الجَبَلَ, (Ṣ,) † *He mounted,* or *ascended, the greater part of the sand,* (Ṣ, Ḳ,) and *of the mountain.* (Ṣ.) [See also 5 in art. جشم.] — تجسّم الأَمْرَ ‡ *i. q.* رَكِبَ مُعْظَمَهُ; (Ṣ, Ḳ, TA;) i. e., *He ventured upon, embarked in,* or *undertook, the main part,* or *bulk, of the affair:* (TḲ:) or *he constrained himself to do it,* or *perform it;* as also تجشّمهُ: (Aboo-Mihjen, Aboo-Turáb, TA:) or both these verbs signify *he took it,* or *imposed it, upon himself,* or *he undertook it, in spite of difficulty* or *trouble* or *inconvenience.* (Aboo-Turáb, TA in art. جشم.) You say also, فُلَانٌ يَتَجَشَّمُ المَجَاشِمَ and ▼ يَتَجَسَّمُ المَجَاسِمَ ‡ [app. meaning *Such a one undertakes, in spite of difficulty* or *trouble* or *inconvenience, those things,* or *affairs, that are causes of difficulty* or *trouble* or *inconvenience;* i. e., *difficult,* or *troublesome,* or *inconvenient things* or *affairs:* supposing the two nouns to be pls. of which the sings. are مَجْسَمَةٌ and مَجْشَمَةٌ, of the measure مَفْعَلَةٌ, like مَبْخَلَةٌ and مَجْبَنَةٌ and مَشَقَّةٌ (originally مَشْقَقَةٌ) &c.]. (TA.)

جِسْمٌ *The body, with the limbs* or *members;* syn. جَسَدٌ; (AZ, Ṣ, Mṣb;) as also ▼ جُسْمَانٌ and جُثْمَانٌ: (AZ, Ṣ, Mṣb:*) or جِسْمٌ and ▼ جُسْمَانٌ are syn. with جَسَدٌ; (Aṣ, Ṣ;) or signify the *whole body and limbs* or *members* of a man, (Ḳ, and T and Mṣb in explanation of جسم,) and of a beast, a camel, and the like, (T, Mṣb,) and of any other species, (Ḳ,) of large make; (T, Mṣb, Ḳ;) and جُثْمَانٌ is syn. with شَخْصٌ [app. as meaning "a person"]; and ▼ جُسْمَانٌ signifies the *whole* جِسْم of a man: (Aṣ, Ṣ:) or جِسْمٌ signifies [*a body,* or *material substance; a solid;*] *a thing having length and breadth and thickness;* so that, when it is cut and divided, no portion thereof ceases to be a جِسْم; whereas a شَخْص [meaning "a person"] ceases to be a شَخْص by its being divided: (Er-Rághib, TA:) *a thing that is capable of being divided in length and breadth and thickness* is called جِسْمٌ طَبِيعِىٌّ, and also, because it is a subject of investigation, or inquiry, in instruction in the mathematical studies, جِسْمٌ تَعْلِيمِىٌّ: (KT:) pl. [of pauc.] أَجْسَامٌ and [of mult.] جُسُومٌ. (Ḳ.) You say, ثَابَ جِسْمُهُ, (M, A, Ḳ, in art. ثوب,) and أَثَابَ جِسْمُهُ, (IḲṭ, M, ib.,) and ثَابَ إِلَيْهِ جِسْمُهُ, (T, M, A, ib.,) ‡ *He became fat, after leanness;* (A;) *his good state of body returned to him;* (M, Ḳ;*) *his condition of body became good, after extenuation; and health,* or *soundness, thereof returned to him.* (T.) And ▼ إِنَّهُ لَنَحِيفُ الجُسْمَانِ [*Verily he is slender, spare,* or *lean, of body*]. (TA.) — الأَجْسَامُ المُخْتَلِفَةُ الطَّبَائِعِ [*The material substances of different natures;* also called الأَجْسَامُ السَّبْعَةُ *the seven material substances,* and الفِلِزَّاتُ; namely,] *the* عَنَاصِر [which are the *four elements, fire, air, earth, and water,*] *and the three products composed of these,* (KT,) which are *minerals, vegetables, and animals.* (Note in a copy of the KT.) الأَجْسَامُ الطَّبِيعِيَّةُ, as used by those who study to discover occult things, signifies *The* عَرْش *and the* كُرْسِىّ: and الأَجْسَامُ العُنْصُرِيَّةُ, *everything beside these two, of the heavens and the* [*elements termed*] أُسْطُقُسَّات *therein.* (KT.)

[جِسْمِىٌّ *Bodily,* or *corporeal.*]

[جِسْمِيَّةٌ *Bodiliness,* or *corporeity.*]

جُسْمَانٌ: see جِسْمٌ, in four places.

جُسْمَانِىٌّ: see جَسِيمٌ.

جُسَامٌ: see what next follows.

جَسِيمٌ *Great; large;* (Ṣ, Mṣb, Ḳ;) as also ▼ جُسَامٌ: (Ṣ, Ḳ:) *big; bulky:* (Mṣb:) fem. with ة: (Ḳ:) pl. جِسَامٌ (Ṣ, Mṣb) [and جُسُمٌ also, like as جُدُدٌ is pl. of جَدِيدٌ]: and *corpulent, large in body,* or *big-bodied;* (Ḳ,* TA;) as also ▼ جُسْمَانِىٌّ, applied to a man. (TA.) You say, هُوَ مِنْ جِسَامِ الأُمُورِ [*It is one of great affairs* or *events*]: and مِنْ جَسِيمَاتِ الخُطُوبِ [meaning the same, or *of great afflictions* or *calamities*]. (TA.) And جُسُمٌ [likewise, or أُمُورٌ جُسُمٌ,] signifies *Great affairs* or *events.* (TA.) جَسِيمُ الأَمْرِ signifies [also] The *bulk,* or the *greater, main, principal,* or *chief, part, of the affair;* and so ▼ أَجْسَمُ الأَمْرِ. (Ṣ.) — [The pl.] جُسُمٌ also signifies † *Intelligent men.* (TA.) — جَسِيمُ الأَرْضِ *Elevated land over which water has risen:* (Ḳ,* TA:) pl. جِسَامٌ. (Ḳ.)

أَجْسَمُ *Greater, larger, bigger,* or *bulkier;* or *greatest, largest, biggest,* or *bulkiest.* (Ṣ, Ḳ.) See also جَسِيمٌ.

مَجَاسِمُ: see 5, last sentence.

جسو: omitted in some copies of the Ṣ.

1. جَسَا, (Ṣ, M, Mṣb, Ḳ,) aor. ـُ, (M, Mṣb, Ḳ,) inf. n. جُسُوٌّ (M, Ḳ) and جَسْوٌ, (M, TA,) *He,* or *it,* (a man, M, TA, or a thing, Mṣb,) *was,* or *became, hard, tough, rigid,* or *stiff:* (M, Mṣb, Ḳ:) [see also جَسَأَ: or, said of a man,] *he was,* or *became, hard, rough, harsh,* or *ungentle; contr. of* لَطُفَ. (Ṣ.) And جَسَتِ اليَدُ, inf. n. جُسُوٌّ and جَسًا, *The hand,* or *arm, was,* or *became, dry,* or *tough,* or *hard,* (Ṣ, TA,) *in the bones, with little flesh;* (TA;) and in the same sense the verb is said of other things. (Ṣ.) — *It* (water) *congealed,* or *froze.* (Ṣ, TA.) — Also, inf. n. جُسُوٌّ, *He* (an old man) *attained the utmost age* [*so that his limbs became rigid*]. (Ṣ, TA.)

3. جاساهُ, (Ḳ,) inf. n. مُجَاسَاةٌ, (TA,) *He treated him,* or *regarded him, with enmity,* or *hostility.* (Ḳ, TA.)

يَدٌ جَاسِيَةٌ *A hand,* or *an arm, dry,* or *tough,* or *hard, in the bones, with little flesh.* (TA.) And دَابَّةٌ جَاسِيَةُ القَوَائِمِ *A beast having tough,* or *hard, legs.* (TA.) And رِمَاحٌ جَاسِيَةٌ *Stiff, rigid,* or *tough, and hard, spears.* (TA.) [See also جَاسِئٌ, in art. جسأ.]

جش

1. جَشَّهُ, (Ṣ, A, Ḳ,) aor. ـُ, inf. n. جَشٌّ, (Ṣ,) *He ground it* (namely wheat, Ṣ, or grain, A) *coarsely;* (Ṣ, A;) as also ▼ أَجَشَّهُ. (Ṣ.) — *He bruised, brayed,* or *pounded, it:* and *he broke it:* (Ṣ, Ḳ:) as also ▼ أَجَشَّهُ. (Ḳ.) — *He beat him,* or *struck him,* with a staff or stick. (Ṣ, Ḳ.)

4: see 1, in two places.

جُشَّةٌ (A, Ḳ, TA) and ▼ جَشَشٌ (TA) *Loudness,* or *vehemence, of voice* or *sound:* (A, Ḳ, TA:) and *a rough sound coming forth from the* خَيَاشِيم [or *air-passages in the nose*], *in which is a hoarseness.* (Ḳ, TA.) You say, ▼ فِى صَهِيلِ الفَرَسِ جَشَشٌ *In the neighing of the horse is a rough sound:* (TA:) which is one of the qualities approved in horses. (IDrd.) And فِى صَوْتِ القَوْسِ جُشَّةٌ عِنْدَ الرَّمْىِ *In the sound of the bow is a roughness of twanging on the occasion of shooting.* (AḤn TA.)

جَشَشٌ: see جُشَّةٌ, in two places.

جَشِيشٌ Wheat *coarsely ground;* as also ▼ مَجْشُوشٌ. (Ṣ.) — And, [as an epithet in which the quality of a subst. predominates,] (TA,) as also ▼ جَشِيشَةٌ, (Ṣ, Ḳ, TA,) *What is coarsely ground,* (Ṣ, Ḳ, TA,) *of wheat &c.,* (Ṣ,) or *of wheat and the like:* (Ḳ, TA:) or the former, *grain when bruised, brayed,* or *pounded, before it is cooked:* and ▼ the latter, *such as is cooked:* but ISd says that this distinction is not of valid authority. (TA.) — Also, the former, *i. q.* سَوِيق [*Meal of parched barley or wheat, coarsely ground, which is made into a kind of gruel*]; (El-Fárisee, Ṣ, Ḳ;) and so ▼ the latter: (A:) or ▼ the latter is the n. un. (El-Fárisee.) You say, اِسْقِنِى ▼ جَشِيشَةً سَوِيقٍ *Give thou me to drink* [*some*] سويق. (A.) Or سويق is not called ▼ جشيشة, but is called جَذِيذَةٌ, q. v. (TA.) — And the former, (Sh, Ḳ,) or ▼ جَشِيشَةٌ, (TA in art. دش,) *Wheat coarsely ground, and put into a cooking-pot, into which some flesh-meat is thrown, or some dates, and then cooked:* (Sh, Ḳ:) also called دَشِيشَةٌ: (TA:) or *a soup made of coarsely bruised wheat.* (TA in art. دش.)

جَشِيشَةٌ: see جَشِيشٌ, in seven places.

أَجَشُّ *Having a rough,* (Ṣ, Ḳ,) or *loud,* or *vehement,* (A,) *voice,* or *sound:* (Ṣ, A, Ḳ:) applied to a man, and a horse, and thunder, (A, Ḳ,) &c. (Ḳ.) You say, رَجُلٌ أَجَشُّ الصَّوْتِ *A man having a* [*rough,* or] *loud,* or *vehement, voice.* (A.) And فَرَسٌ أَجَشُّ, (A,) or أَجَشُّ الصَّوْتِ, (Ṣ, TA,) *A horse in whose neighing is a roughness.* (TA.) And سَحَابٌ أَجَشُّ, (Aṣ,) or أَجَشُّ الرَّعْدِ, (Ṣ,) *Clouds that thunder vehemently.* (Aṣ.) And قَوْسٌ جَشَّآءُ, [جشّاء being the fem. of أجشّ,] *A bow having a rough twanging,* (AḤn, Ḳ,) *when one shoots with it.* (AḤn.) — الأَجَشُّ is also the name of *One of the sounds of which musical modulations are formed,* (Kh, Ḳ,) *which are three in number;* [app. meaning the treble, tenor, and bass, clefs; the last being that to which this term is applied;] *the sound thus called being from the head,* (Kh,) *issuing from the* خَيَاشِيم [or *air-passages in the nose*], *having in it a roughness and hoarseness,* (Kh, Ḳ,) *and followed by a gradual fall* (تَحَدُّر) [*of the voice*] *modulated in accordance to that same sound, and then followed by a sound* [in my original بُوشِي, but I think it probable that this is a mistranscription for بُوحِي, or بُوحَى, or the like, for, though وَشْى might perhaps, by straining a metaphor, be applied to denote a varied sound, its being understood in this sense seems to be forbidden by its being here added] *like the first.* (Kh, TA.) [This explanation is perhaps illustrated by the fact that the bass in the music of the Arabs is often formed of one prolonged note, falling and rising.] — Also جَشَّآءُ, [أَرْضٌ being understood,] *A pebbly plain, fit for palm-trees.* (Ḳ, TA.)

مِجَشٌّ, (Ṣ,) or ↓ مِجَشَّةٌ, (A,) or both, (Ḳ,) *A mill* (Ṣ, Ḳ) *with which* جَشِيش *is ground:* (Ṣ:) or *a small mill with which one grinds coarsely.* (A.)

مِجَشَّةٌ: see what next precedes.

مَجْشُوشٌ: see جَشِيشٌ.

جشأ

1. جَشَأَتْ نَفْسُهُ, (Ṣ, Ḳ,) aor. َ, (Ḳ,) inf. n. جُشُوْءٌ, (Ṣ, Ḳ, KL,) like قُعُودٌ, (TA,) and جَشَآءٌ, (KL, [or جُشَآءٌ, so Golius on the authority of the KL,]) [like جَاشَتْ نَفْسُهُ, and جَاشَتْ,] *His soul* [or *stomach*] *heaved, by reason of grief* or *fright:* (Ṣ, Ḳ; and so in the O; but in one copy of the Ḳ, *by reason of grief* or *joy*: TA:) or [simply] *heaved,* or *rose:* (T in art. ثور:) and *heaved,* or *became agitated by a tendency to vomit;* (Ḳ;) *i. q.* خَبُثَتْ *and* لَقِسَتْ: (Sh, TA:) and جَشَأَتْ إِلَىَّ نَفْسِى *My soul* [or *stomach*] *heaved,* or *became agitated by a tendency to vomit, or became heavy,* (خَبُثَتْ,) *in consequence of pain from something that it disliked.* (ISh, TA.) — جَشَأَ عَنِ الطَّعَامِ *He nauseated food, in consequence of indigestion.* (TA.) — جَشَأَتِ الغَنَمُ *The sheep emitted a sound from their throats.* (Lth, Ḳ.) — جَشَأَتِ الأَرْضُ ‡ *The earth put forth all its plants,* or *herbage:* like as they say, قَآءَتِ الأَرْضُ أُكْلَهَا [lit. "the earth vomited her victuals"]. (TA.) — جَشَأَتِ الرِّيَاضُ بِرُبَّاهَا ‡ [*The meadows,* or *gardens,*] *put forth* [*their good things*]. (TA.) — جَشَأَتِ البِلَادُ بِأَهْلِهَا ‡ [*The countries,* or *towns,* &c.,] *cast forth* [*their inhabitants*]. (TA.) — جَشَأَتِ البِحَارُ بِأَمْوَاجِهَا ‡ [*The seas*] *cast forth* [*their waves*]. (TA.) — Also جَشَأَ said of the sea, ‡ *It rushed on,* (TA,) *grew dark,* (Ḳ, TA,) *and was tumultuous with its waves;* (TA;) and [in the CḲ "or"] *impended over one.* (Ḳ, TA.) And in like manner said of the night, ‡ *It came on suddenly,* (TA,) *grew dark;* (Ḳ, TA;) and [in the CḲ "or"] *impended over one.* (Ḳ, TA.) — جَشَأَتِ الوَحْشُ † *The wild animals made a single leap,* or *spring.* (TA.) — جَشَأَ القَوْمُ † *The people,* or *company of men, went forth* from one country, or town, to another. (Ṣ, Ḳ, TA.) It is said in a trad., جَشَأَتِ الرُّومُ عَلَى عَهْدِ عُمَرَ † *The Greeks rose, and advanced from their country* [*in the time of 'Omar*]. (TA.)

2: see 5.

5. تَجَشَّأَ, (Ṣ,) inf. n. تَجَشُّؤٌ; (Ṣ, Mgh, Ḳ; [in the CḲ, التَّجَشُّ is erroneously put for التَّجَشُّؤُ;]) or تَجَشَّى, inf. n. تَجَشٍّ; (Mṣb;) and ↓ جَشَّأَ, (Ṣ,) inf. n. تَجْشِئَةٌ; (Ṣ, Ḳ;) both signify alike; (Ṣ;) *He eructed,* or *belched;* i. e., *emitted a sound accompanied with wind, from his mouth, on an occasion of satiation of the stomach,* (Mgh, Mṣb,) *intentionally:* (Mgh:) or *it* (the stomach) *emitted wind* (Ḳ, TA) *on an occasion of its impletion with food or drink.* (TA.)

8. اِجْتَشَأَتْهُ البِلَادُ, and اِجْتَشَأَ البِلَادَ, † [*He found the country to disagree with him,* and] *the country disagreed with him.* (Ṣ, Ḳ.)

جَشْءٌ *A light bow:* (Ṣ, Ḳ:) or *a bow that makes a ringing sound:* (Lth, TA:) or *a light rod of the tree called* نَبْع: (Aṣ, Ṣ:) pl. أَجْشَآءٌ, (Ḳ,) anomalous, and asserted by IHsh to be rare, (TA,) and جَشَآتٌ. (Ḳ: in the CḲ, جُشَآتٌ.) — سَهْمٌ جَشْءٌ *A light arrow.* (Yaạḳoob, TA.) ═ *A large number* (IAạr, Ḳ, TA) of men, and of cattle. (IAạr, TA.)

جُشَآءٌ: see جُشْأَةٌ. — Also ‡ *Daybreak:* [or,] accord. to 'Alee Ibn-Ḥamzeh, the *blowing of the wind at daybreak.* (TA.)

جُشَأَةٌ: see جُشْأَةٌ, in two places.

قَوْسٌ جَشْآءُ *A ringing bow.* (TA. [See also جَشَّآءُ, voce أَجَشُّ, in art. جش.])

جُشْأَةٌ *A belch;* i. e., *a sound accompanied with wind, from the mouth, on an occasion of satiation of the stomach;* (Mgh, Mṣb;) a subst. from 5; (Aṣ, Ṣ, Mṣb, Ḳ;) as also ↓ جُشَأَةٌ (Ṣ, Ḳ) and ↓ جُشَآءٌ: (Ḳ: but the first and last of these three words are omitted in some copies of the Ḳ:) or ↓ the second of these three words, accord. to some, is a superlative epithet, signifying *a great,* or *frequent, belcher.* (MF.) — Also † *An invasion* of the night, and of the sea. (Ḳ, TA.) The torrent and the night (السَّيْلُ وَاللَّيْلُ) are called الأَعْمَيَانِ [the two blind things] because their invasion is vehement. (TA.)

جشب

1. جَشُبَ, aor. ُ; and جَشِبَ, aor. َ; (Ḳ;) and جَشُبَ, aor. ُ, inf. n. جَشَابَةٌ; (TA;) said of wheat, or food, (طَعَام,) *It was gross,* or *coarse:* (Ḳ, TA:) *it was badly and coarsely ground:* (TA:) or *it was without seasoning,* or *condiment,* or *anything to render it savoury.* (Ḳ.) — And the first, *It* (a thing) *was thick, gross, big, coarse,* or *rough.* (TA.) — And جَشُبَ, aor. ُ, inf. n. جُشُوبَةٌ, *He* (a man, TA) *was a foul,* or *bad, eater.* (Ḳ, TA.) ═ جَشَبَهُ *He ground it coarsely;* namely, wheat. (Ḳ, TA.) — جَشَبَ اللّٰهُ شَبَابَهُ *God caused his youth,* or *youthful vigour, to pass away:* or *rendered him vile and despicable* (رَدَّاهُ وَأَقْمَأَهُ): or *may God cause* &c. (Ḳ.)

12. اِجْشَوْشَبُوا, accord. to some, or اِخْشَوْشَبُوا, accord. to others, occurs in a trad. of 'Omar; (TA in art. خشب;) [and J says, and so Az accord. to the TA,] the former, if used like the latter, is not improbably correct; but I have not heard it. (Ṣ.) [See art. خشب.]

جَشْبٌ: see جَشِبٌ.

جُشْبٌ The *rinds of pomegranates:* (Ḳ:) of the dial. of El-Yemen. (TA.)

جَشْبٌ (Ṣ, Ḳ) and ↓ جَشِبٌ and ↓ جَشِيبٌ and ↓ مِجْشَابٌ (Ḳ) and ↓ مَجْشُوبٌ, (Ṣ, Ḳ,) applied to wheat, or food, *Gross,* or *coarse:* (Ṣ, Ḳ, TA:) *badly and coarsely ground:* (TA:) or *without seasoning,* or *condiment,* or *anything to render it savoury:* (Ḳ, TA:) جشب [probably جَشْبٌ] signifies also anything *disagreeable in taste, and choking:* (TA:) and *gross,* or *coarse, and dry,* or *tough:* (IAth, TA:) and what is *dry,* or *tough,* of herbage. (TA.) — Also, the first, *A bulky and strong* camel: (ISk, TA:) a *thick-boned* horse. (Ḥam p. 207.)

جَشُوبٌ *A rough,* or *coarse,* (or, as some say, TA,) *short* woman. (Ḳ, TA.)

جَشِيبٌ: see جَشْبٌ. — Also Anything *rough, gross* or *coarse, disagreeable to the taste, and choking.* (Ḳ.) — *A thick, rough,* or *coarse,* garment, or piece of cloth. (Ṣ.) — *A rough,* or *coarse, and old, worn-out,* skin for water or milk. (TA.) — *Rude, uncivil, unkind, rough,* speech or language. (TA.) — And (applied to a man, TA) *A foul,* or *bad, eater.* (Ḳ, TA.)

مُجْشِبٌ *Big,* or *bulky, and courageous, brave,* or *bold.* (IAạr, Ḳ.) ═ [Also, accord. to Golius, as on the authority of Ibn-Maạroof, but in this case probably a mistranscription for مِشْجَبٌ, *A wooden thing upon which clothes are put.*]

مُجَشَّبٌ A man (Sh) *coarse in his means of subsistence.* (Sh, Ḳ.)

مِجْشَابٌ: see جَشْبٌ. — *Thick, gross, big, coarse,* or *rough,* (Ṣ, and Ḥam p. 207,) *and short.* (Ḥam ib.) — مِجْشَابُ البَدَنِ *Thick, gross,* or *big, in body.* (T, TA.)

مَجْشُوبٌ: see جَشْبٌ.

جشر

1. جَشَرَ, aor. ُ, (Aṣ, Ṣ, A, Mgh,) inf. n. جَشْرٌ; (Aṣ, Ṣ, Ḳ;) and ↓ جَشَّرَ, (A,) inf. n. تَجْشِيرٌ; (Ḳ;) *He took,* or *sent, forth* his beasts *to pasture,* (Aṣ, Ṣ, Mgh, Ḳ,) *not to return in the evening:* (Aṣ, Ṣ, Mgh:) [or] *he pastured* his beast *near to the tents* or *houses:* (A:) [or] جَشْرٌ signifies also one's *pasturing* his horses *before his tent* or *house, after their covering:* (Ḳ:) or a people's *taking forth* their horses *and pasturing* them *before their tents* or *houses.* (L.) — And جَشْرٌ and ↓ تَجْشِيرٌ also signify The *leaving* or *neglecting* [a thing]: (Ḳ, TA;) and *dismissing* [it]. (TA.) جَشَرَ القُرْآنَ, meaning *He estranged himself from*

the Ḳur-án, is said of him who has neglected the reading or reciting of it for two months. (L from a trad.) ═ جَشَرَ المَالُ عَنْ أَهْلِهِ *The cattle went forth to the places of pasturage from their owners.* (A.) — جَشَرَ الرَّجُلُ عَنْ أَهْلِهِ ‡ *The man journeyed away from his family,* or *wife.* (A.) — جَشَرَ الصُّبْحُ, (aor. ـُ, Ṣ,) inf. n. جُشُورٌ, (Ṣ, A, Ḳ,) ‡ *The dawn broke,* (Ṣ,) or *rose,* (Ḳ,) or *came forth.* (A.)

2: see 1, in two places.

جَشْرٌ: see جَشَرٌ, in three places.

جَشَرٌ Camels or sheep or goats *pasturing in their place, not returning to their owners* (Aṣ, Ṣ, Ḳ) *at night:* (Ḳ:) or [simply] *not returning to their owners.* (Aṣ, TA.) [See also مُجَشَّرٌ.] — †A people *who pass the night with the camels,* (Aṣ, Ṣ, Ḳ,) *in their place, not returning to their tents* or *houses:* (Aṣ, Ṣ:) *who go forth with their beasts to the place of pasturage, and remain in their place, not returning to the tents* or *houses:* the doing this is not considered as travelling, and therefore is not a legal reason for shortening the ordinary prayers: (A'Obeyd, TA:) and ▼جَشْرٌ signifies the same. (TA, as on the authority of A'Obeyd. [But perhaps this latter is a mistranscription for جُشَّرٌ: see what follows.]) ‡ A man who is *away* (عَزَبٌ, Ḳ, TA) *from his family,* or *wife, with his camels;* (TA;) as also ▼جَشِيرٌ: (Ḳ, TA:) and in like manner the former is applied to a company of men; and so جُشَّرٌ [a pl. of ▼جَاشِرٌ, q. v.]: you say قَوْمٌ جَشَرٌ and جُشَّرٌ. (L, TA.) ═ The *herbs,* or *leguminous plants, of* [*the season,* or *rain, called*] *the* رَبِيع; (L, Ḳ;) as also ▼جَشْرٌ. (L.) And جشر [app. جَشَرٌ or ▼جَشْرٌ] also signifies *A pasture-land in which horses feed.* (TA.)

جَشِيرٌ: see جَشَرٌ. ═ Also *A* [*quiver of the kind called*] وَفْضَة; (Ṣ, Ḳ;) i. e., *a* كِنَانَة; and so جَفِيرٌ; accord. to ISd, *a* [*quiver of the kind called*] جَعْبَة, *of skins, slit in the side in order that the wind may enter it and the feathers may therefore not be eaten:* (TA:) or, accord. to Z, *i. q.* جِرَابٌ. (IAth, TA.) — And *A large* جُوَالِق [or *sack*]: (Ṣ, Ḳ:) pl. [of pauc.] أَجْشِرَةٌ and [of mult.] جُشُرٌ. (TA.)

جَشَّارٌ The *owner* (صَاحِب) *of a pasture-land in which horses feed.* (Ḳ.) You say, "He is the جَشَّار of our camels." (A, TA. [But it seems to be implied in the A that it signifies the same as جَاشِرٌ as explained below.])

جَاشِرٌ One *who takes forth horses and camels to the pasture-land, and remains there:* [see also جَشَّارٌ:] pl. جُشَّارٌ: (TA:) [and جُشَّرٌ is another pl. of the same:] see جَشَرٌ. — Also [the pl.] جُشَّرٌ Camels, and asses, *going whithersoever they will.* (TA.)

جَاشِرِيَّةٌ ‡ *A drink that is taken at daybreak:* (Ṣ, A, Ḳ:) you say, اِصْطَبَحْنَا الجَاشِرِيَّةَ *We drank the morning-draught that is taken at daybreak:* (Ṣ, A:) and it has no verb: (Ṣ:) or it is only *of camels' milk:* (Ḳ:) or it is correctly of general application: or is properly *of wine;* for this is what is most frequently mentioned: and it is also used as an epithet: thus you say شَرْبَةٌ جَاشِرِيَّةٌ. (TA.) — †*A certain kind of food:* (Ḳ, TA:) or *a kind of food eaten at daybreak.* (TA.) — †The [*last part of the night, called the*] سَحَر: (Ḳ:) because near to daybreak. (TA.) — †*Midday:* (Ḳ:) because of the appearance and spreading of its light. (TA.)

مُجَشَّرٌ [A beast] *made to pass the night in the pasture, away from its owner, not brought back in the evening:* (Ḳ,* TA: [see also جَشَرٌ:]) or *not pastured near the water:* (IAạr, TA:) or *that is pastured near to the water.* (El-Mundhiree, TA.) And خَيْلٌ مُجَشَّرَةٌ *Horses pastured* (Ṣ, Ḳ) بِالحِمَى [*in the place of pasturage that is prohibited to the public*]. (Ṣ.)

جشع

1. جَشِعَ, aor. ـَ, inf. n. جَشَعٌ, *He was,* or *became, affected with the most vehement desire, eagerness, avidity, cupidity,* or *hankering,* (Ṣ, O, Ḳ,) and, (O, Ḳ,) as explained by an Arab of the desert to Aṣ, (IDrd,) *with the worst kind thereof,* (IDrd, O, Ḳ,) for eating &c.: (TA:) or, as explained by another Arab of the desert to Aṣ, (IDrd,) *he took his own share, and coveted the share of another:* (IDrd, Ḳ:) and ▼تجشّع signifies the like; (Ṣ;) or *i. q.* تَحَرَّصَ, q. v. (Ḳ.) — جَشَعٌ also signifies The *being impatient on account of separation from an associate.* (TA.) — And The *being frightened, terrified,* or *afraid.* (TA.)

5: see 1.

6. تَجَاشَعَا المَآءَ *They straitened each other in pressing to the water, and* [so I render تَعَاطَشَا] *vied, each with the other, in endeavouring to satisfy their thirst;* (Ḳ;) on the authority of an Arab of the desert. (TA.)

جَشِعٌ part. n. of جَشِعَ, *Affected with the most vehement desire,* &c.: pl. جَشِعُونَ, (Ṣ, Ḳ,) and جَشَاعَى and جُشَعَآءُ and جِشَاعٌ are also pls. [of the same]. (TA.) — الجَشِعُ *The lion.* (TA.) — رَجُلٌ جَشِعٌ بَشِعٌ *A man in whom are combined impatience and fright and a heavy,* or *a heaving, state of the soul.* (TA.)

جَشِيعٌ One *who assumes a false disposition, and that which is not in him.* (TA.)

أَجْشَعُ [comparative and superlative of جَشِعٌ; *More,* and *most, affected with most vehement desire,* &c.]. (TA.)

جشم

1. جَشِمَ الأَمْرَ, aor. ـَ, inf. n. جَشْمٌ (Ṣ, Mṣb, Ḳ) and جَشَامَةٌ, (Mṣb, Ḳ,) *He took,* or *imposed, upon himself the affair,* or *he undertook it, as a task,* or *in spite of difficulty* or *trouble* or *inconvenience;* (Ṣ, Mṣb, Ḳ;) as also ▼تجشّمه, (Ṣ, Mṣb,* Ḳ,) and تَجَسَّمَهُ: (Aboo-Turáb, TA:) or ▼the second and third signify *he constrained himself to do it,* or *perform it:* (Aboo-Miḥjen, Aboo-Turáb, TA in art. جسم:) and ▼تجشّم كَذَا وَكَذَا *He did such and such things against his will, and in spite of difficulty* or *trouble* or *inconvenience.* (TA.) [See also an ex. of ▼تجشّم in art. جسم; conj. 5, last sentence.] جَشِمْتُ إِلَيْكَ عَرَقَ القِرْبَةِ is said to mean [*I have imposed upon myself difficulty* or *trouble* or *inconvenience, in coming to thee,*] *so that I have journeyed, and become in want of the water of the water-skin* in the journey: or the meaning is, *I have suffered, and imposed upon myself, difficulty* or *trouble* or *inconvenience, so that I have sweated like the water-skin:* or by the عَرَق of the قِرْبَة is meant its عَلَق, i. e., its مِعْلَاق, by which it is carried; and the phrase means ▼تَجَشَّمْتُ إِلَيْكَ حَمْلَ القِرْبَةِ [*I have imposed upon myself, in spite of difficulty &c., in coming to thee, the carrying of the water-skin*]; alluding to journeying and its difficulties: (Ḥar p. 511:) [and in like manner,] one says, ▼تَجَشَّمْتُ لَكَ عَرَقَ القِرْبَةِ. (Ṣ in art. عرق.) The sportsman, when he has not taken any game, and has returned disappointed, says, مَا جَشِمْتُ إِلَيْكَ ظِلْفًا [app. meaning *I have not had the trouble of bringing to thee* so much as *a hoof of a gazelle or the like*]. (AZ, TA.) And مَا جَشِمْتُ اليَوْمَ طَعَامًا, i. e., *I have not eaten, to-day, food,* is said on the occasion of the disappointment of any one seeking a thing. (AZ, TA.)

2. جَشَّمَهُ الأَمْرَ, (Ṣ, Mṣb, Ḳ,) inf. n. تَجْشِيمٌ; (Ṣ;) and ▼اجشمهُ; (Ṣ, Mṣb, Ḳ;) *He imposed upon him,* or *made him to undertake, the affair, as a task,* or *in spite of difficulty* or *trouble* or *inconvenience;* or *he ordered, required,* or *constrained, him to do the thing, it being difficult* or *troublesome* or *inconvenient.* (Ṣ, Mṣb, Ḳ.) Hence, (Ṣ, TA,) in a trad. of Zeyd Ibn-'Amr Ibn-Nufeyl, (TA,)

* مَهْمَا تُجَشِّمْنِى فَإِنِّى جَاشِمٌ *

[*Whatever thou impose upon me, in spite of difficulty &c., I undertake it, in spite of difficulty &c.*]. (Ṣ, TA.)

4: see 2.

5: see 1, in six places. — تَجَشَّمْتُ الرَّمْلَ *I mounted,* or *ascended, the greater part of the sand:* some say thus; and some say تَجَسَّمْتُ. (TA.) — تَجَشَّمْتُ فُلَانًا مِنْ بَيْنِ القَوْمِ, *I directed my course,* or *aim, towards such a one,* [*and chose him,* (like تَجَسَّمَتْهُ, q. v.,)] *from among the people,* or *party.* (Abu-n-Naḍr, TA.)

جَشْمٌ *A state of destruction, perdition,* or *death.* (AA, TA.) — See also جُشَمٌ.

جُشُمٌ: see جُشَمٌ. ═ Also *Bad money:* pl. جُشُومٌ. (IKh, TA.)

جَشَمٌ: see جُشَمٌ, in two places. — Also *Fatness.* (AA, Ḳ.)

جَشِمٌ: see جُشَمٌ.

جُشَمٌ *Weight,* or *heaviness;* (Ṣ, Ḳ;) as also ▼جَشَمٌ, (Ḳ,) and ▼جَشْمٌ accord. to the Ḳ, but correctly ▼جُشُمٌ, as in the A and L: (TA:) [and *an unpleasing, and a difficult* or *troublesome* or *an inconvenient, affair:*] a subst. from تَجَشَّمَ كَذَا وَكَذَا explained above: see 1. (TA.) You say, أَلْقَى فُلَانٌ عَلَىَّ جَشْمَهُ (Ṣ,) or ▼جُشَمَهُ, (TA,) *Such a one threw upon me his weight*, or *heaviness:* (Ṣ, TA:) to which Z adds, or *his difficult*, or *troublesome*, or *inconvenient, affair, that he had imposed upon himself*, or *that was imposed upon him.* (TA.) — Also The جَوْف [i. e., the *belly*; or the *chest*; &c.]: or the *breast, with the ribs that contain it:* (Ḳ:) or the *breast* of a camel: (Ṣ, TA:) and the *part* [*of the skin*] *of the camel's breast, and of the rest of the body* thereof, *with which the* [*kind of quiver called*] قرن [i. e. قَرْن] *is covered.* (TA.) You say, غَتَّهُ بِجَشْمِهِ, meaning *He threw his breast upon him.* (TA.)

جُشُمٌ, [app. pl. of ▼جَشِيمٌ, like as جُدُدٌ is pl. of جَدِيدٌ,] *Fat* men: (IAạr, Ḳ:*) and *tall, crafty* or *cunning, and wicked* or *malignant*, men. (IAạr, TA.)

جَشُومٌ A man *who takes*, or *imposes, upon himself*, or *who undertakes, affairs, with energy* or *vigour*, or *in an extraordinary degree, in spite of difficulty* or *trouble* or *inconvenience.* (Mṣb.)

جُشَمٌ, (Ḳ,) or, as in the book of Kr, ▼جُشَمٌ, (TA,) *Thick, gross, coarse, rough, rugged, rude, big*, or *bulky.* (Ḳ.) See also جُشُمٌ.

جَاشِمٌ A man *taking*, or *imposing, upon himself*, or *undertaking, an affair, in spite of difficulty* or *trouble* or *inconvenience.* (Mṣb.) See 2.

الْمُجْشِمُ *The lion.* (Ḳ.)

مَجَاشِمُ: see 5 in art. جسم.

جشن

جَوْشَنٌ and جَوْشَنِيٌّ: see art. جوشن.

جص

2. جَصَّصَ *He plastered* a building *with* جِصّ [or *gypsum*]: (Mgh, Ḳ:) or *he made* a house *therewith:* (Mṣb:) *i. q.* قَصَّصَ, (Ṣ, TA,) which is of the dial. of El-Ḥijáz. (TA.)

جِصٌّ (Ṣ, Mgh, Mṣb, Ḳ) and جَصٌّ, (Ṣ, Mgh, Ḳ,) but the former is the more chaste: (TA, from an Expos. of the Fṣ:) the latter is disapproved by IDrd, and disallowed by ISk; (TA;) and it is said in the Bári', on the authority of AḤát, that the latter is the form used by the vulgar, and the former is that which is correct: (Mṣb:) [*Gypsum; a certain substance*] *with which one builds*, (Ṣ,) or *plasters*; (Mgh;) *well known:* (Mṣb, Ḳ:) arabicized; (Ṣ, Mgh, Mṣb, Ḳ;) because ج and ص do not occur in any Arabic word; (Mṣb;) or, accord. to AZ, there are some [Arabic] words in which they both occur, as has been mentioned in art. اجص: from گَچ, (Mgh, Ḳ, [in the CḲ كَج,]) or, as some say, كَسْج; which are Persian: (TA:) in the dial. of the people of El-Ḥijáz, قَصّ [i. e. قِصّ or قَصّ]. (Lth, TA.) [The n. un. is with ة: see art. قص.]

جَصَّاصٌ *A preparer of* جِصّ [or *gypsum*]. (Ṣ, Ḳ.)

جَصَّاصَاتٌ [pl. of جَصَّاصَةٌ] *Places in which* جِصّ [or *gypsum*] *is made.* (Ḳ.)

جظ

1. جَظَّ, [aor., accord. to rule, ـِ, and inf. n., probably, جَظٌّ,] *He was short and fat.* (IAạr, Ḳ.)

4. اجظّ *He was*, or *became, proud*; or *excessively proud, corrupt, unbelieving*, or *disobedient.* (Sgh, Ḳ.)

جَظٌّ [probably an inf. n. used as an epithet,] applied to a man, (Ṣ,) *Large; big; bulky*; or *large in body, corpulent, and fleshy:* (Ṣ, Ḳ, TA:) or *tall, large in body, a great eater and drinker, who exults, and behaves insolently and ungratefully to God:* (Fr, TA:) occurring in a trad. describing the people of Hell. (Ṣ.)

جعب

1. جَعَبَهُ, (Ṣ, Ḳ,) aor. ـَ, (Ḳ,) inf. n. جَعْبٌ, (TA,) *He prostrated him; he threw him down upon the ground*; (Ṣ, Ḳ, TA;) like جَعَفَهُ; (Ṣ;) as also ▼جَعَّبَهُ, (Ḳ,) inf. n. تَجْعِيبٌ; (TA;) and ▼جَعْبَاهُ, (Ṣ, Ḳ, [in the CḲ erroneously written جَعْبَأَهُ,]) like سَلْقَاهُ from سَلَقَهُ, (Ṣ,) inf. n. جِعْبَآءٌ. (Ṣ, TA [in the latter, in one place, probably by a mistake of a copyist, written جعباة].) — *He inverted it*, or *him; he turned it*, or *him, upside down*, or *over*, or *inside out*; syn. قَلَبَهُ. (Ḳ.) — *He collected it*; (Ḳ;) mostly used in relation to that which is small in quantity, paltry, or inconsiderable. (TA.) — جَعَبَ, (A, Ḳ,) and جَعَّبَ, (A,) *He made a quiver of the kind called* جعبة. (A, Ḳ.)

2: see 1.

5: see 7.

7. انجعب and ▼تجعّب (Ḳ) and ▼تَجَعْبَى (Ṣ, Ḳ) *He became prostrated*, or *thrown down upon the ground.* (Ṣ, Ḳ.)

Q. Q. 1. جَعْبَاهُ: see 1.

Q. Q. 2. تَجَعْبَى: see 7.

جَعْبَةٌ *A* كِنَانَة [or *quiver*] (A, Ḳ) *for arrows* (Ṣ, A, Mgh, Mṣb, Ḳ) *of the kind called* نُشَّاب (Ṣ, A, Mṣb, Ḳ) *and for those that are called* نَبْل *also:* (Ḥam p. 154:) but some make a distinction between جَعْبَة and كِنَانَة: the former, they say, is for نشّاب; and the latter, for نبل: (Mz, MF:) accord. to IDrd, the كنانة is only for نبل, and is of leather: that which is of wood is called جَفِير: and that which is of two pieces [of wood] joined together is called قرن [i. e. قَرْن]: (Ḥam ubi suprà:) accord. to ISh, the جعبة is *round and wide, with a cover on the top, over its mouth:* the وَفْضَة is smaller, and its upper and lower parts are of equal size; whereas the جعبة is *wide in its upper part, and contracted in its lower part*; *wide in its upper part that the feathers of the arrows* [*having ample room*] *may not become detached; for the arrows are put in the quiver with the points downwards:* each of these two kinds is *made of two corresponding pieces of wood:* (TA:) the pl. is جِعَابٌ (Ṣ, A, Mṣb, Ḳ) and جَعَبَاتٌ. (Mṣb.) You say, نَكَبُوا الجِعَابَ وَسَكَبُوا النُّشَّابَ [*They inverted*, or *inclined, the quivers, and poured forth the arrows*]. (A, TA.) And مَعَهُ جَعْبَةٌ فِيهَا بَنَاتُ المَوْتِ [*With him is a quiver in which are the daughters of death*; i. e., *deadly arrows*]. (A, TA.) — Also The *largest of drinking-vessels.* (MF, TA.)

جُعْبُوبٌ A man (Ṣ) *short, and ugly*, or *contemptible*; or *ugly, and small in body:* (Ṣ, Ḳ:) or *weak, and destitute of good:* or *vile*, or *mean, and despicable:* (Ḳ:) or *a low, mean*, or *sordid, and weak* man: pl. جَعَابِيبُ. (TA.)

جِعَابَةٌ The *art of making quivers of the kind called* جِعَاب, pl. of جَعْبَة. (A, Ḳ.)

جِعَابِىٌّ: see what next follows.

جَعَّابٌ [and app. ▼جِعَابِىٌّ also, as seems to be indicated in the Ḳ, where it is mentioned as a surname, but in the CḲ written جَعَّابِىٌّ,] *A maker of quivers of the kind called* جِعَاب. (A, Ḳ.)

مِجْعَبٌ One *who often prostrates*, or *throws down, others*, (صِرِّيعٌ, [in some copies of the Ḳ, erroneously, صَرِيع,]) *but is not himself prostrated*, or *thrown down.* (Ḳ, TA.)

مُتَجَعِّبٌ or مُنْجَعِبٌ (accord. to different copies of the Ḳ) *Dead*, or *dying*; syn. مَيِّتٌ. (Ḳ.)

جعد

1. جَعُدَ, aor. ـُ, inf. n. جُعُودَةٌ (Ṣ, A, Mṣb, Ḳ) and جَعَادَةٌ, (Ḳ,) said of hair, (Ṣ, A, Mṣb, Ḳ,) *It was*, or *became, crisp*, or *curly*, or *twisted, and contracted*; (Mṣb;) *was*, or *became, the contr. of* سَبْط, (Ḳ,) or *of* مُسْتَرْسِل: (Mṣb:) or *was*, or *became, short:* (Kr, Ḳ:) and جَعِدَ, [aor. ـَ,] (Mṣb, TA,) inf. n. جَعَدٌ, (TA,) signifies the same; (Mṣb, TA;) as also ▼تجعّد. (Ḳ.) — †*It became contracted, and compacted in lumps*; (L;) as also ▼تجعّد; (L, Ḳ;*) said of earth, (Ḳ,) or of moist earth. (L.) [The inf. n.] جُعُودَةٌ is also sometimes used in describing the state of the froth, or foam, of a camel's mouth, when it is accumulated. (Ṣ. [See جَعْدٌ.]) — Also, said of a cheek, inf. n. جُعُودَةٌ, † *It was rough*, or *coarse, and short; contr. of* أَسْلَ. (L.)

2. جعّده, (Ṣ, A, Mṣb, Ḳ,) inf. n. تَجْعِيدٌ, (Ṣ, A, Mṣb,) *He crisped*, or *curled*, or *twisted, and contracted, it*; (Mṣb;) *made it the contr. of* سَبْط, (Ḳ,) or *of* مُسْتَرْسِل: (Mṣb:) or *made it short:* (Ḳ:) namely, hair. (Ṣ, A, Mṣb, Ḳ.)

5: see 1, in two places.

جَعْدٌ, applied to hair, (Ṣ, A, Mṣb, Ḳ,) *Crisp*, or *curly*, or *twisted, and contracted*; (Mṣb;) *contr. of* سَبْط, (Ḳ,) or *of* مُسْتَرْسِل: (Mṣb:) or *short.* (Kr, Ḳ.) — Applied to a man, (Ṣ,) *Having hair such as is termed* جَعْد: (Ṣ, Mṣb, Ḳ:) [or]

so جَعْدُ الشَّعَرِ: (A, TA:) fem. with ة: (Ṣ, Mṣb, Ḳ:) pl. جِعَادٌ. (A, Mṣb.) — As an epithet of praise, it has two meanings; namely, † *Compact in limbs, and strong in make; not flabby, nor of slack,* or *incongruous, make;* (L;) or *big,* or *bulky, and compact;* (Ḥam p. 238;) or, as some say, *light,* or *active:* (TA:) and *having crisp,* or *curly, not lank, hair;* because lankness is the prevalent characteristic of the hair of the Greeks and Persians; and crispness, or curliness, is the prevalent characteristic of the hair of the Arabs: but very crisp, or frizzled, or woolly, hair, like that of the Zenj and the Nubians, is disapproved. (L.) — [Hence,] ‡ *Generous; bountiful; munificent;* (T, Ṣ, A, Ḳ;) alluding to a man's being an Arab of generous disposition, because the Arabs are characterized by crisp, or curly, hair. (A.) Aṣ did not know جعد in this sense; but it occurs in many verses of the Anṣár. (T, TA.) — As an epithet of dispraise, it has also two meanings; namely, † *Short, and incongruous in make:* (L:) [*contr. of* سَبْطٌ:] — and ‡ *Niggardly;* (Aṣ, T, Ṣ, L, Ḳ;) as also جَعْدُ اليَدَيْنِ, (Ṣ, Ḳ,) and جَعْدُ الأَصَابِعِ, جَعْدُ الأَنَامِلِ, (Ṣ,) and (A,) or this signifies † *having short fingers,* (Ḳ,) and جَعْدُ البَنَانِ, and جَعْدُ الكَفِّ, (Ḥar p. 96,) and جَعْدُ الجَنَانِ; (A;) *contr. of* [سَبْطُ اليَدَيْنِ, and] سَبْطُ البَنَانِ and سبط اليد [&c.]: (Ḥar ubi suprà:) and *mean; ungenerous; base:* (L:) and جَعْدُ القَفَا ‡ *mean,* or *ignoble, in respect of rank, quality, reputation,* or *the like.* (A, Ḳ.) — A camel *having much fur:* (Ḳ:) or *having crisp,* or *curly,* and *abundant, fur.* (Ṣ.) [Hence,] أَبُو الجَعْدِ a surname of *The camel.* (L.) — † *Soft* moist earth; as also ثَعْدٌ: (Ṣ:) or *moist* earth. (Ḳ.) — † A mess of the kind called حَيْس that is *thick,* (L, Ḳ,) *not flowing;* (L;) as also ▼ مُجَعَّدٌ. (L, Ḳ.) IAạr cites the following words of a poet, accusing a woman of foul conduct:

* وَتَخْلِطُ بِالمَأْلُوطِ حَيْسًا مُجَعَّدًا ▼ *

[*And she mixes thick* حيس *with the food prepared with* أَقِط]; meaning, she confounds men together, and does not select him who is to have intercourse with her. (L.) — † Froth, or foam, *accumulated upon the fore part of the mouth* of a camel. (Ṣ,* L.) And جَعْدُ اللُّغَامِ † A camel *having froth,* or *foam, accumulated upon the fore part of his mouth.* (Ṣ,* L, Ḳ.*) — † A cheek *rough,* or *coarse, and short; not* أَسِيل. (L, Ḳ.) And † A *round* face, *with little* مَلْح [or *beauty*], (Ḳ, TA,) or, as in some copies of the Ḳ, لَحْم [or *flesh*]. (TA.) And قَدَمٌ جَعْدَةٌ ‡ *A short foot;* (A, TA;) characteristic of low origin. (TA.) — It is also applied, in the manner of an intensive epithet, to the plant called صِلِّيَان; and in like manner, with ة, to the plant called بُهْمَى. (TA.) — نَاقَةٌ جَعْدَةٌ † *A she-camel compact in make, and strong.* (TA.)

مُجَعَّدٌ: see جَعْدٌ, in two places.

مُتَجَعِّدٌ Moist earth *contracted, and compacted in lumps.* (L in art. عقد.)

جعر

1. جَعَرَ, aor. ـَ, (Ṣ, Mṣb, Ḳ,) inf. n. جَعْرٌ; (Mṣb;) and ▼ انجعر; (Ḳ;) said of a beast or bird of prey (Ṣ, Mṣb, Ḳ*) having claws, or talons, (Ṣ, Ḳ,*) or a hyena, and a dog, and a cat, (TA,) and metaphorically of a rat or mouse, (Mṣb,) *He voided his dung.* (Ṣ, Mṣb, Ḳ.)

5. تجعّر, (Ṣ, Ḳ,) or تجعّر بِجِعَارٍ, (TA,) *He bound upon his* (i. e. his own) *waist a rope of the kind called* جِعَار. (Ṣ, Ḳ, TA.)

7: see 1.

جَعْرٌ, originally an inf. n., (Mṣb,) The *dung of a beast or bird of prey* (Ṣ, Mṣb, Ḳ) *having claws,* or *talons;* (Ṣ, Ḳ;) as also ▼ جَاعِرَةٌ; (Ḳ;) which is like رَوْثٌ in relation to a horse: (TA:) or the *dung of the hyena:* (A:) [and *of the dog,* and *cat:* see 1:] or *dry dung upon the* مَجْعَر, q. v.: (Ḳ:) or *dung that comes forth dry:* (IAth, TA:) and ‡ *that of the rat or mouse:* (Mṣb:) pl. جُعُورٌ. (Ḳ.) — See also جُعْرُورٌ. — Also *Costiveness.* (TA.)

جُعْرَةٌ *A mark left by the rope called* جِعَار (Th, Ḳ) *upon the waist of a man.* (Th, TA.)

جَعْرَاءُ: see مَجْعَرٌ.

أَبُو جِعْرَانَ [in which the latter word is imperfectly decl. because it is a proper name ending with the augment ان] *The* [*black beetle called*] جُعَل, (Kr, Ḳ, TA,) in a general sense: or, as some say, *a certain species thereof.* (TA.) — And أُمُّ جِعْرَانَ, (Ḳ, TA,) or أُمُّ جِعْرَانَةَ, (so in a copy of the Ḳ,) *The* رَخَمَة [or *female of the vultur percnopterus*]. (Kr, Ḳ.)

جُعْرُورٌ *A bad kind of dates;* (Mṣb, Ḳ;) also metaphorically called جَعْرُ الفَأْرَةِ [*the rat's,* or *mouse's, dung*], because of the bad smell, and the diminutiveness, thereof: (Mṣb:) and you also say تَمْرُ جُعْرُورٍ: (TA:) or *a species of the* دَقَل, *which is the worst kind of dates:* (Ṣ:) or *a species of the* [*kind of palm-tree called*] دَقَل *that bears small things* [or *dates*] *in which is no good.* (Aṣ, TA.) [See عَذْقُ الحُبَيْقِ, in art. حبق.]

جَعْرَى: see مَجْعَرٌ.

جَعَارِ, (Ṣ, A, Ḳ,) like قَطَامِ, (Ḳ,) indecl., with kesr for its termination, because it deviates from its original form, which is جَاعِرَةٌ, and is of the fem. gender, and has the quality of an epithet in which that of a subst. is predominant, so that the thing to which it applies is known by it like as it is known by its proper name; and as it is prevented from being perfectly decl. by two causes, it must be indecl. by reason of three; as we also say with respect to حَلَاقِ, a proper name of death; (Ṣ;) *The she-hyena;* (Ṣ, A, Ḳ;) a name of that animal (Ṣ) because of the abundance of its dung; (Ṣ, A;) as also أُمُّ جَعَارِ, and ▼ جَيْعَرُ, and ▼ أُمُّ جَعْوَرٍ. (Ḳ.) Hence, أَعْيَثُ مِنْ جَعَارِ [*More mischievous than the she-hyena*]: a prov. (A, TA.) And تَيْسِى جَعَارِ (Ḳ) *Be thou like the he-goat in stupidity, O she-hyena;* a prov. applied to a stupid man: (A and TA in art. تيس, q. v.:) or عِيثِى جَعَارِ [*Do mischief, O she-hyena*]; a prov. used in declaring a thing to be vain, or false. (Ḳ.) And

* رُوغِى جَعَارِ وَانْظُرِى أَيْنَ المَفَرّ *

[for المَفَرُّ, *Be afraid, O she-hyena, and look where is a place to which to flee*]: (Ḳ,* TA:) or رُوغِى [i. e. *turn aside, this way and that*]: (Ṣ and TA in art. روغ:) a prov. applied to him who seeks to escape, and cannot: (TA:) or with reference to a coward, and his submissiveness. (Ḳ.) And قُومِى جَعَارِ [*Rise, O she-hyena*]: said to a woman, in reviling her; likening her to a she-hyena. (ISk, TA.)

جِعَارٌ *A certain mark made with a hot iron upon* [*the part called*] *the* جَاعِرَتَانِ: (Ḳ:) accord. to the Tedhkireh of Aboo-'Alee, *one of the marks, so made, of camels.* (Ibn-Ḥabeeb, TA.) ═ *A rope which a man who waters ties to a stake, and then binds upon his waist, when he descends into a well, lest he should fall into it:* (Ṣ:) or *a rope which a drawer of water binds upon his waist,* (Ḳ, TA,) *when he descends into a well,* (TA,) *lest he should fall into the well;* (Ḳ, TA;) *the end being in the hand of another man, who, if he falls, pulls him up with it.* (TA.)

أُمُّ جَعْوَرٍ: } see جَعَارِ.
جَيْعَرُ:

جَاعِرَةٌ: see مَجْعَرٌ. — الجَاعِرَتَانِ *The place of the* [*two marks made by cauterization which are called the*] رَقْمَتَانِ, *in the buttocks* (اِسْت) of an ass: (Ṣ, Ḳ:) or *the places of cauterization in the hinder part, upon the* [*two portions of the thighs called the*] كَاذَتَانِ, of an ass: (TA:) and *the part,* (Ṣ, Ḳ,) or *two parts,* (A,) *which the tail strikes,* (Ṣ, A, Ḳ,) *upon the two thighs* of a horse, (Ṣ, Ḳ,) or of a beast, *where he is cauterized:* (A:) or *the two edges of the haunches projecting over the thighs* [*behind*]; (Aṣ, Ṣ, Ḳ;) i. e., *the two places which the farrier marks, making lines upon them* [*with a hot iron*] (يَرْقُمُهُمَا): or *the heads of the upper parts of the two thighs:* or *the depressed part of the haunch and thigh, in the place of the joint.* (TA.) ═ See also جَعْرٌ.

مَجْعَرٌ The *rump,* or *podex;* or the *anus;* [in the present day, the latter;] syn. دُبُرٌ; (Ṣ, Ḳ;) and ▼ جَعْرَاءُ and ▼ جَعْرَى and ▼ جَاعِرَةٌ the *same;* syn. اِسْتٌ; (Ḳ;) or the last (جاعرة), as some say, *i. q.* حَلْقَةُ الدُّبُرِ. (Ṣ, Ḳ.)

مِجْعَارٌ A man *very,* or *often, costive;* (Ḳ;) as also مِجْعَارُ البَطْنِ. (TA.)

جعس

Q. Q. 1. جَعْمَسَ *He* (a man, TA) *deposited his ordure,* or *excrement, at once:* (Ḳ, TA:) or *in a dry,* or *tough, state.* (TA.) The م is augmentative. (Ṣgh, TA.)

جُعْسٌ, (Ṣ, Ḳ,) and ▼ جُعَسٌ, (TA,) and ▼ جُعْسُوسٌ; (AZ, Ṣ, and Ḳ in art. جعمس,) the first of which is post-classical; (Ṣ, Ḳ;) and the last, the term used by the Arabs [of the classical ages]; (Ṣ;) *Human ordure,* or *excrement;* (AZ, TA in art. جعس;) syn. رَجِيعٌ: (Ṣ, Ḳ:) or the first signifies the *place in which the* جُعْمُوس *falls:* (IDrd, Ḳ:) the م in the last word is augmentative; and its pl. is جَعَامِيسُ. (Ṣ.) You say, رَمَى بِجَعَامِيسِ بَطْنِهِ [*He cast forth the excrements of his belly*]. (Ṣ.)

جُعْمُسٌ: see the next preceding paragraph.

جُعْمُوسٌ; pl. جَعَامِيسُ: see جَعْسٌ.

جُعَامِسٌ A man (TA) *who deposits his ordure,* or *excrement, at once:* (Ḳ:) or *in a dry,* or *tough, state:* as also مُجَعْمِسٌ. (TA.)

مُجَعْمِسٌ: see what next precedes.

جعف

1. جَعَفَهُ, (Ṣ, Ḳ,) aor. ـَ, (Ḳ,) inf. n. جَعْفٌ, (TA,) *He prostrated him; threw him down upon the ground;* (Ṣ, Ḳ;) *cast him upon the ground;* like جَعَبَهُ; (TA;) namely, a man; (Ṣ;) as also اجعفهُ. (Ibn-'Abbád, Ḳ.) — *He pulled it,* or *plucked it, out,* or *up;* (Ṣ, Ḳ;) namely, a thing, (Ṣ,) or a tree; (Ḳ, TA;) *and turned it over,* or *upside down;* (TA;) as also اجتعفهُ. (Ḳ, TA.)

4: see 1.

7. انجعف [*He became prostrated, thrown down upon the ground,* or *cast upon the ground:* see its part. n., below]. — *It* (a thing, Ṣ, or a tree, Ḳ) *became pulled,* or *plucht, out,* or *up.* (Ṣ, Ḳ.)

8: see 1.

جُعَافٌ: see what next follows.

جَاعِفٌ, applied to a torrent, *i. q.* جُحَافٌ [*That carries away everything*]; as also جُعَافٌ: (Ḳ:) and *that overturns everything upon which it comes.* (TA.)

مَجْعَفٌ *A place where one is prostrated,* or *thrown down upon the ground.* (TA.)

مَجْعُوفٌ *Prostrated,* or *thrown down upon the ground;* as also مُنْجَعِفٌ. (TA.)

مُنْجَعِفٌ: see what next precedes.

جعفد

Q. 1. جَعْفَدَةٌ [inf. n. of جَعْفَدَ] a compound word from the phrase جَعَلَنِى ٱللّٰهُ فِدَاكَ [May God make me thy ransom]. (Ibn-Diḥyeh, TA.) [You say, جَعْفَدَهُ, meaning *He said to him* جعلنى الله فداك.]

جعفر

جَعْفَرٌ *A small river; a rivulet, streamlet,* or *brook;* (IAạr, Ṣ, Ḳ;) or *one that is larger than a* جَدْوَل: (Ḳ:) or, as in the Nawádir, *a small river, larger than a* جدول: (TA:) or *a river,* (IDrd, IJ, TA,) absolutely: (TA:) if small, it is a فَلَجٌ: (IDrd, TA:) or (TA, but in the Ḳ "and") *a large, wide river:* (Ibn-El-Ajdábee, Ḳ:) [if so,] bearing two contr. significations: (Ḳ:) or (in some copies of the Ḳ "and") *a full river.* (Ḳ.) — Also (from the last of these significations, TA) ‡ A she-camel *abounding with milk.* (Ḳ, TA.)

جعل

1. جَعَلَ, aor. ـَ, inf. n. جَعْلٌ (Ṣ, Mṣb, Ḳ) and مَجْعَلٌ (Ṣ, TA,) and جِعَالَةٌ and جُعْلٌ (Ḳ) and مَجْعَلٌ, (Ṣ, TA,) *He made* a thing; syn. صَنَعَ; (Mṣb, Ḳ;) but having a more general signification than فَعَلَ and صَنَعَ and their equivalents [as will be shown by what follows]; (Er-Rághib, TA;) and so اجتعل: (Ḳ:) both these verbs signify the same. (Ṣ.) — *He made* a thing of, or from, a thing; as in the saying [in the Ḳur xvi. 74 and xlii. 9], جَعَلَ لَكُمْ مِنْ أَنْفُسِكُمْ أَزْوَاجًا [*He hath made for you, of,* or *from, yourselves, wives*]; and [in the Ḳur xvi. 83] وَجَعَلَ لَكُمْ مِنَ ٱلْجِبَالِ أَكْنَانًا [*And He hath made for you, of the mountains, places of retreat;* as caves, and excavated houses or chambers: so explained by Bḍ]. (TA.) — *He created;* (Ḳ, TA;) *brought into being,* or *existence;* (TA;) as in the saying [in the Ḳur vi. 1], وَجَعَلَ ٱلظُّلُمَاتِ وَٱلنُّورَ [*And hath created,* or *brought into being, the darknesses and the light*]; (Ḳ, TA;) and [in the Ḳur xxi. 31] وَجَعَلْنَا مِنَ ٱلْمَآءِ كُلَّ شَىْءٍ حَىٍّ [*And We have created of water,* or *the seminal fluid, everything living*]; and [in the Ḳur xvi. 80, &c.,] وَجَعَلَ لَكُمُ ٱلسَّمْعَ وَٱلْأَبْصَارَ وَٱلْأَفْئِدَةَ [*And He created for you the ears and the eyes and the hearts*]. (TA.) — *He made,* or *prepared;* as in the saying [in the Ḳur lxv. 2], يَجْعَلْ لَهُ مَخْرَجًا [*He will make,* or *prepare, for him a way of escape,* or *safety*]; and [in the Ḳur lxv. 4] يَجْعَلْ لَهُ مِنْ أَمْرِهِ يُسْرًا [*He will make,* or *prepare, for him an easy state of his circumstances;* i. e., *will make his circumstances,* or *case, easy to him*]. (TA.) — *He made;* meaning *he made to be,* or *become; he constituted; he appointed;* [in which sense it is doubly trans.;] (Ṣ, Ḳ;) as in the saying in the Ḳur [xix. 31], وَجَعَلَنِى نَبِيًّا [*And He hath made me a prophet*]; (Ṣ;) [and in the elliptical phrase, جَعَلَهُ عَلَيْهِ *He made him to be* superintendant, or the like, *over it; set him,* or *appointed him, over it:*] and in the phrase, جَعَلَ ٱلْقَبِيحَ حَسَنًا [*He made that which was bad to be,* or *become, good*]. (Ḳ.) — *He made* a thing *to be in a particular state or condition;* as in the saying [in the Ḳur ii. 20], ٱلَّذِى جَعَلَ لَكُمُ ٱلْأَرْضَ فِرَاشًا [*Who hath made for you the earth to be as a bed*]; and [in the Ḳur lxxi. 15] وَجَعَلَ ٱلْقَمَرَ فِيهِنَّ نُورًا [*And hath made the moon, in them* (the heavens), *to be as a light*]; and so, as some say, in the saying [in the Ḳur xliii. 2], إِنَّا جَعَلْنَاهُ قُرْآنًا عَرَبِيًّا [*Verily we have made it an Arabic Ḳur-án*]. (TA.) — [*He made* a thing *to be in an altered,* or *changed, state* or *condition;* i. e.,] the verb signifies also the changing a thing from its state or condition; as in the saying [in the Ḳur xi. 84 and xv. 74], جَعَلْنَا عَالِيَهَا سَافِلَهَا [*We made their upper part to be their lower part*]; (Ḳ;) and in the words of the Ḳur [lvi. 81], وَتَجْعَلُونَ رِزْقَكُمْ أَنَّكُمْ تُكَذِّبُونَ [*And do ye make* the thanks that ye should render for *your sustenance to be that ye charge with falsehood* the Giver thereof by attributing it to the stars called أَنْوَآء? as expl. by Bḍ and Jel]. (TA.) — *He pronounced* (Er-Rághib, Ḳ) a thing *by a true judgment* or *decision,* (Er-Rághib,) or *as a legal ordinance;* (Ḳ;) as in the saying (of the Legislator, TA), جَعَلَ ٱللّٰهُ ٱلصَّلَوَاتِ ٱلْمَفْرُوضَاتِ خَمْسًا [*God hath pronounced the prayers that are made obligatory to be five*]. (Ḳ.) And *He pronounced* (Er-Rághib, Ḳ*) a thing *by a false judgment* or *decision,* (Er-Rághib,) or *according to his own judgment, heretically;* (Ḳ;) as in the saying [in the Ḳur xv. 91], ٱلَّذِينَ جَعَلُوا ٱلْقُرْآنَ عِضِينَ [*Who pronounced the Ḳur-án to be lies,* or *enchantment,* &c.]. (Er-Rághib, Ḳ.) — *He called,* or *named,* (Ṣ, Mṣb, Ḳ,) a thing; (Mṣb;) as in the saying [in the Ḳur xliii. 18], وَجَعَلُوا ٱلْمَلَائِكَةَ ٱلَّذِينَ هُمْ عِبَادُ ٱلرَّحْمٰنِ إِنَاثًا [*And they have called the angels, who are the servants of the Compassionate, females*]: (Ṣ, Ḳ:) or, as some say, the meaning is, *have described them as,* and *pronounced them to be, females;* like as one says, جَعَلَ فُلَانٌ زَيْدًا أَعْلَمَ ٱلنَّاسِ [*Such a one described Zeyd as,* and *pronounced him to be, the most learned of men*]: or *have held,* or *believed, them to be females;* like as the verb signifies in the saying in the Ḳur [xvi. 59], وَيَجْعَلُونَ لِلّٰهِ ٱلْبَنَاتِ [*And they hold,* or *believe, God to have daughters:* or this may be rendered *and they attribute to God daughters*]. (TA.) You say also, جَعَلْتُ زَيْدًا أَخَاكَ, meaning *I asserted Zeyd to be related to thee* [*as a brother;* or *I called Zeyd thy brother*]. (Ḳ.) — *He thought;* as in the saying, جَعَلَ ٱلْبَصْرَةَ بَغْدَادَ [*He thought El-Baṣrah to be Baghdád*]; (Ḳ;) and so in the saying, جَعَلْتُهُ عَبْدًا فَشَتَمْتُهُ [*I thought him to be a slave, and consequently I reviled him*]. (Ḥam p. 31.) — *He made known,* or *plain,* or *perspicuous;* as in the saying [in the Ḳur xliii. 2, of which one explanation has been given above], إِنَّا جَعَلْنَاهُ قُرْآنًا عَرَبِيًّا [*Verily we have made it known,* &c., *as an Arabic Ḳur-án*]: (Ḳ:) or the meaning is, *we have revealed it* [*as such*]. (TA.) — *He exalted,* or *ennobled;* as in the saying [in the Ḳur ii. 137], جَعَلْنَاكُمْ أُمَّةً وَسَطًا [*We have exalted you,* or *ennobled you, as a nation conforming to the just mean;* or *just,* or *equitable,* or *good*]: (Ḳ:) [or it may be rendered, *we have made you a nation* &c.:] or, as some say, the meaning is, *we have called you,* or *named you, a nation* &c. (TA.) — Also, inf. n. جَعْلٌ, *He put,* or *laid,* a thing; or *put it,* or *laid it, down.* (Ḳ.) And جَعَلَ بَعْضَهُ فَوْقَ بَعْضٍ *He put,* or *threw, one part of it upon another.* (Ḳ.) — *He inserted* a thing into a thing; as in the Ḳur [ii. 18], يَجْعَلُونَ أَصَابِعَهُمْ فِى آذَانِهِمْ [*They insert,* or *put, their fingers into their ears*]. (TA.) — *He put* into the heart, or mind; as in the Ḳur [lvii. 27], وَجَعَلْنَا فِى قُلُوبِ ٱلَّذِينَ ٱتَّبَعُوهُ رَأْفَةً وَرَحْمَةً [*And we put into the hearts of those who followed him pity and com-*

passion]. (TA.) — [*He appointed*, or *assigned*, or *stipulated to give*, or *gave*, wages, pay, or a stipend, &c.]. You say, جَعَلْتُ لَهُ جُعْلًا [*I appointed him, &c., wages, pay, or a stipend*]. (Mgh.) And جَعَلَ لَهُ كَذَا عَلَى كَذَا *He stipulated with him to give him such a thing for* [*doing*] *such a thing*. (Ḳ.) And جَعَلَ [alone] *He gave wages, pay, or a stipend, to another to serve for him in war, i. e., in his stead*. (Mgh.) And أَجْعَلْتُ ♦ لَهُ *I gave to him wages, pay, or a stipend*. (Ṣ,* Mgh.) And أَجْعَلَهُ ♦ جُعْلًا, and أَجْعَلَهُ ♦ لَهُ, *He gave to him wages, pay, or a stipend*. (Ḳ, TA.) And it is said in a trad., جَعَلَ لِقَوْمِهِ مِائَةً مِنَ الإِبِلِ عَلَى أَنْ يُسْلِمُوا [*He gave*, or *stipulated to give, to his people*, or *party, a hundred camels on the condition that they should surrender*]. (Mgh.) — جَعَلَ يَفْعَلُ كَذَا *He set about, began, commenced, took to*, or *betook himself to, doing such a thing*; (Ḳ,* TA;) *he became occupied in doing such a thing*. (TA.) — جَعَلَ is also, sometimes, an intrans. verb included among the verbs of appropinquation (أَفْعَالُ المُقَارَبَةِ); as in the saying,

* وَقَدْ جَعَلْتُ إِذَا مَا قُمْتُ يُثْقِلُنِى *
* ثَوْبِى فَأَنْهَضُ نَهْضَ الشَّارِبِ الثَّمِلِ *

[*And I was beginning to be*, or *at the point of being, in such a state that, when I rose, my garment heavily burdened me, so that I stood up as stands up the intoxicated drinker*]. (Ḳ.) = جَعِلَ, (Ṣ, Ḳ,) aor. ـَ, (Ḳ,) inf. n. جَعَلٌ; (Ṣ;) and ♦ اجعل; (Ḳ;) *It* (water) *had in it many* جِعْلَان, pl. of جُعَلٌ: (Ṣ, Ḳ:) or *had in it dead* جِعْلَان. (Ḳ.) — And جَعِلَ, (TḲ,) inf. n. جَعَلٌ, (IAạr, Ḳ, TḲ,) *He* (a boy, TḲ) *was*, or *became, short and fat*. (IAạr, Ḳ. [In the explanation of الجَعَلُ in the CḲ, القَصِيرُ is erroneously put for القِصَرُ.]) — And *He* (a man, TḲ) *persisted*; or *persisted obstinately*; or *persisted in contention*, or *litigation*; or *contended*, or *litigated*; جَعَلٌ being syn. with لَجَاجٌ. (IAạr, Ḳ.)

3. جاعلهُ, (A, Ḳ,) inf. n. مُجَاعَلَةٌ and جِعَالٌ, (TA,) *He endeavoured to conciliate him by means of a bribe*. (A, Ḳ.* [In the CḲ, رَشَاهُ is erroneously put for رَاشَاهُ.])

4: see جَعَلَ, above, in three places: = and see جَعِلَ, above. = اجعل القِدْرَ *He put down the cooking-pot* (Ṣ, Ḳ) *from the fire* (Ṣ) *with the piece of rag called* جِعَال. (Ṣ, Ḳ.) = أَجْعَلَتْ and ♦ استجعلت said of a bitch, (Ṣ, Ḳ,) and of other animals, (Ḳ,) of any beasts of prey, (Ṣ,) *She desired*, (Ṣ, Er-Rághib,) or *loved*, (Ḳ,) *copulation*: (Ṣ, Ḳ, Er-Rághib:) metonymically used in this sense. (Er-Rághib, TA.)

6. تجاعلوا الشَّىْءَ *They stipulated among themselves to give the thing as wages, pay, or stipend*: (Ḳ:) from جُعْلٌ. (TA.) You say also, تجاعل النَّاسُ بَيْنَهُمْ عِنْدَ البَعْثِ [*The people stipulated among themselves to give wages, or pay, to such of them as should serve as substitutes, on the occasion of being ordered forth to war*]. (TA.)

8. اجتعل: see 1, first sentence. — Also *He took*, or *received*, wages, pay, or a stipend. (Mgh, TA.)

10: see 4.

جَعْلٌ *Short palm-trees*: (Ṣ, Ḳ:) or *shoots*, or *offsets, of palm-trees, cut off from the mother-trees, or plucked forth from the ground, and planted*: or *bad palm-trees*: or *palm-trees that rise beyond the reach of the hand*: (Ḳ:) n. un. with ة: (Ṣ: [in the Ḳ, not so correctly, pl. of جَعْلَةٌ:]) and *palm-trees such as are called* بَعْل [q. v.]. (Ḳ.)

جُعْلٌ *Wages*; *pay*; *a stipend*; or *a thing that is appointed*, or *stipulated, to be given to a man for work*, or *service*; (Ṣ, Mgh, Mṣb,* Ḳ;) of more general import than أُجْرَةٌ and ثَوَابٌ; (TA;) as also ♦ جَعَالَةٌ (Ṣ, Mgh, Mṣb, Ḳ) and (as some say, Mṣb) ♦ جِعَالَةٌ (Aṣ, Mgh, Mṣb, Ḳ) and ♦ جُعَالَةٌ (Mgh, Mṣb, Ḳ) and ♦ جَعِيلَةٌ (Ṣ, Mgh, Mṣb, Ḳ) and ♦ جِعَالٌ (Ḳ) and ♦ جُعَلٌ: (Ḥar p. 134:) pl. جُعُلٌ (TA) and (of جعيلة or جعالة, Mgh) جَعَائِلُ. (Mgh, TA.) Afterwards, (Mgh,) or ♦ جَعَالَةٌ and ♦ جِعَالَةٌ and ♦ جُعَالَةٌ, (Ḳ, TA,) *Wages*, or *pay, or the like, which one gives to a man who goes to war* (Mgh, Ḳ, TA) *as a substitute for the giver*, (Ḳ, TA,) *that he may aid himself thereby to serve in the war*: (Mgh:) pl. of the last three words جَعَائِلُ. (TA.) And جُعْلٌ, (TA in art. رشو,) or ♦ جَعِيلَةٌ, (Ḳ,) *A bribe*. (Ḳ, TA.) And ♦ جَعِيلَةُ الغَرِقِ *What is given*, or *stipulated to be given, to him who dives for goods or for a man drowned*. (TA.)

جَعِلٌ and ♦ جَعَلٌ and ♦ مُجْعِلٌ Water *having in it many* جِعْلَان, pl. of جُعَلٌ: or *having in it dead* جِعْلَان. (Ḳ.) And ♦ أَرْضٌ مُجْعِلَةٌ *A land abounding with* جِعْلَان. (Ḳ.)

جَعَلٌ: see جُعْلٌ.

جِعِلٌ: see جَعِلٌ.

جُعَلٌ [The *species of black beetle called cantharus*;] *a certain insect* (دُوَيْبَّةٌ); (Ṣ, Ḳ;) *a certain black insect, found in moist places*, (TA,) *that rolls along a little ball* [*of dung*] *called* دُحْرُوجَة [*in which it deposits its eggs*]: (Ṣ and Ḳ in art. دحرج:) [see also خُنْفَسَآء: it is strangely explained in the Mṣb as the حِرْبَآء, which is the *male of the* أُمّ حُبَيْن:] pl. جِعْلَانٌ. (Ṣ, Mṣb, Ḳ.) — Hence, as being likened thereto, (TA,) ‡ *A black and ugly and small man*: or one who is *wont to persist*, or *to persist obstinately*, or *to persist in contention* or *litigation*, or *to contend* or *litigate*: and (as some say, TA) i. q. رَقِيبٌ [*a watcher, an observer, &c.*]. (Ḳ, TA.)

جِعَالٌ *A piece of rag with which a cooking-pot is put down* (Ṣ, Ḳ) *from the fire*; (Ṣ;) as also ♦ جِعَالَةٌ and ♦ جُعَالَةٌ: (Ḳ:) pl. جُعُلٌ (Ṣ, TA) and جَعَائِلُ. (TA.) = See also جُعْلٌ.

جَعُولٌ The *young of the ostrich*. (IDrd, Ḳ.)

جَعَالَةٌ: see جُعْلٌ, in three places.

جِعَالَةٌ: } see جُعْلٌ, for each in two places: =
جُعَالَةٌ: } and جِعَالٌ.

جَعِيلَةٌ: see جُعْلٌ, in two places.

جَاعِلٌ [act. part. n. of جَعَلَ] *Giving* [wages, pay, or a stipend: &c.]. (Ḳ.)

مُجْعِلٌ applied to a bitch, (Ṣ, Ḳ,) and to any animal (Ṣ, Ḳ) or beast of prey, (Ṣ,) *Desiring*, (Ṣ,) or *loving*, (Ḳ,) *copulation*. (Ṣ, Ḳ. [See 4.]) = Also, fem. with ة: see جَعِلٌ, in two places.

مُجْتَعِلٌ *Taking*, or *receiving*, [wages, pay, or a stipend.] (Ḳ.)

جعمس

Q. accord. to the Ḳ, but Q. Q. accord. to Ṣgh, جَعْمَسَ: see art. جعس.

جُعْمُوسٌ: }
جُعَامِسٌ: } see art. جعس.
مُجَعْمِسٌ: }

جف

1. جَفَّ, (Ṣ, Mgh, Mṣb,) sec. pers. جَفَفْتَ, (Ḳ,) aor. يَجِفُّ (Ṣ, Mgh, Mṣb, Ḳ) and يَجَفُّ, (Ṣ, Ḳ,) the latter aor. mentioned by AZ, but rejected by Ks; (Ṣ;) and sec. pers. جَفِفْتَ, aor. يَجَفُّ, (Ṣgh, Mṣb, Ḳ,) of the dial. of Benoo-Asad; (Mṣb;) inf. n. جَفَافٌ and جُفُوفٌ, (Ṣ, Mgh, Mṣb, Ḳ,) which are assigned by J and Ṣgh to جَفَّ aor. يَجِفُّ; (TA;) *It* (a thing, Mgh, or a garment, Ṣ, Mṣb, Ḳ, and also said of other things, Ṣ) *was*, or *became, dry*; *it dried*, or *dried up*. (Mgh, Mṣb.) Hence the saying, مَنِ احْتَلَمَ ثُمَّ أَصْبَحَ عَلَى جَفَافٍ *He who experiences an emission of semen in sleep, then rises in the morning with what is on his garment, of the semen, dry*. (Mgh.) And جَفَّ النَّهْرُ, an elliptical phrase, for جَفَّ مَآءُ النَّهْرِ [*The water of the river dried up*]. (Mṣb.) And فُلَانٌ لَا يَجِفُّ لِبْدُهُ ‡ *Such a one does not remit*, or *become remiss, in his work*, or *labour*: (TA:) or *does not cease to go to and fro*. (Ḥar p. 589.) — جَفَّ, inf. n. جُفُوفٌ, said of a man, † *He was*, or *became, silent*; *he did not speak*. (Mṣb.) = جَفَفْتُ الشَّىْءَ إِلَىَّ, aor. أَجُفُّ, inf. n. جَفٌّ, *I collected the thing to me*. (Nawádir of AZ, TA.) — جَفُّوا أَمْوَالَهُمْ *They collected together their camels, and took them away*. (Ṣgh,* Ḳ,* TA.)

2. جفّفهُ, (Ṣ, Mṣb,) inf. n. تَجْفِيفٌ (Ṣ, Mṣb, Ḳ) and تَجِفَّافٌ, (Ḳ,) *He dried it*. (Mṣb, Ḳ.) = جفّف الفَرَسَ, (Ḳ,) inf. n. تَجْفِيفٌ, (Ṣ,) *He clad*, or *attired, the horse with a* تِجْفَاف. (Ṣ, Ḳ.)

5: see R. Q. 2.

8. اجتفّ مَا فِى الإِنَآءِ *He consumed what was in the vessel*; (Ḳ;) i. e., *drank up all of it*; as also اشتفّ. (TA.)

R. Q. 2. تَجَفْجَفَ *It* (a garment, or piece of cloth), *having been moist, dried so far as to retain some moisture:* (Ṣ, Ḳ:) if it has dried entirely, you say of it, قَدْ قَفَّ: (Ṣ:) the verb is originally ▼ تجفّف; the medial ف being changed into ع: it is like تَبْشَّشْ, originally تَبْشَّشْ. (Lth, Ṣ.)

جُفَّ: see جُفَّةٌ.

جُفٌّ *The spathe of the palm-tree;* the *envelope of the* طلع; (AA, A'Obeyd, Ṣ, Ḳ;) as also جُبٌّ; (AA, TA;) or [in other words] the قِيقَآء *of the* طلع; (Ḳ;) i. e., the *envelope that is with the* وَلِيع: (Lth, Ḳ:) or, as some say, the *envelope of the* طلع *when it has become dry:* (TA:) pl. جُفُوفٌ. (A'Obeyd, TA.) [See جُبٌّ.] — *A receptacle such as is termed* وِعَآء, *that is not to be tied round at its mouth.* (Ḳ, TA.) — *An old, worn-out water-skin* or *milk-skin, of which half is cut off and made into a bucket:* (Ṣ, Ḳ:) and sometimes it is *made of the lower part of a palm-tree hollowed out:* (Lth, Ṣ, Ḳ:*) or *a thing that is hollowed out in* (فِى [probably a mistranscription for مِن, i. e. *of*]) *the trunks of palm-trees:* (A'Obeyd, TA:) or *a worn-out milk-skin* or *butter-skin:* (IAạr, TA:) or *a water-skin,* or *milk-skin, of which part is cut off at the fore legs, and in which the beverage called* نَبِيذ *is prepared:* (Ḳt, TA:) or the *lower half of a water-skin* or *milk-skin, made into a bucket:* (IDrd, TA:) or *a thing of camel's skin, like a vessel,* or *like a bucket, in which the rain-water is taken, holding half the quantity of a water-skin or the like.* (TA.) — ‡ *An old man;* (Ḳ;) as being likened to an old, worn-out water-skin or milk-skin: mentioned in the L from El-Hejeree, and by Ṣgh from Ibn-'Abbád. (TA.) — *Anything hollow, such as has something within it, like the nut, and the* مُقْدَة [or *fruit of the* تَنْضُب, &c.: in the CḲ, the مَعِدَة]. (Ibn-'Abbád, Ḳ, TA.) — The *body,* or *substance,* (شَخْص,) of a thing. (TA.) — *An obstruction that one sees between him and the ḳibleh.* (Ibn-'Abbád, Ḳ.) — هُوَ جُفُّ مَالٍ *He is a good manager of cattle,* (Ḳ,) *acquainted with the art of pasturing them, and of collecting them at their proper time in the place of pasture.* (TA.) — See also what next follows.

جَفَّةٌ (Ṣ, Ḳ) and ▼ جُفَّةٌ, (Ṣgh, Ḳ,) but the latter is rare, (Ṣgh,) and ▼ جَفٌّ (Ṣ, Ḳ) and ▼ جُفٌّ, (Ḳ,) *A company of men or people; a collective body thereof:* (Ks, Ṣ, Ḳ:) or *a great number* (Ḳ) *thereof.* (TA.) You say, دُعِيتُ فِى جَفَّةِ النَّاسِ [*I was summoned,* or *invited, among the collective body of people*]. (Ṣ.) And جَآءُوا جَفَّةً وَاحِدَةً (Ṣ, Ḳ) *They came in one collective body.* (Ḳ.) لَا نَفَلَ فِى غَنِيمَةٍ حَتَّى تُقْسَمَ جَفَّةً, (Ṣ, Mgh,) or ▼ جُفَّةً, (Ḳ,) means [*There shall be no gift of spoil*] *until it is divided altogether:* (Ṣ, Mgh, Ḳ:) a saying of Ibn-'Abbás: (Ṣ, Mgh:) accord. to one reading, ▼ عَلَى جُفَّتِهِ, i. e., [until it is divided] *among the collective body of the army first.* (Ḳ. [Golius (here copied by Freytag) appears to have read, لَا نَفَلَ فِى غَنَمٍ جَفَّةً; and hence to have said, of جَفَّةٌ, erroneously, "de pecore non dicitur nisi *totus grex* sit."])

جُفَّةٌ: see what next precedes, in three places. — Also *A great* دَلْو [or *bucket*]. (Ḳ.)

جُفَافٌ *What is dry* of a thing that one has dried. (Ḳ.) You say, اعْزِلْ جُفَافَهُ مِنْ رَطْبِهِ [*Put thou apart what is dry thereof from what is fresh and moist thereof*]. (TA.)

جَفِيفٌ *Dry* herbs or herbage: (Ṣ, Ḳ:) or *dry* leguminous plants or herbs, of the kind that are eaten without being cooked: (TA:) or of this kind and of such as are thick and inclining to bitterness; as also قَفِيفٌ: (TA in art. قف:) or, as some say, ما ضمنت من الريح. (TA in the present art. [But what this means I know not; the verb being evidently mistranscribed.])

جُفَافَةٌ *What has become scattered,* or *strewed, of dry herbage* (حَشِيش) *and of* [*the kind of trefoil called*] قَتّ, (Ṣ, Ḳ, TA,) *and the like.* (TA.)

تِجْفَافٌ *A thing,* (Ṣ, Mgh, Mṣb,) i. e. *a kind of armour,* (IAth, Ḳ,) [*a cataphract,*] *with which a horse is clad,* (Ṣ, IAth, Mgh, Mṣb, Ḳ,) *in war, in the manner of a coat of mail,* (Mgh, Mṣb,) *to defend him from being wounded;* (IAth;) and *sometimes worn by a man, to defend him in war:* (Ḳ:) of the measure تِفْعَالٌ, (Mgh, Mṣb,) the ت being augmentative, (Aboo-'Alee the Grammarian, Ṣ, IJ,) to render the word quasi-coordinate to the class of قِرْطَاسٌ; (IJ;) from جَفَّ, because of its hardness and toughness: (Mgh, Mṣb:) pl. تَجَافِيفُ. (Ṣ, Mgh, Mṣb.) It is said in a trad., أَعِدَّ لِلْفَقْرِ تِجْفَافًا; and one says, اِلْبَسْ لِلْفَقْرِ تِجْفَافًا; [both] meaning, †*Make thou preparation for poverty.* (TA.)

مُتَجَفِّفٌ *Having a* تِجْفَاف *upon his horse.* (Mgh.)

جفا

جَفَا, (Ṣ, Ḳ,) aor. ـُ, (Ḳ,) inf. n. جَفْءٌ, (Ṣ,) *It* (a valley [flowing with water]) *cast forth froth,* or *foam,* (Ṣ, Ḳ,) *and particles of rubbish* or *refuse;* (Ṣ;) as also ▼ اجفأ; (Ḳ;) but this latter is said in the O to be of weak authority. (TA.) And جَفَأَتِ القِدْرُ, and ▼ اجفأت, *The cooking-pot cast forth its froth,* or *foam,* (Ṣ, Ḳ,) *in boiling:* (Ṣ:) or جفأت بزَبَدِهَا *it cast forth its froth,* or *foam:* (Ḥam p. 132:) originally جَفَّت and اجفّت, without ء. (Er-Rághib, TA in art. جفو.) — جَفَأَ الغُثَآءَ عَنِ الوَادِى, (Ḳ,) or جَفَأَ الوَادِىَ, (IAạr, O,) *He* (a man, IAạr, O) *swept off the scum and rubbish of the valley* [after it had flowed, or while it was flowing, with water]. (IAạr, O, Ḳ.) And جَفَأَ القِدْرَ *He cleared off the froth,* or *foam, of the cooking-pot.* (Ḳ, TA.) — Also جَفَأَ القِدْرَ, (Ṣ, Z in the Fáïḳ, TA,) inf. n. as above; (Ṣ, TA;) and ▼ أَجْفَأَهَا; (Z ubi suprà, TA;) but the former is that which is commonly known; (ISd, TA;) the latter is rare; (IAth, TA;) or the latter should not be said, though it occurs in a trad., (Ṣ, TA,*) accord. to one relation; (TA;) *He turned the cooking-pot upside-down,* or *inclined it,* (Ṣ, Z ubi suprà, TA,) *and poured out what was in it:* (Ṣ:) or *he emptied the cooking-pot, and turned it upside-down:* (TA:) and جَفَأَ البُرْمَةَ فِى القَصْعَةِ *He turned the cooking-pot upside-down upon the bowl.* (Ḳ.) — جَفَأَهُ, (Ṣ, Ḳ,) [like حَفَأَهُ,] and جَفَأَ بِهِ الأَرْضَ, (TA,) *He threw him down,* or *prostrated him, on the ground;* (Ṣ, Ḳ, TA;) namely, a man: (Ṣ:) and ▼ اجفأ بِهِ [signifies the same; or] *he threw him,* or *it,* (Ḳ, TA,) *on the ground.* (TA.) — See also 8.

4: see 1, in four places.

8. اجتفأ *He pulled,* or *plucked, up,* or *out,* or *he uprooted,* (Ṣ, Ḳ,) *and threw down,* or *away,* a thing, (Ṣ,) or plants, or herbs, such as are termed بَقْل, (Ḳ,) and trees; (TA;) [but see احتفى;] as also ▼ جَفَأَ, (Ḳ,) aor. and inf. n. as above: (TA:) [or] both signify *he cut* a plant, or herb. (IAạr, Nh.)

جُفَآءٌ *What is cast forth* [*of froth,* or *foam, and particles of rubbish* or *refuse,* (see 1,)] *by a torrent:* (ISk, Ṣ:) the *froth,* or *foam, cast forth by a valley* [*flowing with water*]; and *by a cooking-pot,* (Ḳ, TA,) *in boiling.* (TA.) — Hence, as being likened to the froth, or foam, of the cooking-pot, of which no use is made, (Fr, TA,) *i. q.* بَاطِلٌ [meaning ‡ *A thing that is worthless, useless,* or *unprofitable*]. (Fr, Ḳ, TA.) It is said in the Ḳur [xiii. 18], فَأَمَّا الزَّبَدُ فَيَذْهَبُ جُفَآءً, meaning بَاطِلًا [i. e. ‡ *Now as to the froth,* or *scum, it passeth away as a thing that is worthless,* or *useless,* or *unprofitable*], (Fr, Ṣ, Jel, TA,) and *thrown away.* (Jel.) You say also, ذَهَبَ الزَّبَدُ جُفَآءً, meaning [*The froth,* or *scum, passed away*] *driven from its water.* (TA.) — جُفَآءٌ مِنَ النَّاسِ, occurring in a trad., is explained by IAth as meaning † *The first,* or *foremost, of the men* or *people* (سَرَعَانُهُمْ): but Bkh and Muslim read (instead of جفاء,) أَخِفَّآءُ, pl. of خَفِيفٌ. (TA.) — Also, [like جُفَايَةٌ,] *An empty ship.* (O, Ḳ.)

جفر

1. جَفُرَ *He,* or *it, became wide:* (Ḳ:) or *became inflated,* or *swollen.* (A.) And جَفَرَ جَنْبَاهُ *His* (a kid's, Ṣ and Mṣb, or lamb's, Mṣb) *sides became widened,* or *distended:* (Ṣ, Mṣb:) and ▼ اجفر جَنْبَاهُ [and ▼ انجفر (Ḳ in art. هضر)] *his* (a horse's) *sides became inflated,* or *swollen.* (A.) — *He* (a lamb, Ḳ, and a kid, TA) *became what is termed* جَفْر; as also ▼ تجفّر and ▼ استجفر: (Ḳ:) and ▼ تجفّرت and ▼ استجفرت *she* (a kid) *became a* جَفْرَة. (ISh, TA.) And ‡ *He* (a boy) *became what is termed* جَفْر; as also ▼ تجفّر (TA) and ▼ استجفر: (A:) and this last verb, *he became large in the sides.* (L.) — جَفَرَ عَنِ الضِّرَابِ, (Ṣ, A) (Ṣ,) or عَنِ الإِبِلِ, (A,) aor. ـُ, (Ṣ,) inf. n. جُفُورٌ; (Ṣ, Ḳ;) and ▼ اجفر, inf. n. إِجْفَارٌ; and ▼ اجتفر; and ▼ جفّر, inf. n. تَجْفِيرٌ; (Ḳ;) *He* (a stallion-camel) *ceased,* (Ṣ, Ḳ,) or *abstained,* (A,) *from covering,* (Ṣ, A, Ḳ,) *and avoided it; having indulged in it so much that he was wearied;* (Ṣ;) *and his seminal fluid became little:* (TA:) you say of a ram, رَبَضَ, (Ṣ, A,) not جَفَرَ. (Ṣ.) And

جَفَرَ عَنِ المَرْأَةِ, (IAạr, TA,) and ▼اجفر عَنْهَا, (IAạr, Ḳ,) and ▼اجتفر, and ▼جفّر, (IAạr, TA,) *He* (a man) *abstained from the woman;* (Ḳ;) *he abstained from sexual intercourse with her.* (IAạr, TA.) — جَفَرَ مِنَ المَرَضِ *He recovered from the disease.* (Ḳ, TA.)

2: see 1, in two places. — جفّرهُ الأَمْرُ عَنْهُ *The thing, or affair, cut him off from him, or it.* (IAạr, L.)

4: see 1, in three places. — اجفر also signifies *He cut, abandoned,* or *forsook,* (Ṣ, Ḳ,) another, (Ṣ,) or his companion, or friend, (Ḳ,) *and left off visiting* him. (Ṣ, Ḳ.) And أَجْفَرْتُ مَا كُنْتُ فِيهِ *I left,* or *relinquished, that in which I was occupied.* (Ṣ.) — Also *It* (a thing, TA) *was,* or *became, absent,* or *hidden,* or *concealed,* (Ḳ, TA,) *from one.* (TA.)

5: see 1, in three places.

7: see 1.

8: see 1, in two places.

10: see 1, in three places.

جَفْرٌ *A lamb,* or *kid, whose sides have become widened,* or *distended:* (Mṣb:) or *a lamb,* (IAmb, Mṣb, Ḳ,) and *a kid,* (Ḳ,* TA,) *that has become large, and begun to pasture,* (Ḳ, TA,) *and whose sides have become widened,* or *distended:* (TA:) or *a lamb,* (Ḳ,) or *a kid,* (Ṣ, Mgh, Mṣb, Ḳ,* TA,) *that is four months old,* (Ṣ, Mgh, Mṣb, Ḳ,) *and whose sides have become widened, and that is weaned,* (A'Obeyd, Ṣ,) *and has taken to pasture:* (A'Obeyd, TA:) or *this is sometimes four months, and sometimes five months, after the birth:* or *a young lamb,* and *a kid, after it has been weaned, when six months old:* (IAạr, TA:) pl. [of pauc.] أَجْفَارٌ (Ḳ) and [of mult.] جِفَارٌ (Mṣb, Ḳ) and جِفَرَةٌ: (Ḳ:) fem. with ة: (Ṣ, A, Mgh, Mṣb, Ḳ:) or جَفْرَةٌ signifies *a female kid that has become satiated with leguminous herbs and with shrubs, and is independent of its mother:* (ISh, TA:) IAmb applies it to a female lamb and a female kid; and this is correct, though some say that it is applied to the latter only. (TA.) — *A boy when his belly has become widened,* (A,) or *when his flesh has become swollen out,* (Ḳ,) *and he has begun to eat:* (A, Ḳ:) fem. with ة. (Ḳ.) [See شَادِخٌ; and see also مُطْبِخٌ.] — *A well,* (Mṣb, Ḳ,) or *a wide well,* (Ṣ, A,) *not cased,* or *walled round, within;* (Ṣ, A, Mṣb, Ḳ;) as also ▼جُفْرَةٌ: (R, TA:) or, *of which a portion is cased,* or *walled round, within,* (Ḳ, TA,) *and a portion is not:* (TA:) the former of the masc. gender: pl. جِفَارٌ. (Mṣb.) — [Hence,] فُلَانٌ مُتَهَدِّمُ الجَفْرِ (A, Ḳ*) [*Such a one's well is in a state of demolition;* meaning,] ‡ *such a one has no judgment:* (A:) or *has no intelligence.* (Ḳ.) And إِنَّ جَفْرَكَ عَلَيَّ لَهَارٍ [*Verily thy well is falling in upon me;* meaning] ‡ *thy mischief is coming quickly upon me.* (A, TA.) — [The pl.] جِفَارٌ also signifies [simply] *Wells.* (Ḳ.) — And hence, (TA,) ‡ She-camels *abounding with milk.* (Ḳ, TA.)

جَفْرَةٌ: see جُفْرَةٌ.

جُفْرَةٌ *A round space in the ground:* (Ṣ, Ḳ:) or *a round and wide cavity in the ground:* (L:) pl. جِفَارٌ. (Ṣ.) — Hence, (Ṣ,) The *belly,* or *interior of the body:* (Ṣ:) or the *cavity of the chest:* or *what comprises the belly* [in the TA *the chest*] *and the two sides:* (Ḳ:) or the *place where the ribs curve;* and so in a horse &c.: (TA:) the *middle* of a horse, (Ṣ, Mṣb, Ḳ,) and of a camel: and, as some say, the *middle, and main part,* of anything: and thus, the *main part* of the sea: (TA:) pl. جُفَرٌ and جِفَارٌ, accord. to the Ḳ; but the latter is pl. of جُفْرَةٌ in the sense of "a round cavity." (TA.) — Also [the pl.] جُفَرٌ signifies The *holes that are dug in the ground for props.* (TA.)

جَفِيرٌ *A kind of quiver like the* كِنَانَة, *but wider,* (Lth, Ṣ, TA,) *in which are put many arrows:* (Lth, TA:) or *a* [*quiver of the kind called*] جَعْبَة [q. v.], *of skins, in which is no wood:* or *of wood, in which are no skins;* (Ḳ;) or *in which is no skin;* as in some good lexicons: (TA:) or *of skins, and slit in its side, that the wind may enter it, and the feathers in consequence may not be eaten:* (TA: [see also جَشِيرٌ:]) or the *same as the* جعبة *and the* كنانة: (El-Aḥmar, TA:) or *a quiver for* نَبْل, *wide, of wood.* (Ḥam p. 358.) Hence, لَيْسَ فِى جَفِيرِهِ غَيْرُ زَنْدَيْنِ [*There is not in his quiver* aught *save two pieces of wood for producing fire*]: a prov. applied to him in whom is no good. (Meyd.)

مَجْفَرَةٌ: see مُجْفَرٌ.

مُجْفَرٌ, applied to a horse, (Ṣ, Mṣb, Ḳ,) and with ة applied to a she-camel, (Ṣ,) *Large in the middle:* (Ṣ, Mṣb, Ḳ:) and مُجْفَرُ الجَنْبَيْنِ a horse *inflated,* or *swollen, in the sides.* (A.)

مَجْفَرَةٌ (Ṣ, A, Ḳ) and ▼مَجْفَرٌ (Lḥ, Ḳ) *An impediment to venery;* (Lḥ, A'Obeyd, Ṣ, A, Ḳ;) *and a cause of diminishing the seminal fluid:* (A'Obeyd, TA:) applied to food: (Lḥ, Ḳ:) and such is fasting said to be; (A'Obeyd, Ṣ, Ḳ;) and the sun, (A,* TA,) i. e., sitting in the sun; and such, also, the sleeping between daybreak and sunrise, or in the first part of the day. (TA.)

جفل

1. جَفَلَ, (Ṣ, Mṣb, Ḳ,) aor. ـُ and ـِ, inf. n. جَفْلٌ (Mṣb) and جُفُولٌ, (Mṣb, Ḳ,) *He* (a camel) *took fright,* or *shied, and fled,* or *ran away at random;* or *became refractory, and went away at random;* or *ran away,* or *broke loose, and went hither and thither by reason of his sprightliness:* and ▼اجفل *he* (a bird) *took fright, and flew away;* or *became scared away:* (Mṣb:) or the former, *he* (an ostrich, Ḳ) *hastened,* or *sped,* (Ṣ, Ḳ,) *in his pace,* (TA,) *and went away in the land,* or *country;* as also ▼اجفل; (IDrd, Ḳ;) both, said of an ostrich, mean *he spread his wings, running;* (Ḥam p. 555;) or *spread his wings, and ran quickly,* or *went away at random and swiftly:* (TA:) or جَفَلَتِ النَّعَامَةُ means *the ostrich fled:* (Mṣb:) and ▼اجفل عَنْهُ, said of anything, *he fled from it:* (TA, Ḥam p. 555:) and جَفَلُوا, aor. ـِ, inf. n. جَفْلٌ; (Mṣb;) and ▼اجفلوا (Ṣ, Mṣb) and ▼انجفلوا and ▼تجفّلوا; (Mṣb;) *they* (a company of men) *fled quickly;* (Ṣ, Mṣb;) or the second (Ḳ) and third (Ṣ, Ḳ) signify *they became displaced,* (Ṣ, Ḳ, TA,) *and quickly defeated,* (TA,) *and went away;* (Ṣ, Ḳ, TA;) or these two and the fourth, (TA,) or all the four, (Ḥar p. 373,) *they hastened in defeat and flight:* (TA, and Ḥar ubi suprà:) and جَفَلَتِ الرِّيحُ, (Ḳ,) and ▼اجفلت, (Ṣ, Ḳ,) *the wind was swift* (Ṣ, Ḳ, TA) *in blowing.* (TA.) — جَفَلَ, inf. n. جُفُولٌ, ‡ *It* (hair) *became shaggy,* or *dishevelled, and frouzy,* or *altered in smell, in consequence of its being seldom dressed;* or *dusty and matted, by reason of its being seldom anointed;* (Ḳ, TA;) and *became raised and spread.* (TA.) — جَفَلَ is also trans., signifying *He made* a bird *to take fright, and fly away;* or *he scared* it *away:* its quasi-pass. is ▼اجفل [explained above]; the reverse of the rule commonly obtaining: (Mṣb:) or the former verb, as in the O; not the latter, as in the Ḳ; *he made* a male ostrich *to hasten,* or *speed, in his pace, and to go away in the land,* or *country;* or *made him to spread his wings, and run quickly,* or *go away at random and swiftly:* (TA:) and ▼جفّل *he,* or *it, made* an animal, or animals, *to take fright, and flee,* or *run away at random;* or *scared away* it, or them: (TA:) [and, app., *he frightened;* تَجْفِيل being also said in the TA to be syn. with تَفْرِيغ, which, I think, is evidently a mistranscription for تَفْزِيع.] You say, جفّل ▼ القَنَّاصُ الوَحْشَ [*The sportsman scared away the wild animals*]. (TA.) And أَتَوْهُمْ فَجَفَّلُوهُمْ ▼ عَنْ مَرَاكِزِهِمْ [*They came to them, and scared them,* or *frightened them,* or *made them to flee, away from their stations*]. (TA.) And جَفَلَتِ الرِّيحُ الظَّلِيمَ *The wind put in motion the male ostrich, and drove him away,* or *along:* (Ḳ:) and [in like manner] السَّفِينَةَ [† *the ship*]. (TA.) And جَفَلَتِ الرِّيحُ السَّحَابَ ‡ *The wind smote the clouds, and put them into a state of commotion,* (Ḳ, TA,) *and made them to speed along.* (TA.) And الرِّيحُ تَجْفُلُ الجَهَامَ † *The wind carries away the rainless clouds.* (Mgh. [See also 4.]) Whence, app., (Mgh,) جَفَلَ البَحْرُ سَمَكًا † *The sea cast fish upon the shore;* (Lth, Mgh, Ḳ;) a verb like ضَرَبَ; occurring in a trad., in which it is erroneously said to be أَجْفَلَ. (Mgh.) — Also, (Ḳ,) aor. ـُ, inf. n. جَفْلٌ, (TA,) *He prostrated* a man; *threw* him *down upon the ground.* (Ḳ.) You say, طَعَنَهُ فَجَفَلَهُ, meaning *He thrust him,* or *pierced him,* [with a spear or the like,] *and displaced and prostrated him.* (Mgh.) — *He threw* goods *one upon another.* (IDrd, Mṣb, TA.) — *He,* or *it, overturned,* or *turned upside-down.* (TA.) — Also, aor. ـُ, (Ḳ,) inf. n. جَفْلٌ, (TA,) *He peeled, pared, stripped,* or *scraped off,* a thing; (AZ, Ḳ, TA;) as, for instance, flesh from the bone, and fat from the skin; (AZ, TA;) and so ▼جفّل, (Ḳ,) inf. n. تَجْفِيلٌ: (TA:) *he removed* flesh *from the bone:* (Ḳ:) app. formed by transposition from جَلَفَ. (TA.) — Also, (Mṣb, Ḳ,) aor. ـِ, (Mṣb,) or ـُ, (Ḳ,) *He swept away* mud (Mṣb, Ḳ, TA) from the ground; (TA;) and so ▼جفّل. (Ḳ.) [It seems that Golius found, in a copy of the Ḳ, التِّبْنَ erroneously put for الطِّينَ;

and حَرَقَهُ for جَرَفَهُ; for he has explained the former verb as meaning "*combussit* stramen."]

2: see 1, in five places.

4: see 1, in six places. — You say also, أَجْفَلَتِ الرِّيحُ بِالتُّرَابِ † *The wind carried away the dust; made it to fly away.* (Ṣ.) And اجفل الغَيْمُ *The clouds, or mist, became removed, or cleared off.* (TA.)

5: see 1. — You say of a cock, تجفّل, meaning نَفَشَ بُرَائِلَهُ [i. e., † *He ruffled the feathers around his neck*]. (Ibn-'Abbád, Ḳ, TA.)

7: see 1. — انجفل also signifies ‡ *It went away, or departed;* said of the shade, (Ḳ, TA,) and of the night. (TA.) — *He, or it, became overturned,* or *turned upside-down.* (TA.) — انجفلت الشَّجَرَةُ *The tree, blown upon by a violent wind, became uprooted.* (TA.)

جَفَلٌ: see إِجْفِيلٌ. It is an inf. n. used as an epithet; and means *A people, or party, fleeing quickly;* as also ↓ جَفَالَةٌ. (Mṣb.) — Also *A cloud that has poured forth its water and gone away* (Ṣ, Ḳ) *quickly;* (Ṣ;) because it is then lighter and quicker. (Ḥar p. 373.) — *A ship;* (Ḳ;) because the wind drives it along (تَجْفِلُهَا): (TA:) pl. جُفُولٌ. (Ḳ.) ═ *Ants: black ants:* (Ḳ:) *large black ants:* (TA:) a dial. var. of جَثَلٌ. (Ḳ.)

وَقَعَتْ فِى النَّاسِ جَفْلَةٌ [*Fear fell upon the people;*] *the people feared.* (TA.) ═ شَجَرَةٌ جَفْلَةٌ *A leafy tree; a tree having many leaves.* (Ḳ.) ═ See also what next follows.

جُفْلَةٌ (Ṣ, Ḳ) and ↓ جَفْلَةٌ (TA [there said in one place to be بالفتح, but this is most probably a mistranscription for بالضم,]) *A fleece* of wool: (Ṣ, Ḳ:) [a word used in the sense of] a pass. part. n., like غُرْفَةٌ in the phrase اِغْتَرَفَ غُرْفَةً. (Ṣ.)

دَعَوْتُهُمُ الجَفَلَى, (AZ, Ṣ, Mṣb,* Ḳ,*) and ↓ الأَجْفَلَى, (AZ, Ṣ, Ḳ,*) which latter was unknown to Aṣ, (Ṣ,) *I invited them* to my feast, or food, (AZ, Ṣ, Mṣb, Ḳ,*) *in common,* (AZ, Ṣ, Mṣb,) *without distinction,* (Mṣb,) or *with their company and commonalty.* (Ḳ.) And دُعِىَ فُلَانٌ فِى النَّقَرَى لَا فِى الجَفَلَى, (Akh, Ṣ, Mṣb,*) and ↓ الأَجْفَلَى, *Such a one was invited among the distinguished persons, not among the commonalty.* (Akh, Ṣ.) And دَعْوَةٌ جَفَلَى *A general invitation; contr. of* دَعْوَةٌ نَقَرَى. (Mṣb.) And جَآءَ القَوْمُ ↓ أَجْفَلَةً, and أَزْفَلَةً, (Fr, Ṣ, Ḳ,*) *The people came in a company;* (Fr, Ṣ;) and ↓ بِأَجْفَلَتِهِمْ, and بِأَزْفَلَتِهِمْ, *with their company.* (Fr, Ṣ, Ḳ.) Accord. to some, (Ṣ,) ↓ أَجْفَلَى signifies *A collection,* or *an assemblage,* of any things; (Ṣ, Ḳ;) as also أَزْفَلَى: (Ṣ:) and ↓ جُفَالَةٌ, (Ṣ, Ṣgh, TA,) or ↓ جَفَالَةٌ, (Ḳ,) *a company,* or *an assembly,* (Ṣ, Ṣgh, Ḳ,) of men, (Ṣ, TA,) *going along quickly.* (TA.)

جَفْلَانُ, or جَفْلَانٌ, [whether with or without tenween is not shown,] *Fearful; wont,* or *apt, to take fright and flee,* or *run away at random.* (TA.) [See also جَفَّالٌ.]

جَفَالٌ: see what next follows.

جُفَالٌ *What is cast forth by a torrent,* (Ṣ, Ḳ, TA,) *of rubbish and scum,* or *of rotten leaves mixed with scum;* (TA;) as also ↓ جَفَالٌ, like سَحَابٌ; (TA;) and ↓ جُفَالَةٌ. (Ḳ,* TA.) — *The froth of milk.* (Ḳ.) ═ *Much* (Ḳ) of anything: (TA:) or of wool; as also ↓ جَفِيلٌ: (Ḳ:) or *much wool.* (Ṣ.) The ewe is represented as saying, أُوَلَّدُ رُخَالًا وَأُجَزُّ جُفَالًا وَأُحْلَبُ كُثَبًا ثِقَالًا وَلَمْ تَرَ مِثْلِى مَالًا [*I am delivered of lambs, and I am shorn of much wool, and I am milked of heavy bowlfuls, and thou hast not seen cattle the like of me*]: by أُجَزُّ جُفَالًا is meant *I am shorn* [*of much wool*] *at once;* for nought of her wool falls to the ground until all of it is shorn. (Ṣ.) جفال is applied, by Dhu-r-Rummeh, as an epithet to hair; [meaning *Much,* or *abundant;*] and it is not applied as an epithet to anything save what is much, or abundant. (Ṣ.) Ed-Dejjál [or Antichrist] is described, in a trad., as جُفَالُ الشَّعَرِ *Having much hair:* (TA:) and ↓ جَافِلُ الرَّأْسِ [also] has this meaning. (Ḥam p. 460.)

جَفُولٌ A wind (رِيحٌ) *that smites the clouds, and puts them into a state of commotion;* (Ḳ;) or *that makes them to speed along:* (TA:) a *swift* wind; (TA;) as also ↓ جَافِلَةٌ and ↓ مُجْفِلٌ: (Ṣ, Ḳ:) pl. of the first, (i. e., of جفول,) جُفُلٌ. (Ḳ.) — *Great,* or *large:* so in the phrase جُمَّةٌ جَفُولٌ [*A great, or large, quantity of hair extending beyond the ears*]. (Ḳ.) — An *aged* woman; (Ḳ,* TA;) as also ↓ إِجْفِيلٌ: (Ḳ:) pl. of the former as above. (Ḳ.)

جَفِيلٌ: see جُفَالٌ.

جَفَالَةٌ: see جَفَلٌ.

جُفَالَةٌ: see الجَفَلَى: — and جُفَالٌ. — Also الجُفَالَةُ, (Ḳ,) or جُفَالَةُ القِدْرِ, (Ṣ,) *What one takes from the head* [*of the contents*] *of the cooking-pot with the ladle.* (Ṣ, Ḳ.)

جَفَّالٌ an intensive epithet from جَفَلَ in the first of the senses explained above; i. e., A camel *that takes fright,* or *shies, and flees,* &c., *much,* or *often.* (Mṣb.) [See also جَفْلَانُ.]

جَفَالَةٌ: see الجَفَلَى.

جَافِلٌ part. n. of جَفَلَ in the first of the senses explained above: (Mṣb:) [and in other senses.] — *Hastening,* or *speeding.* (TA.) See جَفُولٌ. — *Disquieted, disturbed, agitated,* or *flurried.* (Ṣ, Ḳ, TA.) — See also جُفَالٌ.

جَيْفَلٌ a name of [*The month*] ذُو القَعْدَةِ, (Ḳ, TA,) in the time of paganism. (TA.)

أَجْفَلَةٌ: see الجَفَلَى, in two places.

أَجْفَلَى: see الجَفَلَى, in three places.

إِجْفِيلٌ *Cowardly,* or *a coward,* (Ṣ, Ḳ, TA,) *that is frightened at everything.* (TA.) A he-ostrich (Ṣ, Ḳ) *that takes fright,* (Ḳ,) *and flees from everything* (Ṣ, Ḳ, TA) *that he sees;* (TA;) as also ↓ جَفَلٌ. (Ḳ.) — *A bow of which the arrow goes far.* (Ḳ.) — See also جَفُولٌ.

مُجْفِلٌ *Turning away,* or *going back,* or *retreating; going away.* (TA.) — See also جَفُولٌ.

مِجْفَلٌ applied to a camel's hump, *Heavy:* [properly, *an instrument of overturning:*] applied as an epithet to a camel's hump that is so heavy as to overturn the animal when, after rolling on the ground, he desires to rise. (TA.)

جفن

1. جَفَنَ نَاقَةً, (Ḳ,) inf. n. جَفْنٌ, (TA,) *He slaughtered a she-camel, and gave her flesh for food* (Ḳ, TA) *to the people,* (TA,) *in bowls* (جِفَان). (Ḳ, TA.)

2. جفّنوا *They made bowls* (جِفَان) [probably meaning *they prepared bowls of food:* accord. to Freytag's Lex., جفّن means "apposuit scutellam;" but he does not name his authority]). (TA.) ═ جفّن and ↓ تجفّن *It* (a grape-vine) *attained to the state of having an* أَصْل [i. e., app., *a stock*]. (TA.)

5: see 2.

جَفْنٌ The *eyelid;* both the *upper* and the *lower:* (Ṣ, Mṣb, Ḳ:) of the masc. gender: (Mṣb:) pl. [of pauc.] أَجْفُنٌ and أَجْفَانٌ and [of mult.] جُفُونٌ. (Ḳ.) — The *upper surface,* and the *lower,* of a cake of bread: both together being called جَفْنَا الرَّغِيفِ. (Lḥ, TA.) — The *scabbard,* or *sheath,* (غِمْد, Ṣ, Ḳ, or غِلَاف, Mṣb,) of a sword: (Ṣ, Mṣb, Ḳ:) [or] the *case,* or *receptacle, in which is* [*put*] *the sword together with its* غمد *and suspensory belt or cord:* (Ṣ voce قِرَابٌ:) [but the former signification only is commonly known:] and [it is said that] ↓ جِفْنٌ signifies the same; (Ḳ;) but this is doubted by IDrd: (M, TA:) pl. [of pauc.] أَجْفَانٌ and [of mult.] جُفُونٌ. (Mṣb.) — The أَصْل [app. here meaning *stock*] *of a grape-vine:* (Ḳ:) or *a grape-vine itself,* in the dial. of El-Yemen; (T, TA;) so called as being imagined to be the receptacle of the grapes: (Er-Rághib, TA:) or *a species of grape:* (ISd, Ḳ:) or the *skin of the grape, in which is the juice:* (IAạr, TA:) or *a climbing shoot of a grape-vine:* (AḤn, TA:) or the *shoots of the grape-vine:* (T, Ṣ, M, Ḳ:) n. un. with ة: (T, Ṣ, M:) or, accord. to IAạr, جَفْنَةٌ is *syn. with* كرعة [app. a mistranscription for كَرْمَةٌ *a single grape-vine*]: or, accord. to some, as ISd says, the *leaves of the grape-vine.* (TA.) [Hence,] مَآءُ الجَفْنِ *The juice of the vine;* (A, TA;) *wine:* (TA:) [or it may originally mean *tears;* then, *rain;* and then, *wine:* for] wine is also called جَفْنُ المَآءِ: and مَآءُ السَّحَابِ means *the clouds.* (TA.) — *A kind of tree, of sweet odour.* (AḤn, Ḳ.) — *A certain plant, of the kind called* أَحْرَار, *that grows in a spreading manner, and, when it dries up, contracts; having grains like the* حُلْبَة [or *fenugreek*]. (AḤn, TA.)

جِفْنٌ: see جَفْنٌ.

جَفْنَةٌ *A* [*bowl of the kind called*] قَصْعَة: (Ḳ:) or *like a* قصعة: (Ṣ:) the *largest kind of* قصعة; (Ks, Ṣ in art. صحف, M;) next to which is the

قصعة [properly so called], which satisfies the hunger of ten [men]; then, the صَحْفَة, which satisfies five; then, the مِئْكَلَة, which satisfies two men, and three; then, the صُحَيْفَة, which satisfies one man: (Ks, Ṣ in art. صحف:) it is peculiarly applied to *a receptacle for kinds of food:* (Er-Rághib, TA:) pl. [of mult.] جِفَانٌ (Ṣ, Mṣb, Ḳ) and جِفَنٌ (Sb, TA) and (of pauc., TA) جَفَنَاتٌ. (Ṣ, Mṣb, Ḳ.) [Hence,] كُفِئَتْ جَفْنَتُهُ [*His bowl was turned upside-down;* meaning] ‡*he was slain;* a phrase similar to هُرِيقَ رِفْدُهُ. (A in art. رفد.) —— †*A small well;* (Ḳ;) as being likened to the جفنة for food. (Er-Rághib, TA.) —— ‡*A generous man:* (Ḳ:) جَفْنَةٌ غَرَّآءُ is an appellation applied to *a generous man who entertains many guests and feeds many:* (IAạr, TA:) he is called جفنة because people are fed in the جفنة, and the epithet غرّآء is added because of the whiteness of the camel's hump in the جفنة. (TA.) —— Also *i. q.* خَمْرَةٌ [meaning *Some wine,* or *a kind of wine:* see also مَآءُ الجَفْنِ, voce جَفْنٌ]. (IAạr, TA.)

جفو

1. جَفَا, (Ḳ,) aor. ـُ, (TA,) inf. n. جَفَآءٌ; and ▼تجافى; *It did not keep,* or *cleave, to its place.* (Ḳ.) You say, جَفَا جَنْبُهُ عَنِ الفِرَاشِ, (Mgh,) and ▼تجافى عَنْهُ, (Ṣ, Mgh,) *His side did not rest,* or *was restless,* or *uneasy, upon the bed;* or *shrank from it;* (Ṣ, Mgh;) and *heaved,* or *rose, from it:* (Mgh:) or the former, *his side did not keep,* or *cleave, to its place upon the bed:* and the latter, *it became withdrawn,* or *removed, from it.* (TA.) And جَفَا السَّرْجُ عَنْ ظَهْرِ الفَرَسِ, (Ṣ, Mṣb, TA,) aor. and inf. n. as above; (Mṣb;) and ▼تجافى; (Ṣ, Mṣb;) *The saddle heaved,* or *rose, from the back of the horse:* (Ṣ, Mṣb:) or *did not keep,* or *cleave, to its place upon his back.* (TA.) —— [Hence,] it is said in a trad. of 'Omar, إِنِّى أَجْفُو عَنْ أَشْيَآءَ مِنَ العِلْمِ *Verily I recoil, shrink,* or *am averse, from some matters of knowledge,* or *science, and ignore them,* or *neglect them.* (Mgh.) —— Also جَفَا, (Mṣb, TA,) aor. as above, (Mṣb,) [and so the inf. n.,] said of a garment, or piece of cloth, *It was thick, coarse,* or *rough:* (Mṣb, TA:) and in like manner said of a reed-pen, *it was thick, coarse,* or *rough, in its nib.* (TA.) —— [And hence, (see جَفَآءٌ, below, and جَافٍ,) *He was,* or *became, thick, gross, coarse, rough,* or *rude, of make:* and, more commonly, *coarse, rough,* or *rude, of nature* or *disposition,* or *in his intercourse and dealings with others; unkind, hard, churlish, uncivil,* or *surly.* (See also 10.)] It is said in a trad., مَنْ بَدَا جَفَا, i. e. [*He who abides in the desert*] *becomes coarse, rough,* or *rude, of nature* or *disposition;* [or *unkind, hard, churlish,* &c.;] by reason of mixing little with men. (TA.) —— جَفَا عَلَيْهِ كَذَا *Such a thing was,* or *became, heavy, onerous, burdensome,* or *oppressive, to him.* (Ḳ.) —— جَفَا جَنْبُهُ عَنِ الفِرَاشِ: and جَفَا السَّرْجُ: see 3. —— جَفَتِ القِدْرُ بِزَبَدِهَا *The cooking-pot cast forth its froth,* or *foam;* as also ▼اجفت; (TA;) [like جَفَأَتْ and اجفأت;] originally without ء. (Er-Rághib, TA.) And جَفَا السَّيْلُ *The torrent drove away* [things in its course]. (Mṣb.) —— جَفَا الرَّجُلَ, aor. as above, *He turned away from the man; avoided him;* or *shunned him:* or *he drove away the man;* from جفا السَّيْلُ, explained above: sometimes meaning, *with hatred.* (Mṣb.) And جَفَاهُ *He withdrew,* or *removed, far,* or *to a distance, from him:* whence the saying of Mohammad Ibn-Sookah, لَمَّا قَلَّ مَالِى جَفَانِى إِخْوَانِى [*When my property became little, my brethren withdrew far from me*]. (TA.) —— And جَفَا مَالَهُ *He did not keep,* or *cleave,* or *hold fast, to his property.* (Ḳ.) —— And جَفَاهُ, (Ṣ, Ḳ,) aor. ـُ, (Ṣ,) inf. n. جَفَآءٌ (Ṣ, Ḳ) and جَفْوٌ, (Ḳ,) *He treated him,* or *behaved towards him, coarsely, roughly, rudely, unkindly, hardly, churlishly, uncivilly,* or *surlily:* (Ṣ, Ḳ:) you should not say جَفَيْتُ. (Ṣ.) The pass. part. n. is ▼مَجْفُوٌّ and ▼مَجْفِىٌّ: (Ṣ, and Ḳ in art. جفى:) the latter formed in accordance with جُفِىَ, in which the و is changed into ى. (Fr, Ṣ.) —— *He did to him what displeased, grieved,* or *vexed, him; did to him what he disliked,* or *hated; did evil to him.* (TA.) So in the trad., مَنْ حَجَّ وَلَمْ يَزُرْنِى فَقَدْ جَفَانِى [*He who performs the pilgrimage* to Mekkeh *and does not visit me,* i. e. does not visit my tomb afterwards, either on his homeward journey or by journeying to El-Medeeneh for that special purpose, *does what displeases* me]. (TA.) —— جَفَتِ المَرْأَةُ وَلَدَهَا *The woman neglected to take care of her child,* or *to pay frequent attention to it.* (TA.)

2. تَجْفِيَةٌ [inf. n. of جفّى] signifies, in Persian, جفاء فرمودن [app. as meaning The *ordering* one *to act,* or *to treat* another, *coarsely, roughly, rudely,* &c.]. (KL. [Accord. to Golius, as on this authority, "*i. q.* جَفَا, sign. *injuria affecit; duriter et inique tractavit:* seu transit. *ejus.*"])

3. جافى جَنْبَهُ عَنِ الفِرَاشِ, and ▼جَفَاهُ, *He caused his side not to rest,* or *caused it to be restless,* or *uneasy, upon the bed;* or *caused it to shrink therefrom:* and *he heaved it, drew it up,* or *raised it,* [*making it to be separated by some space* or *interval*] *from the bed.* (Mgh.) And hence, (Mgh,) جافى عَضُدَيْهِ عَنْ جَنْبَيْهِ (Mgh, TA) (TA) *He put,* or *set, his upper arms apart,* or *remote, from his sides.* (Mgh, TA.) [Thus the Muslim is enjoined to do in prostrating himself in prayer.] And جافى السَّرْجَ عَنْ ظَهْرِ الفَرَسِ, (Ṣ,* Mṣb,) and ▼اجفاهُ, (Ṣ, Ḳ,) and ▼جَفَاهُ, (Ḳ, [said in the TA to be a mistake, but a similar usage of this verb has been mentioned above on the authority of the Mgh,]) *He raised the saddle from the back of the horse:* (Ṣ, Mṣb, Ḳ:) and in like manner, أَجْفَيْتُ ▼ القَتَبَ عَنْ ظَهْرِ البَعِيرِ [*I raised the saddle from the back of the camel*]. (M, TA.)

4. اجفاهُ *He made,* or *caused, him,* or *it, to be,* or *become, distant, remote, far off,* or *aloof.* (TA.) —— See also 3, in two places. —— اجفى المَاشِيَةَ *He fatigued the cattle,* [in some copies of the Ṣ, تَعَبًا is put for أَتْعَبَهَا,] *and did not let them eat,* (AZ, Ṣ, Ḳ, TA,) *nor fed them previously, driving them vehemently.* (TA.) —— أَجْفَتِ القِدْرُ: see 1. —— أَجْفَتِ الأَرْضُ *The land became like the* جُفَآء, or *rubbish and scum cast forth by the torrent of a valley, or by a cooking-pot, in respect of the departure of the good thereof.* (Er-Rághib, TA.)

6. تجافى: see 1, in three places. [Said of a person prostrating himself, or lying down, it means *He drew up his body from that on which he rested.* And تجافى عَنْهُ generally signifies *He,* or *it, receded, withdrew, removed,* or *became remote* or *aloof* or *separated by some space* or *interval, from him,* or *it:* and *he drew away, shrank,* or *flinched, from him,* or *it.*] It is said, of the difference between الذَّبْح and القَتْل, that the former is by cutting the external jugular veins; and the latter, بِإِيقَاعِ الفِعْلِ فِى المَحَلِّ مَعَ التَّجَافِى [*By causing the act to take effect upon the place* thereof *while standing aloof*]; meaning that the قَاتِل strikes from a distance, not knowing whether he will hit the place or not. (Mgh.) And [hence] you say, تجافى لَهُ عَنْ حَقِّهِ † [*He relinquished,* i. e.] *he gave, to him, his right,* or *due.* (TA in art. خمص.) —— *He inclined,* or *declined,* or *turned, from side to side:* and *from right to wrong.* (Ḥar p. 125.) [See 6 in art. دفو.]

8. اجتفاهُ *He removed him,* or *it, from his,* or *its, place.* (Ḳ.)

10. استجفاهُ *He esteemed it* (namely, a bed, &c., Ḳ) جَافٍ, (Ṣ, Ḳ,) i. e. *thick, coarse,* or *rough.* (TA.) —— *He demanded,* or *required, of him that he should do what was displeasing, grievous, vexatious,* or *evil.* (TA.) —— استجفى *He became coarse, rough, rude, unkind, hard, churlish, uncivil,* or *surly.* (KL. [See also 1.])

جَفًا: see جَفَآءٌ.

جَفْوَةٌ *A single act of coarse, rough, rude, unkind, hard, churlish, uncivil,* or *surly, treatment,* or *behaviour.* (TA.) —— See also جَفَآءٌ, in three places.

جِفْوَةٌ: see جَفَآءٌ.

جَفَآءٌ is in make; [signifying *Thickness, grossness, coarseness, roughness,* or *rudeness:*] and in nature, or disposition; (TA;) signifying *coarseness, roughness,* or *rudeness,* (Mgh, Mṣb,) *in one's intercourse and dealings with others;* (Mgh;) *unkindness, hardness, churlishness, incivility,* or *surliness;* a predominant quality of the people of the desert; (Mgh, Mṣb;) from جَفَا said of a garment, or piece of cloth; (Mṣb;) *contr. of* بِرٌّ, (Ṣ,) or *of* صِلَةٌ; (Ḳ;) as also ▼جَفًا, (Ḳ,) accord. to Lth; but Az says that he knew not any one who allowed this latter: (TA:) so, too, ▼جَفْوَةٌ and ▼جِفْوَةٌ, in the sayings فِيهِ جَفْوَةٌ and جِفْوَةٌ [*In him is coarseness, roughness,* or *rudeness,* &c.]: (Ḳ:) and فُلَانٌ ظَاهِرُ الجَفْوَةِ *Such a one is a person in whom coarseness, roughness,* or *rudeness,* &c., *is apparent:* (Ṣ:) but accord. to Lth, جفوة [whether جَفْوَةٌ or جِفْوَةٌ is not shown] denotes a more constant quality than جَفَآءٌ. (TA.) You say also, ▼ بِهِ جَفْوَةٌ, meaning *He is suffering coarseness, roughness,* or *rudeness,* &c. (K.) And أَصَابَتْهُ جَفْوَةُ ▼ الزَّمَنِ ‡[*The roughness,* or *rudeness,* &c., *of time,* or *fortune, smote him*]; and جَفَوَاتُهُ [*its roughnesses,* or *rudenesses,* &c.]. (TA.)

جُفَآءٌ The *rubbish and scum cast forth by the torrent of a valley,* and *by a cooking-pot.* (Er-Rághib, TA.) [See also art. جفا.] — And hence, as being likened to the جُفَآء of the torrent, ‡ The *first,* or *foremost,* of men, or people. (TA.) [But see art. جفا.]

جَافٍ [act. part. n. of 1:] applied to a garment, or piece of cloth, (Mgh, Msb,) and to a bed, &c., (S,* K, TA,) *Thick, coarse,* or *rough.* (Mgh, Msb, TA.) — And [hence] applied to a man, (S, TA,) meaning *Thick, gross, coarse, rough,* or *rude, of make;* and *coarse, rough,* or *rude, of nature* or *disposition; coarse, rough, rude, unkind, hard, churlish, uncivil,* or *surly, in his treatment of,* or *behaviour towards, his companions:* pl. جُفَاةٌ. (TA.) You say also, رَجُلٌ جَافِى الخِلْقَةِ [*A man thick, gross, coarse, rough,* or *rude, of make*]: and جَافِى الخُلُقِ *niggardly and incompliant; coarse, rough,* or *rude,* (K, TA,) *in his intercourse and dealings with others; oppressive when angry and irritated against his companion with whom he sits.* (TA.) And جَافٍ عَنِ المَوْعِظَةِ [*Obdurate against admonition*]. (TA in art. جعظر.)

مَجْفُوٌّ : } see 1.
مَجْفِىٌّ :

جل

1. جَلَّ, aor. يَجِلُّ, (S, Msb, K,) inf. n. جَلَالَةٌ, (S,) or جَلَالٌ, (K, [in the CK, erroneously, جُلَالًا, is put for جَلَالًا,]) or both, (TA, [but see what follows,]) and جُلًّى, (Ham p. 218, see this word below, under جَلِيلٌ,) [in its primary sense, *It was,* or *became, thick, gross, coarse, rough, rugged, rude, big,* or *bulky:* (see جَلِيلٌ:) and then,] *it,* (a thing, Msb,) or *he* (a man, S) *was,* or *became, great;* (S, Msb, K, TA;) [said of a thing, meaning *in size;* and] said of a man, meaning *in estimation, rank,* or *dignity:* (S, TA:) or جَلَالَةٌ signifies *greatness of estimation* or *rank* or *dignity:* but جَلَالٌ, *supreme greatness thereof:* (Er-Rághib, TA:) the latter is an attribute of God only; (As in Ham p. 607, Er-Rághib, TA;) except in few instances: (As ubi suprà:) or it means the *greatness,* or *majesty,* of God: (S, Msb:) or his *absolute independence.* (Bd in lv. 27.) [عَزَّ وَجَلَّ, referring to the name of God expressed or understood, is a phrase of frequent occurrence, meaning, *To Him,* or *to Whom, belong might and majesty,* or *glory and greatness*] — يَجِلُّ عَنِ الإِحَاطَةِ بِهِ [*He is too great to be comprehended within limits*] and يَجِلُّ أَنْ يُدْرَكَ بِالحَوَاسِّ [*He is too great to be perceived by the senses*] are phrases used in speaking of God. (Er-Rághib, TA.) — The saying of El-Ahmar,

* يَا جَلَّ مَا بَعُدَتْ عَلَيْكَ بِلَادُنَا *
* فَابْرُقْ بِأَرْضِكَ مَا بَدَا لَكَ وَارْعُدِ *

[*O, how greatly distant to thee is our country! therefore threaten in thy land as long as it seems fit to thee, and menace*], means مَا أَجَلَّ ▼ ما بعدت [&c.]. (S.) — Also جَلَّ, (S, K,) aor. يَجِلُّ, inf. n. جَلَالَةٌ and جَلَالٌ, (K,) said of a man, (S,) *He became old,* or *advanced in age,* (S, K,) *and firm,* or *sound, in judgment.* (K.) And جَلَّتْ said of a she-camel, *She was,* or *became, old,* or *advanced in age:* (Abu-n-Nasr, S:) and so ▼ تَجَالَّتْ said of a woman. (TA.) — جَلَّتِ الهَاجِنُ عَنِ الوَلَدِ [*The girl married before she had arrived at puberty,* or *the beast covered before she was of fit age,*] *was too young* [*to bear offspring*]: (S:) a prov. (TA.) [Thus the verb bears two contr. significations. See also هَاجِنٌ.] — جَلَّ القَوْمُ, (S, Msb,* K,*) عَنِ البَلَدِ, (S,) or عَنْ مَنَازِلِهِمْ, (K,) aor. ـِ , (Msb, K,) or ـُ [contr. to rule], (S, Sgh,) or both, accord. to Ibn-Málik and others, (TA,) inf. n. جُلُولٌ, (S, K,) [and جَلَآءٌ accord. to the K, but this is an inf. n. of جَلَا], *The people,* or *company of men, went forth,* or *emigrated,* (S, Msb, K,) like جَلَا, (S, K,) *from a country,* or *town,* (Msb,) [or *from their places of abode,*] *to another country,* or *town.* (S, Msb.) — جَلُّوا الأَقِطَ, (K,) [aor., accord. to rule, ـُ,] inf. n. جَلٌّ, (TA,) *They took the main part,* or *portion, of the* [*preparation of milk termed*] اقط. (K.) [See also 5.] — جَلَلْتَ هٰذَا عَلَى نَفْسِكَ *Thou hast brought this as an injury* (جَنَيْتَهُ) *upon thyself.* (K.) — جَلَّ البَعَرَ, (S, K,) aor. ـُ, (S,) inf. n. جَلٌّ (S, K) and جِلَّةٌ, (K,) *He picked up,* (S,) or *collected with his hand,* (K,) *the camels',* or *similar, dung;* (S, K;) and ▼ اجتلّه signifies the same, (S,) or *he picked it up for fuel.* (K.) [See جِلَّةٌ.] — See also 2.

2. جلّل, inf. n. تَجْلِيلٌ, said of a thing, i. q. عَمَّ [as meaning *It included* persons, or things, &c., *in common,* or *generally,* or *universally, within the compass of its influence,* or *effects*]. (S, TA.) So in the phrase سَحَابٌ يُجَلِّلُ الأَرْضَ بِالمَطَرِ [*Clouds that include the land in common,* or *generally,* or *universally, within the compass of their rain;* i. e., *that rain upon the land throughout its general,* or *universal, extent*]: (S, TA:) or, as in the A, *thundering clouds, covering the land with rain.* (TA.) And so in the phrase, جَلَّلَ المَطَرُ الأَرْضَ *The rain included the general,* or *universal, extent of the land within the compass of its fall;* and *covered the land so as not to leave anything uncovered.* (IF, Msb.) — And hence, [in a general sense,] *He covered* a thing. (Msb.) *It* [or *he*] *ascended, rose, mounted, got, was,* or *became, upon,* or *over,* a thing; (Ham p. 45;) as also ▼ تجلّل. (S, K.) — *He clad* a horse (S, K) or beast (K) *with a* جُلّ [or *covering for protection from the cold*]; (S, K;) as also ▼ جَلَّ. (K.)

4. اجلّهُ, (S, K,) inf. n. إِجْلَالٌ, (TA,) [*He made it* جَلِيل, i. e., *thick,* &c.: *contr. of* أَدَقَّهُ: see Ham p. 546. — And hence,] *He magnified him; honoured him;* (K, TA;) as also ▼ تجالّه: (TA:) *he exalted him* (TA) in rank, or station. (S.) It is said in a trad., أَجِلُّوا ٱللّٰهَ يَغْفِرْ لَكُمْ, meaning [*Magnify ye God, and He will forgive you:* or] say ye, يَا ذَا الجَلَالِ وَالإِكْرَامِ [*O Thou who art possessed of greatness,* or *majesty, and bounty*], *and believe in his greatness,* or *majesty:* it is also recited otherwise, with ح; (TA in the present art.;) i. e. أَحِلُّوا ٱللّٰهَ, meaning "Resign yourselves to God;" or "quit ye the danger and straitness of belief in a plurality of Gods, to avail yourselves of the freedom of El-Islám;" (TA in art. حل;) but the former recital is confirmed by another trad., namely, أَلِظُّوا بِيَا ذَا الجَلَالِ وَالإِكْرَامِ [see art. لظ]. (TA in the present art.) [Hence,] مِنْ أَجْلِ إِجْلَالِكَ, and فَعَلْتُ ذٰلِكَ مِنْ إِجْلَالِكَ: see جَلَلٌ. — *He gave him much.* (S.) You say, مَا أَجَلَّنِى وَلَا أَدَقَّنِى (S, TA) *He gave me not much, nor gave he me little:* (S:) or †*he gave me not a camel, nor gave he me a sheep,* or *goat.* (TA.) A poet says, (S,) namely, El-Murrár El-Fak'asee, describing his eye, (TA,)

* بَكَتْ فَأَدَقَّتْ فِى البُكَى وَأَجَلَّتِ *

†*It wept, and shed few tears, and shed many.* (S, TA.) You say also, أَجَلَّ فَرَسَهُ فَرَقًا مِنْ ذُرَةٍ *He gave his horse a large feed of millet.* (TA.) — *He gave him a* جَلِيلَة, i. e., *a she-camel that had brought forth once.* (S, K.) You say, مَا أَجَلَّنِى وَلَا أَحْشَانِى *He gave me not a she-camel that had brought forth once,* (S, K,*) *nor gave he me a young,* or *small, camel.* (S.) — مَا أَجَلَّ : see 1. [You say, مَا أَجَلَّهُ *How great,* &c., *is he,* or *it!*] — اجلّ *He was,* or *became, strong:* — and *He was,* or *became, weak:* thus bearing two contr. significations. (Ibn-'Abbád, K.)

5. تجلّلهُ *He took the greater, main, principal,* or *chief, part of it; the main, gross, mass,* or *bulk, of it;* (S, K;) as also ▼ اجتلّهُ (K) and ▼ تجالّهُ. (Ibn-'Abbád, K. [In the CK, in the explanation of the second and third of these verbs, جِلَالَهُ is erroneously put for جُلَالَهُ.]) — See also 2. — [Hence,] *He sat upon him;* namely, a horse. (KL.) And تجلّل الفَحْلُ النَّاقَةَ (S and K in art. دوم) *The stallion-camel mounted the she-camel.* (TA in that art.)

6. تجالّ i. q. تَعَاظَمَ (S, K) and تَرَفَّعَ. (S.) You say, فُلَانٌ يَتَجَالُّ عَنْ ذٰلِكَ (S, K*) *Such a one exalts himself above that; holds himself above it; disdains it;* or *is disdainful of it;* syn. يَتَرَفَّعُ عَنْهُ, (S,) or يَتَعَاظَمُ; (K;) as also يتجالّ عَلَيْهِ. (TA.) — See also 1. — تجالّهُ : see 4: — and 5.

8: see 5: — and see also 1.

R. Q. 1. جَلْجَلَ [app. *It sounded;* or *made a sound,* or *sounds;* said of a little bell, such as is called جُلْجُل: said also of thunder: and *it sounded vehemently;* or *made a vehement sound,* or *vehement sounds:* and *he threatened:* (see جَلْجَلَةٌ, which seems to be the inf. n. of the verb in these senses:) and,] said of a horse, *he neighed clearly;* or *had a clear neigh.* (K.) — جَلْجَلَهُ,

(Ṣ,) inf. n. جَلْجَلَةٌ, (Ḳ,) *He put it* (a thing, Ṣ) *in motion* (Ṣ, Ḳ) *with his hand.* (Ṣ.) And جلجل القِدَاحَ *He* (a player at the game called المَيْسِر) *moved about* [or *shuffled*] *the gaming-arrows.* (TA.) — *He mixed it.* (Ḳ.) — *He twisted it vehemently,* or *strongly;* namely, the string of a bow or the like. (Ibn-'Abbád, Ḳ.)

R. Q. 2. تَجَلْجَلَ *It was,* or *became, in a state of motion;* or *was put in motion.* (Ḳ.) — *It was,* or *became, agitated* in the mind. (Ḳ,* TA.) — *He sank into the ground.* (Ṣ, Ḳ.) *It sank,* or *became depressed;* syn. تَضَعْضَعَ. (Ḳ.) One says, تَجَلْجَلَتْ قَوَاعِدُ البَيْتِ *The foundations of the house sank,* or *became depressed;* syn. تَضَعْضَعَتْ. (Ṣ.)

جَلٌّ The *sail of a ship:* pl. جُلُولٌ. (Ṣ, Ḳ.) = See also جُلٌّ, in two places: = and جِلٌّ: = and جَلِيلٌ. — Also *Contemptible, mean,* or *paltry:* thus bearing two contr. significations. (Ḳ.)

جُلٌّ The *greater, main, principal,* or *chief, part* of a thing; the *most* thereof; the *main, gross, mass,* or *bulk,* of it; (Ṣ, Mṣb, Ḳ;) as also ▼ جُلَالٌ. (Ḳ.) You say, أَخَذَ جُلَّهُ (Ḳ, TA) and ▼ جُلَالَهُ (Ṣ, Ṣgh, Ḳ) [*He took the greater part of it*]. = *A horse-cloth,* or *covering* (Mṣb, Ḳ) of a horse or similar beast, (Ṣ, Mgh, Mṣb, Ḳ,) *for protection* (Mṣb, Ḳ) *from the cold;* (Mṣb;) as also ▼ جَلٌّ: (Ḳ:) [in Persian جُل:] pl. [of mult.] جِلَالٌ (Ṣ, Mgh, Mṣb, Ḳ) and [of pauc.] أَجْلَالٌ, (Mṣb, Ḳ,) and أَجِلَّةٌ is pl. of جِلَالٌ. (Ṣ, TA.) — The *cover of,* or *a thing with which one covers, a book,* or *volume;* which latter is hence called ▼ مُجَلَّةٌ. (Er-Rághib in TA; but, in this sense, written without any vowel-sign.) = The *place of the pitching and constructing* of a tent or house. (Ḳ.) = Also, (Ṣ, Ḳ,) and ▼ جَلٌّ, (Ḳ,) The *rose,* (AḤn, Ṣ, Ḳ,) the *white* and the *red* and the *yellow;* (AḤn, Ḳ;) *plentiful in the countries of the Arabs,* both *cultivated* and *wild:* (AḤn, TA:) a Persian word, arabicized; (AḤn,* Ṣ, Ṣgh;) from گُل: (Ṣgh, TA:) and the *jasmine:* n. un. with ة. (Ḳ.) = See also جِلٌّ: = and جَلَلٌ.

جِلٌّ: see جَلِيلٌ, in six places. = Also The *stalks of seed-produce* [or *corn*] *when it has been reaped;* (Ṣ, O, Mgh, Ḳ;) as also ▼ جُلٌّ and ▼ جَلٌّ: (Ḳ:) when it has been removed to the place where the grain is trodden out, and has been trodden, and cut by means of the مِدْوَس, it is called تِبْنٌ. (AḤn, Mgh.) And, by amplification, applied to The *stalks remaining upon the field after the reaping.* (Mgh in the present art. and in art. حصد.)

جَلَّةٌ (Ṣ, Mgh, Mṣb, Ḳ) and ▼ جِلَّةٌ and ▼ جُلَّةٌ, (Ḳ,) the second whereof is that which is most known [in the present day], and next the first [which seems to be the most chaste], (TA,) *Camels',* or *sheep's,* or *goats',* or *similar, dung;* syn. بَعَر: (Ṣ, Ḳ:) or *a single lump thereof:* (Mgh, Mṣb, Ḳ:) or *such as has not been broken.* (Ḳ.) [Commonly applied in the present day to *Such dung kneaded with chopped straw and formed into round flat cakes, which are dried in the sun, for fuel.*] You say, إِنَّ بَنِى فُلَانٍ وَقُودُهُمُ الجَلَّةُ [*Verily the sons of such a one, their fuel is the dung of camels* or *sheep* &c.]. (Ṣ.) — Also (metonymically, Mgh) applied to *Human ordure.* (Mgh, Mṣb.)

جُلَّةٌ *A large* [*receptacle made of palm-leaves woven together, such as is called*] قُفَّة, *for dates;* (Ḳ;) *a receptacle* (Ṣ, Mgh, Mṣb, Ḳ) *for dates,* (Ṣ, Mgh, Mṣb,) *made of palm-leaves;* (Ḳ;) [*a thing made of palm-leaves woven together, generally used as a receptacle for dates, but also employed for other purposes, as, for instance, to lay upon the mouth of a watering-trough, where the water is poured in, by way of protection;* see إِزَآءٌ:] pl. جِلَالٌ (Mgh, Mṣb, Ḳ) and جُلَلٌ. (Ḳ.) = See also جَلَّةٌ.

جِلَّةٌ: see جَلَّةٌ: = and جَلِيلٌ; of which it is in most instances a pl.

جَلَلٌ *A great, momentous,* or *formidable, thing, affair, matter, case,* or *event;* as also ▼ جُلَّى (Ṣ, Ḳ, TA) and ▼ جُلَّآءُ: (TA:) or ▼ جُلَّى [as also جَلَلٌ and ▼ جُلَّآءُ] signifies *a hard, difficult, severe,* or *distressing,* and *a great, momentous,* or *formidable, thing,* or *affair,* &c.: (Mṣb:) pl. [of جَلَلٌ,] أَجْلَالٌ; (TA;) and of ▼ جُلَّى, جُلَلٌ. (Ṣ, Ḳ.) El-Ḥárith Ibn-Waạleh says,

* قَوْمِى هُمُ قَتَلُوا أُمَيْمَ أَخِى *
* فَإِذَا رَمَيْتُ يُصِيبُنِى سَهْمِى *
* فَلَئِنْ عَفَوْتُ لَأَعْفُوَنْ جَلَلًا *
* وَلَئِنْ سَطَوْتُ لَأُوهِنَنْ عَظْمِى *

[*My people, they have slain, O Umeymeh,* (أُمَيْمَ being apocopated, for أُمَيْمَةُ,) *my brother; so, if I shoot, my arrow will strike me; and verily, if I forgive, I shall indeed forgive a great thing; but verily, if I assault, I shall indeed weaken my bone:* see Ḥam p. 97]. (Ṣ.) And Beshámeh Ibn-Ḥazn says,

* وَإِنْ دَعَوْتِ إِلَى جُلَّى ▼ وَمَكْرُمَةٍ *
* يَوْمًا سَرَاةَ كِرَامِ النَّاسِ فَادْعِينَا *

[*And if thou invite to a great affair, and a generous act, any day, manly and noble persons, the generous of mankind, invite us*]: (TA:) or جُلَّى is here an inf. n. in the place of جَلَالٌ and جَلَالَةٌ, like رُجْعَى &c. (Ḥam p. 218.) — Also, i. e., جَلَلٌ, *A small,* (Ḳ,) *an easy,* or *a mean, paltry,* or *contemptible, thing, affair, matter, case,* or *event:* (Ṣ, Ḳ, TA:) thus bearing two contr. significations. (Ṣ, Ḳ.) Imra-el-Ḳeys says, on the occasion of his father's having been slain,

* بِقَتْلِ بَنِى أَسَدٍ رَبَّهُمْ * أَلَا كُلُّ شَىْءٍ سِوَاهُ جَلَلْ *

meaning [*By Benoo-Asad's slaying their lord: now surely everything beside it is*] *a mean, paltry,* or *small, matter.* (Ṣ,* TA.) — فَعَلْتُ ذٰلِكَ مِنْ جَلَلِكَ *I did that on account of thee, for thy sake,* or *because of thee;* syn. مِنْ أَجْلِكَ; (Ṣ, Ḳ;*) as also ▼ من جُلِّكَ, (Ḳ,) and من جَلَالِكَ, (Ṣ, Ḳ,) and ▼ من تَجِلَّتِكَ, and ▼ من إِجْلَالِكَ, and من أَجْلِ ▼ إِجْلَالِكَ. (Ḳ.) Jemeel says,

* رَسْمِ دَارٍ وَقَفْتُ فِى طَلَلِهْ *
* كِدْتُ أَقْضِى الغَدَاةَ مِنْ جَلَلِهْ *

meaning [*The remains marking the site of a house, I paused at the relic thereof that was still standing: I almost died, in the early morning,*] *on account of it* (مِنْ أَجْلِهِ), or, as some say, *because of its greatness in my eye.* (Ṣ.) = Accord. to Zj, جَلَلْ is a particle *syn. with* نَعَمْ. (Mughnee.)

جَلَالٌ an inf. n. of جَلَّ. (Ḳ, TA.) — [Hence,] فَعَلْتُ ذٰلِكَ مِنْ جَلَالِكَ : see جَلَلٌ.

جُلَالٌ: see جُلٌّ, in two places: — also, and its fem., with ة, see جَلِيلٌ, in three places: — and see جُلَاجِلٌ.

جَلَالٌ The *deck,* or *part resembling a roof,* of a ship: a sing. word. (Mgh.) — [See جُلٌّ and جُلَّةٌ, of each of which it is a pl.]

جَلِيلٌ, in its primary acceptation, signifies *Thick, gross, coarse, rough, rugged, rude, big,* or *bulky;* applied to a material substance; (Er-Rághib, TA;) opposed to دَقِيقٌ; (Ṣ, Er-Rághib, TA;) as also ▼ جِلٌّ, (Ṣ,) opposed to دِقٌّ: (Ṣ, Ḳ:) [and then,] *great;* (Mṣb, Ḳ;) as also ▼ جِلٌّ and ▼ جَلٌّ (Ḳ) and ▼ جُلَالٌ, (Ṣ, Ḳ,) which is also explained as signifying *large, big, bulky,* or *large in body,* (Ḳ,) and ▼ جُلَّالٌ: fem. جَلِيلَةٌ and ▼ جُلَالَةٌ: (Ḳ:) [also] *great in respect of estimation, rank,* or *dignity:* (Ṣ, TA:) pl. [of pauc.] أَجِلَّةٌ and جِلَّةٌ and [of mult.] أَجِلَّآءُ. (TA.) You say, مَا لَهُ دِقٌّ وَلَا جِلٌّ ▼, i. e., دَقِيقٌ وَلَا جَلِيلٌ [*He has neither slender,* or *fine,* or *small, nor thick,* or *gross,* or *coarse,* &c., or *great*]. (Ṣ.) And ▼ شَجَرٌ جِلٌّ [*Large trees;* or *trees as*] *opposed to* شَجَرٌ دِقٌّ [or *shrubs,* or *bushes*]. (Lth, Mgh in art. بقل.) And ▼ حُلَلٌ جِلٌّ *Thick,* or *coarse,* [*garments,* or *dresses, of the kind called*] حُلَل; opposed to حُلَلٌ دِقٌّ: (Mgh in art. دق:) or the things termed جِلٌّ, of commodities, are *carpets,* and [the *garments called*] أَكْسِيَة [pl. of كِسَآء], and *the like;* (Ḳ;) *contr. of* دِقٌّ; such as the [*cloth called*] حِلْس, and the *mat,* and *the like.* (TA.) And ▼ جُلَالَةٌ signifies *A great she-camel;* (Ṣ, Ḳ;) *big-bodied.* (TA.) You say also, طَحَنَهُ طَحْنًا جَلِيلًا [*He ground it coarsely*]. (Ṣ in art. جش.) الجَلِيلُ, meaning *The great in dignity,* is not applied peculiarly to God: when it is applied to Him, it is because of his creating the great things that are indicative of Him, or because He is too great to be comprehended within limits or to be perceived by the senses. (Er-Rághib, TA.) And قَوْمٌ جِلَّةٌ means *A great people; lords, chiefs,* or *people of rank* or *quality;* (Ḳ;) *a good people;* (TA;) *a people of eminence, nobility, dignity,* or *high rank.* (Ḳ.) — Also *Old,* or *advanced in age,* and *firm,* or *sound, in judgment:*

pl. جِلَّةٌ: (K:) which pl., as meaning *old*, or *advanced in age*, is applied to camels, (S, Sgh, K,) as well as to men. (K.) Hence, in a trad., فَاعْتَرَضَ لَهُمْ إِبْلِيسُ فِى صُورَةِ شَيْخٍ جَلِيلٍ [*And Iblees presented himself to them in the form of an old man advanced in age*]. (TA.) ▼ جِلَّةٌ in the sense last explained above, is also used as a sing., and is applied to the male and the female [of camels]: or signifies *such as is termed* ثَنِيَّة, [i. e., a she-camel *that has entered her sixth year*,] *until she has become a* بَازِل [*in her ninth year*]: or a male camel *that has become a* ثَنِىّ: or it is applied to a she-camel, and ▼ جِلٌّ to a he-camel. (K.) And [the fem.] ▼ جَلِيلَةٌ [used as a subst.] signifies *A she-camel that has brought forth once*: (S, O, K:) and [simply] *a she-camel*; as in the saying, مَا لَهُ جَلِيلَةٌ وَلَا دَقِيقَةٌ *He has neither a she-camel nor a ewe*, or *she-goat*: (S:) or *camels*. (JK and TA in art. دق [q. v., voce دَقِيقٌ].) Also (i. e. ▼ جليلة) *A great palm-tree having much fruit*: pl. جَلِيلٌ; (K;) [or rather this is a coll. gen. n.;] or, accord. to some copies of the K, the pl. is جِلَالٌ. (TA.) ═ Also *i. q.* ثُمَامٌ [*Panicum, or panic grass*]; (S, K;) *a weak plant, with which the interstices of houses are stopped up*: n. un. with ة: (S:) or ▼ جَلِيلَةٌ signifies *a species of* ثُمَام: (TA in art. ثم:) pl. جَلَائِلُ. (S, K.)

جَلِيلَةٌ [used as a subst.]: see the latter part of the next preceding paragraph, in three places.

جُلَّى: see جَلَلٌ, in four places.

جُلَّاءُ: see جَلَلٌ, in two places.

جُلِّىٌّ a rel. n. from جُلٌّ; *A seller of* جِلَال [pl. of جُلّ] *for horses or similar beasts*. (TA.)

جُلَالٌ: see جَلِيلٌ.

جَلَّالَةٌ (S, Mgh, Msb, K) and ▼ جَالَّةٌ (Mgh, Msb) *A cow that repeatedly seeks after filths* [*to eat them*]; (S, K;) the milk of which is forbidden: (S:) *a beast that eats* جِلَّة, meaning *human ordure*; (S, Mgh, Msb;) the flesh of which is forbidden: (Mgh:) pl. [of the former] جَلَّالَاتٌ (Msb) and of the latter جَوَالٌّ; (Mgh, Msb;) the latter pl. occurring in a trad., in which some erroneously substitute for it جَوَّالَاتٌ. (Mgh.)

جُلْجُلٌ [*A little bell, consisting of a hollow ball of copper or brass or other metal, perforated, and containing a loose solid ball*;] *a small* جَرَس [or *bell*]; (Msb, K;) *a thing that is hung to the neck of a horse or similar beast, or to the leg of a hawk*: (Mgh:) pl. جَلَاجِلُ. (S, Mgh, Msb.) You say, فُلَانٌ يُعَلِّقُ الجُلْجُلَ فِى عُنُقِهِ [*Such a one hangs the little bell upon his neck*;] meaning, ‡ *such a one imperils, or endangers, himself*. (TA.) Abu-n-Nejm says,

إِلَّا آمْرَأً يَعْقِدُ خَيْطَ الجُلْجُلِ

[*Except a man who ties the string of the little bell*;] meaning, ‡ *except a bold man, who imperils himself*: AA says that it is a prov., meaning, *except a man who makes himself notorious, so that no one precedes him except a courageous man who cares not for him, and who is stubborn and notorious*. (TA.) — See also جُلَاجِلٌ.

جَلْجَلَةٌ [app. inf. n. of جَلْجَلَ, q. v.;] The *sound*, or *sounding, of a* جُلْجُل, (S,) or *of a* جَرَس [or *bell*]; (TA;) and of thunder: (S, K:) and *vehemence of sound*: and *a threatening* (K, TA) *from behind a thing covering* or *concealing*. (TA.)

جُلْجُلَانٌ *What is* جَلِيل [app. meaning *great in estimation*] of a thing. (Ibn-'Abbád, TA.) ═ Also The *fruit of the* كُزْبَرَة [or *coriander*]: (S, Mgh, K:) and, (Mgh,) accord. to Abu-l-Ghowth, (S,) *sesame*, or *sesamum*, (S, Z, Mgh, TA,) *in its husks, before it is reaped*: (S:) or it signifies also the *grain of sesame* or *sesamum*. (K.) — ‡ The heart's *core* (حَبَّةُ القَلْبِ). (S, Z, K, TA.) You say, أَصَبْتُ جُلْجُلَانَ قَلْبِهِ ‡ [*I hit his heart's core*]. (S.) And اِسْتَقَرَّ ذٰلِكَ فِى جُلْجُلَانِ قَلْبِهِ ‡ [*That rested, or remained, in his heart's core*]. (Z, TA.) And كَلَامٌ خَرَجَ مِنْ جُلْجُلَانِ القَلْبِ إِلَى قَمَعِ الأُذُنِ ‡ [*Speech that came forth from the core of the heart to the meatus of the ear*]. (Z, TA.)

جَلْجَالٌ: see مُجَلْجِلٌ.

جُلَاجِلٌ An ass *that brays clearly*; (S, K;) as also ▼ جُلَالٌ; (El-Moheet, K;) which is in like manner applied to a she-camel. (El-Moheet, TA.) — A boy *light in spirit; brisk, lively*, or *sprightly, in his work*; (K;) as also ▼ جُلْجُلٌ. (Ibn-'Abbád, K.) ═ أَبْثَثْتُهُ جُلَاجِلَ نَفْسِى *I revealed to him what was agitated in my mind*. (Ibn-'Abbád, K,* TA.)

جَالٌّ *Going forth*, or *emigrating, from a country*, or *town, to another country*, or *town*; (Msb;) [as also جَالٍ; (see art. جلو;)] and so جَالَّةٌ, (S, Msb, K,) its pl., (Msb,) applied to a people, or company of men; (S, Msb, K;) originally applied to the Jews who were expelled from El-Hijáz; as also جَالِيَةٌ. (Msb.) — Hence, ▼ جَالَّةٌ, as a subst., meaning The *poll-tax*; (Msb;) as also جَالِيَةٌ, (S and Msb in art. جلو.) You say, أُسْتُعْمِلَ فُلَانٌ عَلَى الجَالَّةِ [*Such a one was employed as collector of the poll-tax*]; like as you say, على الجَالِيَةِ. (S, Msb.) ═ جَالَّةٌ as a fem. epithet used as a subst.: see جَلَّالَةٌ.

جَالَّةٌ (as a subst.): see جَالٌّ; of which it is also pl. and fem.

أَجَلُّ [*Thicker* &c., and *thickest* &c.; see جَلِيلٌ: and] *i. q.* أَعْظَمُ [*more, and most, great* &c.]: (S, TA:) fem. جُلَّى. (Ham. p. 45.) With the article, [as a superlative epithet,] it is applied to God; (S, TA;) and so, by poetic license, الأَجْلَلُ. (TA.)

تَجِلَّةٌ a subst. [signifying The *act of magnifying, or honouring*]; (K, TA;) like تَكْرِمَةٌ. (TA.) — [Hence,] فَعَلْتُ ذٰلِكَ مِنْ تَجِلَّتِكَ, like من إِجْلَالِكَ &c.: see جَلَلٌ.

مَجَلَّةٌ *A* صَحِيفَة [or *book, volume, writing*, or *written paper or the like*,] *in which is science*: (S, K:) and *any book*, or *writing*, (A 'Obeyd, S, K,) is thus called by the Arabs; (A 'Obeyd, S;) as, for instance, that of Lukmán, and one of poetry: (TA:) and so in the phrase used by En-Nábighah (Edh-Dhubyánee, TA) مَجَلَّتُهُمْ ذَاتُ الإِلٰهِ [*Their book is that of God*]: or, as some recite it, he said مَحَلَّتُهُمْ, with حاء, meaning, their abode is one of pilgrimage and of sacred sites. (S, TA.) See جُلٌّ. — [Hence,] *Science*; and the *doctrine*, or *science, of practical law*. (AA, TA.)

مُجَلَّلٌ A horse *clad with a* جُلّ; as also ▼ مَجْلُولٌ; (TA;) which latter is likewise applied to a camel. (Ibn-Abbád, TA.)

سَحَابٌ مُجَلِّلٌ *Clouds that include the land in common*, or *generally*, or *universally, within the compass of their rain*; i. e., *that rain upon the land throughout its general*, or *universal extent*: (S, TA:) or *thundering clouds, covering the land with rain*: (A, TA:) or *clouds in which are thunder and lightning*. (As, TA in art. قصب.) [See also مُجَلْجِلٌ.]

مَجْلُولٌ: see مُجَلَّلٌ. ═ Also Water *into which* جِلَّة [q. v.] *has fallen*. (TA.)

مُجَلْجَلٌ A man *very excellent*, or *elegant, in mind, manners, address, speech, person*, or *the like*; *in whom is no fault*, or *vice*. (K.) — A camel *that has attained his full strength*. (K, TA.) ═ إِبِلٌ مُجَلْجَلَةٌ *Camels having small bells, of the kind called* جُلْجُل, *hung upon them*. (K.)

سَحَابٌ مُجَلْجِلٌ Clouds (سَحَابٌ) *in which is the sound of thunder*: (S, K:* [in the CK, in this instance, erroneously written مُجَلْجَلٌ:]) or *sounding*: (TA:) [see also مُجَلِّلٌ:] and in like manner ▼ جَلْجَالٌ applied to rain. (K, TA.) — *A strong chief*: or [in the CK, "and,"] one *whose voice*, or *fame*, (صَوْت,) *reaches far*: and *bold, vehement in repelling* or *defending, eloquent*, or *able in speech*, (K,) *who subjects himself to peril*, or *danger*. (TA.)

جلب

1. جَلَبَ, (S, A, Mgh, Msb, K,) aor. ـُ and ـِ, (S, Msb, K,) inf. n. جَلْبٌ (S, Mgh, Msb, K) and جَلَبٌ, (S, K,) *He drove*, (A, K,) or *brought, conveyed*, or *transported*, (Mgh,) a thing, (S, A,* Mgh, Msb, K,*) or things, such as camels, sheep, goats, horses, captives, or slaves, or any merchandise, (TA,) *from one place to another*, (A, K,) or *from one country or town to another, for the purpose of traffic*; (Mgh;) as also ▼ اجتلب, (A, K, KL,) and ▼ استجلب. (KL.) And جَلَبْتُ الشَّىْءَ إِلَى نَفْسِى and ▼ اِجْتَلَبْتُهُ signify the same; (S;) i. e. † *I brought, drew, attracted*, or *procured, the thing to myself*. (PS.) [Hence,] ذَا مِمَّا يَجْلِبُ الإِخْوَانَ ‡ [*This is of the things that bring, draw, attract, or procure, brothers*, or *friends*]. (A, TA.) And جَلَبَتْهُ جَوَالِبُ الدَّهْرِ ▼ ‡ [*The calamities of time*, or *of fortune*, or *of fate, brought, drew, or attracted, him, or it*]. (A, TA.)

[Hence also, accord. to some,] لَا جَلَبَ وَلَا جَنَبَ, a trad., explained as meaning, *The owner of cattle shall not be required to drive them*, or *bring them, to the town*, or *country, in order that the collector may take from them the portion appointed for the poor-rate, but this shall be taken at the waters; and when the cattle are in the yards, they shall be left therein, and not brought forth to the place of pasture, for the collector to take that portion*: or, as some say, ولا جنب means, *nor shall one have a horse led by his side, in a race, in order that, when he draws near to the goal, he may transfer himself to it, and so outstrip his fellow*: and other explanations have been given: (Mṣb:) [accord. to some,] لا جلب here means, *they shall not drive*, or *bring, their cattle to the collector of the portions appointed for the poor-rate in the place where he alights, but he shall himself come to their yards and take those portions*: or [جلب here is from the verb جَلَبَ in a sense which will be explained below, and] the trad. relates to horse-racing, and means, *one shall not cause his horse to be followed by a man crying out at it and chiding it; nor shall he have a horse without a rider led by his own horse, in order that, when he draws near to the goal, he may transfer himself to it, and outstrip upon it*: (Mgh:) or الجَلَبُ, which is forbidden, means *the collector's not coming to the people at their waters to take the portions appointed for the poor-rate, but ordering them to drive*, or *bring, their cattle to him*: or it relates to contending for a stake, or wager, and means *the mounting a man upon one's horse, and, when he has drawn near to the goal, following his horse and crying out at it, in order that it may outstrip*; which is a kind of fraud: (Ṣ:) or it is used in both these cases: (A 'Obeyd: [his explanations are virtually the same as those in the Ṣ:]) or the meaning of the trad. [so far as the former clause of it is concerned] is, that *the contributions to the poor-rate shall not be driven*, or *brought, to the waters nor to the great towns, but shall be given in their places of pasture*: or it means, [or rather الجلب means,] *the collector's alighting in a place, and then sending a person*, or *persons, to drive*, or *bring, to him the cattle from their places, that he may take the portion thereof appointed for the poor-rate*: or it [relates to horse-racing, and] means *the sending forth a horse in the race-course, and a number of persons' congregating, and crying out at it, in order that it may be turned from its course*: or *a man's following his horse, and spurring on behind it, and chiding it, and crying out at it*: (Ḳ, TA:) or *the shaking a thing behind a horse that is backward in a race, that it may be urged on thereby, and outstrip*: or *one's riding a horse, and leading behind him another, to urge it on, in contending for a stake*, or *wager*: or *the crying out at a horse from behind, and urging it to outstrip*. (TA. See also 1 in art. جنب.) — جَلَبَ لِأَهْلِهِ *He gained* or *earned; sought* or *sought after* or *sought to gain* [provisions &c.; generally meaning *he purveyed*]; *and exercised art or cunning or skill, in the management of his affairs; for his family*; as also ▼اجلب. (Lḥ, Ḳ.) ═ جَلَبُوا, aor. ـُ and ـِ, (Ḳ,) [inf. n. جَلَبٌ, and perhaps جَلَبَةٌ also;] and ▼جَلَّبُوا; (Ṣ, Ḳ;) and ▼اجلبوا, (Ḳ,) inf. n. إِجْلَابٌ; (Mgh;) [the second of which is the most common;] *They raised cries, shouts, noises, a clamour*, (Ṣ, Mgh, TA,) or *confused cries* or *shouts* or *noises*. (Mgh, Ḳ.*) And جَلَبَ عَلَى فَرَسِهِ, (Ṣ, Mṣb, Ḳ,) aor. ـِ, (Ṣ, Mṣb,) inf. n. جَلْبٌ, (Ṣ,) or جَلَبٌ, (Mṣb,) *He chid*, or *urged on, his horse*; as also ▼جلّب and ▼اجلب; (Ḳ;) the first, rare; the second and third, usual: (TA:) *he cried out at his horse*, (Ṣ, Ḳ,) *from behind him, and urged him to outstrip* [*in a race*], (Ṣ,) aor. ـِ and ـُ; (Ḳ; but this explanation is erased in the copy of the Ḳ in its author's handwriting, as being a repetition; and rightly, accord. to MF; though this requires consideration; TA;) as also ▼اجلب: (Ṣ:) *he urged his horse to run, by striking*, or *goading*, or *by crying out*, or *the like*; as also ▼اجلب: or, as some say, *he led behind his horse that he was riding another horse to urge on the former*, in contending [in a race] for a stake, or wager; as is shown in an explanation of the tradition cited above, لَا جَلَبَ وَلَاجَنَبَ. (TA.) It is said in the Ḳur [xvii. 66], وَأَجْلِبْ▼ عَلَيْهِمْ بِخَيْلِكَ وَرَجِلِكَ *And raise thou confused cries against them*, (Mgh,) or *cry out against them, with thy forces riding and on foot*. (Bḍ. But see another explanation in what follows.) And it is said in a well-known prov., جَلَبَتْ جَلْبَةً ثُمَّ أَمْسَكَتْ *It*, i. e. a cloud (سَحَابَة), *thundered, then refrained from raining*: applied to a coward, who threatens, and then is silent: but accord. to some, it is with ح in the place of ج. (MF. See art. حلب.) [Hence,] جَلَبَ, aor. ـُ and ـِ; and ▼اجلب; *He threatened with evil*; (Ḳ, TA;) followed by an accus. (TA) [or, app., by عَلَى before the object]: or (so in the TA, but in some copies of the Ḳ "and,") *he collected a company, a troop*, or *an army*. (Ḳ, TA.) [It is said that] وَأَجْلِبْ▼ عَلَيْهِمْ, in the Ḳur [xvii. 66], means *And collect thou against them* [*thy forces*], *and threaten them with evil*. (TA. But see another explanation above.) And اجلبوا▼ عَلَيْهِ signifies also *They collected themselves together against him*, (Ṣ, Ḳ,*) *and aided one another*; like احلبوا. (Ṣ.) — جَلَبَ عَلَيْهِ, aor. ـِ, inf. n. جَلْبٌ, *He committed a crime against him; or an offence for which he should be punished*. (Ḳ,* TA.) ═ جَلَبَ, aor. ـُ and ـِ, (Ṣ, Ḳ,) *It* (a wound) *healed*: (Ḳ:) or *it* (an ulcer, Aṣ, or a wound, Ṣ) *became covered with a skin in healing*: (Aṣ, Ṣ:) as also ▼اجلب. (Ṣ, L.) — And *It* (blood) *dried; became dry*; as also ▼اجلب. (Lḥ, Ḳ.) ═ جَلِبَ, aor. ـَ, *It* [app. a company or troop] *assembled*, or *became collected together*. (Ḳ.)

2: see 1, in two places. ═ The inf. n. تَجْلِيبٌ also signifies The act of *bringing together*: or *collecting*. (KL.)

3. [جالب is explained by Golius, as on the authority of the KL, as meaning *He helped*, or *assisted*: but this is a mistake for حالب; for I find مُحَالَبَةٌ explained by يارى كردن in a copy of the KL, and the order of the words there shows that it is not a mistranscription for مجالبة.]

4. اجلب: see 1, in eleven places, in the latter half of the paragraph. ═ Also *His camels brought forth males*; (Ṣ, Ḳ;) because the males that they produce are driven, or brought, from one place to another, and sold; opposed to احلب "his camels brought forth females:" (Ṣ:) and *his camel brought forth a male*. (TA.) أَجْلَبْتَ وَلَا أَحْلَبْتَ *May thy camels bring forth males, and may they not bring forth females*, is a form of imprecation against a man, implying a wish that he may lose the milk [that he would have otherwise]. (TA.) ═ *He aided, helped*, or *assisted*, another. (Ṣ, Ḳ.) [So, too, احلب.] ═ *He put an amulet into a* جُلْبَة [which must therefore signify the *piece of skin* in which an amulet is enclosed, as well as an amulet enclosed in a piece of skin: see مُجْلَبٌ]. (Ḳ.) — اجلب قَتَبَهُ, (Ṣ, Ḳ,) inf. n. إِجْلَابٌ, (T,) *He covered his* قتب [or *camel's saddle*] (Ṣ, Ḳ) *with a* جُلْبَة, i. e., (Ṣ,) *with a piece of fresh, moist skin, which he left upon it until it became dry* [*and tight*]: (Ṣ, Ḳ:*) or *he covered the head of his* قتب *with a piece of kid's*, or *lamb's, skin, and left it to dry upon it*. (T.)

5. [تجلّب rendered by Golius *Clamorem ac murmur excitavit*, as on the authority of the Ḳ, I do not find in that lexicon nor in any other.]

7. انجلب *It* [a camel, sheep, goat, horse, captive, or slave, or a number of camels &c., or any merchandise, (see 1, first sentence,)] *was driven* [or *brought*] *from one place to another* [or *from one country or town to another, for the purpose of traffic*]. (Ḳ.)

8. اجتلب: see 1, first and second sentences. — Also †*He* (a poet) *took*, or *borrowed, from the poetry of another*. (TA.) — And *He sought* or *demanded* [a thing]. (Ḥar p. 44.)

10. استجلبهُ *He sought*, or *demanded*, or *desired, that it* [a camel, sheep, goat, horse, captive, or slave, or a number of camels &c., or any merchandise, (see 1, first sentence,)] *should be driven* [or *brought*] *from one place to another* [*in which he was*, or *from one country or town to another, for sale*]. (Ḳ.) — See also 1, first sentence.

R. Q. 1. جَلْبَبَهُ, (Ḳ,) or جلببهُ جِلْبَابًا, (TA,) inf. n. جَلْبَبَةٌ, the second ب not being incorporated into the first because the word is quasi-coordinate to the class of دَحْرَجَةٌ, (Ṣ,) *He put on him a garment of the kind called* جِلْبَاب. (Ṣ, Ḳ.) Accord. to Kh, the first ب in جلبب is [augmentative] like the و in جَهْوَرَ and دَهْوَرَ: accord. to Yoo, the second is [augmentative] like the ى in سَلْقَى and جَعْبَى. (IJ, TA.)

R. Q. 2. تَجَلْبَبَ, (Ḳ,) and تَجَلْبَبَتْ, (A, Mṣb,) *He*, and *she, put on a garment of the kind called* جِلْبَاب; or *clad himself*, and *herself, therewith*. (A, Mṣb, Ḳ.) And تجلبب بِثَوْبِهِ *He covered himself with his garment*. (Ḥar p. 102.)

جُلْبٌ: see جِلْبٌ. — Also The *blackness* of night; (Ḳ, TA;) and so ▼جِلْبَابٌ. (Ḥar p. 480. [The latter evidently tropical in this sense, and perhaps the former also.])

جِلْبٌ (Ṣ, Ḳ) and ▼جُلْبٌ (Ṣ, L) *A camel's saddle of the kind called* رَحْل, *with what it* contains, or comprises: (Ḳ:) or its cover: (Th, Ḳ:) or its pieces of wood: (Ṣ:) or its curved pieces of wood: (TA:) or its wood, without [the thongs called] أَنْسَاع and other apparatus. (Ḳ, TA.) ═ Also, both words, Clouds, (Ḳ,) or thin clouds, (Ṣ,) in which is no water: (Ṣ, Ḳ:) or clouds appearing, or extending sideways, (مُعْتَرِض,) [in the horizon,] like a mountain [or mountain-range]: (Ḳ, TA:) or a cloud like that which is termed عَارِض [q. v.], but narrower, and more distant, and inclining to blackness: (AZ, TA in art. عرض:) pl. أَجْلَابٌ. (TA.) [See also جُلْبَةٌ.]

جَلَبٌ A thing, or things, driven, or brought, (Ṣ, A, Mgh, Mṣb, Ḳ,) from one country or town to another, (Ṣ, Mgh, Mṣb,) or from one place to another, (A, Ḳ,) for the purpose of traffic; (Mgh;) as horses, &c., (Ḳ,) camels, (TA,) sheep or goats, captives or slaves, (Lth, TA,) or any merchandise: (TA:) and so † جَلَبَةٌ, thus in the handwriting of the author of the Ḳ in his last copy of that work, and mentioned by more than one, (MF, [who adds that it is correct, but SM thinks it a mistake,]) and † جَلِيبَةٌ and † جَلُوبَةٌ: (Ḳ:) [see this last, below:] pl. [of the first] أَجْلَابٌ. (Ḳ.) Hence the prov., النُّفَاضُ يُقَطِّرُ الجَلَبَ The failure of provisions causes the camels, driven, or brought, from one place to another, to be disposed in files for sale. (TA.) — [And, app., Male camels; like جَلُوبَةٌ; because they are driven, or brought, from one place to another, and sold; (see 4;) opposed to حَلَبٌ, q. v.] — Also Persons who drive, or bring, camels and sheep or goats [&c.] from one place or country or town to another, for sale; and so [its pl.] أَجْلَابٌ. (Ṣ.) [In the present day, † جَلَّابٌ signifies One who brings slaves from foreign countries, particularly from African countries, for sale.] ═ Also, (Ṣ, A, Ḳ,) and † جَلَبَةٌ, (Ṣ, A,* Mgh, Ḳ,) [the former an inf. n., and so, perhaps, the latter, but often used as simple substs., the latter more commonly, meaning] Cries, shouts, noises, or clamour: (Ṣ, TA:) or a confusion, or mixture, (A, Mgh, Ḳ,) of cries or shouts or noises, (A, Mgh,) or of crying or shouting or noise. (Ḳ.) — And the former, An assembly of men. (TA.)

جُلْبَةٌ The small piece of skin, (Ṣ,) or the crust, or scab, (A, Ḳ,) that forms over a wound (Ṣ, A, Ḳ) when it heals: (Ṣ, Ḳ:) pl. جُلَبٌ. (A.) — A piece of skin that is put upon the [kind of camel's saddle called] قَتَب. (Ṣ, Ḳ.) [See 4.] — [A piece of skin in which an amulet is enclosed: see 4.] — An amulet upon which is sewed a piece of skin: (Ḳ:) pl. as above. (TA.) — A detached portion of cloud: (Ḳ:) [or] a cloud covering the sky. (IAạr, TA.) [See also جُلْبٌ.] — A piece of land differing from that which adjoins it; a patch of ground; syn. بُقْعَةٌ. (Ḳ.) One says, إِنَّهُ لَفِى جُلْبَةِ صِدْقٍ i. e. فِى بُقْعَةِ صِدْقٍ [app. meaning † Verily he is in a good station or position: see art. بقع]. (TA.) — A detached portion of herbage or pasture. (Ḳ,* TA.) ═ Also Severity, or pressure, of time or fortune; (Ṣ, Ḳ;) like كُلْبَةٌ: (Ṣ:) and hunger: (so in some copies of the Ḳ:) or vehemence of hunger: (so in other copies of the Ḳ:) or severity; adversity; difficulty; trouble: (TA:) and a hard, distressful, or calamitous, year. (Ḳ.)

جَلَبَةٌ: see جَلَبٌ, in two places.

جِلْبَابٌ (Ṣ, A, Mgh, Mṣb, Ḳ, &c.) and † جِلِبَّابٌ; (Ḳ;) the latter mentioned as an ex. of form by Sb, and thought by Seer to be syn. with the former, but not explained by any one except the author of the Ḳ; masc. and fem.; (TA;) A [woman's outer wrapping garment called] مِلْحَفَة: (Ṣ:) or this is its primary signification; but it is metaphorically applied to other kinds of garments: (El-Khafájee, TA:) or a shirt, (Ḳ, TA,) absolutely: or one that envelopes the whole body: (TA:) and a wide garment for a woman, less than the ملحفة: or one with which a woman covers over her other garments, like the ملحفة: or the [kind of head-covering called] خِمَار: (Ḳ:) so in the M: (TA:) or a garment wider than the خمار, but less than the رِدَآء, (Mgh, L, Mṣb,) with which a woman covers her head and bosom: (L:) or a garment shorter, but wider, than the خمار; the same as the مِقْنَعَة: (En-Naḍr, TA:) or a woman's head-covering: (TA:) or the [kind of wrapper called] إِزَار: (IAạr, TA:) or a garment with which the person is entirely enveloped, so that not even a hand is left exposed, (Ḥar p. 162, and TA,) of the kind called مُلَآءَة, worn by a woman: (TA:) or a garment, or other thing, that one uses as a covering: (IF, Mṣb:) pl. جَلَابِيبُ. (Ṣ, Mgh, Mṣb.) — See also جُلْبٌ. — † Dominion, sovereignty, or rule [with which a person is invested]. (Ḳ.)

جُلْبَانٌ and جُلَبَانٌ: see جُلُبَّانٌ, in three places.

جِلِبَّابٌ: see جِلْبَابٌ.

جُلَبَانٌ: see the next paragraph, last sentence.

جُلُبَّانٌ, (Ḳ, TA, in the CḲ جُلُبَّان, and so in the TA in art. خرف,) and without teshdeed, (Ḳ,) [i. e.] † جُلْبَانٌ, (Ṣ, Mṣb,) and, accord. to some, † جُلَبَانٌ also, (Mṣb,) not heard by AḤn from the Arabs of the desert but with teshdeed, though many others pronounce it without teshdeed, and pronounced in the latter manner, he says, it may be a dial. var.; (TA;) [a coll. gen. n.;] A certain plant; (Ḳ;) or a certain grain, or seed, of the kind called قِطْنِيّ [i. e. pulse]; (Mṣb;) the [grain, or seed, called] خُلَّر, which is a thing resembling the مَاش: (Ṣ:) or a dust-coloured, dusky kind of grain or seed, which is cooked; of the colour of the ماش, except in its being of a more dusky shade; but larger: (T, TA:) a certain kind of grain or seed, resembling the ماش, of the kind called قِطْنِيّ, well known: (TA:) [a common kind of vetch, or pea, the common lathyrus, or blue chickling vetch, the lathyrus sativus of Linn., is called in Upper Egypt, and by some of the people of Lower Egypt also, جِلْبَان:] n. un. with ة. (TA.) ═ Also the first, (Ḳ,) and † جُلْبَانٌ, (TA,) and † جِلِبَّانٌ, (MF, on the authority of Ibn-El-Jowzee,) [like جُرُبَّانٌ and جُرْبَانٌ or جِرِبَّانٌ,] A thing like a جِرَاب [or sword-case], of skin, or leather, (Ḳ, TA,) in which is put the sword sheathed, and in which the rider puts his whip and implements &c., and which he hangs upon the آخِرَة or the وَاسِط [see these two words] of the camel's saddle; derived from جُلْبَةٌ meaning "a piece of skin that is put upon a قَتَب:" (TA:) or the case (قِرَاب) of the sword-sheath, or scabbard: (Ḳ:) or جُلُبَّانُ السِّلَاحِ, occurring in a trad., signifies the case (قراب) with its contents: or the sword and bow and the like, which require some trouble to draw forth and use in fight; not such a weapon as the lance. (L, TA.) ═ Also the first, and † جَلَبَّانٌ, (Ḳ, TA,) or † جِلِبَّانٌ, (so in the CḲ,) A clamorous man; or one who makes a confused crying or shouting or noise. (Ḳ, TA.)

جِلِبَّانٌ: see the next preceding paragraph, in two places.

جِلِبَّانَةٌ and جُلُبَّانَةٌ: }
جِلِبْنَانَةٌ and جُلُبْنَانَةٌ: } see جَلَّابَةٌ.

جَلِيبٌ, applied to a male slave, (A, Mgh, Ḳ,) One who is brought from one place or country or town to another [for sale]: (Ṣ, Ḳ:) or one who is brought to the country of the Muslims [for sale]: (Mgh:) pl. جَلْبَى and جُلَبَآءُ. (Ḳ.) It is also applied [in like manner] to a woman: pl. جَلْبَى and جَلَائِبُ. (Lḥ, Ḳ.)

جَلُوبَةٌ A thing that is driven or brought from one place or country or town to another for sale; (T, Ṣ, TA;) such as an aged she-camel, and a he-camel, and a young she-camel such as is called قَلُوص, and any other thing; but not applied to stallion-camels of generous race, that are used for procreation: pl. جَلَائِبُ: or the pl. signifies camels that are brought to a man sojourning at a water, who has not means of carriage; wherefore they put him [and his companions or goods &c.] thereon: (TA:) or جلوبة signifies male camels: [see also جَلَبٌ:] or camels that are laden with the goods or utensils &c. of the people: and it is used alike as pl. and sing. (Ḳ.) See جَلَبٌ, with which it is syn. (Ḳ.)

جَلِيبَةٌ: see جَلَبٌ. — Also † An affected habit or disposition. (Ibn-Abi-l-Ḥadeed, MF.)

جَلَّابٌ: see جَلَبٌ.

جُلَّابٌ Rose-water: an arabicized word, (Ḳ,) from the Persian [گُلْ آبْ]. (TA.)

جَلَّابَةٌ and † مُجَلِّبَةٌ and † جِلِبَّانَةٌ (Ḳ, TA) and † جُلُبَّانَةٌ (CḲ) and † جِلِبْنَانَةٌ and † جُلُبْنَانَةٌ, (Ḳ, TA,) applied to a woman, Clamorous, noisy, very loquacious or garrulous, and of evil disposition: (Ḳ, TA:) or جلبّانة signifies, thus applied, rude and coarse: (TA:) the ل in this word is not a substitute for the ر in جِرِبَّانَةٌ [which has a similar meaning]: for it is from الجَلَبَةُ. (IJ, TA.)

جَالِبٌ (A) and † جَالِبَةٌ (L) and † مَجْلَبَةٌ (Ḥar p. 194 &c.) [all signify] † A cause of bringing or drawing or attracting or procuring of a thing: (Ḥar p. 194, in explanation of the last:) thus مَجْلَبَةُ الدَّمْعِ means † the cause of drawing tears:

(Id p. 15:) pl. of the second, جَوَالِبُ; as in the phrase جَوَالِبُ القَدَرِ † [*the drawing*, or *procuring, causes of destiny*]: (L, TA:) pl. of the third, مَجَالِبُ. (Ḥar p. 430.) You say, لِكُلِّ قَضَآءٍ جَالِبٌ وَلِكُلِّ دَرٍّ حَالِبٌ ‡ [*For every decree of fate there is a drawing*, or *procuring, cause; and for every flow of milk there is a milker*]. (A, TA.) And [hence] the pl. جَوَالِبُ signifies † *Calamities, misfortunes, evil accidents, adversities*, or *difficulties*. (TA.) See an ex. in the first paragraph, near the beginning. — قُرُوحٌ جَوَالِبُ and جُلَّبٌ *Wounds*, or *ulcers, healing*, or *becoming covered with skin in healing*. (Aṣ, TA.)

جَالِبَةٌ: see the paragraph next preceding.

مُجْلِبٌ A person *who puts an amulet into a case of skin*: after which it is sewed upon [the headstall, or some other part of the trappings, of] a horse. (TA.)

مَجْلَبَةٌ: see جَالِبٌ.

مُجَلِّبٌ, applied to thunder, (Ḳ,) and to rain, (TA,) *Boisterous*. (Ḳ, TA.) — مُجَلِّبَةٌ: see جَلَّابَةٌ.

خَرَزَةٌ يَنْجَلِبٌ *A* [i. e. *bead*, or *gem*, or *similar stone*] (T, Ḳ, TA) *used by the Arabs of the desert*, (T, TA,) [or *by the women of the desert, as a charm*,] *for captivating*, or *fascinating, men*; (Ḳ,* TA;) or *for bringing back after flight*; (T, Ḳ;) or *for procuring affection after hatred*: (T, TA:) Az mentions it as a quadriliteral-radical word. (TA.) The Arab women used to say,

* أَخَذْتُهُ بِالْيَنْجَلِبْ * فَلَا يَرِمْ وَلَا يَغِبْ *

* وَلَا يَزَلْ عِنْدَ الطُّنُبْ *

[*I have fascinated him with the yenjelib, and he shall not seek* another, *nor absent himself, nor cease to remain at the tent-rope*]. (Lḥ, TA.)

جلح

1. جَلِحَ, aor. ـَ, inf. n. جَلَحٌ, *He* (a man, Ṣ, L, &c.) *was*, or *became, bald in the two sides of his head*: (Ṣ, Ḳ:) or *in the two sides of the fore part of his head*: (Mṣb:) or *in the fore part of his head*: or *a little more bald than he who is termed* أَنْزَعُ. (L.) [See also جَلَحٌ and أَجْلَحُ.] — جَلِحَتِ الأَرْضُ, inf. n. as above, *The herbage of the land was eaten*; as also جُلِحَت. (TA.) And جَلِحَتِ الشَّجَرَةُ *The branches of the tree were eaten, and it became reduced to its stem*, or *root*. (AḤn, TA.) ═ جَلَحَ المَالُ الشَّجَرَ, aor. ـَ, (Ṣ, Ḳ,) inf. n. جَلْحٌ; (Ṣ;) and † جلّحهُ, inf. n. تَجْلِيحٌ; (TA;) *The cattle ate the trees*: or *ate the upper parts thereof*: (TA:) or *fed upon the upper parts thereof, and peeled them*. (Ṣ, Ḳ.) ═ See also 3.

2: see 1. ═ تَجْلِيحٌ [the inf. n.] signifies also *The acting*, or *advancing, boldly*, (Ḳ,) or *very boldly*: (Ṣ:) or *being bold to do evil or mischief*; and *showing open enmity* or *hostility*: (A:) and *acting with penetrating energy, vigour*, or *effectiveness*, (Ṣ, Ḳ, TA,) in an affair: (TA:) and *going*, or *journeying, vehemently*: (TA:) and *the assaulting*, or *attacking*, (AZ, Ḳ, TA,) of a man, (AZ, TA,) and of an animal of prey. (Ḳ.) See also 3. You say, لَا تُجَلِّحْ عَلَيْنَا يَا فُلَانُ [*Be not bold to do evil* or *mischief*, or *to show open enmity* or *hostility, to us, O such a one*]. (A.) And فِى وَجْهِهِ تَجْلِيحٌ *In his face is* [apparent] *boldness to do evil* or *mischief*, and *a show of open enmity* or *hostility*. (A.) And جلّح عَلَى القَوْمِ *He charged*, or *made an assault* or *attack, upon the people* or *party*. (AZ, TA.) And جلّح تَجْلِيحَ الذِّئْبِ [*He assaulted with the assaulting of the wolf*]. (A.) And جلّح عَلَيْنَا *He came upon us*; or *came down upon us and overcame us*; or *destroyed us*; syn. أَتَى عَلَيْنَا. (ISh, TA.) And جلّح فِى الأَمْرِ *He went at random, heedlessly, without any certain aim* or *object*, or *without consideration, in the affair*; or *pursued a headlong*, or *rash, course therein*. (TA.)

3. مُجَالَحَةٌ [the inf. n.] signifies *The acting openly with another in an affair*: (Aṣ, Ḳ:) and *the showing open enmity* or *hostility with another*. (Ṣ, Ḳ.) You say, جَالَحْتُ الرَّجُلَ بِالأَمْرِ *I acted openly with the man in the affair*. (Ṣ.) And جَالَحَنِى فُلَانٌ *Such a one showed open enmity* or *hostility with me*; as also † جلّح عَلَىَّ. (A.) — Also *The contending with another for superiority in strength*; syn. مُشَادَّةٌ (Ṣ) and مُكَالَحَةٌ. (Ṣ, Ḳ.) You say, جَالَحَنِى فُلَانٌ فَجَلَحْتُهُ † [app. meaning *Such a one contended with me for superiority in strength, and overcame me therein*]. (TA.) — And i. q. مُكَابَرَةٌ [The *contending with another for superiority in greatness*; &c.]. (Ḳ.)

Q. Q. 1. جَلْمَحَ *He shaved* his head: (Fr, Ṣ, Ḳ:) the م is augmentative. (Ṣ.)

جَلَحٌ *Baldness in the two sides of the head*: (Ṣ, Ḳ:) or *in the two sides of the fore part of the head*: (Mṣb:) it is *more than* نَزَعٌ, and *less than* صَلَعٌ, (Ṣ, Mṣb,) which is less than جَلَهٌ: (Mṣb:) or *baldness in the fore part of the head*: or *baldness that is a little more than what is termed* نَزَعٌ. (L.)

جَلَحَةٌ *A part*, or *place, in which is baldness such as is termed* جَلَحٌ. (Ṣ, Mṣb.)

أَرْضٌ جِلِحَانَةٌ *Land that produces no herbage*. (Ḳ.)

جُلَاحٌ A torrent *that carries away everything in its course*. (Ṣ, Ḳ.)

جِلْوَاحٌ *Wide* (Ḳ, TA) *and bare*, or *open*, (TA,) *land*. (Ḳ, TA.)

جُلَّحٌ: see أَجْلَحُ.

جَالِحَةٌ (TA) and جَوَالِحُ (Ṣ, Ḳ [the latter being pl. of the former]) *What flies about in successive portions from the heads of reeds and papyrus-plants* (Ṣ, Ḳ, TA) *and other plants, in the wind*, (TA,) *resembling cotton*; (Ṣ, TA;) and *spiders' webs so flying about*. (TA.) And the latter, *Flakes of snow falling quickly and continuously*. (TA.)

أَجْلَحُ A man *bald in the two sides of his head*: (Ṣ:) or *in the two sides of the fore part of his head*: (Mṣb:) or *in the fore part of his head*: (Mgh, L:) or *a little more bald than he who is termed* أَنْزَعُ: (L:) it signifies more than انزع, and less than أَجْلَى and أَجْلَهُ: (Mgh:) when a man is bald in the sides of his forehead, he is termed انزع; when the baldness is a little more, اجلح; when it extends to the half, or the like, اجلى; and then, اجله: (A 'Obeyd, TA:) the fem. is جَلْحَآءُ: and the pl. جُلْحٌ (L, Mṣb) and جُلْحَانٌ. (L.) — ‡ *Having no horn*; applied to a bull and a he-goat: (A:) and in this sense the fem. is applied to a ewe (T, M, Mṣb) or she-goat, (T, M, A, Mṣb,) and to a cow: (T, M, A:) and in like manner [the pl.] جُلْحٌ is applied to cows or bulls *having no horns*; (Ṣ, TA;) erroneously said in the Ḳ to be جُلَّحٌ, like سُكَّرٌ. (TA.) — † A [woman's camel-vehicle of the kind called] هَوْدَج *that has not a high head* or *top*: (Ibn-Kulthoom, IJ, Ṣ, Ḳ:) or *without a top*: (T:) or one that is *of a square form*: (Aṣ, IJ:) pl. † أَجَالِحُ, (Ṣ, IJ,) like as أَعْزَالٌ is pl. of أَعْزَلُ; a very rare form of pl. of a sing. of the measure أَفْعَلُ. (IJ.) — † A flat roof *not surrounded by a wall or anything else to prevent persons' falling from it*. (IAth, Ḳ.) — قَرْيَةٌ جَلْحَآءُ ‡ *A town having no fortress*: (A, TA:) pl. قُرًى جُلْحٌ: the fortresses being likened to horns. (TA.) — أَكَمَةٌ جَلْحَآءُ † [*A hill*] *not having a pointed summit*. (TA.) And هَضْبَةٌ جَلْحَآءُ † [*A*] *smooth* [*hill*]. (A.) — أَرْضٌ جَلْحَآءُ † *A land in which are no trees*. (TA.) — يَوْمٌ أَجْلَحُ † *A hard, distressing*, or *calamitous, day*; as also أَصْلَعُ. (A, TA.)

أَجْلَاحٌ: see أَجْلَحُ.

إِجْلِيحٌ A plant *of which the upper parts have been eaten*. (TA.)

مَجْلُوحٌ *Eaten*: (Ṣ, Ḳ:) *eaten until nothing of it is left*: (Ṣ:) herbage *so eaten*. (TA.)

مِجْلَحٌ A man (Ṣ) *who eats much*; *a great eater*; *voracious*. (Ṣ, Ḳ.) — See also مِجْلَاحٌ. — سَنَةٌ مُجَلِّحَةٌ *A year of drought, barrenness*, or *dearth*. (TA.) ═ *Insolent and audacious*. (L.) You say, فُلَانٌ وَقِحٌ مُجَلِّحٌ [*Such a one is impudent, insolent, and audacious*]. (A, TA.) — A bold wolf. (TA.)

مِجْلَاحٌ A she-camel (Ṣ) *that bears with hardiness a severe year, preserving her milk*; (Ṣ, Ḳ;) as also † مُجَلِّحَةٌ. (L.) — See also مُجَالِحٌ.

مَجْلُوحٌ A tree *having the head*, or *upper part, eaten*. (L.) — A plant, or tree, *that has been eaten and has grown again*. (TA.)

مُجَالِحٌ i. q. مُكَابِرٌ [*Contending with another for superiority in greatness*; &c.: see its verb, 3]. (Ṣ.) — *The lion*. (Ḳ.) — A she-camel *that yields milk abundantly in winter:* (Ṣ, Ḳ:) or *that crops the twigs of the dry trees in winter, in a year of drought, and becomes fat upon them, and so preserves her milk:* (IAạr, TA:) pl. ♦مَجَالِيحُ: (Ṣ, Ḳ:) or this is pl. of مُجَالِحٌ and ♦مِجْلَاحٌ as epithets applied to a palm-tree and a she-camel *that cares not for the want of rain.* (AḤn, TA.) And مُجَالِحَةٌ A she-camel *that eats the* سَمُر and عُرْفُط, *whether they have leaves upon them or not.* (TA.)

مَجَالِيحُ: see مُجَالِحٌ. — Also Years *of drought that carry off,* or *destroy, the cattle.* (Ṣ, Ḳ.)

جلد

1. جَلَدَهُ, (Ṣ, A, Mgh, &c.,) aor. ـِ, (Mṣb, Ḳ,) inf. n. جَلْدٌ, (Ṣ, Mgh, Mṣb,) *He hit,* or *hurt, his skin;* (Ṣ, Ḳ;) like as you say, رَأَسَهُ, and بَطَنَهُ: (Ṣ:) *he beat his skin:* (Mgh:) *he beat him;* namely, a criminal: (Mṣb:) *he struck him* with a whip, and with a sword: (TA:) *he flogged him* (A, Ḳ) *with a whip,* (Ḳ,) or *with whips:* (A:) جَلَدْتُ is sometimes written and pronounced جَلَتُّ. (MF on the letter د.) You say, جَلَدَهُ الحَدَّ, inf. n. as above, *He inflicted upon him the flogging ordained by the law.* (Ṣ, L.) — جَلَدَتِ الحَيَّةُ *The serpent bit:* (Ḳ:) or, accord. to some, one says of the serpent called أَسْوَدُ, specially, يَجْلِدُ بِذَنَبِهِ [*it strikes with its tail*]. (TA.) — جَلَدَ جَارِيَتَهُ, (Ḳ,) aor. and inf. n. as above, (TA,) ‡ *He lay with his young woman,* or *female slave.* (Ḳ, TA.) [Hence,] جَلَدَ عُمَيْرَةَ [† *i. q.* نَكَحَ اليَدَ], a metonymical phrase: جَلْدُ عُمَيْرَةَ meaning الخَضْخَضَةُ, and التَّدْلِيلُ, also termed الاِسْتِمْنَآءُ بِاليَدِ, and the similar act of a woman is termed الاِعْتِمَارُ: الاِلْطَافُ. (Ḥar p. 572.) — جَلَدَ بِهِ الأَرْضَ *He smote the ground with him;* (TA;) *he threw him down prostrate on the ground.* (A, TA.) And جُلِدَ بِهِ *He fell down* (Ḳ, TA) *upon the ground by reason of much sleepiness;* as also جُلِدَ بِهِ نَوْمًا. (TA.) كُنْتُ أَشْتَدُّ فَيُجْلَدُ بِى, in a trad., means [*I used to exert my strength,* or *energy, but*] *sleep would overcome me so that I fell down.* (L.) — جَلَدَهُ عَلَى الأَمْرِ ‡ *He compelled him against his will to do the thing.* (A, Ḳ.) — يُجْلَدُ بِكُلِّ خَيْرٍ (or, as related by AḤát, يحلد, with ح, TA) ‡ *He is imagined to possess every good quality.* (A, Ḳ.) But the saying of Esh-Sháfi'ee كَانَ مُجَالِدٌ يُجْلَدُ means † *Mujálid used to be pronounced a liar,* (Ḳ, TA,) or *suspected and accused of lying.* (TA.) ═ جُلِدَتِ الأَرْضُ, (Ṣ, L, Mṣb,) the verb being in the pass. form, (Mṣb,) or جَلِدَتْ, (A, Ḳ,) a verb of the same form as فَرِحَ; (Ḳ;) [or both may be correct, like ضُرِبَتْ and ضَرِبَتْ in the same sense;] and ♦أَجْلَدَتْ; (Ḳ;) [but this last I believe to be a mistake for أُجْلِدَتْ, like أُضْرِبَتْ;] *The land was,* or *became, affected,* or *smitten, by hoar-frost,* or *rime.* (Ṣ, A, L, Mṣb, Ḳ.) And جلد البَقْلُ [in the TA جَلِدَ] *The herbs,* or *leguminous plants, were,* or *became, affected,* or *smitten, thereby.* (L, TA.) And ♦أُجْلِدُوا *They* (men) *were,* or *became, affected,* or *smitten, thereby.* (L, Ḳ.) ═ جَلُدَ, aor. ـُ, inf. n. جَلَادَةٌ and جُلُودَةٌ and جَلَدٌ (or this last is a simple subst., L) and مَجْلُودٌ, (an inf. n. like مَحْلُوفٌ and مَعْقُولٌ, (Ṣ, or from تَجَلَّدَ, M in art. عسر,) *He* (a man, Ṣ, L) *was,* or *became, hardy, strong, sturdy,* (Ṣ,* L, Ḳ,*) *and enduring,* or *patient.* (L.)

2. جلّد, (IAạr, T, Ṣ, Mgh,) inf n. تَجْلِيدٌ, (T, Ṣ, Mgh, Ḳ,) *He skinned* a camel (IAạr, T, Ṣ, Mgh, Ḳ) that had been slaughtered: (Ṣ, Ḳ:) one seldom uses سَلَخَ thus [in relation to a camel]. (Ṣ.) — Also *He covered* a thing *with skin;* as, for instance, a pair of socks, or stockings: (Mgh:) and in like manner, [*he bound*] a book: (A, Ḳ:*) and *he clad* a young camel *in the skin of another young camel:* (L: [see جَلَدٌ:]) thus the verb bears two contr. significations. (Mgh.) ═ [*He ordered to be flogged.* (Freytag's Lex.: but without any indication of an authority.)] ═ [*He,* or *it, rendered* a man *hardy, strong, sturdy, and enduring,* or *patient:* so in the present day.]

3. جالدهُ *He contended with him in fight,* whether the fight were with swords or not. (A in art. طرد.) You say, جالدهُ بِالسَّيْفِ, (L,) inf. n. مُجَالَدَةٌ (Ṣ, A, L) and جِلَادٌ, (A, L,) *He contended with him in fight with the sword.* (Ṣ,* A, L.) And جَالَدُوهُمْ بِالسُّيُوفِ *They contended with them in fight with swords.* (A.) And جالدوا بِالسُّيُوفِ, (Ḳ, TA,) and ♦تجالدوا (Ṣ, A, L, and so in the CḲ instead of جالدوا) بالسيوف, (Ṣ,) and ♦اجتلدوا, (Ṣ, A, L,) *They contended, one with another, in fight with swords.* (Ṣ, A,* L, Ḳ.) — [See also حَاضَرَهُ.]

4. اجلدهُ إِلَيْهِ †*He constrained, compelled,* or *necessitated, him to have recourse to,* or *betake himself to, him,* or *it:* (so in some copies of the Ḳ:) or *he made him to stand in need of,* or *to want, him,* or *it.* (AA, L, and so in some copies of the Ḳ and in the TA.) ═ أَجْلَدَتِ الأَرْضُ [or أُجْلِدَتْ]: and أُجْلِدُوا: see 1.

5. تجلّد *He affected hardiness, strength, sturdiness, and endurance,* or *patience; constrained himself to behave with hardiness,* &c. (Ṣ,* A,* L, Ḳ.*) So in the phrase تجلّد لِلشَّامِتِينَ [*He constrained himself to behave with hardiness,* &c., *to those who rejoiced at his misfortune*]. (A, TA.) In the phrase تجلّد عَنْهُ [*He constrained himself to endure with hardiness and patience the loss,* or *want, of him,* or *it*], the verb is made trans. by means of عن because it implies the meaning of تَصَبَّرَ. (L.) — Also *He feigned,* or *made a show of, hardiness, strength, sturdiness, and endurance,* or *patience.* (L.)

6: see 3.

8: see 3. ═ اجتلد مَا فِى الإِنَآءِ, (AZ, TA,) or الإِنَآءَ, (Ḳ,) *He drank all that was in the vessel;* (AZ, Ḳ, TA;) as also اجتلد. (AZ, TA.)

جَلْدٌ (sometimes pronounced جَضْدٌ, Ṣ) and ♦جَلِيدٌ, (applied to a man, Ṣ, A, Mgh,) *Hardy, strong, sturdy,* (Ṣ,* A,* L, Ḳ,*) *and enduring,* or *patient:* (L:) *not* بَلِيد [q. v.]: (Mgh:) pl. [of either] جُلُدٌ, (Ṣ, L, Ḳ,) or جُلْدٌ, (so in some copies of the Ḳ,) and جُلَدَآءُ and أَجْلَادٌ [a pl. of pauc.] (Ṣ, L, Ḳ) and جِلَادٌ. (Ḳ.) And [the fem.] جَلْدَةٌ A *hardy and strong* she-camel; *strong to labour and to journey; that heeds not the cold:* and also *swift:* pl. جَلَدَاتٌ: (L:) and a she-camel *that yields a copious flow of milk:* (Th, TA:) sing. of جِلَادٌ, (Ṣ,) which signifies she-camels *abounding with milk;* as also مَجَالِيدُ, (Ḳ,) pl. of ♦مِجْلَادٌ; (TA;) or she-camels *having neither milk nor young:* (Ḳ:) [see also جَلَدٌ:] or she-camels *that yield the most greasy,* or *unctuous, sort of milk:* and so the sing., جلدة, applied to a ewe or a she-goat. (Ṣ.) جِلَادٌ (pl. of جَلْدَةٌ, TA) is also applied to palm-trees, meaning *Large, hard, hardy,* or *strong:* (Ṣ, Ḳ, TA:) or *such as are not affected by drought.* (TA.) And تَمْرَةٌ جَلْدَةٌ signifies *A tough-skinned, excellent, date;* as also ♦جَلِدَةٌ: and *a hard, compact, date.* (L.)

جِلْدٌ, (Ṣ, Mṣb, Ḳ, &c.,) the only form of the word mentioned by the generality of the lexicographers; (TA;) occurring at the end of a verse with kesr to the second as well as the first letter, ♦جِلِدٌ, agreeably with a license allowed to a poet in such a case, to give to a quiescent letter in a rhyme the same vowel as that which the preceding letter has; (Ṣ;) and ♦جَلَدٌ, (IAạr, Ṣ, Ḳ,) like شِبْهٌ and شَبَهٌ, and مِثْلٌ and مَثَلٌ; but this is said by ISk to be unknown; (Ṣ;) The *skin* of any animal; (Ḳ;) the *integument of the body and limbs* of an animal: (Az, Mṣb:) or the *exterior of the* بَشَرَة [or *upper skin*] of an animal: (Mṣb: [but this is a strange explanation:]) pl. جُلُودٌ (Ṣ, Mṣb, Ḳ) and (sometimes, Mṣb) أَجْلَادٌ [a pl. of pauc.]. (Mṣb, Ḳ.) — [The pl.] أَجْلَادٌ signifies also, and ♦تَجَالِيدُ likewise, The *body and limbs* (Ṣ, A, L) of a man; (Ṣ;) the *whole person,* or *body and limbs,* of a human being; (L, Ḳ;) and his *self:* (L:) so called because enclosed by the skin: pl. of the former, أَجَالِدُ. (L.) You say, مَا أَشْبَهَ أَجْلَادَهُ بِأَجْلَادِ أَبِيهِ *How like are his person and body to the person and body of his father!* (L.) And فُلَانٌ عَظِيمُ الأَجْلَادِ and ♦التَّجَالِيدِ (A, L) *Such a one is large and strong* (L) *in respect of the body and limbs.* (A, L.) And رُدُّوا الأَيْمَانَ عَلَى أَجْلَادِهِمْ *Repeat ye the oaths to the persons, themselves:* occurring in a trad.: said on the occasion of a man's entering among others of whom an oath had been demanded. (L.) — الجِلْدُ also signifies † *The penis.* (Fr, L, Ḳ: but in the CḲ, in this sense, it is written الجَلَدُ.) Agreeably with this explanation, its pl. جُلُود is said by Fr to be used in the Ḳur xli. 20: (L:) or as meaning the *pudenda:* (L, Ḳ:) but ISd holds that this word there means the *skins,* with which, as in manual operations, acts of disobedience are performed. (L.)

جَلَدٌ: see جِلْدٌ. — Also The *skin of a camel,*

or other beast, *with which another beast is clothed:* (L:) the *skin of a young camel, which* (*being stripped off,* Ṣ) *is put over the body of another young camel, in order that the mother of the skinned young one* (*smelling it,* Ṣ) *may conceive an affection for it* [*and suckle it*]: (Ṣ, Ḳ:) or the *skin of a young camel, which is stuffed with panic grass* (ثُمَام), (Ḳ, TA,) *or some other plant,* (TA,) *and put before a she-camel, in order that she may be induced thereby to affect that which is not her young one* [*and so yield her milk*], (Ḳ, TA,) or, *to affect the young one of another.* (L, TA.) ═ A ewe or she-goat *whose young one dies at the time of her bringing it forth;* as also ▼ جَلَدَةٌ: pl. [of the former] جِلَادٌ and [of the latter] جَلَدَاتٌ. (TA.) — Also, [as a coll. gen. n.,] *Great* she-camels, *having neither young ones nor milk;* n. un. with ة: (Ṣ:) [see also جَلَدٌ:] or *great* camels, *among which are no little ones;* (Ḳ;) n. un. with ة: (TA:) and (app. as a quasi-pl. n., TA) sheep or goats, and camels, *having neither young ones nor milk;* (Ḳ;) app. meaning *having no little ones to which they give such:* (Moḥammad Ibn-El-Mukarram, TA:) or she-camels *having no young ones with them, so that they endure patiently the heat and cold:* (Fr, TA:) or she-camels *having no milk, and the young ones of which have gone away from them;* including what are called بَنَاتُ اللَّبُونِ, and such as are above these in age; and also such as are called مَخَاضٌ, and عِشَارٌ, and حِيَالٌ; but when they have given birth to their young, they cease to be termed جَلَدٌ, and are called عِشَارٌ, and لِقَاحٌ: the pl. is أَجْلَادٌ and [pl. pl.] أَجَالِيدُ. (Az, TA.) ═ *Hard ground;* as also ▼ أَجْلَدُ: (Ṣ:) or *hard and level ground;* as also ▼ جَلَدَةٌ: (Ḳ:) or *level but rough ground;* (L;) as also ▼ أَجْلَدُ: (TA:) pl. (of the former, TA) أَجْلَادٌ and (of the latter, TA) أَجَالِدُ. (Ṣ, TA.) You also say أَرْضٌ جَلَدٌ, with fet-ḥ to the ل; (AḤn, TA;) and ▼ جَلْدَةٌ, with a quiescent ل; (Lth, AḤn, TA;) and أرض ▼ جَلَدَةٌ also; and مَكَانٌ جَلَدٌ. (Lth, TA.) ═ Also, as a subst. or an inf. n., (L, [see جَلُدَ,]) *Hardiness, strength, sturdiness,* (Ṣ,* A,* L, Ḳ,*) *and endurance,* or *patience.* (T.)

جُلْدٌ : see جِلْدٌ.

جَلْدَةٌ : see جَلْدٌ.

جِلْدَةٌ a more particular term than جِلْدٌ; (Ṣ, L;) signifying *A piece,* or *portion, of skin.* (L.) — One says also جِلْدَةُ العَيْنِ [app. meaning *The eyelid*]. (TA.) — And قَوْمٌ مِنْ جِلْدَتِنَا *A people,* or *company of men, of ourselves,* and *of our kinsfolk.* (TA.) ═ See also جَلْدٌ.

جَلَدَةٌ : see جَلَدٌ, in three places.

جَلِيدٌ A woman *flogged with a whip;* as also with ة: pl. جَلْدَى and جَلَائِدُ: (Lḥ, L:) the former pl. thought by ISd to be of the former sing.; and the latter, of the latter. (L.) ═ *Hoar-frost,* or *rime;* i. e. *dew that falls* (Ṣ, A, Ḳ) *from the sky* (Ṣ) *upon the ground* (Ṣ, A, Ḳ) *and congeals;* (Ṣ, Ḳ;) also called ضَرِيبٌ and سَقِيطٌ; (Ṣ;) like صَقِيعٌ. (Mṣb.) — *Congealed,* or *frozen, water; ice.* (TA.) ═ See also جَلْدٌ.

جُلَيْدَةٌ [dim. of جِلْدَةٌ] *One of the* طَبَقَات [*coats,* or *tunics,*] *of the eye.* (TA.)

جُلُودِيٌّ [*A dealer in skins;*] a rel. n. from جُلُودٌ, pl. of جِلْدٌ. (TA.)

جَلَّادٌ *One whose office it is to flog others with a whip.* (Mgh.) [In the present day, *An executioner,* in a general sense.]

أَجْلَدُ : see جَلَدٌ, in two places.

تَجَالِيدُ : see جِلْدٌ, in two places.

مِجْلَدٌ *A piece of skin which a wailing woman holds in her hand, and with which she slaps* (Ṣ, Ḳ) *her face* (Ṣ) or *her cheek:* (Ḳ:) pl. مَجَالِدُ; (Kr, Ḳ;) or, as ISd thinks, this is pl. of ▼ مِجْلَادٌ [as syn. with مِجْلَدٌ], for مِفْعَلٌ and مِفْعَالٌ are often interchangeable as measures of words of this kind. (TA.)

مُجَلَّدٌ *Covered with skin:* thus applied to a pair of socks, or stockings, meaning *having skin put upon the upper and lower parts.* (Mgh.) [A book, or portion of a book, *bound:* — and hence, *A volume:* pl. مُجَلَّدَاتٌ.] — A bone *covered only by the skin; having nothing remaining on it but the skin.* (Ḳ.) ═ A horse [*rendered hardy and enduring;*] *that is not frightened by,* (Ḳ,) or *not impatient at,* (Ṣ, and so in some copies of the Ḳ,) *being beaten* (Ṣ, Ḳ) *with the whip.* (TA.) ═ *A certain quantity of a burden,* or *load, of known measure and weight;* (Ḳ;) *six hundred pounds' weight.* (IAạr, TA in art. بهر.)

مُجَلِّدٌ *One who binds books,* or *covers them with skin.* (Ḳ.)

مِجْلَادٌ : see مِجْلَدٌ: ═ and see also جَلْدٌ.

مَجْلُودٌ [*Having his skin hit, hurt,* or *beaten: flogged:* &c.: see also جَلِيدٌ]. ═ أَرْضٌ مَجْلُودَةٌ *Land affected,* or *smitten, by hoar-frost,* or *rime.* (Ṣ, A, L, Mṣb, Ḳ.) ═ مَجْلُودٌ is also an inf. n. of جَلَدَ [q. v.]. (Ṣ, L, Ḳ.)

مُجْتَلَدٌ *A place of contending in fight with swords.* (L from a trad.)

جلس

1. جَلَسَ, (Ṣ, A, Mṣb, Ḳ,) aor. ـِ, (A, Ḳ,) inf. n. جُلُوسٌ (Ṣ, A, Mṣb, Ḳ) and ▼ مَجْلِسٌ, (Ṣ, A, Ḳ,) *He placed his seat,* or *posteriors, upon rugged* [or rather *elevated*] *ground, such as is termed* جَلْسٌ: this is the primary signification: (TA:) [and hence,] *He sat;* i. q. قَعَدَ [when the latter is used in its largest sense]: (Mṣb, and so Ṣ and L and A and Ḳ in art. قعد:) you say, جَلَسَ مُتَرَبِّعًا and قَعَدَ مُتَرَبِّعًا [*He sat cross-legged*]: (Mṣb:) accord. to El-Fárábee and others, *contr. of* قَامَ; and thus it has a more common application than قَعَدَ [when the latter is used in its most proper and restricted sense]: (Mṣb:) but قَعَدَ also signifies the contr. of قَامَ: ('Orweh Ibn-Zubeyr, L in art. قعد:) properly speaking, جَلَسَ differs from قَعَدَ; the former signifying *he sat up;* or *sat after sleeping,* or *prostration,* (Mṣb,) or *after lying on his side;* (B, TA;) and the latter, he sat down; or sat after standing: (Mṣb, B, TA: and see other authorities to the same effect in art. قعد:) for جُلُوسٌ is a change of place from low to high, and قُعُودٌ is a change of place from high to low: and one says, جَلَسَ مُتَّكِئًا, but not قَعَدَ مُتَّكِئًا, meaning [*He sat*] *leaning,* or *reclining, upon one side:* (Mṣb:) but both these verbs sometimes signify *he was,* or *became:* and thus, [it is said,] قَعَدَ مُتَرَبِّعًا and جَلَسَ مُتَرَبِّعًا signify *he was,* or *became, cross-legged:* and جَلَسَ بَيْنَ شُعَبِهَا الأَرْبَعِ in like manner signifies *he was,* or *became,* [*between her four limbs,*] (El-Fárábee, Mṣb,) because the man, in this case, is resting upon his own four limbs. (Mṣb.) [جَلَسَ مَعَهُ and جَلَسَ إِلَيْهِ, like خلا معه and خلا اليه, signify the same; i. e. *He sat with him:* or the latter, *he sat by him;* like "assedit ei."] An instance of the inf. n. مَجْلِسٌ is found in a trad., in which it is said, فَإِذَا أَتَيْتُمْ إِلَى المَجْلِسِ فَأَعْطُوا الطَّرِيقَ حَقَّهُ [*But when ye come to sitting, perform ye the duties relating to the road*]. (TA.) [The trad. commences thus: إِيَّاكُمْ وَالجُلُوسَ عَلَى الطُّرُقَاتِ *Beware ye of sitting on the roads:* and then, after the words before cited, (in which, however, in my copy of the Jámi' eṣ-Ṣagheer, instead of المجلس, I find المَجَالِس, which is pl. of المَجْلِس,) it is added that the duties thus alluded to are the lowering of the eyes, the putting away or aside what is hurtful or annoying, the returning of salutations, the enjoining of that which is good, and the forbidding of that which is evil.] — جَلَسَتِ الرَّخَمَةُ ‡ *The aquiline vulture lay upon its breast on the ground;* syn. جَثَمَتْ: a saying applied to him who is of the seceders. (A, TA.) [See also قَعَدَ.] — جَلَسَ also signifies † *It* (a thing, as, for instance, a plant,) *remained,* or *continued.* (AḤn, TA.) — Also, (aor. ـِ, inf. n. جَلْسٌ, TA,) *He came to* الجَلْس, (TA,) or [*the high country called*] *Nejd:* (T, Ṣ, A, TA:) and in like manner said of a cloud; *it came to Nejd.* (TA.)

3. جَالَسَهُ, inf. n. مُجَالَسَةٌ and جِلَاسٌ, [*He sat with him.*] (TA.) You say, لَا تُجَالِسْ مَنْ لَا تُجَانِسُ [*Sit not with him with whom thou wilt not be congenial*]. (A, TA.) And كَرِيمُ النِّحَاسِ طَيِّبُ الجِلَاسِ [*Generous in origin,* or *disposition; pleasant to sit with;*] is said of a man. (TA.)

4. اجلسهُ [*He seated him; made him to sit:* or *he made him to sit up*]: (Ṣ, Ḳ, TA:) *he gave him place,* or *settled him,* (مَكَّنَهُ,) *in sitting.* (TA.)

6. تجالسوا [*They sat together; one with another;*] (Ṣ, A, TA;) فِي المَجَالِسِ [*in the sitting-places*]. (Ṣ.)

10. استجلسهُ [*He asked him,* or *desired him, to sit:* or *to sit up*]. You say, رَآنِي قَائِمًا فَاسْتَجْلَسَنِي [*He saw me standing, and he asked me,* or *desired me, to sit*]: (A, TA:) but this is at variance with what we have mentioned in the beginning of the art., respecting the distinction [between جَلَسَ and قَعَدَ]. (TA.)

جَلْسٌ *Rugged* ground or land: (Ṣ, Ḳ:) this is the primary signification. (TA.) — [Also, app., *Elevated* ground or land:] a place *elevated and hard:* or, as some say, a tract of land *extending widely.* (Ḥam p. 688.) — [And hence,] الجَلْسُ *What is elevated above the* غَوْر [or *low country*]: (TA:) applied especially to *the country of Nejd.* (T, Ṣ, M, Ḳ.) ═ [Persons *sitting:* or *sitting up:*] a quasi-pl. n., accord. to Sb, or a pl., accord. to Akh, of ↓ جَالِسٌ: said to be used as sing. and pl. and fem. and masc.; but this assertion is of no account: (ISd, L:) or the *people of a* مَجْلِس: (Lḥ, ISd, L, Ḳ:) [↓ جُلُوسٌ is also a pl. of ↓ جَالِسٌ; like as بُكِىٌّ, originally بُكُوىٌ, is of بَاكٍ: or it is an inf. n. used as an epithet: see جَاثٍ:)] you say قَوْمٌ جُلُوسٌ [*a company of men sitting:* or *sitting up*]. (Ṣ.) [See also مَجْلِسٌ.] — Also A woman *who sits in the* فِنَآء [or *court of the house*], *not quitting it:* (Ḳ:) or she who is *of noble rank* (Ḳ, TA) *among her people.* (TA.)

جِلْسٌ: see جَلِيسٌ, in two places.

جَلْسَةٌ *A single sitting:* or *sitting up.* (Mṣb.)

جِلْسَةٌ *A mode* or *manner,* (TA,) *kind,* (Mṣb,) or *state,* (Ṣ, A, Mṣb,) *of sitting:* or *of sitting up.* (Ṣ,* A,* Mṣb, Ḳ.*) You say, هُوَ حَسَنُ الجِلْسَةِ [*He has a good mode,* &c., *of sitting*]. (A, Mṣb, Ḳ.)

جُلَسَةٌ A man (Ṣ) *who sits much; sedentary.* (Ṣ, Ḳ.)

جُلُوسٌ: see جَلْسٌ.

جَلِيسٌ (Ṣ, A, Mṣb, Ḳ) and ↓ جِلِّيسٌ (TA, as found in a copy of the Ḳ, [but this is an intensive form,]) and ↓ جِلْسٌ (Ṣ, A, Ḳ) *A companion with whom one sits:* (A, Mṣb, Ḳ:) fem. of the first with ة: (TA:) and pl. [of the same] جُلَسَآءُ (A, Ḳ) and [irreg., being by rule pl. of جَالِسٌ,] جُلَّاسٌ. (Ḳ.) You say, هُوَ جَلِيسِى and ↓ جِلْسِى [*He is my companion with whom I sit*]; like as you say, هُوَ خِدْنِى and خَدِينِى. (Ṣ.)

جِلِّيسٌ: see جَلِيسٌ.

جَالِسٌ: see جَلْسٌ, in two places. — Also A man, and a cloud, *coming to* [*the high country called*] *Nejd.* (TA.) You say, رَأَيْتُهُمْ يَعْدُونَ جَالِسِينَ *I saw them running, coming to Nejd.* (A, TA.)

مَجْلَسٌ: see 1: — and see مَجْلِسٌ.

مَجْلِسٌ *A sitting-place;* (Ṣ, Mṣb, Ḳ;) as also ↓ with ة; (Fr, Lḥ, Ṣgh, Ḳ;) similar to مَكَانٌ and مَكَانَةٌ: (Ṣgh, TA:) [*a place where persons sit together and converse; a sitting-room:*] *a thing upon which one sits:* (MF:) some make a strange distinction between مَجْلِسٌ and ↓ مَجْلَسٌ, asserting the former to be applied to the *chamber* or *house* (بَيْت) [*in which people sit*]; and the latter, to a *place of honour upon which it is forbidden to sit without permission;* but the former is the only correct form of the two: (MF, TA:) pl. مَجَالِسُ. (Ṣ, Mṣb.) You say, إِرْزُنْ فِى مَجْلِسِكَ and ↓ مَجْلِسَتِكَ [*Be thou grave*] *in thy sitting-place.* (Fr, Ṣgh.) — ‡ The *people of a* مَجْلِس; (Mṣb, TA;) elliptical, for أَهْلُ مَجْلِسٍ: (TA:) *an assembly,* or *a company of men, sitting* [*together*]: (Th, TA:) not well explained as being, with the article ال, syn. with النَّاسُ: (TA:) *persons sitting,* or *sitting up.* (A, TA.) [See also جَلْسٌ.] You say, اِنْفَضَّ المَجْلِسُ † [*The assembly of persons sitting together broke up*]. (Mṣb.) And رَأَيْتُهُمْ مَجْلِسًا *I saw them sitting.* (A, TA.) — † *An oration* or *a discourse,* or *an exhortation,* (خُطْبَةٌ أَوْ عِظَةٌ,) *delivered in a* مَجْلِس; like مَقَامَةٌ. (Mṭr, in the Preface to Ḥar.) — It is also used in the same manner as حَضْرَة and جَنَاب: you say مَجْلِسُ فُلَانٍ [meaning † *The object of resort, with whom others sit and converse, such a one*]; like حَضْرَةُ فُلَانٍ. (Kull p. 146.) [See arts. حضر and جنب. But this usage I believe to be post-classical.] — [Also † *A stool;* meaning, *an evacuation.* So in medical books.]

مَجْلِسَةٌ: see مَجْلِسٌ, in two places.

جلسد

جَلْسَدٌ and الجَلْسَدُ: see art. جسد.

جلف

1. جَلَفَهُ, (Ṣ, Mṣb, Ḳ,) aor. ِ, inf. n. جَلْفٌ, (Ṣ, Mṣb,) *He peeled, pared, stripped,* or *scraped, it off;* (Ṣ, Mṣb, Ḳ, TA;) namely, a thing; (TA;) as, for instance, (Ṣ, TA,) the mud, or clay, (Ṣ, Mṣb, TA,) from the head of a [jar of the kind called] دَنّ. (Ṣ, TA.) You say also, جَلَفَ ظُفْرَهُ عَنْ إِصْبَعِهِ *He stripped off his nail from his finger.* (Lth, TA.) And accord. to some, جَلْفٌ signifies The *scraping off,* or *stripping off, the skin with somewhat of the flesh:* and the act of *pulling,* or *drawing, out,* or *up,* or *off;* or *displacing.* (TA.) — Also i. q. جَرَفَهُ [*He took away, carried away,* or *removed, the whole of it,* or *the greater part of it,* or *much of it;* or *he swept it away*]: (Ḳ:) or, as some say, جَلْفٌ signifies a more intensive and more exterminating action than جَرْفٌ. (TA.) — And *He cut it off;* (Ṣ;) or *pulled it,* or *plucked it, out,* or *up;* or *eradicated,* or *uprooted, it;* (Ḳ;) *and exterminated it;* (Ṣ, Ḳ;) as also ↓ اجتلفهُ. (Ḳ.) — جَلَفَهُ بِالسَّيْفِ *He struck him with the sword:* (Ḳ:) or *he cut,* or *cut a piece from,* or *cut in pieces, his flesh* [*with the sword*]. (A, TA.) — جُلِفَ النَّبَاتُ *The herbage was eaten to the uttermost.* (TA.) — جُلِفَ فِى مَالِهِ جَلْفَةً *He suffered the loss of somewhat of his property,* or cattle. (TA.) ═ جَلِفَ, aor. َ, inf. n. جَلَفٌ and جَلَافَةٌ, † *He was,* or *became, rude in disposition,* or *in make; coarse,* or *churlish.* (Ḳ.)

2. جَلَّفَتْهُ السِّنُونَ *The years of drought,* or *barrenness,* or *dearth, destroyed his cattle.* (Ṣ.) And جَلَّفَتْ كَحْلٌ, (Ṣ,) or كَحْلُ, (Ḳ,) *The year of drought,* or *barrenness,* or *dearth, exterminated the cattle.* (Ḳ.) And ↓ اجتلفت أَمْوَالَهُمْ [*It destroyed their cattle*] is also said of a year of great drought, or barrenness, or dearth. (Ṣ.) And ↓ اجتلفهُ الدَّهْرُ *Time,* or *fortune,* or *misfortune, destroyed his property,* or *cattle.* (TA.)

4. اجلف *He* (a man) *removed the* جِلَاف [or *clay*] *from the head of the* [*jar called*] خُنْبُجَة [i. q. دَنّ]. (IAạr, Ḳ.)

8: see 1; and see also 2, in two places.

جِلْفٌ *A* [*jar of the kind called*] دَنّ: (M, Ḳ:) or *an empty* دَنّ: (AO, Ṣ, Mṣb, Ḳ:) this is said (Ṣ, Mṣb) by AO (Ṣ) to be the primary signification of the word: (Ṣ, Mṣb:) or the *lower part of a* دَنّ *when it is broken:* (ISd, Ṣgh, Ḳ:) and *a* [*receptacle such as is called*] ظَرْف, (AA, Ṣ, Hr, Mṣb, Ḳ, [in the CḲ, erroneously, ظُرُوف,]) and وِعَآء, (AA, Ṣ, Mṣb, Ḳ,) *of any kind,* (AA, Ṣ, Mṣb,) *such as a saddle-bag,* or *pair of saddle-bags, and a sack, in which bread or other food is kept:* (Hr, TA:) pl. [of mult.] جُلُوفٌ (Ṣ, M, Mṣb) and [of pauc.] أَجْلُفٌ and أَجْلَافٌ, which last is rare. (Mṣb.) — Also *A* [*skin of the kind called*] زِقّ *without head and without legs.* (IAạr, Ḳ.) — And *A skinned animal,* (AO, Ṣ,) or *a skinned sheep or goat,* (Ḳ,) *of which the belly has been taken forth,* (AO, Ṣ, Ḳ,) *and the head and legs of which have been cut off;* (Ḳ;) the *body of a skinned sheep or goat, without head and without belly and without legs:* or, as some say, *a body of any kind without a head upon it:* (L:) or *a beast without fat, and without back* [*to bear*], *and without belly to conceive:* (IAth, TA:) or the *skin of a sheep or goat,* and *of a camel:* (Aṣ, Mṣb:) pl. أَجْلَافٌ (Sb, L) and sometimes أَجْلُفٌ: (Sb, TA:) and [it is also said that] أَجْلَافُ الشَّاةِ signifies *the skinned sheep or goat that is without head and without legs and without belly.* (Ṣ, Mṣb.) — Hence, i. e., from اجلاف الشاة, (Ṣ, Mṣb,) ‡ *Rude in disposition* or *in make; coarse,* or *churlish;* (Ṣ, M, Mṣb, Ḳ;) as also ↓ جَلِيفٌ; (Ḳ;) meaning that the person so termed is empty, without intellect: (M, TA:) applied to a Desert-Arab, (Ṣ,) or to an Arab: (so in a copy of the Mṣb:) or it is so applied as though meaning one *with his skin; not having assumed the gentle and soft habits of the people of the towns or villages or cultivated lands;* for when one does this, it is as though he pulled off his skin and clad himself with another: (Mṣb:) or ‡ *stupid, foolish,* or *unsound in intellect;* likened to a skinned sheep or goat because of the weakness of his intellect. (IAth, TA.) — Also *Thick,* or *coarse, dry bread:* or *bread not rendered savoury by anything eaten therewith:* or the *edge* [*of a cake*] *of bread.* (Ḳ.) [See also جِلْفَةٌ.] ═ *A male palm-tree,* (Lth, Ḳ,) *with the spadix of which the female palm-tree is fecundated:* (Lth, TA:) pl. جُلُوفٌ. (TA.) ═ *A certain well-known bird.* (Ḳ.)

جَلْفَةٌ [inf. n. of un. of جَلَفَ, q. v.]. ═ See also جِلْفَةٌ.

جُلْفَةٌ *A part of a skin that is peeled, pared, stripped,* or *scraped, off.* (L, Ḳ.)

جِلْفَةٌ *A broken piece of dry bread,* (Ḳ, TA,) *thick,* or *coarse,* (TA,) *and without anything to* render it savoury: (Ḳ, TA:) pl. جِلَفٌ. (TA.) [See جِلْفٌ, of which it may be regarded as the n. un.] —— A piece of anything: (Ṣgh, Ḳ:) pl. as above. (TA.) —— The portion of a reed for writing that is between its مَبْرًى [or place where the paring is commenced] and its point; as also ↓ جِلْفَةٌ. (Ḳ.)

جُلَافٌ Clay; such as is put upon the head of the [jar called] خُنْبُجَة. [See 4.] (IAạr, Ḳ.)

جَلِيفٌ Peeled, pared, stripped, or scraped, off; as also ↓ مَجْلُوفٌ. (Ḳ.) It is said by some that the last word in the following saying of Ḳeys Ibn-El-Khaṭeem,

* كَأَنَّ لَبَّاتِهَا تَبَدَّدَهَا * هَزْلَى جَرَادٍ أَجْوَافُهُ جُلُفُ *

is pl. of the former in this sense: but accord. to ISk, [the meaning of the verse is, *As though emaciated locusts without heads and without legs occupied the two sides, or the whole, of the part of her breast where the necklace lay*; for he says that] the poet likens the ornaments upon her لَبَّة to locusts without heads and without legs. (TA.) رِجْلٌ جَلِيفَةٌ [*An excoriated leg*]. (TA.) —— جَلِيفَةٌ [or سَنَةٌ جَلِيفَةٌ] *A year that destroys the cattle*; (Ṣ,* Ḳ;) as also ↓ جَالِفَةٌ: (Ṣ, Ḳ:) *any bane*, or *calamity, that destroys the cattle*: pl. جَلَائِفُ and جُلُفٌ and جُلْفٌ. (TA.) You say, أَصَابَتْهُمْ جَلِيفَةٌ عَظِيمَةٌ *A great destruction of their cattle befell them.* (Ṣ, TA.) And سِنُونَ جَلَائِفُ and جُلُفٌ and جُلْفٌ *Years that destroy the cattle.* (Ḳ.) And جَلَائِفُ also signifies *Torrents.* (TA.) ═ See also جِلْفٌ.

جَالِفٌ [act. part. n. of جَلَفَ]. —— جَالِفَةٌ [or شَجَّةٌ جَالِفَةٌ] *A wound of the head that peels off the skin with the flesh*: (Ṣ, Ḳ:) or *that peels off the skin but does not penetrate into the interior.* (Mṣb.) And طَعْنَةٌ جَالِفَةٌ *A spear-wound*, or *the like, that does not penetrate into the interior*; (Ṣ, Ḳ;) opposed to جَائِفَةٌ. (Ṣ.) —— زَمَانٌ جَالِفٌ i. q. جَارِفٌ [*A time*, or *season, that sweeps away*, or *destroys, the cattle*]. (TA.) See also جَلِيفٌ.

مُجَلَّفٌ *Having a portion*, or *portions, taken from its sides.* (Ṣ, Ḳ.) —— *Having somewhat thereof remaining.* (Ṣ, Ḳ.) So explained by Abu-l-Ghowth as occurring in the saying of El-Farezdaḳ,

* وَعَضُّ زَمَانٍ يَا ٱبْنَ مَرْوَانَ لَمْ يَدَعْ *
* مِنَ المَالِ إِلَّا مُسْحَتًا أَوْ مُجَلَّفُ *

i. e., [*And a biting of fortune, O Ibn-Marwán, left not, of the cattle, save*] *such as were destroyed, or they were such as had only a remnant remaining.* (Ṣ.) —— A man (Ṣ) *whose cattle have been destroyed by years of drought*, or *barrenness*, or *dearth.* (Ṣ, Ḳ.) And ↓ قَوْمٌ مُجْتَلَفُونَ *A people*, or *party, whose cattle have been destroyed by a year of drought or the like.* (Ṣ, TA.)

مَجْلُوفٌ: see جَلِيفٌ. —— Also A *skinned* sheep or goat. (L.) —— خُبْزٌ مَجْلُوفٌ *Bread burnt by the oven*, (Ḳ, TA,) *so that its outer parts stick to it.* (TA.)

قَوْمٌ مُجْتَلَفُونَ: see مُجَلَّفٌ.

مُتَجَلِّفٌ *Lean*, or *emaciated*; (Ḳ;) as also مُتَجَرِّفٌ. (TA.)

جلق

جُوَالِقٌ (Ṣ, M, MA, Mgh, Ḳ) and جُوَالَقٌ (IAạr, M, Ḳ) and جِوَالِقٌ (Ḳ) *A sack*; in Pers. جُوَال; (MA, PṢ;) [not so well rendered in the KL by خُرْجِين, which means *a saddle-bag* or *a pair of saddle-bags*, like the Arabic خُرْج;] *a certain kind of* وِعَآء [or *receptacle*], (Ṣ, M, Ḳ,) [*for corn &c.*,] *well known*: (M, Ḳ:) *it has a loop, into which is inserted a stick*, or *piece of wood*, called شِظَاظٌ, (Ṣ and Ḳ in art. شظ,) *this being also inserted into the loop of another* جوالق, *when they are bound upon the camel*: (Ḳ* and TA in that art.:) or *it has two loops, one of which is inserted into the other*, (Ṣ and Ḳ voce قَطَبَ,) [*and then the stick is put through*,] *on the occasion of putting it on a camel*: (TA ibid.:) the word is arabicized; (M, TA;) said to be from كواله, (TA,) or كوالك, (KL,) but correctly from چواله, which is Pers.: (TA:) the pl. is جَوَالِقُ (Ṣ, MA, Mgh, Ḳ) and جَوَالِيقُ, (Sb, Ṣ, MA, Mgh, Ḳ,) the latter occurring in poetry, (TA,) and جُوَالِقَاتٌ (Ṣ, Ḳ) was sometimes used, (Ṣ,) but this is disallowed by Sb. (Ṣ, M.) The saying [of a rájiz], cited by Th,

* أُحِبُّ مَاوِيَّةَ حُبًّا صَادِقًا *
* حُبَّ أَبِي الجُوَالِقِ الجُوَالِقَا *

[*I love Máweeyeh with a true love; with the love of the owner* (lit. *father*) *of the sack for the sack*;] means that the speaker had a vehement love for the food, or wheat, that was in his جوالق. (M.) Another says,

* يَا حَبَّذَا مَا فِي الجَوَالِيقِ السُّودْ *
* مِنْ خُشْكَنَانٍ وَسَوِيقٍ مَقْنُودْ *

[*O, lovely is what is in the black sacks, of biscuit and meal of parched barley sweetened with sugar-candy!*]. (Ṣ.)

جلم

1. جَلَمَهُ, (Ṣ, Mṣb, Ḳ,) aor. ـِ, (Mṣb, Ḳ,) inf. n. جَلْمٌ, (Ṣ, Mṣb,) *He cut it, or cut it off*; (Ṣ, Mṣb, Ḳ;) [like جَرَمَهُ.] —— *He shore it*, or *sheared it*; namely, wool, (Mṣb, Ḳ,) and hair, with the جَلَمَانِ, (Mṣb,) or جَلَمٌ. (TA.) And جَلَمْتُ مِنْهُ *I took* [or *clipped somewhat*] *from it*; namely, wool; like جَرَمْتُ. (Ṣ in art. جرم.) And جَلَمَ الشَّاةَ *He shore, or sheared, the sheep* or *goat.* (Ḥar p. 190.) —— جَلَمَ الجَزُورَ, (Ṣ, Ḳ,) inf. n. as above, (Ṣ,) *He took the flesh that was on the bones of the slaughtered camel*; (Ṣ, Ḳ;) as also ↓ اجتلمهُ. (Ḳ.)

8: see what precedes.

جَلْمٌ The *fat of the* ثَرْب [or *thin integument that covers the stomach and bowels* or *intestines*] *of a sheep* or *goat.* (Ḳ.)

جَلَمٌ The *instrument with which one shears* (Ṣ, Ḳ, TA) *wool and hair*: (TA:) and جَلَمَانِ signifies *a pair thereof*; *a pair of shears*: (Ṣ:) or جَلَمٌ and جَلَمَانِ signify alike, i. q. مِقْرَاضٌ; like as one says مِقْرَاضٌ and مِقْرَاضَانِ, and قَلَمٌ and قَلَمَانِ: and ↓ جَلْمَانٌ is also allowable, as a sing.; (Mṣb;) authorized by Ks; (TA;) and in like manner, قَلْمَانٌ. (Mṣb. [But see مِقْرَاضٌ.]) —— † The *moon*; (Az, Ḳ;) as also ↓ جَيْلَمٌ: (Ḳ:) or the [*new moon, or moon when it is termed*] هِلَال: (Ḳ:) likened to the جَلَم [as meaning the blade of a pair of shears]. (TA.) —— [Hence also, probably,] † *A certain mark of camels, made with a hot iron.* (Ibn-Ḥabeeb, Ḳ.) —— [Hence also,] † The *tick*, or *ticks.* (Ḳ.) —— And, accord. to some, as being likened to these, because of their smallness, (TA,) † The *sheep of Mekkeh*: (A 'Obeyd, TA:) or *certain sheep with long and hairless legs, found at Eṭ-Ṭáïf.* (Ḳ.) And † The *male of the gazelle*, and *of the sheep*: pl. جِلَامٌ. (Ḳ.) And † *A kid*: (Kr, Ḳ:) pl. as above. (Ṣ, TA.)

جَلَمَةٌ The *whole* of a thing; (Ṣ, Ḳ;) as also ↓ جُلْمَةٌ and ↓ جَلْمَةٌ. (Ḳ.) You say, أَخَذْتُ الشَّىْءَ بِجَلَمَتِهِ *I took the thing wholly.* (Ṣ.)

جُلْمَةٌ: see what next precedes.

جَلَمَةٌ *A skinned sheep* or *goat* (Ṣ, Ḳ) *without the intestines and without the legs*, (Ṣ,) [or] *when the shanks and the redundant parts have gone.* (Ḳ.) And the *whole flesh* of a slaughtered camel. (Ṣ.) —— See also جَلَمَةٌ.

جَلْمَانٌ: see جَلَمٌ.

جُلَامَةٌ *Shorn wool.* (Ḳ.)

جُلَّامٌ *Shorn he-goats.* (Ḳ.)

جَيْلَمٌ: see جَلَمٌ.

مَجْلُومٌ *Cut*, or *cut off.* (Mṣb.) —— A *shorn* sheep or goat. (Ḳ,* TA.)

مُجْتَلَمٌ A bone *having the flesh cut off with the* جَلَم. (TA in art. كنت.)

Quasi جلمح

جَلْمَحَ: see Q. Q. 1. in art. جلح.

جلمد

جَلْمَدٌ and ↓ جُلْمُودٌ *Rock*: (Ṣ, Ḳ:) or *a rock*; or *mass*, or *piece, of rock*: (M:) or *a round stone*: (Mgh, Mṣb:) or [*a stone*] *smaller than what is termed* جَنْدَل, *of such a size as that which is thrown with a ballista*: (L:) or *a great stone*: (Ḥar p. 95:) or the latter word, [*a stone*] *like the head of a kid; or less, such as may be carried in the hand by grasping its side but over which the two hands will not meet, with which date-stones &c. are bruised*, or *brayed*: (ISh:) [pl. of the former, جَلَامِدُ; and of the latter, جَلَامِيدُ. Accord. to the Mgh and Mṣb, the م is an augmentative letter; but most of the lexicographers regard it as radical.] —— [Hence,] رَشَحَ جَلْمَدُهُ, said of one known to be a niggard, meaning, † *He gave something.* (Ḥar p. 95.) —— And أَلْقَى عَلَيْهِ جَلَامِيدَهُ † *He threw his weight* (ثِقْلَهُ)

upon him. (Ḳ. [See القى عليه مَتَاقِيلَهُ, voce مِثْقَالٌ.]) — Also جُلْمَدٌ, (L,) or ↑ جُلْمُدٌ, (Ḳ,) *A mass of rock rising out of shallow water.* (IAạr, L, Ḳ.) — And the former, † *A strong* man; and so ↑ جُلْمَدَةٌ: (Ḳ:) or a man *having a strong voice;* and so ↑ جُلْمُدٌ. (L.) ═ Also, جَلْمَدٌ, *Many* camels: (Ṣ:) or camels *composing a large herd:* or camels *advanced in years;* as also ↑ جُلْمُودٌ: (Ḳ:) and sheep *exceeding in number a hundred:* (L, Ḳ:) you say ضَأْنٌ جَلْمَدٌ. (L.) — And *Oxen,* or *cows:* (L:) and the same word, (Ḳ,) or ↑ جَلْمَدَةٌ, (AA, L,) *a single ox* or *cow.* (AA, L, Ḳ.)

جُلْمُدٌ : } see above.
جِلْمِدٌ : }

جَلْمَدَةٌ applied to land (أَرْضٌ) *Stony:* (Ḳ:) and جَلَامِدُ is [its pl.,] like جَرَاوِلُ. (TA.) — See also جَلْمَدٌ, in two places.

جُلْمُودٌ : see جَلْمَدٌ, in two places.

جلنار

جُلَّنَارٌ The *flower of the pomegranate:* an arabicized word, from گُلْنَارْ, (Ḳ,) which is Persian, composed of گُلْ "a flower," and نَارْ "a pomegranate." (MF.) It is said that whoever swallows three grains of it, of the smallest that may be, (Ḳ,) on the condition of his taking them with his mouth from the tree, before their opening, at sunrise, on a Wednesday, (Tedhkiret Dáwood, TA,) will not have ophthalmia in that year. (Ḳ.)

جله

1. جَلِهَ, aor. ـَ, (Ṣ, Mṣb, Ḳ,) inf. n. جَلَهٌ, (Ṣ,* Mṣb, TA,) *He was,* or *became, bald in the fore part of the head:* (Ṣ, Ḳ:) or *in the greater part of his head.* (Mṣb.) [The latter seems to be the correct meaning: see جَلَهٌ below.] — جَلِهَتْ سَاحَتُهُ *His court,* or *yard, was,* or *became, vacant,* or *void.* (JK.) ═ جَلَهَهُ, (Ṣ, Ḳ,) aor. ـَ, (Ḳ,) inf. n. جَلْهٌ, (TA,) *He uncovered* a thing; or *removed* it [from a thing that it covered or concealed]. (Ḳ, TA.) — *He raised* the turban, *while folding it, from the side of his forehead* (عَنْ جَبِينِهِ), (Ḳ, TA,) [like اِجْتَلَى,] *and from the fore part of his head.* (TA.) — *He removed* the pebbles from a place. (Ṣ, Ḳ.) — *He turned back* a person *from a difficult,* or *hard, thing* or *affair.* (Ḳ.)

جَلَهٌ, (JK, Ṣ,) or ↑ جَلَهَةٌ, (Ḳ,) *Baldness in the fore part of the head;* (Ṣ, Ḳ;) *which is the beginning of* صَلَعٌ; *like* جَلَحٌ: (Ṣ:) or *baldness of the greater part of the head;* (Mṣb;) *more than* جَلَحٌ, (JK,) *and more than* صَلَعٌ, which is more than جَلَحٌ. (Mṣb in art. جلح.)

جَلْهَةٌ: see جَلَهٌ. ═ Also The *part that faces one of the brows,* or *brinks,* or *edges, of a valley:* (Ṣ:) or the *side of a valley;* (Ḳ, TA;) the *bank,* or *border, thereof:* (TA:) or *elevated parts in the interior,* or *lower part, of a valley, rising above the water-course, so that, when the valley flows with water, the water does not reach them:* (ISh, TA:) and, some say, the *mouth of a valley:* and some, *a part of a valley uncovered by the torrents, and so made apparent:* (TA:) and [the dual] جَلْهَتَانِ, (JK, M, TA,) or جَلْهَتَا وَادٍ, (Ṣ,) the *two sides,* or *borders, of a valley,* (Ṣ, M, TA,) *when there is in them hardness:* (JK, M, TA:) occurring in a trad., or, as some relate it, ↑ جَلْهَمَتَانِ, with an augmentative م: (TA:) pl. جِلَاهٌ. (Ṣ.) — *A great round rock.* (JK, Ḳ.) — *A large* [*hill, or the like, such as is called*] قَارَة; as also ↑ جَلْهَمَةٌ, with an augmentative م. (TA.) — The *place of alighting and abiding of a people,* or *company of men:* (JK, Ḳ, TA:) and *a yard,* or *wide space, in front,* or *extending from the sides, of their dwelling.* (JK.) — And *A meadow in which water collects and stagnates:* pl. as above. (JK.) ═ *Dates,* (Ḳ, TA,) *of which the stones have been picked out,* (TA,) *macerated and mashed with milk,* (Ḳ,* TA,) *then given to drink to women;* (TA;) *having a fattening property;* (Ḳ, TA;) as also ↑ جَلِيهَةٌ. (Ḳ.)

جَلَهَمَةٌ: see جَلْهَةٌ, in two places.

جَلَيْهَةٌ *A mode of wearing the turban, in which the* جَبِين [or *side of the forehead*] *is uncovered, so that the part where the hair grows is seen.* (JK, Ṣgh.)

جَلِيهَةٌ *A place from which the pebbles have been removed.* (JK, Ṣ, Ḳ.) ═ See also جَلْهَةٌ, last sentence.

أَجْلَهُ *Bald in the fore part of the head;* (Ṣ;) *i. q.* أَجْلَحُ: (TA:) or [it denotes more than the latter; meaning] *bald in the greater part of the head:* fem. جَلْهَآءُ: pl. جُلْهٌ. (Mṣb.) [See جَلَهٌ.] — *Large in the forehead, having the places of growth of the hair receding.* (Ḳ.) — † *A bull having no horn;* (Ks, JK, Ṣ, Ḳ;) like أَجْلَحُ. (Ks, Ṣ.)

مَجْلُوهٌ A tent, or house, or chamber, (بَيْتٌ,) *in which is neither door nor curtain.* (JK, Ḳ.)

جلهق

جُلَاهِقٌ *Bullets,* syn. بُنْدُقٌ, (Ṣ, Mṣb, Ḳ,) or *rounded things,* (En-Naḍr, TA,) *made of clay,* (En-Naḍr, Mṣb, TA,) *which one shoots* [*from a cross-bow*]: (Ḳ:) n. un. with ة: (En-Naḍr, Mṣb, TA:) a Persian word, arabicized; (Mṣb;) in Persian جُلَهْ, meaning "a ball of thread;" pl. جُلَهَا; applied also to "a weaver;" (Ṣ, Ḳ;) i. e., جلها is so applied. (TA.) Hence, قَوْسُ الجُلَاهِقِ [*The cross-bow for shooting bullets*]. (Ṣ, Mṣb.)

Quasi جلهم

جَلْهَمَةٌ: see art. جله.

جلو

1. جَلَا, (Ṣ, Mgh, Mṣb,) [aor. ـُ,] inf. n. جَلَآءٌ, (Mṣb,) *It* (a thing, and † an affair, or a case, Mgh, or † information, or tidings, Mṣb,) *was,* or *became, clear, unobscured, exposed to view, displayed, laid open, disclosed,* or *uncovered,* (Mgh, Mṣb,) لِلنَّاسِ *to men,* or *the people;* (Mṣb;) as also ↑ تَجَلَّى, said of a thing: (Ṣ, Mgh, Mṣb:) *it* († information, or tidings, Ṣ, Mṣb, or † an affair, or a case, Mgh,) *was,* or *became, apparent,* or *plainly apparent, overt, conspicuous, manifest, notorious, plain, obvious,* or *evident,* (Ṣ, Mgh, Mṣb,) لِى *to me,* (Ṣ,) or لِلنَّاسِ *to men,* or *the people.* (Mṣb.) One says, تَجَلَّتِ ↑ الشَّمْسُ *The sun became unobscured,* or *exposed to view, and ceased to be eclipsed.* (TA from a trad.) Er-Rághib says that ↑ التَّجَلِّى is sometimes by the thing itself; as in the phrase [in the Ḳur xcii. 2], وَالنَّهَارِ إِذَا تَجَلَّى [*By the day when it becometh clear,* &c.]: and sometimes, by the case, and the action; as in the saying [in the Ḳur vii. 139], فَلَمَّا تَجَلَّى رَبُّهُ لِلْجَبَلِ [*And when his Lord became manifested to the mountain*]: Zj says that the meaning in this instance is, *appeared,* and so say the Sunnees; El-Ḥasan says that the meaning is, تجلّى بالنور العَرْشِ [*became manifested by light,* the light of *the empyrean*]. (TA.) — جَلَا, [aor. ـُ,] inf. n. جَلَآءٌ (Ṣ, Mgh, Mṣb, Ḳ) and جُلُوٌّ; (Ḳ;) and ↑ اجلى; (Ṣ, Mgh, Mṣb, Ḳ;) *He,* (a man, Mṣb,) or *they,* (a company of men, Mgh, Mṣb,) *went forth,* or *emigrated,* (Ṣ, Mgh, Mṣb,) عَنِ البَلَدِ *from the country,* or *town,* (Ṣ, Mṣb,) and عَنْ أَوْطَانِهِمْ *from their homes:* (Ṣ, Mgh:) [like جَلَّ:] or *they* (a company of men) *dispersed themselves,* or *became dispersed,* عَنِ المَوْضِعِ, and مِنْهُ, *from the place:* (Ḳ:) or جلا means, *in consequence of fear:* and ↑ اجلى, *in consequence of drought:* (AZ, Ḳ:) or ↑ أَجْلَوْا مَنَازِلَهُمْ signifies *they left their place of abode in consequence of fear;* the verb in this case being trans. by itself: but if they have left for some other reason than fear, you say, عَنْ مَنَازِلِهِمْ: (Mṣb:) accord. to IAạr, جلا signifies *he fled, being driven away,* from his home. (TA.) [See also 12.] — جَلِىَ, aor. ـَ, inf. n. جَلًا, *He had that degree of baldness which is termed* ↑ جَلًا; (Ḳ;) i. e. *baldness of the fore part of the head;* (Ṣ, Ḳ;) like جَلَهٌ: (Ṣ:) or *baldness of half of the head;* (Ṣ, Ḳ;) which is the beginning of صَلَعٌ: (Ṣ:) or *baldness less than what is termed* صَلَعٌ. (Ḳ.) And جَلِىَ الجَبِينُ, inf. n. جَلًا, signifies the same as جَلِىَ [*The part above the temple became bald*]. (A 'Obeyd, TA.) ═ جَلَاهُ, [aor. ـُ, inf. n., app., جَلَآءٌ, or perhaps جِلَآءٌ, but the former seems to be indicated by what follows;] (Ṣ, Mgh, Mṣb;) and ↑ جلّاهُ; (MA;) *He made it,* or *rendered it, clear,* or *unobscured; exposed it to view, displayed it, laid it open, disclosed it,* or *uncovered it;* (Ṣ, Mgh, MA;) namely, a thing: (Ṣ, Mgh:) *he made it,* or *rendered it, apparent,* or *plainly apparent, overt, conspicuous, manifest, notorious, plain, obvious,* or *evident;* (Ṣ, Mgh, Mṣb, MA;) namely, † an affair, (Mgh,) or † information, or tidings. (Mṣb.) You say, جَلَا العَرُوسَ, inf. n. جِلْوَةٌ and جَلْوَةٌ (Ṣ, Mṣb, Ḳ) and جُلْوَةٌ (Mṣb, Ḳ) and جَلَآءٌ; (Ḳ;) and ↑ اجتلاها; (Ṣ, Mṣb, Ḳ;) *He displayed the bride,* عَلَى بَعْلِهَا *to her husband:* (Ḳ:) or *he looked at the bride*

displayed: (Ṣ:) and you say also, جُلِيَتْ عَلَى زَوْجِهَا (TA) *She was shown to her husband, and he looked at her displayed:* (Ḥar p. 30:) and جَلَاهَا زَوْجُهَا *Her husband presented,* or *gave, to her* a female slave (Ṣ, Ḳ) or some other thing *at the time of her being displayed to him;* as also ↑ جلّاها: (Ḳ:) and جَلَتِ المَاشِطَةُ العَرُوسَ *The female hairdresser adorned the bride* [*to display her to her husband*]. (TA.) You also say, جَلَا فُلَانٌ الأَمْرَ ‡ *Such a one displayed, discovered, disclosed, revealed,* or *manifested, the affair,* or *case;* as also ↑ جلّاه, and جَلَا عَنْهُ: (Ḳ,* TA:) or جَلَا فُلَانًا الأَمْرَ *he displayed, discovered, &c., to such a one the affair,* or *case;* as also ↑ جلّاه [i. e. جلّاه الأَمْرَ], and جَلَا عَنْهُ [i. e. جلا عنه الأمر or جلا فُلَانًا عَنِ الأَمْرِ]. (So accord. to the CḲ and my MS. copy of the Ḳ. [The reading in the TA is, in my opinion, preferable to the latter.]) And اللّٰهُ يُجَلِّى ↑ السَّاعَةَ ‡ *God will make manifest the hour,* or *time of the resurrection;* or *will make it to appear.* (Ḳ in art. جلى: [but it belongs to the present art.:]) so in the Ḳur vii. 186. (TA.) And هُوَ يُجَلِّى ↑ عَنْ نَفْسِهِ ‡ *He declares,* or *explains, his mind.* (Ṣ.) — جَلَوْتُ السَّيْفَ, inf. n. جِلَآءٌ, (Ṣ, Mṣb, Ḳ, [in the CḲ جَلاء, but it is]) with kesr, (Ṣ, Mṣb,) and جَلْوٌ, (Ḳ,) *I removed,* or *cleared off, the rust from the sword;* (Mṣb;) *I polished,* or *furbished, the sword;* (Ṣ, Ḳ;) and المِرْآةَ *the mirror;* (Ḳ;) and the like; (TA;) [as, for instance,] الفِضَّةَ *the silver;* and so جَلَيْتُهَا. (Ḳ in art. جلى.) And جَلَوْتُ بَصَرِى بِالكُحْلِ [*I cleared my sight with collyrium*]: (Ṣ:) [whence,] جَلَا *He applied collyrium to his eye* or *eyes.* (IAạr, TA.) And جَلَوْتُ هَمِّى عَنِّى ‡ *I removed my anxiety,* or *caused it to depart, from me:* (Ṣ, Ḳ,* TA:*) and ↑ اجلى عَنْهُ الهَمَّ ‡ *He removed,* or *cleared away, from him anxiety.* (Lth, TA.) And جَلَا اللّٰهُ عَنْهُ المَرَضَ ‡ *God removed from him the disease.* (TA.) — جَلَاهُمْ, and ↑ اجلاهم, (Ṣ, Mgh,) or جَلَاهُ, and ↑ اجلاه, (Mṣb, Ḳ,) and ↑ اجتلاه, (Ḳ,) *He,* (a man, Ṣ, Mṣb, or the Sultán, Mgh,) or *it,* (drought, Ḳ,) *caused them,* or *him, to go forth,* or *emigrate;* or *expelled them,* or *him;* or *drove them,* or *him, forth;* (Ṣ, Mgh, Mṣb, Ḳ;) [from their homes, or from his home.] And جَلَا النَّحْلَ, inf. n. جَلَآءٌ, or جِلَآءٌ, (accord. to different copies of the Ḳ,) and جلوة [thus written without any syll. signs]; and ↑ اجتلاها; (TA;) *He smoked* [*out*] *the bees, in order to collect the honey;* (Ḳ;) *he drove away the bees by means of smoke.* (TA.)

2. جلّى: see 1, in six places. — Also, inf. n. تَجْلِيَةٌ and تَجَلِّىٌ, *He* (a hawk, or falcon,) *raised his head, and looked,* (Ḳ, TA,) *seeing the prey:* (TA:) or *he* (a hawk) *closed his eyes, and then opened them, in order to see more clearly.* (Ibn-Ḥamzeh, TA.) — And [hence,] جلّى بِبَصَرِهِ, inf. n. تَجْلِيَةٌ, *He cast his eyes* (Ṣ, Ḳ) *like the hawk looking at the prey.* (Ṣ.) — [جلّى is also mentioned (in Ḥar p. 161), on the authority of Mṭr, as signifying *He,* or *it, outstripped;* from المُجَلِّى meaning "the first of the horses in a race;" but as being not known in this sense on any other authority.]

3. جَالَيْتُهُ بِالأَمْرِ, inf. n. مُجَالَاةٌ, *I acted openly with him in the affair;* as also جَالَحْتُهُ. (Ṣ.)

4. اجلى as an intrans. v.: see 1, in two places. — أَجْلَوْا عَنِ القَتِيلِ *They cleared themselves away,* or *removed, from the slain person.* (Ṣ, Mgh, Mṣb, TA.) — اجلى يَعْدُو *He hastened, running:* (Ḳ:) or *hastened somewhat, running:* (TA:) or اجلى signifies *he became distant,* or *remote, and hastened.* (So accord. to some copies of the Ḳ, where we find وَأَجْلَى بَعُدَ وَأَسْرَعَ instead of وَأَجْلَى يَعْدُو أَسْرَعَ.) — As a trans. v.: see 1, in four places.

5. تجلّى: see 1, in three places: — and see also 7. — تجلّى الشَّىْءَ *He looked at the thing,* (Ḳ in art. جلى,) *standing upon a higher position.* (TA.) [See also 8.]

6. تَجَالَيْنَا *Our states,* or *conditions, became disclosed to each other; the state,* or *condition, of each of us to the other.* (Ṣ.)

7. انجلى *It became removed,* or *cleared away;* said of anxiety, (Ṣ, Ḳ,* TA,) and of an affair [&c.]; as also ↑ تجلّى. (Ḳ,* TA.) You say, انجلى عَنْهُ الهَمُّ *Anxiety became removed,* or *cleared away, from him,* (Ṣ,) كَمَا تَنْجَلِى الظُّلْمَةُ *like as the darkness becomes removed,* or *cleared away.* (TA.)

8. اجتلاه *He looked at him,* or *it.* (Ḳ.) [See also 5.] Hence, اجتلى العَرُوسَ, explained above: see 1. (TA.) — See also 1 in two other places, last two sentences. — اِجْتَلَيْتُ العِمَامَةَ عَنْ رَأْسِى *I raised the turban, while folding it, from the side of my forehead* (عَنْ جَبِينِى): (Ṣ:) [like جَلَيْتُهَا.] — اجتلى *It became polished,* or *furbished;* said of a sword [&c.]. (TA.)

12. اجلولى *He went forth,* or *emigrated, from one country,* or *town, to another.* (IAạr, Ḳ.) [See also 1.]

اِبْنُ جَلَا ‡ *A man who is well known, celebrated,* or *notable;* (Mgh;) *of whom it is said,* جَلَا الأُمُورَ, i. e. *he has made affairs clear, unobscured,* or *manifest;* (Ṣ, Mgh;) or جَلَا أَمْرُهُ, i. e. *his case has become clear, unobscured,* or *manifest:* (Mgh:) or *one whose case is clear, apparent, plainly apparent,* or *manifest;* (Ḳ, TA;) as also ↑ اِبْنُ أَجْلَى: (Ḳ:) applied to a man who is upon an elevated and conspicuous place; and applied by El-Ḥajjáj to himself, as meaning that he was one whom every one knew: (TA:) and also, (Ḳ,) for this reason, (TA,) the name of a certain man, (Ṣ, Ḳ,) well known, (Ḳ,) of the Benoo-Leyth, who was a person of great daring. (TA.) A poet says, (Ṣ,) namely, Soḥeym Ibn-Wetheel Er-Riyáḥee, (TA,)

* أَنَا ٱبْنُ جَلَا وَطَلَّاعُ الثَّنَايَا *
* مَتَى أَضَعِ العِمَامَةَ تَعْرِفُونِى *

[*I am a man well known, celebrated,* or *notable, &c.; and he who rises to eminences,* or *who is accustomed to embark in,* or *surmount, lofty and difficult things: when I put down the turban, ye will know me*]. (Ṣ, TA.) Sb says, (TA,) جلا in this case is a verb in the pret. tense: 'Eesà Ibn-'Omar says that when a man is named ضُرِبَ or قُتِلَ, or the like, the word is imperfectly decl.; and he adduces, in evidence, this verse: others say that جلا may be here without tenween because it is imitative of a phrase, as though the poet said, أَنَا ٱبْنُ الَّذِى يُقَالُ لَهُ جَلَا الأُمُورَ: (Ṣ, TA:) accord. to IB, it is without tenween because it is a verb with its agent [implied in it]. (TA.) — Accord. to some, it signifies ‡ *The daybreak,* or *dawn;* (Ḥar p. 498;) and so ↑ اِبْنُ أَجْلَى: (TA:) accord. to Ḥamzeh, ‡ *the beginning of day:* and accord. to some, ‡ *the moon.* (Ḥar ubi suprà.)

جَلًا: see 1, voce جَلِىَ: — and see جَلَآءٌ.

جِلًا: see جَلَآءٌ.

جِلْوَةٌ *A female slave,* (Ṣ, Ḳ,) or *some other thing,* (Ḳ,) *that is presented,* or *given, by the husband to his bride at the time of her being displayed to him.* (Ṣ,* Ḳ.) One says, مَا جِلْوَتُهَا [*What is her bridal present?*]; and is answered, "Such a thing." (Ṣ.)

جَلَآءٌ *A thing, an affair,* or *a case, that is apparent, manifest, plain,* or *evident.* (Ṣ, Ḳ, TA.) — And *Acknowledgment,* or *confession:* so in the saying of Zuheyr:

* فَإِنَّ الحَقَّ مَقْطَعُهُ ثَلَاثٌ *
* يَمِينٌ أَوْ نِفَارٌ أَوْ جَلَآءُ *

[*For verily the means of deciding the truth are three: an oath, and incongruity* of circumstances, *and acknowledgment,* or *confession*]: (Ṣ:) but Az writes the last word ↑ جِلَآء, with kesr to the ج, as meaning *an evidence,* or *a proof,* and *witnesses;* from مُجَالَاةٌ [inf. n. of 3, q. v.]. (TA.) — جَلَآءُ يَوْمٍ, (Ḳ, TA,) or أَقَمْتُ عِنْدَهُ جَلَآءَ يَوْمِى, (so in some copies of the Ḳ,) [*I remained with him, or at his abode,*] *during the whiteness of my,* or *a, day.* (Zj, Ḳ, TA.) — See also the next paragraph.

جِلَآءٌ: see the paragraph next preceding. — Also, (Ṣ, Mgh, Ḳ,) written by El-Muhellebee ↑ جَلَآءٌ, (TA,) and ↑ جَلًا, which is more correct than the first, (Mgh,) or it is allowable, as also ↑ جِلًا, the former of the last two mentioned on the authority of En-Naḥḥás, (TA,) *Collyrium:* (Ṣ, Ḳ:) or *a particular kind thereof,* (Ḳ, TA,) *that clears the sight;* (TA;) [i. e.] *i. q.* إِثْمِدٌ [*antimony,* or *an ore of antimony*]; (Mgh, TA;) so called because it clears the sight. (Mgh.) — مَا جِلَاؤُهُ *What is his honourable name,* or *surname,* (Ṣ,) or *his good surname,* (Ḳ,) *by which he is addressed?* (Ṣ, Ḳ.)

جَلِىٌّ *Clear, unobscured, exposed to view, displayed, laid open, disclosed,* or *uncovered: apparent,* or *plainly apparent, overt, conspicuous, manifest, notorious, plain, obvious,* or *evident:* (Ṣ, Mṣb, Ḳ, TA:) جَالٍ thus used has not been heard. (Er-Rághib, TA.) It is applied as an epithet to information, or tidings, (Mṣb, TA,) and to analogy, or rule. (TA.) — عَيْنٌ جَلِيَّةٌ *A seeing eye.* (IB, TA.)

جَلِيَّةٌ *Sure information* or *tidings.* (Ṣ.) — أَخْبَرَنِى عَنْ جَلِيَّةِ الأَمْرِ *He informed me of the true,* or *real, state of the affair,* or *case.* (TA.)

دَوَآءٌ جَلَآءٌ [*A medicine that clears the complexion* or *skin*]. (Ḳ voce فُوَّةٌ, &c.)

جِلِّيَانٌ The *act of rendering apparent, open, manifest, plain,* or *evident: rendering clear,* or *unobscured; exposing to view, displaying, laying open, disclosing,* or *uncovering.* (TA.)

جَالٍ *Going forth,* or *emigrating,* from his country, or town: [like جَالٌّ:] and so جَالِيَةٌ, applied to a company of people; [as also جَالَّةٌ;] (Mṣb;) or to *people who have gone forth,* or *emigrated, from their homes;* (Ṣ;) and particularly to those tributaries, (Mgh, Mṣb,) namely, certain Jews, (Mgh,) whom 'Omar expelled from the country of the Arabs; (Mgh, Mṣb;) and afterwards, to *such as have the poll-tax imposed upon them, of the people of the Bible, and of the Magians, though not having emigrated from their homes;* (Mgh;) [i. e.] the *free non-Muslim subjects of a Muslim government;* because they were expelled by 'Omar from Arabia; (Ḳ;) the word being fem. because denoting a جَمَاعَة; (Mgh;) and its pl. is جَوَالٍ. (Mgh, Mṣb.) — Hence, (Mṣb,) ↓ جَالِيَةٌ [as a subst.] is applied to The *poll-tax that is exacted from the persons last mentioned above;* (Ṣ, Mgh, Mṣb;) as also جَالَّةٌ: (Ṣ:) first, in this sense, applied to that which was exacted from the people expelled from Arabia by 'Omar. (Mṣb.) You say, أُسْتُعْمِلَ فُلَانٌ عَلَى الجَالِيَةِ [*Such a one was employed as collector of the poll-tax*]. (Ṣ, Mgh, Mṣb.) = See also جَائِلٌ, in art. جول.

جَالِيَةٌ (as a subst.): see what next precedes.

أَجْلَى *Having that degree of baldness which is termed* جَلًا; i. e. *baldness of the fore part of the head:* or *baldness of half of the head;* (Ṣ, Ḳ;) which is the beginning of صَلَعٌ: (Ṣ:) or *baldness less than what is termed* صَلَعٌ: (Ḳ:) or *baldness of half of the head, and the like:* (A'Obeyd, TA:) fem. جَلْوَآءُ. (Ḳ.) [See أَجْلَحُ.] — *Beautiful,* or *handsome, in face, bald in the sides of the forehead.* (Ḳ.) — جَبْهَةٌ جَلْوَآءُ *A wide forehead.* (Ḳ.) — سَمَآءٌ جَلْوَآءُ † *A cloudless sky:* (Ks, Ṣ, Ḳ:) and لَيْلَةٌ جَلْوَآءُ † *a cloudless, bright, night.* (TA.) — ابْنُ أَجْلَى: see ابْنُ جَلَا, in two places. — Also (i. e. ابن اجلى) † *The lion.* (TA.) = فَعَلْتُهُ مِنْ أَجْلَاكَ, and ↓ إِجْلَاكَ, *I did it on account of thee, for thy sake,* or *because of thee;* syn. مِنْ أَجْلِكَ. (Ḳ.)

فَعَلْتُهُ مِنْ إِجْلَاكَ: see what next precedes.

مَجْلًى sing. of مَجَالٍ, which signifies The *fore parts of the head, which are the* [*first*] *places of baldness:* (Fr, Ṣ:) or *what is seen of the head when one fronts the face.* (TA.)

مُجْلٍ [act. part. n. of 4. Hence,] فَإِمَّا حَرْبٌ مُجْلِيَةٌ وَإِمَّا سِلْمٌ مُخْزِيَةٌ *And either war that shall cause* you *to emigrate, or abasing peace.* (TA.)

المُجَلِّى *The first of the horses in a race.* (Ḳ in art. جلى.)

جلى

Some words are mentioned in the Ḳ under this head belonging to art. جلو, q. v.

جم

1. جَمَّ, (Ṣ, Mgh, Mṣb, Ḳ,) aor. ـِ, inf. n. جَمٌّ, (Mṣb,) said of water, (Ṣ,) &c., (Ṣ, Mṣb,) *It became much,* or *abundant;* (Ṣ, Mṣb;) as also ↓ اِسْتَجَمَّ; said of a thing: (TA:) and, said of water, aor. ـِ (Ṣ, Ḳ) and ـُ, (Ḳ,) the former of which is of the higher authority, (TA,) inf. n. جُمُومٌ, (Ṣ, Mgh, Ḳ,) *it became much,* or *abundant,* (Ṣ, Mgh, Ḳ,) *in the well,* (Ṣ,) *and collected* (Ṣ, Ḳ) *after it had been drawn from;* (Ṣ;) as also ↓ اِسْتَجَمَّ. (Ḳ.) And جَمُّوا *They became many.* (TA.) — جَمَّتِ البِئْرُ, (Ḳ,) aor. ـِ and ـُ, inf. n. جُمُومٌ, (TA,) *The water of the well returned by degrees,* (Ḳ, TA,) *and became much,* or *abundant, and collected;* (TA;) and ↓ اِسْتَجَمَّتْ signifies the same. (Ṣ.) — [Hence, جَمَّتْ مَثَابَةُ جَهْلِهِ, a tropical phrase, explained in art. ثوب.] — جَمَّ الكَيْلُ, inf. n. جَمَامٌ, with ḍamm, *The measure became full* or *filled.* (KL. [But only the inf. n. is there mentioned: so that the verb may be جُمَّ, which see below.]) — جَمَّ الفَرَسُ, (Ṣ, Ḳ,) aor. ـِ and ـُ, (Ṣ, TA,) inf. n. جَمَامٌ (Ṣ, Ḳ) and جَمٌّ; (Ṣ;) and ↓ اِسْتَجَمَّ; (Ṣ;) *The horse abstained from covering,* (Ṣ, Ḳ,) *so that his seminal fluid* (مَاؤُهُ) *collected.* (Ḳ.) — Also, (aor. of the former verb as above, TA, and inf. n. جَمٌّ and جَمَامٌ, Ṣ, Ḳ,) *The fatigue of the horse went away;* (Ṣ;) [*he recovered his strength after fatigue;*] *he recovered from his fatigue, being left unridden;* (Ḳ, TA;) and so ↓ أُجِمَّ [of which see also the pass. form, below]. (M, Ḳ.) — [And hence,] جَمَّ [inf. n. جَمَامٌ, q. v. infrà,] *He rested;* said of a man [as well as of a horse]; (TA;) and so ↓ اِسْتَجَمَّ. (Ḥar p. 324.) — Also, said of a bone, (Ḳ,) inf. n. جَمٌّ, (TA,) *It had much flesh; its flesh became much,* or *abundant.* (Ḳ.) — Also i. q. عَلَا [*He,* or *it, became high,* &c.: perhaps said of water in a well]. (T, TA.) — Also, inf. n. جُمُومٌ, *He rose* (ارْتَفَعَ) *in his pace,* or *going.* (TA.) — Also, (Ṣ, Ḳ,) inf. n. جَمٌّ (TA) and جُمُومٌ, (Ṣ, TA,) said of an event, (Ḳ,) of the arrival of a person, (Ṣ,) *It drew near;* (Ṣ, Ḳ;) *it came to pass:* (Ṣ:) and ↓ أَجَمَّ signifies the same, (Ṣ, Mṣb, Ḳ,) said of an event, of separation from another, (Ṣ, TA,) and of an object of want: (TA:) اَحَمَّ, [q. v.,] with the unpointed ح, in this sense, was not known to Aṣ. (TA.) — And, said of the نَصِىّ, and the صِلِّيَان, [two plants, inf. n. not mentioned,] *It attained to the state of having a* جُمَّة [app. meaning *tuft,* or *flower-bud*]. (TA.) = جُمَّ الكَبْشُ, (TA,) or جَمَّتِ الشَّاةُ, [perhaps a mistranscription for جَمِمَتْ,] aor. ـَ, (Mṣb,) inf. n. جَمَمٌ, (Ṣ, Mṣb, TA,) *The ram,* (TA,) or *sheep* or *goat,* (Ṣ, Mṣb,) *was hornless.* (Ṣ, Mṣb, TA.) = جَمَّهُ, (Ḳ,) [aor. ـُ, accord. to rule,] inf. n. جَمٌّ, (TA,) *He left it* (namely, water [in a well],) *to collect;* as also ↓ أَجَمَّهُ. (Ḳ.) And جُمَّتِ البِئْرُ [*The well was left for its water to collect*]. (TA.) And اِسْتَجَمَّ البِئْرَ *He left the well for some days until its water should collect:* whence the metaphorical phrase, كَانَ يَسْتَجِمُّ مَثَابَةَ سَفَهِهِ, [explained in art. ثوب,] occurring in a trad. (Ḥar p. 68.) — Also *He filled it* (namely, a measure, Ṣ, such as is called مَكُّوك, Ḳ) *so that it had what is termed* جُمَام; and so ↓ أَجَمَّهُ; (Ṣ, Ḳ;) and ↓ جَمَّمَهُ. (Ḳ.) And جُمَّ *It was filled.* (T, TA.) See also جَمَّ, last sentence.

2. جَمَّمَتِ الأَرْضُ *The* جَمِيم [q. v.] *of the land became full,* or *abundant.* (ISh, TA.) — جَمَّمَ [in the CḲ جَمَّمَ] *It* (herbage) *became such as is termed* جَمِيم; as also ↓ تَجَمَّمَ. (Ḳ, TA.) — *He made a* جُمَّة [q. v.] *of his hair.* (Z, TA.) = See also 1, last sentence but two.

4. اَجَمَّ, as an intrans. verb: see 1, in two places. = اَجَمَّهُ: see 1, near the end of the paragraph, in two places. — *He left him unridden, so that he recovered from his fatigue;* namely, a horse. (Ḳ.) And أُجِمَّ *He* (a horse) *was left unridden.* (Ṣ.) — [Hence, *He rested him,* or *gave him rest.*] You say, أَجِمَّ نَفْسَكَ يَوْمًا أَوْ يَوْمَيْنِ [*Rest thyself a day, or two days*]. (Ṣ.) And hence, in a trad. respecting the سَفَرْجَلَة [or quince], تُجِمُّ الفُؤَادَ, i. e. *It rests the heart, and consolidates it, and completes its soundness and liveliness.* (TA.) One says also, إِنِّى لَأَسْتَجِمُّ ↓ قَلْبِى بِشَىْءٍ مِنَ اللَّهْوِ لِأَقْوَى بِهِ عَلَى الحَقِّ [*Verily I relieve my heart with somewhat of diversion, in order that I may become strong thereby for that which is substantial,* or *solid, not vain* or *frivolous*]. (Ṣ.) And اَجَمَّ الأَرْضَ *He gave the land rest from tilling.* (TA in art. بضع.) — *He gave him the* جُمَّة [or *supply of water, that had collected after drawing,*] *of the well.* (Th, TA.) — أُجِمَّ العِنَبُ *The grape-vine had all its branches that were above the ground cut off.* (AḤn, TA.)

5: see 2.

10. اِسْتَجَمَّ, as an intrans. verb: see 1, in five places. — It is said in a trad., مَنْ يُحِبُّ أَنْ يَسْتَجِمَّ لَهُ النَّاسُ قِيَامًا فَلْيَتَبَوَّأْ مَقْعَدَهُ مِنَ النَّارِ, i. e. [*Whoso loveth that men*] *should collect themselves to him, standing* in his presence, and confining themselves to him, [*let him take his sitting-place in the fire* of Hell:] or, accord. as some relate it, يَسْتَخِمُّ; see art. خم; (TA;) and, as some, يَسْتَخِيمُ. (TA in art. خيم, q. v.) — اِسْتَجَمَّتِ الأَرْضُ *The land put forth its plants,* or *herbage,* (Ḳ, TA,) *so that it became like the* [*hair termed*] جُمَّة [i. e. جُمَّة]. (TA.) = As a trans. verb: see 1, near the end of the paragraph, in two

places: — and see 4. — أُسْتُجِمَّتْ جَمَّةُ المَآءِ [*The supply of water that had collected after drawing*] *was drunk.* (TA.)

R. Q. 1. جَمْجَمَ, (Ṣ, TA,) inf. n. جَمْجَمَةٌ, (Ḳ,) *He spoke indistinctly,* (Ṣ, Ḳ,) *not from impotence,* or, accord. to the T, *from impotence;* (TA;) and ▼ تَجَمْجَمَ signifies the same. (Ṣ, Ḳ.) — Also, (TA,) inf. n. as above, (Mgh, Ḳ,) with which مَجْمَجَةٌ is syn., accord. to Ez-Zowzanee, (Mgh,) *He concealed* (Mgh, Ḳ, TA) speech, (Mgh,) or a thing, (Ḳ, TA,) in his bosom. (Mgh, Ḳ, TA.) You say, جمجم شَيْئًا فِى صَدْرِهِ *He concealed a thing in his bosom; did not reveal it.* (TA.) — Also, (TA,) inf. n. as above, (Kr, Ḳ,) *He destroyed,* or *killed,* (Kr, Ḳ, TA,) another or others. (TA.)

R. Q. 2. تَجَمْجَمَ: see R. Q. 1. — Hence, تجمجم عَنِ الأَمْرِ ‡ [*He held back from the thing, not daring to do it;*] *he did not dare to do the thing.* (Ḥam p. 240.)

جَمٌّ, an inf. n. used as an epithet, (Mṣb,) *Much,* or *many;* (Ṣ, Mṣb, Ḳ;) as also ▼ جَمِيمٌ accord. to the copies of the Ḳ, but correctly ▼ جُمَمٌ, as in the L. (TA.) You say مَالٌ جَمٌّ (Mṣb, TA) and ▼ جُمَمٌ (TA) *Much property,* or *many cattle.* (Mṣb.) And it is said in the Ḳur [lxxxix. 21], وَيُحِبُّونَ المَالَ حُبًّا جَمًّا (Ṣ) *And they love property with much love.* (A'Obeyd, TA.) Aboo-Khirásh El-Hudhalee says,

* إِنْ تَغْفِرِ ٱللّٰهُمَّ تَغْفِرْ جَمَّا *

(TA) i. e. [*If thou forgive, O God, Thou forgivest*] *much* sin. (Mgh.) جَمُّ البُعَاقِ, in a trad. respecting prayer for rain, means *Copious, abundant, extensive rain.* (TA in art. بعق.) — The *greater,* or *main, portion* of the ظَهِيرَة [i. e. mid-day, or mid-day in summer, &c.]: and of water; as also ▼ جُمَّةٌ, (as in some copies of the Ḳ,) or ▼ جَمَّةٌ: (so in other copies of the Ḳ, and accord. to the TA: [the former app. the right: if it were the latter, the author of the Ḳ, accord. to a rule which he has seldom neglected, would have added بِالضَّمِّ; as SM has here done, unless his transcriber have thus written by mistake for بِالفَتْحِ:] or جَمٌّ signifies the *water, of a well, that has collected* [*after it has been drawn from*]: and ▼ جَمَّةٌ, the *place in which the water collects:* (Ṣ:) and also, this last, the *water itself:* (TA: [i. e. the *supply of water that has collected after drawing:* see 4, last sentence but one; and see 10, last sentence:]) the pl. (of جَمَّةٌ Ṣ [or of this and of جَمٌّ also]) is جِمَامٌ (Ṣ, Ḳ) and جُمُومٌ. (Ḳ.) — بِئْرٌ جَمَّةٌ: see جَمُومٌ. — جَاؤُوا جَمًّا غَفِيرًا, and الجَمُّ الغَفِيرُ, &c.; see أَجَمٌّ, and art. غفر. — جَمٌّ also signifies *People of the lowest,* or *basest,* or *meanest sort.* (T, TA.) — Also The *measuring to the head of the measure;* [app. an inf. n., of which the verb is ▼ جَمَّ; see 1, last sentence but two; and so ▼ جَمَامٌ and ▼ جُمَامٌ and ▼ جِمَامٌ. (Ḳ.)

جُمَّةٌ: see جَمٌّ, in two places. — Also The *part,* or *place,* of a ship, *in which collects the water that leaks from its* خُرُوز [or *seams:* in the CḲ خُرُوُر]: (Ḳ:) a genuine Arabic word. (TA.) — Also, (Ṣ, Ḳ,) and ▼ جَمَّةٌ, (Ṣ, Ḳ, and Ḥam p. 746,) *A company demanding a bloodwit* (Ṣ, Ḳ) or *an obligation that must be discharged,* (TA, and Ḥam ubi suprà,) or *peace;* or *coming for some other purpose:* (Ḥam:) pl. جمم [probably جُمَمٌ, pl. of the latter, or perhaps of both; or it may be جُمَمٌ, as pl. of both]. (TA.) You say, جَآءَ فِى جُمَّةٍ عَظِيمَةٍ and ▼ جَمَّةٍ عظيمة *He came in a great company* &c. (Ṣ, Ḳ.)

جُمَّةٌ: see جَمٌّ: — and see also جُمَّةٌ, in two places. — Also The *collective mass of the hair of the head,* or the *head of hair,* (مُجْتَمَعُ شَعْرِ الرَّأْسِ, Ṣ, Mgh, Ḳ,) *when more in quantity than what is* termed وَفْرَةٌ: (Ṣ, Mgh:) or the *collective mass of the hair* (مُجْتَمَعُ الشَّعْرِ) *when it hangs down from the head to the lobe of the ear, and to the two shoulder-joints, and more than that;* what does not extend beyond the ears being termed وفرة: (TA:) or the *collective mass* (مُجْتَمَع) *of the hair of the* نَاصِيَة [or *anterior part of the head*]: accord. to some, *what reaches to the two shoulder-joints:* (Mṣb:) in the M it is said to signify the *hair;* and in like manner in the Deewán el-Adab: but ISd adds that it is also said to mean *hair more in quantity than that which is termed* لِمَّة: accord. to IAth, the *hair of the head that falls upon the two shoulder-joints:* in the Muhedhdhib, *what extends beyond the ears:* in the Muḳaddameh of Z, *what extends to the lobe of the ear:* accord. to IDrd, *much hair:* (TA:) [see also لِمَّةٌ and وَفْرَةٌ:] pl. جُمَمٌ (Mṣb, TA) and جِمَامٌ: (TA:) dim. ▼ جُمَيْمَةٌ. (TA.) Hence, رَأَى لُمْعَةً فَغَسَلَهَا بِجُمَّتِهِ, meaning [*He saw a spot, and washed it*] *by a moistening of his* جُمَّة: or *with the water of his* جُمَّة: the prefixed noun being suppressed. (Mgh.) — Also [app. † *A flower-bud;*] the قبة [rendered by Golius "nodosior pars"] *of a plant, from which the flower comes forth.* (KL.) [See an ex. above, voce جَمٌّ. If from جُمَّة applied to hair, it would seem rather to mean *A tuft.*] — One says also, حَذَفَ جُمَّةَ الجَوْزَةِ ثُمَّ أَكَلَهَا ‡ [app. meaning *He threw away the husk of the walnut: then ate it*]. (TA.)

جُمَمٌ: see جَمٌّ, in two places: — and see also جُمَامٌ.

جَمَامٌ *A state of resting;* (Fr, Ṣ, Mṣb, Ḳ;) as also ▼ جَمَامَةٌ: (TA:) particularly of a horse. (Fr, Ṣ, Mṣb.) [See جَمَّ, of which it is an inf. n.] — See also جَمٌّ, last sentence: and see what next follows.

جُمَامٌ *What has collected of the seminal fluid of a horse* [*after his resting from covering*]; as also ▼ جَمَامٌ. (Ḳ.) — Also, and ▼ جَمَامٌ and ▼ جِمَامٌ, (Ṣ, Mgh, Mṣb, Ḳ,) [but see what follows,] The *quantity* [of flour or the like] *that rises above the head* of the [measure termed] مَكُّوك, (Ṣ, Mgh, Ḳ,) *after the filling,* (Mgh,) *exceeding the* طفاف thereof; (Ṣ, Mgh, Ḳ;) as also ▼ جُمَمٌ: (Ḳ:) or the *fill* of a bowl, *without a head:* accord. to ISk, only said of flour and the like: one says, أَعْطَانِى جِمَامَ القَدَحِ دَقِيقًا [*He gave me the bowlful of flour*]: but جمام meaning the "resting" of a horse is with fet-ḥ only: (Mṣb:) or, accord. to Fr, one says ▼ جِمَامُ القَدَحِ مَآءً, with kesr, meaning *the bowlful of water;* and جُمَامُ المَكُّوكِ دَقِيقًا, with ḍamm; and جَمَامُ الفَرَسِ, with fet-ḥ only; and one does not say جُمَامٌ, with ḍamm, except in relation to flour and the like, meaning the *quantity that rises above the head* of the مكّوك, *after the being filled:* one says, أَعْطِنِى جُمَامَ المَكُّوكِ when one puts what the head of the مكّوك will bear, and gives it: (Ṣ, TA:) in the T, it is said that أَعْطِهِ جمامَ المَكُّوكِ means *Give thou him* [the quantity of] *a* مكّوك *without a head:* but [SM says,] I have seen in its margin written that the right meaning is, *the quantity borne by the head of the* مكّوك. (TA.) — See also جَمٌّ, last sentence.

جِمَامٌ: see جُمَامٌ, in three places: — and جَمٌّ, last sentence: — and see also جُمْجُمَةٌ. — It is also a pl. of جَمَّةٌ (Ṣ) [and perhaps of جَمٌّ likewise: (see this latter:)] and of جُمَّةٌ. (TA.)

جَمُومٌ بِئْرٌ (Ṣ, Ḳ) and ▼ جَمَّةٌ (Ḳ) *A well of much water.* (Ṣ, Ḳ.) — فَرَسٌ جَمُومٌ *A horse that, after any run, runs again;* (T, Ṣ, Ḳ;) applied to the female as well as the male: (T, TA:) *a horse that goes much.* (KL.)

جَمِيمٌ *A plant, or herbage, that has grown somewhat, but not attained its full height:* (Ṣ:) or *much,* or *abundant,* herbage: (Ḳ:) or herbage *standing up and spreading:* (AḤn, Ḳ:) or *that has grown up until it has become like the* جُمَّة *of hair:* (TA:) a plant, or herbage, when it first appears in the ground is termed بَارِضٌ; then, بُسْرَةٌ; then, صَمْعَآءُ; then, جَمِيمٌ; and then, [when it is dry,] حَشِيشٌ: (Ṣ in art. بسر:) pl. أَجِمَّآءُ. (Ḳ.) And, with ة, A [plant of the kind termed] نَصِيَّة *that has become half a month old, so that it fills the mouth.* (Ḳ.) — See also جَمٌّ.

جَمَامَةٌ: see جَمَامٌ. — Also The *state of being satiated,* or *satisfied, with food,* and *with drink.* (TA.)

جُمَيْمَةٌ dim. of جُمَّةٌ, q. v. (TA.)

جُمَّى The *bean,* or *beans;* syn. بَاقِلَّى. (AḤn, Ḳ.)

جُمِّىٌّ: see جُمَّانِىٌّ.

جُمَّامٌ: see what next follows.

جُمَّانٌ A measure, (Ṣ,) such as is called مَكُّوك, (Ḳ,) *filled so as to have what is termed* جُمَام; (Ṣ, Ḳ;) as also ▼ جُمَّامٌ: (Ḳ:) [fem. of the former جُمَّى. Hence,] جُمْجُمَةٌ جُمَّى [*A*] *full* [*bowl*]. (Ḳ. In the CḲ جُمْجُمَةٌ.)

جُمَّانِىٌّ, with ن, (Ṣ,) an irreg. rel. n., applied to a man, (Sb, Ṣ,) *Having a long* جُمَّة: (Ṣ, Ḳ:) or *having a great and long* جُمَّة: (Sb, TA:) but if you name a person جُمَّة, the rel. n. formed from it is ▼ جُمِّىٌّ (Sb, Ṣ) only. (Sb, TA.)

جُمْجُمْ : see what next follows.

جُمْجُمَةٌ The *skull*; i. e. the *bone that contains the brain*: (Ṣ, Mṣb, Ḳ:*) or *i. q.* قِحْفٌ [i. e. the *bone above the brain*, or *a separate portion of the skull*, or *a distinct bone of the skull*]: (Ḳ:) or the *bones of the head*; (IAạr, Mgh, TA;) *all of them*; the uppermost of them being the هَامَة; (IAạr, TA;) or the هامة is the جمجمة altogether; (ISh, TA;) and the قحف is said to be a piece of the جمجمة: (TA:) pl. † جُمْجُمٌ, (Ḳ,) [or this (in the CḲ, erroneously, جُمْجَمٌ) is a coll. gen. n.,] and جَمَاجِمُ [is the pl. properly so called, and that which is more commonly known]. (TA.) — Sometimes it is used to signify *A man*; so that one says, خُذُوا مِنْ كُلِّ جُمْجُمَةٍ دِرْهَمًا [*Take ye from every man, or head, a dirhem*]; like as one says, مِنْ كُلِّ رَأْسٍ: (Mṣb:) and وَضَعَ الإِمَامُ الخَرَاجَ عَلَى الجَمَاجِمِ عَلَى كُلِّ جُمْجُمَةٍ كَذَا [*The Imám imposed the tax, or land-tax, upon the heads; upon every head so much*]. (Mgh.) — *A wooden bowl*: (Ṣ, Ḳ:) *a bowl of glass*; as also قِحْفٌ. (Az, TA.) — *A kind of measure for corn or the like.* (Ḳ.) — Also † *Chiefs*, or *lords*, of the Arabs; because the جمجمة is the head, which is the most noble of the members: (TA:) also, (TA,) [the pl.] جَمَاجِمُ has this meaning. (T, Ḳ, TA.) And † *Any sons of a father that are persons of might, or power, and eminence*, or *nobility*: (T, TA:) and [the pl.] جَمَاجِمُ the *tribes* (قَبَائِل) of the Arabs *which comprise* بُطُون, *and in relation to which persons are called*; as Kelb Ibn-Webreh; for when you say كَلْبِىٌّ, you do not need to call the person in relation to any of the بطون: (Ṣ:) or the *tribes* (قبائل) *in relation to which the* بطون *are called*; as also † جُمَّامٌ. (Ḳ.) ═ *A well that is dug in salt ground.* (Ṣ, Ḳ.) ═ *Sixty head of camels.* (IF, IB, TA.) ═ جَمَاجِمُ الحَارِثِ *The piece of wood at the head of which is the ploughshare.* (TA.)

أَجَمُّ [*Greater*, and *greatest, in quantity*, and *in number*, &c.: fem. جَمَّآءُ. Hence,] وَالوَحْىُ أَجَمُّ مَا كَانَ, in a trad. of Anas, means *The revelation being the most that it used to be.* (Sh, TA.) — A bone *having much flesh.* (Ḳ.) You say also اِمْرَأَةٌ جَمَّآءُ العِظَامِ *A woman having much flesh* (Ḳ, TA) *on the bones.* (TA.) And اِمْرَأَةٌ جَمَّآءُ المَرَافِقِ [*A woman having much flesh on the elbows*: or, as seems to be indicated by J, *having no prominence of the elbows*; and if so, from جَمَّآءُ applied to a ewe, in a sense explained in what follows]. (Ṣ.) — جَاؤُوا الجَمَّآءَ الغَفِيرَ, (Ṣ,* Mṣb, Ḳ,) [and جَمَّآءَ غَفِيرًا, &c.,] and † جَمًّا غَفِيرًا, (Ḳ,) [and الجَمَّ الغَفِيرَ, &c.,] *They came all together*, (Ṣ,* Mṣb, Ḳ,) *high and low, none of them remaining behind, and they being many*: (Ṣ, Ḳ, in art. غفر:) see art. غفر. (Ṣ, Ḳ.) ═ *Hornless*, applied to a ram (Mgh, Mṣb, Ḳ) or he-goat; (Mṣb;) and so جَمَّآءُ applied to a ewe (Ṣ, Mgh, Mṣb) or she-goat: (Ṣ, Mṣb:) pl. جُمٌّ. (Mgh, Mṣb.) — And [hence,] ‡ A man *having no spear* (Ṣ, Ḳ, TA) *in war* or *battle*: (Ṣ, TA:) pl. as above. (TA.) The pl. is also applied to horses, (Ṣ,) meaning † *whose owners have no spears*; the spears being regarded as the horses' horns. (Ḥam, p. 90.) — Also † A building *having no* [*acroterial ornaments such as are termed*] شُرَف: (Ṣ:) and the pl., † Mosques *having no* شُرَف (Mgh, TA) *upon them*, (TA,) [i. e.] *upon their walls.* (Mgh.) — † A flat house-top *having no parapet*, or *surrounding wall*. (TA.) — † *Short*; *having no elevation.* (TA.) — † A woman's *anterior pudendum.* (Ḳ.) — And, as being likened thereto, or the reverse may be the case, (TA,) † *A bowl.* (Ḳ.) — Also, the fem., † *Smooth.* (IAạr, Ḳ.) — And hence, because of its smoothness, (IAạr, TA,) † *A helmet*: (IAạr, Ḳ:) to which the epithet غَفِير [q. v.] is applied because it covers the head: but this meaning of "a helmet" was not known to ISd on any other authority than that of IAạr. (TA.)

مَجَمٌّ *A place where water remains*: or *to which it reaches, and where it ends.* (TA.) — † The *breast*, or *bosom*, or *mind*: (Ḳ, TA:) because it is the place in which are collected the knowledge &c. that it retains. (TA.) You say, هُوَ وَاسِعُ المَجَمِّ, i. e. رَحْبُ الذِّرَاعِ وَاسِعُ الصَّدْرِ ‡ [*He is possessed of ample power and might*, and *free from distress of mind* or *from narrowness of mind*]. (IAạr, Ḳ, TA.) And إِنَّهُ لَضَيِّقُ المَجَمِّ ‡ *Verily he is contracted*, or *straitened, in mind* by affairs, or events. (IAạr, TA.)

مَجَمَّةٌ *A thing in which resting is usually known to take place.* (TA.)

مُجَمَّمٌ A boy (IDrd, TA) *having a head of hair such as is termed a* جُمَّة. (IDrd, Ḳ, TA.)

مُجَمِّمَةٌ A woman *who makes her hair to form a* جُمَّة, to make herself like a man: the doing of which is forbidden. (TA.)

جمح

1. جَمَحَ, aor. ـَ, inf. n. جِمَاحٌ (Ṣ, A, Mgh, Mṣb, Ḳ, &c.) and جُمُوحٌ (Ṣ, A, Mṣb, Ḳ) and جَمْحٌ, (Ḳ,) or this last has not been heard, (Mgh,) *He* (a horse) *overcame his rider*, or *gained the mastery over him*, (Ṣ, L, Ḳ,) *running away with him*: (L:) or *broke loose*, or *ran away*, (Mṣb,) *and went at random, without any certain aim, so as not to be turned by anything*: (Mgh, Mṣb:) or *ran so as to have the mastery over his rider*: (Ḥam p. 568:) and جَمَحَ بِرَاكِبِهِ (A, Mgh, Mṣb) *he overcame his rider*, (A, Mgh,) *and ran away so that he could not govern him*: (A:) or *became refractory, so that he overcame his rider*: (Mṣb:) and sometimes, (Mṣb,) this verb also signifies *he was quick*, or *swift*, (A, Mṣb,) *and brisk, lively*, or *sprightly*; denoting in this case a quality that is approved; whereas in the senses before explained it denotes a quality that is disapproved: but in the last sense it is obsolete [unless tropically applied to a man]. (Mṣb.) — [Hence,] † *He* (said of anything [i. e. of a man or any animal]) *went at random*, or *heedlessly, without consideration* or *certain aim, not obeying a guide to the right course.* (TA.) And † *He* (said of a man, Ṣ, L) *hastened*, or *went quickly*, (Ṣ, L, Ḳ,) إِلَيْهِ *to him*, or *it, so that his course was not turned for anything.* (L, TA.) وَهُمْ يَجْمَحُونَ, in the Ḳur [ix. 57], means † *They hastening*, or *going quickly*: (AO, Ṣ, L:) or *hastening so that nothing turns them back, like the horse that is termed* جَمُوح: (Bḍ, Jel:) or *running like horses that overcome their riders and run away so as to be ungovernable by them.* (A.) And جَمَحَ فِى إِثْرِهِ, occurring in a trad., † *He hastened after him*, or *it, so that nothing turned him back.* (L.) — جَمَحَتِ المَرْأَةُ ‡ *The woman went forth from the place where she used to pass the night, in anger, without the permission of her husband.* (Mṣb.) And جَمَحَتْ مِنْ زَوْجِهَا, so in the Ṣ and L &c., but in the Ḳ جَمَحَتْ زَوْجَهَا, [which is evidently a mistake,] (TA,) ‡ *She went forth from the house*, or *tent, of her husband, to her own family, before he divorced her*; (Ṣ, L, Ḳ;) inf. n. جِمَاحٌ. (L, TA.) And جَمَحَتْ إِلَى أَهْلِهَا ‡ *She went to her family without the permission of her husband.* (A.) — جَمَحَتِ السَّفِينَةُ ‡ *The ship quitted her course*, (A, TA,) *and became ungovernable by the sailors*; inf. n. جُمُوحٌ. (TA.) — جَمَحَتِ المَفَازَةُ بِالقَوْمِ ‡ *The desert led the people*, or *party, far away, by reason of its great extent.* (A, TA.) — جَمَحَ بِهِ مُرَادُهُ ‡ *The object of his desire baffled his efforts to attain it.* (A, TA.)

جَمْحَةٌ [*A trick of overcoming the rider, and running away with him*]. You say, دَابَّةٌ سَمْحَةٌ مَا بِهَا جَمْحَةٌ وَلَا رَمْحَةٌ [*A beast submissive*, or *easy*, or *gentle*: *there is not in her a trick of overcoming the rider, and running away with him, nor a trick of kicking*]. (A.)

جَمُوحٌ (T, Ṣ, A, Mgh, L, Mṣb, Ḳ) and † جَامِحٌ (Mgh, Mṣb) A horse *that overcomes his rider*, or *gains the mastery over him*, (Ṣ, A, Mgh, L, Mṣb, Ḳ,) *being refractory*, (Mṣb,) *and runs away with him*, (L,) or *runs away so that his rider cannot govern him*, (A,) or *goes away at random, without any certain aim, so as not to be turned by anything*: (Mgh:) or *that will not bend his head*: (TA:) the former epithet, (T, Mgh, TA,) and the latter, (Mgh, Mṣb,) applied alike to the horse and the mare: (T, Mgh, Mṣb, TA:) and the former has two meanings; one denoting what is a fault, for which the horse may be returned; (T, Mgh, TA;) i. e., *that habitually takes his own way, so that his rider cannot turn him from it*; (T, TA;) or as explained before; (Mgh;) the other meaning being *quick*, or *swift, and brisk, lively, and sprightly*; and this does not imply a fault (T, Mgh, TA) for which he may be returned. (T, TA.) — [Hence,] † Anything [i. e. a man or any animal] *that goes at random*, or *heedlessly, without consideration* or *certain aim, not obeying a guide to the right course*: (TA:) and the former epithet, (Mṣb, Ḳ,) or each of the two, (Ṣ,* A,) ‡ a man *who follows his own natural desire, without consideration, not obeying a guide to the right course of conduct*, (Ṣ, A, Mṣb, Ḳ,) so that he cannot be turned back. (Ṣ, Ḳ.) And ▼جُمَّاحٌ [is pl. of جَامِحٌ, accord. to analogy, and] signifies † Men routed, defeated, or put to flight, in war. (IAạr, Ḳ.) —— مَفَازَةٌ جَمُوحٌ † [A desert that leads one far away, by reason of its great extent: see 1, last sentence but one: likened to a horse that is termed جَمُوحٌ]: occurring in a poem of Dhu-r-Rummeh. (TA in art. نحب.)

جُمَّاحٌ An arrow, (Ṣ, Ḳ,) or a small arrow, (L,) without an iron point, having a round head, with which the art of shooting is learned (Ṣ, L, Ḳ) by a boy: (Ṣ, L:) or one with which boys play, putting upon its head a date or some clay, in order that it may not wound: (L:) or it signifies also a date put upon the head of a piece of wood, with which boys play: (Ḳ:) birds are shot at with it, and knocked down, without being killed, so that the shooter takes them: and it is also called جُبَّاحٌ: (T, TA:) or a boy's arrow, upon the end of which he puts a lump of chewed dates of the size of the عِفَاص [here meaning stopper] of a bottle, in order that it (the end) may go more directly, and be smooth; without feathers, and sometimes without a notch. (AḤn, L.) ═══ Also That [kind of plant] at the extremities of which come forth what resemble ears of wheat, soft, (L, Ḳ,) like foxes' tails, or (L) resembling the heads of the حَلِىّ and the صِلِّيَان and the like: (L, Ḳ:) a coll. gen. n.: n. un. with ة: (L:) pl. جَمَامِيحُ; and in poetry جَمَامِحُ; (L, Ḳ;) the latter allowable only in cases of necessity. (L.) ═══ See also جَمُوحٌ.

جَامِحٌ: see جَمُوحٌ, in two places.

جمد

1. جَمَدَ, aor. ـُ , inf. n. جَمْدٌ and جُمُودٌ, said of water, (Ṣ, M, L, Mṣb, Ḳ,) &c., (Mṣb,) [i. e.,] of anything fluid, or liquid, (Ḳ,) It congealed; concreted; became solid, or contr. of fluid or liquid; froze; syn. قَامَ; (Ṣ, M;) contr. of ذَابَ; (Mṣb, Ḳ;) as also جَمُدَ. (L, Ḳ.) And said of blood, &c., (Ṣ, M,) It congealed, or concreted; syn. قام: (M:) or became dry; dried. (Ṣ.) See also 2. —— Also, inf. n. جُمُودٌ, † He, or it, remained fixed, or stationary. (KL.) You say, مَا زِلْتُ أَضْرِبُهُ حَتَّى جَمَدَ ‡ [I ceased not to beat him until he became motionless]. (A.) —— † [He, or it, was, or became, incapable of growth or increase; lifeless, or dead: see جَامِدٌ.] —— † He was, or became, stupid, dull, wanting in intelligence; inert; not sharp, penetrating, vigorous, or effective, in the performing of affairs; or soft, without strength or sturdiness, and without endurance: see, again, جَامِدٌ.] —— Also, inf. n. جُمُودٌ, ‡ said of a man's state or condition [as meaning, It was, or became, stagnant, or unimproving]. (A.) —— Also جَمَدَتْ, aor. ـُ , inf. n. جُمُودٌ, ‡ She [a camel, &c.,] had little milk. (T, TA.) And جَمَدَتْ عَيْنُهُ ‡ His eye shed few tears: a phrase alluding to hardness of the heart. (Mṣb.) —— Also جَمَدَ, (L, Ḳ,) aor. ـُ , (L,) [inf. n. جُمُودٌ;] and ▼أَجْمَدَ; (A, TA; [in a copy of the A, انجمد, but this is doubtless a mistranscription; see مُجْمِدٌ;]) ‡ He was, or became, niggardly, penurious, or avaricious; (L, A, Ḳ;) as also جَمُدَ [or جَمَدَتْ كَفُّهُ]; (Mṣb;) or جَمَدَتْ يَدُهُ: (A:) and ▼اجمد ‡ he possessed little good: (A, TA:) or جُمُودٌ signifies the refraining, or holding back, from beneficence. (Har p. 149.) —— جَمَدَ لِى عَلَيْهِ حَقِّى ‡ My right, or due, was, or became, incumbent, or obligatory, on him; or established against him; (A, Ḳ,* TA;) as also ذَابَ. (A, TA.) ═══ جَمَدَهُ He cut it, or cut it off. (Ḳ.)

2. جَمَّدَ, inf. n. تَجْمِيدٌ; (Ḳ;) or ▼جَمَدَ; (so in the L;) It (water, and expressed juice, L) was about to congeal, concrete, become solid, or freeze; was at the point of congealing, &c.; expl. by حَاوَلَ أَنْ يَجْمُدَ. (L, Ḳ.) ═══ [And the former, It caused water &c. to congeal.]

4. اجمد: see 1, in two places. —— Also, inf. n. إِجْمَادٌ, He was entrusted with the management of affairs among a people or party [in the game called المَيْسِر: see مُجْمِدٌ]. (T, TA.) ═══ أَجْمَدْتُ عَلَيْهِ حَقِّى ‡ I made my right, or due, incumbent, or obligatory, on him; or established it against him. (A, Ḳ,* TA.)

جَمْدٌ: see جَامِدٌ, in two places.

جُمْدٌ: see جُمُدٌ.

جَمَدٌ pl. [or rather quasi-pl. n.] of جَامِدٌ, q. v. (Ṣ, Mṣb, Ḳ.) —— Also Congealed, or frozen, water; ice: [see also جَمْدٌ, mentioned with جَامِدٌ:] and snow. (Ḳ.) —— See also what next follows.

جُمُدٌ and ▼جُمْدٌ Elevated ground; as also ▼جَمَدٌ: (M, Ḳ:) or a hard, elevated place: (Ṣ, TA:) or جُمُدٌ signifies rugged ground: (TA:) or an elevated, rugged place: (Aṣ, TA:) or a small isolated mountain, not high, sometimes rugged and sometimes soft, and producing trees, only found in rugged land; so called because of its dryness; it is the smallest kind of أَكَمَةٌ, round and small, not extending along the ground, rugged at the top, and producing herbs, or leguminous plants, as well as trees; differing from جُمُودٌ [q. v.]: (ISh, L, TA:) pl. [of mult.] جِمَادٌ (ISh, Ṣ, M, Ḳ) and [of pauc.] أَجْمَادٌ. (Ṣ, M, Ḳ.) —— Also, the first, A stone: pl. جِمَادٌ. (Fr, TA.)

جَمَادٌ † [A thing that does not grow, or increase; that is incapable of growth, or increase; an inorganic thing; as a mineral and the like:] an inanimate thing; a thing that has no soul: [an epithet used as a subst.; or an epithet in which the quality of a subst. is predominant:] pl. جَمَادَاتٌ. (Har p. 13.) [See also جَامِدٌ.] —— ‡ Land (أَرْضٌ) upon which rain has not fallen: (T, Ṣ, Ḳ:) or dry land, upon which no rain has fallen, and in which is nothing: (T, TA:) or land in which is no produce: (A:) or, as some say, rugged land: (L:) or sterile, barren, or unfruitful, land, in which is nothing; as also جِمَادٌ: pl. جُمُدٌ. (AA, L in art. جهد.) —— ‡ A year (سَنَةٌ) in which is no rain: (Ṣ, Ḳ:) or in which is no produce of the earth: (A:) and, in like manner, ▼جَامِدَةٌ a year in which is no herbage, or pasture, no plenty, or fruitfulness, and no rain. (T, TA.) —— ‡ A she-camel having no milk; (Ṣ, M, Ḳ;) and so a ewe or a she-goat: (L:) or having little milk: (T, TA:) and [accord. to some,] a slow she-camel; syn. بَطِيْئَةٌ; (L, Ḳ;) but [this is app. a mistranscription for بَكِيْئَةٌ "having little milk," and] ISd says that the explanation بطيئة does not please him. (TA.) —— See also جَامِدٌ, in two places. ═══ A kind of cloth or garment; as also ▼جِمَادٌ. (Ḳ.)

جَمَادِ, like قَطَامِ, (Ḳ,) or جَمَادِ لَهُ, (Ṣ, A, L,) said with reference to a niggard, (Ṣ, A, L, Ḳ,) in dispraise, (Ḳ,) as an imprecation, meaning ‡ May a stagnant, or an unimproving, state or condition (جُمُودُ الحَالِ) be his lot [or his constant lot]: (A:) or may he not cease to be in a stagnant, or an unimproving, state or condition (لَا زَالَ جَامِدَ الحَالِ). (Ṣ, L.) جَمَادِ is [a proper name,] indecl., with kesr for its termination, because it is transformed from the inf. n., namely, الجُمُودُ, like فَجَارِ, which means الفَجْرَةُ: (Ṣ:) and the contr. of جَمَادِ لَهُ is حَمَادِ لَهُ, (Ṣ,* A,) which denotes praise. (Ṣ.) El-Mutalemmis says,

* جَمَادِ لَهَا جَمَادِ وَلَا تَقُولِى *

* لَهَا أَبَدًا إِذَا ذُكِرَتْ حَمَادِ *

i. e., Say thou جُمُودًا to her, [جُمُودًا,] and say not to her [ever, when she is mentioned,] حَمْدًا and شُكْرًا. (Ṣ.)

جِمَادٌ: see جَمَادٌ, last meaning.

جَمُودٌ: see جَامِدٌ.

جُمُودٌ [app. Elevated tracts,] softer, or more plain, than what is termed جُمُدٌ, and more intermixed with soft, or plain, tracts, sometimes in, or by, that [kind of high ground] which is termed قُفٌّ, and sometimes in, or by, soft, or plain, tracts. (ISh, L, TA.)

جَمِيدُ العَيْنِ: see جَامِدٌ.

جُمَادَى One of the names of the months, (Mṣb, Ḳ,) applied to two of the Arabian months, together called جُمَادَيَانِ, (TA,) and distinguished by the appellations of جُمَادَى الأُولَى and جُمَادَى الآخِرَةُ [the fifth and sixth months of the Arabian year]: (Ṣ, Ḳ:) it is of the measure فُعَالَى, from الجَمْدُ; (Ṣ;) the two months to which it is applied being [said to be] so called because, when the months were named, these two fell in the season of the freezing of water: (ISd, L, Mṣb:) [but this derivation seems to have been invented when the two months thus named had fallen back, into, or beyond, the winter; for when they received this appellation, the former of them evidently commenced in March, and the latter ended in May; therefore I hold the opinion of M. Caussin de Perceval, that they were thus called because falling in a period when the earth had become dry and hard by reason of paucity of rain, from جَمَادٌ, an epithet applied to land upon which rain has not fallen, or from جُمَادَى, an epithet applied to an eye that sheds few tears; which opinion is confirmed by the obvious derivations of the names of other months, صَفَر and رَبِيع and رَمَضَان and شَوَّال:] afterwards, when the lunar months superseded the solar, the same names were retained: (Mṣb:) [see زَمَن, and الجَمْرَة:] جمادى is determinate, (Ḳ,) being a proper name, (TA,) and of the fem. gender: (Mṣb, Ḳ:) if you find it masc., it is because it is made to accord to الشَّهْر: all the other names of the months are masc.: (Fr, IAmb, Mṣb:) the pl. is جُمَادَيَاتٌ, (Fr, L, Ḳ,) agreeably with analogy; and if the form جُمَادٌ [a mistranscription for جَمَائِدُ, like حَبَائِرُ, pl. of حُبَارَى,] were used, it would also be agreeable with analogy. (Fr, L.) The former of these two months is also called جُمَادَى خَمْسَة; and the latter, جُمَادَى سِتَّة; (Ḳ;) which mean, respectively, Jumádà the fifth month and Jumádà the sixth month, from the commencement of the year. (TA.) Lebeed says, [describing a pair of wild asses,]

* حَتَّى إِذَا سَلَخَا جُمَادَى سِتَّةً *
* جَزَآ فَطَالَ صِيَامُهُ وَصِيَامُهَا *

[Until, when they both pass, and come to the end of, Jumádà, completing six months, they satisfy themselves with green pasture so as to be in no need of water, and his and her abstinence from water becomes of long continuance]: thus cited by Bundár; ستّة being in the accus. case as a denotative of state, and by جمادى being meant جمادى الآخرة: or, accord. to IAạr, the poet said ستّة, meaning the six months of winter, which are the months of dew; and Aboo-'Amr Esh-Sheybánee says the like. (MF.) AḤn says that the Arabs applied the name of جمادى to The whole of the winter; [see above;] whether the winter were at the same time as the months so called or not: and Aboo-Sa'eed says the like. (L.) — See also جَامِدٌ.

لَيْلَةٌ جُمَادِيَّةٌ A wintry night. (Aboo-Sa'eed, L.) [See جُمَادَى.]

جُمَادٌ ‡ A sword such that he who is struck with it becomes motionless (يَجْمُدُ): (A, TA:) or a sharp, cutting, sword. (AA, Ḳ.)

جَامِدٌ, applied to water, (Mṣb, Ḳ,) &c., (Mṣb,) [i. e.] anything fluid, or liquid, (Ḳ,) In a state of congelation, concretion, or solidity; freezing; as also ▼ جَمْدٌ; contr. of ذَائِبٌ: (Mṣb, Ḳ:) you say مَآءٌ جَمْدٌ [as well as مَآءٌ جَامِدٌ]: (Mṣb:) or ▼ جَمْدٌ signifies what is congealed, or frozen, of water [&c.]; ice; (Ṣ, A;) contr. of ذَوْبٌ: (Ṣ:) [see also جَمَدٌ:] it is originally an inf. n.: (Ṣ, Mṣb, Ḳ:) [or it is an epithet from جَمَدَ, like ضَخْمٌ from ضَخُمَ:] and ▼ جَمَدٌ is a pl. [or rather a quasi-pl. n.] of جَامِدٌ, (Ṣ, Mṣb, Ḳ,) like as خَدَمٌ is of خَادِمٌ: (Ṣ, Mṣb:) you say, قَدْ كَثُرَ الجَمَدُ [The frozen waters have become many]. (Ṣ.) [Hence,] مُخَّةٌ جَامِدَةٌ A hard piece of marrow. (L.) — † Remaining fixed, stationary, or motionless. (Bḍ and Jel in xxvii. 90.) — † A thing that does not grow, or increase; [incapable of growth, or increase; inanimate;] as stone, in contradistinction to a tree [and an animal]. (Kull.) [See also جُمَادٌ.] You say, لَكَ جَامِدُ هٰذَا المَالِ وَذَائِبُهُ (A, L, Ḳ*) ‡ To thee belongs, or shall belong, what consists of gold and silver [or the like inanimate things], of this property, and what consists of live stock, thereof: (L, Ḳ:) or what consists of stones, thereof, and what consists of trees, thereof: or what is solid, thereof, and what is fluid, or liquid, thereof. (L.) — [Hence its application in lexicology and grammar to † A noun that is not an inf. n. nor derived from an inf. n.; a noun having the quality of a real substantive (اِسْمُ عَيْنٍ), opposed to that which has the quality of an ideal substantive (اِسْمُ مَعْنًى): and † a verb that has but one tense and no inf. n., as لَيْسَ and نِعْمَ &c., opposed (as is said in the TA voce قد) to مُتَصَرِّفٌ: it may be rendered (and so I have rendered it), in these cases, aplastic.] — † Lifeless; dead. (Kull p. 147.) — † Stupid, dull, wanting in intelligence; inert; not sharp, penetrating, vigorous, or effective, in the performing of affairs; or soft, without strength or sturdiness, and without endurance. (TA.) — It is also applied to a man's state, or condition: you say رَجُلٌ جَامِدُ الحَالِ † [A man in a stagnant, or unimproving, state or condition]. (Ṣ, L.) — And to the eye: you say عَيْنٌ جَامِدَةٌ † An eye that sheds no tears; (Ks, Ḳ;) as also ▼ جُمَادَى, (Ks, Ḳ,) and ▼ جَمُودٌ; (Ṣ, Ḳ;) or this last signifies ‡ an eye that sheds few tears. (A.) And رَجُلٌ جَامِدُ العَيْنِ, (A, Ḳ,) and ▼ جَمِيدُ العين, and ▼ جَمَادُ العين, (A,) ‡ A man whose eye sheds few tears; (A;) or whose eye sheds no tears. (Ḳ.) — See also جُمَادٌ. — Also, (L,) and ▼ مُجْمِدٌ, (M, A, Ḳ,) and ▼ جَمَادُ الكَفِّ, (A, Ḳ,) ‡ Niggardly, penurious, or avaricious; (M, A, Ḳ;) niggardly of that which it is incumbent on him to give: (L:) and ▼ مُجْمِدٌ, also, a man of little, or no, good; possessing little, or no, good. (Ḳ.) ═ جَوَامِدُ, (as its pl., IAạr, L,) Limits, or boundaries, or boundary-marks, between lands, (IAạr, L, Ḳ,*) and between two dwellings. (L.)

مُجْمِدٌ: see جَامِدٌ, last sentence but one, in two places. — The person who is entrusted with the management of affairs in a game of chance (قِمَار) [here meaning the game called المَيْسِر]: (Ḳ:) [i. q. ضَرِيبٌ:] or the person entrusted with the management of affairs among a people or party, (T, Ḳ, TA,) who does not take part in the game called المَيْسِر, except that he shuffles the arrows (يَضْرِبُ بِهَا) for the players, and has them placed in his hands, and is confided in with respect to them, and compels him who has incurred an obligation to fulfil it: (L, TA:) or one who takes no part in the game called المَيْسِر, (who is called بَرَمٌ,) but who sometimes shuffles, or deals forth, the arrows, (يُفِيضُ بِهَا,) for the players; so in the following verse of Ṭarafeh:

* وَأَصْفَرَ مَضْبُوحٍ نَظَرْتُ حَوِيرَهُ *
* عَلَى النَّارِ وَاسْتَوْدَعْتُهُ كَفَّ مُجْمِدِ *

[And of many a yellow arrow, changed in colour by fire, I have awaited the sound over the fire, and I have deposited it in the hand of one taking no part in the game but only shuffling, or dealing forth, the arrows for the players]; meaning, I have awaited its sound, which was like an answer proceeding from it, when I straightened it and marked it, over the fire: (Ṣ:) [or, accord. to the EM (p. 105), where we find حِوَارَهُ in the place of حَوِيرَهُ, the meaning is, and of many a yellow arrow, &c., I have awaited the returning and gaining, while we were assembled at the fire, &c.:] or مجمد here means a man taking with both his hands so as not to let anything go forth from them: (AA, TA:) or, accord. to Aṣ, it here means a man entering upon Jumádà, which was in that [the poet's] time a month of cold: (Ṣ, Ḳ:*) or one whose arrow does not gain anything in the game called المَيْسِر: (L:) or a person in whom one confides, and who is tenacious of that which is in his hand or possession, and not to be deceived. (A 'Obeyd, TA.)

مُجْمَدَةٌ A place in which ice is kept. (MA.)

هُوَ مُجَامِدِى He is my neighbour, his house, or tent, adjoining mine. (Ḳ.)

جمر

1: see 2, in two places: ═ and see also 4: — and 5. ═ Also جَمَرَهُ He gave him جَمْر [live, or burning, coals]. (Ḳ.) ═ He put him aside, apart, away, or at a distance. (Th, Ḳ.) — جَمَرَتِ الشَّمْسُ القَمَرَ, aor. ـِ, The sun concealed [or as it were put out] the moon [by its proximity thereto: see اِبْنُ جَمِيرٍ]. (IAạr, TA.) ═ جَمَرَ [said of the moon, It became concealed by its proximity to the sun: see an ex. voce جَمِيرٌ: and see also 4]. ═ Also, (Ḳ,) aor. ـُ, (TA,) He (a horse) leaped while shackled; and so ▼ اجمر. (Ḳ.)

2. جَمَّرَ, inf. n. تَجْمِيرٌ; (Ḳ;) and ▼ جَمَرَ; (Mṣb;) He collected together (Mṣb, Ḳ) a people, and anything. (Mṣb.) — جَمَّرَتْ شَعْرَهَا, inf. n. تَجْمِيرٌ; (Ṣ, A, Ḳ;) and ▼ جَمَرَتْهُ, (Mṣb,) and ▼ أَجْمَرَتْهُ; (Ḳ;) She (a woman) collected together her hair, (Ṣ, A, Mṣb, Ḳ,) and tied it in knots, or made it knotted and crisp, (عَقَدَتْهُ, Ṣ, A, Mṣb,) at the back of her neck; (Ṣ, A, Mṣb, Ḳ;) not letting it hang down loosely: (Ṣ:) or plaited it: (T, TA:) and جَمَّرَ شَعْرَهُ he collected together his hair at the back of his head: (Mgh:) and ▼ اجمرت رأسها she collected together the hair of her head, and plaited it: and ▼ اجمر شَعْرَهُ he disposed his hair in ذَوَائِب [or locks hanging down loosely from the middle of the head to the back, or plaits hanging down]. (TA.) — And جَمَّرَ It (a thing) necessitated a people to unite together. (TA.) — Also, (inf. n. as above, Ṣ,) He (a commander, Aṣ, A) detained the army in the territory of the enemy, (Ṣ, Ḳ,) or on the frontier of the enemy's country, (A,) and did not bring them back (Ṣ, A, Ḳ) from the frontier: (Ṣ:) the doing of which is forbidden: (TA:) or he detained them long on the frontier of the enemy, and did not give them permission to return to their families: (Aṣ, TA:) or he collected them on the frontiers of the enemy, and kept them from returning to their families.

(TA.) ═ See also 4: — and 5. ═ جَمَّرَ الثَّوْبَ, (A, Mgh, Mṣb,) inf. n. as above, (Mgh, Mṣb,) *He fumigated the garment with perfume*; (A,* Mgh, Mṣb;) as also ▼ اجمرهُ: (Mgh, Mṣb, Ḳ:) but the former is the more common. (Mgh.) And جَمَّرَ المَسْجِدَ, (Mgh, TA,) or ▼ اجمرهُ, accord. to different modes of writing the surname of a certain No'eym, i. e., المُجَمِّرُ or المُجْمِرُ, (TA,) [and accord. to different copies of the Ḳ,] *He fumigated the mosque with perfume*: (Mgh:) [or perhaps it may mean *he strewed the ground of the mosque with pebbles*; from جَمْرَةٌ; like حَصَّبَهُ, from حَصْبَةٌ or حَصْبَآءُ or حَصَبَةٌ.] — And جَمَّرَ [for جَمَّرَ لَحْمًا] *He put flesh-meat upon live coals* [*to roast*]. (A.) ═ Also, (A,) inf. n. as above, (Ṣ, A,) *He* (a pilgrim, A) *threw the pebbles* [in the valley of Minè]; (Ṣ;) and so ▼ استجمر. (TA in art. تو.) Hence, يَوْمُ التَّجْمِيرِ [*The day of the throwing of the pebbles*, by the pilgrims, in the valley of Minè]. (A.) [See جَمْرَةٌ.] ═ جَمَّرَ النَّخْلَةَ, (inf. n. as above, A,) *He cut off the heart*, or *pith*, (جُمَّار,) *of the palm-tree*. (Ṣ, A, Ḳ.*)

4. اجمر شَعْرَهُ, and رَأْسَهَا; and أَجْمَرَتْ شَعْرَهَا: see 2. — اجمر الأَمْرُ بَنِى فُلَانٍ *The thing*, or *affair, included the common mass*, (Ḳ,) or *the whole mass*, (TA,) *of the sons of such a one within the compass of its relation or relations, or its effect or effects*, &c. (Ḳ, TA.) — اجمر النَّخْلَ *He computed by conjecture the quantity of the fruit upon the palm-trees, and then reckoned, and summed up the quantity so computed*. (Ḳ.) He who does so is termed ▼ مُجْمِرٌ. (TA.) — اجمر الخَيْلَ *He prepared the horses for racing* &c. *by feeding them with food barely sufficient to sustain them, after they had become fat*, (أَضْمَرَهَا,) *and collected them together*. (Ḳ.) ═ اجمر القَوْمُ عَلَى الشَّىْءِ, (Ṣ,) or على الأَمْرِ; (Ḳ;) and ▼ جَمَّرُوا, (Ḳ,) inf. n. تَجْمِيرٌ; (TA;) and ▼ جَمَرُوا, and ▼ استجمروا; (Ḳ;) *The people, or party, agreed together to do the thing*, (Ṣ, Ḳ,) *and united for it*. (Ḳ.) [See also 5.] ═ اجمر الثَّوْبَ, and المَسْجِدَ: see 2. — اجمر النَّارَ, inf. n. مُجْمَرٌ, *He prepared the fire* [app. *in a* مِجْمَرَة]. (Ṣ,* Ḳ.) ═ اجمر said of a camel, *He had his foot rendered even, so that there was no line between its phalanges*, (Ḳ, TA,) *in consequence of its having been wounded by the pebbles, and become hard*. (TA.) ═ Also, said of a camel, (Ṣ,) and of a man, (TA,) *He hastened*, or *was quick, in his pace*, or *going*; (Ṣ, Ḳ;) and *ran*: (TA:) you should not say اجمز. (Ṣ.) — See also 1. ═ أَجْمَرَتِ اللَّيْلَةُ *The night had its moon concealed by its proximity to the sun*. (Ḳ,* TA.) [See also 1.]

5. تجمّر *It* (a people, or party,) *collected together*; (A, Mgh, TA;) [and] so ▼ جَمَّرُوا; this verb being intrans. as well as trans.: (Mṣb: [see 2:]) and ▼ جَمَرَ *it* (a tribe) *collected together, and became one band*. (Aṣ, TA.) — *It* (an army) *became detained in the territory of the enemy, and was not brought back* (Ṣ, Ḳ) *from the frontier*; (Ṣ;) as also ▼ استجمر. (Ḳ.) ═ See also 10.

8. اجتمر بِالمِجْمَرِ, (Ḳ,) and ▼ استجمر, (AḤn, A, Mgh,) *He fumigated*, or *perfumed, himself with aloes-wood* [or *the like*]. (AḤn, A, Mgh, Ḳ.)

10. استجمر: see 4: — and 5: ═ and 8: ═ and 2. — Also, [and vulgarly ▼ تجمّر,] *He performed the purification termed* اِسْتِنْجَآء *with* جِمَار, (Mgh, Mṣb, Ḳ,) i. e., with *stones*, (AZ, Ṣ, Mṣb,) or *small stones*. (Mgh, TA.)

جَمْرٌ: see what next follows, in two places.

جَمْرَةٌ *A live*, or *burning, coal*; *a piece of smokeless burning fire*: (Mṣb:) or *burning fire*: (Ḳ:) [but the former is the correct explanation:] when cold, [before it is kindled,] it is called فَحْمٌ (TA) [or حَطَبٌ &c.]: and when reduced to powder by burning, رَمَادٌ: (L in art. رمد:) from جَمَرَ "he collected together:" (Mgh:) pl. ▼ جَمْرٌ (Ṣ, Mṣb, Ḳ) [or rather this is a coll. gen. n.] and جِمَارٌ and جَمَرَاتٌ. (Mṣb.) — [Hence,] الجَمْرُ ▼ فِى كَبِدِى ‡ [*Live coals are in my liver*]. (A.) — [Hence also,] الجَمَرَاتُ الثَّلَاثُ † [*The three live coals*; meaning the first three degrees of heat]: the first is in the air; the second, in the earth, or dust; and the third, in the water: [or, accord. to the modern Egyptian almanacs, the first is in the air, and is cold, or cool; the second, in the water, and is lukewarm; and the third, in the earth, or dust, and is hot: the first falling exactly a zodiacal month before the vernal equinox; and each lasting seven days:] whence the saying, كَانَ ذٰلِكَ عِنْدَ سُقُوطِ الجَمْرَةِ † [*That was at the time of the falling of the live coal*]; i. e., when the heat had acquired strength. (TA.) ═ *Any body of men that have united together, and become one band, and that do not form a confederacy with any others*: (Ṣ:) or *a body of men that congregate by themselves, because of their strength and their great valour*; [said to be] from the same word signifying "a live coal:" (Mṣb:) or *any people that endure patiently fighting with those who fight them, not forming a confederacy with any others, nor uniting themselves to any others*: (Lth, TA:) or *a tribe that does not unite itself to any other*: (Ḳ:) or *that comprises three hundred horsemen*, (Ḳ,) or *the like thereof*: (TA:) or *a tribe that fights with a company of tribes*: (TA:) pl. جَمَرَاتٌ. (Ṣ, Mṣb, Ḳ.) You say, بَنُو فُلَانٍ جَمْرَةٌ *The sons of such a one are a people able to defend themselves, and strong*. (TA.) جَمَرَاتُ العَرَبِ is an appellation especially applied to three tribes; namely, Benoo-Ḍabbeh Ibn-Udd, and Benu-l-Ḥárith Ibn-Kaab, and Benoo-Numeyr Ibn-'Ámir; (Ṣ, A, Ḳ;) the first of which became extinguished by confederating with Er-Ribáb, and the second by confederating with Medhhij; the third only remaining [a جمرة] because it formed no confederacy: (Ṣ:) or it is applied to 'Abs and El-Ḥárith and Ḍabbeh; all the offspring of a woman who dreamt that three live coals issued from her فَرْج. (Ṣ, Ḳ.) — Also *A thousand horsemen*. (Ṣ, Ḳ.) One says جَمْرَةٌ كَالجَمْرَةِ [*A troop of a thousand horsemen like the live coal*]. (Ṣ, TA.) ═ *A pebble*: (Ṣ, Ḳ:) or *a stone*: (Mṣb:) or *a small stone* or *pebble*: pl. جِمَارٌ (Mgh, Mṣb, Et-Towsheeḥ, TA) and جَمَرَاتٌ (Mgh, Mṣb.) — Also sing. of جَمَرَاتٌ (Ṣ, Mṣb, Ḳ) and of جِمَارٌ (TA) in the appellations جَمَرَاتُ مِنًى (Mṣb) or جَمَرَاتُ المَنَاسِكِ (Ṣ, Ḳ) and جِمَارُ المَنَاسِكِ, (TA,) which were three in number, (Ṣ, Mṣb, Ḳ,) called الجَمْرَةُ الأُولَى and الجَمْرَةُ الوُسْطَى and جَمْرَةُ العَقَبَةِ, (Ḳ,) at which جَمَرَات [i. e. small pebbles, TA] were cast; (Ṣ, Ḳ;) each of these being *a heap of pebbles, at Minè*, and each two heaps [or rather each heap and that next to it] being about a bow-shot apart: (Mṣb:) accord. to Th, from جَمَّرَهُ "he put him aside, apart, away, or at a distance:" or from أَجْمَرَ "he hastened;" because Adam pelted Iblees in Minè, and he hastened away before him: (Ḳ,* TA:) or from تجمّروا "they collected together:" (Mgh:) or from جَمَرَهُ "he collected it together." (Mṣb.) ═ See also جُمَيْرَةٌ.

جَمَارٌ *An assembly*; *an assemblage*; *a collection*: (Ḳ:) *a people assembled together*. (TA.) — عَدَّ إِبِلَهُ جَمَارًا *He counted*, or *numbered, his camels in one herd*, (Aṣ, TA,) *by looking at their aggregate*. (Aṣ, T voce نَظِيرٌ, q. v.) — جَاؤُوا جَمَارًا, and with tenween, [i. e., app., ▼ جَمَارَى, not, as might be thought at first sight, جَمَارَى,] a form which MF disapproves, though it is said in the TA that his disapproval requires consideration,] *They came all together*, or *all of them*. (Ḳ.)

جَمِيرٌ *A place of assembly of* a people. (Ṣ, Ḳ.) — ابْنَا جَمِيرٍ *The night and the day*: (Ṣ, Ḳ:) so called because of the assembling [of people therein]; like as they are called ابْنَا سَمِيرٍ because people held conversation therein: (Ṣ:) or *the two nights during which the moon becomes concealed by its proximity to the sun*. (TA.) And ابْنُ جَمِيرٍ, (IAạr, Ṣ,) or ▼ ابْنُ جُمَيْرٍ, (Lḥ, Th,) *The moon in the night when it is concealed by its proximity to the sun*: (TA:) or *the moon in the end of the* [*lunar*] *month*; because the sun conceals it (تُوَارِيهِ, i. e. تُجَمِّرُهُ): (IAạr, TA:) or *the dark night*: (Ṣ:) or *the night in which the moon does not rise, either in the first part thereof or in the last*: (TA:) or *the last night of the* [*lunar*] *month*. (Aboo-'Amr Ez-Záhid, TA.) You say, جَآءَنَا فَحْمَةَ ابْنِ جُمَيْرٍ ▼ [*He came to us in the darkest part of the moonless night*, or *of the night in which the moon did not rise*]. (Th, TA.) And لَا أَفْعَلُ ذٰلِكَ مَا جَمَرَ ابْنُ جُمَيْرٍ ▼ [*I will not do that as long as the moon in the end of the lunar month becomes concealed by its proximity to the sun*; i. e., I will never do it]. (Lḥ, TA.) — جَمِيرُ الشَّعَرِ *What is collected together, of the hair, and tied in knots, or made knotted and crisp*. (TA. [See 2.])

ابْنُ جُمَيْرٍ: see جَمِيرٌ, in three places.

جَمِيرَةٌ *A plait of hair*: (T, Mṣb, Ḳ:) and i. q. ذُؤَابَةٌ [app. here meaning *a plait of hair hanging down*; or *a lock of hair hanging down loosely from the middle of the head to the back*]: (TA:) and ▼ جُمَيْرَةٌ *a lock of hair*: (TA:) pl. of the former جَمَائِرُ. (T, Mṣb.)

جَمَارَى: see جَاؤُوا جَمَارًا.

جُمَّارٌ (Ṣ, A, Mgh, Mṣb, Ḳ) and ↓جَامُورٌ (Ḳ) [each a coll. gen. n.] The *heart*, or *pith*, [or *cerebrum*,] *of the palm-tree*, (Ṣ, A, Mṣb, Ḳ, TA,) *that is in the summit of its head, which part is cut off, and its outer portion is stripped off from the pith within it, which is a white substance, like a piece of the hump of a camel, large and soft: it is eaten with honey:* (TA:) *from it come forth the fruit and the branches; and when it is cut off, the tree dies:* (Mṣb:) *the spathe comes forth from it, amid the part whence two branches divide:* (TA:) the *head of the palm-tree; a soft, white substance:* from جَمَرَ "he collected together;" for a similar reason termed كَثَرٌ: (Mgh:) n. un. جُمَّارَةٌ. (A, TA.) [See also قَلْبٌ.] You say, لَهُ سَاقٌ كَالجُمَّارَةِ *He has a shank like a piece of the heart of the palm-tree.* (A.) And الجُمَّارُ فِى خَلَاخِلِهِنَّ ‡ [*Legs like the heart of the palm-tree are within their anklets*]. (A.) Ṣakhr El-Hudhalee says, using a double trope, likening the fresh juicy stalks of the بَرْدِىّ to the pith of the palm-tree, and then applying this expression to the legs of a woman,

* إِذَا عُطِفَتْ خَلَاخِلُهُنَّ غَصَّتْ *
* بِجُمَّارَاتِ بَرْدِىٍّ خِدَالِ *

‡ [*When their anklets are bent*, (for the anklet of the Arab woman is formed of a piece of silver, or other metal, which is bent round so that the two ends nearly meet,) *they are choked*, or *entirely filled up, with plump legs like the pith of the papyrus*]. (A, TA.)

مُجَمَّرٌ: see جَامِرٌ.

جَامُورٌ: see جُمَّارٌ. — Also ‡ *A well-known appertenance of a ship* or *boat*; [i. e., the *head of the mast; a kind of truck, which is made of harder wood than the mast itself.*] (TA.) — And hence, ‡ The *head* [absolutely]: but accord. to Kr, only the vulgar call it so. (TA.)

أَجْمَرُ occurs in a trad., where it is said, دَخَلْتُ المَسْجِدَ وَالنَّاسُ أَجْمَرُ مَا كَانُوا, meaning *I entered the mosque when the people were in their most collected state.* (TA.)

مِجْمَرٌ: see مُجْمَرٌ: — and see also مِجْمَرَةٌ, in two places. — Also, (Ṣ, Ḳ,) and ↓مُجْمِرٌ, (Ḳ,) *A hard* solid hoof: (AA, Ṣ, Ḳ:) and a *hard, strong, compact* camel's foot: or one *that has been wounded by the stones, and become hard.* (TA.)

مُجْمِرٌ: see مُجَمِّرٌ, in two places: — and أَجْمَرَ: النَّخْلُ: — and see also مُجْمِرٌ.

مِجْمَرٌ: see مِجْمَرَةٌ. — Also, (Mgh, Mṣb, Ḳ,) and ↓مُجْمَرٌ, (Ḳ,) *Aloes-wood*, (AḤn, Mgh, Mṣb, Ḳ,) and *the like*, (Mgh,) or *other substance*, (Mṣb,) *with which clothes are fumigated*, (Mgh,) or *with which one perfumes himself by burning it:* (Mṣb:) pl. مَجَامِرُ. (Mgh.)

مِجْمَرَةٌ and ↓مِجْمَرٌ, (Ṣ, Mgh, Mṣb, Ḳ,) which latter is sometimes fem. [like the former], (Ḳ,) or fem. when by it is meant the fire (النَّار), and masc. when meaning the place [of the fire], (TA,) and ↓مُجْمَرٌ, (Ḳ,) *A vessel for fumigation; a censer;* (Mṣb;) *a vessel in which live coals are put*, (Ṣ, Ḳ,) *with incense, or some odoriferous substance for fumigation;* (Ḳ;) *a vessel in which aloes-wood is burned:* it is disapproved, because generally of silver; but not so what is termed مِدْخَنَة: (Mgh:) or ↓مُجْمَرٌ signifies the *thing for which the live coals are prepared:* (Ṣ:) [and مِجْمَرَةٌ also signifies a blacksmith's *fire-place:* (Ḳ in art. كور:)] pl. مَجَامِرُ. (Ṣ.)

مُجَمَّرٌ Flesh-meat *put upon live coals* [*to roast*]. (A.)

مُجَمِّرٌ (Ṣ, Z) and ↓مُجْمِرٌ (TA) One *who collects together his hair, and ties it in knots*, or *makes it knotted and crisp, at the back of his neck, not letting it hang down loosely:* (Ṣ:) or *who plaits the hair of his head.* (TA.) He who does so (while he is a مُحْرِم, TA) is commanded to shave his head. (Ṣ and TA from a trad.) — Also, both the former and ↓ the latter, and ↓جَامِرٌ, which is a possessive epithet, without a verb, One *whose business is to fumigate garments* [&c.] *with perfume.* (TA.)

جمز

1. جَمَزَ, (Ṣ, A, &c.,) aor. ِ, inf. n. جَمْزٌ (Ṣ, Mṣb, Ḳ) and جَمَزَى, (Ḳ,) or the latter is a simple subst., (Mṣb,) said of a camel, (Ṣ, Ḳ,) and of a man, (A, Ḳ,) [*He went at a gentle trot or run;*] *he went a pace quicker than that termed* عَنَق, (Ṣ, A, Mṣb, Ḳ,) *but not so quick as that termed* حُضْر, (Ḳ,) or *not so quick as a vehement* حُضْر; (TA;) *he went the pace with which corpses are conveyed* [*to the tomb;* which, according to the practice prescribed by Mohammad, is a quick pace]: (TA:) or simply, *he went*, or *went along:* (Mṣb:) and *he ran;* syn. عَدَا: (Mgh, Mṣb:) and *he went quickly.* (Mgh, Mṣb, TA.) You say, جَمَزَ بِالجِنَازَةِ *He went a pace quicker than that termed* عَنَق [*with the corpse upon its bier*]. (A.) And جَمَزَ الرَّجُلُ فِى الأَرْضِ *The man went away into, or in, the country* or *land.* (Kr, Ḳ.)

[2. جمّز, if used, *He rode a camel such as is called* جَمَّازٌ or جَمَّازَةٌ. See the act. part. n., below.]

جَمَزَى a subst. from جَمَزَ; [signifying *A gentle trot or run; a pace quicker than that termed* عَنَق, *but not so quick as that termed* حُضْر, or *not so quick as a vehement* حُضْر; &c.] (Mṣb.) You say, هُوَ يَعْدُو الجَمَزَى, (A,) and النَّاقَةُ تَعْدُو الجَمَزَى, and in like manner الفَرَسُ, (Ks, Ṣ,) [*He*, and *the she-camel*, and *the mare* or *horse*, *runs at the pace termed* جَمَزَى.] — See also جَمَّازٌ, in two places.

جَمَّازٌ, applied to a he-camel, (Ṣ, Ḳ,) and جَمَّازَةٌ, applied to a she-camel, (Ḳ,) *That is ridden by the* مُجَمِّز; (Ṣ;) *that goes the pace described above*, [*voce* جَمَزَى *and*] *voce* جَمَزَ: (Ḳ, TA:) [the latter is also said in the TA to be مِن آلَاتِ المحامل; but the correct reading seems to be مِنْ أُولَاتِ المَحَامِلِ; and the meaning, *of those that carry the vehicles called* محامل, pl. of مَحْمِل.] — حِمَارٌ جَمَّازٌ *An ass that leaps, jumps, springs*, or *bounds, quickly:* (Ḳ:) and حِمَارٌ ↓جَمَزَى *a quick ass;* (Ṣ, Ḳ;) or *an ass that leaps, jumps, springs, or bounds, quickly, and is swift;* (TA;) the latter word in this phrase used as a masc. and fem. epithet, though its final letter is a denotative of the fem. gender. (Ḥam p. 277. [See below; and see also حَيَدَى.]) Umeiyeh Ibn-Abee-'Áïdh (Ṣ, TA) El-Hudhalee (TA) says,

* كَأَنِّى وَرَحْلِى إِذَا رُعْتُهَا *
* عَلَى جَمَزَى ↓ جَازِئٍ بِالرِّمَالِ *

[*As though I and my she-camel's saddle, when I frightened her, were upon a swift wild ass satisfied with green pasture, so as to be in no need of water, in the sands*]. (Ṣ, TA.) He likens his she-camel to a wild ass, to which he applies the epithet جمزى, that is, *swift;* meaning, عَلَى حِمَارٍ جَمَزَى. (TA.) Aṣ says that this is the only epithet of the measure فَعَلَى heard by him applied to a male; and that IAạr cited the verse above to him saying حَيَدٍ بِالدِّحَالِ, meaning عَنِ الدِّحَالِ, [i. e., "shying and turning aside from the hollows, narrow at the top but wide below, in the ground:" but this is probably a reading of some in the place of حَيَدَى بِالدِّحَالِ, which ends the next verse, agreeably with what is said in the L in art. حيد:] Az says that عَلَى جَمَزَى may be explained as for عَلَى عَيْرٍ ذِى جَمَزَى, i. e., *upon an ass having the mode of pace termed* جَمَزَى; and نَاقَةٌ وَكَرَى has a similar meaning. (TA.) — See also مُجَمِّزٌ.

جُمَّيْزٌ (Ṣ, Ḳ) and ↓جُمَّيْزَى (Ḳ) [*The sycamore-fig:* and the *sycamore fig-tree: ficus sycomorus;* also called the *Egyptian fig:*] the *male fig;* (Ḳ, TA;) *which is found in the Ghowr*, or *Ghór*, [here meaning *the Valley of the Jordan*,] (TA,) *and is sweet:* (Ḳ, TA:) *this is the yellow: the black makes the mouth bleed:* (TA:) *it is of various colours*, or *kinds*, (أَلْوَان,) (Ḳ,* TA,) *abundant in Syria and in Egypt:* n. un. جُمَّيْزَةٌ: (TA:) [*a fruit*] *resembling the* تِين [or *common fig*]: (Ṣ:) AḤn says, of the kinds of fig is the fig of the جُمَّيْز, *a sweet, moist fig, which has long fruit-stalks, and which is dried in the sun: and there is another species of the* جُمَّيْز, *the fruit of which is like the fig in make, but its leaves are smaller than those of the fig, and its figs are yellow, of a small size, and black: it is found in the Ghowr*, or *Ghór, and is called the male fig: the yellow is sweet: the black makes the mouth bleed: and its fig has no stalk, but cleaves to the wood.* ('Abd-el-Laṭeef, Account of Egypt: White's ed., entitled Abdollatiphi Historiæ Aegypti Compendium: p. 22. See also De Sacy's notes to his transl. of that work, pp. 82—86.) — [الجُمَّيْزَةُ also signifies † *The pudendum muliebre:* opposed to التِّينَةُ as meaning "the anus."]

جُمَّيْزَى: see جُمَّيْزٌ.

جُمَّيْزِىٌّ *A seller of* جُمَّيْز. (TA.)

مُجَمِّزٌ One *who rides the camel called* جَمَّازٌ, (Ṣ,* TA,) or *who rides the she-camel called* جَمَّازَةٌ; (Ḳ,* TA;) as also ↓جَمَّازٌ. (TA.)

جمس

1. جَمَسَ, (A, Mṣb, Ḳ,) aor. ـُ, (Mṣb, TA,) inf. n. جُمُوسٌ (Ṣ, Mṣb, Ḳ) and جَمْسٌ; (TA;) and جَمِسَ, aor. ـَ; (TA;) *It* (grease, Aṣ, Ṣ, A, Mṣb, Ḳ, and clarified butter, and water, A, Ḳ, but جَمَدَ is more commonly said of the last, Ḳ, or جمس is incorrect when said of water, Aṣ, TA) *congealed.* (Aṣ, Ṣ, A, Mṣb, Ḳ.)

جُمْسَةٌ *A tough date:* (IDrd,* M, Ḳ,* TA:) *a date ripening* (Aṣ, Ṣ, Z, Ḳ) *altogether,* (Aṣ, Z, Ḳ, TA,) *but as yet hard, not mellow, or digestible, or easy of digestion:* (Aṣ, Ṣ, Z, Ḳ:) pl. جُمَسٌ. (Aṣ, TA.) [See بُسْرٌ.] — Also *A distinct number, or herd, of camels.* (O, Ḳ.)

جَامِسٌ Grease, (A,) and clarified butter, (TA,) and water, (Ṣ,) or it is improperly applied to the last, (Aṣ, TA,) *in a state of congelation.* (Ṣ, A, Mgh, TA.) — A plant *that has lost its freshness, or juiciness,* (AḤn, Ḳ,) *and become old, and hard,* or *tough.* (AḤn, TA.) — صَخْرَةٌ جَامِسَةٌ *A tough rock,* (TA,) *firm in its place.* (Ḳ, TA.) [In the TA is added مُقْشَعِرَّةٌ: but this is evidently a mistranscription, for مُسْتَقِرَّةٌ, which adds nothing to the explanation.]

جَامُوسٌ [The *buffalo;*] *a kind of* بَقَر; (Mgh, Mṣb, TA;) *well known:* (Ḳ:) n. un. with ة: (Ḳ:) and pl. جَوَامِيسُ: (Ṣ, Mṣb, Ḳ:) an arabicized word, (T, Ṣ, Ḳ,) from the Persian; (T, Ṣ;) originally گَاوْمِيش. (T, Ḳ.)

جمش

1. جَمَشَهُ, (Ṣ, A, Ḳ,) aor. ـُ and ـِ, (TA,) inf. n. جَمْشٌ, (Ṣ,) *He shaved it,* or *removed its hair;* (Ṣ, A, Ḳ;) namely, his pubes; (Ṣ;) or his head. (A, Ḳ.) And جَمَشَتْهُ *She shaved it;* namely, her pudendum; or *removed its hair:* and *it* (نُورَة, q. v.) *removed it;* namely, hair. (A.)

جَمُوشٌ, applied to نُورَة [q. v.], *That removes hair;* (Ṣ, Ḳ;) as also ▼جَمِيشٌ. (Ḳ.) — Applied to a year (سَنَة), † *That shaves off,* (Ṣ,) or *nips, shrinks, shrivels,* or *blasts,* [lit., *burns,* see أَحْرَقَ,] (Ḳ,) *the plants,* or *herbage.* (Ṣ, Ḳ.)

جَمِيشٌ, applied to the pubes, *Shaven,* or *having its hair removed,* (Ṣ, A, Ḳ, TA,) *by means of* نُورَة [q. v.]. (TA.) — Applied to a place, † [*Shorn of its plants,* or *herbage;* or] *having in it no plants,* or *herbage.* (Ṣ, Ḳ.) — See also جَمُوشٌ.

جمع

1. جَمَعَ, (Ṣ, Mgh, Mṣb,) aor. ـَ, (Mgh, TA,) inf. n. جَمْعٌ, (Ṣ, Mgh, Mṣb, Ḳ,) *He collected; brought,* or *gathered, together; gathered up; assembled; congregated; mustered; drew together;* or *contracted;* (Mgh, Er-Rághib, B, Ḳ;*) a thing; (Er-Rághib, Mṣb, B;) *so that the several parts or portions became near together;* (Er-Rághib, B;) or a thing in a scattered, or dispersed, state; (Fr, Ṣ;) and a number of men; (Fr;) as also ▼جمّع; [or this has only an intensive signification;] and ▼اجمع. (TA.) [See also the inf. n., جَمْعٌ, below; and] see 2; and 10. — [جَمَعَ بَيْنَهُمَا *He brought them two together, into a state of union, after separation;* and particularly, *reconciled them; conciliated them:* and *he,* or *it, united, connected,* or *formed a connexion between, them two:* see 3 (last sentence) in art. دنو.] — جَمَعَ عَلَيْهِ ثِيَابَهُ *He put on,* or *attired himself with, his clothes.* (TA.) — جَمَعَتِ الجَارِيَةُ الثِّيَابَ *The girl put on the* دِرْع *and the* خِمَار *and the* مِلْحَفَة; (Ṣ, TA;) i. e., ‡ *became a young woman;* (Ṣ, Ḳ, TA;) *became full-grown.* (TA.) — مَا جَمَعْتُ بِٱمْرَأَةٍ قَطُّ, and عَنِ ٱمْرَأَةٍ, † *I have never gone in to a woman;* or *I have never had a woman conducted to me as my bride.* (Ks, Ḳ.) — فَٱجْمَعُوا — جَمَعَ أَمْرَهُ: see 4. — فَجَمَعَ كَيْدَهُ, and كَيْدَكُمْ: see 4. — [جَمَعَ also signifies *He composed, arranged,* or *settled,* a thing, or an affair; as in the phrase جَمَعَ ٱللّٰهُ شَمْلَهُ: see art. شمل. — Also *It comprised, comprehended,* or *contained.*] — Also *He pluralized* a word; *made it to have a plural,* or *plurals.* (The Lexicons passim.)

2. جمّع, (Fr, Mṣb,) inf. n. تَجْمِيعٌ, (Ḳ,) *He collected; brought,* or *gathered, together; gathered up; assembled; congregated; mustered; drew together;* or *contracted;* [thus I render جَمَعَ, as explained above;] *much; with much,* or *extraordinary, energy,* or *effectiveness,* or *the like; vigorously;* or *well.* (Bḍ in civ. 2; Mṣb, Ḳ.) Thus in the Ḳur [civ. 2], ٱلَّذِى جَمَّعَ مَالًا وَعَدَّدَهُ, (Ṣ,* Bḍ) *Who hath collected much wealth, and hath made it a provision for the casualties of fortune,* or *reckoned it time after time:* (Bḍ:) [or *who hath amassed,* or *accumulated, wealth,* &c.:] or *who hath gained, acquired,* or *earned, wealth,* &c.; thus differing from جَمَعَ, explained above: but it is allowable to say ▼جَمَعَ مَالًا, without teshdeed; (Fr;) and thus it is [generally] read in this passage of the Ḳur. (Bḍ.) See also 1. — جَمَّعَتْ, (TA,) inf. n. as above, (Ḳ,) *She* (a hen) *collected her eggs in her belly.* (Ḳ, TA.) — جَمَّعُوا, (inf. n. as above, Ṣ,) *They were present on the Friday,* (Ṣ, Mgh, Mṣb,) or *with the congregation* [*then collected*], (Mgh,) *and performed the prayers* [*with the congregation*] *on that day.* (Ṣ, Mgh.) — Hence the saying, أَوَّلُ جُمْعَةٍ جُمِّعَتْ فِى الإِسْلَامِ بَعْدَ المَدِينَةِ بِجُوَاثَى [*The first Friday that was observed by the performance of congregational prayer* in the time of *El-Islám, after* the observance thereof in *El-Medeeneh, was in Ju-áthà*]. (TA.)

3. جَامَعَهُ عَلَى أَمْرِ كَذَا, (Ṣ, Ḳ,) inf. n. مُجَامَعَةٌ (TḲ) [and جِمَاعٌ], *He combined with him,* (▼اجتمع مَعَهُ, Ṣ, Ḳ, TA,) *and aided him,* (TA,) *to do such a thing.* (Ṣ,* Ḳ,* TA.) It is said in a trad. of Aboo-Dharr, وَلَا جِمَاعَ لَنَا فِيمَا بَعْدُ i. e. لَا ٱجْتِمَاعَ▼ لَنَا [which may mean *Nor any combining,* or *nor any coming together, for us afterwards:* see 8]. (TA.) — جَامَعَ ٱمْرَأَتَهُ, (Mṣb,) inf. n. مُجَامَعَةٌ (Ṣ, Mṣb, Ḳ) and جِمَاعٌ, (Mgh, Mṣb,) ‡ *He lay with his wife; compressed her.* (Ṣ,* Mgh,* Mṣb, Ḳ.*) [The latter inf. n. is the more common as meaning *Coitus conjugalis,* or the act of *compressing*]. — ٱسْتَأْجَرَ الأَجِيرَ مُجَامَعَةً, and جِمَاعًا, *He hired the hireling for a certain pay every week.* (Lḥ,* TA.)

4. اجمع: see 1. — أَجْمَعْتُ الشَّىْءَ signifies *I put the thing together;* such, for instance, as spoil, or plunder. (Ṣ.) You say, أَجْمَعْتُ النَّهْبَ, meaning *I collected together from every quarter the camels taken as spoil from the people to whom they belonged, and drove them away:* (AHeyth:) or إِجْمَاعٌ signifies [simply] the *driving* of camels *together,* or *collectively.* (Ḳ.) — الإِجْمَاعُ also signifies *The composing and settling a thing which has been discomposed* [*and unsettled*]; *as an opinion upon which one determines, resolves,* or *decides:* (TA:) or جَعْلُ الأَمْرِ جَمِيعًا بَعْدَ تَفَرُّقِهِ, (AHeyth, Ḳ,) i. e. *the determining, resolving,* or *deciding, upon an affair, so as to make it firmly settled,* [*after it had been unsettled in the mind,* or] *after considering what might be its issues, or results, and saying at one time, I will do thus, and at another time, I will do thus.* (AHeyth.) You say, أَجْمَعْتُ الأَمْرَ, (Ks, Ṣ, Mgh,* Mṣb, Ḳ,) and عَلَى الأَمْرِ, (Mgh,* Mṣb, Ḳ,) *I determined, resolved,* or *decided, upon the affair;* (Ks, Ṣ, Mgh,* Mṣb, Ḳ;) as though I collected myself, or my mind, for it; (TA;) as, for instance, a journeying, and a fasting, (Mgh, Mṣb,) and a going forth, and a tarrying or an abiding; (TA;) and in like manner, ▼جَمَعَ أَمْرَهُ *He determined, resolved,* or *decided, upon his affair;* as, for instance, a fasting: (TA:) and أَجْمَعْتُ الرَّأْىَ *I determined,* or *settled, the opinion.* (TA.) You say also, أَجْمِعْ أَمْرَكَ وَلَا تَدَعْهُ مُنْتَشِرًا [*Determine thou,* or *decide, upon thine affair, and do not leave it unsettled*]. (Ṣ.) The saying, in the Ḳur [x. 72], فَأَجْمِعُوا أَمْرَكُمْ وَشُرَكَآءَكُمْ means *Then determine ye,* or *resolve,* or *decide, upon your affair,* (Fr, Ibn-'Arafeh, Bḍ,) *and prepare for it,* (Fr,) or ٱجْعَلُوهُ جَمِيعًا, [which has the former of these meanings, as shown above,] (AHeyth,) *and call ye your companions,* (Fr, Ṣ, Bḍ, Ḳ,) شركاءكم being governed in the accus. case by the verb understood, (Bḍ, TA,) because the verb in the text is not used with شركاء for its object, (Ṣ, Ḳ,) but only the unaugmented verb: (Ṣ:) or the meaning is *then determine ye, with your companions, upon your affair;* (Bḍ, Ḳ;) so says Aboo-Is-ḥáḳ, adding that what Fr says is erroneous: (TA:) or *then determine ye upon your affair and* the affair of *your companions,* for وَأَمْرَ شُرَكَائِكُمْ. (Bḍ.) It is also said that the phrase, in the Ḳur [xx. 67], فَأَجْمِعُوا كَيْدَكُمْ means *Therefore determine ye,* or *resolve,* or *decide, upon your artifice,* or *stratagem:* (TA:) but some read فَٱجْمَعُوا▼ كَيْدَكُمْ, (Bḍ, TA,) meaning *therefore combine ye all your artifice; leave nothing thereof unexerted;* (TA;) and this latter reading is favoured by the phrase ▼فَجَمَعَ كَيْدَهُ [in verse 62 of the same ch.]. (Bḍ.) — Also *The agreeing,* or *uniting, in opinion.* (Ḳ,* TA.) You say, أَجْمَعُوا عَلَى الأَمْرِ meaning *They agreed,* or *were of one mind or opinion, upon,* or *respecting, the affair;* (Mgh, Mṣb;) [and so ▼اجتمعوا عَلَيْهِ;

and عليه ↓تجمّعوا.] — Also *The preparing* [a thing], or *making* [it] *ready*; syn. الإِعْدَادُ. (K, TA. [In the CK, erroneously, الأَعْدَادُ.]) You say, أَجْمَعْتُ كَذَا *I prepared*, or *made ready, such a thing.* (TA.) And أَجْمِعُوا أَمْرَكُمْ *Prepare ye for your affair.* (Fr.) — Also *The binding the teats of a she-camel all together with the* صِرَار, q. v. (K.) You say, اجمع بِالنَّاقَةِ, (S, TA,) and اجمع النَّاقَةَ, (TA,) *He so bound the teats of the she-camel;* (S, TA;) and so أَكْمَشَ بِهَا. (TA.) — Also *The drying* [a thing]; *drying* [it] *up; making* [it] *dry;* syn. التَّجْفِيفُ وَالإِيبَاسُ. (K, TA. [In the CK, erroneously, التَّخْفِيفُ والاِيْبَاسُ.]) Hence the saying of Aboo-Wejzeh Es-Saadee,

* وَأَجْمَعَتِ الهَوَاجِرُ كُلَّ رَجْعٍ *
* مِنَ الأَجْمَادِ وَالدَّمِثِ البَثَاءِ *

i. e. [*And the vehement mid-day-heats*] *dried up every pool left by a torrent* [*of the hard and elevated grounds and of the soft and even ground*]. (TA.) — اجمع المَطَرُ الأَرْضَ *The rain made the whole of the land, both its soft tracts and its hard tracts, to flow:* (K:) and in like manner you say, أَجْمَعَتِ الأَرْضُ سَائِلَةً *The land flowed in its soft tracts* [*as well as in its hard tracts; i. e., in every part*]. (TA.) [See also 10.]

5: see 8, in three places: and see also 4, latter half.

7. انجمع عَنِ النَّاسِ [*He withdrew himself from men*]. (TA in art. قبض.)

8. اجتمع *It* (a thing in a scattered or dispersed state, S, and a number of men, Msb, [and a number of things,]) *became collected, brought together, gathered together, gathered up, assembled, congregated, mustered, drawn together,* or *contracted;* or *it collected, collected itself together, gathered itself together, came together, assembled, congregated, drew itself together, contracted itself; coalesced; combined;* (K, TA;) *so that the several parts or portions became near* [or *close*] *together;* (TA;) as also اِجْدَمَعَ, (K,) with د [substituted for the ت]; (TA;) and ↓تجمّع and ↓تجمّعوا and ↓استجمع signify the same: (Msb, K:) and ↓تجمّعوا signifies *they became collected,* &c., [*from several places,* or] *hence and thence.* (S, K.) [See also 10.] You say also, اجتمع مَعَهُ (Mgh) and بِهِ (Msb) [meaning *He was,* or *became, in company with him; came together with him; met with him; met him; had a meeting,* or *an interview, with him*]. And اجتمع مَعَهُ عَلَى أَمْرِ كَذَا: (S, K:) see 3, first sentence: and see the sentence there next following. And in like manner, ↓تجمّعوا عَلَى فُلَانٍ *They combined, conspired,* or *leagued, together against such a one.* (Ibn-Buzurj, TA in art. ضفر.) [See also اجتمعوا عَلَى الأَمْرِ in 4, latter half.] You also say, اِجْتَمَعَتْ آرَاؤُهُمْ عَلَى الأَمْرِ [*Their opinions agreed together,* or *were in unison, upon,* or *respecting, the affair*]. (Er-Rághib.) And اِجْتَمَعَتْ شَرَائِطُ الإِمَامَةِ *The conditions of the office of Imám occurred together* [or *were combined,* or *they coexisted,* in such a case]; as also ↓اِسْتَجْمَعَتْ. (Msb: [but it is implied in the Mgh that the latter verb in this sense is not of established authority.]) [See a similar ex. voce ارتفع.] — [*He,* or *it, was,* or *became, compact* in make or frame, *compressed, contracted,* or *the like.* — And hence,] *He* (a man) *attained to his full state of manly vigour, and his beard became full-grown.* (K, TA.) The verb is not thus used in speaking of a woman. (S, TA.) — [Hence also,] اجتمع فِي الحَاجَةِ [*He was quick and vigorous* in executing *the needful affair,* or *in* accomplishing *that which was wanted;* as though he compacted his frame, and collected all his energy: see مَشَى مُجْتَمِعًا, below: and see also 10]. (TA in art. كمش.) — [Hence also,] اِجْتَمَعَتِ القِدْرُ *The cooking-pot boiled.* (Z, TA.) — [Hence also, اجتمع said of a thing, or an affair, *It was,* or *became, composed, arranged,* or *settled.*]

10. استجمع كُلَّ مَجْمَعٍ ↓ [*He desired,* or *demanded, the collecting together of every body of soldiers; or he summoned together every body of soldiers*]: said of him who demands, or summons, armies, or military forces. (S, TA.) [But this usage of the verb is perhaps post-classical: for Mtr says,] With respect to the saying of El-Abeewardee,

* شَآمِيَّةٌ تَسْتَجْمِعُ الشَّوْلَ حَرْجَفُ *

[*A north wind, cold and vehement, inviting to collect themselves together the she-camels whose milk has dried up, they having passed seven or eight months since bringing forth,* or *since pregnancy*], it seems that he has compared this verb with the generality of others of the same class, [and so derived the meaning in which he has here used it,] or that he heard it [in that sense] from the people of the cities, or towns, or villages, and cultivated lands. (Mgh.) — استجمع used intransitively is *syn. with* اجتمع, which see in two places, and تجمّع. (Msb, K.) — استجمع السَّيْلُ *The torrent collected itself together from every place.* (S, Mgh, K.) — استجمع الوَادِى *The valley flowed in every place thereof.* (TA.) [See also 4, last signification.] — اِسْتَجْمَعَتْ لَهُ أُمُورُهُ *His affairs,* or *circumstances, all combined in a manner pleasing to him.* (Mgh, K.) — استجمع الفَرَسُ جَرْيًا (S, Mgh, K) *The horse exerted all his force,* or *energy, in running:* (K, TA:) the last word is here in the accus. case as a specificative. (Mgh.) You say also, اِسْتَجْمَعُوا لَهُمْ, meaning *They exerted* [*all*] *their strength, force,* or *energy, for fighting them:* and hence, إِنَّ النَّاسَ قَدْ جَمَعُوا ↓ لَكُمْ [app. meaning *Verily the men,* or *people, have exerted all their strength for fighting you*]. (A, TA.) — استجمع القَوْمُ *The people,* or *company of men, all went away, not one of them remaining;* like as one says of a valley flowing in every place thereof. (TA.) — استجمع البَقْلُ *The herbs,* or *leguminous plants, all dried up.* (TA.)

جَمْعٌ inf. n. of 1. (S, &c.) [Hence,] يَوْمُ الجَمْعِ *The day of resurrection* [when all mankind will be collected together]. (IDrd, K.) — Also, without the article ال, *A name of El-Muzdelifeh* [between 'Arafát and Minè]; (S, Mgh, Msb, K;) determinate, like عَرَفَاتُ: (TA:) so called because people collect themselves there; (S, Msb;) or because Adam there met with Eve (Mgh, Msb) after they had fallen [from Paradise]: (TA:) [or, app., *a name of the tract from 'Arafát to Minè inclusive of these two places:* and hence,] يَوْمُ جَمْعٍ *the day of 'Arafeh* [when the pilgrims halt at Mount 'Arafát]: and أَيَّامُ جَمْعٍ *the days of Minè.* (IDrd, K.) — As an inf. n. used as a subst., properly so termed, (S,* Mgh, Msb,) it also signifies *A collection; a number together; an assembly; a company, troop, congregated* or *collective body, party,* or *group; a mass;* syn. ↓جَمَاعَةٌ, (S, Mgh, L, Msb, K,) *of men;* (S, L, K;) as also ↓مَجْمَعٌ (L, Msb, TA) and ↓مَجْمَعَةٌ (Msb) and ↓مَجْمَعَةٌ (L, TA) and ↓جَمِيعٌ: (O, K:) but ↓جَمَاعَةٌ is also used as signifying *a collection, a number together,* or *an assemblage,* of other things than men; [of beasts, as camels, horses and the like, bulls and cows, and antelopes, gazelles, &c., i. e. *a herd, troop,* or *drove;* of dogs, i. e. *a pack;* of sheep and goats, i. e. *a flock;* of birds, i. e. *a flock* or *bevy;* of bees, and locusts, &c., i. e. *a swarm;*] and even of trees, and of plants; (L, TA;) it signifies *a collection,* or *an assemblage,* or *aggregate,* of any things, *consisting of many* and *of few;* (Msb;) [as also ↓مَجْمَعٌ and ↓مَجْمُوعٌ;] *a number, a plurality,* and *a multitude,* of any things: (TA:) the pl. of جَمْعٌ is جُمُوعٌ. (S, Mgh, Msb, K.) — And particularly, *An army; a military force;* (TA;) as also ↓جَمِيعٌ. (S, K.) Whence the phrase, in a trad., لَهُ سَهْمُ جَمْعٍ, [or, more probably, سَهْمُ جَمْعٍ,] meaning *For him is,* or shall be, the like *of an army's share* of the spoil. (TA.) — Also The *plural* of a thing [or word; i. e. *a proper plural,* according to the grammarians; and also applied by the lexicologists to *a quasi-plural noun,* which the grammarians distinguish by the terms جَمْعٌ لُغَوِىٌّ and اِسْمُ جَمْعٍ]; and so ↓جِمَاعٌ, (S, K,) and ↓جَمِيعٌ, except that this last is what is termed اِسْمٌ لَازِمٌ [app. meaning a subst. which does not govern another as its complement in the gen. case like as جَمْعٌ and جِمَاعٌ do, being thus likened to what is termed فِعْلٌ لَازِمٌ, i. e. an intransitive verb; so that you say of الخِبَآءُ, for instance, الجَمِيعُ الأَخْبِيَةُ *the plural is* الاخبية; for in this manner I always find it used when it has this signification, which is frequently the case in several of the older lexicons, and in some others; not جَمِيعُ الخِبَآءِ الأَخْبِيَةُ]; (TA;) [whereas] you say, [جَمْعُ الخِبَآءِ الأَخْبِيَةُ and] ↓جِمَاعُ الخِبَآءِ الأَخْبِيَةُ, (S, K,) i. e. *the* جَمْع [or *plural*] *of* الخباء *is* الاخبية; (K;) for ↓الجِمَاعُ is *what comprises a number* [of things]. (S, K.) See also this last word below. — And see also the next paragraph, in three places. — The *worst sort of dates;* (S, Mgh, Msb, K;) because they are *collected together and mixed,* (Mgh, Msb,) *from among the dates of fifty palm-trees:* (Mgh:) and afterwards, by predominant usage, [*any*] *bad dates:* (Mgh,* Mṣb:) or *a certain kind of dates* (Ḳ, TA) *mixed together, of several sorts, not in request, and not mixed but for their badness:* (TA:) or it signifies, (Mgh, Ḳ,) or signifies also, (Ṣ, Mṣb,) *palm-trees* (Aṣ, Ṣ, Mgh, Mṣb, Ḳ) *of any kind,* (Aṣ, Mgh, Mṣb,) *growing from the date-stones,* (Ṣ, Ḳ,) *of which the name is unknown.* (Aṣ, Ṣ, Mgh, Mṣb, Ḳ.) — *Red gum;* (Ibn-'Abbád, Ḳ;) [app. because *collected and mixed with gum of lighter colour.*] — The *milk of any camel having her udder bound with the* صِرَار [q. v.]; ([i. e. the *milk that collects in the udder so bound;*] that of any camel not having her udder bound therewith is called فُوَاق;) as also ▼ جَمِيعٌ. (Ḳ.)

جُمْعُ الكَفِّ, (TA, and EM p. 102,) and جَمْعُ الكَفِّ, (Ṣ, Mṣb, Ḳ,) and ▼ جِمْعُ الكَفِّ, (Mṣb, Ḳ, and so in the margin of a copy of the Ṣ, as mentioned in the TA,) and ▼ جِمْعُ الكَفِّ, (Mṣb,) *The fist; the hand clinched;* (Ṣ, Mṣb, Ḳ;) *the hand with the fingers put together and contracted in the palm:* (TA,* and EM ubi suprà:) pl. أَجْمَاعٌ. (Ḳ.) You say, ضَرَبْتُهُ بِجُمْعِ كَفِّى *I beat him,* or *struck him, with my fist.* (Ṣ, Mṣb.*) And ضَرَبُوهُ بِأَجْمَاعِهِمْ *They beat him,* or *struck him, with their* [*clinched*] *hands.* (TA.) And جَآءَ فُلَانٌ بِقُبْضَةٍ مِلْءَ جُمْعِهِ *Such a one came with a quantity in his grasp as much as filled his clinched hand.* (Ṣ, TA.) And جُمْعُ كَفٍّ signifies [also] *The quantity that a hand grasps,* of money &c. (Ḥam p. 778.) — أَخَذْتُ فُلَانًا بِجُمْعِ ثِيَابِهِ, (Ṣ, Mṣb,*) and ▼ بِجِمْعِ ثِيَابِهِ, (Mṣb,) i. e. [*I took,* or *seized, such a one*] *by the part where his garments met together.* (Mṣb.) — أَمْرُهُمْ بِجُمْعٍ, and ▼ بِجِمْعٍ, ‡ *Their affair,* or *case, is concealed,* (Ṣ, Ḳ,) *undivulged by them, and unknown by any one* [*beside them*]. (Ṣ, TA.) — ذَهَبَ الشَّهْرُ بِجُمْعٍ, and ▼ بِجِمْعٍ, *The month passed away wholly; all of it.* (Ḳ, TA.) — هِىَ مِنْ زَوْجِهَا بِجُمْعٍ, (Ṣ, Mgh, Ḳ,) and ▼ بِجِمْعٍ, (Ṣ, Ḳ,) *She is as yet undevirginated,* or *undeflowered,* (Ṣ, Mgh, Ḳ,) *by her husband.* (Ṣ, Mgh.) And طُلِّقَتْ بِجُمْعٍ, or ▼ بِجِمْعٍ, *She was divorced being yet a virgin.* (TA.) And مَاتَتْ بِجُمْعٍ, (Ṣ, Mgh, Mṣb, Ḳ,) and ▼ بِجَمْعٍ, (Ks, Ṣ, Mṣb, Ḳ,) and ▼ بِجِمْعٍ, (Ḳ,) *She died a virgin:* (Mgh, Mṣb, Ḳ:) or it signifies, (Ṣ, Ḳ,) or signifies also, (Mgh, Mṣb,) *she died being with child;* (AZ, Ṣ, Mgh, Mṣb, Ḳ;) *whether suffering the pains of parturition or not:* (AZ:) or *heavy with child:* (Ḳ:) occurring in the first sense, (Mgh, TA,) or, as some say, in the last, (TA,) in a trad., in which it is said that a woman who so dies is a martyr: (Mgh, TA:) it properly signifies *she died with something comprised in her, not separated from her, whether it were a burden in the womb, or her maidenhead:* (Ṣgh:) [the pl. is أَجْمَاعٌ; for] you say, مَاتَتِ النِّسَآءُ بِأَجْمَاعٍ *The women died* [*being virgins:* or] *being with child.* (AZ.) You say also, نَاقَةٌ جُمْعٌ *A she-camel with young.* (TA.) And ▼ اِمْرَأَةٌ جَامِعٌ *A woman with child.* (TA.)

جِمْعٌ: see the next preceding paragraph, in six places.

جُمَعُ pl. of جَمْعَآءُ, fem. of أَجْمَعُ [q. v.].

جُمْعَةٌ is [a subst.] from الاِجْتِمَاعُ, like as [its contr.] فُرْقَةٌ is [a subst.] from الاِفْتِرَاقُ: (Mgh:) and signifies *A state of union, agreement, congruity,* or *congregation:* or *sociableness, socialness, familiarity, companionableness, companionship, fellowship, friendship,* and *amity:* syn. أُلْفَةٌ: as in the saying, أَدَامَ ٱللّٰهُ جُمْعَةَ مَا بَيْنَكُمَا [*May God make permanent the state of union,* &c., *subsisting between you two*]. (Aboo-Sa'eed, Ḳ.) — Hence, (Mgh,) يَوْمُ الجُمُعَةِ, (Ṣ, Mgh, Mṣb, Ḳ,) the original form, (TA,) of the dial. of 'Okeyl; (Mṣb, TA;) and يَوْمُ الجُمْعَةِ, (Ṣ, Mṣb, Ḳ,) the most chaste form, (TA,) of the dial. of El-Ḥijáz; (Mṣb, TA;) and يَوْمُ الجُمَعَةِ, (Mṣb, Ḳ,) of the dial. of Benoo-Temeem; (Mṣb, TA;) and, in consequence of frequency of usage, الجُمْعَةُ alone; (Mgh;) *A well-known day;* (Ḳ;) [*the day of the congregation;* i. e. *Friday;*] formerly called (TA) *the day of* العَرُوبَةُ: (Ṣ, TA:) called يوم الجمعة because of the congregating of the people thereon: (Mṣb:) Th asserts that the first who named it thus was Kaab Ibn-Lu-eí; and he is related to have said that it was thus called because Ḳureysh used to gather themselves together to Ḳuṣeí, [on that day,] in [the building called] دَارُ النَّدْوَةِ: (TA:) accord. to the R, Kaab Ibn-Lu-eí was the first who collected a congregation on the day of العروبة, which was not called الجمعة save since the coming of El-Islám; [or it was not generally thus called before El-Islám; for it is added,] and he was the first who named it الجمعة; for Ḳureysh used to congregate to him on this day, and he used to preach to them, and to put them in mind of the mission of the apostle of God, informing them that he should be of his descendants, and bidding them to follow him and to believe in him: (TA:) or, as some say, it was thus called in the time of El-Islám because of their congregating [thereon] in the mosque: accord. to a trad., the Anṣár named it thus, because of their congregating thereon: (TA:) or it was thus named because God collected thereon the materials of which Adam was created: (I 'Ab:) those who say الجُمَعَةُ regard it as an epithet, meaning that this day collects men much; comparing it to هُمَزَةٌ and لُمَزَةٌ and ضُحَكَةٌ: (TA:) the pl. is جُمَعٌ (Ṣ, Mgh, Mṣb, Ḳ) and جُمُعَاتٌ (Mṣb, Ḳ) and جُمْعَاتٌ (Ṣ, Mgh, Mṣb, Ḳ) and جُمَعَاتٌ; (Mṣb, Ḳ;) of which the last is pl. of جُمَعَةٌ, [as well as of جُمْعَةٌ, accord. to analogy,] but not so جُمَعٌ (AHát) [nor either of the other pls. mentioned above]. — In like manner you say صَلَاةُ الجُمُعَةِ [*The prayer of Friday*], and, in consequence of the frequency of usage, الجُمُعَةُ alone. (Mgh.) — الجُمْعَةُ, with the م quiescent, is also a name for [*The week;* i. e.] *the days of the week* [*collectively*]; of which the Arabs are said, by IAṣr, to have reckoned the Sabbath (السَّبْت [i. e. Saturday]) as the first, though they called Sunday the first of the days. (Mṣb.) — جُمْعَةٌ is also *syn. with* مَجْمُوعَةٌ [meaning Things *collected together;* or *a collection* of things]; (Ḳ;) as in the phrase جُمْعَةٌ مِنْ حَصًى [*a collection of pebbles*]. (TA.) — You say also جُمْعَةٌ مِنْ تَمْرٍ, meaning *A handful of dates.* (Ṣ, Ḳ.)

[جَمْعِىٌّ *Of,* or *relating to, a plural.*]

جُمْعِىٌّ One *who fasts on Friday by himself.* (IAṣr, Th.)

جِمَاعٌ: see جَمْعٌ as signifying "a plural," in three places. [The primary signification seems to be the last there mentioned; where it is said,] الجِمَاعُ is *What comprises a number* [of things]: (Ṣ, Ḳ:) one says, الخَمْرُ جِمَاعُ الإِثْمِ (Ṣ, TA) [i. e. *Wine is what comprises a number of sins:* or] *that in which sin is comprised, and known to be:* the saying is a trad.: (TA:) or جِمَاعُ الإِثْمِ signifies *the plurality* (جَمْع) *of sins.* (Mṣb.) Hence also the saying of El-Ḥasan El-Baṣree, اِتَّقُوا هٰذِهِ الأَهْوَآءَ فَإِنَّ جِمَاعَهَا الضَّلَالَةُ وَمَعَادَهَا النَّارُ [*Beware ye of these natural desires; for what they involve is error, and the place to which they lead is the fire of Hell*]. (TA: in the L, وميعادها.) And it is said in a trad., حَدِّثْنِى بِكَلِمَةٍ تَكُونُ جِمَاعًا, i. e. *Tell me a saying comprising* [virtually] *a plurality of sayings.* (TA.) [See a similar phrase below, voce جَامِعٌ.] — [Hence also,] بُرْمَةٌ جِمَاعٌ *A stone-cooking-pot of the largest size:* (Ks, L:) or قِدْرٌ جِمَاعٌ, and ▼ جَامِعَةٌ, (Ṣ, Ḳ, TA,) *a cooking-pot that comprises a slaughtered camel;* or, accord. to the A, *that comprises a sheep* or *goat:* (TA:) or *a great cooking-pot;* (Ṣ, Ḳ;) as also ▼ جَامِعٌ: (Ṣgh, Ḳ:) pl. [most probably of this last] جُمْعٌ [like as بُزْلٌ is pl. of بَازِلٌ, &c.]. (Ḳ.) — You say also, فُلَانٌ جِمَاعٌ لِبَنِى فُلَانٍ *Such a one is an object of resort for his counsel and authority to the sons of such a one.* (TA.) = [See also 3.]

جَمُوعٌ: see جِمَاعٌ.

جَمِيعٌ *In a state of collection, congregation,* or *union; being together; met together;* [as also ▼ مُجْتَمِعٌ;] *contr. of* مُتَفَرِّقٌ. (Ṣ, Ḳ.) You say قَوْمٌ جَمِيعٌ *A people,* or *number of men, in a state of collection,* &c.; *being together; met together;* syn. ▼ مُجْتَمِعُونَ: (TA:) and in like manner, ▼ إِبِلٌ جَمَاعَةٌ *Camels in a state of collection;* &c. (TA.) — [*All,* or *the whole,* of any things or thing.] See أَجْمَعُ, last sentence. — [As an epithet in which the quality of a subst. is predominant,] *A tribe* [or *any number of men*] *in a state of collection, congregation,* or *union; being together; met together;* syn. ▼ حَىٌّ مُجْتَمِعٌ. (Ṣ, Ḳ.) See also جَمْعٌ, in four places. — A man *compact,* or *compressed,* or *contracted, in make,* or *frame:* (مُجْتَمِعُ ▼ الخَلْقِ;) *strong; who has not become decrepit nor infirm.* (TA.) — رَجُلٌ جَمِيعُ اللَّأْمَةِ *A man having his arms,* or *weapons, collected together.* (TA.) — رَجُلٌ جَمِيعُ الرَّأْىِ, and ▼ مُجْتَمِعُهُ, *A man of right, not disordered* or *unsettled, opinion,* or *judgment,* or *counsel.* (TA.) — جَعَلَ

الأَمْرَ جَمِيعًا بَعْدَ تَفَرُّقِهِ (AHeyth, Ḳ) *He determined, resolved,* or *decided, upon the affair, so as to make it firmly settled,* [*after it had been unsettled in his mind,* or] *after considering what might be its issues, or results, and saying at one time, I will do thus, and at another time, I will do thus.* (AHeyth.)

جَمَاعَةٌ : see جَمْعٌ, in two places.

جَمَّاعٌ and ↓ مِجْمَعٌ [are mentioned together, but not explained, in the TA: the former signifies, and probably, judging from analogy, the latter likewise, as also ↓ جَمُوعٌ, *One who collects much; or who collects many things*]. — إِبِلٌ جَمَّاعَةٌ : see جَمِيعٌ.

جُمَّاعٌ *Anything of which the several component parts are collected, brought, gathered,* or *drawn, together.* (IDrd, Ḳ.) — [Hence,] as an epithet, applied to a woman, it means *Short.* (TA.) — [Hence also,] جُمَّاعُ الثُّرَيَّا *The cluster of the Pleiades:* (IDrd:) or *persons who collect together for the rain of the Pleiades, which is the rain called* الوَسْمِىّ, *looking for the fruitfulness and herbage resulting from it.* (IAar.) — And جُمَّاعُ النَّاسِ *A medley,* or *mixed* or *promiscuous multitude* or *collection, of men,* or *people,* (Ṣ, Mṣb, Ḳ,) *of various tribes;* (Ṣ, Ḳ;) as also جُمَّاعٌ alone: (TA:) or the latter, *people scattered,* or *in a state of dispersion.* (Ḥam p. 302.) — جُمَّاعٌ also signifies The *place* [either properly or tropically] *which comprises the origin* of anything; (Ḳ, TA;) the *source of descent* or *extraction* of people; and hence applied by I'Ab to main tribes from which other tribes are derived; or, as some say, used by him as meaning *various classes of men, such as are termed* أَوْزَاع *and* أَوْشَاب. (TA.) — [And The *main,* or *most essential, part* of a thing. Thus,] جُمَّاعُ جَسَدِ الإِنْسَانِ means *The head of the man.* (TA.) — جُمَّاعُ الثَّمَرِ *The contraction* (تَجَمُّع) *of the envelopes of the flowers of dates, in one place, upon* [*the germs of*] *the fruit, or produce, thereof.* (TA.)

جَامِعٌ [act. part. n. of 1; *Collecting;* &c.] — الجَامِعُ one of the names of God; meaning *The Collector of the created beings for the day of reckoning:* or, as some say, *the Combiner of things of similar natures and of things of contrary natures, in existence.* (IAth.) — *The belly;* [because it collects what passes from the stomach;] of the dial. of El-Yemen. (TA.) — Also, (Mṣb,) or المَسْجِدُ الجَامِعُ, (Ṣ, Ḳ,) [*The congregational mosque;*] *the mosque in which the* [*congregational*] *prayers of Friday are performed;* because it collects the people for a certain time; (Mṣb;) and you may also say, مَسْجِدُ الجَامِعِ, meaning مَسْجِدُ اليَوْمِ الجَامِعِ, (Ṣ, Ḳ,) like as you say الحَقُّ اليَقِينُ and حَقُّ اليَقِينِ, [the latter] as meaning حَقُّ الشَّىْءِ اليَقِينِ ; for it is not allowable to prefix a noun to another of the same meaning except with this kind of subaudition; or, accord. to Fr, the Arabs used to do so because of the difference of the two words themselves: (Ṣ:) or مسجد الجامع is a mistake: (Ḳ:) so says Lth; but all others allow it; for the Arabs prefix a subst. to another signifying the same thing, and also to its epithet, as in the phrases in the Ḳur دِينُ القَيِّمَةِ [ch. xcviii. v. 4] and وَعْدَ الصِّدْقِ [ch. xlvi. v. 15]: (Az, TA:) [pl. جَوَامِعُ.] — مِصْرٌ جَامِعٌ [*A great town comprising a large population; a comprehensive great town*]. (Mṣb in art. مدن [where it is given as the explanation of مَدِينَةٌ]; and Ḳ in art. قرى [where it is less properly given as the explanation of قَرْيَةٌ].) — قِدْرٌ جَامِعٌ and جَامِعَةٌ : see جِمَاعٌ. — اِمْرَأَةٌ جَامِعٌ : see the paragraph commencing with الجَمْعُ ; last signification. — أَتَانٌ جَامِعٌ *A she-ass pregnant when beginning to be so.* (Ṣ, O, Ḳ.) — ↓ جَامِعَةٌ *A* [*collar of the kind called*] غُلّ ; (Ṣ, Ḳ;) because it collects together the two hands to the neck: (Ṣ:) pl. جَوَامِعُ. (TA.) — أَمْرٌ جَامِعٌ *An affair that collects people together:* or, as Er-Rághib says, *a momentous affair, on account of which people collect themselves together;* as though the affair itself collected them. (TA.) [Similar to this is the saying,] الصَّلَاةُ جَامِعَةٌ لِكُلِّ النَّاسِ *Prayer is a collector of all people.* (Mṣb.) — It is said of Mohammad, (Mṣb,) كَانَ يَتَكَلَّمُ بِجَوَامِعِ الكَلِمِ *He used to speak comprehensive but concise language; language conveying many meanings in few words.* (Mṣb, Ḳ. [In the CḲ, الكلم is omitted.]) And hence the saying of 'Omar Ibn-'Abd-el-'Azeez, عَجِبْتُ لِمَنْ لَاحَنَ النَّاسَ كَيْفَ لَا يَعْرِفُ جَوَامِعَ الكَلِمِ, meaning [*I wonder at him who vies with men in endeavouring to show his superiority of intelligence,*] *how it is that he does not* [*know the way to*] *confine himself to conciseness, and abstain from superfluity, of speech.* (TA.) In like manner, (TA,) it is said in a trad., أُوتِيتُ جَوَامِعَ الكَلِمِ, meaning *I have had communicated to me the Ḳur-án,* (Ḳ, TA,) *in which many meanings are comprised in a few words.* (TA.) الجَوَامِعُ مِنَ الدُّعَاءِ, also, signifies *Prayers,* or *supplications, combining petitions for good and right objects of desire with praise of God and with the general prescribed observances proper to the case.* (TA.) You say also, حَمِدْتُ اللّٰهَ بِمَجَامِعِ ↓ المَحَامِدِ *I praised God with words comprising various forms of praise.* (Mṣb.) [See also جِمَاعٌ.] — رَجُلٌ جَامِعٌ *A man who combines such qualities that he is suited to hardship and to easiness of circumstances.* (Aṣ, T in art. ادم.) And رَجُلٌ جَامِعٌ لِلْخَيْرِ (T and M and Ḳ in art. ام) *A man combining all kinds of good qualities.* (TḲ in that art.) — دَابَّةٌ جَامِعٌ *A beast fit for the* إِكَاف *and the* سَرْج [i. e. *for the saddle of either of the kinds thus called*]. (Ṣgh, Ḳ.) — جَمَلٌ جَامِعٌ, and نَاقَةٌ جَامِعَةٌ, (Ḳ,) accord. to ISh, (TA,) *A he-camel, and a she-camel, that fails of putting forth the tooth called* ناب *at the time expected;* expl. by أَخْلَفَا بُزُولًا : but this is not said except after four years: (Ḳ:) so in the copies of the Ḳ; but correctly, accord. to the O and TṢ, this is not said after four years, [app. reckoned from the usual time of بزول, for this is in the ninth year, or, sometimes, in the eighth,] without the exceptive particle. (TA.)

جَامِعَةٌ used as a subst.: see the next preceding paragraph.

أَجْمَعُ [*Collecting, comprising,* or *containing, a greater,* or *the greatest, number* or *quantity; more,* or *most, comprehensive.* Of its usage in a superlative sense, the following are exs.]. إِذَا أَخَذَ شَاهِدَ زُورٍ بَعَثَ بِهِ إِلَى السُّوقِ أَجْمَعَ مَا كَانَ [*When he took a false witness, he sent him to the market when it comprised,* or *contained, the greatest number* of people]: اجمع being here in the accus. case as a denotative of state with respect to the سوق : and the reason why كانت is not here said [instead of كان] is that سوق is sometimes masc. (Mgh.) And اِفْعَلْ مَا هُوَ أَجْمَعُ لِأُصُولِ الأَحْكَامِ [*Do thou that which is most comprehensive in relation to the principles of the ordinances* applying to the case]. (Mṣb in art. حوط.) = [As a simple epithet, *Entire, complete,* or *whole:* fem. جَمْعَآءُ. You say,] بَهِيمَةٌ جَمْعَآءُ *A beast free from defects, entire in all its limbs* or *members, without mutilation, and without cauterization;* (TA;) *a beast from the body of which nothing has gone.* (Ṣ, Ḳ.) — نَاقَةٌ جَمْعَآءُ [may sometimes have the like meaning: or,] accord. to IAar, (TA,) *A she-camel extremely aged,* (Ḳ, TA,) *so that her teeth have become short, and almost gone.* (TA.) = It is also a sing. having the meaning of a pl., (Ṣ, Ḳ,) without any proper sing. of its own: (Ṣ:) its pl. is أَجْمَعُونَ : and its fem. is جَمْعَآءُ : (Ṣ, Ḳ:) and the pl. of this last is جُمَعُ, though by rule it should be formed by the addition of ا and ت to the sing., like as the pl. of أَجْمَعُ is formed by the addition of و and ن; (Ṣ;) the original form from which جُمَعُ is changed being جَمْعَاوَاتٌ ; or it is جَمَاعَى ; it is not جُمْعٌ, because أَجْمَعُ is not an epithet, like as أَحْمَرُ is, of which the pl. is حُمْرٌ ; (L;) for it is determinate, though of the measure of an epithet, which is indeterminate; (AAF;) and though it is in concordance with the noun which precedes it, like an epithet, it is shown to be not an epithet by its not having a broken pl.: (L:) it is a simple corroborative; (Ṣ, Ḳ;) and so are أَجْمَعُونَ and جَمْعَآءُ and جُمَعُ ; not used as an inchoative nor as an enunciative nor as the agent of a verb nor as the objective complement of a verb, like as are some other corroboratives, such as نَفْسُهُ and عَيْنُهُ and كُلُّهُ. (Ṣ.) You say, أَخَذْتُ حَقِّى أَجْمَعَ [*I took my right, or due, all of it,* or *altogether*]: and رَأَيْتُ النِّسْوَةَ جُمَعَ [*I saw the women, all of them,* or *all together*]: the last word in this and similar cases being imperfectly declinable, and determinate, because it is a corroborative of a determinate word: (Ṣ, TA:) and جَاؤُوا أَجْمَعُونَ [*They came, all of them,* or *all together*]: and رَأَيْتُهُمْ أَجْمَعِينَ [*I saw them, all of them,* or *all together*]: and مَرَرْتُ بِهِمْ أَجْمَعِينَ [*I passed by them, all of them,* or *all together*]. (Mṣb.) Fr mentions the phrases, أَعْجَبَنِى القَصْرُ أَجْمَعُ [*The palace pleased me, all of it,* or *altogether*], and الدَّارُ جَمْعَآءُ [*The house, all of it*, or *altogether*], with the accus. case, as denotative of state; but does not allow أَجْمَعُونَ nor جُمَعُ to be used otherwise than as corroboratives: IDrst, however, allows أَجْمَعِينَ to be used as a denotative of state; and this is correct; and accord. to both these ways is related the trad., أَجْمَعُونَ and فَصَلُّوا جُلُوسًا أَجْمَعِينَ [*And pray ye sitting, all of you*, or *all together*]; though some make اجمعين [here] to be a corroborative of a pronoun understood in the accus. case, as though the speaker said, أَعْنِيكُمْ أَجْمَعِينَ [*I mean you, all of you*, or *all together*]: (Ḳ in art. بتع:) or اجمعين in this case is a corruption committed by the relaters in the first age; and he is in error who says that it is in the accus. case as a denotative of state, for corroboratives are determinate, and the denotative of state is literally or virtually indeterminate. (Mṣb.) [Respecting the usage of this corroborative together with others similar to it, see أَبْتَعُ.] You say also, جَاؤُوا بِأَجْمَعِهِمْ, and بِأَجْمُعِهِمْ, with ḍamm to the م, [*They came, all of them*, or *all together*,] (Ṣ, Mṣb, Ḳ,) the latter mentioned by ISk. (Mṣb.) And you say, قَبَضْتُ المَالَ أَجْمَعَهُ [*I took*, or *received, the property, all of it*, or *altogether*]. (Mṣb.) And ▼ جَمِيعٌ, also, is used as a corroborative: (Ṣ, Mṣb:) as in the saying جَاؤُوا جَمِيعًا, meaning *They came, all of them*: (Ṣ:) and قَبَضْتُ المَالَ جَمِيعَهُ, like أَجْمَعَهُ [explained above]: (Mṣb:) and جَمِيعَةٌ occurs as its fem.; but this is extr. (TA.)

مَجْمَعٌ and مَجْمِعٌ, (Ṣ, Mṣb, Ḳ,) the latter anomalous, like مَشْرِقٌ and مَغْرِبٌ &c., (TA,) *A place of collecting, and the like*: (Ṣ, Mṣb,* Ḳ:) [pl. مَجَامِعُ:] [Hence,] مَجْمَعُ البَحْرَيْنِ, in the Ḳur [xviii. 59], means *The place where the two seas meet*. (Bḍ.) And in like manner, where it is said in a trad., فَضَرَبَ بِيَدِهِ مَجْمَعَ بَيْنَ عُنُقِى وَكَتِفِى, [in which مَا seems to have been dropped by the copyist between مجمع and بين,] the meaning is, [*And he struck with his hand*] *the place where my neck and my shoulder-blade meet*. (TA.) [Hence also the phrase مَجَامِعُ المَحَامِدِ, explained above: see جَامِعٌ, near the end of the paragraph. And مَجَامِعُ الأُمُورِ, meaning *The concurrences of affairs*, or *of circumstances*, or *of events*.] — *A place in which people collect, assemble*, or *congregate*: (Mṣb,* TA:) and [in like manner,] ▼ مَجْمَعَةٌ signifies *an assembly-room; a sitting room in which people assemble*: (TA:) [pl. of both مَجَامِعُ.] You say, هٰذَا الكَلَامُ أَوْلَجُ فِى المَسَامِعِ وَأَجْوَلُ فِى المَجَامِعِ [*This language*, or *discourse, is more*, or *most, penetrating into the ears, and more*, or *most, circulating in the places of assembly*]. (TA.) — See also جُمْعٌ, as syn. with جَمَاعَةٌ, in two places; and see 10, first sentence. — [*The whole of anything*, considered as the place in which the several parts thereof are collected: see an instance voce خُفٌّ: and see also مُجْتَمَعٌ.]

أَمْرٌ مُجْمَعٌ, (Ṣ, Ḳ,) and مُجْمَعٌ عَلَيْهِ, (TA,) *An affair determined, resolved*, or *decided, upon*: (Ṣ, Ḳ:) *an affair agreed upon*. (TA.) [The former signification applies to both of the above-mentioned phrases: the latter signification, perhaps, only to the latter phrase.] — خُطْبَةٌ مُجْمَعَةٌ [*A discourse in rhyming prose*, or *the like*,] *in which is no flaw*, or *defect*. (Ibn-'Abbád, Ḳ.)

عَامٌ مُجْمِعٌ *A year of dearth, drought, sterility*, or *unfruitfulness*: (Ks, Ḳ:) because it is an occasion of people's collecting together in the place where herbage, or plenty, is found. (Ks.) And فَلَاةٌ مُجْمِعَةٌ, (Ṣ, TA,) like مُحْسِنَةٌ; (TA;) [in Gol. Lex., erroneously, مُجْمَعَةٌ;] and ▼ مُجَمِّعَةٌ, like مُحَدِّثَةٌ; (TA;) *A desert in which people collect themselves together, not separating themselves, from fear of losing their way, or perishing, and the like*; as though the desert itself collected them. (Ṣ, TA.) And أَرْضٌ مُجْمِعَةٌ, like مُحْسِنَةٌ, *A land of dearth, drought, sterility*, or *unfruitfulness, wherein the camels upon which people journey are not dispersed to pasture*. (TA.)

مُجَمَّعٌ: see جَمَّاعٌ.

مُجَمَّعَةٌ: see مَجْمَعٌ: — and جَمْعٌ, as syn. with جَمَاعَةٌ. — Also *Sands collected together*: (Ḳ:) pl. مَجَامِعُ. (TA.) And *A vacant*, or *void, land, destitute of herbage or vegetable produce, and of water*. (AA, Ḳ.)

مُجَمِّعٌ: see فَلَاةٌ مُجْمِعَةٌ.

مَجْمُوعٌ *Collected; brought*, or *gathered, together; gathered up; assembled; congregated; mustered; drawn together*; [or *contracted*;] (Ṣ, Ḳ, TA;) [*from several places*, or] *hence and thence, although not made as one thing*. (Ṣ, Ṣgh, L, Ḳ.) It is said in the Ḳur [xi. 105], ذٰلِكَ يَوْمٌ مَجْمُوعٌ لَهُ النَّاسُ *That is a day for which mankind shall be collected*. (TA.) — See also جَمْعٌ.

مُجْتَمَعٌ [*A place in which a thing becomes collected, brought together*, or *the like*; or *in which things have become so; where they collect themselves, come together*, or *unite*; or *in which they are comprised*, or *contained; a place in which is a collection of things*]. You say, البَيْضَةُ مُجْتَمَعُ الوَلَدِ [*The egg is that which comprises the young bird*]. (Mgh in art. بيض.) And مُجْتَمَعُ المَوْتِ signifies *the same as* حَوْضُ المَوْتِ, which see, in art. حوض. (TA in that art.) — [Also *The collective mass*, or *whole, of the hair of the head*: (see جُمَّةٌ, in three places:) مُجْتَمَعُ شَعْرِ الرَّأْسِ meaning *the whole head of hair*: see also مَجْمَعٌ.]

مُجْتَمِعٌ: see جَمِيعٌ, in five places. — *A man who has attained to his full state of manly vigour*, (Ṣ, Mgh, TA,) *and whose beard has become full-grown*: (TA:) because at that time his powers have become collected, or because his beard is then full-grown. (Mgh.) [See the verb, 8. And see an ex. in a verse of Suheym Ibn-Wetheel cited in art. دور, conj. 3.] — أَلْقَاهُ مُجْتَمِعًا [*He threw him down gathered together*, or *in a heap*]. (Ṣ and Mṣb and Ḳ in art. كور.) — مَشَى مُجْتَمِعًا *He walked quickly*, (Ḳ, TA,) *with vehemence of motion, and strength of limbs, not languidly*. (TA.)

مُتَجَمَّعُ البَيْدَاءِ *The main part of the desert; the part in which* [*as it were*] *it collects itself*; syn. مُعْظَمُهَا وَمُحْتَفَلُهَا. (TA.)

جمل

1. جَمَلَ, (Ḳ,) aor. ـُ, inf. n. جَمْلٌ, (TA,) *He collected* [a thing, or things]. (Ḳ.) [See also 4.] — Also, (Ṣ, Mgh, Ḳ,) aor. and inf. n. as above, (Ṣ, Mgh,) *He melted* fat; (Ṣ, Mgh, Ḳ;) and so ▼ اجتمل, and ▼ اجمل: (A'Obeyd, Ṣ, Ḳ:) this last was sometimes used: (Ṣ:) the best form is جَمَلَ: (Fr, TA:) accord. to Z, ▼ اجتمل signifies *he made the melted grease of fat to drip upon bread, putting it again over the fire*. (TA. [See also جَمِيلٌ.]) جَمَلَكَ اللّٰهُ, meaning *May God melt thee like as fat is melted*, is a form of imprecation mentioned in a trad., as used by a woman. (TA.) = جَمَلَ الجَمَلَ *He put the he-camel apart from the she-camel that was fit to be covered*. (TA.) = جَمُلَ, aor. ـُ; (Ṣ, Mgh, Mṣb, Ḳ;) and جَمِلَ, aor. ـَ; (Mṣb;) inf. n. جَمَالٌ, (Ṣ, Mgh, Mṣb, Ḳ,*) originally جَمَالَةٌ; (Mṣb;) *He was*, or *became, beautiful, goodly, comely*, or *pleasing*, (Ṣ, M, Mgh, Ḳ,) *in person*, (M, Ḳ,) and *good in action*, or *actions*, or *behaviour*, (M, TA,) or also *in moral character*: (Ḳ:) or *elegant*, or *pretty*; i. e., *delicately*, or *minutely, beautiful*: (Sb, Mṣb:) or *characterized by much goodness, beauty, goodliness, comeliness*, or *pleasingness, in his mind*, or *in his person*, or *in his actions* or *behaviour*; and also, *characterized by much goodness communicated from him to others*. (Er-Rághib, TA.) [See جَمَالٌ, below; and see also جَمِيلٌ.]

2. جمّل, (Ṣ, Ḳ,) inf. n. تَجْمِيلٌ, (Ḳ,) *He*, or *it, embellished*, or *adorned*, another. (Ṣ, Ḳ.) Hence the saying, إِذَا لَمْ يُجَمِّلْكَ مَالُكَ لَمْ يُجْدِ عَلَيْكَ جَمَالُكَ [*If thy wealth do not embellish thee, thy beauty of person, or of moral character, will not suffice thee*]. (TA.) And you say, جَمَّلَ اللّٰهُ عَلَيْهِ, inf. n. as above, meaning, *May God render him beautiful*. (TA.) = *He gave a camel to be eaten*. (Ḳ in art. برقش.) = *He detained* an army *long* [*on the frontier of the enemy*]; (Ḳ, TA:) like جَمَّرَ [q. v.]. (TA.)

3. جامله, (Ḳ,) inf. n. مُجَامَلَةٌ, (Ṣ, TA,) *He coaxed him*, or *wheedled him, with comely behaviour* or *speech* (بِالجَمِيلِ), *not rendering him pure*, or *sincere, brotherly affection*: (ISd, Ḳ:) or *he associated with him in a good manner*: (Ḳ:) or *he treated him with comely behaviour*. (Ṣ, TA.) One says, عَلَيْكَ بِالمُدَارَاةِ وَالمُجَامَلَةِ [*Keep thou to blandishment and coaxing*, &c.]. (TA.)

4. اجمل *He collected* a thing (Mṣb, Ḳ) *without discrimination*, or *distinction*, (Mṣb,) or *from a state of separation*, or *dispersion*. (Ḳ.) [See also 1.] And أُجْمِلَ *It was collected into an aggregate*. (TA.) — *He reduced* a calculation *to its sum; summed it up*: (Ṣ, Ḳ, TA:) and in like manner, *he summed up* a speech, or discourse, *and then analyzed and explained it*. (TA.) —

See also 1. — He made good and large [or liberal]: so in the phrase, اجمل الصَّنِيعَةَ (Ṣ, Ḳ) He made the benefit good and large [or liberal] (Ḳ) عِنْدَ فُلَانٍ [to such a one]. (Ṣ.) — [He acted with goodness, or was good and liberal: and he acted with moderation, or was moderate. You say,] اجمل فِى صَنِيعِهِ [He was good and liberal, or, perhaps, moderate, in his benefit]. (Ṣ.) And اجمل فِى الطَّلَبِ He was moderate, not extravagant, in demanding, or desire. (Mṣb,* Ḳ, TA.) It is said in a trad., أَجْمِلُوا فِى طَلَبِ الرِّزْقِ فَإِنَّ كُلًّا مُيَسَّرٌ لِمَا خُلِقَ لَهُ [Be ye moderate in demanding, or desiring, the means of subsistence, for every one is accommodated to that which is created for him]. (TA.) — اجمل القَوْمُ The people, or company of men, had many camels; or their camels became many. (Ṣ.)

5. تجمّل He beautified, embellished, or adorned, himself. (Ḳ.) — He affected what is جَمِيل [or beautiful, goodly, comely, or pleasing, in person, or in action or actions or behaviour, or in moral character, &c.]. (Ṣ.) You say, تجمّل بِأَكْثَرَ مِمَّا عِنْدَهُ [He affected beautiful, goodly, comely, or pleasing, qualities, more than he possessed]. (TA in art. شبع.) — He was, or became, patient; or restrained himself from impatience; or constrained himself to be patient: (Mgh, TA:) from جَمَال meaning "patience." (Mgh.) Hence the saying,

• وَإِذَا تُصِبْكَ خَصَاصَةٌ فَتَجَمَّلِ •

And when poverty, or straitness, befalls thee, then be patient, or restrain thyself &c. (Mgh in art. خص.) — He ate what is termed جَمِيل, i. e., melted fat. (Ṣ, Ḳ.*)

8. اجتمل: see 1, in two places. — Also He anointed himself with fat. (TA.) — And He ate of a camel. (Ḳ in art. برقش.)

10. استجمل He (a camel) became a جَمَل, (Ṣ, Ḳ,) i. e., such as is termed رَبَاع [or one in his seventh year], (Ṣ,) or such as is termed بَازِل [or one in his ninth year], or, accord. to Z, one that had covered. (TA.)

جَمْلٌ: see جَمَلٌ.

جُمْلٌ: see جُمْلَةٌ and جُمَّلٌ; the latter in two places.

جَمَلٌ (Ṣ, Mgh, Mṣb, Ḳ, &c.) and ↓ جَمْلٌ, (Ḳ,) which latter is so rare that it is said by some to be used only in poetry, in cases of necessity, (MF,) but it is a correct dial. var., (TA,) a word of well-known meaning; (Ḳ;) i. e., [A he-camel; but commonly applied to the camel as a generic term; in like manner as جَامِل is applied to the males and the females; but properly,] the male of the إِبِل; (TA;) the mate of the نَاقَة; (Fr, Ṣ, Mgh;) among camels, corresponding to رَجُل among us; (Sh, Mṣb;) نَاقَة corresponding to مَرْأَة, and بَكْر to غُلَام, and بَكْرَة to جَارِيَة; (Sh, TA;) [in general] peculiarly applied to the male; (Mṣb;) exceptionally to the female, as in the saying شَرِبْتُ لَبَنَ جَمَلِى, (Ḳ,) i. e., I drank the milk of my she-camel; but ISd doubts the correctness of this: (TA:) [as corresponding to رَجُل among us, it signifies a full-grown he-camel:] or it signifies such as is termed رَبَاع [or one in his seventh year]: (Ṣ, ISd, Ḳ:) or such as is termed جَذَع [or one in his fifth year]: (ISd, Ḳ:) or such as is termed بَازِل [or one in his ninth year]: (ISd, Mgh, Mṣb, Ḳ:) or such as is termed ثَنِىّ [or one in his sixth year]: (ISd, Ḳ:) or, accord. to Z, one that has covered: (TA:) [see also بَعِير, and بَكْر, and قَعُود:] pl. [of pauc.] أَجْمَال, (Ṣ, Mgh, Mṣb, Ḳ,) which may be pl. of جَمْل, (TA,) and أَجْمُل (Mṣb) and [of mult.] جِمَال (Ṣ, Mgh, Mṣb, Ḳ) and جُمْل (Ḳ) and جِمَالَة (Mgh, Mṣb, Ḳ) and [quasi-pl. n.] جُمَالَة and جِمَالَة and جَامِل, (Ḳ,) which last is disallowed by some, as will be seen below, (TA,) and [pl. pl.] جِمَالَات, (Ṣ, Mṣb, Ḳ,) which is pl. of جِمَال, (Mṣb, TA,) or it may be pl. of جِمَالَة, (TA,) and جُمَالَات [which see also voce جُمَّل] and (Ḳ) and جَمَائِل, (Ṣ, Ḳ,) pl. of جمالة and جمال, (Ḥam p. 527,) and أَجَامِل. (Ḳ.) One says of camels, when they are males, without any female among them, هٰذِهِ جِمَالَةُ بَنِى فُلَانٍ [These are the he-camels of the sons of such a one]. (ISk, Ṣ. [See also جُمَالَة.]) And they said also جِمَالَانِ [meaning Two herds of camels, thus forming a dual from the pl. جِمَال], like as they said لِقَاحَانِ. (ISd, in TA voce خَيْل.) It is said in a prov., مَا ٱسْتَتَرَ مَنْ قَادَ الجَمَلَ [He does not conceal himself who leads the he-camel]. (TA.) And in another prov., اِتَّخَذَ اللَّيْلَ جَمَلًا †He journeyed all the night. (Ḳ, TA. [See also Freytag's Arab. Prov. i. 230.]) — الجَمَلُ also signifies A certain fish (IAạr, Ḳ) of the sea, (IAạr, TA,) thirty cubits in length: (Ḳ:) or, as some say, جَمَلُ البَحْرِ is the name of a very great fish, also called the بَال, [i. e., the whale,] thirty cubits in length: accord. to some, this, (TA,) or جَمَلُ المَآءِ, (Mgh,) is what is called the كَوْسَج and كُبَع (Mgh, TA) and نُنْغُر, [i. e., xiphias, or sword-fish,] which passes by nothing without cutting it. (TA.) [In the present day, جَمَلُ البَحْرِ is an appellation of The pelican.] — عَيْنُ الجَمَلِ, in the dial. of Egypt, i. q. الشَّاهْ بَلُّوط [The chestnut]. (TA.) — جَمَل signifies also †A woman's husband. (L in arts. اخذ and قيد. See 2 in each of those arts.) — Also ‡ Palm-trees; (Ḳ;) as being likened to the he-camel in respect of their tallness and their bigness and their produce: in some of the copies of the Ḳ, النَّحْل is erroneously put for النَّخْل. (TA.) — See also جُمَّل.

جُمُلٌ: see جُمَّل, in three places.

جُمَّلٌ A company, or congregated body, of men. (ISd, Ḳ.) — See also جُمَّل.

جُمْلَةٌ A strand of a thick rope: pl. [or rather coll. gen. n.] ↓ جُمَل: or many strands of a rope, put together [to compose a cable: see جُمَّل]. (TA, in two places in this art.) — Hence, app., (TA,) The aggregate of a thing; (Ḳ;) the sum, whole, or total; (KL, PṢ;) it implies muchness, or numerousness, and means any aggregate unseparated: (Er-Rághib, TA:) pl. جُمَل. (Ṣ.) [جُمْلَةٌ مِنْ مَالٍ generally means A large sum of money; and in a similar sense جُمْلَة is often used in relation to various things.] It is said in the Ḳur [xxv. 34], وَقَالَ ٱلَّذِينَ كَفَرُوا لَوْ لَا نُزِّلَ عَلَيْهِ ٱلْقُرْآنُ جُمْلَةً وَاحِدَةً, i. e., [And those who disbelieved said, Wherefore was not the Ḳur-án sent down, or revealed, to him] aggregated? (TA:) [or in one aggregate?] or at once? (Bḍ.) [Hence, بِالجُمْلَةِ as meaning Upon the whole; to sum up.] — And hence, in grammar, (TA,) [A proposition; a clause; a phrase; sometimes, a sentence;] a phrase composed of a subject and an attribute, [i. e., composed of an inchoative and an enunciative, (in which case it is termed جُمْلَةٌ ٱسْمِيَّةٌ,) or of a verb and its agent, (in which case it is termed جُمْلَةٌ فِعْلِيَّةٌ,)] (KT, TA,) [&c.,] whether affording a complete sense, as زَيْدٌ قَائِمٌ [Zeyd is standing], or not, as إِنْ يُكْرِمْنِى [If he treat me with honour]. (KT.)

جَمَلَاةٌ: see جَمِيل.

جَمَلُونٌ A building, or structure, in the form of a camel's hump: (TA:) [a ridged roof: so in the present day: pl. جَمَالِين.]

جَمَالٌ inf. n. of جَمُلَ: (Ṣ, Mgh, Mṣb:) [when used as a simple subst., meaning] Beauty, goodliness, comeliness, or pleasingness, syn. حُسْن, (Ṣ, M, Mgh,* Ḳ,) in person, (M, Ḳ,) and goodness in action, or actions, or behaviour, (M, TA,) or also, in moral character: (Ḳ:) or elegance, or prettiness; i. e., delicacy, or minuteness, of beauty: (Sb, Mṣb:) or much goodness, or beauty or goodliness or comeliness, in the mind, or in the person, or in the actions or behaviour; and also, much goodness that is communicated from its possessor to another: (Er-Rághib, TA:) accord. to Aṣ, [when relating to the person,] حُسْن is in the eyes; and جَمَال, in the nose. (TA in art. حسن.) [See also جَمِيل.] One says, جَمَالَكَ أَنْ تَفْعَلَ كَذَا, (IDrd, TA,) or جَمَالَكَ لَا تَفْعَلْ كَذَا, (ISd, Ḳ,) meaning, Keep to that which is most comely for thee to do, and do not thus. (IDrd, ISd, Ḳ. [But see what follows.]) — Also Patience. (Mgh in art. خص.) Aboo-Dhu-eyb says,

• جَمَالَكَ أَيُّهَا القَلْبُ القَرِيحُ •
• سَتَلْقَى مَنْ تُحِبُّ فَتَسْتَرِيحُ •

(Ṣ,* TA, the former of which cites only the first hemistich, and the latter substitutes الجَرِيح for its syn. القَرِيح,) meaning, [Keep thy patience, O thou wounded heart: thou wilt find whom thou lovest, and be at rest: or] keep to thy patience, or thy constraint of thyself to be patient, and thy shrinking from what is foul, and be not impatient in an evil manner. (Ṣ, TA.)

جُمَال: see جَمِيل: — and جُمَالَة.

جَمُولٌ A piece of fat melted. (IAạr, TA.) [See also جَمِيل.] — A fat woman. (IAạr, Ḳ.) — A person, (Ḳ,) or woman, (M,) who melts fat. (M, Ḳ.)

جَمِيلٌ *Melted fat*: (Ṣ, Mgh:) or *melting fat*: or *fat that is melted and collected*: (Ḳ, TA:) or *fat that is melted, and, whenever it drips, made to drip upon bread, and then replaced over the fire* [*that it may drip again*: see جَمَلَ]: (TA:) and ▾ جُمَالَةٌ, also, signifies [the same; or] *melted grease*. (Mgh,* TA.) [See also جَمُولٌ.] ═ Hence, accord. to Abu-l-'Alà, because, when a man becomes fat and in good condition, his جَمَال becomes apparent, (Ḥam p. 155,) as also ▾ جُمَالٌ and ▾ جُمَّالٌ, (Ḳ,) or this last denotes a higher degree of beauty than جَمِيلٌ, (Ṣ, Ṣgh,) and has no broken pl., (TA,) and ▾ أَجْمَلُ, (TA,) *Beautiful, goodly, comely*, or *pleasing*, (Ṣ, M, Mgh, Ḳ,) *in person*, (M, Ḳ,) and *good in action*, or *actions*, or *behaviour*, (M, TA,) or also *in moral character*: (Ḳ:) [like the Greek καλὸς, the Latin *pulcher*, the French *beau*, &c.; and so حَسَنٌ:] or *elegant*, or *pretty*; i. e., *delicately*, or *minutely, beautiful*: (Mṣb:) [or *characterized by much goodness*, or *beauty* or *goodliness* or *comeliness, in his mind*, or *in his person*, or *in his actions* or *behaviour*; and also *characterized by much goodness communicated from him to others*: see جَمَالٌ:] pl. of the first جِمَالٌ: (TA:) fem. جَمِيلَةٌ, (Ṣ, Mgh, Mṣb, Ḳ,) applied to a woman; (Ṣ, Mgh, Mṣb;) as also ▾ جَمْلَآءُ, (Ks, Ṣ, Ḳ,) [said to be] an instance of [the measure] فَعْلَآءُ having no [masc. of the measure] أَفْعَلُ; (TA;) [but see above;] or this is applied to any female as signifying *perfect*, or *complete, in body*. (Ibn-'Abbád, Ḳ.) It is said in a trad., إِنَّ ٱللّٰهَ جَمِيلٌ يُحِبُّ ٱلْجَمَالَ *Verily God is comely in deeds*, (TA,) or *an Abundant Bestower of good things*: *He loveth those who are of the like character*. (Er-Rághib, TA.) And you say, عَامَلَهُ بِٱلْجَمِيلِ [*He treated him with comely*, or *pleasing, behaviour*]. (TA.) And مَاسَحَهُ بِٱلْجَمِيلِ [*He coaxed him*, or *wheedled him, with comely*, or *pleasing, behaviour* or *speech*]. (ISd, Ḳ. [See 3.]) — أَبُو جَمِيلٍ [*The kind of plants called*] ٱلْبَقْلُ; because they embellish by their presence, and render good, the seasoning of food; or because they take away the جَمِيل, i. e., the grease of the flesh-meat, and dry up the food. (Ḥar p. 227.)

جَمَالَةٌ: see the next paragraph.

جُمَالَةٌ: see جُمَّلٌ: ═ and جَمِيلٌ. ═ Also *A herd*, or *distinct number, of camels*; (Ḳ;) mentioned before as a pl. of جَمَلٌ [q. v.]: (TA:) or, *of she-camels among which is no he-camel*; as also ▾ جِمَالَةٌ and ▾ جِمَالَةٌ; (Ḳ;) but this is contradictory to a saying of ISk [respecting جِمَالَةٌ], mentioned above [voce جَمَلٌ; where all these three words are said to be pls. of جَمَلٌ]: (TA:) and also *horses*: pl. ▾ جُمَالٌ, which is extr. [as a pl.; though, in relation to جُمَالَةٌ, it may be a coll. gen. n., forming its n. un. with ة]. (AA, Ḳ.)

جِمَالَةٌ: see what next precedes.

جَمِيلَةٌ *A number of gazelles together*: and *of pigeons*. (Ibn-'Abbád, Ḳ.)

جُمَالِيٌّ applied to a man, (Ṣ, Mṣb, Ḳ,) *Large in make*: (Ṣ, Mṣb:) or *tall in body*: (Mṣb:) or *firm* [*in make*], (Ḳ,) or *big in limbs, complete in make*, (TA,) *like a he-camel*. (Ḳ, TA.) And with ة applied to a she-camel, (Ṣ, Ḳ,) *Resembling a he-camel in greatness of make*: (Ṣ:) or *firm* (Ḳ, TA) *in make*, (TA,) *like a he-camel* (Ḳ, TA) *in greatness of make and in strength*. (TA.)

جُمَّلٌ (Ṣ, Ḳ, &c.) and ▾ جُمَلٌ and ▾ جُمْلٌ (Ḳ) and ▾ جُمُلٌ and ▾ جَمَلٌ (IJ, Ḳ) [*A cable*;] *the rope of a ship*, (Ṣ, Ḳ,) i. e., the *thick rope thereof*, (TA,) *that is also called* قَلْسٌ, (Ṣ, TA,) *consisting of* [*a number of*] *ropes put together*: (Ṣ:) and ▾ جُمَالَةٌ also signifies [the same; or] *a thick rope, because consisting of many strands put together*; pl. جُمَالَاتٌ; (Zj, TA;) which Mujáhid explains as meaning the *ropes of bridges*; but I'Ab, as the *ropes of ships, put together so as to be like the waists of men* [*in thickness*]. (TA.) In all the forms mentioned above, except the last (جمالة), the word is read in the phrase [in the Ḳur vii. 38], حَتَّى يَلِجَ ٱلْجُمَّلُ فِى سَمِّ ٱلْخِيَاطِ [*Until the cable shall enter into the eye of the needle*]: (Ḳ, TA:) I'Ab reads ٱلْجُمَّلُ, (Ṣ, TA,) and so do 'Alee and many others: ▾ جُمَلٌ is pl. [or rather coll. gen. n.] of جُمْلَةٌ, *a strand* of a thick rope; or, accord. to IJ, pl. of جَمَلٌ [q. v.]: the first is explained by Fr as meaning *ropes put together*; but Aboo-Tálib thinks that he meant ▾ جُمَلٌ, without teshdeed. (TA.) ═ حِسَابُ ٱلْجُمَّلِ, (Ṣ, Ḳ,) thought by IDrd to be not Arabic, (TA,) and ▾ ٱلْجُمَلُ, (Ḳ,) but IDrd doubts its correctness, *The calculation by means of the letters* ا, ب, ج, د, *&c.* (TA.)

جَمَّالٌ *An owner*, or *an attendant, of a camel* or *camels*: (KL, TA:* [see also جَامِلٌ:]) and جَمَّالَةٌ *owners*, or *attendants, of camels*; (Ṣ, Ḳ, TA;) similar to خَيَّالَةٌ and حَمَّارَةٌ; (Ṣ, TA;) as the former is to حَمَّارٌ. (TA.) [See an ex. of the latter in a verse cited voce إِذَا.]

جُمَّالٌ: see جَمِيلٌ.

جَامِلٌ [act. part. n. of جَمَلَ. ═ And also part. n. of جَمُلَ]. The Arabs say, أَجْمِلْ إِنْ كُنْتَ جَامِلًا [*Become beautiful, &c., if thou be becoming beautiful, &c.*]: but when they mean the quality [alone], they say, إِنَّهُ لَجَمِيلٌ [*Verily he is beautiful, &c.*]. (Lḥ, TA.) ═ A man *possessing a* جَمَل [or *he-camel*]. (TA. [See also جَمَّالٌ.]) — *A herd*, or *distinct number, of camels*, (Ṣ, Ḳ,* TA,) *males and females*, (TA,) *with their pastors and their owners*: (Ṣ, Ḳ, TA: [also said in the Ḳ to be a pl. of جَمَلٌ: in the CḲ, الجامع is erroneously put for ٱلْجَامِل:]) or a word formed to denote a pl., meaning *camels*, (Ḥam pp. 122 and 490,) *males and females*; (Id p. 122;) derived from جَمَلٌ; (Id. p. 490;) like بَاقِرٌ (Id. ib. and TA) from بَقَرٌ, (Ḥam p. 490,) and كَالِبٌ [from كَلْبٌ]. (TA.) — Also *A great tribe*. (AHeyth, Ḳ.)

أَجْمَلُ [*More*, and *most*, جَمِيل, or *beautiful*, &c.]. (Ṣ, Ḳ.) — See also جَمِيلٌ.

مُجْمَلٌ [pass. part. n. of 4, q. v. — Also, applied to a phrase or the like,] properly, *Including*, or *implying, a number of things, many and unexplained*: (Er-Rághib, TA:) as used by the lawyers, [*confused*, or] *requiring explanation*. (TA.)

مُجَامِلٌ [act. part. n. of 3, q. v. — Also] *One who is unable to answer a question put to him by another person, and therefore neglects it, and bears malice against him for some time*. (TA.)

جمن

جُمَانٌ *Beads made of silver, like pearls*; (Ṣ;) *things in the form of pearls, of silver*; (Ḳ;) one of which is called جُمَانَةٌ, (Ṣ, Ḳ,) pl. جُمَانَاتٌ: (Ḥar p. 181:) or *pearls* (Ḳ, TA) themselves: (TA:) or the first is the proper meaning, and this is metaphorical: (EM p. 161:) [said to be] a Persian word, arabicized. (TA.) Also *A kind of belt* (سَفِيفَة) *woven of leather, in which are beads of every colour, worn by a woman as a* وِشَاح [q. v.]: or *silvered beads*. (Ḳ.)

جمهر

Q. 1. جَمْهَرَ, (T, A, Ḳ,) inf. n. جَمْهَرَةٌ, (A,) *He collected together* (T, A, Ḳ) a thing, (TA,) or earth, or dust, (T, A, TA,) *one part upon another*. (T, TA.) — جمهر ٱلْقَبْرَ, (Ṣ, Mṣb, Ḳ,) inf. n. as above, (Ṣ,) *He collected the earth*, or *dust, over the grave*, (Ṣ, Mṣb, Ḳ,) *not plastering it with clay*, or *mud*, (Ṣ, Ḳ,) *nor making it even*, or *level*. (TA.) — جمهر ٱلْمَتَاعَ *He took the main part of the household-goods*, or *commodities*: and in like manner the verb is used in relation to plants, or herbage. (Kitáb el-Aḍdád, TA.) — جمهر عَلَيْهِ ٱلْخَبَرَ, (Ks, Ṣ, Ḳ,) or لَهُ, (Lth, TA,) or إِلَيْهِ, (AZ, TA,) *He acquainted him with a part of the news*, or *story, and concealed what he desired*, or *meant*: (Ks, Ṣ, Ḳ:) or *he acquainted him with a part of the news*, or *story, incorrectly*, or *not in the proper manner, and omitted what he desired*, or *meant*: (Lth, TA:) or *he acquainted him with a small portion thereof, omitting most of what was necessary for him to know, and relating it in a manner different from the proper way*: (AZ, TA:) and accord. to the Kitáb el-Aḍdád of Abu-ṭ-Ṭeiyib the Lexicologist, it seems to have a contr. signification; for he says that جَمْهَرْتُ لَكَ ٱلْخَبَرَ means, *I acquainted thee with the main part of the news*, or *story*. (TA.)

Q. 2. تَجَمْهَرَ عَلَيْنَا *He held up his head with an assumption of superiority over us; domineered over us*; or *exalted himself above us*. (TA.)

جَمْهَرَةٌ [originally inf. n. of جَمْهَرَ]: see the next paragraph.

جُمْهُورٌ (not جُمْهُورٌ, which is a form of the word mentioned by Et-Tilimsánee, MF) *A quantity of sand rising above what is around it*, (Ṣ, Mṣb, Ḳ,) *and collected together*; (Ṣ;) as also ▾ with ة; (L;) so called from its abundance and height: (Mṣb:) or *a large quantity of sand, heaped up, and extensive*: (Lth, TA:) and ▾ with ة, *sand compacted together, and extending in an oblong form upon the surface of the earth*. (TA.) — The *generality*, or *main part*, of men, or people: (Ṣ, A, Ḳ:) and the *eminent, elevated*, or *noble*, of them: (TA:) and *a great number of people*:

(Mṣb:) pl. جَمَاهِيرُ; (A, Mṣb;) which signifies also *collective bodies* of men. (TA.) You say, هٰذَا قَوْلُ الجُمْهُورِ *This is the saying of the generality,* or *main part.* (A.)—The *generality;* the *greater, main,* or *chief, part;* the *main body, main, gross, mass,* or *bulk;* of anything; (Ḳ;) as also ▼جُمْهُورَةٌ. (W 95.)—Also, (Ḳ,) or ▼جُمْهُورَةٌ, (TA,) *A noble,* or *high-born, woman.* (Ḳ, TA.)

جُمْهُورَةٌ: see جُمْهُور, in three places.

جُمْهُورِىٌّ *An intoxicating beverage:* (AO, Ḳ:) or [*beverage of the kind called*] نَبِيذ *made of grapes, that is three years old:* (Ḳ:) or *i. q.* بُخْتَجٌ; (TA;) which is *expressed juice* [*of grapes*] *cooked* (Mgh voce بختج, and TA) *so as to be reduced to one third,* (Mgh,) *such as is lawful to be drunk:* (TA:) or the *beverage called* بختج *to which what has gone from it has been restored, and which is then cooked, and put into vessels, and becomes very potent:* (AḤn, and Mgh ubi suprà, and TA:) or *juice of grapes cooked until half of it is gone and half remains:* (KL:) called جمهورى because used by most men. (TA.)

جُمَاهِرٌ *Large, big, bulky,* or *corpulent.* (TA.)

مُجَمْهَرَةٌ A she-camel *compact in make;* (Ḳ;) as though she were a جُمْهُور of sand. (TA.)

جن

1. جَنَّهُ, (Ṣ, Mgh, Ḳ,) aor. ـُ, (Mgh, TA,) inf. n. جَنٌّ, (TA,) *It veiled, concealed, hid, covered,* or *protected, him;* (Ṣ, Mgh, Ḳ;) said of the night; (Ṣ, Ḳ;) as also جَنَّ عَلَيْهِ, (Ṣ, Mṣb, Ḳ,) aor. ـُ, (Ṣ, Mṣb,) inf. n. جُنُونٌ, (Ṣ,) or جَنٌّ, (Ḳ,) or both; (TA;) so in the Ḳur vi. 76, meaning *it veiled him, concealed him,* or *covered him, with its darkness;* (Bḍ;) and ▼اجنّهُ: (Ṣ, Mṣb, Ḳ:) or this last signifies *he,* or *it, made,* or *prepared, for him,* or *gave him, that which should veil him, conceal him, &c.:* accord. to Er-Rághib, the primary signification of جَنٌّ is the *veiling,* or *concealing, &c., from the sense.* (TA.) And جُنَّ عَنْهُ means *It* (anything) *was veiled, concealed,* or *hidden, from him.* (Ḳ.)—*He concealed it;* namely, a dead body; as also ▼اجنّهُ: (Ṣ, TA:) or the latter, *he wrapped it in grave-clothing:* (Ḳ:) and *he buried it.* (TA.) And أَجْنَنْتُ ▼ الشَّىْءَ فِى صَدْرِى *I concealed the thing in my bosom.* (Ṣ.) And أَجَنَّتْ ▼ وَلَدًا, (Ṣ,) or جَنِينًا, (Ḳ,) said of a woman, (Ṣ,) or a pregnant female, (Ḳ,) *She concealed* [or *enveloped* in her womb *a child,* or *an embryo,* or *a fœtus*]. (TA.)==جَنَّ, aor. ـِ, inf. n. جَنٌّ, *It* (an embryo, or a fœtus,) *was concealed in the womb.* (Ḳ.)—Also, [inf. n., probably, جَنٌّ and جُنُونٌ and جَنَانٌ, explained below,] *It* (the night) *was,* or *became, dark.* (Golius on the authority of Ibn-Maaroof.)==جُنَّ, (Ṣ, Mṣb, Ḳ,) inf. n. جُنُونٌ (Ṣ, Ḳ) and جِنَّةٌ (Ṣ) and جَنٌّ; (Ḳ;) and ▼اُسْتُجِنَّ, and ▼تَجَنَّنَ, and ▼تَجَانَّ; (Ḳ;) *He* (a man, Ṣ) *was,* or *became,* مَجْنُون [originally signifying *possessed by a* جِنِّىّ, or *by* جِنّ; *possessed by a devil* or *demon;* (see Bḍ li. 39;) and hence meaning *bereft of reason;* or *mad, insane, unsound in mind* or *intellect,* or *wanting therein:* the verbs may generally be rendered *he was,* or *became, possessed;* or *mad,* or *insane*]. (Ṣ, Mṣb, Ḳ.)—جَنَّ الذُّبَابُ, (Ṣ, A, TA,) inf. n. جُنُونٌ, (TA,) † *The flies made much buzzing:* (Ṣ:) or *made a gladsome buzzing* in a meadow. (A, TA.)—جُنَّ النَّبْتُ, inf. n. جُنُونٌ, ‡ *The herbage became tall, and tangled,* or *luxuriant,* or *abundant and dense, and put forth its flowers* or *blossoms:* (Ṣ, TA:) or *became thick and tall and full-grown, and blossomed.* (M, TA.) And جُنَّتِ الأَرْضُ, (Fr, Ḳ,) inf. n. جُنُونٌ, (Ḳ,) ‡ *The land produced pleasing herbage* or *plants:* (Fr, TA:) or *put forth its flowers and blossoms;* as also ▼تَجَنَّنَتْ. (Ḳ, TA.)

2: see 4.

4: see 1, in four places:==and see 8.==Also اجنّهُ *He* (God) *caused him to be,* or *become,* مَجْنُون [originally signifying *possessed by a* جِنِّىّ, or *by* جِنّ; and hence generally meaning *bereft of reason;* or *mad, insane, unsound in mind* or *intellect,* or *wanting therein*]. (Ṣ, Mṣb, Ḳ.) [And so, vulgarly, ▼جَنَّنَهُ, whoever, or whatever, be the agent.]—ما اجنّهُ [*How mad,* or *insane, &c., is he!*] is anomalous, (Th, Ṣ,) being formed from a verb of the pass. form, namely, جُنَّ; (Th, TA;) for of the مَضْرُوب one should not say, مَا أَضْرَبَهُ; nor of the مَسْلُول should one say, مَا أَسَلَّهُ: (Ṣ:) Sb says that the verb of wonder is used in this case because it denotes want of intellect [which admits of degrees]. (TA.)==اجنّ also signifies وَقَعَ فِى مَجَنَّةٍ [app. meaning *He fell into,* or *upon, a place containing,* or *abounding with,* جِنّ]. (TA.)

5: see 1, in two places.—تَجَنَّنَ عَلَيْهِ, and ▼تَجَانَنَ, (Ṣ, Ḳ,) and ▼تَجَانَّ, (Ṣ,) *He feigned himself* مَجْنُون [i. e. *possessed by a* جِنِّىّ, or *by* جِنّ; and hence generally meaning *bereft of reason;* or *mad, insane, &c.;*] *to him;* (Ṣ, Ḳ;) not being really so. (TA.)

6. تَجَانَّ and تَجَانَنَ: see 1:—and see also 5.

8. اجتنّ, (accord. to the Ṣ,) or ▼اُجِنَّ, (accord. to the Ḳ,) *He was,* or *became, veiled, concealed, hidden, covered,* or *protected,* or *he veiled, concealed, hid, covered,* or *protected, himself,* (Ṣ, Ḳ,) عَنْهُ *from him, or it;* (Ḳ;) as also ▼استجنّ. (Ṣ, Ḳ.) You say, استجنّ ▼ بِجُنَّةٍ *He was,* or *became, veiled, &c.,* or *he veiled himself, &c., by a thing whereby he was veiled, &c.* (Ṣ.)

10: see 8, in two places:==and see also 1.==اِسْتِجْنَانٌ is also syn. with اِسْتِطْرَابٌ; (Ṣ, Ḳ;) استجنّهُ meaning استطربهُ, i. e. *He excited him to mirth, joy, gladness,* or *sport.* (TḲ.)

جِنٌّ The *darkness* of night; as also ▼جُنُونٌ and ▼جَنَانٌ, (Ḳ, TA,) the last [written in the CḲ جُنان, but it is] with fet-ḥ: (TA:) or all signify its *intense darkness:* (TA:) or all, the *confusedness of the darkness* of night: (Ḳ:) [all, in these senses, are app. inf. ns.: (see 1:)] the last, ▼جَنَانٌ, also signifies *night* [itself]: (Ḳ:) or [so in copies of the Ḳ, accord. to the TA, but in the CḲ "and,"] the *dense black darkness* of night: (Ṣ, Ḳ:) and ▼جُنُونٌ, the *veiling,* or *concealing,* or *protecting, darkness* of night. (ISk, Ṣ.)—*Concealment:* so in the phrase, لَا جِنَّ بِهٰذَا الأَمْرِ *There is no concealment with this thing.* (Ḳ,* TA.) One of the Hudhalees says,

* وَلَا جِنَّ بِالبَغْضَآءِ وَالنَّظَرِ الشَّزْرِ *

[*And there is no concealment with vehement hatred and the looking with aversion*]. (TA.)==[The *genii;* and sometimes the *angels;*] accord. to some, the *spiritual beings that are concealed from the senses,* or *that conceal themselves from the senses; all of such beings;* (Er-Rághib, TA;) the *opposite of* إِنْس; (Ṣ, Mgh, Mṣb, Er-Rághib, TA;) *thus comprising the angels; all of these being* جِنّ; (Er-Rághib, TA;) thus called because they are feared but not seen: (Ṣ:) or, accord. to others, *certain of the spiritual beings;* for the spiritual beings are of three kinds; the good being the angels; and the evil being the devils (شَيَاطِين); and *the middle kind, among whom are good and evil,* being the جِنّ; as is shown by the first twelve verses of ch. lxxii. of the Ḳur: (Er-Rághib, TA:) or it here means *intelligent invisible bodies, predominantly of the fiery,* or *of the aerial, quality:* or a *species of souls,* or *spirits, divested of bodies:* or *human souls separate from their bodies:* (Bḍ:) or the جِنّ are the *angels* [*exclusively*]; (Ḳ;) these being so called in the Time of Ignorance, because they were concealed, or because they concealed themselves, from the eyes: so, accord. to some, in the Ḳur [xviii. 48], where it is said that Iblees was of the جِنّ: and so, as some say, in the Ḳur [vi. 100], where it is said that they called the جِنّ partners of God: (TA:) but some reject the explanation in the Ḳ, because the angels were created of light, and the جِنّ of fire; and the former do not propagate their kind, nor are they to be described as males and females; contrary to the case of the جِنّ; wherefore it is generally said that in the phrase [in the Ḳur xviii. 48, above mentioned] إِلَّا إِبْلِيسَ كَانَ مِنَ الجِنِّ, what is excepted is disunited in kind from that from which the exception is made, or that Iblees had adopted the dispositions of the جِنّ: (MF, TA:) or, as some say, the جِنّ were *a species of the angels, who were the guardians of the earth and of the gardens of Paradise:* (TA:) ▼جِنَّةٌ, also, signifies the same as جِنٌّ: (Ṣ, Mṣb, Ḳ:) so in the last verse of the Ḳur: (Ṣ:) in the Ḳur xxxvii. 158 meaning the *angels,* whom certain of the Arabs worshipped; (TA;) and whom they called the daughters of God: (Fr, TA:) a single individual of the جِنّ is called ▼جِنِّىٌّ, [fem. with ة:] (Ṣ, TA:) and ▼جَانٌّ, also, is syn. with جِنٌّ: (Mṣb:) or الجَانُّ means *the father of the* جِنّ; (Ṣ, Mgh, TA;) [i. e. any father of جِنّ; for] the pl. is جِنَّانٌ, like حِيطَانٌ pl. of حَائِطٌ: (Ṣ, TA:) so says El-Ḥasan: it is said in the T, on the authority of

AA, that the جَانّ is, or are, *of the* جِنّ: (TA:) or جَانّ is a quasi-pl. n. of جِنّ; (M, Ḳ;) like جَامِلٌ and بَاقِرٌ: (M, TA:) so in the Ḳur lv. 56 and 74: in reading the passage in the Ḳur lv. 39, 'Amr Ibn-'Obeyd pronounced it جَأْنٌ: (TA:) it is related that there were *certain creatures* called the جَانّ, *who were upon the earth, and who acted corruptly therein, and shed blood, wherefore God sent angels who banished them from the earth;* and it is said that these angels became the inhabitants of the earth after them. (Zj, TA.) بَاتَ فُلَانٌ ضَيْفَ جِنٍّ [*Such a one passed the night a guest of* جنّ] means, *in a desolate place, in which was no one that might cheer him by his society* or *converse.* (TA.) The saying of Moosà Ibn-Jábir,

* فَمَا نَفَرَتْ جِنِّى وَلَا فُلَّ مِبْرَدِى *

may mean *And my companions, who were like the* جنّ, *did not flee* when I came to them and informed them, *nor was my tongue, that is like a file, deprived of its sharp edge:* or by his جنّ he means his *familiar* جنّ, such as were asserted to aid poets when difficulties befell them; and by his مبرد, his *tongue:* (Ḥam p. 182 [where other explanations are proposed; but they are far-fetched]:) or by his جنّ he means his *heart;* and by his مبرد, his *tongue.* (Ṣ.) The Arabs liken a man who is sharp and effective in affairs to a جِنِّى and a شَيْطَان: and hence they said, نَفَرَتْ جِنُّهُ, meaning † *He became weak and abject.* (Ḥam ubi suprà.) — The *greater, main,* or *chief, part,* or the *main body,* or *bulk,* of men, or of mankind; as also ↓ جَنَانٌ; (Ḳ;) because he who enters among them becomes concealed by them: (TA:) or the latter means the *general assemblage,* or *collective body,* of men: (IAạr, Ṣ,* TA:) or *what veils, conceals, covers,* or *protects, one,* of a thing. (AA, TA.) — ‡ The *flowers,* or *blossoms,* of plants or herbage. (Ḳ, TA.) — ‡ The *prime,* or *first part,* of youth: (Ṣ, Ḳ, TA:) or the *sharpness,* or *vigorousness, and briskness, liveliness,* or *sprightliness,* thereof. (TA.) You say, كَانَ ذٰلِكَ فِى جِنِّ شَبَابِهِ ‡ *That was in the prime,* or *first part, of his youth.* (Ṣ, TA.) And, أَفْعَلُ ذٰلِكَ الأَمْرَ بِجِنِّ ذٰلِكَ ‡ *I will do that thing in the time of the first and fresh state of that.* (Ṣ, TA.) جِنٌّ may also signify † The *madness,* or *insanity, of exultation,* or *of excessive exultation.* (TA.) And one says, اِتَّقِ النَّاقَةَ فَإِنَّهَا بِجِنِّ ضِرَامِهَا, meaning † *Fear thou the she-camel, for she is in her evil temper on the occasion of her bringing forth.* (TA.) — Also *i. q.* جِدّ [app. جِدّ, as meaning † *Seriousness,* or *earnestness*]; because it is a thing that is an accompaniment of thought, or reflection, and is concealed by the heart. (TA.)

جَنَّةٌ A [*garden, such as is called*] بُسْتَان: (Ṣ, Mgh:) or a *garden,* or *walled garden,* (حَدِيقَة, Mṣb, Ḳ,) *of trees,* or *of palm-trees,* (Mṣb,) or *of palms and other trees:* (Ḳ:) or only *if containing palm-trees and grape-vines;* otherwise, if containing trees, called حديقة: (Aboo-'Alee in the Tedhkireh, TA:) or *any* بستان *having trees by which the ground is concealed:* and sometimes *concealing trees:* (Er-Rághib, TA:) and *palm-trees:* (Ṣ:) or *tall palm-trees:* (Mgh:) or *shadowing trees;* because of the tangling, or luxuriousness of their branches; as though concealing at once what is beneath them: then *a* بستان; because of its dense and shadowing trees: (Bḍ in ii. 23:) or *a* بستان *of palms and other trees, dense, and shadowing by the tangling,* or *luxuriousness, and denseness, of their branches;* as though it were originally the inf. n. of un. of جَنَّهُ, and meaning "a single act of veiling" or "concealing" &c.: (Ksh ib.:) then, with the article ال, [*Paradise,*] *the abode of recompense;* because of the جِنَان therein; (Ksh and Bḍ ib.;) or because the various delights prepared therein for mankind are concealed in the present state of existence: (Bḍ ib.:) [and] hence الجَنَّاتُ [*the gardens* of Paradise], (so in a copy of the Ṣ,) or جَنَّاتُ عَدْنٍ [*the gardens of continual abode*]: (so in another copy of the Ṣ:) [for] the pl. of جَنَّةٌ is جِنَانٌ (Mgh, Mṣb, Ḳ) and جَنَّاتٌ (Mṣb, TA) and أَجِنَّةٌ, but this last is strange. (MF, TA.) [Dim. ↓ جُنَيْنَةٌ, vulgarly pronounced جِنَيْنَة, and applied to *A garden;* as though it were *a little Paradise.*]

جُنَّةٌ *A thing by which a person is veiled, concealed, hidden, covered,* or *protected: an arm,* or *armour, with which one protects himself:* (Ṣ:) *anything protective:* (Ḳ:) or *coats of mail,* and *any defensive,* or *protective, arm* or *armour:* (TA:) pl. جُنَنٌ. (Ṣ.) — *A piece of cloth which a woman wears, covering the fore and hind parts of her head, but not the middle of it, and covering the face, and the two sides of the bosom,* (Ḳ,) or, accord. to the M, *the ornaments* [حَلْى instead of جَنْبَى] *of the bosom,* (TA,) *and having two eye-holes, like the* بُرْقُع. (Ḳ.)

جِنَّةٌ: see its syn. جِنّ: — and جُنُونٌ.

جَنَنٌ *A grave;* (Ṣ, Ḳ;) because it conceals the dead: (TA:) and so ↓ جَنِينٌ, of the measure فَعِيلٌ in the sense of the measure فَاعِلٌ. (Er-Rághib, TA.) — *Grave-clothing;* (Ḳ;) for the same reason. (TA.) — *A garment that conceals the body.* (TA.) [See also جَنَانٌ.] — *A dead body;* (Ṣ, Ḳ;) because concealed in the grave; the word being of the measure فَعَلٌ in the sense of the measure مَفْعُولٌ, like نَفَضٌ in the sense of مَنْفُوضٌ. (TA.)

جُنُنٌ: see جُنُونٌ.

جَنَانٌ: see جِنّ, first sentence, in two places; — and see the same in the latter part of the paragraph. — Also *A garment:* (Ḳ:) or *a garment that conceals one;* as in the saying, مَا عَلَىَّ جَنَانٌ إِلَّا مَا تَرَى [*There is not upon me a garment that conceals me save what thou seest*]. (Ṣ.) [See also جَنَنٌ.] — The حَوِير [or *surrounding adjuncts,* or *appertenances and conveniences,*] (Ḳ, TA) of a house; because concealing the house. (TA.) — The *interior* of a thing that one does not see; (Ḳ;) because concealed from the eye. (TA.) — The *heart;* (T, Ṣ, M, Mṣb, Ḳ;) because concealed in the bosom; (T, M;) or because it holds things in memory: (M, TA:) or *its* رُوع [i. e. the *heart's core,* or the *mind,* or *understanding,* or *intellect*]; (Ḳ;) which is more deeply hidden: (TA:) and (sometimes, TA) the *soul,* or *spirit;* (IDrd, Ḳ;) because the body conceals it: (IDrd, TA:) pl. أَجْنَانٌ. (IJ, Ḳ.) You say, مَا يَسْتَقِرُّ جَنَانُهُ مِنَ الفَزَعِ [*His heart does not rest in its place by reason of fright*]. (TA.) — *A secret and bad action.* (TA. [Before the word rendered "secret" is another epithet, which is illegible.])

جُنَانٌ: see مِجَنٌّ: — and what here next follows.

جُنُونٌ: see جِنٌّ, first sentence, in two places. — Also, inf. n. of جُنَّ; (Ṣ, Ḳ;) [originally signifying *A state of possession by a* جِنِّى, or *by* جِنّ; *diabolical,* or *demoniacal, possession;* and hence meaning] *loss of reason;* or *madness, insanity,* or *unsoundness in mind* or *intellect;* (Mgh;) or *deficiency of intellect:* (Sb, TA:) [it may generally be rendered *possession,* or *insanity:*] ↓ جُنُنٌ is a contraction thereof; (Ṣ, Ḳ;) or accord. to some, an original form: (MF, TA:) and ↓ جِنَّةٌ, also, (an inf. n. and a simple subst., Ṣ,) signifies the same as جُنُونٌ: (Ṣ, Mṣb, Ḳ:) as also ↓ مَجَنَّةٌ, (Ṣ, Ḳ,) and ↓ جُنَانٌ, but this last is vulgar. (TA.) — Also *Persistence in evil;* and *pursuance of a headlong,* or *rash, course.* (Ḥam p. 14.)

جَنِينٌ Anything *veiled, concealed, hidden,* or *covered:* (Ḳ:) applied as an epithet even to rancour, or malice. (TA.) — *Buried; deposited in a grave.* (IDrd, Ṣ.) — *An embryo; a fœtus;* the *child,* or *young, in the belly;* (Ṣ, Mṣb, Ḳ;) [i. e.,] *in the womb:* (Mgh:) pl. أَجِنَّةٌ (Ṣ, Mṣb, Ḳ) and أَجْنُنٌ. (ISd, Ḳ.) — And the former of these pls., *Waters choked up with earth.* (TA.) — See also جَنَنٌ. — Also The *vulva.* (TA.)

جُنَانَةٌ: see مِجَنٌّ.

جَنِينَةٌ, accord. to the copies of the Ḳ, but in the M ↓ جِنِّيَّةٌ, (TA,) *A* [*garment of the kind called*] مِطْرَف, (Ḳ, TA,) *of a round form,* (TA,) *like the* طَيْلَسَان, (Ḳ, TA,) *worn by women:* (TA:) in the T, said to be *certain well-known garments.* (TA.)

جُنَيْنَةٌ: see جَنَّةٌ, last sentence.

جِنِّىٌّ *Of,* or *relating to, the* جِنّ, or جِنَّة. (Ḳ.) — See جِنّ. In the saying,

* وَيْحَكِ يَا جِنِّىَّ هَلْ بَدَا لَكِ *

* أَنْ تَرْجِعِى عَقْلِى فَقَدْ أَنَى لَكِ *

[*Mercy on thee! O Jinneeyeh,* (جِنِّىّ being for جِنِّيَّة,) *doth it appear fit to thee that thou shouldst restore my reason? for the time hath come for thee* to do so], a woman resembling a جِنِّيَّة is meant, either because of her beauty, or in her changeableness. (TA.) — The *tallness,* or *length and height,* of a camel's hump. (TA.)

جِنِّيَّةٌ [fem. of جِنِّىٌّ, q. v.] — See also جَنِينَةٌ.

جَنْجَنٌ and جِنْجِنٌ and ↑جِنْجِنَةٌ (Ṣ, Ḳ) and ↑جُنْجُونٌ (Ḳ) and (as some say, TA) ↑جَنْجَنَةٌ (Ḳ) are sings. of جَنَاجِنُ, which signifies The *bones of the breast:* (Ṣ, Ḳ:) or the *heads of the ribs* of men and of others: (M, TA:) or the *extremities of the ribs, next the sternum.* (T, TA.)

جِنْجِنَةٌ : } see what next precedes.
جُنْجُونٌ : }

جَانٌّ : see جِنٌّ. — Also *A white serpent:* (Lth, Ṣ, Mṣb:) or *a small white serpent:* (Mgh:) or *a great serpent:* (Zj, TA:) or *a species of serpent* (AA, M, Ḳ) *having black-bordered eyes,* (M, Ḳ,) *inclining to yellow,* (M, TA,) *harmless, and abounding in houses:* (M, Ḳ:) pl. جَوَانُّ, (AA, TA,) or جِنَّانٌ. (TA.)

مِنْ أَجْلِ أَنَّكَ كَذَا i. q. أَجِنَّكَ كَذَا [*Because that thou art thus*]; (Ṣ, Ḳ;) from which it is contracted by suppressing the ل and ا, and transferring the kesrah of the ل to the ج. (Ṣ.) A poet says,

* أَجِنَّكِ عِنْدِى أَحْسَنُ النَّاسِ كُلِّهِمْ *

[*Because that thou art in my estimation the goodliest of all mankind*]. (Ṣ.) The مِنْ is omitted as in the phrase فَعَلْتُهُ أَجْلَكَ for مِنْ أَجْلِكَ. (Ks, TA.)

تَجْنِينٌ [an inf. n. used as a simple subst.,] *What is said by the* جِنّ [or *genii*]: or, accord. to Es-Sukkaree, *strange, uncouth speech* or *language, difficult to be understood.* (TA.)

مَجَنٌّ : see مَجْنُونٌ.

مِجَنٌّ *A shield;* (Ṣ, Mgh, Mṣb, Ḳ;) because the owner conceals, or protects, himself with it; (Mgh, Mṣb;) as also ↑مِجَنَّةٌ (Lḥ, Ḳ) and ↑جُنَانٌ and ↑جُنَانَةٌ: (Ḳ:) pl. مَجَانُّ. (Ṣ, Mṣb.) Sb held it to be of the measure فِعَلٌّ, from مجن; but his opinion is opposed by the fact that the word is of the form which is significant of an instrument, by the doubling of the ن, and by the syns. جنان and جنانة. (MF, TA.) It is said in a trad., that the hand [of a thief] shall not be cut off save for the value of a مِجَنّ; which in the time of the Prophet was a deenár, or ten dirhems; for this is the lowest amount for which that punishment is to be inflicted. (Mgh.) You say, قَلَبَ مِجَنَّهُ [*He turned his shield*], meaning ‡ *He dropped shame, and did what he pleased:* or *he became absolute master of his affair,* or *case.* (Ḳ, TA.) And قَلَبْتُ لَهُ ظَهْرَ المِجَنِّ [*I turned towards him the outer side of the shield*], meaning † *I became hostile to him after reconciliation.* (Ḥar p. 265.) — Also *A* [*woman's ornament such as is commonly called*] وِشَاح. (Az, Ḳ.)

مَجَنَّةٌ *A place in which one is veiled, concealed, hidden, covered,* or *protected;* or *in which one veils, conceals, hides, covers,* or *protects, himself.* (Ṣ.) — *A land having in it* جِنّ: (Ṣ:) or *abounding with* جِنّ. (Ḳ.) = See also جُنُونٌ.

مِجَنَّةٌ : see مِجَنٌّ.

مَجْنُونٌ [*Possessed by a* جِنِّىّ, or *by* جِنّ, or *by a devil,* or *demon; a demoniac:* (see Bḍ li. 39:) and hence meaning *bereft of reason;* or *mad, insane, unsound in mind* or *intellect,* or *wanting therein:* (see جُنُونٌ:) it may generally be rendered *possessed;* or *mad,* or *insane:*] part. n. of جُنَّ: (Mṣb:) or anomalously used as pass. part. n. of أَجَنَّهُ: (Ṣ,* Ḳ,* TA:) one should not say ↑مُجَنٌّ: (Ṣ, TA:) [pl. مَجَانِينُ.] — نَخْلَةٌ مَجْنُونَةٌ ‡ *A tall palm-tree:* (Ṣ, Ḳ, TA:) pl. مَجَانِينُ. (Ṣ, TA.) And نَبْتٌ مَجْنُونٌ ‡ *A plant,* or *herbage, that is tangled,* or *luxuriant,* or *abundant and dense, in part, and strong.* (TA.) And أَرْضٌ مَجْنُونَةٌ † *Land producing much herbage, that has not been depastured.* (TA. [See also what next follows.])

أَرْضٌ مُتَجَنِّنَةٌ ‡ *Land having much herbage, so that it extends in every way.* (Ḳ, TA.)

مَنْجَنُونٌ and مَنْجَنِينٌ : see art. منجن.

جنأ

1. جَنَأَ عَلَيْهِ, (Aṣ, Ṣ, L, Ḳ,) aor. ـَ, inf. n. جُنُوْءٌ, (Aṣ, Ḳ,) *He bent down over,* or *fell prostrate upon, him,* or *it;* (Aṣ, Ṣ, L, Ḳ;) namely, a man, (Th, L,) as one does in speaking to another [who is sitting], (Th,) or to shield or protect another from blows, (L,) and as a woman does over a child; (TA;) or a horse, said when a man has bent down to preserve himself [from an arrow &c.]; or a thing; (Ṣ;) as also جَنِئَ, aor. ـَ, inf. n. جَنَأٌ; (Ḳ, TA; [in the CḲ, erroneously, جَنَاء;]) as also ↑اجنأ, (Aṣ, L, Ḳ,) likewise said of a man bending over another to shield or protect the latter; (L, TA;) and ↑جانأ, and ↑تجانأ, (Ṣ, Ḳ,) and ↑اجتنأ. (TA.) — جَنَأَ فِى عَدْوِهِ *He persevered,* or *was fatigued and slow,* (أَلَحَّ, [which has these two different meanings,]) *and bent down, in his running.* (T, TA.) = جَنِئَ, aor. ـَ, (Lth, Ḳ,) inf. n. جَنَأٌ, (Lth, Ṣ,) said of a man, (Ṣ,) *He had a bending forward of the upper part of his back over his breast:* (Lth, Ḳ:) or *was humpbacked:* (Ṣ:) or *he had a bent and humped back:* but Lth denies that جَنَأٌ signifies the *being humpbacked.* (TA.) [See also أَجْنَأُ.]

3 : }
4 : }
6 : } see 1.
8 : }

أَجْنَأُ, applied to a man, (Ṣ,) *Having a bending forward of the upper part of the back over the breast:* (Lth, Ḳ:) or *humpbacked:* (Ṣ:) or *having a bent and humped back;* but see what Lth says, voce جَنِئَ: (TA:) or i. q. أَدْنَأُ and أَقْعَسُ, meaning a man *having a bending in his breast towards his back:* (AA, TA:) accord. to Aṣ, applied to him who has been straight in the back and has then been affected with what is termed جَنَأٌ: it is also applied to an ostrich: fem. جَنْآءُ and جَنْوَآءُ. (TA.) — Also جَنْآءُ *A ewe,* or *a she-goat, having her horns bending backwards;* (Esh-Sheybánee, Ḳ;) and so جَنْوَآءُ. (TA in art. جنو.)

مُجْنَأٌ *A shield:* (Ṣ, Ḳ:) so called because of its being humped, (Ḳ, TA, [in some copies of the former of which, for لَاحَدِيدَ بِهِ, we find لَا حَدِيدَةَ بِهِ, i. e. having no piece of iron in it,]) and on account of its bending form. (TA.)

مُجْنَأَةٌ *A grave.* (Ḳ.) [App. so called because the grave of an Arab of the desert generally has a small oblong humped mound raised over it.]

جنب

1. جَنَبَهُ *He broke his side:* (Ṣ, Ḳ:) or *he hit,* or *hurt, his side.* (TA.) [The aor. of the verb in this sense is probably ـُ, and the inf. n., accord. to the TḲ, is جَنْبٌ.] — *He led him by his side;* (Ṣ, A,* Mṣb, Ḳ;) namely, a horse (Ṣ, A, Mṣb, TA) or the like, (Ṣ, A,) and a captive. (Ṣ, TA.) In this sense, its aor. is ـُ, (A, Mṣb, TA,) and the inf. n. جَنْبٌ (Ṣ, A, Mṣb, Ḳ) and جَنَبٌ. (Ḳ.) Hence, جِنَابٌ : see طَوْعُ الجِنَابِ; which is forbidden (Ṣ, A, TA) in a trad., [in which it is said, لَا جَلَبَ وَلَا جَنَبَ,] (A, TA) relating to horse-racing and to [the collecting of] the poor-rate, (TA,) means [in the former case] *A man's leading, by the side of a horse that he rides in a race, another horse,* (Ṣ, A, Ḳ,) *without a rider,* (TA,) *and when the horse that he rides has become languid and weak,* (Ḳ,) or *when he fears that he will not outstrip upon it,* (Ṣ,) or *when he draws near to the goal,* (A,) *transferring himself to the other,* (Ṣ, A, Ḳ,) *in order that he may outstrip:* (A:) and in relation to the poor-rate, it means *the collector's alighting in the most remote of the places whence the portion appointed for the poor-rate is to be collected, and then ordering that the camels or the like* [*that constitute that portion*] *shall be led to him:* or the *going of the owner of the property to a distance,* [or *aside,* or *out of the way,*] *with his property, so that the collector is obliged to go to a distance in quest of it.* (Ḳ. See more in art. جلب, first paragraph.) — *He placed,* or *put, at a distance,* or *he put,* or *sent, away,* or *far away,* or *far off,* or *he removed far away, alienated,* or *estranged, him,* or *it;* (Ḳ;) *as though he put him,* or *it, aside,* or *as though he walked aside;* as also ↑جانبه. (TA.) And *He pushed, thrust,* or *drove, him,* or *it, away, aside,* or *to a distance.* (Ḳ,* TA.) And جَنَبَهُ الشَّىْءَ, (Ṣ, Ḳ,*) or الشَّرَّ, (Fr, Zj, Mṣb,) aor. ـُ; (Ṣ, Mṣb, Ḳ;) and ↑جنّبهُ, (Fr, Zj, Ṣ, A, Mṣb, Ḳ,) but this has an intensive signification; (Mṣb;) and ↑اجنبهُ; (Fr, Zj, A, Ḳ;) *He put aside,* or *away,* or *he warded off, from him,* (Ṣ,) or *he removed from him,* (Ṣ, Mṣb, Ḳ,) or *removed far from him,* (Mṣb, Ḳ,) *the thing,* (Ṣ, Ḳ,*) or *evil.* (Fr, Zj, A, Mṣb.) It is said in the Ḳur [xiv. 38], وَٱجْنُبْنِى وَبَنِىَّ أَنْ نَعْبُدَ ٱلْأَصْنَامَ [*And put Thou away from me and my sons our worshipping of idols*], (Ṣ,) or, accord. to one reading, ↑وَأَجْنِبْنِى. (TA.) — *He yearned towards, longed for,* or *desired, him,* or *it.* (Ḳ,* TA.) = جَنِبَ بِهِ,

aor. ِ, [*He went aside, apart, out of the way, to a distance,* or *far away, with him,* or *it*: or, like جَنَّبَهُ, in a sense explained above,] *he placed,* or *put, at a distance,* or *he put,* or *sent, away,* or *far away,* or *far off, him,* or *it.* (Ḳ, TA.) —— جَنَبَ فِى بَنِى فُلَانٍ, (Ṣ, Ḳ,*) aor. ُ, inf. n. جَنَابَةٌ; (Ṣ;) and ▼ تجنّب; (so, app., in the TA;) *He alighted,* or *descended and abode,* or *settled, as a stranger, among the sons of such a one.* (Ṣ, Ḳ,* TA.) One says, نِعْمَ القَوْمُ هُمْ لِجَارِ الجَنَابَةِ [*Excellent are the people, they,*] *to the neighbour who is a stranger.* (Ṣ. [See also جُنُبٌ.]) And لَا تَحْرِمْنِى عَنْ جَنَابَةٍ *Do not thou by any means refuse me because of being remote* (Ṣ, A, TA) *in respect of relationship.* (A, TA.) [See also جَنَابَةٌ mentioned below as a subst.] —— جَنَبَتِ الرِّيحُ, (Ṣ, A, Ḳ,) aor. ُ, (TA,) inf. n. جُنُوبٌ; (Ḳ;) and ▼ اجنبت; (TA;) *The wind was,* or *became, such as is termed* جَنُوب [i. e. *south,* or *southerly*]; (Ḳ;) *it blew in the direction of the wind thus called:* (A, TA:) or the former, (Ṣ,) or جُنِبَت, (TA,) *the wind changed,* or *veered, so as to become* جنوب. (Ṣ, TA.) —— [And hence, (see جَنُوبٌ,)] جَنَبَ إِلَيْهِ, (IAar, Ḳ,) or إِلَى لِقَائِهِ, (TA,) aor. ُ; (Ḳ;) and جَنِبَ, aor. َ; (Th, Ḳ;) [inf. n., app., جَنَبٌ, for the verb is said in the Ḳ to be like نَصَرَ and سَمِعَ;] †*He was,* or *became, disquieted by vehement desire to see him,* or *to meet him.* (Ḳ,* TA.) ═ جَنِبَ, aor. َ, (Ṣ,) inf. n. جَنَبٌ, (Ṣ, Ḳ,) *He* (a camel) *limped,* or *halted, by reason of* [*pain in*] *his side:* (Ṣ:) or *he had an affection resembling* ظَلَع [i. e. *limping,* or *halting*], (Ḳ, TA,) *but not the same as this:* (TA:) and, (Ḳ,) or accord. to Aṣ, (Ṣ,) *his lungs clave to his side by reason of vehement thirst:* (Ṣ, Ḳ:) or, accord. to the Arabs of the desert, as ISk says, *he became bent,* or *contorted, by reason of vehemence of thirst:* (Ṣ:) and *he* (a camel) *had a pain in his side from vehemence of thirst.* (TA.) The epithet is ▼ جَنِبٌ; which is applied by Dhu-r-Rummeh to an ass. (Ṣ, TA.) —— جنبت الدَّلْوُ [app. جَنِبَت] *The bucket inclined to one side in consequence of the breaking of one or two of the thongs attaching it to the cross-bars.* (L, TA.) ═ جَنُبَ and جَنِبَ and جُنِبَ are *syn.* with أَجْنَبَ in a sense explained below: see 4. ═ جُنِبَ *He had,* or *became affected by, the disease termed* ذَاتُ الجَنْبِ [or *pleurisy*]: (Ṣ, Mgh, Mṣb:) *he had a complaint of his side.* (Ḳ.) ═ جُنِبُوا *They were,* or *became, affected by the* [*south,* or *southerly, wind called*] جَنُوب. (Ṣ, A, Ḳ.) And also, [in allusion to the fertilizing effect attributed to the wind so called,] *They were,* or *became, affected by that wind in their cattle.* (L, TA.)

2. جَنَّبَهُ: see 1: —— and see also 3. ═ جنّب, inf. n. تَجْنِيبٌ, *He did not send the stallion-camel among his she-camels, nor the ram or he-goat among his ewes or she-goats.* (Ḳ.) —— جنّب القَوْمُ *The milk of the people's camels became little:* (Ṣ:) or *the people's milk ceased;* (Ḳ, TA;) or *became little:* or *the people's camels had no milk:* and جنّب said of a man, *his camels had no milk, nor had his sheep or goats.* (TA.) Hence, عَامُ تَجْنِيبٍ [*A year of little,* or *no, milk*]. (Ṣ, TA.) —— جنّبت الإِبِلُ *The camels, with the exception of one or two, brought forth no young.* (AZ, TA.) *The camels did not conceive, so as to have milk.* (TA.) ═ تَجْنِيبٌ [as an inf. n. of which the verb, if it have one in any of the following senses, is جُنِّبَ,] also signifies *A bending,* or *curving, and tension* [*of the sinews*] (تَوْتِيرٌ), *of the hind leg of a horse;* which is a quality approved: (Ṣ, Ḳ:) or, accord. to AO, *a turning aside of his fore legs in raising them and putting them down:* but accord. to Aṣ, it is *in the hind legs,* and تَجْنِيبٌ is in the back-bone and in the fore legs. (TA.) [See also 2 in art. حنب; and see also مُجَنَّبٌ.]

3. جانبهُ, (A, Ḳ,) inf. n. مُجَانَبَةٌ and جِنَابٌ, (Ḳ,) *He was,* or *became, at,* or *by, his side:* (A, Ḳ:) and *he walked,* or *went, by his side.* (A.) ═ Also *i. q.* بَاعَدَهُ; (A, Ḳ;) i. e. *He was,* or *became,* [*distant, remote, far off,* or *aloof, from him;* or] *apart from him;* or *in a part, quarter,* or *tract, different from that in which he* (the other) *was;* (TA;) thus bearing two contr. significations. (A, Ḳ.) جانبهُ and ▼ تجانبهُ and ▼ تجنّبهُ and ▼ اجتنبهُ all signify the same, (Ṣ, Ḳ,) i. e. *He was,* or *became, distant, remote, far off,* or *aloof,* or *he went,* or *removed,* or *retired,* or *withdrew himself, to a distance,* or *far away,* or *far off,* or *he alienated,* or *estranged, himself,* or *he stood,* or *kept, aloof, from him,* or *it; he shunned,* or *avoided, him,* or *it;* as also ▼ جنّبه (Ḳ) [and ▼ تجنّب مِنْهُ]. You say, جَانِبِ اللِّئَامَ [*Remove thyself far from the mean,* or *ignoble; stand,* or *keep, aloof from them; shun,* or *avoid, them*]. (A.) And لَجَّ فِى جِنَابٍ قَبِيحٍ *He persisted in removing himself to a distance,* or *estranging himself, from his family.* (Ṣ, A, Ḳ. [In two copies of the Ṣ, I find جناب here written with fet-ḥ to the ج; but it is expressly said in the TA to be with kesr.]) —— See also 1.

4. اجنبهُ: see 1, in the former half of the paragraph, in two places. ═ اجنب, (Ṣ, IAth, Mgh, Mṣb, Ḳ, &c.,) inf. n. إِجْنَابٌ; (IAth, TA;) and ▼ جَنِبَ; (IB, Ḳ;) but the former is more common than the latter; and the latter, than the next here following; (IB, TA;) and ▼ جَنُبَ, (Ṣ, Mṣb, Ḳ,) [inf. n. جَنَابَةٌ, agreeably with analogy;] and ▼ جَنَبَ, aor. ُ; (L, TA;) and أُجْنِبَ, and ▼ استجنب, (Ḳ,) and ▼ تجنّب; (L, TA;) *He was,* or *became, in the state of one who is termed* جُنُبٌ; (Ṣ, IAth, Mgh, L, Mṣb, Ḳ;) i. e., *under the obligation of performing a total ablution, by reason of sexual intercourse and discharge of the semen.* (IAth, TA.) لَا يُجْنِبُ, said by I'Ab, of a man, and of a garment, and of the ground, (TA,) and of water, (Mgh, TA,) means ‡ *He,* or *it, will not become polluted* (Mgh, TA) *by the touch of him who is* جُنُب *so that one should need total ablution in consequence of the touching thereof.* (TA.) ═ اجنبوا *They entered upon* [*a time in which blew*] *the* [*south,* or *southerly,*] *wind termed* الجَنُوب. (Ṣ, A, Ḳ.) —— See also 1 in the latter half of the paragraph.

5: see 1: —— and 3, in two places: —— and 4.

6: } see 3.
8: }

10: see 4.

جَنْبٌ, a word of well-known meaning; (Ṣ;) The *side,* or *half,* or *lateral half,* syn. شِقٌّ, (A, Ḳ,) of a man &c.; as also ▼ جَانِبٌ and ▼ جَنْبَةٌ: (Ḳ:) or the *part of a man that is beneath the arm-pit, extending to the flank;* as also ▼ جَانِبٌ, because it is the *side* of the person: (Mṣb:) pl. (of the first, Mṣb) جُنُوبٌ (Mṣb, Ḳ) and [of the same, a pl. of pauc.,] أَجْنَابٌ (CḲ) and [of جَانِبٌ] جَوَانِبُ (Lḥ, ISd, Ḳ, but not in the CḲ) and [app. of جَنْبٌ (like as تَبَائِلُ is a pl. of تَبْلٌ) or of جَنْبَةٌ (like as حَوَائِجُ is pl. of حَاجَةٌ which is originally حَوْجَةٌ) or of both these] ▼ جَنَائِبُ, (M, Ḳ,) which is extr. (M, TA.) [Hence,] قَعَدْتُ إِلَى جَنْبِ فُلَانٍ and الى ▼ جَانِبِ فلانٍ [*I sat by the side of such a one*]: both meaning the same. (Ṣ.) And إِنَّهُ لَمُنْتَفِخُ الجَوَانِبِ ▼ [*Verily he is inflated in the side*]: جوانب being here one of those words which are used in the sing. sense though in the pl. form. (Lḥ, TA.) And أَعْطَاهُ الجَنْبَ [lit. *He gave him the side;* meaning] *he was,* or *became, submissive, manageable, easy,* or *tractable, to him.* (A.) And جَارُ الجَنْبِ *He who cleaves to one, keeping by one's side.* (Ḳ. [Differing from جَارُ الجُنُبِ, q. v. infrà.]) And الصَّاحِبُ بِالجَنْبِ [in the Ḳur iv. 40] *The travelling-companion; the companion in a journey:* (Ṣ, Ḳ:) or *he who is near one;* or *by one's side:* or *the companion in every good affair:* or *the husband:* or *the wife.* (TA.) And ذَاتُ الجَنْبِ, (Ṣ, A, Mgh, Mṣb, Ḳ, &c.,) with which ▼ الجُنَابُ is syn., (Ḳ,) [and sometimes الجَنْبُ, as will be seen in what follows,] *A well-known disease;* (Mgh;) [*the pleurisy;* called by the first of these three appellations in the present day;] *a severe disease, being an inflammatory tumour in the* [*pleura,* or] *membrane within the ribs:* (Mṣb:) or *an ulcer,* or *a purulent pustule, that comes within a man's side:* (Ṣ, TA:) it is *a severe disease in the side:* accord. to El-Hejeree, it is *in either side;* and they assert that *when it is in the left side, the patient perishes:* accord. to ISh, *the* دُبَيْلَة; which is *an ulcer that penetrates into the belly:* or *the ulcer* (دُبَيْلَة and دُمَّل) *that comes forth within the side, and discharges internally; the sufferer from which seldom recovers:* he who suffers from it [and dies in consequence], or, as some say, he who is afflicted by a complaint of the side (absolutely) while warring in the cause of God, is reckoned a martyr: (TA:) [soldiers in a campaign are notoriously more subject to it than persons in most other circumstances; and it is app. for this reason that] it is termed ذَاتُ الصَّنَادِيدِ [the disease of the courageous chiefs]. (A, TA.) ذُو الجَنْبِ, of which ذَاتُ الجَنْبِ is the fem., signifies *Having a complaint of his side by reason of* [*the disease above mentioned,* or *what is termed*] الدُّبَيْلَة. (TA. [See also مَجْنُوبٌ.]) —— A poet says,

* النَّاسُ جَنْبٌ وَالأَمِيرُ جَنْبُ *

[*The people are a side and the prince is a side*]: (Akh, Ṣ, TA:) as though he reckoned the latter equal to all the people. (TA. [This is cited in the Ṣ and TA as though it were an ex. of جنب in the sense here next following: but it seems to be rather an ex. of this word in the sense first explained in the present paragraph.]) — *I. q.* نَاحِيَةٌ [*A side; meaning a lateral, or an outward or adjacent, part or portion, region, quarter, or tract; or a part, region, quarter, or tract, considered with respect to its collocation or juxtaposition or direction, or considered as belonging to a whole; a vicinage, or neighbourhood*]; (Ṣ, Ḳ;) as also ▼ جَانِبٌ (Ṣ, Mṣb, Ḳ) and ▼ جَنَابٌ and ▼ جَنْبَةٌ (Ṣ, Ḳ) and ▼ جَنَبَةٌ (Ṣ) and ▼ جَنَابَةٌ. (L, TA.) It is said that the primary signification of جَنْبٌ is the part of the body mentioned in the beginning of this paragraph, and that its use in the sense of نَاحِيَةٌ is metaphorical, as is the case of يَمِينٌ and شِمَالٌ; but نَاحِيَةٌ is mentioned in the Mṣb as the primary signification of ▼ جَانِبٌ; (MF, TA;) though its primary signification accord. to the Ḳ and ISd seems to be that first mentioned. (TA.) You say, ▼ مَشَوْا جَانِبَيْهِ and ▼ جَنَابَيْهِ and ▼ جَنْبَتَيْهِ and ▼ جَنَابَتَيْهِ [*They walked, or went on foot, on either side of him*]. (A, TA.*) And ▼ مَرُّوا يَسِيرُونَ جَنَابَيْهِ (Ṣ, L) and ▼ جَنْبَتَيْهِ and ▼ جَنَابَتَيْهِ (L, TA) *They went along journeying on either side of him.* (Ṣ, L.) And كُنَّا عَنْهُمْ ▼ جَنَابًا and ▼ جَنَابَيْنِ *We were apart from them* [*on two sides* and *on one side*]. (TA.) And نَزَلُوا فِى جَنَابَاتِ ▼ الوَادِى [*They alighted in the sides of the valley, or in the tracts beside the valley*]. (A.) And ▼ فُلَانٌ لَا يَطُورُ بِجَنَبَتِنَا *Such a one will not approach our quarter:* (Ṣ:) thus accord. to AO; with fet-ḥ to the ن: IJ, however, says, people are wont to say, ▼ أَنَا فِى ذَرَاكَ وَجَنَبَتِكَ [meaning *I am under thy protection and in thy quarter*]; but that the correct expression is ▼ جَنْبَتِكَ, with the ن quiescent. (IB, TA.) The Arabs also said, ▼ الحَرُّ جَانِبَىْ سُهَيْلٍ, meaning † *The heat is on either side of Suheyl* [or *Canopus*: i. e., *during the period next before, and that next after, the auroral rising of Canopus;* which rising began, in central Arabia, at the commencement of the era of the Flight, about the 4th of August, O. S.]: this is the greatest heat. (TA.) One also says, ▼ أَحَاطُوا بِهِ مِنْ جَانِبَيْهِ [meaning *They surrounded him on all his sides; lit., on his two sides*]; dividing the surrounding parts into two, but not meaning that any of these remained vacant. (Expos. of the exs. cited as testimonies by Sb, TA in art. حول.) — Also, [and ▼ جَانِبٌ, which is thus used in the L in art. جنح, and by many authors,] *A part,* or *portion,* of a thing; (L;) the *greater,* or *main,* or *chief, part* or *portion* thereof; *most* thereof; (L, Ḳ;) or *a great part* or *portion* thereof; *much* thereof. (L.) Hence, [or perhaps from جَنْبٌ in the second of the senses assigned to it above, conveying the idea of juxtaposition, and thus of comparison,] هٰذَا قَلِيلٌ فِى جَنْبِ مَوَدَّتِكَ [*This is little in comparison with the magnitude of thy love;* or simply, *in comparison with thy love*]. (TA.) — يَا حَسْرَتَا عَلَى مَا فَرَّطْتُ فِى جَنْبِ اللّٰهِ [in the Ḳur xxxix. 57] means ▼ فى جَانِبِهِ, i. e. † [*O my grief,* or *regret, for my negligence,* or *remissness,*] *in respect of that which is the right,* or *due, of God!* (A, Bḍ, TA,) i. e., (Bḍ,) *in respect of obedience to God!* (Bḍ, Jel:) or, *in respect of* [*the means of attaining*] *nearness to God!* (Fr, TA;) or, *nearness to God in Paradise!* (IAạr, TA:) or, *in respect of the way of God, to which He hath called me!* i. e., the profession of his unity, and the confession of the prophetic office of Moḥammad. (Zj, TA.) The saying of the Arabs, اِتَّقِ اللّٰهَ فِى جَنْبِهِ وَلَا تَقْدَحْ فِى سَاقِهِ [may be rendered † *Fear God in respect of his* (thy brother's) *right,* or *due, and impugn not his honour,* or *reputation:* or] means, accord. to the copies of the Ḳ, لَا تَقْتُلْهُ [*slay him not*], or, as in the L, and in the original draught of the author [of the Ḳ] لا تَغْتَلْهُ [*slay him not clandestinely,* or *on an occasion of inadvertence*], from الغِيلَةُ, *and throw him not into trouble,* or *trial:* (TA:) or, accord. to some, فى جنبه means *in detracting from his reputation,* or *reviling him.* (Ḳ, TA. [See also Freytag's Arab. Prov. i. 240.]) A poet, cited by IAạr, says,

* خَلِيلَىَّ كُفَّا وَٱذْكُرَا ٱللّٰهَ فِى جَنْبِى *

† [*O my two friends, refrain, and be mindful of God in respect of my reputation;* (see also جَانِبٌ;)] meaning, *in detracting from my reputation, or reviling me:* or, accord. to MF, *in my case.* (TA.) And one says, مَا فَعَلْتَ فِى جَنْبِ حَاجَتِى † *What didst thou, or what hast thou done, in the case of the thing that I want?* (L, TA.)

جِنَابٌ: see طَوْعُ الجِنَابِ. = جَنِيبٌ: see جَنْبٌ.

جَنَبٌ also signifies *Short;* (Ḳ;) applied to a man. (TA.)

جَنِبٌ: see جَنَبٌ. — It is also applied as an epithet to a wolf, because he pretends to halt, from guile, or cunning. (L, TA.) — Also *A man who goes aside, or to a distance, from the beaten way, for fear of guests' coming to him for entertainment.* (Ḳ, TA.)

جُنُبٌ, (El-Fárábee, Ṣ, A, Mṣb, Ḳ,) which is sometimes used in the sing. form as pl., and has no fem. form, (TA,) and ▼ جَانِبٌ and ▼ أَجْنَبِىٌّ, (El-Fárábee, Ṣ, Mṣb, Ḳ,) which is said by Az in art. روح to be seldom or never used by the Arabs, but is mentioned by him in its proper art., (Mṣb,) and ▼ أَجْنَبُ, (Az, Ṣ, Mṣb, Ḳ,) are syn., (El-Fárábee, Ṣ, Mṣb, Ḳ,) signifying *A stranger;* (Ḳ;) as also ▼ جَنِيبٌ: (Ṣ:) or *a man who is distant, or remote:* (Mṣb:) or *distant,* or *remote, in respect of relationship:* (Az and Mṣb in explanation of the third and fourth:) [or *not a relation;* as will be seen from what follows:] and ▼ جَانِبٌ [as an act. part. n.] signifies one *alighting,* or *descending and abiding,* or *settling, as a stranger, among a tribe:* (Ṣ:) pl. of the first أَجْنَابٌ, (A, TA,) and of the second جُنَّابٌ, (Ṣ, TA,) and of the fourth أَجَانِبُ. (Mṣb.) الجَارُ الجُنُبُ [occurring in the Ḳur iv. 40] (T, Ṣ, A, Mṣb, Ḳ) and جَارُ الجَنْبِ (TA) *The person who is one's neighbour, but who belongs to another people;* (T, Ṣ, A, Mṣb, Ḳ;) *who is not of one's family nor of one's lineage;* (A;) *who is of another lineage than he of whom he is a neighbour;* (T, TA;) *who is not a relation:* (MF:) or *one who is distant,* or *remote, in an absolute sense:* (TA:) or *the person who is not a relation to another, and who comes to him, and asks him to protect him, and abides with him:* such has the title to respect that belongs to him as neighbour of the other, and to his protection, and as relying upon his safeguard and promise. (TA in art. جور. [Differing from جَارُ الجَنْبِ, q. v. suprà.]) It is said in a trad., هُمْ أَجْنَابُ النَّاسِ *They are the strangers of mankind,* or *of the people.* (TA.) And in another trad., قَالَ لِجَارِيَةٍ هَلْ مِنْ مُغَرِّبَةِ خَبَرٍ قَالَتْ عَلَى جَانِبٍ ▼ الخَبَرُ [*He said to a girl, Is there any news from abroad? She answered,*] *It is for a stranger coming from a journey* [*to give* such *news*]. (TA.) And one says, هُوَ ▼ أَجْنَبِىٌّ مِنِّى [*He is a person not related to me*]. (A.) — Also, ▼ the same four words, (of which only the last is mentioned in this sense in the Ṣ,) *That will not be led; intractable.* (Ḳ.) — جُنُبٌ is also an epithet from الجَنَابَةُ; (Ṣ, Mgh, Mṣb, Ḳ;) signifying A man *under the obligation of performing a total ablution, by reason of sexual intercourse and discharge of the semen:* (IAth, TA: [see 4:]) and is used alike as masc. and fem. (Ṣ, Mgh, Mṣb) and sing. (Ṣ, A, Mgh, Mṣb, Ḳ) and dual (Mṣb, TA) and pl.; (Ṣ, A, Mgh, Mṣb, Ḳ;) being regarded as quasi-coordinate to the class of inf. ns.; for the inf. n., when used as an epithet, must remain, in form, sing. and masc.: (MF in art. عفت:) or one may use the dual form جُنُبَانِ; (Ḳ;) and sometimes they used the pl. أَجْنَابٌ (Ṣ, Mṣb, Ḳ*) and جُنُبُونَ, (Ṣ, Mṣb,) and the fem. pl. جُنُبَاتٌ; (Mṣb) but not جُنُبَةٌ, (Ḳ, TA,) applied to a female. (TA.) It is said in a trad., لَا تَدْخُلُ المَلَائِكَةُ بَيْتًا فِيهِ جُنُبٌ, meaning [*The angels will not enter a house,* or *chamber,* or *tent, in which is*] *one who usually neglects the total ablution when under an obligation to perform it for the cause above mentioned.* (IAth, TA.)

جَنْبَةٌ: see جَنْبٌ, in four places: — and see جَانِبٌ. — Also *Retirement,* or *secession, from others:* (Ḳ, TA:) and in a trad., in which it is enjoined, used as meaning *retirement from women; avoiding the sitting by them, and the approaching the place that they occupy.* (TA.) You say, رَجُلٌ ذُو جَنْبَةٍ *A man of retirement.* (TA.) And نَزَلَ جَنْبَةً *He alighted,* or *descended and abode,* or *settled, in a place aside,* or *apart.* (Ṣ, TA.) And قَعَدَ جَنْبَةً *He* [*sat apart,* or] *retired from others.* (A, TA.) — *The state of being a stranger;* as also ▼ جَنَابَةٌ. (Ḳ. [Both are there mentioned as simple substs.; but the latter is an inf. n.: see جَنَبَ فِى بَنِى فُلَانٍ; and what next follows it: and see also 4.]) Both also signify *Remoteness in respect of relationship.* (TA.) = Also, جَنْبَةٌ, *A piece of skin from the side of a camel,* (Ṣ, L, Ḳ,*) *of which is made a kind of milking vessel* (عُلْبَة), (Ṣ, L,) *larger than the* مِعْلَق, *but smaller than the* جَوْبَة. (L.) = And *Every kind of plant,* (Ṣ,) or *every kind of tree in general,* (K,) *that produces* [*new leaves such as are termed*] رَبْل *in the season of the* صَيْف [which may mean either *summer* or *spring*]: (S, K:) or *every kind of plant that produces leaves in that season without rain:* (TA:) or a name given to *many plants, all of them* عُرُوق [perhaps meaning *resembling roots,* i. e. *straggling,* or *spreading like roots*]; so called because *less than large trees and higher than those that have no root-stock* (أُرُومَة) *in the earth; comprising the* نَصِىّ *and* صِلِّيَان *and* حَمَاط *and* مَكْر *and* حذر [so in the TA, but I do not find it elsewhere, and think it may be a mistranscription for حَزْو, of which حَزْوَة (the name of a certain sour tree) is probably the n. un.,] and دَهْمَاء; *which are smaller than* شَجَر *and superior to* بُقُول: all this has been heard from the Arabs: (T, TA:) or *green and fresh* صِلِّيَان: (TA:) or *what is* [*of a kind*] *between* بَقْل *and* شَجَر; (AHn, K, TA;) *being* [in the TA وهما, but this is evidently a mistake for وَهِىَ,] *of the kind of which the root remains in the winter while the branches perish:* (AHn, TA:) or *herbage of which the root is deep in the earth; such as the* نَصِىّ *and the* صِلِّيَان. (TA voce خَضِرٌ.)

جَنَبَةٌ: see جَنْبٌ, in four places: — and see جَانِبٌ.

جُنَبَةٌ *A thing from which one retires,* or *withdraws himself, to a distance,* or *far away,* or *far off; from which one stands,* or *keeps, aloof.* (K.)

جَنَابٌ: see جَنْبٌ, in five places. [Hence,] كُنَّا عَنْهُمْ جَنَابَيْنِ and جَنَابًا *We were remote,* or *retired, from them;* or *out of their way.* (TA.) — Also, (S, A, Msb, K,) and ▾ جَانِبٌ, (Msb,* TA,) *A court,* or *yard,* or *an open* or *a wide space in front of a house or extending from its sides:* (S, A, K, TA:) and *a place of alighting* or *abode;* or *a settlement,* or *place of settling:* (A:) *a mansion; an abode; a habitation;* or *a place to which a man betakes himself,* or *repairs, for lodging, covert,* or *refuge, in a city or town or village or other place of settled habitations;* syn. رَحْلٌ: (K:) and *a vicinage, neighbourhood,* or *tract adjacent to the place of abode* or *settlement, of a people* or *company of men:* pl. أَجْنِبَةٌ. (S.) You say, أَنَا فِى جَنَابِ زَيْدٍ *I am in the court,* or *yard, of Zeyd;* and *in his place of alighting* or *abode,* or *settlement.* (A, TA.) And فُلَانٌ رَحْبُ الجَنَابِ, (A, TA,) and خَصِيبُ الجَنَابِ, (S, A,) the former meaning *Such a one is possessed of an ample* رَحْل [or *mansion,* &c., as explained above]: (TA:) [and the latter, *such a one is surrounded by a plentiful,* or *fruitful, tract:*] or both mean ‡ *such a one is generous* or *bountiful* [or *hospitable*]. (A.) And فُلَانٌ جَدِيبُ الجَنَابِ (S, TA) [meaning *Such a one is environed by a tract affected with drought,* or *barrenness;* as explained in the S in art. جدب: but generally used tropically, as meaning † *such a one is ungenerous, illiberal,* or *inhospitable*]. And أَخْصَبَ جَنَابُ القَوْمِ [*The neighbourhood of the people,* or *the tract surrounding them, became plentiful,* or *fruitful*]. (S, TA.) And أَجْدَبَ بِنَا الجَنَابُ [*Our neighbourhood,* or *the tract surrounding us, became affected with drought,* or *barrenness*]. (TA from a trad.) — رَجُلٌ لَيِّنُ الجَنَابِ [perhaps a mistranscription for الجَانِبِ] ‡ *A man easy to deal with, compliant,* or *obsequious.* (A.) — الجَنَابُ is also a title often given by writers of letters and the like to any great man to whom others betake themselves, or repair, for protection; and sometimes to God; meaning ‡ *The object of recourse; the refuge; the asylum:* similar to الحَضْرَةُ, q. v., and used in the same manner, i. e., alone, and, without the article, prefixed to the name of the person to whom it is applied, or to a pronoun; but the latter is generally considered as implying greater respect than the former.]

جُنَابٌ *i. q.* ذَاتُ الجَنْبِ: see جَنْبٌ. (K.)

جِنَابٌ *A cord tied to the head and neck of a beast, by which he is led,* or *drawn.* (KL.) [Hence,] فَرَسٌ طَوْعُ الجِنَابِ *A horse easily led;* or *easy to be led; tractable;* [*obedient to the* جناب;] (S, A, K, TA;) as also ▾ طَوْعُ الجَنَبِ. (TA. [See 1, near the beginning.])

جَنُوبٌ, of the fem. gender, and, accord. to Sb, both a subst. and an epithet, [so that one says رِيحٌ جَنُوبٌ, as well as جَنُوبٌ alone and رِيحُ الجَنُوبِ,] (TA,) [The *south wind:* or *a southerly wind:*] the *wind that is opposite to that called the* شَمَال: (S, K:) [consequently, the *wind that blows from the direction of the south pole,* accord. to the S;] the *wind that blows from the direction of the left hand of a person standing opposite to the kibleh* [by which is here meant that corner of the Kaabeh in which is set the Black Stone; which corner is towards the east]: (Th, TA:) or the *wind that blows from the quarter between the place where Canopus rises* [S. 29° E. in central Arabia] *and the place where the same star sets* [S. 29° W. in the same latitude]: ('Omárah, TA:) or *from the quarter between the place where Canopus rises and the place where the sun sets in winter* [W. 26° S. in central Arabia]: (As, TA:) or it is *a hot wind, that blows in every season; blowing from that part of the tract between the quarter whence blows the east wind* (الصَّبَا) *and that whence blows the west wind* (الدَّبُور) *which is next to the place where Canopus rises:* (T, TA:) or the *wind that blows from the quarter between the place where Canopus rises and that where the Pleiades set* [W. 26° N. in central Arabia]: (IAar, K:) [the points whence it usually blows seem to differ somewhat in different parts:] As says that the جنوب is attended by good, and by fecundating influence; and the شمال by drying up [of the earth &c.]: (TA:) accord. to IAar, it is hot in every place, except in Nejd, where it is cold, or cool: (MF:) pl. جَنَائِبُ (T, K) and [of pauc.] أَجْنُبٌ. (T, TA.) — One says, of two persons, when they are on terms of sincere friendship, رِيحُهُمَا جَنُوبٌ † [*Their wind is south,* or *southerly*]; and when they are separated, شَمَلَتْ رِيحُهُمَا † [Their wind has become north, or northerly]. (TA.)

جَنِيبٌ, applied to a horse and a captive, (TA,) *Led by one's side;* as also ▾ مَجْنُوبٌ and ▾ مُجَنَّبٌ: (K:) or you say ▾ خَيْلٌ مُجَنَّبَةٌ, meaning *horses led by the side;* the teshdeed denoting application to many objects: (S, TA:) pl. [of the first, and of جَنِيبَةٌ, q. v., or only of this last,] جَنَائِبُ and [quasi-pl. n.] ▾ جَنَبٌ. (K.) One *walking by the side* of another; (A;) [and] so ▾ جُنَّابٌ. (K.) — Any animal or man that is *obedient, tractable,* or *submissive.* (S, TA.) You say, أَصْبَحَ جَنِيبَهُ *He became compliant to him.* (A.) = See also جُنُبٌ. = Also, applied to a man, [app. *Having a pain in the side;* or *having the pleurisy;* like مَجْنُوبٌ: and hence, or from جَنْبٌ, q. v., irregularly formed,] *as though walking on one side, bent* or *crooked,* مُتَعَقِّفًا: so in the L: in the M and K, on the authority of IAar, مُتَعَقِّبًا [to which I am unable to assign an appropriate meaning, except its modern one of *lagging behind*]: so in the saying of a poet,

* رَبَا الجُوعُ فِى أَوْنَيْهِ حَتَّى كَأَنَّهُ *
* جَنِيبٌ بِهِ إِنَّ الجَنِيبَ جَنِيبُ *

[*Hunger increased in him* (lit. in the two sides of his saddle-bags); *so that he seemed as though he walked on one side, bent thereby; for he who has a pain in his side walks on one side, in that manner*]. (TA.) = Also *An excellent kind of dates,* (K, TA,) *well known;* (TA;) *one of the best kinds of dates.* (Mgh in art. جمع, Msb.)

جَنَابَةٌ: see جَنْبٌ, in four places: and see جَانِبٌ. = See also جُنُبٌ. — Accord. to IAth, its primary signification is *Distance:* and hence it signifies The *state of him who is under the obligation of performing a total ablution, by reason of sexual intercourse and discharge of the semen.* (TA.) — The *sperma genitalis* [itself]. (K. [But in a marginal note in my copy of that work I find this last signification rejected as erroneous.]) = See also the next paragraph.

جَنِيبَةٌ *A led horse or mule or ass;* (S, TA;) *a horse that is led* [*by one's side*], *not ridden:* (Msb:) pl. جَنَائِبُ. (A, TA.) — جَنِيبَتَا البَعِيرِ *The* [*two equal*] *loads on the two sides of the camel.* (K.) — [Hence, app.,] اِتَّقِ ٱللّٰهَ الَّذِى لَا جَنِيبَةَ لَهُ ‡ *Fear thou God, to whom there is no equal.* (A, TA.) — Also, جَنِيبَةٌ, (S,) or ▾ جَنَابَةٌ, (K,) or both, (TA,) *A she-camel that one gives* [or *lends*] *to people,* (S, M, K,) *with money,* (M, TA,) *in order that they may bring corn or other provision for him;* (S, M, K;) also called عَلِيقَةٌ: pl. جَنَائِبُ. (S.) = Also, (Kr, M, K,) and خَبِيبَةٌ, (M, TA,) The *wool of a* ثَنِىّ [or *sheep in its third year*]: (Kr, M, K:) it is better and cleaner than what is termed عَقِيقَةٌ, which is the wool of a جَذَع [or sheep in or before its second year]. (TA.)

[جَنُوبِىٌّ *Of,* or *relating to, the quarter of the wind termed the* جَنُوب; *south,* or *southerly.*]

جَنَائِبُ as an extr. pl.: see جَنْبٌ, first sentence.

جُنَّابٌ: see جَنِيبٌ.

جَانِبٌ; pl. جَوَانِبُ: see جَنْبٌ, in eleven places.

[Hence, لَانَ جَانِبُهُ † *He was*, or *became*, *gentle*, *easy to deal with*, *compliant*, or *obsequious*. And رَجُلٌ لَيِّنُ الجَانِبِ † *A man who is gentle, easy to deal with, compliant*, or *obsequious*; *contr. of* غَلِيظُ الجَانِبِ ; see art. غلظ : and see جَنَابٌ. And] تُزَنُّ بِلِينِ الجَانِبِ † [*She is suspected of easiness*, or *compliance*], (Ḳ in art. لبس,) *towards him who desires of her that he may lie with her*. (TA in that art.) [Hence also,] جَانِبَا الأَنْفِ (CḲ) and ↓ جَنَابَتَا and ↓ جَنْبَتَا and ↓ جَنَبَتَا (Ḳ) *The two sides of the nose*: (Ḳ:) or *the two lines that surround the two sides of the nose* of a doe-gazelle: (Sb, TA:) pl. [of the second, agreeably with analogy,] جَنَائِبُ. (TA.) — See also جَنَابٌ. [It often signifies The *vicinage* or *neighbourhood* of a people &c.: and *a region* or *quarter* or *tract* of a people or country: like ناحية. — The *bank* of a river; and *any bank*, or *steep acclivity*. — And *A limit, bound*, or *boundary*: see a tropical usage of its pl. (جَوَانِب) voce. جَنْو — And عَلَى جَانِبٍ means *Beside, aside*, or *apart*; and so جَانِبًا, and جَانِبٌ مِنْ مَالٍ — فِى جَانِبٍ, in post-classical writings, means *A portion*, and particularly *a large portion, of property*: and جَانِبٌ alone, in the same, *a sum*, and particularly *a large sum, of money*. — The latter, also, in post-classical writings, signifies, like جَنْبٌ, q. v., † *A man's honour*, or *reputation, which should be preserved inviolate*; so used in the Ḳ voce عِرْضٌ, in an explanation of the latter word taken from ʼIAth; *i. q.* نَامُوس and حُرْمَةٌ, as in the TḲ in that case.] ═ *Avoided and despised*. (Ḳ, TA.) — [Hence, perhaps, دَعْ كَذَا جَانِبًا *Let thou*, or *leave thou, such a thing alone*: see an ex. voce أَوْ.] — See also جُنُبٌ, in four places. — And see مُجْتَنَبٌ.

أَجْنَبُ : see جُنُبٌ, in two places.

أَجْنَبِىٌّ : see جُنُبٌ, in three places. — You say also, هُوَ أَجْنَبِىٌّ مِنْ كَذَا (A,) or عَنْ كَذَا, (TA,) ‡ *He has no concern nor acquaintance with such a thing*. (A, TA.)

مَجْنَبٌ (Ṣ, AAF, Ḳ) and ↓ مِجْنَبٌ (AAF, Ḳ) *Much* (A'Obeyd, Ṣ, AAF, Ḳ) of good (A 'Obeyd, Ḳ) and of evil. (Ḳ.) You say, إِنَّ عِنْدَنَا لَخَيْرًا مَجْنَبًا *Verily with us is much good*, and شَرًّا مَجْنَبًا *much evil*. (Ṣ.) And طَعَامٌ مَجْنَبٌ means *Much* [*wheat* or *food*]. (Sh, TA.)

مِجْنَبٌ : see what next follows.

مِجْنَبٌ *A shield*; (Ṣ, A, Ḳ;) because it wards off from its possessor what is displeasing to him; (A, TA;) also with ḍamm to the م [app. ↓ مُجْنِبٌ, act. part. n. of 4]. (Ḳ.) — *A thing by which a person or thing is veiled, concealed*, or *hidden*; *a veil, curtain*, or *covering*; (Ḳ, TA;) for a house, or chamber, or tent. (TA.) — *A thing like a door, upon which the gatherer of honey stands*; (Ḳ, TA;) he being let down [upon it] by means of ropes to [the place of] the honey [in the face of a rock or mountain]. (TA.) — *A thing* (شَبَحٌ [app. here meaning *a wooden implement*]) *resembling a comb without teeth* (Ḳ, TA) *and thin-edged in its lowest part*, (TA,) *with which earth is raised upon*, or *against, the* أَعْضَاد *and* فَلْجَان [*or raised borders of watering-troughs or the like, and streamlets for irrigation*]. (Ḳ, TA. [In the CḲ, الفُلْجان is put for الفَلْجان.]) — *The extreme part of the territory of the foreigners towards that of the Arabs*: (Ṣ, Ḳ:) and the *nearest part of the territory of the Arabs to that of the foreigners*. (Ṣ.) ═ See also مُجْتَنَبٌ.

مُجَنَّبٌ ; and its fem., with ة : see جَنِيبٌ. — Also, the former, (TA,) or ↓ مُجَنِّبٌ, (Ḳ, [but this is said in the TA to be a mistake,]) A horse *wide in the space between the two hind legs*, (Ḳ, TA,) *without what is termed* فَحَجٌ [which is an awkward kind of straddling, with the hocks wide apart]: it is a quality approved. (TA. [See also 2; and see مُجَنَّبٌ.])

مُجَنِّبٌ A man *whose sheep or goats* [&c.] *have few young ones*; [and therefore, *having little milk*;] (TA in art. يسر;) *contr. of* مُيَسِّرٌ. (Ṣ and TA in that art. [See also 2.])

مُجَنِّبَةٌ The *van*, or *fore part*, (Ḳ, TA,) of an army. (TA.)

المُجَنِّبَتَانِ *The right and left wings* of an army: (Ḳ: [Golius has erroneously written مُجَنَّبَتَانِ, and has given J as the authority instead of the Ḳ:]) or مُجَنَّبَةٌ signifies *a portion of an army* (كَتِيبَةٌ) *that takes one of the two sides of a way*: but the former meaning is the more correct. (IAạr, TA.)

مَجْنُوبٌ pass. part. n. of 1 [q. v.]. — See also جَنِيبٌ. — Also *Affected by the disease termed* ذَاتُ الجَنْبِ [or *pleurisy*]: (Ṣ, Mgh, Mṣb, TA:) and said to mean also *having a complaint of his side*, absolutely. (TA.) — And *Affected by the* [*south*, or *southerly*, *wind called*] جَنُوب. (Ṣ, TA.) [And *Affected by that wind in one's cattle*: see 1, last sentence.] سَحَابَةٌ مَجْنُوبَةٌ *A cloud brought by the blowing of that wind*. (Ṣ, A, Ḳ.) The saying of Aboo-Wejzeh,

* مَجْنُوبَةُ الأُنْسِ مَشْمُولٌ مَوَاعِدُهَا *

means *Her familiarity passes away with the* جَنُوب [or *south-wind*], *and her promises pass away with the* شَمَال [or *north wind*]. (IAạr, TA.)

جنث

جِنْثٌ *i. q.* أَصْلٌ [as meaning The *stock*, or *stem*, or the *root*, or *base*, or *lowest part*, of a tree or plant: and the *stock*, or *origin*, of a man]; (Ṣ, Ḳ;) a dial. var. of جِنْسٌ, or a word mispronounced: (Ṣ:) or, as some say, *a root* (of a tree) *of which the* أُرُومَة [i. e. *root-stock*] *is erect in the ground*: or the *part of the* سَاق [i. e. *stock*, or *stem*,] of a tree *that is in the ground, above the* عُرُوق [or *roots properly so called*]: (L, TA:) pl. [of pauc.] أَجْنَاثٌ and [of mult.] جُنُوثٌ. (TA.) One says, فُلَانٌ مِنْ جِنْثِكَ and جِنْسِكَ, meaning مِنْ أَصْلِكَ [i. e. *Such a one is of thy stock*]. (Ṣ.) And إِنَّهُ لَيَرْجِعُ إِلَى جِنْثِ صِدْقٍ [*Verily he traces back his lineage to an excellent stock*, or *origin*]. (TA.)

جنح

1. جَنَحَ, (Ṣ, A, Ḳ, &c.,) aor. َ , (Ṣ, Mṣb, Ḳ,) agreeably with analogy, of the dial. of Temeem, and the most chaste form, (TA,) and ُ , (Ṣ, Mṣb, Ḳ,) of the dial. of Ḳeys, (TA,) and ِ , (Ḳ,) inf. n. جُنُوحٌ, (Ṣ, Mgh, Mṣb, Ḳ,) *He*, or *it, inclined, leant*, or *propended*, (Ṣ, A, Mgh, L, Ḳ,) إِلَيْهِ (L, Mṣb) and لَهُ (L) [*to*, or *towards, it*]; as also ↓ اجتنح, (Ṣ, Mgh, Ḳ,) and ↓ اجنح. (Ḳ [but, by the omission of a و after it, this is made in some copies of the Ḳ to relate to what there follows it].) It is said in the Ḳur [viii. 63], وَإِنْ جَنَحُوا لِلسَّلْمِ فَاجْنَحْ لَهَا (Mgh, L) *And if they incline to peace, incline thou to it*: سلم being here made fem. because syn. with مُصَالَحَة. (L.) You say, جَنَحُوا لِلسَّلْمِ and إِلَيْهِ. (A.) — *He* (a man) *inclined*, or *leant, on one side*; and *leant upon his bow*: as also ↓ اجتنح: and اجتنح عَلَيْهِ *he leant upon him*. (L.) And جَنَحَتْ *She* (a camel lying on her breast) *leant on one side*. (AO, TA.) — *He* (a man) *set about a thing, to do it with his hands, his breast leaning over it*. (T, TA.) — جَنَحَ عَلَى مِرْفَقَيْهِ, inf. n. جُنُوحٌ and جَنْحٌ, *He* (a man) *rested himself upon his elbows, having set them upon the ground or upon a cushion*. (ISh, TA.) — جَنَحَ إِلَيْهِمْ and لَهُمْ *He* [*inclined to them*; or] *followed them and submitted to them*; namely, a sect. (ISh, TA.) — جَنَحَتِ الشَّمْسُ لِلْغُرُوبِ [*The sun inclined to setting*]. (A.) — جَنَحَ, (A, L, Mṣb,) aor. َ , (L, Mṣb,) inf. n. جُنُوحٌ, (Ṣ, L, Ḳ,) said of the night, (Ṣ, A, L, Mṣb, Ḳ,) and of the evening, (A,) and of the darkness, (L,) *It inclined to going*, or *to coming*: (A:) or *it came on*, or *approached*. (Ṣ, L, Mṣb, Ḳ.) — Also, with the same aor. and inf. n., said of a bird, *It contracted its wings to descend*, or *alight, and approached like one falling, and repairing to a place of refuge*. (L.) — جَنَحَتْ said of camels, *They lowered the fore part of the neck* [in running]: or *they went quickly*, or *swiftly*. (TA.) — And, inf. n. جُنُوحٌ, said of a ship (سَفِينَة), *She came to shallow water, and stuck to the ground*, (A, L,) *so as to cease from motion*. (L.) ═ جَنَحَ, inf. n. جُنُوحٌ, *He* (a man) *gave with his* [جَنَاح, or] *hand*. (TA.) ═ جَنَحَهُ, (Ṣ, L,) aor. َ , inf. n. جَنْحٌ, (L,) *He hit*, or *hurt, its* جَنَاح [or *wing*]; (Ṣ, L;) i. e., the جناح of the bird. (Ṣ.) And جَنَحَ فُلَانًا *He hit*, or *hurt, the arm* (جَنَاح) *of such a one*. (Ḳ. [In some copies of the Ḳ, by the omission of a و, this signification is erroneously made to relate to اجنح: so in the copies used by MF and SM, who state that the right verb is جَنَحَ.]) ═ جُنِحَ, (Ṣ, Ḳ,) with ḍamm, (Ṣ,) like عُنِىَ, (Ḳ,) inf. n. جُنُوحٌ, (TA,) *He* (a camel) *had his* جَوَانِح [the ribs so called] *broken by reason of the heaviness of his load*: (Ṣ, Ḳ:) or

he (a camel) *had the first of his ribs broken in the part next the breast.* (TA.) — [جَنِحَ also signifies *He regarded* an act *as a crime,* or *sin.* Thus,] إِنِّى لَأَجْنَحُ أَنْ آكُلَ مِنْهُ, in a trad. respecting the property of the orphan, means *Verily I regard as a crime,* or *sin,* (جُنَاح,) *my eating,* or *devouring,* [aught] *thereof.* (TA.)

[2. جنّحهُ, inf. n. تَجْنِيحٌ, *He furnished it with wings,* or †*the like:* see مُجَنَّحٌ.]

4. اجنح: see 1, first sentence. — اجنحهُ *He made him,* or *it, to incline, lean,* or *propend.* (Ṣ, Ḳ.)

5: see 8.

8. اجتنح: see 1, in three places. — Also *He* (an old man) *leant towards the ground, supporting himself with his hands upon his knees,* by reason of his weakness. (Mgh.) And *He* (a man prostrating himself in prayer) *rested upon his palms, putting his fore arms apart* (*from his sides,* IAth), *not laying them on the ground;* (*so that they became like the wings of a bird;* IAth); as also ▼تجنّح. (Sh, IAth, Mgh, Ḳ.) — اِجْتِنَاحٌ in a she-camel is The *going quickly,* or *swiftly:* (Sh, Ḳ:) or the *going so that her hinder part as it were leans towards her fore part, by reason of her vehement pressing on,* (ISh, Ḳ,*) *by her pushing forward her hind legs towards her breast:* (ISh:) and in a horse, the *running with a uniform leaning on one side.* (A, O, Ḳ.)

10. استجنح *It* (the night) *began.* (L.)

جِنْحُ اللَّيْلِ (Ṣ, A, Mṣb, Ḳ) and ▼جُنْحُهُ (Ṣ, Mṣb, Ḳ) *A part,* or *portion, of the night:* (Ṣ, A, Ḳ:) or *a great,* or *the greater, part thereof:* or *the first part thereof:* or *a part thereof, about the half:* (L:) or *the darkness thereof;* and *its confusedness.* (Mṣb.) كَأَنَّهُ جِنْحُ لَيْلٍ [*As though it were a portion,* &c., *of a night*] is said of a numerous army heavily encumbered. (L.) [See also an ex. in a verse cited voce إِنَّ.]

جِنْحٌ *A side* (Ṣ, Mṣb, Ḳ) of a road. (Ṣ, Mṣb.) — And The *vicinage* or *neighbourhood,* or the *region* or *quarter* or *tract,* and the *shadow* or *shelter* or *protection,* syn. نَاحِيَةٌ and كَنَفٌ, (Ṣ, Ḳ,) as also ▼جَنَاحٌ, (Ḳ,) of a people, or party, or company of men: (Ṣ:) the latter word thus used is tropical. (TA.) You say, بَاتَ بِجِنْحِ القَوْمِ *He passed the night in the vicinage,* &c., *of the people.* (Ṣ, TA.) And ▼أَنَا فِى جَنَاحِهِ ‡ *I am in his shadow, shelter,* or *protection.* (TA.) — See also the next preceding paragraph.

جَنَاحٌ The يَد (Ṣ, Ḳ) [meaning *wing*] of a bird or flying thing; (Ṣ;) i. e., of a bird or flying thing, the *limb that corresponds to the* يد *of a man:* (Mṣb:) and also the يد [i. e. *arm,* sometimes also meaning *hand,* (see a signification of جِنْح,)] of a man: (L, TA:) and (Ḳ) the *upper arm,* or *arm from the shoulder to the elbow:* (Zj, L, Ḳ:) each of these is so called because it is on one side of the body: (L:) and the *armpit:* (Ḳ:) pl. أَجْنِحَةٌ (Ṣ, Mṣb, Ḳ) and أَجْنُحٌ: (IJ, Ḳ:) the sing., though masc., has the latter pl., which properly belongs to a fem. sing. [of this form], because جناح is assimilated to رِيشَةٌ; (IJ;) [or rather, I think, to يَدٌ, which is fem.;] but some assert جناح to be both masc. and fem. (MF.) [Hence,] هُوَ مَقْصُوصُ الجَنَاحِ [*He has the wing clipped;* meaning] ‡ *he is one who lacks strength* or *power* or *ability; he is impotent.* (A, TA.) And خَفَضَ لَهُ جَنَاحَهُ ‡ [*He abased himself to him:* lit. *he lowered to him his wing:* but see an explanation of a similar phrase in the Ḳur, below]. (A.) And رَكِبُوا جَنَاحَىِ الطَّائِرِ, (Fr, L, Ḳ,) in [some of the copies of] the Ḳ, الطَّرِيقِ, (TA,) † *They quitted their homes,* or *accustomed places.* (Fr, L, Ḳ.) And فُلَانٌ فِى جَنَاحَىْ طَائِرٍ ‡ *Such a one is in a state of disquiet, and confounded,* or *perplexed, unable to see his right course.* (L, A.*) And رَكِبَ فُلَانٌ جَنَاحَىِ النَّعَامَةِ ‡ *Such a one employed himself vigorously, labouriously, sedulously,* or *diligently, in an affair;* (A, Ḳ;) *managing well.* (Ḳ.) And نَحْنُ عَلَى جَنَاحِ السَّفَرِ [lit. *We are on the wing of travel;* meaning] ‡ *we are about to travel,* or *journey.* (Ḳ, TA.) And جَنَاحُ الفَرَسِ † *A certain star* [γ] *of Pegasus; one of the four bright stars, in Pegasus, which form a square;* the other three being that at the extremity of the neck, called عَيْنُ الفَرَسِ, [i. e. α of Pegasus,] that called مَنْكِبُ الفَرَسِ, [β of Pegasus,] and the star [α of Andromeda] that belongs to both Pegasus and Andromeda. (Ḳzw.) [And جَنَاحُ سَمَكَةٍ †*The fin of a fish.*] And جَنَاحَا نَصْلٍ †*The two wings,* or *blades, of a spear-head* or *of an arrow-head.* (L.) And جَنَاحُ الرَّحَى † *The wing* (نَاعُور) *of the mill* or *mill-stone.* (L.) And جَنَاحَا عَسْكَرٍ ‡ *The two wings of an army.* (A, TA.) And جَنَاحَا الوَادِى ‡ *The two sides of the valley* (A, L) *down which the water runs, on the right and left.* (L.) And ثَرِيدَةٌ لَهَا جَنَاحَانِ and ▼مُجَنَّحَةٌ بِالعُرَاقِ مِنْ عُرَاقٍ ‡ [*A mess of crumbled bread moistened with broth, having two side-garnishes of bones with some meat remaining upon them*]. (A, TA.) — See also جِنْحٌ, in two places. — Also The *side,* syn. جَانِبٌ. (Ḳ.) So in the saying in the Ḳur [xvii. 25], وَٱخْفِضْ لَهُمَا جَنَاحَ الذُّلِّ, meaning وَأَلِنْ لَهُمَا جَانِبَكَ الذَّلِيلَ ‡ [*And make soft to them* (thy two parents) *thy submissive side;* i. e. *treat them with gentleness and submissiveness:* or the former words may be literally rendered *lower to them the wing of submissiveness;* meaning *be submissive to them*]. (Jel, TA.) — And *A part,* or *portion,* of a thing; as also ▼جُنَاحٌ. (Ḳ.)

جُنَاحٌ *A sin, a crime,* or *an act of disobedience:* (AHeyth, Ṣ, A, IAth, L, Mṣb, Ḳ:) or *an inclining thereto:* (IAth,* L, TA:) and *anxiety, and annoyance* or *molestation* or *hurt, which one is made to bear.* (L, TA.) لَا جُنَاحَ عَلَيْكُمْ, in the Ḳur ii. 235, means, accord. to AHeyth, *There shall be no sin,* or *crime, chargeable upon you:* or, accord. to others, *there shall be no straitening of you.* (L.) — See also جَنَاحٌ, last signification.

جَانِحٌ *Inclining, leaning,* or *propending:* pl. أَجْنَاحٌ, like as أَشْهَادٌ is pl. of شَاهِدٌ. (L, TA.)

جَانِحَةٌ sing. of جَوَانِحُ; (Ṣ, Ḳ;) which latter signifies The *ribs of the breast:* (A:) or the *ribs that are beneath* [those called] *the* تَرَائِب, *of the part next the breast;* (Ṣ, Ḳ;) like the ضُلُوع of the part next the back: (Ṣ:) or the *anterior parts of those ribs;* so called because they incline over the heart: or the *short ribs that are in the anterior part of the breast:* or, of a camel and a horse and the like, the *ribs against which lies the shoulder-blade:* and of a man, the *ribs of the back which are called* دَأْى, *six in number, three on the right and three on the left.* (L.)

مُجْنَحَةٌ *A piece of leather upon the fore part of the camel's saddle, upon which the rider leans with his hands, thus resting himself.* (TA. [See 8.])

مُجَنَّحٌ [*Furnished with wings,* or †*the like*]. — [Hence,] ثَرِيدَةٌ مُجَنَّحَةٌ بِالعُرَاقِ: see جَنَاحٌ. — نَاقَةٌ مُجَنَّحَةُ الجَبِينِ † *A she-camel wide in the* جبين [app. here meaning the *forehead*]. (TA.)

جند

2. جنّد جُنُودًا (Ṣ, A) *He collected,* or *assembled,* جنود [i. e. *armies, military forces,* &c.]. (A.)

5. تجنّد *He took,* or *prepared, for himself, a* جُنْد [i. e. *an army, a military force,* &c.]. (A, TA.)

جُنْدٌ [a coll. gen. n.] *An army; a military force; a legion; a body of troops* or *soldiers;* (L, Ḳ;) *a collected body of men prepared for war:* (Mgh:) *auxiliaries:* (Ṣ, L, Mṣb, Ḳ:) n. un. ▼جُنْدِىٌّ [signifying *a trooper; a soldier; an auxiliary*]; like as رُومِىٌّ is n. un. of رُومٌ: (Mṣb:) pl. of the former, جُنُودٌ (Ṣ, A, Mgh, Mṣb) and [pl. of pauc.] أَجْنَادٌ. (Mgh, Mṣb.) It is said in a trad., ▼الأَرْوَاحُ جُنُودٌ مُجَنَّدَةٌ (Ṣ, A, L) *Souls are troops collected together:* meaning that they are created of two parties, each party agreeing together, and disagreeing with the other party, like two armies opposed, each to the other. (L.) — Also *Any kind,* or *species, of creatures,* or *created things, regarded as alone, by itself,* or *apart from others.* (Ḳ,* TA.) It is said in a prov., إِنَّ لِلّٰهِ جُنُودًا مِنْهَا العَسَلُ *Verily to God belong kinds,* or *species, of created things* [*by means of which He effects his purposes as by armies* or *auxiliaries*], *of which is honey:* (Z, Ḳ:) first said by Mo'áwiyeh, when he heard that El-Ashtar had been given to drink some honey containing poison, in consequence of which he died: and used on the occasion of rejoicing at a misfortune that has befallen an enemy: it occurs in the history of El-Mes'oodee thus; إِنَّ لِلّٰهِ جُنْدًا فِى العَسَلِ. (MF.) — Also, [as a term used in Syria, and afterwards by the Arabs from Syria in Spain,] *A city* [*with its territory;* i. e. *a* *province, or district*]: (Ḳ:) or particularly *a city of Syria* [*with its territory*]: (AO, TA:) pl. أَجْنَادٌ. (TA.) Syria consists of five اجناد; namely, Dimashk [or Damascus], Ḥims [or Emessa], Ḳinnesreen, El-Urdunn [or the Jordan], and Filasteen or Falasteen [or Palestine]: (Ṣ, M, A:) they were thus called because the military forces were thence collected. (A.) [See مِخْلَافٌ.]

جُنْدِىٌّ: see جُنْدٌ. — Also *Of*, or *relating to*, *a* جُنْد *of Syria*. (A.)

جُنْدٌ مُجَنَّدٌ *An army*, or *a military force*, *collected*. (TA.) See also جُنْدٌ.

جندب

جُنْدَبٌ and جُنْدُبٌ and جِنْدَبٌ: see art. جدب.

جندر

جَنْدَرَ: see Q. Q. 1 in art. جدر.

جندل

Q. 1. [جَنْدَلَهُ *He*, or *it*, *made him to cleave to the stones*. Hence,] تَرِبَتْ يَدَاهُ وَجَنْدَلَتْ [*May his arms*, or *his hands*, *cleave to the dust*, or *earth*, *and the stones, by reason of poverty*]. (M in art. ترب.)

جَنْدَلٌ (Ṣ, Ḳ) and جَنَدِلٌ (Ḳ) *Stones*; (Ṣ in art. جدل;) used in the sense of [the pl.] جَنَادِلُ: (Sb, TA:) n. un. جَنْدَلَةٌ: (TA:) or *what a man can lift, of stones*: (Ḳ:) or, as some say, *any stone*: (TA:) or *a mass of stone like a man's head*. (T, TA.) [Hence, تُرْبًا لَهُ وَجَنْدَلًا: see تُرَابٌ.]

جَنَدِلٌ, (Ṣ in art. جدل,) or جُنَدِلٌ, (Kr, Ḳ,) *A place in which are stones* (Kr, Ṣ, Ḳ) *collected together*: (Kr, Ḳ:) but ISd doubts its correctness. (TA.) And أَرْضٌ جَنَدِلَةٌ, (Ḳ,) and sometimes with fet-ḥ, (Ṣgh, Ḳ,) i. e., to the ج, [جَنَدَلَةٌ,] (TḲ,) *A land abounding with stones*. (Ṣgh, Ḳ.)

جُنَادِلٌ *Strong and great*. (Ḳ.)

جنز

1. جَنَزَهُ, aor. ـِ, (A, Mṣb, Ḳ,) and ـُ, (A,) inf. n. جَنْزٌ, (TA,) *He veiled, concealed, hid*, or *covered*, *it*. (A, Mṣb, Ḳ.) — *He collected it*; or *gathered it together* or *up*; (Ḳ;) as also ↓ جنّزه, inf. n. تَجْنِيزٌ. (Ṣgh, TA.) You say also, جُنِزَ الرَّجُلُ, part. n. ↓ مَجْنُوزٌ, *The man was gathered up* [app. as a corpse is gathered up in the grave-clothes]: (TA:) or † *he died*. (Ḥar p. 122.)

2. جَنَّزَهُ, inf. n. تَجْنِيزٌ: see 1. — It is used by El-Ḥasan El-Baṣree as signifying *He put it* (namely a corpse) *upon the bier*. (Ḳ,* TA.)

جِنَازَةٌ and جَنَازَةٌ signify the same, (Mgh, Mṣb, Ḳ,) namely, *A dead person*; *a corpse*: (Ḳ:) but the former is the more chaste: (Mṣb:) or the former signifies the *dead person*; and the latter, the *bier*: (Aṣ, IAạr, Mṣb, Ḳ:) or the former, the *bier*; and the latter, the *dead person*: (Mgh, Mṣb, Ḳ:) so accord. to Th, as related by Aboo-'Amr Ez-Záhid: (Mṣb:) or the former is said by the vulgar to mean the *bier*: (Aṣ, TA:) or the former signifies the *bier with the dead person*: (Ḳ:) or the *dead person upon the bier*: (Ṣ:) and the bier is not so called unless the dead person is upon it: (AAF:) when the dead person is not upon it, it is called سَرِيرٌ and نَعْشٌ: (AAF, Ṣ:) but the vulgar say جَنَازَةٌ, with fet-ḥ; (Ṣ;) which is not allowable: (Lth, Aṣ, Mgh, TA:) or جَنَازَةٌ [so in the TA] signifies the *man*: or the *bier with the man*: (En-Naḍr, TA:) جنازة is derived from جَنَزَهُ, in the first of the senses assigned to it above: (A, Mṣb, TA:) so some assert: (IDrd, TA:) but ISd says, I know not whether this be correct: (TA:) or a جِنَازَة is so called because the clothes are gathered together when the man is upon the bier: ('Abd-Allah Ibn-El-Ḥasan, TA:) some say that it is Nabathean: (TA:) the pl. is جَنَائِزُ. (Ṣ.) The Arabs say, تَرَكْتُهُ جِنَازَةً *I left him a corpse*, or *dead*. (Aṣ, TA.) And ضُرِبَ الرَّجُلُ حَتَّى تُرِكَ جِنَازَةً [so in the TA, *The man was beaten until he was left a corpse*]. (ISh, TA.) And رُمِىَ فِى جِنَازَتِهِ, meaning, *He died*: (Lth, Mgh:) or *he has been carried*, or *lifted, and put* [*into his bier*]: this they say when they give information of the death of a man: (TA:) and they say also, طُعِنَ فِى جِنَازَتِهِ, meaning the same. (Lth, Mgh, TA.) — Also جِنَازَةٌ [or جَنَازَةٌ] † *A wine-skin* [*when emptied*; as though it were a body without a soul]. (Ḳ.) A certain impudent man of the Arabs, 'Amr Ibn-Ḳi'ás, says,

* وَكُنْتُ إِذَا أَرَى زِقًّا صَرِيعًا *
* يُنَاحُ عَلَى جِنَازَتِهِ بَكَيْتُ *

[*And I used, when I saw a wine-skin laid prostrate, its corpse being wailed over, to weep*, or accord. to Th, as is said in the TA in art. بكى, *to sing*]. (TA.) — † *Anything oppressive and grievous*, عَلَى قَوْمٍ *to a people*. (Lth, Ḳ.) — † *A sick person*. (Ṣgh, Ḳ.)

جَنَائِزِىٌّ, [from جَنَائِزُ, pl. of جِنَازَةٌ,] *One who recites* [or *chants the profession of the faith &c.*] *before the dead* [*in a funeral-procession*]. (TA.)

مَجْنُوزٌ: see 1.

جنس

2. جنّسهُ بهِ, (TḲ,) inf. n. تَجْنِيسٌ, from الجِنْسُ, (Ṣ, Ḳ,) [*He made it homogeneous*, or *congenial*, *with it*; or *similar*, or *conformable*, *to it*: expl. in the TḲ, not well, by شاكله; but the inf. n., with tolerable correctness, by ايكى شيئى برى برينه مشاكل قلمق. The usage of the term تَجْنِيسٌ in rhetoric, to signify the use of two or more words completely or partly conformable, is post-classical, like جِنَاسٌ, an inf. n. of جَانَسَ.]

3. جانسهُ, [inf. n. مُجَانَسَةٌ and جِنَاسٌ,] *It was*, or *became*, *homogeneous*, or *congenial*, *with it*; or *similar*, or *conformable*, *to it*; syn. شَاكَلَهُ: (Mgh, Mṣb:) الْمُجَانَسَةُ is from الجِنْسُ. (Ṣ, TA.) You say, هٰذَا يُجَانِسُ هٰذَا *This is homogeneous with this*; syn. يُشَاكِلُهُ: (Mgh, Mṣb:) so says Kh. (Mṣb.) And كَيْفَ يُؤَانِسُكَ مَنْ لَا يُجَانِسُكَ [*How will he be sociable with thee who will not be congenial with thee?*]. (A.) And of a man who has not discrimination nor intelligence, one says, فُلَانٌ يُجَانِسُ البَهَائِمَ وَلَا يُجَانِسُ النَّاسَ [*Such a one resembles the beasts, and does not resemble men*]: (Mgh, Mṣb,* TA:) so says Kh. (Mgh.) But Aṣ says that this usage, (Mgh, Mṣb,) in the first and last of the above-mentioned phrases, (Mṣb,) is post-classical. (Mgh, Mṣb.) The usage of the term جِنَاسٌ by rhetoricians [to signify the complete or partial conformity of two or more words] is post-classical [like تَجْنِيسٌ]. (TA.)

6. تجانس الشَّيْآنِ [*The two things were*, or *became homogeneous, congenial, similar*, or *conformable*,] is a phrase of the scholastic theologians, not [classical] Arabic. (TA.)

جِنْسٌ [*A genus, kind*, or *generical class, comprising under it several species*, or *sorts*; or *comprised under a superior genus, in relation to which it is a species*, or *sort*;] *a* ضَرْبٌ of a thing; (Ṣ;) or of anything; (Mgh, Mṣb;) *any* ضَرْب of a thing; (A, Ḳ;) [as] of men, and of birds, and of the definitions of grammar and of the art of versification, and of things collectively; so accord. to the lexicologists; (ISd, TA;) *a term of more common import than* نوع [which is a *species*, or *sort*]: (Ṣ, A, Mgh, Mṣb, Ḳ:) thus animal is a جنس and man is a نوع, (Mgh, Mṣb,) because the latter is of more particular import than the former, though it is a جنس in relation to what is under it; but the scholastic theologians reverse the case, (Mgh,) for with them جنس is of more particular import than نوع: (Kull p. 139:) thus also camels are a جنس of beasts: (A, Ḳ:) pl. أَجْنَاسٌ [properly a pl. of pauc. but used also as one of mult.] (Mgh, Mṣb, Ḳ) and جُنُوسٌ. (IDrd, Ḳ.) You say, النَّاسُ أَجْنَاسٌ وَأَكْثَرُهُمْ أَنْجَاسٌ [*Men are of several kinds, and most of them are impure*]. (A, TA.) And فُلَانٌ مِنْ جِنْسِكَ, meaning أَصْلِكَ [i. e. *Such a one is of thy stock*]. (Ṣ in art. جنث.) — [Hence, اِسْمُ جِنْسٍ *A generic noun*: and اِسْمُ جِنْسٍ جَمْعِىٌّ *a collective generic noun*.] — أَوْصَى لِجِنْسِهِ signifies *He left by will*, of his property, *to the children of his father*, [or *his kindred by the father's side*,] *exclusively of all relations of the mother*: and so, لِأَهْلِ بَيْتِهِ. (Mgh.) — The assertion, in the Ḳ, that J's saying, on the authority of IDrd, that Aṣ used to say الجِنْسُ as meaning الْمُجَانَسَةُ is a vulgarism, is erroneous, is a matter for consideration; for Aṣ said not this, but [what has been cited above, voce جَانَسَهُ, or] what will be found below, voce مُجَانِسٌ. (TA.)

[جِنْسِىٌّ *Generic*; *generical*.]

[جِنْسِيَّةٌ *Generical quality*.]

مُجَانِسٌ *Homogeneous*; *congenial*; *similar*;

conformable; syn. مُشَاكِلٌ. (Ḳ.) But IDrd asserts that Aṣ used to reject the saying of the vulgar, هٰذَا مُجَانِسٌ لِهٰذَا [*This is homogeneous with this*, &c.], and to say, It is post-classical. (Ṣ.) [See also 3.]

جنف

1. جَنَفٌ (T, Ṣ, Mgh, Mṣb, Ḳ, &c.) and جُنُوفٌ, (Ḳ,) the former being inf. n. of جَنِفَ and the latter of جَنَفَ, (TA,) The act of *inclining*, or *declining* (T, Ṣ, Mgh, Ḳ, TA) in speech and in all affairs: (TA:) and *declining*, or *deviating, from the right course; acting wrongfully, unjustly, injuriously*, or *tyrannically*; (T, Mṣb,* Ḳ, TA;) like حَيْفٌ, which some erroneously assert to be the act of a judge only. (T, TA.) You say, جَنِفَ, (T, Ṣ, Mgh, Mṣb,) aor. ـَ, inf. n. جَنَفٌ; (Ṣ, Mṣb;) [and app. جَنَفَ, aor. ـِ, inf. n. جُنُوفٌ;] and ▼ اجنف; (T, Mṣb, TA;) *He inclined* or *declined* [in speech and in any affair]: (T, Ṣ, TA:) and *he declined*, or *deviated, from the right course; acted wrongfully, unjustly, injuriously*, or *tyrannically*; (T, Mgh, Mṣb, TA;) عَلَيْهِ *against him*. (Mgh.) Hence, in the Ḳur [ii. 178], فَمَنْ خَافَ مِنْ مُوصٍ جَنَفًا (Ṣ, TA,) i. e. [*And he who feareth*, (or, as is said in the Ḳ in art. خوف, *knoweth*,) *from*, or *on the part of, the testator*,] *an inclining* [*to a wrong course*], or *a declining* [*from the right course*]: (TA:) or *a manifest inclining* or *declining*. (Er-Rághib, TA.) You say, جَنِفَ فِى وَصِيَّتِهِ, aor. ـَ; [*He inclined*, &c., *in his testament*;] and so ▼ اجنف. (Ḳ.) And ▼ اجنف signifies *He deviated from that which was right*. (Ḳ.) And اجنف ▼ فِى حُكْمِهِ [*He declined from the right course*, or *acted wrongfully* or *unjustly, in his judgment*]. (TA.) Or ▼ اجنف relates peculiarly to the case of a testament: and جَنِفَ signifies absolutely *He declined*, or *deviated, from that which was right*. (Ḳ.) And you say, جَنِفَ عَنْ طَرِيقِهِ, aor. ـَ; and جَنَفَ, aor. ـِ; inf. n. جَنَفٌ, (Ḳ, TA,) which is of the former verb, (TA,) and جُنُوفٌ, (Ḳ, TA,) which is of the latter verb; *He turned away from his course*, or *way*; *deviated therefrom*. (TA.) — Or [app. a mistake for "and"] جَنَفٌ signifies *Depression* (دُخُولٌ and اِنْهِضَامٌ) *in one of the two sides of the breast*, or *chest*, (الزَّوْر,) *with evenness of the other side*: (Ḳ:) the verb is جَنِفَ: and the part. n. is ▼ جَنِفٌ and ▼ أَجْنَفُ, fem. [of the latter] جَنْفَآءُ. (TA.)

3. جِنَافٌ [an inf. n. of which the verb (جانف) is not mentioned]. You say, لَجَّ فِى جِنَافٍ قَبِيحٍ *He persisted in removing himself to a distance*, or *estranging himself, from his family*; (Aboo-Sa'eed, Ḳ;) like جِنَابٍ قَبِيحٍ. (Aboo-Sa'eed, TA.)

4. اجنف *He committed an act of inclining* [*to a wrong course*], or *declining* [*from the right course*, or *acting wrongfully* or *unjustly* or *injuriously* or *tyrannically*]; like as you say, أَلَامَ, meaning "he did a thing for which he should be blamed." (Ṣ.) See also 1, in five places. — اجنفهُ *He found him to be one who deviated from the right way in his judgment; who acted wrongfully, unjustly, injuriously, or tyrannically, therein*. (Ḳ.)

6. تجانف فى مشيته *He carried himself in a proud and self-conceited manner*, [*affecting an inclining of the body from side to side*,] *in his gait*. (TA.) [And *He inclined on one side in his gait*: said of one who is lame of one leg: see مُخْرُوقٌ.] — تجانف عَنْ طَرِيقِهِ *He affected a declining*, or *deviating, from his course*, or *way*; *he purposely declined*, or *deviated, therefrom*; (Ḳ,* TA;) syn. تَمَايَلَ. (Ḳ.) And in like manner, تجانف إِلَى الشَّىْءِ [*He affected an inclining to the thing*; *intentionally inclined to it*]. (TA.) You say, تجانف لِإِثْمٍ *He inclined to a sin*, (Ṣ, Mgh,) [or *affected an inclining to it*, (see the part. n., below,)] *intending*, or *purposing, the commission of an act of disobedience*. (Mgh.)

جَنَفٌ is [an epithet] like دَنَفٌ, applying to a sing. and a pl., [being] also an inf. n. [of جَنِفَ, q. v.]: Abu-l-'Iyál El-Hudhalee says,

* أَلَّا دَرَأْتَ الْخَصْمَ حِينَ رَأَيْتَهُمْ *
* جَنَفًا عَلَىَّ بِأَلْسُنٍ وَعُيُونِ *

[*Wherefore didst not thou repel the adversaries, when thou sawest them inclining*, or *acting wrongfully, against me with tongues and eyes?*]: or, accord. to one reading, جُنُفًا [which is a pl. of ▼ جَانِفٌ]: (Skr p. 128:) [or, as some say,] جَنَفًا is a pl. [or rather quasi-pl. n.] of ▼ جَانِفٌ, like as رَوَحٌ is of رَائِحٌ: or it may be for ذَوِى جَنَفٍ. (TA.)

جَنِفٌ: see أَجْنَفُ, in two places: — and see also 1, last sentence.

جُنَافِىٌّ One *who carries himself in a proud and self-conceited manner*, (مُخْتَالٌ, [for which Golius, here copied by Freytag, appears to have read مُحْتَالٌ, with the unpointed ح,]) *with an inclining* [*of the body from side to side*]: (Sh, Ḳ:) or one *who affects an inclining* [*of the body from side to side*] (يَتَجَانَفُ) *in his gait, and carries himself in a proud and self-conceited manner therein*. (TA.)

جَانِفٌ: see أَجْنَفُ; and see also جَنَفٌ, in two places.

أَجْنَفُ [fem. جَنْفَآءُ] *Inclining* [*to a wrong course*], or *declining* [*from the right course*], or *acting wrongfully* or *unjustly*, [absolutely, as also ▼ جَنِفٌ and ▼ جَانِفٌ and ▼ مُجْنِفٌ, or peculiarly] in his testament. (Ḳ.) You say, صَادَفَهُ ▼ جَنِفًا فِى حُكْمِهِ [*He found him to be one who deviated from the right way in his judgment; who acted wrongfully, unjustly, injuriously*, or *tyrannically, therein*]. (Ḳ.) And ▼ خَصْمٌ مُجْنِفٌ *An adversary who inclines* [*to a wrong course*], or *declines* [*from the right course*]; (Ḳ;) *who acts wrongfully, unjustly, injuriously*, or *tyrannically*. (TA.) — *Inclining*; [or *pendulous*;] applied to a penis. (TA.) — *Having a bending back*; (Ṣ, Ḳ;) applied to a man. (Ṣ.) — See also 1, last sentence. — *Big*, or *large*; applied to a bowl (قَدَح). (TA.)

مُجْنِفٌ: see أَجْنَفُ, in two places.

غَيْرَ مُتَجَانِفٍ لِإِثْمٍ, in the Ḳur [v. 5], means *Not affecting an inclining to sin*; *intending*, or *purposing, it*. (Mṣb, TA.)

جنق

1. جَنَقُوا, aor. ـُ, (IAạr, Ḳ,) inf. n. جَنْقٌ; (IAạr, TA;) and ▼ جنّقوا, inf. n. تَجْنِيقٌ; (AZ, Ḳ;) *They cast stones of* [or *with*] *the* مَنْجَنِيق: (Ḳ,* TA:) and ▼ مَجْنَقُوا, (Ḳ,) or مَنْجَنِيقًا مَجْنَقُوا [app. meaning *they constructed a* منجنيق], (Lth, TA,) is said by such as hold the م to be radical; (Ḳ;) or it may be that the م is augmentative, and that the verb is [denominative,] similar to تَمَسْكَنَ from مِسْكِينٌ. (TA.) — جَنَقُونَا بِالْمَنْجَنِيقِ *They cast at us with the* منجنيق: mentioned by El-Fárisee on the authority of AZ. (M.) An Arab of the desert, in describing the wars of his people, said, تَارَةً نُجْنَقُ وَأُخْرَى نُرْشَقُ [*One time we are cast at with the* منجنيق, *and another we are shot at with arrows*]: (Ṣ,* M:) also mentioned by El-Fárisee. (M.)

2. جنّقوا المَجَانِيقَ and ▼ مَجْنَقُوهَا [app. *They constructed the engines of the kind called* منجنيق]. (JK.) See also 1.

Q. Q. 1. مَجْنَقُوا: see 1 and 2.

جَنَقٌ The *stones of the* منجنيق. (M, TA.) [But] IAạr says that جنق [app. جُنُقٌ] means The *people who manage the* منجنيق. (TA.)

مُجَيْنِيقٌ dim. of مَنْجَنِيقٌ, q. v. (Sb, Ṣ.)

مَنْجَنُوقٌ: see what next follows.

مَنْجَنِيقٌ (Ṣ, M, Mṣb, Ḳ) and مِنْجَنِيقٌ (Mṣb, Ḳ) and ▼ مَنْجَنُوقٌ (Lth, IAạr, Mṣb, Ḳ) [*A kind of balista*;] *an instrument* [or *engine of war*] *with which stones are cast* (Ṣ, M, Ḳ, TA) *at the enemy*; *made by binding* [*in some manner*] *very tall wooden poles, whereon is placed what is to be cast, which is then struck with a long pole, and so cast to a very distant place*: [this imperfect description (the only one that I have found of this engine) seems to show that it was of a very simple and rude construction:] it is an ancient instrument, anterior to the invention [or use] of gunpowder and cannons by the Christians; and was used by Moḥammad in the siege of Eṭ-Ṭáif; but the first [of the Arabs] who used it in the time of paganism is said to have been Jedheemeh El-Abrash, of the Mulook eṭ-Ṭawáif: (TA:) pl. مَنْجَنِيقَاتٌ (Ṣ, Mṣb, Ḳ) and مَنْجَنُوقَاتٌ (Lth, TA) and مَجَانِيقُ (Sb, Ṣ, Mṣb, Ḳ) and مَجَانِقُ: (Ḳ:) dim. ▼ مُجَيْنِيقٌ: (Sb, Ṣ:) it is fem., (Lth, Ṣ, Mṣb, Ḳ,*) preferably, (Lth, TA,) and in most instances; (Mṣb;) and hence it is also called الأُنْثَى; (M in art. انث;) but sometimes it is masc.: (Mṣb, Ḳ:) and is arabicized, (Ṣ, Mṣb, Ḳ,) from the Persian مَنْ چِى نِيكْ, (Ṣ, Ḳ) i. e. "How excellent am I!" (Ṣ,) or مَنْ چِهْ نِيكْ, i. e. "I, how excellent am I!" (Ḳ;) [Golius asserts it to be from the Greek Μάγγανον; but this is the original of مَنْجَنُونٌ, which has a different meaning: its derivation, however, seems to have been unknown to the Arabs in general; for] some say that مَنْجَنِيقٌ is of the measure مَفْعَلِيل, [so in copies of the Ṣ, but correctly مَنْفَعِيل, as in the Mṣb and TA,] because of the phrase نُجْنِقُ مَرَّةً [see 1]: (Ṣ, Mṣb:*) or, (Mṣb,) accord. to Sb, (Ṣ,) it is of the measure فَنْعَلِيل; (Ṣ, Mṣb;) because مَجَانِيق is its pl. and the dim. is مُجَيْنِيق; and because, if the م were augmentative as well as the ن, there would be two augmentative letters at the beginning of the noun, and this may not be in substs. nor in epithets that are not conformable to augmented verbs; and if the ن were made an essential part of the word, the noun would be a quadriliteral-radical, and augmentative letters are not prefixed to words of this class except nouns conformable to their verbs, like مُدَحْرَج: (Ṣ, TA:) but MF says that these disquisitions are unprofitable; for in his opinion, all the letters of the word are radical, since it is a foreign word. (TA.)

مَنْجَنِيقِىّ [app. *A maker of the kind of engine called* منجنيق; being mentioned as a surname:] a rel. n. from منجنيق. (Ḳ.)

جنك

جُنْكٌ [and جَنْكٌ] *A certain* [*musical*] *instrument, which is beaten like the* عود [or *lute*; namely, the *Persian harp*; two specimens of which are figured in Note 26 to Chap. iii. of my translation of the "Thousand and One Nights"]; mentioned by El-Khafájee in the Shifá el-Ghaleel; and often spoken of: it is an arabicized word; (MF, TA;) from the Persian چنگ: and applied [also] to the دف, [دُفّ or دَفّ, i. e., *tambourine*,] *with which one plays*. (TA.)

جُنْكِىّ [and جَنْكِىّ] *A player on the* جنك *above mentioned*. (TA.)

جنو

جَنًا *i. q.* جَنْأٌ [inf. n. of جَنِئَ, q. v.]. (Ṣgh, Ḳ.)

أَجْنَا, applied to a man, *i. q.* أَجْنَأُ, q. v. (Ṣgh, Ḳ.) And [its fem.] جَنْوَآءُ, *i. q.* جَنْآءُ, (Ṣgh, Ḳ,) applied to a ewe, or a she-goat, meaning *Having her horns bending backwards*. (TA.)

جنى

1. جَنَى الثَّمَرَةَ, (Ṣ, Mgh, Mṣb, Ḳ,) aor. ـِ, (Ṣ, Mṣb,) inf. n. جَنْىٌ, (Ṣ, TA,) or جَنًى, (so accord. to one copy of the Ṣ, and written in the accus. case جنيا in the Ḥam p. 355,) and جِنَايَةٌ, (Ḥam ib., but there without any vowel-signs,) *He gathered, plucked,* or *took from the tree, the fruit*; (Mgh, TA;) *i. q.* ↓ اجتناها (Ṣ, Mṣb, Ḳ) and ↓ تجنّاها: (Ḳ:) and in like manner one says of a thing similar to fruit. (TA.) One says also, جَنَاهُ إِيَّاهَا and جَنَاهَا لَهُ [*He gathered it, plucked it,* or *took it from the tree, for him*]. (A'Obeyd, Ḳ.) And جَنَيْتُكَ أَكْمُؤًا [*I gathered for thee truffles*]. (TA.) And جَنَى ذَهَبًا *He collected gold from its mine*. (TA.) — Aboo-Dhu-eyb uses this verb metaphorically, in the phrase جَنَى العُلَى, meaning ‡ [*He acquired*] *eminence,* or *nobility*. (TA.) — You also say, جَنَى عَلَيْهِ, (Ṣ,) or عَلَى قَوْمِهِ, (Mṣb,) inf. n. جِنَايَةٌ, (Ṣ, Mṣb,) ‡ *He committed,* [*against him,* or] *against his people,* or *party, a crime,* or *an offence for which he should be punished*; (Mṣb;) as also ↓ جاناه [or جَانَى قَوْمَهُ]: (TḲ in art. جر:) [and ‡ *he brought an injury upon him,* or *them*:] and جَنَى الذَّنْبَ عَلَيْهِ, aor. ـِ, inf. n. جِنَايَةٌ, (Ḳ,) with kesr, (TA,) ‡ *He committed the crime, offence,* or *injurious action, against him*; syn. جَرَّهُ إِلَيْهِ: (Ḳ: [see art. جر:]) thus used, also, the verb is metaphorical, from جَنَى الثَّمَرَةَ: (Er-Rághib, TA:) and so in the phrase, جَنَى شَرًّا, meaning ‡ *He brought to pass an evil thing* or *action*. (Mgh.) [See also جِنَايَةٌ, below.] It is said in a trad., لَا يَجْنِى جَانٍ إِلَّا عَلَى نَفْسِهِ † [*An injurer shall not bring injury save on himself*]; meaning that one shall not be prosecuted for an injurious action committed by another, of his relations or of others. (TA.) And a poet says,

جَانِيكَ مَنْ يَجْنِى عَلَيْكَ وَقَدْ *
تُعْدِى الصِّحَاحَ فَتَجْرَبُ الجُرْبُ *

[which may be rendered, † *Thine injurer* whom thou shouldst punish *is he who brings an injury upon thee: but sometimes the mangy* camels *infect the sound* ones *so that these become mangy*; and thus a criminal sometimes brings punishment upon his relations: for] A'Obeyd says that جانيك من يجنى عليك is a prov. applied to the case of a man who is punished for an injurious action; because brothers [sometimes] bring injury upon a man [by occasioning his being punished for an injurious action which they have themselves committed], as the latter hemistich of the verse cited above indicates: but AHeyth says that this prov. means الجَانِى لَكَ الخَيْرَ مَنْ يَجْنِى عَلَيْكَ الشَّرَّ [*The person bringing thee good is he who brings,* or *will bring, upon thee evil*: perhaps intended as a caution; for the Arabs often suspect that a benefactor has some evil intention]: and he cites the following hemistich:

تُعْدِى الصِّحَاحَ مَبَارِكُ الجُرْبِ *

[meaning Sometimes *the places where the mangy camels lie down,* and which afford benefit to other camels, *infect the sound* ones]. (TA. [See also Freytag's Arab. Prov., i. 298.]) You say also, جَنَيْتَ هٰذَا عَلَى نَفْسِكَ [*Thou hast brought this as an injury upon thyself*]. (Ḳ in art. جل.)

3. جَانَى عَلَيْهِ, inf. n. مُجَانَاةٌ, *He accused him of a* جِنَايَة [or *crime*, &c.]. (TA.) — See also 1.

4. اجنى said of a tree, (Ṣ, Ḳ,) or of a palm-tree, (Mṣb,) *It had ripe fruit*: (Ṣ:) or *it attained to the time for the gathering of its fruit*: (Mṣb:) or *it attained to maturity*: (Ḳ:) or, said of a tree, *it had fruit to be gathered and eaten*: and, said of fruit, *it became ripe*: (TA:) and, said of grass, or herbage, *it became abundant*. (KL.) — And اجنت الأَرْضُ *The land had much* جَنًى, (Ṣ, Mṣb, Ḳ,) i. e. *herbage,* and *truffles,* and *the like*. (Ṣ.)

5: see 1. — تجنّى عليه, (Ṣ,* Ḳ,) or تجنّى عليه ذَنْبًا, (TA,) † *He accused him of a crime, an offence,* or *an injurious action, that he had not committed*; (Ṣ, Ḳ;) i. e. *he forged against him the charge of his having committed a crime,* &c., *he being guiltless* [*thereof*]: (TA:) التَّجَنِّى being like التَّجَرُّم. (Ṣ.) You say also, يَتَجَنَّى عَلَيْنَا مَا لَمْ نَجْنِهِ [*He accuses us of committing what we did not commit*]. (Abu-l-'Abbás, TA in art. جرم.)

8: see 1. — اِجْتَنَيْنَا مَآءَ مَطَرٍ *We came to rain-water, and drank it*: (Ḳ:) a phrase mentioned with approval by IAạr, but not explained by him: thought by ISd to have this meaning. (TA.)

جَنًى [in the CḲ جَنْى] *Whatever is gathered,* or *plucked*; as also ↓ جَنَاةٌ: (Ḳ:) or *whatever is gathered,* or *plucked, from trees,* (Ṣ,) *&c.*; (so in a copy of the Ṣ;) as also ↓ جَنَاةٌ: (Ṣ:) so that these two words are of the same class as حَقٌّ and حَقَّةٌ: or the latter of them is a n. un.: (TA:) or the former signifies *what is gathered from trees while fresh*; (Mṣb;) as also ↓ جَنِىٌّ: (Mṣb:) or this last is an epithet applied to fruit, signifying *just gathered* or *plucked*; (Ṣ, Ḳ;) or *gathered,* or *plucked, while fresh*: (TA:) and ↓ مَجْنِىٌّ, also, pl. مَجَانٍ, signifies *fruit gathered* or *plucked*: (Ḥar p. 369:) جَنًى also signifies *fruit* [*ready to be gathered* or *plucked*]; so in the Ḳur lv. 54: (Jel:) and is applied to *fresh ripe dates*: (Fr, Ḳ:) and *grapes*: (TA:) and *truffles,* and *the like*: (Ṣ:) and even *cotton*: (TA:) and *herbage*: (Ṣ:) and *gold,* (Ḳ,) which is collected from its mine: (TA:) and *cowries,* (Ḳ,) as though gathered from the sea: (TA:) and *honey,* (Ḳ,) *when it is gathered*: (TA:) pl. أَجْنَآءٌ (Ḳ) and أَجْنٍ, originally أَجْنُىٌ. (TA.) Hence the saying,

هٰذَا جَنَاىَ وَخِيَارُهُ فِيهِ *
إِذْ كُلُّ جَانٍ يَدُهُ إِلَى فِيهِ *

[*This is what I have gathered, and the best of it is in it; when every gatherer* but myself *has his hand to his mouth*]: or, accord. to one reading, وَهِجَانُهُ فِيهِ (which has the same meaning, TA in art. هجن): a prov., ascribed by Ibn-El-Kelbee to 'Amr Ibn-'Adee El-Lakhmee, the son of the daughter of Jedheemeh: he says that Jedheemeh had ordered the people to gather for him truffles, and some of them ate the best that they found; but 'Amr brought to him the best that he found, and addressed to him these words: and 'Alee is related to have repeated them on an occasion of his entering the government-treasury; meaning that he had not defiled himself with anything of the tribute belonging to the Muslims, but had put it in its places. (TA.)

جَنَاةٌ: see جَنًى, in two places.

جَنِىٌّ: see جَنًى. Also *Dates cut from the tree*. (TA.)

جِنَايَةٌ, primarily, The act of *gathering, plucking,* or *taking from a tree,* fruit: [see 1:] — then, † The *bringing to pass an evil thing,* or *action*: (Mgh, Kull p. 147:) — then, † *Evil,* [*itself*]: — then, † The *doing a forbidden action*: (Kull ib.:) specially used in this last sense; though it has a general application: (Mgh:) — [as a simple subst., it generally signifies] *A crime, an offence,* or *an injurious action, for which one should be punished*:

(Mṣb:) or *an action that a man commits requiring punishment or retaliation to be inflicted upon him in the present world and in the world to come:* (TA:) or *any forbidden injurious action:* (Kull p. 134:) and in the language of the lawyers, especially *a wounding:* and *an amputation*, or *a maiming:* (Mṣb:) and ↓ جِنْيَةٌ signifies the same as جِنَايَةٌ: (Ḥam p. 241:) the pl. of جناية is جِنَايَاتٌ and جَنَايَا; but the latter of these pls. is of rare occurrence. (Mṣb.)

جُنْيَةٌ: see what next precedes. ═ Also *A [garment such as is called]* رِدَآءٌ, (Ḳ, TA,) *of a round form*, (TA,) *made of [the kind of cloth termed]* خَزّ. (Ḳ, TA.)

جَانٍ *A gatherer* of fruit [&c.]: — and also †*A committer of a* جِنَايَة [or *crime*, &c.]: (Ḳ,* TA:) pl. جُنَاةٌ (Ṣ, Ḳ) and جُنَّآءُ (Sb, Ḳ) and [of pauc.] أَجْنَآءُ, which last is extr., (Ṣ, Ḳ,) or doubtful. (Ṣ.) Hence the prov., أَجْنَاؤُهَا أَبْنَاؤُهَا, explained in art. بنى. (Ṣ, TA.) — [Hence also,] الجَانِى †*The wolf.* (IAạr, TA voce جَابٍ, q. v.) ═ Also *i. q.* لَقَّاحٌ; (IAạr, Az, TA;) i. e. *A fecundater of palm-trees.* (Az, TA.)

الجَوَانِى *i. q.* الجَوَانِبُ [pl. of الجَانِبُ]; (Ḳ;) similar to الثَّعَالِى and الأَرَانِى. (TA.)

مَجْنًى: see جَنًى.

مُجْتَنًى *A place of gathering, or plucking,* fruits, &c. (TA.)

جهد

1. جَهَدَ, (Ṣ, A, L, &c.,) aor. ـَ, (Ḳ,) inf. n. جَهْدٌ, (TA,) *He strove, laboured,* or *toiled; exerted himself* or *his power* or *efforts* or *endeavours* or *ability; employed himself vigorously, strenuously, laboriously, diligently, studiously, sedulously, earnestly,* or *with energy; was diligent,* or *studious; took pains,* or *extraordinary pains;* (Ṣ, A, L, Ḳ;) فِى كَذَا *in such a thing;* (Ṣ;) or فِى الأَمْرِ *in the affair;* (A;) as also ↓ اجتهد; (A, Ḳ;) and so ↓ جاهد, with respect to speech and actions: (L:) or جَهَدَ فِى الأَمْرِ, aor. and inf. n. as above, *he did his utmost,* or *used his utmost power* or *efforts* or *endeavours* or *ability, in prosecuting the affair:* (Mṣb:) and ↓ اجتهد and ↓ تجاهد *he exerted unsparingly his power,* or *ability:* (Ṣ, A, Ḳ:) or ↓ اجتهد فِى الأَمْرِ *he exerted unsparingly his power,* or *ability, in the prosecution of the affair, so as to effect his utmost.* (Mṣb.) You say also, اِجْهَدْ جَهْدَكَ فِى هٰذَا الأَمْرِ ‡ *Do thine utmost in this affair:* (Fr, Ṣ, Ḳ:*) but not اُجْهُدْكَ. (Fr, Ṣ.) And ↓ اجتهد رَأْيَهُ ‡ *He took pains,* or *put himself to trouble* or *fatigue, to form a right judgment* or *opinion.* (MA.) And ↓ اِجْتَهَدْتُ رَأْيِى وَنَفْسِى حَتَّى بَلَغْتُ مَجْهُودِى †*I exerted my judgment and my mind so that I attained the utmost of my power,* or *ability.* (T, L.) — جَهَدَ بِهِ *He tried, proved,* or *examined, him,* (L, Ḳ,) عَنِ الخَيْرِ وَغَيْرِهِ [*respecting good* qualities, *&c.*]. (L.) ═ جَهَدَهُ, (Mgh, L, Mṣb,) aor. ـَ, (Mgh, L,) inf. n. جَهْدٌ, (L, Mṣb,) *It,* (an affair, and a disease,) and *he,* (a man,) *affected him severely; harassed, embarrassed, distressed, afflicted, troubled, inconvenienced, fatigued,* or *wearied, him:* (Mṣb:) *it* (disease, L and Ḳ, and fatigue, and love, L) *rendered him lean; emaciated him:* (L, Ḳ:) *he burdened him beyond his power; imposed upon him that which was beyond his power;* as also ↓ اجهدهُ: (Mgh:) and, [as also ↓ اجهدهُ,] *he importuned him, harassed him,* or *plied him hard, in asking, begging,* or *petitioning.* (A.) [Hence,] جُهِدَ, said of a man, *He was severely affected, harassed, embarrassed, distressed, afflicted, troubled, inconvenienced, fatigued,* or *wearied:* (Ṣ, L:) or *was grieved,* or *made sorry* or *unhappy.* (L.) And أَصَابَهُمْ قُحُوطٌ مِنَ المَطَرِ فَجُهِدُوا جَهْدًا شَدِيدًا *Drought befell them, and they consequently became severely distressed.* (Ṣ.) And جُهِدُوا *They were,* or *became, afflicted with drought, barrenness,* or *dearth;* or *with drought, and dryness of the earth.* (L.) And رَجُلٌ يَجْهَدُ أَنْ يَحْمِلَ سِلَاحَهُ, for يَجْهَدُ نَفْسَهُ مِنَ الضَّعْفِ, *A man who imposes upon himself a difficulty,* or *trouble,* or *fatigue,* or *a difficult* or *severe task,* or *who strains,* or *strains himself, in the carrying of his weapons,* or *arms, by reason of weakness.* (Mgh.) And جَهَدَ دَابَّتَهُ and ↓ اجهدها *He jaded, harassed, distressed, fatigued,* or *wearied, his beast; i. q.* ↓ بَلَغَ جَهْدَهَا: (Ḳ:) or *he tasked,* or *plied, his beast beyond his power in journeying,* or *marching,* or *in respect of pace.* (Ṣ, Mṣb.) And أَجْهَدْتُهُ عَلَى أَنْ يَفْعَلَ كَذَا وَكَذَا [*I importuned him,* or *harassed him, to do such and such things*]. (L.) — Also, (Ṣ, Mṣb,) aor. as above, (A,) and so the inf. n., (Mṣb,) ‡ *He deprived it* (namely, milk,) *of its butter,* (Ṣ, A, Ḳ,) *entirely:* (Ṣ, Ḳ:) or *churned it so as to extract its butter and render it sweet and pleasant:* or *mixed it with water:* (Mṣb:) or *diluted it so that it consisted for the most part of water:* and in like manner it is used in relation to broth. (A.) — Hence, (Mṣb,) جَهَدَهَا †*He lay with her; or compressed her:* (L, Mṣb, from a trad.:) or *i. q.* دَفَعَهَا, and حَفَزَهَا [which has a similar meaning]. (L.) — جَهَدَ الطَّعَامَ †*He desired the food eagerly; longed for it;* (Ṣ, Ḳ;) as also ↓ اجهدهُ. (Ḳ.) And جُهِدَ الطَّعَامُ and ↓ أُجْهِدَ †*The food was eagerly desired,* or *longed for.* (Ṣ.) — Also ‡ *He ate much of the food:* (Ṣ, Ḳ:) *he left nothing of it.* (A.) You say also, هٰذَا كَلَأٌ يَجْهَدُهُ المَالُ †*This is herbage,* or *pasture, of which the cattle eat perseveringly.* (AA, TA.) ═ جَهِدَ *It* (a state of life) *was,* or *became, hard, difficult, strait,* or *distressful.* (Ṣ, Ḳ.)

3. جِهَادٌ, inf. n. of جاهد, properly signifies The *using,* or *exerting, one's utmost power, efforts, endeavours,* or *ability, in contending with an object of disapprobation;* and this is of three kinds, namely, a visible enemy, the devil, and one's self; all of which are included in the term as used in the Ḳur xxii. 77. (Er-Rághib, TA.) See also 1, first sentence. You say, جاهد العَدُوَّ, (JK, A, Mgh,) inf. n. as above (JK, Mgh, Ḳ) and مُجَاهَدَةٌ, (JK, Ḳ,) *He fought with the enemy:* (Ḳ:) or *he encountered the enemy, imposing upon himself difficulty* or *distress* or *fatigue,* or *exerting his power* or *efforts* or *endeavours* or *ability,* [or *the utmost thereof,*] *to repel him, his enemy doing the like:* and hence جاهد came to be used by the Muslims to signify generally *he fought, warred,* or *waged war, against unbelievers* and *the like.* (Mgh.) You say also, جاهد فِى سَبِيلِ ٱللّٰهِ, inf. n. جِهَادٌ (Ṣ, Mṣb) and مُجَاهَدَةٌ, (Ṣ,) [*He fought, &c., in the way of God;* i. e., *in the cause of religion.*]

4. اجهد, as trans.: see 1, in six places. — Also *He made,* or *incited,* another, *to strive* or *labour* or *toil, to exert himself* or *his power* or *efforts* or *endeavours* or *ability, &c.;* trans. of 1 in the first of the senses assigned to it above. (JK.) — أُجْهِدَ *He was thrown into a state of difficulty, distress, affliction, trouble, inconvenience,* or *fatigue.* (L.) — اجهد مَالَهُ *He consumed,* or *wasted, and dispersed, his property:* (Ḳ:) or *gave it away, and dispersed it, altogether, here and there.* (En-Naḍr, TA.) ═ As intrans., *He* (an enemy) *strove, laboured,* or *exerted himself, in enmity,* (Ḳ, TA,) عَلَيْنَا *against us.* (TA.) — *He acted with energy,* or *with the utmost energy:* so in the phrases سَارَ فَأَجْهَدَ *He marched,* or *journeyed, and did so with energy,* or *with the utmost energy;* and حَلَفَ بِٱللّٰهِ فَأَجْهَدَ *He swore by God, and did so with energy, &c.:* in which cases one should not say فَجَهَدَ. (Aboo-'Amr Ibn-El-'Alà, L.) — *He took the course prescribed by prudence, precaution, and sound judgment,* فِى الأَمْرِ *in the affair;* syn. اِحْتَاطَ. (L, Ḳ.) — *He became in a state of difficulty, embarrassment, distress, affliction, trouble, inconvenience,* or *fatigue.* (L.) — †*It* (a thing) *became mixed,* or *confused.* (Ḳ.) ═ *He entered upon land such as is termed* جِهَاد: *he went forth into the desert;* and *into the plain,* or *open country.* (JK.) — *It rose up; rose into view; appeared.* (JK.) You say, اجهد لِىَ القَوْمُ *The people,* or *company of men, came within my sight,* or *view;* syn. أَشْرَفُوا. (AA, Ḳ.) And اجهد فِيهِ الشَّيْبُ *Hoariness appeared upon him, and became much:* (TA:) or ‡ *became much, and spread:* (A:) or *became much, and was quick in its progress,* (Ḳ, TA,) *and spread.* (TA.) And أَجْهَدَتْ لَهُ الأَرْضُ *The land became open to him.* (L, Ḳ.*) And in like manner, اجهد لهُ الطَّرِيقُ, (L,) and الحَقُّ, (L, Ḳ,*) *The road,* and †*the truth, became open, apparent, and manifest, to him.* (L, Ḳ.*) And اجهد لَكَ الأَمْرُ †*The thing became,* or *has become, within thy power,* or *reach;* (Aboo-Sa'eed, Ḳ;) *and offered,* or *presented, itself to thee.* (Aboo-Sa'eed, TA.)

6: see 1.

8: see 1, in five places. — اِجْتِهَادٌ as a conventional term means A lawyer's *exerting the faculties* [*of the mind*] *to the utmost, for the purpose of forming an opinion in a case of law* [*respecting a doubtful and difficult point*]: (KT:) the *seeking to form a right opinion:* (KL:) [*investigation of the law,* or the *working out a solution of any difficulty in the law, by means of reason and comparison:* and] the *referring a case proposed to the judge,* [*respecting a doubtful and difficult* *point,*] *from the method of analogy, to the Ḳur-án and the Sunneh.* (L, TA.*)

جُهْدٌ *Power; ability;* as also ▼ جَهْدٌ; (Ṣ, A, IAth, L, Mṣb, Ḳ;) the latter of the dial. of El-Ḥijáz, and the former of other dials.; (Mṣb;) and ▼ مَجْهُودٌ: (A:) جهد in the Ḳur ix. 80 is read both جُهْد and ▼ جَهْد: (Ṣ:) and جُهْدٌ signifies also *labour, toil, exertion, effort, endeavour, energy, diligence, painstaking,* or *extraordinary painstaking:* (L: [see جَهَدَ:]) or ▼ جَهْدٌ has the signification first mentioned above, (Fr, Ṣ, IAth, Mṣb,) and جَهْدٌ, with fet-ḥ, is from اِجْهَدْ جَهْدَكَ, (Fr, Ṣ,) or from جَهَدَ فِى هٰذَا الأَمْرِ, being an inf. n. from this verb, (Mṣb,) and signifies, [as also ▼ مَجْهُودٌ,] *one's utmost; the utmost of one's power* or *ability* or *efforts* or *endeavours* or *energy.* (Fr, Ṣ, IAth, Mṣb, Ḳ.) You say, بَذَلَ الجَهْدَ (Mṣb in art. بلغ, &c.,) and ▼ المَجْهُودَ, (Ṣ, A,) or جَهْدَهُ, (Mgh,) [and ▼ مَجْهُودَهُ,] *He exerted unsparingly his power or ability:* (Mgh:) [or *his utmost power* or *ability* or *efforts* or *endeavours* or *energy;* as shown above.] And بَلَغَ جَهْدَهُ, (A, L,) and ▼ مَجْهُودَهُ, (A,) *He accomplished the utmost of his power* or *ability; did his utmost.* (A, L. [Like جَهَدَ جَهْدَهُ. See also بَلَغَ جَهْدَ دَابَّتِهِ, below.]) And ▼ جُهَيْدَى is syn. with جَهْدٌ; (Ḳ;) as in the saying, لَأَبْلُغَنَّ جُهَيْدَاىَ فِى الأَمْرِ, (JK, TḲ,) i. e. *I will assuredly accomplish the utmost of my power, or ability, in the affair.* (TḲ. [In a copy of the A, جُهَيْدَاكَ; and so in the TA, I believe from that same copy.]) [So, too, is ▼ جُهَادَى; as in the saying,] جُهَادَاكَ أَنْ تَفْعَلَ *The utmost of thy power,* or *ability,* and *the utmost of thy case, is,* or *will be, thy doing* [such a thing]; syn. قُصَارَاكَ [q. v.], (JK, Ḳ,) and غَايَةُ أَمْرِكَ. (TA.) اَلَّذِينَ أَقْسَمُوا بِاللّٰهِ جَهْدَ أَيْمَانِهِمْ, in the Ḳur [v. 58, &c.], means *Who swore by God with the most energetic of their oaths:* (Ḳ,* Jel:) or *the strongest,* or *most forcible, of their oaths;* جهد being originally an inf. n., and in the accus. case as a denotative of state with يَجْهَدُونَ understood before it, or as an inf. n. (Bḍ.)—Also *Difficulty,* or *grievousness; embarrassment, distress, affliction, trouble, inconvenience, fatigue,* or *weariness;* (Ṣ, A, IAth, Mgh, Mṣb, Ḳ;) so accord. to some who say that ▼ جُهْدٌ, with ḍamm, has the first of the significations assigned to it above; (Mṣb;) as also ▼ مَجْهُودٌ: (Mgh:) *a disease,* or *difficulty, that distresses* or *afflicts, a man;* as also ▼ جُهْدٌ. (JK.) Hence, جَهْدُ البَلَآءِ, (Mṣb,) i. e. *A state of difficulty,* or *trouble, to which death is preferred:* or *largeness of one's family,* or *household, combined with poverty.* (L, Ḳ.*) [Hence also,] بَلَغَ جَهْدَ دَابَّتِهِ, [i. e. بَلَغَ مَشَقَّتَهَا,] *i. q.* جَهَدَهَا: see 1. (Ḳ.)—Also *Small provision, upon which a man possessing little property can live* (JK, L) *with difficulty.* (L.) And جَهْدُ المُقِلِّ *What a man who possesses little property can afford to give in payment of the poor-rate required by the law.* (L, from a trad.)

جُهْدٌ: see جَهْدٌ, in five places.==Also *Milk mixed* [*with water:* see مَجْهُودٌ]. (JK.)

جَهَادٌ *Hard land:* (JK, Ṣ:) or *land in which is no herbage:* (TA:) or *hard land in which is no herbage:* (Ḳ:) or *level,* or *even, land:* or *rugged land:* also used as an epithet; so that you say أَرْضٌ جَهَادٌ: (TA:) or *level, smooth land, in which is no hill:* (JK:) or *the most plain and even of land, whether it have produced herbage or not, not having any mountain or hill near it:* and such is what is termed a صَحْرَآء: (ISh, TA:) or *an open tract of land:* (Fr, TA:) or *sterile, barren,* or *unfruitful,* land, *in which is nothing;* as also جَمَادٌ: pl. جُهُدٌ. (AA, L.)==Also The *fruit of the* أَرَاك; (IAạr, Ḳ;) and so جَهَاضٌ. (IAạr, TA.)

مَرْعًى جَهِيدٌ ‡ *Pasture much eaten by cattle.* (Ṣ, A, Ḳ.) And أَرْضٌ جَهِيدَةُ الكَلَإِ ‡ *Land of which the herbage is much eaten by cattle.* (A.)

جُهَادَى: } see جَهْدٌ.
جُهَيْدَى: }

جَاهِدٌ [*Striving, labouring,* or *toiling;* &c.: see 1. Hence,] سَيْرُنَا جَاهِدٌ [*Our journeying is laborious*]. (TA in art. اخو.) And جَهْدٌ جَاهِدٌ [*Intense labour* or *exertion,* or *the like:* or *severe difficulty* or *distress* &c.]: an intensive expression, (Ḳ, TA,) like شِعْرٌ شَاعِرٌ and لَيْلٌ لَائِلٌ. (TA.)—† *Eagerly desiring* [food]; *longing for* [it]: (JK, Ṣ:) pl. أَجْهَادٌ. (JK.)—غَرْثَانُ جَاهِدٌ ‡ *Hungry and greedy, leaving no food.* (A.)

مُجْهَدٌ A man *thrown into a state of difficulty, embarrassment, distress, affliction, inconvenience, trouble,* or *fatigue.* (L.)

هُوَ مُجْهِدٌ لَكَ *He is one who takes the course prescribed by prudence, precaution,* or *sound judgment, for thee;* syn. مُحْتَاطٌ. (L.) And نَصِيحٌ مُجْهِدٌ *A sincere,* or *faithful, and careful, adviser,* or *counsellor.* (L.)—رَجُلٌ مُجْهِدٌ *A man in a state of difficulty, embarrassment, distress, affliction, inconvenience, trouble,* or *fatigue: possessing little property; poor.* (L.)—And *A man whose beast is weak by reason of fatigue.* (L.)

مَجْهُودٌ *Severely affected, harassed, embarrassed, distressed, afflicted, troubled, inconvenienced, fatigued,* or *wearied:* (Ṣ, Mgh, L:) *distressed,* or *afflicted, by disease* or *difficulty:* (JK:) *afflicted with drought, barrenness,* or *dearth;* or *with drought, and dryness of the earth:* (L:) and *angry.* (JK.)—*A hard, difficult, strait,* or *distressful,* state of life. (TA.)—‡ Milk *deprived of its butter* (Ṣ, A) *entirely:* (Ṣ:) or *mixed with water:* (Mṣb:) or *diluted so as to consist for the most part of water;* and in like manner, broth: (A:) or *churned so that its butter is extracted and it is rendered sweet and pleasant:* and used as meaning *eagerly desired,* or *longed for, and drunk without its occasioning disgust, by reason of its sweetness and pleasantness:* (Mṣb:) or *eagerly desired,* or *longed for;* and so food in general: (JK, L:) or *eagerly desired,* or *longed for, and drunk with perseverance, on account of its pleasantness and sweetness.* (L.)==See also جَهْدٌ, in six places.

جهر

1. جَهَرَ, (A, Mṣb, Ḳ,) aor. َ, (Mṣb, Ḳ,) inf. n. جَهْرٌ and جِهَارٌ, (Er-Rághib, TA,) *It* (a thing, A, Mṣb) *was,* or *became, plain, apparent, conspicuous, open,* or *public;* syn. ظَهَرَ, (A, Mṣb,) and بَدَا, (TA,) and عَلَنَ: (Ḳ:) or the radical signification is, *it* (a thing) *was,* or *became, exceedingly plain to be perceived,* either *by the sense of sight* or *by that of hearing.* (Er-Rághib, TA.) [Accord. to some, when relating to what is visible, it is tropical; and when relating to what is audible, proper: but if so, it seems to be so much used in the former sense as to be, in that sense, conventionally regarded as proper. See also جَهْرَةٌ.]==جَهُرَ, aor. ُ, [inf. n., app., جَهَارَةٌ and جُهُورَةٌ,] *He* (a man, TA) *was,* or *became, great,* or *bulky,* (Ḳ, TA,) [and therefore *a conspicuous object,*] *before the eyes of the beholder.* (TA.) [And *He was,* or *became, pleasing,* or *goodly, in aspect:* see جَهَارَةٌ, below.]—Also, (A, Mṣb, Ḳ,) inf. n. جَهَارَةٌ, (A, Mṣb,) *It* (the voice) *rose* [*so as to be plainly heard*]; *was,* or *became, high,* or *loud.* (A, Mṣb,* Ḳ.)—Also, (Ṣ,) inf. n. جَهَارَةٌ, (TA,) *He,* (a man) *was,* or *became, high,* or *loud, of voice.* (Ṣ, TA.)==جَهِرَ, aor. َ, (Mṣb,) inf. n. جَهَرٌ, (Ṣ, Mṣb,) *He* (a man) *was unable to see in the sun.* (Ṣ, Mṣb, TA.) And in like manner said of the eye. (Ḳ.)==جَهَرَهُ, (Mṣb, TA,) inf. n. جَهْرٌ; (TA;) and جَهَرَ بِهِ; (A, Mṣb;) and ▼ اجهرهُ, (A, Mṣb, TA,) [and ▼ اجهر بِهِ;] and ▼ جَهْوَرَهُ; (TA;) *He made it plain, apparent, conspicuous, open,* or *public.* (A, Mṣb, TA.)—جَهَرَ الكَلَامَ, and جَهَرَ بِهِ; (Ḳ;) and ▼ اجهرهُ, inf. n. إِجْهَارٌ; (Ṣ;) and ▼ اجهر بِهِ; (Ḳ;) and ▼ جَهْوَرَ; (TA;) and جَهَرَ بِالقَوْلِ, and بِقِرَآءَتِهِ, (TA,) and بِصَلَاتِهِ, and بِدُعَائِهِ, (Ṣgh, Mṣb, TA,) aor. َ, inf. n. جَهْرٌ and جِهَارٌ; (TA;) and ▼ اجهر بِقِرَآءَتِهِ; (Ṣgh, Mṣb, TA;) *He uttered the speech,* and *the saying,* and *his supplication,* and *his prayer,* and *his recitation, with a plain,* or *an open, voice; openly; publicly:* (Ṣ, Mṣb, Ḳ, TA:) or جَهَرَ بِكَلَامِهِ, (A,) and بِالقَوْلِ, and ▼ جَهْوَرَ; (Ṣ;) and بِقِرَآءَتِهِ; (A;) *he uttered his speech,* and *the saying,* and *his recitation, with a raised,* or *loud, voice; aloud:* (Ṣ, A:) and جَهَرَ الصَّوْتَ *he raised the voice* [*so as to make it plainly heard*]. (Ḳ.)—جَهَرَ بِالمَعَاصِى, and ▼ اجهر, and ▼ جاهر, *He made known the acts of disobedience* that he had committed, *by talking of them:* he who does so is termed ▼ مُجَاهِرٌ بِالمَعَاصِى, and simply ▼ مُجَاهِرٌ. (TA.) And ▼ اجهر مَا فِى صَدْرِهِ *He revealed what was in his bosom.* (A.) And ▼ جَهْوَرَ الحَدِيثَ بَعْدَ مَا هَيْنَمَهُ *He revealed the story after he had concealed it.* (A.) And ▼ اجهر الأَمْرَ *He made the case,* or *affair, notorious.* (TA.)—Also جَهَرَهُ *He discovered it* (Ḳ, TA) *ocularly.* (TA.)—*He saw him* (a man) *without any veil* (Ḳ, TA) *intervening;* (TA;) as also ▼ اجتهرهُ: (Ḳ:) or *he looked towards him,* or *regarded him.* (Ḳ.) You say, مَا فِى الحَىِّ أَحَدٌ

تَجْهَرُهُ عَيْنِى *There is not in the tribe any one whom my eye regards as worthy of notice or respect* by reason of his greatness therein; syn. تَأْخُذُهُ. (TA.) And ▼ اجتهر القَوْمُ فُلَانًا *The people looked towards such a one without any veil intervening between them and him.* (TA.) — *He treated him, or regarded him, with reverence, veneration, respect, or honour:* (Ḳ:) or (TA) *he regarded him as great in his eyes:* (Ḳ, TA:) *he saw him to be great in aspect, or appearance;* (Ṣ;) as also ▼ اجتهره (Ṣ, Ḳ) and ▼ استجهره: (A:) *he was pleased with his beauty, and his form, or appearance, or state of apparel or the like;* as also ▼ اجتهره: (Lḥ,* Ḳ:) or *he pleased him by his beauty and form or appearance* &c.: (A:) or *it pleased him by its beauty;* as also ▼ اجتهره. (TA.) — *He saw it* (an army, Ṣ, A, Ḳ, and a people, TA) *to be numerous in his eyes;* as also ▼ اجتهره. (Ṣ, A, Ḳ.) ═ جَهَرَ البِئْرَ, (Ṣ, Ḳ,) aor. َ, inf. n. جَهْرٌ, (TA,) *He cleared out the well,* (Ṣ, Ḳ,) *and took forth from it the black fetid mud that it contained;* as also ▼ اجتهرها: (Ṣ:) or both signify *he entirely, or nearly, exhausted the well of its water:* (Ḳ:) or the former, *he reached the water of the well,* (Ḳ, TA,) in digging: or so جَهَرَ alone: (TA:) and accord. to Akh, جَهَرْتُ الرَّكِيَّةَ signifies *I cleared out the mud that the water covered in the well, so that the water appeared and became clear.* (Ṣ.) 'Áisheh said, describing her father, ▼ اجتهر دُفُنَ الرَّوَآءِ, lit., *He cleared out the filled-up wells of abundant water so as to make the water well forth;* alluding to his rectifying affairs that had become disordered. (TA from a trad.) ═ جَهَرْنَاهُمْ *We came to them in the morning, at the time called* الصَّبَاح, (Ṣ, A, Ḳ, TA,) *when they were inadvertent.* (Ṣ, Ḳ, TA.) — جَهَرَ الأَرْضَ *He traversed the land* (Ṣ, Ḳ) *without knowledge.* (Ṣ.) ═ جَهَرَ السِّقَآءَ *He shook the milk-skin to make butter,* (Fr, Ṣ, Ḳ,) *and took forth its butter.* (Fr, TA.) ═ جَهَرَتِ الشَّمْسُ المُسَافِرَ *The sun dazzled the eye, and confused the sight, of the traveller;* syn. أَسْدَرَتْ عَيْنَهُ. (Ḳ.)

3. جاهر: see 1. — [Its inf. n.] مُجَاهَرَةٌ signifies The *fighting* [*with any one*] *face to face:* and the *showing open enmity, or hostility, with any one:* and the *reading, or reciting, a thing aloud:* and the *speaking loudly.* (KL.) You say, جاهر بالعَدَاوَةِ, (Mṣb,) inf. n. مُجَاهَرَةٌ (Ṣ, Mṣb) and جِهَارٌ, (Mṣb,) *He showed open enmity or hostility, with another.* (Ṣ,* Mṣb.) And جَاهَرْتُهُمْ بِالأَمْرِ *I acted openly with them in the affair, or case;* syn. عَالَنْتُهُم به. (JK.) [And جاهره *He treated him openly with enmity* &c.] — جَاهَرَهُمْ بِالأَمْرِ, (TA,) inf. n. مُجَاهَرَةٌ and جِهَارٌ, (Ḳ,) [is explained as signifying] *He vied with them, or strove to overcome or surpass them, in the affair, or case.* (Ḳ,* TA.) [But غالبهم, in the TA, and المُغَالَبَةُ, in the Ḳ, are here evidently mistranscriptions for عَالَنَهُم and المُعَالَنَةُ.]

4: see 1, in eight places. — اجهر also signifies *He begat sons goodly in stature* (IAạr, Ḳ) *and in aspect,* (IAạr, TA,) or *in cheeks:* (Ḳ:) or, *a squint-eyed son.* (IAạr, Ḳ.)

6. تَجَاهُرٌ signifies The *showing oneself openly: and acting openly, or being open in one's conduct or converse, with others.* You say,] تَجَاهَرُوا بِالعَدَاوَةِ *They showed open enmity, or hostility, one with another;* syn. تَبَادَوْا بِهَا. (Ṣ in art. بدو.) ═ [And تجاهر *He feigned himself unable to see in the sun:* see the part. n., below.]

8: see 1, in eight places.

10. استجهره: see 1. — Also *He took it forth.* (TA from a trad.)

Q. Q. 1. جَهْوَرَ: see 1, in four places.

جَهْرًا: see جَهْرَةٌ, in two places.

جُهْرٌ: see جَهَارَةٌ, in six places.

جَهُورٌ: see جَهِيرٌ, in two places.

جَهْرَةٌ *A thing that is plain, apparent, conspicuous, open, or public.* (Ḳ.) You say, رَآهُ جَهْرَةً, (Ṣ, A, &c.) *He saw him, or it,* [*plainly,*] *without the intervention of any veil:* (TA:) and ▼ رآه جِهَارًا [signifies the same: or] *he saw him, or it, with exceeding plainness:* (Er-Rághib, TA:) or the former signifies *he saw him, or it, with his eyes, ocularly,* or *before his eyes,* (Ṣ, A, Bḍ in ii. 52, Mṣb,) *without anything intervening:* (Ṣ:) so in the Ḳur. [ii. 52], حَتَّى نَرَى ٱللّٰهَ جَهْرَةً: (Ṣ, Bḍ:) and [some say that] جَهْرَةً is here originally an inf. n. of جَهَرْتُ in جَهَرْتُ بِالقِرَآءَةِ, [like ▼ جَهْرًا,] and metaphorically used in the sense of مُعَايَنَةً: it is in the accus. case as an inf. n.: or it is thus used as a denotative of state relating to the agent or the object: and some read ▼ جَهَرَةً, as an inf. n. like غَلَبَةً, or as pl. of جَاهِرٌ, and as such it is a denotative of state: (Bḍ:) or جَهْرَةً is here from جَهَرْتُ الرَّكِيَّةَ: (Akh, Ṣ:) accord. to Ibn-'Arafeh, it here signifies *unconcealed from us:* (TA:) and in the Ḳur. iv. 152, *ocularly; not concealed from us by anything.* (Ḳ,* TA.) — You say also, كَلَّمَهُ جَهْرَةً [and ▼ جَهْرًا *He spoke to him plainly, with an open voice, aloud, or publicly*]. (Ṣ, TA.) — And لَقِيَهُ نَهَارًا جِهَارًا ▼ and جَهَارًا ▼ [*He met him in the daytime, openly, or publicly*]. (Ḳ.)

جُهْرَةٌ [*A blaze covering the face of a horse:* or the *quality of having such a blaze:*] a subst. from أَجْهَرُ applied to a horse. (TA.) — *A cast in the eye.* (AA, TA. [See also أَجْهَرُ.])

جَهَرَةٌ: see جَهْرَةٌ.

جِهَارًا and جَهَارًا: see جَهْرَةٌ, in three places.

جَهُورٌ: see جَهِيرٌ. — Also, and ▼ مُجْتَهَرٌ, An army *seen to be numerous.* (A.) — And the former, *Bold; daring:* in the Ḳ, erroneously, ▼ جَوْهَرٌ. (TA.)

جَهِيرٌ (in the TA, here, ▼ جَهْرٌ, but in another place, جَهِيرٌ,) *High, loud,* or *vehement,* speech; (Mṣb, Ḳ, TA;) as also ▼ مُجْهَرٌ and ▼ جَهْوَرِىٌّ: (Ḳ:) and so applied to the voice; (Mṣb, TA;) as also ▼ جَهْوَرِىٌّ. (A, TA.) Also, and ▼ مُجْهَرٌ (TA) and جَهِيرُ الصَّوْتِ ▼ (A, TA) and ▼ جَهُورٌ (A) and ▼ جَهْوَرِىٌّ (Ṣ, A) and ▼ جَهْوَرِىُّ الصَّوْتِ, (Ṣ,) A man *having a high, loud,* or *strong voice.* (Ṣ, A, TA.) — A man (Ṣ, A) *of pleasing, or goodly, aspect;* (Ṣ, A, Ḳ;) as also ▼ جَهْرٌ: (Ḳ:) fem. of the former with ة: (Ṣ:) *beautiful:* (Ḳ:) *of goodly aspect, who pleases the beholder by his beauty:* and a face *of goodly, or beautiful, fairness:* (TA:) and ▼ أَجْهَرُ a man (TA) *of goodly aspect,* (Ḳ, TA,) *and of goodly and perfect body.* (AA, Ḳ, TA.) — Also, (Ḳ,) or جَهِيرٌ لِلْخَيْرِ and لِلْمَعْرُوفِ, (A,) *Adapted to, or constituted for, goodness:* (A, Ḳ:) because he who beholds him desires his beneficence: (TA:) pl. جُهَرَآءُ. (A, Ḳ.) ═ Also Milk *not mixed with water:* (Fr, Ṣ, Ḳ:) or *from which the butter has been taken forth.* (TA.)

جَهَارَةٌ [an inf. n. (see جَهُرَ)] *Pleasingness, or goodliness, of aspect;* (Ṣ, A, Ḳ;) as also ▼ جُهُورَةٌ (Ḳ) and ▼ جُهْرٌ: (TA:) [and *a quality pleasing to behold:* for] Abu-n-Nejm says,

* وَأَرَى البَيَاضَ عَلَى النِّسَآءِ جَهَارَةً *

[*And I regard fairness in women as a quality pleasing to behold*]: (Ṣ:) and ▼ جُهْرٌ signifies the *form, or appearance, or the like, and goodliness of aspect,* of a man: (Ḳ:) or *what pleases by its beauty, of the form or appearance or the like,* of a man, *and* his *goodliness of aspect:* (Ṣ:) [and simply *aspect, or outward appearance.*] You say, بَنُونَ ذَوُو جَهَارَةٍ *Sons goodly in stature and in aspect:* (IAạr, TA:) or *in stature and in cheeks:* (Ḳ:) but the former is the more agreeable with authority. (TA.) And مَا أَحْسَنَ ▼ جُهْرَ فُلَانٍ *How goodly is the form, or appearance, or the like, and the beauty of aspect, of such a one!* (Ṣ, A:*) [or simply, *the aspect;* for] you say also, مَا أَسْوَأَ ▼ جُهْرَهُ [*How evil is his aspect!*]. (A.) And رَجُلٌ حَسَنُ الجَهَارَةِ and ▼ الجُهْرِ *A man goodly in aspect.* (TA.) And رَأَيْتُ ▼ جُهْرَهُ فَعَرَفْتُ سِرَّهُ [*I saw his aspect, and so knew his mind*]. (A.)

جُهُورَةٌ: see the next preceding paragraph.

فُلَانٌ عَفِيفُ السَّرِيرَةِ وَالجَهِيرَةِ [*Such a one is chaste in secret conduct and in public behaviour*]. (A.)

جَهْوَرِىٌّ: see جَهِيرٌ, in four places.

جَوْهَرٌ a word of well-known meaning, (Mṣb,) [a coll. gen. n., *Jewels; precious stones; gems; pearls: any kind of jewel, precious stone, or gem:* and also applied (as in the T, M, Mgh, Mṣb, and Ḳ, voce تِبْرٌ, q. v.,) to *native ore:*] *any stone from which is extracted, or elicited, anything by which one may profit:* (Ḳ:) n. un. with ة: (Ṣ:) [pl. جَوَاهِرُ:] it is of the measure فَوْعَلٌ, (Mṣb,) and is from الجَهْرُ signifying a thing's "becoming exceedingly plain to be perceived by the sense of sight:" (Er-Rághib, TA:) or it is of Persian origin, (TA,) arabicized, (Ṣ, TA,) [from كُوهَر,] accord. to most persons. (TA.) — جَوْهَرُ سَيْفٍ *The diversified wavy marks, streaks, or grain, of a sword;* syn. فِرِنْدٌ. (T and Ḳ voce فرند.) — جَوْهَرُ شَىْءٍ [*The essence of a thing;* or *that whereby a thing is what it is; the substance of a thing: the constituent of a thing; the material part thereof;*] *that upon which the natural con-*

stitution of a thing is as it were based; or *of which its natural constitution is made to be;* [or, as IbrD thinks to be meant in the Ḳ, *the collective parts and materials of a thing, of which its natural constitution is moulded;*] expl. by مَا وُضِعَتْ عَلَيْهِ جِبِلَّتُهُ, (Ḳ,) or, as in some Lexicons, [as the JK and the Mṣb,] مَا خُلِقَتْ عَلَيْهِ جِبِلَّتُهُ [which is virtually the same]: (TA:) الجَوْهَرُ and الذَّاتُ and الحَقِيقَةُ and المَاهِيَّةُ are all syn. terms; and the first has other significations; but in the classical language it signifies أَصْلُ المُرَكَّبَاتِ, i. e., الأَصْلُ [*the original of compound things*]; and not what subsists by itself. (Kull.) — [Hence, الجَوْهَرُ الفَرْدُ † *The indivisible atom.*] — In the conventional language of scholastic theology, جَوْهَرٌ signifies ‡ *Substance, as opposed to accident;* in which sense, some assert the word to be so much used as to be, in this sense, conventionally regarded as proper. (TA.) — See also جَهْوَرٌ.

جَوْهَرِىٌّ *A jeweller; a seller of* جَوْهَر [or جَوَاهِر]. (TA.) — [In scholastic theology, † *Of,* or *relating to, substance, as opposed to accident.*]

أَجْهَرُ: see جَهِيرٌ. — Also A man *having the eyeball,* or *globe of the eye, prominent and apparent,* or *large and prominent;* syn. جَاحِظٌ: or *resembling such as is termed* جاحظ: fem. جَهْرَآءُ. (TA.) And this latter, An eye *having the ball,* or *globe, prominent and apparent,* or *large and prominent;* syn. جَاحِظَةٌ: (Ḳ:) or *resembling what is thus termed.* (TA.) — *Having a pretty cast in the eye:* (AA, Ḳ:) fem. as above. (Ḳ.) — *That cannot see in the sun;* (Ṣ, A, Mṣb, Ḳ;) applied to a man, (A, Mṣb,) and to a ram: (Ṣ:) fem. as above: (Ṣ, A, Mṣb, Ḳ:) or *weak-sighted in the sun:* (Lḥ, TA:) or *that cannot see in the daytime;* أَعْشَى signifying "that cannot see in the night:" (TA:) and the fem., a woman *who closes her eyes in the sun.* (A.) — A horse *having a blaze that covers his face:* fem. as above. (Ḳ.) — Also the fem., *Open, bare,* land, *not concealed by anything:* (A:) or *plain* land, *in which are no trees nor hills* (Ḳ, TA) *nor sands:* (TA:) pl. جَهْرَاوَاتٌ. (A, TA.) — And *A company* (Ṣ, Ḳ) *consisting of the distinguished part* (TA) of a people: (Ṣ:) the *more,* or *most, excellent persons* of a tribe. (Ḳ.) You say, [with reference to distinguished persons,] كَيْفَ جَهْرَاؤُكُمْ *How is your company?* (Ṣ.)

مُجْهِرٌ: see مَجْهُورٌ: and see also جَهِيرٌ, in two places.

مِجْهَرٌ (Ṣ, Ḳ) and ↓ مِجْهَارٌ (Ḳ) A man *accustomed to speak with a plain,* or *an open, voice; openly;* or *publicly.* (Ṣ, Ḳ.)

مِجْهَارٌ: see what next precedes.

مَجْهُورٌ بِهِ *Notorious;* applied to a thing: (TA:) and so ↓ مُجْتَهَرٌ applied to a man: (A, TA:) and ↓ مُجْهِرٌ *plain, apparent,* or *conspicuous;* applied to a thing. (TA.) — الحُرُوفُ المَجْهُورَةُ [*The letters that are pronounced with the voice, and not with the breath only; the vocal letters;*] *the letters* (*nineteen in number,* Ṣ) *that are comprised in the saying* ظِلُّ قَوٍ رَبَضٍ إِذْ غَزَا جُنْدٌ مُطِيعٌ: (Ṣ, Ḳ:) opposed to المَهْمُوسَةُ: (TA:) so called [accord. to some] because there is a full stress in the place where any one of them occurs, and the breath is prevented from passing with it until the stress is ended with the passage of the voice. (Sb, Ṣ.) = مَآءٌ مَجْهُورٌ *Water which, having been buried in the earth, has been drawn until it has become sweet.* (TA.) — مَجْهُورَةٌ A well (بِئْرٌ) *cleared out, and cleansed from the black fetid mud which it had contained.* (Ṣ.) — And Wells *frequented* [*and in use*], (Ḳ,) *whether their water be sweet or salt.* (TA.)

مُجَاهِرٌ: see, above, جَهَرَ بِالمَعَاصِى.

مُجْتَهَرٌ: see مَجْهُورٌ: and see also جَهْوَرٌ.

مُتَجَاهِرٌ *Feigning himself* أَجْهَر; as in the saying, cited by Th,

كَالنَّاظِرِ المُتَجَاهِرِ

[*Like the looker that feigns himself unable to see in the sun.*] (TA.)

جهز

1. جَهَزَ عَلَى الجَرِيحِ: see 4, in two places.

2. جَهَّزَ, inf. n. تَجْهِيزٌ, *He fitted out, equipped, furnished,* or *supplied,* a bride, and a traveller, and a corpse, (Ṣ, Mṣb, Ḳ,) and an army, (Ṣ,) *with her,* or *his,* or *their,* جَهَاز [i. e. *requisites, equipage, furniture, accoutrements,* or *apparatus*]: (Ṣ, Mṣb, Ḳ:) *he provided* a warrior *with a beast to ride, and with other requisites for his expedition:* (TA:) *he prepared* him or it. (TA.) You say also, جَهَّزَ عَلَيْهِ الخَيْلَ [*He fitted out the horsemen and sent them against him*]. (Ṣ.) = See also 4.

4. اجهز عَلَى الجَرِيحِ, (Aṣ, JK, Ṣ, Mgh, Mṣb, Ḳ,) inf. n. إِجْهَازٌ; (Mṣb;) and ↓ جَهَزَ عَلَيْهِ, aor. ـَ, (Mṣb, Ḳ,) inf. n. جَهْزٌ; (TA;) *He despatched,* or *he hastened and completed the slaughter of, the wounded man;* (Aṣ, Ṣ, Mgh, Mṣb, Ḳ;) *he made his slaughter sure,* or *certain:* (JK, Ḳ:) and ↓ جَهَّزَ, with teshdeed, signifies the same, but denoting muchness, or frequency, or repetition, of the action, or its application to many objects, and intensiveness: (Mṣb:) or جَهَّزَ على الجريح signifies *he slew the wounded man.* (IDrd, TA.) You should not say, أَجْزَتْ على الجريح, (Ṣ,) or أَجَازَ عليه. (ISd, TA.)

5. تَجَهَّزَ *He fitted out, equipped, furnished,* or *supplied, himself;* or *he* or *it became fitted out* or *equipped* or *furnished* or *supplied; with his* or *its* جَهَاز [or *requisites,* &c.]: (Ḳ:) *he prepared himself.* (Ṣ, Ḳ.) You say, تَجَهَّزْتُ لِأَمْرِ كَذَا, (Ṣ, Ḳ,*) and ↓ اِجْهَازَزْتُ, (Ḳ,) *I prepared myself for such a thing.* (Ṣ, Ḳ.)

11. اِجْهَازَزْتُ لِأَمْرِ كَذَا: see 5.

جَهَازٌ, (Ṣ, Mṣb, Ḳ, &c.,) so accord. to the seven readers in the Ḳur xii. 59 and 70, (Az, Mṣb,) and ↓ جِهَازٌ, (Ṣ, Mṣb, Ḳ,) but the latter is rare, (Mṣb,) or bad, (Az, TA,) or an erroneous pronunciation of the people of El-Baṣrah, (Lth, TA,) The *requisites,* (Mṣb, Ḳ,) *equipments, equipage, furniture, accoutrements,* or *apparatus,* (Mṣb,) of a bride, [i. e. her *paraphernalia,*] and of a traveller, and of a corpse: (Ṣ, Mṣb, Ḳ:) *provisions and other requisites for a traveller:* (Ḥar p. 104:) pl. [of pauc.] أَجْهِزَةٌ, and pl. pl. أَجْهِزَاتٌ. (Ṣ, Ḳ.) — Accord. to some, *Household goods* or *furniture and utensils:* accord. to 'Alee Ibn-'Eesà, *excellent goods that are conveyed from country to country:* and hence the جهاز of the bride: (Ḥar p. 104:) or *excellent goods that are conveyed as merchandise.* (Mgh.) — Also the former, *What is upon a camel that is used for riding* [*consisting of the saddle and its appertenances*]. (Ḳ.) It is said in a prov., ضَرَبَ فِى جَهَازِهِ, with fet-ḥ [to the ج], (Aṣ, Ṣ, Ḳ,) i. e. *He took fright and fled* or *went away at random, and did not return:* (Ḳ:) or it is said of a thing that goes away and does not return: (Aṣ, Ṣ:) originally relating to a camel from whose back the saddle with its apparatus tumbles, falling between his legs, in consequence of which he takes fright and flees or runs away at random, so that he goes away into the land: (Aṣ, Ṣ, Ḳ:) ضَرَبَ signifies *he went;* (سار;) for which in the CḲ is put صار;) and the meaning of the phrase is, *he went stumbling upon his apparatus.* (Ḳ.) In the T it is said, The Arabs say, ضَرَبَ البَعِيرُ فِى جَهَازِهِ, meaning *The camel took fright and ran away at random, beating the ground with his feet so as to throw down the apparatus and load that were upon him.* (TA.) = Also the former, The *pudendum of a woman.* (Ṣ, Ḳ, TA.)

جِهَازٌ: see جَهَازٌ.

فَرَسٌ جَهِيزٌ, (Ṣ, Ḳ,) or جَهِيزُ الشَّدِّ, (AO, TA,) *A horse quick,* or *swift, in running:* (AO, Ṣ:) or the former, *a horse that is light,* or *active.* (Ḳ.) — مَوْتٌ جَهِيزٌ, and ↓ مُجْهِزٌ, *A quick death.* (Ḳ.)

مَوْتٌ مُجْهِزٌ: see what next precedes.

مُجَهَّزٌ *One for whom are prepared travelling-provisions and equipage, that he may perform the pilgrimage for another.* (Mgh.)

مُجَهِّزٌ *One who sends forth traders with excellent goods:* or *who travels with such goods.* (Mgh.) Hence, app., the vulgar term ↓ مُجَاهِزٌ, meaning *A rich merchant.* (Mgh.) — And مُجَهِّزُونَ *Travelling-companions who assist one in the loading of the beasts.* (Mṣb.)

مُجَاهِزٌ: see the next preceding paragraph.

جهش

1. جَهَشَ إِلَيْهِ, (Ṣ, Ḳ,) and جَهِشَ, (Ḳ,) which latter, accord. to IDrd, is the more common, (TA,) aor. ـَ, inf. n. جَهْشٌ (Ṣ, Ḳ) and جُهُوشٌ and جَهَشَانٌ, (Ḳ,) *He betook himself to him by reason of fright* or *fear, seeking protection, and being about,* or *ready, to weep; like as the child betakes himself to his mother by reason of fright* or *fear,* (Aṣ, Ṣ, Ḳ,) *and to his father,* (Aṣ,) *being about,* or *ready, to weep;* (Aṣ, Ṣ;) as also ↓ أَجْهَشَ. (AO, Ṣ, Ḳ.) — جَهَشَتْ نَفْسُهُ, (Ṣ, A,) or جَهَشَتْ إِلَيْهِ نَفْسُهُ, inf. n. جُهُوشٌ; (TA;) and ↓ أَجْهَشَتْ; (Ṣ, A, TA;) *His soul heaved,* (Ṣ, A,)

and he was about, or ready, to weep: (A:) or *his soul heaved, and quitted him.* (TA.) — جَهَشَ لِلشَّوْقِ وَالحُزْنِ *He was ready, or about, to desire, and to mourn:* (IDrd:) and ▼ أَجْهَشَ بِالبُكَآءِ *He was ready, or about, to weep.* (El-Umawee, Ḳ.) — جَهَشَ مِنَ الشَّىْءِ, (AA, Ḳ,) and جَهِشَ, (Ḳ,) aor. ـَ, inf. n. جَهَشَانٌ, (AA, Ḳ,) *He was frightened at the thing, or afraid of it:* (AA, Ḳ:) or *he fled from the thing.* (Ṣgh, Ḳ.) — جَهَشُوا إِلَىَّ فَزِعِينَ *They came quickly to me, frightened.* (A.) — يَجْهَشُ مِنْ أَرْضٍ إِلَى أَرْضٍ *He removes, and goes quickly, from land to land.* (Ḳ.) — جَهَشَ إِلَى القَوْمِ, inf. n. جَهْشٌ, *He came to the people, or company of men.* (TA.)

4: see 1, in three places. = اجهش فُلَانًا *He hastened such a one.* (Ibn-'Abbád, Ḳ.)

جَهْشَةٌ *A flow of tears* (A, Ḳ, TA) *falling one after another on the occasion of* جَهْش [or the *betaking oneself to another by reason of fright* or *fear,* &c.]. (TA.) = *A company* of men; as also ▼ جَاهِشَةٌ: (Ḳ:) or the latter signifies *a party,* and *a multitude,* of men. (TA.)

جَهُوشٌ *Quick; who removes, and goes quickly, from land to land.* (Ḳ.)

جَاهِشَةٌ: see جَهْشَةٌ.

جهض

1: see 4, in two places.

3. جاهضهُ, (Ḳ,) inf. n. جِهَاضٌ, (TA,) *He endeavoured to prevent him,* or *to turn him away* [from a thing]; and *strove to be before him;* syn. مَانَعَهُ and عَاجَلَهُ. (Ḳ.) Hence the saying, in a trad. of Mohammad Ibn-Meslemeh, relating his endeavouring to attack a man on the day of Ohod, فَجَاهَضَنِى عَنْهُ أَبُو سُفْيَانَ *But Aboo-Sufyán endeavoured to turn me away from him* (مَانَعَنِى عَنْهُ), *and put me away* [*from him*]. (TA.)

4. أَجْهَضَنِى *He overcame me* in contending for the thing; as also ▼ جَهَضَنِى. (Ṣ.) You say, اجهضهُ عَنِ الأَمْرِ, *He overcame him in contending for the thing, and made him to quit it,* or *put him away from it;* as also ▼ جَهَضَهُ عَنْهُ, aor. ـَ. (A, Ḳ.) And قُتِلَ فُلَانٌ فَأُجْهِضَ عَنْهُ القَوْمُ *Such a one was slain, and the people were overcome so that he was taken from them.* (Ṣ.) And صَادَ الجَارِحَةُ الصَّيْدَ فَأَجْهَضْنَاهُ عَنْهُ *The beast, or bird, of prey caught the game, and we removed him, and overcame him in contending for that which he had caught.* (Ṣ, A,* Mṣb.) — اجهض is also *syn. with* أَعْجَلَ. (A, Ḳ.) Thus, sometimes, (Ṣ,) أَجْهَضْتُهُ عَنْ كَذَا, (Ṣ,) or عَنِ الأَمْرِ, (Mgh, TA,) signifies *I hurried him so as to prevent him from doing such a thing, or the thing;* (Ṣ, Mgh, TA;) *and made him to quit it.* (Mgh.) — You say also, اجهضهُ عَنْ مَكَانِهِ *He roused him, or made him to rise, from his place.* (A,* TA.) And [elliptically] أَجْهَضْنَا العَدُوَّ *We roused the enemy, or made them to rise, from their places; and removed them therefrom, or made them to quit them.* (Mgh.) = اجهضت *She* (a camel, AZ, Aṣ, Ṣ, A, Ḳ) *cast her young one:* (Ṣ, A:) or *cast her young one with its fur grown,* (Aṣ, Ḳ,) *before it was perfect:* (Aṣ:) or *cast her young one before its form was apparent:* (AZ:) or *in an imperfect state:* (TA:) or اجهضت وَلَدَهَا, inf. n. إِجْهَاضٌ, *she* (a camel, and a woman,) *cast her young one imperfect in form.* (Mṣb.)

جَهْضٌ: see جَهِيضٌ, in two places.

جِهَاضٌ [The *act of* a camel's, and of a woman's, *casting an imperfect fœtus;*] a subst. from اجهضت said of a camel and of a woman. (Mṣb.)

جَهِيضٌ, applied to the young one, or fœtus, of a camel, (Ṣ, A, Mṣb,) and of a woman, (Mṣb,) i. q. ▼ مُجْهَضٌ [pass. part. n. of أَجْهَضَتْ, q. v.]: (Ṣ, A, Mṣb:*) or, (Ḳ,) as also ▼ جَهْضٌ, on the authority of Fr, being, as he says, like حَفْضٌ, in the Ḳ, erroneously, جَهَضٌ, (TA,) *an abortion:* or *a young one perfectly formed, and into which the spirit has been inspired, without its living:* (Ḳ:) or ▼ جَهْضٌ signifies the *young one of a camel that is cast before its form is apparent.* (TA.)

مُجْهَضٌ: see the next preceding paragraph.

مُجْهِضٌ, applied to a she-camel, (Ṣ, Mṣb, Ḳ,) and to a woman, (Mṣb,) act. part. n. of أَجْهَضَتْ [q. v.]; (Ṣ, Mṣb, Ḳ;) as also with ة: (Mṣb:) pl. مَجَاهِيضُ. (Ḳ.)

مِجْهَاضٌ A she-camel *that is accustomed to cast her young* (Ṣ, A) *in an imperfect state.* (TA.)

جهل

1. جَهِلَ; (Ṣ;) and جَهِلَهُ, (Sh, Mṣb, Ḳ,) and جَهِلَ بِهِ (JK) [and مِنْهُ (see جَاهِلٌ)]; aor. ـَ, (Ḳ,) inf. n. جَهْلٌ and جَهَالَةٌ (Ṣ, Mṣb, Ḳ) and جُهُولِيَّةٌ; (TA;) *He was ignorant;* (Ṣ;) *he was characterized by* جَهْل *in any of the senses assigned to this word below:* (TA:) and *he was ignorant of it; he did not know it;* (Sh, JK, Mṣb, Ḳ;) *contr. of* عَلِمَهُ. (Mṣb, Ḳ.) You say, مِثْلِى لَا يَجْهَلُ مِثْلَكَ *The like of me will not be ignorant of the like of thee.* (Sh, TA.) And جَهِلَ عَلَى غَيْرِهِ *He acted in an ignorant or a silly or foolish manner towards another:* and *wrongly.* (Mṣb.) And جَهِلَ فُلَانٌ رَأْيَهُ [i. q. سَفِهَ رَأْيَهُ, *He was ignorant, or silly, or foolish, in his opinion, or judgment*]. (Sh, TA.) And جَهِلَ الحَقَّ *He neglected the truth, or the right, or due;* [or *he ignored it;*] syn. أَضَاعَهُ. (Mṣb.) See also 6. — جَهِلَتِ القِدْرُ ‡ *The cooking-pot boiled vehemently; contr. of* تَحَلَّمَت. (TA.)

2. جهّلهُ, (Mṣb, Ḳ,) inf. n. تَجْهِيلٌ, (Ṣ, Ḳ,) *He attributed to him* جَهْل [or *ignorance,* &c.]. (Ṣ, Mṣb, Ḳ.) — And *He caused him to fall into* جَهْل. (TA.)

3. مُجَاهَلَةٌ The *acting with levity,* and *in an ignorant or a silly or foolish manner, with any one.* (KL.) [You say, جاهلهُ, meaning *He so acted with him.*]

6. تجاهل *He feigned,* or *made a false show of,* جَهْل [or *ignorance,* &c.]. (Ṣ, Ḳ.) And ▼ جَهِلَ عَلَيْهِ *He feigned ignorance* [*to him*]. (Ḳ.)

10. استجهلهُ *He reckoned him,* or *esteemed him,* جَاهِل [or *ignorant,* &c.]. (Ṣ, TA.) — *He,* or *it, excited him to lightness,* or *levity, and unsteadiness.* (Ṣ, Ḳ.) — اِسْتَجْهَلَتِ الرِّيحُ الغُصْنَ ‡ *The wind put the branch into a state of commotion.* (Ḳ, TA.)

جَهْلٌ an inf. n. of 1: *Ignorance; contr. of* عِلْمٌ: (Ṣ, Mṣb,* Ḳ:*) [and *silliness,* or *foolishness:* and *wrong conduct:* (see 1:)] it is of two kinds; namely, simple, which is the *non-existence of knowledge of that which should be known;* and compound, which is *a decisive belief not agreeable with the fact,* or *reality:* so accord. to Ibn-El-Kemál: or, accord. to Er-Rághib, it is of three kinds; namely, the *mind's voidness of knowledge,* which is the primary meaning; and the *believing a thing to be different from what it is;* and the *doing a thing in a manner different from that in which it ought to be done:* or, accord. to El-Ḥarállee, the *proceeding in dubious affairs without knowledge.* (TA.) It is said in a prov., كَفَى بِالشَّكِّ جَهْلًا [*Doubt is sufficient ignorance*]. (Mṣb.) And it is said in a trad., إِنَّ مِنَ العِلْمِ جَهْلًا [*Verily there is,* among the kinds *of knowledge, what is ignorance*]: this is one's learning what is not requisite, and neglecting what is requisite; or a learned man's affecting, or pretending, a knowledge of that which he does not know. (TA.)

الجَاهِلِيَّةُ: see جَهْلَآءُ.

جَهُولٌ: see what next follows.

جَاهِلٌ and ▼ جَهُولٌ, [but the latter has an intensive signification,] *Ignorant:* (Mṣb, Ḳ:) and *silly,* or *foolish, in conduct:* and *wrong in conduct:* (Mṣb:) [*characterized by* جَهْل *in any of the senses assigned to this word above:*] pl. جُهَّالٌ (Ḳ, KL) and جُهَّلٌ and جُهُلٌ and جَهَلَةٌ and جُهَلَآءُ: (Ḳ:) see جَاهِلِىٌّ. The former epithet is mostly used in dispraise: but sometimes not in dispraise; as in the saying [in the Ḳur ii. 274], يَحْسَبُهُمُ الجَاهِلُ أَغْنِيَآءَ i. e., *The ignorant* of their [real] state [*would reckon them possessed of competence*]. (TA.) هُوَ جَاهِلٌ مِنْهُ means هو جاهل بِهِ, (Ḳ, TA,) i. e., *He is ignorant* [*of him,* or *it,* or] *of his,* or *its, state,* or *condition.* (TA.) — الجَاهِلُ *The lion* (Ḳ, TA) *that is ignorant of the prey.* (TA.) [In the CḲ, والجاهِلُ والاَسَدُ, is a mistake for وَالجَاهِلُ الأَسَدُ.]

جَيْهَلٌ: see جَيْهَلَةٌ. = Also *Great* as an epithet applied to a smooth rock (صَفَاةٌ). (Ḳ.)

جَيْهَلَةٌ (IDrd, JK, Ḳ) and ▼ جَيْهَلٌ and ▼ مِجْهَلٌ (IDrd, Ḳ) and ▼ مِجْهَلَةٌ (Ḳ) *A piece of wood with which one stirs live, or burning, coals* (جَمْر), (JK, Ḳ, TA,) or *wine* (خَمْر); (so in some copies of the Ḳ;) of the dial. of El-Yemen. (TA.)

جَاهِلِىٌّ [*A pagan; a pagan Arab; one of those who are called collectively* ▼ الجَاهِلِيَّةُ; and sometimes ▼ الجُهَّالُ, pl. of الجَاهِلُ: and particularly, *a pagan poet;*] *a poet of the first,* or *earliest, of the four classes which are ranked in chronological order; of the class which was succeeded next by the* مُخَضْرَمُون. (Mz, 49th نوع.) [See also شَاهِدٌ.]

الجَاهِلِيَّةُ : see what next precedes. — [Also, or] ▼ الجَاهِلِيَّةُ الجَهْلَآءُ, (JK, Ṣ, Ḳ,) in which the latter word is a corroborative, (Ṣ, Ḳ,) as in لَيْلَةٌ لَيْلَآءُ &c., (Ṣ,) [*The time,* or *state, of ignorance,* or *paganism;* or *of intense ignorance;*] *the time of the* فَتْرَة [or *cessation of the mission of apostles, and of the effacement of the signs of their religion*]. (JK.) One says, كَانَ ذٰلِكَ فِى الجَاهِلِيَّةِ الجَهْلَآءِ [*That was in the time,* or *state, of paganism,* or *of intense ignorance*]. (Ṣ.)

مَجْهَلٌ *An affair,* or *an event,* or *a case,* and *a land,* and *a habit, a property, a quality, a practice,* or *an action, that induces a man to believe a thing to be different from what it is.* (Er-Rághib, TA.) *A desert* (مَفَازَةٌ) *in which are no signs of the way.* (Ṣ.) And أَرْضٌ مَجْهَلٌ *A land in which are no signs of the way:* (TA:) or *in which one will not go aright* (Ḳ, TA) *unless by means of the* [*signs of the way called*] آرَام: (TA:) pl. مَجَاهِلُ, which is the *contr.* of مَعَالِمُ: (TA:) accord. to the Ḳ, it has neither dual nor pl.; but it has both, as 'Iyáḍ and others have affirmed. (MF, TA.)

مَجْهِلٌ : see مَجْهَلَةٌ.

مَجْهَلَةٌ [*A cause of,* or *an incitement to, ignorant, silly, foolish,* or *wrong, conduct;*] *a thing that incites one to* الجَهْل. (Ṣ, Ḳ.) Hence the saying, الوَلَدُ مَجْهَلَةٌ [*Children are a cause of silly,* or *foolish, conduct*]. (Ṣ.)

مَجْهِلَةٌ : see مَجْهَلَةٌ.

مِجْهَالٌ ‡ A she-camel *light, brisk,* or *agile, in her pace,* or *going.* (TA.)

مَجْهُولٌ [*Unknown*]. You say, رَكِبْتُ المَفَازَةَ عَلَى مَجْهُولِهَا [*I ventured upon traversing the desert notwithstanding its unknown character*]. (Ṣ, TA.) — [A man *of unknown origin.* A book *of unknown authorship.* — In grammar, The *passive voice.*] — نَاقَةٌ مَجْهُولَةٌ ‡ *A she-camel that has never been milked:* or *that has no brand upon her:* (Ḳ, TA:) and ‡ *a she-camel that has never conceived.* (Z, TA.)

مُسْتَجْهِلٌ *Making himself like the* جَاهِل [or *ignorant; feigning himself ignorant*]: or *reckoning,* or *esteeming,* جَاهِل. (Ḥar p. 572.)

جهم

1. جَهُمَ, aor. ـُ, inf. n. جُهُومَةٌ (Ṣ, Ḳ) and جَهَامَةٌ, (Ḳ,) *He* (a man) *became frowning,* or *contracted, in face;* or *grinning in face, with a frowning,* or *contraction,* or *a stern, an austere,* or *a morose, look:* (Ṣ:) or *he was,* or *became, coarse,* or *rough, and contracted, and ugly, in face.* (Ḳ.) — Also, said of the pubes, *It was coarse, rough,* or *big.* (TA.) — جَهَمَهُ, (Ṣ, Ḳ,) and جَهِمَهُ, aor. ـَ, in both instances; (Ḳ;) and ▼ تَجَهَّمَهُ, (Ṣ, Ḳ,) and ▼ تَجَهَّمَ لَهُ; (JK, Ḳ;) *He grinned, frowning,* or *looking sternly, austerely,* or *morosely, in his face:* (Ṣ:) or *he met him,* or *regarded him, with a displeasing,* (JK, Ḳ, TA,) *frowning,* or *grinning and frowning,* (TA,) *face:* (JK, Ḳ, TA:) or ▼ تَجَهَّمَ signifies *he showed a sour, a crabbed,* or *an austere, face.* (TḲ in art. عبس.) A poet says,

* وَبَلْدَةٍ تَجَهَّمُ الجَهُومَا *
* زَجَرْتُ فِيهَا عَيْهَلًا رَسُومَا *

(Ṣ,* TA,) i. e. [*Many a region*] *that exhibits to the impotent that which he dislikes,* [or *that frowns upon the weak,* or *impotent, I have chidden therein a strong she-camel that leaves the marks of her footsteps upon the ground.*] (Ṣ.) You say also, الدَّهْرُ يَتَجَهَّمُ ▼ الكِرَامَ ‡ [*Fortune frowns upon the generous*]. (TA.) And تَجَهَّمَنِى ▼ أَمَلِى ‡ [*My object of hope frowned upon me*] is said when one has not attained his object of hope. (TA.)

4. أَجْهَمَتِ السَّمَآءُ *The sky had clouds such as are termed* جَهَام. (Ḳ.)

5 : see 1, in six places.

8. اجتهم *He entered upon,* (Ḳ,) or *journeyed in,* (A, TA,) *the portion of the night termed* جُهْمَة. (A, Ḳ, TA.)

جَهْمٌ *A coarse,* or *rough, and contracted, and ugly,* face; as also ▼ جَهِيمٌ; (Ḳ;) or, as in some of the lexicons, ▼ جَهُومٌ. (TA.) And جَهْمُ الوَجْهِ *Frowning,* or *contracted, in face;* or *grinning in face, with a frowning,* or *contraction,* or *a stern, an austere,* or *a morose, look:* (Ṣ, Mgh:) or *coarse,* or *rough, in face:* (JK, TA:) applied to a man: (JK, Ṣ, Mgh, TA:) and to a lion. (JK, TA.) And [hence,] الجَهْمُ *The lion.* (Ḳ.) — See also جَهُومٌ.

جُهُمٌ : see جَهْمٌ.

جُهْمَةٌ : see جُهْمَةٌ, in two places. — Also *A big cooking-pot.* (Ḳ.)

جُهْمَةٌ and ▼ جُهُمَةٌ, (JK, Ṣ, Ḳ,) both mentioned by Fr, (Ṣ,) *A portion of the night:* (JK:) the *first of the last portions of the night,* (JK, Ṣ, Ḳ, TA,) *extending* [app. *from midnight*] *to near the period a little before daybreak:* (TA:) or the *remaining portion of the darkness of the last part of the night:* (Ḳ:) or the former signifies, [or each,] the *beginning of the night, extending to a fourth part:* or, as some say, the *middle of the night:* (KL:) pl. of the former جُهَمٌ. (JK.) You say, مَضَى مِنَ اللَّيْلِ جُهْمَةٌ and ▼ جُهُمَةٌ [app. meaning *A portion,* or *a portion at the commencement of the latter part, of the night passed*]. (A'Obeyd, TA.) — Also the former, *Eighty camels:* or *the like.* (Ḳ.)

جَهَامٌ *Clouds in which is no water:* (JK, Ṣ, Ḳ:) or *that have poured forth their water* (Ḳ, TA) *with the wind.* (TA.)

جَهُومٌ, applied to a man, (JK, Ṣ,) *Impotent;* (JK, Ṣ, Ḳ;) *weak;* as also ▼ جَهْمٌ. (Ḳ.) See an ex. in the first paragraph.

جَهِيمٌ : see جَهْمٌ.

جهنم

جَهَنَّمُ [*Hell;* or *Hell-fire;*] (T, Ṣ, Ḳ, &c.;) a name of *the fire with which God will punish,* (T, Ṣ,) *in the life to come,* (T,) *his* [*disobedient and unrepentant and unbelieving*] *servants;* (Ṣ;) a proper name of *the abode of punishment:* (Bḍ, ii. 202:) a word rendered quasi-coordinate to the quinqueliteral-radical class by the doubling of the third letter: (Ṣ:) accord. to some, it is an Arabic word, applied to the fire of the world to come because of its depth; [see the last paragraph;] (T, TA;) or originally syn. with النَّارُ [*fire,* or *the fire*]; (Bḍ in ii. 202;) and imperfectly decl. because determinate and of the fem. gender: (T, Ṣ:) accord. to others, it is an arabicized word, (T, Ṣ, Bḍ ubi suprà,) imperfectly decl. because determinate and of foreign origin; (T, TA;) some say, originally Persian; (Ṣ;) others, from the Hebrew كهنام, (TA,) [or as Golius says, גֵּיא הִנֹּם "the Valley of Hinnom," where children were burned alive as sacrifices to the idol Moloch.] — See also جِهِنَّام.

جَهَنَّمِىٌّ [*Of,* or *relating to,* جَهَنَّم, i. e. *Hell,* or *Hell-fire.*]

جِهِنَّامٌ (Ṣ, Ḳ, Ḥam p. 817) and جَهَنَّامٌ [and جُهُنَّامٌ], with each of the three vowels (Ḳ, TA) to the ج, (TA,) [but accord. to the Ḳ it would rather seem to be جِهِنَّامٌ and جُهُنَّامٌ,] and ▼ جَهَنَّمٌ, (Ḳ,) applied to a well (رَكِيَّةٌ, Ṣ, Ḳ, or بِئْرٌ, Ḥam), *Deep;* (Ṣ, Ḳ, Ḥam;) *in which he who falls into it perishes.* (Ḥam.)

جو

جَوٌّ The *air:* (Ḳ:) or [the *atmosphere;* i. e.] *what is between the heaven,* or *sky, and the earth;* (Ṣ, Mgh;) the *air* [or the *region*] *between the heaven,* or *sky, and the earth:* pl. أَجْوَآءٌ. (TA.) فِى جَوِّ السَّمَآءِ, in the Ḳur [xvi. 81], means *In the air between the heaven,* or *sky, and the earth:* (Jel:) or *in the middle of the sky:* (Ḳatádeh, TA:) or *in the air that is remote from the earth.* (Bḍ.) — *A low,* or *depressed, part of the ground;* (M, Ḳ;) as also ▼ جُوَّةٌ: (Ḳ:) or *a wide part of a valley:* (Ṣ, Mṣb:) pl. (in the latter sense, Mṣb, or in the former, Ḳ, TA) جِوَآءٌ (Mṣb, Ḳ) and [of pauc.] أَجْوِيَةٌ. (TA.) It has the latter meaning in the saying of Ṭarafeh, [addressing a lark,]

* خَلَا لَكِ الجَوُّ فَبِيضِى وَاصْفِرِى *

[*The wide part of the valley has become vacant for thee; so lay thine eggs, and whistle*]. (AA, Ṣ.) Az says, I entered, with an Arab of the desert, a [hollow place in which water collected, termed a] دَحْل, in El-Khalṣà, and when we came to the water, he said, هٰذَا جَوٌّ مِنَ المَآءِ لَا يُوقَفُ عَلَى أَقْصَاهُ [*This is an abyss of water: one knows not the utmost extent of it*]. (TA.) — *A vacant, void,* or *desolate, place, in which is no one to cheer by his company.* (Ḥam p. 203.) — The *inside,* or *interior,* of a house or tent; (Ḳ;) and of anything; of the dial. of Syria; also termed ▼ جُوَّةٌ, (TA,) and ▼ جُوَّانِىٌّ; (Ḳ; [in the CḲ

ﻜﺠﻮّاﻧﻴّﻪ is erroneously put for كَجُوَّانِيّهِ;]) in which last, the ا and ن are augmentatives for the purpose of corroboration. (TA.) — [Hence,] أُرِيدُ جُوًّا وَيُرِيدُ بَرًّا *I desire concealment*, or *secrecy*, *and he desires publicity*. (A in art. بر.)

جُوَّةٌ: see جُوٌّ, in two places.

جُوَّانِىٌّ: see جُوٌّ. Hence the saying of Selmán, مَنْ أَصْلَحَ جُوَّانِيَّهُ أَصْلَحَ ٱللّٰهُ بَرَّانِيَّهُ [explained in art. بر]. (TA.) [It generally signifies *Inner*, *inward*, or *interior*; and *secret*, or *private*; opposed to بَرَّانِىٌّ; and is now vulgarly pronounced جُوَّانِى.] It is a rel. n. [irregularly formed] from جُوٌّ signifying "any low, or depressed, part of the ground." (T in art. بر.)

جوأ

1. يَجُوْءُ is a dial. var. of يَجِيْءُ [aor. of جَآءَ and meaning *He comes*, or *will come*]: (Ḳ:) mentioned by Sb, who gives as an ex. أَنَا أَجُوْؤُكَ وَأَنْبُؤُكَ [for انا أَجِيْؤُكَ وأُنْبِئُكَ, *I will come to thee and inform thee*]. (TA.)

جوالق

جِوَالِقٌ and جُوَالَقٌ and جُوَالِقٌ: see art. جلق.

جوب

1. جَابَهُ, (Ṣ,* TA,) aor. يَجُوبُ, (Ṣ, TA,) inf. n. جَوْبٌ (Ṣ, A, Ḳ, TA) and تَجْوَابٌ, (Ḥar p. 336,) *He made a hole in it; or rent*, or *tore, it*; (Ṣ, A, Ḳ, TA;) as also ↓ اجتابهُ: (Ḳ,* TA:) *he made a hole through*, or *in*, or *into, it; perforated, pierced*, or *bored, it*: (TA:) *he cut it*: (Ṣ, A, Ḳ, TA:) *he cut it in like manner as one cuts a* جَيْب [or *an opening at the neck and bosom of a shirt &c.*]: (L, TA:) *he made*, or *cut, a hole in the middle of it; cut a piece out of the middle of it; hollowed it out*; or *excavated it*. (TA.) You say, جاب الصَّخْرَةَ *He made a hole in the rock*; (A, TA;) *perforated, pierced*, or *bored, it*. (TA.) Hence, in the Ḳur [lxxxix. 8], وَثَمُودَ الَّذِينَ جَابُوا الصَّخْرَ بِالوَادِ (Fr, Ṣ, TA) *And Thamood, who made holes in the rocks*, (Fr, TA,) or *cut the rocks*, (Bḍ, Jel,) [or *hollowed them out*,] *and made them dwellings, in the valley*, (Fr, Bḍ, Jel, TA,) i. e., in Wádi-l-Ḳurà. (Bḍ, Jel.) You say also, جاب القَمِيصَ, (Ṣ, A, Ḳ,) aor. يَجُوبُ, [inf. n. جَوْبٌ;] (Ṣ, Ḳ, and Mṣb in art. جيب;) and aor. يَجِيبُ, (Ṣ, Ḳ,) [inf. n., app., جَيْبٌ, originally جَوْبٌ; see a verse cited below, and a remark of Sh thereon;] and ↓ جَوَّبهُ; (A, Ḳ;) *He hollowed out*, or *cut out in a round form, the* جَيْب *of the shirt*: (Ṣ, and Mṣb in art. جيب:) or *he cut the* جَيْب *of the shirt*: (A:) or *he made a* جَيْب *to the shirt*; (Ḳ;) as also جَيَّبَهُ, (Ṣ, and Mṣb in art. جيب,) inf. n. تَجْيِيبٌ. (Ṣ.) And جاب الثَّوْبَ *He cut the garment*, or *piece of cloth*; [or *cut it out*;] as also ↓ اجتابهُ. (A.) And جاب النَّعْلَ, inf. n. جَوْبٌ, *He cut out the sandal*. (TA.) And جاب القَرْنُ [i. e. جاب اللَّحْمَ] *The horn cut the flesh and came forth*. (TA.) — [Hence, also,] جاب, (Ṣ, A, Mṣb, TA,) aor. يَجُوبُ (Ṣ, Mṣb, TA) and يَجِيبُ, (Ṣ, TA,) inf. n. جَوْبٌ; (TA;) and ↓ اجتاب; (Ṣ, A, TA;) ‡ *He traversed*, or *crossed*, (Ṣ, A,* Mṣb, TA,) or *cut through by journeying*, (TA,) a country, (Ṣ, TA,) or a land, (Mṣb,) and a desert, and the darkness: (A,* TA:) and جَوْبٌ signifies likewise the *pouncing down* of a bird. (TA.) A rájiz says,

* بَاتَتْ تَجِيبُ أَدْعَجَ الظَّلَامِ *
* جَيْبَ البِيَطْرِ مِدْرَعَ الهُمَامِ *

† [*She passed the night cutting through the black darkness, like as the tailor cuts through the woollen tunic of the valiant chief, making the opening at the neck and bosom*]: (Ṣ: [but in one copy, instead of جَيْبَ, I here find جَوْبَ; and in art. بطر, شَقَّ:]) and Sh remarks that this [verb تجيب, or the inf. n. جيب,] is not from الجَيْبُ [meaning "the opening at the neck and bosom" of a shirt &c.], because its medial radical is و, and that of الجيب is ى: (TA:) [i. e., جاب, aor. يجيب, is originally جَوَبَ, aor. يَجْوُبُ.] One says also, of news, يَجُوبُ الأَرْضَ مِنْ بَلَدٍ إِلَى بَلَدٍ † *It traverses the earth from country to country*, or *the land from town to town*]. (Ṣ, TA.) And of proverbs, تَجُوبُ البِلَادَ † *They are current in the countries*, or *towns*. (TA.) — It is said in a trad., جِيبَتِ العَرَبُ عَنَّا كَمَا جِيبَتِ الرَّحَا عَنْ قُطْبِهَا † *The Arabs were rent from us, like as the mill-stone is rent from its pivot*; we being in the midst, and they around us. (TA.) ═ جَابَتِ الدَّعْوَةَ: see أَجْوَبُ.

2. جوّب: see 1. — Also, said of the light of the moon, † *It illumined, and rendered clear*, [*by penetrating*,] a dark night. (TA.) ═ جوّب عَلَيْهِ [from جَوْبٌ "a shield"] *He shielded him*. (TA: so accord. to an explanation of the act. part. n.)

3. [جاوبهُ, inf. n. مُجَاوَبَةٌ, *He returned him answer for answer, or answers for answers; held a dialogue, colloquy, conference, disputation*, or *debate, with him; bandied words with him*.] See 6, in two places.

4. اجابهُ, (Ṣ, A, Mṣb, TA,) inf. n. إِجَابَةٌ (Ṣ, Mṣb, Ḳ,* TA) and إِجَابٌ (Ḳ,* TA) and ↓ جَابَةٌ, (Kr, TA,) or this last is a simple subst., (AHeyth, Ṣ, TA,) like طَاعَةٌ and طَاقَةٌ, (Ṣ, A,) used in the place of an inf. n.; (AHeyth, TA;) and ↓ استجابهُ (A, Ḳ, TA) and ↓ استجاب لَهُ and ↓ اِسْتَجْوَبَهُ; (Ḳ, TA;) [for] إِجَابَةٌ and ↓ اِسْتِجَابَةٌ are syn.; (Ṣ, TA;) *He answered him, replied to him, responded to him*, (Mṣb, TA,) either affirmatively or negatively. (Mṣb.) And اجاب قَوْلَهُ *He answered*, or *replied to, his saying*. (Mṣb.) And اجاب عَنْ سُؤَالِهِ (Ṣ, TA) *He answered*, or *replied to, his question*. (TA.) And اجاب دُعَآءَهُ, (Mṣb, TA,*) and ↓ استجاب دُعَآءَهُ, (Ṣ, A, TA,) and ↓ استجاب لَهُ, (Mṣb,) and ↓ استجاب مِنْهُ, (Ḥar p. 307,) said of God, (Ṣ, A, Mṣb, TA,) [*He answered his prayer*;] *He accepted his prayer*; (Mṣb;) *He recompensed his prayer by gift and acceptance*. (TA.) It is said in the Ḳur [ii. 182], أُجِيبُ دَعْوَةَ الدَّاعِ إِذَا دَعَانِ فَلْيَسْتَجِيبُوا ↓ لِى [*I answer the prayer of him who prayeth to me*;] *therefore let them answer me*; (TA;) i. e., *let them answer my call by obedience*, (Jel,) when I call them to belief and obedience: (Bḍ:) accord. to Fr, what is here meant [by the last verb] is تَلْبِيَة [q. v. in art. لبى]: (TA:) [or *let them give me their assent*, or *consent, to my call*; or *let them obey my call*: for you say, اجابهُ إِلَى شَىْءٍ and عَلَى شَىْءٍ, (for the latter of which there is authority in this art. in the TA, but the former is more common,) and] ↓ استجاب لَهُ, *He obeyed him*, or *complied with his desire, in doing a thing*, [or *consented to do it*,] *when summoned*, or *invited, to do it*. (Mṣb.) — اجابت الأَرْضُ † *The land produced plants*, or *herbage*. (Ḥam p. 94.) — دَمْعٌ مُجِيبٌ † *Tears running*, or *flowing*; as though called for and answering the call. (Ḥar p. 71.) ═ The forms أَجْوَبَ and أَجْوِبْ [as verbs of wonder] are not used: therefore you say, مَا أَجْوَدَ جَوَابَهُ and أَجْوِدْ بِجَوَابِهِ [*How good is his answer, or reply!*]; not مَا أَجْوَبَهُ nor أَجْوِبْ بِهِ: nor do you say, هُوَ أَجْوَبُ مِنْكَ [meaning *He is better in answering*, or *replying, than thou*: but see أَجْوَبُ, below]. (Sb, TA.)

6. تجاوبوا i. q. جَاوَبَ ↓ بَعْضُهُمْ بَعْضًا [*They returned one another answer for answer*, or *answers for answers; they answered one another; replied, one to another; held a dialogue, colloquy, conference, disputation*, or *debate, together; bandied words, one with another*]: (Ḳ:) ↓ مُجَاوَبَةٌ and تَجَاوُبٌ both signify i. q. تَحَاوُرٌ. (Ṣ, TA.) In like manner one says of turtle-doves, (A,) of pigeons, of braying camels, and of neighing horses. (TA.) — [Hence,] يَتَجَاوَبُ أَوَّلُ كَلَامِهِ وَآخِرُهُ ‡ *The first and the last parts of his speech correspond*, or *are consistent*. (A, TA.)

7. انجاب [*It* (a garment) *became rent*, or *slit*: see مُنْجَابٌ]. — Said of a cloud, or a collection of clouds, *It cleared away* [*so as to leave an open space*]. (Ṣ, Mṣb.) It is said in a trad., وَٱنْجَابَ السَّحَابُ عَنِ المَدِينَةِ حَتَّى صَارَ كَالإِكْلِيلِ *And the clouds became gathered and drawn together, and cleared away from the city* [*so that they became like a crown*]. (TA.) — [*It* (a place) *was*, or *became, clear, open*, or *unobstructed*.] See جَوْبَةٌ. ═ انجابت *She* (a camel) *stretched forth her neck, to be milked*; (Ḳ;) as though she complied with the desire of her milker to be restrained [for that purpose]: but Fr says that he had not found a verb of this measure from أَجَابَ. (TA.)

8. اجتاب: see 1, in three places. — *He dug* a well. (Ḳ.) And اجتابت, said of a wild cow, *She hollowed out*, or *excavated*, a place to shelter herself from the rain. (TA.) — *He put on*, i. e. *clad himself with*, (T, Ṣ, Ḳ,) a garment, (T,) or a shirt; (Ṣ, Ḳ;) *he entered into* a shirt: and in like manner, † the darkness. (TA.)

10. استجاب and اِسْتَجْوَبَ, inf. n. اِسْتِجَابَةٌ: see 4, nine places.

جَابٌ: see جَأْبٌ, in art. جأب.

جَوْبٌ [an inf. n. (of 1, q. v.,) used in the sense of a pass. part. n. Hence,] a tribe is said to be جَوْبُ أَبٍ as meaning *Cut* [*as it were*] *from one father;* [*sprung from the loins of one father;*] occurring in a trad. (TA.) — *A fire-place;* [so called because hollowed out;] syn. كَانُونٌ. (Ḳ.) — *A large* دَلْو [or *bucket;* because of its hollow form]. (Kr, Ḳ.) — *A shield;* (Ṣ, Ḳ;) as also ▼ جَوْبَةٌ (TA) and ▼ مِجْوَبٌ: (Ḳ:) [see a verse cited voce يَلَبٌ:] pl. of the first أَجْوَابٌ. (TA.) — *A garment like the* بَقِيرَة: [so called because it has a slit in the middle, through which the head is put:] (Ṣ:) or *a woman's shift.* (Ḳ.) — See also جَوْبَةٌ. — [*A kind,* or *sort.*] You say, فُلَانٌ فِيهِ جَوْبَانِ مِنْ خُلُقٍ [*In such a one are two kinds of temper,* or *disposition*]; i. e., he does not remain in one temper, or disposition. (TA.) And Dhu-Rummeh says,

* جَوْبَيْنِ مِنْ هَمَاهِمِ الأَغْوَالِ *

meaning Thou hearest *two kinds of the sounds,* or *voices,* [or *mutterings,*] *of the ghools.* (TA.)

جُيُوبٌ, meaning The [*part called*] طَوْق of a shirt, (see art. جيب,) is, accord. to some, from the root جوب, because the middle of it is cut out: accord. to others, from the root جيب. (TA.)

جَابَةٌ is an inf. n. of أَجَابَ, (Kr, TA,) or a simple subst. (AHeyth, Ṣ, TA) used in the place of an inf. n. (AHeyth, TA. See 4.) Hence, أَسَآءَ سَمْعًا فَأَسَآءَ جَابَةً [*He heard ill, and therefore answered ill*]: (Ṣ, A, Ḳ:) a prov., and therefore not to be rehearsed otherwise than in the original way, as above: [not to be altered by the substitution of إِجَابَةً or إِجَابًا for جَابَةً:] its origin is said to have been this: Sahl [or Suheyl] Ibn-'Amr had an insane son; and a man said to him, أَيْنَ أُمُّكَ, i. e. "Whither is thy tending?" to which he (thinking that he said, أَيْنَ أُمُّكَ ["Where is thy mother?"],) answered, "She is gone to buy flour:" whereupon his father uttered the words of this prov. (TA. [See also Freytag's Arab. Prov. i. 603.]) See also جَوَابٌ. — جَابَةُ المِدْرَى is a dial. var. of جَأْبَةُ المدرى: (Ḳ: [see art. جأب:]) accord. to AO and Sh, it is without ء: accord. to the former, it means A doe-gazelle *when her horn has come forth;* and accord. to the latter, *when her horn has cut the skin and come forth:* (T, TA:) or it means *having smooth horns;* and if so, it has no [known] derivation. (TA.) [See also art. درى.]

جَوْبَةٌ *A depressed place amid the houses of a people, into which the rain-water flows:* (TA:) *a pit, an excavation,* or *a hollow,* (T, Ḳ, TA,) *round and wide:* (T, TA:) *a gap,* or *an opening, in the clouds;* and *in mountains:* and *a clear space* (مَوْضِعٌ يَنْجَابُ ▼) *in a* [*stony tract such as is called*] حَرَّة: (Ṣ:) *a place* (AḤn, Ḳ) *that is clear,* (AḤn,) *plain and smooth,* (AḤn, Ḳ,) *such as is termed* دَارَة, *with few trees, like a round* غَائِط [or *wide and depressed tract*], (AḤn,) *in a tract that is hard,* or *hard and level,* or *level but rough,* (AḤn, Ḳ,) *and such as is of large extent, not in sands nor in a mountain;* so called because [for the most part] clear of trees: (AḤn:) and *an intervening space between houses;* (Ḳ;) as also ▼ جَوْبٌ: (TA:) and *a wide,* or *spacious, and smooth tract, between two lands:* (Ḳ:) *any wide gap,* or *opening: any gap,* or *opening, without buildings:* (TA:) pl. جُوَبٌ (Ṣ, Ḳ) and جَوْبَاتٌ. (TA.) — The former of these pls. also signifies The *pudenda of women;* syn. فُرُوجٌ. (TA.) — See also جَوْبٌ.

جِيبَةٌ *i. q.* جَوَابٌ, q. v. (Ṣ, Ḳ.) So in the phrase, إِنَّهُ لَحَسَنُ الجِيبَةِ [*Verily he is good in respect of answer* or *reply* or *response:* or here it seems rather to signify, agreeably with analogy, *the mode,* or *manner, of answering* or *replying* or *responding*]. (Ṣ.)

جَوَابٌ *An answer, a reply,* or *a response,* (Mṣb, TA,*) to a letter, or writing, and to a saying, or question; and this is either affirmative or negative: (Mṣb:) [accord. to some, it is only after a question or demand; but this is not correct; for it is often a reply to an affirmation:] ▼ جِيبَةٌ [q. v.] is syn. therewith; (Ṣ, Ḳ;) and so are ▼ جَابَةٌ [q. v.] and ▼ مَجُوبَةٌ: (Ḳ:) the pl. of جواب is أَجْوِبَةٌ and جَوَابَاتٌ. (Mṣb.) [Hence, in grammar, حَرْفُ جَوَابٍ *A responsive,* or *replicative, particle.* And جَوَابُ شَرْطٍ *An apodosis;* the *complement,* or *correlative, of a condition;* as أَكْرَمْتُكَ in the saying, إِنْ جِئْتَنِى أَكْرَمْتُكَ; also called جَزَآءٌ, and جَوَابُ جَزَآءٍ. And جَوَابُ قَسَمٍ The *complement of an oath.*] — Also The *sound of a bird pouncing down* from the sky. (TA from a trad.)

جَوَّابٌ [*An excellent well-digger:*] a surname given to Málik Ibn-Kaạb El-Kilábee, (AO, ISk, Ṣ, Ḳ,*) because he dug not a well nor bored a rock without making it to yield water. (AO, ISk, Ṣ.) — † *A traverser of countries; one who travels much.* (TA.) Hence, جَوَّابُ لَيْلٍ سَرْمَدٍ † *One who travels all the night without sleeping.* (TA.) And جَوَّابٌ جَأَّبٌ † *One who traverses the countries and gains wealth.* (TA.) And جَوَّابُ الفَلَاةِ † *The guide of the desert.* (TA.)

الجَائِبُ العَيْنِ *The lion.* (Ḳ.)

خَبَرٌ جَائِبَةٌ ‡ *News that traverses the earth, from country to country,* or *town to town:* (Ṣ, A:*) or *i. q.* طَرِيفَةٌ خَارِقَةٌ [app. a mistranscription for طَرِيفَةٌ خَارِقَةٌ, meaning *recent news that traverses the land*]. (Ḳ.) And [the pl.] جَوَائِبُ † *Tidings from afar.* (Ḳ.) And جَوَائِبُ الأَمْثَالِ † *Current proverbs; such as traverse the countries.* (TA.)

أَجْوَبُ, [see 4,] in the following question, put to Mohammad, (TA,) أَىُّ اللَّيْلِ أَجْوَبُ دَعْوَةً, is either from جُبْتُ الأَرْضَ (K, TA) "I traversed the land," (TA,) and signifies ‡ *More,* or *most, penetrating* to the places whence the answer is imagined to proceed; (Ḳ, TA;) or [it signifies *more,* or *most, quick in being answered,*] from جَابَتِ ▼ الدَّعْوَةُ, of the measure فَعُلَتْ, [i. e., originally جَوُبَتْ,] "the prayer became answered," which, however, is a verb not in use, like as فَقِيرٌ and شَدِيدٌ are imagined to be derived from فَقُرَ and شَدُدَ: (Z, TA:) or it signifies *more,* or *most, quick of answer,* [from أَجَابَ,] and is [anomalous, and] similar to أَطْوَعُ ["more obedient"], from الطَّاعَةُ, [i. e. from أَطَاعَ "he obeyed,"] (M, L, TA,) and to أَعْطَى ["more, or most, excellent in giving," from أَعْطَى "he gave"], and لَوَاقِحُ [pl. of لَاقِحَةٌ a "fecundating" wind, (in the Ḳur xv. 22,) from أَلْقَحَ "he, or it, fecundated"], (M, L, Ḳ, TA,) and the like; (M, L, TA;) and if so, the word is anomalous because a word of the measure أَفْعَل of this kind is not derived from a verb of more than three letters, except in certain cases of deviation from the constant course of speech: (L, TA:) the meaning is, ‡ *What part of the night is that* [*in which prayer most quickly penetrates?* or] *in which prayer is most quick in being answered?* (Mgh:) or *what part of the night is that in which God is most quick in answering prayer?* (L, TA.)

مَجُوبٌ [pass. part. n. of 1, q. v.:] Anything *cut in the middle,* or *of which the middle is cut out;* as also ▼ مُجَوَّبٌ; (T, TA;) and the latter, anything *hollowed out in the middle.* (TA.)

مِجْوَبٌ *An iron instrument with which one cuts* [or *perforates* or *hollows out*]. (Ṣ, TA.) — See also جَوْبٌ.

المُجِيبُ one of the names of God; *The Answerer of prayer; He who recompenses prayer and petition by gift and acceptance.* (TA.)

مَجُوبَةٌ: see جَوَابٌ.

مُجَوَّبٌ: see مَجُوبٌ. — [Hence,] أَرْضٌ مُجَوَّبَةٌ † *A land of which one part has been rained upon* (Ḳ, TA) *and not another.* (TA.)

مِجْوَابٌ *An instrument with which palm-sticks and canes &c. are bored by the maker of cages or crates or the like.* (TA in art. ثعلب.)

مُتَجَاوِبٌ ‡ Speech, or language, *of which the several parts correspond,* or *are consistent.* (A, TA.)

مُنْجَابٌ A garment *rent,* or *slit.* (Ḥam p. 338.)

جوح

1. جَاحَ, (Ṣ, A,) aor. يَجُوحُ, (Ṣ,) inf. n. جَوْحٌ, (Ṣ, Ḳ,) *He* (a man, Ṣ) *extirpated,* or *exterminated,* (Ṣ, A, Ḳ,) a thing: (Ṣ:) *He* (God, Ṣ) *destroyed* (Ṣ, A, Ḳ) a man's property, or cattle, (Ṣ,) *by what is termed* جَائِحَة: (Ṣ, A:) as also ▼ اجاح, (Ṣ, A,) inf. n. إِجَاحَةٌ; (Ḳ;) and ▼ اجتاح, (Ṣ,) inf. n. اِجْتِيَاحٌ. (Ḳ.) You say [also], جَاحَتِ الآفَةُ المَالَ, i. e. الجَائِحَةُ, aor. تَجُوحُهُ, inf. n. as above; and ▼ اجاحته; and ▼ اجتاحته; *The bane,* or *pest,* or *the like, destroyed the property,* or *cattle;* as also جَاحَتْهُ, aor. تَجِيحُهُ, inf. n. جِيَاحَةٌ: (Mṣb:) and one says of anything, such as drought, or dearth, and civil war, or conflict and faction, &c., جاح المَالَ and ▼ اجتاحه, meaning *it extirpated,* or *exterminated, the property,* or *cattle.* (TA.) And جَاحَتْهُمُ الجَائِحَةُ, (Ṣ,) or السَّنَةُ, inf. n. جَوْحٌ and جِيَاحٌ; and ▼ أَجَاحَتْهُمْ; (TA;) and ▼ اِجْتَاحَتْهُمْ; (Ṣ, A, TA;) i. e. [*The calamity, bane, pest,* or *the like,* or *drought,* or *dearth, destroyed* or] *extirpated* [*them,* or] *their property,* or *cattle.*

(TA.) And اجتاح ↑ العَدُوُّ مَالَهُ *The enemy destroyed, or made an end of, his property, or cattle;* syn. أَتَى عَلَيْهِ. (TA.) — Also جاح, aor. يَجُوحُ, inf. n. جَوْحٌ, *He destroyed the property, or cattle, of his relations.* (IAạr, TA.) = And, aor. يَجُوحُ, *He deviated from the road,* or *beaten track.* (IAạr, Ḳ.)

4: see 1, in three places.

8: see 1, in five places.

جَوْحَةٌ: see جَائِحَةٌ.

جَائِحٌ [act. part. n. of 1]. You say آفَةٌ جَائِحَةٌ *A destructive bane* or *pest* or *the like:* (Mṣb:) and سَنَةٌ جَائِحَةٌ *a year of drought* or *dearth* or *sterility:* (Mgh:) pl. جَوَائِحُ. (Mgh, Mṣb.) — [And hence,] The *locust.* (IAạr, T in art. جحى.)

جَائِحَةٌ [fem. of جَائِحٌ, q. v.: and hence, as a subst.,] *A bane, a pest,* (Mṣb,) *a calamity,* or *the like,* (Ṣ, Mṣb, Ḳ,) or *a great calamity,* (Mgh, TA,) *that destroys,* or *extirpates, men's property,* or *cattle;* (Ṣ, Mgh, Mṣb,* Ḳ, TA;) *as drought,* or *dearth;* or *civil war,* or *conflict and faction,* and *the like:* (Ṣ, TA:) *drought,* or *dearth, that destroys,* or *extirpates, men's property,* or *cattle;* as also ↑ جَوْحَةٌ: (Wáṣil, TA:) or *a calamity that befalls a man, and destroys all his property,* or *cattle:* (A 'Obeyd, T:) or *severe drought* or *dearth, that destroys,* or *extirpates, men's property,* or *cattle:* (ISh:) *sometimes it is the effect of large hail; and sometimes, of excessive cold* or *heat:* (T:) or (accord. to Esh-Sháfi'ee, Mgh, Mṣb) *a blast,* or *blight,* or *calamity arising from the atmosphere, that destroys the fruits,* (Mgh, Mṣb, TA,) or *some thereof;* (Mgh;) only relating to fruits: (TA:) pl. جَوَائِحُ. (A, Mgh, Mṣb.) It is said in a trad., أَمَرَ بِوَضْعِ الجَوَائِحِ, which is an elliptical phrase, meaning *He commanded to remit the poor-rate in respect of things affected by a blast,* or *blight,* (Mgh, Mṣb,) namely, property, (Mgh,) or fruits, (Mṣb,) so affected; (Mgh, Mṣb;) i. e., that no portion thereof, (Mgh,) or of the remains thereof, (Mṣb,) should be taken for the poor-rate. (Mgh, Mṣb.)

مُجْتَاحٌ: see what next follows.

مَجُوحٌ Property, or cattle, *destroyed* by a bane, or pest, or the like; as also مَجِيحٌ [belonging to art. جيح], and ↑ مُجْتَاحٌ. (Mṣb.)

مِجْوَحٌ [A person or thing] *that destroys,* or *extirpates, everything.* (Ḳ, TA.)

جود

1. جَادَ, aor. يَجُودُ, inf. n. جُودَةٌ and جَوْدَةٌ, *It* (a thing, Ṣ, or a commodity, an article of household-goods, or the like, Mṣb, and a work, or performance, TA) *was,* or *became,* جَيِّد [i. e. *good, goodly, approvable,* or *excellent;* the verb being the *contr.* of رَدُؤَ, as is implied in the A and Ḳ]: (Ṣ, A, Mṣb, Ḳ:) in this sense, accord. to some, of the class of قَالَ; accord. to others, of the class of قَرُبَ. (Mṣb.) [Also said of a man, meaning *He was,* or *became, excellent,* or *egregious,* in some quality; sometimes, though very rarely, in a quality that is disapproved.] — And جَادَ, (Ṣ, A, Mṣb, Ḳ,) of the class of قَالَ, (Mṣb,) aor. as above, (Ṣ, Mṣb,) inf. n. جُودٌ, (Ṣ, A, Mṣb, Ḳ,) with ḍamm, (Ṣ, Mṣb,) *He was liberal, bountiful, munificent,* or *generous:* (Ḳ:) or *he affected,* or *constrained himself, to be generous:* (Mṣb:) or *he gave without being asked, to preserve the receiver from the ignominy of asking:* (MF:) or *he gave what was meet to him to whom it was meet:* (El-Karmánee, TA:) or *he gave what was meet to him to whom it was meet, not for a compensation;* so that it has a more special signification than أَحْسَنَ. (MF.) You say, جاد بماله [*He was liberal, &c., with his property*]: (Ṣ:) or جاد بالمَالِ *he affected,* or *constrained himself, to be generous with the property.* (Mṣb.) — Hence, (Mṣb,) جاد بِنَفْسِهِ, (Ṣ, Mṣb, Ḳ,) aor. as above, (Ṣ, A,) inf. n. جُودٌ (TA) and جُؤُودٌ, (Ṣ, TA,) ‡ *He gave up his spirit,* (A, Mṣb, TA,) at death; (Ṣ, Mṣb;) like as one gives away his property; said of one in the agony of death: (TA:) and ‡ *he gave away his life,* in war. (Mṣb.) And you say also, جَادَتْ نَفْسُهُ † [*His soul,* or *spirit, resigned itself,* or *departed*]. (Mṣb in art. نفس.) — جاد المَطَرُ, inf. n. جَوْدٌ, *The rain was,* or *became, copious,* or *abundant.* (Ṣ.) And جَادَتِ السَّمَآءُ, (A, Mṣb,) inf. n. جَوْدٌ, with fet-ḥ, *The sky rained.* (Mṣb.) And جَادَتِ العَيْنُ, inf. n. جَوْدٌ and جُؤُودٌ, *The eye shed many,* or *abundant, tears.* (Lḥ, Ḳ.) — جاد said of a horse, (Ṣ, A, L, Mṣb, Ḳ,) aor. as above, (Ṣ,) inf. n. جُودَةٌ (Ṣ, L, Mṣb, Ḳ) and جَوْدَةٌ; (Mṣb, and some copies of the Ḳ;) and ↑ جوّد, (A, L, Ḳ,) inf. n. تَجْوِيدٌ; (TA;) and ↑ اجاد, (L,) and ↑ أَجْوَدَ; (L, Ḳ;) *He became fleet,* or *swift, and excellent,* (L,) فِي عَدْوِهِ (Ṣ, L, Ḳ,*) صَارَ رَائِعًا *in his running.* (A, L, Ḳ.) [See an ex. in a verse cited voce دَامِرٌ, in art. دوم.] — See also 4, in two places. — جاد إِلَيْهِ *He inclined to him,* or *it.* (TA.) = جادهُ *He overcame him in liberality, bounty, munificence,* or *generosity.* (Ḳ.) See 3. — جَادَهُمْ, aor. يَجُودُ, inf. n. جَوْدٌ, *It* (rain) *rained,* or *descended, upon them copiously,* or *abundantly.* (L.) And جِيدُوا *They were rained upon with a copious,* or *an abundant, rain.* (L.) And جِيدَتِ الأَرْضُ, (Ṣ, L, Ḳ,) inf. n. جَوْدٌ; (Aṣ, TA;) and ↑ أُجِيدَتْ; (Ḳ;) *The earth,* or *land, was rained upon with a copious,* or *an abundant, rain:* (Ṣ, L, Ḳ:) or, *so that the moisture of the rain met that of the soil.* (Aṣ, TA.) — جِيدَ, (Ṣ, A, Ḳ,) aor. يُجَادُ, (Ṣ, Ḳ,) inf. n. جُوَادٌ, (Ṣ,* Ḳ,* TA,) † *He* (a man, Ṣ, A) *thirsted,* or *became affected by thirst:* (Ṣ, A, Ḳ:) or *thirsted vehemently:* (accord. to an explanation of جُوَادٌ in the Ḳ:) or *was at the point of death,* or *destruction;* (Ḳ;) as though destruction rained upon him. (TA.) — [Hence,] إِنِّي أُجَادُ إِلَى لِقَائِكَ ‡ *Verily I am affected with a longing desire to meet thee:* (A:) or إِنِّي لَأُجَادُ إِلَيْكَ (Ḳ, TA [in the CḲ, erroneously, لَأَجَادُ]) ‡ *Verily I am affected with a longing desire for thee,* (Ḳ, TA,) i. e., *to meet thee,* (TA,) *and am impelled towards thee:* (Ḳ:) and يُجَادُ إِلَى فُلَانَةَ ‡ *He is affected with longing desire for such a female;* like as you say يُظْمَأُ. (A.) One says also, جَادَهُ الهَوَى ‡ *Love affected him with longing desire,* (شَاقَهُ, L, Ḳ, in the CḲ شَاقَّهُ,) *and overcame him.* (Ḳ.) — [Also, app., جِيدَ, aor. يُجَادُ, inf. n. جُوَادٌ, (as in a sense explained above,) † *He became affected,* or *overcome,* or *distressed,* (see مَجُودٌ,) *by drowsiness,* or *slumber:* for] جُوَادٌ is syn. with نُعَاسٌ: (L, TA:) and you say, جَادَهُ النُّعَاسُ † *Drowsiness,* or *slumber, overcame him;* (L;) as though sleep rained upon him. (TA.)

2: see 4: — and see also 1.

3. جاودهُ *He vied with him,* or *contended with him for superiority, in liberality, bounty, munificence,* or *generosity.* (Ṣ, TA.) You say, جاودهُ ↑ فَجَادَهُ *He vied with him,* or *contended &c., in liberality, &c., and overcame him therein.* (TA.)

4. اجادهُ *He made it good, goodly, approvable,* or *excellent;* (Ṣ, A,* Ḳ;) as also أَجْوَدَهُ, (Ṣ,* Ḳ,) like as they said اطال and اطول, and احال and احول, and اطاب and اطيب, and الان and الين; (Ṣ;) and ↑ جوّدهُ, (Ṣ,* A,) inf. n. تَجْوِيدٌ. (Ṣ.) [Hence,] اجادهُ النَّقْدَ *He gave him the cash,* or *ready money, good.* (Ṣ, Ḳ.) And أَجَدْتُكَ ثَوْبًا *I gave thee a garment,* or *piece of cloth, that was good, goodly,* or *excellent;* or *in a good state.* (A, TA.) — *He gave him a* dirhem, or piece of silver. (Ḳ.) — أُجِيدَتِ الأَرْضُ: see 1. = *He,* or *it, slew him,* or *killed him.* (L.) = اجاد, (inf. n. إِجَادَةٌ, Mṣb,) *He said, gave utterance to, uttered,* or *expressed, what was good, approvable,* or *excellent;* *he said,* or *did, well,* or *excellently;* مِنْ قَوْلٍ أَوْ فِعْلٍ (Mṣb;) أَتَى بِالجَيِّدِ (L, Mṣb, Ḳ) as also أَجْوَدَ; (L;) and ↑ جاد, inf. n. جُودَةٌ. (L.) [You say, قَالَ فَأَجَادَ *He said, and said well:* and فَعَلَ فَأَجَادَ *He did, and did well.*] And اجاد فِى عَمَلِهِ, and اجود, *He did well,* or *excellently, in his work.* (L.) — Said of a horse, and اجود likewise: see 1. — Also *He had with him a horse such as is termed* جَوَاد [i. e. *fleet,* or *swift, and excellent*]: (Ṣ:) or *he became possessed of such a horse;* (A, Ḳ;) as also اجود. (Ḳ.) — أَجَادَتْ *She brought forth a child,* or *children, of liberal, bountiful,* or *generous, disposition.* (A.) And اجاد بالوَلَدِ *He begot the child,* or *children, of liberal, bountiful,* or *generous, disposition;* (Ḳ;) and in like manner, جاد ↑ به أَبَوَاهُ [*His two parents so engendered him*]. (TA.)

5. تجوّد *He chose what was good, goodly, approvable,* or *excellent, among all things.* (Ḥam p. 299.) *He affected nicety,* or *refinement; he was,* or *became, nice, exquisite, refined,* or *scrupulously nice and exact;* or *he chose what was excellent,* or *best, to be done; and exceeded the usual bounds;* فِى صَنْعَتِهِ *in his work of art,* or *his manufacture;* syn. تَنَوَّقَ. (A, TA.) And تجوّد وَبَالَغَ فِى مَطْعَمِهِ وَمَلْبَسِهِ [*He was dainty, nice, exquisite, refined,* or *scrupulously nice and exact;* or *he chose what was excellent,* or *best; and exceeded the usual bounds; in his food and his*

apparel]; (JK and Ḳ in art. نوق;) *he was studious of his diet and apparel, always eating exquisite food and wearing sumptuous clothing.* (TḲ in that art.) ═ تجوّده : see 10. —— تَجَوَّدْتُهَا لَكَ *I chose, or selected, the best, or most excellent,* (الأَجْوَدَ ▼,) *thereof for thee.* (TA.)

6. تجاودوا *They considered* [or tried] *which of them had the best argument, or plea, or allegation:* (Ḳ, TA:) so says Aboo-Sa'eed on the authority of an Arab of the desert. (TA.) And يَتَجَاوَدُونَ الحَدِيثَ *They consider, or see,* [or try,] *which of them will be best in narration, or talk, or discourse.* (A.) —— [Also *They vied, or contended together for superiority, in liberality, bounty, munificence, or generosity.*]

10. استجادهُ *He reckoned it, or esteemed it, good, goodly, approvable, or excellent:* (Ṣ:) or *he found it to be so:* (Ḳ:) or *he desired, or sought, that it might be so,* (A, Ḳ,) *and chose it, or selected it;* (A;) as also ▼ تَجَوَّدَهُ. (A.) You say also, اِسْتَجْوَدَ رَأْيَهُ [*He esteemed his judgment, or opinion, good:* or *found it to be so*]. (TA in art. جزل.) —— *He desired, or sought, or demanded, his liberality, bounty, munificence, or generosity.* (Ḳ.) —— *He desired, or sought, that he* (a horse) *might be such as is termed* جَوَاد [i. e. *fleet, or swift, and excellent*]. (Ḳ.) ═ استجاد *It came or happened, well.* (KL.)

جَوْدٌ *Copious, or abundant,* rain; (Ṣ, L, Ḳ;) as also ▼ جَائِدٌ: (Ṣ:) or rain *that thoroughly irrigates everything:* (M:) or rain *that is not exceeded:* (M, L, Ḳ:) accord. to some, who observe that the phrase, mentioned by Sb, أَخَذَتْنَا بِالجَوْدِ وَفَوْقَهُ [*Thou hast assailed us with a storm* of reproach or the like *not to be exceeded, and with that which is above it,*] is one of hyperbole and reproach. (M, L.) It is an inf. n. thus used as an epithet [and therefore applicable without variation to a fem. as to a masc. n., and to a dual and a pl. as to a sing. n.]: (L:) and is also pl. [or rather a quasi-pl. n.] of جَائِدٌ, (Ṣ, L, Ḳ,) like as صَحْبٌ is of صَاحِبٌ. (Ṣ, L.) You say مَطَرٌ جَوْدٌ [*A copious, or an abundant, rain; &c.*]: (L:) and سَحَابَةٌ جَوْدٌ [*A cloud yielding a copious, or an abundant, rain; &c.*]: (IAạr, L:) and هَاجَتْ لَنَا سَمَآءٌ جَوْدٌ [*A copious, or an abundant, rain, &c., became stirred up for us*]: (Ṣ, Ḳ:*) and you also say, [contr. to the usage mentioned above, or as though جَوْدٌ were an epithet from جَادَ, and this originally جَوُدَ, like ضَخْمٌ from ضَخُمَ, but used as a subst.,] مَطْرَتَانِ جَوْدَانِ [*Two showers of rain, copious, or abundant, &c.*]. (Ṣ, Ḳ.) ▼ تَجَاوِيدُ, [app. signifying the same as جَوْدٌ used as a pl.,] occurring in the following verse of Sakhr El-Ghei,

* يُلَاعِبُ الرِّيحَ بِالعَصْرَيْنِ قَسْطَلُهُ *
* وَالوَابِلُونَ وَتَهْتَانُ التَّجَاوِيدِ *

[*Its dust makes sport with the wind in the morning and evening, or night and day, and so do the violent showers of big drops, and the pouring of copious, or abundant, rains, &c.*], (L, Ḳ,*) is a pl. having no sing.; (Ḳ;) or it may be so, like تَعَاجِيبُ and تَعَاشِيبُ and تَبَاشِيرُ; or it may be pl. of تَجْوَادٌ [an inf. n.]. (L.) You say also, أَصَابَتْهُ تَجَاوِيدُ ▼ المَطَرِ [*Copious showers of rain fell upon him, or it*]. (A.) —— See also جَوَادٌ.

جَوْدَةٌ [an inf. n. of 1, (q. v.,) in two senses; as also جُودَةٌ: and an inf. n. of un., signifying] *A single affection of thirst; a thirsting.* (Ṣ, Ḳ.) —— See also جُوَادٌ.

جَادِيٌّ *Saffron.* (Ṣ, Ḳ.)

جَوَادٌ, used alike as masc. and fem., (Ṣ, Ḳ,) *Liberal, bountiful, munificent, or generous:* (Ṣ,* Ḳ:) or one *who affects, or constrains himself, to be generous:* (Mṣb:) or *who gives without being asked, to preserve the receiver from the ignominy of asking:* (MF:) or *who gives what is meet to him to whom it is meet:* (El-Karmánee, TA:) or *who gives what is meet to him to whom it is meet, not for a compensation;* so that it has a more special signification than مُحْسِنٌ: (MF:) pl. [of pauc., masc.,] أَجْوَادٌ and (of mult., TA) جُوْدٌ, (Ṣ, A, Ḳ,) like as قُذُلٌ is pl. of قَذَالٌ, but the و is made quiescent because it is an unsound letter, (Ṣ,) [in some copies of the Ḳ جُوُدٌ,] and أَجَاوِدُ, (Ṣ, Ḳ,) contr. to analogy, (TA,) or أَجَاوِيدُ, [reg., as pl. of أَجْوَادٌ,] (A,) and جُودَآءُ (Ṣ, Ḳ) and جُودَةٌ, (CḲ, [in some copies of the Ḳ omitted,]) or جَوْدَةٌ, or جُوُدَةٌ, [written in the latter manner in a MS. copy of the Ḳ,] with ة added to the [proper] pl. form [جُوْدٌ or جُوُدٌ], accord. to the doctrine of Sb: (TA:) جُوْدٌ is used as a fem. pl., (Ṣ, Mṣb,) and is like نُوْرٌ pl. of نَوَارٌ. (Ṣ.) —— Also, applied alike to the male and the female, (Ṣ,) *A courser; a fleet, or swift, and excellent,* horse; (L;) a horse *fleet, or swift, in running;* or *excellent in running,* or *in the motion of his legs;* as also ▼ جَوْدٌ: (Bḍ in xxxviii. 30:) or *that outstrips others:* (Jel ib.:) *i. q.* رَائِعٌ: (Ṣ, L, Ḳ:) pl. جِيَادٌ, (Ṣ, A, Bḍ, L, Msb, Ḳ,) which by rule should be جِوَادٌ, like طِوَالٌ, but this latter form has not been heard from the Arabs; (L;) or جِيَادٌ is pl. of جَوْدٌ, or of جَيِّدٌ; (Bḍ ubi suprà;) and جَوَادٌ has also for its pl. أَجْيَادٌ, [a pl. of pauc., and irregular, or this is pl. of جَيِّدٌ, and therefore, though irregularly, retains the ى substituted for و,] (Ṣ, L,) and أَجْوَادٌ, [also a pl. of pauc., but agreeable with rule, or this is pl. of جَوْدٌ,] (L,) and أَجَاوِيدُ (Ṣ, L) is pl. of أَجْوَادٌ. (L.) Hence, أَقْبَلَ جَوَادًا † *He came on, or advanced, like a horse that is termed* جواد: and سِرْتُ إِلَيْهِ جَوَادًا † *I went to him, or it, like a horse that is so termed.* (Mgh in art. غذ.) You say also, عَدَا عَدْوًا جَوَادًا *He ran a long run.* (A, TA.) And سِرْنَا عُقْبَةً جَوَادًا, and عُقْبَتَيْنِ جَوَادَيْنِ, and عُقَبًا جِيَادًا (Ṣ, A) and أَجْوَادًا, (A, TA,) *We journeyed a long march* or *stage, and two long marches* or *stages, and long marches* or *stages.* (Ṣ, A, TA.)

جُوَادٌ [accord. to the TA inf. n. of جِيدَ, which see in two places,] †*Thirst:* (Ṣ, Ḳ:) or *vehemence of thirst.* (Ḳ.) —— Also, [accord. to the Ḳ ▼ جُودَةٌ, but this is corrected in the TA,] † *Drowsiness,* or *slumber.* (TA.)

جَوْدٌ : see جَائِدٌ.

جَيِّدٌ, originally of the measure فَيْعِلٌ, (Ṣ, Mṣb,) as the Baṣrees say, i. e. جَيْوِدٌ, (Mṣb, TA,) the و being changed into ى because of its being meksoor and preceded by ى, and the augmentative ى being then incorporated into it; (TA;) or, as the Koofees say, of the measure فَيْعَلٌ, like عَيْطَلٌ &c., because there is found no sound word of the measure فَيْعِلٌ except صَيْقِلٌ, a woman's name, and the unsound is accorded to the sound; or, as others say, of the measure فَعِيلٌ, [and so I find in one copy of the Ṣ,] originally جَوِيدٌ, the kesreh of the و being, accord. to them, suppressed because difficult of pronunciation, and the quiescent و and ى thus coming together, [the latter receives the rejected kesreh, and] the و is changed into ى and incorporated into the [augmentative] ى; (Mṣb;) *Good, goodly, approvable, or excellent; contr. of* رَدِىٌّ; (A, Ḳ;) applied to a thing, (Ṣ,) or a commodity, an article of household-goods, or the like, (A, Mṣb,) and a work, or performance: (TA:) pl. جِيَادٌ (Ṣ, A, Mṣb, Ḳ) and جِيَادَاتٌ, (Ḳ,) the latter a pl. pl., [i. e. pl. of جِيَادٌ,] (TA,) and جَيَائِدُ, (Ṣ, Ḳ,) with hemz, [and, accord. to some,] contr. to analogy. (Ṣ.) [It is also applied to a man, meaning *Excellent,* or *egregious,* in some quality; sometimes, though very rarely, in a quality that is disapproved.]

أَجْوَدُ [*Better,* and *best; more,* and *most, goodly* or *approvable* or *excellent*]: see 5. —— [*More,* and *most, liberal, bountiful, munificent,* or *generous.* Hence,] أَجْوَدُ مِنْ حَاتِمٍ [*More liberal, &c., than Ḥátim*]: a prov. (Meyd.) —— [*More,* and *most, fleet,* or *swift, and excellent;* relating to a horse. Hence,] أَجْوَدُ مِنَ الجَوَادِ المُبَرِّ [*More fleet, &c., than the courser that surpasses others*]: a prov. (Meyd.)

تَجَاوِيدُ : see جَوْدٌ, in two places.

مَجُودٌ A field, or garden, *rained upon:* (A:) [or *rained upon copiously,* or *abundantly.*] And أَرْضٌ مَجُودَةٌ *Land rained upon with a copious,* or *an abundant, rain.* (Ṣ, L, Ḳ.) —— † A man (Ṣ, A) *affected with thirst:* (Ṣ, A, Ḳ:) [or, *with vehement thirst:* (see جُوَادٌ:)] or *at the point of death,* or *destruction.* (Ḳ.) —— And [hence,] ‡ *Affected with longing desire.* (L.) —— Also † *Overcome by drowsiness,* or *slumber:* (TA:) or *distressed by drowsiness,* or *slumber, &c.* (Lḥ, L.)

مُجِيدٌ : see مِجْوَادٌ. —— Also A man *possessing a horse such as is termed* جَوَاد [i. e. *fleet,* or *swift, and excellent*]: pl. مَجَاوِيدُ [by rule pl. of مِجْوَادٌ, q. v.]. (A, TA.) —— حَتْفٌ مُجِيدٌ ‡ *Present death.* (Ḳ, TA.)

مِجْوَادٌ One *who says, utters,* or *expresses,* or *who does,* (Ḳ, TA,) *much,* or *often,* (TA,) *what is good, goodly, approvable,* or *excellent;* (Ḳ, TA;) as also ▼ مُجِيدٌ: (TA:) [or rather the latter is a simple, not an intensive, epithet:] the former is applied to a poet, (Ṣ, A, Ḳ,) as syn. with the latter, (Ḳ,) or as meaning *who says,* or *utters, much,* or *often, what is good,* or *excellent:*

(Ṣ:) and both are applied to a workman, or an artificer: pl. of the former مَجَاوِيدُ. (A.)

جوذاب

جُوذَابٌ : see art. جذب.

جور

1. جَارَ, aor. يَجُورُ, (TA,) inf. n. جَوْرٌ, (Ṣ, A, Ḳ,) *He declined*, or *deviated, from the right course;* (Ṣ, A;) and so جَارَ عَنِ القَصْدِ: (A:) *he wandered from the right way:* (TA:) *he pursued a wrong course:* (Ḳ:) or *he left the right way in journeying:* and *it* (anything) *declined.* (TA.) You say also, جارَ عَنِ الطَّرِيقِ *He declined*, or *deviated, from the road*, or *way.* (Ṣ, Mgh, Mṣb.) — And جار, (Ṣ, Mgh, Mṣb,) aor. as above, (Mṣb,) and so the inf. n., (Mgh, Mṣb, Ḳ,) *He acted wrongfully, unjustly, injuriously*, or *tyrannically*, (Ṣ,* Mgh, Mṣb, Ḳ,) عَلَيْهِ *against him*, (Ṣ, TA,) فِى حُكْمِهِ *in his judgment*, (Mṣb,) or فِى الحُكْمِ *in judgment.* (Ṣ, TA.) — جَارَتِ الأَرْضُ ‡ *The plants*, or *herbage, of the land grew tall:* (A, TA:) and so جَارَت. (TA.) = See also 10.

2. جوّرهُ, (Ṣ, A, Ḳ,) inf. n. تَجْوِيرٌ, (Ṣ,) *He attributed*, or *imputed, to him*, or *charged him with*, or *accused him of, wrongful, unjust, injurious*, or *tyrannical, conduct;* (Ṣ, Ḳ;) *contr. of* عَدَّلَهُ. (A.) = *He prostrated him* (Ṣ, Ḳ) by a blow, (Ṣ,) or by a thrust of a spear or the like; from جار "he, or it, declined;" (A;) like كَوَّرَهُ. (Ṣ.) — *He threw it down*, (TA,) and *overturned it;* (Ḳ, TA;) namely, a building, and a tent, &c.: (TA:) *he took it to pieces;* namely, a tent. (A.)

3. جاورهُ, inf. n. مُجَاوَرَةٌ and جِوَارٌ (Ṣ, Mṣb, Ḳ) and ♦ جُوَارٌ, (Ṣ, M, and some copies of the Ḳ,) or the last is a simple subst., (Mṣb,) and ♦ جَوَارٌ, (M, and so in some copies of the Ḳ instead of جُوَار,) of which forms the second (جِوَار) is more chaste than the third (Ṣ, TA) and than the fourth, as relating to the verb in the sense here following, though some disapprove of it, and assert the third and the fourth to be more chaste; (TA;) *He became his* جار [or *neighbour*]; (Ḳ;) *he lived in his neighbourhood*, or *near to him:* (Mṣb, TA:) or *he lived in a dwelling contiguous to his.* (Mṣb.) — Also جاورهُ, (TA,) inf. n. جِوَارٌ, (Ḳ,) and ♦ جُوَارٌ is said to be a quasi-inf. n., and more chaste than جِوَار as relating to the verb in the sense here following; (TA;) *He bound himself to him by a covenant to protect him.* (Ḳ, TA.) — And جاور بَنِى فُلَانٍ, and فى بنى فلان, inf. n. مُجَاوَرَةٌ and جِوَارٌ, *He protected himself by a covenant with the sons of such a one;* from مُجَاوَرَة signifying the "living near." (TA.) — And جاور, inf. n. مُجَاوَرَةٌ, *i. q.* اِعْتَكَفَ فِى مَسْجِدٍ [*He confined himself in a mosque, or place of worship, during a period of days and nights, or at least during one whole day, fasting from daybreak to sunset, and occupying himself in prayer and religious meditation, without any interruption by affairs distracting the mind from devotion and not pressing*]. (Ṣ, Ḳ.) But جاور بِمَكَّةَ, and بِالمَدِينَةِ, signifies absolutely *He abode in Mekkeh*, and *El-Medeeneh;* not necessarily implying conformity with the conditions of اِعْتِكَاف required by the law [though generally meaning *for the purpose of study:* and so in the neighbourhood of the great collegiate mosque called the Azhar, in Cairo: so that the term ♦ مُجَاوِرٌ means *a student* of Mekkeh &c.]. (TA.)

4. اجارهُ, (Ṣ, A, &c.,) inf. n. إِجَارَةٌ (Mgh, Ḳ) and ♦ جَارَةٌ, (Kr, Ḳ,) [or the latter is rather a quasi-inf. n., like طَاعَةٌ from أَطَاعَهُ,] *He protected him; granted him refuge;* (Ḳ;) *preserved, saved, rescued*, or *liberated, him;* (Ṣ, A, Mṣb, Ḳ;) from (مِنْ) wrongful, unjust, injurious, or tyrannical, treatment; (Ṣ, Ḳ;) from punishment; (Ṣ, A;) or from what he feared: (Mṣb:) *he aided him; succoured him; delivered him from evil:* the أ having a privative effect. (Mgh.) It is said of God, يُجِيرُ وَلَا يُجَارُ عَلَيْهِ *He protects, but none is protected against him.* (TA.) And in the Ḳur [lxxii. 22], قُلْ إِنِّى لَنْ يُجِيرَنِى مِنَ ٱللّٰهِ أَحَدٌ *Verily none will protect me against God.* (TA.) — اجار المَتَاعَ *He put the household-goods*, or *commodities, into the repository*, (Ḳ, TA,) *and so preserved them from being lost.* (TA.) — It is said [of God] in a trad., يُجِيرُ بَيْنَ البُحُورِ *He makes a division between the seas, and prevents one from mixing with another and encroaching upon it.* (TA.)

5. تجوّر *He became prostrated;* (Ṣ;) *he fell down;* (Ḳ;) by reason of a blow. (Ṣ, TA.) — *It* (a building, TA) *became thrown down*, or *demolished.* (Ḳ.) — *He* (a man, TA) *laid himself down on his side* (Ḳ) upon his bed. (TA.)

6. تَجَاوَرُوا and ♦ اِجْتَوَرُوا (Ṣ, Ḳ) are syn., (Ṣ,) signifying *They became mutual neighbours; they lived near together:* (Ḳ,* TA:) the [radical] و in the latter verb remaining unaltered because this verb is syn. with one in which the و must preserve its original form on account of the quiescence of the preceding letter, namely, تجاوروا, (Ṣ, TA,) and to show that it is syn. therewith: but اِجْتَارُوا also occurs. (TA.) — [Also *They bound themselves by a covenant to protect one another.*]

8: see 6.

10. استجار and ♦ جَارَ, (Ḳ,) the latter like جَارَ as syn. with مُسْتَجِيرٌ, (TA,) *He sought, desired*, or *asked, to be protected; to be granted refuge; to be preserved, saved, rescued*, or *liberated.* (Ḳ.) And استجارهُ *He desired him*, or *asked him, to preserve, save, rescue, or deliver, him*, (Ṣ, A, Mṣb,) مِنْ فُلَانٍ *from such a one.* (Ṣ.) And استجار بهِ *He had recourse to him for refuge, protection*, or *preservation; he sought his protection.* (TA.)

♦ جَارٌ *A neighbour; one who lives near to another;* (Ṣ, Mgh, Mṣb, Ḳ;) *one who lives in the next tent* or *house:* (IAạr, Th, T, Mṣb:) pl. [of mult.] جِيرَانٌ (Mṣb, Ḳ) [and جِوَارٌ (a pl. not of unfrequent occurrence, and mentioned by Freytag as used by El-Mutanebbee,)] and [of pauc.] جِيرَةٌ and أَجْوَارٌ; (Ḳ;) like قَاعٌ, pl. قِيعَانٌ and قِيعَةٌ and أَقْوَاعٌ, the only similar instance: (TA:) fem. with ة. (Mgh.) الجَارُ ذُو القُرْبَى [in the Ḳur iv. 40] is *The relation*, or *kinsman, who is abiding in one's neighbourhood:* or *who is abiding in one town or district or the like while thou art in another, and who has that title to respect which belongs to nearness of relationship:* (TA:) or *the near neighbour:* (Bḍ, Jel:) or *the near relation:* (Jel:) or *he who is near, and connected, by relationship* or *religion.* (Bḍ.) جَارُ الجُنُبِ: and الجَارُ الجُنُبُ and جَارُ الجَنْبِ: see art. جنب. جَارٌ نَقِيعٌ *A stranger* [*who has become one's neighbour*]. (TA.) — *A person whom one protects from wrongful, unjust, injurious*, or *tyrannical, treatment.* (Ṣ, Mgh, Mṣb, Ḳ.) — *One who seeks*, or *asks, protection* (Mṣb, Ḳ) of another: جَارُكَ signifying *he who seeks thy protection.* (TA.) — *A protector;* (A, Mgh, Mṣb, Ḳ;) *one who protects another from that which he fears;* (Mṣb;) *one who grants refuge*, or *protects*, or *preserves.* (AHeyth.) هُمْ ♦ جَارَةٌ مِنْ ذٰلِكَ الأَمْرِ *They are protectors from that thing*, is a phrase mentioned by Th, respecting which ISd says, I know not how this is, unless the sing. be supposed to be originally ♦ جَائِرٌ, so as to have a pl. of the measure فَعَلَةٌ [as جَارَةٌ is originally جَوَرَةٌ]. (TA.) — *An aider*, or *assister.* (IAạr, Mṣb, Ḳ.) — *A confederate.* (IAạr, Mṣb, Ḳ.) — A woman's *husband.* (Mṣb, Ḳ.) — A man's *wife;* (Mṣb;) as also ♦ جَارَةٌ: (Ṣ, M, A, Mgh, Mṣb, Ḳ:) or the latter, the *object of his love:* (M:) and the latter also, a woman's *fellow-wife;* (Mgh, Mṣb, TA;) so called because the term ضَرَّةٌ is disliked, (Mgh, Mṣb,) as being of evil omen. (Mgh.) — *A partner who has not divided with his partner:* so in the trad. الجَارُ أَحَقُّ بِصَقَبِهِ [explained in art. صقب]; as is shown by another trad. (Az, Mṣb.) — *A partner*, or *sharer*, (Mṣb, Ḳ,) *in immoveable property, such as land and houses*, (Mṣb, TA,) and *in merchandise*, (Ḳ, TA,) *whether he divide the property with the other or not*, (Mṣb,) or *whether he be partner in the whole or only in part.* (TA.) — *One who divides with another.* (IAạr, Ḳ.) — ‡ The فَرْج [or *pudendum*] *of a woman:* and ‡ the *anus;* as also ♦ جَارَةٌ. (IAạr, Ḳ, TA.) — The *part* (IAạr, Ḳ) *of the sea-shore* (IAạr) *that is near to the places where people have alighted and taken up their abode.* (IAạr, Ḳ.)

جَوْرٌ, an inf. n. used as an epithet, (TA,) *i. q.* ♦ جَائِرٌ; (Ḳ, TA;) i. e. *Declining*, or *deviating, from the right course:* and *acting wrongfully, unjustly, injuriously*, or *tyrannically:* (TA:) pl. [of the latter], applied to men, ♦ جَوَرَةٌ, (Ḳ,) in which the و remains unaltered contr. to rule, (TA,) and ♦ جَارَةٌ, (A, Ḳ,) as in all the copies of the Ḳ, but some substitute for it, as a correction, ♦ جَوَرَةٌ, [found in a copy of the A,] which, however, requires consideration, (TA,) and جَائِرُونَ. (Ḳ.) You say طَرِيقٌ جَوْرٌ *A road*, or *way, deviating from the right course.* (TA.) And هُوَ جَوْرٌ عَنْ طَرِيقِنَا *He is declining*, or *deviating, from our way.* (TA.) — Also, for ذُو جَوْرٍ, meaning *Wronged*, or *unjustly treated, by the judge.*

(Mgh from a trad.) — عِنْدَهُ مِنَ المَالِ الجَوْرُ ‡ *He possesses, of property, an extraordinary abundance.* (A, TA.) See also جَوْرٌ.

جَارَةٌ : see جَارٌ, in three places: — and جَوْزٌ : — and see also 4.

جُوْرَةٌ and جَوْرَةٌ : see جَوْرٌ.

إِنَّهُ لَحَسَنُ الجِيرَةِ *Verily he is good in respect of the mode,* or *manner, of* جِوَار [i. e. *living as a neighbour,* or *binding himself by covenant to protect others*]. (TA.)

جُورٌ A rain *accompanied by vehement thunder:* (K:) or *by a vehement sound of thunder:* (S:) or a *copious* rain; as also جَارٌ and جُوْرٌ ; (K in art. جَارٌ ;) and, accord. to Aṣ, جُوَارٌ : (TA:) and an *exceedingly great* torrent. (TA. [In this last sense written in a copy of the A ▼جَوْرٌ, and there said to be tropical.]) See جُوَارٌ : and see also art. جر. — You say also بَازِلٌ جُورٌ (S) [app. meaning *A camel nine years old that brays loudly:* or] *hard and strong:* and بَعِيرٌ جُورٌ *a bulky camel.* (TA.)

جُوَارٌ : see 3. — Also The *part of the exterior court* or *yard* of a house *that is coextensive with the house.* (K,* TA.) — *Abundant and deep water.* (K.) Whence ▼جُورٌ applied to rain. (TA.) — *Ships:* a dial. var. of جَوَارٍ ; on the authority of Ṣá'id, (K,) surnamed Abu-l-'Alà: (TA:) said in the K to be strange; but similar instances are well known. (MF.)

جِوَارٌ : see 3, in two places. — Also, and ▼جُوَارٌ, or the latter is only an inf. n., The *covenant between two parties by which either is bound to protect the other.* (TA.)

جُوَارٌ : see what next precedes. — [Also a pl. of جَارٌ.]

جَائِرٌ : see جَوْرٌ. — Also ‡ *Wide and big;* applied to a [bucket of the kind called] غَرْبٌ : and so, with ة, applied to a [skin of the kind called] قِرْبَةٌ. (A, TA.)

مُجُورٌ [as meaning *Thrown down,* or *overturned,*] occurs in the following prov.:

* يَوْمٌ بِيَوْمِ الحَفَضِ المُجَوَّرِ *

[*A day for a day of the household-goods* (or, accord. to the TA, *the hair-cloth tent*) *thrown down,* or *overturned*]: applied in the case of rejoicing at a calamity befalling another: a man had an aged paternal uncle, and used continually to go into the latter's tent, or house, and throw down his household-goods, one upon another; and when he himself grew old, sons of a brother of his did to him as he had done to his paternal uncle; wherefore he said thus, meaning, this is for what I did to my paternal uncle. (K.)

مُجَاوِرٌ : see 3, last sentence.

جوز

1. جَازَ المَوْضِعَ, (S, K,) or المَكَانَ, (A, Mgh, Mṣb,) aor. يَجُوزُ, (S, Mṣb,) inf. n. جَوَازٌ (S, Mṣb, K) and جَوْزٌ and جُؤُوزٌ and مَجَازٌ, (K,) *He went,* or *passed, in,* or *along, the place, and left it behind;* (Mgh, K;) [whether this be meant for one signification or two, does not appear; but in either case it is evident that one signification is *he passed through,* or *over,* or *along, and beyond, the place;* and this signification is of frequent occurrence;] as also جَازَ بِهِ ; (K;) and ▼اجازه ; (Mgh;) and ▼جاوزه, (Mgh, K,) inf. n. جِوَازٌ ; (K, TA; in the CK جُوَازٌ ;) and ▼تجاوزه ; (Mgh;) lit., *he traversed,* or *crossed, its* جَوْز, i. e., *middle, and passed through it:* (Mgh:) or *he went,* or *passed, in,* or *along, the place;* (Aṣ, S, A, Mṣb, TA;) as also جاز به, and ▼جاوزه, (TA,) and ▼اجازه, (A,) and ▼اجتازه : (S: [so it appears from its being said that اِجْتِيَازٌ is syn. with سُلُوكٌ :]) and in like manner, الطَّرِيقَ *the road:* (TA:) جاوز ▼ الموضعَ and جازهُ signify the same: (TA:) or ▼اجازه (Aṣ, S, Mṣb, K) and ▼جاوزه and ▼تجاوزه (A) signify *he left it behind him,* (Aṣ, S, A, K,) *and traversed,* or *crossed, it;* (Aṣ, S, A, Mṣb;) and ▼جاوزه and جاوز ▼ به also signify *he left it behind.* (TA.) You say, جُزْتُ خِلَالَ الدِّيَارِ, which is like جُسْتُ [*I passed amid,* or *among, the houses:* (see the remarks on the letter ز :) or *I went to and fro amid,* or *among, the houses,* in a hostile attack upon them: or *went round about them*]. (Ibn-Umm-Kásim, TA.) And جُزْتُ بِكَذَا, i. e., اِجْتَزْتُ ▼ به [*I passed by, and beyond, such a thing*]. (TA.) And جاز عَلَيْهِ *He passed by him,* or *it;* syn. مَرَّ بِهِ, and اِمْتَرَّ بِهِ and عَلَيْهِ. (M and K in art. مر.) And جَازَهُ *He passed,* or *crossed, over it.* (L.) جاز and ▼اجاز are syn. [in this last sense]. (TA.) You say, أَعَانَكَ اللّٰهُ عَلَى إِجَازَةِ ▼ الصِّرَاطِ (A, TA) *May God aid thee* [*to pass,* or *cross, over,* or] *to pass along, and to leave behind thee, the Ṣiráṭ.* (TA.) And it is said in a trad. respecting the Ṣiráṭ, فَأَكُونُ أَنَا وَأُمَّتِى أَوَّلَ مَنْ يُجِيزُ ▼ عَلَيْهِ [*And I, with my people, shall be the first who will pass over it*]: يجيز being here syn. with يَجُوزُ. (TA.) — جُزْتُ جاز الدِّرْهَمُ فَتَجَوَّزْتُهُ ▼ : — الشَّىْءَ إِلَى غَيْرِهِ : see 3. — [*The piece of money passed,* or *was current, and he accepted it as current:* in the TA written جاز الدرهم فتجوزه, and without any syll. signs; but that the reading which I have adopted is right appears from what immediately follows:] a poet says,

* دَرَاهِمُ مِنْهَا جَائِزَاتٌ ▼ وَزُيَّفُ *

[*Pieces of money whereof there are current and bad*]: and Lḥ mentions the saying, لَمْ أَرَ النَّفَقَةَ تَجُوزُ بِمَكَانٍ كَمَا تَجُوزُ بِمَكَّةَ [*I have not seen money for expenses pass away in a place as it passes away in Mekkeh*]: ISd says, He has not explained it, but I think that the meaning is تَنْفُقُ. (TA.) — جاز الشَّىْءُ, inf. n. جَوَازٌ, *The thing was,* or *became, allowable; it passed for lawful:* as though it kept the middle (جَوْز) of the road. (TA.) You say, جَازَ البَيْعُ, and النِّكَاحُ, (A, Mgh,) and العَقْدُ وَغَيْرُهُ, (Mṣb,) [*The sale,* and *the marriage,* and *the contract, or other thing, was,* or *became, allowable;* or] *passed as right, sound, valid,* or *good* [*in law:*] (Mṣb:) or *had effect.* (Mgh.) [And جاز لَهُ أَنْ يَفْعَلَ كَذَا *It was allowable to him to do so.* And يَجُوزُ أَنْ يَكُونَ كَذَا *It may be so;* or *such a thing may be.*] — جَازَهُ in the sense of اجازه : see 4, second sentence, in two places.

2: see 4, in nine places.

3. جاوزه and جاوز به, inf. n. جِوَازٌ : see 1, in six places. — جاوز الحَدَّ, and القَدْرَ, inf. n. مُجَاوَزَةٌ ; and so ▼تجاوز, alone; *He exceeded,* or *transgressed, the proper bound,* or *limit,* or *measure; acted extravagantly, exorbitantly,* or *immoderately: he,* or *it, was,* or *became, excessive, extravagant, exorbitant,* or *immoderate.* (The Lexicons &c. passim.) — جَاوَزْتُ الشَّىْءَ إِلَى غَيْرِهِ, (S, Mṣb*) *I passed from the thing* [*to another thing*]; (Mṣb;) as also تَجَاوَزْتُهُ ▼ ; (S, Mṣb;) *i. q.* جُزْتُهُ ▼. (S.) — جاوز عَنْ ذَنْبِهِ : see 6. — [Hence, app.,] كَانَ مِنْ خُلُقِى الجِوَازُ *It was of my disposition to be easy,* or *facile, in selling and demanding.* (TA from a trad.) — جاوز به : see 4, in two places.

4. اجاز and اجازه : see 1, in six places. — اجازه *He made him to go,* or *pass along;* as also جَازَهُ ▼ : (TA:) *he made him to pass through,* or *over,* or *along and beyond:* (S, IF, Mṣb, K;) as also [جاوز ▼ به, as will be shown by an ex. below, and ▼جوّزه, and] جَازَهُ, for which we find جاوزه incorrectly substituted in the K. (TA.) A rájiz says,

* خَلُّوا الطَّرِيقَ عَنْ أَبِى سَيَّارَةْ *

* حَتَّى يُجِيزَ سَالِمًا حِمَارَهْ *

[*Leave ye the road to Aboo-Seiyárah until he make his ass to pass through,* or *over, safely*]. (S.) And it is said in the Kur [vii. 134, and x. 90], وَجَاوَزْنَا ▼ بِبَنِى إِسْرَائِيلَ البَحْرَ [*And we made the Children of Israel to pass through the sea*]. (TA.) You say also جَوَّزَ ▼ لَهُمْ إِبِلَهُمْ, inf. n. تَجْوِيزٌ, *He led for them their camels one by one until they passed.* (K.) — [*He made it to pass,* or *be current;* as also جوّزه ▼ : as in the following phrases.] جَعَلْتُهُ جَائِزًا *i. q.* أَجَزْتُ عَلَى اسْمِهِ [*I made his name to pass,* or *be current,* by stamping money with it]: (ISk, S, TA:) and ضَرَبْتُ [*I coined,* or *minted,* money *in his name*]. (ISd, TA.) And جَوَّزَ ▼ الضَّرَّابُ الدَّرَاهِمَ, inf. n. تَجْوِيزٌ, [*The coiner,* or *minter,*] *made the dirhems,* or *pieces of money, to pass,* or *be current.* (Mgh.) — *He made it, or held it, to be allowable,* or *to pass for lawful; he allowed it,* or *permitted it;* (S, K, TA;) as also جَوَّزَهُ ▼ : (S, TA:) syn. سَوَّغَ : (S, K:) and syn. of إِجَازَةٌ, [the inf. n. of the former verb,] إِذْنٌ. (K, TA: omitted in the CK.) You say, اجاز لَهُ مَا صَنَعَ, (S, K,*) and جَوَّزَ ▼ له, (S,) *He made,* or *held, what he did to be allowable,* &c. (S, K.) And هٰذَا مِمَّا لَا يُجَوِّزُهُ ▼ العَقْلُ [*This is of the things which reason will not allow*]. (A, TA.) — [*He granted him the authority or degree of a licentiate in some one or more of the various departments of learning, for the instruction of others therein;*] *he* *granted him a license with respect to the matters that he had related and heard [from other learned men, to teach the same].* (TA.) You say also, اجاز لفُلَانٍ جَمِيعَ مَسْمُوعَاتِهِ مِنْ مَشَائِخِهِ [*He granted him a license with respect to all the matters which he had heard from his sheykhs, to teach the same to others*]. (TA.) The licentiate is termed ↑مُجَازٌ: and the matters which he relates are termed ↑مُجَازَاتٌ. (TA.) — اجاز البَيْعَ, (A, Mgh, Ḳ,) and النِّكَاحَ, (A, Mgh,) and العَقْدَ, (Mṣb,) *He* (the judge, A, Mgh) *made the sale,* (A, Mgh, Ḳ,) and *the marriage,* (A, Mgh,) and *the contract,* (Mṣb,) *to have effect; he executed* or *performed it;* (Mgh, Mṣb Ḳ;) لَهُ *for him:* (Ḳ:) *he decreed it.* (Mgh.) And [in like manner] اجاز رَأْيَهُ, and ↑جوّزه, *He made his judgment,* or *opinion, to have effect; he executed* or *performed it.* (Ḳ.) Hence the saying, in a trad. of Aboo-Dharr, قَبْلَ أَنْ يُجِيزُوا عَلَىَّ, i. e., *Before they slay me, and execute your order upon me.* (TA.) ═ أَجَازَنِى (Ṣ, Ḳ*) ‡ *He gave me water for,* (Ṣ,) or *he watered* [*for me*], (Ḳ,) *my land,* or *my beasts.* (Ṣ, Ḳ.) And ↑جوّز إِبِلَهُ, (Ḳ,) inf. n. تَجْوِيزٌ, (TA,) *He watered his camels.* (Ḳ.) And اجاز الوَفْدَ *He gave to the party who came as envoys, or the like, the quantity of water sufficient to pass therewith from one watering-place to another.* (TA.) And أَجَازَهُ مَآءً يَجُوزُ بِهِ الطَّرِيقَ † *He gave him water wherewith to travel the road.* (A.) And أَجِزْنِى مَآءً *Give thou me some water that I may go my way, and pass from thee.* (Aboo-Bekr, TA.) — Hence, (Aboo-Bekr, TA,) اجازه بِجَائِزَةٍ, (Aboo-Bekr, TA,) and اجازه بِجَائِزَةٍ سَنِيَّةٍ, (Ṣ, A,) † *He* (the Sultán) *gave him a gift,* or *present,* (Aboo-Bekr, TA,) and *he gave him a gift,* or *present, of high estimation.* (Ṣ, A.*) Or the origin of the expression was this: Ḳaṭan the son of 'Owf, of the tribe of Benoo-Hilál-Ibn-'Ámir-Ibn-Ṣaṣa'ah, gave the government of Fáris to 'Abd-Allah Ibn-'Abbás; and El-Aḥnaf passing by him with his army on an expedition to Khurásán, he waited for them upon a bridge, and said, أَجِيزُوهُمْ [*Make ye them to pass over*]; and he began to mention the lineage of each man and to give him according to his rank: (Ṣ:) or from the fact that a certain commander, having a river between him and an opposing force, said, مَنْ جَازَ هٰذَا النَّهْرَ فَلَهُ كَذَا [*Whoso passeth this river shall have such a thing*]; and whenever one passed over, he received a جَائِزَة. (TA.) You say also, أَجَازَهُ, meaning † *He gave him.* (TA.) And it is said in a trad., أَجِيزُوا الوَفْدَ بِنَحْوِ مَا كُنْتُ أُجِيزُهُمْ بِهِ *Give ye to the party who come as envoys, or the like, a similar* جَائِزَة *to that which I used to give them.* (TA.)

5. تجوّز اللَّيْلُ *The darkness of the night cleared away.* (A.) ═ تجوّز فِى صَلَاتِهِ *He relaxed,* or *remitted, in his prayer;* (Ṣ, A, Mgh, Mṣb, Ḳ, TA;) and so in other things; (A;) *and abridged it; and was quick in it:* said to be from "الجَوْزُ" "the act of traversing, and going, or passing along:" (TA:) or *did less than was sufficient in it.* (Mṣb.) — Hence, تجوّز فِى أَخْذِ الدَّرَاهِمِ, (A, Mgh,) or تجوّز الدَّرَاهِمَ, (Ḳ,) *He accepted the dirhems,* or *pieces of money, as current; did not reject them:* (A, Mgh:) see 1: or *he accepted them as they were,* or *notwithstanding what was in them:* (Lth, TA:) or *he accepted them notwithstanding what was intermixed with them,* (Ḳ, TA,) [*of bad money,*] *concealed therein, and notwithstanding their fewness.* (TA.) In the phrase التَّجَوُّزُ بِدُونِ الحَقِّ [*The accepting less than what was due*], the inf. n. is made trans. by means of ب because it implies the meaning of الرِّضَا [which is made trans. by the same means]. (Mgh.) ↑تَجَاوَزَ also occurs in the sense of تَجَوَّزَ in a trad. of Ibn-Rawáḥah: هٰذَا لَكَ وَتَجَاوَزْ فِى القَسْمِ *This is thine,* or *for thee, and be thou remiss,* or *not extreme, in,* or *with respect to, the division:* and is allowable, though we have not heard it. (Mgh.) You say also, تَجَوَّزَ فِى هٰذَا الأَمْرِ مَا لَمْ يَتَجَوَّزْ فِى غَيْرِهِ *He bore patiently,* or *with silence and forgiveness, and with feigned neglect,* or *connivance, in this affair,* or *case, what he did not so bear in another.* (Ḳ,* TA.) — See also 6, in three places. ═ تجوّز فِى كَلَامِهِ *He made use of a trope,* or *tropes, in his speech.* (Ṣ, Ḳ.) [See مَجَازٌ, below.]

6. تجاوزهُ: see 1, first sentence: and see also 3. ═ تجاوز i. q. أَفْرَطَ, [i. e., جاوز الحَدَّ, explained above,] فِيهِ *in it,* or *with respect to it.* (Ḳ. See 3.) — تجاوز عَنْهُ, (Ṣ, A, Mgh, Mṣb,) and ↑تجوّز; (Ṣ, A, Mgh;) and تجاوز عَنْ ذَنْبِهِ, (A, Ḳ,) and ↑تجوّز, and ↑جاوز; (Ḳ;) *He* (God, Ṣ, A, or a man, Mṣb) *passed him by,* or *over, without punishing him;* or *forgave him;* (Ṣ, A, Mgh, Mṣb;) namely, an evil-doer; (A, Mgh, Mṣb;) and *He passed by,* or *over, without punishing,* or *forgave, his sin* or *offence.* (A, Ḳ.*) You say, اَللّٰهُمَّ تَجَاوَزْ عَنِّى, and ↑تَجَوَّزْ عَنِّى, *O God, pass me by,* or *over, without punishing me;* or *forgive me.* (Ṣ, A.) تجاوز عَنْهُ, followed by a noun in the accus. case, also signifies *He forgave him* a thing. (L.) And the same alone, *He feigned himself neglectful of it; he connived at it.* (Ḳ.) — [Also, this last phrase alone, *He transcended it.*] — تَجَاوَزَ فِى القَسْمِ: see 5.

8. اجتازهُ: and اجتاز بِهِ: see 1.

10. استجازهُ *He asked,* or *demanded, of him permission.* (Ḳ,* TA.) — *He asked,* or *demanded, of him* [*the authority or degree of a licentiate; i. e.,*] *a license with respect to the matters that he had related and heard* [*from other learned men, to teach the same*]. (TA.) [See 4.] ═ ‡ *He asked,* or *demanded, of him* (Ṣ, Ḳ) *water for,* (Ṣ,) or *to water* [*for him*], (Ḳ,) *his land,* or *his beasts.* (Ṣ, Ḳ.) ═ *He approved it.* (Ḥar p. 326.)

جَوْزٌ The *middle* (Ṣ, Ḳ) of a thing, (Ḳ,) or of anything; (Ṣ;) [as, for instance,] of a desert, (A,) and of a camel, (TA,) and of the night: (A, TA:) and the *main part* of a thing, (Ḳ,) or of the night: (TA:) pl. أَجْوَازٌ; (Sb, Ṣ, A;) beside which it has no other. (Sb.) ═ [*The walnut;* or *walnuts;*] *a well-known fruit,* (Ḳ,) *which is eaten:* (Mṣb:) a Persian word, (Ṣ,) arabicized; (Ṣ, Mṣb, Ḳ;) originally كَوْزْ: (Mgh, Mṣb, Ḳ:) n. un. جَوْزَةٌ: (Ṣ, TA:) pl. جَوْزَاتٌ: (Ṣ, Ḳ, TA: in the CḲ جَوْزَانٌ:) *the tree thereof abounds in the land of the Arabs, in the province of El-Yemen, where it bears fruit and is cultivated; and in the Sarawát* (السَّرَوَات) *are trees thereof, which are not cultivated: the wood thereof is characterized by hardness and strength.* (AḤn, TA.) — جَوْزُ بَوَّا, (Ḳ,) or جَوْزُ بُوَى, with the short alif, as heard from the physicians, in Persian كَوْزْ بُويَا, (Mgh, under the letter ب,) [vulgarly called جَوْزُ الطِّيبِ, *The nutmeg;*] *a certain medicine;* (Ḳ;) *it is of the size of the gall-nut* (عَفْص), *easily broken, with a thin coat,* (Mgh, TA,) *having a pleasant odour,* (Mgh,) or *a pleasant and sharp odour; and the best kind is the red, with a black coat, and heavy:* (TA:) *it is good for the* [*affection of the face termed*] لَقْوَة, *strengthens the stomach and heart, and removes cold.* (Mgh.) — جَوْزُ مَاثِلٍ [*The datura stramonium,* or *thorn-apple;*] also *a certain medicine;* (Ḳ;) *having the property of producing torpor; resembling the* جَوْزُ القَىْءِ (see what follows); *having upon it small, thick thorns; and its seed is like that of the* أُتْرُجّ [or *citron*]. (TA.) — جَوْزُ القَىْءِ [*Nux romica;*] also *a certain medicine,* (Ḳ,) *having a power similar to that of the white* خَرْبَق [or *hellebore*]. (TA.) — جَوْزُ الهِنْدِ [*The cocoa-nut;*] *what is commonly called the* نَارَجِيل. (TA.)

جَوْزَةٌ: see جَائِزَةٌ, in four places. ═ Also n. un. of جَوْزٌ [q. v.].

جِيزَةٌ: see جَائِزَةٌ.

الجَوْزَآءُ *A certain constellation* (نَجْمٌ); (Ṣ;) *a certain sign of the Zodiac;* (Ḳ;) [namely, *Gemini;*] *said to cross the* جَوْز (i. e. the *middle,* TA) *of the sky;* (Ṣ, TA;) for which reason it is [asserted to be] thus called. (TA.) — Also i. q. الجَبَّارُ [*The constellation Orion*]: (A and Ḳ in art. جبر:) *it has three very bright stars disposed obliquely in the midst thereof, called by the Arabs* فَقَارُ الجَوْزَآءِ, نِطَاقُ الجَوْزَآءِ, *and* النَّظْمُ. (Ḥar p. 456.)

↑جَوَازٌ † The *act of watering,* or *giving to drink:* (Ṣ:) or *a single watering of,* or *giving drink to,* camels. (TA.) [See also جَائِزَةٌ.] A rájiz says,

* يَا صَاحِبَ ٱلْمَآءِ فَدَتْكَ نَفْسِى *
* عَجِّلْ جَوَازِى وَأَقِلَّ حَبْسِى *

[*O master of the water (may my soul be thy ransom) hasten the watering of my camels, and make my detention little*]. (TA.) — ‡ *The water with which beasts are watered,* or *with which seed-produce is watered:* (AA, Ṣ, Ḳ:) [and] *water which is given one that he may travel with it the road.* (A, Mgh.) [See also جَائِزَةٌ.] — Hence, (Mgh,) † *The traveller's pass,* (A, Mgh, Ḳ,) *given him to prevent any one's offering opposition to him:* (A, Mgh:) pl. أَجْوِزَةٌ. (A, TA.) ═ *The office,* or *authority, of a guardian and affiancer.* (TA.)

جَائِزٌ [act. part. n. of جَازَ, in all its senses]. — *Passing*, or *current*, money. (Mgh.) See an ex. above, voce جَازَ. [And hence,] جَوَائِزُ الأَشْعَارِ, and الأَمْثَالِ, (K, TA,) for the former of which we find, in some copies of the K, الشِّعْرِ, which is incorrect, (TA,) *Verses*, or *poems*, and *proverbs, current from country to country*, or *from town to town*. (K, TA.) — Applied to a contract, [and a sale and a marriage, *Allowable*; *passing for lawful*;] *passing as right, sound, valid*, or *good* [*in law*]; *having effect*. (Msb.) ═ [The *beam* of a house, or chamber, *upon which rest the* عَوَارِض, or *rafters*;] *that upon which are placed the extremities of the pieces of wood in the roof of a house or chamber*; (AO, TA;) the *palm-trunk*, (S,) or *piece of wood, which passes across between two walls*, (K,) *called in Persian* تِير, (S, K,) *which is the* سَهْم *of the house or chamber*: (S:) pl. [of pauc.] أَجْوِزَةٌ, (S, CK, TA,) in [some of] the copies of the K, incorrectly, أَجْوُزٌ, (TA,) [and both these are given in the CK,] and [of mult.] جُوزَانٌ (S, K) and جِيزَانٌ (CK, but omitted in my MS. copy of the K and in the TA,) and جَوَائِزُ. (Seer, K.)

جَائِزَةٌ † *A draught of water*; (S, K;) as also † جُوزَةٌ: (K:) or † the latter signifies *a single watering*, or *giving of water to drink*; (S, K; [see an ex. in art. اذن, conj. 2;]) or *such as a man passes with from one person to another*: and † both signify the *quantity of water with which the traveller passes from one watering-place to another*; as also † جِيزَةٌ. (TA.) It is said in a prov., لِكُلِّ جَابِهٍ جَوْزَةٌ ثُمَّ يُؤَذَّنُ, i. e., † *For every one that comes to us for water is a single watering*, or *giving of water to drink*; *then he is repelled from the water*: or, as in the M, *then his ear is struck*, to indicate to him that he has nothing more than that to receive from us. (TA.) — Hence, (A, Mgh,) accord. to Aboo-Bekr, (TA,) [but see 4,] † *A gift*, or *present*: (Aboo-Bekr, S, Mgh, K:) pl. جَوَائِزُ. (S, A, Mgh.) — Hence also, (Mgh,) ‡ *Kindness and courtesy*: (K:) or *kindnesses and courtesy shown to those who come to one as envoys or the like*: (Mgh:) or *provisions for a day and a night given to a guest at his departure after entertainment for three days*. (Mgh, TA.) It is said in a trad., الضِّيَافَةُ ثَلَاثَةُ أَيَّامٍ وَجَائِزَتُهُ يَوْمٌ وَلَيْلَةٌ وَمَا زَادَ فَهُوَ صَدَقَةٌ, meaning, [The period of] *the entertainment of a guest is three days*, during the first of which the host shall take trouble to show him large kindness and courtesy, and on the second and third of which he shall offer him what he has at hand, not exceeding his usual custom; *then he shall give him that wherewith to journey for the space of a day and a night*; *and what is after that shall be as an alms and an act of favour*, which he may do if he please or neglect if he please. (TA.)

مَجَازٌ *A way, road*, or *path*, (S, K, TA,) *which one travels from one side* [or *end*] *to the other*; (K, TA;) as also † مَجَازَةٌ. (TA.) You say, جَعَلَ فُلَانٌ ذٰلِكَ الأَمْرَ مَجَازًا إِلَى حَاجَتِهِ † *Such a one made that thing a way to the attainment of his want*. (S, TA.) مَجَازَةُ نَهْرٍ signifies *A bridge*. (A.) And † مَجَازَةٌ alone [also] signifies *A road* (طَرِيقَةٌ) *in a* سَبْخَة [or *salt tract*]. (K.) — *A privy*, or *place where one performs ablution*; syn. مُتَبَرَّزٌ. (TA.) ═ *A trope*; *a word*, or *phrase, used in a sense different from that which it was originally applied to denote, by reason of some analogy, or connexion, between the two senses*; as, for instance, أَسَدٌ, properly signifying "a lion," applied to "a courageous man;" (KT, &c.;) *what passes beyond the meaning to which it is originally applied*; (TA;) [being of the measure مَفْعَلٌ in the sense of the measure فَاعِلٌ;] contr. of حَقِيقَةٌ. (K.) [This is also called مَجَازٌ لُغَوِىٌّ, and مَجَازٌ لُغَةً; to distinguish it from what is termed مَجَازٌ عُرْفًا, and مَجَازٌ عُرْفِىٌّ, which is *A word*, or *phrase, so little used in a particular proper sense as to be, in that sense, conventionally regarded as tropical*; as, for instance, دَابَّةٌ in the sense of "a man," or "a human being;" it being commonly applied to "a beast," and especially to "a horse" or "a mule" or "an ass."] A حَقِيقَة, when little used, becomes what is termed مَجَازٌ عُرْفًا. (Mz, 24th نوع.) The مَجَاز is either what is termed اِسْتِعَارَةٌ [i. e. *a metaphor*] (as أَسَدٌ used as meaning "a courageous man"), or مَجَازٌ مُرْسَلٌ [*a loose trope*] (as يَدٌ used as meaning "a benefit," "benefaction," "favour," or "boon"). (KT, &c.) [مَجَازٌ also signifies *A tropical meaning*.]

مُجَازٌ: and مُجَازَاتٌ: see 4, in the middle of the paragraph.

مُجِيزٌ *A commissioned agent* of another; *an executor appointed by a will*; syn. وَكِيلٌ, and وَصِىٌّ; because he executes what he is ordered to do: so in the conventional language of the people of El-Koofeh: (Mgh:) or *a slave who has received permission to traffic*. (Mgh, K.) — The *guardian and affiancer* [of a woman]; syn. وَلِىٌّ. (K.) You say, هٰذِهِ ٱمْرَأَةٌ لَيْسَ لَهَا مُجِيزٌ [*This is a woman who has no guardian and affiancer*]: and Shureyh is related to have said, إِذَا أَنْكَحَ ٱلْمُجِيزَانِ فَالنِّكَاحُ لِلْأَوَّلِ [*When the two guardians and affiancers give a woman in marriage, the marriage is the former's*]. (TA.) — The *manager of the affairs of an orphan*. (K.)

مَجَازَةٌ: see مَجَازٌ, in three places. ═ أَرْضٌ مَجَازَةٌ (S, A) *A land containing trees of the* جَوْز [or *walnut*]: (S:) or *a land* (in the K, مَكَان [*a place*], which is wrong, TA) *abounding with* جَوْز. (A, K.)

[مَجَازِىٌّ *Tropical*.]

مُجْتَازٌ *Going*, or *passing along*. (K.) — One *who travels*, or *penetrates, along a road*. (K.) — One *who loves to hasten*, or *outstrip*. (K, TA.)

جوس

1. جَاسَ, aor. يَجُوسُ, (S, TA,) inf. n. جَوْسٌ, (A, K,) *He sought for*, or *after*, (Zj, S, A, K,) a thing, (Zj, A, K,) or news, or tidings, (S,) *with the utmost of his endeavour*. (Zj, A, K.) — [Hence,] جَاسُوا خِلَالَ الدِّيَارِ, (S, A,) in the Kur [xvii. 5], (TA,) inf. n. as above, (S, A, K,) *They went through the midst of the houses* (تَخَلَّلُوهَا), *and sought for what was in them, as a man seeks for news*, or *tidings*; as also † اجتاسوا: (S:) or *they went to and fro among the houses*, in a sudden attack: (A, K:) or the meaning in the Kur is, *they slew you amid your houses*; and جَاسُوا signifies the same; *going and coming*: (Fr, TA:) and, inf. n. as above, (K,) and جَوَسَانٌ, (S, K,) *they went round about* (Zj, S, A, K) *by night*, (S,) *among the houses*, (Zj, A, K,) *looking if any remained whom they had not slain*, (Zj, TA,) or *doing mischief*: (A:) and † اِجْتَاسَ signifies the same as جَوَسَانٌ, (K, TA,) the *going round about by night*. (TA.) — جِيسَ also signifies *It* (anything) *was trodden*: جَوْسٌ is said to be like دَوْسٌ: (TA:) and you say, جَآءَ فُلَانٌ يَجُوسُ النَّاسَ *Such a one came stepping over the people*; syn. يَتَخَطَّاهُمْ. (A, TA.*) And accord. to A'Obeyd, جُسْتُهُ and جَشْتُهُ both signify *I came into it, and trod it*; (خَالَطْتُهُ وَوَطِئْتُهُ;) meaning, any place. (TA.) You say also, جَاسَهُمُ الأَسَدُ *The lion trod upon them*: or *came into the midst of them, and did mischief among them*. (TA; [in which the inf. n. of the verb in this sense is said to be جَوْسٌ and جَوْسَانٌ: but the latter is probably a mistranscription for جَوَسَانٌ.])

8: see 1, in two places.

جَوَّاسٌ One *who treads upon everything*: or *who comes into the midst of a people, and does mischief among them*. (TA.) — Hence, (TA.) The *lion*. (K, TA.) And in like manner it is applied to a man. (TA.)

جوسق

جَوْسَقٌ i. q. قَصْرٌ [*A palace*: or *a pavilion*, or *kind of building wholly or for the most part isolated, sometimes on the top of a larger building*, i. e., *a belvedere, and sometimes projecting from a larger building, and generally consisting of one room if forming a part of a larger building*]: (S, K, and Ham p. 823:) or *a fortress*; syn. حِصْنٌ: (M, IB, TA:) or [*a building*] *resembling a* حِصْن: (M:) an arabicized word, (Lth, JK, S, M,) from the Persian كُوشك: (TA:) its primary meaning is *a* حِصْن *in a state of demolition*; and *a ruined* قَصْر: pl. جَوَاسِقُ and جَوَاسِيقُ; the latter formed by giving fulness of sound to the kesreh, or by poetic license. (Ham ubi suprà.) There were, in the Karáfeh, [the great burial-ground of the Egyptian metropolis,] numerous قُصُور, i. e., what are called جواسق, having belvederes (مَنَاظِر) and gardens: but most of the جواسق were without gardens and without a well; being lofty belvederes: all of them were called قُصُور. (El-Makreezee's "Khitat," ii. 453.)

جوش

جُؤْشُوشٌ جَوْشٌ The *breast*; (S, A, K;) as also جُؤْشُوش (S) and جَوْشَنٌ: (S, A:) and [particularly] *that of a man*; as also ▼جُوشٌ: (K:) and the *middle* of a man. (K.) — Also †The *fore part* (صَدْر) of the night; like جَرْشٌ; (S;) and so ▼جُوشٌ: (TA:) or the *middle* thereof; (K;) as also جَوْزٌ: (AA:) and *a great portion* thereof: or *of the latter part* thereof: (K:) or *a portion of the latter part* thereof: (TA:) or *from the time of a quarter to a third* thereof. (T, TA.) You say, مَضَى جَوْشٌ مِنَ اللَّيْلِ †[*A portion of the fore part, &c., of the night passed*]. (S.)

جُوشٌ: see above, in two places.

جوشن

جَوْشَنٌ The *breast*: (IDrd, S, Mgh, K:) or the *broad part of the middle of the breast*: and the *fore part of the body* of a locust. (TA.) — *A coat of mail*: (S, Mgh, K:) or *mail with which the breast and the* حَيْزُوم [*or parts adjoining the breast*] *are clad*. (M, TA.) — The *middle* of the night: (S, K:) and (so in the S, but in the K "or") the *fore part* thereof: (S, K:) pl. جَوَاشِنُ. (S.) You say, مَضَى جَوْشَنٌ مِنَ اللَّيْلِ *An early portion of the night passed*: (S:) or [simply] *a portion of the night*. (M, TA.) It is a dial. var. of جَوْشٌ, [in the first of the meanings explained above, and also as relating to the night,] although augmented. (TA.)

جَوْشَنِىٌّ *A manufacturer of coats of mail*. (K.)

جوع

1. جَاعَ, (S, Msb, K,) aor. يَجُوعُ, (S,) inf. n. جَوْعٌ, (Msb, K,) or جُوعٌ, (S, so in two copies,) or this is a simple subst., (Msb, TA,) and مَجَاعَةٌ, (S, K,) *He was, or became, hungry*; or *empty in the belly*; (TA;) *contr. of* شَبِعَ. (S, K, TA.) [See also جُوعٌ below.] — [Hence,] جَاعَ إِلَيْهِ, (K,) or جَاعَ إِلَى لِقَائِهِ, (M, TA,) ‡*He desired*, (M, K,*) *and longed*, (K,) [*as though hungering*,] *to meet with him*; like عَطِشَ. (M, K.*) And جَاعَ إِلَى مَالِهِ ‡*He longed for his property*. (AZ.)

2: see 4, in two places.

4. اجاعهُ, (S, Msb, K,) inf. n. إِجَاعَةٌ; (Msb;) and ▼جوّعهُ, (S, Msb, K,) inf. n. تَجْوِيعٌ; (Msb;) *He constrained him to be hungry, or empty in the belly*: (S,* K, TA:) or *he debarred him from food and drink*. (Msb.) It is said in a prov., أَجِعْ كَلْبَكَ يَتْبَعْكَ [*Make thy dog to be hungry and he will follow thee*]; (S, K;) meaning †*constrain thou the ignoble to have recourse to thee, by want, in order that he may continue by thee*; (K,* TA;) for if he be in no need of thee, he will leave thee: and for أَجِعْ, some say ▼جَوِّعْ. (TA.)

5. تجوّع *He made himself hungry, or empty in the belly, intentionally, or purposely*. (S, K.) You say, تَجَوَّعْ لِلدَّوَآءِ and تَوَحَّشْ لِلدَّوَآءِ, [*Make thyself hungry, or make thy stomach empty of food and beverage*, (see art. وحش,) or] *abstain thou from eating the full quantity of food, for the purpose of taking medicine*. (TA.)

10. استجاع *He showed hunger*. (KL, PS.) — [Hence,] اِسْتِجَاعَةُ العِلْمِ †*The being insatiable of knowledge*. (TA.)

جُوعٌ, a subst., (Msb, TA,) signifying *Hunger*; or *emptiness of the belly*; (TA;) *contr. of* شِبَعٌ; (S, K, TA;) as also ▼مَجَاعَةٌ, [properly an inf. n.,] and ▼مَجْوَعَةٌ, (K, TA,) and ▼مَجُوعَةٌ. (TA.) You say, جُوعًا لَهُ وَنُوعًا [May God decree *hunger* to *him*]: accord. to Sb, an instance of inf. ns. in the accus. case by reason of a verb understood: it is a form of imprecation: and the latter noun may not be put before the former, because it is a corroborative to it: (TA:) or, accord. to some, نُوعٌ means "thirst." (S, &c., in art. نوع.) And ▼عَامُ مَجَاعَةٍ and ▼مَجْوَعَةٍ (S, K) and ▼مَجُوعَةٍ (TA) *A year in which is hunger, or emptiness of the belly*: (K, TA:) and ▼عَامُ المَجَاعَةِ and ▼المَجُوعَةِ [*the year of hunger, &c.*]: (Msb:) pl. مَجَائِعُ (K) and مَجَاوِعُ: as in the phrases أَصَابَتْهُمُ المَجَاوِعُ [*Cases of hunger, &c., befell them*] and وَقَعُوا فِى المَجَاوِعِ [*They fell into cases of hunger, &c.*]. (TA.) And ▼الرَّضَاعَةُ مِنَ المَجَاعَةِ, meaning *The sucking which occasions interdiction* [of marriage with the woman whose milk is sucked and certain of her relations] *is that consequent upon hunger* which is stopped by the milk in the time of infancy of the child; not when the child's hunger is only to be stopped by solid food. (Mgh.) [See also 1 in art. رضع.] And it is said in a prov., سَمِنَ كَلْبٌ بِجُوعِ أَهْلِهِ, i. e. [*The dog's becoming fat is*] *by reason of* [*the hunger of his owners* occasioned by] murrain befalling the camels; (K,* TA;) his owners falling into hunger and distress and leanness: (TA:) or كلب was the name of a certain man, who was an object of fear, wherefore he was asked for a pledge, and he pledged his family: then obtaining possession of the camels, or cattle, of the people to whom he had pledged his family, he drove them away, and left his family: (K, TA:) some relate this prov. differently, saying [سَمِنَ كَلْبٌ "a dog," or "Kelb," "became fat," and] بِبُؤْسِ أَهْلِهِ ["by reason of the distress of his owners," or "his family"]. (TA.) [See Freytag's Arab. Prov. i. 615.]

جَوْعَةٌ *A single temporary affection of hunger*. (S, TA.) *A state of destitution and hunger* of a tribe. (TA.)

جَوْعَانُ: see the next paragraph.

جَائِعٌ and ▼جَوْعَانُ, (Msb, K, TA,) but not جَيْعَانُ, [as the vulgar say,] for this a mistake, (TA,) *Hungry*; or *empty in the belly*: (K,* TA:) or *debarred from food and drink*: (Msb:) the fem. [of the former] is جَائِعَةٌ and [of the latter] جَوْعَى: (Msb, K:) and the pl. [of the former] is جُوَّعٌ (S, Msb, K) and جُيَّعٌ, with the و changed into ى, (L,) and [of the latter, or perhaps of both,] جِيَاعٌ (S, K) and جَاعَةٌ (K* in art. سوع) and جَيَاعَى [with the و changed into ى contr. to rule, if this be not a mistake for جِيَاعٌ, in which the و is changed into ى by rule]. (Msb.) You say جَائِعٌ نَائِعٌ; the latter word being an imitative sequent; (TA;) or, accord. to some, signifying "thirsty." (S, &c., in art. نوع.) — رَجُلٌ جَائِعُ القِدْرِ ‡*A man whose cooking-pot is not full*. (TA.) — اِمْرَأَةٌ جَائِعَةُ الوِشَاحِ ‡*A woman slender in the* [*waist*, or] *belly*. (K,* TA.) [See art. وشح.]

مَجَاعٌ [The *space in which one becomes hungry*]. You say, هُوَ مِنِّى عَلَى قَدْرِ مَجَاعِ الشَّبْعَانِ, i. e., عَلَى قَدْرِ مَا يَجُوعُ الشَّبْعَانُ [*He, or it, is distant from me as far as the space in which he who is satiated with food becomes hungry*]: (O, K:*) and in like manner, عَلَى قَدْرِ مَعْطَشِ الرَّيَّانِ [as far as the space in which he who is satisfied with drink becomes thirsty]. (Z, TA.)

مَجَاعَةٌ: see جُوعٌ, in four places.

مَجْوَعَةٌ and مَجُوعَةٌ: see جُوعٌ, in five places.

مُسْتَجِيعٌ A man (S) *who always shows himself*, or *is seen, to be hungry*: (S, A, O, K:) or, accord. to Aboo-Sa'eed, *who is always eating one thing after another*. (Sgh, L.)

جوف

1. جَوِفَ The *being* [*hollow*, or] *wide and hollow within*: (PS:) or the *being empty, vacant*, or *void*: an inf. n. of which the verb is of the class of تَعِبَ [i. e. جَافَ, originally جَوِفَ, like خَافَ, sec. pers. جِفْتَ, aor. يَجَافُ]: (Msb:) the *being wide, spacious*, or *ample*: (K:) the inf. n., or source, whence شَىْءٌ أَجْوَفُ. (S.) [See also 10.] — جَافَهُ, [aor. يَجُوفُ,] inf. n. جَوْفٌ, *It reached his* جَوْف [or *inside*, or *interior*, &c.]. (TA.) *It* (medicine) *entered his* جَوْف. (TA. [See also 8.]) And جَافَتْهُ الجِرَاحَةُ *The wound reached his* جَوْف. (Msb.) — طَعَنَهُ فَجَافَهُ, and ▼اجافهُ, *He pierced him and pierced his* جَوْف: (Mgh, Msb:) and ▼جَوَّفَهُ, inf. n. تَجْوِيفٌ, *he pierced him in his* جَوْف. (TA.) جُفْتُهُ بِالطَّعْنَةِ, and ▼أَجَفْتُهُ الطَّعْنَةَ, *I made the spear-wound, or the like, to reach his* جَوْف. (Ks, A'Obeyd, S, K.) جَافَ الصَّيْدَ *He made the arrow to enter the* جَوْف *of the object of the chase*. (TA.)

2. تَجْوِيفٌ The *making* [a thing] *hollow*, or *empty in the middle*. (KL, PS.) You say, جَوَّفَهُ, inf. n. تَجْوِيفٌ, [*He made it hollow; hollowed it out*;] *he made it to have a* جَوْف. (Msb.) And of a thing that is مُجَوَّفٌ, (S, K,) i. e. أَجْوَفُ, (S,) you say, فِيهِ تَجْوِيفٌ [*In it is a hollowing out*; meaning *a hollow*, in which sense تجويف has a pl., namely, تَجَاوِيفُ]. (S, K.) — See also 1.

4: see 1, in two places. — اجاف البَابَ ‡*He shut*, or *closed, the door*. (S, K, TA.) Hence, in a trad., وَأَجِيفُوا الأَبْوَابَ وَأَطْفِئُوا المَصَابِيحَ [*And shut ye the doors, and extinguish the lamps*]. (TA.)

5. تجوّف *It was, or became, hollow, or empty within*. (KL.) — تجوّفهُ: see 8. — تَجَوَّفَتِ الخُوصَةُ

العَرْفَجِ *The leaf was in the* جَوْف [or *inside*] *of the* [*plant called*] عرفج, *not having yet come forth.* (Ṣ.)

8. اجتافهُ *He entered its* جَوْف [or *inside*, or *interior*; *he entered into the midst of it*]; as also ▼تجوّفهُ. (Ṣ, Ḳ.) [See an ex. in a verse of Lebeed, voce أَصْلٌ: and see also 1.]

10. استجاف and اِسْتَجْوَفَ *It* (a thing) *became wide, spacious,* or *ample.* (Ṣ, Ḳ.) [See also 1, first sentence.] ═ استجافهُ *He found it* (a place) *to be* أَجْوَف [i. e. *hollow*, or *empty within*; or *wide, spacious,* or *ample*]. (O, L, Ḳ.)

جَوْفٌ [*A hollow; an interior empty, vacant,* or *void, space;*] *a vacancy*: pl. أَجْوَافٌ: this is the primary signification: then it was used in relation to a thing capable of being occupied and of being unoccupied; so as to be applied in the sense next following. (Mṣb.) —— The *inside,* or *interior,* (Mṣb, KL,) of a house [&c.]. (Mṣb.) —— [The *midst,* or *middle,* of a thing.] —— *A low,* or *depressed,* (Ṣ, Ḳ, TA,) *and wide,* (TA,) *tract,* or *portion, of land,* or *ground*: (Ṣ, Ḳ, TA:) *what is wider than the* شِعْب; *the* [*water-courses termed*] تِلَاع, *and the valleys, flow into it; and it has* جِرَفَة [or *abrupt, water-worn, banks*]: *sometimes it is wider than a valley, and deeper*: and *sometimes it is a plain,* or *soft, tract, that retains water*: and *sometimes it is completely round, so that it retains water*: accord. to IAạr, it signifies *a valley*: or, as some say, the *interior* (بَطْن) *of a valley.* (TA.) —— The *belly,* or *abdomen,* of a man: (Ṣ, Ḳ:) or, accord. to ISd, the *interior of the belly*: and the *part upon which close the shoulder-blades and the upper arms and the ribs and the two flanks* (الصُّقْلَانِ): (TA:) the *chest,* or *thorax*; i. e., the *part of the body that is separated from the* بَطْن [or *belly,* or *abdomen,*] *by the* حِجَاب [or *diaphragm,* or *midriff*]; *containing the heart and its appertenances*: (Zj in his "Khalḳ el-Insán:") pl. as above. (TA.) See also جَائِفٌ. It is one of the words that are not used adverbially except with prepositions. (Sb, TA.) It is said in a trad., لَا تَنْسَوُا الجَوْفَ وَمَا وَعَى [*Forget not ye the* جوف *and what it hath collected*]; meaning what enters into it, of food and beverage: but some say that الجوف here means *the belly and the* فَرْج [or *vulva,* or *pudendum muliebre*], together, which are also called ▼الأَجْوَفَانِ: and some say that the meaning is, *the heart and what it hath retained, and kept in memory,* of the knowledge of God. (A 'Obeyd, TA.) —— الأَجْوَافُ is also applied by the people of El-Ghowr (Ḳ) and of El-Yemen (TA) to *The tents* (فَسَاطِيط) *of their* عُمَّال [or *governors,* or *collectors of the poor-rates*]. (Ḳ.) —— جَوْفُ اللَّيْلِ الآخِرُ, occurring in a trad., means † *The last third of the night*: [or] *the fifth of the sixths of the night*: (Ḳ:) not the half, as some assert. (TA.)

جَوْفِىٌّ : see أَجْوَفُ.

جُوفِىٌّ : see أَجْوَفُ. ═ Also, and without teshdeed, (Ṣ, Ḳ,) [app. meaning, when with the article ال, written and pronounced الجُوفِى,] in the accus. case جُوفِيًا, by poetic license, (Ṣ,) *A species of fish*; and so ▼جُوَافٌ. (Ṣ, Ḳ.)

جُوفَانٌ The *penis of an ass*: (El-Muärrij, Ḳ:) and *of a man.* (TA.)

جُوَافٌ : see جُوفِىٌّ.

جَائِفٌ *Reaching the* جَوْف. (Mṣb.) [Hence,] جِرَاحَةٌ جَائِفَةٌ, (Ṣ, Mgh, Ḳ, &c.,) or طَعْنَةٌ جَائِفَةٌ, (Mṣb,) *A spear-wound,* or *the like, that reaches the* ▼جَوْف, (Ṣ, Mgh, Mṣb, Ḳ,) by which is here meant [the *interior of the body* or *head,* or], accord. to IAth, *any vital part,* as *the belly* and *the brain*: (TA:) and sometimes, *that penetrates into the* جوف: (A 'Obeyd, Ṣ, Mgh:) and *that passes through* also: (A 'Obeyd, Ṣ:) and said to be *such as is in the pit between the collar-bones,* and *in the pubes*; but not in the neck, nor in the throat, nor in the thigh, nor in the leg: (Mgh:) not if it reaches the interior of the bone of the thigh: (Mṣb:) opposed to جَالِفَةٌ. (Ṣ in art. جلف.) —— Hence, جَائِفَةٌ is applied to ‡ *A great fault* or *imperfection* or *vice.* (TA from a trad.) —— تَلْعَةٌ جَائِفَةٌ *A deep* [*water-course,* &c.: see تلعة]: pl. جَوَائِفُ. (Ḳ, TA. [In the CḲ, قَصِيرَةٌ is erroneously put for قَعِيرَةٌ.]) —— جَوَائِفُ النَّفْسِ *The deep recesses of the* جَوْف [or *chest*] *in the places where the soul has its seat*; expl. by مَا تَقَعَّرَ مِنَ الجَوْفِ فِى مَقَارِّ الرُّوحِ. (L, Ḳ.) So in the phrase, used by El-Farezdaḳ, وَرَدَّ النَّفْسَ بَيْنَ الجَوَائِفِ [*And he drove back the soul into the midst of the deep recesses of the chest*]: (L, TA:) but some read بَيْنَ الشَّرَاسِيفِ. (TA.) —— الجَائِفُ [*The cephalic vein*;] *a vein that runs along the upper arm to the* [*cartilage called*] نَغْض *of the shoulder-blade*; it is the فَلِيق. (TA.)

أَجْوَفُ *Having a* جَوْف; (TA;) [i. e.,] *hollow,* or *empty within*; (KL, PṢ;) *having in it a* تَجْوِيف [or *hollowing out,* meaning *a hollow*], (Ṣ,) and so ▼مُجَوَّفٌ: (Ṣ, Ḳ: [but the latter is more properly rendered *hollowed,* or *hollowed out*:]) *empty, vacant,* or *void*: (Mṣb:) *wide, spacious,* or *ample*; (Ṣ, Ḳ;) as also ▼مُسْتَجَافٌ, (Ṣ, TA,) and ▼جُوفِىٌّ, with ḍamm, (Ḳ,) thus correctly written, being a rel. n. altered from the original form, like سُهْلِىٌّ and دُهْرِىٌّ, (Ṣgh, TA,) but meaning *wide in the* جَوْف [or *belly,* &c.], written by J [in the Ṣ] ▼جَوْفِىٌّ, with fet-ḥ: (TA:) *great in the* جَوْف; (TA;) as also ▼مِجْوَفٌ; (AO, Ṣ, Ḳ;) each applied to a man: (TA:) [fem. جَوْفَاءُ:] pl. جُوفٌ. (TA.) You say لُؤْلُؤٌ أَجْوَفُ, and ▼مُجَوَّفٌ, [*Hollow, and hollowed, pearls*; or] both signify the same. (TA.) And قَنَاةٌ جَوْفَاءُ *An empty* [or *a hollow*] *cane,* or *reed*: (Ḳ:) and in like manner, شَجَرَةٌ [*a tree*]; (Ṣ, Ḳ;) *having a* جوف. (Ṣ.) And دَلْوٌ جَوْفَاءُ *A wide,* or *an ample, bucket*: (Ḳ:) and دِلَآءٌ جُوفٌ *wide,* or *ample, buckets*: (Ṣ:) and قِدْرٌ جَوْفَاءُ *a wide, capacious, cooking-pot.* (Ḥam p. 719.) And الأَجْوَفُ *The lion that is great in the* جَوْف [or *belly,* &c.]. (Ḳ.) And الأَجْوَفَانِ *The belly and the* فَرْج [or *vulva,* or *pudendum muliebre*]; (Ṣ, Ḳ;) because of their width. (TA.) See also جَوْفٌ. Hence the trad., إِنَّ أَخْوَفَ مَا أَخَافُ عَلَيْكُمُ الأَجْوَفَانِ [*Verily what I most fear for you are the belly and the vulva*]. (TA.) —— ‡ A *cowardly* man; as also ▼مِجْوَفٌ, and ▼مُجَوَّفٌ; the last explained in the Ḳ as meaning *having no heart*: pl. [of the first] جُوفٌ. (TA.) —— A horse *white in the* جَوْف [or *belly*] *as far as the part where the sides terminate,* whatever be the colour of the rest of him; (AO, TA;) as also ▼مُجَوَّفٌ. (TA.) [See also مُجَوَّفٌ.] —— In the conventional language of the science of inflection, † [A *hollow* word; i. e.] a word *having an infirm letter for its medial radical*; (Ḳ, TA;) as قَالَ and بَاعَ. (TA.)

مُجَافٌ ‡ A *shut,* or *closed,* door. (TA.)

مَجُوفٌ : see أَجْوَفُ.

مِجْوَفٌ : see أَجْوَفُ, in two places.

مُجَوَّفٌ : see أَجْوَفُ, in three places. —— Also A beast whose بَلَق [q. v.] *reaches up to his belly*: (Aṣ, Ṣ, Ḳ:) or a horse *whose* بَلَق *reaches to his sides* is said to be مُجَوَّفٌ بَلَقًا. (AA, TA.) [See also أَجْوَفُ, last meaning but one.] —— And an epithet applied to the bird called صُرَد, because it is *white in the belly.* (Mgh and Mṣb in art. صرد.)

مُسْتَجَافٌ : see أَجْوَفُ.

جول

1. جَالَ, (Ṣ, Ḳ, &c.,) aor. يَجُولُ, (Ṣ,) inf. n. جَوْلٌ (Ṣ, Ḳ) and جُوْلٌ (Ḳ) and جَوَلَانٌ (Az, Ṣ, ISd, Z, Ṣgh) and جُؤُولٌ (ISd, Ḳ) and جِيلَانٌ, (Ibn-'Abbád, Ḳ, TA,) in some copies of the Ḳ جِيَلَانٌ; (TA;) and in like manner, ▼اجتال and ▼انجال; (Ṣ, Ḳ;) *He went round,* or *about,* or or *round about*; as also ▼جوّل, inf. n. تَجْوَالٌ: (Ḳ:) or جوّل signifies *he went round,* or *about,* or *round about, much,* or *often*; agreeably with what Sb says of the measure تَفْعَالٌ; but accord. to the O, تَجْوَالٌ is an inf. n. of جال. (TA.) You say, جال فِى البِلَادِ *He went about,* or *round about, in the countries,* or *districts, not remaining fixed,* or *settled*: (Mṣb:) and ▼جوّل البِلَادَ, (T, TA,) or جوّل فِى البِلَادِ, (Ṣ,) inf. n. تَجْوِيلٌ, (T, TA,) or تَجْوَالٌ, (Ṣ,) *he went about,* or *round about, much,* or *often, in the countries,* or *districts.* (T, Ṣ, TA.) And جال فِى المَيْدَانِ, aor. as above, inf. n. جَوْلَةٌ and جَوَلَانٌ, *He* (a horse) *traversed the sides,* or *lateral parts* or *tracts, of the horse-course*; which are termed أَجْوَالٌ, pl. of جُولٌ. (Mṣb.) And جال فِى الحَرْبِ, inf. n. جَوْلَةٌ, *He wheeled round,* or *about, in battle.* (Ḳ.) And فَرَّ لِلْجَوَلَانِ ثُمَّ عَادَ لِلْقِتَالِ [*He fled, to wheel round,* or *about, and then returned to the fight*]. (Mṣb in art. كر.) And جالوا, (Mṣb,) or ▼تَجَاوَلُوا, (Ṣ, Ḳ,) means جَالَ بَعْضُهُمْ عَلَى بَعْضٍ, (Ṣ, Mṣb,

Ḳ,) i. e., *They assailed*, or *assaulted, one another*, (TA,) فِى الحَرْبِ [*in battle*]; (Ṣ, Mṣb, Ḳ;) [and so, app., ↑ جَاوَلُوا, inf. n. مُجَاوَلَةٌ :] and كَانَتْ بَيْنَهُمْ ↑ مُجَاوَلَاتٌ (Ṣ, Ḳ*) *There were between them mutual* [*assailings, or assaults, and*] *defendings.* (Ibn-'Abbád, TA.) And جَالَ القَوْمُ, inf. n. جَوْلَةٌ, *The company of men were routed, defeated*, or *put to flight*, (اِنْكَشَفُوا,) *and then returned to the fight*, or *charged*, or *assaulted.* (Ḳ.) And أَصَابَ المُسْلِمِينَ جَوْلَةٌ *Defeat befell the Muslims:* a metonymy; used only in relation to the favourites of God; from الجَوَلَانُ. (Mgh.) — جَالُوا فِى الضَّلَالَةِ, (Ṣgh, TA,) or إِلَى الضَّلَالَةِ, (A, TA,) † *They became excited to lightness*, or *levity, and unsteadiness, and carried away, and driven*, (Ṣgh, TA,) or *they became fascinated, and turned away*, (A, TA,) by the influence of devils, (A, Ṣgh, TA,) from their religion, (Ṣgh, TA,) or from the right course, (A, TA,) *to error.* (A, Ṣgh, TA.) — جَالَ التُّرَابُ, (ISd, Ḳ,) inf. n. جَوْلٌ, (TA,) *The dust went away, and rose;* as also ↑ انجال : (ISd, Ḳ:) or the latter signifies *became removed*, or *cleared away* [by the wind]; syn. اِنْكَشَطَ. (T, TA.) — يَجُولُ فِى صَدْرِى أَنْ أَفْعَلَهُ ‡ [*It is revolved in my bosom*, or *mind, that I should do it*]. (TA.) — See also 4, in two places. ═ جَالَ الشَّىْءَ, (Ḳ,) inf. n. جَوْلٌ, (TA,) *He chose*, or *selected, the thing.* (Ḳ.) You say, جُلْتُ هٰذَا مِنْ هٰذَا *I chose*, or *selected, this from this.* (AA, Ṣ.) And ↑ اجْتَلْتُ مِنْهُمْ جَوْلًا *I chose*, or *selected, from them* [*a choice portion*]; (Ṣ, Ḳ,* TA;) and *separated some of them from others.* (TA.) And ↑ اجتال مِنْ مَالِهِ جَوْلًا, and جَوَالَةً, *He chose*, or *selected, from his property*, or *cattle, a choice portion.* (TA.)

2: see 1, in two places.

3. جَاوَلُوا, inf. n. مُجَاوَلَةٌ : and كَانَتْ بَيْنَهُمْ مُجَاوَلَاتٌ : see 1.

4. اجالهُ, (Mṣb, Ḳ,) and اجال بِهِ, (Ḳ,) inf. n. إِجَالَةٌ, (Ṣ,) *He, or it, made*, or *caused, him*, or *it, to go, move*, or *turn round*, or *about*, or *round about; to circle*, or *revolve;* (Ṣ, Mṣb, Ḳ;) as also ↑ جَالَ بِهِ. (Zj, Ḳ.) One says in the game called المَيْسِر, [see this word,] أَجِلِ السِّهَامَ [*Turn thou round about*, i. e., *shuffle, the arrows* in the رِبَابَة]. (Ṣ, TA.) And اجال السِّهَامَ بَيْنَ القَوْمِ *He moved about the arrows*, [i. e., *shuffled them* in the رِبَابَة,] (Az, ISd, TA,) *and then distributed them among the people*, or *party.* (Az, TA.) And اجال سَيْفَهُ *He brandished, flourished*, or *played with, his sword, turning it round about.* (Mṣb.) And الرِّيحُ تَجُولُ بِالتُّرَابِ, (Lth, TA,) and بِالحَصَى, (Ḳ, TA,) [*The wind makes the dust*, and *the pebbles, to turn round about, to circle*, or *to revolve.*] — أَجَالُوا الرَّأْىَ فِيمَا بَيْنَهُمْ ‡ *They turned about*, or *revolved*, [*in their minds*, *the idea*, or *opinion, respecting the matter that was between them.*] (TA.) And اجالوا الفِكَرَ † [*They turned about*, or *revolved, thoughts, ideas, schemes*, or *contrivances, in their minds*]. (Jel in ix. 48.) — [اجال خَمْسَهُ فِى وِعَائِهِ, in the 7th Maḳámeh of Ḥar, (p. 76 of the sec. ed.,) is explained in a MS. of that work as meaning ادخل, (De Sacy's Chrest. Ar., sec. ed., p. 185,) i. e., *He inserted:* but the proper meaning is, *he turned about*, or *round about, his five* fingers *in his bag.*] — ↑ أَجِلْ جَائِلَتَكَ ‡ *Accomplish*, or *finish, the affair in which thou art engaged.* (M, Ḳ, TA.)

6: see 1.

7: see 1, in two places.

8: see 1, first sentence. ═ اجتالهم *He turned them from their course.* (Ḳ.) *He* (the devil) *caused them to leave*, or *forsake, the right way.* (T, TA.) اِجْتَالَتْهُمُ الشَّيَاطِينُ عَنْ دِينِهِمْ *The devils excited them to lightness*, or *levity, and unsteadiness, so that they turned away from their religion*, to error; i. e., *they carried them away and drove them* [*from their religion*]. (Ṣgh, TA.) [See also 10.] ═ See also 1, last two sentences.

10. اِسْتَجَالَتْهُمُ الشَّيَاطِينُ *The devils turned them from the right course, to error; fascinated them so that they turned with them.* (A, TA.) And استجالهُ الشَّىْءُ *The thing excited him to lightness*, or *levity, and unsteadiness.* (TA.) [See also 8.] — استجالتِ الخَيْلُ مَا مَرَّتْ بِهِ *The horses removed*, or *displaced, that by which they passed.* (O, TA.) — أُسْتُجِيلَ الرَّبَابُ *The* رباب [or *white clouds*] *were driven together after a state of dispersion, and became ready to rain:* (M, TA:) or it means جَاءَتْهُ الرِّيحُ فَاسْتَجَالَتْهُ, i. e., *the wind came to them, and removed them*, or *displaced them, and dissundered them, and drove them away.* (TA.) — اِسْتَجَلْنَا الجَهَامَ ‡ *We saw the rainless clouds going about*, or *round about, in the horizon*, (A, TA,) or *in the sky.* (TA.)

جَالٌ : see جُولٌ, in five places: — and مِجْوَلٌ.

جَالٌ : see جَائِلٌ.

جَوْلٌ : see جَوَلَانٌ, in two places. ═ Also *A large army*, or *military force*, or *troop of horse:* (Ṣgh, Ḳ:) pl. جُوُولٌ. (Ṣgh, TA.) *A herd of camels:* and *a troop of* خَيْل [meaning *horses* or *horsemen*]: as also ↑ جُولٌ in both these senses: (Ḳ:) or the latter, which is also explained in the Ḳ as signifying *a herd* of camels, and *a flock* of ostriches and of sheep or goats, is pl. of the former: (TA:) or the former signifies *thirty* [*horses or horsemen*]: or *forty:* (Ḳ:) or *less:* or *more:* (TA:) or the *choice*, or *best, of camels:* and [in like manner] ↑ جَوَالَةٌ signifies the *choice and best;* as in the saying, أَخَذَ جَوَالَةَ مَالِهِ [*He took the choice and best of his cattle*, or *property*]. (Ḳ. [See also 1, last two sentences: and see جَوَلَانٌ.]) And *Many great sheep or goats.* (Ḳ.) — Also *A male mountain-goat that is old*, or *advanced in age:* (M, Ḳ:) pl. أَجْوَالٌ. (M, TA.)

جُولٌ The *wall* [*that surrounds the interior*] *of a well:* accord. to A 'Obeyd, *every side of a well, from its top to its bottom:* and ↑ جَالٌ signifies the same: (Ṣ:) or the former, the *side*, or *lateral part*, (M, Mṣb, Ḳ,) *of a well*, and *of a grave*, and *of the sea*, and *of a mountain;* as also ↑ جَالٌ (M, Ḳ) and ↑ جِيلٌ : (Ḳ, TA; in the CḲ جُمْلٌ:) or the *surrounding parts* [or *sides*] of a grave: (M, TA:) and ↑ جَالَا الوَادِى *the two sides of the water of the valley:* and ↑ جَالَا البَحْرِ *the two shores of the sea*, or *great river:* (T, TA:) pl. [of pauc.] أَجْوَالٌ, (Az, Ṣ, Mṣb, Ḳ,) pl. of جُولٌ and جَالٌ, (TA,) and [of mult.] جِوَالٌ and جُوَالَةٌ; (so in copies of the Ḳ, and in the M, but in some copies of the Ḳ جُوَالٌ and جُوَالَةٌ;) and أَجَاوِلُ is pl. of أَجْوَالٌ. (TA.) Also, as in the T and the Moḥeeṭ, (TA,) The *portion of rock that is at the bottom of the water*, (Ḳ, TA.) *upon which is the casing of the well; so that if it quit its place, the well falls to ruin:* this is [said to be] the primary meaning of the word: and hence the saying, هٰذَا مَآءٌ لَا يُدْرَكُ جُولُهُ [*This is water of which the rock beneath it is not to be reached*]. (TA.) — [And from this word as signifying the casing of a well, or the portion of rock above mentioned,] ‡ *Intelligence;* (Ṣ, Ḳ, TA;) *judgment, and intelligence*, or *full intelligence*, or *intelligence to which one has recourse;* (T, TA;) *understanding of the heart;* (TA;) and *resolution*, or *fixed purpose of mind;* (Ṣ, M;) and *prudence:* (T, TA:) العَزْمُ in the Ḳ is erroneously put for الحَزْمُ. (TA.) You say, of a man, مَا لَهُ جُولٌ ‡ *He has not intelligence and judgment*, or *fixed purpose of mind, to withhold him*, or *protect him;* like the جول of a well; (Ṣ, M,* TA;*) because a well, when cased with stone or the like, is stronger. (TA.) And رَجُلٌ لَهُ زَبْرٌ وَجُولٌ ‡ *A man having judgment and intelligence*, or *full intelligence*, or *intelligence to which recourse is had;* whose جول *does not become demolished:* and in like manner, هُوَ مَزْبُورٌ مَا فَوْقَ الجُولِ مِنْهُ وَصُلْبٌ مَا تَحْتَ الزَّبْرِ لَيْسَ لِفُلَانٍ مِنَ الجُولِ : and in the contr. case, جُولٌ ‡ *Such a one has not intelligence nor prudence;* i. e., *his* جول *is demolished, therefore one is not sure that the* زبر [*that rests upon it*] *may not also fall:* and لَيْسَ لَهُ جُولٌ, and ↑ جَالٌ, ‡ *He has not prudence.* (T, TA.) — فَعَلْتُهُ مِنْ جُولِهِ *I did it on account*, or *for the sake*, or *because, of him*, or *it.* (Ibn-'Abbád, TA.) ═ See also جَوْلٌ: ═ and جَوَلَانٌ.

جِيلٌ : see جُولٌ.

جَوَلَانٌ *Dust;* as also ↑ جَوْلٌ and ↑ جُولٌ, (Ḳ,) both mentioned by Az, (TA,) and ↑ جَيْلَانٌ, (Ḳ,) mentioned by ISd: [or] all signify *dust which the wind makes to turn about or round about, to circle, or to revolve, upon, or from, the surface of the earth.* (TA.) And *Small pebbles which the wind makes to turn about or round about, to circle, or to revolve;* (Ḳ, TA;) as also ↑ جَوْلٌ and ↑ جَيْلَانٌ. (TA.) ═ جَوْلَانٌ : see أَجْوَلُ.

جَوَلَانُ الهُمُومِ ‡ *The first*, or *beginning*, [lit. *the revolving*, (see 1,)] *of anxieties.* (Ibn-'Abbád, Ḳ, TA. [In the CḲ, erroneously, جُوْلَانُ.]) You say, فِى قَلْبِهِ جَوَلَانُ الهُمُومِ ‡ *In his heart are revolving anxieties.* (A, TA.) ═ جَوْلَانُ المَالِ *The small*, or *young, and bad, of cattle:* (Fr, Ṣ, Ḳ:) so in the M and O; but in a copy of the M, written جَوَلَانٌ; which is app. a mistake. (TA.) Accord. to Ibn-'Abbád, *The choice*, or *best, of cattle:* the contr. of what is said by Fr. (TA. [See also جَوْلٌ.])

جَيْلَانٌ: see جَوْلَانٌ, in two places. — جَيْلَانٌ: see أَجْوَلُ.

جَوْلَانِىٌّ: see أَجْوَلُ. — Also ‡ A man *whose benefits are common to the near and the distant;* (Ḳ, TA;) *whose benefits go round to every one.* (Ṣgh, TA.)

جَيْلَانِىٌّ: see أَجْوَلُ.

جَوِيلٌ *What the winds sweep away* (AḤn, M, Ḳ) *and round about,* (AḤn, M,) *of fragments of plants and of the fallen leaves of trees;* (AḤn, M, Ḳ;) as also ▼ جَائِلٌ. (M, TA.)

جَوَالَةٌ: see جَوْلٌ.

جَوَائِلُ أَمْرٍ † *The turns* (دَوَائِر) *of an affair,* or *event.* (TA.)

جَوَّالٌ One *who goes about,* or *round about, much,* or *often, in the countries,* or *districts,* (Mṣb, TA,) *not remaining fixed,* or *settled;* (Mṣb;) as also ▼ جَوَّالَةٌ [but in a more intensive sense, meaning *who does so very much,* or *very often*]. (TA.) — A horse *having a flexible head:* (TA:) and ▼ أَجْوَلِىٌّ a *swift* horse, *that turns about howsoever one turns him.* (Ḳ,* TA.)

جَوَّالَةٌ: see the next preceding paragraph.

جَائِلٌ: see جَوِيلٌ. — Also *Rainless clouds going round about.* (A, TA.) — And, applied to a [woman's ornament of the kind termed] وِشَاح, and to a camel's belly-girth, *Loose; not tight; unsteady;* as also ▼ جَالٌ. (T, TA.) [Hence,] ‡ اِمْرَأَةٌ جَائِلَةُ الوِشَاحَيْنِ *A woman slender in the waist.* (Z, TA.)

جَائِلَةٌ *An affair in which one is engaged.* (M, Ḳ.) See 4, last sentence.

أَجْوَلُ [*More,* and *most, wont to go round,* or *about,* or *round about; to circle,* or *revolve;*] is from the first of the verbs in this art.: and hence the prov., أَجْوَلُ مِنْ قُطْرُبٍ [*More wont to go about,* or *round about,* or *more restless, than a* قطرب; a certain animalcule, or insect, that is constantly moving about: see art. قطرب]. (Ḥar p. 661.) — Also, [as meaning † *More,* and *most, circulating,*] applied to language, or discourse. (TA in art. جمع.) [See an ex. voce مَجْمَعٌ.] — يَوْمٌ أَجْوَلُ, and ▼ جَيْلَانِىٌّ, and ▼ جَوْلَانِىٌّ, (Lḥ, M, Ḳ,) and ▼ جَوْلَانٌ, and ▼ جَيْلَانٌ, (M, Ḳ,) *A day of much dust* (T, M, Ḳ) *and wind:* (T, TA:) from جَوْلٌ signifying "dust." (TA.)

أَجْوَلِىٌّ: see جَوَّالٌ.

مَجَالٌ *A place in which one goes round,* or *about,* or *round about:* (TA:) [*a field of battle: a circus:*] *a place of exercise* for horses. (Ḥar p. 16.) — [Hence] one says, لَمْ يَبْقَ مَجَالٌ فِى الأَمْرِ ‡ [*There remained not any scope in the affair,* or *case*]. (TA.)

مِجْوَلٌ *A certain garment for women,* (M, Ḳ,) *doubled, and sewed together at one of its two sides, and having an opening made to it at the neck and bosom; in which a woman goes about:* (M, TA:) or *for a young girl;* (Ḳ;) the دِرْع being for a woman: (TA:) *a small garment in which a girl goes about:* (Ṣ:) or *a garment which a girl wears before she is made to keep herself behind,* or *within, the curtain, and in which she goes about:* (Z, TA:) accord. to IAạr, *i. q.* صُدْرَةٌ. (TA.) Imra-el-Ḳeys says,

إِلَى مِثْلِهَا يَرْنُو الحَلِيمُ صَبَابَةً
إِذَا مَا ٱسْبَكَرَّتْ بَيْنَ دِرْعٍ وَمِجْوَلِ

[*At the like of her the staid would fixedly gaze with tenderness of desire, when she has become of erect and justly-proportioned stature, between* such as wears *a woman's shirt and* such as wears *a young girl's garment*]. (Ṣ,* TA.) — *A woman's anklet.* (Ibn-'Abbád, Ḳ.) — *An amulet, a phylactery,* or *charm of the kind termed* عُوذَة. (IAạr, Ḳ.) — *A crescent of silver in the middle of the necklace termed* قِلَادَة. (IAạr, Ḳ.) — *Silver* [itself]. (Th, Ḳ.) — *A good,* or *sound,* دِرْهَم [or *silver coin*]. (IAạr, Ḳ.) — *A shield;* (Ṣ, O, Ḳ;) sometimes used in this sense; (Ṣ, O;) as also ▼ جَالٌ. (Ibn-'Abbád, TA.) — *A large wooden bowl.* (IAạr, TA.) — *A white* ثَوْب [or *piece of cloth*] *that is put upon the hand of him to whom the players at the game called* المَيْسِر *commit the arrows* [*to be shuffled and distributed, in order that he may not be able to distinguish them by the feel,*] *when they have collected themselves.* (ISd, Ḳ,* TA.) [For the same purpose, a piece of thin skin was also used: see رِبَابَة.] — *A pool of water left by a torrent;* because the water goes round about in it. (IF, TA.) — *A wild ass.* (IAạr, Ḳ.)

مُسْتَجَالٌ [pass. part. n. of 10, *Turned from the right course,* &c.:] *excited to lightness,* or *levity, and unsteadiness:* (TA:) *being bereft of his reason,* or *intellect.* (AA, TA.)

جوم

جَامٌ *A vessel,* (Ḳ,) or فَاثُور [i. e. *a basin,* or *a table,* or *a tray used as a table*], (IAạr, TA,) *of silver:* (IAạr, Ḳ, TA:) or *a white dish* or *tray, of glass* or *of silver:* (Mgh:) or *a vessel* (ظَرْف) *of glass:* (Ḥar p. 200:) [a Persian word, i. e. جام, arabicized; or] a genuine Arabic word: (TA:) pl. [of pauc.] أَجْؤُمٌ, with ء, (IAạr, Ḳ,) and أَجْوَامٌ, and (as some say, IAạr, TA, [of mult.,]) جُومٌ, (Ḳ,) with ḍamm, (TA, in the CḲ جَوم,) and [of pauc. or mult.] جَامَاتٌ: (IAạr, Ḳ:) but IB says that جَامٌ is pl. of ▼ جَامَةٌ, as is also جَامَاتٌ: [instead of which he should rather have said that جام is a coll. gen. n. of which the n. un. is جامة, (though this requires consideration, as the former is commonly used as a sing.,) and that the pl. of the latter is جامات:] that its dim. is ▼ جُوَيْمَةٌ: and that it, i. e. جام, is of the fem. gender. (TA.)

جَامَةٌ: } see above.
جُوَيْمَةٌ: }

جون

1. جَانَ, (Ḳ, TA, [in the CḲ, erroneously, جَانٌ,]) inf. n. جَوْنٌ, (TA,) *It* (the face) *became black.* (Ḳ.)

جَوْنٌ *White:* and *black:* (Ṣ, Mṣb, Ḳ:) thus bearing two contr. significations: (Ṣ:) and ▼ جُونِىٌّ, also, has the latter signification: (IAth, TA in art. جوت:) or جَوْنٌ signifies *black tinged over with red:* (T, M, TA:) and *black intermixed with red;* the *colour of the* قَطَا: (T, TA:) and also *red:* (Ḳ:) or *of a pure red colour:* (TA:) and, applied to a horse and a camel, *of the colour termed* أَدْهَم, (Ṣ, Ḳ,) *intensely black:* (Ṣ:) every camel, and every wild ass, seen from a distance, is of this colour: fem. with ة: (T, TA:) and, applied to a plant, or herbage, *green,* (Ḳ,) or *intensely green,* (TA,) *inclining to blackness:* (Ḳ, TA:) pl. جُونٌ; (Ṣ, TA;) like as صُتْمٌ is of صَتْمٌ, (Ṣ,) and وُرْدٌ of وَرْدٌ. (M, TA.) You say also, الشَّمْسُ جَوْنَةٌ *The sun is characterized by what is termed* جُونَة: (Ṣ:) or *is intensely glistening and clear.* (Az, TA.) [See also جَوْنَة below.] See also جُونِىٌّ. Accord. to ISk, أَبُو الجَوْنِ means *The white man:* opposed to أَبُو البَيْضَآءِ meaning the negro. (TA in art. بيض.) — Also † *Day:* (AO, Ṣ, Ḳ:) pl. as above. (Ḳ.) So in the saying,

غَيَّرَ يَا بِنْتَ الحُلَيْسِ لَوْنِى
مَرُّ اللَّيَالِى وَٱخْتِلَافُ الجَوْنِ

[*The passing of the nights, and the alternating of the day, have changed, O daughter of El-Ḥoleys, my colour*]. (AO, Ṣ.) — And, accord. to certain of the lawyers, metaphorically, ‡ The *light:* and the *darkness.* (Mṣb.) — And accord. to IAạr, † The فرق [app. فَرَق, meaning *day-break*]. (TA.) — الجَوْنَانِ *The two extremities of the bow.* (Fr, Az, Ḳ.)

جَوْنَةٌ The *sun;* (Ḳ;) [i. e.] the *sun's disc;* because it becomes black [or of a blackish colour tinged with red] at setting; (Ṣ;) or it may be because of its whiteness and clearness; but it is said to be only applied to the sun when it is setting; opposed to غَزَالَةٌ; as observed by MF: (TA:) [see also جَوْنٌ:] the *sun* is also called ▼ جُوَنَةٌ, (Ḳ,) because of its becoming black [or of a blackish colour tinged with red] at setting. (TA.) — *A* [*jar such as is called*] خَابِيَة: (IAạr, TA:) or *a* خابية *smeared with tar,* or *pitch.* (Ṣ.) [See an ex. in a verse of Lebeed cited in art. دكن.] See also جُونَة. — And *A bucket* (دَلْو) *that has become black.* (IAạr, TA.) — And *i. q.* فَحْمَةٌ [which may here mean either *A piece of charcoal,* or the *blackness* of night or the like]. (IAạr, Ḳ.) — And *i. q.* أَحْمَرُ [perhaps as a subst., meaning *A red thing*]. (Ḳ.) — See also جُونِىٌّ.

جُونَةٌ The *quality* [i. e. *colour*], in horses, *denoted by* [*the epithet*] جَوْنٌ; like غُبْسَة and وُرْدَة; (Ṣ;) in horses, *i. q.* دُهْمَة: (Ḳ:) and in the sun, also, the *quality denoted by* جَوْنَة [as fem. of جَوْنٌ, q. v.]: and *blackness;* as in the saying, لَا أَفْعَلُهُ حَتَّى تَبْيَضَّ جُونَةُ القَارِ [*I will not do it until the blackness of pitch*, or *tar*, *become white*]: but if you say ▼ جُونَةُ القَارِ, the meaning is the خَابِيَة [*smeared with tar, or pitch*]. (Ṣ.) ═ *A small basket* (سُلَيْلَة), (Ḳ,) or سَفَط, (Ḳ in art. جان,) *of a round form*, (TA,) *that is with the sellers of perfumes*, (Ṣ, Ḳ,) *used for containing their perfumes*: (Ḳ in art. جان:) called in Persian شِيشَهْ دَانْ [*a receptacle for bottles* or *the like*]: (KL:) originally with ء: (Ḳ:) or sometimes pronounced with ء: (Ṣ:) El-Fárisee approved the suppression of the ء: (M, TA:) pl. جُوَنٌ. (Ṣ, M, Ḳ.) [See also رُبْعَة.] ═ *A small mountain.* (Ḳ.)

جُونَة: see جُونَة. — Also *A cooking-pot*; (Ḳ;) because it is black. (TA.) — And *A she-camel such as is termed* دَهْمَآء [*of an intense, or a dark, gray colour, without any admixture of white*]; from جَان said of the face. (Ḳ.)

جُونِىٌّ: see جُونٌ. — Also *A species of the kind of bird called* قَطًا, (Ṣ, Ḳ,) *black in the belly and wings, larger than the* [*species called*] كُدْرِىّ, *one of the former species being equal to two of the latter*: (Ṣ, TA:) or, accord. to ISk, the قطا compose two species; one called جُونِىّ and كُدْرِىّ; and the other, غَطَاط; and the former is *dusky*, or *dingy*, or *of a hue inclining to black and dust-colour*, (أَكْدَر,) *in the back, black in the inner side of the wing, yellow in the throat, short in the legs, having in the tail two feathers longer than the rest of the tail*: (T, TA:) or, as some say, the كُدْرِيَّة and جُونِيَّة are one of the two species of the قطا, and the other is the غطاط; and the former are *short in the legs, yellow in the necks, black in the primary feathers of the wings, of a white hue tinged with red* (صُهْب) *in the tertials*: (TA voce غطاط, q. v.:) [but see كُدْرِىّ: the جونىّ is described by De Sacy, on the authority of the book entitled دُرَّة المُنْتَقَاة من عجائب المخلوقات وغرائب الموجودات, thus: "le djouni a les barbes internes des ailes et les pennes primaires noires; il a la gorge blanche, ornée de deux colliers, l'un jaune et l'autre noir; son dos est d'un gris cendré, moucheté, mêlé d'un peu de jaune: on appelle cette espèce djouni, parce que sa voix ne rend pas un son clair et sonore, mais qu'elle fait entendre seulement une sorte de gargouillement dans le gosier:" (Chrest. Arabe, 2nd ed., ii. 369:)] it is stated in the handwriting of Aṣ, on the authority of the Arabs, that جُونِىّ, applied to the قطا, is with ء; app. meaning that it was pronounced جُؤْنِىّ: (M, TA:) a single bird of this species is termed جُونِيَّة: (Ṣ:) and you say also ▼ قَطَاةٌ جُونَةٌ, with fet-ḥ: (TA:) [but جُونِىّ seems to be also used as a n. un., like رُومِىّ: for it is said that] جُونٌ is pl. [or rather coll. gen. n.] of جُونِىّ, like as تَمْرٌ is of تَمْرَةٌ. (Ḥam p. 605.)

جوه

1. جَاهَهُ بِالْمَكْرُوهِ, (Ṣ,) or بِمَكْرُوهٍ, (Ḳ,) [aor. يَجُوهُ,] inf. n. جَوْهٌ, (Ṣ,) i. q. جَبَهَهُ بِهِ [*He met him*, or *encountered him, with*, or *he said to him*, or *did to him*, or *he accused him, to his face, of, that which was*, or *a thing that was, disliked*, or *hated*]. (Ṣ, Ḳ.) And [in like manner,] جَاهَهُ بِشَرٍّ *He encountered him with evil* [*speech* or *conduct*]; or *confronted him therewith*: whence لَا جُهْتَ, i. e. *Mayest thou not be encountered with evil*; said in chiding a he-camel: see جَاهِ below. (TA.) A woman of Ghaṭafán, being chidden by her son, and being asked why she did not reply against him, said, أَخَافُ أَنْ يَجُوهَنِى بِأَكْثَرَ مِنْ هٰذَا, meaning *I fear that he would encounter me with more than this.* (JK.)

[2. جوّه and ▼ اجاه are rendered by Golius *Ad dignitatem evexit: spectabilem reddidit*: as on the authority of the Ṣ: but in my copies of the Ṣ the two verbs are اوجه and وجّه; and belong to art. وجه, though mentioned in the present art.]

[4: see 2.]

5. تجوّه *He magnified himself*; or *was*, or *became, proud, haughty*, or *disdainful*: or *he affected rank, station*, or *dignity, not possessing it.* (TA.)

جَاهٌ *Rank, station*, or *dignity*, (Ṣ, Ḳ, TA, and JK in art. وجه,) *with*, or *in the estimation of, the Sultán*; (JK, TA;) as also ▼ جَاهَةٌ, (Ḳ,) on the authority of Lḥ, (TA, as from the Ḳ, [but not in my copies,]) or, accord. to Ṣgh, of Ks.: the former word [probably arabicized from the Persian جَاه; but] said to be formed by transposition from وَجْهٌ; this being first changed to جَوْهٌ; then, to جَوَهٌ; and then, to جَاهٌ: or, accord. to Lḥ, it is not from وَجْهٌ, but from جُهْتُ [app. جُهْتُ, first pers. sing. of جَاهَ, q. v.]; though he does not explain what is جُهْتُ. (TA.) You say, فُلَانٌ ذُو جَاهٍ [*Such a one is possessed of rank, station*, or *dignity*]. (Ṣ.) And لِفُلَانٍ جَاهٌ فِيهِمْ *To such a one belongs rank, station*, or *dignity, among them.* (Aboo-Bekr, TA.) The dim. of جَاهٌ [or of جَاهَةٌ] is ▼ جُوَيْهَةٌ. (TA.) ═ جَاهِ, indecl., with kesr for its termination; and, accord. to Aṣ, sometimes, جَاهٍ, with tenween; (Ṣ;) or جَاهِ جَاهِ, (JK, Ḳ,) and جَاهٍ جَاهٍ, and جَاهَ جَاهَ, (Lḥ, Ḳ,) and جُوه ▼ جُوهِ, or جُوهٍ جُوهٍ, (accord. to different copies of the Ḳ, but) indecl., with kesr for the termination, [i. e. جُوهِ جُوهِ, or جُوهِ جُوهِ,] mentioned in the M, (TA,) and ▼ جَاهِ لَا جُهْتَ, (IDrd, TA, [see 1,]) ejaculations used for chiding a he-camel, not a she-camel: (Aṣ, JK, IDrd, Ṣ, Ḳ:) or one says to a she-camel, عَاجِ and جَاهِ. (A 'Obeyd, TA in art. عوج: [or perhaps there is an omission here: I think it more likely that what A 'Obeyd said was that one says to a she-camel عَاجِ, and to a he-camel جَاهِ.])

جُوهٌ The *face*, or *countenance*; syn. وَجْهٌ; as also جِيهٌ: (Lḥ, Ḳ:) the latter with kesr: (TA:) so in the saying, نَظَرَ بِجُوهِ سَوْءٍ and بِجِيهِ سَوْءٍ [*He looked with an evil face* or *countenance*]. (Lḥ, Ḳ.) ═ جُوهِ جُوهِ, or جُوهٍ جُوهٍ: see the next preceding paragraph.

جَاهَةٌ: }
جُوَيْهَةٌ: } see جَاهٌ.

تَجَاهَ and تِجَاهَ and تُجَاهَ: see art. وجه.

جوهر

جَوْهَرٌ: }
جَوْهَرِىٌّ: } see art. جهر.

جوى

1. جَوِىَ, (Ṣ, Ḳ,) aor. يَجْوَى, (TA,) inf. n. جَوًى, (Ḳ,) *He* (a man, Ṣ) *was*, or *became, affected with what is termed* ▼ جَوًى, (Ṣ, Ḳ,) meaning *ardour*: and *violence of amorous desire*; or *of grief*, or *sorrow*: (Ṣ:) or *inward love*: (M, Ḳ:) and *grief*, or *sorrow*: (Ḳ, and so in a copy of the Ṣ:) and *ardour*: and *violence of love*; or *of grief*, or *sorrow*: (Ḳ:) [see حُبٌّ:] — and also as meaning *consumption*; or *an ulcer in the lungs*: and *long continuance*, or *oppressiveness, of disease*: and *a disease in the chest*: (Ḳ:) or *any inward disease during which one does not find food to be wholesome*: (TA:) part. n. ▼ جَوٍ; (Ṣ, Ḳ;) fem. جَوِيَةٌ. (TA.) — You say also, جَوِيَتْ نَفْسِى, meaning *I found the country*, or *town, to disagree with me.* (Ṣ.) And جَوِيَتْ نَفْسُهُ مِنْهُ and عَنْهُ [*He found it to disagree with him*: a meaning indicated, but not expressed]. (Ḳ.) See also 8. — And جَوِيَتِ الأَرْضُ *The land stank.* (TA.)

8. اجتواهُ *He disliked residing in it*, namely, a country, or town, *even if in the enjoyment of ease and plenty*: (Ṣ:) or *he disliked it*, (Ḳ, TA,) *and found it to disagree with him*; (TA;) as also ▼ جَوِيَهُ: (Ḳ, TA:) or *he disliked it*, namely, a city, *and found it to be insalubrious*: or, as AZ says, *he disliked it*, namely, a country, *even if it agreed with him in respect of his body*: and he says, in his Nawádir, that اِجْتِوَآءٌ signifies the *yearning towards*, or *longing for, home, and disliking a place, even if in the enjoyment of ease and plenty*: and *disliking it without yearning towards*, or *longing for, home*: and also the *not finding the food nor the beverage in a land to be wholesome*; *but not when one likes the residing in it but its food and beverage do not agree with him.* (TA.) ═ And اجتوى *His heart was burnt by the fire of enmity.* (Ḥam p. 219.)

جَوًى: see 1: — and see what next follows, in two places.

جَوٍ part. n. of 1, q. v.; (Ṣ, Ḳ;) and ▼ جَوًى signifies the same, being an inf. n. used as an epithet. (Ḳ.) — Hence, (Ṣ,) the former, (Ṣ, Ḳ,) as also ▼ the latter, (Ḳ, and so in a copy of the Ṣ,) signifies *Stinking* water; (Ḳ;) or water *that has become altered and stinking.* (Ṣ, TA.) — أَرْضٌ جَوِيَةٌ and ▼ جَوِّيَةٌ *A land that disagrees with one.* (Ḳ.)

جُوِّىٌّ *Contracted in the bosom*, (Ḳ, TA,) *by reason of disease therein*, (TA,) *so that his tongue cannot explain for him*, (Ḳ,) *or so that his tongue can hardly, or not at all, explain for him.* (TA.)

جَوٍ: see أَرْضٌ جَوِيَةٌ. —

جى

3. جَايَاهُ, inf. n. مُجَايَاةٌ, *He faced him, fronted him, was opposite to him,* or *was over against him*: a dial. var. of جَابَاهُ. (IAạr, Ḳ.) You say also, مَرَّ بِى مُجَايَاةً *He passed by me being in front,* or *opposite.* (TA.)

جيأ

1. جَآءَ, aor. يَجِىْءُ, inf. n. مَجِىْءٌ (Ṣ, Mṣb, Ḳ, &c., [the most common form, but] deviating from the general rule [respecting inf. ns. of this class], for the inf. n. of a verb of the form فَعَلَ having its aor. of the form يَفْعِلُ is [accord. to the general rule, if commencing with an augmentative م,] مَفْعَل, though some words, beside مَجِىْءٌ, deviate from this rule by being of the measure مَفْعِل, [for مَجِىْءٌ is originally مَجْيِئٌ,] as مَعِيش and مَكِيل and مَحِيد and مَسِير and مَصِير and مَمِيل and مَقِيل and مَزِيد and مَعِيل and مَبِيع and مَحِيض and مَحِيص, Ṣ,* TA) and جَيْئَةٌ (Ṣ, Ḳ, of the form of an inf. n. of un., but used as an inf. n. in an absolute sense, like رَجْفَةٌ and رَحْمَةٌ, Ṣ, TA) and جَىْءٌ, (Ḳ,) *He,* or *it, came;* or *was,* or *became, present;* syn. أَتَى; (Ṣ, Ḳ;) or حَضَرَ, said of a man [&c.]; (Mṣb;) or حَصَلَ [meaning *it came, came to pass, happened, took place, betided, befell,* or *occurred; it resulted; it ensued;* &c.]; and it is used in relation to ideal, as well as real, substantives; so that إِذَا جَآءَ نَصْرُ ٱللّٰهِ [*When the assistance of God shall come* (in the Ḳur cx. 1)] is [not a figurative but] a proper phrase. (Er-Rághib, TA.) Sb mentions, on the authority of certain of the Arabs, هُوَ يَجِيكَ [for هو يَجِيْئُكَ *He comes,* or *will come, to thee*], with the hemzeh suppressed: (TA:) and he also mentions يَجُوءُ as a dial. var. of يَجِىْءُ. (Id. in art. جوأ, q. v.) [As shown above,] جَآءَ is used intransitively and transitively. (Mṣb, MF.) You say, جَآءَ زَيْدٌ *Zeyd came;* or *was,* or *became, present.* (Mṣb.) And جِئْتُ مَجِيْئًا حَسَنًا [*I came with a good coming;* or *in a good manner*]. (Ṣ.) And جِئْتُ زَيْدًا *I came to Zeyd.* (Mṣb.) And sometimes one says, جِئْتُ إِلَيْهِ meaning *I went* [as well as *I came*] *to him,* or *it.* (Mṣb.) And جِئْتُ مِنَ ٱلْبَلَدِ [*I came from the town,* or *country*]: and مِنَ ٱلْقَوْمِ, meaning مِنْ عِنْدِ ٱلْقَوْمِ [*from the presence of the people,* or *company of men*]. (Mṣb.) And جَآءَ ٱلْغَيْثُ *The rain* [*came,* or] *descended.* (Mṣb.) And جَآءَ أَمْرُ ٱلسُّلْطَانِ *The order,* or *command, of the Sulṭán came,* or *arrived.* (Mṣb.) And جِئْتُ بِهِ (Ṣ, Mṣb, Ḳ) and أَجَأْتُهُ↓, both signifying the same, (Ṣ, Ḳ,) [*I came with him,* or *it;*] *I brought him,* or *it, with me.* (Mṣb.) And ٱلْحَمْدُ لِلّٰهِ ٱلَّذِى جَآءَ بِكَ [*Praise be to God who brought thee*]; and ٱلْحَمْدُ لِلّٰهِ إِذْ جِئْتَ [*Praise be to God because,* or *that, thou camest,* or *hast come*]; but not ٱلْحَمْدُ لِلّٰهِ ٱلَّذِى جِئْتَ: (Ṣ, TA:) and [in like manner] you say, ٱلْحَمْدُ لِلّٰهِ إِذْ كَانَ كَذَا; but not ٱلْحَمْدُ لِلّٰهِ ٱلَّذِى كَانَ كَذَا unless you say بِهِ or مِنْهُ or عَنْهُ [after ٱلَّذِى]. (ISk, TA.) [Hence, جَآءَ بِوَلَدٍ *He begot a child,* or *children;* like أَتَى بِوَلَدٍ. And جَآءَتْ بِهِ *She brought him forth; gave birth to him;* like أَتَتْ بِهِ. And جَآءَ بِمَعْنًى *It* (a word) *conveyed,* or *imported, a meaning.*] — [جَآءَ بِشَىْءٍ also signifies *He brought to pass, did, executed, performed,* or *effected, a thing:* and *he said, gave utterance to,* or *uttered, a thing:* like أَتَى بِهِ in both these senses.] And جَآءَ كَذَا *He did thus,* or *such a thing.* (TA.) Hence, [in the Ḳur xix. 28,] لَقَدْ جِئْتِ شَيْئًا فَرِيًّا (TA) [*Verily, O Mary, thou hast done*] *a thing hitherto unknown; a thing deemed strange.* (Bḍ. [See another ex. voce إِمْرٌ, likewise from the Ḳur.]) And جِئْتُ شَيْئًا حَسَنًا *I did a good thing.* (Mṣb.) And جَآءَ بِٱلْبَدِيعِ *He produced a new saying,* or *new poetry, not after the similitude of anything preceding.* (TA in art. بدع.) And جَآءَ جَرْيًا بَعْدَ جَرْىٍ (Ḳ in art. تأر) or [more commonly] جاء بِجَرْىٍ بَعْدَ جَرْىٍ (M in that art.) [*He* (a horse) *performed,* or *fetched, run after run*]. — جَآءَ is also *syn. with* صَارَ, like أَتَى; as in the saying, جَآءَ ٱلْبِنَآءُ مُحْكَمًا *The building became,* or *came to be, firm, strong,* or *compact.* (Kull p. 11.) [And hence the phrase,] مَا جَآءَتْ حَاجَتَكَ, (M, Ḳ,) thus in all the copies of the Ḳ, with the noun in the accus. case; i. e. *What became,* or *has become, thy want?* syn. مَا صَارَتْ; (M, Ḳ;) or *What was thy want?* syn. مَا كَانَتْ: (Er-Radee, TA:) ما being here an interrogative, and the [implied] pronoun [in the verb] being made fem. because its predicate is fem.: but some say حَاجَتُكَ, in the nom. case, [as it is in the CḲ, meaning *What did,* or *has, thy want become?*] regarding حاجتك as the subject of جاءت, and ما as the predicate of this verb. (TA.) — See also 3.

3. مُجَايَاةٌ [inf. n. of جَايَأَ] signifies The act of *facing,* or *fronting; being opposite,* or *over against:* (IAạr, Ḳ:) and the act of *coinciding;* as also جِيَآءٌ↓. (AZ, Ḳ.) You say of a man, جَايَأَنِى مِنْ قُرْبٍ *He faced me, fronted me, was opposite to me,* or *was over against me, at a short distance.* (TA.) And مَرَّ بِى مُجَايَأَةً *He passed by me being in front,* or *opposite.* (TA.) And جَايَأْتُ فُلَانًا *I coincided with such a one in his coming.* (TA.) And لَوْ جَاوَزْتَ هٰذَا ٱلْمَكَانَ لَجَايَأْتَ ٱلْغَيْثَ *Hadst thou passed beyond this place, thou hadst met with rain,* or *coincided with rain in its coming.* (TA.) — جَاآنِى فَجِئْتُهُ↓, [so in copies of the S, and in copies of the Ḳ, as from the Ṣ, but in the TA, as from the Ṣ, جَاءَأَنِى, and said to be with two hemzehs, though this is evidently wrong,] aor. أَجِيْئُهُ, the former verb of the measure فَاعَلَنِى, (Ṣ,) is [said to be] a mistake for جَايَأَنِى فَجِئْتُهُ, since the former verb has an infirm letter [ى] for its medial radical and ء for its final, not the reverse, (Ṣgh, Ḳ,) [therefore] what J says is not allowable unless it be an instance of transposition; (IB, TA;) but what is given by F [and Ṣgh as the correct form] is that which is accord. to rule, and what J says is that which has been heard from the Arabs, as ISd has pointed out; (TA;) [and rule is not to be regarded when it is contr. to classical usage;] the meaning is, *He vied with me,* or *strove to surpass me, in frequency of coming, and I surpassed him therein.* (Ṣ, Ḳ.)

4. أَجَآءَهُ *He made him,* or *it, to come.* (Kull p. 11.) — [Hence,] أَجَآءَهُ i. q. جِئْتُ بِهِ: see 1. (Ṣ, Ḳ.) — أَجَآءَهُ إِلَيْهِ *I compelled him, constrained him,* or *necessitated him, to have recourse,* or *betake himself, to it;* (Fr, Ṣ, Ḳ;) or *made him to want it,* or *be in need of it:* (Ṣ:) in the dial. of Temeem, أَشَآءَهُ. (TA in art. شيأ.) It is said in a prov., شَرٌّ مَا يُجِيْئُكَ إِلَى مُخَّةِ عُرْقُوبٍ [*It is an evil thing that compels thee to have recourse to the marrow of a hock*]; for, as Aṣ says, the عرقوب contains no marrow, and only he who cannot obtain any [other] thing is made to want it. (Ṣ.) And it is said in the Ḳur xix. 23, فَأَجَآءَهَا ٱلْمَخَاضُ إِلَى جِذْعِ ٱلنَّخْلَةِ *And the motion of the child in her womb compelled her to betake herself to the trunk of the palm-tree.* (Bḍ.)

جِيْئَةٌ [*A coming;*] a subst. from جَآءَ, (Ṣ, Ḳ,) of the measure فِعْلَةٌ, with kesr to the ج. (Ṣ.)

جَيْئٌ and جِيَآءٌ: see what next follows.

جَيَّآءٌ, (Ḳ,) mentioned by Sb as an extr. word, (TA,) [but regularly formed, of the measure فَعَّالٌ,] and جَيَّاءٌ↓, also written جَآءٌ, (Ḳ,) with the ى changed into hemzeh, (TA,) and جَيِّئٌ↓, (Ḳ,) [originally جَيْوِئٌ, of the measure فَيْعِلٌ, denoting intensiveness, in the CḲ written جَايِئٌ,] mentioned by IJ as anomalous, *A frequent comer.* (TA.) One says, إِنَّهُ لَجَيَّآءٌ بِخَيْرٍ *Verily he is a frequent bringer of good.* (TA.)

[جَاءٍ, originally جَايِئٌ, then جَائِئٌ, then جَائِى, and then جَآءٍ, *Coming;* act. part. n. of 1.]

جيب

1. جَابَ, aor. يَجِيبُ: see 1 in art. جوب, in two places.

2. جَيَّبَ, inf. n. تَجْيِيبٌ: see 1 in art. جوب.

جَيْبٌ The طَوْق [or *opening at the neck and bosom*] (Ḳ) of a shirt (Ṣ, Ḳ) and the like; (Ḳ;) as, for instance, of a coat of mail: (TA:) or the *opening* of a shirt *at the uppermost part of the breast:* (Mṣb, MF:) or the *opening* in a garment *for the head to be put through:* or *such an opening as a sleeve and a* طَوْق: (MF:) pl. [of mult.] جُيُوبٌ, (Mṣb, Ḳ,) also pronounced جِيُوبٌ, (TA,) [like بِيُوتٌ for بُيُوتٌ,] and [of pauc.] أَجْيَابٌ: (Mṣb:) this is said to be its proper art., (Ḳ, TA,) not جوب, because its pl. is جُيُوبٌ. (TA.) [The Arabs often carry things within the bosom of the shirt &c.; and hence the word is now applied by them to *A pocket.*] — † The *heart;* the *bosom.* (Ḳ.) So in the saying, هُوَ نَاصِحُ ٱلْجَيْبِ † [*He is pure,* or *sincere, of heart* or *bosom*]: (Ḳ:) or *trusty, trustworthy,* or *faithful.* (Ṣ. [See also art. نصح.]) A poet says,

* وَخَشَّنْتَ صَدْرًا جَيْبُهُ لَكَ نَاصِحُ

† [*And thou hast exasperated a bosom the heart of which was faithful to thee*]. (TA.) You say also, هُوَ دَنِسُ الجَيْبِ † [*He is a person of foul heart*]. (A in art. دنس.) And رَدَعَ جَيْبَهُ عَنْهُ † *He cleared his heart, or bosom, of it.* (Ḳ in art. ردع, q. v.) — جَيْبُ الأَرْضِ † *The place of entrance of the land, or country:* (Ḳ:) pl. جُيُوبٌ. (TA.)

جِيبَةٌ: see art. جوب.

يَاقُوتٌ مُجَيَّبٌ *Hollowed* [*sapphires*]: occurring in a trad. describing the banks of the river of Paradise: but accord. to one reading, it is لُؤْلُؤٌ مُجَوَّفٌ; accord. to another, مُجَيَّبٌ or مُجَوَّبٌ; and accord. to another, مُجَوَّبٌ or مُجَيَّبٌ. (TA.)

جيح

1. جَاحَتِ الآفَةُ المَالَ, aor. تَجِيحُهُ, inf. n. جِيَاحَةٌ, [perhaps a mistranscription for جَائِحَةٌ,] *i. q.* جَاحَتْهُ of which the aor. is تَجُوحُهُ: see art. جوح. (Mṣb.) And جَاحَهُمُ اللّٰهُ, inf. n. جَيْحٌ and جَائِحَةٌ, the latter an inf. n. like عَافِيَةٌ, *God afflicted them with a calamity.* (TA.)

مُجِيحٌ *i. q.* مُجُوحٌ: see art. جوح. (Mṣb.)

جيد

1. جَيِدَ, (Lḥ, L,) or جَادَ, of the same class as تَعِبَ, aor. يَجَادُ, (Mṣb,) inf. n. جَيَدٌ, (Lḥ, Ṣ, A, L, Mṣb, Ḳ,) *He had a long neck:* (A, Mṣb, Ḳ:) or *a long and beautiful neck:* (Ṣ, L:) or *a slender and long neck.* (L, Ḳ.)

جِيدٌ, of the measure فِعْلٌ, (Sb, Akh,) or it may be originally of the measure فُعْلٌ, (Sb, TA,) *The neck:* (Ṣ, L, Mṣb, Ḳ:) said by Sh to be used only in praise; and عُنُقٌ, in dispraise; the use of the former in the Ḳur cxi. being ironical; (TA;) but accord. to Esh-Shihąb, the contr. is often the case: (MF:) generally applied to that of a woman: (L:) or the *part of the neck upon which the necklace lies:* or *its fore part:* (L, Ḳ:) pl. [of pauc.] أَجْيَادٌ (Ṣ, L, Mṣb, Ḳ) and [of mult.] جُيُودٌ. (L, Ḳ.) لَيِّنَةُ الأَجْيَادِ means A female *soft in respect of the neck;* as though the term جِيدٌ applied to each distinct part of the neck, and the pl. denoted the whole neck. (L.)

جَيْدَانَةٌ: see أَجْيَدُ.

جَيِّدٌ: see art. جود.

أَجْيَدُ *Having a long neck:* (A, Mṣb, Ḳ:) or *having a long and beautiful neck:* (Ṣ, L:) or *having a slender and long neck:* (L, Ḳ:) or it is not applied to a man: (T, TA:) fem. جَيْدَآءُ, (Ṣ, L, Mṣb, Ḳ,) with which ▼ جَيْدَانَةٌ is syn.; (Ḳ;) or this signifies *having a beautiful neck:* (L:) pl. جُودٌ [originally جُيْدٌ]. (Ṣ, A, Ḳ.) And عُنُقٌ أَجْيَدُ *A long and beautiful,* or *slender and long, neck.* (L.)

جير

2. جَيَّرَ *He plastered* a watering-trough or tank *with* جَيَّار. (TA.)

جَيْرِ, with kesr to the ر, (Ṣ, Mughnee, Ḳ, &c.,) like أَمْسِ; (Mughnee;) and جَيْرَ, like أَيْنَ; and sometimes جَيْرُ; (Mughnee, Ḳ;) or this, where it occurs, is for جَيْرِ إِنَّ, and is properly written جَيْرُنَّ إِنَّ in the sense of نَعَمْ being a corroborative of جَيْرِ, and its hemzeh and sheddeh and final vowel being here suppressed: (Mughnee:) a form of oath, (Ṣ, Ḳ,) or put in the place of an oath, (IAmb, TA,) meaning *Verily,* or *truly;* syn. حَقًّا: (Ṣ, Ḳ:) or a responsive particle, (Mughnee,) meaning *yes;* syn. نَعَمْ [which is most approved as responsive to an interrogation], (Mughnee, Ḳ,) or أَجَلْ [which is most approved as responsive to an affirmation]; (Sharḥ et-Tes-heel, Ḳ;) not a noun in the sense of حَقًّا, for were it so it would be an inf. n.; nor in the sense of أَبَدًا, for were it so it would be an adv. n. of time; and if it were a noun it would be decl., and would admit the article ال, and would not have إِنَّ for a corroborative, nor have لَا opposed to it, as it has in the saying,

* إِذَا تَقُولُ لَا ٱبْنَةُ العُجَيْرِ *
* تَصْدُقُ لَا إِذَا تَقُولُ جَيْرِ *

[*When the daughter of El-'Ojeyr says* لا, *she speaks truly: not when she says* جير]: (Mughnee:) or it is a verbal noun, meaning *I know;* syn. أَعْرِفُ; as is mentioned by Ibn-Abi-r-Rabeeạ, and by Er-Radee on the authority of 'Abd-El-Ḳáhir. (MF.) You say, جَيْرِ لَا آتِيكَ *Verily,* or *truly,* [&c.,] *I will not come to thee.* (Ṣ.) And جَيْرِ لَا أَفْعَلُ ذٰلِكَ *Verily,* or *truly,* [&c.,] *I will not do that.* (Ḳ,* TA.) And لَا جَيْرِ لَا أَفْعَلُ *No, verily,* or *truly,* [&c.,] *I will not do* [that]. (Ḳ.)

جِيرٌ *Gypsum;* syn. جِصٌّ. (IAạr, TA.) [In modern Arabic, *Lime:* see also what next follows.]

جَيَّارٌ *Quick lime,* and the *mixtures thereof, with which are plastered watering-troughs* or *tanks, and baths;* syn. صَارُوجٌ: (Ṣ, Ḳ:) *quick lime and gypsum mixed with ashes:* (IAạr, TA:) or *quick lime* alone. (TA. [See also جِيرٌ.]) = [*A lime-burner:* so in the present day: see أَتُّونٌ.] = *Heat in the chest, by reason of rage* or *hunger;* as also ▼ جَائِرٌ: (Ṣ, Ḳ:) or *cough,* or *the like.* (Ḥam p. 56.) It is app. of the measure فَعَّالٌ; or it may be of the measure فَيْعَالٌ; or فَوْعَالٌ [originally جَوْيَارٌ]. (IJ, TA.) — *Strength,* or *vehemence.* (TA.)

جَائِرٌ: see جَيَّارٌ.

مُجَيَّرٌ A watering-trough or tank *made small:* or *made deep:* or *plastered with gypsum.* (Ḳ.)

جيش

1. جَاشَ, (Ṣ, A, Mṣb, Ḳ,) aor. يَجِيشُ, (Ṣ, Mṣb, Ḳ,) inf. n. جَيْشٌ (Mṣb, Ḳ) and جَيَشَانٌ (T, Ḳ) and جُيُوشٌ, (Ḳ,) said of a cooking-pot (T, Ṣ, A, Mṣb, Ḳ) &c., (T, A, Ḳ,) *It boiled,* or *estuated:* (T, Ṣ, A, Mṣb, Ḳ:) or *began to do so, not yet boiling* or *estuating;* this latter being said by some to be the correct meaning. (IB, L, TA.) — ‡ *It* (the sea) *estuated* (A,* Ḳ, TA) with the waves, (A, TA,) *so that it was unnavigable.* (TA.) — † *It* (a valley) *flowed with much water, its water,* or *waves, rising high.* (Ṣ, Ḳ.*) — † *It* (a water-spout, or pipe,) *poured forth water:* (TA:) and [in like manner] you say of the eye, جَاشَتْ, meaning, † *it flowed,* or *overflowed, with tears.* (Ḳ.) — † *He* (a horse) *reared, and became excited.* (TA.) — جَاشَتْ نَفْسُهُ, (Ṣ, Ḳ,) or إِلَيْهِ نَفْسُهُ, (A,) ‡ *His soul* [or *stomach*] *heaved;* or *became agitated by a tendency to vomit;* syn. غَثَتْ; (Ṣ, Ḳ, TA;) or فَارَتْ; (T in art. ثور; see ثَارَتْ نَفْسُهُ in that art.;) *as though what was in his belly rose to his fauces:* (TA:) or *his soul* [*as it were*] *turned round,* [i. e., *he became giddy,*] *with the tendency to vomit:* (Ṣ, Ḳ:) as also ▼ تَجَيَّشَتْ; (Ḳ;) which occurs in the former of these senses in a trad.: (TA:) and جَاشَتْ نَفْسُهُ also signifies *his soul* [or *stomach*] *heaved, by reason of grief* or *fright;* (Ḳ;) [as also جَأَشَتْ;] or when this last signification is meant, you say جَشَأَتْ: (Ṣ:) and *his* (a coward's) *soul purposed flight:* or *was frightened:* as also جَأَشَتْ, in either of these two senses: (TA:) and جَاشَتْ إِلَيْهِ النَّفْسُ *his heart quitted its place by reason of fear.* (EM p. 79.) — جَاشَ صَدْرُهُ ‡ *His bosom boiled with wrath,* or *rage.* (TA.) You say also, صَدْرُهُ يَجِيشُ عَلَيَّ بِالغِلِّ ‡ [*His bosom boils against me with rancour, malevolence, malice,* or *spite;* or *with latent rancour,* &c.]. (A.) — جَاشَ الهَمُّ فِي الصَّدْرِ *Anxiety boiled in the bosom:* and in like manner, جَاشَتِ الغُصَّةُ فِي الصَّدْرِ † [*Choking wrath* or *rage boiled in the bosom*]. (T, TA.) — جَاشَتِ الحَرْبُ بَيْنَهُمْ (A, L) ‡ *War,* or *the war,* [*boiled,* or *raged,* or] *began to boil* [or *rage,*] *between them.* (L, TA.) — In the following words of a poet, cited by IAạr,

* قَامَتْ تَبَدَّى لَكَ فِي جَيْشَانِهَا *

† [*She arose, showing herself to thee*] *in her strength and youth,* [تَبَدَّى being for تَتَبَدَّى, جيشان, [the inf. n.,] meaning as rendered above, is with sukoon [to the ى] by poetic license. (ISd, TA.)

2. جَيَّشَ, [from جَيْشٌ,] *He collected,* or *assembled, armies,* or *military forces.* (Ṣ.) And جَيَّشَ جَيْشًا [*He collected,* or *assembled, an army,* or *a military force*]. (A.)

5. تَجَيَّشَتْ نَفْسُهُ: see 1. = تَجَيَّشُوا, from جَيْشٌ, *They became collected,* or *assembled, as an army,* or *a military force:* or *they formed themselves into an army, or a military force.*]

10. اسْتَجَاشَ, [from جَيْشٌ,] *He demanded,* or *summoned, armies,* or *military forces,* مِنْ مَحَلِّ كَذَا *from such a place.* (A.) And اسْتَجَاشَهُ *He demanded of him an army,* or *a military force.* (Ṣ.)

جَاشٌ: see جَأْشٌ, in art. جأش, in two places; and see الجَائِشَةُ, below.

جَيْشٌ *An army; a military force:* (A, Ḳ:) or *a body of men in war:* (TA:) or *men going to war* or *for some other purpose:* (T, Ḳ, TA:) pl. جُيُوشٌ. (Ṣ, A, Mṣb, Ḳ.)

جَيْشَةٌ *A single rising,* or *heaving,* or *the like:* pl. جَيْشَاتٌ: hence the phrase جَيْشَاتُ الأَبَاطِيلِ [app. meaning *The risings of false* or *vain things in the mind, or the like*]. (TA.)

مِرْجَلٌ جَيَّاشٌ [*A cooking-pot boiling,* or *boiling much*]. (A.) — فَرَسٌ جَيَّاشٌ † *A horse that rears and is excited when thou puttest him in motion with thy heel.* (Ḳ,* TA.)

الجَائِشَةُ *The soul;* syn. النَّفْسُ; (Ḳ;) [as also ▼ الجَاشُ, sometimes written الجَأْشُ, without ء;] mentioned by some in art. جأش. (TA.)

Quasi جيع

جَيَاعَى and جُيَّعٌ: see جَائِعٌ, in art. جوع.

جيف

1. جَافَتِ الجِيفَةُ, aor. تَجِيفُ; (Ḳ;) and ▼ جَيَّفَتْ, (Ṣ,* Ḳ,) inf. n. تَجْيِيفٌ; (Ṣ;) and ▼ اجْتَافَتْ, (Ḳ,) [and ▼ تَجَيَّفَتْ, Golius, as from the Ḳ, but not found by me in any copy thereof,] and ▼ انْجَافَتْ; (TA;) *The dead body stank,* or *became stinking.* (Ṣ,* Ḳ, TA.)

2. جَيَّفَ *He became a stinking dead body.* (Mgh, KL.) — See also 1.

5:
7: } see 1.
8:

جِيفَةٌ [*A carcass,* or *corpse,* i. e.] *a dead body* [of a beast or a man], *that has become stinking;* (Ṣ, Mgh, Ḳ;) or, as some say, *in a general sense;* [*whether stinking or not:*] (TA:) or, of beasts and cattle, *an animal that has died a natural death, or been killed otherwise than in the manner prescribed by the law, and has become stinking:* (Mṣb:) [and the *corpse of a man:* (see جَيَّافٌ:)] pl. [of mult.] جِيَفٌ (Ṣ, Mgh, Mṣb, Ḳ) and [of pauc.] أَجْيَافٌ. (Ṣ, Ḳ.) [Hence,] جِيفَةُ لَيْلٍ قُطْرُبُ نَهَارٍ *One who sleeps all the night, and labours all the day.* (TA from a trad. [See also art. قطرب.]) IDrd mentions this word in art. جوف, holding the ى to be originally و. (TA.)

جَيَّافٌ *A rifler,* or *ransacker, of graves;* (Ḳ, TA;) because he removes the [grave-] clothes from the corpses, and takes them; or, as some say, because of the stinking nature of his act. (TA.)

جيل

جِيلٌ *A nation, people, race, tribe,* or *family of mankind;* (Ṣ, Mṣb, Ḳ;) *such as the Turks, and the Greeks,* (Ṣ, TA,) *and the Chinese:* (TA:) pl. أَجْيَالٌ (M, Mṣb, TA) and جِيلَانٌ. (M, TA.) — And *A generation of men.* (TA.) — See also جُولٌ.

جَيَلٌ: see جَيْأَلٌ, in art. جأل.

جَيْلَانُ الحَصَى *Small pebbles which the wind makes to turn about* or *round about, to circle,* or *to revolve:* (Ṣ, Ḳ:) but this belongs to art. جول. (TA.) — يَوْمٌ جَيْلَانٌ: see أَجْوَلُ, in art. جول.

يَوْمٌ جَيْلَانِيٌّ: see أَجْوَلُ, in art. جول.

جيم

2. جَيَّمَ جِيمًا *He wrote a* ج. (Ḳ.)

جِيمٌ The *letter* ج: masc. and fem.: (T, Ḳ,* TA:) pl. أَجْيَامٌ and جِيمَاتٌ. (TA.)

Quasi جيه

جِيهٌ: see جُوهٌ, in art. جوه.

ح

The sixth letter of the alphabet: called حَآءٌ [and حَا (respecting which latter see the letter ب). It is one of the letters termed مَهْمُوسَة, or non-vocal, i. e. pronounced with the breath only, without the voice; and of those termed حَلْقِيَّة i. e. faucial, or guttural, for] the place of its utterance is in the fauces; and were it not for a hoarse aspiration with which it is pronounced, it would resemble ع: next after it [with respect to the place of utterance] is ه: [ع having the lowest place of utterance; then ح; and then ه:] and ح and ه are never consociated in any uncompounded word of the which the letters are all radicals, because of the mutual nearness of their places of utterance: they occur together in حَيْهَل; but this is only a compound word in the classical language; and as the name of a certain kind of tree it is a post-classical word. (Kh, L.) ═ [It is often put for حِينَئِذٍ. ═ As a numeral, it denotes *Eight.*]

حا

حَا and حَآءٌ: see the letter ح, and see arts. حوأ and حى.

حب

1. حَبَّ [signifying *He, or it, was, or became, loved, beloved, an object of love, affected, liked, or approved,* is originally حَبُبَ or حَبِبَ]. You say, حَبَّ إِلَىَّ هٰذَا الشَّىْءُ, [aor. ـِ or ـُ,] inf. n. حُبٌّ, *This thing was, or became, an object of love to me.* (Ḳ. [The meaning is there indicated, but not expressed. In the CḲ, الشَّىْءَ is erroneously put for الشَّىْءُ.]) And حَبِبْتُ إِلَيْهِ *I became loved, beloved, or an object of love, to him:* [said to be] the only instance of its kind except شَرُرْتُ and لَبُبْتُ. (Ḳ.) And مَا كُنْتَ حَبِيبًا وَلَقَدْ حَبِبْتَ, with kesr, *Thou wast not loved, and thou hast become loved.* (Ṣ.) See also 5. — حَبَّ, formed from حَبُبَ, by making the former ب quiescent and incorporating it into the latter, is also a verb of praise [signifying *Beloved, lovely, pleasing, charming, or excellent, is* he, or it]; (TA;) and so حُبَّ, [which is more common,] formed from the same, by incorporating the former ب into the latter after transferring the ḍammeh of the former to the ح. (ISk, Ṣ, TA.) A poet says,

* وَزَادَهُ كَلَفًا فِى الحُبِّ أَنْ مَنَعَتْ *
* وَحَبَّ شَيْئًا اِلَى الاِنْسَانِ مَا مُنِعَا *

[*And her denying increased his devotion in love: for lovely, as a thing, to man, is that which is denied*]. (TA.) And Sá'ideh says,

* هَجَرَتْ غَضُوبُ وَحُبَّ مَنْ يَتَجَنَّبُ *
* وَعَدَتْ عَوَادٍ دُونَ وَلْيِكَ تَشْعَبُ *

[*Ghadoob hath forsaken thee, (and lovely is the person who withdraweth far away,) and obstacles in the way of thy drawing near have occurred to separate thee and her*]. (Ṣ, TA.) [See also حَبَّذَا, below.] — [Both are also verbs of wonder.] You say, حَبَّ بِفُلَانٍ, (Aṣ, Ṣ, and so in copies of the Ḳ,) and حُبَّ, (I'Aḳ p. 236, [where both forms are mentioned as correct,] and so in the CḲ,) *How beloved, or lovely, &c., is such a one* (Aṣ, Ṣ, Ḳ) to me! (Aṣ, Ṣ.) [See also 4.] A'Obeyd and Fr read this حَبَّ, saying that it means حَبُبَ بفلان, and that the former ب is rendered quiescent by the suppression of its ḍammeh, and incorporated into the latter. (Ṣ,* TA.) ═ See also 4, in two places. ═ Also حَبَّ, [aor., accord. to analogy, ـِ,] *He stood still, stopped, or paused.* (Ḳ.) ═ And حُبَّ, with ḍamm, *He was fatigued, or tired.* (Ḳ.)

2. حَبَّبَهُ إِلَىَّ *He, or it, [rendered him, or it, an object of love, lovely, or pleasant, to me;] made me to love, affect, like, approve, or take pleasure in, him, or it.* (Ḳ.) You say, حَبَّبَهُ إِلَىَّ إِحْسَانُهُ [*His beneficence made him an object of love to me*]. (A, TA.) And حَبَّبَ ٱللّٰهُ إِلَيْهِ الإِيمَانَ [*God made faith lovely to him*]. (A, TA.) And حُبِّبَ إِلَىَّ بِأَنْ تَزُورَنِى [*Thy visiting me hath been made pleasant to me*]. (A, TA.) ═ حَبَّبَ الدَّوَآءَ [*He formed the medicine into pills, or little clots or balls: see its quasi-pass., 5*]. (Ḳ in حثر, &c.) ═ And حَبَّبَ *He filled* a water-skin &c. (AA, TA.) ═ See also 5.

3. مُحَابَّةٌ, (Ṣ,) or مُحَابَبَةٌ, (Ḳ,) and حِبَابٌ signify the same [as inf. ns. of حَابَّ]. (Ṣ, Ḳ.) [You say, حَابَّ بَعْضُهُمْ بَعْضًا *They loved, affected, liked, approved, or took pleasure in, one another.*] And حَابَّهُ *He acted, or behaved, in a loving, or friendly, manner with him, or to him.* (A, TA.) — See also 4.

4. أَحَبَّهُ, (Ṣ, A, Mṣb, Ḳ,) inf. n. إِحْبَابٌ; (KL;) and ↓ حَبَّهُ, (Ṣ,) first pers. حَبَبْتُهُ, (Mṣb, Ḳ,) aor. ـِ, which is anomalous, (Ṣ, Mṣb, Ḳ,) the regular aor. being ـُ, which is unused, (Mṣb,) [said to be] the only instance of a trans. verb whose second and third radical letters are the same having the measure يَفْعِلُ as that of its aor. without having also the measure يَفْعُلُ, (Ṣ,) and therefore by some disapproved, as not chaste, and disallowed by Az, though he allows the pass. form حُبَّ, (TA,) inf. n. حُبٌّ, (Ḳ,) or this is a simple subst., (Mṣb,) and حِبٌّ; (Ḳ;) and [↓ حَبِبَهُ] first pers. حَبِبْتُهُ, aor. ـَ; and ↓ حَابَّهُ, inf. n. حِبَابٌ, of the dial. of Hudheyl; (Mṣb;) and ↓ اسْتَحَبَّهُ; (Mṣb, Ḳ;) signify the same; (Ṣ, Mṣb, Ḳ;) *He loved, affected, liked, approved, or took pleasure in, him, or it:* (A, Ḳ, and KL in explanation of the first and last:) *he held him, or esteemed him, as a friend:* (KL in explanation of the first and last:) or اِسْتِحْبَابٌ signifies the *esteeming* [a person or thing] *good:* (Ṣ:) and the *preferring, or choosing,* [a person or thing,] as also إِحْبَابٌ: (KL:) and اسْتَحَبَّهُ عَلَى غَيْرِهِ means *he loved, or esteemed, him, or it, above another, or others; preferred him, or it, to another, or others.* (Ḳ, A,* TA.) مَا أَحْبَبْتُ ذٰلِكَ, in the dial. of the tribe of Suleym, is for مَا أَحْبَبْتُهُ [*I loved not, or liked not, that*]; like ظَنْتُ for ظَنَنْتُ, and ظَلْتُ and ظِلْتُ for ظَلِلْتُ. (Lḥ, TA.) [أُحِبُّ أَنْ يَكُونَ كَذَا may be rendered *I would that it were thus, or that such a thing were.*] It is said of Ohod, in a trad., هُوَ جَبَلٌ يُحِبُّنَا وَنُحِبُّهُ, meaning *It is a mountain whose inhabitants love us, and whose inhabitants we love:* or it may mean we love the mountain itself, because it is in the land of people whom we love. (IAth, TA.) And one says فِى سَاعَةٍ يَحِبُّهَا الطَّعَامُ for يُحَبُّ فِيهَا [*In an hour, or a time, in which food is loved, or liked*]. (TA.) — مَا أَحَبَّهُ إِلَىَّ i. q. حَبَّ بِهِ [*How beloved, lovely, pleasing, charming, or excellent, is he, or it, to me!*]; (Aṣ, Ṣ, Ḳ,* TA;) and so أَحْبِبْ إِلَىَّ بِهِ. (A, TA.) [De Sacy, in his Gram. Ar., sec. ed., ii. 221, mentions the saying, مَا أَحَبَّ الْمُؤْمِنَ لِلّٰهِ وَمَا أَحَبَّهُ إِلَى ٱللّٰهِ, as meaning *How greatly does the believer love God! and how great an object of love is he to God!*] ═ أَحَبَّ, (Ṣ, Ḳ,) inf. n. as above, (Ṣ,) also signifies *He* (a camel) *kneeled and lay down, and would not spring up:* (Ḳ:) or *was restive:* or *kneeled and lay down:* (Ṣ:) or *was afflicted by a fracture, or disease, and would not move from his place until cured,* or *remained there until he died:* (AZ, Ṣ, Ḳ:) or *became jaded:* (TA: [agreeably with this last explanation the act. part. n. is rendered in the Ṣ and Ḳ on the authority of Th:]) or *was at the point of death,*

by reason of violent disease, and therefore kneeled and lay down, and could not be roused. (AHeyth, TA.) Accord. to AO, أَحْبَبْتُ حُبَّ الخَيْرِ عَنْ ذِكْرِ رَبِّى [in the Ḳur xxxviii. 31] means *I have stuck to the ground, on account of my love of the horses,* [lit., *of good things,*] *and so been diverted from prayer,* until the time of prayer has passed: (TA:) by الخير is meant الخَيْل. (Jel.) ═ Also *He became in a state of recovery from his disease.* (Ḳ.) ═ And *It* (seed-produce) *had, bore,* or *produced, grain.* (Ṣ, Ḳ.)

5. تحبّب *He manifested,* or *showed, love,* or *affection,* (Ṣ, Ḳ,) إِلَيْهِ *to him.* (Ṣ.) تحبّب and ▼حَبَّبَ are both *syn. with* تودّد. (TA.) — [Also, app., *He became,* or *made himself, an object of love* or *affection* to him: see مُحَبَّب, said to be syn. with مُتَحَبِّب.] ═ *He became swollen,* or *inflated, like a jar* (حُبّ), from drinking. (A, TA.) — *It* (a water-skin &c.) *became full.* (AA, TA.) — *He began to be satiated with drink.* (Ḳ.) — *He* (an ass &c.) *became filled with water:* (Ṣ:) and ▼حَبَّبَ also is used in this sense, but ISd doubts its correctness: (TA:) one says, شَرِبَتِ الإِبِلُ حَتَّى حَبَّبَتْ *The camels drank until they were satiated.* (Ṣ.) ═ تحبّب الجَلِيدُ كَاللُّؤْلُؤِ الصِّغَارِ [*The hoar-frost formed into grains like small pearls*]. (TA in art. صاب.) — تحبّب الرِّيقُ عَلَى الأَسْنَانِ [*The saliva formed,* or *collected, in little bubbles upon the teeth*]. (Az, TA.) — تَقَطَّعَ اللَّبَنُ وَتَحَبَّبَ [*The milk became decomposed, and formed little clots of curd*]. (Ṣ in art. بحثر.) — تحبّب الزُّبْدُ [*The butter formed into little clots,* when first appearing in the milk or cream]. (Ṣ and Ḳ in art. ثمر.) The verb is also used in like manner in relation to honey, (Ḳ in art. حثر,) and دِبْس, (Ṣ in that art.,) and medicine. (TA in that art. [See also 2.]) — تحبّب الجِلْدُ [*The skin broke out with pimples,* or *small pustules:* so in the language of the present day: see حَبٌّ.] (TA in art. حثر.)

6. تحابّوا *They loved,* or *affected,* or *liked, one another.* (Ṣ, A,* Ḳ.*)

10: see 4. ═ اِسْتَحَبَّتْ كَرِشُ المَالِ *The stomachs of the cattle,* or *camels &c., retained the water* [*that they had drunk*], *and the time between the two waterings thereof became long,* or *became lengthened.* (Ḳ.) This is at the conjunction of [the periods of] الجَبْهَة and الطَّرْف [the ninth and tenth of the Mansions of the Moon, which, in central Arabia, about the commencement of the era of the Flight, took place on the 12th of August, O. S., (see مَنَازِلُ القَمَرِ, in art. نزل,)] when Canopus rises with them. (TA. [الصَّرْفَة is there put for الطَّرْف; but evidently by a mistake of a copyist. There is also another mistake, though a small one, in the foregoing passage: for Canopus rises, in central Arabia, after الطرف, and before الجبهة; and rose aurorally, in that latitude, about the commencement of the era of the Flight, on the 4th of August, O. S.])

حَبْ حَبْ and حَبِ حَبِ *A cry by which a he-camel is chidden, to urge him on.* (TA voce حَوْبِ, q. v.)

حَبٌّ, (Ṣ, Mṣb, Ḳ,) a [coll.] gen. n., (Mṣb,) n. un. حَبَّةٌ; (Ṣ, Mṣb, Ḳ;) *Grain* of wheat, barley, lentils, rice, &c.: (Az, TA:) accord. to Ks, only of wheat and barley: (TA:) or *wheat &c. while in the ears* or *other envelopes:* (Mṣb:) [but applied also to various other *seeds;* among which, to beans, (as in the Mgh in art. بقل,) and peas and the like; and *kernels;* and] the *stones* of grapes, dates, pomegranates, and the like: (Mgh voce عَجَمٌ:) by some it is applied even [to *berries;* as, for instance,] to *grapes:* you say حَبَّةٌ مِنْ عِنَبٍ, as well as مِنَ البُرِّ, and مِنَ الشَّعِيرِ, and the like: (TA:) [and hence, to *beads:* (see حَبٌّ:)] the pl. (of حَبٌّ, Mṣb) is حُبُوبٌ (Ṣ, Mṣb, Ḳ) and حُبَّانٌ, like تُمْرَانٌ, (Ḳ,) pl. of تَمْرٌ; (TA;) and (of حَبَّةٌ, Mṣb) حَبَّاتٌ (Mṣb, Ḳ) and حِبَابٌ, [or this is pl. of حَبٌّ also,] like كِلَابٌ as pl. of كَلْبَةٌ [and of كَلْبٌ]: (Mṣb:) and حَبٌّ is also called [by lexicologists, but not by grammarians,] a pl. of حَبَّةٌ. (TA.) — [Hence,] *Seed-produce,* whether small or large. (TA.) — And الحَبَّةُ الخَضْرَآءُ (Ṣ, Ḳ) *i. q.* البُطْمُ [*The fruit of the terebinth-tree,* or *pistacia terebinthus* of Linn. (Delile, Flor. Æg. no. 936.)] (Ḳ.) — And الحَبَّةُ السَّوْدَآءُ (Ṣ, Ḳ) *i. q.* الشُّونِيزُ [*The black aromatic seed of a species of nigella*]. (Ḳ.) [But see art. سود. And for other similar terms, see the latter word of each.] — And حَبُّ المُزْنِ and حَبُّ الغَمَامِ and حَبُّ قُرٍّ *Hail.* (Ṣ. [See a metaphorical usage of the first of these in a verse cited voce أَنَّ.]) — [Hence likewise,] حَبٌّ also signifies *Pimples,* or *small pustules:* [so in the present day: and any *similar small extuberances:* a coll. gen. n.: n. un. with ة.] (Ṣ and Ḳ* in art. حثر.)

حُبٌّ *Love; affection;* syn. وُدٌّ, (A,) or وِدَادٌ; (Ḳ;) *inclination of the nature,* or *natural disposition, towards a thing that pleases,* or *delights;* (Kull p. 165;) *contr. of* بُغْضٌ: (Mgh, TA:) حُبٌّ and ▼حِبٌّ and ▼حُبَّةٌ [this being said in the Ṣ to be syn. with حُبٌّ and in the Ḳ to be syn. with مَحَبَّةٌ, and it is used as an inf. n. in an ex. cited voce دَاحَ in art. دوح,] and ▼حُبَابٌ (Ṣ, Ḳ) and ▼حِبَابٌ (Ḳ) and ▼مَحَبَّةٌ (Ṣ) signify the same; (Ṣ, Ḳ;) i. e., as above. (Ḳ.) The degrees of حُبّ are as follow: first, هَوًى, the "inclining of the soul, or mind;" also applied to the "object of love itself:" then, عَلَاقَةٌ, "love cleaving to the heart;" so termed because of the heart's cleaving to the object of love: then, كَلَفٌ, "violent, or intense, love;" from كُلْفَةٌ signifying "difficulty, or distress, or affliction:" then, عِشْقٌ, ["amorous desire;" or "passionate love;"] in the Ṣ, "excess of love;" and in the language of the physicians, "a kind of melancholy:" then, شَغَفٌ, "ardour of love, accompanied by a sensation of pleasure;" like لَوْعَةٌ and لَاعِجٌ; the former of which is "ardour of love;" and the latter, "ardent love:" then, جَوًى, "inward love;" and "violence of amorous desire," or "of grief, or sorrow:" then, تَتَيُّمٌ, "a state of enslavement by love:" then تَبْلٌ, "lovesickness:" then, وَلَهٌ, "distraction, or loss of reason, in love:" and then, هُيَامٌ "a state of wandering about at random in consequence of overpowering love." (Kull ubi suprà.) [Accord. to the Mṣb, it is a simple subst.: but accord. to the Ḳ, an inf. n.; and hence,] حُبًّا لِمَا أَحْبَبْتُمْ, meaning أُحِبُّ حُبًّا [*I love with loving,* i. e. *much, what ye have loved*]. (Ḥar p. 186.) Hence the phrase, نَعَمْ وَحُبَّةً وَكَرَامَةً [*Yea; and with love and honour* will I do what thou requirest: or *for the sake of* the *love and honour* that I bear thee: or حُبَّة may be here used for حُبًّا to assimilate it in termination to كرامة: see what follows]. (Ṣ, TA.) Hence also the saying of Abu-l-'Aṭà Es-Sindee,

* فَوَاللّٰهِ مَا أَدْرِى وَإِنِّى لَصَادِقٌ *
* أَدَآءٌ عَرَانِى مِنْ حُبَابِكِ ▼ أَمْ سِحْرُ *

[*And by God, I know not* (*and indeed I am speaking truth*) *whether disease have befallen me in consequence of love of thee, or enchantment*]: (Ṣ, TA:) but IB says that the reading best known is ▼ مِنْ حِبَابِكِ; and that حباب, here, may be an inf. n. of حَابَبْتُهُ; or it may be pl. of حُبٌّ, like as عِشَاشٌ is of عُشٌّ; (TA;) or it may be an inf. n. of حَبَبْتُهُ: some also read ▼مِنْ حَبَابِكِ, with fet-ḥ to the ح, said to mean *on account of the love of thee, and of the main amount thereof:* (Ḥam p. 26:) and some read مِنْ جَنَابِكِ "from thy part" [or "from thee"]. (TA.) — See also حَبِيبٌ. ═ Also a Persian word, arabicized, (AḤát, Ṣ, Mṣb,) from خُنْب, (AḤát, TA,) [or خُبْ or خُنْبْ,] *i. q.* خَابِيَةٌ, (Ṣ, Mṣb,) *A jar,* (Ḳ, MF,) *whether large or small, used for preparing wine:* (MF:) or *a large jar:* (Ḳ:) or *one for water:* (IDrd, TA:) or the *four pieces of wood upon which is placed a two-handled,* or *two-eared, jar:* (Ḳ, TA: [in the CḲ, by a misplacement of words, this last signification is assigned to حُبَاب:]) pl. [of pauc.] أَحْبَابٌ (Ḳ) and [of mult.] حِبَابٌ and حِبَبَةٌ. (Ṣ, Mṣb, Ḳ.) From this last signification is [said to be] derived the phrase حُبًّا وَكَرَامَةً, [pronounced حُبًّا وَكَرَامَةً,] lit. *A jar-stand and a cover* will I give thee, or the like], كرامة signifying the "cover" of a jar, (Ḳ, TA,) whether of wood or of baked clay. (TA.) [If this be the true derivation, the phrase may have originated from a person's asking of another the loan or gift of a jar, and the latter's replying "Yea; and I will give thee a jar-stand and a cover;" meaning "I will do what thou requirest, and more:" but this phrase is now, and perhaps was in early times, generally used, agreeably with the more common significations of the two words, in the sense assigned above to the phrase حُبَّةً وَكَرَامَةً.]

حِبٌّ: see حُبٌّ: — and حَبِيبٌ, in four places: ═ and حُبَّةٌ. ═ Also, and ▼حِبَابٌ, [but the latter is doubted by the author of the TA, and thought to be perhaps syn. with حُبٌّ in the sense of مُحِبٌّ, and in the L it is said to be syn. with حِبٌّ, but in what sense is not explained,] *An ear-ring* [*formed*] *of one* حَبَّة [or *bead*]. (Ḳ.)

حَبَّةٌ n. un. of حَبٌّ [q. v.]. (Ṣ, Mṣb, Ḳ.) [Hence,] جَابِرُ بْنُ حَبَّةَ a name of †*Bread*. (ISk, Ṣ.) — See also حِبَّةٌ, in two places. — [*A grain*; meaning the *weight of a grain of barley*;] *a well-known weight*. (Ḳ.) — *A* [*small*] *piece*, or *portion*, *of a thing*. (Ṣ, Ḳ.) — حَبَّةُ القَلْبِ *The heart's core*; (AA, TA;) *the black*, or *inner, part of the heart*; or i. q. ثَمَرَتُهُ; (Ṣ, A, Ḳ;) which is that [same thing]: (Ṣ:) or *a black thing in the heart*: (Ḳ:) or *the black clot of blood that is within the heart*: (T, TA:) or *the heart's blood*. (Ḳ.) You say, أَصَابَتْ فُلَانَةُ حَبَّةَ قَلْبِهِ [*Such a woman smote his heart's core*]. (A, TA.) ═ *A want*: or *an object of want*; *a needful*, or *requisite, thing*: syn. حَاجَةٌ. (Ḳ.)

حُبَّةٌ: see حُبٌّ, in two places: — and حَبِيبٌ. — [It is also used in a pl. sense.] You say, هُوَ مِنْ حُبَّةِ نَفْسِى [*He is of the beloved of my soul*]. (TA voce حُمَّةٌ.) — And حُبَّتُكَ also signifies *What thou lovest to receive as a gift*, or *to have*. (Ḳ.) You say, اِخْتَرْ حُبَّتَكَ *Choose thou what*, or *whom, thou lovest*; as also ▼ مَحَبَّتَكَ. (TA.) ═ Also *A grape-stone*: sometimes without teshdeed; (Ḳ;) i. e. حُبَةٌ. (TA.)

حِبَّةٌ, a pl., [or rather quasi-pl. n.,] The *seeds of desert-plants that are not used as food*; pl. حِبَبٌ: (Ṣ:) or *seeds of herbs*, or *leguminous plants*, (بُقُول,) and *of odoriferous plants*: (Ḳ:) or *of the latter only*; (Ks, Az, TA;) and one of such seeds is called ▼ حَبَّةٌ; (Az, TA;) or حِبَّةٌ; the coll. n. being ▼ حِبٌّ: (Mṣb:) or *different seeds of every kind*: or the *seeds of the herbage called* عُشْب: or *all seeds of plants*: sing. the same, and ▼ حَبَّةٌ: or this signifies *everything that is sown*: and حِبَّةٌ, the *seed of everything that grows spontaneously, without being sown*: or *a small plant growing among the kind of herbage called* حَشِيش: (Ḳ:) and *dry herbage, broken in pieces, and heaped together*: (Aboo-Ziyád, Ḳ:) or *dry herbs* or *leguminous plants*: (Ḳ:) or the *seeds of wild herbs* or *leguminous plants*, and *of those of the kind called* عُشْب, *and their leaves, that are scattered and mixed therewith*; such as the قُلْقُلَان and بَسْبَاس and ذُرَق and نَفَل and مُلَّاح, and all kinds of those herbs or leguminous plants that are eaten crude, and those that are thick, or gross, and bitterish: upon these seeds and leaves, cattle, or camels &c., pasture and fatten in the end of [the season called] the صَيْف. (T, TA.)

حَبَبٌ: see حَبَابٌ. — Also, (Ṣ, Ḳ,) and ▼ حِبَبٌ, (Ḳ,) *A beautiful arrangement of the teeth in regular rows*. (Ṣ, Ḳ.) — And *Streaks of saliva on the teeth*. (TA.) — And (both accord. to the Ḳ, but the latter only accord. to the TA,) The *saliva that flows over the teeth, or collects in the mouth, in little bubbles*. (T, Ḳ, TA.)

حِبَبٌ: see حَبَابٌ: — and حَبَبٌ.

حَبَابٌ: see حُبٌّ. — حَبَابُكَ *Thine utmost*: (Mṣb:) or *the utmost of thy power*: (Ṣ:) or *the utmost of thy love*: or, *of thine endeavour* (جَهْدُكَ [like غَنَامَاكَ and قُصَارَاكَ and حُمَادَاكَ and جُهَادَاكَ and نُعَامَاكَ]). (Ḳ. [In the CḲ جُهْدُكَ.]) You say, حَبَابُكَ كَذَا, (Ḳ,) and حبابك أَنْ تَفْعَلَ ذٰلِكَ, (Ṣ, Mṣb,* TA,) and حبابك أَنْ يَكُونَ ذٰلِكَ, (TA,) *Thine utmost*, (Mṣb,) or *the utmost of thy power*, (Ṣ,) or *of thy love*, or *of thine endeavour*, (Ḳ,) *will be such a thing*, (Ḳ,) and *thy doing that*, (Ṣ, Mṣb,* TA,) and *that event's taking place*. (TA.) ═ Also, and ▼ حَبَبٌ and ▼ حِبَبٌ, The *main body*, the *mass*, or *bulk*, or *greater part* or *portion*, of water, (Ṣ, Ḳ,) and of sand, (Ḳ,) and of [the beverage called] نَبِيذ: but it is said that the third word applies particularly to water: (TA:) or the first signifies the *streaks*, or *lines*, of water, (Aṣ, Ḳ, TA,) *resembling variegated work*: (Aṣ, TA:) or the *waves* of water *that follow one another*: (TA:) or the *bubbles* (Ṣ, A, Ḳ) of water, (Ṣ, Ḳ,) or of wine, (A, TA,) *that float upon the surface*; (Ṣ, A, Ḳ;) as also the second (AḤn, A) and the third: (AḤn, TA:) [it is a coll. gen. n., in this sense, of which the n. un. is with ة:] accord. to IDrd, حَبَبُ المَآءِ and حَبَابُ المَآءِ signify تَكَسُّرُهُ [app. meaning the *ripple*, or *broken surface, of water*, such as is seen when it is slightly fretted by wind, and when it flows over uneven ground]. (TA.) طِرْتَ بِعُبَابِهَا وَفُزْتَ بِحَبَابِهَا, in a trad. of 'Alee, relating to Aboo-Bekr, is explained as meaning *Thou hast outrun others, and attained to the place where the flood of El-Islám collects, and reached the first* [*springs*] *thereof, and drunk the purest of it, and become possessor of its excellencies*: [this is the only explanation of it that I have found:] but it is also otherwise explained. (IIr and others, TA in art. عب.) — حَبَابٌ also signifies ‡ *Dew-drops*; (A;) the *dew* (IAth, Ḳ) *that is on trees &c. in the evening*. (IAth, TA.) It is said in a trad., of the inhabitants of Paradise, that their food shall turn into a sweat like حَبَابُ المِسْكِ, by which is meant *Musky dew*: or, perhaps, *musky bubbles*. (IAth, TA.)

حُبَابٌ: see حُبٌّ, in two places: — and حَبِيبٌ. ═ Also The *serpent*: (Ṣ, IAth, Ḳ:) or *a serpent not of a malignant species*: (TA:) and the name of *a devil*, (Ṣ, Ḳ,) accord. to some; (Ṣ;) but said to be so only because a serpent is called شَيْطَان. (A'Obeyd, Ṣ, TA.) — And a pl. [or rather coll. gen. n.] of which the sing. [or n. un.] is حُبَابَةٌ [accord. to the CḲ حَبَابَةٌ], meaning *A certain black aquatic insect* or *small animal*. (Ḳ.) ═ أُمُّ حُبَابٍ ‡ *The present world*; (Ḳ, TA;) metonymically used in this sense. (TA.)

حِبَابٌ: see حُبٌّ, in two places: ═ and حِبٌّ.

حَبِيبٌ A person *loved, beloved, affected, liked*, or *approved*; (Ṣ,* A, Mṣb,* Ḳ;) as also ▼ مَحْبُوبٌ and ▼ مُحَبٌّ, (Ṣ, Mṣb, Ḳ,) of which two the former is generally used for the latter, (Ṣ, Ḳ, TA,) in like manner as are used مَزْكُومٌ and مَحْزُونٌ and مَجْنُونٌ and مَكْزُوزٌ and مَقْرُورٌ, each of which has its proper verb of the measure فُعِلَ, (TA,) and ▼ حِبٌّ (Ṣ, Ḳ) and ▼ حُبَابٌ and ▼ حُبَّةٌ, which last is also applied to a female, and has for its pl. حُبَبٌ: (Ḳ:) the fem. of حَبِيبٌ is with ة; (Mṣb, Ḳ;) and so is that of ▼ مَحْبُوبٌ, (Ḳ, TA,) [and that of ▼ مُحَبٌّ,] and that of ▼ حِبٌّ: (TA:) the pl. of حَبِيبٌ is أَحِبَّآءُ, instead of حُبَبَآءُ, which would be the reg. pl. but for the repetition of the ب; and the pl. of حَبِيبَةٌ is حَبَائِبُ: (Mṣb:) the pl. of ▼ حِبٌّ is أَحْبَابٌ [a pl. of pauc.] and حِبَّانٌ (Ḳ) and حُبُوبٌ (MF) and حِبَبَةٌ and حُبَّانٌ and ▼ حُبٌّ, which last is rare (عَزِيزٌ) [as a pl.], or is a quasi-pl. n. (Ḳ.) Though ▼ مُحَبٌّ is uncommon, it occurs in the following verse of 'Antarah:

* وَلَقَدْ نَزَلْتِ فَلَا تَظُنِّى غَيْرَهُ *
* مِنِّى بِمَنْزِلَةِ المُحَبِّ المُكْرَمِ *

[*And thou hast taken (and imagine not otherwise), in respect of me*, i. e. of my heart, *the place of the beloved, the honoured*; or *become in the condition of the beloved, &c.*]. (T, TA.) — Also, (IAar, KL, TA,) and ▼ حِبٌّ, (Ḳ, KL,) A person *loving*; *a lover*; *a friend*; (KL;) i. q. ▼ مُحِبٌّ: (IAar, Ḳ, TA:) [fem. of each with ة:] the pl. of the first (i. e. حبيب) is أَحْبَابٌ (TA) [and أَحِبَّةٌ and أَحِبَّآءُ, mentioned by Golius as from the Ṣ, but not in my copies of the Ṣ: both, however, are correct: the former, the more common: the latter, a pl. of pauc.]. You say اِمْرَأَةٌ ▼ مُحِبٌّ and ▼ مُحِبَّةٌ لِزَوْجِهَا [*A woman loving to her husband*]. (Fr, Ṣ, Ḳ.*) — أَبُو حَبِيبٍ *The kid*. (Ḥar p. 227.) — الحَبِيبَةُ: see مُحَبٌّ.

حُبَاحِبٌ, (Ḳ,) or أَبُو حُبَاحِبٍ, (Ṣ,) [*A kind of fire-fly*;] *a fly that flies in the night*, (Ḳ,) *resembling fire*, (Ṣ,) *emitting rays like a lamp*: (Ḳ:) AḤn says that حباحب and ابو حباحب were both unknown to him, and that nothing respecting them had been heard by him from the Arabs; but that some people asserted the insect thus called to be the يَرَاع, *a moth that, when it flies by night, no person not knowing it would doubt to be a spark of fire*: Aboo-Ṭálib says, as on the authority of Arabs of the desert, that حباحب is the name of *a flying thing longer than the common fly, and slender, that flies between sunset and nightfall, resembling a spark of fire*: (TA:) or, accord. to Aṣ, it is *a flying thing, like the common fly, with a wing that becomes red; when it flies appearing at a distance like a lighted piece of fire-wood*. (Ḥar p. 500.) نَارُ الحُبَاحِبِ (Ṣ, Ḳ) and نَارُ أَبِى حُبَاحِبٍ and simply الحُبَاحِبُ (Ṣ) mean *The fire of the fly above mentioned*: or *of El-Ḥobáḥib* or *Aboo-Ḥobáḥib*: (TA:) [for] El-Ḥobáḥib, (Ṣ,) or Aboo-Ḥobáḥib, (Ḳ,) is said to have been a niggardly man, who never lighted any but a faint fire, fearing to attract guests, so that his fire became proverbial. (Ṣ, Ḳ.) El-Kumeyt says, describing swords,

* يَرَى الرَّاؤُونَ بِالشَّفَرَاتِ مِنْهَا *
* كَنَارِ أَبِى حُبَاحِبَ وَالظُّبِينَا *

[*The beholders see, in the sides of the blades thereof, and the extremities, the semblance of the fire of the fire-fly*]: (Ṣ:) here the poet has made حباحب imperfectly decl., regarding it as a fem. [proper] name [of the fly above mentioned]. (TA.) Or نار الحباحب (Ṣ, Ḳ) and simply الحباحب (Ṣ) signify *The fire that is struck by a horse's hoofs:* (Fr, Ṣ:) or *the sparks of fire that are made to fly forth in the air by the collision of stones:* or *the sparks that fall from the pieces of wood that are used for producing fire* [by means of *friction*]: (Ḳ:) or they are derived from حَبْحَبَةٌ, (IAạr, Ḳ,) signifying "weakness," (IAạr, TA,) [and their meaning is *faint fire.*] — أُمُّ حُبَاحِبٍ *A flying insect resembling the* [*species of locust called*] جُنْدُب, (Ḳ,* TA,) *spotted with yellow and green:* when people see it, they say, بُرْدِى يَا حُبَاحِبُ [*Spread forth thy wings* (بُرُودَكَ), *O ḥobáḥib*]; whereupon it spreads its two wings, which are adorned with red and yellow. (TA.)

حَبَّذَا, meaning حَبِيبٌ, as in the phrase حَبَّذَا الأَمْرُ [*Loved, beloved, affected, loved,* or *approved, is the thing,* or *affair;* or *lovely, charming,* or *excellent, is it*]; (Ḳ;) and in حَبَّذَا زَيْدٌ [*Loved, beloved, &c., is Zeyd*]; (Ṣ;) is composed of حَبَّ, (Sb, Fr, Ṣ, Ḳ,) a verb of praise, in the pret. form, invariable, originally حَبُبَ, (Fr, Ṣ,) and ذَا, (Sb, Fr, Ṣ, Ḳ,) its agent, (Ṣ,) which together constitute it a single word, (Sb, Ṣ, Ḳ,) a noun, (Sb, Ḳ,) or occupying the place of a noun, (Ṣ,) governing the noun [particularized by praise] that follows it in the nom. case; (Sb, Ṣ, Ḳ;) the place that it occupies in construction making it virtually in the nom. case as an inchoative, and the noun that follows it being its enunciative: (Ṣ:) [but see what follows.] It is used in the same manner as a prov.; (Sb, Ḳ;) [i. e., it is not altered to agree in number or gender with the noun particularized by praise, which follows it;] remaining the same when used in the dual and pl. and fem. sense; so that one says, حبذا زيد and الزَّيْدَانِ and الزَّيْدُونَ and هِنْدٌ and أَنْتِ and أَنْتُمَا and أَنْتُمْ [&c.]; (Ibn-Keysán, TA;) and حبذا امرأةٌ, not حَبَّذِهِ المَرْأَةُ; (Sb, Ṣ, Ḳ;*) which shows that the noun that follows it may not be regarded as a substitute for ذا: (Ṣ:) [but see what follows.] It is allowable, but bad, to say, زَيْدٌ حَبَّذَا. (TA.) [There are, however, various opinions respecting حبذا and the noun that follows it.] Some hold that حبذا is a noun, composed of حَبَّ and ذَا, and is an inchoative, and that the noun particularized by praise is its enunciative; or that the former is an enunciative, and the latter an inchoative, reversing the usual order: others hold that حَبَّ is a verb in the pret. form; and ذا, its agent; and that the noun particularized by praise may be an inchoative, of which حبذا is the enunciative; or it may be an enunciative of which the inchoative is suppressed, so that حبذا زيد is for حبذا هُوَ زَيْدٌ [*Loved,* or *beloved, &c., is this person: he is Zeyd*], or حبذا المَمْدُوحُ زَيْدٌ [*loved, &c., is this person: the person praised is Zeyd*]: others hold that حبّذا is a pret. verb, composed of حَبَّ and ذَا, and that the noun following it is its agent; but this is the weakest of opinions: one also says, in dispraise, لَاحَبَّذَا زَيْدٌ. (I'Aḳ p. 235.)

حَابٌّ An arrow *that falls* [*in the space*] *around the butt:* pl. حَوَابُّ. (Ḳ.)

أَحَبُّ [*More,* and *most, loved, beloved,* &c. You say, هٰذَا أَحَبُّ إِلَيَّ مِنْ ذَاكَ *This is more an object of love, affection, liking,* or *approval,* or *is more lovely, charming,* or *pleasing, to me than that.* And هُوَ أَحَبُّهُمْ إِلَيَّ *He is the most beloved of them to me.*]

مُحِبٌّ: see حَبِيبٌ, in three places. — المُحِبَّةُ and ▼المَحْبُوبَةُ and ▼المُحَبَّبَةُ and ▼الحَبِيبَةُ are epithets of *El-Medeeneh.* (Ḳ.)

مُحَبٌّ, and its fem.: see حَبِيبٌ, in three places.

مَحَبَّةٌ: see حُبٌّ. — Also *A cause of love* or *affection:* (Jel in xx. 39:) [pl. مَحَابُّ, like مَعَاشُ pl. of مَعِيشَةٌ, &c.] You say, أُوتِىَ فُلَانٌ مَحَابَّ القُلُوبِ [*Such a one was gifted with* qualities that are *the causes of the love of hearts*]. (A, TA.)

مُحِبَّةٌ: see حِبَّةٌ.

مُتَحَبِّبٌ إِلَى النَّاسِ *i. q.* مُتَحَبِّبٌ [see 5]. (A, TA.) — مُحَبَّبٌ: see المُحَبَّبَةُ.

مَحْبُوبٌ: see حَبِيبٌ, in two places. — المَحْبُوبَةُ: see مُحَبٌّ. — أُمُّ مَحْبُوبٍ a surname of *The serpent.* (Ḳ.) [See also حُبَابٌ.]

حبر

1. حَبَرَهُ, (Ṣ, Mṣb, TA,) aor. ـُ, (Mṣb,) inf. n. حَبْرٌ; (Ṣ, Mṣb, TA;) and ▼حَبَّرَهُ, (TA,) inf. n. تَحْبِيرٌ; (Ṣ, Ḳ, TA;) or the latter has an intensive signification; (Mṣb;) *He made it beautiful, beautified it,* (Ṣ, Ḳ, TA,) or *adorned it,* or *embellished it,* (Mṣb,) and *made it plain;* (TA;) namely, handwriting, and poetry, &c., (Ṣ, Ḳ, both in relation to the latter verb, and TA in relation to both verbs,) such as language, or speech, and science, (Ṣ, TA,) and pronunciation, and a recitation; meaning, with respect to the last, the voice [with which he recited]. (TA.) — Also حَبَرَهُ, (Ṣ, A, L, Mṣb, but in the Mṣb "or," not "also,") aor. ـُ, inf. n. حَبْرٌ (Ṣ, Mṣb) and حَبْرَةٌ; (Ṣ;) and ▼احبرهُ; (Ḳ;) and in an intensive sense ▼حبّرهُ; (Mṣb;) *He,* (God, A,) or *it,* (a thing, or an affair or event, Ṣ, L,) *made him happy, joyful,* or *glad;* (Ṣ, A, L, Mṣb, Ḳ;) *affected him with a happiness, joy,* or *gladness, that made his face to shine,* or *of which the mark,* or *sign,* (حَبَار, i. e. أَثَر,) *appeared upon his countenance;* (Bḍ in xliii. 70, in explanation of the pass. form of the first of these verbs;) *he made him to enjoy a state of ease and plenty;* and *treated him with honour:* (Lth and Ṣ in explanation of the pass. form of the first verb as used in the Ḳur xxx. 14:) or *treated him with extraordinary honour.* (Bḍ in xliii. 70, and TA.) [حُبِرَ, properly signifying *He was made happy,* &c., may be used as meaning *he was,* or *became, happy,* &c.; like سُرَّ; and حُبُورٌ, and its syns. mentioned with it below, may be regarded as its inf. ns. Golius, app. from his finding حَبَرَ explained in the KL as an inf. n. meaning The *being happy,* &c., (شَادْ شُدَنْ,) assigns to حَبَرَ, as on the authority of that lexicon, the meaning of "hilaris lætusque fuit;" but I have not found this verb in any Arabic work.] — حَبَرَ جِلْدُهُ *His skin was beaten so that there remained the mark of the beating.* (Ḳ.) — حَبِرَ الجُرْحُ, (Ṣ, Ḳ,) aor. ـَ, (Ḳ,) inf. n. حَبَرٌ, (Ṣ,) *The wound broke out afresh:* (Ṣ, Ḳ:) or *became healed, but left scars.* (Ks, Ṣ, Ḳ.) — حَبِرَتْ أَسْنَانُهُ, aor. ـَ, (Ṣ, A,* Mṣb, Ḳ,) inf. n. حَبَرٌ, (Ṣ, Mṣb,*) *His teeth became of a yellow colour mingled with the whiteness:* (Ḳ:) or *became yellow;* (A, Mṣb;) syn. قَلِحَتْ. (Ṣ.) [See also حِبِرٌ.]

2. حبّرهُ: see 1, in two places. — Also, inf. n. تَحْبِيرٌ, *He pared it well;* namely, an arrow. (TA.)

4. احبرهُ: see 1. — احبر بِهِ *He,* or *it, left a mark upon him,* or *it.* (TA.) And احبرتِ الضَّرْبَةُ جِلْدَهُ and بِجِلْدِهِ *The blow made a mark,* or *marks, upon his skin.* (TA.)

حَبَرٌ: see حُبُورٌ, in two places: — and حِبِرٌ, in two places: — and حِبْرٌ. — Also حَبْرٌ and ▼حِبْرٌ; (Ṣ, A, Mṣb, Ḳ, &c.;) but Aṣ says, I know not whether it be the former or the latter: (Ṣ:) IAạr says both: A 'Obeyd says that some of the lawyers say the former; and some, the latter; (TA;) and that in his opinion it is the former: (Ṣ, TA:) AHeyth, that it is the former only: (TA:) Th mentions the former only: (Mṣb:) Fr says it is the latter only: (TA:) and the latter is [said to be] the more chaste because the pl. is of the measure أَفْعَال, and not فُعُول: (Ṣ, TA:) [but a pl. of the latter measure is also mentioned:] *A learned man* (Aṣ, Ṣ, Mṣb, Ḳ) *of the Jews:* (Ṣ, A:) or *whether he be a Christian or Jewish or Subean subject of a Muslim government, who pays a poll-tax for his freedom and toleration, or one who, having been such, has become a Muslim:* or *one skilled in the beautifying of language:* (A 'Obeyd, Ṣ:) or *a good,* or *righteous, man:* (Kaạb, Ḳ, TA:) pl. (of the former, Mṣb) حُبُورٌ, (Mṣb, Ḳ,) [but this is seldom used,] and (of the latter, Mṣb) أَحْبَارٌ. (IDrst, Ṣ, A, Mṣb, Ḳ, &c.)

حِبْرٌ *Ink,* syn. مِدَادٌ, (Mṣb,) and نِقْسٌ, (Ḳ,) *with which one writes:* (Ṣ, Mṣb:) so called because it is one of the means of beautifying writings; (Moḥammad Ibn-Zeyd, TA;) or because it beautifies, and makes plain, handwriting; (Hr, TA;) or because of the marks that it leaves: (Aṣ, TA:) pl. [of pauc.] أَحْبَارٌ (IDrst, TA) and [of mult.] حُبُورٌ. (TA.) — *I. q.* وَشْىٌ [The *variegation,* or *figuring,* of cloth or of a garment; or *a kind of variegated,* or *figured, cloth or garment*]: (IAạr, Ḳ:) pl. حُبُورٌ. (Ḳ,* TA.) [See also حِبَرَةٌ.] — *A mark,* or *sign, of the enjoyment of ease and plenty:* (Aṣ, Ṣ, Ḳ: [in one copy of the Ṣ, and in the CḲ, for أَثَرُ النِّعْمَةِ, I find, erroneously, أَثَرُ النَّعْمَةِ:]) and [hence,] *beauty;* (Aṣ, Ṣ, A, Ḳ;)

beauty of aspect; or a beautiful and pleasing aspect, that satisfies the eye by its comeliness: (Aṣ, Ṣ, TA:) *colour; complexion:* (Fr, IAạr, Ṣ, TA:) pl. أَحْبَارٌ (Ṣ) and حُبُورٌ. (Ḳ,* TA.) One says, إِنَّهُ لَحَسَنُ الحِبْرِ وَالسِّبْرِ *Verily he is beautiful, and of goodly appearance:* (Aṣ, Ṣ:) or *of beautiful complexion.* (IAạr.) And ذَهَبَ حِبْرُهُ وَسِبْرُهُ *His colour,* or *complexion,* (Fr, Ṣ,) or *beauty,* (A,) *and goodliness of form* or *aspect, departed:* (Fr, Ṣ, A:) from the saying, جَآءَتِ الإِبِلُ حَسَنَةَ الأَحْبَارِ وَالأَسْبَارِ [*The camels came beautiful in colours and in appearances*]. (Fr, Ṣ, A.*) One says also, فُلَانٌ حَسَنُ الحَبْرِ ▼ وَالسَّبْرِ; where حبر seems to be the inf. n. of حَبَرْتُهُ "I made him, or it, beautiful." (Ṣ.) — Also, (Ṣ, Ḳ,) and ▼ حَبْرٌ (TA) and ▼ حَبَرٌ (Ḳ) and ▼ حُبَارٌ (Ṣ, Ḳ) and ▼ حِبَارٌ, (A, Ḳ,) *A mark,* or *trace,* (Ṣ, A, Ḳ,) of beating, (A,) or of a blow that has not brought blood, or of a healed wound, (TA,) or of work, or labour: (A, TA:) pl. of the first [or second] حُبُورٌ (Yaạḳoob, Ṣ, Ḳ) and [of the first and third, accord. to analogy,] أَحْبَارٌ; (TA;) and of the fourth حَبَارَاتٌ, (Yaạḳoob, Ṣ, TA,) it having no broken pl. (TA.) One says, بِهِ حُبُورٌ *Upon him are marks* [of beating, &c.]. (Ṣ.) And بِجِلْدِهِ حِبَارُ ▼ الضَّرْبِ *Upon his skin is the mark of beating.* (A.) And بِيَدِهِ حِبَارُ ▼ العَمَلِ *Upon his hand is the mark of work, or labour.* (A.) — See also حِبْرٌ. — And see حُبُورٌ. — Also, [like the Hebrew חָבֵר, and the Chaldee חֲבַר,] *A like; an equal; a fellow.* (Ḳ.) — See also حَبْرٌ.

حَبِرٌ: see حُبُورٌ: — and حِبْرٌ: — and حَبِرَةٌ.

حَبِيرٌ: see حَبِيرٌ.

حُبْرَةٌ: see حَبِيرٌ.

حِبِرٌ, (Mṣb, Ḳ,) the only subst. of this form beside إِبِلٌ, (Mṣb,) [and a few rare dial. vars.,] and ▼ حِبْرٌ (Ḳ) and ▼ حَبْرٌ (A, Ḳ) and ▼ حِبِرَةٌ (Ṣ, Mṣb, Ḳ) and ▼ حَبْرَةٌ (A, Ḳ,) and ▼ حُبْرَةٌ; (Ḳ;) or حِبِرٌ, without ة, [as also حِبْرٌ and حَبْرٌ,] is a pl. [or rather a coll. gen. n.], (Ṣ,) and with ة it is said to be a n. un.; (Mṣb;) *A yellowness that mingles with the whiteness of the teeth;* (Ḳ;) *a yellowness of the teeth;* (Sh, A, Mṣb;) *what is termed* قَلَحٌ *in the teeth:* (Ṣ:) or قَلَحٌ is when they become green: and when the crust increases so as to encroach upon the gums, and to make the roots of the teeth to appear, this is what is termed حَفَرٌ and حَفْرٌ: (Sh, Mṣb, TA:) pl. حُبُورٌ. (Ḳ.)

حَبْرَةٌ: see حُبُورٌ, in three places. — Also *Extraordinariness* (مُبَالَغَةٌ) *in a thing that is described as beautiful.* (Ḳ.) [See 1.] — *A musical performance,* or *concert, instrumental* or *vocal* or *both,* (سَمَاعٌ,) *in Paradise;* (Zj, Ḳ;) agreeably with which signification Zj explains [the verb in] the verse of the Ḳur [xxx. 14, or xliii. 70]: (TA:) and *any sweet melody.* (Ḳ.) — See also حِبِرٌ.

حُبْرَةٌ: see حِبِرٌ.

حَبُرَةٌ: see حُبُورٌ: — and see also the next paragraph, in two places.

حِبَرَةٌ (Ṣ, Mgh, Mṣb, Ḳ) and ▼ حَبْرَةٌ (Ḳ) *A [garment of the kind called]* بُرْدٌ, (Ṣ, Mgh,) or *a sort of* بُرْد, (Ḳ,) *of the fabric of El-Yemen,* (Ṣ, Mgh, Ḳ,) *striped* (مُنَمَّرٌ [or this word, q. v., may perhaps signify *spotted*]); (TA;) *a kind of garment of the fabric of El-Yemen, of cotton* or *linen, striped* (مُخَطَّطٌ): (Mṣb:) pl. حِبَرٌ and حِبَرَاتٌ (Ṣ, Mgh, Mṣb, Ḳ) and حِبْرٌ and حِبْرَاتٌ: (TA:) [or rather ▼ حِبَرٌ and ▼ حِبْرٌ are coll. gen. ns.] Accord. to Lth, (Az, Mgh, TA,) حبرة is not a place, nor a known thing, but only signifies وَشْىٌ [see حَبِيرٌ]; (Az, Mgh, Mṣb, TA;) and one says بُرْدُ حِبَرَةٍ (Mṣb, TA) and بُرُودُ حِبَرَةٍ, (TA,) and بُرْدٌ حِبَرَةٌ (Mgh, Mṣb, TA) and بُرُودٌ حِبَرَةٌ, (Mgh, TA,) like as one says ثَوْبُ قِرْمِزٍ, the word قرمز signifying a certain dye. (Az, Mṣb, TA.) [The term ▼ حَبَرَة is now applied in Egypt to *A lady's outer covering of silk, black for the married,* and *white for the unmarried, worn in riding and walking abroad; the former worn also by concubine slaves.* See also حَبِيرٌ.]

حَبَرَةٌ: see حِبَرَةٌ.

حِبْرِيٌّ *A seller of ink.* (Ḳ.) ▼ حَبَّارٌ, also, is mentioned as having the same signification; and some say that analogy is a sufficient authority for it: but it is disallowed by F. (TA.)

حِبَرِيٌّ, not ▼ حَبَّارٌ, (Ḳ,) or the latter is allowable on the ground of analogy, (MF,) *A seller of the garments called* حِبَرٌ. (Ḳ.) [See حِبَرَةٌ.]

حُبْرُورٌ (Mṣb, Ḳ) and ▼ حِبْرِيرٌ and ▼ حُبُرْبُرٌ and ▼ حُبْرُبُورٌ and ▼ يَحْبُورٌ [in the CḲ يُحْبُورٌ] and ▼ حُبُورٌ (Ḳ) The *young one of the* حُبَارَى: (Mṣb, Ḳ:) pl. حَبَارِيرُ and حَبَابِيرُ. (Ḳ.) [See also يَحْبُورٌ below.]

حِبْرِيرٌ: حُبُرْبُرٌ: حُبْرُبُورٌ:	see what next precedes.

حَبَارٌ: see حِبْرٌ. — Also The هَيْئَة [i. e. *form,* or *aspect,* or *the like,* or *goodliness of form* or *aspect,*] of a man. (Aboo-Ṣafwán, Lḥ.)

حِبَارٌ: see حِبْرٌ, in three places.

حُبُورٌ and ▼ حِبْرٌ, (Ṣ, Ḳ,) or ▼ حِبْرٌ, with kesr, (Mṣb,) and ▼ حَبِرٌ, which last occurs in a verse of El-'Ajjáj, for حِبْرٌ, [by poetic license,] (Ṣ,) and ▼ حُبْرَةٌ (A, Ḳ) and ▼ حَبَرَةٌ, (Ḳ,) *Happiness, joy,* or *gladness:* (Ṣ, Mṣb, Ḳ:) or the first signifies *cheerfulness;* i. e. *pleasure,* or *delight, and dilatation of the heart, which has a visible effect in the aspect:* (TA voce سُرُورٌ:) and the same word (IAth) and ▼ حَبْرَةٌ (Az, IAth, Ḳ) and ▼ حَبْرٌ, (Ḳ,) *a state of ease and plenty;* syn. نَعْمَةٌ: (IAth, Ḳ: [in the CḲ and in a MS. copy of the Ḳ, erroneously, نِعْمَةٌ:]) or *a state of complete,* or *perfect, ease and plenty:* (Az:) and *ampleness of the circumstances of life.* (IAth.) [See 1. Hence the saying,] كُلُّ حَبْرَةٍ ▼ بَعْدَهَا عَبْرَةٌ [*After every state of happiness, or joy,* &c., *is a tear*]. (A.)

حَبِيرٌ *A [garment of the kind called]* بُرْد, *variegated,* (مُوَشًّى,) (Ḳ,) [i. e.] *striped.* (TA.) One says بُرْدٌ حَبِيرٌ and بُرْدُ حَبِيرٍ. (TA.) [See also حِبَرَةٌ. Hence the saying,] لَبِسَ حَبِيرَ الحُبُورِ وَاسْتَوَى عَلَى سَرِيرِ السُّرُورِ ‡ [*He clad himself with the mantle of cheerfulness, and seated himself firmly upon the couch of happiness*]. (A.) — Also, applied to a garment, or piece of cloth, *New:* (Ṣ, Ḳ:) and *soft and new;* (Ḳ, TA;) applied to the same; (TA;) and so ▼ حِبْرٌ; (Ḳ;) which also signifies a *soft* thing: (TA:) pl. of the former حُبُرٌ. (Ḳ.) — And *Clouds;* syn. سَحَابٌ: (Ṣ:) or *clouds spotted* (مُنَمَّرٌ); (Ḳ;) *in which one sees what resembles* تَنْمِير, *by reason of the abundance of their water;* but Er-Riyáshee disapproves of this. (TA.)

حُبَارَى [a word respecting which J says,] its alif [written ى] is not the fem. alif nor the alif of quasi-coördination; [as F says of the alif of قَبَعْثَرَى, though he finds fault with J for saying thus of the alif of حُبَارَى; (see أَلِفُ التَّكْثِيرِ, in art. ا;)] the name [says J] being only composed with it, so that it is as it were a part of the word itself, which is imperfectly decl. when determinate and when indeterminate; i. e., without tenween: (Ṣ:) but its alif is the fem. alif; for were it not so, it would be perfectly decl.; (Ḳ;) and J says that it is imperfectly decl.: (TA:) and his saying that the alif is [as it were] a part of the word itself is a strange expression, for which it would be difficult to give an answer, and which therefore requires not exorbitance: but "it is sufficient excellence for a man that his faults may be counted:" (M:) [*A species of bustard;*] *a certain bird,* (Ṣ, Mṣb, Ḳ,) *well known, of the form of the goose, with a dust-colour upon its head and belly, and the back and wings of which are for the most part of the colour of the quail;* (Mṣb;) or it is *a long-necked bird, of an ash-colour, of the form of the goose, with a beak somewhat long, and that is preyed upon, but does not itself prey:* Az says that *it does not drink water, and that it lays its eggs in distant sands:* [the truth is, that it drinks seldom: the male bird has a pouch, extending from beneath the tongue to the breast, said to be large enough to contain seven quarts of water; and it has been supposed by some that he fills this with water for the supply of himself and his mate:] and Az further says, We used, when we journeyed, to proceed in the mountains of Ed-Dahnà, and sometimes we picked up in one day between four and eight of its eggs: *it lays four eggs, of a bluish colour, more delicious in taste than those of the domestic hen and than those of the ostrich:* and others say that *it brings its food from a greater distance than any other bird; sometimes from a distance of many days' journey:* also, that *it is constantly provided with a thin excrement, or dung, which it voids upon the hawk when pursued by the latter; thus saving itself, by preventing the hawk from continuing its flight, and, as some say, causing its feathers to drop off:* whence the prov., أَسْلَحُ مِنْ حُبَارَى: [see art. سلح:] (TA:) حُبَارَى is applied alike to the male and the female, and used as sing. and pl.: (Ṣ, Ḳ:) but it has pl. forms, (TA,) namely, حُبَارَيَاتٌ (Ṣ, Mṣb, Ḳ, TA) and حُبَارَاتٌ: (TA:) accord. to Sb, it has not حَبَارٍ, [in the TA incorrectly written حَبَارِى, as though it had the article ال prefixed to it, or were prefixed to another noun,] nor حَبَائِرُ, [though both of these are mentioned as pls. of it in several of the grammars of the Arabs,] in order to distinguish between حُبَارَى and nouns of the measures فَعْلَاةٌ and فِعَالَةٌ and the like. (TA.) It is said in a prov.,

وَكُلُّ شَىْءٍ قَدْ يُحِبُّ وَلَدَهْ *
حَتَّى الحُبَارَى وَتَطِيرُ عَنَدَهْ *

[*And everything certainly loves its offspring: even the bustard; and it flies by its side*]: (Ṣ, Mgh:*) [in the TA, وَيَدِفُّ عَنَدَهْ:] it flies by the side of its young one to teach it to fly before its wings have grown, because of its stupidity: (TA:) the حبارى is thus specially mentioned because it is proverbial for stupidity, and, notwithstanding its stupidity, loves its offspring, and teaches it to fly. (Ṣ, Mgh.) Another prov. is, فُلَانٌ مَيِّتٌ كَمَدَ الحُبَارَى [*Such a one is dying with the concealed grief of the bustard*]: because the حبارى moults with other birds, but its new feathers are slow in coming: so when the other birds fly, it is unable to do so, and dies of concealed grief. (TA.) [See also حُبْرُورٌ, and يَحْبُورٌ.]

حُبَارٌ: see حِبْرَى — and حَبْرَى.

حُبُورٌ: see حُبْرُورٌ.

حَابُورٌ *A sitting-place,* or *a company sitting together,* (مَجْلِسٌ,) *of unrighteous persons* [or *revellers*]: (Ṣ, Ḳ:) from حَبَرَهُ "it made him happy," &c. (Ṣ.)

مَحْبَرَةٌ, (Mṣb, Ḳ,) which is the most approved form, (Mṣb, TA,) and ▼مِحْبَرَةٌ, (Ṣ, Mṣb,) because it is an instrument, (Mṣb, TA,) a correct form, though said in the Ḳ to be incorrect, (TA,) and ▼مَحْبُرَةٌ (Mṣb, Ḳ) and ▼مَحْبِرَةٌ, (Ḳ,) the last used by poetic license, (TA,) *The place,* (Ṣ, Ḳ,) or *earthern pot,* or *glass bottle,* (TA,) *in which ink is put:* (Ṣ, Ḳ, TA:) pl. مَحَابِرُ. (Mṣb.) ═ Also, the first of these words, *A thing,* or *things, in which happiness, joy,* or *gladness, is usually found:* such are women said to be. (TA from a trad.) [*A cause of happiness, joy,* or *gladness;* agreeably with analogy: of the same class as مَبْخَلَةٌ and مَجْبَنَةٌ.]

مِحْبَرَةٌ:
مَحْبُرَةٌ:
مَحْبِرَةٌ:
} see the next preceding paragraph.

مُحَبَّرٌ A man (T) *having his skin marked by the bites of fleas.* (T, Ḳ.) — *An arrow well pared.* (Ḳ.)

يَحْبُورٌ, applied to a man, [*Very happy, joyful, glad,* or *cheerful;*] of the measure يَفْعُولٌ from الحُبُورُ: (Ṣ:) *a soft, tender,* or *delicate,* man: pl. يَحَابِيرُ. (AA, TA.) ═ *A certain bird:* or *the male of the* حُبَارَى: or *its young one.* (Ḳ.) See حُبْرُورٌ.

حبس

1. حَبَسَهُ, (Ṣ, A, Mṣb, Ḳ,) aor. ـِ, (Mṣb, Ḳ,) inf. n. حَبْسٌ (Ṣ, A, Mgh, Mṣb, Ḳ) and مَحْبِسٌ, (Lth, Sb, Ḳ,) *He confined, restricted, limited, kept in, prevented from escape, kept close, kept within certain bounds* or *limits, shut up, imprisoned, held in custody, detained, retained, arrested, restrained, withheld, debarred, hindered, impeded,* or *prevented, him* or *it; contr. of* خَلَّاهُ; (Ṣ, TA;) syn. مَنَعَهُ, (A, Mgh, Mṣb, Ḳ, TA,) and أَمْسَكَهُ; (TA;) as also ▼احتبسهُ: (Ṣ, Ḳ:*) and i. q. ضَبَطَهُ (Sb, TA in this art.) or ضَبَطَ عَلَيْهِ (TA in art. ضبط) [*he took, held,* or *retained, him* or *it, strongly, vehemently,* or *firmly;* &c.]. You say, لَا يُحْبَسُ دَرُّكُمْ meaning, لَا تُحْبَسُ ذَوَاتُ الدَّرِّ [*Your milch animals shall not be confined,* or *restrained from pasturing*]. (TA.) And حَبَسَ المِلْكَ عَلَيْهِ †[*He confined,* or *restricted, the property to him,* by will or otherwise]. (Mgh in art. وقف.) And حَبَسَ نَفْسَهُ عَلَى كَذَا [*He confined,* or *restricted, himself to such a thing*]. (Ṣ and Ḳ voce تَحَبَّسَ.) And حَبَسَهُ عَنْ وَجْهِهِ [*He restrained,* or *withheld, him from his course, purpose,* or *object*]. (Ṣ in art. الت; &c.) And حَبَسَهُ عَنْ حَاجَتِهِ [*He withheld,* or *debarred, him from the thing that he wanted*]. (Ḳ in art. بيت; &c.) — [Hence,] حَبَسَهُ, (IDrst, Mgh, Mṣb, Ḳ,) inf. n. حَبْسٌ; (TA;) and ▼احبسهُ, (Ṣ, IDrst, Mgh, Mṣb, Ḳ, [in one copy of the Ṣ, and in one of the A, ▼احتبسهُ, which is perhaps allowable,]) inf. n. إِحْبَاسٌ; (TA;) and ▼حبّسهُ, (IDrd, Mgh,) inf. n. تَحْبِيسٌ; (IDrd, TA;) فِى سَبِيلِ ٱللّٰهِ; (Ṣ, IDrst, A, Mgh;) ‡ *He bequeathed it,* or *gave it,* (namely, a horse, Ṣ, IDrst, A, Mgh, Ḳ,) *unalienably,* (Ṣ, IDrst, Mgh, Mṣb, Ḳ,) *to be used in the cause of God,* or *religion;* (Ṣ, IDrst, A, Mgh, Ḳ;) i. e., *to the warriors, to ride it in war against unbelievers and the like:* (TA:) it is said that the chaste forms are ▼احبسهُ and ▼حبّسهُ: (TA:) or the latter of these two is sometimes used; (Mgh;) but has an intensive signification [or is applied to several objects]: (Mṣb:) حَبَسَهُ is said to be a bad form; (TA;) it is used by the vulgar, but is allowable: ▼احبسهُ is used in preference, to signify the bequeathing or giving of horses and other articles of property that are forbidden to be [afterwards] sold or given, to distinguish between that which is so forbidden and that which is not: (IDrst, TA:) the reverse is the case with respect to وَقَفَهُ and أَوْقَفَهُ and وَقَّفَهُ; for the first of these three is the most chaste, and the last of them is disapproved and rare: (TA:) ▼حبّس شَيْئًا, inf. n. تَحْبِيسٌ, signifies † *He made a thing to remain in itself unalienable,* (Ḳ,* TA,) *not to be inherited nor sold nor given away,* (TA,) *assigning the profit arising from it to be employed in the cause of God,* or *religion.* (Ḳ, TA.) Moḥammad is related to have said to 'Omar, respecting some palm-trees belonging to the latter, (Mgh, TA,) which he (the latter) desired to give in charity, (TA,) حَبِّسْ ▼الأَصْلَ وَسَبِّلِ الثَّمَرَةَ ‡ *Make thou the property itself to remain unalienable,* (Mgh, TA,) *in perpetuity,* (Mgh,) *not to be inherited nor sold nor given away,* (TA,) *and assign thou the profit arising therefrom to be employed in the cause of God,* or *religion.* (Mgh, TA.) [See حَبِيسٌ.]

2: see 1, in four places.

3. حابس صَاحِبَهُ, (Ḳ,) inf. n. مُحَابَسَةٌ, (TḲ,) i. q. حَبَسَ [*He confined his companion,* or *restricted him,* &c.]: or [more probably, and agreeably with analogy,] *he confined his companion,* &c., *the latter doing the same to him.* (TḲ.)

4: see 1, in three places.

5. تحبّس عَلَى كَذَا *He confined, restricted, limited,* or *withheld, himself* (حَبَسَ نَفْسَهُ) *to such a thing.* (Ṣ, Ḳ.) تحبّس عَنِ الرُّكْبَانِ *He held back from the riders.* (TA.) تحبّس فِى الأَمْرِ [*He withheld himself,* or *held back, in,* or *respecting, the affair*]. (TA in art. حوز.)

7: see 8.

8. احتبس quasi-pass. of حَبَسَهُ; *He,* or *it, was,* or *became, confined, restricted, limited,* &c.; and *he confined, restricted, limited,* &c., *himself;* (Ṣ, A, Ḳ;) [as also ▼انحبس; but this latter is probably post-classical.] — Said of urine [as meaning *It became suppressed*]. (Ṣ and Mṣb in art. حقب; &c.) ═ احتبسهُ syn. with حَبَسَهُ, which see, in two places. (Ṣ, Ḳ.) — Also *He appropriated it to himself; restricted it to his own special possession:* (A, TA:) or *he made,* or *constituted, it* (اِتَّخَذَهُ) *what is termed* حَبِيس. (TA.)

حَبْسٌ *A place of confinement, restriction, imprisonment,* or *the like; a prison; a jail;* (A, Mṣb;) as also ▼مَحْبِسٌ, (Lth, A, TA,) which is also an inf. n.; (Lth, TA;) or, accord. to analogy, ▼مَحْبَسٌ: (Sb, TA:) pl. of the first, حُبُوسٌ; (Mṣb;) and of the second [and third], مَحَابِسُ. (A.) ═ See also حِبْسٌ.

حُبْسٌ a contraction of حُبُسٌ, which is pl. of حَبِيسٌ [q. v.]. (IAth, TA.)

حِبْسٌ *A dam constructed of wood or stones, in a channel of water, to confine the water,* (Ṣ, Ḳ,) *that people may drink from it and water their beasts;* (Ṣ, TA;) as also ▼حَبْسٌ: (El-'Ámiree, Ḳ:) pl. أَحْبَاسٌ (Ṣ, TA) and حِبَاسٌ: (Meyd, in Golius:) or *a dam by which the water-course of a valley is obstructed, in any place where it is confined:* (TA:) or *stones put in the mouth of a river* or *rivulet* or *the like, preventing the overflowing of the water:* (IAar, TA:) or *a* مَصْنَعَة *for water;* [i. e. *a thing like a* حَوْض, or *watering-trough for beasts &c., in which the rain-water is collected;*] (Ṣ;) as also ▼حَابِسٌ: (TA:) or *a thing like a* مَصْنَعَة *for water:* (AA, Ḳ:) pl.

أَحْبَاس: (AA, TA:) and ▼ حَبَاسَةٌ and ▼ حِبَاسَةٌ signify the same as حِبْسٌ: or, accord. to Lth, the حباسات in a piece of land are *what surround a [portion of ground such as is called]* دَبْرَةٌ, *which is the same as a* مَشَارَةٌ, *in which the water is confined until they are full, when it is made to flow to other parts:* (TA:) or a حِبْسٌ is *what is surrounded by dams* [or *by ridges of earth*] *which confine,* or *retain, the water* [*for irrigation*]; as also مَشَارَةٌ and دَبْرَةٌ. (R, TA in art. شور.) —— Also *Water collected, and having no supply to increase it:* (Ibn-'Abbád, Ķ:) thus called by the name of that by which it is confined. (TA.)

حُبْسَةٌ a subst. from اِحْتِبَاسٌ [signifying *A state of confinement, restriction, limitation, &c.*]: you say, الصَّمْتُ حُبْسَةٌ [*Speechlessness is a state of restriction*]. (Ş, TA.) —— [Hence,] *A difficulty of utterance which prevents one's speaking distinctly;* (A;) *a difficulty of speech,* (Mbr, Ķ,) *and hesitation,* (Mbr, TA,) *when one desires to speak;* (Mbr, Ķ;) *a hesitation in speech.* (Mşb.)

حَبِيسٌ i. q. ▼ مَحْبُوسٌ, [pass. part. n. of 1,] *Confined; restricted; limited;* &c. (TA.) —— ‡ Anything *bequeathed,* or *given, unalienably,* (Lth, Mgh, Mşb,) *for the sake of God;* whether an animal or land or a house; (Mgh;) as also ▼ مَحْبُوسٌ and ▼ مُحَبَّسٌ and ▼ مُحْبَسٌ: (Mşb:) pl. of the first حُبُسٌ, (Mgh, Mşb,) and, by contraction, حُبْسٌ: (Mşb:) حَبِيسٌ is used as a sing. and as a pl.: (Mşb:) it is of the measure فَعِيلٌ in the sense of the measure مَفْعُولٌ; and is sometimes used in the place of the pass. part. n. of حَبَّسَ: (TA:) it is also particularly applied to a horse *bequeathed,* or *given, unalienably, to be used in the cause of God,* or *religion;* (Ş, A,* Mgh, Ķ;) i. e., *to the warriors, to ride it in war against unbelievers and the like;* (TA;) as also ▼ مُحَبَّسٌ (Ş, Mgh, Ķ) and ▼ مَحْبُوسٌ: (Ķ:) and حُبُسٌ, (Ş,) or حُبْسٌ, (Ķ,) to what is, or are, *bequeathed,* or *given, unalienably,* (Ş, Ķ,) *not to be sold nor inherited,* (TA,) of palm-trees, or vines, &c., (Ķ,) as land, and anything that is a source of profit, (TA,) *itself to remain unalienable, and the profit arising therefrom to be employed in the cause of God,* or *religion:* (Ķ, TA:) but the حُبُس which Mohammad is related to have made common property were *what the pagan Arabs bequeathed,* or *gave, unalienably, for* (عَلَى [so in the TA, and this I regard as the true reading, rather than مِن, which is the reading in the Mgh and L,]) *the* [*camels called*] سَوَائِب *and* بَحَائِر, *and such as was called* حَام: (Mgh, L, TA:) Hr, in the Ghareebeyn, gives the reading حُبْس, which, says IAth, if correct, is a contraction of حُبُس. (TA.) [From حَبِيس, used as a subst., has been formed, app. in post-classical times, the pl. أَحْبَاس: see De Sacy's "Chrest. Ar.," sec. ed., vol. i. p. 189.] ▼ حَبِيسَةٌ, also, [used as a subst.,] signifies † *A thing that is bequeathed,* or *given, unalienably, in the way of beneficence:* and its pl. is حَبَائِس. (TA.)

حَبَاسَةٌ and حِبَاسَةٌ: pl. حِبَاسَاتٌ: see حِبْسٌ.

حَبِيسَةٌ: see حَبِيسٌ, last sentence.

[حَبَّاسٌ *A jailer.*]

حَابِسٌ [act. part. n. of حَبَسَ; *Confining; restricting; limiting;* &c.]: pl. حُبَّسٌ. (IAth, TA.) [Hence,] حَابِسُ الفِيلِ *The Restrainer of the Elephant:* an epithet applied to God; alluding to the case of Abrahah. [See Ķur ch. cv.] (TA.) And زِقٌّ حَابِسٌ *A skin that retains the water* [&c.]. (TA.) And كَلَأٌ حَابِسٌ *Herbage that is abundant, and retaining the water.* (TA.) —— See also حِبْسٌ. = Also *i. q.* مَحْبُوسٌ, or ذُو حَبْسٍ. (Ham p. 188.)

مَحْبِسٌ and مَحْبَسٌ: see حَبْسٌ. —— Also, the latter, [or both,] The *manger,* or *stable, of a beast.* (TA.)

مُحْبَسٌ: see حَبِيسٌ, in two places.

مُحَبَّسٌ: see حَبِيسٌ.

مَحْبُوسٌ: see حَبِيسٌ, in three places.

إِبِلٌ مُحْتَبِسَةٌ *Camels that remain at the house;* syn. دَاجِنَةٌ: as though they were restrained from pasturing. (TA.)

حبش

1. حَبَشَ لَهُ, (Ķ,) aor. ِ, (TĶ,) inf. n. حَبْشٌ and حُبَاشَةٌ; (Ķ, TĶ;) or حَبَشَ لَهُ حُبَاشَةً; (Ş;) [whence it appears probable that the author of the Ķ is in error in regarding حُبَاشَةٌ as an inf. n.;] *He collected for him something;* as also ▼ حَبَّشَ, inf. n. تَحْبِيشٌ: (Ş, Ķ:) and ▼ تحبّشه and ▼ احتبشه likewise signify *he collected it.* (TA.) You say also, ▼ حَبَّشَ قَوْمَهُ, inf. n. تَحْبِيشٌ, *He collected his people.* (Ş.) And حَبَشَ لِعِيَالِهِ, inf. n. حَبْشٌ, *He gained,* or *earned, and collected, for his family,* or *household;* like هَبَشَ; as also ▼ احتبش. (TA.)

2: see 1, in two places.

4. احبشت بِوَلَدِهَا *She brought forth her child like an Abyssinian* (حَبَشِيّ) *in colour.* (Ş.)

5. تحبّشوا *They collected themselves together,* (Ş,* A, TA,) عَلَيْهِ *against him;* as also تهبّشوا. (TA.) = تحبّشه: see 1.

8: see 1, in two places.

الحُبُشُ: see the next paragraph.

الحَبَشُ, (Ş, A, Mşb, Ķ,) a coll. gen. n., (Mşb,) and ▼ الحُبُشُ, (A, MF,) or this is a pl., and the former is also said to be an anomalous pl., (TA,) and ▼ الحَبَشَةُ, (Ş, A, Mşb, Ķ,) also said to be an anomalous pl., (TA,) and wrong with respect to rule, (T, M,) having no sing. of the measure فَاعِل, (M,) for they did not use حَابِش as a sing. thereof, like فَاسِق as sing. of فَسَقَة, (T,) but الحَبَشَة became used as a dial. var., (T, Mşb,) commonly obtaining, for الحَبَش, (Mşb,) and is allowable in poetry in cases of necessity, (T,) and ▼ الأَحْبُش, (IDrd, Ķ,) also used as syn. with الحَبَش, (IDrd,) or it is pl. of الحُبْش, with ḍamm, not a sing. as it seems to be from the mention of it in the Ķ, (MF,) and ▼ الأُحْبُوش, (A, TA,) and الحُبْشَان, (A,) which is a pl. (IDrd, Ş, Ķ) of الحَبَش, (IDrd,) like as حُمْلَان is pl. of حَمَل, (Ş,) and الحُبُوش, (A,) [also a pl.,] and ▼ الحَبِيش, which is also a pl., (TA,) [or rather a quasi-pl. n.,] and الأَحَابِش, which is likewise a pl., (Ķ,) app. of أُحْبُش, (TA,) and الأَحَابِيش, (A,) [which is pl. of أُحْبُوش,] *A certain race of the blacks;* (Ş, A, Mşb, Ķ, &c.;) [namely, *the Abyssinians;* who, however, are not properly called "blacks:"] one of whom is called حَبَشِيّ. (A, Mgh, Mşb.) The dim. of حَبَش is حُبَيْش. (Mşb.)

الحَبَشَةُ: see الحَبَشُ. —— It also signifies *The country of the* حُبْشَان [or *Abyssinians*]: (Ķ:) a proper name applied thereto. (TA.)

حَبَشِيّ: fem. with ة. For the latter, see حَبَشِيّ.

حَبَشِيّ a rel. n. from الحَبَشَة; (TA;) [signifying *Of,* or *belonging to,* or *relating to, Abyssinia* or *the Abyssinians.*] —— [*An Abyssinian;*] *one of the race called* الحَبَش. (A, Mgh, Mşb.) —— حَبَشِيَّة (Ķ) and ▼ حُبْشِيَّة (A, Ķ) *A black,* (A,) or *an intensely black,* (Ķ,) she-camel. (A, Ķ.) —— الحَبَشِيّ مِنَ النَّمْلِ *The black ant.* (M in art. دلم.)

الحَبِيشُ: see الحَبَشُ.

حُبَيْش dim. of حَبَش, q. v. (Mşb.) —— Also *A certain well-known bird;* [the *Numidia;* which comprises the species commonly called the *Guinea-hen,* and *pintado:* so applied in the present day:] the word is thus, [without the article ال, apparently as a proper name, and] in the dim. form, like كُعَيْت and كُمَيْت: (Ş, TA:) it is strangely omitted in the Ķ. (TA.)

حُبَاشَةٌ *What is collected,* (Ş,* and TA in art. هبش,) of men, and of property; as also هُبَاشَةٌ: (TA ubi suprà:) pl. حُبَاشَات. (Ş, and TA ubi suprà.) —— *A company,* or *body, of men, not of one tribe;* (Ş, Ķ;) like هُبَاشَةٌ; (TA;) as also ▼ أُحْبُوش and أَحَابِيش; (Ş;) or as also ▼ أُحْبُوشَةٌ, (Ķ, TA,) of which the pl. is أَحَابِيش; (TA:) the pl. of حباشة in this sense is as above. (TA.)

الأُحْبُشُ: see الحَبَشُ.

أُحْبُوشٌ: see حُبَاشَةٌ. —— الأُحْبُوشُ: see الحَبَشُ: accord. to some, it signifies *Any company,* or *body, of men;* because, when they are collected together, they are [in their general hue] black. (TA.)

أُحْبُوشَةٌ: see حُبَاشَةٌ.

حبط

1. حَبِطَ, aor. َ, inf. n. حَبَطٌ, (As, Ş, Ķ, &c.,) *He* (a beast, As, Ş, or a camel, ISd, Ķ) *ate much,* (Ş,) or *had pain in his belly from pasture which he found unwholesome,* or *from eating much of herbage,* (ISd, Ķ,) *so that he became swollen,* or *inflated, thereby* (Ş, ISd, Ķ) *in his belly,* (Ş,) *and there would not come forth from him* (Ş, ISd, Ķ) *what was in it,* (Ş,) or *anything;*

(ISd, Ḳ;) *he did not void either thin dung or urine, his belly being bound:* (Az:) or *he* (a sheep, or goat, ISk, Ṣ) *became swollen,* or *inflated, in his belly, in consequence of eating* [*the herb called*] ذُرَق, (ISk, Ṣ, Ḳ,*) *which is the* حَنْدَقُوق [i. e. the herb *lotus, melilot,* or *bird's-foot-trefoil*]: (ISk, Ṣ:) or *he* (a beast) *lighted upon good pasturage, and ate immoderately, so that he became swollen,* or *inflated, and died:* (Z, IAth:) or, in speaking of a horse, you do not say, حَبِطَ الفَرَسُ, but حَبِطَ قُصْرَى الفَرَسِ, or خَاصِرَتُهُ, or مَوْقِفُهُ, because it means that *the horse's belly became swollen,* or *inflated:* (ISd, Z, L:) you say also, حَبِطَ بَطْنُهُ *his belly became swollen,* or *inflated, so that he died:* (Az, TA:) or *his* (a man's) *belly became swollen,* or *inflated, by food &c.:* (Mbr, TA in art. حبطأ:) and حبط is also said of the skin, meaning *it became swollen,* or *inflated.* (TA.) [See also Q. Q. 3; and see حَبَطٌ below.] —— Hence, app., i. e. from حبط said of the belly, (Az, TA,) or it is from this verb said of a beast, (Z, IAth, TA,) حَبِطَ عَمَلُهُ, (Az, Ṣ, Mṣb, Ḳ, &c.,) aor. ـَ; (Az, Mṣb, Ḳ;) and حَبَطَ, aor. ـِ; (AZ, Az, Mṣb, Ḳ;) the latter, says Az, heard by AZ from an Arab of the desert, but I have not heard it on any other authority; (TA;) inf. n. حَبْطٌ, (Az, Ṣ, Ḳ, [but in the Mṣb it seems to be indicated that it is حَبَطٌ,]) with the ب quiescent, (Az, Ṣ,) thus differing from the inf. n. of حبط said of the belly, (Az, TA,) and حُبُوطٌ, (Az, Ṣ, Mṣb, Ḳ,) which latter, accord. to AZ, is the inf. n. of حَبَطَ like ضَرَبَ; (T, TA;) ‡ *His work,* or *deed, became null,* or *void,* or *of no account; it went for nothing; it perished;* (Az, Mṣb, TA;) for like as he of whom one says حَبِطَ بَطْنُهُ perishes, so does the work, or deed, of the hypocrite: (Az, TA:) or *it became ineffective of reward; its reward became annulled.* (Ṣ, Ḳ.) And hence also, (Z, TA,) حَبِطَ دَمُهُ, aor. ـَ, (Z, Mṣb, Ḳ, TA,) but not حَبَطَ also, as is implied in the Ḳ, (TA,) and in this case the inf. n. is حَبَطٌ, (Mṣb,* TA,) with the ب movent, (TA,) ‡ *His blood* (the blood of one slain, Ḳ) *went for nothing; unretaliated, and uncompensated by a mulct.* (Mṣb, Ḳ, TA.) —— حبط said of the water of a well, *i. q.* أَحْبَطَ, q. v. (TA.) —— Said of a wound, (Ṣ, Ibn-'Abbád, Ḳ,) aor. ـَ, (Ḳ,) inf. n. حَبَطٌ, with fet-ḥ to the ب, (Ṣ, Ḳ,) *It had scars remaining after having healed:* (Ibn-'Abbád, Ḳ:*) or *it broke open again;* or *became recrudescent;* syn. عَرِبَ [which has the signification given above on the authority of Ibn-'Abbád as well as what follows it] and نُكِسَ. (Ṣ.) [See also حَبَطٌ below.]

4. [احبطهُ seems to signify, in its primary acceptation, *He made him,* (namely a beast,) or *it,* (the belly,) *to be in the state termed* حَبَطٌ, which see below. —— And hence,] احبط عَمَلَهُ ‡ *He* (God, Ṣ, Ḳ, or a man, Mṣb) *made his work,* or *deed, to become null,* or *void,* or *of no account; to go for nothing; to perish;* (Mṣb, Ḳ,* TA;) *to be ineffective of reward;* or *he annulled its reward.* (Ṣ.) So it signifies in the Ḳur [xxxiii. 19, &c.]: and you say, إِنْ عَمِلَ عَمَلًا صَالِحًا أَتْبَعَهُ مَا يُحْبِطُهُ وَإِنْ أَرْسَلَ كَلِمًا طَيِّبًا أَرْسَلَ خَلْفَهُ مَا يُحْبِطُهُ ‡ [*If he do a good deed, he makes to follow it that which annuls it; and if he send forth good words, he sends forth after them that which annuls them*]. (TA.) And hence also, (Z, TA,) احبط الدَّمَ ‡ *He made the blood to go for nothing; unretaliated, and uncompensated by a mulct.* (Mṣb, Ḳ,* TA.*) —— احبطهُ الضَّرْبُ *The beating made a mark* or *scar,* or *marks* or *scars, upon him.* (TA.) == احبط مَآءُ الرَّكِيَّةِ, (Ḳ,) inf. n. إِحْبَاطٌ, (AA, Ṣ,) *The water of the well went away, and did not return* (AA, Ṣ, Ḳ) *as it was;* (AA, Ṣ;) as also ▼حَبِطَ, aor. ـَ. (TA.) —— احبط عَنْ فُلَانٍ *He turned away from, avoided, shunned, and left, such a one.* (IDrd, Ḳ.)

Q. Q. 3. اِحْبَنْطَى *He* (a man, TA) *was,* or *became, swollen,* or *inflated, in his belly:* (Ḳ, TA:) *he* (a man) *was short and bigbellied:* (Ṣ:) *he* (a man) *was,* or *became, filled with wrath,* or *rage;* or *by repletion of the belly;* as also اِحْبَنْطَأَ: from حَبَطٌ. (TA.) [See 1.]

حَبَطٌ [inf. n. of حَبِطَ, q. v.:] A beast's *having the belly swollen,* or *inflated, so that what is in it does not come forth, in consequence of eating much:* (Ṣ:) or *pain in the belly,* of a camel, *from pasture which he finds unwholesome,* or *from herbage of which he has eaten much, so that he becomes swollen,* or *inflated, therefrom,* (ISd, Ḳ,) *in his belly,* (TA,) *and nothing comes forth from him:* (ISd, Ḳ:) or *a swelling,* or *inflation, of the belly,* (Ḳ,) or a beast's *having the belly swollen,* or *inflated,* (ISk, Ṣ,) *from eating* [*the herb called*] ذُرَق: (ISk, Ṣ, Ḳ:) [see 1:] and *a swelling in the udder* or *other thing:* (Ḳ:) or, accord. to the M, the *slightest swelling in the udder:* or, as some say, *swelling,* or *inflation, wherever it be, from disease or other cause.* (TA.) It is said in a trad., إِنَّ مِمَّا يُنْبِتُ الرَّبِيعُ مَا يَقْتُلُ حَبَطًا أَوْ يُلِمُّ [*Verily, of what the* (rain, or season, called) ربيع *causes to grow, is what kills by inflation of the belly, or nearly does so*]. (Ṣ, TA.) —— The *scars,* or *marks, of a wound,* or *of whips, upon the body, after healing:* or the *swollen scars,* or *marks,* (*of whips,* TA,) *not lacerated:* when mangled and bleeding, they are termed عُلُوب [pl. of عَلْبٌ]: (Ḳ:) the *excrescent flesh upon the scars of wounds.* (Ṣgh.)

حَبِطٌ part. n. of حَبِطَ; A camel [or other beast *having his belly swollen,* or *inflated, so that what is in it does not come forth, in consequence of eating much:* or] *having pain in the belly, from pasture which he finds unwholesome,* or *from herbage of which he has eaten much, so that he is swollen,* or *inflated, therefrom,* [*in his belly,*] *and nothing comes forth from him:* (Ḳ:) [see حَبَطٌ:] pl. حَبَاطَى (Ḳ) and حَبِطَةٌ. (M, TA.) You say also فَرَسٌ حَبِطُ القُصْرَى *A horse swollen,* or *inflated, in the flanks.* (TA.)

حُبَاطٌ The *disease in which the belly is swollen,* or *inflated, from eating* [*the herb called*] ذُرَق: (Ḳ:) or, as Az says, accord. to some, it is with the pointed خ, from التَّخَبُّطُ signifying "the being in a state of commotion, agitation, convulsion, tumult, or disturbance." (TA.)

حُبَيْطٌ: } see حَبَنْطَى.
حُبَيْطِىٌّ: }

حَبَنْطًى, with tenween, and حَبَنْطَأٌ, the ن and the ا [which latter is written in the former word ى] being added to render the word quasi-coordinate to سَفَرْجَلٌ, (Ṣ, TA,) the derivation being from حَبِطَ, (TA,) A man *short and bigbellied;* (Ṣ, TA;) as also حَبَنْطَاةٌ and ▼ مُحْبَنْطٍ: (Ṣ:) [see the last of these words below:] or *filled with wrath,* or *rage;* or *by repletion of the belly;* (Ḳ;) as also حِبِنْطًى and حَبَنْطَاةٌ: (Ks, Lḥ:) and this last, a woman *short, ugly, and bigbellied;* (Ḳ;) also related with ء [i. e. حَبَنْطَأَةٌ, or, as it is written in the L, حَبَنْطَآءَةٌ, but this I think a mistranscription]. (TA.) When you form the dim., you may reject the ن, and change the ا [which is the final letter] into ى, so that [the dim. becomes originally حُبَيْطِىٌ, for which, accord. to a well-known rule,] you say ▼حُبَيْطٍ, with kesr to the ط, and with tenween; for the ا is not to denote the fem. gender, that the letter preceding it should be with fet-ḥ, as in [حُبَيْلَى and بُشَيْرَى] the dims. of حُبْلَى and بُشْرَى: you may also retain the ن, and reject the ا; saying ▼حُبَيْنِطٌ: and thus you may do in the case of any noun having two letters added for the purpose of quasi-coordination: you may also put a compensation for the letter rejected in either place, or not: if you put a compensation in the former instance, you say ▼حُبَيْطِىٌّ, with teshdeed to the ى, and with kesr to the ط; and in the latter instance, you say ▼حُبَيْنِيطٌ. (Ṣ, O, TA.)

حُبَيْنِطٌ: } see حَبَنْطَى.
حُبَيْنِيطٌ: }

مُحْبَنْطٍ and مُحْبَنْطِئٌ A man, or child, *swollen,* or *inflated, in his belly:* (TA:) or *filled with anger:* (AZ, TA:) or *who becomes angry, deeming a thing slow* or *tardy* or *late:* (IAth, TA:) or *refraining as one who seeks or desires, not as one who refuses:* (TA:) or the former, *becoming angry;* and the latter, *swollen,* or *inflated:* (IB, TA:) or the former, *deeming a thing slow* or *tardy* or *late;* and the latter, *bigbellied:* and the latter also signifies *cleaving to the ground.* (TA.) See also حَبَنْطَى.

حبق

1. حَبَقَ, (Ṣ, Mṣb, Ḳ,) said of a goat, (Lth, TA,) or mostly said of the camel and of the goat, (Ḳ,) and sometimes of a man, (TA,) or حَبَقَتْ, said of a she-goat, (Mṣb,) aor. ـِ, (Ṣ, Mṣb, Ḳ,) inf. n. حَبْقٌ (Ṣ, Ḳ) and حَبِقٌ (Mṣb, Ḳ) and حُبَاقٌ, (Ḳ,) *He,* or *she, broke wind.* (Ṣ, Mṣb, Ḳ.) —— [Hence,] يَحْبِقُونَ عَلَى فُلَانٍ ‡ *They revile such a one; and act in an ignorant,* or *a silly,* or *foolish, and a wrong manner towards him.* (TA.)

حَبِقٌ: see حَبْقٌ.

حَبَقٌ [The *mentha pulegium* of Linn., or *penny-royal*; so generally called in the present day, in Egypt and other countries; accord. to Golius, applied by the Moors and Egyptians to *ocimum* (i. e. *basil*), which, he says, the Easterns call حبق النبطى; but he should have said الحَبَقُ النَّبَطِىُّ, which see below;] *a certain plant of sweet odour*, (Ḳ,) *of sharp flavour, the leaves whereof are like those of the* خِلَاف [q. v.]; *of which one kind grows in the plains, and another on the mountains; not depastured*; (TA;) *called in Persian* الفُوتَنْج, (Ḳ, in the CḲ الفُوتَنَج,) or الفُوذَنْج, (Ṣ,) or پُودِينَه: (TA:) AḤn says, on the authority of an Arab of the desert, that *it is a cause of diminishing the seminal fluid; that the horse rolls upon it and it diminishes his seminal fluid; and it is put into the pillow which is placed beneath the head of a man and it diminishes his seminal fluid*: (TA:) *it resembles the sweet-smelling plant called the* نَمَّام [q. v., in the CḲ, erroneously, ثُمام]; (Ḳ,* TA;) *and grows abundantly by water*: (TA:) [a coll. gen. n.: n. un. with ة: and] pl. حِبَاقٌ. (IKh, TA.) — حَبَقُ المَآءِ and حَبَقُ التِّمْسَاحِ [*Mentha aquatica*, or *water-mint*,] الفُوتَنْجُ النَّهْرِىُّ; (Ḳ;) so called because it grows upon the sides of rivers, and because the crocodile eats of it much. (TA.) — حَبَقُ الفِيلِ, or حَبَقُ القَنَا, [*Marjoram, sweet marjoram*,] المَرْزَنْجُوشُ. (Ḳ.) — حَبَقُ الرَّاعِى [*Common artemisia*, or *mugwort*,] البِرِنْجَاسِفُ [or البَرَنْجَاسِفُ]. (Ḳ, TA: in the CḲ البِرَنْجَاسِف.) — حَبَقُ البَقَرِ [*Chamomile*] البَابُونَجُ. (Ḳ.) — حَبَقٌ الشُّيُوخِ [*Marum*; so called in the present day;] المَرْوُ; (Ḳ;) also called رَيْحَانُ الشُّيُوخِ. (TA.) — الحَبَقُ الكَرْمَانِىُّ and الحَبَقُ الصَّعْتَرِىُّ [*Basil-royal*] الشَّاهَسْفَرَمُ [from the Persian شَاهْ سِفَرَمْ or شَاهْ سِپَرَمْ &c.]; (Ḳ, TA; in the CḲ الشاهَسْفَرَمُ;) *which is the Sulṭán of the* رَيَاحِين; *also called* الرَّيْحَانُ المُطْلَقُ; *and which is sown in houses*. (TA.) — الحَبَقُ القَرَنْفُلِىُّ [*Common clinopodium*, or *wild basil*,] الفَرَنْجَمُشْكُ; (Ḳ, TA; in the CḲ الفرَنْجَمِشك;) [a word of Persian origin,] meaning the musk of the Franks. (TA.) — الحَبَقُ النَّبَطِىُّ, i. e. رَيْحَانُ الحَمَاحِمِ [which is *Garden-basil*: الحَمَاحِمُ is said in the Ḳ, art. حمر, to be الحَبَقُ البُسْتَانِىُّ, with wide leaves; also called الحَبَقُ النَّبَطِىُّ]. (TA.) — حَبَقُ تُرُنْجَانٍ [*Melissa, citrago, balm-mint*, or *balm-gentle*,] البَاذِرَنْجبويه. (TA.) — البَقْلُ المَكِّىُّ الحَبَقُ الرَّيْحَانِىُّ *What is eaten of* [see art. مقل]. (Ḳ.)

حَبْقٌ, (Ṣ, O, L, TA,) in the Ḳ, erroneously, حِبق, (TA,) *Emission of wind from the anus, with a sound*; (Ṣ, O, L, Ḳ, TA;) mostly used in relation to camels and sheep or goats; (Ḳ;) accord. to Lth, in relation to goats; but sometimes used in relation to human beings; a simple subst., as well as an inf. n.; (TA;) as also ▼ حُبَاقٌ (Ḳ) and ▼ حَبِقٌ. (TA.)

حَبْقَةٌ *A single emission of wind from the anus, with a sound*: (Ḳ:) or *a slight emission thereof*. (IDrd, TA.)

يَا حَبَاقِ is said to a female slave, [in reviling her, meaning *O thou stinking one!*] (Ḳ,) like as one says to her يَا دَفَارِ. (TA.)

حُبَاقٌ: see حَبْقٌ.

عَذْقُ الحُبَيْقِ, (Aṣ, Ṣ, Mṣb,) or, accord. to Málik Ibn-Anas, عَذْقُ ٱبْنِ الحُبَيْقِ, (Mṣb,) and لَوْنُ الحُبَيْقِ, (Ṣ, and TA in art. جعر,) or عَذْقُ حُبَيْقٍ, (Ḳ, in the CḲ عِذْقُ حُبَيْقٍ,) *A sort of* دَقَل, *of bad quality*: (Aṣ, Ṣ:) or *dates such as are termed* دَقَل; (Mṣb, Ḳ;) *dust-coloured, small, and somewhat long; of bad quality*: (Aṣ:) so called because of their badness; (Mṣb;) or so called in relation to [a man named] Ibn-Ḥobeyḳ. (TA.) It is said in a trad., نَهَى عَنْ لَوْنَيْنِ مِنَ التَّمْرِ الجُعْرُورِ وَلَوْنِ الحُبَيْقِ [*He* (Moḥammad) *forbade two sorts of dates; the* جعرور *and* لون الحبيق]: (Ṣ:) or نَهَى عَنِ الجُعْرُورِ وَعَذْقِ الحُبَيْقِ: (Mṣb:) meaning, in the case of the poor-rate. (Ṣ, Mṣb.)

حبك

1. حَبَكَهُ, aor. ـِ (Ṣ, Ḳ) and ـُ, (Ḳ,) inf. n. حَبْكٌ, (Ṣ, Ḳ,) *He bound it*, or *tied it*; and *made it fast*, or *firm*: (Ḳ: [see also 2:]) *he made it well*: (TA:) *he wove it well*, (Ṣ, Ḳ, TA,) *and firmly*, or *compactly*; (TA;) namely, a piece of cloth: (Ṣ, Ḳ, TA:) *he made the effect of the work therein to be beautiful*; i. e., in a piece of cloth: and ▼ احتبكه signifies the same: (Ḳ:) or this latter, *he made it* (i. e. anything) *firm*, or *compact*; and *made it well*. (IAạr, Ṣ, Mṣb.) It is said of 'Áisheh, in a trad., كَانَتْ تَحْتَبِكُ ▼ تَحْتَ الدِّرْعِ فِى الصَّلَاةِ *She used to bind the* إِزَار [or *waist-wrapper*], *and make it fast, beneath the shift, in prayer*; (Ṣ;) from حُبْكَةٌ, q. v.: (TA:) or كَانَتْ فِى الصَّلَاةِ تَحْتَبِكُ ▼ بِإِزَارٍ فَوْقَ القَمِيصِ *she used, in prayer, to bind an* ازار *over the shirt*. (Mṣb.) [It is said that] ▼ اُحْتِبَاكٌ is also *syn. with* اِحْتِبَآءٌ, on the authority of Aṣ: (Ṣ:) [i. e., that] احتبك is *syn. with* احتبى: (Mṣb:) [and that] احتبك بإزاره signifies احتبى, (Ḳ,) or احتبى بِهِ وَشَدَّهُ إِلَى يَدَيْهِ: so says Aboo-'Obeyd, as on the authority of Aṣ: but Az says that this is a mistake: that what Aṣ said was, that الاحتياك, with ى, is syn. with الاحتباء, as ISk relates. (TA.) One says also, حَبَكْتُ الحَظِيرَةَ بِقَصَبَاتٍ كَمَا تُحْبَكُ عُرُوشُ الكَرْمِ بِالحِبَالِ [*I bound the enclosure for cattle with canes*, or *reeds*, (or perhaps we should read بِقُضْبَانٍ, i. e. *with twigs*,) *like as the trellises of the grape-vine are bound with cords*: see also the last sentence of this paragraph]. (Az, TA.) — [In the present day, حَبَكَ also signifies *He sewed* the leaves of a book: and *he bound* a book.] ═ حَبْكٌ also signifies The act of *cutting*: and *smiting* [or *severing*] *the neck*. (Ḳ.) One says, حَبَكَهُ بِالسَّيْفِ, aor. ـِ and ـُ, inf. n. حَبْكٌ, (IAạr, TA,) *He struck him*, or *smote him, upon his middle*, or *waist, with the sword*: or *he cut the flesh* [or *his flesh*] *above the bones* [*with the sword*]: (TA:) or *he smote* [or *severed*] *his neck with the sword*: or *he smote him with the sword*. (IAạr, TA.) And حَبَكَ عُرُوشَ الكَرْمِ *He cut the trellises of the grape-vine*. (TA. [But this has another meaning, explained above.])

2. حبّك, (A, TA,) inf. n. تَحْبِيكٌ, (Sh, Ḳ,) *He made firm*, or *fast*, (Sh, A, Ḳ,) a knot. (A, TA. [See also 1.]) ═ *He striped*, or *wove with stripes*, (A, Ḳ,) a [garment of the kind called] كِسَآء. (A, TA.)

5. تحبّك *He bound*, or *tied, the* حُبْكَة, i. e. the حُجْزَة: [see حُبْكَة, below:] (Ḳ:) or *i. q.* تَلَبَّبَ بِثِيَابِهِ [*he raised*, or *tucked up, his clothes*; or *girded himself, and raised*, or *tucked up, his clothes*; &c.]. (IDrd, Ḳ.) And تحبّكت بِنِطَاقِهَا *She* (a woman) *bound*, or *tied, her* نِطَاق [q. v.] *upon her waist*. (IDrd, Ḳ.)

8: see 1, in four places; and see حُبْكَةٌ.

الحُبُكُ and الحِبِكُ and الحَبَكُ and ذَاتِ الحُبُكِ and الحُبْكُ and الحِبْكُ (TA) and الحِبَكُ (Bḍ in li. 7] and الحِبُكُ and الحبك (TA) are various readings in the Ḳur [li. 7]: الحُبْكُ is a contraction of الحُبُكُ, of the dial. of Benoo-Temeem: الحِبْكُ is a contraction of الحِبِكُ: الحِبَكُ is as though its sing., or n. un., were حِبْكَةٌ: الحَبَكُ is as though its sing. were حَبَكَةٌ: الحُبُكُ is the common reading, and is pl. of حِبَاكٌ [q. v.] or of حَبِيكَةٌ: الحِبُكُ is of a form unused [in any other instance]: (TA:) الحِبَكُ is like النِّعَمِ [as though its sing. were حِبْكَةٌ]: (Bḍ:) الحِبُكُ is affirmed to be a mixture of two dial. vars.: الحِبِكُ is of a rare measure, like إِبِلٌ &c. (TA.)

حُبْكَةٌ *i. q.* حُجْزَةٌ [i. e. The *part of the* إِزَار (or *waist-wrapper*) *where it is tied round the waist; which part is folded*, or *doubled*]: (Sh, Ḳ:) whence ▼ الاِحْتِبَاكُ, meaning "the binding, or tying, the ازار:" or the *folds of the* حُجْزَة, *let down, before the wearer, for the purpose of his carrying anything therein*. (TA.) And *An* ازار [*itself*]; as also ▼ حِبَاكٌ. (Ḥam p. 37.) And *A cord, or rope, which one binds on the waist*: (Ḳ:) and ▼ حِبَاكٌ [also] signifies *a cord, or rope, or an* ازار, *or other thing, with which the waist is bound*; pl. حُبَكٌ: whence the saying, فُلَانٌ حُبُكُ النِّطَاقِ, meaning ‡*Such a one prepared himself to go away*; or *applied himself exclusively and diligently to an affair*. (Ḥar p. 160.) And The *thong* (القِدَّةُ [in the CḲ, erroneously, القُدَّةُ]) *that connects the head to the* [*pieces of wood called*] فَرَاصِيف, *of the* [*camel's saddle called*] قَتَب, (Ḳ, TA,) *and of the* [*saddle called*] رَحْل; (TA;) as also ▼ حِبَاكٌ. (Ḳ.) Pl. (of the former, TA) حُبَكٌ and (of the latter, TA) حُبُكٌ. (Ḳ.)

حِبَاكٌ: see حُبْكَةٌ, in three places. — Also *An enclosure for cattle* (حَظِيرَةٌ), [*made*] *with canes*, or *reeds*, (بِقَصَبَاتٍ, [or perhaps we should read

بِقُضْبَانٍ, i. e. *with twigs*,]) *put crosswise, and then bound, or tied*: (Az, TA:) or *pieces of wood put together like a* حَظِيرَة, *and then bound in the middle with a cord, or rope, that joins them together.* (Lth, TA.) —— The حِفَاف [i. e. *selvages, or the like*,] of a garment, or piece of cloth. (Z, TA.) —— The *black threads with which are sewed the borders, or extremities,* of a [cloth of the kind called] لِبْد. (Ibn-'Abbád, TA.) —— *A streak, or line*, (طَرِيقَة,) in sand and the like; as also ↑ حَبِيكَة: pl. of the former حُبُك; and of the latter ↑ حَبَائِك: (Ṣ:) or حُبُك, the pl. of حِبَاك, signifies the *ridges* of sand [*that are formed by the wind*]; (Ḳ;) the *ripples* (دَرَج) of sand, and of water, when moved by the wind; pl. of حِبَاك and of ↑ حَبِيكَة: (Az, TA:) [i. e.] حُبُكُ المَآءِ signifies المُتَكَسِّرُ مِنْهُ [*the ripples of water*]: and so حُبُكُ الشَّعْرِ الجَعْدِ [*the rimples, or wavy forms, of crisp hair, appearing as though it were crimped*]: (Ḳ:) [and the like of other things: this is what is meant by the following passage:] Fr says, الحُبُكُ تَكَسُّرُ كُلِّ شَىْءٍ كَالرَّمْلِ إِذَا مَرَّتْ بِهِ الرِّيحُ السَّاكِنَةُ وَالمَآءِ القَائِمِ إِذَا مَرَّتْ بِهِ الرِّيحُ وَدِرْعُ الحَدِيدِ لَهَا حُبُكٌ أَيْضًا وَالشَّعْرَةُ الجَعْدَةُ تَكَسُّرُهَا حُبُكٌ: (Ṣ:) [respecting the حُبُك of a coat of mail, here mentioned, see what follows: in like manner,] ↑ حَبِيك (T, Ḳ) and ↑ حَبَائِك and حُبُك, all as pls. of ↑ حَبِيكَة, [or rather ↑ حَبِيك is a coll. gen. n.,] signify the *streaks* of locks of hair; (Ḳ;) or of a helmet; (T, Ḳ; [in the CḲ, البَيْضَة is erroneously put for البَيْضَة;]) and likewise of sand, *such as are made by the wind*: (T, TA:) the حُبُك of the sky, (Ṣ, Ḳ,) sing. ↑ حَبِيكَة, (Ḳ,) are the *tracks of the stars*: (Ṣ, Ḳ:) and ↑ حَبَائِك signifies also *streaks, or tracks, in the sky*: and the *heavens*; because in them are the paths of the stars: and حُبُك, the *streaks of a mountain*: (TA:) and حُبُكُ دِرْعٍ, *the rows of rings of a coat of mail*: (TḲ in art. حرشف:) [in a passage in the Ṣ, cited above, it seems to be implied that it means *the rimples, or folds, thereof*:] or *the scales of silver with which a coat of mail is ornamented*; likened to the scales on the back of a fish, by their being termed the حَرْشَف of a coat of mail: (TA in art. حرشف:) and حِبَاكُ الحَمَامِ, *the blackness of the part above the wings of the pigeon.* (Ibn-'Abbád, A, Ḳ.) The phrase رَأْسُهُ حُبُكٌ, in a description of Ed-Dejjál [or Antichrist], means *The hair of his head is rimpled* (مُتَكَسِّر) *by reason of crispness*; like stagnant water, and sand, when the wind blows upon them, and they in consequence thereof become rippled (يَتَجَعَّدَانِ); *and marked with streaks*: or, as some say, it is ↑ مُحَبَّكُ الشَّعْرِ, as in the Ḳ, meaning the same; (TA;) or *crisp-haired*: (Ḳ:) or حُبُكُ الشَّعْرِ, (IDrd, Ḳ,* TA,) meaning the same: (TA:) or إِنَّ شَعْرَهُ حُبُكٌ حُبُكٌ: (Ṣ:) or رَأْسُهُ حُبُكٌ حُبُكٌ. (TA.) In the phrase, in the Ḳur [li. 7], وَالسَّمَآءِ ذَاتِ الحُبُكِ, it is said that الحبك means *the tracks of the stars*, (Ṣ, Er-Rághib, TA,) *and the milky way*: or *ideal tracks*: (Er-Rághib, TA:) or *streaks of clouds*: (TA:) or *beautiful* طَرَائِق [which is generally understood to mean, in this instance, *streaks*, or the like; but may also be rendered *stages, one above another*, to the number of seven]: (Zj, TA:) or *structures*, or *construction*: (Mujáhid, TA:) or *beautiful construction.* (I 'Ab, TA.) See also the paragraph, above, commencing with ذَاتُ الحُبُك.

حَبِيك and ↑ مَحْبُوك *Bound, or tied*; *made fast, or firm*: (Ḳ, TA:) *made well*: *woven well*: (TA:) *made beautiful in the effect of the work therein*: applied to a piece of cloth: (Ḳ, TA:) and the former, [app. as meaning *firmly*, or *well, made*,] to a bow-string also. (TA.) —— For the former, see also حِبَاك, in two places.

حَبِيكَة and its pl. حَبَائِك: see حِبَاك, in seven places.

[حَبَّاك, in the present day, signifies *A sewer of the leaves of books*: *a binder of books*: and also *an ornamental sewer*: and *a maker of the kind of lace called* شَرِيط.]

مُحَبَّك *Striped*; applied to a [garment, or particularly to one of the kind called] كِسَآء. (A, TA.) —— مُحَبَّكُ الشَّعْرِ: see حِبَاك, in the latter part of the paragraph.

مَحْبُوك: see حَبِيك. —— [Hence,] *A strong horse*; (Ḳ;) *firm, or compact, in make*: (TA:) or *strong in make*; applied to a horse &c. (Ṣ.) And دَابَّةٌ مَحْبُوكَةٌ *A beast having a well-knit frame.* (Sh, TA.) And مَحْبُوكُ المَتْنِ وَالعَجُزِ *Even, and high, in the back and rump.* (Lth, TA.)

حبل

1. حَبَلَهُ, (Ḳ,) aor. ِ, inf. n. حَبْلٌ, (TA,) *He bound, tied, or made fast, him, or it, with a rope, or cord.* (Ḳ, TA.) —— [Hence,] حَبْلٌ signifies [also] † The *making a covenant.* (KL.) —— And † The *obtaining* أَمَان [i. e. *a promise*, or *an assurance, of security* or *safety*]. (KL.) —— And The *placing a snare for game.* (KL.) And The *catching game with, or in, a snare.* (KL.) You say, حَبَلَ الصَّيْدَ, (Az, ISd, Mṣb, Ḳ,) aor. ُ, (Mṣb,) inf. n. حَبْلٌ; (Mṣb, TA;) and ↑ احتبلهُ, (Az, Ṣ, ISd, Mṣb, Ḳ,) and ↑ تحبّلهُ; (TA;) *He took, or caught, the game with the* حِبَالَة [or *snare*]: (Az, Ṣ, ISd, Mṣb, Ḳ:) or *he set up the* حِبَالَة *for the game.* (ISd, Ḳ.) And حَبَلَتْهُ الحِبَالَةُ *The snare* [*caught him*, or] *clung to him*: and hence, قَذًى حَبَلَتْهُ عَيْنُهُ ‡ [*Motes which his eye caught*]; a metaphorical phrase, used by Er-Rá'ee; the eye being likened to the snare; and the motes, to game. (TA.) And حُبِلَ عَنِ البَرَاحِ † [*He was prevented, as by a snare, or by a rope, from quitting his place*]. (TA.) And احْتَبَلَهَا ↑ زَوْجُهَا [app. meaning † *Her husband entrapped her*: or *laid a snare for her*]. (TA.) And ↑ احتبلهُ المَوْتُ ‡ [*Death ensnared him*; or *took him*]. (ISd, Z, TA.) And حَبَلَتْهُ فُلَانَةُ ‡ *Such a woman smote his heart with her love*; [or *captivated him*;] as also ↑ احْتَبَلَتْهُ. (TA.) [And accord. to the CḲ, حَبْلٌ also signifies the same as مُدَاهَنَةٌ [i. e. † The *endeavouring to conciliate*; &c.]: but the reading in the TA, and in my MS. copy of the Ḳ, is دَاهِيَة: which, however, occurs afterwards in the Ḳ as a meaning of حِبْل and of حَبْل.] —— حَبِلَتْ, (Ṣ, Mgh, Mṣb, Ḳ,) aor. َ, (Mṣb, Ḳ,) inf. n. حَبَلٌ, (Ṣ, Mgh, Mṣb, Ḳ, TA, [in the CḲ حَبْل,]) said of a woman, (Ṣ, Mgh, Mṣb,) and of any female beast, (Mṣb,) *She was, or became, pregnant*: (Ṣ, Mṣb, Ḳ:) حَبَلٌ and حَمْلٌ signifying the same: (AO, Ṣ, ISd, Ḳ:*) or the former applies only to human beings; and the latter, to others. (Mṣb, TA.) You say وَقْتُ حَبَلِ أُمِّهِ بِهِ [*The time of his mother's being pregnant with him*]. (Ṣ.) —— [Hence,] حَبَلٌ signifies also ‡ The *being full.* (ISd, Ḳ, TA.) You say, حَبِلَ مِنَ الشَّرَابِ and المَآءِ, aor. َ, (Ḳ,) inf. n. حَبَلٌ, (Ḳ,* TḲ,) ‡ *He became full of beverage, or wine, and of water,* (Ḳ, TA,) *and his belly became swollen* [*therewith, like that of a pregnant woman*]. (TA.) —— And ‡ The *being angry.* (Ḳ,* TA.) You say, حَبِلَ فُلَانٌ ‡ *Such a one became angry.* (TḲ.)

2. حَبَّلَ الزَّرْعُ, inf. n. تَحْبِيلٌ, (M, A, Ḳ, [in the CḲ, and in my MS. copy of the Ḳ, erroneously, الزَّرْعَ,]) ‡ *The seed-produce shot forth* (قَذَفَ) *one part thereof upon another, or parts thereof upon others*: (M, Ḳ, TA:) or *the ears of the seed-produce* [or *corn*] *became compacted and filled with the grain.* (A, TA.)

4. احبل العِضَاهُ *The* [*trees called*] عضاه [*produced their* حُبْل, or حَبَل; or] *scattered their blossoms, and organized and compacted their fruit* [i. e. *their pods with the seeds therein*]; expl. by تَنَاثَرَ وَرْدُهَا وَعَقَدَ [meaning عَقَدَ الثَّمَرَ]: (A, O, Ḳ:) from الحُبْلَة [q. v.], like عَلَّفَ from العُلَّف. (AA, O, TA.) ══ احبلهُ, (Ṣ, Ḳ,) inf. n. إِحْبَالٌ, (TA,) *He fecundated it*; syn. أَلْقَحَهُ. (Ṣ, Ḳ.)

5: see 1.

8: see 1, in four places.

حَبْلٌ i. q. رَسَنٌ [as meaning *A rope, or cord*]; (Ṣ;) *a certain thing well known*; (Mṣb;) *a thing with which one ties, binds, or makes fast, a beast &c.*; syn. رِبَاطٌ: (M, Ḳ:) and i. q. رَسَنٌ [as meaning *a halter*]; (M, Mṣb, Ḳ;) as in the Ḳur cxi. 5; (TA;) and so ↑ مُحَبَّلٌ: (M, Ḳ:) in the former sense, the pl. [of pauc.] is أَحْبُلٌ (Ṣ, M, Ḳ) and أَحْبَالٌ (M, Ḳ) and [of mult.] حِبَالٌ (Ṣ, M, Mṣb, Ḳ) and حُبُولٌ (M, Ḳ) and حِبَالَةٌ (L voce جُرُوح) [and حُبُولَةٌ, agreeably with a usage of the Arabs, which is, to add ة to any pl. of the measure فِعَالٌ or of that of فُعُولٌ, (see حِجَارَة,)] and ↑ حَبَائِلُ, which is anomalous, as in the phrase حَبَائِلُ اللُّؤْلُؤِ [*cords of pearls*], occurring in a trad.; or this is a mistranscription for جَنَابِذُ, (Ḳ, TA, [in the CḲ حَنَائِذُ,]) with ج [and ن] and ذ: (TA:) and in the latter sense, the pl. is حُبُولٌ. (M, Mṣb, Ḳ.) In a trad. in which it is said that a man's hand is to be cut off for his stealing a

حَبْل, the حبل of a ship may be meant. (Mgh in art. بيض.) — [Hence, †A bond; cause of union; or link of connexion:] connexion with another by the bond of love or friendship or the like; (Ṣ, Ḳ, TA;) pl. حِبَالٌ: (TA:) mutual connexion by such a bond. (ISd, Mṣb, Ḳ.) You say, وَصَلَ فُلَانٌ حَبْلَ فُلَانٍ † Such a one married his daughter to such a one. (Ḥar p. 223.) And هُوَ يَحْطِبُ فِى حَبْلِ فُلَانٍ † He aids such a one in seeking, or demanding, a woman in marriage. (TA.) And it is said in a trad., إِنَّ بَيْنَنَا وَبَيْنَ الْقَوْمِ حِبَالًا وَنَحْنُ قَاطِعُوهَا † Verily there is between us and the party a connexion by the bond of love or friendship or the like, and we are severing it. (TA.) You say also, إِنَّهُ لَوَاسِعُ الْحَبْلِ ‡ Verily he is large, or liberal, in disposition; [or in the scope of his friendship;] and ضَيِّقُ الْحَبْلِ ‡ narrow therein. (TA.) — ‡ A covenant, or compact: (Ṣ, Mṣb, Ḳ, TA:) ‡ a covenant, or an obligation, by which one becomes responsible for the safety, or safe-keeping, of a person or thing: (Ḳ, TA:) and ‡ a promise, or an assurance, of security, or safety; (A 'Obeyd, Ṣ, Mṣb, Ḳ, TA;) such as a man, desiring to make a journey, used [and still uses] to take from the chief of a tribe: (A 'Obeyd, TA:) pl. حِبَالٌ. (TA.) You say, كَانَتْ بَيْنَهُمْ حِبَالٌ فَقَطَعُوهَا ‡ There were between them covenants, and obligations whereby they were responsible for one another's safety, and they broke them. (TA.) And it is said in the Ḳur [iii. 108], إِلَّا بِحَبْلٍ مِنَ ٱللّٰهِ وَحَبْلٍ مِنَ النَّاسِ ‡ Unless [they have] a covenant from God and a covenant from men: (Ibn-'Arafeh, TA:) for the unbeliever requires a covenant from God, which consists in his being of those who have a revealed scripture without which he cannot retain his religion nor enjoy protection, and a covenant granted to him by men. (Er-Rághib, TA.) And it is also said in the Ḳur [iii. 98], وَٱعْتَصِمُوا بِحَبْلِ ٱللّٰهِ i. e. ‡ [And hold ye fast] by the covenant of God: (TA:) or ‡ the means of approach, or access, unto God; i. e. the Ḳur-án, and the Prophet, and intelligence, &c., which are the means of obtaining the protection of God; for حَبْل is metaphorically applied to ‡ any means of access to a thing: (Er-Rághib, TA:) or these words of the Ḳur mean ‡ and follow ye the Ḳur-án, and abstain from schism. (A 'Obeyd, TA.) And in like manner, the saying of Ibn-Mes'ood, عَلَيْكُمْ بِحَبْلِ ٱللّٰهِ, means ‡ Keep ye to the Book of God; for it is a security for you, and a covenant, against the punishment of God. (A'Obeyd, TA.) — ‡ An elongated, or extended, tract of sand, (T, Ṣ, M, Mgh, Ḳ,) collected together, abundant, and high: (T, TA:) or حَبْلٌ مِنَ الرَّمْلِ means a long, extended, tract of sand, collected together, and elevated: (Mṣb:) [or simply a long, or long and elevated, tract of sand; likened to a rope, as is indicated in the Mgh:] pl. حِبَالٌ. (TA.) — [†A long, creeping, or twining, stalk or shoot or branch; likened to a rope or cord: pl. حِبَالٌ: often occurring in descriptions of plants by AḤn and others.] — See also حَبَلَة. — الحَبْلُ † The وَرِيد; [a name applied to each of the two carotid arteries, and sometimes to each of the two external jugular veins;] also called حَبْلُ الوَرِيدِ; a vein between the windpipe and the [two sinews called the] عِلْبَاوَانِ; (Fr, TA;) a certain vein in the neck, (Ṣ,) or in the حَلْق. (Mṣb.) — † The عَاتِق [or part between the shoulder-joint and the neck]: (Ḳ:) or الحَبْلُ, (Ḳ,) or حَبْلُ العَاتِقِ, (TA,) signifies the طَرِيقَة [app. here meaning, as it does in some other instances, oblong muscle] that is between the neck and the head of the shoulder-blade: or a sinew between the neck and the shoulder-joint: (Ḳ:) or حَبْلُ العَاتِقِ signifies a bond, or ligament, between the عاتق and the neck; (T, Mṣb, TA;) or between the neck and the shoulder-joint: (Lth, TA:) or certain sinews. (Ṣ.) — † A certain vein, or nerve, (عِرْق,) in the fore arm, (Ḳ, TA,) extending from the wrist until it becomes concealed in the shoulder-joint: (TA:) or حَبْلُ الذِّرَاعِ is [a vein, or nerve,] in the arm: (Ṣ:) or حِبَالُ الذِّرَاعَيْنِ signifies the sinews that appear upon the two fore arms; and in like manner, those of a horse. (TA.) One says, هُوَ عَلَى حَبْلِ ذِرَاعِكَ, (Ṣ, TA,) a prov., (Ṣ,) meaning ‡ He, or it, is near to thee: (T, Ṣ, Ṣgh:) or within thy power, or reach; or possible, or practicable, to thee; or easy to thee. (ISd, Z, TA.) — Also, (Ḳ,) or حَبْلُ الفَقَارِ, (TA,) † A certain vein, or nerve, (عِرْق,) in the back, (Ḳ, TA,) extending from the beginning thereof to its end. (TA.) — الحِبَالُ فِى السَّاقِ, (Ḳ,) or حِبَالُ السَّاقَيْنِ, (M,) † The sinews of the two shanks. (M, Ḳ.) — الحِبَالُ فِى الذَّكَرِ, (Ḳ,) or ▼ حَبَائِلُ الذَّكَرِ, (M,) † The veins (عُرُوق) of the penis. (M, Ḳ.) — الحَبْلُ also signifies The station of the horses collected for a race, before they are let go. (Ḳ.) [Probably it was marked by an extended rope; and for that reason was thus called.] ═ Also Heaviness; weight, or weightiness; ponderousness; syn. ثِقْلٌ. (Az, Ḳ.)

حِبْلٌ: see حَبْلَة.

حِبْلٌ A calamity, or misfortune; (Ṣ, Ḳ;) as also ▼ حُبُولٌ: (Ḳ:) pl. حُبُولٌ. (Ṣ, Ḳ.) ISd cites as an ex. the saying of El-Akhṭal,

* وَكُنْتُ سَلِيمَ القَلْبِ حَتَّى أَصَابَنِى *
* مِنَ اللَّامِعَاتِ المُبْرِقَاتِ حُبُولُ *

[And I was sound of heart until calamities befell me from the resplendent females, exhibiting their beauty]. (TA.) — رَجُلٌ حِبْلٌ † A learned, sagacious, intelligent man. (IAạr, Ḳ.*) [And حِبْلٌ also signifies † Very intelligent, or very cunning. Pl. أَحْبَالٌ.] You say, إِنَّهُ لَحِبْلٌ مِنْ أَحْبَالِهَا, meaning ‡ Verily he is one who possesses much intelligence, or much cunning: and verily he is a gentle manager of cattle. (ISd, Ḳ, TA.)

حَبَلٌ: see حَبَلَة. ═ It is also an inf. n.; i. e., of حَبِلَتْ. (Ṣ, Mgh, Mṣb, Ḳ.) ═ And it is also a simple subst.: (Ḳ, TA: [in the CḲ, وَاسْمٌ جَمْعٌ is erroneously put for وَٱسْمٌ:]) [i. e.] it also signifies The fœtus in the womb: (Mgh:) pl. أَحْبَالٌ. (Ḳ.) It is said in a trad., نَهَى عَنْ حَبَلِ ▼ الحَبَلَةِ, (Ṣ, Mgh,) or نَهَى عَنْ بَيْعِ حَبَلِ الحَبَلَةِ, (Mṣb, Ḳ,) i. e. He forbade the selling of the offspring of the offspring (Ṣ, Mṣb, Ḳ) in the belly (Mṣb, TA) of the she-camel &c.; (Mṣb;) [i. e.,] the offspring of the fœtus (A 'Obeyd, Ṣ, Mṣb) in the belly of the she-camel [&c.]; (A 'Obeyd, Mṣb;) [i. e.,] what the fœtus will bring forth, if it be a female; (Mgh;) the ة in الحبلة being the sign of the fem. gender; (A 'Obeyd, Mgh, Mṣb;) or a sign of intensiveness of the signification: (IAmb, TA:) for the Arabs in the Time of Ignorance used to sell the offspring of the offspring in the bellies (T, M, Mṣb, TA) of pregnant beasts, (T, Mṣb,) or of sheep or goats: (M, TA:) or the meaning is, what is in the belly of the she-camel: (A 'Obeyd, Esh-Sháfi'ee, Ḳ:) or the produce of the grape-vine before it has attained to maturity: (M, Ḳ:) but Suh disapproves of this last explanation, as a mistake occasioned by the ة in الحبلة. (TA.) — † Anything that is in another thing: thus, for instance, the pearl is the حَبَل of the oyster-shell; and the wine is the حَبَل of the glass bottle. (A, TA.) ═ ‡ Fulness; (ISd, Ḳ, TA; [see حَبِلَ:]) as also ▼ حُبَالٌ. (IAạr, Ḳ.) — ‡ Anger: (Ḳ, TA:) ‡ anger and grief; as in the saying بِهِ حَبَلٌ ‡ In him is anger and grief: (Az, ISd, Ḳ, TA:) from the same word as meaning the "pregnancy" of a woman. (Az, TA.) ═ حَبَلْ حَبَلْ A cry by which sheep or goats are chidden. (Ṣgh, Ḳ.)

حُبْلَة: see حَبَلَة.

حُبْلَةٌ The fruit, or produce, of the [kind of trees called] عِضَاه, (Ṣ, Ḳ,) in general: (Ḳ:) or the pod, or receptacle of the seeds, of the سَمُر and سَلَم; [so accord. to AZ; as appears from a comparison of passages in art. بل in the T and TA;] that of other [trees of the kind called] عضاه being termed سِنْفَة: (TA:) or the fruit, or produce, of the سَمُر, resembling the [species of kidney-bean called] لُوبِيَآء; (IAạr, TA;) or of the سَلَم and سَمُر and سَيَال, (M, Ḳ,) which is a curved thing, containing small black grains, resembling lentils: (M, TA:) or, accord. to AO, a species of tree; as is the سَمُر: (Az, TA:) pl. ▼ حُبْلٌ, [or rather this is a coll. gen. n.,] and [the proper pl. is] حُبَلٌ. (Ḳ.) Hence, in a trad. of Saad, وَمَا لَنَا طَعَامٌ إِلَّا الحُبْلَةُ وَوَرَقُ السَّمُرِ [We having no food except the حبلة and the leaves of the سمر]. (Ṣ, TA.) — A kind of ornament worn by women, (Ṣ, Ḳ, TA,) fashioned in the form of the fruit thus called, (TA,) and put upon necklaces, (Ṣ, TA,) used in the Time of Ignorance. (Aṣ, TA.) — A certain herb, (بَقْلَة, ISd, Ḳ,) sweet, or pleasant, of the herbs termed ذُكُور: so says ISd: and in one place he says, a certain tree which [the lizards termed] ضِبَاب eat. (TA.) — See also what next follows.

حَبَلَةٌ (M, A, Ḳ) and ▼ حَبْلَةٌ, (M, A,) or ▼ حُبْلَةٌ, (Ḳ,) † A grape-vine; (M, A, Ḳ;) its branches being likened to ropes, or cords: (A, TA:) or a stock of a grape-vine: (Ḳ:) the first of these words has the latter signification (Mgh, TA) accord. to Aṣ: (TA:) or it signifies a stock of a grape-vine having its branches spread upon its *trellises:* (TA:) or the first and second signify *a branch of a grape-vine:* (Ṣ:) or, accord. to Lth, حبلة [thus in the TA, without any syll. sign,] signifies *a grape-vine:* and also *a* طاق [app. here meaning *an arch*] of the branches of a grape-vine: so in the T: (TA:) and ↓حَبَلٌ and ↓حُبْلٌ [are coll. gen. ns., and] signify *grape-vines.* (Ḳ.) — حَبَلَةُ عَمْرٍو *A sort of grapes of Eṭ-Ṭáïf, white, and pointed at the extremities.* (TA.) = See also حَبْلٌ: = and see what next follows.

حُبْلَى *Pregnant;* (Ṣ, Mgh, Mṣb, Ḳ;) as also ↓حَابِلَةٌ; (Ḳ;) and ↓حُبْلَانَةٌ also occurs in the same sense: (ISd, Ḳ:) applied to a woman, (Ṣ, Mgh,) or, accord. to AZ, to any animal having a nail, (Ṣ,) or to any beast, as, for instance, a sheep, or goat, and a cat: (Mṣb:) pl. of the first حَبَالَى (Ṣ, Mgh, Mṣb, Ḳ) and حُبْلَيَاتٌ (Mṣb, Ḳ) and حَبَالَيَاتٌ, (Ṣ, TA,) which last is pl. of حَبَالَى: (TA:) and the pl. of حَابِلَةٌ is ↓حَبَلَةٌ, (Ḳ,) which is extr. (TA.) One says, اللَّيْلُ حُبْلَى لَسْتَ تَدْرِى مَا تَلِدُ † [*The night is pregnant: thou knowest not what it will bring forth*]: meaning that the events of the night are not to be trusted. (TA.) — See also حُبْلَانَةٌ.

حُبْلِىٌّ and ↓حُبْلَوِىٌّ and ↓حُبْلَاوِىٌّ *Of, or relating to, one that is* حُبْلَى, i. e. *pregnant.* (Ṣ, Ḳ.)

حُبْلَانَةٌ: see حُبْلَى. — [Hence,] حَبْلَانُ ‡ *Full* [of beverage, or wine, and of water; see حَبِلَ]; as also ↓حُبْلَانٌ: fem. of the former حَبْلَى; and of the latter ↓حُبْلَى [which is anomalous]: (AḤn, ISd, Ḳ, TA:) and ↓أَحْبَلُ a man *full of beverage or wine.* (Z, TA.) — And حَبْلَانُ ‡ *Angry;* (Ḳ, TA;) *full of anger;* عَلَى فُلَانٍ *against such a one:* (TA:) fem. with ة. (Ibn-'Arafeh, Ḳ, TA.)

حُبْلَانٌ: see the next preceding paragraph. [By rule, it should be with tenween, like عُرْيَانٌ, and should form its fem. with ة.]

حُبْلَوِىٌّ: / حُبْلَاوِىٌّ: } see حُبْلِىٌّ.

حُبَالٌ: see حَبْلٌ. = Also † *Much hair.* (Az, TA.)

حَبُولٌ: see حِبْلٌ.

حَبِيلُ بَرَاحٍ † One *who stands in his place like the lion, not fleeing:* (Ṣ:) or ‡ *courageous:* (Ḳ, TA:) and an appellation given to ‡ *a lion;* (Ḳ, TA;) as though he were prevented, as by a snare, or by a rope, from quitting his place; not quitting it, by reason of his boldness. (TA.)

حِبَالَةٌ (Ṣ, Mṣb, Ḳ) and ↓أُحْبُولَةٌ (Lth, Mṣb, Ḳ) and ↓أُحْبُولٌ (Lth, Ḳ) *A snare;* or *thing by means of which one takes, catches,* or *snares, game,* or *wild animals,* or *birds;* (Ṣ, M, Ḳ;) *of whatever kind it be;* (M, TA;) *a* شَرَك, and *the like:* (Mṣb:) or حبالة peculiarly applies to the *cord* (حَبْل) *of him who takes, catches,* or *snares, game or the like:* (Er-Rághib, TA:) pl. of the first حَبَائِلُ, (Mṣb, TA,) and of the second [and third] أَحَابِيلُ. (Mṣb.) It is said in a prov., خَشِّ ذُؤَالَةَ بِالحِبَالَةِ [*Frighten thou the wolf to catch him with the snare*]; ذؤالة meaning the wolf: applied to him whose threatening is not cared for: i. e., threaten another than me; for I know thee. (Meyd, TA.) — [Hence,] النِّسَآءُ حَبَائِلُ الشَّيْطَانِ † [*Women are the snares of the Devil*]. (TA.) And حَبَائِلُ المَوْتِ † *The causes of death.* (Ḳ.) And هُوَ حِبَالَةُ الإِبِلِ † *He is one who takes good care of the camels, so that they do not escape from him.* (TA.) — For the pl. حَبَائِلُ, see also حَبْلٌ, in two places; in the first sentence, and near the end of the paragraph.

حَابِلٌ One *who binds, ties,* or *makes fast, a rope,* or *cord.* (TA.) Hence, (TA,) يَا حَابِلُ اذْكُرْ حَلًّا, a prov., (Ḳ, TA,) meaning *O binder,* or *tyer, of the rope, bear in mind* the time of *untying.* (TA.) — *The setter of the snare* (حِبَالَة) *for game;* (Ṣ, TA;) as also ↓مُحْتَبِلٌ. (TA.) It is said in a prov., اخْتَلَطَ الحَابِلُ بِالنَّابِلِ (Ṣ) † *The setter of the snare became confounded with the shooter of the arrows:* (TA in art. خلط:) or, in this instance, (Ṣ,) الحابل signifies *the warp;* and النابل, *the woof.* (Ṣ, Ḳ.) And in another prov., ثَارَ حَابِلُهُمْ عَلَى نَابِلِهِمْ † *They kindled mischief among themselves:* (Ḳ, TA:) الحابل [properly] signifying *the owner of the* حِبَالَة; and النابل, *the shooter with* نَبْل, or *the owner of* نَبْل: i. e., their case became confused: and sometimes it is applied to a party whose case has become turned from its proper state, and who become roused, or stirred up, one against another. (Az, TA.) One says also, حَوَّلَ حَابِلَهُ عَلَى نَابِلِهِ † *He turned it upside down.* (Ḳ.) And اجْعَلْ حَابِلَهُ عَلَى نَابِلِهِ, and حَابِلَهُ نَابِلَهُ, †*Turn thou it upside down.* (TA.) — ‡ *An enchanter.* (Ṣgh, Ḳ, TA.) = A [lizard of the kind called] ضَبّ *that feeds upon the* حُبْلَة [q. v.]; (Ṣ, M, Ḳ;) and so a gazelle. (TA.) = حَابِلَةٌ: see حُبْلَى.

حَابُولٌ *A rope* [*in the form of a hoop*] *by means of which one ascends palm-trees;* (Ṣ, M, Ḳ;) *made of bark, or of* [*the fibres of the palm-tree called*] لِيف. (Ḥar pp. 544-5.)

أَحْبَلُ: see حَبْلَانُ, voce حُبْلَانَةٌ.

أُحْبُولٌ and أُحْبُولَةٌ: see حِبَالَةٌ.

مَحْبَلٌ The *time of pregnancy:* (Ḳ:) [or the *time of* one's *mother's pregnancy:* for] you say, كَانَ ذٰلِكَ فِى مَحْبَلِ فُلَانٍ *That was in the time of such a one's mother's being pregnant with him.* (Ṣ, TA.) So in the saying of El-Mutanakhkhil El-Hudhalee:

* لَا تَقِهِ المَوْتَ وَقِيَّاتُهُ * خُطَّ لَهُ ذٰلِكَ فِى المَحْبَلِ *

[*His possessions by means of which he preserves himself shall not preserve him from death: that was written for him in the time when his mother was pregnant with him:* or the last word is المَهْبِلِ: so in the TA in arts. هبل and وقى: see what here follows, in the next sentence]: or the meaning is that here following. (TA.) — [The *register of God's decrees;* which is called] the *first writing:* (ISd, Ḳ:) but in the verse cited above, the last word, accord. to some, is ↓المَحْبِلِ, (TA,) which means المَهْبِل, (Ḳ, TA,) and this is the reading best known, signifying *the place of gestation in the womb.* (TA.)

مَحْبِلٌ: see what next precedes.

مُحَبَّلٌ: see حَبْلٌ, first sentence. — Also Hair *crisped,* or *twisted and contracted:* so accord. to the Ḳ; in which is added, شِبْهُ الجُمَّلِ; but the right reading is شِبْهُ الحَبْلِ [*like the rope* or *cord*]: or *having its locks twisted like ropes* or *cords:* [thus many Ethiopian races, and some of the Arab women, twist their hair, like cords; and thus, generally, did the ancient Egyptians:] or, accord. to the M, *i. q.* مَضْفُورٌ [meaning *plaited,* or *twisted*]. (TA.)

مَحْبُولٌ A wild animal *caught,* or *entangled, in a* حِبَالَة [or *snare*]: (Ṣ:) or one *for which a* حبالة *has been set, though he may not as yet have fallen into it:* and ↓مُحْتَبَلٌ [in the CḲ erroneously مُحْتَبِل] one *that has fallen into it,* (ISd, Ḳ,) *and been taken.* (ISd, TA.)

مُحْتَبَلٌ: see what next precedes. — Also [The *place of the hobble;* i. e.] ‡ the *pastern* of a beast: (T, TA:) or the *pasterns* of a horse: (Ṣ, Ḳ:) originally used in relation to a bird caught in a snare. (A, TA.)

مُحْتَبِلٌ: see حَابِلٌ.

حبن

1. حَبِنَ, aor. ـَ; (Ṣ, Ḳ;) and حُبِنَ; inf. n. (of the former, TA) حَبَنٌ and (of the latter, TA) حَبْنٌ; (Ḳ;) *He* (a man) *had the dropsy;* as also ↓احتبن: (KL:) *he had a disease in the belly, whereby it became large and swollen.* (Ḳ.) — [Hence,] حَبِنَ عَلَيْهِ, aor. ـَ, (Ḳ,) inf. n. حَبَنٌ, (TA,) ‡ *He became filled with anger against him.* (Ḳ, TA.)

4. احبنهُ [*It caused him, or his belly, to become large and swollen*]: said of a disease [app. dropsy] that has befallen one; or of much eating. (TA.)

8: see 1.

حَبْنٌ The *tree called* دِفْلَى [q. v.]; as also ↓حَبِينٌ. (Ḳ.)

حِبْنٌ and ↓حِبْنَةٌ *i. q.* دُمَّلٌ [all which are applied in the present day to *A boil*]: (Ḳ:) and [*small swellings or pustules, of the kind termed*] خُرَاج, (Ḳ,) *like* دُمَّل: (Ṣ Ḳ:) or *a thing that comes upon the body,* or *person, generating pus,* or *thick purulent matter, and swelling:* pl. [of the former] حُبُونٌ. (Ḳ.) = Also, the former, *An ape,* or *a monkey;* syn. قِرْدٌ. (Kr, Ḳ.)

حَبَنٌ The *dropsy;* (Ṣ;) *a disease in the belly, whereby it becomes large and swollen.* (Ḳ.) — The *yellow water* [*of the blood;* i. e. the *serum:* a superabundant effusion of which, in the body, constitutes dropsy]. (TA.)

حِبْنَةٌ: see حِبْنٌ.

حَبِينٌ: see أَحْبَنُ: = and see also حَبْنٌ.

أُمُّ حُبَيْنٍ *A certain small beast or reptile,* (Ṣ, Ḳ,) *well known;* (Ḳ;) *the* عَظَايَة: (Mgh:) or *a species of the* [*kind of lizards termed*] عِظَآء *of stinking odour:* (Mṣb:) so called because of the largeness of its belly; from أَحْبَنُ [q. v.]: also called ▼ حُبَيْنَةٌ; (Ṣ, Mṣb, Ḳ;) and sometimes the article ال is prefixed to it, (Ṣ, Mṣb, Ḳ,) so that it is called أُمُّ الحُبَيْنِ, (Ṣ, Mṣb,) by poetic license: (TA:) it is *of the form of the* حِرْبَآء [or *chameleon*], *broad in the breast, and large in the belly:* (TA:) or, accord. to some, (TA,) it is the *female of the* حِرْبَآء: (Ṣ and Mṣb and Ḳ in art. حرب, and TA in the present art.:) accord. to Az, it is *a small reptile resembling the* [*kind of lizard called*] ضَبّ: (Mṣb:) or, as some say, *a certain reptile of the size of a man's hand:* or, accord. to Ibn-Ziyád, *a dust-coloured reptile, with four legs, and of the size of a frog that is not large; and when the children hunt it, they say to it,*

* أُمَّ الحُبَيْنِ انْشُرِى بُرْدَيْكِ * انَّ الأَمِيرَ نَاظِرٌ إِلَيْكِ *

[*Umm-el-Ḥobeyn, spread forth thy two wings: verily the commander is looking at thee*]: *they hunt it until fatigue overcomes it, when it stops, standing upright upon its two hind legs, and spreads forth two wings that it has, of the same dust-colour; and when they hunt it further, it spreads forth wings that were beneath those two wings, than which nothing more beautiful in colour has been seen, yellow and red and green and white, in streaks, one above another, very many; and when it has done this, they leave it: no offspring of it is found; nor any genital organ:* (TA:) the appellation أُمُّ حُبَيْنٍ is determinate, like ابْنُ عِرْسٍ and ابْنُ آوَى; (Ṣ, Mṣb;) and [so is ▼ حُبَيْنَةُ,] like أُسَامَةُ; (Ṣ;) but determinate as a generic appellation: (Ṣ, Mṣb:) the suppression of the article does not render it indeterminate; which is contr. to rule: (Ṣ, Ḳ:) the pl. is أُمَّهَاتُ حُبَيْنَاتٍ, [which is strange,] and أُمَّاتُ حُبَيْنٍ. (Mṣb.) — The Arabs say, in one of their imprecations, صَبَّ اللّٰهُ عَلَيْكَ أُمَّ حُبَيْنٍ مَاحِضًا meaning † [*May God pour upon thee*] *the night.* (Ibn-Buzurj, TA in art. محض.)

حُبَيْنَةُ: see the next preceding paragraph, in two places.

أَحْبَنُ *Having the dropsy;* (Ṣ, Mgh, Mṣb;) as also ▼ مَحْبُونٌ (KL) [and ▼ حَبِينٌ; so in the Lex. of Golius; and so in the present day]: *having a disease in the belly, whereby it becomes large and swollen:* (Ḳ:) fem. حَبْنَآءُ, (Ṣ, Ḳ,) applied to a woman: (Ṣ:) pl. حُبْنٌ. (TA.) — Hence, (TA,) the fem., ‡ *Big-bellied;* (Ḳ, TA;) applied to a woman. (TA.) And † A foot (قَدَمٌ) *having much flesh in the* بَخْصَة [app. here meaning the *pulpy portion of the sole*]; (Ḳ;) *as though it were swollen.* (TA.) And † A pigeon (حَمَامٌ) *that does not lay eggs:* pl. حُبْنٌ. (Ḳ.)

مَحْبُونٌ: see the next preceding paragraph.

مُحْبَئِنٌّ † *Angry.* (Ḳ.)

حبو

1. حَبَا, (Mṣb, Ḳ,) [aor. يَحْبُو,] inf. n. حُبُوٌّ, (Ḳ,) *He, or it* (a thing, Mṣb, TA), *was,* or *became,* or *drew, near.* (Mṣb, Ḳ.) And hence, (TA.) حَبَوْتُ لِلْخَمْسِينَ *I was,* or *became,* or *drew, near to fifty* [years]; (Ṣ, ISd, TA;) [as also حبوت الخمسين; for] IAạr says that حَبَاهَا and حَبَا لَهَا both have this signification. (TA.) — حَبَتِ الأَضْلَاعُ إِلَى الصُّلْبِ *The ribs joined to the backbone;* (Ḳ;) and in like manner, with the same meaning, one says of the entrails: and *the ribs were near to the backbone.* (TA.) And حَبَتِ الشَّرَاسِيفُ, (Ḳ,) inf. n. as above, (TA,) i. e. [*The extremities of the ribs, projecting over the belly,*] *were long, so that they were near one another.* (Ḳ.) And حَبَا المَسِيلُ *The water-course, or channel of a torrent, became* [*contracted,*] *so that one part thereof was near to another.* (Ḳ.) — حَبَا, (Ṣ, Mgh, Mṣb, Ḳ,) aor. يَحْبُو, (Mgh, Mṣb,) inf. n. حَبْوٌ, said of a child, (Ṣ, Mgh, Mṣb, Ḳ,) before he stands; (Lth, TA;) as also حَبِىَ, aor. يَحْبَى, inf. n. حَبْىٌ, which, however, is rare; (Mṣb;) *He crept,* or *crawled,* [or *dragged himself along,*] *upon his posteriors;* (Mgh;) or so حبا عَلَى آسْتِهِ: (Ṣ:) or *he went along upon his posteriors, protruding his chest:* (Ḳ:) or *went along on four* [or, as we say, *on all fours*]: in this last sense it is used by the lawyers. (Mgh.) And, said of a man, *He went along upon his hands,* or *arms, and his belly:* (Ḳ:) or *upon his hands,* or *arms, and his knees:* or *upon his posteriors:* or *upon his elbows and knees:* (TA:) [or *he crept,* or *crawled:* for] you say, مَا جَآءَ إِلَّا حَبْوًا, meaning *He came not save creeping,* or *crawling:* and مَا نَجَا فُلَانٌ إِلَّا حَبْوًا [*Such a one escaped not save creeping,* or *crawling*]. (TA.) Also, said of a camel having his fore shank bound up to his arm, *He crept,* or *crawled, along:* [or *he dragged himself along on the ground:*] and, said of a camel, *he lay down, and crept,* or *crawled,* [or *dragged himself along,*] *by reason of fatigue:* or, as some say, *being constrained to ascend a difficult tract of sand, he protruded his chest, and then crept,* or *crawled.* (TA.) — [Hence,] said of an arrow, *It glided along the ground, and then hit the butt:* (Ṣ:) or so حبا إِلَى الغَرَضِ. (Mṣb.) — And حبا المَالُ, (Ḳ,) inf. n. حَبْوٌ, (TA,) *The cattle clave to the ground, motionless, by reason of emaciation.* (Ḳ.) — And حَبَتِ السَّفِينَةُ, (Ḳ,) inf. n. حَبْوٌ, (TA,) *The ship ran.* (Ḳ.) — حَبَا لَهُ *It* (a thing) *presented itself,* or *its breadth,* or *width,* or *its side, to him,* or *it;* syn. اِعْتَرَضَ, (Ḳ,) or عَرَضَ; (Mgh;) as do, for instance, waves to a ship; (TA;) and as clouds, like a mountain, before they cover the sky. (Ṣ.) And حبا الرَّمْلُ, aor. يَحْبُو, inf. n. حُبُوٌّ, *The sands rose up, extending sideways* (مُعْتَرِضًا): (TA:) or *extended widely.* (IAạr, TA.) — حَبَاهُ, (Ṣ, Mṣb, Ḳ,) aor. as above, (TA,) inf. n. حَبْوٌ (TA) and حَبْوَةٌ, (Ṣ, TA,) or this is a simple subst., (Ḳ,) and the inf. n. is حِبَآءٌ, (Mṣb,) or this last also is a simple subst., (Ṣ,* Ḳ,) *He gave him* (Ṣ, Mṣb, Ḳ) a thing (Mṣb) *without any compensation* (Mṣb, Ḳ) *and without* [*receiving*] *any favour,* or *benefit:* or *in a general sense.* (Ḳ. [See also حِبَآءٌ below.]) You say, حَبَاهُ كَذَا and بِكَذَا *He* (God, or a man,) *gave him such a thing without* [*receiving*] *any favour,* or *benefit, and without requital.* (Ḥam pp. 327 and 654.) — And also, (Ḳ,) aor. as above, inf. n. حِبَآءٌ, (TA,) *He denied him, refused him,* or *refused to give him;* (Ḳ, TA;) on the authority of IAạr only. (TA.) Thus the verb bears two contr. significations. (Ḳ.) — حبا مَا حَوْلَهُ *He defended, protected,* or *guarded, what was around him;* (Aṣ, Ṣ, Ḳ;) as also ▼ حَبَّاهُ, inf. n. تَحْبِيَةٌ. (Ṣ, Ḳ.) J cites as an ex. of the former verb, from a poem of Ibn-Aḥmar, the phrase لَمْ يَحْبُهَا فَحْلٌ [as though meaning *A stallion did not defend them*]; referring to she-camels: but accord. to AḤn, it means *did not regard them;* being occupied with himself. (TA.) — You say also, فُلَانٌ يَحْبُو قَصَاهُمْ and يَحُوطُ قَصَاهُمْ [*Such a one fights in their defence;* or *defends them in a distant quarter:* but generally meant ironically: see 1 in art. حوط]: both signify the same. (Abu-l-'Abbás, TA.)

2: see 1.

3. حَابَاهُ, (Mṣb, Ḳ,) inf. n. مُحَابَاةٌ (Mṣb, Ḳ, KL) and حِبَآءٌ, (Ḳ,) *He vied,* or *contended, with him in giving.* (KL.) — *He aided him,* or *assisted him: he treated him,* or *behaved towards him, with partiality; was partial towards him:* and *inclined towards him:* (Ḳ:) *he treated him in an easy and a gentle manner.* (Mṣb.) — حَابَاهُ فِى البَيْعِ, (Ṣ, MA,) inf. n. مُحَابَاةٌ, (Ṣ, Mgh, KL,) *He abated the price,* or *payment, to him in selling:* (MA, KL, PṢ:) or *he treated him in an easy and a gentle manner therein:* (TḲ:) from حِبَآءٌ signifying "a gift." (Mgh.)

4. رَمَى فَأَحْبَى *He shot, and made his arrow to fall short of the butt* (IAạr, Ḳ) *and then to leap so as to hit the butt.* (IAạr, TA.)

5: see what next follows.

8. احتبى *He drew together and confined his back and his shanks* (Ṣ, Mgh, Mṣb, and Ḥar p. 179) *with his* رِدَآء, (Ṣ,) or *with a garment,* or *piece of cloth,* or *with some other thing,* (Mgh, Mṣb, and Ḥar ubi suprà,) *when sitting, to be like him who is leaning* [*his back against a wall*]: (Ḥar ubi suprà:) *he drew his legs against his belly with a garment,* or *piece of cloth, confining them therewith, together with his back, and binding it,* or *making it tight, upon them, so as to preserve him from falling,* [*when he sat,*] *like a wall:* (IAth, TA:) and ▼ تحبّى signifies the same: (TA:) or احتبى بِالثَّوْبِ *he inwrapped himself with the garment:* or *he drew together and confined his back and his shanks with a turban or the like:* (Ḳ:) for the Arabs not having walls in their deserts to lean against in their assembling, the man used to set up his knees in his sitting, and put against them a sword, or surround them [and his back] with a piece of cloth, or knit his hands, or arms, together upon them, and rest against them; this standing him in stead of leaning. (Ḥar ubi suprà.) The doing this in one garment is forbidden, in a trad., lest, by accident,

what decency requires to be concealed should become exposed. (IAth, TA.) You say also, احتبى بيديه [*He confined his legs against his belly with his hands, or arms, in sitting, to support himself by so doing*]. (Ṣ, Mṣb.*) [See also قُرْفُصَآءُ.] الاِحْتِبَآءُ with the sword is practised on the occasions of making a covenant for mutual protection, or war, or appointing a chief, and the like; because the sword may be wanted in these cases. (Ḥam p. 711.)

حَبِىٌّ : see حَبًا.

حَبْيَةٌ *A grape*: (Ḳ:) or *grapes when they first grow, from the berry, not from planting*: (TA:) pl. حُبًى. (Ḳ.)

حِبْوَةٌ : see حِبَآءٌ.

حُبْوَةٌ a subst. from اِحْتَبَى, (Yaạkoob, Ṣ, Ḳ,) as also ▼ حِبْوَةٌ (Ṣ, Mṣb, Ḳ) and حُبْيَةٌ (Ḳ) and ▼ حِبَآءٌ and ▼ حُبَآءٌ: (Ks, Ḳ:) meaning [The *act denoted by* اِحْتَبَى; i. e. اِحْتِبَآءٌ: and also] *a turban, or piece of cloth, or some other thing. with which a man performs what is termed* الاِحْتِبَآء: (Ḥar p. 179:) pl. حُبًى (Yaạkoob, TA) and حِبًى. (Yaạkoob, Ṣ, TA.) [See an ex. from a trad. voce نَمِرَةٌ: and see also a verse of El-Farezdaḳ cited voce حَلَّ.] Hence, حَلَّ حُبْوَتَهُ and عَقَدَ حُبْوَتَهُ mean † *He rose*, or *stood up*, and † *He sat*. (Ḥar p. 179. The former phrase is also mentioned in the Ṣ.) And the saying, الحُبَى حِيطَانُ العَرَبِ [*The things used for the purpose of* اِحْتِبَآء *are the walls of the Arabs*: see 8]. (TA.) And the saying, in a trad. of El-Aḥnaf (when he was asked in a time of war, "When is forbearance?"), ▼ عِنْدَ الحُبَآءِ [*On the occasion of* اِحْتِبَآء]; meaning that forbearance is to be approved in peace, not in war. (TA.) الحبوة on Friday, when the Imám is reciting the khuṭbeh, is forbidden; because الاِحْتِبَآء induces sleep, and exposes the purity of the worshipper to be annulled. (TA.) ═ See also حِبَآءٌ.

حِبْوَةٌ : see حُبْوَةٌ: ═ and see also حِبَآءٌ.

حُبَآءٌ; see حُبْوَةٌ, in two places.

حِبَآءٌ (Ṣ, Mgh, Ḳ) a subst. from حَبَاهُ "he gave him without any compensation" &c., (Ḳ,) as also ▼ حُبْوَةٌ (Mṣb, Ḳ) and ▼ حَبْوَةٌ and ▼ حِبْوَةٌ; (Ḳ;) all held by Lḥ to be inf. ns.: (TA:) or meaning *A gift*. (Ṣ, Mgh.) And the first, The *dowry of a woman* or *wife*. (TA.) ═ See also حُبْوَةٌ, in two places.

حَبِىٌّ : see حَابٍ. — Also *A collection of clouds*; syn. سَحَابٌ; because it creeps along; or from حَبَا meaning عَرَضَ, wherefore it is also called عَارِضٌ: (Mgh:) or applied to a collection of clouds as meaning *that presents itself*, or *its breadth*, or *width*, or *its side*, or *extends sideways*, (Ṣ, Ḥam p. 785, and EM p. 51,) *heaped up*, (EM,) *in the tracts of the horizon*, (Ḥam,) *like a mountain, before it covers the sky*; (Ṣ, EM;) as also ▼ حَبًا; (Ṣ;) so called because near to the earth, (Ṣ, Ḥam,) as though creeping, or crawling, like a child; or from حَبَا; like as سَحَابٌ is from سَحَبَ, (Ḥam,) or from سَحْبُ أَهْدَابِهِ: (TA:) or, as also ▼ حُبِىٌّ, *a collection of clouds overpeering* (يُشْرِفُ, in [some of] the copies of the Ḳ, erroneously, يشرق, TA) *from the horizon upon the earth*: or *heaped up, one part above another*. (Ḳ, TA.)

حُبِىٌّ : see the next preceding paragraph.

حَابٍ *Near*; applied to a thing of any kind. (Ṣ.) [Hence,] حَابِى الضُّلُوعِ *Having the heads of the ribs connected* [*by means of the cartilages*], *one with another*. (Az, TA.) And إِنَّهُ لَحَابِى الشَّرَاسِيفِ *Verily he is protuberant in the two sides*. (Ṣ.) — *Having the shoulder-joints elevated to*, or *towards, the neck*; (Ḳ;) applied to a man, and likewise to a camel. (TA.) ═ An arrow *that creeps along* (Ḳt, Ḳ) *upon the ground* (Ḳt) *to the butt*, (Ḳt, Ḳ,) *having fallen short of it*: (Ḳt:) or an arrow *that glides along the ground, and then hits the butt*: pl. حَوَابٍ. (Mṣb.) Hence the saying, in a trad., إِنَّ حَابِيًا خَيْرٌ مِنْ زَاهِقٍ, i. e. An arrow *such as is termed* حَابٍ, though weak, having hit the butt, *is better than one that goes beyond the butt* by its vehemence of passage, and its force, not having hit it: meaning, by the two arrows, one who attains the truth, or right, or a part thereof, though weak; and another who goes beyond it, and far from it, though strong. (TA.) ═ A thing *presenting itself*, or *its breadth*, or *width*, or *its side*; as also ▼ حَبِىٌّ; (Ḳ;) as in the saying of El-'Ajjáj, describing a [vessel such as is called] قُرْقُور,

* فَهْوَ إِذَا حَبَا لَهُ حَبِىٌّ *

i. e. [*So it*,] *when waves present themselves*, or *their breadth*, &c., *to it*. (TA.) [Hence,] رَمْلٌ حَابٍ *Overpeering sands presenting themselves*, or *their breadth*, &c. (TA.) And جَبَلٌ حَابٍ *A heavy, overpeering mountain*. (TA.) — Also *A certain plant*: (Ḳ:) so called because of its height. (TA.) And حَابِيَةٌ A tract of sand (رَمْلَةٌ), (Ḳ, TA,) *elevated and overpeering*, (TA,) *producing that plant*. (Ḳ, TA.)

حبى

1. حَبَى, aor. يَحْبِى, inf. n. حَبْىٌ : see 1 in art. حبو.

حُبْيَةٌ : see حُبْوَةٌ.

حُبِىٌّ and حَبِىٌّ : see art. حبو.

حت

1. حَتَّهُ, (A, Mgh, Mṣb, Ḳ, &c.,) aor. ـُ, (Mgh, Mṣb,) inf. n. حَتٌّ, (Ṣ, Mgh, Mṣb,) *He scraped it*, or *rubbed it, off*, (Az, Mgh, Mṣb, TA, and Ḥam p. 310,) or *rubbed it and scraped it off*, (A, Ḳ,) or *scraped it off by little and little*, (Az, TA,) namely, a thing, (TA,) as, for instance, blood, (A, TA,) or semen, (Ṣ, A,) or something dry, (Ḥam ubi suprà,) from a garment, (Ṣ, A, Ḥam, TA,) or the like, (Ṣ, Ḥam,) *with the hand*, or *with a stick*, or *piece of wood*, (Mgh, Ḥam,) or *with the end of a stone* or *of a stick* or *piece of wood*. (Az, Mṣb.) And حَتَّ الوَرَقَ, (A, Mṣb,) inf. n. as above, (Ṣ, Mṣb,) *He removed the leaves* [*by rubbing* or *scraping*], (Mṣb,) from a branch, (Ṣ,) or from trees. (A.) — Hence, حَتَّ اللّٰهُ مَالَهُ, (A, TA,) inf. n. as above, (TA,) ‡ *God destroyed*, or *may God destroy, his property*: (A:) or *God caused his property to pass away, and so reduced him to poverty*; or *may God cause* &c. (TA.) — And حَتَّهُ عَنِ الشَّىْءِ, (A,* TA,) aor. and inf. n. as above, (TA,) ‡ *He repelled him, drove him back*, or *turned him back, from the thing*. (A,* TA.) — حَتَّهُ مِائَةَ دِرْهَمٍ ‡ *He payed him hastily a hundred dirhems*. (A, TA.*) And حَتَّهُ مِائَةَ سَوْطٍ ‡ *He inflicted upon him hastily a hundred lashes with a whip*. (Ṣ, A.) — حَتَّ الشَّىْءَ i. q. حَطَّهُ [*He put the thing; put it down*; &c.]. (Ḳ.) ═ See also 6.

4. احتّ *It* (the kind of tree called أُرْطًى) *dried*, or *dried up*. (Ḳ.)

6. تحاتّ *It became rubbed and scraped off*; as also ▼ انحتّ: (Ḳ:) *it* (a thing) *became scattered, strewn*, or *dispersed*; or *became so by degrees, part after part*; syn. تَنَاثَرَ. (Ṣ.) And تحاتّ الوَرَقُ, (A, TA,) or تَحَاتَّتْ; (Ḳ;) and ▼ انحتّ, (A,) or انحتّت; (Ḳ;) and ▼ حَتَّتْ, (Ḳ,) [aor., app., ـِ, the verb being intrans.,] inf. n. حَتٌّ; (TA;) and ▼ تَحَتْحَتَتْ; (Ḳ;) *The leaves became rubbed and scraped off*: (A:) or *fell* (Ḳ) from the branch &c.: or *fell successively, one after another*. (TA.) And تحاتّت الشَّجَرَةُ *The tree shed its leaves, one after another*. (Mṣb.) And انحتّ ▼ شَعَرُهُ عَنْ رَأْسِهِ *His hair fell off from his head*. (TA.) And تحاتّت أَسْنَانُهُ *His teeth fell out, one after another*. (TA.) — [Hence,] تحاتّت ذُنُوبُهُ †*His sins fell from him*. (TA, from a trad.)

7: see 6, in three places.

R. Q. 1. حَتْحَتَ : see حَتْحَتَةٌ.

R. Q. 2: see 6.

حَتِّ (indecl., with kesr for its termination, TA) *A cry by which birds are chidden*. (Ḳ.)

حَتٌّ : see حُتَاتٌ. — Also *Dead*; [as though strewn upon the ground, in fragments;] applied to locusts (جَرَاد): pl. أَحْتَاتٌ; (Ḳ;) its only pl. (MF.) [Hence, app.,] تَرَكُوهُمْ حَتًّا بَتًّا and حَتًّا فَتًّا ‡ *They destroyed them*. (A, TA.) — *Dates* (تَمْرٌ) *not sticking together*. (Ḳ,* TA.) [See also حُتٌّ.] — ‡ *A fleet*, or *swift*, horse; (Ṣ, A;) *as though he scraped the ground*; (A;) *light in pace, and wide in step*: pl. as above: (Ṣ:) or *a fleet*, or *swift, and excellent* horse; (Ḳ;) *that runs swiftly and much, or that furrows the ground much with his feet*: (TA:) also *generous and high-bred* (كَرِيمٌ عَتِيقٌ [app. as applied to a horse]): (Ḳ:) and *a fleet*, or *swift*, camel: (Ḳ:) *a quick-paced and light-paced* camel; as also ▼ حَتْحَتٌ: (TA:) and *a male ostrich*. (Ḳ.) The Hudhalee says, (Ṣ,) namely, El-Aạlam, (TA,)

* على حَتِّ البُرَايَةِ زَمْخَرِيِّ السَّـ *
* ـسَوَاعِدِ ظَلَّ فِى شَرْيٍ طِوَالِ *

(Ṣ, TA:) he likens himself, says Aṣ, in his running and fleeing, to a male ostrich, as is shown by what precedes this verse: (Ṣ:) by حتّ البراية is said to be meant حَتٌّ عِنْدَ البُرَايَةِ, i. e. ‡ *fleet*, or *swift, when emaciated by journeying*; the subst. براية being said to be put for the inf. n. بَرْى: (A,* L:) some of the Baṣrees say that the poet means a camel; but Aṣ disapproves of this, because to that which he here describes he has before applied the epithet هِجَفّ: ISd says that in his opinion he likens his horse or his camel to a male ostrich, because of the epithet هجفّ, preceding, and because neither the horse nor the camel eats the colocynth, but this plant is cropped by ostriches; شرى meaning the colocynth: IJ says that شرى here signifies a tree of which bows are made; and the poet means that if the trees thus called are tall, they conceal him, and he is the more lonesome, or sad; and that if they were short, his eye would range freely, and he would be pleased, and would run gently. (L.) — See also حَتَّى, near the end of the paragraph.

حُتٌّ, applied to سَوِيق, *i. q.* مَلْتُوتٌ [i. e. *Moistened*, or *stirred about, with water*, &c.]. (Ḳ.) [See also حُتٌّ.]

حَتَتٌ *A disease that affects trees, in consequence of which their leaves fall off.* (TA.)

حتّة [app. حِتَّةٌ, as pronounced in the present day; pl. حِتَتٌ; now applied to *A bit* of anything; properly, *a paring*, or *scraping*;] *a piece of peel* or *bark* or *crust* or *the like*. (TA.)

حَتَاتٌ *Cries, shouts, noises*, or *clamour*; or *a confusion*, or *mixture, of cries* or *shouts* or *noises*, or *of crying* or *shouting* or *noise*. (Ḳ.)

حُتَاتٌ *What is rubbed and scraped off*; or *what becomes scattered, strewn*, or *dispersed*; or *what becomes so by degrees, part after part*; of a thing. (Ṣ, TA.) A word of this class generally ends with ة; (TA;) [as قُلَامَةٌ and نُجَارَةٌ &c.; but ▼ حُتَاتَةٌ seems to be also used in the same sense: and hence the phrase,] مَا فِى يَدِي مِنْهُ حُتَاتَةٌ, so in the A, but in the Ḳ ▼ حَتٌّ, (TA,) *There is not aught of it in my hand.* (A, Ḳ, TA.) — Also *A disease that attacks the camel, so that he becomes emaciated, and his flesh and fat and colour change, and his hair falls off.* (TA.)

حَتُوتٌ A palm-tree *of which the full-grown unripe dates fall off and become scattered, one after another*; as also ▼ مِحْتَاتٌ: (Ḳ:) and the latter, a tree *that scatters its leaves.* (TA.)

حُتَاتَةٌ: see حُتَاتٌ.

حَتَّى is a particle, used in three senses: (Mughnee:) it is a particle denoting the end of an extent; (Mughnee, Ḳ;) which is its predominant meaning; (Mughnee;) asserted by some to be always its meaning: (TA:) and denoting a cause, or motive: and syn. with إِلَّا as an exceptive; (Mughnee, Ḳ;) which last is the rarest, and is mentioned by few. (Mughnee.) — It is used as a preposition governing the gen. case, in the same manner as إِلَى (Ṣ, Mughnee) in respect of meaning and government, (Mughnee,) denoting the end of an extent; (Ṣ;) [signifying *To, till, until*, or *to the time of*;] but the word that it so governs must be a noun properly so called, not a pronoun except in a case of poetic license; and must signify the last part, or portion, of what is signified by that which precedes حَتَّى, as in the saying, أَكَلْتُ السَّمَكَةَ حَتَّى رَأْسِهَا [*I ate the fish*, even *to its head*]; or must be prefixed to the word signifying that part, or portion, as in the saying, [in the Ḳur xcvii. last verse,] سَلَامٌ هِيَ حَتَّى مَطْلَعِ الفَجْرِ [A night of *peace*, or of *salutation*, is *it, until the time of the rising of the dawn*]. (Mughnee.) It is also followed by a manṣoob aor., as in سِرْتُ حَتَّى أَدْخُلَهَا [*I journeyed until I entered it*, ادخلها here virtually meaning دَخَلْتُهَا], أَنْ being here understood after حَتَّى, and the ان together with the verb being rendered in grammatical analysis by an inf. n. governed in the gen. case by حَتَّى [so that حتّى ان ادخلها means حَتَّى دُخُولِي إِيَّاهَا]: this is one of the cases in which حَتَّى differs from إِلَى; for one may not say, سِرْتُ إِلَى أَدْخُلَهَا [with أَنْ understood after إلى]: and in the same sense it is used in the phrase, [in the Ḳur xx. 93,] حَتَّى يَرْجِعَ إِلَيْنَا مُوسَى [*Until Moses return to us*]. (Mughnee.) — It is also syn. with كَيْ, denoting a cause or motive [of action &c., signifying *To the end that, in order that*, or *so that*], as in the saying, أَسْلِمْ حَتَّى تَدْخُلَ الجَنَّةَ [*Become a Muslim, to the end that*, or *in order that*, or *so that, thou mayest enter Paradise*]; being in this case, likewise, followed by a manṣoob aor. (Mughnee.) — It is also used [as a preposition virtually governing the gen. case, أَنْ being understood after it,] in the sense of إِلَّا, meaning *Except*, or *unless*, likewise followed by a manṣoob aor., as in the following verse:

* لَيْسَ العَطَآءُ مِنَ الفُضُولِ سَمَاحَةً *
* حَتَّى تَجُودَ وَمَا لَدَيْكَ قَلِيلُ *

[*The giving of superfluities is not liberality*: (giving is not liberality) *except*, or *unless*, (or here we may also say *until*,) *thou be bountiful when little is in thy possession*]. (Mughnee.) — It is also a conjunction, like وَ, [signifying *And*, or rather *even*,] (Ṣ, Mughnee,) but on three conditions: first, that the word following it and conjoined by it be a noun properly so called, not a pronoun: secondly, that this noun signify a part, or portion, of what is signified by that which precedes حَتَّى, as in قَدِمَ الحُجَّاجُ حَتَّى المُشَاةُ [*The pilgrims arrived: even those on foot*], and أَكَلْتُ السَّمَكَةَ حَتَّى رَأْسَهَا [*I ate the fish: even its head*]: thirdly, that the noun following it and conjoined by it denote either the greatest or the least [literally or figuratively] of what are included in the signification of the noun that precedes حَتَّى, as in مَاتَ النَّاسُ حَتَّى الأَنْبِيَآءُ [*Men have died: even the prophets*], and زَارَكَ النَّاسُ حَتَّى الحَجَّامُونَ [*The people visited thee: even the cuppers*]. (Mughnee.) — It is also used as an inceptive particle, (Ṣ, Mughnee,) preceding a nominal proposition, (Mughnee,) as in the following verse (of Jereer [so in a copy of the Ṣ]):

* فَمَا زَالَتِ القَتْلَى تَمُجُّ دِمَآءَهَا *
* بِدِجْلَةَ حَتَّى مَآءُ دِجْلَةَ أَشْكَلُ *

[*And the slain ceased not to emit their blood into the Tigris, so that the water of the Tigris was of a mixed colour consisting of red and white*]: (Ṣ, Mughnee:) and preceding a verbal proposition, of which the verb is a pret., as in the phrase, [in the Ḳur vii. 93,] حَتَّى عَفَوْا وَقَالُوا [*So that they became numerous, and said*]: (Mughnee:) and preceding a marfooʼ aor., as in the phrase, [in the Ḳur ii. 210,] حَتَّى يَقُولُ الرَّسُولُ [*So that the Apostle said*, or, as in the Ṣ, *so that* this was the case: *the Apostle said*], accord. to him who reads يَقُولُ; (IHsh, in De Sacy's "Anthol. Gramm. Ar.," p. 82 of the Arabic text;) others reading يَقُولَ, which, as well as يَقُولُ, here means قَالَ. (Jel.) — [Respecting the cases in which the manṣoob aor. is used after حَتَّى, and those in which the marfooʼ aor. is used, the following observations are made.] When حَتَّى precedes a future, the latter is manṣoob, by reason of أَنْ understood before it, as in the saying, سِرْتُ إِلَى الكُوفَةِ حَتَّى أَدْخُلَهَا [*I journeyed to El-Koofeh until I entered it*: see above]: (Ṣ:) it is not manṣoob unless the verb is a future: if it is future with respect to the time of speaking, it must be manṣoob, as in حَتَّى يَرْجِعَ إِلَيْنَا مُوسَى [cited above]: if the verb is future with respect to what precedes, only, it may be manṣoob, as in حَتَّى يَقُولَ الرَّسُولُ [mentioned above]; or it may be marfooʼ; but not unless [somehow] denoting a present time; (Mughnee;) and if present with respect to the time of speaking, it must be marfooʼ, as in the saying, سِرْتُ إِلَى الكُوفَةِ حَتَّى أَدْخُلُهَا [*I have journeyed to El-Koofeh so that now I am entering it*]; (Ṣ, Mughnee;*) but if not really present, it is not marfooʼ unless denoting a past event as though it were present, as in حَتَّى يَقُولُ الرَّسُولُ [explained above]; nor may it be marfooʼ unless denoting an effect of what precedes it; so that you may not say, سِرْتُ حَتَّى تَطْلُعُ الشَّمْسُ, nor مَا سِرْتُ حَتَّى أَدْخُلُهَا unless the ما is regarded as prefixed to the entire affirmative phrase that follows it, nor هَلْ سِرْتَ حَتَّى تَدْخُلُهَا; nor may it be marfooʼ unless it denote a complement to what precedes it, so that you may not say سَيْرِى حَتَّى أَدْخُلُهَا, as the inchoative would in this case be without an enunciative. (Mughnee.) — [The following verse of El-Farezdaḳ is cited in the Mughnee as an ex. of حَتَّى used as an inceptive particle:

* فَوَا عَجَبًا حَتَّى كُلَيْبٌ تَسُبُّنِى *
* كَأَنَّ أَبَاهَا نَهْشَلٌ وَمُجَاشِعُ *

and it is there added that يَسُبُّنِى النَّاسُ must be understood in this verse before حَتَّى: but I rather think that حَتَّى is here a conjunctive particle, and that the meaning is, *And alas, my wonder!* the people revile me: *even Kuleyb revile me, as though their father were Nahshal and* as though he were *Mujáshiʻ*.] — It should be observed that حَتَّى may be used in three different ways in some sentences: thus you may say, using it as a preposition (in the sense of إِلَى), أَكَلْتُ السَّمَكَةَ حَتَّى رَأْسِهَا [*I ate the fish, even to its head*]; and, using it as a conjunction, حَتَّى رَأْسَهَا [*even its head*]; and, using it as an inceptive particle, حَتَّى رَأْسُهَا [for حَتَّى رَأْسُهَا مَأْكُولٌ *so that its head was eaten*]. (Mughnee.) — It is said in the Ḳ that حَتَّى renders makhfooḍ and marfooạ and manṣoob; and that therefore Fr said, أَمُوتُ وَفِى نَفْسِى مِنْ حَتَّى شَىْءٌ ["I shall die with something respecting حَتَّى (remaining unsettled) in my mind"]: this is said on the supposition that حَتَّى deviates in government from an established rule, which is, that a particle which governs one part of speech governs that only; those, for instance, which render manṣoob and mejzoom governing verbs only, and the particles that govern nouns governing none but nouns: but the truth is, that حَتَّى governs only the gen. case; a marfooạ noun or aor. after it would be so without it, as حَتَّى in this case is only an inceptive particle; and a manṣoob aor. after it is rendered so by أَنْ understood, so that حَتَّى in this case virtually governs a noun in the gen. case [as has been shown above]: therefore the remark quoted above from the Ḳ is faulty, and confounds things that should be distinguished. (MF, TA.) — Some say that حَتَّى is [a noun] of the measure فَعْلَى, from ♦ الحَتُّ *the ceasing from a thing, having finished it*; or *the becoming unoccupied by a thing*; like شَتَّى from الشَّتُّ: but Az disapproves of this, because, were it so, the pronunciation termed إِمَالَة would be allowable in its case, and it is not so: حَتَّى, he says, is a particle, not a noun, nor a verb. (TA.) — حَتَّامَ is originally حَتَّى مَا [*Till when? until when?* or *how long?*]: the ا of ما is elided because the expression is used interrogatively, like as it is when any preposition is prefixed to it if used interrogatively, as in بِمَ and فِيمَ and عَمَّ. (Ṣ.) — In the dial. of Hudheyl, عَتَّى is said for حَتَّى. (L.)

مَا تَرَكُوا إِلَّا رِمْدَةً حَتَّانَ, or حَتَّانٍ, (as in different copies of the Ḳ in art. رمد,) *They left not of them so much as thou mightest rub thy hands therewith and then blow it away in the wind after rubbing it off.* (Ḳ ubi suprà.)

حَتْحَتٌ: see حَتٌّ.

حَتْحَتَةٌ ‡ *Quickness,* (Ḳ, TA,) *and haste,* in anything. (TA.) [App. an inf. n., of which the verb is ♦ حَتْحَتَ.] Hence the prov., شَرُّ السَّيْرِ الحَتْحَتَةُ ‡ [*The worst pace is that which is quick and hasty*: but in Freytag's "Ar. Prov." (i. 654,) الحَقْحَقَةُ]. (TA.)

حَتْحَاتٌ i. q. حَثْحَاثٌ [*Quick,* or *swift,* &c.]. (Ḳ.)

حَتُوتٌ: see مِحَتَّاتٌ.

حتد

1. حَتَدَ بِمَكَانٍ, aor. ـِ, (Ṣ, L, Ḳ,) inf. n. حَتْدٌ, (L,) *He remained, stayed, abode,* or *dwelt, in a place,* (Ṣ, L, Ḳ,) *and became fixed,* or *settled.* (Ṣ, L.) = حَتِدَ, aor. ـَ, (L, Ḳ,) inf. n. حَتَدٌ, (L,) *It* (anything, L, Ḳ) *was pure,* (L,) or *pure in origin.* (Ḳ.)

2. حَتَّدَهُ, inf. n. تَحْتِيدٌ, *He chose it for its purity and excellence.* (Ḳ.)

حَتِدٌ Anything (L, Ḳ) *pure,* (L,) or *pure in origin.* (Ḳ.)

مَحْتِدٌ *Origin*; syn. أَصْلٌ; (IAạr, Ṣ, L, Ḳ;) app. *in respect of race,* or *lineage,* only, as several of the lexicologists have expressly asserted; (MF;) as also مَحْفِدٌ (Ṣ, L) and مَحْقِدٌ and مَحْكِدٌ: (IAạr, L:) pl. مَحَاتِدُ. (A.) You say, فُلَانٌ مِنْ مَحْتِدِ صِدْقٍ, (Ṣ,) or فِى مَحْتِدِ صِدْقٍ, (A,) [*Such a one is of a good,* or *an excellent, origin.*] And هُوَ كَرِيمُ المَحْتِدِ [*He is generous in respect of origin*]. (A, L.) — Also *Nature; natural,* or *native, disposition, temper,* or *the like.* (L, Ḳ.) You say, of a man who has done an act of kindness and reverted from it, رَجَعَ إِلَى مَحْتِدِهِ *He returned to his natural disposition.* (L.)

حتر

حِتَارٌ The *circuit, rim,* or *surrounding edge,* of a thing; *what surrounds* [*the whole* of] a thing: (Ṣ, Ḳ:) pl. حُتُرٌ. (Ṣ.) — The *hoop* of a sieve [and the like]. (TA.) — The *anus*; syn. حَلْقَةُ الدُّبُرِ; (Ḳ;) or such is the meaning of حِتَارُ الدُّبُرِ, and شَرْجُهُ: (Mgh in art. شرج:) and the *extremities of its skin*; i. e. the *place where the external skin and the extremities of the* خَوْرَانِ [or *rectum*] *meet*: or the *edges of the* دُبُر [or *anus*]: (TA:) or the *part between the anus and the anterior pudendum*: or the *line between the two testicles.* (Ḳ.) — حِتَارُ الأُذُنِ *The circuit of the edges of the gristles of the ear.* (TA.) — حِتَارُ العَيْنِ *The edges of the eyelids, which meet when the eye is closed*: (TA:) or the زِيق *of the eyelid,* (Ḳ accord. to some copies, [and this is the right reading, meaning *its edge,* زِيق being here used tropically, its proper signification being the "part" of a shirt "that surrounds the neck,"] as is said in the TA,) *internally*: in most copies of the Ḳ رِيقُ الجَفْنِ, [in the CḲ رِيْقُ الجَفْنِ,] with راء [not زاى]. (TA.) — حِتَارُ الظُّفُرِ *The part of the flesh which is around the nail.* (TA.)

حتف

1. حَتَفَ, aor. ـِ, is a verb mentioned by IḲooṭ and IḲṭṭ and others, as derived from حَتْفٌ signifying "death," though J says that no verb is formed from this word; as IF and Meyd and Az also assert: (MF, TA:) and حُتُوفٌ is its inf. n., as well as pl. of حَتْفٌ: [accord. to SM, it is intrans., signifying *He died*; for he says,] hence the saying in the A, المَرْءُ يَسْعَى وَيَطُوفُ وَعَاقِبَتُهُ الحُتُوفُ [*Man labours, and goes about: and his end is dying*]. (TA.) [But see what follows.] = IF says that no verb is formed from حَتْفٌ signifying "death;" and so, after him, J; and Az says that he had heard no such verb: but IḲooṭ mentions حَتَفَهُ, aor. ـِ, inf. n. حَتْفٌ, as signifying *He,* or *it, killed him*; or *caused him to die.* (Mṣb.)

حَتْفٌ *Death*: (Ṣ, Mṣb, Ḳ, &c.:) pl. حُتُوفٌ. (Ṣ, Ḳ.) You say, مَاتَ حَتْفَ أَنْفِهِ, (A'Obeyd, Ṣ, Mgh, Mṣb, Ḳ,) and حَتْفَ فِيهِ, but this is rare, and حَتْفَ أَنْفَيْهِ, (Ḳ,) which may mean مَنْخِرَيْهِ, or it may mean أَنْفِهِ وَفِيهِ, the انف being made predominant, (TA,) *He died upon his bed*; (A'Obeyd, Mgh, Mṣb, Ḳ;) [*a natural death*;] *respiring until he yielded his last breath*; (Mṣb;) *not from slaughter, nor beating,* (A'Obeyd, Ṣ, Mṣb, Ḳ,) *nor drowning, nor burning,* (A'Obeyd, Ṣgh, Mṣb, Ḳ,) *nor by a wild beast, &c.*: (A'Obeyd, TA:) the nose is particularized as meaning that the spirit passes forth from it with the breath; or because they imagined that the spirit of the sick man passes forth from his nose, and that of the wounded man from his wound; (IAth, Ḳ;) or because the spirit passes forth from the mouth and the nose, and the latter of these is made predominant: (O, TA:) and حتف is put in the accus. case in the manner of an inf. n. (TA.) This phrase was used in the time of paganism, by Es-Semow-al: (Mṣb, TA:) or, accord. to some, in the phrase attributed to him, the right reading is not مات حتف انفه, but مَاتَ فِى فِرَاشِهِ; and the former was first used by Moḥammad. (Ḥam p. 52.) It is said of a human being: (Ṣ, Mgh:) and then of any animal when it dies without any accidental cause: (Mgh:) hence, of a fish that has died, and floats upon the water. (Mṣb, TA.) It is said in a trad. of 'Ámir Ibn-Fuheyreh,

* وَالمَرْءُ يَأْتِى حَتْفُهُ مِنْ فَوْقِهِ *

[*And man, his death comes from above him*]: meaning that his caution and his cowardice will not repel from him the decree of death when it befalls him: originally said by 'Amr Ibn-Mámeh. (L, TA.) = حَيَّةٌ حَتْفَةٌ [*A dead serpent*] is a phrase like امْرَأَةٌ عَدْلَةٌ: (Z, TA:) the latter word is here an epithet. (Z, Ḳ.)

حتك

1. حَتَكَ, aor. ـِ, inf. n. حَتْكٌ and حَتَكَانٌ, *He walked with short steps, and quickly*; (Ṣ, Ḳ;) said of a man (T, Ṣ) &c.; like رَتَكَ, except that this is said peculiarly of the camel: (T, TA:) and ♦ تَحَتَّكَ signifies the same; (ISd, Ḳ;) or *he walked with a moving,* or *shaking, of his limbs, and with short steps.* (TA.) — لَا أَدْرِى أَيْنَ حَتَكُوا, (Ḳ,) or عَلَى أَيِّ وَجْهٍ حَتَكُوا, (Ṣ,) *I know not whither,* or *in what direction, they went,* or *have gone.* (Ṣ, Ḳ.) = حَتَكَهُ, (Ḳ,) aor. ـِ, inf. n. حَتْكٌ, (TA,) *He scraped it up*; or *searched,* or *sought, for it,* or *after it, in the dust,* or *earth*; namely, a thing; syn. بَحَثَهُ. (Ḳ, TA. [In the CḲ, نَحَتَهُ.]) — *He* (an ostrich, Ḳ, and any bird, TA) *dug it up,* or *hollowed it out,* (Ḳ, TA,) with his wings; (TA;) namely, sand, (Ḳ, TA,) and pebbles. (TA.)

5: see 1.

حَتَكٌ: see حَوْتَكَةٌ.

حَتَكَةٌ A man *despised and little in the eyes of others.* (Az, TA.)

حَتَكَى: see حَوْتَكَةٌ.

حَاتِكٌ *Slow, or short in step, and lacking strength or power.* (Az, TA.)

حَوْتَكٌ and ✝حَوْتَكِىٌّ *Short, and lean, or emaciated, and small in body, or slender in the bones,* (Az, Ṣ, Ḳ,) *and short in step;* (Az, TA;) applied to a man and to an ass: (TA:) or the former signifies anything *short:* (AZ, Th, TA:) or *small in body, and mean,* or *ignoble:* (Az, TA:) and ✝the latter, anything *small,* or *young:* (Ḥam p. 631:) and also, the latter, a man *that eats vehemently.* (Ibn-'Abbád, Ḳ.)

حَوْتَكَةٌ The *manner of walking of him who is short;* as also ✝حَتَكَى. (Ibn-'Abbád, Ḳ.) = Also sing. of حَوَاتِكُ, (Ibn-'Abbád, TA,) which signifies *Ill-fed* beasts. (Ibn-'Abbád, Ḳ.) — The same pl. also signifies *Young ostriches:* (Ṣ, Ḳ:) the *little ones of ostriches;* as also ✝حَتَكٌ. (Ibn-'Abbád, Ḳ.)

حَوْتَكَانٌ *Young children.* (Ibn-'Abbád, TA.)

حَوْتَكِىٌّ: see حَوْتَكٌ, in two places.

حَوْتَكِيَّةٌ *A certain kind of turban, worn by the Arabs:* (Sh, Ḳ:) said by some to be so called after a man named حَوْتَك, who wore it. (TA.)

حتم

1. حَتَمَهُ, aor. ـِ, (Ḳ,) inf. n. حَتْمٌ, (Ṣ, TA,) *He made it, or rendered it, firm; or established it, or settled it, firmly;* namely, a thing, or an affair. (Ṣ, Ḳ, TA.) — *He decreed it; ordained it; pronounced it; or decided it judicially:* (TA:) and so حَتَمَ بِهِ. (TḲ.) [See حَاتِمٌ.] — *He necessitated it; or made it, or rendered it, necessary, requisite,* or *unavoidable;* syn. أَوْجَبَهُ: (Ṣ, Ḳ,* TA:) and so حَتَمَ بِهِ. (Mṣb.) See حَاتِمٌ. You say, حَتَمْتُ عَلَيْهِ الشَّىْءَ *I made it, or rendered it, necessary,* &c., *for him to do,* or *to suffer, the thing.* (Ṣ. [Golius, reading أَوْحَيْتُ for أَوْجَبْتُ, the explanation in the Ṣ, has rendered حَتَمَ as signifying "inspiravit:" and the next explanation given by him, i. e. "indicavit," is a misprint for "judicavit."]) Or حَتَمَ عَلَيْهِ الأَمْرَ, aor. and inf. n. as above, signifies *He made the thing,* or *affair,* or *event, to be absolutely,* or *decisively,* or *irreversibly, necessary, requisite,* or *unavoidable, for him to do,* or *to suffer.* (Mṣb.) حُتُومٌ, which is pl. of حَتْمٌ, may also be an inf. n. of حَتَمَ, and is likewise pl. of حَاتِمٌ. (TA.)

5. تحتّم *He made a thing to be necessary,* or *unavoidable.* (Ḳ,* TA.) — [Hence, perhaps,] تحتّم لِفُلَانٍ بِخَيْرٍ *He wished such a one good:* or *he augured good for him.* (Ḳ.) — See also 7. = [*It* (a thing that was eaten) *was soft, yielding, crummy,* or *easily broken.*] It is said in the Ṣ that التَّحَتُّمُ signifies البَشَاشَةُ: but in a marginal note it is stated that there is an omission in this explanation, the right rendering being هشاشة الشَّىْءِ المَأْكُولِ. (TA.) One says, هُوَ ذُو تَحَتُّمٍ [*It is soft, yielding, crummy,* or *easily broken:* explained in the Ḳ by هَشَاشٌ, which is evidently a mistranscription for هَشَاشٌ, syn. with هَشٌّ]. (Ṣ, Ḳ.) And هُوَ غَضُّ المُتَحَتَّمِ [*It is fresh, juicy, sappy,* or *moist, in its soft, yielding, crummy,* or *crumbling, nature;* مُتَحَتَّمٌ being a regular inf. n.] (Ṣ, Ḳ.) — *It* (a ثُؤْلُول [or wart] when it had become dry) *crumbled,* or *broke into small bits.* (TA.) And *It* (a glass vessel) *broke in pieces, one part upon another.* (TA.) — *He ate a thing that was soft, yielding, crummy,* or *easily broken, in his mouth.* (Lth, Ḳ.) — *He ate the* حُتَامَة, i.e., the *food that remained upon the table,* or *what fell from it during eating,* (Ḳ, TA,) *of the crumbs of bread &c.* (TA.) It is said in a trad. that he who eats and does this will enter Paradise. (TA.) = تحتّم لِكَذَا *He was, or became, cheerful, brisk, lively,* or *sprightly, by reason of such a thing.* (Ḳ.)

7. انحتم *It* (a thing, or an affair, or event,) *was, or became, absolutely,* or *decisively,* or *irreversibly, necessary, requisite,* or *unavoidable;* as also ✝تحتّم. (Mṣb.)

Q. Q. 4. اِحْتَامَّ *He cut* or *cut off* [a thing]. (Ḳ.)

حَتْمٌ [inf. n. of 1: when used as a simple subst.,] *A decree, an ordinance, a sentence,* or *a judicial decision;* (Ṣ, Ḳ;) accord. to some, *adapted to a particular case:* (TA:) pl. حُتُومٌ. (Ṣ, Ḳ.) It is said in the Ḳur [xix. 72], كَانَ عَلَى رَبِّكَ حَتْمًا مَقْضِيًّا [*It is* imposed by Himself *upon thy Lord* as *a decree judicially decided*]. (TA.) And Umeiyeh Ibn-Abi-ṣ-Ṣalt says,

عِبَادُكَ يُخْطِئُونَ وَأَنْتَ رَبٌّ
بِكَفَّيْكَ المَنَايَا وَالحُتُومُ

[*Thy servants sin: and Thou art a Lord: in thy hands are the decrees and the judicial decisions*]. (Ṣ.) — [As a term of the law, it is sometimes used instead of قَضَآءٌ as opposed to أَدَآءٌ.] = Also [an inf. n. used as an epithet, signifying] *Necessary, requisite,* or *unavoidable; that must inevitably be done* [or *suffered*]: (TA:) or *indissoluble and irreversible:* as in the phrase قَضَآءٌ حَتْمٌ [*an indissoluble and irreversible decree* or *ordinance* or *sentence* or *judicial decision*]. (Mṣb in art. جزم.) — And [hence,] *Death.* (MA.) = *Pure; free from admixture; genuine:* formed by transposition from مَحْتٌ. (Ḳ.) One says, هُوَ الأَخُ الحَتْمُ *He is the pure,* or *genuine, true, brother.* (TA.)

حُتْمَةٌ *Blackness;* (Ḳ;) as also ✝حَتَمَةٌ. (TA.)

حَتَمَةٌ: see what next precedes. = Also *A flask* (قَارُورَةٌ) *broken into small pieces.* (Ḳ.)

حُتَامَةٌ The *food remaining upon the table:* (Ṣ, Ḳ:) or *what has fallen from it during eating,* (Ḳ,) *of the crumbs of bread, &c.* (TA.)

حُتُومَةٌ *Acidity, sourness,* or *pungency;* syn. حُمُوضَةٌ. (Ḳ.)

حَاتِمٌ *A judge;* or one *who decrees, ordains,* or *pronounces a judicial decision:* (Ṣ,* Ḳ,* TA:) pl. حُتُومٌ, (Ḳ,) like as شُهُودٌ is pl. of شَاهِدٌ. (TA.) — [Hence,] because, as they [the Arabs] hold, or assert, it necessitates separation, or departure, يُوجِبُ ✝يَحْتِمُ بِالفِرَاقِ (Ṣ, Mṣb, TA,) i. e. (Mṣb,) by its croaking, (Mṣb, TA,) الحَاتِمُ signifies *The* غُرَاب [a name applied to any species of *crow;* and here, app., particularly to the *raven*]: (Mṣb:) or *the black* غُرَاب: (Ṣ, Ḳ:) and [*the bird called*] غُرَابُ البَيْنِ, [see art. بين,] *which is red in the beak and legs:* (Ḳ:) said by Lḥ to be *that which is fond of plucking out its feathers, and which is held to be of evil omen.* (TA.) — And [hence,] حَاتِمٌ signifies also *Unlucky,* or *inauspicious.* (TA.) — See also أَحْتَمُ.

حَنْتَمٌ, of the measure فَنْعَلٌ, signifies *A green jar* (جَرَّةٌ خَضْرَآءُ, meaning خَزَفٌ أَخْضَرُ): it is also applied to anything [*of the colour termed*] أَسْوَد: and الأَخْضَرُ is, with the Arabs, أَسْوَدُ [which may mean either that green is, with the Arabs, termed اسود, or that الاخضر is, with the Arabs, black; but the former appears to be the right meaning: see أَسْوَدُ]. (Mṣb.) [See أَخْضَرُ: and see also what here follows.]

أَحْتَمُ Anything (TA) *black;* (Ḳ, TA;) as also ✝حَاتِمٌ. (TA.) [See also what next precedes.]

حتى

حَتِّىٌّ The سَوِيق [or *meal of what has been parched,* or perhaps *of what has been dried in the sun,*] *of the* مُقْل [or *fruit of the Theban palm,* or *cucifera Thebaica*]: (Ṣ, Ḳ:) or *what is rasped, of the* مُقْل, *when it has become ripe, and is then eaten:* (AḤn, TA:) also, (Ḳ,) or as some say, (TA,) the [*fruit called*] مُقْل, (Ḳ, TA,) *itself:* (TA:) or *what is bad thereof:* or *what is dry thereof.* (Ḳ.) [See an ex. in a verse cited voce دُرّ.] — The *refuse* (ثُفْل) *and skins* (قُشُور) *of dates:* (Ḳ:) [like حُتًّا and حُتِّىّ.] — The *scaly substances* (قِشْر [app. meaning *bits of the wax*]) *of honey,* or *of honey in the wax.* (Th, Ḳ.) — *I. q.* دِمْنٌ [*Dung of beasts, compacted together;* &c.] (Az, Ḳ, TA. [In the CḲ, الزِّمْنُ is put for الدِّمْنُ.]) — The *apparatus* (مَتَاع) *of the* [*kind of basket, made of palm-leaves, called*] زَبِيل: or *its* عَرْق [meaning the *suspensory, by which it is carried:* see this word, which also means the "suspensory" of a water-skin]; (Ḳ;) *its* كِتَاف [or *cord by which it is carried, being attached*] *in its* شَفَة [or *edge,* lit. *lip, and* app., as is commonly the case, *passed through a loop-shaped handle in the opposite edge, so that the two opposite edges are drawn together when it is carried:* كِتَاف originally signifying "a rope with which one's arms or hands are tied together behind his back"]. (TA.) — The مَتَاع [or *furniture and utensils,* &c.,] of a house or tent. (TA.) — And *What is bad of spun thread.* (TA.)

حُتًّى: see art. حت.

حث

1. حَثَّهُ, aor. ـُ, inf. n. حَثٌّ, *He hastened him,*

or *hurried him uninterruptedly*, or *in any manner*. (TA.) And حَثَّهُ عَلَيْهِ, (Ṣ, A, Mṣb, Ḳ,) aor. ُ, (Mṣb,) inf. n. حَثٌّ, (Ṣ, Mṣb, Ḳ,) with which are syn. ▼ حُثْحُوثٌ [app. a quasi-inf. n.] and ▼ حِثِّيثَى [an intensive inf. n.]; and ▼ اسْتَحَثَّهُ; and ▼ حَثَّثَهُ, (Ṣ, Ḳ,) inf. n. تَحْثِيثٌ; (Ṣ;) and ▼ احْتَثَّهُ; (A, Ḳ;) and ▼ احثّهُ; (Ḳ;) and ▼ حَثْحَثَهُ; (Ṣ, A, Ḳ;) *He incited, excited, urged, or instigated, him to it, or to do it*, (Ṣ, A, Mṣb, Ḳ,) namely, a thing; (Ṣ, Mṣb;) syn. حَرَّضَهُ; (Mṣb;) or حَضَّهُ; (Ṣ, A, Ḳ;) or حَثٌّ is used in relation to pace, or journeying, or marching; and حَضٌّ in other cases: so says El-Ḥareeree, on the authority of Kh. (TA.) You say, حَثَّ دَابَّتَهُ, and ▼ حَثْحَثَهَا, [*He incited, or urged, his beast*,] بِالسَّوْطِ [*with the whip*]. (A.) And حَثَّ الفَرَسَ عَلَى العَدْوِ, and ▼ اسْتَحَثَّهُ, *He made the horse to go quickly, or in a brisk or sprightly manner; or urged him to run by striking him with the foot, or by beating him.* (Mṣb.) — See also R. Q. 1.

2: see 1.

4: see 1.

6. تَحَاثُّوا, (Ṣ, A,) inf. n. تَحَاثٌّ, (Ḳ,) *They incited, excited, urged, or instigated, one another.* (Ḳ, TA.) You say, لَا يَتَحَاثُّونَ عَلَى طَعَامِ المِسْكِينِ *They do not incite, excite, urge, or instigate, one another to feed the needy.* (Ṣ, TA.) And التَّقْوَى أَصْلُ مَا تَحَاثَّ النَّاسُ عَلَيْهِ [*Piety is the principal, or best, thing to which men incite one another.*] (A, TA.)

8. احتثّ *He was, or became, incited, excited, urged, or instigated.* (Ṣ, Ḳ.) — See also 1. This verb is both trans. and intrans. (Ḳ.)

10: see 1, in two places.

R. Q. 1. حَثْحَثَ: see 1, in two places. — Also *He moved about* [a thing]; or *put* [it] *in motion*, or *into a state of commotion*; (Ḳ;) [and so ▼ حَثَّ, as is implied in the M, voce فَدَّ, where it is used as meaning *it* (a bird) *moved, or flapped*, its wings.] You say, حَثْحَثَ المِيلَ فِى العَيْنِ *He moved about the collyrium-style in the eye.* (A, TA.) And [hence,] حَثْحَثُوا ذٰلِكَ الأَمْرَ ثُمَّ تَرَكُوهُ † *They stirred up* (حَرَّكُوا) *that affair, and then left it*, or *abandoned it.* (TA.) — *He scattered about* the utensils, or furniture, of his house, or tent; as also عَثْعَثَ. (TA in art. عث.) — Also, inf. n. حَثْحَثَةٌ, *It was, or became, in a state of commotion*: (L:) or *in a state of consecutive motion.* (TA.) *It* (lightning) *was, or became, in a state of commotion*, (Ḳ, TA,) as some say, (TA,) *in the clouds.* (Ḳ, TA.) — *It* (rain, and hail, and snow,) *appeared and went away, without pouring down.* (L.) — *He* (a man) *slept.* (TA.)

حُثٌّ *Small particles of straw.* (Ṣ, Ḳ, TA.) — *Anything bruised, brayed, or broken into small particles.* (L.) — *Coarse sand*: (Aṣ, Ṣ:) or *what is* مُتَرَقْرِق, (Ḳ,) probably, [says SM,] a mistake for مُدَقْوَق [*broken into small particles*], agreeably with an explanation of حُثٌّ in the L, for, as to مترقرق, [he adds,] I have not found it in any book, (TA, [meaning that he had not found a signification assigned to it that is appropriate here,]) *of sand*, and *of earth, or dust*: or *what is dry, or firm, or hard, and coarse, of sand.* (Ḳ, TA.) — *Bread without any seasoning, or condiment, to render it pleasant, or savoury.* (A'Obeyd, Ṣ, Ḳ.) — سَوِيقٌ حُثٌّ, (Ṣ, Ḳ,) as also عُثٌّ, (TA in art. عث,) [*Meal of parched barley or wheat*] *not moistened, or stirred about, with water &c.*; expl. by غَيْرُ مَلْتُوتٍ; (Ṣ, and in a similar manner in the Ḳ;) and in like manner حُثٌّ is applied to collyrium (كُحْل), and to musk: or سويق *not finely ground.* (TA.) [See also حُتٌّ.] — تَمْرٌ حُثٌّ *Dates not sticking together.* (IAar, TA.) [See also حَتٌّ.]

حَثَاثٌ and ▼ حِثَاثٌ *Sleep*; as also ▼ حَثْحَاثٌ and ▼ حُثْحُوثٌ. (TA.) You say, مَا ذُقْتُ حِثَاثًا *I tasted not sleep.* (TA.) And مَا اكْتَحَلْتُ حَثَاثًا and حِثَاثًا *I slept not*: (Ṣ, Ḳ:) Aṣ says the latter; but A'Obeyd asserts the former to be the more correct: (Ṣ:) Th mentions both. (TA.) And مَا كَحَلْتُ عَيْنِى بِحِثَاثٍ *I have not anointed my eye with sleep.* (TA.) And مَا جَعَلْتُ فِى عَيْنِى حِثَاثًا [*I have not put any sleep into my eye*]; (A, TA;) meaning *I have been very wakeful.* (TA.) Or حثاث signifies *Light, or little, sleep.* (IDrst, TA.) It is related on the authority of an Arab of the desert that it signifies *A little collyrium*: and on the authority of El-Fihree, that it is *syn.* with بَرُودٌ, i. e. كُحْل [*collyrium*]. (TA.) — نَوْمٌ حَثَاثٌ or حِثَاثٌ *Little sleep.* (TA.) [See also حَثِيثٌ.]

حِثَاثٌ: see the next preceding paragraph.

حَثُوثٌ: see what next follows, in two places.

حَثِيثٌ, in the sense of حَاثَّةٌ, A woman *inciting, exciting, urging, or instigating.* (TA.) — And in the sense of مَحْثُوثَةٌ, A woman *incited, excited, urged, or instigated.* (TA.) [So, too, a man.] — And [hence,] A *sharp* man, *quick in his affair; as though his soul incited him*; as also ▼ حَثُوثٌ. (TA.) — And *Quick, or swift*; as also ▼ حَثُوثٌ and ▼ حَثْحَاثٌ and ▼ حُثْحُوثٌ: (Ḳ:) pl. of the first حِثَاثٌ. (L.) You say فَرَسٌ حَثِيثُ السَّيْرِ [*A horse quick, or swift, of pace*]. (A.) And مَضَى حَثِيثًا (A, Mṣb) *He passed, or went away, quickly, or swiftly.* (Mṣb.) And وَلَّى حَثِيثًا *He retreated quickly and eagerly.* (Ṣ.) And قَرَبٌ ▼ حَثْحَاثٌ i. e. [*A*] *quick* [*night-journey to water*], *in which is no flagging*: (Ṣ:) or [*a*] *hard* [*night-journey to water*]; as also حَذْحَاذٌ and تَحْتَاحٌ. (TA.) And ▼ خِمْسٌ حَثْحَاثٌ i. e. [*A journey in which the second and third and fourth days are without water*,] *in which is no flagging*; as also حَذْحَاذٌ: or *long and fatiguing, in which is no flagging*; as also قَعْقَاعٌ. (TA.) And ▼ حَيَّةٌ حَثْحَاثٌ *A serpent that is in constant motion.* (TA.) — نَوْمٌ حَثِيثٌ *Light sleep.* (IDrst, TA.) [See also حَثَاثٌ.]

حِثِّيثَى: see 1.

حَثْحَاثٌ: see حَثَاثٌ: — and see also حَثِيثٌ, in four places.

حُثْحُوثٌ: see 1: — and see also حَثَاثٌ: — and حَثِيثٌ.

فَرَسٌ جَوَادُ المَحَثَّةِ *A horse that runs repeatedly when incited* [*to do so*]. (Ṣ.)

حثرم

حُثْرُمَةٌ *Thickness of the lip.* (Ḳ.) — See also what next follows.

حِثْرِمَةٌ The دَائِرَة [or *small protuberance*] *in the middle of the upper lip*, (Ṣ, Ḳ,) *beneath the nose*; (Ḳ;) also termed خِثْرِمَةٌ, with خ as well as ح; (Ṣ and Ḳ in art. خثرم;) or, accord. to Aboo-Ḥátim Es-Sijzee, حُثْرُمَةٌ; (TA;) and حِثْرِبَةٌ: (Ḳ in art. حثرب:) when somewhat long, it is termed بُظَارَةٌ [q. v.]; (Ṣ in art. بظر;) and when it is so, the man is termed أَبْظَرُ. (Ṣ in the present art.) — Also The أَرْنَبَة [i. e. the *lobule*, or *lower portion, of the nose*]: or the *extremity of this*: (IAar, Ḳ:) accord. to IDrd, [▼ حُثْرُمَةٌ,] with fet-ḥ. (TA.)

حُثَارِمٌ A man (TA) *having a thick lip.* (Ḳ,* TA.)

حثف

حَثْفٌ and حِثْفٌ dial. vars. of حَتْفٌ and حِفْتٌ: (O, Ḳ:) pl. أَحْثَافٌ. (TA.)

حثل

حَثْلٌ: see what follows, in two places.

حُثَالَةٌ [The *grain of the weed called*] زُوَان, and *the like*, (M, Ḳ,) *of what is worthless*, (M, TA,) *found in wheat*, (M, Ḳ,) *and thrown away*; (M, TA;) said by Lḥ to be *somewhat grosser than dust, or earth, and than what are termed* دُقَاق [q. v.]: (TA: [see also حُصَالَةٌ:]) and *bits, or particles, that fall off, or are pared off*, (Ḳ, TA,) *from dates, and barley, and the like*: (TA:) or *what falls of the husks of barley, and rice, and of the skin of dates; and of everything of which bits, or particles, fall off, or are pared off; when it is picked for the purpose of removing what is bad*: (Ṣ:) the *refuse* of قَرَظ [or leaves of the mimosa flava] *after picking to remove what is bad*: (TA:) *dregs* of oil (Ṣ, TA) &c.; (TA;) the *thick, or turbid, portion that remains at the bottom of that which is clear*: (Mṣb in art. ثفل:) *broken pieces of straw*: (TA in art. حثر:) *what is worthless*: (Ḳ:) *what is bad*, and *what remains*, of wheat: (Lḥ, TA:) *what is bad* (Az, Ṣ, Ḳ) of dates, (Az, TA,) or, (Ṣ, Ḳ,) app., (Ṣ,) of anything; (Ṣ, Ḳ;) as also ▼ حُثْل. (Ḳ.) [Hence,] † The *refuse* of men or mankind; the *bad*, or *evil*, thereof; (T, TA;) as also ▼ حُثْل. (T.)

حثم

حَثْمٌ: see what next follows.

حَثْمَةٌ *A hill, or rising ground*; (Mṣb;) as also

↑ حَثْمَةٌ: (Az, as heard from the Arabs; and TA:) or *a red* [*hill such as is termed*] أَكَمَة: (Ṣ:) or *a small red* أَكَمَة: (Ḳ:) or *a black* أَكَمَة *consisting of stones:* (M, Ḳ:) as also ↑ حَثَمَةٌ: (Ḳ:) or, as some say, *an elevated road:* (Mṣb:) pl. حِثَامٌ (Ḳ,* TA) and حَثَمَاتٌ: (TA:) and ↑ حَثْمٌ [as a coll. gen. n.] signifies *elevated roads.* (TA.)

حَثَمَةٌ: see حَثْمَةٌ, in two places.

حثو and حثى

1. حَثَا التُّرَابَ and حَثَى التراب, (Ṣ, Mgh, Mṣb, Ḳ,) the latter of which is the more usual, or more supported by authority, (TA,) aor. يَحْثُو and يَحْثِى, (Ṣ, Mṣb, Ḳ,) inf. n. حَثْوٌ and حَثْىٌ (Ṣ, Mgh, Mṣb, Ḳ) and تَحْثَاءٌ, (Ṣ,) *He poured dust* (Mṣb, TA) *with his hand;* (Mṣb;) *threw it,* or *cast it;* (TA;) or *seized it with his hand and threw it;* (Mgh, Mṣb;) عَلَيْهِ [*upon him,* or *it*], (Ḳ,) and فِى وَجْهِهِ [*in his face*]: (Ṣ, Mṣb:) الحَثْىُ and الهَيْلُ both signify *the pouring of dust;* but the former, *not without* [*first*] *raising* it; and the latter, the sending forth, or down, without raising. (Ḥam p. 477.) [See an ex. in a verse cited in art. اى, conj. 6.] Accord. to IAth, the saying, in a trad., أُحْثُوا فِى وُجُوهِ المَدَّاحِينَ التُّرَابَ means † [*Repel ye with*] *disappointment* [*those who praise much*]: but some make it to have its overt meaning [of *throw ye dust in the faces of those who praise much*]. (TA.) The verb is also used in relation to water; as in the saying, ↑ يَكْفِيهِ أَنْ يَحْثُوَ ثَلَاثَ حَثَوَاتٍ, (Mṣb,) and كَانَ يَحْثِى ثَلَاثَ حَثَيَاتٍ, (TA, [and the like is said in the Mgb,]) i. e. † [*It is sufficient for him to throw,* and *he used to throw,*] *three handfuls* [of water]. (Mṣb, TA.) — حَثَى said of a jerboa means *He went so deep into his hole that he could not be dug out, the direction of his hole being unknown, and it being seen to be filled with dust,* or *earth, even with the rest of the ground.* (TA in art. حفر.) The jerboa throws up (يَحْثُو) the dust, or earth, from his hole. (IAạr, TA.) — حَثَوْتُ لَهُ also means † *I gave him something little in quantity,* or *paltry.* (Ṣ, Ḳ.) — You say also, حَثَا التُّرَابُ or حَثَى, aor. يَحْثُو and يَحْثِى; (Ḳ;) so in the copies of the Ḳ, but correctly يَحْثَا, [or rather يَحْثَى,] which is extr., like جَبَا [or جَبَى,] aor. يَجْبَا [or يَجْبَى,] and قَلَا [or قَلَى], aor. يَقْلَا [or يَقْلَى]; i. e. *The dust became poured;* or *thrown,* or *cast.* (TA.)

4. أَحْثَاهَا and احثى الأَرْضَ [*He turned,* or *threw, up the earth,* or *ground, searching for what was in it:*] both signify the same. (TA in art. حوث.) And أَحْثَتِ الخَيْلُ البِلَادَ *The horses bruised,* or *crushed, the regions* [*with their hoofs*]; as also أَحَاثَتْهَا. (Ḳ.)

10. اِسْتَحْثَوْا *They threw,* or *cast, dust, each in another's face.* (TA.)

حَثًا (Ṣ, ISd, TA) and حَثًى (Ḳ, TA) *Dust poured;* or *thrown,* or *cast:* (ISd, Ḳ, TA:) or *dust being poured;* or *being thrown,* or *cast:* dual حَثَوَانِ and حَثَيَانِ. (TA.) — Also, (Ḳ,) written in both ways as above, (TA,) The *skins* (قُشُور) *of dates;* (Ḳ;) and the *bad thereof:* (TA:) [like حَثِىٌّ:] pl. of حَثَاةٌ: (Ḳ:) [or rather this is a n. un., and what is called the pl. is a coll. gen. n.;] like حَصَاةٌ and حَصًا. (TA.) — And *Straw* (تِبْنٌ [in the CḲ. erroneously, تِين]): (Ḳ, TA:) or *broken pieces thereof;* (Lḥ, Ṣ, Ḳ;) i. e., *of straw:* (Lḥ, Ṣ:) or *straw* (تِبن [in the CḲ, again, تين]) *separated from the grain.* (Ḳ.)

حَثْىٌ *What is raised with the hand* [*to be poured,* or *to be thrown,* or *cast, therefrom; of dust;* and also † *of water*]; (Ḳ, TA;) or, as in some copies of the Ḳ, *with the two hands:* and so [↑ حَثْوَةٌ and] ↑ حَثْيَةٌ; [or rather these signify *a single handful of dust,* and † *of water, raised to be poured* or *thrown;*] pl. [حَثَوَاتٌ and] حَثَيَاتٌ. (TA.) See 1.

حَثْوَةٌ; pl. حَثَوَاتٌ: } see حَثْىٌ.
حَثْيَةٌ; pl. حَثَيَاتٌ: }

أَرْضٌ حَثْوَاةٌ *Land abounding in dust:* (Ṣ, Ḳ:) but IDrd says that it is asserted to be not of established authority. (TA.)

حَاثِيَاءُ *One of the holes of the jerboa;* (TA;) *like* نَافِقَاءُ: (Ḳ:) or the *dust,* or *earth, of the jerboa,* (IAạr, Ḳ,* TA,) *which he throws up with his legs from his hole:* (IAạr, TA:) pl. حَوَاثٍ [like as نَوَافِقُ is pl. of نَافِقَاءُ]. (IB, TA.)

مَحْثُوٌّ (Ḳ) and ↑ مَحْثِىٌّ (ISd, TA) Dust *poured;* or *thrown,* or *cast.* (Ḳ,* TA.) ↑ يَا لَيْتَنِى المَحْثِىُّ عَلَيْهِ [*O would that I were he upon whom dust is poured,* or *thrown:* (in Freytag's Arab. Prov. ii. 920, المَحْثِىّ:)] said on the occasion of wishing to be in the condition of him of whom the honourable treatment that he receives is concealed, and of whom the contemptuous treatment that he receives is made apparent: originating from the fact that a man was sitting with a woman, and a man attached to her approached; so, when she saw him, she threw dust in his [the former's] face, to show this man who was sitting with her that he should not draw near to her, lest their case should become known. (TA.)

مَحْثِىٌّ: see what next precedes, in two places.

حج

1. حَجَّ, aor. ـُ, (Ṣ, A, Mgh, Mṣb,) inf. n. حَجٌّ, (Ṣ, Mgh, Mṣb, Ḳ,) *He repaired,* or *betook himself, to,* or *towards,* syn. قَصَدَ, (Ṣ, A, Mgh, Mṣb, Ḳ,) a person (Ṣ, A, Mgh) [or place], in an absolute sense: or to, or towards, an object of reverence, veneration, respect, or honour: or, accord. to Kh, *he repaired,* or *betook himself, much,* or *frequently, to,* or *towards,* an object of this kind: and also *he repaired to, betook himself to,* or *visited,* a person: (TA:) and *he went to,* or *visited,* a person *repeatedly,* or *frequently.* (ISk, T, Ṣ, Mgh, Ḳ.*) You say also, حَجَّ بَنُو فُلَانٍ فُلَانًا *The sons of such a one continued long going repeatedly to visit such a one.* (Ṣ.) — Hence, (Ṣ, Mgh, Mṣb,) aor. and inf. n. as above, (Ṣ,) and inf. n. حِجٌّ also, (Sb, L,) or this is a simple subst., (Ṣ, Mṣb, Ḳ,) by a conventional usage, (Ṣ,) or predominantly, (Mgh,) or by restriction of its usage in the law, (Mṣb,) *He repaired to Mekkeh,* (Ṣ, Ḳ,) or *to the Kaạbeh,* (Mgh, Mṣb,) *to perform the religious rites and ceremonies of the pilgrimage;* (Ṣ, Mgh, Mṣb, Ḳ;) or *for the purpose of the* عُمْرَة [q. v.; but this latter meaning is very rare: the usual meaning is, *he performed the pilgrimage to Mekkeh and Mount 'Arafát, with all the rites and ceremonies prescribed to be observed at, and between, those two places*]: (Mṣb:) or *he repaired to the House* [*of God, at Mekkeh,*] *and performed the actions prescribed for that occasion by the law of the Ḳur-án and the Sunneh.* (L.) [See حَجٌّ, below.] You say also, حَجَّ البَيْتَ, aor. ـُ, inf. n. حَجٌّ, (T, Ṣ,) and ↑ احتجّهُ, (El-Hejeree, TA,) *He performed the pilgrimage to the House* [*of God, at Mekkeh*]; (T, Ṣ;) because people repair to it every year. (T, TA.) And حَجُّوا مَكَّةَ [*They performed the pilgrimage to Mekkeh*]. (A.) And مَا حَجَّ وَلٰكِنَّهُ دَجَّ *He did not repair to Mekkeh to visit the House of God,* (Aboo-Ṭálib, Az,) or *for the performance of the rites and ceremonies of the pilgrimage,* (Mṣb,) *but he journeyed for mercantile purposes.* (Aboo-Ṭálib, Az, Mṣb. [See also art. دج.]) And hence, accord. to some, لَجَّ فَحَجَّ, a prov., which see below. (TA.) — Also, (TA,) inf. n. حَجٌّ, (Ḳ,) *He came,* or *arrived.* (Ḳ, TA.) You say, حَجَّ عَلَيْنَا فُلَانٌ *Such a one came to us.* (TA.) — Also, [aor., accord. to rule, as above,] inf. n. حَجٌّ, *He shaved* [his head; as one does on completing the performance of the rites and ceremonies of the pilgrimage: see حَجٌّ, below]. (TA.) — Also, (IAạr, A, &c.,) aor. ـُ, inf. n. حَجٌّ, (TA,) *He probed* a fracture of the head, (Ḳ,) or a wound, (A, TA,) *with a* مِحْجَاج, (A, Ḳ,) or مِيل, (TA,) *for the purpose of curing it:* (TA:) or *he probed* a wound *to know its depth:* (IAạr, TA:) or *he examined* a cleft in the head *to know whether there were in it bone or blood:* (ISh, TA:) or *he dressed and cured* a wound in the head reaching to the brain: or *he poured boiled clarified butter upon* a fracture of the head, in consequence of which the blood was mixed with the brain, *until the blood appeared, which he took away with a little cotton:* (TA:) or حَجِيجٌ, inf. n. حَجٌّ, signifies *he probed a fracture of his head for the purpose of curing it:* (Ṣ:) or *he made a perforation in the bone* [*of his broken head*] (قَدَحَ فِيهِ) *with an iron instrument; it being broken so that the brain was befouled with blood, and pulled off the skin that had dried up, and then cured it, so that it closed up with a* [*new*] *skin:* it relates to a wound reaching to the brain. (L.) — Also, aor. ـُ, inf. n. حَجٌّ, *He cut out and extracted* a bone *from a wound.* (TA.) — Also, (A, Mṣb,) aor. ـُ, (Mṣb, TA,) inf. n. حَجٌّ, (Ḳ,) *He overcame* another *in,* or *by, an argument, a plea, an alle-*

gation, a proof, an evidence, or a testimony. (A, Mṣb, Ḳ.) See 3. It is said in a prov., لَجَّ فَحَجَّ (Ṣ, TA) *He was pertinacious in litigation, dispute, or altercation, and overcame therein* [as is implied in the Ṣ, and expressed in the TA]: or *he persevered until he performed the pilgrimage* [not having intended to do so when he set out: see Freytag's "Arab. Prov." ii. 452]. (TA.) ═ Also, (TA,) [aor., accord. to rule, ـِ ,] inf. n. حَجٌّ; (Ḳ;) and ▼ حَجْحَجَ, (Ḳ,) inf. n. حَجْحَجَةٌ; (TA;) *He refrained, forbore,* or *abstained,* (Ḳ, TA,) عَنْ شَىْءٍ *from a thing.* (TA.) [See also the latter verb below.]

3. حَاجَّهُ, (Ṣ, A, Mgh, Mṣb,) inf. n. مُحَاجَّةٌ (A, Mṣb, TA) and حِجَاجٌ, (TA,) *He contended with him in,* or *by, an argument, a plea, an allegation, a proof, an evidence,* or *a testimony.* (Ṣ, Mgh, Mṣb, TA.) You say, حَاجَّهُ فَحَجَّهُ *He contended with him in,* or *by, an argument, &c., and he overcame him therein,* or *thereby.* (Ṣ, A,* Mgh, Mṣb.) — [And hence, حَاجَّ *He pleaded in a lawsuit.*]

4. أَحَجَّهُ *He sent him to perform the pilgrimage to Mekkeh, and the religious rites and ceremonies thereof.* (Ṣ, Mṣb, Ḳ.)

6. تَحَاجٌّ [inf. n. of تَحَاجُّوا] The *contending, one with another, in a litigation, a dispute,* or *an altercation;* (Ṣ, Ḳ;) the *adducing arguments, pleas, allegations, proofs, evidences,* or *testimonies, one with another.* (KL.)

8: see 1. ═ [احتجّ بِشَىْءٍ *He adduced,* or *urged,* or *defended himself by adducing* or *urging, a thing as an argument, a plea, an allegation, a proof, an evidence,* or *a testimony.*] You say, اِحْتَجَّ عَلَى خَصْمِهِ بِحُجَّةٍ شَهْبَآءَ [*He argued against his adversary with a strong,* or *a difficult, argument, plea, &c.*]. (A.)

R. Q. 1. حَجْحَجَ, inf. n. حَجْحَجَةٌ: see 1, last signification. — Also *He retired,* or *drew back;* or *did so in fear:* (Ṣ, Ḳ:) or *he lacked power,* or *ability.* (TA.) One says, حَمَلُوا عَلَى القَوْمِ حَمْلَةً ثُمَّ حَجْحَجُوا *They made a single charge,* or *assault, upon the party, and then retired,* or *drew back;* or *drew back in fear:* (Ṣ, TA:) or *lacked power,* or *ability.* (TA.) — *He refrained from saying what he desired,* or *was about, to say;* (Ṣ, Ḳ;) like مَجْمَجَ: (Ṣ:) or *he did not reveal,* or *manifest, what was in his mind.* (M, TA.) It is said in one of the provs. of Meyd, نَفْسَكَ بِمَا تُحَجْحِجُ أَعْلَمُ *Thou thyself knowest better than others* [*what thou refrainest from uttering,* or] *what is in thy mind.* (TA.) — *He remained, stayed, abode,* or *dwelt,* (Ḳ, TA,) بِمَكَانٍ *in a place; not quitting it;* as also ▼ تَحَجْحَجَ. (TA.)

R. Q. 2. تَحَجْحَجَ: see what next precedes.

حَجٌّ and ▼ حِجٌّ, the former an inf. n., and the latter a simple subst., (Ṣ, Mṣb, Ḳ,) or the latter also is an inf. n., (Sb, L,) [both used as substs.,] The *pilgrimage to Mekkeh,* (Ṣ, Ḳ,) or *to the Kaabeh,* (Mṣb,) *to perform the religious rites and ceremonies prescribed to be observed on that occasion:* (Ṣ, Mṣb, Ḳ:) Ks makes no difference between these two words: some say that the former is employed to signify the *religious rites and ceremonies of the pilgrimage* because they follow the repairing to Mekkeh, or because they are completed by shaving [the head], or because people continue long going to and fro to perform them: accord. to Az, it signifies the *performance of the religious rites and ceremonies of the pilgrimage of one year;* and some say ▼ حِجٌّ and ▼ حِجَّةٌ: (TA:) or this last signifies *a single pilgrimage, for the performance of its appointed religious rites and ceremonies;* deviating from rule; (Ṣ, Mgh, Mṣb, Ḳ;) for by rule it should be ▼ حَجَّةٌ, (Ṣ, Mgh, Ḳ,) which, Th says, has not been heard from the Arabs: (Mgh, Mṣb:) Ks says that ▼ حَجَجْتُ حِجَّةً and رَأَيْتُ رِؤْيَةً are the only deviations from the model of فَعَلْتُ فَعْلَةً in all the language of the Arabs: but El-Athram and others are related to have said, We have not heard from the Arabs حَجَجْتُ حِجَّةً nor رَأَيْتُ رِؤْيَةً; they saying only ▼ حَجَجْتُ حَجَّةً: (L, TA:) whence it appears that ▼ حَجَّةٌ and ▼ حِجَّةٌ were both used: (TA:) the pl. of the latter is حِجَجٌ: (Mgh, Mṣb:) so in the saying, نَذَرَ خَمْسَ حِجَجٍ [*He made a vow to perform five pilgrimages*]. (Mgh.) Hence, ▼ ذُو الحِجَّةِ (Ṣ, Mgh, Mṣb) and ▼ ذُو الحَجَّةِ, (Mṣb, TA,) which latter is said by Ḳz and 'Iyáḍ and Ibn-Ḳurḳool to be the more common, (TA,) [or, accord. to Fei, the contr. is the case, for he says,] some pronounce it in the latter manner, (Mṣb,) [*The last month of the Arabian calendar;*] *the month of the pilgrimage;* (Ṣ, Mgh, Mṣb;) so called because the pilgrimage to Mekkeh, and the religious rites and ceremonies thereof, are performed in it: (TA:) pl. ذَوَاتُ الحِجَّةِ: (Ṣ, Mṣb:) they did not say ذَوُو الحِجَّةِ agreeably with the singular. (Ṣ.) [Hence also,] ▼ وَحِجَّةِ اللّٰهِ لَا أَفْعَلُ [*By the pilgrimage* which is the ordinance *of God, I will not do* this or that thing]: a form of oath used by the Arabs. (Ṣ, Ḳ.) What is commonly termed الحَجُّ is sometimes termed الحَجُّ الأَكْبَرُ [*The greater pilgrimage*]: العُمْرَةُ [q. v.] being termed الحَجُّ الأَصْغَرُ [*the minor pilgrimage*]. (Kull p. 168.) — See also حَاجٌّ.

حِجٌّ: see حَجٌّ, in two places: — and see also حَاجٌّ.

حَجَّةٌ: see حَجٌّ, in five places. ═ Also, (IAạr, Ḳ,) and ▼ حُجَّةٌ, (Ṣ, Ḳ,) the former of which is the word commonly known, (IAạr in a marginal note in a copy of the Ṣ,) and ▼ حَاجَّةٌ, which is a subst. like كَاهِلٌ and غَارِبٌ, (L,) The *lobe of the ear.* (Ṣ, L, Ḳ.) — And the first, The *bore,* or *perforation, of the lobe of the ear.* (AA, TA.) — And *A bead,* or *a pearl, that is hung in the ear;* (Ḳ;) sometimes called ▼ حَاجَّةٌ. (IDrd, TA.)

حُجَّةٌ *A mode* [*of argument or the like*] *by which one overcomes in a litigation, dispute,* or *altercation;* so called because recourse is had to it (لِأَنَّهَا تُحَجُّ), i. e. (تُقْصَدُ): (T, TA:) *that by which one rebuts,* or *repels, an adversary in a litigation, dispute,* or *altercation: an argument; a plea; an allegation:* [it may be true or false: see Ḳur xlii. 15, and xlv. 24:] (TA:) *a proof; an evidence; a testimony:* (Ṣ, Mṣb, Ḳ:) [*a title; a voucher:* often thus used in the present day:] also applied to a person; like ثَبْتٌ; (A and Mgh and TA in art. ثبت;) [as in the saying, مَنْ حَفِظَ حُجَّةٌ عَلَى مَنْ لَمْ يَحْفَظْ *He who preserves in his mind* a word, or an authority, &c., *is an evidence against him who does not;* occurring often in the larger lexicons, expressing the superior authority of hearsay, or usage, over analogy &c.; and in the saying,] أَنْتَ حُجَّةٌ عَلَى نَفْسِكَ [*Thou art an evidence against thyself*]; a phrase mentioned by Akh: (Ṣ in art. بصر:) [also, *an excuse:*] pl. حُجَجٌ (A, Mṣb) and حِجَاجٌ. (TA.)

حِجَّةٌ: see حَجٌّ, in four places. — Also *A year:* (Ṣ, Mṣb, Ḳ:) pl. حِجَجٌ. (Ṣ, A, Mṣb.) You say, أَقَمْتُ عِنْدَهُ حِجَّةً [*I stayed at his abode a year*], and ثَلَاثَ حِجَجٍ كَوَامِلَ [*three complete years*]. (A.) ═ See also حَجَّةٌ.

حُجُجٌ: see حَجِيجٌ, in two places: — and see also حَجَاجٌ.

حَجَاجٌ and ▼ حِجَاجٌ The *surrounding bone* of the eye, (Mṣb, TA,) *upon* [*the upper part of*] *which grows the eyebrow;* (TA;) the *bone that surrounds the cavity* of the eye, *upon* [*the upper part of*] *which grows the hair of the eyebrow:* (ISk, TA:) it is said in a trad. that a female hyena and her young ones were within the حِجَاج of the eye of an Amalekite: (TA:) or the [*supra-orbital*] *bone upon which grows the hair of the eyebrow;* (Ṣ, Ḳ;) the *bone that projects over the cavity* of the eye: (IAmb, Mṣb:) or the *upper bone, beneath the eyebrow:* (TA:) of the masc. gender: (Mṣb:) pl. [of pauc.] أَحِجَّةٌ (Ṣ, Mṣb) and [of mult.] ▼ حُجُجٌ, deviating from a general rule, accord. to which a sing. of the measure to which this belongs does not assume this form of pl. because the reduplication is disapproved: also, by poetic license, حَوَاجِجُ, contr. to rule, for حَوَاجُّ. (TA.) The expression فِى حَجَا حَاجِبٍ ضَمِرٍ is used by poetic license for فِى حَجَاجِ حَاجِبٍ ضَمِرٍ. (TA.) — [Hence,] both words also signify ‡ The *upper limb of the disk* (i. q. حَاجِب) of the sun, appearing when it begins to rise. (A, Ḳ, TA: but in the A, only the latter form of the word is given.) — Also, [hence,] both words, ‡ *A side.* (A,* Ḳ.) You say, مَرُّوا بِحِجَاجَىِ الجَبَلِ ‡ *They passed by the two sides of the mountain.* (A.)

حِجَاجٌ: see the paragraph next preceding.

حَجِيجٌ A man *upon whom the operation termed* حَجٌّ (the probing of a fracture of the head, &c.,)

has been performed; (Ṣ, L;) as also ▼ مَحْجُوجٌ. (L.) And A fracture of the head *that has been medically treated*, or *cured*:—and also *A certain mode of medical treatment*, or *curing*, of such a fracture. (Aṣ, TA.)—▼ حُجُجٌ (pl. of حَجِيجٌ, TA) signifies *Probed* wounds. (Ḳ.)—And ▼ this same pl., Roads *much furrowed* [by the feet of beasts or men] (مُحَفَّرَةٌ): (L, Ḳ:) but it is uncertain whether its sing., if it have any, be حَجِيجٌ or حِجَاجٌ. (MF.)═Also *i. q.* ▼ مُحَاجٌّ as act. part. n. of حَاجَّ: so in the phrase, أَنَا حَجِيجُهُ *I am he who will overcome him by arguments*, or *proofs*, or *the like*: occurring in a trad. relating to Ed-Dejjál. (TA.)═See also حَاجٌّ.

حَجَّاجٌ *A frequent performer of the pilgrimage to Mekkeh, and of the religious rites and ceremonies ordained for that occasion*: the ا in this word, as in other epithets of the same measure, does not [regularly] admit of imáleh; but when it is used as a proper name, it admits this, agreeably with rule: some pronounce its ا with imáleh even when it is in the nom. or accus. case, contr. to rule. (TA.)

حَاجٌّ act. part. n. of 1; *Repairing*, or *betaking himself*, *to* [a person or place]. (Mṣb.)—And hence, (Ṣ, Mṣb,) A man *repairing to Mekkeh*, (Ṣ, Ḳ,) or *to the Kaabeh*, (Mṣb,) *to perform the religious rites and ceremonies of the pilgrimage*; (Ṣ, Mṣb, Ḳ;) or *for the purpose of the* عُمْرَة: (Mṣb: [but see 1:]) [*a pilgrim* of Mekkeh; or *one who has performed the pilgrimage* of Mekkeh: see what follows:] as also ▼ حَاجِجٌ, (Ṣ, Ḳ,) the original form, sometimes used by poetic license: (Ṣ:) pl. حُجَّاجٌ and ▼ حَجِيجٌ (Ṣ, A, Mṣb, Ḳ) and حُجٌّ; (Ṣ, Ḳ;) or rather the second of these is a quasi-pl. n., a kind of noun which, as well as the coll. gen. n., is often called by the lexicographers a pl., though not so called by the grammarians: (MF:) حَاجٌّ is also used as a pl., syn. with حُجَّاجٌ, like as سَامِرٌ is with سُمَّارٌ: (Mgh:) it may be considered as a gen. n., and is sometimes a quasi-pl. n., like جَامِلٌ and بَاقِرٌ; (TA;) as is also ▼ حِجٌّ; signifying *a company of pilgrims* of Mekkeh; or *pilgrims, collectively*; (ISk, L;) and likewise ▼ حَجٌّ. (So in a marginal note in a copy of the Ṣ.) The fem. is ▼ حَاجَّةٌ: pl. حَوَاجُّ: (Ṣ, Ḳ:) you say حَوَاجُّ بَيْتِ ٱللّٰهِ when they have performed the pilgrimage; but when they have not yet performed it, [being in the act of performing it,] you say حَوَاجُّ بَيْتَ ٱللّٰهِ, in which latter case you would say حَوَاجٌّ were not this word imperfectly decl.; [and in like manner, حَاجُّ بَيْتِ ٱللّٰهِ, and حَاجٌّ بَيْتَ ٱللّٰهِ;] like as you say ضَارِبُ زَيْدٍ أَمْسِ, and ضَارِبٌ زَيْدًا غَدًا. (Ṣ.) [▼ حَاجِّىٌّ, as a n. un. of حَاجٌّ, considering the latter as a coll. gen. n., like رُومٌ, of which the n. un. is رُومِىٌّ, is commonly used by the Turks and Persians as signifying *a pilgrim* of Mekkeh: but I have not found it so used in any classical Arabic work.] You say, أَقْبَلَ ٱلْحَاجُّ وَٱلدَّاجُّ *The company of pilgrims* to Mekkeh, *and of men travelling for mercantile purposes, came*. (TA. [See also art. دج.]) And لَمْ يَتْرُكْ حَاجَّةً ▼ وَلَا دَاجَّةً *He left not a company of pilgrims* to Mekkeh (جَمَاعَةً حَاجَّةً), *nor a company of their followers*, or *dependents*. (TA from a trad. [See also arts. دج and دوج.])═Also *Overcoming in* [or *by*] *an argument*, or *a plea*, or *the like*. (Mgh.)

حَاجَّةٌ: see حَاجٌّ, in two places:═and see also حُجَّةٌ, in two places.

حَاجِجٌ:
حَاجِّىٌّ: } see حَاجٌّ.

هُوَ أَحَجُّ مِنْهُ *He is one who overcomes in* [or *by*] a حُجَّة [i. e. *an argument*, &c.,] *more than he*. (Mgh.)

مَحَجَّةٌ *A road*, or *way*: (Mgh, TA:) or the *middle* of a road; (M, voce جُرْجُمَة;) the *beaten track*, or *part* of a road *along which one travels*; (T, TA;) the *main part, and middle*, of a road; syn. جَادَّةٌ: (Ṣ, Mṣb:) pl. مَحَاجُّ. (A, TA.)—[Hence,] اِجْعَلِ ٱلْأَمْرَ مَحَجَّةً وَاحِدَةً †*Make thou the affair*, or *case*, [*uniform*, or] *one uniform thing*. (Fr, TA in art. بأج.)

مِحْجَاجٌ *A surgeon's probe*. (Ṣ, A, Ḳ.)═A man *much addicted to litigation, dispute*, or *altercation*. (Ṣ, Ḳ.)

مَحْجُوجٌ A man *repaired to*. (Ṣ.)═See also حَجِيجٌ.═Also A man *overcome in* [or *by*] *a* حُجَّة [i. e. *an argument*, &c.]. (A,* Mgh.)

مُحَاجٌّ: see حَجِيجٌ.

ضَرْبٌ مُحَجْحِجٌ *A blow that is feeble, and falling short*. (IAạr, TA.)

حجأ

1. حَجَأَ بِهِ, aor. ـَ, *He rejoiced in it*, or *at it*; namely, a thing, or an affair or event; (Ṣ, Ḳ;) as also حَجِئَ بِهِ, aor. ـَ: (Ḳ:) or the latter, *he was*, or *became, attached to it, and tenacious of it*; (Ṣ, Ḳ;) as also حَجِىَ بِهِ, without ء; and ▼ تَحَجَّأَ بِهِ: (Ṣ:) or *he laid*, or *kept, hold upon it, and clave to it*; (Fr, Ḳ;) as also حَجِىَ بِهِ, without ء; (Fr;) and حَجَا بِهِ, (TA,) and ▼ تَحَجَّأَ بِهِ. (Fr, Ḳ.)—حَجَأَ عَنْهُ كَذَا, aor. ـَ, *He kept*, or *withheld, from him such a thing*. (Ḳ.)

5: see above, in two places.

8. اِحْتَجَأَ بِهِ *He had recourse to him for protection*. (TA in art. حجر.)

حَجِئٌ بِهِ *Attached to, and tenacious of, him*, or *it*. (Fr, Ṣ.)—إِنَّهُ لَحَجِئٌ إِلَى بَنِى فُلَانٍ *Verily he is betaking himself for refuge*, or *protection, to the sons of such a one*. (AZ, Ḳ.*)—هُوَ حَجِئٌ بِكَذَا *He is adapted* or *disposed, apt, meet, suited, suitable, fitted, fit, competent*, or *proper, for such a thing*; or *worthy of it*: (Lḥ, Ḳ:) a dial. var. of حَجِىٌّ: (Lḥ:) you say, هُمَا حَجِئَانِ, and هُنَّ حَجَايَا. (TA.)

مَحْجَأٌ *A refuge; a place to which one has recourse for refuge* or *protection*. (Lḥ, Ḳ.)

حجب

1. حَجَبَهُ, (Ṣ, A, Mṣb,) aor. ـُ, (Mṣb,) inf. n. حَجْبٌ, (Mgh, Mṣb,) *He*, or *it, prevented, hindered, debarred*, or *precluded, him*, or *it*: (Mgh, Mṣb:) *he*, or *it, precluded him*, or *it*; i. e. *prevented him*, or *it, from entering*. (Ṣ, A.) [Hence,] الإِخْوَةُ يَحْجُبُونَ الأُمَّ عَنِ الثُّلُثِ [*Brothers* of a person deceased *preclude the mother from* receiving *the third* of the inheritance]. (Ṣ, A.)—Also, (A, Ḳ,) aor. as above, (TA,) inf. n. حَجْبٌ and حِجَابٌ, (Ḳ,) *He*, or *it, veiled, concealed, hid, covered*, or *protected, him*, or *it*; (A, Ḳ;) as also ▼ حَجَّبَهُ. (Ḳ.)—[And *It intervened* between two things. Thus the diaphragm is described as] جِلْدَةٌ تَحْجُبُ بَيْنَ الفُؤَادِ وَالبَطْنِ [*A piece of skin that intervenes between the heart and the belly*]. (A.)—[And *He held the office of* حَاجِب, i. e. *door-keeper*, or *chamberlain*. You say,] فُلَانٌ يَحْجُبُ لِلْأَمِيرِ *Such a one holds the office of* حاجب *to the prince, governor*, or *commander*. (A, TA.)

2: see 1.

5: see 8.

8. احتجب [*He*, or *it, became prevented, hindered, debarred*, or *precluded*: *he became secluded*; or *he secluded himself*:] *he*, or *it, became veiled, concealed, hidden, covered*, or *protected*; as also ▼ تحجّب. (Ḳ, TA.) You say, احتجب المَلِكُ عَنِ النَّاسِ [*The King secluded*, or *concealed, himself*, or *became secluded* or *concealed, from the people*]. (Ṣ, A.) And احتجبت الشَّمْسُ فِى السَّحَابِ ‡[*The sun became concealed*, or *concealed itself, in the clouds*]. (A, TA.)—[Hence, app.,] احتجبت المَرْأَةُ بِيَوْمٍ مِنْ تَاسِعِهَا, (Ḳ,) or مِنْ يَوْمٍ مِنْ تَاسِعِهَا, and (TA,) [as though meaning *The woman secluded herself from* the commencement of *a day of her ninth* month of pregnancy:] said of a pregnant woman, (TA,) when a day has passed of her ninth [month, during which it was probably a custom for a woman to seclude herself in the house or tent]. (Ḳ, TA.)

10. استحجبهُ *He appointed him to the office of* حَاجِب [i. e. *door-keeper*, or *chamberlain*]. (Ṣ, Ḳ.)

حَجْبٌ: see حَجَبَةٌ.═Also [*The windpipe*;] the *passage of the breath*. (Ḳ.)

حِجْبٌ *A hill*; syn. أَكَمَةٌ: (Ḳ:) or *a lofty* أَكَمَة. (TA.)

حِجَابَةٌ: see حَجَبَةٌ.

حَجَبَةٌ The *head* [or *crest*] *of the hip* or *haunch* (Ṣ, A) [of a man, (see حَرْقَفَةٌ,) and] of a horse;

(A;) i. e. *each of the* حَجَبَتَانِ, *which project above, or beyond, the* خَاصِرَتَانِ [*or two flanks*]: (Ṣ:) or the dual signifies the *two edges of the hip or haunch, that project above, or beyond, the* خَاصِرَة [*or flank*]: (Ḳ:) or the *two bones above the pubes, that project above, or beyond, the soft parts of the belly, on the right and left:* (Zj in his "Khalḳ el-Insán,"* and Ḳ:) or the *heads of the two hip-bones or haunch-bones, next the* حَرْقَفَتَانِ [q. v.]; pl. [or rather coll. gen. n.] ↓حَجَبٌ, and pl. of pauc. حَجَبَاتٌ: (TA:) and in a horse, the *parts of the two hips, or haunches, that project above, or beyond, the* [*skin called*] صِفَاق, *of the belly.* (Ḳ.)

حِجَابٌ [*A thing that prevents, hinders, debars, or precludes:*] *a thing that veils, conceals, hides, covers, or protects;* (Ṣ, Mṣb, Ḳ, TA;) because it prevents seeing, or beholding: (Mṣb:) *a thing,* (A, Ḳ,) or *body,* (Mṣb,) *that intervenes* (A, Mṣb, Ḳ) *between two things,* (A, Ḳ,) or *between two bodies;* which is [said to be] the primary signification; (Mṣb;) [*a partition, a bar, a barrier,* or *an obstacle:*] and sometimes applied to ideal things: (Mṣb:) pl. حُجُبٌ. (A, Mṣb, Ḳ.) You say, ضُرِبَ الحِجَابُ عَلَى النِّسَآءِ [*The veil,* or *curtain, was put,* or *let down, over the women*]. (A, TA.) And لَهُ دَعَوَاتٌ تَخْرِقُ الحُجُبَ [*He has prayers that rend the veils*]. (A, TA.) And مَا لِدَعْوَةِ المَظْلُومِ حِجَابٌ [*There is no veil,* or *obstacle, to the prayer of the wronged*]. (A, TA.) It is said in a trad., مَنِ اطَّلَعَ الحِجَابَ وَاقَعَ مَا وَرَآءَهُ [*He who gets sight and knowledge of the veil falls into that which is behind it*]: i. e., when a man dies, he falls into what is behind [one of] the two veils, that of Paradise and that of Hell: (ISh, TA:) or, accord. to some, اِطِّلَاعُ الحِجَابِ signifies the *stretching out the head* [*and looking over the veil*]; for he who examines into a thing stretches out his head to see what is behind the veil, or covering. (TA.) And in another trad., a saying of Mohammad, (TA,) إِنَّ ٱللّٰهَ يَغْفِرُ لِلْعَبْدِ مَا لَمْ يَقَعِ الحِجَابُ [*Verily God forgiveth the servant* (his creature) *as long as the precluding event shall not have happened*]: الحجاب here meaning the dying in the belief in a plurality of gods: (Ḳ,* TA:) as though one were precluded from true belief by death. (TA.) One says also, العَجْزُ حِجَابٌ بَيْنَ الإِنْسَانِ وَمُرَادِهِ [*Inability is a bar between man and his desire*]. (Mṣb.) And المَعْصِيَةُ حِجَابٌ بَيْنَ العَبْدِ وَبَيْنَ رَبِّهِ [*Disobedience is a bar between the servant and his Lord*]. (Mṣb.) — [Hence, in the present day, †*A written charm* or *amulet;* generally worn in a case (called بَيْت حِجَاب) suspended, on the right side, by a string passing over the left shoulder, or on some other part of the person: pl. of mult. حُجُبٌ, and of pauc. حِجَابَاتٌ and أَحْجِبَةٌ.] — [Hence also,] *A thin piece of flesh,* (Ḳ,) *resembling a piece of skin,* (TA,) *in the interior of the body, between the two sides, intervening between the lungs and the* قصب [in the Ḳ القَصَب, but this is evidently a mistranscription for القُصْب the *lower intestines;* for the حجاب is the *diaphragm,* or *midriff*]: (Ḳ, TA:) and حِجَابُ القَلْبِ, (A,) or حِجَابُ الجَوْفِ, (Ṣ,) signifies [*the same;* as also حِجَابُ الكَبِدِ; (see قَلْبٌ;)] *what intervenes between the heart and the rest of the* جوف; (Ṣ;) *the piece of skin that intervenes between the heart and the belly:* (A, TA:) or حِجَابُ القَلْبِ signifies *a certain fat that clothes the heart:* (ÁHeyth, TA in art. شغف:) [or it signifies, or signifies also, *the septum cordis:* see قَلْبٌ:] pl. حُجُبٌ. (A, TA.) Hence the saying, ‡هَتَكَ الخَوْفُ حِجَابَ قَلْبِهِ [*Fear rent open his midriff:* or *his septum cordis*]. (A, TA.) — † The *horizon:* [because it terminates the view:] so in the phrase, تَوَارَتْ بِٱلْحِجَابِ † *It* (the sun) *became concealed by the horizon;* occurring in the Ḳur [xxxviii. 31], and in a trad. (TA.) — ‡ *A mountain:* (A:) or *an elevated part of a mountain.* (Ḳ.) You say, قَعَدَ فِي ظِلِّ الحِجَابِ ‡ *He sat in the shade of the mountain.* (A.) — † The *place where a* [*stony tract such as is called*] حَرَّة *ends.* (Ḳ.) — † *A tract of sand uniformly continuous, and long.* (Ḳ.) — † The *light* of the sun: or the *tract,* or *side,* of the sun: (Ḳ, TA:) or [like حَاجِبٌ] *a side,* or *part,* of the sun. (TA.)

حِجَابَةٌ, (Ḳ,) or ↓حِجْبَةٌ, (Ṣ,) The *office of door-keeper* [or *chamberlain*]. (Ṣ,* Ḳ.) — And the former, The *office of door-keeper and guardian of the Kaabeh.* (TA.)

حَاجِبٌ, an epithet in which the quality of a subst. predominates, (TA,) *A door-keeper;* (Mṣb, Ḳ;) so called because he prevents persons from entering: (Mṣb:) [*a chamberlain:*] pl. حُجَّابٌ (Ṣ, Mṣb, Ḳ) and حَجَبَةٌ. (Mṣb, Ḳ.) And حَجَبَةُ البَيْتِ [*The door-keepers and guardians of the Kaabeh:* see حِجَابَةٌ]. (TA.) — *Each of the two bones over the eyes, with the hair and flesh upon them:* (IF, Mṣb, Ḳ:) or the *eyebrow;* the *hair growing on either of those bones:* (AZ, Ḳ:) so called because it precludes the rays of the sun from the eye: (TA:) of the masc. gender: (Lḥ, TA:) pl. حَوَاجِبُ. (Ṣ, Mṣb, Ḳ.) One says, إِنَّهُ لَمُزَجَّجُ الحَاجِبِ [*Verily he has the eyebrow made narrow and long, by the removal of redundant hairs;* or *made narrow and long and arched;* or *lengthened with antimony*]. (Lḥ, TA.) — [Hence, as being likened thereto,] † The *piece of wood that is over the lintel of a door-frame.* (Az, TA.) [See عَتَبَةٌ.] — ‡ The *edge,* (A,) or *side,* (Ḳ,) or *upper limb of the disk, that appears when it begins to rise,* (T, TA,) or the *first part that appears,* (Mgh,) of the sun, (T, A, Mgh, Ḳ,) and of the moon: (T, TA:) likened to the حاجب (A, Mgh) of the face (Mgh) of a man: (A:) and حَوَاجِبُ الشَّمْسِ *the sides of the sun.* (Ṣ.) You say, بَدَا حَاجِبُ الشَّمْسِ, (T, A, TA,) and القَمَرِ, (T, TA,) ‡ *The upper limb of the disk,* (T, TA,) or *the edge,* (A,) *of the sun appeared,* (T, A, TA,) and *of the moon.* (T, TA.) — † The *edge* of anything. (Ḳ.) A woman said to a man who was eating of the middle of a round cake of bread, كُلْ مِنْ حَوَاجِبِهَا ‡ *Eat of its edges,* (Aṣ, TA,) or *its sides.* (A.) — † [*The beginning* of the dawn.] You say, لَاحَتْ حَوَاجِبُ الصُّبْحِ ‡ *The beginnings of the dawn appeared.* (A, TA.)

مُحْتَجَبٌ: see what next follows.

مَحْجُوبٌ [pass. part. n. of 1]. You say مَلِكٌ مُحْتَجِبٌ (A) and ↓مُحْتَجَبٌ (Ṣ) and ↓مَحْجُوبٌ (A) and ↓مُحَوْجَبٌ (TA) [*A king secluded,* or *concealed,* from the people]. And اِمْرَأَةٌ مَحْجُوبَةٌ *A woman veiled, or concealed by a curtain or the like.* (TA.) And هُوَ مَحْجُوبٌ عَنِ الخَيْرِ [*He is debarred from good.*] (A, TA.) — † *Blind;* (Ṣ, Ḳ;) and so ↓مُحَوْجَبٌ. (TA.)

مُحْتَجَبٌ: see the next preceding paragraph.

مُحَوْجَبٌ: see مَحْجُوبٌ, in two places.

حجر

حَجَرَ, aor. ُ, (ISd, TA,) inf. n. حَجْرٌ (ISd, Mgh, Ḳ) and حُجْرٌ and حِجْرٌ and حُجْرَانٌ and حِجْرَانٌ, (ISd, Ḳ,) *He prevented, hindered, withheld, restrained, debarred, inhibited, forbade, prohibited,* or *interdicted,* (ISd, Mgh, Ḳ,) عَلَيْهِ *from him,* or *it:* (ISd, TA:) [or عليه is here a mistranscription for عَنْهُ: for] you say, لَا حَجْرَ عَنْهُ, meaning *There is no prevention,* &c., *from him,* or *it:* (TA:) and حَجَرَ عَلَيْهِ, aor. ُ, inf. n. حَجْرٌ, (Ṣ, A,* Mṣb,) *He* (a ḳádee, or judge, Ṣ, A) *prohibited him* (a young or a lightwitted person, TA) *from using,* or *disposing of, his property according to his own free will:* (Ṣ, A, Mṣb, TA:) or حَجَرَ عَلَيْهِ فِي مَالِهِ *he* (a ḳádee) *prevented,* or *prohibited, him from consuming,* or *wasting,* or *ruining, his property.* (Mgh.) — See also 5: — and 8.

2. حَجَّرَهُ: see 5. — حَجَّرَ حَوْلَ أَرْضِهِ [*He made a bound,* or *an enclosure, around his land*]. (A. [Perhaps from what next follows; or the reverse may be the case.]) — حَجَّرَ عَيْنَ البَعِيرِ, (Mṣb,) inf. n. تَحْجِيرٌ, (Ṣ, L,) *He burned a mark round the eye of the camel with a circular cauterizing instrument:* (Ṣ, L, Mṣb:) and حَجَّرَ عَيْنَ الدَّابَّةِ, and حَوْلَهَا, [i. e. حَوْلَ عَيْنِهَا, like as is said in the A,] *he burned a mark round the eye of the beast.* (L.) — حَجَّرَ البَعِيرُ *The camel had a mark burned round each of his eyes with a circular cauterizing instrument.* (Ḳ. [Perhaps this may be a mistake for حُجِّرَ البَعِيرُ: or for حَجَّرَ البَعِيرَ, meaning *he burned a mark round each of the eyes of the camel* &c.: but see what follows.]) — حَجَّرَ القَمَرُ, (Ṣ, Ḳ,) inf. n. as above, (Ḳ,) *The moon became surrounded by a thin line, which did not become thick:* (Ṣ, Ḳ:) and (Ṣ [in the Ḳ "or"]) *became surrounded by a halo in the clouds.* (Ṣ, Ḳ.)

5. تَحَجَّرَ عَلَيْهِ *He straitened him,* (Ḳ, TA,) and *made* [a thing] *unlawful to him,* or *not allowable.* (TA.) And تَحَجَّرَ مَا وَسَّعَهُ ٱللّٰهُ *He made strait to himself what God made ample.* (A.) And تَحَجَّرْتَ عَلَيَّ مَا وَسَّعَهُ ٱللّٰهُ *Thou hast made strait and unlawful to me what God has made ample.* (Mgh.) And تَحَجَّرَ وَاسِعًا *He made strait what was ample:* (Mṣb:) or *he made strait what God made ample, and made it to be*

peculiar to himself, exclusively of others; as also ▼ حُجْرَة and ▼ حُجُرَة. (TA.) ═ See also 8: ═ and 10. — [Hence, perhaps,] تحجّر للبُرْءِ *It* (a wound) *closed up, and consolidated, to heal.* (TA from a trad.)

8. احتجر, (TA,) or احتجر حُجْرَةً, (Ṣ, Mṣb,) and ▼ استحجر and ▼ تحجّر, (Ḳ,) *He made for himself a* حُجْرَة [i. e. *an enclosure for camels*]. (Ṣ, Mṣb, Ḳ.) — And hence, (Mṣb,) احتجر الأَرْضَ, (Mgh, Mṣb, Ḳ,) and ▼ تحجّرها, (TA,) *He placed a land-mark to the land,* (Mgh, Mṣb, Ḳ,) *to confine it,* (Mgh, Mṣb,) *and to prevent others from encroaching upon it.* (Mgh, TA.) — احتجر به *He sought protection by him,* (A,* Ḳ,) as, for instance, by God, مِنَ الشَّيْطَانِ *from the devil.* (A.) ═ احتجر اللَّوْحَ *He put the tablet in his* حِجْر [or *bosom*]. (Ḳ.)

10. استحجر: see 8. ═ Also *It* (clay) *became stone:* (TA:) or *became hard; as when it is made into baked bricks:* (Mgh:) or *became hard like stone;* (A, Mṣb;) as also ▼ تحجّر. (A.) — † *He became emboldened* or *encouraged,* or *he emboldened* or *encouraged himself,* (Ḳ, TA,) عَلَيْهِ *against him.* (TA.)

Q. Q. 1. حَنْجَرَهُ *He slaughtered him by cutting his throat* [*in the part called the* حَنْجَرَة]. (Ḳ in art. حنجر.)

حَجْرٌ: see حِجْرٌ, in three places. ═ Also, and ▼ حِجْرٌ, (Ṣ, A, Mgh, Mṣb, Ḳ, TA,) [the latter of which I have found to be the more common in the present day,] and ▼ حُجْرٌ, (Ḳ, [but this I have not found in any other lexicon, and the TA, by implication, disallows it,]) The حِضْن; (Mgh, Mṣb, Ḳ;) [i. e. the *bosom;* or *breast;* agreeably with explanations of حِضْن in the Ḳ: or] the *part beneath the armpit, extending to the flank;* (Mgh, Mṣb;) [agreeably with other explanations of حِضْن;] *of a man or woman:* (Ṣ, A, Mgh, Mṣb, Ḳ:) pl. حُجُورٌ. (Ṣ, Mṣb.) Hence the saying, (Mgh,) فُلَانٌ فِى حَجْرِ فُلَانٍ † *Such a one is in the protection of such a one;* (AZ, T, Mgh, Mṣb;) as also ▼ فى حُجْرَتِهِ. (TA.) And نَشَأَ فى حِجْرِهِ ▼ and حُجْرِهِ † *He grew up in his care and protection.* (Ḳ.) — Also ▼ حِجْرٌ (T, Ḳ) and حَجْرٌ (T, TA) [The *bosom* as meaning] the *fore part of the garment;* or the *part, thereof, between one's arms.* (T, Ḳ.) — See also حُجْرَة: — and مَحْجِرُ العَيْنِ. ═ Also *An extended gibbous tract of sand.* (Ḳ.)

حُجْرٌ: see حِجْرٌ, in three places: ═ and حُجُرٌ: — and مَحْجِرُ العَيْنِ.

حِجْرٌ (Ṣ, A, Mgh, Mṣb, Ḳ) and ▼ حُجْرٌ (Ṣ, Mgh, Mṣb, Ḳ) and ▼ حَجْرٌ, (Ṣ, Ḳ,) of which the first is the most chaste, (Ṣ,) and ▼ مَحْجُورٌ (Ṣ, Ḳ) and ▼ حَاجُورٌ (Ḳ) [and ▼ مُحْجُورٌ], *Forbidden, prohibited, unlawful, inviolable,* or *sacred.* (Ṣ, A, Mgh, Mṣb, Ḳ.) Each of the first three forms occurs in different readings of the Ḳur vi. 139. (Ṣ.) You say, هٰذَا حِجْرٌ عَلَيْكَ *This is forbidden,* or *unlawful, to thee.* (A.) In the time of paganism, a man meeting another whom he feared, in a sacred month, used to say, ▼ حِجْرًا مَحْجُورًا, meaning *It is rigorously forbidden* to thee [to commit an act of hostility against me] in this month: and the latter, thereupon, would abstain from any aggression against him: and so, on the day of resurrection, the polytheists, when they see the punishment, will say to the angels, thinking that it will profit them: (Lth, Ṣ:*) but Az says that I'Ab and his companions explain these words [occurring in the Ḳur xxv. 24] otherwise, i. e., as said by the angels, and meaning, the joyful annunciation is forbidden to be made to you: and accord. to El-Ḥasan, the former word will be said by the sinners, and the latter is said by God, meaning it will be forbidden to them to be granted refuge or protection as they used to be in their former life in the world: but Az adds, it is more proper to regard the two words as composing one saying: (TA:) and the latter word is a corroborative of the former, like مَائِتٌ in the expression مَوْتٌ مَائِتٌ. (Bḍ.) The same words in the Ḳur xxv. 55 signify *A strong mutual repugnance,* or *incongruity;* as though each said what one says who seeks refuge or protection from another: or, as some say, *a defined limit.* (Bḍ.) A man says to another, "Dost thou so and so, O such a one?" and the latter replies حِجْرًا, or ▼ حُجْرًا, or ▼ حَجْرًا, meaning [I pray for] *preservation, and acquitment,* from this thing; a meaning reducible to that of prohibition, and of a thing that is prohibited. (Sb.) The Arabs say, on the occasion of a thing that they disapprove, ▼ حُجْرًا لَهُ, with ḍamm, meaning, *May it be averted.* (Ṣ.) — Ḥomeyd Ibn-Thowr says,

فَهَمَمْتُ أَنْ أَغْشَى إِلَيْهَا مَحْجِرًا ▼
وَلَمِثْلُهَا يُغْشَى إِلَيْهِ المَحْجَرُ ▼

meaning, *And I purposed doing to her a forbidden action: and verily the like of her is one to whom that which is forbidden is done.* (Ṣ, Ḳ.) ▼ مَحْجَرٌ is also explained as signifying حُرْمَةٌ; [app. meaning *a thing from which one is bound to refrain, from a motive of respect* or *reverence;*] and to have this meaning in the verse above. (Az.) — Also, the first of these words, *Any* حَائِط [i. e. *garden,* or *walled garden of palm-trees,*] *which one prohibits* [*to the public*]. (Ṣ.) — And الحِجْرُ *That* [*space*] *which is comprised by* [*the curved wall called*] *the* حَطِيم, (Ṣ, A, Mgh, Ḳ,) *which encompasses the Kaabeh on the north* [or rather *north-west*] *side;* (Ṣ, A, Ḳ;) *on the side of the spout:* (Mgh:) or *the* حطيم [*itself*], *which encompasses the Kaabeh on the side of the spout.* (Mṣb.) [It is applied to both of these in the present day; but more commonly to the former.] — Also, حِجْرٌ, The *anterior pudendum* of a man and of a woman; and so ▼ حُجْرٌ: (Ḳ, TA:) the latter the more chaste. (TA.) — *A mare;* the *female of the horse:* (Ṣ, A, Mṣb, Ḳ:) and *a mare kept for breeding;* (A;) as though her womb were forbidden to all but generous horses: (T:) but in the latter sense the sing. is scarcely ever used; though its pl., the first of the following forms, (as well as the second, A,) is used to signify mares kept for breeding: (Ḳ:) ▼ حِجْرَةٌ, as a sing., is said by F and others to be a barbarism: it occurs in a trad.; but perhaps the ة is there added to assimilate it to بَغْلَةٌ, with which it is there coupled: (MF:) the pl. [of pauc.] is أَحْجَارٌ (Mṣb, Ḳ) and [of mult.] حُجُورٌ (A, Mṣb, Ḳ) and حُجُورَةٌ. (Ḳ.) A poet says,

إِذَا خَرِسَ الفَحْلُ وَسْطَ الحُجُورِ
وَصَاحَ الكِلَابُ وَعَقَّ الوَلَدْ

When the stallion, seeing the army and the gleaming swords, *is mute in the midst of the mares kept for breeding,* and does not look towards them, *and the dogs bark* at their masters, because of the change of their appearances, *and children behave undutifully* to their mothers whom fear diverts from attending to them. (A.) — *Relationship* [*that prohibits marriage*]; *nearness with respect to kindred.* (Mṣb, Ḳ.) — *Understanding, intelligence, intellect, mind,* or *reason:* (Ṣ, A, Mṣb, Ḳ:) so in the Ḳur lxxxix. 4: (Ṣ, Bḍ:) thus called because it forbids that which it does not behoove one to do. (Bḍ.) One says, فِى ذٰلِكَ عِبْرَةٌ لِذِى حِجْرٍ *In that is an admonition to him who possesses understanding,* &c. (A.) ═ See also حَجْرٌ, in three places.

حَجَرٌ [*A stone;* explained in the Ḳ by صَخْرَةٌ; but this means "a rock," or "a great mass of stone" or "of hard stone"]; (Ṣ, Ḳ, &c.;) so called because it resists, by reason of its hardness; (Mgh;) and ▼ أُحْجُرٌ signifies the same: (Fr, Ḳ:) pl. (of pauc., of the former, Ṣ) أَحْجَارٌ (Ṣ, Mgh, Ḳ) and أَحْجُرٌ (Ḳ) and (of mult, Ṣ) حِجَارٌ and [more commonly] حِجَارَةٌ, (Ṣ, Ḳ,) which last is extr. [with respect to rule], (Ṣ,) or agreeable with a usage of the Arabs, which is, to add ة to any pl. of the measure فِعَالٌ or of that of فُعُولٌ, as in the instances of ذِكَارَةٌ and فِحَالَةٌ and ذُكُورَةٌ and فُحُولَةٌ. (AHeyth.) And (metonymically, TA) ‡ *Sand:* (IAạr, Ḳ:) pl. أَحْجَارٌ. (TA.) — [Hence,] أَهْلُ الحَجَرِ *The people of the desert, who dwell in stony and sandy places:* occurring in a trad., coupled with أَهْلُ المَدَرِ. (TA.) — الحَجَرُ الأَسْوَدُ, and simply الحَجَرُ, *The* [*Black*] *Stone of the Kaabeh.* (Ḳ, TA.) El-Farezdaḳ applies to it, in one instance, the pl. الأَحْجَارُ, considering the sing. as applicable to every part of it. (TA.) — One says, فُلَانٌ حَجَرُ الأَرْضِ, meaning † *Such a one is unequalled.* (TA.) And رُمِىَ فُلَانٌ بِحَجَرِ الأَرْضِ ‡ *Such a one has had a very sagacious and crafty and politic man made to be an assailant against him.* (Ḳ,* TA.) El-Aḥnaf Ibn-Ḳeys said to 'Alee, when Mo'áwiyeh named 'Amr Ibn-El-'Áṣ as one of the two umpires, قَدْ رُمِيتَ بِحَجَرِ الأَرْضِ فَاجْعَلْ مَعَهُ ٱبْنَ عَبَّاسٍ فَإِنَّهُ لَا يَعْقِدُ عُقْدَةً إِلَّا حَلَّهَا † *Thou hast had a most exceedingly sagacious and crafty and politic man made to be an assailant against thee: so appoint thou with him Ibn-'Abbás; for he will not tie a knot but he shall untie it:* meaning *one that shall stand firm like a stone upon the ground.* (L from a trad.) One says also, رُمِىَ فُلَانٌ بِحَجَرِهِ, meaning ‡ *Such a one was coupled* [or *opposed*] *with his like:* (A:) [as though he had a stone suited to the purpose of knocking him down cast at him.] —— لِلْعَاهِرِ الحَجَرُ, occurring in a trad., means † *For the fornicator*, or *adulterer*, *disappointment, and prohibition*: accord. to some, it is meant to allude to *stoning*; [and it may have had this meaning in the first instance in which it was used;] but [in general] this is not the case; for every fornicator is not to be stoned. (IAth, TA.) [See also art. عهر.] —— الحَجَرُ *Gold*: and *silver*. (Ḳ.) Both together are called الحَجَرَانِ. (Ṣ.)

حَجِرٌ [*Stony; abounding with stones*]. You say أَرْضٌ حَجِرَةٌ [so in several copies of the Ḳ; in the CḲ حَجَرَةٌ;] *Land abounding with stones*; as also ↓ حَجِيرَةٌ and ↓ مُتَحَجِّرَةٌ. (Ḳ.)

حُجْرٌ The *flesh surrounding the nail*. (Ḳ.)

حَجْرَةٌ *A severe year, that confines men to their tents, or houses, so that they slaughter their generous camels to eat them.* (L in art. نبت, on a verse of Zuheyr.) ═ *A side*; *an adjacent tract* or *quarter*; (ISd, Ḳ;) as also ↓ حُجْرَةٌ: (EM p. 281:) pl. of the former ↓ حَجْرٌ, [or rather this is a coll. gen. n., of which the former is the n. un.,] and حَجَرَاتٌ (Ṣ, Ḳ) and ↓ حَوَاجِرُ: (Ḳ:) the last is mentioned by ISd as being thought by him to be a pl. of حَجْرَةٌ in the sense above explained, contr. to analogy. (TA.) Hence, حَجْرَةُ قَوْمٍ *The tract* or *quarter adjacent to the abode of a people*. (Ṣ.) And حَجْرَتَا الطَّرِيقِ *The two sides of the road*. (TA.) And حَجْرَتَا عَسْكَرٍ *The two sides of an army*; (A, TA;) *its right and left wings*. (TA.) And قَعَدَ حَجْرَةً *He sat aside*. (A.) And سَارَ حَجْرَةً *He journeyed aside, by himself*. (TA.) And ↓ مَحْجَرًا is also said to signify the same, in the following ex.: تَرْعَى مَحْجَرًا وَتَبْرُكُ وَسَطًا *She* (the camel) *pastures aside, and lies down in the middle*. (TA.) It is said in a prov., يَرْبِضُ حَجْرَةً وَيَرْتَعِى وَسَطًا *He lies down aside, and pastures in the middle*: (Ṣ:) or فُلَانٌ يَرْعَى وَسَطًا وَيَرْبِضُ حَجْرَةً *Such a one pastures in the middle, and lies down aside*: (TA:) applied to a man who is in the midst of a people when they are in prosperity, and when they become in an evil state leaves them, and lies down apart: the prov. is ascribed to Gheylán Ibn-Muḍar. (IB.) Imra-el-Ḳeys says, [addressing Khálid, in whose neighbourhood he had alighted and sojourned, and who had demanded of him some horses and riding-camels to pursue and overtake a party that had carried off some camels belonging to him (Imra-el-Ḳeys), on Khálid's having gone away, and returned without anything,]

* فَدَعْ عَنْكَ نَهْبًا صِيحَ فِى حَجَرَاتِهِ *
* وَلٰكِنْ حَدِيثًا مَا حَدِيثُ الرَّوَاحِلِ *

[*Then let thou alone spoil by the sides of which a shouting was raised: but relate to me a story. What is the story of the riding-camels?*]: hence the prov., المُلْكُ لِلّٰهِ

* وَدَعْ عَنْكَ نَهْبًا صِيحَ فِى حَجَرَاتِهِ *

[*Dominion belongeth to God: then let thou alone &c.*]; said with reference to him who has lost part of his property and after that lost what is of greater value. (TA.) [And hence the saying,] قَدِ ٱنْتَشَرَتْ حَجْرَتُهُ † *His property has become large, or ample*. (Ṣ.) —— See also حَجْرٌ.

حُجْرَةٌ *An enclosure* (حَظِيرَة) *for camels*. (Ṣ, Ḳ.) —— [And hence,] The حُجْرَة of a house; (Ṣ;) [i. e.] *a chamber* [in an absolute sense, and so in the present day]; syn. بَيْتٌ: (Mṣb:) or *an upper chamber*; syn. غُرْفَةٌ: (Ḳ:) pl. حُجَرٌ and حُجُرَاتٌ (Ṣ, Mṣb, Ḳ) and حُجَرَاتٌ and حُجْرَاتٌ. (Z, Mṣb, Ḳ.) —— See also حَجْرَةٌ.

حُجُرٌ: see حُجْرَةٌ.

حُجْرِىٌّ and حِجْرِىٌّ *A right, or due; a thing, or quality, to be regarded as sacred, or inviolable*; (Ḳ;) *a peculiar attribute*. (TA.)

أَرْضٌ حَجِيرَةٌ: see حَجِرٌ.

حَاجِرٌ The *part of the brink* (شَفَة) *of a valley that retains the water*, (Ṣ, Ḳ,) *and surrounds it*; (ISd;) as also ↓ حَاجُورٌ: pl. of the former حُجْرَانٌ. (Ṣ, Ḳ.) *High land* or *ground, the middle of which is low*, or *depressed*; (Ḳ;) as also ↓ مَحْجِرٌ: (TA:) and ↓ مَحَاجِرُ [pl. of the latter] *low places in the ground, retaining water*. (A.) *A fertile piece of land, abounding with herbage, low*, or *depressed, and having elevated borders, upon which the water is retained*. (AḤn.) *A place where water flows*, or *where herbs grow, surrounded by high ground*, or *by an elevated river*. (T, TA.) *A place where trees of the kind called* رِمْث *grow*; *where they are collected together*; and *a place which they surround*: (M, Ḳ:) pl. as above. (Ḳ.) —— *A wall that retains water between houses*: so called because encompassing. (TA.)

حَاجُورٌ: see حَجْرٌ —— and حَاجِرٌ. —— Also *A refuge*; *a means of protection* or *defence*: analogous with عَاثُورٌ, which signifies "a place of perdition:" whence,

* وَقَالَ قَائِلُهُمْ إِنِّى بِحَاجُورٍ *

And their sayer said, Verily I lay hold on that which will protect me from thee *and repel thee* from me; مُتَمَسِّكٌ being understood. (TA.)

حَوَاجِرُ: see حَجْرَةٌ.

حُنْجُرَةٌ and ↓ حُنْجُورٌ, (Ṣ, Ḳ,) each with an augmentative ن, (Ṣ, Mṣb,) [The *head of the windpipe*; consisting of *a part*, or the *whole*, *of the larynx*: but variously explained; as follows:] the *windpipe*; syn. حُلْقُومٌ: (Ṣ, Ḳ:) or the former [has this meaning, i. e.], the *passage of the breath*: (Mgh, Mṣb:) or the *extremity of the* حلقوم, *at the entrance of the passage of the food and drink*: (Bḍ in xxxiii. 10:) or [the *head of the larynx*, *composed of the two arytenoides*;] *two of the successively-superimposed cartilages of the* حلقوم (طَبَقَانِ مِنْ أَطْبَاقِ الحُلْقُومِ), *next the* غَلْصَمَة [or *epiglottis*], *where it is pointed*: or the *inside*, or *cavity, of the* حلقوم: and so ↓ حُنْجُورٌ: (TA in art. حنجر:) or ↓ the latter is *syn. with* حَلْقٌ [q. v.]: (Mṣb:) pl. حَنَاجِرُ. (Ḳ.)

حُنْجُورٌ: see the next preceding paragraph, in three places. —— Also *A small* سَفَط [or *receptacle for perfumes and the like*]. (Ḳ.) —— And *A glass flask* or *bottle* (قَارُورَة), (Ḳ, TA,) *of a small size*, (TA,) *for* ذَرِيرَة [q. v.]. (Ḳ, TA.)

أُحْجُرٌ: see حَجَرٌ.

مَحْجِرٌ: see حَاجِرٌ, in four places. —— Also, (Ṣ,) or ↓ مَحْجَرٌ and ↓ مَحْجِرٌ, (Ḳ,) The *tract surrounding a town* or *village*: (Ṣ, Ḳ:) [pl. مَحَاجِرُ.] Hence the مَحَاجِر of the kings (أَقْيَال) of El-Yemen, which were *Places of pasturage*, whereof each of them had one, *in which no other person pastured his beasts*: (Ṣ, Ḳ:) the محجر of a قَيْل of El-Yemen was his *tract of land into which no other person than himself entered*. (T.) —— See also حَجْرَةٌ. —— And see مَحْجِرُ العَيْنِ.

مَحْجِرٌ (Ṣ, Ḳ) and ↓ مَحْجَرٌ (Ḳ) *A garden surrounded by a wall*; or *a garden of trees*; syn. حَدِيقَةٌ: (Ṣ, Ḳ:) or *a low*, or *depressed, place of pasture*: (T, TA:) or *a place in which is much pasture, with water*: (A,* TA:) pl. مَحَاجِرُ. (Ṣ, A.) See also حَاجِرٌ for the former word and its pl.: and see مَحْجِرُ العَيْنِ. —— مَحْجِرُ العَيْنِ (Ṣ, Ḳ, &c.) and ↓ مَحْجِرُهَا (TA) and ↓ مَحْجَرُهَا (Ḳ) and simply المحجر (Mṣb, TA) and ↓ المَحْجَرُ (Ḳ) and ↓ المُحْجَرُ, which occurs in a verse of El-Akhṭal, (IAạr,) [*The part which is next below*, or *around, the eye, and which appears when the rest of the face is veiled by the* نقاب *or the* بُرْقُع:] *that part* [*of the face, next below the eye*,] *which appears from out of the* [*kind of veil called*] نِقَاب (T, Ṣ, A, Mṣb, Ḳ) of a woman (A, Mṣb, Ḳ) and of a man, *from the lower eyelid*; and sometimes *from the upper*: (Mṣb:) or *the part that surrounds the eye* (Mṣb, Ḳ) *on all sides*, (Mṣb,) *and appears from out of the* [*kind of veil called*] بُرْقُع: (Mṣb, Ḳ:) or *the part of the bone beneath the eyelid, which encompasses the eye*: (TA:) and محجر العين means also *what appears from beneath the turban of a man when he has put it on*: (Ḳ: [accord. to the TA, the turban itself; but this is a meaning evidently derived from a mistranscription in a copy of the Ḳ, namely, عِمَامَتُهُ for عِمَامَتِهِ:]) also مَحْجِرُ الوَجْهِ *that part of the face against which the* نقاب *lies*: and المحجر *the eye* [*itself*]: (T, TA:) the pl. of محجر is مَحَاجِرُ. (A, Mṣb.)

مُحْجَرٌ: see مَحْجِرٌ: —— and see also مَحْجِرٌ, in two places.

مَحْجُورٌ عَلَيْهِ, for which the doctors of practical law say مَحْجُورٌ only, omitting the preposition and the pronoun governed by it, on account of the frequent usage of the term, A person *prohibited* [by a ḳáḍee] *from using*, or *disposing of, his property according to his own free will*: (Mṣb:) or *prohibited from consuming*, or *wasting*, or *ruining, his property*. (Mgh.) —— See also حِجْرٌ, in two places.

أَرْضٌ مُتَحَجِّرَةٌ: see حَجِرٌ.

حجز

1. حَجَزَهُ, aor. ِ (Ṣ, Ḳ) and ُ, (Ḳ,) inf. n. حَجْزٌ (Ṣ, Ḳ) and حِجَازَةٌ and ↓حِجِّيزَى, (Ḳ,) [or the last is rather a quasi-inf. n. of تَحَاجَزُوا, and, accord. to some, it is of an intensive form,] *He, or it, prevented, hindered, impeded, withheld, restrained,* or *debarred, him,* or *it;* syn. مَنَعَهُ, (Ṣ, Ḳ,) and كَفَّهُ. (Ḳ.) It is said in a trad., وَلِأَهْلِ القَتِيلِ أَنْ يَحْجُزُوا الأَدْنَى فَالأَدْنَى *And it is for the family of the slain person to prevent the nearest* [in relationship] *and then the nearest* [after him] from retaliating the slaughter. (TA.) And one says, كَانَتْ بَيْنَ القَوْمِ رِمِّيَّا ثُمَّ صَارَتْ إِلَى ↓حِجِّيزَى *There was a shooting of arrows* or *the like* [or *a great shooting* &c.] *between the people; then they withheld themselves* [or *withheld themselves much*] *from each other:* (Ṣ, L:) which is a prov. (TA.) It is also said in another prov., مَا يُحْجَزُ فُلَانٌ فِى العِلْمِ ‡ *Such a one's case* [with respect to knowledge or science] *cannot be concealed.* (A, TA.) — حَجَزَ بَيْنَهُمَا, (A, Mṣb, Ḳ,) aor. ِ (Mṣb, Ḳ) and ُ, (Ḳ,) inf. n. حَجْزٌ, (Mṣb,) *He, or it, intervened as,* or *made,* or *formed, a separation, a partition, a fence, a barrier,* or *an obstacle,* or *obstruction, between them two; he separated,* or *parted, them;* (Mṣb, Ḳ;) i. e., two things; (Mṣb;) or two persons or things facing, or opposite to, each other. (Az, A, TA.)

2. سَطْحٌ لَمْ يُحَجَّزْ بِجِدَارٍ [*A flat roof that has not been fenced round with a wall to prevent persons falling from it*]. (Ḳ in art. جلح.)

3. حَاجَزُوا عَدُوَّهُمْ, (A,) inf. n. مُحَاجَزَةٌ, (Ṣ, Ḳ,) *They reciprocally prevented their enemy from fighting with them, or reciprocally abstained from fighting with them, and made peace with them;* syn. كَافُّوهُمْ, (A,) and syn. of the inf. n. مُمَانَعَةٌ, (Ṣ, Ḳ, TA,) and مُسَالَمَةٌ. (TA.) It is said in a prov., إِنْ أَرَدْتَ المُحَاجَزَةَ فَقَبْلَ المُنَاجَزَةِ [*If thou desire the reciprocal prevention of fighting, and the making of peace, let it be before fighting*]: (Ṣ, TA:) or المُحَاجَزَةُ قَبْلَ المُنَاجَزَةِ [*The reciprocal prevention of fighting, and the making of peace, should be before fighting*]. (A.) [See also art. نجز.]

4: see 8.

6. تَحَاجَزَا *They prevented one another,* or *mutually abstained, from fighting,* [*and made peace, one with another;* (see also 3;)] syn. تَمَانَعَا: (Ṣ,* Ḳ, TA:) said of two troops. (Ṣ.) — تحاجز القَوْمُ *The people separated themselves, one from another; they left, forsook,* or *relinquished, one another;* as also ↓انحجزوا and ↓احتجزوا. (TA.) = Also تحاجز القَوْمُ *The people took,* or *took hold of, one another by the* حُجَز [pl. of حُجْزَةٌ]; *took hold of one another's* حُجَز: (TA:) [or *sought aid, or refuge and protection, one of another:* see حُجْزَةٌ.]

7. انحجز quasi-pass. of حَجَزَهُ in the first of the senses explained above; (TA;) *He, or it, was, or became, prevented, hindered, impeded, withheld, restrained,* or *debarred; he withheld,* or *restrained, himself; he refrained, forbore,* or *abstained.* (Ṣ, Ḳ, TA.) — انحجز عَنْهُ *He left, forsook,* or *relinquished, it.* (TA.) — See also 6. = See also 8.

8. احتجز بِهِ *He, or it, was, or became, defended,* or *he defended himself, by it;* syn. اِمْتَنَعَ. (TA.) [See also another explanation in what follows.] — احتجزا *They two were,* or *became, separated,* or *parted, each from the other.* (TA.) See also 6. = احتجزهُ *He carried it in his* حُجْزَة: (A, Ḳ:*) like as اِحْتَضَنَهُ signifies "he carried it in his حِضْن." (A.) — احتجز بِإِزَارِهِ *He tied his* ازار *upon* [or *around*] *his waist;* (Ṣ, Mgh, Mṣb, Ḳ;) *he made the two ends thereof to meet, and tied it* upon his waist; (A, TA;) *he wrapped it round his waist.* (TA.) — Hence, احتجز بِالحِرَارِ وَالجِبَالِ *It was encompassed by the stony tracts called* حرار [pl. of حَرَّةٌ], *and by the mountains.* (Mgh.) = احتجز *He,* (a man, TA,) or *it,* (a party of people, Ṣ,) *came to the province called El-Ḥijáz;* (Ṣ, Ḳ;) as also ↓انحجز; (ISk, Ṣ, Ḳ;) and ↓احجز, (Ḳ,) inf. n. إِحْجَازٌ. (TA.)

حَجْزٌ One *who abstains from what is unlawful and indecorous.* (TA.) [See also حُجْزَةٌ.]

حُجْزٌ: see حُجْزَةٌ, in two places.

حُجْزَةٌ *A severe year, that confines men to their tents or houses, so that they slaughter their generous camels to eat them.* (L in art. نبت, on a verse of Zuheyr.)

حُجْزَةُ الإِزَارِ *The place* [or *part*] *of the* ازار *where it is tied* [*round the waist*]; (Ṣ, Mṣb, Ḳ, TA;) *the place where the end of the* ازار *is folded,* or *doubled, in wrapping it round:* (Lth, TA:) and [in like manner] حُجْزَةُ السَّرَاوِيلِ [*the tuck,* or *doubled upper border, of the trousers, through which passes the waist-band,* i. e., *the band or string that binds them round the waist;*] *the part of the trousers in which is the* تِكَّة [or *waist-band*]; (Ṣ;) *the place of the* تِكَّة: (Ḳ:) pl. حُجَزٌ (Mṣb, TA) and حُجُزَاتٌ and حُجَزَاتٌ: (TA:) and hence حُجْزَةٌ is applied to the *garment called* إِزَار *itself;* as also ↓حُجْزٌ; of which latter the pl. is حُجُزٌ, with two dammehs, [app. contracted into حُجْزٌ,] and pl. pl. حُجُوزٌ: Z says that ↓حُجْزٌ and حُجْزَةٌ signify the same. (TA.) — Hence, شِدَّةُ الحُجْزَةِ ‡ *Patience,* (Ḳ, TA,) and *hardiness.* (TA.) One says, هُوَ شَدِيدُ الحُجْزَةِ ‡ *He is patient in difficulty.* (TA.) And it is said in a trad. of 'Alee, when he was asked respecting the Benoo-Umeiyeh, هُمْ أَشَدُّنَا حُجَزًا, or حُجْزَةً, accord. to different relations, ‡ *They are the most patient of us in difficulty.* (TA.) — You say also رَجُلٌ طَيِّبُ الحُجْزَةِ, (A, TA,) and كَرِيمُ الحُجْزَةِ, and كَرِيمُ الحُجْزِ, [app. a contraction of الحُجُزِ,] (TA,) ‡ *He is one who abstains from what is unlawful and indecorous* [especially *with respect to women*]; like طَيِّبُ الإِزَارِ. (TA.) — And هُوَ نَاتِئُ الحُجْزَةِ ‡ *He is full in the flanks:* the being so is a fault. (Ḳ, TA.) — You also say, أَخَذَ بِحُجْزَتِهِ, meaning ‡ *He sought aid of him:* (A, TA:) or *he had recourse to him for refuge and protection.* بِسَبَبٍ مِنْهُ (TA.) And أَخَذَ بِحُجْزَةِ اللّٰهِ, i. e., [meaning, † *He laid hold of a means of obtaining access, or nearness, to God*]: said of Mohammad, in a trad. (TA.) And هٰذَا كَلَامٌ آخِذٌ بَعْضُهُ بِحُجَزِ بَعْضٍ ‡ *This is language disposed in regular order, consecutively.* (A, TA.) It is said in a trad., إِنَّ الرَّحِمَ أَخَذَتْ بِحُجْزَةِ الرَّحْمٰنِ [*Verily* الرَّحِمُ *is connected with* الرَّحْمٰن]: said to mean, that the name of الرحم is derived from الرحمن; so that it is as though it attached itself thereto, and laid hold of the middle thereof. (IAth, TA.)

حِجْزَةٌ *A mode, manner,* or *form, of tying the* إِزَار. (TA.)

حَاجِزٌ: see حِجَازٌ.

حَجَازَيْكَ *Separate thou,* or *part thou, the people;* (Ṣ, A, Ḳ; and T in art. دول, on the authority of IAar;) *time after time:* (Ḳ:) app. meaning, *without intermission:* (TA:) or it may mean *withhold thyself.* (IAar, T in art. دول.)

حِجِّيزَى: see 1, in two places.

حَاجِزٌ *A thing intervening, as a separation, a partition, a fence, a barrier,* or *an obstacle,* or *obstruction, between two other things,* (TA,) or *between two things facing,* or *opposite to, each other;* (Az, A, TA;) as also ↓حِجَازٌ. (Az, A, TA.) Hence the province called El-Ḥijáz is thus named, because it forms a separation between Nejd and the Ghowr, or Ghór; (Ṣ, Mgh, TA;) or between Nejd and the Saráh; (Mṣb, Ḳ;) or between the Ghowr, or Ghór, and Syria (Mgh, Mṣb, TA) and the Bádiyeh; (Mgh, TA;) or between Nejd and Tihámeh; (Ḳ;) or because the Ḥirár [or certain stony tracts] separate it from the high part of Nejd; (Az, TA;) or because it is encompassed by the Ḥirár and the mountains, (Mgh, Mṣb,*) or by the five Ḥirár, namely, the Ḥarrah of Benoo-Suleym and that of Wáḳim and that of Leylà and that of Showrán and that of En-Nár. (Aṣ, Ḳ.) — Also sing. of حَجَزَةٌ, (Ḳ,) which latter signifies *Wrongers,* or *wrongdoers, who prevent one from obtaining his right:* (Ṣ:) or *persons who defend men, one from another, and decide between them justly.* (Az, TA.) In the Ḳ, the signification of "wrongers, or wrongdoers," is combined with the contrary explanation given by Az. (TA.) It is said in a trad. of Ḳeyleh, أَيَعْجِزُ ابْنُ هٰذِهِ أَنْ يَنْتَصِفَ مِنْ وَرَاءِ الحَجَزَةِ [*Is the son of this woman unable to obtain his right in the absence of the wrongdoers who prevent his doing so?*]: (Ṣ:) or [according to one relation] أَيُلَامُ ابْنُ ذِهْ أَنْ يَفْصِلَ الخُطَّةَ مِنْ وَرَاءِ الحَجَزَةِ [*Is the son of this woman to be blamed for deciding the affair in the absence of those who defend men, one from another, and decide between them justly?*]: by "the son of this woman" she means her own son: she says, if he suffer a wrong, and allege for himself that which repels from him the wrong, [without having recourse to the judge,] he is not to be blamed. (TA.) [See also خُطَّةٌ.]

مَحْجُوزٌ *Hit, or hurt, in the place* [or *part of the body*] *where the* إِزَار *is tied.* (Ḳ.)

مُحْتَجَزٌ The *place* [or *part of the body*] *where the* إِزَار *is tied.* (Ḳ.)

مُحْتَجِزٌ *Having his waist bound* [*with the* إِزَار]: and with ة, a woman *having her* مِئْزَر *bound upon* [*that part of her person which is termed*] *the* عَوْرَة. (TA.)

حجف

3. حاجفهُ *He opposed him and repelled him,* [app. *with a* حَجَفَة, (see the act. part. n., below,)] *being opposed and repelled by him.* (Ṣ.)

حَجَفٌ: see what next follows, in two places.

حَجَفَةٌ *A shield made of skins,* (Ṣ, Ḳ, TA,) or, as some say, *of the skins of camels, cut out in a round form,* (TA,) *without wood and without sinews;* (Ṣ, Ḳ, TA;) as also دَرَقَةٌ: (Ṣ:) or *of skins sewed one over another;* as also دَرَقَةٌ: (ISd:) or *a small shield made of two* [*pieces of*] *skins sewed together, one over the other:* (Mṣb:) pl. ▼حَجَفٌ, (Ṣ, Mṣb, Ḳ,) [or rather this is a coll. gen. n.,] and [the pl. is] حَجَفَاتٌ. (Mṣb.) The saying of the rájiz, (Ṣ,) namely, Su-r-edh-Dhi-b, (TA,)

* بَلْ جَوْزِ تَيْهَآءَ كَظَهْرِ الحَجَفَتْ *

means رُبَّ جَوْزِ تَيْهَآءَ [i. e. *Many a middle of a desert in which one loses his way, like the back of a shield such as is called* حجفة]: and [by الحجفت is meant الحجفة; for] some of the Arabs, in pausing upon a ة, make it ت: thus they say, هٰذَا طَلْحَتْ, and خُبْزُ الذُّرَتْ: (Ṣ, Ṣgh:) these are Ṭeiyi. (Ṣgh, TA.) — ‡ The *breast,* or *chest;* (Abu-l-'Omeythil, Ḳ:) as being likened to a shield: (Abu-l-'Omeythil, TA:) pl. [or coll. gen. n.] ▼حَجَفٌ. (Abu-l-'Omeythil, Ḳ.)

مُحَاجِفٌ *Fighting with another, having a* حَجَفَة: (Ṣ, Ḳ:) *opposing reciprocally.* (Ḳ.)

حجل

1. حَجَلَ, aor. ِ and ُ, inf. n. حَجَلَانٌ (Ṣ, Ḳ) and حَجْلٌ, (Ḳ,) *He walked having his legs shackled:* (Ṣ:) or *he raised one leg, and went slowly on the other leg:* (M, Ḳ:) or *he went with short steps, like him who has his legs shackled:* (Ḥam p. 221:) and *he raised one leg, and hopped on the other:* (TA:) it is said of a bird: (Ṣ:) and it means, (Ṣ, Ḳ,) in like manner, (Ṣ,) as also ▼حجّل, (TA,) *he leaped in going;* (Ṣ, Ḳ, TA;) said of a crow, or raven; (Ḳ, TA;) *as leaps* (يَحْجُلُ) *the camel that is hocked* [*in one leg*] *upon three legs, and the boy upon one leg or upon two.* (Ṣ.) ═ حَجَلَتْ عَيْنُهُ, aor. ُ, inf. n. حُجُولٌ; (Ḳ;) and ▼حجّلت, (Aṣ, Ṣ, Ḳ,) inf. n. تَحْجِيلٌ; (Aṣ, Ṣ;) *His eye sank,* or *became depressed, in his head;* (Aṣ, Ṣ, Ḳ;) said of a man, and of a camel, and of a horse: (TA:) and ▼حَوْجَلَ, alone, signifies the same; (Ibn-'Abbád, Ḳ;) said of a man. (Ibn-'Abbád, TA.) ═ حَجَلَ بَيْنَهُ وَبَيْنَهُ, inf. n. حَجْلٌ, *An obstacle was made to intervene between him, or it, and him, or it.* (Ḳ.)

2. [حجّل, inf. n. تَحْجِيلٌ, originally, *He ornamented* a woman, or her legs, *with anklets:* and *he shackled* a man, or a man's legs: see حَجَلَ. — And hence,] حُجِّلَتْ قَوَائِمُهُ, inf. n. تَحْجِيلٌ, said of a horse, *His legs were white in the lower parts, the whiteness extending* [*upwards*] *beyond the pasterns but not extending beyond the knees and hocks;* because they [the lower parts of the leg] are the places of the احجال, i. e., the anklets, and the shackles. (Ṣ, TA.) [See تَحْجِيل explained as a simple subst., below.] — [Hence also,] حَجَّلَتْ بَنَانَهَا *She* (a woman) *coloured the dye of her fingers,* or *of the extremities of her fingers.* (Ḳ, TA.) In the copies of the T, لَوَّثَتْ is put in the place of لَوَّنَتْ, app. by a mistake. (TA.) — [Hence also,] تَحْجِيلٌ in the وُضُوء signifies The *washing a portion of the* عَضُد [or *upper arm,* perhaps a mistake for the ذِرَاع, or *fore arm,*] *and a portion of the shank, while washing the hand and foot.* (Mṣb.) — [Hence also,] حُجِّلَ القِرَى, (TA,) inf. n. as above, (Ḳ,) ‡ *A little milk, as much as the measure of the* تَحْجِيل *of a horse, was poured into the bowl for the guest,* or *guests, and then the bowl was filled up with water;* this being done in a case of dearth, or drought, and want of milk: (Ḳ,* TA:) or, accord. to Aṣ, it means *the bowl for the guest,* or *guests, was concealed in the* حَجَلَة, through niggardliness, in order that the owners might drink its contents. (TA.) — [Hence also, as تَحْجِيل renders a horse conspicuous,] حَجَّلَ فُلَانٌ أَمْرَهُ † *Such a one made his case,* or *affair, notorious,* or *public.* (TA.) — See also 1, first sentence. ═ حَجَّلَهَا, inf. n. as above, *He made for her a* حَجَلَة: (M, Ḳ:) or *he brought her,* or *put her, therein.* (O, Ḳ.) — [And hence حجّل signifies also *He concealed* a thing *in the* حَجَلَة: see above.] ═ See also 1, second sentence.

4. احجل البَعِيرَ *He loosed the camel's shackle from his left fore leg, and fastened it upon the right:* (Ṣ, O, Ḳ:) or, accord. to the M, *he loosed it from his right fore leg, and fastened it upon the left.* (TA.)

Q. Q. 1. حَوْجَلَ: see 1.

حَجْلٌ: see what next follows.

حِجْلٌ and ▼حَجْلٌ (Ṣ, Mgh, Mṣb, Ḳ) and ▼حِجِلٌ (Ṣgh, Ḳ) and ▼حِجِّلٌ (Ḳ) *An anklet;* or *a pair of anklets;* syn. خَلْخَالٌ: (Ṣ, Mgh, Mṣb, Ḳ:) and the first and second (Ṣ, Mgh, Mṣb, Ḳ) and third, as some say, (Ḳ,) by a metaphor, (Mṣb,) ‡ *a shackle;* or *a pair of shackles,* or *hobbles;* syn. قَيْدٌ: (Ṣ, Mgh, Mṣb, Ḳ:) and † *the two rings of the* قَيْد: (Ḳ:) pl. [of pauc.] أَحْجَالٌ (Ṣ, Mgh, Mṣb, Ḳ) and [of mult.] حُجُولٌ. (Mgh, Mṣb, Ḳ.) You say, فِى سَاقَيْهَا حِجِلٌ ▼ [or حِجْلٌ &c.] *Upon her legs are anklets.* (TA.) And القُيُودُ حُجُولُ الرِّجَالِ وَالحُجُولُ قُيُودُ رَبَّاتِ الحِجَالِ, i. e. *Shackles are the anklets of men; and anklets are* [*for the mistresses of the curtained canopies,* i. e.,] *for women.* (TA.) And خَرَجَ يَجُرُّ رِجْلَيْهِ وَيُطَابِقُ فِى حِجْلَيْهِ [*He went forth dragging his legs, and hobbling in his shackles*]. (TA.) And [hence] فَرَسٌ بَادٍ حُجُولُهُ i. q. مُحَجَّلٌ [q. v.]. (TA.) ═ Also, the first, *Whiteness:* (M, Ḳ:) pl. أَحْجَالٌ. (Ḳ.)

حَجَلٌ [The *partridge;* or *partridges;* comprising several species, of which those most commonly known appear to be identical with the *Barbary partridge* and the *Greek partridge;* both *red-legged:* accord. to Forskål, ("Descr. Animal.," pp. vii. and 11,) applied both to this bird, *tetrao perdix,* and also to the *phasianus meleagris:*] *a well-known bird;* (Mṣb;) *i. q.* قَبْجٌ: (ISh, Ṣ:) or *the male of the* قَبْج: (Ḳ:) or the *females of the* يَعَاقِيب [pl. of يَعْقُوبٌ, q. v.]: (Lth:) *also called* [دَجَاجُ البَرِّ] دجاج البر: *there are two species;* [نَجْدِىٌّ] نجدى *of Nejd*] *and* [تِهَامِىٌّ] تهامى *of Tihámeh*]: *the former species is* أَخْضَرُ [here meaning *of a dark,* or *an ashy, dust-colour*], *with red feet* [or *legs*]; *the latter, of the former colour intermixed with white:* but نجدى is found used for the male; and غرغرة and بنت السعدى, for the female: (Dmr, cited by Freytag:) a single bird of the kind is called ▼حَجَلَةٌ: (Ṣ, Mṣb, Ḳ:) حَجَلٌ is a pl., as also حِجْلَانٌ and ▼حِجْلَى; (Ṣ;) or [rather] حَجَلٌ is a coll. gen. n., (Mṣb, Ḳ,) and the pl., (Mṣb,) or quasi-pl. n., (Ḳ,) is ▼حِجْلَى; (Mṣb, Ḳ;) which is the only instance of its kind except ظِرْبَى: (Ṣ, Ḳ: in a copy of the Mṣb ظِرْبِى:) *its flesh is of moderate temperament,* (Ḳ, TA,) *more delicate than that of the* دُرَّاج *and that of the* فَوَاخِت, *and very fattening:* (TA:) *the swallowing half a mithkál of its liver is good for the epilepsy; and the introduction of its gall-bladder into the nose once in every month sharpens the intellect greatly, and strengthens the sight:* (Ḳ:) *its flesh is good for the dropsy, benefits the stomach, and increases the venereal faculty.* (Ibn-Seenà, TA.) — Also, (Ṣ,) or ▼حَجَلَةٌ, of which حَجَلٌ is pl., (Ḳ,) or حَجَلَةٌ is n. un. of حَجَلٌ, [which is a coll. gen. n.,] (Ṣ,) The *young offspring of camels; the little ones thereof.* (Ṣ, Ḳ.) — دِبِى حَجَلٍ *A certain game* (Fr, Ḳ) *of the Arabs of the desert.* (Fr.) ═ See also حَجَلَة.

حِجِلٌ: } see حِجْلٌ, in three places.
حِجِّلٌ: }

حَجَلَةٌ [*A kind of curtained canopy* or *alcove* or *the like, prepared for a bride;*] *a thing like a* قُبَّة: (M, Ḳ:) and *a place,* (Ḳ,) or *a tent,* or *pavilion,* or *chamber,* (بَيْت,) (Ṣ,) *adorned with cloths* (Ṣ, Ḳ) *and with raised couches* (Ṣ) *and with curtains, for a bride:* (Ṣ, Ḳ:) or *the curtain of the bride, within a* بَيْت [meaning *tent,* or *pavilion,* or *chamber*]: (Mgh:) pl. حِجَالٌ (Ṣ, Mgh, Ḳ) and [coll. gen. n.] ▼حَجَلٌ. (Ḳ.) [See أَرِيكَةٌ, and مِنَصَّةٌ.] ═ See also حَجَلٌ, in two places.

حِجْلَى: see حَجَلٌ, in two places.

حَجْلَآءُ, applied to a ewe, (Ṣ,* Ḳ,* TA,) *Whose fore and hind shanks are white,* (Ṣ, Ḳ, TA,) *and the rest of her black:* so in the M and O. (TA.) [See also خَدَمَةٌ, voce أَخْدَمُ.]

حَجِيلٌ A horse *that is* مُحَجَّلٌ [q. v.] *in three legs.* (Fr, Ḳ.)

حَاجِلٌ [part. n. of حَجَلَ] has for its pl. حُجَّلٌ, which is applied by Jereer to crows or ravens [as meaning *Leaping in going, as though shackled*]. (TA.) [The fem. pl.] حَاجِلَاتٌ is also applied to camels, (Ṣ, Ḳ,) meaning *That have been smitten in their legs,* (Ṣ,) or *that have been hamstrung,* (Ḳ,) *and in consequence walk not on all of their legs.* (Ṣ, Ḳ.)

حَوْجَلَةٌ (Ṣ, Ḳ, &c.) and حَوْجَلَّةٌ, (M, Ḳ,) like حَوْصَلَةٌ and حَوْصَلَّةٌ, and دَوْخَلَةٌ and دَوْخَلَّةٌ, &c., (TA,) *A flask,* or *bottle;* syn. قَارُورَةٌ: (Ḳ:) or *a small* قارورة *with a wide head,* (Ṣ, M, O,) [*the head*] *resembling a* سُكُرُّجَة *and the like:* (M, TA:) or *a* قارورة *large in the lower part:* (Ḳ:) or *one like the* قَوَارِير *of* [*the kind of perfume called*] ذَرِيرَة: (TA:) pl. حَوَاجِلُ and حَوَاجِيلُ; (M, Ḳ;) in the latter of which, the ى may be inserted by poetic license, or as a substitute for one of the لs in حوجلّة. (M, TA.) [See also حَوْقَلَةٌ.]

تَحْجِيلٌ [inf. n. of 2, q. v.: and also used as a simple subst., signifying] *Whiteness in the legs of a horse,* (Ṣ, Ḳ,) *all of them;* (Ḳ;) or *in three of the legs;* (Ṣ;) *in the two hind legs and a fore leg;* (Ḳ;) or *in a hind leg and the two fore legs;* (TA;) or *in the two hind legs* (Ṣ, Ḳ) *only;* (Ḳ;) or *in one hind leg only;* (Ḳ;) *but not in the two fore legs alone, nor in one fore leg without the other, unless with the two hind legs,* (AO, Ṣ, Ḳ, TA,) or *with one hind leg;* (AO, Ṣ, TA;) *whether little or much, so that it extends* [*upwards*] *beyond the pastern but not beyond the knee and hock.* (Ṣ.) — Also *A whiteness in a she-camel's teats, occasioned by the* صِرَار [q. v.]. (Ḳ.) — And, accord. to ISk and the Ḳ, *A certain mark made with a hot iron upon a camel:* but Ṣgh says that the right word is تَحْجِينٌ, with ن. (TA.)

مُحَجَّلٌ *Wearing* أَحْجَال, i. e. *anklets;* [or *adorned therewith;*] applied to a woman [without ة because men do not wear anklets]: if applied to a man, *shackled.* (Ḥam p. 238.) — [And hence,] applied to a horse, (Ṣ, Mgh, Mṣb, Ḳ,) *Having what is termed* تَحْجِيلٌ, as explained in the first sentence of the paragraph next preceding; (Ṣ, Ḳ;) as also ↓مَحْجُولٌ: (Ḳ:) *white in the place of the anklet,* and *above that;* wherefore the horse is thus termed: (Ḥam p. 53:) *having his legs,* (Mgh, Mṣb,) *all four,* (Mgh,) *white; the whiteness extending* [*upwards*] *beyond the pasterns,* (Mgh, Mṣb,) *to a third,* (Mgh,) or *to half,* (Mgh, Mṣb,) or *thereabout,* (Mṣb,) or *to two thirds,* (Mgh,) *of the shank.* (Mgh, Mṣb.) When the whiteness is in all the four legs, he is termed مُحَجَّلٌ أَرْبَعٍ: when in the two hind legs, مُحَجَّلُ الرِّجْلَيْنِ: when in one of the hind legs, extending [upwards] beyond the pastern, مُحَجَّلُ الرِّجْلِ اليُمْنَى or اليُسْرَى: when in three legs, exclusive of a hind leg or of a fore leg, مُحَجَّلٌ ثَلَاثٍ مُطْلَقُ يَدٍ or رِجْلٍ: when in the fore leg and hind leg of one side, مُمْسَكُ الأَيَامِنِ مُطْلَقُ الأَيَاسِرِ or مُمْسَكُ الأَيَاسِرِ مُطْلَقُ الأَيَامِنِ: when on opposite sides, whether little or much, مَشْكُولٌ. (Ṣ.) Hence, in a trad., أُمَّتِى الغُرُّ المُحَجَّلُونَ يَوْمَ القِيَامَةِ مِنْ آثَارِ الوُضُوْءِ † [*My followers will be those having a whiteness on the forehead and on the wrists and ankles, on the day of resurrection, from the effects of the ablution for prayer*]. (TA.) [Hence also, because the horse that is مُحَجَّل is conspicuous,] رَكِبَ الشَّادِخَةَ المُحَجَّلَةَ † *He committed a bad and notorious deed.* (Ṣ in art. شدخ, q. v.) And the saying of El-Jaadee, satirizing Leylà El-Akhyaleeyeh,

فَقَدْ رَكِبَتْ أَمْرًا أَغَرَّ مُحَجَّلَا

† [*For she has committed a glaring, notorious deed*]. (Az, TA.) And يَوْمٌ أَغَرُّ مُحَجَّلٌ † *A day bright and beaming with happiness and cheerfulness.* (Ḥar p. 377.) — Also *A she-camel's udder having a whiteness in the teats, occasioned by the* صرار [q. v.]. (Ḳ.) ═ *A woman who keeps,* or *cleaves, to the* حِجَال [pl. of حَجَلَةٌ]: and in like manner, a man; meaning † *one who keeps much,* or *habitually, to the company of women.* (Ḥam p. 238.)

مَحْجُولٌ: see مُحَجَّلٌ.

حجم

1. حَجَمَ, said of the breast of a woman or girl: see 4. ═ حَجَمَ, (ISk, Ṣ, Ḳ,) aor. ـُ and ـِ, inf. n. حَجْمٌ, (Ḳ,) *He* (a child, ISk, Ṣ, TA) *sucked* (ISk, Ṣ, Ḳ) the breast of his mother. (ISk, Ṣ, TA.) — [Hence,] حَجَمَهُ, aor. ـُ, inf. n. حَجْمٌ, (Ṣ, Mgh, Mṣb,) *He scarified him* [*and drew blood from him with the* مِحْجَمَة; i. e. *he cupped him*]; (Mṣb;) *he performed upon him the operation of the* حَجَّام, (Mgh,) or, *of the* حَاجِم, (Ṣ,) *who sucks the mouth of the* مِحْجَمَة [*to draw the blood*]. (Az, TA.) — And [hence,] حَجَمَتْهُ الحَيَّةُ † *The serpent bit him.* (TA.) And حَجَمَتِ الفُحُولُ العَيْرَ ‡ *The stallions bit the ass.* (TA.) And حَجَمَ العَظْمَ, (TA,) inf. n. as above, (Ḳ,) † *He ate off the flesh from the bone.* (Ḳ, TA.) ═ حَجَمَ البَعِيرَ, (Ṣ, Mṣb, TA,) aor. ـُ, (Ṣ, TA,) inf. n. as above, (TA,) *He put a* حِجَام *upon the mouth* [or *muzzle*] *of the camel,* when excited, *in order that he might not bite;* (Ṣ, TA;) [*he muzzled the camel;*] *he bound the camel's mouth* [or *muzzle*] *with a thing.* (Mṣb.) — Hence, (Ḥar p. 474,) حَجَمْتُهُ عَنِ الشَّىْءِ, (Ṣ, Mṣb,* TA,) and عَنْ صَاحِبَتِهِ, (TA,) inf. n. as above, (Ḳ,) *I made him to refrain, forbear, abstain,* (Ṣ, TA,) or *go back;* (Mṣb;) or *I withheld him,* or *restrained him;* (Ṣ, Ḳ, TA;) *from the thing,* (Ṣ, TA,) and *from his female companion.* (TA.) And حَجَمَ طَرْفَهُ عَنْهُ *He turned away his eye,* or *eyes, from him,* or *it.* (TA.) And ↓أَحْجَمْتُهُ عَنْ حَاجَتِهِ is also mentioned, as meaning *I withheld him from the object of his want;* or *prevented him from attaining it.* (TA.)

2. حجّم: see 4. ═ Also, inf. n. تَحْجِيمٌ, ‡ *He looked hard:* (Ḳ, TA:) and so, accord. to Az جَحَّمَ (TA.)

4. احجم *It* (the breast of a woman or girl) *was,* or *became, swelling, prominent,* or *protuberant;* (Mgh, Ḳ;) as also ↓حَجَمَ, inf. n. حُجُومٌ: (Ḳ:) or *was,* or *became, round, and swelling, prominent,* or *protuberant;* as also ↓حجّم: (A, TA:) properly, *became such as to have what is termed* حَجْمٌ, (Mgh, TA,) meaning *projection, protrusion, prominence,* or *protuberance,* and *elevation:* (Mgh:) or, as some say, *became such that the suckling might suck it.* (TA.) — أَحْجَمَتْ لِلْمَوْلُودِ ‡ *She* (a woman) *suckled the new-born child for the first time.* (Ḳ, TA.) ═ احجم عَنْهُ *He refrained, forbore, abstained,* or *desisted, from it;* (Ṣ, Ḳ;) quasi-pass. of حَجَمْتُهُ, like as أَكَبَّ is of كَبَبْتُهُ; which are extr. of their kind; (Ṣ;) and اجحم عنه signifies the same; (Ṣ in art. جحم;) but is a rare dial. var.: (Ḥar p. 95:) or *he drew back from it,* or *him, in awe,* or *fear:* (Ḳ, TA:) or *he receded,* or *drew back, from it;* namely, a thing, or an affair: (Mṣb:) and أَحْجَمْتُ عَنِ القَوْمِ *I dreaded,* or *feared, the people,* or *party, and returned, and left them, after I had desired to go to them.* (AZ, Mṣb.) — احجم also signifies *He* (a man) *advanced,* or *went forward;* and so اجحم: both of these verbs thus having two contr. meanings. (MF.) ═ See also 1, last sentence.

8. احتجم *He performed the operation of cupping:* (MA, KL, PṢ:) [or rather *he had that operation performed upon him;* or *had blood drawn from him by that operation;* a quasi-pass. verb, like افتصد and اكتوى and احترق and countless others:] or *he sought,* or *demanded, the performance of that operation* [*upon himself*]. (Ḳ, TA.) One says, اِحْتَجَمْتُ مِنَ الدَّمِ [app. meaning *I had some of the blood drawn from me by cupping*]. (Ṣ.) ═ *He* (a camel) *was withheld,* or *restrained,* or *prevented, from biting* [*by being muzzled*]. (TA.)

حَجْمٌ *A rising, protuberant,* or *prominent, part* of a thing: (Ṣ:) *a projection, protrusion, prominence,* or *protuberance;* (Ṣ, Mgh;) and *a rising,* or *an elevation:* (Mgh:) or the *part* of a thing *that one feels beneath his hand:* (El-Ghooree, Mgh:) or the *part* of a thing *that one feels projecting, protruding, prominent,* or *protuberant, beneath his hand:* pl. حُجُومٌ. (Ḳ.) One says, لَيْسَ لِمِرْفَقِهِ حَجْمٌ *There is no projection, protrusion, prominence,* or *protuberance, to his elbow.* (Ṣ.) Lth says that حَجْمُ العَظْمِ means One's *perceiving the feel of the bones* [or *bone*] *behind the skin:* thus explaining it after the manner of explaining inf. ns.: and ISd says, I know not whether it be in his opinion an inf. n. or a simple subst. (TA.) And Lth says that الحَجْمُ means One's *perceiving the feel of a thing beneath a garment:* [and that] one says, مَسِسْتُ بَطْنَ الحُبْلَى فَوَجَدْتُ حَجْمَ الصَّبِىِّ فِى بَطْنِهَا [which plainly means, *I felt the belly of the pregnant woman, and perceived the bulging of the child in her belly*]. (Mgh, TA.) It is said [of a woman's garment] in a trad., لَا يَصِفُ حَجْمَ عِظَامِهَا [lit.

‡ *It does not describe the projecting of her bones*]: meaning, the garment does not stick to her body so as to tell what projects of her bones: it is made to be a describer by way of comparison. (IAth, TA.) —— [In post-classical works it is often used as signifying *Bulk, bigness,* or *magnitude.*]

حِجَامٌ [*A muzzle for a camel,* commonly *a net of cord;*] *a thing that is put upon the mouth,* or *muzzle, of a camel,* (Ṣ, Ḳ,) *when he is excited,* (Ṣ,) *to prevent his biting.* (Ṣ, Ḳ.)

حَجُومٌ *Gentle, nice,* or *skilful, in operation;* syn. رَفِيقٌ [in the CḲ, erroneously, رَقِيق] as an epithet applied to a حَاجِم [or cupper]; as also ▼ مِحْجَمٌ. (Ḳ.) —— ‡ The فَرْج [meaning *vulva,* or *vagina,*] *of a woman.* (Ḳ, TA.)

حِجَامَةٌ The *craft,* or *art, of the* حَاجِم or حَجَّام [i. e. *cupper*]: (Ṣ,* Mgh, Mṣb, Ḳ:) and the *operation that he performs.* (Ṣ,* TA.)

حَجَّامٌ *A cupper;* syn. مَصَّاصٌ [lit. *one who is in the habit of sucking*]; (Ḳ;) *one who performs the operation termed* حَجْم; (Ṣ, Mgh;) *one who scarifies* [*and draws blood with the* مِحْجَمَة]; (Mṣb;) as also ▼ حَاجِمٌ; (Ṣ, Mṣb, Ḳ;) but the former is an intensive epithet [denoting frequency or habit]: (Mṣb:) so called because he sucks the mouth of the مِحْجَمَة. (Az, TA.) Hence the prov. أَفْرَغُ مِنْ حَجَّامِ سَابَاطَ [*More unoccupied than the cupper of Sábáṭ* (the name of a place)]: for military forces used to pass by the man to whom it relates, and he used to cup them on credit, in consequence of want of employment, waiting for payment until they should return. (Ṣ: more fully explained in the Ḳ in art. سبط.)

حَاجِمٌ: see what next precedes.

حُوجَمٌ: see what next follows.

حُوجَمَةٌ The *red rose:* (Ḳ:) or [rather] *a single red rose:* (A 'Obeyd, Ṣ:) pl. [or rather coll. gen. n.] ▼ حُوجَمٌ. (A 'Obeyd, Ṣ, Ḳ.)

مَحْجَمٌ, (Mgh, Mṣb,) or ▼ مَحْجَمَةٌ, (TA,) The *place to which the* مِحْجَمَة *is applied,* (Mgh, Mṣb,* TA,) *in the neck:* (Mgh, TA:) pl. مَحَاجِمُ. (Mgh, Mṣb.)

مِحْجَمٌ The *scarifying instrument of the* حَجَّام. (IAth, TA.) —— See also مِحْجَمَة. ═ And see حَجُوم.

مَحْجَمَةٌ: see مَحْجَم.

مِحْجَمَةٌ and ▼ مِحْجَمٌ The قَارُورَة [or *glass vessel,* or *the like,*] *of the* حَجَّام; (Az, Ṣ, Mgh, Mṣb;) the *thing with which the* حَجَّام *operates;* (Ḳ;) the *utensil in which the blood produced by the operation of* حِجَامَة *is collected by sucking:* (IAth, TA:) pl. مَحَاجِمُ. (TA.)

مُحْجِمٌ A man (TA) *who draws back much,* or *often, in awe,* or *fear.* (Ḳ, TA.)

مَحْجُومٌ A breast *sucked.* (TA.) —— A man [*cupped,* or] *operated upon by the* حَاجِم. (Ṣ.) ═ A camel [*muzzled,* or] *having a* حِجَام *put upon his mouth* [or *muzzle*] *in order that he may not bite.* (Ṣ.)

حجن

1. حَجَنَهُ, aor. ـِ, (Ḳ, TA, [in the CḲ ـُ,]) inf. n. حَجْنٌ, (TA,) *He bent it,* or *made it crooked* [or *hooked*]; namely, a stick, or branch, or slender piece of wood; as also ▼ حَجَّنَهُ, (Ḳ,) inf. n. تَحْجِينٌ. (TA.) —— † *He marked him* (i. e. a camel) *with the brand of the* مِحْجَن, which is *a line with a crooked,* or *hooked, end, like the stick called* مِحْجَن; inf. n. as above. (TA.) —— *He drew it,* or *pulled it,* [or *hooked it,*] (Ṣ, Ḳ,) *towards himself* (Ṣ) *with the* مِحْجَن; as also ▼ احتجنهُ. (Ṣ, Ḳ.) —— ‡ *He turned him away* (Ḳ, TA) عَنِ الشَّىْءِ *from the thing.* (TA.) ═ حَجِنَ عَلَيْهِ and بِهِ, aor. ـَ, (Ḳ,) inf. n. حَجَنٌ, (TA,) *He was,* or *became, avaricious, tenacious,* or *niggardly, of it;* (Ḳ;) like حَجِىَ بِهِ. (TA.) —— حَجِنَ بِالدَّارِ *He remained, stayed, dwelt,* or *abode, in the house.* (Ḳ.)

2: see 1.

4. احجن, said of the ثُمَام [or panic grass], *It put forth its* خُوص; (A 'Obeyd, Ṣ, Ḳ;* [in the Ḳ, *its* خُوصَة;]) [i. e.] *its leaves appeared.* (TA.)

5. تحجّن *It was,* or *became, crooked,* [or *hooked,*] or *curved:* (T, Ḳ:) said of a thing that is termed أَحْجَن. (T.)

8: see 1. —— [Hence,] احتجن المَالَ ‡ *He drew the property,* or *camels &c., together* (Ṣ, Ḳ, TA) *to himself,* (Ṣ, TA,) *and took,* or *took possession of, it,* or *them.* (Ṣ, Ḳ, TA.) And احتجنهُ ‡ *He took possession of it* (i. e. a portion of land), *exclusively of others.* (TA from a trad.) And احتجنهُ لِنَفْسِهِ دُونَ أَصْحَابِهِ ‡ *He appropriated it* (a thing) *to himself, exclusively of his companions.* (T, TA.) And احتجن مَالَ غَيْرِهِ † *He took away,* and *stole, the property of another.* (TA.) —— Also † *He put the property,* or *camels &c., into a good,* or *right, state, and drew together what had become scattered thereof.* (TA.) —— And احتجن عَلَيْهِ † *He straitened him.* (TA.)

حَجَنٌ *Crookedness,* [or *hookedness,*] or *curvature;* (Ṣ, Ḳ;) as also ▼ حُجْنَةٌ. (Ḳ.) —— See also حُجْنَة.

حَجِنٌ: see أَحْجَنُ.

حُجْنَةٌ: see حَجَنٌ. —— Also *Crispness* [or *recurvation*] *in the extremities* of hair. (T, TA. [See أَحْجَنُ.]) —— *A place of crookedness* or *curvature* (ISd, TA) of a staff or stick. (TA.) —— The *hook in the head* of a spindle, (Ṣ,* Ḳ,* TA,) *with which the thread is caught preparatively to the twisting thereof.* (TA.) —— The خُوصَة, (Ḳ,) or خُوص, (Ṣ,) [i. e.] the *leaves* [or *blades*], (TA,) *of* ثُمَام [or panic grass]; (Ṣ, Ḳ, TA;) as also ▼ حَجَنَةٌ. (Ḳ.) And ▼ حُجَنٌ [of which ▼ حُجَنَةٌ is the n. un., if not a mistranscription of حُجْنٌ,] *Tender,* or *soft, shoots, that grow upon the sides of the stalks of the* ثُمَام *and the* ضَعَة [which is said to be a species of ثُمَام]. (TA.) And حُجَنٌ, pl. of حُجْنَةٌ, The *fruit-stalks of grapes.* (TA.) ═ Also *A thing,* or *portion of a thing, that one has drawn and appropriated to oneself.* (TA.)

حَجَنَةٌ: see the next preceding paragraph, in two places.

حَجُونٌ *Sluggish, lazy,* or *indolent:* (Ḳ:) from حَجِنَ بِالدَّارِ [q. v.]. (TA.) —— غَزْوَةٌ حَجُونٌ ‡ *A hostile,* or *hostile and plundering, expedition, in which the party feigns to be going in one direction, and then turns to another:* (A, Ḳ,* TA:) or *far-extending.* (Ṣ, Ḳ.) And سِرْنَا عُقْبَةً حَجُونًا ‡ *We journeyed a long stage.* (Ṣ, TA.)

أَحْجَنُ *Crooked,* [*hooked,*] or *curved:* fem. حَجْنَآءُ: pl. حُجْنٌ. (Ḥam p. 403.) You say, الصَّقْرُ أَحْجَنُ المِنْقَارِ *The hawk is crooked* [or *hooked*] *in the bill.* (TA.) And صَقْرٌ أَحْجَنُ المَخَالِبِ *A hawk having crooked* [or *hooked*] *talons.* (Ṣ, TA.) And أَنْفٌ أَحْجَنُ [*A hooked nose;*] *a nose having the tip approaching the mouth,* and, Az adds, *having its* نَاشِرَتَان [or *two alæ*] *receding in an ugly manner.* (TA.) And أُذُنٌ حَجْنَآءُ *An ear having one* [app. *the upper*] *of its two extremities turning towards the forehead, downwards:* or *having its edges turning towards the other ear, in the direction of the forehead:* (M, Ḳ:) in either case, *curving.* (M, TA.) And شَعْرٌ أَحْجَنُ ‡ *Hair that is crisp,* or *curly, in its extremities:* or, accord. to AZ, *wavy hair:* (T:) or *hair that is recurvate at its extremities* (مُعَقَّف), *and intermingling:* (M:) or *hair forming a succession of rimples* (مُتَسَلْسِل), *pendulous, wavy, and crisp,* or *curling, in the extremities;* as also ▼ حَجِنٌ. (Ḳ.)

تَحْجِينٌ †*A crooked,* [or *hooked,*] or *curved, brand,* or *mark made with a hot iron* [*upon a camel*]: (Ḳ:) [originally inf. n. of 2; but in this sense,] a subst., properly speaking, like تَنْبِيتٌ and تَبْيِينٌ. (TA. [See also مِحْجَن.])

مِحْجَنٌ *A crooked,* [or *hooked,*] or *curved, staff* or *stick;* as also ▼ مِحْجَنَةٌ: (Ḳ:) or *a stick,* (IAth, Mgh, Ḥam p. 403,) or *staff,* (IAth, Ḥam,) or *piece of wood,* (Mṣb,) *with a crooked,* or *hooked, head,* (IAth, Mgh,) or *crooked at the end;* (Mṣb;) *like the* صَوْلَجَان: (Ṣ, Mgh, Mṣb, Ḥam:) *one draws towards him with it the extremities* [*of the branches*] *of trees, and the like:* (Ḥam ubi suprà:) or *a stick with a crooked,* or *curved, end, being naturally so on the tree on which it has grown;* distinguished from a صولجان, the end of which is crooked, or curved, artificially: (T:) or, accord. to AZ, *any stick with a curved head:* (Mṣb:) or it signifies also *anything bent,* or *crooked:* (Ḳ:) pl. مَحَاجِنُ. (Mṣb, TA.) The appellation صَاحِبُ المِحْجَنِ [*The owner of the crooked stick or staff*] was given to a certain man who, in the Time of Ignorance, used to sit in the highway, and take with his محجن one thing after another, of the goods of the passers-by; and if any one were cognizant of his doing, he excused himself, saying that the thing had caught to his محجن. (TA.)

You say, فُلَانٌ لَا يَرْكُضُ المِحْجَنَ [lit. *Such a one will not kick the crooked stick* or *staff*], meaning † *such a one is of no use*, or *stands one in no stead*: the saying originating from the fact that a مِحْجَن is put between the hind legs of the camel, and if he be inert, or wanting in vigour, he will not kick it; but if he be sharp in spirit, he will kick it and go on. (TA.) And you say, إِنَّهُ لَمِحْجَنُ مَالٍ, meaning † *Verily he is one who puts the cattle into a good state, and pastures and manages them well.* (TA.) Also † *A brand*, or *mark made with a hot iron, upon a camel, in the form of a line with a crooked*, or *hooked, end, like the stick so called.* (TA. [See also تَحْجِينٌ.]) — And The [*hooked*] *bill* of a bird; because of its crookedness. (TA.)

مِحْجَنَةٌ: see the next preceding paragraph.

مَحْجُونٌ A camel *marked with the brand termed* مِحْجَن. (TA.)

حجو

1. حَجَا, (Ṣ, Ḳ,) aor. يَحْجُو, (Ṣ, TA,) inf. n. حَجْوٌ, (TA,) *He thought*, or *opined*: (Ṣ, TA:) or *he thought*, or *opined*, a thing, *and, doing so, claimed* it (ادّعاهُ), *not being certain of it*: (Ḳ:) [or *he thought* it *and asserted* it; as appears from a verse here following, cited in the TA as an ex. of the meaning explained as above in the Ḳ:] and ▾ تَحَجَّى بِظَنِّهِ, and تحجّى alone, *he thought*, or *opined*, a thing, *not being certain of it*. (T, TA.) You say, أَنَا أَحْجُو بِهِ خَيْرًا *I think*, or *opine, that there is good in him.* (Ṣ, TA.) And Aboo-Shembel says, respecting Aboo-'Amr Esh-Sheybánee,

* قَدْ كُنْتُ أَحْجُو أَبَا عَمْرٍو أَخَا ثِقَةٍ *
* حَتَّى أَلَمَّتْ بِنَا يَوْمًا مُلِمَّاتُ *

[*I used to think and assert Aboo-'Amr to be a trustworthy person, until, one day, misfortunes befell us*]. (TA.) — حَجَا القَوْمَ كَذَا وَكَذَا, (Ṣ, Ḳ*) *He repaid, requited, compensated*, or *recompensed, the people*, or *party*, [*with*, or *for, such and such things*,] syn. جَزَاهُمْ; [so in my copies of the Ṣ, and in the Ḳ; but Freytag gives, as the reading found by him in the Ṣ, حزاهم, as though meaning *he computed them by conjecture to be such and such in number*; which is agreeable with what follows;] and *he thought them to be so*. (Ṣ.) ═ *He directed himself*, or *his course* or *aim, to*, or *towards*, him, or it: (Az, TA:) and ▾ تحجّى الشَّىْءَ *he directed himself*, or *his course* or *aim, to*, or *towards, the thing*. (Ṣ, TA.) ═ Also, (Ḳ,) inf. n. as above, (TA,) *He kept* a secret: (Ḳ:) or *he concealed* it: (AZ, TA:) [and ▾ حاجى has a similar meaning; for] you say, لَا مُحَاجَاةَ عِنْدِى فِى كَذَا, i. e. *There is no concealment with me in respect of such a thing*; as also لَا مُكَافَأَةَ. (TA.) — لَا يَحْجُو إِبِلَهُ, said of a pastor, *He does not*, or *will not, keep and tend*, or *pasture and defend, his camels*. (TA.) One says also of a pastor whose sheep or goats [and camels] are lost by him, and become dispersed, مَا يَحْجُو غَنَمَهُ وَلَا إِبِلَهُ [*He does not keep his sheep or goats, nor his camels, from being lost and becoming dispersed*]. (TA.) — سِقَآءٌ لَا يَحْجُو المَآءَ *A skin that will not hold*, or *retain, water*. (TA.) — مَا حَجَوْتُ مِنْهُ شَيْئًا *I did not keep*, or *retain in my memory, of it, aught*; as also مَا حَجَوْتُ. (Ks, TA.) ═ Also, (Ḳ,) inf. n. as above, (TA,) *He*, or *it, prevented*, or *withheld*. (Ḳ, TA.) [See حِجًا.] ═ حَاجَيْتُهُ فَحَجَوْتُهُ: see 3. ═ حَجَا بِالمَكَانِ, (Ṣ, Ḳ,) inf. n. as above, (Ḳ,) *He remained, stayed, dwelt*, or *abode, in the place*, (Ṣ, Ḳ, TA,) *and became fixed* [*therein*]; (TA;) as also ▾ تحجّى بِهِ. (Ṣ, Ḳ.) And حَجَا, (IAạr, Ḳ,) inf. n. as above, (IAạr,) *He stopped*, or *paused*. (IAạr, Ḳ.) — And حَجَا بِهِ, [inf. n. as above, (see Ḥam p. 523,)] *He was*, or *became, tenacious*, or *avaricious, of it*, namely, a thing; (Ṣ, Ḳ;) as also ▾ تحجّى بِهِ. (TA.) And حَجِىَ بِهِ, (Fr, Ṣ, Ḳ,) aor. ـَ, inf. n. حَجًى [or حَجًا], (TḲ,) *He was*, or *became attached to it, and clave to it*; (Fr, Ṣ, Ḳ;) as also ▾ تحجّى بِهِ; and حَجِئَ بِهِ; (Fr, Ṣ;) and تحجّأ بِهِ. (Fr, TA.) ▾ تحجّى also signifies *He kept to the* حَجَا, or *place of bending* or *turning of a valley*. (TA.) And ▾ تَحَجَّيْتُ بِهٰذَا المَكَانِ *I preceded you*, or *outwent you, to this place, and clave to it before you*. (Ṣ, TA.) And [hence,] ▾ تحجّى بِهِمْ is said to mean † *He was first*, or *foremost*, or *quick, to blame them*. (TA.) — حَجِىَ, (Ḳ, TA,) aor. ـَ, (TA,) is also [said to be] *syn. with* عَدَا, (Ḳ,) *He ran*; (TḲ;) thus bearing two contr. significations: (Ḳ:) but this requires consideration. (TA.) — [Golius, as on the authority of the KL, assigns to it also the meaning "Hilaris et lubens fuit:" but in this sense it is said in the KL to be with ء for its last radical letter: see art. حجأ.]

3. ▾ حَاجَيْتُهُ فَحَجَوْتُهُ, (T, Ṣ, Ḳ,) inf. n. of the former مُحَاجَاةٌ and حِجَآءٌ, (Ḳ,) *I contended*, or *vied, with him in intelligence* or *skill and knowledge*, or *in endeavouring to show my superiority in intelligence &c.*, (فَاطَنْتُهُ,) *and I overcame him therein*; (Ḳ;) from حِجًى [or حِجًا] meaning "intelligence." (Ḥar p. 9.) [And hence, *I tried him with an enigma* or *enigmas, and overcame him therein*: (see 6:) or] *I proposed to him an enigma* [&c.]: (T, TA:) [or *I contended with him in proposing an enigma* or *enigmas*, &c.]: i. e. دَاعَيْتُهُ فَدَعَيْتُهُ: (Ṣ:) so in the handwriting of Aboo-Zekereeyà, and in that of Aboo-Sahl, [and in my two copies of the Ṣ,] but in [some of the copies of] the Ṣ دعيته. (TA.) And حُوجِىَ بِهِ [*He was tried with it as an enigma to be explained by him*; or *he had it proposed to him as an enigma*]. (TA.) — مُحَاجَاةٌ signifies [also] The *asking a thing* of one *much, so as to weary*; as also مُدَاعَاةٌ. (KL.) — And حِجَآءٌ, (Ḳ, TA,) [accord. to the CḲ حَجًا, but correctly] like كِتَابٌ, (TA,) An *engaging in conflict*, or *fight*. (Ḳ, TA.) — See also 1.

4. مَا أَحْجَاهُ, and أَحْجِ بِهِ, *How well adapted* or *disposed*, or *how apt, meet, suited, suitable, fitted, fit, competent*, or *proper*, or *how worthy, is he!* (Ṣ, Ḳ, TA:) verbs of wonder having no corresponding verb of the measure فَعُلَ. (TA.) You say, مَا أَحْجَاهُ بِذٰلِكَ *How well adapted* or *disposed*, &c., *is he for that!* (Ṣ, TA.)

5: see 1, in eight places. — You say also, تحجّى لَهُ, meaning *He knew it*, or *understood it, readily*, or *with quickness of intelligence*. (AHeyth, TA.)

6. تَحَاجَوْا *They tried one another with enigmas*: [or *proposed enigmas, one to another*: or *contended, one with another, in proposing an enigma*, or *enigmas*: (see 3:)] from حِجًى [or حِجًا] meaning "intelligence." (Ḥar p. 189.) You say, يَتَحَاجَوْنَ بِأُحْجِيَّةٍ [*They try one another with an enigma*: or *contend, one with another, in proposing an enigma*]: (Ṣ, TA:) التَّحَاجِى signifying التَّدَاعِى. (TA.)

8. احتجى *He found out*, or *discovered*, (أَصَابَ,) that with which he was tried as an enigma to be explained by him. (TA.)

حَجًا or حَجًى *A side, region, quarter*, or *tract*: (Ṣ, Mṣb, Ḳ, TA:) and *an extremity*: (TA:) pl. أَحْجَآءٌ. (Ṣ, Mṣb.) — *A refuge*; *a place to which one has recourse for refuge*, or *protection*; as also ▾ مَحْجًى (Lḥ, TA) and مَحْجًا. (Lḥ, Ḳ in art. حجأ.) — *Elevated ground*. (TA.) — *A place of bending* or *turning* of a valley. (TA.) — *Anything by which one is veiled, concealed*, or *protected*; (Mgh, Mṣb, TA;) as also ▾ حِجًا. (Mgh, TA.) [Hence applied to *A parapet* on the top of a house; as is indicated in the Mgh and TA.] ═ *Bubbles upon water, arising from the drops of rain*: pl. [or rather coll. gen. n.] of ▾ حَجَاةٌ: (Ṣ, Ḳ: [in the CḲ, erroneously, حِجاةٌ:]) the latter word, which is like حَصَاةٌ, is explained by Az as signifying *a bubble that rises upon water, like a flask*; and as having for its pl. حَجَوَاتٌ: and the same word (حجاة) signifies also *a pool of water*, itself, *such as is left by a torrent*. (TA.) ═ See also حِجًى. ═ حَجَاءِ is also used, by poetic license, for حَجَاجِ, q. v. (TA in art. حج.)

حَجٍ: see حِجًى. ═ إِنَّهُ لَحَجِىٌّ إِلَى بَنِى فُلَانٍ means لَاجِئٌ لَهُمْ [i. e. *Verily he is betaking himself for refuge*, or *protection, to the sons of such a one*; like لَحَجِئٌ: but by rule it should be لَحَجٍ; and thus, perhaps, it is correctly written]. (AZ, TA.)

حِجًا or حِجًى *Intelligence*, or *understanding*; (Ṣ, Mṣb, Ḳ;) *sagacity*, or *skill and knowledge*: (Ḳ:) [said to be] from حَجَا meaning "he, or it, prevented, or withheld;" because it prevents, or withholds, a man from doing that which is bad, or corrupt. (TA.) ═ And *i. q.* مِقْدَارٌ [*Quantity, measure*, &c.]: pl. أَحْجَآءٌ. (Ḳ.) ═ See also حَجًا. ═ [Also said by Golius, as on the authority of J, to signify "an enigma," as though syn. with أُحْجِيَّةٌ; but I have not found it in this sense in any copy of the Ṣ, nor in any other lexicon.]

حَجَاةٌ: see حَجًا.

حَجْوَى a subst. signifying مُحَاجَاةٌ [i. e. *A*

contention in intelligence &c.; and particularly, *in trying with an enigma* or *enigmas; in proposing an enigma* or *enigmas;* or simply the *proposal of an enigma; see 3*]; (T, TA;) the subst. denoted by حَاجَيْتُهُ فَحَجَوْتُهُ; (K;) [or rather by the verb حَاجَيْتُ, agreeably with the foregoing explanation from the T;] as also ▼حُجًيَّا, (Ṣ, K, TA,) with a ḍammeh, (K, TA,) and with teshdeed of the ى; (TA; [in some copies of the K erroneously written ▼حُجْيًا;]) and ▼أُحْجِيَّةٌ: (Ṣ:) or ▼حُجَيًّا is the dim. of حَجْوَى: (T, TA:) and ▼أُحْجِيَّةٌ signifies *a saying of which the meaning differs from the letter,* as also ▼أُحْجُوَّةٌ, (K,) but احجيّة is preferable, (T, TA,) and كَلِمَةٌ ▼مُحْجِيَةٌ, (K,) like مُحْسِنَةٌ; (TA; [in the CK, erroneously, مُحْجِيَةٌ;]) [i. e.] ▼أُحْجِيَّةٌ signifies *an enigma; a riddle;* (MA, PṢ, TK;) and so ▼حُجَيًّا; (PṢ;) or *a question put to one with the view of causing him to make a mistake;* (TA;) and is from حِجًى [or حِجًا] meaning "intelligence," because مُحَاجَاةٌ is like a vying, or contending, in intelligence: (Ḥar p. 9:) the pl. of ▼أُحْجِيَّةٌ is أَحَاجِىُّ and أَحَاجٍ, (MA, and Ḥar ubi suprà,) agreeably with a general rule relating to words of its measure, as أُثْفِيَّةٌ and أُمْنِيَّةٌ. (Seer, in Ḥar ubi suprà.) One says, ▼حُجَيَّاكَ مَا كَانَ كَذَا وَكَذَا [*The question of contention with thee in trying thine intelligence by an enigma,* or *the enigma proposed to thee, is, What is,* or *was, such a thing, and such a thing?*]: it means *a certain game, and a question put with the view of causing one to make a mistake:* A'Obeyd says, It is *like their saying, Disclose what is in my hand, and thou shalt have such a thing.* (Ṣ.) One says also, أَنَا ▼حُجَيَّاكَ فِى هٰذَا, meaning مَنْ يُحَاجِيكَ [i. e. *I am he who contends with thee in intelligence,* or *in trying with an enigma,* &c., *respecting this thing*]. (Ṣ.) And بَيْنَهُمْ ▼أُحْجِيَّةٌ يَتَحَاجَوْنَ بِهَا [*Between them is a contention in intelligence,* or *in proposing enigmas,* or *between them is an enigma, with which they try one another*]. (Ṣ.)

هُوَ حَجِىٌّ بِهِ *He is adapted, disposed, apt, meet, suited, suitable, fitted, fit, competent,* or *proper, for it;* or *worthy of it;* as also ▼حَجٍ, (Ṣ, M, K, TA,) of which the pl. is حَجُونَ; (TA;) and ▼حَجًى, (Ṣ, M, K, TA,) which last has no fem. nor dual nor pl. form, (Ṣ, M,) retaining the same form as fem. and dual and pl.: (M:) and in like manner you say, إِنَّهُ لَمَحْجَاةٌ ▼أَنْ يَفْعَلَ ذٰلِكَ *Verily he is adapted,* &c., *to do that;* (Ṣ, K;*) and إِنَّهُمْ لَمَحْجَاةٌ; and إِنَّهَا لَمَحْجَاةٌ. (Ṣ.) — حَجِىٌّ is also syn. with حَرِيصٌ [*Vehemently desirous, eager,* &c.]. (KL.)

حُجَيًّا: see حَجْوَى, in five places. — Respecting this word in the phrase حُجَيَّاكَ حَجٍ, see حَاجَةٌ (in art. حوج), near the end of the paragraph.

أَحْجَى *More,* and *most, adapted, disposed, apt, meet, suited, suitable, fitted, fit, competent, proper,* or *worthy:* (TA, Ḥam p. 523:) so in the saying, هُوَ أَحْجَى أَنْ يَكُونَ كَذَا [*He is more,* or *most, adapted,* &c., *to be thus;* or *worthy of being thus*]: (TA:) and فُلَانٌ أَحْجَى بِكَذَا [*Such a one is more,* or *most, adapted,* &c., *for such a thing;* or *worthy of it*]. (Ḥam ubi suprà.) Hence, in a trad., مَعَاشِرُ هَمْدَانَ أَحْجَى حَىٍّ بِالْكُوفَةِ [*The companies of Hemdán are the most worthy tribe in El-Koofeh*]: or, as some say, the meaning is, *the most intelligent tribe.* (TA.)

أُحْجُوَّةٌ: see حَجْوَى.

أُحْجِيَّةٌ: see حَجْوَى, in five places.

مَحْجِىٌّ: see حَجًا.

مُحْجٍ *Niggardly, tenacious,* or *avaricious.* (K.) — كَلِمَةٌ مُحْجِيَةٌ: see حَجْوَى.

مَحْجَاةٌ: see حَجِىٌّ.

حد

1. حَدَّ, (A, Mgh, Mṣb,) aor. ـُ, (Mgh, TA,) inf. n. حَدٌّ, (Ṣ, Mgh, Mṣb, K,) *He,* or *it, prevented, hindered, impeded, withheld, restrained, debarred, inhibited, forbade, prohibited,* or *interdicted:* (Ṣ, A, Mgh, Mṣb, K, TA:) this is the primary signification: (Mgh:) and *he repelled, turned away,* or *averted,* (L, K, TA,) evil [or the like], and also a person from a thing, good or evil. (L.) You say, حَدَّ الرَّجُلَ عَنِ الأَمْرِ *He prevented,* or *hindered,* and *withheld,* or *restrained, the man from the thing,* or *affair.* (L.) And حَدَدْتُ فُلَانًا عَنِ الشَّرِّ *I prevented,* or *hindered, such a one from* [*falling into*], or *preserved him from, evil.* (L.) And قَدْ حَدَّ ٱللّٰهُ ذٰلِكَ عَنَّا [*God hath forbidden us that*]. (Ṣ.) And اَللّٰهُمَّ ٱحْدُدْهُ (T, A, L) *O God, prevent him from hitting the mark:* said with reference to a man shooting, or casting a missile weapon, or the like. (T, L.) And حُدَّ *He* (a man) *was prevented,* or *withheld, from obtaining good fortune, success,* or *what he desired* or *sought.* (L.) And حَدَّ ٱللّٰهُ عَنَّا شَرَّ فُلَانٍ *May God repel,* or *avert, from us, the evil,* or *mischief, of such a one.* (L.) — [Hence,] حَدَّهُ, (Ṣ, L, Mṣb,) aor. ـُ, (L,) inf. n. حَدٌّ, (L, Mṣb, K,) *He inflicted upon him the castigation,* or *punishment, termed* حَدّ; (Ṣ, L;) *he inflicted upon him* (namely, a criminal or an offender [against the law],) *a castigation,* or *punishment, that should prevent him from returning to his crime* or *offence, and that should prevent others from committing such a crime or such an offence:* (K,* TA:) *he inflicted upon him a flogging.* (Mṣb.) — حَدَّ شَيْئًا مِنْ غَيْرِهِ, aor. ـُ, (L,) inf. n. حَدٌّ; (L, K;) and ▼حَدَّدَهُ; (L;) *He distinguished,* or *separated by some mark or note,* or *marks or notes, a thing from another thing.* (L, K.*) And حَدَّ الدَّارَ, aor. and inf. n. as above; (Ṣ, Mṣb;) and ▼حَدَّدَهَا, inf. n. تَحْدِيدٌ; (Ṣ;) *He distinguished the house from the parts adjoining it, by mentioning* [or *defining*] *its limits.* (Mṣb.) — [And hence, حَدَّ in logic, inf. n. حَدٌّ, †*He defined* a word; as also ▼حَدَّدَ, inf. n. تَحْدِيدٌ.] — حَدَّ, (L, Mṣb, K,) aor. ـِ, (L, Mṣb,) inf. n. حَدٌّ; (L;) and ▼حَدَّدَ, (Ṣ, L, Mṣb, K,) [which is more common,] inf. n. تَحْدِيدٌ; (Ṣ;) and ▼احدّ, (Ṣ, L, K,) which is the form preferred by Lḥ, (L,) inf. n. إِحْدَادٌ; (Ṣ;) and ▼استحدّ; (Aṣ, Ṣ, L;) *He edged,* or *sharpened,* a knife, (L, K,) a blade, (Ṣ,) a sword, (L, Mṣb,) or anything blunt, (L,) [and *pointed,* or *made sharp-pointed,* an arrow-head or the like,] *with a stone* or *file.* (L, K.) — [And hence,] حَدَّ بَصَرَهُ إِلَيْهِ, aor. ـِ; (Lḥ, L;) and ▼احدّهُ, (L,) or احدّ النَّظَرَ اليه; (Ṣ, Mṣb;) and ▼حَدَّدَهُ; (K in art. لتا, &c.;) ‡*He looked sharply at him,* or *it;* (L;) or *intently,* or *attentively.* (Mṣb.) — حَدَّتْ, (Ṣ, Mgh, L, K,) or حَدَّتْ عَلَى زَوْجِهَا, (Mṣb,) aor. ـُ and ـِ, inf. n. حِدَادٌ (Ṣ, Mgh, L, Mṣb, K) and حَدٌّ; (L, K;) and ▼احدّت, (Aṣ, Ṣ, A, Mgh, L, Mṣb, K,) inf. n. إِحْدَادٌ; (Mgh, Mṣb;) the former the more common in the language of the Arabs, but the latter preferred by the early grammarians, (Fr, TA,) and the only form known to Aṣ, (Ṣ,) who rejected the former; (Mṣb;) *She* (a woman) *abstained from the wearing of ornaments,* (A'Obeyd, Ṣ, A, Mgh, L, Mṣb, K,) *and the use of perfumes,* (L,) *and dye for the hands* &c., (Ṣ, Mgh,) *because forbidden such things,* or *because she forbade herself,* (Mgh,) *and put on the garments of mourning,* (A,) *after the death of her husband,* (Ṣ, Mgh,) or *on account of the death of her husband,* (A'Obeyd, A, Mṣb,) *for the period called* العِدَّة: (K:) or *she mourned for her husband, and put on the garments of mourning, and abstained from the wearing of ornaments, and the use of dye for the hands* &c. (L.) The epithets applied to a woman in this case are ▼حَادٌّ (Ṣ, L, Mṣb, K) and ▼مُحِدٌّ (Ṣ, A, Mgh, L, Mṣb, K) and ▼مُحِدَّةٌ also, but the first [always] without ة, (Mṣb,) or both more chaste without ة. (TA.) — حَدَّ, aor. ـِ, inf. n. حِدَّةٌ; (Ṣ, L, Mṣb, K;) and ▼احتدّ; (L, K;) [and app. ▼انحدّ, q. v.;] *It* (a sword, Ṣ, Mṣb, and a knife, L, K, [or the like,] and a canine tooth, L) *was,* or *became,* [*edged,* or] *sharp,* or *pointed.* (Ṣ, L, Mṣb, K.) — [And hence,] حَدَّ, aor. ـِ, inf. n. حِدَّةٌ, ‡*He was,* or *became, sharp* [or *effective*] *in respect of eloquence,* and *of intellect,* or *understanding,* and *of anger.* (L.) And حَدَّ عَلَيْهِ, aor. ـِ, (Ṣ, L, K,) inf. n. حِدَّةٌ and حَدٌّ, (Ṣ, L,) ‡*He became excited against him by sharpness,* or *hastiness, of temper; by irascibility, passionateness,* or *angriness;* (Ks, Ṣ, L, K;) as also ▼احتدّ عَلَيْهِ: (TA:) and حَدَّ عَلَيْهِ, aor. as above, inf. n. حَدَدٌ; (L, K;) and ▼حَادَّ, (accord. to some copies of the K,) and ▼احتدّ, (Ṣ, [in which it is not followed by عليه,] A, L, K,) and ▼استحدّ; (L, K;) ‡*he was angry with him;* (Ṣ,* A, L, K;) but Az remarks upon the last of these verbs as not heard from the Arabs of classical times in this sense: (L:) and ▼تحدّد بِهِمْ †*he became exasperated by them:* syn. تحرّش. (AZ, L.)

2. حدّد as a trans. v.: see 1, in five places.

—حدّد بَلَدًا *He repaired*, or *betook himself, to the limits*, or *boundaries, of a country*, or *town*. (L.) And حدّد إِلَيْهِ and لَهُ *He repaired*, or *betook himself, to him*, or *it*. (K.) ═ As an intrans. v., inf. n. تَحْدِيدٌ, *It* (seed-produce) *was late in coming forth because of the lateness of rain*, (K, TA,) *and then came forth* [pointed,] *without forking*, or *shooting forth into separate stalks* or *stems*. (TA.) — حدّد عَلَيْهِ : see 1.

3. أَرْضُنَا تُحَادُّ أَرْضَكُمْ *Our land borders upon*, or *is conterminous with, your land*; syn. تُتَاخِمُهَا. (K in art. تخم.) — [And hence,] حادّهُ, (L, K,) inf. n. مُحَادَّةٌ, (S,) ‡ *He acted towards him with reciprocal anger and enmity* (L, K) *and opposition* or *contrariety* or *repugnance*, (S, K,) *contending with him*, (TA,) *and refusing to do what was incumbent on him*: (S:) like شَاقَّهُ : as though meaning he became in the حَدّ, i. e. the side, region, quarter, or tract, in which was (or opposite to that in which was, Zj) his enemy; like as شاقّه means he became in the شِقّ, i. e. the side, or quarter, in which was [or opposite to that in which was] his enemy: (L:) and ↓ تحادّه, (TA,) inf. n. تَحَادٌّ, (S,) signifies the same. (S, TA.)

4: see 1, in three places.

5: see 1, last sentence.

6: see 3.

7. انحدّ *It was*, or *became, slender*. (TA in art. ابر.) — See 1, latter part.

8: see 1, latter part, in three places.

10. استحدّ as a trans. v.: see 1. ═ Also ‡ *He shaved* (S, Mgh, K) *his pubes* (S, Mgh) *with* [*a razor of*] *iron*: (Mgh, K:) derived from حَدِيدٌ. (Mgh.) — See also 1, last sentence.

حَدْ, for أَحَدْ, in the phrase يَا حَدْ رَآهَا : see أَحَدْ, in art. احد.

حَدٌّ *Prevention, hinderance, an impediment, a withholding, restraint, a debarring, inhibition, forbiddance, prohibition*, or *interdiction*; (S, Mgh, L, Msb, K, TA;) as also ↓ حَدَدٌ : (S, L, K:) and, both words, *a repelling*, or *an averting*. (K. [See 1.]) A poet says, (S,) namely, Zeyd Ibn-'Amr Ibn-Nufeyl, (TA,)

* لَا تَعْبُدُنَّ إِلٰهًا غَيْرَ خَالِقِكُمْ *
* وَإِنْ دُعِيتُمْ فَقُولُوا دُونَهُ حَدَدُ ↓ *

[*Ye shall by no means worship any deity except your Creator; and if ye be invited* to do so, *say ye, There is an impediment in the way of it*, or *a prohibition against it*]. (S, TA.) And one says, ↓ دُونَ مَا سَأَلْتَ عَنْهُ حَدَدٌ (A,* L) *There is an impediment*, or *a prohibition, in the way of that respecting which thou hast asked*. (L.) And لَا حَدَدَ ↓ عَنْهُ *There is nothing to prevent*, or *hinder, one from it*. (L. [But this admits of another meaning, as will be seen, under the word حَدَدٌ, below.]) — [Hence,] *A restrictive ordinance*, or *statute*, of God, *respecting things lawful and things unlawful*: pl. حُدُودٌ. (L.) The حُدود of God are of two kinds: first, those *ordinances prescribed to men* (T, Mgh, L) *respecting eatables and drinkables and marriages &c.; what are lawful thereof and what are unlawful*: (T, L:) the second kind, *castigations*, or *punishments, prescribed*, or *appointed, to be inflicted upon him who does that which he has been forbidden to do*; (T, Mgh, L;) as the حَدّ of the thief, which is the cutting off of his right hand for stealing a thing of the value of a quarter of a deenár or more; and that of the fornicator or fornicatress, which is flogging with a hundred stripes and banishment for a year; and that of the adulterer or adulteress, which is stoning; and that of the person who [falsely] charges an honest or a married woman with adultery, which is flogging with eighty stripes [as is also that of the person who has committed the crime of drunkenness]: (T, L:) the first kind are called حدود because they denote limits which God has forbidden to transgress: the second, because they prevent one's committing again those acts for which they are appointed as punishments; (T, Mgh, L;) or because the limits thereof are determined: (Mgh:) the latter kind of حدّ is also explained as being *that* [*castigation*, or *punishment*,] *which prevents the criminal from returning to his crime, and prevents others from committing his crime*. (L, K.*) لَوْ رَأَيْتَهُ عَلَى حَدّ, in a saying of 'Omar, means *Hadst thou seen him engaged in an affair requiring the infliction of the* حدّ. (Mgh.) — *A bar, an obstruction, a partition*, or *a separation*, (S, A, Mgh, L, Msb,* K,) *between two things*, (S, A, L, K,) or *between two places*, (Mgh,) [or *between two persons*,] *to prevent their commixture*, or *confusion*, or *the encroachment of one upon the other*: (L:) an inf. n. used as a subst.: (Mgh:) pl. حُدُودٌ. (L.) — *A limit*, or *boundary*, of a land or territory: pl. as above. (L.) [Hence, جَاوَزَ الحَدَّ † *He*, or *it, exceeded the proper, due*, or *common, limit; was excessive, immoderate, beyond measure, enormous, inordinate*, or *exorbitant*.] — [And hence, in logic, † *A definition*.] It is applied by the learned to the حَقِيقَة of a thing, [or *that by being which a thing is what it is*,] because it is [a term] collective and restrictive. (Mgh.) — The *end, extremity*, or *utmost point*, of a thing: (S, L, K:) pl. as above. (L.) — [† The *point*, or *verge*, of an event.] The saying مُسْلِمَةٌ مَوْقُوفَةٌ عَلَى حَدِّ مُحَرَّمٍ means † *A Muslimeh brought to the point*, or *verge, of being subjected to an infidel's lying with her*: and in like manner, مُسْلِمٌ مَوْقُوفٌ عَلَى حَدِّ كُفْرٍ † *A Muslim brought*, by beating or slaughter, *to* [*the point*, or *verge, of*] *denying God*. (Mgh.) — The *edge*, or *extremity of the edge*, (S, L,) and *point*, (L,) of anything, (S, L,) as of a sword, a knife, a spear-head, and an arrow: (L:) the *part* of a sword [&c.] *with which one cuts*: (MF:) pl. as above. (L.) — See also حِدَّةٌ, in four places. — [And hence, app.,] *Arms*, or *weapons*; as in the phrase ذَوُو حَدٍّ [*Possessors of arms* or *weapons*: or this may mean ‡ *persons endowed with valour*]. (Ham p. 143.) — *A side, region, quarter*, or *tract*. (L.) — † *Station, standing, rank, condition*, or *the like*; syn. مَرْتَبَةٌ. (KL.) — [† *A case*: as when a noun is said to be فِى حَدِّ الرَّفْعِ *in the nominative case*. — And † *A class*, or *category*: as when a verb is said to be مِنْ حَدِّ ضَرَبَ *of the class*, or *category, of* ضَرَبَ.] — [‡ *A quarter* of the year.] You say, أَقَامَ حَدَّ الرَّبِيعِ ‡ *He remained, stayed*, or *abode, during the quarter of the* ربيع. (A.) ═ See also مَحْدُودٌ.

حُدٌّ : see مَحْدُودٌ.

حُدَّةٌ *A small quantity* of water or milk &c. *remaining in a vessel or skin*; syn. صُبَّةٌ and صُبَابَةٌ. (K.)

حِدَّةٌ [*Sharpness* of a sword, a knife, or the like: see 1]. — [And hence,] ‡ *Sharpness*, or *hastiness, of temper; irascibility, passionateness*, or *angriness*; (Ks, S, A, L, K;) as also ↓ حَدٌّ : (Ks, S, L, K:) ‡ *sharpness* [or *effectiveness*] *in respect of eloquence*, and *of intellect* or *understanding*, and *of anger*: (L:) ‡ *sharpness, penetrating energy, vigorousness, effectiveness, and brishness, in the performance of affairs*; and also, *in matters of religion, with ambition to attain what is good*: from حَدٌّ as signifying the "edge" of a sword [&c.]: (L:) and ↓ the latter word, [or rather both,] ‡ a man's *sharpness, penetrating energy*, or *vigour, in the exercise of courage; his mettle*; (L;) his *valour*, or *valiantness, in war*. (S, A, L, K.) You say, ↓ إِنَّهُ لَبَيِّنُ الحَدِّ ‡ *Verily he is one who displays sharpness like that of a knife*. (L.) — حِدَّةٌ and ↓ حَدٌّ, as denoting a quality of anything, are syn. (K.) [Both signify † *Sharpness; vehemence; force*; and *strength*: and] both, † the *force*, or *strength*, of wine and the like; syn. سَوْرَةٌ; (Msb and K, in explanation of the former, [which is the more common,] in art. سور;) meaning شِدَّةٌ; (MF;) and صَلَابَةٌ. (S and L in explanation of the latter in the present art.) [Also, the former, † *Pungency; acridness*.]

حَدَدٌ : see حَدٌّ, first four sentences. — You say also, مَا لِى عَنْ هٰذَا الأَمْرِ حَدَدٌ, (S, A,*) and مَا لِى عَنْهُ مُحْتَدٌّ ↓, (K,) and ↓ مُحَدٌّ, (K, TA,) with damm, of the same measure as مُكْرَمٌ, (TA,) or ↓ مَحَدٌّ, (so in the CK,) *I have no way of avoiding*, or *escaping, this thing*. (S, A, K.) And مَا أَجِدُ مِنْهُ مُحْتَدًّا ↓ وَلَا مُلْتَدًّا *I find not any way of avoiding, nor any way of escaping, it*. (S.) ═ Also, (L,) and ↓ مَحْدُودٌ, (Msb,) *Prevented, hindered, impeded, withheld, restrained, debarred, inhibited, forbidden, prohibited*, or *interdicted*. (L, Msb.) You say, هٰذَا أَمْرٌ حَدَدٌ *This is a forbidden*, or *prohibited, thing; a thing unlawful to be done*, or *committed*. (S. [See also what follows.]) And حَدَدًا أَنْ يَكُونَ هٰكَذَا (S,* A, L) *Forbidden be it that it should be so*: like as you say, مَعَاذَ ٱللّٰهِ قَدْ حَدَّ ٱللّٰهُ ذٰلِكَ عَنَّا. (S, A,* L.) أَمْرٌ حَدَدٌ also signifies *A disallowed, and vain*, or *false, thing* or *affair*. (L.) And دَعْوَةٌ حَدَدٌ *A vain*, or *false, pretension*. (S, L, K.)

حَدَادِ, like قَطَامِ, [indecl., a proper name, for الحَادَّةُ, fem. act. part. n. of حَدَّ; like فَجَارِ for الفَاجِرَةُ; and hence, for يَا حَادَّةُ;] occurring in the phrase, حَدَادِ حُدِّيهِ [*O averter, avert him*, or *it*]: said [with respect] to him whose aspect, or countenance, thou dislikest. (A,* Ḳ.) — [It is also a proper name for الحَدُّ; like فَجَارِ for الفَجْرَةُ or الفُجُورُ; as in the following hemistich:]

حَدَادِ دُونَ شَرِّهَا حَدَادِ

[May there be *an impediment in the way of her evil*, or *mischief: an impediment*]. (L.) — حَدَادُكَ: see the next paragraph.

حُدَادُكَ أَنْ تَفْعَلَ كَذَا — حُدَادُ: see حَدِيدٌ, (Ḳ, TA,) with ḍamm, (TA,) or ▼حَدَادُكَ, (so in a MS. copy of the Ḳ and in the CḲ,) *The utmost of thy power*, or *of thine ability*, [*will be*] *thy doing such a thing*; and *the end of thy case*; syn. قُصَارَاكَ, (Ḳ,) [or قُصَارُكَ,] and مُنْتَهَى أَمْرِكَ. (TA.)

حِدَادٌ The *black garments of mourning* [*worn by a widow*]. (Ṣ, A, Mgh, L.)

حَدِيدٌ i. q. ▼مُحَادٌّ. (A.) You say, فُلَانٌ حَدِيدُ فُلَانٍ *Such a one is the close*, or *next*, *neighbour of such a one*; meaning that the house of the former is next by the side of that of the latter; (A,* L;) or that the land of the former is adjacent to that of the latter. (Ṣ, L.) And هُوَ حَدِيدِى فِى الدَّارِ, i. e. ▼مُحَادِّى [*He is my next neighbour in respect of house*]. (A.) And دَارِى حَدِيدَةُ دَارِهِ, and ▼مُحَادَّتُهَا (L, Ḳ,) or ▼مُحَادَّةٌ لِدَارِهِ, (A,) *My house is close*, or *next*, or *adjoining*, *to his house*; meaning that the limit of the former is like that of the latter. (L, Ḳ.*) — Also, (Ṣ, L, Mṣb, Ḳ,) used as masc. and fem. without ة, and also as fem. with ة, (L,) and ▼حَادٌّ, (Ṣ, L, Mṣb,) but this is disapproved by IKh, (TA,) though allowed by some as agreeable with analogy, (MF,) and ▼حُدَادٌ, (Aṣ, L, Ḳ,) and ▼حُدَّادٌ, (AA, Ṣ, L, Ḳ,) [*Edged*, or *sharpened*; or] *sharp*; applied to a sword, (Ṣ, Mṣb,) a knife, (L, Mṣb, Ḳ,) [and the like: and *pointed*, or *sharp-pointed*:] pl. [of the first] حِدَادٌ, (Ṣ, L, Ḳ,) masc. and fem.; (L;) and حَدِيدَاتٌ and حَدَائِدُ, (L, Ḳ,) fem. (L.) And نَابٌ حَدِيدٌ and حَدِيدَةٌ *A sharp canine tooth*: (L, Ḳ:) حُدَادٌ thus applied has not been heard. (L.) — [Hence,] رَجُلٌ حَدِيدٌ ‡ *A man who is sharp* [or *effective*] *in respect of eloquence*, and *of intellect* or *understanding*, and (as also ▼مُحْتَدٌّ, Ṣ) *of anger*: pl. أَحِدَّآءُ and أَحِدَّةٌ and حِدَادٌ. (L, Ḳ.) And أَلْسِنَةٌ حِدَادٌ † *Sharp tongues*. (Ṣ.) And رَجُلٌ حَدِيدُ النَّاظِرِ ‡ [*A man who looks sharply*, or *boldly*;] *a man not suspected of evil*, *so that he should cast down his eyes*. (L.) فَبَصَرُكَ اليَوْمَ حَدِيدٌ [in the Ḳur l. 21] means † *And thy sight*, or *intellect*, *to-day*, *is*] *sharp*, or *piercing*; so that thou perceivest therewith what thou didst not know, or what thou deemedst improbable, in thy life on earth: (Jel:) or *thy judgment*, *to-day*, *is penetrating*. (L.) [Hence also,] رَائِحَةٌ حَدِيدَةٌ (L) and ▼حَادَّةٌ (L, K) ‡ *A sharp*, or *pungent*, *odour*. (L, Ḳ.) And نَاقَةٌ حَدِيدَةُ الجِرَّةِ ‡ *A she-camel whose cud has a pungent odour*; (Ḳ, TA;) which is a quality approved. (TA.) — حَدِيدٌ also signifies [*Iron*;] *a certain substance*, (L,) *well known*; (Ṣ, L, Ḳ;) so called because of its resistance: (Ṣ, L:) ▼حَدِيدَةٌ is a more particular term, (Ṣ,) signifying *a piece thereof*; (L;) [and *an instrument*, or *implement*, *thereof*:] pl. حَدَائِدُ (Ṣ, L, Ḳ) and حَدَائِدَاتٌ; (Ṣ, L;) the latter (which is erroneously written in the Ḳ حَدِيدَاتٌ, TA) is a pl. pl., (L,) sometimes occurring in poetry. (Ṣ.) It is said in a prov.,

إِنَّ الحَدِيدَ بِالحَدِيدِ يُفْلَحُ

Verily iron with iron is cloven, or *cut*. (Ṣ and Ḳ in art. فلح.) And in another, تَضْرِبُ فِى حَدِيدٍ بَارِدٍ [*Thou beatest upon cold iron*]: applied in relation to him who hopes for that of which the attainment is remote, or improbable; and to him in whom is nothing to be hoped for. (Ḥar p. 633.) — Also † *Like iron in hardness*: applied in this sense to solid hoofs. (Mgh.)

حَدَادَةٌ One's *wife*. (Sh, Ḳ.)

حِدَادَةٌ The *office of a door-keeper*. (Mṣb.) — The *art of a blacksmith*, or *worker in iron*. (Mgh.) [The *art of a maker of coats of mail*.]

حَدِيدَةٌ: see حَدِيدٌ.

حُدَّى: see حِدَأَةٌ, in art. حدأ.

حَدَّادٌ *A door-keeper*: (Ṣ, A, Mgh, L, Mṣb, Ḳ:) so called because he prevents men from entering. (Mgh, L.) — *A keeper of a prison*: (Ṣ, Mgh, Ḳ:) because he prevents persons from going out, or because he works the iron of the shackles. (Ṣ. [See what follows.]) — The *person who inflicts the punishment termed* حَدّ: so in the saying, أُجْرَةُ الحَدَّادِ عَلَى السَّارِقِ [*The pay of the inflicter of the* حدّ *is to be imposed upon the thief*]; or, as some say, the meaning here is, *the keeper of the prison*, because, in general, he has the charge of the amputation; but the former meaning is the more probable, and more obvious. (Mgh.) — *A seller of wine*; *a vintner*: because he withholds his wine until he obtains for it a price that contents him: so in the following verse of El-Aạshà:

فَقُمْنَا وَلَمَّا يَصِحْ دِيكُنَا
إِلَى جَوْنَةٍ عِنْدَ حَدَّادِهَا

[*And we arose, when our cock had not yet crowed, to a wine-jar smeared with pitch, in the possession of its seller*]. (Ṣ, L.) — *A blacksmith*; *a worker in iron*. (Mgh, L, Ḳ.) *A maker of coats of mail*. (TA.)

حُدَّادٌ: see حَدِيدٌ.

حَدْحَدٌ *Short* (L, Ḳ) *and thick*: an epithet applied to a man. (L.)

حَادٌّ; fem. with ة: see حَدِيدٌ, in two places. — See also 1, voce حَدَّتْ.

أَحَدُّ [*More*, and *most*, *sharp*: &c.] — You say, هُوَ مِنْ أَحَدِّ الرِّجَالِ ‡ *He is of the most sharp*, or *hasty*, *in temper*, or *of the most irascible*, *passionate*, or *angry*, *of men*. (A, TA.)

مُحَدٌّ, or مُحِدٌّ: see حَدَّدَ.

مُحِدٌّ and مُحِدَّةٌ: see 1, voce حَدَّتْ.

مَحْدُودٌ: see حَدَّدَ. — Also A man (L) *denied*, or *refused*, *good*, or *prosperity*; *prevented*, or *withheld*, *from obtaining good*; (T, L, Ḳ;) and so ▼حُدٌّ, with ḍamm, (Ḳ,) or ▼حَدٌّ; (as in the L;) the latter heard only from Lth: (T, TA:) *withheld from good fortune &c.*; (Ṣ, L;) *withheld from sustenance*; *contr. of* مَجْدُودٌ: (Mgh:) and *withheld from evil*. (L, Ḳ.)

مُحَادٌّ and مُحَادَّةٌ: see حَدِيدٌ, in four places.

مُحْتَدٌّ: see حَدِيدٌ: — and see also حَدَّدَ, in two places.

حدأ

حَدَأٌ: see حَدَأَةٌ: — and see also حِدَأَةٌ, in two places.

حِدَأٌ: see حِدَأَةٌ, in three places: — and see also what next follows.

حَدَأَةٌ (Aṣ, Ṣ, Ḳ) and ▼حِدَأَةٌ, but the former is the more chaste, (TA,) *A double-headed* فَأْس [i. e. *hoe*, or *adz*, or *axe*]: (Aṣ, Ṣ, Ḳ:) [a kind of فأس used in the present day is a hoe with two heads, one at each end of the handle:] or the *head of a* فأس: and the *head of an arrow*: (Ḳ:) pl. of the former ▼حَدَأٌ (Aṣ, Ṣ, Ḳ) [or rather this is a coll. gen. n.] and حِدَآءٌ, (Ḳ, TA, [in the CḲ حِدَاءٌ,]) mentioned by AO and Aṣ and A'Obeyd; (TA;) and the pl. of حِدَأَةٌ is ▼حِدَأٌ (TA) [or rather this, like حَدَأٌ, is a coll. gen. n.]. — See also the next paragraph, in two places.

حِدَأَةٌ, (Ṣ, Mṣb, Ḳ,) or ▼حِدَأٌ, [but see what follows,] sometimes pronounced ▼حَدَأٌ, (Mgh,) [The *kite*; vulgarly called حِدَايَة;] *a certain bird*, (Ṣ, Mgh, Ḳ,) *well known*; (Ṣ, Ḳ;) *a certain noxious bird*; (Mṣb;) *surnamed* أَبُو الخَطَّافِ *and* أَبُو الصَّلْتِ; (TA;) *that preys upon large field-rats* (جِرْذَان): (Mgh, TA:) J and Ṣgh say that the word should not be pronounced ▼حَدَأَةٌ; but AḤei mentions this pronunciation on the authority of [some of] the Arabs; and accord. to IAạr and IAmb, the فأس [see above] and this bird were sometimes called alike ▼حَدَأَةٌ and ▼حَدَأٌ: the more approved pronunciation of the name of the bird, however, is with kesr [i. e. حِدَأَةٌ]: the pl. is ▼حِدَأٌ (Ṣ, Mṣb, Ḳ) and حِدَآءٌ, (Ḳ,) both extr., (TA,) [or rather the former is a coll. gen. n.,] and حِدْآنٌ: (Mṣb, Ḳ:) and the following are variations of the name of this bird: حُدَّى, and حُدَيَّا, (TA,) the latter said by AḤát to be an erroneous form of the word, used by the people of El-Ḥijáz, (Mgh, TA,) and ▼حُدَيْئَةٌ, app. a dim., for ▼حُدَيْأَةٌ, also pronounced حُدَيَّةٌ, (TṢ, TA,) and حُدَوٌ, occurring in a trad. in conjunction with أَفْعَو [for أَفْعَى], (Mgh, TA,) of the dial. of the people of Mekkeh. (TA in art. حدو.)

Hence the saying, حِدَأً حِدَأً وَرَآءَكِ بُنْدُقَة, (Ṣ, Ḳ, TA,) for which the vulgar say, حِدَا حِدَا, (Ṣ,) [accord. to some, meaning *O kite, O kite, a bullet is behind thee:* accord. to others, *O Ḥidà, O Ḥidà, Bunduḳah is behind thee:*] Esh-Sharḳee (Ibn-El-Ḳuṭámee, TA) says, (Ṣ,) حِدَأً and بُنْدُقَة were two tribes, descendants of حِدَأُ بْنُ نَمِرَةَ and سَعْدُ العَشِيرَةِ, بُنْدُقَةُ بْنُ مَظَّةَ (Ṣ, Ḳ,*) and both of; (Ṣ, TA;) the former in El-Koofeh, and the latter in El-Yemen: the former attacked the latter, and obtained spoil from them; and then the latter attacked the former, and destroyed them: (TA:) and hence this saying: (Ṣ, Ḳ, TA:) or حِدَأً is here an apocopated form of حِدَأَةً: (Ṣ, Ḳ:) so says ISk: (Ṣ:) and AO says that by it is here meant the bird [i. e. the kite]; and by بندقة, the thing with which one shoots [from a cross-bow, namely, a bullet]; and the prov. is used to caution a person: accord. to Ibn-El-Kelbee, it is applied to him who esteems himself cunning in an affair, and is outwitted therein by another: accord. to the A, to him who is threatened with an evil near at hand. (TA.) ═ حِدَأَةٌ also signifies The سَالِفَة (meaning the *fore part*, TA, [or the *fore part from beneath the ear to the middle of the collar-bone*,]) *of the neck of a horse:* (Aṣ, Ḳ:) pl. حِدَآءٌ. (Aṣ, TA.) ═ See also حِدَأَةٌ.

حُدَيْئَةٌ and حُدَيْئَةٌ: see the next preceding paragraph.

حدب

1. حَدِبَ, aor. ـَ, inf. n. حَدَبٌ; (Ṣ,* A, Mgh,* Mṣb, Ḳ;) and ▾احدب, and ▾تحادب, (Ḳ,) and ▾احدودب; (Ṣ, Ḳ;) *He* (a man, Mṣb) *was,* or *became, humpbacked;* (Mgh, Mṣb;) *he had a prominent,* or *protuberant, back, and a hollow,* or *receding, chest* (A,* Ḳ) *and belly:* (Ḳ:) [accord. to the Mṣb, from حَدَبٌ signifying "elevated ground;" but the reverse is indicated in the A:] and *it* (the back) *was,* or *became, humped,* or *protuberant;* (Ṣ, A;*) as also ▾انحدب. (KL.) — And the first, †*It* (a thing) *rose,* or *grew up* or *out, high:* (KL:) [*it was,* or *became, gibbous,* or *convex;* as also ▾احدودب.] — حَدِبَ عَلَيْهِ, (Ṣ, A, Ḳ,) inf. n. as above; (KL, TA;) and ▾تحدّب; (Ṣ, A, Ḳ;) ‡*He was,* or *became, affectionate, favourable,* or *kind, to him.* (Ṣ, A,* Ḳ, KL, TA.) And حَدِبَتْ عَلَى وَلَدِهَا, (Ḳ,* TA,) inf. n. as above; (TA;) and ▾تحدّبت; (Ḳ;) ‡*She* (a woman) *applied herself constantly to the care of her child,* or *children, after the loss of her husband, not marrying again.* (Ḳ, TA.) ═ حَدَبَ عَنْهُ, aor. ـُ, inf. n. حَدْبٌ, *He repelled from him, and defended him.* (MF, TA.)

2. تَحْدِيبٌ [inf. n. of حدّب] The act of *elevating,* or *raising high, the back.* (KL.) — [And, accord. to Golius, as on the authority of the KL, The *making* a thing *gibbous,* or *convex:* but this meaning which the word has in the present day, I do not find in my copy of the KL.]

4. احدبهُ *He* (God) *rendered him humpbacked.* (Ṣ.) — †*He,* or *it, rendered him affectionate, favourable,* or *kind.* (KL.)

5: see 1, in two places. — تحدّب بِهِ *He,* or *it, clung,* or *clave, to it.* (Ḳ, TA.)

6: see 1.

7: see 1.

9: see 1.

12: see 1, in two places. — Also †*It* (sand) *was,* or *became, curved,* or *winding;* or *curved,* or *winding, and long.* (Ḳ.)

حَدَبٌ ‡*High,* or *elevated, ground;* so in the Ḳur xxi. 96; (Ṣ, A, Mṣb;) as also ▾حَدَبَةٌ; and so حَدَبٌ مِنَ الأَرْضِ: (A:) or *rugged and high ground:* (T, Ḳ:) pl. حِدَابٌ (Ṣ) [and app., accord. to the TA, أَحْدَابٌ also, a pl. of pauc.]. And حَدَبُ الرَّمْلِ ‡ *Sand brought by the wind,* [or *blown together,*] *and elevated.* (A, TA.) And hence, as being likened to such sand, (IAạr, TA,) حَدَبُ البُهْمَى ‡ *What is scattered, and heaped up, of* [*the species of barley-grass called*] بهمى. (IAạr, Ḳ, TA.) And حَدَبُ المَآءِ † *The elevated waves of water:* (T, TA:) or *the rolling over of water, volume over volume:* (Ḳ, TA:) or *the rolling of water in waves.* (TA.) And حَدَبُ الغَدِيرِ † *The motion and waves of the pool of water left by a torrent.* (IAạr, TA.) And حَدَبُ السَّيْلِ ‡ *The rise,* or *swell, and abundance, of the torrent.* (A, TA.) — †*A slope in a declivity;* expl. by حُدُورٌ فِي صَبَبٍ, as in the correct copies of the Ḳ, and in the L; in some copies of the Ḳ حدوب; (TA;) [in the CḲ حُدُورٌ;] as the حَدَب of waves (in some copies of the Ḳ, of the wind, TA, [an evident mistranscription, الريح for الموج,]) and of sand. (Ḳ.) — †*A mark left upon the skin;* (Aṣ, Ḳ;) such as the [*weal* or] *swelling and thickness produced by beating.* (Aṣ, TA.) — ‡The *intenseness of the cold* of winter. (A, Ḳ.) ═ *A certain plant:* or the [*plant called*] نَصِىّ. (Ḳ.)

حَدِبٌ: see أَحْدَبُ. — Also ‡ *Affectionate, favourable,* or *kind.* (A, TA.) You say, هُوَ حَدِبٌ عَلَى أَخِيهِ ‡ *He is affectionate, &c., to his brother.* (A.) ═ أَرْضٌ حَدِبَةٌ *A land abounding with the plant called* حَدَب. (Ḳ.)

حَدَبَةٌ *A hump on the back.* (Az, Ṣ, A, Mgh.) — See also حَدَبٌ.

حَدَابِ, like قَطَامِ, (Ḳ,) indecl., (TA,) †*A year of drought, barrenness,* or *dearth:* (Ḳ:) or *a year of severe drought.* (TA.)

حُدَيْبَآءُ: see what next follows.

أَحْدَبُ *Humpbacked;* (Ṣ, Mgh, Mṣb;) *having a prominent,* or *protuberant, back, and a hollow,* or *receding, chest and belly;* (Ḳ;) and ▾حَدِبٌ signifies the same: (Sb, Ṣ, Ḳ:) fem. of the former حَدْبَآءُ: (Mṣb:) and pl. حُدْبٌ. (Mṣb, TA.) ابْنَةٌ ▾حُدَيْبَآءُ (dim. of حَدْبَآءُ), meaning *A little humpbacked daughter,* occurs in a trad. (TA.) — Hence, آلَةٌ حَدْبَآءُ, (see a verse of Kaab Ibn-Zuheyr, voce آلَةٌ, in art. اول,) †*A gibbous bier:* (A,* TA:) or (as used in that verse) it means *a distressing state,* or *condition:* or *an elevated apparatus.* (TA.) And رَمْلَةٌ حَدْبَآءُ † [*A gibbous tract of sand*]. (ISh, Ḳ in art. دبح, &c.) And نَاقَةٌ حَدْبَآءُ, (Ṣ, A,) or دَابَّةٌ حدبآءُ, (Ḳ,) ‡*A she-camel,* (Ṣ, A,) or *a beast,* (Ḳ,) *the prominent parts of whose hips,* (Ṣ, A, Ḳ,) *and the bone of whose back,* (TA,) *appear,* (Ṣ, A, Ḳ,) *by reason of her leanness.* (A, TA.) And حَدْبَآءُ حِدْبِيرٌ and حِدْبَارٌ are expressions used in the same sense: (L, TA:) pl. حُدْبٌ حَدَابِيرُ. (Ṣ, L, TA.) — الأَحْدَبُ is the name of *A vein* (عِرْق) *penetrating into,* or *lying within, the bone* (عَظْم [app. a mistranscription for عَضَلَة the *upper portion*]) *of the fore-arm.* (Ḳ.) — أَمْرٌ أَحْدَبُ (A) and خُطَّةٌ حَدْبَآءُ (A, TA) ‡*A difficult affair:* (A, TA:) and أُمُورٌ حُدْبٌ (A, TA) and حُدْبُ الأُمُورِ (Ḳ) ‡*difficult affairs;* (A, Ḳ, TA;) sing. حَدْبَآءُ [for خُطَّةٌ حَدْبَآءُ or the like]. (Ḳ.) And سَنَةٌ حَدْبَآءُ ‡ *A severe, cold year.* (A, TA.) [Hence,] وَسِيقٌ أَحْدَبُ †*A quick driving.* (TA.) — [Hence, also,] الأَحْدَبُ [used as a subst.] †*Vehemence, severity, difficulty,* or *distress;* syn. الشِّدَّةُ. (Ḳ.) ═ [Also †*More,* and *most, affectionate, favourable,* or *kind.*] أَحْدَبُهُمْ عَلَى المُسْلِمِينَ, said of Aboo-Bekr, in a trad. of 'Alee, means †*The most affectionate, favourable,* or *kind, of them, to the Muslims.* (TA.)

حدث

1. حَدَثَ, (Ṣ, A, Mgh, Mṣb, Ḳ,) aor. ـُ, (Mgh, Mṣb,) inf. n. حُدُوثٌ (Ṣ, Mgh, Mṣb, Ḳ) and حَدَاثَةٌ, (A, Ḳ,) *It was new,* or *recent;* contr. of قَدُمَ: (Ṣ,* A, Ḳ:) *it* (a thing) *came into existence; began to be; had a beginning; began,* or *originated; existed newly, for the first time, not having been before:* (Ṣ, Mgh, Mṣb, TA:) but when mentioned with قَدُمَ, it is written حَدُثَ, with ḍamm to the د, (Ṣ, Mgh, Ḳ,) as in the saying, أَخَذَنِى مَا قَدُمَ وَمَا حَدُثَ, (Ṣ,) or أَخَذَهُ الخ, (A, Mgh,) meaning *Old and new anxieties and thoughts* [*came into my mind,* or *his mind,* or *overcame me,* or *him*]; (TA;) or *old and new griefs* or *sorrows;* (Mgh;) the former saying occurring in a trad.: (TA:) the verb is not thus in any other case [in this sense]. (Ṣ.) You say, حَدَثَ بِهِ عَيْبٌ *A vice,* or *fault,* or *the like, originated in him,* or *it, not having been before.* (Mṣb.) And حَدَثَ أَمْرٌ *An affair,* or *event, originated:* (Mgh:) or *happened,* or *came to pass.* (Ṣ.) حُدُوثٌ is of two kinds: حُدُوثٌ زَمَانِىٌّ, which is A thing's *being preceded by non-existence:* and حُدُوثٌ ذَاتِىٌّ, which is a thing's *being dependent upon another for its existence.* (KT.) — حَدَاثَةٌ and حُدُوثَةٌ, [as inf. ns. of which the verb, if they have one, is, accord. to analogy, حَدُثَ,] relating to a man, signify The *being young;* or [as simple substs.] *youthfulness.* (ISd, Ḳ.)

2. حدّثهُ [*He told him,* or *related to him, something; he discoursed to him,* or *talked to him:* see also 5]. You say, حدّثهُ الحَدِيثَ, (L,) and حدّثهُ بِهِ, (A,* L,) inf. n. تَحْدِيثٌ, a word of well-known meaning, (Ṣ,) *He told him,* or *related to him, the story,* or *narrative,* or *tradition.* (L.)

[And حدّث *He related traditions of Mohammad:* and حدّث عن فُلَانٍ *he related such traditions heard, or learned, from such a one:* the verb in this sense being an Islámee term.] — [Hence,] تَرَكْتُ البِلَادَ تُحَدِّثُ † *I left the countries, or towns, resounding with a buzzing, or confused noise.* (Th, ISd.)

3. حادث سَيْفَهُ, (TA,) inf. n. مُحَادَثَةٌ, (Ṣ, Ḳ,) *He polished his sword;* (Ṣ,* Ḳ,* TA;) [*as though he made it new by doing so;*] as also ↓ احدثهُ, (TA,) inf. n. إِحْدَاثٌ. (Ḳ.) — Hence, حَادِثُوا هٰذِهِ القُلُوبَ بِذِكْرِ ٱللّٰهِ فَإِنَّهَا سَرِيعَةُ الدُّثُورِ † *Polish and cleanse ye these hearts by the remembrance of God, like as the sword is polished:* [*for they quickly become sullied:*] a trad. of El-Ḥasan. (TA.) = مُحَادَثَةٌ and ↓ تَحَادُثٌ, words of well-known meaning, (Ṣ,) are syn.: (Ḳ:) [but the former generally relates to two persons: the latter, to more than two:] you say, حادث صَاحِبَهُ [*He talked, or conversed in words, with his companion*]: (A:) and حادثوا and ↓ تحادثوا [*They talked, or conversed in words, together, or one with another*]. (TḲ.)

4. احدثهُ (Ṣ, A, Mṣb, TA) and ↓ استحدثهُ (A) *He* (God, Ṣ, or a man, Mṣb) *brought it into existence, caused it to be, made it, produced it, effected it, or did it, newly, for the first time, it not having been before; began it, or originated it; invented it; innovated it.* (Ṣ, Mṣb, TA.) [Hence,] احدث أَمْرًا [*He brought to pass an event*]. (Ḳur lxv. 1.) And احدث حَدَثًا *He originated an innovation* [see حَدَثٌ]. (TA.) — See also 3. — Also احدث, (Ṣ, L, Mṣb, Ḳ,) inf. n. إِحْدَاثٌ, (Mṣb,) from الحَدَثُ, (Ṣ,) † *He voided his ordure; or broke wind:* (L, Ḳ:) it has both these meanings: (L:) or *he did a thing that annulled his state of legal purity.* (Mṣb.) [See حَدَثٌ.] — And ‡ *He committed adultery, or fornication:* (Ḳ, TA:) and in like manner one says of a woman [احدثت]. (TA.)

5. تحدّث [*He talked; conversed in words; told, or related, stories, or narratives*]. (Ṣ.) And تحدّث بِهِ [*He talked of it; told it; related it*]; (Ṣ, A, Mṣb, Ḳ;) namely, a حَدِيث, (Mṣb,) or what is termed أُحْدُوثَة. (Ṣ, Ḳ.) And يَتَحَدَّثُ إِلَى النِّسَآءِ [*He talks to women*]. (Ṣ, A.*) [See also 2.] — It is said in a trad., يَبْعَثُ ٱللّٰهُ السَّحَابَ فَيَضْحَكُ أَحْسَنَ الضَّحِكِ وَيَتَحَدَّثُ أَحْسَنَ الحَدِيثِ ‡ [*God shall send the clouds, and they shall laugh with the best laughing, and talk with the best talking*]: the talking here mentioned, says IAth, is said to mean thundering; and the laughing, lightning; thundering being likened to talking because it announces rain, and its near coming: or by laughing may be meant the smiling of the earth, and the appearing of the flowers or blossoms; and by talking, the talking of men in describing and mentioning the plants or herbage: this figure of speech is termed مَجَازٌ تَعْلِيقِىٌّ, and is one of the most approved kinds of مجاز. (TA.)

6: see 3, in two places.

10: see 4. — You say also, استحدث خَبَرًا *He found new tidings or information:* (Ṣ:) or *he gained, or acquired, tidings or information.* (A.)

حِدْثٌ and ↓ حُدْثٌ and ↓ حِدِّيثٌ and رَجُلٌ حِدْثٌ (Ḳ) and ↓ مُحْدَثٌ (L) *A man of many stories or narratives,* (L, Ḳ,) *and who relates them well:* (L:) or رَجُلٌ حُدْثٌ ↓ and ↓ حِدِّيثٌ signify *a man who relates stories, or narratives, well:* and رَجُلٌ حِدِّيثٌ ↓ signifies *a man of many stories or narratives;* (Ṣ, A, El-Wáʼee;) but is used by the vulgar to signify *a man who relates stories, or narratives, well.* (El-Wáʼee, TA.) And you say رَجُلٌ حِدْثُ مُلُوكٍ *A man who is a companion of kings in talk* (Ṣ, A, Ḳ) *and in their nocturnal conversations:* (Ṣ:) and حِدْثُ نِسَآءٍ *one who talks to women;* (Ṣ, A;) or *who talks with women.* (Az, TA in art. تبع.) And ↓ هُوَ حِدِّيثُهُ [*He is his story-teller*]. (A.)

حَدَثٌ *A novelty, or new thing; an innovation; a thing not known before:* and particularly *relating to El-Islám* [i. e. *to matters of religious doctrine or practice or the like*]: (Mgh:) [and so ↓ أَمْرٌ مُحْدَثٌ; for] مُحْدَثَاتُ الأُمُورِ (pl. of ↓ مُحْدَثٌ, TA) signifies *innovations of people of erroneous opinions,* (Mṣb, TA,) *inconsistent with the doctrines, or practices, of the just of preceding times:* or *what is not known in revealed scripture, nor in the Sunneh, nor in the general conventional tenets of the doctors of the law:* and حَدَثٌ, [in like manner,] *an innovation that is disapproved, not agreeable with custom, or usage, and not known in the Sunneh.* (TA.) ↓ آوَى مُحْدَثًا, occurring in a trad., means *He entertained an innovation;* [i. e. *he embraced, or held, it;*] or *he was content, or pleased, with it;* or *he bore it patiently:* or, as some say, it is آوَى مُحْدِثًا ↓, meaning *he entertained, or harboured in his dwelling, a criminal, or an offender, and protected him from retaliation.* (TA.) — Also *i. q.* ↓ حَادِثَةٌ and ↓ حَدَثَانٌ [in some copies of the Ṣ ↓ حِدْثَانٌ] and ↓ حُدْثَى [signifying *An accident, an event, a hap, or a casualty:* and generally *an evil accident or event, a mishap, a misfortune, a disaster, a calamity, or an affliction*]: (Ṣ:) [the most common of these words is ↓ حَادِثَةٌ; and its pl., حَوَادِثُ, is more common than the sing.:] the pl. of حَدَثٌ is أَحْدَاثٌ. (TA.) أَحْدَاثُ الدَّهْرِ and ↓ حَوَادِثُهُ (A, Ḳ) and ↓ حِدْثَانُهُ, (Ḳ,) or, as is said by Fr and others, this last is ↓ حَدَثَانُهُ, (TA,) signify *The accidents, or casualties, of time or fortune;* or *the evil accidents, or calamities, of time or fortune.* (A, Ḳ.) ↓ حَوَادِثُ occurs used as a sing., said to be put by poetic license for ↓ حَدَثَانٌ: and this latter is also used [as a pl.] for حَوَادِثُ: so say Az and AAF: and it is said to be a noun in the sense of نَوَائِبُ الدَّهْرِ and حَوَادِثُ الدَّهْرِ: accord. to Fr, the Arabs say, [using it as a pl.,] أَهْلَكَتْنَا الحَدَثَانُ [*The accidents, or evil accidents, of time, or fortune, destroyed us*]: some say الحَدَثَانِ, making it dual of حَدَثٌ, and meaning thereby *the night and day;* like as they say [in the same sense] المَلَوَانِ and الجَدِيدَانِ &c. (TA.)

— [Hence] حَدَثٌ is a term applied by Sb to The مَصْدَر [or *infinitive noun*]; because all مصادر are [significant of] accidents [considered as subsisting in, or proceding from, agents]: and the pl. which he assigns to it in this sense is أَحْدَاثٌ. (TA.) — † The *voiding of ordure;* or the *breaking of wind;* syn. إِبْدَآءٌ: (Ḳ:) or *legal impurity that forbids, or prevents, one's performing prayer &c.:* (KT:) or *a state annulling legal purity:* pl. أَحْدَاثٌ. (Mṣb.) [See 4.] — *I. q.* وَلِىٌّ † [The *rain following that called the* وَسْمِىّ]: (L:) or الأَحْدَاثُ [pl. of الحَدَثُ] signifies *the rains of the commencement,* or *first part, of the year.* (Ḳ.) — *Young,* applied to a man, (A,* L, Mṣb,*) and to a horse or an ass or the like, and a camel, and, accord. to IAạr, to a mountain-goat: (L:) pl. أَحْدَاثٌ (A, L, Mṣb,) and حُدْثَانٌ. (L.) You say رَجُلٌ حَدَثٌ, (Th, Ṣ, L, &c.,) and ↓ حَدِيثٌ, حَدَثُ السِّنِّ, (Th, Ṣ, A, Mṣb, Ḳ,) and السِّنِّ, (IDrd, Ḳ, [but this is by some disallowed, as will be seen below,]) *A young man:* (Ṣ, L, Mṣb, Ḳ:) and in the pl. sense you say غِلْمَانٌ أَحْدَاثٌ and رِجَالٌ أَحْدَاثٌ [pls. of حَدَثٌ], (Ṣ,) and حُدْثَانٌ and حُدْثَانُ السِّنِّ, [or these, as is implied above, are not allowable,] and حُدَثَآءُ السِّنِّ [pl. of ↓ حَدِيثٌ]. (ISd, TA.) J says, [in the Ṣ,] if you mention the سِنّ, you say حَدِيثُ السِّنِّ ↓ [lit. *Young of tooth*]: and IDrst says, the vulgar say, هُوَ حَدَثُ السِّنِّ, like as you say حديث السِّنِّ; but it is a mistake; for حَدَثٌ is an epithet applied to the man himself, and is originally an inf. n.; one should not apply it as an epithet to the سِنّ nor to the ضِرْس nor to the ناب; but ↓ حَدِيثٌ is an epithet applied to anything *recent.* (TA.)

حِدْثٌ : } see حَدَثٌ, first sentence; each in
حُدْثٌ : } two places.

حُدْثَى : see حَدَثٌ.

حِدْثَى : see what next follows.

حِدْثَانٌ The *first,* or *beginning,* or *commencement,* of a state, or a case, or an affair; (Ṣ, A, Mgh, Ḳ;) as also ↓ حَدَاثَةٌ: (Ṣ, Mgh, Ḳ:) and its *freshness;* which is also a signification of both these words. (Ṣ, Mgh.) So in the saying, اِفْعَلْ ذٰلِكَ الأَمْرَ بِحِدْثَانِهِ and ↓ بِحَدَاثَتِهِ [*Do thou that thing while it is in its first and fresh state*]. (Ṣ, Mgh.*) One says also, أَتَيْتُهُ فِى حِدْثَانِ شَبَابِهِ and ↓ حِدْثَى شبابه and ↓ حَدِيثِ شبابه † *I came to him in the beginning, or first period, of his youth.* (Aboo-'Amr Esh-Sheybánee, TA.) And it is said in a trad., addressed to 'Áisheh, لَوْلَا حِدْثَانُ قَوْمِكِ بِالكُفْرِ لَهَدَمْتُ الكَعْبَةَ وَبَنَيْتُهَا, (Mgh,* TA,) or, as some relate it, ↓ حَدَاثَةُ قومك, which means the same, (Mgh,) i. e. *Were it not for the shortness of the period that has elapsed since thy people were in the state of infidelity, I would pull down the Kaabeh, and build it* [anew]. (TA.) — See also حَدَثٌ, in two places.

حَدَثَانٌ, used as a sing. and as a pl.: see حَدَثٌ, in three places.

حَدِيثٌ *New, recent;* (Ḳ;) contr. of قَدِيمٌ: (Ṣ:) *having,* or *having had, a beginning; existing newly, for the first time, not having been before;* as also ↓ حَادِثٌ: (Mṣb:) *brought into existence, caused to be, made, produced,* or *done, newly, for the first time, not having been before; begun,* or *originated; invented; innovated;* as also ↓ مُحْدَثٌ. (TA.) —— See حَدَثٌ, last two sentences, in four places. And see حِدْثَانٌ. You say also, هُوَ حَدِيثُ عَهْدٍ بِالإِسْلَامِ *He is,* or *was, recently become a Muslim.* (Mṣb.) And حَدِيثُو عَهْدٍ بِكُفْرِهِمْ, (TA,) or بِالجَاهِلِيَّةِ, or حَدِيثٌ عَهْدُهُمْ, (Mgh,) Men *lately in their state of infidelity* [or *in the state of paganism* or *ignorance*]; *who have but recently ceased to be in their state of infidelity* [&c.]. (TA.) === Also *i. q.* خَبَرٌ [*Information; a piece of information; intelligence; an announcement; news,* or *tidings; a piece of news; an account; a narration,* or *narrative; a story;* &c.]; (Ṣ, Ḳ;) employed to signify little and much; (Ṣ;) and ↓ حِدِّيثَى signifies the same: (Ḳ:) or *a thing,* or *matter, that is talked of, told,* or *narrated, and transmitted:* (Mṣb:) [and *talk,* or *discourse:*] and [in like manner] ↓ أُحْدُوثَةٌ signifies *a thing that is talked of, told,* or *narrated:* (Ṣ, Ḳ:) or this last signifies *a wonderful thing:* (IB, TA:) it has been asserted, says MF, that there is no difference between احدوثة and حديث in usage, and in denoting *what is good* and *what is evil;* in contradiction to such as say that the former peculiarly signifies *that* [*kind of story*] *in which there is no profit nor any truth;* such as amatory stories, and the like fictions of the Arabs: Fr asserts it to signify peculiarly *a laughable and an absurd story;* differing from حديث: and Ibn-Hishám El-Lakhmee, in his Expos. of the Fṣ, says that it is only used to denote *what is bad,* or *evil:* but Lb replies against him, in his Expos., that it is sometimes used to denote *what is good;* as in a saying mentioned by Yaạḳoob, which see below: (TA:) the pl. of حَدِيثٌ is أَحَادِيثُ, contr. to analogy, (Ṣ, Ḳ,) said by Fr to be pl. of ↓ أُحْدُوثَةٌ, and then used as pl. of حديث, (Ṣ,) but IB says that this is not the case; (TA;) and حِدْثَانٌ and حُدْثَانٌ are also pls. of حديث, (Ḳ, TA,) sometimes occurring; the latter, rare. (TA.) You say, سَمِعْتُ حَدِيثًا حَسَنًا (TA) and ↓ حِدِّيثَى حَسَنَةً (Ṣ, A,* TA) [*I heard a good story* or *narrative* &c.]; both meaning the same. (TA.) And اِنْتَشَرَ لَهُ فِى النَّاسِ أُحْدُوثَةٌ ↓ حَسَنَةٌ [*A good story of him became spread abroad among the people*]: a saying mentioned by Yaạḳoob in his "Iṣláḥ." (TA.) And ↓ أُحْدُوثَةٌ مَلِيحَةٌ [*A pretty story*], and أَحَادِيثُ مِلَاحٌ [*pretty stories*]. (A.) And ↓ قَدْ صَارَ فُلَانٌ أُحْدُوثَةً [‡ *Such a one has become the subject of a story,* or *of a wonderful story:* and in like manner, as is said in the A, صَارُوا أَحَادِيثَ: there said to be tropical]. (IB, TA.) —— Hence the حَدِيث of the Apostle of God: (Mṣb:) [i. e.] حَدِيثٌ also signifies *A narration of a* مُحَدِّث: (L:) [meaning حَدِيثٌ نَبَوِىٌّ, i. e. *a tradition,* or *narration, relating,* or *describing, a saying* or *an action* &c. *of Mohammad:*] this word and خَبَرٌ both signify *a tradition that is traced up to Mohammad,* or *to a Ṣaḥábee,* or *to a Tábi'ee:* (TA in art. رق:) or حديث is applied to *what comes from the Prophet:* خَبَرٌ, to what comes from another than the Prophet; or from him or another: and أَثَرٌ, to what comes from a Companion of the Prophet; but it may also be applied to a saying of the Prophet: (Kull p. 152:) the word in this sense, i. e. the حديث of the Prophet, has for its pl. only أَحَادِيثُ; and therefore Sb mentions it in the category of those words which have pls. anomalously formed; such as عَرُوضٌ, pl. أَعَارِيضُ; and بَاطِلٌ, pl. أَبَاطِيلُ. (TA.) [الحَدِيثَ written at the end of a quotation of a part of a trad. is for اِقْرَإِ الحَدِيثَ *Read the tradition.*] —— حَدِيثٌ قُدْسِىٌّ [*A holy tradition* or *narration*] means *what God has told to his prophet by inspiration,* or *by a dream,* or *in sleep, and the prophet has told in his own phraseology:* the Ḳur-án is esteemed above this, because [it is held that] its words also were revealed: (KT:) *that of which the words are from the apostle, but the meaning is from God, by inspiration,* or *by a dream,* or *in sleep.* (Kull p. 288.)

حَدَاثَةٌ: see حِدْثَانٌ, in three places. [Hence,] حَدَاثَةُ السِّنِّ ‡ *Youth; the first period of life.* (TA.)

حُدَّاثٌ: see مُحَدِّثٌ.

حِدِّيثٌ: see حَدَثٌ, in three places.

حِدِّيثَى: see حَدِيثٌ, in two places.

حَادِثٌ: see حَدِيثٌ, first sentence.

حَادِثَةٌ; and its pl., حَوَادِثُ: see حَدَثٌ, in four places.

أَحْدَثُ *More,* and *most, new,* or *recent:* fem. حُدْثَى; as in the phrase اِمْرَأَتِى الحُدْثَى, occurring in a trad., *My wife who was more,* or *most, recently married.* (TA.)

أُحْدُوثَةٌ: see حَدِيثٌ, in five places.

مُحْدَثٌ: see حَدِيثٌ: —— and see also حَدَثٌ, in two places. —— Also, applied to a poet, *i. q.* مُوَلَّدٌ [*A post-classical author:* itself a post-classical term]. (Mz 49th نوع.) [And المُحْدَثُونَ *The moderns;* or *people of later times;* opposed to القُدَمَآءُ.]

مُحْدِثٌ: see حَدَثٌ.

مُحَدَّثٌ A *true,* or *veracious,* man: (Ḳ:) a man *of true opinion:* (Ṣ:) *of true conjecture:* (A, TA:) *inspired; into whose mind a thing is put, and who tells it conjecturally and with sagacity; as though he were told a thing, and said it:* occurring in a trad.: (TA:) such was 'Omar. (A, TA.)

مُحَدِّثٌ *A teller,* or *relater, of stories, narratives,* or *traditions:* [and particularly *a relater of,* or *one skilled in, the traditions of Mohammad:*] ↓ حُدَّاثٌ in the sense of مُحَدِّثُونَ, signifying a company of men *telling,* or *relating, stories* &c., is an anomalous pl., formed by assigning it to the same predicament as words of similar meaning, of which سُمَّارٌ, pl. of سَامِرٌ, is an ex. (L.) See also حَدَثٌ.

أَرْضٌ مَحْدُوثَةٌ † *Land upon which the rain called* حَدَثٌ *has fallen.* (L.)

حدج

1. حَدَجَهُ, (Ṣ, A, Ḳ,*) aor. ِ, inf. n. حَدْجٌ (Ṣ, Ḳ) and حِدَاجٌ, (TA,) *He bound the* حِدْج *upon him,* i. e., upon the camel; (Ṣ, A, Ḳ;) as also ↓ احدجهُ: (Ḳ:) or *he bound upon him the* حِدَاجَة, i. e., *the* [*saddle called*] قَتَب *and its apparatus;* (Az, TA;) *which apparatus consists of the* بِدَادَانِ, *with the two girths called the* بِطَانَانِ *and the* حَقَب, *without which a camel is not* [*said to be*] مَحْدُوج. (Sh, TA.) [See حِدْج.] Accord. to J, حَدَجَ also signifies *He bound* loads, or burdens, *and divided* them *into camel-loads:* (TA:) but this is a meaning that was unknown to the Arabs. (Az, TA.) J cites as an ex. the words of El-Aạshà,

* أَلِلْبَيْنِ تُحْدَجُ أَحْمَالُهَا *

[*Is it for separation that her loads are bound* &c.?]: but he adds that, accord. to one reading, the poet said أَجْمَالُهَا: and this [SM says] is the right reading. (TA.) —— [Hence, حَدَجَ is used to signify ‡ *He betook himself to warring for the sake of the religion.*] 'Omar is related to have said, حِجَّةٌ هٰهُنَا ثُمَّ احْدِجْ هٰهُنَا حَتَّى تَفْنَى, meaning Perform *one pilgrimage,* then ‡ *betake thyself to warring for the sake of the religion until thou become old and weak,* or *die;* احدج literally signifying *bind the* حداجة *upon the camel.* (Az, TA.) —— [Hence also,] حَدَجَهُ, (TA,) inf. n. حَدْجٌ, (Ḳ,) ‡ *He imposed upon him in a sale.* (Ḳ, TA.) You say, حَدَجْتُهُ بِبَيْعِ سَوْءٍ (A, TA) ‡ *I imposed upon him with a bad sale,* and بِمَتَاعِ سَوْءٍ *with bad merchandise.* (TA.) The person imposed upon is likened to a camel upon which a حداجة is bound. (Az, TA.) —— And حَدَجْتُهُ بِمَهْرٍ ثَقِيلٍ ‡ *I imposed upon him a heavy dowry, by deceit and fraud.* (A, TA.) === Also, aor. ِ, inf. n. حَدْجٌ, *He cast* حَدَج [or *unripe and hard colocynths,* or *small colocynths,* or *small and green colocynths* or *melons,*] *at him.* (A, TA.) —— Hence, (A, TA,) حَدَجَهُ بِسَهْمٍ, (Ṣ, A,) inf. n. حَدْجٌ, (Ḳ,) ‡ *He shot at him with an arrow.* (Ṣ, A, Ḳ.) And حَدَجَهُ بِعَصًا, inf. n. حَدْجٌ, ‡ *He beat him,* or *struck him, with a staff,* or *stick.* (Ibn-El-Faraj, Ḳ,* TA.) —— [Hence also,] حَدَجَهُ بِالتُّهَمَةِ, inf. n. حَدْجٌ, ‡ *He cast suspicion upon him.* (Ḳ,* TA,* TḲ.) And حَدَجَهُ بِذَنْبِ غَيْرِهِ (Ṣ, A) ‡ *He accused him of the crime,* or *offence, of another,* (Ṣ, TA,) and *put it upon him.* (TA.) And حَدَجَهُ بِبَصَرِهِ, (Ṣ, A,) aor. ِ, inf. n. حَدْجٌ (Ṣ, TA) and

حُدُوجٌ; and ▾ حدّجهُ, inf. n. تَحْدِيجٌ; (TA;) *He cast his eyes at him;* (Ṣ, TA;) as also حَدَجَ إِلَيْهِ بَصَرَهُ: or *he looked intently, and sharply, at him:* or *he looked at him with a look which he* [the latter] *suspected and disliked:* (TA:) but حَدْجٌ in looking may be unattended by alarm, or fear: (Az, TA:) ▾ تَحْدِيجٌ is like تَحْدِيقٌ, (Ṣ,) syn. therewith: (Ḳ:) and also signifies the *looking intently, after alarm,* or *fear.* (TA.) — Also حَدَجَ, aor. ـِ, inf. n. حُدُوجٌ, † *He* (a horse) *looked at the figure of a man,* or *the like, seen from a distance,* or *heard a sound, and raised his ears, and directed his eyes, towards it.* (TA.)

2. حدّجهُ, inf. n. تَحْدِيجٌ: see 1, in two places.

4: see 1, first sentence. = أَحْدَجَتْ شَجَرَةُ الحَنْظَلِ *The colocynth-plant bore,* or *produced, fruit such as is termed* حَدَج. (Ṣ.)

حِدْجٌ *A certain thing upon which the women of the Arabs of the desert ride; not a* رَحْل *nor a* هَوْدَج: (Lth, TA:) *a certain vehicle,* or *thing to ride upon, for women,* (Az, Ṣ, A, Ḳ,) *like the* مِحَفَّة, (Az, Ṣ, Ḳ,) *and like the* هَوْدَج; (Az, TA;) as also ▾ حِدَاجَةٌ: (Ṣ, A, Ḳ:) pl. of the former حُدُوجٌ and أَحْدَاجٌ (Ṣ, A, Ḳ) and حُدُجٌ; (AAF, TA;) and pl. of the latter حَدَائِجُ: (Yaḳoob, Ṣ, A:) Az, however, says that ISk makes no difference between the حِدْج and the ▾ حِدَاجَة, though there is a difference between them accord. to the Arabs, as will be seen from what follows: Sh says that حِدْج is a name given to *a* هَوْدَج *bound upon a* قَتَب [or *small kind of camel's saddle*] *when it is bound upon the camel at once with all its apparatus:* he also says that ▾ حِدَاجَة is a name given to the *apparatus composed of the* أَبِدَّة, [pl. of بِدَاد, q. v.,] *which are also called* مَثَانِى القَتَبِ, [*and which are appertenances of the* قتب,] *when they are filled, and drawn together, and bound, and tied to the* قتب: [and he shows, in his explanation of the verb حَدَجَ, that *this apparatus comprises the* قتب *and* بِدَادَان *with the two girths called the* بِطَان *and the* حَقَب: this is what is meant in the Ḳ by the saying that ▾ الحِدَاجَةُ also signifies الأَدَاةُ:] Aboo-Saʼeed El-Kilábee says that ▾ حداجة signifies the *apparatus* (اداة) *of the* قتب: and Az says that it signifies the قتب *with its apparatus.* (TA.) — Also *A load,* or *burden.* (Ṣ, Ḳ.) — And [its pl.] حُدُوجٌ, *Camels with their* رِحَال [or *saddles*]. (TA.)

حَدَجٌ [a coll. gen. n.] The *colocynth,* or *colocynths, when unripe and hard:* (TA:) or *when become hard;* (Ṣ, TA;) *before becoming yellow:* (TA:) or *small colocynths:* (A:) or the *colocynth* or *colocynths,* and the *melon* or *melons,* (M, Ḳ,) *while small and green, before becoming yellow,* (M,) or *while continuing succulent,* or *fresh,* or *green:* (Ḳ:) or [more correctly] the *melon* or *melons;* and the *colocynth,* or *colocynths, while continuing succulent,* or *fresh,* or *green:* (T:) n. un. with ة. (Ṣ.)

حِدَاجَةٌ: see حِدْجٌ, in five places.

حدر

1. حَدَرَ, aor. ـُ (M, Mṣb, Ḳ, &c.) and ـِ, (M, Ḳ,) inf. n. حُدُورٌ (T, Ṣ, M, Mṣb, Ḳ) and حَدْرٌ, (T, M, Ḳ,) *He made to descend,* or *to go down* or *downwards* or *down a declivity; sent, let,* or *put, down,* or *from a higher to a lower place* or *position;* (T, Ṣ, M, A, Mṣb, Ḳ;) as also ▾ احدر: (Mṣb:) [or this latter is not chaste; for, accord. to J,] one says, حَدَرَ السَّفِينَةَ *he lowered the ship;* or *sent it to a lower place,* (Ṣ,) or *from a higher to a lower part of a river;* (A;) but one should not say, احدرها. (Ṣ.) You say also, حَدَرَ الحَجَرَ مِنَ الجَبَلِ *He rolled down the stone from the mountain.* (A.) — حَدَرَتْهُمُ السَّنَةُ ‡ *Dearth, scarcity,* or *drought, made them to descend* [*from the desert*]; *brought them to a descent;* (T, Ṣ;) *brought them,* (TA,) or *brought them down,* or *made them to descend,* (A,) *to the towns,* or *villages.* (A, TA.) — حَدَرَ اللِّثَامَ عَنْ حَنَكِهِ *He turned down the* لثام [or *muffler*] *from the part beneath his chin.* (TA.) — حَدَرَ الدَّمْعَ, aor. ـُ and ـِ, inf. n. حُدُورٌ and حَدْرٌ, *He shed,* or *let fall, tears;* as also ▾ حدّرهُ. (TA.) And العَيْنُ تَحْدُرُ الدَّمْعَ, (A, Ḳ,*) and تَحْدِرُهُ, inf. n. حَدْرٌ, (Ḳ,) ‡ *The eye sheds,* or *lets fall, tears;* (A;) or *flows with tears.* (Ḳ.) And الدَّمْعُ يَحْدُرُ الكُحْلَ ‡ [*The tears make the collyrium to flow down*]. (A.) — حَدَرَ الدَّوَآءُ بَطْنَهُ, (A,) aor. ـُ, (TA,) inf. n. حَدْرٌ, (Ḳ,) ‡ *The medicine made his belly to discharge itself.* (A, Ḳ.) [And ▾ حدّر الطَّمْثَ † *It* (a medicine) *caused the menstrual flux to descend:* see مُحَدِّرٌ.] = حَدَرَهُ, (T, Ṣ, Mgh, Ḳ,) aor. ـُ and ـِ, (Ḳ,) inf. n. حَدْرٌ; (Ṣ, Mgh, Ḳ;) and ▾ احدرهُ, (T, Ṣ, A, Ḳ,) inf. n. إِحْدَارٌ; (Ḳ;) ‡ *He made* the skin *to swell,* (T, Ṣ, A, Mgh, Ḳ,) *and to become thick,* (A,) by beating. (T, Ṣ, A, Mgh.) = حَدَرَ الثَّوْبَ, (A, Ḳ,) aor. ـُ and ـِ, inf. n. حَدْرٌ; (Ḳ;) and ▾ احدرهُ, (Ṣ, Ḳ,) inf. n. إِحْدَارٌ; (Ḳ;) ‡ *He twisted the unwoven warp,* (Ḳ,) or *the extremities of the unwoven warp,* (Ṣ, A,) *of the garment,* or *piece of cloth;* (Ṣ, A, Ḳ;) *like as is done with the ends of* [*garments of the kind called*] أَكْسِيَة [pl. of كِسَآء]: (Ṣ:) because its length is thus diminished. (A.) = See 7. — [Hence,] حَدَرَ فِى القِرَآءَةِ, (Ṣ, Mgh, Mṣb, Ḳ,*) and فِى الأَذَانِ, (Ṣ, Mgh, Mṣb,) and فِى الإِقَامَةِ, (Mṣb,) aor. ـُ (Ṣ, Mgh, Mṣb, Ḳ) and ـِ, (Ḳ,) inf. n. حَدْرٌ; (Ṣ, Mgh, Mṣb, Ḳ;) and ▾ حدّر, inf. n. تَحْدِيرٌ; (Ḳ;) and حَدَرَ القِرَآءَةَ, (A, Mṣb,) and الأَذَانَ, and الإِقَامَةَ; (Mṣb;) ‡ *He hastened,* or *was quick, in the reading,* or *recitation,* (Ṣ, A, Mgh, Mṣb, Ḳ,*) and *in the call to prayer,* (Ṣ, Mgh, Mṣb,) and *in the* [*form of words called the*] اقامة; (Mṣb;) and *he hastened the reading,* or *recitation,* &c. (Mṣb.) = حَدَرَ and حَدُرَ, inf. n. [of the latter, accord. to analogy,] حُدُورَةٌ, *It* (a bow-string) *was thick and strong.* (TA. [See also حَادِرٌ.]) — And [hence, app.,] ‡ *It* (a boy) *was,* or *became, such as is termed* حَادِر [q. v.]: (TA:) [or] حَدُرَ, aor. ـُ; (Lth, Aṣ, Ṣ, A, Ḳ;) and حَدَرَ, aor. ـِ; (ISd, Ḳ;) inf. n. [of the former] حَدَارَةٌ (A, Ḳ) and حَدْرٌ; (Ṣ, Ḳ;) ‡ *he was,* or *became, compact in make,* (Aṣ, Ṣ, Ḳ,) *and thick:* (TA:) or *short and fleshy:* (A:) and *he was,* or *became, fat, with thickness,* (Ḳ, TA,) *and shortness.* (TA. [See حَادِر.]) — And حَدَرَ, (T, Ṣ, A, Ḳ,) aor. ـُ (T, Ṣ, Ḳ) and ـِ, (Ḳ,) inf. n. حُدُورٌ (T, Ṣ, A) and حَدْرٌ; (Ḳ;) and ▾ احدر, inf. n. إِحْدَارٌ; and ▾ حدّر, inf. n. تَحْدِيرٌ; (Ḳ, TA;) or the first form only; (T;) ‡ *It* (the skin) *became swollen,* (T, Ṣ, TA,) as also ▾ انحدر, (Ṣ, Ḳ,) by reason of beating: (T, Ṣ, TA:) or *became swollen and thick,* by reason thereof. (A, Ḳ.) — حَدُرَتِ العَيْنُ, inf. n. حَدَارَةٌ, † *The eye was,* or *became, large and wide:* (Mṣb:) *was,* or *became, beautiful.* (TA.)

2: see 1, in four places.

4: see 1, in four places. = Also احدر الثَّوْبَ †*He sewed the garment,* or *piece of cloth, the second time, after the* [*slight sewing termed*] مَلّ, or شَلّ. (Ṣ.)

5. تحدّر الدَّمْعُ (Ṣ, Ḳ*) and ▾ تحادر (A) *The tears descended gently,* or *little by little.* (Ṣ, A, Ḳ.*) And رَأَيْتُ المَطَرَ يَتَحَادَرُ عَلَى لِحْيَتِهِ *I saw the rain descending and dropping upon his beard.* (TA.)

6: see 5, in two places.

7. انحدر *He,* or *it, descended; went down, downwards, down a declivity,* or *from a higher to a lower place* or *position:* (Ṣ, A, Mṣb, Ḳ:) and [in like manner] ▾ حَدَرَ, inf. n. حَدْرٌ, (TA,) or حُدُورٌ, (A,) *he went down,* or *descended, a declivity.* (A, TA.) [Hence,] انْحَدَرْتُ إِلَى البَصْرَةِ *I went down to El-Baṣrah.* (Ṣ.) — Also *He journeyed,* or *went, towards El-ʼIráḳ,* and *Syria,* and *ʼOmán:* opposed to أَصْعَدَ, which signifies "he journeyed, or went, towards Nejd, and El-Ḥijáz, and El-Yemen:" (ISk, on the authority of ʼOmárah, TA in art. صعد:) or the former, *he journeyed,* or *went, towards El-ʼIráḳ:* and the latter, "he journeyed, or went, towards the Ḳibleh:" (Aboo-Ṣakhr, T, TA ubi suprà:) and ▾ مُنْحَدَرٌ is used as an inf. n. of the former; like as مُصْعَدٌ is of the latter: (T, TA ubi suprà:) also, the former verb, *he returned from any town or country:* and the latter, "he commenced a journey or the like, in any direction." (Ibn-ʼArafeh, TA ubi suprà.) — Also, said of a place, *It sloped down.* (Mṣb.) = See also 1, last sentence but one.

حَدْرٌ: see حُدُورٌ, in two places.

حَدْرَةٌ *A single thread, of the threads of a* [*garment of the kind called*] كِسَآء. (TA.) [See حَدَرَ الثَّوْبَ.] = عَيْنٌ حَدْرَةٌ (Aṣ, T, Ṣ, Mṣb, Ḳ) and ▾ حُدُرَّى (Ḳ) †*An eye compact and hard:* (Aṣ, T, Ṣ:) or *thick and hard:* (Ḳ:) or *wide and large and projecting:* (T:) or *large and wide:* (Mṣb:) or *large:* (Ḳ:) or *wide:* (TA:) or *sharp-sighted.* (Ḳ.)

حَدْرَةٌ *A herd* of camels, (Ṣ, Ḳ,) *like,* or *about, a* صِرْمَة, (Ṣ,) which is [as some say] from ten to forty: when they amount to sixty, they are termed a صدعة: (TA:) *a flock* of sheep or goats. (Lḥ, TA.) — See also حُدُورَةٌ.

حَدْرَآءُ: see حَدُورٌ — عَيْنٌ حَدْرَآءُ †*A beautiful eye.* (TA.)

حُدْرَى: see حُدْرَةٌ.

حَدُورٌ (Ṣ, A, Mṣb, Ḳ) and ▼حَدَرٌ (Ṣ, Ḳ) and ▼حَدْرَآءُ, (T, Ḳ,) of the same measure as صَفْرَآءُ, (T,) [in the CḲ, erroneously, حُدَرَآءُ,] and ▼أُحْدُورٌ (Ḳ) and ▼حَادُورٌ (Ḳ) and ▼مُنْحَدَرٌ (Ṣ, Ḳ) [which is of frequent occurrence] and ▼مُنْحُدُرٌ and ▼مُنْحَدِرٌ, or ▼مَنْحَدِرٌ, or ▼مُنْحَدُرٌ, (as in different copies of the Ḳ, the last of these being the third form given in the CḲ,) *A declivity,* or *declivous place; a place sloping down; a slope; a place of descent,* or *by which one descends:* (Ṣ, A, Mṣb, Ḳ:) a حدور is *at the foot of a mountain,* and *in any place.* (TA.) You say, هَبَطْنَا فِى حَدُورٍ صَعْبَةٍ [*We descended a difficult declivity*]. (A.) And كَأَنَّمَا يَنْحَطُّ فِى حَدَرٍ▼ [*As though he were descending a declivity*]: (Ṣ:) occurring in a trad. (TA.)

حَدُورَةٌ: see what next follows.

حُدُورَةٌ and ▼حَدُورَةٌ and ▼حَادُورَةٌ ‡*A flow,* or *flowing, of tears from the eye.* (Lḥ, ISd, Ḳ, TA.) — Also the first, (Ṣ,) so accord. to the M, &c., (TA,) or ▼حُدْرَةٌ, (Ḳ,) *Multitude, and congregation.* (Ṣ, M, Ḳ.) You say حَىٌّ ذُو حُدُورَةٍ *A tribe numerous and congregated.* (Ṣ, M.)

حَادِرٌ A rope *strongly twisted:* a bow-string *strong and full.* (TA. [See also 1.]) — A *thick* spear. (TA.) And كُعُوبٌ حَوَادِرُ *Thick and round knots,* or *joints,* of a spear. (TA.) — A cake of bread (رَغِيفٌ) *complete:* or *having thick edges.* (TA.) — ‡A man *compact in make:* (Ṣ:) a boy *short and fleshy:* (A:) a youth *thick and compact:* (TA:) or *full of fat and flesh, with softness,* or *thinness, of skin:* (Lth, Az:) a boy *full in body, and of great force:* (Th:) or a boy *full of youthful vigour;* as also حَادِرَةٌ: [but this is an intensive epithet:] (Lth, Az:) or a *fat* boy: (Ḳ:) or a boy *fat, thick, and compact in make:* (ISd:) or *goodly,* or *beautiful:* (ISd, Ḳ:) pl. حَدَرَةٌ. (TA.) Also the fem., حَادِرَةٌ, †A *thick,* or *bulky,* she-camel. (T in art. زنب.) And the same, †*Bulky in the shoulder-joints.* (IB.) And حَوَادِرُ [the pl. fem.] †*Compact and bulky* camels or the like. (TA.) — †Anything *full of moisture, and of beautiful make.* (TA.) And حَادِرَةُ الْعَيْنَيْنِ †A she-camel *having full eyes:* (Ṣ:) or *having eyes full of fat, equal, and beautiful.* (TA.) — A tribe *congregated.* (TA.) — A *lofty* mountain. (TA.) — See also الحَيْدَرَةُ.

حُنْدُرٌ and ▼حُنْدُورَةٌ (Ṣ, Ḳ) and ▼حُنْدُورٌ (Ḳ) and ▼حِنْدِيرٌ (Th, Ḳ) and ▼حِنْدُورَةٌ and ▼حِنْدِيرَةٌ and ▼حِنْدَوْرٌ and ▼حِنْدَارَةٌ, (Ḳ,) of which ▼حِنْدِيرَةٌ is the most approved form, (TA,) The *black* of the eye. (Ṣ, Ḳ.) One says, هُوَ عَلَى حُنْدُرِ عَيْنِهِ and ▼حُنْدُورَةِ عَيْنِهِ (Ṣ, Ḳ) and ▼حُنْدُورِ عَيْنِهِ and ▼حِنْدِيرَةِ عَيْنِهِ (TA) †*He is deemed burdensome,* or *troublesome, by him, so that he cannot look at him by reason of hatred.* (Ṣ, Ḳ.) And جَعَلْتُهُ عَلَى حِنْدِيرَةِ عَيْنِى▼ and ▼حُنْدُورَةِ عَيْنِى †*I made him, or it, a conspicuous object,* or *a thing in full view, of my eye.* (Ṣ, Ḳ.) Several lexicographers mention these forms in art. حندر, regarding the ن as a radical letter, as it should not be held to be augmentative, when occupying the second place in a word, unless on strong evidence. (TA.)

حَيْدَرٌ and الحَيْدَرُ: see what next follows.

الحَيْدَرَةُ †*The lion;* (Ṣ, Ḳ;) as also ▼الحَيْدَرُ, (Ḳ,) and ▼حَيْدَرٌ, without ال, (TA,) and ▼الحَادِرُ: (Ḳ:) or *the lion that is, among other lions, like the king among men;* (IAạr;) because of the thickness of his neck, and the strength of his fore legs. (Th, TA.) — Also حَيْدَرَةٌ ‡*Destruction,* or *perdition;* (AZ, Ḳ;) and so ▼حَادُورٌ: (Ḳ:) or *a severe calamity;* as though it were a lion in its severity. (A.)

حَادُورٌ: see حَدُورٌ. — Also *An ear-ring;* syn. قُرْطٌ: (Ṣ, Ḳ:) pl. حَوَادِيرُ. (TA.) — ‡*A laxative medicine;* (A, Ḳ,* TA;) *contr. of* عَاقُولٌ. (A.) — See also الحَيْدَرَةُ.

حَادُورَةٌ: see حُدُورَةٌ.

حُنْدُورٌ and حِنْدَوْرٌ: see حُنْدُرٌ, in three places.

حِنْدِيرٌ: }
حِنْدَارَةٌ: } see حُنْدُرٌ.

حُنْدُورَةٌ and حِنْدُورَةٌ and حِنْدَوْرَةٌ: see حُنْدُرٌ, in six places.

حِنْدِيرَةٌ: see حُنْدُرٌ, in three places.

أَحْدَرُ †*More, most,* or *very, fat and thick.* (TA.)

أُحْدُورٌ: see حَدُورٌ.

مُحَدِّرٌ لِلطَّمْثِ †[*Emmenagogue*]. (Ḳ in arts. نجد and جزر &c.)

مُنْحَدَرٌ and مُنْحُدُرٌ and مُنْحَدِرٌ, or مَنْحَدِرٌ, or مُنْحَدُرٌ: see حَدُورٌ: — and for the first, see also 7.

حدس

1. حَدَسَ, [aor., app., ـِ and ـُ,] inf. n. حَدْسٌ, *He threw, cast,* or *shot.* (TA.) You say, حَدَسْتُ بِسَهْمٍ *I shot an arrow.* (Ṣ.) And حَدَسْتُهُ بِكَذَا *I threw, cast,* or *shot, at him with such a thing.* (A.) — Hence, حَدْسُ الظَّنِّ *The conjecturing without evidence or proof.* (TA.) You say, حَدَسَ, (Ṣ, A, Mṣb, Ḳ,) aor. ـِ (Ṣ, Mṣb, Ḳ) and ـُ, (Ḳ,) inf. n. حَدْسٌ, (Ṣ, A, Mṣb, Ḳ,) *He opined:* (Ṣ, A, Ḳ:) or *he formed a confirmed opinion:* (Mṣb:) *he formed a surmise, or an opinion;* or *he spoke conjecturally,* or *surmising:* (Ṣ, A, Ḳ:) *he surmised respecting the meanings of speech* or *language,* (A, Ḳ,) and *things.* (Ḳ.) And هُوَ يَحْدِسُ *He says a thing according to his opinion.* (Ṣ, TA.) And بَلَغَنِى عَنْ فُلَانٍ أَمْرٌ وَأَنَا أَحْدِسُ فِيهِ *A thing has been told me of such a one, and I speak of it conjecturally,* or *surmising.* (TA.) And حَدَسْتُ عَلَيْهِ ظَنِّى *I formed my opinion of it, not being certain of it;* as also نَدَسْتُ. (TA.) And حَدَسَ الكَلَامَ عَلَى عَوَاهِنِهِ *He spoke without anything to guide him, and without caution.* (TA.) And حَدَسَ الشَّىْءَ *He computed by conjecture the quantity, measure,* or *the like, of the thing.* (A.) And قَالَهُ بِالحَدْسِ [generally meaning *He said it conjecturally,* or *surmising:* but also meaning] *he said it by means of intuition.* (A, TA.) [حَدْسٌ is also explained in the A as signifying نَظَرٌ خَافٍ: in the TA نَظَرٌ خَفِىٌّ: both app. meaning *An obscure,* or *an occult, mode of judging* of a thing.] — حَدْسٌ is also *syn. with* قَصْدٌ, (Ḳ,) used transitively, (T, Ḳ,) [app. signifying The *aiming at* a thing,] *by,* or *with, whatever thing it be;* [app. meaning *by any mental operation;*] *by opinion,* or *by judgment,* or *by intelligence* or *cunning sagacity.* (TA.) — حَدَسَ فِى الأَرْضِ, (El-Umawee, Mṣb,) aor. ـِ, (El-Umawee, TA,) inf. n. حَدْسٌ, (Ṣ,) *He went away,* (Ṣ, Mṣb, TA,) or *journeyed,* (TA,) *into,* or *in,* or *through, the country,* or *land, without guidance:* (Ṣ, Mṣb, TA:) or simply *he went away into,* or *in, the country,* or *land;* as also عَدَسَ. (El-Umawee, TA.) — [Hence, app., the phrase حَدَسَ فِى صَدْرِى شَىْءٌ, which seems to mean *A thing came at random into my mind.* See هَجَسَ.] — Also حَدَسَ, inf. n. as above, *He went in a right course,* or *direction:* (TA:) or *in one regular, uniform,* or *constant, course:* (O, Ḳ:) or, accord. to Az, *not in one regular, uniform,* or *constant, course.* (TA.) — And حَدَسَ فِى السَّيْرِ, (Mṣb,) inf. n. حَدْسٌ, (Ḳ,) *He hastened,* or *was quick, in pace,* or *in journeying.* (Mṣb, Ḳ.)

5. تحدّس الأَخْبَارَ, (Ṣ, Ḳ,) and عَنِ الأَخْبَارِ, (AZ, Ṣ, A, Ḳ,) *He sought to learn the news,* or *tidings, without others' knowing of him;* (AZ, Ṣ, M, A, Ḳ;) as also تندّس عنها, and توجّس: (AZ, TA:) or *he sought for,* or *inquired respecting, the news,* or *tidings, in order to know what others knew not.* (A.)

حَدَّاسٌ One *who opines,* or *conjectures, much;* syn. ظَنَّانٌ. (TA.)

حدق

1. حَدَقَهُ (Ḳ, TA) بِعَيْنِهِ, aor. ـِ, inf. n. حَدْقٌ, (TA,) *He looked at it.* (Ḳ, TA.) It is said in a trad., فَحَدَقَنِى القَوْمُ بِأَبْصَارِهِمْ *And the people,* or *party, cast the blacks of their eyes at me.* (TA.) And حَدَقَهُ *He,* or *it, hit,* or *hurt, the black of his eye.* (Ḳ.) — حَدَقَ, (Ḳ, TA,) aor. ـِ, (TA,) inf. n. حُدُوقٌ, said of one that is dying (مَيِّت), *He opened his eyes, and moved his eyelids,* or *twinkled with his eyes.* (Ḳ.) You say, رَأَيْتُ المَيِّتَ يَحْدِقُ *I saw him that was dying open his eyes,* &c. (TA.) — See also 4.

2. حدّق, (Mgh,) or حدّق النَّظَرَ, (Mṣb,) inf. n. تَحْدِيقٌ, (Ṣ, Mgh, Mṣb, Ḳ,) *He looked hardly,* or *intently,* (Ṣ, Mgh, Mṣb, Ḳ,) *and rolled the black*

of the eye, (Ḥar p. 221,) إِلَيْهِ *at him:* (Mgh, Mṣb:) and ↓ حَدْلَقَةٌ, with an augmentative ل, is like تَحْدِيقٌ; [the verb of which it is the inf. n., namely,] ↓ حَدْلَقَ, said of a man, signifying *he rolled the black of his eye in looking.* (Ṣ.)

4. احدقوا به; (Ṣ, Mgh, Mṣb, Ḳ;) and ↓ حَدَقُوا بِهِ, (Ṣ, Mṣb, Ḳ,) aor. ـِ, (Mṣb, Ḳ,) inf. n. حَدْقٌ; (TḲ;) and ↓ احدودقوا به; (Ṣgh, Ḳ;) *They surrounded, encompassed,* or *encircled, him,* or *it;* (Ṣ, Mgh, Mṣb;) namely, a man, (Ṣ,) or a town or the like: (Mṣb:) or *they went round* or *round about, circuited,* or *compassed, him,* or *it.* (Ḳ.) You say of anything, احدق به as meaning *It surrounded, encompassed,* or *encircled, it;* (TA;) as, for instance, a house [or a wall] surrounds a garden. (Mgh.) Thus you say, عَلَيْهِ شَامَةٌ سَوْدَآءُ قَدْ أَحْدَقَ بِهَا بَيَاضٌ [*Upon him is a black mole which whiteness has surrounded*]. (TA.) And أَحْدَقَتْ بِهِ المَنِيَّةُ ‡ *Death encompassed him.* (TA.) — أَحْدَقُوا بِهِ الأَحْدَاقَ, a phrase used by El-Ḥareeree, means *They made the blacks of the eyes to surround him.* (Ḥar p. 186.) — احدقت الرَّوْضَةُ, (Ḳ,) or احدقت الروضة عُشْبًا, (Zj, TA,) i. e. [*The meadow*] *became a* حَدِيقَة [q. v.], (Zj, Ḳ,) [*by producing herbs such as are termed* عشب; for] without عشب it is a روضة. (Zj, TA.)

12: see 4.

Q. Q. 1. حَدْلَقَ, and its inf. n. حَدْلَقَةٌ: see 2.

حَدَقٌ: see حَدَقَةٌ, in three places. — Also The [*fruit of the*] بَاذِنْجَان [q. v.: accord. to Forskål (Flora Ægypt. Arab., p. 47), *solanum cordatum*]: (IAạr, Az, Ḳ:) [a coll. gen. n.:] n. un. with ة: likened to the blacks of the eyes of the [species of antelope called] مَهًا: in the handwriting of 'Alee Ibn-Ḥamzeh, written حذق, with the dotted ذ; but this is not known. (TA.)

حَدَقَةٌ (Ṣ, Mṣb, Ḳ, &c.) and ↓ حُنْدُوقَةٌ and ↓ حِنْدِيقَةٌ, (Ḳ,) but IDrd doubts the correctness of this, (TA,) The *black of the eye;* (IDrd, Mṣb, Ḳ;) i. e. the *round part in the middle of the eye;* (TA;) the *greater black* of the eye; (Ṣ, TA;) the smaller being the نَاظِر [or pupil], in which is what is termed إِنْسَانُ العَيْنِ, for it is like a mirror, in which one facing it sees his figure; (TA;) [the *part, of the eye, that is surrounded by the white;* the *iris, together with the whole of the anterior chamber of the eye;*] *what is in the middle of the white of the eye:* (Zj in his "Khalḳ el-Insán:") or, as some say, externally, the *black of the eye;* and internally, *its* خَرَزَة [or *lens*]: (TA:) [and sometimes the *eye,* absolutely; as, for instance, in a phrase mentioned voce ذَآب, in art. ذوب:] pl. ↓ حَدَقٌ, (Ṣ, Mṣb, Ḳ) [or rather this is a coll. gen. n.,] and [the pl. is] أَحْدَاقٌ [a pl. of pauc.] (Ḳ) and حَدَقَاتٌ (Mṣb) and حِدَاقٌ; (Ṣ, Mṣb, Ḳ;) which last is applied by Aboo-Dhu-eyb to the حَدَقَة *together with what surrounds it.* (TA.) [Hence,] ↓ تَكَلَّمْتُ عَلَى حَدَقِ القَوْمِ *I spoke while the people,* or *party, looked at me.* (TA.) And ↓ رُمَاةُ الحَدَقِ *Those who hit the mark in throwing* or *shooting.* (TA.) And نَزَلُوا فِى مِثْلِ حَدَقَةِ البَعِيرِ ‡ *They alighted,* or *alighted and abode, in a tract abounding with herbage;* likened to the حدقة of the camel because this is plentifully supplied with moisture. (TA from a trad.)

حَدِيقَةٌ *A walled garden; a garden surrounded by a wall:* (Ṣ, Mṣb, TA:) of the measure فَعِيلَةٌ in the sense of the measure مَفْعُولَةٌ: (Mṣb:) or *any round piece of land surrounded by a fence* or *the like,* or *by elevated land:* (TA:) or *whatever is surrounded by building:* (Ḳ:) and, by extension of its application, *a garden, though without a wall:* (Mṣb:) or *a meadow,* or *garden,* (رَوْضَة,) *having trees:* (Ṣ, Ḳ:) or *a garden of palm-trees and of other trees,* (Zj, IDrd, Ḳ,) *dense and luxuriant,* (Zj, IDrd, TA,) *and,* as some say, *fruit-bearing:* (TA:) or *a garden* (جَنَّة) *of palm-trees and grape-vines:* (TA:) or *a distinct collection of palm-trees:* (Ḳ:) or *a plot of seed-produce:* (Kr, TA:) or *a hollow in a valley, that retains water:* and *any depressed place in a valley, that retains water, though water be not in its bottom:* (TA:) pl. حَدَائِقُ. (Ṣ, Mṣb, Ḳ.) Hence the saying, وَرَدَ عَلَىَّ كِتَابُكَ فَتَنَزَّهْتُ فِى بَهْجَةِ حَدَائِقِهِ ‡ [*Thy letter has come to me, and I have recreated myself in contemplating the beauty of its garden-like phrases*]. (TA.)

حُدْلُقَةٌ (mentioned in the Ḳ in art. حدلق) *A large* حَدَقَة [or *black of the eye*]: (Ṣ, Ḳ:) a meaning which shows the ل to be augmentative: (TA:) or *some part of the body that is unknown:* (Ḳ:) one says, أَكَلَ الذِّئْبُ مِنَ الشَّاةِ الحُدْلُقَةَ [*the wolf ate, of the sheep,* or *goat, the* حدلقة]: A 'Obeyd says, it is some part of the body thereof, but I know not what it is: (Ṣ:) or the *eye:* (Ṣ, Ḳ:) so says Lḥ; (Ṣ;) and so Kr: (TA:) Aṣ heard an Arab of the desert, of the Benoo-Saad, say that it means its غَلْصَمَة [or *epiglottis*]. (IB, TA.) — عَيْنٌ حُدْلُقَةٌ *An eye of which the ball,* or *globe, is prominent,* or *large and prominent;* or *of which the black is prominent;* syn. جَاحِظَةٌ. (TA.)

حَدَوْلَقٌ *Short and compact.* (IDrd, Ḳ.)

حُنْدُوقَةٌ: } see حَدَقَةٌ.
حِنْدِيقَةٌ: }

حَنْدَقُوقٌ and its vars. (mentioned in this art. by J and Ṣgh): see art. حندق.

مُحْدَقٌ *A difficult,* or *distressing, affair* or *event; in consequence of which men look hardly,* or *intently.* (TA.)

حدلق

For several words mentioned under this head in some of the Lexicons, see art. حدق.

حدم

1. حَدَمَتْهُ الشَّمْسُ, and النَّارُ, aor. ـِ, inf. n. حَدْمٌ, *The sun,* and *the fire, was,* or *became, vehemently hot upon him.* (Mṣb.)

4: } see what next follows.
5: }

8. احتدمت النَّارُ, and احتدم الحَرُّ, in the Ḳ, erroneously, ↓ أَحْدَمَتْ, *The fire,* and *the heat, burned,* or *burned fiercely:* (Ḳ,* TA:) and احتدمت النَّارُ *the fire flamed,* or *blazed:* (Ṣ, Mgh, Ḳ:) and in like manner, [as meaning *it became inflamed,* or *made to flame* or *blaze,*] the verb is said of anything: (T, TA:) or احتدمت النَّارُ, and احتدم النَّهَارُ, *the fire,* and *the day, was,* or *became, vehemently hot:* (Mṣb:) and احتدم, said of a day, (AZ, TA,) and of heat, (Ṣ in art. حمد,) signifies the same as احتمد, (AZ, Ṣ ubi suprà, TA,) from which it is formed by transposition. (Ṣ ubi suprà.) And احتدم *He suffered vehement heat* from the sun, and from fire. (Mṣb.) And احتدمت القِدْرُ *The cooking-pot boiled vehemently.* (AZ, TA.) — [Hence,] احتدم الشَّرَابُ ‡ *The wine,* or *beverage, estuated,* or *fermented;* syn. غَلَى. (Mgh, TA.) And احتدم صَدْرُ فُلَانٍ غَيْظًا (Ṣ) ‡ *The bosom of such a one burned with anger, wrath,* or *rage.* (TA.) And احتدم عَلَيْهِ غَيْظًا ‡ *He burned against him with anger, wrath,* or *rage;* (Ḳ, TA;) as also ↓ تحدّم. (Ḳ.) And احتدم الدَّمُ ‡ *The blood became intensely red, so as to be* [*nearly*] *black;* (Ṣ, Mṣb, Ḳ, TA;) and *became vehemently burning.* (Mṣb. [See also the part. n., below.])

حَدْمٌ and ↓ حَدَمٌ The *vehement burning* of fire, (Ḳ,) and of heat: or, accord. to the T, حدم [app. حَدْمٌ] signifies the *vehement heating* of a thing by the sun, and by fire: [see 1:] accord. to AZ, زَفْرُ النَّارِ signifies "the flaming, or blazing, of fire;" and شَيْقُهَا and حَدْمُهَا and حَدَمُهَا and كَلْحَبَتُهَا have all one meaning [app. the third of the meanings assigned below to حَدَمَة]. (TA.)

حَدَمٌ: see what next precedes.

حَدَمَةٌ *Fire:* and the *sound* of fire: (Ḳ:) or the *sound of the flaming,* or *blazing,* of fire; (Fr, Ṣ;) as also حَمَدَةٌ. (Fr, TA.) — † The *sound of the belly* of the serpent, (Ḳ,) or, as some say, of the species of serpent termed أَسْوَد: (TA:) or the *rustling sound of the skin* of the serpent, *caused by rubbing one part thereof against another;* as though it were the confused and continued sound of a thing مُحْتَدِم [i. e. flaming, or boiling]. (AḤát, TA.) — † The *purring* of a cat: likened to the sound of flaming, or blazing. (TA.)

قِدْرٌ حَدِمَةٌ *A cooking-pot quickly boiling; contr. of* صَلُودٌ: (Fr, Ṣ, A, TA:) in the Ḳ, erroneously, حَدَمَة, like فَرِحَة. (TA.)

مُحْتَدِمٌ A day *vehemently hot.* (Ṣ.) — † Blood *intensely red, inclining to blackness:* or, as some say, *intensely burning.* (Mgh.)

حدو

1. حَدَا الإِبِلَ, (Ṣ, Mgh,) or حَدَا بِالإِبِلِ, (Mṣb,) or both, (Ḳ,) aor. يَحْدُو, (Mṣb,) inf. n. حَدْوٌ (Ṣ, Mgh, Ḳ) and حُدَآءٌ (Ṣ, Ḳ) and حِدَآءٌ, *He drove the camels;* (Ṣ, Mgh, Ḳ;) and *chid them:* (Ḳ:) [and ↓ احتداها app. has the former signification:] and *he sang to them:* (Ṣ:) or *he urged,* or *excited, the camels by singing to them,* which

is termed † حُدَآءٌ: (Mṣb:) or حَدَا لَهَا signifies *he sang to them.* (Mgh.) The Arabs in driving their camels used commonly to sing verses of the kind termed رَجَز. (TA in art. رجز.) [It is said that] † حُدَآءٌ originated from the fact of a Desert-Arab's beating his young man, or boy, and biting his fingers; whereupon he went along saying وَىْ دَىْ, meaning يَا يَدَىَّ ["O my two hands!"]; and the camels went on at his cry; therefore his master bade him keep to it: (Ḳ in art. دى:) so says IAṣr. (TA in that art. [Other (similar) accounts of its origin are mentioned by MF in remarking on this passage of the Ḳ.]) حَدَا signifies also *He raised his voice with* [*the singing termed*] الحُدَآء. (Ḥar p. 576.) [And *He breathed short (anhelavit), and sent forth a voice* or *sound.* (Golius, from a gloss in the KL.)] — You say also of the north wind, تَحْدُو السَّحَابَ, i. e. † *It drives along the clouds.* (Ṣ.) — And حَدَاهُ عَلَى كَذَا † *He urged him, incited him,* or *put him in motion* or *action, to do such a thing.* (Mṣb, TA.) — And حَدَاهُ † *It followed it;* namely, the night the day; (Ḳ;) as also † احتداهُ: (AḤn, Ḳ:) and so the [wild] he-ass his she-asses; and anything any other thing. (TA.) Hence the saying, لَا أَفْعَلُهُ مَا حَدَا اللَّيْلُ النَّهَارَ † *I will not do it as long as the night follows the day*]. (TA.) — See also 5.

4: see 5.

5. تحدّاهُ, in its primary sense, is from الحُدَآءُ, and signifies *He* (a driver of camels) *vied, competed,* or *contended for superiority, with him,* or *emulated him,* (namely, another driver,) *in driving camels,* or *urging them by singing to them; each of them desiring to elicit the ability of the other in doing so.* (Z, TA.) — [And hence,] *He vied,* or *competed, and contended for superiority, with him,* (Ṣ, Ḳ,* mentioned in the Ḳ in art. حدى,) *in an action* or *a work* [*of any kind*]. (Ṣ.) You say, تَحَدَّيْتُ النَّاسَ القُرْآنَ *I sought to elicit the ability of the people in order that it might be known which of us was most skilled in reading* or *reciting* [*the Ḳur-án*]: it means like as does the saying of a person contending with others for the superior glory of his people, "Bring ye a people like my people, or like one of them." (Mṣb.) It is said of Mohammad, in a trad., تَحَدَّى العَرَبَ بِالقُرْآنِ [*He vied,* or *contended, with the Arabs by means of the Ḳur-án*]. (TA.) And one says, تَحَدَّى صَاحِبَهُ القِرَآءَةَ and الصِّرَاعَ, [*He vied,* or *contended, with his companion in reading,* or *reciting,* and *in wrestling,*] in order that it might be seen which of them was the superior reader or reciter, and the superior wrestler. (TA.) — Also *He aimed at it; made it his object; sought, endeavoured after, pursued,* or *endeavoured to reach* or *attain* or *obtain, it; intended* or *purposed it;* (AA, and Ḳ* and TA in art. حدى;) as also † حَادَاهُ, (AA, TA,) and † احداهُ. (Ṣgh, and Ḳ ib.) Hence the saying of Mujáhid, كُنْتُ أَتَحَدَّى القُرْآنَ فَأَقْرَأُ [*I used to aim at reading,* or *reciting, the Ḳur-án, and so to read,* or *recite*]. (AA, TA.)

6. تَحَادَتِ الإِبِلُ *The camels urged on one another.* (Ḳ.)

8: see 1, in two places.

لَا أَفْعَلُهُ حَدَا الدَّهْرِ *I will not do it ever;* (Ḳ;) *as long as the night follows the day.* (TA. [See 1.])

حَدْوَآءُ *The north wind;* (Ṣ, Ḳ;) because it drives along the clouds: the masc. form, أَحْدَى, is not used. (Ṣ.)

حُدَآءٌ: see 1, in two places.

حُدُوٌّ: } see حِدَأَةٌ, in art. حدأ.
حُدَىٌّ: }
حُدَيَّةٌ: }

حُدَيَّا *A vying,* or *competition, and contention for superiority.* (Ḳ. [There mentioned in art. حدى; but belonging to the present art., (see 5,) like as حُجَيَّا belongs to art. حجو.]) — [*One who vies,* or *competes, and contends for superiority.*] You say, أَنَا حُدَيَّاكَ بِهٰذَا الأَمْرِ [*I am he who vies,* &c., *with thee in this affair*], meaning *come forth to me as an adversary, by thyself alone,* (T, Ṣ,* Ḳ,* TA,) *and compete,* or *contend, with me* [*in this affair*]. (T, TA.) — And هُوَ حُدَيَّاهُمْ *He aims at them; makes them his object; seeks, endeavours after, pursues,* or *endeavours to reach,* or *attain, them.* (TA.) — And هٰذَا حُدَيَّا هٰذَا *This is the like,* or *like in form, of this.* (Aṣ, TA.) — And حُدَيَّا النَّاسِ *One of the men* or *people.* (Kr, Ḳ.) — See also حِدَأَةٌ, in art. حدأ.)

حَدَّآءٌ: see what next follows.

حَادٍ *Driving* or *a driver* [of camels; or *urging* or *exciting* them, or *one who urges* or *excites* them, *by singing* to them: see 1]: (Mgh:) pl. حُدَاةٌ. (TA.) You say رَجُلٌ حَادٍ and † حَدَّآءٌ [which latter is an intensive epithet]. (Ḳ.) — It is also applied to a [wild] he-ass, as meaning *Driving before him his she-asses.* (Ṣ,* TA.) He is said to be حَادِى ثَلَاثٍ [*A driver before him of three* she-asses], (Ṣ, TA,) and حَادِى ثَمَانٍ [*a driver before him of eight* she-asses]. (TA.) — حَادِى النَّجْمِ [lit. *The driver,* or *urger, of the asterism*] means الدَّبَرَانُ [i. e. †*the Hyades;* or *the five chief stars thereof;* or *the brightest star thereof, α of Taurus*]: (TA:) and so حَادِى النُّجُومِ [lit. *the driver,* or *urger, of the stars*]. (Ṣ voce مِجْدَحٌ.) — الحَوَادِى [pl. of the fem. الحَادِيَةُ] means †*The hind legs;* because they follow the fore legs. (Ḳ.) And †*The latter* or *hinder,* or *the last;* or *hindmost, parts* or *portions* of anything. (Az, TA.) — حَادٍ is also the act. part. n. of حَدَا as syn. with تَحَدَّى; and thus means *Aiming* at a thing; &c. (AA, TA.) — حَادِىَ عَشَرَ, &c.: see art. وحد.

إِحْدَى fem. of أَحَدٌ: see art. احد.

أُحْدُوَّةٌ and بَيْنَهُمْ أُحْدِيَّةٌ *Among them is in use a certain kind of* حُدَآء. (Lḥ, Ḳ.)

حدى

For several words mentioned in the Ḳ under this head, see art. حدو.

حذ

1. حَذَّ, aor. ـُ, (L, Mṣb,) inf. n. حَذٌّ, (L, Mṣb, Ḳ,) *He cut,* or *cut off,* a thing: (Mṣb:) or *cut,* or *cut off, quickly;* (IDrd, A, L;) as also هَذَّ: (A:) or *cut off quickly and utterly:* (L:) حَذٌّ is *syn. with* جَذٌّ. (Ḳ.)

حَذٌّ † *Quickness in speech and actions.* (TA.)

حُذَّةٌ *A piece of flesh-meat;* (A,* L, Ḳ;) as also حُزَّةٌ. (A, L.)

حَذَذٌ, an inf. n. having no verb, (Az, L,) *Lightness of the tail:* (Ṣ, L, Ḳ:) *lightness of the hair of the tail of* a horse: *shortness of the tail of* an ass; and *of* a bird of the kind called قَطًا; or *lightness of the feathers of the tail* thereof; or †*lightness and quickness of the flight* thereof: (L:) and *lightness of the beard.* (L.) — † *Lightness of hand.* (Ṣ. [See أَحَذُّ.]) † *Quickness:* or *quickness and lightness* or *agility.* (L.)

حُذْحُذَةٌ and حُذْحُذٌ A *short* woman. (L.)

قَرَبٌ حَذْحَاذٌ † *A quick* [*night-journey to water*]; (Ṣ, L, Ḳ;) like حَثْحَاثٌ: (Ṣ, A, L:) and *far-extending,* or *distant;* as also † حُذَاحِذٌ. (L.)

And خِمْسٌ حَذْحَاذٌ (L) and † أَحَذُّ (Ḳ) [*A journey of five days whereof the second and third and fourth are without water*] *in which is no languor,* or *flagging;* (L;) and *quick.* (Ḳ.)

حُذَاحِذٌ: see what next precedes.

أَحَذُّ *Deprived* of an arm, or a hand, &c., *by amputation:* fem. حَذَّآءُ: [pl. حُذٌّ:] thus أَحَذُّ اليَدِ means *having the arm,* or *hand, amputated:* and [hence] † *not having the means of acquiring eminence,* or *nobility.* (L.) [Hence, also,] أَصُولُ بِيَدٍ حَذَّآءَ, in a trad. of 'Alee, means † *Shall I assault the enemy with a short arm, that will not attain to that which I desire?* or, accord. to one reading, it is جَذَّآءَ, meaning *amputated:* alluding to his companions' falling short of their duty, and keeping back from the war. (L in arts. حذ and جذ.) [Another meaning of أَحَذُّ اليَدِ will be found below.] — A *light-tailed* camel: (Ṣ, L:) a horse *light,* or *scanty, in the hair of the tail:* (A, L:) or *having it* (*the tail,* Mṣb) *cut off:* (A, Mṣb:) a *short-tailed* ass. (L.) قَطَاةٌ حَذَّآءُ *A bird of the kind called* قَطًا *having a light,* or *scanty, tail; having a tail of which the feathers have become light,* or *scanty;* (Ṣ, L;) *having few feathers in the tail:* (A:) or *short-tailed:* (L:) or † *light and quick in flight.* (A,* L.) And لِحْيَةٌ حَذَّآءُ *A light,* or *scanty, beard.* (L.) — *Lean, lank,* or *light of flesh;* or *slender;* or *lank in the belly.* (Ḳ.) — *Smooth;* (Kh, Mṣb;) a thing *to which nothing clings,* or *attaches.* (Mṣb,* TA.) وَلَّتِ الدُّنْيَا حَذَّآءَ, in a خُطْبَة of 'Otbeh Ibn-Ghazwán, means ‡ *The world hath retired, passing away quickly,* (A,* L, Ḳ,*) *nothing clinging to it* [*so as to retard it*]: (L, Ḳ:) or *the people thereof not clinging to aught of it* [*so as to retard it*]: (A:) or *quickly; its latter part being cut off.* (Az, L.) — A sword *quick in cutting.* (A.) — ‡ *Light-handed;* (Ṣ, A, L, Ḳ;) *quick-handed;* (L;) i. e. *thievish;* or *quick in* *getting*, or *attaining:* (A:) [and so أَحَذُّ اليَدِ, as is implied in the Ṣ and L.] El-Farezdaḳ uses the expression أَحَذُّ يَدِ القَمِيصِ, (Ṣ, L,) meaning thereby أَحَذُّ اليَدِ. (L.) نَاقَةٌ حَذَّآءُ † *A she-camel quick in pace.* (A.) — قَلْبٌ أَحَذُّ † *A sharp, quick, active mind.* (L.) — سَيْرٌ أَحَذُّ ‡ *A very quick, unpleasant journey* or *pace.* (A.) See also حَذْحَاذٌ. — أَمْرٌ أَحَذُّ ‡ *An affair,* or *event, quick in passing:* (L:) or *decisive and quick:* (TA:) or *severe and abominable;* (A, L, Ḳ;) *without a parallel:* (A:) or *as though it eluded every one, so that he could not attain to it, and was not sufficient for it:* (A:) pl. حُذٌّ. (L, Ḳ.) — يَمِينٌ حَذَّآءُ ‡ *An oath which a man takes quickly:* (Ṣ, Ḳ:) or *an abominable, severe oath, whereby one decides a right,* or *due:* (A, L:) some say جَذَّآءُ. (Ṣ.) It is said in a prov., تَزَبَّدَهَا حَذَّآءَ ‡ *He swallowed it* [i. e. *took it,* namely, an oath, *hastily,*] *like as one swallows butter.* (TA.) — عَزِيمَةٌ حَذَّآءُ ‡ *An energetic and effective resolution, from which one does not turn to anything.* (A.) — حَاجَةٌ حَذَّآءُ ‡ *A light want, quickly accomplished.* (A, L.) — رَحِمٌ حَذَّآءُ ‡ *A bond of relationship* [*cut,* or *severed,* or] *not made close by affection;* expl. by إِذَا لَمْ تُوصَلْ: (Fr, Ṣ, Ḳ:) as also جَذَّآءُ [q. v.]. (Fr, Ṣ.) — أَحَذُّ أَبَدُّ: see بَدٌّ.

حذر

1. حَذِرَ, aor. ـَ, inf. n. حَذَرٌ; and ▼ احتذر; (Mṣb, Ḳ;) *He was cautious, wary,* or *vigilant; was on his guard; took care;* (Ḳ, TA;) *was in a state of preparation;* (Mṣb;) *was in fear; feared.* (TA.) [You say, حَذِرَ مِنْ أَمْرٍ and ▼ احتذر منه *He was cautious, &c., of a thing,* or *an event.* And حَذِرَ عَلَيْهِ مِنْ كَذَا and ▼ احتذر *He was cautious, &c., for him, of such a thing.* And both verbs are also trans.: for you say,] حَذِرَهُ, (Ṣ, A, Mṣb,) aor. and inf. n. as above; (Ṣ;) and ▼ احتذرهُ, (TA,) and ▼ حَاذَرَهُ; (A;) *He was cautious of it; guarded,* or *was on his guard, against it;* (Ṣ, A;) *prepared, prepared himself,* or *was in a state of preparation, against it;* (TA;) *feared it.* (Mṣb, TA.) [And حَذِرَ أَنْ يَفْعَلَهُ and ▼ احتذر *He was cautious of doing it; or he feared doing it.*] And حَذِرَ الشَّىْءُ فَحَذِرَهُ *The thing was an object of fear, and so he feared it.* (Mṣb.) And حُذِرَ المَوْتُ [*Death was an object of fear*]: and ▼ حاذر المَوْتَ [*He feared death*]. (A.)

2. تَحْذِيرٌ [The *cautioning* another; *putting* him *on his guard; making* him *to be cautious* or *wary* or *vigilant, to be on his guard, to take care,* or *to be in a state of preparation;*] the *making to fear,* or *be in fear.* (Ṣ, TA.) [You say, حَذَّرَهُ مِنْ أَمْرٍ *He cautioned him against a thing.* And the verb is also doubly trans.: you say,] حَذَّرَهُ الأَمْرَ [*He cautioned him against,* or *made him to fear, the thing,* or *event*]. (TA.) And أُحَذِّرُكَهُ [*I caution thee against him,* or *it*]. (Ḳ.) And it is said in the Ḳur [iii. 27 and 28], يُحَذِّرُكُمُ اللّٰهُ نَفْسَهُ *God maketh you to fear Himself.* (TA.)

3. حَاذَرَا, (TḲ,) inf. n. مُحَاذَرَةٌ (Ṣ, Ḳ) and حِذَارٌ, (Ṣ,) *They two were cautious,* or *in fear, each of the other; were on their guard,* or *in a state of preparation, each against the other.* (TḲ.) حِذَارٌ is *syn. with* مُحَاذَرَةٌ, (Ṣ,) and مُحَاذَرَةٌ is between two. (Ḳ.) — See also 1, in two places.

8: see 1, in five places.

11. احذارّ *He was angry,* (Ḳ,) *and prepared himself to do mischief,* (TA,) *and drew himself together* (تَقَبَّضَ): so in some copies of the Ḳ and in other lexicons: or *became enraged* (تَغَيَّظَ): so in other copies of the Ḳ. (TA.)

حِذْرٌ: see what next follows, in two places.

حَذَرٌ and ▼ حِذْرٌ, (Ṣ, A, Mṣb, Ḳ,) the former an inf. n., (Ṣ, Mṣb,) and the latter a simple subst., (Mṣb,) *Caution, wariness, vigilance, guard,* or *care;* (Ṣ, A, Ḳ;) as also ▼ مَحْذُورَةٌ; (Ḳ;) or *a state of preparation;* (Mṣb;) or *fear;* (Mgh,* TA;) and so ▼ مَحْذُورَةٌ: (Ṣ, Mṣb, Ḳ:) [pl. أَحْذَارٌ.] You say, أَخَذَ حِذْرَهُ ▼ *He took care; was cautious,* or *vigilant.* (Bḍ in iv. 73 and 103.) And هُوَ ٱبْنُ أَحْذَارٍ ‡ *He is a son* [i. e. *a person*] *of resolution,* or *determination, and caution,* or *wariness.* (Ṣ, Ḳ.) And حَذَرَكَ زَيْدًا: see حَذَارِ. — أَبُو حَذَرٍ *The male chameleon:* (Ḳ:) because of its frequent changes. (TA.)

حَذُرٌ, and its pl.: see what next follows.

حَذِرٌ (Ṣ, A, Mṣb, Ḳ) and ▼ حَذُرٌ (Ṣ, A, Ḳ) *Cautious; wary; vigilant; on his guard; careful;* (Ṣ;) or *in a state of preparation;* (Mṣb;) as also ▼ حَاذِرٌ; (Ṣ, A, Mṣb;) and *fearful:* (Ṣ:) or *very cautious* or *wary* or *vigilant* or *careful;* as also ▼ حَاذُورَةٌ (Ḳ) and ▼ حِذْرِيَانٌ: (A, Ḳ:) or this last signifies *very fearful and cautious* &c.: (Ṣ:) pl. of the first حَذِرُونَ and حَذَارَى. (Ṣ, Ḳ.) Sb cites, as an ex. of حَذِرٌ used transitively,

* حَذِرٌ أُمُورًا لَا تُخَافُ وَآمِنٌ *
* مَا لَيْسَ مُنْجِيهِ مِنَ الأَقْدَارِ *

[*Cautious,* or *very cautious, of things not to be feared, and trusting in that which will not save him from the decrees of destiny*]: but this is extr.; for an epithet of the measure فَعِلٌ is not [regularly] trans., so as to govern an objective complement. (Ṣ, TA.) In the Ḳur xxvi. 56, some read ▼ حَاذِرُونَ; and some, حَذِرُونَ; and ▼ حَذُرُونَ: حاذرون signifying *in a state of preparation;* (Zj, Ṣ;) or *fully equipped with arms:* (Sh:) and حذرون, *in a state of fear;* (Ṣ;) or *in a state of preparation;* (TA;) or *in a state of preparation with the accoutrements of war;* (Ibn-Mes'ood;) or *cautious,* or *vigilant.* (Zj.)

حِذْرِيَةٌ *A rugged piece of ground:* (Ṣ, Ḳ:) or the *top of a mountain, when it is hard and rugged, but level:* (Aboo-Kheyreh:) and *rough ground:* (TA:) and *a rugged* [*hill such as is termed*] أَكَمَةٌ; as also ▼ حِذْرِيَآءُ: (Ḳ:) pl. حَذَارَى and حَذَارٍ. (Ṣ.) ═ Also *The* عِفْرِيَة [or *feathers of the back of the neck*] *of a cock:* (Ṣ, Ḳ:) pl. as above. (Ḳ.)

حِذْرِيَآءُ: see what next precedes.

حِذْرِيَانٌ: see حَذِرٌ.

الحُذُرَّى, [like البُذُرَّى,] *What is false, vain,* or *ineffectual;* syn. البَاطِلُ. (Ḳ.)

حَذَارِ [an imperative verbal noun] meaning *Beware; be cautious, wary,* or *on thy guard;* or *take care.* (Ṣ, A, Ḳ.) The poet (Abu-n-Nejm, TA) says,

* حَذَارِ مِنْ أَرْمَاحِنَا حَذَارِ *

Beware of our spears: beware. (Ṣ, A.) And you say, سُمِعَتْ حَذَارِ فِى عَسْكَرِهِمْ [The cry "*Beware*" *was heard in their army*]. (TA.) When the word is repeated, the second is sometimes with tenween: (Ḳ:) but this is only in poetry, when required by the metre, as in the following verse, cited by Lḥ:

* حَذَارِ حَذَارٍ مِنْ فَوَارِسِ دَارِمِ *
* أَبَا خَالِدٍ مِنْ قَبْلِ أَنْ تَتَنَدَّمَا *

[*Beware, beware of the horsemen of Dárim, O Aboo-Khálid, before that thou repent*]. (TA.) You say also, حَذَارِكَ, [with the ك of allocution,] meaning *Beware thou.* (Lḥ, TA.) And when you caution a person [against another], ▼ حَذَرَكَ زَيْدًا [*Beware thou of Zeyd*]; (Ḳ,* TA;) and حَذَارَيْكَ زَيْدًا, (Ḳ,) which latter means *Beware thou, and beware thou again, of Zeyd:* both then [also] being verbal nouns. (TA.)

حَذِيرٌ [*A cautioner*]. You say, أَنَا حَذِيرُكَ مِنْهُ i. e. مُحَذِّرُكَ [*I am thy cautioner against him,* or *it*]; (TA;) or أُحَذِّرُكَهُ [*I caution thee against him,* or *it*]: (Ḳ:) known to Aṣ as heard only from Lth. (TA.) [See also what next follows.]

حُذَارِيَاتٌ *Persons who make others to fear:* (Ḳ:) or rather, as others than F explain it, مُنْذِرُونَ [*cautioners,* or *warners,* &c.]. (TA.)

حَاذِرٌ and حَاذِرُونَ: see حَذِرٌ.

حَاذُورَةٌ: see حَذِرٌ.

أَحْذَرُ [*More, and most, cautious, wary, vigilant, careful,* or *fearful*]. You say, أَحْذَرُ مِنَ الغُرَابِ *More fearful* [or *cautious,* &c.] *than the raven:* a prov. (Mgh.)

مَحْذُورٌ A thing *that is feared.* (Mṣb.) One says, وَقَاكَ اللّٰهُ كُلَّ مَحْذُورٍ [*May God preserve thee from everything that is feared*]. (A.)

مَحْذُورَةٌ *A calamity that is feared,* or *regarded with caution:* (Ḳ:) or *a troop of horse making a hostile attack,* or *incursion, upon a people:* or *i. q.* صَيْحَةٌ [app. as meaning *a hostile attack,* or *incursion, when it comes upon a tribe suddenly,* or *unexpectedly;* or it may here mean *a punishment,* or *chastisement;* or *a crying-out,* which is the primary signification]: (A:) and *war.* (Ḳ.) — See also حَذَرٌ, in two places.

حذف

1. حَذَفَهُ, (Ṣ,* Mgh, Mṣb, Ḳ, &c.,) aor. ـِ, (Mgh, Mṣb, Ḳ,) inf. n. حَذْفٌ, (Ṣ, Mgh, Mṣb,) *He cut*

it off: (Mgh, Mṣb:) or he cut it [so as to lessen it] at its extremity; (TA;) he cut off somewhat from the extremity of it; he curtailed it; as, for instance, the tail of a beast: (Lth, TA:) and he made it to fall; dropped it; rejected it. (Ṣ, Mgh, Mṣb, Ḳ.) One says, حَذَفْتُ مِنْ شَعَرِى, (Ṣ,) or مِنْ شَعَرِهِ, (Mṣb, Ḳ,*) and مِنْ ذَنَبِ الدَّابَّةِ, (Ṣ, Mṣb,) [شَيْئًا being understood,] I took, or cut off, [somewhat] from my hair, [or his hair,] and from the tail of the beast; (Ṣ, Ḳ;*) I clipped it. (Mṣb.) And حَذَفَ الشَّعَرَ [He clipped the hair]: said of a cupper. (TA.) And ↓ احتذف الثَّوْبَ He cut off a piece from the garment, or cloth. (TA.) And حَذَفْتُ رَأْسَهُ بِالسَّيْفِ I cut off a portion of his head with the sword: (IF, Mṣb:) I struck his head with the sword and cut off a portion of it. (Ṣ.) — Also, inf. n. as above, †He elided it, struck it off or out, or rejected it; namely, a letter, [and a syllable,] from a word: (MA, PṢ:) he omitted it. (MA.) [† He suppressed it; namely, a word of a proposition or sentence.] And حَذَفَ السَّلَامَ, (Ḳ,) inf. n. as above, (TA,) ‡ He made the salutation to be light [of utterance], and concise; (Ḳ, TA;) i. e., the salutation in prayer. (TA.) And حَذَفَ فِى قَوْلِهِ, (Mṣb,) and فِى الأَذَانِ, and القِرَآءَةِ, aor. and inf. n. as above, (Mgh,) ‡ He was concise, (Mgh, Mṣb,) and quick, in his saying, (Mṣb,) and in the call to prayer, and the recitation, or reading. (Mgh.) — حَذَفَهُ بِالعَصَا He struck him, or beat him, with the staff, or stick: (TA:) and he cast, or threw, at him the staff, or stick. (Ṣ, Ḳ, TA.) It is said in a prov. of the Arabs, mentioned by Sb, إِيَّاكُمْ وَأَنْ يَحْذِفَ أَحَدُكُمُ الأَرْنَبَ, i. e. [Beware ye] lest any one of you cast at, or shoot, the hare: because this animal is of evil omen. (TA. [But the reading there given is ابّاى: an evident mistranscription.]) Or حَذَفَهُ, inf. n. as above, signifies He struck, or he cast at, or shot, him, or it, from one side. (Lth, TA.) — [Hence,] حَذَفَ فُلَانًا بِجَائِزَةٍ † He gave such a one a gift. (Z, Ḳ.) — And حَذَفَ بِهَا † He broke wind. (Ibn-'Abbád, TA.) — حَذَفَ فِى مِشْيَتِهِ He moved about his side and his hinder parts (in the CḲ he moved about his hinder parts and his shoulder-joint) in his gait: and (in the CḲ "or") he went with short steps. (En-Naḍr, Ḳ, TA.)

2. حذّفهُ, [inf. n. تَحْذِيفٌ,] He clipped it much: he took, or cut off, from its lateral parts, whatever it was, so as to make it even: (Mṣb:) he (a workman, or an artificer,) made it (a thing) becomingly even; as though he cut off from it whatever required to be cut off, so that it became free from everything unseemly, and was nicely, neatly, or properly, trimmed. (A, TA.) تَحْذِيفُ الشَّعَرِ [or الرَّأْسِ] signifies The cutting of the hair so as to form a طُرَّة [q. v.], by taking from its sides so as to make it even [with the cut portion over the forehead]; (T, Mgh;) as is done by, or to, a girl: (Mgh:) or تحذيف الرأس is a custom of women, consisting in the removing of the hair from [the sides of] the head as far as a line upon the side of the face made by putting one end of a string, or thread, upon the top of the ear, and the other end upon the angle of the جَبِين [or part above the temple]: (Mṣb:) accord. to En-Naḍr, the تحذيف of the طُرَّة is the making a [طُرَّة such as is termed] سُكَيْنِيَّة, [i. e., after the fashion of Sukeyneh the daughter of El-Hoseyn, as is shown in the Ṣ and Ḳ &c. in art. سكن,] like as do the Christians. (L, TA.) — Also, inf. n. as above, ‡ He prepared it; or put it into a right, or good, state; and made it; or made it skilfully, or well. (Ṣ, Ḳ, TA.)

8: see 1.

حَذَفٌ Small, black sheep or goats, (Ṣ, Mṣb, Ḳ,) of El-Hijáz; (Ṣ, Ḳ;) or of Jurash, (ISh, Ḳ, TA,) of El-Yemen, small, with short, or short and fine, wool or hair, (ISh, TA,) without tails and without ears: (ISh, Ḳ, TA:) or the young ones of sheep or goats, in general: and metaphorically applied to ‡ gazelles: (TA:) n. un. with ة. (Ṣ, Mṣb.) — A certain kind of bird: (Ṣgh, Ḳ:) or small بَطّ [or ducks]: (Ḳ:) like [or likened to] the sheep, or goats, thus called: it is said by IDrd to be not a genuine Arabic word. (TA.) — The small زَاغ [or rook], which is eaten; (Lth, Ḳ;) the small black birds of the crow-kind, called زِيغَان [pl. of زاغ], which are eaten: n. un. with ة. (ISh, TA.) — The leaves of seed-produce, (L,) or of grain. (O, Ḳ.)

حذفة A piece cut off from a garment. (L, TA.)

حُذُفَّةٌ Short: applied to a woman: (Ṣgh, Ḳ:) and to a ewe. (Ṣgh.)

أُذُنٌ حُذَفَةٌ An ear that is as though it were clipped, or cut off. (Ḳ, TA.)

حُذَافَةٌ Clippings, or what one cuts off, of a hide, (Lh, Ṣ, Ṣgh, Ḳ,) &c.: (Ṣ, Ṣgh, Ḳ:) or what is cut off, of a thing, and thrown away. (TA.) — [Hence,] one says, مَا فِى رَحْلِهِ حُذَافَةٌ ‡ There is not in his travelling-utensils any food: (Ṣ, Ṣgh, Ḳ:) or any small quantity of food &c. (Z, TA.) And أَكَلَ الطَّعَامَ فَمَا تَرَكَ مِنْهُ حُذَافَةً † [He ate the food, and left not of it anything]. (ISk, Ṣ.) And احْتَمَلَ رَحْلَهُ فَمَا تَرَكَ مِنْهُ حُذَافَةً † [He took up and carried away his travelling-apparatus, and left not of it anything]. (ISk, Ṣ.) Accord. to the companions of A'Obeyd, the word is حذاقة, with ق; but this is disallowed by Sh; and is wrong. (Az, TA.)

الحُذَّافَةُ The anus, or the podex; syn. الإِسْتُ. (Ḳ.)

حَاذِفٌ [act. part. n. of 1]. — You say, هُمْ مَا بَيْنَ حَاذِفٍ وَقَاذِفٍ, i. e. [They are partly, or in part,] beating with the staff, or stick, and [partly, or in part,] pelting with stones; [or some beating &c., and the others pelting &c.] (TA in the present art.; and Ṣ and TA in art. قذف, but without مَا before بين.)

رَجُلٌ مُحَذَّفُ الكَلَامِ ‡ A man chastened, good, free from every fault, in respect of speech: and you say also مُحَذَّفَةُ الكلام; in which the ة is added to give intensiveness to the signification: the latter occurring in a trad. (TA.)

حذفر

حِذْفَارٌ (Ṣ, Ḳ) and ↓ حُذْفُورٌ (Ḳ) A side; or a lateral, or an adjacent, part; (Ṣ, Ḳ, TA;) of a thing: and the upper, or uppermost, part of a thing: (Ṣ:) pl. حَذَافِيرُ. (Ṣ, Ḳ.) You say, أَخَذَهُ بِحَذَافِيرِهِ and بِحِذْفَارِهِ and ↓ بِحُذْفُورِهِ He took it altogether: (Ḳ, TA:) or with its lateral parts: or with its higher, or highest, parts. (TA.) And أَعْطَاهُ الدُّنْيَا بِحَذَافِيرِهَا He gave him the world, or worldly goods, altogether: (Ṣ, TA:) or with the lateral parts thereof: or with the higher, or highest, parts thereof. (TA.) And جَآؤُوا بِحَذَافِيرِهِمْ They came altogether. (TA.)

حُذْفُورٌ: see above, in two places. — Also A numerous congregation or assemblage. (Ḳ.) — And An eminent, an elevated, or a noble, man: (Ḳ, TA:) pl. as above. (TA.) — And the pl., (as some say, TA,) Persons prepared for war. (Ḳ.) Whence the phrase, أُشْدُدْ حَذَافِيرَكَ Prepare thyself (Ḳ, TA) for war, &c. (TA.)

حذق

1. حَذَقَهُ, aor. ـِ, inf. n. حَذْقٌ (Ṣ, Ḳ, TA, in the CḲ حِذْق) and حَذَاقَةٌ, (Ḳ, TA, in the CḲ حِذاقة,) He cut it; (Ṣ, Ḳ;) namely, a rope, (Ṣ,) or a thing: (Ḳ:) or he stretched it, or extended it, to cut it with a reaping-hook and the like, (Ḳ, TA,) so that there remained not of it anything. (TA.) — حَذَقَ الرِّبَاطُ يَدَ الشَّاةِ The bond made an impression upon the fore leg of the sheep, or goat, (IDrd, Ḳ,) by cutting. (IDrd.) — حَذَقَ فَاهُ, (IDrd, Ṣ, Ḳ,) inf. n. حَذْقٌ, (Ṣ,) said of vinegar, (IDrd, Ṣ, Ḳ,) and of milk [when sour], and of the beverage called نَبِيذ, and the like, (TA,) ‡ It stung, bit, or burned, his mouth, by its strength and sharpness, (IDrd, Ṣ, Ḳ, TA,) and contracted it. (Ḳ.) — And حَذَقَ, (Ṣ, Mṣb, Ḳ,) aor. ـِ, (Mṣb, Ḳ,) inf. n. حُذُوقٌ, said of vinegar, (Ṣ, Mṣb, Ḳ,) and of milk, (TA,) ‡ It was, or became, sour, (Ṣ, Mṣb, Ḳ, TA,) in the utmost degree, (Mṣb,) so that it burned the tongue. (Mṣb, TA.) — حَذَقَ القُرْآنَ, and (so in the Ṣ, but in the Ḳ "or") العَمَلَ, aor. ـِ; and حَذِقَ, aor. ـَ; (Ṣ, Ḳ;) or حَذِقَ فِيهِ, and حَذَقَ; (TA;) inf. n. (of both, Ṣ) حِذْقٌ and (of the former, Ṣ) حَذَاقٌ and حِذَاقٌ and حَذَاقَةٌ (Ṣ, Ḳ) and حَذْقٌ and حِذَاقَةٌ; or ↓ this last is a simple subst.; (Ḳ;) ‡ He (a boy) was, or became, skilled in the Kur-án, and the work; (Ṣ;) or learned the whole of it, and was, or became, skilled in it: (Ḳ, TA:) from الحَذْقُ signifying "the act of cutting." (Z, TA.) You say, هٰذَا يَوْمُ حِذَاقِهِ This is the day of his finishing [the learning or reciting] of the Kur-án. (Ṣ, Ḳ.) And حَذَقَ فِى صَنْعَتِهِ, [and بِهَا also,] aor. ـِ; and حَذِقَ, aor. ـَ; † He was, or became, skilled in his art, or habitual work or occupation, and knew its abstrusities and niceties. (Mṣb.)

2. تَحْذِيقٌ, [inf. n. of حذّقهُ † He, or it, made, or rendered, him skilful,] from الحِذْقُ, rests upon analogy, not upon the authority of hearsay. (Mgh.)

4. احذقهُ † *It* (the heat) *rendered it sour, so that it burned the tongue;* namely, vinegar. (TA.)

5. تحذّق عَلَيْنَا † *He feigned,* or *made a show of, skilfulness to us.* (TA.) And ▼ حَذْلَقَ, (Ṣ, Ḳ, mentioned in the latter in art. حذلق,) with an augmentative ل, (Ṣ,) inf. n. حَذْلَقَةٌ, (A, TA,) ‡ *He feigned,* or *made a show of, skilfulness, and* [in some copies of the Ḳ "or"] *laid claim to more than he possessed;* as also ▼ تَحَذْلَقَ: (Ṣ, Ḳ, TA:) or ▼ حَذْلَقَةٌ signifies † the *employing oneself,* or *using art* or *artifice, with skilfulness, cleverness,* or *ingeniousness:* and تحذلق ▼ فِى كَلَامِهِ † *he feigned,* or *made a show of, skilfulness, cleverness,* or *ingeniousness, in his speech.* (L.) You say, ▼ فِيهِ حَذْلَقَةٌ and ▼ تَحَذْلُقٌ ‡ [*In him is a quality of feigning,* or *making a show of, skilfulness,* &c.]. (A, TA.)

7. انحذق *It* (a rope) *was,* or *became, cut.* (Ḳ, TA.) Hence the saying of the poet,

* يَكَادُ مِنْهُ نِيَاطُ القَلْبِ يَنْحَذِقُ *

[*The suspensory of the heart is near to becoming severed in consequence thereof*]. (TA.)

Q. Q. 1. حَذْلَقَ, inf. n. حَذْلَقَةٌ: see 5, in three places. — Also, [perhaps originally,] *It was,* or *became, sharpened.* (TA.)

Q. Q. 2. تَحَذْلَقَ: see 5, in three places.

حُذْقَةٌ *A piece,* or *portion cut off,* of a rope: pl. حِذَاقٌ and حُذَاقٌ; as in the phrase, تَرَكْتُ الحَبْلَ حِذَاقًا and حُذَاقًا [*I left the rope in pieces*]. (Ḳ.) [See also what next follows.]

حَذِيقٌ (Ṣ, Ḳ) and ▼ مَحْذُوقٌ (Ḳ) *Cut:* (Ṣ, Ḳ,* TA:) pl. أَحْذَاقٌ. (Lḥ, TA.) One says حَبْلٌ أَحْذَاقٌ *A rope altogether worn out; as though it were cut:* (Lḥ, Ḳ,* TA:) every part of it being termed حَذِيقٌ. (Lḥ, TA.)

مَا عِنْدَهُ حُذَاقَةٌ ‡ *He has not aught of food.* (Ḳ, TA.) [See also حُذَافَةٌ, with ف.]

حِذَاقَةٌ: see 1.

حُذَاقِىٌّ *Sharpened:* applied to a knife: (Ibn-'Abbád, Ḳ:) and ▼ حِذْلَاقٌ signifies the same, applied to a thing [of any kind]. (TA.) — See also حَاذِقٌ. — ‡ A man *chaste,* or *eloquent, of tongue;* (Ṣ, Ḳ,* TA;) *perspicuous in language.* (Ṣ.) — *A young ass;* syn. جَحْشٌ. (Ḳ.)

حِذْلَاقٌ † A *loquacious* man; *who affects to be commended for,* or *glories in, that which he does not possess.* (TA.)

حُذَاقِىٌّ: see حِذْلَاقٌ.

حَاذِقٌ *Cutting,* or *sharp:* (Ṣ, Ḳ:*) applied to a knife [&c.]. (Ṣ.) — ‡ Wine (شَرَاب) *that has attained to its full maturity* [*of strength, so that it stings, bites,* or *burns, the tongue;* see 1]: (AḤn, TA:) likewise applied to vinegar [in the same sense; or as meaning *sour,* or *sour in the utmost degree, so that it burns the tongue;* see again 1]; as also ▼ حُذَاقِىٌّ. (TA.) — ‡ *Skilled,* or *skilful, and thoroughly learned,* [*so as to know abstrusities and niceties,* (see 1,)] in an art, or a habitual work or occupation, and in the Ḳur-án: pl. حُذَّاقٌ. (TA.) You say, فُلَانٌ فِى صَنْعَتِهِ حَاذِقٌ بَاذِقٌ (Ṣ, TA) ‡ *Such a one is skilled,* or *skilful,* &c., *in his art,* or *habitual work* or *occupation;* (TA;) using the latter word as an imitative sequent. (Ṣ, TA.) — ‡ *Bad, evil, wicked, mischievous,* or *the like;* syn. خَبِيثٌ. (TA.)

حَذِيقٌ: see مَحْذُوقٌ.

مُتَحَذْلِقٌ *Feigning,* or *making a show of, skilfulness, cleverness,* or *ingeniousness:* or *desiring to exceed his due bounds.* (L.) [See 5.]

حذلق

For several words mentioned under this head in the Ḳ, see art. حذق.

حذم

1. حَذَمَهُ, (Ṣ, Mṣb, Ḳ,) aor. ـِ, (Mṣb, Ḳ,) inf. n. حَذْمٌ, (Ṣ, Mṣb,) *He cut it,* or *cut it off,* (Ṣ, Mṣb, Ḳ,) *in any manner:* (TA:) or *hastily,* or *quickly.* (Ḳ.) — And [*He did it quickly:* or] *he was quick in it;* [as also حَذَمَ فِيهِ;] i. e., in any action. (Ṣ, Mṣb.) You say, حَذَمَ فِى مَشْيِهِ, (Mṣb,) and فِى قِرَآءَتِهِ (Ṣ, Ḳ,) *He was quick* [*in his walking,* or *going,* and *in his reading,* or *reciting*]: (Ṣ, Mṣb, Ḳ:) and so in other things: (Ḳ:) by "other things" being here meant walking and the like; for حَذْمٌ, of which the verb is حَذَمَ, signifies the *being quick* in walking, app. *with a stretching out of the arms backwards:* (TA:) and *a light, an active,* or *an agile, walking.* (Ṣ, TA.) [See also حَذَمَانٌ, below.] Hence, (Mgh, Mṣb, TA,) 'Omar said, (Ṣ, TA,) to the مُؤَذِّن of Jerusalem, (TA,) إِذَا أَذَّنْتَ فَتَرَسَّلْ وَإِذَا أَقَمْتَ فَاحْذِمْ, (Ṣ, A, Mgh, Mṣb, TA,) i. e. [*When thou chantest the* أَذَان, *be moderate; not quick: and when thou chantest the* إِقَامَة,] *cut short the lengthening of thine utterance;* meaning, *be quick* in the اقامة: (Aṣ, Mgh, TA:) another reading is said to have been given by Z, namely, [فَاخْذِمْ,] with خ. (TA.)

حَذْمٌ The *flight of a bird that has its wings clipt,* (Ḳ, TA,) *as the pigeon and the like.* (TA.)

حِذْمٌ A *sharp,* or *cutting,* sword; as also ▼ حِذْيَمٌ, with kesr to the ح, (Ḳ,) and with fet-ḥ to the ى; (TA;) or ▼ حَذْيَمٌ. (So in two copies of the Ṣ.)

حُذَمٌ and ▼ حُذَمَةٌ *Short in stature and in step;* (Ḳ;) applied to a man: (TA:) and the latter is likewise fem.; (Ṣ, Ḳ, TA;) applied to a woman as meaning *short in stature.* (Ṣ, TA.)

حُذُمٌ *Swift* hares. (IAạr, Ḳ. [See also حُذَمَةٌ.]) — And *Skilful* thieves. (IAạr, Ḳ.)

حُذَمَةٌ; see حُذَمٌ. — One says of the female hare, حُذَمَةٌ لُذَمَةٌ تَسْبِقُ الجَمْعَ بِالأَكَمَةِ, i. e. *Swift, persistent in running, she outstrips the company* pursuing her *upon the hill.* (TA.)

حَذَمَانٌ *Quickness* in walking or going: (Ḳ:) accord. to Aboo-'Adnán, *a pace of the kind termed* زَمِيلٌ, *exceeding what is termed* مَشْىٌ. (TA.) — And *Slowness* (Ḳ, TA) in walking or going: so says Aboo-'Adnán on the authority of Khálid Ibn-Jembeh. (TA.) Thus it bears two contrary meanings. (Ḳ.)

حُذَامٌ an epithet applied to a slave: so in the saying, اِشْتَرَى عَبْدًا حُذَامَ المَشْىِ *He bought a slave slow* [*in gait*], *lazy,* (Ḳ, TA,) *devoid of good:* so says Khálid Ibn-Jembeh. (TA.)

حَذْيَمٌ: see حِذْيَمٌ.

حِذْيَمٌ *Skilful* (Ḳ, TA) in a thing. (TA.) — See also حِذْمٌ.

حذو

1. حَذَا النَّعْلَ, [aor. يَحْذُو,] inf. n. حَذْوٌ and حِذَآءٌ, *He measured the sandal,* or *sole,* or *made it according to a measure;* (Ḳ;) and *cut it* (T, Ḳ) *according to a pattern:* (T, TA:) or *he made the sandal,* or *sole,* لِى *for me.* (Mgh.) And حَذَا النَّعْلَ بِالمِثَالِ *He cut the sandal,* or *sole, by the pattern.* (Mgh.) And حَذَا النَّعْلَ بِالنَّعْلِ, (Ṣ, Mṣb, Ḳ,) inf. n. حَذْوٌ, (Ṣ,) *He measured the sandal,* or *sole, by the sandal,* or *sole;* or *made it according to the measure thereof;* (Ṣ, Mṣb, Ḳ;) and *cut it according to the pattern and measure thereof:* (Mṣb:) and in like manner, حَذَا القُذَّةَ بِالقُذَّةِ *He measured the feather for an arrow by the feather for an arrow;* or *made it according to the measure thereof.* (Ḳ.) Hence the prov., حَذْوَ القُذَّةِ بِالقُذَّةِ (TA,) [meaning † *With like for like*]. (Ṣ, TA.) And the saying, جَزَيْتُهُ حَذْوَ النَّعْلِ بِالنَّعْلِ, meaning † *I requited him* [*with like for like*]. (Ḥar p. 43. [See also Freytag's Arab. Prov. i. 345.]) One says also, هُوَ جَيِّدُ الحِذَآءِ *He is good in respect of proportion,* or *conformation.* (TA.) And [in like manner] a beast is said to be حَسَنُ الحِذَآءِ *Goodly,* or *beautiful, in respect of proportion,* or *conformation.* (TA in art. حذى.) — [And hence,] حَذَا حَذْوَ زَيْدٍ † [*He did as Zeyd did;*] *he did the deed of Zeyd.* (Ḳ.) And ▼ احتذى مِثَالَهُ, (Ṣ, Ḳ,) or عَلَى مِثَالِهِ, (T, TA,) ‡ *He followed,* or *imitated, his example* (T, Ṣ, Ḳ, TA) *in his affair,* or *case:* (TA:) or احتذى بِهِ *he followed,* or *imitated, his example in his affairs:* (Mṣb:) syn. اقتدى. (T, Ṣ, Mṣb, Ḳ.) — See also 3, in three places. — حَذَاهُ نَعْلًا *He put on him* (namely a man [i. e. *on his foot*]) *a sandal;* as also ▼ احذاه: (Ḳ:) the former is authorized by Aṣ, but the latter is disallowed by him: Az explains the former by حمله على نعل [as though meaning *he gave him a sandal upon which to walk;* like as حَمَلَهُ عَلَى دَابَّةٍ for means "he gave him a beast upon which to ride"]: and accord. to J, (TA,) أَحْذَيْتُهُ نَعْلًا signifies *I gave him a sandal.* (Ṣ, TA.) — حَذَا زَيْدًا, (Ḳ,) inf. n. حَذْوٌ, (TA,) *He gave to Zeyd.* (Ḳ, TA. [See also 4 in art. حذى.]) — حَذَا مِنْهُ حِذْوَةً *He cut off from it a piece* of flesh-meat. (TA.) — حَذَا الجِلْدَ, aor. يَحْذُو, *i. q.* قَوَّرَهُ [*He cut a piece out of the skin,* generally meaning *in a round form*]. (TA.) — حَذَا لِسَانَهُ, (Ḳ,) aor.

as above, inf. n. حَذْوٌ, (TA,) ‡ *It* (wine, or beverage,) *bit his tongue;* (AHn, Ḳ, TA;) a dial. var. of حَذَى, aor. يَحْذِى, which is the word well known. (AHn, TA.) —— حَذَا التُّرَابَ فِى وُجُوهِهِمْ i. q. حَثَاهُ [i. e. *He poured with his hand, threw,* or *cast, the dust in their faces*]. (IAth, Ḳ.)

3. حَاذَاهُ, (Ṣ, Mgh, Mṣb, Ḳ,) inf. n. مُحَاذَاةٌ (Mṣb, TA) and حِذَآءٌ, (Mṣb,) *He,* (Mgh, Mṣb,) and *it,* (Mgh,) *was,* or *became, over against,* or *opposite to, him,* or *it;* (Ṣ, Mgh, TA;) *faced,* or *fronted, him,* or *it;* (TA;) syn. آزَاهُ, (Mṣb,* Ḳ,) and قَابَلَهُ; (TA;) as also ▼حَذَاهُ, aor. يَحْذُو, (Mgh, Mṣb,) inf. n. حَذْوٌ: (Mṣb:) [and] the latter, (Ṣ,) or each, (Ḥar p. 43,) signifies *he sat over against,* or *opposite to, him,* or *it.* (Ṣ, and Ḥar ibid.) مَا يَحْذُو ▼رَأْسَهَا مِنَ الشَّعَرِ وَلَا يَسْتَرْسِلُ means مَا يُحَاذِيهِ [i. e. *What is against,* or *opposite to, her head, of the hair, and does not hang down* below it]. (Mgh. [See another ex. in an explanation of آخِرَةٌ.]) One also says, أَتَيْتُ أَرْضًا قَدْ حَذَا ▼بَقْلُهَا عَلَى أَفْوَاهِ غَنَمِهَا *I came to a land the herbs of which were opposite to the mouths of its sheep,* or *goats, not rising beyond them.* (Sh, TA.) And تَحْذُ ▼بِحِذَآءِ هٰذِهِ الشَّجَرَةِ *Be thou over against,* or *opposite to, this tree.* (TA.) —— حَاذَتْهَا, said of a girl, *She matched her,* namely, another girl; *she was,* or *became, her match, fellow,* or *equal;* syn. تَارَبَتْهَا. (A and TA in art. ترب.) —— [Hence مُحَاذَاةٌ signifying *A conformity, a mutual resemblance,* or *a correspondence, with regard to sound, of two words occurring near together;* like اِزْدِوَاجٌ &c.: see art. زوج.]

4: see 1, in two places. —— احذاهُ also signifies *He gave him* a thing. (TA.) [See حِذْوَةٌ; and see 4 in art. حذى.]

5: see 3.

[6. تَحَاذَوْا *They were,* or *became, over against,* or *opposite to, each other; they faced,* or *fronted, each other.* —— And *They matched each other; each of them was,* or *became, the match, fellow,* or *equal, of the other.*] —— التَّحَاذِى in selling and buying: see التَّرَاوُضُ. [This, perhaps, may belong to art. حذى.]

8. احتذى *He wore,* or *put on, a sandal,* or *sandals.* (Ṣ, TA.) Hence the saying, خَيْرُ مَنِ احْتَذَى النِّعَالَ [*The best of those who have worn sandals*]. (TA.) A rájiz says,

* كُلُّ الحِذَآءِ يَحْتَذِى الحَافِى الوَقِعْ *

[*The barefooted whose sole is hurt by the rugged ground and stones will put on any sandal:* a prov.]. (Ṣ. [See Freytag's Arab. Prov. ii. 317.]) —— See also 1.

10. استحذاهُ *He asked him to give him a sandal* [or *a pair of sandals*]. (Ṣ,* TA.)

حِذَةٌ: see حِذَآءٌ, in three places. —— You say also, جَآءَ الرَّجُلَانِ حِذَتَيْنِ *The two men came together, side by side.* (TA.) [See also حِذْيَةٌ, in art. حذى.]

حَذْوٌ [originally an inf. n. of 1, q. v.]: see حِذَآءٌ, in three places.

حِذْوَةٌ: see حِذَآءٌ: —— and see also حُذَاوَةٌ.

حِذْوَةٌ *A gift;* (Ḳ, and Ḥam p. 596;) as also ▼حُذْيَا; (TA;) or حُذْيَا: (Mgh, and Ḥam ubi suprà:) or *a portion that is given of spoil.* (Ṣ.) [See also حُذْيَا, in art. حذى.] —— And *A piece,* (Ḳ,) or *small piece,* (TA,) *of flesh-meat:* (Ḳ, TA:) or *a piece of flesh cut lengthwise;* or so حِذْيَةٌ, accord. to some. (Mgh.) [See also حِذْيَةٌ, in art. حذى.] —— See also حِذَآءٌ, in three places.

حِذَآءٌ *A sandal,* or *sole;* syn. نَعْلٌ; (Ṣ, Mṣb, TA;) vulgarly called حذوة: (TA [there written without any syll. signs]:) pl. أَحْذِيَةٌ. (Mṣb.) —— † The *sole* of a camel's foot and of the hoof of a horse (Ṣ, Mṣb, TA) or similar beast; (Mṣb;) as being likened to a نَعْل. (TA.) Hence the saying in a trad., (Ṣ,* Mṣb,* TA,) respecting a stray she-camel, (Mṣb, TA,) مَعَهَا حِذَاؤُهَا وَسِقَاؤُهَا (Ṣ, Mṣb, TA) † *With her are her foot,* with which she may defend herself from the smaller beasts of prey, *and* [*her stomach,* which, by her replenishing it, affords her means of long] endurance of the want of water: (Mṣb:) [or] it means that she has ability to traverse the land and to go to water. (TA.) —— †*A wife;* because she is مَوْطُوءَةٌ, like the نَعْل. (TA.) —— Also, [originally an inf. n.,] i. q. إِزَآءٌ [The *front,* as meaning the *part, place,* or *location, that is over against, opposite, facing, fronting,* or *in front*]. (Ṣ, Ḳ.) One says, جَلَسَ بِحِذَائِهِ [*He sat in the place over against, opposite to, facing,* or *fronting, him;* or simply *he sat over against,* &c.]. (Ṣ.) And هُوَ حِذَآءَكَ and ▼حِذْوَتَكَ and ▼حِذَتَكَ [in the CḲ, erroneously, حِذْيَتَكَ and حَذْوَتَكَ] and ▼مُحَاذَاتَكَ [*He, or it, is over against, opposite to, facing,* or *fronting, thee*]; (Ḳ;) and ▼هُوَ حُذَيَّاكَ; i. e. بِإِزَائِكَ. (Ḳ in art. حذى.) And دَارِى حِذَآءَ دَارِهِ (Ṣ) and ▼حِذْوَةَ دارِه (Ṣ, Ḳ) and ▼حُذْوَةَ داره (Ṣ) and ▼حِذَةَ داره (Ṣ, Ḳ) and ▼حَذْوَ داره and ▼حِذْوَةَ داره and ▼حَذْوَ داره (Ḳ) i. e. إِزَآءَهَا [*My house is over against,* &c., *his house*]; (Ḳ;) and [so] بِحِذَآءِ داره. (Mṣb.) And ▼رَفَعَ يَدَيْهِ حَذْوَ أُذُنَيْهِ and حِذَآءَ اذنيه [*He raised his hands over against,* or *opposite to, not higher than, his ears*]: (Mgh,* Mṣb:) both are correct expressions. (Mgh.) [Hence, عَلَى حِذَآءٍ وَاحِدٍ *Corresponding to,* or *matching, one another; uniformly disposed.*] —— See also art. حذى.

حُذَاوَةٌ and حِذَاوَةٌ and ▼حُذْوَةٌ *What falls, and is thrown away, of skins, when they are pared and cut:* whence the saying in a trad. respecting the bridal furniture of Fáṭimeh, that one of her two beds was stuffed with حذوة of the makers of sandals. (TA.)

حُذْيَا [perhaps belonging to this art., like as حُجْيَا and حُذْيَا belong to arts. حجو and حدو]: see حِذْوَةٌ, and art. حذى: —— and see also حِذَآءٌ.

حَذَّآءٌ *A maker of sandals:* whence the prov.,

* مَنْ يَكُ حَذَّآءً تَجِدْ نَعْلَاهُ *

[*He who is a maker of sandals, his pair of sandals is good*]. (TA. [Freytag (Arab. Prov. ii. 665) gives it thus: مَنْ يَكُنْ أَبُوهُ حَذَّآءً تَجِدْ نَعْلَاهُ *He whose father is a maker of sandals, his pair of sandals is good.*])

حَاذٍ A man *wearing a sandal* [or *a pair of sandals*]. (TA.)

هُوَ مُحَاذَاتَكَ: see حِذَآءٌ.

حذى

1. حَذَتِ الشَّفْرَةُ النَّعْلَ, [aor. and inf. n. as below,] *The shoemaker's knife cut the sandal,* or *sole.* (Ṣ.) And حَذَى يَدَهُ *He cut,* or *cut off, his arm,* or *hand,* (T, Ṣ, Ḳ,) *with a knife.* (Ṣ.) And حَذَى أُذُنَهُ *He cut off a piece of his ear.* (TA.) —— حَذَى الجِلْدَ, aor. يَحْذِى, *He,* or *it, wounded the skin.* (TA.) And حَذَى الإِهَابَ, (Ḳ,) inf. n. حَذْىٌ, (TA,) *He rent,* or *tore, much,* or *he made many holes in, the skin,* or *hide.* (Ḳ, TA.) —— حَذَى لِسَانَهُ, (Ṣ, Mgh, Ḳ,) and فَاهُ (Ṣ) or فَمَهُ, (TA,) aor. as above, (Ṣ, Mgh, Mṣb,) and so the inf. n., (Ṣ, TA,) ‡ *It,* namely, vinegar, and wine, or beverage, (Ṣ, Mgh,) and [sour] milk, (Mgh, Ḳ,) &c., (Ḳ,) *bit his tongue,* (Ṣ, Mgh, Ḳ,) and *his mouth;* (Ṣ, TA;) *had a burning effect upon it, like cutting.* (Mgh.) —— حَذَى فُلَانًا بِلِسَانِهِ ‡ *He spoke evil of such a one, saying of him what would grieve him if he heard it; reviled, vilified,* or *vituperated, him;* (Ḳ, TA;) [*as though*] *he cut him with his tongue.* (TA.) A man who does so is termed ▼مِحْذَآءٌ. (Ḳ, TA.) —— See also what next follows.

4. احذاهُ; (Ṣ, Mgh, Ḳ;) as also ▼حَذَاهُ, first pers. حَذَيْتُ; (Mgh;) *He gave him a thing,* or *somewhat:* (Mgh:) *he gave him a portion,* or *share,* of spoil. (Ṣ, Mgh, Ḳ.) [See also 4 in art. حذو.] —— [Hence,] أَحْذَيْتُهُ طَعْنَةً ‡ [*I gave him a thrust, piercing thrust,* or *stab;*] *I thrust, pierced,* or *stabbed, him.* (Lḥ, TA.)

6. تَحَاذَى القَوْمُ فِيمَا بَيْنَهُمْ ‡ *The party,* or *company of men, divided among themselves* (Ḳ, TA) water *equally;* like تصافنوا. (TA.) —— See also 6 in art. حذو.

حِذْىٌ: see حِذْيَةٌ.

حُذْيَةٌ, with ḍamm, The *diamond with which stones are cut and bored.* (TA.)

حِذْيَةٌ *A piece cut lengthwise* (Aṣ, Ṣ, Ḳ) of flesh-meat: (Aṣ, Ṣ:) or *a small piece* (Ḳ) thereof. (TA.) [See also حِذْوَةٌ, in art. حذو.] Hence the saying in a trad., إِنَّمَا فَاطِمَةُ حِذْيَةٌ مِنِّى يَقْبِضُنِى مَا يَقْبِضُهَا †[*Fáṭimeh is only a piece of me: what distresses,* or *grieves, her, distresses,* or *grieves, me*]. (TA.) —— جَآءَا حِذْيَتَيْنِ *They two came side by side:* (Ḳ, TA:) and so جَآءَا حِذَتَيْنِ, mentioned in art. حذو. (TA.) —— *A gift;* as also ▼حِذْىٌ (TA) and ▼حُذْيَا, (Mgh, and Ḥam p. 596,) or ▼حُذَيَّا, (TA in art. حذو,) and حِذْوَةٌ. (Ḳ in that art., and Ḥam ubi suprà.) [See also حُذْيَا.]

حُذْيَا: see what next precedes. —— Also *A share,*

or *portion, of spoil;* (Ṣ, Ḳ;) and so ↑ حُذْيًا and ↑ حِذَايَةٌ (Ḳ) and ↑ حَذِيَّةٌ (Ṣ, Ḳ) and حِذْوَةٌ. (Ṣ.)

حِذَآءٌ The *gathering of the crop of grapes:* or the *time thereof:* syn. قِطَافٌ. (Ḳ.) ═ See also art. حذو.

حُذَايَةٌ: } see حُذْيَا.
حَذِيَّةٌ: }

حُذْيَا *A gift,* or *present, for bringing good news.* (Ḳ, TA.) See also حَذِيَّةٌ. You say, أَخَذَهُ بَيْنَ الحُذْيَا وَالخُلْسَةِ *He took it as something between a gift and a thing carried off by force.* (ISd, Ḳ.) —— See also حُذْيَا. ═ هُوَ حُذَيَّاكَ: see حِذَآءٌ, in art. حذو.

مِحْذًى *A large,* or *broad, knife;* or *such as is used by a shoemaker* or *maker of sandals.* (TA.)

مِحْذَآءٌ: see 1.

حر

حَرَّ, sec. pers. حَرِرْتَ, aor. ـَ; (Ṣ, A, Mṣb, Ḳ;) and حَرَّ, sec. pers. حَرَرْتَ, aor. ـِ and ـُ; inf. n. حَرٌّ and حُرُورٌ (Ṣ, Mṣb, Ḳ) and حَرَارَةٌ, (Ṣ, Ḳ,) or this is a simple subst., (Mṣb,) and حِرَّةٌ; (TA;) and ↑ احرّ, (Ṣ, Ḳ,) a dial. var. heard by Ks, (Ṣ,) and mentioned by Zj and IḲṭṭ; (TA;) *It* (a day, Ṣ, A, Mṣb, Ḳ, and food, Mṣb) *was,* or *became, hot;* (A, Mṣb, Ḳ;) or *very hot.* (TA.) And حَرَّتِ النَّارُ, sec. pers. حَرِرْتِ, aor. ـَ, *The fire burned up, and became fierce* or *hot.* (Mṣb.) —— See also 10. —— حَرَّ, sec. pers. حَرِرْتَ, aor. ـَ, inf. n. حَرَّةٌ, *He* (a man, Ṣ) *thirsted; was,* or *became, thirsty.* (Ṣ, Ḳ.) Lḥ mentions حَرِرْتَ يَا رَجُلُ, aor. ـَ, inf. n. حِرَّةٌ [perhaps a mistake for حَرَّةٌ] and حَرَارَةٌ: [app. in the same sense:] ISd says, I think he means [from] الحَرّ, not الحُرِّيَّةُ. (TA.) And حَرُورٌ [an inf. n. of the same verb] signifies The liver's *becoming dry from thirst* or *grief.* (TA.) ═ حَرَّ, sec. pers. حَرِرْتَ, aor. ـَ, (Ṣ, A,* Mgh, Mṣb, Ḳ,) inf. n. حَرَارٌ, (Ṣ, Mgh, Mṣb, Ḳ,) *He* (a slave, Ṣ) *became free:* (Ṣ, A, Mgh, Mṣb, Ḳ:) and ↑ تحرّر in the same sense is agreeable with analogy. (Mgh.) —— And حَرَّ, sec. pers. and aor. as above, inf. n. حُرِّيَّةٌ, *He* (a man) *was free-born,* or *of free origin.* (Ṣ.) ═ حَرَّ, [sec. pers. حَرَرْتَ,] aor. ـُ, inf. n. حَرٌّ, *He heated water* (A,* Ḳ) &c. (A.) ═ حَرَّ, aor. ـُ, *He cooked* [*what is termed*] حَرِيرَة: (Ḳ:) and حَرَّتْ *she made* حَرِيرَة. (A.) Hence, in a trad., ذُرِّى وَأَنَا أَحُرُّ لَكِ *Sprinkle thou* the flour, *and I will make* of it حَرِيرَة *for thee.* (TA.)

2. حرّر, inf. n. تَحْرِيرٌ, *He freed, liberated,* or *emancipated,* a slave. (A, Mgh, Mṣb.) And حَرَّرَ رَقَبَةً *He freed a neck* [i. e. *a slave*]. (Ṣ, Ḳ.) —— Also *He set apart* a child *for the worship of God and the service of the mosque or oratory:* (Ṣ, TA:) or *he devoted* him *to the service of the church as long as he should live, so that he could not relinquish it while he retained his religion.* (TA.) —— Also, inf. n. as above, ‡ *He made* a writing &c. *accurate,* or *exact;* (Ṣ, Ḳ;) *he made* a writing *beautiful,* or *elegant, and free from defects, by forming its characters rightly, and rectifying its faults:* (A:) *he wrote* a writing *well,* or *elegantly, and accurately,* or *exactly;* (TḲ;) *he wrote well,* or *elegantly:* (KL:) and *he made* an account, or a reckoning, *accurate, without mistake, and without omission, and without erasure.* (TA.) [And simply ‡ *He wrote* a letter &c.]

4. احرّ: see 1. —— Also *His* (a man's) *camels became thirsty.* (Ṣ, Ḳ.) ═ Also *He* (God) *made* a man's liver *to become dry by reason of thirst* or *grief.* (TA.) And *He made* a man's bosom *thirsty;* as in the saying, used by the Arabs in cursing a man, مَا لَهُ أَحَرَّ ٱللّٰهُ صَدْرَهُ [*What aileth him? May God make his bosom thirsty*]: or the meaning is هَامَتَهُ [app. here used as signifying the bird called هَامَة, in the form of which the soul was believed to issue from a slain man, and to call incessantly for drink until the slaughter of the slayer]. (TA.)

5: see 1.

10. استحرّ (Ṣ, Ḳ) and ↑ حَرَّ (Ṣ, TA) ‡ *It* (slaughter) *was,* or *became, vehement,* (Ṣ, Ḳ,) *and great in extent;* (TA;) and the same is said of death. (TA.) ═ استحرّها *He asked,* or *desired, of her* [*that she should make what is termed*] حَرِيرَة. (A.) [See 1, last signification.]

حِرٌ: see حِرٌ, below; and see also art. حرح.

حِرِىٌّ: see art. حرح.

حَرٌّ *Heat; contr. of* بَرْدٌ; (Ṣ, A, Mgh, Mṣb, Ḳ;) as also ↑ حَرَارَةٌ, (Ṣ,* Mṣb,* Ḳ,) *contr. of* بُرُودَةٌ; (Ṣ;) and ↑ حُرُورٌ (Ṣ,* Mṣb,* Ḳ) and ↑ حِرَّةٌ: (TA:) [see 1, first sentence:] pl. [of the first] حُرُورٌ and ↑ أَحَارِرُ; (Ḳ;) the latter anomalous, both as to its measure and in the non-incorporation of the first ر into the second: it is mentioned on the authority of AZ and others; but IDrd doubts its correctness; and the author of the Wá'ee mentions أَحَارٌّ as a pl. form, but apparently to avoid contrariety to rule: the pl. of ↑ حَرَارَةٌ as a simple subst., or as an inf. n., but more probably as the former, is حَرَارَاتٌ. (TA.) —— † *A burning of the heart, from pain* and *wrath* and *distress* or *affliction* or *trouble* or *fatigue.* (TA.) [See also حَرَارَةٌ.] —— † *Difficulty,* or *severity,* of work. (TA.) ═ See also حَارٌّ: ═ and حَرَّةٌ: ═ and سَاقُ حُرٍّ, voce حُرٌّ.

حُرٌّ *Free, ingenuous,* or *free-born; contr. of* عَبْدٌ: (Ṣ, A, Mgh, Mṣb, Ḳ:) fem. حُرَّةٌ: (Ṣ, Mgh, Mṣb, Ḳ:) pl. masc. أَحْرَارٌ (Mṣb, Ḳ) and حِرَارٌ; (IJ, Ḳ;) not حَرَارٌ, as some say; nor is حِرَارٌ an inf. n. as well as a pl., as others say: (MF:) pl. fem. حَرَائِرُ, (Mṣb, Ḳ,) contr. to analogy, and, as Suh says, the only instance of the kind except شَجَرٌ مَرَائِرُ as pl. of شَجَرَةٌ مُرَّةٌ; for the [regular] pl. of فُعْلَةٌ is فُعَلٌ; but حُرَّةٌ has this form of pl. because it is syn. with كَرِيمَةٌ and عَقِيلَةٌ [as will be seen in what follows]; and مُرَّةٌ, because it means خَبِيثَةُ الطَّعْمِ. (Mṣb.) Omar said to the women who used to go forth to the mosque, لَأَرُدَّنَّكُنَّ حَرَائِرَ [lit. *I will assuredly make you to become free women*]; meaning *I will assuredly make you to keep to the houses:* for the curtain is lowered before free women; not before slave-women. (TA.) [See also حُرِّيَّةٌ.] —— ‡ *Generous, noble,* or *well-born;* like as عَبْدٌ is used to signify "ignoble," or "base-born:" (Mgh:) and so the fem. حُرَّةٌ; (Ṣ, Mgh, Ḳ;) applied to a woman; (TA;) and to a she-camel: (Ṣ:) and so the masc. applied to a horse. (Ḳ, TA.) [Hence,] ‡ بَاتَتْ بِلَيْلَةِ حُرَّةٍ [*She passed a virgin's night*] is said of her whose husband has not been able to devirginate her (Ṣ, A, Ḳ) in the night when she has been first brought to him: (TA:) because the حُرَّة is modest and repugnant: (Ḥar p. 418:) in the contr. case one says, بِلَيْلَةِ شَيْبَآءَ: (Ṣ, L:) and one says also بِلَيْلَةٍ حُرَّةٍ and بِلَيْلَةٍ شَيْبَآءَ. (TA.) [And hence,] لَيْلَةُ حُرَّةٍ and لَيْلَةٌ حُرَّةٌ signify also † *The first night of the* [*lunar*] *month:* (Ḳ:) its last night is called لَيْلَةُ شَيْبَآءَ and لَيْلَةٌ شَيْبَآءُ. (TA.) You say also وَجْهٌ حُرٌّ ‡ [app. meaning *An ingenuous countenance*]. (A.) —— ‡ *Generous,* or *ingenuous, in conduct:* as in the saying of Imra-el-Ḳeys,

لَعَمْرُكَ مَا قَلْبِى إِلَى أَهْلِهِ بِحُرّْ

[*By thy life, my heart is not generous in conduct to its,* or *his, companion*]; meaning that it is averse therefrom, and inclines to another. (Az, TA.) [Hence,] سَحَابَةٌ حُرَّةٌ ‡ *A cloud bountiful with rain;* (A;) or *abounding with rain.* (Ṣ, Ḳ.) —— ‡ A *good* deed or action. (Ḳ, TA.) You say, مَا هٰذَا مِنْكَ بِحُرٍّ ‡ *This is not good,* or *well, of thee.* (Ṣ, A.) —— † Anything *good,* or *excellent;* as poetry, &c. (TA.) You say كَلَامٌ حُرٌّ ‡ [app. meaning *good,* or *excellent, speech* or *language*]. (A.) —— ‡ *Good* earth, or clay, and sand: (Ḳ, TA:) or earth, or clay, *in which is no sand:* (Ṣ, A:) and sand *in which is no earth* or *clay:* (Ṣ:) or sand *that has good herbage:* (A:) you say رَمْلَةٌ حُرَّةٌ; (Ṣ, A;) and the pl. is حَرَائِرُ: (Ṣ:) or sand *in which is no mixture of any other thing:* (Mṣb: [accord. to which, this is the primary meaning of the word, whence the meaning of "free," i. e. the "contr. of عَبْدٌ:" but accord. to the A and TA, it is tropical:]) and أَرْضٌ حُرَّةٌ ‡ *land in which is no salt earth:* (A:) or *in which is no sand:* as applied to *that upon which no tithe is levied,* it is post-classical. (Mgh.) —— ‡ The *middle,* (Ṣ, A, Ḳ,) and *best part,* (TA,) of sand, (Ṣ, Ḳ, TA,) and of a house. (Ṣ, A, TA.) —— † The *best* of anything; (Ḳ, TA;) as, for instance, of fruit. (TA.) —— Also sing. of أَحْرَار in the term أَحْرَارُ البُقُولِ, (TA,) which means ‡ *Herbs,* or *leguminous plants, that are eaten without being cooked;* (Ṣ, A;) as also ↑ حُرِّيَّةُ البُقُولِ: (A:) or *such as are slender and succulent;* and ذُكُورُ البُقُولِ means "such as are thick and rough:" (AHeyth:) or the former are *such as are slender and soft;* and the latter, "such as are hard and thick:" (TA in art. عشب:) or the former are such as are slender and sweet; and the latter, "such as are thick, and inclining to bitterness:" (TA in art. ذكر:) or the former are such as are rough; and these are three, namely, النَّفَل and القَفْعَآءُ and الحُرَيْثُ: or الحُرُّ is applied to a plant of the kind called النَّجِيل, growing in salt grounds. (TA.) —— حُرُّ الوَجْهِ ‡ What appears of the face: (Ḳ, TA:) or what appears of the elevated part of the cheek; (Ṣ;) [i. e.] the ball, or most prominent place, of the cheek; (W p. 28;) and ↓ الحُرَّةُ signifies [the same, or] the elevated part of the cheek: (TA:) or the former is what fronts one, of the face: or the four tracks of the tears, from each corner of each eye. (TA.) One says, لَطَمَهُ عَلَى حُرِّ وَجْهِهِ ‡ [He slapped him on the ball of his cheek]. (Ṣ, TA.*) ==== The young one of a gazelle. (Ṣ, Ḳ.) —— The young one of a serpent: (Ṣ, Ḳ:) or of a slender serpent: or it is a slender serpent, like the جَانّ, of a white colour: or a white serpent: or a serpent, absolutely. (TA.) —— The young one of a pigeon: (Ṣ, Ḳ:) or the male thereof. (TA.) —— سَاقُ حُرٍّ [is said to signify] The male of the قَمَارِىّ [or kind of collared turtle-doves of which the female is called قُمْرِيَّة (see قُمْرِىّ)]: (Ṣ, Mṣb, Ḳ:) Ḥomeyd Ibn-Thowr says,

وَمَا هَاجَ هٰذَا الشَّوْقَ إِلَّا حَمَامَةٌ

دَعَتْ سَاقَ حُرٍّ تَرْحَةً وَتَرَنُّمَا

[And nothing excited this desire but a pigeon (see حَمَامٌ) that called ساق حرّ, sorrowing and warbling]: or, accord. to IJ, the right reading is

دَعَتْ سَاقَ حُرٍّ فِى حَمَامٍ تَرَنَّمَا

[that called ساق حرّ among other pigeons, warbling]: but some say that الساق is the pigeon; and حر, its young one: or ساق حرّ is the cry of the قمارى, and is an onomatopœia: accord. to Aboo-'Adnán, it is ↓ ساق حَرٍّ, and means the warbling of the pigeon: and Ṣakhr El-Gheí makes it a compound, and indecl.; using the phrase, تُنَادِى سَاقَ حُرَّ [she calls ساق حرّ]: on which IJ observes, Aṣ says, ساق حرّ is thought to mean the young one of the bird; but it is her cry: and he (IJ) adds, the fact that the poet [Ṣakhr] does not make it decl. is an evidence of the correctness of the assertion of Aṣ; for, were it decl., he would have said سَاقَ حُرٍّ if it consisted of two nouns whereof the former was prefixed to the other so as to govern it in the gen. case, or ساق حُرًّا if it were a compound; as it is indeterminate: and its being made decl. by Ḥomeyd does not show it to be not significant of a sound; for sometimes an expression significant of a sound consists of two nouns whereof the former is prefixed to the latter so as to govern it in the gen. case, like خَازِ بَازِ. (M, MF, TA.)

حِرٌ (Mṣb, Ḳ) and ↓ حِرٌّ (Ṣ, Mgh, Mṣb, and Ḳ, in art. حرح) The vulva, or pudendum, of a woman: (Mṣb, Ḳ:) the former a dial. var. of the latter; (Ḳ;) originally حِرْحٌ [q. v.]. (Mṣb.)

حَرَّةٌ A stony tract, of which the stones are black (Ṣ, Mgh, Mṣb, Ḳ) and worn and crumbling, (Ṣ, Ḳ,) as though burned with fire: (Ṣ:) or a hard and rugged tract of ground, strewn with black and worn and crumbling stones, as though they were rained down: (TA:) or a level tract abounding with stones, over which it is difficult to walk, and hard: (IAạr:) or one [whereof the stones are] black above and white beneath: accord. to AA, of a round form: such as is oblong, not wide, is termed كُرَاعٌ: (TA:) pl. ↓ حَرٌّ, (Ḳ,) or rather this is a coll. gen. n., (MF,) and حِرَارٌ (Ṣ, A, Mgh, Mṣb, Ḳ) and حَرَّاتٌ and حَرُّونَ, (Ṣ, Ḳ,) with و and ن like أَرَضُونَ, (Yoo, Ṣ,) to which it is made like because it is fem., as أَرْضٌ is, (Yoo,) and ↓ أُحَرُّونَ, (Ṣ, Ḳ,) as though the sing. were أَحَرَّةٌ, (Yoo, Sb, Ṣ,) though this sing. is not used; (Yoo;) or as though its sing. were أَحَرّ, accord. to Th, who app. means that this place is hotter than others. (TA.)

الحُرَّةُ: see حُرٌّ —— حُرَّةُ الذِّفْرَى ‡ The part of the protuberance behind the ear where the earring swings about: (Ṣ, Ḳ:*) or it is an epithet, signifying beautiful and smooth and long in the protuberance behind the ear; applied to a woman and to a she-camel. (TA.) —— الحُرَّتَانِ is also said to signify The two ears. (TA.) One says, حَفِظَ ٱللّٰهُ كَرِيمَتَيْكَ وَحُرَّتَيْكَ (A, TA) i. e. ‡ [May God preserve thy two eyes and] thy two ears. (TA.) ==== Chamomile, or chamomile-flowers; syn. البَابُونَج. (TA.)

حِرَّةٌ: see حَرٌّ —— Also A heat, or burning, in the throat: when it increases, it is termed حَرُورَةٌ. (TA.) [See also حَرَارَةٌ.] —— Thirst: (Ṣ, A:) or the heat and burning of thirst: (IDrd:) it may be said that it is with kesr [instead of fet-ḥ (see 1)] for the purpose of its being assimilated in form to قِرَّةٌ, with which it occurs. (Ṣ, Ḳ.) One says, رَمَاهُ ٱللّٰهُ بِالحِرَّةِ تَحْتَ القِرَّةِ (A, Ḳ) May God afflict him by thirst with cold: and بِالحِرَّةِ وَالقِرَّةِ by thirst and cold. (TA.) And أَشَدُّ العَطَشِ حِرَّةٌ عَلَى قِرَّةٍ The most severe of thirst is thirst in a cold day. (Ṣ.) And حِرَّةٌ تَحْتَ قِرَّةٍ Thirst in a cold day: (ISd:) a prov., applied to him who makes a show of the contrary of that which he conceals; (TA;) or who makes a show of friendship while he conceals hatred. (Meyd.)

حَرَارٌ: see حُرِّيَّةٌ.

حَرُورٌ, of the fem. gender, (Mṣb,) A hot wind, (Mṣb,) in the night or in the day; (AA, Fr, Mṣb;) as also سَمُومٌ: (AA, Mṣb:) or the former is a hot wind in the night, and sometimes in the day; (AO, Ṣ, Ḳ;) and the latter, a hot wind in the day, and sometimes in the night: (AO, Ṣ:) or the former, a hot wind in the night; like the latter in the day: (Ṣ:) or the former, in the day; the latter being in the night; accord. to Ru-beh, as said to AO: (Mṣb:) pl. حَرَائِرُ. (A.) —— The heat of the sun: (Ḳ:) or heat [absolutely]: (ISd:) constant heat: (Ḳ:) the fire of Hell: (Th, Ḳ:) pl. as above. (TA.) In the Ḳur [xxxv. 20], وَلَا ٱلظِّلُّ وَلَا ٱلْحَرُورُ means Nor shade nor heat: (ISd:) or nor Paradise nor Hell: (Th:) or nor the people of truth, who are in the shade of truth, nor the people of falsehood, who are in constant heat, night and day. (Zj.)

حُرُورٌ: see حَرٌّ.

حَرِيرٌ Heated by wrath &c.; as also ↓ مَحْرُورٌ: (Ṣ, Ḳ:) fem. of each with ة; the former being with ة because it is syn. with حَزِينَةٌ [afflicted with grief or sorrow]: or حَرِيرَةٌ signifies affected with grief or sorrow, and having the liver burned [thereby]: (TA:) or heated in the bosom: (Az, TA:) and its pl. is حَرِيرَاتٌ. (Az, Ṣ, TA.) ==== Silk; syn. إِبْرِيسَمٌ: (Mṣb:) or dressed silk; syn. ابريسم مَطْبُوخٌ: (Mgh, Mṣb:) and a garment, or stuff, made thereof: (Mgh:) or stuff wholly composed of silk: or of which the woof is silk: (Mgh, from the Jemạ et-Tefáreeḳ:) n. un. with ة; (Mṣb;) meaning one of the garments, or pieces of stuff, called حَرِير. (Ṣ, Ḳ.)

حَرَارَةٌ: see حَرٌّ, in two places. —— Also I. q. حَرْوَةٌ as used in the saying, إِنِّى لَأَجِدُ لِهٰذَا الطَّعَامِ حَرْوَةً فِى فَمِى, (Ṣ, TA,) meaning Verily I find that this food has a burning effect, or a pungency, in my mouth. (TA.) It signifies A burning in the mouth, from the taste of a thing: and in the heart, from pain: and hence one says, وَجَدَ حَرَارَةَ السَّيْفِ, and الضَّرْبِ, and المَوْتِ, and الفِرَاقِ, [He felt the burning effect of the sword, and of beating, and of death, and of separation.] (IDrst, TA.) [See also حِرَّةٌ.] ==== See also حُرِّيَّةٌ.

حُرُورَةٌ: see حُرِّيَّةٌ.

حَرِيرَةٌ n. un. of حَرِيرٌ [q. v.]. (Mṣb.) ==== Also A kind of soup of flour and grease or gravy: (TA:) or flour cooked with milk, (Ṣ, Ḳ,) or with grease or gravy: (Ḳ:) it is of flour, and خَزِيرَةٌ is of bran: (Sh:) [when a mess of this kind is thickest,] it is عَصِيدَةٌ; then, نَجِيرَةٌ; then, حَرِيرَةٌ; then, حَسْوٌ. (IAạr.) [See also نَفِيتَةٌ.]

حَرُورِىٌّ: see the next paragraph.

الحَرُورِيَّةُ and حَرُورِيَّةٌ: see حُرِّيَّةٌ ==== A sect of the heretics, or schismatics; (خَوَارِج [q. v.];) so called in relation to Ḥaroorà (حَرُورَآءُ), a certain town (Az, Ṣ, A, Mgh, Mṣb) of El-Koofeh, (Az, Mgh, Mṣb,) from which it is distant two miles; (TA;) because they first assembled there (Az, Ṣ, Mgh, Mṣb) and professed the doctrine that government belongs only to God: (Az, Ṣ, Mgh:) they dived so deeply into matters of religion that they became heretics; and hence the appellation is applied also to any who do thus: (Mgh, Mṣb:) they consisted of Nejdeh and his companions, (Ḳ,) and those holding their tenets: (TA:) they were also called المُبَيِّضَةُ, because their ensigns in war were white: (T voce المُحَمِّرَةُ:) a man of this sect is called ↓ حَرُورِىٌّ; (Ṣ, Ḳ;) and a woman, as well as the sect collectively, حَرُورِيَّةٌ: (Mgh, Mṣb:) which also signifies the *quality of belonging to this sect.* (Ṣ,* Ḳ,* TA.)

حَرِّىٌّ A camel *that pastures in a stony tract such as is termed* حَرَّةٌ. (Ṣ, Ḳ.)

حُرِّيَّةٌ The *state, or condition, of freedom; contr. of slavery;* as also † حُرُورِيَّةٌ (Ṣ, A, Mṣb, Ḳ) and † حَرُورِيَّةٌ, (Ṣ, Mṣb, Ḳ,) of which two the latter is the chaste form, (Mgh,) or it is more chaste than the former, which is the regular form, (MF,) and † حَرَارٌ, (Ṣ, A, Mṣb, Ḳ,) not حِرَارٌ, (TA,) and † حُرُورَةٌ (Ḳ, TA [in the CḲ حَرُورَةٌ]) and † حَرَارَةٌ. (TA.) — *Free persons, collectively.* (Mgh.) [See حُرٌّ.] — ‡ The *eminent, elevated,* or *noble persons* of the Arabs, (Ḳ, TA,) and of the foreigners. (TA.) You say, هُوَ مِنْ حُرِّيَّةِ قَوْمِهِ *He is of the noble ones of his people:* (A:) or *of the choicest, best, or most excellent, of his people.* (TA.) — † *Sandy, soft earth,* (Ḳ, TA,) *good, and fit to produce plants or herbage.* (TA.) — حُرِّيَّةُ البُقُولِ : see حُرٌّ.

حَرَّانُ *Thirsty:* (Ṣ, A, Ḳ:) or it has an intensive signification, as will be shown by what follows: (TA:) fem. حَرَّى: pl. (masc. and fem., TA) حِرَارٌ (Ṣ, TA) and حَرَارَى and حُرَارَى. (TA.) One says حَرَّانُ يَرَّانُ جَرَّانُ. (TA.) It is said in a trad., فِى كُلِّ كَبِدٍ حَرَّى أَجْرٌ, meaning *For* the giving of drink to *any liver that is dried up by thirst from intense heat,* there shall be *a recompense:* and in another, فِى كُلِّ كَبِدٍ † حَارَّةٍ أَجْرٌ. (IAth, TA.) — [See also a tropical use of this word in a verse cited in art. حسب, conj. 2.]

حَارٌّ *Hot:* (Mṣb:) a *very hot* day, and food. (A.) IAạr says, I do not say † يَوْمٌ حَرٌّ. (TA in art. قر.) [This seems to imply that some allow it; and it is common in the present day. See جُوْمٌ.] — See an ex. of its fem., حَارَّةٌ, in the next preceding paragraph. — †*Difficult, troublesome, distressing, fatiguing,* or *severe* work. (Ḳ, TA.) El-Ḥasan, when [his father] 'Alee ordered him to flog El-Weleed the son of 'Oḳbeh for drinking wine, in the days of 'Othmán, said, وَلِّ حَارَّهَا مَنْ تَوَلَّى قَارَّهَا † *Set thou over what is evil thereof him who has superintended what is good thereof:* (Mgh:) or *set thou over what is difficult of the affair him who has superintended what is profitable thereof:* (Mṣb:) meaning that only he should undertake the infliction of the flogging who superintends the profitable affairs of government. (Mgh.) — جَآءَ فُلَانٌ حَارًّا مُخُّهُ, and حَارَّ العِظَامِ, ‡ *Such a one came in a plump, or fat, state; contr. of* بَارِدًا مُخُّهُ, and بَارِدَ العِظَامِ. (A and TA in art. برد.)

أَحَرُّ [*Hotter:* and *hottest*]. — أَحَرُّونَ : see حَرَّةٌ. — هُوَ أَحَرُّ حُسْنًا مِنْهُ † *He is more delicate* [or *more free from defects*] *in goodliness,* or *beauty, than he.* (Ḳ, TA.)

أَحَارِرُ : see حُرٌّ, first sentence.

مُحِرٌّ A man *whose camels are thirsty.* (Ṣ.)

مُحَرَّرٌ *Freed from slavery; emancipated.* (TA.) — A child *devoted by the parent to the service of a church.* (TA.) [See also 2.]

مَحْرُورٌ : see حَرِيرٌ.

حرب

1. حَرَبَهُ, (Ṣ, A, Ḳ,) aor. ُ, inf. n. حَرْبٌ, (Ṣ, Ḳ,) *He despoiled him of his wealth, or property;* or *plundered him;* (Ṣ, A, Ḳ;) *leaving him without anything.* (Ṣ.) — [Hence,] حُرِبَ, (A, Mgh, Mṣb,) or حُرِبَ مَالَهُ, (Ṣ,) *He was,* or *became, despoiled,* or *plundered,* (Ṣ, A, Mgh, Mṣb,) *of his wealth,* or *property,* (Ṣ,) or *of all his wealth,* or *property;* as also حَرِبَ, (Mgh, Mṣb,) aor. َ, (Mṣb,) inf. n. حَرَبٌ. (Mgh, Mṣb.) You say, مَا لَهُ حُرِبَ وَحُرِبَ : see art. جرب. (TA.) And حُرِبَ دِينَهُ † *He was despoiled of his religion; was rendered,* or *became, an unbeliever.* (TA.) — [And hence,] حَرِبَ, aor. َ, inf. n. حَرَبٌ, *He said* وَا حَرَبَا, or وَا حَرَبَاهُ : [see حَرَبٌ, below.] (TA.) — And حَرِبَ, (Ṣ, A, Ḳ,) aor. َ, (Ḳ,) inf. n. حَرَبٌ, (A, TA,) ‡ *He* (a man, Ṣ, A) *was, or became, angry,* (A,) or *violently angry.* (Ṣ, Ḳ.) And *i. q.* كَلِبَ [meaning † *He was, or became, affected with canine madness:* see حَرِبٌ]. (Ḳ.) And † *He* (an enemy) *was,* or *became, like a lion;* as also † استحرب. (TA.)

2. حرّب, inf. n. تَحْرِيبٌ, *He sharpened* a spear-head. (Ṣ, Ḳ.) — ‡ *He angered:* (Ṣ, A:) or *angered violently:* (Ḳ:) and *he provoked,* or *exasperated.* (Ṣ, Ḳ, TA.) And it is said to signify † *He acquainted* a person *with a thing that angered him:* but where it is said to have this meaning, it is accord. to one reading with ج and hemzeh [in the places of ح and ب]. (TA.)

3. حاربهُ, (Ṣ, A, Mṣb, Ḳ,) inf. n. مُحَارَبَةٌ (Mṣb, Ḳ) and حِرَابٌ, (Ḳ,) *He waged,* or *contended in, war with him; warred, or battled, with him.* (Ṣ,* A, Mṣb, Ḳ.) See also 6. — *He was, or became, hostile, or an enemy, to him.* (Ṣ,* TA.) — *He disobeyed Him;* namely, God. (TA.)

4. احرب الحَرْبَ *He excited, provoked,* or *stirred up, war.* (Ḳ.) — احربهُ *He guided him to spoil, or plunder; guided him,* or *showed him the way, to obtain spoil, or plunder, of an enemy;* (Ṣ, Ḳ, TA;) *acting as a spy.* (TA.) — *He found him to be despoiled,* or *plundered, of his wealth, or property,* or *of all his wealth,* or *property.* (TA.)

6. تحاربوا and † احتربوا (Ṣ, A, Ḳ) and † حاربوا (Ṣ) *They waged, or contended in, war, one with another; warred, or battled, one with another.* (Ṣ, A, Ḳ.)

8: see 6. — أُحْتُرِبَ *It was all plundered, taken, or carried off.* (Ḥar p. 313.)

10: see 1, last meaning.

حَرْبٌ *War, battle, fight,* or *conflict;* (Mṣb, TA;) *contr. of* سِلْمٌ; (TA;) consisting, first, in *shooting arrows, one at another;* then, in *thrusting, one at another, with spears;* then, in *combating one another with swords;* and then, in *grappling and struggling together:* (Suh, TA:) it is [generally] fem.; (Ṣ, L, Mṣb;) but its dim. is † حُرَيْبٌ, without ة, (Kh, Ṣ, L, Mṣb,) contr. to rule, (L, Mṣb,) like ذُرَيْعٌ, and قُوَيْسٌ, and فُرَيْسٌ in a fem. sense, (L,) because originally an inf. n. [of which the verb (حَرَبَ) seems not to have been used as meaning "he waged, or contended in, war"], (El-Mázinee, Ṣ,) or in order that it may not be confounded with the dim. of حَرْبَةٌ: (Mṣb:) Seer makes its origin to be the epithet حَرْبٌ, which, however, is originally an inf. n.: (L:) sometimes it is masc.; (IAạr, Mbr, Ṣ, Mṣb, Ḳ;) but this is extr.: (L:) the pl. is حُرُوبٌ. (Ṣ, Ḳ.) You say, وَقَعَتْ بَيْنَهُمْ حَرْبٌ [*War happened between them*]. (Ṣ.) And قَامَتِ الحَرْبُ عَلَى سَاقٍ *The war,* or *battle, became vehement, so that safety from destruction was difficult of attainment.* (Mṣb.) And making it masc., as meaning قِتَالٌ, you say حَرْبٌ شَدِيدٌ *A vehement fight* or *battle.* (Mṣb.) [Hence,] اِبْنُ حَرْبٍ *A warrior:* (Er-Rághib, TA in art. بنى:) and اِبْنُ الحَرْبِ [*the warrior;* or] *he who suffices for war, and who defends.* (Mṣb in that art.) And دَارُ الحَرْبِ *The country, or countries, of the unbelievers,* (Mṣb,) or *of* [*those called by the Muslims*] *the polytheists,* (Ḳ,) *between whom and the Muslims there is not peace.* (Mṣb, Ḳ.) In the saying of Aboo-Ḥaneefeh, كَانَتْ مَكَّةُ إِذْ ذَاكَ حَرْبًا, the meaning is دَارَ حَرْبٍ [*Mekkeh was at that time a place of which the people were at war with the Muslims*]. (Mgh.) = It is also an epithet; originally an inf. n. (L.) You say رَجُلٌ حَرْبٌ, (Ḳ, TA,) [in the CḲ حَرِبٌ, but it is] like عَدْلٌ, (TA,) *A man vehement in war, and courageous;* as also † مِحْرَبٌ and † مِحْرَابٌ: (Ḳ:) or † مِحْرَبٌ signifies *a man of wars;* (Ṣ;) or *a man of war,* as also † مِحْرَابٌ; and *a known, experienced warrior.* (TA.) [Being originally an inf. n.,] حَرْبٌ as an epithet is used in the same form as masc. and fem. and sing. and pl.: (Ḳ:) so that one says اِمْرَأَةٌ حَرْبٌ and قَوْمٌ حَرْبٌ, (TA,) as also † قَوْمٌ مَحْرَبَةٌ. (Ṣ, Ḳ.) — Also *An enemy,* (Ṣ, Ḳ,) whether, or not, actually at war. (Ḳ.) So in the saying, أَنَا حَرْبٌ لِمَنْ حَارَبَنِى [*I am an enemy to him who wars with me, or who is an enemy to me*]. (Ṣ.) And فُلَانٌ حَرْبُ فُلَانٍ *Such a one is the enemy of such a one.* (TA.) Some hold that حَرْبٌ is a pl. [or rather a quasi-pl. n.] of حَارِبٌ or مُحَارِبٌ. (TA.)

حَرَبٌ inf. n. of حَرِبَ. (A, Mgh, Mṣb.) وَا حَرَبَا is an ejaculation expressive of grief, lamentation, or regret, [meaning *Alas, my spoliation!* or *my loss!* or *my grief!*] (ISd, Mgh, TA,) used in an absolute manner, like وَا أَسَفَا, (ISd, TA,) or يَا أَسَفَا, (Mgh,) from حَرَبَهُ "he despoiled him of his wealth, or property:" (Ḳ:) [or from حَرِبَ, q. v.:] or it originated from the fact that Ḥarb the son of Umeiyeh, when any one died, used to ask his family what they required to expend on the occasion, and used to supply them therewith; (TA;) and when he himself died, the people of Mekkeh and its neighbourhood bewailed him, saying, وَا حَرْبَا, (Th, Ḳ,* TA,) or وَا حَرْبَاهُ, (TA,) [*Alas for Ḥarb!*] and then they changed the expression to وَا حَرَبَا, (Th, Ḳ,) or وَا حَرَبَاهُ, and it became used in the case of bewailing any person who was dear, and in the cases of other calamities: but this account of the origin did not please ISd. (TA.) —— Also *Perdition, destruction*, or *death*. (Ḥar p. 158.)

حَرِبٌ: see حَرِيبٌ. ═ Also ‡ *Angry*: (A:) or *violently angry*: (Ṣ, Ḳ:) applied to a man and to a lion. (Ṣ, A.) And *i. q.* كَلِبٌ [meaning *Affected with canine madness*]: pl. حَرْبَى, (Ḳ,) syn. with كَلْبَى, but unknown to Az in this sense except in one instance. (TA.)

حَرْبَةٌ [*A dart*, or *javelin*;] *a certain weapon* (Ḳ) *resembling a spear*, (Mṣb,) *but smaller*, (TA,) *having a wide head*; (Aṣ, TA;) *not reckoned among* رِمَاح: (IAṣr, TA:) dim. ▼حُرَيْبَةٌ: (Mṣb:) pl. حِرَابٌ. (Ṣ, A, Mṣb, Ḳ.) You say, أَخَذُوا الحِرَابَ لِلْحِرَابِ [*They took the darts*, or *javelins, for contending in war*, or *battle*]. (A.) ═ *A thrust, stick*, or *stab*. (Ḳ.) —— *Spoliation*. (Ḳ.) —— *Corruptness of religion*. (Ḳ.) ═ حَرْبَةُ *a name of Friday*; (Ḳ, TA;) accord. to the Námoos, because it is a time for warring with oneself: (TA:) pl. حَرْبَاتٌ and حَرَبَاتٌ. (Ḳ.)

حِرْبَةٌ *A mode*, or *manner, of war, battle, fight*, or *conflict*. (Ḳ.)

حِرْبَاءُ [The *male chameleon*;] the *male of what is called* أُمُّ حُبَيْنٍ; (Ṣ, Mṣb, Ḳ; [but see the latter appellation in art. حبن;]) *a well-known animal*: (TA:) or *a certain reptile, like the* عَظَايَة, (Ḳ,) said to be *larger than this latter*, (Mṣb,) *somewhat larger*, (Ṣ,) *that turns itself*, (Ṣ, Mṣb,) or *its head*, (Ḳ,) *towards the sun*, (Ṣ, Mṣb, Ḳ,) *turning with the sun as the sun turns, and assuming various colours* (Ṣ, Mṣb) *by reason of the heat of the sun*: (Ṣ:) Az describes it as *a reptile resembling in form what is called* سَامّ أَبْرَص, *with four legs, slender head*, [which is not correct as applied to the chameleon,] *and striped back; that all the day looks towards the sun*; and he adds that *its flesh is impure, and the Arabs never eat it*: (TA:) [accord. to Freytag, the word, thus applied, is said (but I know not on what authority) to be from حُرْبًا, meaning حافظ الشمس (guardian of the sun):] the fem. is with ة: (Ṣ:) and the pl. حَرَابِىُّ. (Ṣ, Mṣb.) [The word حرباء is used in passages cited in the TA as masc. and fem.; whence it seems that it may be written حِرْبَاءَةٌ as well as حِرْبَاءٌ.] The Arabs used the expression حِرْبَاءُ تَنْضُبٍ or تَنْضُبَ, like ذِئْبُ غَضًا: (Ṣ:) [the latter word in each of these cases being the name of a tree:] the former is proverbially applied to a prudent man; because the حرباء does not quit the first branch but to leap upon the second. (TA.) The phrase انْتَصَبَ العُودُ فِى الحِرْبَاءِ is used, by inversion, for انتصب الحرباء فى العود [*The male chameleon stood erect upon the branch*]: for it stands erect upon stones, and upon the roots or trunks of trees, looking towards the sun, and declines as the sun declines. (TA.) —— Also ‡ The *back*: or *its flesh*: (Ḳ:) or حِرْبَاءُ المَتْنِ means *the flesh along either side of the backbone*: (TA:) or this, (TA,) or الحرباء, (Ḳ,) *the ridge of the backbone*: (Ḳ, TA:) or حَرَابِىُّ المَتْنِ (Ṣ, L, TA) signifies *the portions of flesh*, (Ṣ,) or *the flesh*, (L, TA,) *along either side of the backbone*: (Ṣ, L, TA:) the sing. is حِرْبَاءٌ; likened to the حرباء [or male chameleon] of the desert, and therefore tropical: Kr says that the sing. of حَرَابِىُّ الظُّهُورِ is حِرْبَاءٌ accord. to rule; showing that it has no known sing. on the authority of hearsay. (L, TA.) ═ The *nails*, (Ṣ,) or *a nail*, (Ḳ,) *of a coat of mail*: (Ṣ, Ḳ:) or the *head of a nail in a ring of a coat of mail*: (Ḳ:) pl. as above. (TA.) ═ And *Rugged ground*: (Ḳ:) or *rugged and hard ground*; accord. to Th; but the word commonly known is حِزْبَاءٌ, with záy. (TA.) [This meaning has been supposed to be assigned in the Ḳ to مُحْرِبَةٌ; but the TA shows that such is not the case.]

حَرِيبٌ and ▼مَحْرُوبٌ (Ṣ, Mgh, Mṣb, Ḳ) and ▼حَرِبٌ (MF) *Despoiled of his wealth*, or *property*; *plundered*; (Ṣ, Mgh, Mṣb, Ḳ, MF;) *and left without anything*: (Ṣ, Mgh, Mṣb:) pl. (of the first, TA) حَرْبَى and حُرَبَاءُ. (Ḳ.) And حَرِيبَةٌ and ▼مَحْرُوبَةٌ A woman *deprived of her child*, or *children*. (TA.) And ▼محروب † *Despoiled of his religion*; *rendered*, or *become, an unbeliever*. (TA.)

حُرَيْبٌ dim. of حَرْبٌ, q. v.

حَرَابَةٌ: see what next follows.

حَرِيبَةٌ (Ṣ, A, Ḳ) and ▼حَرَابَةٌ (A) *Wealth*, or *property, of which one is despoiled*, or *plundered*: (A, Ḳ:) a man's property is not so called until he has been despoiled of it: (TA:) or (Ḳ, but in the A "and") *wealth*, or *property, by means of which one lives*, or *subsists*: (Ṣ, A, Ḳ:) pl. of the former [and of the latter also accord. to analogy] حَرَائِبُ. (TA.)

حُرَيْبَةٌ dim. of حَرْبَةٌ, q. v.

حَرَّابَةٌ *A troop of plunderers*. (TA.)

حَارِبٌ [act. part. n. of حَرَبَ]. —— It occurs in a trad. as signifying *One who strips people forcibly of their clothes*. (TA.)

مِحْرَبٌ and مِحْرَبَةٌ: see حَرْبٌ, in three places.

أَرْضٌ مَحْرَبَةٌ (Ṣ, Ḳ, in the CḲ مُحْرِبَةٌ) *A land containing*, (Ṣ,) or *abounding with*, (Ḳ,) *animals of the kind called* حِرْبَاء [i. e. *male chameleons*]. (Ṣ, Ḳ.)

المِحْرَبُ and ▼المُتَحَرِّبُ *The lion*. (Ḳ, TA.)

مِحْرَابٌ: see حَرْبٌ, in two places. ═ Also The *upper end of a sitting-room*, (Mṣb, and so accord. to an explanation of the pl. مَحَارِيبُ, in the Ṣ, on the authority of Fr,) or *of a house*, or *tent*, or *chamber*; (Ḳ;) the *chief*, or *most honourable, sitting-place*; (AO, L, Mṣb, Ḳ;*) whence, in a trad., كَانَ يَكْرَهُ المَحَارِيبَ [*he used to dislike the uppermost*, or *chief, sitting-places in rooms*]: (L:) the *place where kings and chiefs and great men sit*: (Mṣb:) *a high place*: (Aṣ, Hr, TA:) *a* [*chamber of the kind called*] غُرْفَة: (Ṣ, Mṣb, Ḳ:) the *highest chamber in a house*: *a chamber to which one ascends by stairs*: (Zj, TA:) *a king's closet*, or *private chamber, into which he retires alone, out of the way of the people*: (Ḳ:) *a* [*pavilion*, or *building of the kind called*] قَصْر: (Aṣ, TA:) the *station of the Imám in a mosque*: (Ḳ:) the مِحْرَاب [or *niche* which shows the direction of the ḳibleh] *of a mosque*; from the same word as signifying the "upper end of a sitting-room;" (Fr, Ṣ, Mṣb;) or, as some say, because the person praying wars with the devil and with himself by causing the attention of his heart: (Mṣb:) the *highest place in a mosque*: (Zj, TA:) the *ḳibleh*: (L, TA:) *a mosque*, or *place of worship*; so in the Ḳur xix. 12: (Ṣ, L:) *a place of assembly*. (Aṣ, TA.) مَحَارِيبُ بَنِى إِسْرَائِيلَ means *The places of worship of the Children of Israel*, (T, Ḳ,) *in which they used to assemble for prayer*, (T, TA,) or *in which they used to sit*; (Ḳ;) as though they sat therein to consult respecting war. (TA.) [See also مَذْبَحٌ.] —— *I. q.* أَجَمَةٌ, (Ḳ,) meaning The *haunt of a lion*. (TA.) —— The *neck* of a beast. (Lth, Ḳ, TA.)

مَحْرُوبٌ and مَحْرُوبَةٌ: see حَرِيبٌ, in three places.

المُتَحَرِّبُ: see المِحْرَبُ.

حرث

1\. حَرَثَ, aor. ـِ (Ṣ, Mṣb, Ḳ) and ـُ, (Ḳ,) inf. n. حَرْثٌ, (Ṣ, A, Mṣb, Ḳ,) *He gained, acquired*, or *earned*, (Ṣ, A, Ḳ,) wealth; (Ṣ;) as also ▼احترث: (Az, TA:) *he collected* wealth. (Ṣ, A, Mṣb, Ḳ.) —— *He sought, sought after*, or *sought to gain, sustenance*; and *laboured diligently*; لِعِيَالِهِ *for his family*; as also ▼احترث: (TA:) *he worked*, or *laboured*, for the goods of the present world, (Az, TA,) and ‡ for those of the world to come. (Az, A, TA.) You say, أُحْرُثْ لِآخِرَتِكَ ‡ *Labour for thy good in the world to come*. (A, TA.) And it is said in a trad., أُحْرُثْ لِدُنْيَاكَ كَأَنَّكَ تَعِيشُ أَبَدًا (Ṣ, TA) *Labour for thy good in the present world as though thou wert to live for ever*: and, in continuation, وَاعْمَلْ لِآخِرَتِكَ كَأَنَّكَ تَمُوتُ غَدًا ‡ *and work for thy good in the world to come as though thou wert to die to-morrow*. (TA.) —— Also حَرَثَ, (T, Ṣ, Mgh, Mṣb, Ḳ,) aor. ـِ and ـُ, (Ḳ,) inf. n. حَرْثٌ (T, Mgh, Mṣb, Ḳ) and حِرَاثَةٌ; (TA;) and ▼احترث; (T, Ṣ;) *He sowed*; (T, Ṣ, Ḳ;) *he cast seed upon the ground*: (T, TA:) [accord. to Bḍ (xlii. 19), this is the primary signification: see حَرْثٌ, below:] and the former verb, *he tilled*, or *cultivated*, land, either *by sowing* or *by planting*: (TA:) or *he ploughed up* land *for sowing*: (Mgh, Mṣb:) or *he ploughed* land; because the doing so is a means of gain. (Ḥam p. 70.) And the former verb, *He ploughed up* the ground *by much walking upon it*; as also ▼احرث. (TA.) —— Also, the former verb, (L, Ḳ,) aor. ـِ and ـُ, (Ḳ,) inf. n. حَرْثٌ, (A, L, Ḳ,) *He took*, or *had, four wives together*. (A, L, Ḳ.) —— *Immoderatè inivit*: (A, Ḳ:) *multùm inivit*. (IAṣr, L.) And حَرَثَ ٱمْرَأَتَهُ *Multùm inivit mulierem suam*. (IAṣr, L.) —— ‡ *He emaciated*, or *rendered lean*, (IAṣr, Ṣ, A, Ḳ,) a beast, (Ḳ,) or a camel, (IAṣr, TA,) or a she-camel, (IAṣr, Ṣ, A,) and a horse, (IAṣr, TA,) *by journeying* (IAṣr, Ṣ, A, Ḳ) *thereon*; (IAṣr, Ṣ, Ḳ;) as also ▼احرث, (so in the

A and L and TA, and in some copies of the Ṣ, in this art., and so in the Ṣ and L and Ḳ in art. لهد,) or ▼ احترث. (So in some copies of the Ṣ in the present art.) — ‡ *He stirred* a fire, (Ṣ, A, Ḳ,) *and made it to burn up,* (TA,) with the مِحْرَاث. (A, TA.) — ‡ *He examined, looked into, scrutinized,* or *investigated:* (Ḳ, TA:) app. in an absolute sense: but accord. to some of the leading lexicologists, *he examined, looked into, scrutinized,* or *investigated,* and *studied,* the book, or the Ḳur-án: (TA:) *he studied* the Ḳur-án: (Ṣ:) or *he studied* the Ḳur-án *long, and meditated upon* it. (A, TA.) — † *He called to mind* a thing, or an affair, *and became excited thereby:* [for ex.] Ru-beh says,

* وَالقَوْلُ مَنْسِيٌّ إِذَا لَمْ يُحْرَثِ *

[*And the saying is forgotten if it be not called to mind so as to produce excitement*]. (TA.) — † *He applied himself to the study of* الفِقْه [i. e. *the law*]; or *he learned the science so called.* (Ḳ.)

4: see 1, in two places.

8: see 1, in four places.

حَرْثٌ *Gain, acquisition,* or *earning;* (Jel in xlii. 19;) as also ▼ حَرِيثَةٌ; of which the pl. is حَرَائِثُ: (Ḳ:) and *recompense,* or *reward.* (Bḍ and Jel in xlii. 19, and TA. [Accord. to Bḍ, in the place here referred to, this is from the same word as meaning "seed-produce: but the reverse seems to be the case accord. to the generality of the lexicologists.]) مَنْ كَانَ يُرِيدُ حَرْثَ الآخِرَةِ, in the Ḳur xlii. 19, means † *Whoso desireth the reward,* or *recompense,* (Bḍ,) or *the gain,* i. e. *reward,* or *recompense,* (Jel,) [*of the world to come.*] — *A lot, share,* or *portion.* (TA.) — *Worldly goods.* (TA.) — † *Seed-produce:* (Ṣ,* Ḳ,* TA:) ‡ *what is grown,* or *raised, by means of seed,* and *by means of date-stones,* and *by means of planting:* (Mgh:) an inf. n. used as a proper subst.: (Mgh, Mṣb:) pl. حُرُوثٌ. (Mṣb.) — † *A place ploughed for sowing;* (Mgh, Mṣb;) as also ▼ مَحْرَثٌ, (Mṣb,) pl. مَحَارِثُ: (Mgh, Mṣb:) or *land prepared for sowing:* (Jel in ii. 66:) and it is said to signify also *a plain,* or *soft, place;* perhaps because one ploughs in it. (Ḥam p. 70.) [Being originally an inf. n., it is also used in a pl. sense.] It is said in the Ḳur ii. 223, نِسَاؤُكُمْ حَرْثٌ لَكُمْ (Mgh, Mṣb) ‡ *Your wives,* or *women, are unto you things wherein ye sow your offspring:* (Bḍ, Jel:) they are thus likened to places that are ploughed for sowing. (Mgh, Mṣb.) — [And hence,] ‡ *A wife;* as in the saying, كَيْفَ حَرْثُكَ ‡ [*How is thy wife?*]. (A, TA.) — *A road,* or *beaten track,* or the *middle of a road, that is much trodden* [*as though ploughed*] *by the hoofs of horses or the like.* (Ḳ,* TA.) — [*A ploughshare:* so in Richardson's Pers. Ar. and Engl. Dict., ed. by Johnson; and so, app., in the Munjid of Kr, voce عُقَابٌ.]

حَرِيثَةٌ: see حَرْثٌ. — The pl., حَرَائِثُ, also signifies † Camels *emaciated by travel:* (El-Khaṭṭábee, Ḳ:) originally applied to horses: of camels you [generally] say, أَحْرَفْنَاهَا ["we rendered them lean"], with ف; and نَاقَةٌ حَرْفٌ means "a lean she-camel." (El-Khaṭṭábee, TA.)

حَرَّاثٌ *A sower, plougher, tiller,* or *cultivator, of land;* (Ṣ, TA;) as also ▼ حَارِثٌ [pl. حُرَّاثٌ]: (KL:) *a plougher of land for sowing.* (Mṣb.) — *One who eats much; a great eater.* (IAạr, TA.)

حَارِثٌ *A collector of property.* (Mṣb.) — الحَارِثُ, (Ḳ, [also written الحٰرِثُ, in the CḲ, erroneously, الحَرَثُ,]) as a generic proper name, (MF,) and أَبُو الحَارِثِ, (Ṣ, Ḳ,) the latter the better known, (TA,) *The lion:* (Ṣ, Ḳ:) because he is the prince of beasts of prey, and the strongest to acquire. (Ḥar p. 662.) — See also حَرَّاثٌ.

مَحْرَثٌ: see حَرْثٌ.

مَحْرُوثَةٌ: see أَرْضٌ مُحْرَثَةٌ.

مُحْرَثٌ: see what next follows.

مِحْرَاثٌ The *thing* (i. e. the *piece of wood,* or the *wooden thing,* TA) *with which the fire is stirred* (Ṣ, A, Ḳ) *in the* [*kind of oven called*] تَنُّور; (Ṣ;) as also ▼ مِحْرَثٌ: (Ḳ:) and مِحْرَاثُ النَّارِ *the shovel* (مِسْحَاة) *with which the fire is stirred.* (TA.) [Hence,] مِحْرَاثُ الحَرْبِ † *That which* [or *he who*] *stirs up,* or *excites, war.* (TA.) — [In the present day, it signifies *A plough:* and (like حَرْثٌ) *a ploughshare.*]

أَرْضٌ مَحْرُوثَةٌ and ▼ مُحْرَثَةٌ *Ground ploughed up by people's treading much upon it.* (T, TA.)

حرج

1. حَرِجَ, aor. َ, inf. n. حَرَجٌ, *It* (a number of things) *became collected together:* and, necessarily, *became close, strait,* or *narrow:* (so accord. to an explanation of the inf. n. by Er-Rághib, in the TA:) said of anything, *it was,* or *became, close, strait,* or *narrow.* (KL.) One says of dust, حَرِجَ إِلَى حَائِطٍ, or سَنَدٍ, *It rose,* (Lth, Az, TA,) *in a narrow place,* (TA,) *and became collected* [*against a wall,* or *an acclivity* or *the like*]. (Lth, Az, TA.) — حَرِجَ صَدْرُهُ, aor. and inf. n. as above, † *His bosom became strait,* or *contracted;* (Ṣ, A, Mgh, Mṣb, TA;) *not expanded,* or *dilated, by reason of what was good.* (TA.) And حَرِجَ alone, aor. and inf. n. as above, † *He became disquieted, and contracted in bosom:* and † *he became in doubt; he doubted;* because doubt disquiets the mind. (So accord. to explanations of the inf. n. by Er-Rághib, in the TA.) — Also حَرِجَ, aor. and inf. n. as above, [† *He became straitened,* or *in difficulty:* and particularly, *by the commission of a sin,* or *crime:* (see حَرِجٌ, below:) and hence, simply,] † *he committed a sin, a crime,* or *an act of disobedience for which he deserved punishment.* (Mṣb.) — Also *He looked, and was unable to move from his place by reason of fear and rage.* (T, TA.) And حَرِجَتِ العَيْنُ, (Ṣ, A, Ḳ,) aor. َ, (Ḳ,) inf. n. as above, (TA,) ‡ *The eye became dazzled,* (حَارَتْ, Ṣ, Ḳ, TA,) or *sank in its socket,* (غَارَتْ,) *and its vision became straitened:* (A, TA:) or *it did not turn about,* nor *wink, by reason of intent gazing.* (TA.) — Also, (Ṣ, A, Ḳ,) aor. as above, (Ḳ,) and so the inf. n., (Ṣ, Ḳ,) ‡ *It was,* or *became, forbidden,* or *prohibited,* (Ṣ, A, Ḳ,) and *attended with straitness,* or *difficulty.* (A.) So in the saying, حَرِجَ عَلَىَّ ظُلْمُكَ ‡ *The wronging of thee is forbidden,* or *prohibited, to me.* (Ṣ, TA.) And حَرِجَ عَلَيْهِ السُّحُورُ ‡ *The meal termed* سحور *became forbidden,* or *prohibited, to him,* (A, TA,) namely, a man fasting, *and attended with difficulty,* (A,) *by reason of the straitness of the time thereof.* (TA.) And حَرِجَتِ الصَّلَاةُ ‡ *Prayer became forbidden,* or *prohibited,* (A, and TA as from the Ḳ, [but not found by me in the copies of the Ḳ,]) عَلَيْهَا *to her* [by reason of legal impurity, as is shown in the A]. (A, TA.) — حَرِجَ إِلَيْهِ ‡ *He betook himself,* or *had recourse, to him,* or *it, for protection from a strait,* or *difficulty.* (TA.) And حَرِجَ إِلَى كَذَا وَكَذَا † *He betook himself to such and such things.* (TA.)

2. حرّجهُ, (TA,) inf. n. تَحْرِيجٌ, (Ṣ, Ḳ,) † *He made it strait,* or *difficult;* (Ṣ, Ḳ, TA;) and *forbade it to be violated;* namely, a right. (TA.) — حرّج عَلَى حَيَّةٍ † *He said to a serpent,* [by way of warning, lest it should be a Jinnee,] *Thou wilt be in a strait if thou return to us; therefore blame us not if we reduce thee to a strait by pursuing and driving away and killing.* (TA from a trad.)

4. احرجهُ *He made him to betake himself to a narrow,* or *confined, place;* and so أَجْحَرَهُ and أَحْرَدَهُ. (TA.) And *He made him* (a dog or a beast of prey) *to betake himself to a narrow,* or *confined, place, and then attached him.* (TA.) [Hence,] احرجهُ إِلَيْهِ † *He constrained him to betake himself,* or *have recourse, to him,* or *it.* (Ṣ, A, Ḳ.) And احرجهُ إِلَى كَذَا وَكَذَا † *He made him to betake himself to such and such things:* (TA:) or *he,* or *it, caused him to want such and such things.* (AA, TA in art. دمج.) — ‡ *He caused him to fall into a strait,* or *difficulty:* (A, TA:) *he straitened him; reduced him to a strait,* or *difficulty.* (TA.) — † *He made him,* or *caused him, to fall into a sin, a crime,* or *an act of disobedience for which he deserved punishment.* (Ṣ, Ḳ, TA.) — أَحْرَجْتُ الصَّلَاةَ † *I made,* or *pronounced, prayer to be forbidden,* or *prohibited.* (Ḳ.) = احرج كَلْبَهُ, (A,) or احرجهُ مِنْ صَيْدِهِ, (Aṣ, TA,) *He gave to his dog a portion of his prey.* (A.)

5. تحرّجهُ † *He made it strait,* or *difficult, to himself.* (TA.) = And تحرّج ‡ *He put away,* or *cast away, from himself, sin,* or *crime;* (TA;) *he shunned, avoided,* or *kept aloof from, sin,* or *crime;* (Mgh;) *he did a deed whereby he shunned, avoided,* or *kept aloof from, sin,* or *crime;* (Mṣb, TA;) syn. تَأَثَّمَ. (Ṣ, A, Mgh.) And تحرّج مِنْهُ ‡ *He shunned, avoided,* or *kept aloof from, it, as a sin,* or *crime.* (A,* Mgh.) [See تَحَنَّثَ.]

حَرْجٌ: see حَرَجٌ, in two places.

حِرْجٌ : see حَرَجٌ. ═ Also *The dog's portion of the prey, or game;* (Ṣ, A, Ḳ;) such as the head and the shanks and the belly: (TA:) *what is thrown to the dog, of the prey, or game, that he has taken:* (Az, TA:) or *a piece of flesh:* pl. أَحْرَاجٌ. (TA.) ═ And *A cowry;* syn. وَدَعَةٌ: (Ṣ, A, Ḳ:) pl. أَحْرَاجٌ (Ṣ, A) and أَحْرِجَةٌ (T, TA) and حِرَاجٌ; (TA;) the second, [as also the first,] a pl. of pauc.: (T, TA:) or *cowries* (وَدَع) *which are hung upon the necks of dogs.* (Aṣ, TA.) — And *A dog's collar* [*of cowries*]: (TA:) or *a collar* [*of cowries*] *for any animal.* (T, TA.)

حَرَجٌ [inf. n. of 1, q. v.:] ‡ *Straitness; a strait,* or *difficulty.* (A,* TA.) — ‡ *A sin, a crime,* or *an act of disobedience for which one deserves punishment;* syn. إِثْمٌ; (Ṣ, Mṣb,* Ḳ;) as also ▼حُرْجٌ: (Yoo, Ṣ, Ḳ:) or the *straitness* [*which is the consequence*] *of sin* or *crime.* (A, Mgh.) — [Hence,] لَا حَرَجَ i. q. لَا بَأْسَ [*There is,* or *will be, no harm* in thy doing this or that]; and لَا إِثْمَ [*there is,* or *will be, no sin,* or *crime*]. (IAth, TA.) ═ See also حَرِجٌ, in six places. — Also, applied to a she-camel, ‡ *Lean, lank, light of flesh, slender,* or *lank in the belly;* (Ṣ, Ḳ;) as also ▼حُرْجُوجٌ, (Ṣ, A,) accord. to AZ, (Ṣ,) and ▼حُرُوجٌ: (A:) or ▼حُرْجُوجٌ signifies, so applied, *lean,* &c., as above, *and sharp-spirited:* (Ḳ:) or this last, (Ḳ,) and حَرَجٌ and ▼حُرُوجٌ, (TA,) *fat,* (Ḳ, TA,) *large-bodied,* (TA,) *and long* [lit. *long upon the face of the ground,* as distinguished from tall]: or *strong:* (Ḳ, TA:) and حَرَجٌ signifies also, (Ḳ,) or, as some say, and so do ▼حُرْجُوجٌ and ▼حُرْجُجٌ and ▼حُرُجٌ, (Ṣ,) so applied, *long* [lit. *long upon the face of the ground*]: (Ṣ:) and some allow ▼حِرْجِيجٌ in the sense of ▼حُرْجُوجٌ; (TA;) which last is originally ▼حُرْجُجٌ, which is originally ▼حُرُجٌ: (Ṣ:) the pl. of ▼حُرْجُوجٌ (Ṣ) and of ▼حِرْجِيجٌ (L) is حَرَاجِيجُ. (Ṣ, L.) ═ See also حَرَجَةٌ, in three places. ═ Also *A thing composed of pieces of wood,* (Aṣ, Ṣ, Ḳ,) *bound together,* (Aṣ, Ṣ,) *in which dead bodies are carried;* (Aṣ, Ṣ, Ḳ;) *sometimes put over the bier of a woman:* (Ṣ:) accord. to the T, the حرج of a bier is a شِجَار, [i. e. the *frame-work of a* هَوْدَج,] *which is constructed of wood, and put over the bier of a corpse:* accord. to ISd, the حرج is *a vehicle for women and men, which has no head.* (TA.) See also نَعْشٌ, in two places.

حَرَجٌ and ▼حَرِجٌ *A strait, narrow, confined,* or *close,* place: (TA:) or *strait, narrow, confined,* or *close, in the utmost degree:* (Zj, T:) or *a strait, narrow, confined,* or *close,* place, *abounding with trees,* (Ṣ, Ḳ,) *and impenetrable to the pasturing animals:* (Ṣ:) and ▼حَرِيجٌ, also, applied to a place, signifies the same as حَرَجٌ. (TA.) — صَدْرٌ حَرَجٌ (Ṣ, Mṣb, TA) and ▼حَرِجٌ, (Ṣ, A, TA,) like فَرَدٌ and فَرِدٌ, and وَحَدٌ and وَحِدٌ, and دَنَفٌ and دَنِفٌ, (Ṣ,) *A bosom strait,* or *contracted;* (A, Mṣb, TA;) *not expanded,* or *dilated, by reason of what is good.* (TA.) يَجْعَلْ صَدْرَهُ ضَيِّقًا حَرَجًا or ▼حَرِجًا, accord. to different readings, [in the Ḳur vi. 125,] (Ṣ,) is explained by I 'Ab as meaning *He will make his bosom strait,* †*impenetrable to wisdom.* (TA.) — Also حَرِجٌ and ▼حَرَجٌ A man *having a strait,* or *contracted, bosom, which does not expand,* or *dilate, by reason of what is good:* the former has a dual and a pl.; but the latter has only the sing. form, because it is [properly, or originally,] an inf. n.: Zj says that the former is a part. n., and that by the latter is meant ذُو حَرَجٍ. (TA.) — And the former, †*One who fears,* or *dreads, to venture upon an affair.* (TA.) — And † *That seldom,* or *never, withdraws from fight:* (Ḳ:) *that will not be put to flight,* as though it were difficult for him to find an excuse for being put to flight. (TA.) — And *Committing a sin, a crime,* or *an act of disobedience for which he deserves punishment;* (Mṣb;) and so ▼حَارِجٌ, which is thought by ISd to be after the manner of a rel. n., because it has no corresponding verb [of which it may be regarded as the part. n.; the regular part. n. being حَرِجٌ, as حَرِجَ is intrans.]. (TA.) — Also †*Abstaining from sin,* or *crime;* and so ▼حَرِجٌ and ▼مُتَحَرِّجٌ. (TA.) [Thus bearing two contr. significations. See 5.] — Also, and ▼حَرَجٌ, ‡ *Forbidden,* or *prohibited:* so in the phrase, ظُلْمُكَ حَرِجٌ عَلَىَّ and حَرَجٌ ‡ [*The wronging of thee is forbidden,* or *prohibited, to me*]. (A.)

حَرَجَةٌ ‡ *A wood,* or *collection of trees;* (Ṣ, Ḳ, TA;) so called because of their closeness: or *dense and tangled trees:* (TA:) or *a thicket,* or *collection of dense and tangled trees, of the kind called* سَلَم, *into which no one can penetrate;* (AHeyth, Az, TA;) or *of the* سَمُر *and* طَلْح *and* عَوْسَج *and* سَلَم *and* سِدْر; or *of the* سِدْر *and olive and other trees:* or *a place in a wood where trees are dense and tangled, extending as far as a stone's throw:* and also *a tree which the pasturing animals cannot reach:* (TA:) pl. ▼حَرَجٌ (Ṣ, Ḳ) [or rather this is a coll. gen. n., of which حَرَجَةٌ is the n. un.,] and حَرَجَاتٌ (Ṣ, A) and حِرَاجٌ (Ṣ) and [of pauc.] أَحْرَاجٌ: (A, TA:) or حَرَجٌ signifies *a place in which is a collection of trees, and where they are close together.* (A.) — Also ‡ *A collection of camels:* (Ṣ, Ḳ, TA:) *a hundred camels:* (ISd, TA:) pl. [or rather coll. gen. n.] ▼حَرَجٌ. (Ḳ.)

حُرْجُجٌ : see حَرَجٌ, in two places.

حُرْجُوجٌ : see حَرَجٌ, in five places.

حِرْجِيجٌ : see حَرَجٌ, in two places.

حِرَاجُ الظَّلْمَاءِ, (Ḳ,) or حِرَاجٌ مِنَ الظَّلَامِ, (A, TA,) and حِرَاجٌ مِنَ الظَّلْمَاءِ, (TA,) ‡ *Dense darkness.* (A,* Ḳ, TA.)

حُرُوجٌ : see حَرَجٌ, in two places.

حَرِيجٌ : } see حَرِجٌ.
حَارِجٌ :

مُحَرَّجٌ A dog *having a collar of cowries;* (Ṣ, Ḳ;) from حِرْجٌ: (Ṣ:) *having cowries upon his neck.* (Aṣ, TA.)

حَلَفَ فُلَانٌ بِالْمُحَرِّجَاتِ ‡ *Such a one swore by the three divorces* [*which render the wife absolutely forbidden to the husband*]: (A:) or *by the oaths that rendered his scope strait,* or *narrow.* (Ḥar p. 178.)

مُتَحَرِّجٌ : see حَرِجٌ.

حرح

1. حَرَحَهَا, aor. ـَ, *He hit,* or *hurt, her* (a woman's) حِرْح or حِر [i. e. *vulva,* or *pudendum*]. (Ḳ.)

حِرٌ, (Ṣ, Mgh, Mṣb, Ḳ,) similar in form to يَدٌ and دَمٌ, (Mṣb,) and حِرٌّ (AHeyth, Az, Mṣb) and حِرَةٌ; (Ḳ;) the first of which is the most common; and the last, strange; (TA;) or the first is sometimes used instead of the second; (Mṣb;) *The vulva,* or *pudendum, of a woman:* originally ▼حِرْحٌ, (Ṣ, Mgh, Mṣb, Ḳ,) as is shown by the form of its broken pl., (Ṣ, Mgh, Mṣb,) and by that of its dim.; for in the formation of the broken pl. and of the dim., a word is restored to its original state: the final ح in the original is elided, and ر is put in its stead, and incorporated into the medial radical letter; and sometimes the word is used without any letter of compensation: thus are formed حِرٌّ and حِرٌ: (Mṣb:) the pl. is أَحْرَاحٌ, (Ṣ, Mgh, Mṣb, Ḳ,) its only broken pl., (TA,) and حِرُونَ, (Ṣ, Ḳ,) which latter is like لِدُونَ and مِئُونَ, pls. of defective nouns [لِدَةٌ and مِئَةٌ]: (Ṣ:) dim. ▼حُرَيْحٌ. (Mṣb.)

حِرْحٌ : see the paragraph next preceding.

حَرِحٌ A man *loving,* (L, TA,) or *fond of,* (Ḳ,) *the* أَحْرَاح [or *pudenda*] *of women:* (L, Ḳ:) an epithet after the manner of a rel. n. (Sb, TA.) — See also حِرِىٌّ.

حِرَحِىٌّ : see what next follows.

حِرِىٌّ, and (if you will, Ṣ) ▼حِرَحِىٌّ, (Ṣ, Ḳ,) in which latter the medial radical letter is with fet-ḥ, as in the rel. ns. formed from يَدٌ and غَدٌ, namely, يَدَوِىٌّ and غَدَوِىٌّ; (Ṣ;) [*Of,* or *belonging to,* or *relating to, the vulva,* or *pudendum, of a woman;*] rel. ns. of حِرٌ [and حِرْحٌ]: and (if you will, Ṣ) you may say ▼حَرِحٌ, like سَتِهٌ. (Ṣ, Ḳ.)

حُرَيْحٌ : see حِرٌ, above.

مَحْرُوحَةٌ A woman *hit,* or *hurt, in her* حرح or حِر [i. e. *vulva,* or *pudendum*]. (Ḳ.)

حرد

1. حَرَدَ, (Ṣ, A, Mṣb, Ḳ,) aor. ـِ, (Ṣ, Ḳ,) inf. n. حَرْدٌ, (Ṣ, Mṣb,) *He tended, repaired, betook himself,* or *directed himself* or *his course* or *aim, to* or *towards; made for* or *towards; aimed at; sought, pursued, desired,* or *intended;* (him, or it; IAạr, Ḳ;) syn. قَصَدَ. (IAạr, Ṣ, A, Mṣb, Ḳ.) Agreeably with this explanation, some render the

words of the Ḳur [lxviii. 25], وَغَدَوْا عَلَى حَرْدٍ قَادِرِينَ. (Ṣ.) You say to a man, قَدْ حَرَدْتُ ↓حَرْدَكَ *I have tended, repaired, &c., to, or towards, thee;* like أَقْبَلْتُ قِبَلَكَ and قَصَدْتُ قَصْدَكَ (Fr, Ṣ,* L) (Fr, L.) A rájiz says, (Ṣ,) namely, Ḥassán, (so in a copy of the Ṣ,)

* أَقْبَلَ سَيْلٌ جَآءَ مِنْ أَمْرِ ٱللّٰهْ *
* يَحْرِدُ حَرْدَ ٱلْجَنَّةِ ٱلْمُغِلَّهْ *

[*A torrent advanced, that came by the command of God, tending to the fruitful garden*]. (Ṣ.) ═ Also, aor. ـِ, (Ḳ,) inf. n. حَرْدٌ, (Ṣ, L,) *He prevented, hindered, impeded, withheld, restrained, debarred, inhibited, forbade, prohibited,* or *interdicted;* (IAạr, Ṣ, Ḳ;) and so ↓حرّد, (L, Ḳ,) inf. n. تَحْرِيدٌ. (TA.) Agreeably with this explanation, also, some render the words of the Ḳur cited above: from حَارَدَتْ said of she-camels, meaning "they became scanty in their supplies of milk." (Ṣ.) ═ Also, aor. ـِ, (Ṣ, L, Ḳ,) or ـُ, (AZ, Ṣ, L,) inf. n. حُرُودٌ; (Ṣ, Ḳ;) [and app. ↓تحرّد and ↓انحرد; (see حَرِيدٌ;)] *He* (a man) *separated himself from others;* (Ḳ;) *he left,* or *abandoned,* or *forsook, his people, and removed from them;* (AZ, Ṣ;) *he retired from his people, and alighted,* or *took up his abode, in a place by himself.* (Ṣ.) ═ حَرِدَ, (Sb, Ṣ, Mṣb, Ḳ,) aor. ـَ, (Mṣb, Ḳ,) and حَرَدَ, aor. ـِ, (L, Ḳ,) inf. n. حَرْدٌ, (Sb, Aṣ, T, IDrd, Ṣ, Mṣb, &c.,) so says Aboo-Naṣr Aḥmad Ibn-Ḥátim, companion of Aṣ, (Ṣ,) and حَرَدٌ, (T, Ṣ, Mṣb,) this latter form of the inf. n. sometimes used, accord. to ISk, (Ṣ,) and this is the form heard by AZ and AO and Aṣ from the Arabs of chaste speech, (TA,) but both forms are chaste, (IAạr, TA,) though the former is the more common, (IAạr, Mṣb,) *He was,* or *became, angry:* (Ṣ, Mṣb, Ḳ, &c.:) *he was,* or *became, exasperated* (تحرّش) *by one who angered him, and desired to kill him.* (T, L.) And حَرِدَ عَلَيْهِ (A, L) and حَرَدَ (L) *He was angry with him.* (A, L.) ═ حَرِدَ, (Ṣ, Mṣb, Ḳ,) aor. ـَ, (Ṣ, Ḳ,) inf. n. حَرَدٌ, (Ṣ, Mgh, Mṣb,) *He* (a camel) *had the disease termed* حَرَدٌ [q. v.]: (Ḳ:) *he had the tendons,* or *sinews, of one of his fore legs relaxed by the cord whereby the fore shank is sometimes bound up to the arm,* or *had them in that state naturally,* (Ṣ, Mgh, Mṣb,*) *so that he shook his fore legs,* (Ṣ,) or *so that he beat the ground* [*with the fore leg*], (Mgh, Mṣb,) *in walking,* or *going:* (Ṣ, Mgh, Mṣb:) or *he* (a camel) *had the tendon,* or *sinew, of his arm broken, so that his fore leg became lax, and he never ceased to shake it:* the tendon, or sinew, breaks only in the outer side of the arm, and it [the arm] seems, when the camel walks or is in motion, as though it stretched, by reason of his raising it so high from the ground, and by reason of its laxness: (ISh, TA:) or *he* (a beast) *raised his legs very high, in walking,* or *going, and put them down in their place, by reason of his being very short in his step.* (L.) — Also, aor. and inf. n. as above, *He* (a man) *was oppressed by the weight of his coat of mail, so that he was unable to stretch himself out in walking.* (Ḳ.) — And, with the same aor. and inf. n., *It* (a bowstring) *had one or more of the several portions of which* (*by their being twisted together*) *it was composed longer than others.* (Ḳ.)

2. حرّد: see 1. ═ Also, (T, L, Ḳ,) inf. n. تَحْرِيدٌ, (Ḳ,) *He twisted* a rope *so tightly that the strands formed knots, and overlay one another:* (T, L:) and *he rolled* a rope *in twisting it* (أَدْرَجَ فَتْلَهُ) *so that it became round.* (AḤn, L, Ḳ.) [See also the pass. part. n., below.] — And, (Ḳ,) inf. n. as above, (Ṣ, Ḳ,) *He crooked, curved,* or *bent,* a thing, (Ṣ, Ḳ,) *in the form of an arch.* (Ṣ.) — See also حُرْدِىٌّ. [It seems to be implied in the L, that one says حرّد حَائِطَ القَصَبِ, meaning *He bound a* حُرْدِىّ (q. v.) *upon the fence of reeds,* or *canes,* of a fold for sheep &c.] ═ Also, (Ḳ,) inf. n. as above, (T, Ḳ,) *He* (a man) *betook himself,* or *repaired, for covert,* or *lodging, to a* [*house,* or *hut, such as is called*] كُوخ, (T, Ḳ,) *with a gibbous roof.* (Ḳ.)

3. حَارَدَتْ, (Ṣ, A, Ḳ,) inf. n. حِرَادٌ, (Ṣ,) *She* (a camel) *was,* or *became, scanty in her supply of milk:* (Ṣ, A, Ḳ:) or *ceased to yield milk,* or *to have milk in her udder.* (Ḳ.) — [Hence,] ‡ *She* (a woman) *ceased to have milk in her breasts.* (L.) — And ‡ *It* (a بَاطِيَة or other vessel) *ceased to have wine,* or *beverage, in it.* (L.) — And ‡ *It* (a year, سَنَةٌ,) *was one of little rain.* (Ṣ, A, Ḳ.) — And حارد ‡ *He* (a man) *was about to give, and then refrained.* (A.) — And حَارَدَتْ حَالِى ‡ *My state,* or *condition, became changed, so as not to be known,* or *so as to be displeasing.* (A.)

4. احردهُ *He separated,* or *set apart,* (Ḳ,) and *removed,* (TA,) *him,* or *it.* (Ḳ, TA.)

5: see 1.

7: see 1. — [Also,] *It* (a star) *darted down.* (Ḳ.)

حَرْدٌ *i. q.* قَصْدٌ: whence the phrase, قَدْ حَرَدْتُ حَرْدَكَ: see 1. ═ *Anger;* [as also ↓حَرَدٌ: see 1:] so in the prov., تَمَسَّكْ بِحَرْدِكَ حَتَّى تُدْرِكَ حَقَّكَ *Retain,* or *persist in, thine anger until thou obtain thy right.* (TA.) *Rancour,* or *enmity which one retains in the heart, watching for an opportunity to indulge it.* (El-Ḳálee, MF.) ═ See also حَرِيدٌ.

حُرْدٌ The مَبْعَر [i. e. the *intestine,* or *gut, containing the* بَعْر, or *dung,*] of a camel, (Aṣ, Ṣ, Ḳ,) male or female; (Ḳ;) as also ↓حُرْدَةٌ: (Aṣ, Ḳ:) pl. حُرُودٌ. (Aṣ, Ṣ.) — An *intestine,* or *a gut:* (T:) pl. as above: (IAạr:) [or] أَحْرَادٌ signifies the *intestines,* or *guts,* of camels; and is probably a pl. of حُرْدٌ, like حُرُودٌ, as the مَبَاعِر and the أَمْعَآء are nearly alike. (L.) Accord. to Lth [and the Ḳ], حُرْدٌ signifies *A piece of a camel's hump:* but this is a mistake: it means (as explained above) an intestine, or a gut. (T.)

حَرَدٌ: see حَرْدٌ. ═ Also *A certain disease in the legs of camels,* (Ḳ, TA,) *occasioning them, in walking,* or *going, to shake their legs, and to beat the ground with them much:* (TA:) or *a certain disease in their fore legs;* (Ḳ, TA;) *not in the hind legs; caused by the cord whereby the fore shank is sometimes bound up to the arm:* (TA:) or *an aridity in the tendons,* or *sinews, of one of the fore legs, occasioned by that cord,* (Ḳ, TA,) *when the animal is young and recently weaned,* (TA,) *in consequence of which he beats the ground with his fore legs,* (Ḳ, TA,) or [*strikes*] *his breast* [*therewith*], *in walking,* or *going:* (TA:) the disease thus called is casual; [or generally so; (see حُرْدِيَّة;)] not natural. (T.) [See حَرِدَ.]

حَرِدٌ: see حَرِيدٌ: ═ and حَارِدٌ: ═ and أَحْرَدُ, in two places. ═ Also A rope *uneven in its strands.* (AḤn, TA.) A bow-string *having one or more of the several portions of which* (*by their being twisted together*) *it is composed longer than others.* (Ḳ.) [See also مُحَرَّدٌ.] ═ A man *in want,* or *needy.* (Yoo, on the authority of an Arab of the desert.)

حُرْدَةٌ: see حُرْدٌ.

حُرْدِىٌّ *A bundle of reeds,* or *canes, which is laid upon the rafters,* or *pieces of wood,* (called رَوَافِدُ, IAạr, L,) *of a roof:* (IAạr, Mgh, Mṣb:) [the reeds, or canes, which are thus used in the construction of a roof are tied together in small bundles, each of which I have generally found to consist of about five or six: over them is added a coat of plaster:] pl. حَرَادِىُّ: a Nabathæan word: (Ṣ, Mgh, Mṣb, Ḳ:) arabicized: (Ṣ:) you should not say حُرْدِىّ. (ISk, Ṣ, Mgh.) — Also, (L, Ḳ,) and ↓حُرْدِيَّةٌ, (Mgh, L, Ḳ,) The *girdle* (حِيَاصَة, Mgh, L, Ḳ, TA, in the CḲ حِيَاطَة) *of a fold for sheep, &c.* (حَظِيرَة), *which is bound upon the fence* (حَائِط) *of reeds,* or *canes,* (Mgh, L, Ḳ,) *crosswise:* (Mgh, L:) accord. to IDrd, Nabathæan. (L.) You say, ↓حَرَّدَهُ, inf. n. تَحْرِيدٌ. (L.) — Also ↓حُرْدِيَّةٌ, (Lth, Mṣb,) in the 'Eyn هُرْدِيَّةٌ, (Mgh,) but this latter is disallowed by ISk, (Mṣb,) *Reeds,* or *canes, which are connected, in a bent form, with the arched branches* (طَاقَات) *of a grape-vine,* (Lth, Mgh, Mṣb,) *and upon which the shoots of the vine are let fall.* (Mgh.) — Also حُرْدِىٌّ, with ḍamm, [irregularly formed from حُرْدٌ, unless it be a mistake for حَرْدِىٌّ,] A man *having wide,* or *capacious, intestines* [like those of the camel]. (L, TA.)

حُرْدِيَّةٌ: see what next precedes, in two places.

حَرْدَانُ: see حَرِيدٌ: ═ and حَارِدٌ.

حَرُودٌ (Ṣ, A, Ḳ) and ↓مُحَارِدٌ (A, Ḳ) and ↓مُحَارِدَةٌ (Ḳ, TA, but omitted in some copies of the Ḳ) A she-camel *yielding little milk:* (Ṣ, A, Ḳ:) or *ceasing to yield milk,* or *to have milk in her udder.* (Ḳ.)

حُرُودٌ and ↓حَرَائِدُ, (Ḳ, TA,) or ↓حَرَادِيدُ, (so in a MS. copy of the Ḳ and in the CḲ,) The *prominent edges of a rope:* (Ḳ: [in a MS. copy of the Ḳ and in the CḲ, for حَبْل is erroneously put جَبَل:]) or the former, *knots,* and *parts overlying one another, in a rope, in consequence of the strands' being twisted very tightly.* (Az, on the authority of Arabs of his time.) — Also the former, pl. of حُرْدٌ [q. v.]. (Aṣ, Ṣ.)

حَرِيدٌ A man *who separates himself from others;*

as also ▼ حَرِدٌ and ▼ حَرْدٌ and ▼ حَارِدٌ and ▼ مُتَحَرِّدٌ (Ḳ) and ▼ حَرْدَانُ : (L:) fem. حَرِيدَةٌ, not حَرْدَى : (L:) or a man *who has left*, or *abandoned*, or *forsaken, his people, and removed from them:* (AZ, Ṣ:) or a *sole*, or *single*, man: (Aṣ, Ṣ:) and ▼ مُنْحَرِدٌ signifies *solitary*, in the dial. of Hudheyl: (Aṣ, Ṣ:) pl. (of the first, Ṣ) حُرَدَاءُ (Ṣ, Ḳ) and (of the second, TA) حِرَادٌ. (Ḳ.) You say, حَلَّ حَرِيدًا *He alighted and abode aside*, or *apart, from the people.* (A.) And حَىٌّ حَرِيدٌ *A tribe that separates itself from others,* (Ḳ, TA,) *not mixing with them when departing and alighting,* (TA,) *either on account of its might or on account of its smallness of number* (Ḳ, TA) *and its meanness of condition.* (TA.) And كَوْكَبٌ حَرِيدٌ (Ṣ, A) and ▼ مُنْحَرِدٌ (Ṣ) *A solitary star.* (Ṣ.) Aboo-Dhu-eyb says,

* كَأَنَّهُ كَوْكَبٌ فِى الجَوِّ مُنْحَرِدُ ▼ *

[*As though it were a solitary star in the region between the heaven and the earth*]: but AA reads [منجرد,] with ج, explaining it in the same sense; and saying that the poet means سُهَيْل [or Canopus]. (Ṣ.) [See also 7.] And they say, كُلُّ قَلِيلٍ فِى كَثِيرٍ حَرِيدٌ [*Everything little among much*, or *small in number among great in number, is solitary*]. (AZ, Ṣ.)

حُرَيْدَاةٌ *A tendon*, or *sinew, that is in the place of the cord whereby the fore shank is sometimes bound up to the arm, occasioning a beast to be what is termed* أَحْرَدُ, (Ḳ,) i. e., *to shake one of his fore legs in walking*, or *going:* sometimes this is natural. (TA.) [See حَرَدٌ.]

حَرَائِدُ : } see حُرُودٌ.
حَرَادِيدُ : }

حَارِدٌ : see حَرِيدٌ. — Also, (Ṣ, A, Ḳ,) and ▼ حَرِدٌ (A, Ḳ) and ▼ حَرْدَانُ, *Angry:* (Ṣ, A, Ḳ:) *exasperated* (مُتَحَرِّشٌ) *by him who has angered him, and desirous of killing him:* (T, L:) or the first, *compact in make, strong, feared,* or *dreaded; whom, by reason of* [*his*] *disdainfulness* (عزة [i. e. عِزَّة]) *one thinks to be angry.* (Ḥam p. 300.) أَسَدٌ حَارِدٌ *An angry lion:* pl. حَوَارِدُ. (Ṣ, A.)

أَحْرَدُ A camel (or a beast, L) *having the disease*, or *fault, termed* حَرَدٌ; (Ṣ, Mgh, L, Mṣb, Ḳ;) as also ▼ حَرِدٌ: (Ḳ:) fem. of the former حَرْدَاءُ. (Ṣ.) — A man *oppressed by the weight of his coat of mail, and unable to stretch himself out in walking;* (T, TA;) [and] so ▼ حَرِدٌ. (Ḳ.) — ‡ *Niggardly; mean; sordid.* (Ḳ, TA.) And أَحْرَدُ اليَدَيْنِ † *Close-fisted*, or *niggardly.* (T.)

مُحَرَّدٌ A rope *plaited so that it has prominent edges, by reason of its distortion.* (Ṣ, L. [See also 2; and see حَرِدٌ.]) And A bow-string *strongly twisted, having one or more of its strands*, or *the several portions of which (by their being twisted together) it is composed, appearing over*, or *above, others;* as also مُعَجَّرٌ. (L.) — *Crooked, curved*, or *bent*, (Ṣ, Ḳ,) [*in the form of an arch:* see 2:] applied to anything. (Ṣ.) — A room *in which are* [*bundles such as are called*] حَرَادِىّ *of reeds*, or *canes*, (Ṣ, L.) *laid across* [*over the rafters of the roof*]. (L;) as also مُحَرَّدَةٌ applied as an epithet to a room of the kind called غُرْفَة: (Ṣ, L:) and the former word, (Ḳ,) used as a subst., (TA,) signifies as above. (Ḳ, TA.) — Also, (Ḳ,) or بَيْتٌ مُحَرَّدٌ, (Aṣ, Ṣ, A,) *A house* [or *hut*] *with a gibbous roof, such as is termed* كُوخ. (Aṣ, Ṣ, A,* Ḳ.*)

مُحَارِدٌ and مُحَارِدَةٌ : see حُرُودٌ.

مُتَحَرِّدٌ : see حَرِيدٌ.

مُنْحَرِدٌ : see حَرِيدٌ, in three places.

حردن

حِرْدَوْنٌ : see the next article.

حرذن

حِرْذَوْنٌ (Ṣ, L, Ḳ) and حِرْذَوْنٌ : (L, Ḳ:) some say the former; and some, the latter: (Mṣb:) *A certain small reptile* (دُوَيْبَّة): some say, the *male of the* [*kind of lizard called*] ضَبّ: (Ṣ, Ḳ:) accord. to Aṣ and IDrd, and several others, *a reptile* (دَابَّة) *of which the real character is unknown;* and therefore, by several authors, said to be *a reptile* (دَابَّة) *of the reptiles* (دَوَابّ) *of the deserts:* (Mṣb:) in the O (Mṣb) and in the L, (TA,) it is said to be *a small reptile* (دُوَيْبَّة) *resembling the* حِرْبَاء [or *chameleon*], (Mṣb, TA,) *of a beautiful kind,* (TA,) *diversified with various colours and spots, found in the region of Egypt; and* [it is said that] *it* (*the male thereof*, Mṣb) *has two penes* (نِزْكَانِ), *like as the* ضَبّ *has:* (Mṣb, TA:) accord. to some, the ن is augmentative: accord. to others, radical: the pl. is حَرَاذِينُ [or حَرَاذِين]. (Mṣb.) — Also, the former, The [*kind of lizard called*] عَظَاءَة: (Th, Seer, TA:) not what is here, immediately before, described. (TA.) — [In the present day, Any *lizard.*] — And A she-camel *that is ridden until no strength remains in her:* (TA.)

حرز

1. حَرُزَ, aor. ـُ, (Ḳ,) inf. n. حَرَازَةٌ and حِرْزٌ, (TA,) *It* (a place, TA) *was*, or *became, fortified, strong*, or *protected against attack.* (Ḳ, TA.) — حَرِزَ, aor. ـَ, *He was very pious*, or *abstinent from unlawful things.* (Ṣgh, Ḳ.) — حَرَزَهُ : see 4, in three places.

2. حَرَّزَهُ : see 4, in two places.

4. احرزه, inf. n. إِحْرَازٌ, *He kept, preserved*, or *guarded, it; he took care of it;* (TA;) as also ▼ حَرَزَهُ, (Ḳ,) aor. ـِ, (TḲ,) inf. n. حَرْزٌ; (TA;) or the latter is formed by substitution of a letter from حَرَسَهُ: (Ḳ:) or the former signifies *he put it in a* حِرْز [q. v.]; (Mgh, Mṣb;) and so ▼ the latter: (TA:) and the former, *he preserved it from being taken.* (TA.) You say, أَحْرَزَهُ فِى وِعَائِهِ [*He kept*, or *preserved, it in his*, or *its, receptacle*]. (A.) And أَحْرَزْتُ المَتَاعَ *I put the goods into the* حِرْز. (Mṣb.) And حَرِّزُوا ▼ أَنْفُسَكُمْ *Preserve ye*, or *guard ye, yourselves:* (A:) [or *do so strenuously;* for it is said that] حَرَّزَهُ, inf. n. تَحْرِيزٌ, signifies *he took extraordinary pains in keeping, preserving*, or *guarding, it.* (Ḳ.) You say also أَحْرَزَتْ فَرْجَهَا *She* (a woman, TA) *guarded her pudendum;* (Ḳ, TA;) as though she put it in an inaccessible حِرْز. (TA.) And احرز المَكَانُ الرَّجُلَ *The place protected the man; afforded him refuge;* as also ▼ حَرَّزَهُ, (Ḳ,) inf. n. تَحْرِيزٌ. (TA.) — *He made it firm*, or *strong.* (KL.) [*He fortified it*, or *protected it against attack:* see حَرُزَ.] — *He drew, collected*, or *gathered, it together;* (Mṣb, TA;) as also ▼ حَرَزَهُ, [aor. ـِ,] inf. n. حَرْزٌ. (TA.) Hence, (Mṣb,) أَحْرَزَ قَصَبَ السَّبْقِ *He grasped*, or *clutched, the winning-canes; he got them for himself:* (Mṣb:) ‡ *he outstripped; outran;* or *won the race.* (A, TA. See قَصَبٌ.) [Hence also,] أَحْرَزَ الأَجْرَ *He took, received*, or *got possession of, the recompense, reward, hire, pay*, or *wages;* syn. حَازَهُ. (Ḳ.) Whence the prov., أَحْرَزْتُ نَهْبِى وَأَبْتَغِى النَّوَافِلَ [*I have gained my spoil, and I seek the superabundant gain*]: originally said by Aboo-Bekr: he used to perform the prayer called الوتر in the beginning of the night, and to say these words; meaning, that he had performed his وِتْر, and was safe from its escaping his observance, and that he had gained his recompense for it; and if he awoke in the night, would perform the supererogatory prayers. (TA.) You say also, أَحْرَزَ الخَطَرَ [*He won the bet*]. (A in art. خطر.)

5. تحرّز مِنْهُ : see 8.

8. احترز *He prepared himself; he was*, or *became, in a state of preparation.* (Mṣb in art. حذر.) — احترز مِنْهُ, and ▼ تحرّز منه, *He guarded against it; was cautious of it;* syn. تَوَقَّاهُ, (Ṣ,) or تَوَقَّى مِنْهُ, (Ḳ,) and تَحَفَّظَ مِنْهُ; (A, Mṣb;) namely, a thing; (Ṣ, Mṣb;) or an enemy: (A:) as though he put himself into a حِرْز to secure himself therefrom. (TA.)

10. اِسْتَحْرَزَ *It was*, or *remained*, [or *was preserved*,] *in the* [or *in a*] حِرْز [or *place of custody*, &c.]. (A.)

حِرْزٌ *A place that is fortified, strong*, or *protected against attack:* (Ṣ, Mgh, Ḳ:) or *a place in which a thing is kept, preserved*, or *guarded; a place of custody or protection:* (Mṣb:) or *a place or other thing that protects a man:* or *a place or other thing that is held in one's possession* (حِيز), or *to which one betakes himself for refuge or protection:* (TA:) pl. أَحْرَازٌ. (Mṣb, TA.) You say, هُوَ فِى حِرْزٍ لَا يُوصَلُ إِلَيْهِ *He is in a place of protection to which there is no access.* (TA.) And هَتَكَ السَّارِقُ الحِرْزَ [*The thief broke into the place of custody*]. (A.) — [Hence,] *An amulet*, or *a charm, bearing an inscription, which is hung upon a person to charm him against the evil eye &c.;* syn. تَعْوِيذٌ, (Ṣ,) or عُوذَةٌ: (A, Ḳ:) pl. as above. (A.) — *A share*, or *portion:*

pl. as above: you say, أَخَذَ حِرْزَهُ *He took*, or *received, his share*, or *portion*. (A, TA.)

حَرِيزٌ A place *fortified, strong*, or *protected against attack*; (A, TA;) as also ▼ مُحْرَزٌ. (TA.) You say, حِرْزٌ حَرِيزٌ (Ṣ, Mṣb, TA) *A strong fortified place*: (TA:) the latter word is a corroborative. (Mṣb.) [See also حَارِزٌ. Hence,] لَا حَرِيزَ مِنْ بَيْعٍ [*There is nothing kept from sale*]: (A, TA:) a prov.; (TA;) meaning, if thou give me a price that I approve, I will sell to thee. (A, TA.) [Hence also,] حَرَائِزُ [a pl.] Camels *that are not sold, because of their preciousness*. (Ḳ.) And فُلَانٌ حَرِيزٌ مِنْ هٰذَا *Such a one is a person who keeps aloof from*, or *shuns, this*. (A.) — A recompense or the like, *taken, received*, or *got possession of*; as also ▼ مُحْرَزٌ. (TA.)

حَارِزٌ occurs in a trad., in a form of prayer; اَللّٰهُمَّ اجْعَلْنَا فِى حِرْزٍ حَارِزٍ, meaning *O God, place us in a protecting asylum*. (TA.)

مُحْرَزٌ: see حَرِيزٌ, in two places.

حرس

1. حَرَسَهُ, (Ṣ, A, Mgh, Mṣb, Ḳ,) aor. ِ (Ṣ, Mṣb, TA) and ُ, (TA,) inf. n. حِرَاسَةٌ (Ṣ, Mgh, Ḳ) and حَرْسٌ, (Ḳ,) or ▼ the former is a simple subst., (Mṣb,) and the latter is an inf. n. only on the authority of analogy, though often used by [the Ḥanafee Imám] Moḥammad, (Mgh,) *He guarded, kept, preserved*, or *took care of, him* or *it*: (Ṣ, Mgh, Mṣb:) [and so ▼ حارسهُ; or *he guarded him, being guarded by him*: see 3 in art. ربأ.] You say, حَرَسَهُ اللّٰهُ مِنَ الْبَلَآءِ [*May God guard him from trial*, or *affliction*]: and أَدَامَ حِرَاسَتَهُ [*May He continue the guarding of him*]. (A.) — حَرَسَ, aor. ِ, (Mṣb, Ḳ,) inf. n. حَرْسٌ, (Mṣb,) ‡ *He stole*; (Mṣb, Ḳ;) [ironically used in this sense; see حَارِسٌ;] as also ▼ احترس: (Ḳ:) or the latter, *he stole a sheep or goat by night*: (Ṣ:) or both, *he stole* camels and sheep or goats *by night, and ate* them: (TA:) or the latter, *he stole* [a sheep or the like] *from the mountain*: (El-Fárábee, Mṣb:) or *he took*, (Sh, TA,) or *stole*, (TA,) a thing *from the place of pasturage*. (Sh, TA.) You say also, حَرَسَنِى شَاةً, (A, Mgh,) and ▼ أَحْرَسَنِى, (A, TA,) [but the latter is perhaps a mistranscription for ▼ اِحْتَرَسَنِى,] ‡ *He stole from me a sheep or goat*. (Mgh.)

3: see above.

4: see 1, last signification.

5. تحرّس منهُ, and ▼ احترس منه, *He guarded*, i. e., *guarded himself, against him*; syn. تحفّظ منه. (Ṣ, Mṣb, Ḳ.)

8: see 5: — and see حَرَسَ, in two places.

حَرَسٌ: } see حَارِسٌ: for the former, in two places.
حَرَسِىٌّ: }

حَرَاسَةٌ: see 1. — فُلَانٌ يَأْكُلُ الحَرَاسَاتِ *Such a one eats stolen things*: (A:) or *steals the sheep or goats of people, one after another, and eats of them*. (TA.)

حَرِيسَةٌ *What is guarded, kept, preserved*, or *taken care of*. (Mṣb.) — ‡ *A thing stolen*: (Ḳ:) or *a sheep*, or *goat, that is stolen by night*: (Ṣ:) of the measure فَعِيلَةٌ in the sense of the measure مَفْعُولَةٌ: (TA:) hence, حَرِيسَةُ الجَبَلِ (Ṣ) *a sheep*, or *goat, that is overtaken by the night before its return to its nightly resting-place, and is stolen from the mountain*: (Mṣb:) or *a sheep*, or *goat, that is stolen, of those that are guarded*, or *kept, in the mountain*: or, as some say, from حَارِسٌ applied ironically to a thief: (Mgh:) pl. حَرَائِسُ. (Ṣ, Ḳ.) Hence the saying, (TA,) لَا قَطْعَ فِى حَرِيسَةِ الجَبَلِ ‡ [*There shall be no amputation* of the hand *for the sheep*, or *goat, that is stolen by night from the mountain*]. (A, TA.) IF says that there are two explanations of the expression حريسة الجبل: some make it to signify *theft*, or *the thing stolen*, (السَّرِقَةُ,) *itself*: others make the meaning to be, that there shall be no amputation for [stealing] *what is guarded*, or *kept, in the mountain*, because it is not a place well protected: ISk says that الحَرِيسَةُ signifies السَّرِقَةُ. (Mṣb.) — *A wall of stones, made for sheep*, or *goats*, (Ḳ,) *to guard them*. (TA.)

حَارِسٌ *Guarding, keeping*, or *preserving*; *a guardian*, or *keeper*: (Ṣ, Mgh, Mṣb:) pl. ▼ حَرَسٌ (Mgh, Mṣb, Ḳ) [or this is rather a quasi-pl. n.] and حُرَّاسٌ (Mṣb, Ḳ) and [pl. of pauc.] أَحْرَاسٌ. (Ḳ.) ▼ حَرَسٌ also signifies The *guards* of a Sulṭán; (Ṣ,* Mṣb, Ḳ,* TA;) and so حُرَّاسٌ: (Ṣ, Ḳ:) the former is thus used as a gen. n.: (Ṣ, Mṣb:) and the n. un. is ▼ حَرَسِىٌّ: (Ṣ, Mṣb, Ḳ:) you do not say حَارِسٌ unless you mean to denote thereby the signification of guarding, or keeping, without the quality of a gen. n. (Ṣ, Mṣb.) — Also ‡ *A thief*; used in this sense ironically; (A, Mgh, TA;) because they found guardians to be thieves; (A, TA;) and so ▼ مُحْتَرِسٌ: (TA:) pl. of the former, حُرَّاسٌ. (A.)

مُحْتَرَسٌ [pass. part. n. of 8]. You say, مُحْتَرَسٌ مِنْ مِثْلِهِ وَهُوَ حَارِسٌ [*From such as he does one guard himself, whereas he is a guardian*]: a prov.: (Ṣ, Ḳ:) alluding to him who finds fault with a bad man when he is himself worse than he: (Ḳ:) or to him who is intrusted with the guarding of a thing when one is not secure from his being unfaithful with respect to it. (TA.) [See also Freytag's Arab. Prov., ii. 706.]

مُحْتَرِسٌ: see حَارِسٌ, last signification.

حرش

1. حَرَشَ الضَّبَّ, (Ṣ, A, Ḳ,) aor. ُ, inf. n. حَرْشٌ (Ṣ, Ḳ) and تَحْرَاشٌ, (Ḳ,) *He hunted*, or *sought to capture or catch*, or *captured* or *caught, the* [*lizard called*] ضَبّ; syn. صَادَهُ; (Ṣ, A, Ḳ;) *by moving about his hand at its hole*, (Ṣ, Ḳ,) *at the entrance thereof*, (Ḳ,) *in order that it might imagine it to be a serpent, and put forth its tail to strike it, whereupon he would seize it*; (Ṣ, Ḳ;) as also ▼ احترشهُ: (A, Ḳ:) or, as also ▼ احترشهُ, and ▼ تحرّشهُ, and ▼ تحرّش بهِ, *he traced its hole, and made a noise with his staff*, or *stick, at it, and inserted the end of this into the hole, and the* ضَبّ, *hearing the sound, thought it to be a beast desiring to come in upon it, so it came backwards upon its feet and hinder part, fighting, and striking with its tail, whereupon the man hastened with it, and seized it firmly by its tail, and it was unable to escape from him*. (TA.) And hence, *He hunted*, or *sought to capture*, or *captured, the* ضَبّ *in any manner*. (Ḥam p. 61.) Hence also the saying, لَهُوَ أَخْبَثُ مِنْ ضَبٍّ حَرَشْتَهُ [*Verily he is worse than a* ضَبّ *which thou hast hunted*]: for sometimes the ضَبّ scents [its pursuer], and circumvents [him], and cannot be caught. (TA.) And hence the prov., alluding to one's discoursing to a learned man with the desire of instructing him, أَتُعَلِّمُنِى بِضَبٍّ أَنَا حَرَشْتُهُ [*Dost thou acquaint me with a* ضَبّ *which I have captured?*]. (A'Obeyd, Az.) Hence also the prov., هٰذَا أَجَلُّ مِنَ الحَرْشِ [*This is a greater matter than the hunting*, or *capturing, of the* ضَبّ]: (M, A, Ḳ:) originating in one of their fables, to the effect that a ضَبّ said to its young one, "O my little son, beware thou of الحَرْش:" and the young one heard, one day, the fall of a digging-implement upon the mouth of the hole; so he said, "O my father, is this الحَرْش?" to which his father answered, "O my little son, this is a greater matter than الحَرْش:" (M, Ḳ:*) and it became a prov., which is applied to him who fears a thing and falls into that which is more severe. (M.) [Hence also the saying,] ▼ اِحْتَرَشَ ضَبَّ العَدَاوَةِ بَيْنَهُمْ ‡ [*He roused the rancour of enmity between them*]. (TA.) — حَرَشَهُ, (Ṣ, Ḳ,) aor. ُ, (Ḳ,) inf. n. حَرْشٌ (Ṣ, Ḳ) and تَحْرَاشٌ, (Ḳ,) signifies also *He scratched him with the nails*; or *wounded him in the outer skin*; (Ṣ, Ḳ;) and so خَرَشَهُ, with خ. (Ṣ.) — Both also signify *It* (a fly) *bit him*. (TA in art. خرش.) — And حَرَشَ البَعِيرَ *He scratched*, or *rubbed, the* غَارِب [or *withers*] *of the camel* with his staff, or stick, *to make him go*. (TA.) — And *He scratched*, or *rubbed, the camel so as to abrade the upper skin, and make it bleed*; whereupon it is smeared with هِنَآء [or tar]; as also حَرَّشَهُ. (TA.) = حَرَّشَ بَيْنَ القَوْمِ; &c.: see 2, in two places. = حَرِشَ, aor. َ, inf. n. حَرَشٌ, *He deceived, beguiled*, or *circumvented*; syn. خَدَعَ: and ▼ احترش signifies the same; or nearly the same; i. e. *he endeavoured to deceive, beguile*, or *circumvent*; syn. of the inf. n. خِدَاعٌ. (TA.)

2. حَرَّشَ بَيْنَ القَوْمِ, inf. n. تَحْرِيشٌ; (Ṣ,* A,* Ḳ,* TA;) and ▼ حَرَشَ بَيْنَهُمْ, (A, TA,*) inf. n. حَرْشٌ; (TA;) *He excited discord, dissension, disorder, strife, quarrelling*, or *animosity, between*, or *among, the people*; (Ṣ,* A,* Ḳ,* TA;) and (so in the Ṣ, but in the Ḳ "or") بَيْنَ الكِلَابِ *between*, or *among, the dogs*; (Ṣ, Ḳ;) and البَهَائِمِ *the beasts*; *exciting*, or *provoking, them, one against another*; *as is done with camels, and rams, and cocks, &c.*; the doing of which is forbidden in a trad.; (TA;) or حَرْشٌ and تَحْرِيشٌ signify one's *inciting* a man, and a lion, *to attack*

his adversary; (TA;) and مُحَارَشَةٌ and حِرَاشٌ [inf. ns. of ▼حَارَشَ] are syn. with تَحْرِيشٌ [in the last of the senses above]; as also مُهَارَشَةٌ and هِرَاشٌ: (TA in art. هرش, q. v.:) you say, حَرَّشَهُ [and ▼حَارَشَهُ, meaning, *he incited him,* &c.; or rather, *he exasperated him;* app. from حَرْشٌ or or حُرْشَةٌ, signifying "roughness"]. (AZ, Ṣ in art. ذأر.) — [Hence, app.,] تَحْرِيشٌ also signifies The *mentioning a thing that renders reproof necessary.* (TA.)

3. حارش الضَّبُّ الأَفْعَى *The* ضبّ *fought with the viper, the latter desiring to come in upon him.* (TA.) — See also 2.

4. احرش الهِنَآءُ البَعِيرَ [app. originally signifying *The tar made the camel to scratch:* and hence meaning,] *the tar made the camel to break out with small pustules;* syn. بَثَّرَهُ: (Ḳ:) or *excoriated him, and made him to bleed.* (Ibn-'Abbád.)

5. تَحَرَّشَهُ and تحرّش بِهِ: see حَرَشَ الضَّبَّ. — تحرّش [is also quasi-pass. of 2. You say,] تحرّش بِهِمْ [*He became exasperated by them*]. (AZ, L in art. حد, in explanation of the phrase تَحَدَّدَ بِهِمْ.) [See also حَرِدَ.]

8. احترشهُ: see حَرَشَ الضَّبَّ, in three places. — See also حَرِشَ.

حَرْشٌ *A mark,* or *trace;* syn. أَثَرٌ: (Ṣ, Ḳ:) by poetic license written ▼حَرَشٌ: (Ṣ:) or *a mark upon the back:* or *a mark of a blow* or *beating, upon a camel, which has healed, but upon which no hair nor fur grows:* or, as heard by Az, from more than one of the Arabs of the desert, *a gall,* or *sore, on the back, which has healed,* or *become covered with a skin in healing:* or *a scar of a gall,* or *sore, on the back:* (TA:) pl. حِرَاشٌ. (Ṣ, TA.)

حَرْشٌ *Roughness, harshness,* or *coarseness;* as also ▼حُرْشَةٌ: (Ḳ:) or *roughness,* &c., *of the skin.* (Ṣ.) [App., it has no verb: see حَرِشٌ, voce أَحْرَشُ.] — See also حَرْشٌ.

أَحْرَشُ: see حَرِشٌ.

حَرَشٌ: see حُرْشَةٌ.

حَارِشُ ضِبَابٍ *A hunter,* or *catcher, of* [*lizards of the kind called*] ضِبَاب [pl. of ضَبٌّ]: (Ṣ, A:) pl. حَرَشَةٌ. (A.)

أَحْرَشُ Anything *rough, harsh,* or *coarse;* as also ▼حَرِشٌ, on the authority of AḤn, and thought by Az to be a possessive epithet, [meaning *having roughness,* &c., from حَرْشٌ or حُرْشَةٌ,] because he had not heard any verb belonging to it: (TA:) or the former is applied to a ضَبّ, signifying *rough;* (Ṣ, Ḳ;) or *rough in the skin,* (A, TA,) *as though notched,* or *serrated:* (TA:) and in like manner, its fem., حَرْشَآءُ, to a serpent (حَيَّة), signifying *rough;* (Ḳ;) or *rough in the skin:* (Ṣ, TA:) and the masc. to a deenár, signifying *rough* (Ṣ, A, Ḳ) *by reason of its newness;* (A, Ḳ;) *good, rough, recently coined; having upon it the roughness of the stamp:* pl. حُرْشٌ (TA) [and حُرْشٌ]: and to a camel, signifying *whose galls,* or *sores, on his back have healed,* or *become covered with a skin in healing:* (Az, as heard by him from more than one of the Arabs of the desert:) and the fem., above mentioned, is applied to a she-camel, signifying, *having the mange,* or *scab,* (Ḳ, TA,) *and not smeared* [*with tar*]; (TA;) she being so called because of the roughness of her skin: (Az, TA:) and to a نُقْبَة [or scab], signifying *having small pustules,* (Ṣ,) *not smeared* [*with tar*]. (Ṣ, A.)

حرص

1. حَرَصَهُ, aor. [ـِ and] ـُ, (TḲ,) inf. n. حَرْصٌ, (T, Ṣ, Ḳ, &c.,) *He rent it;* or *clave it:* (T, Ṣ, Ḳ, &c.:) and *he,* or *it, stripped off, scraped off, rubbed off, abraded,* or *otherwise removed, its superficial part, peel, rind,* or *the like:* (T, Ḳ, &c.:) the former, (Az,) or the latter, (Az, Er-Rághib, B,) is the primary signification. (TA.) You say, حَرَصَ القَصَّارُ الثَّوْبَ, (Ṣ, A, Mgh, Mṣb,) aor. ـِ (Ṣ, Mṣb, TA) and ـُ, (Mṣb,) inf. n. حَرْصٌ, (Mṣb, TA,) *The beater and washer and whitener rent the garment,* or *piece of cloth,* (A, Mgh, Mṣb, TA,) *in beating it:* (Mgh:) or *made holes in it by beating it:* (Ṣ, TA:*) or *made holes and rents in it thereby:* or *abraded its superficial part thereby.* (TA.) And حَرَصَ المَرْعَى *The pasture,* or *herbage, became entirely consumed, nothing of it being left;* (IF, Ḳ;) as though it were stripped off from the surface of the land. (TA.) — حَرَصَ عَلَيْهِ, (T, Ṣ, A, Mṣb, Ḳ,*) aor. ـِ (T, Ṣ, Mṣb, Ḳ) and ـُ, mentioned by IḲṭṭ and the author of the Iḳtitáf, (MF,) inf. n. حِرْصٌ; (Mṣb;) and حَرِصَ, aor. ـَ, (T, Mṣb, Ḳ,) inf. n. حَرَصٌ; (Mṣb; [but it seems to be indicated in the Ḳ, by its being said that the pret. is like سَمِعَ, that it is حَرْصٌ;]) the latter form of the verb, however, is bad; (T, TA;) though El-Ḳurṭubee is asserted by MF to have said that the former is of weak authority; which is clearly wrong, as the readers of the Ḳur are generally agreed in using it; (TA;) *He desired it vehemently, eagerly, greedily, very greedily,* or *with avidity; he hankered after it; he coveted it; he desired it excessively,* or *inordinately:* (Ṣ,* Ḳ,* TA:) or *he desired it culpably;* namely, worldly good: (Mṣb:) said by Az to be derived from the first of the significations mentioned in this art., or, as is also said by Er-Rághib, and in the B, from the second thereof; because, says Az, he who does so [may be said by a figure of speech to be one who] scratches off the outer skin of the faces of men by his so doing: but MF regards this as improbable; and says that most of the lexicologists hold the last to be the primary signification, and others to be taken from it: (TA:) حَرَصَ عَلَيْهِ, aor. ـِ, inf. n. حِرْصٌ, also signifies *he strove,* or *laboured; exerted himself; took pains,* or *extraordinary pains; to acquire, obtain,* or *attain, it:* (Mṣb, TA:*) and ▼احترص signifies the same as حَرَصَ [*he desired vehemently,* &c.]: (Ḳ:) and *he strove,* or *laboured; exerted himself; took pains,* or *extraordinary pains; to acquire, obtain,* or *attain, a thing.* (AA, Ḳ,* TA.) You say, لَا حَرَصَ ٱللّٰهُ مَنْ حَرَصَ [*May God not guard him who desires inordinately,* or *culpably*]. (A.) Aboo-Dhu-eyb makes حَرَصْتُ trans. by means of بِ, in the following verse:

* وَلَقَدْ حَرَصْتُ بِأَنْ أُدَافِعَ عَنْهُمُ *
* فَإِذَا المَنِيَّةُ أَقْبَلَتْ لَا تُدْفَعُ *

[*And verily I had purposed to defend them by repelling aggression from them; but lo, the decree of death came: it was not to be repelled*]: meaning, هَمَمْتُ. (TA.) — [Also *He was excessively solicitous,* or *careful, and fearful, respecting him;* and *excessively pitiful,* or *compassionate, to him.* See حِرْصٌ and حَرِيصٌ.]

3. حارص عَلَى الأَمْرِ *He kept,* or *applied himself, constantly,* or *perseveringly, to the thing,* or *affair.* (TA in art. حفظ.)

4. مَا أَحْرَصَكَ عَلَى الدُّنْيَا [*How vehemently* or *inordinately* or *culpably desirous art thou,* or *how greedy* or *covetous art thou, of worldly goods!*] (A.)

5. إِنَّهُ لَيَتَحَرَّصُ غَدَآءَهُمْ وَعَشَآءَهُمْ *Verily he watches for the time of* (يَتَحَيَّنُ) *their morning-meal and their evening-meal:* (Ḳ:) from حِرْصٌ as meaning "vehemence of desire," &c. (TA.)

8: see حَرَصَ عَلَيْهِ.

حِرْصٌ a subst. from حَرَصَ عَلَيْهِ, (Mṣb,) *Vehemence of desire; eagerness; avidity; cupidity; hankering; greediness; vehement greediness,* or *avidity, for an object sought after; covetousness; excessive,* or *inordinate, desire;* (TA;) or i. q. جَشَعٌ: (Ṣ, Ḳ:) or *culpable desire* for worldly good: (Mṣb:) [the different opinions respecting its derivation have been mentioned above, voce حَرَصَ:] also the *act of striving,* or *labouring; exerting oneself; taking pains,* or *extraordinary pains; to acquire, obtain,* or *attain, a thing.* (Mṣb, TA.) قُرِنَ الحِرْصُ بِالحِرْمَانِ [*Greediness,* &c., *is coupled with prohibition* of the object thereof] is a saying of the Arabs. (TA.) — Also *Excessive solicitude,* or *care, and fear,* respecting a person; and *excessive pity* or *compassion.* (TA.) [See حَرِيصٌ.]

حَرْصَةٌ *A rent* in a garment or piece of cloth. (TA.) — See also حَارِصَةٌ, in two places.

حَرِيصٌ A garment, or piece of cloth, *rent* [in being beaten and washed and whitened: see 1]. (Ḳ.) — *Vehemently desirous; eager; greedy; hankering; vehemently greedy; covetous; excessively,* or *inordinately, desirous:* (Ṣ,* A,* Ḳ,* TA:) [or *culpably desirous:* see حَرَصَ عَلَيْهِ:] fem. with ة: (TA:) pl. حِرَاصٌ, (A, L, Ḳ, [in the CḲ, erroneously, حُرَّاصٌ,]) which is masc. (A, L) and fem., (L, TA,) and حُرَصَآءُ, (Ḳ,) which is masc., (TA,) and حَرَائِصُ, fem. (TA.) — Also *Excessively solicitous,* or *careful, and fearful,* respecting a person; and *excessively pitiful,* or *compassionate.* (TA.) حَرِيصٌ عَلَيْكُمْ, in the Ḳur [ix. 129], means [*Vehemently desirous,* &c.,] *of profiting,* or *benefiting, you:* or *excessively soli-*

citous, or careful, and fearful, respecting you; and *excessively pitiful, or compassionate, to you.* (TA.)

حَرِيصَةٌ : see the next paragraph, in three places.

حَارِصَةٌ *A wound in the head* (شَجَّةٌ) *by which the skin is cleft* (Ṣ, A, Mgh, Mṣb, Ḳ) *a little;* (Ṣ, Ḳ;) the first [in degree] of شِجَاج [pl. of شَجَّةٌ]; (TA;) as also ↓ حَرْصَةٌ (Ṣ, Ḳ) and ↓ حُرْصَةٌ: (TA:) or, accord. to IAạr, ↓ حَرِيصَةٌ is *syn. with* شَجَّةٌ, q. v. (Az.) [See also دَامِعَةٌ.] — Also, and ↓ حَرِيصَةٌ, *A cloud* (سَحَابَةٌ) *the rain of which,* (Ṣ, A, Ḳ,) *falling vehemently,* (A,) *removes the surface of the ground.* (Ṣ, A, Ḳ.) One says, رَأَيْتُ العَرَبَ حَرِيصَةً عَلَى وَقْعِ الحَرِيصَةِ↓ [*I saw the Arabs vehemently desirous of the descent of the cloud of which the rain should fall vehemently so as to remove the surface of the ground*]. (A, TA.)

أَحْرَصُ [comparative and superlative of حَرِيصٌ; i. e., *More,* and *most, vehemently desirous,* &c.]. It is said in the Ḳur [ii., 90], وَلَتَجِدَنَّهُمْ أَحْرَصَ النَّاسِ عَلَى حَيَاةٍ, meaning, *And thou wilt assuredly find them the most vehement of men in desire,* or *the most excessive of men in greediness, of life.* (TA.)

حِمَارٌ مُحَرَّصٌ *An ass much lacerated by the bites of other asses.* (A.)

أَرْضٌ مَحْرُوصَةٌ *Land depastured and trodden.* (TA.)

حرض

1. حَرِضَ, (Ṣ, Mṣb, Ḳ,) aor. ـَ, (Mṣb, Ḳ,) inf. n. حَرَضٌ, (Mṣb,) *His stomach became in a corrupt,* or *disordered, state:* (Ḳ:) or *he* (a man) *became in a corrupt,* or *disordered, state, and sick,* or *diseased, so as to defile himself in his clothes:* [see حَرَضٌ, below:] or *he became emaciated* (lit. *dissolved*) *by grief,* or *by excessive love:* (Ṣ:) or *he became at the point of death:* (Mṣb:) and *he suffered protracted disquietude of mind,* and *disease;* as also حَرَضَ, aor. ـُ: (Ḳ:) and حَرُضَ, aor. ـُ and ـِ, inf. n. حُرُوضٌ (Ḳ) and حَرَضٌ, (TA,) *he became heavily oppressed by disease;* or *constantly affected thereby so as to be at the point of death:* (Ḳ:) or this last form of the verb signifies *he died,* or *perished.* (TA.) — [Hence, app.,] حَرُضَ, and حَرَضَ, aor. ـُ, inf. n. حُرُوضٌ, as in the L; not حَرِضَ, as in the Ḳ; † *He was,* or *became, low, base, mean,* or *sordid; unable to rise from,* or *quit, his place;* a signification given in the Ḳ to حَرِضَ: or *low, base, mean,* or *sordid; possessing no good:* (TA:) [but of the correctness of one of the two forms here mentioned on the authority of the L, the author of the TA expresses a doubt: app. with respect to the latter of them; for it is said,] حَرُضَ, inf. n. حَرَاضَةٌ and حُرُوضَةٌ and حُرُوضٌ, also signifies *he* (a man, TA) *was,* or *became, low, base, mean,* or *sordid, and bad, corrupt,* or *vicious, and neglected,* or *forsaken;* (Ḳ, TA;) as also حَرِضَ. (Ḳ: but only the former, حَرُضَ, is given in this sense in the TA.) — حَرَضَ as a trans. v.: see 4, in two places. — حَرَضَ, aor. ـُ, also signifies *He picked up from the ground* إِحْرِيض [or *safflower*]. (O, Ḳ.)

2. حرّضهُ: see 4. — Also, inf. n. تَحْرِيضٌ, *He rendered him free from,* or *rid him of,* حَرَض [q.v.]; like as قَذَّيْتُهُ signifies "I rid him of what is termed قَذًى." (TA.) [Thus it bears two contr. significations.] — And, [hence, perhaps,] (ISd, A, &c.,) inf. n. as above, (Ṣ, ISd, A, &c.,) *He excited, incited, urged,* or *instigated, him,* (Zj, Ṣ, ISd, Ḳ,) *and roused him to ardour,* (Ṣ,) عَلَى القِتَالِ *to fight,* (Zj, Ṣ,) or عَلَى الشَّىْءِ *to do the thing,* (A,* Mṣb,) *in order that he might be known to be such as is termed* حَارِض *if he held back from it:* (Zj:) so in the Ḳur [iv. 86 and] viii. 66: (Zj:) or *he excited, incited, urged,* or *instigated, him to apply himself constantly,* or *perseveringly,* to fight: (TA:) [see 3:] and أَحْرَضَهُ↓ عَلَى الشَّىْءِ, inf. n. إِحْرَاضٌ, signifies the same as حرّضهُ. (TṢ.) — حرّض, inf. n. as above, *He had a* حُرْضَة, i. e., *a person entrusted with the office of turning about,* or *shuffling, the gaming-arrows of the players.* (TṢ.) — *He employed the portion of his property set apart for traffic in* حُرْض [q. v.], (IAạr, Ḳ,) i. e. أُشْنَان. (TA.) — *He dyed* a garment, or piece of cloth, *with* إِحْرِيض [q. v.]. (IAạr, Ḳ.)

3. حارض, (Ibn-'Abbád,) inf. n. مُحَارَضَةٌ, (Ibn-'Abbád, Ḳ,) *He contended with another in shuffling,* or *playing with, gaming-arrows.* (Ibn-'Abbád, Ḳ.) [See حُرْضَةٌ.] — حارض عَلَى العَمَلِ, (Lḥ,) inf. n. as above, (Lḥ, Ḳ,) *He applied himself constantly,* or *perseveringly, to work:* (Lḥ, Ḳ:) and على القِتَالِ *to fight.* (Lḥ.)

4. احرضهُ *It* (disease, A, TA) *pressed heavily upon him;* or *clave to him constantly: it caused him to be at the point of death;* as also ↓ حَرَضَهُ: *it corrupted,* or *disordered, his body, so that he became on the brink of death.* (TA: [in which this last signification is said to be tropical: but accord. to the A, it is evidently not so.]) *It* (food) *caused him to be sick,* or *diseased.* (A.) *It* (love, AO, Ṣ) *corrupted,* or *disordered, him.* (AO, Ṣ, Ḳ.) — ‡ *He corrupted, vitiated, marred,* or *destroyed, it;* namely, a thing; as also ↓ حرضهُ: (A:) and *he annulled it; rendered it null,* or *void.* (TA.) You say also, حَرَضَ↓ نَفْسَهُ, aor. ـُ, (Ḳ,) inf. n. حَرْضٌ, (TA,) ‡ *He corrupted,* or *vitiated,* or *destroyed, himself,* or *his own soul:* (Ḳ,* TA:) and احرض نَفْسَهُ †*he destroyed himself,* or *his own soul,* by telling a lie. (TA.) And سُوْءُ حَمْلِ الفَاقَةِ يُحْرِضُ الحَسَبَ, occurring in a saying of Aktham Ibn-Ṣeyfee, means † *The ill-bearing of poverty annuls the grounds of pretension to respect.* (TA.) — احرضهُ عَلَى الشَّىْءِ: see 2. — احرض †*He* (a man) *begat evil offspring.* (Ṣ, Ḳ.)

حَرْضٌ: see حَرَضٌ, last sentence: — and see also what here follows.

حُرْضٌ, (Mgh,) or ↓ حُرُضٌ, (Mṣb,) or both, (Ṣ, Ḳ,) the former mentioned by Sb, but in some of the copies of his book written with fet-ḥ (حَرْضٌ↓), *i. q.* شَجَرُ الأُشْنَانِ [The *trees,* or *plants, from which potash is obtained;* the *kind of plants called kali,* or *glasswort,* &c.]; *which are of the kind called* نَجِيل: (Az, TA:) Aboo-Ziyád says that *what is termed* حُرْض *is slender in the extremities* (دُقَاقُ الأَطْرَافِ), *but its tree is large, being sometimes used for shade, and affords firewood, and it is that with which people wash clothes;* and he adds, we have not seen any حُرْض purer or whiter than some which grows in El-Yemámeh, in a valley thereof called جَوُّ الخَضَارِمِ: (TA:) *i. q.* أُشْنَانٌ [q. v.]; (Ṣ, A, Mgh, Mṣb, Ḳ;) *with which the hands are washed after food.* (TA. [But see حُرَّاض.]) So in the Ḳur [xii. 85], accord. to one reading, (Ḳ,) the reading of El-Ḥasan El-Baṣree, (Ṣgh,) ↓ حَتَّى تَكُونَ حُرُضًا, (Bḍ,) meaning *Until thou be like* اشنان *in dryness;* as explained in the Ḳ, except that نُحُولًا is there erroneously put for قُحُولًا: (TA:) but Es-Suddee disapproved of this reading. (Ṣgh.) — Also حُرْض, [and app. حُرُض also,] *i. q.* جِصّ [or *Gypsum*]. (TA.)

حَرَضٌ *Corruptness in the body,* and *in the intellect,* (Ibn-'Arafeh, A, Ḳ,) and † *in one's course of conduct,* or *tenets.* (Ibn-'Arafeh, Ḳ.) [See 1.] — *A man in a corrupt,* or *disordered, state, and sick,* or *diseased,* (Ṣ, Ḳ,) *so that he defiles himself* (يُحْدِثُ) [but in some copies of the Ṣ this word is omitted]) *in his clothes;* (Ṣ;) as also ↓ حَارِضَةٌ and ↓ حَارِضٌ and ↓ حَرِضٌ; (Ḳ;) ↓ which last also signifies a man *having his stomach in a corrupt,* or *disordered, state;* and *suffering protracted disquietude of mind, and disease:* (TA:) also the first, (حَرَضٌ,) *weary,* or *fatigued:* (Ḳ:) and *at the point of death;* (Mṣb, Ḳ;) as also ↓ حَارِضٌ; (Ḳ;) which last also signifies one *near to dying,* or *to perishing;* and *having his body corrupted,* or *disordered, by disease, so as to be at the point of death,* and so ↓ حَرِضٌ; (TA;) and [in like manner] ↓ مُحْرَضٌ signifies *dying,* or *perishing, from disease, being neither living so as to be an object of hope, nor dead so as to be an object of despair:* (T, TA:) حَرَضٌ also signifies *emaciated* (lit. *dissolved*) *by grief,* or *by excessive love;* (AA accord. to the Ṣ, or AO accord. to the TA, and Ḳ;) as also ↓ مُحْرَضٌ, (Ṣ,) or ↓ مُحَرَّضٌ: (Ḳ:) and *heavily oppressed by disease;* or *constantly affected thereby so as to be at the point of death:* so in the Ḳur xii. 85: (Ḳ:) [in the CḲ, حُرُضًا is here erroneously put for مَرَضًا:] or it there signifies *heavily pressed upon by disease;* or *affected by constant disease:* (AZ:) or *extremely aged;* or *old and weak:* (Ḳatádeh:) and anything *withering:* (TA:) [the following observation, which is inserted in the Ṣ after the first of the significations here given of حَرَضٌ used as an epithet, and in the Ḳ after a later signification which is said to be tropical, applies to it, when so used, in all its senses:] it is employed alike as sing. as pl. (Fr, Ṣ, Ḳ) and masc. (Fr) and fem.; (Fr, Ḳ;) being originally an inf. n.: (Fr, Mṣb:) or, like every inf. n. used as an epithet, it is for ذُو followed by the inf. n., and therefore has no dual nor pl.

form: (Zj:) but some of the Arabs use ▼ حَارِضٌ as an epithet applied to a male, and ▼ حَارِضَةٌ as applied to a female; and these have duals and pls.: (Fr:) and sometimes حَرَضٌ has pls.; namely أَحْرَاضٌ; (K;) which is also pl. of حَرِضٌ and of حَارِضٌ; or, accord. to the L, it is allowable as a pl. of حَرِضٌ, in the place of the more common pl. حَرِضُونَ; (TA;) and ▼ حُرْضَانٌ; (K;) which is more approved; (TA;) and حِرَضَةٌ. (K: [this last being expressly said in the TA to be thus written, but in the CK it is written حَرَضَة.]) — Also, applied to a man, (A,) ‡ *Possessing no good;* (A, K;) like ▼ حَارِضَةٌ, (TA,) which latter is explained by Aṣ as signifying a man *in whom is no good:* (T, TA:) or the former, one *whose good is not hoped for, nor his evil feared:* (K:) and a *bad* man: (K:) and *low, base, mean,* or *sordid; unable to rise from,* or *quit, his place;* as also ▼ حَرِيضٌ and ▼ حَرِضٌ and ▼ مُحْرَضٌ, (K, [this last, in the CK, written مُحَرَّض,]) or ▼ مُحْرَضٌ, (TA,) and ▼ إِحْرِيضٌ: (K:) or *low, base, mean,* or *sordid; in whom is no good:* (TA:) and [in like manner] ▼ حَارِضٌ signifies *bad, corrupt,* or *vitious, and neglected,* or *forsaken;* (K;) and so ▼ مَحْرُوضٌ, (TA,) and ▼ حِرْضَةٌ, of which the pl. is حِرَضٌ; (TA;) ▼ مَحْرُوضٌ also signifying *made,* or *asserted, to be low, base, mean,* or *sordid;* (K, TA;) and so ▼ حَارِضٌ, and ▼ حِرْضَةٌ; and this last signifying also *having in him no good:* (TA:) and حَرَضٌ likewise signifies one *who does not take to himself arms, nor fight:* (Lth, K:) its pl. is أَحْرَاضٌ (A, TA) and ▼ حُرْضَانٌ: (TA:) both these pls. signify *weak* men, *who will not fight:* (S:) and the former of them is explained as signifying the *lowest, basest,* or *meanest, sort* of mankind: and men *corrupt in their course of conduct,* or *tenets:* also the latter of them as signifying men *who know not the place of their chief:* and ▼ حَارِضٌ, of which the fem. is with ة, signifies a *stupid* man. (TA.) — Also, applied to a she-camel, *Lean,* or *emaciated:* (K, TA:) and ▼ حُرْضَانٌ, so applied, *vile:* and *perishing,* or *dying;* in which sense it is likewise applied to a male camel. (TA.) — Also, applied to language, or speech, † *Bad;* (K;) and so, by poetic license, ▼ حُرُضٌ; or this, accord. to Sgh, is a dial. var.: (L, TA:) and *perishing:* pl. أَحْرَاضٌ. (TA.)

حَرِضٌ: see حَرَضٌ, in three places, near the beginning: — and again in the latter half of the paragraph.

حُرُضٌ: see حُرْضٌ, in two places.

حُرْضَةٌ The *person called* أَمِينُ مُقَامِرِينَ; (O, K;) [i. e.] the *man who turns round about,* or *shuffles, the arrows* [*in the* رِبَابَة], or *who deals them forth,* (الَّذِى يَضْرِبُ بِالقِدَاحِ, S, or يُفِيضُ, A,) *for the players in the game called* القِدَاح, (المَيْسِر, (S, A,) *in order that he may eat of their meat* [*without having contributed to pay for the slaughtered camel*]: (A:) like him who is termed بَرَم, (S, A,) always a low, or mean, person, (S,) an object of dispraise: (A:) called thus because of his lowness, or meanness. (L.) — Also *One who does not purchase flesh-meat, nor eat it unless he find it in the possession of another person.* (AHeyth, Az.)

حِرْضَةٌ: see حَرَضٌ, latter half, in two places.

حُرْضَانٌ: see حَرَضٌ, (of which it is a syn. and a pl.,) latter half, in three places.

حَرِيضٌ: see حَرَضٌ, in the latter half of the paragraph.

حَرَّاضٌ *One who burns* حُرْض [*kali,* or *glasswort,* &c.] *for* قِلْى [or *potash*]; (K; [in the CK, لِلْقِلْي is erroneously put for لِلْقِلْى;]) *one who makes a fire upon* حُرْض *for the purpose of procuring from it* قِلْى; (S;) i. e. for the dyers; and ▼ إِحْرِيضٌ also signifies *one who makes a fire upon* أُشْنَان [or حُرْض]: it is said that [plants of the kind called] حَمْض are burned, in their fresh state, and then water is sprinkled upon their ashes, which in consequence are compacted, and become قِلْى [q. v.]. (TA.) — Also *One who makes a fire upon masses of hard stone for the purpose of preparing thence* نُورَة [or *quick lime*], or جِصّ [which is *gypsum*]. (S, K.)

حَرَّاضَةٌ *A place in which* أُشْنَان [or حُرْض] *is burned* [*for making potash*]. (TA.) — Also *A place for the preparing, by fire, of* [*quick lime,* (see حَرَّاضٌ,) or] *gypsum.* (TA.)

حَارِضٌ and حَارِضَةٌ: see حَرَضٌ, from near the beginning to near the end.

إِحْرِيضٌ: see حَرَضٌ, latter half: — and see also حَرَّاضٌ. — Also *Safflower;* syn. عُصْفُرٌ; (S, A, K;) a general name thereof: or عُصْفُر *that is put into cooked flesh-meat:* or the *grain thereof.* (TA.)

مُحْرَضٌ: see حَرَضٌ, in three places.

مِحْرَضَةٌ, with kesr, *A vessel for* حُرْض; (S, K;) *made of wood, or of brass, and the like;* (TA;) *i. q.* أُشْنَانَةٌ: (A:) pl. مَحَارِضُ. (A, TA.)

مُحْرَضٌ : } see حَرَضٌ; for each in two places.
مَحْرُوضٌ : }

حرف

1. حَرَفَ الشَّيْءَ عَنْ وَجْهِهِ, (AO, S, Msb, K,) aor. ـِ, (Msb,) or ـُ, (K,) inf. n. حَرْفٌ, (S, Msb,) *He turned the thing from its proper way,* or *manner:* (K:) or *altered it therefrom:* (Msb:) and ▼ حَرَّفَهُ, inf. n. تَحْرِيفٌ, has this latter meaning: (K,* TA:) or has an intensive signification of this kind. (Msb.) ▼ تَحْرِيفُ الكَلِمِ عَنْ مَوَاضِعِهِ signifies *The altering words from their proper meanings:* (S,* TA:) and agreeably with this explanation, the verb is used in the Kur iv. 48, &c.: (TA:) or تحريف signifies the *perverting* of language: (Msb:) or the *altering* a word *in form;* as in writing بَرَّدَ for بُرْدٌ; or *vice versâ:* (KT:) [and the *mistranscribing* a word in any manner: commonly used in this sense in the lexicons &c.: or the *altering* a word *by substituting one letter, or more, for another,* or *others.* See also صَحَّفَ.] — See also 7. — حَرَفَ لِعِيَالِهِ, (Msb, K,) aor. ـِ, (Aṣ, S, K,) or ـُ, (Msb,) *He earned* or *gained* [*subsistence*], or *laboured to do so, for his family,* or *household,* (Aṣ, S, Msb, K,) *from this and that quarter;* (Aṣ, S;) as also ▼ احترف: (Mgh,* Msb, TA:) and ▼ احترف بِيَدَيْهِ [*he earned,* or *gained, with his hands*]: and ▼ تحرّف لعياله *he applied himself to earn* or *gain* [*subsistence*] *for his family,* or *household, by means of any,* or *every, art* or *craft:* (TA:) and ▼ احرف *he laboured,* or *sought gain* or *sustenance, for his household,* or *family;* expl. by كَدَّ عَلَى عِيَالِهِ. (IAar, K.) — حَرَفَ عَيْنَهُ, inf. n. حَرْفَةٌ, (K,) not an inf. n. of un., (TA,) *He applied collyrium to his eye* (K, TA) *with the* [*style called*] مِيل. (TA.) — حُرِفَ فِى مَالِهِ, inf. n. حَرْفَةٌ, *He suffered the loss of somewhat of his property.* (Lḥ, K.)

2: see 1, in two places. — [Hence,] طَاعُونٌ يُحَرِّفُ القُلُوبَ [*A pestilence*] *causing the hearts* [of those witnessing its effects] *to turn away, and be aloof:* (K:) occurring in a trad.: or, accord. to one relation, يُحَوِّفُ القلوب, (TA,) i. e., turning the hearts from confidence, and inclining them to removal and flight. (K and TA in art. حوف.) — تَحْرِيفُ القَلَمِ *The nibbing the writing-reed obliquely;* (S,* K,* TA;) *making the right tooth of the nib higher* [i. e. *longer*] *than the left.* (TA.) You say also, حَرَّفَ القَطَّةَ [*He made the nibbing oblique*]. (TA.) And حرّف السِّكِّينَ فِى حَالِ القَطِّ [*He turned the knife obliquely in nibbing*]. (TA.) — See also 7. — تَحْرِيفٌ also signifies The *putting in motion,* or *into a state of commotion;* syn. تَحْرِيكٌ. (TA.) — قَالَ بِيَدِهِ فَحَرَّفَهَا كَأَنَّهُ يُرِيدُ القَتْلَ, in a trad., means [*He made a sign with his hand,*] *and imitated with it the cutting of a sword with its edge.* (TA.)

3. حُورِفَ *He was debarred from the means of subsistence;* because he of whom this is said is aloof (بِحَرْفٍ) from the means of subsistence. (Mgh.) And حُورِفَ كَسْبُ فُلَانٍ *Such a one was made to experience difficulty* (S, TA) *in his buying and selling, and was straitened* (TA) *in his means of subsistence;* as though his means of subsistence were turned away from him: (S, TA:) or *he had his gain, or earnings, turned away from him.* (Msb.) It is said in a trad. of Ibn-Mes'ood, مَوْتُ المُؤْمِنِ عَرَقُ الجَبِينِ تَبْقَى عَلَيْهِ البَقِيَّةُ مِنَ الذُّنُوبِ فَيُحَارَفُ بِهَا عِنْدَ المَوْتِ, i. e. [*The death of the believer is accompanied with sweating of the side of the forehead: some sins remain chargeable against him, and*] *he is made to experience difficulty by them* [*in dying*], in order that his sins may be diminished. (S.) — مُحَارَفَةٌ has also a meaning like مُفَاخَرَةٌ: Sá'ideh says,

* فَقَدْ عَلِمُوا فِى الغَزْوِ كَيْفَ نُحَارِفُ *

[*And they certainly know, in warfare, how we vie for superiority in glory:* or] accord. to Skr, it means *how we deal with them;* as when one says *to a man, What is thy* حِرْفَةٌ (i. e. *thine occupation*) *and thy lineage?* (TA:) [or the meaning may be *how we requite*; for] ═ حارفهُ بِسُوْءٍ signifies *He requited him for evil* (Ḳ, TA) *that he had done.* (TA.) And it is said in a trad., إِنَّ ٱلْعَبْدَ لَيُحَارَفُ عَنْ عَمَلِهِ ٱلْخَيْرَ أَوِ ٱلشَّرَّ, i. e. [*Verily the servant*] *shall be requited* [*for his deed; the good* I mean, *or the evil*]. (IAạr, TA.) And ▼ احرف also signifies *He requited* for good or evil. (IAạr, Ḳ.) ═ مُحَارَفَةٌ signifies also The *measuring* a wound *with the* مِحْرَاف, i. e. the *probe.* (Ḳ,* TA.)

4. احرف: see 1. —— Also, (inf. n. إِحْرَافٌ, Mṣb,) *His* مال [or *cattle*] *increased, and became in a good state* or *condition.* (AZ, Ṣ, Mṣb, Ḳ.) One says, جَآءَ بِالحِلْقِ وَالإِحْرَافِ, meaning *He came with,* or *brought, much cattle.* (AZ, Ṣ. [See حِلْقٌ.]) ═ *He emaciated,* or *rendered lean,* a she-camel: so says Aṣ: others say احرث. (Ṣ.) [See حَرْفٌ: and see حَرِيثَةٌ.] ═ See also 3, last sentence but one.

5: see 7: —— and see also 1.

7. انحرف [*It became turned,* or *altered, from its proper way,* or *manner*; quasi-pass. of 1 in the first of the senses explained above: and] *he turned aside*; (Az, Ṣ, Mgh, Mṣb, Ḳ;) as also ▼ تحرّف; (Az, Ṣ, Mgh, Ḳ;) and ▼ احرورف; (Az, Ṣ, Ḳ;) and ▼ حَرَفَ, inf. n. حَرْفٌ; (TA;) عَنْهُ *from it.* (Az, Ṣ, Mṣb, TA.) [Hence,] one says, انحرف مِزَاجُهُ [*His temperament,* or *constitution, became disordered*]; as also ▼ حُرِّفَ, [app. a mistranscription for حُرِّفَ,] inf. n. تَحْرِيفٌ. (TA.) [And انحرف عَلَيْهِ *He turned against him, with enmity,* or *anger.*] And انحرف إِلَيْهِ *He turned to,* or *towards, him,* or *it.* (TA.)

8: see 1, in two places.

12: see 7.

حَرْفٌ The *extremity, verge, border, margin, brink, brow, side,* or *edge,* (Ṣ, Mgh,* Ḳ, TA,) of anything; (Ṣ, Ḳ;) as, for instance, the *side* of a river or rivulet, and of a ship or boat, (TA,) and of the notch of an arrow; (Mṣb;) and the *edge* of a sword: (L, TA:) pl. [of mult.] حُرُوفٌ, and [of pauc.] أَحْرُفٌ. (TA.) Hence, (Ṣ,) [*A point, a ridge, a brow,* and *a ledge,* of a mountain:] the *pointed, sharp,* or *edged, summit* of a mountain: (Ṣ, Mṣb, Ḳ:) *a projecting portion in the side* of a mountain, *in form like a small* دُكَّان [i. e. *bench*] or *the like:* and *a portion in the summit* of a mountain, *having a thin edge,* or *ridge, rising above the upper part of the back:* (Sh, TA:) pl. (of the word thus used in relation to a mountain, TA) حِرَفٌ; (Fr, Ṣ, Mṣb, Ḳ;) accord. to Fr, (Mṣb,) the only instance of the kind except طِلَلٌ as pl. of طَلٌّ. (Mṣb, Ḳ.) [Hence, also,] *A nib,* of a writing-reed, *obliquely cut:* so in the phrase قَلَمٌ لَا حَرْفَ لَهُ, in the Ṣ and Ḳ in art. جزم, *a writing-reed not having a nib obliquely cut.* (TA in that art. [See 2 in the present art.]) And حَرْفَا الرَّأْسِ *The two lateral halves of the head.* (TA.) [Hence, also, the phrase] فُلَانٌ عَلَى حَرْفٍ مِنْ أَمْرِهِ [and بِحَرْفٍ مِنْهُ (see 3, first sentence,)] *Such a one is* [*standing*] *aloof with respect to his affair,* (عَلَى نَاحِيَةٍ مِنْهُ, ISd, TA,) [*in suspense,*] *waiting, and looking to the result, if he see, in regarding it from one side, what he likes;* (TA;) *turning from it if he see what does not please him.* (ISd, TA.) The saying, in the Ḳur xxii. 11, وَمِنَ ٱلنَّاسِ مَنْ يَعْبُدُ ٱللّٰهَ عَلَى حَرْفٍ means *And of men is he who serves God standing aloof with respect to religion, in a fluctuating state, like him who is in the outskirts of the army, who, if sure of victory and spoil, stands firm, and otherwise flees:* (Kṣh, Bḍ:*) or the meaning is, *who serves God in doubt,* or *suspense,* (Zj, Ḳ, Jel,) *being unsteady like him who alights and abides upon the* حَرْف [i. e. *point,* or *ridge,* or *brow,*] *of a mountain:* (Jel:) or *in a state of disquietude respecting his case;* (Ibn-'Arafeh, Ḳ;) i. e. *not entering into the religion firmly,* or *steadily:* (Ḳ:) or *who serves God in one mode of circumstances;* i. e. *when in ample circumstances, and not when straitened in circumstances;* (Az, Ṣ, Ḳ;) as though good fortune and plenty were one side, and an evil state were another side: (Az, TA:) [hence,] حَرْفٌ sometimes signifies *a mode,* or *manner,* and *a way.* (Mṣb.) —— *A letter of the alphabet:* pl. حُرُوفٌ: (Ṣ, Mṣb, Ḳ:) the letters being thus called because they are the extremities of the word [and of the syllable]. (Kull.) The saying of the lawyers, تُبْطَلُ الصَّلَاةُ بِحَرْفٍ مُفْهِمٍ [*Prayer is made null by a significant letter*] means only by an imperative of a verb of which the first and last radical letters are infirm; such as قِ from وَقَى, and فِ from وَفَى, and the like. (Mṣb.) —— As a grammatical term, †[*A particle*; i. e.] *what is used to express a meaning, and is not a noun nor a verb:* every other definition of it is bad: (Ḳ:) pl. حُرُوفٌ. (Mṣb, &c.) —— And ‡*A word* [absolutely: often used in this sense in lexicons &c.]. (Kull.) —— *A dialect, an idiom,* or *a mode of expression, peculiar to certain of the Arabs:* pl. [of pauc.] أَحْرُفٌ: so in the saying (of Moḥammad, TA) نَزَلَ ٱلْقُرْآنُ عَلَى سَبْعَةِ أَحْرُفٍ *The Ḳur-án has been revealed according to seven dialects,* of the dialects of the Arabs: (A'Obeyd, Az, IAth, Ḳ:) or this means, *according to seven modes,* or *manners,* (Mgh, Mṣb,) *of reading:* whence فُلَانٌ يَقْرَأُ بِحَرْفِ ٱبْنِ مَسْعُودٍ *Such a one reads in the manner of reading of Ibn-Mes'ood.* (Mgh.) ═ Applied to a she-camel, †*Lean,* or *light of flesh;* or *lean, and lank in the belly;* (Ṣ, Ḳ;) *and firm, strong,* or *hardy;* likened to the حَرْف of a mountain; (Ṣ;) or to the حرف of a sword, (Z, O, TA,) in respect of her leanness, or thinness, and her sharpness and effectiveness in pace; (Z, TA;) or to a letter of the alphabet, meaning the letter ا, in respect of her leanness: (TA:) or *excellent,* or *high-bred,* or *strong and light and swift, sharp and effective in pace, rendered lean by journeyings;* likened to the حرف of a sword: (L:) or *emaciated:* (Ṣ, Ḳ:) so Aṣ used to say: (Ṣ:) but this is inconsistent with Dhu-r-Rummeh's description of a she-camel by the epithets جُمَالِيَّةٌ حَرْفٌ سِنَادٌ: (TA:) [see حَرِيثَةٌ:] or [in the CḲ "and"] *great; big; of great size;* (Ḳ, TA;) likened to the حرف of a mountain: (TA:) it is applied only to a she-camel: one may not say جَمَلٌ حَرْفٌ. (IAạr, TA.)

حُرْفٌ and ▼ حُرْفَةٌ (Ṣ, Ḳ) and ▼ حِرْفَةٌ (Mgh, Ḳ) and ▼ حِرَافٌ (TA) *Ill-fatedness; privation of prosperity;* or the *being denied prosperity;* syn. حِرْمَانٌ [as inf. n. of حُرِمَ]: (Ḳ, TA:) *lack of good fortune, so that one has no increase of his cattle* or *other property:* (Ṣ:) *debarment from the means of subsistence.* (Mgh.) Hence the saying of 'Omar, لَحِرْفَةُ ▼ أَحَدِهِمْ أَشَدُّ عَلَيَّ مِنْ عَيْلَتِهِ, (Ṣ, Ḳ,) or, accord. to one reading, ▼ لَحُرْفَةُ, (TA,) [*Verily the ill-fatedness of any one of them is more distressing to me than his poverty:*] i. e., the supplying the wants of the poor man is easier to me than the making the bad to thrive: or the meaning is, *the want of the means of gaining subsistence by any one of them, and grief on that account, is more distressing to me than his poverty:* so in the Nh. (TA.) ═ الحُرْفُ *A certain grain, resembling* الخَرْدَل [or *mustard*]; (Az, Mṣb, TA;) called by the vulgar, (AḤn, TA,) or in the dial. of El-'Irák, (TA in art. رشد,) حَبُّ الرَّشَادِ, (AḤn, Ṣ, Ḳ,) or الرَّشَادُ: (Mṣb:) n. un. with ة, (TA,) applied to *a single grain thereof.* (Mṣb.) [See art. رشد.] Hence حِرِّيفٌ [q. v.]. (Ṣ, Mṣb.)

حُرْفَةٌ: see حُرْفٌ, in two places.

حِرْفَةٌ *A craft,* or *handicraft,* (Ṣ, Ḳ, TA,) *by which one gains his subsistence; a mode,* or *manner, of gain; any habitual work* or *occupation* of a man; because he turns (يَنْحَرِفُ, Ḳ, i. e. يَمِيلُ, TA) to it; (Ḳ, TA;) a subst. from اِحْتَرَفَ: (Mgh, Mṣb:) pl. حِرَفٌ. (TA.) ═ See also حُرْفٌ, in two places.

حُرْفِيٌّ *A seller of* الحُرْف, i. e. حَبّ الرَّشَاد. (Ḳ.)

حِرَافٌ: see حُرْفٌ.

حَرِيفٌ *A fellow-worker,* syn. مُعَامِلٌ, (Ṣ, Mgh, Mṣb, Ḳ,) *in one's craft* or *ordinary occupation:* (Ḳ:) and *an associate:* (KL:) pl. حُرَفَآءُ. (Mṣb.) —— It is mostly used by foreigners as meaning *A companion in drinking:* and by most of the Turks, as implying vituperation; [like our term "fellow;"] so that when any one of them addresses another by this epithet, he is angry. (TA.)

حَرَافَةٌ The *quality,* or *property,* of *burning,* or *biting, the tongue; acritude.* (Ṣ, Mṣb, TA.)

حِرِّيفٌ, from الحُرْف, *Burning,* or *biting, to the tongue:* (Ṣ, Mṣb, TA:) it is applied in this sense to an onion, and to other things: one should not say حَرِيفٌ. (Ṣ, TA.)

مَحْرِفٌ *A place to which to turn away,* or *back,* from a thing. (AO, Ṣ, Ḳ.) So in the saying, مَا لِي عَنْ هٰذَا ٱلْأَمْرِ مَحْرِفٌ [*I have no place to which to turn away,* or *back, from this thing*]. (AO, Ṣ, Ḳ.*) —— Also, and ▼ مُحْتَرَفٌ, *A place in which a man earns or gains* [*subsistence*], or *labours to do so, and employs himself as he pleases,* or *follows his various pursuits.* (Ḳ.)

مُحْرِفٌ A man *whose property increases, and becomes in a good state* or *condition;* or *whose cattle increase* &c. (Ṣ, Mṣb.)

مِحْرَفٌ : } see مِحْرَافٌ.
مِحْرَفَةٌ :

مُحَرَّفٌ [pass. part. n. of 2, q. v. —] One whose *property has gone.* (TA.) — A *writing-reed nibbed obliquely; having the right tooth of the nib higher* [i. e. *longer*] *than the left.* (TA.)

مُحَرِّفُ القُلُوبِ, applied to God, *The Turner,* or *Incliner, of hearts:* or *the Mover of hearts:* (TA:) or *the Remover of hearts.* (Fr, TA voce مُحَرِّك, q. v.)

مِحْرَافٌ (Ṣ, L, Ḳ) and ▼ مِحْرَفٌ, (L, TA,) or ▼ مِحْرَفَةٌ, (Akh, TA,) *A probe with which the depth of a wound is measured:* (Ṣ, L, Ḳ:) pl. of the first مَحَارِيفُ; and [of the second, or,] accord. to Akh, of the last, مَحَارِفُ. (TA.)

مُحَارَفٌ *Prevented,* or *withheld, from obtaining good; withheld from good fortune,* or *from sustenance; denied,* or *refused, good,* or *prosperity; lacking good fortune; having no increase of his cattle or other property;* (Ṣ, Mgh,* Ḳ;*) *contr. of* مُبَارَكٌ: (Ṣ:) or *having his gain,* or *earnings, turned away from him:* (Mṣb:) or *who obtains not good from a quarter to which he betakes himself:* or *scanted in his means of subsistence:* or *who works not,* or *labours not, to earn,* or *gain:* or *who earns,* or *gains, with his hands, but not enough for the support of himself and his household* or *family:* (TA:) مُحَارِفٌ and مُجَارَفٌ are dial. vars. thereof. (TA in art. جرف.)

مُحْتَرِفٌ : see مُحْرِفٌ.

مُحْتَرِفٌ *A handicraftsman; a worker with his hands.* (Ṣ, TA.)

مُتَحَرِّفًا لِقِتَالٍ, in the Ḳur [viii. 16], means *Turning away for the purpose of returning to fight:* the doing which is one of the stratagems of war. (Mgh, Mṣb.*)

حرق

1. حَرَقَهُ, aor. ـِ, inf. n. حَرْقٌ : see 4. ═ حَرَقَهُ, (Ṣ, Ḳ,) aor. ـُ, (TA,) inf. n. حَرْقٌ, (Ṣ,) *He filed it:* and *he rubbed one part of it with another.* (Ṣ, Ḳ.) — And hence, (Ṣ,) حَرَقَ نَابَهُ, aor. ـُ and ـِ, (Ṣ, Ḳ,) inf. n. as above, (TA,) *He ground his dog-tooth, so that it made a grating sound:* (Ṣ, Ḳ:) when said of a stallion-camel, denoting threatening: and, accord. to IDrd, when the like is said of a she-camel, it is asserted to denote a consequence of fatigue. (TA.) And ▼ حَرَّقَ الأَسْنَانَ (Ḳ and TA in art. وهم) *He grated the teeth.* (TA in that art.) One says, فُلَانٌ يَحْرُقُ عَلَيْكَ الأُرَّمَ غَيْظًا (Ṣ, A*) *Such a one grinds together the* ارّم [or *teeth,* or *molar teeth,* (as the word is generally understood to mean in this case, but other meanings are assigned to it,)] *at thee* [*in anger,* or *rage*], *like one filing:* (A, TA:) or, as some say, الأُرَّمَ [*the canine teeth*]: and the verb is also used without the objective complement, because the meaning is understood. (Ḥam p. 115.) IDrd makes the act to be that of the canine tooth; saying, حَرَقَ نَابُ البَعِيرِ, meaning *The canine tooth of the camel made a grating sound.* (TA.) AHát also mentions the saying, فُلَانٌ يَحْرُقُ نَابَهُ عَلَيَّ [*Such a one's canine tooth makes a grating sound at me*]: and Zuheyr uses the phrase يَحْرُقُ نَابُهُ عَلَيْهِ. (Ḥam p. 286.) — حَرْقٌ also signifies The *act of eating to the uttermost.* (IAạr, TA.) ═ حَرِقَ *He* (a man) *was,* or *became, evil in disposition.* (TA.) ═ حَرَقٌ, as an inf. n., [i. e. of حَرِقَ,] signifies A garment's, or cloth's, *being burnt by beating* [*with too much violence*]. (KL.) — And The *springing forth,* or *shooting forth, vehemently,* of lightning. (KL.) ═ حَرِقَ شَعَرُهُ, (Ṣ, Ḳ,) aor. ـَ, (Ḳ;) inf. n. حَرَقٌ, (TA,) *His hair fell off piecemeal.* (Ṣ, Ḳ.) [And حَرِقَتِ النَّاصِيَةُ *The forelock* of the horse *became thin,* or *scanty:* for it is said that] الحَرَقُ in relation to the ناصية is like السَّفَا. (TA.) And حَرِقَتِ اللِّحْيَةُ *The beard was,* or *became, shorter upon the chin than upon the two sides of the face.* (TA.) ═ حُرِقَ, aor. ـُ, inf. n. حَرْقٌ, *His* حَارِقَة [q. v.] *became cut,* or *severed:* said of a man: in speaking of a camel, حُرِقَ, like عُنِيَ, is more commonly used than حَرِقَ. (TA.)

2. حَرَّقَهُ, inf. n. تَحْرِيقٌ : see 4. — تحريق also signifies Fire's *making a mark,* or *impression, upon a thing.* (TA.) — حَرَّقَ الإِبِلَ, said of pasturage, (Ḳ,) [particularly] of what is termed حَمْض, (Ṣ,) *It made the camels thirsty.* (Ṣ, Ḳ.) ═ See also 1.

3. حَارَقَهَا, (Ḳ,) inf. n. مُحَارَقَةٌ, (Ṣ,) *He lay with her* (Ṣ, Ḳ) [عَلَى الحَارِقَةِ, i. e.] *on the side.* (Ḳ.)

4. أَحْرَقَتْهُ النَّارُ, inf. n. إِحْرَاقٌ, (Mṣb,) [*The fire burned him.*] And احرقه بِالنَّارِ (Ṣ, Mṣb, Ḳ) [*He burned him,* or *it, with fire*]: this phrase, and حَرَقَهُ بالنار, aor. ـِ, (Ḳ,) inf. n. حَرْقٌ, (TA,) signify the same; as also ▼ حرّقه : (Ḳ:) or this last [signifies *he burned him,* or *it, much,* or *frequently,* or *repeatedly;* for it] denotes muchness, or frequency, or repetition, of the action. (Ṣ, Mṣb, TA.) — [Hence, احرقه † *It pained him;* or *caused him burning pain:* said of beating, or a blow; and of a galling, or chafing; and of fever, passionate desire, rage or anger, hunger, &c.] And أَحْرَقَنَا فُلَانٌ † *Such a one afflicted, distressed, annoyed, molested,* or *hurt, us.* (TA.) And احرقهُ بِاللِّسَانِ † *He blamed, upbraided,* or *reproached, him; detracted from his reputation.* (Mṣb.) And احرق البَرْدُ الكَلَأَ [† *The cold nipped, shrunk, shrivelled,* or *blasted, the herbage;* like أَنْضَجَ, q. v.; and like the Lat. "ussit," and "adussit:" comp. Virgil, Georg. i. 93, "Boreæ penetrabile frigus adurat:" and Lucan, iv. 52, "Urunt montana nives:" and Ecclesiasticus, xliii. 20 and 21, "When the cold north wind bloweth, and the water is congealed into ice, it abideth upon every gathering together of water, and clotheth the water as with a breastplate: it devoureth the mountains, and burneth the wilderness, and consumeth the grass as fire"]: (Ṣ and Ḳ voce حَسّ :) and [in like manner] احرق النَّبَاتَ is said of heat, and of cold, and of a wind, and of other banes, or causes of mischief or harm. (TA.) And احرقهُ † *He,* or *it, destroyed,* or *caused to perish, him,* or *it.* (TA.) — You say also, أَحْرِقْ لَنَا فِي هٰذِهِ القَصَبَةِ نَارًا *Give thou,* or *bring thou, to us, upon this cane, some fire.* (IAạr, TA.) ═ Also احرق *He made,* or *prepared, what is termed* حَرِيقَة. (Ḳ.)

5: see 8. — [Hence,] هُوَ يَتَحَرَّقُ جُوعًا † [*He burns with hunger*]: like يَتَضَرَّمُ. (TA.)

8: احترق [*It burned,* or *became burnt,*] بِالنَّارِ [*with fire*]: and ▼ تحرّق [*it burned,* or *became burnt, much,* or *frequently,* or *repeatedly*]: each is a quasi-pass.; (Ṣ, Mṣb, Ḳ, TA;) [the former, of احرق or حَرَقَ ; and the latter, of حَرَّقَ.] — [Hence,] one says of a horse, يَحْتَرِقُ فِي عَدْوِهِ [† *He is fiery, ardent,* or *vehement, in his running*]. (Ṣ.) And احترق النَّبَاتُ [† *The plant,* or *plants,* or *herbage, became nipped, shrunk, shrivelled,* or *blasted:* see 4]: this is said of a consequence of heat, and of cold, and of a wind, and of other banes, or causes of mischief or harm. (TA.) And احترقت الفِضَّةُ † *The silver became black.* (Ḥar p. 114.) And احترق † *He,* or *it, perished.* (TA.)

حُرْقٌ : see حَرَقٌ, in two places.

حَرِقٌ † *An angry man.* (TA.)

حَرَقٌ [*A burning* by means of fire;] a subst. (Mgh, Mṣb) from الإِحْرَاقُ, (Mgh,) [i. e.] from إِحْرَاقُ النَّارِ : (Mṣb:) or *fire,* (Ṣ, Mṣb, Ḳ,) itself; (Mṣb;) [the *fire of a burning house &c.;*] as also ▼ حَرِيقٌ (Mgh) and ▼ حَارِقَةٌ : (Ḳ:) or the *flame of fire.* (IAạr, Th, Mgh, Ḳ.) The first is meant in the saying, ضَالَّةُ المُؤْمِنِ حَرَقُ النَّارِ [*The stray-beast of the believer is* a cause of *the burning of fire*]: (Mgh:) or it here signifies *the flame of fire:* a trad., meaning that if any one takes the stray-beast of a believer to possess it, his doing so will bring him to the flame of the fire [of Hell]. (Az, Mgh, TA.) And hence, (Mgh,) الحَرَقُ شَهَادَةٌ, (Mgh, TA,) i. e. [*Burning,* or] *fire,* [or *flame, is* a cause of one's receiving the reward of *martyrdom:*] occurring in another trad. (TA.) You say also فِي حَرَقِ اللّٰهِ *In the fire of God.* (Ṣ.) And أَلْقَى اللّٰهُ الكَافِرَ فِي ▼ حَارِقَتِهِ, i. e. [*May God cast the unbeliever*] *into his fire.* (TA.) — *A burn,* (Ṣ,) or *a mark of burning,* (Ḳ,) *in a garment,* or *piece of cloth, from the beating* (Ṣ, Ḳ) *of the washer, and whitener, and the like;* (Ḳ;) and so, sometimes, ▼ حُرْقٌ : (Ṣ:) or the former, *a hole thus caused in a garment,* or *piece of cloth;* (IAạr, Mgh, TA;) and so, sometimes, ▼ the latter; which also signifies *a hole caused by fire, in a garment,* or *piece of cloth.* (Mgh.)

حَرِقٌ A cloud *lightening vehemently.* (Ṣ, Ḳ.) — *Sharp;* as though *having the quality of burning;* applied to an iron head or blade of an arrow or a spear or sword &c.; (TA;) and so ▼ حُرُقَةٌ and ▼ حُرَاقَةٌ and ▼ حَارُوقَةٌ, applied to swords. (Ḳ.) ═ See also حَرِقُ الشَّعَرِ. — حَرِقٌ *Having the hair falling off piecemeal:* (Ṣ, Ḳ:) and حَرِقُ الجَنَاحِ has a similar meaning; (Ṣ, TA;) i. e. [*having the feathers of the wing falling off piecemeal:* or] *short in the wing:* or *having it cut off.* (TA.) And رِيشٌ حَرِقٌ *Feathers falling off, and becoming scattered, by degrees.* (TA.) And لِحْيَةٌ حَرِقَةٌ *A beard that is shorter upon the chin than upon the two sides of the face.* (TA.) — Also, حَرِقٌ, A man *having the extremities much chapped:* (Ḳ:) so some say. (TA.) — See also مَحْرُوقٌ.

حَرْقَةٌ: see what next follows.

حُرْقَةٌ [A state of *burning;*] a subst. from اِحْتَرَقَ; as also ▾حَرِيقٌ. (Ṣ, Ḳ.) Thus the latter means in the Ḳur [lxxxv. 10], ▾وَلَهُمْ عَذَابُ الحَرِيقِ [*And for them* shall be *the punishment of burning:* as in other passages in the Ḳur]. (TA.) — †*A burning such as a man experiences from the taste of a thing in which is heat, or from love, or grief;* (TA;) and *such as is experienced in the eye from ophthalmia,* and *in the heart from pain:* (Lth, TA:) *heat;* as in the phrase, فِى جَوْفِهِ حُرْقَةٌ [†*In his belly,* or *chest, is heat*]; and so ▾حُرَقَةٌ and ▾حَرِيقَةٌ. (Ḳ.)

حُرَقٌ: see حُرَقَةٌ.

حُرْقَانٌ *A rubbing together of the thighs.* (Ṣ, Ḳ.)

حُرَاقٌ: see حَرَاقٌ, in two places. — †A horse *that runs much:* (Ḳ:) or حُرَاقُ العَدْوِ a horse *that is fiery, ardent,* or *vehement,* (يَحْتَرِقُ,) *in his running.* (Ṣ.) — †*Very salt* water; (Ṣ, Ḳ;) as also ▾حُرَّاقٌ: (Ḳ:) as though it burned the fauces of the drinker: (TA:) or *such as is exceeded* [*in saltness*] *by nothing; that makes the urine of the camels to burn;* as also قُعَاعٌ. (IAạr, TA.) — Also, (Ṣ, Ḳ, &c.,) and ▾حُرَاقَةٌ (Ṣ, Mgh, Ḳ) and ▾حُرَّاقٌ, (Ḳ,) or this is vulgar, (O, TA,) and ▾حُرَّاقَةٌ, or this is incorrect, (Ḳ,) or vulgar, (Ṣ, O,) and ▾حُرُوقٌ and ▾حَرُوقٌ (Fr, O, Ḳ) and ▾حَرُوقَةٌ, (Fr, Ṣ, O, Ḳ) [*Tinder;* i. e.] *a thing,* (Ṣ, Ḳ,) or *burnt rag,* (AḤn, ISd, TA,) *into which fire falls when it is struck:* (AḤn, Ṣ, ISd, Ḳ, TA:) or *what remains of burnt cloth:* (Mgh:) [and *any substance used for receiving fire that is struck;* as, for instance, the pith of the عُشَر.]

حَرَاقٌ, applied to fire, (نَارٌ,) *That burns everything;* as also ▾حُرَاقٌ: (Aboo-Málik, TA:) *that spares, or leaves, nothing.* (IAạr, Ḳ.) — †*A man that spoils, mars, destroys, or consumes, everything;* (IAạr, Ḳ;) *sparing nothing; like the fire thus termed;* (IAạr, TA;) as also ▾حُرَاقٌ. (Ḳ.) In some copies of the Ḳ, مَنْ يُفْسِدُ فِى كُلِّ شَىْءٍ; but correctly, without فى. (TA.) — رَمْىٌ حِرَاقٌ †*A vehement throwing* or *casting* or *shooting.* (Ḳ.)

حُرُوقٌ: see حُرَاقٌ.

حَرُوقٌ: see the next paragraph.

حَرِيقٌ: see حُرَقٌ: — and see also حُرْقَةٌ, in two places. — *Heat,* or †*cold,* or *a wind,* or *some other cause of mischief* or *harm, that burns,* or †*nips, shrinks, shrivels, or blasts,* (يُحْرِقُ,) *herbage.* (TA.) — Also i. q. ▾مُحْرَقٌ, [i. e. *Burnt,*] (Mgh, Mṣb,) and so ▾مَحْرُوقٌ: (TA:) pl. of the first حَرْقَى; like قَتْلَى and جَرْحَى, pls. of قَتِيلٌ and جَرِيحٌ. (Mgh.) Thus, in a trad., الحَرِيقُ شَهِيدٌ [*The burnt is a martyr*]: (Mgh:) or ▾الحَرِقُ, i. e. *he who falls into fire, and takes fire and burns.* (TA.) — The *grating sound* of the dog-tooth by reason of anger, or rage; as also ▾حُرُوقٌ. (TA.)

حُرَاقَةٌ: see حُرَاقٌ.

حَرِيقَةٌ: see حَرُوقَةٌ.

حَرِيقَةٌ: see حُرْقَةٌ. — Also, (Yaạḳoob, Ṣ, Ḳ,) and ▾حَرُوقَةٌ, (Ḳ,) *A kind of food,* (Ḳ,) *thicker than what is termed* حَسَآء; (Yaạḳoob, Ṣ, Ḳ;) *like* نَفِيتَةٌ: (Ṣ:) or *water,* (Ḳ,) i. e. *hot water,* (TA,) *upon which a little flour is sprinkled, and which swells,* or *becomes inflated, in boiling,* (Ḳ, TA,) *and becomes of a whitish dust-colour: it is licked up with the tongue: and is also called* نَفِيتَةٌ: they made use of it in hard and dear times, and when the cattle were lean, and when the season was severe: (TA:) or *it was made by sprinkling flour upon water or fresh milk until it swelled, and became* [*like*] *what is termed* حساء: a man used to satisfy his household with it when fortune overcame him: and it is also called نفيتة: (ISk, Az, TA:) pl. حَرَائِقُ. (Ṣ.) One says, وَجَدْتُ بَنِى فُلَانٍ مَا لَهُمْ عَيْشٌ إِلَّا الحَرَائِقُ [*I found the sons of such a one having no means of subsistence other than the messes of the kind called* حرائق]. (Ṣ.)

حَرُوقَةٌ: see حُرَاقٌ.

حُرَّاقٌ: see حُرَاقٌ, in two places: — and see also مَحْرُوقٌ, in two places.

حَرُوقٌ: see حُرَاقٌ.

حَرَّاقَةٌ *A kind of ship,* (Lth, Ṣ, Ḳ,*) [*built*] *at El-Baṣrah,* (Ḳ,) *in which are engines for throwing fire upon the enemy at sea, or on a large river:* (Lth, Ṣ, Ḳ:) accord. to some, *such an engine itself:* (ISd, TA:) accord. to the A, [*a bark;*] *a light-going ship:* (TA:) [it is often used in this last sense in post-classical works:] pl. حَرَّاقَاتٌ (Ḳ) [and حَرَارِيقُ]. — Also the former pl., The *places of those who fry* [*meat &c.*], and *of the makers of charcoal:* (Lth, Ḳ:) of the dial. of the people of El-Baṣrah. (Lth, TA.)

حُرَّاقَةٌ: see حُرَاقٌ: — and see also حُرَاقٌ.

حِرِّيقَى The *act of copulation upon the side.* (Z, TA.) [See 3.]

حَارِقَةٌ: see حَرِقٌ, in two places. — الحَارِقَتَانِ *The heads* [*of the bones*] *of the two thighs, in the two hips:* or *two sinews in the two hips:* (Ṣ, Ḳ:) when these are severed, the man walks upon the extremities of his toes, and cannot do otherwise: when one so walks by choice, you say that he is مُكْتَامٌ, part. n. of اِكْتَامَ: (IAạr, TA:) the حارقة is also explained as being the *sinew that connects the thigh and the hip:* or the *sinew that connects the head* [*of the bone*] *of the thigh and that* [*of the bone*] *of the upper arm, which turn in the* صَدَفَة [or *socket*] *of the hip and of the shoulder-blade:* when it is severed, it never unites: or *a sinew in the* حُرْبَة [or *socket of the hip*], *that suspends* [*the bone of*] *the thigh to the hip, and by means of which the man walks:* it is said that when the حارقة is displaced, the man becomes lame. (TA.) — Also, the sing., The *side of the body.* (AHeyth, TA.)

حَارُوقَةٌ: see حَرِقٌ.

مُحْرَقٌ: see حَرِيقٌ.

المُحَرِّقُ *A certain idol, of Bekr Ibn-Wáïl,* (Ḳ,) *which was in Selmán.* (TA.)

مَحْرُوقٌ: see حَرِيقٌ. — *Having his* حَارِقَة [q. v.] *severed;* (Ṣ, TA;) as also ▾حَرِقٌ; which latter is [said to be] the more common: (TA:) [but this I doubt:] or, as some say, (Ṣ,) *having his hip dislocated:* (Ṣ, Ḳ:) [pl. of the latter, deviating from rule, ▾حُرَّاقٌ, occurring in a verse below.] The rájiz says, (Ṣ,) namely, Aboo-Moḥammad El-Ḥadhlamee, (TA,) describing a pastor, (Ṣ,)

* يَظَلُّ تَحْتَ الفَنَنِ الوَرِيقِ *
* يَشُولُ بِالمِحْجَنِ كَالمَحْرُوقِ *

[*He continues,* or *continues during the day, beneath the leafy branch, raising the crooked-headed stick, like the* محروق]: i. e. he stands upon one leg, stretching himself up towards the branches, and drawing them to him with the محجن, and shaking off their leaves for the camels: (Ṣ, TA:) or he stands upon the extremities of his toes, [see حَارِقَةٌ,] in order to reach the branch and bend it to his camels. (ISd, TA. But see another meaning of the last word, below.) And another says,

* هُمُ الغِرْبَانُ فِى حُرُمَاتِ جَارٍ *
* وَفِى الأَدْنَيْنَ حُرَّاقُ ▾ الوُرُوكِ *

[*They are* like *the crows in respect of the sacred rights of a neighbour; and in respect of inferiors,* like *those who are dislocated in the hips,* or *who have the sinews of the hip-joints severed*]: i. e., when a neighbour having a sacred right to respect alights among them, they are like the crow, which loaths not the gall on the back nor that which is unclean; and in wrongful treatment of their inferiors, like the محروق, who walks with an inclining of the body (يَمْشِى مُتَجَانِفًا); and they abstain from aiding and defending them. (Ṣ, TA.) — Accord. to Ibn-'Abbád, in the saying of the rájiz cited above, it means (TA) The *iron instrument with which one roasts meat;* syn. سَفُّودٌ. (Ḳ, TA.)

حرقد

حِرْقِدٌ: see what follows.

حَرْقَدَةٌ The عُقْدَة *of the* حُنْجُور [meaning the *head of the windpipe;* app. called its عقدة, or "knob," because its anterior portion forms a protuberance, that which we term "pomum Adami:" see غَلْصَمَة]: (Ṣ, L, Ḳ:) pl. حَرَاقِدُ. (L.) — And The *root of the tongue;* (IAạr, L;) as also ▾حِرْقِدٌ. (IAạr, L, Ḳ.)

حرقص

حُرْقُوصٌ *A certain insect, resembling the flea,* (Ṣ, A, Ḳ,) *to which, sometimes, there grow two wings, and then it flies;* (Ṣ;) or, as some say, *larger than the flea;* (TA;) or *like the tick;* (A;) accord. to Lth, *a certain variegated insect* (دُوَيْبَة مُجَزَّعَة); (TA;) *the* حُمَة [i. e. *venom,* or *sting,*] *of which is like that of the hornet;* (A, Ḳ;) *which sticks to men, and bites,* or *stings;* (تَلْدَغُ;) *and to which the extremities of whips are likened;* (A;) or, accord. to Az, *it has no* حُمَة *when it bites; but its bite occasions much pain,* [*though*] *it has no venom* (سُمّ) *like that of hornets:* (TA:) or (Ḳ) *a certain small insect,* (IDrd, TA,) *resembling the tick, that sticks to men:* (IDrd, Ḳ, TA:) or it is *smaller than the* [*black beetle called*] جُعَل; (ISk, TA;) or, accord. to the M, *a thing like a small pebble, speckled a little with red or yellow, but its prevailing colour is black; which collects, and enters beneath men, and in their groins,* or *armpits,* or *the like, and bites them; and rends the skins in which water or milk is kept;* or, as in the T, *a certain small insect,* (TA,) *which makes holes in the skins wherein water or milk is kept, and* (as Az heard the Arabs of the desert to assert, TA) *enters into the pudenda of girls;* (Ḳ, TA;) *and is of the same kind as* جِعْلَان [pl. of جُعَل], *but smaller; black, speckled with white:* (TA:) because of its entering into the فَرْج of the virgin girl, it is called عَاشِقُ الأَبْكَارِ: (IB, TA:) pl. حَرَاقِيصُ. (Ḳ.) — Also The *stone of a green unripe date.* (AA, Ḳ.)

حرقف

Q. 1. حَرْقَفَ الأَتَانَ *He* (an ass) *took hold upon the* حَرَاقِف [pl. of حَرْقَفَة] *of the she-ass* [*with his fore legs*]. (Ibn-'Abbád, Sgh, Ḳ.)

حَرْقَفَةٌ The *bone of the* حَجَبَة, which is the *head* [or *crest*] *of the hip* or *haunch:* (Ṣ, Ḳ:) or the *head of the upper part of the hip* or *haunch:* (TA in art. اكر:) or [the dual] حَرْقَفَتَانِ signifies the *place where the head of each thigh unites with,* or *meets, the hip* or *haunch,* (Zj in his "Khalḳ el-Insán," and TA,*) *externally:* (TA:) [see an explanation of حَجَبَة, in which a distinction is made between this latter term and حَرْقَفَة: a distinction is also made between them by Zj in his work cited above, as a reference to حجبة will show:] pl. حَرَاقِفُ (Ṣ, Ḳ) and حَرَاقِيفُ. (TA.) One says, المَرِيضُ إِذَا طَالَتْ ضَجْعَتُهُ دَبِرَتْ حَرَاقِفُهُ [*The sick man, when his lying on the side is of long continuance, his* حراقف *become galled;* i. e. *the exterior prominent regions of his hip-joints*]. (Ṣ.)

حُرْقُوفٌ An *emaciated* beast, or horse or the like; (Ṣ, Ḳ, TA;) i. e., *whose* حَرَاقِيف *are apparent.* (TA.)

حرك

1. حَرَكَ, aor. ِ, (IḲṭṭ, Mṣb, Ḳ,) inf. n. حَرَكٌ, (IḲṭṭ, Mṣb, MF,) or حَرْكٌ, with fet-ḥ, (Ḳ,) as in the 'Eyn and O, but disallowed by MF, (TA,) [and probably transcribed from some lexicon in which, as is often the case, "with fet-ḥ" relates to the medial radical letter,] and حَرَكَةٌ, (Ḳ,) or this is an inf. n. of un.; (Mṣb; [but in general usage it is not thus restricted;]) and ♦ تَحَرَّكَ [which is the more common]: (Ṣ, Mṣb, Ḳ:) [both signify the same; *It,* or *he, moved; was,* or *became, in a state of motion, commotion,* or *agitation; shook, shook about, wabbled, tottered, waggled, wagged,* or *nodded:* or the latter verb, more properly, *it,* or *he, was put,* or *it put itself,* or *he put himself, in a state of motion, commotion,* or *agitation:* and the latter also signifies *he became active;* said of a growing child, and of a young gazelle &c.: (see حَرِكٌ:)] the former is the *contr.* of سَكَنَ; (Mṣb, Ḳ;) and the latter is *quasi-pass. of.* حَرَّكَهُ. (Ṣ, Mṣb, Ḳ.) — حَرَكَهُ, (AA, Ṣ, Ḳ,) aor. ُ, inf. n. حَرْكٌ, (Ṣ,) *He hit,* or *hurt, his* (a man's, AA, Ḳ) حَارِك. (AA, Ṣ, Ḳ.) Accord. to Fr, حَرَكَ حَارِكَهُ *He cut his* حَارِك. (TA.) And accord. to AZ, حَرَكَهُ بِالسَّيْفِ, inf. n. حَرْكٌ, *He smote* [meaning *severed*] *his neck with the sword.* (TA.) — حَرِكَ, aor. َ, (IAạr, Ḳ,) inf. n. حَرَكٌ, (TḲ,) *He was,* or *became, incapable of coition* (IAạr, Ḳ) *with women.* (IAạr, TA.)

2. حَرَّكَهُ, [inf. n. تَحْرِيكٌ,] *He moved; put in motion; put in a state of motion, commotion,* or *agitation; moved about; agitated, stirred,* or *shook; it,* or *him.* (Ṣ, Mṣb, Ḳ.) [Hence,] حَرَّكَ أَحْشَائِى [*It agitated my bowels*] is like the phrase حَرَّكَ مِنِّى [*it agitated me*], said by one who has been agitated by reason of an event or affair. (Ḥam p. 183.) — [*He made it* (a letter) *movent;* i. e., *made it to be immediately followed by a vowel; contr. of* سَكَّنَهُ.] — *He urged him* (a camel) *to go, but he went not.* (Ibn-'Abbád, Z.)

5. تَحَرَّكَ: see 1. — [Also *It* (a letter) *was,* or *became, movent;* i. e., *immediately followed by a vowel; contr. of* سَكَنَ.]

حَرِكٌ, applied to a boy, *Light, active, agile, brisk, lively,* or *sprightly; and sharp,* or *quick, in intellect.* (Ṣ, Ḳ.)

حَرَكَةٌ *Motion; commotion; agitation; contr. of* سُكُونٌ; (Ṣ, Mṣb, Ḳ;*) and so ♦ حَرَاكٌ, (Ṣ, Mṣb, Ḳ,) as in the phrase, مَا بِهِ حَرَاكٌ [*There is not in him any motion*]: (Ṣ, Ḳ:) and, accord. to El-Khafájee, حِرَاكٌ also; but this is disallowed by MF. (TA.) [The first, accord. to the Mṣb, is an inf. n. of un.: but see 1.] — [Also *Activity:* often used in this sense in the classical language, and in the present day. — And A letter's *having a vowel immediately following:* and *a vowel* itself.]

حُرْكُكَةٌ *i. q.* حَرْقَفَةٌ [q. v.]: (Ṣ: [in the Ḳ حُرْقُوفٌ, which is evidently a mistake:]) pl. حَرَاكِكُ and حَرَاكِيكُ, (Ṣ, Ḳ,) meaning the *heads,* (Ṣ, TA,) or *extremities,* (TA,) *of the two hips,* or *haunches, that are next the ground when one sits:* (Ṣ, TA:) in the latter pl., which is extr., the ى may be inserted for euphony, because of the double ك. (TA.)

حُرْكُوكٌ: see حَارِكٌ.

حَرَاكٌ: see حَرَكَةٌ.

حَرِيكٌ *Incapable of coition;* (IAạr, Ḳ;) applied to a man and to a horse. (IAạr, TA in art. عجز.) — One who is *weak in the waist, so that, when he walks, he is as though he were plucking up himself* [or *his feet*] *from the ground:* (IDrd, Ḳ:*) fem. with ة. (Ḳ.) A man *weak in the* حَرَاكِيك [pl. of حُرْكُكَة, q. v.]. (TA.)

حَارِكٌ The كَاهِل [or *withers*]; (Ṣ;) as also ♦ حُرْكُوكٌ: (Ḳ:) and the *branches of the two shoulder-blades* of a horse: (Ṣ:) or the *upper part of the* كاهل (Ḳ) of a horse: (TA:) or *a bone projecting from the two sides thereof,* (Ḳ, TA,) *bordered by the two branches of the two shoulder-blades:* (TA:) or the *place of growth of the lowest part of the mane, next the back, upon which he who mounts lays hold:* (Ḳ:) or حَارِكَانِ signifies the *place where the two shoulder-blades meet.* (Mṣb.)

مُحَرَّكٌ The *upper extremity of the neck,* (AZ, Ḳ, TA,) *at the joint of the head.* (TA.)

مُحَرِّكُ القُلُوبِ *The Turner* [or *Mover*] *of hearts;* [applied to God;] occurring in a trad., in which some read in its stead مُحَرِّفُ القُلُوبِ, meaning "The Remover of hearts:" (Fr, TA:) Abu-l-'Abbás says that the former is preferable. (TA.)

مِحْرَاكٌ *A thing,* (Ṣ,) or *piece of wood,* (Ḳ,) *with which a fire is stirred.* (Ṣ, Ḳ.) — *A style with which a receptacle for ink is stirred.* (Lth, TA.)

مُحْتَرِكٌ One *who keeps,* or *cleaves, to the* حَارِك *of his camel.* (Ibn-'Abbád, Ḳ.)

حرم

1. حَرُمَ, (Ṣ, Mṣb, Ḳ,) aor. ُ, (Ḳ,) inf. n. حُرْمٌ, (Mṣb, Ḳ) and حُرُمٌ (Mṣb) and حُرْمَةٌ (IḲooṭ, Ṣ, Mṣb) and حُرُمَةٌ (IḲooṭ, Mṣb) and حَرَامٌ, (Mṣb, Ḳ,) *It* (a thing, Ṣ, Mṣb) *was,* or *became, forbidden, prohibited,* or *unlawful,* (Mṣb,) عَلَيْهِ *to him.* (Ṣ, Ḳ.) And حَرُمَتِ الصَّلَاةُ, (Ṣ, Mṣb, Ḳ,) inf. n. حُرْمٌ (Ṣ, Ḳ) and حُرُمٌ (Ḳ) and حُرُومٌ; (Az, TA;) and حَرِمَت, (Ṣ, Mṣb, Ḳ,) aor. َ, inf. n. حَرَمٌ [in the CḲ حَرْم] and حَرَامٌ; (Mṣb, Ḳ, TA;) *Prayer was,* or *became, forbidden, prohibited,* or *unlawful,* (Mṣb,) عَلَيْهَا *to her;* (T, Ṣ, Ḳ;) namely, a woman (T, Ṣ, Ḳ) menstruating. (Ṣ.) And حَرُمَ السُّحُورُ عَلَى الصَّائِمِ [*The meal before day-break was,* or *became, forbidden to the faster*]. (Ḳ.) And حَرُمَتِ المَرْأَةُ عَلَى زَوْجِهَا, aor. ُ, inf. n. حُرْمٌ and حَرَامٌ, [*The woman was,* or *became, forbidden to her husband.*] (Az, TA.) — [Also *It* (a place, a possession, a right, an office or a function, a quality, a command or an ordinance, &c.,) and *he,* (a person,) *was,* or *became, sacred,* or *inviolable,* or *entitled to reverence, respect,* or *honour;* whence several applications of its part. n. حَرِيمٌ, q. v.] — حَرَمَهُ الشَّىْءَ, aor. ِ; (Ṣ, Mṣb, Ḳ;) and حَرِمَهُ الشىءَ, aor. َ; (Ḳ;) inf. n. حِرْمٌ (Ṣ, Mṣb, Ḳ) and حِرِمٌ and حَرِمَةٌ (Ḳ) and حِرْمَةٌ and حِرْمَانٌ (Ṣ, Mṣb, Ḳ) and حَرِيمَةٌ (Ṣ, Ḳ) and حَرِيمٌ (Ḳ) and مَحْرُمَةٌ; (Ḥar p. 69;) and ♦ احرمه الشىءَ, (Ṣ, Mṣb, Ḳ,) but this last is of weak authority; (Ḳ;) *He denied him,* or *refused him, the thing;* (Ṣ, Ḳ;) *he refused to give him the*

thing: (TA:) *he rendered him hopeless of the thing:* (PṢ:) accord. to the T, حَرْمٌ signifies the act of *denying* or *refusing* [a thing]; and حِرْمَةٌ is the same as حِرْمَانٌ; (TA;) which signifies [also the *denying*, or *refusing*, a thing; or] the *rendering unprosperous*, or *unfortunate*; (KL;) [and frequently, as inf. n. of the pass. v. حُرِمَ, the *being denied prosperity; privation of prosperity; ill-fatedness:* see its syn. حُرْفٌ.] — حَرِمَتْ, aor. ـَ, inf. n. حِرَامٌ; (Ḳ;) and ↓ استحرمت; (Ṣ, Ḳ;) said of a female cloven-hoofed animal, *She desired the male:* (Ṣ, Ḳ:) accord. to El-Umawee, (Ṣ,) likewise said of a she-wolf and of a bitch: (Ṣ, Ḳ:) and sometimes also said of a she-camel: but mostly of a ewe or she-goat. (TA.) — حَرِمَ, aor. ـَ, (Ṣ, Ḳ,) inf. n. حَرَمٌ, (Ṣ,) accord. to AZ and Ks, (Ṣ,) *He was overcome in contending for stakes*, or *wagers, in a game of hazard*, (Ṣ, Ḳ,) *not having himself overcome therein.* (Ḳ.) — Also حَرِمَ, aor. ـَ, (Ḳ,) inf. n. حَرَمٌ, (TA,) *He persisted;* or *persisted obstinately;* or *persisted in contention, litigation*, or *wrangling;* or *he contended, litigated*, or *wrangled.* (Ḳ.)

2. حرّمهُ, inf. n. تَحْرِيمٌ, (Ṣ, Mṣb, Ḳ,) said of God, (Ḳ,) and of a man, (Ṣ, Mṣb,) *He forbade it, prohibited it*, or *made it unlawful*, (Ṣ, Mṣb, Ḳ,*) عَلَيْهِ *to him;* (Ṣ;) as also ↓ احرمهُ, (Ṣ,* Mṣb, Ḳ,) inf. n. إِحْرَامٌ. (Ṣ.) The saying اَللّٰهُ أَكْبَرُ at the commencement of prayer is termed تَكْبِيرَةُ التَّحْرِيمِ [*The* تكبيرة *of prohibition*], because it prohibits the person praying from saying and doing anything extraneous to prayer: and it is also termed ↓ تَكْبِيرَةُ الإِحْرَامِ, meaning *the* تكبيرة *of entering upon a state of prohibition* by prayer. (TA.) It is said in a trad., of Ibn-'Abbás, إِذَا حَرَّمَ الرَّجُلُ ٱمْرَأَتَهُ فَهِىَ يَمِينٌ يُكَفِّرُهَا [*When the man declares his wife to be forbidden to him, it is an oath, which he must expiate*]: for the تَحْرِيم of a wife and of a female slave may be without the intention of divorce. (TA.) And حَرَّمْتُ الظُّلْمَ عَلَى نَفْسِى, occurring in another trad., [lit. *I have forbidden myself wrongdoing*, said by Moḥammad,] means *I am far above wrongdoing.* (TA.) تَحْرِيمٌ [as the inf. n. of حُرِّمَ] means The *being refractory*, or *untractable;* [as though forbidden to the rider;] whence مُحَرَّمٌ [q. v.] applied to a camel. (TA.) — [Also *He made*, or *pronounced, it*, or *him, sacred*, or *inviolable*, or *entitled to reverence* or *respect* or *honour;* whence المُحَرَّمُ applied to the حَرَم of Mekkeh, &c.:] *he*, or *it, made him*, or *it, to be reverenced, respected*, or *honoured.* (KL.) — *He bound it hard;* namely, a whip. (KL.) — *He tanned it incompletely* [*so that it became*, or *remained, hard*]; namely, a hide. (KL.) — See also 4, in two places.

4. احرم, [inf. n. إِحْرَامٌ,] *He entered upon a thing* [or *state* or *time*] *that caused what was before allowable*, or *lawful, to him to be forbidden*, or *unlawful.* (Ṣ,* Mṣb. [See also 5.]) And hence, (Ṣ, Mṣb,) *He purposed entering upon the performance of the* حَجّ or *the* عُمْرَة: (Mṣb:) or *he* (the performer *of the* حَجّ or *the* عُمْرَة) *entered upon acts whereby what was allowable*, or *lawful, to him became forbidden*, or *unlawful;* (Ḳ, TA;) as venereal intercourse, and the anointing of oneself, and wearing sewed garments, and hunting and the like: (TA:) you say, احرم بِالحَجِّ and بِالعُمْرَةِ, because what was allowable to the person became forbidden; as the killing of objects of the chase, and [venereal intercourse with] women. (Ṣ.) And *He entered into the* حَرَم, i. e. *Mekkeh* or *El-Medeeneh*, (Ḳ, TA,) or *the sacred territory of either of those cities:* (TA:) or *he entered into a sacred*, or *an inviolable, state;* or *into a state of security* or *safety*, (Ṣ, Ḳ, TA,) *being assured by a compact*, or *bond, that he should not be attacked* [&c.]: (TA:) or it signifies, (Ḳ,) or signifies also, (Ṣ,) *he entered upon a sacred month;* (Ṣ, Mṣb, Ḳ;) and so ↓ حرّم, (Ḳ, TA, [in the CḲ حَرُمَ,]) inf. n. تَحْرِيمٌ. (TA.) And *He entered* [as a subject] *into the covenanted state of security of the government of the Khaleefeh.* (TA.) 'Omar said, الصِّيَامُ إِحْرَامٌ [*Fasting is a state of prohibition*], because the faster is prohibited from doing that which would break his fast. (Sh, TA.) And الرَّجُلُ يُحْرِمُ فِى الغَضَبِ, a saying of El-Ḥasan, means *The man swears in anger*, because he becomes prohibited thereby (لِتَحْرِيمِهِ ↓ بِهِ) [from doing, or refraining from, a thing]. (TA.) See also 2, second sentence. — احرم عَنْهُ *He refrained from it* [as though he were prohibited from doing it]. (El-Mufaḍḍal, TA.) — احرمهُ: see 2, first sentence. — See also 1. — Also *He overcame him in contending for stakes*, or *wagers, in a game of hazard;* (AZ, Ks, Ṣ, Ḳ;) and so ↓ حرّمهُ, (Ḳ,) inf. n. تَحْرِيمٌ. (TA.)

5. تحرّم [*He became in a state of prohibition*]: see 4. [Thus it is similar to 4 in the first of the senses assigned to this latter above. Like as you say, احرم بِالحَجِّ and بِالعُمْرَةِ, so] you say, تحرّم بِالصَّلَاةِ [*He became in a state of prohibition by prayer;* i. e.] *he pronounced the* تَكْبِير [or تَكْبِيرَة, also termed تَكْبِيرَةُ الإِحْرَامِ, and تَكْبِيرَةُ التَّحْرِيمِ, (see 2,)] *for prayer; he entered upon prayer.* (MA.) — [Also *He protected*, or *defended, himself.*] You say, تحرّم مِنْهُ بِحُرْمَةٍ, meaning تمنّع and تحمّى [*He protected*, or *defended, himself*] بِذِمَّةٍ [*by a compact*, or *covenant, whereby he became in a state of security* or *safety*, or *by a promise*, or *an assurance, of security* or *safety*]; (Ḳ;) or بِصُحْبَةٍ [*by companionship*]; or بِحَقٍّ [*by a right*, or *due*]. (TA.) And تحرّم بِصُحْبَتِهِ [*He protected*, or *defended, himself by his companionship:* or, as explained in the PṢ, *he sought protection*, or *security, by his companionship*]. (Ṣ.) — Also [*He was*, or *became, entitled to reverence, respect*, or *honour;* or] *he possessed what entitled him to reverence, respect*, or *honour.* (KL.)

8. احترمهُ *He held him in reverence, respect*, or *honour; he reverenced, respected*, or *honoured, him.* (MA.) [See حُرْمَةٌ. Golius and Freytag explain اِحْتَرَمَ as meaning "Dignitate et præsidio venerabilis fuit:" but it is the pass., اُحْتُرِمَ, that has this meaning; or rather, *he was held in reverence*, &c.; *was reverenced*, &c.]

10. استحرم [*He deemed himself in a state of prohibition*]. It is said in a trad., of Adam, اِسْتَحْرَمَ بَعْدَ مَوْتِ ٱبْنِهِ مِائَةَ سَنَةٍ لَمْ يَضْحَكْ [*He deemed himself in a state of prohibition, after the death of his son, a hundred years, not laughing*]: from أَحْرَمَ signifying "he entered into a sacred, or an inviolable, state." (TA.) — استحرمت, said of a female cloven-hoofed animal, &c.: see 1.

حِرْمٌ: see حَرَمٌ.

حُرْمٌ The *state of* إِحْرَام (Az, Ṣ, Ḳ) *on account of the performance of the* حَجّ or *the* عُمْرَة; (Az, TA;) as also ↓ حِرْمٌ. (Ḳ in art. حل. [See 4 in the present art.]) Hence the saying, فَعَلَهُ فِى حُلِّهِ وَحُرْمِهِ, and ↓ فِى حِلِّهِ وَحِرْمِهِ, *He did it when he was free from* احرام *and when he was in the state of* احرام. (Ḳ in art. حل.) And hence the saying of 'Áisheh, respecting Moḥammad, كُنْتُ أُطَيِّبُهُ لِحُلِّهِ وَحُرْمِهِ, i. e. [*I used to perfume him when he was free from* احرام *and*] *when he was in the state of* احرام: (Ṣ, Mṣb:*) or *when he became free from* احرام *and when he* performed the ablution and *desired to enter upon the state of* احرام *for the* حَجّ or *the* عُمْرَة. (Az, TA.) [حُرْمُكَ in copies of the Ḳ, explained as meaning نِسَاؤُكَ وَمَا تَحْمِى, is a mistranscription for حُرَمُكَ: see حُرْمَةٌ.]

حِرْمٌ: see حُرْمٌ, in two places. — See also حَرَامٌ, in two places. — وَحِرْمٌ عَلَى قَرْيَةٍ أَهْلَكْنَاهَا أَنَّهُمْ لَا يَرْجِعُونَ, (Ṣ,* Ḳ,* TA,) in the Ḳur [xxi. 95], (TA,) thus read by some, (Ṣ, TA,) means وَاجِبٌ [i. e. *It is a necessary lot of* the people of *a town that we have destroyed that they shall not return*] (Ṣ, Ḳ, TA) to their present state of existence: (TA:) so explained by Ks, (Ṣ, TA,) and by I 'Ab and Fr and Zj: (TA:) some read ↓ حَرُمَ: (Bḍ:) the people of El-Medeeneh read ↓ حَرَامٌ; meaning *forbidden;* and accord. to this reading and meaning, لا is redundant: (TA:) [or حَرَامٌ in this instance is syn. with وَاجِبٌ, like حِرْمٌ; for it is said that] the explanation of Ks is confirmed by the saying of 'Abd-er-Raḥmán Ibn-Jumáneh [in the TA حمانة, app. for جُمَانَة,] El-Muḥáribee, a Jáhilee,

* فَإِنَّ حَرَامًا لَا أَرَى ٱلدَّهْرَ بَاكِيًا *

* عَلَى شَجْوِهِ إِلَّا بَكَيْتُ عَلَى عَمْرِو *

[*For it is a necessary thing that I should not ever see one weeping for his sorrow but I should weep for 'Amr*]. (TA.)

حَرَمٌ: see حَرَامٌ, with which it is sometimes syn., like as زَمَنٌ is with زَمَانٌ. (Ṣ, Mṣb, TA.) — [Hence,] الحَرَمُ *The* حَرَم [or *sacred territory*] *of Mekkeh*, (Lth, Az, Mṣb,* Ḳ,) *upon the limits of which were set up ancient boundary-marks* [*said to have been*] *built by Abraham;* (Az, TA;) also called حَرَمُ ٱللّٰهِ and حَرَمُ رَسُولِ ٱللّٰهِ (Ḳ) and ↓ المُحَرَّمُ: (Lth, Ḳ:) also *the* حَرَم *of El-Medeeneh:* (Mṣb:) [and *Mekkeh* itself: and *El-Medeeneh* itself:] and الحَرَمَانِ [*the sacred territory* *of Mekkeh and that of El-Medeeneh:* and] *Mekkeh* [itself] *and El-Medeeneh* [itself]: pl. أَحْرَامٌ: (Ḳ:) and حَرَمُ ٱللّٰهِ is also applied to *Mekkeh* [itself]. (Ṣ.) — See also حَرِيمٌ, in two places.

حُرُمٌ: see حَرَامٌ, with which it is syn. (TA.) Zuheyr says,

* وَإِنْ أَتَاهُ خَلِيلٌ يَوْمَ مَسْأَلَةٍ *
* يَقُولُ لَا غَائِبٌ مَالِى وَلَا حَرِمُ *

[*And if a friend come to him, on a day of solicitation, he says, My cattle are not,* or *my property is not, absent, nor forbidden,* or *refused*]: (Ṣ, IB, TA:) [in the Ṣ, this is cited as an ex. of حَرِمٌ as syn. with حِرْمَانٌ, which is an inf. n. of حَرَمَهُ, q. v.: but] IB says that حرم means مَمْنُوعٌ: (TA:) يقول in this verse is marfooạ though commencing an apodosis, because meant to be understood as put before [in the protasis], accord. to Sb; as though the poet said, يَقُولُ إِنْ أَتَاهُ خَلِيلٌ: accord. to the Koofees, it is so by reason of ف understood. (Ṣ, TA.)

حُرْمَةٌ The *state of being forbidden, prohibited,* or *unlawful:* (KL:) [and *of being sacred,* or *inviolable; sacredness,* or *inviolability:* (see حَرُمَ, of which it is an inf. n.:)] and the *state of being revered, respected,* or *honoured.* (KL.) See also مَحْرَمٌ. — Also, (Az, Mgh, Mṣb, Ḳ,) and ▾ حُرُمَةٌ, (Mgh, Ḳ,) and ▾ حُرَمَةٌ, (Ḳ,) *Reverence, respect,* or *honour;* (Az, Ḳ, TḲ;) a subst. from اِحْتِرَامٌ, (Mgh, Mṣb,) like فُرْقَةٌ from اِفْتِرَاقٌ; (Mṣb;) and ▾ مَحْرَمٌ signifies the same; but properly, *a place of* حُرْمَة: (Mgh:) pl. of the first حُرُمَاتٌ and حُرَمَاتٌ and حُرْمَاتٌ, like غرفات pl. of غُرْفَةٌ. (Mṣb.) When a man has relationship [to us], and we regard him with bashfulness, we say, لَهُ حُرْمَةٌ [*Reverence, &c., is due to him;* or *is rendered to him*]. (Az, TA.) And we say, لِلْمُسْلِمِ عَلَى ٱلْمُسْلِمِ حُرْمَةٌ [*Reverence, &c., to the Muslim is incumbent on the Muslim*]. (Az, TA.) — Also *A thing that should be sacred,* or *inviolable;* (Ṣ, Mṣb, Ḳ;) and so ▾ مَحْرُمَةٌ and ▾ مَحْرَمَةٌ (Ṣ, Mṣb) and ▾ مَحْرَمٌ: (Mṣb:) as, for instance, a man's *honour,* or *reputation:* (TḲ:) *a thing which one is under an obligation to reverence, respect,* or *honour* [*and defend*]: (Jel in ii. 190:) *a thing of which one is under an obligation to be mindful, observant,* or *regardful:* (Bḍ ibid.:) [*everything that is entitled to reverence, respect, honour,* or *defence, in the character and appertenances of a person: a thing that one is bound to do,* or *from which one is bound to refrain, from a motive of reverence, respect,* or *honour:* (see the next sentence:) and *any attribute that renders the subject thereof entitled to reverence, respect,* or *honour:*] the pl. of حُرْمَةٌ is حُرُمَاتٌ (Bḍ and Jel ubi suprà, and TA) [and حُرَمَاتٌ and حُرْمَاتٌ, as above,] and حُرَمٌ; (Mṣb;) and that of ▾ مَحْرَمٌ [and ▾ مَحْرُمَةٌ and ▾ مَحْرَمَةٌ] is مَحَارِمُ; (Mṣb;) and مَحْرُمَاتٌ and مَحْرَمَاتٌ [also] are pls. of ▾ مَحْرُمَةٌ and ▾ مَحْرَمَةٌ. (Aṣ, Ṣ.) حُرُمَاتُ ٱللّٰهِ means [*The inviolable ordinances and prohibitions of God:* or] *the ordinances of God, and other inviolable things:* (Bḍ and Jel* in xxii. 31:) or *what it is incumbent on one to perform, and unlawful to neglect:* (Zj, Ḳ:) or *all the requisitions of God relating to the rites and ceremonies of the pilgrimage and to other things:* (Ksh in xxii. 31:) or *the* حَرَم [or *sacred territory*] *and the requisitions relating to the pilgrimage:* (Bḍ ubi suprà:) or *the requisitions relating to the pilgrimage* in particular: (Ksh ubi suprà:) or *the Kaạbeh and the sacred mosque and the sacred territory and the sacred month and the person who is in the state of* إِحْرَام: (Ksh and Bḍ ibid.:) or *the inviolability* (حُرْمَة) *of the sacred territory and of the state of* إِحْرَام *and of the sacred month:* (TA:) or *Mekkeh and the pilgrimage and the* عُمْرَة, *and all the acts of disobedience to God which He has forbidden:* (Mujáhid, TA:) or [simply] *the acts of disobedience to God.* ('Aṭà, TA.) — And [hence, because it should be regarded as sacred, or inviolable,] *i. q.* ذِمَّةٌ [*A compact,* a *covenant,* or *an obligation;* and particularly *such as renders one responsible for the safety,* or *safe-keeping, of a person* or *thing,* or *for the restoration of a thing,* or *for the payment of a sum of money, &c.;* or *by which one becomes in a state of security* or *safety:* and simply *responsibility,* or *suretiship:* and *security,* or *safety; security of life and property; protection,* or *safeguard; a promise,* or *an assurance, of security, safety, protection,* or *safeguard; indemnity;* or *quarter:* or *an obligation, a duty,* or *a right,* or *due, that should be regarded as sacred,* or *inviolable,* or *the nonobservance of which is blameable*]. (Ḳ.) — And [hence also] A man's حُرَم [i. e. his *wives,* or *women under covert,*] and his *family:* (Ṣ:) and [in like manner the pl.] حُرَمٌ, accord. to the Ḳ حُرْمٌ, but correctly like زُفَرٌ, (TA,) a man's *wives,* or *women* [*under covert*], (Ḳ, TA,) and his *household,* or *family,* (TA,) and *what he protects,* or *defends;* as also مَحَارِمُ, of which the sing. is ▾ مَحْرُمَةٌ and ▾ مَحْرَمَةٌ: (Ḳ, TA:) and hence حُرْمَة is applied by the vulgar to signify *a wife.* (TA.) [In Ḥar, p. 377, a man's حُرْمَة is said to mean his حُرَم and his family: and in p. 489, a man's حُرَم is said to mean his family and his wives and those whom he protects, or defends. See also حَرِيمٌ.] — Also *A share, portion,* or *lot;* syn. نَصِيبٌ. (Ḳ.)

حِرْمَةٌ (Ḳ) and ▾ حَرَمَةٌ (Lḥ, Ṣ, Ḳ) The *desire* of a female cloven-hoofed animal, (Ḳ,) or of a ewe, or she-goat, (Ṣ,) and of a she-wolf and of a bitch, (Ḳ,) *for the male:* (Ṣ, Ḳ:) حَرَمَةٌ in ewes, or she-goats, is like ضَبَعَةٌ in she-camels, and حِنَآءٌ in ewes. (Ṣ.) It is also used, in a trad., in relation to male human beings. (Ḳ.) It is said in a trad., respecting those whom the hour [of the resurrection] shall overtake, تُبْعَثُ عَلَيْهِمُ ٱلحِرْمَةُ وَيُسْلَبُونَ ٱلْحَيَآءَ, i. e. *Venereal desire* [*shall be made to befall them, and they shall be bereft of shame*]. (Ṣ.)

حَرَمَةٌ: see what next precedes.

حُرُمَةٌ }
حُرَمَةٌ } see حُرْمَةٌ.

حَرْمَى, applied to a female cloven-hoofed animal, (Ḳ,) or to a ewe, or she-goat, (Ṣ,) and to a she-wolf and to a bitch, (Ḳ,) *Desiring the male:* pl. حِرَامٌ and حَرَامَى, (Ṣ, Ḳ,) like عِجَالٌ and عَجَالَى, (Ṣ,) or the latter pl. is حُرَامَى; (so accord. to some copies of the Ḳ [like عُجَالَى];) as though its masc., if it had a masc., were حَرْمَانُ. (Ṣ.) — حَرْمَى وَٱللّٰهِ means the same as أَمَا وَٱللّٰهِ [*Verily,* or *now surely, by God*]; (Ḳ;) as also حَزْمَى وَٱللّٰهِ. (Ḳ in art. حزم.)

حِرْمِىٌّ, applied to a man, *Of,* or *belonging to, the* حَرَم: fem. حِرْمِيَّةٌ. (Ṣ, Mṣb, TA.) [In the TA it is said that Mbr mentions two forms of the epithet حرمية as applied to a woman: it does not specify what these are; but one seems to be حُرْمِيَّةٌ, for he says that it is from the phrase وَحُرْمَةِ ٱلْبَيْتِ "by the sacredness of the House" of God.] Az says, on the authority of Lth, that when they applied the rel. n. from ٱلحَرَم to anything not a human being, [as, for instance, to a garment, or piece of cloth,] they said ▾ ثَوْبٌ حَرَمِىٌّ: (Mṣb:) [but] they also said سِهَامٌ حِرْمِيَّةٌ, (Ṣ,) or حَرَمِيَّةٌ, (Mṣb,) meaning *Arrows of the* حَرَم: (Ṣ, Mṣb:) and حِرْمِيَّةٌ [also, or قَوْسٌ حِرْمِيَّةٌ,] meaning *A bow made of a tree of the* حَرَم. (Ḥam p. 284.) — Also *A man of the* حَرَم *whose food was eaten by a pilgrim, and in whose clothes this pilgrim performed his circuiting round the Kaạbeh:* and *a pilgrim who ate the food of a man of the* حَرَم, *and performed his circuiting round the Kaạbeh in this man's clothes:* each of these was called the حِرْمِىّ of the other: every one of the chiefs of the Arabs who imposed upon himself hardship, or strictness, in his religious practices had a حِرْمِىّ of the tribe of Ḳureysh; and when he performed the pilgrimage, would not eat any food but that of this man, nor perform his circuiting round the Kaạbeh except in this man's clothes. (TA.)

حَرَمِىٌّ: see the next preceding paragraph.

حَرَامٌ *Forbidden, prohibited,* or *unlawful:* and *sacred,* or *inviolable;* as in the phrases ٱلْبَيْتُ ٱلْحَرَامُ [*the Sacred House* of God (i. e. the Kaạbeh)] and ٱلْمَسْجِدُ ٱلْحَرَامُ [*the Sacred Mosque* of Mekkeh] and ٱلْبَلَدُ ٱلْحَرَامُ [*the Sacred Town* or *Territory*]: (Mṣb:) *contr. of* حَلَالٌ; (Ṣ;) as also ▾ حَرَمٌ (Ṣ, Mṣb) and ▾ حِرْمٌ (Ṣ, Mṣb, Ḳ) and ▾ حُرُمٌ [q. v.] (TA) [and in its primary sense ▾ حَرِيمٌ] and ▾ مُحَرَّمٌ: (Ṣ, Mgh, Mṣb:) the pl. [of حَرَامٌ, agreeably with analogy,] is حُرُمٌ; (Ḳ;) and مَحَارِمُ also is a pl. of حَرَامٌ, contr. to rule, (TA,) and signifies *things forbidden by God.* (Ḳ.) See also حَرَمٌ. — حَرَامُ ٱللّٰهِ لَا أَفْعَلُ, (as in some copies of the Ṣ,) or حَرَامَ ٱللّٰهِ لا افعل, (as in other copies of the Ṣ and in the Ḳ,) is a saying like يَمِينُ ٱللّٰهِ لَا أَفْعَلُ, or يَمِينَ ٱللّٰهِ لا افعل: (Ṣ, Ḳ:) it may mean a declaration that the wife or the female slave shall be forbidden [to him who utters it], without the intention of divorcing [thereby the former, or of emancipating the latter; so that it may be rendered, according to the two different readings, I imprecate upon myself, or that which I imprecate upon myself is, *what is forbidden of God*, if I do it: *I will not do* such a thing: in like manner, عَلَيَّ الحَرَامُ is often said in the present day]. (TA. [See 2.]) — ابْنُ حَرَامٍ [*An illegitimate son*: and *a disingenuous*, or *dishonest, person*.] — شَهْرٌ حَرَامٌ [*A sacred month*]: (Mṣb:) pl. حُرُمٌ. (Ṣ, Mṣb, Ḳ.) الأَشْهُرُ الحُرُمُ [*The sacred months*] (Ṣ,* Mṣb, Ḳ) were four; namely, ذُو القَعْدَةِ and ذُو الحِجَّةِ and المُحَرَّمُ and رَجَبٌ; (Ṣ, Mṣb, Ḳ;) three consecutive, and one separate: (Ṣ, Mṣb:) in these the Arabs held fight to be unlawful; except two tribes, Khath'am and Ṭeiyi; unless with those who held these months as profane. (Ṣ, TA.) — حَرَامٌ applied to a man signifies *Entering into the* حَرَم [or *sacred territory of Mekkeh* or *of El-Medeeneh*, or *Mekkeh* or El-*Medeeneh* itself]; and is applied also to a woman; and to a pl. number: (TA:) or i. q. ↓ مُحْرِمٌ (Ṣ, Mṣb) as meaning [*in*, or *entering upon, the state of* إِحْرَام: i. e. *entering upon the performance of those acts of the* حَجّ, *or of the* عُمْرَة, *whereby certain things before allowable*, or *lawful, to him became forbidden*, or *unlawful*; (see 4;) or] *purposing to enter upon the performance of the* حَجّ or *the* عُمْرَة: (Mṣb:) as also ↓ حِرْمٌ: you say, أَنْتَ حِلٌّ and انت حِرْمٌ [*Thou art one who has quitted his state of* إِحْرَام and *thou art in*, or *entering upon, the state of* احرام]: (TA:) the pl. of حَرَامٌ thus applied is حُرُمٌ: (Ṣ, Mṣb:) the fem. of ↓ مُحْرِمٌ is with ة; and the pl. masc. مُحْرِمُونَ; and the pl. fem. مُحْرِمَاتٌ. (Mṣb.) — See another meaning voce حِرْمٌ.

حَرِيمٌ: see حَرَامٌ.

حَرُومٌ A she-camel *that does not conceive when covered*. (AA, Ḳ. [In the CḲ, مُقْتَاطَة is erroneously put for مُغْتَاطَة.])

حَرِيمٌ: see حَرَامٌ. — [Hence,] The *appertenances*, or *conveniences*, (حُقُوق and مَرَافِق, Ṣ, Mṣb, Ḳ,) *that are in the immediate environs*, (Ṣ, Mṣb,) of a thing, (Mṣb,) or of a well &c., (Ṣ,) or *that are adjuncts* [or *within the precincts*] of a house; (Ḳ;) because it is forbidden to any but the owner to appropriate to himself the use thereof: (Mṣb:) or, of a well, the *place where is thrown the earth that has been dug out*, (Ḳ, TA,) *and the walking-place on either side*; in the case of a well dug in a waste land that has no owner, said in a trad. to be forty cubits: (TA: [but see بَدِىٌّ:]) and of a river, or rivulet, or canal, the *place where the mud is thrown out, and the walking-place on each side*: (TA:) and of a house, the *interior part upon which the door is closed*: (Ibn-Wásil El-Kilábee, TA:) or the *interior part*, or *middle*, (قَصَبَة,) thereof: (T, TA:) [and particularly the *women's apartments*, and the *portion that is forbidden to men who are not related to the women within the prohibited degrees of marriage*:] and the *court of a mosque*: (T, TA:) [and in general,] *a place which it is incumbent on one to defend* [*from intrusion*]: (Ḥam p. 492:) *a thing that one protects, and in defence of which one fights*; [and particularly, like حُرْمَة as used by the vulgar, a man's *wife*; and also his *female slave*; or *any woman under covert*; and, like حُرَمٌ, pl. of حُرْمَة, as used in the classical language, his *wives*, or *women under covert, and household*;] as also ↓ حِرْمٌ: pl. حُرُمٌ, (Ḳ,) the pl. of حَرِيمٌ; (TA;) and أَحْرَامٌ, (Ḳ,) which is the pl. of ↓ حِرْمٌ. (TA.) — *A partner, copartner*, or *sharer*. (Ḳ.) — *A friend*: so in the saying, فُلَانٌ حَرِيمٌ صَرِيحٌ *Such a one is a genuine*, or *sincere, friend*. (TA.) — The *garment of the* مُحْرِم, (Ṣ, Ḳ,) [*which he wears during the performance of the* حَجّ or *the* عُمْرَة;] called by the vulgar ↓ إِحْرَامٌ and ↓ حِرَامٌ. (TA.) — The *clothes which the* مُحْرِمُونَ *used to cast off*, (Ṣ,* Ḳ, TA,) *when, in the time of paganism, they performed the pilgrimage to the House* [*of God, at Mekkeh*], namely, *those that were upon them when they entered the* حَرَم [*or sacred territory*], (TA,) *and which they did not wear* (Ḳ, TA) *as long as they remained in the* حَرَم: (TA:) for the Arabs used to perform their circuiting round the House naked, with their clothes thrown down before them during the circuiting; (T, Ṣ, TA;) they saying, "We will not perform the circuiting round the House in clothes in which we have committed sins, or crimes:" and the woman, also, used to perform the circuiting naked, except that she wore a رَهْط of thongs. (TA.) A poet says,

* كَفَى حَزَنًا مَرِّى عَلَيْهِ كَأَنَّهُ *
* لَقًى بَيْنَ أَيْدِى الطَّائِفِينَ حَرِيمُ *

[*Sufficiently grievous is my passing by him as though he were a thing thrown away, a cast-off garment of a* مُحْرِم, *before those performing the circuiting* round the Kaabeh]. (Ṣ.)

حَرِيمَةٌ *Anything eagerly desired*, or *coveted, that escapes one, so that he cannot attain it*. (Ṣ.) And حَرِيمَةُ الرَّبِّ *That which the Lord denies to whomsoever He will*. (Ḳ.)

حَارِمٌ *Denying, refusing*, or *refusing to give*. (TA.) — هُوَ بِحَارِمِ عَقْلٍ, (so in the copies of the Ḳ,) or مَا هُوَ بِحَارِمِ عَقْلٍ, (so in the TA,) means *He has intellect*, or *intelligence*: (Ḳ:) a phrase mentioned, and thus explained, by AZ: and so بِعَارِمِ عَقْلٍ. (TA.) [The right reading is evidently that given in the TA.]

إِحْرَامٌ inf. n. of 4. — See also حَرِيمٌ.

مُحْرِمٌ: see حَرَامٌ, with which it is syn. (Ṣ, Mgh, Mṣb.) [And see an ex. voce حَدٌّ.] — See also حُرْمَةٌ, in three places. — Also *A female relation whom it is unlawful to marry*: (T, Mṣb:) [and *such a male relation* likewise:] and رَحِمٌ مُحْرِمٌ *relationship that renders it unlawful to marry*. (Ḳ.) You say, هِىَ لَهُ مُحْرِمٌ [*She is a relation to him such as it is unlawful for him to marry*]: and هُوَ لَهَا مُحْرِمٌ and هُوَ مُحْرِمٌ مِنْهَا (Mgh) and هُوَ ذُو مُحْرِمٍ مِنْهَا *he is one whom it is unlawful for her to marry*, (Ṣ,) and ذُو رَحِمٍ مُحْرِمٍ and ذُو رَحِمٍ مُحْرِمٌ, applying محرم as an epithet to رحم and to ذو; (Mgh, Mṣb;) and ذُو حُرْمَةٍ ↓ فِى القَرَابَةِ: (Ḥam p. 669:) and in the case of a woman, ذَاتُ رَحِمٍ مُحْرِمٍ. (Mṣb.) — مَحَارِمُ اللَّيْلِ ‡ *The fearful places of the night*, (IAạr, Ṣ, Ḳ, TA,) *which the coward is forbidden to traverse*. (IAạr, Ṣ, TA.) [See also مَخَارِمُ, pl. of مَخْرِمٌ.]

مُحْرِمٌ: see حَرَامٌ, in two places: *Contr. of* مُحِلٌّ: and as such signifying [also] one *with whom it is unlawful to fight*: (Ṣ:) or, as such, *whom it is unlawful to slay*: (TA in art. حل:) and, as such also, one *who has a claim*, or *covenanted right, to protection*, or *safeguard*. (Ṣ in art. حل.) Er-Rá'ee says,

* قَتَلُوا ٱبْنَ عَفَّانَ الخَلِيفَةَ مُحْرِمًا *

(Ṣ,) meaning [*They slew* ('Othmán) *Ibn-'Affán, the Khaleefeh*,] *while entitled to the respect due to the office of Imám and to the* [*sacred*] *city and to the* [*sacred*] *month*: for he was slain [in El-Medeeneh and] in [the month of] Dhu-l-Ḥijjeh. (Ḥam p. 310.) And one says, إِنَّهُ لَمُحْرِمٌ عَنْكَ *Verily he is one whom it is unlawful for thee to harm*: (Ḳ:) or *for whom it is unlawful to harm thee*: (IAạr, Th:) or *whom it is unlawful for thee to harm and for whom it is unlawful to harm thee*. (Az, TA.) And مُسْلِمٌ مُحْرِمٌ *A Muslim is secure, as to himself and his property, by the respect that is due to El-Islám*: or *a Muslim refrains from the property of a Muslim, and his honour*, or *reputation, and his blood*. (TA.) — One *who is at peace with another*. (IAạr, Ḳ.) — One *who is in the* حَرِيم *of another*. (Ḳ.) You say, هُوَ مُحْرِمٌ بِنَا *He is in our* حَرِيم. (TA.) — *Fasting*, or *a faster*: because the faster is prohibited from doing that which would break his fast. (TA.) — And, for a like reason, *Swearing*, or *a swearer*. (TA.)

مُحْرِمَاتٌ and مَحْرُمَةٌ and مُحْرِمَةٌ, pl. مَحَارِمُ (Ḳ) and مُحْرِمَاتٌ: (Aṣ, Ṣ:) see each voce حُرْمَةٌ, in four places.

مُحَرَّمٌ [*Forbidden, prohibited*, or *made unlawful*: and *made*, or *pronounced, sacred*, or *inviolable*, or *entitled to reverence* or *respect* or *honour*]. It is said in a trad., أَمَا عَلِمْتَ أَنَّ الصُّورَةَ مُحَرَّمَةٌ, i. e. [*Knowest thou not that the face is*] *forbidden to be beaten? or that it has a title to reverence or respect or honour?* (TA.) — المُحَرَّمُ *The first of the months* (Ṣ, Mṣb, Ḳ,* TA) *of the year* (Mṣb) *of the Arabs* [*since the age of paganism*]; (TA;) the article ال being prefixed because it is originally an epithet; but accord. to some, it is not prefixed to the name of any other month; or, accord. to some, it may be prefixed to صفر and شوّال: (Mṣb:) and [*in the age of paganism, the seventh month, also called*] شَهْرُ ٱللّٰهِ الأَصَبُّ, (Ḳ, TA,) [الاصبّ being app. a dial. var. of الأَصَمّ,] i. e. رَجَبٌ; [for] Az says, the Arabs used to call the month of رَجَب, in the age of paganism, الأَصَمّ and المُحَرَّم; and he cites the saying of a poet,

* أَقَمْنَا بِهَا شَهْرَىْ رَبِيعٍ كِلَاهُمَا *
* وَشَهْرَىْ جُمَادَى وَٱسْتَحَلُّوا المُحَرَّمَا *

[*We stayed in it during the two months of Rabeea, both of them, and the two months of Jumádà; and they made El-Moharram to be profane;* app. by postponing it, as the pagan Arabs often did]: the Arabs called it thus because they did not allow fighting in it [unless they had postponed it]: (TA:) the pl. is مُحَرَّمَاتٌ (Mṣb, Ḳ) and مَحَارِمُ and مَحَارِيمُ. (Ḳ.) — See also مُحْرَمٌ. — مُحَرَّمٌ applied to a camel means *Refractory, or untractable:* (TA:) [or,] thus applied, [like عَرُوضٌ, q. v.,] *submissive in the middle part,* [but] *difficult to be turned about,* [i. e. *stubborn in the head,*] *when turned about:* (Ḳ: [in the CḲ, الذَّلُولُ الوَسْطُ is erroneously put for الذَّلُولُ الوَسَطِ: in my MS. copy of the Ḳ, الذَّلُولُ الوَسَطِ:]) and with ة, a she-camel *not broken,* or *not trained:* (TA:) or *not yet completely broken* or *trained:* (Ṣ, TA:) and مُحَرَّمَةُ الظَّهْرِ a she-camel that is *refractory,* or *untractable; not broken,* or *not trained:* in this sense heard by Az from the Arabs. (TA.) — ‡ A skin *not tanned:* (Ḳ:) or *not completely tanned:* (Ṣ:) or *tanned, but not made soft, and not thoroughly done.* (TA.) — ‡ A *new whip:* (Ḳ:) or a whip *not yet made soft.* (Ṣ, A, TA.) — ‡ An Arab of the desert *rude in nature or disposition, chaste in speech, that has not mixed with people of the towns* or *villages.* (TA.) — † The *part* of the nose *that is soft in the hand.* (Ḳ.)

مَحْرُومٌ *Denied,* or *refused, a gift:* (Mṣb,* TA:) or *denied,* or *refused, good,* or *prosperity:* (Az, Ḳ:) in the Ḳur lxx. 25, (I'Ab, Ṣ,) [it has this latter, or a similar, meaning;] *i. q.* مُحَارَفٌ [q. v.]; (I'Ab, Ṣ, Ḳ;) *who hardly,* or *never, earns,* or *gains, anything:* (Ḳ:) or *who does not beg, and is therefore thought to be in no need, and is denied:* (Bḍ:) and *who has no increase of his cattle* or *other property:* (Ḳ:) opposed to مَرْزُوقٌ: (Az, TA:) accord. to some, *who has not the faculty of speech,* like the dog and the cat &c. (Ḥar p. 378.) — *Held in reverence, respect,* or *honour; reverenced, respected,* or *honoured;* and so ▼ مُحْتَرَمٌ. (KL. [But the latter only is commonly known in this sense.])

مَحَارِمُ an anomalous pl. of حَرَامٌ, q. v.: (TA:) — and pl. of مَحْرَمَةٌ and مَحْرُمَةٌ: (Ḳ:) — and also of المُحَرَّمُ. (Ḳ.)

مَحَارِيمُ a pl. of المُحَرَّمُ. (Ḳ.)

مُحْتَرَمٌ [erroneously written in the Lexicons of Golius and Freytag مُحْتَرِمٌ]: see مَحْرُومٌ.

حرن

1. حَرَنَ, aor. ـُ; (Ṣ, Mgh, Mṣb, Ḳ;) and حَرُنَ; (Ṣ, Mṣb, Ḳ;) inf. n. حِرَانٌ (Mgh, Mṣb, Ḳ) and حُرُونٌ, (Mgh, Mṣb,) or the former is a simple subst., and the latter is the inf. n., (Ṣ,) and حُرَانٌ; (Ḳ;) said of a horse (Ṣ, Mgh, Mṣb, Ḳ) or similar beast, (Mṣb, Ḳ,) *He was,* or *became, restive,* or *refractory, and, when vehemently running, stopped:* (Ṣ:) or *stopped, and was restive,* or *refractory:* (Mgh:) or *stopped when one desired to call into action his power of running:* said peculiarly of a solid-hoofed animal: (M, Ḳ:) or, accord. to Lḥ, one says also حَرَنَتِ النَّاقَةُ, meaning *the she-camel stood still, and would not move from her place:* and حِرَانٌ is used by AO in relation to a she-camel. (TA.) And حَرَنَ, inf. n. حُرُونٌ, *i. q.* تَأَخَّرَ [*He went back* or *backwards, drew back, receded,* &c.]. (Aṣ, TA.) — حَرَنَ بِالمَكَانِ, inf. n. حُرُونَةٌ, [perhaps a mistranscription for حُرُونٌ,] *He kept,* or *clave, to the place, and did not quit it.* (TA.) — حَرَنَ فِى البَيْعِ ‡ *He did not exceed nor fall short in selling.* (Ṣ, Ḳ, TA.) — حَرَنَ القُطْنَ *He separated and loosened the cotton* [*by means of a bow and a kind of wooden mallet, by striking the string of the bow with the mallet*]; syn. نَدَفَهُ. (Ḳ.)

حِرَانٌ a subst. from حَرَنَ said of a horse [or similar beast; i. e. *Restiveness,* or *refractoriness,* &c.]: (Ṣ:) or an inf. n. (Mgh, Mṣb, Ḳ.)

حَرُونٌ an epithet applied to a horse (Ṣ, Mgh, Mṣb, Ḳ) or similar beast, (Mṣb, Ḳ,) [meaning *Restive,* or *refractory,* &c.,] from حَرَنَ (Ṣ, Mgh, Mṣb, Ḳ) or حَرُنَ: (Ṣ, Mṣb, Ḳ:) pl. حُرُنٌ, with two dammehs. (TA.) — Also An animal of the chase *that does not quit the higher,* or *highest, part of the mountain.* (Ṣ, Ḳ.)

مِحْرَنٌ The *wooden implement* (*a kind of mallet*) *with which, together with a bow, by striking with the former the string of the latter, cotton is separated and loosened;* syn. مِنْدَفٌ. (Ḳ.)

مُحْرَانٌ *Honey:* (Ḳ:) pl. مَحَارِينُ. (So in the TA, as from the Ḳ.) — See also the pl. below.

مَحَارِنُ: see what follows.

مَحَارِينُ (Ṣ, Ḳ) and ▼ مَحَارِنُ (Ṣ [in which the latter occupies the first place, the former occurring in an ex.,]) ‡ The *bees that stick to the honey, and are extracted with the* مَحَابِض, (Ṣ, Ḳ, TA,) or *wooden implements with which the honey itself is extracted:* (TA:) or the *bees that stick in the hive, and are with difficulty extracted:* or the *bees that die in the honey:* (T, TA:) sing. ▼ مِحْرَانٌ. (Ḳ.) — And The *pods of cotton.* (Ḳ.)

حرو

حَرْوَةٌ *A burning* (M, Ḳ) *which a man experiences* (M) *in the fauces* (الحَلْق) *and the chest and the head, by reason of anger, wrath,* or *rage, and of pain.* (M, Ḳ.) — *Acritude* (Ṣ, Ḳ) of food, (Ṣ,) or *in the taste* of mustard (Ḳ, TA) and the like; (TA;) as also ▼ حَرَاوَةٌ. (Ṣ, Ḳ.) You say, إِنِّى لَأَجِدُ لِهٰذَا الطَّعَامِ حَرْوَةً and ▼ حَرَاوَةً *Verily I find that this food has an acrid quality,* (Ṣ,) or *a burning quality.* (TA.) [See also حِرَّةٌ, and حَرَارَةٌ.] And one says, ▼ لِهٰذَا الكُحْلِ حَرَاوَةٌ فِى العَيْنِ [*This collyrium has a burning effect in the eye*]. (TA.) — *A disagreeable odour, that has a sharpness,* or *pungency,* (M, Ḳ,) *in the* خَيَاشِيم [or *air-passages of the nose*]. (M.)

حَرَاوَةٌ: see above, in three places.

حرى

1. حَرَى, aor. يَحْرِى, (Ṣ, Ḳ,) inf. n. حَرْىٌ, (Ṣ,) *It* (a thing Ṣ) *decreased, diminished,* or *waned,* (Ṣ, Ḳ, TA,) *after increase;* (TA;) as does, for instance, the moon. (Ṣ, TA.) [See an ex. in a verse cited in art. است.] — حَرَى أَنْ يَكُونَ ذٰلِكَ *i. q.* عَسَى [*May-be,* or *may-hap,* &c., *that will be*]. (TA.) — حَرِىَ بِكَذَا *He was,* or *became, adapted, disposed, apt, meet, suited, suitable, fitted, fit, competent,* or *proper, for such a thing;* or *worthy of it.* (MA.) — حَرَّاهُ: see 5.

4. احراهُ *It* (time) *caused it* (a thing, Ṣ) *to decrease, diminish,* or *wane.* (Ṣ, Ḳ.) — مَا أَحْرَاهُ, and أَحْرِ بِهِ, *How well adapted* or *disposed,* or *how apt, meet, suited, suitable, fitted, fit, competent,* or *proper,* or *how worthy, is he!* (Ṣ, Ḳ.) [You say, مَا أَحْرَاهُ بِذٰلِكَ *How well adapted* or *disposed,* &c., *is he for that!*]

5. تحرّى signifies قَصَدَ الحَرَى; i. e. *He sought,* or *repaired to, the vicinage, quarter, tract,* or *region, of a people:* this is said to be the primary signification: (Mgh:) and تَحَرَّاهُ *he sought,* or *repaired to, his vicinage,* &c.; (قَصَدَ حَرَاهُ;) as also ▼ حَرَّاهُ, aor. يُحَرِّى: (TA:) *he aimed at it; made it his object; sought, endeavoured after, pursued,* or *endeavoured to reach* or *attain* or *obtain, it; intended* or *purposed it;* namely, a thing. (Ṣ, Mgh, Mṣb, Ḳ, TA.) Hence, in the Ḳur [lxxii. 14], فَأُولٰئِكَ تَحَرَّوْا رَشَدًا *Those have aimed at,* or *sought,* &c., *a right course.* (Ṣ, TA.) And تَحَرَّيْتُ مَرْضَاتَهُ *I aimed at,* or *sought,* &c., *his approval.* (Mgh.) And the trad., تَحَرَّوْا لَيْلَةَ القَدْرِ فِى العَشْرِ الأَوَاخِرِ *Seek ye the Night of* القدر *in the last ten* [nights of Ramadán]. (TA.) — Also *He sought what was most meet, suitable, fit, proper,* or *deserving,* (Ṣ, Mgh, Mṣb, Ḳ,) *to be done,* (Ṣ, Ḳ,) *of two things,* (Mgh, Mṣb,) *according to the opinion predominating in his mind,* (Ṣ,) فِى الأَمْرِ [*in the affair,* or *case*]: (Mṣb:) or *he sought,* or *endeavoured, and strove in seeking, and deciding upon, the singling out of a thing, by deed* and *by word.* (TA.) — And *He tarried, waited,* or *paused in expectation,* بِالمَكَانِ *in the place.* (Ṣ, Ḳ.)

حَرًى The *vicinage, quarter, tract,* or *region,* (Aṣ, T, Ṣ, IAth, Mgh, Ḳ,) of a man, (Aṣ, T, IAth,) or of a people; (Mgh;) the *environs* (Aṣ, T, Ṣ) of a man, (Aṣ, T,) or of a house; (Ṣ;) and ▼ حَرَاةٌ signifies the same: (Ṣ, Ḳ:) and [it is said that] the former signifies also the *place of the eggs* of an ostrich: (Ṣ, Ḳ:) and *a covert,* or *hiding-place, among trees,* of a gazelle: (Ḳ,* TA:) Lth says that it signifies the *place of laying eggs* of the ostrich; or the *covert,* or *lodging-place,* of the gazelle: but this is false; for with the Arabs the word signifies as explained above on the authority of Aṣ; and the حرى of the place of laying eggs of the ostrich, and of the covert of the gazelle, is the *environs* thereof: (T, TA:) pl. أَحْرَآءٌ. (Ḳ.) You say, اِذْهَبْ فَلَا أَرَيَنَّكَ بِحَرَاىَ and ▼ بِحَرَاتِى [*Go thou, so that I may by no means see thee in my vicinage,* &c.]. (Ṣ.) And لَا تَطُرْ حَرَانَا *Approach not thou our environs.* (Ṣ.) And نَزَلْتُ بِحَرَاهُ and بِعَرَاهُ [*I alighted,* or *descended and abode, in his vicinage,* &c.]. (Ṣ.) ═ See also حَرِىٌّ, in six places.

حَرٍ : see حَرِىٌّ, in four places.

حَرَاةٌ : see حَرًى, in two places.

حَرًى *Adapted, disposed, apt, meet, suited, suitable, fitted, fit, competent, proper,* or *worthy;* as also ▼ حَرٍ and ▼ حَرِىٌّ, which last has no dual nor pl., (Ṣ, Mṣb, Ḳ,) nor a fem. form, remaining unaltered, because it is [originally] an inf. n.; [see a verse cited voce نَقْرٌ;] or, accord. to Lḥ, one may say حَرَيَانِ, because Ks has related that some of the Arabs dualize what they do not pluralize: (TA:) the pl. of حَرِىٌّ is حَرِيُّونَ and أَحْرِيَآءُ; (Ṣ, Mṣb;) and the pl. of حَرِيَّةٌ is حَرِيَّاتٌ and حَرَايَا: the pl. of ▼ حَرٍ is أَحْرَآءٌ (Ṣ, TA) and حَرُونَ; and the pl. of حَرِيَةٌ is حَرِيَاتٌ. (TA.) You say, إِنَّهُ لَحَرِىٌّ بِكَذَا, and ▼ لَحَرٍ, and ▼ لَحَرًى, (Ḳ, TA,) *Verily he,* or *it, is adapted,* &c., *to such a thing;* or *worthy of such a thing.* (TA.) And هُوَ حَرِىٌّ أَنْ يَفْعَلَ ذٰلِكَ, and ▼ حَرٍ, and ▼ حَرًى, *He is adapted,* &c., *to do that:* (Ṣ, Mṣb:) and انّه لَمَحْرًى ▼ ان يفعل, (Lḥ, Ḳ, [in some copies of the Ḳ, erroneously, لَمَحْرِىٌّ,]) and ▼ لَمَحْرَاةٌ, (Ḳ,) which last has no dual nor pl. nor fem. form, like مَخْلَقَةٌ [q. v.] and مَظِنَّةٌ: (TA:) and هٰذَا الأَمْرُ مَحْرَاةٌ ▼ لِذٰلِكَ [*This thing,* or *affair, is adapted,* &c., *to that*]. (Ṣ.) And hence the phrase, بِالْحَرَى ▼ أَنْ يَكُونَ ذَالِكَ (Ṣ,* Ḳ) *It is suitable, fit,* or *proper, that that should be.* (PṢ.) [But this phrase, in the present day, means *Rather that should be.* And hence, ▼ كَمْ بِالْحَرَى *How much rather.*] One says also, of a man who has attained to fifty [years], ▼ فَحَرًى, meaning *He is adapted,* &c., *to attain all that is good.* (Th, TA.) And one says إِنَّهُ لَحَرِىٌّ as meaning *Verily it is probable;* or *likely to happen* or *be,* or *to have happened* or *been;* as also لَخَلِيقٌ. (TA in art. خلق.)

حِرِىٌّ : see art. حرح.

حَارٍ masc. of حَارِيَةٌ, (M, TA,) which is an epithet applied to a viper (أَفْعَى); (Ṣ, M, Ḳ;) meaning *That has decreased in its body by reason of age;* and it is *the worst,* or *most malignant* or *noxious, that is:* (Ṣ:) or *that has become old, and has wasted in its body, and whereof there remains not save its head and its breath* (نَفَسُهَا) [in the CḲ نَفْسُهَا]) *and its poison:* (M, Ḳ:) dim. ▼ حُوَيْرٍ. (TA.) One says, رَمَاكَ ٱللّٰهُ بِأَفْعَى حَارِيَةٍ [meaning †*May God smite thee with* an evil like *a viper wasted by age*]. (Ṣ.)

حُوَيْرٍ : see what next precedes.

أَحْرَى *More,* and *most, adapted, disposed, apt, meet, suited, suitable, fitted, fit, competent,* or *proper;* or *more,* and *most, worthy,* or *deserving.* (Ṣ,* Mgh, Mṣb, Ḳ.) A thing is said to be أَحْرَى بِالِٱسْتِعْمَالِ [*More,* or *most, meet,* &c., *to be done*]. (Ṣ, Ḳ.)

مَحْرًى : see حَرِىٌّ.

مَحْرَاةٌ : see حَرِىٌّ, in two places.

حز

1. حَزَّهُ, (Ṣ, A, Mṣb,) aor. ـُ, inf. n. حَزٌّ, (Ṣ, Mṣb, Ḳ,*) *He cut it;* (Ṣ, A, Ḳ;) namely, his, or its, head; (A;) as also ▼ اِحْتَزَّهُ: (Ṣ, A, Ḳ:) or *he cut it* (namely, a thing, or flesh-meat,) *without separating; made an incision in it:* or *he cut it with labour:* (TA:) and *he notched it;* or *made a notch in it;* namely, a piece of wood. (Ṣ, Mṣb, TA.) It is said in a prov., حَزَّتْ حَازَّةٌ مِنْ كُوعِهَا [*A woman cutting cut a part* of the skin *of the extremity of the bone of her fore arm next the thumb:* nearly the same as another prov. حَلَأَتْ حَالِئَةٌ عَنْ كُوعِهَا]: alluding to a people's being occupied by their own affair so as to be diverted thereby from attending to other things. (AZ, Ḳ.) You say also, حَزَّ فِى رَأْسِ ٱلْقَوْسِ *He made a notch in the head of the bow.* (A.) — [Hence the saying,] الإِثْمُ مَا حَزَّ فِى قَلْبِكَ ‡[*Sin is that which makes an impression upon thy heart, causing thee to waver lest it be an act of disobedience because of thy not being easy respecting it*]. (A. [See حَزَّازٌ, below; and see also حَكَّ, and حَاكَ.]) Of anything making an impression in, or upon, the bosom, and causing one to waver or scruple, you say حَزَّ. (Ṣ.)

2. [حَزَّزَهُ, inf. n. تَحْزِيزٌ, *He cut it,* or *notched it, much,* or *in many places; he made notches in it; he made it serrated; he jagged it.* You say,] حَزَّزَ أَسْنَانَهُ, (Ṣ, Ḳ,) inf. n. تَحْزِيزٌ, (TA,) *He made his teeth serrated, and sharpened their extremities,* to make them like those of a young person. (Ṣ, Ḳ, TA.) [See also تَحْزِيزٌ, below.]

5. تَحَزَّزَ *It was cut much,* or *in many places,* or *into many pieces:* (Ṣ, Ḳ:) [*it was notched much,* or *in many places; was made serrated; was jagged.*]

8. اِحْتَزَّهُ : see 1.

حَزٌّ *A notch,* or *an incision,* (Ṣ, A, Mṣb, Ḳ,) in a thing; (Ṣ, Ḳ;) as, [for instance,] in a piece of wood, and a tooth-stick (سِوَاك), and a bone, (TA,) and a bow: (A, TA:) and in like manner ▼ مَحَزٌّ, *the notch of a bow, into which the ring of the string falls:* (A and Ḳ, voce كُظْرٌ:) [or the former is a coll. gen. n.:] n. un. ▼ حَزَّةٌ; (Ṣ, TA;) [for which, in the Ṣ and L and Ḳ, voce طَرِيدَةٌ, we find ▼ حُزَّةٌ, perhaps a dial. var.] You say, رَدَّ الوَتَرَ إِلَى حَزِّهَا *He put back the bow-string to its notch* (A, TA) in the head of the bow. (TA.) For the explanation of حَزُّ ٱلكَرَاكِرِ, see كِرْكِرَةٌ. ═ *A time; a particular time;* [*a nick of time:*] (Ṣ, Ḳ:) and ▼ حَزَّةٌ signifies the same; and also *a particular state* or *condition.* (A, TA.) You say, هٰذِهِ حَزَّةُ ▼ مَجِىْءِ فُلَانٍ [*This is the time of the coming of such a one*]. (A.) And كَيْفَ جِئْتَ فِى هٰذِهِ الحَزَّةِ ▼ [*How hast thou come at this time,* or *in this state?*]. (A.) And لَقِيتُهُ عَلَى حَزَّةٍ ▼ مُنْكَرَةٍ [*I met him at an evil time,* or *in an evil condition*]. (A.)

حَزَّةٌ : see حَزٌّ, in five places.

حُزَّةٌ : see حَزٌّ. ═ Also *A piece* of flesh-meat *cut off lengthwise:* (Ṣ, Ḳ:) or only of liver: (Ḳ, TA:) not of a camel's hump, nor of flesh, or other thing: (TA:) or it signifies also *a piece cut off* of anything, such as a melon &c.: used in this sense by the people of Syria: (TA:) pl. حُزَزٌ. (Mṣb.) — ‡The *neck:* (Ṣ, Mṣb, Ḳ:) accord. to some: (Mṣb:) so termed metaphorically. (Ṣ.) So in a trad., أَخَذَ بِحُزَّتِهِ *He took hold of his neck.* (Ṣ, TA.) ═ حُزَّةُ السَّرَاوِيلِ i. q. حُجْزَتُهُ: (IAạr, Az, Ṣ, Mṣb, Ḳ:) but disallowed by Aṣ. (TA.)

حَزَازٌ *Scurf* of the head: ▼ n. un. with ة [signifying *a particle,* or *flake, thereof*]. (Ṣ, Ḳ.)

حَزِيزٌ *Rugged ground:* (IDrd:) or *a rugged and extended place:* (Ṣ, Ḳ:) or *a place abounding with rugged stones like knives:* or *hard ground in a tract abounding with pebbles:* (TA:) or *rugged and hard ground with a slight elevation:* (ISh:) or *depressed ground:* (TA:) pl. [of pauc.] أَحِزَّةٌ and [of mult.] حِزَّانٌ (Ṣ, Ḳ) and حُزَّانٌ and حُزُزٌ, (Ḳ, TA,) with two ḍammehs, (TA,) or حُزَزٌ. (CḲ.)

حَزَازَةٌ : see حَزَازٌ : ═ and see also حَزَّازٌ.

حَزَّازٌ Food *that becomes acid in the stomach,* (Ḳ, TA,) *by reason of its badness, and so* [*as it were*] *cuts* (يَحُزُّ) *into the heart.* (TA.) Hence the saying, أَنْتَ أَثْقَلُ مِنَ الحَزَّازِ [*Thou art heavier,* or *more difficult to be borne, than the food that becomes acid in the stomach,* &c]. (AHeyth, on the authority of Abu-l-Ḥasan El-Aạrábee.) — [And hence, app.,] *Anything that makes an impression in,* or *upon, the heart,* or *bosom, causing one to waver or scruple;* lit., that cuts into the heart, and scrapes in the bosom; expl. by مَا حَزَّ فِى القَلْبِ, حَكَّ فِى الصَّدْرِ (A'Obeyd, Ṣ, Ḳ,) and (Ḳ;) as also ▼ حُزَّازٌ (Ḳ) [and ▼ حَازَّةٌ, as will be seen below]: and *pain in the heart, arising from wrath* &c.; as also ▼ حُزَّازٌ, (A'Obeyd, Ṣ,) and حَزَازَةٌ; (A'Obeyd, Ṣ, Ḳ;) of which last the pl. is حَزَازَاتٌ. (A'Obeyd, Ṣ.) [See also حَزْحَزَةٌ.] It is said in a trad., الإِثْمُ حَوَازُّ القُلُوبِ, (Ṣ, Mgh, and Ḳ* in art. حوز,) i. e., *Sin is those things that make an impression upon hearts,* (Lth, Mgh, MṢ, and Ḳ ubi suprà,) *like as cutting, or notching, makes an impression upon a thing,* (TA,) *and that cause one to suspect that they may be acts of disobedience, by reason of uneasiness respecting them,* (Mgh,) or *to waver respecting them, lest they should be so, for that reason,* (Ḳ,) or *to be uneasy in heart respecting them:* (MṢ:) حوازّ being pl. of ▼ حَازَّةٌ, (Mgh, Ḳ,) like as دَوَابُّ is pl. of دَابَّةٌ: (Mgh:) Sh. reads حَوَازُ القُلُوبِ, which he explains as meaning, "what overcomes hearts, (يَغْلِبُ عَلَيْهَا, i. e., مَا يَحُوزُهَا,) so that they commit that which is not incumbent:" (Mgh, and Ḳ* and TA ubi suprà:) but the former reading is the more common: (Mgh, TA:) and some read حَزَّازٌ; and some, حَوَّازٌ. (TA in art. حوز.) [See also حَزَّازُ الصَّخْرِ and حَزَّازُ الحَجَرِ ـــ.] [قَصَّ: see بَيْقُ الحَجَرِ, in art. بيق.]

حُزَّازٌ : see حَزَّازٌ, in two places.

بِهِ حَازٌّ, said of a camel, *He has an incision*, or *a cut, in the edge of the callous protuberance upon his breast, produced by his elbow, which makes it bleed:* if it does not make it bleed, it is termed مَاسِحٌ : (Ṣ, Ḳ:) or حَازٌّ is *a cut*, or *an incision, in the arm, penetrating through the skin, to the flesh, opposite the callous protuberance upon the breast of a camel;* also termed عَرْكٌ : (El-'Adebbes El-Kinánee:) or *a cut*, or *an incision, in the said protuberance:* it is a subst., like فَاغِطٌ and نَاكِتٌ. (TA.)

حَازَّةٌ ; pl. حَوَازُّ : see حَزَّازٌ, in two places.

حَزْحَزَةٌ *A pain in the heart, arising from fear or from physical suffering:* (Ḳ:) pl. حَزَاحِزُ. (TA.) [See also حَزَّازٌ.]

تَحْزِيزٌ [see 2.—] The *being cut*, or *notched, much*, or *in many places; being serrated*, or *jagged, like the teeth of the* مِنْجَل : and sometimes this is in the edges, or extremities, of the human teeth. (TA.) You say, فِى أَسْنَانِهِ تَحْزِيزٌ (Ṣ, A, Ḳ) *In his teeth is a serration, and a sharpness of the extremities* [such as is seen in the teeth of young persons]; syn. أُشُرٌ ; (Ṣ, Ḳ;) *the like of the serration of the teeth of the* مِنْجَل. (A.) — Also The *marks of cutting* or *notching*. (TA.)

مَحَزٌّ *A place of cutting* [or *notching*]. (TA.) You say, قَطَعَ فَأَصَابَ المَحَزَّ (A, TA) *He cut, and hit the place of cutting.* (TA.) And تَكَلَّمَ وَأَشَارَ فَأَصَابَ المَحَزَّ ‡ [*He spoke, and indicated*, or *advised, and hit upon the right thing*]: (A, TA:) [app. alluding to the right place of incision of the حَرْكَرَة, which is a nice and difficult operation: see حَرْكَرَة.] — [*A notched*, or *small hollowed, place, made by cutting* or *otherwise*. — *A groove*, or *the like*. Occurring in the Ḳ, voce بَكَرَة, and in art. حصر, &c.] — See also حَزٌّ, first signification.

حزب

1. حَزَبَهُ, (Ṣ, A, Mgh, Mṣb, Ḳ,) aor. ـُ, (Mgh, Mṣb,) inf. n. حَزْبٌ, (Ḳ, TA,) *It* (an event) *befell him:* (Ṣ, A, Mgh, Mṣb, Ḳ:) and *it became severe to him; distressed him;* or *pressed severely*, or *heavily, upon him:* or *it straitened him*, or *overcame him*, (Ḳ, TA,) *suddenly*, or *unexpectedly*. (TA.)

2. حَزَّبَ, (A, Ḳ,) inf. n. تَحْزِيبٌ, (Ḳ,) *He collected, congregated*, or *assembled*, people: (TA:) *he collected*, or *formed*, people *into* أَحْزَاب, (A, Ḳ,) i. e. *parties, classes, bodies, divisions*, or *the like*. (A.) — ‡ *He divided* the Ḳur-án *into* أَحْزَاب, (Ṣ, A, Mgh, TA,) meaning *set portions for particular acts of prayer, &c.;* the doing of which is forbidden. (Mgh.) [But it may also be used as meaning † *He divided* the Ḳur-án *into sixtieth portions.*]

3. حَازَبَهُ *He was*, or *became, of the number of his partisans*, or *party:* (TA:) *he helped*, or *aided, him.* (A.) — See also 5.

5. تَحَزَّبُوا *They became* [or *formed themselves into*] أَحْزَاب, (A, Mṣb, Ḳ,) i. e. *parties, classes, bodies, divisions*, or *the like;* (A;) as also ▼ حَازَبُوا : (Ḳ:) *they collected themselves together*, (Ṣ, Mgh,* TA,) against (عَلَى) others. (Mgh.)

حَزْبٌ and ▼ حُزَابَةٌ *A severe*, or *distressing, event:* or *one that straitens*, or *overcomes*, (Ḳ, TA,) *suddenly*, or *unexpectedly*. (TA.)

حِزْبٌ, in its primary acceptation, *A party*, or *company of men, assembling themselves on account of an event that has befallen them* (لِأَمْرٍ حَزَبَهُمْ): (Ksh and Bḍ in v. 61:) [and then, in a general sense,] *an assembly, a collective body*, or *company, of men:* (IAạr, A, Mgh, L, Ḳ:) *a party, portion, division*, or *class*, (Ṣ, A, L, Mṣb, Ḳ, TA,) *of men:* (L, Mṣb, TA:) the *troops*, or *combined forces, of a man;* (Ḳ, TA;) *his party, partisans*, or *faction, prepared*, or *ready, for fighting and the like:* (TA:) the *companions*, (Ṣ, Ḳ,) *sect*, or *party in opinions* or *tenets*, (Ḳ,) *of a man:* (Ṣ, Ḳ:) *any party agreeing in hearts and actions, whether meeting together or not:* (El-Moajam, TA:) pl. أَحْزَابٌ. (Ṣ, A, Mgh, Mṣb, L, Ḳ.) And the pl., with the article, *Those people who leagued together to wage war against Mohammad:* (Ḳ:) or *the parties that combined to war with the prophets.* (Ṣ.) And in the Ḳur xl. 31, *The people of Noah and 'Ád and Thamood, and those whom God destroyed after them*, (Ḳ, TA,) *as the people of Pharaoh.* (TA.) And يَوْمُ الأَحْزَابِ [*The day of the combined forces;*] *the day* [or *war*] *of the moat* (الخَنْدَق). (Mgh, Mṣb, TA.) — *I. q.* وِرْدٌ, (Ṣ, Mgh, Mṣb, Ḳ, TA,) either in its proper sense, *A turn*, or *time, of coming to water:* or in the sense next following, which is tropical. (TA.) — ‡ *A set portion of the Ḳur-án*, (A, Mgh, L, TA,) and *of prayer*, (Mgh, L, TA,) *&c.*, (Mgh,) *of which a man imposes upon himself the recital* (A, Mgh, TA) *on a particular occasion*, (Mgh,) or *at a particular time;* (TA;) *a set portion of prayer*, and *of recitation* [*of the Ḳur-án*], *&c., which a person is accustomed to perform:* (Mṣb:) pl. as above. (Mgh.) You say, قَرَأَ حِزْبَهُ مِنَ القُرْآنِ ‡ [*He recited his set portion of the Ḳur-án*]. (A.) And كَمْ حِزْبُكَ ‡ [*How much is thy set portion* of the Ḳur-án?]. (A.) — [Also † *A sixtieth portion of the Ḳur-án.*] — † *A portion, share*, or *lot*, (Mṣb, TA,) of wealth, or property: or perhaps a mistranscription for جِزْبٌ ; since IAạr says that حِزْبٌ signifies "a company of men;" and جِزْبٌ, "a portion, share, or lot." (TA.) = *A weapon*, or *weapons, of war;* syn. سِلَاحٌ ; (M, A, Ḳ, TA;) i. e. آلَةُ حَرْبٍ. (TA.) = See also what next follows.

حِزْبَاءُ, (Ṣ,) or ▼ حِزْبٌ and حِزْبَاءَةٌ, (Ḳ, TA,) *Rugged ground:* (Ṣ, Ḳ:) or *very rugged ground:* (TA:) or the first signifies *hard, elevated ground:* (Ḥam p. 664:) and the last, *a most rugged tract of* [*high ground such as is termed*] قُفّ, *slightly elevated, in another hard* قُفّ ; (ISh, TA;) or *a rugged, elevated place:* (TA:) the first is a pl.; (Ḳ;) [or rather a coll. gen. n., of which the last is the n. un.; i. e.,] the last is a more special term than the first; (Ṣ;) and the pl. is حَزَابِ, (Ṣ, in copies of the Ḳ حَزَابِى,) like صَحَارٍ, originally حَزَابِىُّ ; (Ṣ, TA;) and also explained as signifying *extended, rugged, narrow places.* (TA.)

حَزَابٌ *Thick, coarse, rude*, or *bulky, and short;* as also ▼ حِنْزَابٌ : (Ṣ:) *thick, coarse, rude*, or *bulky, and inclining to shortness;* as also ▼ حَزَابِيَةٌ, (Ṣ, Ḳ,) in which the ى is for the purpose of quasi-coordination to the quadriliteral-radical class, as in عَلَانِيَةٌ and قَبَاعِيَةٌ from عَلَنٌ and قَبْعٌ, (Ṣ,) and ▼ حِنْزَابٌ ; (Ḳ;) applied to a man, (Ṣ, TA,) and to an ass: (TA:) and ▼ حَزَابِيَةٌ also signifies *thick, coarse, rude*, or *bulky*, applied to a camel, and to a pubes; and *hardy, strong*, or *sturdy*, applied to an ass. (TA.) = Also pl. of حِزْبَاءَةٌ. (Ṣ.)

حَزِيبٌ : see حَازِبٌ.

حُزَابَةٌ : see حَزْبٌ.

حَزَابِيَةٌ : see حَزَابٌ, in two places.

حَازِبٌ and ▼ حَزِيبٌ A *severe*, or *distressing*, event: pl. [app. of either word] حُزُبٌ, (Ḳ,) or, accord. to MF, حُزْبٌ ; and pl. of the former word حَوَازِبُ. (TA.) — Also, the former, *What falls to one's lot*, of work. (TA.)

حِنْزَابٌ, in which the ن is said by some to be augmentative, and by others to be radical: (TA:) see حَزَابٌ, in two places. — Also The *carrot of the land* (جَزَرُ البَرِّ): [this would rather seem to mean the *wild carrot*, but for what here follows:]) the carrot of the sea (جَزَرُ البَحْرِ) is called قُسْطٌ. (Ṣ.) [See also art. حنزب.] — The *cock*. (Ḳ.) — *A species of* [*the birds called*] قَطًا. (Ḳ.) [See also art. حنزب.]

حُنْزُوبٌ *A certain plant* [app. that called حِنْزَابٌ, mentioned above: see art. حنزب].

حَيْزَبُونٌ *An old woman:* (Ṣ, TA:) or [*an old woman*] *in whom is no good:* (TA:) or *a cunning*, or *crafty, old woman*. (Ḥar p. 76.) The ن is augmentative, as it is in زَيْتُونٌ. (TA.)

حزر

1. حَزَرَهُ, aor. ـِ and ـُ, inf. n. حَزْرٌ (Ṣ, M, Mṣb, Ḳ) and مَحْزَرَةٌ, (Th, Ḳ,) *He computed*, or *determined, its quantity, measure, size, bulk, proportion, extent, amount, sum*, or *number:* (Ṣ, Mgh,* Mṣb, Ḳ:) [more commonly,] *he computed by conjecture its quantity* or *measure &c.;* syn. خَرَصَهُ, (Ṣ, Ḳ,) and قَدَّرَهُ بِالحَدْسِ ; (M;) *he took its quantity* or *measure &c. by the eye.* (TḲ.) [*He conjectured it;* and so ▼ حَزَّرَهُ, inf. n. تَحْزِيرٌ : perhaps post-classical: whence عِلْمُ التَّحْزِيرِ *The science of divination.*] You say, حَزَرَ النَّخْلَ *He computed by conjecture the quantity of the fruit upon the palm-trees.* (A, Mṣb.) And حَزَرْتُ قِرَاءَتَهُ عِشْرِينَ آيَةً ‡ *I computed his recitation*, or *reading, to be twenty verses* [of the Ḳur-án]. (A.) And

حَزَرْتُ قُدُومَهُ يَوْمَ كَذَا ‡ *I computed his arrival to be on such a day.* (A.) And اِحْزِرْ نَفْسَكَ هَلْ تَقْدِرُ عَلَيْهِ ‡ *Measure thyself, whether thou be able to do it.* (A.) ═ حَزَرَ, (Ṣ, M, Ḳ,) aor. ِ , inf. n. حَزْرٌ and حُزُورٌ, (M,) *It* (milk, Ṣ, M, Ḳ, and beverage of the kind called نَبِيذ, Ṣ, Ḳ) *became sour,* or *acid.* (Ṣ, M, Ḳ.) It is said in a prov., عَدَا القَارِصُ فَحَزَرَ [explained in art. قرص]. (A.) — Hence, (TA,) ‡ *It* (a man's face) *was,* or *became,* [*sour,* i. e.] *frowning, contracted, stern, austere,* or *morose.* (Ḳ, TA.)

2: see above.

حَزْرَةُ المَالِ *The better,* or *best, of cattle* or *other property;* (Ṣ, A, Mgh, Mṣb, Ḳ;) as also ▼ حَزِيرَةُ المَالِ: or the latter signifies *property to which the heart clings:* (TA:) the term حزرة is applied alike to what is masc. and what is fem.: (AO, Mṣb:) the pl. is حَزَرَاتٌ (Ṣ, Mṣb, Ḳ) and حَزْرَاتٌ, as though the sing. were an epithet: (Mṣb:) it is applied to the better or best of property because the owner of such property always, when he sees it, computes its quantity or number in his mind: accord. to Aboo-Sa'eed, حَزَرَاتُ الأَمْوَالِ signifies *those kinds of property which their owners love:* accord. to AO, *choice kinds of property.* (TA.) You say also, هٰذَا حَزْرَةُ نَفْسِى, meaning *This is the best of what I have:* and of such property the collector of the poor-rate is forbidden to take. (Ṣ, Mgh,* TA.) Also, هِىَ حَزْرَةُ قَلْبِهِ *It is the choice thing of his heart.* (AO.)

حَزْرَآءُ *Sour* [milk such as is termed] ضَرْبَةٌ. (Ḳ. [In some copies of the Ḳ, ضربة, which SM thinks to be a mistake for صربة. See also حَازِرٌ.])

حَزْوَرٌ (TA) and ▼ حَزْوَرَةٌ (Ṣ, Ḳ) and ▼ حِزْوَارَةٌ (Ḳ) *A small hill:* (Ṣ, Ḳ, TA:) or the first, *rugged ground:* (TA:) and the second, *stony ground:* (Abu-ṭ-Ṭeiyib:) pl. [of the first and second] حَزَاوِرُ (Ṣ, Ḳ) and حَزَاوِرَةٌ (Ḳ, mentioned by Abu-ṭ-Ṭeiyib as pl. of the second,) and [of the third] حَزَاوِيرُ. (Ḳ.) — Also حَزْوَرٌ and ▼ حَزَوَّرٌ, (Ṣ, A, Ḳ,) [the latter the more common,] † *A strong* boy; (Ḳ;) one *that has attained to youthful vigour,* or *the prime of manhood:* (TA:) or a boy *who has become strong,* (Ṣ, A,) *and has served:* (Ṣ:) or one *who has nearly attained the age of puberty, and has not had commerce with a woman* (ولم يفعل [app. meaning وَلَمْ يَفْعَلْ بِٱمْرَأَةٍ]): (Yaạḳoob, Ṣ:) or one *who has attained the age of puberty, and has become strong:* (Mgh, TA, in explanation of the latter word:) or a boy *who has nearly attained the age of puberty;* so called, accord. to several authors, as being likened to a hill: or one *who has fully attained that age:* (TA:) or, accord. to Aṣ and El-Mufaḍḍal, a *young* boy, who *has not attained the age of puberty:* and sometimes, one *who has attained that age, and become strong in body, and has borne arms:* and this is the right explanation: (Az:) and a *strong* man: and, contr., a *weak* man: (AḤát, Ḳ:) or, accord. to some of the lexicologists, when applied to a boy, or young man, it signifies *strong:* and when applied to an old man, *weak:* (Abu-ṭ-Ṭeiyib:) pl. حَزَاوِرَةٌ (Ṣ, TA) and حَزَاوِرُ. (TA.)

حَزْوَرَةٌ: see حَزْوَرٌ.

حَزِيرَةُ المَالِ: see حَزْرَةُ المَالِ.

حَزِيرَانُ *The name of a month, in Greek;* [*the Syrian month corresponding to June, O.S.;*] (Ṣ, Ḳ;) [*next*] *before* تَمُّوز. (Ṣ.)

حَزَوَّرٌ: } see حَزْوَرٌ.
حِزْوَارَةٌ: }

حَازِرٌ A man *computing,* or *who computes, by conjecture, the quantity* or *number* [&c.] of a thing or things. (Ṣ, TA.) [See 1.] ═ Applied to milk, and to the beverage called نَبِيذ, *Sour,* or *acid:* (Ṣ, Ḳ:) or, applied to milk, it means *more than* حَامِض: (TA:) or *i. q.* حَامِز: (IAạr, TA:) or more than حامز. (TA in art. حمز.) — And hence, (TA,) applied to a face, ‡ [*Sour,* i. e.] *frowning, contracted, stern, austere,* or *morose.* (Ḳ, TA.) [See also what follows.]

مَحْزُورٌ, (Ḳ, TA,) in some copies of the Ḳ مُحْزَوِّرٌ, (TA,) [in the CḲ مُحْزَوِّر,] ‡ *Angered;* (Ḳ, TA;) and *having a frowning, contracted, stern, austere,* or *morose face.* (TA.) [See also what next precedes.]

حزق

1. حَزَقَ, (Ṣ, Ḳ,) aor. ِ , inf. n. حَزْقٌ, (Ḳ,) *He tied,* or *bound,* a thing (Ṣ, Ḳ) with a rope: (Ṣ:) *he tied,* or *bound,* it *strongly* with a rope. (TA.) — *He bound,* or *bound round,* a foot, or a man. (Ḳ, accord. to different copies.) — *He strung* a bow. (TA.) — *He pulled,* or *drew,* a bond, and a bow-string, *hard,* or *vehemently.* (Ḳ.) — *He,* or *it, squeezed, pressed, compressed,* or *pinched,* a thing. (Ḳ.) You say of a tight boot, حَزَقَ رِجْلَهُ *It compressed,* or *pinched, his foot.* (Ḳ.) — *He straitened; made strait,* or *narrow.* (TA.) — حَزَقُوا بِهِ *They surrounded, encompassed,* or *encircled, him,* or *it.* (TA.) ═ *He broke wind:* (Ḳ:) said of an ass. (TA.) Hence the saying of 'Alee, in disparagement of the schismatics, حَزْقٌ عَيْرٍ حَزْقٌ عَيْرٍ; meaning "the case is not as ye assert:" said by El-Mufaḍḍal to be a proverbial phrase, applied in relation to a man who relates a piece of information that is not complete nor realized. (TA.)

4. احزقهُ, (Ḳ,) inf. n. إِحْزَاقٌ, (TA,) *He prevented, withheld, debarred,* or *forbade, him,* (Az, Ḳ,) عَنْهُ *from it.* (TA.)

5. تحزّق *i. q.* تجمّع [*It became collected, brought together,* &c.]. (TA.)

7. انحزق *i. q.* انضمّ [*It became drawn, collected,* or *gathered, together;* or *drawn and joined,* or *adjoined;* &c.]. (TA.)

حِزْقٌ and ▼ حِزْقَةٌ and ▼ حَازِقَةٌ and ▼ حَزِيقٌ and ▼ حَزِيقَةٌ (Ṣ, Ḳ) and ▼ حِزَاقَةٌ (Ḳ) *A collection* (Ṣ, Ḳ) of men, and of birds, and of palm-trees, &c.: pl. of the second حِزَقٌ. (Ṣ.)

حَزْقٌ [*Niggardliness,* or *tenaciousness, of that which is in one's hands*]: see حُزُقٌّ.

حِزْقَةٌ: see حِزْقٌ: — and see also حَزِيقَةٌ.

حُزُقٌ: see the next paragraph.

حُزُقٌّ and ▼ حُزُقَّةٌ *Short:* or *short in step by reason of the weakness of his body:* (Ḳ:) or *short, and short in step:* (Ṣ:) or *narrow in judgment:* (Ḳ:) so Aṣ, in explanation of the latter word; applied to a man and to a woman: (TA:) or this signifies *short:* and *short and ugly:* and the former, *narrow in power and judgment, and avaricious, niggardly,* or *tenacious:* (Sh, T, TA:) and the latter, (AO, TA,) or both, (Ḳ,) *large-bellied and short, and, in walking, turning about his buttocks;* as also ▼ أُحْزُقَّةٌ and ▼ حُزُقَةٌ: [the last in the CḲ without teshdeed:] or [in the CḲ "and"] the first two words, and ▼ حُزُقٌ and ▼ حُزُقَةٌ, a *short* man *who is short in step by reason of his shortness* or *of the weakness of his body:* or a man *niggardly,* or *tenacious, of that which is in his hands;* and the subst. [signifying the quality thus denoted] is ▼ حَزْقٌ: (Ḳ:) also *evil in disposition,* (IAạr, Ḳ,) *and niggardly:* (IAạr, TA:) and *straitened in circumstances:* (Sh, Ḳ:) or الحُزُقَّةُ [in the CḲ الحُزُقَةُ] signifies [sometimes] *a sort of game;* (Ḳ;) as in a trad., in which it is said of some girls, تَعِبْنَ الحُزُقَّةَ [*They played at the game of* الحزقة]. (TA.) حُزُقَّهْ حُزُقَّهْ تَرَقَّ عَيْنَ بَقَّهْ (Ṣ, TA) is a saying of the Arabs, (Ṣ,) explained as meaning *Short in step by reason of thy weakness, short in step* &c., *climb up,* [*O eye of a gnat* or *musquito;*] and was said, as is related in a trad., by the Prophet, in dancing El-Ḥasan and El-Ḥoseyn; whereupon the child would climb up until he placed his feet upon the Prophet's chest: (TA:) حُزُقَّهْ is for يَا حُزُقَّةُ, or أَنْتَ حُزُقَّةٌ: (IAth, TA:) and تَرَقَّ means اِرْقَ, from رَقِيتُ فِى الدَّرَجَةِ: (Ṣ:) and عَيْنَ بَقَّهْ is an allusion to smallness of the eye, (IAth, TA, and Ḥar p. 619,) as being likened to the eye of the gnat or musquito; or denotes smallness of person. (Ḥar.)

حُزُقَةٌ: see حُزُقٌّ, in two places.

حُزُقَّةٌ: see حُزُقٌّ.

حِزَاقٌ *Anything with which one ties, binds,* or *makes fast.* (TA.) [The meaning of بِالضَّرُورَةِ assigned by Golius to حِزَاقًا is a mistake, occasioned by his misunderstanding a passage in the Ḳ, where it is said of a woman that she used حِزَاقًا by poetic license (لِلضَّرُورَةِ) for حَازُوقًا, a proper name of a man.]

حَزِيقٌ: see حِزْقٌ: — and see also حَزِيقَةٌ.

حِزَاقَةٌ: see حِزْقٌ.

حَزِيقَةٌ: see حِزْقٌ. — Also, (Ḳ,) and ▼ حِزْقَةٌ, (TA,) *A part,* or *portion,* (Ḳ, TA,) [of a swarm] of locusts; (TA;) as also حِرْقَةٌ; (Ḳ and TA in art. حرق;) or of anything; (Ḳ, TA;) even of wind: (TA:) pl. of the former حَزَائِقُ and حِزَقٌ (Ḳ, TA [in the CḲ حِزَقٌ, which is pl. of حِزْقَةٌ,]) and [coll. gen. n.] ▼ حَزِيقٌ. (Ḳ.) — And *i. q.* حَدِيقَةٌ [*A walled garden;* &c.]: (Ḳ:) or the *like of a* حديقة. (Ibn-'Abbád, TA.)

حَارِقٌ One who is *pinched by a tight boot*: (Ṣ, Ḳ:) of the measure فَاعِلٌ in the sense of the measure مَفْعُولٌ. (Ḳ.) One says, لَا رَأْىَ لِحَارِقٍ [*No counsel, or advice, is possible to one who is pinched by a tight boot*]. (Ṣ, TA.) [See also حَاقِنٌ.]

حَارِقَةٌ: see حَزِقٌ.

أُحْزُقَةٌ: see حَزِقٌ.

إِبْرِيقٌ مَحْزُوقُ العُنُقِ *A narrow-necked ewer.* (A, Nh, Ḳ.)

مُتَحَزِّقٌ *Very niggardly* or *tenacious* or *avaricious*. (Ṣ, Ḳ.)

حزم

1. حَزَمَهُ, (Ṣ, Mṣb, Ḳ,) aor. ـِ, (Ḳ,) inf. n. حَزْمٌ, (Ṣ,) *He bound it*, or *tied it*; (Ṣ, Ḳ;) namely, a thing: (Ṣ:) or *he made it a* حُزْمَة [q. v.]. (Mṣb.) —— حَزَمَ الدَّابَّةَ (Ṣ, Mṣb) or الفَرَسَ, (Ḳ,) aor. as above, (Mṣb,) and so the inf. n., (Mgh, Mṣb,) *He bound the beast* [or *horse*] *with the* حِزَام [or *girth*]; (T,* Ṣ,* Mgh,* Mṣb;) and *with a rope*; (T, TA;) or *he bound the* حِزَام *of the* [*beast* or] *horse*. (Ḳ.) —— [And hence,] حَزَمَ رَأْيَهُ, [aor. and] inf. n. as above, † *He made his judgment, opinion*, or *counsel, firm*, or *sound*. (Mṣb.) It is said in a prov., قَدْ أَحْزِمُ لَوْ أَعْزِمُ [*Certainly I make firm* my determination *if I determine* upon doing a thing]; meaning I know الحَزْم [i. e. *prudence*, or *discretion*, and *precaution*], though I do not practise it. (IB, TA. [See also Freytag's Arab. Prov. ii. 262.]) ═ [Hence, also,] حَزُمَ, (Ṣ, Ḳ,) aor. ـُ, (Ḳ,) inf. n. حَزَامَةٌ (Ṣ, Ḳ*) and حُزُومَةٌ, (Ḳ,* TḲ,) but this latter is not of established authority, (TA,) and حَزْمٌ, (CḲ,* TḲ, [or this is probably a simple subst. in relation to حَزُمَ,]) *He possessed the quality of* حَزْم [explained below, as meaning *prudence*, or *discretion*, &c.]. (Ṣ, Ḳ.) ═ حَزِمَ, aor. ـَ, inf. n. حَزَمٌ, (Ṣ, Ḳ,) *He was*, or *became, choked*, (Ḳ,) or *he had what resembled a choking*, (Ṣ,) *in his chest*. (Ṣ, Ḳ.)

4. احزمهُ *He made for him*, or *put to him*, [namely, a horse, as is implied in the Ḳ,] *a* حِزَام [or *girth*]. (Ḳ.)

5. تحزّم and ↓احتزم, (Ṣ, Ḳ,) [said of a horse, as is implied in the Ḳ, and of a man,] *He became furnished with a* حِزَام [i. e. *girth*, or *girdle*]: (Ḳ:) [or, said of a man, *he became girt*; or *he girded himself*;] or *i. q.* تَلَبَّبَ, meaning *he bound his waist with a rope* [or *girdle*]. (Ṣ.) It is said in a trad., نَهَى أَنْ يُصَلِّىَ الرَّجُلُ حَتَّى يَحْتَزِمَ ↓ [*He forbade that the man should pray unless he were girt*, or *unless he girded himself*]. (TA.) ═ تحزّم فِى أَمْرِهِ *He acted with prudence*, or *discretion, and precaution, in his affair*, or *case*. (TA.)

8. احتزم: see 5, in two places: —— and see حَزْمٌ. —— Also *It was*, or *became, inwrapped*. (Ḥam p. 614.)

12. اِحْزَوْزَمَ, (Ḳ,) from الحَزْمُ; like اِعْشَوْشَبَ, from العُشْبُ; (TA;) *It* (a place) *was*, or *became, rough*, or *rugged*: (Ḳ:) or *elevated*. (TA.) —— *It was*, or *became, collected together, and compacted*, or *compact*. (Ḳ.) —— *He* (a man) *was*, or *became, big*, or *large, in the belly, without being full*. (Ḳ, TA.)

حَزْمٌ [*Prudence*, or *discretion*, and *precaution*;] *sound management of one's affair* or *case*, (Ṣ, Ḳ,) and *taking the sure course therein*, (T, Ṣ, Ḳ,) and *precaution, that it may not become beyond the power of management*: (TA:) said in a trad. to consist in evil opinion: and in another, in the asking counsel of people of judgment and obeying them: (TA:) or *good judgment*: (Mgh:) or *strength*, [or *firmness of mind* or *of judgment*, (see حَازِمٌ,)] and *sound management*: (Ḥam p. 33:) the first part thereof said by Aktham Ibn-Ṣeyfee to be consultation: (Ḥam ibid:) from the same word as signifying the act of "binding the حِزَام," (Mgh,) or from this word as signifying the act of "binding with the حِزَام," and "with the rope:" (T, TA:) and ↓حُزْمَةٌ signifies the same; as in the saying, إِنَّ الوَحَاةَ مِنْ طَعَامِ الحُزْمَةِ [*Verily quickness is of the food of prudence*, &c.], a prov., mentioned by Ibn-Kethweh, alluding to people's collecting themselves together and aiding one another, when they act with quickness, or sharpness, and vigour; and said in praise of him who thus acts. (TA.) You say, أَخَذَ بِالحَزْمِ (TA) and [sometimes] فِى الحَزْمِ (Ḳ in art. حوط) [*He took the course prescribed by prudence, discretion, precaution*, or *good judgment*; *he used precaution*: and, like أَخَذَ بِالثِّقَةِ, *he took the sure course* in his affair]. ═ *Elevated ground*; as also ↓أَحْزَمُ and ↓حَيْزُومٌ: (Ḳ:) or this last signifies *rough*, or *rugged, ground*: (Yz, IB, Ḳ:) and حَزْمٌ is [*ground*] *more elevated than what is termed* حَزْنٌ: (Ṣ:) or *more rough*, or *rugged, than what is termed* حَزْنٌ: (Ḥam p. 45:) or *elevated ground*, or *rugged and elevated ground, that is girt* (↓اُحْتُزِمَ) *by a torrent*: or *rugged ground, having many stones, which are more rugged and rough and scabrous than those of the* أَكَمَة, *but the top of which is broad and long, extending to the length of two leagues, and three, and less than that, which the camels do not ascend except by a road that it has*: accord. to Yaạkoob, the م is a substitute for the ن of حَزْنٌ: pl. حُزُومٌ. (TA.)

حَزَمٌ [in a horse (see أَحْزَمُ)] *Largeness*, or *fulness, of the sides*, or *of the sides and belly and flank*; *contr. of* هَضَمٌ. (Ṣ.)

حُزْمَةٌ: see حَزْمٌ.

حُزْمَةٌ *A bundle*, or *what is bound round*, (Ḳ, TA,) of firewood &c.: (Ṣ:) pl. حُزَمٌ. (Mṣb, TA.)

حَزْمَى وَٱللّٰهِ *i. q.* أَمَا وَٱللّٰهِ; (Ḳ;) as also حَرْمَى وَٱللّٰهِ [q. v.]. (Ḳ in art. حرم.)

حُزُمَّةٌ *Short*; (Ḳ;) applied to a man. (TA.)

حِزَامٌ [The *girth* of a horse and the like; and the *girdle* of a man;] the *thing with which one girths*, or *girds*; as also ↓حِزَامَةٌ and ↓مِحْزَمٌ and ↓مِحْزَمَةٌ: (Ḳ:) pl. حُزُمٌ, (Mṣb, Ḳ, TA, [in the CḲ حُزْمٌ,]) i. e., pl. of حِزَامٌ, (Mṣb, TA,) [and أَحْزِمَةٌ is pl. of pauc. of the same:] the pl. of مِحْزَمَةٌ [and مِحْزَمٌ] is مَحَازِمُ. (TA.) [J says,] The حزام of the beast is well known: and hence the saying, جَاوَزَ الحِزَامُ الطُّبْيَيْنِ [*The girth passed beyond the two teats*]; (Ṣ;) meaning † *the affair*, or *case, became distressing*, and *formidable*. (Ḳ in art. طبى.) —— Hence, also, The حزام [or *swaddling-band*] *of a child in his cradle*. (Ṣ.) —— [And hence, also,] أَخَذَ حِزَامَ الطَّرِيقِ ‡ *He took the middle*, and *main part*, or *beaten track, of the road*. (TA.)

حَزِيمٌ: see حَازِمٌ: ═ and see also حَيْزُومٌ, in two places.

حِزَامَةٌ: see حِزَامٌ.

حَزَّامٌ *A binder of paper into bundles*: in [the dial. of] Má-wará-en-Nahr. (TA.)

حَازِمٌ (Ṣ, Ḳ) and ↓حَزِيمٌ (Ḳ) *Possessing the quality of* حَزْم [explained above, as meaning *prudence*, or *discretion*, and *precaution*; or *good judgment*; &c.]: or *intelligent*; *discriminating*, or *discerning*; *possessing firmness*, or *soundness, of judgment*, or *knowledge, and skill in affairs*, or *experience and good judgment*; *using precaution in affairs*: (TA:) pl. (of the former, TA) حَزَمَةٌ (Ḳ, TA [in the CḲ, erroneously, حُزْمَةٌ,]) and حُزَّامٌ and حُزُمٌ and حُزَّمٌ and [of pauc.] أَحْزَامٌ; (TA;) and (of حَزِيمٌ, TA) حُزَمَاءُ. (Ḳ.)

حَيْزُومٌ and ↓حَزِيمٌ The *breast*, or *chest*: (Ḳ:) or the *middle thereof*; (Ṣ, Ḳ;) and the *part which the* حزام [i. e. *girth* or *girdle*] *embraces*, (Ṣ, TA,) *where the heads of the* جَوَانِح [or *ribs of the breast*] *meet, above the lower extremity of the sternum, opposite the* كَاهِل [or *uppermost third portion of the backbone*]: (TA:) the *part of the breast which is the place of the* حزام: (Ḥam p. 704, in explanation of the latter word:) and the former word, the *part that surrounds the back and the belly*: or the *ribs of* [*the part where lies*] *the heart*: and the *part of the side of the breast on the right and left of the* حُلْقُوم [or *windpipe*]; (Ḳ;) the two parts thus described being called حَيْزُومَانِ: (TA:) pl. of the former حَيَازِيمُ; (TA;) and of the latter أَحْزِمَةٌ [a pl. of pauc.] (Kr, Ḳ) and حُزُمٌ [a pl. of mult.]. (Ḳ.) One says, أُشْدُدْ حَيْزُومَكَ لِهٰذَا الأَمْرِ, and حَيَازِيمَكَ, i. e. † *Dispose and subject thyself to this affair*, or *case*; meaning *prepare thyself for it*: and شَدَّ حَزِيمَهُ↓ [† *He disposed and subjected*, or *prepared, himself*]: (TA:) or شَدُّ الحَيَازِيمِ is an expression denoting, by way of similitude, *patient endurance* of that which has befallen one. (Ḥam p. 163.) —— And the former, † The *breast* [or *bows*] *of a ship* or *boat*. (MA.) ═ حَيْزُومُ [so in my copies of the Ṣ, imperfectly decl., app. regarded as of foreign origin, (not الحَيْزُومُ as is implied in the Ḳ,)] the name of *One of the horses of the angels;* (Ṣ;) *the horse of Gabriel:* (Ḳ:) accord. to some, [حيزون,] with ن in the place of the م. (TA.) ═ See also حَزْمٌ.

أَحْزَمُ [*More, and most, prudent, discrete, or cautious*]. Hence the prov., أَحْزَمُ مِنْ حِرْبَآءٍ [*More prudent, or cautious, than a chameleon*]. (Meyd. [See Freytag's Arab. Prov. i. 309.]) ═ Also, applied to a horse, (Ṣ,) *Large, or full, in the sides, or in the sides and belly and flank; contr. of* أَهْضَمُ. (Ṣ, Ḳ.) — And, applied to a camel, (TA,) *Large in the* حَيْزُوم: (Ḳ:) or *large in the place of the* حِزَام [or *girth*]. (T, TA.) — See also حَزْمٌ.

أَحْزَامٌ *i. q.* أَحْزَابٌ [pl. of حِزْبٌ]: (Ḳ:) the م is a substitute for the ب. (TA.)

مَحْزِمٌ, of a beast, The *part upon which lies the* حِزَام [or *girth*]. (Ṣ.)

مِحْزَمٌ : } see حِزَامٌ.
مِحْزَمَةٌ :

حزن

1. حَزِنَ, (Ṣ, Mṣb, Ḳ,) aor. ـَ, (Mṣb, Ḳ,) inf. n. حَزَنٌ; (Mṣb, TA;) and ▼ تَحَزَّنَ, and ▼ احتزن, (Ṣ, Ḳ,) and ▼ تحازن; (Ḳ;) *He was, or became, affected with* حُزْن [q. v.; i. e. *he grieved, mourned, or lamented; or was sorrowful, sad, or unhappy;* &c.; لَهُ and عَلَيْهِ *for him* or *it*]. (Ṣ, Mṣb, Ḳ.) لَا تَحْزَنْ and لَا تَحْزَنُوا, in the Ḳur [ix. 40, &c., and iii. 133], do not denote a prohibition of getting حُزْن; for حُزْن does not come by the will of man: the real meaning is *Do not thou, and ye, that which engenders* حُزْن; *do not thou, and ye, acquire* حُزْن. (Er-Rághib. [But this requires consideration; or, rather, is not in every case admissible.]) ═ حَزَنَهُ, (Ṣ, Mṣb, Ḳ,) aor. ـُ, (Mṣb, TA,) inf. n. حُزْنٌ; (Ḳ;) and ▼ احزنهُ; *He,* (another person, Ṣ,) or *it,* (an affair, or an event, or a case, Mṣb, Ḳ,) *caused him to be affected with* حُزْن [which see below; i. e. *grieved him; or caused him to mourn or lament, or to be sorrowful or sad or unhappy;* &c.]: (Ṣ, Mṣb, Ḳ:) accord. to Yz, (Ṣ,) the former is of the dial. of Ḳureysh; and the latter, of the dial. of Temeem: (Ṣ, Mṣb:) and so say Th and Az: (Mṣb, TA:) but the former is said to be the more approved: (TA:) or, accord. to AZ, the aor. of the former is used, but not the pret.; (Mṣb, TA;) and when the act is ascribed to God, the latter verb is used: Z, also, says that what is well known in usage is the employing the pret. of the latter and the aor. of the former: (TA:) or احزنهُ signifies *he made him* حَزِين [q. v.]; and حَزَنَهُ, [in some copies of the Ḳ ▼ حَزَّنَهُ,] *he made* حُزْن *to be in him:* (Sb, Ḳ:) or حَزَنَهُ, *it caused him to fall into* حُزْن. (TA.) ═ حَزُنَتِ الأَرْضُ, (TA,) inf. n. حُزُونَةٌ; (Ṣ, TA;) and ▼ استحزنت; (TA;) *The ground was, or became, rough,* (TA,) or *rugged and hard.* (Ṣ.) — One says also of a beast that is not easy to ride upon, يَحْزُنُ المَشْىَ ‡ [*He is rough in pace*]: and فِيهِ حُزُونَةٌ ‡ [*In him is roughness* in pace]. (TA.)

2. يَقْرَأُ بِالتَّحْزِينِ *He reads, or recites, with a slender* [*and plaintive*] *voice.* (Ṣ, Ḳ.) — See also 1.

4. احزنهُ: see 1. ═ احزن بِنَا المَنْزِلُ *The place of alighting, or abode, was, or became, rough, or rugged and hard, with us: or made us to be on rugged ground.* (TA.) — And احزن *He was, or became, in a tract such as is termed* حَزْن [i. e. *rugged,* or *rugged and hard:* opposed to أَسْهَلَ]. (Ḳ.) [And hence,] احزنوا † *They used roughness* with men: opposed to اسهلوا. (TA in art. سهل.)

5: see 1. — تحزّن عَلَيْهِ *He expressed pain, grief, or sorrow, or he lamented, or moaned, for, or on account of, him, or it;* syn. تَوَجَّعَ. (Ḳ.)

6: } 8: } 10: } see 1.

حَزْنٌ *Rugged* (Ṣ, Mṣb, Ḳ) *and hard* (Ṣ) *ground:* (Ṣ, Mṣb, Ḳ:) or *rugged high ground:* (TA: [see also حَزْمٌ:]) good land, though hard, is not thus termed: (ISh:) pl. حُزُونٌ: (Mṣb, TA:) and ▼ حَزْنَةٌ signifies the same as حَزْنٌ: (Ḳ:) so too, as some say, does ▼ حُزُنٌ, with two dammehs; or, as others say, this is a pl. of حَزْنٌ: and you say also ▼ أَرْضٌ حَزْنِيَّةٌ [meaning the same as حَزْنٌ, or *land of a rugged,* or *rugged and hard,* or *rugged and high, kind*]. (TA.)

حُزْنٌ and ▼ حَزَنٌ, (Lth, Ṣ, Ḳ,) the former said by AA to be used when the nom. or gen. case is employed, and the latter when the accus. is employed; (TA;) or the former is a simple subst., and the latter an inf. n.; (Mṣb;) *Grief, mourning, lamentation, sorrow, sadness, or unhappiness; contr. of* سُرُورٌ: (Ṣ, TA:) or *i. q.* هَمٌّ: (Ḳ:) or [هَمٌّ, accord. to common usage, is for some evil that is expected to happen; whereas] حُزْنٌ is *grief arising on account of an unpleasant event that has happened, or on account of an object of love that has passed away;* and is *the contr. of* فَرَحٌ: (El-Munáwee, TA:) or *a roughness in the spirit, occasioned by grief:* (Er-Rághib, TA:) pl. أَحْزَانٌ; (Ḳ;) [properly a pl. of pauc.; but] it has no other pl. (TA.) [Hence,] عَامُ الحُزْنِ *The year* [*of mourning;*] *in which died Khadeejeh and Aboo-Tálib:* (IAạr, Th, Ḳ:) so called by Moḥammad. (IAạr, Th.) الحَمْدُ لِلّٰهِ الَّذِى أَذْهَبَ عَنَّا الحَزَنَ▼, in the Ḳur [xxxv. 31], is said to mean [*Praise be to God, who hath dispelled from us*] *the anxiety* (هَمّ) *of the morning and evening meals:* or *all grieving anxiety of the means of subsistence:* or *the grief of punishment:* or *of death.* (TA.)

حَزَنٌ: see حُزْنٌ, in two places.

حَزُنٌ : } see حَزِينٌ.
حَزِنٌ :

حُزُنٌ : } see حَزْنٌ.
حَزْنَةٌ :

حَزْنَةٌ *Rugged mountains:* pl. حَزْنٌ. (Aṣ, Ṣ, Ḳ.) — And [hence,] the pl., † *Difficulties, hardships, or distresses.* (TA.)

حَزْنِىٌّ A camel *that pastures in a tract such as is termed* حَزْن. (Ṣ, TA.) — أَرْضٌ حَزْنِيَّةٌ: see حَزْنٌ.

حَزْنَانُ *Affected with vehement, or intense,* حُزْن [i. e. *grief, mourning,* &c.]; as also ▼ مِحْزَانٌ. (Ḳ,* TA.)

حَزُونٌ A sheep, or goat, (شَاةٌ,) *evil in disposition.* (Ṣ, Ḳ.)

حَزِينٌ (Ṣ, Mṣb, Ḳ) and ▼ حَزِنٌ (Ṣ, Ḳ) and ▼ حَزُنٌ (Ḳ) *Grieving, mourning, lamenting, sorrowful, sad, or unhappy:* (Ṣ, Mṣb,* Ḳ:* [see also مَحْزُونٌ; with which, accord. to the Ḳ, all seem to be syn.; and with which the first may be regarded as properly syn. on the ground of analogy, being from حَزِنَ, not from حَزَنَ:]) pl. (of the first, TA) حِزَانٌ and حُزَنَآءُ (Ḳ, TA) and حَزَانَى. (Ḳ voce ضَرِيسٌ.) — صَوْتٌ حَزِينٌ *A soft* or *gentle, easy, slender, plaintive, and melodious, voice.* (TA.) — مَالِكٌ الحَزِينِ *A certain bird.* (TA. [See art. ملك.])

حُزَانَةٌ A man's *family,* or *household, for whose case he suffers grief and anxiety.* (Ṣ, Ḳ.) [And simply One's *family,* or *household.*] One says, فُلَانٌ لَا يُبَالِى إِذَا شَبِعَتْ خِزَانَتُهُ أَنْ تَجُوعَ حُزَانَتُهُ [*Such a one cares not, when his store-room is full, that his family, or household, suffer hunger*]. (A, TA.) ═ *A prior right which the Arabs enjoy over the foreigners, on their first arrival* [*in the territory of the latter*], *with respect to the houses and lands:* (M, Ḳ:) or *a condition which the Arabs used to impose upon the foreigners in Khurásán, when they took a town, or district, pacifically, that when the soldiery* [*of the former*] *passed by them, singly or in companies, they should lodge them, and entertain them, and supply them with provisions for their march to another district.* (Az, TA.)

حَيْزُونٌ: see حَيْزُومٌ, in art. حزم.

مَحْزَنٌ: see مَحْزُونٌ.

مُحْزِنٌ [*Grieving,* or *causing to mourn* or *lament,* &c.,] is applied to an event, or a case; and also, but not حَازِنٌ, to a voice. (TA.)

مِحْزَانٌ: see حَزْنَانُ.

مَحْزُونٌ *Grieved;* or *caused to mourn* or *lament,* or *to be sorrowful* or *sad* or *unhappy;* (AA, Ṣ, Ḳ;) as also ▼ مَحْزَنٌ. (Ḳ.) — مَحْزُونُ اللِّهْزِمَةِ *Rough in the* لهزمة [app. meaning the *angle of the lower jaw,* or the *flesh on that part*]: and *having the* لهزمة *hanging down,* [*by the relaxation of its muscle,*] *in consequence of grief.* (TA.)

حزو and حزى

1. حَزَا الشَّىْءَ, (Aṣ, Ṣ,) aor. يَحْزُو, (Ṣ,) inf. n. حَزْوٌ; (Aṣ, TA;) and حَزَى الشَّىْءَ, aor. يَحْزِى, (Ṣ,) [inf. n. حَزْىٌ;] *He computed,* or *determined, the quantity, measure, or the like, of the thing;* (Ṣ;) *he computed by conjecture the quantity,* &c.

(Aṣ, Ṣ.) You say, حَزَيْتُ النَّخْلَ, (Ṣ, Mṣb,) inf. n. حَزْىٌ; and حَزَوْتُهُ, inf. n. حَزْوٌ; *I computed by conjecture the quantity of the fruit upon the palm-trees.* (Mṣb.) In the copies of the Ḳ, خَرَصَهُ, meaning حَزَّى النَّخْلَ تَحْزِيَةً, is erroneously put for حَزَى النَّخْلَ حَزْيًا, the words of Aṣ. (TA.) — حَزَا, (Lth, Ḳ,) aor. يَحْزُو, (Lth,) inf. n. حَزْوٌ; (Ḳ;) and حَزَى aor., يَحْزِى, (Lth, Ḳ,) inf. n. حَزْىٌ; (Ḳ;) and ▼تَحَزَّى, (Lth, Ḳ,) inf. n. تَحَزٍّ and تَحَزُّوٌ; (Ḳ;) *He divined:* (Lth, Ḳ, TA:) and particularly *from the flight,* or *cries, &c., of birds.* (Ḳ.) And حَزَا, (AZ, Ḳ,) aor. يَحْزُو, (TA,) inf. n. حَزْوٌ; (AZ;) and حَزَى, aor. يَحْزِى; (TA;) *He chid* birds, *to make them rise and fly,* or *drove* them *away, and divined from their flight,* or *cries, &c.:* (AZ, Ḳ,* TA:) and *drove* them: (Ḳ,* TA: but the error in the Ḳ, mentioned above, makes the verb in this case also, and in the next following, to be حَزَّى, and the inf. n. to be تَحْزِيَةٌ: TA:) [for] when a crow croaks in front of a man, and he desires to obtain some object of want, he says, "It is good," and goes forth: when it croaks behind his back, he says, "This is evil," and does not go forth: and when a thing passes by him from the direction of his right hand, he augurs good from it; but if from the direction of his left hand, he augurs evil from it. (AZ, TA.) ═ حَزَا الشَّخْصَ, aor. يَحْزُو, (Ṣ,) inf. n. حَزْوٌ; (TA;) and حَزَى الشخص, (Ṣ, Ḳ,) aor. يَحْزِى; (Ṣ;) said of the سَرَاب [or mirage], *It raised* [*to the eye*] *the figure of a man or other thing seen from a distance:* (Ṣ, Ḳ, TA:) or, accord. to IB, the correct phrase is حزى الآلَ [which may mean that one should say, حَزَى الآلُ الشَّخْصَ, or حَزَى السَّرَابُ الآلَ: see آل in art. اول]. (TA.)

5: see 1.

حَزَّآءٌ: see what follows.

حَازٍ One *who computes by conjecture the quantity of fruit upon palm-trees.* (Mṣb, TA.) — *A diviner:* (Lth, TA:) one *who examines the members, and the moles of the face, divining* [*from them*]. (Ṣ, Ḥar p. 170.) One says, عَلَى الحَازِى هَبَطْتَ [meaning † *Thou hast lighted upon him who is knowing*]. (Ḥar ib.) — *An astronomer,* or *astrologer;* as also ▼حَزَّآءٌ. (TA.) — The pl. is حُزَاةٌ and حَوَازٍ. (TA.)

حس

1. حَسَّ, (first pers. حَسَسْتُ, or حَسِسْتُ, and حَسِيتُ, and حَسَيْتُ, and حَسْتُ,) as syn. with أَحَسَّ: see 4, throughout. — حَسَّ لَهُ, first pers. حَسَسْتُ, aor. حَسِ; (Ṣ, Ḳ;) and first pers. حَسِسْتُ, [aor. حَسَ;] (Yaạḳoob, Ṣ, Ḳ;) inf. n. حِسٌّ (Ṣ, Ḳ) and حَسٌّ, (Ḳ,) or the latter is inf. n. of both verbs, but the former is a simple subst., (TA,) *He was,* or *became, tender,* or *compassionate, towards him:* (Ṣ, Ḳ:) [lit., *he felt for him:*] ISd says that, for رَقَقْتُ لَهُ, the explanation of حسست له, he found in the book of Kr the verb written with ف and ق; but that the former [?] is the right: (TA:) or *he lamented for him;* (Aboo-Málik, TA;) as also ▼تَحَسَّسَ لَهُ. (Ḳ,* TA.) The Arabs say, إِنَّ العَامِرِىَّ لَيَحِسُّ لِلسَّعْدِىِّ *Verily the 'Ámiree is tender, or compassionate, towards the Saạdee;* because of the relationship that subsists between them. (TA.) And Abu-l-Jarráḥ El-'Oḳeylee said, مَا رَأَيْتُ عُقَيْلِيًّا إِلَّا حَسَسْتُ لَهُ *I have not seen an 'Oḳeylee without my feeling tenderness, or compassion, towards him;* (Ṣ, TA;) for the same reason. (TA.) [See حَاسَّةٌ, second signification.] ═ حَسَّهُ, [aor. حَسُ, inf. n. حَسٌّ,] *He made his* حِسّ [i. e. *sound, or motion,*] *to cease.* (Bḍ in iii. 145.) — Hence, (Bḍ, ubi suprà,) حَسَّهُمْ, aor. حَسُ, (Ṣ, A, Bḍ, Mṣb,) inf. n. حَسٌّ, (Mṣb, Ḳ,) [and app., accord. to Fr, حِسٌّ, or this may be a simple subst.,] *He slew them:* (Bḍ, Mṣb, Ḳ:) and *extirpated them:* (Ḳ:) or *he slew them quickly:* (A, TA:) or *he extirpated them by slaughter:* (Aboo-Is-ḥáḳ, Ṣ:) or *he slew them with a quick and extirpating slaughter:* or *with a vehement slaughter:* (TA:) it occurs in the Ḳur iii. 145: (Ṣ, A:) accord. to Fr, حِسٌّ, [so in the TA,] in this instance, is the *act of slaying and destroying.* (TA.) [Hence, حُسَّ app. signifies *It was utterly destroyed, so that nothing was left in the place thereof;* and so ▼احتسّ; for it is said that] حِسٌّ and اِحْتِسَاسٌ, with respect to anything, signify أَنْ لَا يُتْرَكَ فِى المَكَانِ شَىْءٌ. (TA.) Also حَسَّهُمْ, aor. as above, *He trod them under foot, and despised them.* (TA.) And حَسَّ البَرْدُ الجَرَادَ *The cold killed the locusts.* (Ṣ.) — [Hence,] الجَرَادُ تَحُسُّ الأَرْضَ *The locusts eat the herbage of the land.* (TA.) And حَسَّ البَرْدُ الكَلَأَ, (Ṣ, Ḳ,) or الزَّرْعَ, (A,) aor. حَسُ, (Ṣ, TA,) inf. n. حَسٌّ, (TA,) ‡ *The cold nipped, shrunk, shrivelled,* or *blasted,* (lit., *burned,* أَحْرَقَ, q. v.,) *the herbage,* (Ṣ, Ḳ,) or *the seed-produce.* (A.) ═ حَسَّ الدَّابَّةَ, (Ṣ, TA,) or حَسَّهَا بِالمِحَسَّةِ, (A,) aor. حَسُ, (Ṣ,) inf. n. حَسٌّ, (Ṣ, Ḳ,) ‡ *He curried the beast; removed the dust from it with the* مِحَسَّة. (Ṣ, A, Ḳ, TA.) [See a proverb, voce حَشٌّ.] Hence the saying of Zeyd Ibn-Ṣooḥán, on the day of the battle of the Camel, when he was carried off from the field, about to die, اُدْفِنُونِى فِى ثِيَابِى وَلَا تَحُسُّوا عَنِّى تُرَابًا ‡ [*Bury ye me in my clothes, and*] *shake not off any dust from me.* (Ṣ, TA.)

2. حَسَّسْتُ الشَّىْءَ [a mistake in the CḲ for حَسَسْتُ]: see 4.

4. احسّ الشَّىْءَ, (Ṣ, Mṣb, Ḳ,) inf. n. إِحْسَاسٌ, (Mṣb,) *He perceived,* or *became sensible of,* (وَجَدَ,) *the* حِسّ (i. e. *motion or sound,* TA) *of the thing:* (Ṣ, Ḳ, TA:) *he knew the thing by means of* [*any of*] *the senses:* (IAth, accord. to his explanation of الإِحْسَاس as signifying العِلْمُ بِالحَوَاسِّ:) *he ascertained the thing as one ascertains a thing that is perceived by the senses:* (Bḍ in iii. 45:) *he knew the thing;* or *he perceived it by means of any of the senses;* syn. عَلِمَ بِهِ, (Mṣb,) and عَلِمَهُ, and عَرَفَهُ, and شَعَرَ بِهِ; (TA;) as also ▼حَسَّهُ, aor. حَسُ, inf. n. حَسٌّ and حِسٌّ and حَسِيسٌ; (TA, in explanation of the saying in the Ḳ that حَسَسْتُ الشَّىْءَ [in the CḲ erroneously ▼حَسَّسْتُ] signifies the same as أَحْسَسْتُهُ;) and in like manner احسّ بِهِ is syn. with شَعَرَ بِهِ; (L, Mṣb;) and so is ▼حَسَّ بِهِ, aor. حَسُ, inf. n. حِسٌّ (L, Mṣb) and حَسٌّ and حَسِيسٌ; (L;) or حِسٌّ is a simple subst.: (M, L:) accord. to Fr, إِحْسَاسٌ is syn. with وُجُودٌ; and Zj says that the meaning of احسّ is عَلِمَ and وَجَدَ: (TA:) or احسّ signifies *he perceived,* or *became sensible of;* syn. وَجَدَ: and *he thought,* or *opined:* (Akh, Ṣ, Ḳ:) and *he saw;* syn. أَبْصَرَ: and *he knew:* (Ḳ:) and is trans. in these senses by itself, (Akh, Ṣ,) or by means of بِ: (TA:) and أَحْسَسْتُ بِهِ signifies *I knew it certainly; was certain,* or *sure, of it;* (Ṣ;) as also ▼حَسِسْتُ بِهِ, (Ṣ, Ḳ,) with kesr; (Ḳ;) speaking of news, or tidings: (Ṣ:) and for أَحْسَسْتُ, some say أَحْسَيْتُ, (Ṣ, Mṣb, Ḳ,) changing the [second] س into ى; (Ṣ, Mṣb;) and أَحَسْتُ, (T, Ṣ, L, Ḳ,) with a single س, which is an extr. kind of contraction, (Ṣ, Ḳ,) but made in all other similar cases, where the last radical letter is quiescent; (Sb, L, TA;) and [thus] for أَحْسَسْنَ, we find أَحَسْنَ; (Ṣ;) and in like manner, for ▼حَسِسْتُ, (Ṣ, Ḳ,) or ▼حَسَسْتُ, (Mṣb,) some say حَسِيتُ, (T, Ṣ, M, L,) and حَسَيْتُ, (M, L, Mṣb, Ḳ, [in the CḲ حَسَيْتُ, which is the modern vulgar form,]) and حَسْتُ. (Abu-l-Ḥasan, IAạr, T, M, L.) You say, احسّهُ بِعَيْنِهِ [*He perceived him, or it, with his eye*]; (Ibn-Zekereeyá, TA in art. جس;) and so ▼حَسَّهُ. (Ṣgh, TA ibid.) [In the present art. in the TA, it is said that حَسَّهُ بالنصل is syn. with أَحَسَّهُ: but بالنصل is evidently a mistranscription for بِالبَصَرِ.] And it is said in the Ḳur [iii. 45], فَلَمَّا أَحَسَّ عِيسَى مِنْهُمُ الكُفْرَ *And when Jesus ascertained their unbelief as one ascertains a thing that is perceived by the senses:* (Bḍ:) or *opined it;* or *perceived it,* or *became sensible of it:* (Akh, Ṣ:) or *saw it:* (Lḥ, TA:) or *knew it,* or *perceived it sensibly.* (Mṣb.) And in the same [xix. last verse], هَلْ تُحِسُّ مِنْهُمْ مِنْ أَحَدٍ *Dost thou see of them any one?* (TA.) You say also, هَلْ أَحْسَسْتَ الخَبَرَ *Hast thou known the news?* (TA.) And أَحْسَسْتُ بِالخَبَرِ, and حَسِيتُ بِهِ, and ▼حَسِسْتُ بِهِ, and أَحْسَيْتُ بِهِ, *I knew the news certainly.* (Ṣ.) And أَحْسَسْتُ الخَبَرَ, and أَحَسْتُهُ, and ▼حَسِيتُهُ, and حَسْتُهُ, *I knew somewhat of the news.* (T, L, TA.) And مَا أَحْسَسْتُ بِالخَبَرِ, and مَا أَحَسْتُ بِهِ, and ▼مَا حَسِيتُ بِهِ, and مَا حَسْتُ بِهِ, *I knew not aught of the news.* (T, L, TA.)

5. تحسّس *He listened* to the discourse of people: (El-Ḥarbee, Ḳ:) accord. to Aboo-Mo'ádh, it is similar to تَسَمَّعَ and تَبَصَّرَ: (TA:) or *he sought repeatedly,* or *time after time, to know* a thing, *by the sense* (الحاسّة) [of hearing &c.]: (Ḥar p. 678:) or (so accord. to the TA, but in the Ḳ "and,") *he sought after,* (Mṣb, Ḳ,) or *sought after repeatedly,* or *time after time,* (Mṣb,) news, or tidings, of a people, in a good cause; (Ḳ;) تجسّس signifying the doing the same in an evil cause: A'Obeyd says, you say, تَحَسَّسْتُ الخَبَرَ

and تَحَسَّسَهُ; and Sh says that تَنَدَّسَهُ is similar to it; and IAạr, that تَحَسَّسْتُ الخَبَرَ and تَبَحَّثْتُهُ [but this is app. a mistranscription for تَبَحَّثْتُهُ] signify the same. (TA.) You say also, تحسّس مِنَ الشَّىْءِ *He asked, or inquired, after news, or tidings, of the thing.* (Ṣ, TA.) And تحسّس فُلَانًا, and مِنْ فُلَانٍ, *He inquired, or sought for information, respecting such a one;* as also تجسّس: or the former signifies *he sought after him for himself;* and the latter, "he sought after him for another." (TA, art. جس.) The passage in which it occurs in the Ḳur xii. 87, has been differently interpreted, accord. to the several explanations here given. (TA.) ═ See also 7.

7. انحسّ † *It became pulled out* or *up* or *off; became eradicated,* or *displaced; fell,* or *came, out;* syn. اِنْقَلَعَ: (Ṣ, Ḳ:) *it fell; fell off;* or *fell continuously, by degrees,* or *one part after another;* syn. تَحَاتَّ, (Ṣ, Ḳ,) and تَسَاقَطَ: (A, TA:) *it broke in pieces:* (TA:) said by Az to be a dial. var. of انْحَتَّ. (TA.) You say, انحسّت أَسْنَانُهُ (Ṣ, TA) ‡ *His teeth fell,* or *came, out,* (انقلعت,) *and broke in pieces.* (TA.) And انحسّ شَعَرُهُ ‡ *His hair fell off continuously.* (A.) And in like manner, تَحَسَّسَتْ ▼ أَوْبَارُ الإِبِلِ, (TA,) and تَحَسْحَسَتْ ▼, (Ḳ, TA,) ‡ *The fur of the camels fell off continuously, and became scattered.* (Ḳ,* TA.)

8. احتسّ: see حَسَّهُ.

R. Q. 1. حَسْحَسَ لَهُ: see حَسَّ لَهُ.

R. Q. 2. تَحَسْحَسَتْ أَوْبَارُ الإِبِلِ: see 7.

حِسٌّ [accord. to some, a subst. from 1, q. v., in several senses explained above; but accord. to others, an inf. n. As a simple subst., it is often used as syn. with حَاسَّةٌ in the first of the senses assigned to the latter below; i. e., *A sense; a faculty of sense;* as, for instance, in the Ḳ in art. سمع. — Hence الحِسُّ المُشْتَرَكُ: see art. شرك]. ═ *A sound:* (Ḳ:) or *a low, faint, gentle,* or *soft, sound;* as also ▼ حَسِيسٌ: (Ṣ, Mgh, Mṣb, TA:) [in the present day it often signifies the *voice of a man or woman;* and particularly *when soft:*] *a plaintive voice or sound, in singing* or *weeping,* and *such as that of a lute;* syn. رَنَّةٌ: (TA:) or ▼ حَسِيسٌ, [or both,] the *sound by which a thing is perceived:* (Bḍ in xxi. 102:) and the former, *motion:* (Ḳ:) and the *passing* of anything *near by one, so that he hears it without seeing it;* as also ▼ حَسِيسٌ. (Ḳ, TA.) It is said in the Ḳur [xxi. 102], لَا يَسْمَعُونَ حَسِيسَهَا ▼ *They shall not hear its low sound:* (Ṣ:) [meaning, that of hell-fire:] or *the sound by which it shall be perceived:* (Bḍ:) or *the motion of its flaming.* (TA.) And in a trad., فَسَمِعَ حِسَّ حَيَّةٍ *And he heard the motion,* and *the sound of the passing along, of a serpent.* (TA.) And you say, مَا سَمِعَ لَهُ حِسًّا وَلَا جَرْسًا *He heard not any motion, nor any sound, of him,* or *it.* (TA.) It may refer to a man and to other things; as, for instance, wind. (TA.) ═ *A pain which attacks a woman after childbirth,* (Ṣ, A, Ḳ,) *in the womb:* (A:) or the *pain of childbirth, when the latter is* [*first*] *felt:* but the former meaning is confirmed by a trad. (TA.) — حِسُّ الحُمَّى *The commencement of fever, when the latter is* [*first*] *felt;* (TA;) as also ▼ حُسَاسُهَا: (Lḥ, TA:) or حِسٌّ signifies *a touch,* or *slight affection,* of fever, *at its very commencement.* (TA.) ═ *Cold that nips, shrinks, shrivels,* or *blasts,* (lit., *burns,* يُحْرِقُ, for which, in the TA, is substituted يَقْطَعُ,) *the herbage.* (Ṣ, Ḳ.) [See also حَاسَّةٌ.]

حَسَاسٌ, with fet-ḥ, *Perception by means of any of the senses;* syn. وُجُودٌ. (L, TA.) Hence the proverb, لَا حَسَاسَ مِنَ ٱبْنَىْ مُوقِدٍ [*There is no perceiving of the two sons of the place of the kindling of fire*]: (L, TA:) for they say that two men used to kindle a fire in a road, and, when people passed by them, to entertain them as guests; and a party passing by when they had gone, a man said these words. (L.) And hence the saying, ذَهَبَ فُلَانٌ فَلَا حَسَاسَ بِهِ *Such a one has gone, and there is no perceiving him:* or *there is no perceiving his place.* (TA.)

حُسَاسُ الحُمَّى: see حِسٌّ.

سَنَةٌ حَسُوسٌ *A year of severe sterility and drought;* (Ṣ, Ḳ;*) *in which is little good fortune;* (TA;) as also ▼ حَاسُوسٌ: (Ḳ:) or *a year that consumes everything.* (TA.)

حَسِيسٌ: see حِسٌّ, in four places. ═ *Slain; killed.* (Ṣ, Mṣb.)

حِسِّىٌّ [*Relating to sense; sensible,* or *perceptible by sense;*] *opposed to* مَعْنَوِىٌّ. (Kull p. 101 &c.)

حَسَّاسٌ *Having strong perception:* an epithet applied in this sense to the devil. (TA.) — رَجُلٌ حَسَّاسٌ لِلْأَخْبَارِ *A man having much knowledge of news.* (Mṣb.) — القُوَّةُ الحَسَّاسَةُ [*The sensitive faculty*]. (Er-Rághib, TA in art. حى.)

حَاسُوسٌ *One who searches for news* or *tidings;* (TA;) like جَاسُوسٌ: (Ḳ, TA:) or the former relates to good, and the latter to evil. (Ḳ, TA.) ═ *Unfortunate; unlucky;* (IAạr, Ḳ;) as also ▼ مَحْسُوسٌ; (Lḥ, TA;) applied to a man. (IAạr, Ḳ.) — See also حَسُوسٌ.

حَاسَّةٌ sing. of حَوَاسُّ, (A, Mṣb, Ḳ,) which signifies The five *senses;* (Ṣ, Mṣb;) the *hearing,* the *sight,* the *smell,* the *taste,* and the *touch:* (Ṣ, Mṣb, Ḳ:) these are the external: the internal are also five; but authors disagree respecting the seats thereof. (TA.) [See also حِسٌّ.] — [*A feeling;* as in the saying,] أَطَّتْ لَهُ مِنِّى حَاسَّةُ رَحِمٍ [*A feeling of relationship,* or *consanguinity, pleaded for him on my part*]. (Aboo-Málik, TA.) — [*An organ of sense;* as when you say,] العَيْنُ حَاسَّةُ الرُّؤْيَةِ [*the organ of the sense of sight*]. (Ṣ in art. عين.) ═ حَاسَّةٌ [also signifies † *A thing that destroys, consumes,* or *injures, herbage* or *the like.* Hence,] حَوَاسُّ الأَرْضِ *Cold* and *hail* [in one copy of the Ṣ *heat*] and *wind* and *locusts* and *the beasts* [*that pasture*]: (Ṣ, Ḳ:) these also being five. (Ṣ.) You say, أَصَابَتِ الأَرْضَ حَاسَّةٌ † *Cold smote the land:* (Lḥ, TA:) the ة is to denote intensiveness. (TA.) [See also حِسٌّ, last signification.] And أَصَابَتْهُمْ حَاسَّةٌ ‡ *Injury befell them* (Ṣ, TA) from cold (Ṣ, A, TA) or some other cause. (Ṣ.) حَاسَّةٌ also signifies † *Wind that removes the dust into the pools of water left by torrents, and fills them, so that the moist earth dries up.* (AḤn, TA.) And † *Locusts eating the herbage of the land.* (TA.) You say also, مَرَّتْ بِالقَوْمِ حَوَاسُّ, meaning † *Severe years passed over the people.* (Lḥ, TA.)

البَرْدُ مَحَسَّةٌ لِلْكَلَإِ, (Ṣ, TA,) and لِلنَّبَاتِ, (A, TA,) ‡ *The cold is a cause of nipping, shrinking, shrivelling,* or *blasting,* (lit., *of burning,* see 1,) *to the herbage.* (Ṣ, TA.)

مِحَسَّةٌ ‡ *A currycomb;* syn. فِرْجَوْنٌ; (Ṣ, A,* Ḳ;) it is *an instrument of iron, having teeth like the* مُشْط. (TA in art. نمص.)

مَحْسُوسٌ pass. part. n. of 1, q. v. — [As a subst. it means *A thing perceived by any of the senses; an object of sense:* pl. مَحْسُوسَاتٌ.] — الطَّرِيقُ المَحْسُوسَةُ [lit., *The way that is sensibly perceived;* app., *the milky way in the sky:* or] *the tract in the sky along which* (مِنْهَا) *the* [*wandering*] *stars* [or *planets*] *take their courses.* (TA voce المَجَرَّةُ.) ═ أَرْضٌ مَحْسُوسَةٌ † *Land smitten by locusts.* (TA.) ═ See also حَاسُوسٌ.

حسب

1. حَسَبَهُ, (Ṣ, A, Mgh, &c.,) aor. ـُ, (Ṣ, Mgh, Mṣb, &c.,) inf. n. حَسْبٌ (Ṣ, A, Mgh, Mṣb, Ḳ) and حُسْبَانٌ (Ṣ, Mgh, Mṣb, Ḳ) and حِسْبَانٌ (Ḳ) and حِسَابٌ, (Ṣ, Ḳ,) which is generally an inf. n. of this verb, but sometimes of حَاسَبَ, (TA,) and حِسَابَةٌ (Ṣ, Ḳ) and حِسْبَةٌ, (Mṣb, Ḳ,) or this is like قِعْدَةٌ and رِكْبَةٌ, [denoting a mode, or manner,] as in a verse of En-Nábighah cited below, (Ṣ,) and حُسْبَةٌ, which is of rare occurrence, (MF, TA,) *He numbered, counted, reckoned, calculated,* or *computed, it;* (Ṣ, A, Mgh, Mṣb, Ḳ;) namely, property [&c.]. (A, Mgh, Mṣb.) You say, مَنْ يَقْدِرُ عَلَى عَدِّ الرَّمْلِ وَحَسْبِ الحَصَى [*Who can count the sands, and number the pebbles?*]. (A.) And أَلْقِ هٰذَا فِى الحَسْبِ [*Throw thou this into the reckoning*]; i. e., *into what thou hast reckoned.* (A.) وَالشَّمْسُ وَالقَمَرُ بِحُسْبَانٍ, in the Ḳur [lv. 4], means *And the sun and the moon* [run their courses] *according to a* [certain] *reckoning;* or *through a series of mansions* [or *constellations*], the bounds of which they do not transgress: (TA:) or بحسبان alludes to the numbers of the months and years and all other times: [but properly speaking,] حسبان is here an inf. n.: (Zj, TA:) or, accord. to Akh, a pl. of حِسَابٌ; (Ṣ, TA;) and so says AHeyth: or, accord. to some, it is here a proper subst., signifying the *firmament.* (TA.) حُسْبَانًا in the Ḳur vi. 96 is held by Akh to be for بِحِسَابٍ, بِحُسْبَانٍ, meaning [as in the phrase quoted above, from the Ḳur lv. 4, accord. to the first explanation]. (TA.) And حِسَابُكَ عَلَى اللّٰهِ signifies حُسْبَانُكَ عَلَى اللّٰهِ [*On*

God be it to reckon with thee: see also حَسِيبُكَ ٱللّٰهُ]. (TA.) Az says that the reckoning in buying and selling is termed حِسَابٌ because one knows thereby what is sufficient. (TA.) وَٱللّٰهُ سَرِيعُ الحِسَابِ, in the Ḳur [ii. 198, &c., *God is quick in reckoning*], signifies that his reckoning is necessary, or of necessity, and that his reckoning with one person does not divert Him from reckoning with another. (TA.) And يَرْزُقُ مَنْ يَشَآءُ بِغَيْرِ حِسَابٍ, in the Ḳur [ii. 208, &c., *He supplieth whom He willeth, without reckoning*], means, *without sparing,* or *scanting;* as when a man expends without reckoning: but the phrase is variously explained, as meaning *without appointing for any one what is deficient:* or *without fearing that any one will call Him to account for it:* or *without the receiver's thinking that He will bestow upon him,* or *without his reckoning upon the supply;* so that it may be from حَسِبَ "he thought," or from حَسَبَ "he reckoned." (L, TA.) The saying, cited by IAạr,

* يَا جُمْلُ أُسْقَاكِ بِلَا حِسَابَهْ *

as related by J [in the Ṣ], but correctly أُسْقِيتِ, (TA,) means [*O Juml, mayest thou be given rain*] *without reckoning,* and *without measure.* (Ṣ.) An instance of حِسْبَةٌ as similar to قِعْدَةٌ and رِكْبَةٌ occurs in the saying of En-Nábighah,

* فَكَمَّلَتْ مِائَةً فِيهَا حَمَامَتُهَا *
* وَأَسْرَعَتْ حِسْبَةً فِى ذٰلِكَ العَدَدِ *

[*And she completed a hundred, in which was her pigeon; and she was quick in the mode of computing that number*]. (Ṣ.) — حَسِبَهُ كَذَا, [a verb of the kind termed أَفْعَالُ القُلُوبِ, having two objective complements, the former of which is called its noun, and the latter its enunciative,] aor. ـَ and ـِ; (Ṣ, Mṣb, Ḳ;) the former the more approved, (TA,) of the dialects of all the Arabs except Benoo-Kinaneh; the latter aor. being peculiar to the dial. of this tribe, (Mṣb,) and contr. to analogy, (Ṣ, Mṣb,) for by rule it should be ـَ [only]; and حَسِبَ is the only verb of the measure فَعِلَ having both يَفْعِلُ and يَفْعَلُ as the measures of its aor. except نَعِمَ and يَئِسَ and بَئِسَ [and وَهِلَ and وَحِرَ and يَبِسَ and وَلِهَ and وَغِرَ mentioned by Ibn-Málik (with the preceding) cited in the TA voce وَرِثَ]; but eight verbs having an unsound letter for the first radical have kesreh to the medial radical in the pret. and aor., viz., وَرِىَ and وَرِثَ and وَرِعَ and وَرِمَ and وَثِقَ and وَلِىَ and وَمِقَ and وَفِقَ; (Ṣ;) inf. n. حِسْبَانٌ (Ṣ, Mgh, Mṣb, Ḳ) and مَحْسِبَةٌ and مَحْسَبَةٌ (Ṣ, Ḳ) and حِسَابٌ; (TA; [but see what follows;]) *He* [*counted, accounted, reckoned,* or *esteemed,* meaning] *thought,* or *supposed, him,* or *it, to be so.* (Ṣ, Mgh, Mṣb, Ḳ.) You say, حَسِبْتُهُ صَالِحًا [*I counted him,* or *thought him, good,* or *righteous*]. (Ṣ.) And حَسِبْتُ زَيْدًا قَائِمًا [*I thought Zeyd to be standing*]. (Mṣb.) And مَا كَانَ فِى حِسْبَانِى كَذَا [*Such a thing was not in my thought*]: you should not say فِى حِسَابِى, (Ḳ,) unless you mean thereby *it was not included in my reckoning,* or, by amplification of the sense, *I did not think it.* (MF.) — حَسُبَ, aor. ـُ, (Ṣ, Mṣb, Ḳ,) inf. n. حَسَابَةٌ (Ṣ, Ḳ) and حَسَبٌ, (Mṣb, Ḳ,) *He was,* or *became, characterized,* or *distinguished, by what is termed* حَسَبٌ *as explained below* [i. e. *grounds of pretension to respect* or *honour;* &c.]. (Ṣ, Mṣb, Ḳ.)

2. حَسَّبَهُ, inf. n. تَحْسِيبٌ: see 4. — Also *He placed a pillow for him; supported him with a pillow;* (Ṣ, Ḳ;) *seated him upon a* حُسْبَانَة, or مِحْسَبَة. (TA.) — And hence, *He honoured him.* (L.) — *He buried him:* (TA:) or *buried him in stones:* [see حَسَبٌ:] or *buried him wrapped in grave-clothing:* namely, a dead person. (Ḳ, TA.) — Nuheyk El-Fezáree says, (Ṣ, TA,) addressing 'Ámir Ibn-Eṭ-Ṭufeyl, (TA,)

* لَتَقَيْتَ بِالوَجْعَآءِ طَعْنَةَ مُرْهَفٍ *
* حَرَّانَ أَوْ لَثَوَيْتَ غَيْرَ مُحَسَّبِ *

(Ṣ, TA) *Thou wouldst have avoided, by turning thy hinder part, the thrust* [*of a thin, thirsty weapon*], *or thou wouldst have taken thy resting-place* (TA) *not honoured,* or *not shrouded,* (Ṣ, TA,) or *not pillowed:* غير محسب being variously rendered: one person prefers the meaning *not buried:* Az says that the signification of burial in stones and that of wrapping in grave-clothes, assigned to the verb, were unknown to him; and that غير محسب signifies *not supported with a pillow.* (TA.)

3. حَاسَبَهُ, inf. n. مُحَاسَبَةٌ (Ṣ, TA) and sometimes حِسَابٌ, which is also an inf. n. of حَسَبَ, or, accord. to Th, it seems to be a quasi-inf. n., (TA,) [*He reckoned with him.*] And حَاسَبَهُ عَلَيْهِ [*He called him to account for it*]. (TA.)

4. أَحْسَبَهُ, (Th, Ṣ, Ḳ,) inf. n. إِحْسَابٌ, (TA,) *He gave him what sufficed,* or *satisfied, him,* مِنْ كُلِّ شَيْءٍ *of everything:* (Th, TA:) *he contented him:* (Ḳ:) or *he gave him what contented him;* as also حَسَّبَهُ: (Ṣ:) and both verbs, inf. n. of the latter تَحْسِيبٌ, *he gave him to eat and drink until he was satisfied:* (Ḳ:) and the former, [or both,] *he gave him until he said* حَسْبِى [*It is sufficient for me*]. (AZ, Ṣ.) You say also, أَعْطَى فَأَحْسَبَ *He gave, and* †*gave much:* (Ṣ:) and اِحْتَسَبْتُهُ, [if not a mistranscription for أَحْسَبْتُهُ,] ‡ *I gave him much.* (A, TA.) — Also *It* (a thing, Ṣ, Mṣb,) *sufficed him:* (Ṣ, A, Mṣb:) *he sufficed him.* (TA.) You say, مَرَرْتُ بِرَجُلٍ أَحْسَبَكَ مِنْ رَجُلٍ, and بِرَجُلَيْنِ أَحْسَبَاكَ [مِنْ رَجُلَيْنِ], and بِرِجَالٍ أَحْسَبُوكَ [مِنْ رِجَالٍ], *I passed by a man sufficient for thee as a man,* i. e., *supplying to thee the place of any other* [by his excellent qualities], and *by two men* &c., and *by men* &c. (Ṣ.) [The verb here is rendered, in grammatical analysis, by its act. part. n. See also حَسْبُ.]

5. تَحَسَّبَ ‡ *He sought,* or *sought leisurely and repeatedly, to learn news:* (A, Ḳ, * TA:) *he sought after news:* (Ḳ, * TA:) *he inquired,* or *asked, respecting news;* (Ṣ, Ḳ, * TA; [in the CḲ, اسْتَخْبَرَ is erroneously put for اِسْتَخْبَرَ;]) of the dial. of El-Ḥijáz: (TA:) *he searched after news as a spy.* (A'Obeyd, TA.) It is said in a trad., accord. to one reading, كَانُوا يَجْتَمِعُونَ فَيَتَحَسَّبُونَ الصَّلَاةَ ‡ *They used to assemble, and endeavour to ascertain the time of prayer:* but the common reading is يَتَحَيَّنُونَ. (TA.) — Also *He reclined upon a pillow.* (Ḳ.)

8. اِحْتَسَبَ [for اِحْتَسَبَ أَجْرًا *He reckoned upon a reward:* or] *he sought a reward* [from God in the world to come]. (TA.) وَيَرْزُقْهُ مِنْ حَيْثُ لَا يَحْتَسِبُ, in the Ḳur lxv. 2, means [*And He will supply him with the means of subsistence*] *whence he does not reckon,* or *expect; whence does not occur to his mind.* (Bḍ, Jel.) And مَنْ صَامَ رَمَضَانَ إِيمَانًا وَٱحْتِسَابًا, in a trad., *Whoso fasteth during Ramaḍán, believing* in God and his Apostle, *and* [*reckoning upon a reward,* or] *seeking a reward* from God. (Mgh, * TA.) You say also, اِحْتَسَبَ بِكَذَا أَجْرًا عِنْدَ ٱللّٰهِ (Ṣ, Ḳ) *He reckoned upon obtaining,* [or *he sought,*] *by such a thing,* or *such an action, a reward from God:* (PṢ:) or *he prepared,* or *provided, such a thing, seeking thereby a reward from God.* (Ḳ.) And اِحْتَسَبَ عِنْدَ ٱللّٰهِ خَيْرًا *He prepared,* or *provided, in store for himself, good,* [i. e. *a reward,*] *with God.* (A, Mgh.) And اِحْتَسَبَ الأَجْرَ عَلَى ٱللّٰهِ *He laid up for himself, in store, the reward, with God, not hoping for the reward of the present life;* اِحْتِسَابُ الأَجْرِ relating only to an action done for the sake of God. (Mṣb.) [Hence,] اِحْتَسَبَ وَلَدَهُ, (A, Mgh,) or ٱبْنَهُ, (Mṣb,) or ٱبْنًا لَهُ, or بِنْتًا, (Ṣ, Ḳ, *) is said when one has lost by death an adult child or son or daughter; (Ṣ, A, Mgh, Mṣb, Ḳ;) meaning *He prepared,* or *provided, in store for himself, a reward, by his patience on the occasion of his being afflicted and tried by the death of his adult child:* (Mgh, * TA:) when a man has lost by death a child not arrived at the age of puberty, you say of him, اِحْتَسَبَ افْتَرَطَهُ. (Ṣ, A, Mṣb, Ḳ.) [Hence also,] اِحْتَسَبَ عَمَلَهُ [*He reckoned upon,* or *prepared for himself, a reward by his deed:* or] *he did his deed seeking a reward from God in the world to come.* (L, TA.) — اِحْتَسَبْتُ بِالشَّيْءِ *I included the thing in a numbering,* or *reckoning;* or *made account of it; accounted it a matter of importance.* (Mṣb.) And فُلَانٌ لَا يُحْتَسَبُ [for لا يحتسب به] *Such a one is made no account of; is not esteemed,* or *regarded, as of any account,* or *importance.* (A, TA.) — اِحْتَسَبْتُ عِنْدَهُ means اِكْتَفَيْتُ [*I was,* or *became, sufficed,* or *contented, thereat,* or *with him,* or *at his abode*]. (A, TA.) [And IbrD thinks that the verb has the same signification in the phrase اِحْتَسَبْتُ عَلَيْهِ بِالمَالِ, quoted in the TA from the A; holding عليه to be here used in the sense of عَنْهُ; so that the meaning is *I was,* or *became, sufficed, so as to have no need of him,* or *it, by the property:* but I doubt whether this phrase be correctly transcribed.] — اِحْتَسَبَ also signifies اِنْتَهَى [*He abstained,* or *desisted;* app. as one sufficed, or contented]. (Ḳ.) — And اِحْتَسَبَ عَلَيْهِ كَذَا *He disapproved and disallowed his doing,* or *having done, such a*

thing; (Ṣ, Ḳ;*) namely, a foul deed: (TA:) whence the appellation ▼ مُحْتَسِبٌ. (Ḳ.) And accord. to some, اِحْتَسَبَ ٱللّٰهُ عَلَيْهِ means *He said, May God take,* or *execute, vengeance upon him;* or *punish him;* for his evil deeds. (Ḥar p. 371. [See حَسِيبٌ.]) [In the present day, احتسب عَلَيْهِ is used as meaning *He prayed for aid against him by saying,* حَسْبُنَا ٱللّٰهُ *God is,* or *will be, sufficient for us.*] — You say also, اِحْتَسَبَ فُلَانًا, (Ḳ,) or اِحْتَسَبَ مَا عِنْدَهُ, (A,) meaning ‡ *He endeavoured to learn what such a one had* [*in his mind,* or *in his possession*]. (A, Ḳ,* TA.) — See also 4.

9. اِحْسَبَّ *He* (a camel) *was,* or *became, of a white colour intermixed with red* (Ṣ, TA) *and with black.* (TA.)

حَسْبٌ *Sufficiency.* (Ḳ voce هَبْ.) — حَسْبُ is a [prefixed] noun (Ṣ) [syn. with كَفْى, as is implied in the Ḳ voce قَطْ; or] syn. with كَافِى; (Mṣb;) or [virtually] meaning كَفَى [as a pret. in the sense of an emphatic aor.]; (Ṣ, Ḳ;) or يَكْفِى: (TA:) Sb says that it is used to denote the being sufficed, or content. (TA.) You say, حَسْبُكَ دِرْهَمٌ [and بِحَسْبِكَ دِرْهَمٌ, in which latter the ب is redundant; meaning *Thy sufficiency,* or *a thing sufficing thee, is a dirhem;* a phrase which may be used in two ways; as predicating of what is sufficient, that it is a dirhem; and as predicating of a dirhem, that it is sufficient; in which latter case, بحسبك is an enunciative put before its inchoative, (as also حَسْبُكَ,) so that the meaning is, *a dirhem is a thing sufficing thee,* i. e. *a dirhem is sufficient for thee;* as is shown in a marginal note in my copy of the Mughnee, in art. ب; or, accord. to the Ṣ and Ḳ, *a dirhem suffices thee:* accord. to Bḍ (iii. 167), بحسبك means أَحْسَبَهُ, مُحْسِبُكَ, and كَافِيكَ, from meaning كَفَاهُ; and is shown to have this meaning by its not importing a determinate signification in consequence of its being a prefixed noun with its complement in the saying, هٰذَا رَجُلٌ حَسْبُكَ *This is a man sufficing thee*]. (Ṣ, Mṣb, Ḳ.) You say also, حَسْبُكَ ذٰلِكَ *That is,* or *will be,* [or *let that be,*] *sufficient for thee.* (TA.) And حَسْبُكَ ٱللّٰهُ, in the Ḳur viii. 65, *God is,* or *will be, sufficient for thee.* (Fr, TA. See also حَسِيبُكَ ٱللّٰهُ.) And حَسْبُكَ بِصَدِيقِنَا [*A person sufficing thee is our friend*]; in which the ب is added to denote emphatic praise. (Fr, TA in art. ب.) In the saying, هٰذَا رَجُلٌ حَسْبُكَ مِنْ رَجُلٍ *This is a man sufficing thee as a man,* i. e. *supplying to thee the place of any other* [by his excellent qualities], (Ṣ, Ḳ,) and مَرَرْتُ بِرَجُلٍ حَسْبِكَ مِنْ رَجُلٍ *I passed by a man sufficing thee as a man,* (TA,) حسبك is an expression of praise, referring to the indeterminate noun [رجل]; because, in its case, [what is originally (see below)] an inf. n. (فَعْلٌ) [under which term lexicologists, but not grammarians, include the مَصْدَر]) is rendered, in grammatical analysis, by another word, [i. e., by an act. part. n.,] as though one said مُحْسِبٌ لَكَ, or كَافٍ لَكَ. (Ṣ. [Thus حسبك in these two instances is a صِفَة, i. e. an epithetic phrase; and من رجل is a تَمْيِيز, i. e. a specificative phrase.]) When the noun to which حسبك refers is determinate, you put حسب in the accus. case, as a حَال, i. e. a denotative of state; as in the saying, هٰذَا عَبْدُ ٱللّٰهِ حَسْبَكَ مِنْ رَجُلٍ *This is 'Abd-Allah; being one sufficing thee as a man.* (Ṣ. [Here من رجل is, as before, a specificative phrase.]) [See also 4, the corresponding verb.]) حسب, in this manner, is used alike as sing. and dual and pl.; (Ṣ, Ḳ;) being [originally] an inf. n. (Ṣ.) It is also used alone, [as a prefixed noun of which the complement is understood,] as in the phrase زَيْدٌ حَسْبُ, without tenween, for حَسْبِى or حَسْبُكَ [&c., meaning *Zeyd is sufficient for me* or *for thee* &c.]; like as one says, جَآءَنِى زَيْدٌ لَيْسَ غَيْرُ, for لَيْسَ غَيْرُهُ عِنْدِى. (Ṣ. [That is, حَسْبُ, when thus used, is subject to the same rules as غَيْرُ and قَبْلُ and بَعْدُ &c. when so used.]) — See also حَسَبٌ, in three places. — Also, (TA,) and ▼ حِسْبَةٌ, (Ḳ,) *Burial* of the dead: (TA:) or *burial* of the dead *in stones* [app. meaning *in a grave cased with stones*]: or *burial* of the dead *wrapped in grave-clothes:* like تَحْسِيبٌ. (Ḳ. [See 2.])

حَسَبٌ i. q. ▼ مَحْسُوبٌ; (Ṣ, Ḳ;) of the measure فَعَلٌ in the sense of the measure مَفْعُولٌ, like نَفَضٌ in the sense of مَنْفُوضٌ; (Ṣ;) *Numbered, counted, reckoned, calculated,* or *computed.* (Ṣ, Ḳ.) — *A number counted.* (L.) — *Amount, quantity,* or *value.* (L.) Sometimes, (S, L, Ḳ,) by poetic license, (Ṣ,) and in prose, (L,) ▼ حَسْبٌ. (Ṣ, L, Ḳ.) You say, اَلْأَجْرُ بِحَسَبِ مَا عَمِلْتَ, and ▼ بِحَسْبِ, *The recompense is,* or *shall be, according to the amount,* or *quantity,* or *value, of thy work.* (L.) And يُجْزَى ٱلْمَرْءُ عَلَى حَسَبِ عَمَلِهِ *The man is,* or *shall be, paid according to the amount,* or *quantity, of his work.* (Mṣb.) And عَلَى حَسَبِ مَا أَسْدَيْتَ إِلَىَّ شُكْرِى لَكَ [and ▼ حَسْبَمَا (for عَلَى حَسَبِ مَا)] *According to the amount,* or *value, of the benefit,* or *benefits, that thou hast conferred upon me are my thanks to thee.* (L.) And لِيَكُنْ عَمَلُكَ بِحَسَبِ ذٰلِكَ *Let thy deed,* or *work, be correspondent to the quantity,* or *number, of that:* or *adequate,* or *equivalent, to that.* (Ṣ.) And هٰذَا بِحَسَبِ ذَا *This is equal in number* or *quantity,* or *is equivalent, to that.* (Ḳ.) And مَا أَدْرِى مَا حَسَبُ حَدِيثِكَ, i. e. ما قَدْرُهُ [app. *I know not what is the value of thy story*]. (Ks, Ṣ.) And أَحْسَنْتُ إِلَيْهِ حَسَبَ ٱلطَّاقَةِ and عَلَى حَسَبِ ٱلطَّاقَةِ *I benefited him according to the measure of ability.* (Mgh.) — Also [*Grounds of pretension to respect* or *honour, consisting in any qualities* (*either of oneself or of one's ancestors*) *which are enumerated,* or *recounted, as causes of glorying:* and hence signifying *nobility; rank* or *quality; honourableness,* or *estimableness, from whatever source derived:*] originally, (MF,) *what one enumerates,* or *recounts, of the deeds,* or *qualities, in which his ancestors have gloried:* (Ṣ, A, Mgh,* Ḳ, MF:) secondly, *what one enumerates,* or *recounts, of his own deeds,* or *qualities, in which he glories:* thirdly, *what one enumerates,* or *recounts, of any deeds,* or *qualities, that are causes of his glorying, of whatever kind they be:* (MF:) or the *memorable deeds,* or *qualities, of one's ancestors;* and *one's own deeds,* or *qualities, in which he glories;* because they were enumerated, or recounted, by the Arabs in contending, or disputing, for glory; (T, Mṣb,* TA;) the latter consisting in *such qualities as courage, and good disposition, and liberality:* (Mṣb:) or *what are enumerated,* or *recounted, of generous actions,* or *qualities:* (Mṣb:) or *good actions,* or *conduct, of oneself,* and *of one's ancestors:* (Sh, Mgh:) or *generosity,* or *nobility, of actions* or *conduct:* (IAạr, Ḳ:) or *righteous, virtuous,* or *good, actions* or *conduct:* (Ḳ:) or *good disposition:* (TA:) or *religion;* (Ṣ, Mṣb, Ḳ;) *piety;* because true nobility consists in religion or piety: (MF:) or *wealth;* (Ṣ, Ḳ;) because it serves in lieu of true nobility: (TA:) in this sense, and in the sense next preceding, it has no corresponding verb: (TA:) or *state,* or *condition;* [i. e. *good state* or *condition;*] syn. بَالٌ [i. q. حَالٌ]: (Ḳ:) or *intellect,* or *understanding:* (MF:) and a man's *relations, consisting of his children and others:* pl. أَحْسَابٌ. (Az, Mgh.) Accord. to ISk, (Ṣ, Mṣb,) حَسَبٌ and كَرَمٌ may pertain to him who has not noble ancestors; but not شَرَفٌ nor مَجْدٌ. (Ṣ, Mṣb,* Ḳ.) حَسَبٌ is also used elliptically, (Mgh, TA,) [in the sense of حَسِيبٌ, q. v.,] for ذُو حَسَبٍ, (TA,) and for ذَوُو حَسَبٍ. (Mgh.) — اِشْتَرَى بِالْحَسَبِ *He bought* a thing *in an honourable manner with respect to himself and the seller:* حسب, here, is said to be from حَسَّبَهُ "he honoured him;" or from حُسْبَانَةٌ "a small pillow" [because him for whom you put a pillow you honour: see 2]. (TA.)

حُسْبَةٌ, in a camel, *A colour in which are whiteness and redness* (Ḳ, TA) *and blackness:* (TA:) in a man, [*a reddish colour such as is termed*] شُقْرَة *in the hair of the head:* (Ḳ:) and also in a man, (Ḳ, TA,) and in a camel, (TA,) *whiteness and redness produced by a whiteness of the skin arising from disease and infecting the hair* [*so as to turn it red*]: (Ḳ, TA:) accord. to IAạr, *blackness inclining to redness.* (TA.) — Also *Leprosy.* (Ḳ.)

حِسْبَةٌ [originally The act of *numbering, counting,* &c.: or *a mode,* or *manner, of numbering,* &c.: see 1. —] A subst. from اِحْتَسَبَ أَجْرًا; (Ṣ, Mṣb, Ḳ;) syn. with اِحْتِسَابٌ (A) [as meaning *A reckoning upon,* or *seeking,* or *preparing* or *providing,* or *laying up for oneself in store, a reward in the world to come*]. You say, فَعَلَهُ حِسْبَةً [*He did it reckoning upon,* or *seeking,* &c., *a reward in the world to come*]. (A, TA.) — هُوَ حَسَنُ ٱلْحِسْبَةِ *He is good in respect of managing, conducting, ordering,* or *regulating,* (Ṣ, A, Mṣb, Ḳ,) *and examining,* or *judging,* (Mṣb,) *and sufficing,* (A,) فِى ٱلْأَمْرِ *in the affair.* (Ṣ, A, Mṣb.) This is not from اِحْتِسَابُ ٱلْأَجْرِ; for احتساب الاجر relates only to an action done for the sake of God. (Mṣb.) — *A reward,* or *recompense:* pl. حِسَبٌ. (Ṣ, Ḳ.) — [The *office of the* مُحْتَسِب.] — See also حَسْبٌ, last sentence.

حُسْبَانٌ: see حِسَابٌ. — Also *A punishment.* (Ṣ, Ḳ.) — *A calamity; an affliction with which a man is tried.* (Aboo-Ziyád, Ḳ.) —— *Evil; mischief.* (Aboo-Ziyád, Ḳ.) —— *Locusts.* (Aboo-Ziyád, Ṣ, Ḳ.) —— *Dust:* or *smoke:* syn. عَجَاجٌ. (Ḳ.) —— *Fire.* (TA.) This, and each of the five significations next preceding, and that next following, have been assigned to the word as used in the Ḳur xviii. 38. (TA.) See also حُسْبَانَةٌ. —— *Small arrows,* (Mgh, Mṣb, Ḳ,) or *short arrows,* (Ṣ,) *which are shot from Persian bows:* (Mgh, Mṣb:) said by IDrd to be, in this sense, post-classical: (TA:) or *arrows which a man shoots in the hollow of a reed,* or *cane; drawing the bow, he discharges twenty of them at once, and they pass by nothing without wounding it, whether it be an armed man or another object; they come forth like rain, and scatter among the people:* (ISh, TA:) or *small arrows, with slender heads, in the hollow of a reed,* or *cane, which, when discharged, come forth like a shower of rain, and scatter, and pass by nothing without wounding it:* (Az, Mṣb:) or *iron-headed arrows, like large needles, slender, but somewhat long, and without edges* [*to the heads*]: (Th, TA:) n. un. with ة. (Ṣ, Mgh, Mṣb, Ḳ.) ═ It is also said to signify The *circumference* of a mill-stone: —— and hence, in the Ḳur lv. 4, [see 1, above,] to mean The [*revolving*] *firmament.* (El-Khafájee, MF.)

حُسْبَانَةٌ n. un. of حُسْبَانٌ [q. v.]. (Ṣ, Mgh, &c.) —— Also *A thunderbolt;* syn. صَاعِقَةٌ: (Ḳ:) and حُسْبَانٌ ♦, [of which it is the n. un.,] *thunderbolts;* syn. صَوَاعِقُ. (Bḍ and Jel in xviii. 38.) —— *A hailstone;* syn. بَرَدَةٌ. (Ḳ. [In some copies of the Ḳ بُرْدَةٌ.]) —— *A cloud.* (Ḳ.) ═ *A small ant.* (Ḳ.) ═ *A small pillow;* (Ṣ, Ḳ;) and so مِحْسَبَةٌ ♦: (Ḳ:) or this signifies *a pillow of skin,* or *leather.* (TA.)

حِسَابٌ and حُسْبَانٌ ♦ [*A numbering, counting, reckoning, calculation,* or *computation:* see 1:] both signify the same: (Ṣ:) or the latter is pl. of the former, (Ṣ, Ḳ, TA,) accord. to Akh (Ṣ, TA) and AHeyth and others, when the former signifies *what is numbered; &c.;* [*a number;* or *quantity;*] and the former has also for a pl. [of pauc.] أَحْسِبَةٌ. (TA.) You say, رَفَعَ العَامِلُ حِسَابَهُ and حُسْبَانَهُ [*The agent presented his reckoning,* &c.]. (A.) Hence, حِسَابُ الجُمَّلِ and الجُمَلِ: see art. جمل. [And حِسَابُ عَقْدِ الأَصَابِعِ *The numbering, counting,* or *reckoning, with the fingers.*] And يَوْمُ الحِسَابِ [*The day of reckoning;* i. e., of the final judgment]. (Ḳur xxxviii. 15, &c.) —— حِسَابٌ also signifies The *reckoning,* or *enumerating,* or *recounting, of causes of glorying;* or *of memorable,* or *generous, actions* or *qualities.* (Mṣb.) —— And ‡ *A great number* of men: (A, L, Ḳ:) of the dial. of Hudheyl. (L.) —— And † A *sufficing* thing, (Ṣ, Ḳ,) and gift, (Ṣ, Ḳ, and Bḍ in lxxviii. 36,) as also حُسَابٌ ♦: (Bḍ ib.:) or a *large* gift: (Jel ib.:) or a gift *according to one's works.* (Bḍ ib.)

حَسِيبٌ *A reckoner,* or *taker of accounts:* [see also حَاسِبٌ:] or *a sufficer,* or *giver of what is sufficient;* (Ḳ, TA;) from أَحْسَبَ, of the measure فَعِيلٌ in the sense of the measure مُفْعِلٌ. (TA.) It has the former of these significations, or the latter, in the phrase, كَفَى بِٱللّٰهِ حَسِيبًا [*God is sufficient as a reckoner,* or *as a giver of what sufficeth*], (Fr, Ḳ, TA,) in the Ḳur [iv. 7, and xxxiii. 39]: (TA:) and so in the Ḳur iv. 88. (TA.) —— [Hence,] حَسِيبُكَ ٱللّٰهُ, (Ṣ, Ḳ,) in the L حَسْبُكَ ♦ ٱللّٰهُ, (TA,) [both of which phrases are used in the present day in the sense here following,] *May God take,* or *execute, vengeance upon thee;* or *punish thee:* (Ṣ, L, Ḳ:) meaning an imprecation though literally predicatory. (IAmb, Ḥar p. 371.) [See also حُسْبَانُكَ عَلَى ٱللّٰهِ, voce حَسْبٌ.] ═ Also *Characterized,* or *distinguished, by what is termed* حَسَبٌ *as explained above* [i. e. *grounds of pretension to respect or honour;* &c.]: (Ṣ, Ḳ:) *generous, liberal, honourable,* or *noble:* (Mṣb:) *bountiful,* or *munificent:* and *having a numerous household:* (Az, Mgh:) pl. حُسَبَآءُ. (A, Ḳ.)

حِسَابٌ: see حُسَابٌ.

حَاسِبٌ [act. part. n. of 1; *Numbering, counting,* &c.:] *a reckoner; an accountant:* [see also حَسِيبٌ:] pl. حُسَّبٌ and حُسَّابٌ (TA) and حَسَبَةٌ. (A.)

أَحْسَبُ, (Ṣ, Ḳ,) fem. حَسْبَآءُ, (TA,) A camel *of a colour in which are whiteness and redness* (Ṣ, Ḳ, TA) *and blackness:* (TA:) a man *in the hair of whose head is* [*a reddish colour such as is termed*] شُقْرَةٌ: (Ṣ, Ḳ:) a man, (Ḳ,) and a camel, (TA,) *whose skin has become white by reason of disease, and whose hair is infected* [*and turned red*] *in consequence thereof, so that he has become white and red:* (Ḳ:) accord. to Sh, *that has no* [*distinct*] *colour; of whom,* or *of which, one says, I think so, and I think so.* (TA. [The latter clause of this explanation (in the TA الذى يقال احسب كذا واحسب كذا) I have rendered conjecturally; supposing فيه to have been omitted by a copyist, after يقال.]) —— Also *A leper.* (Lth, T, Ḳ.) —— And † A *mean, avaricious,* man. (Ṣ, TA.)

إِبِلٌ مُحْسِبَةٌ *Camels that have much flesh and fat:* (TA:) or محسبة has two meanings; from حَسَبٌ signifying "nobility;" [i. e. *noble* camels;] and from إِحْسَابٌ; i. e. *satisfying, with their milk, their owners and the guest.* (IAṭr, TA.)

مِحْسَبَةٌ: see حُسْبَانَةٌ.

مُحَسَّبٌ: see 2.

مَحْسُوبٌ: see حَسْبٌ, first sentence.

مُحْتَسِبٌ [The *inspector of the markets and of the weights and measures &c.*] is an appellation derived from اِحْتَسَبَ, as shown above: see this verb. (Ḳ.) You say, فُلَانٌ مُحْتَسِبُ البَلَدِ [*Such a one is the inspector of the markets &c. of the town*]: you should not say مُحْسِبُ. (Ṣ.)

حسد

1. حَسَدَهُ الشَّىْءَ and حَسَدَهُ عَلَى الشَّىْءِ, (Ṣ, A, Mṣb, Ḳ,) aor. ـُ and ـِ, (Ṣ, Ḳ,) the latter form of aor. used by some, (Akh, Ṣ,) the former being that which commonly obtains, (TA,) inf. n. حَسَدٌ (Akh, Ṣ, A, Mṣb, Ḳ [in the CḲ حَسْدٌ]) and حَسْدٌ, but the former is more common, (Mṣb,) and حُسُودٌ and حَسَادَةٌ (Ṣ, Ḳ) and حَسِيدَةٌ; (CḲ;) and حَسَّدَهُ ♦, [which probably has an intensive signification,] (Ḳ,) inf. n. تَحْسِيدٌ; (TA;) *He envied him for the thing,* or *envied him the thing,* meaning a blessing, or a cause of happiness; i. e. *he disliked that he should possess it, and wished that it might depart from him* [*and be transferred to himself*]: (Mṣb:) or *he wished,* or *regarded him with a wish, that the thing,* meaning as above, *might depart from him:* (A:) or *he wished that he might be deprived of the thing,* meaning as above, or an excellence: (Ḳ: [in the CḲ, for يُسْلَبَهُمَا, is put يَسْلُبَهُمَا:]) or *he wished that the thing,* meaning a blessing, or a cause of happiness, (Ṣ, Ḳ,) or an excellence, (Ḳ,) *might become transferred from him* (another) *to himself.* (Ṣ, Ḳ.) —— And حَسَدَهُ عَلَى شَجَاعَتِهِ وَنَحْوِهَا *He wished that he possessed such as his* (another's) *courage, and the like, without wishing that the other should be deprived of it;* the verb in this case being syn. with غَبَطَ; and implying admiration. (Mṣb.) —— حَسَدَنِى ٱللّٰهُ إِنْ كُنْتُ أَحْسُدُكَ (M, Ḳ) is a saying of the Arabs, mentioned by Lḥ, strange and abominable, (M,) meaning *May God punish me for my envy if I envy thee.* (M, Ḳ.)

2: see 1.

4. صَحِبْتُهُ فَأَحْسَدْتُهُ *I associated with him and found him to be envious.* (A.)

6. تحاسدوا *They envied* (حَسَدُوا) *one another.* (Ṣ, A,* Ḳ.)

حَسَدٌ *Envy;* or the *wishing that a blessing,* or *a cause of happiness, may depart from its possessor* (Ṣ, A) *and become transferred to oneself.* (Ṣ.) [See 1.]

حَسُودٌ *Envious:* (Mṣb, Ḳ:) used also as a fem. epithet without ة: (TA:) pl. حُسُدٌ. (Ḳ.)

حَاسِدٌ *Envying:* (Ṣ, Mṣb, Ḳ:) pl. حَسَدَةٌ (Ṣ, A, Mṣb, Ḳ) and حُسَّادٌ (Mṣb, A, Ḳ) and حُسَّدٌ. (A, Ḳ.)

المَحْسَدَةُ مَفْسَدَةٌ [*That which is a cause of envy is a cause of corruption, or evil*]. (A.)

مَحْسُودٌ *Envied.* (Ṣ, A, Mṣb.)

حسر

1. حَسَرَهُ, aor. ـِ (Ṣ, Mṣb, Ḳ) and ـُ, (Mgh, Mṣb, Ḳ,) inf. n. حَسْرٌ (Ṣ, Mṣb, Ḳ) and حُسُورٌ, (TA,) *He removed it, put it off, took it off,* or *stripped it off,* (Mgh, Ḳ, TA,) عَنْ شَىْءٍ *from a thing which it covered* or *concealed.* (TA.) حَسَرَ is said of anything as meaning *It was removed, put off, taken off,* or *stripped off,* from a thing which it covered or concealed. (A.) You say, حَسَرَ كُمَّهُ عَنْ ذِرَاعِهِ *He removed his sleeve from his fore arm.* (Ṣ, A.) And simply حَسَرَ عَنْ ذِرَاعِهِ *He uncovered his fore arm.* (Mṣb.) And حَسَرَ عِمَامَتَهُ عَنْ رَأْسِهِ *He removed,* or *took off, his turban from his head.* (A.) And حَسَرَتْ دِرْعَهَا, (A, Mṣb,) aor. ـِ, (Mṣb,) *She* (a woman) *took off her shift* (A, Mṣb) عَنْ جَسَدِهَا *from her body:* (A:) and خِمَارَهَا *her head-covering.* (Mṣb.) —— [Hence,] حَسَرَتِ الرِّيحُ السَّحَابَ ‡ [*The wind removed the clouds* from the

sky]. (A.) And حَسَرَ قِنَاعَ الهَمِّ عَنِّى ‡ [*He, or it, removed the covering of anxiety from me*]. (A.) —— Also, (K,) inf. n. حَسْرٌ, (TA,) *He peeled a branch of a tree.* (K, TA.) —— And *He swept a house or chamber.* (K, TA.) —— And حَسَرُوهُ, aor. ـُ, inf. n. حَسْرٌ and حُسُورٌ, ‡ *They begged of him and he gave them until nothing remained in his possession.* (TA.) ═ حَسَرَ, (Ṣ, A, K,) aor. ـِ and ـُ, (TA,) inf. n. حَسْرٌ (Ṣ, TA) and حُسُورٌ; (TA;) and ▼ احسر, (Ṣ, K,) inf. n. إِحْسَارٌ; and ▼ حسّر, inf. n. تَحْسِيرٌ; (TA;) *He,* (a man, Ṣ, A,) and *it,* (a journey, TA,) *tired, fatigued,* or *jaded,* (Ṣ, A, K,) a beast, (A, TA,) or a camel: (Ṣ:) and *he drove* a camel *until he tired, fatigued,* or *jaded, him.* (K.) And حَسَرَتِ الدَّابَّةُ *The beast was fatigued so that it was left to remain where it was.* (AHeyth.) —— And حَسَرَ, aor. ـُ, †*It* (the distance to which it looked, and the indistinctness of the object,) *fatigued* the eye. (TA.) And حَسَرَ البَصَرُ مِنْ طُولِ النَّظَرِ ‡ [*The eye was fatigued by the length of looking:* see a similar meaning of حَسِرَ and حَسُرَ, below]. (A.) ═ See 7, with which حَسَرَ is syn. —— [Hence,] حَسَرَ, (ISk, A, Mgh, Mṣb,) aor. ـُ, (TA,) ‡ *It* (water) *sank and disappeared;* or *became low;* or *retired:* (ISk, A, Mgh:) *it sank and disappeared,* or *retired, from its place:* (Mṣb:) properly, *it became removed from the shore:* (Mgh:) and *it* (the sea, or great river,) *sank,* or *retired,* from (عَنْ) El-'Irák, and from the shore, *so that the ground which was beneath the water appeared:* (TA:) you do not say, in this sense, ▼ انحسر. (Az. [But this latter is sometimes used, as, for instance, in the Mṣb art. جزر.]) Hence, in a trad., كُلْ مَا حَسَرَ عَنْهُ البَحْرُ وَدَعْ مَا طَفَا عَلَيْهِ [*Eat thou that from which the sea retires, and leave what floats upon it*]. (Mgh.) ═ حَسِرَ, aor. ـَ, (Ṣ, A, K,) inf. n. حُسُورٌ (Ṣ, A) and حَسَرٌ; (TA;) and حَسُرَ, aor. ـُ, (A, K,) inf. n. حَسَرٌ; (TA;) and ▼ استحسر, (Ṣ, K,) and ▼ تحسّر; (Ṣ;) *He* (a camel, Ṣ, or a beast, A) *became tired, fatigued,* or *jaded,* (Ṣ, K, TA,) *by travel:* (TA:) [or] the last signifies *he* (a camel) *fell down from fatigue.* (Ḥam p. 491.) [Hence,] it is said in a trad., ▼ أُدْعُوا اللّٰهَ وَلَا تَسْتَحْسِرُوا † *Supplicate ye God, and be not weary:* and a similar instance occurs in the Kur xxi. 19. (TA.) —— [Hence also,] حَسَرَ, aor. ـِ, (Ṣ, K,) or ـُ, (Mṣb,) inf. n. حُسُورٌ; (Ṣ, Mṣb, K;) and حَسِرَ, aor. ـَ; (A;) ‡ *It* (the sight) *was,* or *became, dim, dull,* or *hebetated;* (Ṣ, Mṣb, K;) and *it failed;* (Ṣ, K;) [or *became fatigued;*] by reason of length of space [overlooked], (Ṣ, Mṣb, K,) and the like; (Ṣ, Mṣb;) or by long looking. (A.) ═ حَسِرَ عَلَيْهِ, aor. ـَ, inf. n. حَسَرٌ (Ṣ, Mṣb, K) and حَسْرَةٌ, (Ṣ, K,) or the latter is a simple subst., (Mṣb,) and حَسَرَانٌ, (TA,) *He grieved for it, or at it;* or *regretted it; he felt,* or *expressed, grief, sorrow,* or *regret, on account of it;* syn. تَلَهَّفَ; (Mṣb, K;) as also ▼ تحسّر: (Ṣ, K:) or the former, *he grieved for it,* or *regretted it,* (تَلَهَّفَ عَلَيْهِ, Ṣ, A, or نَدِمَ عَلَيْهِ, TA,) namely, a thing that had escaped him, *most intensely.* (Ṣ, A,* TA.) [See حَسْرَةٌ.]

2. حسّر, inf. n. تَحْسِيرٌ: see 1. —— Also *He despised* another: *he annoyed,* or *vexed,* him: (K:) *he drove him away.* (TA.) —— *He caused him to experience,* or *fall into, grief,* or *regret:* (Mgh, Mṣb, K:) or *to grieve for,* or *to regret, most intensely, a thing that had escaped him.* (Ṣ.) ═ حَسَّرَتِ الطَّيْرُ, (Ṣ,) inf. n. as above; (Ṣ, K;) and ▼ تحسّرت, (A, TA,) and ▼ انحسرت; (TA;) *The birds moulted; shed their feathers.* (Ṣ, A, K,* TA.)

4: see 1. ═ Also احسر القَوْمُ *The people,* or *party, experienced fatigue.* (TA.)

5. تحسّر *It* (the plumage of a bird, A, and the fur, or soft hair, of a camel, Ṣ, K) *fell off;* (Ṣ, A, K;) when relating to the fur, or soft hair, of a camel, [said to be] *by reason of fatigue;* (K;) but this restriction is not necessary; for its falling off is sometimes occasioned by diseases; though it may be said that the former cause is the more common. (TA.) You say also, تحسّر الوَبَرُ عَنِ البَعِيرِ *The fur, or soft hair, fell off from the camel:* and in like manner one says of the plumage from the birds: (A:) and of the hair from the ass. (TA.) See also 2. —— تَحَسَّرَتْ بَيْنَ يَدَيْهِ [*She uncovered herself,* or *her head and forehead,* or *her head,* or *her face, before him:* (see حَاسِرٌ:) or] *she sat before him with her face uncovered.* (TA from a trad.) ═ See also 1, in two places.

7. انحسر *It became removed, put off, taken off,* or *stripped off,* from a thing which it covered or concealed; (Ṣ, A, Mgh, Mṣb;) as also ▼ حَسَرَ, (K,) which occurs in poetry, (TA,) inf. n. حُسُورٌ. (K.) [See also 5.] —— *It* (the darkness) *became removed,* or *cleared away;* (A, Mṣb;) عَنْهُ [*from him,* or *it*]. (A.) —— See also 1: —— and 2.

10: see 1, in two places.

حَسِيرٌ: see حَسِرٌ.

حَسْرَةٌ *Grief,* or *regret;* syn. تَلَهُّفٌ, (Mṣb, K,) and تَأَسُّفٌ, (Mṣb,) or نَدَامَةٌ, (Jel in ii. 162 and viii. 36 and xxxix. 57,) or نَدَمٌ and غَمٌّ: (Bḍ in viii. 36:) or *intense lamentation* or *expression of pain* or *of grief* or *of sorrow;* syn. شِدَّةُ التَّأَلُّمِ: (Jel in vi. 31 and xxxvi. 29:) or *most intense grief* or *regret* (أَشَدُّ النَّدَمِ, Ṣ, or أَشَدُّ التَّلَهُّفِ, Zj) *for a thing that has escaped one,* (Ṣ,) so that he who feels it is like a beast that is tired, or fatigued, or jaded, (حَسِير,) and of no use: (Zj in xxxvi. 29 of the Kur:) pl. حَسَرَاتٌ. (Mṣb.) You say, يَا حَسْرَتَا عَلَيْهِ [*O my grief,* or *regret,* &c., *for it!*] (A.)

حَسْرَانٌ: see what next follows.

حَسِيرٌ *Tired, fatigued,* or *jaded,* (Ṣ, K,) *by much travel;* (TA;) applied to a camel, (Ṣ, K,) alike to the male and the female; and so ▼ حَاسِرٌ and حَاسِرَةٌ, applied to a horse or the like: (TA:) and ▼ مَحْسُورٌ a camel *fatigued,* or *jaded; emaciated by fatigue,* or *made to exert himself beyond his strength in a journey:* (Ḥam p. 208:) pl. of the first حَسْرَى. (Ṣ, K.) —— ‡ Sight that is *dim, dull,* or *hebetated,* and *failing,* by reason of length of space [overlooked] (Ṣ, Mṣb, K, TA) and the like; (Ṣ, Mṣb;) as also ▼ مَحْسُورٌ; (Ṣ, K;) or [*fatigued*] by long looking. (A.) —— Also, (Ṣ, K,) and ▼ حَسِرٌ and ▼ حَسْرَانُ, (TA, [but whether the latter be with or without tenween is not shown,]) *Grieving,* or *regretting:* (K:) or *grieving,* or *regretting, most intensely,* on account of a thing that has escaped one. (Ṣ, TA.)

حَاسِرٌ A man *having no* مِغْفَر [or *covering for the head, made of mail,* &c.,] (Ṣ, K,) *upon him;* (Ṣ;) *nor a coat of mail;* (Ṣ, K;) *contr. of* دَارِعٌ; (Mgh;) *nor a helmet upon his head;* (TA;) *contr. of* مُقَنَّعٌ: (Mgh:) or *having no* جُنَّة [or *defensive covering,* &c.]: (K:) a man *having no turban on his head:* (TA:) a man *having his head uncovered:* (A:) pl. حُسَّرٌ, and pl. pl. حُسَّرُونَ; the latter a form used by one of the poets; the former pl. applied to *foot-soldiers in war,* because they uncover their arms and legs, or because they have not upon them coats of mail nor helmets; occurring in this sense in a trad. (TA.) Also, without ة, *A woman who has taken off her shift from her person:* (ISd, Mṣb, TA:) *who has taken off her clothes from her person:* *who has uncovered her head and her fore arms:* *who has taken off her head-covering:* and, with ة, a woman *having her face uncovered:* pl. حُسَّرٌ and حَوَاسِرُ. (TA.) ابْنُوا المَسَاجِدَ حُسَّرًا, in a trad. of 'Alee, means *Build ye mosques,* or *oratories, with bare walls, with no* شُرَف [or *acroterial ornaments* or *crestings*]. (TA.) ═ See also حَسِيرٌ.

مَحْسَرٌ ‡ The *internal,* or *intrinsic, state* or *quality,* (Ṣ, A, K,) of a person; (Ṣ, A;) as also ▼ مَحْسِرٌ: (K:) and the latter, [or both,] the *nature,* or *natural disposition.* (K, TA.) You say, فُلَانٌ كَرِيمُ المَحْسَرِ ‡ *Such a one is generous,* or *noble, in respect of his internal,* or *intrinsic, state* or *quality:* (Ṣ, A:) or ▼ المَحْسِرِ, meaning as above: or *in respect of his nature,* or *natural disposition:* or *face,* or *countenance.* (TA.)

مَحْسِرٌ *The face,* or *countenance:* (K:) [or *a part, of the person, that is uncovered:*] the pl., مَحَاسِرُ, signifies the *parts, of the person* of a woman, *that are exposed to view;* namely, the *face, arms, and legs.* (Az.) You say امْرَأَةٌ حَسَنَةُ المَحَاسِرِ [*A woman beautiful in respect of the parts, of the person, that are exposed to view*]. (A.) —— [Hence, ‡ *An elevated, plain tract, bare of herbage* or *trees*]. You say أَرْضٌ عَارِيَةُ المَحَاسِرِ ‡ *Land bare of herbage:* (A:) and in like manner, فَلَاةٌ عارية المحاسر *a desert without any covering of trees;* its محاسر meaning its *elevated and plain tracts of ground that are uncovered by plants* [or *trees*]. (T, TA.) —— See also مَحْسَرٌ, in two places.

مِحْسَرَةٌ *An instrument for sweeping; a broom,* or *besom.* (Ṣ, K.)

مُحَسَّرٌ: see حَسِيرٌ. —— Also *Annoyed; vexed:* and *despised:* (Ṣ, K:) applied to a man. (Ṣ.) It is said in a trad. that the companions of a man who is to come forth in the end of time, to be called أَمِيرُ الغَضَبِ, or, as some say, أَمِيرُ العُصَبِ,

shall be مُحْسُورُونَ, (TA,) meaning *despised*; (Ṣ, TA;) i. e. *annoyed*, or *vexed, and caused to grieve* or *regret*, or *to grieve* or *regret most intensely*: or *driven away*, or *outcasts, and fatigued*; from حَسَرَ signifying "he fatigued" a beast. (TA.)

مَحْسُورٌ [pass. part. n. of حَسَرَهُ; *Removed; put, taken*, or *stripped, off*: &c. — And hence,] ‡ *A man who has given all that he had, so that nothing remains in his possession*: thus it is said to mean in the Ḳur xvii. 31. (TA.) — See also حَسِيرٌ.

حسك

1. حَسَكَ عَلَيَّ, (Ṣ, Ḳ,) aor. ـَ, (Ḳ,) inf. n. حَسَكٌ, (Ṣ,) ‡ *He bore rancour, malevolence, malice*, or *spite, against me*; and *enmity*: (Ṣ:) or *he was angry with me*. (Ḳ, TA.)

2. حسّك, inf. n. تَحْسِيكٌ, † *He was*, or *became, niggardly, tenacious*, or *avaricious*. (Sh, IAth.)

4. أَحْسَكَتِ النَّفَلَةُ *The plant of the species termed* نَفَل [q. v.] *put forth a* حَسَكَة, i. e. *a prickle*. (TA.)

حَسَكٌ [is applied, app. in the classical language, as it is in the present day, to *Various species of thistle*, and *other prickly plants*: also to the *heads of thistles* and *the like*: and particularly to the *caltrop*, or *tribulus*: and hence the explanations here following:] the حَسَك [or *prickly heads*] *of the* [*plant called*] سَعْدَان: (Ṣ:) or *a certain herb*, (Aboo-Ziyád, Mgh, TA,) *inclining to yellowness*, (Aboo-Ziyád, TA,) *having* [*a head of*] *prickles of rounded form*: (Aboo-Ziyád, Mgh, TA:) *seldom, or never, does any one walk upon it, when it has dried up, without putting upon his feet boots or sandals: and the ants transport its produce* [or *heads*] *to their nests*: (Aboo-Ziyád, TA:) *a certain plant, the produce* [or *head*] *whereof (which is rough* [or *prickly*], TA) *clings to the wool of sheep*, (Ḳ,) *and to the fur of camels, in their places of pasturing*: (TA:) *its leaves are like those of purslane* (الرِّجْلَة), *or narrower, and at its leaves are compact and hard prickles, having three forks*: [hence it seems to be *a species of three-horned caltrop*: or for "three," we should perhaps read *four*: (see another application of the word in what follows:)] *the drinking* [*of an infusion*] *thereof has the effect of crumbling the stone of the kidneys and bladder; and the drinking of the expressed juice of its leaves is good for the venereal faculty, and for difficulty in the discharge of urine, and for the bite of vipers; and the sprinkling it in the dwelling kills fleas*: (Ḳ:) also, accord. to Aboo-Naṣr, the *produce of the* [*plant called*] نَفَل: (TA:) and *sharp, hard prickles* or *thorns*: (TA in art. مسك:) n. un. with ة: (Ṣ, Mgh:) which some hold to apply to *any fruit*, or *produce*, of a plant, *that is of the kind termed* عُقْدَة [i. e. *forming a compact and roundish head*]; and hence, to the *pod of the cotton-plant*: and it also signifies *a prickle*, or *thorn*. (TA.) [Hence,] إِنَّهُ لَحَسَكَةٌ ‡ *Verily he is rough*. (A, TA.) And إِنَّهُمْ لَحَسَكٌ أَمْرَاسٌ † *Verily they are strong*: and of one person you say حَسَكَةٌ مَرِسٌ (TA.) And هُوَ حَسَكَةٌ مَسَكَةٌ ‡ *He is courageous*: (Ḳ and TA in art. مسك:) and of a number of persons you say حَسَكٌ مَسَكٌ. (TA in that art., q. v.) [See also حَسِكٌ.] — See also حَسِيكَةٌ. — Also [† *Caltrops*, as meaning] *a kind of instrument used in war*, (Ṣ, Ḳ,) *made like the* حَسَك *mentioned in the first sentence of this paragraph*, (Ṣ,) or *like the prickles of the* حَسَك, (Ḳ,) *of iron*, (Ṣ, Ḳ,) or *of canes*, (Ḳ,) and sometimes *of wood*, (TA,) *and cast*, (Ḳ,) or *set up*, (TA,) *around the army*, (Ḳ, TA,) *in the ways of the horses*. (TA.) — And ‡ *Rancour, malevolence, malice*, or *spite*; and *enmity*; (Ḳ, TA;) as also حَسَكَةٌ, (Ḳ,) and حَسِيكَةٌ and حُسَاكَةٌ. (Ṣ, Ḳ. [The last in the CḲ written حَسَاكَة; but expressly said in the TA to be with ḍamm, and so written in copies of the Ṣ and Ḳ.]) You say, فِي صَدْرِهِ عَلَيَّ حَسِيكَةٌ and حُسَاكَةٌ [*In his bosom is rancour*, &c., *against me*]. (Ṣ.)

حَسِكٌ † *Affected with rancour, malevolence, malice*, or *spite*; and *enmity*: (TA:) or *angry*. (Ḳ.) You say, إِنَّهُ لَحَسِكُ الصَّدْرِ عَلَى فُلَانٍ † *Verily he is affected with rancour*, &c., *of the bosom against such a one*. (TA.) — حَسِكٌ مَرِسٌ † *Courageous* [*and strong*]; *not to be attempted* [*in fight*]. (A, TA.) [See also حَسَكٌ.]

حَسَكَةٌ: see حَسَكٌ, (of which it is properly the n. un.,) in four places.

حَسِيكَةٌ: see حَسَكٌ.

حُسَاكَةٌ: see حَسَكٌ, last two significations.

حَسِيكَةٌ: see حَسَكٌ, last two significations. — Also, (Ṣ, IF, Ḳ,) and حُسْكُكٌ, (Ḳ,) so accord. to Az, on the authority of Lth, but in the 'Eyn, and also in the Moḥeeṭ, as Ṣgh says, حُسَّكٌ, which (SM says) is probably a mistranscription, (TA,) The *hedge-hog*: ('Eyn, Ṣ, Ḳ:) or *a large hedge-hog*. (TA.)

حسل

8. احتسل *He hunted, caught, snared*, or *entrapped, the* [*young lizards termed*] حُسُول, pl. of حِسْل. (O, Ḳ.*)

حِسْلٌ The *young one of the* [*kind of lizard called*] ضَبّ, (AZ, Ṣ, Mgh, Ḳ,) *when it first comes forth from its egg*: (AZ, Ṣ, Ḳ:) it is next called غَيْدَاقٌ; then, مُطَبِّخٌ; then, خُضَرِمٌ; and then, ضَبٌّ: (Ṣ and L voce مُطَبِّخ: [but see this word:]) pl. [of pauc.] أَحْسَالٌ (Ḳ) and [of mult.] حُسُولٌ (Ṣ, Ḳ) and حِسْلَانٌ, with kesr, and حِسَلَةٌ, (Ḳ, TA,) with kesr and then fet-ḥ. (TA. [In the CḲ حَسَلَة.]) [Hence,] أَبُو الحِسْلِ, (Ṣ,) or أَبُو حِسْلٍ, and أَبُو حُسَيْلٍ, (Ḳ,) *The* [*lizard called*] ضَبّ. (Ṣ, Ḳ.) [Hence also,] لَا آتِيكَ سِنَّ الحِسْلِ, i. e. *I will not come to thee ever*, (Ṣ, Ḳ,) *until thy death*: (Ṣ:) because the tooth of the حسل does not fall out: (Ṣ, Ḳ:) a prov. (Ṣ.)

حُسَيْلٌ: see حِسْلٌ [of which it is the dim.].

حسم

1. حَسَمَهُ, (Ṣ, Mṣb, Ḳ,) aor. ـِ, (Mṣb, Ḳ,) inf. n. حَسْمٌ, (Mgh, Mṣb,) *He cut it*; or *cut it off*: (Ṣ, Mṣb, Ḳ:) *he cut it off entirely*. (Mgh, Mṣb.) — Hence, حَسَمَ العِرْقَ: (Ṣ:) you say, حَسَمَ العِرْقَ, (Ḳ,) inf. n. as above, (TA,) *He cut the vein, and then cauterized it to prevent the flow of the blood*: (Ḳ:) or this is an elliptical expression, originally حَسَمَ دَمَ العِرْقِ, meaning *he stopped the flow of blood from the vein by cauterization*. (Mṣb.) And hence, (Mgh,) اِقْطَعُوهُ ثُمَّ آحْسِمُوهُ, (Ṣ, Mgh,) or اِقْطَعُوا يَدَهُ ثُمَّ آكْوُوهَا [*Cut ye off his hand, then cauterize it*], (Ṣ,* Mgh,* TA,) in order that the blood may stop. (Ṣ, Mgh, TA.) You say also, حَسَمْتُ الدَّابَّةَ, meaning *I cauterized the beast by successive operations*. (Bḍ in lxix. 7.) — [Hence, also,] حَسَمَ الدَّآءَ, (Ḳ,) inf. n. as above, (TA,) *He stopped the disease by a remedy*. (Ḳ.) And حَسَمَتْهُ أُمُّهُ الرَّضَاعَ, and الغِذَآءَ, *His mother stopped his sucking*, and *his food*: (TA:) and حُسِمَ رِضَاعُهُ [*His sucking was stopped*]; (Ḳ;) said of a child. (TA.) And حَسَمَ فُلَانًا الشَّىْءَ, (Ḳ,) inf. n. as above, (TA,) *He prevented such a one from attaining the thing*. (Ḳ.) And أَنَا أَحْسِمُ عَلَى فُلَانٍ الأَمْرَ *I cut off from him the thing, so that he cannot attain aught thereof*. (TA.) See also حُسُومٌ, below. — You say also, حَسَمَهُمْ, aor. ـِ, inf. n. حُسُومٌ, *It caused them to pass away, come to an end, cease, perish*, or *come to nought*. (Zj, TA.) See, again, حُسُومٌ, below. — حَسَمَ فِى العَمَلِ, (TḲ,) inf. n. حُسُومٌ, (Yoo, Ḳ,) *He strove, laboured, toiled*, or *exerted himself, and wearied himself, in work*. (Yoo, Ḳ, TḲ.)

7. انحسم *It was*, or *became, cut*, or *cut off*: (Ṣ, Mṣb, Ḳ:) [or *it was*, or *became, cut off entirely*: see 1, of which it is the quasi-pass.]

حُسَامٌ *A sword*; because it cuts that upon which it comes: (Mṣb:) or *a sharp sword*; (Ṣ, Ḳ, TA;) and in the same sense applied to a [knife such as is termed] مُدْيَة: (TA:) and (so in the Ṣ, but in the Ḳ "or") the *edge of a sword, with which one strikes*. (Ṣ, Ḳ.) — لَيْلَةٌ حُسَامٌ *A lasting night*: (Ḳ:) or *a night of lasting evil*, especially. (TA.)

حُسُومٌ *Unluckiness*, or *inauspiciousness*. (Ṣ,* Ḳ,* TA.) Some explain it thus in the passage here following. (Ṣ,* TA.) — سَخَّرَهَا عَلَيْهِمْ سَبْعَ لَيَالٍ وَثَمَانِيَةَ أَيَّامٍ حُسُومًا, in the Ḳur lxix. 7, means *He sent it* (the wind) *upon them by force*, (Jel,) or *made it to prevail against them by his power*, (Bḍ,) [*seven nights and eight days*] *consecutively*; (T, Ṣ, Bḍ, Ḳ, Jel;) an expression taken from the repetition of the act of cauterization, (T, Bḍ, Jel.) i. e. the act of the حَاسِمٌ; (Jel;) whence this word (حاسم) is applied to anything *made consecutive*; (T, TA;) and حُسُومٌ is its pl.: (T, Bḍ, TA:) or حُسُومٌ signifies the *making consecutive*. and, as an epithet, *consecutive, and continuous from first to last*: (Fr, TA:) or, accord. to some, الأَيَّامُ الحُسُومُ means *the days that are consecutive with evil* especially; and such ISd thinks to be the meaning: (TA:) or اللَّيَالِى الحُسُومُ means *the nights that cut off good, or prosperity*, (تَحْسِمُ الخَيْرَ,) *from their people*: (Ṣ,* Ḳ: [and the like is said by Bḍ in lxix. 7:]) or حُسُومًا in the Ḳur means *causing* them *to pass away, come to an* *end, cease, perish,* or *come to nought:* (Zj, T:) or it may be an inf. n., meaning *for the purpose of cutting off:* or an inf. n. of a verb meant to be understood, as a denotative of state; i. e. ▼تَحْسِمُهُمْ حُسُومًا [agreeably with the explanation of Zj]; and this is confirmed by the reading with fet-ḥ [i. e. حَسُومًا, though this is a very rare form of inf. n.]: (Bḍ:) you say أَيَّامٌ حُسُومٌ, (Ḳ,) in which case the latter word is an inf. n. used as an epithet, meaning *cutting off,* or *preventing, good,* or *prosperity;* (TA;) and أَيَّامُ حُسُومٍ, which has a similar meaning. (Ḳ, TA.)

حَاسِمٌ; pl. حُسُومٌ: see this latter word.

مَحْسَمَةٌ *A cause,* or *means, of cutting off,* or *stopping;* (T, Ḳ, TA;) syn. مَقْطَعَةٌ. (T, TA.) So in the saying, هٰذَا مَحْسَمَةٌ لِلدَّاءِ *This is a cause,* or *means, of cutting off,* or *stopping, the disease.* (Ḳ,* TA.) And hence, (TA,) عَلَيْكُمْ بِالصَّوْمِ فَإِنَّهُ مَحْسَمَةٌ لِلْعِرْقِ وَمَذْهَبَةٌ لِلْأَشَرِ, a trad., (T, Ṣ, TA,) meaning [*Keep ye to fasting, for it is*] *a cause,* or *means, of stopping venereal intercourse,* [*and a cause,* or *means, of dispelling exultation,* or *excessive exultation, and resting the mind upon things agreeable with natural desire:*] (TA:) or *an impediment to venery, and a cause of diminishing the seminal fluid, and of stopping venereal intercourse* or *passion,* &c. (T, TA.)

مَحْسُومٌ A child (TA) *whose sucking is stopped:* (Ḳ, TA:) and *whose food is stopped.* (TA.) And A child *badly fed.* (Ṣ, Ḳ.) Hence the prov., وَلْغُ جُرَيٍّ كَانَ مَحْسُومًا [*The lapping of a little puppy that had been badly fed*]: said on the occasion of a greedy person's taking much of a thing that he had not been able to obtain, and that he had become able to obtain; or in ordering one to take much when able. (TA. [See Freytag's Arab. Prov. ii. 817; where another reading is given, namely, مَحْشُومًا in the place of مَحْسُومًا, as well as the reading here given.])

حسن

1. حَسُنَ, (Ṣ, Mgh, Mṣb, Ḳ, &c.,) which may also be written and pronounced حَسْنَ, with the ḍammeh suppressed, (Ṣ,) and حَسَنَ, (Ḳ,) aor. ـُ, (TA,) inf. n. حُسْنٌ (Ṣ,* Mṣb, Ḳ,* TA) and حُسْنَى, (Ḥam p. 657, and Bḍ in ii. 77,) *He,* or *it* (a thing, Ṣ, Mṣb), *had,* or *possessed, the quality termed* حُسْنٌ [which see below; i. e., *was,* or *became, good,* or *goodly,* (generally the latter,) *beautiful, comely,* or *pleasing,* &c.; and ▼تحسّن often signifies the same, as in the phrase تحسّن عِنْدَهُ *it was,* or *became, good,* &c., *in his estimation*]: (Ṣ, Ḳ, TA:) and [in like manner] ▼أَحْسَنَ زَيْدٌ means *Zeyd became possessed of* حُسْن. (Mughnee in art. ب.) — One may not say حُسْنَ, transferring the ḍammeh of the س to the ح and making the former letter quiescent, except in one case; because it is [virtually, together with its agent expressed or implied, in this case,] a predicate: [see I 'Aḳ p. 234:] this is allowable only in the case of a verb of praise or dispraise; حُسْنَ, in respect of the transference of the medial vowel, being likened to نِعْمَ and بِئْسَ, which are originally نَعِمَ and بَئِسَ: and thus one does in all verbs like these two in meaning: a poet says,

* لَمْ يَمْنَعِ النَّاسُ مِنِّى مَا أَرَدْتُ وَمَا *
* أُعْطِيهِمُ مَا أَرَادُوا حُسْنَ ذَا أَدَبَا *

[*Men have not withheld from me what I have desired, nor do I give them what they have desired: good,* or *very good, is this as a mode of conduct!*]: meaning حَسُنَ هٰذَا أَدَبًا. (Ṣ, TA.) You say also, حَسُنَ زَيْدٌ, [meaning *Good,* or *goodly,* &c., or *very good* &c., *is Zeyd!* or] meaning ▼أَحْسِنْ بِهِ [i. e. *how good,* or *goodly,* &c., *is Zeyd!* as also ▼مَا أَحْسَنَهُ]. (B, TA in art. ب.)

2. حسّنهُ, (Ṣ, Ḳ,) inf. n. تَحْسِينٌ, (Ṣ,) *He made it,* or *rendered it,* حَسَن [i. e. *good,* or *goodly,* (generally the latter,) *beautiful, comely,* or *pleasing,* &c.]; (Ḳ;) *he beautified, embellished,* or *adorned, it;* (Ṣ, TA;) as also ▼احسنهُ. (TA.) You say, ▼أَحْسَنَ الحَلَّاقُ رَأْسَهُ *The shaver beautified,* or *trimmed, his head.* (TA.) And الَّذِى ▼أَحْسَنَ كُلَّ شَىْءٍ خَلَقَهُ [*Who hath made good,* or *goodly, everything that He hath created*], in the Ḳur [xxxii. 6], means حَسَّنَ خَلْقَ كُلِّ شَىْءٍ [*hath made good,* or *goodly, the creation of everything*]. (TA.) — [See also تَحْسِينٌ.] — And see 10.

3. إِنِّى أُحَاسِنُ بِكَ النَّاسَ (Ṣ, TA) *Verily I contend with men for thy superiority in* حُسْن [i. e. *goodness,* or *goodliness,* &c.]. (TA.) [حَاسَنَ followed by an accus. is rendered by Golius, as on the authority of J, who gives no explanation of it, "Bene tractavit et egit."]

4. احسن as an intrans. v.: see 1. — Also *He did that which was* حَسَن [meaning *good, comely,* or *pleasing; he acted well*]; (Mṣb;) *he did a good deed:* (Er-Rághib, TA:) [for] إِحْسَانٌ is the *contr. of* إِسَآءَةٌ: (Ḳ:) it differs from إِنْعَامٌ in being to oneself and to another; whereas the latter is only to another: (TA:) and it surpasses عَدْلٌ, inasmuch as it means the *giving more than one owes,* and *taking less than is owed to one;* whereas the latter means the giving what one owes, and taking what is owed to one. (Er-Rághib, TA.) You say, أَحْسَنْتُ إِلَيْهِ and بِهِ [*I acted,* or *behaved, with goodness, well,* or *in a good* or *comely* or *pleasing manner, towards him; did good to him; benefited him; conferred a benefit,* or *benefits, upon him*]: both signify the same: (Ṣ, TA:) and hence, in the Ḳur [xii. 101], قَدْ أَحْسَنَ بِى إِذْ أَخْرَجَنِى مِنَ السِّجْنِ meaning إِلَىَّ [i. e. *He hath acted well towards me, when he brought me forth from the prison*]: (AHeyth, Az:) or, accord. to some, the verb in this case is made to import the meaning of لَطَفَ [which is trans. by means of ب, i. e. *He hath acted graciously with me*]. (Mughnee in art. ب.) — الإِحْسَانُ is also explained as meaning الإِخْلَاصُ [i. e. *The being sincere,* or *without hypocrisy;* or *the asserting oneself to be clear of believing in any beside God*]; which is a condition of the soundness, or validity, of الإِيمَان and الإِسْلَام together: and as denoting *watchfulness, and good obedience:* and as meaning *the continuing in the right way, and following the way which those* [*of the righteous*] *who have gone before have trodden;* this last being said to be the meaning in the Ḳur ix. 101. (TA.) — As a trans. v.: see 2, in three places. — احسنهُ also signifies ‡*He knew it:* (Ṣ, Ḳ, TA:) [or] *he knew it well;* (Er-Rághib, Mṣb;) and so احسن بِهِ, as in the saying, هُوَ يُحْسِنُ بِالعَرَبِيَّةِ †*He knows well the Arabic language.* (MA.) Hence the saying of 'Alee, قِيمَةُ المَرْءِ مَا يُحْسِنُهُ ‡[*The value of the man is what he knows,* or *knows well*]. (TA.) النَّاسُ أَبْنَآءُ مَا يُحْسِنُونَ is another saying of 'Alee, meaning ‡*Men are named,* or *reputed, in relation to what they know, and to the good deeds that they do.* (TA.) — مَا أَحْسَنَهُ and أَحْسِنْ بِهِ: see 1, last sentence. You say also, ▼مَا أُحَيْسِنَهُ [i. e. *How very good,* or *goodly,* &c., *is he!*]; using the dim. form; like مَا أُمَيْلِحَهُ [q. v.]. (Ṣ and Ḳ in art. ملح.) — Also *He* (a man, IAạr) *sat upon a high hill,* or *heap, of sand, such as is termed* حَسَن. (IAạr, Ḳ.)

5. تحسّن: see 1. — Also *i. q.* تَجَمَّلَ [i. e. *He beautified, embellished,* or *adorned, himself:* and *he affected what is beautiful, goodly,* or *comely, in person,* or *in action* or *actions* or *behaviour,* or *in moral character,* &c.]. (TA.) [تَحَسَّنَتْ, said of a woman, occurs, in the former sense, in the Ṣ and Ḳ in art. رعد, and in the TA in art. نقط, &c.] — دَخَلَ الحَمَّامَ فَتَحَسَّنَ *He entered the hot bath and was shaven.* (TA.)

6. تحاسن [*He affected to be* حَسَن (i. e. *good, goodly, beautiful, comely,* &c.), *not being really so*]. (A in art. صبح. [See 6 in that art.])

10. استحسنهُ *He counted, accounted, reckoned,* or *esteemed, him,* or *it,* حَسَن [i. e. *good, goodly, beautiful, comely, pleasing,* &c.; *he approved, thought well of,* or *liked, him,* or *it*]; (Ṣ, Ḳ;) as also ▼حسّنهُ, inf. n. تَحْسِينٌ. (Ḥar p. 594.) Hence the saying, صَرْفُ هٰذَا ٱسْتِحْسَانٌ وَالمَنْعُ قِيَاسٌ [*The making this word perfectly declinable is approvable, but the making it imperfectly declinable is agreeable with analogy*]. (TA.)

حُسْنٌ (Ṣ, Ḳ, &c.) and ▼حُسُنٌ, which is of the dial. of El-Ḥijáz, and ▼حَسَنٌ, (MF, TA,) *Goodness,* or *goodliness,* [generally the latter,] *beauty, comeliness,* or *pleasingness;* contr. of قُبْحٌ: (Ṣ:) *i. q.* جَمَالٌ: (Ḳ:) but accord. to Aṣ, [when relating to the person,] حُسْن is in the eyes, and جَمَال is in the nose: (TA:) *symmetry;* or *just proportion of the several parts of the person, one to another:* (Kull:) or *anything, moving the mind, that is desired,* or *wished for; such as is approved by the intellect;* and *such as is approved by natural desire;* and *such as is approved by the faculty of sense:* in the common conventional language, mostly applied to *what is approved by the sight:* in the Ḳur, mostly to *what is approved by mental perception:* it is in accidents as well as in substances: (Er-Rághib, TA:) the pl. is ▼مَحَاسِنُ, (Ṣ, Ḳ,) like مَلَامِحُ pl. of لَمْحَةٌ, and مَشَابِهُ pl. of شِبْهٌ, &c., (Ḥar p. 9,) contr. to rule, (Ṣ, Ḳ,) as though pl. of ▼مَحْسَنٌ or ▼مُحْسِنٌ: (Ṣ accord. to different copies:) or, accord. to Lḥ and Eth-Tha'álibee, مَحَاسِنُ has no proper sing. (TA.) وَقُولُوا لِلنَّاسِ حُسْنًا, in the Ḳur [ii. 77], means *And say ye to men a saying having in it goodness* (قَوْلًا ذَا حُسْنٍ): or حُسْنًا may mean حَسَنًا: (Zj, TA:) and some read here حَسَنًا: and some, حُسُنًا, accord. to the dial. of El-Ḥijáz: and some, ▼حُسْنَى, as an inf. n., like بُشْرَى: (Bḍ:) but AḤát and Zj disallow this; the former saying that حُسْنَى is like فُعْلَى [as fem. of أَفْعَلُ denoting the comparative and superlative degrees], and therefore should have the article ال. (TA.) وَوَصَّيْنَا الإِنْسَانَ بِوَالِدَيْهِ حُسْنًا, in the Ḳur [xxix. 7], means [in like manner] *And we have enjoined man to do to his two parents what is good* (مَا يَحْسُنُ حُسْنًا): (TA:) and here [also] some read حَسَنًا; and some, إِحْسَانًا. (Bḍ.) [See another ex. of a similar kind, from the Ḳur xviii. 85, voce إِمَّا, near the beginning of the paragraph.] — سِتُّ الحُسْنِ [*The convolvulus cairicus* of Linn.; abundant in the gardens of Cairo;] *a certain plant that twines about trees and has a beautiful flower.* (TA.) — See also حَسَنٌ.

حَسَنٌ *Having*, or *possessing, the quality termed* حُسْن [which see above; *good*, or *goodly*, (generally the latter,) *beautiful, comely, pleasing* or *pleasant*, &c.]; (Mṣb, Ḳ, TA;) either *intrinsically*, as when applied to belief in God and in his attributes; or *extrinsically*, as when applied to war against unbelievers, for this is not good in itself: said to be the only epithet of its measure except بَطَلٌ: (TA:) and ▼حَسِينٌ signifies the same, (IB, Ḳ,) because from حَسُنَ, like عَظِيمٌ and كَرِيمٌ from عَظُمَ and كَرُمَ, (IB, TA,) and ▼حُسَانٌ, (Ḳ,) but this is an intensive epithet, [signifying *very good* or *goodly* &c.,] (IB, TA,) and ▼حُسَّانٌ, (Ḳ,) also an intensive epithet, (Ṣ, IB,) and ▼حَاسِنٌ, (Ḳ,) [properly signifying *being*, or *becoming, good* or *goodly* &c.,] cited by Lḥ as used in a future sense, (TA,) and ▼مُحَسَّنٌ as applied to a face: (Ḳ:) the fem. is حَسَنَةٌ, and ▼حَسْنَآءُ, applied to a woman, (Ṣ, Mṣb, Ḳ,) though the corresponding masc. of this latter, namely, ▼أَحْسَنُ, is [said to be] not used (Ṣ, Ḳ) as applied to a man [in the sense of حَسَنٌ], (Ṣ,) [but the phrase هُوَ أَحْسَنُهُمْ وَجْهًا as meaning حَسَنُهُمْ وَجْهًا is mentioned in the Ṣ in art. بيض, (see بَيَاضٌ, and see also the pl. أَحَاسِنُ in what here follows,)] and ▼حُسَّانَةٌ: (Ṣ, Ḳ:) the pl. masc. is حِسَانٌ, (Mṣb, Ḳ,) pl. of حَسَنٌ used as an epithet; but when حَسَنٌ is used as a [proper] name, its pl. is حَسَنُونَ; (Mṣb;) and حِسَانٌ may also be pl. of حَسِينٌ; (TA;) and حُسَّانُونَ, (Sb, Ḳ,) pl. of ▼حُسَّانٌ, which has no broken pl.: (Sb:) and أَحَاسِنُ القَوْمِ means حِسَانُهُمْ [*the good*, or *goodly*, &c., *of the party*, or *company of men*]: (Ḳ:) the pl. fem. is حِسَانٌ, (Ḳ,) like the masc., pl. of حَسَنَةٌ, and the only instance of its kind except عِجَافٌ, pl. of عَجْفَآءُ. (TA.) You say رَجُلٌ حَسَنٌ بَسَنٌ [*A man very good* or *goodly* &c.], using بسن as an imitative sequent [for the purpose of corroboration]. (Ṣ.) — [حَدِيثٌ حَسَنٌ *A tradition of good authority*; generally applied to one transmitted in the first instance by two or more relaters. — Also meaning *Good, comely, good-humoured, pleasing*, or *pleasant, discourse* or *talk*.] — الحَسَنُ *The bone that is next to the elbow*; as also ▼الحُسْنُ: (Ḳ:) or *the extremity of the bone of the upper half of the arm next the shoulder-joint*, because of the abundance of flesh that is upon it; the extremity of that bone next the elbow being called القَبِيحُ: (TA in art. قبح:) or *the upper part of that bone*; the lower part thereof being called القبيح. (Fr, TA in that art.) — *A kind of tree, of beautiful appearance*, (Ḳ, TA,) *also called the* أَلَآء, *that grows in rows upon a hill, or heap*, (كَثِيب,) *of sand*; so called because of its beauty; whence the كثيب is called نَقَا الحَسَنِ: thus described by Az, on the authority of 'Alee Ibn-Ḥamzeh. (TA.) — [And hence, perhaps,] حَسَنٌ signifies also *A high* كَثِيب [or *hill*, or *heap, of sand*]: (IAạr, Ḳ:) whence it is used as a [proper] name of a boy. (IAạr, TA.) ═ See also حُسْنٌ, first sentence.

الحُسَنُ : see أَحْسَنُ.

حُسُنٌ : see حُسْنٌ, first sentence.

حِسْنَةٌ *A ledge* (رَيْدٌ) *projecting from a mountain*: pl. حِسَنٌ. (Ḳ.)

حَسَنَةٌ fem. of حَسَنٌ [q. v.]. (Ṣ, Mṣb, Ḳ.) — Also, [used as a subst., or as an epithet in which the quality of a subst. is predominant, *A good act* or *action*;] *an act of obedience* [*to God*; often particularly applied to *an alms-deed*]: (Ksh and Bḍ in iv. 80:) and the *reward* [*of a good action*]: (Er-Rághib, TA:) *a good, benefit, benefaction, boon*, or *blessing*: (Ksh and Bḍ ibid.:) *contr. of* سَيِّئَةٌ [in all these senses]: (Ṣ, Ḳ:) as contr. of this latter word, it signifies *any rejoicing*, or *gladdening, good* or *benefit* &c. *that betides a man in his soul and his body and his circumstances*: (Er-Rághib, TA:) pl. حَسَنَاتٌ: (Ḳ, and Ḳur vii. 167, &c.:) it has no broken pl. (TA.) Hence, in the Ḳur iv. 80, it means *Abundance of herbage*, or *of the goods, conveniences, and comforts, of life; ampleness of circumstances; and success*: and سَيِّئَةٌ there means the contr. of these. (Er-Rághib, TA.) In the Ḳur xi. 116, الحَسَنَاتُ is said to mean *The five daily prayers*, as expiating what has been between them. (TA.) — As an epithet, [fem. of حَسَنٌ,] it is applied to an accident as well as to a substance. (Er-Rághib, TA.)

حُسْنَى : see حُسْنٌ, and أَحْسَنُ; the latter, in three places.

حَسْنَآءُ : ⎫
حُسَانٌ : ⎬ see حَسَنٌ.
حَسِينٌ : ⎭

حُسَيْنٌ [dim. of حَسَنٌ. — Also] *A high mountain*: whence it is used as a [proper] name of a boy. (TA.)

حُسَيْنَى One's *utmost*, [or rather one's *best*,] or the *utmost of* one's *power* or *ability* or *deed* or *case*: so in the saying, حُسَيْنَاهُ أَنْ يَفْعَلَ كَذَا [*His utmost*, or *best*, &c., *is*, or *will be, the doing such a thing*]: and ▼حُسَيْنَاؤُهُ means the same. (Ḳ,* TA.)

حُسَيْنَآءُ : see what next precedes. ═ Also *A kind of tree, with small leaves*. (Ḳ.)

حُسَّانٌ; and its fem., with ة: see حَسَنٌ, in three places.

حَاسِنٌ : see حَسَنٌ. — [Hence,] الحَاسِنُ *The moon*. (AA, Ṣ.)

أَحْسَنُ, fem. حَسْنَآءُ, pl. أَحَاسِنُ: see حَسَنٌ. — الأَحْسَنُ denotes the comparative and superlative degrees [of حَسَنٌ]; as in the phrase هُوَ الأَحْسَنُ [*He, or it, is the better*, and *best*; or *the more*, and *most, goodly* or *beautiful* or *comely* &c.]: (Ḳ:) ▼الحُسْنَى is the fem.; as in the phrase الأَسْمَآءُ الحُسْنَى *The best names*; those of God; which are ninety and nine: (Jel in vii. 179:) it signifies *the contr. of* السُّوأَى: (Ṣ, Ḳ:) the pl. of الأَحْسَنُ is الأَحَاسِنُ. (Ḳ.) In the saying, in the Ḳur [vi. 153 and xvii. 36], وَلَا تَقْرَبُوا مَالَ اليَتِيمِ إِلَّا بِالَّتِى هِىَ أَحْسَنُ [*And approach ye not the property of the orphan*, to make use of it,] *except by that act which is best* to be done with it, the meaning is, such an act as the taking care of it, and increasing it: (Bḍ:) or, as some say, the meaning is, the taking, of his property, what will [suffice to] conceal those parts of one's person that should not be exposed, and stay one's hunger. (TA.) [The fem.] ▼الحُسْنَى is applied to accidents only: not to substances. (Er-Rághib, TA.) It means also, [as an epithet in which the quality of a subst. predominates, *That which is better*, and *that which is best*. And hence,] *The good final* or *ultimate state* or *condition* [appointed for the faithful]: (Ḳ:) so, it is said, in the Ḳur xli. 50. (TA.) And *The view, or vision, of God*; (Ḳ;) accord. to some: but it is said that in the Ḳur x. 27, it means *Paradise*; and زِيَادَةٌ, which there follows it, means the view, or vision, of the face of God. (TA.) And *Victory*: and *martyrdom*: (Th, Ḳ:) whence, [in the Ḳur ix. 52,] إِحْدَى الحُسْنَيَيْنِ [*one of the two best things*]; (Ḳ;) *victory or martyrdom*. (Ksh, Bḍ, Jel.) And *The saying* لَا إِلٰهَ إِلَّا اللهُ. (Jel in xcii. 6 and 9.) The pl. of ▼الحُسْنَى is الحُسْنَيَاتُ and ▼الحُسَنُ, (Ḳ, [the latter like رُجَعٌ pl. of رُجْعَى, but misunderstood by Freytag as syn. with المَحَاسِنُ, which next follows it in the Ḳ,]) neither of which is used without the article ال. (TA.)

مَا أُحَيْسِنَهُ : see 4, last sentence but one.

تَحْسِينٌ a subst. of the measure تَفْعِيلٌ; (Ḳ;) or rather an inf. n. used as a subst.; (TA;) pl. تَحَاسِينُ: whence كِتَابُ التَّحَاسِينِ (Ḳ) [*Caligraphy*; or] *deliberate, orderly, and regular writing*; (TḲ;) [or *close and compact writing, without spaces, or gaps, and without elongation of the letters*;] *contr. of* المَشْقُ. (Ḳ. [See كِتَابٌ مَشْقٌ.])

مُحَسَّنٌ : see حُسْنٌ, and مَحَاسِنُ.

حُسْنٌ : see مُحْسَنٌ.

مُحْسِنٌ *Doing*, or *who does, that which is* حَسَن [meaning *good, comely*, or *pleasing*]; (Ḳ, TA;) as also ▼ مِحْسَانٌ: (Ḳ:) or the latter [is an intensive epithet, meaning *doing*, or *who does, much that is good, comely*, or *pleasing*: or] means *constantly doing that which is* حَسَن. (TA.) — إِنَّا نَرَاكَ مِنَ الْمُحْسِنِينَ, in the Ḳur xii. 36, means ‡ *Verily we see thee to be of those who know*, or *know well, the interpretation of dreams*: (Ksh, Bḍ, TA:*) or † *of those endowed with knowledge*: or *of the doers of good to the prisoners*: (Ksh, Bḍ:) or *of those who aid the weak and the sufferer of wrong, and visit the sick*. (TA.)

مَحْسَنَةٌ [*A cause of good*: pl., app., ▼ مَحَاسِنُ; like as مَسَاوٍ, originally مَسَاوِئُ, is said to be pl. of مَسَآءَةٌ, originally مَسْوَأَةٌ]. You say, هٰذَا الطَّعَامُ مَحْسَنَةٌ لِلْجِسْمِ [*This food is a cause of good*, i. e. *beneficial, to the body*]. (Ṣ.)

مُحْسَنٌ : see حُسْنٌ.

مُحْسِنٌ : see مِحْسَانٌ.

مَحَاسِنُ The *beautiful places* [or *parts*] *of the body*: (Ḳ:) accord. to some, (TA,) the sing. is ▼ مَحْسَنٌ: or it has no sing.: (Ḳ:) the former opinion is disapproved by ISd.: the latter is the opinion of the grammarians and of the generality of the lexicologists: and therefore, says Sb, the rel. n. is ▼ مَحَاسِنِىٌّ; for if مَحَاسِنُ had a sing., it would be restored to the sing. in forming the rel. n. (TA.) You say, فُلَانَةُ كَثِيرَةُ الْمَحَاسِنِ *Such a woman has many beautiful places* [or *parts*] *of the body*. (TA.) And مَحَاسِنُ الوَجْهِ وَمَسَاوِيهِ [*The beauties of the face, and its defects*]: (Ḳ in art. لمح:) [for] مَحَاسِنُ signifies the *contr. of* مَسَاوٍ. (Ṣ.) — [As contr. of مَسَاوٍ, it signifies also *Good qualities of any kind*: and also *good actions*; like حَسَنَاتٌ: agreeably with an explanation in the KL, نيكوئيها.] — See also حُسْنٌ: — and مَحْسَنَةٌ.

مَحَاسِنِىٌّ : see the next preceding paragraph.

حسو

1. حَسَا, (Ṣ, Mṣb, Ḳ,) aor. يَحْسُو, (Mṣb,) inf. n. حَسْوٌ, (Ṣ, Mṣb,) and some say that حُسْوَةٌ also is an inf. n., (Mṣb,) [but this is properly an inf. n. of un.,] *He* (a man) *supped*, or *sipped*, or *drank by little and little*, (Ḳ,) soup, or broth, (Ṣ, Ḳ,) or سَوِيق, and the like; (Mṣb;) as also ▼ احتسى (Ṣ, Ḳ) and ▼ تحسّى, (Ḳ,) or the last means, *in a leisurely manner*. (Sb, Ṣ.) You say also, حَسَوْتُ حَسْوَةً وَاحِدَةً [*I supped*, or *sipped, one sup*, or *sip*]. (Ṣ.) And حَسَوْتُ as meaning *I drank* [or *supped* or *sipped*] *what is termed* حَسْو and حَسَآء. (ISk, TA.) [Hence,] ▼ اِحْتَسَوْا كَأْسَ الْمَنَايَا † [*They sipped the cup of death*; lit., *deaths*]: and ▼ اِحْتَسَوْا أَنْفَاسَ النَّوْمِ † [*They sipped the draughts of sleep*; meaning *they took naps*]. (TA.) — One says also of a bird, حَسَا الْمَآءَ, (Mṣb, Ḳ,) aor. as above, (Mṣb,) inf. n. حَسْوٌ, (Mṣb, Ḳ,) like as one says of a man, شَرِبَ; (TA;) [*He sipped the water*:] one should not say, in this case, شَرِبَ. (Mṣb, Ḳ.) Hence the prov., (Mṣb,) يَوْمٌ كَحَسْوِ الطَّيْرِ [*A sleeping like the sipping of the bird*]; i. e., *of short duration*; (Ṣ, M, Mṣb, TA;) likened, in its quick ending, to a bird's swallowing water: (Mṣb:) in the copies of the Ḳ, يَوْمٌ [*a day*]; and so in [some copies of] the Ṣ, and in the A. (TA.) And the saying, نِمْتُ نَوْمَةً كَحَسْوِ الطَّيْرِ *I slept* [*a sleep like the sipping of the bird*; meaning,] *a short sleep*. (T, Mṣb.*)

2: see 4, in two places.

3. [حاساهُ *He supped*, or *sipped, with him* soup, or broth, &c.] You say, حَاسَيْتُهُ كَأْسًا مُرَّةً [*I supped*, or *sipped, with him a bitter cup*]. (TA.)

4. أَحْسَيْتُهُ المَرَقَ, (Ṣ, Ḳ,) inf. n. إِحْسَآءٌ, (TA,) *I made him*, or *gave him, to sup*, or *sip, the soup*, or *broth*; (Ṣ,* Ḳ;) as also ▼ حَسَّيْتُهُ, (Ḳ,) inf. n. تَحْسِيَةٌ. (TA.) It is said in a prov.,

* لِمِثْلِهَا كُنْتُ أُحَسِّيكَ ▼ الحُسَى *

[*For the like thereof I used to give thee the mouthfuls of soup to sup*, or *sip*;] meaning † *for the like of this case I used to act with goodness to thee*. (A, TA. [See also Freytag's Arab. Prov. ii. 437; where we read أُحَسِّيهَا.])

5: see 1.

6. تحاسوا (TA) *They supped*, or *sipped*, [soup, or broth, &c.,] *one with another*. (KL.)

8: see 1, in three places. — [Hence also,] احتسى سَيْرَ الفَرَسِ, and الجَمَلِ, and النَّاقَةِ, † *He elicited*, or *exacted, the utmost pace*, or *power of going on, of the horse*, and *of the he-camel*, and *of the she-camel*. (TA.)

حَسًا : see حَسَآءٌ.

حَسْوٌ, inf. n. of 1. — See also حَسَآءٌ.

حَسْوَةٌ *A single act of supping* or *sipping*; (Ṣ,* Ḳ;) as also ▼ حُسْوَةٌ; but the former is the more chaste: (Ḳ:) some say that these are dial. vars., like نَقْبَةٌ and نُقْبَةٌ, and جَرْعَةٌ and جُرْعَةٌ: but accord. to Yoo, the former denotes the act, and the latter is the [proper] subst. (TA.) See also what next follows.

حُسْوَةٌ *A sup*, or *sip*; i. e. *a small quantity of what is supped*, or *sipped*: (Ḳ:) or *as much as is supped*, or *sipped*, (يُحْسَى,) *at once*: (Ṣ:) or *a mouthful of what is supped*, or *sipped*; and some say that ▼ حَسْوَةٌ is a dial. var.; but others, that this is an inf. n. [of un.]: (Mṣb:) pl. حُسًى [for an ex. of which see 4] (Mṣb, TA) and حُسُوَاتٌ or حُسْوَاتٌ or حُسَوَاتٌ (Mṣb) and [of pauc.] أَحْسِيَةٌ and أَحْسُوَةٌ [in some copies of the Ḳ, erroneously, أُحْسُوَةٌ], and pl. pl. أَحَاسِى [or rather أَحَاسٍ], (Ḳ,) which ISd thinks to be rather a pl. of ▼ حَسَآءٌ, contr. to rule. (TA.) You say, فِى الإِنَآءِ حُسْوَةٌ [*In the vessel is a sup, or sip*]. (Ṣ, Mṣb.) See also حَسْوَةٌ.

حَسَآءٌ *A well-known kind of food*; (Ṣ;) *soup*; i. e. *what is supped*, or *sipped*; (Ḳ;) *thin cooked food*, (Sh, IAth, Mṣb,) *that is supped*, or *sipped*, (IAth, Mṣb,) such as is prepared for one who has a complaint of his chest, (Sh,) *made of flour and water and oil or grease, and sometimes sweetened*: (IAth:) also called ▼ حَسْوٌ (Sh, Ṣ, Mṣb, Ḳ) and ▼ حَسِيَّةٌ (Sh, Ḳ) and ▼ حَسًا and ▼ حَسُوٌّ; (Ḳ;) the last two, the latter of which is like the inf. n., mentioned by IAạr, but regarded by ISd as of doubtful authority. (TA.) See also حُسْوَةٌ.

حَسُوٌّ : see what next precedes. — Also One *who sups*, or *sips, much*: (Ṣ, Ḳ:) an epithet applied to a man. (Ṣ.)

حَسِيَّةٌ : see حَسَآءٌ.

[حَاسٍ act. part. n. of 1. Hence,] حَاسِى الذَّهَبِ [lit. The *supper*, or *sipper, of gold*;] a surname of Ibn-Judh'án, because he had a vessel of gold from which he supped, or sipped. (Ṣ, CḲ.)

مَحْسًى [The *mouth*; lit. the *place of supping*, or *sipping*]. One says of him who is short, هُوَ قَرِيبُ المَحْسَى مِنَ المَفْسَى [*He has the mouth near to the anus*]. (TA.)

حسى

1. حَسِيتُ الخَبَرَ : حَسَى حِسْيًا : see 8. — حَسِيتُ الخَبَرَ *i. q.* حَسِسْتُ [i. e. *I knew the news certainly*; or *knew somewhat of the news*; see 4 in art. حس]; as also ▼ أَحْسَيْتُ الخَبَرَ. (Ṣ, TA.) And حَسِىَ مَا فِى نَفْسِهِ *He knew what was in his mind*; as also ▼ احتساهُ. (Ḳ.) One says also, هَلْ ▼ أَحْسَيْتَ مِنْ فُلَانٍ شَيْئًا, meaning *Hast thou found*, or *discovered*, [or *learned*,] *anything from such a one?* (Az, TA.)

4: see 1.

8. احتسى *He dug out the sand from a* حِسْى *to procure the water beneath*: (Ṣ:) *he dug out the earth for the water to come forth*: (TA:) and احتسى حِسْيًا (T, Ḳ) *he fetched out, by digging, the water of a* حِسْى; so as heard by Az from more than one of Benoo-Temeem: (TA:) or *he dug a* حِسْى; as also ▼ حَسَاهُ. (Ḳ.) — [Hence,] † *He asked*, or *sought, information, news*, or *tidings*. (TA.) — See also 1, in two places.

حَسًى : see what next follows.

حِسْىٌ (T, Ṣ, Ḳ) and ▼ حَسْىٌ (Aḥmad Ibn-Yaḥyà, AAF, Ḳ) and ▼ حَسًى, accord. to the Ḳ, but this is unknown, and the correct word is حِسًا, [or rather ▼ حِسًى,] mentioned by IAạr, (TA,) *Water which the earth imbibes from sand* [*above it*], *and which, when it reaches what is hard, is arrested thereby: one digs out the sand from over it, and draws it forth*: (Ṣ:) or *accumulated sand, beneath which is hard rock, so that, when the sand is rained upon, it imbibes the water of the rain, which, reaching the rock beneath, is arrested thereby, and the sand prevents the heat of the sun from drying up the water; wherefore, when the heat is vehement, the upper portion of the sand is dug out from over the water, and it wells *forth, cold and sweet, and is taken by little and little:* (Az, TA:) or *soft*, or *plain, ground, in which water remains and collects:* or *rugged ground, over which is sand, that collects the rain-water; so that whenever a bucketful is drawn forth, another collects:* (K, TA:) so in the M: (TA:) pl. [of pauc.] أَحْسَآءٌ (S, K) and [of mult.] حِسَآءٌ: (K:) أَحْسَآءٌ is syn. with كِرَارٌ. (S.) [See also حَشْرَجٌ.] — حِسْىٌ also signifies *A small quantity of water;* and so ▼حَسَآءٌ. (Th, TA.)

حَسَى : }
حِسَى : } see حِسْىٌ.
حَسَآءٌ : }

حش

1. حَشَّ, (Msb, K,) aor. حَشَّ, [contr. to general rule in the case of an intrans. v. of this class, unless the sec. pers. of the pret. be حَشُشْتَ, which seems to be not improbable,] inf. n. حَشٌّ, (Msb,) *It* (a plant, or herbage, Msb, or a shoot of a palm-tree cut off from the mother-tree, or plucked forth from the ground, and planted, K) *dried*, or *dried up.* (Msb, K.) [Accord. to my copy of the Msb, the same is said of a well; but I incline to think that بِئْر is here written by mistake for تِبْن (meaning straw) or some similar word.] You say also, حَشَّ الوَلَدُ, (IAar, S, A, K,) aor. and inf. n. as above, (TA,) or inf. n. حُشُوشٌ, (IAar,) and, as some say, حُشَّ; (A 'Obeyd, S;) and ▼اِسْتَحَشَّ; (TA;) *The child*, or *young one*, (S, A, K,) of a she-camel, (IAar,) *dried up* in the belly, (S, A, K,) or womb, (TA,) the time of the birth having been exceeded. (TA.) And حَشَّتِ اليَدُ, (A, K,) and حُشَّتْ, (Yoo,) and ▼احشّت, (S, K,) and ▼استحشّت, (Yoo, K,) *The arm*, or *hand, dried up;* (S, A;) *and became unsound in its veins* or *ducts, and so rendered motionless;* syn. شَلَّتْ: (S, K:) or, as some say, *became slender and small.* (TA.) ═ حَشَّهُ, (S, Msb, K,) aor. حَشَّ, [in this case agreeable with general rule,] inf. n. حَشٌّ, (Msb, TA,) *He cut it*, namely, حَشِيش [or dry herbage]: (S, Mgh, Msb, K:) and *he collected it;* as also ▼احتشّهُ: (TA:) or the former has the former signification [only]; and ▼ the latter signifies *he sought it, and collected it.* (S, K, TA.) You say also, حَشَّ لِبَعِيرِهِ *He collected dry herbage* (حَشِيش) *for his camel.* (TA in art. بقل.) And حَشَّ عَلَى دَابَّتِهِ *He cut dry herbage* (حشيش) *for his beast.* (TA.) And حَشَّ عَلَى غَنَمِهِ *He beat the branches of the trees so that its leaves became scattered* [*for*, or *upon, his sheep* or *goats*]; like هَشَّ. (TA.) — Also, (S, K,) aor. as above, (S, A, K,) and so the inf. n., (TA,) *He threw to him* (namely a horse) حَشِيش [or *dry herbage*]; (S, K;) *he fed him therewith.* (A, TA.) Az says, I have heard the Arabs say to a man حُشَّ فَرَسَكَ [*Feed thou thy horse with dry herbage*]. (TA.) Hence the prov., أَحُشُّكَ وَتَرُوثُنِى [*I feed thee with dry herbage and thou dungest upon me*]: (S, A, K:) and if it were said with س, [أُحِسُّكَ, "I curry thee,"] it would not be strange: (S:) applied to him who does evil to one who does good to him: (Az, K:) or to any one to whom a benefit has been done and who requites it with the contrary thereof, or is not grateful for it nor profits by it: and thus the prov. is related in the T and S and M and A [and K]; but by 'Abd-es-Selám El-Basree, أَحُشُّكَ وَتَرُوثُنِينِى. (TA.) — Hence, (A,) حَشَّ النَّارَ, (S, A, K,) aor. and inf. n. as above, (S,) and Az adds بِالْحَطَبِ, (TA,) ‡ *He kindled the fire;* or *made it to burn*, or *to burn fiercely;* (S, A, K;) *and fed it with firewood, like as one feeds a beast with* حَشِيش: (A, TA:) or *he collected to it what was scattered of the firewood:* (TA:) and *he stirred it.* (K.) — And حَشَّ الحَرْبَ, aor. and inf. n. as above, ‡ *He kindled, and excited*, or *provoked, war*, or *the war.* (TA.) — And حَشَّ فُلَانًا ‡ *He improved*, or *made good, the condition*, (A, K,) or *property*, (O,) *of such a one.* (A, O, K.) — And حَشَّ مَالَهُ ‡ *He multiplied his property*, or *made it to be much*, (A, K,*) by [adding to it] the property of another: (A:) or حَشَّ بِهِ مَالًا *he put property into*, or *among, his property:* (Skr:) or *he strengthened him with property.* (El-Báhilee.) — And حَشَّ سَهْمَهُ, (S, A, O,) aor. and inf. n. as above, (TA,) ‡ *He feathered his arrow:* (A, O:) or *stuck the feathers upon the sides of his arrow:* (S:) or *mounted them upon his arrow.* (TA.)

4. احشّ *It* (herbage) *became in such a state that it might be cut* (ISh, K) *and gathered*, (TA,) *being dried up.* (ISh.) — أَحَشَّتِ اليَدُ: see حَشَّتْ. — Also احشّت *She* (a woman, S and K, and a camel, TA) *had her child*, or *young one, dried up in her belly.* (S, K.) ═ أَحَشَّ اللّٰهُ يَدَهُ [*May God make his arm*, or *hand, to dry up;* or *to become unsound in its veins* or *ducts, and so rendered motionless;*] is a form of imprecation used by the Arabs. (TA.) — احشّ فُلَانًا *He cut* (K) *and collected* (TA) حَشِيش [or *dry herbage*] *with such a one;* (K;) as though he helped him in doing so. (TA.)

8. احتشّهُ: see حَشَّهُ, in two places.

10. اِسْتَحَشَّ الوَلَدُ; and اِسْتَحَشَّتِ اليَدُ: see حَشَّ; and حَشَّتْ.

حَشٌّ (S, Mgh, Msb, K) and ▼حُشٌّ, (S, Msb, K,) but the former is the more common, (Msb,) and ▼حِشٌّ, (K,) *A garden:* (El-Fárábee, S, Mgh, Msb, K:*) or *a garden of palm-trees:* (AHát, Msb:) pl. حِشَّانٌ (S, Msb) and حُشَّانٌ. (Msb.) — Hence, ‡ *A privy;* (El-Fárábee, S, A, Mgh, Msb, K;) likewise called بَيْتُ الحَشِّ or ▼الحُشُّ: (Msb:) because they used to ease themselves in the gardens: (S, Mgh, Msb, K:) then, when they made privies, they applied thus this appellation: (Msb:) and in like manner, ▼مَحَشٌّ; but accord. to the Abridgment of the 'Eyn., this is proper, not tropical: (Msb:) or this last, also written ▼مِحَشٌّ, signifies the same; (TA;) or *a place in which human ordure has become collected:* (K:) the pl. of حَشٌّ as applied to a privy is حُشُوشٌ (S, Mgh, K) and حُشُّونَ. (Ibn-'Abbád, K.) ═ See also مَحَشَّةٌ.

حُشٌّ: see حَشٌّ, in two places: ═ and see حَشِيشٌ.

حِشٌّ: see حَشٌّ.

حُشَاشَةٌ: see مَحَشٌّ: ═ and see حُشَاشَةٌ.

مَحَشٌّ: see حِشَاشٌ.

حُشُوشٌ جَنِينُهَا [*Having her fœtus dried up in her womb*]. (L from a verse of Ibn-Mukbil.)

حَشِيشٌ *Dry herbage;* (Msb;) *dry pasture*, or *fodder:* (El-Fárábee, S, Mgh, Msb, K:) of the measure فَعِيلٌ in the sense of the measure فَاعِلٌ: (Msb:) what is fresh is not so called: (S, Msb:) but عُشْبٌ is applied to what is fresh and what is dry: this, says ISd, is the opinion of the generality of the lexicologists: some [he adds] assert that حشيش is *green pasture* or *herbage*, as well as *dry:* but he says that this is not correct; [and the like is said in the Msb;] for this word is properly applied to denote dryness and contraction: ISh says that it is applied to all *herbs*, or *leguminous plants, fresh* as well as *dry;* as also عَلَفٌ and خَلًى: Az says that when they use it unrestricted, the Arabs mean thereby حَلِىّ, [which is the *herb called* نَصِىّ *when it has become dry and white*,] in particular; and that this is the best kind of fodder; that horses thrive upon it, and it is one of the best pastures for camels, or for camels and sheep and goats; a good supply in years of scarcity: (TA:) or it signifies *cut herbage* or *pasture;* and is of the measure فَعِيلٌ in the sense of the measure مَفْعُولٌ: (Msb:) the n. un. is with ة, signifying *a fascicle*, or *wisp, of* حَشِيش: (TA:) [and sometimes *a herb of any kind:* the pl. is حَشَائِشُ.] — [It is also applied, in the present day, to *Hemp*, used for its intoxicating property; both *fresh* and *dry:* app. what is termed حَشِيشُ الحَرَافِيشِ in the K, voce بَنْجٌ, q. v.: and also termed حَشِيشَةُ الفُقَرَآءِ: see De Sacy's "Chrest. Arabe," sec. ed., vol. i. pp. 210—283. — حَشِيشَةُ السُّلْطَانِ: see خَرْدَلٌ.] — حَشِيشٌ also signifies *A child*, or *young one, that has dried up in the belly of its mother;* (Mgh, Msb, TA;) and so ▼حُشٌّ and ▼مَحْشُوشٌ and ▼أُحْشُوشٌ: (TA:) or حُشٌّ [and the rest], *a child*, or *young one, that perishes in the belly of its mother.* (K.) It is said in a trad., فَأَلْقَتْ حَشِيشًا *And she cast forth a child*, or *young one, dried up.* (Mgh.) And you say, أَلْقَتْ وَلَدَهَا حَشِيشًا *She* (a camel) *cast forth her young one dried up.* (Msb.)

حُشَاشَةٌ The [*last*] *remains of the spirit* (S, A,* Msb, K) *in the heart*, (TA,) [or *of life;*] *in a sick man*, (S, Msb, K,) and *in one who is wounded;* (K;) as also ▼حُشَاشٌ, (S, Msb, K,) the ة being sometimes elided. (Msb.) — And †*Any remains*, or *relic.* (TA.) You say, مَا بَقِىَ مِنَ المُرُوَّةِ إِلَّا حُشَاشَةٌ تَتَرَدَّدُ فِى أَحْشَآءِ مُحْتَضَرٍ ‡ [*There remained not, of manliness, save a last relic going to and fro*, or *wavering, in the entrails of one at the* *point of death*]. (A, TA.) And مَا بَقِىَ مِنَ الشَّمْسِ إِلَّا حُشَاشَةُ نَازِعٍ ‡ [*There remained not, of the sun, save a last departing relic*]. (A, TA.)

[حَشَائِشِىٌّ *One skilled in the knowledge of herbs*: so in modern works. — Accord. to Golius, as on the authority of the KL, but not in my copy of that work, *A collector of hay; a forager.*]

حُشَّاشٌ, [pl. of ▼ حَاشٌّ,] *Cutters,* or *cutters and collectors, of* حَشِيش [or *dry herbage*]: (TA:) or *seekers and collectors thereof.* (Ṣ.) ═ See also مِحَشٌّ.

حَاشٌّ: see its pl. حُشَّاشٌ.

أُحْشُوشٌ: see حَشِيش, last signification.

مَحَشٌّ, (Ṣ, A, TA,) or ▼ مِحَشٌّ, (Ḳ, [but this seems to be a mistake occasioned by the accidental omission of وَالمَحَشُّ, as is indicated by the addition of وَيُكْسَرُ shortly after, referring to the word in a sense different from that which is here next mentioned,] *A place,* (Ṣ,) or *land,* (Ḳ,) *in which is much* حَشِيش [or *dry herbage*]; (Ṣ, Ḳ;) as also ▼ مَحَشَّةٌ: (Ḳ:) or *a place in which one cuts* حشيش: (A:) and the first (مَحَشٌّ) *a place in which are much pasture,* or *herbage, and wealth,* or *good things.* (Ḳ.) You say, هٰذَا مَحَشٌّ صِدْقٍ, meaning *This is a* [*good*] *region abounding in* حَشِيش. (TA.) And إِنَّكَ بِمَحَشِّ صِدْقٍ فَلَا تَبْرَحْهُ *Verily thou art in a place abounding in good things, therefore do not quit it*: so in some copies of the Ṣ; and accord. to this explanation, the word is tropically used: in other copies of the Ṣ, *in a place abounding in* حَشِيش. (TA.) — See also مِحَشٌّ. — Also the former, *A thing in which* حَشِيش *is put*; and so ▼ مِحَشٌّ; but the former is the more chaste; (A 'Obeyd, Ṣ, Ḳ;) and ▼ مَحَشَّةٌ, (Ḳ,) and ▼ مِحَشَّةٌ, which is more chaste; so in some copies of the Ḳ; (TA;) and ▼ حُشَّاشٌ, like غُرَابٌ; of which the pl. is أَحِشَّةٌ: (TA:) the first two of these words are applied to *a woollen* كِسَآء [q. v.] *in which* حَشِيش *is put*: (IAth:) and ▼ حِشَاشٌ, with kesr, signifies *a* [*sack of the kind called*] جُوَالِق *in which is* حَشِيش. (Ḳ.) — See also مِحَشٌّ.

مُحِشٌّ A woman, (Ṣ, Ḳ,) and a she-camel, (TA,) *whose child,* or *young one, dries up in her belly.* (Ṣ, Ḳ, TA.) — An arm, or a hand, (يَدٌ,) *drying up*; or *becoming unsound in its veins or ducts, and so rendered motionless*: or *becoming slender and small.* (TA.)

مِحَشٌّ *An instrument with which* حَشِيش [or *dry herbage*] *is cut*; (A 'Obeyd, Ṣ;) as also ▼ حُشَّاشٌ, like رُمَّانٌ: (TA:) or *a plain* [i. e. *not serrated*] مِنْجَل [or *reaping-hook*] *with which* حَشِيش *is cut*; as also ▼ مَحَشٌّ; but the former is the more chaste; (Ḳ;) or, accord. to the L, the latter is the better. (TA.) ═ See also مَحَشٌّ, in two places. ═ Also *An iron instrument with which a fire is stirred*; and so ▼ مِحَشَّةٌ: (Ṣ, Ḳ:) pl. مَحَاشُّ. (A.) — [Hence, ‡ *A kindler, an exciter,* or *a provoker, of war*: or] *a courageous man.* (Ḳ.) Of such one says, نِعْمَ مِحَشُّ الكَتِيبَةِ ‡ [*Excellent is the exciter of the army,* or *troop*]. (Ṣ, A.) And مِحَشُّ حَرْبٍ signifies ‡ *A kindler and an exciter of war*: (Ḳ, TA:) or *a conductor of war.* (Ḥam p. 14.) You say, هُمْ مَحَاشُّ الحُرُوبِ ‡ *They are the kindlers and exciters of wars.* (A.) ═ See also مَحَشٌّ.

مِحَشَّةٌ: see مِحَشٌّ, in two places. ═ Also ‡ The *podex*: or *anus*: (Ṣ, Mgh, Mṣb,* Ḳ:*) and so ▼ حُشٌّ: (TA:) pl. of the former مَحَاشُّ; (Ṣ, Mgh, Ḳ;) and of the latter حُشُوشٌ: (TA:) the former also occurs written with س. (Ṣ, Mgh.)

مَحَشَّةٌ: see مَحَشٌّ: ═ see also مِحَشٌّ. — Also *A staff,* or *stick*: or *a rod, wand,* or *twig.* (TA.)

مَحْشُوشٌ: see حَشِيش, last signification.

حشب

4. احشبهُ *He angered him.* (Ḳ.)

8. احتشبوا *They collected themselves together; congregated.* (El-Muarrij, Ḳ.)

حِشْبٌ: see what next follows.

حَشِيبٌ *A thick, coarse,* or *rough, garment* or *piece of cloth*; (Aboo-Semeyda' El-Aạrábee, Ḳ;) as also جَشِيبٌ and ▼ حِشْبٌ. (TA.) ═ See also حَوْشَبٌ.

حَشِيبِىٌّ: see what next follows.

حَوْشَبٌ The *fetlock-joint* (مَوْصِلُ الوَظِيفِ) *in the pastern* (رُسْغ) *of a beast*: (Ṣ, Ḳ:) or, (Ḳ,) as also ▼ حَشِيبٌ and ▼ حَشِيبِىٌّ, (so in the TA,) *a bone in the inside of the hoof, between the tendons* (عَصَب) *and the* وَظِيف [or *shank*; app. the *lower pastern-bone*]: (Ḳ:) or the *contents* (حَشْو) *of the hoof*: (AA, TA:) or *a small bone, like a* سُلَامَى [or *finger-bone*, a description aptly applying to either of the pastern-bones, the upper of which seems to be here meant], *at the extremity of the* وَظِيف, *between the head thereof and the place where the hoof is set on,* (Aṣ, Ṣ, Ḳ,) *entering into the* جُبَّة: (Aṣ, Ṣ: [see this last word (جُبَّة), to which various significations are assigned; here said in the TA to be that which contains the حوشب and دَخِيس (both of which words seem to be syn.), between, or amid, the flesh and the tendons:]) or the *bone of the* رُسْغ [or *pastern*]: (T, Ḳ:) or a name applied to *each of the two bones of the pastern* (رسغ) *of a horse.* (TA.) ═ *Lean, and lank in the belly.* (Ḳ.) — And *Big-bellied*: or *big in the sides*: (TA:) or *swollen,* or *inflated, in the sides*: (Ṣ, Ḳ:) or *swollen in the belly, and short*: (Skr p. 57: [see an ex. in a verse cited voce مُجْرٍ in art. جرو:]) bearing two contr. significations: (Ḳ:) fem. with ة: (TA:) pl. حَوَاشِبُ. (Skr, Ṣ.) ═ The *male hare*: (Ḳ,* TA:) and [so in the Ḳ; but accord. to the TA, "or"] the *calf.* (Ḳ.) Also, accord. to the Ḳ, the "male fox:" but this is a mistake, occasioned by the occurrence of the words حَوْشَب and قَعْنَب together in a verse: the latter of these two signifies the "male fox." (TA.) ═ *A company of men*; as also ▼ حَوْشَبَةٌ: (El-Muarrij, Ḳ:*) *a large number of men collected together.* (TA.)

حَوْشَبَةٌ: see what next precedes.

حشد

1. حَشَدَ, aor. ـِ (A, Mṣb, Ḳ) and ـُ, (Mṣb, Ḳ,) inf. n. حَشْدٌ, (A, Mṣb, Ḳ,) *He collected together* (A, Mṣb, Ḳ) people, or a company of men. (A, Mṣb.) — [Hence,] بِتُّ فِى لَيْلَةٍ تَحْشُدُ عَلَىَّ الهُمُومَ ‡ [*I passed a night that brought anxieties crowding together upon me*]. (A, TA.) ═ حَشَدُوا, (Ṣ, A, Mṣb,) aor. ـِ, inf. n. حَشْدٌ, (Ṣ,) or حُشُودٌ; (A;) and ▼ احتشدوا, and ▼ تحشّدوا; (Ṣ, A;) *They collected themselves together,* or *assembled,* (Ṣ, A, Mṣb,) *and came round about* (حَفُّوا [but see what follows]) *aiding one another*: (A:) or حَشَدُوا signifies *they were prompt,* or *active,* [instead of حَفُّوا, in the Ḳ, I read خَفُّوا, as in the L,] *in aiding one another*: or *they complied quickly, when called,* or *summoned*: (L, Ḳ:) the verb is thus generally used in relation to a collective number: seldom in relation to one: (L:) or *they collected themselves together,* or *assembled, for one thing* or *affair*; as also ▼ احشدوا, and ▼ احتشدوا, and ▼ تحاشدوا. (L, Ḳ.) And حشدوا عَلَيْهِ, (L,) and ▼ تحاشدوا عَلَيْهِ, *They collected themselves together, aiding one another against him.* (A.) And احتشد ▼ القَوْمُ لِفُلَانٍ *The people,* or *party, collected themselves together to such a one, and prepared, equipped,* or *furnished, themselves* [for action]. (TA.) — حَشَدُوا لَهُ *They combined for him,* [or *on his account,*] *and took pains,* or *exerted themselves, in treating him with courtesy and honour*; and so حَفَلُوا لَهُ. (Fr, L.) And *They showed honour, and gave a hospitable entertainment, to him*; namely, a guest. (L.)

4: see 1.

5: see 1.

6: see 1, in two places.

8: see 1, in three places. — احتشد لَهُ فِى كَذَا *He prepared himself for him* [*in such a case,* or *in such an affair*]. (A.) — احتشد فِى الضِّيَافَةِ, and للضِّيَافَةِ *He strove, laboured,* or *exerted himself, in,* and *for, the entertainment of a guest* or *guests.* (A.)

حَشْدٌ, (Ṣ, Ḳ,) originally an inf. n., (Ṣ,) and ▼ حَشَدٌ, (Ḳ,) *An assembly,* or *a collected* or *congregated body* or *party,* (Ṣ, Ḳ,) of men. (Ṣ.)

حَشَدٌ: see what next precedes.

حَشِدٌ One *who does not leave* [unemployed] *any endeavour or aid or property that he possesses*; as also ▼ مُحْتَشِدٌ; (L, Ḳ;) and ▼ حَاشِدٌ, pl. حُشُدٌ. (L.) — وَادٍ حَشِدٌ *A valley similar to land such as is termed* حَشَادٌ: (Ḳ:) *a valley which a small and an inconsiderable quantity of water causes to flow.* (TA.)

أَرْضٌ حَشَادٌ *Land that does not flow with water save in consequence of much rain*: (Ṣ:) or *in consequence of* [*lasting, or continuous, rain, such as is termed*] دِيمَة: (Ḳ:) or *that flows with water in consequence of the least rain;* (ISk, M, Ḳ;) as also زَهَّادٌ and شَحَّاحٌ and نَزِلَةٌ: (ISk:) or *such as is the quickest to flow with water:* (AA in a marginal note in a copy of the Ṣ:) or حَشَّادٌ signifies a water-course *of which the ground is hard, quickly flowing with water, having many minor water-courses* (شِعَاب) *pouring into its bed and uniting one with another.* (ISh.)

حَاشِدٌ [A people *collecting themselves together,* or *assembling,* &c. (see 1)]: pl. حُشُدٌ, occurring in a trad., followed by وُفُدٌ. (L.) — جَآءَ فُلَانٌ حَافِلًا حَاشِدًا and ↑مُحْتَفِلًا مُحْتَشِدًا *Such a one came* [*full of energy,*] *prepared, furnished, equipped,* or *accoutred.* (Ṣ.) See also حَشِدٌ.

مَحْشُودٌ A man *with whom is an assembly,* or *a collected body,* or *party, of men:* (L:) or one *to whom others collect themselves together; who is served,* or *waited on:* (A:) or *obeyed by others,* (Ṣ, Ḳ,) *among his people,* (TA,) *and whom they are prompt* [instead of يَحِفُّونَ, in most of the copies of the Ḳ, I read يَخِفُّونَ, as in others and in the Ṣ,] *to serve,* (Ṣ, Ḳ,) *and to whom they collect themselves together.* (TA.)

مَحَاشِدُ *Places where people are collected to go forth:* or it is a pl. of حَشْدٌ, contr. to rule, like مَشَابِهُ [pl. of شَبَهٌ] and مَلَامِحُ [pl. of لَمْحَةٌ]. (L from a trad.) [See مَخَاطِبُ.]

مُحْتَشِدٌ: see حَشِدٌ, and حَاشِدٌ.

حشر

1. حَشَرَ, aor. ـِ and ـُ, (Ṣ, Mṣb, Ḳ,) the former of which aors. is found in the seven readings of the Ḳur, (Mṣb,) inf. n. حَشْرٌ, (Ṣ, Mṣb, Ḳ,) *He congregated,* or *collected together,* (Ṣ, Mṣb, Ḳ,) men: (Ṣ, Mṣb:) or *he congregated* them, or *collected* them *together, and drove* them: (Mṣb, TA:) *he made* them *to go forth, collected together, from one place to another:* (Bḍ in lix. 2:) *he,* or *it, compelled* them *to emigrate:* (Ḳ,* TA: [in the CḲ الخَلَآءُ is put by mistake for الجَلَآءُ, the explanation of the inf. n.:]) and [simply] *he drove* towards a place or quarter. (TA.) Hence يَوْمُ الحَشْرِ ‡ [*The day of congregation,* &c.; meaning] *the day of resurrection:* (Ṣ,* TA:) [see also مَحْشَرٌ:] and سُورَةُ الحَشْرِ ‡ [*The Chapter of the Compulsion to emigration;* which is the fifty-ninth chapter of the Ḳur-án]. (TA.) It is said by most of the expositors of the Ḳur that the wild animals and other beasts, and even the flies, will be collected together (تُحْشَرُ) for retaliation; and they cite a trad. on this subject. (TA.) So in the Ḳur [lxxxi. 5], وَإِذَا الوُحُوشُ حُشِرَتْ *And when the wild animals shall be collected together,* (Bḍ, Jel,) from every quarter, (Bḍ,) after resurrection; (Jel;) or *raised to life,* (Bḍ,) for the purpose of their retaliating, one upon another; after which they shall return to dust: (Bḍ, Jel:) or the meaning is, *shall die,* (Az, Ṣ,) in the present world; accord. to some: (Az:) and thus says 'Ikrimeh, (Ṣ, TA,) on the authority of I'Ab, (TA,) as is related by Sa'eed Ibn-Masrooḳ: (Ṣ, TA:) but accord. to some, the two meanings are nearly the same; for each denotes collection. (TA.) حَشْرٌ also signifies The *going forth with a people fleeing* or *hastening* or *dispersing themselves in war;* when used absolutely. (TA.) — حَشَرَتْهُمُ السَّنَةُ, aor. ـِ and ـُ, (Lth,) inf. n. حَشْرٌ, (Ḳ,) ‡ *The year of dearth destroyed their camels and other quadrupeds;* because it causes the owners to collect themselves from the various quarters to the cities or towns: (Lth:) or *it caused them to go down to the cities or towns:* (A:) or *it distressed them;* app., because of their collecting themselves together from the desert to the places of settled abodes: (Abu-ṭ-Ṭeiyib:) and حَشَرَتِ السَّنَةُ مَالَ فُلَانٍ *The year of dearth destroyed the camels &c. of such a one.* (Ṣ, Ḳ.*) — حَشَرَهُ, (Ṣ, A,) inf. n. حَشْرٌ, (Ṣ, Ḳ,) ‡ *He made it* (a spear-head, Ṣ, A) *thin,* or *slender:* (Ṣ, A, Ḳ:) *he made it* (a spear-head, and a knife,) *sharp,* or *pointed, and thin,* or *slender:* (TA:) *he made it small, and thin,* or *slender:* (Th:) *he pared it;* namely, a stick: (TA:) *he pared it, and made it sharp,* or *pointed.* (Ṣ.)

7. انحشروا ‡ *They* (people) *became collected together* from the desert to the places of settled abodes. (Abu-ṭ-Ṭeiyib.)

حَشْرٌ ‡ Anything *thin,* or *slender,* or *elegant.* (TA.) You say أُذُنٌ حَشْرٌ ‡ *A thin,* or *an elegant, ear;* (Lth, ISk, Ṣ, A, Ḳ;) as though it were pared, (Lth, Ṣ,) and made sharp: (Ṣ:) or *small, elegant, and round:* (Lth:) or *thin at the end:* (Th:) or *sharp-pointed:* (TA:) and the epithet is the same for the dual also and the pl.: (Ḳ:) [J says that] it does not admit the dual form nor the pl., because it is originally an inf. n., and the expression above mentioned is like مَآءٌ غَوْرٌ and مَآءٌ سَكْبٌ: but اذن حَشْرَةٌ is sometimes said: (Ṣ:) and the pl. حُشُورٌ occurs in a verse of Umeiyeh Ibn-Abee-'Áïdh: (TA:) and you also say اذن ↑مَحْشُورَةٌ. (TA.) حَشْرٌ is also applied in the same sense as an epithet to other things. (Ṣ.) You say قُذَّةٌ حَشْرٌ ‡ *A thin,* or *an elegant, feather of an arrow;* (Lth, Ṣ, A, Ḳ;) as though it were pared: (Lth:) or *sharp-pointed.* (TA.) Also سِنَانٌ حَشْرٌ ‡ *A thin,* or *slender, spear-head:* (Ṣ, Ḳ:) or *sharp,* or *sharp-pointed:* and سِكِّينٌ حَشْرٌ in like manner: and حَرْبَةٌ حَشْرَةٌ: (TA:) and سَهْمٌ حَشْرٌ, and سِهَامٌ حُشْرٌ: like جَوْنٌ and جُونٌ, and وَرْدٌ and وُرْدٌ: (Akh, Ṣ:) or سَهْمٌ حَشْرٌ signifies *an arrow having straight,* or *even, feathers;* and so سهم ↑مَحْشُورٌ; and ↑حَشِرٌ, of the same measure as كَتِفٌ, an arrow *having good feathers attached to it.* (TA.) You also say بَعِيرٌ حَشْرُ الأُذُنِ ‡ *A camel having a thin,* or *an elegant, ear.* (TA.)

حَشِرٌ: see حَشْرٌ.

حَشَرَةٌ and حَشَرَاتٌ, (Ḳ,) each being a coll. n. without a sing.; (TA;) or the former is sing. of the latter; (Ṣ, Mṣb;) *Any small animals that creep or walk upon the earth;* (Ṣ, Mgh, Mṣb, Ḳ;) *as jerboas and hedgehogs and lizards of the kind called* ضَبّ *and the like:* (TA:) or the former, (Mṣb,) or latter, (Mgh,) is applied to *rats or mice, and jerboas, and lizards of the kind above mentioned,* (Mgh, Mṣb,) *collected together:* (Mṣb:) or *any venomous* or *noxious reptiles* or *the like, such as scorpions and serpents;* syn. هَوَامّ; (Aṣ, Ḳ;) as also أَحْرَاشٌ and أَحْنَاشٌ. (Aṣ.) — Also the former, *Whatever is captured, snared, entrapped, hunted,* or *chased, of wild animals* or *the like, birds,* and *fish, &c.;* (Ḳ;) *whether small or great:* (TA:) or the *great thereof:* or *what is eaten thereof:* (Ḳ:) thus in all the copies of the Ḳ; but the pronoun [in the latter case] does not refer to the animals &c. above mentioned: it is expressly said in the T and M that the word signifies *whatever is eaten of herbs,* or *leguminous plants, of the earth, such as the* ذُعَاع *and* فَثّ. (TA.)

حَاشِرٌ One *who congregates,* or *collects together,* people. (TA.) With the article ال, applied to Mohammad; (Ṣ, Ḳ;) because he collects people after him (Ṣ, IAth) and to his religion. (IAth.) — *A collector of spoils:* (El-Ḥulwánee, Mgh:) and [its pl.] حُشَّارٌ signifies *collectors of the tithes and poll-tax.* (TA.)

مَحْشَرٌ (Ṣ, Ḳ) and مَحْشِرٌ (Ḳ) *A place of congregation:* (Ṣ, Ḳ:) a term used when people are collected together to a town or country, and to an encampment, and the like. (TA.) Hence, يَوْمُ المَحْشَرِ [*The day of the place of congregation;* meaning the day of judgment]. (TA.)

مَحْشُورٌ; and its fem., with ة: see حَشْرٌ.

حشرج

Q. 1. حَشْرَجَ, (TA,) inf. n. حَشْرَجَةٌ, (Ṣ, Ḳ, TA,) *He rattled in the throat, in dying: he made his breath,* or *spirit,* (نَفَسَهُ, or نَفْسَهُ, accord. to different copies of the Ṣ and Ḳ,) *to reciprocate:* (Ṣ, Ḳ:) said also of the chest: or *he made the sound of his breath to reciprocate in his throat,* or *fauces, without uttering it with his tongue.* (TA.) Also said of an ass, *He made his voice to reciprocate in his throat:* (Ṣ, Ḳ, TA:) or *uttered his voice from his chest.* (TA.)

حَشْرَجٌ *Water that is beneath the ground, unperceived, in the wide water-channels that contain small pebbles, and which, when one has dug to the depth of a cubit, gushes forth abundantly:* waters of this description are called by the Arabs أَحْسَآء [pl. of حِسْىٌ] and كِرَارٌ [pl. of كَرٌّ] and حَشَارِجُ: and *sweet water, of the water of a* حِسْى: (Az, TA:) or *water that runs, clear and shallow, over pebbles,* or *over small pebbles:* (TA:) *what is termed* حِسْىٌ, *among pebbles:* (ISk, Ṣ, Ḳ:) or *what is termed* حِسْىٌ, *having pebbles in it:* (Ḳ accord. to the TA:) or *what resembles that which is termed* حِسْىٌ, *in which waters collect:* or *a small,* or *round, hollow,* or *cavity, in a mountain, in which water becomes clear,* (Az, Ḳ, TA,) *after collecting:* (Az, TA:) or *water in a small,* or *round, hollow,* or *cavity, in a mountain.* (A.) — *Soft foraminous stones* (كَذَّان) *of the ground:* n. un. with ة. (Ḳ.) — *A small,* (A, TA,) or *thin,* (Ḳ,) *and clean,* (TA,) كُوز [or *mug*], (A, Ḳ, TA,) *in which water is cooled,* (A,) *of the manufacture of El-Ḥeereh.* (Ḳ.) — The *cocoa-nut.* (Kr, TA.)

حَشْرَجَةٌ inf. n. of حَشْرَجَ [q. v.].—[The rattles;] the voice of a sick person reciprocated in the throat, or fauces. (A.)

حشف

1. حَشَفَ, said of a she-camel's dug, *Its milk became drawn up* or *withdrawn* or *withheld*, or *it went away, from it.* (IDrd, L, TA. [See also 4 and 10.])

2. حَشَّفَ عَيْنَهُ, inf. n. تَحْشِيفٌ, *He* (a man, TA) *contracted his eyelids, and looked through the interstices of their lashes.* (IDrd, Ḳ.)

4. احشف, said of a she-camel's udder, *It became contracted, and like an old worn-out water-skin* or *milk-skin.* (TA. [See also 1 and 10.])—احشفت النَّخْلَةُ *The palm-tree bore dates such as are termed* حَشَف. (Ṣ, Mgh, Mṣb.)

5. تحشّف *He wore old and worn-out clothing,* (O, L, KL, TA,) *such as is termed* حَشِيف: (O, L, TA:) in the copies of the Ḳ, erroneously, ▼استحشف. (TA.)

10. استحشف, said of an udder, (Jm, Ḳ,) *It became contracted:* (Jm:) or *became dried up and contracted.* (Ḳ. [See also 1 and 4.]) And استحشفت الأُذُنُ *The ear became dried up* (Mgh, Mṣb, Ḳ) *and contracted.* (Ḳ.) And استحشف الأَنْفُ *The cartilage of the nose became dried up from want of natural motion.* (Mṣb.)—See also 5.

حَشْفٌ *Dry bread.* (Ḳ.)

حَشَفٌ The *worst kind of dates;* (Ṣ, Mgh, Mṣb, Ḳ;) *that dry up without ripening, so that they have no flesh:* (Mṣb:) or *dates without firmness, having no stones;* (Ḳ;) like شِيصٌ: (TA:) or *dry,* or *tough, bad dates;* (Ḳ;) for when they dry up, they become hard and bad, without taste and without sweetness: (TA:) or *of which the lower portion has become bad and rotten, while in its place:* (IAạr, TA in art. حشو:) n. un. with ة. (Mṣb.) [Hence,] أَحَشَفًا وَسُوْءَ كِيلَةٍ, a prov., (Ṣ, Meyd, O,) meaning Dost thou combine *the worst of dates and bad measure?* applied to him who combines two bad qualities. (Meyd, O.)—A *worn-out* udder; (Ṣ, Ḳ;) as also ▼حَشِفٌ: (Ḳ:) or an udder *of which the milk has dried up, so that it has become contracted.* (EM p. 67.)—*A thing that is lean,* and *dry,* or *withered.* (KL.)

حَشِفٌ: see حَشَفٌ.—تَمْرٌ حَشِفٌ *Dates having many such as are termed* حَشَف. (TA.)

حَشَفَةٌ The *head* [or *glans*] *of the penis:* (TA:) or the *part of the penis,* (Ṣ, Ḳ,) [i. e.] the *part of the head of the penis,* (Mgh,) *that is above* [i. e. *beyond*] *the place of circumcision:* (Ṣ, Mgh, Ḳ:) [accord. to the latter explanation, somewhat more than the glans:] the mulct for the cutting off of which is the whole price of blood. (TA.)

حَشِيفٌ *Old, and worn-out:* applied to clothing or a garment. (Ṣ, Ḳ, TA.)

نَخْلَةٌ مِحْشَافٌ [*A palm-tree that bears dates such as are termed* حَشَف]. (Ṣ and L voce مِعْزَارٌ.)

مُتَحَشِّفٌ A man *clad in old and worn-out clothing* [*such as is termed* حَشِيف]: (Ṣ, TA:) a man *in evil condition; slovenly in his person; threadbare, shabby,* or *mean, in the state of his apparel:* or *dried up, and shrivelled:* or *having his garment tucked up.* (TA.)

حشك

1. حَشَكَتِ الدِّرَّةُ, aor. ـِ, inf. n. حَشْكٌ and حُشُوكٌ; for the former of which, حَشَكٌ is used by poetic license: (Ṣ:) or حَشُكَت, aor. ـُ, (TA,) inf. n. حَشَكٌ (Ḳ, TA) and حُشُوكٌ: (TA:) *The flow of milk became full:* (Ṣ:) or *became vehement in the udder:* or *collected quickly therein:* (Ḳ, TA:) but accord. to Lth, حَشْكٌ and ▼حَشَكٌ are like نَفْضٌ and نَفَضٌ, and نَقْضٌ and نَقَضٌ; the former being an inf. n., and the latter a subst. [in the proper sense of the term, app. signifying *milk collected,* or *collected quickly, in the udder*]. (TA.)—[Hence,] حَشَكَتِ السَّحَابَةُ, (Ḳ,) aor. ـِ, inf. n. حَشْكٌ, (TA,) ‡ *The cloud had much water.* (Ḳ, TA.) And حَشَكَتِ السَّمَآءُ, aor. ـِ, inf. n. حَشْكٌ, †*The sky let fall a rain such as is termed* حَشْكَة. (AZ, Ṣ.) And حَشَكَ الوَادِى, †*The valley poured* (دَفَعَ) *with water.* (TA.)—And حَشَكَتِ النَّخْلَةُ ‡ *The palm-tree bore much fruit.* (Yaạḳoob, Ṣ, Ḳ, TA.)—And حَشَكَ القَوْمُ, (Ṣ, Ḳ,) inf. n. حَشْكٌ, or, accord. to Th, حَشَكٌ, (TA,) † *The people collected themselves together,* or *assembled.* (Th, Ṣ, Ḳ.)—حَشَكَتْ لَبَنَهَا, inf. n. حَشْكٌ and حُشُوكٌ, *She* (a camel) *collected her milk.* (Ḳ.)—حَشَكَ النَّاقَةَ, (Ṣ, Ḳ,) aor. ـُ, (Ḳ,) inf. n. حَشْكٌ, (TA,) *He left milking the she-camel until her milk collected* (Ṣ, Ḳ, TA) *in her udder.* (TA.)

حَشَكٌ: see 1.

حَشْكَةٌ † *A rain exceeding such as is termed* بَغْشَة; like غَبْيَةٌ and حَفْشَةٌ. (Ṣ).

جَآءُوا بِحَشَكَتِهِمْ, (Ḳ,) or, as in the Moḥeeṭ, جَآءَ فُلَانٌ بِحَشَكَتِهِمْ, (TA,) means † [*They came, or such a one came,*] *with their company.* (Ḳ, TA.)

حِشَاكٌ, (IDrd, Ṣ, Ṣgh,) thus correctly written, like كِتَابٌ, but in [most of] the copies of the Ḳ like سَحَابٌ, (TA, [in the CḲ like سَحَابَةٌ,]) *A piece of wood which is put in the mouth of a kid,* (Ṣ, Ḳ,) *across,* (Ṣ,) *and tied* (Ṣ, Ḳ) *at the back of his neck,* (Ṣ,) *to prevent him from sucking:* (Ṣ, Ḳ:) also called شِبَامٌ. (IDrd, Ṣ.)

حَشُوكٌ A she-camel *collecting milk in her udder* (Ṣ, Ḳ) *quickly.* (Ṣ.)

حَاشِكٌ [act. part. n. of حَشَكَ]. You say شَاةٌ حَاشِكَةٌ *A sheep, or goat, abounding with milk.* (TA in art. لجب.) [And hence,] نَخْلَةٌ حَاشِكٌ ‡*A palm-tree bearing much fruit.* (Yaạḳoob, Ṣ, Ḳ.)—Also † *Consecutive,* or *uninterrupted.* (Ibn-'Abbád, Ḳ.)

حشم

1. حَشِمَ, aor. ـَ, (Mṣb, Ḳ,) inf. n. حَشَمٌ, (Mṣb,) *He was,* or *became, angry;* (Mṣb, Ḳ;) as also ▼احتشم: (Mṣb:) or the latter signifies *he became angered.* (TA.)—And *He was,* or *became, confounded and stupified by shame;* or *ashamed and confounded* or *stupified, and remained speechless and motionless.* (Mṣb.) See also 8.—حَشَمَهُ, aor. ـِ, inf. n. حَشْمٌ; (Mṣb;) or حَشَمَهُ, aor. ـُ; (Ḳ;) *He angered him;* (Mṣb, Ḳ;) as also ▼احشمهُ, (IAạr, Ṣ, Mṣb, Ḳ,) and ▼حَشَّمَهُ. (Ḳ.) And, accord. to AZ (Ṣ) and El-Fárábee, (Mṣb,) حَشَمَهُ, (Ṣ, Mṣb, Ḳ,) aor. ـِ and ـُ, (Ḳ,) *He annoyed him,* (Ṣ, Mṣb, Ḳ,) *and said to him what he disliked,* (Ḳ,) *and angered him;* (Ṣ, Mṣb;) namely, a man sitting with him. (Ṣ, Mṣb, Ḳ.) An Arab of chaste speech is related to have said, ذٰلِكَ مِمَّا ▼يُحْشِمُ بَنِى فُلَانٍ, meaning *That is of the things that anger the sons of such a one.* (Ṣ.)—Accord. to IAạr, (Ṣ,) حَشَمَهُ signifies *He caused him to become confounded and stupified by shame;* or *to become ashamed and confounded* or *stupified, and to remain speechless and motionless;* (Ṣ, Ḳ;) as also ▼احشمهُ: (Ṣ, Mṣb, Ḳ:) and both signify *it caused him to be affected with shame, shyness,* or *bashfulness;* or *to shrink;* as in the saying, to one shrinking from food, مَا الَّذِى حَشَمَكَ, or ▼أَحْشَمَكَ, *What caused thee to be affected with shame, &c.?* (TA.)—حَشَمَ, aor. ـِ, inf. n. حُشُومٌ, *He became fat,* or *in a good condition of body, after leanness.* (Ḳ.) And حَشَمَتِ الدَّابَّةُ فِى أَوَّلِ الرَّبِيعِ, (Ḳ,) aor. ـِ, inf. n. حَشْمٌ, (TA,) or حُشُومٌ, (TḲ,) *The beast obtained somewhat of the* [*herbage called*] ربيع, *in the beginning thereof, and became fat, and in good condition, and large in the belly,* (Ḳ, TA,) *and goodly:* (TA:) or, as En-Naḍr says, حَشَمَتِ الدَّوَابُّ *the beasts became in good condition.* (Ṣ.)—مَا حَشَمَ مِنْ طَعَامِنَا *He ate not of our food* (Ḳ, TA) aught. (TA.)—مَا حَشَمَ الصَّيْدَ *He hit not, or obtained not, or found not, the game,* or *object of the chase.* (Ḳ.)—حَشِمَ, inf. n. حَشَمٌ (TA) and حُشُومٌ, (Ḳ,) *He was,* or *became, fatigued, tired,* or *wearied.* (Ḳ, TA.) The Arabs say, الحُشُومُ يُورِثُ الحُشُومَ *Labour,* or *toil, occasions fatigue.* (Ýoo, TA.)

2: see 1.

4: see 1, in four places.

5: see 8.—You say also, هُوَ يَتَحَشَّمُ المَحَارِمَ *He guards against things forbidden.* (TA.)

8. احتشم: see 1.—Also *He felt,* or *had a sense of,* or *was moved* or *affected with, shame,* or *shyness,* or *bashfulness.* (Mṣb.) احتشم مِنْهُ (Ṣ, Mgh, Ḳ) and عَنْهُ, (Ḳ,) and احتشمهُ, (Ṣ, Mgh,) or this last is not allowable except when مِنْ is meant to be understood, (TA,) signify the same; (Ṣ, Mgh;) i. e. *He was ashamed of it,* or *abashed at him;* or *was ashamed to do it,* or *shy of doing it:* (Mgh, Ḳ:) or it signifies, (Mgh,) or signifies also, (Ḳ,) *he shrank from it,* or *him:* (Mgh, Ḳ:) or, as some say, thus used it is vulgar; for حِشْمَة, with the Arabs, is only anger: (Mgh:) but IB cites, from Kutheiyir,

إِنِّى مَتَى لَمْ يَكُنْ عَطَاؤُهُمَا
عِنْدِى بِمَا قَدْ فَعَلْتُ أَحْتَشِمُ

as meaning [*Verily I, when the gift of them two*

in my possession is not for what I have done,] am *ashamed*, or *abashed*: and in a trad. of 'Alee, respecting the thief, occurs the saying, إِنِّى لَأَحْتَشِمُ أَنْ لَا أَدَعَ لَهُ يَدًا, meaning *Verily I am ashamed not to leave him a hand*; and *I shrink from it*. (TA.) [And ↓ حَشَمَ app. signifies the same; for,] accord. to Aṣ, (TA,) [the inf. n.] حُشُومٌ signifies The act of *shrinking*. (Ḳ.) You say also, إِنِّى لَأَتَحَشَّمُ ↓ مِنْهُ *Verily I abstain from it*, or *refrain from it, to shun blame*, or *through disdain and pride; disdain*, or *scorn, it*; (أَتَذَمَّمُ مِنْهُ;) *and am ashamed of it*. (Ḳ.) ═ Also *He was*, or *became, master of many* حَشَم [or *dependents* &c.] *and servants*. (KL.)

حَشَمٌ A man's *special dependents, consisting of his family and slaves* [and *others*], or *his neighbours, who are angry on his account* (Ḳ, TA) *when an event befalls him*; (TA;) as also ↓ حُشْمَةٌ; (Yoo, TA;) in the Ḳ, erroneously, حَشَمَةٌ; (TA;) and أَحْشَامٌ; (Ḳ;) which IAạr thinks to be pl. of حَشَمٌ used in a sing. sense; (TA;) [for] this word is applied to *one* [*of such persons*] as well as to a pl. number: (Ḳ:) you say, هٰذَا الغُلَامُ حَشَمٌ لِى [*This young man*, or *slave, is a dependent of mine*]: (IAạr, TA:) or حَشَمٌ signifies, (ISk, Mgh, Mṣb,) or signifies also, (Ḳ,) a man's *relations and household*; (ISk, Mgh, Mṣb, Ḳ;) or his *servants*; (Ṣ, Mṣb;) and *those who are angry on his account* (ISk, Ṣ, Mgh, Mṣb) *when an event befalls him*; (Mgh, Mṣb;) for which reason they are thus called: (Ṣ:) or a man's *followers*; and *those on whose account he should be angry*: (Ḥam p. 614:) or the حَشَم of a man are *those who are angry on his account*; or *those on whose account he is angry*: (Ḥar p. 164:) accord. to ISk, (Mṣb,) it is a word having a pl. signification, and having no proper sing.: (Mgh, Mṣb:) but some say that it has for its pl. أَحْشَامٌ: (Mgh:) and accord. to the Ḳ, ↓ حُشَمَآءُ [in the CḲ حُشْماء] signifies *neighbours* and *guests*; as though it were pl. of حَشِيمٌ, like as كُرَمَآءُ is of كَرِيمٌ: but [perhaps this should be أَحْشَامٌ; for] we find in the M, هٰؤُلَآءِ أَحْشَامِى, meaning *These are my neighbours*, and *my guests*: (TA:) and ↓ حُشُمٌ, with two dammehs, signifies *slaves*; (IAạr, TA;) or, as some say, *followers, whether slaves or free persons*. (TA.) ═ Also *An object of desire* or *quest*; syn. طَلِبَةٌ [in the CḲ طَلَبَةٌ]; and so ↓ حُشُومٌ. (Ḳ.) You say, هُوَ حَشَمُهُ *It is his object of desire* or *quest*. (TḲ.)

حُشُمٌ Persons *having*, or *possessing*, (ذَوُو, as in the explanation of IAạr, for which ذُو is erroneously substituted in the copies of the Ḳ, TA,) *consummate shame, shyness, bashfulness*, or *pudency*. (IAạr, Ḳ, TA.) — See also حَشَمٌ.

حُشْمَةٌ (in the Ḳ, erroneously, حَشَمَةٌ, TA): see حَشَمٌ — Also [in the CḲ, erroneously, حَشْمَةٌ] *A woman*, or *a wife*; syn. مَرْأَةٌ. (Ḳ, TA.) ═ *I. q.* ذِمَامٌ [app. as meaning *protection, safeguard*, or *security of life and property*]. (Yoo, Ḳ.) So in the phrase, لَهُ الحُشْمَةُ [*Protection, &c., is due to him*]. (Yoo, TA.) — *Relationship*. (Ḳ.) So in the phrase, فِيهِمْ حُشْمَةٌ [*Among them is relationship*]. (TA.) ═ See also حِشْمَةٌ.

حِشْمَةٌ *Anger*. (Aṣ, Ṣ, Mgh, Mṣb, TA.) — And *Shame, shyness, bashfulness*, or *pudency*; (Ṣ, Mṣb, Ḳ;) and *a shrinking* (Lth, Mgh, Ḳ, TA) *from one's brother in a place of eating, and in seeking*, or *requesting, a thing that one wants*. (Lth, Mgh, TA.) It has been asserted, (Mgh, Mṣb, TA,) on the authority of Aṣ, (Mṣb, TA,) that it signifies only "anger:" (Mgh, Mṣb, TA:) but several authors have refuted this assertion, by showing that it occurs in trads. as meaning "shame." (MF, TA.) — Also The *act of annoying a person sitting with one, and saying to him what he dislikes*; and so ↓ حُشْمَةٌ. (Ḳ.)

حُشَمَآءُ, in the CḲ حُشْماء: see حَشَمٌ.

حُشُومٌ: see حَشَمٌ, last sentence but one. ═ It is also an inf. n. of 1. (Ḳ.)

حَشِيمٌ, (Ṣ, Ḳ,) in some of the copies of the Ṣ حِشِّيمٌ, which is app. a mistake, (TA,) [thus I find it in one of my copies of the Ṣ,] *i. q.* ↓ مُحْتَشَمٌ; (Ṣ, Ḳ;) i. e. *Regarded with reverence, veneration, respect, honour, awe*, or *fear*; (TA;) applied to a man. (Ṣ.)

حَاشِمٌ A man *being*, or *becoming, fat*, or *in a good condition of body, after leanness*. (TA.)

مَحْشُومٌ *Angered*. (TA.) [But it is implied in the Ṣ that it signifies *Confounded and stupified by shame*; or *ashamed and confounded* or *stupified, and remaining speechless and motionless*.] A poet says,

لَعَمْرُكَ إِنَّ قُرْصَ أَبِى خُبَيْبٍ

بَطِىْءُ النُّضْجِ مَحْشُومُ الأَكِيلِ

[*By thy life, verily the round cake of bread of Aboo-Khubeyb is slow in becoming thoroughly baked: the eater is angered*, or *confounded and stupified by shame, &c.*]. (Ṣ, TA.)

مُحْتَشَمٌ: see حَشِيمٌ.

إِنَّهُ لَمُحْتَشِمٌ بِأَمْرِى *Verily he is grieved and disquieted* (مُغْتَمٌّ) *by my affair*, or *case*. (AA, TA.)

حشو

1. حَشَا, (Ṣ, Mgh, Mṣb, TA,) aor. يَحْشُو, (Mṣb, TA,) inf. n. حَشْوٌ, (Ṣ, Mgh, Mṣb, Ḳ,) *He filled*, (Ḳ, TA,) or *stuffed*, (KL, PṢ,) a pillow, or cushion, [and a garment, (see حَشْوٌ, below,)] &c., (Ṣ, Mgh,* Mṣb, Ḳ,) with a thing, (Ḳ,) with cotton, (Mṣb, TA,) and the like. (TA.) [And *He stuffed* a lamb, or a fowl, and a vegetable, &c., with rice &c.] — Hence, حَشَا الغَيْظَ, aor. and inf. n. as above, ‡ [*He stuffed wrath* into a man's bosom: see an ex. in a verse cited in the first paragraph of art. حفل:] and حُشِىَ الرَّجُلُ غَيْظًا وَكِبْرًا ‡ [*The man was stuffed with wrath and pride*], and حُشِىَ الرَّجُلُ بِالنَّفْسِ and حُشِىَ النَّفْسَ † [*The man was stuffed with pride*, or *self-magnification*, or *with disdain*, or *scorn*]. (TA.) — [Hence also,] صِغَارُ الإِبِلِ تَحْشُو الكِبَارَ † *The young camels enter*, or *occupy the spaces, among the old ones*. (TA.) — [رَسَمَ كِتَابًا وَلَمْ يَحْشُهُ, a phrase occurring in the 1st نَوْع of the Mz, means † *He sketched out a book, and did not fill it up*.] — حَشَاهُ [also signifies *He foisted it* into a thing. — And] *He hit*, or *hurt, his* حَشًا [q. v., like أَحْشَاهُ]. (Ḳ.) You say, حَشَاهُ سَهْمًا, inf. n. as above, *He hit*, or *hurt, his* حَشًا [*with an arrow*]. (TA.)

3. مَا أَجْلَهُ وَلَا حَاشَاهُ *He gave him not a* جَلِيلَة [i. e. *a she-camel that had brought forth once*] *nor* حَاشِيَة [i. e. *small*, or *young, camels*]: (Ḳ:) or ↓ مَا أَجَلَّنِى وَلَا أَحْشَانِى *He gave me not a she-camel that had brought forth once nor gave he me a young*, or *small, camel*. (Ṣ in art. جل.)

4: see 3.

5: see 8. — تَحَشَّى فِى بَنِى فُلَانٍ † *He became received among the sons of such a one, and harboured, protected*, or *lodged, by them*. (TA in art. حشى [but belonging to the present art.].)

7: see 8. — انحشى صَوْتٌ فِى صَوْتٍ [app. † *A sound became blended in a sound*], and حَرْفٌ فِى حَرْفٍ [*a letter in a letter*]: mentioned by Az. (TA in art. حشى [but app. belonging to the present art.].)

8. احتشى *It* (a thing) *became filled* [or *stuffed*; as also ↓ انحشى]. (Ḳ.) And in like manner you say of a man, احتشى مِنَ الطَّعَامِ *He became filled* [or *stuffed*] *with food*. (TA.) And احْتَشَتِ الرُّمَّانَةُ بِالحَبِّ *The pomegranate became filled with the grains*, or *seeds*. (TA.) — احْتَشَتْ *She* (a مُسْتَحَاضَة) *stuffed her vulva* (نَفْسَهَا) *with the* [*rags termed*] مَقَارِم [in the CḲ, erroneously, مَقَارِمِ], (Ḳ, TA,) *and the like*: and in a similar sense احتشى is used as said of a man having the [disorder termed] إِبْرِدَة. (TA.) And احْتَشَتْ بِالكُرْسُفِ (Ṣ, Mgh, TA) and الكُرْسُفَ (Mgh, TA) *She* (a حَائِض, Ṣ, Mgh) *stuffed her vulva with cotton*, (Mgh, TA,) to arrest the blood. (Ṣ.) — احْتَشَتْ حَشِيَّةً and بِحَشِيَّةٍ *She* (a woman) *wore a* حَشِيَّة; (IAạr, Ḳ;) as also ↓ تَحَشَّتْ [alone]. (Az, TA in art. حشى.) A poet says,

لَا تَحْتَشِى إِلَّا الصَّمِيمَ الصَّادِقَا

[*She will not wear any stuffing but that which is genuine and true*]: meaning that she will not wear حَشَايَا because the largeness of her posteriors renders it needless for her to do so. (IAạr, TA.)

حِشَةٌ, pl. حِشُونَ : see وَحْش.

حَشًا The *contents of the belly*: (Ḳ:) or *a bowel*, or *an intestine, into which the food passes from the stomach*; syn. مِعًى: (Mṣb:) pl. أَحْشَآءٌ: (Mṣb, Ḳ:) and ↓ حُشْوَةٌ and ↓ حِشْوَةٌ signify the *bowels*, or *intestines*; [like أَحْشَآءٌ;] syn. أَمْعَآءٌ: (Mṣb:) or these are called ↓ حُشْوَةُ البَطْنِ and ↓ حِشْوَتُهُ: (Ṣ, TA:) or حشوة signifies *all that is in the belly except the fat*; so accord. to Az and Esh-Sháfi'ee: or, accord. to Aṣ, the *place of the food, comprising the* أَحْشَآء *and the* أَقْصَاب: (TA:) [see also مَحْشًى:] الحَشَا is the name of *all the places of the food*: (Zj in his "Khalḳ el-Insán:") [see also, for other meanings, its dial. var. حَشًى, in art. حشى:] the word belongs to this art. and to art. حشى; the dual being حَشَوَانِ and حَشَيَانِ. (TA.) ═ *A side,* (Mṣb, TA,) *region, quarter,* or *tract.* (Mṣb.) You say, أَنَا فِى حَشَا فُلَانٍ *I am in the quarter and protection of such a one:* pl. as above. (Ḥar p. 61.) [See, again, حَشًى, in art. حشى.]

حَشْوٌ, like the inf. n., (TA,) *Stuffing;* (PṢ;) [i. e.] *what is put into a pillow, or cushion, &c.:* (Ḳ, TA:) and [hence] *cotton:* and the *seeds used for seasoning food,* [and the *rice &c.,*] *with which the belly of a lamb is stuffed:* pl. ▼مَحَاشٍ, deviating from rule. (TA.) — ‡ The *soul* of a man. (Ḳ, TA.) — † [*A parenthesis;*] *a redundant part,* or *portion, of speech,* or *of a sentence,* (Ḳ, TA,) *upon which nothing is syntactically dependent.* (TA. [See Ḥar pp. 85 and 86.]) — †[*A digression.*] — † The *portion* of either hemistich of a verse *that is comprised between the first and last foot.* (KT, &c.) — † *Small,* or *young, camels,* (Ṣ, Ḳ,) *among which are no great,* or *old, ones;* (Ṣ, TA;) as also ▼حَاشِيَةٌ: (Ṣ, Ḳ:) so called because they enter, or occupy the spaces, among the latter; or because they go against the sides of the latter: (TA:) accord. to ISk, (Ṣ,) ▼ابْنُ الحَاشِيَتَانِ signifies [*the camel termed*] ابْنُ المَخَاضِ *and* [*that termed*] ابْنُ اللَّبُونِ: (Ṣ, and Ḳ in art. حشى:) the pl. [of حَاشِيَةٌ] is ▼حَوَاشٍ. (TA.) It is said in a trad. respecting the poor-rate, خُذْ مِنْ حَوَاشِى ▼أَمْوَالِهِمْ, i. e., accord. to IAth, † *Take thou of the small,* or *young, of their camels; such as those termed* ابن المخاض *and* ابن اللبون. (TA. [But see another explanation of this saying voce حَاشِيَةٌ in art. حشى.]) — And حَشْوٌ and ▼حَاشِيَةٌ signify also † The *like* of mankind; (Ṣ;) [i. e.] حَاشِيَةٌ signifies ‡ the *lower* or *lowest, baser* or *basest, meaner* or *meanest, sort,* or the *rabble,* or *refuse,* of mankind, or of the people; (TA in art. حشى, and Ḥar p. 61;) as also حَشْوٌ [which is of frequent occurrence in this sense]; (KL;) and ▼حِشْوَةٌ; (Ṣ, TA;) *such as servants and the like.* (Ḥar ubi suprà, in explanation of حاشية. [See also this word in art. حشى.]) You say, ▼جَآءَ فُلَانٌ مَعَ حَاشِيَتِهِ † *Such a one came with those who were in his quarter and protection:* but this may be from حَشًا signifying "a region, quarter, or tract;" servants and followers being in the quarter and protection of their master. (Ḥar ubi suprà.) And فُلَانٌ مِنْ حِشْوَةِ ▼بَنِى فُلَانٍ † *Such a one is of the lower* or *lowest,* &c., *of the sons of such a one.* (Ṣ.) — See also حُشْوَةٌ. ═ Also *A stuffed garment.* (Mgh.)

أَرْضٌ حَشَاةٌ ‡ *Black land, in which is no good.* (Ḳ, TA.)

حُشْوَةٌ and حِشْوَةٌ: for each, see حَشًا, in two places: — and for the latter, see also حَشْوٌ, in two places. — You say also, مَا أَكْثَرَ حُشْوَةَ أَرْضِهِ and حِشْوَةَ ارضه, i. e. ▼حَشْوُهَا and دَغَلُهَا [app. meaning ‡ *How many are the thickets,* or *the like,* that obstruct the tracts *of his land!*]. (Lḥ, Ḳ, TA.)

حَشِىٌّ Herbage *that has become dry in its lower part, and rotten:* (IAạr, Ḳ:) or *dry:* (Aṣ, Ṣ, Ḳ:) like خَشِىٌّ [q. v.]. (Ṣ, TA.)

حَشِيَّةٌ *A stuffed bed:* (Ḳ:) pl. حَشَايَا. (TA.) ['Antarah says that a saddle was to him what the حَشِيَّة, or stuffed bed, is to others: see EM p. 229.] — Also, (Ḳ,) and ▼مِحْشًى, (Ṣ, Ḳ,) *A pillow,* (Ḳ,) or *the like,* (Ṣ,) *with which a woman makes her posteriors* (Ṣ, Ḳ) or *her body* (Ḳ) *to appear large:* (Ṣ, Ḳ:) pl. of the former as above; (TA;) and of the latter مَحَاشٍ. (Ṣ, TA. [In the Ṣ, it is only said of the former that it is the sing. of حَشَايَا.]) — [Also the former, The *pad* of a رَحْل (or camel's saddle): see مُرَبَّطَةٌ.]

حَاشِيَةٌ, and its dual and pl.: see حَشْوٌ, in six places. — See also art. حشى.

مَحْشًى The *place of the food in the belly.* (Ḳ.) [See also حَشًا, and مَحْشَاةٌ.]

مِحْشًى : see حَشِيَّةٌ.

مَحْشَاةٌ [The *rectum;*] the *lowest of the places of the food,* (Aṣ, TA,) [i. e.] the *portion of the intestines which is the lowest of the places of the food,* (IAth, TA,) *leading* [*immediately*] *to the place of egress;* (Aṣ, TA;) in a beast, *i. q.* مَبْعَرٌ: (TA: [explained in the Ḳ in art. حش, to which it does not belong:]) pl. مَحَاشٍ. (IAth, TA.) Hence, إِيَّاكُمْ وَإِتْيَانَ النِّسَآءِ فِى مَحَاشِيهِنَّ فَإِنَّ كُلَّ مَحْشَاةٍ حَرَامٌ. (TA.)

مِحْشَاةٌ *A coarse* [*garment of the kind called*] كِسَآء, (Aṣ, Ṣ, TA,) *that abrades the skin:* (TA:) pl. مَحَاشٍ. (Aṣ, Ṣ.) [But accord. to some, a garment of this kind is called مِحْشًا or مِحْشَآء.]

[مَحْشُوٌّ and مَحْشِىٌّ *Filled,* or *stuffed.*]

مَحَاشٍ pl. of مِحْشًى, (Ṣ, TA,) and of مَحْشَاةٌ, (IAth, TA,) and of مِحْشَاةٌ, (Aṣ, Ṣ,) and irreg. pl. of حَشْوٌ, q. v. (TA.)

حشى

1. حَشِىَ, [aor. يَحْشَى,] inf. n. حَشًى, *He* (a man walking quickly, and one speaking with sharpness, or hastiness, of temper, TA) *breathed short,* or *unintermittedly; panted for breath;* or *was out of breath.* (Ṣ, Ḳ, TA.) ═ See also 3.

2. حَشَّى, (TA,) inf. n. تَحْشِيَةٌ, (KL, TA,) *He made a* حَاشِيَة *to a garment,* or *piece of cloth.* (KL.) — And [hence,] † *He wrote a note* or *comment,* or *notes* or *comments, upon the margin,* or *margins, of a book:* (KL, TA:) in this sense it is vulgar [or post-classical]. (TA.) ═ [Also, app., *He,* or *it, caused* one *to breathe short* or *unintermittedly, to pant for breath,* or *to be out of breath:* see مُحَشِّيَةُ الكِلَابِ, below.]

3. حَاشَاهُ مِنْهُمْ, (IAmb, Ḳ, &c.,) inf. n. مُحَاشَاةٌ, (KL,) *He set him aside as excluded from the description of them;* [*excluded him from them;*] *did not include him among them:* (IAmb, TA:) from حَشًى meaning the "side" of a thing: (Az, TA:) *he excepted him from them;* as also ▼تَحَشَّاهُ. (Lḥ, Ḳ, TA.) You say, شَتَمْتُهُمْ وَمَا حَاشَيْتُ مِنْهُمْ أَحَدًا and ▼مَا تَحَشَّيْتُ, i. e. [*I reviled them and*] *I did not say* [*of any one of them*] حَاشَى لِفُلَانٍ [*Far is such a one* from being included among those of whom I speak! or from any cause of reproach!]; or *I did not except, of them, any one.* (Lḥ, TA. [The same saying is repeated in the TA, in the supplement to this art. of the Ḳ, but with the substitution of حشيت for حاشيت; perhaps by a mistake of the copyist: or it may be inferred from a statement of Fr and IAmb, which will be found below, that one also says ما ▼حَشَيْتُ.]) And En-Nábighah [Edh-Dhubyánee] says, (Mbr, Ṣ,)

* وَلَا أَرَى فَاعِلًا فِى النَّاسِ يُشْبِهُهُ *

* وَمَا أُحَاشِى مِنَ الأَقْوَامِ مِنْ أَحَدِ *

[*And I shall not see an actor among mankind resembling him; and I do not except, of the companies of men, any one*]: which shows حَاشَى to be a verb perfectly inflected. (Mbr, Ṣ, Mughnee.) And hence the trad. of the Prophet, قَالَ أُسَامَةُ أَحَبُّ النَّاسِ إِلَىَّ مَا حَاشَى فَاطِمَةَ, i. e. [*He said, Usámeh is the most beloved of men to me:*] *he did not except Fátimeh:* ما being here a negative; not, as Ibn-Málik imagines, supposing this clause to be of the words of the Prophet, the ما which, with a verb following it, conveys the meaning of an inf. n. (Mughnee.) — حَاشَى also denotes exception as a particle, or as a verb (Ṣ, Mughnee) having but one tense and no inf. n., (Mughnee,) according as it governs a gen. or an accus. case. (Ṣ.) You say, ضَرَبْتُهُمْ حَاشَى ▼زَيْدٍ [*I beat them, except Zeyd*], using it as a particle; and ضربتهم حَاشَى زَيْدًا, using it as a verb. (Ṣ, I 'Aḳ* p. 169.) And sometimes one says, قَامَ القَوْمُ مَا حَاشَى زَيْدًا [*The party stood, except Zeyd*]; (Mughnee, I 'Aḳ p. 169;) like مَا خَلَا; though Ibn-Málik [like Sb] disallows it: (I 'Aḳ:) this being shown to be allowable by the saying of the poet,

* رَأَيْتُ النَّاسَ مَا حَاشَى قُرَيْشًا *

* فَإِنَّا نَحْنُ أَفْضَلُهُمْ فَعَالَا *

[*I have seen mankind, except Kureysh, that we are the most excellent of them in conduct*]. (Mughnee, I 'Aḳ p. 170.) The agent of حاشى [used as a verb having but one tense and no inf. n.] is a pronoun implied, relating to the inf. n. of the preceding verb, or to its act. part. n., or to a portion of what is meant by the [preceding] general noun: so that when one says, قَامَ القَوْمُ حَاشَى زَيْدًا, the meaning is, [*The party stood,* but] *their standing,* or *the stander of them,* or *a portion of them, was apart from Zeyd.* (Mughnee.) [In the Ḳ, its use as a verb of this kind is not mentioned; but it is there said to govern the gen. case, like حَتَّى.] — It also denotes remoteness from imperfection or the like, or freedom therefrom; as in the phrase ▼حَاشَى لِلّٰهِ [*I ascribe unto God remoteness from every imperfection* or *the like,* or *freedom therefrom;* generally implying wonder or admiration, so that it may be rendered, *how* *far, or how free, is God from every imperfection!*]: (Mughnee:) or this phrase means مَعَاذَ ٱللّٰهِ [*I seek protection by God*; or, as often used by late writers, and in the present day, *God forbid!*]; as also ▼حَاشَ لِلّٰهِ; (Ṣ, Ḳ;) the former being the original expression: (Ṣ:) it occurs, read in both these ways, in the Ḳur xii. 31 and 51; where it implies wonder at the power of God as manifested in the creation of such a person as Joseph. (Ksh, Bḍ.) Mbr and IJ and the Koofees hold that it is a verb; and that, in the [latter] verse the meaning is, Joseph *hath kept aloof from disobedience for the sake of God*; but such an interpretation as this is not suitable in the case of [the former verse,] حَاشَ لِلّٰهِ مَا هٰذَا بَشَرًا: the truth is, that it is a noun, syn. with التَّنْزِيه or البَرَآءَة, [accord. to different copies of the Mughnee, meaning تَنْزِيهًا or بَرَآءَةً,] as is shown by another reading, ▼حَاشًا لِلّٰهِ, with tenween, like بَرَآءَةً لِلّٰهِ; and by the reading of Ibn-Mes'ood, ▼حَاشَ ٱللّٰهِ, like مَعَاذَ ٱللّٰهِ: some assert that it is a verbal noun, meaning أَتَبَرَّأُ [*I assert myself to be free*, or *clear*, to God], or تَبَرَّأْتُ [*I have asserted myself* &c.]; but its being decl. in some dials. contradicts this. (Mughnee.) One says also, ▼حَاشَاكَ and ▼حَاشَى لَكَ [*Far art thou from being included among those of whom I speak! or from any cause of reproach! or the like*]; both meaning the same. (Ṣ, Ḳ.) And some of the Arabs say ▼حَشَى, (Fr, IAmb, TA,) dropping the ا. (IAmb, TA.) — Sb says, (Ṣ, Mughnee,) with most of the Baṣrees, (Mughnee,) that it is only a particle governing the gen. case, (Ṣ, Mughnee,) syn. with إِلَّا; (Mughnee;) because, if it were a verb, it would be allowable to make it a connective to مَا, like خَلَا; and this he asserts to be not allowable. (Ṣ.) Mbr says that it is sometimes a verb; as is shown by the verse of En-Nábighah cited above; and by the saying حَاشَى لِزَيْدٍ, because a particle cannot be made to precede immediately another particle; and because it suffers elision, as in the phrase حَاشَ لِزَيْدٍ, seeing that elision takes place in nouns and verbs only, exclusively of particles: (Ṣ, Mughnee*:) and IJ and the Koofees say the like: but Mbr also holds, in common with El-Jarmee and El-Mázinee and Zj and Akh and AZ and Fr and Aboo-'Amr Esh-Sheybánee, that it is often a particle governing the gen. case, and seldom a trans. verb having but one tense and no inf. n., because syn. with إِلَّا. (Mughnee.) = See also 5, in two places.

5. تَحَشَّى [*He went aside, apart*, or *aloof*, or *he removed, withdrew*, or *retired to a distance*,] is from الحَاشِيَةُ; like تَنَحَّى from النَّاحِيَةُ: (TA:) and [▼تحاشى has a similar meaning:] you say, تَحَاشَيْتُ مِنْ كَذَا *I kept aloof from, shunned*, or *removed myself far from, such a thing*; from الحَشَا [or الحَشَى] meaning النَّاحِيَةُ. (Ḥar p. 194.) — And تحشّى مِنْهُ; (IAạr, Ḳ;) and ▼حاشى مِنْهُ, (Ḥar p. 294,) inf. n. مُحَاشَاةٌ; (KL;) *He abstained*, or *refrained*, (IAạr, Ḳ, KL, Ḥar,) *from him*, (IAạr, Ḳ, Ḥar,) or *it*, (IAạr, KL,) *through disdain and pride*; or *he disdained*, or *scorned*, or *was ashamed of, him*, or *it*. (IAạr, Ḳ, Ḥar.) — And لَا يَتَحَشَّى, (El-Báhilee, TA,) and ▼لَا يُحَاشِى, (Ḥar p. 294,) *He will not care, mind, heed*, or *regard*. (El-Báhilee, Ḥar, TA.) = See also 3, in two places, first and second sentences.

6: see 5.

حَشَى, for حَاشَى: see 3.

حَشًى, [in some copies of the Ṣ written with a final ا, (see حَشًا in art. حشو,)] *The contents of the ribs*; or *what the ribs enclose*: (Ṣ:) or *the contents of the belly, below the* حِجَاب [or *diaphragm*], *consisting of the liver and the spleen and the* كَرِش [or *stomach*, properly that of a ruminant, but also applied to that of a man,] *and what succeeds to this*: or the *portion between the shortest rib, which is in the extremity of the side, and the hip*, or *haunch*: or the *exterior of the belly*: (Ḳ: [in the CḲ, او ظاهِرِ البَطْنِ is erroneously put for او ظاهِرُ البَطْنِ:]) and, accord. to the copies of the Ḳ, the حِضْن: [in the CḲ, والحِضْنِ:] but correctly, the خَصْر [or *waist*]: (TA:) the dual is حَشَيَانِ, (Az, TA,) and the dual of حَشًا is حَشَوَانِ: (TA in art. حشو:) and the pl. [of both these sings.] is أَحْشَآءٌ [generally meaning the *bowels*, or *intestines*]. (Ṣ, TA.) [It is often used as meaning The *belly*: and the *waist*.] You say رَجُلٌ خَمِيصُ الحَشَى *A man lank in the belly*. (Ṣ and Ḳ in art. خمص.) And هُوَ لَطِيفُ الحَشَى *He is slender in the waist*: and of a woman, هِىَ ضَامِرَةُ الحَشَى: and of women, هُنَّ ضَوَامِرُ الأَحْشَآءِ. (TA.) — *A side, quarter, region*, or *tract*: (Ṣ, Ḳ:) *a vicinage*, or *neighbourhood*: *shelter*, or *protection*. (Z, Ḳ,* TA.) You say, أَنَا فِى حَشَاهُ *I am in his quarter, vicinage*, or *protection*: (Z, Ḳ:) [see also حَشًا, in art. حشو:] and ▼هٰؤُلَآءِ حَاشِيَتُهُ *These are in his quarter*, or *vicinage, and shadow*, or *protection*. (TA.) = Also inf. n. of حَشِىَ [q. v.]. (Ṣ,* Ḳ.)

حَشٍ A man *having a complaint of his* حَشًى [q. v.]. (Az, TA.) — Also, and ▼حَشْيَانُ, *Breathing short*, or *unintermittedly*; *panting for breath*; or *out of breath*; (Ṣ, Ḳ, TA;) applied to a man; (Ṣ;) in walking quickly, and in speaking with sharpness, or hastiness, of temper: (TA:) fem. [of the former] حَشِيَةٌ and [of the latter] حَشْيَا, (Ḳ,) of the measure فَعْلَى. (TA.)

حَشْيَانُ: see what next precedes.

حَاشَ ٱللّٰهِ and حَاشًا لِلّٰهِ and حَاشَ لِلّٰهِ: see 3.

حَاشَى as a particle, and as a noun; as in حَاشَى زَيْدٍ and لِلّٰهِ, and حَاشَاكَ and حَاشَى لَكَ: see 3, in four places.

حَاشِيَةٌ [The *selvage*, or *selvedge*, i. e.] the *side*, (Mgh, Mṣb, TA,) or *each of the two sides that have no unwoven extremity*, (M, TA,) or *each of the two long sides that have at their two ends the unwoven threads*, (T, TA,) of a garment, or piece of cloth: (Ṣ, T, M, Mgh, Mṣb, Ḳ:) and in like manner, of other things; (Ḳ, TA;) as, for instance, †the *side* of a tract of herbage, and of the mirage; and †the *extremity* of the مَقَام [app. meaning مَقَام إِبْرٰهِيم in the Temple of Mekkeh]: (TA:) pl. حَوَاشٍ. (Ṣ, Mgh, Mṣb.) — †The *margin* of a book or writing. (TA.) — ‡*A writing*, or *writings, upon the margin* [or *margins*] *of a book*: (TA:) [i. e. *a marginal note*, or *marginal notes*; *an annotation*, or *a comment, scholium*, or *gloss*: and hence, *a series of annotations, comments, scholia*, or *glosses*; *a commentary on particular words and passages of a book*; distinguished from a شَرْح, which is an exposition, explanation, or interpretation, in the form of a running commentary, comprising the entire text of the work which it expounds.] — †*An indefinite portion* of property, or of cattle: (Mṣb:) [or *a portion from the side*; not selected: for, accord. to some,] the saying, خُذْ مِنْ حَوَاشِى أَمْوَالِهِمْ means †*Take thou from the sides of their collections of cattle*, [for the poor-rate,] without choosing [the best]: this meaning being from حَاشِيَةٌ as used in relation to a garment, or piece of cloth, meaning the "side." (Mgh. [But see another explanation of this saying voce حَشْوٌ.]) — حَاشِيَةُ النَّسَبِ †*The collateral relation* or *relations*; *such as the paternal uncle, and his son*: as though from حاشية as meaning the "side" of a garment, or piece of cloth. (Mṣb.) — حَاشِيَةُ رَجُلٍ ‡*A man's family and dependents*, (Ḳ, TA,) *who are under his protection*. (TA.) — See also حَاشِيَةٌ and its dual and pl. voce حَشْوٌ, in six places. — And see حَشًى (in the present art.), last sentence but one. — رَجُلٌ رَقِيقُ الحَوَاشِى ‡*A man gentle, gracious*, or *courteous, to his associates*. (TA.) — عَيْشٌ رَقِيقُ الحَوَاشِى ‡*A plentiful, easy, life*. (Ṣ, TA.) — جَمْعٌ كَثِيفُ الحَوَاشِى ‡*A congregated body crowding close together, side against side*. (Ḥar p. 294.)

مُحْشِيَةُ الكِلَابِ [*She that causes the dogs to pant for breath*;] applied to *the hare*; meaning that the dogs run after her until they are out of breath. (ISk, Ṣ. [In one copy of the Ṣ, erroneously, مُحَشِّيَة.])

حص

1. حَصَّهُ, aor. ـُ, (TA,) inf. n. حَصٌّ, (A, Ḳ,) *He*, or *it, shaved it off*; namely, hair. (A, Ḳ, TA.) You say also, حَصَّتِ البَيْضَةُ رَأْسَهُ (Ṣ, A) [*The helmet rubbed off his hair*: or] *rendered his hair scanty*. (Ṣ.) — *He cut off from it*, either with the مِشَارَة, [a word for which I do not find any apposite meaning, and which is perhaps a mistranscription,] or with the shears: (Er-Rághib, as quoted in the TA:) whence, accord. to some, the word حِصَّة. (TA.) — حَصُّوا بَيْنَهُمْ رَحِمًا ‡*They cut*, or *severed, a tie of relationship between them*. (TA.) — جَآءَتْ سَنَةٌ فَحَصَّتْ كُلَّ شَىْءٍ ‡[*There came a dearth*, or *drought*, or *a year of drought, and*] *it did away with*, or *consumed*, or *destroyed, everything*. (TA, from a trad.) — حَصَّ الجَلِيدُ النَّبْتَ †*The hoar-frost*, or *rime, nipped, shrunk, shrivelled*, or *blasted*, (lit. *burned*, أَحْرَقَ, q. v.,) *the plant*, or *plants*, or *herbage*: (AḤn:) a dial. var. of حَسَّ, q. v. (TA.) = حَصَّ, quasi-pass. of حَصَّهُ in the first of the senses

explained above: see 7, in two places. ═ حَصَّنِى مِنَ المَالِ كَذَا, (A, Mgh,* Mṣb, Ḳ,*) aor. ـُ, (A, Mgh, Mṣb,) *Such a thing became my portion of the property:* (A,* Ḳ:) or *came to me, and became my portion:* (Mgh:) or *came to me as my portion.* (Mṣb.) ═ حَصَّ, aor. ـِ, [contr. to rule, by which it should be ـَ, the verb being intrans., unless the sec. pers. pret. be حَصِصْتَ,] inf. n. حَصٌّ, (Ṣ, TA,) with which حُصَاصٌ, q. v., is syn., (Ṣ, &c.,) *He ran vehemently and quickly:* (Ṣ, TA:) and ▼ حَصْحَصَ, (TA,) inf. n. حَصْحَصَةٌ, (Ṣ, Ḳ,) *he was quick* (Ṣ, Ḳ, TA) *in going,* (TA,) *and in journeying* or *pace.* (Ṣ, TA.) ═ حَصَّ is also *syn. with* ▼ حَصْحَصَ in all its meanings; like كَبَّ and كَبْكَبَ, and كَفَّ and كَفْكَفَ. (Er-Rághib.)

2: see R. Q. 1, in two places.

3. حَاصَصْتُهُ الشَّىْءَ *I divided with him the thing, each of us allotting to himself his portion.* (TA.) See also 6.

4. أَحْصَصْتُهُ *I gave him his portion, lot,* or *share:* (Ṣ, Mṣb, Ḳ:) or *his portion, lot,* or *share, of food and beverage &c.* (TA.) And احصصت القَوْمَ *I gave the company of men their portions, lots,* or *shares.* (A.)

6. تحاصّوا *They* (namely, creditors,) *divided property among themselves in portions, lots,* or *shares;* (Ṣ,* Mgh, Mṣb, Ḳ;*) as also ▼ حاصّوا, (Ḳ,) inf. n. مُحَاصَّةٌ (Ṣ, TA) and حِصَاصٌ; (TA;) *every one of them taking his portion.* (TA.)

7. انحصّ quasi-pass. of حَصَّهُ in the first of the senses explained above; *It* (hair) *became shaven off;* as also ▼ حَصَّ, [sec. pers., app., حَصِصْتَ, and aor. يَحَصُّ,] inf. n. حَصِيصٌ [and app. حَصَصٌ, q. v. infrà]; or, as some say, حص [perhaps a mistake for حَصَصٌ or حَصِيصٌ] signifies the hair's *going from the head by shaving* or *by disease:* (TA:) and the former verb, *it* (the hair) *went* from the head; (Ḳ, TA;) *became removed,* or *stripped off;* (TA;) *fell off, and became scattered, by degrees;* (Ṣ, TA;) as also ▼ حَصَّ: (TA:) the former is also said of the plumage of a bird; (A;) and of the foliage of trees; in the last of the senses mentioned above: (TA:) and [in like manner] ▼ تَحَصْحَصَ *it* (fur, and the nap of cloth,) *became removed,* or *stripped off.* (IAạr.) You say also, انحصّت اللِّحْيَةُ *The beard became short, its hair breaking off in pieces.* (TA.) And انحصّ الذَّنَبُ *The tail became cut off.* (Ḳ.) It is said in a prov., أَفْلَتَ وَانْحَصَّ الذَّنَبُ [*He escaped, but the tail became cut off*]: applied to him who has been at the point of destruction, and then escaped: (Ḳ:) or alluding to the coward's escape from destruction after being at the point thereof: related to have been said by Mo'áwiyeh, on the occasion of the safe return of an ambassador whom he had sent to the King of the Greeks, appointing for him a threefold bloodwit [if he should be slain] on the condition of his proclaiming the call to prayer on entering his court; which he did; whereupon the King's generals, who were with him, sprang forward to slay him; but he forbade them, and sent him back furnished with requisites for his journey. (A 'Obeyd.) [See also Freytag's Arab. Prov., ii. 201.] You also say, انحصّ رَأْسُهُ [*His head shed, or lost, its hair: or part thereof*]: (A:) and ▼ تَحَصْحَصَ الحِمَارُ, and البَعِيرُ, *the ass's,* and *the camel's, hair fell off.* (TA.)

R. Q. 1. حَصْحَصَ, inf. n. حَصْحَصَةٌ: see 1, last two sentences. — The inf. n. also signifies The *walking of him who is shackled.* (Ḳ, TA.) ═ *He strove,* or *laboured; exerted himself; took pains,* or *extraordinary pains;* or *exceeded the usual bounds;* in his affair. (Abu-l-'Abbás, TA.) ═ *He* (a camel) *fixed,* or *made firm* or *steady,* his knees, *in order to rise* (Ṣ, Ḳ*) *with the load;* and his stifle-joints: (Ṣ:) or *lay down upon his breast, with folded legs.* (TA.) ═ Hence, as some say, آلْآنَ حَصْحَصَ الحَقُّ, in the Ḳur [xii. 51], meaning, *Now the truth hath become established:* or, as others say, it is from حِصَّةٌ, and means, *now hath the portion of truth become distinct from that of falsehood:* (TA:) or *now hath the truth become distinct, apparent,* or *manifest,* (Ṣ, Mṣb, Er-Rághib, TA,) *after concealment,* (TA,) or *by the coming to light of that which was concealed in the mind.* (Er-Rághib, TA.) You say, حَصْحَصَ الشَّىْءُ *The thing became distinct, apparent,* or *manifest,* (Kh, Ṣ, Ḳ,) *after having been concealed;* (Kh;) as also ▼ حَصَّصَ, inf. n. تَحْصِيصٌ: (Ḳ:) and some read ▼ حُصِّصَ الحَقُّ in the Ḳur ubi suprà. (TA.) One should not say حُصْحِصَ in this sense; (TA;) nor تَحَصْحَصَ. (Ez-Zejjájee.)

R. Q. 2. تَحَصْحَصَ: see 7, in two places.

حِصَّةٌ *A portion of a sum:* (Er-Rághib:) and used to signify *a portion, lot,* or *share,* (Ṣ, Mṣb, Er-Rághib, Ḳ,) of food, and of beverage, and of land, &c.: (TA:) accord. to some, from حَصَّهُ signifying "he cut off from it:" (TA:) pl. حِصَصٌ. (A, Mṣb, Ḳ.)

حَصَصٌ *Paucity,* or *scantiness, of the hair of the head;* (Ṣ, Ḳ;) and *of the fetlock of a horse:* (TA:) also *shortness of the beard, when its hair breaks off in pieces:* (TA:) and the *state of one suffering from a protracted disease, whose hair does not grow long.* (TA.)

حُصَاصٌ *Mange,* or *scab:* (Ibn-'Abbád, Ḳ:) because the hair falls off in consequence of it. (TA.) ═ *Vehemence of running,* (Aṣ, Ṣ, Mgh, Ḳ,) of an ass, (Mgh,) and *quickness thereof:* (Aṣ, Ṣ:) [see 1, last sentence but one:] or, accord. to 'Áṣim Ibn-Abi-n-Nujood, (Ṣ,) an ass's *straightening and erecting the ears, and moving about,* or *wagging, the tail, and running:* (Ṣ, Ḳ:) accord. to some, (Ṣ,) *an emission of wind from the anus, with a sound;* (Ṣ, Mgh, Ḳ;) as A'Obeyd says, in relation to a trad. in which that which it signifies is attributed to the devil as the effect of his hearing the call to prayer; but he adds that the saying of 'Áṣim is more pleasing to him; and it is also the saying of Aṣ, or like it. (Ṣ.)

حَصِيصٌ i. q. ▼ مَحْصُوصٌ [*Shaven off*]; applied to hair: (Ḳ:) or it is a subst. applied to that hair [*which is shaven off*]. (TA.) [See also حَصِيصَةٌ:] — and see أَحَصُّ.

حَصِيصَةٌ *What is collected from shaving or plucking out.* (TA.) [See also حَصِيصٌ.] — Also The *hair and fur of the ear, whether shaven off or not:* or, as some say, *hair and fur in general:* but the former explanation is more known. (TA.) — And *What is above the* أَشْعَر [or *part next the hoof* (in the CḲ erroneously written شَعْر)] *of the horse;* (Ibn-'Abbád, Ḳ;) i. e., *of the hair that surrounds the hoof:* so called because of the paucity of that hair. (Ibn-'Abbád.)

قَرَبٌ حَصْحَاصٌ *A laborious,* (Ḳ,) *quick night-journey to water, in which is no flagging;* (Aṣ, Ṣ, Ḳ;) like حَثْحَاثٌ: (Ṣ:) or *such as is far-extending,* or *long:* and سَيْرٌ حَصْحَاصٌ *a quick journey,* or *pace;* like حَثْحَاثٌ. (TA.)

حَاصَّةٌ *A disease in consequence of which the hair gradually falls off and becomes scattered:* (Ṣ, A, Ḳ:) or *a disease that takes away the hair:* (IAth:) or *that takes away the hair entirely.* (A 'Obeyd.) ═ بَيْنَهُمْ رَحِمٌ حَاصَّةٌ i. q. ▼ مَحْصُوصَةٌ; (Ḳ;) ‡ *Between them is a tie of relationship* which they have *severed,* or *cut;* not treating one another with the affection due to it; (TA;) [so that it is an act. part. n. in the sense of a pass. part. n.;] as also ▼ رَحِمٌ حَصَّاءُ: (A, TA:) or the meaning is, ذَاتُ حَصٍّ [*having a severing;* so that it is a possessive epithet]. (Ḳ.)

أَحَصُّ [*Having the hair shaven off,* or *rubbed off,* or *fallen off,* either *wholly* or *partly*], applied to the head; pl. حُصٌّ: (A:) a man *having little hair upon the head:* (Ṣ, Ḳ:) or a man *having no hair;* (Mgh;) a man *whose hair has all gone;* fem. حَصَّآءُ, applied to a woman: (Et-Tirmidhee:) also, [a man] *having no hair upon his breast:* and a man *suffering from a protracted disease, whose hair does not grow long:* (TA:) and a horse *having little hair in the fetlock,* and *in the tail;* which is a fault; (TA;) as also ▼ حَصِيصٌ; (Ḳ,* TA;) on the authority of IDrd: (TA:) and the fem., a she-camel *having no fur upon her:* and the masc., a tail *having no hair upon it:* and ▼ مَحْصُوصٌ applied to the back of the neck, *of which the hair has been shaven off* (TA.) You say also رَجُلٌ أَحَصُّ اللِّحْيَةِ *A man whose beard has become short, its hair having broken off in pieces:* and لِحْيَةٌ حَصَّآءُ *a beard that has become short in like manner.* (TA.) And طَائِرٌ أَحَصُّ الجَنَاحِ (Ṣ, A, Ḳ) *A bird having little plumage in the wing:* (Ḳ:) or *whose plumage of the wing has gradually fallen off and become scattered:* (Ṣ,* TA:) pl. as above. (Ṣ.) — ‡ A sword *having in it, or upon it, no* أُثْر [or *diversified wavy marks, streaks,* or *grain*]. (Ḳ, TA.) — Applied to a man, (A,) ‡ *Unlucky;* (AZ, A, Ḳ;) *unpropitious; in whom is no good:* (AZ, A:) and the fem., applied to a woman, also signifies ‡ *unlucky;* (Ḳ, TA;) *in whom is no good.* (TA.) And hence, (A,) or because they keep pace together in their prices (يَتَمَاشَيَانِ أَثْمَانُهُمَا) until they grow old and weak, when their prices become diminished and they die, (Ṣ,) الأَحَصَّانِ signifies ‡ *The slave and the ass.* (Ṣ, A, Ḳ.) —— ‡ [A man] *who cuts, or severs, the tie of relationship.* (TA.) —— رَجُلٌ حَصَّآءُ: see حَاصَّةٌ. —— سَنَةٌ حَصَّآءُ (Ṣ, A, Ḳ) ‡ *A sterile year, in which is no good:* (Ṣ, Ḳ:) or *a year of drought, in which is little herbage:* or *a year in which is no herbage.* (TA.) —— يَوْمٌ أَحَصُّ ‡ *A day intensely cold.* (TA.) It was said to a man of the Arabs, "Which of the days is the most cold?" and he answered, الأَحَصُّ الأَزَبُّ; (TA;) the former meaning, ‡ *The day whose sun rises* (Ḳ, TA) *the horizon being red,* (TA,) *and its sky* (سَمَآؤُهُ), accord. to the copies of the Ḳ, but correctly *its north wind* (شَمَالُهُ), (TA,) *being clear,* (Ḳ, TA,) *and such that a touch is not felt by reason of the cold;* and it is *that in which there are no clouds, and of which the cold does not abate:* and the latter meaning, *the day in which blows the wind called* النَّكْبَآء, *driving along clouds in which is no water, wherein no sun rises, and in which is no rain.* (TA.) Z says, (TA,) it was said to one of them, "Which of the days is the coldest?" and he answered, الأَحَصُّ الوَرْدُ وَالأَزَبُّ الهَلُوفُ, i. e., *The clear,* [*in which the horizon is red,*] *and the cloudy, in which blows the wind called* النَّكْبَآء. (A, TA.) —— رِيحٌ حَصَّآءُ ‡ *A wind that is clear, without dust.* (Ḳ, TA.)

مَحْصُوصٌ: see حَصِيصٌ; and أَحَصُّ; and حَاصَّةٌ.

حصب

1. حَصَبَهُ, (Ṣ, A, Mṣb, Ḳ,) aor. ـِ (Ṣ, A, Mṣb) and ـُ, (Mṣb,) inf. n. حَصْبٌ, (Mṣb, TA,) *He threw at him,* or *pelted him with, pebbles,* (Ṣ, A,* Ḳ,) or *small pebbles.* (Mṣb.) And hence, in a general sense, *He pelted him.* (Ḥar p. 234.) And حَصَبَتِ الرِّيحُ بِالحَصْبَآءِ [*The wind cast,* or *drove along,* or *tore up, the pebbles,* or *small pebbles*]. (A.) —— Also, (A,) or ▼حَصَّبهُ, inf. n. تَحْصِيبٌ, (Ṣ,) or both, (Mṣb, Ḳ,) but the latter has an intensive signification, (Mṣb,) *He spread pebbles in it,* (A, Ḳ,) namely, a mosque, (A,) or a place; (Ḳ;) *he strewed it,* namely, a mosque, (Ṣ, Mṣb,) &c., (Mṣb,) *with pebbles,* (Ṣ,) or *with small pebbles.* (Mṣb.) —— حَصَبَ بِهِ النَّارَ *He threw it* (anything) *into the fire.* (AO, Ṣ.) —— حَصَبَ النَّارَ *He threw* حَصَب [or *firewood, &c.,*] *into the fire.* (A.) [Also,] inf. n. as above, *He kindled the fire,* or *made it to blaze* or *flame,* with حَصَب. (TA.) ═══ حَصَبُوا عَنْهُ ‡ *They hastened from him, or it, in flight.* (A, TA.) —— حَصَبَ عَنْ صَاحِبِهِ † *He turned away from his companion;* as also ▼احصب. (Ḳ.) —— حَصَبَ فِى الأَرْضِ † *i. q.* ذَهَبَ فِيهَا [which has two meanings: *he went away in,* or *into, the country,* or *land:* and *he discharged his excrement:* the former seems to be here meant]. (Ṣ.) ═══ حَصِبَ; (Ḳ; [in a copy of the A حُصِبَ, but this is probably a mistranscription, as appears to be indicated by its being there added that the part. n. is مَحْصُوبٌ;]) and حَصَبَ, aor. ـِ, (Ḳ,) inf. n. حَصْبٌ, (KL,) or حَصَبٌ; (TḲ, and indicated in the Ḳ;) [and app. ▼حُصِّبَ also; (see مَحْصُوبٌ;)] *He broke forth with* حَصْبَة [i. e. *measles,* or *spotted fever*]. (Ḳ, KL.) The second of these verbs signifies as above, said of a person's skin. (Ṣ.)

2. حَصَّبَهُ: see 1. ═══ Also حَصَّبَ, (T, TA,) inf. n. تَحْصِيبٌ, (T, Mgh, Ḳ,) *He* (a pilgrim) *slept* [or *stopped to sleep*] *in El-Moḥaṣṣab* (▼المُحَصَّب), (T, Mgh,* Ḳ,) which is the name of the way between the mountains opening upon the part called الأَبْطَح, (T, Ḳ,) between Mekkeh and Minè, (T, Mṣb,) so called from the pebbles in it, (T, TA,) and also called ▼الحَصْبَآء, (Mṣb,) *for an hour,* or *a short time,* (سَاعَةً,) *of the night,* (T, Mgh, Ḳ,) *in returning from Minè to Mekkeh:* (T, Mgh,* TA:) this was formerly done in imitation of Moḥammad; but it is said to be voluntary; not obligatory. (T, TA.) Also *He slept at that place after going forth from Mekkeh.* (TA.) ▼المُحَصَّبُ is also the name of the place where the pebbles are cast in Minè; (Aṣ, Ṣ, A, Mgh, Mṣb, Ḳ;) also called ▼حِصَابٌ. (TA.) ═══ حُصِّبَ: see 1.

4. احصب, (Ṣ, A, Ḳ,) inf. n. إِحْصَابٌ, (TA,) *He* (a horse, Ṣ, A, or other beast &c., TA) *struck up the pebbles* in his running. (Ṣ, A, Ḳ.) —— See also 1.

6. تحاصبوا *They pelted one another with pebbles.* (A, Ḳ.)

حَصَبٌ *Stones;* as also ▼حَصْبَةٌ, n. un. ▼حَصَبَةٌ, which is extr. [as n. un. of حَصْبَة, but not of حَصَب]. (Ḳ.) —— *A stone that is thrown;* like نَفَضٌ in the sense of مَنْفُوضٌ. (TA.) —— *Firewood,* (Ḳ,) in a general sense; (TA;) in the dial. of El-Yemen: (Fr, TA:) or *what is thrown into a fire,* (A 'Obeyd, Ṣ, Ḳ,) *of firewood* and *of other things;* (TA;) in the dial. of Nejd: (Fr, TA:) or *firewood prepared for fuel:* (Mṣb:) or *firewood with which a fire is lighted;* firewood not being so called until it is thus used. (Ḳ.) حَصَبُ جَهَنَّمَ, in the Ḳur [xxi. 98], signifies, in the Abyssinian language, accord. to 'Ikrimeh, *The firewood* [or *fuel*] *of Hell.* (TA.)

حَصِبٌ [*Pebbly*]. You say أَرْضٌ حَصِبَةٌ and ▼مَحْصَبَةٌ (T, Ṣ, A, Ḳ) *A land containing,* (T, Ṣ,) or *abounding with,* (A, Ḳ,) *pebbles.* (T, Ṣ, A, Ḳ.) And ▼مَكَانٌ حَاصِبٌ *A place containing pebbles.* (TA.) —— See also حَاصِبٌ.

حَصْبَةٌ [*A single throwing of pebbles*]. —— [Hence, app., because immediately following the day of the last throwing of pebbles in the Valley of Minè,] لَيْلَةُ الحَصْبَةِ *The night* [*next*] *after the days called* أَيَّامُ التَّشْرِيقِ [which are *the 11th and 12th and 13th of Dhu-l-Ḥijjeh*]. (Ḳ.) —— See also حَصَبٌ. ═══ Also, (Ṣ, A, Mṣb, Ḳ,) and ▼حَصَبَةٌ, (Ṣ, Mṣb, Ḳ,) and (sometimes, Ṣ) ▼حَصِبَةٌ, (Ṣ, Ḳ,) [*Measles,* or *spotted fever;*] *a certain cutaneous eruption:* (Ṣ, A, Mṣb, Ḳ:) by some, [contr. to general authority,] said to be *small-pox.* (Mṣb.)

حَصَبَةٌ: see حَصَبٌ, and حَصْبَآءُ: ═══ and see also حَصْبَةٌ.

حَصِبَةٌ: see حَصْبَةٌ.

حَصْبَآءُ *Pebbles:* (Ṣ, A, Ḳ:) or *small pebbles:* (Mṣb:) accord. to Sb, a quasi-pl. n.: (TA:) sing. ▼حَصَبَةٌ. (Ḳ.) —— See also 2.

حِصَابٌ: see 2.

حَاصِبٌ [*A thrower,* or *pelter, of stones*]. You say, هُوَ حَاصِبٌ لَيْسَ بِصَاحِبٍ [*He is a pelter of stones* (app. meaning *a calumniator*): *he is not a friend*]. (A, TA.) [Hence also,] حَاصِبٌ, (Ṣ, Ḳ,) or رِيحٌ حَاصِبٌ, (A,) *A violent wind that raises the pebbles;* (Ṣ, A;*) as also ▼حَصِبَةٌ: (Ṣ:) or *a wind that bears along the dust* (Ḳ, TA) *and pebbles:* (TA:) and *a wind casting down pebbles from the sky:* or *a wind that tears up the pebbles.* (TA. [See the Ḳur liv. 34, &c.]) —— And hence, † *A punishment from God.* (TA.) —— Dust *containing pebbles.* (IAạr, TA.) See also حَصِبٌ. —— *Clouds* (سَحَابٌ) *casting down snow and hail:* (Ḳ:) or *clouds* (سحاب), because of their casting down snow and hail. (TA.) —— *Pebbles* [*borne*] *in the wind.* (ISh, TA.) You say, كَانَ يَوْمُنَا ذَا حَاصِبٍ [*Our day was one in which pebbles were blown about by the wind*]. (TA.) —— *Small particles of snow and hail scattered about.* (Ḳ.) —— *A large number of men on foot.* (Az, TA.)

مَحْصَبَةٌ: see حَصِبٌ.

مُحَصَّبٌ: see مَحْصُوبٌ: ═══ and see also 2, in two places.

مَحْصُوبٌ *Affected with the cutaneous eruption termed* حَصْبَة [i. e. *measles,* or *spotted fever*]; (A, Ḳ;) as also ▼مُحَصَّبٌ. (TA.)

حصد

1. حَصَدَ, (Ṣ, A, Mgh, &c.,) aor. ـُ and ـِ, inf. n. حَصْدٌ (Ṣ, Mgh, Mṣb, Ḳ) and حَصَادٌ (Lḥ, Mgh, Ḳ) and حِصَادٌ, (Lḥ, Ḳ,) *He reaped,* or *cut* (A, Mgh, Ḳ) *with the* مِنْجَل, (Ḳ,) seed-produce, (Ṣ, A, Mgh, Mṣb, Ḳ,) and plants, or herbage; (Ṣ,* Ḳ;) originally used in relation to seed-produce; (TA;) as also ▼احتصد. (Ḳ.) —— [Hence,] حَصَدَهُمْ بِالسَّيْفِ ‡ [*He reaped,* or *mowed, them down with the sword*]: (A:) *he slew them:* or *he exerted his utmost power* or *ability,* or *exceeded the ordinary bounds, in slaying them:* (TA:) *he exterminated them* [*with the sword*]. (Mṣb, TA.) —— And مَنْ زَرَعَ الشَّرَّ حَصَدَ النَّدَامَةَ ‡ [*He who sows evil reaps repentance*]. (A.) ═══ حَصِدَ, (L,) [aor. ـَ,] inf. n. حَصَدٌ, (L, Ḳ,) *It* (a rope, and a bow-string,) *was,* or *became, strongly twisted,* and *firmly,* or *compactly, made;* (L, Ḳ;*) as also ▼استحصد: (Ṣ,* A, L, Ḳ:*) and *it* (a coat of mail) *was,* or *became,* [*close in its rings,* (see حَصِدٌ,) *and*] *firm, and compactly made.* (L, Ḳ.*)

4. احصد *It* (seed-produce) *attained to the proper time for its being reaped;* as also ▼استحصد: (Ṣ, Mgh, Mṣb, Ḳ:) or the latter, † *it invited the act of reaping.* (TA.) ═══ *He twisted* a rope (Ṣ, Ḳ) *firmly.* (TA.) —— [*He made firm,* in a general sense. (Golius as from the KL; but not in my copy of that work.)]

8: see 1.

10: see 4: ═ and see also 1. — [Hence,] † *It* (the affair, or state, of a people) *became established*, or *settled, firmly, soundly, thoroughly*, or *well*. (TA.) — *It* (a people, or party) *collected together*, or *assembled, and rendered mutual aid.* (Ṣ, Ḳ.) — *He was*, or *became, angry*: (Ḳ:) or *violently angry*. (TA.)

حَصَدٌ: see حَصِيدٌ.

أَحْصَدُ and ▼ مُحْصَدٌ (Ṣ, Ḳ) and ▼ حَبْلٌ حَصِدٌ and ▼ مُسْتَحْصِدٌ (Ḳ) *A rope strongly twisted, and firmly, or compactly, made*: (Ṣ, Ḳ:) and وَتَرٌ أَحْصَدُ *a bow-string strongly twisted*: (TA:) and ▼ دِرْعٌ حَصْدَآءُ *a coat of mail close in its rings, compact and strong.* (L, Ḳ.)

حَصَادٌ The *time*, or *season, of reaping*; as also ▼ حِصَادٌ. (Ḳ, TA. [In the CḲ, each is erroneously made to be with ة.]) One says, [also, using each as an inf. n.,] هٰذَا زَمَنُ الحَصَادِ, (Ṣ, A,) or أَوَانُ الحَصَادِ, (Mṣb,) and ▼ الحِصَادِ, (Ṣ, Mṣb,) [*This is the time*, or *season, of reaping*: for] both are also inf. ns. of حَصَدَ in the first of the senses explained above. (Lḥ, Ḳ.) ═ See also حَصِيدٌ. — Also *What remains upon the ground*, of seed-produce, *among the lower parts of the stalks of that which has been reaped*; and so حَصَائِدُ, pl. of ▼ حَصِيدٌ and ▼ حَصِيدَةٌ. (Mgh.) — And *What falls off, and becomes scattered*, of seeds of wild leguminous plants when they dry up. (L.) — And The *fruit*, or *produce*, of any tree. (L.)

حِصَادٌ: see حَصَادٌ, in two places.

حَصِيدٌ *Reaped* seed-produce; (Ṣ, A, Mgh, Mṣb, Ḳ;) as also ▼ حَصِيدَةٌ (Ṣ, Mgh, Ḳ) and ▼ مَحْصُودٌ and ▼ حَصَدٌ (Ṣ, Mṣb, Ḳ) and ▼ حَصَادٌ, which last is originally an inf. n.: (Mgh:) pl. of the first, (A, Mgh,) and of the second, (Mgh,) حَصَائِدُ. (A, Mgh.) And حَبُّ الحَصِيدِ [see Ḳur l. 9] *Grain that is reaped*: (L:) *grain of wheat* and *of barley* and *of anything that is reaped*; as though for حَبُّ النَّبْتِ الحَصِيدِ: (Zj:) or *grain of reaped wheat*. (Lth.) — See also حَصَادٌ. — It is also said to signify Seed-produce *torn up and carried away by the wind.* (L.) — Also † *Slain* [or *mowed down*] *with the sword, like seed-produce reaped.* (Jel in xxi. 15.) — حَصَائِدُ أَلْسِنَتِهِمْ, occurring in a trad., means ‡ *The words that their tongues utter, and* [*as it were*] *cut off, against others*; (Ṣ, A,* L;) being words wherein is no good: the tongue being likened to a reaping-hook; and the words that it utters, to reaped seed-produce: حصائد, here, is pl. of ▼ حَصِيدَةٌ. (L.) — See also what next follows.

حَصِيدَةٌ: see حَصِيدٌ, in two places: — and حَصَادٌ. — Also The *lower parts of seed-produce, which the reaping-hook does not reach.* (Ḳ.) — Also *A place of seed-produce*: (Ḳ:) or ▼ حَصِيدٌ has this signification; such a place being so called because it is reaped: حَصِيدَةٌ, accord. to Az, signifies *a field of which all the produce has been reaped*: the pl. is حَصَائِدُ. (L.) — *A place of reaping.* (Mṣb.)

حَاصِدٌ *A reaper*: pl. حَصَدَةٌ and حُصَّادٌ. (Ḳ.)

أَحْصَدُ; fem. حَصْدَآءُ: see حَصِدٌ, in three places. — Also شَجَرَةٌ حَصْدَآءُ *A tree abounding with leaves* [and therefore *compact*]. (Ḳ.)

مُحْصَدٌ: see حَصِدٌ. — [Hence,] مُحْصَدُ الرَّأْيِ ‡ A man *whose judgment*, or *opinion, is well*, or *rightly, directed*, (Ṣ, Ḳ,) *and sound*, or *firm*. (TA.) ═ *What has dried up while standing* [of seed-produce]. (Ḳ.)

مُحْصِدٌ Seed-produce *that has attained to the proper time*, or *season, for its being reaped*; as also ▼ مُسْتَحْصِدٌ. (Mgh, Mṣb.)

مِحْصَدٌ *A reaping-hook*, syn. مِنْجَلٌ, (Ṣ, Ḳ,) *with which seed-produce is cut.* (TA.)

مَحْصُودٌ: see حَصِيدٌ.

مُسْتَحْصِدٌ: see مُحْصِدٌ: ═ and see حَصِدٌ. — [Hence,] رَأْىٌ مُسْتَحْصِدٌ ‡ *Sound*, or *firm, judgment* or *opinion*. (TA.)

حصر

1. حَصَرَهُ, (Ṣ, A,) aor. ـُ (Ṣ, Ḳ) and ـِ, (Ḳ,) inf. n. حَصْرٌ, (Ṣ, Ḳ,) *He*, or *it, straitened him*; (Ṣ, A, Ḳ;) so in the Ḳur ix. 5; (TA;) *and encompassed*, or *surrounded, him*. (Ṣ, A.) You say حَصَرُوهُ, (Ṣ, Mṣb,) or حَصَرُوا بِهِ, (Ḳ,) aor. ـُ, (Ṣ, Mṣb,) inf. n. حَصْرٌ, (Mṣb,) *It* (a hostile party, ISk, Ṣ, Mṣb, or a people, Ḳ) *encompassed him*, or *surrounded him*, (Mṣb, Ḳ,) *and prevented him from going to his business*: (Mṣb:) or *straitened him, and encompassed* or *surrounded him*; as also ▼ حَاصَرُوهُ, inf. n. مُحَاصَرَةٌ and حِصَارٌ. (ISk, Ṣ.) The ▼ مُحَاصَرَة of an enemy is well known. (Ḳ.) You say ▼ حَاصَرَهُمُ العَدُوُّ, inf. ns. as above, [*The enemy besieged*, or *beset, them*;] and بَقِينَا فِى الحِصَارِ أَيَّامًا *We remained in the state of siege some days*; or *in the place of confinement*; and حُوصِرُوا مُحَاصَرَةً شَدِيدَةً [*They were besieged*, or *beset, vehemently*]. (A.) — Also حَصَرَهُ, (Ṣ, A, Ḳ, &c.,) aor. ـُ (Mgh, Ḳ) and ـِ, (Ḳ,) inf. n. حَصْرٌ, (A, Mgh, Ḳ,) *He*, (Akh, Ṣ, A,) *or it*, (Ṣ,) *confined, kept close, imprisoned, detained, retained, restrained, withheld*, or *prevented, him*; (AO, Aboo-'Amr Esh-Sheybánee, Akh, Ṣ, A;) as also ▼ أَحْصَرَهُ: (Aboo-'Amr Esh-Sheybánee, Ṣ:) or a distinction should be made between these two forms, as will be seen in what follows. (TA.) And *It* (a hostile party, and a disease, ISk, Th, Mṣb) *detained, restrained, withheld*, or *prevented, him* (ISk, Th, Mṣb, Ḳ) *from journeying* &c.; (Ḳ;) as also ▼ احصره: (AO,* ISk, Th, Mṣb, Ḳ:) or the latter signifies *it* (disease) *prevented him from journeying*, or *from a thing that he desired*: so in the Ḳur ii. 192: (ISk, Ṣ:) or [more properly] *it* (disease, or urine, [&c.,]) *made him to restrain himself*: (Akh, Ṣ, Ḳ:) or إِحْصَارٌ signifies the *being prevented from attending the religious rites and ceremonies of the pilgrimage, by disease*, or *the like*: (IAth:) or أُحْصِرَ is said when a man is turned back from a course which he desired: and حُصِرَ, when he is confined, or restrained, or the like: (Yoo:) or, accord. to Fr, the Arabs say, of him whom fear or disease prevents from accomplishing his pilgrimage or his عُمْرَة [q. v.], (Mgh,* TA,) and of any one that is not forcibly constrained, as by imprisonment, or by enchantment or the like, (TA,) ▼ أُحْصِرَ: and of him who is imprisoned or restrained by a Sultán, or by one who overpowers, حُصِرَ: this distinction is observed by them: (Mgh,* TA:) but if you mean that the constraining power of the Sultán is a preventing cause, and you do not refer to the act of the agent, it is allowable for you to say, قَدْ أُحْصِرَ الرَّجُلُ: and if you say of him whom pain or disease makes to restrain himself, that the disease, or fear, restrains him, it is allowable for you to say, حُصِرَ: or, as Aboo-Is-ḥáḳ the Grammarian says, the correct rule, accord. to the lexicologists, is, that one says of him whom fear and disease prevent, ▼ أُحْصِرَ: and of him who is confined or restrained by another, حُصِرَ: and thus it is because he who refrains from conducting himself freely in an affair restrains himself: and thy saying حَصَرْتُهُ means that thou hast restrained him; not that he has restrained himself: so that it is allowable to say in this case [when you do not mention the agent], ▼ أُحْصِرَ. (TA.) [Accord. to Z,] حُصِرَ عَنْهُ and دُونَهُ [lit. *He was withheld from it*] is said when a man is ashamed at a thing, and leaves it, or abstains from it, or when he is unable to effect a thing, or finds his wish difficult of attainment. (A. [See also حَصِرَ, in what follows, in this paragraph.]) حَصَرْتُ الغُرَمَآءَ فِى المَالِ means حَصَرْتُ قِسْمَةَ المَالِ فِى الغُرَمَآءِ [*I restricted the division of the property among the creditors*]: for the prevention is not against them, but against others, from their being sharers with them in the property: the phrase is inverted, like أَدْخَلْتُ القَبْرَ المَيِّتَ. (Mṣb.) — Also حَصَرَهُ, (Ḳ,) aor. ـُ, inf. n. حَصْرٌ, (TA,) *He took the whole of it*; (Ḳ;) [*appropriated it to himself exclusively*;] *acquired it; took it to himself.* (TA.) — And حُصِرَ, (Ṣ, A, Mgh, Ḳ,) and ▼ أُحْصِرَ, (Ṣ, A, Ḳ,) or حُصِرَ عَلَيْهِ خَلَاؤُهُ, بِغَائِطِهِ, and ▼ أُحْصِرَ, (Ks,) or حَصِرَ, aor. يَحْصَرُ, inf. n. حَصْرٌ [and حَصَرٌ, or this latter is a simple subst.], (Ibn-Buzurj,) *He* (a man, Ṣ, A) *suffered suppression of the fæces*, or *constipation of the bowels*: (Ks, Ibn-Buzurj, Ṣ, A, Mgh, Ḳ:) [distinguished from أُسِرَ: (see حُصْرٌ:) or] حُصِرَ عَلَيْهِ بَوْلُهُ signifies *he suffered suppression of his urine.*. (Ibn-Buzurj.) ═ حَصِرَتْ, [intrans.,] with fet-ḥ [to the ص], and ▼ أَحْصَرَتْ, *She* (a camel) *had a narrow orifice to the teat.* (Ṣ.) And حَصُرَ, aor. ـُ; and حَصِرَ, aor. ـَ; and ▼ أَحْصَرَ, (Ḳ,) or أُحْصِرَ; (so in the TA;) *It* (the orifice of her teat) *was*, or *became, narrow.* (Ḳ,* TA.) — And حَصِرَ, aor. ـَ, inf. n. حَصَرٌ, *He was*, or *became, unable to express his mind, to say what he would, to find words to express what he would say*; *he faltered in speech*; (Ṣ, Mgh, Ḳ, Expos.

of the "Mufaṣṣal" of Z;) *by reason of shame and confusion of mind, or other* [*accidental*] *cause;* wherein, only, it differs from عَيِىَ. (Expos. of the "Mufaṣṣal" of Z.) And also, (Mṣb, Ḳ,) or حَصِرَ فِى القِرَآءَةِ, (Ṣ,) *He faltered,* or *became impeded, and was unable to proceed, in reading,* or *recitation.* (Ṣ, Mṣb, Ḳ.) And حَصِرَ aor. ـَ, *He was ashamed, and cut short, as though the affair straitened him like as the prison straitens the prisoner.* (TA.) And حَصِرَ عَنْهُ *He became impeded, and was unable to do it.* (Ṣ.) And حَصِرَ عَنِ المَرْأَةِ, aor. ـَ, [inf. n. حَصْرٌ,] *He abstained from sexual intercourse with the woman,* (Ḳ, TA,) *though able to enjoy it:* (TA:) or حَصِرَ عَنْ أَهْلِهِ, (Ṣ,) or عَنِ النِّسَآءِ, (Az,) *he was prevented by impotence from having sexual intercourse* (Az, Ṣ) *with his wife,* (Ṣ,) or *with women.* (Az. [See حَصُورٌ.]) — Also حَصِرَ, (Mgh, TA,) or حَصِرَ صَدْرُهُ, (Ṣ, Mṣb, TA,) aor. ـَ, (Mṣb,) inf. n. حَصَرٌ, (Ṣ, Mṣb, Ḳ,) *He became straitened in his bosom; his bosom became straitened.* (Ṣ, Mgh, Mṣb, Ḳ,* TA.) In the Ḳur [iv. 92], أَوْ جَآؤُوكُمْ حَصِرَتْ صُدُورُهُمْ أَنْ يُقَاتِلُوكُمْ means عَنْ قِتَالِكُمْ [*Or who come to you, their bosoms being contracted so that they are incapable of fighting you;* or *their bosoms shrinking from fighting you*]: (TA:) Akh and the Koofees allow that the pret. here may be a denotative of state; but Sb does not allow this use of the pret. unless with قَدْ; and he makes حصرت صدورهم to be an imprecation [meaning *may their bosoms become contracted*]: (Ṣ:) accord. to Fr, the Arabs say, أَتَانِى فُلَانٌ ذَهَبَ عَقْلُهُ, meaning قد ذهب عقله: Zj says, Fr makes حصرت a denotative of state; but it is not so unless with قد: Th says that if قد be understood, it approximates to a denotative of state, and becomes like a noun; and some read حَصِرَةً صُدُورُهُمْ: AZ does not allow this use of the pret. [as a denotative of state] unless preceded by و or قد. (TA.) — And حَصِرَ, alone, *He vomited.* (Mgh.) — And *He became affected with a disease,* or *malady,* by a thing. (TA.) — Also, (Ṣ, Ḳ,) aor. ـَ, inf. n. حَصَرٌ, (Ḳ,) *He was,* or *became, niggardly, tenacious, penurious, or avaricious.* (Ṣ, Ḳ.) One says, شَرِبَ القَوْمُ فَحَصِرَ عَلَيْهِمْ فُلَانٌ *The party drank, and such a one was niggardly to them,* (AA, Ṣ, L,) *not expending upon those who drank with him.* (L.) — [Hence,] حَصِرَ بِالسِّرِّ *He concealed the secret;* (Ḳ;) *refrained from divulging it.* (TA.) ═ حَصَرَ البَعِيرَ, aor. ـُ and ـِ, (TA,) inf. n. حَصْرٌ, (Ḳ,) *He bound a* حِصَار, (Ḳ, TA,) or *a* مِحْصَرَة, (TA,) *upon the camel;* (Ḳ, TA;) as also ↓ احتصره: (Ṣ, Ḳ, TA:) and *he made for,* or *put to, the camel a* حِصَار: as also ↓ احصره. (TA.)

3: see 1, in three places.

4: see 1, in eleven places.

7. انحصر *He,* or *it, was,* or *became, restrained, withheld,* or *prevented.* (KL.)

8: see 1, last sentence.

حُصْرٌ (Ṣ, Mgh, Ḳ, &c.) and ↓ حُصُرٌ (A, and Expositions of the Fṣ) *Suppression of the fæces;* or *constipation of the bowels:* (Yz, Aṣ, Ṣ, A, Mgh, Ḳ:) suppression of the urine is termed أُسْرٌ: (Yz, Aṣ, Mgh:) or حُصْرٌ signifies also *suppression of the urine,* like أُسْرٌ. (Ibn-Buzurj.)

حَصَرٌ [inf. n. of حَصِرَ, q. v., passim. — Also] *Suppression of the flow of milk of a camel, from a heaviness,* or *heaving, of the stomach, or a tendency to vomit;* and *unwillingness to yield a flow of milk.* (TA.)

حَصِرٌ A man *unable to express his mind; to say what he would; to find words to express what he would say;* (Mgh, TA;) *by reason of shame and confusion of mind, or other* [*accidental*] *cause:* (TA: [see حَصِرَ:]) and one who is *impeded, and unable to proceed, in reading,* or *recitation:* (Mṣb, TA:) and so ↓ حَصِيرٌ and ↓ مَحْصُورٌ, in both these senses. (TA.) — *Contracted in the bosom; having the bosom contracted;* (Mgh, TA;) as also ↓ حَصِيرٌ and ↓ حَصُورٌ. (Ḳ.) In the Ḳur iv. 92, some read حَصِرَةً صُدُورُهُمْ [*Their bosoms being contracted*]. (TA. [See 1, latter part.]) — *Affected with vomiting.* (Mgh.) — *Niggardly, tenacious, penurious,* or *avaricious;* (Ḳ;) as also ↓ حَصِيرٌ and ↓ حَصُورٌ: (Ṣ, Ḳ:) and ↓ حَصِيرٌ one *who will not drink wine, by reason of niggardliness:* (Ḳ:) and ↓ حَصُورٌ one *who will not expend upon those who drink with him:* (L:) and one *who* [*by reason of niggardliness*] *does not take part in the game called* المَيْسِر. (Suh.) — Also, (Ṣ,) or حَصِرٌ بِالأَسْرَارِ, (A,) and ↓ حَصُورٌ [alone], (Ḳ,) *A strict concealer of secrets:* (Ṣ:) or [simply] *a concealer of secrets.* (A, Ḳ.) — حَصِرَةُ الشَّخْبِ A she-camel *whose flow of milk is suppressed.* (TA.)

حُصُرٌ: see حُصْرٌ.

حَصْرَآءُ *Impervia coëunti* mulier; syn. رَتْقَآءُ. (A, Ḳ.)

حُصْرِىٌّ [and حُصُرِىٌّ, which latter is now the more common,] *A maker,* or *seller, of* حُصْر [or حُصُر, i. e. *mats,* pl. of حَصِير]. (Ibn-Khillikán, p. 19 of vol. i. of De Slane's ed.)

حَصَارٌ: see the next paragraph.

حِصَارٌ: see حَصِيرٌ. — [*A fortress; a fort; a castle.*] ═ Also, (Ṣ, Ḳ,) and ↓ حَصَارٌ, (Ḳ,) *A kind of pillow, cushion, or pad, which is put upon a camel, and of which the hinder part is raised so that it is made like the* آخِرَة *of a camel's saddle, the fore part being stuffed so that it is made like the* قَادِمَة [or rather واسط or وَاسِطَة] *of a camel's saddle, and which is ridden upon;* and so ↓ مِحْصَرَةٌ: (Ḳ:) or *a kind of saddle upon which those who break, or train, beasts ride:* or *a* [*piece of stuff of the kind called*] كِسَآء, *which is thrown upon the back of the camel, behind the rider:* (TA:) or ↓ مِحْصَرَةٌ (Ḳ) and حِصَارٌ (TA) signify *a small* [*saddle of the kind called*] قَتَب, (Ḳ, TA,) *which is bound upon a camel, and upon which is thrown the apparatus of the rider.* (TA.)

حَصُورٌ One *who has no sexual intercourse with women,* (Ṣ, Mgh, Ḳ,) *though able to have it,* (Ḳ,) *abstaining from them from a motive of chastity, and for the sake of shunning worldly pleasures:* (TA:) or *who is prevented from having it,* (Ḳ, TA,) *by impotence:* (TA:) or *who does not desire them,* (IAạr, A, Mṣb, Ḳ,) *nor approach them:* (IAạr, Ḳ:) applied also to a horse, i. q. عِنِّين. (IAạr, TA in art. عجز.) In the Ḳur [iii. 34], applied to John the Baptist. (TA.) — *Castrated;* (Ḳ;) *having the penis and testicles amputated.* (TA.) — *Very fearful* or *cautious; who abstains,* or *refrains, from a thing through fear.* (Ḳ.) — See also حَصِرٌ, in four places. — Also A she-camel *having a narrow orifice to the teat.* (Ṣ, Ḳ.)

حَصِيرٌ: see مَحْصُورٌ, in two places: — and see حَصِرٌ, in four places. — Also *A king:* (Ṣ, A, Ḳ:) because he is secluded: (Ṣ, A:) or because he prevents those who have access to him. (TA.) ═ *A prison;* (Ṣ, A, Mgh, Mṣb, Ḳ;) as also ↓ حِصَارٌ. (TA.) So [accord. to some] in the Ḳur xvii. 8. (Ṣ, ISd.) ═ *A mat woven of reeds* [or *of rushes*] (Mṣb, Ḳ) or *of palm-leaves;* (IDrd and Ḳ voce تَذْرُعٌ, &c.;) syn. بَارِيَّةٌ; (Mṣb, Ḳ;) vulgarly ↓ حَصِيرَة: (Mṣb:) or *a thing woven,* [سَفِيفَة, in the L and TA erroneously written سقيفة,] *made of* بَرْدِىّ [or *papyrus*] and *of* أَسَل [or *rushes*], *and then spread upon the ground like a carpet:* (TA:) pl. حُصُرٌ (Mṣb, TA) and, by contraction, حُصْرٌ. (TA.) Hence the prov., أَسِيرٌ عَلَى حَصِيرٍ [*A captive upon a mat*]. (TA.) And بَنَاتُ الحَصِيرِ *Bugs;* syn. بَقٌّ. (TA in art. بق.) — *Anything woven.* (Ḳ.) — *A garment,* or *piece of cloth, ornamented and variegated, which, when spread out, captivates hearts in a manner peculiar to it, by its beauty.* (Ḳ.) So, accord. to some, in the trad. of Ḥodheyfeh, تُعْرَضُ الفِتَنُ عَلَى القُلُوبِ عَرْضَ الحَصِيرِ [expl. in art. عرض, conj. 1]. (B.) — *A bed;* or *a thing spread to lie upon;* as though it were a mat: so, accord. to El-Ḥasan, in the Ḳur xvii. 8, referred to above. (TA.) — *A sitting-place;* syn. مَجْلِس: (Ḳ, and so in two copies of the A:) MF thinks it to be a mistake for مَحْبِس [*a prison,* or *place of confinement*]. (TA.) — The *surface of the ground:* (Mṣb, Ḳ:) whence, accord. to some, it is applied to that which is spread upon the ground [i. e. a mat]: (TA:) pl. [of pauc.] أَحْصِرَة and [of mult.] حُصُرٌ. (Ḳ.) — *Water.* (Ḳ.) [Perhaps because its surface, when rippled by the wind, is likened to a thing woven: see نَسْج.] — The *diversified wavy marks, streaks,* or *grain,* (فِرِنْد,) of a sword, (Ḳ, TA,) *resembling the tracks of ants:* (TA:) or its حَصِيرَانِ are its *two sides.* (Ḳ,* TA.) — *A road,* or *way.* (IAạr, Ḳ.) — *A row* of men, and of other things. (Ḳ.) — *A certain vein extending across upon the side of a beast, towards the belly:* (Ḳ:) so, accord. to some, in the trad. of Ḥodheyfeh mentioned above: (TA:) or *a portion of flesh so situate;* (Ḳ;) i. e., *from the shoulder-blades to the flank;* as also ↓ حَصِيرَة, explained in the Ḳ as *a portion of flesh lying across in the side of a horse, which one sees when he is made lean by scanty food:* (TA:) or the former signifies the *sinew that* is between the part called the صِفَاق and the part where the false ribs end; (Ḳ, TA;) which is the end of the side: (TA:) or the part that is between the vein that appears in the side of the camel and horse, lying across, and what is above it, to the part where the side terminates: (Aṣ, Ṣ:) or the حصير of the side is what appears of the upper parts of the ribs. (Ibn-Es-Seed.) — Also The side itself. (Az, Ṣ, Ḳ.) Hence the phrase, دَابَّةٌ عَرِيضُ الحَصِيرَيْنِ A beast having wide sides. (A, TA.) And أَوْجَعَ ٱللّٰهُ حَصِيرَيْهِ [May God make his sides to ache; meaning] may he be severely beaten. (A, TA.) A certain elegant scholar says, أَثَّرَ حَصِيرُ الحَصِيرِ فِى حَصِيرِ الحَصِيرِ The mat of the prison made marks upon the side of the king. (MF.)

حَصِيرَةٌ: see حَصِيرٌ, in two places. — Also A place in which dates are dried: (Ṣ, Ḳ:) or, accord. to Az, it is with ض. (TA.)

مَحْصُورٌ: see مَحْصَرٌ.

مِحْصَرَةٌ: see حِصَارٌ, in two places.

مَحْصُورٌ Straitened: [encompassed, or surrounded:] besieged, or beset, in a fortress. (TA.) Confined, kept close, imprisoned, detained, retained, restrained, withheld, or prevented; (Akh, Ṣ, TA;) as also ▼حَصِيرٌ. (Ibn-Es-Seed.) Detained, restrained, withheld, or prevented, from journeying &c.; as also ▼حَصِيرٌ and ▼مُحْصَرٌ: (TA:) [or this last signifies made to restrain himself: see 1.] See also حَصِرٌ. — Suffering suppression of the fæces, or constipation of the bowels: (Ibn-Buzurj, Mgh, Ḳ:) [distinguished from مَأْسُورٌ: (see حُصْرٌ:) or] it also signifies suffering suppression of the urine. (Ibn-Buzurj.) — A camel having upon him [or furnished with] a حِصَار. (Ḳ.)

حصرم

Q. 1. حَصْرَمَ, [inf. n. حَصْرَمَةٌ,] He braced his bow, making the string tight, or tense. (Ṣ, Ḳ.) — He twisted a rope strongly. (Ḳ.) — He filled (AḤn, Ḳ) a vessel, (AḤn, TA,) or a skin: (Ḳ:) or he filled a skin so that it became strait [or tense]. (TA.) — He pared and shaped a reed for writing. (Ḳ.) — [Golius has added the signification "Excitavit," as on the authority of the KL; but in my copy of that work, I find, as the inf. n. of the verb having this signification, حَثْحَثَةٌ, which immediately follows the significations of حَصْرَمَةٌ, and hence appears to have been omitted in the copy of the KL used by Golius.]

Q. 2. [تَحَصْرَمَ app. signifies It (a grape) became in the state in which it is termed حصرم. And hence,] تَزَبَّبَ قَبْلَ أَنْ يَتَحَصْرَمَ [app. meaning † He did the latter part of an affair before the first; as when a man writes a book before he has qualified himself by preparatory study]: a prov. (TA.) — تحصرم said of butter [in the process of formation] means It became dissundered, or separated [into clots], by reason of intense cold; and did not coalesce; as also تحضرم. (TA.)

حُصْرُمٌ: see the next paragraph.

حِصْرِمٌ The first of grapes, (Ṣ, Mgh, Mṣb, Ḳ,) such as are crude and sour, (Mgh,) as long as they remain sour, (Mṣb,) or as long as they remain green: (Ḳ:) or grapes when hard: (Az, TA:) or, accord. to AḤn, grapes when they have become organized and compacted: or, as he says on one occasion, [the n. un.] حِصْرِمَةٌ signifies a grape when it germinates. (TA.) The rubbing of the body in the bath with bruised, or pulverized, حصرم dried in the shade prevents the origination of [the cutaneous disorder termed] حَصَف in the year in which this is done, and strengthens the body, and cools it. (Ḳ.) — Dates, or fruit, (تَمْرٌ, or ثَمَرٌ, accord. to different copies of the Ḳ, the latter being the reading in the M, TA,) not yet ripe. (M, Ḳ, TA.) — Fruit plucked from the tree called the مَظّ, (Ḳ,) i. e. the wild pomegranate. (TA.) — In the "Jemʼ et-Tefáreeḳ," it is said to signify Grape-stones: but this requires consideration. (Mgh.) — What is lean, dry, or withered, (syn. حَشَفٌ,) of anything. (AZ, Mṣb, Ḳ.) — And hence, (Mṣb,) † A niggardly man; (ISk, Ṣ, Mṣb, Ḳ;) narrow in disposition; as also ▼مُحَصْرَمٌ (Ṣ, TA) and ▼مُتَحَصْرِمٌ: (Ḳ,* TA:) or ▼مُحَصْرَمٌ means having little, or no, good. (TA.) — Short; (Ḳ;) and فَاحِش [app. as meaning evil in disposition]; as also ▼حُصْرُمٌ. (TA.) — And An iron [hooked] instrument with which the bucket is extracted from a well; (Ḳ;) also called عُوذَقٌ. (TA.)

حَصْرَمَةٌ [inf. n. of Q. 1, q. v.] — Also Niggardliness, tenaciousness, or avarice. (Ḳ, TA.)

[حِصْرِمِىٌّ Omphacine. (Golius, on the authority of Meyd.)]

حِصْرِمِيَّةٌ Soup made [or flavoured] with unripe grapes or dates, or with the juice thereof. (MA.)

مُحَصْرَمٌ Butter dissundered, or separated [into clots], by reason of intense cold; not coalescing; (Ḳ;) [as also ▼مُتَحَصْرِمٌ: see Q. 2.] — A scanty, or small, gift. (TA.) — Anything straitened, or scanted. (TA.) — See also حِصْرِمٌ, in two places. — شَاعِرٌ مُحَصْرَمٌ i. q. مُخَضْرَمٌ, (Ḳ,) which is the more common term; meaning A poet that lived in the time of paganism and in that of El-Islám. (TA.)

مُتَحَصْرِمٌ: see مُحَصْرَمٌ: — and حِصْرِمٌ.

حصف

1. حَصُفَ, aor. ـُ, (Ṣ, Ḳ,) inf. n. حَصَافَةٌ, (Ṣ, TA,) [app., in its primary and proper sense, It was, or became, compact, or firm and close; said of a rope, and of a web, and the like: see its part. n., حَصِيفٌ; and see also 10. — And hence,] ‡ He (a man, TA) was, or became, firm, or sound, in intellect or sense; (Ṣ, Ḳ, TA;) or strong in intellect, and good in judgment. (TA.) — حَصِفَ, aor. ـَ, inf. n. حَصَفٌ, It (the skin, Ṣ, TA, or the body, or person, Mṣb) was, or became, affected with dry mange or scab: (Ṣ, Ḳ, TA:) or broke out with small pustules, (Mṣb, TA,) like the small-pox, (Mṣb,) generating matter, or thick purulent matter, and not becoming large; sometimes coming forth in the soft parts of the belly in the days of heat. (TA.)

4. احصف, (Ḳ,) inf. n. إِحْصَافٌ, (Ṣ,) He twisted a rope firmly, strongly, or compactly. (Ṣ, Ḳ.) And احصف نَسْجَهُ [He made his weaving, or his web, compact; or firm and close; or close in texture, and strong: see the pass. part. n. below]: said of a weaver. (TA.) — [Hence,] ‡ He established, or settled, or he did, performed, or executed, an affair, firmly, solidly, soundly, or well. (Ṣ, Ḳ, TA.) — ‡ He (a man, and a horse,) passed, went, or went along, quickly, or swiftly: (Ṣ, Ḳ:) or raised the pebbles in his running: (Ṣgh, Ḳ:) or went with short steps, but quickly: (ISk, Ḳ:) or he (a horse) ran quickly, with short steps, at the utmost rate of the pace termed حُضْر. (AO, TA.) — احصفهُ الحَرُّ, inf. n. as above, The heat caused pustules [such as are termed حَصَف] to come forth upon his body, or person. (TA.)

10. استحصف It (a thing, Ṣ, TA) was, or became, firm, strong, compact, sound, or free from defect: (Ṣ, Ḳ, TA:) properly said of a rope, as meaning it was, or became, firmly, strongly, or compactly, twisted: and — tropically said of judgment [as meaning ‡ it was, or became, firm, strong, sound, or good]: and of an affair [as meaning ‡ it was, or became, established, or settled, or done, performed, or executed, firmly, solidly, soundly, or well]. (TA. [See also حَصُفَ.]) — ‡ It (the vulva) was, or became, narrow, and firm, or tough, on the occasion of جِمَاع. (Ḳ, TA.) The quality thus denoted is approved; and the woman possessing it is termed ▼مُسْتَحْصِفَةٌ. (TA.) — ‡ It (time, or fortune,) pressed hardly, or severely, (Ṣ, Ḳ, TA,) عَلَيْهِ upon him. (Ṣ, TA.) — † It (a company of men) collected, or congregated. (TA.)

حَصَفٌ Dry mange or scab: (Ṣ, Ḳ:) or small pustules, [like the small-pox, (see حَصِفَ,)] that generate matter, or thick purulent matter, and do not become large; sometimes coming forth in the soft parts of the belly in the days of heat. (TA.)

حَصِفٌ: see حَصِيفٌ. — Also part. n. of حَصِفَ. (Mṣb.)

حَصِيفٌ Anything firm, strong, compact, sound, and free from defect. (TA.) A garment, or piece of cloth, compactly, or firmly and closely, woven: (TA:) or dense; concealing [what is within it]: (Kf, TA:) and ▼مُحْصَفٌ [in like manner] signifies dense and strong. (TA.) — [Hence,] ‡ Firm, or sound, in intellect or sense; (Ṣ, Ḳ, TA;) and ▼حَصِفٌ signifies the same; and firm in judgment; [or possessing firmness, or soundness, of intellect and judgment; for] it is a possessive epithet: and the former is said to signify strong in intellect, and good in judgment. (TA.) حَصِيفُ العُقْدَةِ, occurring in a letter of ʼOmar, means † [Firm] in judgment, and in the management, conducting, ordering, or regulating, of affairs. (TA.)

مُحْصَفٌ: see حَصِيفٌ. — [Hence,] بَيْنَهُمَا حَبْلٌ مُحْصَفٌ ‡ Between them two is a firm tie of brotherhood. (TA.)

مُحْصِفٌ A horse *that goes in the manner denoted by the verb* احصف; (Ḳ;) as also ▼مِحْصَفٌ (Ṣ, Ḳ) and ▼مِحْصَافٌ; (Ḳ;) [or] the last is applied to a she-camel: (Ṣ:) the fem. of the first is with ة. (TA.)

مِحْصَفٌ : } see what next precedes.
مِحْصَافٌ : }

مُسْتَحْصِفٌ A *narrow* فَرْج [or vulva]. (Ṣ.) See also 10.

حصل

1. حَصَلَ, (Mṣb, Ḳ, &c.,) aor. ُ, (TA,) inf. n. حُصُولٌ (Mṣb, Ḳ, &c.) and مَحْصُولٌ, (Ḳ,) like مَعْقُولٌ and مَعْسُورٌ and مَيْسُورٌ, (TA,) [*It was, or became, produced, educed, extracted, taken forth, or fetched out*; as gold or silver from the stone of the mine, and the kernel from the shell, and wheat from the straw: (see 2:)] *it came out, it became apparent:* (KL:) *it was, or existed, or came into being or existence; it became realized; syn. with the complete* [i. e. *attributive*] كَانَ: (Mṣb in art. كون:) [*it presented itself: it was, or became, prepared, or ready: it became attained, obtained, gotten, or acquired:*] *it came, came to pass, happened, took place, betided, befell, or occurred;* said of an event; *syn. with* وَقَعَ, (TA in art. وقع,) which is also *syn. with the complete* [or *attributive*] كَانَ; (Mṣb in art. كون;) likewise *syn. with* جَآءَ: (Er-Rághib, TA in art. جيأ:) [*it resulted;* and particularly *as a sum;* and *as a product;* and *as a quotient: it ensued: it arose, originated, proceeded, came, supervened, or accrued:* in which senses, also, it is *syn. with the attributive* كَانَ, and *with* جَآءَ, followed by مِنْ:] *it remained, and continued, when the rest had gone, or passed away;* (Ḳ, TA;) relating to a reckoning, and to an action, and the like: (TA:) and *i. q.* ثَبَتَ and وَجَبَ; as in the saying, حَصَلَ لِى عَلَيْهِ كَذَا [*Such a thing, or sum, was, or became, or proved to be, binding, obligatory, or incumbent, on him to render as a debt to me*]. (Mṣb.) ═ حَصِلَ, [aor. َ,] inf. n. حَصَلٌ, *He* (a horse) *had a complaint of his belly from eating the earth of the herbage:* (Ṣ:) or حَصِلَتِ الدَّابَّةُ, aor. َ, (M, Ḳ,) inf. n. as above, (TA,) *the beast ate earth,* (M, Ḳ,) or *pebbles,* (Ḳ,) *and they remained in its inside,* (M, Ḳ,) *fixed:* (M:) or حَصَلٌ signifies a horse's *taking into the mouth earth from the herbs, some of which earth, collecting in his belly, kills him:* and the horse so killed is said to be ▼حَصِلٌ: (T, TA:) or a camel's *having pebbles* [*which he has swallowed*] *remaining in the omasum, so as not to come forth in the cud when he ruminates;* and when this is the case, they sometimes kill: or a young camel's *eating earth, and in consequence not ejecting the cud;* which sometimes kills it. (TA.) —— Said of a boy, it signifies وَقَعَ الحَصَى (O) or وَقَعَتِ الحَصَاةُ (Ḳ) فِى أُنْثَيَيْهِ (O, Ḳ) [app. meaning *The stones, or the stone, fell, or descended, in his scrotum:* Freytag, following the TḲ, in which فى انثيه is considered (I know not on what authority) as meaning فِى مَثَانَتِهِ, renders it "laboravit lapidibus in vesica urinæ orientibus"].

2. حَصَّلَ, inf. n. تَحْصِيلٌ, a trans. verb; (Ṣ, Mṣb;) i. e. trans. of حَصَلَ, primarily signifying, accord. to IF, (Mṣb,) *He produced, educed, extracted, took forth,* or *fetched out,* gold [or silver] from the stone of the mine; (Mṣb, Er-Rághib, TA;) and in like manner, the kernel from the shell; and [the grain of] wheat from the straw: (Er-Rághib, TA:) *he made* a thing *apparent;* (Az, Er-Rághib, TA;) as, for instance, the kernel from the shell; and the حَاصِل [or result] of a computation: (Er-Rághib, TA:) [*he brought into being, or existence; he realized:*] *he prepared,* or *made ready:* (PṢ:) *he separated, discriminated,* or *distinguished,* (Az, Ḳ,) what remained and continued, when the rest had gone, or passed away: (Ḳ: [in the CḲ, مَا يُحَصَّلُ is erroneously put for مَا يَحْصُلُ:]) *he perceived* a thing: *he attained,* or *obtained,* a thing: syn. أَدْرَكَ [in both these senses: and also as meaning *he overtook*]: (Abu-l-Baḳá, TA:) *he took,* or *got,* or *acquired,* advantage, or profit; (KL;) *i. q.* أَخَذَ, and حَازَ: (B and TA in art. اخذ:) *he collected:* (Az, Er-Rághib, TA:) and [hence, app.,] تَحْصِيلُ كَلَامٍ signifies *The reducing a sentence,* or *the like, to its* ▼مَحْصُول [here meaning its *essential import,* or its *sum and substance*]: (Ṣ, TA:) and حَصَّلَ الكَلَامُ كَذَا [*The sentence, or speech, comprehended,* or *comprised, within its scope, such a thing*]. (Mṣb in explanation of تَضَمَّنَ.) وَحُصِّلَ مَا فِى الصُّدُورِ, in the Ḳur [c. 10], means *And what is in the breasts,* or *minds,* [of men] *shall be made apparent:* (Az, Er-Rághib, TA:) or *discriminated:* (Az, Bḍ, TA:) or *collected,* (Fr, Az, Bḍ, Er-Rághib, TA,) *in the registers.* (Bḍ.) ═ See also 4, in two places.

4. احصل النَّخْلُ; (Ṣ, Ḳ;) and ▼حَصَّلَ, inf. n. تَحْصِيلٌ; (Ḳ;) *The palm-trees had* حَصَل; i. e., *dates that had not yet become hard,* (Ṣ, Ḳ,) *and of which the* ثَفَارِيق [or *bases*] *had not yet appeared;* (Ṣ;) or *dates that had become hard and round:* and also, *had* حَصَل as meaning *spadixes* (طَلْع) *that had become yellow:* (Ḳ:) or احصل البَلَحُ *the dates came forth from their* ثَفَارِيق, *small:* and ▼حَصَّلَ *they became round.* (TA.) —— احصل القَوْمُ *The people had unripe, or ripening, dates appearing upon their palm-trees.* (TA.)

5. تحصّل *It became collected, and remained,* or *continued.* (Ḳ, TA.)

Q. Q. 1. حَوْصَلَ *He* (a bird, Ṣ) *filled his* حَوْصَلَة [i. e. *stomach,* or *crop*]. (Ṣ, Ḳ.) You say [to a bird], حَوْصِلِى وَطِيرِى [*Fill thy stomach,* or *crop, and fly*]. (Ṣ.)

حَصْلٌ: see what next follows: —— and see حُصَالَةٌ.

حَصَلٌ (Ṣ, Ḳ) and ▼حَصْلٌ, (M, Ḳ,) the latter used by poetic license, (ISd, TA,) *Dates before they have become hard,* (Ṣ, Ḳ,) *and before their* ثَفَارِيق [or *bases*] *have appeared;* n. un. حَصَلَةٌ: (Ṣ:) or *when they have become hard and round.* (IAạr, Ḳ.) And The *spadix of the palm-tree* (طَلْع) *when it has become yellow.* (Ḳ.) Also, the former, *What fall, and become scattered, of the produce of a palm-tree, green and fresh, like small green beads.* (Aboo-Ziyád, TA.) —— See also حُصَالَةٌ.

حَصِلٌ : see حَصِلٌ.

حَصِيلٌ *A certain plant.* (Ṣ, M, O, Ḳ.)

حُصَالَةٌ *What remains, of grain, in the place where it has been trodden out, after the removal* [*of the bulk*] *of the grain:* (Ṣ, O:) or, as also ▼حَصَلٌ (Ḳ, TA) and ▼حَصْلٌ, (Ḳ,) *what remains, of barley and wheat, in the place where it has been trodden out, after the bad thereof has been removed:* and *what comes forth from wheat, and is thrown away, such as* [*the weed called*] زُؤَان, (Ḳ, TA,) *and* دنقة [i. e. دُنْقَة or دَنَقَة] *and the like:* or *what comes forth from barley and wheat, and is thrown away, when it is somewhat grosser than dust,* or *earth, and than what are termed* دُقَاق [q. v.]: (TA:) or the *remains of wheat in the sieve, after the sifting, with what are mixed therewith;* as also خُصَالَةٌ; but the former word is the more known. (JK and TA in art. خصل.) [See also حُثَالَةٌ.]

حَصِيلَةٌ : see حَاصِلٌ.

حُصَّالَةٌ : see حَوْصَلَةٌ.

حَاصِلٌ (T, Ṣ, M, Mṣb, Ḳ, KL) and ▼حَصِيلَةٌ (Ṣ, Ḳ, &c.) and ▼مَحْصُولٌ (Ṣ, Mṣb, Ḳ) [and ▼مُحَصَّلٌ] *Produce;* or *what is produced, educed, extracted, taken forth,* or *fetched out: what is made apparent: profit, advantage, gain,* or *acquisition:* (KL in explanation of the first word [but applying to all]:) [the *result* of a thing:] *a remain, remainder, remaining portion, remnant, relic, residue,* or the *remains,* of a thing; (Ṣ;) *what remains, and continues,* of anything, *when the rest has gone,* or *passed away:* (Ḳ:) it is of a reckoning, or computation, and of actions, and the like: (T, M, TA:) pl. of the second حَصَائِلُ. (Ṣ, TA.) The first also particularly signifies *What is cleared,* or *purified,* of silver [and of gold] *from the stone of the mine.* (TA.) [And The *produce,* or *net produce,* of land &c.; of anything that is a source of revenue; as also the third. The *result* of an arithmetical process; the *sum,* the *product,* and the *quotient.* The *sum,* or *sum and substance,* or *essential import,* of a sentence or the like; as also the third (see 2) and the fourth. And the *result, end, conclusion, event, issue, ultimate consequence* or *effect,* or *ultimity,* of anything.] ═ See also حَوْصَلَةٌ.

حَوْصَلٌ *A depressed place where water rests in a meadow, where the herbage is the latest to dry up:* whence the ▼حَوْصَلَة of a bird, as being the resting-place of what it eats. (Az, TA.) —— The *place where water rests,* or *remains, in the furthest part of a watering-trough* or *tank;* (Ḳ;) as also ▼حَوْصَلَةٌ. (ISd, Ḳ.) —— See also حَوْصَلَةٌ. —— Also A sheep or goat *large in the part of the belly above the navel.* (M, Ḳ.) ═ *A certain plant.* (TA.)

حَوْصَلٌ The [*plant called*] بَاذَنْجَان [q. v.] (Ḳ.)

حَوْصَلَةٌ: see حَوْصَلٌ, in two places. — The حَوْصَلَة of a bird (Ṣ, Mṣb, Ḳ) is [The *stomach; the triple stomach, consisting of the crop, or craw, the second stomach, and the gizzard,* or *true stomach:* and often, particularly, the *first of these three:* see جِرِّيَّة and جِرْبَة:] *that which, to a bird, is like the* مَعِدَة *to a man;* (Ḳ;) also called حَوْصَلَّة ↓ (Mṣb, Ḳ) and حَوْصَلَاء ↓ and حَوْصَل ↓: (Ḳ:) and of an animal having a cloven hoof or a خُفّ, i. q. مَصَارِين [q. v.]: (Az, TA:) pl. حَوَاصِلُ. (Ṣ, TA.) — Hence the حَوَاصِل [i. e. † *Store-rooms,* or *magazines,*] of kháns: [also meaning † the *cells* of prisons:] of which the sing. is حَوْصَلَة: not, as the vulgar say, حَاصِل ↓. (TA.) — Also, the sing., The *lower part of the belly, as far as the pubes,* (Ḳ, TA,) of a man, (TA,) and of any animal: (Ḳ, TA:) or the *place where the fæces collect, below the navel:* or the *part between the navel and the pubes.* (TA.) — نَاقَةٌ ضَخْمَةُ الحَوْصَلَةِ *A she-camel big in the belly.* (TA.)

حَوْصَلَاء: see حَوْصَلَة.

حَوْصَلَّة: see حَوْصَلَة. — Also *A thing resembling a* حُقَّة [q. v.], *made of baked clay;* vulgarly called حُصَّالَة ↓. (TA.)

حَاصِلٌ: see مُحَصَّل.

مُحَصِّلٌ One who *clears,* or *purifies,* silver [and gold] *from the stone of the mine.* (TA.) And مُحَصِّلَةٌ A woman *who separates* (تُحَصِّلُ) *the earth of the mine* [*for the purpose of extracting the gold or silver*]. (Ṣ, Ḳ.)

مَحْصُولٌ: see حَاصِلٌ: and see also 2.

مُحَوْصَلٌ (Ḳ) and مُحَوْصِلٌ, (Ḳ, TA,) or مُحْصَوْصِلٌ ↓, (so in my MS. copy of the Ḳ,) or مُحْصَوْصَلٌ, (so in the CḲ,) One who is *protuberant in his lower part* [*of the belly*], *next his navel, like her who is pregnant:* (Ḳ:) so in the M. (TA.)

مُحْصَوْصِلٌ, or مُحْصَوْصَلٌ: see what next precedes.

حصن

1. حَصُنَ, (Mgh, Mṣb, Ḳ,) aor. ـُ, (Ḳ, TA,) inf. n. حَصَانَةٌ, (Ṣ, Mgh, Mṣb, TA,) said of a حِصْن [or fortress], (Ṣ,) or of a place, (Mgh, Mṣb, TA,) i. q. مَنُعَ, [for which the CḲ erroneously substitutes وَمَنَعَ, after كَكَرُمَ, as though the verb were حَصَنَ and حَصُنَ,] (Ḳ, TA,) [i. e.] *It was,* or *became,* مَنِيع [meaning *inaccessible,* or *unapproachable,* or *difficult of access*]; *it was,* or *became, unattainable, by reason of its height;* (Mṣb;) *it was fortified,* or *protected against attack, so that one could not gain access to what was within it.* (Mgh.) — Hence, (Mgh,) حَصُنَتْ, (Ṣ, Mṣb, Ḳ,) aor. ـُ, (Ḳ,) inf. n. حُصْنٌ (Ṣ, Mgh, Ḳ) and حِصْنٌ (Ḳ) and حَصَانَةٌ; (Ṣ,* Mṣb,* TA;) and أَحْصَنَتْ ↓, (Ṣ, Mgh, Ḳ,) inf. n. إِحْصَانٌ; (Mgh;) and تَحَصَّنَتْ ↓; (Ḳ;) said of a woman, *She was,* or *became, continent,* or *chaste;* or *she abstained from what was not lawful nor decorous;* syn. عَفَّتْ: (Ṣ, Mgh, Mṣb, Ḳ:) or *she was,* or *became, married;* or *she had a husband;* (Ḳ;) as also حَصَنَتْ and أُحْصِنَتْ ↓: (Ḥam p. 101, in which حَصَنَتْ is likewise mentioned in this sense:) or أُحْصِنَتْ ↓ signifies *she was,* or *became, pregnant;* (Ḳ;) as though pregnancy protected her from a man's going in to her. (TA.) ═ حَصَّنَهُ, inf. n. حَصْنٌ, *He preserved,* or *guarded, him,* or *it, in places inaccessible,* or *unapproachable,* or *difficult of access, as in a* حِصْن [or *fortress*]. (TA.) [See also 2 and 4.]

2. حَصَّنَهُ, [inf. n. تَحْصِين;] and احصنهُ ↓; *He made it,* or *rendered it, inaccessible,* or *unapproachable,* or *difficult of access;* (Mṣb, Ḳ;) *he made it,* or *rendered it, unattainable, by reason of its height;* (Mṣb;) *he fortified it,* or *protected it against attack, so that one could not gain access to what was within it;* (Mgh;) namely, a place. (Mgh, Mṣb.) حَصَّنْتُ القَرْيَةَ *I built a wall around the town,* or *village.* (Ṣ.) — For the former verb, see also 4. ═ And see 5.

4. احصنهُ: see 2. — Also *He,* (God,) or *it,* (a coat of mail [or the like],) *protected,* or *defended, him.* (Fr, Mgh,* TA.) — العِفَّةُ تُحْصِنُ مِنَ الرِّيبَةِ [*Continence,* or *chastity, preserves from suspicion,* or *evil opinion*]. (Mgh.) — الَّتِى أَحْصَنَتْ فَرْجَهَا, in the Ḳur [xxi. 91 and lxvi. last verse], means *Who preserved her pudendum from that which is unlawful* or *indecorous;* (Zj, Mgh,* TA;) *who abstained from what is unlawful* or *indecorous;* or *was continent,* or *chaste.* (Mṣb.) — احصن المَرْأَةَ *He* (her husband) *caused the woman to abstain from that which is unlawful* or *indecorous, or to be continent or chaste;* (Ṣ, Mgh, Ḳ;*) as also حَصَّنَهَا ↓. (Ḳ.) And احصنهُ التَّزَوُّجُ [*Marriage caused him to abstain from that which is unlawful &c.*]. (Ḳ.) — [Also *He married the woman;* i. e. *gave her in marriage.*] See أُحْصِنَتْ above, in the first paragraph. In the Ḳur iv. 30, some read فَإِذَا أُحْصِنَّ, meaning *And when they are married.* (Ṣ, TA.) And a poet says,

* أَحْصَنُوا أُمَّهُمْ مِنْ عَبْدِهِمْ *
* تِلْكَ أَفْعَالُ القَزَامِ الوَكَعَهْ *

i. e. *They married* [*their mother to their slave: such are the deeds of the mean, the base*]. (Ṣ.) ═ أَحْصَنَتْ, intrans.: see 1, in two places. — In the Ḳur iv. 30, some read فَإِذَا أَحْصَنَّ; and accord. to Ibn-Mes'ood, this, said of female slaves, means *And when they are Muslimehs.* (TA.) Accord. to Aboo-Ḥaneefeh, الإِحْصَانُ in a case of stoning involves six conditions; *The being a Muslimeh, and free, and of sound intellect, and of the age of puberty, and validly married, and having had her marriage consummated:* and in a case of charging with adultery, *the being a Muslimeh, and free, and of sound intellect, and of the age of puberty, and continent,* or *chaste.* (Mgh.) — And أَحْصَنَ *He* (a man, Ṣ, Mṣb) *married,* or *took a wife.* (Ṣ, Mṣb, Ḳ.) With the lawyers, إِحْصَانٌ means The act of *coïtus conjugalis in a case of valid marriage;* and accord. to Esh-Sháfi'ee, *by a free man who has attained to puberty,* and *in the case of a free woman who has attained to puberty,* among the Muslims and the believers in a plurality of gods; meaning, *in a case of valid marriage.* (Mṣb.)

5. تحصّن, said of the enemy, (Ṣ, TA,) [*He fortified himself:* or] *he entered the* [or *a*] حِصْن [or *fortress*]: or *protected himself by it:* or *took it, or made it, as a place of abode.* (TA.) — And hence, *He guarded,* or *protected, himself in any way.* (TA.) — See also 1. — Also *He* (a horse, TA) *became a* حِصَان, (Ḳ,) i. e. a *stallion,* or *fit to cover:* (TA voce رَاحَ:) or *affected to be so:* (Az, TA:) [and so حَصُنَ ↓ or حَصَنَ; for] a horse in this case is said to bear evidence of التَّحَصُّن and التَّحْصِين. (Ṣ,* Ḳ, TA.)

حِصْنٌ [*A fortress; a fort; a fortified place;*] *a place of which the interior is inaccessible;* (Ḳ;) *any place that is fortified, or protected against attack, so that one cannot gain access to what is within it;* (Mgh;) *a place that is unattainable, by reason of its height;* (Mṣb;) *a fortified city:* (TA:) pl. [of mult.] حُصُونٌ (Ṣ, Mṣb, Ḳ) and حِصَنَةٌ and [of pauc.] أَحْصَانٌ. (Ḳ.) [Hence,] أَبُو الحُصَيْنِ: see أَبُو الحِصْنِ, below. [Hence, also,] خَيْلُ العَرَبِ حُصُونُهَا ذُكُورُهَا وَإِنَاثُهَا ‡ [*The horses of the Arabs are their* حُصُون; *the males thereof and the females thereof*]. (TA.) A man said to 'Abd-Allah Ibn-El-Ḥasan, "My father has left the third of his property for the حُصُون:" and he replied "Buy thou horses:" so in the A: in the M, "Buy thou therewith horses, and mount [men] on them [to fight] in the cause of God." (TA.) — [Hence, also,] ‡ *Arms.* (Ḳ, TA.) You say, جَآءَ يَحْمِلُ حِصْنًا ‡ *He came bearing arms.* (TA.) — Also The [*new moon;* or the *moon when it is termed*] هِلَال: in the Ḳ, الهِلَاكُ is erroneously put for الهِلَال. (TA.)

حَصْنَآء: see what next follows.

حَصَانٌ (Ṣ, Mgh, Mṣb, Ḳ) and حَاصِنٌ ↓ (Ṣ, Mgh, Ḳ) and حَاصِنَةٌ ↓ (Ḳ) and حَصْنَآءُ ↓ (Ṣ, Ḳ) and مُحْصِنَةٌ ↓ (Th, Ṣ, Mgh, Mṣb, Ḳ) and ↓ (Th, Ṣ, Mṣb, Ḳ,) applied to a woman, *Continent,* or *chaste;* or *abstaining from what is not lawful nor decorous,* (Th, Ṣ, Mgh, Mṣb, Ḳ,) or *from that which induces suspicion* or *evil opinion:* (Sh and TA in explanation of the first of these epithets:) or *married; having a husband:* (Ḳ:) or حَصَانٌ has both of these significations: (Ḥam p. 101:) and accord. to Th, (Ṣ,) مُحْصَنَةٌ ↓, with fet-ḥ only, has the latter signification; (Ṣ, Mgh,* Mṣb;*) and means *caused to be continent or chaste, or to abstain from that which is unlawful* or *indecorous,* by her husband: (Mgh: [and the same is implied in the Ṣ:]) and this epithet is also applied to a woman *emancipated:* and to one *having become a Muslimeh:* (Az, TA:) [certain particular applications of مُحْصَنَة ↓ have been implicatively shown above: see 4:] the pl. of حَصَان is حُصُن

and حَصَانَاتٌ: and the pl. of ▼حَاصِنٌ and ▼حَاصِنَةٌ is حَوَاصِنُ and حَاصِنَاتٌ: the former of which (حواصن) also signifies *pregnant*, (Ḳ,) applied to women: (TA:) the pl. محصنات, in the first instance of its occurrence in ch. iv. [verse 28] of the Ḳur, is read by all ▼مُحْصَنَات, (A 'Obeyd, TA,) meaning *having husbands*; (A 'Obeyd, Mgh, Mṣb, TA;) because when such women are made captives, their marriage-tie is cut: but in other instances, some read thus, understanding it in the sense last explained; and others read ▼مُحْصِنَات, as meaning *that have become Muslimehs*: (A 'Obeyd, TA:) in the Ḳur iv. 29, it means *free* women: and in the Ḳur v. 7, *continent*, or *chaste*, women: (Mgh: [in the Mṣb, it is said to have the latter of these last two meanings in iv. 29, and the former of them in v. 7:]) ▼مُحْصَنَات is the more common in the language of the Arabs. (Fr, TA.) —— حَصَانٌ also signifies *A pearl*, or *a large pearl*; syn. دُرَّةٌ: (Ḳ:) because it is protected in the interior of the shell that contains it. (TA.)

حِصَانٌ A *generous*, or *high-bred*, horse, (Mṣb, Ḳ,) *of whose seed one is niggardly*: (Ḳ:) or a *male* horse: (Mgh, Ḳ:) or this latter is a secondary meaning, originating from frequency of usage: (Ṣ, Mṣb:) or a *stallion* horse; or one *fit to cover*: (TA voce رَاحَ:) the حِصَان is so called because he preserves his rider: (TA:) or because his back is like the حِصْن to his rider; (Mgh, Mṣb;) wherefore horses are called حُصُون: (Mgh:) or because one is niggardly of his seed, so that he is not made to cover any but a generous mare: (Ṣ, Mgh, Mṣb:) pl. حُصُنٌ. (Mgh, Mṣb, Ḳ.)

حَصِينٌ, applied to a place, (Mṣb, TA,) or to a حِصْن [or fortress], (Ṣ,) *Inaccessible*, or *unapproachable*, or *difficult of access*; syn. مَنِيعٌ; (Mṣb, Ḳ;*) [*unattainable, by reason of its height; fortified*, or *protected against attack, so that one cannot gain access to what is within it*; (see حَصُنَ, of which it is the part. n.;)] a building *that protects him who has recourse to it for refuge*. (Sb, TA.) —— هُوَ الحَصِينُ أَنْ يُرَامَ, meaning ▼أَحْصَنُ مِنْ أَنْ يُرَامَ: see (near its end) the first paragraph in art. ال. —— دِرْعٌ حَصِينٌ (Ḳ) and حَصِينَةٌ (Sh, Ḳ) *A coat of mail firmly, strongly*, or *compactly made*: (Ḳ:) or *trusty*, or *trusted in, having the rings* [for الخلق in the L and TA, I read الحَلَق,] *near together; such that weapons produce no effect upon it*: (Sh, L, TA:) so called because it is [as] a حِصْن to the body. (Er-Rághib, TA.)

حُصَيْنٌ dim. of حِصْنٌ. —— [Hence,] أَبُو الحُصَيْنِ *The fox*; syn. الثَّعْلَبُ; (Ṣ, Ḳ;) so called because of his protecting himself from causes of harm by his acuteness; (Ḥar p. 663;) as also ▼أَبُو الحِصْنِ. (M, Ḳ.)

حَاصِنٌ and حَاصِنَةٌ: see حَصَانٌ, in four places. —— The latter also signifies A man's *wife*: and so حَاضِنَةٌ. (TA.)

أَحْصَنُ [*More, and most, strongly fortified*, or *protected against attack*]. (TA in art. لوم.) See also حَصِينٌ.

مُحْصَنٌ Wheat *stored up*. (TA in art. عسى.) —— A man *caused, by marriage, to abstain from that which is unlawful* or *indecorous*, or *to be continent* or *chaste*; expl. by قَدْ أَحْصَنَهُ التَّزَوُّجُ. (Ḳ.) And A man *married*, or *having a wife*; (Ṣ, Mṣb, Ḳ;) as also ▼مُحْصِنٌ: (Zj, Mṣb:) the former anomalous [if أُحْصِنَ be not allowable as meaning, like أَحْصَنَ, "he married," or "took a wife:" but see 1 and 4]. (Ṣ, Mṣb.) For the fem., مُحْصَنَةٌ, and its pl., مُحْصَنَاتٌ, see حَصَانٌ, in four places.

مُحْصِنٌ: see مُحْصَنٌ: and for the fem., مُحْصِنَةٌ, and its pl., مُحْصِنَاتٌ, see حَصَانٌ, in three places.

مِحْصَنٌ *A lock*; syn. قُفْلٌ. (Ḳ.) —— The *piece of iron that extends upwards upon the nose of the horse, having its base in the* كِعَامَة, which is the *iron thing that embraces*, or *clasps*, (تَلْتَقِمُ,) *the muzzle of the horse*. (IDrd in his book on the Saddle and Bridle, pp. 8-9.) [Jac. Schultens, as cited in Freytag's Lex., explains it as *Ferramentum quoddam in fræno equi et frænum ipsum*.] —— *A* [*basket of the kind called*] زَبِيل: (Ḳ:) one should not say مِحْصَنَةٌ. (TA.)

حصى and حصو

1. حَصَا, (Ṣ, TA,) [aor. يَحْصُو,] inf. n. حَصْوٌ, *i. q.* مَنَعَ [as meaning *He denied*, or *refused*]: (Ṣ, Ḳ:) doubly trans., as in the phrase حَصَوْتَنِى حَقِّى [*Thou deniedst me*, or *refusedst me, my right*, or *due*]. (Ṣ, TA.) ═ حَصَيْتُهُ, (Ḳ, TA,) inf. n. حَصْىٌ, (TḲ,) *I cast a pebble*, or *small stone, at him*, or *it*: (TA:) or *I smote him*, or *it, with a pebble*, or *small stone*. (Ḳ.) ═ حَصِيَتِ الأَرْضُ, (Ḳ,) aor. تَحْصَى, (TA,) [inf. n. حَصًى,] *The land*, or *ground, abounded with pebbles*, or *small stones*. (Ḳ.) ═ حُصِىَ [*He had a stone in his bladder* or *kidneys*: or] *he had a hardening of the urine in the bladder so that it became like a pebble*, or *small stone*: (Lth, Ḳ:) said of a man: part. n. ▼مَحْصِىٌّ. (Lth, TA.)

4. احصاه, (inf. n. إِحْصَآءٌ, TA,) *He numbered, counted, calculated, reckoned*, or *computed, it*: (Ṣ, Mṣb, Ḳ:) or *he reached the last number thereof*: (Ḥam p. 297:) or *he collected it into an aggregate by numbering*: from حَصًى; because they used pebbles in numbering like as we use the fingers; (Er-Rághib;) or because they used to divide a thing upon the pebbles, and when nothing remained they said, أَحْصَيْنَا, i. e. *We have come to the pebbles*; or because they used to reckon the spoils, and to divide among themselves, and then to take the pebbles, and put marks upon them, and when they ended the reckoning, and came to the marks, they said, أَحْصَيْنَا: (Ḥam ubi suprà:) or *he retained it in his memory*: (Ḳ,* TA:) or *he understood it*. (Ḳ.) —— *He registered*, or *recorded, its number*: (Ksh and Bḍ in xvi. 18, and TA:) and [simply] *he registered it*, or *recorded it*. (Bḍ and Jel in xxxvi. 11 and lxxviii. 29.) —— *He was able to number it*, and *to register it* or *record it*: (TA:) and *he was able to do it*. (Mṣb, TA.) —— *He comprehended it*, or *knew it altogether*: (TA:) or [simply] *he knew it*. (Mṣb.) You say of God, أَحْصَى كُلَّ شَىْءٍ بِعِلْمِهِ *He hath comprehended everything by his knowledge, nothing escaping Him thereof, small or great*. (TA.) And you say [of a man], أَحْصَى عِلْمَهُ [*He had*, or *attained, a comprehensive and complete knowledge of it*]. (Ḳ in art. حوط. [In the CḲ, erroneously, عَلِمَهُ.])

حَصًى *Pebbles*, or *small stones*: (Ḳ:) *things that one throws, like the dung of sheep* or *goats*: (ISh, TA:) n. un. حَصَاةٌ: (Ṣ, Mṣb, Ḳ:) pl. حَصَيَاتٌ (Ṣ, Ḳ) and حُصِىٌّ (Ḳ) and حِصِىٌّ, thus, also, in [some of] the copies of the Ḳ. (TA.) [Hence,] طَرْقُ الحَصَى [*Pessomancy*; also termed الضَّرْبُ بِالحَصَى;] *a certain kind of divination*. (Ḥar p. 655.) And بَيْعُ الحَصَاةِ *The bargaining by one of the two persons' saying, When I throw the pebble to thee, the sale is binding*, or *settled*, or *concluded*: or *by his saying, I sell to thee, of the commodities, that upon which thy pebble shall fall when thou throwest it*: or *I sell to thee the portion of the land extending to the place which thy pebble shall reach*: all of which practices are forbidden. (TA. [See also 3 in art. نبذ.]) And حَصَاةُ القَسْمِ *A pebble which is put into a vessel, into which is then poured as much water as will cover it; this being done when they are journeying and have but little water, which they thus apportion*. (Ḳ in art. قسم.) —— حَصَى لُبَانٍ is a vulgar term for عَسَلُ اللُّبْنَى, (Ḳ in art. عسل,) i. e. المَيْعَةُ [generally applied to *Storax*, or *styrax*], *which is sometimes used for fumigation*. (TA in art. لبن.) [In the present day, pronounced حَصَى لِبَان, and applied to *Frankincense*: or *the coarser tears thereof; manna thuris*. Also to *The officinal rosemary; rosmarinus officinalis*.] ═ *A number*: (Ṣ, Ḳ:) or *a great number*; (Ḳ;) as being likened to the pebbles in multitude. (TA.) You say, نَحْنُ أَكْثَرُ مِنْهُمْ حَصًى *We are more than they in number*. (Ṣ.)

حَصَاةٌ n. un. of حَصًى [q. v.]. (Ṣ, Mṣb, Ḳ.) —— Also [*A stone in the bladder* or *kidneys*: or] *a hardening of the urine in the bladder so that it becomes like a pebble*, or *small stone*. (Ḳ.) —— حَصَاةُ مِسْكٍ *A hard piece that is found in the vesicle of musk*: (Ṣ, TA:) or *any piece of musk*. (Lth, TA.) ═ *Intelligence*, or *understanding*, (Ṣ, Ḳ,) and *judgment*, or *mental perception*; (Ḳ;) as also أَصَاةٌ. (TA.) You say, فُلَانٌ ذُو حَصَاةٍ *Such a one possesses intelligence*, or *understanding*, (Ṣ, TA,) and *judgment*, or *mental perception*: (TA:) [or] *is prudent*, or *discreet*, and *wont to conceal his secret*. (Aṣ, TA.) Or حَصَاة was used by the Arabs not as meaning intelligence, or understanding; but as meaning *Gravity, staidness*, and *forbearance*, or *clemency*. (Ḥar p. 537.) And حَصَاةُ اللِّسَانِ means *Gravity*, or *staidness, in respect of the tongue*. (TA.) ═

The *act of numbering, counting, calculating, reckoning,* or *computing:* a subst. from الإِحْصَآءُ. (Az, IB, TA.)

نَهْرٌ حَصَوِىٌّ : } see أَرْضٌ مَحْصَاةٌ.
أَرْضٌ حَصِيَةٌ :

حَصِىٌّ *Possessing full, sound, and strong, intelligence* or *understanding.* (Ḳ,* TA.)

حَصَاوِىٌّ Bread *made upon the pebbles:* but this is a vulgar term. (TA.)

المُحْصِى one of the [ninety and nine] best names of God; *He who comprehendeth everything by his knowledge; nothing escaping Him thereof, small or great.* (TA.)

مَحْصِىٌّ : see 1, last sentence.

أَرْضٌ مَحْصَاةٌ *A land containing pebbles,* or *small stones:* (Ṣ:) or *abounding therewith;* (Ḳ;) as also ▼ أَرْضٌ حَصِيَةٌ : and in like manner, نَهْرٌ ▼ حَصَوِىٌّ *a river,* or *rivulet,* or *the like, abounding therewith.* (TA.)

حض

1. حَضَّهُ, (Ṣ, A, Mṣb, Ḳ,) aor. ـُ, (Mṣb,) inf. n. حَضٌّ (Ṣ, Mṣb, Ḳ) and ▼ حُضٌّ, (IDrd, Ḳ,) or this latter is a simple subst., (Ṣ, Ḳ,) and ▼ حِضِّيضَى, (Ḳ,) or this also is a simple subst., (Ṣ, TA,) and ▼ حُضِّيضَى, (Ḳ,) or this also is a simple subst., and is the only instance of the measure فُعِّيلَى; and of these last two, the former is the more approved; (TA;) [both of these, accord. to some, have an intensive signification;] *He excited, incited, urged,* or *instigated, him;* syn. حَثَّهُ; (T, Ṣ, A, Ḳ;) or حَمَلَهُ; (Mṣb;) *and roused him to ardour;* (Ḳ;) عَلَيْهِ *to do it;* (Ḳ;) i. e. عَلَى الأَمْرِ *to do the thing;* (Mṣb;) or عَلَى الخَيْرِ *to do good;* (T, A;) or عَلَى القِتَالِ *to fight;* (Ṣ;) or it is used as relating to pace, and to driving, and to any other thing; (M, TA;) and El-Ḥareeree says, on the authority of Kh, that حَثَّهُ is used peculiarly in relation to pace, or journeying, or marching: (TA in art. حث:) and ▼ حَضَّضَهُ, (T, Ṣ, Ḳ,) inf. n. تَحْضِيضٌ, (T, Mṣb,) signifies the same; (Ḳ;) or has a more intensive signification; (Mṣb;) or *i. q.* حَرَّضَهُ [which is equivalent to حَضَّهُ with the addition given above from the Ḳ]; (T, Ṣ;) and this is used as relating to fighting: (T, TA:) or تَحْضِيضٌ signifies the *asking,* or *requiring, with urgency.* (Mughnee voce أَلَا.) It is said in the Ḳur [lxxxix. 19], accord. to the reading of the people of El-Medeeneh, وَلَا يَحُضُّونَ عَلَى طَعَامِ المِسْكِينِ; and accord. to that of El-Ḥasan, تَحُضُّونَ; [*Nor do they,* or *ye, excite* one another *to feed the poor,* or *destitute*]: (TA:) or *nor do they,* or *ye, command to feed the poor,* &c.: (Fr, TA:) and there are other readings thereof: see 3 and 6. (TA.) The grammarians apply to the particles هَلَّا and أَلَّا and لَوْلَا and لَوْمَا the term حُرُوفُ ▼ التَّحْضِيضِ [*Particles of exciting*]; and say that when they are followed by a future, they denote exciting to an action, and seeking or desiring or demanding the performance of it; and when followed by a preterite, reproof for not doing a thing. (Mṣb.)

2. حَضَّضَهُ, inf. n. تَحْضِيضٌ : see 1, in two places.

3. حَاضَّهُ, (TḲ,) inf. n. مُحَاضَّةٌ, (Ṣ, Ḳ,) *He excited him,* &c., as above, *being excited,* &c., *by him.* (Ṣ, Ḳ, TḲ.) In the Ḳur ubi suprà, among the various readings are these two: وَلَا يُحَاضُّونَ and وَلَا تُحَاضُّونَ, meaning, accord. to Fr, *Nor are they,* or *ye, mindful.* (TA.)

6. تَحَاضُّوا *They excited,* &c., *one another.* (Ṣ, Ḳ.) In the Ḳur ubi suprà, some read وَلَا تَحَاضُّونَ (Fr, Ṣ, TA) *Nor do ye excite one another.* (Fr.)

حُضٌّ, said to be a subst.: see 1.

حَضِيضٌ *A depressed piece* (قَرَار) *of ground at the place where a mountain ends;* (Ṣ;) or *a depressed piece* (قرار) *of ground at,* or *by, the* سَفْح [i. e. *foot,* or *bottom,* or *lowest part,*] *of a mountain;* or *in,* or *at, the lower,* or *lowest, part thereof,* (فِى أَسْفَلِهِ,) *the* سفح *being behind the* حضيض; *the* حضيض *being in the part next the* سفح, *and the* سفح *being* دُونَ ذٰلِكَ [which seems to be virtually a repetition, signifying *behind that part,* or it may mean *above that part*]: (TA:) or *i. q.* سَفْحٌ: (Ḳ in art. سفح:) or *a depressed piece* (قرار) *of ground:* (A:) or *a tract* (حَجْز) [in the CḲ, erroneously, حَجَر]), and *a depressed piece* (قرار), *in land* or *ground:* (Ḳ:) and *any low piece* (سَافِل) *of land* or *ground:* (TA:) and the *ground* [in an absolute sense]: (Ṣ:) pl. [of pauc.] أَحِضَّةٌ and [of mult.] حُضُضٌ. (Ḳ.) It has the last of the above-mentioned significations in a certain trad., in which it is related that a present being brought to Moḥammad, he, not finding anything upon which to put it, said, ضَعْهُ بِالحَضِيضِ, meaning بِالأَرْضِ; [i. e. *Put thou it upon the ground;*] adding, for I am only a servant; I eat as the servant eateth. (Ṣ.)

حُضِّيضَى and حِضِّيضَى, said to be substs.: see 1.

حضأ

1. حَضَأَ النَّارَ, (Ṣ, Ḳ,) aor. ـَ, (Ḳ,) *He lighted,* or *kindled, the fire;* or *made it to burn, burn up, burn brightly* or *fiercely, blaze,* or *flame:* (Ṣ, Ḳ:) or *he opened it,* (Ḳ,) i. e. *stirred it,* (TA,) *in order that it might blaze,* or *flame;* as also ▼ احتضأها. (Ḳ.) It is also without ء. (Ṣ. [See art. حضو.]) — حَضَأَتِ النَّارُ *The fire became lighted* or *kindled;* or *it burned, burned up,* &c. (Ḳ.)

8: see above.

أَبْيَضُ حَضِىءٌ, (Ḳ,) so accord. to the correct copies of the Ḳ, but in some copies حَضِىٌّ, (TA,) *Very white.* (Ḳ.)

مِحْضَأٌ *A stick,* or *piece of wood, with which a fire is stirred;* as also مِحْضَاةٌ; (Ṣ, Ḳ;) the latter accord. to those who pronounce the verb without ء. (Ṣ.)

حضر

1. حَضَرَ, aor. ـُ; (Ṣ, A, Mṣb, Ḳ, &c.;) and حَضِرَ, (AA, Kh, Lth, Fr, Ṣ, Mṣb, Ḳ, &c.,) aor. as above, (Kh, Lth, Fr, Az, Ṣ, Mṣb, &c.,) not ـَ, as is implied in the Ḳ; but the latter form of the pret. is disallowed by some; (MF;) and, with its aor., is an instance of the intermixture of dialects; (Mṣb;) and is like فَضِلَ, aor. ـُ, and نَعِمَ, aor. ـُ; which are said by IḲooṭ to be the only instances of the kind; (MF;) inf. n. حُضُورٌ (Ṣ, Mṣb, Ḳ) and حَضَارَةٌ; (Ḳ;) and ▼ احتضر, and ▼ تحضّر; (Ḳ;) *He was,* or *became, present; contr. of* غَابَ: (Ṣ, Ḳ:) *he came after having been absent.* (Mṣb.) — حَضَرَتِ الصَّلَاةُ, (Lth, A, L, Mṣb,) and, as the people of El-Medeeneh say, حَضِرَت, but all say تَحْضُرُ, (Lth, L,) originally حَضَرَ وَقْتُ الصَّلَاةِ, ‡*The time of prayer came,* or *arrived.* (Mṣb.) — [حَضَرَ also signifies † *He,* or *it, was,* or *became, ready,* or *prepared.* See 4; and see also حَاضِرٌ.] = حَضَرَهُ, (AA, Fr, A, Mgh, Ḳ, &c.,) and حَضِرَهُ, (AA, Fr, &c.,) aor. and inf. ns. as above; (TA;) and ▼ احتضره, (Mgh, TA,) and ▼ تحضّره; (Ḳ;) *He was, or became, present with him; attended him; came into his presence; came to him:* (Ḳ, &c.:) and *he was,* or *became, present at it,* or *in it;* namely, a place. (Mgh.) One says, حَضَرَتِ القَاضِىَ امْرَأَةٌ, (Az,) and حَضَرَت, (Fr, Ṣ,) and حَضَرَ, in which the ت is elided because القاضى intervenes between the verb and امرأة, (Sh,) but the first is the most approved, (Az,) [*A woman came into the presence of,* or *presented herself before,* or *came to, the judge.*] And حَضَرْتُ مَجْلِسَ القَاضِى, aor. ـُ, inf. n. حُضُورٌ, *I was present at,* or *attended, the court of the judge.* (Mṣb.) [And حَضَرَ دَرْسًا *He attended a lecture.*] And حَضَرُوا المِيَاهَ *They stayed,* or *dwelt, by the waters.* (Ṣ. [See حَاضِرٌ.]) — أَعُوذُ بِكَ رَبِّ أَنْ يَحْضُرُونِ [in the Ḳur xxiii. 100] means [*I seek thy protection, O my Lord,*] *from their* (the devils') *bringing evil upon me:* (Ṣ:) or [*from their being present with me:* or] *hovering around me.* (Ksh, Bḍ.) — الجِنُّ تَحْضُرُ اللَّبَنَ, (Ṣ, Ḳ,) or ▼ تَحْتَضِرُهُ, (T, TA,) † [*The jinn, or genii, come to, and taint, the milk.*] — حُضِرَ, (A,) and ▼ اُحْتُضِرَ, (A, Mgh, Ḳ,) *i. q.* حَضَرَهُ المَوْتُ, (A, Ḳ,) i. e. ‡[*He was visited by the angel of death;*] *he became at the point of death; in the agony of death;* as also ▼ اِحْتَضَرَهُ المَوْتُ: (Mṣb:) or *he was visited by death,* or *by the angels of death;* meaning *he died:* (Mgh:) or ▼ اُحْتُضِرَ means *he died a youth.* (Ṣ and TA voce أُجْزِرَ, q. v.) — حَضَرْنَا عَنْ مَآءِ كَذَا ‡ *We removed from such a water.* (Ḳ, TA.) — حَضَرْتُ الأَمْرَ ‡ *I was present at the affair,* or *event.* (A.) — حَضَرْتُ الأَمْرَ بِخَيْرٍ ‡ *I formed a right opinion,* or *judgment, respecting the thing,* or *affair.* (A.) — حَضَرَهُ الهَمُّ, and ▼ تحضّره, and ▼ احتضره, ‡ [*Anxiety befell him.*] (Ṣ, A.) — حَضَرَنِى كَذَا † *Such a thing occurred to my mind.* (Mṣb.) And قُولُوا مَا يَحْضُرُكُمْ †[*Say ye what is in your minds;* or] *what is ready with you.* (TA from a trad.) = حَضَرَ,

(Mṣb,) inf. n. حِضَارَةٌ, (AZ, Ṣ, Ḳ,) or حَضَارَةٌ, (Aṣ, Ṣ, A,) or both, (Mṣb,) [see بدا, the contr. of حَضَرَ, in art. بدو,] *He resided, dwelt,* or *abode, in a region, district,* or *tract, of cities, towns,* or *villages, and of cultivated land;* (Ṣ, Mṣb, Ḳ;) [as also ▼تحضّر: or this latter signifies *he became an inhabitant of such a region, district,* or *tract:*] you say ▼بَدَوِىٌّ يَتَحَضَّرُ [*an inhabitant of the desert who becomes an inhabitant of a region, district,* or *tract, of cities* &c.]; and [contr.] حَضَرِىٌّ يَتَبَدَّى. (A.) [See also 8.]

3. حَاضَرْتُهُ, (A, TA,) inf. n. مُحَاضَرَةٌ, (TA,) *I witnessed it; saw it,* or *beheld it, with my eye.* (A, TA.) — مُحَاضَرَةٌ between people is One's *giving to another the answer,* or *reply, that presents itself to him:* and حاضر الجَوَابَ signifies *He gave the answer,* or *reply, readily,* or *presently.* (Ḥar p. 180.) — حَاضَرْتُهُ, (Ṣ,) inf. n. as above, (Ḳ,) [also] signifies *I sat with him, with my knee to his knee, each of us sitting upon his knees,* in contending or disputing, (جَاثَيْتُهُ, Ṣ, Ḳ,*) *in the presence of the Sulṭán:* (Ṣ, Ḳ:) the meaning is similar to that of مُغَالَبَةٌ and مُكَاثَرَةٌ, (Ṣ,) or مُكَابَرَةٌ [which seems to be the right reading]. (TA.) — [And حاضرهُ بِكَذَا *He disputed, debated,* or *bandied words, with him respecting such a thing.*] — And حاضرهُ بِحَقِّهِ, (Lth, TA,) inf. n. as above, (Lth, Ḳ,) *He contended,* or *disputed, with him for his* (the latter's) *right,* or *due, and overcame him, and went off with it:* (Lth, Ḳ:) and مُجَالَدَةٌ, also, [which is one of the explanations assigned to مُحَاضَرَةٌ in the Ḳ,] is syn. with مُحَاضَرَةٌ as the inf. n. of the verb in this sense [unless it be a mistranscription for مُجَادَلَةٌ, which I think not improbable]. (TA.) — Also حَاضَرْتُهُ, (Ṣ, A,) inf. n. as above, (Ḳ,) *I ran with him:* (Ṣ, Ḳ:) or *I vied,* or *contended, with him in running;* syn. عَادَيْتُهُ; from الحُضْرُ. (A.)

4. احضرهُ, (Ṣ, A, Ḳ,) [inf. n. إِحْضَارٌ,] *He caused him,* (Ṣ, A,) or *it,* (Ḳ,) *to be present; he brought him,* or *it.* (Ṣ, Ḳ.) [It is also doubly trans.] You say, احضرهُ إِيَّاهُ *He caused him,* or *it, to be present with him, to attend him, to come into his presence,* or *to come to him;* or *he brought him,* or *it, to him.* (Ḳ.) And طَلَبْتُ فُلَانًا فَأَحْضَرَنِيهِ صَاحِبُهُ [*I demanded such a one, and his companion caused him to come to me,* or *brought him to me*]. (A.) [Hence,] أَحْضِرْ ذِهْنَكَ ‡[*Summon thine intellect; have thy wits about thee*]. (A.) — Also †*He made it ready,* or *prepared it;* syn. أَعَدَّهُ. (TA in art. عد.) — احضر, (Ṣ,) inf. n. إِحْضَارٌ; (Ṣ, A, Ḳ, &c.;) and ▼احتضر; (Ṣ;) *He* (a horse, Ṣ, Ḳ, and a man, Kr) *ran;* syn. عَدَا: (Ṣ:) or *rose in his running;* [app. meaning *trotted;*] syn. ارْتَفَعَ فِى عَدْوِهِ. (Ḳ.)

5: see 1, in five places.

8: see 1, in seven places. — [احتضر also signifies *He came to a region, district,* or *tract, of cities, towns,* or *villages, and of cultivated land.* See مُحْتَضَرٌ, voce حَاضِرٌ; and see also حَضَرَ, last signification.] — See also 4.

10. استحضرهُ *He desired,* or *demanded, his presence.* (A.) [*He desired,* or *required,* or *requested, that he,* or *it, should come,* or *be brought.*] — *He made him* (a horse) *to run;* syn. أَعْدَاهُ. (Ṣ.)

حَضْرٌ The *intruding uninvited at feasts.* (IAạr, Ḳ.)

حُضْرٌ (Az, Ṣ, Ḳ) and [in poetry] ▼حُضُرٌ (Ḥam p. 277) and ▼حِضَارٌ (Az, TA) *A run,* or *running;* syn. عَدْوٌ: (Ṣ:) or the *rising* of a horse *in running;* [app. meaning *trotting;*] syn. ارْتِفَاعُ فَرَسٍ فِى عَدْوِهِ: (Ḳ:) or *vehement running.* (Ḥam p. 277, in explanation of حُضُرٌ.) It is said in a trad., أَقْطَعَ ٱبْنَ ٱلزُّبَيْرِ حُضْرَ فَرَسِهِ بِأَرْضِ ٱلْمَدِينَةِ [*He assigned to Ibn-Ez-Zubeyr the extent of his horse's run in the land of El-Medeeneh*]. (TA.)

حَضَرٌ: see حَضْرَةٌ. — Also, (Ṣ, A, Mṣb, Ḳ,) and ▼حَاضِرَةٌ (Ṣ, A, Ḳ) and ▼حِضَارَةٌ and ▼حَضَارَةٌ, (Ḳ,) [or the last two are app. only inf. ns. of حَضَرَ as contr. of بَدَا,] *A region, district,* or *tract, of cities, towns,* or *villages, and of cultivated land;* (Ṣ;) *contr. of* بَدْوٌ (Ṣ, A, Mṣb) *and* بَادِيَةٌ: (Ṣ, Ḳ:) pl. [of the second] حَوَاضِرُ. (A.) You say, هُوَ مِنْ أَهْلِ الحَضَرِ (A) and ▼الحَاضِرَةِ (Ṣ, A) and الحَوَاضِرِ (A) *He is of the people of the region,* or *regions,* &c., *of cities, towns,* or *villages, and of cultivated land;* (Ṣ, A;*) *contr. of* مِنْ أَهْلِ البَادِيَةِ. (Ṣ.) — And the first signifies also *Residence at home;* contr. of سَفَرٌ. (M and Ḳ in art. سفر.)

حَضِرٌ: see حَضْرٌ: — and حَاضِرٌ.

حَضِرٌ One *who intrudes uninvited at feasts; a smell-feast; a spunger;* (TA;) *one who watches for the time of* (يَتَحَيَّنُ) *the feeding of others, in order that he may attend it;* as also ▼حَضُرٌ, (Ḳ,) and ▼حُضُرٌ. (IAạr, Ḳ, TA.) — A man *unfit for journeying:* (T, Ṣ:) or one *who does not desire journeying:* or i. q. حَضَرِىٌّ. (Ḳ.)

حَضُرٌ: see حَضِرٌ: — and حُضْرٌ.

حَضْرَةٌ, originally an inf. n., signifying *Presence:* and afterwards applied to signify ‡*a place of presence* [as also the several forms occurring in the following phrases]. (MF.) You say, كَلَّمْتُهُ بِحَضْرَةِ فُلَانٍ, (Ṣ, A,* Mṣb,) and كَانَ ذٰلِكَ بِحَضْرَتِهِ (Ḳ,* TA,) and ▼حُضْرَتِهِ and ▼حِضْرَتِهِ (Ṣ, Ḳ) and ▼حَضَرِهِ (Yaạkoob, Ṣ, Mṣb, Ḳ) and ▼حَضَرَتِهِ (Ḳ) and ▼مَحْضَرِهِ, (Ṣ, A, Mṣb, Ḳ,) all syn. expressions, (Ḳ,) meaning ‡[*I spoke to him,* and *that was* or *happened,*] *in the presence,* i. e. *the place of presence, of such a one.* (Ṣ, A, Mṣb.) And ▼فُلَانٌ حَسَنُ الحَضْرَةِ (Ṣ, M, A, Ḳ) and ▼الحِضْرَةِ (Ṣ, M) ‡*Such a one is a person whose presence is attended by good.* (Ḳ.) And غَطِّ إِنَاءَكَ بِحَضْرَةِ الذُّبَابِ ‡[*Cover thy vessel in the presence of the flies,* lest they taint it.] (A, TA. [Or perhaps this is a mistranscription, for يَحْضُرُهُ الذُّبَابُ, meaning, if thou do not, *the flies will come to it,* and taint it.]) — It is also applied as a title, by writers of letters and the like, to any great man with whom people are wont to be present; [and sometimes to God; and meaning ‡*The object of resort;*] as in the phrase, الحَضْرَةُ العَالِيَةُ تَأْمُرُ بِكَذَا ‡[*The exalted object of resort commands such a thing*]. (MF.) [It is similar to الجَنَابُ; but is generally considered as implying greater respect than the latter. It is often prefixed to the name of the person to whom it is applied, or to a pronoun: as حَضْرَةُ فُلَانٍ ‡*The object of resort, such a one:* and حَضْرَتُكَ ‡*The object of resort, thyself.*] — Also ‡The *vicinity* of a thing, (T, A,) and of a man. (Ṣ. [So accord. to two copies of the Ṣ; but الرَّجُلِ is there an evident mistranscription, for الرَّحْلِ, "of the house," or "abode:" see what follows.]) You say, كُنْتُ بِحَضْرَةِ الدَّارِ ‡*I was in the vicinity of,* or *near to, the house.* (T, A.) And كُنَّا بِحَضْرَةِ مَآءٍ ‡*We were by a water.* (TA from a trad.) And بِحَضْرَةِ المَآءِ ‡*In the vicinity of,* or *near to, the water.* (A.) — Also The فِنَآء of a رَجُل. (Ṣ. [So accord. to two copies of the Ṣ; where it is said, حَضْرَةُ الرَّجُلِ قُرْبُهُ وَفِنَآؤُهُ: but the right reading is evidently الرَّحْلِ: so that the second of the two meanings thus explained is, The *court,* or *yard, in front,* or *extending from the sides,* of a house, or an abode.]) — And ‡*Apparatus for building, such as baked bricks, and gypsum-plaster:* so in the saying, جَمَعَ الحَضْرَةَ يُرِيدُ بِنَآءَ دَارٍ ‡[*He collected the apparatus, such as the baked bricks, &c., desiring to build a house*]. (A.) — See also حَضَرٌ.

حُضْرَةٌ, حِضْرَةٌ, حَضَرَةٌ: see حَضْرَةٌ, in five places.

حَضَرِىٌّ *An inhabitant of a region, district,* or *tract, of cities, towns,* or *villages, and of cultivated land;* (Ṣ, A,* Mṣb;) opposed to بَدَوِىٌّ. (Ṣ, A.) [See also حَضِرٌ.]

حَضَارِ [an imperative verbal n.] *Be thou present.* (A.) — Also *A certain star,* (Ṣ, Ḳ,) *upon the right hoof of Centaurus:* upon his other fore leg is الوَزْنُ. (Ḳzw.) It is said, حَضَارِ وَالوَزْنُ مُحْلِفَانِ [*Ḥaḍári and El-Wezn are two causes of swearing*]: they are two stars that rise before Canopus (Suheyl); and when either of them rises, it is thought to be Canopus, because of their resemblance to it: (AA, Ṣ:*) they are termed محلفان because of the disagreement of their beholders when they rise; one swearing that the one rising is Canopus, and another swearing that it is not. (AA, TA.) Th says that it is *a dim, distant, star;* and cites this verse:

* أَرَى نَارَ لَيْلَى بِالعَقِيقِ كَأَنَّهَا *
* حَضَارِ إِذَا مَا أَعْرَضَتْ وَفُرُودُهَا *

I see the fire of Leylá, in El-'Akeek, dim in the distance, *as though it were Ḥaḍári, when it appears, with its Furood,* which are dim stars around Ḥaḍári. (TA.) — حَضَارٌ: see what next follows.

حِضَارٌ (Ṣ, Ḳ) and † حَضَارٌ (Ḳ) *White:* (Sh, T:) or *excellent and white:* (Ṣ, Ḳ:) or *red:* (Ḳ:) but this requires consideration: (TA:) applied to camels, and to a single camel: (Ṣ, Ḳ:) or *having no sing.* (Ḳ.) And the former, *A she-camel combining strength with excellence of pace:* (El-Umawee, T, Ṣ, Ḳ:) but Sh says that he had not heard it used in this sense; and that it only signifies "white," as applied to camels. (TA.) —— See also حَضْرٌ.

حَضِيرٌ ‡ One *who always forms right opinions,* or *judgments, respecting things,* or *affairs.* (A.) —— See also حَضِيرَةٌ, in three places.

حَضَارَةٌ and حِضَارَةٌ: see حَضْرٌ.

حَضِيرَةٌ The *collective body* of a people: (Fr, Ḳ:) so in the following ex., (Fr,) from a poem of Selmà El-Juhaneeyeh, in which she bewails the death of her brother As'ad, and celebrates his praises: (Ṣ:)

* يَرِدُ المِيَاهَ حَضِيرَةً وَنَفِيضَةً *

نفيضة signifying the same: (Fr:) [so that the meaning is, *Coming to the waters in a collective and congregated body*:] or the former signifies waters *by which people are dwelling,* or *staying;* and the latter, "by which there is not any one:" (IAạr, Sh:) or the former, people *dwelling,* or *staying, by the waters;* and the latter, men "going before an army as scouts, or explorers:" (Aṣ:) but what IAạr says, mentioned above, is better: (Az:) or the former, *a company of seven,* or *eight, men;* and the latter, "one;" and also men "who explore a place thoroughly:" (A'Obeyd:) or the former, *a company of four,* or *five, men,* (Ṣ, Ḳ,) *engaged in a warring and plundering expedition:* (Ṣ:) or *seven:* (TA:) or *eight:* or *nine:* (Ḳ: in some copies of the Ḳ "seven;" but the former is the right reading: TA:) or *ten:* or *a company of men not more than ten* (نَفَرٌ) *with whom one goes on a warring and plundering expedition:* (Ḳ:) or, accord. to AAF and the M and Ḳ, the *foremost,* or *preceding, portion of an army:* and accord. to IB, نفيضة signifies "a party sent to a place to discover whether there be there an enemy or any cause of fear:" (TA:) pl. حَضَائِرُ. (Ṣ.) —— *A place where dates are dried:* (El-Báhilee, ISk, Az, Mgh, Mṣb, Ḳ:) because frequented: pl. as above. (Mgh.) [See also حَضِيرَةٌ and حَظِيرَةٌ.] —— Also, (Ṣ,) or † حَضِيرٌ, (Ḳ, TA,) *What collects in a wound,* (Ṣ, Ḳ,) *of thick purulent matter.* (Ṣ.) —— And the former *What collects in the membrane that encloses the fœtus, of the* [*fluid called*] سُخْد, (Ṣ,) and *the like.* (TA.) You say, أَلْقَتِ الشَّاةُ حَضِيرَتَهَا *The ewe,* or *she-goat, ejected her* سُخْد *and water and blood, after having brought forth.* (Ṣ.) —— And *What a woman emits after childbirth and* [*after*] *the stopping of her blood:* and † حَضِيرٌ is its pl. [or a coll. gen. n.]. (Ḳ. [Or, accord. to some copies of the Ḳ, and the TA, The *stopping of her blood,* or *its ceasing to flow,* is a signification distinct from what precedes it.)] —— And *What a she-camel emits after bringing forth:* or, accord. to AO, the *membrane that envelops the fœtus.* (TA.) —— And (Ḳ, TA, [in the CḲ "or"]) † the latter signifies *Thick blood which collects in the membrane that encloses the fœtus.* (Ḳ,* TA.)

حَاضِرٌ A man *present:* (A, Ḳ:) pl. [حَاضِرُونَ and] حُضَّرٌ and [more commonly] حُضُورٌ, (Ṣ, Ḳ,) which last is originally an inf. n. (Ṣ.) You say, فَعَلْتُهُ وَفُلَانٌ حَاضِرٌ *I did it such a one being present.* (A.) And هُوَ مِنْ حَاضِرِى المَلِكِ [*He is of those who are in the presence of the king*]. (A.) —— So, too, applied to a time: and you say, وَقْتُهَا حَاضِرٌ, for الصَّلَاةُ حَاضِرَةٌ, ‡ *The time of prayer is come.* (Mṣb.) —— [Also † *Ready,* or *prepared:* often used in this sense in the lexicons &c., as in modern Arabic. See 4.] You say, قُولُوا مَا هُوَ حَاضِرٌ عِنْدَكُمْ † *Say ye what is ready with you* [or *in your minds*]. (TA.) And هُوَ حَاضِرٌ بِالجَوَابِ ‡ [*He is ready with the answer,* or *reply*]; and بِالنَّوَادِرِ ‡ [*with rare words* or *expressions*]; (A;) as also † حَضِرٌ: (TA:) which latter word, alone, signifies a man *having the quality of perspicuity of speech, and intelligence;* syn. ذُو البَيَانِ وَالفِقْهِ. (Ḳ.) —— *A visiting angel:* and † حَاضِرَةٌ is applied to *a class* or *company* [*of visiting angels*]. (TA.) —— One *coming to a region, district,* or *tract, of cities, towns,* or *villages, and of cultivated land;* contr. of بَادٍ; (Ṣ, Ḳ;) as also † مُحْتَضِرٌ. (Ṣ.) —— A man *staying, residing, dwelling,* or *abiding,* بِمَوْضِعٍ *in a place.* (Ṣ.) —— [A man, or people,] *staying,* or *dwelling, by water;* (Ṣ,* TA;) contr. of بَادٍ: (Ḳ:) pl. حُضُورٌ (TA) and حَضَرَةٌ and حُضَّارٌ: (Ṣ:) one says, مَا عَلَى المَآءِ حَاضِرٌ [*There is not any one staying,* or *dwelling, by the water*]: and هٰؤُلَآءِ قَوْمٌ حُضَّارٌ and مَحَاضِرُ [which is pl. of † مَحْضَرٌ, a syn. of حَاضِرٌ in this sense; i. e. *These are a people staying,* or *dwelling, by water*]: (Ṣ:) or حَاضِرٌ signifies any people *that have alighted and taken up their abode by a constant source of water, and do not remove from it in winter nor in summer, whether they have alighted and taken up their abode in towns or villages, and cultivated land, and houses of clay, or pitched their tents by the water, and remained there, and sustained their beasts with the water and herbage around them:* (TA:) or حَىٌّ حَاضِرٌ, without ة, signifies *a tribe that has alighted and is abiding by a constant source of water:* (T, TA:) and † حَاضِرَةٌ and حَاضِرُونَ, *a people staying,* or *dwelling, by waters; alighting there in the hottest time of summer:* when the weather becomes cool, they migrate from the constant sources of water, and go into the desert, seeking the vicinity of herbage; and then they are termed بَادِيَةٌ and بَادُونَ. (T in art. بدو.) —— Also *A great tribe* (Ṣ, Ḳ) or *company of men:* (TA:) or *a tribe,* (ISd,) or *company of men,* (Mgh,) *when staying, or dwelling, in the abode which is their place of assembling;* (ISd, Mgh;) as also † حَاضِرَةٌ. (Mgh.) One says حَاضِرُ طَيِّئٍ *The great tribe of Ṭeiyi.* (Ṣ.) It is a pl., (Ṣ,) or coll. n., (ISd,) [or quasi-pl. n.,] like سَامِرٌ and حَاجٌّ (Ṣ, ISd) for سُمَّارٌ and حُجَّاجٌ. (Ṣ.) —— Also, of the measure فَاعِلٌ in the sense of the measure مَفْعُولٌ, (TA,) *A place where people are present;* or *where people stay,* or *dwell, by waters:* syn. مَكَانٌ مَحْضُورٌ: one says, نَزَلْنَا حَاضِرَ بَنِى فُلَانٍ [*We alighted and took up our abode,* or *sojourned, at the place where the sons of such a one were present;* or *were staying,* or *dwelling, by waters*]. (El-Khaṭṭábee.) [See also مَحْضَرٌ.]

حَاضِرَةٌ: see حَاضِرٌ, in three places: —— and see حَضَرٌ, in two places.

مَحْضَرٌ *A place where people are present,* or *assembled.* (Ḳ,* TA.) See also حَضْرَةٌ. —— *A place to which people return* (مَرْجِعٌ [here a n. of place, agreeably with analogy,]) *to the waters,* (Ṣ, Ḳ,) or *to the constant sources of water;* (T, TA;) contr. of مَبْدًى: (T and Ṣ in art. بدو:) a place to which one goes (مَذْهَبٌ) in search of herbage is called مُنْتَجَعٌ; and every such place is called مَبْدًى, of which the pl. is مَبَادٍ: *watering-places* (مَنَاهِل) are called مَحَاضِرُ [pl. of مَحْضَرٌ] because of the congregation and presence of men at them. (T, TA.) [See also حَاضِرٌ, last signification.] —— [*People present,* or *assembled; an assembly:* so in the present day.] —— *A people dwelling,* or *staying, by waters:* (Ḳ,* TA:) [pl. مَحَاضِرُ:] see حَاضِرٌ. —— The *record of a ḳáḍee* (or *judge*), *in which his sentence is written,* syn. سِجِلٌّ: (Ṣ, Ḳ:) or *what is written when a person brings a charge against another:* when the latter makes his reply, and proves it, it [the writing] is [called] سِجِلٌّ; and when judgment is given, تَوْفِيقٌ. (Kull p. 352.) This is thought by MF to be a recent conventional term; but it has been heard from the Arabs [of the classical times], and is mentioned by ISd and others. (TA.) —— Also *A signature* (خَطٌّ) *that is written at the end of the record of the signatures* (خُطُوط) *of the witnesses, in testimony of the truth of the contents of what precedes.* (Ḳ. [In the CḲ, وَاقِعَة is erroneously put for وَاقِعَة; and خُطُوط, for خُطُوط.]) But this is a recent conventional application. (MF, TA.) —— فُلَانٌ حَسَنُ المَحْضَرِ † *Such a one is a person who speaks well of the absent.* (Ṣ.)

مِحْضِيرٌ: see مِحْضَارٌ.

مَحْضُورٌ [pass. part. n. of حَضَرَهُ]. [Hence,] اللَّبَنُ مَحْضُورٌ, (Ṣ, A, Ḳ,) and † مُحْتَضَرٌ, (Ṣ, A,) فَغَطِّ إِنَآءَكَ, (Ṣ,) ‡ *Milk is much subject to taint,* or *much tainted;* [lit.] *come to* [*and tainted;* i. e.,] by the jinn, or genii, (Aṣ, T, Ṣ, Ḳ,) and beasts, &c.; (Aṣ, T;) [*therefore cover thou thy vessel.*] And (in like manner [one says], Ḳ) الكُنُفُ مَحْضُورَةٌ † [*Privies are haunted* by jinn, or genii]. (Ṣ, Ḳ.) It is said in a trad., إِنَّ هٰذِهِ الحُشُوشَ مُحْتَضَرَةٌ † † [*Verily these privies are haunted* by jinn]. (TA.) And in another trad., إِنَّهَا مَشْهُودَةٌ مَحْضُورَةٌ *Verily it* (the prayer of daybreak) *is attended* by the angels of the night and the day. (TA.) —— Also, (Mṣb,) and † مُحْتَضَرٌ, (Mgh, Mṣb,) ‡ *At the point of death; in the agony of death:* (Mṣb:) [*visited by death;* or *by the angel,* or *angels, of death:* (see 1:)] or the latter, *near to death.* (Mgh.)

مِحْضِيرٌ, applied to a horse, (Ṣ, A, Ḳ, &c.,) and to a mare, (Ṣ, M,) *That runs much,* or *vehemently;* syn. كَثِيرُ العَدْوِ, (Ṣ,) or شَدِيدُ الحُضْرِ; (M;) as also ↓ مِحْضَارٌ, applied without ة to a mare; (M;) or this latter is not allowable; (Ṣ, Ḳ;) or is of weak authority: (Ḳ:) pl. [of both] مَحَاضِيرُ. (A.)

مُحْتَضَرٌ: see مَحْضُورٌ, in three places. Also †*A man afflicted by demoniacal possession,* or *insanity,* or *madness.* (TA.) كُلُّ شِرْبٍ مُحْتَضَرٌ, in the Ḳur liv. 28, *Every share of the water shall be come unto* in turn, means, the water shall be come to by the people on their day, and by the she-camel on her day: (Jel:) or it means, the people shall come to their shares of the water, and the she-camel shall come to her share thereof. (Ḳ.)

مُحْتَضِرٌ: see حَاضِرٌ.

حضن

1. حَضَنَ الصَّبِىَّ, (Ḳ,) aor. ـُ, (TA,) inf. n. حَضْنٌ and حِضَانَةٌ, *He put the child in his* حِضْن [i. e. *under his arm,* or *in his bosom*]: or *he nourished him, reared him, fostered him, brought him up,* (Ḳ, TA,) *and took care of him;* (TA;) as also ↓ احتضنهُ. (Ḳ, TA.) And حَضَنَتْ وَلَدَهَا, (Ṣ, Mgh,) aor. ـُ, (Mgh, TA,) inf. n. حِضَانَةٌ, (Mgh,) said of a woman, (Ṣ, Mgh, TA,) *She put her child in her* حِضْن, *and* [*thus*] *carried him* [*under her arm,*] *on one of her two sides:* (TA:) it has a similar meaning to the phrase next following: (Ṣ:) or it means *she had charge of her child, and carried him, and reared him,* or *fostered him.* (Mgh.) —— حَضَنَ بَيْضَهُ, (Ṣ, Mgh, Mṣb, Ḳ,) and عَلَى بَيْضِهِ, (TA,) aor. ـُ, (Ṣ, Mṣb,) inf. n. حَضْنٌ (Mgh, Mṣb) and حِضَانَةٌ (Mṣb, Ḳ) and حِضَانٌ and حُضُونٌ, (Ḳ,) said of a bird, (Ṣ, Mgh, Mṣb, Ḳ,) *He pressed,* or *compressed, his eggs* (Ṣ, Mṣb) *to himself,* (Ṣ,) *beneath his wing,* (Ṣ, Mṣb,) or *beneath his two wings;* (so in some copies of the Ṣ;) *he sat upon his eggs, protecting them with his two sides* (بِحِضْنَيْهِ); (Mgh;) *he brooded upon his eggs to hatch them:* (Ḳ:) as also ↓ احتضن. (KL.) —— حَضَنَ بَيْضَةً تَحْتَ دَجَاجَةٍ لَهُ حَتَّى أَفْرَخَتْ, meaning *He put an egg beneath a hen belonging to him, and made her to sit* [or *brood*] *upon it* [*until it became hatched*], if remembered to have been heard [from any of the Arabs of pure speech], is a tropical usage of the verb, like as when one says "The Emeer built the city:" otherwise, it is correctly [↓ حَضَّنَ,] with teshdeed. (Mgh.) —— حَضَنَهُ عَنْ كَذَا, inf. n. حَضْنٌ and حِضَانَةٌ, ‡ *He made him to turn away, withdraw,* or *retire, from such a thing, and had it to himself exclusively;* (Ṣ, Ḳ, TA;) as though he put him aside from it, or by its side: *he excluded him from participation in it;* in which sense ↓ أَحْضَنَهُ مِنْهُ is disapproved: (TA:) *he impeded him,* or *debarred him, from it.* (ISd, TA.) It is related in a trad. of Ibn-Mes'ood that, when he made his will, he said, وَلَا تُحْضَنُ زَيْنَبُ عَنْ ذٰلِكَ, meaning †*And Zeyneb* (his wife) *shall not be precluded from looking into that and executing it;* namely, his will: or *shall not be precluded from it, nor shall any matter* [*relating to it*] *be decided without her.* (TA.) And you say also, حَضَنَهُ عَنْ حَاجَتِهِ *He withheld him from the object of his want;* as also ↓ احتضنهُ. (Ṣ, ISd, Ḳ.) And حَضَنَ مَعْرُوفَهُ, (Ḳ,) and حَدِيثَهُ, (TA,) عَنْ جِيرَانِهِ, (Ḳ,) and مَعَارِفِهِ, (TA,) inf. n. حَضْنٌ, (Ḳ,) †*He turned his beneficence,* (Ḳ, TA,) and *his discourse,* (TA,) *from his neighbours,* (Ḳ, TA,) and *his acquaintances, to others:* on the authority of Lḥ. (TA.) ═ حَضِنَتْ, aor. ـَ, inf. n. حَضَانٌ, (Ḳ,) or this is a simple subst., (A 'Obeyd, TA,) *She* (a ewe [or goat], and a camel, and a woman,) *had one of her teats,* or *breasts, larger than the other.* (Ḳ.) [See حَضُونٌ.]

2: see 1.

3. فُلَانٌ يُحَاضِنُ النِّسَآءَ [*Such a one indulges himself with women in mutual embracing* or *pressing to the bosom*]. (IAạr, TA in explanation of the epithet عُقَرَةٌ, q. v.)

4. احضن الطَّائِرَ البَيْضَ *He made the bird to sit* [or *brood*] *upon the eggs.* (Mṣb.) —— أَحْضَنَهُ مِنْهُ: see 1. —— احضن بِحَقِّى ‡ *He went away with,* or *took away, my right,* or *due;* (Ḳ, TA;) as though he put it by his side. (TA.) —— احضن الرَّجُلَ, (AZ, Ṣ, Ḳ,*) and احضن بِهِ, (Ḳ,) i. q. أَزْرَى بِهِ † [*He held him in little,* or *light,* or *mean, estimation,* or *in contempt;* &c.]. (AZ, Ṣ, Ḳ.)

[6. تحاضنوا *They embraced one another,* or *pressed one another to the bosom.* See also 3.]

8. احتضنهُ *He put it* (a thing) *in his* حِضْن [i. e. *under his arm,* or *in his bosom*]: (Ṣ, Mṣb:) *he took it up, and put it in his* حِضْن, *like as a woman takes up her child, and carries him* [*in her* حضن or] *on one of her two sides.* (TA.) —— See also 1, in three places.

حُضْنٌ: see what next follows.

حِضْنٌ *The part beneath the armpit,* (Ṣ, Mgh, Mṣb, Ḳ,) *extending to the* كَشْح [or *flank*]: (Ṣ, Mṣb, Ḳ:) or the *bosom,* or *breast;* syn. صَدْر: [الصُّدُر in the CḲ should be الصَّدْرُ:] and the *upper arms with what is between them:* (Ḳ:) and ↓ مُحْتَضَنٌ signifies the same: (Ṣ:) pl. of the former أَحْضَانٌ (Mṣb, Ḳ*) [and accord. to Freytag's Lex. حُضُونٌ also]. —— The *side* of a thing, (Ṣ, Ḳ,) and of a man: (Mgh:) the *lateral,* or *adjacent, part* of a thing: pl. أَحْضَانٌ. (Ṣ,* Ḳ.) حِضْنَا المَفَازَةِ means *The two borders* [*the nearer border and the further*] *of the desert.* (M, TA.) And حِضْنَا اللَّيْلِ †*The two sides* [or *first and last portions*] *of the night.* (TA.) And [as the حِضْن of a man or woman is often a place of concealment,] one says, مَا زَالَ يَقْطَعُ أَحْضَانَ اللَّيْلِ ‡[meaning *He ceased not to traverse the shades of the night*]. (TA.) عَلَيْكُمْ بِالحِضْنَيْنِ, in a trad. of 'Alee, means [*Keep ye to*] *the two wings of the army.* (TA.) You say also, أَخَذَ فُلَانٌ حَقَّهُ عَلَى حِضْنِهِ, i. e. *Such a one took his right,* or *due, by force.* (TA.) —— Also ‡*The quantity that is carried in the* حِضْن. (A.) —— Also, (Ṣ, Ḳ,) and ↓ حُضْنٌ, (Ḳ,) The *hole,* or *den,* or *subterranean habitation,* of the hyena: (Ṣ, Ḳ:) or the *place of hunting,* or *of capture,* of the hyena. (IB, TA.) —— And, both these words, The *circuit,* or *surrounding part,* of a mountain: or its *base;* or *lower,* or *lowest, part.* (Ḳ.) Accord. to Az, حِضْنَا الجَبَلِ means *The two lateral,* or *adjacent, parts of the mountain.* (TA.)

حَضَنٌ *Ivory:* (ISk, Ṣ, Ḳ:) the *tusk of the elephant.* (T, TA.)

حَضَانٌ The *state,* or *condition,* of a ewe, or she-goat, (Ṣ, TA,) and of a she-camel, and of a man in respect of his testicles, and of the pudendum muliebre, (TA,) *denoted by the epithet* حَضُونٌ. (Ṣ, TA.) [See also حَضِنَتْ.]

حَضُونٌ A ewe, and a she-camel, and a woman, *having one of her teats,* or *breasts, larger than the other:* (Ḳ:) or, applied to a ewe or she-goat, i. q. شَطُورٌ; i. e. *having one of her teats longer than the other:* (Ṣ:) or a she-camel, and a she-goat, *of which one of her* طُبْيَان [meaning either *two mammæ* or *two teats*] *has gone.* (A' Obeyd, TA.) —— Also A man *having one of his testicles larger than the other.* (Ḳ.) —— And A pudendum muliebre *having the edge of one of its labia majora* (i. e. *having one of its* شُفْرَان) *larger than the other.* (Ḳ.)

حَضَانَةٌ and حِضَانَةٌ [The *office,* or *occupation, of carrying and rearing* or *fostering a child:* the latter, accord. to the Ḳ and the Mgh, is an inf. n.: (see 1, first two sentences:) but accord. to Fei,] each is a subst. from حَاضِنٌ applied to a man, and حَاضِنَةٌ applied to a woman. (Mṣb.)

حَاضِنٌ A man *who has the charge of* [*carrying and*] *rearing,* or *nourishing,* or *fostering, a child:* (Mṣb,* TA:) and حَاضِنَةٌ A woman *who has the charge of a child,* (Ṣ, Mgh, Mṣb,* Ḳ, TA,) *who carries him,* (Mgh,) *and takes care of him,* (TA,) *and rears,* or *nourishes,* or *fosters, him:* (Ṣ, Mgh, TA:) pl. of the former حُضَّانٌ (TA) [and حَضَنَةٌ (as in a phrase below), agreeably with a general rule: and pl. of the latter, also agreeably with a general rule, حَوَاضِنُ]. —— [Hence,] هُوَ مِنْ حَضَنَةِ العِلْمِ, ‡ i. e. علمته [a mistranscription for عَلَمَتِهِ, meaning *He is of the servants of learning,* or *science*]. (TA.) —— حَمَامَةٌ حَاضِنٌ (Mgh, Mṣb, TA) and حَاضِنَةٌ (Mṣb) *A pigeon sitting* [or *brooding*] *upon its eggs, protecting them with its two sides;* (Mgh;) or *pressing,* or *compressing, its eggs beneath its wing.* (Mṣb.) —— [Hence,] سُفْعٌ حَوَاضِنُ [pl. of حَاضِنَةٌ] ‡ *Three stones for supporting a cooking-pot, cleaving to the ground,* (Ḳ, TA,) *with the ashes.* (TA.) —— حَاضِنَةٌ also signifies A man's *wife:* and so حَاضِنَةٌ. (TA.) —— And a palm-tree (نَخْلَةٌ) *having short racemes:* (Kr, Ḳ:) or one *of which the racemes have come forth, and quitted their spathes, and are short in their fruit-stalks.* (AḤn, Ḳ.)

مَحْضَنٌ and مَحْضِنٌ The *place in which a bird broods upon its eggs to hatch them:* (K:) pl. مَحَاضِنُ. (TA.) See also what next follows.

مِحْضَنَةٌ *A shallow bowl, made of clay, for the pigeon* (K, TA) [*to lay its eggs therein, and*] *to brood therein upon its eggs:* (TA:) مَحَاضِنُ [is its pl.], accord. to rule, pl. of ▼مَحْضَنٌ [&c., and] signifies the *places, in pigeon-towers, in which the pigeons lay their eggs.* (Mgh.) [See what next precedes.]

مُحْتَضَنٌ: see حِضْنٌ.

حضو

1. حَضَا النَّارَ, (K,) first pers. حَضَوْتُ, (S,) inf. n. حَضْوٌ, (K,) *He lighted,* or *kindled, the fire;* or *made it to burn, burn up, burn brightly* or *fiercely, blaze,* or *flame:* (S:) or *he stirred the live coals of the fire after they had become* [*partially*] *extinguished.* (K. [حَمَدَ, there, seems to be a mistake for خَمَدَ.]) It is also with ء. (S, TA. [See art. حضأ.])

مِحْضًى i. q. كُورٌ [as meaning either *A blacksmith's fire-place,* or the *skin with which he blows his fire*]. (K.)

مِحْضَاةٌ *A stick,* or *piece of wood, with which a fire is stirred;* as also مِحْضَأٌ; the latter accord. to those who pronounce the verb with ء. (S.)

حط

1. حَطَّهُ, (S, Msb, TA,) aor. ـُ, (Msb, TA,) inf. n. حَطٌّ, (Msb, K, TA,) *He put it down,* syn. وَضَعَهُ, (K,* TA,) and أَنْزَلَهُ, (Msb, TA,) *from a high to a lower place;* (Msb;) namely a load, or any other thing from a back; (TA;) a camel's saddle, (S, Msb, TA,) or other thing; (Msb;) a horse's saddle; and a bow; (S, TA;) as also ▼اِحْتَطَّهُ. (K,* TA.) You say, حَطَّ الأَحْمَالَ عَنِ الدَّوَابِّ *He put down the loads from the beasts.* (L.) And حَطَّ عَنْهُ وِزْرَهُ [*He put down from him his heavy burden:* or ‡*his heavy burden of sin*]: (S,* K,* TA:) if a real load be intended, this is proper; but if an ideal thing, it is tropical; (TA;) [as when you say,] حُطَّ عَنَّا ذُنُوبَنَا [‡*Put Thou down from us* the burden of *our sins*]. (K.) And حَطَّ رَحْلَهُ [lit. *He put down his camel's saddle;*] meaning ‡*he stayed,* or *abode.* (TA.) And حَطَّ alone, [elliptically,] †*He alighted;* or *alighted and abode;* (S, TA;) فِى مَكَانٍ *in a place.* (TA.) — *He threw it down;* namely a thing. (TA.) [*He paid it down;* namely money.] — *He made it to descend,* or *to go down,* or *downwards,* or *down a declivity,* (S, K,* TA,) *from above to below.* (K, TA.) Imra-el-Keys says, [describing a horse in motion,]

كَجُلْمُودِ صَخْرٍ حَطَّهُ السَّيْلُ مِنْ عَلِ *

[*Like a mass of rock which the torrent has made to descend from above*]. (TA.) — [†*He lowered him,* or *degraded him.*] — حَطَّ مِنَ الثَّمَنِ كَذَا ‡*He abated of the price so much;* syn. أَسْقَطَ; (Mgh, Msb;) لَهُ *to him.* (Msb.) And حَطَّ مِنْهُ حَطِيطَةً وَافِيَةً ‡[*He made a large abatement of it*]. (TA.) [See حَطِيطَةٌ.] And حَطَطْتُ مِنَ الدَّيْنِ ‡*I abated* [somewhat] *of the debt.* (Msb.) [See also 10.] — حَطَّهُ is also syn. with حَتَّهُ: as in a trad. where it is said of Mohammad, that he sat by a dry branch of a tree, فَقَالَ بِيَدِهِ وَحَطَّ وَرَقَهَا [*And he made a sign with his hand, and removed its leaves*]; meaning, *scattered its leaves.* (AA, TA.) And so in the phrase, الزُّبْدُ يُحَطُّ عَنِ اللَّبَنِ [*The froth is removed,* or *skimmed off, from the milk*]. (TA.) — حَطَّ السِّعْرُ, aor. ـِ, (L, TA, [but I think it more probable that the aor. is ـُ, agreeably with analogy, like as some say of قَطَّ, which, accord. to Sh, signifies the same, or nearly so, but accord. to others, the contr. of حَطَّ,]) inf. n. حَطٌّ and حُطُوطٌ, (L, K,) ‡*The price was,* or *became, low,* or *cheap;* (L, K, TA;) *it abated;* (Fr, TA in art. قط;) and so ▼انحطّ. (Fr ubi suprà, S,* TA.) — حَطَّ البَعِيرُ, (S, K,) فِى السَّيْرِ (S,) inf. n. حِطَاطٌ, (S, K,) ‡*The camel bore upon his nose-rein* (S, K) *towards one side,* (K,) *in going;* (S;) as also ▼انحطّ: (K:) or, accord. to AA, انحطّت النَّاقَةُ فِى سَيْرِهَا signifies †*the she-camel was quick in her pace;* (S, TA;) and so حَطَّتْ: (TA:) and حِطَاطٌ also signifies †*vehement running.* (TA.) [From what next follows, it seems that the verb in this sense is of the measure فَعِلَ, aor. ـَ.] — The poet 'Amr-Ibn-El-Ahtam uses the phrase, حَطِّى فِى هَوَاىَ, meaning †*Rest thou upon my love, and incline my way.* (TA.) And it is said in a trad., فَحَطَّتْ إِلَى الشَّابِّ †*And she inclined with her heart towards the youth.* (TA.) — حَطَّ فِى عِرْضِ فُلَانٍ †*He launched forth into,* or *pressed on in, the reviling of such a one.* (TA.) — حَطَّ فِى الطَّعَامِ ‡*He ate the food;* (K, TA;) as also ▼حطّط, (Sgh, K,) inf. n. تَحْطِيطٌ: (TA:) or *he ate much of the food.* (A, TA.) — حَطَّ الجِلْدَ, inf. n. حَطٌّ, *He polished,* or *smoothed, the skin;* and *figured,* or *decorated, it;* (K, TA;) and *ruled it,* or *made lines upon it;* (TA;) *with the* مِحَطّ. (K, TA.)

2: see 1; last sentence but one.

7. انحطّ, quasi-pass. of حَطَّهُ; *It was,* or *became, put down;* (S,* TA;) said of [a load, or any other thing from a back;] a camel's saddle; a horse's saddle; (TA;) [&c.] — *He,* or *it, descended; went down,* or *downwards,* or *down a declivity; from above to below.* (S,* TA.) You say, انحطّ فِى حَدُورٍ [*He descended a declivity*]. (S and TA in art. حدر, from a trad.) — [†*He became lowered,* or *degraded.*] — Said of a price: see 1. — Said of a camel: see 1. — Also †*He,* or *it, went back,* or *backward: went away: perished.* (TA.) — See also R. Q. 1.

8: see 1; first signification.

10. استحطّهُ وِزْرَهُ *He asked him to put down from him* [*his heavy burden:* or ‡*his heavy burden of sin*]: (K, TA:) if a real load be intended, this is proper; but if an ideal thing, it is tropical. (TA.) — اِسْتَحَطَّنِى مِنَ الثَّمَنِ, (S, Msb,) or مِنْ ثَمَنِهِ, (K,) شَيْئًا, (S, K,) or كَذَا, (Msb,) ‡*He asked, demanded,* or *desired, of me an abatement, a lowering, a diminution,* or *a lessening, of somewhat,* or *so much, of the price.* (S,* Msb, K.) — استحطّ †*He deserved to be lowered,* or *degraded.* (Har p. 258.)

R. Q. 1. حَطْحَطَ i. q. ▼انحطّ [but in what sense is not pointed out]: (Ibn-'Abbád, K:) said of a thing. (Ibn-'Abbád, TA.) — *He was,* or *became, quick,* (IDrd, K,) in his walk, or going, and in his work. (IDrd, TA.)

حِطَّةٌ [*A petition for the putting down of a heavy burden from one:* or, ‡*of the heavy burden of sin:* or merely *a putting down thereof:*] a subst. from استحطّهُ وِزْرَهُ, explained above; as also ▼حِطِّيطَى. (K.) It is said in the Kur [ii. 55, and vii. 161], وَقُولُوا حِطَّةٌ; i. e. they were told to say حِطَّةٌ for the purpose of asking thereby for the putting down of their heavy burdens from them, and they should be put down from them: (Ibn-Isráeel, TA:) the meaning is, *And say ye, Put Thou down from us our heavy burdens,* (S, TA,) or ‡*our sins:* (Ibn-'Arafeh, K:) or [*and say ye,*] *Our petition is* حِطَّةٌ; i. e. *that Thou wouldest put down from us our sins:* (Aboo-Is-hák, K:) or, accord. to some, حِطَّةٌ is a word which the children of Israel were commanded to say; and if they said it, their heavy burdens, or sins, were put down: (S, TA:) accord. to IAar, it is the saying لَا إِلٰهَ إِلَّا اللّٰهُ: (TA:) or it means *forgiveness:* (I'Ab:) or our affair is *an alighting and abiding* in this town: (Bd in ii. 55:) and there is another reading, وَقُولُوا حِطَّةً, which is explained in two ways; either by making the verb to govern the noun, as though he had said, *and say ye a saying which shall put down from you your heavy burdens,* or ‡*sins;* or by making the noun to be in the accus. case as an inf. n. meaning *supplicating* and *petitioning* [*that God may put down from you your heavy burdens,* or ‡*sins*]; i. e. اُحْطُطْ اللّٰهُمَّ أَوْزَارَنَا حِطَّةً: (TA:) but they changed this saying, (Fr, Sgh, K,) using for it a Nabathean expression; (Fr, TA;) saying حِطَّى سُمْقَاثَا, i. e. "red wheat," (Sgh, K,) accord. to Es-Suddee and Mujáhid; or, accord. to IAar, حِنْطَةٌ شُمْقَاثَا, i. e. "good wheat." (Sgh, TA.) You say also, سَأَلَهُ ▼الحِطِّيطَى, i. e. الحِطَّةَ [*He asked of him the putting down of his heavy burden from him:* or ‡*his sin*]. (TA.) And it is said in a trad., that when God tries a person with a trial in his body, هُوَ لَهُ حِطَّةٌ, i. e. *It is to him a mode of putting down from him his sins.* (TA.) الحِطَّةُ is also *A name of the month of Ramadán,* in the Gospel, or some other book: (Az, K,* TA:) because it puts down somewhat of the sin of him who observes the fast thereof. (Az, TA.) — Also ‡*A decrease,* or *state of diminution, in respect of rank,* or *station:* (TA:) [or *low,* or the *lowest, rank,* or *station:* for] ▼الحِطَطُ, (K, TA,) which is its pl., (TA,) is explained as signifying, (K, TA,) on the authority of IAar, (TA,) مَرَاكِبُ السَّفَلِ, or correctly مَرَاتِبُ السِّفَلِ: (K, TA:) the latter [meaning ‡*the ranks,* or *stations, of the lowest,* or *meanest, of mankind,*] is the right reading, as verified by Az. (TA.)

حَطَاطِ : see حِطَّةٌ, last sentence. ═ Also ‡ *Smooth, or sleek, bodies*; (IAạr, Ḳ, TA;) as though polished with the مِحَطَّة. (TA.)

حَطُوطٌ *A declivity, or declivous place; a place of descent, or by which one descends*: (Ṣ, TA:) *a hill* (أَكَمَةٌ) *that is difficult of descent*: or, accord. to IDrd, *a difficult* أَكَمَة. (TA.) ═ † An *excellent, swift* she-camel; (Ṣ, Ḳ;) as also ↓ مُنْحَطَّةٌ. (TA.)

كَعْبٌ حَطِيطٌ ‡ *An ankle-bone covered with flesh and fat so that it is not apparent.* (TA.)

حَطِيطَةٌ ‡ *What is abated, or diminished, of a price*: (Mgh, Mṣb, Ḳ, TA:) pl. حَطَائِطُ. (TA.) You say, الحَطِيطَةُ كَذَا وَكَذَا مِنَ الثَّمَنِ ‡ [*The portion that is abated is so much and so much of the price*]. (Ṣ.) And طَلَبَ مِنِّى حَطِيطَةً ‡ [*He sought, or demanded, of me an abatement of price*]. (TA.)

حِطِّيطَى : see حِطَّةٌ; first sentence; and again, in the latter half of the paragraph.

سِعْرٌ حَاطٌّ ‡ *A low, or cheap, price*; as also ↓ مَحْطُوطٌ. (TA.)

أَحَطُّ † *Smooth in the* مَتْنَانِ [or *two portions of flesh and sinew next the back-bone, on each side*]. (IAạr, Ḳ.) [See also مَحْطُوطٌ.]

مَحَطٌّ [*A place where loads, &c., are put down.* — And hence,] † *A place where one alights, or alights and abides*; (Ṣ, TA;) as also ↓ مَحَطَّةٌ: pl. [of either] مَحَاطُّ and [of the latter] مَحَطَّاتٌ. (TA.) — [Hence also,] مَحَطُّ سُفُنٍ [*A place for unloading of ships*]. (Ṣ and Ḳ in art. فرض.) — [Hence also,] هٰذَا مَحَطُّ الكَلَامِ ‡ [*This is the meaning, or intent, of the saying, or sentence*: used in this sense in the present day]. (TA.)

مِحَطٌّ (Ṣ, O, Ḳ) and ↓ مِحَطَّةٌ (Ḳ) *An instrument of iron*, (O, Ḳ,) or *of wood, prepared* (Ḳ, in some copies of which, for مُعَدَّةٌ, we find مُعَدَّلَةٌ, which is a mistake, TA,) *for the polishing, or smoothing, of skins*, (O, Ḳ,) *to make them soft and beautiful*; (O;) and *for figuring, or decorating, them*; (Ḳ;) [and *for ruling them*, or *making lines upon them*: see 1; last sentence:] or *an instrument with which tattooing is performed*: or *an iron instrument used by sewers of boots &c., with which they figure, or decorate, the leather*: (Ṣ:) or *an instrument* (T, A, TA) *with a pointed extremity*, (T, TA,) *used by binders of books*, (T, A, TA,) *and by others*. (A, TA.)

مَحَطَّةٌ : see مَحَطٌّ.

مِحَطَّةٌ : see مِحَطٌّ.

مَحْطُوطٌ [originally, *Put down*; applied to a load, &c. See 1. — And hence,] applied to a price: see حَاطٌّ. ═ Leather *polished*, or *made smooth* [&c.: see 1; last sentence]. (TA.) — ‡ A sword *made thin*. (TA.) — أَلْيَةٌ مَحْطُوطَةٌ ‡ [*A rump*] *without* مَأْكَمَة [or *protuberant buttocks*]; (Ḳ, TA;) *as though it were smoothed* (حُطَّتْ) *with the* مِحَطّ. (TA.) And جَارِيَةٌ مَحْطُوطَةُ المَتْنَيْنِ ‡ *A girl having the* مَتْنَانِ [or *two portions of flesh and sinew next the back-bone, on each side*,] *extended* [or *long*]; *as though they were smoothed with the* مِحَطّ: (TA:) or *having them extended* [or *long*] *and even* (Ṣ, TA) *and beautiful*. (Az, TA.) And مَحْطُوطَةُ الكَشْحِ may mean † A woman *whose flank is as though it were smoothed with the* مِحَطّ: but it is better explained as meaning *elevated*, or *protuberant, in the posteriors, and depressed in the flank*. (Ḥam p. 141.) [See also أَحَطُّ.]

مُنْحَطٌّ ‡ A shoulder-joint (Ḳ, TA) *neither high nor low*, (TA,) *of the most beautiful kind*. (Ḳ, TA.) ═ See also حَطُوطٌ.

حطأ

1. حَطَأَ *He cast*, or *threw*; syn. رَمَى [and, like this, trans. by means of بِ]. (Ḳ.) [Hence,] حَطَأَ بِهِ الأَرْضَ, (AZ, Ṣ, Ḳ,) aor. ـَ, (Ḳ,) inf. n. حَطْءٌ, (Ṣ,) *He cast him*, or *threw him, down prostrate upon the ground*: (AZ, Ṣ, Ḳ:) or, *with violence*: one says, اِحْتَمَلَهُ فَحَطَأَ بِهِ الأَرْضَ [*He raised him upon his back, and threw him down violently upon the ground*]. (Lth, TA.) And حَطَأَتْ بِوَلَدِهَا *She* (a woman) *cast forth her child from her womb*. (TA in art. زكب.) And حَطَأَتِ القِدْرُ بِزَبَدِهَا *The cooking-pot cast forth its froth*, or *scum*, (Ṣ, TA,) in boiling. (TA.) And حَطَأَ بِسَلْحِهِ *He cast forth his excrement*, or *ordure*. (Ṣ.) And حَطَأَ [alone], aor. ـَ and ـِ, *He cast forth his excrement, or ordure, at once*, (Ḳ, TA,) *quietly*, or *gently*. (TA.) And حَطَأَ, (Ḳ,) or حَطَأَ بِهَا, (Ṣ,) *He broke wind, with a sound*. (Ṣ, Ḳ.) — حَطَأَ بِهِ عَنْ رَأْيِهِ *He turned him back from the opinion, or judgment, that he had formed*. (IAth, Ḳ.) — حَطَأَهُ *He struck him*, or *slapped him, on the back with his open hand*: (Ṣ, Ḳ:) or *struck him gently on the back with the palm of his hand*: (Ṣ in art. لطع:) or *slapped him on the back, or between the shoulders, or on the upper part of the side, or on the chest*; (Khálid Ibn-Jembeh, TA;) or *on the back of the head*: (AZ, TA:) and *he struck him*, or *beat him*, (Sh, Ḳ,*) *with his hand*; (Sh, TA;) but it is said that the meaning is, *he struck him on the back of the neck*: (TA:) it also occurs without ء. (TA.) — حَطَأَهَا *He lay with her*. (Ṣ, Ḳ.*)

حَطْأَةٌ *A slap on the back* [&c. (see the verb)] *with the open hand*. (Ṣ, TA.) It also occurs without ء, written حَطْوَةٌ. (TA.)

حُطَئٌ, applied to a man, *Low, ignoble, mean*, or *sordid*. (AZ, Ṣ, Ḳ.) You say also حُطَئٌ بُطَئٌ, using the latter word as an imitative sequent. (Ṣ.)

حُطَيْئَةٌ, applied to a man, *Ugly*; or *contemptible*; or *ugly in aspect and small in body*: (Th, Ṣ, Ḳ:) or *short*. (Ṣ, Ḳ.)

حطب

1. حَطَبَ, (Ṣ, A, Ḳ,) aor. ـِ, (Ḳ,) inf. n. حَطْبٌ, (TA,) *He collected* حَطَب, (Ṣ, A, Ḳ,) i. e. *firewood*; (A, Ḳ;) as also ↓ احتطب: (Ṣ, A, Ḳ:) [and] so حَطَبَ الحَطَبَ, (Mgh, Mṣb,) aor. and inf. n. as above; and ↓ احتطب. (Mṣb.) — [Hence,] حَطَبَ فِى حَبْلِهِ, (A,) or فِى حَبْلِهِمْ, (Ḳ,) [lit. *He collected firewood in his*, or *their, rope*; meaning] ‡ *he aided*, or *assisted*, (A, Ḳ,) *him*, (A,) or *them*. (Ḳ.) One says, إِنَّكَ تَحْطِبُ فِى حَبْلِهِ وَتَمِيلُ إِلَى هَوَاهُ ‡ [*Verily thou aidest him, and inclinest to his desire*]. (A, TA.) — And حَطَبَ عَلَيْهِ بِخَيْرٍ (A, Mgh) ‡ *He brought to him* خَيْر, (Mgh,) meaning *wealth*, or *property*. (A.) — And فُلَانٌ يَحْطِبُ عَلَى فُلَانٍ ‡ *Such a one incites, urges*, or *instigates*, [others] *against such a one*. (Ḥar p. 209.) — And حَطَبَ بِهِ ‡ *He calumniated him*, or *slandered him*. (A, Mgh, Mṣb, Ḳ. [See حَطَبٌ, below.]) ═ حَطَبَهُ, (Ṣ, Ḳ,) aor. ـِ, (TA,) *He collected firewood for him*: (Ḳ:) or *he brought firewood to him*: (Ṣ, Ḳ:) as also ↓ احتطب لَهُ. (TA.) You say, فُلَانٌ يَحْطِبُ رُفْقَاءَهُ وَيَسْقِيهِمْ [*Such a one collects firewood for, or brings firewood to, his companions, and gives them to drink*]. (A, TA.) — Also *He collected for him* a thing *as firewood*. (TA.) — حَطَبُوا كَرْمَهُمْ, inf. n. حَطْبٌ, [and app. also حِطَابٌ, which see below, or this may be a simple subst.,] ‡ *They cut off the* حَطَب *of their grape-vine*; (A;) i. e. *the dry portions thereof, that were of no use but for fire*. (Mgh.) ═ حَطِبَ and ↓ احطب *It*, (a place, Ḳ,) or *he*, (a man, TA,) *abounded with firewood*. (Ḳ,* TA.) ═ حَطُبَ *He was, or became, lean, or meagre*. (A.)

4. احطب: see 1. — Also † *It* (a grape-vine) *attained to the time for having its* حَطَب [q. v.] *cut off from it*: (Ṣ:) or ‡ *attained to the time of bearing grapes and for the cutting off of what required to be cut off*; as also ↓ استحطب: (A:) or the latter, *it required to have its upper, or uppermost, portions cut off*, (Ḳ,) or *somewhat of those portions*. (TA.)

8. احتطب: see 1, in three places. — Also † *It* (rain) *tore up the roots, or stems, of the trees* [as though it made firewood of them]. (Ḳ.) — † *He* (a camel, TA) *fed upon the small and dry parts of the branches*. (Ḳ, TA.) This a camel does only by reason of soundness, and of surpassing strength. (TA.) — احتطب عَلَيْهِ فِى الأَمْرِ *i. q.* احتقب [app. meaning † *He took upon himself a burden, or a responsibility, in the affair*; as though he bound it like a bundle of firewood upon his back]. (Ḳ.)

10: see 4.

حَطَبٌ *Firewood*; the *portions of trees that are prepared for fuel*: (M, A, Ḳ:) pl. أَحْطَابٌ. (Mṣb.) — And hence, because enmity is kindled with it like as fire is with حَطَب [properly so called], (Ḥar p. 209,) ‡ *Calumny*, or *slander*; (A, Mgh, TA;) also termed الحَطَبُ الرَّطْبُ. (TA.) The former is said to have this meaning in the Ḳur cxi. 4: (Mgh, TA:) or to mean *thorns*, which the woman there spoken of is related to have been in the habit of throwing in the way of Mohammad. (TA.) You say, يَحْمِلُ الحَطَبَ بَيْنَ النَّاسِ, meaning ‡ *He goes about* *with calumny,* or *slander, among,* or *between, the people.* (A.) — Also ‡ The *prunings* of a grape-vine; (Ṣ,* A;) and so ▼ حِطَابٌ: (A:) the *dry portions* of a grape-vine *that are of no use but for fire:* (Mgh:) ▼the latter is explained by IṢh as meaning the *prunings that are cut off each year from the upper,* or *uppermost, portions* of a grape-vine. (TA.)

حَطِبٌ, applied to a man, (Ṣ,) [and app. to cattle, (see حَشِبٌ,)] † *Very lean* or *meagre;* as also ▼ أَحْطَبُ: (Ṣ, Ḳ:) or the latter means *unlucky,* or *unfortunate;* syn. مَشْؤُومٌ; (Ḳ;) in some copies of the Ḳ مَوْسُومٌ; (TA;) and its fem. is حَطْبَآءُ. (Ḳ.)

حِطَابٌ [app. an inf. n., see 1,] † The *pruning of a grape-vine until reaching the part in which the sap runs.* (Ḳ.) — See also حَطَبٌ, in two places.

حَطِيبٌ, (Ṣ, Mṣb, Ḳ,) fem. with ة, (Ḳ,) A place, (Ṣ, Mṣb, Ḳ,) or land, (Ḳ,) *abounding with firewood.* (Ṣ,* Mṣb, Ḳ.)

حَطُوبَةٌ *A handful,* or *bundle, of firewood.* (Ḳ,* TA.)

حَطَّابٌ: see حَاطِبٌ. — Also † A camel *that feeds upon the small and dry parts of branches:* (Ḳ, TA:) fem. with ة. (TA.) [See 8.]

حَاطِبٌ *A collector of firewood:* (Mṣb, TA:) and so ▼ حَطَّابٌ, (A, Mṣb,) but in an intensive sense: (Mṣb:) or the latter, *one who collects and sells firewood:* (TA:) and حَطَّابَةٌ *a company of collectors of firewood:* (Ṣ, Mgh:) and إِمَآءٌ حَوَاطِبُ [pl. of حَاطِبَةٌ] *female slaves who collect firewood.* (A, TA.) — [Hence,] هُوَ حَاطِبُ لَيْلٍ ‡ *He is one who confuses in his speech,* (A, Ḳ,) and *in his affair:* (TA:) or *one who speaks what is bad and what is good:* (Ṣ, TA:) like him who collects firewood by night; (TA;) for this person sees not what he collects in his rope, (Ṣ, TA,) so he collects bad and good. (TA.) Az says, one who acts unjustly towards himself is likened to a collector of wood by night, because the latter may chance to put his hand upon a viper, and be bitten by it: and so is one who does not restrain his tongue, but censures others; by doing which a man sometimes occasions his own death. (TA.) It is said in a prov., المِكْثَارُ حَاطِبُ لَيْلٍ ‡ *The loquacious is* [like] *a collector of firewood by night.* (A 'Obeyd, TA.)

أَحْطَبُ: see حَطِبٌ.

مِحْطَبٌ † *A pruning-knife.* (Ḳ, TA.)

مُحَاطِبَةٌ † A she-camel *that eats dry thorns.* (Ṣ, Ḳ.)

مُحْتَطِبٌ † *A small creeping thing that goes along upon the ground with pieces of wood,* or *stick, clinging to it.* (Ḥam p. 207.)

حطم

1. حَطَمَ, aor. ـِ, inf. n. حَطْمٌ, *It broke,* or *became broken, in pieces;* as also ▼ انحطم (Mṣb) and ▼ تحطّم: (TA:) or these two, (Ṣ, Ḳ,) or [correctly] the former [only], (TA,) *it broke,* or *became broken:* (Ṣ, Ḳ, TA:) or they are peculiarly said of that which is dry, or tough; (Ḳ, TA;) as a bone and the like. (TA.) — [Hence,] حَطِمَتِ الدَّابَّةُ † *The beast became aged* [*and emaciated and weak,* or *broken with age:* see حَطِمٌ, below]. (Ṣ.) — And † *The beast had a disease* (*termed* حَطَمٌ) *in his legs.* (TA.) = حَطَمَهُ, (Ṣ, Mṣb, Ḳ,) aor. ـِ, (Mṣb, Ḳ,) inf. n. حَطْمٌ, (Ṣ, Mṣb, Ḳ,) *He broke it:* (Ṣ, Ḳ:) or it applies peculiarly to that which is dry, or tough; (Ḳ, TA;) as a bone and the like: (TA:) as also ▼ حطّمه, (Ḳ,) inf. n. تَحْطِيمٌ: (Ṣ:) or the latter signifies *he broke it in pieces,* (Ṣ,) and so the former; (Mṣb;) or the latter, *he broke it much.* (Mṣb, TA.) — *He,* or *it, crushed it,* or *bruised it;* as, for instance, a lion, that which he devours; and as a camel and a sheep or goat, the ground with his feet or hoofs, and the trees and herbs in eating them; and as the wind, that upon which it blows [vehemently]. (TA.) It is said in a trad., رَأَيْتُ جَهَنَّمَ يَحْطِمُ بَعْضُهَا بَعْضًا [*I saw Hell-fire, one part thereof crushing another:* or, as though *pressing upon another;* from what next follows]. (TA.) One says of people crowding together, يَحْطِمُ بَعْضُهُمْ بَعْضًا [*They crush, bruise,* or *press upon, one another*]. (TA.) And of a vehement driver, يَحْطِمُ المَالَ [*He bruises the cattle,* or *camels &c.*]. (A, TA.) — One says also, لَا تَحْطِمْ عَلَيْنَا المَرْتَعَ, meaning ‡ *Spoil not thou to us the pasturage by pasturing upon it.* (TA.) — And حَطَمَ فُلَانًا أَهْلُهُ ‡ *His family rendered such a one a broken old man;* as though they loaded him with their burdens. (TA.) And حَطَمَتْهُ السِّنُّ (Ṣ) † *Age rendered him infirm.* (TA.)

2: see 1.

5: see 1. You say also, تحطّم البَيْضُ عَنِ الفِرَاخِ [*The eggs broke in pieces so as to disclose the young birds*]. (TA.) And تَحَطَّمَتِ الأَرْضُ يُبْسًا *The ground,* or *earth, crumbled by reason of excessive dryness.* (TA.) And تحطّم النَّاسُ *The people crowded together, crushing, bruising,* or *pressing upon, one another.* (TA.) And ▼ انحطم النَّاسُ عَلَيْهِ *The people pressed together,* or *crowded, upon it,* or *him.* (ISd, TA.) — And تحطّم عَلَيْهِ غَيْظًا ‡ *He became inflamed with wrath,* or *rage, against him.* (Ḳ,* TA.)

7: see 1: — and 5.

حَطْمٌ inf. n. of حَطَمَ [q. v.]. (Mṣb.) — Also *A certain disease in the legs of a beast.* (Ḳ.)

حَطِمٌ A thing (Mṣb) *breaking in pieces of itself.* (Ṣ, Mṣb, Ḳ.) — † A horse *broken by age:* (Ṣ:) or a horse *weak by reason of leanness and old age:* (Az, TA:) or an *aged* beast. (Mṣb.)

حُطَمٌ One *who breaks the ranks on the right and left;* and حَطَّامُ الصُّفُوفِ [signifies the same]. (TA.) — See also حُطَمَةٌ, in five places.

حُطْمَةٌ: see حُطَمٌ.

حَطْمَةٌ The *crowding, thronging,* or *pressing,* of men; and their *pushing one another.* (TA.) — The *tide* (دُفْعَة) of a torrent; like طَحْمَة. (Ṣ.) — The *havoc* of a lion among cattle. (TA.) — ‡ *Dearth, drought,* or *sterility;* or *a year of dearth,* &c.; (Ṣ, Ḳ, TA;) because it breaks (تَحْطِمُ) everything; (TA;) as also ▼ حُطَمَةٌ and ▼ حَاطُومٌ: (Ḳ:) or this last is not used except as meaning *continual dearth* &c. (TA.) [See also the last of these words below.]

حُطْمَةٌ: see what next precedes.

حِطْمَةٌ *What is broken in pieces,* or *what one breaks,* [accord. to different copies of the Ḳ, the former accord. to the reading in the TA,] of a thing that is dry, or tough; (Ḳ, TA;) as also ▼ حُطَامَةٌ: (Ḳ:) pl. of the former حِطَمٌ: whence صَعْدَةٌ حِطَمٌ [meaning *a spear,* or *spear-shaft, broken in pieces,* as is indicated in the TA], in which the term حِطْمَةٌ is regarded as applying to every portion. (Ḳ,* TA.) [See حُطَامٌ.]

حُطَمَةٌ *A vehement fire,* (Ḳ,) *that breaks in pieces everything that is cast into it.* (TA.) Hence, (Ṣ, TA,) الحُطَمَةُ a name of *Hell,* (Ḳ,) or of *Hell-fire:* (Ṣ, Ḳ:) or, as some say, *the fourth stage of Hell:* (Ḥar. p. 347:) or *a gate of Hell.* (Ḳ.) — ‡ A man *who eats much;* (Ṣ, TA;) as also ▼ حُطَمٌ; *who breaks everything in eating:* (Ḥar p. 580:) and the latter, and ▼ حُطَمٌ, an *insatiable* man. (TA.) — † *A large number of camels,* (T, Ṣ, Ḳ,) and *of sheep* or *goats:* (T, Ḳ:) because they break, or crush, (T, Ṣ, TA,) the herbage, (T, TA,) or everything, (Ṣ, TA,) or the ground with their feet or hoofs, and the trees and herbs in eating them. (TA.) — Also, and ▼ حُطَمٌ, (Ṣ, Ḳ,) ‡ A pastor *having little mercy upon the cattle;* (Ṣ, TA;) or *who acts injuriously towards them;* (Ḳ, TA;) *causing them to crush,* or *bruise, one another;* (Ṣ, Ḳ, TA;) or as though he crushed, or bruised, them by his vehement driving: (A, TA:) or the former signifies a pastor *who does not allow his beasts to avail themselves of the plentiful pasturages, nor let them disperse themselves in the pasturage:* and ▼ the latter, one *who is ungentle,* or *rough;* as though he broke, or crushed, or bruised, them when driving them or pasturing them: and ▼ سَوَّاقٌ حُطَمٌ signifies a man *who drives beasts vehemently, crushing them,* or *bruising them, by reason of his vehement driving;* but it is used by way of comparison, as meaning ‡ *cunning and versatile.* (TA.) Hence, شَرُّ الرِّعَآءِ الحُطَمَةُ [*The worst of pastors is the ungentle, who causes the beasts to crush,* or *bruise, one another*]: (Ṣ, Ḳ:) accord. to the Ṣ, a prov.: accord. to Ṣgh and the Ḳ, not a prov., but a trad.: but many of the trads. are reckoned among provs.: it is applied to him who governs, or manages, ill. (MF, TA.) Hence also what is related in a trad. of 'Alee, that Ḳureysh, when they saw him in war, or battle, used to say, احْذَرُوا الحُطَمَ ▼ احْذَرُوا الحُطَمَ [*Beware ye of the rough one! Beware ye of the rough one!*]. (TA.)

حُطَمِيَّاتٌ *Coats of mail;* so called from a maker thereof named حُطَمَةٌ: or *such as break the swords:* or *such as are heavy and wide:* (Ḳ:) the first of which explanations is the most probable. (TA.)

حُطَامٌ *What is broken in pieces,* of a thing that

is dry, or tough. (Ṣ, Ḳ. [In the CḲ, by the accidental omission of وَعُقْرُوبٌ, this signification and the next here following, from the Ḳ, are assigned to صَعْدَةٌ حِطَمٌ. See حُطَمَةٌ, which, accord. to some copies of the Ḳ, is syn. with حُطَامٌ in the sense explained above.]) And *Fragments* of eggs; (A, TA;) or of an egg-shell; so in a verse of Eṭ-Ṭirimmáḥ: (TA:) or the *shell* of the egg. (Ḳ.) — [See a tropical usage of it in an ex. cited, from a trad., voce ثُمَامٌ.] — حُطَامُ الدُّنْيَا †*The frail*, or *perishing, goods*, or *possessions, of the present world*: accord. to Z, from حُطَام signifying the "fragments" of eggs: (TA:) or [simply] *the goods of the present world*. (TA in art. عرض.)

حَطُومٌ The *lion*, (Ḳ,) *that crushes*, or *bruises, everything that he devours*; (TA;) as also ▼حُطَامٌ and ▼مِحْطَمٌ. (Ḳ.) And A wind (رِيحٌ) *that crushes everything*. (TA.)

حَطِيمٌ *Herbage remaining from the preceding year*: (Lḥ, Ḳ:) because dry, and broken in pieces. (Lḥ, TA.) — الحَطِيمُ *The* حِجْر [q. v.] (Mṣb, Ḳ) *of Mekkeh*, (Mṣb,) [i. e.] *of the Kaabeh*; (Ḳ;) *which is excluded from the Kaabeh*; said in the M to be *of the part next the spout*; and in the T, to be *that in* [or rather *over*] *which is the spout*: so called because it was left broken when the House was raised: or because the Arabs used to throw in it, or upon it, the clothes in which they performed their circuitings, and it remained until it became broken by length of time: (TA:) or *the wall of the* حِجْر *of the Kaabeh*; (I'Ab, Ṣ, Ḳ;) *the wall over which is the spout of the Kaabeh*; (Ḥam p. 710;) *the wall that* [*partly*] *encloses the* حِجْر *of the Kaabeh, on the western* [or rather *north-western*] *side*: (Ḥar p. 389:) or *the part between the angle* [*of the Black Stone*] *and* [*the well of*] *Zemzem and the Maḳám* [*-Ibráheem*] *and*, some add, *the* حِجْر: or *from the Maḳám to the door*: (Ḳ:) or *the part between the black angle and the door and the Maḳám, where the people crowd together to offer up their supplications, so that they crush*, or *bruise*, or *press upon, one another*: (Ḳ,* TA:) and there the pagans used to confederate. (Ḳ.)

حُطَامَةٌ: see حُطَمَةٌ.

حُطَامٌ: see حَطُومٌ: and حُطَمٌ.

حَاطُومٌ: see حُطَمَةٌ. — Also ‡*A digestive*; syn. هَاضُومٌ. (Ḳ, TA. [In the CḲ, erroneously, حاضوم.]) It is implied in the Ḳ that this is also a signification of حُطَمَةٌ and حُطَمَةٌ; which it is not. (TA.) One says, نِعْمَ حَاطُومُ الطَّعَامِ البِطِّيخُ ‡[*Excellent*, or *most excellent, is the digestive of food, the melon*, or *water-melon*]. (A, TA.)

مِحْطَمٌ: see حَطُومٌ.

حظ

حَظَّ, sec. pers. حَظِظْتَ, (Ṣ, Ḳ,) aor. يَحَظُّ, (Ṣ,) inf. n. حَظٌّ, (Ḳ,) *He was*, or *became, fortunate*, or *possessed of good fortune*, (Ṣ, Ḳ,) فِى الأَمْرِ *in the affair*; (Ḳ;) as also ▼أَحَظَّ: (Ḳ, TA:) or ▼the latter signifies *he was*, or *became, rich, wealthy*, or *opulent*; or *in a state of*, or *possessed of, competence*, or *sufficiency*; *in no need*; *without wants*; or *with few wants*. (O, TṢ.) Lth says, I have not heard any verb from حَظٌّ: but it has a verb, transmitted from the Arabs, which Lth did not know, and had not heard: and AHeyth says, in writing to Ibn-Buzurj, هُمْ يُحْظُونَ بِهِمْ, and يُجَدُّونَ بِهِمْ, (Az, TA,) meaning *They become possessed of good fortune, and riches*, or *competence*, or *sufficiency*. (L in art. جد.)

4: see 1, in two places. — أَحْظَيْتُهُ عَلَيْهِ [*I preferred him above him*] may be from the same root as the other words of this art.; the second ظ being changed into ى; [like as أَمْلَيْتُ is, accord. to some, formed from أَمْلَلْتُ;] or it may be from حُظْوَةٌ. (TA.)

حَظٌّ *Fortune*; or particularly *good fortune*; syn. جَدٌّ (Ṣ, Nh, Mṣb, Ḳ) and بَخْتٌ: (Nh:) and *a share, portion*, or *lot*: (Ṣ, Mṣb, Ḳ:) or particularly *a share, portion*, or *lot, of something good* or *excellent*: (Lth, Ḳ:) some of the people of Ḥimṣ say حَنْظٌ; but when they form a pl., they return to the original, saying حُظُوظٌ: and the ن is regarded by them as a nasal sound, not as a radical letter: and in like manner they do in the case of every word having a doubled letter, such as رُزٌّ and أُتْرُجٌّ, saying رُنْزٌ and أُتْرُنْجٌ: (Lth, L:) the pl. (of pauc., Ṣ) is أَحُظٌّ (Ṣ, Ḳ) and (of mult., Ṣ) حُظُوظٌ (AZ, Ṣ, Mṣb, Ḳ) and حُظُوظَةٌ (Ibn-'Abbád, Ḳ) and حِظَاظٌ (IJ, Ḳ) and حُظٌّ (AZ, Ḳ) and ▼أَحَاظٍ, (Ṣ, L, Ḳ, [in the CḲ, erroneously, أَحاظٌ,]) irregularly, as though it were pl. of أَحْظٍ, (Ṣ, L,) or it is regularly formed from أَحُظّ, which latter is [irregular, being] originally أَحْظُظٌ, [which is the original form of the pl. of pauc. mentioned above,] (IB,) and another pl. is ▼حِظَآءٌ, (L, [and so in the TA as from the Ḳ, but in several copies of the Ḳ ▼حُظَّآءٌ, which is of one of the forms of quasi-pl. ns.,]) also irregular, being formed [from حِظَاظٌ] by a change of the second ظ [into ى, and then into ء]. (L.) — See also حَظِيظٌ.

حِظَآءٌ and حُظَّآءٌ: see حَظٌّ.

حَظِّىٌّ, or حَظِىٌّ: see what next follows.

حَظِيظٌ and ▼حَظٌّ (Ṣ, Ḳ) and ▼حَظِّىٌّ, as a relative n., accord. to [most of] the copies of the Ḳ, or as a defective word, [i. e. ▼حَظِىٌّ, with a single ظ, as we find it in the CḲ,] accord. to Az, who says that it is originally حَظٌّ, (TA,) and ▼مَحْظُوظٌ, (AA, Ṣ, Mṣb, Ḳ,) *Fortunate*; or *possessed of good fortune*; (Ṣ, Mṣb, Ḳ;) *possessing a good share* (حَظّ) *of the means of subsistence*: (TA:) or the first, accord. to Fr, *possessing competence*, or *sufficiency*; or *rich*, or *wealthy*, or *opulent*: (TA:) the pl. [accord. to analogy of حَظِيظٌ] is أَحِظَّآءُ. (So in the L: [in the TA written أَحْظَآءُ, which I think a mistake, though it seems to be there implied that it is pl. of حَظٌّ, and if so, we must suppose it to be originally أَحْظَاظٌ, like as حِظَآءٌ, a pl. of the subst. حَظٌّ, if correct, is originally حِظَاظٌ.])

فُلَانٌ أَحَظُّ مِنْ فُلَانٍ *Such a one is more fortunate than such a one*. (Ṣ,* Mṣb,* TA.)

أَحَاظٍ said to be an irreg. pl. of حَظٌّ, q. v.

مَحْظُوظٌ: see حَظِيظٌ.

حظر

1. حَظَرَهُ, (Mṣb, Ḳ,) and حَظَرَ عَلَيْهِ, (Ḳ,) aor. ـُ, (Mṣb,) inf. n. حَظْرٌ (Ṣ, A, Mṣb) and حِظَارٌ, (TA,) *He forbade it*; *prohibited it*; *interdicted it*. (Ṣ, A, Mṣb, Ḳ.) The Arabs say, لَا حِظَارَ عَلَى الأَسْمَآءِ *There is no prohibition against names*; i. e., no one is forbidden to be named, or to name himself, as he pleases. (TA.) — حَظَرَ الشَّىْءَ عَلَيْهِ *It* (anything intervening) *debarred the thing from him*. (L.) And حُظِرَ عَلَيْهِ كَذَا *Such a thing was debarred from him*, by something intervening. (A.) — Also حَظَرَهُ, (Mṣb, Ḳ,) inf. n. حَظْرٌ, (Mgh,) *He took it to*, or *for, himself*; (Mgh, Mṣb, Ḳ, TA;) as though he withheld it from others. (TA.) — And حَظَرَ, (Ḳ,) aor. ـُ, inf. n. حَظْرٌ, (TA,) *He confined* cattle *in a* حَظِيرَة. (Ḳ.) And حَظَرَ عَلَى نَعَمِهِ *He confined his cattle in a* حِظَار. (Az.) — And حَظَرَ, (Ḳ,) [aor. app. as above,] inf. n. حَظْرٌ; (TA;) ▼احتظر; (A, Mṣb, Ḳ;) *He made a* حَظِيرَة: (A, Mṣb, Ḳ:) or the former, (Mgh,) or ▼احظر inf. n. إِحْظَارٌ, (TA,) *he made a* حظيرة *for another*: and ▼احتظر *he made a* حظيرة *for himself*. (Mgh, TA.)

2. [حَظَّرَ, inf. n. تَحْظِيرٌ, app. signifies *He made a limit of separation*, or *the like*. For] زَمَنُ التَّحْظِيرِ (used as an era, Mgh) points to what 'Omar did, in dividing Wádi-l-Ḳurà among the Muslims and Benoo-'Odhrah, after the expulsion of the Jews: (Mgh, Ḳ, TA:) as though he assigned to every one a limit of separation. (TA.)

4: see 1.

8: see 1, in two places. — Also احتظر بِهِ †*He protected*, or *defended, himself by means of him*, or *it*. (TA.)

حَظَرٌ *Trees with which a* حَظِيرَة *is made*. (A, Ḳ.) — And *Fresh thorns*. (Ḳ.) وَقَعَ فِى الحَظِرِ الرَّطْبِ †*He fell into that to which he was not equal*, (Ḳ, TA,) is a prov., originating from the fact of the Arabs' collecting fresh thorns, and making of them enclosures, into which a man sometimes falls so that he becomes caught therein. (TA.) And جَآءَ بِالحَظِرِ الرَّطْبِ ‡*He came with*, or *brought, a large number of cattle, and of men*: or *an odious lie*: (Ḳ:) [or *calumny*, or *slander*, and *falsehood*:] is said of a calumniator, or slanderer, and liar, who kindles by his calumnies the fire of enmity, and makes it to burn up. (A.) And أَوْقَدَ فِى الحَظِرِ الرَّطْبِ ‡*He uttered calumny*, or *slander*: (Ḳ:) or *he went about with calumny*, or *slander, and foul conduct*. (TA.)

حِظَارٌ (Sh, T, Ḳ) and حَظَارٌ (T, Ḳ) *i. q.* حَائِطٌ

[*A wall, or wall of enclosure, &c.*]: (Sh, T, Ḳ:) and *anything intervening between a person and a thing, or between two things, and forming a barrier, an obstruction, a partition,* or *a fence.* (TA.) See also the next paragraph, in two places.

حَظِيرَةٌ *An enclosure* of a thing, *of wood,* or *of canes* or *reeds:* (Ḳ:) [*a kind of pen:*] *an enclosure for camels,* (Ṣ, Mgh,) *made of trees, to protect them from the cold and wind;* (Ṣ;) as also ▼حِظَارٌ: (Ṣ, Ḳ:) *an enclosure for sheep or goats, &c., made of trees, to confine and protect them:* pl. حَظَائِرُ and حِظَارٌ: (Mṣb:) Az heard the Arabs apply the term ▼حَظَارٌ, with fet-ḥ [to the ح], to *a wall made of trees placed one upon another to form a protection for camels or sheep or goats from the cold of the north wind in winter.* (TA.) The pl. حَظَائِرُ is met. applied, by the poet El-Marrár Ibn-Munḳidh, to ‡ [*Enclosures of*] *palm-trees.* (TA.) [Hence,] حَظِيرَةُ القُدْسِ † *Paradise:* (Ḳ:) occurring in a trad. (TA.) And هُوَ نَكِدُ الحَظِيرَةِ ‡ *He is a person of little good,* or *of no good:* (Ṣ, Ḳ:) or *niggardly, tenacious, penurious,* or *avaricious.* (A.) — Also *A place in which dates are dried:* (Ḳ:) of the dial. of Nejd: as also حَضِيرَةٌ and حَصِيرَةٌ. (TA.)

مَحْظُورٌ *Forbidden; prohibited; interdicted:* (Ṣ, Mgh, Ḳ:) *confined to one class of men, exclusively of others;* thus in the Ḳur xvii. 21. (Ḳ.)

مُحْتَظَرٌ: see what next follows.

مُحْتَظِرٌ *A maker of a* حَظِيرَة. (Ṣ, Mṣb.) In the Ḳur liv. 31, some read, كَهَشِيمِ المُحْتَظِرِ; and others, ▼المُحْتَظَرِ: (Ṣ, TA:) the former meaning *Like the dry fragments of plants,* or *trees, which the maker of a* حظيرة *collects:* the latter, *like the dry fragments of plants,* or *trees, of a* حظيرة. (TA.)

حظل

1. حَظَلَهُ, (Mṣb,) or حَظَلَ عَلَيْهِ, (Ṣ, Ḳ, TA,) aor. ِ (Ṣ, Mṣb, Ḳ) and ُ, (Ḳ,) inf. n. حَظْلٌ (Ṣ, Mṣb, Ḳ) and حَظَلَانٌ and حِظْلَانٌ, (Ḳ,) *He forbade, prohibited,* or *interdicted, him,* (Ṣ, Mṣb, Ḳ, TA,) like حَظَرَهُ, (Mṣb, TA,*) or particularly (TA) *from free action, and motion,* (Ṣ, Ḳ, TA,) *and walking,* (Ḳ,) [or *walking out,*] or *somewhat from walking.* (TA.) حَظْلٌ also signifies A man's *regarding* his wife *with jealousy, and forbidding* her, or *preventing* her, *from free action, and from walking* [*out*]: (TA:) and يَحْظُلُ, *he straitens, and withholds, restrains,* or *debars:* (Fr, IAạr, TA:) or يَحْظُلُهَا, *he prevents her,* or *restrains her, from appearing* [*in public*]. (TA.) — حَظَلَ المَشْىَ, (Ṣ, Ḳ,) aor. ِ, (Ṣ,) inf. n. حَظَلَانٌ, *He checked,* or *restrained, somewhat of his walking:* and the inf. n. signifies *the walking of him who is angry.* (Ṣ, Ḳ.) El-Marrár Ibn-Munḳidh says,

* وَحَشَوْتُ الغَيْظَ فِى أَضْلَاعِهِ *
* فَهُوَ يَمْشِى حَظَلَانًا كَالنَّقِرْ *

[*And I stuffed wrath within his ribs, so that he walks checking somewhat his pace, like the* نَقِر]; (ISk, Ṣ;) i. e., like the *ram that has a vein,* or *nerve, twisted in his hock;* so that he somewhat checks his walk. (TA.) — حَظِلَ, aor. َ, signifies also *He walked on one side, by reason of some complaint:* (Az, TA:) and حَظَلَانٌ, a man's *being lame.* (TA.) — حَظِلَتْ, aor. َ, (M, Ḳ,) inf. n. حَظَلٌ, (TA,) *She* (a ewe or a goat) *limped,* or *was slightly lame, and her colour changed, in consequence of a tumour in her udder:* (M, Ḳ:) or, said of a she-camel, and of a ewe or she-goat, *her udder became swollen, and her milk became bad.* (AḤei, TA.) = حَظِلَ, (Ṣ, Ḳ,) aor. َ, (Ḳ,) inf. n. حَظَلٌ, (TḲ,) *He* (a camel) *ate much of* حَنْظَل [or *colocynths*]: (Ṣ, Ḳ:) or *became sick from eating* حنظل: (AḤei, TA:) but seldom does he eat them. (TA.)

4. احظل *It* (a place) *abounded with* حَنْظَل [or *colocynths*]. (R, TA.)

حَظِلٌ *A parsimonious* man, *who reckons with his family,* or *calls them to account, for what he expends upon them;* as also ▼حَظَّالٌ (Ṣ, M, Ṣgh, Ḳ) and ▼حَظُولٌ: (M, Ḳ:) which last also signifies [simply] *niggardly,* or *avaricious.* (TA.) = A camel *eating much of* حَنْظَل [or *colocynths*]: (Ṣ, Ḳ:) or *that eats* حنظل: (Mṣb:) or *that pastures upon* حنظل, *and becomes sick in consequence thereof:* (AḤn, TA:) but it is seldom that he eats them: (TA:) pl. حَظَالَى. (Ṣ, Ḳ.)

حِظْلَانٌ The *quality of parsimony, and reckoning with one's family,* or *calling them to account, for what one expends upon them.* (Ṣ, Ḳ.)

حَظُولٌ: see حَظِلٌ. — Also A ewe, or she-goat, *that limps,* or *is slightly lame, and changed in colour, in consequence of a tumour in her udder:* (M, TA:) or a she-camel, and a ewe or she-goat, *having her udder swollen, and her milk bad.* (AḤei, TA.)

حَظَّالٌ: see حَظِلٌ.

حَاظِلٌ [act. part. n. of حَظَلَ; *Forbidding,* &c.: —] accord. to Az, *Walking on one side, by reason of some complaint:* and accord. to AḤei, *flagging in his walking, by reason of pain* or *anger.* (TA.)

حَنْظَلٌ, accord. to some derived from حَظِلَ, the last verb in the first paragraph of this art.: see art. حنظل. (TA.)

حظو

1. حَظِيَتْ عِنْدَ زَوْجِهَا, (Ṣ, Ḳ,* TA,) aor. َ, (Ḳ,) inf. n. حُظْوَةٌ and حِظْوَةٌ and حِظَةٌ, (Ṣ, TA,) *She was,* or *became, fortunate,* or *happy, with her husband; near to his heart; in favour with him,* or *beloved by him;* (Ḳ,* TA;) as also ▼احتظت: and حَظِىَ هُوَ عِنْدَهَا [*he was,* or *became, fortunate,* or *happy, with her;* &c.]; as also ▼احتظى. (Ḳ,* TA.) And حَظِىَ عِنْدَ النَّاسِ, aor. and inf. n. as above, *He was,* or *became, in favour with,* or *beloved by, and in high estimation with, the people,* or *men.* (Mṣb.) And حَظِىَ عِنْدَ الأَمِيرِ and ▼احتظى بِهِ [*He was,* or *became, in favour, and high estimation,* or *an occupant of a high place, with the prince,* or *commander:*] both signify the same. (Ṣ, TA.) And حَظِىَ بِكَذَا *He was,* or *became, fortunate by means of such a thing.* (MA.) [In the vulgar dial., *He acquired,* or *obtained, such a thing.*] = حَظَا, aor. يَحْظُو, (Ḳ,) inf. n. حَظْوٌ, (TA,) *He went in a gentle,* or *leisurely, manner, such as is termed* حُظْلًا. (Ḳ.)

4. احظاهُ *It* [or *he*] *caused him* [*to be fortunate* or *happy, to be in favour* or *to be beloved,* or] *to occupy a high place* or *rank* [*in the estimation of another* or *others*]. (Ḥar p. 379.) — [*He favoured him,* بِكَذَا *with such a thing:* for] احظى also signifies تَفَضَّلَ trans. by means of عَلَى. (Ḥar p. 687.) — And أَحْظَيْتُهُ عَلَى فُلَانٍ *I preferred him above such a one.* (Ṣ, TA.) [See also 4 in art. حظ.]

8: see 1, in three places.

حِظَةٌ: } see حُظْوَةٌ.
حِظْوٌ, or حَظْوٌ: }

حَظٌ: see حُظْوٌ.

حِظًى, or حِظًا; see حُظْوَةٌ.

حَظْوَةٌ: see حُظْوَةٌ. = Also, (Ṣ, Ḳ,) and ▼حُظْوَةٌ (Ḳ) and ▼حِظْوَةٌ, (MF, TA,) *A small arrow,* (Ṣ, Ḳ,) *a cubit in length,* (Ṣ,) *with which children play,* (Ḳ,) *and with which they learn to shoot:* (TA:) and *any rod,* or *twig, growing upon the stock* (أَصْل) *of a tree, that has not yet become strong:* (Ḳ:) pl. (in both senses, TA) حِظَآءٌ and حَظَوَاتٌ. (Ṣ, Ḳ.) The dim. ▼حُظَيَّةٌ signifies *Such an arrow having no head:* the pl. is حُظَيَّاتٌ: (Ṣ:) and [hence,] إِحْدَى حُظَيَّاتِ لُقْمَانَ *One of the* [*small headless*] *arrows of Luḳmán,* the son of 'Ád, is a prov., applied to him who is known for evil conduct, and from whom proceeds (Ṣ, Ḳ) something, (Ṣ,) or some good act. (Ḳ. [See Freytag's Arab. Prov. i. 52.])

حُظْوَةٌ and ▼حِظْوَةٌ (Ḳ) and ▼حَظْوَةٌ (Th, MF) and ▼حِظَةٌ, (Ḳ,) [all, except the third, said to be inf. ns. of حَظِىَ and حَظِيَتْ,] *A state of fortunateness* or *happiness; nearness to the heart; a state of favour, of being beloved,* or *of being in high estimation;* (see 1;)] *high rank* or *standing, in the estimation of another* or *others;* (Ḳ, TA;) and *ideal nearness:* or *rank, station,* or *dignity, and advancement in the favour of a man of power* or *authority, and the like:* (TA:) and *a good share of the means of subsistence:* (Ḳ:) pl. حُظًا and حِظَآءٌ: (Ḳ:) and ▼حِظًى [or حِظًا] signifies *the same as* حظوة; (IAmb, TA;) or the *same as* ▼حُظْوٌ, (so in some copies of the Ḳ, in art. حظى,) or ▼حِظْوٌ, (so in other copies of the Ḳ and in the TA,) mentioned by Ṣgh, on the authority of Fr, (TA,) i. e. the *same as* حَظٌّ [*good fortune,* &c.]: (Ibn-Buzurj, Ḳ:) pl. أَحْظٍ, and pl. pl. أَحَاظٍ. (Ḳ.) Accord. to AZ, one says, إِنَّهُ لَذُو حُظْوَةٍ فِيهِنَّ and عِنْدَهُنَّ [*Verily he is a possessor of fortunateness,* &c., *among them* and *in their estimation*; i. e., among those women and in the estimation of those women]; and he adds that one does not say this except in relation to a state subsisting between men and women: (TA:) and the mullà 'Alee, in his "Námoos," [an Expos. of the Ḳámoos,] says that حظوة seems to apply peculiarly to the case of a woman, as it does in the common conventional language: but it is of common application, agreeably with the explanations in the Ḳ, as is expressly asserted on the authority of Th and others. (MF.) ══ See also حَظْوَةٌ.

حِظْوَةٌ : see حُظْوَةٌ : ══ and see also حَظْوَةٌ.

حَظِىٌّ part. n. of حَظِىَ, (Mṣb,) [*Fortunate* or *happy*,] *in favour with*, or *beloved by, and in high estimation with, others*; (Ṣ,* Mṣb, TA;*) *occupying a high place* or *rank* [*in the estimation of another* or *others*]; (Ṣ, TA;) and ↓ حَظٍ signifies the same: (Ḥar p. 623:) fem. ↓ حَظِيَّةٌ, (Ṣ, Mṣb, Ḳ,) applied to a woman *in favour with*, or *beloved by, and in high estimation with, her husband*; (Mṣb;) pl. حَظَايَا. (Ṣ, TA.) You say, ↓ هِىَ حَظِيَّتِى [*She is my favourite*], and إِحْدَى حَظَايَاىَ [*one of my favourites*]. (Ṣ, TA.) For حَظِيَّةٌ, the vulgar say, erroneously, ↓ مَحْظِيَّةٌ; [meaning thereby *A concubine*; in which sense حَظِيَّةٌ is used by late writers;] and making the pl. مَحَاظِى, which is also wrong. (TA.) Hence the prov., إِلَّا حَظِيَّةٌ ↓ فَلَا أَلِيَّةٌ, (Ṣ, Ḳ, TA,) explained in art. الو.—الحَظِىُّ also is applied to *The eighth of the horses that are started together in a race*. (Ḥam p. 46.)

حَظِيَّةٌ : see what next precedes, in three places.

حُظَيَّةٌ : see حُظْوَةٌ.

حُظَيًّا *A certain gentle*, or *leisurely, manner of going*. (Ḳ.)

هُوَ أَحْظَى مِنْهُ *He is nearer to him*, [or *more in favour with him, more beloved and esteemed by him*,] *and more fortunate* or *happy* [*with him*]. (TA.) [It may also mean *He is nearer*, or *more in favour*, &c., *than he*.]

مَحْظِيَّةٌ : see حَظِىٌّ.

حف

1. حَفُّوهُ, (Ksh and Bḍ in xviii. 31,) or حَفُّوا حَوْلَهُ, (Ṣ, Ḳ,) or بِهِ, (Mṣb, and W p. 153, [and so in the present day, because syn. with احاطوا به and استداروا به and اطافوا به, &c.,]) but the verb is properly trans. by itself, (W ibid.,) aor. ـُ, (Ṣ,) inf. n. حَفٌّ (Ṣ, TA) and حِفَافٌ; (TA [accord. to a meaning there assigned to it];) and ↓ حَفَّفُوا and ↓ احتفّوا; all signify the same; (Ḳ, TA;) *They went round about, circuited, compassed*, or *surrounded, it*, or *him*. (Ṣ, Ksh, Bḍ, Mṣb, TA.) You say, حَفَّ القَوْمُ بِالبَيْتِ, [or rather حَوْلَ البَيْتِ,] *The company of men went round about the House* [called the House of God, i. e. the Kaabeh]. (Mṣb.) And it is said in a trad., فَيَحُفُّونَهُمْ بِأَجْنِحَتِهِمْ *And they circuit round about them with their wings*. (TA.) And in a prov., مَنْ حَفَّنَا أَوْ رَفَّنَا فَلْيَقْتَصِدْ, i. e. *Whoso goes round about us, and minds*, or *manages, our affairs*, (Ḳ, TA,) *and treats us with honour*; (TA;) or [in the Ḳ "and"] *serves us*, (Ṣ, Ḳ,) *and guards us, defends us*, or *takes care of us, and regards us*, or *behaves towards us, with benevolence and solicitude*; (Ṣ, TA;) or [in the Ḳ "and"] *praises us*; (A 'Obeyd, Ḳ, TA;) *let him* [*act moderately, and*] *not exceed the due bounds*, (A 'Obeyd, Ḳ, TA,) *but speak truth*. (A 'Obeyd, TA.) Hence the saying, (Ḳ,) مَا لَهُ حَافٌّ ↓ وَلَا رَافٌّ (Ṣ, Ḳ) [*He has not any who goes round about him, and minds*, or *manages, his affairs*, &c.]. And ذَهَبَ مَنْ كَانَ يَحُفُّهُ وَيَرُفُّهُ (Ṣ, Ḳ) [*He went away*, or *has gone away, who used to go round about him*, &c.; or] *who used to give to him, and bring him corn* or *food*: (TA:) [for] حَفَّهُ signifies also *he gave to him*. (Mṣb.) And هُوَ يَحُفُّ وَيَرُفُّ *He stands and sits*: and *he acts as a sincere*, or *faithful, adviser, and with benevolence and solicitude*. (Aṣ, (TA.) [See also art. رف.] One says, of persons in want, حَفَّتْهُمُ الحَاجَةُ, (Ṣ, Ḳ,) aor. ـُ, (Ṣ,) inf. n. حَفٌّ, (TA,) ‡ [*Want beset*, or *encompassed, them*; or *has beset*, &c.;] and ↓ هُمْ قَوْمٌ مَحْفُوفُونَ ‡ [*They are persons beset*, or *encompassed*, by want]. (Ṣ, Ḳ, TA.)—حَفَّهُ بِالشَّىْءِ, aor. ـُ, (Ṣ, O, Ḳ,) *He surrounded it*, or *him, with the thing*; (Ḳ, TA;) as, for instance, a هَوْدَج with pieces of cloth; (Ṣ, O;) and so ↓ حفّفهُ, inf. n. تَحْفِيفٌ. (Ṣ.) It is said in the Ḳur [xviii. 31], حَفَفْنَاهُمَا بِنَخْلٍ *We made them*, namely, the two gardens, *to be surrounded by palm-trees*; (Ksh, Bḍ;) *We made palm-trees to encompass their* أَحِفَّة, (Ḳ,) i. e., their *sides*. (TA.) And you say, حَفَفْتُهُ بِهِمْ *I surrounded it*, or *him, with them*. (Ksh and Bḍ in xviii. 31, and TA.) And حُفَّتِ الجَنَّةُ بِالمَكَارِهِ (TA) a trad., meaning † *Paradise is encompassed by things that one dislikes to do*: these being likened to a wall, through which alone one can enter Paradise. (Gloss in a copy of the "Jámi' eṣ-Ṣagheer" of Es-Suyootee.) ══ حَفَّ شَارِبَهُ, (Ṣ, Mṣb, Ḳ,) and رَأْسَهُ, (Ṣ, Ḳ,) and اللِّحْيَةَ, (M,) aor. ـُ, (Ṣ,) or ـِ, (M, IB, TA,) [the former contr. to rule, and disapproved by IB,] inf. n. حَفٌّ, (Ṣ, M, TA,) *He cut*, or *clipped*, (Ṣ, M, Mṣb, Ḳ,) *his mustache*, (Ṣ, Mṣb, Ḳ,) and *the hair of his head*, (Ṣ, Ḳ,) and *the beard*, (M,) *much*, or *short*, or *to the utmost degree*. (Ṣ, Mṣb, Ḳ.)—حَفَّتْ وَجْهَهَا (Ṣ, Mgh, Mṣb, Ḳ) مِنَ الشَّعَرِ, (Ṣ, Ḳ,) aor. ـُ, (Ṣ, Mṣb,) inf. n. حَفٌّ (Ṣ, Mṣb, Ḳ) and حِفَافٌ; (Ṣ, Ḳ;) and ↓ احْتَفَّتْ; (Ṣ, Ḳ;) said of a woman; (Ṣ, Mgh, Mṣb, Ḳ;) *She plucked out the hair of her face*: (Mgh:) or *she embellished her face by removing the hair thereof*: (Mṣb:) or *she scraped off the hair of her face* (Ḳ, TA) *with a razor*: (TA:) and ↓ احْتَفَّتْ *she ordered another to pluck out the hair of her face with two threads*: (Ḳ,* TA:) so some say: and ↓ أَحَفَّتْ, inf. n. إِحْفَافٌ, signifies the same as احتفّت. (TA.) ══ حَفَّ رَأْسُهُ, aor. ـِ, inf. n. حُفُوفٌ, *His head remained long without ointment*, (Aṣ, Ṣ, Ḳ,) *and its hair was shaggy, matted, frouzy*, or *dusty*: (TA:) and حَفَّتِ اللِّحْيَةُ, aor. and inf. n. as above, *The beard was shaggy, matted, frouzy from long want of ointment*, or *dusty*. (M, TA.) El-Kumeyt says, describing a wooden peg or stake, (Ṣ, L,) long neglected, (L,)

* وَأَشْعَثَ فِى الدَّارِ ذَا لِمَّةٍ *
* يُطِيلُ الحُفُوفَ فَلَا يَقْمَلُ *

[*And a wooden peg* or *stake, in the dwelling, having a head of battered and pendent fibres, long neglected, but not lousy*: the fibres being likened to hair; and (as is said in the TA in art. شعث, where this verse is cited, but with ذِى in the place of ذا,) the term اشعث being used to signify a wooden peg or stake because its head is bruised, or battered, and separated, so that the parts do not cohere]. (Ṣ, L.)—حَفَّتِ الثَّرِيدَةُ *The* ثريدة [or *mess of crumbled bread moistened with broth*] *became dry in its upper part* [*by reason of paucity of broth*], *and cracked open in several places*. (TA.) [See the part. n., حَافٌّ.]—[The inf. n.] حُفُوفٌ signifies The *being dry, without grease*. (TA.)—And حَفَّ بَطْنُهُ *His* (a man's) *belly became dry in consequence of his not having eaten greasy food nor flesh-meat*. (TA.)—حَفَّتِ الأَرْضُ, (Mṣb, Ḳ,) aor. ـِ, (Mṣb, TA,) inf. n. حُفُوفٌ, (TA,) *The earth*, or *land, dried up*: (TA:) or *its plants*, (Mṣb,) or *its herbs*, or *leguminous plants*, (Ḳ,) *dried up*, (Mṣb, Ḳ,) *for want of water*. (TA.)—حَفَّ سَمْعُهُ, (IAar, Ḳ,) inf. n. حُفُوفٌ, (IAar, TA,) † *His hearing went away entirely*. (IAar, Ḳ.) ══ حَفَّ, (Ṣ, Ḳ,) aor. ـِ, (Ṣ,) inf. n. حَفِيفٌ, (Ṣ, Ḳ, KL,) *He* (a horse) *made a sound*, (Ṣ, Ḳ, KL,) *such as is termed* دَوِىّ [i. e. *a confused and continued sound*], (Ṣ,) *with his fore and hind feet*, (KL,) *in his running*, (Ṣ, Ḳ,) or *in going along*. (KL.) Said also of violent rain, *It made a* [*pattering*] *sound*. (Aṣ, TA.) And of a viper, *It made a* [*rustling*] *sound with its skin*: فَحَّ, inf. n. فَحِيحٌ, signifies "it made a sound to proceed from its mouth:" (Aboo-Kheyreh, Ḳ:) or حَفَّتْ, inf. n. as above, said of the female of the [kind of serpents called] أَسَاوِد, *she made a* [*rustling*] *sound with her skin by rubbing one part thereof with another*. (L.) And in like manner it is said of a tree, meaning *It made a* [*rustling*] *sound* (Ḳ, TA) *by the blowing of the wind upon its branches*. (TA.) And of a bird, meaning *It made a* [*rustling*] *sound* (Ḳ, TA) *with its wing* [or *wings*]: (TA:) and ↓ حَفْحَفَ signifies the same, said of the wing of a bird; and likewise, of a hyena, (IDrd, Ḳ,) as also خَفْخَفَ. (TA.) [Hence,] said of the [beetle called] جُعَل, [because of the humming that it makes in flying,] *It flew*. (TA.)

2: see 1, in two places:—see also 4. ══ Also حفّف, inf. n. تَحْفِيفٌ, ‡ *He* (a man, TA) *was in a state of embarrassment*, or *distress, and his property became little*: (Ḳ, TA:) from حَفَّتِ الأَرْضُ "the earth, or land, dried up." (TA.) حفّف وَجْهَهُ occurs in a trad. [app. in the same sense]. (TA.)

4. أَحَفَّتْ, said of a woman: see 1. — أَحْفَفْتُ رَأْسِى *I made my head to remain long without ointment* [*so that the hair became shaggy, matted, frowsy,* or *dusty*]. (Aṣ, Ṣ, Ḳ.) — [Hence, app.,] أَحْفَفْتُهُ ‡ *I spoke evil of him.* (Ibn-'Abbád, Ḳ, TA.) — أَحْفَفْتُ الفَرَسَ *I urged the horse* (Ṣ, O, L, Ḳ) *to run vehemently* (O, Ḳ) *so as to cause him to make a sound such as is termed* دَوِىّ [i. e. *a confused and continued sound*] (Ṣ, O, L, Ḳ) *in his running,* [*with his feet,* (see حَفِيف,)] (Ṣ, L,) or *in his belly:* (O, Ḳ:) the former is probably the right meaning. (TA.) — أَحْفَفْتُ الثَّوْبَ *I wove the piece of cloth with the* حَفّ, i. e. the مِنْسَج; as also ▼حَفَّفْتُهُ, (Ḳ, TA, [in the CḲ حَفَفْتُهُ,]) inf. n. تَحْفِيف. (TA.)

8. احتفّوا: see 1, first sentence. — احتفّ بِهِ *He,* or *it, became encompassed,* or *surrounded, by it:* and hence, *became in the midst of it.* (Ḥar p. 445.) — اِحْتَفَّتْ, said of a woman: see 1, in two places. — احتفّ النَّبْتَ *He cut the herbage;* syn. جَزَّهُ: (so in some copies of the Ḳ, and in the TḲ:) or حَزَرَهُ [*he computed by conjecture its quantity*]: (so in other copies of the Ḳ, and in the TA:) mentioned by Ṣgh: in some copies of the Ḳ, حَزَّزه [*he jagged it*]: in one, جوّره, which is a mistake. (TA.) — اِحْتَفَّتِ الإِبِلُ الكَلَأَ *The camels ate the herbage:* or *obtained some of it.* (TA.) — And احتفّ *He ate up entirely* what was in the cooking pot: like as اشتفّ signifies "he drank up entirely" what was in the vessel. (Ṣ.)

10. استحفّ أَمْوَالَهُمْ *He took the whole of their possessions* (Ḳ, TA) in an incursion into the territory of an enemy. (TA.)

R. Q. 1. حَفْحَفَ: see 1, last sentence but one. — Also ‡ *He* (a man, TA) *was,* or *became, straitened in his means of subsistence.* (IAạr, Ḳ, TA.)

حَفٌّ: see حَفَّةٌ, in three places. — [It is said, accord. to the KL, to signify also *What is called in Persian* زين كوهه, app. meaning *a saddle-bow:* but this signification, if correct, is probably post-classical.] — Also, and ▼حَفَفٌ and ▼حِفَافٌ, *A time,* or *season:* (L:) or *i. q.* أَثَرٌ [*a track,* &c.]. (Ḳ.) You say, جَآءَ عَلَى حَفِّ ذٰلِكَ, and ▼حَفَفِهِ, and ▼حِفَافِهِ, (L, Ḳ,) *He,* or *it, came in the time,* or *season, of that:* (L:) or the meaning is عَلَى أَثَرِهِ [lit. *in the track thereof;* and hence, *after,* or *near after, that*]. (Ḳ.) — فُلَانٌ حَفٌّ بِنَفْسِهِ *Such a one is busied with,* or *anxious about, himself.* (TA.)

حَفَّةٌ *i. q.* مِنْوَالٌ; i. e. The *web-beam of a loom;* the *wooden thing* [or *roller*] *upon which the weaver winds the web,* or *piece of cloth* [*as it is woven*]: ▼حَفٌّ signifying the مِنْسَج [which generally means the *weaver's loom;* but explained in the TḲ as meaning here the *stay of a weaver's loom;* in the KL, said to be *what is called in Persian* كار چوب, but this is the حَفَّة, to which the same explanation is assigned in the KL]: (Ṣ, Ḳ:*) so accord. to Aṣ: [for] Aboo-Sa'eed [i. e. Aṣ] says, the حَفَّة is the مِنْوَال; and it should not be called the ▼حَفّ; for the حَفّ is the مِنْسَج: (Ṣ, O:) [the former is also applied to the *yarn-beam, upon which the yarn is rolled:* see نِيرٌ:] in the L, it is said that the حَفَّة of the weaver is the *wide piece of wood with which he arranges the woof between* [*the threads of*] *the warp:* or, as some say, the *three canes:* and some say that it is ▼حِفَّةٌ, with kesr: and it is said to be the *thing with which the weaver strikes, like a sword:* and the ▼حَفّ is the *cane that comes and goes* [or *goes to and fro;* app. meaning the *shuttle*]: Az says, thus it is with the Arabs: and its pl. [the pl. of حَفّ] is حُفُوفٌ. (TA.) One says, مَا أَنْتَ بِحَفَّةٍ وَلَا نِيرَةٍ [*Thou art neither a* حفّة *nor a* نيرة]; the نيرة being the transverse piece of wood: alluding to him who neither profits nor harms; meaning that he is good for nothing. (TA.) [See also a similar saying voce نِيرٌ.] — Also *What camels have eaten,* or *obtained,* (اِحْتَفَّتْ,) *of herbage.* (TA.) — See also حَفَفٌ.

حِفَّةٌ: see حَفَّةٌ.

حَفَفٌ The *verge* of an event, or affair. (Ḳ,* TA.) You say, هُوَ عَلَى حَفَفِ أَمْرٍ *He is on the verge of an event,* or *affair.* (TA.) — See also حَفٌّ, in two places. — Also, (Aṣ, Ṣ, Ḳ,) and ▼حُفُوفٌ, (Ḳ,) ‡*An evil state,* or *condition, of life;* and *paucity of property;* (Aṣ, Ṣ, Ḳ, TA;) as though one were placed aloof (فِى حَفَفٍ, i. e. جَانِبٍ,) from the means of subsistence: (Er-Rághib, TA:) or the former signifies *straitness of the means of subsistence;* (IDrd, TA;) and so ▼the latter: (TA:) or the former, *a* [*bare*] *sufficiency of the means of subsistence:* (Lḥ, TA:) or *a state in which the family,* or *household, is proportionate to the provisions:* (Th, TA:) it is coupled with ضَفَفٌ: and is said to signify *straitness;* the latter signifying "paucity of food with numerousness of the eaters thereof;" or, as some say, "food proportionate to the household:" (TA:) or the former signifies *a state in which the eaters are proportionate to the property;* and the latter, "a state in which the eaters are more than proportionate to the property:" (Abu-l-'Abbás, TA:) or the former, *want;* and the latter, "paucity [of property]:" (IAạr, TA:) or both signify the same. (TA.) One says, مَا رُئِىَ عَلَيْهِمْ حَفَفٌ وَلَا ضَفَفٌ *There was not seen upon them a trace of want.* (Ṣ.) And أَصَابَهُمْ مِنَ العَيْشِ حَفَفٌ and ضَفَفٌ and قَشَفٌ, *Straitness of the means of subsistence befell them.* (Aṣ, TA.) And مَا عِنْدَ فُلَانٍ إِلَّا حَفَفٌ مِنَ المَتَاعِ *There is not with such a one aught save a scanty supply of the necessaries of life.* (TA.) And هٰذِهِ حَفَّةٌ مِنْ مَالٍ, or مَتَاعٍ, *This is a scanty supply of the necessaries of life, not exceeding the wants of its people* or *owners.* (TA.)

حِفَافٌ A *side* (Ṣ, Ḳ) of a thing; حِفَافَا شَىْءٍ signifying *the two sides of a thing:* (Ṣ:) pl. أَحِفَّةٌ. (Ḳ.) — *A border* of hair *remaining around the head of one who has become bald:* (Ṣ, Ḳ:*) pl. as above. (Ṣ, Ḳ.) Dhu-r-Rummeh says, (Ṣ, TA,) describing bowls [of food], (TA,)

* لَهُنَّ إِذَا أَصْبَحْنَ مِنْهُمْ أَحِفَّةٌ *
* وَحِينَ يَرَوْنَ اللَّيْلَ أَقْبَلَ جَائِيَا *

meaning *They,* i. e. the bowls, *have a party of them surrounding them* [*when they are set in the beginning of the day, and when they see the night, that it has advanced, coming on*]. (Ṣ, TA.) And you say, قَوْمُهُ أَحِفَّةٌ بِهِ *His people are surrounding him.* (TA.) — حِفَافُ الرَّمْلِ *The place where the sand ends:* pl. as above. (TA.) — كَانَ الطَّعَامُ حِفَافَ مَا أَكَلُوا *The food was proportionate to what they ate.* (TA.) — See also حَفٌّ, in two places.

حُفُوفٌ an inf. n. [See حَفَّ رَأْسُهُ, &c.]. — See also حَفَفٌ, in two places.

حَفِيفٌ The دَوِىّ [or *confused and continued sound*] (Ṣ, O, Ḳ) [*of the feet*] of a horse *in running,* (Ṣ,) or *of the belly* of a horse *in running vehemently:* (O, Ḳ:) the former is probably the right meaning: (TA: [see 1 and 4:]) the *sound of the feet* of camels *when going a vehement pace:* (TA:) the [*pattering*] *sound* of violent rain: (Aṣ, TA:) the [*rustling*] *sound of the skin* of a serpent, (L, Ḳ,) *caused by rubbing one part thereof with another:* (L:) the [*rustling*] *sound of the wing* [or *wings*] of a bird: (Ṣ, TA:) the [*rustling*] *sound* of a tree agitated by the wind: the [*rustling,* or *murmuring,*] *sound* of the wind, in, or upon, anything by [or through] which it passes: *a plaintive sound,* or *moaning:* the [*murmuring,* or *quivering,*] *sound* of the flaming, or blazing, of fire; and the like: (TA:) the [*rushing*] *sound* of a stone thrown by a مَنْجَنِيق: the [*whizzing*] *sound* of a penetrating or transpiercing arrow [app. in its passage through the air: see a verse cited voce ذُلَّةٌ]: (TA:) the *humming,* or *buzzing,* (دَوِىّ,) of bees. (Ṣ and Ḳ in art. دوى.) The saying, cited by IAạr,

* أَبْلِغْ أَبَا قَيْسٍ حَفِيفَ الأَثَأَبَهْ *

is explained by him as meaning [*Tell thou Aboo-Keys*] *that he is weak in intellect;* as though he were the حفيف of the tree called أَثْأَبَة when it is agitated by the wind: some say that it means [*tell thou Aboo-Keys that*] *I will threaten him and agitate him like as the wind agitates this tree;* but ISd says that this is nought. (TA.) — *Dry* herbage; as also ▼حَفِيفٌ. (TA.)

حُفَافَةٌ *Hair plucked out:* or *what has fallen of hair plucked out.* (TA.) — *Remains of straw, and of* [*the trefoil,* or *dry trefoil, called*] قَتّ. (Ibn-'Abbád, Ḳ.)

حَفْحَفَةٌ [inf. n. of حَفْحَفَ]. — See فَهِرَ.

حَفَّانٌ A *full* vessel: (Ḳ:) or a vessel *nearly filled to* [*the top of*] *each side:* (TA:) or a vessel *of which the contents, measured therein, reach to* [*the top of*] *each side.* (Ṣ, Ḳ.) — The *young ones* of an ostrich; *male* and *female:* (Ṣ, Ḳ:) or, accord. to ISd, *females* only: (MF, TA:) n. un. with ة. (Ṣ, Ḳ.) — The *feathers,* or *plumage,* of the ostrich. (TA.) — The *young ones* of camels: (TA:) sometimes these are thus termed:

(Ṣ in art. حفن:) [app. as being likened to those of the ostrich:] or *such camels as are under* [i. e. *younger than*] *those termed* حِقَاق: (TA:) n. un., applied to a male and a female, as above. (Ṣ in art. حفن.) —— *Servants:* (Ṣ, Ḳ:) as though likened to the young ones of the ostrich. (TA.)

حَافٌّ *Going round about, circuiting, compassing,* or *surrounding.* (Ṣ, Mṣb, Ḳ.) It is said in the Ḳur [xxxix. last verse], وَتَرَى المَلَائِكَةَ حَافِّينَ مِنْ حَوْلِ العَرْشِ (Zj, Ṣ, Ḳ*) *And thou shalt see the angels surrounding the* عرش: (Zj, TA:) or *surrounding the sides thereof:* (Ṣgh, Ḳ:) or *going round about on either side thereof.* (Er-Rághib, TA.) —— مَا لَهُ حَافٌّ وَلَا رَافٌّ: see 1. ═ سَوِيقٌ حَافٌّ [*Meal of parched barley*] *not moistened with water* or *with clarified butter* or *the like.* (Lth, Ḳ.) [خُبْزٌ حَافٌّ, in the present day, means *Dry bread;* i. e. *bread without anything savoury.*] And هُوَ حَافُّ المَطْعَمِ *He is one whose food is dry.* (TA.) ═ See also حَافٌّ, in art. حوف.

مَحْفُوفٌ [*Encompassed,* or *surrounded*]. You say, هُوَ مَحْفُوفٌ بِخَدَمِهِ [*He is encompassed,* or *surrounded, by his servants*]. (TA.) —— هُمْ قَوْمٌ مَحْفُوفُونَ: see 1.

مِحَفَّةٌ, with kesr; (Ṣ, Ṣgh, Mṣb, Ḳ;) in the "Mesháriḳ" of 'Iyáḍ said to be [مَحَفَّةٌ,] with fet-ḥ, (MF,) *A vehicle of the kind used for women, like the* هَوْدَج, (Ṣ, Mṣb, Ḳ,) *except that it has no* قُبَّة [or *dome-like,* or *tent-like, top*], (Ṣ, Ḳ,) *which the* هودج *has:* (Ṣ:) or *a camel's saddle* (رَحْل) *surrounded* (يُحَفُّ [*with pieces of cloth* (see 1) *upon a wooden frame*]), *upon which a woman rides:* accord. to IDrd, so called because the [frame of] wood [with the pieces of cloth attached thereto] surrounds on all sides the sitter upon it. (TA.)

هَوْدَجٌ مُحَفَّفٌ بِالدِّيبَاجِ [*A* هودج *hung round with silk brocade*]. (TA.)

حفث

حَفِثٌ (Ṣ, Ḳ) and حِفْثٌ and ↑ حَفِثَةٌ (Ḳ) [*A certain portion* or *appertenance*] *of the stomach of a ruminant animal, that which has* طَرَائِق [meaning either *furrows* or *streaks,* but more probably the former], *as though it,* or *they,* (كَأَنَّهَا,) *were the coverings* (أَطْبَاق) *of the feces in the stomach:* (Az, L:) or *that* [*part*] *which has coverings* (أَطْبَاق [or probably this signifies here *folds, one above another,*]) *at the lower part of the stomach of a ruminant, towards the side of the latter, from which the feces of the stomach never pass forth:* [app. meaning the *third stomach,* or *omasum;* commonly called the *manyplies,* because of its many plies, or folds, and strata super strata; and by some, the *millet;* from which the food, being already ruminated, does not pass out again to the mouth, as it does from the first and second stomachs:] it pertains to the camel, and to the sheep and goat, and oxen; or, accord. to IAạr, [only] to the sheep and goat: (L:) the حَفِث *of the stomach of a ruminant;* (Ṣ;) *i. q.* قِبَةٌ, (Ṣ, Ḳ,) or فِبَةٌ: (TA:) or *that which is with the stomach of a ruminant, and which resembles it:* (T, TA:) or *that which has* طَرَائِق, *by the side of which is the* قبة, another thing, which has not طرائق: it is called حَفِث and حَثِف and حِفْث and حِثْف and فَحِث and, as some say, فِتْح and بُحْف: (AA, TA:) pl. أَحْفَاثٌ. (Ḳ.) —— Also the first, *A certain great kind of serpent, resembling a* جِرَاب [or *traveller's provision-bag*]. (Ḳ.)

حَفِثَةٌ: see above.

حَفَافِيَةٌ *Big, bulky,* or *corpulent.* (Ḳ.)

حُفَّاثٌ *A certain kind of serpent, that blows, but does not hurt:* (Ṣ:) *a kind of serpent larger than that called* حَفِث, (Ḳ, TA,) *speckled with black and white, party-coloured; that eats herbs,* or *dry pasture, and threatens, but does not hurt any one:* (TA:) or, accord. to Sh, *a bulky serpent, with a large head, red, speckled with white and black, resembling that called* الأَسْوَد, *but not the same as this latter; if one irritate it, its jugular vein becomes distended:* accord. to ISh, it is *larger than that called* الأَرْقَم, *but is speckled with black and white in the same manner as this latter:* pl. حَفَافِيثُ. (Az, TA.) [Hence,] اِحْرَنْفَشَ حُفَّاثُهُ ‡ *His external jugular veins* (أَوْدَاجُهُ [likened to serpents]) *became distended by rage,* or *anger.* (TA.)

حفد

1. حَفَدَ, [aor. ـِ, as appears from what follows,] inf. n. حَفْدٌ and حَفَدَانٌ (Ṣ, A) and حُفُودٌ, (A,) *He* (a camel, Ṣ, A, and an ostrich, Ṣ) *was quick,* or *went quickly;* (Ṣ, A;) *was continuous in his course* or *pace:* and some say that ↑ احفد is syn. with حَفَدَ, meaning *he went quickly:* (Ṣ:) accord. to A 'Obeyd, احفد, said of an ostrich, is syn. with حَفَدَ, inf. n. حَفْدٌ: and it is said that حَفَدَ, inf. n. حَفَدَانٌ, signifies *he went a pace such as is termed* خَبَب, *quicker than that of walking:* (L:) or حَفَدَ, inf. n. حَفْدٌ (TA) and حَفَدٌ and حَفَدَانٌ; and ↑ احفد, inf. n. إِحْفَادٌ; *he went a pace less quick than that termed* خَبَب. (Ḳ, TA.) —— And حَفَدَ, (A, L, Mṣb, Ḳ,) aor. ـِ, (L, Mgh, Mṣb, Ḳ,) inf. n. حَفْدٌ (T, Ṣ, Mgh, Mṣb, Ḳ) and حَفَدَانٌ; (Ḳ;) and ↑ احفد, inf. n. إِحْفَادٌ; (Mṣb;) and ↑ احتفد; (A, Ḳ;) ‡ *He was quick* (Ṣ, A, Mṣb) *in an affair, and active, agile,* or *prompt, in performing it:* (A:) or *he was quick in service:* (Mgh:) or *he was active, agile,* or *prompt, in work;* and *quick:* (Ḳ:) or *he was active, agile,* or *prompt,* in service and in work: (T:) or *he was quick* therein. (L.) Hence, (Mgh,) وَإِلَيْكَ نَسْعَى وَنَحْفِدُ, (Ṣ, Mgh,* L, Mṣb,) in a form of supplication, (Ṣ, L, Mṣb,) which is uttered standing, termed دُعَآءُ القُنُوتِ, means *And we are quick in working for Thee and in serving Thee:* (L:) or *quick to obey Thee:* (Mṣb:) or *we work for Thee by obeying Thee:* (Mgh:) [for] —— حَفَدَ, (A, L, Mṣb, Ḳ,) aor. ـِ, (L,) inf. n. حَفْدٌ, (L, Mṣb,) also signifies ‡ *He served* (A, L, Mṣb, Ḳ) a person: (A:) [I have marked this, and the significations explained in the second sentence above, as tropical on the authority of the A: but] accord. to A 'Obeyd, the primary signification of this verb is *he served and worked.* (L.)

4: see 1, in three places. ═ احفدهُ *He made him,* or *incited him,* (namely, a camel, Ṣ, A,) *to go quickly,* (Ṣ, A, Ḳ,) *with a continuous course* or *pace.* (Ṣ.)

8: see 1.

حَفَدٌ *A pace less quick than that termed* خَبَب. (Ḳ.) [See 1.] ═ See also حَافِدٌ.

حَفِيدٌ: see حَافِدٌ.

حَفَّادٌ A camel *that goes quickly, with a continuous course* or *pace.* (Ṣ.)

حَافِدٌ sing. of حُفَّادٌ (L) [and of أَحْفَادٌ, a pl. of pauc.,] and of حَفَدَةٌ, (Ṣ, A, L, Mṣb,) which last signifies ‡ *Assistants, helpers,* or *auxiliaries;* and *any who work,* or *labour, in obedience to orders, and strive together in quickness;* (Ibn-'Arafeh;) *whatever serve thee and work for thee and assist thee;* (El-Ḥasan;) *assistants, helpers,* or *auxiliaries,* and *servants;* (Ṣ, A, Mgh, Mṣb;) as also ↑ حَفَدٌ, which is likewise a pl. [or rather a quasi-pl. n.] of حَافِدٌ; (Ḳ, TA;) [and حُفَّادٌ:] and also, (A, Mgh, Mṣb, Ḳ,) hence, (A, Mgh, Mṣb,) as some say, (Ṣ,) a man's *grandchildren;* (Ṣ, Mgh, Mṣb, Ḳ;) because they are like servants while young: (Mṣb:) or *sons' children:* (A:) or *a son's children:* (Mgh:) likewise pl. of حَافِدٌ: (Ṣ:) and ↑ حَفِيدٌ, which is said in the Ḳ to be syn. with حَفَدَةٌ as meaning "grandchildren," is a sing., of which حَفَدَةٌ and حُفَدَآءُ [and أَحْفَادٌ] are pls., (TA,) and signifies *a grandchild:* (L, TA:) [it is vulgarly applied to *a son's son;* and سِبْط, to a daughter's son:] or حَفَدَةٌ signifies a man's *children:* (CḲ:) or his *daughters;* (Ḳ;) by which, as some say, are meant those *who serve their parents in the house:* (TA:) or his *children and grandchildren who serve him;* accord. to Zirr and 'Ikrimeh; but this is contradicted by 'Abd-Allah Ibn-Mes'ood and others: (L:) or *such relations as are termed* أَصْهَار: ('Abd-Allah Ibn-Mes'ood, L, Ḳ:) or *such as are termed* أَخْتَان: (Fr:) or one's *wife's sons by her former husband.* (Eḍ-Ḍaḥḥáḳ.)

مَحْفِدٌ *Origin,* syn. أَصْلٌ, (Ṣ, Ḳ,) of a man; (Ṣ;) or in a general sense; (L;) *i. q.* مَحْتِدٌ and مَحْفَدٌ and مُحْفَدٌ. (IAạr.) —— And The *base,* or *lower part,* (أَصْل,) *of a camel's hump:* (IAạr, Yaạḳoob, Ṣ, M, Ḳ:) or the *hump itself.* (TA.)

مَحْفُودٌ A man *served,* or *waited on, by others;* (Ṣ, A, Ḳ;) and *obeyed:* (A:) one *whom his companions serve and honour, and whom they hasten to obey.* (TA.)

مُحْتَفِدٌ A sword *quick in cutting.* (Ṣ, Ḳ.) مُحْتَفِدُ الوَقْعِ [meaning A sword *quick in falling*] occurs in a verse of El-Aạshà describing a sword, accord. to one reading: but Az says that the right reading is مِحتفل, with ل. (L.)

حفر

1. حَفَرَ, (Ṣ, A, Ḳ, &c.,) aor. ـِ, (Mṣb, Ḳ,) inf. n. حَفْرٌ, (Mgh, Mṣb,) *He dug, excavated,* or *hollowed out,* the ground, or earth; (KL, PṢ, &c.;) *he cleared out* a thing, (Ḳ,) as one does the ground; (Ṣ, Mṣb, Ḳ;) and a well; (the Lexicons passim;) and a river; (A, Mgh;) with a مِحْفار; (A;) or with an iron implement; (Ḳ;) and ↓احتفر signifies the same. (Ṣ, A, Ḳ.) And حَفَرَ عَلَيْهِ, and حَفَرَهُ, and ↓احتفرهُ, *He dug for him,* (namely, a lizard of the kind called ضَبّ, or a jerboa,) *to fetch him forth.* (A, TA.) — [*He burrowed.*] — † *It* (a torrent) *furrowed* a valley. (Mṣb.) [See also 5.] — ‡ *Inivit* feminam: (IAạr, Mṣb, Ḳ:) the action being likened to that of a man digging a river. (IAạr.) — هٰذَا غَيْثٌ لَا يَحْفِرُهُ أَحَدٌ ‡ *This is a rain of which no one knows the utmost extent.* (Ḳ,* TA.) — حَفَرَ ثَرَى زَيْدٍ ‡ *He searched into the affair,* or *case, of Zeyd,* (A, Ḳ,) *and became acquainted with it.* (Ḳ.) — And حَفَرَ, (Ṣ, A, Ḳ,) aor. as above, (Ṣ,) and so the inf. n., (Ṣ, A,) † *He,* or *it, emaciated,* or *rendered lean:* (Ṣ, Ḳ:) *it* (a copious flow of milk, TA) *emaciated* a she-goat: (Ḳ, TA:) ‡ *he* (a young camel) *rendered* his mother *flabby in flesh by much sucking.* (A.) There is no pregnant animal that pregnancy does not emaciate, except the camel: (Ṣ, A:) she fattens in pregnancy. (Ṣ.) — حفر رَوَاضِعَ *He* (a child) *shed his* [or *milk-teeth*]. (Ḳ, TA.) [See also 4.] — حَفَرَتْ, or حُفِرَتْ, رَوَاضِعُ المُهْرِ, (accord. to different copies of the A,) ‡ *The milk-teeth of the colt became in a wabbling,* or *loose, state, previously to their falling out;* because, when they have fallen out, their sockets become hollow. (A.) [See 4.] — حَفَرَتِ الأَسْنَانُ, aor. ـِ, (Ṣ, Mgh, Mṣb, Ḳ,) inf. n. حَفْرٌ; (Ṣ, Mṣb;) and حَفِرَت, aor. ـَ, (Ṣ, Mgh, Mṣb, Ḳ,) inf. n. حَفَرٌ, in the dial. of Benoo-Asad, (Ṣ, Mṣb,) and this is the worse of these two forms, (Ṣ,) and حُفِرَ; (El-Wá'ee;) and حُفِرَت; (Ḳ;) ‡ *The teeth became affected with what is termed* حَفْر [q. v. infrà] or حَفَر: (Ṣ, Mṣb, Ḳ:) or *became unsound:* (Mgh:) and حَفِرَ فُوهُ and حُفِرَ ‡ *his teeth cankered.* (A.) IDrst says, in the Expos. of the Fṣ, that حَفَرَ, aor. ـِ, inf. n. حَفْرٌ, is trans.; and that the cause of حَفْر of the teeth, [or the agent of the verb حَفَرَ,] is old age, or the continuance of a yellow incrustation, [or tartar,] or some kind of canker that effects them: but that the verb in the phrase حَفِرَتْ سِنُّهُ, aor. ـَ, inf. n. حَفَرٌ, is intrans. (MF.) [The truth probably is, that the former verb is both trans. and intrans., and hence حَفَرَتِ الأَسْنَانُ; and that the latter is intrans. only.] — And حَفِرَ, aor. ـَ, † *It was,* or *became, in a bad, corrupt,* or *unsound, state.* (Az.)

3. حافر, (A,) inf. n. مُحَافَرَةٌ, (TA,) *He* (a jerboa) *went deep into his hole;* (A;) *so deep that he could not be dug out.* (TA.)

4. احفر فُلَانًا بِئْرًا *He assisted such a one to dig a well.* (Ḳ.) — احفر الصَّبِىُّ, (Ḳ,) inf. n. إِحْفَارٌ, (TA,) ‡ *The child shed his two upper and lower central incisors:* سَقَطَتْ لَهُ الثَّنِيَّتَانِ العُلْيَيَانِ وَالسُّفْلَيَانِ: so in the Ḳ: and to these words we find added, in some copies of the Ḳ, لِلْإِثْنَاءِ وَالإِرْبَاعِ; and then, وَالمُهْرُ سَقَطَتْ ثَنَايَاهُ وَرَبَاعِيَاتُهُ; but in some good and corrected copies, we read, after السفليان, thus, والمهر للاثناء والارباع سقطت ثناياه ورباعياته: to which, in some lexicons, [as in the Ṣ, though the explanation which follows is there different,] after والارباع, is added وَالقُرُوحِ. (TA. [This is evidently the right reading; and therefore I follow it in an explanation in what is here immediately subjoined.]) — احفر المُهْرُ لِلْإِثْنَاءِ وَالإِرْبَاعِ ‡ *The colt shed his central incisors, or nippers, and each of the teeth immediately next to these:* (Ḳ: see what next precedes:) or احفر المُهْرُ لِلْإِثْنَاءِ وَالإِرْبَاعِ وَالقُرُوحِ *the colt shed his milk teeth* (رَوَاضِع), [*the central pair, the second pair, and the third pair, in each jaw,*] *and grew others:* (Ṣ:) or احفر المهر, [inf. n. إِحْفَارٌ,] signifies, *the colt had his milk-teeth in a wabbling, or loose, state, previously to their falling out;* because, when they have fallen out, their sockets become hollow: (A:) or *the colt had his lower and upper central pairs of nippers, of his milk-teeth, in a wabbling, or loose, state:* this is during a period extending from thirty months, at the earliest, to three years: then the teeth fall out: then a lower and an upper central pair of nippers grow in the place of the milk-nippers which have fallen out, after three years; and the epithet مُبْدِئٌ is applied to the colt; and the epithet ثَنِىٌّ is [also] then applied to him, and continues to be until [again it is said of him] يُحْفِرُ, meaning, *he has his lower and upper pairs of nippers, of his milk-teeth, in a wabbling, or loose, state:* then these fall out, when he has completed four years: then the term إِبْدَآءٌ is [again] applied to him; [i. e., he is again termed مُبْدِئٌ;] and he is, and ceases not to be, termed رَبَاعٍ, until [it is said of him] يُحْفِرُ لِلْقُرُوحِ [in the TA, تُحْفِرُ القُرُوحِ, which is an evident mistake,] meaning, *he has his two corner nippers* [*in each jaw*] *in a wabbling, or loose, state:* this is when he has completed five years: then the term إِبْدَآءٌ is applied to him as before described: then he is [also said to be] قَارِحٌ. (TA from the "Kitáb el-Kheyl" of AO.) [See also 1.]

5. تحفّر ‡ *It* (a torrent) *made hollows in the ground.* (A.) [See also 1.]

8: see 1, first and second sentences.

10. استحفر *He asked,* or *desired,* [another] *to dig* a well, or pit, and a rivulet, or canal. (KL.) — استحفر النَّهْرُ *It was time for the river,* or *rivulet, or canal, to be dug* [or *cleared out*]. (Ṣ.)

حَفْرٌ: see حَفَرٌ, in two places; and حَفِيرٌ. — Also † *Emaciation,* or *leanness.* (Kr.) [See 1.] — Also, and ↓حَفَرٌ, (Az, Ṣ, Mṣb, Ḳ,) the latter of the dial. of the Benoo-Asad, and the worse of the two forms, (Ṣ,) said by IḲt to be a bad form, (TA,) and by ISk to be a vulgar mispronunciation, which is attributed to his not having heard the dial. of the Benoo-Asad, (Mṣb,) ‡ *A scaling* (سُلَاق) *in the roots of the teeth:* (Yaḳkoob, Ṣ, Ḳ:) or *a rottenness,* or *an unsound state, of the roots of the teeth,* (Ṣ, Mṣb,) *by reason of a scaling of those parts:* (Mṣb:) or *what adheres to the teeth, externally and internally:* (Az:) or *an erosion of the roots of the teeth by a yellow incrustation between those parts and the gum, externally and internally, pressing upon the bone so that the latter scales away if it be not quickly removed:* (Sh:) or *a cankering of the teeth:* (A:) or *a yellowness upon the teeth:* (IDrd, IKh, Ḳ:) or حَفْرٌ signifies *a pimple,* or *small pustule, in the gum of a child.* (El-Wá'ee.) [See 1: and see also حَبْرٌ.]

حَفَرٌ *A well that is widened* (Ḳ, TA) *beyond measure;* (TA;) as also ↓حَفْرٌ (Ḳ) and ↓حَفِيرٌ and ↓حَفِيرَةٌ. (TA.) — See also حَفِيرٌ. — The *earth that is taken forth from a hollow, cavity, pit, or the like, that is dug in the ground;* (Ṣ, Ḳ;) like هَدَمٌ: (Ṣ:) [see also حَفِيرَةٌ:] or *what is dug,* or *excavated;* like عَدَدٌ and خَبَطٌ and نَفَضٌ in the senses of مَعْدُودٌ and مَخْبُوطٌ and مَنْفُوضٌ: (Mṣb:) or *a place that is dug,* (Az, Ṣ, Mṣb,) *like a moat or a well;* (Az, Mṣb;) as also ↓حَفْرٌ: (TA:) pl. أَحْفَارٌ, (Mṣb, Ḳ,) and pl. pl. أَحَافِيرُ. (Ḳ.) — See, again, حَفِيرٌ. — And see حُفْرَةٌ.

حُفْرَةٌ *What is dug, excavated, hollowed out,* or *cleared out,* (Mṣb, Ḳ,) *in the ground;* (Mṣb;) [i. e. *a hollow, cavity, pit, hole, trench, ditch,* or *furrow, dug,* or *excavated, in the ground:* and *any hollow,* or *cavity, in the ground, whether made by digging* or †*natural: a burrow:*] as also ↓حَفِيرَةٌ, (Mgh, Mṣb, Ḳ,) which is of the measure فَعِيلَةٌ in the sense of the measure مَفْعُولَةٌ: (Mṣb:) pl. of the former حُفَرٌ; (Ṣ, Mṣb;) and of the latter حَفَائِرُ. (Mṣb.) — See also حَفِيرٌ.

حَفِيرٌ is of the measure فَعِيلٌ in the sense of the measure مَفْعُولٌ [meaning *Dug, excavated, hollowed out, or cleared out, in the ground*]. (TA.) [Hence,] رَكِيَّةٌ حَفِيرَةٌ *A newly-dug well;* as also ↓حَفَرٌ. (TA.) — See also this last word. — Also, (IAạr, Ṣ, A, Ḳ,) and ↓حَفِيرَةٌ and ↓حَفْرٌ, (A,) [or ↓حَفَرٌ, q. v., and ↓حُفْرَةٌ, as is shown by an explanation of its pl. (حُفَرٌ) in the Ḥam p. 562,] *A grave.* (IAạr, Ṣ, A, Ḳ.)

حَفِيرَةٌ: see حَفَرٌ: — and حُفْرَةٌ: — and حَفِيرٌ. — Also *What is dug out of a mine.* (Mgh.)

حَفَّارٌ *A grave-digger.* (Ḳ.)

حَافِرٌ, [*Digging: a digger.* — And hence,] The حافر of a beast, (دَابَّة, Ṣ, Ḳ,) i. e., of a horse, or mule, or ass; (TA;) [namely, the *hoof; a solid hoof;*] as though it dug the ground by reason of the vehemence of its tread upon it; (Mṣb;) a subst., like كَاهِلٌ and غَارِبٌ: (TA:) pl. حَوَافِرُ. (Ṣ, A, Ḳ.) — [Hence, by a synecdoche,] خُفٌّ وَحَافِرٌ ‡ *Camels and horses.* (Mgh in art. خف.)

—— حَافِرٌ is also applied to ‡The *foot of a man*, (Ṣ, TA,) when it is meant to be characterized as ugly. (TA.) —— اَلنَّقْدُ عِنْدَ الحَافِرَةِ▾, (Ṣ, A, Ḳ,) and الحَافِرِ, (A, Ḳ,) is a prov., (Ṣ,) meaning, ‡*The payment in ready money is on the occasion of the first sentence spoken* (Yaḳḳoob, T,* Ṣ, Ḳ) *by the seller, when he says "I have sold to thee"* [such a thing]. (T.) The origin of the saying was this: horses were the most excellent (Ḳ) and precious (TA) of the things that they possessed; and they used not to sell them on credit: a man used to say the words above to another; meaning that its hoof should not remove until he received its price: (Ḳ:) and he who says عند الحافرة (since he makes الحافر to mean the beast, الدَّابَّة, itself, and since its use in this sense is frequent without the mention of ذَات [prefixed to it],) subjoins to it the sign [ة] of the fem. gender to show that ذَات الحَافِرِ is meant by this name. (TA.) Or they used to say this on the occasion of racing and betting: and the meaning is, *when the horse's hoof first falls upon the dug ground* [at the goal]: (Abu-l-'Abbás, Az, Ḳ:) ▾حَافِرَةٌ, (Abu-l-'Abbás,) or حَافِرٌ, (Ḳ,) signifying *dug ground*; (Abu-l-'Abbás, Ḳ;*) *ground that is dug by a horse's feet*; (Ḥar p. 653;) like as one says مَآءٌ دَافِقٌ, meaning مَدْفُوقٌ. (TA.) Lth says that the saying means, when thou buyest it, thou dost not quit thy place until thou payest ready money. (TA.) This was its origin: then it came to be so often said as to be used with reference to any priority. (Ḳ.) —— [Thus,] ▾حَافِرَةٌ signifies ‡*The original state or constitution of a thing; that wherein it was created:* and the *returning in a thing, so that the end thereof is brought back to its beginning.* (Ḳ.) It is said in the Ḳur [lxxix. 10], أَئِنَّا لَمَرْدُودُونَ فِى الحَافِرَةِ▾, i. e., ‡*Shall we indeed be restored to our first state?* (Ṣ:) i. e., *to life?* (Fr:) or *to the present world, as we were:* (IAạr:) or *to our first creation, after our death.* (TA.) IAạr cites the following verse:

* أَحَافِرَةً▾ عَلَى صَلَعٍ وَشَيْبٍ *
* مَعَاذَ ٱللّٰهِ مِنْ سَفَهٍ وَعَارِ *

meaning ‡*Shall I return to my first state, wherein I was in my youth*, when I indulged in amatory conversation, and silly and youthful conduct, *after hoariness, and baldness of the fore part of my head?* [*I beg God to preserve me from lightwittedness and shameful conduct.*] (Ṣ.) One says also, رَجَعَ إِلَى حَافِرَتِهِ▾, (A,) and حَافِرِهِ, (TA,) ‡*He became old and decrepit:* (A, TA:) [*as though he returned to his first state; or became in a state of second childishness.*] And اِلْتَقَوْا فَٱقْتَتَلُوا عِنْدَ الحَافِرَةِ▾ (Ṣ, A, Ḳ) and الحَافِرِ (A) ‡*They met, and fought one another at the first of their meeting.* (Ṣ, Ḳ.) And فَعَلَ كَذَا عِنْدَ الحَافِرَةِ▾ and الحَافِرِ ‡*He did so at the first, without delay.* (TA.) And رَجَعَ عَلَى حَافِرَتِهِ▾ ‡*He returned by the way by which he had come:* (T, Ṣ:) or *by which he had come forth.* (Ḳ.)

حَافِرَةٌ : see حَافِرٌ, in nine places.

مِحْفَرٌ (Ḳ) and ▾مِحْفَارٌ (A, Ḳ) and ▾مِحْفَرَةٌ (Ḳ) *A spade;* syn. مِسْحَاةٌ: (Ḳ:) *an implement for digging* (A, Ḳ, TA) *of the same kind as a* مسحاة: (TA:) pl. of the first [and last] مَحَافِرُ. (Ḥam p. 665.)

مِحْفَرَةٌ : see what next precedes.

طُرُقٌ مُحْفَرَةٌ [app. *Roads much furrowed* by the feet of beasts or men: see حَجْحَجَ]. (L and Ḳ in art. حج.)

مِحْفَارٌ : see مِحْفَرٌ.

مَحْفُورٌ [*i. q.* حَفِيرٌ as meaning *Dug*: see the latter.] —— فَمُ فُلَانٍ مَحْفُورٌ [and أَسْنَانُهُ مَحْفُورَةٌ] ‡*The teeth of such a one are affected with what is termed* حَفْرٌ or حَفَرٌ. (Ṣ, TA.) And صَبِىٌّ مَحْفُورٌ †*A child having a pimple, or small pustule, in the gum.* (El-Wá'ee.)

فُلَانٌ أَرْوَغُ مِنْ يَرْبُوعٍ مُحَافِرٍ *Such a one is more elusive than a jerboa that goes so deep into his hole that he cannot be dug out.* (A, TA.)

حفز

1. حَفَزَهُ, aor. ـِ, (Ṣ, Ḳ,) inf. n. حَفْزٌ, (Ṣ,) *He hastened, or hurried, or incited, him, or it, from behind, either by driving or otherwise:* this is the primary signification. (TA.) You say, حَفَزَهُ عَنِ الأَمْرِ, (Ḳ,) aor. and inf. n. as above, (IDrd, TA,) *He hastened, or hurried, and urged, him away from the thing* or *affair.* (IDrd, Ḳ,* TA.) —— *He pushed him, or it, from behind.* (Ṣ, Ḳ.) [Hence,] حَفَزَ اللَّيْلُ النَّهَارَ, (Ḳ,) aor. as above, (Ṣ,) and so the inf. n., (TA,) †*The night urged on the day.* (Ṣ, Ḳ, TA.) —— *He put in motion,* and *disturbed,* or *removed, him,* or *it.* (Mgh.) —— *He thrust him, or pierced him,* بِالرُّمْحِ *with the spear.* (Ṣ, Ḳ.) —— [Hence,] حَفَزَ signifies also † *Inivit feminam.* (Ṣgh, Ḳ.) —— حَفَزُوا عَلَيْنَا الخَيْلَ وَالرِّكَابَ *They poured upon us* [*the horses and the camels with their riders*]. (Shujáạ El-Aạrábee, TA.)

5: see 8, in two places.

8. احتفز *He urged, or pressed forward, and strove,* in his gait, or pace; (IAạr, Ḳ;) [and so ▾تحفّز: see الدَّوَالِيكَ, in art. دلك; and دَوَالَيْكَ, in art. دول.] —— *He sat upright, not in an easy posture;* syn. اِسْتَوْفَزَ; as also ▾تحفّز. (Ḳ.) [See the part. n., below.] —— *He drew himself together* (تَضَامَّ) *in his prostration and sitting.* (Ḳ.) It is said in a trad. of 'Alee, إِذَا صَلَّتِ المَرْأَةُ فَلْتَحْتَفِزْ *When the woman prays, let her draw herself together in her sitting and prostration,* (Ṣ, Mgh,* TA,) *and not put her arms apart from her sides, like the man.* (Ṣ, TA.) —— *He settled himself in a sitting posture upon his buttocks:* (En-Naḍr, Ḳ:) or *upon his knees, as though he would rise:* (TA:) or *he was uneasy, and raised himself, being vexed, or disquieted by grief:* (IAth:) or *he desired to rise and to lay violent hands upon a thing,* while sitting. (TA.)

مُحْتَفِزٌ *Hasting;* (TA;) *sitting upright, not in an easy posture,* (مُسْتَوْفِزٌ, Ṣ, Mgh, TA,) *desiring to rise, not sitting firmly upon the ground.* (TA.)

حفش

حَفَشٌ : see what follows.

حِفْشٌ *A receptacle for spindles:* (Ṣ, Ḳ:) *a* سَفَط [or *receptacle woven of palm-leaves, in which are stowed perfumes and similar things, of the apparatus of women*]; (Ḳ; [in the CḲ, السَّفَط is erroneously put for السَّقَط;]) *a* سَفَط *in which perfumes are kept:* (TA:) *a* دُرْج, (Ḳ,) i. e. *a small receptacle of the kind called* سَفَط, *in which a woman keeps her perfumes and apparatus,* or *implements;* (TA in art. درج;) *a woman's* دُرْج: (Mgh:) pl. as below. (TA.) —— ‡*A small* بَيْت [or *tent*]; (A'Obeyd, Ṣ, Mgh;) as being likened to a woman's دُرْج: (Mgh:) or *a very small* بَيْت [or *tent*], (Ḳ, TA,) *having a low roof;* so called because of its narrowness; and accord. to some, ▾حَفْش, and ▾حَفَش: (TA:) or *a very small* بيت *of* [*goats'*] *hair,* (Ḳ, TA,) *of the* بُيُوت [or *tents*] *of the Arabs of the desert:* so says Kh: pl. as below. (TA.) It is said in a trad., دَخَلَتْ حِفْشًا, meaning †*She entered a small* بيت [or *tent*]. (Mgh.) And in another trad., هَلَّا قَعَدَ فِى حِفْشِ أُمِّهِ, (Ṣ,) or هلّا جَلَسَ الخ, (Mgh,) i. e. †[*Wherefore did he not sit*] *in the small* بيت *of his mother?* (Ṣ, Mgh:) or, accord. to some, the meaning is, عِنْدَ حِفْشِ أُمِّهِ [i. e., *by the receptacle for the spindles, &c., of his mother*]: (Ṣ:) or, as some say, *by the* فَرْج *of his mother;* for (TA) —— It also signifies †The فَرْج [*pudendum muliebre*]. (Ḳ.) —— And *Any vessels that are held in little account,* (Lth, Ḳ, TA,) *that are used as receptacles in the house, or tent, for perfumes and the like,* (Lth, TA,) *such as glass flasks, or bottles, &c.:* (Lth, Ḳ, TA:) and *an old and worn-out thing,* (Ḳ, TA,) *that is of no use:* (TA:) and *a large, old and worn-out, sack* (Ḳ, TA) *of* [*goats'*] *hair:* (TA:) pl. (in all the senses of the word, TA) أَحْفَاشٌ (Ḳ, TA) and حِفَاشٌ: (TA:) or أَحْفَاشُ البَيْتِ means *the furniture and utensils of the house or tent;* and *the paltry articles thereof, or such as are held in little account.* (Aboo-Sinán, Ḳ.) —— And أَحْفَاشُ الأَرْضِ means *The* [*lizards called*] ضِبَاب, *and the hedgehogs,* (Ḳ, TA,) *and the jerboas,* (TA,) *of the earth:* (Ḳ, TA:) not the same as أَحْنَاش: so says Aboo-Ziyád. (TA.)

حَفَشٌ : see above.

حفظ

1. حَفِظَهُ, (Ṣ, Mgh, Mṣb, Ḳ,) aor. ـَ, (Ḳ,) inf. n. حِفْظٌ, (Ṣ, Mgh, Mṣb,) *He kept it, preserved it, guarded it, protected it,* or *took care of it;* (Ṣ, Ḳ;) namely, a thing; (Ṣ;) *he prevented it from perishing,* or *becoming lost;* (Mgh, Mṣb;) namely, a thing, (Mgh,) or property &c.; (Mṣb;) and hence the saying, (Mgh,) حِفْظٌ is the *contr.* of نِسْيَان; (M, Mgh;) i. e., it signifies the *taking*

care, being careful; (M;) *being mindful, regardful, attentive,* or *considerate:* (M, Ḳ:) [see also 5:] and ▼ احتفظ به signifies the same as حَفِظَهُ. (Ṣ, Mṣb.) [Hence,] you say, حَفِظَ المَالَ *He kept and tended,* or *pastured and defended, the camels* or *the like.* (Ḳ.) [And حَفِظَ حُرْمَةَ صَاحِبِهِ *He was regardful of everything entitled to reverence, respect, honour, or defence, in the character and appertenances of his companion,* or *friend.*] And حَفِظَ السِّرَّ *He kept the secret.* (TA.) [And حَفِظَ يَمِينَهُ *He kept his oath:* but this has also another meaning, as will be seen below.] And حَفِظَ القُرْآنَ *He kept,* or *retained, the Ḳur-án in his mind,* or *memory; got it, knew it,* or *learned it, by heart.* (Ṣ,* Mṣb, Ḳ.) [See also 5.] And حَفِظَ عَنْ فُلَانٍ [*He learned by heart from such a one:* and, followed by an accus. case, the same; or *he retained in his memory, as learned,* or *heard, from such a one;* or *he remembered to have heard from such a one*]. (TA &c. passim.) And one says of God, قَدْ حَفِظَ عَلَى خَلْقِهِ وَعِبَادِهِ مَا يَعْمَلُونَ مِنْ خَيْرٍ أَوْ شَرٍّ [*He hath preserved from oblivion, for,* or *against, his creatures and his servants, what they do of good or evil*]. (TA.) — Also *He kept it from being used,* or *employed, on,* or *for, ordinary, mean,* or *vile, occasions,* or *purposes.* (Mgh, Mṣb.) You say, فُلَانٌ يَحْفَظُ نَفْسَهُ وَلِسَانَهُ *Such a one keeps himself and his tongue from ordinary, mean,* or *vile, employment, in that which does not concern him.* (Mgh.) And hence the saying in the Ḳur [v. 91], وَٱحْفَظُوا أَيْمَانَكُمْ, accord. to one of the modes of interpreting it; i. e. *And keep ye your oaths from being used,* or *uttered, on,* or *for, ordinary, mean,* or *vile, occasions,* or *purposes;* agreeably with what is said in ii. 224 of the Ḳur, where ordinary and frequent swearing by God is forbidden. (Mgh.) [Another meaning of which this phrase is susceptible has been shown above.]

2. حَفَّظْتُهُ الكِتَابَ *I incited him,* or *urged him,* [or *made him,*] *to commit to memory,* or *learn by heart, the book:* (Ṣ:) and [in like manner,] ▼ أَحْفَظْتُهُ الحَدِيثَ *I made him to retain the narration,* or *tradition, in his mind,* or *memory;* or *to know it,* or *learn it, by heart.* (TA in art. زكت.)

3. مُحَافَظَةٌ The *defending* of those persons, or things, that are sacred, or inviolable, or that one is bound to respect or honour, and to defend, (Ḳ, TA,) on the occasions of wars; (TA;) as also حِفَاظٌ. (Ḳ, TA.) You say, حافظ حُرِيمَهُ *He defended his wife,* or *wives,* or *the like.* (TḲ.) [And hence,] you say, إِنَّهُ لَذُو حِفَاظٍ, and ذُو مُحَافَظَةٍ, meaning *Verily he is disdainful,* or *scornful.* (Ṣ, TA.) — The *being mindful, watchful, observant,* or *regardful:* (Ṣ, and TA in art. رعى:) [see also 5:] or the *keeping, attending,* or *applying oneself, constantly, perseveringly,* or *assiduously,* (Ḳ, TA,) to a thing, or an affair. (TA.) You say, حافظ عَلَى الأَمْرِ, (TA,) or على الشَّىْءِ, inf. n. محافظة, (Mṣb,) *He kept, attended,* or *applied himself, constantly,* &c., *to the thing,* or *affair.* (TA.) And hence the saying in the Ḳur [ii. 239], حَافِظُوا عَلَى الصَّلَوَاتِ *Perform ye the prayers in their proper times:* or, accord. to Az, *keep ye, attend ye,* or *apply yourselves, constantly,* or *perseveringly, to the performance of the prayers in their proper times.* (TA.) — حِفَاظٌ, is also explained as signifying The *being mindful,* or *observant, of a covenant,* and the *keeping,* or *fulfilling, of a promise, with forgiveness, and holding fast to love* or *affection.* (TA.)

4: see 2. — احفظهُ, (Ṣ, Ḳ, TA,) and احفظهُ حِفْظَةً, inf. n. إِحْفَاظٌ, [*He made him to conceive what is termed* حِفْظَة, or حَفِيظَة;] *he angered him; made him angry:* (Ṣ, Ḳ, TA:) and in the same sense it is said of a speech, or word: (TA:) or only *he angered him by evil,* or *foul, speech,* (Ḳ, TA,) and *making him to hear what he disliked,* or *hated.* (TA.)

5. تحفّظ *He guarded himself;* syn. اِحْتَرَزَ, (Ḳ, TA,) or تَحَرَّزَ, (Mṣb,) and تَحَرَّسَ, and اِحْتَرَسَ, (Ṣ and Mṣb and Ḳ in art. حرس,) مِنْهُ *from him,* or *it,* (Ṣ in art. حرس, &c.,) or عَنْهُ. (TA.) *He was,* or *became, careful, mindful, attentive,* or *considerate;* (TA;) *watchful, vigilant,* or *heedful;* (Ṣ, O, L, TA;) in affairs, and speech, and to avoid a slip, or fault; as though he were cautious, or careful, or fearful, of falling. (L, TA.) [See also 1, and 3.] — [In the last of the senses explained above, it is also trans.: you say, تحفّظ أَمْرَهُ *He was careful, mindful,* &c., *of his affair,* or *case:* see Bḍ in xxxiii. 52.] — تَحَفَّظْتُ الكِتَابَ *I learned the book by heart, one part,* or *thing, after another.* (Ṣ, TA.) [See also حَفِظَ القُرْآنَ, in the first paragraph.]

8. احتفظ به: see 1. — احتفظهُ لِنَفْسِهِ, (Ḳ,) and احتفظ به لنفسه, (TA,) *He appropriated it, took it,* or *chose it, to,* or *for, himself.* (Ḳ, TA.) — احتفظ [*He conceived,* or *became affected with, what is termed* حِفْظَة, or حَفِيظَة;] *he became angered,* or *angry:* (Ṣ, Ḳ:) or *he became angered by evil,* or *foul, speech.* (Ḳ.)

10. استحفظهُ, (Ṣ, Ḳz, Ṣgh, Mṣb, Ḳ,) followed by إِيَّاهُ, (Ḳ,) or الشَّىْءَ, (Ḳz, Mṣb,) or مَالًا, or سِرًّا, (Ṣgh,) [but in the Ṣ, nothing follows it,] *He asked him to keep, preserve, guard,* or *take care of,* or *to preserve from perishing* or *becoming lost,* or *to be careful of,* or *mindful of,* or *attentive to,* (Ṣ, Ṣgh, Mṣb, Ḳ,) *it,* (Ṣ, Ḳ,) or *the thing,* (Mṣb,) or *property,* or *a secret:* (Ṣgh:) or *he placed the thing with him for him to keep it, preserve it, guard it,* or *take care of it,* &c.: (Ḳz:) or *he intrusted him with the thing; intrusted it to him;* or *gave it to him in trust,* or *as a deposite.* (Mṣb.) It is said in the Ḳur [v. 48], بِمَا ٱسْتُحْفِظُوا مِنْ كِتَابِ ٱللّٰهِ, meaning *By that which they have been required to keep,* &c., *of the Book of God:* (Mṣb:) or *by that with which they have been intrusted, of the Book of God.* (Mṣb, TA.)

حِفْظٌ inf. n. of 1 [q. v.]. (Ṣ, Mgh, Mṣb.) — See also حَافِظٌ, last sentence but one.

حِفْظَةٌ: see حَفِيظَةٌ.

رَجُلٌ حُفَظَةٌ *A man of much* حفظ [app. meaning *retention in the mind,* or *memory:* see 1]. (Ṣgh.)

حَفِيظٌ: see حَافِظٌ, in seven places: — and see مَحْفُوظٌ.

حَفِيظَةٌ The *defence of those persons,* or *things, that are sacred,* or *inviolable,* or *that one is bound to respect* or *honour, and to defend;* a subst. from 3, in the first of the senses mentioned above: (Ḳ, TA:) pl. حَفَائِظُ. (TA.) Hence the saying, تَنْقُضُ الأَحْقَادَ, الحَفَائِظُ تُذْهِبُ الأَحْقَادَ, (TA,) or (Ṣ,) [*The acts of defending those whom one is bound to respect or honour, and to defend, put away,* or *annul, rancorous feelings;*] i. e., when thou seest thy relation, or kinsman, wronged, thou defendest him, though rancour be in thy heart. (Ṣ, TA.) — Also, and ▼ حِفْظَةٌ, *Indignation,* and *anger,* (Ṣ, Ḳ, TA,) *by reason of violence,* or *injury, done to something which one is bound to honour* or *respect, and to defend,* or *of wrong done to a relation,* or *kinsman, in one's neighbourhood,* or *of the breach of a covenant.* (TA.) It is said in a prov., المَقْدِرَةُ تُذْهِبُ الحَفِيظَةَ [*Power to revenge dispels anger,* or *indignation,* &c.]; meaning that it is incumbent to forgive when one has power [to revenge]. (A, TA.) — *An amulet,* or *a charm, bearing an inscription, which is hung upon a child, to charm against the evil eye* &c. (TA.)

حَافِظٌ and ▼ حَفِيظٌ *Keeping, preserving, guarding,* or *taking care of,* a thing; or *a keeper, preserver,* &c.: *keeping and tending,* or *pasturing and defending,* camels or the like; or *a keeper and tender* thereof: (Ḳ:) *keeping* a secret [and an oath]: (TA:) *keeping,* or *retaining,* the Ḳur-án [&c.] *in the mind,* or *memory; knowing* it, or *learning* it, *by heart:* (Ḳ:) *intrusted with a thing,* (Ḳ, TA,) *to keep it, preserve it, guard it,* or *take care of it:* (TA:) [*careful, mindful, attentive,* or *considerate:* (see 1:)] and the latter, *a keeper,* or *person mindful, of the ordinances prescribed by God:* (Bḍ and Jel in l. 31:) pl. of the former حَفَظَةٌ and حُفَّاظٌ: (Mṣb, Ḳ:) the latter pl. particularly applied to persons *endowed with a faculty of retaining in the mind what they have heard, and seldom forgetting what they learn by heart.* (TA.) You say, ▼ فُلَانٌ حَفِيظُنَا عَلَيْكُمْ i. e. حَافِظُنَا [*Such a one is our keeper over you*]. (TA.) It is said in the Ṣ that ▼ حَفِيظٌ is syn. with ▼ مُحَافِظٌ; [but this seems to be a mistranscription for حَافِظٌ;] and hence (it is there added) the saying in the Ḳur [vi. 104, and xi. 86], ▼ وَمَا أَنَا عَلَيْكُمْ بِحَفِيظٍ [*And I am not a defender,* or *a watcher,* or, as I rather think, *a keeper, over you*]. (TA.) You say also, رَجُلٌ حَافِظٌ لِدِينِهِ وَأَمَانَتِهِ وَيَمِينِهِ [*A man who is a keeper,* &c., *of his religion and his deposite and his oath*]; and ▼ حفيظ likewise: (Mṣb:) but حافظ لِيَمِينِهِ signifies also *who keeps his oath from being used,* or *uttered, on,* or *for, ordinary, mean,* or *vile, occasions,* or *purposes.* (Mgh.) And رَجُلٌ حَافِظٌ

الْعَيْن *A man whom sleep does not overcome:* (Lḥ, Ḳ:) because the eye guards the person when sleep does not overcome it. (TA.) ▼ الحَفِيظُ is also a name of God; meaning [*The Preserver of all things;*] *He from whose preservation nothing is excluded,* (Ḳ,* TA,) *not even a thing of the weight of a* ذَرَّة [q. v.], (TA,) *in the heavens, nor on the earth;* (Ḳ, TA;) *who preserves from oblivion, for, or against, his creatures and his servants, what they do of good or evil; who preserves the heavens and the earth by his power, and whom the preservation of both does not burden.* (TA.) And الحَفَظَةُ is an appellation of *The recording angels, who write down the actions of the sons of Adam,* or *mankind;* (Ṣ, Ḳ;) as also الحَافِظُونَ. (Ḳ.) ▼ حَفِيظٌ is sometimes trans.; as in the saying, هُوَ حَفِيظٌ عِلْمَكَ وَعِلْمَ غَيْرِكَ [*He knows by heart thy science, and the science of others beside thee*]. (TA.) [القُوَّةُ الحَافِظَةُ, and simply الحَافِظَةُ, signify *The retentive faculty of the mind; retentiveness of mind;* or *memory;* as also ▼ الحِفْظُ, for حِفْظُ القَلْبِ.] — حَافِظٌ also signifies ‡ *A distinct and direct road;* (En-Naḍr, Ḳ, TA;) *not one that is apparent at one time and then ceases to be traceable.* (En-Naḍr, TA.)

مُحْفِظَاتٌ *Things that anger a man, when he has his kinsman, or neighbour, slain.* (TA.) — And مُحْفِظَاتُ رَجُلٍ *A man's women and others whom he protects, and for whose defence he fights* [*when required to do so:* because they occasion his being angered when they are injured]. (TA.)

مَحْفُوظٌ *Kept, preserved, guarded,* or *taken care of, because of the high estimation in which it is held;* as also ▼ حَفِيظٌ. (TA.) It is said in the Ḳur [lxxxv. 21 and 22], بَلْ هُوَ قُرْآنٌ مَجِيدٌ فِى لَوْحٍ مَحْفُوظٍ (TA) [*Nay, it is a glorious Ḳur-án,* written *upon a tablet preserved*] from the devils and from the alteration of anything thereof: (Jel:) or, accord. to one reading, مَحْفُوظٌ, this epithet being thus made to relate to the Ḳur-án. (TA.) — [Hence, as an epithet in which the quality of a subst. predominates, and then as a subst.,] *A young child;* in the dial. of Mekkeh; as a term of good omen: pl. مَحَافِيظُ. (TA.) — [Also *Kept,* or *retained, in the mind,* or *memory; known,* or *learned, by heart.* Hence the phrase,] عَرَضَ مَحْفُوظَاتِهِ عَلَى فُلَانٍ *He showed the things which he kept,* or *retained, in his mind,* or *memory,* or *which he knew,* or *had learned, by heart, to such a one.* (TA.)

مُحَافِظٌ: see حَافِظٌ.

حفل

1. حَفَلَ, aor. ـِ, inf. n. حَفْلٌ and حُفُولٌ and حَفِيلٌ, said of water, and of milk (Ḳ) in the udder (TA) [or breast], *It collected;* as also ▼ تحفّل and ▼ احتفل. (Ḳ, TA. [In the CḲ, احْتَفَلَهُ is erroneously put for احْتَفَلَ.]) And حَفْلٌ signifies The *collecting of water,* i. e. *its becoming collected, in its* مَحْفِل, meaning its *place of collecting.* (TA.) — حَفَلَ الدَّمْعُ, (M, Ḳ,) inf. n. حَفْلٌ, (TA,) *The tears became copious.* (M, Ḳ, TA.) In some copies of the Ḳ, نثر is here erroneously put for كَثُرَ. (TA.) — حَفَلَ الوَادِى بِالسَّيْلِ *The valley brought the torrent so as that it filled its sides;* as also ▼ احتفل: (Ḳ:) or the latter signifies *the valley became filled by the torrent:* (Ṣ:) or احتفل الوادى *the valley became full, and flowed.* (Mṣb.) — حَفَلَتِ السَّمَآءُ, (Ṣ, M, Ḳ,) inf. n. حَفْلٌ, (Ṣ,) *The sky rained vehemently:* (Ḳ:) or *the rain fell profusely.* (Ṣ,* M, TA.) — حَفَلَتْ, said of a woman, *She collected the milk in her breasts.* (TA.) And of camels, one says, (Ḳ in art. شكر,) حَفَلَتْ مِنَ الرَّبِيعِ [*They abounded in milk, or had their udders full, from the herbage called* ربيع]. (Ṣ and Ḳ in that art. [See حَافِلٌ.]) — حَفَلَ القَوْمُ, (Ṣ, Mṣb, Ḳ,) aor. ـِ, inf. n. حَفْلٌ; (Mṣb;) and ▼ احتفلوا; (Ṣ, Mṣb, Ḳ;) *The people,* or *party, collected themselves together* (Ṣ, Mṣb, Ḳ) in a sitting-place: (Mṣb:) *they collected themselves together, aiding one another,* or *for one thing* or *affair;* syn. احتشدوا. (Ṣ.) And حَفَلُوا لَهُ *They combined for him,* [or *on his account,*] *and took pains,* or *exerted themselves, in treating him with courtesy and honour;* as also حَشَدُوا لَهُ. (Fr, L in art. حشد.) — See also 8. ═ حَفَلَهُ: see 2. ═ حَفَلْتُ بِفُلَانٍ *I managed,* or *conducted, the affair,* or *affairs, of such a one.* (Mṣb.) And ▼ احتفل بِالأُمُورِ *He managed,* or *conducted, affairs,* or *the affairs, well.* (IDrd, Ḳ.) — حَفَلْتُ كَذَا, aor. ـِ, *I cared for, minded, heeded,* or *regarded, such a thing.* (Ṣ.) And مَا حَفَلَ بِهِ, and مَا حَفَلَهُ, aor. ـِ, (M, Ḳ,) inf. n. حَفْلٌ; (TA;) and ▼ ما احتفل بِهِ [and لَهُ; so in the T and TA in art. ربأ]; *He did not care for, mind, heed,* or *regard, it,* or *him.* (M, Ḳ.) And لَا تَحْفِلْ بِهِ *Do not thou care for, mind, &c., it,* or *him.* (Ṣ.) And لَا تَحْفِلْ بِأَمْرِهِ *Do not thou ask his affair, nor be disquieted thereby.* (Mṣb.) ═ حَفَلَهُ, aor. ـِ, (Ṣ,) inf. n. حَفْلٌ, (TA,) *He,* or *it, made it,* or *rendered it, clear, unobscured, apparent, plainly apparent,* or *conspicuous;* (Ṣ, TA;) as, for instance, black hair the colour of a pearl, increasing [in appearance] its whiteness; (TA;) *exposed it to view; displayed it;* syn. جَلَاهُ [perhaps here signifying also *he polished it*]. (Ṣ, TA.)

2. حفّله, (Mgh,* Mṣb, Ḳ,) inf. n. تَحْفِيلٌ; (TA;) and ▼ حَفَلَهُ, (Ḳ,) inf. n. حَفْلٌ; (TA;) *He collected it,* or *caused it to collect;* (Mgh, Mṣb, Ḳ;) namely, water, (Ḳ,) and milk (Mgh, Mṣb, Ḳ) in the udder (Mgh) of a ewe or she-goat, (Mgh, Mṣb,) or of a she-camel, or of a cow, in order to deceive the purchaser, that he might increase the price. (Mgh.) Hence one says, حفّل الشَّاةَ, (Ṣ,* Mṣb, Ḳ,*) or البَقَرَةَ, or النَّاقَةَ, (TA,) inf. n. as above, (Ṣ, Ḳ,) *He abstained from milking the ewe* or *she-goat,* (Ṣ, Mṣb, Ḳ,*) or *the cow,* or *the she-camel,* (TA,) *for some days, in order that the milk might collect in her udder, for sale,* (Ṣ, Ḳ,*) or *until the milk collected in her udder.* (Mṣb.) The Prophet forbade the doing this. (Ṣ.) ═ *He adorned him,* or *it.* (Ḳ,* TA.)

5. تحفّل: see 1, first signification. — Also, said of a sitting-place, *It abounded with company; had many persons in it.* (ISd, Ḳ.) ═ *He adorned himself;* (Ḳ;) as also ▼ احتفل. (TA.) And تَحَفَّلِى لِزَوْجِكِ *Adorn thyself that thou mayest be in favour with thy husband.* (TA.) And العَرُوسُ ▼ تَحْتَفِلُ *The bride adorns herself.* (TA.) — *It was,* or *became, clear, unobscured, apparent, plainly apparent, conspicuous, exposed to view,* or *displayed;* (Ṣ;) as also ▼ احتفل: (Ṣ, Ḳ:*) each is quasi-pass. of حَفَلَهُ as explained in the last sentence of the first paragraph in this art.: (Ṣ:) the latter, said of a road, means *It was,* or *became, apparent.* (Aṣ, Ḳ.)

8. احتفل: see 1, in five places: ═ and 5, in three places. ═ اِحْتِفَالٌ also signifies The *exceeding the usual,* or *ordinary,* or *the just,* or *proper, bounds,* or *degree; acting egregiously,* or *immoderately,* or *extravagantly; striving,* or *labouring; exerting oneself,* or *one's power* or *efforts* or *endeavours* or *ability;* or *the like;* syn. مُبَالَغَةٌ; and so حَفِيلٌ [an inf. n. of ▼ حَفَلَ]. (M, Ḳ.) — And احتفل, said of a horse, *He showed his rider that he had attained his utmost speed of running, and yet had some remaining power.* (AO, Ḳ.)

حَفْلٌ *A company* of men; as in the saying, عِنْدَهُ حَفْلٌ مِنَ النَّاسِ [*With him,* or *at his abode, is a company of men*]: originally an inf. n. (Ṣ.) — جَمْعٌ حَفْلٌ and ▼ حَفِيلٌ [which latter is also originally an inf. n. (see 1 and 8)] *A numerous company.* (Ḳ.) ═ ذُو حَفْلٍ, (Ḳ,) and ▼ ذو حَفْلَةٍ, (Ṣ, Ḳ,) and ▼ حَفِيلٌ, (Ḳ,) فِى أَمْرِهِ, (TA,) *A man who exerts himself,* or *his power* or *efforts* or *endeavours* or *ability,* or *who takes pains* or *extraordinary pains, in that which he sets about.* (Ṣ, Ḳ, TA.) And أَخَذَ لِلْأَمْرِ ▼ حَفْلَتَهُ *He strove,* or *laboured; exerted himself,* or *his power* or *efforts* or *endeavours* or *ability;* or *took pains* or *extraordinary pains; in the affair.* (Ṣgh, Ḳ.)

حِفْلٌ: see حُفَالَةٌ.

حَفْلَةٌ: see حَفْلٌ, in two places: and see also جَاؤُوا بِحَفِيلَتِهِمْ, below.

دَعَاهُمُ الحَفَلَى and ▼ الأَحْفَلَى dial. vars. of الجَفَلَى and الأَجْفَلَى, (M, Ḳ,) which are more common; meaning *He invited them with their company.* (M, TA. [See art. جفل.])

حُفَالٌ Milk *collected.* (IAạr, Ḳ.) — *A great company.* (IAạr, Ḳ.)

حَفُولٌ: see حَافِلٌ. — Also, applied to a woman, *Beautiful, goodly,* or *comely;* syn. جَمِيلَةٌ: (Ibn-'Abbád, TA:) pl. حَفَائِلُ, or, as some say, حَوَافِلُ. (TA.)

حَفِيلٌ: see حَفْلٌ, in two places.

حُفَالَةٌ The *bad,* or *vile,* of anything: (Aṣ, Ṣ:) of wheat, *what comes forth and is thrown away;* [like حُثَالَةٌ;] (TA;) and ▼ حِفْلٌ [in like manner] signifies the حُثَالَة of wheat: (AA, TA:) also, the former, *what is thin,* of the dregs of oil (Ḳ, TA) and perfume, (TA,) and of the froth of milk: (CḲ:) or it signifies also the *froth* of milk: (ISd, Ḳ, TA:) and the *worthless* of mankind; *those in whom is no good;* (Aṣ, Ṣ;) like حُثَالَةٌ; (Aṣ, Ṣ, Ḳ;) as in the saying هُوَ مِنْ حُفَالَتِهِمْ [*He is of the worthless of them*]. (Aṣ, Ṣ.)

جَاؤُوا بِحَفِيلَتِهِمْ *They came, all of them,* or *all together:* (M, Ḳ:) in the O, ▼ بِحَفْلَتِهِمْ. (TA.) — كَانَ حَفِيلَةُ مَا أَعْطَى دِرْهَمًا *The utmost amount that he gave was a dirhem.* (TA.)

حَافِلٌ A valley, and a small water-course (شُعْبَةٌ), *flowing with a copious torrent.* (Ṣ.) An udder *full of milk:* (Ṣ:) or *having much milk:* pl. حُفَّلٌ (Ḳ) and حَوَافِلُ also: (Ḥar p. 131:) it has also the latter meaning applied to a ewe or she-goat; (Ḳ;) pl. حُفَّلٌ: (TA:) and so have حَافِلَةٌ and ▼ حَفُولٌ applied to a she-camel. (Ḳ.) مَدَامِعُ حُفَّلٌ *Copious flowings of tears.* (TA.)

الحَفَلَى : see دَعَاهُمُ الأَحْفَلَى.

مَحْفِلٌ *A place of collecting of water.* (TA.) — *A place of assembling,* or *congregating,* (T, Ṣ, Mṣb, Ḳ,) of a people; (Ṣ, Mṣb;) as also ▼ مُحْتَفَلٌ: (Ṣ, Ḳ:) or *a place of assembling,* or *congregating, of many persons:* (El-Ámidee, MF:) or *a place in which is an assembly,* or *congregation:* (El-Munáwee, TA:) and *a sitting-place:* (T, TA:) pl. مَحَافِلُ. (Mṣb.) — [And The *elevated platform for the* مُبَلِّغُون *in a mosque;* also (in Egypt) called دَكَّةٌ, vulg. دِكَّة, it is surrounded by a low railing or parapet, and generally supported by small columns.]

مُحَفَّلَةٌ A ewe, or she-goat, *left unmilked* (Ṣ, Mṣb) *for some days, in order that the milk may collect in her udder, for sale,* (Ṣ,) or *until the milk has collected in her udder:* (Mṣb:) or a ewe, or she-goat, or a she-camel, or a cow, *whose milk has been made to collect in the udder,* in order to deceive the purchaser, that he may increase the price: (Mgh:) originally مُحَفَّلٌ لَبَنُهَا. (Mṣb.)

مُحَافِلٌ *Contending for superiority in number &c.* (TA.) — هُوَ مُحَافِظٌ عَلَى حَسَبِهِ مُحَافِلٌ *He is one who preserves his nobility,* or *honourableness.* (Az, Ḳ.)

مُحْتَفَلٌ : see مَحْفِلٌ. — Also The *most fleshy part* of the flesh of the thigh and shank. (TA.) — And The *main part* of an affair: (TA:) [and likewise of a place, or tract, or region; for] مُحْتَفَلُ البَيْدَاءِ signifies *the main part of the desert;* syn. مُتَجَمَّعُهَا and مُعْظَمُهَا. (TA in art. جمع.)

حفن

1. حَفَنَهُ, (Ṣ,) aor. ـِ, (PṢ, TḲ,) [or ـُ, as in a phrase following,] inf. n. حَفْنٌ, (M, Ḳ,) *He took it* (a thing) *with the palms of his two hands and with the fingers put together* [*so as to make the two hands like a bowl*]: (M, Ḳ:) or *he scooped it up,* or *out,* (جَرَفَهُ,) *with both his hands:* (Ṣ, Ḳ:) said only of what is dry, as flour, and sand, and the like. (Ṣ.) — حَفَنَ القَوْمَ *He gave to every one of the party a* حَفْنَة. (TA.) — حَفَنَ لَهُ, (Mṣb,) or حَفَنَ لَهُ حَفْنَةً, (Ṣ,) aor. ـُ, (Mṣb,) inf. n. حَفْنٌ, (Mṣb, Ḳ,) *He gave to him a small quantity.* (Ṣ, Ḳ.) — حَفَنَ المَآءَ عَلَى رَأْسِهِ *He threw the water upon his head with his two hands* [*put together so as to be like a bowl*]. (IAạr, TA.)

8. احتفنهُ (Ṣ, Ḳ) لِنَفْسِهِ (Ṣ) *He took it* (a thing) *for himself.* (Ṣ, Ḳ.) — احتفن مِنْهُ ‡ *He took much of it.* (A, TA.) — احتفن الشَّجَرَ † *He pulled up the trees from the ground.* (Ḳ.) And احتفن الرَّجُلَ † *He uprooted the man:* (Az, Ṣ:) [or] احتفنهُ signifies ‡ *he put his hands, or arms, beneath his knees, and took him* [*by that part,* i. e.] *by the inner side of the knee, and then carried him,* or *carried him off* or *away.* (Ḳ, TA.)

حَفْنٌ The *act of turning the feet as though one were throwing the dust* (كَأَنَّهُ يَحْثُو) *with them, when walking.* (Ḳ.)

حَفْنَةٌ, (Ṣ, Mgh, Mṣb, TA, and so, accord. to the TA, in the Ḳ,) or ▼ حُفْنَةٌ, (so in copies of the Ḳ,) [but the former is that which is commonly known,] *A handful:* (Mgh, Ḳ:) or the *quantity that fills the two hands* [*when they are put together so as to be like a bowl*]; (Ṣ, Mṣb;) of wheat [or the like]: (Ṣ:) pl. حَفَنَاتٌ. (Ṣ, Mṣb.) Hence, (in the saying of Aboo-Bekr, TA,) إِنَّمَا نَحْنُ حَفْنَةٌ مِنْ حَفَنَاتِ ٱللّٰهِ, (Ṣ, TA,) i. e., ‡ *We shall be but little,* on the day of resurrection, *like a* حفنة, in the estimation *of God;* (TA;) meaning we shall be but a small thing in comparison with the dominion and the mercy of God. (Ṣ, TA.) — Also The *hand* (كَفّ) itself. (Ḥar p. 296.) — See also what next follows.

حُفْنَةٌ *A hollow, cavity, trench,* or *the like, dug,* or *excavated, in the ground,* (Ṣ, Ḳ, TA,) *wherever it be;* or *excavated by a torrent in rugged ground, in the channel of the water:* (TA:) and a [*hollow, or cavity, in the ground, such as is termed*] نُقْرَة, (ISk, Ḳ,) *having in it water, and in its bottom pebbles and earth;* (ISk, TA;) as also ▼ حَفْنَةٌ: (Ḳ:) *a well,* or *pit:* (KL:) pl. of the former حُفَنٌ; (Ṣ, Ḳ;) which is explained by Sh as meaning *small round hollows or cavities, in which rain-water stagnates, excavated by the water, in the form of pools.* (TA.) — See also حَفْنَةٌ.

حَفَّانٌ, n. un. with ة, belongs to art. حف, q. v. (Ṣ, Ḳ.)

مِحْفَنٌ, applied to a man, (TA,) i. q. كَثِيرُ الحَفْنِ [i. e. One *who takes much with the palms of his two hands and with the fingers put together:* or *who scoops up, or out, much, with both his hands:* see 1, first sentence]. (ISd, Ḳ.)

حفى and حفو

1. حَفِىَ, (Ṣ, Mgh, Mṣb,) aor. ـَ, (Ṣ, Mṣb,) inf. n. حَفَآءٌ, (ISk, Zj, IB, Mgh, Mṣb,) like سَلَامٌ, (Mṣb,) with fet-ḥ, not حِفَآءٌ, as in the Ṣ, (IB,) [for] this latter is a simple subst., (Mṣb,) *He walked barefoot, without sandal and without boot;* (ISk, Zj, Ṣ, IB, Mgh, Mṣb;) as also ▼ احتفى: (Ḳ:) [but Mṭr says,] this latter, occurring in the sense of the former in a trad. of 'Omar, I have not found [elsewhere]. (Mgh.) And حَفِىَ مِنْ نَعْلَيْهِ and خُفِّهِ, inf. n. حُفْوَةٌ and حِفْيَةٌ and حِفَايَةٌ, [or all these are simple substs., *He was,* or *became, bare of his sandals* and *his boot.*] (TA.) — And حَفِىَ, inf. n. حَفًا, (Ṣ, Mgh, Mṣb, Ḳ, &c.,) *His foot,* i. e. a man's, (Ṣ, Mṣb, Ḳ,) and a camel's, (Ḳ,) or *his solid hoof,* (Ṣ, Ḥam p. 476,) or this also, (Ḳ,) *was,* or *became, attenuated* [*in the sole*], (Ṣ, Mṣb, Ḳ,) or *chafed, abraded,* or *worn,* (Ḥam ubi suprà, and PṢ,) by much walking or treading or travel: (Ṣ, Mṣb, Ḥam:) or *it* (the foot of a man [and the hoof of a horse or the like]) *became attenuated* [&c.] *by much walking* or *treading* or *travel:* (Mgh:) and حَفِىَ said of a horse, *his hoof became abraded,* or *grazed.* (Ṣ.) ═ حَفِىَ بِهِ, inf. n. حَفَاوَةٌ (Ṣ, Mgh, Ḳ) and حِفَاوَةٌ and حِفَايَةٌ and تَحْفَايَةٌ, (Ḳ, TA,) with kesr; (TA; [in the CḲ تَحْفَايَةٌ; both extr.;]) and ▼ تحفّى بِهِ, (Ṣ, Ḳ,) and ▼ احتفى بِهِ; (Ḳ;) *He showed him much honour,* (Ṣ, Mgh, Ḳ,) *and kindness,* or *goodness and affection and gentleness, and regard for his circumstances;* (Ṣ;) *regarded him,* or *behaved towards him, with affection,* or *benevolence and solicitude;* (Mgh;) *and manifested joy,* or *pleasure;* (Ḳ;) and *asked,* or *inquired, much respecting his state,* or *condition;* (Ṣ, Ḳ;) and *was careful,* or *solicitous, respecting his case:* (Ṣ:) or حَفِىَ عَنْهُ signifies *he asked,* or *inquired, much respecting him:* (Ḥar p. 284:) and ▼ احفى *he exceeded the usual bounds in making much inquiry respecting another, and in carefulness,* or *solicitude, respecting his case:* (Ḥar p. 348:) and ▼ تَحَفٍّ signifies the *speaking and meeting in a good,* or *pleasing, manner;* and *exceeding the usual bounds in returning a salutation,* and *in asking,* or *inquiring:* accord. to Zj, حَفِىَ بِهِ, inf. n. حُفْوَةٌ, means *he acted towards him with kindness,* or *goodness and affection and gentleness, and regard for his circumstances:* accord. to Aṣ, حَفِىَ بِهِ, aor. ـَ, inf. n. حَفَاوَةٌ, *he employed himself in the accomplishment of that which he needed,* or *required, and made his abode pleasant:* (TA:) and حَفَاوَةٌ signifies also the *being importunate, pressing, urgent,* or *persevering,* (Ḳ, TA,) *in asking.* (TA.) Hence the prov., مَأْرُبَةٌ لَا حَفَاوَةٌ [or مَأْرَبَةٌ, and مَأْرُبَةٌ, explained in art. ارب]. (Ṣ, Ḳ.) — حَفِيتُ, or حَفَيْتُ, إِلَيْهِ بِالوَصِيَّةِ, (accord. to different copies of the Ṣ,) *I exceeded the usual bounds to him in the injunction,* or *charge:* mentioned by A'Obeyd: (Ṣ:) and ▼ تحفّى اليه *he exceeded the usual bounds to him in the injunction,* or *charge.* (TA.) ═ حَفَا بِهِ, (Ḳ,) and حَفَاهُ, (TA,) inf. n. حَفْوٌ, *He* (God) *honoured him.* (Ḳ, TA.) — And حَفَاهُ *He* (a man) *gave to him.* (Ḳ.) — And *He denied him,* or *refused him,* or *refused to give him;* (IAạr, Ḳ;) inf. n. حَفْوٌ. (IAạr.) Thus it bears two contr. significations. (Ḳ.) You say, أَتَانِى فَحَفَوْتُهُ *He came to me and I denied him,* &c. (IAạr.) And حَفَوْتُهُ مِنْ كُلِّ خَيْرٍ, aor. ـُ, inf. n. حَفْوٌ, *I debarred him from,* or *prevented him from obtaining, all good.* (Aṣ, Ṣ.) ═ حَفَا شَارِبَهُ : see 4.

3. حافاهُ, (inf. n. مُحَافَاةٌ, TA,) *He disputed with him in words;* or *did so vehemently,* or *obstinately.* (AZ, Ṣ, Ḳ.*)

4. احفى, said of a man, *His beast became attenuated,* or *chafed,* or *abraded, in* [*the sole of*] *the hoof.* (Ṣ.) ═ احفاهُ *He* (God) *caused him to be barefooted, without sandals or boots:* and hence احفى also signifies *he made his feet bare of the sandals* or *boots.* (TA.) — *He made his foot,* i. e. a man's, and *his hoof,* i. e. a horse's or the like, *to become attenuated,* [or *chafed,* or *worn, in the sole,*] by much walking or treading or travel. (Ṣ.) ═ *He repeated it;* namely, a question. (Ḳ.) — *He importuned, pressed,* or *urged, him, and harassed,* or *molested, him in so doing:* (Lth, Ḳ:) or *he asked him, and pressed him much in seeking* or *demanding:* (TA:) *he importuned him* in asking or questioning: (Mṣb:) *he harassed,* or *wearied, him, and went to the utmost length in asking him,* or *questioning him.* (TA.) See also 1. إِحْفَآءٌ signifies also The *going to the utmost length in disputing.* (Ṣ, TA.) It is in asking respecting a thing, and in seeking or demanding, and in seeking or demanding from another: you say, احفى فى المَسْأَلَةِ, [&c.,] and ↓ تحفّى فيها, *He exceeded the usual bounds in the question,* or *in asking.* (Ḥam p. 80.) — And hence, (Ḥam p. 80,) احفى شَارِبَهُ *He clipped his mustache to the utmost degree;* (Ṣ, Ḥam;) *clipped it closely:* (Ṣ:) or *he clipped it much;* (Mgh, Mṣb, Ḳ;) as also ↓ حَفَاهُ. (Ḳ.) — احفى فَمَهُ *He examined to the utmost his* [app. a horse's] *teeth.* (TA.) — أَحْفَيْتُهُ *I incited,* or *urged, him to inquire respecting,* or *to investigate, the news,* or *information,* (Ḳ, TA,) *to the utmost.* (TA.) ═ أَحْفَيْتُ بِهِ i. q. أَزْرَيْتُ [*I despised him;* or *I brought against him the imputation of a fault,* or *the like, desiring to involve him in confusion thereby;* &c.]. (Ḳ.)

5. تحفّى, inf. n. تَحَفٍّ: see 1, in three places: and see also 4. — Also i. q. اِجْتَهَدَ [*He strove,* or *laboured,* &c., فِى أَمْرٍ *in an affair*]. (Ḳ.) — And i. q. اِحْتَفَلَ [app. as meaning *He practised artifice,* or *the like*]. (Ḳ.)

6. تَحَافَيْنَا إِلَى السُّلْطَانِ *Each of us preferred a complaint against the other to the Sultán,* (Ḳ, TA,) *and he referred our case to the judge* (الحَافِى, i. e. القَاضِى). (TA.)

8. احتفى: see 1, first sentence. ═ And احتفى بِهِ: see 1. ═ احتفى also signifies *He pulled up herbs,* or *leguminous plants, from the ground* (T, Mgh, Ḳ) *with the ends of his fingers, because of their shortness and paucity;* (T, Mgh;) as also اِحْتَفَأَ, (T, Mgh, Ḳ,) accord. to one reading of a trad. in which it occurs; and احتفأ, accord. to another reading of the same; but both these are disallowed by Aboo-Sa'eed. (T, Mgh.) — *He uprooted,* or *extirpated,* anything. (Aboo-Sa'eed, T.) — احتفى القَوْمُ المَرْعَى *The people,* or *party, fed their animals upon the pasturage and left not aught of its herbage:* the subst. [denoting the *act of doing this*] is ↓ حفوة [app. حُفْوَةٌ]. (TA.)

10. استحفى *He asked,* or *inquired after, news,* or *tidings,* (A, Ḳ,) *exceeding the usual bounds in doing so.* (A.)

حَفًا: see حِفْوَةٌ.

حَفٍ, applied to a man, (Ṣ, Mgh, Mṣb, Ḳ,) and a camel, (Ḳ,) and a horse or the like, (Zj, Ṣ, Ḳ, Ḥam p. 476,) *Having his foot,* (Ṣ, Mgh, Mṣb, Ḳ,) or *hoof,* (Zj, Ṣ, Ḳ, Ḥam,) *attenuated* [*in the sole*], (Zj, Ṣ, Mgh, Mṣb, Ḳ,) or *chafed, abraded,* or *worn,* (Ḥam, PṢ,) by much walking or treading or travel; (Ṣ, Mgh, Mṣb, Ḥam;) as also ↓ حَافٍ. (Ḳ.)

حَفْوَةٌ: see 8.

حُفْوَةٌ: see what next follows.

حِفْوَةٌ (Ṣ, Ḳ) and ↓ حُفْوَةٌ (Ḳ) and ↓ حِفْيَةٌ and ↓ حِفَايَةٌ, (Ṣ, Ḳ,) all simple substs., (Ḳ,) [but all except the second mentioned in one place in the TA as inf. ns.,] and ↓ حِفَآءٌ, (Ṣ, Mṣb,) a simple subst., differing from the inf. n., which is حَفًا, (Mṣb,) The *state of being barefoot, without sandal and without boot:* (Ṣ, Mṣb:) or the *state of having the foot,* or *hoof, attenuated* [*in the sole,* or *chafed, abraded,* or *worn,* by much walking or treading or travel]: (Ḳ:) [but it seems that this is a mistake, and that] the word denoting the latter meaning is ↓ حَفًا [an inf. n.: see 1]. (ISk, Zj, Ṣ, Mgh, Mṣb.)

حِفْيَةٌ: } see حِفْوَةٌ.
حِفَآءٌ: }

حَفِىٌّ *Showing much honour,* (Mgh, Ḳ,) *and manifesting joy* or *pleasure* [*at meeting another*]; as also ↓ حَافٍ; (Ḳ;) *and behaving with affection,* or *benevolence and solicitude;* (Mgh;) *showing kindness,* or *goodness and affection and gentleness, and regard for the circumstances of another:* (Lth, TA:) and *asking,* or *inquiring, much respecting another's state,* or *condition;* as also ↓ حَافٍ: (Ḳ:) *going to the utmost length in asking,* or *inquiring:* (Ṣ:) and [hence, perhaps,] *knowing in the utmost degree:* (Ṣ, Ḳ:) and *importunate, pressing,* or *urgent, in asking,* or *inquiring:* (Ḳ:) pl. حُفَوَآءُ. (Fr, Ḳ.) You say, هُوَ حَفِىٌّ بِهِ *He is one who behaves with affection,* or *benevolence and solicitude, to him; who shows him much honour* [&c.]. (Mgh.) إِنَّهُ كَانَ بِى حَفِيًّا, in the Ḳur [xix. 48], means *Verily He is gracious* [*to me*]: (Zj:) or *knowing* [*with respect to me*] *and gracious* [*to me*], *answering my prayer when I pray to him:* (Fr:) or *mindful, regardful,* or *considerate, of me.* (TA.)

حِفَايَةٌ: see حِفْوَةٌ.

حَافٍ *Walking barefoot, without sandal and without boot:* (ISk, Zj, Ṣ, Mgh, Mṣb:) pl. حُفَاةٌ. (Mgh, Mṣb.) — See also حَفٍ. ═ And see حَفِىٌّ, in two places. ═ Also *A judge;* syn. قَاضٍ. (Ḳ.)

حق

1. حَقَّ, aor. ـِ (Ṣ, Mṣb, Ḳ, &c.) and ـُ, (IDrd, Mṣb, Ḳ,) [the latter irregular,] inf. n. حَقَّةٌ (Ḳ, TA) and حَقٌّ (IDrd, TA) and حُقُوقٌ, (TA,) i. q. صَارَ حَقًّا [i. e., accord. to the primary meaning of حَقٌّ, as explained below, on the authority of Er-Rághib, *It was,* or *became, suitable to the requirements of wisdom, justice, right* or *rightness, truth,* or *reality* or *fact;* or *to the exigencies of the case*]: (TA:) *it was,* or *became, just, proper, right, correct,* or *true; authentic, genuine, sound, valid, substantial,* or *real; established,* or *confirmed, as a truth* or *fact:* and *necessitated, necessary, requisite,* or *unavoidable; binding, obligatory, incumbent,* or *due:* syn. وَجَبَ; (T, Ṣ, Mṣb, Ḳ, &c.;) and ثَبَتَ: (Mṣb, TA:) *it was,* or *became, a manifest and an indubitable fact* or *event;* as explained by IDrd in the Jm; (TA;) *it happened, betided,* or *befell, surely, without doubt* or *uncertainty.* (Ḳ.) It is said in the Ḳur xxxvi. 6, لَقَدْ حَقَّ القَوْلُ عَلَى أَكْثَرِهِمْ, i. e. *The saying,* "I will assuredly fill Hell with genii and men together," [Ḳur xi. 120 and xxxii. 13,] (Bḍ,) or *the sentence* of punishment, (Jel,) *hath become necessitated* [*as suitable to the requirements of justice,* or *as being just* or *right,*] *to take effect upon the greater number of them;* syn. وَجَبَ, (Jel, TA,) and ثَبَتَ. (TA.) And this, namely, ثَبَتَ, is the meaning of the verb in the phrase, حَقَّ عَلَيْكَ القَضَآءُ [*The sentence was,* or, emphatically, *is, necessitated as suitable to the requirements of justice to take effect upon thee;* or *it was,* or *is, necessary, just,* or *right, that the sentence should take effect upon thee*]. (TA.) [In like manner,] one says, يَحِقُّ عَلَيْكَ أَنْ تَفْعَلَ كَذَا *It is necessary for thee* [*as suitable to the requirements of wisdom* or *justice* or *the like*], or *incumbent on thee,* or *just* or *proper* or *right for thee, that thou shouldst do such a thing.* (TA.) [Thus one says,] الحَقِيقَةُ مَا يَحِقُّ عَلَيْكَ أَنْ تَحْمِيَهُ [*The* حقيقة *is that which it is necessary for thee* &c., or *that which it behooveth thee, that thou shouldst defend it,* or *protect it*]. (Ṣ,* Ḳ.) Accord. to Sh, the Arabs said, حَقَّ عَلَىَّ أَنْ أَفْعَلَ ذٰلِكَ and حُقَّ: but accord. to Fr, when you say حَقَّ, you say عَلَيْكَ; and when you say حُقَّ, you say لَكَ. (TA.) [Accordingly] one says, حُقِقْتَ أَنْ تَفْعَلَهُ and حُقَّ لَكَ أَنْ تَفْعَلَ ذَا: both mean the same: (Ks, Ṣ, Ḳ:) [i. e., each has one, or the other, or both, of the meanings next following:] or the former means *It was,* or, emphatically, *is, rendered* حَقّ [or *suitable to the requirements of wisdom* or *justice* &c.] *for thee,* or *necessary for thee,* or *incumbent on thee,* or *just* or *proper* or *right for thee,* [or *it behooved* or *behooves thee,*] *that thou shouldst do,* or *to do, this,* or *that:* and [the latter, or] حُقِقْتَ بِأَنْ تَفْعَلَ, *Thou wast,* or, emphatically, *art, rendered* حَقِيق [or *adapted, disposed, apt, meet, suited, suitable, fitted, fit, proper, competent,* or *worthy,*] *that thou shouldst do,* or *to do, this,* or *that*]: (A, TA:) and in like manner, حُقَّتْ signifies in the Ḳur lxxxiv. 2 and 5: (Bḍ, Jel:*) or حُقِقْتَ بِأَنْ تَفْعَلَ may mean *thou wast,* or *art, known by the testimony of thy circumstances to be* حَقِيق &c. (A, TA.) And مَا كَانَ يَحُقُّكَ أَنْ تَفْعَلَهُ

[virtually] means the same as مَا حَقَّ لَكَ [best rendered in this case *It did not behoove thee to do it*]. (TA.) One says also, حُقَّ أَنْ تَفْعَلَ [*It was*, or, emphatically, *is*, *necessary* &c. *that thou shouldst do* or *to do* such a thing]: but they did not say, حَقَقْتَ أَنْ تَفْعَلَ. (Fr, TA.) — But حَقَّ عَلَىَّ أَنْ تَفْعَلَ هٰذَا means *Thy doing so distressed*, or *hath distressed*, or *afflicted*, *me*; or, emphatically, *distresses*, or *afflicts*, *me*; like عَزَّ عَلَىَّ. (Ṣ and Ḳ and TA in art. عز.) And in like manner, [حَقَّ مَا أَنَّكَ ذَاهِبٌ [or حَقَّ أَنَّكَ ذَاهِبٌ] means عَزَّ مَا أَنَّكَ ذَاهِبٌ [*It is distressing to me that thou art going away*]. (TA in art. عز.) And لَعَزَّ مَا is used in the same manner as لَحَقَّ مَا q. v. (A and TA in art. عز.) You say also, حَقَّتِ الحَاجَةُ *Want befell*, or *betided*, or *happened*, *and was severe*, or *distressing*: (Mṣb, TA:) [which is said to be] from the phrase, حَقَّتِ القِيَامَةُ, aor. ـِ, *The resurrection included*, or *shall include*, *within its sphere* [*all*] *the created beings*. (Mṣb.) — حَقَّتْ, aor. ـِ, (Ḳ,) inf. n. حِقَّةٌ (Ṣ,* Mṣb, Ḳ,) and حِقٌّ, (Ḳ,) or, accord. to ISd, it should rather be حِقَاقَةٌ and حُقُوقَةٌ, because حِقَّةٌ is used as an epithet, [as will be seen below,] and the inf. n. in a case like this, by rule, should differ from the epithet, (TA,) *She* (a camel) *became a* حِقّ, or حِقَّة; i. e., *entered the fourth year*: (Ḳ:) and ▼أَحَقَّ, inf. n. إِحْقَاقٌ, *he* (a camel) *became a* حِقّ: because, so they say, he is then fit to be laden: (Mṣb:) and ▼احقّت *she* (a young camel) *completed three years*; (Aboo-Málik, Ḳ;) *became a* حِقَّة; (Ibn-'Abbád, Ḳ;) like حَقَّتْ. (TA.) You say, هُوَ حِقٌّ بَيِّنُ الحِقَّةِ [*He is a* حِقّ, *bearing evidence of being such*]: (Ṣ:) and هِىَ حِقٌّ (Ḳ) and حِقَّةٌ (Mṣb, Ḳ) بَيِّنَةُ الحِقَّةِ [*she is a* حِقّ or حِقَّة, *bearing evidence &c.*]: (Mṣb:) [a phrase] to which a parallel is scarcely known, (Mṣb,) or to which there is no parallel (Ḳ) except أَسَدٌ بَيِّنُ الأَسَدِ [a lion bearing evidence of being like a lion in boldness]. (TA.) — حِقٌّ [as inf. n. of حَقَّتْ] also signifies *A she-camel's overpassing the days* [*corresponding to those*] *in which she was covered* [*in the preceding year*]: (Ḳ:) or her *completing* [*the time of*] *her pregnancy*; as also ▼اِسْتِحْقَاقٌ. (TA.) — And حَقَّتْ and ▼احقّت and ▼استحقّت *She* (a camel) *became fat*. (TA. [See also 8, last signification.]) — حَقَّهُ, (Ḳ,) aor. ـُ, inf. n. حَقٌّ, (TA,) *He*, or *it*, *rendered it* [*suitable to the requirements of wisdom, justice, rightness, truth, or reality* or *fact*; or *to the exigencies of the case*; (see the first of the significations in this art.;) or] *necessary*, *requisite*, or *unavoidable*; *binding*, *obligatory*, *incumbent*, or *due*; or *just*, *proper*, or *right*; syn. أَوْجَبَهُ; (Ḳ;) [whence حَقَّ لَكَ أَنْ تَفْعَلَ ذَا, explained above;] as also ▼حقّقهُ (Ḳ) and ▼احقّهُ; (Ṣ, Ḳ;) which last some explain by صَيَّرَهُ حَقًّا [meaning as above; or *he rendered it true*;] or صَيَّرَهُ حَقًّا لَا شَكَّ فِيهِ [*he rendered it true, so that there was no doubt respecting it*]; as also حَقَّقَهُ, inf. n. حَقٌّ: and حَقَّهُ signifies also *he established it so that it became true and undoubted in his estimation*: (TA:) or حَقَّهُ signifies, (Ṣ, Mṣb,) or signifies also, (Ḳ,) *he assured*, or *certified*, *himself of it*; *he ascertained it*; *he was*, or *became*, *sure*, or *certain*, *of it*; (A 'Obeyd, Ṣ, Mṣb,* Ḳ;*) and so ▼تحقّقهُ (A 'Obeyd, Ṣ, Ḳ) and ▼احقّهُ: (Ṣ, Mṣb:*) or *he pronounced it*, or *held* or *believed it*, *to be established as a necessary truth* or *fact*; as also ▼احقّهُ: and ▼حقّقهُ has a similar, but intensive, signification: (Mṣb:) or ▼احقّهُ signifies *he established it as true*; or *he judged*, or *decided*, *it to be so*: (TA: [*contr. of* أَبْطَلَهُ: see an ex., from the Ḳur viii., voce أَبْطَلَ:]) and ▼حقّقهُ, inf. n. تَحْقِيقٌ, signifies صَدَّقَهُ [as meaning *he verified it*, or *proved it to be true* or *veritable*; or *he found it to be true* or *veritable*; both of which significations are of very frequent occurrence]; (Ṣ, Ḳ;) as also حَقَّهُ, inf. n. حَقٌّ: and accord. to IDrd, ▼حقّقهُ signifies [also] صَدَّقَ قَائِلَهُ [*he proved*, or *found*, or *pronounced*, *the sayer of it to be true*]: and حقّق is also said to signify *he said, "This thing is the truth;"* like صَدَّقَ. (TA.) You say, حَقَقْتُ عَلَيْهِ القَضَآءَ, aor. ـُ, inf. n. حَقٌّ, *I necessitated the sentence* [*as suitable to the requirements of justice*] *to take effect upon him*; or *necessitated* [*as suitable &c.*] *the taking effect of the sentence upon him*; syn. أَوْجَبْتُهُ; as also ▼أَحْقَقْتُهُ, inf. n. إِحْقَاقٌ. (TA.) And أُحِقَّ عَلَيْكَ القَضَآءُ *The sentence was*, or, emphatically, *is*, *necessitated* [*as suitable to the requirements of justice*] *to take effect upon thee*; syn. أُثْبِتَ. (TA.) And حَقَقْتُ حَذَرَهُ, (Ṣ, Ḳ,) or حِذْرَهُ, (so in one copy of the Ṣ,) aor. and inf. n. as above, (Ṣ,) [*I rendered his caution*, or *fear*, *necessary*; or *justified it*; meaning] *I did that of which he was cautious*, or *that which he feared*; (Ṣ, Ḳ;) as also ▼أَحْقَقْتُ حذره: (Ṣ:) or, accord. to Az, the latter only is right. (TA.) And حَقَقْتُ ظَنَّهُ; (Ks, TA;) and ▼حَقَّقْتُهُ, (Ks, Ṣ, TA,) inf. n. تَحْقِيقٌ: (Ṣ:) both signify the same; (Ks, TA;) i. e. صَدَّقْتُ; (Ṣ;) which means *I found his opinion to be true*; (Ksh and Bḍ and Jel, in xxxiv. 19;) or *proved it to be true*: (Ksh, ibid.:) and so تَحْقِيقُ قَوْلِهِ *his saying*: (Ṣ:) and تَحْقِيقٌ signifies [also] the *strengthening*, or *confirming*, a saying; or *making* it *strong*, or *firm*. (KL.) And أَنَا أَحُقُّ لَكُمْ هٰذَا الخَبَرَ *I will know*, or *ascertain*, *the truth*, or *real nature*, *of this piece of news* or *information*, *for you*. (TA.) And أَظُنُّهُ وَلَا أَحُقُّهُ [*I think it, but I do not know the truth of it*, or *am not certain of it*]. (T in art. إِيَّا; &c.) And حَقَقْتُ العُقْدَةَ, [written in the TA without any syll. signs, so that it may be either thus or ▼حَقَّقْتُهَا; but it is most probably the former, as the quasi-pass. is not تحقّقت, but انحقّت: it signifies lit. *I made the knot right*, or *sure*; meaning] ‡ *I tied*, or *made fast*, or *tightened*, *the knot*; (Ibn-'Abbád, TA;) or *I tied*, or *tightened*, *firmly the knot*. (A, TA.) — [*He*, or *it*, *rendered him* حَقِيق, i. e. *adapted*, *disposed*, *apt*, *meet*, *suited*, *suitable*, *fitted*, *fit*, *proper*, or *worthy*, to do a thing &c.; whence بِأَنْ تَفْعَلَ, or حَقِقْتَ أَنْ تَفْعَلَهُ, explained above. — *It was necessary for him*, or *incumbent on him*, or *just* or *proper* or *right for him*, or *it behooved him*, to do a thing &c.; whence مَا كَانَ يَحُقُّكَ أَنْ تَفْعَلَهُ, explained above.] — Also, (Ṣ, Ḳ,) aor. ـُ, (Ḳ,) inf. n. حَقٌّ, (TA,) *He overcame him in disputing*, or *contending*, *for a right*, or *due*; (Ṣ, Ḳ,* TA;) and so ▼احقّهُ, (Ḳ,) inf. n. إِحْقَاقٌ, mentioned by Az on the authority of Ks, but, he adds, disallowed by A 'Obeyd. (TA.) See 3. — Also *He* (a man) *came to him*, namely, another man; (A 'Obeyd, Ṣ, Ḳ;) and so ▼احقّهُ. (A 'Obeyd, Ṣ.) [Hence, app.,] حَقَّتْنِى الشَّمْسُ *The sun reached me*. (TA.) And لَا يَحِقُّ مَا فِى هٰذَا الوِعَآءِ رِطْلًا [app. يَحُقُّ] *What is in this receptacle* [*does not reach*, or *amount*, *to a pound*; i. e.,] *does not weigh a pound*. (TA.) — حَقَّ الطَّرِيقَ, (Ḳ,) aor. ـُ, (TA,) inf. n. حَقٌّ, (TḲ,) *He went upon the* حَاقّ *of the road*; (Ḳ;) i. e. *the middle of it*: the doing of which is forbidden, in a trad., to women. (TA.) And حَقَّ فُلَانًا, (Ḳ,) aor. and inf. n. as above, (TA,) *He beat*, or *struck*, *such a one in*, or *upon*, *the* حاقّ *of his head*; (Ḳ;) i. e. *the middle of it*: (TA:) or *in*, or *upon*, *the* حُقّ *of his* كَتِف; i. e. *the small hollow upon the head of his shoulder-blade*: (Ḳ:) or, as some say, *the head of the upper arm*, *in which is the* وَابِلَة. (TA.)

2. حقّقهُ, inf. n. تَحْقِيقٌ: see حَقَّهُ, above, in six places. [Hence تَحْقِيقُ الهَمْزَةِ *The uttering of the hemzeh with its true*, or *proper*, *sound*; opposed to تَخْفِيفُهَا. Hence also] صَبَغْتُ الثَّوْبَ صَبْغًا تَحْقِيقًا *I dyed the garment*, or *piece of cloth*, *with a saturating dyeing*. (TA.) And تَحْقِيقٌ signifies also The *weaving* a garment, or piece of cloth, *strongly*, or *firmly*. (KL.) — حقّق فِى أَمْرِهِ *He was serious*, or *in earnest*, *in his affair*; *contr. of* هَزَلَ. (L in art. جد.)

3. حاقّهُ, (Ṣ, Ḳ,) inf. n. حِقَاقٌ and مُحَاقَّةٌ, (TA,) *He disputed*, *litigated*, or *contended*, *with him*, (Ṣ, Ḳ,) *each of them laying claim to a right*, or *due*: (Ṣ, TA:) the verb is mostly used in the third person. (TA.) You say, حَاقَّنِى وَلَمْ يُحَاقَّنِى فِيهِ أَحَدٌ [*He disputed, &c., with me, and no one had disputed, &c., with me respecting it*]. (TA.) [But] you say also, ▼حَاقَقْتُهُ فَحَقَقْتُهُ *I disputed*, *litigated*, or *contended*, *with him for a right*, or *due*, *and I overcame him in doing so*. (TA.) And إِنَّهُ لَنَزِقُ الحِقَاقِ ‡ *Verily he is one who disputes*, or *litigates*, or *contends*, *respecting small things*. (Ṣ, Ḳ, TA.) And مَا لَهُ فِيهِ حَقٌّ وَلَا حِقَاقٌ, i. e. [*He has no right*, or *due*, *to exact*, *in respect of him*, or *it*, *nor any cause of*] *disputing*, or *litigating*, or *contending*. (Ṣ.) And it is said respecting women, (Ḳ,) in a trad. of 'Alee, (TA,) إِذَا بَلَغْنَ نَصَّ الحِقَاقِ فَالعَصَبَةُ أَوْلَى, or الحَقَائِقِ: (Ḳ:) accord. to some, الحقاق here means the same as المُحَاقَّة: accord. to others, it properly signifies the camels thus called: and so الحقائق;

this [likewise] being a pl. of ↑ حِقَّةٌ; or it is pl. of ↑ حَقِيقَةٌ. (TA. [See art. نص; in which this trad. is more fully, but somewhat differently, cited; and fully explained.]) ═ [Also, app., *He acted seriously*, or *in earnest, with him* in an affair: see 3 in art. جد: and see also 2 above, last signification.]

4. أَحَقَّ, [inf. n. إِحْقَاقٌ,] *He spoke truth; said what was true:* [very common in this sense; *contr. of* أَبْطَلَ:] or *he revealed*, or *manifested*, or *showed, a truth*, or *a right* or *due:* or *he laid claim to a right,* (or *to a thing,* TA) *and it was,* or *became, due to him.* (Mṣb.) ═ See also حَقَّ, as an intrans. verb, in three places; relating to camels. — أَحَقَّ القَوْمُ *The people's cattle became fat.* (TA.) And أَحَقَّ القَوْمُ مِنَ الرَّبِيعِ *The people's cattle became fat by means of the* [herbage *called*] ربيع. (AḤn,* ISd, TA.) ═ As a trans. verb: see حَقَّهُ, in nine places. You say also, أَحْقَقْتُ الأَمْرَ, inf. n. as above, ‡ *I did, performed,* or *executed, the affair in a firm, solid, sound,* or *good, manner;* or *put it into a firm, solid, sound,* or *good, state.* (TA.) — رَمَى فَأَحَقَّ الرَّمِيَّةَ ‡ *He cast,* or *shot, and killed on the spot the animal at which he cast,* or *shot.* (Ibn-'Abbád, Z, Ḳ,* TA.) — أَحَقَّتْ إِبِلُنَا رَبِيعًا and ↑ استحقّت ربيعا † *Our camels found* [*herbage such as is termed*] ربيع *full-grown, and pastured upon it.* (TA.)

5. تحقّق [*It was,* or *became,* or *proved to be, a truth, a reality,* or *a fact.*] [Hence,] تحقّق عِنْدَهُ الخَبَرُ *The information was,* or *proved, true, right, correct,* or *valid, in his estimation.* (Ṣ, Ḳ.*) ═ تحقّقهُ: see حَقَّهُ.

6. تَحَاقَّ is syn. with تَخَاصُمٌ; and ↑ اِحْتِقَاقٌ, with اِخْتِصَامٌ; [*The disputing, litigating,* or *contending, together;*] (Ṣ, Ḳ;) [for] تَحَاقُّوا and اِخْتَصَمُوا signify the same; (Ḳ in art. خصم;) [or rather] the meaning of [تحاقّ and] ↑ احتقاق is [the *disputing,* &c., *together for a right,* or *due;*] *each one's,* or *every one's, saying,* "*The right is mine,*" and "*with me;*" or *demanding his right,* or *due.* (TA.) One says, تَحَاقُّوا [*They disputed,* &c., *together for a right,* or *due*]. (TḲ.) And ↑ اِحْتَقَّا *They two disputed,* &c., (Ḳ, TA,) *each of them demanding his right,* or *due.* (TA.) And ↑ احتقّ فُلَانٌ وَفُلَانٌ [*Such a one and such a one disputed,* &c., *together for a right,* or *due*]. (Ṣ.) One does not say of a single person [تحاقّ nor] ↑ احتقّ; like as one does not say of one only [تخاصم nor] اختصم. (Ṣ.)

7. اِنْحَقَّتِ العُقْدَةُ ‡ *The knot became tied,* or *made fast,* or *tightened.* (Ibn-'Abbád, Ḳ, TA.)

8: see 6, throughout. ═ اِحْتَقَّتْ بِهِ الطَّعْنَةُ † *The thrust,* or *piercing, killed him:* (AA, Ḳ:) or ‡ *went right,* or *directly, into him:* (Aṣ, TA:) or † *penetrated into his belly,* or *inside:* (L, TA:) or *hit,* or *struck, the socket,* or *turning-place, of his hip,* which is termed its حُقّ. (Ḳ,* TA.) One says, رَمَى فُلَانٌ الصَّيْدَ فَاحْتَقَّ بَعْضًا وَشَرَمَ بَعْضًا † *Such a one shot,* or *cast, at the objects of the chase, and killed some, and wounded some so that they escaped:* (Ṣ:) or *pierced into the bellies,* or *insides, of some, and wounded the skin of some without so piercing.* (L.) ═ احتقّهُ إِلَى كَذَا *He kept him,* or *held him, back,* or *retarded him,* [*until such a time,* or *such an event,*] *and straitened him.* (TA.) ═ احتقّ الفَرَسُ *The horse became lean,* or *light of flesh;* or *slender, and lean;* or *lean, and lank in the belly.* (Ṣ, Ḳ, TA.) — And احتقّ المَالُ *The cattle became fat:* (Ḳ: [see also the last meaning of 1 as an intrans. verb:]) but in the A and O and L, احتقّ القَوْمُ *the people's cattle became fat, and their fatness ended,* or *attained the extreme point.* (TA.)

10. استحقّهُ *He demanded it as his right,* or *due.* (TA.) [And hence,] *He had a right,* or *just title* or *claim, to it; he was,* or *became, entitled to it; he deserved it,* or *merited it;* syn. اِسْتَوْجَبَهُ: (Ṣ, Mṣb, Ḳ:) or these two verbs are nearly the same; (TA;) [the former meaning *he was,* or *became, adapted, disposed, apt, meet, suited, suitable, fitted, fit, competent,* or *proper, for it;* which is the most proper meaning of the phrase صَارَ حَقِيقًا بِهِ, as well as of the verb استحقّ; but this verb has also the former of these two meanings.] When a man purchases a house, and another lays claim to it, and establishes a just evidence of his claim, and the judge decides for him according to his evidence, one says of him, قَدِ ٱسْتَحَقَّهَا عَلَى المُشْتَرِى [*He has a right to it in preference to the purchaser*]; meaning that he is to possess it in preference to the purchaser. (TA.) And of a camel such as is termed حِقّ one says, استحقّ أَنْ يُرْكَبَ [*He was,* or *has become, fit to be ridden*], (Ḳ,) and أَنْ يُحْمَلَ عَلَيْهِ [*to be laden*]: (Ṣ, Mṣb:) and استحقّ الضِّرَابَ [*He was,* or *has become, fit for covering*]. (L, Ḳ.) — [Hence, *It* (an action, and anything,) *deserved it, merited it,* or *required it.*] And استحقّ إِثْمًا *He did what necessitated sin;* (Ksh and Bḍ and Jel in v. 106;) [*was guilty of a sin;*] and *deserved its being said of him that he was a sinner;* (Ksh ibid.;) i. q. اِسْتَوْجَبَهُ. (TA.) And استحقّوا *They committed sins for which he who should punish them would be excusable, because they deserved punishment;* like أَوْجَبُوا, and أَعْذَرُوا, and اِسْتَلَاطُوا. (IAạr, TA in art. لوط.) — استحقّت إِبِلُنَا رَبِيعًا: see 4, last sentence. — استحقّت النَّاقَةُ لَقَاحًا *The she-camel conceived,* or *became pregnant;* and استحقّ لَقَاحُهَا [signifies the same]. (TA.) — See also 1, as an intrans. verb, last two sentences.

R. Q. 1. حَقْحَقَ, inf. n. حَقْحَقَةٌ, *He went the pace,* or *in the manner, termed* حَقْحَقَةٌ; (TA;) which means *a pace,* or *manner of going, in which the beast is made to exert himself to the very utmost, and which is the most fatiguing to the* ظَهْر [meaning the *camel that is ridden,* or the *beast that carries one*]: (Ṣ, Mgh, Ḳ:) or *a journeying in the beginning,* or *first part, of the night;* (Lth, Ṣ, Ḳ;) which is forbidden: (Lth, Ṣ, TA:) or, as some say, the *fatiguing a while, and abstaining a while:* (Lth, TA:) but Az says that Lth is not correct in either of his explanations of this word: (TA:) or *an obstinate persisting in journeying:* or *an obstinate persisting in journeying until the camel that one is riding perishes* or *breaks down:* (Ḳ:) or, accord. to Az, the correct meaning, confirmed by what the Arabs said, is the *making the camel to go on, and urging him to that which fatigues him, and that which is beyond his power, until he breaks down with his rider:* or, accord. to IAạr, the *jading of the weak* [*beast*] *by hard journeying.* (TA.) It is related in a trad., that Mutarrif Ibn-Esh-Shikhkheer said to his son, when he took extraordinary pains in religious exercises, (Ṣ, TA,) and was immoderate therein, خَيْرُ الأُمُورِ أَوْسَاطُهَا وَالحَسَنَةُ بَيْنَ السَّيِّئَتَيْنِ (TA,) وَشَرُّ السَّيْرِ الحَقْحَقَةُ [*The best of affairs,* or *actions,* or *cases, are such of them as are between two extremes; and the good action is between the two things; and the worst kind of journeying is that in which the beast is made to exert himself to the very utmost,* &c.]: (Ṣ, TA:) meaning, pursue thou the middle course in religious exercises, and burden not thyself, lest thou become disgusted; for the best of works is that which is continued, though it be small. (TA.)

حَقٌّ *contr. of* بَاطِلٌ [used as a subst. and as an epithet or act. part. n.]: (Ṣ, Mṣb, Ḳ:) or, as an inf. n. [and used as a simple subst.], *contr. of* بُطْلَانٌ; and as an act. part. n., and a simple epithet, *contr. of* بَاطِلٌ. (Kull.) [As a subst.,] its primary signification is *Suitableness to the requirements of wisdom, justice, right,* or *rightness, truth, reality,* or *fact;* or *to the exigencies of the case;* as the suitableness of the foot of a door in respect of its socket, for turning round rightly: (Er-Rághib, TA:) [and particularly] the *suitableness* of a judgment, and of what involves, or implies, a judgment, [i. e., of a saying, and a religion, and a persuasion, or the like, (as will be shown by one of the explanations of its meanings as an epithet,)] *to reality* or *fact;* and the *suitableness* of reality or fact *to a judgment:* (Kull:) [the *state,* or *quality,* or *property, of being just, proper, right, correct,* or *true; justness, propriety, rightness, correctness,* or *truth; reality,* or *fact;* the *state,* &c., *of being established,* or *confirmed, as a truth* or *fact; of being necessary, requisite,* or *unavoidable; of being binding, obligatory, incumbent,* or *due:* (as shown above: see 1, first sentence:)] and *existence* in relation to substances, absolutely: and *everlasting existence* [in relation to God]: (Kull:) pl. حُقُوقٌ and حِقَاقٌ: it has no pl. of pauc. (TA.) As an act. part. n. and a simple epithet, it is applied to a judgment [as meaning] *suitable to reality* or *fact;* and to a saying, and a religion, and a persuasion, considered as involving, or implying, such a judgment: (Kull:) to that which is *suitable to the requirements of wisdom, justice, right,* or *rightness;* as when one says that every act of God is حَقّ: to a belief, in a thing, *suitable to the reality of the case;* as when one says that belief in the resurrection is حَقّ: and to an action, and a saying, *accordant to what is requisite* or *obligatory, in quality and measure and time;* as when one says that the action of another is حَقّ, and that his saying is حَقّ: (Er-Rághib, TA:) [thus it signifies *just, proper, right, correct,* or *true; authentic, genuine, sound,* *valid, substantial,* or *real; established,* or *confirmed, as a truth* or *fact:* and *necessary, requisite,* or *unavoidable:* and *binding, obligatory, incumbent,* or *due:*] also the *necessarily-existing by his own essence* [applied to God; as an epithet of Whom it has other meanings assigned to it by some, as will be seen below]: and anything *existing,* of an objective kind: (Kull:) *existing as an established fact,* or *truth,* (Ḳ, TA,) *so as to be undeniable.* (TA.) In the saying, هٰذَا عَبْدُ ٱللّٰهِ الحَقَّ لَا البَاطِلَ [*This is 'Abd-Allah, truly; not falsely*], the article ال is prefixed as it is in the phrase, أَرْسَلَهَا العِرَاكَ; but sometimes it is dropped, so that one says حَقًّا لَا بَاطِلًا. (Sb, TA.) And in the phrase, لَحَقُّ لَا آتِيكَ, a form of oath, the nom. case is used without tenween; but when the ل is dropped, one says, حَقًّا لَا آتِيكَ: (Ṣ, TA:) [the latter means *Truly I will not come to thee:* the former seems to be best explained by what here follows:] accord. to the A, لَحَقُّ لَا أَفْعَلُ is originally لَحَقُّ ٱللّٰهِ لَا أَفْعَلُ [*The truth,* or *existence, of God* is that by which I swear, *I will not do* such a thing]; the affixed noun [ٱللّٰه] being suppressed, and meant to be understood. (TA.) الحَقُّ مَعِى الحَقُّ بِيَدِى [*The right is mine*] and الحَقُّ عَلَيْكَ [*The right is with me* and *The right is against thee,* which last is often used as meaning *thou art in fault,* or *in the wrong,*] are said by one disputing, or contending, for a thing. (TA.) [And in like manner one says الحَقُّ بِيَدِكَ and الحَقُّ مَعَكَ as meaning *Thou art in the right,* and الحَقُّ عَلَىَّ as meaning *I am in the wrong.*] One says also, كَانَ ذٰلِكَ عِنْدَ حَقِّ لَقَاحِهَا, and ▼ حِقِّ لَقَاحِهَا ‡ *That was on the occasion of the establishment of the fact of her conception,* or *pregnancy.* (Ṣ, A, Ḳ,* TA.) And هٰذَا العَالِمُ حَقُّ العَالِمِ [like هٰذَا العَالِمُ جِدُّ العَالِمِ,] *This is the learned* man, *the extremely learned* man. (Sb, TA.) And حَقُّ عَلِيمٍ means *Very* [or *extremely*] *knowing.* (Ḥam p. 139.) [Respecting the expressions حَقُّ اليَقِينِ and الحَقُّ اليَقِينُ, see art. يقن.] — [From the primary and general signification, explained in the first sentence of this paragraph, are deduced several particular meanings here following.] — *Equity,* or *justice.* (Ḳ.) — [The *right mode,* or *manner, of acting* or *being.*] — *Veracity* (Ḳ) *in discourse.* (TA.) — *Prudence.* (Ḳ, TA.) — [*A right,* or *due, of any kind: a just claim: a desert,* or *thing deserved: anything that is owed;* as *a fee, hire,* or *pay,* and *a price: a duty; an obligation:*] the sing. of حُقُوقٌ. (Ṣ, Ḳ.) [You say, هٰذَا حَقِّى *This is my right,* or *due,* &c. And هٰذَا حَقٌّ لِى *This is a right,* or *due, belonging to me;* or *a thing due,* or *owed, to me:* or *this is a duty to me.* And هٰذَا حَقٌّ عَلَىَّ *This is a right,* or *due, the rendering of which is binding, obligatory,* or *incumbent, on me:* or *this is my duty.* And hence, حَقُّ الطَّرِيقِ *The duty that relates to the road:* see art. طرق.] ▼ حَقَّةٌ is a more particular, or peculiar, or special, term. (Ṣ, Ḳ.) You say, ▼ هٰذِهِ حَقَّتِى [*This is my particular,* or *peculiar,* or *special, right* or *due* &c.: but it is explained as] meaning حَقِّى. (Ṣ.) And ▼ هٰذِهِ حِقَّتِى *This is my just,* or *necessary,* or *incumbent, right* or *due* &c. (Ḳ.) — *A share,* or *portion;* as in the saying, أَعْطِ كُلَّ ذِى حَقٍّ حَقَّهُ *Give thou to every one to whom belongs a share,* or *portion, his share,* or *portion, that is appointed,* or *assigned, to him.* (TA.) — *Property: a possession.* (Ḳ.) — [*An appertenance.* Hence the pl.] حُقُوقٌ signifies The مَرَافِق [or *appertenances,* or *conveniences, such as the privy and the kitchen and the like,*] of a house. (Mṣb, TA.) — [*A necessary,* or *requisite, thing.*] — *A thing,* or *an event, that is decreed,* or *destined.* (Ḳ, TA.) It is said to have this meaning in the Ḳur [xv. 8], in the words, مَا نُنَزِّلُ المَلَائِكَةَ إِلَّا بِالحَقِّ [*We send not down the angels save with that which is decreed,* or *destined*]: (TA:) or, as some say, it means here *revelation:* (Ksh, Bḍ:) or *punishment.* (Ksh, Bḍ, Jel.) — [And hence,] *Death.* (Ḳ.) So accord. to some in the Ḳur [l. 18], where it is said, وَجَآءَتْ سَكْرَةُ المَوْتِ بِالحَقِّ [*And the confusion of the intellect by reason of the agony of death shall come with death:* but other and obvious meanings are assigned to it in this instance]. (TA.) — [As an epithet,] الحَقُّ is one of the names of *God:* or one of the epithets applied to Him: (Ḳ:) meaning *the Really-existing; whose existence and divinity are proved to be true:* (IAth, TA:) or *the Creator according to the requirements of wisdom, justice, right,* or *rightness.* (Er-Rághib, TA.) — It is also applied to *The Ḳur-án.* (Ḳ.) — And to [*The religion of*] *El-Islám.* (Ḳ.) — See also حَقِيقٌ, in two places. — And see حَاقٌّ, in two places.

حُقٌّ: see حُقَّةٌ. — Also The *breast,* or *mamma,* of an old woman. (TA.) — *A tuber* of a truffle. (TA.) — The *small hollow upon the head* of the shoulder-blade: (Ḳ:) or, as some say, the حُقّ of the shoulder-blade is the *head of the upper arm, in which is the* وَابِلَة: (TA:) or this latter is another signification of حُقّ. (Ḳ.) — The *head,* (Ḳ,) or *lower part of the head,* (TA,) of the hip, *in which is the thigh-bone;* (Ḳ, TA;) the *socket,* or *turning-place,* of the hip. (TA.) — The *socket,* or *turning-place, of the foot* of a door. (TA.) You say, لَقِيتُهُ عِنْدَ حُقِّ بَابِ المَسْجِدِ, meaning *I met him,* or *found him, near to the mosque:* and لَقِيتُهُ مِنْ حَاقِّ ▼ المَسْجِدِ [app. means the same]. (TA.) — See also حَاقٌّ, in two places. — Also The *web* of a spider. (Az, Ḳ.)

حِقٌّ A camel *three years old,* (Ṣ, Mgh,) *that has entered the fourth year:* (Ṣ, Mgh, Mṣb:) or a camel *entering the fourth year:* (Ḳ:) so called because fit to be laden (Ṣ, Mṣb) and made use of; (Ṣ;) or because fit to be ridden; or because fit for covering: (Ḳ:) the female is termed ▼ حِقَّةٌ, (Ṣ, Mgh, Mṣb, Ḳ,) and حِقٌّ also: (Ṣ, Ḳ:) the pl. (of حِقٌّ, Mṣb) is حِقَاقٌ (Ṣ, Mgh, Mṣb, Ḳ) and (of حِقَّةٌ, Mṣb) حِقَقٌ, (Mṣb, Ḳ,) and the pl. pl., (Ḳ,) i. e. pl. of حِقَاقٌ, (Ṣ,) is حُقُقٌ, (Ṣ, Ḳ,) and sometimes حَقَائِقُ, (Ṣ, TA,) or this is a pl. of حِقَّةٌ. (TA: see 3.) Or [so in the Ḳ, but it should rather be "and,"] حِقٌّ signifies A she-camel *whose teeth have fallen out by reason of extreme age.* (Ḳ.) — One says, رَأَيْتُهَا وَهِىَ حِقَّةٌ as meaning † [*I saw her when she was*] *like a she-camel termed* حِقَّة *in bigness.* (TA.) — And [the pl.] حِقَاقٌ is applied to The *young ones* of trees: (TA:) and particularly of the [species of mimosa termed] عُرْفُط: (Ḳ, TA:) as being likened to the camels termed حِقَاق. (TA.) = Also ‡ The *time of year in which a she-camel was covered in the preceding year;* (Ṣ, TA;) and so ▼ حِقَّةٌ: (TA:) or the *usual period of* her *gestation.* (L in art. نضج.) You say, أَتَتِ النَّاقَةُ عَلَى حِقِّهَا ‡ *The she-camel arrived at the time of year in which she had been covered in the preceding year:* (Ṣ, TA:) and اتت على ▼ حِقَّتِهَا signifies the same; or *she completed her period of gestation, and overpassed by some days the time of year in which she had been covered in the preceding year, to complete the formation of the fœtus.* (TA.) And جَازَتِ الحِقَّ *She* (a camel) *overpassed the year without bringing forth.* (Aṣ, Ṣ.) [See also the last sentence but one in the explanations of 1 as an intrans. verb.] — حِقٌّ: see حَقٌّ: كَانَ ذٰلِكَ عِنْدَ حِقِّ لَقَاحِهَا —

حَقَّةٌ: see حَقٌّ, in two places: — and حَقِيقَةٌ, also in two places: — and حَاقَّةٌ.

حُقَّةٌ *A receptacle of wood,* (Ḳ, TA,) or *of ivory,* or *of some other material proper to be cut,* or *shaped out;* (TA;) *a receptacle for perfume;* (Ḥar p. 518;) [generally *a small round box,* used for unguents and perfumes &c.; and applied also to *a small cocoa-nut used as a box* for snuff &c.;] *a thing well known:* (Ṣ:) [also *a receptacle for wine:* (see تَامُورٌ, in art. امر:)] pl. ▼ حُقٌّ, [or rather this is a coll. gen. n., as is indicated in the TA, and it is now used as a sing., like حُقَّةٌ,] and حُقَقٌ, (Ṣ, Ḳ,) which latter is pl. of حُقَّةٌ, (ISd, TA,) and حِقَاقٌ (Ṣ, Ḳ) and حُقُوقٌ and [of pauc.] أَحْقَاقٌ, (Ḳ,) which three are pls. of حُقٌّ. (TA.) — And ‡ *A woman;* (Ḳ, TA;) as being likened thereto. (TA.) = See also حَاقَّةٌ.

حِقَّةٌ: see هٰذِهِ حِقَّتِى, voce حَقٌّ. = See also حِقٌّ, in three places.

حَقَقٌ, in a horse, The quality of *not sweating:* (Ṣ,* Ḳ:) which is a fault. (TA.) — And, in a horse also, The *putting down the hind hoof in the place* [*that has just before been that*] *of the fore hoof:* (Ṣ,* Ḳ:) which is also a fault. (Ḳ.) [See أَحَقُّ.]

حُقَّقٌ [app. pl. of the act. part. n. حَاقٌّ, like نُزَّلٌ pl. of نَازِلٌ, &c.,] Persons *who have recently known,* or *been acquainted with, events,* or *affairs, good* and *evil.* (TA.) — And Persons *establishing a claim* or *claims.* (TA.)

حَقِيقٌ *Adapted, disposed, apt, meet, suited, suitable, fitted, fit, proper, competent,* or *worthy;* syn. خَلِيقٌ, (Sh, Ṣ, Mgh, Mṣb, Ḳ,) and جَدِيرٌ; (Ḳ;) as also ▼ حَقٌّ, (Ibn-'Abbád, Ḳ,) and [some say] ▼ مَحْقُوقٌ: (Sh, Ṣ, Mgh, Ḳ:) حَقِيقٌ is said to be of the measure فَعِيلٌ in the sense of the measure مَفْعُولٌ; but accord. to the A, it is not so, because its fem. is with ة; but is from the supposed verb حَقُقَ, and is like خَلِيقٌ from خَلُقَ, and جَدِيرٌ from جَدُرَ: and ▼ مَحْقُوقٌ signifies [properly] *rendered adapted* &c.: (TA:) the pl. of حقيق is أَحِقَّآءُ; and that of ▼ محقوق is مَحْقُوقُونَ. (Ṣ.) You say, هُوَ حَقِيقٌ بِهِ (Sh, Ṣ, Mṣb, Ḳ) and ▼ مَحْقُوقٌ بِهِ (Sh, Ṣ, Ḳ) and ▼ حَقٌّ بِهِ (Ibn-'Abbád, Ḳ) [*He is adapted, &c., for it*; or *worthy of it*]. And to a woman, أَنْتِ حَقِيقَةٌ ▼ مَحْقُوقَةٌ لِذٰلِكَ (A, TA) and حَقِيقَةٌ لِذٰلِكَ and بِكَذَا [*Thou art adapted, &c., for such a thing* and *for that thing*; or *worthy of it*]. (TA.) And أَنْتَ حَقِيقٌ بِأَنْ تَفْعَلَ (A, Mgh) and ▼ مَحْقُوقٌ (A) [*Thou art adapted, &c., for thy doing* such a thing; or *worthy of doing* it]. And هُوَ حَقِيقٌ أَنْ يَفْعَلَ كَذَا [*He is adapted, &c., for his doing such a thing*; or *worthy to do it*]; (Ṣ;) in which case, ان is for بِأَنْ. (Mgh.) [And حَقِيقٌ بِكَذَا also signifies *Having a right*, or *just title* or *claim, to such a thing; entitled to such a thing.*] It is said in the Ḳur [vii. 103], حَقِيقٌ عَلَى أَنْ لَا أَقُولَ عَلَى اللّٰهِ إِلَّا الْحَقَّ, meaning I am *disposed* [not] *to say* [of God aught save] *the truth*: or, as some say, I am *vehemently desirous* [that I should not say &c.]; for, accord. to Aboo-'Alee, أَنَا حَقِيقٌ عَلَى كَذَا means *I am vehemently desirous of such a thing*: but one reading, that of Náfi', is حَقِيقٌ عَلَيَّ أَنْ لَا أَقُولَ, *It is binding*, or *obligatory*, or *incumbent, on me* [*that I should not say*]. (TA.)

حَقِيقَةٌ The *essence* of a thing as meaning *that by being which a thing is what it is*; [or *that in being which a thing consists*;] as when we say that a rational animal is the حقيقة of a human being: (KT:) or *that by being which a thing is what it is, considered with regard to its reality*, is termed حَقِيقَةٌ: considered with regard to its individuality, هُوِيَّةٌ: and without regard thereto, مَاهِيَّةٌ: (KT, TA:) *the ultimate and radical constituent* of a thing. (Mṣb, TA.) — [Also The *essence* of a thing as meaning the *property* or *quality*, or the *aggregate of properties or qualities, whereby a thing is what it is*; the *essential property* or *quality*, or the *aggregate of the essential properties or qualities*, of a thing; *that which constitutes the particular and distinguishing nature* of a thing or of a genus or species; i. q. ذَاتِيَّةٌ: and] the *truth, reality*, or *true* or *real nature* or *state* [or *circumstances* or *facts*, the *very nature*, and the *gist*, and the *pith, marrow*, or *most essential part*], of a case, or an affair: pl. حَقَائِقُ: see 3. (TA.) One says, بَلَغَ حَقِيقَةَ الأَمْرِ *He arrived at* [the knowledge of] *the truth, reality*, or *true* or *real nature* or *state* [&c.], *of the case*, or *affair*. (TA.) And حَقِيقَةُ الأَمْرِ signifies ▼ الحَقَّةُ; (Ṣ, Ḳ;) as also ▼ الحَاقَّةُ. (TA.) Hence the saying, لَمَّا عَرَفَ ▼ الحَقَّةَ مِنِّى هَرَبَ [*When he knew the truth, reality*, or *true* or *real nature* or *state* &c., *of the case*, or *affair, from me, he fled*]. (Ṣ, TA.) And لَمَّا رَأَى ▼ الحَاقَّةَ مِنِّى هَرَبَ [*When he saw the truth*, &c.]. (TA.) [حَقِيقَةً is often used as meaning *In truth*, or *truly*; *in reality*, or *really*; and *in fact*.] You say also, عَرَفْتُهُ حَقِيقَةَ المَعْرِفَةِ [*I knew it with reality of knowledge*]. (Mṣb in art. كنه.) And حَقِيقَةُ الإِيمَانِ means *Genuine belief* or *faith*; *reality of belief* or *faith*. (TA.) [And you say, هٰذَا شَىْءٌ لَا حَقِيقَةَ لَهُ *This is a thing having no reality*.] — [Also *A word*, or *phrase, used in its proper* or *original*, or *in a proper* or *an original, sense*;] *that which is constantly used according to its original application*; or a name for *that whereby is meant what it was* [*originally*] *applied to denote*; (TA;) *contr. of* مَجَازٌ: (Ṣ, Ḳ:) of the measure فَعِيلَةٌ in the sense of the measure فَاعِلَةٌ, from حَقَّ الشَّىْءُ, signifying ثَبَتَ: the ة is affixed for the conversion of the word from an epithet to a subst.: (TA:) [pl. as above]. [It is also called حَقِيقَةٌ لُغَوِيَّةٌ, and حَقِيقَةٌ لُغَةً; to distinguish it from what is termed حَقِيقَةٌ عُرْفِيَّةٌ, and حَقِيقَةٌ عُرْفًا, which is *A word*, or *phrase, so much used in a particular tropical sense as to be, in that sense, conventionally regarded as proper*; as, for instance, عَدْلٌ in the sense of "just;" it being properly an inf. n.] A مَجَازٌ, when much used, becomes what is termed حَقِيقَةٌ عُرْفًا. (Mz 24th نوع.) حَقِيقَةٌ means also *A proper* (opposed to a tropical) *signification*.] — الحَقِيقَةُ also signifies ‡ *That which, or those whom, it is necessary for one, or it behooveth one, to defend, or protect*, (Ṣ, L, Ḳ, TA,) *of the people of one's house*, (L,) or *such as the wife, and the female neighbour, and property, &c.*: (Ḥam p. 181:) pl. as above. (L.) You say, فُلَانٌ حَامِى الحَقِيقَةِ ‡ [*Such a one is the defender*, or *protector, of that which*, or *those whom, it is necessary, &c., to defend*, or *protect*]. (Ṣ, TA.) [See also ذِمَارٌ. And see an ex. of this signification, or of the next, in a verse cited in p. 288.] — Also † *The banner*, or *standard*: (Ṣ, Ḳ, and Ḥam ubi suprà:) this being included in the preceding meaning. (Ḥam.) — And † *That which is sacred*, or *inviolable*; *that which one is under an obligation to respect*, or *honour*. (TA.)

[حَقِيقِىٌّ rel. n. of حَقِيقَةٌ, *Essential*, &c.]

حَقَّانِىٌّ [*Of*, or *relating to*, الحَقّ as meaning *justness, propriety, rightness, correctness*, or *truth*; &c.: and hence *just, proper*, &c.; like حَقٌّ when used as an epithet: and *of*, or *relating to*, الحَقّ as meaning *God*:] a rel. n. from الحَقّ, like رَبَّانِىٌّ from الرَّبّ. (TA.)

قَرَبٌ حَقْحَاقٌ [*A night-journey to water*] *made with labour* or *exertion* or *haste*; (Ḳ;) as also قَهْقَاهٌ and هَقْهَاقٌ; and so ▼ مُحَقْحِقٌ. (TA.) [See R. Q. 1.]

حَاقٌّ *i. q.* صَادِقٌ [as used in the phrases صَادِقُ الحَلَاوَةِ and صَادِقُ الحُمْلَةِ, &c.: see art. صدق]: so in the phrase حَاقُّ الجُوعِ [*Vehement hunger*]: (Ḳ:) occurring in a trad. of Aboo-Bekr: but accord. to one reading, it is حَاقُ الجُوعِ, without teshdeed to the ق, from حَاقَ بِهِ البَلَآءُ, inf. n. حَيْقٌ and حَاقٌ, "trial, or trouble, beset him;" and means *the besetting of hunger*: or it may mean حَائِقُ الجُوعِ [*besetting hunger*]. (TA.) One says also, رَجُلٌ حَاقُّ الرَّجُلِ and ▼ حَاقَّةُ الرَّجُلِ *A man perfect in manliness*: and حَاقُّ الشُّجَاعِ and ▼ حَاقَّةُ الشُّجَاعِ *perfect in courage*. (Ḳ, *TA.) And Az relates that he heard an Arab of the desert say, of a mark of mange, or scab, that appeared upon a camel, هٰذَا حَاقُّ صُمَادِحِ الجَرَبِ [*This is a most sure*, or *a truth-telling, evidence of genuine mange*, or *scab*]. (TA.) — Also The *middle* of the head; (Ṣ, Ḳ;) as also ▼ حَقٌّ: (Ḳ:) and of the back of the neck; as also ▼ حُقٌّ: (TA: [thus the latter is there written, in this instance, with ḍamm:]) and of the eye: (TA:) and of a road: (Ḳ, *TA:) and of winter. (Ṣ.) One says, سَقَطَ عَلَى حَاقِّ رَأْسِهِ (Ṣ, Ḳ) and ▼ حَقِّ رَأْسِهِ (Ḳ) *He fell upon the middle of his head*: (Ṣ, Ḳ:) and عَلَى حَاقِّ القَفَا and ▼ حُقِّ القَفَا *upon the middle of the back of the neck*. (TA.) And أَصَابَ حَاقَّ عَيْنِهِ *He*, or *it, hit the middle of his eye*. (TA.) And رَكِبَ حَاقَّ الطَّرِيقِ *He went upon the middle of the road*. (Ḳ, *TA.) And جِئْتُهُ فِى حَاقِّ الشِّتَآءِ *I came to him in the middle of winter*. (Ṣ.) And لَقِيتُهُ مِنْ حَاقِّ المَسْجِدِ: see حَقٌّ. — هُوَ فِى حَاقٍّ مِنْ كَذَا *He is in straitness by reason of such a thing*. (TA.)

حَاقَّةٌ: see حَقِيقَةٌ, in two places. [In the sense in which it is there explained, its pl. is حَوَاقُّ; and so in other senses; agreeably with analogy: see the second of the sentences here following.] — Also *A severe calamity* or *affliction, the happening of which is fixed*, or *established*; and so ▼ حَقَّةٌ; (Ḳ;) which signifies also, [according to another explanation,] like ▼ حُقَّةٌ, [simply,] *a calamity*; or *a great, formidable, terrible*, or *momentous, thing*, or *event*: (Az, Ḳ:) and حَاجَةٌ حَاقَّةٌ *a want that befalls*, or *happens, and is severe*, or *distressing*. (Mṣb.) — And الحَاقَّةُ [in the Ḳur lxix. 1 and 2] means *The resurrection*: (Ṣ, Mṣb, Ḳ:) because in it shall be [manifest] the true natures (حَوَاقّ) of things, or actions; or because in it shall be [or shall happen (Bḍ)] severe calamities (حَوَاقُّ الأُمُورِ); (Fr, Ṣ, Bḍ, Ḳ;) namely, the reckoning and the recompensing: (Bḍ:) or because in it things shall be surely known (Bḍ, Jel) which are denied; namely, the raising of the dead, and the reckoning, and the recompensing: (Jel:) or because including within its sphere [all] the created beings. (Mṣb. [Several other reasons are assigned; but these which I have mentioned appear to be the most generally approved.]) — See also حَاقٌّ, in two places.

أَحَقُّ [comparative and superlative of حَقِيقٌ]. You say, هُوَ أَحَقُّ بِكَذَا [*He is more, and most, adapted, disposed, apt, meet, suited, suitable, fitted, fit, proper*, or *competent, for such a thing*; or *more, and most, worthy*, or *deserving, of it*: and *he has a better, and the best, right to such a thing*; or *a more just*, and *the most just, title* or *claim* *to it; or he is more, and most, entitled to it*]: this phrase is used in two senses: first, as denoting the possession of an exclusive right or title, i. e., without the participation of another; as when you say, زَيْدٌ أَحَقُّ بِمَالِهِ *Zeyd is entitled to his property exclusively of any other person:* secondly, as denoting the possession of a right or title in participation with another person, but in a superior degree; as in the saying, الأَيِّمُ أَحَقُّ بِنَفْسِهَا مِنْ وَلِيِّهَا, (Mṣb,) i. e. *The woman that has not a husband and is not a virgin* [*is more entitled to dispose of herself than is her guardian*]; (Mgh in art. ايم;) meaning that they participate [in the right], but that her right is the stronger: (Mṣb:) a saying of Mohammad, in which the ايم is opposed to the بِكْر, for it is added that the بكر is to be asked her permission: but one reading substitutes الثَّيِّبُ for الايم. (Mgh ubi suprà.) In the saying, in the Ḳur [v. 106], لَشَهَادَتُنَا أَحَقُّ مِنْ شَهَادَتِهِمَا, it may be formed from اِسْتَحَقَّ by rejection of the augmentative letters, so that the meaning is, [*Verily our testimony is*] *more deserving of being accepted* [*than the testimony of them two*]: or it may be from حَقَّ الشَّىْءُ signifying ثَبَتَ, and so mean *more true,* or *valid.* (TA.) ═ Applied to a horse, *That does not sweat.* (Ṣ, Ḳ.) — And, likewise thus applied, *That puts down his hind hoof in the place* [*that has just before been that*] *of his fore hoof.* (Ṣ,* Ḳ.) [See حَقْحَقٌ.]

مُحِقٌّ *Speaking truth; saying what is true;* (Mṣb;) *contr. of* مُبْطِلٌ: (Ḳ:) or *revealing,* or *manifesting,* or *showing, a truth,* or *a right* or *due:* or *laying claim to a right* [or *to a thing* (see 4)] *which is,* or *becomes, due to him.* (Mṣb.)

مُحَقَّقٌ, [in the CḲ, erroneously, مُحْقَّق,] applied to speech, or language, ‡ *Sound,* or *compact,* (Ṣ, Ḳ, TA,) *and orderly.* (TA.) — And, applied to a garment, or piece of cloth, ‡ *Firmly,* or *compactly, woven,* (Ṣ, Ḳ, TA,) *and figured with the form of* حُقَق [pl. of حُقَّةٌ, q. v.]. (TA.)

[مُحَقِّقٌ is often used as meaning *A critical judge in matters of literature.*]

مَحْقُوقٌ: see حَقِيقٌ, in six places.

مَحَاقٌّ, applied to cattle, *Such as have not brought forth, nor been milked* (لَمْ يُحْلَبْنَ [in the CḲ, erroneously, لَمْ يُحْلِبْنَ]), *in the next preceding year:* (Ibn-'Abbád, Ḳ:) or *whose first and second milkings are of biestings.* (AHát, TA.)

طَعْنَةٌ مُحْتَقَّةٌ (in [some of] the copies of the Ḳ, erroneously, مُحْتَقِقَةٌ, TA) *A thrust,* or *piercing, in which is no swerving from the right direction.* (Ṣ, A, O, L, Ḳ.)

حَقْحَاقٌ: see مُحَقْحِقٌ.

حقب

1. حَقِبَ, (Ṣ, A, Mṣb, Ḳ,) aor. َ, (A, Mṣb, Ḳ,) inf. n. حَقَبٌ, (Mṣb, TA,) *It* (a camel's urine) *became suppressed:* and, elliptically, (Mṣb,) *he* (a camel) *suffered suppression of his urine,* (Ṣ, Mṣb,) or *had difficulty in staling,* (A, Ḳ,) *in consequence of the pressure of his* حَقَب [or *hind girth*] *upon his sheath,* (Ṣ, A, Ḳ,) which sometimes kills the beast; as also ▼احقب. (TA.) And حَقِبَتْ *She* (a camel) *suffered suppression of her milk in consequence of the pressure of the* حَقَب *upon her udder.* (A.) — [Hence,] said of rain, (IAạr, L, Mṣb, Ḳ,) &c., (Ḳ,) ‡ *It was delayed;* (L, Mṣb;) *was withheld;* (IAạr, L, Ḳ;) as also ▼احقب. (TA.) And of a year (عَام), ‡ *Its rain was withheld.* (Ṣ, R, A.) And of a mine, ‡ [*It ceased to yield;* or] *nothing was found in it;* as also ▼احقب. (Ḳ, TA.) And of a gift, or benefit, ‡ *It became little,* or *ceased.* (TA.) And of an affair, ‡ *It became perverted, marred,* or *disordered, and impeded.* (L.) ═ See also 8.

4. احقب *He girded* a camel *with a* حَقَب. (Ṣ.) — *He made* a person *to ride behind him on the same beast;* (A, TA;) as also ▼استحقب, (A,) or ▼احتقب, q. v. (TA.) ═ See also 1, in three places.

8. احتقب *He bound a* حَقِيبَة *behind* [*on his camel* or *horse*]; (Az, TA;) as also ▼استحقب: (Ḥam p. 289:) *he bound* anything *behind the* [*camel's saddle called*] رَحْل or قَتَب: (Ḳ:) *he put on,* or *conveyed, a* حقيبة *behind him on his horse* [or *camel*]; as also ▼حَقَّبَ: (Mṣb:) *he carried,* or *conveyed,* a thing *behind him* [*on his beast*]; as also ▼استحقب: and *he made a* person *to ride behind him on the* حقيبة. (TA.) See also 4. — [Hence,] احتقب and ▼استحقب (Ṣ, A, Ḳ) ‡ *He bore,* or *took upon himself the burden of,* (Ṣ, A,) a thing, (Ṣ,) good, or evil; (A;) syn. اِحْتَمَلَ: (Ṣ, A:) and *laid* it *up for the future;* (A, Ḳ;) namely, good, or evil: (A:) for a man [as it were] bears his actions, and lays them up for the future [to be rewarded or punished for them]. (TA.) And hence, احتقب الإِثْمَ (Ṣ, Mṣb) † [*He bore,* or *took upon himself the burden of, the sin;* or] *he committed the sin:* as though it were a thing perceived by the senses, which he bore or carried [behind him]: (Mṣb:) or as though he collected it into a mass, and conveyed it behind him [as a حقيبة]. (Ṣ.)

10: see 4 and 8; the latter in three places.

حُقْبٌ, (A, Mṣb, Ḳ,) or ▼حُقُبٌ, (Ṣ,) or the latter also, (A, Mṣb, Ḳ,) *i. q.* دَهْرٌ; (Ṣ, A, Mṣb, Ḳ, and Bḍ in xviii. 59;) [as meaning] *A long time:* (Bḍ ib.:) and the former, (Ṣ,) or both, (A, Mṣb,* Ḳ,) *eighty years;* (Ṣ, A, Mṣb, Ḳ, and Bḍ ubi suprà;) as some say: (Mṣb and Bḍ:) or *more:* (Ṣ, A, Ḳ:) or, as some say, *seventy:* (Bḍ:) and *a year;* (A, Ḳ;) as also ▼حِقْبَةٌ: (Ṣ, A, Ḳ:) or *years:* (A, Ḳ:) pl. of the former حِقَابٌ [a pl. of mult.], (Ṣ, TA,) and of the latter, (Ṣ, TA,) or former, (Mṣb,) or of both, (TA,) أَحْقَابٌ (Ṣ, A, Mṣb, Ḳ) and أَحْقُبٌ [both pls. of pauc.]. (Az, Ḳ.)

حَقَبٌ *A camel's hind girth;* the *girth that is next to the flank:* (A, Ḳ:) or *a rope with which a camel's saddle is bound to his belly,* (Ṣ, A, Mṣb, Ḳ,) *next to the sheath of his penis, in order that the fore girth may not draw it forward* (Ṣ, TA) *nor hurt him,* (TA,) or *in order that the saddle may not shift forward to his withers:* (Mṣb:) pl. أَحْقَابٌ. (Mṣb.) — And *A cord with which the* حَقِيبَة *is bound.* (ISh, TA.) — See also حِقَابٌ. ═ In excellent she-camels, *Smallness,* or *slenderness, of the flanks, with tenseness,* or *firmness, of the skin of those parts:* a quality approved. (Az, TA.)

حُقُبٌ: see حُقْبٌ.

حُقُبٌ: see حُقْبٌ.

حِقْبَةٌ *A period of time,* (A, Mṣb, Ḳ,) *undefined:* (A, Ḳ:) accord. to some, *i. q.* حُقْبٌ: (Mṣb:) see this latter: pl. حِقَبٌ (Ṣ, Ḳ) and حُقُوبٌ. (Ḳ.)

حِقَابٌ *A thing to which a woman hangs ornaments, and which she binds upon her waist;* as also ▼حَقَبٌ: (Ḳ:) *an ornamented thing which a woman binds upon her waist:* (Ṣ:) accord. to Az, *like the* بَرِيم, *except that the latter has different-coloured threads.* (TA.) — *A thread,* or *string, that is bound upon the waist of a child to avert the evil eye.* (Az, Ḳ.) — The *whiteness that appears at the root of the nail.* (Ḳ.)

حَقِيبَةٌ *A bag,* or *receptacle,* (A, TA,) *in which a man puts his travelling-provisions:* (TA;) and *any other thing that is conveyed behind a man* [*on his beast*]: (A [accord. to which this is a proper signification]:) *what the rider conveys behind him:* (MF [accord. to whom this is a tropical signification, from the same word in the last of the senses mentioned below]:) *what is borne, of goods or utensils or the like, upon the horse, behind the rider:* (Mṣb [accord. to which, also, this is tropical]:) *anything that is bound at the hinder part of the* [*camel's saddle called*] رَحْل *or of the* [*saddle called*] قَتَب: (Ḳ:) *what is put behind the* رَحْل: they used to put the coats of mail behind their رِحَال, in the [receptacles called] عِيَاب, that they might put them on in case of war: (Ḥam p. 458:) *a thing like a* بَرْذَعَة, [*a covering for a camel's back,*] *of two kinds;* namely, *that of the* [*cloth called*] حِلْس, *which is hollowed out, so as to admit the upper part of the camel's hump;* and *that of the* [*saddle called*] قَتَب, *which is behind:* ISh says that *it* (the حقيبة) *is placed upon the hinder part of the camel, beneath the two hinder curved pieces of wood of the* قَتَب: (TA:) *a* رِفَادَة [*or kind of pad, or stuffed thing,*] *placed at the hinder part of the* قَتَب: (Ḳ:) pl. حَقَائِبُ. (Ṣ, A.) You say, مَلَأَ حَقِيبَتَهُ [*He filled his* حقيبة]. (A.) And أَرْدَفَهُ خَلْفَهُ عَلَى الحَقِيبَةِ *He made him to ride behind him on the* حقيبة. (TA.) — [Hence,] ‡ *A thing* [*of an ideal kind*] *that one takes upon himself,* or *lays up for the future* [*to be rewarded or punished for it*]. (A.) You say, اِحْتَمَلَ حَقِيبَةَ سُوْءٍ ‡ [*He took upon himself a burden of evil:* as though he bound it behind him: see 8]. (A, TA.) And البِرُّ خَيْرُ حَقِيبَةٍ ‡ [*Piety is the best thing that one can take upon himself, and lay up for the future* to profit thereby]. (A, TA.) — [Hence also, accord. to the A, which I follow in marking this signification as tropical, but accord. to the Mṣb and to MF it is the primary signification,] ‡The *hinder parts*, or *posteriors*, (A, Mṣb, MF, TA,) of a woman, (A, Mṣb,) and of a man: (TA:) pl. as above. (Mṣb.) So in the phrase نُفُجُ الحَقِيبَةِ ‡*Large*, (A,) or *prominent*, (TA,) *in the posteriors*. (A, TA.)

حَاقِبٌ A camel *suffering suppression of his urine:* (Mṣb:) and ▼حَقِبٌ [signifies the same; or] a camel *having difficulty in staling, in consequence of the pressure of his* حَقَب [or *hind girth*] *upon his sheath*, which sometimes kills him. (A, TA.) And the former, A man *who is caused to hurry by the issuing of his urine:* (Mṣb:) or *who requires to go to the privy* (Mṣb, TA) *for the discharge of his urine*, (Mṣb,) [or *to evacuate his bowels*,] *and does it not until he suffers constipation:* (Mṣb, TA:) or one *suffering constipation*. (Mṣb.) [See an ex. voce حَاقِنٌ.]

أَحْقَبُ A wild ass *having a whiteness in the belly:* (Ḳ:) or *white in the part where the hind girth* (حَقَب) *would be placed:* (A, Ḳ:) the former is the more approved meaning: (TA:) or *a wild ass;* so called because white in the flanks: (Ṣ:) fem. حَقْبَآءُ: (Ṣ, A:) pl. حُقْبٌ. (A.) — Also حَقْبَآءُ *A* قَارَة [or *small isolated mountain*], (Ṣ, Ḳ,) *slender*, (TA,) *rising high into the sky*, (Ṣ, Ḳ,) *of which the flanks*, or *middle parts*, (الحَقْوَانِ,) *are enveloped by the mirage* (السَّرَاب, so in the Ḳ accord. to the TA), or *by dust* (التُّرَاب, accord. to the CḲ and a MS. copy of the Ḳ): or حَقْبَآءُ, (Ḳ,) or قَارَةٌ حَقْبَآءُ, (TA,) signifies *a* قارة *having, in its middle part, dust of a whitish hue* (أَعْفَر), *with* بُرْقَة [app. meaning *a mixture of blackness and whiteness*] *of the rest*. (Ḳ, TA.)

مُحْقَبٌ *Made to ride behind another on the same beast*. (Ṣ.) — *Bound upon the* [حَقِيبَة or] حَقَائِب. (Ḥam p. 289.) — The *fox:* (Ḳ:) so called because of the whiteness of his belly. (TA.)

مُحْقِبٌ One *who makes* another *to ride behind him on the same beast*. (Ḳ.) — Hence, in a trad., المُحْقِبُ النَّاسَ دِينَهُ †*He who makes his religion to follow that of others, without evidence, proof, or consideration*. (TA.)

حقد

1. حَقَدَ عَلَيْهِ, (Ṣ, A, L, Mṣb, Ḳ,) aor. ـِ; and حَقِدَ عليه, aor. ـَ; inf. n. (of the former, Ṣ, or of both, L, TA) حِقْدٌ, (Ṣ, L, Mṣb,* Ḳ,) or this is a simple subst., (L,) and (of the former, L, or of both, TA) حَقْدٌ (L, Ḳ) and (of the latter, Ṣ, L) حَقَدٌ (Ṣ, L, Ḳ) and [app. of both, or perhaps this is a simple subst.,] حَقِيدَةٌ; (Ḳ;) and ▼تحقّد; (L, Ḳ;) [and ▼احتقد; (see حَاقِدٌ;)] *He bore rancour, malevolence, malice, or spite*, (ضِغْن,) *against him;* (Ṣ, L;) *he hid enmity, and violent hatred, against him in his heart;* (Mṣb;) *he retained*, or *held fast, enmity against him in his heart, watching for an opportunity to indulge it or exercise it* (A, L, Ḳ) *speedily*. (A.) [Accord. to the A, this signification is from the one here next following.] — حَقِدَ, (IAạr, A, L,) or حَقَدَ, (Ḳ,) inf. n. حَقْدٌ; (L;) and ▼احقد, (IAạr, A, L,) and ▼احتقد, (Ḳ,) or both these augmented forms; (TA;) *It* (a mine) *ceased to yield anything;* (IAạr, L, Ḳ;) or *yielded nothing:* (A:) and *it* (rain) *withheld itself*. (IAạr, L, Ḳ.) — حَقِدَتْ, (Ḳ, TA,) inf. n. حَقَدٌ; (TA;) or حَقَدَتْ; (CḲ;) or ▼احقدت; (so in a MS. copy of the Ḳ;) *She* (a camel) *became full of fat*. (Ḳ.)

4. احقدهُ *He*, (a person, Ṣ, L,) or *it*, (a thing, L,) *caused him to bear rancour, malevolence, malice, or spite*, (ضِغْن,) [against another;] (Ṣ, L;) [*caused him to hide enmity, and violent hatred, in his heart;* (see 1;)] *caused him to retain*, or *hold fast, enmity in his heart, watching for an opportunity to indulge it or exercise it* [*speedily*]. (L, Ḳ.) — احقد, intrans.: see 1, in two places. — احقدوا *They sought to obtain something from a mine and found it not*. (Ṣ, Ḳ: copied by J from a book, but not heard by him.)

5: see 1.

6. تحاقدوا [*They bore mutual rancour, malevolence, malice, or spite; they hid mutual enmity, and violent hatred, in their hearts;* (see 1;)] *they retained*, or *held fast, mutual enmity in their hearts, watching for opportunities to indulge it or exercise it speedily*. (A.)

8: see 1, in two places.

حِقْدٌ (Ṣ, A, L, Mṣb, Ḳ) and ▼حَقِيدَةٌ (L, Ḳ) *Rancour, malevolence, malice*, or *spite;* syn. ضِغْنٌ; (Ṣ, A, L;) *concealment of enmity, and violent hatred, in the heart;* (Mṣb;) *retention of enmity in the heart, with watchfulness for an opportunity to indulge it or exercise it* [*speedily:* see 1]: (L, Ḳ:) pl. of the former أَحْقَادٌ [a pl. of pauc.] (Ṣ, A, L, Mṣb, Ḳ) and حُقُودٌ (A, L, Ḳ) and (of the latter, L) حَقَائِدُ. (L, Ḳ.)

حَقِدٌ: see حَاقِدٌ.

حَقُودٌ (Ṣ, A, L, Ḳ). A man [*very rancorous, malevolent, malicious, or spiteful; wont to hide enmity, and violent hatred, in his heart;*] *wont often*, or *to a great degree, to retain*, or *hold fast, enmity in his heart, watching for opportunities to indulge it or exercise it* [*speedily*]. (L, Ḳ.)

حَقِيدَةٌ: see حِقْدٌ.

حَاقِدٌ (A, L) and ▼حَقِدٌ (Ḥam p. 516) [*Bearing rancour, malevolence, malice, or spite; hiding enmity, and violent hatred, in the heart;*] *retaining*, or *holding fast, enmity in the heart, and watching for an opportunity to indulge it or exercise it* [*speedily*]; applied to a man; (L;) and in like manner to a heart, as also ▼مُحْتَقِدٌ. (A.)

مَحْقِدٌ *Origin;* syn. أَصْلٌ; (IAạr;) i. q. مَحْتِدٌ (Ḳ) and مَحْفِدٌ and مَحْكِدٌ. (TA.)

مَحْقُودٌ عَلَيْهِ A man [*against whom rancour, malevolence, malice, or spite, is borne; against whom enmity, and violent hatred, are hidden in the heart;*] *against whom enmity is retained in the heart*, or *hearts, with watchfulness for an opportunity to indulge it or exercise it speedily*. (A.)

مُحْتَقِدٌ: see حَاقِدٌ.

حقر

1. حَقُرَ, aor. ـُ, inf. n. حَقَارَةٌ (Ṣ, A, Mṣb, Ḳ) and حُقْرٌ; (TA;) and حَقَرَ, aor. ـِ, inf. n. حُقْرٌ and حُقْرِيَّةٌ; (Ḳ, TA;) *He*, or *it, was*, or *became, contemptible, despicable, mean, paltry, abject, ignominious, base*, or *vile*, (Ṣ, A, Mṣb, Ḳ,) and *held of no weight* or *worth*. (Mṣb.) And حَقُرَ فِي عَيْنِي, (TA,) or حَقَرَ, (so in a copy of the A,) and ▼تحاقر, (Ḳ,) *He*, or *it, was*, or *became, contemptible, despicable, mean, paltry, &c., in my eye*. (Ḳ,* TA.) And ▼تَحَاقَرَتْ إِلَيْهِ نَفْسُهُ *He* (lit. *his spirit*, or *soul*, or *his own self*,) *became contemptible, &c., in his own estimation*. (Ṣ, TA.) You say also, حَقْرًا لَهُ وَعَقْرًا [*May he be contemptible, or despicable, and beget no children*]: (A, TA:) a form of imprecation. (TA.) And حَقْرٌ نَقْرٌ [in which the latter word is an imitative sequent, or a corroborative: see حَقِيرٌ]. (TA.) And حَقُرْتَ وَنَقُرْتَ *Thou hast become very contemptible, despicable, mean, paltry, &c.;* syn. صِرْتَ حَقِيرًا نَقِيرًا. (Ḳ.) — حَقَرَهُ, (Ṣ, A, Mṣb, Ḳ,) aor. ـِ, (Mṣb, Ḳ,) inf. n. حَقْرٌ; (Ḳ, TA;) and ▼حقّره, (A,) inf. n. تَحْقِيرٌ; (Ḳ;) and ▼احتقره, (Ṣ, A, Mṣb, Ḳ,) and ▼استحقره; (Ṣ, A, Ḳ;) *He contemned*, or *despised, him*, or *it; held him*, or *it, to be contemptible, despicable, mean, paltry, abject, ignominious, base*, or *vile*, (Ṣ, A, Mṣb, Ḳ, TA,) and *of no weight* or *worth*. (Mṣb.) [Hence,] مَنْ حَقَرَ حَرَمَ [*He who is contemned*, or *despised, is prohibited*, or *debarred*, from what is good]: a prov. (A.)

2. حقّره, inf. n. تَحْقِيرٌ: see 1. — Also *He*, or *it, made him to be contemned* or *despised, contemptible* or *despicable, mean, paltry, abject, ignominious, base*, or *vile*. (TA.) — Also, (A, Ḳ,) inf. n. as above, (Ṣ, A, Ḳ,) said of a noun, (A,) and of speech, (Ḳ,) i. q. صَغَّرَهُ [*He made it* (namely, a noun,) *diminutive in form:* and *he used the diminutive form in it;* namely, speech].

6: see 1, in two places.

8: see 1.

10: see 1.

حُقْرٌ: see حَقَارَةٌ.

حُقْرَةٌ, a simple subst., *Contempt*. (Mṣb.)

حُقْرِيَّةٌ: see حَقَارَةٌ.

حَقِيرٌ *Contemned* or *despised, contemptible* or *despicable, mean, paltry, abject, ignominious, base*, or *vile*, (Ṣ, A, Mṣb,) and *held of no weight* or *worth;* (Mṣb;) as also ▼حَيْقُورٌ and ▼حُقَّرٌ: (Ḳ, TA:) or these two signify *weak:* or *of mean*, or *ignoble, origin*. (Ḳ.) You say also حَقِيرٌ نَقِيرٌ, (A,) using the latter word as an imitative sequent, (Ṣ and Ḳ in art. نقر,) or as a corroborative. (TA.)

حَقَارَةٌ (Ṣ, A, Mṣb, Ḳ) and ▼حُقْرٌ and ▼حُقْرِيَّةٌ, (Ḳ,) all of which are inf. ns., (TA,) and ▼حُقَارَةٌ and ▼حِقَارَةٌ (Ḳ) and ▼مَحْقَرَةٌ, (Ṣ, Ḳ,) *Contemptibleness, despicableness, meanness, paltriness, abjectness, ignominiousness, baseness*, or *vileness:* (Ṣ, A, Mṣb, Ḳ:) [or the last rather signifies a *cause of contemptibleness* &c.; being similar to مَجْبَنَةٌ and مَبْخَلَةٌ &c.:] you say, هٰذَا الأَمْرُ مَحْقَرَةٌ ▼ بِكَ i. e. حَقَارَةٌ [or rather *This thing is a cause of contemptibleness* &c. *to thee*]. (Ṣ.)

حُقَارَةٌ: } see حَقَارَةٌ.
حِقَارَةٌ: }

حَاقِرٌ act. part. n. of حَقَرَهُ, *Contemning, despising,* &c. (A.) You say, هُوَ حَاقِرٌ نَاقِرٌ [using the last word as an imitative sequent, or a corroborative: see حَقِيرٌ]. (A.)

حَيْقُرٌ: } see حَقِيرٌ.
حَيْقَرٌ: }

الحَاقُورَةُ *The Fourth Heaven.* (Ḳ.)

مَحْقَرَةٌ: see حَقَارَةٌ, in two places.

مُحَقَّرَاتٌ *Small sins;* syn. صَغَائِرُ: (Ṣ, Ḳ:) but this is an application proper to the law; for the [pagan] Arabs knew not صَغَائِر nor كَبَائِر: or, accord. to the writers on strange words, it signifies *actions which a man contemns,* or *despises,* or *holds in light estimation, even if great sins.* (MF.)

حقط

حَيْقُطٌ: see what follows.

حَيْقُطَانٌ, (IDrd, Ṣ, Ḳ,) and حَيْقَطَانٌ, but the former is the more chaste, (IDrd,) and the latter is mentioned by none but IDrd, (IKh,) and ▼ حَيْقُطٌ, (Ḳ,) The [bird called] دُرَّاج [i. e. *attagen, francolin, heath-cock,* or *rail:* but see this last Arabic word]: (Ḳ:) or the *male of the* دُرَّاج: (Ṣ, Ḳ:) but IF says, I do not think it correct: (TA:) fem. حَيْقُطَانَةٌ. (Ḳ.)

حقف

1. حَقَفَ, aor. ـِ, inf. n. حُقُوفٌ, *It* (a thing) *was,* or *became, curved, bent,* or *winding.* (Mṣb.) حُقُوفٌ also signifies A gazelle's *being in the condition denoted by* [the part. n.] حَاقِفٌ *in the phrase* ظَبْيٌ حَاقِفٌ, which see below. (Ibn-'Abbád, Ḳ.)

12. احقوقف, said of sand, and of the هِلَال [or moon when, being near the sun, it shows a narrow rim of light], (Ṣ, Ḳ,) and of the back, (Ḳ,) or of anything, such as the back of a camel, (L,) *It was curved, bent,* or *winding,* (Ṣ, L, Ḳ,) *and long.* (L, Ḳ.)

حِقْفٌ *A curving,* or *winding, tract of sand:* (Ṣ, Mgh, Mṣb, Ḳ:) pl. [of pauc.] أَحْقَافٌ (Ṣ, Mṣb, Ḳ) and [of mult.] حِقَافٌ (Ṣ, Ḳ) and حُقُوفٌ, (O, L, Ḳ,) and pl. pl. حَقَائِفُ and حِقَفَةٌ, (Ḳ,) the former of these two being pl. of أَحْقَافٌ or of حِقَافٌ, (L, TA,) but the latter of them, accord. to the Ó, is a pl., not a pl. pl.: (TA:) or *a heap,* or *hill, of sand, that has assumed a bowed form:* (IDrd, TA:) or *a curving,* or *winding, thin tract of sand:* (Ḥar p. 51:) or *a great, round tract of sand:* (Ibn-'Arafeh, Ḳ:) or *an oblong, elevated tract of sand:* (Fr, Ḳ:) or الأَحْقَافُ applies [particularly] to *certain oblong tracts of sand in the region of Esh-Shihr:* (Ḳ,* TA:) accord. to J, [in the Ṣ,] to *the country of 'Ád.* (TA.) —— Also The *lower,* or *lowest, part of a tract of sand,* and *of a mountain,* and *of a wall.* (O, L, Ḳ.)

حَاقِفٌ A gazelle [*lying*] *bent together,* or *curled,* (A 'Obeyd, Ṣ, Mgh, Mṣb, Ḳ,) *like a* حِقْف *of sand,* (Az, Ṣgh, Ḳ,) *in his sleep,* (A 'Obeyd, Ṣ, Ḳ,) or *in consequence of a wound &c.:* (Mṣb:) or *lying upon his breast on a* حِقْف *of sand:* (IAạr, Ḳ:) or, as some say, *upon,* or *at, the lower,* or *lowest, part of a* حِقْف. (Mgh.)

أَحْقَفُ A camel *lank in the belly.* (ISh, Ḳ.)

حقل

3. مُحَاقَلَةٌ [inf. n. of حَاقَلَ] The *selling of seed-produce in the ear for the grain of wheat:* (Ṣ, M, Mgh, Ṣgh, Mṣb, Ḳ:) or the *selling of seed-produce while in growth, before it appears to be in a good state:* or the *making a bargain,* or *contract, with another, for labour upon land, on the condition of his receiving a third,* or *a fourth,* or *less,* or *more, of the produce:* or the *hiring of land for the grain of wheat:* (M, Mgh, Ṣgh, Ḳ:) or the *buying of seed-produce while in growth for the grain of wheat.* (Mgh.)

4. احقل *It* (seed-produce) *became such as is termed* حَقْل, q. v.: (Ṣ, Ḳ:) or *put forth its head.* (Mgh.) —— احقلت الأَرْضُ *The land became in the condition of having what is termed* حَقْل. (Ḳ,* TA.)

Q. Q. 1. حَوْقَلَ, inf. n. حَوْقَلَةٌ and حِيقَالٌ, for which latter some say حَوْقَالٌ, *He* (an old man) *became aged, and languid in respect of the venereal faculty:* (Ṣ:) or حَوْقَلَةٌ signifies the *lacking ability to exercise the venereal faculty,* (Ḳ,) accord. to Az, *on the occasion of one's having his bride brought to him.* (TA.) —— حَوْقَلَةٌ also signifies The *being weary, and weak.* (Ḳ.) —— The state of *sleeping.* (Ḳ.) —— The act, or state, of *retrograding,* or *declining;* syn. إِدْبَارٌ. (Ḳ.) —— An old man's *resting,* or *staying, his hands upon his waist.* (Ḳ.) —— The *walking quickly, and with short steps.* (Ḳ.) ═══ And حَوْقَلَهُ, (TA,) inf. n. حَوْقَلَةٌ, (Ḳ, TA,) *He impelled,* or *repelled, him,* or *it;* syn. دَفَعَهُ. (Ḳ,* TA.) ═══ حَوْقَلَةٌ [inf. n. of حَوْقَلَ] also signifies The *saying* لَا حَوْلَ وَلَا قُوَّةَ إِلَّا بِاللّٰهِ; (TA in art. حول;) i. q. حَوْلَقَةٌ. (Ḳ in that art.)

حَقْلٌ *Seed-produce when its leaves have branched forth,* (Ṣ, M, Mṣb, Ḳ,) *and become apparent and numerous,* (M, Ḳ,) *before its stalks have become thick:* (Ṣ:) or *when its shoots have come forth near together:* or *as long as it is green:* (M, Ḳ:) or *of which the head has come forth.* (Lth, L in art. فرخ.) —— And *Land such as is termed* قَرَاح; i. e. *land in which are no trees:* (Mṣb:) or *good* قَرَاح: n. un. with ة: (Ṣ:) or *a good* قَرَاح *in which one sows;* as also with ة: (Ḳ:) or *a place that has never been sown;* and so, accord. to some, with ة: (TA:) whence, (Ḳ,) it is said in a prov., (Ṣ,) لَا تُنْبِتُ البَقْلَةَ إِلَّا الحَقْلَةُ [*Nothing but the good seed-plot produces the herb*]: (Ṣ, Ḳ:) said to be applied to a base saying proceding from a base man: [or it means, as the father, so is the son: (Freytag's Arab. Prov. ii. 516:)] but ISd says that الحقلة is not known; and he thinks that the ة is added in order to make it accord with البقلة; or that it means *a portion of what is termed* حَقْل: (TA:) the pl. is حُقُولٌ. (Mṣb.)

حَاقِلٌ *A tiller,* or *cultivator, of land.* (TA.)

حَوْقَلٌ An *old* man, *who is languid in respect of the venereal faculty:* or an *aged* man, absolutely: and a man who is *weary.* (TA.) —— See also what next follows.

حَوْقَلَةٌ *A soft,* or *flaccid, penis;* (Ṣ, Ḳ;) as also ▼ حَوْقَلٌ: (TA:) or the latter signifies [simply] *a penis:* (Ḳ:) the former, accord. to Abu-l-Ghowth, signifies the *penis of an old man who is languid in respect of the venereal faculty:* some of those of post-classical times pronounce it with ف. (Ṣ.) —— *A flask,* or *bottle,* (قَارُورَةٌ,) *with a long neck, used by the water-carrier:* (Ḳ:) app. formed by substitution [of ق for ج] from حَوْجَلَةٌ. (TA.)

مَحْقَلَةٌ sing. of مَحَاقِلُ, (TḲ,) which signifies *Places of seed-produce;* syn. مَزَارِعُ. (Ḳ: but in some copies, in the place of المَحَاقِلُ and المَزَارِعُ, we find المَحَاقِلُ; and المَزَارِعُ.) [See an ex. voce مَحْقَنَةٌ.]

حقن

1. حَقَنَهُ, aor. ـُ and ـِ, (Ḳ,) inf. n. حَقْنٌ, (TA,) i. q. حَبَسَهُ [as meaning *He confined it; kept it in; prevented it from escape; retained, restrained,* or *withheld, it*]; (Ḳ;) as also ▼ احتقنه, (as in some copies of the Ḳ,) or ▼ احقنه; (as in other copies and in the TA;) but see, in what follows, what is said of this last in the Ṣ. (TA.) —— حَقَنَ اللَّبَنَ (Ṣ, Mgh, Ḳ) فِى السِّقَاءِ, (Ḳ,) aor. ـُ, (Ṣ,) inf. n. as above, (TA,) *He collected the milk in the skin,* (Ṣ, Mgh,) *and poured fresh milk upon that which was curdled,* or *thick,* or *upon that which was churned:* (Ṣ:) or *he poured the milk into the skin,* [*and kept it therein*] *that its butter might come forth.* (Ḳ.) And حَقَنَ المَآءَ فِى السِّقَاءِ, aor. ـُ, inf. n. as above, *He collected the water in the skin.* (Mṣb.) —— حَقَنَ البَوْلَ, (Ks, Ṣ, M,) or بَوْلَهُ, (Mgh, Mṣb,) *He kept in,* or *retained,* (M, Mgh, Mṣb,) *and collected,* (Mgh, Mṣb,) *the urine,* (M,) or *his urine:* (Mgh, Mṣb:) one should not say ▼ احقنه; (Ks, Ṣ, M;) nor should one say [of the urine] حَقَنَنِى هُوَ. (M.) —— حَقَنَ دَمَهُ, (Ṣ, Mgh, Mṣb, Ḳ,) and حَقَنَ لَهُ دَمَهُ, (TA from a trad.,) † *He prevented,* or *forbade, the shedding of his blood,* (Ṣ, Mgh, TA,) and *the slaying him;* (TA;) [*he spared his blood,* or *forbore to shed it;*] i. e., (Mgh,) *he saved him* (Mgh, Ḳ) *from slaughter* (Ḳ) *when it had become lawful to slay him;* from حَقَنَ اللَّبَنَ; (Mgh;) contr. of هَدَرَهُ; as though he collected his blood in him, and did not pour it forth. (Mṣb.) —— حَقَنَ مَآءَ وَجْهِهِ † *He preserved* [*the lustre of his face;* meaning his

honour, or reputation]. (TA.) — حَقَنَ المَرِيضَ (Ṣ,* Mgh, Mṣb, Ḳ*) He administered to the sick person what is termed حُقْنَة, i. e. [a clyster,] a medicine put into a مِحْقَنَة; (Mgh;) he conveyed medicine into the inside of the sick person by his anus (مِنْ مَخْرَجِهِ) with the مِحْقَنَة. (Mṣb.) See also 8.

4. احقن He collected different sorts of milk [in a skin, old and fresh,] to become good. (Ḳ.) — See also 1, in two places.

5. تَحَقَّنَتِ الإِبِلُ The camels became full in their insides. (TA.)

8. احتقن as a trans. v.: see 1. — Also It (blood) collected in the inside in consequence of a spear-wound, or stab, or the like, penetrating thereinto. (TA.) — And He (a man) administered to himself, or had administered to him, a حُقْنَة [or clyster]: (Ṣ,* Mgh, Mṣb:) or he (a sick man) made use of the حُقْنَة in consequence of suppression of his urine. (Ḳ.) The saying اِحْتَقَنَ الصَّبِىُّ بِلَبَنِ أُمِّهِ [meaning The child had its mother's milk administered to it as a clyster] is far-fetched: and أُحْتُقِنَ is not allowable: the right expression is ↓ حُقِنَ, or عُولِجَ بِالحُقْنَةِ. (Mgh.) — And اِحْتَقَنَتِ الرَّوْضَةُ The روضة [i. e. meadow, or garden,] had its sides elevated عَلَى سَائِرِهَا [above the rest of it]: so says AḤn: in the Ḳ, على سَرَارِهَا [above the depressed, or the best, or most fruitful, part of it: in the CḲ, سِرارِها]. (TA.)

حَقِنٌ: see حَاقِنٌ.

حُقْنَةٌ A pain in the belly: pl. أَحْقَانٌ. (IAạr, Ḳ.)

حُقْنَةٌ The administration of a medicine to a sick person by his anus; (TA;) [i. e. the administration of a clyster;] the conveyance of a medicine to the inside of a sick person by his anus with the مِحْقَنَة. (Mṣb.) — And hence, (Mṣb,) [A clyster;] a medicine so administered to a sick person: (Ṣ, Mgh, Mṣb, Ḳ:) pl. حُقَنٌ. (Mṣb.) — And, by extension of the meaning, The tube of a مِحْقَنَة. (Mgh.)

حَقِينٌ Confined, kept in, prevented from escape, retained, restrained, or withheld; as also ↓ مَحْقُونٌ. (Ḳ.) — And, as a subst., (Ṣ,) Milk collected in a skin, (Ṣ, IF, Mṣb,) when fresh milk has been poured upon that which has become curdled, or thick, or upon that which has been churned. (Ṣ, TA.) Hence the prov., أَبَى الحَقِينُ العِذْرَةَ; i. e. العُذْرَ; [The milk collected in the skin, &c., has disallowed the excuse;] (A 'Obeyd, Ṣ, TA;) applied to him who excuses himself when having no [real] excuse: (TA:) originally said by a man who asked some people to give him milk to drink, upon their excusing themselves while they had milk which they had collected in a skin; meaning, this حقين belies you. (A'Obeyd, TA.)

حَاقِنٌ A man keeping in, or retaining, and collecting, his urine; from حَقِينٌ meaning "milk collected in a skin:" (IF, Mṣb:) [suffering from retention of the urine:] having urine that distresses [by its quantity]: (Ṣ, TA:) having much urine retained and collected: (Mgh:) and ↓ حَقِنٌ signifies the same. (TA.) Hence the saying, (Mgh, TA,) in a trad., (TA,) لَا رَأْىَ لِحَاقِنٍ وَلَا حَاقِبٍ وَلَا حَازِقٍ (Ṣ,* Mgh, TA*) No counsel, or advice, is possible to one who has much urine retained and collected, nor to one suffering suppression of the feces, nor to one who is pinched by a tight boot. (Mgh, TA.*) — [Collecting grease, or melted grease, in a skin.] You say, أَنَا مِنْهُ كَحَاقِنِ الإِهَالَةِ † I am a person skilled in it; because such a one does not collect the اهالة in a skin until he knows that it has cooled, lest the skin should be burnt. (Ḳ.) — † A moon a little after or before the change (هلال) having its two extremities elevated, and its back decumbent. (Ḳ.) Hence the saying, هِلَالٌ وَاقِفٌ خَيْرٌ مِنْ هِلَالٍ حَاقِنٍ ‡ [A هلال erect (here meaning nearly erect) is better, or more auspicious, than a هلال decumbent]. (A, TA.) [See also أَدْفَقُ.]

الحَاقِنَةُ The stomach; (Ḳ;) an epithet in which the quality of a subst. is predominant; because it retains, or collects, the food: (TA:) and the pit between each collar-bone and what is termed حَبْلُ العَاتِقِ [explained in art. حبل, q. v.]; (AA, Ṣ, Ḳ;*) the two together being termed [the] حَاقِنَتَانِ: (AA, Ṣ:) the pit of each collar-bone: (T, TA:) the part between the collar-bone and the neck: (TA:) or the lower part of the belly: (Ṣ, Ḳ:) pl. الحَوَاقِنُ. (T, TA.) [See also الذَّاقِنَةُ.] Hence, (Ḳ,) it is said in a prov., لَأُلْحِقَنَّ حَوَاقِنَكَ بِذَوَاقِنِكَ, (Ṣ, Ḳ,) i. e. [I will assuredly conjoin] the lower part of thy belly with the upper part thereof: or the part of thy belly that retains, or collects, the food, with the lower part of thy belly, and with thy knees. (TA.) The ذَاقِنَة is [also said to be] the extremity of the حُلْقُوم [or windpipe]: and hence the saying attributed to 'Áisheh, that the Prophet died [with his head] between her حَاقِنَة and her ذَاقِنَة. (Ṣ.)

مِحْقَنٌ A skin in which milk is collected; fresh milk being poured into it upon that which is curdled, or thick, or upon that which has been churned: (Ṣ:) or a skin into which milk is poured, [and in which it is kept,] that its butter may come forth. (Ḳ.) — And A funnel with which the milk is poured into the skin: (Az, Ḳ:*) or a funnel which is put into the mouth of a skin, into which one then pours wine, and water. (TA.)

[مَحْقَنَةٌ app. sing. of مَحَاقِنُ as used in the following saying; like as مَحْقَلَة is sing. of مَحَاقِل.] بَارَكَ ٱللّٰهُ فِى مَحَاقِلِكُمْ وَمَحَاقِنِكُمْ May God bless your seed-produce and your progeny. (TA.)

مِحْقَنَةٌ The thing with which a حُقْنَة [or clyster] is administered; (Mgh, Mṣb;) being a leathern pouch, furnished with a tube. (Mgh.)

مِحْقَانٌ One who retains his urine, and, when he discharges it, discharges much: (Ṣ, Ḳ:) applied by ISd peculiarly to a camel. (TA.)

مَحْقُونٌ: see حَقِينٌ.

حقو

1. حَقَاهُ, [aor. يَحْقُوهُ,] inf. n. حَقْوٌ, He, or it, hit, or hurt, his حَقْو [i. e. waist, or flank]. (Ḳ.) — Also, said of water, It reached up to his حَقْو. (Fr, TA.) — حُقِىَ He (a man, Ṣ) was, or became, affected with the pain of the belly termed حَقْوَة: (Ṣ, Ḳ:) and, inf. n. حَقًا, [which, as well as the part. n. حَقٍ, suggests that the verb is حَقِىَ, like فَرِحَ, though it is said in the CḲ to be like عُنِىَ,] he had a complaint of his حَقْو: (CḲ, but wanting in MS. copies of the Ḳ:) or ↓ تحقّى has this latter meaning. (So in a copy of the Ḳ.)

5: see above.

حَقٍ Having a complaint of his حَقْو. (CḲ, but wanting in MS. copies of the Ḳ.)

حَقْوٌ The waist; syn. خَصْرٌ: (Ṣ:) [or] the place (in the side, TA) where the [waist-wrapper called] إِزَار is bound; (Ṣ, Mṣb, Ḳ,* TA;) i. e. (Mṣb) the flank; syn. خَاصِرَةٌ, (A 'Obeyd, Mṣb, TA,) or كَشْحٌ: (Ḳ:) there are two [parts], called together حَقْوَانِ: (A'Obeyd, TA:) and the [waist-wrapper itself that is called] إِزَار; (Ṣ, M, Mṣb, Ḳ;) because it is bound upon the [part of the body called] حَقْو; (Mṣb,* TA;) as also ↓ حِقْوٌ; (M, Ḳ;) which MF asserts to be also a var. of حَقْو as syn. with كَشْح; of the dial. of Hudheyl: and in the Ḳ is here added, or the place of the tying thereof; as also ↓ حَقْوَةٌ and ↓ حِقَاءٌ; [the latter written in the CḲ حَقاء; but said in the TA to be like كِتَاب;] whereas this is the primary signification: (TA:) pl. (of pauc., Ṣ, TA) أَحْقٍ, (Ṣ, Mṣb, Ḳ,) originally أَحْقُوٌ, (Ṣ, TA,) and أَحْقَآءٌ (Ḳ) and (of mult., Ṣ, TA) حُقِىٌّ, (Ṣ, Mṣb, Ḳ,) [originally حُقُوٌّ,] of the measure فُعُولٌ, (Ṣ,) like فُلُوسٌ, (Mṣb,) and حِقَآءٌ, (Mṣb, Ḳ,) which is pl. of حَقْو and of حَقْوَة, as well as syn. with the latter in a sense pointed out above. (TA.) Hence, عَاذَ بِحَقْوِهِ ‡ He had recourse to him for refuge, protection, or preservation. (TA.) [See an ex. in a verse cited voce سَمْعٌ.] — ‡ The part of an arrow which is the place of the feathers: (Ḳ:) or the slender part of the hinder portion, next the feathers; (Ṣ, TA;) or, as in the A, below the feathers. (TA.) — ‡ A rugged place elevated above a torrent; (M, Ḳ, TA;) pl. حِقَآءٌ: (Ḳ:) or † any place which a water-course reaches: (Aṣ, TA:) and ↓ حَقْوَةٌ signifies † the like of the [elevated piece of ground termed] نَجْوَة, but higher, upon which the animals of prey preserve themselves from the torrent; and its pl. is حِقَآءٌ. (TA.) Accord. to En-Nadr, † An elevated piece of ground; an acclivity; pl. أَحْقٍ and حُقِىٌّ: accord. to Z, † the سَفْح [or lowest part, or face, &c.,] of a mountain: (TA:) also ‡ [each of] the two sides of a ثَنِيَّة [or long mountain traversed by a road; or a part of a mountain that requires one, in traversing it, to ascend and descend; &c.]: (Ḳ, TA:) Lth says that when you look at the head of a ثَنِيَّة of a mountain, you see its prominence to have what are termed حَقْوَانِ. (TA.)

حَقْوٌ : see حِقْوٌ.

حَقْوَةٌ : see حَقْوٌ, in two places. ▬ Also *A pain of the belly,* (Ṣ, M, Ḳ,) *which affects a man,* (Ṣ, M,) *from eating flesh-meat,* (M, Ḳ,) *occasioning diarrhœa;* (M;) or *occasioning an inflation in the* حَقْوَانِ [or *two flanks*]; (T;) and so ♦ حِقَآءٌ : (M, Ḳ:) or *i. q.* هَيْضَةٌ [generally meaning *cholera*]. (TA in art. طسأ.) — And *A certain malady in camels, in consequence of which the belly is rent by the* [*affection of the lungs termed*] نُحَاز, [*which occasions violent coughing,*] (Ḳ, TA,) *and the animal voids not the urine nor dung,* (Ḳ in art. ققأ,) *often, also, having the veins and flesh choked with blood, and becoming swollen,* or *inflated, often to such a degree that the stomach bursts in consequence thereof.* (TA in that art.) The word is mostly used in relation to a human being. (TA.)

حِقَآءٌ : see حَقْوٌ, with which it is syn. in one sense pointed out above; and of which it is also a pl., as well as of حَقْوَةٌ. — Also The *cord,* or *the like, with which the horse-cloth is bound upon the belly of the horse when he is made to run a heat or two heats and then covered over to make him sweat and to reduce his fat,* [see 1 in art. حنذ,] *by way of preparing him for racing or the like.* (TA.) ▬ See also حَقْوَةٌ.

مَحْقُوٌّ (Ṣ, Ḳ) and مَحْقِىٌّ (Ḳ) applied to a man, (Ṣ,) *Affected with the pain of the belly termed* حَقْوَةٌ. (Ṣ, Ḳ.) And *Having a complaint of his* حَقْو. (CḲ, but wanting in MS. copies of the Ḳ.)

حك

1. حَكَّهُ, aor. ُ, (Ṣ, Mṣb,) inf. n. حَكٌّ, (Ṣ, Mgh, Mṣb, Ḳ,) [*He scratched, scraped, rubbed, grated, chafed,* or *fretted, it:* or] *he scraped off, abraded,* or *otherwise removed, its superficial part:* (Mgh, Mṣb:) حَكٌّ signifies the act of *scratching:* (KL:) or the *making a body to pass upon another body with collision:* (Ḳ:) [as meaning scratching and the like,] it is with the nail, and with the hand, &c. (TA.) مَا حَكَّ ظَهْرِى مِثْلُ يَدِى [*Nothing has scratched my back like my hand*] is a prov., meaning that one should abstain from relying upon others: and the same meaning is intended in the following verse:

مَا حَكَّ جِلْدَكَ مِثْلُ ظُفْرِكَ *

فَتَوَلَّ أَنْتَ جَمِيعَ أَمْرِكَ *

[*Nothing has scratched thy skin like thy nail: so manage thou thyself all thine affair*]. (Ḥar pp. 432 et seq.) The saying, in a trad., إِذَا حَكَكْتُ قُرْحَةً دَمَّيْتُهَا [lit. *When I scratch a sore, I make it bleed,*] means ‡ *when I desire an object, I attain it.* (TA.) — [Hence,] حَكَّ فِى صَدْرِى, and ♦ احكّ, and ♦ احتكّ, (Ḳ,) the first whereof, which is mentioned by IDrd preceded by the negative مَا, is the most approved, (TA,) ‡ *It wrought,* or *operated, in,* or *upon, my mind:* (Ḳ, TA:) said of a suggestion of the devil, that comes into one's mind. (TA.) Or حَكَّ فِى صَدْرِهِ كَذَا, aor. ُ, means † *Such a thing occurred to his mind as a thing outweighed in probability,* or *a matter of suspicion.* (Mṣb.) And you say مَا حَكَّ فِى صَدْرِى † *It did not make an impression upon my mind.* (Ḥar p. 648.) It is said in a trad., الإِثْمُ مَا حَكَّ فِى صَدْرِكَ † *Sin is that which makes an impression upon thy mind, and induces a suspicion that it is an act of disobedience, because the mind is not dilated thereby.* (Mgh. [See also حَاكَ, in arts. حوك and حيك; and see حَزَّ.]) You say also, مَا حَكَّ فِى صَدْرِى مِنْهُ شَىْءٌ ‡ *Nothing thereof was unsettled, so as to be doubtful, in my mind.* (Ṣ, TA.) And مَا حَكَّ فِى صَدْرِى كَذَا ‡ *Such a thing did not cause dilatation* [or *pleasure*] *in my mind.* (Ṣ, Ḳ, TA.) ▬ See also 8. ▬ حَكِكَتِ الدَّابَّةُ, aor. َ, (Kr, Ḳ,) a verb of an unusual form, with the reduplication distinct, like لَحِحَتْ in the phrase لَحِحَتْ عَيْنُهُ, &c., (TA,) *The beast had its hoof worn away at the edges.* (Ḳ,* TA.)

2. حكّك, inf. n. تَحْكِيكٌ, *He scratched* [&c.] *well* [or *much*]. (KL.)

3. حَاكَّهُ, (TA,) inf. n. مُحَاكَّةٌ (Ṣ, Ḳ, KL) and حِكَاكٌ, (TA,) ‡ *He emulated, rivalled,* or *imitated, him;* [originally, I suppose, *in scratching,* or *the like;*] (Ḳ, KL, TA;) the inf. n. being syn. with مُبَارَاةٌ; (Ḳ, TA;) or like مُبَارَاةٌ. (Ṣ.) — حَاكَّ الشَّرَّ (Ḳ) ‡ *He produced,* or *effected,* or *brought to pass, evil,* or *mischief.* (TḲ.)

4: see 8: — and see also 1.

5. فُلَانٌ يَتَحَكَّكُ بِى *Such a one rubs,* or *scratches, himself against me;* syn. يَتَمَرَّسُ بِى: (Ṣ: so in two copies:) or ‡ *becomes exasperated by me;* syn. يَتَحَرَّشُ بِى: (TA:) *and addresses,* or *applies, himself to do evil,* or *mischief, to me.* (Ṣ, Ḳ, TA.) لَقَدْ تَحَكَّكَتِ العَقْرَبُ بِالأَفْعَى † *The scorpion has addressed itself to do evil,* or *mischief, to the viper,* is a prov., applied to him who contends with his superior in strength and power, and does evil to him. (Ḥar p. 478.)

6. تَحَاكَّا [*They scratched, scraped, rubbed, grated, chafed,* or *fretted, each other;* or] *their two bodies became in collision, and each of them scratched, &c.,* (حَكَّ,) *the other.* (Ḳ.) — هٰذَا أَمْرٌ تَحَاكَّتْ فِيهِ الرُّكَبُ, and ♦ احتكّت, ‡ *This is a case in which the knees are in contact,* and *in collision,* is a saying by which is meant equality of station or rank, or the sitting together upon the knees in contending for superiority in glory or excellence or nobility. (TA.) — [تحاكّ also signifies *It became scraped off,* or *rubbed off, by degrees;* the verb in this sense being similar to تساقط &c.: see حُكَاكَةٌ.]

8. احتكّ بِهِ *He scratched, scraped,* or *rubbed, himself* (حَكَّ نَفْسَهُ) *against it;* (Ṣ, Ḳ;) as the mangy or scabby [camel] does against a piece of wood. (TA.) — احتكّ رَأْسِى *My head induced me,* or *caused me, to scratch it;* (دَعَانِى إِلَى حَكِّهِ;) [i. e. *it itched;*] as also ♦ اِسْتَحَكَّنِى and ♦ أَحَكَّنِى and ♦ حَكَّنِى; (Ḳ;) though this last is held by IB to be erroneous: (TA:) and in like manner one says of all the other members. (M, TA.) — See also 1: — and 6. — احتكّ حَافِرُهُ مِنْ كَثْرَةِ السَّيْرِ [*His hoof became chafed, abraded,* or *worn, by much travel*]. (Ḥam p. 476.)

10: see 8.

حِكَّةٌ ‡ *Doubt* (Ḳ, TA) in religion &c.; (TA;) as also ♦ حُكَّةٌ: (AA, TA:) because it makes an impression (يَحُكُّ) upon the mind. (TA.) ▬ حِكُّ شَرٍّ, explained in the Ḳ, as also ♦ حِكَاكُ شَرٍّ, by the words يُحَاكُّهُ كَثِيرًا, means ‡ *A producer of much evil,* or *mischief:* (TḲ:) it is a tropical phrase: and in like manner one says حِكُّ ضِغْنٍ ‡ [*a producer of much rancour, malevolence, malice,* or *spite*]: and حِكُّ مَالٍ ‡ [*a producer of much wealth*]. (TA.)

حِكَّةٌ [*An itching;*] a subst. from اِحْتَكَّ as used in the phrase احتكّ رَأْسِى [q. v.]; as also ♦ حُكَاكٌ. (Ḳ.) — And The جَرَب [i. e. *mange,* or *scab*]: (Ṣ, Ḳ:) or it differs from the latter; and is said to be the *dry* جَرَب: (MF:) or *anything that one scratches; as the* جَرَب *and the like:* (Mgh:) [in the present day particularly applied to the *itch:*] *a certain cutaneous disease;* said in the medical books to be *a thin humour, causing swelling, originating beneath the skin, not accompanied with pus, but with what resembles bran, and quick in passing away.* (Mṣb.) — And hence † *Lice.* (Mgh.) ▬ See also حِكٌّ.

حَكَكٌ *A wearing away at the edges* in a beast's hoof. (Ḳ,* TA.) ▬ *A gait in which is commotion, like the gait of a short woman who moves about her shoulder-joints.* (Ibn-'Abbád, L, Ḳ.) ▬ *Soft,* or *uncompact, white stones:* (Ṣ:) or *a kind of white stone, like marble,* (Ḳ, TA,) *more soft,* or *uncompact, than marble, but harder than gypsum:* n. un. with ة: (TA:) or, with ة, *ground in which are soft,* or *uncompact, stones, like marble:* (ISh, TA:) or, accord. to ADḳ, ♦ حُكَكَاتٌ, with ḍamm, and then fet-ḥ, signifies *ground in which are white stones, resembling* أَقِط, *that break into many pieces; and such is only in low land,* (TA.)

حُكُكٌ ‡ *Evil,* or *mischievous, persons.* (IAạr, Ḳ, TA.) — And ‡ *Such as are importunate in demanding things wanted.* (IAạr, Ḳ, TA.)

حُكَكَاتٌ : see حَكَكٌ.

حُكَاكٌ *A thing that is rubbed,* or *grated,* (حُكَّ,) *upon another thing, so as to produce* حُكَاكَةٌ. (IDrd, TA.) — *I. q.* بُورَقٌ [q. v.]. (Ṣgh, Ḳ.) ▬ See also حِكَّةٌ.

حِكَاكٌ [*A thing against which a beast rubs,* or *scratches, himself*]. The Arabs say, فُلَانٌ جِذْلُ حِكَاكٍ خُشِّعَتْ عَنْهُ الأُبَنُ † [*Such a one is a rubbing-post from which the knots have become worn down*]; meaning that he is so pruned, or trimmed, [figuratively speaking,] that nothing is cast at him but it glances off from him, and recoils. (TA.) [See مُحَكَّكٌ.] ▬ حِكَاكُ شَرٍّ : see حِكٌّ.

حَكِيكٌ i. q. ▼ مَحْكُوكٌ [i. e. *Scratched, scraped, rubbed,* &c.; and particularly *worn by rubbing* or *friction;*] applied to a كَعْب [app. as meaning an ankle-bone, or rather the skin upon that bone]: and *having the edges worn away;* syn. نَحِيتٌ, (Ṣ,) or مَنْحُوتٌ; (Ḳ;) applied to a solid hoof; (Ṣ, Ḳ, TA;) as also ▼ أَحَكُّ: (Ḳ, TA:) and كُلُّ نَحِيتٍ خَفِىٍّ [so in copies of the Ḳ: in the CḲ the last word in this explanation is خَفِىٌّ: but I doubt not that the right reading is حَفِىٍّ, with the unpointed ح; and that the meaning of the whole is, whatever (i. e. whatever foot) is *worn by rubbing* or *friction; that has become attenuated,* or *chafed, by much walking* or *treading;* agreeably with the explanation that follows]: the subst. is حَكَكٌ: and you say, حَكِكَتِ الدَّابَّةُ. (Ḳ.) And A horse *having the hoofs much worn* (مُنْحَتُ الحَوَافِرِ, IDrd, Ḳ, in the CḲ الحَافِرِ) *by the erosion of the ground, so as to be attenuated.* (IDrd, TA.)

حُكَاكَةٌ *What falls from a thing* عِنْدَ الحَكِّ [i. e. *on the occasion of scratching, scraping, rubbing, grating,* &c.]. (Ṣ, Ḳ.) And *What is scraped,* or *rubbed,* or *grated,* (مَا حُكَّ,) *between two stones, and then used as a collyrium for ophthalmia:* (Ḳ:) or *what is scraped off,* or *rubbed off, by degrees,* (مَا تَحَاكَّ,) *between two stones, when one of them is rubbed with the other, for medicine and the like.* (TA.)

[حَكَّاكٌ *A lapidary.*]

حَكَّاكَةٌ ‡ *A thing that makes an impression upon hearts:* pl. حَكَّاكَاتٌ: (IAth, TA:) or the pl. signifies ‡ [*suggestions of the devil* or *of the mind, whereby the mind is disturbed; such as are termed*] وَسَاوِسُ: (Ḳ, TA:) *things that make an impression* (تَحُكُّ) *upon the heart, and are dubious to a man:* such are sins said to be. (TA.)

حَاكَّةٌ *A tooth:* (Ṣ, Ḳ:) thus called because it rubs, or grates, (تَحُكُّ,) either its fellow or what one eats: an epithet in which the quality of a subst. predominates. (TA.) So in the saying, مَا بَقِيَتْ فِى فِيهِ حَاكَّةٌ [*There remained not in his mouth a tooth*]. (Ṣ.) The Arabs also say, مَا فِيهِ حَاكَّةٌ وَلَا تَاكَّةٌ, meaning *There is not in him,* or *it, a grinder* (ضِرْسٌ) *nor a dog-tooth.* (Aboo-'Amr Ibn-El-'Alà, TA.)

أَحَكُّ: see حَكِيكٌ. — Also A man (TA) *having no* حَاكَّة, i. e., *no tooth, in his mouth.* (Ḳ,* TA.)

مَا أَنْتَ مِنْ أَحْكَاكِهِ *Thou art not of his,* or *its, men:* (Ibn-'Abbád, Ḳ:) [app. meaning thou art not the man to cope with him, or to accomplish it.]

مِحَكٌّ [*A touchstone; the stone upon which pieces of money* &c. *are rubbed to try their quality;*] the *stone of the* نَقَّادُون. (Ḥar p. 66.) [It is commonly called in the present day مِحَكَّة: which also signifies *a stone for rubbing the soles of the feet,* &c.: and *a rasp.*]

الجِذْلُ المُحَكَّكُ [*The rubbing-post;* i. e.] *the thing that is set up in the place where camels lie down, at their watering-place, for the mangy camels to rub against it.* (Ṣ, Ḳ.) Hence the saying of El-Ḥobáb Ibn-El-Mundhir El-Anṣáree, (Ṣ,) أَنَا جُذَيْلُهَا المُحَكَّكُ وَعُذَيْقُهَا المُرَجَّبُ [see جِذْلٌ,] meaning I am he by means of whose counsel, or advice, and forecast, relief is sought: (Ṣ, Ḳ:*) or it has another meaning, preferred by Az, i. e., that the sayer was one who had been strengthened by experience, who had experienced and known affairs, and been tried, or proved, by them, and found to be one who bore up against difficulty, strong and firm, such as would not flee from his adversary: or the meaning is, I am, exclusively of [the rest of] the Anṣár, a rubbing-post for him who would oppose me, and with me should the stubborn be coupled: the dim. form is here used for the purpose of aggrandizement. (TA.) [See also حِكَاكٌ.]

مَحْكُوكٌ : see حَكِيكٌ.

حكأ

1. حَكَأَ : see what next follows.

4. احكأ, (Sh, Ṣ, Ḳ,) inf. n. إِحْكَآءٌ; (TA;) and ▼ حَكَأَ, aor. ـَ , (Ḳ,) inf. n. حَكْءٌ; (TA;) and ▼ احتكأ; (Ḳ;) *He tightened* a knot; (Ṣ, Ḳ;) *made* it *firm:* (Sh, TA:) as also احكى (Ṣ) and حَكَى. (Ṣ and Ḳ in art. حكى.) [See a verse of 'Adee Ibn-Zeyd cited voce صُلْبٌ.] — See also 8.

8. احتكأت العُقْدَةُ *The knot became tight,* or *firm.* (Sh, TA.) And احتكأ العِقْدُ فِى عُنُقِهِ *The necklace became fast upon his neck.* (Sh, TA.) — [Hence,] احتكأ الشَّىْءُ فِى صَدْرِى † *The thing became established in my mind, so that I did not doubt respecting it:* and احتكأ الأَمْرُ فِى نَفْسِى † *The matter became established in my mind.* (TA.) [See also اِحْتَكَى; and see اِحْتَكَّ in the first paragraph of art. حك.] You say also, سَمِعْتُ أَحَادِيثَ وَمَا احتكأ فِى صَدْرِى مِنْهَا شَىْءٌ (in the CḲ ▼ مَا احكأ) † *I heard stories, and nothing of them was unsettled in my mind.* (Ḳ,* TA.) And لَوِ احتكأ لِى أَمْرِى لَفَعَلْتُ كَذَا † *Had my case been manifest to me at the first, I had done thus.* (L, TA.) — See also 4.

حكر

1. حَكَرَ i. q. احتكر, q. v. (A.) — Also حَكَرَهُ, aor. ـِ, inf. n. حَكْرٌ, *He wronged him; acted wrongfully,* or *injuriously; towards him;* (T, Ḳ,* TA;) *and detracted from his reputation,* or *impugned his character;* (T, TA;) *acted,* or *behaved, towards him with bad fellowship,* (T, Ḳ,* TA,) and *with difficulty,* or *hardness, and perverseness:* (TA:) and *he brought upon him distress,* or *trouble, and harm,* or *injury, in his intercourse with him, and his ways of life.* (T, TA.) The epithet applied to him who does so is ▼ حَكِرٌ, [not a reg. part. n., but] a kind of relative epithet. (T, TA.) You say, فِيهِ حَكَرٌ *In him is difficulty,* or *hardness, and perverseness, and a quality of bad fellowship.* (A.) — حَكِرَ, aor. ـَ, (TA,) inf. n. حَكَرٌ, (Ḳ, TA,) *He was obstinate,* or *persistent,* or *persistent in contention,* (Ḳ,* TA,) *and difficult,* or *hard;* (TA;) and *kept a thing to himself, not allowing any one to share with him in it.* (Ḳ,* TA.) The part. n. is ▼ حَكِرٌ. (TA.)

2. حكّر أَرْضَ فُلَانٍ, inf. n. تَحْكِيرٌ, as used by the people of Egypt, [*He made the land of such a one to be a* حِكْر; i. e.] *he* [*enclosed, and*] *debarred others from building upon,* [*or otherwise making use of,*] *the land of such a one.* (El-Maḳreezee's Khiṭaṭ, ii. 114.) [A post-classical phrase: see حِكْرٌ.]

3. حاكرهُ, (TḲ,) inf. n. مُحَاكَرَةٌ, (A, Ḳ,) *He contended, litigated,* or *wrangled, with him.* (A, Ḳ, TḲ.)

5: see 8, in two places.

8. احتكر *He withheld,* (A, Mgh, Mṣb, Ḳ,) or *collected and withheld,* (Ṣ, M,) wheat, (Ṣ, M, A, Mgh, Mṣb,) and the like, of what is eaten, (M, TA,) *waiting for a time of dearness;* (Ṣ, M, A, Mgh, Mṣb, Ḳ;) as also ▼ حَكَرَ, (A,) and ▼ تحكّر: (Ḳ:) *he bought* wheat *and withheld* it *in order that it might become scarce and dear.* (TA.) And ▼ تحكّر فِى بَيْعِهِ *He waited, and watched,* [*for a time of dearness,*] *in his selling.* (ISh.) [This last verb is perhaps not transitive.]

حُكْرَةٌ : see حُكْرٌ.

حِكْرٌ, as meaning مَا يُجْعَلُ عَلَى العَقَارَاتِ وَيُحْبَسُ [app. a mistranscription for مَا يُحْكَرُ مِنَ العقارات ويحبس, which expresses the correct signification, in the dial. of Egypt, i. e. *What is enclosed, of lands,* or *of lands and houses,* or *of lands and palm-trees* &c., *and debarred from others, so that they may not build upon it nor otherwise make use of it*], is a post-classical term: (TA:) [pl. أَحْكَارٌ. See also حَاكُورَةٌ.]

حُكْرٌ and ▼ حُكُرٌ (Ḳ) and ▼ حُكْرَةٌ (TA) *What is withheld,* (Ḳ,) [or *collected and withheld,* (see 8,)] of wheat, and the like, of what is eaten, (TA,) *in expectation of its becoming dear.* (Ḳ.) — See also the last of these words.

حَكِرٌ One *who withholds* a thing, *and keeps* it *to himself:* (A:) one *who withholds* (Ḳ, TA) [or *collects and withholds*] wheat, and the like, of what is eaten, (TA,) *in expectation of dearness:* (Ḳ, TA:) one *who ceases not to withhold his merchandise when the market is full of people and of goods for sale, that it may be sold for much* [بِالكَثِيرِ: in the L and Ḳ بالكسر, which is evidently a mistranscription]. (L, TA.) — See also 1, in two places.

حُكُرٌ : see حُكْرٌ.

حُكْرَةٌ *Collection and retention:* this is the primary signification. (Er-Rághib.) — The *withholding,* (A, Mgh, Mṣb, Ḳ,) or *collecting and withholding,* (Ṣ,) *wheat,* (Ṣ, A, Mgh, Mṣb,) *and the like, of what is eaten,* (TA,) *waiting for a time of dearness;* (Ṣ, A, Mgh, Mṣb, Ḳ;) a subst. from اِحْتِكَارٌ; (Mṣb, Ḳ;) as also ▼ حُكْرٌ and ▼ حُكُرٌ. (Mṣb.) You say that the trade of such a one is الحُكْرَةُ. (A.) — Accord. to some, The *selling* a thing without knowing its measure or weight. (TA.) — And A collection, or an aggregate. (TA.) — See also حَكَرٌ.

حَاكُورَةٌ A piece of land retained and enclosed by its proprietor (تحكر [i. e. تُحْكَرُ]) for sowing [and planting] trees [&c.], near to the houses and abodes: of the dial. of Syria. (TA.) [See also حِكْرٌ.]

حكل

1. حَكَلَ: see what next follows.

4. احكل عَلَىَّ الخَبَرُ The information was dubious, confused, or vague, to me; (Zj, Ṣ, Ḳ;) as also ▾ حَكَلَ, (Ḳ,) and ▾ احتكل: (Ṣ,* Ḳ,* TA:) and احكل الأَمْرُ (Mṣb) and ▾ احتكل The affair, or case, was dubious, or confused, (Mṣb, TA,) عَلَيْهِ to him. (TA.)

8. احتكل: see 4, in two places. — Also He learned a foreign language after Arabic. (Fr, Ḳ.)

حُكْلٌ, [app. a pl., of which the sing. is not used; as though its sing. were أَحْكَلُ; like حُرْسٌ, pl. of أَحْرَسُ;] of animals, (TA,) Having no voice, or sound, to be heard; (Ṣ, Ḳ, TA;) like the ذَرّ [or young ones of ants], (Ḳ,) and نَمْل [or ants in general]: or the mute of birds and beasts. (TA.) — الحُكْلُ is also [said to be] a name of Solomon: (Ḳ:) so, accord. to Lth, in the rejez of Ru-beh, where he says,

* لَوْ أَنَّنِى أُوتِيتُ عِلْمَ الحُكْلِ *

[Were I gifted with the knowledge of Solomon]. (TA.)

حُكْلَةٌ A barbarousness, or vitiousness, in speech or utterance; or an impediment therein; (Ṣ, Mṣb, Ḳ;) so that one does not speak distinctly. (Ṣ.) You say, فِى لِسَانِهِ حُكْلَةٌ [In his speech, or utterance, is a barbarousness, &c.]. (Ṣ.)

حُلْكَةٌ: see حُكْلَةٌ.

حَكِيلَةٌ A mispronunciation; or a word mispronounced; syn. لُثْغَةٌ. (TA.)

حكم

1. حَكَمَهُ, (Ṣ, Ḳ,) [aor. ـُ,] inf. n. حُكْمٌ, (Mṣb, Ḳ, [in the TḲ حَكْمٌ,]) in its primary acceptation, (Mṣb,) He prevented, restrained, or withheld, him (Ṣ, Mṣb, Ḳ) from acting in an evil, or a corrupt, manner; as also ▾ احكمهُ: (Ḳ:) and (Ḳ) from doing that which he desired; as also ▾ احكمهُ; and ▾ حَكَّمَهُ, (Ṣ, Ḳ,) inf. n. تَحْكِيمٌ: (Ṣ:) and حُكُومَةٌ [is another inf. n. of حَكَمَهُ, and], accord. to Aṣ, primarily signifies the turning a man back from wrongdoing. (TA.) Ibráheem En-Nakha'ee is related to have said, ▾ حَكِّمِ اليَتِيمَ كَمَا تُحَكِّمُ وَلَدَكَ, meaning Restrain thou the orphan from acting in an evil, or a corrupt, manner, and make him good, or virtuous, as thou restrainest thine offspring &c.: and of every one whom thou preventest, or restrainest, or withholdest, from doing a thing, thou sayest, [حَكَمْتَهُ and] ▾ حكّمته and ▾ احكمته: or, accord. to Aboo-Sa'eed Ed-Dareer, as related by Sh, the forementioned saying of En-Nakh'ee means let the orphan decide respecting his property, when he is good, or virtuous, as thou lettest thine offspring &c.; but this explanation is not approved. (Az, TA.) And Jereer says,

* أَبَنِى حَنِيفَةَ أَحْكِمُوا ▾ سُفَهَآءَكُمْ *
* إِنِّى أَخَافُ عَلَيْكُمْ أَنْ أَغْضَبَا *

[O sons of Haneefeh, restrain your lightwitted ones: verily I fear for you that I may be angry]: (Ṣ, TA:) i. e., restrain and prevent them from opposing me. (TA.) You say, also, ▾ احكمهُ عَنِ الأَمْرِ He made him to turn back, or revert, from the thing, or affair. (Ḳ.) — حَكَمَ الفَرَسَ, and ▾ احكمهُ, and ▾ حكّمهُ, He pulled in the horse by the bridle and bit, to stop him; he curbed, or restrained, him. (TA.) And حَكَمَ الدَّابَّةَ, (Ṣ,) or الفَرَسَ, (Ḳ,) inf. n. حَكْمٌ; (Ṣ; [so in my two copies of that work;]) and ▾ أَحْكَمَهَا, (Ṣ,) or احكمهُ; (Ḳ;) He put a حَكَمَة [q. v.] to the bit of the beast, or horse. (Ṣ,* Ḳ.) — And ▾ حكّم الحَوَادِثَ † [He controlled events: see مُحَكَّمٌ]. (MF.) — حَكَمْتُ عَلَيْهِ بِكَذَا originally signifies I prevented, restrained, or withheld, him from doing, or suffering, any other than such a thing, so that he could not escape it. (Mṣb.) [Hence it means I condemned him to such a thing; as, for instance, the payment of a fine or of a debt, and death.] And hence, (Mṣb,) حَكَمَ, (Ṣ, Mgh, Mṣb, Ḳ,) aor. ـُ, (Ṣ, Ḳ,) inf. n. حُكْمٌ (Ṣ, Mgh, Mṣb, Ḳ) and حُكُومَةٌ, (Ḳ,) He judged, gave judgment, passed sentence, or decided judicially, بَيْنَهُمْ between them, (Ṣ, Mṣb, Ḳ, TA,) and لَهُ in his favour, and عَلَيْهِ against him. (Ṣ, TA.) And حَكَمَ عَلَيْهِ بِالأَمْرِ He decided judicially the thing, or affair, or case, against him. (Ḳ, TA.) And حَكَمَ لَهُ عَلَيْهِ بِكَذَا [He awarded by judicial sentence in his favour, against him (i. e. another person), such a thing]. (Mgh.) [And حَكَمَ عَلَيْهِ He exercised judicial authority, jurisdiction, rule, dominion, or government, over him. And حَكَمَ بِكَذَا He ordered, ordained, or decreed, such a thing.] — حَكَمَ عَنِ الأَمْرِ He turned back, or reverted, from the thing, or affair. (IAạr, Az, Ḳ.) — حَكُمَ, (Ṣ, MA, TA,) with damm to the ك, (Ṣ,) like كَرُمَ, (TA,) [not حَكَمَ as in the Lexicons of Golius and Freytag,] inf. n. حُكْمٌ (KL, MA) and حِكْمَةٌ, (MA,) He was, or became, such as is termed حَكِيمٌ [i. e. wise, &c.]. (Ṣ, KL, MA, TA.) — And حكم, inf. n. حكم, [so in the TA, without any syll. signs, app. حَكُمَ, inf. n. حُكْمٌ,] is said of a man, signifying He reached the utmost point, or degree, in its meaning (فِى مَعْنَاهُ [i. e., app., in what is the radical meaning of the verb, namely, in judging; like قَضُوَ]); in praising, not in dispraising. (TA.)

2. حكّمهُ, inf. n. تَحْكِيمٌ: see 1, in five places. — Also [He made him judge; or] he committed to him the office of judging, giving judgment, passing sentence, or deciding judicially; (Mgh, Mṣb;) or he ordered him to judge, give judgment, pass sentence, or decide judicially; (Ḳ;) or he allowed him to judge, &c.; (TA;) فِى الأَمْرِ in the affair, or case. (Ḳ.) And حَكَّمْتُهُ فِى مَالِى I gave him authority to judge, give judgment, pass sentence, or decide judicially, respecting my property. (Ṣ, TA.) — Hence, حَكَّمَتِ الخَوَارِجُ The [schismatics called the] خوارج asserted that judgment (الحُكْمُ) belongs not to any but God. (Mgh.) ▾ تَحَكُّمُ الحَرُورِيَّةِ, in the Ḳ, erroneously, تَحْكِيمُ الحَرُورِيَّةِ, (TA,) signifies The assertion of the [schismatics called] حرورية that there is no judgment (حُكْمَ) but God's, (Ḳ, TA,) and that there is no judge (حَكَمَ) but God. (TA.)

3. حاكمهُ إِلَى الحَاكِمِ, (Ḳ,) inf. n. مُحَاكَمَةٌ, (Ṣ,) He summoned him to the judge, and litigated with him, (Ṣ, Ḳ, TA,) seeking judgment: and he made a complaint of him to the judge; or brought him before the judge to arraign him and litigate with him, and made a complaint of him. (TA.) And حَاكَمْنَاهُ إِلَى اللّٰهِ We summoned him to the judgment of God [administered by the Ḳádee]. (TA.) بِكَ حَاكَمْتُ, occurring in a trad., is said to mean I have submitted the judgment [of my case] to Thee, and there is no judgment but thine; and by Thee [or thy means or aid] I have litigated in seeking judgment and in proving the falseness of him who has disputed with me in the matter of religion. (TA. [The past tense, here, is perhaps used as a corroborative present.])

4: see 1, in seven places. The saying of Lebeed, describing a coat of mail,

* أَحْكَمَ الجِنْثِىُّ مِنْ عَوْرَاتِهَا *
* كُلَّ حِرْبَآءٍ إِذَا أُكْرِهَ صَلّ *

is explained as meaning Every nail repelled the sword from its interstices: [when it was struck with force, it made a clashing sound:] or, as some say, [the right reading is الجِنْثِىُّ and كُلَّ, (as in the Ṣ in arts. جنث and صل,) and, accord. to some, صَنْعَتِهَا in the place of عوراتها, (as in the Ṣ and M in art. صل,) and] the meaning is, the manufacturer thereof made firm, or strong, every nail [of its interstices, or of its fabric: &c.]: احكم in this case signifying أَحْرَزَ [agreeably with the explanation here next following]. (TA.) — احكمهُ, (Ṣ, Mgh, Mṣb, Ḳ,) inf. n. إِحْكَامٌ, (TA,) i. q. أَتْقَنَهُ [He made it, or rendered it, (namely, a thing, Ṣ, Mgh, Mṣb,) firm, stable, strong, solid, compact, sound, or free from defect or imperfection, by the exercise of skill; he made it firmly, strongly, solidly, compactly, so that it was firmly and closely joined or knit together, soundly, thoroughly, skilfully, judiciously, or well; he so constructed, constituted, established, settled, arranged, did, performed, or executed, it; he put it into a firm, solid, sound, or good, state, or on a firm, solid, sound, or good, footing: and he knew it, or learned it, soundly, thoroughly, or well; see 1, last sentence, in art. حنك]. (Mṣb, Ḳ.) Hence, in the Ḳur [xi. 1], كِتَابٌ أُحْكِمَتْ آيَاتُهُ (TA) i. e. [A book whereof the verses are rendered valid] by arguments and proofs; (Bḍ;) or by command

and prohibition, and the statement of what is lawful and unlawful: (TA:) or *disposed in a sound manner,* (Ksh, Bḍ,) *with respect to the words and meanings,* (Bḍ,) *like a building firmly and orderly and well constructed:* (Ksh:) or *prevented from being corrupted* (Ksh, Bḍ) *and from being abrogated:* (Bḍ:) or *made to be characterized by wisdom,* (Ksh, Bḍ,) *as comprising the sources of speculative and practical wisdom.* (Bḍ.) And hence one says of a man such as is termed حَكِيم, [i. e. wise, &c.,] قَدْ أَحْكَمَتْهُ التَّجَارِبُ [*Tryings have rendered him firm,* or *sound, in judgment*]. (TA.) — [Hence, أُحْكِمَ عَنْ كَذَا *It was secured from such a thing:* see مُحْكَمٌ.] — [إِحْكَامٌ is also often used as the inf. n. of the pass. verb, signifying The *being firm,* &c.; or *firmness,* &c.: see مِرَّةٌ.] — See also حَكَمَةٌ.

5. تَحَكَّمَ فِيهِ *He did* [or *decided*] *according to his own judgment, or did what he judged fit, respecting it,* or *in it:* (Mṣb:) or *he had authority to judge, give judgment, pass sentence,* or *decide judicially, respecting it;* (Ḳ, TA;) as also ↓احتكم فيه: (Ṣ, Ḳ:) each is quasi-pass. of حَكَّمَهُ; the former regular, and the latter irregular: (TA:) or the former signifies *he pretended to have authority to judge,* &c. (KL.) You say, ↓احتكم عَلَىَّ فِى مَالِى *He had authority over me to judge,* &c., *respecting my property.* (Ṣ.) — See also 2.

6. تحاكموا إِلَى الحَاكِمِ *They summoned one another to the judge,* [*seeking judgment,* (see 3,)] *and litigated;* as also ↓احتكموا إِلَيْهِ. (Ṣ, TA.)

8: see 5, in two places: — and 6: — and 10.

10. استحكم *He* (a man) *refrained from what would injure him in his religion and his worldly concerns.* (Aboo-'Adnán, TA.) — Also quasi-pass. of أَحْكَمَهُ (Ṣ, Mgh, Mṣb, Ḳ) as signifying أَتْقَنَهُ; (Mṣb, Ḳ;) [*It was,* or *became, firm, stable, strong, solid, compact, firmly and closely joined* or *knit together, sound,* or *free from defect* or *imperfection, by the exercise of skill; firmly, strongly, solidly, compactly, soundly, thoroughly, skilfully, judiciously,* or *well, made* or *constructed* or *constituted* or *established* or *settled* or *arranged* or *done* or *performed* or *executed:* and, said of a quality or faculty &c., *it was,* or *became, firm, strong, sound, free from defect* or *imperfection, established,* or *confirmed:*] and, said of an affair, or a case, *it was,* or *became, in a firm, solid, sound,* or *good, state,* or *on a firm, solid, sound,* or *good, footing;* as also ↓احتكم. (TA.) ═ استحكم عَلَيْهِ الأَمْرُ *The thing,* or *affair, became confused and dubious to him;* syn. اِلْتَبَسَ: so in the A. (TA. [But this seems to require confirmation.])

حُكْمٌ [inf. n. of 1, q. v.,] originally signifies *Prevention,* or *restraint.* (Mṣb.) — And hence, (Mṣb,) *Judgment,* or *judicial decision:* (Ṣ, Mṣb, Ḳ, TA:) or *judgment respecting a thing, that it is such a thing,* or *is not such a thing, whether it be necessarily connected with another thing,* or *not:* (TA:) [whence,] in logic, [what our logicians term *judgment;* i. e.] the *judging a thing to stand to another* [*thing*] *in the relation of an attribute to its subject, affirmatively* or *negatively;* or the *perception of relation* or *non-relation:* (Kull:) or it properly signifies *judgment with equity* or *justice:* (Az, TA:) and ↓حُكُومَةٌ signifies the same; (Ḳ, TA;) originally, accord. to Aṣ, the *restraint* of a man from *wrongdoing:* (TA:) [each, though an inf. n., being used as a simple subst., has its pl.:] the pl. of the former is أَحْكَامٌ, (Ḳ,) [properly a pl. of pauc., but] its only pl. form: and the pl. of the latter is حُكُومَاتٌ. (TA.) You say, هُوَ يَتَوَلَّى الحُكُومَاتِ↓ وَيَفْصِلُ الخُصُومَاتِ [*He presides over the affairs of judgment, and decides litigations*]. (TA.) And it is said in a trad., إِنَّ مِنَ الشِّعْرِ لَحُكْمًا, meaning *Verily, of poetry, there is* that which is *true judgment:* so says Er-Rághib: or, as others say, *profitable discourse, such as restrains from, and forbids, ignorant and silly behaviour;* i. e., [what contains] exhortations and proverbs profitable to men: or, the right reading is, as some relate it, ↓لَحِكَمًا [i. e. *wisdom,* &c.]: (TA:) or حِكَمًا [pl. of حِكْمَةٌ]. (So in a copy of the "Jámi' eṣ-Ṣagheer" of Es-Suyooṭee.) — [The *exercise of judicial authority; jurisdiction; rule; dominion;* or *government.* See also حُكُومَةٌ.] — *An ordinance; a statute; a prescript; an edict; a decree;* or *a particular law;* like قَضَآءٌ. Hence the phrase حُكْمُ العَادَةِ *According to custom* or *usage;* properly, *according to the ordinance of custom* or *usage.* — *A rule* in grammar &c.; as when one says, حُكْمُ الفَاعِلِ الرَّفْعُ or أَنْ يُرْفَعَ, i. e. *The rule applying to the case of the agent is that it be put in the nom. case;* and حُكْمُهُ حُكْمُ كَذَا, or كَحُكْمِ كَذَا, i. e. *The rule applying to it is* the same as *the rule applying to such a thing,* or *like the rule applying to such a thing.* — It may often be rendered *Predicament:* (thus the last of the foregoing exs. may be rendered *Its predicament is* the same as *the predicament of such a thing,* or *like the predicament of such a thing:*) and حُكْمًا, or فِى الحُكْمِ, *predicamentally,* or *in respect of predicament;* and *virtually;* as distinguished from لَفْظًا (literally), and حَقِيقَةً (really), and the like.] — Also *Knowledge of the law in matters of religion.* (TA.) — See also حِكْمَةٌ, in two places. It is a more general term than حِكْمَةٌ; for all حِكْمَة is حُكْم, but the reverse is not the case. (Er-Rághib, TA.)

حَكَمٌ: see حَاكِمٌ, in two places; and مُحَكَّمٌ. [Hence,] الحَكَمُ [*The Judge*] is one of the names of God. (TA.) — A man *advanced in age* (Ḳ, TA) *to the utmost degree.* (TA.) ═ See also حَكَمَةٌ.

حِكْمَةٌ [properly, or primarily,] signifies *What prevents,* or *restrains, from ignorant behaviour:* (Mgh:) [in its most usual sense, which is *wisdom,* agreeably with explanations here following,] it is derived from حَكَمَةٌ, signifying a certain appertenance of a beast, [a kind of curb,] because it prevents its possessor from having bad dispositions: (Mṣb:) it means *knowledge;* or *science;* (Ṣ, Ḳ;) as also ↓حُكْمٌ: (Ṣ, TA:) or [generally] *knowledge of the true natures of things, and action according to the requirements thereof;* and therefore it is divided into intellectual and practical: or *a state,* or *quality, of the intellectual faculty:* this is the theological حِكْمَة: in the Ḳur xxxi. 11, by the حِكْمَة given by God to Luḳmán, is meant the *evidence of the intellect in accordance with the statutes of the law:* (TA:) in the conventional language of the learned, it means the *perfecting of the human mind by the acquisition of the speculative sciences, and of the complete faculty of doing excellent deeds, according to the ability possessed:* (Bḍ on the passage of the Ḳur above mentioned:) or it means the *attainment of that which is true,* or *right, by knowledge and by deed:* so that in God it is the *knowledge of things, and* the *origination thereof in the most perfect manner:* and, in man, the *knowledge and doing of good things:* or it means *acquaintance with the most excellent of things by the most excellent kind of knowledge:* (TA:) [and in the modern language, *philosophy:* pl. حِكَمٌ:] see حُكْمٌ. — Also *Equity,* or *justice,* (Ḳ, TA,) *in judgment* or *judicial decision;* and so ↓حُكْمٌ. (TA.) — And i. q. حِلْمٌ; (Ḳ, TA;) i. e. [*Forbearance,* or *clemency,* or] the *management of one's soul and temper on the occasion of excitement of anger:* which, if correct, is nearly the same as equity or justice. (TA.) — And *Obedience of God:* and *knowledge in matters of religion, and the acting agreeably therewith:* and *understanding:* and *reverential fear; piety; pious fear;* or *abstinence from unlawful things:* and the *doing,* or *saying, that which is right:* and *reflection upon what God has commanded, and doing according thereto.* (TA.) — And [*Knowledge of*] *the interpretation of the Ḳur-án, and saying that which is right in relation to it:* so in the Ḳur ii. 272. (TA.) — And The *gift of prophecy,* or the *prophetic office;* (Ḳ, TA;) and *apostleship:* so in the Ḳur ii. 252 and iii. 43 and xxxviii. 19: (TA:) or in the [first and] last of these instances it means — The *Book of the Psalms* [of David]: or, as some say, *any saying,* or *discourse, agreeable with the truth:* (Mgh:) and it also means [in other instances] the *Book of the Law of Moses:* (TA:) and the *Gospel:* and the *Ḳur-án:* (Ḳ:) because each of these comprises what is termed الحِكْمَةُ المَنْطُوقُ بِهَا, i. e. *the secrets of the sciences of the law and of the course of conduct;* and الحِكْمَةُ المَسْكُوتُ عَنْهَا, i. e. *the secrets of the science of the Divine Essence.* (TA.)

حَكَمَةٌ [*A kind of curb for a horse;*] *a certain appertenance of a beast:* so called because it renders him manageable, or submissive, to the rider, and prevents him from being refractory and the like; (Mṣb;) or because it prevents him from vehement running: (TA:) it is the *appertenance* of the لِجَام [or bridle] *that surrounds the* حَنَك [or *part beneath the chin and lower jaw*]: the Arabs used to make it *of untanned thong* or *of hemp;* because what they aimed at was courage, not finery: (Ṣ:) or the *appertenance* of the لجام *that surrounds the* حَنَكَان [which word app. here means the *two jaws*] *of the horse, and in which are* [*attached*] *the* عِذَارَان [or *two side-pieces of the headstall, that lie against the two cheeks*]: (Ḳ:) or *a ring which surrounds the* مَرْسِن [or *part of the nose which is the place of the halter*] *and the* حَنَك [or *part beneath the chin and lower jaw*], *of silver* or *iron* or *thong:* (IDrd in his Book on the Saddle and Bridle:) or *a ring which is upon* (فِى) *the mouth of the horse:* (ISh, TA:) pl. حَكَمَاتٌ (Ṣ, TA) and [coll. gen. n.] ↓ حَكَمٌ. (TA.) Zuheyr says, describing horses,

* قَدْ أُحْكِمَتْ ↓ حَكَمَاتِ القِدِّ وَالأَبَقَا *

meaning قَدْ أُحْكِمَتْ بِحَكَمَاتِ القِدِّ وَبِحَكَمَاتِ الأَبَقِ [*That had been curbed with curbs of untanned thong, and with curbs of hemp*]: (Ṣ, TA:) or, accord. to Abu-l-Ḥasan, [the meaning is *that had been furnished with curbs* &c.; for he says that] احكمت is here made trans. because it implies the signification of قُلِّدَتْ: (TA:) some relate the hemistich thus:

* مُحْكُومَةً ↓ حَكَمَاتِ القِدِّ وَالأَبَقَا *

[*furnished with curbs of untanned thong, and hemp*]. (Ṣ, TA.) — † The *chin* of a sheep (Ṣ, Ḳ) or goat. (Ṣ.) — And, of a man, ‡ The *fore part of the face:* (Ḳ, TA:) or, as some say, the *lower part of the face:* a metaphorical term from the حَكَمَة of the لِجَام: (TA:) or [in some copies of the Ḳ "and"] ‡ his *head:* [accord. to the CḲ, or the *fore part of the head* of a man:] and ‡ his *state,* or *condition:* and ‡ *rank,* and *station.* (Ḳ, TA.) You say, رَفَعَ ٱللّٰهُ حَكَمَتَهُ ‡ *God exalted,* or *may God exalt, his head,* or *his state,* or *condition,* and *his rank,* and *station:* because the stooping of the head is a characteristic of the low, or abject. (TA.) And لَهُ عِنْدَنَا حَكَمَةٌ ‡ *He has rank in our estimation.* (TA.) And فُلَانٌ عَالِى الحَكَمَةِ ‡ [*Such a one is elevated in respect of rank,* or *station.*] (TA.) = [See also حَاكِمٌ, of which it is a pl.]

حَكِيمٌ *Possessing knowledge* or *science;* [in its most usual sense,] *possessing* حِكْمَة [as meaning *wisdom*]; (Ṣ, TA; [see also أَحْكَمُ الحَاكِمِينَ;]) [*wise; a sage:* and in the modern language, *a philosopher:* and particularly *a physician:*] one *who performs,* or *executes, affairs firmly, solidly, soundly, thoroughly, skilfully, judiciously,* or *well;* (Ṣ, IAth;) so that it is, in this sense, of the measure فَعِيلٌ in the sense of the measure مُفْعِلٌ: (IAth, TA:) one *who executes well, and firmly, solidly,* &c., *the niceties of arts:* (TA:) [pl. حُكَمَآءُ.] الحَكِيمُ [as meaning *The All-wise*] is one of the names of God. (TA.) — الذِّكْرُ الحَكِيمُ, applied to the Ḳur-án, means [*The admonition*] *that decides judicially in your favour and against you:* or *that is rendered free from defect* or *imperfection; in which is no incongruity, nor any unsoundness.* (TA.)

حُكُومَةٌ an inf. n. of حَكَمَ [q. v.]: (Ḳ:) [and used as a simple subst.; pl. حُكُومَاتٌ:] see حُكْمٌ, in two places. — Also [*Judicial authority; authority to judge, give judgment, pass sentence,* or *decide judicially,* فِى أَمْرٍ *respecting an affair,* or *a case;*] a subst. from اِحْتَكَمَ and تَحَكَّمَ; and so ↓ أُحْكُومَةٌ. (Ḳ, TA.)

حَاكِمٌ *One who judges, gives judgment, passes sentence,* or *decides judicially; a judge; an arbiter, arbitrator,* or *umpire;* (Ṣ,* Mṣb, Ḳ, TA;) *between people:* (Mṣb, TA:) [*one who exercises judicial authority, jurisdiction, rule, dominion,* or *government; a ruler,* or *governor:*] and ↓ حَكَمٌ signifies the same: (Ṣ, Mgh, Mṣb, Ḳ:) the حاكم between people is so called because he restrains from wrongdoing: (Aṣ, TA:) the pl. is حُكَّامٌ (Mṣb, Ḳ) and حَكَمَةٌ, meaning *judges,* [&c.,] (TA,) and حَاكِمُونَ is allowable. (Mṣb.) It is said in a prov., فِى بَيْتِهِ يُؤْتَى الحَكَمُ ↓ [*In his house the judge is to be come to*]. (Ṣ. [See Freytag's Arab. Prov. ii. 204.]) الحَاكِمُ [as meaning *The Supreme Judge*] is one of the names of God. (TA.) See also the next paragraph. = [The pl.] حَكَمَة also signifies *Mockers, scoffers,* or *deriders.* (TA. [The ح in this case seems to be a substitute for ه: see art. هكم.])

أَحْكَمُ الحَاكِمِينَ ↓ [*The most qualified to judge of those who judge:* or] *the most knowing and most just* [*of them*]: (Bḍ and Jel in xi. 47, where it is applied to God:) or it may mean *the wisest of those who possess attributes of wisdom;* supposing حَاكِمٌ to be [a possessive epithet] from الحِكْمَةُ, like دَارِعٌ from الدِّرْعُ. (Bḍ.)

أُحْكُومَةٌ: see حُكُومَةٌ.

مُحْكَمٌ [pass. part. n. of أَحْكَمَهُ;] applied to a building [&c.,] *Made,* or *rendered, firm, stable, strong, solid, compact,* &c.; *held to be secure from falling to pieces.* (KT.) — And hence, A passage, or portion, of the Ḳur-án *of which the meaning is secured* (أُحْكِمَ) *from change, and alteration, and peculiarization, and interpretation not according to the obvious import, and abrogation.* (KT.) And سُورَةٌ مُحْكَمَةٌ *A chapter* of the Ḳur-án *not abrogated.* (Ḳ.) And الآيَاتُ المُحْكَمَاتُ, [see Ḳur iii. 5, where it is opposed to آيَاتٌ مُتَشَابِهَاتٌ,] *The portion commencing with* قُلْ تَعَالَوْا أَتْلُ مَا حَرَّمَ رَبُّكُمْ [Ḳur vi. 152], *to the end of the chapter:* or *the verses that are rendered free from defect or imperfection, so that the hearer thereof does not need to interpret them otherwise than according to their obvious import; such as the stories of the prophets;* (Ḳ;) or *so that they are preserved from being susceptible of several meanings.* (Bḍ in iii. 5.) And المُحْكَمُ *The portion of the Ḳur-án called* المُفَصَّل [q. v.]; because nought thereof has been abrogated: or, as some say, *what is unequivocal,* or *unambiguous;* because its perspicuity is made free from defect, or imperfection, and it requires nothing else [to explain it]. (TA.)

[مَحْكَمَةٌ *A place of judging; a tribunal; a court of justice.*]

مُحَكَّمٌ فِى نَفْسِهِ [One *who is made to judge respecting himself:* and particularly] one *who is given his choice between denial of God and slaughter, and chooses slaughter.* (Mgh.) In a trad., in which it is said, إِنَّ الجَنَّةَ لِلْمُحَكَّمِينَ [*Verily Paradise is for the* مُحَكَّمُون], (Ṣ,) لِلْمُحَكِّمِينَ ↓, (Ṣ, Ḳ,) or, as some read, لِلْمُحَكِّمِينَ (Ḳ,) denotes a people of those who are called أَصْحَابُ الأُخْدُودِ, who were given their choice between slaughter and the denial of God, and chose the remaining constant to El-Islám, with slaughter: (Ṣ, Ḳ:) or المحكّمون means *those who fall into the hand of the enemy, and are given their choice between* [*the profession of*] *belief in a plurality of Gods, and slaughter, and choose slaughter.* (IAth, TA.) — المُحَكَّمُ occurring in a poem of Ṭarafeh, (Ṣ,) or this is a mistake, and the right reading is ↓ المُحْكَمُ, (Ḳ,) An *old man, tried,* or *proved, and strengthened by experience in affairs;* (Ṣ, Ḳ;) *to whom* حِكْمَة [or *wisdom,* &c.,] *is attributed:* (Ṣ:) or both are correct, like مُجَرَّبٌ and مُجَرِّبٌ, as several authors have allowed; the former meaning one *whom events have controlled* (حَكَّمَتْهُ الحَوَادِثُ), *and tried,* or *proved;* and the latter, one *who has controlled* (حَكَّمَ), *and experienced, events.* (MF.)

مُحَكِّمٌ, and its pl. مُحَكِّمُونَ: see مُحَكَّمٌ. — المُحَكِّمَةُ is an appellation applied to the [schismatics called the] خَوَارِج because they disallowed the judgment of the ↓ حَكَمَانِ [or two judges], (Ṣ,) namely, Aboo-Moosà El-Ash'aree and 'Amr Ibn-El-'Áṣ, (Ḳ, TA,) and said that judgment (الحُكْمُ) belongs not to any but God. (Ṣ.)

فَرَسٌ مَحْكُومَةٌ *A horse* [*furnished with a* حَكَمَة; or] *having a* حَكَمَة *upon his head.* (Az, TA.) See حَكَمَةٌ.

مُتَحَكِّمٌ A judge *who judges without evidence:* and one *who judges in the way of asking respecting a thing with the desire of bringing perplexity,* or *doubt, and difficulty, upon the person asked.* (Ḥar p. 97.)

حكو

1. حَكَوْتُهُ, (Ṣ, Mṣb, Ḳ,) aor. ـُ, (Mṣb, Ḳ,) *i. q.* حَكَيْتُهُ; (Ṣ, Mṣb, Ḳ;) mentioned by AO. (Ṣ.) See art. حكى.

رَجُلٌ حَكَوِىٌّ *A man who relates stories,* or *tales, and extraordinary things:* but this is a vulgar word. (TA.)

حكى

1. حَكَيْتُ الحَدِيثَ, aor. ـِ, (Ḳ,) inf. n. حِكَايَةٌ; (TA;) and حَكَوْتُهُ, aor. ـُ; (Ḳ;) [*I related, recited, rehearsed, narrated,* or *told, the narrative, story, tradition,* &c.:] both signify the same. (Ḳ.) And حَكَيْتُ عَنْهُ الكَلَامَ, inf. n. as above; (Ṣ, Ḳ;) and حَكَوْتُهُ; (AO, Ṣ;) *i. q.* نَقَلْتُهُ [meaning *I transmitted, quoted,* or *transcribed, from him, the speech, discourse, saying, sentence,* or *the like; reported, related, recited, rehearsed, narrated, told,* or *mentioned, the speech,* &c., *as from him*]. (Ḳ.) — حَكَيْتُ الشَّىْءَ, aor. and inf. n. as above, *I did the like of the thing;* as, for instance, another person's art: *I imitated it;* or *emulated it:* as also حَكَوْتُهُ, aor. ـُ. (Mṣb.) One is related to have said, لَا أَحْكُو كَلَامَ رَبِّى,

meaning *I will not imitate*, or *emulate*, *the words of my Lord*. (ISk, Mṣb.) And you say, حَكَيْتُ فُلَانًا; and ▼حَاكَيْتُهُ, (Ḳ,) inf. n. مُحَاكَاةٌ; (TA;) [*I imitated such a one;*] *I did the deed of such a one:* (Ḳ:) or *I said the like of the saying of such a one, exactly;* (Ḳ,* TA;) *not exceeding it:* (TA:) and حَكَيْتُ فِعْلَهُ, and ▼حَاكَيْتُهُ, *I did the like of his deed:* (Ṣ:) but in relation to what is bad, foul, or unseemly, the latter verb is most used. (TA.) — And حَكَيْتُهُ and ▼حَاكَيْتُهُ *I resembled him*, (Ḳ,) and *it*. (TA.) You say, فُلَانٌ يَحْكِى الشَّمْسَ حُسْنًا; and ▼يُحَاكِيهَا, inf. n. as above; *Such a one resembles the sun in beauty*. (Ṣ.) — حَكَى and ▼احكى *He tightened* a knot; (Ṣ, Ḳ;) and *made it strong:* (IḲṭṭ, TA:) dial. vars. of حَكَأَ and أَحْكَأَ. (Ṣ.)

3: see 1, in four places.

4: see 1, last sentence. — احكى عَلَيْهِمْ i. q. أَبَرَّ [*He overcame them;* &c.]. (Ṣgh, Ḳ.)

8. احتكى أَمْرِى *My affair was*, or *became, in a firm, solid, sound*, or *good, state*. (Ḳ.) — احتكى ذٰلِكَ فِى صَدْرِى *That came into my mind*. (Fr, TA.) [See also اِحْتَكَأَ; and see اِحْتَكَّ in the first paragraph of art. حك.]

اِمْرَأَةٌ حَكِىٌّ *A calumnious woman;* (Ḳ;) *one who repeats the sayings of others in a calumnious manner*. (TA.)

حِكَايَةٌ inf. n. of 1. (Ṣ, Mṣb, Ḳ.) — [Used as a simple subst., it signifies *A narrative, story*, or *tale:*] pl. حِكَايَاتٌ. (TA.) — As a conventional term in grammar, [meaning *Imitation*, or *conforming*, in language,] it is of two kinds: حِكَايَةُ جُمْلَةٍ [*The imitation of a proposition*, or *phrase*], and حِكَايَةُ مُفْرَدٍ [*The imitation of a single word*]: the former is of two sorts; one of which is termed حِكَايَةُ مَلْفُوظٍ [*The imitation of a thing uttered*], whereof قَالُوا الحَمْدُ لِلّٰهِ [They said, "Praise be to God," (Ḳur vii. 41,)] is an ex.; and the other, حِكَايَةُ مَكْتُوبٍ [*The imitation of a thing written*], whereof قَرَأْتُ عَلَى فَصِّهِ مُحَمَّدٌ رَسُولُ اللّٰهِ [I read upon his ring-stone, "Moḥammad is the Apostle of God,"] is an ex.: and the latter kind, also, (that of a single word,) is of two sorts; one of which is with the interrogative أَىّ or مَنْ, as when, to one who says, رَأَيْتُ رَجُلًا [pronounced رَجُلَا, I saw a man], you say, أَيًّا [What (man)?], or مَنًا [Whom?], and the like, which sort is agreeable with a common rule; the other sort being without an interrogative, as the saying of one of the Arabs, to one who said, هَاتَانِ تَمْرَتَانِ [These are two dates], دَعْنَا مِنْ تَمْرَتَانِ [Abstain thou from troubling us about two dates], which sort is anomalous. (El-Ashmoonee's Expos. of the Alfeeyeh of Ibn-Málik.) [حِكَايَةُ صَوْتٍ signifies *A word imitative of a sound; an onomatopœia:* see اِسْمُ صَوْتٍ, in art. سمو.]

حل

1. حَلَّ, (Ṣ, Mgh, Mṣb, Ḳ,) aor. ـُ, inf. n. حَلٌّ, (Ṣ, Mgh, Mṣb,) *He untied*, or *undid*, (Ḳ, TA,) or *opened*, (Ṣ,) a knot: (Ṣ, Mgh, Mṣb, Ḳ:) this is the primary signification. (Er-Rághib, TA.) Hence, يَا عَاقِدُ ٱذْكُرْ حَلًّا [*O tyer of a knot, bear in mind the time of untying:* or يا حَابِلُ *O binder*, or *tyer, of the rope:* (see art. حبل:)] (Ṣ, TA:) or, as IAạr relates it, يا حَامِلُ [*O loader*]: a prov., applied to the consideration of results; because a man may tie a load too tightly; and when he desires to untie it, may injure himself and his camel. (TA. [See also a similar saying below, in the second paragraph.]) And الشُّفْعَةُ كَحَلِّ العِقَالِ [*The right of preemption is like the untying of the cord with which a camel's fore shank and arm are bound together*]: meaning that it is accomplished as quickly and easily as the عقال is untied: the explanation that it passes away quickly, like the camel when his عقال is untied, is improbable. (Mgh, Mṣb.*) And hence the saying [in the Ḳur xx. 28], وَٱحْلُلْ عُقْدَةً مِنْ لِسَانِى † [*And loose Thou an impediment of*, or *from, my tongue*]. (Er-Rághib, TA.) The pass. is pronounced by some حُلَّ, and by others حِلَّ: thus in the saying of El-Farezdaḳ,

* فَمَا حِلَّ مِنْ جَهْلٍ حُبَى حُلَمَائِنَا *
* وَلَا قَائِلُ المَعْرُوفِ فِينَا يُعَنَّفُ *

[*And the garments of our forbearing men by which they support themselves in sitting by binding them*, or *making them tight, round the shanks and back are not loosed through ignorance, nor is the accepter of the benefit, among us, reproached*]; the kesreh of the first ل [in the original form حُلِلَ] being transferred to the ح: but Akh heard it pronounced in this instance حُلَّ; and some, he says, in this word, and in others like it, as رُدَّ and شُدَّ, only impart to the ḍammeh somewhat of the sound of kesreh, by the pronunciation termed إِشْمَام. (Ṣ.) — [*He*, or *it, dissolved, melted*, or *liquefied*, a thing; as also ▼حلّل, inf. n. تَحْلِيلٌ, often said of a medicine as meaning *it acted as a dissolvent*.] حَلَّ, (M, Ḳ,) inf. n. as above, (TA,) meaning *It was dissolved, melted*, or *liquefied*, is said of anything congealed, frozen, or solid. (M, Ḳ.) — [† *He solved* a problem, or riddle, &c. — † *He analyzed* a thing; as also ▼حلّل, inf. n. as above.] حَلَّ اليَمِينَ: see 2. — From حَلُّ الأَحْمَالِ عِنْدَ النُّزُولِ [*The untying, unbinding*, or *loosing, of the loads on the occasion of alighting*], حَلَّ, inf. n. حُلُولٌ, came to be used alone as meaning نَزَلَ [i. e. *He alighted;* or *descended and stopped* or *sojourned* or *abode* or *lodged* or *settled;* and simply *he took up his abode;* or *he abode, lodged*, or *settled;* in a place]. (Er-Rághib, TA.) You say, حَلَّ المَكَانَ (Ṣ, Ḳ) or بِالْبَلَدِ, (Mṣb,) and حَلَّ بِالْمَكَانِ (Ṣ, Mgh,* Ḳ) or البَلَدَ, (Mṣb,) aor. ـُ (Ṣ, Mṣb, Ḳ) and ـِ, (Ḳ,) both of which forms of the aor. are mentioned by Ibn-Málik, (TA,) inf. n. حُلُولٌ (Ṣ, Mgh, Mṣb, Ḳ) and حَلٌّ (Ṣ, Ḳ) and حَلَلٌ, which is extr., (Ḳ,) and مَحَلٌّ (Ṣ, TA) and مَحِلٌّ; (TA;) and ▼احتلّ بِهِ (Ṣ,* Ḳ) and ▼احتلّهُ; (Ḳ;) meaning نَزَلَ بِهِ [*He alighted*, or *descended and stopped* &c. (as above), *in the place* or *the country* or *town* &c.]. (Mṣb, Ḳ, TA.) And in like manner, حَلَّ القَوْمَ and حَلَّ بِالقَوْمِ (Ṣ, ISd, TA) and ▼احتلّهم and ▼احتلّ بهم (ISd, TA) [*He alighted*, or *descended and stopped* &c., *at*, or *in, the abode of the people* or *party*]; and حَلَّ إِلَى القَوْمِ signifies the same. (TA.) And حُلَّ المَكَانُ *The place was alighted in*, or *taken as an abode;* (TA;) *was inhabited*. (Ḳ.) [Hence, in philosophy, حُلُولٌ signifies † *Temporary* or *separable*, and *permanent* or *inseparable, indwelling* or *inbeing:* and حَلَّ بِهِ or فِيهِ, † *It had*, or *became in the condition of having, such indwelling* or *inbeing in it*. And حَلَّ بِهِ or فيه is often said of joy and grief and the like, meaning † *It took up its abode in him*. And حَلَّ مَحَلَّ ▼كَذَا † *It took*, or *occupied, the place of such a thing*.] — And hence, (TA,) حَلَّ الهَدْىُ, (Ṣ, Mṣb, Ḳ,) aor. ـِ, (Ṣ, Mṣb,) inf. n. حِلَّةٌ and حُلُولٌ, (Ṣ, Ḳ,) ‡ *The beasts for sacrifice arrived at the place where it was lawful for them to be sacrificed*, (Ṣ, Ḳ, TA,) or *at the place in which they should be sacrificed*. (Mṣb.) — حَلَّ said of a punishment has for its aor. ـُ and ـِ, and the inf. n. is حُلُولٌ: (Mṣb:) [but it is said that] حَلَّ العَذَابُ, aor. ـُ, means † *The punishment alighted*, or *descended:* and حَلَّ, aor. ـِ, † *it was*, or *became, due*, or *necessitated by the requirements of justice to take effect:* it is said in the Ḳur [xx. 83], فَيَحِلَّ عَلَيْكُمْ غَضَبِى † [*lest my anger alight upon you*, or *befall you*]; or فَيَحِلَّ عليكم † [*lest it become due to you*]; accord. to different readings: (Ṣ, O:) or when you say, حَلَّ بِهِمُ العَذَابُ, [you mean † *The punishment alighted upon them*, or *befell them;* and] the aor. is ـُ only: and when you say, حَلَّ عَلَىَّ, or لَكَ, [you mean † *It became due to me*, or *to thee;* and] the aor. is ـِ: أَنْ يَحِلَّ عَلَيْكُمْ غَضَبٌ مِنْ رَبِّكُمْ [in the Ḳur xx. 89,] means † *that anger should alight upon you*, or *befall you, from your Lord* [accord. to those who read thus instead of يَحِلَّ; but the latter is the common reading]. (TA.) You say also, حَلَّ أَمْرُ اللّٰهِ عَلَيْهِ, aor. ـِ, inf. n. حُلُولٌ, meaning †[*The threatened punishment of God*] *was*, or *became, due to him*, or *necessitated by the requirements of justice to take effect upon him*. (Ḳ.) And حَلَّ حَقِّى عَلَيْهِ, (Mṣb,* Ḳ,) aor. ـِ, (Mṣb, Ḳ,) inf. n. حُلُولٌ, (Mṣb,) or مَحِلٌّ, (Ḳ,) ‡ *My right*, or *due, was*, or *became, a thing the rendering of which was obligatory*, or *incumbent, on him*. (Mṣb,* Ḳ, TA.) And حَلَّ الدَّيْنُ, (Ṣ, Mṣb, Ḳ,) aor. ـِ, inf. n. حُلُولٌ, (Ṣ, Mṣb,) † *The debt was*, or *became*, or *fell, due;* (Ḳ,* TA;) *its appointed term*, or *period, ended*, (Mṣb, TA,) *so that the payment of it became due*. (TA.) And حَلَّ عَلَيْهِ الدَّيْنُ † *The payment of the debt became obligatory on him*. (Mgh.) — حَلَّ, aor. ـِ, inf. n. حِلٌّ (Ṣ, Mgh, Mṣb, Ḳ) and حَلَالٌ, (Ṣ,) ‡ *It* (a thing, Ṣ, Mgh, Mṣb) *was*, or *became, lawful, allowable*, or *free;* (Ṣ, TA;) لَكَ *to thee:* (Ṣ:) contr. of حَرُمَ: (Mṣb, Ḳ:) a metaphorical signification, from حَلُّ العُقْدَةِ ["the untying of the knot"]. (TA.) Hence the saying, الزَّوْجُ أَحَقُّ

بِرَجْعَتِهَا مَا لَمْ تَحِلَّ لَهَا الصَّلَاةُ † [*The husband is entitled to taking her back to the marriage-state as long as prayer is not lawful to her*]. (Mgh.) And the saying, in a trad., لَمَّا رَأَى الشَّمْسَ قَدْ وَقَبَتْ قَالَ هٰذَا حِينُ حِلِّهَا, i. e. [*When he saw that the sun had set, he said,*] *This is the time of its becoming lawful*; meaning the prayer of sunset. (TA.) — [حَلَّ, aor. ـِ, inf. n. حِلٌّ, † *He was, or became, free from*, or *quit of*, *an obligation*, or *responsibility*.] You say, جَعَلَهُ فِى حِلٍّ مِنْ قِبَلِهِ † [*He made him*, or *pronounced him, free from*, or *quit of, obligation*, or *responsibility, with respect to him*]. (TA.) [And أَنْتَ فِى حِلٍّ مِنْ كَذَا † *Thou art free from*, or *quit of, obligation*, or *responsibility, with respect to such a thing*.] And حَلَّ said of the مُحْرِم, (Ṣ, Mṣb,) or حَلَّ مِنْ إِحْرَامِهِ, (Ḳ,) inf. n. حِلٌّ, (Ṣ,) or حَلَالٌ, (Mṣb, Ḳ,) or both; (TA;) and ↓ احلّ, (Ṣ, Mṣb, Ḳ,) and ↓ تحلّل; (Bḍ and Jel in ii. 192;) ‡ *He quitted his state of* إِحْرَام: (Mṣb, Ḳ:) this, also, is a metaphorical signification, from حَلَّ العُقْدَةَ. (TA.) [Hence,] فَعَلَهُ فِى حِلِّهِ وَحِرْمِهِ, and فِى حِلِّهِ ↓ وحُرْمِهِ, † *He did it when he was free from* إِحْرَام *and when he was in the state of* احرام. (Ḳ.) And شُهُورُ الحِلِّ, (Ṣ,) or أَشْهُرُ الحِلِّ: (Ḳ:) see حِلٌّ, below. And حَلَّتِ المَرْأَةُ, (Ṣ, Ḳ,) [aor. ـِ,] inf. n. حِلٌّ and حُلُولٌ, (TA,) ‡ *The woman quitted her* [*period termed*] عِدَّة: (Ṣ, Ḳ:) this, too, is a metaphorical signification, from حَلَّ العُقْدَةَ: (TA:) or حَلَّتْ لِلزَّوَاجِ † *she became free from any obstacle to marriage, as, for instance, by having accomplished the* عِدَّة. (Mṣb.) [And † *The woman became free from the marriage tie, by the death of her husband*, or *by divorce*.] You say, أَنْتِ فِى حِلٍّ مِنِّى † *Thou art divorced from me*. (TA.) And حَلَّتِ اليَمِينُ † *The oath* [*became discharged*; and thus,] *proved true*. (Mṣb.) — حَلَّ, aor. ـِ, inf. n. حَلٌّ, † *He ran*. (TA.) ═ حَلَّ بِهِ: see 4. ═ حَلَّ, sec. pers. حَلِلْتَ, aor. ـَ, inf. n. حَلَلٌ, *He* (a man) *had a pain in his hips and* [in the CḲ "or"] *his knees*. (Ḳ.) [See also حَلَلٌ, below.]

2. حلّلهُ: see 1, in two places: — and see also 4, in four places. — حَلَّلَ اليَمِينَ, (Ṣ, Mgh, Mṣb, Ḳ,) inf. n. تَحْلِيلٌ (Mgh, Ḳ) and ↓ تَحِلَّةٌ, (Ṣ,* Mgh, Ḳ,) and ↓ تَحِلٌّ, which is anomalous, (Ḳ,) or ↓ تَحِلَّةٌ is a simple subst., (Mṣb,) [and] so is ↓ حِلٌّ, (Ḳ, TA,) † *He expiated the oath*: (Ḳ:) or † *he freed the oath from obligation by making an exception*, or *saying* إِنْ شَآءَ اللّٰهُ, or *by expiation*: (Mgh, Mṣb:) or † *he did that whereby he became free from violating*, or *failing of keeping, the oath*; [generally meaning *he made an exception in the oath*, or *he expiated it*;] as also ↓ تحلّل: (Mṣb:) and ↓ تحلّل فِى يَمِينِهِ † *he made an exception*, or *said* إِنْ شَآءَ اللّٰهُ, *in his oath*, (Ṣ, Mgh, Ḳ,) *immediately*: (TA:) and ↓ تحلّل مِنْهَا † *he became quit of it by expiation*, (Mgh, TA,) or *by a violation of it requiring expiation*, or by *making an exception*, or *saying* ان شاء الله, *in it*. (TA.) One says to him who goes to a great length in threatening, or him who exceeds the due bounds in what he says, ↓ حِلًّا أَبَا فُلَانٍ, meaning † *Make thou an exception*, or *say* ان شاء الله, *O father of such a one*, in thine oath; regarding him as a swearer: and in like manner one says, ↓ يَا حَالِفُ ٱذْكُرْ حِلًّا † [*O swearer, bear in mind* the *making an exception*, or *saying* ان شاء الله]. (Ṣ,* TA. [See a similar saying in the second sentence of this art.]) In the saying لَأَفْعَلَنَّ كَذَا ↓ إِلَّا حِلُّ ذٰلِكَ أَنْ أَفْعَلَ كَذَا, [the particle] الّا is syn. with لٰكِنْ; and the meaning is said to be, † [*I will assuredly do such a thing: but*] *the annulling of the obligation*, or *the expiation, of* [*that*] *my asseveration* (تَحْلِيلَةُ, or ↓ تَحِلَّةُ قَسَمِى,) *shall be my doing such a thing*. (TA.) One says also, ↓ فَعَلْتُهُ تَحِلَّةَ القَسَمِ, meaning † *I did it* only enough *to annul the obligation of*, or *to expiate, the oath*; not exceeding therein the ordinary bounds. (Ṣ, Mṣb.) It is said in a trad., لَا يَمُوتُ لِلْمُؤْمِنِ ثَلَاثَةُ أَوْلَادٍ فَتَمَسَّهُ النَّارُ إِلَّا ↓ تَحِلَّةَ القَسَمِ meaning † [*Three children of the believer shall not die and the fire* of Hell *touch him*] *save* enough *to annul the obligation of*, or *to expiate, the oath* that is implied in the saying in the Ḳur [xix. 72], "There is not any of you that shall not come to it." (A'Obeyd, Ṣ, TA.) Hence تَحْلِيلٌ came to be applied to anything in which the ordinary bounds were not exceeded. (Ṣ, Mṣb.) One says, ضَرَبْتُهُ تَحْلِيلًا, (Ṣ, TA,) or ضَرْبًا تَحْلِيلًا, (Ḳ,) meaning † *I beat him moderately; not exceeding the ordinary bounds*. (Ḳ,* TA.) And Kaab Ibn-Zuheyr says, speaking of the feet of a she-camel, وَقْعُهُنَّ الأَرْضَ تَحْلِيلُ, meaning *Their falling on the ground is without vehemence*. (Ṣ.) [In like manner, also,] ↓ تَحِلَّةُ القَسَمِ is descriptive, by way of comparison, of littleness; as is ↓ تَحِلَّةُ اليَمِينِ: (Mgh:) or of anything occupying little time: (TA:) and ↓ إِلَّا تَحِلَّةَ القَسَمِ, in the trad. cited above, means † [*slightly*, or] *with a slight touch*. (Mgh.) A poet says,

أَرَى إِبِلِى عَافَتْ جَدُودَ فَلَمْ تَذُقْ

بِهَا قَطْرَةً إِلَّا ↓ تَحِلَّةَ ↓ مُقْسِمِ

† [*I see my camels loathed* the water of *Jadood, so that they did not taste in it a drop save sparingly*]. (Ṣ.) — حَلَّلَ مَا بِهِ مِنَ الدَّآءِ, inf. n. تَحْلِيلٌ, † *He, or it, removed what was in him, of disease*. (Ḥar p. 231.) ═ حلّلهُ الحُلَّةَ *He clad him with the* حُلَّة. (TA.)

3. حالّهُ *He alighted*, or *descended and stopped* or *sojourned* or *abode* or *lodged* or *settled, with him*; and simply *he took up his abode, lodged*, or *settled, with him*; syn. حَلَّ مَعَهُ. (Ḳ.) You say, يُحَالُّهُ فِى دَارٍ وَاحِدَةٍ [*He takes up his abode, lodges*, or *settles, with him in one house*]. (Ṣ.) And, of a woman, تُحَالُّ زَوْجَهَا فِى فِرَاشٍ [*She takes her place with her husband in a bed*]. (Mgh.)

4. احلّهُ *He made him to alight*, or *descend and stop* or *sojourn* or *abide* or *lodge* or *settle*; and simply *he made him to take up his abode, to lodge*, or *to settle*; syn. أَنْزَلَهُ; (Ṣ, Ḳ;) as also ↓ حلّلهُ, and ↓ حَلَّ بِهِ: (Ḳ:) said also of a place [as though meaning *it invited him to alight*, &c.]. (ISd, TA.) So in the phrases احلّهُ المَكَانَ and ↓ حلّلهُ المَكَانَ, and بِالْمَكَانِ, *He made him to alight*, or *descend and stop* &c., *in the place*. (Ḳ.) — احلّ بِنَفْسِهِ † [*He caused punishment* (العُقُوبَة being understood) *to alight*, or *descend, upon himself*; or] *he did what necessitated*, or *he deserved, punishment*. (Ṣ, Ḳ.) — احلّهُ اللّٰهُ عَلَيْهِ † *God necessitated it, as suitable to the requirements of justice, to take effect upon him*; namely, his threatened punishment (أَمْرَهُ). (Ḳ,* TA.) — And احلّهُ ‡ *He* (God, Mṣb and Ḳ, and a man, Ṣ, Mṣb) *made it lawful, allowable*, or *free*; as also ↓ حلّلهُ, (Ṣ, Mgh, Mṣb, Ḳ, TA,) inf. n. تَحْلِيلٌ and ↓ تَحِلَّةٌ. (Ṣ.) Hence, [in the Ḳur ii. 276,] أَحَلَّ ٱللّٰهُ البَيْعَ † *God has made selling to be lawful*, or *allowable, giving the choice to practise it or abstain from it*. (Mṣb.) And hence also, أَحْلَلْتُ لَهُ الشَّىْءَ † *I made*, or *have made, lawful, allowable*, or *free, to him, the thing*. (Ṣ.) And أَحْلَلْتُ المَرْأَةَ لِزَوْجِهَا † *I made*, or *have made, the woman lawful to her husband*. (Ṣ.) — And أَحْلَلْتُهُ and ↓ حَلَّلْتُهُ † *I made him*, or *pronounced him, free from*, or *quit of, obligation*, or *responsibility, with respect to what was between me and him*. (Ḥam p. 446.) And ↓ تحلّلهُ † *He made him*, or *pronounced him, free from*, or *quit of, obligation*, or *responsibility, with respect to himself*. (TA.) — أَحِلُّوا ٱللّٰهَ يَغْفِرْ لَكُمْ, or أَحِلُّوا, with ح, accord. to different readings of a trad.: see 4 in art. حل. ═ احلّ as an intrans. verb: see 1, near the end of the paragraph. Also † *He entered upon* [*any of*] *the profane months*. (Ṣ, Ḳ.) And † *He went forth to the* حِلّ: (Ṣ, Ḳ:) or *he became in the* حِلّ; which means the *region without the* حَرَم [or *sacred territory*]: (Mṣb:) or *he became free from*, or *quit of, an obligation* [*of any kind*] *that was upon him*. (Ṣ, Ḳ.) — It is said in a trad., أَحِلَّ بِمَنْ أَحَلَّ بِكَ, meaning † *Whoso quits the state of* إِحْرَام, *and makes it lawful to attack thee, and fights with thee, do thou so with him, though thou be in the state of* احرام: or it means, *if a man make lawful to him what is unlawful to him, as thy honour or reputation, and thy property, repel him from thyself in such a way as thou canst*. (Ṣgh, TA.) — احلّت † *She* (a ewe or goat) *secreted milk in her udder without bringing forth*: (Ṣ, O:) or † *she*, (a ewe or goat, Ḳ, and a camel, TA,) *after her milk had become scanty, or had dried up, yielded her milk abundantly in consequence of her having eaten the* [*herbage termed*] رَبِيع: in which case she is said to be ↓ مُحِلٌّ. (Ḳ.) And احلّت عَلَى وَلَدِهَا † *She* (a camel) *yielded her milk abundantly to her young one*. (ISd, TA.)

5. تحلّل *It passed away by becoming dissolved, melted*, or *liquefied*. (KL.) [And تحلّل إِلَيْهِ † *It became reduced by analysis to it*: occurring in this sense in the TA, art. قطع, in two places.] — † *It* (a disease) *went away by degrees.* (Ḥar p. 231.) — See also 1, near the end of the paragraph. [Hence,] † *He became* ▼ حَلَال, meaning *he finished his prayer.* (Ḥar p. 348.) — تحلّل فِى يَمِينِهِ : and تحلّل مِنْهَا : see 2. — تحلّل السَّفَرُ بِالرَّجُلِ † [*The journey caused the man to fall sick after it; or*] *the man fell sick after arriving from the journey.* (ISd, Ḳ.) — تحلّلهُ : see 4.

7. انحلّت العُقْدَةُ *The knot became untied,* or *undone,* (Ḳ, TA,) or *opened.* (Ṣ.) — [And انحلّ *It became dissolved, melted,* or *liquefied.* — † *It* (a problem, or riddle, &c.) *became solved.* — † *It* (a thing) *became analyzed.* — † *He relaxed;* or *became free from self-restraint.*] — انحلّت اليَمِينُ † *The oath became freed from obligation* [*by an exception made in it, or by expiation*]. (Mṣb.)

8. احتلّ : see 1, in four places.

10. استحلّهُ † *He reckoned it, accounted it, esteemed it,* or *deemed it, lawful, allowable,* or *free:* (Ṣ, O:) [and consequently, *he profaned, desecrated,* or *violated, it;* i. e., a thing that should be regarded as sacred, or inviolable:] or *he took it as,* or *made it, lawful, allowable,* or *free:* or *he asked him to make it so to him.* (Ḳ.)

R. Q. 1. حَلْحَلَهُمْ *He removed them,* (Ṣ, Ḳ,) or *unsettled them, from their place,* (Ṣ,) or *from their places, and put them in motion.* (Ḳ.) — حَلْحَلَ بِالنَّاقَةِ, (Ṣ,) or بِالإِبِلِ, (Ḳ,) *He said to the she-camel,* (Ṣ,) or *to the camels,* (Ḳ,) حَلْ, (Ṣ, Ḳ,) or حَلِ حَلِ. (Ḳ.)

R. Q. 2. تَحَلْحَلَ عَنْ مَكَانِهِ *He removed from his place;* or *quitted it.* (Ṣ.) And تَحَلْحَلُوا *They removed from their places, and became in motion,* (Ḳ, TA,) *and went away.* (TA.)

حَلْ (Ṣ, Ḳ) and حَلِ, the latter used in the case of connexion with a following word, (Ṣ,) or حَلِ حَلِ, (Ḳ,) *A cry by which a she-camel is chidden,* like as a male camel is by the cry حَوْبِ: (Ṣ:) or *a cry by which camels are chidden;* (Ḳ,* TA;) *but only female camels;* as also حَلِى. (TA.)

حَلٌّ *Oil of sesame,* or *sesamum.* (Ṣ, Ḳ.)

حُلٌّ : see 1, near the end of the paragraph.

حِلٌّ an inf. n. of 1 [q. v.], in several senses. (Ṣ, Ḳ, &c.) — [Hence,] أَشْهُرُ الحِلِّ, (Ṣ,) or شُهُورُ الحِلِّ, (Ḳ,) [*The profane months;* i. e. *all the months except those termed* الأَشْهُرُ الحُرُمُ : see حَرَامٌ.] — Also a simple subst. from حلّل اليَمِينَ : see 2, in four places. — See also حَلَالٌ, in seven places. — [Hence,] الحِلُّ † *The region that is without the* حَرَم [*or sacred territory*]. (Ṣ, Mṣb, Ḳ.) — See also حَالٌّ. — Also *A butt; an object of aim, at which one shoots or throws.* (Ḳ.)

حَلَّةٌ : see مَحَلٌّ. — *A large basket,* (Ḳ,) or *a thing of the form of a large basket,* (Sgh, TA,) *of reeds,* or *canes,* (Sgh, Ḳ, TA,) *in which wheat is put:* so in the conventional language of the people of Baghdád: (Sgh, TA:) but in that of Egypt, *a copper cooking-pot:* (TA:) [pl. حِلَلٌ.] — The *direction* (جِهَة, and قَصْد,) of a thing; as also ▼ حِلَّةٌ : (Ḳ:) as when you say حلّة الغَوْرِ *in the direction of the Ghowr;* syn. قَصْدَهُ. (Sb, TA.) — فِيهِ حَلَّةٌ *In him is weakness, and languor;* as also ▼ حِلَّةٌ. (M.)

حُلَّةٌ [*A dress consisting of*] *an* إِزَار [i. e. *a waist-wrapper*] *and a* رِدَاء [*or wrapper for the whole body*], (Ṣ, M, Mgh, Ḳ,) or *a* بُرْد [*which is another kind of wrapper for the whole body*], or *some other garment:* (M, Ḳ:) only applied to *a dress consisting of two garments* (Ṣ, M, Nh, Mṣb, Ḳ) *of one kind:* (Nh, Mṣb:) or *either of the two garments by itself:* or *a* رِدَاء *and a shirt, completed by a turban;* or *a good garment;* but not so called when upon a man; for in this case it means *two garments,* or *three:* or *any good new garment that is worn, thick* or *coarse,* or *fine* or *thin:* (TA:) or *a lined garment:* (Ḳ:) but with the Arabs of the desert it means [*a dress consisting of*] *three garments,* i. e. *a shirt and an* إِزَار *and a* رِدَاء : (TA:) pl. حُلَلٌ (Mṣb) [and حِلَالٌ, as below]: accord. to A'Obeyd, حُلَلٌ means بُرُود *of El-Yemen,* (Ṣ, TA,) *from various places;* and a garment of this kind is asserted to be meant in a trad. in which it is said that the best kind of grave-clothing is the حُلَّة : it is also said that حُلَلٌ is applied to the *and* جِبَر *and* خَزّ *and* قَزّ *and* قُوهِىّ *and* وَشْى *and* خَرِير مَرْوِىّ. (TA.) — Also † *A weapon,* or *weapons:* (Sgh, Ḳ:) pl. حُلَلٌ and حِلَالٌ. (Ḳ.) You say, لَبِسَ حُلَّتَهُ, meaning † *He wore,* or *put on, his weapon,* or *weapons.* (Sgh, TA.) — Also [like إِزَار] ‡ *A wife.* (TA.)

حِلَّةٌ *A mode,* or *manner, of* حُلُول [i. e. *alighting, taking up one's abode, lodging,* or *settling*]. (Ḳ.) — See also حَالٌّ, in three places. — And see مَحَلٌّ, in two places. — Also ‡ *A collection of* بُيُوت [i. e. *tents,* or *houses,*] (Mṣb, Ḳ) *of men;* (Ḳ;) as also ▼ مَحَلَّةٌ : (Ḥar p. 333:) or (Ḳ) *a hundred thereof,* (Mṣb, Ḳ,) and *more:* pl. حِلَالٌ. (Mṣb.) — † *A sitting-place,* or the *people thereof;* syn. مَجْلِس : [or] † *a place of assembly:* pl. as above. (Ḳ.) — See also حَلَّةٌ, in two places.

حَلَلٌ *A laxness in the legs* of a beast: or *in the tendons,* or *sinews,* (Ḳ,) *and weakness in the* نَسَا [q. v.], (TA,) *with laxness of the hock:* or it is peculiar to camels: (Ḳ:) and signifies *a weakness in the* عُرْقُوب [i. e. *hock,* or *hock-tendon,*] of a camel; (Fr, Ṣ, O;) or *in each* عرقوب of a camel: (M, TA:) if in the knee, it is termed طَرَقٌ. (Fr, Ṣ.) — And *Paucity of flesh in the posteriors and thighs;* or *smallness and closeness of the buttocks;* or *paucity of flesh in the thighs;* syn. رَسَحٌ ; (Ḳ;) in a woman. (TA.) — And *Pain in the hips,* or *haunches, and the knees,* in a man. (Ḳ.)

حَلَالٌ ‡ *Lawful, allowable,* or *free; contr. of* حَرَامٌ ; (Ṣ, Mṣb, Ḳ;) as also ▼ حِلَالٌ (Ḳ) and ▼ حِلٌّ (Ṣ, Mgh, Mṣb, Ḳ) and ▼ حَلِيلٌ : (Ḳ:) a metaphorical signification, from حَلُّ العُقْدَةِ ["the untying of the knot"]: according to some, meaning what is *not forbidden;* and therefore *including what is disapproved and what is not disapproved:* accord. to others, that *for which one is not punishable.* (TA.) [Hence, مَالٌ حَلَالٌ † *Wealth lawfully acquired.* And ابْنُ حَلَالٍ † *A legitimate son:* and *an ingenuous,* or *honest, person.*] And الحُلْوُ الحَلَالُ ‡ *Language in which is nothing that induces doubt,* or *suspicion:* (Ḳ, TA:) and *the man in whom is nothing that induces doubt,* or *suspicion.* (TA in art. حلو.) And هُوَ حِلٌّ لَكَ ‡ *It is lawful, allowable,* or *free, to thee.* (TA.) And هُوَ حِلٌّ ▼ بِلٌّ, meaning the same; (Ṣ;) or حِلٌّ ▼ وَبِلٌّ ; (Ḳ;) and هِىَ لِشَارِبٍ حِلٌّ ▼ وَبِلٌّ : (TA:) see art. بل. — ‡ A man *who has quitted his state of* إِحْرَام ; (Ṣ, Mṣb, Ḳ;) as also ▼ مُحِلٌّ, (Mṣb,) and ▼ حِلٌّ, (Ṣ, Mṣb,) and حِلٌّ ▼ مِنَ الإِحْرَامِ ; (Ṣ, TA;) or this last signifies one *who has not become in that state:* (TA:) by rule one should say حَالٌّ, which is not used in this sense. (Ḳ.) — ‡ A woman *free from any obstacle to marriage, as, for instance, by having accomplished the* عِدَّة. (Mṣb.) — See also 5.

حِلَالٌ : see حَلَالٌ.

حَلِيلٌ : see حَلَالٌ. — Also *A fellow-lodger,* or *fellow-resident,* of another, *in one house:* fem. with ة. (Ṣ.) — And hence, (TA,) *A husband:* (Ṣ, Mṣb, Ḳ:) and with ة *a wife;* (Ṣ, Mgh, Mṣb, Ḳ;) as also without ة : (Ḳ:) or they are so called because each occupies a place, in relation to the other, which none beside occupies: (Mṣb:) [but there may be two or more wives to one man:] accord. to some, they are so called because the husband is lawful to the wife, and the wife to the husband: but the word, [or rather each word,] thus applied, is ancient: not a law-term: the pl. is حَلَائِلُ. (TA.) — Also *A neighbour:* (Mṣb:) fem. with ة. (Ṣ.) — And *A guest.* (Mṣb.)

حَلَّالٌ † *One who solves astronomical problems.* (TA.)

حُلَّانٌ : see تَحِلَّةٌ. — دَمُهُ حُلَّانٌ † *His blood goes for nothing;* or *is of no account.* (Ḳ.)

حَلْحَالٌ [*The act of chiding a she-camel,* or *she-camels, by the cry* حَلْ : a subst. from حَلْحَلَ ; like زَلْزَالٌ from زَلْزَلَ : or] a subst. derived from حَلْ, or حَلِ. (TA.)

حُلَاحِلٌ *A grave, staid,* or *sedate, chief:* (Ṣ:) or one who is *grave, staid,* or *sedate, in his sitting-place;* *a chief among his kinsfolk:* (TA:) or *a courageous chief:* or *a portly man, characterized by much manly virtue:* or *grave, staid,* or *sedate, with a forbearing,* or *clement, disposition:* applied to a man: (Ḳ:) never to a woman: (TA:) and ▼ مُحَلْحَلٌ signifies the same: (Ḳ:) or the former, *a chief with whom men often alight,* or *abide:* (Ḥar p. 69:) pl. حَلَاحِلُ. (Ṣ.)

حَالٌّ [*Untying, undoing,* or *opening,* a knot:]

act. part. n. of حَلَّ in the phrase حَلَّ العُقْدَةَ. (Mṣb.) — [And hence, (see 1,)] *Alighting, or descending and stopping or sojourning or abiding or lodging or settling; or simply taking up one's abode; or abiding, lodging, or settling; in a place;* syn. نَازِلٌ; (Ḳ, TA;) as also ↑ حِلٌّ; occurring in the Ḳur xc. 2: (TA:) pl. of the former حُلُولٌ and حُلَّالٌ and حُلَّلٌ, (Ḳ,) and quasi-pl. n. ↑ حِلَّةٌ; (TA [in which it is in one place called a pl. (not a quasi-pl. n.) of حَالٌّ];) and the pl. of حِلَّةٌ is حِلَالٌ. (TA.) You say حَىٌّ حُلُولٌ *A tribe that is [abiding] in one place.* (Ḥam p. 171.) And ↑ قَوْمٌ حِلَّةٌ *A people, or party, alighting, &c.,* (Ṣ, Mṣb, Ḳ,) *and comprising a numerous company:* and in like manner, ↑ حَىٌّ حِلَّةٌ, (Ṣ,) and حَىٌّ حِلَالٌ, (Ṣ, TA,) *a numerous tribe [alighting, &c.].* (TA.) [See also نَفْرٌ.] — Hence, الحَالُّ المُرْتَحِلُ † *He who completes the reading, or reciting, of the Kur-án, and then immediately recommences it;* likened to him who travels much, and does not come to his family: or *the warrior who does not return from his warring.* (TA.) — دَيْنٌ حَالٌّ † *A debt of which the appointed term, or period, is ended;* (Mṣb;) *a debt falling due;* (TA;) contr. of مُؤَجَّلٌ. (Mgh.) — See also مُحَلَّلٌ.

أَحَلُّ *Having what is termed* حَلَلٌ [q. v.]: fem. حَلَّآءُ: and pl. حُلٌّ, applied to horses, (Ḳ, TA,) and to camels, and to wolves: (TA:) a camel *having a weakness in the* عُرْقُوب [i. e. *hock,* or *hock-tendon*]: (Fr, Ṣ:) and *having a laxness in his legs:* it is discommended in everything, except the wolf. (Ṣ.) — The fem., applied to a woman, signifies *Having little flesh in the posteriors and thighs; or having small and close buttocks; or having little flesh in the thighs.* (TA.)

إِحْلِيلٌ (Ṣ, Mṣb, Ḳ) and ↑ تَحْلِيلٌ (Ḳ) *The orifice through which the urine passes forth* (Ṣ, Mṣb, Ḳ) *from the penis of a man:* (Ḳ:) and *the orifice through which the milk passes forth from the breast* (Ṣ, Mṣb, Ḳ) and *from the udder.* (Ṣ, Mṣb.)

تَحِلٌّ: see 2, near the beginning.

تَحِلَّةٌ: see 2, in nine places: and see also 4. — Also † *A thing with which an oath is expiated;* (Ḳ;) [and so ↑ حُلَّانٌ; as in the saying,] أَعْطِهِ حُلَّانَ يَمِينِهِ † *Give thou to him that with which he may expiate his oath.* (ISd, Ḳ.)

تَحْلِيلٌ: see إِحْلِيلٌ.

مَحَلٌّ *A place where a person or party alights, or descends and stops or sojourns or abides or lodges or settles; a place of alighting, or descending and stopping &c.; or simply where one takes up his abode, abides, lodges, or settles;* (Ṣ, Mṣb, TA;) as also ↑ مَحِلٌّ: (Mṣb:) and ↑ مَحَلَّةٌ (Ṣ, Mṣb, Ḳ) and ↑ حِلَّةٌ, (Ṣ,) or ↑ حَلَّةٌ, (Ḳ,) signify *a [particular, or special,] place of alighting, or descending and stopping &c.,* (Ṣ, Mṣb, Ḳ,) *of a people or party:* (Ṣ, Mṣb:) the pl. of مَحَلٌّ is مَحَالٌّ; and the pl. of مَحَلَّةٌ is مَحَلَّاتٌ. (TA.) You say, هُوَ فِى حِلَّةِ ↑ صِدْقٍ, i. e. فِى مَحَلَّةِ ↑ صِدْقٍ, [*He is in a good, or an excellent, place of alighting, &c.*]. (Ṣ.) — [Hence, † *A place,* in a general sense. Thus in the phrase, حَلَّ مَحَلَّ كَذَا: see 1. And in the phrases, used in grammar, مَحَلُّهُ الرَّفْعُ † *Its place in construction is that of the nominative case;* and مَرْفُوعٌ مَحَلًّا † Virtually *in the nominative case by reason of the place which it occupies* in construction; and the like.] — [Hence, also,] a term applied by Ks to † *An adverbial noun of place or time.* (T voce ظَرْفٌ.) — [Hence, also, † *A person, considered as one in whom some quality has place.*] You say, هُوَ مَحَلٌّ لِأَنْ يُقَالَ فِيهِ إِنَّهُ لَخَيْرٌ وَعَسَى أَنْ يَفْعَلَ خَيْرًا † [*He is a person fit, or proper, for one's saying of him, Verily he is good, and may-be he will do good*]. (A and TA in art. ان.) ═ It is also an inf. n. (Ṣ, TA. [See 1.])

مَحِلٌّ: see مَحَلٌّ. — Also The *lawful place of slaughter of a beast for sacrifice;* (Ṣ;) accord. to some, *to the pilgrim on the general day of sacrifice, and to the performer of the* عُمْرَة *on the day of his entering Mekkeh;* or, as others say, *to him who is in the state of* إِحْرَام. (TA.) — And The *term, or period, of falling due* of a debt. (Ṣ, TA.) ═ It is also an inf. n. (Ḳ, TA. [See 1.])

مُحَلٌّ: see مُحَلَّلٌ.

مُحِلٌّ [*Making* one *to alight,* or *descend and stop* &c.]. [Hence,] المُحِلَّتَانِ † *The cooking-pot and the hand-mill:* and المُحِلَّاتُ *the cooking-pot and the hand-mill and the bucket and the knife and the axe and the instrument for striking light* (قَدَّاحَة, Ṣ, or زَنْد, Ḳ) *and the water-skin* (Ṣ, Ḳ) *and the bowl:* (Ḳ:) for he who has with him these things alights, or abides, wheresoever he will; but he who has not must be near to persons from whom he may borrow some one or more thereof. (Ṣ.) [Hence, also,] تَلْعَةٌ مُحِلَّةٌ † A تلعة [q. v.] *comprising one* بَيْت [or tent], or *two.* (O, Ḳ.) — † [*Making* a debt *to fall due.*] The Arabs used to say, when they saw the new moon, لَا مَرْحَبًا بِمُحِلِّ الدَّيْنِ وَمُقَرِّبِ الآجَالِ † [*No welcome be to that which makes the debt to fall due, and makes near the appointed periods!*]. (TA.) — See also حَلَالٌ. — Also † One *with whom it is lawful to fight:* (Ṣ in art. حرم:) or *whom it is lawful to slay:* (TA:) contr. of مُحْرِمٌ, in the former sense, (Ṣ ubi suprà,) or in the latter sense. (TA.) — † One *who has no claim, or covenanted right, to protection, or safeguard, or respect;* (Ṣ, TA;) contr. of مُحْرِمٌ, in this sense also. (Ṣ.) — † A man *who violates that which is sacred:* or *who does not hold that there is any sacredness pertaining to the sacred month.* (Ḳ.) — See also 4, last sentence but one.

مَحَلَّةٌ: see مَحَلٌّ, in two places: — and see also حِلَّةٌ.

مُحَلَّلٌ: see مِحْلَالٌ. Also Any water *at which camels have abode, and which they have consequently rendered turbid.* (Ḳ.) ═ مُحَلَّلٌ لَهُ A man *whose remarriage to his wife whom he has trebly divorced has been made lawful to him by her having been married to, and divorced by, another man;* (Mgh,* TA;) as also ↑ مُحَلٌّ لَهُ (Mgh) and ↑ مَحْلُولٌ لَهُ (TA) and ↑ حَالٌّ, (El-Kurkhee, Mgh,) or this last signifies [properly] one *whose wife is lawful to him.* (TA.) ═ مُحَلَّلٌ also signifies A thing *little in quantity.* (Ḳ.)

مُحَلِّلٌ † A man *who marries a woman that has been trebly divorced,* (Ṣ, Mṣb, Ḳ, TA,) *on the condition of his divorcing her after consummation of the marriage,* (TA,) *in order that she may become lawful to [be remarried to] the former husband.* (Ṣ, Mṣb, Ḳ, TA.) — In a case of racing, † *He that intervenes between two contending for a stake or stakes,* (Ṣ,) or the *third horse in a contest for a stake or stakes;* (Ḳ;) *if he outstrip, he takes [the stake or stakes]; and if he be outstripped, he is not fined:* (Ṣ, Ḳ:) the case is this: two men lay two stakes; and then another comes, and starts his horse with the two others, without [laying] a stake; if one of the first two outstrip, he takes the two stakes, and this is lawful because of the third; but if the مُحَلِّل outstrip, he takes the two stakes; and if he be outstripped, there is no fine for him: he must be a horse of which one is sure that he may outstrip; otherwise it is termed قِمَار: and he is also called دَخِيل: (TA:) the مُحَلِّل in racing is so called because he makes lawful the contest for a stake or stakes, which had otherwise been unlawful. (Mṣb.)

مِحْلَالٌ A place, (Ṣ,) or a meadow (رَوْضَةٌ), (Ḳ,) and a land (أَرْضٌ), (TA,) and a house (دَارٌ), (Mgh and Mṣb in art. اتى,) *in which people alight, or descend and stop, or abide, much,* or *often;* (Ṣ, Ḳ, TA, and Mgh and Mṣb ubi suprà;) as also ↑ مُحَلَّلٌ applied to a place: (Ṣ, TA:) or *chosen as a place of alighting, &c.:* or, accord. to ISd, *that makes [or invites] people to alight, &c., in it much,* or *often;* because a word of the measure مِفْعَالٌ has only the meaning of an act. part. n.: and, as some say, a meadow and a land are only thus called if *abounding with herbage wholesome to the cattle.* (TA.)

مَحْلُولٌ: see مُحَلَّلٌ.

مُحَلْحَلٌ: see حُلَاحِلٌ.

حلا

1. حَلَاهُ, aor. ـُ ; and ↑ أَحْلَاهُ; *He applied the collyrium called* حُلَاءَة *and* حُلُوء *to his eyes:* (Ḳ:) or, accord. to AZ, ↑ أَحْلَاهُ, inf. n. إِحْلَآءٌ, signifies, *he rubbed for him powder from two stones, and applied their powder as a collyrium to his eyes when they were diseased:* (TA:) and accord. to ISk, حَلَأَ لَهُ حَلُوءًا signifies *he rubbed for him a stone upon another stone, then put the powder [thus obtained] upon the palm of his hand, and rubbed off with it the rust of a mirror,* [see صَدَأٌ and صَدًا, the mirror being of bronze, or other metallic substance,] *then applied it as a collyrium*

to his eyes. (Ḳ,* Ṣ.) ═ حَلَأَهُ, inf. n. حَلْءٌ, *He flogged him* with a whip. (Ṣ.) —— And, as also ↓حَلَّأَهُ, *He struck him* with a sword, (Ṣ, Ḳ,) or a staff or stick. (TA.) —— حَلَأَ بِهِ الأَرْضَ *He threw him down on the ground, prostrate:* (Ḳ:) like جَلَأَ به الارض, which, accord. to Az, is a dial. var. of حلأ. (TA.) —— حَلَأَهَا ‡ *He lay with her; or compressed her.* (Ḳ, TA.) ═ حَلَأَهُ, (Ṣ, Ḳ,) and ↓حَلَّأَهُ, (Ḳ,) and ↓احلأه, (TA,) *He gave him* money. (AZ, Ṣ, Ḳ.) [Hence,] مَا حَلَئْتُ مِنْهُ بِطَائِلٍ [*I gained not*, or *derived not, any great profit from him*, or *it*]. (T.) [See also 1 in art. حلى.] ═ حَلَأَ الجِلْدَ, aor. َ, inf. n. حَلْءٌ and حَلَأَةٌ, *He* (a currier) *shaved the hide;* (Ṣ, Ḳ;) i. e., *removed what remained of the flesh.* (Ḳ.) —— Hence the prov., حَلَأَتْ حَالِئَةٌ عَنْ كُوعِهَا [*A woman shaving a hide grazed the skin of the extremity of the bone of her fore arm next the thumb:* see also حَزَّ]: for the dexterous woman sometimes hurries, and so grazes the skin of her wrist-bone. (Ṣ.) The prov., however, is differently explained: see حَالِئَةٌ. (TA.) —— حَلَأْتُ الصُّوفَ, inf. n. حَلْءٌ; as also حَلَتُّهُ; *I tore the wool* from the sheep. (Lḥ, TA in art. حلت.) ═ حَلِئَ الأَدِيمُ, inf. n. حَلَأٌ, *The hide had in it what is called* تِحْلِئٌ. (Ṣ.) —— حَلِئَ *He had pustules* (حَلَأ, for which is put in the Ḳ تِحْلِئ) *upon his lips after a fever.* (TA.) And حَلِئَتِ الشَّفَةُ *The lip broke out with pustules after an illness;* (Ṣ, Ḳ;) as also حَلِيَت. (T.)

2: see 1, in two places. ═ حلّأ, inf. n. تَحْلِئٌ and تَحْلِئَةٌ, *He drove away, and debarred,* (camels or other animals, Ṣ, or people, TA,) from the water. (Ṣ, Ḳ.) حَلَّيْتُم occurs in a trad. for حَلَّأْتُم, like قَرَيْتُ for قَرَأْتُ, contr. to analogy; it being a rule not to change hemzeh into ى unless the next preceding letter is meksoor. (TA.) ═ حَلَّأَ السَّوِيقَ, inf. n. تَحْلِئَةٌ; as also ↓احلأ; *He sweetened the* سويق [or *mess made of the meal of parched barley*]: but hemzeh does not properly belong to this verb; for it is from الحَلْوَآءُ. (Fr, Ṣ, Ḳ.) [See 2 in art. حلو.]

4: see 1, in three places: ═ and see also 2.

حَلَأٌ *Pustules breaking out upon the lips after a fever.* (Ṣ, Ḳ.) [See also حَلًا, in art. حلى.]

حُلَأَةٌ: see حَلُوءٌ.

حِلَأَةٌ *A land abounding with trees:* (Ḳ:) or *the name of a certain place,* (Ḳ,) *intensely cold;* (TA;) as also حِلَآءَةٌ. (Ḳ.)

حُلَأَةٌ and ↓حَلُوءٌ *What is rubbed between two stones, to be applied as a collyrium* (Ṣ, Ḳ) *for a pain in the eyes:* (TA:) [but see the verb, in the explanations of which this collyrium seems to be more correctly described:] or حَلُوءٌ is *a stone which a person with diseased eyes uses as a remedy:* (Ḳ:) or, accord. to ISk, *a stone that is rubbed upon, and then used as a collyrium;* [i. e., its powder is so used.] (TA.) ↓حَلُوءٌ تُحَكُّ بِالذَّرَارِيحِ [*A powder for the eyes, that is rubbed together with cantharides,*] is a prov., applied to him whose words are fair, and whose actions are foul. (TA.) —— حُلَأَةٌ also signifies *That which a currier shaves off from the inner side of a hide.* (Ṣ, Ḳ.)

حَلُوءَةٌ: see the next preceding paragraph.

حَالِئَةٌ *A malignant serpent,* (Sh, Ḳ,) *the action of which, in poisoning him whom it bites, is like that of the oculist who rubs powder* [*from two stones*] *for him who has diseased eyes, and applies it to them.* (Sh.) [Hence, accord. to some, the prov. above mentioned, as is stated (but without explanation) in the TA.]

تِحْلِئٌ and ↓تِحْلِئَةٌ The *hair on the surface of a hide, and its dirt, and blackness:* (Ḳ:) or *what is pared off from the back of a hide.* (Lḥ, TA in art. بشر.) —— Also *What the knife spoils, of a hide, in the process of shaving it.* (Ṣ, Ḳ.) —— رَجُلٌ تِحْلِئَةٌ ‡ *A heavy,* or *dull,* or *troublesome, man,* (TA,) *who sticks to another* [*like dirt*], *and vexes him.* (Ḳ.)

تِحْلِئَةٌ: see the next preceding paragraph.

مِحْلَأٌ: see what next follows.

مِحْلَأَةٌ *A currier's knife, used for shaving the inner surface of the hide:* (Ḳ:) and ↓مِحْلَأٌ the *iron instrument, or stone, with which one shaves off the* تِحْلِئ *of a hide, and with which one skins.* (TA voce مِحْسَرٌ, q. v.)

حلب

1. حَلَبَ, (Ṣ, Mgh, Mṣb, Ḳ,) aor. ُ (Ṣ, Mṣb, Ḳ) and ِ, (Ḳ,) inf. n. حَلْبٌ (Ṣ, A, Mṣb, Ḳ) and حَلَبٌ (A, Mgh, Mṣb, Ḳ) and حِلَابٌ; (Ḳ;) and ↓احتلب; (Ṣ, Ḳ;) *He milked* (TA) a she-camel (Ṣ, Mgh, Mṣb, TA) &c., (Mṣb,) a ewe, a she-goat, and a cow: (TA:) *he drew forth the milk in the udder:* (A, Ḳ: [see also 10:]) and حَلَبَ اللَّبَنَ [*he drew the milk from the udder*]. (Ṣ and Ḳ in art. رثأ, &c.) حَلَبْتَ بِالسَّاعِدِ الأَشَدِّ [*Thou hast milked with the strongest fore arm*] is a prov., meaning thou hast asked aid of him who will perform thine affair, or thy want: (TA:) or, accord. to A'Obeyd, حَلَبْتُهَا بالشد *I have milked her* &c., meaning I have taken it by force when I could not by gentle means. (TA in art. شد.) And حُلِبَتْ صُرَامِ, (TA,) or صُرَامُ, (Ṣ, and some copies of the Ḳ, in art. صرم,) *The last milk was,* or *has been, drawn from the udder,* is another prov., used in a case when an evil has attained its crisis: (TA:) or it means †*the excuse reached,* or *has reached, its utmost point:* (AO, Ṣ and Ḳ in art. صرم:) or *the she-camel that had* [*little or*] *no milk was milked,* or *has been milked;* denoting a calamity. (IB, TA in art. صرم.) And إِنْ كُنْتَ كَاذِبًا فَحَلَبْتَ قَاعِدًا *If thou lie, mayest thou milk sitting;* i. e., lose thy camels, and become an owner only of sheep or goats, and thus, after having milked camels standing, milk sheep or goats sitting: this, also, is a proverbial saying, like the following: مَا لَهُ حَلَبَ قَاعِدًا وَأَصْطَبَحَ بَارِدًا [*What aileth him?*] *May he* [be reduced to] *milk* a sheep or goat *sitting, and drink cold* water, not hot milk. (TA.) And حَلَبَ الدَّهْرَ أَشْطُرَهُ †*He experienced good fortune and evil,* is another prov. (TA. [See this and other exs. in art. شطر.]) So, too, حَلَبَتْ ↓حَلْبَتَهَا ثُمَّ أَقْلَعَتْ (TA [but not there explained]) [lit. *She performed her act of milking,* and *then desisted:* but, as explained in Freytag's Arab. Prov. (i. 343 and 281), meaning †*It* (a cloud, سَحَابَة,) *sent forth a fine rain,* and *then ceased:* and some read أَحْلَبَتْ ↓حَلَبَةً, meaning the same: see another reading voce جَلَبَ]. —— [Hence, حَلَبَهُ †*He mulcted him:* see an ex. voce قَشَّ: and see حَلَبٌ.] —— [Hence also,] حَلَبَ, (A, Ḳ,) aor. ُ, (A,) inf. n. حَلْبٌ, (TA,) ‡ *He set himself upon his knees, in the posture of the milker:* (A:) *he sat on his knees;* (Ḳ;) or *on his knee: he sat on his knee* in eating, or in milking a ewe or she-goat: *he kneeled.* (TA.) You say, أَحْلُبْ وَكُلْ ‡[*Kneel thou like the milker, and eat*]: (A:) it denotes a lowly [and becoming] mode of sitting while eating. (TA.) لَيْسَ فِي كُلِّ حِينٍ أَحْلُبُ فَأَشْرَبُ ‡[*Not at every time is it said, kneel thou, and understand*] is a prov. applied in the case of anything that is forbidden, or denied: AA says that الحلب signifies *the act of kneeling;* and الشرب *the understanding* a thing: and one says to a stupid person, أَحْلُبْ ثُمَّ آشْرَبْ *Kneel thou; then understand:* one says also, ليس كل حين احلب فاشرب [in Freytag's Arab. Prov., ii. 437, thus: لَيْسَ كُلَّ حِينٍ أَحْلُبُ فَأَشْرَبُ, and explained as meaning *Not every time permits to milk and then to drink:* i. e. not every time aids thee in performing a work; therefore thou shouldst act prudently, and not expend thy wealth without rule and measure.] (TA.) ═ حَلَبَ الرَّجُلَ, aor. ُ, *He milked for the man.* (Ṣ.) —— حَلَبَهُ and ↓احلبه *He assigned to him, to be milked by him,* a ewe or she-goat, and a she-camel: (Ḳ:) or the latter, *he assigned to him what he should milk.* (Ṣ.) ═ حَلَبُوا, (Ḳ,) aor. ُ, (TA,) inf. n. حَلْبٌ and حُلُوبٌ, *They assembled,* or *collected themselves together, from every quarter.* (Ḳ, TA.) [See also 4.] ═ حَلِبَ, aor. َ, *It* (hair &c.) *was,* or *became, black.* (Ḳ.)

3. حالبه *He milked with him.* (Ḳ.) —— See also 4. —— حَالَبْتُهَا, inf. n. مُحَالَبَةٌ, *She vied with her in patience during milking.* (L.)

4. احلب أَهْلَهُ, (Ṣ,) inf. n. إِحْلَابٌ (Ḳ) and إِحْلَابَةٌ, (Ṣ, Ḳ,) [which latter see also below,] *He milked for his family, while he was in the place of pasturage, and then sent to them the milk there drawn by him:* (Ṣ, Ḳ:) or *he conveyed to his tribe what had been milked while the camels were in the places of pasturage, and had been collected to the quantity of a camel-load.* (TA.) —— See also 1, in three places. [In the last of those instances, the verb, as explained in the Ḳ, is doubly trans.; and hence,] —— احلبه is also used as meaning † *He gave him* a thing. (TA.) —— Also *He assisted him to milk,* or *in milking.* (Ṣ, A, Mgh, Ḳ.) —— And hence, (A, Mgh,) as also ↓حالبه, (Ṣ,) by extension, (A,) in a general sense, (Mgh,) ‡*He*

assisted him, or *aided him*. (Ṣ, A.) And احلب غَيْرَ قَوْمِهِ †*He entered among a party*, or *people, not his own, and aided some of them against others.* (TA.) And احلبوا †*They aided their companions.* (TA.) †*They assembled*, or *collected themselves together, from every quarter, to render aid*, عَلَيْهِ *against him*; (Ṣ;) like اجلبوا; (Ṣ in art. جلب;) as also ↓استحلبوا: (TA:) †*they collected themselves together from every quarter for war &c.*: (Az, TA:) †*they assembled from every quarter*, عَلَيْهِ *against him.* (TA. [See also 1.]) ═ احلب *His camels brought forth females*: opposed to اجلب "his camels brought forth males." (Ṣ, A, Ḳ.) One says, أَأَحْلَبْتَ أَمْ أَجْلَبْتَ *Have thy camels brought forth females, or have they brought forth males?* (M, Ḳ.) See also أَجْلَبَ.

5. تحلّب *It flowed*; (Ṣ, A, Ḳ, KL;) [or *oozed*, or *exuded*;] said of milk; (KL;) and ‡of water; (A;) and ‡of sweat, (Ṣ, A, Ḳ,) as also ↓انحلب; (Ṣ;) and ‡of moisture, or dew. (L.) — ‡*It* (one's body) *flowed*, عَرَقًا *with sweat*: and in like manner, the eye [with tears]; (Ḳ;) and the mouth [with saliva]; (A, Ḳ;) as also ↓انحلب. (Ḳ.) — †*He sweated.* (TA.) — It is also said of the [tribute termed] فَىْء [as meaning †*It flowed in*; or *was collected*: see حَلَبٌ]. (TA.)

7: see 5, in two places.

8: see 1, first sentence.

10. استحلب *He drew forth* milk. (Ṣ, A, Ḳ.*) — [Hence,] اِسْتَحْلَبَتِ الرِّيحُ السَّحَابَ (A, TA) ‡*The wind drew forth a shower of fine rain from the clouds*; or *caused them to send forth fine rain.* (TA.) [And اِسْتَحْلَبَهُ فِى فِيهِ †*He sucked it in his mouth so as to draw forth its moisture* or *what dissolved thereof*: see an ex. voce مُرٌّ.] — نَسْتَحْلِبُ الصَّبِيرَ نَسْتَخِيرُ, occurring in a trad., means †[*We desire*, or *look for, a shower of rain from the white clouds*]. (TA.) — See also 4.

حَلْبٌ: see حَلَبَةٌ.

حَلَبٌ is an inf. n.: (Ṣ, A, Mṣb, Ḳ: see 1:) — and also signifies *Milk drawn from the udder*; (Ṣ, A,* Mgh, Ḳ;) or so لَبَنٌ حَلَبٌ; (Mṣb;) and so ↓حَلِيبٌ; (Ṣ, A, Ḳ;) or لَبَنٌ حَلِيبٌ; (Mṣb;) and ↓حِلَابٌ: (TA:) or (Ḳ, TA, in the CḲ "and") ↓حَلِيبٌ signifies [*fresh milk*, i. e.] *milk of which the taste has not become altered*; (Ḳ, TA;) and حَلَبٌ is thought by ISd to be used in this sense. (TA.) — [Hence,] ‡The [*tax called*] جِبَايَة: (A:) or the *kind of* جباية (Ṣ, Ḳ) *that is similar to the* صَدَقَة *and the like*, (Ḳ,) *whereof the assessment is not certain*, or *defined*: (Ṣ, Ḳ:) pl. أَحْلَابٌ. (A, TA.) The pl. also means †*Profits*, or *advantages*, such as accrue to a commander, or governor. (TA in art. رضع.) — ‡*An evil result*: so in the saying, ذَاقُوا حَلَبَ أَمْرِهِمْ ‡[*They tasted the evil result of their affair*, or *action*]. (A.) — مَا لَهُ لَا حَلَبَ وَلَا جَلَبَ, mentioned by IAạr, but not explained by him, (TA,) is said to be a form of imprecation [meaning *What aileth him? May he have neither she-camels nor he-camels*]; (Ḳ;) and this is the opinion generally held: (TA:) but some say that there is no reason for this [assertion; holding the meaning to be, *he has neither she-camels nor he-camels*; the former لا being redundant: see 4; and see also جَلَبٌ]. (Ḳ.) ═ Also The *covering, exterior part, peel*, or *the like*, (syn. قِشْر,) of anything. (Kr, TA.)

حُلُبٌ [a pl. of which the sing. is not mentioned,] *Black*; as applied to animals. (Ḳ. [See also حُلْبُوبٌ.]) — And *Intelligent*; as applied to men. (Ḳ.)

حَلْبَةٌ [*A single act of milking*:] see 1. — [*A time of milking*. And hence,] الحَلْبَتَانِ *The morning and evening*; (IAạr, Ḳ;) because they are the two milking-times. (TA.) — [†*A fine rain*; or *a shower of fine rain*: pl. حَلَبَاتٌ: the sing. occurring in the TA in art. هضب, and the pl. in the same and in the Ṣ in that art.: see also 1 in the present art.] ═ *A number of horses started together for a wager*: (Ḳ:) *horses assembled from every quarter for a race*, (Ṣ, Mgh, Mṣb, Ḳ,) *not from one stable*, (Ṣ, Ḳ,) or *not from one quarter*: (Mṣb:) or *horses that come from every quarter to aid*: (A: [but this is probably a false rendering, occasioned by an omission, which has combined portions of explanations of two words:]) pl. حَلَائِبُ, (Mṣb, Ḳ,) because the sing. has the meaning of ↓حَلِيبَةٌ, (Mṣb,) [as pl. of حَلْبَةٌ irreg., and حِلَابٌ and حَلَبَاتٌ. (TA.) You say, جَآءَتِ الفَرَسُ فِى آخِرِ الحَلْبَةِ *The mare came among the last of the horses* [*in the race*]. (Mṣb.) And فُلَانٌ سَابِقُ الحَلَائِبِ †[*Such a one is the winner in races*, or *in contests*]. (TA.) — And *A race-ground.* (A.) You say, فُلَانٌ يَرْكُضُ فِى كُلِّ حَلْبَةٍ مِنْ حَلَبَاتِ المَجْدِ ‡[*Such a one urges on in every scene of glorious contest*]. (A, TA.)

حُلْبَةٌ [*Fenugreek*; *trigonella fœnumgræcum* of Linn.;] *a certain grain*, (Ṣ, Mgh, Mṣb,) *well known*, (Ṣ, Mgh,) *which is eaten*; also pronounced ↓حُلُبَةٌ: (Mṣb:) *a certain plant*, (AḤn, Ḳ,) *having a yellow grain, used medicinally; and made to germinate* [*in a vessel of water*], *and eaten*; (AḤn, TA;) *useful as a remedy for diseases of the chest, for cough, asthma, phlegm, and hæmorrhoids, for giving strength to the back, for the liver and the bladder, and as a stimulant to the venereal faculty*, (Ḳ,* TA,) *alone or compounded*; *and a common article of food of the people of El-Yemen*: pl. [or rather coll. gen. n.] ↓حُلْبٌ. (TA.) — The [*plant otherwise called*] عَرْفَج. (AḤn, Ḳ.) — The قَتَاد [or *tragacantha*]. (AḤn, Ḳ.) — The *leaves of the* عِضَاه *when they have become harsh and dry, and dusty* or *dust-coloured, and when its branches and thorns have become thick*: (TA:) or it is [*what is in a similar state*] *of the fruit of the* عضاه: (IAth, TA:) the word is sometimes pronounced ↓حُلُبَةٌ. (TA.) — The *kind of food called* فَرِيقَة, (Ḳ, TA,) *which is given to women when childbearing*; (TA;) as also ↓حُلُبَةٌ. (Ḳ.) ═ *A pure black colour.* (Ḳ.) [See حُلْبُوبٌ.]

حُلُبَةٌ: see the next preceding paragraph, in three places.

حَلَبَى: see حَلُوبٌ, in two places.

حَلَبَاةٌ A female slave *who kneels by reason of indolence.* (TA.)

حَلْبَاةٌ; and its pl. حَلَبَاتٌ: see حَلُوبٌ, in three places.

حُلْبُوبٌ *Black* hair &c. (T, Ḳ. [See also حُلُبٌ.]) And أَسْوَدُ حُلْبُوبٌ *Intensely black.* (Ṣ.)

حَلَبُوتٌ:
حَلَبُوتَى:
حَلَبَانَةٌ:
} see each in two places voce حَلُوبٌ.

حِلَابٌ: see حَلَبٌ: — and مِحْلَبٌ. ═ It is also a pl. of حَلْبَةٌ, as shown above. (TA.)

حَلُوبٌ and ↓حَلُوبَةٌ (of which the latter is the more common, TA) A she-camel *that is milked*; (Ḳ;) both signify alike: (TA:) or the former is an epithet, signifying as above; and the latter is a subst., signifying the *animal that is milked*; (Ṣ,* A, Mgh, Mṣb, TA;) though some say the reverse: or sometimes the former is used for the latter, meaning *a milch camel*, &c.: accord. to Meyd, the latter signifies *a she-camel that is milked for the guest, and for the people of the tent or house*: (TA:) the former is used by some as a sing., and by others in a pl. sense: (IB, TA:) and [in like manner] the latter is applied to a single she-camel or ewe or she-goat, and to more: (Ḳ:) the pl. (of the latter, TA) is حَلَائِبُ (Ṣ, Ḳ) and حُلُبٌ; (Ḳ;) and حُلْبٌ, supposed to be a contraction of حُلُبٌ, also occurs as a pl. epithet applied to ewes and to she-goats. (Lḥ, TA.) You say حَلُوبَةٌ تُثْمِلُ وَلَا تُصَرِّحُ *A milch camel that gives much froth in her milk, and does not give pure*, or *clear, milk*: a prov., applied to him who promises much, but performs little. (Meyd, TA.) And دَرَّتْ حَلُوبَةُ المُسْلِمِينَ †[*The milch camel of the Muslims has yielded a copious supply of milk*] is said when the dues of the government-treasury are in a good state. (IAạr, Suh, TA.) ↓حَلَبَانَةٌ, also, signifies A she-camel *having milk*; (IAạr, Ṣ, Ḳ;) *that is milked*; a *milch* camel; (A, Ḳ;) like حَلُوبٌ; (TA;) and so ↓حَلْبَاةٌ (IAạr, Ḳ) and ↓حَلَبُوتٌ (ISd, Ḳ) and ↓حَلَبَى and ↓حَلَبُوتَى, (Ḳ,) like as they said رَكَبَى and رَكَبُوتٌ and رَكْبَاةٌ and رَكَبَانَةٌ (TA) and رَكَبُوتَى: (Ḳ:) or *fit to be milked*: (Ṣ and TA voce رَكُوبٌ:) and حَلُوبٌ and the rest of the foregoing epithets, except حَلَبُوتٌ, [which I nevertheless believe to be perfectly syn. with them, like as خَلَبُوتٌ is syn. with خَلَّابٌ accord. to the Ṣ,] are also mentioned as having an intensive signification. (TA.) You say ↓نَاقَةٌ حَلَبَانَةٌ رَكَبَانَةٌ (A, Ḳ) and ↓حَلْبَاةٌ رَكْبَاةٌ (TA) [and ↓حَلَبُوتَى رَكَبُوتَى and ↓حَلَبَى رَكَبَى and ↓حَلَبُوتٌ رَكَبُوتٌ] (Ḳ) *A she-camel that is milked and ridden*: (A, Ḳ:) or *that yields abundance of milk and that is submissive to be ridden.* (TA.) AZ mentions ↓نَاقَةٌ حَلَبَاتٌ, the latter word in the pl. form; as also نَاقَةٌ رَكَبَاتٌ. (TA. [But in each case I think that the ت is a mistake for ة.]) ═ See also حَالِبٌ. — [Hence,] هَاجِرَةٌ حَلُوبٌ † [*A summer-midday*] *that draws forth the sweat.* (Ḳ.)

حَلِيبٌ : see حَلَبٌ, in two places. — Also ‡ *A beverage* [*of the kind termed* نَبِيذ,] *prepared from dates.* (Ḳ, TA.) — And † *Fresh blood.* (Ḳ.)

حِلَابَةٌ [or perhaps حُلَابَةٌ, like عُصَارَةٌ &c., † *An exuding fluid*]. (AḤn, TA voce نفط, q. v.)

حَلُوبَةٌ : see حَلُوبٌ.

حَلِيبَةٌ : see حَلَبَةٌ: — and see what next follows.

حَلَائِبُ used as a pl. of حَلَبَةٌ [q. v.], because the latter has the meaning of ▼ حَلِيبَةٌ. (Mṣb.) — Also *Companies, assemblies,* or *troops.* (Ḳ.) — And The *sons of the paternal uncle:* (Ḳ:) or a man's *assistants,* or *auxiliaries, consisting of the sons of the paternal uncle in particular.* (TA.)

حُلَّبٌ *A certain plant,* (Ṣ, Ḳ,) *that grows in the hot season, in the plains and on the sides of valleys, cleaving to the ground so as almost to be buried in it, not eaten by the camels, but only by the sheep or goats,* (TA,) *and by the gazelles:* (Ṣ, TA:) *it increases the milk, and fattens; and gazelles are snared* [*while pasturing*] *upon it* (تحتبل عليها): (TA:) hence the expressions تَيْسٌ ذُو حُلَّبٍ and تَيْسُ الحُلَّبِ [*a buck-gazelle that feeds upon the* حلّب]: (Ṣ:) *it is a curling herb, of a dusty colour inclining to green, that spreads upon the ground; when a piece of it is cut off, a milky fluid flows from it:* (Aṣ, Ṣ:) AḤn says, *it is a plant that spreads upon the ground, evergreen, having small leaves, with which they tan:* Aboo-Ziyád says, *it is included among what are termed* الخِلْفَةُ, *and is a tree that expands over the ground, cleaving thereto, intensely green, growing most when the heat becomes great:* and he adds, on the authority of Arabs of the desert, that *it lies upon the ground, having small and bitter leaves, and a root penetrating deep into the earth, and small twigs: it is of the kind of plants termed* رِبْحَةٌ. (TA.)

حُلَّبِيٌّ A skin for water or milk *tanned with* [*the leaves of*] *the* حُلَّب; (Ṣ, Ḳ;) as also ▼ مَحْلُوبٌ. (Ḳ.)

يَوْمٌ حَلَّابٌ † *A dewy day.* (Sh, Ḳ.)

حَالِبٌ *A milker;* (Ṣ, Ḳ;) as also ▼ حَلُوبٌ; (Ḳ;) but the latter has an intensive signification: (TA:) pl. حَلَبَةٌ. (Ṣ, A.) You say, هُمْ حَلَبَةُ الإِبِلِ [*They are the milkers of the camels*]. (A.) And شَتَّى تَؤُوبُ الحَلَبَةُ [*Separately the milkers return*]: (Ṣ, A:) for when they assemble to milk their camels, each occupies himself with milking his own, and then they return, one after another; (Ṣ, TA;) or they water them together, and return separately to their abodes, where each one milks: (TA:) a prov., (Ṣ, A, TA,) relating to the manners of men in assembling and separating: (TA:) you should not say الحَلَبَةُ. (Ṣ.) IKṭṭ gives it differently, thus: حَتَّى تَؤُوبَ الحَلَبَةُ [*Until the milkers return*]: but the former reading is that commonly known. (IB, TA.) لَيْسَ لَهَا رَاعٍ وَلٰكِنْ حَلَبَةٌ [*They* (i. e. camels) *have not a pastor, but milkers*] is another prov., applied to a man who asks thine aid, and whom thou aidest, but on whose part there is no aid. (TA. [That is, You ask aid of one to whom you render no aid. See also Freytag's Arab. Prov. ii. 427.]) — [Hence,] الحَالِبَانِ ‡ [*The two spermatic ducts;*] *two veins, or ducts, which supply the penis with* [*the spermatic*] *fluid;* whence the phrase, دَرَّ حَالِبَاهُ, meaning ‡ *his penis became erect:* (A, TA:) † *two veins, or ducts, in the kidneys:* (Zj in his "Khalḳ el-Insán:") or † *two veins, or ducts,* (Ṣ, TA,) *of a green colour,* (TA,) *on either side of the navel:* (Ṣ, TA:) accord. to some, † *two veins, or ducts, within the two horns.* (TA. [But I think that, in this instance, القرنين is a mistranscription for العرنين, meaning *the nose:* see what follows.]) حَوَالِبُ [is the pl., and] signifies ‡ The *sources* [*whence flows the milk*] of the udder: (A, TA:) and ‡ the *sources whence flow the tears* of the eye: (A, Ḳ:) and ‡ the *sources* of a spring, (A,) or of a well: (Ḳ:) and حَوَالِبُ الأَسْهَرَيْنِ † *the veins, or ducts, that excern the mucus from the nose, and the spermatic fluid from the penis.* (AA, T. [But see art. سهر.])

إِحْلَابٌ : see what next follows.

إِحْلَابَةٌ *Milk which a man draws for his family, while he is in the place of pasturing, and then sends to them:* (A, Ḳ:) or *milk that remains over and above what fills the skin:* (Ḳ:) or *what remains over and above the contents of the skin when the pastor brings the skin on the occasion of his conducting his camels to water and it contains milk;* this being the احلابة of the tribe: or *milk which people collect, to the quantity of a camel-load, while their camels are in the place of pasturing, and convey to the tribe;* as also ▼ إِحْلَابٌ, pl. أَحَالِيبُ; whence the phrases, قَدْ جَآءَ بِإِحْلَابَيْنِ, and بِثَلَاثَةِ أَحَالِيبَ *He has come with two camel-loads of milk collected while the camels were in the pasture, and with three such loads:* when, in the case of milking ewes or goats or cows, people do thus, one says of them, جَآؤُوا بِإِمْخَاضَيْنِ, and بِثَلَاثَةِ أَمَاخِيضَ. (TA. [See also 4.])

تَحْلَبَةٌ and تُحْلَبَةٌ and تِحْلِبَةٌ and تَحْلُبَةٌ and تَحْلِبَةٌ (Ḳ) and تُحْلُبَةٌ and تَحْلَبَةٌ and تُحْلَبَةٌ (AḤei, TA) and تِحْلَابَةٌ (Ḳ) A ewe, or she-goat, *from whose udder somewhat* [*of milk*] *has issued before her being mounted by the ram:* (Ḳ:) and a she-camel *that emits,* or *yields, milk before conception:* (Seer, TA:) or you say, accord. to Ks, عَنْزٌ تَحْلِبَةٌ, or تُحْلَبَةٌ, [accord. to different copies of the Ṣ,] meaning *a she-goat from whose udder some milk has issued before she has been mounted by the ram:* and accord. to AZ, عَنَاقٌ تَحْلِبَةٌ, or تُحْلَبَةٌ, [accord. to different copies of the Ṣ,] and تِحْلِبَةٌ, and تُحْلُبَةٌ, *a young she-goat that is milked before she conceives.* (Ṣ.)

مَحْلَبٌ *A place of milking.* (Mṣb.) ═ [Also The *prunus mahaleb* of Linn.; *a small kernel of the stone of a wild cherry, much esteemed by the Egyptians,* (*and by the Arabs in general,* E. W. L.,) *and employed by them in many diseases, as a beckic and carminative; brought from Europe:* (Rouyer, in the "Descr. de l'Égypte," xi. 452 of the 8vo. ed.:)] *a kind of odoriferous tree:* (A:) *a certain tree having a grain* (حَبّ [which may mean *a kernel*]) *that is put into perfumes and aromatics;* (Mṣb,* TA;) *the perfume in which it is incorporated being termed* ▼ مَحْلَبِيَّةٌ: so say IDrst and others: AḤn says that he had not heard of its growing anywhere in the country of the Arabs: accord. to Aboo-Bekr Ibn-Ṭalḥah, *a tree having a grain* (حَبّ) *like that of the* رَيْحَان [which is likewise used in medicine, called بِزْرُ الرَّيْحَانِ, i. e. the seed of the ocimum basilicum, or common sweet basil]: accord. to Aboo-'Obeyd El-Bekree, the [*tree called*] أَرَاك: (TA:) [J says,] حَبُّ المَحْلَبِ *is an aromatic medicine, the place whereof is* المَحْلَبِيَّةُ, (Ṣ,) which is a town (بَلَد) near El-Móṣil: (Ḳ, TA:) IKh calls it *a kind of perfume:* some say it is *the grain of the* خِرْوَع [or *castor-oil-plant*]: others, that the محلب is *the fruit,* or *produce, of the kind of tree termed* شَجَرُ اليُسْرِ, *which the Arabs call* الأُسْر: IDrd says that it is *the grain with which one perfumes;* calling the grain by the name of محلب: (TA:) *the best is the white, pearly, and clear.* (Ibn-Seenà, book ii. p. 210.) Accord. to IDrst, this word is originally an inf. n., and حبّ المحلب and شجرة المحلب mean حبّ الحلب and شجرة الحلب. (TA. [IbrD informs me that it is a custom of some of the Arabs, previously to their milking, to chew some محلب, and to anoint with it the teat of the animal.]) ═ *Honey.* (Ḳ.)

مُحْلِبٌ [One *who assists in milking.* — And hence, in a general sense,] ‡ *An aider,* or *assistant:* (Ṣ, Ḳ:) or *an aider,* or *assistant, not belonging to the party,* or *people, whom he aids:* if of that party, or people, the aider is not so called, accord. to the T. (TA. [But see 4.])

مِحْلَبٌ (Ṣ, A, Mṣb, Ḳ) and ▼ حِلَابٌ (A, Mṣb, Ḳ) *A milking-vessel; a vessel into which one milks;* (Ṣ, A, Mṣb, Ḳ;) *made of the skin of a camel's side,* or *of other skin:* (MF:) *a vessel into which ewes are milked.* (Az, TA.)

مَحْلَبِيَّةٌ : see مَحْلَبٌ.

مَحْلُوبٌ Milk *drawn from the udder.* (Ṣ,* Ḳ, &c.) ═ See also حُلَّبِيٌّ.

حلت

حِلْتِيتٌ : see what follows.

حِلْتِيتٌ ; (Ṣ, Ḳ;) for which you should not say حِلْتِيثٌ, with ث; sometimes written حِلْتِيتٌ; (Ṣ; as in one copy; but in another, and in that from which SM quotes, حِلْتِيثٌ;) and ▼ حِلْتِيتٌ; (Ḳ;) [*Assa,* or *asa:* of which there are two kinds; حِلْتِيتٌ مُنْتِنٌ, or *assa fœtida;* and حِلْتِيتٌ طَيِّبٌ, or *assa dulcis:*] the *gum of the* أَنْجُدَان: (Ṣ, Ḳ:)

ISd says, حلتيت is an Arabic or an arabicized word: [and is the name of *a certain plant*:] I have not heard that it grows in Arabia; but *it grows between Bust and the country of El-Keekán: it is*, he says, *a plant that lies prone upon the ground, and from the middle of it there comes forth, and rises high, a reed, or cane, at the head whereof is a knob* (كُعْبُرَة): it is also, he adds, the name of the *gum that comes forth at the roots of the leaves of that reed, or cane: the people of the part above mentioned*, he says, *cook the plant thus called, and eat it; and it is not a plant that remains during the winter.* (TA.) In the T, Az states that حلتيت is said, on the authority of Lth, to be the same as انجرد [app. a mistake for أُنْجُرَة; or for أُنْجُذَان, from the Persian أَنْگُزَد, signifying *assa fœtida*]; but, he adds, the word that I remember to have heard as the same as انجرد is خلتيت, with خ; and I do not think it to be genuine Arabic. (TA, here and in art. خلت.)

حلج

1. حَلَجَ القُطْنَ, (Ṣ, A, Mṣb, Ḳ,) aor. ـِ (Ṣ, Mṣb, Ḳ) and ـُ, (Ṣ, Ḳ,) inf. n. حَلْجٌ, *He separated*, or *cleared, the cotton from its seeds, with the wooden implement termed* مِحْلَج: (Mṣb:) or *he separated and loosened the cotton* (نَدَفَهُ) *with the* مِحْلَاج *upon the* مِحْلَج, (TA,) or *upon the* مِحْلَجَة. (A.) [Golius describes the operation thus: "e nucibus parum excerptum, inter ferramenta duo (quorum superius, axiculo seu specillo simile, ceu torno vertitur, inferius quiescit, simul autem inter sese arcte cohærent,) attrahitur et à semine duriore segregatur:" but see مِحْلَاج and مِحْلَج.] — [Hence,] حَلَجَ signifies also ‡ *He made* a cake of bread *round* (A, Ḳ) [*by rolling it*] with the مِحْلَاج. (A.) — And ‡ *He mixed, and stirred about, and beat*, what is termed تَلْبِينَة, or هَرِيسَة. (A, TA.) — And ‡ *He beat* another with a staff, or stick. (A.) — ‡ *He twisted* a rope. (A, TA.)

5. تحلّج السَّحَابُ †*The clouds became in a state of commotion, and lightened.* (Ḳ.) — مَا تَحَلَّجَ ذٰلِكَ فِى صَدْرِى ‡ *That* [thing, or affair,] *did not become agitated to and fro in my mind, so that I should doubt respecting it.* (TA.) And مَا تَحَلَّجَ فِى صَدْرِى مِنْهُ شَىْءٌ ‡ [*Nothing of it was doubtful in my mind;*] *I doubted not respecting aught of it.* (A.) You say, دَعْ مَا تَحَلَّجَ فِى صَدْرِكَ and مَا تَخَلَّجَ ‡ [*Leave thou that which has been doubtful in thy mind*]. (Lth, TA.) تحلّج فى صَدْرِى and تخلّج mean ‡ *I doubted respecting it:* (Aṣ, TA:) or both mean nearly the same. (Sh, TA.) The saying of 'Adee, (Ḳ,) [or,] correctly, the saying of the Prophet to 'Adee Ibn-Zeyd, لَا يَتَحَلَّجَنَّ فِى صَدْرِكَ طَعَامٌ ضَارَعْتَ فِيهِ النَّصْرَانِيَّةَ (TA,) or لا يَحْتَلِجَنَّ ▼, [accord. to different copies of the Ḳ, in the CḲ ضارعْتَ فيه النَّصْرانِيَّةُ,] means ‡ *Let not aught* [of doubt] *enter thy heart* on account of it; [i. e., *on account of food in respect of which thou hast resembled* those who follow *the Christian religion;*] i. e., it is clean. (Sh, Ḳ, TA.) Accord. to IAth, this is from حَلْج ▼, signifying *motion*, and *commotion*, or *agitation:* and it is also related with خ [in the place of ح], meaning the same. (TA.)

8: see 5.

حَلْجٌ: see 5, last sentence.

سَحَابٌ حَلُوجٌ †*Lightning clouds.* (Ḳ.)

حَلِيجٌ and مَحْلُوجٌ ▼ Cotton *upon which the operation signified by the verb* حَلَجَ *has been performed;* (Ṣ, Ḳ;) cotton *separated*, or *cleared, from its seeds.* (Mṣb.) — For the former, see also حَلِيجَة.

حِلَاجَةٌ The *art*, or *business, of performing the operation, upon cotton, signified by the verb* حَلَجَ. (Ḳ.)

حَلِيجَةٌ †*Milk in which are dates:* (Ḳ:) *milk in which dates have been macerated, steeped*, or *soaked;* (Ṣ, TA;) *and which is sweet:* (TA:) or *clarified butter* [*poured*] *upon pure milk:* or *dregs squeezed from a butter-skin* (عُصَارَةُ نِحْىٍ) [or perhaps the latter word is a mistranscription for نَخِىّ, and, if so, the meaning is the *expressed juice of the species of dates termed* نَخِىّ]: (Ṣ, Ḳ:) and the *expressed juice of* حِنَّآء: and *some fresh butter milked upon:* (Ḳ:) pl. حُلُجٌ: accord. to the T, حُلُجٌ signifies *dates with milk:* (TA:) and accord. to Kr, حَلِيجٌ ▼, without ة, is a name given to *dates milked upon, and then mashed with the hand.* (ISd, TA.)

حَلَّاجٌ One *whose business is to perform the operation, upon cotton, signified by the verb* حَلَجَ. (Ṣ, Ḳ.)

مِحْلَجٌ The *thing on which is performed the operation, upon cotton, signified by the verb* حَلَجَ; (Ṣ, Ḳ;) as also مِحْلَجَةٌ ▼: (Ṣ, A, Ḳ:) it is *of wood* or *of stone.* (TA.) — See also مِحْلَاجٌ. — Also †The *axis* (مِحْوَر) *of the sheave of a pulley.* (Ḳ.)

مِحْلَجَةٌ: see مِحْلَجٌ.

مِحْلَاجٌ The *thing with which is performed the operation, upon cotton, signified by the verb* حَلَجَ: (Ṣ, A, Ḳ:) or مِحْلَجٌ ▼ is the name of the *wooden implement* [*with which that operation is performed*, or] *with which cotton is separated*, or *cleared, from its seeds.* (Mṣb.) — Also †*An implement of wood*, (Ḳ,) or *of stone*, (TA,) *with which a cake of bread is expanded;* (Ḳ;) *a rolling-pin for dough:* (A, TA:) pl. مَحَالِجُ and مَحَالِيجُ. (TA.) — And ‡*A bull's horn.* (A, TA.)

حَلِيجٌ: see مَحْلُوجٌ.

حلزن or حلز

حَلَزُونٌ [The *snail;*] *a certain creeping thing* (دَابَّةٌ), (Ḳ in art. حلز,) or *a small creeping thing*, (Ṣ and Ḳ in art. حلزن,) *that is found upon the* [*kind of tree called*] رِمْث: (Ṣ, Ḳ:) or *of the kind called* أَصْدَاف [i. e. *of the testaceous kind*, or *shell-kind:* applied in the present day to the *snail with its shell*, and to *any kind of spiral shell*]: (Ḳ:) it is *a kind of worm, having a shell within which it conceals itself:* (TA in art. حلزن:) *its flesh is good for the stomach, and for the wound of the mad dog, and for dissolving hard tumours, and curing ulcers; its shell, burnt, clears the mange*, or *scab, and the* [*species of leprosy termed*] بَهَق [q. v.], *and the teeth; and the application of it externally draws forth the* سَلَا [or perhaps it should be سُلَّآء (or *prickles of the palm-tree*)] *from within the flesh, and, mixed with vinegar, stops bleeding from the nose:* (Ḳ in art. حلزن:) Az agrees with the author of the Ḳ in mentioning this as a triliteral-radical word; but Aṣ and J hold the ن to be a radical letter, (TA in art. حلز,) and so do Lth and A'Obeyd. (TA in art. حلزن.)

حلس

1. حَلَسَ البَعِيرَ, aor. ـِ (Sgh, L, Ḳ) and ـُ, (L,) inf. n. حَلْسٌ; (TA;) and احلسهُ ▼, (Ṣ, Ḳ, &c.,) inf. n. إِحْلَاسٌ; (TA;) *He clad*, or *covered, the camel with a* حِلْس [q. v.]; (Ṣ, Ḳ, &c.;) *put upon him a* حِلْس. (Sh.) — حَلَسَتِ السَّمَآءُ, (T, Ḳ,) inf. n. حَلْسٌ, (TA,) ‡ *The sky rained continually;* as also احلست ▼: (Ḳ:) or *rained a fine and continual rain;* (T;) and so ▼ the latter. (T, Ṣ, A, Ḳ.)

4: see 1, in three places: — and see 10, in two places.

10. استحلسهُ *He made it to be as a* حِلْس. (TA.) — So the verb signifies in the phrase استحلس فُلَانٌ الخَوْفَ [in the CḲ فُلانًا الخَوْفُ] (TA) ‡ *Such a one relinquished not fear.* (Mgh,* Ḳ, TA.) — استحلس اللَّيْلُ بِالظَّلَامِ ‡ *The night became dense with darkness.* (A, TA.) — استحلس النَّبْتُ ‡ *The herbage covered the land with its abundance* (Aṣ, Ṣ, Ḳ, TA) *and tallness;* (Z, TA;) as also احلس ▼. (Ḳ.) And أَحْلَسَتِ ▼ الأَرْضُ ‡ *The land became altogether green* [*as though covered with a* حِلْس: see the part. n. below]: (Sh, TA:) or, as also استحلست, *became clad with sprouting herbage:* or *became green, with erect herbage.* (TA.)

حِلْسٌ *A piece of cloth* (كِسَآء), (Ṣ, A, Mgh, Mṣb, Ḳ,) *of thin texture*, (Ṣ, TA,) *which is put on the back of a camel*, (Ṣ, A, Mgh, Mṣb, Ḳ,) *beneath the* بَرْذَعَة, (Ṣ, A, Mgh, Ḳ,) or *beneath the* رَحْل; (Mṣb;) *a piece of hair-cloth used as a covering for a horse or the like:* (A:) or *anything that is next the back of the camel or other beast, beneath the saddle, in the place of the* مِرْشَحَة, *being beneath the felt cloth:* (TA:) and *a* [*piece of cloth* of the kind called] كِسَآء, (Ṣ,* A, Mgh, Ḳ,) or *a piece of hair-cloth*, (A,) or *the like*, (TA,) or *a carpet*, (IAṭr, Mṣb,) *that is spread in a house or tent*, (Ṣ, A, Mgh, Mṣb, Ḳ,) *beneath the best of the pieces of cloth:* (Ṣ, Mgh,

Ḳ:) and ▼ حَلَسٌ signifies the same, in both applications: (A'Obeyd, Ṣ, Ḳ:) pl. [of pauc.] أَحْلَاسٌ (Ṣ, Mṣb, Ḳ) and [of mult.] حُلُوسٌ (Ḳ) and حِلَسَةٌ. (Fr, Ṣgh, Ḳ.) — [Hence,] فُلَانٌ مِنْ أَحْلَاسِ الخَيْلِ ‡ *Such a one is of those who train and manage horses and are constantly upon their backs.* (TA.) And نَحْنُ أَحْلَاسُ الخَيْلِ ‡ *We are acquirers of horses and constantly upon their backs.* (Ṣ.) — هُوَ حِلْسُ بَيْتِهِ † أُمُّ الحِلْسِ *The she-ass.* (Ṣ, Ḳ.) — ‡ *He is one who does not quit his place* [or *house* or *tent*]: (Ḳ:) said [generally] in dispraise; meaning, that he is not fit for anything but to keep to the house or tent. (Az, TA.) [But it does not always imply dispraise; for] it is said in a trad., (Ṣ,) كُنْ حِلْسَ بَيْتِكَ, (Ṣ, A,) or كُنْ حِلْسًا مِنْ أَحْلَاسِ بَيْتِكَ, (TA,) ‡ *Keep thou to thy house* or *tent;* (A;) *quit not thou thy house* or *tent:* (Ṣ:) meaning, in a case of sedition. (TA.) You say also, فُلَانٌ مِنْ أَحْلَاسِ البِلَادِ, and حِلْسٌ بِهَا, ‡ *Such a one does not quit the country,* by reason of his love of it: and this is said in praise; meaning, that he is a person of might and strength, and that he does not quit it, not caring for debt nor for dearth or drought, waiting until the country be fruitful. (Az, TA.) And فُلَانٌ كَالحِلْسِ المُلْقَى [*Such a one is like the castaway* حلس] meaning, † *is one who stands in no stead when an event presses heavily upon him, or oppresses him suddenly:* and, accord. to El-Marzooḳee, هُوَ كَالحِلْسِ, as meaning † *He is one who does not sit a horse well; is not a horseman.* (Ḥam p. 143.) And هٰذَا مِنْ أَحْلَاسِ فُلَانٍ † *This is not of the implements,* or *apparatus,* or *the like, of such a one.* (Ḥam ibid.) — حِلْسٌ مِنَ النَّاسِ ‡ *A great one of men;* syn. كَبِيرٌ; (Ḳ, TA;) because he keeps to his place of abode, not quitting it: but [SM adds] I have seen, in the Moḥeeṭ, this expression explained by كَثِيرٌ [*a multitude of men*]; and Ṣgh explains it as meaning *a company of men.* (TA.) — هُوَ حِلْسُهَا [app., † *He is the careful and skilful manager of it, constantly attending to it*]: accord. to Fr, this expression, and هُوَ ٱبْنُ بَجْدَتِهَا, and ٱبْنُ بُعْثُطِهَا, and سُرْسُورُهَا, and سَفِيرُهَا, and سِمْسَارُهَا, all signify the same. (TA.) — رَفَضْتُ فُلَانًا وَنَفَضْتُ أَحْلَاسَهُ ‡ *I have forsaken,* or *abandoned, such a one.* (A, TA.) — الحِلْسُ *The fourth of the arrows used in the game called* المَيْسِر; (A'Obeyd, Ṣ, Ḳ;) as also ▼ الحَلِسُ: (IF, Ḳ:) *it has four notches, and four portions assigned to it if it be successful, and the forfeiture of four portions if unsuccessful.* (Lḥ, TA.)

حَلَسٌ : } see حِلْسٌ.
الحَلِسُ : }

أَرْضٌ مُحْلِسَةٌ ‡ *Land covered with abundant herbage, as though with a* حِلْس: (Ḳ, TA:) or *altogether green.* (Sh, TA.)

حلف

1\. حَلَفَ, aor. ـِ, inf. n. حَلْفٌ and حَلِفٌ (Ṣ, Mṣb, Ḳ) and حِلْفٌ (Ḳ) and مَحْلُوفٌ, (Ṣ, Ḳ,) like مَجْلُودٌ and مَعْقُولٌ and مَعْسُورٌ, (Ṣ,) and مَحْلُوفَةٌ (Lth, Ḳ) [and مَحْلُوفَآءُ, like مَشْعُورَآءُ, as will be seen from what follows], *He swore.* (Ṣ.) You say, حَلَفَ بِٱللّٰهِ [*He swore by God*]. (Mṣb.) [And حَلَفَ إِنَّهُ كَذَا *He swore it was so.* And حَلَفَ لَهُ عَلَى كَذَا *He swore to him to do such a thing.*] And حَلَفَ يَمِينًا (T in art. ثنى, &c.) and عَلَى يَمِينٍ (El-Jámi' eṣ-Ṣagheer voce مَنْ, &c.) [*He swore an oath*]. And لَا وَمَحْلُوفَائِهِ لَا أَفْعَلُ [*No, by the swearing it,* (meaning *no, I swear it,*) *I will not do* such a thing]. (Ibn-Buzurj, Ḳ.*) And أَحْلِفُ مَحْلُوفَةً بِٱللّٰهِ, meaning مَحْلُوفَةً بِٱللّٰهِ, i. e. [*I swear*] *an oath* [*by God*]. (Lth, Ḳ.) Accord. to IAth, the primary signification of حَلْفٌ is The act of *confederating,* or *making a compact* or *confederacy, to aid,* or *assist;* and *making an agreement:* [but this meaning is afterwards said in the TA to be tropical:] when the object of this, in the time of paganism, was to aid in sedition or the like, and in fighting, and incursions into the territories of enemies, it was forbidden by Moḥammad: when the object was to aid the wronged, and for making close the ties of relationship, and the like, he confirmed it. (TA.)

2: see 4, in three places.

3\. حَالَفَهُ عَلَى كَذَا *He swore with him respecting,* or *to do, such a thing.* (TA.) — Also, (Ṣ,* Ḳ,* TA,) inf. n. مُحَالَفَةٌ and حِلَافٌ, (TA,) ‡ *He united with him in a confederacy, league, compact,* or *covenant,* (Ṣ, Ḳ, TA,) [*respecting,* or *to do, such a thing.*] — And حَالَفَهُ ‡ *He clave, clung, kept,* or *held fast, to it:* (Ḳ, TA:) see a verse of Aboo-Dhu-eyb in art. خلف, voce خَالَفَ. (TA.) You say, حَالَفَ بَثَّهُ, and حُزْنَهُ, ‡ *He clave to his grief,* or *sorrow.* (TA.) — مُحَالَفَةٌ also signifies † *The establishing a brotherhood.* (TA.) It is said in a trad., حَالَفَ بَيْنَ قُرَيْشٍ وَالأَنْصَارِ, † *He established a brotherhood between Ḳureysh and the Assistants.* (Ṣ, TA.)

4\. احلفهُ, (Ṣ, Mṣb, Ḳ,) inf. n. إِحْلَافٌ; (Mṣb;) and ▼ حلّفهُ, (Ṣ, Mṣb, Ḳ,) inf. n. تَحْلِيفٌ; (Mṣb, Ḳ;) and ▼ استحلفهُ; all signify the same; (Ṣ, Mṣb, Ḳ;) [*He made him to swear:* and ▼ the last, *he asked him,* or *required him, to swear:* and *he conjured him,* or *adjured* him; as is shown in the M in art. بلو; (see 8 in that art. in the present work;) and so ▼ the second; as is shown in the explanation of the phrase أَعْمُرَكَ ٱللّٰهَ أَنْ تَفْعَلَ in the Ḳ and TA:] said [for instance] of a judge. (TA.) You say, حلّفهُ ▼ and ▼ استحلفهُ بِٱللّٰهِ مَا فَعَلَ and احلفه [*He made him to swear by God he did not,* or *had not done,* such a thing]. (TA.) — [Hence,] أَحْلَفَ الغُلَامُ *The boy passed the time when he had nearly attained to puberty:* (Ḳ:) so says Lth; adding that some say, قَدْ أَحْلَفَ: and this Z mentions also, and he adds, *so that it was doubted whether he had attained to puberty:* but Az says that أَحْلَفَ الغُلَامُ in this sense is a mistake; and that it means only *he nearly attained to puberty; so that those who looked at him differed in opinion; one saying and swearing that he had attained to puberty, and another saying and swearing the contrary.* (TA.) — And أَحْلَفَتِ الحَلْفَآءُ *The* حلفاء *attained to maturity.* (IAạr, Ḳ.) [By الحلفاء would seem to be here meant the clamorous female slave: for when this word means a kind of grass, the ا is not that which denotes the fem. gender, but is a letter of quasi-coordination, if its n. of un. be حَلْفَآءَةٌ, as in the Mṣb: but accord. to Sb, it is in this sense sing. and pl.; and as pl., it is fem.; and in a description of it by Aboo-Ziyád, cited by AḤn, it is made fem.]

6\. تحالفوا عَلَى كَذَا *They swore, one to another, respecting,* or *to do, such a thing;* as also ▼ احتلفوا. (TA.) — And تحالفوا ‡ *They confederated;* or *united in a confederacy, league, compact,* or *covenant.* (Ṣ, Ḳ, TA.) And تحالفا † *They two united in a confederacy &c. that their case should be one in respect of aiding and defending.* (Mṣb.) And تحالفا بِالأَيْمَانِ أَنْ يَكُونَ أَمْرُهُمَا وَاحِدًا † *They two united in a confederacy &c., by oaths, that their case should be one.* (Lth, TA.)

8: see 6.

10: see 4, in three places.

حِلْفٌ † *A confederacy, league, compact,* or *covenant,* (Ṣ, Mṣb, Ḳ,) *between persons;* (Ṣ, Ḳ;) as also ▼ حِلْفَةٌ: (Mṣb:) because it is not concluded, or ratified, but by swearing. (ISd, TA.) — † *Friendship;* or *true,* or *sincere, friendship.* (Ḳ.) — ‡ *A confederate* of another; *one who unites in a confederacy, league, compact,* or *covenant;* (TA;) as also ▼ حَلِيفٌ: (Ṣ, Mṣb, Ḳ, TA:) or *a friend,* or *sincere friend, who swears to his companion that he will not act unfaithfully with him:* (Ḳ:) or a friend, or true friend, is thus called because he so swears; as also ▼ حَلِيفٌ: (TA:) pl. of the former أَحْلَافٌ; (Ṣ,* Ḳ;) and of the latter حُلَفَآءُ. (TA.) By the احلاف are meant, in a poem of Zuheyr, Asad and Ghaṭafán; because they united in a confederacy to aid each other; and the same appellation is applied to a party of Thaḳeef; (Ṣ, Ḳ;) and to six tribes of Ḳureysh, namely, 'Abd-ed-Dár, Kaạb, Jumaḥ, Sahm, Makhzoom, and 'Adee: (Ḳ:) and ▼ الحَلِيفَانِ to Benoo-Asad and Ṭeiyi, (Ṣ, O, Ḳ,) or Asad and Ghaṭafán; (ISd, TA;) and Fezárah and Asad also (Ṣ, Ḳ) are termed حَلِيفَانِ. (Ṣ.)

حَلْفٌ: see حَلْفَآءُ.

حَلْفَةٌ *An oath.* (Mṣb, TA.*) You say, حَلَفَ حَلْفَةً, (TA,) and ▼ مَحْلُوفَةً, i. e. *He swore an oath;* (Lth, Ḳ;) and ▼ حَلَفَ أُحْلُوفَةً [which means the same]: (Lḥ, TA:) this last word is of the measure أُفْعُولَةٌ from الحَلِفُ. (Ḳ.) — See also حِلْفٌ.

حِلْفَةٌ: see حَلْفَآءُ.

حَلَفَةٌ: see حَلْفَآءُ. — أَرْضٌ حَلِفَةٌ *Land abounding with* [*the kind of grass called*] حَلْفَآء; as also ▼ محلفة [app. مُحْلِفَةٌ]: (TA:) or *producing* حلفاء. (AḤn, TA.)

حَلْفَآءُ [app. حَلْفَآءَةٌ accord. to some, and حَلْفَآءُ accord. to others, (in the CḲ, erroneously, حُلَفَا,)], (Ṣ, Mṣb, Ḳ, &c.,) in measure like حَمْرَآءُ, [and if so, حَلْفَآءُ, but see what follows,] (Mṣb,) and ♦ حَلِفٌ, (Akh, Ḳ,) [A kind of high, coarse grass; called by the botanists poa multiflora, and poa cynosuroides;] a certain plant, (Ṣ, Mṣb, Ḳ,) [growing] in water, (Ṣ,) well known, (Mṣb,) of those termed أَغْلَاث : (TA:) Aboo-Ziyád says of the حلفا that it seldom grows anywhere but near to water or to the bottom of a valley; and is long, or tall, (سلبة,) rough to the touch; seldom, or never, does any one lay hold upon it, for fear of his hand being cut; sometimes camels and sheep or goats eat a little of it; and it is much liked by oxen: (AḤn, TA:) [a coll. gen. n.:] n. un. ♦ حَلَفَةٌ, (Ṣ, Ḳ,) accord. to AZ, (Ṣ,) or Aboo-Ziyád, and AḤn, (TA,) and ♦ حَلِفَةٌ, (Ṣ, Ḳ,) accord. to Aṣ, (Ṣ,) and حَلْفَاةٌ, (Mṣb, Ḳ,) like صَحْرَاةٌ: (Ḳ: [in the CḲ like صَخْرَةٌ, and omitted in my copy of the TA:]) [this last n. un. requires that the coll. gen. n. should be حَلْفَآءُ: (see 4, last sentence:) but] Sb says that حلفا is sing. and pl.: [see شَجْرٌ:] (TA:) [as pl., it is fem.; and it is made fem. in the description by Aboo-Ziyád, cited above:] sometimes it has حَلَافِىُّ for pl.: and its dim. is ♦ حُلَيْفِيَةٌ. (O, TA.) أَنَا الَّذِى فِى الحَلْفَآءِ, occurring in a trad., means † I am the lion; because that beast repairs to the places where the حلفا grows: and [hence,] كَأَنَّهُ أَخُو الحَلْفَآءِ means † As though he were the lion. (TA.) —— حَلْفَآءُ also signifies A clamorous female slave: (IAạr, Ḳ:) pl. حُلُفٌ. (Ḳ.)

حَلِيفٌ : see حِلْفٌ, in three places. —— † Whatever cleaves, clings, keeps, or holds fast, to another thing, is termed its حَلِيف : whence one says, فُلَانٌ حَلِيفُ الجُودِ † [Such a one cleaves to liberality], &c. (TA.) You say also, هُوَ حَلِيفُ السَّهَرِ, meaning ‡ He is sleepless. (TA.) —— حَلِيفُ اللِّسَانِ ‡ Sharp-tongued; (Ṣ, Z, Ḳ;) chaste, or eloquent, in speech; (Ṣ;) who conforms to the desire of his companion, as though he were a confederate. (Z, TA.) —— حَلِيفُ الغَرْبِ, in a poem of Sa'ideh Ibn-Ju-eiyeh, (Skr, Ḳ,* TA,) means ‡ A sharp spear-head, (Ḳ,) or a spear with a sharp head: (Skr, TA:) or it means a brisk, lively, or sprightly, horse. (Skr, Ḳ.) Az says, سِنَانٌ حَلِيفٌ means ‡ A sharp spear-head: and I think that it is termed حليف because the sharpness of its point is likened to the sharpness of the points of [the grass called] حَلْفَآء. (TA.)

حَلَافَةٌ † Sharpness, in anything. (TA.)

وَادٍ حَلَافِىٌّ A valley that produces [the grass called] حَلْفَآء. (Ṣgh, Ḳ.)

حُلَيْفِيَةٌ : see حَلْفَآء.

حَلَّافٌ and حَلَّافَةٌ : see what next follows.

حَالِفٌ [Swearing:] and ♦ حَلَّافٌ that swears much, or often; and so ♦ حَلَّافَةٌ [but in a more intensive sense]. (TA.)

مَا أَحْلَفَ لِسَانَهُ ‡ How sharp-tongued is he, (Ḳ,* TA,) and how chaste, or eloquent, in speech! (TA.)

أُحْلُوفَةٌ : see حِلْفَةٌ.

مُحْلِفٌ ‡ Anything respecting which one doubts, so that people swear respecting it; (ISd, L, Ḳ, TA;) so called because it occasions swearing: (ISd, TA:) such is also termed مُحْنِثٌ. (L.) [Hence,] ‡ A boy of whom one doubts whether he have attained to puberty. (IAạr, TA.) [And hence] it is said, حَضَارِ وَالوَزْنُ مُحْلِفَانِ ‡ [Haḍári and El-Wezn are two causes of swearing]: these are two stars: the reason of the saying is that which is explained in art. حضر, voce حَضَارِ. (Ṣ, Ḳ.) Hence, also, كُمَيْتٌ مُحْلِفَةٌ, (Ṣ,) or كميت مُحْلِف, (Ḳ,) i. e. ‡ [A bay] not of a clear hue; (Ṣ, TA;) between that termed أَحْوَى and that termed أَحَمّ: accord. to the Ḳ, of a clear hue; but this is the meaning of غَيْرُ مُحْلِف. (TA.) A poet says, (Ṣ,) namely Hubeyreh Ibn-'Abd-Menáf El-Yarboo'ee, also called, after his mother, Ibn-El-Kelḥabeh, (IB,)

كُمَيْتٌ غَيْرُ مُحْلِفَةٍ وَلٰكِنْ
كَلَوْنِ الصِّرْفِ عُلَّ بِهِ الأَدِيمُ

[A bay not of a dubious hue, but like the colour of the صرف (q. v.) with which the hide is dyed a second time]; i. e., of a clear hue, so that one does not swear that she is otherwise than such: (Ṣ, L:) accord. to IAạr, not requiring her owner to swear that he has seen her like in generousness: but the former is the right meaning. (L.) Also نَاقَةٌ مُحْلِفَةٌ ‡ A she-camel respecting the fatness of which one doubts. (TA.)

محلفة [app. مَحْلَفَةٌ]: see حِلْفَةٌ.

مَحْلُوفَةٌ : see حِلْفَةٌ.

حلق

1. حَلَقَ رَأْسَهُ, (Ṣ, Ḳ,) and شَعَرَهُ, (Ṣ, M, Mṣb,) aor. ـِ, (Ṣ, Mṣb, Ḳ,) inf. n. حَلْقٌ (Ṣ,* M, Mṣb, Ḳ) and حِلَاقٌ (Ṣ,* Mṣb, Ḳ*) and تَحْلَاقٌ, (Ṣ,* Ḳ,) He removed the hair of his head [with a razor, or shaved his head], (Ḳ,) [and he shaved off his hair;] as also ♦ احتلقهُ; (Ṣ, Ḳ;) and ♦ حَلَّقَهُ, (Ḳ,) inf. n. تَحْلِيقٌ: (TA:) or the latter verb has an intensive signification, (O, Mṣb,) and applies to many objects, (Ṣ, Mṣb,) as in the phrase, حَلَّقُوا رُؤُوسَهُمْ [they shaved their heads]: (Ṣ:) and you say also, حَلَقَ مِعْزَهُ [he shore his goats]; but not save in the case of sheep: (Ṣ:) [for] الحَلْقُ with respect to the hair of human beings and of goats is like الجَزُّ with respect to wool. (M, TA.) [Hence,] إِنَّ رَأْسَهُ لَجَيِّدُ الحِلَاقِ [Verily his head is well shaven]. (Ṣ, Ḳ.*) And يَوْمُ تَحْلَاقِ اللِّمَمِ [The day of the shaving off of the locks termed لمم]; which was a day fought by Teghlib (Ṣ, Ḳ) against Bekr Ibn-Wáïl; (Ṣ;) because their [i. e. Teghlib's] distinctive sign was shaving (الحَلْق), (Ṣ, Ḳ,) on that day. (Ṣ.) —— عَقْرًا حَلْقًا, or ♦ عَقْرَى حَلْقَى, (Ṣ, Ḳ,*) is an expression occurring in a trad.: (Ṣ:) the latter is rare; or is an incorrect variation of the relaters of traditions: (Ḳ:) A'Obeyd says, it is عَقْرًا حَلْقًا, for which the relaters of traditions say ♦ عَقْرَى حَلْقَى; and the original form and meaning is عَقَرَهَا اللّٰهُ وَحَلَقَهَا, (Ṣ,) or عَقَرَهَا اللّٰهُ عَقْرًا وَحَلَقَهَا حَلْقًا, (TA,) i. e., [accord. to A'Obeyd,] May God wound her body, and afflict her with pain in her حَلْق [or fauces]: (Ṣ, Ḳ:*) but this explanation is not valid: accord. to the T, it is a form of imprecation uttered against a woman, [not in earnest, though denoting a degree of displeasure,] meaning may she be bereft of her husband, or become a widow, so that she shall shave off her hair: and Az says that ♦ عَقْرَى حَلْقَى means she is unlucky [to others] and annoying: ISd says, it is said to mean she is unlucky [to others]; but I am not sure of it. (TA.) Accord. to Aboo-Naṣr (Ṣ, TA) Aḥmad Ibn-Ḥátim, (Ṣ,) one says on the occasion of an event at which one wonders, خَمْشَى ♦ عَقْرَى حَلْقَى, as though [meaning May she who has occasioned this, scratch and wound her face, and shave off her hair:] from الحَلْق [the act of shaving] and العَقْر [the act of wounding] and الخَمْش syn. with الخَدْش [the act of scratching]: (Ṣ, TA:*) and he cites this verse:

أَلَا قَوْمِى أُولُو عَقْرَى وَحَلْقَى ♦
لِمَا لَاقَتْ سَلَامَانُ بْنُ غَنْمِ

(TA, and so in some copies of the Ṣ,) meaning [Now surely] my people have women who have wounded and scratched their faces and shaven off their hair [on account of what the tribe of Selámán Ibn-Ghanm has experienced]: so, says IB, IḲṭṭ relates this verse, and so Hr in the Ghareebeyn: but ISk, thus:

أَلَا قَوْمِى إِلَى عَقْرَى وَحَلْقَى

[and so I find it in one copy of the Ṣ:] and IJ explains it by saying that عقرى وحلقى originally denotes the case of a woman who, when some one honourable in her estimation has been smitten, or wounded, takes a pair of sandals, and beats with them her head, and wounds or scratches it, and shaves off her hair; and the poet means, my people have come to the condition of wounded, or scratched, and shaven, women. (TA.) [Fei says,] حَلْقًا لَهُ وَعَقْرًا is a form of imprecation, meaning May God afflict him with pain in his حَلْق [or fauces], and wound his body: but the relaters of traditions say ♦ عَقْرَى حَلْقَى, with the fem. alif, making them act. part. ns.; [the former meaning, accord. to one of the explanations given above, an unlucky woman to others, though this is doubtful; and] the latter meaning a woman annoying her people: (Mṣb:) or both these words are inf. ns., like دَعْوَى. (TA in art. عقر. [See more in that art.]) —— They said also, بَيْنَهُمْ أَحْلِقِى وَقُومِى [Among them is heard the saying, Shave, O woman, and arise]; i. e. among them is trial, or trouble, and distress, affliction, calamity, or adversity: and يَوْمُ أَحْلِقِى وَقُومِى [A day of the saying Shave, &c.; i. e., of trial, &c.]. (TA.) —— Also حَلَقَ الشَّىْءَ, aor. ـِ, inf. n. حَلْقٌ, He peeled the thing; or stripped off, or otherwise removed, its superficial part: or he peeled, stripped, pared, scraped, or rubbed, off the thing: syn. قَشَرَهُ. (TA.) —— And حَلَقَ † He, or it, destroyed; and cut off entirely, like as the razor does hair. (TA.) —— And, aor. as above, † He (a man) pained, or caused to suffer pain. (IAạr,

TA.) ═ حَلَقَهُ, (Ṣ, Ḳ,) aor. ُ (Ḳ) and ِ, (TA,) *He hit, or hurt, his* حَلْق [or *fauces*]; (Ṣ, Ḳ;) a verb similar to رَأَسَهُ and عَضَدَهُ and صَدَرَهُ, meaning "he struck his head" and "his upper arm" and "his breast:" and *He* (God) *afflicted him with pain in his* حَلْق; as explained in a phrase mentioned above. (Ṣ.) — And ‡ *He filled it,* namely, a watering-trough or tank, (Ḳ, TA,) *up to its* حَلْق [q. v.]; (TA;) as also ↓ احلقهُ. (Ṣgh, Ḳ.) ═ حَلَقَ الشَّىْءَ i. q. قَدَّرَهُ [*He made the thing according to a measure*; &c.]; (Ḳ;) like خَلَقَهُ [q. v.], with the pointed خ. (TA.) ═ حَلَقَ الضَّرْعُ, aor. َ, [so in the TA, app. a mistranscription for ُ, since neither the medial nor final radical letter is faucial,] inf. n. حُلُوقٌ, † *The udder rose to the belly, and became contracted:* — and also † *The udder contained much milk:* (Kr, ISd, TA:) thus it has two contr. meanings. (TA.) [See the part. n. حَالِقٌ.] ═ حَلِقَ, aor. َ, *He* (a man) *suffered pain: or had a complaint of his* حَلْق [or *fauces*]. (IAạr, TA.)

2. حلّق, inf. n. تَحْلِيقٌ: see 1, first sentence. ═ حلّقهُ حَلْقَةً *He clad him with a* حلقة [or *coat of mail, &c.*]. (TA.) — حلّق حَلْقَةً *He turned* [or *drew*] *a circle.* (TA.) — [Hence, perhaps,] حلّق عَلَى ٱسْمِ فُلَانٍ [if, as I suppose, originally meaning *He drew a line round the name of such a one*;] ‡ *he cancelled the stipend, or pay, or allowance, of such a one.* (TA.) — [حلّق الإِبِلَ *He branded the camels with a mark in the form of a ring:* see the pass. part. n.] — حلّق بِإِصْبَعِهِ *He bent his finger round like a* حَلْقَة [or *ring*]. (TA.) — حلّق said of the moon, *It had a halo around it*; (Ḳ,* TA;) as also ↓ تحلّق. (Ḳ.) — Said of a bird, inf. n. as above, ‡ *It soared in its flight,* (Ṣ, Ḳ, TA,) *and circled in the air.* (TA.) — Said of the نَجْم, (Ḳ,) meaning the Pleiades (الثُّرَيَّا), (T in art. فغر,) † *It was, or became, high:* (Ḳ:) or *it became overhead.* (T ubi suprà: see فَغَرَ.) It is said that تَحْلِيقُ الشَّمْسِ, in the former part of the day, means † *The sun's rising high from the east:* and in the latter part of the day, *the sun's going down:* but Sh says, I know not التحليق except as meaning *the being, or becoming, high.* (TA.) — حلّق بِبَصَرِهِ إِلَى السَّمَآءِ † *He raised his eyes towards the sky.* (TA.) — حلّق ضَرْعُ النَّاقَةِ, inf. n. as above, † *The she-camel's milk became drawn up* [and consequently *her udder* also] (IDrd, Ḳ) *to her belly.* (IDrd, TA.) And accord. to ISd, حلّق اللَّبَنُ † *The milk* [*became drawn up, or withdrawn,* i. e.,] *went away.* (TA.) And حلّق is said of the water in a drinking-trough, meaning † *It became little in quantity; and went away.* (TA.) — حَلَّقَتْ عُيُونُ الإِبِلِ ‡ *The eyes of the camels sank, or became depressed, in their heads.* (AA, Ḳ, TA.) — حلّق البُسْرُ, inf. n. as above, † *The ripening dates became ripe* [*as far as the* حَلْق, i. e.,] *to the extent of two thirds:* (AḤn, Ḳ:) and ↓ حَلْقَنَ signifies the same; or *they began to be ripe* (Ḳ in art. حلقن) *next the base;* (TA in that art.;) as also ↓ حَلْقَمَ. (TA in art. حلقم.) — حلّق بِهِ ‡ *It* (a draught of [milk and water such as is termed] صُوَاح) *caused his belly to become inflated.* (Ibn-'Abbád, Ḳ, TA.) — حلّق بِالشَّىْءِ إِلَيْهِ *He threw the thing to him.* (Ḳ.)

4: see 1, near the end.

5. تحلّقوا *They sat in rings, or circles.* (Ṣ, Ḳ.) The doing thus before prayers [in the mosque] is forbidden. (TA.) — See also 2.

7. انحلق شَعْرُهُ [*His hair came off*; as though it were shaven]. (Ḳ voce مُتَقَوِّبٌ.)

8: see 1, first sentence.

Q. Q. 1. حَلْقَمَهُ *He cut, or severed, his* حُلْقُوم [q. v. voce حَلْقٌ]. (Mṣb. See also art. حلقم.) ═ حَلْقَمَ and حَلْقَنَ: see 2. ═ حَوْلَقَ, (TA,) inf. n. حَوْلَقَةٌ, (Ṣ,) *He said* لَا حَوْلَ وَلَا قُوَّةَ إِلَّا بِٱللّٰهِ [see art. حول:] so says ISk: (Ṣ:) others say حَوْقَلَ. (IAth, TA.)

حَلْقٌ [The *fauces:* and hence, by a synecdoche, the *throat,* or *gullet,* i. e. the *œsophagus*:] the *place of the* غَلْصَمَة [or *epiglottis*]; and the *place of slaughter in an animal:* (AZ, TA:) or the *fore part of the neck:* (Zj in his "Khalḳ el-Insán:") or the *passage of,* or *place by which pass, the food and drink, into the* مَرِىْء [or *œsophagus*]: (TA:) or i. q. ↓ حُلْقُومٌ: (Ṣ, Mṣb, Ḳ:) [but] the latter is the *windpipe;* the *passage of the breath;* (Zj ubi suprà, Az, Mṣb;) *which has branches branching from it into the lungs,* [namely, *the bronchi, consisting of two main branches, which divide into smaller and smaller,*] *called the* قَصَب: (Zj ubi suprà, and Mṣb:) [this word (حلقوم), however, as well as the former, is sometimes applied to the *throat,* or *gullet:* but the former (حلق) generally signifies the *fauces;* and the latter (حلقوم), the *windpipe:* (see another explanation of the latter word in art. حلقم, from the M:) a morsel of food, or the like, is commonly said to stick in the حلق, but not in the حلقوم:] حَلْقٌ is of the masc. gender: (Mṣb:) and its pl. is حُلُوقٌ, (Ṣ, Mṣb,) and sometimes حُلُقٌ; (Mṣb;) or حُلْقٌ, which is extr.; and pl. of pauc. أَحْلَاقٌ; (TA;) and أَحْلُقٌ is allowable [as a pl. of pauc.] on the ground of analogy; but it has not been heard from the Arabs: (Mṣb:) ↓ حُلْقُومٌ is of the measure فُعْلُومٌ, (TA,) the م being augmentative, (Mṣb,) accord. to Kh; but of the measure فُعْلُولٌ accord. to others: (TA:) and its pl. is حَلَاقِيمُ, and, by contraction, حَلَاقِمُ. (Mṣb.) — ‡ The *part through which the water runs* of a watering-trough or tank, and of a vessel: pl. حُلُوقٌ. (TA.) — And [the pl.] حُلُوقٌ signifies ‡ The *water-courses,* and *valleys,* of a land; and the *narrow,* or *strait, places,* of a land, (Ḳ, TA,) and of roads. (TA.) — حَلْقُ الجَوِّ [app. † *The upper region of the air:* see 2, as said of a bird, &c.]. (Z, TA.) — The حَلْق of a date is † The *part at the extremity of two thirds* thereof: or *a part near to the base* thereof. (TA.) ═ *Unluckiness* [*to others*]. (IAạr, Ḳ.) Hence, [accord. to some,] عَقْرًا حَلْقًا [explained above: see 1]. (TA.)

حَلْقٌ The *state of being bereft of a child by death;* syn. ثُكْلٌ [in the CḲ, erroneously, شُكْل]. (Ḳ, TA.) So in the prov., لِأُمِّكَ الحَلْقُ [*May bereavement of her child befall thy mother*]: or, accord. to the A, it means *shaving of the head* [on account of such, or a similar, bereavement]. (TA.)

حَلْقٌ ‡ *Numerous cattle:* (Ṣ, Ḳ:) because the herbage is cropped by them like as hair is shaven or shorn. (Ḳ.) You say, جَآءَ فُلَانٌ بِالحَلْقِ وَالإِحْرَافِ (Ṣ) *Such a one came with,* or *brought, much cattle.* (AZ, Ṣ in art. حرف.) ═ The *seal-ring* (IAạr, Ṣ, Ḳ) *that is on the hand* [or *finger*], or *in the hand,* (IAạr, TA,) *of a king:* (IAạr, Ṣ, Ḳ:) or *a seal-ring of silver, without a* فَصّ [or *gem set in it*]. (ISd, Ḳ.) [Hence,] أُعْطِىَ فُلَانٌ الحَلْقَ *Such a one was made prince, or governor, or commander.* (TA.)

حَلَقٌ: see حَلْقَةٌ. — Also Camels *branded with the mark termed* حَلْقَة; (Ḳ;) and so ↓ مُحَلَّقَةٌ. (Ṣ, Ḳ.)

حَلْقَةٌ [*A single act of shaving*]. One says to a beloved child, when he belches, حَلْقَةً وَكَبْرَةً وَشَحْمَةً فِى السُّرَّةِ, i. e. *May thy head be shaven time after time,* (Ibn-'Abbád, Ḳ,*) *so that thou mayest grow old,* (Ibn-'Abbád, TA,) [*and acquire fat at the navel:*] or *mayest thou be preserved so as to have thy head shaven, and to grow old.* (A, TA.) ═ As meaning *A ring;* i. e. *anything circular;* as a حلقة of iron, and of silver, and of gold; (TA;) a حلقة of a coat of mail, &c.; (Mgh;) the حلقة of a door; and a حلقة of people; (Ṣ, Ḳ;) in this last instance meaning *a ring* of people; (Mṣb, TA;) it is also with fet-ḥ to the ل; i. e. ↓ حَلَقَةٌ; (Ṣ, Mgh, Mṣb, Ḳ;) mentioned by Yoo, on the authority of Aboo-'Amr Ibn-El-'Alà, (Ṣ, Mṣb,) and with kesr; (Ḳ;) i. e. ↓ حَلِقَةٌ; mentioned by Fr and El-Umawee, as of the dial. of Belḥárith Ibn-Kaab; accord. to the O; or ↓ حِلَقَةٌ, accord. to the L: (TA:) or there is no such word as ↓ حَلَقَةٌ, (Ṣ, Ḳ,) in chaste speech, (TA,) except as pl. of حَالِقٌ; (Ṣ, Ḳ;) accord. to Aboo-'Amr Esh-Sheybánee; (Ṣ;) or it is a dial. var. of weak authority; (Ḳ;) accord. to Th, allowed by all, though of weak authority; (Ṣ;) or it is used by poetic license: (Mgh:) Lḥ says that the حلقة of a door is حَلْقَة and ↓ حَلَقَة; Kr says the same of the حلقة of a company of men; Lth says that it is the former in this case, but that some say the latter; A 'Obeyd prefers the latter in the case of a حلقة of iron, but allows the former; and prefers the former in the case of a حلقة of people, but allows the latter; and Abu-l-'Abbás prefers the former in both cases, but allows the latter: (L:) the pl. is ↓ حَلَقٌ, (Ṣ, Mṣb, Ḳ,) which is anomalous in relation to حَلْقَة, (Ṣ, Mṣb,) or [rather] a quasi-pl. n., (TA,) but regular in relation to حَلَقَة, (Mṣb, TA,) [as a coll. gen. n.,] like قَصَبٌ in relation to قَصَبَة; (Mṣb;) and, (Ḳ,) accord. to Aṣ, (Ṣ,) حِلَقٌ, (Ṣ, Ḳ,) as pl. of حَلْقَة meaning a حلقة of men and of iron, (TA,) like بِدَرٌ (Ṣ, Ḳ) pl. of بَدْرَة, and قِصَعٌ pl. of قَصْعَة; (Ṣ;) or this is a regular pl. of حِلَقَة; (TA;) and حَلَقَاتٌ, (AA,

Yoo, Ṣ, Ḳ,) which is pl. of حَلْقَةٌ; (TA;) and حِلَقَاتٌ, (Ḳ,) which is pl. of حِلْقَةٌ; (TA;) and حِلَاقٌ in relation to a company of men. (TA.) You say, اِنْتَزَعْتُ حَلْقَتَهُ [lit. *I pulled off his ring*], meaning, (app., Ibn-'Abbád,) † *I outwent him,* or *preceded him.* (Ibn-'Abbád, Ḳ.) And كَالْحَلْقَةِ المُفْرَغَةِ [*Like the solid and continuous ring*]: a prov., applied to a company of men united in words and action. (TA.) And ضَرَبُوا بُيُوتَهُمْ حِلَاقًا *They pitched their tents in one series,* (Ḳ, TA,) *so as to form a ring* [or *rings*]: the last word being a pl. of حَلْقَةٌ or of حِلْقَةٌ. (TA.) And it is said in a trad., نُهِىَ عَنِ الحِلَقِ قَبْلَ الصَّلَاةِ, i. e. *Rings* of men [sitting in the mosque *before prayer are forbidden*]. (TA.) — [Hence,] حَلْقَتَا الرَّحِمِ ‡ [*The two rings of the womb*]: one of these is *the mouth of the vulva, at its extremity;* [*the meatus of the vagina:*] and the other is *that which closes upon the* مَآء [or *seminal fluid*] *and opens for the menstrual discharge;* [*the os uteri:*] (Ḳ:) or, as some say, the other is *that whence the urine is emitted;* [*the meatus urinarius:* but the former is the right explanation: and hence] one says, وَقَعَتِ النُّطْفَةُ فِى حَلْقَةِ الرَّحِمِ ‡ *The seminal fluid fell into the entrance of the womb.* (TA.) [Hence also,] حَلْقَةُ الدُّبُرِ † *The anus;* syn. حِتَارَةٌ and شَرْجَةٌ. (Mgh in art. شرج.) [See also خَاتَمٌ, last sentence but two.] — حَلْقَةٌ also signifies *A brand upon camels,* (Ḳ, TA,) *of a round form, like the* حلقة [or *ring*] *of a door.* (TA.) — And *A coat of mail:* [because made of rings:] (Ḳ:) or *coats of mail:* (Ṣ, Mgh:) or *arms,* or *weapons, in general,* (M, Mgh, Mṣb,) *and coats of mail, and the like.* (M, TA.) It is said in a trad., إِنَّكُمْ أَهْلُ الحَلْقَةِ وَالحُصُونِ [*Verily ye are people of the coat of mail, &c., and of fortresses*]. (TA.) — And *A rope.* (Ḳ, TA.) — And, of a vessel, (AZ, Ḳ,) and of a watering-trough, (AZ,) ‡ The *portion that remains vacant after one has put in it somewhat* (AZ, Ḳ) *of food or beverage, up to the half;* the *portion that is above the half* being thus called: (AZ:) [or] of a watering-trough, ‡ the *fulness;* or *less than that.* (Aboo-Málik, Ḳ.) One says, وَفَّيْتُ حَلْقَةَ الحَوْضِ and الإِنَآءِ ‡ [*I filled up the* حلقة *of the watering-trough* and *of the vessel*]. (AZ, TA.)

حِلْقَةٌ: see حَلْقَةٌ.

حَلَقَةٌ: see حَلْقَةٌ, in three places.

حَلِقَةٌ: see حَلْقَةٌ.

حَلْقَى: see 1, in six places.

حَلْقِىٌّ [*Of,* or *relating to, the* حَلْق; *faucial; guttural*]. الحُرُوفُ الحَلْقِيَّةُ [*The faucial,* or *guttural, letters*] are six; namely, ء and ه, to which are appropriated the furthest part of the حَلْق; and ع and ح, to which are appropriated the middle thereof; and غ and خ, to which are appropriated the nearest part thereof. (TA.)

بُسْرٌ حُلْقَانٌ † *Ripening dates that have become ripe as far as the* حَلْق; which is said by some to be *near the base:* (TA:) or *that have begun to be ripe* (Ḳ in art. حلقن) *next the base;* (TA in that art.;) and so ▼ رُطَبٌ مُحَلْقِمٌ; and a single date in that state is termed ▼ رُطَبَةٌ حُلْقَامَةٌ: (Ḳ in art. حلقم:) or *ripening dates that have become ripe to the extent of two thirds;* as also ▼ مُحَلْقِنٌ, (Ṣ, Ḳ,) and ▼ مُحَلِّقٌ, (Ḳ, TA,) like مُحَدِّثٌ: (TA:) [in the CḲ مُحَلَّقٌ, like مُعَظَّمٌ:]) and the last signifies, (Ḳ,) accord. to Ibn-'Abbád, (TA,) dates *partly ripe* (Ḳ, TA) *and partly unripe:* (TA:) n. un. with ة: (Ṣ, Ḳ:) such dates are also termed ▼ حَوَالِيقُ, held by ISd to be a kind of rel. n., [as though pl. of حَالِقَةٌ,] though the reason of the insertion of the ى in this word, he says, was unknown to him: (TA:) and ▼ رُطَبٌ حُلْقَانِىٌّ: (TA from a trad.:) the pl. of مُحَلِّقٌ is مَحَالِيقُ. (TA.)

حُلْقُومٌ: see حَلْق, in two places.

رُطَبَةٌ حُلْقَامَةٌ: } see حُلْقَانٌ.
رُطَبٌ حُلْقَانِىٌّ: }

حَلَاقِ, (Ṣ, Ḳ,) indecl., with kesr for its termination, because changed from its original form, which is حَالِقَةٌ, of the fem. gender, and an epithet in which the quality of a subst. is predominant; (Ṣ;) ‡ *Death* (Ṣ, Ḳ, TA) *that peels* [*people*] *off;* (TA;) as also حَلَاقٌ, (Ḳ,) allowed by Ibn-'Abbád; and, accord. to the Tekmileh, ▼ حِلَاقٌ also. (TA.) One says, سُقُوا بِكَأْسِ حَلَاقِ ‡ [*They were given to drink the cup of death*]. (ISd, TA.) [See also جَعَارِ.]

حُلَاقٌ *Pain in the* حَلْق [or *fauces*]. (Ṣ, Ḳ.)

حِلَاقٌ: see حَلَاقِ.

رَأْسٌ حَلِيقٌ *i. q.* ▼ مَحْلُوقٌ [*A shaven head*]: (ISd, TA:) and شَعَرٌ حَلِيقٌ [*hair shaven off*]: (AZ, Ṣ:) and لِحْيَةٌ حَلِيقٌ [*a beard shaven off*]; not حَلِيقَةٌ: (AZ, Ṣ, Ḳ:) and عَنْزٌ ▼ مَحْلُوقَةٌ [*a shorn she-goat*]. (AZ, Ṣ.) The pl. of حَلِيقٌ is حَلْقَى [and حِلَاقٌ]. (TA.)

حُلَاقَةٌ *Shorn hair* of a goat. (Ṣ, Ḳ.)

حَلَّاقٌ: see what next follows.

حَالِقٌ [*Shaving:* and] *a shaver;* (Ṣ, TA;) and *a shearer* of goats: (T, TA:) pl. حَلَقَةٌ: (T, Ṣ, Ḳ:) and ▼ حَلَّاقٌ is syn. with حَالِقٌ; (TA;) [or has an intensive signification, or denotes frequency of the action.] The saying لَا تَفْعَلْ ذٰلِكَ أُمُّكَ حَالِقٌ means [*Do not thou that:*] *may* God cause *thy mother* to be bereft of her child so that she shall *shave off her hair.* (Ṣ.) And حَالِقَةٌ occurs in a trad. as an epithet applied to a woman cursed by Mohammad; (TA;) meaning *One who shaves off her hair in the case of an affliction:* (Ḳ, TA:) or *who shaves her face for the sake of embellishment.* (TA.) It is also applied to a wound on the head (شَجَّةٌ) *That scrapes off the skin from the flesh.* (TA in art. دمغ.) — ‡ *Sharp;* applied to a knife: (TA:) and so ▼ حَالُوقَةٌ; applied to a sword; and also to a man. (Ibn-'Abbád, Ḳ.) [Hence, perhaps,] فُلَانٌ حَالِقٌ إِلَىَّ بِعَيْنِهِ † *Such a one is looking at me intently,* or *sharply;* as also ▼ مُحَلِّقٌ. (T, TA in art. زنر.) — † *Quick,* or *swift; and light, active,* or *agile.* (TA.) — † *Lean,* or *light of flesh; slender, and lean;* or *lean, and lank in the belly.* (TA.) — Accord. to A'Obeyd and the Ḳ, it means *An udder:* and accord. to the Ḳ, it means also *full:* (TA:) but it is an epithet applied to an udder; and thus applied, it has this latter meaning, i. e. ‡ *full;* (T, Ṣ, TA;) so ISd thinks; (TA;) as though the milk in it reached to its حَلْق: (Ṣ, TA:) or *big, so that it rubs off the hair of the thighs by reason of its bigness:* (TA:) and it has also the contr. meaning; (T, TA;) *raised* (IAạr, T, Kr, ISd, TA) *towards the belly,* (Kr, ISd, TA,) *and contracted,* (T, Kr, ISd, TA,) *so that its milk has become scanty,* (IAạr, T, TA,) or *has gone away:* (Kr, ISd, TA:) pl. حُلُقٌ and حَوَالِقُ (Ṣ, TA) and حَلَقَةٌ. (TA. [The last is mentioned as pl. of حالق in the latter sense.]) Accord. to Aṣ, أَصْبَحَتْ ضَرَّةُ النَّاقَةِ حَالِقًا means † *The she-camel's udder became nearly full.* (TA.) And one says نَاقَةٌ حَالِقٌ meaning *A she-camel having much milk:* (TA:) or *having great abundance of milk, and a large udder:* and إِبِلٌ ▼ مُحَلِّقَةٌ *camels having much milk:* (En-Naḍr, TA:) and the pl. of حالق is حَوَالِقُ and حُلُقٌ. (TA.) — ‡ *A high mountain,* (Ṣ, Ḳ, TA,) *rising above what surrounds it, and without vegetable produce:* or, as some say, a mountain *having no vegetable produce;* as though it were shaven, or shorn; of the measure فَاعِلٌ in the sense of the measure مَفْعُولٌ: but Z says that it is from حَلَّقَ, said of a bird: (TA:) and *a high,* or *an overtopping* or *overlooking, place.* (Ṣ.) One says also, هَوَى مِنْ حَالِقٍ, meaning † *He fell from a high to a low place.* (Ḥar p. 37.) And its pl. حُلُقٌ signifies † The *vacant spaces between heaven and earth.* (TA.) = ‡ *Unlucky* (Ḳ, TA) *to a people;* as though peeling them; and so ▼ حَالِقَةٌ, accord. to the copies of the Ḳ; but correctly ▼ حَالُوقَةٌ, as in the O and Tekmileh. (TA.) = *A tendril,* or *twining portion,* of a grape-vine, (Ṣ, Ḳ, TA,) and of a colocynth and the like, (TA,) *hanging to the shoots:* (Ṣ, Ḳ, TA:) because it has a circular form, like a حَلْقَة [or ring]. (T, TA.)

حَالِقَةٌ [an epithet (being fem. of حَالِقٌ q. v.) in which the quality of a subst. predominates] ‡ *A year of drought, barrenness,* or *dearth:* so in the saying, وَقَعَتْ فِيهِمْ حَالِقَةٌ لَا تَدَعُ شَيْئًا إِلَّا أَهْلَكَتْهُ ‡ [*A year of drought, &c., happened among them, not leaving anything without its destroying it*]. (TA.) — And الحَالِقَةُ ‡ *The cutting,* or *abandoning,* or *forsaking, of kindred,* or *relations;* syn. قَطِيعَةُ الرَّحِمِ; (Khálid Ibn-Jenebeh, Ḳ, TA;) *and mutual wronging, and evil-speaking:* (Khálid Ibn-Jenebeh, TA:) or *that which destroys, and utterly cuts off, religion;* like as the razor utterly cuts off hair: occurring in a trad., in which البَغْضَآءُ [i. e. vehement hatred] and الحَالِقَةُ are termed the disease of the nations (دَآءُ الأُمَمِ). (TA.) — See also حَالِقٌ, last sentence but one.

حَالُوقَةٌ : see حَالِقٌ, fifth sentence, and last sentence but one.

حُلْقَانٌ : see حَوَالِقُ.

مِحْلَقٌ *A razor;* (K;) the *instrument of shaving.* (TA.) — [Hence,] كِسَآءٌ مِحْلَقٌ (S, K) †*A very rough* [*garment of the kind called*] كِسَاء; (K, TA;) as though it shaved off the hair, (S, K,) by reason of its roughness: pl. مَحَالِقُ. (S.)

المُحَلَّقُ *The place of the shaving of the head, in* [*the valley of*] *Minà.* (Lth, K.) ═ مُحَلَّقَةٌ, applied to camels: see حَلْقٌ.

مُحَلِّقٌ : see حُلْقَانٌ: — and حَالِقٌ, in two places. — Also A vessel *less than full.* (K.) — †*Lean,* or *emaciated;* applied to sheep or goats. (Ibn-'Abbád, K.) — فَلَاةٌ مُحَلِّقٌ †*A desert in which is no water.* (TA.)

مَحْلُوقٌ : see حَلِيقٌ, in two places.

مُحَلْقِمٌ : } see حُلْقَانٌ.
مُحَلْقِنٌ : }

حلقم

Q. 1. حَلْقَمَهُ, (S, K,) inf. n. حَلْقَمَةٌ, (TA,) *He cut,* or *severed, his* حُلْقُوم [or *windpipe*]; (S, K;) accord. to the K, meaning *his* حَلْق; but see the explanations of حلقوم below: (TA:) *he slaughtered him in the manner termed* ذَبْح. (TA.) [Mentioned in the Msb in art. حلق.] ═ حَلْقَمَ الرُّطَبُ [or rather البُسْرُ] †*The dates began to be ripe next the base;* as also حَلْقَنَ, in which the ن is asserted by Yaakoob to be a substitute for م. (TA.)

Q. 3. اِحْلَنْقَمَ *He left,* or *forsook, food;* expl. by تَرَكَ الطَّعَامَ. (K.)

حُلْقُومٌ The *windpipe,* or *passage of the breath;* (T, Mgh, TA;) by the cutting, or severing, of which, and of the مَرِىء [or œsophagus] and وَدَجَان [or two external jugular veins], the lawful slaughtering of an animal is completed: (T, TA:) accord. to the S and K, [and to the Msb, in art. حلق, though it is there correctly and fully explained as meaning the *windpipe,*] i. q. حَلْقٌ: but in the M it is explained [agreeably with general usage] as the *passage of the breath, and of coughing, from the* جَوْف [or *chest*]; *consisting of a series of successively-superimposed cartilages* (أَطْبَاقُ غَرَاضِيفَ), *before which, in the exterior of the throat, is nothing but skin; having its lower extremity in the lungs, and its upper extremity at the root of the tongue: from it pass forth the breath and the wind and the saliva and the voice:* [see also another explanation voce حَلْقٌ, from Zj in his "Khalk el-Insán," and the Msb:] pl. حَلَاقِمُ and حَلَاقِيمُ. (TA.) Accord. to some, the م is augmentative: accord. to others, radical. (TA.) — [Hence,] they say, نَزَلْنَا مِثْلَ حُلْقُومِ النَّعَامَةِ, meaning †*We alighted in a strait,* or *confined, place.* (TA.) And حَلَاقِيمُ البِلَادِ means †*The strait,* or *confined, parts of the country,* or *of countries:* (Mgh:) or *the lateral,* and *extreme, parts thereof.* (TA.)

حُلْقَامَةٌ : see what follows.

رُطَبٌ مُحَلْقِمٌ [or rather بُسْرٌ محلقم] †*Dates that have begun to be ripe next the base;* (K;) [or *that have become ripe to the extent of two thirds;* (see حُلْقَانٌ in art. حلق;)] as also مُحَلْقِنٌ: (TA:) and ▾ رُطَبَةٌ حُلْقَامَةٌ [or rather بُسْرَةٌ حلقامة (in the CK حِلْقَامَةٌ)] is applied to a single date in this sense. (K,* TA.)

حلقن

Q., or Q. Q., 1. حَلْقَنَ : see 2 in art. حلق.

حُلْقَانٌ, n. un. with ة : }
حُلْقَانِىٌّ : } see art. حلق.
مُحَلْقِنٌ : }

حلك

1. حَلِكَ, aor. ـَ, inf. n. حُلُوكَةٌ (S, Sgh, TA) and حُلُوكٌ; (Sgh, TA;) and حَلَكَ, aor. ـُ, (K, TA,) inf. n. حَلَكٌ; (K,* TK;) the former verb strangely overlooked by F; (TA;) *It* (a thing, S) *was,* or *became, intensely black;* (S, Sgh, K;) as also ▾ اِحْلَوْلَكَ. (S, TA.)

12: see above.

حُلْكٌ : see حُلَكَةٌ.

حَلَكٌ *Blackness:* (S:) or *intense blackness;* as also ▾ حُلْكَةٌ; (K;) *like the colour of the crow,* or *raven:* (TA:) [for] حَلَكُ الغُرَابِ signifies *the blackness of the crow,* or *raven;* as in the saying, أَسْوَدُ مِثْلُ حَلَكِ الغُرَابِ [*black like the blackness of the crow,* or *raven*]: (S:) or it means *its* حَنَك [q. v.]: (K:) or if you say مثل حَنَكِ الغراب, you mean its beak; (S;) or *the blackness of its feathers:* (Er-Rághib, TA in art. حنك:) or they said أَسْوَدُ مِنْ حَلَكِ الغُرَابِ [*blacker than the* حلك *of the crow,* or *raven*]: or من حَنَكِ الغراب: an Arab of the desert, being asked by Fr whether he said the latter or the former, answered that he never said the former: Umm-El-Heythem, being asked by AHát the same, answered that she said the former, and never the latter; (TA;) and she explained the former as meaning its *two jaws and the part around them;* adding that [the saying that it means] its beak is nought: and IDrd is related to have disallowed the saying من حنك الغراب: (TA in art. حنك:) accord. to AZ, الحَلَكُ means *the colour;* and الحَنَكُ, the beak: some say that the ن in the latter is a substitute for the ل in the former; but others deny this. (TA.) In the saying of a poet,

* مِدَادٌ مِثْلُ حَالِكَةِ ▾ الغُرَابِ *

[*Ink like the* حالكة *of the crow,* or *raven*], cited by Th, حالكة may be a dial. var. of حَلَكٌ: or it may mean its *feathers;* its خَافِيَة or its قَادِمَة or *other feathers.* (TA.)

حُلْكَةٌ : see حَلَكٌ: ═ and see also حُلَكَةٌ. ═ Also i. q. حُكْلَةٌ: (K:) formed from the latter by transposition: so in the saying, فِى لِسَانِهِ حُلْكَةٌ [*In his speech is a barbarousness,* or *a vitiousness,* or *an impediment,* &c.]. (TA.)

حُلَكَةٌ, (S, Msb,) or ▾ حَلَكَةٌ, (ISd, K,) and ▾ حُلُكَةٌ (S, Msb, K) and ▾ حُلْكَةٌ (K) and ▾ حُلَكَآءُ, and ▾ حُلَكَى, (IDrd, K,) or ▾ حَلَكَى, (L,) and ▾ حُلَّكَى, (IDrd, K,) *A species of the* [*kind of lizard called*] عَظَآءَة: (S, Msb, K: [in the CK, erroneously, عِظاء:]) or (S, K [but in the Msb *which is*]) *a small reptile,* (S, Msb, K,) *resembling a fish, of a blue* [or *greyish*] *colour, and glistening,* (Msb,) or *smooth, and having a mixture of whiteness and redness,* (TA in art. نقى,) *that dives into the sand,* (S, Msb, K,) *like as the aquatic bird dives into the water; the Arabs call it* بَنَاتُ النَّقَا, *because it dwells in the sand-hills;* (Msb;) *and* شَحْمَةُ النَّقَا; (TA in art. نقى;) *and they liken to it the fingers* (بَنَان) *of girls, because of their softness,* or *suppleness:* it is also called حُكَلَةٌ, which is app. formed by transposition: (Msb:) ▾ حُلَكٌ [is the coll. gen. n., or quasi-pl. n.; for it] signifies شَحْمُ النَّقَا. (L in art. عوج.) ═ For the first of these words, see also حَالِكٌ, in two places.

حُلَكَى : }
حَلَكَةٌ : }
حُلُكَةٌ : } see حُلَكَةٌ.
حَلَكَآءُ : }
حُلَكَآءُ : }
حُلَّكَى : }

حُلْكُوكٌ : }
حَلَكُوكٌ : } see حَالِكٌ.
حُلْكُلِكٌ : }

حَالِكٌ *Intensely black;* as also ▾ مُحْلَوْلِكٌ (K) and ▾ حُلْكُوكٌ and ▾ حُلْكُلِكٌ and ▾ حَلَكُوكٌ and ▾ مُحْلَنْكِكٌ and ▾ مُسْتَحْلِكٌ (Ibn-'Abbád, K) and ▾ حُلَكَةٌ. (TA.) You say أَسْوَدُ حَالِكٌ and حَالِكٌ; both meaning the same; (S;) i. e. *Black that is intensely black.* (TA in art. حنك.) And إِنَّهُ لَحُلَكَةٌ *Verily he,* or *it, is intensely black.* (TA.)

حَالِكَةٌ : see حَلَكٌ.

مُحْلَنْكِكٌ : }
مُحْلَوْلِكٌ : } see حَالِكٌ.
مُسْتَحْلِكٌ : }

حلم

1. حَلَمَ, (S, Msb, K, [in the CK, erroneously, حَلُمَ,]) aor. ـُ, inf. n. حُلْمٌ (Msb, TA) and حُلُمٌ, of which the former is a contraction, (Msb,) [both used also as simple substs.,] *He dreamed,* or *saw a dream* or *vision* (S, Msb, K) فِى نَوْمِهِ (K) *in his sleep;* (S,* Msb, K;) as also ▾ احتلم, (S, ISd, Msb, K,) and ▾ انحلم, (ISd, K,) and ▾ تحلّم. (K.) You say, حَلَمَ بِهِ, (S, K, [in the CK, again, erroneously, حَلُمَ,]) and عَنْهُ, (K,)

and عَنْهُ ▼تَحَلَّمَ, (TA,) and حَلَمَهُ also, (Ṣ,) *He dreamed, or saw a dream* or *vision, of it:* (Ṣ, Ḳ:) or *he saw it in sleep.* (M, Ḳ.) And حَلَمَ بِالمَرْأَةِ *He* (a man) *dreamed in his sleep that he was compressing the woman.* (TA.) — [Hence,] حُلْمٌ and ▼اِحْتِلَامٌ signify [*The dreaming of*] *copulation in sleep:* (Ḳ:) and the verbs are حَلَمَ and ▼اِحْتَلَمَ. (TA.) And [hence,] both signify *The experiencing an emission of the seminal fluid;* properly, *in dreaming;* and tropically if meaning, *without dreaming, whether awake or in sleep,* or by extension of the signification. (TA.) And hence, (Mgh,) حَلَمَ, (Mgh, Mṣb,) aor. ـُ, inf. n. حُلْمٌ; (Mgh;) and ▼اِحْتَلَمَ; (Mgh, Mṣb;) *He* (a boy) *attained to puberty,* (Mṣb,) [or] *to virility.* (Mgh, Mṣb.) ═ حَلُمَ, with ḍamm [to the ل], inf. n. حِلْمٌ, (Ṣ, Mṣb, Ḳ,) [*He was,* or *became, forbearing,* or *clement;*] *he forgave and concealed* [*offences*]: or *he was,* or *became, moderate, gentle, deliberate, leisurely in his manner of proceeding* or *of deportment* &c., *patient* as meaning *contr. of hasty, grave, staid, sedate,* or *calm;* (Ṣ, Ḳ;) and † *intelligent:* (Ḳ:) or *he managed his soul and temper on the occasion of excitement of anger.* (TA.) [See حِلْمٌ below.] You say, حَلُمَ عَنْهُ and ▼تَحَلَّمَ [*He treated him with forbearance,* or *clemency,* &c.]: both signify the same. (TA.) And يَحْلُمُ عَمَّنْ يَسُبُّهُ [*He treats with forbearance,* or *clemency,* &c., *him who reviles him*]. (TA in art. حمل.) ═ حَلِمَ, aor. ـَ, (Ḳ,) inf. n. حَلَمٌ, (TA,) *He* (a camel) *had* [*upon him*] *many ticks, such as are termed* حَلَم. (Ḳ.) — Also the same verb, (Ṣ, Ḳ,) with the same inf. n., (Ṣ,) *It* (a hide, or skin,) *had in it worms, such as are termed* حَلَم, (Ṣ, Ḳ, TA,) *whereby it was spoilt and perforated,* (Ṣ, TA,) *so that it became useless.* (TA.) A poet says, (Ṣ,) namely, El-Weleed Ibn-'Okbeh, TA,)

* فَإِنَّكَ وَالكِتَابَ إِلَى عَلِيٍّ *
* كَدَابِغَةٍ وَقَدْ حَلِمَ الأَدِيمُ *

[*For verily thou, as to the letter,* or *writing, to 'Alee, art like a woman tanning when the hide has become spoilt and perforated by worms*]: (Ṣ, TA:) he was urging Mo'áwiyeh to contend in battle with 'Alee, [as though] saying to him, Thou labourest to rectify a matter that has become completely corrupt, like this woman who tans the hide that has become perforated and spoilt by the حَلَم. (TA.) [The latter hemistich of this verse is a prov.: see Freytag's Arab. Prov. ii. 346.] ═ حَلَمَهُ, (Ḳ,) inf. n. حَلْمٌ, (TA,) *He plucked the* حَلَم *from it;* [app., accord. to the Ḳ, the worms thus called from a hide, or skin;] as also ▼حلّمه: (Ḳ:) or, accord. to Az, *he took from him,* namely, a camel, *the* [*ticks called*] حَلَم. (TA.)

2. حلّمه, (Ṣ, Mgh, Mṣb, Ḳ,) inf. n. تَحْلِيمٌ (Ṣ, Ḳ) and حِلَّامٌ, like كِذَّابٌ, (Ḳ,) signifies جَعَلَهُ حَلِيمًا [i. e. *He made him to be forbearing,* or *clement,* &c.; or *he pronounced him to be so;* or *he called him so;* or *he held,* or *believed,* or *thought, him to be so*]: (Ṣ, Ḳ:) or *he enjoined him* الحِلْم [i. e. *forbearance,* or *clemency,* &c.]: (Ḳ:) or *he attributed to him* الحِلْم. (Mgh, Mṣb.) ═ حلّم [so in the TA, evidently حَلَّمَ, (see 5, its quasi-pass.,)] also signifies *It fattened* a lamb, or kid; said of sucking. (TA.) — And *He filled* a skin. (TA.) ═ See also 1, last sentence.

4. احلمت *She* (a woman) *brought forth* حُلَمَآء [i. e. *children that were forbearing,* or *clement,* &c.]. (Ḳ.)

5. تحلّم: see 1, first and second sentences. — Also *He affected,* or *pretended, to dream,* or *see a vision in sleep:* whence, in a trad., مَنْ تَحَلَّمَ مَا لَمْ يَحْلُمْ [*He affected,* or *pretended, to have dreamed that which he did not dream*]. (TA.) And *He asserted himself falsely to have dreamed,* or *seen a vision in sleep.* (TA.) And تحلّم الحُلْمَ *i. q.* اِسْتَعْمَلَهُ [*He feigned the dream;* or *made use of it as a pretext*]. (Ḳ.) ═ *He affected,* or *endeavoured to acquire,* (تَكَلَّفَ) [*the quality termed*] الحِلْم [i. e. *forbearance,* or *clemency,* &c.]. (Ṣ, Ḳ.) A poet says,

* تَحَلَّمْ عَنِ الأَدْنَيْنَ وَاسْتَبْقِ وُدَّهُمْ *
* وَلَنْ تَسْتَطِيعَ الحِلْمَ حَتَّى تَحَلَّمَا *

[*Endeavour thou to treat with forbearance the meaner sort of people, and preserve their love; for thou wilt not be able to be forbearing unless thou endeavour to be so*]. (Ṣ.) — See also حَلَمَ. — [Hence,] تَحَلَّمَتِ القِدْرُ عَنْهُ ‡ *The cooking-pot ceased to boil; contr. of* جَهِلَتْ. (TA in art. جهل.) — See also 6. ═ *It became fat;* said of the [kind of lizard called] ضَبّ; (L in art. ملح;) and likewise of cattle: (Ḳ:) [or] *it became fat and compact;* said of a child, and of the ضَبّ: (Ṣ:) [or] *it began to be fat;* said of a child, and of the ضَبّ, (Ḳ,) and of the jerboa, and of the قُرَاد [or tick]; in the Ḳ, erroneously, جَرَاد. (TA.) — تَحَلَّمَتِ القِرْبَةُ *The skin became full.* (TA.)

6. تحالم *He made a show of having* الحِلْم [i. e. *forbearance, or clemency,* &c.], *not having it;* (Ṣ, TA;*) and ▼تحلّم [in like manner] signifies [sometimes] *he made a show of* الحِلْم; expl. by أَظْهَرَ الحِلْمَ. (TA in art. فصح.)

7: see 1.

8: see 1, in four places.

حِلْمٌ: see حُلْمٌ.

حُلْمٌ an inf. n. of حَلَمَ; as also ▼حُلُمٌ. (Mṣb.) — And *A dream,* or *vision in sleep;* (Ṣ, Ḳ;) as also ▼حُلُمٌ: (Ḳ:) accord. to most of the lexicologists, as well as F, *syn. with* رُؤْيَا: or it is specially *such as is evil;* and رؤيا is the contr.: this is corroborated by the trad., الرُّؤْيَا مِنَ اللّٰهِ وَالحُلْمُ مِنَ الشَّيْطَانِ [*The* رؤيا *is from God, and the* حلم *is from the Devil*]: (MF:) and by the phrase, in the Ḳur [xii. 44 and xxi. 5], أَضْغَاثُ أَحْلَامٍ [*The confused circumstances of dreams,* or *of evil dreams*]: but each is used in the place of the other: (TA:) أَحْلَامٌ is the pl. (Ḳ.) — أَحْلَامُ نَائِمٍ [lit. *The dreams of a sleeper;*] *a kind of thick cloths,* or *garments,* (IKh, Z, TA,) *striped, of the people of El-Medeeneh.* (Z, TA.)

حِلْمٌ [*Forbearance; clemency;*] *the quality of forgiving and concealing* [*offences*]: (Mṣb:) or *moderation; gentleness; deliberateness; a leisurely manner of proceding,* or *of deportment,* &c.; *patience,* as meaning *contr. of hastiness: gravity; staidness; sedateness; calmness:* syn. أَنَاةٌ: (Ṣ, Ḳ:) or *these qualities with power* or *ability* [*to exercise the contrary qualities*]; expl. by أَنَاةٌ *and* سُكُونٌ *with* قُدْرَةٌ *and* قُوَّةٌ: (Kull p. 167:) or *the management of one's soul and temper on the occasion of excitement of anger:* (TA:) or *tranquillity on the occasion of emotion of anger:* or *delay in requiting the wrongdoer:* (KT:) it is described by the term ثِقَلٌ, or *gravity;* like as its contr. [سَفَهٌ] is described by the terms خِفَّةٌ and عَجَلٌ, or levity, or lightness, and hastiness: (TA in art. رجح:) also † *intelligence;* (Ḳ;) which is not its proper signification, but a meaning assigned because it is one of the results of intelligence: and ▼حَلَمٌ, with fet-ḥ, is likewise said to have this last meaning; but this requires consideration: (TA:) the former is one of those inf. ns. that are [used as simple substs., and therefore] pluralized: (ISd, TA:) the pl. [of pauc.] is أَحْلَامٌ and [of mult.] حُلُومٌ. (Ḳ.) Hence, in the Ḳur [lii. 32], أَمْ تَأْمُرُهُمْ أَحْلَامُهُمْ بِهٰذَا, (Ḳ,) said to mean † *Do their understandings enjoin them this?* (TA.) And أُولُو الأَحْلَامِ, occurring in a trad., means † *Persons of understanding.* (TA.)

حَلَمٌ: see حَلَمَةٌ, in two places.

حَلِمٌ A camel *having* [*upon him*] *many ticks, such as are called* حَلَم. (Ḳ.) And A camel *spoilt by the abundance of those ticks that were upon him.* (TA.) — Also A hide, or skin, *spoilt and perforated by* [*the worms termed*] حَلَم: and ▼حَلِيمٌ, [in like manner,] a hide, or skin, *spoilt by the* حَلَم *before it is stripped off.* (TA.) And عَنَاقٌ حَلِمَةٌ A she-kid *whose skin has been spoilt by the* حَلَم; (Ḳ,* TA;) as also ▼تَحْلِمَةٌ, of which the pl. is تَحَالِمُ: (Ḳ:) the pl. of حَلِمَةٌ is حِلَامٌ. (TA.)

حُلُمٌ: see حُلْمٌ, in two places. — Also *A* [*dream of*] *copulation in sleep.* (Ḳ.) Hence, بَلَغَ الحُلُمَ *He attained to puberty,* or *virility,* in an absolute sense. (TA.) It is said in the Ḳur [xxiv. 58], وَإِذَا بَلَغَ الأَطْفَالُ مِنْكُمُ الحُلُمَ فَلْيَسْتَأْذِنُوا [*And when your children attain to puberty,* or *virility, they shall ask permission* to come into your presence]. (TA.) [And hence,] أَضْرَاسُ الحُلُمِ, (also called أَضْرَاسُ العَقْلِ, TA in art. ضرس,) [*The teeth of puberty,* or *wisdom-teeth,*] so called because they grow after the attaining to puberty, and the completion of the intellectual faculties: (Ṣ, L, Mṣb, all in art. نجذ:) they are *four teeth that come forth after the* [*other*] *teeth have become strong.* (TA in art. ضرس.)

حَلَمَةٌ *A small tick:* (Ḳ:) or *a large tick;* (Ṣ, Mgh, Mṣb, Ḳ;) like عَلٌّ; (Ṣ;) and said to be *like the head* [or *nipple, when small,*] *of a woman's breast:* (Mṣb:) or *a tick in the last stage of its growth;* for at first, when small, it is called قَمْقَامَةٌ; then, حَمْنَانَةٌ; then, قُرَادٌ; and then, حَلَمَةٌ: (Aṣ, TA:) the pl., (Ṣ,) or [rather] coll. gen. n., (Mgh, Mṣb,) is ▼حَلَمٌ. (Ṣ, Mgh, Mṣb.) — And hence, as being likened thereto, (Mgh,) †The *head* [or *nipple, when small,*] *of a woman's breast,* (T, Ṣ, Mgh,) *in the middle of the* سُعْدَانَة [or *areola*]; (T, TA;) in like manner called قُرَادٌ: (Mgh:) the *little thing rising from the breast of a woman:* (TA:) the حَبَّة [or *small extuberance like a pimple*] *upon the head of the breast of a woman:* (Mṣb:) the ثُؤْلُول [or *small excrescence*] *in the middle of the breast of a woman:* (Ḳ:) and the *head* [or *nipple*] *of each of the two breasts of a man:* (Mṣb:) the two together are termed حَلَمَتَانِ: (Ṣ:) the protuberant piece of flesh is termed حَلَمَة as being likened in size to a large tick. (Mṣb.) — Also *A certain worm, incident to the upper and lower skin of a sheep or goat,* (Aṣ, Ṣ,) *in consequence of which, when the skin is tanned, the place thereof remains thin:* (Ṣ:) or *a certain worm, incident to skin, which it eats, so that, when the skin is tanned, the place of the eating rends:* pl. [or rather coll. gen. n.] ▼حَلَمٌ. (Ḳ.) = And *A species of plant;* (Ṣ, Ḳ;) accord. to Aṣ, *also called* يَنَمَةٌ: (Ṣ:) Aṣ is also related to have said that it is *a plant of the kind termed* عُشْب, *having a dusty hue, a rough feel, and a red flower:* another says that *it grows in Nejd, in the sands, has a blossom, and roughish leaves, and thorns resembling the nails of a man;* and that *the camels suffer adhesion of the spleen to the side, and their young are cast,* [for وتزل احباكها (an evident mistranscription in the TA), I read وَتَزِلُّ أَحْبَالُهَا,] *when they depasture it from the dry branches:* accord. to AḤn, it is [*a plant*] *less than a cubit* [*in height*], *having a thick,* or *rough, leaf, and branches, and a flower like that of the anemone, except that it is larger, and thicker,* or *rougher:* accord. to the Ḳ, it signifies also the *tree* [or *plant*] *called* سَعْدَان; which is one of the most excellent kinds of pasture: but Az says, it has nothing in common with the سعدان, which is a herb having round [heads of] prickles; whereas the حلمة *has no prickles, but is a well-known kind of* جَنْبَة; and I have seen it: (TA:) [Dmr, accord. to Golius, describes it as "*a herb less than the arnoglossa*" (or *arnoglossum*), "*whitening in the leaves, and downy.*"]

حَلِيمٌ *Having* حِلْم [i. e. *forbearance,* or *clemency,* &c.; *forbearing,* or *clement,* &c.]: (Mgh, Mṣb, Ḳ:) pl. حُلَمَآءُ and أَحْلَامٌ. (Ḳ.) In the Ḳur xi. 89, it is said to be used by way of scoffing [or irony]. (TA.) اَلْحَلِيمُ is one of the names of God; meaning [*The Forbearing,* or *Clement,* &c.; or] *He Whom the disobedience of the disobedient does not flurry, nor anger against them disquiet, but Who has appointed to everything a term to which it must finally come.* (TA.) — حَلِيمَةٌ مُغْتَاظَةٌ ‡[lit. *Calm, angry;* or *the like;* because what it contains is sometimes still and sometimes boiling;] is an appellation given to *a stone cooking-pot.* (A and TA in art. غيظ.) = *A fat* camel: (Ṣ:) or a camel *becoming fat.* (ISd, Ḳ.) ISd says, I know not any unaugmented verb belonging to it in this sense. (TA.) = And *Coming fat.* (ISd, Ḳ.) = See also حَلِمٌ.

حَالِمٌ originally signifies ▼مُحْتَلِمٌ [i. e. *Dreaming:* and particularly *dreaming of copulation:* and *experiencing an emission of the seminal fluid in dreaming*]. (Mgh.) — Hence used in a general sense, (Mgh,) meaning One *who has attained to puberty,* or *virility;* (AHeyth, Mgh, Mṣb, TA;) as also ▼مُحْتَلِمٌ. (Mṣb, TA.)

حَالُومٌ *A sort of* أَقِط [q. v.; i. e. *a certain preparation of dried curd*]: (ISd, Ḳ:) or *milk that is made thick, so that it becomes like fresh cheese;* (Ṣ, Ḳ;) but this it is not: (Ṣ:) a word of the dial. of Egypt. (TA.)

أَحْلَامٌ *Bodies;* syn. أَجْسَامٌ. (ISd, Ḳ.) ISd says, I know not any sing. of it [in this sense]. (TA.) = It is also pl. of حُلْمٌ: = and of حِلْمٌ: = and of حَلِيمٌ. (Ḳ.)

تَحْلِمَةٌ: see حَلِمٌ.

مُحْتَلِمٌ: see حَالِمٌ, in two places.

حلو

1. حَلَا, aor. ـُ; (Ṣ, Mṣb, Ḳ;) and حَلِيَ, aor. ـَ; and حَلُوَ, aor. ـُ; (Ḳ;) inf. n. حَلَاوَةٌ (Ṣ, Mṣb, Ḳ) and حَلْوٌ and حُلْوَانٌ; (Ḳ;) *It* (a thing, Ṣ, Mṣb) *was,* or *became, sweet;* (Ṣ, Ḳ;) as also ▼احلولى; (Ṣ, Ḳ;) or this latter signifies *it was,* or *became, very sweet.* (TA.) And in like manner, حَلَا لِى [*It was,* or *became, sweet to me*]. (Mṣb.) And حَلَوَتِ الفَاكِهَةُ, aor. ـُ, inf. n. حَلَاوَةٌ, [*The fruit was,* or *became, sweet.*] (TA.) — حَلِىَ بِعَيْنِى (Ṣ, Mṣb, Ḳ) and فِى عَيْنِى (Ṣ) and بِصَدْرِى (Ṣ, Mṣb) and فِى صَدْرِى (Ṣ) and بِقَلْبِى, (Ḳ,) aor. ـَ; (Ṣ, Mṣb, Ḳ;) and حَلَا, aor. ـُ; (Ṣ, Ḳ;) inf. n. حَلَاوَةٌ (Ṣ, Mṣb, Ḳ) and حُلْوَانٌ; (Ḳ;) *He,* (Ṣ,) or *it,* (Mṣb,) *pleased me,* or *excited my admiration;* (Ṣ, Mṣb;) *was goodly,* or *beautiful, in my estimation:* (Mṣb:) or one says, حَلِىَ فِى عَيْنِى (Lth, Aṣ, Ṣ, Ḳ*) or فِى صَدْرِى, (Aṣ, TA,) aor. ـَ, inf. n. حُلْوَانٌ; (Lth, TA;) and حَلَا فِى فَمِى, (Lth, Aṣ, Ṣ, Ḳ,*) aor. ـُ, inf. n. حَلْوٌ (Lth, TA) and حَلَاوَةٌ. (TA.) تَحْلَى بِهِ العَيْنُ is an inverted phrase, used by a poet, for يَحْلَى بِالعَيْنِ. (Ṣ.) It has been said that حَلِىَ فِى عَيْنِى (ISd, and Ḳ in art. حلى) and فِى صَدْرِى (ISd) is from الحَلْىُ, (ISd, Ḳ,) meaning the thing that is worn, because it signifies *It was beautiful in my eye, like the* حَلْى; not from حَلَاوَةٌ; (ISd, and TA in art. حلى;) but this saying is not valid, nor approved. (TA in the present art.) Accord. to Lḥ, one says, بِقَلْبِى and فِى عَيْنِى, and حَلِيَتِ المَرْأَةُ بِعَيْنِى and فِى قَلْبِى, aor. ـَ, inf. n. حَلَاوَةٌ; and حَلَتْ, aor. ـُ, inf. n. حَلَاوَةٌ; [*The woman was pleasing in my eye, and in my mind,* or *heart.*] (TA.) — مَا يَمُرُّ وَمَا يَحْلُو ‡ *He is not bitter* at one time, *and he is not sweet* at another. (IAạr, Ḳ, TA.) [In some copies of the Ḳ, وَلَا يَحْلُو.]) And أَمُرُّ وَأَحْلُو ‡ *I am bitter* at one time, *and I am sweet* at one time. (IAạr, M in art. مر.) [See also 4.] = حَلِىَ الشَّىْءَ, aor. ـَ, (Ḳ,) inf. n. حَلَاوَةٌ; (TḲ;) and ▼استحلاهُ, (Ṣ, Mṣb, Ḳ,) from الحَلَاوَةُ, like استجادهُ from الجَوْدَةُ, (Ṣ,) [the most common form,] and ▼تحلّاهُ, (Ḳ,) and ▼احلولاهُ, (Ṣ, Ḳ,) which is the only trans. verb of its measure except one other, in the phrase اِعْرَوْرَيْتُ الفَرَسَ; (Ṣ;) all signify the same; (Ḳ;) *He esteemed the thing sweet* [both properly and metaphorically as is indicated in the TA]: (Mṣb in explanation of the second of these verbs, and TḲ in explanation of all of them:) and [in like manner] ▼أَحْلَيْتُهُ *I found it to be sweet:* (Ṣ, Ḳ:) or this signifies *I made it sweet:* (Ḳ:) or it has both of these significations: (Ṣ:) you say, أَحْلَيْتُ, هٰذَا المَكَانَ ▼اِسْتَحْلَيْتُهُ, meaning [*I esteemed,* or *found, this place to be sweet,* or *pleasant*]. (TA.) And حَلِيَتْهُ العَيْنُ [*The eye esteemed,* or *found, him,* or *it, to be pleasing,* or *goodly,* or *beautiful*]. (IAạr, TA in art. حلى.) — حَلِىَ مِنْهُ بِخَيْرٍ, (Ḳ,) aor. ـَ; (TA;) and حَلَا, (Ḳ,) aor. ـُ; (TA;) *He found,* or *experienced,* or *he got,* or *obtained, from him,* or *it, good.* (Ḳ.) [See also 1 in art. حلى.] = حَلَاهُ الشَّىْءَ, (Mṣb,* Ḳ,) or حَلَاهُ كَذَا مَالًا, (Ṣ,) aor. ـُ, (Ṣ, Mṣb,) inf. n. حَلْوٌ (Ṣ, Ḳ [in the CḲ حُلْو]) and حُلْوَانٌ, (Ṣ,) *He gave him* (Ṣ, Mṣb, Ḳ) *the thing,* (Ḳ,) or *such a piece of property,* [*as a gratuity,*] *for something that he had done; not as hire,* or *pay,* or *wages.* (Ṣ.) — You say also حَلَوْتُ meaning *I bribed* another; *gave him a bribe.* (TA.) — And ▼لَأَحْلُوَنَّكَ حُلْوَانَكَ *I will assuredly give thee thy requital.* (IAạr, Ḳ.) — حُلْوَانٌ also signifies A man's *taking for himself a portion of the dowry of his daughter:* an act for which the Arabs used to reproach him who did it. (Ṣ, Mṣb.) You say, حَلَاهُ, inf. n. حَلْوٌ [in the CḲ حُلْو, and حُلُوٌّ also,] and حُلْوَانٌ, *He gave him in marriage his daughter,* or *his sister,* (Ḳ, TA,) or *any woman,* (TA,) *for a certain dowry, on the condition that he should assign to him a certain portion of it.* (Ḳ, TA.) = حَلَوْتُ المَرْأَةَ *i. q.* حَلَيْتُهَا, i. e. *I assigned,* or *gave, to the woman* حَلْى [or *ornaments*]. (Ṣ.)

2. حلّاهُ, (Ṣ, Ḳ, [erroneously written in the CḲ without the sheddeh,]) inf. n. تَحْلِيَةٌ, (Ḳ,) *He made it sweet;* (Ṣ, Ḳ;) namely, food; (Ṣ;) or a thing; (Ḳ;) as also ▼احلاهُ: (Ṣ:) and so حَلَّأَهُ, which is anomalous: (Ḳ:) sometimes they said, حَلَّأْتُ السَّوِيقَ [*I sweetened the meal of parched barley,* or *the mess made thereof*]; pronouncing with hemz that which is not [properly] with hemz: (Ṣ:) this is said by Lth to be a mistake on their part. (TA.) — [Hence,] حَلَّيْتُ الشَّىْءَ فِى عَيْنِ صَاحِبِهِ [*I made the thing to seem pleasing,* or *goodly,* or *beautiful, in the eye of its possessor*]. (Ṣ.)

3. حَالَيْتُهُ, (inf. n. مُحَالَاةٌ, TḲ,) ‡ *I jested,* or *joked, with him.* (Ṣ, Ḳ, TA.)

4: see 2. — Hence, (TA,) مَا أَمَرَّ وَمَا أَحْلَى

† *He said not anything:* (Ṣ, TA:) or the meaning is similar to that of the phrase next following. (TA.) مَا يُمِرُّ وَمَا يُحْلِى † *He says not a bitter thing nor a sweet thing:* and *he does not a bitter thing nor a sweet thing.* (Ḳ.) [See a similar phrase near the middle of the first paragraph.] — See also another signification in the first paragraph. — مَا أَحْلَاهُ [*How sweet, &c., is it!*] is said by some to be an instance of a verb having a dim. form; so that you say, ▼ مَا أُحَيْلَاهُ [*How very sweet, &c., is it!*]; like مَا أُمَيْلِحَهُ [q. v.]. (TA in art. ملح.)

5. تحلّاهُ: see 1.

6. تَحَالَتْ *She* (a woman) *affected,* or *made a show of, sweetness, and self-conceitedness.* (Ṣ.)

8. احتلى لِنَفَقَةِ ٱمْرَأَتِهِ, and لِمَهْرِهَا, *He exercised art,* or *ingenuity, for* [*the purpose of procuring*] *the expenses of his wife,* and *her dowry:* one says, إِحْتَلِ فَتَزَوَّجْ [*Exercise thou art, &c., and marry*]. (TA.)

10: see 1, in two places. — استحلاهُ also signifies *He sought* [*to elicit*] *its,* or *his, sweetness.* (TA.)

12. احلولى: see 1, first sentence. — [Hence,] said of a man, *He was,* or *became, sweet in disposition.* (IAạr.) — احلولاهُ: see 1.

حَلًا A medicine *mixed,* or *moistened, with water* or *the like.* (Ḳ.)

حُلْوٌ *Sweet; contr. of* مُرٌّ; (Ṣ, Ḳ;) i. e., in the mouth: and in like manner, in the eye [meaning *pleasing,* or *goodly,* or *beautiful:* see 1]: (TA:) applied also to a saying, and to an action: (Ḳ:) fem. with ة. (Mṣb.) And الحُلْوُ الحَلَالُ † *Language in which is nothing that induces doubt,* or *suspicion:* (Ḳ and TA in art. حل:) and *the man in whom is nothing that induces doubt,* or *suspicion.* (TA in the present art.) And حُلْوٌ, also, applied to a man, † One *who is excited to briskness, liveliness,* or *sprightliness,* (يُسْتَخَفُّ,) *and is esteemed pleasing,* or *goodly,* or *beautiful, in the eye;* (Ḳ,* TA;) as also ▼ حُلْوٌ: (IAạr, Ḳ:) the fem. is حُلْوَةٌ: the pl. masc. حُلْوُونَ, and pl. fem. حُلْوَاتٌ: (Ḳ:) there is no broken pl., masc. or fem. (TA.)

حلو *A small* حَفّ [q. v., in the CḲ, erroneously, خُفّ,] *with which one weaves:* (Ḳ, TA:) or *the wooden thing which the weaver turns round:* [app. meaning the *yarn-beam, upon which the yarn is rolled;* termed حَفَّةٌ:] the poet Shemmákh likens the tongue of a braying [wild] ass to a حلو that has slipped from the back of a loom. (TA.)

حُلْوَاةٌ: see حَلْوَى.

أَحْلَى: see حَلْوَى.

حَلْوَاءُ and ▼ حَلْوَى, (Ṣ, Mṣb, Ḳ, &c.,) the latter mentioned by Aṣ, of the fem. gender, (TA,) [*Sweetmeat;* as also ▼ حَلَاوَةٌ; (see حَلَّآء, below;) this last and حَلْوَى used in this sense in the present day;] *an eatable,* (T, Ṣ, M, Mṣb,) *well known,* (Ḳ,) *prepared with sweetness;* (T, M, Mṣb;) said to be peculiarly applied to *such as is prepared with art* [as distinguished from such as is naturally sweet]: (TA:) the حلواء mentioned in a trad. is said to be that which is termed مَجِيع [made of dates kneaded with milk]: (MF, TA:) the pl. of حَلْوَى is حَلَاوَى, with fet-ḥ to the و. (Mṣb.) — The former is also applied by some to *Fruit;* syn. فَاكِهَةٌ: (T, TA:) or both, (Ḳ,) or the former, (TA,) to *sweet fruit.* (Ḳ, TA.) — See also حُلَاوَةٌ.

حُلْوَانٌ is a subst. [as well as an inf. n.], signifying *A gift:* (Mṣb:) [*a gratuity:* so in the present day:] the *hire,* or *pay,* of a broker; (Lḥ, Ḳ;) and of a diviner, (Aṣ, Ṣ,* Mṣb,* Ḳ,) for divination, (Aṣ,) which is forbidden in a trad.: (Ṣ, Mṣb:) and *a requital;* see 1. (IAạr, Ḳ.) Also The *dowry,* or *nuptial gift,* of a woman: (Mṣb, Ḳ:) [or *a portion thereof which the father or guardian of the bride used, in some cases, to take for himself;* see 1:] or *a gift to a woman in consideration of having her as a wife during a certain fixed period;* (Ḳ;) according to a practice obtaining in Mekkeh: (TA:) or *a gift of the nature of a bribe.* (Ḳ.)

حُلْوٌ: see حُلْوٌ. — نَاقَةٌ حُلْوَةٌ: see the next paragraph.

قَوْلٌ حَلِىٌّ *A saying sweet in the mouth.* (Ḳ.) — نَاقَةٌ حَلِيَّةٌ (Lḥ, M, Ḳ) and ▼ حُلْوَةٌ, (Ḳ,) the latter is the original form [but app. obsolete], (Lḥ, M,) *A she-camel eminent,* (Lḥ, M,) or *perfect,* (Ḳ,) *in pleasingness,* or *goodliness,* or *beauty:* (Lḥ, M, Ḳ:) or *pleasing in appearance and pace.* (TA in art. علو.)

حُلَاوَةُ القَفَا: see حُلَاوَةٌ.

حَلَاوَةٌ *Sweetness; contr. of* مَرَارَةٌ. (TA.) [See 1, of which it is an inf. n.] — See also حَلْوَآءُ. — أَرْضٌ حَلَاوَةٌ *Land that produces herbs,* or *leguminous plants, of the kind termed* ذُكُور [q. v.]. (Ḳ.) — See also what next follows.

حُلَاوَةُ القَفَا (T, Ṣ, Mṣb, Ḳ) and ▼ حَلَاوَةُ القفا, (IAth, Ḳ,) but this is said by Ks to be unknown, (TA,) and ▼ حِلَاوَةُ القفا (IAth) and ▼ حُلَاوَآءُ القفا, (Ḳ, TA,) with ḍamm, mentioned by Lḥ, (TA, [in the CḲ حَلَاوَة,]) and ▼ حُلْوَآءُ القفا (Ṣgh, Ḳ) and ▼ حُلَاوَى القفا and ▼ حَلَاوَآءُ القفا, (Ṣ, Ḳ,) *The middle of the back of the neck:* (T, Ṣ, Mṣb, Ḳ:) or, as some say, *the* [*small protuberance termed*] فَأْس *of* [or rather *above*] *the back of the neck.* (T.) — حُلَاوَةٌ is also said to signify the same as حُلَآءَةٌ, i. e. *What is rubbed between two stones, to be used as a collyrium.* (TA. [See the latter of these two words in art. حلأ.])

حِلَاوَةُ القَفَا: see the next preceding paragraph.

حُلَاوَى *A certain plant:* (Ṣ:) or *a certain small tree,* (Ḳ,) *of the kind termed* جَنْبَة, *evergreen:* (TA:) and, (Ḳ,) or, as some say, (TA,) *a certain thorny plant,* (Ḳ,) *having a yellow flower, and small round leaves like those of the* سَذَاب [or *rue*]: (TA:) *a species of plant found in the desert:* (T, TA:) pl. حُلَاوَى, (Ḳ,) like the sing., (TA,) and, (Ḳ,) or, as some say, (TA,) حُلَاوِيَاتٌ. (Ḳ.) It has been said that the sing. is حُلَاوِيَةٌ, like رَبَاعِيَةٌ: but Az says that this was not known by him: Aṣ mentions, as of the measure فُعَالَى, the words خُزَامَى and رُخَامَى and حُلَاوَى; each the name of a plant. (TA.) — حُلَاوَى القَفَا: see حُلَاوَةٌ.

حُلَاوَآءُ القَفَا: see حُلَاوَةٌ.

حَلَّآءٌ *A maker and seller of* حَلَاوَة [or *sweetmeat*]. (TA.)

[أَحْلَى *More,* and *most, sweet,* both properly and metaphorically:] ▼ حُلْوَى is [its fem.,] the *contr. of* مُرَّى: you say, خُذِ الحُلْوَى وَأَعْطِهِ المُرَّى [*Take thou the sweeter,* or *sweetest, and give to him the bitterer,* or *bitterest*]. (Ṣ.)

مَا أُحَيْلَاهُ: see 4, last sentence.

حلى

1. حَلَيْتُ المَرْأَةَ, aor. ـِ, inf. n. حَلْىٌ, *I assigned,* or *gave, to the woman* حُلِىّ [or *ornaments*]; and so حَلَوْتُهَا. (Ṣ.) [See also 2.] — حَلِيَتْ, (Ṣ, Mṣb, Ḳ,) aor. ـَ, (Ḳ,) inf. n. as above, (Mṣb, Ḳ,) *She* (a woman) *acquired an ornament,* or *ornaments:* (Ḳ:) or *she wore an ornament,* or *ornaments;* as also ▼ تحلّت: (Mṣb, Ḳ:) or the former signifies *she had an ornament,* or *ornaments:* (Ṣ, Ḳ:) and ▼ the latter, *she adorned herself* with an ornament, or ornaments: (Ṣ, Mgh,* TA:) or *she made for herself an ornament,* or *ornaments.* (Mṣb, TA.) — لَمْ يَحْلَ مِنْهُ بِطَائِلٍ *He gained not,* or *derived not, from him,* or *it, any great profit, advantage,* or *benefit:* the verb is not used in this sense except in negative phrases; (Ṣ, TA;) and is from الحَلْىُ and الحِلْيَةُ; because the mind reckons an ornament as an acquisition: not from حلو. (TA.) [But an affirmative phrase, with the verb حَلِىَ used in a similar sense, is mentioned in the Ḳ in art. حلو: see 1 in that art. See also 1 in art. حلأ.] — See also حَلِى, below. — حَلِيَتِ الشَّفَةُ: see حَلِئَت, in art. حلأ.

2. حلّى المَرْأَةَ, (Ṣ, Mṣb, Ḳ,) inf. n. تَحْلِيَةٌ, (Ḳ,) *He decked the woman with an ornament,* or *ornaments:* (Ṣ, Mṣb, Ḳ: [see also 1, first sentence: and in like manner, السَّيْفَ *the sword:*]) or *he made for her an ornament,* or *ornaments:* or *he described her:* (Ḳ:) or you say also حَلَّيْتُ الرَّجُلَ as meaning *I described the* حِلْيَة [i. e. *quality,* or *qualities* or *attributes,* or *state* or *condition,*] *of the man:* (Ṣ:) and you say also, حلّاهُ عَلَيْهِ [*he described it to him*]. (L in art. سملج.) The verb in the first of these senses is doubly trans., as in the saying in the Ḳur [xviii. 30, &c.], يُحَلَّوْنَ فِيهَا مِنْ أَسَاوِرَ مِنْ ذَهَبٍ [*They shall be decked therein with bracelets of gold*]. (TA.) — حَلَّيْتُمْ for حَلَّأْتُمْ: see 2 in art. حلأ.

5. تَحَلَّتْ: see 1, in two places. — [Hence,] تحلّى فُلَانٌ بِمَا لَيْسَ فِيهِ *Such a one affected that which was not in him.* (TA.) — تحلّاهُ *He knew his* [حِلْيَة, i. e.] *quality,* or *qualities* or *attributes,* or *state* or *condition.* (TA.)

حَلًا *Pustules breaking out in the mouths of children* [app. *after a fever*; like أُحَلًا]. (Kr, M.)

حَلْىٌ *An ornament* (Mgh, Mṣb,* Ḳ) *of a woman*, (Ṣ, Mgh, Mṣb,) *of moulded metal*, or *of stones*, (Ḳ,) or *of gold*, or *of silver*, and some say, or *of jewels*, or *gems*: (Mgh:) pl. حُلِىٌّ (Ṣ, Mgh, Mṣb, Ḳ) and حِلِىٌّ, also, because of the ى, like عِصِىٌّ: (Ṣ, TA:) or حَلْىٌ is a pl. [or coll. gen. n.], and its sing. [or n. un.] is ▼حَلْيَةٌ: (Ḳ:) so says AAF: (TA:) ▼حِلْيَةٌ, also, signifies the same as حَلْىٌ; (Ḳ; [in the CḲ حَلْىٌ;]) and particularly, (Ḳ,) the *ornament*, or *ornaments*, (حَلْىٌ, Ḳ, or زِينَةٌ, Mgh and Mṣb,) *of gold* or *silver*, (Mgh,) of a sword, (Ṣ, Mgh, Mṣb, Ḳ,) as also ▼حَلَاةٌ, (Ḳ,) or of a lamp, and of other things: (Mgh:) accord. to Lth, حَلْىٌ signifies *any* حِلْيَة [i. e. *ornament*, or *ornaments*,] *with which one decks a woman or a sword and the like*: but accord. to others, only *of a woman*; and one says only ▼حِلْيَةٌ in relation to other things, to a sword [for instance], and the like: (TA:) the pl. of حِلْيَةٌ is حِلًى and حُلًى; (Ṣ, Ḳ;) or, accord. to IF, it has no pl. (Mṣb.)

حَلِى *A cry by which a she-camel is chidden*; as also حَلِ and حَلْ; like as a he-camel is by the cry حَوْبِ and حَوْبَ &c.: (TA voce حَوْبِ:) or *by which female camels are chidden*; as also حَلْ, and, when in connexion with a following word, حَلِ. (TA voce حَلْ, in art. حل.) One says, in chiding the she-camel, حَلِى لَا ▼حَلِيتِ [*On! mayest thou not gain any great benefit*: like as one says, in chiding the he-camel, حَبْ لَا مَشِيتَ, &c.]. (TA in the present art.)

حَلَاةٌ: }
حَلْيَةٌ: } see حَلْىٌ.

حِلْيَةٌ: see حَلْىٌ, in two places. — Also The *quality*, or the *aggregate of the attributes* or *qualities*, or the *state* or *condition*, (Ṣ, Mgh, Mṣb, Ḳ,) and the *make*, and *form*, (Ḳ,) and the *appearance in respect of colour*, or *complexion*, &c., (Mgh,) of a man: (Ṣ, Mgh:) pl. حِلًى and حُلًى. (Mgh, Mṣb, TA.) *A description of the face*, or *countenance*, of a man. (TA.) — الحِلْيَةُ in a trad. respecting الوُضُوْء means [*The mode of ablution termed*] التَّحْجِيلُ. (TA. [See 2 in art. حجل.])

حَلِىٌّ *Dry, dried up*, or *tough*. (TA.) — *What has become dry* (Ṣ, Ḳ) *and white* (Ḳ [but see نَصِىٌّ]) *of the* [*plant called*] نَصِىّ, (Ṣ, Ḳ,) *and* سَبَط: Az says, *it is one of the best kinds of pasture of the people of the desert for camels and horses; and when its produce appears, it resembles eared corn*: accord. to Lth, it is *every plant that resembles corn, or seed-produce, in its manner of growth*; but this, says Az, is a mistake: (TA:) n. un. with ة: (Ḳ:) and pl. أَحْلِيَةٌ. (Ṣ.) — The *pole, or long piece of wood*, [app. *of a plough*,] *that is between the two bulls*: of the dial. of El-Yemen. (TA.)

حَلِيَّةٌ: see حَالٍ.

حُلَّيَا *A certain plant*. (Ḳ.) — And *A certain food of the Arabs*, (Ṣgh, Ḳ,) *in which dates are rubbed and pressed* [or *mashed*] *with the hand*. (Ṣgh.)

حَالٍ (Ḳ) and حَالِيَةٌ (Ṣ, Ḳ) A woman *having an ornament* or *ornaments*; (Ṣ, Ḳ;) as also ▼حَلِيَّةٌ: (Ṣ:) or *wearing an ornament* or *ornaments*; as also ▼مُتَحَلِّيَةٌ: (Ḳ:) pl. حَوَالٍ. (Ṣ.) — And [hence,] حَالِيَةٌ means ‡ Trees *having leaves and fruit*. (TA.)

مُحَلًّى *Ornamented*: applied to a sword [&c.]. (Ṣ.) — [*Described*.]

مُتَحَلِّيَةٌ: see حَالٍ.

حم

1. حَمَّ, (Ṣ, Ḳ,) sec. pers. حَمِمْتَ, aor. ـَ, inf. n. حَمٌّ, (TA,) [or perhaps this should be حُمُومٌ,] *It* (water) *became hot*. (Ṣ, Ḳ, TA.) — حَمِمْتُ, aor. ـَ, (Ḳ,) inf. n. حَمَمٌ, (Ṣ, Ḳ,) *I was*, or *became*, أَحَمّ, signifying *black*; (Ṣ, Ḳ; [accord. to the latter of which, and accord. to El-Hejeree, this epithet also signifies *white*; but it appears from the TA that the former only is here meant; and the verb seems primarily to signify *I became rendered black by heat*;]) as also ▼اِحْمَوْمَيْتُ [originally اِحْمَوْمَمْتُ, or from حَمَى, q. v.], and ▼تَحَمَّمْتُ, (Ḳ, [omitted in the TA,]) and ▼تَحَمْحَمْتُ. (Ḳ, TA: the last, in the CḲ, written تَحْمَحَمْتُ.) — حَمَّ الجَمْرُ, sec. pers. حَمِمْتَ, aor. ـَ, inf. n. حَمَمٌ, *The live coals became black, after their flaming had ceased*, or *after they had become extinguished*: (Mṣb:) or حَمَّتِ الجَمْرَةُ, (Ṣ, Ḳ,) sec. pers. as above, (TA,) aor. ـَ, *the live coal became a piece of charcoal*, (Ṣ, Ḳ,) or *of ashes*. (Ṣ.) — حَمَّهُ, (Ṣ, Ḳ,) aor. ـُ, (Ṣ,) inf. n. حَمٌّ, (TA,) *He heated it*, namely, water, (Ṣ, Ḳ, TA,) *with fire*; (TA;) as also ▼أَحَمَّهُ, (Ṣ, Ḳ,) and ▼حَمَّمَهُ. (Ḳ.) You say, أَحِمُّوا ▼ لَنَا المَآءَ, (TA,) or مِنَ المَآءِ, (Ṣ,) *Heat ye for us the water*, or *some of the water*. (Ṣ, TA.) — *He heated it*; *kindled fire in it*; *filled it with firewood, to heat it*; or *heated it fully with fuel*; namely, an oven. (Ḳ,* TA.) — حَمَّ الأَلْيَةَ, (Ṣ,) or الشَّحْمَةَ, (Ḳ,) aor. ـُ, (Ṣ,) inf. n. حَمٌّ, (TA,) *He melted* [*the fat of a sheep's tail*, or *the piece of fat*]. (Ṣ, Ḳ.) — حَمَّ نَفْسَهُ: see 4. — حُمَّ *He* (a man, Ṣ) *was*, or *became, fevered*, or *affected with fever*; or *he had*, or *was sick of, a fever*: (Ṣ, Mgh, Mṣb, Ḳ:) or one says [of himself], حُمِمْتُ حُمَّى, (Ḳ, TA, [in the CḲ, erroneously, حَمَمْتُ,]) حُمَّى being held by ISd to be an inf. n. like بُشْرَى and رُجْعَى; (TA;) and the simple subst. [also] is حُمَّى: (Ḳ:) [or the inf. n. is حَمٌّ; for] you say, حُمِمْتُ حَمًّا; and the simple subst. is حُمَّى. (L.) And حُمَّ عَلَى طَعَامٍ *He had a fever from eating* [certain] *food*. (Ḳ,* TA.) And حَمَّ, [app. حُمَّ,] inf. n. حُمَامٌ, said of a camel, *He had a fever*. (TA. [See حُمَامٌ, below.]) — حَمَّهُ said of an affair, an event, or a case: see 4. — حَمَّ ارْتِحَالَ البَعِيرِ, (Fr, Ṣ, Ḳ,) aor. ـُ, (Ṣ,) *He hastened the going*, or *departure, of the camel*. (Fr, Ṣ, Ḳ.) — حَمَّ لَهُ كَذَا, and ▼أَحَمَّ, *He* (God) *decreed*, or *appointed, to him*, or *for him, such a thing*. (Ḳ, TA.) And حُمَّ, (Ṣ, Ḳ,) inf. n. حَمٌّ, (Ḳ,) or حُمُومٌ, (Ḥar p. 347,) *It* (a thing, Ṣ, or an event, Ḳ) *was decreed*, or *appointed*; (Ṣ, Ḳ;) as also ▼أُحِمَّ. (Ṣ.) And حُمَّ لَهُ ذٰلِكَ *That was decreed*, or *appointed, to him*, or *for him*. (Ḳ.) — حَمَّ حَمَّهُ, (Ṣ, Ḳ,) aor. ـُ, (Ṣ,) i. q. قَصَدَ قَصْدَهُ [like أَمَّ أَمَّهُ, q. v.; حَمَّ in this sense being a dial. var. of أَمَّ, as also أَبَّ]. (Ṣ, Ḳ.) — See also 4 as an intrans. v.

2. حَمَّمَهُ: see 1. — Also, (Ṣ, Mṣb, Ḳ,*) inf. n. تَحْمِيمٌ, (Mṣb,) *He blackened* (Ṣ, Mṣb, Ḳ) *his* (a man's, Ṣ) *face*, (Ṣ, Ḳ,) or *it*, one's face, (Mṣb,) *with charcoal*. (Ṣ, Mṣb, Ḳ.) [Hence,] حُمِّمَ وَجْهُ الزَّانِى *The face of the fornicator*, or *adulterer, was blackened* [*with charcoal*]. (Mgh. [See 2 in art. جبه.]) — [Using the verb intransitively,] you say also, حَمَّمَ رَأْسُهُ *His head became black after shaving*: (Ṣ, Mgh, TA:) [i. e.] *the hair of his head grew* [*again*] *after it had been shaven*. (Ḳ.) And hence, حَمَّمَ بِالمَآءِ, said of the hair, *It was rendered black by the water*: because the hair, when shaggy, or dishevelled, in consequence of its being seldom dressed or anointed, becomes dusty; and when it is washed with water, its blackness appears. (TA.) And حَمَّمَ الغُلَامُ *The boy's*, or *young man's, beard appeared*. (Ḳ.) And حَمَّمَ الفَرْخُ *The young bird's plumage came forth*: (Ṣ, Ḳ:) or its *down*. (TA.) And حَمَّمَتِ الأَرْضُ *The herbage of the land appeared, of a green hue inclining to black*. (Ḳ.) — حَمَّمَ امْرَأَتَهُ, (Ṣ, Ḳ,) inf. n. تَحْمِيمٌ (Mgh, TA) [and تَحِمَّةٌ], *He gave a present to his wife after divorce*: (Ṣ, M, Ḳ:*) the explanation in the Ḳ, مَتَّعَهَا بِالطَّلَاقِ, should be, as in the [Ṣ and] M, مَتَّعَهَا بِشَىْءٍ بَعْدَ الطَّلَاقِ. (TA.) The verb is doubly trans., as meaning أَعْطَى: so in the phrase, حَمَّمَهَا خَادِمًا سَوْدَآءَ *He gave her, after divorce, a black female slave*: or this may be for حَمَّمَهَا بِهَا. (TA.) [Hence,] ثِيَابُ التَّحِمَّةِ *The clothing with which a man attires his wife when he gives her a gift after divorce*. (Ḳ, TA.)

3. حَامَّهُ, inf. n. مُحَامَّةٌ, i. q. قَارَبَهُ [app. as meaning *He approached*, or *drew near to, him*, or *it*]. (Ḳ.) And حَامَمْتُهُ, (inf. n. as above, Ḳ,) *I desired*, or *sought, to obtain from him*, or *I demanded of him, something*. (El-Umawee, Ṣ, Ḳ.)

4. أَحَمَّهُ as syn. with حَمَّهُ and حَمَّمَهُ: see 1, in two places. — Also *He washed him* (namely, another man,) *with* حَمِيم [i. e. *hot water*]. (Ṣ.) And أَحَمَّ نَفْسَهُ *He washed himself with cold water*, (Ḳ,) accord. to IAạr: but accord. to others, *with hot water*; as also ▼حَمَّ نَفْسَهُ: and حُمُومٌ [is an inf. n. of حَمَّ, and] signifies the *washing oneself*; but is of a vulgar dialect. (TA.

[See also 10.]) — *He* (God) *caused him to have,* or *be sick of, a fever.* (Ṣ, Mṣb, Ḳ.) — *It* (an affair, an event, or a case,) *rendered him anxious, disquieted him,* or *grieved him;* syn. أَهَمَّهُ; (Ṣ, Ḳ;) as also ↓ حَمَّهُ. (Ḳ.) And أُحِمَّ *He* (a man) *was affected with confusion, perplexity, fear, impatience, disquietude,* or *agitation,* and *anxiety,* or *grief.* (TA.) ═ *He* (God) *rendered him,* or *caused him to be,* أَحَمّ, (Ṣ, Ḳ,) i. e. *black.* (Ṣ.) ═ *He caused it to draw near,* or *approach.* (Mṣb.) ═ أَحَمَّتِ الأَرْضُ *The land had fever in it:* (Ṣ, Ḳ:) or *had much fever in it.* (TA.) ═ احمّ *It drew near,* or *approached;* (Ṣ, Mṣb, Ḳ;) as also ↓ حُمَّ, [in the Ḥam p. 530, written حَمَّ,] aor. ـُ, inf. n. حَمّ: (Mṣb:) *it was,* or *became, present:* (Ḳ:) *its time drew near,* or *came;* as also اُحِمَّ: so says Ks; and thus this last verb is explained by Aṣ; but he knew not احمّ in this sense. (Ṣ, TA.) You say, أَحَمَّتِ الحَاجَةُ and أُحِمَّتْ *The object of want became near;* (ISk, TA;) and both are mentioned by Fr. (Ṣ.) And احمّ قُدُومُهُمْ and أُحِمَّ *Their coming drew near.* (Fr, TA.) The Kilábeeyeh says, احمّ رَحِيلُنَا فَنَحْنُ سَائِرُونَ غَدًا [*Our departure has drawn near, and we are going to-morrow*]: and احمّ رحيلنا فنحن سائرون اليَوْمَ [Our departure is determined upon, and we are going to-day]; meaning we have determined upon our going to-day. (TA.) ═ أُحِمَّ لَهُ كَذَا; and أُحِمَّ: see 1, near the end of the paragraph.

5: see 1: ═ and see also 10.

8. احتمّ *He was,* or *became, anxious, disquieted,* or *grieved,* syn. اهتمّ, (Ṣ, TA,) لَهُ *for him;* as though for one near and dear to him: (TA: [see حَمِيم:]) or *he was,* or *became, anxious, disquieted,* or *grieved, and sleepless:* (Ḥam p. 90:) or *he was,* or *became, anxious, disquieted,* or *grieved, by night:* (Ḳ, and Ḥam ibid.:) اِحْتِمَام differing from اِهْتِمَام in being [often] by day: (Ḥam p. 433:) and *he slept not by reason of anxiety, disquietude,* or *grief.* (Ḳ.) And احتمّت العَيْنُ *The eye was,* or *became, sleepless, without pain.* (Ḳ.) Also احتمّ لِفُلَانٍ *He was,* or *became, sharp, hasty,* or *irascible, towards such a one.* (TA.)

10. استحمّ *He washed himself with hot water:* (Ṣ, Mṣb, Ḳ: or accord. to some copies of the Ḳ, استحمّ بِالحَمِيمِ has this meaning:) this is the primary signification: (Ṣ:) then applied, (Ṣ, Mṣb,) by reason of frequency of usage, (Mṣb,) to mean *he washed himself with any water.* (Ṣ, Mṣb. [See also 4.]) — *He entered the* حَمَّام [or *hot bath*]: (Mgh, TA:) ↓ تحمّم [in this sense] is not of established authority. (Mgh.) — *He sweated:* (Ṣ, Ḳ:) said of a man, (TA,) and of a horse (Ṣ, TA) or similar beast. (TA.)

12: see 1, second sentence.

R. Q. 1. حَمْحَمَ, [inf. n. حَمْحَمَةٌ,] *He* (a horse) *uttered his cry,* [or *neighed,*] *when desiring fodder;* as also ↓ تَحَمْحَمَ: (Ṣ:) accord. to Az, حَمْحَمَةٌ is app. a word imitative of the cry of the horse when he desires fodder; or when he sees his master to whom he has been accustomed, and behaves familiarly towards him: (TA:) or it signifies a horse's *uttering a cry with a kind of yearning sound, in order that his master may feel tenderness for him;* as also ↓ تَحَمْحُمٌ: (EM p. 250:) or, of a بِرْذَوْن [or hack, or the like,] the *uttering of a cry* [or *neighing*] *such as is not loud;* and of a horse [of good breed], the *uttering of a cry not so loud as the* صَهِيل [or *usual neighing*]: (Lth, TA:) or, of the بِرذون, the *uttering of a cry when desiring the barley:* (Ḳ,* TA:) and the عَزّ, or عِزّ, [accord. to different copies of the Ḳ, but each is app. a mistranscription, for عِيّ as meaning *faltering of the voice* or *cry.*] of the horse, *when falling,* or *stopping, short in neighing, and seeking self-help* [*to finish it*]; as also ↓ تَحَمْحُمٌ: (Ḳ:) and the bull's *uttering a cry with the desire of leaping the cow.* (Az, Ḳ.)

R. Q. 2. تَحَمْحَمَ: see 1, second sentence: ═ and see also R. Q. 1, in three places.

حَمّ: see حَامِم, throughout.

حَمْ: see art. حمو.

حُمّ, [in the CḲ, erroneously, حَمّ,] The *vehemence,* or *intenseness, of the heat* of the ظَهِيرَة [or midday in summer]. (Ḳ, TA.) You say, أَتَيْتُهُ حُمَّ الظَّهِيرَةِ [*I came to him during the vehemence of the heat of the midday in summer*]. (TA.) — The *main,* or *chief, part* of a thing; (Ḳ;) and so ↓ حُمَّة in the phrase حُمَّةُ الحَرِّ [*the main,* or *chief, part of the heat*]. (Ṣ, TA.) — See also حَمِيمَة. — The *remains of the* أَلْيَة [or *tail of a sheep*] *after the melting* [*of the fat*]: n. un. with ة: and *what is melted thereof:* (Ṣ:) or the *part of the* الية *of which one has melted the grease,* (Aṣ, T, Ḳ,) *when no grease remains in it;* (Aṣ, T, TA;) and *of fat:* n. un. with ة: or *what remains of melted fat:* (Ḳ:) accord. to Az, the correct explanation is that of Aṣ: but he adds, I have heard the Arabs call thus *what is melted of the hump of a camel:* and they called the hump الشَّحْم. (TA.) — *Property,* or *cattle and the like;* and *goods, commodities,* or *household-furniture and utensils.* (Sh, TA.) ═ مَا لَهُ سَمٌّ وَلَا حَمٌّ غَيْرُكَ, (Ṣ,) or مَا لَهُ حَمٌّ ولا سَمٌّ, (Ḳ,) and حُمٌّ ولا سُمٌّ, (Ṣ,) or سَمٌّ ولا حَمٌّ, (Ḳ,) and ↓ حُمٌّ ولا رُمٌّ, and ↓ حُمٌّ ولا زَمٌّ, (TA,) *He has no object in his mind except thee;* syn. هَمٌّ: (Ṣ, Ḳ,* TA: [see also art. سم:]) or ما له حمّ ولا سمّ, (Ḳ,) or حمّ ولا رمّ, (TA,) means *he has neither little nor much.* (Ḳ, TA.) — And مَا لِي مِنْهُ حُمٌّ, (Ṣ,) or عُنْهُ, (Ḳ,) and ↓ حُمٌّ, (Ṣ, Ḳ,) and رُمٌّ, and رُمٌّ, (TA,) *I have not any means,* or *way, of separating myself from it,* or *of avoiding it.* (Ṣ, Ḳ,* TA.)

حُمٌّ: see حُمٌّ, in three places.

حَمَّةٌ *A hot spring,* (IDrd, Ṣ, Mgh, Ḳ,) *by means of which the diseased seek to cure themselves.* (IDrd, Ṣ, Ḳ.) In a trad., (Ṣ, TA,) the learned man (العَالِم) is said to be like the حَمَّة, (Ṣ, Mgh, TA,) to which the distant resort, and which the near neglect. (TA.)

حُمَّةٌ: see حُمَى: — and see also حُمّ. — Also The *vehemence,* and *main force,* of the movements of two armies meeting each other. (TA from a trad.) — The *sharpness* of a spear-head. (TA.) — The *venom,* or *poison,* of the scorpion: (TA:) a dial. var. of حُمَة, (Ḳ,) accord. to IAạr; but others allow not the teshdeed, [and among them J,] and assert the word to be originally حُمْوٌ (TA.) — *A decreed,* or *predestined, case of separation:* (Ṣ, Ḳ:) and *of death;* (TA;) as also ↓ حِمَامٌ: (Ṣ, Ḳ:) you say حِمَامُ المَوْتِ, and الحِمَامُ alone as in a verse cited voce عَتَبَ [q. v.]: (TA:) the pl. of حُمَّةٌ is حُمَمٌ and حِمَامٌ. (Ḳ.) ═ *Blackness;* (Ṣ, TA;) the *colour denoted by the epithet* أَحَمّ [q. v.]: (Ṣ, Ḳ:) *a colour between* دُهْمَة [or *blackness*] and كُمْتَة [or *a blackish red*], *inferior* [*in depth, or brightness,*] *to what is termed* حُوَّة [app. as meaning *redness inclining to blackness*]. (M, Ḳ.) — The *black sediment of clarified butter,* and *the like, in the bottom of the skin.* (TA.) ═ Also i. q. حُبَّةٌ: so in the phrases فُلَانٌ حُمَّةُ نَفْسِى [*Such a one is the beloved of my soul*] (Az, TA) and هُوَ مِنْ حُمَّةِ نَفْسِى [*He is of the beloved of my soul*]: and the م is said to be a substitute for ب. (TA.) [See also أَحَمّ, which is used as syn. with أَحَبّ.]

حِمَّةٌ: see حَمِيم, in two places. ═ Also *Death;* or the *decreed term of life:* (Ḳ:) pl. حِمَم. (TA.)

حُمَمٌ *Charcoal:* (Ṣ, Mgh, Ḳ:) or *cold charcoal:* (TA:) or *burnt wood* and *the like:* (Mṣb:) or *charcoal that does not hold together:* (Mṣb in explanation of the n. un. in art. قبس:) and *ashes:* and *anything burnt by fire:* (Ṣ, TA:) n. un. with ة: (Ṣ, Mṣb, Ḳ:) which is tropically applied to ‡*live coals* [or *a live coal*]. (Mṣb.) [Hence] the n. un. is also used as meaning † *Blackness of complexion.* (TA from a trad. of Luḳmán Ibn-'Ád.) And جَارِيَةٌ حُمَمَةٌ means † *A black girl* or *female slave.* (TA. [See also أَحَمّ.])

حَمَامٌ [The *pigeon,* both *wild* and *domestic,* but more properly the former; and sometimes not strictly confined to denote the pigeon-kind:] *a certain wild bird, that does not keep to the houses; well-known:* (ISd, Ḳ:) or *any collared,* or *ringed, bird;* (Ṣ, Mṣb, Ḳ;) so with the Arabs; *such as the* فَوَاخِت *and the* قَمَارِىّ *and the* سَاقُ حُرّ *and the* قَطَا *and the* وَرَاشِين *and the like,* (Ṣ, Mṣb,) and the *domestic* [*pigeons*] (الدَّوَاجِن), also, (El-Umawee, Ṣ, Mṣb,) that are taken into houses for the purpose of producing their young ones; (El-Umawee, Ṣ;) to which last alone the term is applied by the vulgar: accord. to Ks, it is the *wild* [*species*]; and the يَمَام is that which keeps to the houses: accord. to Aṣ, the latter is the حَمَام وَحْشِىّ [or *wild pigeon*]; *a species of the birds of the desert:* (Ṣ, Mṣb:) or, accord. to Esh-Sháfi'ee, حَمَام signifies *any kind of bird that drinks in the manner denoted by the verb* عَبَّ, [i. e. *continuously,*] *and cooes; including the* قَمَارِىّ *and* وَرَاشِين *and* فَوَاخِت; *whether it be, or be not, collared,* or *ringed; domestic* or *wild:* (Az, TA:) *the flesh thereof strengthens the venereal faculty, and increases the seminal fluid and the blood; the putting it, cut open while alive, upon the place stung by a scorpion, is a proved cure; and the blood stops bleeding from the nose:* (Ḳ:) the n. un. is with ة; (Ṣ, Mṣb;) which is applied to the male and the female: (Ṣ, Mṣb, Ḳ:) and in like manner, حَمَامٌ, because the ة is added to restrict to unity, not to make fem.: (Ṣ:) but to distinguish the masc., you may say, رَأَيْتُ حَمَامًا عَلَى حَمَامَةٍ, i. e. *I saw a male* [*pigeon*] *upon a female* [*pigeon*]: (Zj, Mṣb:) accord. to ISd and the Ḳ, however, حَمَامٌ should not be applied to the [single] male: (TA:) in a verse of Ḥomeyd Ibn-Thowr, cited voce حُرٌّ, by the n. un. is meant *a* قُمْرِيَّة: the pl. of حمامة is حَمَامٌ, (Ṣ,) [or rather this is the coll. gen. n.,] and حَمَائِمُ (Ṣ, Ḳ) and حَمَامَاتٌ: (Ṣ:) and sometimes حَمَامٌ is used as a sing.: [so in an ex. above: and] Jirán-el-'Owd says,

* وَذَكَّرَنِى الصِّبَا بَعْدَ التَّنَائِى *
* حَمَامَةُ أَيْكَةٍ تَدْعُو حَمَامَا *

[*And a female pigeon of a thicket, calling a male pigeon, reminded me of youth, after estrangement*]: a poet also says,

* حَمَامَا قَفْرَةٍ وَقَعَا فَطَارَا *

[*Two pigeons of a desert tract alighted and flew away*]: and El-Umawee cites, as an ex. of حَمَام applied to the domestic [pigeons],

* قَوَاطِنًا مَكَّةَ مِنْ وُرْقِ الحَمِى *

[*Inhabiting Mekkeh, of the pigeons of a white colour inclining to black*]; by الحمى [or rather it should be written الحَمَا] meaning الحَمَام. (Ṣ.)

حُمَامٌ The *fever* (حُمَّى) *of camels;* (Ṣ;) as also ↓ حُمَّةٌ: (TA:) or *of all beasts,* (Ḳ, TA,) *including camels:* (TA:) accord. to ISh, when camels eat date-stones, [which are often given to them as food,] they are [sometimes] affected with حُمَام and قُمَاح; the former of which is *a heat affecting the skin, until the body is smeared with mud, or clay, in consequence of which they forsake the abundant herbage, and their fat goes away; and it continues in them a month, and then passes away.* (Az, TA.) — حُمَامُ قُرٍّ *The disease termed* مُومٌ, *which affects men.* (TA.) — See also حَمِيمٌ. = *A noble chief,* or *lord:* (Ḳ:) thought by Az to be originally هُمَامٌ. (TA.)

حِمَامٌ: see its syn. حُمَّةٌ; of which it is also a pl. (Ḳ.)

حَمِيمٌ The قَيْظ [or *summer:* or the *most vehement heat of summer, from the auroral rising of the Pleiades* (*at the epoch of the Flight about the 13th of May O.S.*) *to the auroral rising of Canopus* (*at the same period about the 4th of August O.S.*): or *vehemence of heat*]: (Ṣ, Ḳ:) or *a period of about twenty nights, commencing at the* [*auroral*] *rising of* الدَّبَرَان [*at the epoch of the Flight about the 26th of May O.S.*]. (AZ, T voce نَوْءٌ.) — *Live coals with which one fumigates.* (IAạr, Sh.) — *Hot water;* (T, Ṣ, ISd, Mgh, Mṣb, Ḳ;) as also ↓ حَمِيمَةٌ: (Ṣ, ISd, Ḳ:) or so مَآءٌ حَمِيمٌ (Mṣb:) pl. حَمَائِمُ; (Ḳ;) i. e. pl. of حَمِيمٌ, accord. to IAạr; but accord. to ISd, of حَمِيمَةٌ. (TA.) — And *Cold water:* (Ḳ:) or *cold,* applied to water: so, accord. to IAạr, in the saying of a poet,

* وَسَاغَ لِىَ الشَّرَابُ وَكُنْتُ قِدْمًا *
* أَكَادُ أَغَصُّ بِالمَآءِ الحَمِيمِ *

[*And wine has become easy to swallow to me, whereas I used, in old time, nearly to be choked with cold water*]: (Az, TA:) thus bearing two contr. significations. (Az, Ḳ.) — The *rain that comes in the time of vehement heat;* (Ṣ;) or *after the heat has become vehement,* (M, Ḳ,) because it is hot; (M;) or *in the* صَيْف [or *summer*], *when the ground is hot.* (TA.) — ‡ *Sweat;* (Az, Ṣ, A, Ḳ;) as also ↓ حَمَّةٌ: (Az, A, Ḳ:) and ↓ حُمَام is said to signify the *sweat of horses.* (Ḥam p. 92.) One says, (to a person who has been in the bath, A, TA,) طَابَ حَمِيمُكَ and ↓ طَابَتْ حِمَّتُكَ, meaning *May thy sweat be good,* or *pleasant;* (Az, A, Ḳ;) and consequently, *may God make thy body sound,* or *healthy:* (A, TA:) or the former may mean as above, or *may thy bathing be good,* or *pleasant:* (IB:) one should not say, ↓ طَابَ حَمَامُكَ, (Ḳ, TA,) though MF defends it. (TA.) = *A relation,* (Lth, Ṣ, Ḳ,) *for whose case one is anxious* or *solicitous,* (Ṣ,) or *whom one loves and by whom one is beloved:* (Lth, Ḳ:) or *an affectionate,* or *a compassionate, relation, who is sharp,* or *hasty, to protect his kinsfolk:* or *an object of love; a person beloved:* (TA:) or *a man's brother;* his *friend,* or *true friend;* because anxious, or solicitous, for him: (Ḥam p. 90:) and ↓ مُحِمٌّ signifies the same: the pl. [of حميم] is أَحِمَّآءُ: and sometimes حَمِيمٌ is used as a pl., and as fem.; (Ḳ;) as well as sing. and masc. (TA.) — الحَمِيمُ بِالحَاجَةِ *He who devotes himself to obtain the object of want; who is solicitous for it.* (TA.) A poet says,

* وَلَا يُدْرِكُ الحَاجَاتِ إِلَّا حَمِيمُهَا *

[*And none will attain the objects of want but he who devotes himself to obtain them; who is solicitous for them*]. (IAạr, TA.)

حَمَامَةٌ n. un. of حَمَامٌ [q. v.]. (Ṣ, Mṣb.) — [Hence, app.,] † *A woman:* or *a beautiful woman.* (Ḳ, TA. [In the CḲ, only the latter.]) = The *middle* of the breast or chest. (Ḳ, TA.) The قَصّ [or *breast,* or *head of the breast,* or *pit at the head of the breast,* or *middle of the breast,* or the *sternum,*] of a horse. (Ḳ.) The *callous protuberance upon the breast* of a camel. (Ḳ.) — The *sheave of the pulley of a bucket.* (Ḳ.) — The *ring of a door.* (Ḳ.) — The *clean court* of a قَصْر [or palace, &c.]. (Ḳ.) = See also the next paragraph.

حَمِيمَةٌ: see حَمِيمٌ. — Also *Heated milk.* (Ḳ.) = Also, (Ṣ, Ḳ,) as well as ↓ حَمٌّ, (Ḳ, TA, [in the CḲ, erroneously, حُمٌّ,]) sing. of حَمَائِمُ signifying ‡ *Such as are held in high estimation, precious,* or *excellent,* or the *choice,* or *best,* (Ṣ, Ḳ, TA,) of cattle or other property, (Ṣ,) or of camels: (Ḳ:) and accord. to Kr, the sing. is used as a pl. in this sense: (ISd, TA:) ↓ حَمَامَةٌ, likewise, signifies the *choice,* or *best,* of cattle or other property; and so ↓ حَامَّةٌ, of camels: (Ḳ:) or you say إِبِلٌ حَامَّةٌ, meaning *excellent,* or *choice, camels.* (Ṣ.)

حُمَمَةٌ; accord. to the Ḳ, حُمُمَاتٌ, but this is the pl.; (TA;) *A live coal;* syn. جَمْرَةٌ: (Ḳ, TA:) or *redness;* syn. حُمْرَةٌ: (CḲ, and so in a MS. copy of the Ḳ:) [in Freytag's Lex., the pl. is explained as meaning *rednesses* of the skin; and so ↓ حُمَامَى.]

حُمَامَى: see what next precedes.

حَمَّامِىٌّ *One who flies pigeons* (حَمَام), *and sends them* [*as carriers of letters*] *to various towns* or *countries.* (TA.)

حُمَّى, (Ṣ, Ḳ, &c.,) a subst. from حُمَّ, (Lḥ, L, Ḳ,) imperfectly decl., because of the fem. alif [which terminates it], (Mṣb,) *A fever; a disease by which the body becomes hot:* from الحَمِيمُ: said to be so called because of the excessive heat; whence the trad., الحُمَّى مِنْ فَيْحِ جَهَنَّمَ [*Fever is from the exhalation of Hell*]: or because of the sweat that occurs in it: or because it is of the signs of الحِمَام [i. e. the decreed, or predestined, case of death]; for they say, الحُمَّى رَائِدُ المَوْتِ [*Fever is the messenger that precedes death*], or بَرِيدُ المَوْتِ [*the messenger of death*], or بَابُ المَوْتِ [*the gate of death*]: (TA:) and ↓ حُمَّةٌ signifies the same: (Ḳ, TA:) pl. of the former حُمَّيَاتٌ. (Mṣb.)

حُمَّآءُ: see حُمَامٌ.

حَمَّامٌ [*A hot bath;*] *a certain structure,* (Ṣ,) *well known;* (Mṣb;) so called because it occasions sweating, or because of the hot water that is in it; accord. to ISd, derived from الحَمِيمُ; (TA;) *i. q.* دِيمَاس: (Ḳ:) of the masc. gender, (Mgh, Ḳ,) and fem. also, (Mgh,) generally the latter; (Mṣb;) but some say that it is a mistake to make it fem., (MF, TA,) though IB cites a verse in which a fem. pronoun is asserted to refer to a حمّام: (TA:) pl. حَمَّامَاتٌ; (Ṣ, Mgh, Ḳ;) accord. to Sb, [not because the sing. is fem., but] because, though masc., it has no broken pl. (TA.) See also حَمِيمٌ.

حَمَّامِىٌّ The *owner* [or *keeper*] *of a* حَمَّام [or *hot bath*]. (Mgh.)

حُمْحُمٌ: see أَحَمُّ.

حِمْحِمٌ: see أَحَمُّ, in two places.

حَامَّةٌ The خَاصَّةٌ [or *particular,* or *special, friends,* or *familiars*], (Ṣ, Ḳ,) consisting of the *family and children* (Ḳ) *and relations,* (TA,) of a man. (Ḳ.) You say, كَيْفَ الحَامَّةُ وَالعَامَّةُ [*How are the particular,* or *special, friends, &c., and the common people?*]. (Ṣ.) And هٰؤُلَآءِ حَامَّةُ الرَّجُلِ *These are the relations of the man.* (Lth, Ṣ.) [See حُمَّةٌ, and أَحَمُّ.] — See also حَمِيمَةٌ. — Also

i. q. عَامَّةٌ. (Ḳ.) [It would seem that this signification might have been assigned to it in consequence of a misunderstanding of the words in the Ṣ, وَالحَامَّةُ الخَاصَّةُ يُقَالُ كَيْفَ الحَامَّةُ وَالعَامَّةُ : but accord. to the TĶ, one says, جَاؤُوا حَامَّةً, meaning عَامَّةً, i. e. *They came generally*, or *universally*.]

آلُ حٰمٓ and آلُ حَامِيمَ, ذَوَاتُ حَامِيمَ (Ḳ,) or ▾ and ذَوَاتُ حٰمٓ, (Ṣ,) آل being prefixed in this case in like manner as in آلُ فُلَانٍ, (Fr, Ṣ,) *Certain chapters of the Ḳur-án* (Ṣ, Ḳ) *commencing with* حاميم [or حٰمٓ], (Ḳ,) [namely, *the fortieth and six following chapters*,] *called by Ibn-Mes'ood* دِيبَاجُ القُرْآنِ: (Ṣ:) one should not say حَوَامِيمُ: (Ḳ:) this is vulgar: (Ṣ:) but it occurs in poetry. (Ṣ, Ḳ.) — Also, (Ḳ,) accord. to I'Ab, ▾ حٰمٓ is *One of the names of God*; (Mgh;) or it is *the most great name of God*; (Ḳ;) occurring in a trad., in which it is said, إِنْ بُيِّتُّمْ فَقُولُوا حٰمٓ لَا يُنْصَرُونَ, meaning *If ye be attacked by night, say ye* حٰمٓ; and when ye say this, *they shall not be made victorious*: (Mgh:) or the meaning is, [*say ye*] *O God, they shall not be made victorious*; not being an imprecation; for were it so, it would be لَا يُنْصَرُوا: (IAth, TA:) or it is an oath; (Mgh, Ḳ;) and the meaning of the trad. is, [say ye] *By God, they shall not be made victorious*: but حٰمٓ is not among the numbered names of God: it has therefore been deemed preferable to understand it as here meaning the seven chapters of the Ḳur-án commencing therewith: (Mgh:) or it is *an abbreviation of* الرَّحْمٰنُ, wanting the letters الرن to complete it: (Zj, Ḳ:) or, as some say, it means [حُمَّ مَا هُوَ كَائِنٌ, i. e.] قُضِيَ مَا هُوَ كَائِنٌ [*What is taking place has been decreed*]. (Az, TA.) It is imperfectly decl. because determinate and of the fem. gender; or because it is of a foreign measure, like قَابِيلُ and هَابِيلُ, (Ksh, Bḍ,) and determinate. (Ksh.)

أَحَمُّ *Black*; (Ṣ, Ḳ;) applied to anything; as also ▾ يَحْمُومٌ, (Ḳ,) and ▾ حُمَاحِمٌ, (Aṣ, Ḳ,) or this signifies *intensely black*, (Ṣ,) and ▾ حُمْحُمٌ, (Ḳ,) which IB explains as *a black hue of dye*: (TA:) [the fem. of the first is حَمَّآءُ: and the pl. حُمٌّ: and] the pl. of ▾ the second is يَحَامِيمُ, and by poetic license يَحَامِمُ. (Sb, TA.) You say, رَجُلٌ أَحَمُّ *A black man*. (Ṣ.) And رَجُلٌ أَحَمُّ المُقْلَتَيْنِ *A man having black eyes*. (TA.) And كُمَيْتٌ أَحَمُّ [*A blackish bay horse*]: pl. كُمْتٌ حُمٌّ; which are the strongest of horses in skin and hoofs. (Ṣ.) And ▾ شَاةٌ حُمَاحِمٌ *A black sheep or goat*. (TA.) And لَيْلٌ أَحَمُّ *Black night*. (TA.) — [Hence,] الحَمَّآءُ *The anus* (سَافِلَة, Ṣ, or اِسْت, Ḳ) of a human being: (Ṣ:) pl. حُمٌّ. (Ṣ, Ḳ.) — And أَحَمُّ *An arrow before it has been furnished with feathers and a head*; syn. قِدْحٌ. (Ḳ.) — حَمَّآءُ applied to a lip (شَفَة) and to a gum (لِثَة) means *Of a colour between* دُهْمَة *and* كُمْتَة. (M, TA. [See حُمَّةٌ.]) — Accord. to some, (TA,) أَحَمُّ also signifies *White*: thus having two contr. meanings. (Ḳ, TA.) — Also *A more, or most, particular, or special, and beloved*, friend or the like. (Az, TA. [See حُمَّةٌ, and حَمِيمٌ, and حَامَّةٌ.])

مُحَمٌّ: see مَحَمَّةٌ: — and see also حَمِيمٌ.

مِحَمٌّ i. q. قُمْقُمَةٌ; (Mgh, Mṣb;) i. e. *A vessel of copper* [or *brass*], *in which water is heated*, (KL, and Mṣb in art. قم,) *having a long and narrow neck*: (KL:) or *a small* قُمْقُم [here meaning the same as قُمْقُمَة], *in which water is heated*. (Ṣ.)

مَحَمَّةٌ, applied to food [&c.], (TA,) *Any cause of fever*; *or a thing from the eating of which one is affected with fever*: (Ḳ,* TA:) such, for instance, the eating of fresh ripe dates is said to be. (TA.) And أَرْضٌ مَحَمَّةٌ (Ṣ, M, Ḳ) and ▾ مُحِمَّةٌ, (M, Ḳ,) mentioned by AAF, but not known by the lexicologists except as agreeable with analogy, [see its verb, 4,] (M, TA,) *A land in which is fever*: (Ṣ, Ḳ:) or *in which is much fever*. (Ḳ.)

مَحْمُومٌ *Fevered*, or *affected with fever*, or *sick of a fever*. (Ṣ, Mgh, Mṣb, Ḳ.) — Applied to water, *like* مَحْمُودٌ [q. v.]. (Az, TA.) — *Decreed*, or *appointed*. (Ṣ, TA.)

مُحَامٌّ *Keeping constantly, firmly, steadily, steadfastly*, or *fixedly*, عَلَى أَمْرٍ *to an affair*. (AZ, Ḳ.*)

مُسْتَحَمٌّ, (TA,) or مُسْتَحَمَّةٌ, (Mgh,) *A place in which one washes with hot water*. (Mgh,* TA.)

يَحْمُومٌ: see أَحَمُّ, in two places. — Also *Smoke*: (Ṣ, M, Ḳ:) or *black smoke*: (Bḍ in lvi. 42:) or *intensely black smoke*. (Jel ibid. and TA.) — *A black mountain*: (Ḳ:) or *a certain black mountain in Hell*. (TA.) — The *canopy*, or *awning*, *that is extended over the people of Hell*: so, as some say, in the Ḳur lvi. 42. (TA.) — *A certain bird*: (Ḳ:) so called because of the blackness of its wings. (TA.) — نَبْتٌ يَحْمُومٌ *A plant, or herbage, green, full of moisture, and black*. (TA.)

حمأ

1. حَمَأَ البِئْرَ, (Ṣ, Ḳ,) aor. ـَ, (Ḳ,) inf. n. حَمْءٌ, (Ṣ,) *He cleansed the well of its* حَمْأَة [or *black, fetid mud*]. (Ṣ, Ḳ.) In the T, this signification is assigned to ▾ أَحْمَأَ البِئْرَ; and the signification assigned below to the latter is given to حمأ: but Aṣ says that he does not find this authorized by usage. (TA.) — حَمِئَ, aor. ـَ, inf. n. حَمْءٌ and حَمَأٌ, *It* (water) *was*, or *became, mixed with black, fetid mud, and so rendered turbid*, (Ḳ, TA,) *and altered in odour*. (TA.) And حَمِئَتِ البِئْرُ, inf. n. حَمَأٌ, *The well had in it black* [*fetid*] *mud* (Ṣ, Mṣb) *in abundance*: (Ṣ:) and the like is also said of a spring. (TA.) — حَمِئَ عَلَيْهِ *He was angry with him*; (El-Umawee, Ṣ, Ḳ;) as also حَمِيَ. (Lḥ, TA.)

4. أَحْمَأَ البِئْرَ, (ISk, Ṣ, Ḳ,) inf. n. إِحْمَآءٌ, (ISk, Ṣ,) *He threw* حَمْأَة [or *black, fetid mud*] *into the well*. (ISk, Ṣ, Ḳ.) — See also 1.

حَمْءٌ and حَمٌ and حَمُو [only used as a prefixed n. governing the gen. case (see حَمٌ in art. حمو)] and حَمًا (Ṣ, Mṣb, Ḳ) and ▾ حَمَأٌ (Ḳ) *A woman's husband's father*; (IF, Mṣb, Ḳ;) and *a man's wife's father*: (IF, Mṣb:) or *a man's wife's father or wife's brother or wife's paternal uncle*: (M, Mṣb:) or *any one of a woman's husband's* [*male*] *relations*, (Ṣ, O, Ḳ,) as the *brother* and the *father* (Ṣ, O) and the *paternal uncle*; (O, TA;) and *of a man's wife's relations*: (Ḳ:) pl. أَحْمَآءٌ: (Ṣ, Ḳ:) the fem. is حَمَاةٌ. (TA.) — See also حَمَاة.

حَمَأٌ: see حَمْءٌ: — and see also what next follows.

حَمْأَةٌ (Ṣ, Mṣb, Ḳ) and ▾ حَمَأٌ, (Ṣ, Ḳ,) but accord. to some, the latter is the pl. [or rather quasi-pl. n.] of the former; and sometimes, by poetic license, it is written ▾ حَمْأَةٌ; (TA;) *Black mud*: (Ṣ, Mṣb:) or *black fetid mud*. (Ḳ.) — Also, the former, *A certain plant*, (Ḳ,) *that grows in Nejd, in the sands, and in plain, or soft, land*. (TA.)

حَمِئُ العَيْنِ A man *of an evil eye*, (Fr, Ḳ,) *who injures with his eye him whom he sees and admires*: no verb belonging to it has been heard. (Fr, TA.) — حَمِئَةٌ A well (بِئْر) or a spring (عَيْن) *foul with black, fetid mud*. (TA.)

حمد

1. حَمِدَهُ, aor. ـَ, inf. n. حَمْدٌ (Ṣ, L, Mṣb, Ḳ) and مَحْمَدٌ and مَحْمِدٌ (L, Ḳ) and مَحْمَدَةٌ (Ṣ, L, Ḳ) and مَحْمِدَةٌ; (L, Ḳ, and so in a copy of the Ṣ;) the last of these inf. ns. [and the third also] extr.; (L;) or the last is an inf. n. and the last but one signifies "a praiseworthy quality," or "a quality for which one is praised;" (El-Fenáree, MF;) or the last may be a simple subst.; (Ḥar p. 302;) *He praised, eulogized*, or *commended, him*; *spoke well of him*; *mentioned him with approbation*; (Akh, Ṣ, L, Mṣb;) عَلَى كَذَا *for such a thing*; (L, Mṣb;) *contr. of* ذَمَّهُ: (Ṣ, L:) accord. to IAmb, formed by transposition from مَدَحَ: (marginal note in a copy of the MṢ:) but it is of less common application than the latter verb; (Mṣb in art. مدح;) signifying *he praised him*, &c., *for something depending on his* (the latter's) *own will*: thus, the describing a pearl as clear is not حَمْد, but it is مَدْح: (Kull p. 150:) or i. q. شَكَرَهُ: (Lḥ, Ḳ:) but it differs [sometimes] from this; (Mṣb;) for شُكْر is only on account of favour received; whereas حَمْد is sometimes because of favour received, (Th, Az, Mṣb,) and sometimes from other causes; (Th;) [and thus] the latter is of more common application than the former; (Ṣ;) therefore you do not say, شَكَرْتُهُ عَلَى شَجَاعَتِهِ; but you say, حَمِدْتُهُ على شجاعته *I praised him, &c., for his courage*. (Mṣb.) حمد also implies admiration: and it implies the magnifying, or honouring, of the object thereof; and lowliness, humility, or submissiveness, in the person who offers it; as in the saying of the afflicted, الحَمْدُ لِلّٰهِ *Praise be to God*; since in this case there is no worldly blessing, favour, or

benefit. (Mṣb.) This last phrase is generally pronounced as it is written above: but some of the Arabs are related to have pronounced it الحَمْدَ لِلّٰهِ, putting the former word in the accus. case as the absolute complement of the verb أَحْمَدُ understood: and others, الحَمْدِ لِلّٰهِ; assimilating the final vowel of the former word to the vowel immediately following it: and others, الحَمْدُ لُلّٰهِ; assimilating the first vowel in لله to the vowel immediately preceding it: Zj, however, disapproves of the latter two modes of pronouncing it: some of them also said, بَدَأْتُ بِالحَمْدُ لِلّٰهِ, meaning *I began with* the saying *Praise be to God.* (L.) [See also حَمْدٌ below.] You say, أَحْمَدُ إِلَيْكَ اللّٰهَ *I praise God* (Az, A,* L, Ḳ) *to thee, or in thy presence:* (L:) or *with thee:* (Kh, Az:) or *I praise to thee God's benefits, and his blessings, or favours; or I praise to thee God's blessings, or favours, and discourse to thee of them.* (L.) And حَمِدَ لَهُ أَمْرًا ‡ *He approved of a thing for him.* (L, Ḳ.*) And حَمِدَ إِلَيْهِ أَمْرًا ‡ *He approved of a thing for him, and commanded, or enjoined, him to do it.* (L.) And جَاوَرْتُهُ فَمَا حَمِدْتُ جِوَارَهُ ‡ [*I became his neighbour, and did not approve of being so*]. (A.) See also 4. — Also, (aor. and inf. n. as above in the beginning of this art., Ḳ,) *He recompensed, or requited, him: he gave him, or paid him, his due.* (L, Ḳ.) ═ حَمِدَ عَلَيْهِ, aor. ـَ, (L, Ḳ,*) inf. n. حَمَدٌ, (TA,) *He was angry with him.* (L, Ḳ.)

2. حمّد, inf. n. تَحْمِيدٌ, has a more intensive signification than حَمِدَ; (Ṣ;) [*He declared the praises of* God: or] *he praised* God *much, with good forms of praise* (بِالمَحَامِدِ الحَسَنَةِ): (T, L:) or *repeatedly; or time after time.* (L, Ḳ.) تحميد [used as a simple subst.] has a pl., namely, تَحَامِيدُ. (A.) [See an ex. voce خَاتَمٌ, in the latter part of the paragraph.]

4. احمد *He* (a man, Ṣ) *came to a state, or result, such as was praised, or commended, or approved;* properly, *his affair, or case, came to such a state or result:* (Ṣ, L, Ḳ:) or (so in the Ḳ, but in the L "and") *he did, or said, that for which he should be praised, or commended; or that which was praiseworthy, or commendable;* (A, L, Ḳ;*) *contr. of* أَذَمَّ. (A.) And احمد أَمْرُهُ † *His affair, or case, was, or became, praiseworthy, or approvable, in his estimation:* (Ḳ:) or احمد أَمْرَهُ (as in the L) *he esteemed his affair, or case, praiseworthy, or approvable.* (L [agreeably with what next follows].) ═ احمدهُ *He found him* (a man, A, L) [or *it*] *to be such as is praised, commended, or approved; or praiseworthy, commendable, or approvable;* (Ṣ, A, L, Mṣb;) *contr. of* أَذَمَّهُ: (TA in art. ذم:) *he made it manifest that he was worthy of praise, eulogy, commendation, or approbation:* (L:) *he approved of his action, and his course of conduct, or his tenet or tenets, and did not expose it, or them, to others.* (Ḳ.) And أَحْمَدْتُ صَنِيعَهُ ‡ [*I found his action to be praiseworthy, or commendable, or approvable*]. (A.) And احمد الأَرْضَ ‡ *He approved the land as a dwelling-place:* (A:) or *he found the land to be such as is praised, commended, or approved;* as also † حَمِدَهَا; (L, Ḳ;) but the former verb is the more chaste in this sense. (L.) And احمد مَوْضِعًا ‡ *He found a place to be such as is praised, commended, or approved, and convenient, or suitable, so that he approved it as a dwelling-place, or for its pasture.* (Ṣ, L.)

5. تحمّد *He affected, or made a show of,* (تَكَلَّفَ,) *praise.* (A.) You say, وَجَدْتُهُ مُتَحَمِّدًا † مُتَشَكِّرًا [*I found him affecting, or making a show of, praise and thanks*]. (A.) — *He praised himself.* (KL.) [Golius assigns this meaning to † احتمد, as on the authority of the KL; but it is not assigned to this verb in my copy of the KL.] — فُلَانٌ يَتَحَمَّدُ النَّاسَ [app. a slight mistranscription, for لِلنَّاسِ, i. q. إِلَى النَّاسِ, as in an ex. in the next sentence but one,] *Such a one pretends to men, or shows them, that he is praiseworthy,* بِجُودِهِ *for his liberality.* (L.) — تحمّد عَلَيْهِ *He reproached him for a favour, or benefit, which he* (the former) *had bestowed, or conferred; or recounted his gifts, or actions, to him;* syn. امْتَنَّ. (Ṣ, L, Ḳ.) One says, مَنْ أَنْفَقَ مَالَهُ عَلَى نَفْسِهِ فَلَا يَتَحَمَّدْ بِهِ عَلَى النَّاسِ [*Whoso expends his property upon himself, he shall not reproach men therewith as for favours, or benefits, bestowed*]: (Ṣ, A:) or فلا يتحمّد به إِلَى النَّاسِ [*he shall not pretend to men that he is praiseworthy on account of it*]: a prov., meaning that a man is not praised for his beneficence to himself, but for his beneficence to others. (L.)

6. تحامدوا ‡ [*They praised, or commended,* a thing, *one to another*]. You say, الرِّعَآءُ يَتَحَامَدُونَ الكَلَأَ ‡ [*The pastors praise, or commend, one to another, the herbage*]. (A.)

8. احتمد: see 5. ═ Said of heat, [*It burned, or burned fiercely; or was, or became, vehement:*] formed by transposition from احتدم. (Ṣ.)

10. اِسْتَحْمِدِ اللّٰهَ إِلَى خَلْقِهِ بِإِحْسَانِهِ إِلَيْهِمْ وَإِنْعَامِهِ عَلَيْهِمْ [so I find it written, as though meaning *Demand thou, of his creatures, the praising of God, by reason of his beneficence to them, and his bounty to them:* but I think that we should read اِسْتَحْمَدَ اللّٰهُ, and that the meaning is, *God hath demanded praise of his creatures by his beneficence, &c.*]. (A.)

حَمْدٌ *Praise, eulogy, or commendation;* &c. (Ṣ, &c. [For further explanations of this word, and respecting the phrase الحَمْدُ لِلّٰهِ and its variations, see 1: and see also شُكْرٌ.]) سُبْحَانَكَ اللّٰهُمَّ وَبِحَمْدِكَ, said by a person praying, means [*I extol, or celebrate, or declare, thy remoteness, or freedom, from every impurity, or imperfection,* &c., *O God,* (see art. سبح,)] *and I begin with praising Thee;* أَبْتَدِئُ being understood: (Az, L, Mṣb:) or by بحمدك is meant الحَمْدُ لَكَ *praise be to Thee:* and nearly the same is said in explanation of the phrase in the Ḳur [ii. 28], نُسَبِّحُ بِحَمْدِكَ, that by بحمدك is meant حَامِدِينَ لَكَ: [see, again, art. سبح:] or by the expression وبحمدك is meant, accord. to Aboo-'Othmán El-Mázinee, *and by praising Thee I extol thy remoteness, or freedom, from every impurity,* &c.; سَبَّحْتُكَ being understood: or the و is redundant, as it is in the phrase, رَبَّنَا وَلَكَ الحَمْدُ [*O our Lord, praise be to Thee*], in which the و is sometimes omitted: or, accord. to Aboo-'Amr Ibn-El-'Alà, the و is corroborative, as in the phrase, وَهُوَ لَكَ for هُوَ لَكَ. (Mṣb.) لِوَآءُ الحَمْدِ بِيَدِي يَوْمَ القِيَامَةِ [*The standard of praise shall be in my hand on the day of resurrection* (said by Moḥammad)] means that he shall be singularly distinguished by praise, or praising, on that day. (L.) — See حَمَادِ: — and حَمَادَاكَ. ═ See also حُمَيْدٌ. ═ It is also said to signify *The young one of the kind of bird called* قَطًا: so in the prov.,

حَمْدُ قَطَاةٍ يَشْتَهِي الأَرَانِبَ *

A young one of a kaṭà desires to make the hares its prey: applied to a weak man who desires to insnare a strong one. (Meyd, TA.) ═ See also what next follows.

حَمَدَةٌ *The sound of the flaming, or blazing, of fire;* (Ṣ, Ḳ;) as also حَدَمَةٌ [from which it is formed by transposition: see 8: and † حَمْدٌ app. signifies the same: see حَدْمٌ]. (TA.)

حُمَدَةٌ: see حَمَادِ.

حَمَادِ لَهُ *Praise, and thanks, be to him:* (Ṣ, L, Ḳ:) i. e., to such a one: (Ṣ, L:) *contr. of* جَمَادِ لَهُ [q. v.]. (Ṣ and A in art. جمد.) حَمَادِ is indecl., with kesr for its termination, because it deviates from its original, which is the inf. n. [† الحَمْدُ]: (Ṣ, L:) [i. e.,] it is [a quasi-inf. n., (see اِسْمٌ مَصْدَرٌ in art. صدر,) being] a proper name for [المَحْمَدَةُ as syn. with الحَمْدُ]. (Sharḥ Shudhoor edh-Dhahab.)

حَمُودٌ: see what next follows.

حَمِيدٌ and † مَحْمُودٌ (Ṣ, A, L, Ḳ) and † حَمُودٌ (as in copies of the Ḳ, but this seems to be an intensive epithet,) *Praised, eulogized, or commended; spoken well of; mentioned with approbation; approved; such as is praised,* &c.; *praiseworthy, laudable; commendable, or approvable:* (Ṣ, L, Ḳ: [in which, as well as in numberless exs., all these significations are clearly indicated, though not so clearly explained; the Arabic words to which they apply exactly agreeing with the Latin "laudatus," which means both "praised" and "praiseworthy:"]) the fem. of the first is with ة, (L, Ḳ,) because the signification, though properly that of a pass. part. n., nearly agrees with that of an act. part. n.: (L:) you say, [هِيَ حَمِيدَةٌ *She is praised,* &c.; and] أَفْعَالُهُ حَمِيدَةٌ ‡ [*His actions are praised,* &c.]. (A.) † حَمْدٌ, also, [originally an inf. n., like its contr. ذَمٌّ,] used as an epithet applied to a man, is *syn. with* مَحْمُودٌ; (Ḳ;) and as an epithet applied to a woman, *syn. with* مَحْمُودَةٌ, (TA,) as is also حَمْدَةٌ: (Ḳ, TA:) and you likewise say مَنْزِلٌ حَمْدٌ (Ḳ) and مَنْزِلَةٌ حَمْدَةٌ (Lḥ) † *A place where one alights, sojourns, or abides, such as is praised, or approved,* (Ḳ, TA,) and *convenient, or suitable.*

(TA.) الحَمِيدُ, meaning *He who is praised*, or *praiseworthy, in every case*, is an epithet applied to God; one of the names termed الأَسْمَآءُ الحُسْنَى. (L.) المَقَامُ المَحْمُودُ ♦ [mentioned in the Ḳur xvii. 81] means †*The station in which its occupant shall be praised* by all creatures [on the day of resurrection] because of his being quickly reckoned with, and relieved from long standing: or it is the station of the intercessor. (L.)

حَمْدَكَ ♦ حَمَادَاكَ أَنْ تَفْعَلَ كَذَا (Ṣ, L, Ḳ*) and (L) *The utmost of thy power*, or *of thine ability*, [or *the utmost of thy praiseworthy actions*, (see an ex. of the pl. in what follows,) *will be*] *thy doing such a thing;* syn. مَبْلَغُ جَهْدِكَ, (L,) or قُصَارَاكَ, (Ṣ, L,) and غَايَتُكَ: (Ṣ, L, Ḳ:) and in like manner, حَمَادِى *The utmost of my power*, &c. (Ḳ.) حَمَادَيَاتُ النِّسَآءِ غَضُّ الطَّرْفِ, said by Umm-Selemeh, means *The utmost of the praiseworthy qualities of women is the lowering of the eye.* (L.)

حَمَّادٌ (TA) and ♦ حُمَدَةٌ (A, Ḳ) A man (TA) *who praises* things *much; a great*, or *frequent, praiser:* (A, Ḳ, TA:) or the latter, a man *who praises* things *much and extravagantly.* (Ṣ.) You say, إِنَّهُ لَحَمَّادٌ لِلّٰهِ *Verily he is one who praises God much*, or *repeatedly*, or *time after time.* (L, Ḳ.)

العَوْدُ أَحْمَدُ is a prov., (Ṣ,) meaning †*Repetition is more attributive of praise* (أَكْثَرُ حَمْدًا): (Ṣ, A, Ḳ:) for generally you do not desire to return to a thing save after experience, or knowledge, [and approbation,] thereof: [the act of returning, therefore, implies praise:] or the meaning is, when one begins a kind act, he attracts praise to himself; and *when one repeats, he gains more praise for himself:* or احمد is from the pass. part. n., and the meaning is, the beginning is praised, or praiseworthy; and *repetition is more deserving of being praised.* (Ḳ.) [See Freytag's Arab. Prov. ii. 130.]

مَحْمَدَةٌ (Ṣ, Mgh) and مَحْمِدَةٌ (Mgh) †[*A cause of praise, commendation*, or *approval; a praiseworthy, commendable*, or *approvable, quality* or *action;*] *a thing for which one is*, or *is to be, praised, commended*, or *approved:* (Mgh:) [see 1, first sentence:] *contr. of* مَذَمَّةٌ: (Ṣ:) [pl. مَحَامِدُ.] You say, هٰذَا طَعَامٌ لَيْسَتْ عِنْدَهُ مَحْمِدَةٌ, with kesr to the second م, †[*This is food in which is no approvable quality;*] *the eating of which is not approved.* (A.) — [The pl.] مَحَامِدُ signifies [also] †*Forms of praise.* (Mṣb in art. جمع; &c.) [See 2.]

مُحَمَّدٌ A man *praised much*, or *repeatedly*, or *time after time:* (L, Ḳ:) *endowed with many praiseworthy qualities.* (Ṣ, L.)

مَحْمُودٌ: see حَمِيدٌ, in two places.

يَوْمٌ مُحْتَمِدٌ *A day intensely*, or *vehemently, hot:* (Ḳ:) as also مُحْتَدِمٌ [from which it is formed by transposition: see 8]. (TA.)

مُتَحَمِّدٌ: see 5.

حمدل

Q. 1. حَمْدَلَ, (Mṣb and TA in art. بسمل,) inf. n. حَمْدَلَةٌ, (Ḳ,) *He said* الحَمْدُ لِلّٰهِ: (Ḳ, and Mṣb and TA ubi suprà:) a word of the kind termed مَنْحُوتٌ. (TA.)

حمر

1. حَمَرَ, (Ṣ, Ḳ,) aor. ِ, (Ṣ,) inf. n. حَمْرٌ, (TA,) *He pared* a thong; *stripped* it *of its superficial part:* (Ṣ, Ḳ:) or *he* (a sewer of leather or of skins) *pared* a thong *by removing its inner superficial part, and then oiled it*, previously to sewing with it, so that it became easy [to sew with; app. because this operation makes it to appear of a red, or reddish, colour]. (Yaạkoob, Ṣ.) — And [hence,] *He pared*, or *peeled*, anything; *divested* or *stripped it of its superficial part, peel, bark, coat, covering, crust*, or *the like:* and ♦ حَمَّرَ, inf. n. تَحْمِيرٌ, signifies the same in an intensive degree, or as applying to many objects; syn. قَشَّرَ. (TA.) — Also, (Ṣ, Ḳ,) aor. and inf. n. as above, (Ṣ,) *He skinned* a sheep [and thus made it to appear red]. (Ṣ, Ḳ.) — *He shaved* the head [and thus made it to appear red, or of a reddish-brown colour, the common hue of the Arab skin]. (Ḳ.) And حَمَرَتِ المَرْأَةُ جِلْدَهَا [*The woman removed the hair of her skin*]. (TA.) The term حَمْرٌ is [also] used in relation to soft hair, or fur, (وَبَر,) and wool. (TA.) — حَمَرَهُ بِالسَّوْطِ *He excoriated him* (قَشَرَهُ) *with the whip.* (TA.) — حَمَرَ الأَرْضَ, aor. and inf. n. as above, *It* (rain) *removed the superficial part of of the ground.* (TA.) — حَمَرَهُ بِاللِّسَانِ †*He galled him* (قَشَرَهُ) *with the tongue.* (TA.) — حَمِرَ, aor. َ, (Lth, Ṣ, Ḳ,) inf. n. حَمَرٌ, (Lth, Ṣ,) *He* (a horse) *suffered indigestion from eating barley:* or *the odour of his mouth became altered*, or *stinking*, (Ḳ, TA,) *by reason thereof:* (TA:) or *he became diseased from eating much barley*, (Lth,) or *he suffered indigestion from eating barley*, (Ṣ,) *so that his mouth stank:* (Lth, Ṣ:) and in like manner one says of a domestic animal [of any kind]: part. n. ♦ حَمِرٌ. (TA.) — حَمِرَ عَلَىَّ, (Sh, Ḳ,*) aor. and inf. n. as above, (Sh,) *He* (a man) *burned with anger and rage against me.* (Sh, Ḳ.*) — حَمِرَتِ الدَّابَّةُ, (Ḳ,) aor. and inf. n. as above, (TA,) [*The horse*] *became like an ass in stupidity, dulness*, or *want of vigour, by reason of fatness.* (Ḳ.)

2. حمّر, inf. n. تَحْمِيرٌ: see 1. — Also *He cut* [a thing] *like pieces*, or *lumps, of flesh-meat.* (Ḳ.) — *He dyed* a thing *red.* (Mṣb.) — [*He wrote with red ink.* — See also تَحْمِيرٌ, below.] — *He called* another *an ass; saying, O ass.* (Ḳ.) — *He rode a* مُحَمِّر; i. e. *a horse got by a stallion of generous race out of a mare not of such race; or a jade.* (A, TA.) — *He spoke the language*, or *dialect, of Ḥimyer;* (Ṣ, Ḳ;) as also ♦ تَحَمْيَرَ. (Ḳ.)

4. احمر *He* (a man, TA) *had a white child* (وَلَدٌ أَحْمَرُ) *born to him.* (Ḳ.) — *He fed* a beast *so as to cause its mouth to become altered in odour*, or *stinking*, (Ḳ, TA,) *from much barley.* (TA.)

5. تحمّر *He asserted himself to be related to* [*the race of*] *Ḥimyer:* or *he imagined himself as though he were one of the Kings of Ḥimyer:* thus explained by IAạr. (TA.)

7. انحمر مَا عَلَى الجِلْدِ [*What was upon the skin became removed*]: said of hair and of wool. (TA.)

9. احمرّ, (Ṣ, Mṣb, Ḳ,) inf. n. احْمِرَارٌ, (Ḳ,) *It became* أَحْمَر [or *red*]; (Mṣb, Ḳ;) as also ♦ احمارّ: (Ḳ:) both these verbs signify the same: (Ṣ:) or the former signifies *it was red, constantly, not changing from one state to another:* and ♦ the latter, *it became red, accidentally, not remaining so;* as when you say, جَعَلَ يَحْمَارُّ مَرَّةً وَيَصْفَارُّ أُخْرَى *He*, or *it, began to become red one time and yellow another.* (TA.) [It is also said that] every verb of the measure افْعَلَّ is contracted from افْعَالَّ; and that the former measure is the more common because [more] easy to be pronounced. (TA.) — احمرّ البَأْسُ ‡*War*, or *the war, became vehement*, or *fierce:* (Ṣ, A, IAth, Mṣb, Ḳ:) or *the fire of war burned fiercely.* (TA.)

11: see 9, in two places.

Q. Q. 2. تَحَمْيَرَ: see 2. — Also *He* (a man, TA) *became evil in disposition.* (Ḳ.)

حَمِرٌ, applied to a horse &c.: see حَمِرَ. — Also A man *burning with anger and rage:* pl. حَمِرُونَ. (Sh.)

حُمَرٌ (incorrectly written, by some physicians and others, ♦ حُمَّرٌ, with teshdeed, MF) and ♦ حُومَرٌ (which is of the dial. of the people of 'Omán, a form disallowed by MF, but his disallowal requires consideration, TA) The *tamarind-fruit:* (Ḳ:) *it abounds in the Saráh* (السَّرَاة) *and in the country of 'Omán, and was seen by AḤn in the tract between the two mosques* [*of Mekkeh and El-Medeeneh*]: *its leaves are like those of the* بَلْخِىّ *called* خِلَاف: AḤn says, *people cook with it: its tree is large, like the walnut-tree; and its fruit is in the form of pods, like the fruit of the* قَرَظ. (TA.) — Also, the former word, *Asphaltum*, or *Jews' pitch; bitumen Judaicum;* syn. قَفْرٌ يَهُودِىّ. (Ibn-Beyṭár: see De Sacy's "Abd-allatif," p. 274.) — See also حُمْرٌ.

حُمْرَةٌ [*Redness;*] *a well-known colour;* (Mṣb, Ḳ;) the *colour of that which is termed* أَحْمَرُ: (Ṣ, A:) it is in animals, and in garments &c.; and, accord. to IAạr, in water [when muddy; for it signifies *brownness*, and *the like:* but when relating to complexion, *whiteness:* see أَحْمَرُ]. (TA.) — الحُمْرَةُ [*Erysipelas:* to this disease the term is evidently applied by Ibn-Seenà, in vol. ii. pp. 63 and 64 of the printed Arabic text of his قانون; and so it is applied by the Arabian physicians in the present day:] *a certain disease which attacks human beings, in consequence of which the place thereof becomes red;* (ISk, TA;) *a certain swelling, of the pestilential kind;* (T, Ḳ;) *differing from phlegmone.* (Ibn-Seenà ubi suprà.) — ذُو حُمْرَةٍ *Sweet:* applied to fresh ripe dates. (Ḳ.) — See also حُمْرٌ.

حُمَارَةٌ: see حُمَرَى.

حَمْرَآءُ [originally fem. of أَحْمَرُ, q. v.]: see حَمَارَّةٌ.

حِمِرٌ *Violent* rain, (Ṣ,) *such as removes the superficial part of the ground.* (Ṣ, Ḳ.) — *A severe* night-journey to water. (TA.) = *The most copious portion* of rain; and *violence* thereof. (TA.) — † The *violence, vehemence,* or *intenseness,* of anything; as also ↓ حِمِرَّةٌ and ↓ حُمْرَةٌ. (TA.) — See also حَمَارَّةٌ, in two places. — Also The *evil,* or *mischief,* of a man. (Ḳ.)

حِمِرَّةٌ: see the next preceding paragraph.

حِمَارٌ [The *ass;*] the *well-known braying quadruped;* (TA;) *i. q.* عَيْرٌ; (Az, Ṣ;) applied to the *male;* (Mṣb;) both *domestic* and *wild:* (Az, Ḳ:) the former is also called حِمَارٌ أَهْلِىٌّ; (Mṣb;) and the latter, حِمَارٌ وَحْشِىٌّ, (Ḳ,) and حِمَارُ الوَحْشِ, and ↓ يَحْمُورٌ: (Ṣ, Ḳ:) أَتَانٌ is the appellation applied to the *female;* and sometimes ↓ حِمَارَةٌ: (Ṣ, Mṣb, Ḳ:*) pl. [of pauc.] أَحْمِرَةٌ and [of mult.] ↓ حَمِيرٌ [more properly termed a quasi-pl. n.] and حُمُرٌ (Ṣ, Mṣb, Ḳ) and حُمْرٌ (Ṣ) and حُمُورٌ and ↓ مَحْمُورَآءُ, (Ḳ,) the last [a quasi-pl. n.] of a very rare form [of which see instances voce شَىْءٌ], (TA,) and حُمُرَاتٌ, (Ṣ, Ḳ,) which is said to be a pl. of حُمُرٌ. (TA.) — [Hence,] مُقَيِّدَةُ الحِمَارِ † *A stony tract, of which the stones are black and worn and crumbling, as though burned with fire;* syn. حَرَّةٌ: because the wild ass is impeded in it, and is as though he were shackled. (TA.) — And [hence,] بَنُو مُقَيِّدَةِ الحِمَارِ † *Scorpions:* because they are generally found in a حَرَّة. (TA. [See an ex. in verses cited voce زِمِّخٌ.]) = *A piece of wood in the fore part of the* [*saddle called*] رَحْل, (Ḳ, TA,) *upon which a woman* [*when riding*] *lays hold:* and *in the fore part of the* [*saddle called*] إِكَاف: and, accord. to Aboo-Sa'eed, the *stick upon which* [*the saddles called*] أَقْتَاب [pl. of قَتَبٌ] *are carried.* (TA.) — The *wooden implement of the polisher, upon which he polishes iron* [*weapons &c.*]. (Lth, Ḳ.*) — *Three pieces of wood,* (T, Ḳ,) or *four,* (T,) *across which is placed another piece of wood; with which one makes fast a captive.* (T, Ḳ. [The last words of the explanation are يُؤْسَرُ بِهَا.]) — حِمَارُ الطُّنْبُورِ [*The bridge of the mandoline;*] *a thing well-known.* (TA.) — حِمَارُ قَبَّانٍ [*The wood-louse;* so called in the present day;] *a certain insect;* (Ṣ, Ḳ;) *a certain small insect,* (Mṣb, TA,) *that cleaves to the ground,* (TA,) *resembling the beetle, but smaller,* (Mṣb,) *and having many legs:* (Mṣb, TA:) *when any one touches it, it contracts itself like a thing folded.* (Mṣb.) The حمار قبّان is also called حِمَارُ البَيْتِ; app. because its back resembles a قُبَّة. (TA in art. قب, q. v.) — حِمَارَانِ *Two stones,* (Ṣ, Ḳ,) *which are set up,* (Ṣ,) *and upon which is placed another stone,* (Ṣ, Ḳ,) *which is thin,* (TA,) *and is called* عَلَاة, (Ṣ,) *whereon* [*the preparation of curd called*] أَقِط *is dried.* (Ṣ, Ḳ.) — الحِمَارَانِ *The two bright stars* [α and δ] *in Cancer.* (Ḳzw.)

حَمِيرٌ Anything *pared,* or *peeled; divested,* or

stripped, of its superficial part, peel, bark, coat, covering, crust, or *the like;* as also ↓ مَحْمُورٌ. (TA.) [See 1.] — Also, and ↓ حَمِيرَةٌ, *i. q.* أُشْكُزٌ, i. e. *A thong,* or *strap,* (Ṣ, Ḳ,) *white, and having its outside pared,* (Ṣ,) *in a horse's saddle,* (Ḳ,) or *with which horses' saddles are bound,* or *made fast:* (Ṣ:) so called because it is pared. (TA.) = See also حِمَارٌ.

حَمِيرَةٌ: see حَمِيرٌ.

حِمَارَةٌ: see حِمَارٌ. — Also *A great,* (Ḳ,) or *great and wide,* (TA,) *mass of stone,* or *rock:* (Ḳ:) and *stones set up around a watering-trough* or *tank, to prevent its water from flowing forth:* (Ṣ:) and *a stone,* (Ḳ,) or *stones,* (Ṣ,) *set up around the booth in which a hunter lurks:* (Ṣ, Ḳ:) but J should have said that حَمَائِرُ signifies *stones:* that حِمَارَةٌ is the sing.: that this latter signifies *any wide stone:* and the pl., *stones that are set round a watering-trough* or *tank, to prevent the water from overflowing:* (IB:) and حَمَائِرُ المَآءِ signifies *four large and smooth masses of stone at the head of the well, upon which the drawer of the water stands.* (TA in art. علق.) Also, the sing., *A wide stone that is put upon a trench* or *an oblong excavation, in the side of a grave, in which the corpse is placed:* (Ḳ:) or *upon a grave:* (TA:) pl. as above. (Ḳ.) — *A piece of wood in the* [*woman's vehicle called*] هَوْدَج. (Ḳ.) — *Three sticks,* or *pieces of palm-branches, having their* [*upper*] *ends bound together and their feet set apart, upon which the* [*vessel of skin called*] إِدَاوَة *is hung, in order that the water may become cool.* (TA.) And its pl., حَمَائِرُ, *Three pieces of wood bound together* [*in like manner*], *upon which is put the* وَطْب [*or milk-skin*], *in order that the* [*insect called*] حُرْقُوص *may not eat it.* (TA.) — حِمَارَةُ القَدَمِ, (Ḳ,) or ↓ حمارّة القدم [thus, without any vowel-sign written], with teshdeed to the ر, (IAth,) *The elevated,* or *protuberant, part of the foot, above the toes* (Ḳ, TA) *and their joints,* where the foot of the thief is directed, in a trad., to be cut off. (TA.)

[حِمَارِىٌّ *Of, or relating to, asses; asinine.*]

حِمَارِيَّةٌ [*Asinineness*]. (A in art. حطب.)

حُمَيْرَةٌ: see حَمِيرٌ.

حُمَيْرَآءُ dim. of حَمْرَآءُ, fem. of أَحْمَرُ, q. v.

الحِمْيَرِيَّةُ *The language,* or *dialect, of* [*the race of*] *Ḥimyer,* who had words and idioms different from those of the rest of the Arabs. (TA.)

حَمَّارٌ: see what next follows.

حَمَارَّةٌ, (Ṣ, Ḳ, &c.,) a word of a rare form, of which the only other instances are said to be حَبَالَّةٌ and زَرَافَّةٌ and زَعَارَّةٌ and سَبَارَّةٌ and صَبَارَّةٌ and عَبَالَّةٌ, (TA,) and sometimes ↓ حَمَارَةٌ, without teshdeed, in poetry, (Ṣ, Ḳ,) and in prose also, as is said by Lḥ and others, (TA,) ‡ The *intenseness of heat* (Lth, Ks, Ṣ, A, Ḳ) of summer; (Lth, Ks, Ṣ, A;) and so ↓ حَمْرَآءُ; (TA;) which also signifies the same in relation to the noon, or summer-noon; (Ḳ;) and ↓ حَمْرِىٌّ, (Az, TA in art. بيض,) and ↓ حِمِرٌ: (TA:) or the *most intense heat* of summer; (TA;) as also ↓ حِمِرٌّ: (Ḳ, TA:) and sometimes, though rarely, used in relation to winter [as signifying the *intenseness of cold;* like صَبَارَّةٌ]: (TA:) pl. [or rather coll. gen. n.] ↓ حَمَارٌّ. (Ṣ.) = See also حَمَارَةٌ, last sentence.

حُمَّرٌ and ↓ حُمَرٌ, (Ṣ, Mṣb, Ḳ,) the former of which is the more common, (Ṣ, Mṣb,) [coll. gen. ns.,] *A kind of bird,* (Ṣ, Mṣb, Ḳ,) *like the sparrow:* (Ṣ, Mṣb:) accord. to Es-Sakháwee, the *lark;* syn. قُبَّر [q. v.]: and حُمَّرَةٌ is said in the Mujarrad to be an appellation applied by the people of El-Medeeneh to the [*bird commonly called*] بُلْبُل; as also نُغَرَةٌ: (Mṣb:) حُمَّرَةٌ and حُمَرَةٌ are the ns. of un.: (Ṣ, Mṣb, Ḳ:) pl. حُمَّرَاتٌ (Ṣ, TA) [and حُمَرَاتٌ]. = See also حَمِرٌ.

حَمَّارٌ: see حَمَّارَةٌ. — Also *A seller of asses.* (TA.)

حَمَّارَةٌ, [a coll. gen. n.,] *Owners,* or *attendants, of asses* (Ṣ, Ḳ, TA) *in a journey;* (Ṣ, TA;) as also ↓ حَامِرَةٌ: (Ḳ:) n. un. ↓ حَمَّارٌ (Ṣ, TA) and ↓ حَامِرٌ. (TA.) = See also مُحَمِّرٌ, in two places.

حَامِرٌ: see حَمَّارَةٌ.

حُومَرٌ: see حُمَّرٌ.

حَامِرَةٌ: see حَمَّارَةٌ.

أَحْمَرُ [*Red:* and also *brown,* or *the like:*] a thing *of the colour termed* حُمْرَةٌ: (Mṣb, Ḳ:) it is in animals, and in garments &c.; and, accord. to IAṭr, in water [when muddy]: and so ↓ يَحْمُورٌ: (Ḳ:) fem. of the former حَمْرَآءُ: (Mṣb:) pl. حُمْرٌ and حُمْرَانٌ: (Ḳ:) or when it means *dyed with the colour termed* حُمْرَة, the pl. is حُمْرٌ (Ṣ, Mṣb) and حُمْرَانٌ; for you say ثِيَابٌ حُمْرٌ and حُمْرَانٌ [*red garments*]: (TA:) but if you apply it as an epithet to a man, [in which case it has other meanings than those explained above, as will be shown in what follows,] the pl. is أَحَامِرُ (Ṣ) and حُمْرٌ: (TA:) or if it means *a thing having the colour termed* حُمْرَة, the pl. is أَحَامِرُ, because, in this case, it is a subst., not an epithet. (Mṣb.) ↓ أُحْمُرِىٌّ also signifies the same as أَحْمَرُ: (Ḥam p. 379:) or, as some say, it has an intensive sense. (TA voce حَرُورِيُّونَ.) It is said in the Ṣ, in art. دك, that حَمْرَاوَاتٌ is a pl. of حَمْرَآءُ, like as دَكَّاوَاتٌ is of دَكَّآءُ; but it is not so. (IB in that art.) — Applied to a camel, *Of a colour like that of saffron when a garment is dyed with it so that it stands up by reason of* [*the thickness of*] *the dye:* (TA:) or *of an unmixed red colour;* (Aṣ, Ṣ in art. كمت, and TA;) and so the fem. when applied to a she-goat. (TA.) It is said that, of she-camels, the حَمْرَآء is the most able to endure the summer midday-heat; and the وَرْقَآء, to endure night-journeying; and that the صَهْبَآء is the most notable and the most beautiful to look at: so said Aboo-Naṣr En-Na'ámee: and the Arabs say that the best of camels are the حُمْر and the صُهْب. (TA.) [Hence,] حُمْرُ النَّعَمِ signifies † *The high-bred,* or *excellent, of camels:* and is proverbially applied to *anything highly prized, precious, valuable,* or *excellent.* (Mgh, Mṣb.) —— Applied to a man, (AA, Sh, Az,) *White* (AA, Sh, Az, Ḳ) *in complexion;* (Az;) because أَبْيَض might be considered as of evil omen [implying the meaning of leprosy]: (AA, Sh:) or, accord. to Th, because the latter epithet, applied to a man, was only used by the Arabs as signifying "pure," or "free from faults:" but they sometimes used this latter epithet in the sense of "white in complexion," applied to a man &c.: (IAth:) fem., in the same sense, حَمْرَآءُ: the dim. of which, ▼ حُمَيْرَآءُ, occurs in a trad., applied to 'Áïsheh. (Ḳ,* TA.) So, accord. to some, in the trad., بُعِثْتُ إِلَى الأَحْمَرِ وَالأَسْوَدِ, (TA,) i. e. *I have been sent to the white and the black;* because these two epithets comprise all mankind: (Az, TA:) [therefore, by the former we should understand the white and the red races; and by the latter, the negroes: but some hold that by the former are meant the foreigners, and] by the latter are meant the Arabs. (TA.) One says also, [when speaking of Arabs and more northern races,] أَتَانِى كُلُّ أَسْوَدَ مِنْهُمْ وَأَحْمَرَ, meaning *Every Arab of them, and foreigner, came to me:* and one should not say, in this sense, أَبْيَض. (AA, Aṣ, Ṣ.) الحَمْرَآءُ, also, is applied to *The foreigners* (العَجَمُ) [*collectively*]; (Ṣ, A, Ḳ;) because a reddish white is the prevailing hue of their complexion: (Ṣ:) or *the Persians and Greeks:* or *those foreigners mostly characterised by whiteness of complexion; as the Greeks and Persians.* (TA.) You say, لَيْسَ فِى الحَمْرَآءِ مِثْلُهُ *There is not among the foreigners* (العَجَم) *the like of him.* (A.) And accord. to some, الأَحْمَرُ وَالأَبْيَضُ means *The Arabs and the foreigners.* (TA.) الحَمْرَآءُ [so in the TA, but correctly أَبْنَآءُ الحَمْرَآءِ,] is an appellation applied to *Emancipated slaves:* and اِبْنُ حَمْرَآءِ العِجَانِ, meaning *Son of the female slave,* is an appellation used in reviling and blaming. (TA.) —— Also ‡ A man *having no weapons with him:* pl. حُمْرٌ (A, Ḳ) and حُمْرَانٌ. (Ḳ.) —— الحُسْنُ أَحْمَرُ means *Beauty is in* الحُمْرَة [app. *fairness of complexion;* i. e. *beauty is fair-complexioned*]: (TA:) or † *beauty is attended by difficulty;* i. e. he who loves beauty must bear difficulty, or distress: (IAth:) or the lover experiences from beauty what is experienced from war. (ISd, Ḳ.) —— الأَحْمَرُ *A sort of dates:* (Ḳ:) so called because of their colour. (TA.) —— الأَحْمَرُ وَالأَبْيَضُ *Gold and silver.* (TA.) And الأَحْمَرَانِ *Flesh-meat and wine;* (Ṣ, A, Ḳ;) said to destroy men: (Ṣ:) so in the saying, نَحْنُ مِنْ أَهْلِ الأَسْوَدَيْنِ لَا الأَحْمَرَيْنِ *We are of the people of dates and water, not of flesh-meat and wine:* (A:) or *the beverage called* نَبِيذ *and flesh-meat.* (IAṭr.) Also *Wine and* [*garments of the kind called*] بُرُود. (Sh.) And *Gold and saffron;* (Az, ISd, Ḳ;) said to destroy women; i. e. the love of ornaments and perfumes destroys them: (Az:) or these are called الأَصْفَرَانِ; (AO, TA;) and milk and water, الأَبْيَضَانِ; (TA;) and dates and water, الأَسْوَدَانِ. (A, TA.) And الأَحَامِرَةُ *Flesh-meat and wine and* [*the perfume called*] الخَلُوق: (Ṣ, Ḳ:) or *gold and flesh-meat and wine;* as also الأَحَامِرُ: (TA in art. خضر:) or *gold and saffron and* الخَلُوق. (ISd, TA.) —— المَوْتُ الأَحْمَرُ † *Slaughter;* (L, Ḳ;) because it occasions the flowing of blood: (TA:) and [so in the L, but in the Ḳ "or"] ‡ *violent death:* (Ṣ, A, L, Ḳ:) or *death in which the sight of the man becomes dim by reason of terror, so that the world appears red and black before his eyes:* (A 'Obeyd:) or it may mean † *recent, fresh, death;* from the phrase next following. (Aṣ.) —— وَطْأَةٌ حَمْرَآءُ ‡ *A new,* or *recent, footstep,* or *footprint:* opposed to دَهْمَآءُ. (Aṣ, Ṣ, A.) —— سَنَةٌ حَمْرَآءُ ‡ *A severe year;* (Ṣ, Ḳ;) because it is a mean between the سَوْدَآء and the بَيْضَآء: or *a year of severe drought;* because, in such a year, the tracts of the horizon are red: (TA:) when الجَبْهَةُ [the tenth Mansion of the Moon (see مَنَازِلُ القَمَرِ in art. نزل)] breaks its promise [of bringing rain], the year is such as is thus called. (AḤn.) —— See also حَمْرَآءُ voce حَمَارَّةٌ. —— جَآءَ بِغَنَمِهِ حُمْرَ الكُلَى, and, in like manner, سُودَ البُطُونِ, ‡ *He brought his sheep,* or *goats, in a lean,* or *an emaciated, state.* (A,* TA.)

أُحْمُرِىٌّ: see أَحْمَرُ.

تَحْمِيرٌ [an inf. n. (of حَمَّرَ) used as a subst.] *A bad kind of tanning.* (Ḳ. [For دِبْغ in the CḲ, I read دَبْغ, as in other copies of the Ḳ.])

مِحْمَرٌ *i. q.* مِحْلَأٌ; (Ḳ; in the CḲ مِحْلاء;) i. e. The *iron instrument,* or *stone, with which one shaves off the hair and dirt on the surface of a hide, and with which one skins.* (L, TA. [But for the last words of the explanation in those two lexicons, يَنشف به, I read يُنْتَقُ بِهِ.]) —— Also, (Ṣ, TA,) in the Ḳ, [and in a copy of the A,] مُحَمَّرٌ, which is a mistake, (TA,) A horse *got by a stallion of generous,* or *Arabian, race, out of a mare not of such a race;* or *not of generous birth;* or *a jade;* syn. هَجِينٌ; (Ṣ, A, Ḳ;) in Persian, پَالَانِى; (Ṣ, Ḳ;) as also ▼ حَمَّارَةٌ: (Ḳ:) or a horse *of mean race, that resembles the ass in his slowness of running:* and a *bad* beast: (TA:) pl. مَحَامِرُ (Ṣ, A, TA) and مَحَامِيرُ: (TA:) and accord. to the T, ▼ حَمَّارَةٌ signifies [not as it is explained above, as a sing., but] *i. q.* مَحَامِرُ; and Z explains it as an epithet applied to horses, signifying *that run like asses.* (TA.) —— Also An *ignoble,* or a *mean,* man: (Ḳ,* TA:) and a man *who will not give unless pressed and importuned.* (Ḳ,* TA.)

المُحَمِّرَةُ *A sect of the* خُرَّمِيَّة, *who opposed the* مُبَيِّضَة (Ṣ, Ḳ) *and the* مُسَوِّدَة: (TA:) a single person thereof was called مُحَمِّرٌ: (Ṣ, Ḳ:) *they made their ensigns red, in opposition to the* مسوّدة *of the Benoo-Háshim;* and hence they were thus called, like as the حَرُورِيَّة were called المُبَيِّضَة because their ensigns in war were white. (T.)

مَحْمُورٌ: see حَمِيرٌ.

مَحْمُورَآءُ: see حِمَارٌ.

يَحْمُورٌ The *wild ass:* see حِمَارٌ: (Ṣ, Mgh, Ḳ:) or *a certain kind of wild animal:* (Mgh:) [the *oryx;* to which the name is generally applied; and so in Hebrew: see also بَقَرُ الوَحْشِ, in art. بقر:] *a certain beast* (Ḳ, TA) *resembling the she-goat.* (TA.) —— And *A certain bird.* (Ḳ.) —— See also أَحْمَرُ.

حمز

1. حَمَزَ, aor. ـِ, inf. n. حَمْزٌ, *It* (milk) *was,* or *became, sour,* [*so as to burn,* or *bite, the tongue;* and so, app., حَمُزَ, inf. n. حَمَازَةٌ; (see حَمْزٌ, below;) or] *in a less degree than such as is termed* حَازِرٌ. (TA.) —— [Hence, app.,] حَمُزَ, inf. n. حَمَازَةٌ, † *He* (a man) *was,* or *became, strong, robust, sturdy,* or *hardy.* (Ṣ,* Ḳ,* TA.) —— حَمَزَ اللِّسَانَ, aor. ـِ, *It* (beverage, or wine,) *stung,* or *bit, the tongue:* (Ṣ, Ḳ:) or *it* (milk, and نَبِيذ,) *burned the tongue by its strength and sharpness.* (Mgh.) —— And حَمَزَهُ, aor. ـِ, (TA,) inf. n. حَمْزٌ, (Ḳ,) *He took it, seized* or *grasped it, contracted it,* or *drew it together;* syn. قَبَضَهُ, (Ḳ,* TA,) and ضَمَّهُ. (TA.) See حَمُوزٌ. You say, حَمَزَتِ الكَلِمَةُ فُؤَادَهُ ‡ *The saying contracted his heart,* (Lḥ, A, TA,) *and grieved him,* (Lḥ, TA,) or *pained him.* (TA.) —— Also, (A, TA,) aor. as above, (TA,) and so the inf. n., (Ḳ,) ‡ *He sharpened it;* (A, Ḳ,* TA;) namely, an iron instrument, (TA,) an arrow-head or the like. (A.) So in the dial. of Hudheyl. (TA.)

حَمْزٌ *Acritude* of a thing; *a quality,* or *property, like that of burning,* or *biting,* (Ṣ,* Ḳ,* TA,) *such as the taste of mustard:* (TA:) and [in like manner] ▼ حَمَازَةٌ signifies the *quality,* or *property, of burning, and sharpness; as in beverage,* or *wine:* (TA:) and ▼ حَمْزَةٌ *a sourness* in milk, *with a biting of the tongue;* (A;) or *a sourness* in milk, *less than that of milk which is termed* حَازِرٌ. (TA.)

حَمْزَةٌ: see what next precedes.

حَمُوزٌ, applied to the beverage termed نَبِيذ, [app. when it is in a state of fermentation,] *Digestive.* (Fr, TA.) —— إِنَّهُ لَحَمُوزٌ لِمَا حَمَزَهُ ▼ means *Verily he is one who keeps,* or *guards,* or *takes care of, prudently,* or *effectually, what he has collected together.* (Ḳ.) [To the explanation in the Ḳ (ضَابِطٌ لِمَا ضَمَّهُ) is added in the TA, وَمُحْتَمِلٌ لَهُ, app. a mistranscription for وَمُسْتَقِلٌّ بِهِ *and one who manages it well.*]

رَجُلٌ حَمِيزُ الفُؤَادِ, and ▼ حَامِزُهُ, † *A strong-hearted man:* (Ṣ, TA:) or *a man who is active, sharp* or *quick in intellect, clever, ingenious,* or *acute in mind,* (Ḳ, TA,) *and strong-hearted.* (TA.) —— And ▼ the latter, † *A man contracted in heart.* (TA.)

حَمْزٌ: see حَمَازَةٌ.

حَامِزٌ *Sour,* (TA,) *and burning,* or *biting, to the tongue,* or *acrid;* (A, TA;) applied to beverage, or wine: (A:) and milk *that bites the tongue:* (A:) or milk, and نَبِيذ, *that burns the tongue by its strength and sharpness.* (Mgh.)

You say also رُمَّانَةٌ حَامِزَةٌ *A pomegranate in which is sourness.* (A, Ḳ.) — هَمٌّ حَامِزٌ † *Intense,* or *severe, anxiety.* (TA.) And حُزَازٌ حَامِزٌ (Ṣ, TA) † *A wringing,* or *poignant and burning, pain in the heart, such as arises from wrath &c.* (TA.) — See also حَمِيزٌ, in two places.

أَفْضَلُ الأَعْمَالِ أَحْمَزُهَا ‡ *The most excellent of deeds is the strongest,* or *most powerful:* (Ṣ, Ḳ,* TA:) or *the most painful* (A, Mgh, TA) *and distressing:* (Mgh, TA:) from حَامِزٌ, applied to milk and to نَبِيذ, signifying "that burns the tongue by reason of its strength:" (Mgh:) a trad., related by I'Ab; (Ṣ, TA;) said by Mohammad. (TA.) You say also, فُلَانٌ أَحْمَزُ أَمْرًا مِنْ فُلَانٍ ‡ *Such a one is in harder,* or *more difficult, circumstances* (أَشَدُّ أَمْرًا) *than such a one:* (TA:) or [*more*] *contracted in circumstances.* (ISk, TA.)

رَجُلٌ مَحْمُوزُ البَنَانِ *A man strong in the tips,* or *ends, of the fingers:* (Ṣ, Ḳ, TA:) occurring in a verse of Aboo-Khirásh: (Ṣ:) but [SM says,] what I read in a poem of that author is مَحْمُوزُ القِطَاعِ, meaning, *hard and pointed in the arrowheads.* (TA.)

حمس

1. حَمِسَ, aor. ـَ, (Ṣ, A, Ḳ,) inf. n. حَمَسٌ (Ṣ) and حَمَاسَةٌ, (Ḥam p. 2,) *He was,* or *became, hard, firm, strong, strict,* or *rigorous,* in religion, and in fight, (Ṣ, A, Ḳ,) and in courage, (TA,) and in an affair. (Ḥam p. 2.) [See also 5.] — ‡ *It* (an affair, or a case, TA) *was,* or *became, severe, rigorous, distressful,* or *afflictive:* (Ḳ, TA:) and ‡ *it* (war, or the clamour thereof, الوَغَى,) *was,* or *became, hot,* (A, TA,) or *vehement.* (TA.) — حَمُسَ, aor. ـُ, inf. n. حَمَسٌ, *He* (a man) *was,* or *became, courageous.* (Sb, TA.)

5. تحمّس *He acted,* or *behaved, with forced hardness, firmness, strictness,* or *rigour,* (Ṣ, A, Mgh,) in his religion. (A, Mgh, Ḳ.) — *He* (a man) *feigned disobedience;* syn. تَعَاصَى. (Ṣ, TA.) — *He protected,* or *defended, himself,* (syn. تَحَرَّمَ,) بِهِ *by means of him.* (Sh, TA.)

6. تحامسوا *They vied with, strove to surpass,* or *contended for superiority with, one another in strength,* (تَشَادُّوا,) *and fought one another.* (TA.)

حَمِسٌ: see أَحْمَسُ, in three places.

حَمَاسٌ *Hardness; firmness; strength: defence: conflict.* (TA.) [See also حَمَاسَةٌ.]

حَمِيسٌ *Vehement.* (TṢ, Ḳ.) So in the saying of Ru-beh,

* لَاقَيْنَ مِنْهُ حَمَسًا حَمِيسَا *

[*They experienced from it vehement strength*]: (TṢ, TA:) or, as Az says, *strength and courage.* (TA.) — See also أَحْمَسُ, in two places.

حَمَاسَةٌ *Courage:* (Ṣ, Ḳ, TA:) *defence: conflict.* [See also حَمَاسٌ.]

أَحْمَسُ *Hard, firm, strong, strict,* or *rigorous,* in religion, and in fight, (Ṣ, Ḳ,) and in courage; (TA;) as also ▼ حَمِسٌ: (Ṣ, Ḳ:) pl. of the former, حُمْسٌ. (Ḳ.) — Hence, *A pious man, who carefully abstains from unlawful things:* because he exceeds the usual bounds in matters of religion, and is hard to himself; as also ▼ مُتَحَمِّسٌ. (TA.) — Sing. of الحُمْسُ, (Mgh,) which latter is an epithet applied to *The tribes of Ḳureysh* (Ṣ, A, Ḳ) *and Kinάneh* (Ṣ, Ḳ) *and Jedeeleh,* (Ḳ,) i. e. *Jedeeleh of Ḳeys, consisting of* [*the tribes of*] *Fahm and 'Adwán the two sons of 'Amr the son of Ḳeys the son of 'Eylán, and the Benoo-'Ámir Ibn-Ṣaạṣa'ah,* (AHeyth, TA,) *and their followers in the Time of Ignorance;* (Ḳ;) or *to Ḳureysh and their coreligionists;* (Mgh;) because of the hardships which they imposed upon themselves in matters of religion, (Ṣ, A, Mgh, Ḳ,) as well as in courage, (TA,) for they used not to enjoy the shade in the days of Minè, nor to enter the houses by their doors, (Ṣ, Mgh, TA,) while they were in the state of إِحْرَام, (TA,) nor to clarify butter, nor to pick up [dung such as is called] جِلَّة, (Ṣ, L,) or بَعْر, (TA,) [for fuel,] and they dwelt in the Ḥaram, (AHeyth, TA,) and did not go forth in the days of the مَوْسِم to 'Arafát, but halted at El-Muzdelifeh, (AHeyth, Mgh, TA,) saying, "We are the people of God, and we go not forth from the Ḥaram:" (AHeyth, TA:) or they were thus called because they made their abode in the Ḥaram: (Ṣgh, TA:) or because they betook themselves for refuge to the حَمْسَآء (الحَمْسَآء), which is *the Kaạbeh,* so called because its stones are *white inclining to blackness:* (Ḳ:) the Benoo-'Ámir were of the حُمْس, though not of the inhabitants of the Ḥaram, because their mother was of the tribe of Ḳureysh: the term الأَحَامِسُ also, [pl. of ▼ حَمِسٌ or of ▼ حَمِيسٌ,] is applied to *those of the Arabs whose mothers were of the tribe of Ḳureysh.* (TA.) — Also *Courageous;* (Sb, Ṣ, Ḳ;) and so ▼ حَمِيسٌ and ▼ حَمِسٌ: (Ḳ:) pl. [of the first, masc. only,] أَحَامِسُ and [masc. and fem.] حُمْسٌ and [of the second or third] أَحْمَاسٌ. (TA.) — الأَحَامِسُ is also said to be applied to *The tribe of Ḳureysh:* or, accord. to some, to *the Benoo-'Ámir,* because descendants of Ḳureysh: the former is said by IAạr. (TA.) — Hence, (A, TA,) وَقَعَ فِى هِنْدِ الأَحَامِسِ, (A, TṢ, Ḳ,) or لَقِىَ هِنْدَ الأَحَامِسِ, (L,) ‡ *He fell into distress* (A, L) *and trial:* (A:) or *into calamity:* (Ḳ:) or *he died:* (Ḳ:) or the latter phrase has this last meaning. (ISd, A, and TA in art. هند.) هِنْدٌ was the name of a courageous people of the Arabs. (A, TA.) — عَامٌ أَحْمَسُ, (Ṣ, A, Ḳ,) and سَنَةٌ حَمْسَآءُ, (Ḳ,) ‡ *A severe year.* (Ṣ, A, Ḳ.) They say also سِنُونَ أَحَامِسُ ‡ *Severe years:* (Ḳ:) the masc. form [of the epithet] being used because by سنون is meant أَعْوَام; or the epithet being used after the manner of a subst.: (ISd, TA:) and سِنُونَ حُمْسٌ signifies the same: (Ḳ:) or the latter, *years of hunger.* (Az, TA.) — نَجْدَةٌ حَمْسَآءُ † *Vehement* [*courage,* or *fight,* &c.]. (TA.) — مَكَانٌ أَحْمَسُ ‡ *A hard place:* (Ṣ, Ḳ:) or *a rugged and hard place:* (A:) pl. أَمْكِنَةٌ حُمْسٌ. (Ḳ.) You say also أَرْضٌ أَحَامِسُ, with the pl., meaning, ‡ *A sterile, barren,* or *unfruitful, and narrow, land:* (A:) or *a land in which is no herbage nor pasturage nor rain nor anything.* (TA.) And أَرَضُونَ أَحَامِسُ ‡ *Sterile, barren,* or *unfruitful, lands.* (Ṣ, L.)

مُتَحَمِّسٌ: see أَحْمَسُ, second signification.

حمش

1. حَمِشَ, (A, Ḳ,) aor. ـَ, inf. n. حَمَشٌ and حُمُوشٌ, (Ḳ,) *He* (a man) *became slender in the shanks.* (A, Ḳ.) — حَمِشَ عَظْمُ سَاقِهِ, aor. ـَ; (Mṣb;) and حَمِشَتْ قَوَائِمُهُ, or حَمُشَتْ; (accord. to different copies of the Ṣ;) and حَمِشَتِ السَّاقُ, inf. n. حَمَشٌ; (so in a copy of the A;) or حَمُشَتِ السَّاقُ, aor. ـُ; (Ḳ;) and حَمُشَتْ, (Lḥ, A, Ḳ,) aor. ـُ, inf. n. حُمُوشَةٌ (A, Ḳ) and حَمَاشَةٌ, (TA,) *The small bone of his shank,* (Mṣb,) and *his legs,* (Ṣ,) and *the shank* (A, Ḳ) of a woman, (A,) *became slender.* (Ṣ, A, Mṣb, Ḳ.) And the like is said, metaphorically, of the whole body. (TA.) You say also, استحمش ▼ الوَتَرُ *The bow-string became slender:* and its being so is better. (TA.)

10: see the last sentence above.

حَمْشٌ: }
حَمِشٌ: } see the next paragraph, throughout.
حَمِيشٌ: }

أَحْمَشُ السَّاقَيْنِ *Slender in the shanks,* applied to a man; (Ṣ, Ḳ;) as also ▼ حَمْشُ السَّاقَيْنِ, (Ṣ, Mgh, Mṣb, Ḳ,) and ▼ حَمِيشُهُمَا: (TA:) and so ▼ حَمْشَةُ السَّاقَيْنِ, applied to a woman. (A.) And ▼ حَمْشُ الخَلْقَةِ ‡ *Slender in make,* applied to a man. (TA.) أَحْمَشُ also signifies *Slender,* applied to the small bone of the shank: (Mṣb:) and so حَمْشَآءُ [the fem.], and ▼ حَمْشَةٌ, and ▼ حَمِيشَةٌ, applied to the shank (سَاق), and to the fore arm (ذِرَاع), and to the legs (قَوَائِم): and [the pls.] حُمْشٌ (TA) and حِمَاشٌ, (Ḳ,) applied to shanks (سُوق): (Ḳ, TA:) and ▼ حَمِشٌ, and ▼ حَمْشٌ, and ▼ مُسْتَحْمِشٌ, applied to a bow-string; (Ḳ,* TA;) the last on the authority of Ibráheem El-Ḥarbee; (TA;) and each of the last three epithets with ة added, applied to bow-strings. (Ḳ,* TA.) You say also ▼ لِثَةٌ حَمِيشَةٌ *A gum having little flesh:* (Ḳ:) or *a thin gum.* (TA.)

مُسْتَحْمِشٌ: see أَحْمَشُ.

حمص

حِمَّصٌ and حِمِّصٌ; (Ṣ, Mṣb, Ḳ;) the former preferred by Th, (Ṣ, TA,) and by the Koofees, (Mṣb, TA,) and the only word of that form except قِنَّبٌ and قِلَّفٌ and قِنَّفٌ and جِلَّبٌ; (Fr, TA;) the latter alone allowed by Mbr, (Ṣ,) and this alone mentioned by Sb, (TA,) and preferred by the Baṣrees, (Mṣb, TA,) and said by Mbr to be the only word of this form except جِلِّزٌ, meaning "short," and جِلِّقٌ, the name of a place in Syria, (Ṣ, TA,) but IÁar did not know this latter form of the word; (Az, TA;) [The *cicer arietinum; or chick-peas;*] *a certain grain,* (Ṣ, Mṣb, Ḳ,) *well known,* (Mṣb, Ḳ,) *of the description termed* القَطَانِيّ: (AḤn:) n. un. حِمَّصَةٌ and حِمِّصَةٌ: (TA:) *it is white, and red, and black, and of a sort called* كُرْسُنِيّ [or كَرْسَنِيّ?]; *and is also wild, and cultivated in gardens: the wild sort is the hotter, and the more contracted; the nutriment of the garden-sort is the better; and the black is the most powerful in its operations:* (the Minháj, TA:) *it is flatulent, lenitive, diuretic, having the property of increasing the seminal fluid and the carnal appetite and the blood:* (Ḳ:) Hippocrates says that *it has in it two substances, which quit it by cooking; one of them salt, or saline, which is lenitive; and the other sweet, which is diuretic; and it clears away spots in the skin, and beautifies the complexion, and is beneficial for hot tumours, and its oil is serviceable for the ringworm, or tetter; and its meal, for the fluid of foul ulcers; and the infusion thereof, for toothache, and for swelling of the lip; and it clears the voice:* (TA:) *it also strengthens the body and the penis;* (Ḳ;) *wherefore it is given as fodder to the stallions of horses and the like, and of camels;* (TA;) *on the condition of its being eaten not before* [other] *food nor after it, but in the midst thereof;* (Ḳ;) or, correctly, as in the Minháj, *it should be eaten between two meals.* (TA.)

حمض

1. حَمَضَ, aor. ـُ; and حَمُضَ, (Ṣ, A, Mṣb, Ḳ,) aor. ـُ, (Ṣ,) or ـَ, (Ḳ,) or both; (TA;) and حَمِضَ, aor. ـَ; (Ḳ;) inf. n. [of the first] حُمُوضَةٌ (Ṣ, Mṣb, Ḳ) and [of the second or third] حَمْضٌ, (as in some copies of the Ṣ and of the Ḳ,) or حَمَضٌ; (as in other copies of the Ṣ and of the Ḳ;) said of a thing, (Ṣ, A, Mṣb,) or the third is said particularly of milk, (Ḳ,) *It was,* or *became,* حَامِض [i. e. *acid, sour, sharp* or *biting to the taste, pungent,* or *in taste like vinegar* or *like sour milk:* see حُمُوضَةٌ below]; (Ṣ, Mṣb, Ḳ;) as also ▼حمّض, inf. n. تَحْمِيضٌ. (TA.) You say, جَآءَنَا بِإِدْلَةٍ مَا تُطَاقُ حَمْضًا, or حَمَضًا, (accord. to different copies of the Ṣ,) *He brought us some thick and very sour milk, not to be endured by reason of sourness.* (Ṣ.) = [Hence, or from حَمْضٌ, q. v. infrà,] حَمَضَتِ الإِبِلُ, (Aṣ, Ṣ, A, Ḳ,) aor. ـُ, (Aṣ, Ṣ, TA,) inf. n. حُمُوضٌ (Aṣ, Ṣ, Ḳ) and حمض; (Ḳ;) and ▼احمضت; (A, TṢ, Ḳ;) *The camels pastured upon* حَمْض [q. v.]; (Aṣ, Ṣ, A;) or *ate it.* (Ḳ.) — [And hence, because camels become weary of eating حَمْض,] حَمَضْتُ عَنْهُ † *I disliked him, or it.* (Ṣgh, Ḳ.) — And [because camels are eager for حَمْض after eating long of خُلَّة,] حَمَضْتُ بِهِ † *I eagerly desired him, or it.* (Ṣgh, Ḳ.)

2. حمّض, inf. n. تَحْمِيضٌ: see 1, first signification. = [It seems to be also syn. with تحمّض, q. v.: for, —] said of a man, it signifies أَتَى المَرْأَةَ فِي دُبُرِهَا, as though he shifted from the better of the two places to the worse thereof, by reason of preposterous desire: (TA:) as also ▼احمض: opposed to أَخَلَّ [q. v.]. (TA in art. خل.) — تَحْمِيضٌ also signifies † تَقْلِيلٌ (Ṣ, TA) in جِمَاع. (TA.) — Also † The *giving,* or *doing, little of a thing.* (Ṣ, Ḳ.) You say, حَمَّضَ لَنَا فُلَانٌ فِي القِرَى † *Such a one gave,* or *did, little to us in entertaining.* (Ṣ.) = حَمَّضْتُ الإِبِلَ: see 4. — حمّضه عنه: see 4.

4. احمضت الأَرْضُ *The land became abundant in* حَمْض [q. v.]. (Ṣ.) — احمض القَوْمُ *The people, or company of men, lighted on,* or *found,* حَمْض. (TA.) — احمضت الإِبِلُ *i. q.* حَمَضَت, q. v. (A, TṢ, Ḳ.) — [And hence,] احمض القَوْمُ ‡ *The people, or company of men, launched into,* or *entered upon, cheering discourse.* (A, TA.) I 'Ab used to say to his companions, أَحْمِضُوا ‡ [*Launch ye forth,* or *enter upon, cheering discourse*]; (A, TA;) whereupon they would begin to recite poetry, and to relate the memorable conflicts of the Arabs; (A;) because they then entered into traditions and stories of the Arabs, being weary of the interpretation of the Ḳur-án, [like camels betaking themselves to the pasture termed حَمْض when weary of that termed خُلَّة.] (TA.) [And in like manner,] إِحْمَاضٌ also means †The *changing from seriousness to jesting* or *joking.* (Ḥar p. 10.) — See also 2. [And see 5.] = احمضتُ الإِبِلَ; (Ṣ, Ḳ;) or ▼حَمَّضْتُهَا, inf. n. تَحْمِيضٌ; (ISk;) *I pastured the camels upon* حَمْض. (ISk, Ṣ, Ḳ.) — [And hence, as camels are pastured upon حَمْض after they have pastured for a time upon خُلَّة,] احمضه عنه, and ▼حمّضه, ‡ *He shifted him from it* [to another thing]. (TA.)

5. تحمّض [app. signifies, in its primary acceptation, *He* (a camel) *betook himself to the pasture termed* حَمْض after eating for a time of that termed خُلَّة. (See also 1 and 4 and 2.) — And hence,] ‡ *He shifted from one thing to another thing.* (TA.) — [Hence also,] one says to a man when he comes threatening, أَنْتَ مُخْتَلٌّ فَتَحَمَّضْ ‡ [*Thou art disordered in temper, therefore sooth thyself*]: (Ṣ, A:) from خُلَّة and حَمْض. (Ṣ.) [See also خُلِّيّ, in art. خل.]

حَمْضٌ [a coll. gen. n.] *A kind of plant in which is saltness,* (A, Mṣb,) *which camels eat as though it were fruit, and after which they drink:* (A:) other plants are termed خُلَّة: (Mṣb:) or *what is salt and bitter, of plants;* (Ṣ, Ḳ;) *such as the* رِمْث *and the* أَثْل *and the* طَرْفَآء *and the like:* (Ṣ:) what is sweet is called خُلَّة: (Ṣ, Ḳ:) or *any kind of plant that is salt,* or *sour, rising upon* [*several*] *stems, and having no* [*single*] أَصْل [or *stock*]: (M [as cited in the L, but I doubt whether the passage be correctly transcribed]:) or *any salt, or sour, kind of trees; having a juicy and quivering leaf, which, when squeezed, bursts forth with water; and having a pungent,* or *strong, odour; that cleanses the garment and the hand when they are washed with it; such as the* نَجِيل *and the* خِذْرَاف *and the* إِخْرِيط *and the* رِمْث *and the* قِضَة *and the* قُلَّام *and the* طَرْفَآء *and the* حُرُض *and the* رُغْل *and the* هَرْم *and the like:* (Lḥ:) or *any plant that does not dry up in the* رَبِيع [or *spring*], *but endures the hot season, having in it saltness; when camels eat it, they drink upon it; and when they do not find it, they become thin and weak:* (Lth, T:) the Arabs say that the خُلَّة is the bread of camels, and the حَمْض is their fruit, (Ṣ, A, Mṣb, Ḳ,*) or, as some say, their flesh-meat; (Ṣ;) or their خَبِيص: (TA in art. خل:) and they say that flesh-meat is the حَمْض of men: (TA:) the n. un. is with ة: (Mgh:) and the pl. is حُمُوضٌ. (Ṣ, Ḳ.) [In Isaiah xxx. 24, the word rendered "clean" in our authorized version is thought by some to mean "salt" or "sour."] — Hence the saying,

* جَاؤُوا مُخِلِّينَ فَلَاقَوْا حَمْضَا *

‡ *They came eagerly desiring evil,* or *mischief, and found him who cured them of that which affected them:* which is like the saying of Ru-beh,

* وَنُورِدُ المُسْتَوْرِدِينَ حَمْضَا *

‡ *And him who cometh to us seeking to do evil,* or *mischief, we cure of his disease:* for camels, when they are satiated with خُلَّة, eagerly desire حَمْض [to cure them of the effect of the former]. (TA. [See also خُلِّيّ, in art. خل.]) — Hence, also, by way of comparison, حَمْض is applied to ‡ *Evil,* and *war:* and خُلَّة, to *ease,* or *repose;* freedom from trouble or inconvenience, and toil or fatigue; or tranquillity; and ampleness of circumstances: (T and TA in art. خل:) and the former, to *death:* and the latter, to life. (Ḥam p. 315.) — نَفْسٌ حَمْضَةٌ and فُؤَادٌ حَمْضٌ mean † *A mind that takes fright at a thing, and shrinks from it, at first hearing it.* (TA.)

حَمْضَةٌ † *Eager desire* for a thing. (Ṣ, Ḳ.) It is said in a trad., الأُذُنُ مَجَّاجَةٌ وَلِلنَّفْسِ حَمْضَةٌ; (Ṣ, TA;) and in another, لِلْأُذُنِ مَجَّةٌ وللنفس حمضة; (TA;) [both meaning the same;] † *The ear is wont to reject* what it hears, not retaining it, when one is exhorted to do a thing, or forbidden to do it, *while the mind has eager desire* to hear: (IAth:) or *the ear retains not* all that that it hears, *while having eager desire* for what it deems elegant, of extraordinary matters of discourse and speech. (Az.) This usage of the word is taken from the eager desire of camels for حَمْض when they have become weary of خُلَّة. (Ṣ.)

حَمْضِيَّةٌ, and إِبِلٌ حَمْضِيَّةٌ, and بَعِيرٌ حَمْضِيٌّ: see حَمِضَةٌ. — حَمِيضَةٌ and أَرْضٌ حَمْضِيَّةٌ: see حَامِض.

حُمُوضَةٌ [*Acidity; sourness;* the *quality of being sharp* or *biting to the taste; pungency;*] the *taste of that which is termed* حَامِض. (Ṣ, Ḳ.) [See 1.] الحُمُوضَةُ is also explained as signifying *That which bites the tongue; as the taste of vinegar, and of milk such as is termed* حَازِر: which is extr., [if it be meant thereby that the word is thus used as an epithet to qualify a subst., or as an epithet in which the quality of a subst.

is predominant, but I rather think that it is a loose way of explaining it as an inf. n. used as a simple subst.,] for [the measure] فُعُولَةٌ does not belong [save] to inf. ns. (TA: [in which the word إِلَّا is evidently omitted by an oversight in transcription, and therefore has been supplied by me in rendering the passage.])

أَرْضٌ حَمِيضَةٌ *Land abounding with* حَمْض; (ISh, Ḳ;) as also ▾ أرض مُحْمِضَةٌ; (Ṣ;) and ▾ ارض حَمْضِيَّةٌ: (TA:) pl. of the first, أَرَضُونَ حُمُضٌ, (as in some copies of the Ḳ,) or حَمْضٌ: (as in other copies of the same, and in the TA:) and حُمُوضٌ [which seems to be another pl. of the first of these epithets] is explained as signifying land *possessing* حَمْض. (TA.)

حُمَّاضٌ [*Sorrel*; or particularly the *rose-flowered sorrel*; more commonly called in the present day حُمَيْض;] *a certain plant having a red flower*; (Ṣ;) *a herb*, or *leguminous plant, of the kind termed* ذُكُور, *having a produce*, or *fruit, red like blood*; (Ḥam p. 823;) *a certain herb* (Ḳ, TA) *growing in the mountains, of herbs of the* [*season called*] رَبِيع, (TA,) *the leaves of which are like those of the* هِنْدِبَآء, (Ḳ, TA,) *large and broad*; (TA;) *it is acid*, (Ḳ, TA,) *intensely so*; *its flower is red, and its leaves are green*: (TA: [in which is here added ويتناوس فى ثمره مثل حَبّ الرمّان, app. for وَيَتَنَوَّسُ الخ; meaning that it waves much to and fro when blown by the wind, and describing its fruit as containing what resemble the grains of the pomegranate:]) *it is pleasant to the taste*; (Ḳ, TA;) *and is eaten by men, but in small quantity*: AḤn and Aboo-Ziyád say, *it grows very tall, and has a wide leaf, and a red flower, which, when it is near to drying up, becomes white*: and Aboo-Ziyád says, *in our mountain-country it is abundant; and is of two species; one of these two is acid*, [but] *pleasant to the taste*; (TA;) *and one species thereof is bitter*; (Ḳ, TA;) *in the lower parts of each, when they are full grown, is a redness; and the seeds and leaves of the acid species are used medicinally*: Az says, *it is a wild herb*, or *leguminous plant, that grows in the days of the* [*season called*] رَبِيع, *in the channels of water, and has a red flower, and is of the herbs*, or *leguminous plants, which are termed* ذُكُور: IB says, *the places of its growth are the small channels of water, and the places to which valleys take their courses; and in it is acidity: sometimes, also, the people of settled habitations make it to grow in their gardens, and water it and sustain it so that it does not dry up in the time when the wild herbs*, or *leguminous plants, dry up*: it is also said in the Minháj that *it is both wild and growing in gardens*; that *the wild is called* سَلِق, [but this name is commonly applied to bete,] *and in all of this there is not acidity: the garden-kind resembles the* هندبآء, *and in this is acidity, and an excessive viscous moisture: the best is the acid, garden-kind*: here ends the quotation from the Minháj: (TA:) *each species*, (Ḳ, TA,) *the bitter and the pleasant*, or *the garden-kind and the wild*, (TA,) *is good for thirst, and for inflammation arising from yellow bile; and strengthens the bowels; and allays heaving of the stomach, and hot palpitation, and tooth-ache; and is good for the black* [or *livid*] *jaundice*; (Ḳ,* TA;) *and, when cooked, and applied externally, for the leprosy; and for the ringworm* (قُوبَآء); *and for glandular swellings in the neck, so much so that it is said to do good to him who has these even when hung upon the neck: with vinegar, also, it is good for the mange*, or *scab; and it is astringent; and puts a stop to malacia* [so I render شَهْوَةُ الطِّين, lit. "the longing for clay"]: *its seeds are cold in the first degree, and have an astringent property, particularly when fried*: (TA:) *they say that if these be hung, in a purse, upon a woman's left upper arm, she will not become pregnant as long as they remain upon her*: (Ḳ,* TA:) *they are also good for the sting of scorpions; and if some of the seeds be swallowed before the scorpion's stinging, its stinging will not hurt*. (TA.) ═ Also *What is in the interior of the* [*kind of citron called*] أُتْرُجّ: (A, Ḳ:) n. un. with ة: (A:) *it is cold and dry in the third degree; used as a liniment, it removes freckles and the like, and clears the complexion; and it suppresses* (يَقْمَعُ) *the yellow bile; and gives appetite for food; and is good for hot palpitation; and made into a beverage, it sweetens the odour of the mouth; and is good for looseness arising from yellow bile; and is suitable for those who are fevered*. (TA.) [In the present day, in Egypt, this name is applied to *A species of citron*, itself, *with a conical apex, and very acid pulp*.]

حُمَيْضَى *A certain plant*: not from حُمُوضَة. (TA.)

حُمَّاضِيَّةٌ *A confection composed of* حُمَّاض *of the* أُتْرُجّ. (TA.)

حَامِضٌ [*Acid; sour; sharp* or *biting to the taste; pungent; having a taste like that of vinegar* or *like that of sour milk*; see حُمُوضَة;] (Ṣ, Mṣb, Ḳ;) applied to milk (TA) and other things; (Mṣb;) and ▾ مُحْمِضٌ signifies the same, applied to a grape. (TA.) — [Hence,] رَجُلٌ حَامِضُ الفُؤَادِ † *A man whose heart*, or *mind, is altered and bad*, (O, Ḳ,) فِى الغَضَبِ *in anger*. (O.) And فُلَانٌ حَامِضُ الرِّئَتَيْنِ † *Such a one is in a loathing state of mind*; syn. مُرُّ النَّفْسِ. (Ṣ.) ═ إِبِلٌ حَامِضَةٌ *Camels pasturing upon* حَمْض; (Ṣ;) or *eating it*; (Ḳ;) or *pasturing upon* حَمْض *after pasturing upon* خُلَّة: (ISk:) pl. حَوَامِضُ: (Ṣ, Ḳ:) and ▾ إِبِلٌ حَمْضِيَّةٌ *Camels staying among* حَمْض; (Aṣ, Ṣ, Ḳ;) as also ▾ حَمْضِيَّةٌ, contr. to rule: (TA:) and بَعِيرٌ ▾ حُمْضِىٌّ *a camel eating* حَمْض. (TA.)

مَحْمَضٌ and مُحْمِضٌ, (Ṣ, Ḳ,) the latter on the authority of A'Obeyd, (Ṣ,) *A place in which camels pasture upon* حَمْض. (Ṣ, Ḳ.*)

حَمِيضَةٌ: see أَرْضٌ مُحْمِضَةٌ.

حَامِضٌ: see مُحْمِضٌ.

لَبَنٌ مُسْتَحْمِضٌ *Milk slow in thickening*. (Ibn-'Abbád, Ḳ.)

حمق

1\. حَمُقَ, aor. ـُ; and حَمِقَ, aor. ـَ; (T, Ṣ, Mgh, Mṣb, Ḳ, &c.;) inf. n. (of the former, Ṣ) حَمَاقَةٌ, (Ṣ, Ḳ,) or this is a simple subst., (Mṣb,) and (of the latter, Ṣ) حُمْقٌ (Ṣ, Mgh,* Mṣb,* Ḳ) and حُمُقٌ; (Ṣ, Ḳ;) *He was*, or *became, foolish*, or *stupid*; i. e., *unsound in intellect* or *understanding*; (T, Mgh, Mṣb;) and *stagnant*, or *dull, therein*;* (T, Mgh;) or *deficient*, or *defective, therein*; (IF, Mgh;) or *he had little*, or *no, intellect* or *understanding*; (Ṣ, Ḳ;) as also ▾ انحمق and ▾ استحمق (Ḳ) and ▾ تحمّق. (TA.) [See حُمْقٌ, below.] One says to a man, تِيسِى, and إِحْمَقِى, [as though he were a she-hyena, or a woman,] when he speaks foolishly, or stupidly, or says what is not like anything. (AZ, TA voce تَأْسَ.) — حَمُقَتِ السُّوقُ, (Ṣ, M, Ḳ,) with ḍamm; (so in two copies of the Ṣ;) or, as in [some copies of] the Ṣ, حَمِقَت; (TA;) and ▾ انْحَمَقَتْ; (Ṣ, Mgh, Ḳ;) ‡ *The market was*, or *became, stagnant*, or *dull, with respect to traffic*. (Ṣ, M, Mgh, Ḳ, TA.) And حَمُقَتْ تِجَارَتُهُ ‡ *His merchandise was*, or *became, unsaleable*, or *difficult of sale*, or *in little demand*. (TA.) — حَمِقَ, aor. ـَ, inf. n. حَمَقٌ, † *His beard became light*, or *scanty*. (Mṣb.) ═ حُمِقَ *He had the disease termed* حُمَاق. (TA.)

2\. حمّقهُ, inf. n. تَحْمِيقٌ, *He ascribed to him* حُمْق [i. e. *foolishness*, or *stupidity*, &c.]. (Ṣ, Ḳ.) — حَمَّقَتْهُ الهَجْعَةُ *The light sleep in the first part of the night rendered him like the* أَحْمَق [i. e. *foolish*, or *stupid*, &c.]: so says IKh; and he cites, from a poet, the phrase حَمَّقَتْهُ بِهَجْعَةٍ; in which the ب is redundant, and the noun occupies the place of one in the nom. case. (TA.) — [And hence, if correct,] حَمَّقَ *He drank wine*: (Ḳ:) or *he became intoxicated, so that his reason departed*: thus explained by IAmb; but disallowed by Ez-Zejjájee. (IB, TA.)

3\. حامقهُ *He aided him in his* حُمْق [i. e. *foolishness*, or *stupidity*, &c.]. (Ṣ.)

4\. احمقهُ *He found him to be* أَحْمَق [i. e. *foolish*, or *stupid*, &c.]. (Ṣ, Ḳ.) [See also 10.] — احمق بِهِ *He mentioned him*, or *spoke of him, as characterized by* حُمْق [i. e. *foolishness*, or *stupidity*, &c.]. (TA.) — احمقت *She* (a woman) *brought forth a child that was* أَحْمَق [i. e. *foolish*, &c.]; (Ṣ;) or *brought forth* حَمْقَى [i. e. *foolish children*]. (Ḳ.) ═ مَا احمقهُ [*How foolish*, or *stupid*, &c., *is he!*] an expression of wonder. (TA.)

5: see 1.

6\. تحامق *He affected* حَمَاقَة [i. e. *foolishness*, or *stupidity*, &c.; meaning *he feigned it*]. (Ṣ.)

7\. انحمق: see 1. — Also *He acted in the manner of the* حَمْقَى [i. e. *foolish*, or *stupid*, &c.]; (Ḳ;) and so ▾ استحمق. (Lth, T, Mgh, Ḳ.) — *He* (a man, TA) *was*, or *became, abject, humble*, or *submissive*, (Ḳ, TA,) *and impotent to do*, or *accomplish, a thing*. (TA.) — ‡ *It* (a garment) *became old, and worn out*. (Ṣ, Mgh, Ḳ, TA.) —

† *It* (food, or wheat,) *became cheap.* (Az, TA.) —— انحمقت السُّوقُ : see 1.

10. استحمق : see 1: — and 7. — استحمقهُ *He counted, accounted, or esteemed, him* أَحْمَق [i. e. *foolish, or stupid,* &c.]: (Ṣ, Mgh, TA:) or *he found him to be so*; like احمقهُ. (TA.)

حُمْقٌ *Foolishness*, or *stupidity*; i. e. *unsoundness in the intellect* or *understanding*; (T, Mgh, Mṣb;) and *stagnancy*, or *dulness, therein*; (T, Mgh;) or *deficiency*, or *defectiveness, therein*; (IF, Mgh;) or *paucity*, or *want, thereof*; and ▼ حُمُقٌ and ▼ حَمَاقَةٌ signify the same: (Ṣ, Ḳ:) [all are said to be inf. ns.; but the last, accord. to the Mṣb, is a simple subst.: (see 1:)] the proper and primary signification of حُمْقٌ is [said to be] the *putting a thing in a wrong place, with knowledge of its being bad* [to do so]. (TA.) [Hence,] نَوْمَةُ الحُمْقِ *The sleep after* [*the period of the afternoon called*] *the* عَصْر; when no one sleeps except one who is intoxicated, or one who is insane, or unsound in mind. (Ḥar p. 223. [See also خُرْقٌ and خُلُقٌ.]) — And *Deceit*; or *a deception.* (TA.) — [It is said that] الحُمْقُ also signifies *Wine*: (Z, Ḳ:) as being a cause of حُمْق; like as wine is called إِثْم as being a cause of إِثْم: (Z, TA:) but Ez-Zejjájee disallows this: and [it is also said that] ▼ الحُمَيْقَةُ signifies the same, because wine occasions حُمْق to its drinker. (TA.)

حَمِقٌ : see أَحْمَقُ. — Also † *Having a scanty beard.* (IDrd, Ḳ.)

حُمُقٌ : see حُمْقٌ.

حَمْقَان : see أَحْمَقُ.

حُمَاقٌ (Ṣ, Ḳ) and حَمَاقٌ (ISd, Ḳ) and ▼ حُمَيْقَى (AZ, Ḳ) and ▼ حُمَيْقَاءُ (IDrd, Ḳ) The جُدَرِىّ [or *small-pox*]: (Ḳ:) or the *like thereof*, (Ṣ, Ḳ,) *which attacks a human being*, (Ṣ,) *and spreads in a scattered manner upon the body*, or *person*: (Ḳ:) accord. to Lḥ, *a certain thing that comes forth upon children.* (TA.)

حُمَيْقٌ a contracted dim. of أَحْمَقُ; or dim. of حَمِقٌ: [the dim. form being app. used in this case to denote enhancement of the signification: (see also حُمَيْقَةٌ:)] so in the prov., (TA,) عَرَفَ حُمَيْقٌ جَمَلَهُ [*A very foolish*, or *stupid, man knew his camel*]; i. e. he knew thus much, although أَحْمَق: or, as some relate it, عَرَفَ حُمَيْقًا جَمَلُهُ, i. e. *his camel knew* him, [namely, *a very foolish*, or *stupid, man*,] and emboldened himself against him; or it means that he knew his quality: (Ḳ, TA:) it is applied to the case of excessive familiarity with men: (TA:) or to him who deems a man weak, and is therefore fond of annoying, or molesting, him, (Ḳ, TA,) and ceases not to act wrongfully towards him: or, as some say, [حميق is here a proper name; and] this person had a camel with which he was familiar, and he made an attack upon him. (TA.) [See Freytag's Arab. Prov. ii. 85.]

حَمَاقَةٌ : see حُمْقٌ.

حُمُوقَةٌ : see أُحْمُوقَةٌ.

حُمَيْقَى : see حُمَاقٌ.

حُمَيْقَاءُ : see حُمْقٌ: — and حُمَاقٌ.

حُمَيْقَةٌ : } see what next follows.
حُمُوقَةٌ : }

حُمَّيْقَةٌ, (Ḳ,) but in the Tekmileh with teshdeed to the ى, and with kesr to the same, [app. ▼ حُمَيِّقَةٌ,] (TA,) and ▼ حُمُوقَةٌ, (Ḳ,) *Foolish*, or *stupid*, (أَحْمَق,) *in the utmost degree.* (Ibn-'Abbád, Ḳ, TA.) [It seems to be implied in the Ḳ that أُحْمُوقَةٌ signifies the same: but see this word below.]

أَحْمَقُ (Ṣ, Mgh, Mṣb, Ḳ, &c.) and ▼ حَمِقٌ (Ṣ, Mṣb) and ▼ حَمْقَانٌ [whether with or without tenween is not shown] (Sb, TA) *Foolish*, or *stupid*; i. e. *unsound in intellect* or *understanding*; (T, Mgh, Mṣb;) and *stagnant*, or *dull, therein*; (T, Mgh;) or *deficient*, or *defective, therein*; IF, Mgh;) or *having little*, or *no, intellect* or *understanding*: (Ṣ, Ḳ:) fem. of the first حَمْقَاءُ; (Ṣ, Mṣb;) and of the second حَمِقَةٌ: (TA:) pl. of the first, applied to men and to women, (Ṣ, Ḳ,) حُمْقٌ, (so in two copies of the Ṣ,) or حُمُقٌ, with two dammehs, (Ḳ,) and حَمْقَى and حَمَاقَى (Ṣ, Ḳ) and حُمَاقَى (Ṣgh, Ḳ) and حِمَاقٌ. (Ibn-'Abbád, Ḳ.) Accord. to some, أَحْمَقُ is from the phrase اِنْحَمَقَتِ السُّوقُ: and accord. to some, from the phrase لَيَالٍ مُحْمِقَاتٌ, because the احمق deceives one at first by what he says. (TA.) The sounds of wailing for the dead, and trilling, or quavering, in playing, are termed أَحْمَقَانِ because of the حُمْق of the person from whom they proceed. (Mgh.) — البَقْلَةُ الحَمْقَآءُ (Ṣ, Ḳ) and بَقْلَةُ الحَمْقَاءِ, (Ḳ,) the latter for بَقْلَةُ الحَبَّةِ الحَمْقَاءِ, *i. q.* الرِّجْلَةُ [*Garden purslane*]; (Ṣ, ISd, Ḳ;) which is the name applied to it by the vulgar; (ISd, TA;) the chief of herbs, or leguminous plants: called by those names because exuding mucilage (مُلَعَّبَة), so that it is likened to the أَحْمَق whose slaver is flowing: IDrd says, they assert that it is so called because it grows in the tracks of men, so that it is trodden upon; and in water-courses, so that the water uproots it: IF says that it is so called because of its weakness: and it is said that some persons, hating 'Áisheh, called it بَقْلَةُ عَائِشَةَ; but this is one of their fanciful assertions; for such was its name in the time of utter paganism: so says Ṣgh. (TA.) — [أَحْمَقُ also signifies *More*, and *most, foolish*, or *stupid*, &c. Hence,] it is said in a prov., أَحْمَقُ مِنْ رِجْلَةٍ [*More foolish*, or *stupid, than a plant of garden-purslane*: explained by what precedes]. (TA.) [See also another prov. voce ثَمَانُونَ.] And in a trad., أَحْمَقُ الحُمْقِ الفُجُورُ [*The most foolish of foolishness*, or *the most stupid of stupidness, is vice*, or *immorality*, or *unrighteousness*]. (A in art. كيس.)

أُحْمُوقَةٌ is from الحُمْقُ, like أُحْدُوثَةٌ from الحَدِيثُ, and أُعْجُوبَةٌ from العَجَبُ: (TA:) it signifies *An action*, or *a deed, of those that are done by the* حَمْقَى [i. e. *foolish, or stupid, persons*]; (Mgh;) [*a foolish*, or *stupid, action* or *deed*:] it is like ▼ حُمُوقَةٌ, which means *an action, a practice*, or *a habit, in which is* حُمْق [i. e. *foolishness*, or *stupidity*, &c.]. (TA.) One says, وَقَعَ فُلَانٌ فِى أُحْمُوقَةٍ [*Such a one fell into the commission of a foolish*, or *stupid, action*, &c.]. (TA.) [See حُمَيْقَةٌ.]

مُحْمِقٌ and مُحْمِقَةٌ (Ṣ, Ḳ) A woman *who brings forth a child that is* أَحْمَق [i. e. *foolish*, &c.]; (Ṣ;) or *who brings forth* حَمْقَى [i. e. *foolish children*]: (Ḳ:) or, accord. to IDrd, the latter has this signification; but the former signifies a man *who begets* حَمْقَى; and he does not allow its application to a woman. (TA.) — المُحْمِقَاتُ ‡ *The nights* [*that make a fool of one*; i. e.] *during the whole of which the moon is above the horizon but intercepted by clouds*; so that one imagines that he has arrived at the time of morning; (A, O, Ḳ, TA;) because he sees light, but sees not the moon: derived from الحُمْقُ. (TA.) One says, غَرَّنِى غُرُورَ المُحْمِقَاتِ ‡ [*He, or it, deceived me with the deceiving of the nights thus called*]. (TA.) And you say, سِرْنَا فِى لَيَالٍ مُحْمِقَاتٍ ‡ [*We journeyed during such nights*]; because the rider therein thinks that he has arrived at the time of morning until he becomes weary. (TA.)

مِحْمَاقٌ A woman who is *accustomed to bring forth* حَمْقَى [i. e. *foolish children*]. (Ṣ, Ḳ.)

مَحْمُوقٌ A man [or child] *affected with* حُمَاق [q. v.]. (A 'Obeyd, Ṣ.)

حمل

1. حَمَلَهُ, aor. ـِ, inf. n. حَمْلٌ (Ṣ, Mgh, Mṣb, Ḳ, &c., in some copies of the Ṣ حِمْلٌ) and حُمْلَانٌ, (Mgh, Ḳ,) *He bore it, carried it, took it up and carried it, conveyed it*, or *carried it off* or *away*, (MA,) عَلَى ظَهْرِهِ (Ṣ, MA,) *upon his back*, or عَلَى رَأْسِهِ *upon his head*; (MA;) and ▼ احتملهُ signifies the same: (Mṣb, Ḳ:) or the latter is used in relation to an object inconsiderable and small in comparison with that in relation to which the former is used; as in the saying of En-Nábighah, (TA,)

إِنَّا ٱقْتَسَمْنَا خُطَّتَيْنَا بَيْنَنَا •
فَحَمَلْتَ بَرَّةَ وَٱحْتَمَلْتُ ▼ فَجَارِ •

[*Verily we have divided our two qualities between us, and thou hast borne* as thy share *goodness, and I have borne* as my share *wickedness*]. (TA in the present art., and Ṣ and TA &c. in arts. بر and فجر.) Hence, in the Ḳur [xx. 100], فَإِنَّهُ يَحْمِلُ يَوْمَ القِيَامَةِ وِزْرًا [*He shall bear, on the day of resurrection, a heavy burden*]. (TA.) Hence also, in the Ḳur [vii. 189], حَمَلَتْ حَمْلًا خَفِيفًا [*She bore a light burden*]; (Ṣ, TA;) i. e., [as some say,] the seminal fluid. (TA.) Hence also, in the Ḳur [xxix. 60], وَكَأَيِّنْ مِنْ دَابَّةٍ لَا تَحْمِلُ رِزْقَهَا [*And how many a beast is there that does not bear its sustenance!*], meaning, † *does not provide its sustenance*, but is sustained by God. (TA.) يَحْمِلُ الحَطَبَ [lit. *He carries firewood*],

(A in art. حطب,) or الحَطَبَ الرُّطَبَ [*juicy, or fresh, firewood*], (Er-Rághib, TA,) means ‡ *he goes about with calumny*, or *slander*. (A in art. حطب, and Er-Rághib* and TA.*) — حَمَلَهُ عَلَى الدَّابَّةِ, (Mṣb, TA,) aor. ـِ, (TA,) inf. n. حَمْلٌ, (Mṣb, TA,) [*He carried him*, or *mounted him*, (namely, a man, Mṣb) *upon the beast*; as also ↓ اِحتملهُ.] And حَمَلَهُ [alone] *He gave him a beast upon which to ride*. (T, TA. [See Ḳur ix. 93.]) أَحْمَلَهُ is not used in this sense. (T, TA.) — See also 4. — حَمَلَتِ المَرْأَةُ, aor. ـِ, (Ḳ,) inf. n. حَمْلٌ, (TA,) ‡ *The woman became pregnant*, or *conceived*: (Ḳ, TA:) and حَمَلَتْ وَلَدَهَا *She became pregnant with*, or *conceived, her child*: (Mṣb:) one should not say, حَمَلَتْ بِهِ; or this is rare; (Ḳ;) or one should not say this, but it is frequently said; (IJ, TA;) [for] as حَمَلَتْ is syn. with عَلِقَتْ, (Mṣb, TA,) and the latter is trans. by means of بِ, the former is thus made trans., (TA,) therefore one says, حَمَلَتْ بِهِ فِى لَيْلَةِ كَذَا وَفِى مَوْضِعِ كَذَا, meaning *She became pregnant with him*, or *conceived him, in such a night, and in such a place*. (Mṣb.) حَمَلَتْ is also said of a ewe or she-goat, and of a female beast of prey, [and app. of any female,] accord. to IAạr; meaning † *She was*, or *became, in the first stage of pregnancy*. (TA.) — حَمَلَتِ الشَّجَرَةُ, inf. n. حَمْلٌ, † *The tree* [*bore*, or] *produced*, or *put forth, its fruit*. (Mṣb.) — حَمَلَ بِدَيْنٍ, and بِدِيَةٍ, inf. n. حَمَالَةٌ, † [*He bore*, or *took upon himself, the responsibility*, or *he was*, or *became, responsible, for a debt*, and *a bloodwit*:] (Mṣb:) [for] حَمَلَ بِهِ, aor. ـِ, inf. n. حَمَالَةٌ, signifies كَفَلَ. (Ṣ,* Ḳ.) And حَمَلَ الحَمَالَةَ and ↓ تحمّلها † [*He was*, or *became, responsible for the bloodwit*, or *debt* or *the like*]: both signify the same: (Ṣ, TA:) and ↓ تحمّل بِهِ † *He took it upon himself*, or *became responsible*, or *answerable, for it*: (Mṣb in art. كفل:) and ↓ تحمّل مُعْظَمَهُ † *He took*, or *imposed, upon himself*, or *undertook, the main part of it*: (Jel in xxiv. 11:) and ↓ احتمل الأَمْرَ † *He took*, or *imposed, upon himself*, or *undertook, the thing*, or *affair*; *he bore*, or *took upon himself, the burden thereof*. (L in art. قلد.) You say, حَمَلَ قَوْمٌ عَنْ قَوْمٍ دِيَةً, or غَرَامَةً, (Ḳ, TA,) or (TA,) † [*A party bore*, or *took upon itself, for a party, the responsibility for a bloodwit*, or *a debt* or *the like*;] as also ↓ تحمّل. (Ṣ.) [And حَمَلَ عَنْ فُلَانٍ لِفُلَانٍ كَذَا † *He bore*, or *took upon himself, for such a one, the responsibility, to such a one, for such a thing*.] And ↓ تحمّل حَمَالَةً بَيْنَ قَوْمٍ † *He bore*, or *took upon himself, the responsibility for the bloodwits between people*, in order to make peace between them, when war had occurred between them, and men's blood had been shed. (TA, from a trad.) — حَمَلَ ظُلْمًا † [*He made himself chargeable with wrongdoing*]. (Ḳur xx. 110.) — حَمَلَ الأَمَانَةَ: see أَمَانَةٌ: accord. to some, it means † *He took upon himself*, or *accepted, the trust*: accord. to others, *he was unfaithful to it*: and ↓ احْتَمَلَهَا means the same.] — حَمَلْتُ إِدْلَالَهُ: see 8. — حَمَلَ عَنْهُ: see 8. — حَمَلَ فُلَانٌ الحِقْدَ عَلَى فُلَانٍ † *Such a one* [*bore* or] *concealed in his mind rancour, malevolence, malice*, or *spite, against such a one*. (TA.) And فُلَانٌ لَا يَحْمِلُ, i. e. يُظْهِرُ غَضَبَهُ [which may be meant as the explanation of لا يحمل, i. e. † *Such a one shows* (or *will not conceal*) *his anger*; and thus SM understood it; or as the explanation of يحمل alone, i. e. *such a one will not show his anger*]: (Az, TA:) [for] حَمَلَ الغَضَبَ, (Ḳ,) aor. ـِ, inf. n. حَمْلٌ, (TA,) means ‡ *he showed*, or *manifested, anger*. (Ḳ, TA.) And hence, it is said, is the saying, in a trad., إِذَا بَلَغَ المَآءُ قُلَّتَيْنِ لَمْ يَحْمِلْ خَبَثًا, i. e. † [*When the water amounts to the quantity of two vessels of the kind called* قُلَّة,] *impurity does not appear in it*: (O, Ḳ,* TA:) or the meaning is, † *it does not admit the bearing of impurity*: for one says, فُلَانٌ لَا يَحْمِلُ الضَّيْمَ, i. e. † *such a one refuses to bear*, or *submit to, and repels from himself, injury*. (Mṣb.) You say also, حَمَلَ مِنْ ذٰلِكَ أَنَفًا † *He conceived, in consequence of that, disdain*, or *scorn, arising from indignation and anger*. (TA in art. انف, from a trad.) — حَمَلَ الحَدِيثَ † [*He bore in his memory, knowing by heart, the tradition*, or *narrative*, or *story*; and in like manner, القُرْآنَ *the Ḳur-án*]. (Mṣb in art. روى.) — حَمَلَ فُلَانًا, and ↓ تحمّل بِهِ and عَلَيْهِ, † *He relied upon such a one* in intercession, and in a case of need. (TA.) — حُمِلَ عَلَى النَّاقَةِ † *The she-camel was covered by a stallion*. (M in art. صمد.) — حَمَلَ عَلَيْهِ [as syn. with حَمَّلَهُ]: see 2, in three places. — حَمَلَ عَلَى دَابَّتِهِ فَوْقَ طَاقَتِهَا فِى السَّيْرِ † [*He tasked his beast beyond its power in journeying*, or *marching*, or *in respect of pace*]. (Ṣ in art. جهد.) And حَمَلَ عَلَى نَفْسِهِ فِى السَّيْرِ † *He jaded*, or *fatigued, himself*, or *tasked himself beyond his power, in journeying*, or *marching*. (Ṣ, TA.) [See also 6.] — حَمَلَ عَلَيْهِ فِى الحَرْبِ, inf. n. حَمْلَةٌ [which is properly an inf. n. of un.], (T, Ṣ,) † *He charged*, or *made an assault* or *attack, upon him in war*, or *battle*. (TA.) — حَمَلْتُ عَلَى بَنِى فُلَانٍ † *I made mischief*, or *I excited disorder, disagreement, dissension*, or *strife, between*, or *among, the sons of such a one*. (AZ, Ṣ.) — حَمَلَهُ عَلَى الأَمْرِ, aor. ـِ, † *He incited, excited, urged, instigated, induced*, or *made, him to do the thing*, or *affair*. (ISd, Ḳ.) — حَمَلَ لَفْظًا عَلَى لَفْظٍ آخَرَ, aor. ـِ, inf. n. حَمْلٌ, a phrase often used in lexicology and grammar, † *He made*, or *held, a word*, or *an expression, to accord in form*, or *in meaning*, or *syntactically, with another word*, or *expression*. One says, يُحْمَلُ عَلَى الأَكْثَرِ † *It* (a word) *is made to accord in form with* those words with which it may be compared that constitute *the greater number*: thus one says of رَحْمَانُ, which is made to accord in form with words of the measure فَعْلَانُ, though it has not a fem. of the measure فَعْلَى, in preference to فَعْلَانٌ, because words of the measure فَعْلَانُ are more numerous than those of the measure فَعْلَانٌ. And يُحْمَلُ عَلَى نَقِيضِهِ † *It* (a word) *is made to accord in form with its contrary in meaning*: thus عِجَافٌ, an anomalous pl. of أَعْجَفُ, is made to accord in form with سِمَانٌ, a regular pl. of سَمِينٌ. And يُحْمَلُ عَلَى المَعْنَى † *It* (a word) *is made to accord syntactically with its meaning*: and يُحْمَلُ عَلَى اللَّفْظِ † *It is made to accord syntactically with its grammatical character*: the former is said when, in a sentence, we make a masc. word fem., and the contrary, because the meaning allows us to substitute a fem. syn. for the masc. word, and a masc. syn. for the fem. word: for ex., it is said in the Ḳur vi. 78, فَلَمَّا رَأَى الشَّمْسَ بَازِغَةً قَالَ هٰذَا رَبِّى "And when he saw the sun rising, he said, This is my Lord:" here (by saying الشمس بازغة) is first made to accord syntactically with its grammatical character (تُحْمَلُ عَلَى اللَّفْظِ); and then (by saying هٰذَا instead of هٰذِهِ) it is made to accord syntactically with its meaning (تُحْمَلُ عَلَى المَعْنَى), which is الجِرْمُ or the like: this is allowable; but the reverse in respect of order is of weak authority; because the meaning is of more importance than the grammatical character of the word. (Collected from the Kull pp. 156 and 157, and other works.)] — حَمَلَهُ أَحْسَنَ مَحْمَلٍ † [*He put the best construction upon it*; namely, a saying: محمل being here an inf. n.]. (TA in art. ابو.) — حَمَلَهُ عَلَى النَّاسِخِ † *He attributed it to*, or *charged it upon, the copyist*; namely, a mistake. حُمِلَ عَلَى النَّاسِخِ, said of a mistake, occurs in the Ḳ in art. ربع. — حَمَلَ شَيْئًا عَلَى آخَرَ, in logic, means † *He predicated a thing of another thing*.] — See also حُمْلَانٌ.

2. حَمَّلَهُ الشَّىْءَ, (Mṣb,) and الرِّسَالَةَ, (Ṣ, TA,) inf. n. تَحْمِيلٌ, (TA,) *He made him*, or *constrained him, to bear* or *carry* [*the thing*, and *the message*; and in like manner, ↓ حَمَلَ عَلَيْهِ الشَّىْءَ]. (Ṣ, Mṣb,* TA.) [And حَمَّلَهُ, alone, *He loaded him*; namely, a camel, &c.] You say also, حَمَّلَهُ الأَمْرَ ↓ فَتَحَمَّلَهُ, inf. n. of the former تَحْمِيلٌ and حِمَّالٌ, like كِذَّابٌ, [which is of the dial. of El-Yemen], and of the latter verb تَحَمُّلٌ and تِحِمَّالٌ [like تِكِلَّامٌ &c.], (Ḳ,) † *He imposed upon him the affair, as a task*, or *in spite of difficulty* or *trouble* or *inconvenience, and he undertook it, as a task*, &c. (Mṣb in art. كلف.) And حَمَّلْتُهُ أَمْرِى فَمَا ↓ تَحَمَّلَ † [*I imposed upon him my affair, as a task*, &c., *but he did not undertake it*]. (TA.) It is said in the Ḳur [xxiv. 53], فَإِنَّمَا عَلَيْهِ مَا حُمِّلَ وَعَلَيْكُمْ مَا حُمِّلْتُمْ † [*Upon him rests only that which he has had imposed upon him; and upon you, that which ye have had imposed upon you*]: i. e., upon the Prophet rests the declaring of, that which has been revealed to him; and upon you, the following him as a guide. (TA.) And رَبَّنَا وَلَا ↓ تَحْمِلْ عَلَيْنَا إِصْرًا كَمَا حَمَلْتَهُ عَلَى الَّذِينَ مِنْ قَبْلِنَا رَبَّنَا وَلَا تُحَمِّلْنَا مَا لَا طَاقَةَ لَنَا بِهِ † [*O our Lord, and do not Thou impose upon us a burden, like as Thou imposedst it upon those before us: O our Lord, and do not Thou impose upon us that which we have not power to bear*]: (Ḳur ii. last verse:) or, accord. to one reading, تُحَمِّلْ, which has an intensive signification [when followed by على]. (Bḍ.) — حَمَّلَهُ ذَنْبًا † *He charged him*

with a crime, or an offence: see a verse of En-Nábighah cited voce عَرٌّ.]

3. حاملهُ [*He bore with him* a burden]. You say, of a Wezeer, حَامَلَ المَلِكَ أَعْبَآءَ المُلْكِ † [*He bore with the King the burdens of the regal office*]. (A in art. وزر.) [See also 4.] — Also † *He requited him;* namely, a man: or, accord. to AA, مُحَامَلَةٌ signifies the *requiting with beneficence.* (TA.)

4. احملهُ *He helped him to bear,* or *carry,* (T, Ṣ,) *that which he was bearing,* or *carrying:* (T, TA:) or you say, احملهُ الحِمْلَ *he helped him to bear,* or *carry, the load,* or *burden:* and ↓ حَمَّلَهُ, i. e. فَعَلَ ذٰلِكَ بِهِ [*he did that with him*]. (M, O, Ḳ.) [See also 3.] = أَحْمَلَتْ *She* (a woman, Ṣ, Ḳ, and a camel, Ṣ) *yielded her milk without being pregnant.* (Ṣ, Ḳ.)

5. تحمّل *He took upon himself the bearing,* or *carrying, of loads,* or *burdens:* this is the primary signification. (Ḥar p. 48.) — [Hence, †*He burdened himself with,* or *he became,* or *made himself, chargeable with,* or *he bore,* or *took upon himself, the burden of,* a sin, or crime, or the like; as also ↓ احتمل:] you say احتمل إِثْمًا meaning تحمّلهُ. (Jel in iv. 112 and xxxiii. 58.) And تحمّل غُرْمًا † *He took,* or *imposed, upon himself a debt,* or *fine.* (MA.) — [And hence, likewise, several other significations:] see 2, in two places: — and 8: — and 1, in six places. — Also He *bound the load,* or *burden,* [or *the loads,* or *burdens, on the saddle,* or *saddles,* or *on the beast,* or *beasts;*] (Ḥar p. 48;) and ↓ احتمل signifies [the same, or] *he put,* or *placed, the load,* or *burden,* [or *the loads,* or *burdens,*] *on the saddle,* [or *saddles,* or *on the beast,* or *beasts.*] (Ḥar p. 556.) — [And hence,] تحمّلوا and ↓ احتملوا † *They went away, departed,* or *journeyed.* (Ṣ, TA.)

6. تحامل عَلَيْهِ [*He bore, bore his weight, pressed,* or *pressed heavily, upon it,* or *him*]. You say, تَحَامَلَ عَلَى رَأْسِ رُمْحِهِ مُعْتَمِدًا عَلَيْهِ لِيَمُوتَ [*He bore, bore his weight, pressed,* or *pressed heavily, upon the head of his spear, leaning upon it, in order that he might die*]. (Mgh in art. ركز.) And تَحَامَلْتُ عَلَيْهِ كَالعَاصِرِ [*I pressed,* or *pressed heavily, upon it, like the squeezer of* fruit &c.]. (Mṣb in art. همز.) — [Hence,] † *He wronged him;* or *treated him wrongfully,* or *unjustly.* (Ṣ, Mgh, and Ḥar p. 80.) And it is asserted that one says, تحامل الزَّمَانُ عَنْ فُلَانٍ meaning † *Time,* or *fortune, turned from such a one, and took away his property:* and تحامل إِلَيْهِ † *It became favourable to him.* (Ḥar ibid.) — [Also] † *He imposed upon him,* or *tasked him with, that which he was not able to bear,* or *to do.* (M, O, Ḳ.) And تحامل عَلَى نَفْسِهِ, (Ṣ, O,) or تحامل فِى الأَمْرِ and بِالأَمْرِ, (M, Ḳ,) † *He imposed upon himself,* or *tasked himself with,* or *constrained himself to do, the thing,* or *affair, notwithstanding difficulty,* or *trouble,* or *inconvenience,* (Ṣ, M, O, Ḳ,) *and fatigue.* (M, TA.) And تَحَامَلْتُ فِى المَشْىِ † *I constrained myself to walk, notwithstanding difficulty,* or *trouble,* or *inconvenience, and fatigue:* whence, رُبَّمَا يَتَحَامَلُ الصَّيْدُ وَيَطِيرُ, i. e. † *Sometimes the game will constrain itself* to fly, *notwithstanding difficulty,* &c., *and will fly.* (Mgh.) [See also two similar phrases in the first paragraph.] — ↓ مُتَحَامَلٌ is used as its inf. n., and also as a noun of place: using it as an inf. n., you say, مَا فِى فُلَانٍ مُتَحَامَلٌ i. e. تَحَامُلٌ † [*There is not, in such a one, wrong-doing,* &c.]: and using it of a place, هٰذَا مُتَحَامَلُنَا † [*This is our place of wrong-doing,* or *wrong-treatment,* &c.]. (Ṣ, TA.)

7. انحمل عَلَى الأَمْرِ † *He was,* or *became, incited, excited, urged, instigated, induced,* or *made, to do the thing,* or *affair.* (ISd, Ḳ.)

8. احتمل *He raised* a thing *upon his back.* (Ḥar p. 41.) — See also 1, in five places: and see 5, in three places. — † *He bore, endured,* or *sustained.* (KL.) You say, اِحْتَمَلْتُ مَا كَانَ مِنْهُ † [*I bore,* or *endured, what proceeded from him,* or *what he did* or *said,* or] *I forgave what proceeded from him, and feigned myself neglectful of it.* (Mṣb.) And ↓ حَمَلْتُ إِدْلَالَهُ and اِحْتَمَلْتُ † [*I bore,* or *endured, his presumptuousness occasioned by his confiding in my love*]. (Ṣ.) And احتملهُ † [*He bore with, endured, suffered,* or *tolerated, him;* or] *he bore,* or *endured, his annoyance,* or *molestation,* (احتمل أَذَاهُ,) *and feigned himself neglectful of what proceeded from him, and did not reprove him.* (Ḥar p. 41.) And احتمل † *He was forbearing,* or *clement; he acted with forbearance,* or *clemency; he treated with forbearance,* or *clemency, him who reviled him:* (TA:) *he forgave an offence;* as also ↓ تحمّل: (Ḥar p. 637:) and ↓ حَمَلَ عَنْهُ ‡ *he treated him with forbearance,* or *clemency.* (Ḳ, TA.) [And احتمل النِّعْمَةَ † *He bore wealth;* or *he had,* or *exercised, the quality of doing so;* generally meaning, *in a becoming,* or *proper, manner;* but also *absolutely,* as is shown by the phrase] سُوْءُ احْتِمَالِ النِّعْمَةِ † [*The bearing of wealth ill,* or *in an evil manner*]. (Er-Rághib voce بَطِرَ.) And احتمل الصَّنِيعَةَ † *He bore the benefit as a badge, and was thankful,* or *grateful, for it.* (ISd, Ḳ.) — [In lexicology, said of a word or phrase or sentence, † *It bore, admitted,* or *was susceptible of,* a meaning, a sense, or an interpretation: and, elliptically, † *it bore, admitted,* or *was susceptible of, two,* or *more, different meanings, senses,* or *interpretations; it was equivocal.*] In the conventional language of the lawyers, and the Muslim theologians [and men of science in general], (Mṣb,) it is used, (Kull,) or may be used, (Mṣb,) as importing supposition, and admissibleness, or allowableness; and thus used, it is intrans.: and also as importing necessary implication, and inclusion; and thus used, it is trans.: you say, يَحْتَمِلُ أَنْ يَكُونَ كَذَا † [*It is supposable,* or *admissible,* or *allowable, that it may be thus;* or simply *it may be thus;* as also يُحْتَمَلُ, which is often used in this sense]: and اِحْتَمَلَ الحَالُ وُجُوهًا كَثِيرَةً † [*The case necessarily implied,* or *included, many (possible) modes,* or *manners of being;* or *admitted of being put,* or *explained,* or *understood, in many ways;* or *bore many kinds of interpretation*]. (Mṣb, Kull.) — احتملهُ الغَضَبُ † *Anger disquieted,* or *flurried, him.* (Mj, TA.) And أُحْتُمِلَ [alone] † *He was disquieted,* or *flurried, by anger:* (T, TA:) or, accord. to the Mj and M and O; but accord. to the Ḳ, followed by لَوْنُهُ; (TA;) † *he was angry, and his colour changed.* (Ḳ, TA.) — [اِحْتَمَلَتْ *She* (a woman) *used* a drug, or the like, *in the manner of a suppository in the vagina:* so in the present day: and so in the Ḳ, on the words نَفْطٌ قُنْبِيطٌ and &c.] — احتمل *He bought what is termed* حَمِيلٌ, i. e. *a thing* [in the CḲ للشّيْءِ is put for للشَّىْءِ] *carried from one country or town to another* (Ḳ, TA) *among a party of captives.* (TA.)

10. سَأَلْتُهُ أَنْ يَحْمِلَنِى signifies اِسْتَحْمَلْتُهُ [i. e. *I asked him to carry me,* or *to give me a beast on which to ride*]. (Ṣ.) — استحملهُ نَفْسَهُ † *He imposed upon him his wants and affairs.* (M, Ḳ.)

R. Q. 1. حَوْمَلَ *He carried water.* (Ibn-'Abbád, Ḳ.)

حَمْلٌ [inf. n. of 1, q. v. — ‡ *Gestation:* see an ex. voce إِنْىٌ. — And hence,] † The *young that is borne in the womb* (M, Ḳ) of any animal; (M, TA;) and † the *fruit* of a tree, (IDrd, Ṣ, M, Mṣb, Ḳ,) as also ↓ حِمْلٌ: (IDrd, Ṣ, M, Ḳ:) or the former, † the *thing that is in a belly,* or *on the head of a tree:* (ISk, Ṣ, M, Mgh, Ḳ:) and ↓ the latter, *a thing borne,* or *carried,* (Mṣb, Ḳ,) *on the back;* [i. e. *a load,* or *burden;*] (Mṣb;) the *thing that is on the back* or *on the head:* (ISk, Ṣ, M, Mgh, Ḳ:) or the former, † *a burden that is borne internally;* as the *young in the belly,* and the *water in the clouds,* and the *fruit in the tree* as being likened to the حَمْل of the woman: and ↓ the latter, *a burden that is borne externally;* as the *thing that is borne on the back:* (Er-Rághib, TA:) or [when applied to fruit] the former signifies *a fruit that is internal:* and ↓ the latter, *a fruit that is external:* (M, Ḳ:) or the former, *fruit of a tree when large,* or *much:* and ↓ the latter, *fruit when not large,* or *when not much and large:* (Ḳ accord. to different copies:) this is the saying of AO, mentioned in the T, in art. شمل, where, in the copies of the T, is found ما لم يكثر, not ما لم يكبر: (TA:) and the former also occurs as meaning *a burden that requires, for the carrying it, a beast or the hire of a porter:* (Mgh:) the pl. [of pauc.] of the latter (Mgh, Mṣb, Ḳ) and of the former (Ḳ) is أَحْمَالٌ (Ṣ, Mgh, Mṣb, Ḳ) and [the pl. of mult.] (of the former, Ḳ,* TA) حِمَالٌ (Ḳ) and (of the latter, Mṣb) حُمُولٌ (Mṣb, Ḳ) and حُمُولَةٌ. (Ṣ, M, Mgh, Ṣgh.) Hence, (in a trad., TA) هٰذَا الحِمَالُ لَا حِمَالُ خَيْبَرَ † [*This is the fruit: not the fruit of Kheyber*]: meaning that it is the fruit of Paradise; and that it does not fail, or come to an end. (M, Ḳ.) — See also what next follows.

حِمْلٌ: see حَمْلٌ, in five places. — حُمُولٌ, (Ṣ, M, Ḳ,) as pl. of حِمْلٌ, (M, Ḳ,) and of ↓ حَمُولٌ also, (Ḳ,) signifies likewise [*Vehicles of the kind called*] هَوَادِج [pl. of هَوْدَجٌ], (M, Ḳ,) *whether having in them women or not:* (M, TA:) or † *camels upon which are* هوادج, (AZ, Ṣ, M,

O, Ḳ,) *whether there be in them women or not:* (AZ, Ṣ, O:) it is not applied to camels unless they have upon them هوادج. (M, TA.) — See also مَحْمِلٌ, and حَمُولَةٌ.

حَمَلٌ *A lamb;* i. e. the *young one of the ewe in the first year;* (Mgh, Mṣb;) *i. q.* بَرَقٌ; (Ṣ;) or خَرُوفٌ [explained in the Ḳ in art. خرف as the *male young one of the sheep-kind; or such as has pastured, and become strong*]: (Ḳ, and Ṣ and Mṣb in art. خرف:) or *such as is termed* جَذَعٌ, [i. e. *a year old,* or *from six to ten months,*] *of the young of the sheep-kind;* and *less than this* [*in age*]: (ISd, Ḳ:) accord. to Er-Rághib, it signifies مَحْمُولٌ [*borne,* or *carried*]; and the young of the sheep-kind is particularly called thus because borne, or carried, on account of its impotence, and of the nearness of the time when its mother was pregnant with it: (TA:) pl. حُمْلَانٌ (Ṣ, M, Mgh, Ṣgh, Mṣb, Ḳ) and أَحْمَالٌ. (M, Ḳ.) — [Hence,] الحَمَلُ † [*The sign Aries;*] *a certain sign of the zodiac;* (Ḳ;) *the first of the signs of the zodiac;* (Ṣ;) *the constellation comprising, first, the* شَرَطَانِ, *which are its two horns; then, the* بُطَيْنٌ; *then, the* ثُرَيَّا. (T, TA.) One says, مُطِرْنَا بِنَوْءِ الحَمَلِ and بنوء الطَّلِيِّ † [*We were,* or *have been, given rain by the auroral setting of Aries:* so the pagan Arabs used to say: see نَوْءٌ; and see مَنَازِلُ القَمَرِ, in art. نزل]. (TA.) One says also, هٰذَا حَمَلٌ طَالِعًا † [*This is Aries, rising*]; suppressing the ال, but making the noun to remain determinate; and thus one does in the case of every name of a sign of the zodiac, preserving the ال or suppressing it. (TA.) — حَمَلٌ signifies also ‡ *Clouds containing much water:* (M, Ḳ, TA:) or *black clouds:* (T, TA: [see also حَوْمَلٌ, below:]) or, as some say, the *rain* [*supposed to be given*] *by the* نَوْء [see above] *of* الحَمَل. (TA.)

حَمْلَةٌ † *A charge,* or *an assault* or *attack, in war,* or *battle.* (T, Ḳ.)

حُمْلَةٌ: see what next follows.

حِمْلَةٌ and ▼ حُمْلَةٌ *Carriage from one* دار [app. here meaning *country,* or *town,* or *the like,*] *to another.* (Ḳ.)

حُمْلَانٌ an inf. n. of حَمَلَ [q. v.]. (Mgh, Ḳ.) — Also *A beast upon which a present is borne.* (M, Mgh, O, Ḳ.) — *Hire for that which is borne,* or *carried.* (Lth, Mgh, TA.) — And, as a conventional term (Mgh, O, Ḳ) of the صَاغَة [or workers in gold and silver], (Ṣgh, Ḳ,) *Adulterating alloy* (غِشّ) *that is added to dirhems,* or *coin* (يُحْمَلُ عَلَى الدَّرَاهِمِ). (Mgh, Ṣgh, Ḳ.) — Also pl. of حَمَلٌ [q. v.]. (Ṣ, M, &c.)

حَمَالٌ or حِمَالٌ: see حَمَالَةٌ.

حَمُولٌ † *Forbearing,* or *clement.* (M, Ḳ.)

حَمِيلٌ *i. q.* ▼ مَحْمُولٌ [*Borne, carried, taken up and carried, conveyed,* or *carried off* or *away*]. (Mṣb, Ḳ.) — Hence, (Mṣb,) The *rubbish,* or *rotten leaves, and scum, that are borne of a torrent.* (Ṣ, Mṣb, Ḳ.*) — *A thing* [شَىْءٌ, accord. to copies of the Ḳ and the TA, but accord. to the CḲ سَبْىٌ, agreeably with the next of the explanations here following,] *that is carried from one country or town to another* (Ḳ, TA) *among a party of captives.* (TA.) — *A captive;* because carried from one country or town to another. (Mṣb.) — *One who is carried a child from his country, not born in* [*the territory of*] *El-Islám:* (Ṣ, O:) or *one who is carried from his country to the country of El-Islám:* or *a child with a woman who carries it, and says that it is her son:* or *any relation,* or *kinsman, in the territory of the enemy:* (Mgh:) or *one that is carried from the territory of the unbelievers to that of El-Islám, and who is therefore not allowed to inherit without evidence:* (Th, TA:) or *a child in the belly of his mother when taken from the land of the unbelievers.* (Ḳ.) — *A foundling,* or *child cast out by his mother, whom persons carry off and rear:* (Ḳ:) in some copies of the Ḳ, فَيُرَبُّونَهُ is erroneously put for فَيُرَبُّونَهُ. (TA.) — † *One whose origin,* or *lineage, is suspected;* or *who claims for his father one who is not;* or *who is claimed as a son by one who is not his father;* syn. دَعِىٌّ. (Ṣ, Mṣb, Ḳ.) — † *A stranger:* (Ḳ:) as being likened to [the حميل of] the torrent, or to the child in the belly. (Er-Rághib, TA.) — † *One who is responsible,* or *a surety,* (Ṣ, Mṣb, Ḳ,) for (ب) a debt or a bloodwit; as also ▼ حَامِلٌ: (Mṣb:) because he bears [or is burdened with] the obligation, together with him upon whom the obligation properly rests. (TA.) — † *What is withered and black* of the ثُمَام and وَشِيج (Ḳ, TA) and ضَعَة and طَرِيفَة. (TA.) — † The [*thong called*] شِرَاك [*of a sandal*]. (O, Ḳ.) In one copy of the Ḳ, الشريك is put in the place of الشراك. (TA.)

حَمَالَةٌ *A bloodwit,* (Ṣ, Ḳ, TA,) or *a debt, an obligation,* or *a responsibility, that must be paid, discharged,* or *performed, taken upon himself by a person,* (Ṣ, TA,) or *taken upon themselves by a party of men,* (Ḳ, TA,) *for others;* (Ṣ, Ḳ, TA;) as also ▼ حَمَالٌ, accord. to the T and M; or ▼ حِمَالٌ, accord. to the Ḳ: (TA:) or *a responsibility which one takes upon himself* for a debt or a bloodwit: pl. حَمَالَاتٌ: (Mṣb:) the pl. of حَمَلٌ is حمال. (Ḳ.)

حِمَالَةٌ The *occupation,* or *business, of a porter,* or *carrier of burdens.* (M, Ḳ.) — Also said to be sing. of حَمَائِلُ, and syn. with مِحْمَلٌ, which see, in two places.

حَمُولَةٌ *A camel,* or *horse,* or *mule,* or *an ass, upon which burdens are borne:* (Mgh, Mṣb:) and sometimes applied to *a number of camels:* (Mṣb:) *camels that bear burdens:* and *any beast upon which the tribe carries,* namely, *an ass* or *other animal;* (Ṣ;) or *a beast upon which people carry,* namely, *a camel,* and *an ass,* and *the like;* (Ḳ;) *whether the loads be thereon or not:* (Ṣ, Ḳ:) or *such as are able to bear:* (Az, TA:) or particularly applied to *such as have on them the loads;* as also ▼ حَمُولٌ: (ISd, TA:) accord. to the T, *not including asses nor mules:* applied to *one* and to *more than one:* (TA:) a word of the measure فَعُولٌ receives the affix ة when it has the meaning of a pass. part. n. (Ṣ, TA.) — Also, accord. to the Ḳ, The *loads,* or *burdens,* themselves: but this, accord. to the Ṣ and M [and Mgh] and Ṣgh, is [حُمُولَةٌ, a pl. of حِمْلٌ,] with ḍamm [to the ح]. (TA.)

حَمِيلَةٌ † *i. q.* كَلٌّ and عِيَالٌ: so in the saying, هُوَ حَمِيلَةٌ عَلَيْنَا † [*He is a burden upon us; one whom we have to support*]. (O, Ḳ.) — Also said to be sing. of حَمَائِلُ, and syn. with مِحْمَلٌ, q. v.

حَمَائِلُ: see مِحْمَلٌ, in two places.

حَمَّالٌ *A porter,* or *carrier of burdens.* (Mṣb, Ḳ.) — حَمَّالَةُ الحَطَبِ [is applied in the Ḳur cxi. 4 to a woman, lit. meaning *The female carrier of firewood:* and as an intensive epithet is applied to a man, as meaning] ‡ *The calumniator,* or *slanderer.* (TA.)

حَامِلٌ [*Bearing, carrying, taking up and carrying, conveying,* or *carrying off* or *away;*] act. part. n. of 1 having for its object what is borne on the back [&c.]: (Mṣb:) fem. with ة: (Ṣ, Mṣb:) pl. masc. حَمَلَةٌ: (Ṣ, TA:) and pl. fem. حَامِلَاتٌ. (TA.) Hence, حَمَلَةُ العَرْشِ [*The bearers of the* عرش, or *empyrean,* held by the vulgar to be *the throne of God*]. (Ṣ, TA.) And the phrase فَالْحَامِلَاتِ وِقْرًا [in the Ḳur li. 2, lit. *And the bearers of a load,* or *heavy load:*] meaning † *the clouds.* (TA.) — Applied to a woman, ‡ *Pregnant;* (Ṣ, Mgh, Mṣb, Ḳ, &c.;) as also حَامِلَةٌ: (Ṣ, Mṣb, Ḳ:) the former as being an epithet exclusively applied to a female: the latter as conformable to its verb, which is حَمَلَتْ; (Ṣ, Mṣb;) or as being used in a tropical [or doubly tropical] manner, meaning pregnant in past time or in future time; (Mṣb;) or as a possessive epithet [meaning *having a burden in the womb*]: (TA:) [see an ex. of the latter in a verse cited in the first paragraph of art. مخض:] accord. to the Koofees, the former, not being applied to a male, has no need of the sign of the fem. gender: but the Baṣrees say that this [rule] does not uniformly obtain; for the Arabs say رَجُلٌ أَيِّمٌ and اِمْرَأَةٌ أَيِّمٌ, and رَجُلٌ عَانِسٌ and اِمْرَأَةٌ عَانِسٌ; and that, correctly speaking, حَامِلٌ and طَالِقٌ and حَائِضٌ and the like are epithets masc. in form applied to females, like as رَبْعَةٌ and رَاوِيَةٌ and خُجَأَةٌ are epithets fem. in form applied to males. (Ṣ.) It is also applied to a she-camel [and app. to any female] in the same sense. (Mgh.) — Applied to trees (شَجَر), † *Bearing fruit:* (TA:) fem. with ة. (Ḳ.) — See also حَمِيلٌ. — [Respecting this epithet, and the phrases حَامِلُ الأَمَانَةِ and مُحْتَمِلُ الأَمَانَةِ, see also أَمَانَةٌ, last sentence but one.] — حَمَلَةُ القُرْآنِ † [*Those who bear in their memory the Ḳur-án, knowing it by heart*]. (Ṣ, TA.)

حَوْمَلٌ Clouds (سَحَابٌ) *black by reason of the abundance of their water.* (O, Ḳ.) [See also حَمَلٌ.] — A *clear* torrent. (Ḳ.) — The *first* of anything. (Ḳ.)

حَامِلَةٌ fem. of حَامِلٌ [q. v.]. (Ṣ, Mṣb.) —

حَوَامِلُ is its pl.: and signifies The *legs*; (M, Ḳ;) because they bear the man. (TA.) —— And The *sinews*, or *tendons*, of the foot and of the fore arm; (M, Ḳ;) and the [*veins called the*] رَوَاهِش thereof. (M, TA. [See also الوَرِيدُ.]) —— See also مَحْمِل.

مَحْمِلٌ [of which the primary signification is *A place of bearing* or *carrying*], (Ṣ, Mgh, Mṣb, Ḳ,) or ♦ مِحْمَلٌ [which primarily signifies *An instrument for bearing* or *carrying*], (M, Mgh,) or the latter is allowable, (Mṣb,) The [*kind of vehicle called*] هَوْدَج; (Mṣb;) as also ♦ حِمْلٌ: (M, Ḳ:) or the *large* هودج *termed* حَجَّاجِىٌّ: (Mgh:) or *a pair of dorsers*, or *panniers*, or *oblong chests*, (شُقَّانِ,) *upon a camel, in which are borne two equal loads*, (Ḳ,) [*and which, with a small tent over them, compose a* هودج;] *first made use of by El-Ḥajjáj Ibn-Yoosuf Eth-Thakafee*: (TA:) *one of the* مَحَامِل *of the pilgrims*: (Ṣ:) مَحَامِل being the pl. (Ḳ.) Hence, ♦ مَحَامِلِىٌّ *A seller of* مَحَامِل. (Ḳ.) [What is now particularly termed the محمل (vulgarly pronounced مَحْمَل) of the pilgrims is an ornamented هودج, which is borne by a camel, but without a rider, and is regarded as the royal banner of the caravan; such as is described and figured in my work on the Modern Egyptians. (See also مَحَارَةٌ, in art. حور.)] Its application to ‡ The *camel that bears the* محمل is tropical. (Mgh.) [See also حِمْل. The assertion that it signifies also the silk covering that is sent every year for the Kaabeh is erroneous. This covering is sent from Cairo, with the baggage of the chief of the Egyptian pilgrim-caravan.] —— Also مَحْمِلٌ, (Ḳ,) or ♦ مِحْمَلٌ, (M,) *A basket* (زِنْبِيل) *in which grapes are carried to the place where they are to be dried*; and so ♦ حَامِلَةٌ. (Ḳ.) —— One says also, مَا عَلَى فُلَانٍ مَحْمِلٌ † *There is no ground of reliance upon such a one*; syn. مُعْتَمَدٌ: (Ṣ:) or *no relying*, or *reliance*: (MA:) or *no ground* (lit. *place*) *for imposing upon such a one the accomplishment of one's wants*. (M, TA.) And مَا عَلَى البَعِيرِ مَحْمِلٌ مِنْ ثِقَلِ الحِمْلِ † [*There is no ground of reliance, or no relying, upon the camel, by reason of the heaviness of the load.*] (TA.)

مُحْمِلٌ A woman, (Ṣ, M, Ḳ,) and a she-camel, (Ṣ, M,) *who yields her milk without being pregnant*. (Ṣ, M, Ḳ.)

مِحْمَلٌ: see مَحْمِلٌ, in two places. —— The عِلَاقَة of a sword (Ṣ, Mṣb,* Ḳ) &c.; (Mṣb;) i. e. its *suspensory thong* [or *cord* or *shoulder-belt*], *by which the wearer hangs it upon his neck*; (Ṣ, TA;) as also ♦ حِمَالَةٌ (Ṣ, Mṣb, Ḳ) and ♦ حَمِيلَةٌ: (IDrd, Ḳ:) and the ♦ حِمَالَة of the bow is similar to that of the sword: the wearer throws it upon his right shoulder, and puts forth his left arm from it, so that the bow is on his back: (AḤn, TA:) the pl. of مِحْمَل is مَحَامِلُ: (Az, Mṣb:) and that of حِمَالَة, (Ṣ, Mṣb,) or of حَمِيلَة, (Kh, TA,) is ♦ حَمَائِلُ; (Kh, Ṣ, TA;) or, accord. to Aṣ, حَمَائِل has no proper sing., its sing. being only مِحْمَل. (Ṣ, TA.) —— Dhu-r-Rummeh applies it to ‡ The *root of a tree*; (Ṣ, Ḳ;) likening this to the محمل of a sword. (Ṣ.) —— مَحَامِلُ الذَّكَرِ and ♦ حَمَائِلُهُ † *The veins in the root and skin of the penis*. (M, Ḳ.)

نَاقَةٌ مُحَمَّلَةٌ *A she-camel heavily burdened*, or *overburdened*. (TA.)

مَحْمُولٌ: see حَمِيل. —— Also ‡ *A fortunate man*: from the riding of beasts such as are termed فُرْه, (Ḳ,* TA,) i. e. brisk, sharp, and strong. (TA in art. فره.) —— [In logic, † *A predicate*: and † *an accident*: in each of these senses *contr.* of مَوْضُوع.]

مَحْمُولَةٌ *A dust-coloured wheat*, (Ḳ, TA,) *like the pod of the cotton-plant*, (TA,) *having many grains*, (Ḳ, TA,) *and large ears, and of much increase, but not approved in colour nor in taste*: so in the M. (TA.)

مُحَامِلٌ † One *who is unable to answer thee; and who does it not, to preserve thine affection*. (TA.)

مَحَامِلِىٌّ: see مَحْمِلٌ.

مُحْتَمِلُ الأَمَانَةِ: see أَمَانَةٌ, last sentence but one.

مُتَحَامِلٌ: see 6, last sentence.

شَهْرٌ مُسْتَحْمِلٌ *A month that brings people into difficulty, or distress*; (Ḳ, TA;) *that is not as it should be*. (TA.) Such is said by the Arabs to be the case إِذَا نَحَرَ هِلَالٌ شَمَالًا [app. meaning *when a new moon faces a north-east wind*]. (TA.)

حملق

Q. 1. حَمْلَقَ, (Ṣ, Ḳ,) inf. n. حَمْلَقَةٌ, (Ḥar p. 273,) said of a man, (Ṣ, TA,) and of a lion, (TA,) *He opened his eyes, and looked hard*: (Ṣ, Ḳ:) or *he opened his eyes*: and حملق إِلَيْهِ *He looked at him, or it*: or *he looked hard at him, or it*. (TA.)

حِمْلَاقُ العَيْنِ (Ṣ, M, Ṣgh, Ḳ) and حُمْلَاقُهَا and ♦ حُمْلُوقُهَا (M, Ḳ) *The inner part of the eyelids, that is blackened by the collyrium*: or *the portions of the white of the globe of the eye that are covered by the eyelids*: (Ṣ, Ḳ:) or *the red inner part of the eyelid, the redness of which is seen when it is turned out for the application of the collyrium*: (L, Ḳ:) or *what cleaves to the eye, of the place of the collyrium, internally*: (M, Ḳ:) or *the sides of the globe of the eye*: or *the part of the skin of the eyelid that is next to the globe of the eye*: (TA:) pl. حَمَالِيقُ; (Ṣ, Ḳ;) which some explain as signifying the *portions of the flesh of the eyelids that are next to the globe of the eye*. (TA.) One says, جَآءَ فُلَانٌ مُتَلَثِّمًا لَا يَظْهَرُ مِنْ حُسْنِ وَجْهِهِ إِلَّا حَمَالِيقُ حَدَقَتَيْهِ [*Such a one came wearing a* لِثَام; *nothing appearing of the beauty of his face except the inner edges of his eyelids*, &c.]. (Ṣ.) —— حَمَالِيقُ المَرْأَةِ signifies *The part*, or *parts, upon which close the two edges*, or *borders, of the labia majora of the vulva of the woman*. (T, TA.)

حُمْلُوقٌ: see the next preceding paragraph.

مُحَمْلِقٌ Eyes *having around their globes a whiteness unmixed with blackness*: [it would seem to be a mistranscription for مُحَمْلَقَةٌ; but perhaps it is an epithet applied to a man having eyes of this description; for it is immediately added,] whence عَيْنٌ مُحَمْلَقَةٌ [app. meaning *an eye having around it such a whiteness*]. (TA.)

حمن

حَمْنٌ: see what next follows.

حَمْنَانٌ *Small* قِرْدَان [or *ticks*]; as also ♦ حَمْنٌ: n. un. with ة: (Ḳ:) accord. to Aṣ, the first of the قُرَاد is termed قَمْقَامَةٌ, when very small; then it is termed حَمْنَانَةٌ; then, قُرَادٌ; then, حَلَمَةٌ; then, عَلٌّ; and then, طِلْحٌ. (Ṣ.) —— Also *A sort of grapes of Eṭ-Ṭáïf*, (Ḳ,) *black inclining to redness*, (TA,) *of which the berries are small* (Ḳ) *and few*: (TA:) or the *small berries that are between the large berries*, (Ḳ,) so in the M, (TA,) *among the grapes*. (Ḳ.)

أَرْضٌ مُحْمِنَةٌ and مُحَمَّنَةٌ *A land abounding with the small* قِرْدَان *termed* حَمْنَان. (Ḳ.)

حمو

حَمٌ, (Ṣ, Mgh, Mṣb, Ḳ,) like أَبٌ, originally حَمَوٌ, as is shown by its pl., which see below, (Ṣ,) and ♦ حَمًا, (Ṣ, Mgh, Mṣb, Ḳ,) and ♦ حَمْوٌ, (Ṣ, Mṣb, Ḳ,) like أَبُو, (Ṣ, Mṣb,) used only as a prefixed n. governing the gen. case, except in poetry, (Ṣ, TA,) and ♦ حَمُوٌ, (Ḳ,) and حَمْءٌ [mentioned in art. حمأ, as well as حَمَأٌ], (Ṣ, Mgh, Mṣb, Ḳ,) A woman's *husband's* [*male*] *relation*, (Ṣ, Mgh, Mṣb, Ḳ,) whoever he be, (Ṣ, Mṣb,) such as his *father*, (Ṣ, Mṣb, Ḳ,) and *brother*, (Ṣ, Mṣb,) and *paternal uncle*, (Mṣb,) &c.: (TA:) the fem. is ♦ حَمَاةٌ, (Ḳ,) signifying a woman's *husband's mother*; and having no dial. var.: (Ṣ, Mṣb:) and the pl. is أَحْمَآءٌ: (Ṣ, Mgh, Ḳ:) and ♦ حَمُو الرَّجُلِ, حَمْو الرَّجُلِ, (M, Mṣb,) or ♦, (Ḳ,) signifies *the man's wife's father*, (IF, M, Mṣb, Ḳ,) or *wife's brother*, or *wife's paternal uncle*; (M, Mṣb, Ḳ;) so that حمو applies to a relation on either side, like صِهْرٌ: (Mṣb:) and أَحْمَآءٌ means a husband's *people*: (Mgh:) or the أَحْمَآء are peculiarly of the wife; (Ḳ;) and the أَخْتَان are of the man [or husband]: (TA:) or [in other words] the احماء are [the woman's *relations*] *on the side of the husband*; (Aṣ, TA;) and the اختان are [the husband's relations] on the side of the wife; (Aṣ, Ṣ, Mṣb, TA;) and those of both sides are included by the term [أَصْهَار, pl. of] صِهْرٌ: (Aṣ, Ṣ, TA:) and أَحْمَآءُ فُلَانَةَ is said to mean *such a woman's husband's people*. (IB, TA.)

حُمَةٌ: see art. حمى.

حَمًا :
حَمْوٌ : } see حَمٌ, above, in three places.

حَمُوٌ: see حَمٌ, in two places. —— حَمُو الشَّمْسِ

[in the CK حَمْوُ] *The heat of the sun:* (Ks, Ṣ,* Ḳ:) as also حَمْيًا. (Ks, Ṣ.)

حَمَاةٌ: see حَمْو. ═ Also The *muscle* (عَضَلَة) *of the* سَاق [which means the *shank* of a human being, and, properly speaking, the *corresponding part* (commonly called the *thigh*) of a horse and the like]: (Ṣ, Ḳ:) *an elevated piece of flesh in the inner side of the* ساق: (Lth:) Aṣ says, in the ساق of the horse are the حَمَاتَانِ, which are the *two pieces of flesh that are in the side of the* ساق, *appearing like two sinews, in the outer side and the inner side:* (Ṣ, TA:) or, accord. to ISh, they are the *two elevated pieces of flesh in the half of each* ساق, *in the outer side:* or, accord. to ISd, the *two compact pieces of flesh in the upper portion of the outer side of each* ساق: (TA:) pl. حَمَوَاتٌ. (Ṣ, Ḳ.)

حَمْوَةٌ: see حَمَى الكَلَأَ, in the second sentence of art. حمى.

حِمْوَةٌ: see حَمَيْتُ المَرِيضَ, in art. حمى.

حِمَوَانِ an irregular dual of حِمًى. (Ks, Ṣ and TA in art. حمى, q. v.)

حُمُوَّةٌ: see حُمَيًّا, in art. حمى.

حمى

1. حَمَاهُ, (Ṣ, Mgh, Ḳ,) aor. ـِ, (Ḳ,) inf. n. حِمَايَةٌ (Ṣ, Mgh, Ḳ [but said in the Mṣb to be a simple subst., though afterwards there mentioned as an inf. n.,]) and حَمْىٌ and مَحْمِيَةٌ, (Ḳ,) *He prohibited it,* or *interdicted it;* or *he protected it, defended it,* or *guarded it, from,* or *against, encroachment, invasion,* or *attack.* (Ṣ,* Mgh, Ḳ,* TA.) You say, حَمَى الكَلَأَ, inf. n. حَمْىٌ and حِمْيَةٌ and حِمَايَةٌ and حِمْوَةٌ, [the last irreg.,] *He prohibited,* or *interdicted,* &c., *the herbage,* or *pasture.* (Ḳ,* TA.) And حَمَى المَكَانَ مِنَ النَّاسِ, aor. ـِ, inf. n. حَمْىٌ and حِمْيَةٌ [and حِمَايَةٌ, though here, in the Mṣb, said to be a simple subst.], *He prohibited,* or *interdicted, the place;* or *he protected, defended,* or *guarded, it; from the people* [*in general*]: (Mṣb:) and, accord. to IB, احماهُ ▼ signifies the same as حَمَاهُ: (TA:) or المكان ▼ احمى signifies *he made the place to be what is termed* حِمًى, (Ṣ, Mṣb, Ḳ,) *not to be approached* (Mṣb, Ḳ) *nor ventured upon,* or *attempted:* (Mṣb:) or it signifies, (Ḳ,) or signifies also, (Mṣb,) *he found it to be what is termed* حِمًى: (Mṣb, Ḳ:) or الحِمَى ▼ احمى signifies *he made the* حمى *to be refrained from by people, and to be acknowledged as a* حمى: and حَمَاهُ, *he prohibited,* or *interdicted, it;* or *he protected it, defended it,* or *guarded it, from,* or *against, encroachment, invasion,* or *attack:* (AZ:) accord. to Suh, in the R, احماهُ ▼ is of weak authority; but both these verbs are chaste. (TA.) [Hence,] حَمَى ظَهْرَهُ [*He prohibited,* or *interdicted, his back* to be used for bearing a rider or any burden], said of a stallion-camel when he is termed حَامٍ, q. v. (Fr, Ṣ, Ḳ.) You say also, حَمَاهُ الشَّىْءَ and حَمَاهُ مِنَ الشَّىْءِ [*He protected, defended,* or *guarded, him from the thing*]. (TA.) And عَنْهُ ▼ حَامَيْتُ, inf. n. مُحَامَاةٌ and حِمَآءٌ, (Ṣ, Ḳ,) *I protected, defended,* or *guarded, him.* (Ḳ.) One says, الضَّرُوسُ تُحَامِى ▼ عَنْ وَلَدِهَا [*The biting she-camel defends her offspring*]. (Ṣ.) And فُلَانٌ ▼ احمى عِرْضَهُ [*Such a one defended his honour,* or *reputation*]. (TA.) And حَمَيْتُ القَوْمَ, inf. n. حِمَايَةٌ, *I aided* [*and defended*] *the people,* or *party.* (Mṣb.) And حَمَيْتُ المَرِيضَ (Ṣ, Mṣb, Ḳ) الطَّعَامَ, (Ṣ,) or مَا يَضُرُّهُ, (Ḳ,) inf. n. حِمْيَةٌ (Ṣ, Mṣb) and حِمْوَةٌ, (Ṣ, TA,) [the latter irreg.,] *I prohibited,* or *interdicted, the sick man,* (Ḳ,) or *ordered him to abstain,* (PṢ,) *from the food,* (PṢ,) or *from what would injure him.* (Ḳ.) ═ حَمِىَ, said of the day, and of an oven, (Ṣ,) and حَمِيَتْ, said of the sun, and of fire, aor. ـَ, (Ḳ,) inf. n. حَمْىٌ (Ṣ, Ḳ) and حُمِىٌّ (Ḳ) and حُمُوٌّ [originally حُمُوىٌ], (Lḥ, Ḳ,) *It was,* or *became, vehemently hot.* (Ṣ, Ḳ.) And حَمِىَ المِسْمَارُ, inf. n. حَمْىٌ and حُمُوٌّ, *The iron nail was,* or *became, hot.* (Ḳ.) And حَمِيَتِ الحَدِيدَةُ *The piece of iron was,* or *became, vehemently hot by means of fire.* (Mṣb.) — حَمِىَ الوَطِيسُ [lit. *The oven became vehemently hot;*] means † *the war,* or *fight, became vehement;* (Ṣ and Ḳ in art. وطس;) and is used as a prov., relating to a severe case or event. (Aṣ, TA in that art.) — حَمِىَ الفَرَسُ, inf. n. حِمًى [and app., accord. to the TA, حَمْىٌ also], *The horse was,* or *became, hot, and sweated.* (Ḳ.) — حَمِيتُ فِى الغَضَبِ [*I was,* or *became, hot in anger*], inf. n. حَمْىٌ. (Lḥ, TA.) And غَضَبًا ▼ احتمى [*He became hot by reason of anger*]. (A in art. لغد.) And حَمِيتُ عَلَيْهِ, accord. to El-Umawee, [حَمِئْتُ,] with hemz, *I was,* or *became, angry with him.* (Ṣ, TA.) And حَمِىَ أَنْفُهُ *He became vehemently angry,* or *enraged.* (IAth, TA in art. انف.) And حَمِىَ عَنْهُ, (Ṣ,) or مِنْهُ, (Ḳ,) or both, (TA,) aor. ـَ, (Ḳ,) inf. n. حَمِيَّةٌ (Ṣ, Mgh,* Mṣb,* Ḳ) and مَحْمِيَةٌ, (Ṣ, Mgh,* Ḳ,) *i. q.* أَنِفَ [*He disdained it; scorned it;* &c.]; (Ṣ, Mgh,* Mṣb,* Ḳ;) *he was ashamed,* and *he disdained,* or *scorned, to do it.* (Ṣ, TA.) And حَمِىَ مِنْ ذٰلِكَ أَنَفًا *He was seized,* or *affected, thereat,* or *by reason of that, with disdain, scorn,* or *indignation.* (TA, from a trad.) And حَمِىَ also signifies *He refused to bear, endure,* or *tolerate, wrongful treatment.* (TA.) ═ See also 4.

3: see 1, in two places. — حَامَيْتُ عَلَى ضَيْفِى *I exerted myself for my guest* [*in paying honour to him, and entertaining him*]. (Ṣ, Ḳ.)

4. احمى: see 1, in five places. ═ Also *He made* the sun, and fire, *to be vehemently hot;* said of God: (Lḥ, Ḳ:) and in like manner, a piece of iron; said of a man: (Mṣb:) [or] *he heated* an iron nail, (ISk, Ḳ,) and a piece of iron, (ISk, Ṣ,) &c., in the fire: (ISk:) one should not say ▼ حَمَى in this sense; (ISk, Ṣ, Mṣb, TA;) app., in chaste speech; for otherwise one does say, حَمَى الشَّىْءَ فِى النَّارِ, meaning *He put the thing into the fire* [*and so heated it*]. (TA.) And احمى المِيسَمَ and احمى عَلَيْهِ *He kindled fire upon the branding-iron* [*and so heated it*]. (Mgh.) — [Hence,] احماهُ عَلَى القِتَالِ [*He excited him to ardour for fight*]. (Ṣ in art. حرض; &c.)

5: see 8.

6. تحاماهُ النَّاسُ *Men guarded against, were cautious of, and kept aloof from,* or *shunned,* or *avoided, him,* or *it.* (Ṣ, Ḳ.)

8. احتمى *He protected, defended,* or *guarded, himself,* [or *he became protected,* &c.,] from a thing. (KL.) — And *He* (a sick man, Ḳ) *refrained, forbore,* or *abstained,* (Ḳ, KL,) مِنَ الطَّعَامِ [*from food,* or *the food*], (Ṣ,) or مِمَّا يَضُرُّهُ [*from what would injure him*]; (TA;) as also ▼ تحمّى. (Ḳ.) أَحْتَمِيَا occurs at the end of a verse, preserving the original form, [for أَحْتَمِىَ,] accord. to a dial. of certain of the Arabs. (Ṣ.) ═ احتمى غَضَبًا: see 1.

12. احموى *It* (a thing, such as the night, and a collection of clouds,) *was,* or *became, black.* (Lth, Ḳ.) [See also the part. n., مُحْمَوٍّ, below: and see the second sentence of the first paragraph of art. حمر.]

حُمَةٌ The *venom,* or *poison,* (Lth, Lḥ, Ṣ, Ḳ,) and *hurt,* (Ṣ,) of a scorpion, (Lth, Ṣ,) and of anything that stings or bites: (Lth:) originally حُمْوٌ or حُمْىٌ: (Ṣ:) and IAṭr mentions حُمَّةٌ [q. v. in art. حم]. (TA.) — And The *sting* of the hornet, (Lth, Ḳ,) and of the scorpion, (Lth, IAth,) and the like, (Lth,) and of the serpent; (Ḳ;) because the venom comes forth from it: (IAth:) so applied by the vulgar: (Lth:) pl. حُمَاتٌ and حُمًى. (Ḳ.) — *Vehemence* of cold. (Ḳ,* TA.)

حَمْىُ الشَّمْسِ: see حَمْوٌ, in art. حمو.

حَمِى وَٱللّٰهِ [or حَمَا واللّه] *i. q.* أَمَا واللّه q. v. (Ṣgh, Ḳ.) ═ الحَمِىُّ [or الحَمَا] for الحَمَامِ: see حَمَام, in art. حم.

حِمًى A thing *prohibited,* or *interdicted;* (Ṣ, Ḳ;) as also ▼ حِمَآءٌ and ▼ حِمْيَةٌ; (Ḳ;) and *not to be approached:* (Ṣ:) [and, as an epithet in which the quality of a subst. is predominant,] *a place of herbage,* or *pasture,* (Lth, Mgh, Mṣb,* TA, and Ḥam p. 589,) *and of water,* (Ḥam ibid.,) *prohibited to the people,* [i. e. *to the public,*] (Lth, Mgh, and Ḥam ubi suprà,) *so that they may not pasture their beasts in it,* (Lth, Mgh,) *nor approach it,* (Mgh, Mṣb,) *nor venture upon it:* (Mṣb:) it was a custom of the noble among the Arabs, in the Time of Ignorance, when he alighted in a district [that pleased him], among his kinsfolk, to incite a dog to bark, and to prohibit for his own special friends or dependents the space throughout which the bark of the dog was heard, so that none else should pasture his beasts there; while he shared with the people in the other places of pasture, around it: but the Prophet forbade this: (Esh-Sháfi'ee, TA:) he said, "There shall be no حمى except for God and for his Apostle;" (Esh-Sháfi'ee, Ṣ, Mgh, TA;) meaning, except for the horses employed in war against the unbelievers and for the camels taken for the poor-rate: (Esh-Sháfi'ee, Mgh, TA:) afterwards, the term was applied in a general sense: (Esh-Sháfi'ee, TA:) the pl. is أَحْمَآءٌ (Ṣ and Ḳ in art. حجر) and أَحْمِيَةٌ: (Ḥam p. 496:) and the dual is حِمَيَانِ and حِمَوَانِ; (ISk, Ṣ, Mṣb, TA;) the latter irreg., (TA,) heard by Ks, but he preferred the former. (Ṣ.) You say, هٰذَا شَىْءٌ حِمًى *This is a thing prohibited,* or *interdicted; not to be approached.* (Ṣ.) And كَلَأٌ حِمًى *Herbage,* or *pasture, that is prohibited,* or *interdicted.* (Ḳ.) ═ See also حِمَآءٌ.

حِمْيَةٌ: see the next preceding paragraph. ═ Also an inf. n. of حَمَى المَكَانَ [q. v.]: (Mṣb:) and of حَمَيْتُ المَرِيضَ [q. v.]. (Ṣ, Mṣb.) ═ Also The *practising abstinence;* (PṢ in art. ازم;) [especially the *abstaining from things injurious in a case of sickness;*] the *abstaining,* or *desisting, from eating.* (TA in that art.)

حِمَآءٌ: see حِمًى ═ حِمَآءٌ لَكَ or حِمَآءً لَكَ *i. q.* فِدَآءٌ لَكَ [May such a person, or thing, be *a ransom for thee!*] or فِدَآءً لَكَ [meaning فُدِيتَ, i. e. *mayest thou be ransomed!* or فَدَاكَ, i. e. *may* such a one *ransom thee!*]. (Ṣ, accord. to different copies.) [And in like manner,] the Arabs said, لَكَ الفِدَى وَالحِمَى ✻ [*Ransom, or ransoming, be for thee!*] pronouncing the former noun with the short alif when thus coupling it with الحِمَى. (El-Ḳálee, TA in art. فدى.)

حَمِىٌّ A sick man *prohibited,* or *interdicted, from what would injure him,* (IAạr, Ḳ,) *of food and drink.* (IAạr.) — *Protected, defended,* or *guarded* (Ṣ,* Mgh, Ḳ, TA) from evil, &c. (TA.) 'Ásim Ibn-Thábit El Anṣáree was called حَمِىُّ الدَّبْرِ [*The protected by hornets,* or *by the swarm of bees*], (Ṣ, Mgh,) because his corpse was protected from his enemies by large hornets, (Ṣ in art. دبر,) or by a swarm of bees. (Mgh.) ═ One *who will not bear, endure,* or *tolerate, wrongful treatment.* (Ḳ.) And حَمِىُّ الأَنْفِ A man *who refuses to submit to wrongful treatment.* (TA. [See also أَنِفٌ.])

حَمِيَّةٌ an inf. n. of حَمِىَ: (Ṣ, Ḳ: [see حَمِىَ عَنْهُ:]) *Disdain, scorn,* or *indignation;* and *anger;* syn. أَنَفَةٌ, (Mgh, Mṣb, TA,) and غَيْرَةٌ; (TA;) because a means of protection: (Mgh:) *care of what is sacred,* or *inviolable,* or *of what one is bound to respect, or honour, and to defend, and of religion, to avoid suspicion.* (KT.)

حُمَيَّا The *vehemence of anger;* and the *commencement* [or *outburst*] *thereof:* (Ḳ:) *spirit,* and *anger;* as in the saying, إِنَّهُ لَشَدِيدُ الحُمَيَّا [*Verily he is vehement in spirit,* and *in anger*]. (TA.) — The *assault* of wine *upon the head;* or its *rush into the head:* (Ḳ:) or the *beginning of* its *assault upon,* or *rush into, the head:* (Ṣ:) and its *force,* or *vehemence:* or its *intoxicating operation:* or its *overpowering influence upon the head;* (Ḳ;) or *upon the drinker:* (Lth, TA:) or the *creeping* [*of the fumes*] of wine [*through the drinker*]. (A 'Obeyd, TA.) One says, سَارَتْ فِيهِ حُمَيَّا الكَأْسِ, meaning [*The fumes of*] *the cup of wine mounted into his head.* (TA.) And حُمَوَّةٌ [originally حُمَوْيَةٌ] signifies The *assault,* or *attack,* of pain. (Ṣ, TA.) — The *prime,* and *sprightliness,* of youth; (Ḳ:) and the *flush,* or *impetuosity,* (سَوْرَة,) thereof. (TA.) You say, فَعَلَ ذٰلِكَ فِى حُمَيَّا شَبَابِهِ *He did that in the flush,* or *impetuosity,* (سورة,) *and sprightliness, of his youth.* (TA.) ═ هُوَ حَامِى ✻ الحُمَيَّا *He is the protector, defender,* or *guarder* [*from encroachment*], *of that which he possesses, and of which he has the superintendence,* or *management.* (Ṣ, Ḳ.)

حَامٍ [act. part. n. of حَمَى. And hence,] *A stallion-camel that has prohibited,* or *interdicted, his back* [to be used for bearing a rider or any burden]; (حَمَى ✻ ظَهْرَهُ; Fr, Ṣ, Ḳ;) *that is not ridden,* (Fr, Ṣ, Mgh,) *nor shorn of any of his fur;* (Fr, Ṣ;) *that is left at liberty, not made any use of,* (Ḳ,) *nor debarred from pasturage* (Fr, Ṣ, Mgh, Ḳ) *nor from water:* (Ḳ:) he is *one that has long continued with a people:* (Ṣ:) or *whose offspring's offspring has conceived:* (Fr, Ṣ, Mgh:) or *that has covered a certain number of times,* or *ten times:* (Ḳ:) it is mentioned in the Ḳur [v. 102]. (Ṣ, Mgh. [See also بَحِيرَةٌ.]) — الحَامِى is also an appellation applied to *The lion;* and so ✻ المُحْمِىُّ; (Ḳ;) in the Tekmileh, الحامى and ✻ المُحْمِى [app. المُحْمِى, as in a copy of the Ḳ]. (TA.) — You say also, فُلَانٌ حَامِى الحَقِيقَةِ [*Such a one is the protector,* or *defender, of that which,* or *those whom, it is necessary for him,* or *incumbent on him, to protect,* or *defend*]; like حَامِى الذِّمَارِ [q. v. in art. ذمر]; and حَامِى ✻ الحُمَيَّا [explained in the next preceding paragraph]: pl. حُمَاةٌ and [coll. gen. n.] ✻ حَامِيَةٌ: (Ṣ:) this last word signifies *a company,* or *party, protecting,* or *defending, their companions,* (Ḳ,) or *themselves:* (TA:) and also a man who is *a protector,* or *defender, of his companions* (Ḳ) *in war:* (TA:) or *a strenuous protector and defender* of a party; for the ة is to give intensiveness to the signification: (Mgh:) and you say, هُوَ عَلَى حَامِيَةِ القَوْمِ, meaning *He is the last of those who protect, or defend, the party in their going away* (Ḳ) *and in their state of defeat.* (TA.) ═ [*Hot:* or *vehemently hot.*] You say حَدِيدَةٌ حَامِيَةٌ *A piece of iron vehemently hot by means of fire.* (Mṣb.) And قِدْرُ القَوْمِ حَامِيَةٌ تَفُورُ *The people's cooking-pot is hot, boiling:* meaning † *the people are mighty, strong,* or *invincible, and vehemently impetuous in valour.* (TA.)

حَامِيَةٌ: see حَامٍ. — Also *A great,* or *wide,* or *great and wide, mass of stone,* (حِجَارَةٌ, Ḳ accord. to the TA,) or the *stones,* (حِجَارَة, so in some copies of the Ḳ,) *with which a well is cased:* (Ḳ:) pl. حَوَامٍ: (TA:) or the latter signifies the *stones &c. with which a well is cased, to protect its sides from becoming dirty and disordered:* (Ḥam p. 62:) or *great and heavy stones:* and also *large masses of rock which are placed in the last parts of the casing* [*of a well*] *if it falls out through age: they dig out hollows, and build them therein, so that they suffer not the earth to come near to the casing, but repel it:* (ISh:) and *all the stones* [*of the casing*] *of a well, matching one another, none of them larger than another.* (AA.) — The *circuit of the solid hoof:* (Ḥam p. 62:) or [the dual] حَامِيَتَانِ signifies the *part on the right and left of the toe of the solid hoof:* (AO, Ṣ:) or [the pl.] حَوَامٍ signifies the *right and left edges of the solid hoofs;* (Aṣ, TA;) *between them are* [*the*] نُسُور [or *frogs*], *like hard date-stones:* (Aboo-Dáwood, TA:) or the *right and left sides of the solid hoof.* (Ḳ.) — [The pl. also signifies The *sides of a mountain.* (Freytag, from the Deewán of Jereer.)] — Also, the sing., *i. q.* أُثْفِيَّةٌ [i. e. *Any one of the three stones on which the cooking-pot is placed*]: (AA, Ḳ:) pl. as above. (TA.) — مَضَيْتُ عَلَى حَامِيَتِى means *I went my own way.* (Ṣgh, Ḳ.)

هُوَ أَحْمَى أَنْفًا مِنْ فُلَانٍ (Ṣ, TA) *He is more resistive than such a one.* (TA.)

المُحْمِى: }
المُحْمِىُّ: } see حَامٍ.

مُحْمَوْمٌ *Black;* applied to such a thing as the night, and a collection of clouds: or, applied to the latter, *heaped up, and black.* (Lth.)

حن

1. حَنَّ, aor. ـِ, inf. n. حَنِينٌ, *He was,* or *became, affected with* [*a yearning, longing,* or *desire,* or] *an intense emotion of grief* or *of joy;* as also ✻ اِسْتَحَنَّ and ✻ تَحَانَّ. (Ḳ.) [See an instance of its denoting an emotion of joy voce حَنَّانٌ.] You say, حَنَّ إِلَيْهِ, aor. and inf. n. as above, *He,* or *his soul, yearned towards, longed for,* or *desired, him,* or *it.* (Ṣ.) And حَنَّ إِلَى وَطَنِهِ *He yearned towards, longed for,* or *desired, his home.* (TA.) And تَحِنُّ إِلَى زَوْجِهَا الأَوَّلِ [*She yearns towards her first,* or *former, husband*]. (TA.) And حَنَّتْ, inf. n. as above, *She* (a woman) *yearned towards, longed for,* or *desired, her child,* or *children.* (Mṣb.) So, too, one says of a she-camel, meaning *She yearned towards, longed for,* or *desired, her home,* or *her young one;* and in like manner, of a pigeon: but in most instances it means *she* (a camel) *yearned with a cry,* or *uttered a cry when yearning* [or *a yearning cry* or *the cry produced by yearning*], towards her young one or her companions: or *she uttered a cry with emotion* after her young one: in its primary sense, *she reiterated her* [*yearning*] *cry after her young one:* but when you say, حَنَّ قَلْبِى إِلَيْهِ, you mean *My heart yearned towards, longed for,* or *desired, him,* or *it,* without the uttering of a cry or sound. (TA.) They said also, لَا أَفْعَلُهُ حَتَّى يَحِنَّ الضَّبُّ فِى أَثَرِ الإِبِلِ الصَّادِرَةِ [*I will not do it until the lizard called* ضَبّ *yearn after the camels returning from the water;* meaning *I will never do it*]: this is only a prov.; for the ضبّ has no حَنِين, nor does it ever go to the water. (TA.) [And حَنَّ العُودُ, aor. and inf. n. as above, † *The lute produced plaintive sounds:* or *excited lively emotions of sadness,* or *of mirth:* see حَنَّانٌ.] And in like manner one says of a musical reed: see

مُنْقَلِب.] And حَنَّتِ القَوْسُ, (Ḳ,) aor. as above, (Ṣ,) and so the inf. n., (TA,) †*The bow* [*twanged,* or] *made a sound* (Ḳ, TA) when its string had been pulled and then let go. (Ṣ.) And حَنَّتِ الطَّسْتُ إِذَا نُقِرَتْ ‡ [*The brazen basin rang when it was knocked,* or *pecked*]. (TA.) And حَنَّتِ السَّارِيَةُ [*The mast creaked,* or *made a creaking sound*]. (TA in art. صر.) And حَنَّ قِدْحٌ لَيْسَ مِنْهَا † [*An arrow of those used in the game called* المَيْسِر *produced a sound: it was not of them*]: a prov., applied to a man who ascribes to himself a false origin, or who arrogates to himself that to which he has no relation: by the قدح is meant one of the arrows of the ميسر; for when this is not of the same substance as the others, and is made to vibrate, it produces a sound different from the sounds of the others, and is known thereby. (TA. [See also Freytag's Arab. Prov. i. 341.]) And حَنَّتِ الرِّيحُ and ▼ استحنّت † [*The wind made a plaintive,* or *moaning,* or perhaps *a shrill, sound; made a sound like the* حَنِين *of camels:* see حَنُونٌ]: both signify the same. (TA.) [See also حَنِينٌ, below.] — حَنَّ عَلَيْهِ, (Ṣ, Mṣb, TA,) aor. as above, (Ṣ, Mṣb,) inf. n. حَنَانٌ (Ṣ, Mṣb, Ḳ*) and حَنَّةٌ (Mṣb, TA*) and حَنٌّ, (Ḳ,* TA,) *He was merciful, compassionate,* or *pitiful, towards him,* or *it;* (Ṣ, Mṣb, Ḳ,* TA;) as also ▼ تحنّن: (Ṣ, Ḳ:) *he was,* or *became, favourably inclined towards him,* or *it;* (Mṣb;) and so ▼ تحنّن: (TA:) *he was,* or *became, affectionate,* or *pitiful,* or *compassionate, towards him;* (Ḳ,* TA;) as also ▼ حَنْحَنَ. (IAạr, Az, Ḳ.) And تَحَنَّنَتْ ▼ عَلَى وَلَدِهَا *She* (a camel, and a ewe or goat,) *became favourably inclined,* or *compassionate, towards her young one.* (Lḥ, TA.) — See also 2. — حَنَّ عَنِّى, aor. ـُ, means صَدَّ; (Ṣ;) i. e. *He turned away from me, avoided me,* or *shunned me:* so that it is anomalous; for by rule the aor. should be ـِ; and it is not mentioned among the exceptions [to the rule applying to a case of this kind]. (MF, TA.) [But it appears from what here follows that صَدَّ may perhaps be here meant to be understood in its trans. sense.] — حَنَّهُ, (Ḳ,) [aor. ـُ, as is shown below,] inf. n. حَنٌّ, (TA,) signifies صَدَّهُ and صَرَفَهُ [*He turned him,* or *it, away,* or *back*]. (Ḳ, TA.) You say, حُنَّ عَنِّى شَرَّكَ, inf. n. حَنٌّ, *Turn thou away,* or *back, from me thy evil,* or *mischief.* (Ḳ.) And مَا تَحُنُّنِى شَيْئًا مِنْ شَرِّكَ *Thou dost not turn away,* or *back, from me aught of thy evil,* or *mischief.* (Ṣ.) — [حُنَّ, inf. n. حَنٌّ, app. *He was,* or *became, possessed by a demon,* or *by one of the tribe* or *kind* or *class termed* الحِنّ; and hence, *he was,* or *became, mad,* or *insane:* for] حُنَّ is *syn. with* جُنُونٌ; (TA as from the Ḳ; [but not in the CḲ nor in my MS. copy of the Ḳ;]) whence مَحْنُونٌ applied to a man [as meaning مَجْنُونٌ]. (TA.)

2. حَنَّنَهُ عَلَى غَيْرِهِ, accord. to modern usage, and perhaps classical also, *He,* or *it, caused him to be merciful, compassionate, pitiful,* or *favourably inclined, towards another.*] — حَمَلَ فَحَنَّنَ *He charged,* or *made an assault* or *attack, and was cowardly, and retreated.* (Ḳ, TA.) — مَا حَنَّنَ عَنِّى *He did not turn away from me; did not leave,* or *relinquish, me.* (TA.) [And ▼ احنّ, or ▼ حَنَّ, seems to have a similar meaning: for] you say, أَثَرٌ لَا يُحِنُّ عَنِ الجِلْدِ *A mark that does not go away from the skin:* or, accord. to Th, who does not explain it, it is يَحِنُّ. (TA.) — حَنَّنَتِ الشَّجَرَةُ *The tree blossomed,* or *flowered:* (Ḳ:) and in like manner one says of a herb. (TA.)

4. احنّ القَوْسَ *He made the bow to* [*twang,* or] *give a sound,* [by pulling, and then letting go, the string.] (Ḳ.) — And احنّ *He* (a man, TA) *did wrong, committed a mistake,* or *missed* [the object of his aim]. (Ḳ.) — See also 2.

5: see 1, in three places.

6: see 1, first sentence.

10: see 1, in two places. — استحنّهُ الشَّوْقُ إِلَى وَطَنِهِ [*Longing for his home affected him with intense emotion*]. (IB, TA.)

R. Q. 1. حَنْحَنَ: see 1.

الحِنُّ *A tribe of the* جِنّ [or *genii*], (Ṣ, Ḳ,) *that were before Adam;* (TA;) *of which are black dogs:* (Ḳ:) or the *lowest,* or *meanest, sort of the* جِنّ: (Ḳ:) or the *weak ones thereof:* (IAạr, Ḳ:) or the *dogs thereof:* (Fr, TA:) or *certain creatures between the* جِنّ *and mankind.* (Ṣ, Ḳ.)

حَنَّةٌ: see حَنَانٌ. — It is said in a prov., لَا تَعْدَمُ نَاقَةٌ مِنْ أُمِّهَا حَنَّةً, and ▼ حَنِينًا, meaning [*The she-camel will not be without*] *likeness* [*to her mother*]: and one says of a man who resembles another man, and of any one who resembles his father and his mother, لَا تَعْدَمُ أَدْمَاءُ مِنْ أُمِّهَا حَنَّةً [*A female camel of the colour termed* أُدْمَة (i. e. *white,* or *very white,* &c.,) *will not be without likeness to her mother*]. (TA. [See also Freytag's Arab. Prov. ii. 497.]) — The [*grumbling cry termed*] رُغَاء of a camel. (Ṣ, Ḳ.) — *A man's wife.* (Ṣ, Ḳ.) — See also what next follows.

حِنَّةٌ: see حَنَانٌ. — Also *i. q.* جِنَّةٌ; (Ṣ, Ḳ;) as also ▼ حُنَّةٌ: (Ḳ:) so in the phrase, بِهِ حِنَّةٌ [*In him is demoniacal possession,* or *madness,* or *insanity:* see الحِنُّ]. (Ṣ.)

حَنَانٌ *Mercy, compassion,* or *pity:* (Ṣ, Ḳ:) *tenderness of heart;* (Ḳ;) which is the same; (TA;) as also ▼ حِنَّةٌ, with kesr; (Kr, TA;) for which the vulgar say ▼ حَنِيَّةٌ: (TA:) and ▼ حَنَّةٌ [in like manner] signifies *affection,* and *compassion.* (Az, TA.) وَحَنَانًا مِنْ لَدُنَّا, in the Ḳur [xix. 14], respecting which I 'Ab is related by 'Ikrimeh to have said, I know not what is الحَنَانُ, means *And mercy from us.* (Ṣ, TA.) The Arabs say, حَنَانَكَ يَا رَبِّ and حَنَانَيْكَ [I beg *thy mercy, O my Lord*]: both signify the same; i. e. رَحْمَتَكَ: (Ṣ:) the latter is the expression commonly used: (A 'Obeyd, in a marginal note in a copy of the Ṣ:) or [rather] the latter means *have mercy on me time after time,* and *with mercy after mercy:* (Ḳ,* TA:) it is a dualized inf. n., of which the verb is not expressed; like لَبَّيْكَ and سَعْدَيْكَ: (TA:) or it means [*let thy mercy be continuous to me;*] *whenever I receive mercy and good from Thee, let it not cease, but be conjoined with other mercy from Thee:* (ISd, TA:) the dual form is not to be understood as restricting the signification to duality: (Suh, TA:) the word is not used in this form otherwise than as a prefixed noun: (Sb, TA:) but sometimes they said حَنَانًا, in the sing., without prefixing it. (ISd, TA.) They said also, سُبْحَانَ اللّٰهِ وَحَنَانَيْهِ, meaning [*I extol,* or *celebrate,* or *declare, the absolute purity,* or *perfection,* or *glory, of God,*] *and I beg his mercy;* like as they said, سُبْحَانَ اللّٰهِ وَرَيْحَانَهُ. (TA.) And حَنَانَ اللّٰهِ as meaning مَعَاذَ اللّٰهِ [*I seek the protection,* or *preservation, of God*]. (Ḳ.) — Also *i. q.* رِزْقٌ [*Means of subsistence,* &c.]: and بَرَكَةٌ [*a blessing; any good that is bestowed by God; prosperity,* or *good fortune; increase;* &c.]. (Ḳ.) — *A quality inspiring reverence* or *veneration* or *respect* or *honour:* (El-Umawee, Ḳ:) *gravity, staidness,* or *sedateness.* (Ḳ.) One says, مَا تَرَى لَهُ حَنَانًا *Thou seest him not to possess any quality inspiring reverence* &c. (El-Umawee, TA.) — *Evil,* or *mischief, long continuing.* (Ḳ.)

حَنُونٌ A wind (رِيحٌ) [*that makes a plaintive,* or *moaning,* or perhaps *a shrill, sound;*] *that makes a sound like the* حَنِين *of camels.* (Ṣ, Ḳ, TA.) — A woman *who marries from a motive of tenderness,* or *compassion, for her children,* (Ḳ, TA,) *when they are young,* (TA,) *in order that the husband may maintain them.* (Ḳ, TA.)

حَنِينٌ an inf. n. of 1: (Ṣ, Mṣb, Ḳ:) *A yearning, longing,* or *desire;* (Ṣ, Ḳ;) *a yearning,* or *longing, of the soul:* (Ṣ:) or the *expression of pain arising from yearning* or *longing* or *desire:* (Ḥam p. 538:) *violence of weeping:* and *a lively emotion:* or the *sound produced by such emotion, proceeding from grief,* or *from joy:* (Ḳ:) or *a sound proceeding from the bosom on the occasion of weeping:* حَنِين is from the nose: (TA:) or the former is [*a sound*] *without weeping and without tears:* if with weeping, it is termed خَنِينٌ: (R, TA:) or the former is *a yearning,* or *longing,* or *desire, with affection,* or *pity,* or *compassion;* as when one speaks of the حنين of a woman and of a she-camel for her young one: and sometimes this is *accompanied with a sound,* or *cry;* wherefore it is explained as *a sound,* or *cry, indicating yearning* or *longing* or *desire, and affection* or *pity* or *compassion:* and sometimes it is confined to the form; as in the case of the حنين [or *leaning,* or *inclining,*] of the trunk of a palm-tree [which is mentioned in a trad.]: (Er-Rághib:) the حنين of the she-camel is her *cry in her yearning towards her young one:* (Ṣ:) or her *yearning towards her young one with a cry,* and *without a cry;* (Lth, TA;) mostly the former: originally, her *reiterating her* [*yearning*] *cry after her young one.* (TA.) You say also, رِيحٌ لَهَا حَنِينٌ كَحَنِينِ الإِبِلِ † [*A wind that has a plaintive,* or *moaning,* or perhaps *a shrill, sound, like the* حنين *of camels*]. (Ṣ, Ḳ.*) — See also حَنَّةٌ. — And الحَنِينُ, and ▼ جُنَيْنٌ and الجُنَيْنُ, two names of [*The* *months called*] جُمَادَى الأُولَى and الآخِرَة : (K:) or حَنِينٌ is a name of جمادى الاولى, like a proper name; as also الحَنِينُ : (M, TA:) or the name by which the tribe of 'Ád called جمادى الآخرة : (Ibn-El-Kelbee, in TA voce مُؤْتَمِر : see شَهْرٌ :) or, accord. to Fr and El-Mufaddal, the Arabs used to call this month ▼ حُنَيْنٌ : (T, TA:) pl. [of pauc.] أَحِنَّةٌ and [of mult.] حُنُونٌ and حَنَائِنُ. (K.)

حُنَيْنٌ : see what next precedes.

كَلْبٌ جِنِّىٌّ *A dog of the tribe of the* جِنّ *called* الجِنّ. (TA.)

حَنَانٌ : see حَنِينٌ.

حَنَّانٌ One *who yearns towards, longs for,* or *desires, a thing,* (K,) and *inclines to it.* (TA.) [Hence,] حَنَّانَةٌ A woman *who remembers a former husband with yearning* (الحَنِين) *and grieving,* or *moaning,* (K, TA,) *in tenderness for her children, when they are young, that the husband may maintain them;* like أَنَّانَةٌ : or *who yearns towards her former husband, and inclines to him:* or *who yearns towards her child,* or *children, by her husband who has separated from her:* (TA:) or a woman *who yearns towards her former husband, and grieves for him:* or *who marries, having been divorced, and yearns towards him who has divorced her.* (Har p. 569.) And † *A bow;* (K;) [because of the sound made by the twanging of its string;] accord. to AHn, as a proper name; but ISd holds it to be, when thus applied, an epithet in which the quality of a subst. is predominant: (TA:) or a bow *that* [*twangs,* or] *makes a sound* (S, K) *when its string has been pulled and then let go.* (S.) And عُودٌ حَنَّانٌ ‡ [*A lute that produces plaintive sounds:* or] *that excites lively emotions of sadness,* or *of mirth.* (TA.) And سَحَابٌ حَنَّانٌ † *Clouds that have* [or *produce*] *a* حَنِين [or *moaning sound,* by their thunder heard from a distance,] *like the* حَنِين *of camels.* (TA.) And سَهْمٌ حَنَّانٌ † *An arrow that produces a sound when thou triest its sonorific quality by turning it round between thy fingers:* (AHeyth, K, TA: [in the CK, نَقَرْتَهُ is erroneously put for نَقَّرْتَهُ :]) or *that produces a sound when it is turned round* (أُدِيرَ [or أُدِرَ]) *with the ends of the fingers upon the thumbs, by reason of the excellence and compactness of its wood.* (TA. [See دَرَّ السَّهْمُ, in art. در.]) And خِمْسٌ حَنَّانٌ ‡ *i. q.* بَائِصٌ [*A hurrying,* or *hard, journey in which the camels are watered only on the first and fifth days:* (in the CK and a MS. copy of the K, erroneously, نابض :)] (K, TA) i. e. (Aç, TA) *in which there is a* حَنِين [or *yearning* of the camels] *by reason of its quickness;* (Aç, K, TA;) or *in which the camels yearn* [*towards their accustomed places*] (تَحِنُّ) *by reason of fatigue.* (A, TA.) And طَرِيقٌ حَنَّانٌ ‡ *A conspicuous road,* (S, K, TA,) *in which the old camel becomes joyous* (يَحِنُّ, i. e. يَنْبَسِطُ): or, accord. to the A, *a road in which there is* [*heard*] *a* حَنِين [or *yearning cry*] *of the camels;* like طَرِيقٌ نَهَّامٌ meaning a road in which is [heard] a نَهِيم [or chiding] of camels. (TA.) — One *who shows favour,* or *presents a favourable aspect, to him who turns from him,* or *shuns him.* (K.) — *Merciful,* or *having mercy.* (S.) [Hence,] الحَنَّانُ a name of God; (K;) meaning *The Merciful* (Aboo-Is-hák, Az, IAth, K) *to his servants.* (IAth, TA.)

حِنَّانٌ *i. q.* حِنَّآءٌ [*Lawsonia inermis,* or *Egyptian privet,* mentioned in art. حنأ]; (K;) a dial. var. of the latter: (Fr, Th, TA:) and حُنَّانٌ is said to be a pl.; (TA in the present art.;) i. e. of حِنَّآءٌ, anomalously; or a dial. var. thereof. (TA in art. حنأ.) [See also what next follows.]

حُنُونٌ *i. q.* فَاغِيَةٌ [The *flower of the* حِنَّآء]: or the *flower of any tree* (K) and *plant:* n. un. with ة. (TA.) [See also what next precedes.]

حَنِينٌ and الجِنِّينُ : see جِنِّينٌ.

حَانٌّ *Yearning, longing,* or *desiring:* (S:) or *being affected with an intense emotion of grief* or *of joy.* (K.) — [Hence, the fem.] حَانَّةٌ signifies *A she-camel;* [because of her yearning towards her young one;] (S, K;) as also ▼ مُسْتَحَنٌّ, (as in some copies of the S,) or ▼ مُسْتَحِنٌّ, (as in other copies of the S and in the K,) [both of which may be correct, as استحنّ is both trans. and intrans.:] or مُسْتَحَنٌّ signifies one who is *affected with intense emotion by longing for his home* (الَّذِى اِسْتَحَنَّهُ الشَّوْقُ إِلَى وَطَنِهِ). (IB, TA.) One says, مَا لَهُ حَانَّةٌ وَلَا آنَّةٌ *He has not a she-camel nor a sheep,* or *goat.* (S, TA.) [See also آنٌّ.] AZ mentions the saying, مَا لَهُ حَانَّةٌ وَلَا جَارَّةٌ, as meaning *He has not camels that yearn* [towards their young ones] (تَحِنّ) *nor such as carry goods,* or *furniture and utensils, and wheat,* or *food.* (TA.)

مَحْنُونٌ, applied to a man, (S,) *i. q.* مَجْنُونٌ [properly *Possessed by a* جِنِّىّ; and hence, *mad,* or *insane*]: (S, K: [see الجِنّ :]) or *i. q.* مَصْرُوعٌ [as meaning *affected with epilepsy*]: (K:) or one *who is affected with epilepsy* (يُصْرَعُ) *and then revives for a time.* (AA, TA.)

مُسْتَحَنٌّ, or مُسْتَحِنٌّ : see حَانٌّ.

حنأ

1. حَنَأَ, aor. ـَ, *It* (a place) *became green, and tangled,* or *luxuriant,* or *abundant and dense, in its herbage,* or *plants.* (K.) ═ حَنَأَتْ يَدَهَا : see 2. — حَنَأَهَا *He lay with her.* (K.)

2. حَنَّأَ, inf. n. تَحْنِئٌ and تَحْنِئَةٌ *He dyed* (AZ, S, K) his head, (AZ, TA,) or his beard, (S,) *with* حِنَّآء : (AZ, S, K:) and حَنَّأَتْ يَدَهَا *She* (a woman) *dyed her hand therewith;* as also ▼ حَنَأَتْهَا, aor. ـَ. (Msb.)

5. تَحَنَّأَ *It* (his head, or his beard,) *was dyed with* حِنَّآء. (AHn, K.)

حِنَّآءٌ [*A certain plant*] *well known;* (S, K;) [the *Lawsonia inermis,* or *Egyptian privet;*] *used for dyeing the extremities* [i. e. *the hands and feet and head*]: (TA:) [in the present day, the plant itself is called تَمْرُ الحِنَّآءِ, (vulgo تَمْر حِنّا,) and its leaves, used for dyeing the hands &c., are called حِنَّآء :] accord. to some, it is the pl. of حِنَّآءَةٌ; [or rather a coll. gen. n., of which حِنَّآءَةٌ is the n. un.;] but it is generally asserted that حِنَّآءَةٌ is a more special word than حِنَّآء, [as in the S and Msb,] and not the sing. of the latter: (TA:) pl. حُنَّانٌ. (K.) حُنَّانٌ is said to be an anomalous pl. of حِنَّآء; or a dial. var. of the latter, and not a pl.: (TA:) and حِنَّانٌ is a dial. var. of حِنَّآءٌ. (Fr, Th, TA in art. حن.)

حِنَّائِىٌّ *A seller of* حِنَّآء : pl. حِنَّائِيُّونَ. (K.)

أَخْضَرُ حَانِئٌ *Very green; intensely green.* (K, TA.)

حنب

1. حَنِبَ : see the next paragraph.

2. حَنَّبَهُ, inf. n. تَحْنِيبٌ, *It* (old age, TA) *bent him down.* (K,* TA.) [See its quasi-pass., 5.] — حَنَّبَ أَزَجًا *He built firmly a structure of the kind called* أَزَج, *and made it curved.* (K.) — تَحْنِيبٌ [as an inf. n. of which the verb, if it have one in any of the following senses, is حُنِّبَ,] also signifies *A bending,* or *curving, and tension* [*of the sinews*] (تَوْتِيرٌ), *of the backbone* (صُلْب) *and fore legs of a horse:* (Aç, S:) or *a convexity in the shank* (وَظِيف) *of each of the fore legs of a horse,* (K,* TA,) *not being a great curvature,* (TA,) *and in the backbone* (صُلْب): (K:) it is a quality indicative of strength: (Az, TA:) تَحْنِيبٌ (with ح) is [the same] in the hind legs: (Aç, S, K:) or it [i. e. تحنيب as indicated in the K and by an explanation of مُحَنَّبٌ, but the same explanation is also given to مُجَنَّبٌ,] is *width in the space between the hind legs, without what is termed* فَحَج [i. e. *straddling*], or فَجَج [i. e. the *having the fore parts of the feet near together and the heels distant,* or *having the legs wide apart* (like فَحَج), or *having the thighs* or *the middles of the legs wide apart*]; (so accord. to different copies of the K;) which is a quality approved: (TA:) or *a curving in the* سَاقَانِ [here app. meaning the *hind legs,* or rather the *hind shanks;* see مُحَنَّبٌ]: as also حَنَبٌ [an inf. n. of which the verb, if it have one, is app. ▼ حَنِبَ]: (K:) or *a curving of the ribs.* (TA.) [See تَجْنِيبٌ: and see also مُحَنَّبٌ.]

5. تحنّب *He* (a man, S) *was,* or *became, crooked, curved,* or *bent.* (S, K.) — [And hence,] تحنّب عَلَيْهِ ‡ *He was,* or *became, affected with compassion for him.* (K, TA.)

حُنْبَاةٌ : see what follows.

مُحَنَّبٌ An old man *bent* (K, TA) *with age.* (TA.) — A horse *characterized by what is termed*

قَحْنِيبٌ: (K:) [see 2:]) accord. to A 'Obeyd, *wide in the space between the hind legs, without* فَحَجٍ, or فَحْجٍ: (so accord. to different copies of the S: [see these terms explained above, conj. 2:]) an epithet of commendation: (S:) accord. to ISh, *having curved bones:* the mare, he says, is termed ✝حَنْبَآءُ: and this latter, accord. to As, is an epithet applied to a mare *curved in the* سَاقَانِ [or *shanks*] *of the fore legs:* or, accord. to IAar, *in* [*the shanks of*] *the hind legs:* or, as he says in another place, *curved in the* ساق [here app. meaning the *hind shank*]. (TA.) [See also مُجَنَّبٌ.]

حنبل

حَنْبَلٌ, [mentioned in the S and Msb in art. حبل,] applied to a man, (S, Msb,) *Short:* (S, Msb, K:) and, (K,) as some say, (Msb,) *large in the belly,* (Az, ISd, Msb, K,) *and short:* (Az, ISd, Msb:) and [in the CK "or"] *fleshy;* (ISd, K;) as also ✝حِنْبَالٌ. (K.) — *A fur-garment:* (Az, S, K:) or *an old and worn-out fur-garment.* (ISd, K.) — *An old and worn-out boot.* (ISd, K.) — The *sea;* as also ✝حِنْبَالَةٌ. (ISd, K.)

حِنْبَالٌ: see above. — Also, (T, O, TA,) and ✝حِنْبَالَةٌ, (T, O, K,) [but the latter has a more intensive signification,] *Loquacious; a great talker.* (T, O, K.)

حِنْبَالَةٌ: see حَنْبَلٌ: — and see also حِنْبَالٌ.

حنت

حَانُوتٌ and حَانُوتِيٌّ: see arts. حون and حنو.

حنتم

حَنْتَمٌ *A green* جَرَّة [or *jar*], (S, K,) to which some add, *inclining to redness:* (TA:) or *wine-jars,* (A 'Obeyd, Nh,) *glazed,* or *varnished, green,* (Nh,) *which used to be carried to El-Medeeneh, with wine in them:* (A 'Obeyd, Nh:) the use of which, for preparing نَبِيذ therein, is forbidden in a trad., because it quickly became potent in them, by reason of the glazing, or varnish; or, as some say, because they used to be made of clay kneaded with blood and hair; but the former is the right reason: afterwards applied to *any jars,* or *pottery:* (Nh:) thus some explain it as a sing.; (MF;) and the pl. is حَنَاتِمُ: (Az, TA:) others, as a pl. [or coll. gen. n.], of which the sing. [or n. un.] is with ة: (MF:) some say that the ن is augmentative: so says the author of the Msb: others, that it is radical. (TA.) [See art. حتم.] — *Black clouds;* (Az, K;) as also [the pl.] حَنَاتِمُ: (Az, S, K:) because, with the Arabs, السَّوَادُ is [used for] خُضْرَةٌ: (S: [see أَسْوَدُ; and see also حَنْتَمٌ in art. حتم:]) or as being likened to حَنَاتِم (meaning jars) filled [with water]: (Az, TA:) n. un. with ة. (K.) — The *colocynth-plant;* (K, TA;) because of its intense greenness: n. un. with ة. (TA.)

حنث

1. حَنِثَ فِى يَمِينِهِ, (S, A, Msb, K,*) aor. ـَ, (Msb, K,) inf. n. حِنْثٌ (S,* A, Msb, K*) and حَنَثٌ, (TA,) *He violated,* or *broke,* or *failed of performing, his oath:* (A, Msb, K, TA:) *he was untrue in his oath:* (S, TA:) *he committed a sin,* or *crime, in his oath.* (TA.) Also *He retracted,* or *revoked, his oath.* (TA.) — And the verb alone, *He said what was not true.* (Khálid Ibn-Jembeh, TA.) — *He inclined from what was false to what was true:* or *from what was true to what was false.* (K.) — [And app. † *He committed a sin,* or *crime:* a meaning which seems to be indicated in the K: see حِنْثٌ, below.]

2. حَنَّثَهُ *He pronounced him* (جَعَلَهُ) *a violator,* or *non-performer, of his oath:* (Msb:) [a verb similar to أَثَّمَهُ and فَسَّقَهُ &c.]

4. أَحْنَثْتُهُ (S, K) فِى يَمِينِهِ (S) *I made him to violate,* or *break,* or *fail of performing, his oath.* (S, K.*) — احنث فُلَانٌ قَسَمَ فُلَانٍ *Such a one assented not,* or *consented not, to the conjurement of such a one; contr. of* أَبَرَّ. (T and TA in art. بر.)

5. تَحَنَّثَ † *He put away,* or *cast away, from himself* الحِنْث, i. e. *sin,* or *crime:* (ISd, Towsheeh, TA:) it is said that there are only six verbs of the measure of this signifying the putting away, or casting away, from oneself a thing, [i. e. the thing denoted by the root,] which are تحنّث and تأثّم and تحوّب and تحرّج and تنجّس and تهجّد: (TA:) † *he did a work whereby to escape from sin,* or *crime:* (IAar, Msb:) † *he applied himself to acts,* or *exercises, of devotion;* (S, A, Msb, K;) accord. to Ez-Zuhree and the K, during numerous nights; but it has been shown by the Expositors of El-Bukháree, and others, that this addition is taken from words following the verb in the explanation of a trad., and has nothing to do with the meaning of the verb itself: (MF, TA:) † *he relinquished* [*the worship of*] *idols;* (S, K;) like تَحَنَّفَ; (S;) and it may be that the ث in this verb is interchangeable with ف, (A,) or a substitute for ف. (TA.) You say also, تَحَنَّثَ مِنْ كَذَا ‡ *He abstained from such a thing as a sin,* or *crime;* syn. تأثّم, (S, A, K,) and تحرّج. (A.) And تَحَنَّثَ بِأَفْعَالٍ † *He sought to bring himself near unto God,* or *to advance himself in God's favour, by works* [of devotion &c.]. (TA.)

حِنْثٌ The *violation,* or *failure of performance, of an oath:* (S, A, K, TA:) [an inf. n. used as a simple subst.:] pl. أَحْنَاثٌ; as in the saying, عَلَيْهِ أَحْنَاثٌ كَثِيرَةٌ [*He is accountable for,* or *chargeable with, many violations,* or *failures of performance, of oaths*]. (TA.) — ‡ *A sin,* a *crime, an offence,* or *an act of disobedience.* (S, A, Msb, K.) So in the saying [in the Kur lvi. 45], وَكَانُوا يُصِرُّونَ عَلَى الحِنْثِ العَظِيمِ ‡[*And they persisted in great sin*]; (A;) meaning belief in a plurality of gods. (Bd, Jel.) And hence, (Bd in lvi. 45,) بَلَغَ الحِنْثَ ‡ *He* (a boy) *attained to the age when he was punishable for sin:* (A,* Bd:) or *attained to* [*the age when he became punishable for*] *disobedience and* [*rewardable for*] *obedience:* (S:) or *attained to the age when the pen* [*of the recording angels*] *began to register his acts of obedience and of disobedience:* *attained to the age of puberty:* *attained to manhood.* (TA.) [Hence also,] أَوْلَادُ الحِنْثِ † *The children of adultery* or *fornication:* occurring in a trad., accord. to one reading: accord. to another reading, اولاد الخُبْثِ. (L.)

حَانِثٌ *Violating,* or *failing to perform, his oath;* or *a violator,* or *non-performer, of his oath.* (Msb.)

مَحَانِثُ: see مَحْنَثٌ.

مَحْنَثٌ A thing *respecting which people differ, and which admits of its being regarded in two different ways;* as also مُخْلِفٌ. (L.) [So called because it may make one to be untrue in an oath.]

مَحَانِثُ, accord. to some, having no sing.; but accord. to others, its sing. is ✝مَحْنَثٌ, and this is app. the truth, and is agreeable with analogy; (MF;) *Places* [i. e. *occasions*] *of falling into the violation of an oath* or *oaths,* or †*into sin* or *crime;* expl. by مَوَاقِعُ حِنْثٍ (TA) and مَوَاقِعُ إِثْمٍ. (K.)

حنجر

حَنْجَرَةٌ, here mentioned in the K: see art. حجر, in which I have mentioned it as Q. Q., like the two words here following, which are mentioned in the latter art. in the S and K &c.

حُنْجُرَةٌ: } see art. حجر.
حُنْجُورٌ: }

حندر

حُنْدُرٌ and حُنْدُورٌ &c.: see art. حدر.

حندس

Q. 1. حَنْدَسَ: see what next follows.

Q. 2. تَحَنْدَسَ *It* (the night) *was,* or *became, dark:* (K:) or *intensely dark;* (TA;) as also ✝حَنْدَسَ. (Ham p. 140.) — And *He* (a man) *fell down;* and *was weak:* (Sgh, K:) mentioned by Sgh in art. حدس. (TA.)

حِنْدِسٌ *Darkness:* (K:) or *intense darkness:* (Ham p. 140:) pl. حَنَادِسُ. (K.) You say, سَرَوْا فِى حِنْدِسِ اللَّيْلِ [*They journeyed in the darkness,* or *intense darkness, of the night*]: and فِى حَنَادِسِ الظُّلَمِ [*in the darkness,* or *intense darkness, of the nights called* الظُّلَم, q. v.]. (A, in art. حدس.) Z derives it from الحَدْسُ meaning نَظَرٌ خَافٍ. (TA. [See art. حدس.]) — Also, as an epithet applied to night, (S, K,) *Dark:* (K:) or *intensely dark;* (S;) and so ✝مُحَنْدِسٌ. (Ham p. 140.) You say لَيْلَةٌ حِنْدِسَةٌ and لَيْلٌ حِنْدِسٌ. (TA.) — [Hence,] الحَنَادِسُ *Three nights* (K, TA) *of the lunar month,* (TA,) *next after the*

أَسْوَدُ حِنْدِسٌ [Hence also,] (K.) — [q. v.]. ظُلْمٌ a phrase like أَسْوَدُ حَالِكٌ [*Intensely black*]. (L.)

حُنْدُوسٌ a term applied by the people of the East to *Dirhems cut in pieces*; which pieces they use for change, and in almsgiving: so explained by Esh-Shereeshee. (Ḥar p. 81.)

حِنْدِسٌ: see مُحَنْدِسٌ.

حندق

حُنْدُوقَةٌ : } see حَدَقَةٌ, in art. حدق.
حِنْدِيقَةٌ :

حَنْدَقُوقٌ, (Ṣ, Ḳ, &c.,) mentioned by J and Ṣgh in art. حدق; but IB says that it belongs to the present art., the ن being a radical letter; and thus Sb mentions it as an epithet, in a sense explained below; (TA;) and حِنْدَقُوقٌ (Sh, Ḳ) and حُنْدُقُوقٌ (TA) and ▼حَنْدَقُوقَى, (Ḳ,) allowed by Sh, but disallowed by J, (TA,) and ▼حَنْدَقُوقِي and ▼حُنْدُقُوقَى and ▼حِنْدَقُوقَى (Ḳ) and ▼حَنْدَقُوقِى; (TA;) [The *herb lotus, melilot, sweet trefoil*, or *bird's-foot-trefoil*; so in the present day;] *a certain plant*; (Ṣ;) *a certain herb*, or *leguminous plant*, (Ḳ,) *resembling fresh*, or *green*, قَتّ [q. v.]; (TA;) *i. q.* ذُرَق: (Ṣ, Ḳ:) a Nabathæan name, arabicized. (Ṣ.) ═ Also the first, A *tall* man, *incompact*, or *incongruous, in make*, (Ibn-Es-Serráj, Ḳ,) *like him who is* مَجْنُون [or *insane*]; (Ibn-Es-Serráj, TA;) or, as some say, *like the* أَحْمَق [or *foolish*, or *stupid*, &c.]: (TA:) or it signifies also *i. q.* أَحْمَق. (Ḳ.) And One *who turns about the eyes*; or *who does so much*, or *frequently*. (AO, Az, Ḳ.) [But in this sense it belongs to art. حدق.]

حَنْدَقُوقَى and حَنْدَقُوقِى &c.: see the paragraph immediately preceding.

حنذ

1. حَنَذَ, aor. ـِ, inf. n. حَنْذٌ (Ṣ, L, Ḳ, &c.) and تَحْنَاذٌ, [which latter is an intensive form,] *He roasted* a kid, &c.: (M, L:) or *he roasted* flesh-meat *with heated stones*, (T, A, L,) [*in a hole dug in the ground*, (see حَنِيذٌ,)] *in order to cook it thoroughly*: (A:) or *he roasted* a sheep or goat, and *put upon it heated stones to cook it thoroughly*: (Ṣ, L, Ḳ:) or *he roasted* flesh-meat *by covering it over in fire, or burying it therein*: (AZ, L:) or *he roasted* it *so that it dripped*: (M:) or *he roasted* it *without overdoing it* [*so that the moisture dripped from it*]: or *he scalded* a kid &c., *so as to remove its hair*. (L.) [See also حَنِيذٌ; and see 4.] — حَنَذَتْهُ الشَّمْسُ, (A, L, Ḳ,) aor. ـِ, inf. n. حَنْذٌ, (L,) ‡ *The sun burned him*, (A, L, Ḳ,) namely, a traveller, *and pained his brain*. (Ḳ.) — حَنَذَ الفَرَسَ, (Ṣ, A, L, Ḳ,) aor. ـِ, (Ṣ, L, Ḳ,) inf. n. حَنْذٌ (Ṣ, L) and حِنَاذٌ, (A, L,) ‡ *He urged on the horse*, (A, Ḳ,) *and made him to run a heat or two heats*, (Ṣ, L, Ḳ,) *and then threw upon him coverings*, (Ṣ, A, L, Ḳ,) *one above another*, (L,) *to the number of five or six*, (T, L,) *in the sun*, (Ṣ, Ḳ,) *in order that he might sweat*, (T, Ṣ, A, L, Ḳ,) *to reduce his fat, and prevent his breathing hard*. (T, L.) To a horse to which this is done you apply the epithets ▼حَنِيذٌ and ▼مَحْنُوذٌ. (Ṣ, A, L, Ḳ.) If the horse do not sweat, you say of him, كَبَا. (Ṣ, L.) — حَنَذَ لَهُ: see 4.

4. احنذ *He cooked* flesh-meat *thoroughly*. (L.) — [See also 1] — Also, inf. n. إِحْنَاذٌ, ‡ *He put much mixture* [*of water*] *into the beverage* or *wine* (شَرَاب): or, as some say, the contr., i. e. *he put little thereof*: (Ḳ, TA:) accord. to Fr, (L,) *i. q.* أَعْرَقَ and أَخْفَسَ: (T, L:) AHeyth disallows its being syn. with these two verbs: (L:) but in the M it is said that ▼حَنَذَ لَهُ, aor. ـِ, means † *he made the water for him little, and the beverage*, or *wine, much*; as also احفس: (L, TA:) accord. to Fr, إِذَا سَقَيْتَ فَأَحْنِذْ means أَخْفِسْ, i. e. † [*when thou givest to drink*,] *make the water little, and the* نَبِيذ *much*; (T, L, TA;) or this phrase means عَرِّقْ شَرَابَكَ, i. e. *pour into thy beverage*, or *wine, a little water*: (Ṣ, L:) in the A it is said that إِذَا سَقَيْتَهُ فَأَحْنِذْ لَهُ means ‡ [*when thou givest him to drink*,] *give him to drink pure* [*wine*] *that shall burn his inside*. (TA.)

10. استحنذ ‡ *He lay upon his side in the sun*, (Ḳ, TA,) *and threw clothes over him there*, (TA,) *in order that he might sweat*. (Ḳ, TA.) You say, اِسْتَحْنَذْتُ فِى الشَّمْسِ ‡ *I endeavoured to make myself sweat by throwing clothes over me in the sun*. (A.)

حَنْذٌ: see حَنِيذٌ. ═ Also † The *intenseness*, or *violence, and burning*, of heat. (Ṣ, L.)

حَنْذَةٌ † *Intense*, or *violent, heat*. (Ḳ.)

حِنْذِيذٌ † A man (TA) *that sweats much*. (Ḳ, TA.)

حَنَاذِ, like قَطَامِ, [a proper name, indecl., with kesr for its termination,] † *The sun*. (Ḳ.)

حِنَاذٌ † *Heat*: so in the phrase ▼حِنَاذٌ مُحْنِذٌ † *Burning heat*. (L.)

حَنِيذٌ *Roasted* flesh-meat; as also ▼مَحْنُوذٌ and ▼حَنْذٌ, which last is an inf. n. used as an epithet: (L:) or *roasted with heated stones*, (T, A, L,) *in order to be cooked thoroughly*: (A:) or a sheep or goat *roasted, and having heated stones put upon it to cook it thoroughly*; (Ṣ, L, Ḳ;) as also ▼مَحْنُوذٌ; (TA;) which Ibn-'Arafeh explains as meaning *roasted by means of heated stones put upon it, so thoroughly as to fall off from the bones*: or حَنِيذٌ, he says, signifies *roasted by means of hot stones so as to drip*: (L:) or *hot, roasted*, flesh-meat, *of which the moisture drips*: (Sh, L, Ḳ:*) this is said by Az to be the best explanation that has been given of it: (TA:) or *roasted* flesh-meat *not overdone*: (L:) or *roasted by being buried in the fire*: (AZ, L:) or [roasted flesh-meat] *for which one has dug a hole in the ground in which it is then covered over* [*with fire or heated stones*], agreeably with a well-known practice of the Arabs of the desert: (Fr, L:) or *roasted in a hole dug in the ground, heated stones being put upon it*. (Ḥar p. 20.) [See also مَرْمُوضٌ.] — See also 1. — Also † *Heated* water: (Ḳ:) or *hot* water. (Sh, T, L.) — † *A kind of oil*. (Ḳ.) — † *A perfumed preparation of* خِطْمِىّ [or *marsh-mallows*] *and the like, for washing the head*. (Ḳ,* TA.)

شَرَابٌ مُحْنَذٌ † *Beverage*, or *wine, mixed with much water*. (IAạr, TA.) [But see 4.]

مُحْنِذٌ: see حِنَاذٌ.

مَحْنُوذٌ: see حَنِيذٌ, in two places: — and see also 1.

حنزب

حِنْزَابٌ An ass *of middling make* (مُقْتَدِرُ الخَلْقِ). (Ḳ.) — Also *Short and strong*: or *broad*: (Ḳ:) or *short and broad*; applied to a man: (Th, TA:) and *thick, coarse, rude*, or *bulky*, (Ḳ, TA,) *and short*. (TA.) [See حَزَابٍ, in two places.] — The *cock*. (Ḳ.) — Also, and ▼حُنْزُوبٌ, *A flock of* [*the kind of birds called*] قَطًا: (Ḳ:) or the *male of the* قطا. (TA.) [See حنزاب in art. حزب.] — And both these words, The *carrot of the land* (جَزَرُ البَرِّ) [see حنزاب in art. حزب]: (Ḳ:) n. un. of the former with ة: that of the latter is not known to have been in use. (TA.) — This is [said to be] the proper place of these two words; [the ن being regarded as a radical letter;] (Ḳ, TA;) not art. حزب. (TA.)

حُنْزُوبٌ: see above: and see also art. حزب.

حنش

1. حَنَشَ, (Ṣ, A, Mṣb, Ḳ,) aor. ـِ, (Mṣb, Ḳ,) *He hunted, sought to catch or capture*, or *caught or captured*, (Ṣ, A, Mṣb, Ḳ,) such animals as are hunted &c., (Ṣ, Mṣb, Ḳ,) or such as are termed أَحْنَاش, pl. of حَنَشٌ; (A;) as also ▼احنش. (TA.) — حَنَشَتْهُ الحَيَّةُ *The serpent bit him*. (A, TA.)

4: see above.

حَنَشٌ *Anything that is hunted, or caught or captured, of birds or flying things, and of* [or, accord. to the CḲ, this word "of" should be omitted here, as well as where it next occurs,] *what are termed* الهَوَامّ [*venomous or noxious reptiles or the like, such as scorpions and serpents*], (Ṣ, A, Mgh,* Mṣb, Ḳ,) *and of what are termed* حَشَرَاتُ الأَرْضِ, (Ḳ,) *such as the hedgehog, and the* [*lizards of the kinds called*] ضَبّ *and* وَرَل, *and the* [*rat called*] جُرَذ, *and the common rat or mouse, and the serpent*: (TA:) or *any animal whose head resembles that of the serpent*, (Lth, Mgh, Mṣb,* Ḳ,) *of chameleons and of* [*the lizards called*] سَوَامّ أَبْرَص *and the like*: (Lth, Mgh,* Mṣb:*) or *any creeping thing, of beasts and of birds or flying things*: (Kr, TA:*) and the *serpent*: (Ṣ, Kr, A, Mgh, Ḳ:) or the *viper*: (Ṣ:) or *a kind of white and thick serpent, like the* ثُعْبَان, or *larger*; or the *black kind thereof*: (TA:) or *a serpent that blows, but does not hurt*: (Ḥam p. 626:) and the *common fly*: (Ibn-'Abbád, A, Ṣgh, Ḳ:) pl. أَحْنَاشٌ (Ṣ, A, Mgh, Ḳ) and حِنْشَانٌ. (A, TA.)

مَحْنُوشٌ *Stung*, or *bitten*, *by what is termed* حَنَش. (IAạr, Ḳ.)

حنط

1. حَنَطَ, (Ḳ,) aor. ـِ, (TḲ,) inf. n. حُنُوطٌ, *It* (seed-produce) *attained to the time for its being reaped;* as also ↓ احنط. (Ḳ.) — *It* (a tree of the kind called رِمْث) *became mature, and its leaves became white;* as also ↓ احنط: (Ṣ:) or *it became white and mature,* (Ḳ, TA,) *and there came forth upon it a dust-coloured fruit, and what resembled pieces of glue appeared upon its tops;* (TA;) as also حَنِطَ, aor. ـَ; (Ḳ, TA;) and ↓ احنط: (TA:) [the last, though omitted in the Ḳ, seems to be the most common:] or *its colour became white inclining to yellowness, and its odour sweet:* (IAth:) Az relates, on the authority of IAạr, that one says, أَوْرَسَ الرِّمْثُ, and ↓ احنط; like غَضِبَ العَرْفَجُ: and one says, of the رِمْث, when it first breaks out for its leaves to come forth, قَدْ أَقْمَلَ; and when it has increased little by little, قَدْ أَدْبَى; and when its greenness has increased, بَقَلَ: and when it has become white and mature, حَنَطَ: (TA:) or ↓ احنط is said of a tree, and of a herb, meaning *its fruit became mature;* and so حَنَطَ, inf. n. حُنُوطٌ. (AḤn.) — *It* (leather) *became red.* (Ṣ, Ḳ.) [The inf. n. of the verb in this sense is not mentioned.] — See also 2.

2. حنّطهُ, inf. n. تَحْنِيطٌ; (Ṣ, TA;) in the Ḳ, ↓ حَنَطَهُ, aor. ـِ; which is a mistake; (TA;) *He prepared him* (a dead person [i. e. *for burial*]) [and also *it* (grave-clothing)] *with* حُنُوط [q. v.]; (Ṣ, Ḳ;) as also ↓ احنطهُ. (Ḳ.) And [hence,] ↓ أُحْنِطَ [lit. *He was prepared for burial with* حنوط, is used to signify] *he died.* (Ḳ.)

4. احنط: see 1, in five places. — أَحْنَطَهُ: and أُحْنِطَ: see 2. — The former also signifies *He,* or *it, made him,* or *it, to bleed: made him,* or *it, to be bloody;* or *smeared, befouled,* or *defiled, him,* or *it, with blood: it* (blood) *befouled,* or *defiled, him,* or *it.* (IAạr.)

5. تحنّط *He* (a dead person) *was,* or *became, prepared* [*for burial*] *with* حُنُوط. (Ḳ.) — Also, or تحنّط بِالحَنُوطِ, (Ṣ,) *He* (a man) *made use of* حنوط *for himself, in his clothing:* (Ṣ,* TA:) so in a trad.: meaning, on his going forth to battle; as though desiring thereby to prepare himself for death, and to induce himself to endure the fight with patience. (TA.)

10. استحنط [lit. *He desired to be prepared for burial with* حَنُوط: and hence meaning] *he* (a man, Fr) *emboldened himself,* or *became emboldened, to encounter death, holding his life in light estimation.* (Fr, Ḳ.)

حِنْطَةٌ *Wheat;* and the *grain of wheat;* syn. بُرٌّ (Ṣ, Mṣb, Ḳ) and قَمْحٌ and طَعَامٌ; (Mṣb;) of the first three of which words, بُرّ is the most chaste; (Ṣ in art. بر;) the *well-known grain called* بُرّ: (TA:) *chewed, and applied as a poultice, it is good for the bite of a dog:* (Ḳ:) or, correctly, *what is chewed thereof disperses humours; but for the bite of a dog, it is coarsely pounded, and put upon the bite;* as is said by the author of the "Minháj:" and one of its well-known properties is this; that *when it is put upon a piece of heated iron, and powdered, and ringworms* (قَوَابِى) *are smeared with the moisture thereof, it removes them:* (TA:) pl. حِنَطٌ. (Ṣ, Ḳ.)

حِنْطِىٌّ *An eater of much* حِنْطَة [or *wheat*], *in order that he may grow fat.* (Ḳ.) — Accord. to Aboo-Naṣr and Aboo-Sa'eed, (TA,) *Inflated,* or *swollen;* syn. مُنْتَفِخٌ. (Ḳ, TA.)

حِنَاطٌ: see what next follows.

حَنُوطٌ (Ṣ, IAth, Mṣb, Ḳ) and ↓ حِنَاطٌ (IAth, Mṣb, Ḳ) [*Perfume such as is termed*] ذَرِيرَة: (Ṣ:) or *odoriferous substances* (IAth, Mṣb, Ḳ) *of any kind* (Ḳ) *that are mixed* (IAth, Mṣb, Ḳ) *for a corpse,* (Mṣb, Ḳ,) *in particular,* (Mṣb,) or *for grave-clothes and for the bodies of the dead, consisting of* ذَرِيرَة, *or musk, or ambergris, or camphor, or other substance, namely, Indian cane, or sandal-wood, bruised:* derived from حَنَطَ said of the رِمْث, signifying that its colour became white inclining to yellowness, and its odour sweet: (IAth:) the term حنوط is applied to *anything with which a corpse is perfumed, consisting of musk and* ذَرِيرَة *and sandal-wood and ambergris and camphor, and other things that are sprinkled upon it for the purpose of perfuming it and drying up its moisture.* (Mṣb.)

حِنَاطَةٌ The *trade of the* حَنَّاط [q. v.]. (Ṣ, Ḳ.)

حَنَّاطٌ *A seller of* حِنْطَة [or *wheat*]; (Ṣ, Mgh, Mṣb, Ḳ;) as also ↓ حَنَّاطِىٌّ; (Ḳ;) a rel. n. from the former. (Mṣb.) [The pl.] حَنَّاطُونَ is explained by the lawyers as signifying *Persons who transport wheat* (حنطة) *from the ship to the houses.* (Mgh in art. نقل.)

حَنَّاطِىٌّ: see the next preceding paragraph.

حَانِطٌ *A possessor of* حِنْطَة [or *wheat*]: (Ḳ:) or *one who possesses much thereof.* (Ṣgh, Ḳ.) [A possessive epithet, like لَابِنٌ and تَامِرٌ.] And قَوْمٌ حَانِطُونَ *A people whose seed-produce has attained to the time for its being reaped:* [in this sense also] a possessive epithet. (TA.) — Also, [act. part. n. of حَنَطَ, or,] accord. to Sh and ISd, an act. part. n. of أَحْنَطَ, as applied to the رِمْث, contr. to analogy, meaning [*Mature and*] *having its leaves become white;* as also ↓ مُحْنِطٌ: (TA:) and, applied to a tree, and a herb, *having its fruit mature.* (AḤn.) Also, accord. to Sh, i. q. وَارِسٌ, in the phrase حَانِطُ الغَضَى [app. meaning *What is putting forth its leaves, of trees of the kind called* غَضًى]: but accord. to Ibn-'Abbád and the Ḳ, the *fruit of the kind of tree called* غَضًى. (TA.) — Also *Red* leather. (Ṣ, TA.) And أَحْمَرُ حَانِطٌ *Intensely red:* (IF, Ḳ:) because wheat (الحِنْطَة) is called الحَمْرَآءُ. (IF.)

مُحْنِطٌ: see حَانِطٌ.

حنظل

Q. 1. حَنْظَلَتِ الشَّجَرَةُ *The tree became bitter in its fruit* [*like the* حَنْظَل]. (AḤei, TA.)

حَنْظَلٌ [The *colocynth; cucumis colocynthis;*] *a certain bitter plant;* (Mṣb;) [and *its fruit;*] *well known;* (Ḳ;) i. q. شَرْىٌ: (Ṣ:) n. un. with ة: (Ṣ, Mṣb, Ḳ:*) [accord. to Freytag (who refers to Avic. p. 175, and Sprengel. hist. rei herb. vol. i. p. 269,) applied also to the *momordica elaterium,* or *cucumis prophetarum:*] *there is a male species, and a female; the former fibrous; the latter soft,* or *easily broken, white, and easy to swallow:* (TA:) *the choice sort of it is the yellow;* (Ḳ;) or, accord. to the "Ḳánoon" of the Ra-ees [Ibn-Seenà, from which the description of its properties and uses, in the Ḳ and TA, is, with some slight variations, taken], *the white, very white, and soft; for the black and the hard are bad, and it is not plucked until it becomes yellow, and the greenness has completely gone from it;* (TA:) *its pulp attenuates the thick phlegmatic humour that flows upon the joints* (Ḳ, TA) *and tendons,* (TA,) *when swallowed* (Ḳ, TA) *in the dose of of twelve ḳeeráts,* (TA,) *or used in the manner of a clyster: it is beneficial for melancholy, and epilepsy, and the* [*sort of doting termed*] وَسْوَاس, *and alopecia* (دَآءُ الثَّعْلَب), *and elephantiasis* (الجُذَام), (Ḳ, TA,) *and* [*the disease of the tumid leg, termed*] دَآءُ الفِيل; *for these three used by rubbing; and for the cold* نِقْرِس [i. e. *arthritis,* or *gout*], (TA,) *and for the bite of vipers, and the sting of scorpions, especially its root;* (Ḳ, TA;) *for this last being the most beneficial of medicines; a drachm of its root, administered to an Arab stung by a scorpion in four places, being said to have cured him on the spot: that which is plucked green relaxes* [*the bowels*] *excessively, and produces excessive vomiting:* so in the "Ḳánoon:" (TA:) *it is also beneficial for the tooth-ache, by fumigating with its seeds; and for killing fleas, by sprinkling what is cooked thereof; and for the sciatica, by rubbing with what is green thereof:* (Ḳ, TA:) *its root is cooked with vinegar, and one rinses the mouth with it for the tooth-ache; and the vinegar is cooked in it in hot ashes: when cooked in olive-oil, that oil, being dropped* [*into the ear-hole*], *is beneficial for ringing in the ears: it is beneficial also for the moist and flatulent colic: and sometimes it attenuates the blood: administered as a suppository in the vagina, it kills the fœtus:* (TA:) *when the plant bears a single fruit, this is very deadly.* (Ḳ, TA.) [See also هَبِيد.] Accord. to [many of] the leading authorities among the Arabs, (TA,) the ن in this word is augmentative; (Mṣb, TA;) because of their saying, حَظِلَ البَعِيرُ, meaning "the camel became sick from eating حَنْظَل;" and J and Ṣgh [and Fei and others] have mentioned it in art. حظل: but ISd says that this is not an evidence of its being radically triliteral; and that حَظِلَ is like ضَفْنَةٌ (as an epithet applied to a woman) from الضَّفَابِيسُ, which must be acknowledged to be radically quadriliteral. (TA.)

حنف

1. حَنَفَ, aor. ـِ, (Ḳ,) inf. n. حَنْفٌ, (TḲ,) *He, or it, inclined,* or *declined.* (Ḳ, TA.) You say, حَنَفَ إِلَيْهِ (TA) and ▼ تحنّف اليه (Ḳ) *He inclined to it.* (Ḳ, TA.) And حَنَفَ عَنْهُ and ▼ تحنّف عنه *He declined from it.* (TA.) — حَنِفَ, aor. ـَ, (Mṣb, Ḳ,) inf. n. حَنَفٌ; (Mṣb;) and حَنُفَ, aor. ـُ; (Ḳ;) *He had that kind of distortion which is termed* حَنَفٌ *as explained below.* (Mṣb, Ḳ.)

2. حنّفهُ, (Ḳ,) or حنّف رِجْلَهُ, (Ṣ,) inf. n. تَحْنِيفٌ, (Ḳ,) *He rendered him,* (Ḳ,) or *his leg,* or *foot,* (Ṣ,) أَحْنَف. (Ṣ, Ḳ.)

5. تحنّف: see 1, in two places. — [Hence,] *He did according to the* حَنِيفِيَّة; (Ṣ, Ḳ;) i. e. the *law of Abraham,* which is the *religion of El-Islám:* (TA:) or *he became circumcised:* or *he turned away from the worship of idols;* (Ṣ, Ḳ;) *and became,* or *made himself, a servant of God;* or *applied,* or *devoted, himself to religious services* or *exercises.* (Ṣ.) [See تَحَنَّثَ.]

حَنَفٌ, originally, *A natural wryness:* and particularly *an inversion of the foot, so that the upper side becomes the lower:* so says IDrd; (Mgh;) or *a crookedness in the leg,* or *foot;* (Ṣ, O, Ḳ;) i. e., (Ṣ, O, but in the Ḳ "or") *a turning of one of the great toes towards the other:* (Ṣ, O, Ḳ:) or [*a distortion that causes*] *one's walking on the outer part of the foot, on the side in which is the little toe:* (Ḳ: [and so accord. to an explanation of ▼ أَحْنَفُ by IAạr cited in the Ṣ:]) or *an inclining* [app. *inwards*] *in the fore part of the foot.* (Lth, Ḳ.) — Accord. to Ibn-'Arafeh and the Ḳ, it signifies also *A right state* or *condition* or *tendency;* and accord. to the former, the epithet ▼ أَحْنَفُ is applied to him who has a wry leg, or foot, only by way of presaging a right state: but Er-Rághib explains حَنَفٌ better, as signifying *an inclining, from error, to a right state* or *tendency.* (TA.)

حَنَفِيَّةٌ The *persons called in relation to the Imám Aboo-Ḥaneefeh* [*because they hold his tenets*]; as also ▼ أَحْنَافٌ. (TA.) حَنَفِىٌّ [is its n. un.: and] signifies [also] one *who is of the religion of Abraham.* (Mgh. [See also حَنِيفٌ.]) — *A* مِيضَأَةٌ; [by which is here meant *a vessel with a tap, for the purpose of ablution,* such as is often used in a private house; and *a fountain,* i. e. *a tank with taps, for the same purpose,* in a mosque; because persons of the persuasion of Aboo-Ḥaneefeh must perform the ablution preparatory to prayer with running water, or from a tank or the like at least ten cubits in breadth and the same in depth;] but this application is post-classical. (TA.) — سُيُوفٌ حَنَفِيَّةٌ, (L, Ḳ,* TA,) or ▼ حَنِيفِيَّةٌ, (so accord. to the CḲ,) or حَنْفِيَّةٌ, (so in a MS. copy of the Ḳ,) *Certain swords, so called in relation to El-Aḥnaf Ibn-Ḳeys;* because he was the first who ordered to make them: by rule it should be أَحْنَفِيَّةٌ. (Lth, L, Ḳ.)

حَنِيفٌ *Inclining to a right state* or *tendency:* (Er-Rághib, TA:) or *right,* or *having a right state* or *tendency;* (Akh, Ṣ, TA;) thus applied in like manner as أَعْوَرُ is applied to a crow: (Ṣ:) [and particularly] *inclining, from one religion, to another:* (Ḥam p. 358:) or *inclining, from any false religion, to the true religion:* (Mgh:) or *inclining in a perfect manner to El-Islám, and continuing firm therein:* (Ḳ:) and any one *who has performed the pilgrimage:* (Aṣ, Ḳ, TA:) so say I 'Ab and El-Ḥasan and Es-Suddee; and Az says the like on the authority of Eḍ-Ḍaḥḥák: (TA:) or one *who is of the religion of Abraham,* (Ḳ, TA,) *in respect of making the Sacred House* [*of Mekkeh*] *his ḳibleh, and of the rite of circumcision:* (TA:) [and] *a Muslim;* (Ṣ, Mgh, Mṣb;) because he inclines to the right religion: (Mṣb:) but in this last sense, it is a conventional term of the professors: (Mgh:) [or,] accord. to AO, the worshipper of idols, in the Time of Ignorance, called himself thus; and when El-Islám came, they thus called the Muslim: accord. to Akh, it was applied in the Time of Ignorance to him *who was circumcised, and who performed the pilgrimage to the* [*Sacred*] *House;* because the Arabs in the Time of Ignorance held nothing of the religion of Abraham except circumcision and that pilgrimage: accord. to Ez-Zejjájee, it was applied in the Time of Ignorance to him *who made the pilgrimage to the* [*Sacred*] *House and performed the ablution on account of* جَنَابَة *and was circumcised;* and when El-Islám came, it was applied to the Muslim, because of his turning from the belief in a plurality of gods: (TA:) also one *who devotes himself to religious exercises;* or *applies himself to devotion:* (Mṣb:) its predominant application is to Abraham: (Mgh:) pl. حُنَفَآءُ. (AO, TA.) — [Hence,] حَسَبٌ حَنِيفٌ *Recent* [*grounds of pretension to respect* or *honour*]; *of the time of El-Islám; not old.* (TA.) — *Short.* (Ḳ.) — *A maker of sandals.* (Ḳ.)

حُنَيْفٌ: see أَحْنَفُ.

حَنِيفِيَّةٌ, accord. to Th and Zj, *An inclining* to a thing: but ISd says that this explanation is nought. (TA.) — The *law of Abraham;* which is the *religion of El-Islám:* also termed مِلَّةٌ حَنِيفِيَّةٌ. (TA.) — See also حَنَفِيَّةٌ.

أَحْنَفُ *Having that kind of distortion which is termed* حَنَفٌ *as explained above;* (Ṣ, Mṣb, Ḳ;) applied to a man: (Ṣ, Mṣb:) and so [the fem.] حَنْفَآءُ applied to a leg or foot: (Ḳ:) accord. to IAạr, one *who walks on the outer part of his foot,* (Ṣ,) or *of his feet,* (Mṣb,) *on the side in which is the little toe:* (Ṣ:) or *who has one of his great toes turning towards the other:* (Mgh:) its abbreviated dim. is ▼ حُنَيْفٌ. (Mṣb.) See حَنَفٌ, in two places. — Also حَنْفَآءُ, *A curved staff* or *stick;* in the dial. of Syria. (TA.) — *A bow;* (Ḳ;) because of its curved shape. (TA.) — *A razor;* (Ḳ;) for the same reason. (TA.) — The *chameleon.* (Ḳ.) — The *tortoise.* (Ḳ.) — *A certain marine fish, also called* أَطُومٌ. (Ḳ.) — *A certain tree.* (IAạr, Ḳ.) — ‡ *A changeable female slave, at one time lazy and at another brisk.* (IAạr, Ḳ.)

أَحْنَافٌ: see حَنَفِيَّةٌ.

حنق

1. حَنِقَ, (Ṣ, Mṣb, Ḳ,) aor. ـَ, (Mṣb, Ḳ,) inf. n. حَنَقٌ (Ṣ,* Mṣb, Ḳ) and حَنْقٌ, (Ḳ,) *He was,* or *became, angered,* or *enraged:* (Ṣ, Mṣb, Ḳ:) or *vehemently angered* or *enraged:* (Ḳ, Ḥam p. 29:) عَلَيْهِ *against him:* (Ṣ:) or it may be from the signification of "cleaving," or "sticking;" as though meaning † *rancour, malevolence, malice,* or *spite, clave to his bosom.* (Ḥam p. 29. [See also حَنَقٌ, below.]) — [Hence,] لَا يَحْنَقُ عَلَى جِرَّتِهِ and مَا يَحْنَقُ عَلَى جِرَّةٍ: [thus written in the TA; though it seems to be implied by the manner in which the two sayings are there introduced, in this art., that the verb in each case is ▼ يُحْنِقُ, from أَحْنَقَ in the first of the senses assigned to it below:] see جِرَّةٌ.

2: see 4.

4. احنق ‡ *He bore rancour, malevolence, malice,* or *spite,* [*cleaving to him, so as*] *not to be relaxed;* or *he hid, in his heart, enmity, and violent hatred, not to be relaxed.* (Ḳ, TA.) See also 1. — *It* (the back-bone, or the back, الصُّلْبُ) *clave to the belly* [*by reason of leanness*]. (Ḳ.) — *It* (a camel's hump) *became lean and thin.* (Ṣ.) — *He* (an ass, Ḳ, or, as some say, a camel, and a horse or a mule or an ass, TA) *became lean,* or *light of flesh,* or *slender, and lean,* or *lean, and lank in the belly, in consequence of much covering:* (Ḳ, TA:) or, said of a horse or other animal, (A, TA,) or of a camel, (Ḥar p. 173,) *his belly clave to his back-bone,* or *back, by reason of leanness.* (A, Ḥar p. 173.) — And *He* (a camel) *became fat; had much fat:* thus it has two contr. meanings. (Az, TA.) — *It* (seed-produce [meaning corn]) *spread forth the awn,* or *beard, of its ears, after they had formed, at the head, what resembled little compact balls;* (Ḳ, TA;) as also ▼ حنّق. inf. n. تَحْنِيقٌ. (Ibn-'Abbád, Ḳ.) IAạr says that you say of seed-produce [or corn] قَنْبَعَ ثُمَّ أَحْنَقَ ثُمَّ مَدَّ لِلْحَبِّ أَعْنَاقَهُ ثُمَّ حَمَلَ الدَّقِيقَ, i. e. *Its ears had, at the head, what resembled little compact balls: then the extremities of its awn,* or *beard, appeared: then the internodal portions of its culms appeared: then* [*it bore farina;* or] *it increased, and* [*its heads*] *became like the heads of birds.* (TA.) — *He angered,* (Ṣ, Mṣb, Ḳ,) or *enraged,* another. (Ṣ, Mṣb.) — *He made* a beast *lean,* or *light of flesh.* (Ḥam p. 29.)

حَنَقٌ *Anger,* or *rage:* (Ṣ, Ḳ:) or *vehemence of anger* or *rage:* (M, Ḳ, Ḥam p. 29:) or † *anger,* or *rage, that cleaves to one:* (Ḥar p. 173:) or *rancour, malevolence, malice,* or *spite:* (Ḥar p. 568:) pl. حِنَاقٌ. (Ṣ, Ḳ.)

حَنِقٌ (Ṣ, Mṣb, Ḳ) and ▼ حَنِيقٌ (ISd, O, Ḳ) *Angered,* or *enraged;* (Ṣ, O, Mṣb, Ḳ;) as also ▼ مُحْنَقٌ: (Ṣ, Mṣb:) or *vehemently angered* or *enraged.* (Ḳ.)

حَنِيقٌ: see حَنِقٌ: — and مُحْنِقٌ.

مُحْنَق : see حَنِق.

مُحْنِق *Lean*, or *light of flesh*; or *slender, and lean*; or *lean, and lank in the belly*: (AHeyth, T, TA:) and so the pl. مَحَانِيقُ applied to camels: (Ṣ, Ḳ:) or the former, applied to an ass, (Ṣ,) or a camel, (ISd, TA,) *lean*, &c., as above, (Ṣ, ISd, TA,) *in consequence of much covering*, (Ṣ,) or *from lust*, or *hunger*: and so مَحَانِيقُ applied to horses; as though they imagined the sing. to be مِحْنَاقٌ: (ISd, TA:) or the sing., applied to a horse, (A,) and to an ass, (TA,) *lean*, &c., as above, *so that the belly cleaves to the back*: and so the pls. مَحَانِقُ and مَحَانِيقُ applied to horses: (A, TA:) or مَحَانِيقُ signifies *fat*; as also حُنُقٌ [pl. of ▼حَنِيقٌ]; (Ḳ, TA;) both applied to camels. (TA.)

حنك

1. حَنَكَ الصَّبِىَّ, (Ṣ, Mṣb, Ḳ,) aor. ـُ and ـِ, inf. n. حَنْكٌ; (Mṣb;) and ▼حنّكهُ, (Ṣ, Mṣb, Ḳ,) inf. n. تَحْنِيكٌ; (Mgh, Mṣb;) *He chewed some dates*, or *some other thing* (Ṣ, Mgh, Mṣb, Ḳ) *of a similar kind*, (Mṣb,) *and rubbed therewith the* حَنَك [i. e. *palate*, or *soft palate*,] *of the child*. (Ṣ, Mgh, Mṣb, Ḳ.) — And حَنَكَ الفَرَسَ, aor. as above, (Ṣ, Ḳ,) and so the inf. n., (Ṣ,) *He put a rope in the mouth of the horse*; (Ṣ, Ḳ;) held by ISd to be derived from الحَنَكُ, though it is said that this is not the case; (TA;) as also ▼احتنكهُ; (Ṣ, Ḳ;) which signifies accord. to Yoo *he put a rope in his mouth and led him*: and thus Ibn-'Arafeh explains the saying of Iblees, in the Ḳur [xvii. 64], لَأَحْتَنِكَنَّ ذُرِّيَّتَهُ إِلَّا قَلِيلًا, i. e. † *I will assuredly lead to obey me his progeny, except a few*. (TA. [But see 8.]) — And [hence,] حَنَكَتْهُ السِّنُّ, inf. n. حَنْكٌ and حَنَكٌ, (Ḳ,) ‡ *Age rendered him firm*, or *sound, in judgment, by means of experience*: (TḲ:) or *experiences rendered him firm*, or *sound, in judgment*; (Ḳ, TA;) as also ▼حنّكتهُ, (Zj, Ṣ, Ḳ,) inf. n. تَحْنِيكٌ; (TA;) and ▼احنكتهُ, (Zj, Ṣ, Ḳ,) and ▼احتنكتهُ: (Ḳ:) this is said to be the case when the wisdom-tooth (سِنُّ العَقْلِ) grows forth: and accord. to Lth, حَنَكَتْهُ السِّنُّ signifies *his teeth called* أَسْنَانُ العَقْلِ [*the wisdom-teeth*] *grew forth*. (TA.) And حَنَكَتْهُ الأُمُورُ ‡ *Affairs did to him what is done to the horse by putting the rope in his mouth*; i. e., *rendered him experienced and submissive*: or *trained*, or *disciplined, and reformed*, or *improved, him*; as also ▼حنّكته. (TA.) And حَنَكَهُ الدَّهْرُ ‡ *Time*, or *fortune, tried*, or *proved, him, and taught him, and rendered him expert*, or *experienced, and well informed*, or *firm*, or *sound, in judgment*. (IAạr, TA.) — And حَنَكَ الشَّىْءَ, (Ṣ, Ḳ,) inf. n. حَنْكٌ, (TA,) ‡ *He understood the thing, and knew it soundly, thoroughly*, or *well*; syn. فَهِمَهُ وَأَحْكَمَهُ; (Ṣ, Ḳ, TA;) like لَقِفَهُ, inf. n. لَقْفٌ. (TA.)

2. حنّكهُ, inf. n. تَحْنِيكٌ, *He rubbed his* حَنَك [i. e. *palate*, or *soft palate*,] (Ḳ, TA) *so as to make it bleed*: (TA:) or *he stuck a piece of wood*, or *stick, into his* (a beast's) *upper* حَنَك, or *the extremity of a horn, so as to make it bleed*; because of something happening therein. (Az, TA.) — See also 1, in three places. — Also *He turned the piece of cloth* [*forming part of the grave-clothing*] *beneath his* (a corpse's) حَنَك, i. e., *the part beneath his chin*. (Mgh.) [See also المُحَنَّكُ, below.]

4: see 1. — Also احنكهُ عَنِ الأَمْرِ *He turned him back*, or *away, from the affair*. (Ḳ,* TA.)

5. تحنّك *i. q.* تَلَحَّى; (Ṣ;) i. e. *He turned* [*a portion of*] *the turban beneath his* حَنَك [here meaning *the part beneath his chin and lower jaw*]. (Ṣ, Ḳ.) = See also 8.

8. احتنك الجَرَادُ الأَرْضَ ‡ *The locusts ate what was upon the land*; (Ṣ, Ḳ, TA;) and *consumed*, or *made an end of, its herbage*: (Ṣ:) or *gained the mastery over the land with the* حَنَك [here meaning the *mouth*], *and ate* [*the produce of*] *it, and extirpated it*: (Er-Rághib, TA:) derived from الحَنَكُ, by which is sometimes meant "the mouth," and "the beak." ('Ináyeh, MF.) And احتنك البَعِيرُ الصِّلِّيَانَةَ † *The camel pulled up by the roots the* [*plant called*] صِلِّيَانة. (Az, TA.) And احتنك [for احتنك النَّبْتَ † *He cropped the herbage*] is said of a young gazelle. (Ḳ voce شَصَرٌ, q. v.) And احتنكهُ † *He took his* (a man's) *property*; (ISd, Ḳ;) as though he ate it with the حَنَك. (ISd, TA.) And † *He took it entirely*; *took the whole of it*; namely, what another possessed. (ISd, TA.) And ‡ *He gained the mastery over him, or it*; *got him*, or *it, in his power*. (Ḳ, TA.) Accord. to Akh, لَأَحْتَنِكَنَّ ذُرِّيَّتَهُ, in the Ḳur [xvii. 64, cited, and explained on the authority of Ibn-'Arafeh, above], means ‡ *I will assuredly extirpate his progeny*; and *I will assuredly incline them* [*to obey me*]: (TA:) or, accord. to Fr, ‡ *I will assuredly gain the mastery over his progeny*. (Ṣ, TA.*) — See also 1, in three places. = Also احتنك [and ▼تحنّك, the latter found by Reiske in this sense, as mentioned in Freytag's Lex.,] ‡ *He* (a man) *was*, or *became, firm*, or *sound, in judgment*, [*by means of experience*:] (Ṣ, TA:) or *experienced and submissive, like the horse in whose mouth the rope has been put*. (TA.)

10. استحنك † *He* (a man, TA) *ate vehemently*, (Ṣgh, Ḳ,) or *strongly and vehemently*, (T, TA,) *after eating little*, (Ṣgh, Ḳ,) or *after eating feebly and little*. (T, TA.) = اِسْتَحْنَكَتِ العِضَاهُ † *The* [*trees called*] عضاه *were*, or *became, pulled up by the roots*. (Ḳ.)

حَنْكٌ : see حُنْكَةٌ, in two places.

حِنْكٌ : see حُنْكَةٌ.

حَنَكٌ The *part beneath the chin* [*and lower jaw*], (Ṣ, Mgh,) of a man &c.: (Ṣ:) or [the *palate*, or *soft palate*;] the *interior of the upper part of the inside of the mouth*, (Ḳ, TA,) of a man and of a beast: (TA:) and the *lower part, from the extremity of the fore part of the two jaws*, (Ḳ,) *below these*: (TA:) or the *roof of the upper part of the mouth*, (Zj in his "Khalḳ el-Insán," El-Ghooree, Mgh, TA,) *from which depends the* لَهَاة [or *uvula*]: (Zj ubi suprà:) and also applied to the *two jaws*: (TA:) or, accord. to IAạr, the حنك is the *lower part of the mouth*, [*beneath the lower jaw*,] and the فقم is the upper part: and the حَنَكَانِ are the *upper and the lower*: (Th, Az, Mgh, TA:) but حَنَك is scarcely ever applied to the *upper* alone: [this art., however, shows instances in which it is thus applied:] (Az, TA:) it is masc.: (Mṣb:) pl. أَحْنَاكٌ, (Mṣb, Ḳ,) which is its only pl. form. (TA.) Sometimes, [as is often the case in modern Arabic,] The *mouth* is meant thereby. ('Ináyeh, MF.) And The *beak*: (Ṣ, 'Ináyeh:) حَنَكُ الغُرَابِ signifying *the beak of the crow*, or *raven*: or *the blackness thereof*: (Ḳ:) or *the blackness of its feathers*: (Er-Rághib, TA:) [whence the saying,] أَسْوَدُ مِثْلُ حَنَكِ الغُرَابِ, (Ṣ,) or مِنْ حَنَكِ الغُرَابِ; respecting which see حَلَكٌ. (TA.) — And ‡ *A party of men seeking after herbage in a district*, or *country, to pasture* [*their animals*] *upon it*: (Ḳ, TA:) pl. أَحْنَاكٌ. (TA.) You say, مَا تَرَكَ الأَحْنَاكُ فِى أَرْضِنَا شَيْئًا, meaning ‡ *The parties of men passing* [*in search of herbage left not in our land anything*]. (TA.)

حُنُكٌ : see حُنْكَةٌ, in two places: = and see also حَنِيكٌ. — Also † *Intelligent*; applied to a woman; (Ḳ;) and, as some say, with ة: (TA:) and to a man: (Ḳ:) so says Fr: (TA:) and pl. of ▼حَنِيكٌ, which signifies the same; (TA;) as does also ▼مَحْنُوكٌ. (IAạr, TA.) — *Eaters*: applied to men. (TA.)

حُنْكَةٌ (Lth, Ṣ, Ḳ) and ▼حُنْكٌ, (Lth, Ḳ,) or ▼حِنْكٌ, (Ḳ,) and ▼حُنُكٌ, (Lth, TA,) ‡ *Firmness*, or *soundness, of judgment*, (Ṣ, Ḳ, TA,) *produced by experience*: (Ḳ, TA:) or *age and experience*, (Lth, TA,) *and knowledge*, or *skill, in affairs*: (TA:) or *experience, and good judgment*: (W p. 176:) or *mature, sound*, or *right, judgment*. (MA.) They say, هُمْ أَهْلُ الحُنْكِ and ▼الحُنُكِ and ▼الحُنْكَةِ ‡ *They are people of age and experience* [&c.]. (Lth, TA.) = Also the first, (Ṣ, Ḳ,) and ▼حِنَاكٌ, (Ḳ,) [or] the latter is pl. of the former, (A 'Obeyd, Ṣ,) [or is also pl. of the former,] *A thong*, (قِدَّةٌ, A 'Obeyd, Ṣ, Ḳ, [in the CḲ قُدَّةٌ,]) or *a piece of wood*, (Ḳ,) *which conjoins the* [*pieces of wood called*] عَرَاصِيف, (so in two copies of the Ṣ,) or غَرَاضِيف, (Ḳ, TA, [in the CḲ الغَراضِيفُ,]) *of the* [*saddle called*] رَحْل: so in the T. (TA.)

حِنَاكٌ *A bond for the neck, with which a captive is bound*: *whenever it is pulled, it goes against*, or *hurts, his* حَنَك [i. e., *the part beneath the chin and lower jaw*]. (TA.) — You say also أَخَذَ بِحِنَاكِ صَاحِبِهِ, meaning *He laid hold upon the* حَنَك [or *part beneath the chin and lower jaw*], *and the* لَبَب [or *part between the collar-bones*], *of his companion, and then dragged him to him*. (TA.) — See also المُحَنَّكُ: — and see حُنْكَةٌ.

حَنِيكٌ ‡ *A man rendered firm*, or *sound, in judgment, by means of experience*; (Ḳ, TA;) as also ▼مُحَنَّكٌ and ▼مُحْتَنَكٌ (Ṣ, Ḳ) and ▼مُحْتَنِكٌ (Ḳ) and ▼مُحْتَنَكٌ (TA) and ▼حُنُكٌ, q. v.: (Fr, Ḳ:) or محنك [i. e. either ▼مُحَنَّكٌ or ▼مُحْنَكٌ], accord.

to Lth, signifies a man *whom the management of affairs has rendered experienced so that nothing that he does is despised*: and † مُحْتَنِكٌ, a man *whose intellect and age have reached the utmost degree* [*of maturity*]. (TA.) — Also † *An old man.* (IAạr, TA.) — And † *Niggardly, tenacious,* or *avaricious.* (AA, TA.) — And حَنِيكَةٌ *A good eater;* applied to a دَابَّة [or beast]; (Ḳ;) to a she-camel, and to a sheep or goat. (TA.)

أَسْوَدُ حَانِكٌ i. q. حَالِكٌ, (Ṣ, Ḳ,) i. e. *Black that is intensely black.* (TA.)

أَحْنَكُ (Ṣ, Ḳ) in the saying هٰذَا البَعِيرُ أَحْنَكُ الإِبِلِ *This camel is the most voracious of the camels,* (Ṣ,) or in the phrase أَحْنَكُ البَعِيرَيْنِ *the more voracious of the two camels,* (Ḳ,) and أَحْنَكُ الشَّاتَيْنِ *the more voracious of the two sheep or goats,* (TA,) is anomalous, because one does not [regularly] use a word of this kind denoting a natural attribute: (Ṣ, Ḳ:) and it has no verb; (Sb, TA;) like أَبْرَحُ. (L in art. برح.)

مُحْنَكٌ: see حَنِيكٌ, in two places.

المِحْنَكُ and † المِحْنَاكُ, (Ḳ,) the former, only, mentioned by IDrd, (TA,) signify الخَيْطُ الَّذِى يُحْنَكُ بِهِ (Ḳ [so in the CḲ, app. meaning *The string with which the lower jaw of a corpse is tied up:* in a MS. copy of the Ḳ, يُحْنَكُ; as though the meaning were, *the string that is used as a halter, put in a horse's mouth:* but the former I regard as the right reading: in the TA, يحنك, without any syll. signs].)

مُحَنَّكٌ: see مَحْنُوكٌ: — and see also حَنِيكٌ, in two places.

مَحْنُوكٌ A child *whose* حَنَك [i. e. *palate,* or *soft palate,*] *has been rubbed with some chewed dates,* or *some other thing* (Ṣ, Mṣb, Ḳ) *of a similar kind;* (Mṣb;) as also † مُحَنَّكٌ. (Ṣ, Mṣb, Ḳ.) — See also حَنَكٌ.

مُحْتَنِكٌ and مُحْتَنَكٌ: see حَنِيكٌ, in three places.

حنو and حنى

1. حَنَاهُ, (Ḳ,) first pers. حَنَوْتُ, (Ṣ, Mṣb,) aor. ـُ, (Mṣb,) inf. n. حَنْوٌ, (Ḳ, [in the CḲ, حناء is erroneously substituted, as another inf. n., for حَنَّاهُ, another form of the verb,]) *He bent it,* or *curved it;* (Ṣ, Mṣb, Ḳ;) namely, a branch, or stick, or piece of wood, (Ṣ, Mṣb,) and his back; (Ṣ;) as also حَنَاهُ, (Ḳ in art. حنى,) first pers. حَنَيْتُ, (Ṣ, Mṣb,) aor. ـِ, inf. n. حَنْىٌ, (Mṣb,) or حِنَايَةٌ; (Ḳ in art. حنى;) but the former verb is the better known; (ISd, TA in that art.;) and † حَنَّاهُ, (Ḳ and TA in this art.,) inf. n. تَحْنِيَةٌ. (Ḳ in art. حنى.) You say also, لَمْ يَحْنِ ظَهْرَهُ, as meaning *He did not bend his back for the act of* رُكُوع [in prayer]. (TA.) And حَنَاهُ الدَّهْرُ *Time bent him,* by reason of age. (Mṣb.) And فُلَانٌ مِمَّنْ لَا تُحْنَى عَلَيْهِ الأَصَابِعُ [*Such a one is of those at the mention of whom the fingers are not bent;* (see a portion of the first paragraph of art. ثنى, commencing with the second sentence;)] i. e. *such a one is not counted among the brethren.* (Th, TA.) And [hence] a poet says,

* بَرَكَ الزَّمَانُ عَلَيْهِمُ بِجِرَانِهِ *
* وَأَلَحَّ مِنْكَ بِحَيْثُ تُحْنَى الإِصْبَعُ *

[*Time,* or *fortune, bore upon them with its weight, and was pressing in its exactions from thee, so that the finger was bent*]; meaning that it took the best, those that were counted [with the fingers]. (IAạr, TA.) — حَنَا قَوْسًا, inf. n. حَنْوٌ, (Ḳ,) *He strung,* or *braced the string of, a bow;* because, in doing so, one bends it: (TA:) or *he made a bow.* (Ḳ.) — حَنَا يَدَهُ *He twisted his arm,* or *hand;* (Ḳ in this art.;) as also حَنَى يَدَهُ, aor. ـِ, inf. n. حِنَايَةٌ. (Ḳ in art. حنى.) — حَنَا also signifies *He peeled* a branch, or stick; or *stripped* it *of its bark;* (ISd, TA in art. حنى;) and so حَنَى; (ISd and Ḳ in that art.;) but the former is the better known. (ISd, TA.) — حَنَتْ عَلَى وَلَدِهَا, (Ṣ, Mṣb, Ḳ,) aor. ـُ, (Ṣ, Mṣb,) inf. n. حُنُوٌّ, (Ṣ, Ḳ;) and حَنَتْ, aor. ـِ; (Mṣb;) *She* (a woman) *undertook the care and maintenance of her children, and did not marry again, after* [*the loss of*] *their father:* (AZ, Ṣ:) or, (Mṣb, Ḳ,) as also † احنت, (Hr, Ḳ,) *she acted affectionately towards her children,* (Mṣb, Ḳ,) *and did not marry again, after* [*the loss of*] *their father.* (Mṣb.) And حَنَوْتُ عَلَيْهِ *I was affectionate, kind,* or *compassionate, towards him.* (Ṣ.) And † تحنّى عَلَيْهِ, also, *He was affectionate, kind,* or *compassionate, towards him;* like تحنّن. (Ṣ.) Accord. to IAạr, حَنَا عَلَى قَرَابَتِهِ and † احنى and † حنّى all signify the same as رَتَمَ [evidently, I think, a mistranscription for رَحِمَ; i. e. *He compassionated his relations*]. (TA.) — حَنَتْ, (Ṣ, M, TA,) aor. ـِ, (TA,) inf. n. حُنُوٌّ, (M, TA,) said of a ewe, and of a wild cow, *She desired the male,* (Ṣ, M, TA,) *and submitted herself to him.* (M, TA.)

2: } see 1; for each in two places.
4: }

5: see 7: — and see also 1, near the end of the paragraph.

7. انحنى *It* (a thing, Ṣ, as, for instance, a branch, or stick, TA) *was,* or *became, bent,* or *curved;* (Ṣ, Ḳ;) as also † تحنّى: and the latter, *it* (a حِنْو, q. v.) *was,* or *became, crooked.* (TA.) [Also *He bent,* or *bowed, himself.*] And انحنى مِنَ الكِبَرِ [*He became bent,* or *bowed, by reason of age*]. (Mṣb.)

حَنْوٌ: see what next follows.

حِنْوٌ (Ṣ, Mgh, Ḳ) and † حَنْوٌ (Ḳ) *Anything in which is a bending, curving,* or *crookedness,* (Ḳ,) *and the like;* (TA;) of the body, (Ḳ,) such as the *bone beneath the eyebrow,* (Ḳ,* TA,) and the *jaw-bone,* and *a rib;* and of other things, such as the [*high ground termed*] قُفّ, and the [*tract of sand termed*] حِقْف, (Ḳ,) and the *bend of a valley* [like مَحْنِيَة &c.]: (TA:) see مَحْنِيَة: and *any curved piece of wood,* (Ḳ,) such as *any of the curved pieces of wood of the horse's saddle,* and *of the* [*camel's saddle called*] قَتَب, (Ṣ,* CḲ, TA,) and *of the* [*camel's saddle called*] رَحْل: (CḲ, TA:) *each of* [*the curved pieces of wood called*] *the* قَرَبُوسَانِ, the *anterior* and the *posterior, of the horse's saddle:* (Mgh:) and *a bending, curving,* or *crookedness,* of anything: whence, حِنْوُ الجَبَلِ [*the bend of the mountain*]: (Ṣ:) pl. [of pauc.] أَحْنَآءٌ (Ṣ, Mgh, Ḳ) and [of mult.] حُنِىٌّ, [in the CḲ, erroneously, حَنِىٌّ,] and حِنِىٌّ [originally حُنُوٌّ, first altered to حُنِىٌّ, and then, because of the kesreh, to حِنِىٌّ]. (Ḳ, TA.) — [The dual] حِنْوَانِ signifies *The two curved pieces of wood, with a net upon them, by means of which wheat is conveyed to the heap that is collected together in the place where it is trodden out.* (Ḳ.) — And the pl. أَحْنَآءٌ, i. q. جَوَانِبُ [as meaning † *Sides, regions, quarters,* or *the like;* and *limits, bounds,* or *boundaries*]; like أَعْنَآءٌ. (Ṣ.) The saying, أَزْجُرْ أَحْنَآءَ طَيْرِكَ means نَوَاحِيَهُ, [i. e. † *Straiten thou the limits,* or *bounds, of thy levity,* (alluding to the restraining of birds from flight,)] *on the right and left, and before and behind;* by طيرك being meant خِفَّتَكَ and طَيْشَكَ. (Ṣ, TA.) — أَحْنَآءُ الأُمُورِ † *Dubious affairs:* (Ḳ,* TA:) or, as some say, it means أَطْرَافُهَا وَنَوَاحِيهَا [† *the ends, and limits,* or *bounds, of affairs*]. (TA.)

حَنْوَةٌ *A stooping of the head, and bowing of the back,* in prayer. (TA.)

حِنَآءٌ The *desire of a ewe, and of a wild cow, for the male.* (Ṣ.) [See also حِرْمَةٌ.]

حَنِىٌّ: see what next follows.

حَنِيَّةٌ *A bow;* (T, Ṣ, Ḳ;) so called because it is bent: (T:) pl. † حَنِىٌّ, (Ṣ,* Ḳ,) [or this is a coll. gen. n., of which حَنِيَّةٌ is the n. un.,] and حَنَايَا, (T, Ṣ, Ḳ,) or this may be pl. of حَنِىٌّ. (TA.) — Also applied to *A* [*bowed,* or *curved,*] *structure* or *building.* (M, TA.)

حِنَايَةٌ *A state of bending,* or *curving.* (Ḳ.) Hence the saying, respecting a man in whose back is a bending, إِنَّ فِيهِ لَحِنَايَةً يَهُودِيَّةً [*Verily in him is a Jewish bending.*] (TA.)

حَانٍ [*Bending,* or *curving*]. — [Hence,] حَانِيَةٌ, applied to a ewe, (Ḳ,) and to a she-camel, (TA,) *That twists her neck, without any disease;* (Ḳ;) as also † حَنْوَآءُ, applied to a ewe or she-goat: (TA:) and sometimes, the former, *from a disease.* (TA.) — And حَانِيَةٌ, applied to a woman, *That undertakes the care and maintenance of her children,* (AZ, Ṣ, TA,) or *acts affectionately towards them,* (TA,) *and does not marry again, after* [*the loss of*] *their father:* (AZ, Ṣ, TA:) pl. حَوَانٍ and حَانِيَاتٌ; the latter applied in a similar sense to she-camels. (TA.) — And حَانٍ, applied to a ewe, (Aṣ, Ṣ, M,) and to a wild cow, (Ṣ,) *Desiring the male,* (Aṣ, Ṣ, M,) *and submitting herself to him:* (M:) and حَانِيَةٌ, (Ḳ,) applied to a ewe, (TA,) *vehemently desirous of the male.* (Ḳ.)

حَانَاةٌ: see what next follows.

حَانُوتٌ (Mgh and Ḳ in this art., and Mṣb in art. حون, and mentioned also in the Ḳ in arts. حنت and حين) *A shop;* (Mṣb, Ḳ;) as also ▾ حَانَاةٌ and ▾ حَانِيَةٌ, (Ḳ,) but this last was unknown to Sb: (TA:) [see دُكَّانٌ, in art. دكن:] and particularly the *house* [or *shop*] *of a vintner, in which wine is sold;* as also حَانَةٌ [mentioned in art. حون]; (Mṣb, TA;) called by the people of El-'Irák مَاخُور: (TA:) the first of these words (i. e. حانوت) is said to be of the measure فَعَلُوتٌ, (AAF, Mṣb, TA,) like رَهَبُوتٌ and مَلَكُوتٌ; the و [of the root] being changed into ا, as in طَالُوت and جَالُوت and the like: (Mṣb:) or originally فَعَلُوتٌ, and then فَلَعُوتٌ, like طَاغُوت: (Mgh:) or it is originally حَانُوَةٌ, like تَرْقُوَةٌ: (Mgh, Mṣb: [but the former is said in the Mgh to be the right opinion:]) or it is originally of the measure فَاعُولٌ; (M, Mṣb;) so says El-Fárábee; (Mṣb;) from حَنُوتٌ; what is so called being likened to the building termed حَنِيَّةٌ; the ت being a substitute for و, as is said by AAF, (M,) and being originally ة, changed into ت because of the quiescent letter before it: (Mṣb:) Az says that the ت is augmentative: (TA:) the word is fem. and masc.: (Mgh, Mṣb, TA:) accord. to Zj, when masc., meaning بَيْتٌ: (Mṣb:) pl. حَوَانِيتُ. (Mṣb, TA.) — Also The *vintner* himself: (Ḳ in art. حنت:) [like حَانِىٌّ &c.]

حَانِيَةٌ [fem. of حَانٍ, q. v.] — The pl. الحَوَانِى [as a subst.] signifies *The longest of the ribs;* (Ḳ;) *four in number; two on either side.* (TA.) — And حَوَانِى الهَرَمِ *The benders of the back of the old man, that cause him to fall prostrate.* (TA. [But what this means I know not.]) — See also حَانُوتٌ.

حَانِىٌّ rel. n. of حَانِيَةٌ, as also ▾ حَانَوِىٌّ; (TA;) or of حَانَةٌ; (Mṣb;) and some say that ▾ حَانَوِىٌّ is rel. n. of حَانَاةٌ: (TA:) and ▾ حَانُوتِىٌّ is rel. n. of حَانُوتٌ; applied to a man [as meaning *A shopman,* or the *occupant of a shop:* and particularly, as also the first and second, *a vintner:* see also حَانُوتٌ, last sentence; and see the next paragraph]: (Mṣb:) or the Arabs did not say ▾ حَانُوتِىٌّ. (Fr, TA in art. حنت.)

حَانِيَّةٌ *Wine;* (Ṣ and Ḳ in art. حين, and Ḳ in the present art.;) a rel. n. from حَانَةٌ [mentioned in art. حون]: (Ṣ, TA:) or *vintners;* (Ḳ in this art.;) [from حَانِىٌّ, q. v.; or] a rel. n. from حَانِيَةٌ. (TA.)

حَانَوِىٌّ : } see حَانِىٌّ; for each in two places.
حَانُوتِىٌّ :

أَحْنَى الظَّهْرِ A man *having a* [*bent, bowed,* or] *humped back:* fem., applied to a woman, حَنْوَآءُ and حَنْيَآءُ: (Ṣ:) the former fem. also applied, in the same sense, to a she-camel. (Ḳ.) — See also the former fem. voce حَانٍ. — فُلَانٌ أَحْنَى النَّاسِ ضُلُوعًا عَلَيْكَ †*Such a one is the most affectionate, kind,* or *compassionate, of men towards thee.* (Ṣ.)

مَحْنَاةٌ : } see what next follows.
مَحْنُوَةٌ :

مَحْنِيَةٌ *A bend,* or *place of bending,* of a valley; (Ṣ, Ḳ;) *below the acclivity;* (TA;) as also ▾ مَحْنُوَةٌ and ▾ مَحْنَاةٌ: (Ḳ:) pl. مَحَانٍ; (Ṣ;) with which أَحْنَآءُ [pl. of ▾ حِنْوٌ] is syn: (TA:) and ▾ مُنْحَنَى الوَادِى [in like manner] signifies *the* [*bending*] *part of the valley that is below the acclivity.* (TA.) — Also, [in one copy of the Ḳ written مَحْنِيَّةٌ,] *A bending,* or *curving, tract* of ground, (Sb, ISd, Ḳ,) whether of sand or of any other kind: (Sb, ISd:) Sb says that the ى is substituted for و, because it is from حَنَوْتُ; which shows that he did not know حَنَيْتُ. (ISd.) And مَحْنِيَةُ رَمْلٍ *The part of a tract of sand over which winds* [*a portion such as is termed*] *a* حِقْف. (TA.) — Also *A milking-vessel* (عُلْبَة, q. v.,) *made of the skins of camels* [or rather *of a piece of the skin of a camel*]: *sand is put into a piece of the skin, which is then suspended, and dries, becoming like a bowl:* (Ḳ:) it is more convenient (أَرْفَقُ) to the pastor than other kinds. (TA.)

مَحْنُوٌّ and ▾ مَحْنِىٌّ *Bent,* or *curved.* (Mṣb.)

مَحْنِىٌّ: see what next precedes.

مَحْنِيَةٌ: see مُنْحَنَى الوَادِى.

حو

1. حَوِىَ [originally حَوِوَ]: see what next follows.

9. اِحْوَوَى, (Aṣ, Ṣ, Ḳ,) [like اِرْعَوَى, originally اِحْوَوَّ, of the measure اِفْعَلَّ, then اِحْوَوَوَ, and then اِحْوَوَى] aor. يَحْوَوِى, inf. n. اِحْوِوَآءٌ, said of a horse, (Aṣ, Ṣ,) *He was,* or *became, of the colour termed* حُوَّةٌ [q. v. infrà]; as also ▾ اِحْوَاوَى, (Aṣ, Ṣ, Ḳ,) [originally اِحْوَاوَّ, of the measure اِفْعَالَّ, then اِحْوَاوَوَ, and then اِحْوَاوَى,] aor. يَحْوَاوِى, inf. n. اِحْوِيوَآءٌ, (Aṣ, Ṣ,) or, accord. to ISd, correctly, اِحْوِيَّآءٌ, because the ى changes the و [after it] into ى, as it does in أَيَّامٌ [which is originally أَيْوَامٌ]; (TA;) and ▾ اِحْوَوَّى, (ISd, Ḳ,) [accord. to the pronunciation of the Koofees, as will be seen below, originally اِحْوَوَّوَ,] said by IB to be found thus written in some of the copies of the book of Aṣ, [that entitled كِتَابُ الفَرَسِ,] but to be a mistake, because it is agreed that there is not in the language a verb ending with three letters of the same kind except اِبْيَضَضَ [for اِبْيَضَّ]; (TA;) and ▾ حَوِىَ, (Aṣ, Ṣ, Ḳ,) like رَضِىَ, (Ḳ,) [originally حَوِوَ,] aor. يَحْوَى, inf. n. حُوَّةٌ, (Aṣ, Ṣ,) or حَوًى; (Ḳ;) this last verb mentioned by Aṣ as used by some of the Arabs. (Ṣ.) And ▾ اِحْوَاوَتِ الأَرْضُ *The land was,* or *became, green;* as also ▾ اِحْوَوَتْ; (Ḳ;) [or the latter is correctly اِحْوَوَتْ:] IJ says that ▾ اِحْوَاوَتْ is of the measure اِفْعَالَّتْ, [originally اِحْوَاوَوَتْ, then اِحْوَاوَتْ,] and that the Koofees say اِحْوَوَّتْ and ▾ اِحْوَوَتْ; but ISd says that their usage is wrong, for the Arabs say اِحْوَوَى, like اِرْعَوَى, and do not say اِحْوَوَّ. (TA.)

11. اِحْوَاوَى : } see 9; for each in three places.
13. اِحْوَوَّى :

حُوَّةٌ [*A brown colour;*] *redness inclining to blackness:* (Aṣ, Ṣ, Ḳ:) or *a colour intermixed with* [*the blackish red termed*] كُمْتَة, *like the rust of iron:* (Ṣ:) or *blackness inclining to greenness.* (Ḳ.) In the lip, [The *brownish colour termed*] سُمْرَة; (Ṣ;) [i. e.] *a colour resembling* [*that termed*] اللَّمَى *and* اللَّعَس: (T, TA:) or *a blackness* in the lips; which is approved. (Ḥam p. 386.)

أَحْوَى *Of the colour termed* حُوَّة [q. v. suprà]: and also *black:* (Ḳ:) or *black by reason of* [*intense*] خُضْرَة [by which may be here meant either *greenness,* or *dark,* or *ashy, dust-colour*]: (TA:) applied to a horse, *i. q.* كُمَيْتٌ [i. e. *bay*] *overspread with blackness;* (TA;) or *red in the back;* (En-Naḍr, TA;) or *more yellow than, but nearly the same as, such as is termed* أَحْمَر, so that one swears, of such a horse, that he is أَحْمَر: (AO, TA: [see مُدَمًّى:]) applied to a camel, *whose* خُضْرَة [here meaning *dark,* or *ashy, dust-colour*] *is intermixed with blackness and yellowness:* (Ṣ:) applied to a man, *having* [*a brownish colour such as is termed*] سُمْرَة *in the lip;* (Ṣ;) or *having a blackness in the lips,* which is approved; (Ḥam p. 386;) fem. حَوَّآءُ, applied to a woman, (Ṣ,) and also to a lip (شَفَةٌ) as meaning *red inclining to blackness:* (Ḳ:) applied to a plant, *inclining to blackness by reason of its intense greenness;* (Ḳ;) and such is the softest of plants: (TA:) the pl. is حُوٌّ; occurring in a trad., in which the best of horses are said to be those thus termed: (TA:) the dim. of أَحْوَى is ▾ أُحَيْوٍ, in the dial. of him who says أُسَيْوِدُ [instead of أُسَيِّدُ, dim. of أَسْوَدُ]; but there is a difference of opinion as to the form with idghám: 'Eesà Ibn-'Omar says ▾ أُحَيُّ, making it perfectly decl., which Sb pronounces a mistake: 'Amr Ibn-El-'Alà, or Aboo-'Amr Ibn-El-'Alà, (accord. to different copies of the Ṣ,) says ▾ أُحَيِّ, after the manner of أُحَيْوٍ, which, also, Sb disallows: Yoo says ▾ أُحَىُّ, and this, says Sb, is the regular and right form. (Ṣ.) [Hence,] بَكْرَةٌ حَوَّآءُ *A sheave of a pulley formed of black wood.* (TA.) And نَمْلٌ حُوٌّ *Red ants; called* نَمْلُ سُلَيْمَانَ. (TA.) فَجَعَلَهُ غُثَآءً أَحْوَى, in the Ḳur [lxxxvii. 5], means, accord. to Fr, *And hath made it* (the herbage mentioned before) *dried up, black by reason of oldness:* or it may mean *and hath made it to become* غُثَآء [or *decayed,* or *dried-up, leaves and stalks,*] after it has been green. (TA.)

أَحْوِىٌّ rel. n. of أَحْوَى. (TA.)

أُحَىُّ and أُحَيِّ and أُحَيُّ: see أَحْوَى.

أُحَيْوٍ: see أَحْوَى.

حوا

حَآءٌ *The name of the letter* ح, q. v.; as also حَا: (Ṣ and Ḳ and TA in باب الألف اللّيِّنة:) it is called حَا in a case of pause, and حَآءٌ when made a noun: and when it is not called a letter, [i. e. when one does not prefix to it the word حَرْف,] it is [properly] fem.: its dim. is حُيَيَّةٌ, meaning *a* ح *written small*, or *indistinct*: (Lth, TA ubi suprà:) and its pl. is أَحْوَآءٌ and أَحْيَآءٌ and حَاآتٌ. (TA ubi suprà.)

حَائِىٌّ and حَاوِىٌّ and حَوِىٌّ [a mistranscription for حَوَوِىٌّ, like نَوَوِىٌّ and تَوَوِىٌّ,] rel. ns. of حَآءٌ and حَا the names of the letter ح. (B, TA ubi suprà.)

حوب

1. حَابَ, (Mṣb, Ḳ,) sec. pers. حُبْتَ, (Ṣ,) aor. يَحُوبُ, (Ṣ, Mṣb,) inf. n. حَوْبٌ (Ṣ, Mṣb, Ḳ) and حَوْبَةٌ and حِيَابَةٌ, (Ṣ, Ḳ, accord. to one copy of the Ḳ حِيَابٌ,) and حِيبَةٌ (TA) and حُوبٌ; (Ḳ;) or this last is a simple subst.; or, as some say, it and حَوْبٌ are two dial. vars.; that with ḍamm, of the dial. of El-Ḥijáz; and that with fet-ḥ, of the dial. of Temeem; (Mṣb;) accord. to Zj, that with ḍamm signifies "sin, or crime;" and that with fet-ḥ, the "act" of a man; [i. e. the "act of committing a sin, or crime;"] (TA;) *He sinned; committed a sin*, or *crime*; *did what was unlawful*; (Ṣ, Mṣb, Ḳ;) بِكَذَا [*by such a thing*]. (Ṣ, Ḳ.) —— Also, aor. as above, [inf. n. not mentioned,] *He*, or *it*, *became in an evil condition*, or *state*. (TA.) —— *He slew* [another]: of the dial. of the tribe of Asad. (TA.) ═ حَوْبٌ also signifies The act of *chiding* a male camel [by the cry حَوْبِ]. (Lth, TA.) [See also 2.]

2. حَوَّبَ بِالإِبِلِ, (Ṣ, Ḳ,*) inf. n. تَحْوِيبٌ, (Ḳ,) *He chid the camels* (Ṣ, Ḳ) *by the cry* حَوْبِ حَوْبِ. (Ṣ.) [See also 1.]

4. أَحْوَبَ *He pursued a course that led him to sin*, or *crime*. (Ḳ, TA.) ═ مَا أَحْوَبَهُ for مَا أَحْيَبَهُ: see 4 in art. حب.

5. تَحَوَّبَ *He abstained from, shunned*, or *avoided, sin*, or *crime*; *put it away from himself*: (A 'Obeyd, Ṣ, Ḳ, TA:) *he applied himself to acts*, or *exercises, of devotion*; *became devout*, or *a devotee*. (IJ, TA.) Here the form تَفَعَّلَ is deprived of the radical signification, as in the cases of the syn. words تَأَثَّمَ and تَحَنَّثَ; though its property is oftener to confirm the radical signification. (TA. [See تحنّث.]) You say, تحوّب مِنْ كَذَا *He abstained from such a thing as a sin*, or *crime*. (A 'Obeyd, Ṣ, TA. [See also another explanation below.]) —— *He humbled himself* in his prayer, or supplication. (TA.) —— *He expressed pain, grief*, or *sorrow*; *lamented*, or *complained*. (Ṣ, Ḳ,* TA.) And تحوّب مِنْ كَذَا *He was enraged, and expressed pain* or *grief* or *sorrow*, or *lamented*, or *complained, by reason of such a thing*. (TA. [See another explanation above.]) —— *He cried out, expressing pain* or *grief* or *sorrow*, or *lamenting*, or *complaining*: *he cried aloud*, or *vehemently*, in prayer, or supplication. (TA.) *He wept, in impatience*, or *sorrow, and with loud crying*: and sometimes, in a general sense, *he cried out*, or *aloud*, (TA.) —— *He* (a jackal) *cried*, or *howled*: because his cry is like that of a person expressing pain or grief or sorrow, or lamenting, or complaining, as though he were writhing from the pain of hunger or beating. (Ṣ, TA.)

حَبِ and حَبْ: } see حَوْبِ, in five places.
حَابِ and حَابْ: }

حُوبٌ: see حَابٌ.

حَوْبِ and حَوْبُ and حَوْبَ (Ṣ, Ḳ) and ↓ حَابِ (Ḳ) *A cry used for chiding a camel*: (Ṣ:) or *a cry by which a male camel is chidden*, (Lth, IAth, Ḳ,) *to urge him on*; (Lth, TA;) like as a she-camel is by the cry حَلْ and حَلِ and حَلِى: the first form (حَوْبِ) is that used by the Arabs [in general]; but the other forms are allowable: حَوْبْ حَوْبْ also occurs, with the ب quiescent; and حَوْبًا حَوْبًا occurs in a trad., in the same sense: also, حَبِ لَا مَشَيْتَ, and ↓ حَبِ and ↓ حَابِ and ↓ حَابِ [*O! mayest thou not walk*, or *mayest thou not be rightly directed*; حب &c. being syn. with حَوْبِ, and followed by an imprecation]. (TA.) Hence, حَوْبُكَ هَلْ يُعْتَمُ بِالسَّمَارِ *Urge on! Should a delay be made in bringing milk much diluted with water?* i. e., if thou entertain with milk much diluted with water, wherefore tardiness? a prov., applied to him who delays the fulfilment of his promise, and then gives little. (MF.)

حَوْبٌ: see حُوبٌ, in two places: ═ and see also حَوْبَةٌ, in four places. —— Also *Grief*, or *sorrow*: and *loneliness*, or *solitariness*: and so ↓ حُوبٌ, in both these senses. (Ḳ.) —— *Difficulty, distress, trouble*, or *fatigue*; syn. جَهْدٌ. (Ḳ. [That جهد is to be thus understood here is indicated in the TA.]) —— *Pain*. (Ḳ.) ═ A *difficult* road. (TA.) ═ *A kind*, or *sort*: and *a mode*, or *manner*. (Ḳ, TA.) You say, سَمِعْتُ مِنْ هٰذَا حَوْبَيْنِ *I heard*, or *have heard, of this, two kinds*, or *modes*: and رَأَيْتُ مِنْهُ حَوْبَيْنِ *I saw*, or *have seen, of it, two kinds*, or *modes*. (TA.) ═ *A he-camel*: (Ḳ:) or *a bulky he-camel*: so called from the cry حَوْبِ, by which he is urged; like as a mule is called عَدَسٌ: (Lth, TA:) or it signifies originally a he-camel, and hence, from its frequency of usage, the cry حوب by which he is urged. (Ḳ,* TA.)

حُوبٌ (Ṣ, A, Mṣb, Ḳ) and ↓ حَوْبٌ, (Mṣb,* Ḳ,) said by some to be two dial. vars., (Mṣb, [see 1, first sentence,]) and ↓ حَابٌ (Ṣ, Ḳ) and ↓ حُوبَةٌ (A 'Obeyd, Ḳ) and ↓ حَوْبَةٌ (A 'Obeyd, TA) and ↓ حَابَةٌ (Ḳ) and ↓ حِيبَةٌ, (TA,) *Sin*, or *crime*: or *a sin*, or *a crime*: (Ṣ, A, Mṣb, Ḳ:) accord. to A 'Obeyd, the first and second signify *any sin* or *crime*; (TA;) [as also, app., حَابٌ;] and حوبة [i. e. حُوبَةٌ and حَوْبَةٌ, the former particularly mentioned in the Mṣb, and app. حَابَةٌ also], *a single sin* or *crime*: (Mṣb, TA:) accord. to Fr, حُوبٌ signifies *great sin*, or *a great sin*: accord. to Ḳatádeh, *wrong, injustice*, or *tyranny*: thus in the Ḳur iv. 2; where El-Ḥasan read ↓ حَوْبًا instead of حُوبًا. (TA.) One says, رَبِّ تَقَبَّلْ تَوْبَتِى وَاغْسِلْ ↓ حَوْبَتِى (T, TA) i. e. [*O my Lord, accept my repentance, and wash away*] *my sin*, or *crime*. (A 'Obeyd, TA.) El-Mukhabbal Es-Saadee says,

فَلَا تُدْخِلَنَّ الدَّهْرَ قَبْرَكَ حَوْبَةً ↓
يَقُومُ بِهَا يَوْمًا عَلَيْكَ حَسِيبُ

[*Then introduce not thou, ever, into thy grave, a sin with which a reckoner*, or *taker of vengeance, may one day rise up against thee*]. (TA.) ═ حُوبٌ also signifies *Perdition, destruction*, or *death*. (Ḳ.) [Hence, app.,] ابْنَةُ حوبٍ *A quiver*; syn. كِنَانَةٌ. (TA. [The vowel of the ح is not indicated.]) —— *Disease*. (Ḳ.) —— *A trial, a trouble*, or *an affliction*. (Ḳ.) You say, هٰؤُلَآءِ عِيَالُ أَبِى حُوبٍ [*These are the family of the father of trouble*; i. e., *of one who is in trouble*]. (TA.) —— See also حَوْبٌ. ═ And see حَوْبَةٌ.

حَابَةٌ: see حُوبٌ.

حَوْبَةٌ: see حُوبٌ, in three places. ═ Also *Maternal tenderness of heart*. (Ḳ.) —— *Anxiety*; (Ṣ, Ḳ;) and so ↓ حِيبَةٌ. (TA.) —— *Want; poverty; indigence*; (Ṣ, Ḳ;) as also ↓ حِيبَةٌ and ↓ حَوْبٌ. (Ḳ.) You say, in prayer, إِلَيْكَ أَرْفَعُ حَوْبَتِى i. e. [*To Thee I make known*] *my want*. (TA from a trad.) And أَلْحَقَ ٱللّٰهُ بِهِ الحَوْبَةَ *May God bring upon him want*, or *poverty*, or *indigence*. (Ṣ,* TA.) [And hence,] ↓ ابْنُ حَوْبٍ *A man oppressed by difficulty, trouble, distress*, or *adversity*; *a man in need*: i. e. *any man in such a state*. (IAạr, TA.) And ↓ عِيَالُ ٱبْنِ حَوْبٍ [*The family of a man oppressed by difficulty, &c.*]. (TA.) —— *A state*, or *condition*; as also ↓ حِيبَةٌ: (Ḳ:) but only used in speaking of an evil state; as in the phrases, بَاتَ بِحَوْبَةِ سَوْءٍ and ↓ بِحِيبَةِ سَوْءٍ *He passed the night in an evil state* or *condition*. (TA.) —— [Hence also, for ذُو حَوْبَةٍ, and ذَاتُ حَوْبَةٍ, and ذَوُو حَوْبَةٍ,] *A weak man*; (AZ, Ṣ, Ḳ;) as also ↓ حُوبَةٌ: (Ḳ:) and *a weak woman*: (TA:) and *weak persons*: (Ṣ:) and [*a man who can neither profit nor harm*; or] *a man having neither good nor evil*: (Ṣ:) pl. حُوَبٌ. (AZ, Ṣ.) It is said in a trad., اِتَّقُوا ٱللّٰهَ فِى الحَوْبَاتِ, for ذَوَاتِ الحَوْبَاتِ, i. e. *Fear ye God with respect to the needy women, who cannot do without some one to maintain them, and to take constant care of them*. (TA.) And you say, إِنَّ لِى حَوْبَةً أَعُولُهَا *Verily I have a weak family to maintain*. (Ṣ.) —— *A person whom one is under an obligation to respect*, or *honour*, or *defend, and who may be subjected to loss*, or *ruin*, [*if abandoned*,] *such as a mother*, or *sister*, or *daughter*, or *any other female relation within the prohibited degrees of marriage*; as also ↓ حِيبَةٌ: (ISk, Ṣ:) *any such relation whom it is sinful to subject to loss*, or *ruin, by abandoning her*: (A 'Obeyd, TA:) or *a mother*: (Ḳ:) by some explained peculiarly as having this meaning: (A 'Obeyd, TA:) and *a* *wife*; or *a concubine*; (K;) because both require to be maintained: (TA:) and, as also ▼ حَوْبٌ, The *father and mother*: and *a sister*: and *a daughter*. (K.) You say, لِى فِى بَنِى فُلَانٍ حَوْبَةٌ and ▼ حِيبَةٌ (ISk, S, K*) and ▼ حُوبَةٌ (K) *I have, among the sons of such a one, a female relation such as any of those above specified*: (ISk, S:) or *one to whom I bear relationship on the side of the mother*: (K:) or *a relation within the prohibited degrees of marriage*. (AZ, TA.) — *A sacred*, or *an inviolable, right* of a person, *which it would be sinful to disregard*; as in the saying, فَعَلْتُهُ بِحَوْبَةِ فُلَانٍ [*I did it for the sake of the sacred*, or *inviolable, right of such a one*]. (A.) — *A horse*, or *similar beast*; syn. دَابَّةٌ: (K:) for this, also, cannot do without some one to take constant care of it, and to sustain it. (TA.) = The *middle of a house*. (K.) Perhaps the ب in this instance is a substitute for م. (TA.)

حُوبَةٌ: see حُوبٌ: = and see also حَوْبَةٌ, in two places. = حُوبَةٌ مِنَ الأَرْضِ *A bad tract of land*; as also ▼ حِيبَةٌ. (TA.)

حِيبَةٌ: see حُوبٌ: = and see also حَوْبَةٌ, in six places: = and حُوبَةٌ.

حَوْبَآءُ The *soul*; syn. نَفْسٌ; (AZ, S, K;) as also ▼ حُوبٌ: (AZ, K:) or the *soul whose seat is in the heart*; syn. رُوحُ القَلْبِ [also called the *animal soul*, رُوح حَيَوَانِىّ: see art. روح]: AHei asserts, in a disquisition on the heart, that this word is formed by transcription form حَبْوَآءُ: (TA:) pl. حَوْبَاوَاتٌ. (S, K.) You say, حَرَسَ ٱللّٰهُ حَوْبَآءَكَ [*May God guard*, or *preserve, thy soul*]. (A.) — [Also] The *body*, or *person*; in Persian تَن. (KL.)

حَائِبٌ *Slaying*; or *a slayer*: of the dial. of the tribe of Asad. (TA.)

أَحْوَبُ, as an epithet applied to a man, *More*, or *most*, or *very, sinful*, or *criminal*. (S, TA. [This meaning is implied, but not expressed.])

مُحَوِّبٌ, (K,) or, accord. to some, مُحَوَّبٌ, (MF,) and ▼ مُتَحَوِّبٌ, (K,) A man *whose wealth passes away from him, and then returns*. (K.)

مُتَحَوِّبٌ: see what next precedes.

حوت

1. حَاتَ عَلَى الشَّىْءِ, aor. يَحُوتُ, (S,) inf. n. حَوْتٌ and حَوَتَانٌ, (K,) † *He went*, or *circuited, round about the thing*; said of a bird, (S, K, TA,) going, or circuiting, round about water [like a fish (حُوت) in water]; (TA;) and of a wild animal: (K:) as also حات به. (TA.)

3. حاوته ‡ *He strove*, or *endeavoured, to turn him*, or *to entice him to turn*, syn. رَاوَغَهُ, (S, L, A, &c.,) in the K رَاغَمَهُ, but the former is the right explanation, (TA,) عَنْ شَىْءٍ *from a thing*, as, for instance, his family, and his property; (S, TA;) ‡ *he strove*, or *endeavoured, to beguile him*. (A.) You say, ظَلَّ يُحَاوِتُنِى بِخُدْعَةٍ ‡ *He passed the day striving*, or *endeavouring, to turn me*, or *entice me, by guile, like as the fish* (حُوت) *does in the water*. (A.) — † *He strove*, or *contended, with him, to repel him*. (K.) — † *He consulted him*, or *consulted with him*. (K.) — † *He talked with him, consulting*, or *making promises, in the case of a sale*. (K.)

حُوتٌ *Fish*: (M, A, K:) or [rather] *a fish*: (S, TA:) or *a great fish*; *any great fish*: (M, Msb, TA:) of the masc. gender: (Msb:) pl. [of mult.] حِيتَانٌ (S, Msb, K) and حِوَتَةٌ and [of pauc.] أَحْوَاتٌ. (K.) — [Hence,] صَاحِبُ الحُوتِ (in the Kur lxviii. 48) [a surname of *The prophet*] *Jonas*. (Bd, Jel.) — [Hence also,] الحُوتُ † [*The constellation Pisces*;] *a certain sign of the Zodiac*. (S, K.) And الحُوتُ الجَنُوبِىُّ † *The constellation Piscis Australis*. (Kzw &c.) And فَمُ الحُوتِ † *The bright star* [*α, called by European astronomers Fomalhaut*,] *in the mouth of Piscis Australis*. (Kzw &c.)

حُوتِىٌّ [*Of*, or *relating to*, or *like, a fish*, or *great fish*]. You say, هُوَ حُوتِىُّ الاِلْتِقَامِ [*He is like a fish, or great fish, in swallowing*]. (A, TA.)

حَيُّوتٌ The *male of the* حَيَّة [or *serpent*]. (A.) [But the proper place of this is art. حيو.]

حوث

حَوْثُ a dial. var. of حَيْثُ, (S, K,) of the dial. of Teiyi, (Lh, IHsh, K,) or of that of Temeem: (L:) some of the Arabs say حَوْثُ, like as some say حَيْثَ; (Ks, Lh, TA;) and some say حَوْثِ: (Mughnee and TA in art. حيث:) حَوْثُ is the original form; (ISd, TA;) but حَيْثُ is the more chaste of the two forms, and that used in the Kur-án; though both forms are good. (AZ, TA.) See art. حيث.

حوج

1. حَاجَ, (S, Msb, K,) aor. يَحُوجُ, (S, Msb,) inf. n. حَوْجٌ; (S, K;) and ▼ احتاج, (S, K,) inf. n. اِحْتِيَاجٌ; (K;) and ▼ أَحْوَجَ; (S, Msb, K;) [the second of which is the most common;] the last, irreg. [for by rule it should be أَحَاجَ]; (MF;) are syn.; (S, Msb,* K;) and حَاجَ, aor. يَحِيجُ, inf. n. حَيْجٌ, signifies the same; (M, TA;) *He*, or *it, wanted, needed*, or *required*. (KL, TA.) You say, حاج إِلَيْهِ, (M, TA,) and ▼ احتاج اليه, and احتاجه, *He*, or *it, wanted, needed*, or *required, him*, or *it*. (TA.) [And in like manner, ▼ احتاج أَنْ يَفْعَلَ كَذَا *He wanted, needed, required*, or *found it necessary, to do*, or *that he should do, such a thing*.] And ▼ أُحْتِيجَ إِلَى فُلَانٍ [*Such a one was wanted*, or *needed*]. (JK in art. خل.) — Also حاج, aor. and inf. n. as above, *He desired, sought*, or *sought after*. (TA.) — And حاج and ▼ احتاج *He was*, or *became, poor*, or *in poverty* or *want* or *need*. (TA.)

2. حَوَّجَ بِهِ الطَّرِيقُ, inf. n. تَحْوِيجٌ, i. q. عَوَّجَ [*The road led him aside*]. (K.) — حَوَّجْتُ لَهُ, (K,) inf. n. as above, (TA,) *I forsook my way in*, or *in respect of, love of him*. (K.)

4. أَحْوَجَ: see 1. = It is also transitive. (Msb.) You say, أَحْوَجَهُ إِلَيْهِ غَيْرُهُ [*Another made him to want*, or *be in need of, him*, or *it*]. (S.) And أَحْوَجَنِى إِلَيْكُمْ زَمَانُ السَّوْءِ [*Evil time*, or *evil fortune, made me to be in need of thee*]. (A.) And لَا أَحْوَجَنِى ٱللّٰهُ إِلَى فُلَانٍ [*May God not cause me to want such a one*]. (A.) And أُحْوِجْتُ إِلَيْهِ [*I was caused to want him*, or *it*]. (A.) And أَحْوَجَهُ ٱللّٰهُ إِلَى كَذَا [*God caused him to want such a thing*]. (Msb.)

5. تحوّج *He sought an object of want*, or *what he wanted*; (K;) or *one object of want after another*. (TA.) And خَرَجَ يَتَحَوَّجُ *He went forth seeking*, or *seeking leisurely*, or *time after time, what he wanted, of the means of his subsistence*. (A, TA.) And تحوّج إِلَى الشَّىْءِ *He wanted the thing, and desired it*. (L, TA.)

8: see 1, in five places. — Also احتاج إِلَيْهِ *He inclined to him*. (K.)

حَاجٌ: see حَاجَةٌ. = Also *A certain kind of thorny plant* or *tree*. (S, K.) See art. حيج.

حَوْجٌ *Safety*; *freedom from evil, harm*, or *the like*: so in the phrase حَوْجًا لَكَ [God grant *safety to thee*]: (K:) said to a person stumbling. (TA.)

حُوجٌ *Poverty*. (K.)

حَاجَةٌ (S, K, &c.) and ▼ حَائِجَةٌ, (AA, IDrd, &c.,) of which latter the former is a contraction, (Kh in the 'Eyn, TA,) or the latter word is unused [except as an epithet, as will be seen below], and ▼ حَوْجَآءُ, (S, K, &c.,) are syn. words, of which the meaning is well known: (S, K, TA:) *Want*; *need*; *necessity*, or *necessitude*; *exigency*: (TA:) [whence,] اِبْنُ حَاجَةٍ *one who is constantly in want*, or *need*: (Har p. 143:) حَاجَةٌ is a more general term than فَقْرٌ; or each of these terms is more general than the other in some respects, and more particular in others: (TA:) and the former signifies also *a thing wanted, needed*, or *required*; *an object of want, of need*, or *of exigence*; *a want*; *a needful*, or *requisite, thing, affair*, or *business*: (A, TA:) [and *a thing to be done, an affair*, or *a business*:] pl. ▼ حَاجٌ, [or rather this is a coll. gen. n., of which حَاجَةٌ is the n. un.,] and حَاجَاتٌ, [which is of more frequent occurrence,] (S, Msb, K,) and حِوَجٌ, [which is of rare occurrence,] (S, K,) and حَوَائِجُ, (S, Msb, K,) which is anomalous, (S, K,) and was disapproved by As, and pronounced by him to be post-classical, but he disapproved it only because of its being anomalous, (S,) and he is said to have retracted his assertion of its being post-classical, (TA,) for it is of frequent occurrence in the [classical] language of the Arabs, (S,) in their verses and in the traditions: (IB, TA:) it seems as though formed from the sing. حَائِجَةٌ, (S, K,) which some assert to have been not used; or, accord. to some, it may be pl.

of حَوْجَآءُ, changed from the regular form of حَوَائِجُ, [originally حَوَاجِىُ,] like صَحَارٍ, by putting the [elided] ى before the ج, agreeably with what is often done in the language of the Arabs. (TA.) You say, قَضَى حَاجَتَهُ [*He accomplished his want*]: (TA:) a phrase which signifies [also] *he did his business*; meaning *he eased nature.* (ISk, TA.) And خُذْ حَاجَتَكَ مِنَ الطَّعَامِ [*Take what thou wantest*, or *requirest, of the food*]. (A.) And فِى نَفْسِى حَاجَةٌ and ↓ حَائِجَةٌ and ↓ حَوْجَآءُ [*In my mind is a want*]. (AA, TA.) And لِى عِنْدَ فُلَانٍ حَاجَةٌ [*I want a thing of such a one*]. (TA.) [When the thing wanted, or not wanted, is mentioned, or referred to by a pronoun, the subst. denoting it, or the pronoun referring to it, is preceded by إِلَى, as in the sayings لِى حَاجَةٌ إِلَى كَذَا *I have a want of such a thing*, and مَا لِى إِلَيْهِ حَاجَةٌ *I have not any want of it*, and مَا حَاجَتُكَ إِلَيْهِ *What is* the reason of *thy want of it*, or *thy wanting it?*] You say also, مَا بَقِىَ فِى صَدْرِهِ حَوْجَآءُ ↓ وَلَا لَوْجَآءُ إِلَّا قَضَاهَا, *There remained not in his bosom a want but he accomplished it.* (TA.) [The dim. of حَوْجَآءُ is ↓ حُوَيْجَآءُ: whence the saying,] مَا لِى فِيهِ حَوْجَآءُ ↓ وَلَا لَوْجَآءُ وَلَا حُوَيْجَآءُ ↓ وَلَا لُوَيْجَآءُ *I have no want*, [*nor any little want*,] *with respect to him*, or *it*. (Lḥ, Ṣ, Ḳ.) [See also حَوْجَآءُ, below.] In the phrase خُذْ حُجَيَّاكَ [*Seek the little thing that thou wantest*], it seems that the second and third radical letters of the latter word [originally حُوَيْجَآءَكَ] have been transposed [and that حُجَيْوَآءَكَ has been then changed into حُجَيَّاكَ]. (IDrd, AAF, TA.) — مَا تَرَكْتُ مِنْ حَاجَةٍ وَلَا دَاجَةٍ *I left not any act of disobedience to which I was enticed.* (TA from a trad. [But see دَاجَةٌ in art. دوج.])

حَوْجَآءُ: see حَاجَةٌ, in four places: — and see حَائِجَةٌ. — You say also, مَا فِى صَدْرِى بِهِ حَوْجَآءُ وَلَا لَوْجَآءُ *There is not in my bosom any doubt respecting him*, or *it*. (Th, Ṣ, Ḳ.) And لَيْسَ فِى أَمْرِكَ حُوَيْجَآءُ ↓ وَلَا لُوَيْجَآءُ وَلَا رُوَيْغَةٌ [*There is not any doubt in*, or *in respect of, thine affair*, or *thy case*]. (Ṣ.) — And كَلَّمْتُهُ فَمَا رَدَّ عَلَىَّ حَوْجَآءَ وَلَا لَوْجَآءَ, like the phrase فَمَا رَدَّ عَلَىَّ سَوْدَآءَ وَلَا بَيْضَآءَ, i. e. *I spoke to him, and he returned me not a bad word nor a good one.* (ISk, Ṣ, Ḳ.)

حُوَيْجَآءُ: see حَاجَةٌ, in two places: — and see حَوْجَآءُ. — You say also, خُذْ حُوَيْجَآءَ مِنَ الأَرْضِ, *Take thou a different and winding road of the land.* (Ḳ.)

حَائِجَةٌ: see حَاجَةٌ, in two places. — You say also حَاجَةٌ حَوْجَآءُ and ↓ حَاجَةٌ حَائِجَةٌ *A great want; a thing much wanted.* (TA.)

مُحْوِجٌ A man *in want, needy, indigent*, or *poor*; (Mṣb, TA;) [as also ↓ مُحْتَاجٌ:] the pl., by rule, should be مُحْوِجُونَ because it is an epithet applied to a rational being; but the pl. used is مَحَاوِيجُ: some, however, reject this, and assert that it has not been heard [as a classical word]: (Mṣb:) ISd thinks that محاويج is pl. of ↓ مِحْوَاجٌ, if the latter word have been used: (TA:) the vulgar say مَحَاوِجُ [as pl. of مُحْوِجٌ], meaning مُحْتَاجُونَ. (Mgh.)

مِحْوَاجٌ: and مُحْتَاجٌ: see the next preceding paragraph.

حود

1. حَادَ, aor. يَحُودُ, i. q. حَادَ, aor. يَحِيدُ, explained in art. حيد. (Ḳ.)

3. تُحَاوِدُهُ الحُمَّى [like تُعَاوِدُهُ] *The fever returns to him repeatedly*, or *time after time.* (Ḳ.) — هُوَ يُحَاوِدُنَا بِالزِّيَارَةِ *He visits us* [*repeatedly*, or *time after time*,] *during the space of several days.* (TA.)

حوذ

1. حَاذَ الإِبِلَ, aor. يَحُوذُ, (Ṣ, A, L,) inf. n. حَوْذٌ, (Ṣ, L, Ḳ,) *He drove the camels quickly*; (Ṣ, L, Ḳ;) as also ↓ أَحْوَذَهَا, (Ṣ, L,) inf. n. إِحْوَاذٌ: (Ḳ:) or *violently*; (M, L;) like حَازَهَا, inf. n. حَوْزٌ: (L:) or *roughly*: (B:) or *he drove the camels to water*; like حازها. (A. TA.) — Also *He collected the camels together to drive them.* (L.) — And حُذْتُ الإِبِلَ and حِذْتُهَا, *I mastered*, or *gained the mastery over, the camels*: two forms of the verb mentioned by Zj and IKṭṭ and others, as coordinate to قَالَ and خَافَ. (MF, TA.) And حاذ الحِمَارُ أُتُنَهُ *The he-ass gained the mastery over his she-asses, and collected them together*; like حازها: (L:) [and so أَحْوَذَ ↓ جَانِبَيْهَا:] Lebeed says,

* إِذَا اجْتَمَعَتْ وَأَحْوَذَ جَانِبَيْهَا *
* وَأَوْرَدَهَا عَلَى عُوجٍ طِوَالِ *

[*When they became collected together, and he gained the mastery over their flanks*, or] *drew them together so that not one of them escaped him*, [*and brought them to the watering-place*, galloping *upon long crooked legs*; for] by عوج he means [crooked] legs. (Ṣ, L.) — And [hence,] حَاذَهُ, aor. and inf. n. as above; (L;) and ↓ اسْتَحْوَذَ عَلَيْهِ, (Ṣ, A, L, Ḳ,*) and استحاذ; (Ṣ, L;) *He overcame, mastered*, or *gained the mastery over, him*, or *it*: (Ṣ, A, L, Ḳ:) [like حَازَهُ.] You say, استحوذ ↓ عَلَى كَذَا *He mastered such a thing; gained the mastery over it; gained possession of it.* (L.) استحوذ ↓ عَلَيْهِمُ الشَّيْطَانُ [in the Ḳur lviii. 20] means *The devil hath overcome them*, or *hath gained the mastery over them*: (Ṣ, L:) or *hath gained the mastery over their hearts*: (Th, L:) or *hath gained the mastery over them, and inclined them to that which he desired of them*: (Mṣb:) or *drove them, having gained the mastery over them.* (B.) And أَلَمْ نَسْتَحْوِذْ ↓ عَلَيْكُمْ [in the Ḳur iv. 140], *Did we not acquire the mastery over your affairs, and gain possession of your affection?* (Ṣ, L:) or *did we not gain the mastery over you by befriending and aiding you?* (Aboo-Is-ḥáḳ, L:) or *did we not overcome you, and have it in our power to slay you?* (Bḍ.) AZ says that in all verbs coordinate to استحوذ, the original letters of the root may be preserved: that the Arabs say اسْتَحْوَبَ and اسْتَجَابَ, and اسْتَصْوَبَ and اسْتَصَابَ: and that their doing so is agreeable with a rule constantly obtaining with them. (Ṣ.) The grammarians say that he who says حَاذَ, aor. يَحُوذُ, says only استحاذ; and he who says أَحْوَذَ, says in like manner استحوذ. (L.) — Also حاذ, aor. يَحُوذُ, (L,) inf. n. حَوْذٌ, (L, Ḳ,) *He guarded, kept, kept safely, protected, took care of*, or *minded*, [a person, or thing;] syn. حَاطَ, (L,) inf. n. حَوْطٌ. (L, Ḳ.) And حاذ عَلَيْهِ, (L,) inf. n. حَوْذٌ; (Ḳ;) and ↓ أَحْوَذَ, inf. n. إِحْوَاذٌ; (TA:) *He was observant*, or *regardful, of it*; (L, Ḳ;) as, for instance, of prayer: from حاذ الإِبِلَ signifying "he collected the camels together to drive them." (L.) — And حاذ, *He did* a thing, or an affair, *firmly, soundly, thoroughly*, or *well*. (L.) — Also, aor. and inf. n. as above; and ↓ أَحْوَذَ, inf. n. إِحْوَاذٌ; and أَحْوَذَ السَّيْرَ; *He journeyed hard; went a hard*, or *vehement, pace*. (L.)

4. أَحْوَذَ, inf. n. إِحْوَاذٌ: see 1, in four places. — احوذ ثَوْبَهُ *He gathered together his garment*, (L, Ḳ,) *and drew it to him.* (L.) — احوذ القِدْحَ [in some copies of the Ḳ القَدَحَ] *He* (a workman, صَانِعٌ [in some copies of the Ḳ صَائِغٌ]) *made the arrow light, by scraping*, or *paring, it*: (L, Ḳ:) a phrase used by Lebeed, in describing the arrow termed المَنِيح. (L.)

10. اسْتَحْوَذَ and استحاذ: see 1, in four places.

الحَاذُ (A, L, Mṣb) and حَاذُ المَتْنِ (Ṣ, L, Ḳ) *The part of the back* of a horse *that is beneath the* لِبْد [*or saddle-cloth*]; (Ṣ, A, L, Mṣb, Ḳ;) i. e. *the middle thereof*; (Mṣb;) as also الحَالُ (Sh, A, L) and حَالُ المَتْنِ: (Ṣ, L:) also *the line along the middle of the back* (M, L) of a man; (L;) but الحال is more approved in this sense: (M, L:) and الحاذ signifies also *the back* of a man; (L, Ḳ;) like الحال: pl. أَحْوَاذٌ. (L.) [Hence,] خَفِيفُ الحَاذِ ‡ A man *light of back*; (Ṣ, A, L, Mṣb;) as also خَفِيفُ الحَالِ; (A;) meaning *having little property*: and also *having a small family to maintain*: (L:) or *having little property and a small family to maintain*: (L, Ḳ:) like خَفِيفُ الظَّهْرِ. (A, L, Mṣb.) And كَيْفَ حَاذُكَ and حَالُكَ †[lit. *How is thy back?* meaning, *thy state*, or *condition?*]. (Sh, L.) حَاذٌ and ↓ حَاذَةٌ both signify † *A state*, or *condition*; like حَالٌ and حَالَةٌ, which are the more approved words. (TA.) You say, هُمَا بِحَاذَةٍ ↓ وَاحِدَةٍ † *They both are in one state, or condition.* (Ḳ.) — الحَاذَانِ *The parts of the backs of the two thighs* (of a camel, A) *against which the tail falls*, (Ṣ, A,* L, Ḳ,) *on this side and on that*: (L:) or *the parts of the two thighs of a beast that face one when he stands behind it*: and *two portions of flesh in the outer sides of the two thighs of a man* &c.: (ISd, L:) or *the hinder parts of the two* *thighs*: pl. أَحْاذٌ. (Ham p. 443.) They say, أَنْقَعُ اللَّبَنِ مَا وَلِيَ حَاذَيِ النَّاقَةِ [*The most thirst-quenching of milk is that which is next to the hinder parts of the two thighs of the she-camel*]: i. e., when it is fresh-drawn, without her having been previously sucked by a young one. (TA. [But the first word, there, is انفع, which I regard as a mistranscription.])

حَاذَةٌ: see الحَاذُ, in two places.

حَوِيذٌ: see أَحْوَذِىٌّ.

طَرْدٌ أَحْوَذُ *A quick hunting.* (L.)

أَحْوَذِىٌّ *Quick in journeying*, or *in pace; one who goes a journey of ten nights in three.* (L.) And hence, ‡ *Quick in everything that he undertakes: quick, sharp, and active in affairs:* (L:) *active and skilful:* (Ḳ:) *active in a thing by reason of his skilfulness:* (AA, Ṣ, L:) applied [as meaning *active by reason of expertness*] to the wing of a bird of the kind called قَطًا, by a poet, (Ṣ, L,) namely, Ḥomeyd Ibn-Thowr: (Ṣ:) *quick in his affairs, who prosecutes them,* or *carries them on, well:* (L:) one *who prosecutes,* or *carries on, affairs in the best manner, by reason of his knowledge thereof:* (A:) one *who manages things skilfully, well,* or *thoroughly:* (Mṣb:) *ready,* or *prompt, in affairs, who masters them, and to whom nothing is out of his way,* or *sphere,* or *compass;* (Aṣ, Ṣ, L, Ḳ;) as also ↓ حَوِيذٌ: (L,* Ḳ:) one *who overcomes,* or *masters.* (L.) And أَحْوَزِىٌّ signifies the same. (Ṣ and Ḳ &c. in art. حوز.) — It is applied by a poet to thick water (مَآءٌ مِنَ الكُثْرَةِ) as meaning † *Quick in moving the bowels.* (Ṣ, L.)

حور

1. حَارَ, aor. يَحُورُ, (Ṣ,) inf. n. حَوْرٌ and حُؤُورٌ (Ṣ, Ḳ) and حُورٌ, a contraction of the form next preceding, used in poetry, in case of necessity, (TA,) and مَحَارٌ (Ṣ, Ḳ) and مَحَارَةٌ (Ḳ) and حَوْرَةٌ, (TA,) *He,* or *it, returned,* (Ṣ, L, Ḳ,) إِلَى شَىْءٍ *to a thing,* and عَنْهُ *from it.* (L.) — [Hence,] حار عَلَيْهِ *It* (a false imputation) *returned to him* [who was its author; or *recoiled upon him*]. (TA, from a trad.) — And حَارَتِ الغُصَّةُ *The thing sticking in the throat, and choking, descended;* as though it returned from its place. (TA.) — [And حار, inf. n. حَوْرٌ and حُورٌ, *He returned from a good state to a bad.*] You say, حَارَ بَعْدَ مَا كَانَ (TA on the authority of 'Áṣim, and so in a copy of the Ṣ,) *He returned from a good state after he had been in that state:* (A 'Obeyd, Ṣ,* TA:) so says 'Aṣim: (TA:) or حار بعد ما كَارَ (TA, and so in copies of the Ṣ,) *He became in a state of defectiveness after he had been in a state of redundance:* (TA:) or it is from حار, inf. n. حَوْرٌ, *He untwisted* his turban: (Zj, TA:) and means † *He became in a bad state of affairs after he had been in a good state.* (TA. [See حَوْرٌ, below.]) — حَارَ وَبَارَ *He became in a defective and bad state.* (TA. [Here بار is an imitative sequent; (see حَائِرٌ;) as is also يَبُورُ in a phrase mentioned below.]) — حار, aor. as above, (Mṣb,) inf n.

حَوْرٌ (Ṣ, A, Mṣb, Ḳ) and حُورٌ (Ṣ, A, Ḳ) and مَحَارَةٌ (Ṣ) and مَحَارٌ, (M and TA in art. اول,) *It decreased,* or *became defective* or *deficient.* (Ṣ,* A,* Mṣb, Ḳ.* [See also حَوْرٌ, below.]) — Also, inf. n. حُؤُورٌ (TA) and حُورٌ, (Ṣ, Ḳ,) *He perished,* or *died.* (Ṣ,* Ḳ,* TA.) — Also, aor. يَحُورُ, inf. n. حَوْرٌ, *He,* or *it, became changed from one state,* or *condition, into another:* and *it became converted into* another thing. (TA.) — مَا يَحُورُ فُلَانٌ وَلَا يَبُورُ *Such a one does not increase nor become augmented* [in his substance] (Ibn-Hánì, Ḳ*) is said when a person's being afflicted with smallness of increase is confirmed. (Ibn-Hánì, TA.) ═ حار, (TḲ,) inf. n. حَوْرٌ, (Ḳ,) *He was,* or *became, confounded,* or *perplexed, and unable to see his right course;* syn. تَحَيَّرَ. (Ḳ,* TḲ.) [See also art. حير.] ═ See also 2. ═ حَوِرَ, aor. ـَ, inf. n. حَوَرٌ; (Ḳ;) and حَوِرَتْ, aor. and inf. n. as above; (Mṣb;) and ↓ احورّ, (Ḳ,) inf. n. اِحْوِرَارٌ; (TA;) and احورّت; (Ṣ, Ḳ;*) *He,* (a man, Ḳ, TA,) and *it,* (an eye, Ṣ, Mṣb, Ḳ,* TA,) *was,* or *became, characterized by the quality termed* حَوَرٌ *as explained below.* (Ṣ, Mṣb, Ḳ, TA.)

2. حوّرهُ, inf. n. تَحْوِيرٌ, *He made him,* or *it, to return.* (Zj, Ḳ.) — *He* (God) *denied him,* or *prohibited him from attaining, what he desired,* or *sought; disappointed him; frustrated his endeavour,* or *hope;* (Ḳ, TA;) *and caused him to return to a state of defectiveness.* (TA.) ═ حوّر, inf. n. as above, *He whitened* clothes, or garments, (Ṣ, Mṣb,) and wheat, or food: (Ṣ:) and ↓ حار, (Ḳ,) aor. يَحُورُ, inf. n. حَوْرٌ, (TA,) *he washed and whitened* a garment, or piece of cloth; (Ḳ;) but حوّر is better known in this sense. (TA.) — حوّر عَيْنَ البَعِيرِ, (inf. n. as above, TA,) *He burned a mark round the eye of the camel with a circular cauterizing-instrument,* (Ṣ, Ḳ,*) on account of a disorder: because the place becomes white. (TA.) ═ [*He prepared* skins such as are called حَوَرٌ: a meaning indicated, but not expressed, in the TA. — And app. *He lined* a boot *with such skin:* see مُحَوَّرٌ.] ═ Also, (inf. n. as above, TA,) *He prepared* a lump of dough, *and made it round,* (Ṣ, Ḳ,) *with a* مِحْوَرٌ, (TA,) *to put it into the hole containing hot ashes in which it was to be baked:* (Ṣ, Ḳ:) *he made it round with a* مِحْوَرٌ. (A.)

3. حاورهُ, (A, Mgh, Mṣb,) and حاورهُ الكَلَامَ, (TA in art. رجع, &c.,) inf. n. مُحَاوَرَةٌ (Ṣ, Mgh, Ḳ) and حِوَارٌ, (A, Mgh,) *He returned him answer for answer,* or *answers for answers; held a dialogue, colloquy, conference, disputation,* or *debate, with him;* or *bandied words with him;* syn. جَاوَبَهُ, (Ṣ, and Jel in xviii. 35,) and رَاجَعَهُ الكَلَامَ, (A, Mgh, Mṣb,) or رَاجَعَهُ فِى الكَلَامِ, (Bḍ in xviii. 32,) or, of the inf. n., مُرَاجَعَةُ النُّطْقِ. (Ḳ.) And حاورهُ *He vied,* or *competed, with him,* or *contended with him for superiority, in glorying,* or *boasting,* or *the like;* syn. فَاخَرَهُ. (Jel. in xviii. 32.)

4. احار [*He returned* a thing]. You say, طَحَنَتْ فَمَا أَحَارَتْ شَيْئًا *She ground, and did not return* (مَا رَدَّتْ) *anything* of the flour [app. for the loan of the hand-mill: see حُورٌ, below]. (Ṣ, Ḳ.) — احار الغُصَّةَ *He swallowed the thing sticking in his throat and choking him;* [as though he returned it from its place: see 1: see also 4 in art. حير: and see an ex. voce مُشْقَرٌ.] (TA.) And فُلَانٌ سَرِيعُ الإِحَارَةِ *Such a one is quick in swallowing:* [said to be] from what next follows. (Meyd, TA.) — احار, (Ṣ, Ḳ, &c.,) inf. n. إِحَارَةٌ, (TA,) *He returned* an answer, or a reply. (Mṣb, TA.) You say, كَلَّمْتُهُ فَمَا أَحَارَ إِلَىَّ جَوَابًا *I spoke to him, and he did not return to me an answer,* or *a reply.* (Ṣ, A,* Mṣb,* Ḳ.*) And in like manner, مَا أَحَارَ بِكَلِمَةٍ [*He did not return a word* in answer, or in reply]. (TA.) ═ احارت *She* (a camel) *had a young one such as is called* حُوَار. (Ḳ.)

6. تحاوروا, (Mṣb, Ḳ, &c.,) inf. n. تَحَاوُرٌ, (Ṣ, Ḳ,) *They returned one another answer for answer,* or *answers for answers; held a dialogue, colloquy, conference, disputation,* or *debate, one with another;* or *bandied words, one with another;* syn. تَجَاوَبُوا, (Ṣ, Ḳ,) and تَرَاجَعُوا, (Jel in lviii. 1,) or تَرَاجَعُوا فِى الكَلَامِ, (Mṣb, Ḳ,) or تَرَاجَعُوا الكَلَامَ. (Bḍ in lviii. 1.) [And *They vied,* or *competed,* or *contended for superiority, one with another, in glorying,* or *boasting,* or *the like:* see 3.]

9. احورّ, (Ṣ, Ḳ, &c.,) inf. n. اِحْوِرَارٌ, (Ḳ,) *It* (a thing, Ṣ, Mṣb, and the body, TA, and the part around the eye, A, and bread, Ṣ, or some other thing, TA) *was,* or *became, white.* (Ṣ, A, Mṣb, Ḳ.) — See also 1, last sentence.

10. استحارهُ *He desired him to speak* [or *to return an answer* or *a reply; he interrogated him*]. (Ṣ, Ḳ.) And استحار الدَّارَ *He desired the house to speak* [*to him; he interrogated the house;* as a lover does in addressing the house in which the object of his love has dwelt]. (IAạr.)

حَوْرٌ inf. n. of حَارَ. (Ṣ, A, Mṣb, Ḳ.) [Hence,] نَعُوذُ بِاللّٰهِ مِنَ الحَوْرِ بَعْدَ الكَوْنِ, (TA on the authority of 'Áṣim, and so in a copy of the Ṣ,) a trad., (TA,) meaning *We have recourse to God for preservation from decrease, or defectiveness, after increase,* or *redundance:* (Ṣ:) or مِنَ الحَوْرِ بَعْدَ الكَوْرِ, (TA, and so in copies of the Ṣ,) meaning as above: (Ṣ, TA:) or † *from a bad state of affairs after a good state;* from حَوْرٌ signifying the "untwisting" a turban: (TA:) or *from returning and departing from the community* [*of the faithful*] *after having been therein;* [from حَارَ "he untwisted" his turban, and] from كَارَ "he twisted" his turban upon his head. (Zj, TA. [See also كَوْرٌ.]) ↓ حَوْرٌ فِى مَحَارَةٍ ↓, (Ṣ, Ḳ,) and حُورٌ, (Ḳ,) *Deficiency upon deficiency,* (Ṣ, Ḳ,) and *return upon return,* (TA,) is a prov., applied to him whose good fortune is retiring; (Ṣ, Ḳ;) or to him who is not in a good state; or to him who has been in a good state and has become in a bad state: (Ḳ:) or the saying is, فُلَانٌ حَوْرٌ فِى مَحَارَةٍ ↓ [*Such a one is suffering deficiency upon deficiency:* حَوْرٌ being used in the sense of حَائِرٌ, like بَوْرٌ in the sense of بَائِرٌ]: so heard by IAạr; and said by him to be applied in the case of a thing not in a good state; or to him who has been in a good state and has become in a bad state. (TA.) One says also, الْبَاطِلُ فِى حُوْرٍ *What is false, or vain, is waning and retreating.* (TA.) And إِنَّهُ فِى حُورٍ ▼ وَبُورٍ, (Ḳ,) or حُورٍ بُورٍ, (Ḳ in art. حير,) *Verily he is engaged in that which is not a skilful nor a good work or performance:* (فِى غَيْرِ صَنْعَةٍ وَلَا إِجَادَةٍ): so in the L: in the Ḳ, for اجادة is put إِتَاوَة [which is evidently a mistake]: TA:) or *he is in a bad state, and a state of perdition:* (TA in art. حير:) or *in error.* (Ḳ. [See also بُورٌ: and see بَائِرٌ, in art. بور; where it is implied that بور is here an imitative sequent of حور.]) And ذَهَبَ فُلَانٌ فِى الحَوَارِ ▼ وَالبَوَارِ *Such a one went away in a defective and bad state.* (L, TA.) — See also حُوَيرٌ. = *What is beneath the* [part called] كَوْر *of a turban.* (Ḳ.) = *The bottom* of a well or the like. (Ḳ.) — Hence, (TA,) هُوَ بَعِيدُ الحَوْرِ † *He is intelligent;* (Ḳ;) *deep in penetration.* (TA.)

حُورٌ: see حَوَرٌ, in two places. = Also [app. *A return of flour for the loan of a hand-mill;* like عُقْبَةٌ (a subst. from أَعْقَبَ) signifying some broth which is returned with a borrowed cooking-pot:] a subst. from احارت in the phrase طَحَنَتْ فَمَا أَحَارَتْ شَيْئًا [q. v. suprà]. (Ṣ, Ḳ.)

حَوَرٌ *Intense whiteness of the white of the eye and intense blackness of the black thereof,* (Ṣ, Mṣb, Ḳ,) *with intense whiteness,* or *fairness, of the rest of the person:* (Ḳ:) or *intense whiteness of the white of the eye and intense blackness of the black thereof, with roundness of the black, and thinness of the eyelids, and whiteness,* or *fairness, of the parts around them:* (Ḳ:) or *blackness of the whole* [*of what appears*] *of the eye, as in the eyes of gazelles* (AA, Ṣ, Mṣb, Ḳ) *and of bulls and cows:* (AA, Ṣ:) and this is not found in human beings, but is attributed to them by way of comparison: (AA, Ṣ, Mṣb, Ḳ:) Aṣ says, I know not what is الحَوَرُ in the eye. (Ṣ.) — Also [simply] *Whiteness.* (A.) = *Red skins, with which* [*baskets of the kind called*] سِلَال *are covered:* (Ṣ, Ḳ:) [a coll. gen. n.:] n. un. with ة: (Ṣ:) pl. حُورَانٌ: (Ḳ, TA: in the CḲ حُوَرَانٌ:) or (so in the TA, but in the Ḳ "and") *a hide dyed red:* (Ḳ, TA:) or *red skins, not* [*such as are termed*] قَرَظِيَّة: pl. أَحْوَارٌ: (AḤn:) or *skins tanned without* قَرَظ: or *thin white skins, of which* [*receptacles of the kind called*] أَسْفَاط *are made:* or *prepared sheep-skins.* (TA.) [In the present day, pronounced حُور, applied to *Sheep-skin leather.*] = *A certain kind of tree:* the people of Syria apply the name of حَوَر to the *plane-tree* (دُلْب); but it is حَوْرٌ, with two fet-ḥahs: in the account of simples in the Ḳánoon [of Ibn-Seenà], it is said to be *a certain tree of which the gum is called* كهربا: (Mgh:) [by the modern Egyptians (pronounced حُور) applied to the *white poplar:*] *a certain kind of wood, called* البَيْضَآء, (Ḳ,) *because of its whiteness.* (TA.) = الحَوَرُ *The third star,* [ε,] *that next the body, of the three in the tail of Ursa Major.* (Mir-át ez-Zemán, &c. [In the Ḳ it is incorrectly said to be the third star of بَنَاتُ نَعْشٍ الصُّغْرَى. See القَائِدُ, in art. قود.])

حَارَةٌ [*A quarter of a city or town; generally consisting of several narrow streets, or lanes, of houses, and having but one general entrance, with a gate, which is closed at night; or, which is the case in some instances, having a by-street passing through it, with a gate at each end:*] *a place of abode of a people, whereof the houses are contiguous:* (Mṣb:) *any place of abode of a people whereof the houses are near* [*together*]: (Ḳ in art. حير:) *a spacious encompassed tract* or *place;* syn. مُسْتَدَارٌ مِنْ فَضَآءٍ: (A:) pl. حَارَاتٌ. (A, Mṣb.)

حِيرَةٌ: see حُوَيرٌ.

حَوْرَآءُ fem. of أَحْوَرُ [q. v.]. — Also *A round,* or *circular, burn, made with a hot iron;* (Ḳ;) [*around the eye of a camel;* (see 2;)] so called because its place becomes white. (TA.)

حُوَرْوَرَةٌ: see حَوَارِيَّةٌ, under حَوَارِىٌّ.

حَوَارٌ: see حُوَيرٌ: = and see حَوْرٌ.

حُوَارٌ, (Ṣ, Ḳ, &c.,) and sometimes with kesr [▼حِوَارٌ], (Ḳ,) but this latter is a bad form, (Yaạḳoob,) *A young camel when just born:* (T, Ḳ:) or *until weaned;* (Ṣ, Ḳ;) i. e. *from the time of its birth until big and weaned;* (TA;) when it is called فَصِيلٌ: (Ṣ:) fem. with ة: (IAạr:) pl. (of pauc., Ṣ) أَحْوِرَةٌ and (of mult., Ṣ) حُورَانٌ and حِيرَانٌ. (Ṣ, Ḳ.) [Its flesh is insipid: see a verse cited as an ex. of the word مَسِيخٌ.] — [Hence,] عَقْرَبُ الحِيرَانِ *The scorpion of winter;* because it injures the حُوَار, (Ḳ, TA,) i. e. the young camel. (TA.)

حِوَارٌ: see حُوَيرٌ: = and see also حُوَارٌ.

حُوَيرٌ (Ṣ, Ḳ) and ▼حُوَيرَةٌ, (Ṣ, and so in some copies of the Ḳ,) or ▼حُوَرْوَرَةٌ, (so in other copies of the Ḳ and in the TA,) and ▼حَوَارٌ (Ṣ, Ḳ) and ▼حِوَارٌ (Ḳ) and ▼مَحُورَةٌ (Ṣ, Ḳ, TA, in the CḲ مَحْوُرَةٌ) and ▼مَحْوَرَةٌ and ▼مُحَاوَرَةٌ [originally an inf. n. of 3] and ▼حِيرَةٌ (Ḳ) and ▼حَوْرٌ, (TA,) *An answer; a reply.* (Ṣ, Ḳ.) You say, مَا رَجَعَ إِلَىَّ حُوَيرًا, &c., *He did not return to me an answer,* or *a reply.* (Ṣ.) [See a verse of Ṭarafeh cited voce مُمْهَدٌ.]

حُوَيرَةٌ, or حُوَرْوَرَةٌ: see what next precedes.

حَوَارِىٌّ *One who whitens clothes,* or *garments, by washing and beating them.* (Ṣ, M, Mṣb, Ḳ.) Hence its pl. حَوَارِيُّونَ is applied to The *companions* [i. e. *apostles* and *disciples*] *of Jesus,* because their trade was to do this. (Ṣ, M, Mṣb.) [Or it is so applied from its bearing some one or another of the following significations.] — *One who is freed and cleared from every vice, fault,* or *defect:* [or] *one who has been tried,* or *proved, time after time, and found to be free from vices, faults,* or *defects;* from حَارَ "he returned." (Zj, TA.) — *A thing that is pure,* or *unsullied:* anything *of a pure,* or *an unsullied, colour:* and hence, — *One who advises,* or *counsels,* or *acts, sincerely, honestly,* or *faithfully:* (Sh:) or *a friend;* or *true,* or *sincere, friend:* (TA:) or *an assistant:* (Ṣ, Mṣb, Ḳ:) or *a strenuous assistant:* (TA:) or *an assistant of prophets:* (Ḳ:) or *a particular and select friend and assistant of a prophet:* and hence the pl. is applied to the *companions of Mohammad* also. (Zj.) — *A relation.* (Ḳ.) — And حَوَارِيَّةٌ *A white,* or *fair,* woman; (A;) as also ▼حُوَرْوَرَةٌ; (T, Ḳ;) and so ▼حَوْرَآءُ, without implying حَوَر of the eye: (TA:) pl. of the first حَوَارِيَّاتٌ: (A:) or this pl. signifies women *of the cities* or *towns;* (Ḳ;) so called by the Arabs of the desert because of their whiteness, or fairness, and cleanness: (TA:) or women *clear in complexion and skin;* because of their whiteness, or fairness: (TA:) or women *inhabitants of regions, districts,* or *tracts, of cities, towns,* or *villages, and of cultivated land:* (Ksh and Bḍ in iii. 45:) or [simply] *women;* because of their whiteness, or fairness. (Ṣ.)

حُوَّارَى *White,* applied to flour: (A,* Ḳ:) such is the best and purest of flour: (Ḳ, TA:) and in like manner applied to bread: (A:) or *whitened,* applied to flour; (Ṣ;) and, in this latter sense, to any food. (Ṣ, Ḳ.) [See also سَمِيدٌ: and see مُحَوَّرٌ.]

رَجُلٌ حَائِرٌ بَائِرٌ *A man in a defective and bad state:* (Ṣ, TA:) or *perishing,* or *dying.* (Ṣ.) [See the same phrase in art. حير: see also حُورٌ: and see بَائِرٌ, in art. بور; where it is said that بائر is here an imitative sequent of حائر.] = See also مَحَارَةٌ.

أَحْوَرُ, (Ḳ,) applied to a man, (TA,) *Having eyes characterized by the quality termed* حَوَر *as explained above:* (Ḳ:) and so حَوْرَآءُ, [the fem.,] applied to a woman: (Ṣ, Mṣb, Ḳ:*) pl. حُورٌ. (Ṣ, Ḳ.) And حُورُ العِينِ, applied to women, *Having eyes like those of gazelles and of cows.* (AA, Ṣ.) Az says that a woman is not termed حَوْرَآء unless *Combining* حَوَر *of the eyes with whiteness,* or *fairness, of complexion.* (TA.) See also حَوَارِيَّةٌ, under حَوَارِىٌّ. — طَرْفٌ أَحْوَرُ *An eye of pure white and black.* (A.) — الأَحْوَرُ *A certain star:* (Ṣ, Ḳ:) or (Ḳ) *Jupiter.* (Ṣ, Ḳ.) = Also ‡ *Intellect:* (ISk, Ṣ, Ḳ:) or *pure,* or *clear, intellect;* like an eye so termed, of pure white and black. (A.) So in the saying, مَا يَعِيشُ بِأَحْوَرَ ‡ [*He does not live by intellect:* or *by pure,* or *clear, intellect*]. (ISk, Ṣ, A.)

أَحْوَرِىٌّ A man (TA) *white,* or *fair,* (Ṣ, Ḳ,) *of the people of the towns or villages.* (TA.) [See also حَوَارِىٌّ; of which the fem. is applied in like manner to a woman.]

مَحَارٌ: see مَحَارَةٌ, in two places.

مِحْوَرٌ *The pin of wood, or,* as is sometimes the case, *of iron, on which the sheave of a pulley turns;* (Ṣ;) *the iron* [*pin*] *that unites the bent piece of iron which is on each side of the sheave of a pulley, and in which it* [the محور] *is inserted, and the sheave itself:* and *a piece of wood which unites* (تَجْمَعُ) *the sheave of a large pulley* [app. *with what is on each side of the latter*; for it seems to mean here, also, the *pivot*]: (Ḳ:) some say that it is so called because it turns round, returning to the point from which it departed: others, that it is so called because, by its revolving, it is polished so that it becomes white: (Zj:) pl. مَحَاوِرُ. (A.) One says, قَلِقَتْ مَحَاوِرُهُ, meaning ‡ *His circumstances,* (A,) or *affair,* or *case,* (Ḳ,) *became unsettled:* (A, Ḳ:) from the state of the pin of the sheave of a pulley when it becomes smooth, and the hole becomes large, so that it wabbles. (A.) — Also *A thing* (Ḳ) *of iron* (TA) *upon which turns the tongue of a buckle at the end of a waist-belt.* (Ḳ.) — And *An iron instrument for cauterizing* [app. *of a circular form*: see 2]. (Ḳ.) — And The *wooden implement* (Ṣ, Ḳ) *of the baker,* or *maker of bread,* (Ṣ,) *with which he expands the dough,* (Ḳ,) *and prepares it, and makes it round, to put it into the hot ashes in which it is baked:* (TA:) so called because of its turning round upon the dough, as being likened to the مِحْوَر of the sheave of a pulley, and because of its roundness. (T.)

مَحَارَةٌ: see حَوْرٌ, in two places. — Also *A place that returns* [*like a circle*]: or *in which a return is made* [*to the point of commencement*]. (Ḳ.) — *A mother-of-pearl shell; an oyster-shell:* (Ṣ, IAth, Mṣb, Ḳ:) or the *like thereof, of bone:* (Ṣ, Ḳ:) pl. مَحَاوِرُ and [coll. gen. n.] ▼مَحَارٌ. (L.) — And hence, *A thing in which water is collected;* as also ▼حَائِرٌ. (IAth.) — [Hence also,] *An oyster* [itself]; expl. by دَابَّةٌ فِى الصَّدَفَيْنِ. (L in art. محر.) — The *cavity of the ear;* (Ḳ;) i. e. the *external, deep, and wide, cavity, around the ear-hole;* or the صَدَفَة [or *concha*] of the ear. (TA.) — The *part of the shoulder-blade called its* مَرْجِع [q. v.]: (Ṣ, Ḳ:) or the *small round hollow that is in that part of the shoulder-blade in which the head of the humerus turns.* (TA.) — The *small round cavity of the hip:* and the dual signifies the *two round heads* [?] *of the hips, in which the heads of the thighs turn.* (TA.) — The *palate;* syn. حَنَكٌ: and without ة, i. e. ▼مَحَارٌ, the same, of a man: and, this latter, the *place,* in a beast, *where the farrier performs the operation termed* تَحْنِيك: (TA:) or the former signifies the *upper part of the mouth* of a horse, *internally:* (IAạr, TA:) or the *inner part of the palate:* (Abu-l-'Omeythil, TA:) or, [which seems to be the same,] the *portion of the upper part of the mouth which is behind the* فَرَاشَة [or فَرَاش]: and the *passage of the breath to the innermost parts of the nose:* (TA:) or مَحَارَةُ الحَنَكِ signifies the *part* [*of the palate*] *which is a little above the place where the farrier performs the operation termed* تحنيك. (Ṣ.) — The *part between the frog and the extremity of the fore part of a solid hoof.* (Abu-l-'Omeythil, Ḳ.) *What is beneath the* إِطَار [q. v., app. here meaning the اطار of the hoof of a horse or the like]. (TA.) And The مَنْسِم [i. e. *toe,* or *nail,* &c.,] *of a camel.* (TA.) — *A thing resembling* [*the kind of vehicle called*] *a* هَوْدَج; (Ḳ;) pronounced by the vulgar [مَحَّارَة,] with teshdeed: pl. مَحَارَاتٌ (TA) [and مَحَائِرُ, which is often applied in the present day to the *dorsers,* or *panniers,* or *oblong chests, which are borne, one on either side, by a camel, and, with a small tent over them, compose a* هودج]: the [*ornamented* هودج *called the*] مَحْمِل [vulgarly pronounced مَحْمَل] *of the pilgrims* [*which is borne by a camel, but without a rider, and is regarded as the royal banner of the caravan; such as is described and figured in my work on the Modern Egyptians*]. (Mṣb.) — *I. q.* خَطّ [*A line,* &c.]. (Ḳ.) — And *i. q.* نَاحِيَةٌ [*A side, region, quarter, tract,* &c.]. (Ḳ.)

مُحْوَرَةٌ and مَحْوَرَةٌ: see حُوَيْرٌ.

مُحْوَرُ القِدْرِ *The whiteness of the froth,* or *of the scum, of the cooking-pot.* (Ṣ.) — جَفْنَةٌ مُحْوَرَةٌ, [in the copies of the Ḳ, erroneously, مُحَوَّرَةٌ,] *A bowl whitened by* [*containing*] *camel's hump,* (Ṣ, L, Ḳ,) or *its fat.* (A.)

مُحَوَّرٌ Dough *of which the surface has been moistened with water, so that it is shining.* (TA.) [See also 2.] — أَعْيُنٌ مُحَوَّرَاتٌ, in a verse of El-'Ajjáj, *Eyes of a clear white* [*in the white parts*] *and intensely black in the black parts.* (Ṣ.) — A boot *lined with skin of the kind called* حَوَرٌ. (Ḳ.)

مُحَوِّرٌ *A possessor of* [*flour,* or *bread, such as is termed*] حُوَّارَى. (TA.)

مُحَاوَرَةٌ: see حُوَيْرٌ.

حوز

1. حَازَهُ, (Ṣ, A, Mṣb,) aor. يَحُوزُهُ, (Mṣb,) inf. n. حَوْزٌ and حِيَازَةٌ; (Ṣ, A, Mṣb, Ḳ;) as also حَازَهُ, [aor. يَحِيزُهُ,] inf. n. حَيْزٌ; (Mṣb;) *He drew, collected,* or *gathered, it together;* (Ṣ, A, Mṣb, Ḳ;) and so ▼احتازهُ, (TA,) inf. n. إِحْتِيَازٌ; (Ḳ;) and ▼حَوَّزَهُ, inf. n. تَحْوِيزٌ: (TA:) *he drew, collected,* or *gathered, it together* (namely, property or wealth &c., TA) *to himself;* (Ṣ, A, Mṣb;) as also ▼احتازهُ, (Ṣ,) and ▼احتازهُ لِنَفْسِهِ, (A, TA,) and حَازَهُ إِلَيْهِ, and ▼احتازهُ اليه. (TA.) You say, عَلَيْكَ بِحِيَازَةِ المَالِ *Take thou to the collecting of wealth.* (A, TA.) — حَازَهُ, aor. يَحُوزُهُ, (TA,) inf. n. حَوْزٌ, (Ḳ, TA,) *He had, held,* or *possessed, it; had it,* or *held it, in his possession; had, took, got, obtained,* or *acquired, possession,* or *occupation, of it;* (AA, Ḳ,* TA;) [المَلْكُ, given as an explanation of the inf. n. in the CḲ, is a mistake for المِلْكُ;]) *he took,* or *received, it; he had it,* or *took it, to,* or *for, himself.* (AA, TA.) [See حَوْزَةٌ, below. Hence, *It comprehended, comprised,* or *embraced, it.*] — حَازَ الأَرْضَ, inf. n. حَوْزٌ, *He took for himself the land, and marked out its boundaries, and had an exclusive right to it.* (TA: but only the inf. n. is there mentioned.) — حَازَ, aor. يَحُوزُ, also signifies [*He* or] *it overcame, conquered,* or *mastered,* [a thing,] as in an instance in art. حز, voce حَزَّازٌ: (Sh, Ḳ:) [as also حَاذَ.] — Also, (A, TA,) inf. n. حَوْزٌ, (Ḳ,) ‡ *He compressed* a woman: (A,* Ḳ,* TA:) [as though he mastered her.] — حَازَ الحِمَارُ أُتُنَهُ *The he-ass gained the mastery over his she-asses, and collected them together;* as also حَاذَهَا. (L in art. حوذ.) — حَازَ الإِبِلَ, aor. يَحُوزُهَا, (Ṣ, A, Mṣb,) inf. n. حَوْزٌ, (Ṣ, Ḳ,) *He drove the camels gently;* (Ṣ, Mṣb, Ḳ;) as also حَازَهَا, aor. يَحِيزُهَا, (Ṣ, Mṣb,) inf. n. حَيْزٌ; (Ṣ, TA;) and ▼حوّزها. (TA.) Also *He drove the camels vehemently;* (Ḳ;) and so حازها, aor. يَحِيزُهَا, (TA in art. حيز,) inf. n. حَيْزٌ: (Ḳ in art. حيز:) thus bearing two contr. significations: (Ḳ:) [as also حَاذَهَا:] you say [also] ▼أَحِزْهَا, [unless this be a mistranscription for حُزْهَا,] meaning, *Drive thou them vehemently.* (TA.) Also *He drove the camels to water;* (A;) and so ▼حوّزها; (Ṣ, A;) [and حَاذَهَا:] or ▼حوّزها, (Aṣ, Ṣ, Ḳ,) inf. n. تَحْوِيزٌ, (Ḳ,) signifies *he drove them during the first night to water,* (Aṣ, Ṣ, Ḳ,) *it being distant from the pasture:* (Aṣ, Ṣ:) because in that night they are driven gently. (TA.) [See also حَوْزٌ, below.] — حَازَ الشَّىْءَ *He removed the thing from its place; put it away; placed it at a distance.* (Sh, TA.)

2. حَوَّزَهُ: see 1, first sentence: — and حَوَّزَ الإِبِلَ: see 1, in three places.

4. أَحِزِ الإِبِلَ: see 1.

5. تَحَوَّزَ *He,* or *it, writhed,* or *twisted, about,* (Ḳ, TA,) *and turned over and over;* (TA;) as also ▼تحيّز: (Ḳ:) or *was restless,* or *unquiet, not remaining still, upon the ground.* (Lth, TA.) You say, تَحَوَّزَتِ الحَيَّةُ, and ▼تَحَيَّزَت, *The serpent writhed,* or *twisted, about.* (Both in the Ṣ; and the latter in the Ḳ in art. حيز.) And مَا لَكَ تَتَحَيَّزُ▼ تَحَيُّزَ الحَيَّةِ, and تَتَحَوَّزُ تَحَوُّزَ الحَيَّةِ, *Wherefore dost thou writhe about like the writhing about of the serpent?* the latter verb, accord. to Sb, is of the measure تَفَيْعَلَ, from حُزْتُ الشَّىْءَ. (Ṣ.) — *He removed, withdrew,* or *retired to a distance,* (A'Obeyd, Ṣ, Ḳ,) *and drew back,* (Ṣ,) عَنْهُ [or مِنْهُ] *from him* or *it;* (TA;) as also ▼تحيّز; (A'Obeyd, Ṣ;) and ▼انحاز. (A.) You say, دَخَلَ عَلَيْهِ فَمَا تَحَوَّزَ لَهُ عَنْ فِرَاشِهِ *He went in to him and he did not move for him from his bed,* or *mattress.* (TḲ.) And El-Ḳaṭámee says, (Ṣ, TA,) describing an old woman of whom he sought hospitality, and who eluded him, (TA,)

* تَحَيَّزُ▼ مِنِّى خَشْيَةً أَنْ أُضِيفَهَا *
* كَمَا ٱنْحَازَتِ▼ الأَفْعَى مَخَافَةَ ضَارِبِ *

She (this old woman) *retires and draws back from me for fear of my alighting at her abode as a guest* [*like as the viper turns away in fear of a beater*]: or, as some relate the verse, تَحَوَّزُ. (Ṣ.) — *He tarried,* or *loitered: he was slow in rising;* as also تحوّس: *he desired to rise, and it was tedious to him to do so;* as also ▼تحيّز. (TA.) AA says, تَحَوَّزَ تَحَوُّزَ الحَيَّةِ, [as though meaning, *He was slow in rising like as the rising of the serpent is slow:* for he adds,] and it is slow in rising when it desires to rise. (Ṣ.)

6. تحاوز الفَرِيقَانِ *The two parties,* or *divisions,*

turned away, each from the other, (Ṣ, Ḳ,) in war or battle. (Ṣ.)

7 انحاز القَوْمُ *The company of men left their appointed station,* (Ṣ, Ḳ, TA,) *and place of fighting,* (TA,) *and turned away to another place.* (Ṣ,* Ḳ,* TA.) You say also, انحاز عَنْهُ *He turned away from him:* (Ṣ, Ḳ:) and انحاز إِلَيْهِ *he turned to,* or *towards, him; and he joined himself to him.* (Ḥar pp. 122 and 320.) You say of friends, انحازوا عَنِ العَدُوِّ, and حَاصُوا; [*They turned away from the enemy;*] and of enemies, انْهَزَمُوا, and وَلَّوْا مُدْبِرِينَ. (Ṣ, TA.) Or انحاز signifies *He separated himself from others that he might be with those who were fighting.* (Aboo-Is-ḥáḳ, TA.) And انحاز الرَّجُلُ إِلَى القَوْمِ signifies the same as ↓ تحيّز إِلَيْهِم [*The man turned, removed, withdrew,* or *retired,* or *he joined himself, to the company of men*]. (Mṣb.) See 5, in two places. — انحاز عَلَى الشَّىْءِ [for عن, in the TA, I have substituted على, as the former is apparently a mistranscription] *He drew himself together, and fell to the thing;* expl. by ضَمَّ بَعْضَهُ عَلَى بَعْضٍ وَأَكَبَّ عَلَيْهِ. (TA.)

8. احتازهُ: see حَازَهُ, in four places, first sentence.

Q. Q. 2. تَحَيَّزَ, [originally تَحَيْوَزَ,] of the measure تَفَيْعَلَ, (Sb, Ṣ, TA,) [from حَيِّزٌ, originally حَيْوِزٌ,] *He turned aside to a* حَيِّز [or place, &c.]. (Mgh.) You say also تحيّز المَالُ [*The property,* or *the camels or the like,*] *became drawn, collected,* or *gathered, together;* or *drew, collected,* or *gathered, themselves together; to a* حَيِّز. (Mṣb.) — See also 5, throughout; and see 7.

حَوْزٌ inf. n. of 1 [q. v.]. — فِى حَوْزِهِ: see حَوْزَةٌ. — *A place of which a man takes possession,* (TA,) *and around which a dam* (مُسَنَّاة) *is made:* (Ḳ, TA:) pl. أَحْوَازٌ. (TA.) — حَوْزُ الدَّارِ: see حَيِّزٌ. — لَيْلَةُ الحَوْزِ *The first night during which camels repair towards the water* (Aṣ, Ṣ, Ḳ) *when it is distant from the pasture:* (Aṣ, Ṣ:) because they are driven gently that night: but when their faces are turned towards the water and they are left to pasture that night, the night is called لَيْلَةُ الطَّلَقِ. (TA.) One says to a man, when he holds back respecting an affair, دَعْنِى مِنْ حَوْزِكَ وَطَلَقِكَ † [*Let me alone and cease from this and that discursion of thine*]. (TA.) And one says also, طَوَّلَ عَلَيْنَا فُلَانٌ بِالحَوْزِ وَالطَّلَقِ قَبْلَ القَرَبِ † [*Such a one was prolix,* or *tedious, to us with this and that discursion before coming to the point*]. (TA.) — حَوْزٌ is also used as an epithet; though properly an inf. n.: you say, سَوْقٌ حَوْزٌ [*A gentle driving:* or *a vehement driving*]. (TA.)

حَوْزَةٌ i. q. حَيِّزٌ, as pointed out in two places below. (Ṣ, Mṣb, &c.) — [Hence,] † *A thing that is in one's possession* or *occupation; a thing that is one's property:* so in the saying of a certain woman, وَأَحْمِى حَوْزَةَ الغَائِبِ † *And I guard from encroachment the property of the absent:* meaning her فَرْج, which was the property of her husband by the marriage-contract: whence it appears that, if this saying be the only ground upon which Az has asserted that one of the significations of حَوْزَةٌ is the فرج of a woman, [as is also said in the Ḳ,] his assertion requires consideration; for a woman's فرج is her own when she has no husband; and when she is married, it is her husband's property. (L, TA.) You say also, صَارَ فِى حَوْزَتِهِ, and ↓ فِى حَوْزِهِ, [and ↓ فِى حَيِّزِهِ,] *It became in his possession,* or *occupation.* (L, TA.) And فُلَانٌ مَانِعٌ حَوْزَتَهُ † *Such a one defends,* or *guards, from encroachment,* or *invasion,* or *attack, what is in his* حَيِّز [or *place;* meaning, *in his possession* or *occupation*]. (TA.) In like manner, a poet says,

* حَمَى حَوْزَاتِهِ فَتُرِكْنَ قَفْرًا *

He guarded from encroachment his tracts of pasture-land [*so that they were left deserted*]. (Fr, TA.) And it is said in a trad., فَحَمَى حَوْزَةَ الإِسْلَامِ ‡ *And he defended,* or *protected,* or *guarded, from encroachment,* or *invasion,* or *attack, the limits,* [meaning, *what the limits comprised,* i. e., *the territory,*] *and the tracts,* or *regions, of El-Islám* [meaning, *of the Muslims*]. (TA.) حَوْزَةُ المُلْكِ signifies [in like manner] بَيْضَتُهُ [i. e. † *The seat of regal power:* or *the heart,* or *principal part, of the kingdom*]. (Ṣ, Ḳ.) — † *Nature;* or *natural disposition, temper,* or *other quality* or *property;* (Ḳ, TA;) *whether good or evil.* (TA.)

حَيِّزٌ, (Ṣ, Mgh, Mṣb,) of the measure فَيْعِلٌ, (Mgh, Mṣb,) from الحَوْزُ, (Ṣ,* Mgh,) as signifying "the drawing, collecting, or gathering, together," (Mgh,) originally حَيْوِزٌ, (TA,) and also contracted into حَيْزٌ, (Ṣ, Mṣb, TA,) like هَيِّنٌ and هَيْنٌ, and لَيِّنٌ and لَيْنٌ; (Ṣ, TA;) [*The continent,* or *container,* or *receptacle,* of anything; like بَيْضَةٌ; as also ↓ حَوْزَةٌ, q. v.:] *any place in which a thing is:* (Mgh:) in scholastic theology, the *imaginary portion of space occupied by a thing having extent, as a body;* or *by a thing not having extent, as an indivisible atom:* in philosophy, the *inner surface of a container, which is contiguous* [*in every part*] *to the outer surface of the thing contained:* and [hence,] الحَيِّزُ الطَّبِيعِىُّ [*the proper natural place of a thing;*] *that in which the nature of a thing requires it to be.* (KT.) — *A quarter, tract, region,* or *place, considered relatively,* or *as part of a whole;* or *a part,* or *portion, of a place;* syn. نَاحِيَةٌ; (Ṣ, Mgh, Mṣb;) as also ↓ حَوْزَةٌ: (Ṣ, Mṣb, Ḳ:) so the authors on practical law mean by حَيِّز; *such,* for instance, *as a room,* or *an apartment, of a house:* (Mgh:) pl. أَحْيَازٌ, (Ṣ, Mṣb, TA,) which is extr., (TA,) being from the contracted form [حَيْز]: (Mṣb:) by rule it should be أَحْوَازٌ, (Az, Mṣb, TA,) like أَمْوَاتٌ, pl. of مَيِّتٌ [and مَيْتٌ]: (Az, TA:) or by rule [if from the uncontracted form حَيِّز] it should be حَيَائِزُ, with hemz, accord. to Sb; or حَيَاوِزُ, with و, accord. to Abu-l-Ḥasan. (TA.) حَيِّزُ الدَّارِ, (Ṣ, Mṣb, TA,) as also حَوْزُ الدَّارِ, (TA,) signifies *What is annexed to the house,* (Ṣ, TA,) or *appertains thereto,* (Mṣb,) *of the* مَرَافِق (Ṣ, Mṣb, TA) *and* مَنَافِع (TA) *and* نَوَاحٍ; (Mṣb;) [i. e., *of the conveniences thereof, such as the privy and the kitchen and the like, and other parts or apartments;*] such are termed collectively أَحْيَازُ الدَّارِ; (Mṣb;) and each part or apartment (نَاحِيَة), by itself, is termed حَيِّز. (TA.) — [Hence the saying,] أَنَا فِى حَيِّزِهِ وَكَنَفِهِ ‡ [*I am in his quarter and protection*]. (A, TA.) — [And hence also the saying,] فِى حَيِّزِ التَّوَاتُرِ ‡ *In the manner, and place, of* [*that kind of transmission which is termed*] التواتر [which is "transmission by such a number of persons as cannot be supposed to have agreed to a falsehood:" as explained in the Mz, 3rd نوع]. (Mgh.) — And صَارَ فِى حَيِّزِهِ: see حَوْزَةٌ. — [And عَلَى حِيَزِهِ *By himself* or *itself.*]

الإِثْمُ حَوَّازُ القُلُوبِ: see حَزَّازٌ, in art. حز.

أَوْ مُتَحَيِّزًا إِلَى فِئَةٍ, in the Ḳur [viii. 16], signifies *Or turning aside to a different company of the Muslims:* (Mgh, Mṣb:*) or the meaning is, *or separating themselves from others to betake themselves to* [*a different company of*] *those engaged in fighting.* (Aboo-Is-ḥáḳ, TA.) The original form of مُتَحَيِّز is مُتَحَيْوِز. (TA.)

قِطْعَةٌ مِنَ الأَرْضِ مُسْتَحِيزَةٌ [*A portion of the earth,* or *of land, comprehended within certain limits*]. (M and Ḳ in art. بلد.)

حوش

1. حَاشَ الصَّيْدَ, (Ṣ, A, Ḳ,) aor. يَحُوشُ, (Ṣ,) inf. n. حَوْشٌ and حِيَاشٌ, (TA,) *He came around the chase,* or *game, to turn it towards the snare;* (Ṣ, A, Ḳ;) as also ↓ أَحَاشَهُ and ↓ أَحْوَشَهُ, (Ṣ, Ḳ,) inf. n. إِحَاشَةٌ and إِحْوَاشٌ. (TA.) — حُشْتُ عَلَيْهِ الصَّيْدَ *I aided him to hunt,* or *catch, the chase,* or *game;* as also ↓ أَحَشْتُهُ عليه, and ↓ أَحْوَشْتُهُ عليه, and أَحْوَشْتُهُ إِيَّاهُ, on the authority of Th: (TA:) and حَاشَ عَلَيْهِ الصَّيْدَ *He scared the chase,* or *game, towards him, and drove and collected it to him;* as also ↓ احاشهُ. (TA.) — حَاشَ الذِّئْبُ الغَنَمَ *The wolf drove along the sheep or goats.* (TA.) — حَاشَ الإِبِلَ *He collected together, and drove, the camels.* (Ṣ, Ḳ.) — حَاشَهُ, inf. n. حَوْشٌ, also signifies [simply] *He collected it; drew it together.* (TA.) [See also 2.] — هُوَ يَحُوشُ الطَّعَامَ, (A,) inf. n. حَوْشٌ, (Ḳ,) *He eats from the sides of the food so as to consume it:* (A, Ḳ:) from IF. (TA.) — [See also 7.]

2. حوّش, (TA,) inf. n. تَحْوِيشٌ, (Ḳ,) *He collected several things:* or *collected much.* (Ḳ,* TA.) [See also 1.]

3. حاوش البَرْقَ *He turned aside from the place of the rain of the lightning, whichever way it turned.* (Ibn-'Abbád, Ḳ.) — Hence, (TA,) حَاوَشَهُ, (A,) inf. n. مُحَاوَشَةٌ, (TA,) *He circumvented him:* or *he endeavoured to induce him to turn,* or *incline,* or *decline;* or *endeavoured to turn him by deceit,* or *guile:* syn. دَاوَرَهُ: (A,

TA:*) in war, and in litigation or contention [&c.]. (TA.) You say, ظَلِلْتُ أُحَاوِشُهُ وَأُحَاوِلُهُ حَتَّى فَعَلَ [*I continued during the day to circumvent him, or to endeavour to induce him to turn, &c., and to delude him, or act towards him with artifice, like a fish in the water, until he did* what I desired: see also what next follows]. (A.) — [And hence,] حَاوَشْتُهُ عَلَيْهِ *I excited, incited, urged, or instigated, him to do it.* (Ibn-'Abbád, A,* Ṣgh, Ḳ.) [It is indicated in the A that in the ex. immediately preceding this, أُحَاوِشُهُ may also be rendered agreeably with this explanation.]

4. احاش الصَّيْدَ, and أَحْوَشَهُ: see 1, in five places.

5. تحوّش القَوْمُ عَنِّى *The people, or company of men, removed, withdrew, or retired to a distance, from me.* (Ṣ, Ḳ.*) And تحوّش عَنِ القَوْمِ *He removed, &c., from the people, or company of men.* (TA.) — تحوّشت مِنْ زَوْجِهَا *She became forlorn of her husband;* syn. تَأَيَّمَتْ. (Ṣgh, Ḳ.) — تحوّش *He felt, or had a sense of, or was moved with, shame, or shyness, or bashfulness.* (AA, Ḳ.)

6: see 8.

7. انحاش عَنْهُ *He took fright, and fled from him; or was averse from him; and shrank from him;* (Ṣ,* Ḳ;) and *was frightened at him;* and *was moved by him.* (TA.) [In the TA it is here added, that this verb is quasi-pass. of الحَوْشُ in the sense of النِّفَارُ; but this seems to indicate that a copyist has written النفار by mistake for الإِنْفَارُ, which is a syn. of the inf. n. of 1 in a sense explained above: so that انحاش signifies *He became scared,* or the like.] Hr mentions this verb in art. حيش; but it belongs to the present art. (IAth.) You say, زَجَرَهُ فَمَا انْحَاشَ لِزَجْرِهِ *He chid him* (meaning a wolf or other animal) *but he did not take fright and flee, &c., at his chiding.* (TA.) And مَا يَنْحَاشُ فُلَانٌ مِنْ شَىْءٍ, (Ṣ, A,*) and لِشَىْءٍ, and مِنْ فُلَانٍ, (TA,) *Such a one is not moved by, and does not care for, or regard, anything,* (Ṣ, A, TA,) and *such a one.* (TA.) — انحاشت الإِبِلُ *The camels became collected together.* (Har p. 130.)

8. احتوش القَوْمُ الصَّيْدَ, (Ṣ, Mṣb, Ḳ,) and, more commonly, بِالصَّيْدِ, (Mṣb,) *The people, or company of men, encompassed, or surrounded, the chase, or game:* (Mṣb:) or *scared it, one, or one party, to another:* (Ṣ, Ḳ:) the و remaining here unchanged as it does in اِجْتَوَرُوا. (Ṣ.) And احتوشوا فُلَانًا (A, TA) *They encompassed, or surrounded, such a one:* (A:) or *they made such a one to be in the midst of them;* (TA;) as also احتوشوا عَلَيْهِ, (Ṣ, Ḳ,) [and احتوشوا حَوَالَيْهِ, (M and O in art. حول,)] and ‡ تحاوشوه بَيْنَهُمْ, (Ḳ,) or تحاوشوه. (TA.) — Hence the phrase احتوش الدَّمُ الطُّهْرَ † [*The blood invaded from every quarter the state of pureness*]; as though the blood encompassed the pureness, and enclosed it on either side. (Mṣb.) [Alluding to the collecting of the blood about the uterus previously to menstruation.]

حَاشَ لِلّٰهِ i. q. تَنْزِيهًا لِلّٰهِ. One should not say حَاشَ لَكَ, but حَاشَاكَ, and حَاشَى لَكَ. (Ṣ, Ḳ.) [See these phrases explained in art. حشى.]

حَوْشٌ *A thing resembling* [*the kind of enclosure, made of trees* or *of wood, &c., for camels* or *sheep* or *goats, called*] *a* حَظِيرَة: a word of the dial. of El-'Irák. (Ṣgh, Ḳ.) — Applied by the people of Egypt to The *court* (فِنَآء) *of a house:* (TA:) [and to *any court,* or *enclosure, surrounded by dwellings* or *the like,* or *by these and walls,* or *by walls alone:* pl. of pauc. أَحْوَاشٌ, and of mult. حِيشَانٌ.]

حُوشٌ: and الحُوشُ: see the next paragraph, in four places.

حُوشِىٌّ *Wild; untamed; undomesticated; uncivilized; unfamiliar;* syn. وَحْشِىٌّ. (Ṣ, Mṣb.) — Applied to a man, ‡ *Wild; uncivilized; unfamiliar;* (A;) *unsociable; that does not mix with others.* (Ṣ, A.) — Applied to a camel, or other [animal], *Wild:* (Ḳ:) [or] the epithet thus applied is tropical; (A, TA;) and what are thus called, (Ḳ,) or الإِبِلُ الحُوشِيَّةُ [*the camels termed* حوشيّة], (Ṣ, A, Mṣb,) are so named from ‡ الحُوشُ, the appellation of *certain stallions of the camels of the jinn,* or *genii,* which covered some of the she-camels of Arabs, (IḲt, Ṣ, A, Mṣb, Ḳ,) as they assert, (Ṣ, Ḳ,) namely, of the she-camels of Mahrah, (Ḳ,) meaning the Benoo-Mahrah-Ibn-Ḥeydán, (TA,) and the offspring were *the camels called* النَّجَائِبُ المَهْرِيَّةُ, (Mṣb, TA,) which scarcely ever become tired; and the like of this is said by AHeyth: (TA:) it is also said that ‡ الحُوشُ, (Ṣ, Ḳ,) from which the epithet above mentioned, thus applied, is a rel. n., (TA,) is *the country of the jinn,* (Ṣ, Ḳ,) *beyond the sands of Yebreen, which no man inhabits:* (Ṣ:) or an appellation of *certain sons of the jinn,* whose country is called بِلَادُ الوَحْشِ by Ru-beh: (TA:) or it is like الحُوشُ: (Mṣb:) or إِبِلٌ حُوشِيَّةٌ means *camels of the jinn:* or *wild camels;* (TA;) as also ‡ حُوشٌ: (Ṣ:) or *camels not completely broken* or *trained, because of their unyielding spirit.* (TA.) — Hence, (A,) رَجُلٌ حُوشِىُّ الفُؤَادِ, (A,) or ‡ حُوشُ الفُؤَادِ, (Ṣ, Ḳ,) ‡ *A man acute, or sharp, in intellect.* (Ṣ,* A, Ḳ,* TA.*) — You say also, كَلَامٌ حُوشِىٌّ ‡ *Strange, uncouth, unusual, extraordinary,* or *unfamiliar, speech; such as is difficult to be understood;* (Mṣb, Ḳ, TA;) i. q. وَحْشِىٌّ; (Ṣ, A;) [opposed to فَصِيحٌ:] and in like manner, لَفْظَةٌ حُوشِيَّةٌ *a word,* or *phrase, that is strange, uncouth, unusual, &c.;* as also لَفْظَةٌ وَحْشِيَّةٌ, and غَرِيبَةٌ, and شَارِدَةٌ; all opposed to لَفْظَةٌ فَصِيحَةٌ. (Mz, 13th نوع.) — And لَيْلٌ حُوشِىٌّ ‡ *A night that is dark* (A, Ḳ) *and terrible.* (A, TA.)

حُوشِيَّةٌ [*Wildness;* and *the like;* the *quality of that which is termed* حُوشِىٌّ:] ‡ *unsociableness of disposition;* or the *quality of not mixing with others;* in a man. (Ṣ.)

مُحْتَوَشٌ *Encompassed,* or *surrounded.* (Mṣb.)

حوص

1. حَاصَهُ, (A, TA,) aor. يَحُوصُ, (A, Ḳ,) inf. n. حَوْصٌ (Ṣ, A, Mgh, Ḳ) and حِيَاصَةٌ, (A, TA,) *He sewed it,* or *sewed it up* or *together;* (Ṣ,* A, Mgh,* Ḳ, TA;) namely, a garment, or piece of cloth: (A, TA:) or *he sewed it with stitches far apart:* (IB, TA:) or *he sewed it without a patch;* and only said of a skin, or of skin, and of a camel's foot. (TA.) Hence the prov., إِنَّ دَوَآءَ الشَّقِّ أَنْ تَحُوصَهُ [*Verily the remedy for the rent is that thou sew it up*]. (A, Ḳ.) [And hence the saying,] لَأَطْعَنَنَّ فِى حَوْصِهِمْ (Ṣ, A) ‡ *I will assuredly mar what they have repaired:* (A:) or *I will assuredly make a hole in what they have sewed up, and I will assuredly mar what they have repaired.* (Ṣ, IB.) Or لَأَطْعَنَنَّ فِى حَوْصِكَ ‡ *I will assuredly circumvent thee,* (لَأَكِيدَنَّكَ [so in copies of the Ḳ, and in the TA, but in the CḲ, erroneously, لَأَكِيدَنَّكَ,]) *and I will assuredly labour for thy destruction.* (AZ, Ḳ.) And it is said in a prov., طَعَنَ فِى حَوْصٍ أَمْرٍ لَيْسَ مِنْهُ فِى شَىْءٍ, (Ḳ,) and حُوصٍ أَمْرٍ, and ‡ حُوصَى أَمْرٍ, (Yoo, Ḳ,) ‡ *He laboured at that which he could not do well, and tasked himself to do that with which he had no concern.* (ISh, Ḳ.) Or طَعَنْتَ فِى حَوْصِ أَمْرٍ لَسْتَ مِنْهُ فِى شَىْءٍ ‡ *Thou hast spoken of a thing with which thou hast no concern.* (A.) You say also, قَبْلَ أَنْ أَدْخُلَ فِى حَوْصِ النَّاسِ أَطْمَعُ فِى ضَرْبِهِمْ ‡ *Before I penetrate into the case of the people, and test them, I am ambitious of smiting them.* (A.) And مَا طَعَنْتَ فِى حَوْصِهِ ‡ *Thou hast not attained the object of thy desire* [*with respect to him* or *it:* or perhaps the right reading is حَوْصِكَ, in which case the addition which I have made should be omitted]. (IB, as quoted in the L and TA.) — حُصْتُ عَيْنَ البَازِى, (Ṣ,) or عَيْنَ الصَّقْرِ, (A,) aor. as above, and inf. n. of both the forms mentioned above, (Ṣ,) [*He seeled the falcon,* or *hawk; closed its eyes by running a thread through their lids.*] — حَاصَ سِقَآءَهُ *He closed a rent in his skin for water or milk by inserting into it two pieces of stick or wood;* not having a سِرَاد with which to sew it. (TA.) — حَوْصٌ also signifies The *making a coarctation between two things;* (Ṣ, Ḳ;) and so حِيَاصَةٌ. (Ḳ.) You say حَاصَ بَيْنَهُمَا [*He contracted the space between them;* meaning two things]. (TḲ.) = حَوِصَتْ عَيْنُهُ, (A, Mṣb,) aor. تَحْوَصُ, inf. n. حَوَصٌ, (Mṣb,) *His eye was,* or *became, narrow,* or *contracted, in its outer angle.* (A, Mṣb.) [But see حَوَصٌ below.] And حَوِصَ, (Ṣ, Ḳ,) aor. يَحْوَصُ, (TA,) said of a man, (Ṣ,) *He had, in his eye,* or *eyes, what is termed* حَوَصٌ [as explained below]. (Ṣ, Ḳ.) [See also حَوِصَتْ عَيْنُهُ.]

3. هُوَ يُحَاوِصُ فُلَانًا *He looks at such a one*

from the outer angle of his eye, concealing [*his doing*] *that.* (Ṣ.) [See also 6: and see خَاوَصَ.]

6. تخاوص إِلَى الشَّمْسِ *i. q.* اِسْتَوْضَحَهَا [app. *He blinked at the sun; or looked at it contracting his eyelids*]. (A in art. وضح.) [See also 3: and see تَخَاوَصَ.]

8. احتاصت *Her* (a camel's) *vulva was,* or *became, impervious to the virga,* or *nervus, of the stallion:* you should not say حَاصَتْ. (TA.) And احتاصت رَحِمَهَا دُونَ الفَحْلِ *Her* (a camel's) *vulva was rendered impervious to the virga,* or *nervus, of the stallion, by her having a ring tied over it,* i. e., over her vulva. (L, Ḳ,* TA.*) — احتاص † *He was prudent, and guarded himself.* (Ṣgh, Ḳ.)

خُوص *syn. with* حَوَص: see 1.

حَوَص *Narrowness,* or *contraction, of the outer angle of the eye,* (Ṣ, Mṣb, and so in some copies of the Ḳ,) or *eyes,* (as in some copies of the Ḳ and in the TA,) *as though they were sewed up;* (TA;) or *one of them:* (Ḳ:) or *narrowness,* or *contraction, of one of the eyes,* (Lth, Ṣ, Mgh, TA,) *exclusively of the other:* (Lth, Mgh, TA:) or, as Az says, accord. to all of them, *narrowness,* or *contraction, in both of the eyes:* (Mgh, TA:) or *in the slit of the eye.* (TA.) — See also أَحْوَص.

حُوصَى *syn. with* حُوص *and* حَوَص: see 1.

حِوَاص *A wooden implement with which one sews.* (Fr, Ṣgh, Ḳ.)

حِيَاصَةٌ, originally حِوَاصَةٌ, (Ḳ, &c.,) *A strip,* or *thong, of leather, in the girth:* or *a long strip,* or *thong, of leather:* (TA:) or *a strip,* or *thong, of leather, with which the girth of a horse's saddle is tied:* (Ḳ:) or *the girth of a beast.* (T, TA.) This is the primary signification: and sometimes it is used to signify — *Anything with which a man binds his waist:* [particularly, in modern usage, *a kind of girdle, zone,* or *waist-belt, which is fastened round the waist with a buckle or clasp; worn by men and by women; and when worn by wealthy women, generally adorned with jewels &c., and having two plates of silver or gold, also generally jewelled, which clasp together:* in earlier post-classical times, it is described as being *of silver,* and *of gold:* (see Dozy's "Dict. des Noms des Vêtements chez les Arabes," pp. 145-7:) pl. حَوَائِص:] of the dial. of Syria. (TA.) — [See also another application of this word voce خُرْدِيّ.]

حَائِص, applied to a she-camel, *Whose vulva is impervious to the virga,* or *nervus, of the stallion;* (Ṣ, TA;) *like* رَتْقَآء *applied to a woman;* (Fr, Ṣ, Ḳ;) as also حَائِصَةٌ and ▼ مُحْتَاصَةٌ: (TA:) and ▼ حَوْصَآء, applied to the same, † *having a narrow,* or *contracted, vulva:* (IAạr, TA:) and ▼ مُحْتَاصَةٌ, applied to the same, also signifies *having her vulva rendered impervious to the virga,* or *nervus, of the stallion, by its having a ring tied over it.* (L, Ḳ,* TA.*)

أَحْوَص A man *having in his eye,* or *eyes, what is termed* حَوَص [as explained above]: (Ṣ, A, Mgh, Mṣb, Ḳ:) or *having a narrowness,* or *contraction, in his eyes:* (Az, TA:) fem. حَوْصَآءُ: (Ṣ, Mṣb:) pl., when it is used as an epithet, حُوص; but when it is used as a [proper] name, أَحَاوِص: (Mṣb:) and ▼ حَوَص, for ذُوو حَوَص, is used as syn. with حُوص, meaning *having small eyes.* (IAạr.) The fem. is also applied to the eye itself, (A, TA,) signifying, *Narrow in its slit, whether it be sunken or prominent.* (TA.) — For another application of the fem., see حَائِص. — بِئْرٌ حَوْصَآءُ ‡ *A narrow well.* (A, TA.*)

مُحْتَاصَةٌ: see حَائِص, in two places.

حوض

1. حَاضَ المَآءَ, (A, Ḳ,) aor. يَحُوضُ, inf. n. حَوْض, (TA,) *He collected the water:* (A, Ḳ:) and, as also ▼ حَوَّضَهُ, inf. n. تَحْوِيض, *he guarded it,* or *took care of it:* (TA:) and ▼ the latter, *he made for it a* حَوْض [q. v.], or *place in which to collect.* (TA.) — Also حَاضَ, (Ṣ, TA,) or حَاضَ حَوْضًا, (A, Ḳ, [unless by this be meant that حَوْض is the inf. n.,]) aor. as above, (Ṣ, TA,) and so the inf. n., (Ṣ,) *He made a* حَوْض; (Ṣ, A, Ḳ;) as also ▼ حَوَّضَ, inf. n. as above; (TA;) and ▼ احتاض, inf. n. اِحْتِيَاض. (Th, TA.) You say, ▼ حَوَّضَ لِإِبِلِهِ [*He made a* حَوْض *for his camels*]: and also ▼ تَحَوَّضُوا حِيَاضًا [*They made* حِيَاض, pl. of حَوْض]. (A.)

2: see 1, in four places. — أَنَا أُحَوِّضُ حَوْلَ ذٰلِكَ الأَمْرِ ‡ *I have within my compass,* or *power, and care, that thing,* or *affair;* expl. by أَدُورُ حَوْلَهُ: (Ṣ, A, O, L, Ḳ:*) like أُحَوِّطُ: mentioned by Yaạḳoob: from مُحَوَّض, explained below: (Ṣ:) in the Ḳ, [هٰذَا is put in the place of ذٰلك, and] لَكَ is erroneously put for حول. (TA.) You say also, فُلَانٌ يُحَوِّضُ حَوْلَ فُلَانَةَ ‡ *Such a man has within his power and care such a female,* (يَدُورُ حَوْلَهَا,) *and toys, dallies, wantons,* or *holds amorous converse, with her.* (A, TA.)

5: see 1.

8: see 1.

10. اِسْتَحْوَضَ *It* (water) *collected,* or *became collected:* (Ṣ:) or *made for itself a* حَوْض. (O, L, Ḳ.)

حَوْض [*A watering-trough* or *tank, for beasts &c., generally constructed of stones cemented and plastered with mud, and made by the mouth of a well; and any similar receptacle for water;*] *a place in which water collects,* or *is collected:* (Mṣb,* TA:) accord. to some, from حَاضَتِ المَرْأَةُ; (Ḳ, TA;) [see art. حيض;] because the water flows to it; for, says Az, the Arabs put و in the place of ى, and ى in that of و: (TA:) accord. to others, from حَاضَ المَآءَ, explained above: (Ḳ, TA:) and ▼ مُحَوَّض signifies the same: (TA:) pl. [of pauc.] of the former, أَحْوَاض and [of mult.] حِيَاض, (Ṣ, Mṣb, Ḳ,) originally حِوَاض, (Mṣb;) and حِيضَان. (TA; and in a copy of the Ṣ in the place of حِيَاض, which is the form given in other copies.) — حَوْضُ الرَّسُولِ [*The pool of the Apostle,* meaning Moḥammad;] *that of which the Apostle's people will be given to drink on the day of resurrection:* [or] *i. q.* الكَوْثَر, q. v. (TA.) AZ mentions the saying سَقَاكَ ٱللّٰهُ بِحَوْضِ الرَّسُولِ (A,* TA) and مِنْ حَوْضِهِ (TA) [*May God give thee to drink from the pool of the Apostle*]. — حَوْضُ الحِمَارِ is an expression of revilement, signifying مَهْزُومُ الصَّدْرِ ‡ [lit. *Depressed in the breast,* or *bosom;* app. meaning *narrow-minded;* or *illiberal;* or *niggardly*]. (Ṣgh, Ḳ.) — حَوْضُ المَوْتِ ‡ *i. q.* مُجْتَمَعُهُ [app. meaning *The place where death is met; where the draught thereof is drunk*]: so termed by way of simile: pl. as above. (TA.) — اِنْصَبَّ عَلَيْهِمْ حَوْضُ الغَمَامِ and حِيَاضُهُ ‡ [*The reservoir of the clouds,* and *the reservoirs thereof, poured forth upon them*]. (A, TA.) — مَلَأَ حَوْضَ أُذُنِهِ بِكَثْرَةِ كَلَامِهِ ‡ *He filled the concha* (صَدَفَة) *of his ear with the abundance of his speech.* (A, TA.)

مُحَوَّض *A thing like a* حَوْض, *made to a palm-tree, that it may imbibe therefrom;* (Ṣ, Ḳ;) *a thing that is made around a tree, in the form of the* شَرَبَة, q. v. (M, TA.) See also حَوْض.

حوط

1. حَاطَ بِهِ, aor. يَحُوطُ: see 4, in three places. — حَاطَهُ, (Ṣ, Mṣb, Ḳ,) aor. as above, (Ṣ, Mṣb,) inf. n. حَوْط (Ṣ, Mṣb, Ḳ) and حِيطَة and حِيَاطَة, (Ṣ, Ḳ, TA, [the second and third, in the CḲ, erroneously, with fet-ḥ to the ح, the former of them being expressly said in the Ṣ and TA, and the latter also in the TA, to be with kesr, and both being shown in the Ṣ to be originally with و, i. e. حِوْطَة and حِوَاطَة,]) and حِيَاط is used in poetry for the last of these; (TA;) and ▼ حَوَّطَهُ, (Ḳ,) inf. n. تَحْوِيط; (TA;) and ▼ تَحَوَّطَهُ; (Ḳ, TA; [omitted in the CḲ;]) *He guarded, kept, kept safely, protected,* or *took care of, him,* or *it;* (Ṣ, Mṣb, Ḳ, TA;) *he defended him,* or *it;* (TA;) *he paid frequent attention to him,* or *it;* (Ḳ, TA;) *he minded,* or *was regardful of, the things that were for his,* or *its, good.* (TA.) You say, لَا زِلْتَ فِي حِيَاطَةِ ٱللّٰهِ *Mayest thou not cease to be in the protection of God.* (TA.) And مَعَ فُلَانٍ حِيطَةٌ لَكَ *There is with such a one compassion and affection for thee:* you should not say عَلَيْكَ. (Ṣ.) And أَحُوطُ عِرْضِي [*I guard,* or *defend,* or *take care of, my honour,* or *reputation*]. (TA.) And هُوَ يَتَحَوَّطُ أَخَاهُ *He takes care of,* or *pays frequent attention to, his brother; and undertakes,* or *superintends,* or *manages, his affairs.* (TA.) And حَاطَهُمْ قَصَاهُمْ and بِقَصَاهُمْ *He fought in their defence.* (TA.) [But this is generally meant ironically.] When an affliction befalls thee, and thy brother does not guard thee, or defend thee, and does not aid thee, one says [to thee], حَاطَكَ القَصَآءُ [so in the TA, app. a mistranscription for القَصَا or القَصَآءَ, with which, however, it is nearly syn.,] which is used ironically; i. e. *He guarded thee,* or *defended thee,* in

a distant quarter; meaning, ‡*he did not guard thee,* or *defend thee;* for he who guards, or defends, his brother, draws near to him, and supports him, or aids him. (A, TA.) [See also 1 in art. حبو.] You say also, حَاطُونَا القَصَا, (Ḳ,) or القَصَا, (TḲ,) [both are said to be correct in the TA in art. قصو, on the authority of Ibn-Wellád,] in some of the copies of the Ḳ with ق and ض, and in some with ق and ص, the latter unpointed, and so in [a copy of] the A, (TA,) ‡*They retired to a distance from us, they being around us, and we not being distant from them, had they desired to come to us.* (Ḳ, TA.) And حُطْنِى القَصَا ‡*Retire thou to a distance from me;* (Ibn-Wellád, and Ḳ in art. قصو;) as also القَصَآءَ. (Ibn-Wellád, and TA in that art.) And لَأَحُوطَنَّكَ القَصَا وَلَأَغْزُوَنَّكَ بِالعَصَا, in each case with the short ا, meaning *I will assuredly leave thee, and not go near thee;* [*and I will assuredly go against thee to fight thee with the staff.*] (Ks, TA in art. قصو.) حُطْ حُطْ means *Take thou care of the tie of kindred, and preserve it.* (IAạr, Ḳ.*) It also signifies *Deck thou the boys* (الصِّبْيَةَ [in the CḲ الصَّبِيَّةَ *the girl*]) *with the* حَوْط [*for preservation from the evil eye*]. (IAạr, Ḳ.) And حُوطُوا غُلَامَكُمْ *Deck ye your boy with the* حَوْط. (AA.) — حَاطَ الحِمَارُ عَانَتَهُ, (Ṣ,* Mṣb, Ḳ,) aor. as above, (Ṣ, Mṣb,) inf. n. حَوْطٌ, (Mṣb,) *The* [wild] *he-ass collected,* or *drew together,* (Ṣ,* Mṣb, Ḳ,*) *and guarded,* or *took care of,* (TA,) *his* عَانَة [app. meaning his *herd of wild asses:* or the phrase may mean *the he-ass drew towards himself,* or *compressed, and guarded, his she-ass:* Freytag here renders عانة by "pubem;" and Golius, by "veretrum"]. (Ṣ, Mṣb, Ḳ.)

2. حَوَّطَ حَوْلَهُ, inf. n. تَحْوِيطٌ, *He surrounded it by some such thing as earth, so as to make this to encompass it.* (Mṣb.) And حَوَّطَ كَرْمَهُ, inf. n. as above, *He built a* حَائِط [or *wall*] *around his vine.* (Ṣ.) — Hence, أَنَا أُحَوِّطُ حَوْلَ ذٰلِكَ الأَمْرِ ‡*I have within my compass,* or *power, and care, that thing,* or *affair;* [like أُحَوِّضُ, q. v.;] syn. أَدُورُ. (Ṣ, TA.) [Hence also, حَوَّطَ عَلَيْهِ, in the present day, is used to signify †*He monopolized it.* See also 4.] — حَوَّطَ حَائِطًا, (Ḳ,) inf. n. as above, (TA,) *He made a* حائط [meaning either a *walled garden* or a *wall;* app. a *wall of enclosure*]; (Ḳ, TA;) as also ↓ احاطهُ. (IDrd, TA.) — See also 1.

3. حاوط فُلَانًا ‡*He endeavoured to induce such a one to turn, or incline;* or *endeavoured to turn him by deceit,* or *guile;* (دَاوَرَهُ;) *in a matter that he desired of him, and which he refused him:* (Ḳ:) as though each of them were guarding, or taking care of, (يَحُوطُ,) the other. (Ḳ: and so in the A, in illustration of what next follows.) حَاوِطْهُ فَإِنَّهُ يَلِينُ لَكَ ‡*Endeavour thou to induce him to turn, or incline;* or *endeavour thou to turn him by deceit,* or *guile;* [*for he will relent to thee;*] syn. دَاوِرْهُ. (A, TA.)

4. احاط بِهِ and ↓ حَاطَ بِهِ signify the same [i. e. *It,* or *he, surrounded, encompassed, environed, enclosed,* or *hemmed in, it,* or *him*]. (TA.) You say, احاط القَوْمُ بِالبَلَدِ, inf. n. إِحَاطَةٌ; and ↓ حَاطُوا بِهِ; *The people surrounded, encompassed, environed, encircled,* or *beset, the sides of the town.* (Mṣb.) And احاطت الخَيْلُ بِفُلَانٍ, (Ṣ, TA,) and ↓ حَاطَتْ بِهِ, (TA,) and ↓ احتاطت بِهِ, (Ṣ,) *The horses,* or *horsemen, surrounded, encompassed, environed, encircled,* or *beset, such a one.* (Ṣ, TA.) [And احاطوا به مِنْ جَانِبَيْهِ, meaning *They surrounded him on all his sides;* lit. *on his two sides:* see جَنْبٌ.] — It is said in the Ḳur [xvii. 62], إِنَّ رَبَّكَ أَحَاطَ بِالنَّاسِ †*Verily thy Lord hath men in his grasp,* or *power:* (Bḍ, TA:) or †*hath destroyed them;* meaning Ḳureysh. (Bḍ.) You say also, أُحِيطَ بِفُلَانٍ, meaning †*Such a one was destroyed:* or †*his destruction drew near.* (TA.) And hence the saying in the Ḳur [xviii. 40], وَأُحِيطَ بِثَمَرِهِ †*And its fruit became smitten by that which destroyed and spoiled it:* (TA:) or †*his possessions became destroyed:* from أَحَاطَ بِهِ العَدُوُّ [*the enemy surrounded him*]. (Bḍ.) [Hence also, in the same, ii. 75,] وَأَحَاطَتْ بِهِ خَطِيئَتُهُ †*And over whom his sin hath gained the mastery, affecting all the circumstances of his case, so that he hath become as though he were entirely encompassed thereby:* (Bḍ:) or †*who hath died in the belief of a plurality of Gods.* (TA.) You also say, احاط بِهِ الأَمْرُ †*The thing beset him on every side, so that he had no place of escape from it.* (TA.) And احاط عَلَيْهِ †*He took it entirely to himself, debarring others from it:* [see also 2.] (TA in art. شرب.) — احاط بِهِ عِلْمًا, (Ḳ,) or احاط بِهِ عِلْمُهُ, (Ṣ, Mṣb, TA,) and احاط بِهِ عِلْمُهُ, (Ṣ, TA,) ‡[*He comprehended it,* or *knew it altogether, in all its modes* or *circumstances;*] *he knew it extrinsically and intrinsically;* (Mṣb;) or *he attained the utmost particular thereof, and had a comprehensive and complete knowledge thereof:* or *he attained everything* [*relating to it*], *and the utmost knowledge thereof.* (Ḳ, accord. to different copies. [In the CḲ, أَحْصَى عِلْمَهُ is put, erroneously, for أَحصى عِلْمَهُ.]) It is said in the Ḳur [xxvii. 22], أَحَطْتُ بِمَا لَمْ تُحِطْ بِهِ ‡*I have known in all its circumstances, or modes, that which thou hast not so known.* (TA.) And you say also, عَلِمَهُ عِلْمَ إِحَاطَةٍ ‡*He knew it in all its circumstances,* or *modes; nothing of them escaping him.* (TA.) — See also 2.

5: see 1, in two places.

8. احتاط: see 4. — Also ‡*He took the course prescribed by prudence, precaution,* or *good judgment; he used precaution; he took the sure course;* (Ṣ,* Ḳ,* TA;) لِنَفْسِهِ *for himself;* (Ṣ, TA;) [and مِنَ الشَّىْءِ *against the thing:*] *he sought the most successful means, and took the surest method;* لِلشَّىْءِ *for* [*the accomplishment,* or *attainment, of*] *the thing.* (Mṣb.) The subst. [denoting the abstract signification of the inf. n., اِحْتِيَاطٌ,] is حِيطَةٌ, (Mṣb,) i. e. ↓ حَيْطَةٌ and ↓ حِيطَةٌ, (Ḳ, TA,) which latter is originally حِوْطَةٌ, (TA,) [and is also an inf. n. of 1,] and ↓ حَوْطَةٌ. (Ḳ, TA.) Some hold احتياط to belong to art. حيط. (Mṣb.) You say also ↓ استحاط فِى الأُمُورِ [meaning in like manner †*He took the course prescribed by prudence, &c., in affairs,* or *in the affairs:* as is shown below: see مُسْتَحَاطٌ]. (TA.)

10: see 8.

حَوْطٌ *A twisted string of two colours, black and red,* (IAạr, Ḳ,) *called* بَرِيم, (IAạr,) *upon which are beads and a crescent of silver, which a woman binds upon her waist,* [*and which is bound upon a boy,* (see 1,)] *in order that the evil eye may not smite her* [or *him*]: (IAạr, Ḳ:) and also the *crescent above mentioned;* as well as the string with it. (TA.) [See also تَحْوِيطَةٌ.]

حَوْطَةٌ:
حَيْطَةٌ: } see 8.
حِيطَةٌ:

حُوَاطٌ: see what next follows.

حُوَاطَةٌ *An enclosure* (حَظِيرَة) *made for wheat:* (Ṣ, Ḳ:) or it signifies *a thing which one soon quits,* or *relinquishes,* or *from which one soon abstains;* and so ↓ حُوَاطٌ, as occurring [accord. to one relation] in a verse cited voce عَرَسَ. (L.)

حَيِّطٌ, [originally either حَيْوِطٌ or حَوْيِطٌ,] like سَيِّدٌ, A man *who guards, protects,* or *defends,* (يَحُوطُ,) *his family and his brethren.* (TA.)

[حَوَّاطٌ *A monopolizer:* so in the present day.]

حُوَّاطُ أَمْرٍ ‡*The undertakers, superintendents,* or *managers, of an affair.* (Ḳ, TA.) [See a verse cited voce عَرَسَ.]

حَائِطٌ *A wall.* (Mṣb,* Ḳ, TA:) or *a wall of enclosure:* (Mṣb,* TA:) or *one that surrounds a garden:* (Mgh:) [often applied to a *fence* of wood, or sticks, or of reeds, or canes:] so called because it surrounds what is within it; (TA;) but it is a subst., like سَقْفٌ and رُكْنٌ, though implying the meaning of surrounding: (IJ, TA:) or it is an act. part. n., from حَاطَ: (Mṣb:) pl. حِيطَانٌ, (Ṣ, Mṣb, Ḳ,) in which the و is changed into ى because of the kesreh before it, (Ṣ,) but by rule it should be حُوطَانٌ; (Sb, Ḳ;) and حِيَاطٌ. (IAạr, Ḳ.) — And hence, (Mgh,) *A garden,* (Mgh, Mṣb, Ḳ,) in general: or *a garden of palm-trees, surrounded by a wall:* (TA:) pl. حَوَائِطُ. (Mṣb, TA.)

اِفْعَلِ الأَحْوَطَ †*Do thou that which is most comprehensive in relation to the principles of the ordinances* [*applying to the case*], (مَا هُوَ أَجْمَعُ لِأُصُولِ الأَحْكَامِ,) *and furthest from the sophistications of interpretations not according to the obvious meanings.* (Mṣb.) And هٰذَا أَحْوَطُ †*This is more,* or *most, conducive to put* [*one*] *in a position of* اِحْتِيَاط [*or taking the course prescribed by prudence, precaution,* or *good judgment;* &c.: see 8]. (Mgh.) The word أَحْوَطُ is from the phrase حَاطَ الحِمَارُ عَانَتَهُ; not from الاِحْتِيَاط; because the أَفْعَل of excess is not formed from a verb of five letters: (Mṣb:) or it is anomalous, like أَحْصَرُ from الاِحْتِصَارُ. (Mgh.) [It may be rendered *More*, or *most, prudent*: or *more*, or *most, sure*.]

تَحُوطُ and التَّحُوطُ &c.: see what next follows.

تُحِيطُ and ▼ تَحُوطُ (ISk, TṢ, Ḳ) and تَحِيطُ and تُحِيطُ and ▼ يَحِيطُ (TṢ, Ḳ) and ▼ التَّحُوطُ and التَّحِيطُ (L, Ḳ) [and ▼ تَحُوِّطُ and ▼ تَحُوطُ (mentioned, with the third and fourth, in Freytag's Arab. Prov., ii. 803, as on the authority of Fr,)] ‡ *The year of dearth, scarcity, or straitness, that destroys the beasts*, (Fr, Ḳ,* TA,) or *men*: (A, TA:) تَحُوطُ being from حَاطَ بِهِ in the sense of أَحَاطَ; or it is used as a term of good omen; accord. to the A. (TA.) You say, وَقَعُوا فِى تُحِيطَ, &c., [the last word being a noun imperfectly decl., ‡ *They fell into the affliction of a year of dearth*, &c.] (ISk, TA.)

تَحْوِيطَةٌ *A thing that is hung upon a boy to repel the evil eye*: of the dial. of El-Yemen. (TA.) [See also حُوطٌ.]

مُحَاطٌ *A place behind the camels or other beasts and the people [to whom they belong], surrounding and protecting them*: (Ḳ:) some say that أَرْضٌ مُحَاطٌ signifies *land surrounded by a wall*: if not so surrounded, it is termed ضَاحِيَةٌ. (TA.)

فُلَانٌ مُحَاطٌ بِهِ † *Such a one is slain; is destroyed*. (TA.)

مُحِيطٌ [act. part. n. of 4; *Surrounding, encompassing*, or *enclosing*: &c.] — It is said in the Ḳur [lxxxv. 20], وَٱللّٰهُ مِنْ وَرَائِهِمْ مُحِيطٌ † *And God, behind them, includeth them altogether within his power; not one shall escape Him*. (TA.) And again, [xi. 85,] عَذَابَ يَوْمٍ مُحِيطٍ † *The punishment of a day which shall beset on every side so that there shall be no place of escape from it*: (TA:) or *of a destructive day*; meaning the day of resurrection: or *the punishment [of a day] of extermination*: the epithet, which is that of the punishment, being applied to the day because it includes it. (Bḍ.) And again, [ii. 18,] وَٱللّٰهُ مُحِيطٌ بِالْكَافِرِينَ, explained by Mujáhid as signifying † *And God will collect together the unbelievers* on the day of resurrection. (TA.)

كَرْمٌ مُحَوَّطٌ *A vine having a wall built around it*. (Ṣ.)

هُوَ مُحْتَاطٌ فِى أَمْرِهِ and ▼ مُسْتَحِيطٌ فِيهِ [*He is taking the course prescribed by prudence, precaution, or good judgment; or using precaution; or taking the sure course; or seeking the most successful means, and taking the surest method; in his affair*: see 8]. (TA.)

مُسْتَحِيطٌ: see what next precedes.

يَحِيطٌ: see تُحِيطُ.

حوف

1. حَافَهُ, [aor. يَحُوفُ,] inf. n. حَوْفٌ, *He, or it, was in*, or *on, the* حَافَة [or *side*] *of it*; i. e., of a thing. (TA.) — *He visited him*. (TA.) — See also 2.

2. حَوَّفَهُ, (Ḳ,) inf. n. تَحْوِيفٌ, (TA,) *He put him, or it, on the* حَافَة, (Ḳ, TA,) i. e., the *side*. (TA.) — *It* (the [rain called] وَسْمِىّ) *surrounded it*; namely, a place; (Ḳ;) as though it took its حَافَات [or sides]. (TA.) — سُلِّطَ عَلَيْهِمْ طَاعُونٌ يُحَوِّفُ القُلُوبَ, occurring in a trad., means [*A pestilence was made to overcome them*] *turning the hearts* [of those witnessing its effects] *from confidence, and inclining them to removal and flight therefrom*; (Ḳ,* TA;) from حَافَةٌ signifying the "side" of a place: (TA:) but some relate it otherwise, saying ▼ يَحُوفُ, like يَقُولُ: (Ḳ:) and some say يُحَرِّفُ; and thus A'Obeyd affirms it to be. (TA. [See 2 in art. حرف.])

5. تَحَوَّفَهُ *He took its* حَافَة [or *side*]: and *he took from its* حَافَة: as also تَحَيَّفَهُ: (TA:) or *he took by little and little from it*, namely, a thing, (Ṣ, Ḳ,) or *from its* حافات [or *sides*]; (Ṣ in art. حيف;) as also [تخوّفه and] تحيّفه. (Ṣ and Ḳ* in that art.) Abd-Allah Ibn-'Ajlán En-Nahdee says, (TA,) or some other poet, (L in art. خوف,)

* تَحَوَّفَ الرَّحْلُ مِنْهَا تَامِكًا قَرِدًا *
* كَمَا تَحَوَّفَ عُودَ النَّبْعَةِ السَّفَنُ *

[*Her saddle abraded from a long and high, compact hump; like as the piece of skin used for smoothing arrows has abraded from the rod of the tree called* نبعة]. (TA.) [See also 5 in art. خوف, where another reading of this verse is given.]

حَافٌ: see حَافَةٌ. — *A certain vein, of a green colour*, or *of a dark*, or *an ashy, dust-colour*, (أَخْضَرُ,) *beneath the tongue*: (O, Mṣb:) there are two veins of this description, called the حَافَانِ: (Ḳ:) accord. to some, the ف is with teshdeed. (TA.)

حَوْفٌ: see حَافَةٌ, first and last significations. — Also *A skin that is slit [in several places, into a number of thongs or strips,] of a form like that of the* إِزَار [or *waist-wrapper*], *worn [round the hips] by one who is in menstruis, and by boys*; (Ṣ, Ḳ;) i. e. *a* رَهْط: (Ṣ:) pl. أَحْوَافٌ: (TA:) accord. to IAạr, *a skin that is slit into thongs or strips*: or, as he says on one occasion, a وَتْر, i. e., (TA,) *a* نُقْبَة *of skin or leather, slit into strips of the width of four fingers each*, (IAạr, Ḳ, TA,) or *a span*, (IAạr, TA,) *worn by a young girl before she has attained to puberty*, (IAạr, Ḳ, TA,) and *worn by her in menstruis*; of the dial. of El-Hijáz; in the dial. of Nejd called رَهْط: (IAạr, TA:) or *red skin or leather, cut in the form of thongs or strips, upon which are put [ornaments of the kind termed]* شَذْر; *worn by a girl over her garments*: (Ḳ:) [see also حَوْف:] accord. to IAth, i. q. بَقِيرَة; i. e. *a garment without sleeves*. (TA.) — *A thing*, (Ḳ,) i. e. *a kind of vehicle in which a woman rides upon a camel*, (TA,) *resembling, but not the same as, the* هَوْدَج: (Ḳ:) of the dial. of the people of El-Ḥowf [in 'Omán], and of the people of Esh-Shiḥr. (TA.)

حَافَةٌ *A side* of anything; originally حَوَفَةٌ; (Mṣb;) and so ▼ حَوْفٌ: (TA:) حَافَتَانِ signifying the *two sides* of a valley (Ṣ, Mṣb, Ḳ) &c.; (Ḳ;) i. e., of any other thing: (TA:) pl. حَافَاتٌ (Mṣb, Ḳ) and حِيفٌ, which is irregular, and حِيَفٌ, which is regular, and حَوَائِفُ, which is extr., like حَوَائِجُ, and changed by transposition to حَوَافٍ, (TA in art. حيف,) and [coll. gen. n.] ▼ حَافٌ. (Ḥam p. 150.) It is said in a trad., عَلَيْكُنَّ بِحَافَاتِ الطَّرِيقِ [*Keep ye* (O women) *to the sides of the road*]. (TA.) — Also, of the دَوَائِس [or bulls or cows that tread wheat or other grain] (Ḳ, TA) in the كُدْس [or wheat &c. collected together in the place where it is trodden out], (TA,) *Such as is*, or *are, at the extremity, exceeding the others in going round*. (Ḳ, TA.) — *Want*: (Ḳ:) and *hardness*, or *difficulty, of life*; (Ḳ, TA;) as also ▼ حَوْفٌ. (TA.)

مُحَوَّفٌ *Bordered with herbage*. (TA in art. عبد.)

حوق

1. حَاقَ, aor. يَحُوقُ, (Ṣ,) inf. n. حَوْقٌ, (Ṣ, Ḳ,) *He swept* a house, or chamber. (Ṣ.) — *He rubbed and smoothened* (Ḳ, TA) a thing. (TA.) — *He sharpened* a spear-head. (Ḥam p. 177.) — حَاقَ بِهِ, (TḲ,) [aor. as above,] inf. n. حَوْقٌ, (Ḳ,) [like حَاقَ بِهِ, aor. يَحِيقُ, inf. n. حَيْقٌ,] *It surrounded, encompassed, encircled*, or *beset, him*, or *it*. (Ḳ, TḲ.)

حَوْقٌ: see حُوقٌ, in two places. — Also i. q. حَوْقَلَةٌ [app. as meaning *A soft*, or *weak, penis, such as that of an old man*]. (TA.)

حُوقٌ The [*corona* or] *surrounding edges of the glans of the penis*; (Ṣ, Ḳ;) as also ▼ حُوُقٌ, (Ibn-'Abbád, Ḳ,) which is a rare dial. var., (TA,) and ▼ حُوَّقٌ: (Ibn-'Abbád, TA:) or ▼ حَوْقٌ, (Ḳ, TA,) with fet-ḥ, (TA,) [in the CḲ حُوق,] signifies *a roundness in the penis*. (Th, Ḳ.)

حُوَّقٌ: see what next precedes.

حُوَاقَةٌ *Sweepings*. (Ṣ, Ḳ.) — And i. q. قُمَاشٌ [(written in the TA with س, which is evidently a mistake, a result of an oversight,) What is *bad* of anything; or *what is collected hence and thence*; or *small particles*, or *fragments*, of anything; or *small rubbish*, or *broken particles of things, on the surface of the ground*]. (Ks, TA.)

أَحْوَقُ A penis (TA) *having a large glans*; as also ▼ مُحَوَّقٌ. (Ḳ, TA.) And فَيْشَلَةٌ حَوْقَاءُ (Ḳ) and كَمَرَةٌ حَوْقَاءُ (TA) *A large* (Ḳ, TA) *and prominent* (TA) *glans of a penis*. (Ḳ, TA.)

مَحُوقٌ [*Swept*. — And hence, † *Shaven*.] It is said in a trad., يَسْجُدُونَ مَحُوقَةً رُؤُوسُهُمْ † *They prostrate themselves having the middle of their heads shaven*: the removal of the hair from that part being likened to sweeping. (TA.) [Hence also] أَرْضٌ مَحُوقَةٌ † *Land having little*, (Ḳ,) or *very little*, (TA,) *herbage*; *by reason of paucity of rain*; (Ḳ, TA;) as though it were swept. (TA.) — *Rubbed and smoothened*; as also ▼ مَحِيقٌ (Ḳ, TA) and ▼ مَحْوُوقٌ. (TA.)

مَحْوُوقٌ: see مَحِيقٌ.

مِحْوَقَةٌ *A broom; a thing with which one sweeps.* (Ṣ, Ḳ.)

أُحْوُقٌ : see مُحْوَقٌ.

مُحْوَقٌ : see مَحْيُوقٌ.

حوقل

حَوْقَلَ, &c.: see art. حقل.

حوك

1. حَاكَ الثَّوْبَ, aor. يَحُوكُ, inf. n. حَوْكٌ (Ṣ, Mṣb, Ḳ) and حِيَاكٌ (Ḳ) and حِيَاكَةٌ, (Ṣ, Ḳ,) [which last see also below,] *He wove the piece of cloth.* (Ṣ, Ḳ.) [See also art. حيك.] —— Hence, حَاكَ الشِّعْرَ, aor. as above, inf. n. حَوْكٌ, ‡ *He wove poetry.* (Mbr, TA.) —— And المَطَرُ يَحُوكُ الأَرْضَ, inf. n. حَوْكٌ, ‡ [*The rain frets the surface of the ground; makes rippling lines upon it.*] (TA.) —— حَاكَ الشَّىْءُ فِى صَدْرِى, (Ḳ,) inf. n. حَوْكٌ, or, accord. to Az, the aor. of the verb in this case is يَحِيكُ, (TA,) [and the inf. n. is حَيْكٌ, (see art. حيك,)] † *The thing became fixed in my mind.* (Ḳ.) [See also حَكَّ.]

5. تَحَوَّكَ بِالثَّوْبِ, (Az in art. حبك,) and ▼احتاك به, (Aṣ, ISk, Ḳ in art. حيك,) *i. q.* اِحْتَبَى به, (Aṣ, ISk, Ḳ in art. حيك, and TA,) i. e. *He drew his legs against his belly with the garment, or piece of cloth, confining them therewith, together with his back, and binding it, or making it tight, upon them, so as to preserve him from falling,* [*when he sat,*] *like a wall.* (IAth, TA in art. حبو.) [See 1 in art. حبك.]

8: see 5.

حَوْكٌ inf. n. of 1 [q. v.]. (Ṣ, Mṣb, Ḳ.) —— [Hence,] ذَا عَلَى حَوْكِ ذَا † *This is like this in age, and in aspect, appearance, or person.* (A, TA.) And نَاسٌ لَيْسَ عَلَيْهِمْ حَوْكَةُ ▼ قُرَيْشٍ † *Men who do not resemble Ḳureysh.* (A, TA.)

حَوْكٌ as used in the saying هٰؤُلَآءِ حَوْكُ سَوْءٍ, meaning † *These are children that are lean, or emaciated, or slender in the bones, and small in the body,* has no sing. form. (O, TA.)

حَوْكَةٌ : see حَوْكٌ.

حِيَاكَةٌ The *art of weaving.* (Lth, Mṣb.*) [See also 1, of which it is said in the Ṣ and Ḳ to be an inf. n.]

حَائِكٌ *A weaver:* [fem. with ة:] (Ṣ, Ḳ:) pl. masc. حَاكَةٌ and حَوَكَةٌ, (Ṣ, Mgh, Mṣb, Ḳ,) the former regular, (TA,) and the latter anomalous, (Ṣ, TA,) like خَوَنَةٌ [pl. of خَائِنٌ]; (Ṣ in art. خون;) and pl. fem. حَوَائِكُ. (Ṣ, Ḳ.)

مَحَاكَةٌ *A place of weaving.* (Ṣ, Ḳ.)

تَرَكْتُهُمْ فِى مَحْوَكَةٍ ‡ *I left them* [engaged] *in fight.* (Ḳ, TA.)

حول

1. حَالَ, (Ṣ, Mgh, Mṣb, Ḳ, &c.,) aor. يَحُولُ, (Mṣb, Er-Rághib,) inf. n. حَوْلٌ and حُؤُولٌ (Ḳ, Er-Rághib) [and حَوَلَانٌ], *It* (a thing) *became altered, transmuted,* or *changed,* (Ṣ,* Mgh, Mṣb, Er-Rághib, TA,) *from its state,* or *condition,* (Ṣ, Mgh,) or *from its natural state* or *condition,* and *its constitution;* as also ▼استحال; (Mṣb;) *i. q.* ▼تحوّل; (Ḳ;) which [here] signifies [as above; or] *it became altered, transmuted,* or *changed,* whether *essentially* or *substantially,* or *in respect of predicament* [or *state* or *condition*], or *by saying;* (Er-Rághib, TA;) and ▼احال signifies the same: (Ḳ:) or this last is said of a man, meaning تَحَوَّلَ ▼ مِنْ شَىْءٍ إِلَى شَىْءٍ [*he shifted,* or *turned, from one thing to another*]: (TA:) and hence, (TA,) *he became a Muslim:* (IAạr, Ḳ, TA:) for in this case one says of a man, تَحَوَّلَ ▼ عَمَّا كَانَ يَعْبُدُ إِلَى الإِسْلَامِ [*he turned from that which he was worshipping to El-Islám*]. (IAạr, TA.) —— [Hence, also,] حَالَ, inf. n. مَحَالٌ and حِيلَةٌ, *i. q.* احتال, q. v. (Ḥam p. 652.) —— And حَالَ and ▼استحال *It* (anything) *shifted,* or *removed,* or *went,* or *became shifted* or *transferred;* syn. تحوّل: or *it moved;* syn. تحرّك: so accord. to different copies of the Ḳ: or, accord. to the O, the former verb has both of these significations: (TA:) or it has the latter of these significations, said of a شَخْص [i. e. a man, or person, or the figure of a thing seen from a distance]: (Ṣ:) or both verbs signify *it* (anything) *became altered,* or *changed,* (M, Ḳ,) *from straightness,* or *evenness,* (Ḳ,) *to crookedness,* or *unevenness.* (M, Ḳ.) You say, حَالَتِ القَوْسُ *The bow became crooked* (Ḳ, TA) *in the portion between the part grasped by the hand and the curved extremity;* or *in the curved extremity:* (TA:) or *reverted from the state into which it was brought by pressure* [*with the* ثِقَاف], *and became crooked in the portion between the part grasped by the hand and the curved extremity;* as also ▼استحالت: (Ṣ, O:) and in like manner, استحالت ▼ الأَرْضُ عَنِ الاِسْتِوَآءِ إِلَى العِوَجِ [*The ground became altered,* or *changed, from evenness to unevenness*]: (Ṣ:) or [simply] ▼استحالت *it became uneven:* (Mṣb:) [and أَحَالَتِ ▼ الارض (Ḳ in art. صمت) app. signifies the same:] or ▼استحال signifies *it* (a thing) *was disposed,* or *was about, to become altered,* or *changed.* (Er-Rághib, TA.) And حَالَ لَوْنُهُ *Its colour altered,* or *changed, and became black.* (Ṣ.) And حال *It* (a thing) *shifted from its way,* or *manner,* or *direction.* (TA.) And حَالَ وَتَرُ القَوْسِ *The string of the bow shifted from its place* on the occasion of shooting: and حَالَتِ القَوْسُ وَتَرَهَا [*the bow shifted from its string.*] (TA.) And حَالَ مِنْ مَكَانِهِ, inf. n. حِوَلٌ, (O, TA,) or this is a simple subst., (Ṣ, M, Ḳ,) *He,* or *it, shifted,* or *removed, from his,* or *its, place.* (O, TA. [See حِوَلٌ, below.]) And حال إِلَى مَكَانٍ آخَرَ *i. q.* ▼تحوّل [i. e. *He,* or *it, shifted,* or *removed,* or *became shifted* or *transferred, to another place*]. (Ṣ.) And حَالَ عَنِ العَهْدِ, inf. n. حُؤُولٌ, *i. q.* انقلب [i. e. *He withdrew,* or *receded, from the covenant, compact, agreement,* or *engagement*]. (Ṣ.) —— حَالَ فِى مَتْنِ فَرَسِهِ, inf. n. حُؤُولٌ, *He leaped, and rode, upon the back of his horse;* as also ▼احال: (Ṣ:) or حال فِى ظَهْرِ دَابَّتِهِ *he leaped, and seated himself firmly, upon the back of his beast;* as also ▼احال: (Ḳ, TA:) and حَالَ عَلَى الفَرَسِ, (TA,) inf. n. حَوْلَةٌ, (Ḳ,* TA,) *he seated himself firmly upon the horse.* (Ḳ,* TA.) —— حَالَ صَبُوحُهُمْ عَلَى غَبُوقِهِمْ, *Their morning-draught and their evening-draught became one,* is said of people suffering from drought, and scarcity of milk. (TA.) —— حَالَ, (Mgh, Mṣb, Ḳ,) aor. as above, inf. n. حَوْلٌ, (Mṣb,) said of a year (حَوْلٌ), (Mgh, Ḳ,) *It passed:* (Mṣb:) or *it revolved and passed:* (Mgh:) or *it became complete.* (Ḳ.) You say, حَالَ عَلَيْهِ الحَوْلُ, (Ṣ, Ḳ,) inf. n. حَوْلٌ and حُؤُولٌ, (Ḳ,) *The year passed over him,* or *it;* [or *he,* or *it, became a year old;*] (Ṣ, Ḳ;) as also ▼احال. (Ṣ.) And حَالَ, alone, (Mṣb, TA,) and ▼احال and ▼أَحْوَلَ, (Mṣb,) *The year passed over it.* (Mṣb, TA.) And ▼احال (Ṣ, Ḳ) and ▼أَحْوَلَ (Ṣ, TA) and ▼احتال (Ḳ) *A year passed over it;* [or *it became a year old;*] (Ṣ, Ḳ, TA;) said of wheat, or food, and of other things: (Ṣ, TA:) and so حالت and ▼احالت and ▼أَحْوَلَتْ said of a دار [or house]: (Ṣ:) or حَالَتِ الدَّارُ and ▼احالت and ▼أَحْوَلَتْ and حِيلَ بِالدَّارِ *years passed over the house:* (Ḳ:) or *the house became altered,* or *changed, and years passed over it:* and in like manner one says, أَعَامَتْ and أَشْهَرَتْ. (TA.) And حال said of a boy, *A year passed over him;* [or *he became a year old;*] (Ṣ;) as also ▼احول. (Ḳ.) And ▼احال بِالمَكَانِ (Ks, Ṣ, Mṣb, Ḳ) and ▼أَحْوَلَ (Ks, Ṣ, Ḳ) *He remained, stayed, abode,* or *dwelt, a year in the place:* (Ks, Ṣ, Mṣb, Ḳ:) or, as some say, *a long time.* (TA.) —— حَالَتْ, inf. n. حِيَالٌ (Ṣ, Mṣb, Ḳ) and حِيَالَةٌ and حُؤُولٌ (Ḳ) and حُولٌ; (Ṣ;) and ▼احالت, and ▼حوّلت; (Ḳ;) said of a she-camel, (Ṣ, Ḳ,) &c., (Ḳ,) *She did not conceive,* or *become pregnant, during a year,* or *two years,* or *some years:* (Ḳ:) or *she, having been covered by the stallion, did not become pregnant:* (Ṣ, Ḳ:) or, said of a woman, and of a she-camel, *she did not become pregnant.* (Mṣb.) And حالت, (Ṣ, Mgh, Mṣb, TA,) and ▼احالت, (Mgh,) said of a palm-tree (نَخْلَة), *It bore one year, and not another year:* (Mgh, TA:) or *did not bear,* (Ṣ, Mṣb,) *having been fecundated.* (Ṣ.) —— حَالَ الشَّىْءُ بَيْنِى وَبَيْنَكَ, (Ṣ, Er-Rághib,) or بَيْنَ الشَّيْئَيْنِ, (Mgh,* Ḳ,) inf. n. حَيْلُولَةٌ, [originally حَيْوُولَةٌ,] (Mgh, Mṣb,) like كَيْنُونَةٌ [&c.], (Mgh,) and حُؤُولٌ (Mgh) and حَوْلٌ, (Er-Rághib, TA,) *The thing intervened as a separation, a partition, a fence, a barrier,* or *an obstacle,* or *obstruction,* (Ṣ, Ḳ, Er-Rághib, TA,) *between me and thee,* (Ṣ, Er-Rághib, TA,) or *between the two things.* (Ḳ.) You say, حال النَّهْرُ بَيْنَنَا *The river intervened as a separation,* or *an obstacle, between us, preventing conjunction,* or *communication.* (Mṣb.) And حال الشَّىْءُ دُونَ الشَّىْءِ [*The thing intervened as an obstacle in the way to the thing*]. (Ṣ voce اِعْتَرَضَ.) It is said in the Ḳur [viii. 24], وَاعْلَمُوا أَنَّ اللّٰهَ يَحُولُ بَيْنَ المَرْءِ وَقَلْبِهِ [*And know ye that God interveneth,* or *interposeth, between the man and his heart,* or *secret thoughts,* or *desire*]: indicating that He turns him from his desire: (Er-Rághib, TA:) or that He possesses his heart, and

turns it as He wills: (O, TA:) or that a man cannot believe nor disbelieve unless it be God's will: (Jel:) or, as some say, that God destroys a man; or reduces him to the vilest condition of life, in order that he may not know, after knowing, anything. (Er-Rághib, TA. [See other remote interpretations in the Ksh, and the Expos. of Bd.]) And in the same [xxxiv. 53], وَحِيلَ بَيْنَهُمْ وَبَيْنَ مَا يَشْتَهُونَ [*And an obstacle shall be made to intervene between them and that which they shall eagerly desire*]. (TA.) — And حال الشَّىْءُ *The thing poured out, or forth.* (TA.) [See also 4.] ═ حَوِلَتْ عَيْنُهُ, (Ṣ, Ḳ,) and حَالَتْ, inf. n. تَحَالُ, (Ḳ,) the latter, of the dial. of Temeem, accord. to Lth, (TA,) [but see what follows,] inf. n. حَوَلٌ; (Ṣ, M, Mgh, Ḳ;) and ▼ احولّت, (Ṣ, Ḳ,) inf. n. ▼ إِحْوِلَال; (Ḳ;) *His eye* [*squinted*; i. e.] *had the white apparent at the outer angle, and the black next the inner angle:* (M, Ḳ:) or *had the black turning towards the nose:* (Lth, M, Ḳ:) or *had one of the two blacks turned towards the nose, and the other towards the temple:* (Mgh:) or *had its black next the outer angle:* or *his eye was as though it looked towards the* حِجَاج [or *supraorbital bone*]: or *had the black inclining towards the outer angle:* (M, Ḳ:) the first of which meanings is that commonly known: but some say that حَالَتْ signifies it was turned from its proper state: or it is anomalous: (TA:) the epithet applied to the man is ▼ أَحْوَلُ, (Ṣ, Mgh, Ḳ,) and ▼ حَوِلٌ: and that applied to the eye is [the fem. of the former of these, i. e.] ▼ حَوْلَاءُ: (Ḳ:) the pl. of which, and of the masc., is حُولٌ. (Ḥar p. 412.)

2. حوّلهُ, inf. n. تَحْوِيلٌ, *He altered it, transmuted it, or changed it,* whether *essentially,* or *substantially,* or *in respect of predicament* [or *state or condition*], or *by saying;* (Er-Rághib, TA;) [as also ▼ احالهُ.] — [Hence, *He turned it over, or about, in his mind, considering what might be its results, and so managed it;* namely, an affair; like قَلَّبَهُ.] You say, رَجُلٌ بَصِيرٌ بِتَحْوِيلِ الأُمُورِ [*A man who is knowing, skilful,* or *intelligent, in turning affairs over, or about, in his mind, &c.*]. (Ṣ, TA.) And ▼ احال رَأْيَهُ فِى الأَمْرِ *He altered, or changed, his opinion respecting the thing,* or *affair.* (MA.) — *He shifted it, removed it,* or *transferred it, from one place to another:* (Ṣ,* Mgh, O, Mṣb, TA:) or حوّلهُ إِلَيْهِ *he shifted it, removed it,* or *transferred it, to it,* or *him:* (Ḳ:) and ▼ احالهُ signifies the same. (Mṣb.) [Hence,] حوّل الرِّدَآءَ, (Mṣb,) inf. n. as above, (Mgh,) *He put the right side of the* [*garment called*] رداء *on the left:* (Mgh:) or *he shifted each extremity of the* رداء *to the place of the other.* (Mṣb.) — *He transferred,* or *transcribed, what was in it,* namely, a book, or writing, *to another, without doing away with the original form.* (TA.) — *He made it,* or *pronounced it to be,* مُحَالٌ [i. e. *absurd, inconsistent, self-contradictory, unreal,* or *impossible*]. (Ḳ. [See also 4.]) — حوّل عَيْنَهُ: see 4. ═ See also 5, in two places. — حَوَّلَتْ said of a she-camel &c., *i. q.* حَالَتْ: (Ḳ:) see 1.

3. حاولهُ, (Ṣ, M, Ḳ,) inf. n. مُحَاوَلَةٌ (M, Ḳ, KL) and حِوَالٌ, (M, Ḳ, TA, [in the CḲ, erroneously, حَوال,]) *He desired it:* (Ṣ, KL:) *he sought it:* (M, Ḳ, KL:) or *he sought it by an artful contrivance* or *device;* or *by artful,* or *skilful, management; by turning over,* or *revolving, thoughts, ideas, schemes,* or *contrivances, in his mind, so as to find a way of attaining his object;* syn. طَلَبَهُ بِحِيلَةٍ, (A,) or بِٱلْحِيلَةِ. (Ḥar p. 326.) Aboo-Ḥeiyeh En-Numeyree says,

وَمَنْ يُحَاوِلُ شَيْئًا فِى فَمِ الأَسَدِ

[*And who will seek to get a thing in the mouth of the lion?*] (Freytag's Arab. Prov. ii. 714.) And one says, حاول مِنْهَا الجِمَاعَ [*He desired,* or *sought, of her, copulation, using blandishment,* or *artifice, for that purpose*]. (T in art. رود.) بِكَ أُحَاوِلُ, occurring in a trad., means بك أُطَالِبُ [app. *By means of Thee I seek,* or *demand,* that which I want]. (Az, M, A, TA.) One says also, حَاوَلْتُهُ أَنْ يَفْعَلَ الأَمْرَ [*I sought,* or *endeavoured, to induce him to do the thing*], and أَنْ يَتْرُكَهُ [*to leave it*]. (A in art. دور.) And, of water, حاول أَنْ يَجْمُدَ (L and Ḳ in art. جمد) † *It was about to congeal,* or *freeze; was at the point of congealing,* or *freezing.* (TḲ in that art.) — حَاوَلْتُ لَهُ بَصَرِى *I looked sharply,* or *intently, at him; I cast my eyes at him.* (ISd, Ḳ.)

4. احال, as an intrans. v.: see 1, in eighteen places. — تَجَنَّبَ رَوْضَةً وَأَحَالَ يَعْدُو [*He withdrew from a meadow, and set to running,*] is a prov., meaning he forsook abundance of herbage, or of the goods and conveniences and comforts of life, and preferred to it straitness, or difficulty. (Ṣ.) — احال عَلَيْهِ بِالسَّوْطِ *He set upon him with the whip,* (Ṣ, MA, Ḳ,) يَضْرِبُهُ [*striking him*]: (Ṣ:) or *he desired to strike him with the whip:* or *he struck him with the whip:* (MA:) and أَحَلْتُهُ بِالسَّوْطِ, and بِالرُّمْحِ, [if احلته be not a mistranscription for أَحَلْتُ عَلَيْهِ, in the MS. from which I take this, as it may be inferred to be from what here precedes and follows,] *I aimed at him with the whip,* and *with the spear, and set upon him with it:* whence the saying, of him who has struck one at the point of death, and killed him, يُحِيلُ المَوْتَ عَلَى الضَّرْبِ, i. e. † *He makes death* [*as it were*] *to hang upon, and cleave to, striking;* like as the spear is made to cleave to the مُحَال عَلَيْهِ, who is the person thrust, or pierced. (Mṣb.) El-Farezdaḳ says, (Ṣ, TA,) addressing Hubeyreh Ibn-Ḍamḍam, (TA,)

وَكُنْتَ كَذِئْبِ السَّوْءِ لَمَّا رَأَى دَمًا
بِصَاحِبِهِ يَوْمًا أَحَالَ عَلَى الدَّمِ

i. e. [*And thou wast like the wicked wolf: when he saw blood upon his companion, one day,*] *he set upon the blood.* (Ṣ, TA.) — [Hence, perhaps,] حال عَلَيْهِ *He reckoned him,* or *esteemed him, weak.* (Ḳ.) — احال اللَّيْلُ *Night poured upon the earth;* (Ḳ;) and *came on.* (TA.) [See also 1, last sentence but one.] — احال بِفُلَانٍ الخُبْزُ *The bread fattened such a one;* and in like manner one says of anything by which one becomes fat. (AA, TA.) — And احال *He did,* or *said, what was* مُحَال [i. e. *absurd, inconsistent, self-contradictory, unreal,* or *impossible*]. (Ṣ, Ṣgh, Ḳ.*) — And *His camels did not conceive,* or *become pregnant,* (AA, Ṣ, Ḳ,) *during a year,* or *two years,* or *some years,* (Ḳ,) *having been covered.* (Ṣ.) ═ As a trans. v.: see 2, in three places. — احال الغَرِيمَ *He referred the creditor, from himself,* [*for the payment of what was due to him,*] *to another.* (M, Ḳ.) And أَحَلْتُهُ بِدَيْنِهِ *I transferred his debt* [i. e. the debt due to him from me] *by making another person than myself responsible for it.* (Mṣb.) And أَحَلْتُ زَيْدًا بِمَا كَانَ لَهُ عَلَىَّ عَلَى رَجُلٍ *I referred,* or *turned over, Zeyd, for the payment of what was due to him from me, to a certain man, transferring the responsibility for the debt to the latter:* in which case, I am termed ▼ مُحِيلٌ; and Zeyd is termed ▼ مُحَالٌ; and the other man, ▼ مُحْتَالٌ عليه, ▼ مُحَالٌ عَلَيْهِ, and ▼ حَوِيلٌ; and the property, ▼ مُحَالٌ بِهِ: (Mgh:) and ▼ حَيِّلٌ, [originally حَيْوِلٌ or حَوِيلٌ,] also, is applied to him to whom the reference is made; and to him who accepts the reference; both together being termed حَيِّلَانِ. (Aboo-Sa'eed, TA.) And you say, احال عَلَيْهِ بِدَيْنِهِ [*He referred* a person *to him for the payment of his debt*]. (Ṣ.) And احال [alone] *He transferred the debt for which he was responsible to the responsibility of another.* (Ḥar p. 59.) And أَحَلْتُ الأَمْرَ عَلَى زَيْدٍ [*I turned over the affair to Zeyd;*] *I made the performance of the affair to be required restrictively of Zeyd.* (Mṣb.) — احال عَلَيْهِ المَآءَ *He poured out the water upon it:* (Ḳ:) or احال المَآءَ مِنَ الدَّلْوِ *he poured forth the water from the bucket, and turned over the latter.* (Ṣ.) — احال اللّٰهُ عَلَيْهِ الحَوْلَ [*God made the year to pass over him,* or *it*]: (Lḥ, TA:) or احال اللّٰهُ الحَوْلَ *God made the year complete.* (Ḳ, TA.) — احال الرَّجُلُ إِبِلَهُ العَامَ [*The man made his she-camels to pass the year without becoming pregnant;* or] *the stallion did not cover the man's she-camels during the year.* (Lḥ, TA.) — أَحْوَلَ عَيْنَهُ, (Ks, Lḥ, Ṣ,) or أَحَالَهَا, and ▼ حوّلها, (Ḳ,) *He made his eye to be* حَوْلَآءَ [i. e. *squinting, &c.*]. (Ks, Lḥ, Ṣ, Ḳ.) [See 1, last sentence.] — احال كَلَامَهُ *He made his speech* مُحَال [i. e. *absurd, inconsistent,* or *self-contradictory*]. (Ṣ. [See also 2.]) ═ مَا أَحْوَلَ حِيَلَكَ *How fluctuating,* and *shifting,* and *varying, are thy evasions, wiles, artifices,* or *artful contrivances* or *devices!* (Ḥar p. 309.) — And مَا أَحْوَلَهُ *How surpassing is he in the practice of evasions, shifts, wiles, artifices,* or *artful contrivances* or *devices;* or *in turning over,* or *revolving, thoughts, ideas, schemes,* or *contrivances, in his mind, so as to find a way of attaining his object!* (Fr, Ṣ, Ḳ;) as also ما أَحْيَلَهُ. (Fr, Ṣ in art. حيل, and Ḳ.)

5. تحوّل: see 1, first sentence, in three places. [Hence,] تحوّل مُبْتَدِعًا [*He turned innovator*]. (O and Ḳ in art. بدع.) — *It shifted,* or *removed,*

or went, or became shifted or transferred, (Ṣ, Mṣb, Ḳ,) from one place to another, (Ṣ,) or from its place; (Mṣb;) as also ↑ حوّل (Ṣ,* Mṣb, Ḳ*) [and حال and استحال, as shown above: see 1, near the beginning.] You say, تحوّل مِنْ مَكَانه *It shifted, &c., from its place.* (Mṣb.) And تحوّل عَنْهُ *It shifted from it to another.* (Ḳ.) And تحوّل إِلَى مَكَانٍ آخَرَ i. q. حَالَ, q. v. (Ṣ.) And حَوَّلَتِ ↑ المَجَرَّةُ *The Milky Way became in the midst of the sky;* which it does in the summer, (Sh, Ḳ, TA,) when the season of heat comes on. (Sh, TA.) — See also 8, in two places. — Also, (Ṣ, Ḳ,) or تحوّل حَالًا, (TA,) *He carried a bundle upon his back.* (Ṣ, Ḳ, TA.) And تحوّل الكِسَآءَ *He put a thing in the* [*garment called*] كساء, *and then carried it on his back.* (M, Ḳ.) — تحوّلهُ بِالمَوْعِظَةِ *He sought to avail himself of the state in which he might be rendered prompt, or willing, to accept admonition.* (AA, Ḳ.)

8. احتال عَلَيْهِ بِالدَّيْنِ [meaning *He was referred, or turned over, to him for the payment of the debt*] is from الحَوَالَةُ. (Ṣ, TA.) You say, احتال زَيْدٌ بِمَا كَانَ لَهُ عَلَىَّ عَلَى رَجُلٍ *Zeyd was referred, or turned over, for the payment of what was due to him from me, to a certain man, to whom the responsibility for the debt was transferred.* (Mgh.) — احتال said of a year; see 1. — احتال (Ṣ, MA, Mṣb, Ḳ, KL) and ↑ تحوّل (Ṣ, Ḳ) and تحيّل (Ḳ) signify the same, (Ṣ, Ḳ,) from الحِيلَة [q. v.]; (Ṣ;) and ↑ حَالَ, (Ḥam p. 652,) inf. n. مَحَال and حِيلَة, (Ḥam ib. and Ḳ,*) also signifies the same as احتال; (Ḥam ib.;) which means *He practised* حيلة [i. e. *an evasion or elusion, a shift, a wile, an artifice,* or *an artful contrivance or device, a machination, a trick, a plot, a stratagem,* or *an expedient,* &c.]: (MA, KL:) or *he exercised art, artifice, cunning, ingenuity,* or *skill, and excellence of consideration or deliberation, and ability to manage according to his own free will,* (Ḳ, TA,) *with subtilty:* (TA:) or *he sought* الحِيلَة i. e. [*means of evading,* or *eluding, a thing,* or *of effecting an object, by*] *the exercise of art, artifice, cunning, ingenuity,* or *skill, in the management of affairs; by the turning over,* or *revolving, thoughts, ideas, schemes,* or *contrivances, in his mind, so as to find a way of attaining his object.* (Mṣb.) You say, احتال فِى الأَمْرِ and ↑ تحوّل [&c., *He practised an evasion or elusion, &c., in the affair*]. (Ḳ.) [And احتال عَلَيْهِ *He practised an artifice,* or *an artful contrivance or device,* &c., *against him.* And احتال لعياله *He exercised art, artifice, cunning, ingenuity, or skill, in the management of affairs, for his family,* or *household.*] — احْتَوَلُوهُ *They encompassed,* or *surrounded, him;* or *made him to be in the midst of them.* (M, O, Ḳ.)

9. احولّت عَيْنُهُ: see 1, last sentence. — [احولّت الأَرْضُ: see 11.]

10. استحال: see 1, in six places. — Also *It* (speech, or language, Ṣ, Mṣb, or a thing, TA) *became* مُحَال [i. e. *absurd, inconsistent, self-contradictory, unreal,* or *impossible*]. (Ṣ, Mṣb, TA.) — استحالهُ *He looked at it,* (Ṣ, M, Ḳ,) namely, a شَخْص [i. e. a man, or person, or the figure of a thing seen at a distance], (Ṣ,) *to see if it moved:* (Ṣ, M, Ḳ:) as though he sought, or desired, its motion and change. (TA.) And استحال الجَهَامَ *He looked at* [*the waterless clouds,* or *the clouds that had poured forth their water, to see if they changed* or *moved*]. (TA.) — *He reckoned it* مُحَال [i. e. *absurd, inconsistent,* &c.: see above]. (KL.)

11. احوالّت الأَرْضُ, (Ḳ, TA, [in the CḲ احولّت,]) inf. n. اِحْوِيلَال, (TA,) ‡ *The land became green, and its herbage stood erect,* or *became strong and erect.* (Ḳ, TA.) [See حُوَلَآءُ.]

حَالٌ The *state, condition,* or *case,* (صِفَة,) of a thing; [considered as subject to change;] (Mṣb, Er-Rághib, TA;) as also ↑ حَالَةٌ: (Mṣb:) or the *quality,* or *manner of being,* and *state,* or *condition,* of a man, (Ḳ, TA,) *in respect of good or evil;* (TA;) as also ↑ حَالَةٌ: (Ḳ:) or the *particular case,* or *predicament,* of a man &c., *in respect of changing events, in the soul and the body and the acquisitions:* (Er-Rághib, TA:) and in the coventional language of the logicians, *a fleeting,* or *quickly-transient, quality, such as accidental heat and cold and moisture and dryness;* as also ↑ حَالَةٌ: (TA:) *anything changing:* (Ḥam p. 288:) the *time in which one is;* (Lth, Ḳ;) [the *present time;*] the *end of the past,* and the *beginning of the future:* and as a conventional term, [in grammar, the *present tense:* and ‡ the *future:* and also] *a denotative of state* of the agent or of the objective complement; [the former termed حَالٌ مِنَ الفَاعِلِ; and the latter, حَالٌ مِنَ المَفْعُولِ; and each said to be مَنْصُوبٌ عَلَى الحَالِ, i. e. *put in the accus. case as a denotative of state,* unless expressed by a complete proposition;] as [قَائِمًا] in the phrase زَيْدٌ فِى الدَّارِ قَائِمًا [Zeyd is in the house, standing], and in ضَرَبْتُ زَيْدًا قَائِمًا [I beat Zeyd standing]: (Ibn-El-Kemál, TA:) it is fem., like حَالَةٌ; (Mṣb;) and masc.; (Mṣb, Ḳ;) but mostly fem.: (TA:) the pl. is أَحْوَالٌ and أَحْوِلَةٌ, (Ḳ,) [both properly pls. of pauc., but the former often used as a pl. of mult., and often signifying *circumstances,*] the latter anomalous: (TA:) the pl. of ↑ حَالَةٌ is حَالَاتٌ: (TA:) or ↑ حالة is the n. un. or sing. of حَالٌ and أَحْوَالٌ [and حَالَاتٌ], used in relation to a man. (Ṣ, O.) You say حَالٌ حَسَنٌ and حَسَنَةٌ ↑ حَالَةٌ [*A good state* or *condition* &c.; as also ↑ حَسَنَةٌ]. (Mṣb.) And أَحْوَالُهُ ↑ حَالَاتُ الدَّهْرِ and *The changes,* or *vicissitudes, of time* or *fortune.* (Ḳ.) [And اِفْعَلْهُ حَالًا and فِى الحَالِ *Do thou it now,* or *immediately.* And عَلَى كُلِّ حَالٍ *In any case:* a phrase of frequent occurrence. The phrase قَالَ لِسَانُ الحَالِ † *The tongue of the case said,* (often used by late writers,) means *the case seemed to say.*] — *A load,* or *burden:* (Ḥam p. 299:) [whence, perhaps, خَفِيفُ الحَالِ (which see in what follows) as meaning † *having a small family to maintain:*] and hence, (Ḥam ib.,) *a bundle,* or *bundle of clothes,* (عِكَارَة,) *which is carried on the back* (Ṣ, Ḥam ib.) *by a man:* (Ṣ:) or *a thing that a man carries on his back,* (ISd, O, Ḳ,) *whatever it be.* (ISd, TA.) — *A* [*garment of the kind called*] كِسَآء *in which one collects,* or *seeks and collects, dry herbage:* (O, Ḳ:) or *a garment,* or *piece of stuff, of which two ends are tied in a knot behind the flanks, and the other two ends over the head; in which one collects dry herbage;* also called شِكْبَانٌ. (TA in art. شكب.) — *A child's go-cart, by means of which he practises walking;* (Ṣ, Ḳ*, TA;) *resembling a small* عَجَلَة; (Ṣ;) also called دَرَّاجَةٌ [q. v.]. (Ṣ, TA.) — The *part* of a horse *that is the place of the* لِبْد [or *saddle-cloth*]: or the *line along the middle of the back:* (Ḳ, TA:) or حَالُ مَتْنِ الفَرَسِ signifies *the middle of the back of the horse; the place of the* لبد. (Ṣ.) [See also its syn. حَاذٌ.] خَفِيفُ الحَالِ signifies the the same as خَفِيفُ الحَاذِ, (A in art. حوذ,) which means ‡ A man *light of back;* (Ṣ, A, L, Mṣb, all in art. حوذ;) i. e. *having little property:* and also *having a small family to maintain;* (L in that art.;) or *having little property and a small family to maintain;* (L and Ḳ in that art.;) like خَفِيفُ الظَّهْرِ. (A, L, Mṣb, all in that art.) — *Black mud:* (Ṣ, Ḳ:) from حَالَ "it became altered, or changed." (TA.) It is said in a trad. that the حال of El-Kowthar is musk: (TA:) and in another, that Gabriel took of the حال of the river [Nile] and put it into the mouth of Pharaoh; (Ṣ, TA;) but here it has the meaning next following. (TA.) *Black fetid mud;* syn. حَمْأَة. (Ḳ, and Ḥam p. 288.) And *Soft earth.* (Ḳ, and Ḥam ib.) — And hence, ‡ *Weakness,* and *softness.* (Ḥam ib.) — *Stinking flesh-meat.* (Ḥam ib.) — *Hot ashes* (IAạr, Ḳ, and Ḥam ib.) — The *leaves of the* سَمُر [*acacia,* or *mimosa, gummifera,*] *beaten and shaken off into a garment,* or *piece of cloth.* (Ḳ.) — *Milk.* (M, Ḳ.) — In the dial. of Hudheyl, (IAạr, TA,) *A wife.* (IAạr, Ḳ.)

حَوْلٌ *A year;* (Ṣ, Mṣb, Ḳ, Er-Rághib, &c.;) so termed in consideration of its changing, and of the revolution of the sun in its places of rising and setting; (Er-Rághib, TA;) or as being the period in which [certain] plants attain their complete strength: (El-Ḥarállee, TA:) and *even if it has not passed;* because it will be [properly speaking] a حَوْل: an inf. n. used as a simple subst.: (Mṣb:) pl. [of pauc., but also used as a pl. of mult.,] أَحْوَالٌ (M, Mṣb, Ḳ) and [of mult.] حُؤُولٌ and حُوُولٌ, (M, Ḳ,) the former with ء and the latter with و. (TA.) — *Strength, power, might,* or *force;* syn. قُوَّةٌ; (Ṣ, TA;) as also ↑ حِيلَةٌ [originally حِوْلَةٌ] (TA) and ↑ حَوْلَةٌ, (Ḳ, TA,) or this last is a n. un. from حَوْلٌ: (TA:) it is *in the soul and the body and the acquisitions:* and hence the saying, in a trad., لَا حَوْلَ وَلَا قُوَّةَ إِلَّا بِٱللّٰهِ العَلِىِّ العَظِيمِ [*There is no strength nor power but in, or by means of, God, the High, the Great*]: (Er-Rághib, TA:) or it here signifies *motion;* [see also حَوْلَةٌ;] and the meaning is, *there is no motion nor power,* or *ability, but by the will of God*: (AHeyth, TA:) or, as some say, the meaning is, *there is no* حول [i. e. *changing*, or *turning*, or *receding*,] from disobedience, *nor* قوّة [i. e. *power*] to obey, *but by the accommodating*, or *disposing, of God*. (Mṣb.) — See also حِيلَةٌ. — حَوْلَ شَيْءٍ *The lateral*, or *adjacent, part to which a thing may shift*, or *remove*: (Er-Rághib, TA:) or *the environs* [*of a thing*]. (Mṣb.) You say, قَعَدُوا حَوْلَهُ, (Ṣ, Ḳ,*) or قَعَدْنَا حَوْلَهُ, the noun being in the accus. case as an adv. n. of place, i. e., [*They sat*, or *we sat, around him*, or *it*, or] *in his*, or *its, environs*; (Mṣb;) and ▼حَوَالَهُ, and حَوْلَيْهِ, (Ṣ, Ḳ,) dual of حَوْل, (TA,) and ▼حَوَالَيْهِ, (Ṣ, Mṣb, Ḳ,) dual of حَوَال, (TA,) and أَحْوَالَهُ, (ISd, Ḳ,) pl. of حَوْل, and used to give intensiveness to the meaning: (ISd, TA:) but you should not say حَوَالِيهِ. (Ṣ, Ṣgh.) And اَللّٰهُمَّ حَوَالَيْنَا ▼ وَلَا عَلَيْنَا [*O God*, be Thou *around us* as our protector, *and not against us*,] occurs in a trad. respecting prayer. (TA.) It is said, in the Expos. of the exs. cited as testimonies by Sb, that one sometimes says ▼حَوَالَيْكَ and حَوْلَيْكَ, meaning *Around thee, in every direction*; dividing the surrounding parts into two; like as one says, أَحَاطُوا بِهِ مِنْ جَانِبَيْهِ, not meaning that any of the surrounding parts remained vacant. (TA.) [See also an ex. voce دُنْدُنَ.]

حَوَل inf. n. of حَوِلَتْ عَيْنُهُ. (Ṣ, Ḳ. [See 1, last sentence.]) — See also حَائِلٌ.

حِوَل: see حُوَّل: — and see also 1, last sentence.

حُوَل: see حُوَّل: — and حَائِلٌ: — and حِيَال.

حِوَل *Removal* from one place to another, *in a passive sense*; a subst. from تَحَوَّلَ: (Ṣ, O, Ḳ:) and *in an active sense*; a subst. from حَوَّلَهُ; (Ḳ;) accord. to ISd, it is the latter [only]; (TA;) as also ▼حُوَيل. (Ḳ.) Hence, in the Ḳur [xviii. 108], لَا يَبْغُونَ عَنْهَا حِوَلًا [*They shall not desire removal from it*]: (Ṣ, M, O, Ḳ:) or, as some say, it here means ▼حِيلَةٌ; i. e. *they shall not* [*desire to*] *practise*, or *seek to practise, any evasion from it* to another abode. (TA.) — See also حِيلَةٌ [of which it is said to be both a syn. and a pl.]. — And see حَوْلَةٌ. — Also *A furrow*, or *trench, in the ground, in which palm-trees are planted in a row*. (ISd, Ḳ.)

حَالَةٌ; pl. حَالَاتٌ: see حَالٌ, in seven places.

حَوْلَةٌ *Motion*, or *removal*, (تَحَوُّلٌ or تَحَرُّكٌ, accord. to different copies of the Ḳ, the former being the reading in the TA,) and *change of state*. (Ḳ.) [See also حَوْل. And see حَالَ عَلَى الفَرَسِ, of which it is the inf. n.] — See also حِيلَةٌ. — And see حَوْل, as meaning قُوَّةٌ.

حُولَةٌ: see حُوَّل: — and حِيلَةٌ. — Also *A wonder*, or *wonderful thing*: pl. حُوَل. (Ḳ: [but probably this should be حُوَل, as below.]) — [It is also used as a pl., signifying *Wonders*.] You say, هٰذَا مِنْ حُولَةِ الدَّهْرِ *This is of the wonders of the age*, or *of time*, or *fortune*; as also مِنْ ▼حُولَانِهِ, and ▼حِولِهِ [pl. of ▼حِيلَةٌ], and ▼حَوْلَانِهِ. (Ḳ, TA: the last, in one copy of the Ḳ, ▼حَوْلَانِهِ, and in the CḲ ▼حُوَلانه.) — *An evil*, or *abominable, event* or *accident*; (Ḳ,* TA;) *a calamity*, or *misfortune*: pl. حُوَل: as in the saying, هُوَ حُولَةٌ مِنَ الحُوَلِ *It is a calamity of calamities*. (Ṣ, TA.) It is also used as an epithet; so that one says, جَآءَ بِأَمْرٍ حُولَةٍ [*He did*, or *brought to pass, an evil*, or *abominable, thing*]. (M, TA.)

حِيلَةٌ, (Ṣ, M, Mṣb, Ḳ, &c.,) originally حِوْلَةٌ, (Er-Rághib, TA,) [said to be an inf. n., (see 8,)] and ▼حَوْل (Ṣ, M, Ḳ) and حَيْل and ▼حِوَل, (M, Ḳ,) which is also a pl. of the first, (Ḳ,) and حُوَيْل ▼ حُولَةٌ (Ḳ) and ▼حُولَةٌ (Ks, TA) and ▼مَحَالَةٌ (M, Ḳ) and ▼حَائِلَةٌ (Ḥam p. 652) and ▼ (Ṣ, Ḳ) and ▼مَحَال, (M, Ḳ,) [said to be an inf. n., (see 8,)] and ▼مَحِيلَةٌ, (Ṣgh, TA,) *i. q.* اِحْتِيَال and تَحَوُّل and تَحَيُّل; (M, Ḳ; [see 8;]) [or *A mode*, or *manner, of changing from one state to another*, or *of shifting from one thing to another*; حِيلَةٌ being of the measure فِعْلَةٌ from حَالَ, like جِلْسَةٌ &c. from جَلَسَ &c.; or from تَحَوُّل as syn. with حَال; (see what follows;) *a mode*, or *means, of evading* or *eluding a thing*, or *of effecting an object*; *an evasion* or *elusion, a shift, a wile, an artifice*, or *artful contrivance* or *device, a machination, a trick, a plot, a stratagem*, or *an expedient*;] *a means of effecting one's transition from that which he dislikes to that which he likes*; (KT, in explanation of the first word;) *art, artifice, cunning, ingenuity*, or *skill*, (Mṣb, Ḳ,) *in the management of affairs*; i. e. the *turning over*, or *revolving, thoughts, ideas, schemes*, or *contrivances, in the mind, so as to find a way of attaining one's object*; (Mṣb;) and *excellence of consideration* or *deliberation*; and *ability to manage according to one's own free will*, (Ḳ, TA,) *with subtilty*: (TA:) accord. to Er-Rághib, حِيلَةٌ signifies *a means of attaining to some state concealedly*; and it is mostly used of that in which is sin, or offence, or disobedience; but sometimes of that in the exercise of which is wisdom; and hence God is described as شَدِيدُ ▼المِحَالِ, meaning *strong in attaining, concealedly from men, to that in which is wisdom*: accord. to Abu-l-Baḳà, it is from التَّحَوُّل; because by it one shifts from one state to another, by a species of forecast, and gentleness, or delicacy, so as to change a thing from its outward appearance: (TA:) the pl. of حِيلَةٌ is حِيَل and حِوَل [which latter is the most common form, and also, as well as حِوَل, said to be syn. with the sing.,] and حِيلَاتٌ. (Ḳ.) One says, لَا حِيلَةَ لَهُ [*He has no mode*, or *means, of evading* &c.]. (TA.) [And مَا بِيَدِي حِيلَةٌ *I have no mode*, or *means, of evading* &c.]. And ▼اَلْمَرْءُ يَعْجِزُ لَا مَحَالَةَ [*Man becomes impotent: there is no avoiding it*]. (Ṣ.) ▼لَا بُدَّ مِنْهُ means لَا مَحَالَةَ [*There is no avoiding it*, or *escaping it*]. (Ṣ,* Ḳ.) One says, اَلْمَوْتُ آتٍ ▼لَا مَحَالَةَ [*Death comes: there is no avoiding it*]. (Ṣ.) See also حَوْل. — And see حُولَةٌ: — and حَوْل. — [عِلْمُ الحِيَلِ *The science of mechanics*.]

حُولَةٌ: see حَوْل, in two places.

حَوْلِيٌّ A solid-hoofed animal *in his first year*: (Ṣ, O:) or a solid-hoofed animal, &c., *a year old*; *a yearling*: (Ḳ:) it is applied in this sense to a camel: and also to a plant: (TA:) and so ▼مُحْوِلٌ and ▼مُحِيلٌ applied to wheat, or food, &c.: (Ṣ, O:) and ▼مُحْوِلٌ applied to a boy-child: (Ḳ:) or, as some say, this signifies *in the state of childhood*; not limited to a year old: (TA:) the fem. of حَوْلِيٌّ is حَوْلِيَّةٌ: pl. [masc. حَوْلِيُّ الغَضَا; and] fem. حَوْلِيَّاتٌ. (Ṣ, Ḳ.) حَوَالِيُّ *Young trees of the kind called* غَضًا. (TA.)

حُوَلِيٌّ: see حَوْل.

حُوَلَآءُ and حِوَلَآءُ, the latter like عِنَبَآءُ and سِيَرَآءُ, which are the only other words of this measure, (Ṣ, Ḳ,) accord. to Kh, (Ṣ,) [The *membrane that encloses the she-camel's fœtus in the womb*;] to the she-camel, *like the* مَشِيمَة (Ḳ, TA) to the woman; (TA;) i. e., (Ḳ,) *a skin* (Ṣ, Ḳ) *of a dark*, or *an ashy, dust-colour* (خَضْرَآء), *full of water*, (Ḳ,) *which comes forth with the fœtus, containing* أَغْرَاس [pl. of غِرْس, q. v.], *and having lines*, or *streaks, which are red, and of a dark*, or *an ashy, dust-colour* (خُضْر): (Ṣ, Ḳ:) or *it comes forth after the fœtus, in the first* سَلًى; *and is the first thing that comes forth therefrom*: so says ISk: and the word is sometimes used in relation to a woman: (TA:) or, accord. to AZ, the *water that comes forth upon the head of the fœtus*, [i. e.] *at the birth*: (Ṣ:) or *a membrane like a large bucket, which is full of water, and bursts when it falls upon the ground: then comes forth the* سَلَى; *and a day, or two days, after that, the* صَاءَة. (TA.) Hence the saying, نَزَلُوا فِي مِثْلِ حِوَلَآءِ, (Ḳ,) and فِي مِثْلِ حُوَلَآءِ النَّاقَةِ, (TA,) ‡ *They alighted amid abundance of water and green herbage*. (Ḳ,* TA.) And رَأَيْتُ أَرْضًا مِثْلَ الحُوَلَآءِ ‡ *I saw land having dark green herbage*. (TA.) — See also حُولَةٌ.

حُولَانٌ and حَوْلَانٌ and حُوَلَانٌ: see حُولَةٌ.

حُوَلَلٌ: see حَائِلٌ, in four places.

حَوَلْوَلٌ: see حُوَّل, in two places.

حَوَالٌ The *changing*, or *varying*, of time, or fortune. (Ḳ.) — حَوَالَهُ, and حَوَالَيْهِ, and حَوَالَيْنَا, and حَوَالَيْكَ: see حَوْل.

حِوَالٌ: see حَائِلٌ, in two places.

حِيَالٌ [in the CḲ, erroneously, حَيال] The *front* of a thing, as meaning the *part, place*, or *location, that is over against, opposite, facing, fronting*, or *in front*; syn. قُبَالَة. (Ḳ, and Mgh in art. حيل.) You say, قُمْتُ حِيَالَهُ *I stood in front of him*; *in the part, place*, or *location, that was over against him, opposite to him*, &c.; syn. قُبَالَتَهُ. (Mṣb in art. حيل.) And قَعَدَ حِيَالَهُ

and بِحِيَالِهِ *He sat in front of him, over against him, opposite to him, facing* or *fronting him;* syn. بِإِزَائِهِ. (Ṣ, Ḳ.) And هٰذَا حِيَالَ كَلِمَتِكَ *This is opposite to thy saying;* syn. مُقَابَلَةَ; in the accus. case, as an adv. n. of place: thus related by IAạr from the Arabs: but one may also say حِيَالُ كَلِمَتِكَ [*the opposite of thy saying*], making the phrase to consist of an inchoative and an enunciative: so says ISd. (TA.) It is originally with و [in the place of the ى]. (Ṣ, O.) —— عَلَى حِيَالِهِ [*By himself* or *itself; independently*]. You say, أَعْطِ كُلَّ وَاحِدٍ مِنْهُمْ عَلَى حِيَالِهِ *Give thou to every one of them by himself;* syn. بِٱنْفِرَادِهِ; (Ṣ in art. وحد;) or عَلَى حِدَتِهِ. (Mgh in art. حيل.) And فَعَلْتُ كُلَّ شَيْءٍ عَلَى حِيَالِهِ *I did everything by itself;* syn. بِٱنْفِرَادِهِ. (Mṣb in art. حيل.) ═ Also *A string that is tied from the camel's* بِطَان [or *belly-girth*] *to his* حَقَب [or *hind girth*], *to prevent the* حقب *from going against the sheath of his penis:* (Ḳ:) so, too, in the M: but in the O, as on the authority of AA, ▼ حِوَلٌ, like صِرَدٌ, signifies the *string that is between the* حقب *and the* بطان. (TA.)

حَوِيلٌ: see 4, in the latter half of the paragraph. —— [Hence,] One who is *responsible,* or *answerable.* (Ḳ.) —— And *A witness.* (Ḳ.) ═ See also حَوْلٌ: —— and حِيلَةٌ. —— Also [*Desire: or a seeking: or a seeking by an artful contrivance or device, or by artful or skilful management, to find a way of attaining an object:*] a subst. from حَاوَلَهُ. (Ṣ, O, Ḳ.)

حَوَالَةٌ The *effecting a transition of one river,* or *rivulet, to another.* (M, Ḳ.) [This is what is meant by the حوالة in المُزَارَعَة, mentioned in the Mgh, as "customary in the cases of certain plants, as rice, and the بَاذِنْجَان, and in planting."] —— The *transfer of a claim,* or *of a debt, by shifting the responsibility from one person to another:* (Mgh:) the *transfer of a debt by shifting the responsibility of him who transfers it to him to whom it is transferred:* (KT:) [*a reference made by a debtor, of his creditor, to a debtor of the former, for the payment of what is owed by the former to the latter: an order for the payment of a debt, or of a sum of money, given by one person, upon another, to a third person:* so in the present day:] a subst. (Ṣ, Ḳ) from أَحَلْتُهُ بِدَيْنِهِ, (Ṣ,) or from أَحَالَ عَلَيْهِ بِدَيْنِهِ, (Mṣb,) or from أَحَالَ الغَرِيمَ. (Ḳ.) [See 4.] —— *A responsibility; accountableness.* (Ḳ.)

حَوَالَىْ and حُوَالَىْ: see the next paragraph.

حُوَّلٌ, applied to a man, signifies بَصِيرٌ بِتَحْوِيلِ الأُمُورِ [i. e. *Knowing, skilful,* or *intelligent, in turning affairs over, or about, in his mind, considering what may be their results, and so managing them*]; (Ṣ, TA;) as also حُوَّلٌ قُلَّبٌ, (TA, and so in copies of the Ṣ,) and ▼ حُوَّلِىٌّ قُلَّبٌ, (TA, and so in a copy of the Ṣ,) and ▼ حُوَّلِىٌّ قُلَّبِىٌّ, and ▼ حَوَّالٌ; (TA;) and ▼ حُوَلَةٌ signifies مُحْتَالٌ [i. e. one *who exercises art, artifice, cunning, ingenuity,* or *skill, and excellence of consideration* or *deliberation, and ability to manage according to his own free will, with subtilty;* &c.; see the verb (8) of which مُحْتَالٌ is the part. n.]: (Ṣ:) or حُوَلٌ or حُوَّلٌ and ▼ حُوَلَةٌ and ▼ حُوَلَةٌ and ▼ حُولَةٌ and ▼ حُوَلِىٌّ, [in the CḲ, erroneously, حَوَلِى,] like سُكَّرِىٌّ, [in the CḲ like سَكْرِىٌ,] and ▼ حَوَالِىٌّ and ▼ حُوَالِىٌّ and ▼ حُوَلْوَلٌ, signify شَدِيدُ الاحْتِيَالِ [i. e. one *who exercises great art, artifice,* &c.]: (Ṣgh, Ḳ:) all of these forms are mentioned by ISd, except حُوَلَةٌ and حُوَلِىٌّ: (TA:) accord. to some, حُوَّلٌ قُلَّبٌ signifies *experienced,* or *expert, in affairs;* or one *who has been tried,* or *proved, and strengthened by experience in affairs:* (Ḥar p.312:) and ▼ حَوِلٌ signifies the same as حُوَّلٌ; (Ḥam p. 34;) *having much* حِيلَة [i. e. *art, artifice,* &c.]: accord. to analogy, it should be [حَالٌ,] like مَالٌ and صَاتٌ as epithets applied to a man: (Idem pp. 530 and 531:) حَيَّالٌ, also, [in like manner,] signifies صَاحِبُ حِيلَةٍ [i. e. one *who exercises art, artifice,* &c., as above]; and so ▼ حَيْلِىٌّ [from حِيَلٌ, pl. of, or syn. with, حِيلَةٌ]: (TA:) and ▼ حُوَلْوَلٌ [mentioned above (in the CḲ, erroneously, in this instance, حَواوِل)] signifies also *cunning,* or *intelligent,* or *skilful and knowing;* and *quick and sharp* or *vigorous* or *effective;* syn. مُنْكَرٌ كَمِيشٌ; (Ḳ, TA;) applied to a man. (TA.)

حَيَّالٌ: see 4, in the latter half of the paragraph.

حُوَّلِىٌّ: see حُوَّلٌ, in three places.

حَوَّالٌ: see حُوَّلٌ.

حَائِلٌ *Altering,* or *being transmuted,* or *changing;* or *altered,* or *transmuted,* or *changed;* [*in any manner;* and particularly] *in colour;* (Ḳ, TA;) *and becoming,* or *become, black;* applied to a bone, and any other thing. (TA.) —— Anything [*shifting,* or *moving,* or] *that has shifted,* or *moved, in* (فِى) [app. a mistranscription for مِنْ *from*]) *its place.* (TA.) —— A she-camel, and any female, *not conceiving,* or *not becoming pregnant, during a year,* (M, Ḳ,) or *two years,* (Ḳ,) or *some years:* (M, Ḳ:) or a she-camel *not pregnant* (Ṣ, Mṣb, Ḳ) *after having been covered by the stallion;* (Ṣ, Ḳ;) because denoting a change from what is usual; (Er-Rághib, TA;) and in like manner applied to a ewe, or she-goat; (TA;) and to a woman: (Mṣb:) pl. حِيَالٌ and حُوَلٌ (Ṣ, Ḳ) and حُوَّلٌ, (Ḳ,) and ▼ حُولَلٌ (M, Ḳ) is a quasi-pl. n.: (M, TA:) [whence,] حَائِلٌ حُولٌ and ▼ حُولَلٌ used as intensive epithets: or حَائِلٌ signifies *not conceiving in one year,* (Ḳ, TA,) *when she has been covered:* (TA:) and حَائِلٌ حُولٍ and ▼ حُولَلٍ, *not conceiving during two years;* (Ḳ;) *not conceiving in the first year after having been covered, nor in the next following year;* like عَائِطٌ عُوطٍ and عِيطٍ and عُوطَطٍ: (Ṣ in art. عوط:) one says also, لَقِحَتْ عَلَى حُولٍ and ▼ حُولَلٍ [app. meaning *She conceived after having failed to do so for two years;* for it seems that in this case حول and حولل are inf. ns., or that the latter is a subst. having the sense of an inf. n.: see 1, and see also عُوطَطٌ]: (TA:) and ▼ مُحْوِلٌ signifies the same as حَائِلٌ. (Ḳ.) Also A palm-tree (Mṣb, Ḳ) *that bears one year, and not another year:* (Ḳ:) or *not bearing.* (Mṣb.) —— The *female young one of a camel, at the time of her birth:* the male is termed سَقْبٌ: (Ṣ, Ḳ:) pl. حُولٌ and حَوَائِلُ. (TA.) One says, نُتِجَتِ النَّاقَةُ حَائِلًا حَسَنَةً [*The she-camel brought forth a beautiful female young one*]. (Ṣ.) And لَا أَفْعَلُ ذٰلِكَ مَا أَرْزَمَتْ أُمُّ حَائِلٍ [*I will not do that as long as a mother of a female young camel utters her gentle yearning cry*]. (Ṣ.) ═ Also, (Lth, Mgh, O, TA,) and ▼ حِوَالٌ (Lth, Ḳ) and ▼ حُوَلٌ and ▼ حَوَلٌ, (Ḳ,) *A thing that intervenes as a separation, a partition, a fence, a barrier,* or *an obstacle,* or *obstruction,* between two other things. (Lth, Mgh,* O, Ḳ.) One says, هٰذَا حِوَالٌ ▼ بَيْنَهُمَا, i. e. حَائِلٌ [*This is a thing intervening as a separation,* &c., *between them two*]; like حِجَازٌ and حَاجِزٌ. (Lth, O, TA.)

حَائِلَةٌ: see حِيلَةٌ.

أَحْوَلُ; and its fem. حَوْلَآءُ: see 1, last sentence. —— أَحْوَلُ مِنْ بَوْلِ الجَمَلِ [*More wry than the urine of the he-camel*]: because it does not come forth straight, but [backwards, and] inclining to one side: a prov. (TA.) —— هُوَ أَحْوَلُ مِنْكَ *He is one who has more* حِيلَة [meaning *art, artifice, cunning, ingenuity,* or *skill, in the management of affairs,* &c.,] *than thou;* (Fr, Ṣ, Ḳ;) as also ▼ أَحْيَلُ. (Ḳ.) And هُوَ ▼ أَحْيَلُ النَّاسِ [*He is the most artful, cunning, ingenious,* or *skilful, of men*]; originally أَحْوَلُ. (MF in art. رود: see أَرْوَدُ.)

أَحْيَلُ: see the next preceding paragraph, in two places.

تَحَاوِيلُ الأَرْضِ means أَنْ تُعْطِلَ حَوْلًا وَتُصِيبَ حَوْلًا, (O, Ḳ,) i. e. *The leaving the land unsown one year, and sowing it another year:* whereby the land is strengthened. (TḲ.)

مَحَالٌ: see حِيلَةٌ, in two places: —— and see also مَحَالَةٌ, in three places.

مُحَالٌ; and مُحَالٌ عَلَيْهِ; and مُحَالٌ بِهِ: see 4, in the latter half of the paragraph. —— مُحَالٌ also signifies [*Absurd; inconsistent; self-contradictory;*] *comprising two contradictories;* as when one speaks of one body in two places in one case [or time]: (Er-Rághib, TA:) *that cannot be conceived as existing in reality:* (TA:) i. q. بَاطِلٌ [as meaning *untrue,* or *unreal*]; (Mṣb, TA;) *impossible; that cannot be:* (Mṣb:) *perverted; turned from its proper way* or *manner of being;* (Ḳ;) applied to speech; (Ṣ, Ḳ;) as also ▼ مُسْتَحِيلٌ. (Ḳ.)

مُحْوِلٌ: see مُحِيلٌ: —— and see also حُوَّلِىٌّ, in two places.

مُحِيلٌ: see 4, in the latter half of the paragraph. —— Also A woman *that brings forth a boy next after a girl; or the reverse:* and in like manner applied to a she-camel; as also ↑ مُحْوِلٌ and ↑ مُحَوِّلٌ: (Ks, Ṣgh, Ḳ:) and accord. to some, ↑ مُتَحَوِّلٌ [if not a mistranscription for مُحْوِلٌ or مُحَوِّلٌ] signifies a she-camel *that brings forth one year a male, and another year a female.* (TA.) — See also حُولِىٌّ.

مَحَالَةٌ: see حِيلَةٌ, in four places. — Also *A machine* (مَنْجَنُونٌ, Lth, Ḳ) *over which* [*passes the rope whereby*] *water is drawn*: (Lth, TA:) and, (Ḳ,) or as some say, (TA,) *a great* بَكَرَة [or *sheave of a pulley*], (Ṣ in art. محل, and Ḳ,) *by means of which camels draw water*: (Ṣ ubi suprà, and TA:) [see سَانِيَةٌ:] pl. مَحَاوِلُ and [coll. gen. n.] ↑ مَحَالٌ. (Ḳ.) — The *vertebræ*; as also ↑ مَحَالٌ: (Ḳ: [in the CḲ, الفَقَار is erroneously put for الفَقَار:]) or the latter has this meaning; and the former signifies *a single vertebra*: and the م may be radical: (M, TA:) pl. مَحَالَاتٌ. (T in art. ملح.) — The *middle* (وَاسِط, as in the M and O; in the Ḳ, erroneously, وَاسِطَة, TA) *of the back*; (M, O, Ḳ;) as also ↑ مَحَالٌ: but accord. to some, the م is radical. (TA.)

مَحِيلَةٌ: see حِيلَةٌ.

مُحَوِّلٌ: see حَائِلٌ: — and مُحِيلٌ.

مِحْوَالٌ A man *who says much that is* مُحَال [or *absurd, inconsistent, self-contradictory, unreal,* or *impossible*]. (Lth, Ḳ.*)

مُحْتَالٌ: see 4, in the latter half of the paragraph. — أَرْضٌ مُحْتَالَةٌ ‡ *Land upon which rain has not fallen.* (TA.) — See also حُوَّلٌ.

مُتَحَوَّلٌ [pass. part. n. of تَحَوَّلَهُ. — Also] an inf. n. of تَحَوَّلَ. (Ḥam p. 503.) — And *A place to which one shifts, removes,* or *becomes transferred.* (Idem ib.)

مُتَحَوِّلٌ: see مُحِيلٌ.

رِجْلٌ مُسْتَحَالَةٌ *A leg that is crooked in the two extremities of its shank.* (M, O, TA.) In the Ḳ, رِجْلٌ is erroneously put for رِجْلٌ, and سَاقَيْهِ for سَاقَهَا. (TA.) And ↑ قَوْسٌ مُسْتَحِيلَةٌ (Ṣ, Ḳ) and مُسْتَحَالَةٌ (Ḳ) *A bow that is crooked* (Ṣ, Ḳ, TA) *in the portion between the part grasped by the hand and the curved extremity, or in its curved extremity.* (TA.) And ↑ أَرْضٌ مُسْتَحِيلَةٌ *Uneven ground*: (Ṣ, TA:) or *i. q.* مُسْتَحَالَةٌ, (Ḳ,) which means land *that has been left* [*unsown,* or *uncultivated,*] *a year,* or *years.* (M, Ḳ.)

مُسْتَحِيلٌ; fem. with ة: see the paragraph next preceding, in two places: — and see also مُحَالٌ. — Also *Full.* (Ḳ.)

حولق

Q. Q. 1. حَوْلَقَ, inf. n. حَوْلَقَةٌ: see Q. Q. 1 in art. حلق.

حوم

1. حَامَ, (Ṣ, Mṣb, Ḳ,) حَوْلَ المَآءِ, (Ṣ, Mṣb,) or عَلَى الشَّىْءِ, (Ḳ,) aor. يَحُومُ, (Ṣ,) inf. n. حَوْمَانٌ (Ṣ, Mṣb, Ḳ) and حَوْمٌ, (Ṣ, Ḳ,) said of a bird, or flying thing, (Ṣ, Mṣb, Ḳ,) &c., (Ṣ,) *It went,* [or *hovered,*] or *circled,* (Ṣ, Mṣb, Ḳ,) *round about the water,* (Ṣ, Mṣb,) *by reason of thirst,* (TA,) or *round about the thing*: (Ḳ:) and in like manner حَامَتْ, said of camels. (Ḳ, TA.) — [Hence,] it is said in a trad., مَنْ حَامَ حَوْلَ الحِمَى يُوشِكُ أَنْ يَقَعَ فِيهِ, meaning † *He who approaches acts of disobedience is near to falling therein.* (Mṣb.) And you say, هُوَ يَحُومُ حَوْلَ غَرَضٍ لَهُ ‡ [*He hovers about an object of aim,* or *desire, that he has*]. (TA.) And حَامَ فُلَانٌ عَلَى الأَمْرِ, inf. n. حَوْمٌ and حِيَامٌ and حُؤُومٌ, (Ḳ, TA,) the second with kesr [to the ح, and therefore with ى in the place of و, but written in the CḲ with fet-ḥ], (TA,) ‡ *Such a one desired,* or *sought, the thing.* (Ḳ, TA.) And حَامَ عَلَى قَرَابَتِهِ ‡ *He affected,* or *inclined towards, his relations, like one going round about the water.* (TA.)

2. حَوَّمَ فِى الأَمْرِ *i. q.* اِسْتَدَامَ ‡ [app. as meaning † *He paused, and acted with deliberation,* or *in a patient* or *leisurely manner,* or *he waited in expectation, in the affair,* or *case,* like one who hovers about a thing]. (Ḳ, TA.)

جَيْشُ حَامٍ [lit. *The army of Ham,* the son of Noah; meaning, because of its blackness,] ‡ *the night.* (TA.)

حَوْمٌ *A large herd of camels,* (Ṣ, M, Ḳ,) *amounting to a thousand*: (M, Ḳ:) or *an indefinite number*; (Ḳ,* TA;) i. e. *many*: a quasi-pl. n.; or, as some say, a pl. (TA.)

حَوْمٌ, applied to wine, (TA,) *That circulates* [or *produces a sensation of circular motion*] *in the head.* (Ḳ.) — Accord. to Aṣ, so applied, it signifies *Much in quantity.* (TA.)

حَوْمَةٌ The *main part* or *portion,* (Ṣ, Ḳ,) of water, (Ṣ,) or of the sea, (Ḳ,) and of sand, and of a fight, &c.: (Ṣ, Ḳ:) the *part where is most water,* of the sea; (Ḥam p. 329, and TA;) the *deepest part* thereof: (TA:) and in like manner, of a watering-trough: (Ḥam ibid., and TA:) or *a copious,* or *deep, part* of water: (Lḥ, TA:) and hence, the *vehemence of war* or *fight*: (Ḥam p. 329:) or *a place of fight*; because the opposing parties go round about it: (Ḥam p. 492:) or the *most vehement part* [or the *thickest*] of a fight: (Ḳ,* TA:) pl. حَوْمَاتٌ. (Ḥam p. 329.) [See also حَوْبَةٌ, last signification.]

حَامِىٌّ *A black,* or *negro,* boy or young man, (Ṣ, Ḳ,) or slave: (Ṣ:) so termed in relation to حَام [or Ham, the son of Noah], the father of the blacks, or negroes. (Ṣ, Ḳ.)

حَائِمٌ *Thirsty, and going,* [or *hovering,*] or *circling, round about the water*: (Ḥam p. 753:) and hence, (Id. p. 317,) whatever is *thirsty*: (Id. ibid., and Ḳ:) [fem. with ة: pl. masc. حُوَّمٌ:] pl. fem. حَوَائِمُ. (Ḥam p. 317.) You say إِبِلٌ حُوَّمٌ *Thirsty camels that go round about the water*: (Aṣ, TA:) or, as also إِبِلٌ حَوَائِمُ, *thirsty camels*: (Ḳ:) or *very thirsty camels.* (TA.) And it is said in a trad., respecting the prayer for rain, اَللّٰهُمَّ ٱرْحَمْ بَهَائِمَنَا الحَائِمَةَ *O God, have mercy upon our beasts that are going about the water and not finding any to which they may come to drink.* (TA.) You say also هَامَةٌ حَائِمَةٌ *A thirsty head*: (TA:) or *a head of which the brain is thirsty.* (T, TA.) — Also ‡ *Desiring,* or *seeking*: pl. حُوَّمٌ, (Ḳ, TA,) [in the CḲ حُوَّم, but it is] like سُكَّرٌ. (TA.)

حون

حَانَةٌ *A place where wine is sold*; (Mṣb and TA in this art., and Ṣ and Ḳ in art. حين;) the *place of the vintner*; (Kr;) the ↑ حَانُوتُ [or *shop*] *of the vintner*: pl. حَانَاتٌ: (Ṣ, Mṣb:) thought by AḤn to be from the Persian خَانَهْ. (TA.)

حَانَاةٌ: see art. حنو.

حَانُوتٌ, mentioned in the Ḳ in arts. حنت and حين and حنو: see حَانَةٌ, above; and see also art. حنو [to which several words allied to this are to be regarded as belonging]. It is masc. and fem.: and is [said to be] originally حَانُوَةٌ, like تَرْقُوَةٌ; and its pl. is حَوَانِيتُ: (Ṣ:) or, accord. to IB, it is originally حَنَوُوتٌ; then, حَوَنُوتٌ; and then, حَانُوتٌ. (TA in art. حين.)

حَانِيَةٌ: } see art. حنو.
حَانِىٌّ: }

حَانِيَّةٌ *Wine*; (Ṣ and Ḳ in art. حين, [in the CḲ, erroneously, حَانِئَة,] and Ḳ in art. حنو;) a rel. n. from حَانَةٌ: (Ṣ, TA:) or *vintners*; (Ḳ in art. حنو;) [from حَانِىٌّ; or] a rel. n. from حَانِيَةٌ. (TA in art. حنو.)

حوى

1. حَوَاهُ, aor. ـِ, (Ṣ, Mṣb, Ḳ,) inf. n. حَىٌّ, (Ṣ,) or حَوَايَةٌ, (Mṣb,) or both, (Ḳ,) *He collected it*; *brought it, drew it,* or *gathered it, together*; (Ṣ, Mṣb, Ḳ;) as also ↑ تَحَوَّاهُ: (TA:) *he grasped it*; *got,* or *gained, possession of it*: (Mṣb, Ḳ:) and ↑ اِحْتَوَاهُ signifies the same; (Ṣ, Mṣb, Ḳ;) as also ↑ اِحْتَوَى عَلَيْهِ: (Mṣb, Ḳ:) or this last, *he took,* or *got, possession of it*; *took it, got it,* or *held it, within his grasp,* or *in his possession*: (Ṣ:) and حَوَاهُ signifies also *he possessed it*: (Mṣb:) and, said of a place [&c., as also ↑ اِحْتَوَى عليه and ↑ اِحْتَوَاهُ], *it comprised, comprehended,* or *contained, it.* (TA.) — *He turned it round, made it to turn round,* or *wound it.* (Ḥar p. 230. [See حَىٌّ in art. حى.]) — حَوَّى حَوِيَّةً *He made a vehicle for a woman such as is called* حَوِيَّةٌ. (TA.) — حَوِىَ [originally حَوِوَ]: see art. حو.

2. تَحْوِيَةٌ [inf. n. of حَوَّى] signifies The act of *drawing together,* or *contracting*: and the state of *drawing together,* or *contracting*; or *being drawn together,* or *contracted*; as also تَحَوٍّ [inf. n. of ↑ تَحَوَّى]: (Lḥ, Ḳ:) or, in the opinion of ISd, تَحْوِيَةٌ has the former signification; and ↑ تَحَوٍّ, the latter. (TA.) A woman of the tribe of Kelb, being asked what she did in the rainy night, answered, أُحَوِّى نَفْسِى [*I draw myself*

together]. (Lḥ, TA.) — كَانَتْ تُحَوِّى وَرَآءَهُ بِعَبَاءَةٍ أَوْ كِسَآءٍ, occurring in a trad. of Ṣafeeyeh, means *She used to wind an* عباءة *or a* كساء *round the hump of the camel, behind him,* and then to ride upon it. (IAth. [See حَوِيَّةٌ.])

5. تَحَوَّى, (Ṣ, Ḳ,) inf. n. تَحَوٍّ, (Ḳ,) *It assumed a round, or circular, form;* or *coiled itself:* (Ḳ:) or *it gathered itself together, and coiled itself,* or *assumed a round, or circular, form.* (Ṣ.) You say, تَحَوَّتِ الحَيَّةُ (Ṣ, Ḳ*) *The serpent gathered itself together, and wound, or coiled, itself;* (TA;) whence, as some say, the word حَيَّةٌ: (Ḳ:) and in like manner one says of the intestines. (Ḳ.) — See also 2, in two places.

6: see 1.

8: see 1, in four places. — احتوى حَوِيًّا *He made a small watering-trough,* or *tank, for his camels.* (TA.)

حَوًى, (Az, IB, TA,) or † حَوِيَّةٌ, (Ḳ,) The *having,* or *assuming, a round,* or *circular,* or *coiled, form;* or *roundness* or *circularity* [or the *state of being coiled*]; of anything; (Az, Ḳ, TA;) as, for instance, of a serpent; (Az, IB, TA;) and of certain asterisms, which appear regularly disposed in a round, or circular, form. (Az, TA.) [See 5.]

حِوَآءٌ *A place that comprises, comprehends,* or *contains, a thing:* for instance, the belly [or womb] of the mother is a حواء to the child [or fœtus]. (TA.) — *A collection of tents* (بُيُوت), *near together;* as also † مُحْوًى: (Ḳ:) or *a collection of tents* (بيوت) *of people, made of camel's fur:* (Ṣ:) or *a collection of tents* (بيوت) *of people, at a water:* (TA:) pl. أَحْوِيَةٌ: (Ṣ, TA:) and † محتوى [app. مُحْتَوًى] and † محوى [app. مُحْوًى] signify *a place in which the tents* (بيوت) *of a tribe are collected together:* (Lth, TA:) the last is of the dial. of El-Yemen, where it is applied to *a few small tents collected together in a tract,* or *region, of fruitful,* or *productive, land:* (TA:) and its pl. is محاوى [or, more probably, مَحَاوٍ; with the article, المَحَاوِى]. (Lth, TA.) *A tent of* [*goats'*] *hair, and of* [*camels'*] *fur, of the Arabs of the desert.* (KL.)

حَوِىٌّ *Possessing, after deserving,* or *after becoming entitled.* (IAạr, Ḳ.) — *A small watering-trough,* or *tank,* (Ḳ, TA,) *which a man makes for his camel; called also* مَرْكُوّ. (TA.) [See also the next paragraph.]

حَوِيَّةٌ: see حَوًى. — Also *A winding,* or *circling,* or *coiled, gut* or *intestine;* and so † حَاوِيَةٌ and † حَاوِيَآءُ: (Ḳ:) [also called قِتْبٌ:] or the [*kind of gut, or intestine, termed*] دُوَّارَةٌ *in the belly of the sheep* or *goat;* as also † حَاوِيَةُ البَطْنِ and حَوِيَّةُ البَطْنِ: (IAạr:) † حَاوِيَةٌ and † حَاوِيَآءُ البَطْنِ all signify the same: (Ṣ:) the pl. is حَوَايَا; (Ḳ;) or this is the pl. of حَوِيَّةٌ, signifying the أَمْعَآء [or *intestines into which the food passes from the stomach*]; and the pl. of

حَاوِيَةٌ and حَاوِيَآءُ is [said to be] حَوَاوٍ; (Ṣ;) but IB says that this latter pl. is not held to be allowable by Sb, and that حَوَايَا is the pl. of all the three sings., [originally] of the measure فَعَائِلُ as pl. of the first, and فَوَاعِلُ as pl. of the second and third [though in these two cases it should be by rule حَوَاوٍ]: AHeyth says that حَوَايَا as pl. of حَاوِيَةٌ is like زَوَايَا as pl. of زَاوِيَةٌ: and ISk mentions الحَاوِيَاتُ as pl. of الحَاوِيَةُ and الحَاوِيَآءُ, [in the latter case like قَاصِعَاتٌ as pl. of قَاصِعَآءُ,] and explains it as signifying بَنَاتُ اللَّبَنِ [app. meaning *the small guts,* or *intestines, in which originate the lacteals*]. (TA.) — Also sing. of حَوَايَا signifying *Winding excavations* or *hollows, which the rain fills, and in which it remains a long time because the soil at the bottom thereof is cohesive and hard, retaining the water:* the Arabs call them [also] أَمْعَآء, likening them to the حَوَايَا of the belly: accord. to AA, the pl. signifies *i. q.* مَسَاطِحُ [pl. of مِسْطَحٌ], *made by collecting earth and stones upon smooth and hard rock, to confine thereby water:* accord. to IB, on the authority of IKh, *wells that are dug in the district of Kelb, in hard ground, whereby is confined the water of the torrents, which they drink throughout the year:* accord. to Naṣr, *a certain construction with masses of rock in the form of a pool, on the way to Et-Teghlibeeyeh, near Ood:* accord. to ISd, the sing. signifies *a smooth and hard rock which is surrounded with stones and earth, in which water collects.* (TA.) — Also *A* [*garment of the kind called*] كِسَآء, *stuffed* [*with* ثُمَام *or the like*], *which is wound round the hump of the camel;* (Ṣ, Ḳ;*) *i. q.* سَوِيَّةٌ, *except that the former is only for camels, and the latter is sometimes for other animals:* pl. حَوَايَا: (Ṣ:) a subst. from 2 in the last of the senses assigned to it above: (IAth:) [the same is app. meant by what here follows:] *a certain thing that is prepared for a woman to ride upon.* (TA. [But it was also used by men.]) 'Omeyr Ibn-Wahb El-Jumaḥee said, on the day of Bedr, when he computed the number of the companions of the Prophet, رَأَيْتُ الحَوَايَا عَلَيْهَا المَنَايَا [meaning *I saw the* حوايا *with the men of courage upon them*]: (Ṣ:) [for] مَنِيَّةٌ [sing. of منايا] means † "a man of courage upon his saddle." (TA.)

حُوَيَّةٌ [a dim. of حَيَّةٌ]: see what next follows.

رَجُلٌ حَوَّآءٌ *A man who collects serpents* (حَيَّات); and so † حَاوٍ: [which latter, vulgarly pronounced حَاوِى, is also now applied to *a serpent-charmer;* and *a juggler who performs various tricks with serpents &c.:*] (Ḳ in art. حى:) or the latter, (Ṣ in that art.,) or both, (T in that art.,) *an owner of serpents;* (T, Ṣ;) and so حَاىٍ: (T:) the pl. of حَاوٍ is حُوَاةٌ. (TA.) — [Hence,] الحَوَّآءُ وَالحُوَيَّةُ † *The constellations Serpentarius and Serpens.* (Ḳzw.)

حَاوٍ: see what next precedes.

حَاوِيَةٌ: see حَوِيَّةٌ, in three places.

حَاوِيَآءُ: see حَوِيَّةٌ, in two places.

حَاوِىٌّ: see حَائِىٌّ, in art. حوأ.

حَيَّةٌ [meaning *A serpent*] is said by some to be from تَحَوِّى, because what is so termed gathers itself together, and winds, or coils, itself; (ISd,* Ḳ,* TA;) and to be originally حَوْيَةٌ; (TA in art. حى;) and their opinion is strengthened by the forms and meaning of the words حَوَّآءٌ and حَاوٍ (TA) [and by the form and meaning of the word مَحْوَاةٌ]: or the حيّة is so called because of the length of its life (لِطُولِ حَيَاتِهَا). (Ḳ. [See the next art.])

أَحْوَى: } see art. حو.
أُحْوِىٌّ: }

محوى [app. مُحْوًى]: see حِوَآءٌ.

أَرْضٌ مَحْوَاةٌ *A land abounding with* حَيَّات [or *serpents*]: (TA:) or *containing serpents;* as also مَحْيَاةٌ. (Ibn-Es-Sarráj, Ṣ in art. حى.)

مَحْوِىٌّ pass. part. n. of حَوَاهُ. (Mṣb.)

مُحْوًى: see حِوَآءٌ.

محتوى [app. مُحْتَوًى]: see حِوَآءٌ.

حى or حى and حيو:

1. حَيِىَ, (Ṣ, Mgh, Mṣb, Ḳ,) and حَىَّ, (Ṣ, Ḳ,) which latter is the more common, (Fr, Ṣ,) [like مَلَّ, originally مَلِلَ,] dual حَيِيَا, and حَيَّا, (Fr,) pl. حَيُوا, (Fr, Ṣ,) like خَشُوا, (Ṣ,) and حَيُّوا, (Fr, Ṣ,) as some say, (Ṣ,) aor. يَحْيَا, (Ṣ, Mṣb, Ḳ,) and يَحَىُّ, (Fr, Ḳ,) [like يَمَلُّ, originally يَمْلَلُ,] occurring in poetry, but improper, and disallowed by the Baṣrees, (Fr,) inf. n. حَيَاةٌ [q. v. infrà], (IB, Mgh, Mṣb,) or حَيَآءٌ, (Ḳ,) and حِىٌّ and حَيَوَانٌ, (IB,) which last has an intensive signification, like its contr. مَوَتَانٌ, (Mṣb,) *He,* or *it, lived;* or *was,* or *became, in the state termed* حَيَاة, explained below. (Ṣ, Ḳ.) [The inf. n. حَيَوَانٌ (q. v. infrà) suggests the supposition that حَيِىَ may be originally حَيِوَ; but I find no authority for this supposition; and if it be the case, this verb presents the only instance of a root of which the medial radical letter is ى and the final و.] — حَيُوا, said of a people, or company of men, † *They were, or became, in good condition:* (AA, Ṣ:) or *they were,* or *became, fat, by having the means of subsistence,* بَعْدَ هُزَالٍ [*after leanness*]. (AZ. [See also 4.]) — You say also, حَيِيَتِ النَّارُ, inf. n. حَيَاةٌ and حَيَآءٌ, † [*The fire was,* or *became, alive,* or *burning,*] like as you say, مَاتَتْ. (AḤn.) — حَيِىَ الطَّرِيقُ † *The road,* or *way, was,* or *became, apparent,* or *distinct.* (Ḳ.) One says, إِذَا حَيِىَ لَكَ الطَّرِيقُ فَخُذْ يَمْنَةً † [*When the road,* or *way, becomes apparent,* or *distinct, to thee, take to the right*]. (TA.) — حَيِىَ and حَيِىَ مِنْهُ, aor. يَحْيَا, inf. n. حَيَآءٌ: see 10, in two places. — حَىَّ الخَمْسِينَ: see 2.

2. حَيَّاهُ, inf. n. تَحِيَّةٌ, (Mgh, Mṣb, Ḳ,) in its primary sense, *i. q.* ▼ اَحْيَاهُ. (Mgh.) You say, حَيَّاكَ ٱللّٰهُ, (Ṣ, Ḳ,) meaning *May God preserve thee alive; prolong thy life;* or *make thee to continue in life;* syn. أَبْقَاكَ; (Fr, Selemeh Ibn-'Áṣim, Ḳ;) as also ▼ أَحْيَاكَ; (Selemeh Ibn-'Áṣim, TA;) or عَمَّرَكَ: (Aboo-'Othmán El-Mázinee, Mgh, TA:) or *may God save thee;* or *make thee to be free from evil,* or *harm,* or *the like:* (Fr, TA:) or *may God make thee to have dominion:* (Fr, Ṣ, Ḳ:) or *may God honour thee, and benefit thee.* (Ḥam. p. 489.) And حَيَّاكَ ٱللّٰهُ وَبَيَّاكَ [explained in art. بى]. (TA.) And حَيَّا ٱللّٰهُ وَجْهَكَ [*May God preserve thy face:* see مُحَيًّا]. (Ḥam p. 23.) — Also *He said to him* حَيَّاكَ ٱللّٰهُ, explained above: (Mgh:) originally, *he prayed for his life:* and then, *he prayed for him,* absolutely: (Mṣb:) *he saluted him;* (Ḳ, TA;) and so حَيَّاهُ تَحِيَّةَ ٱلْمُؤْمِنِ: (Lḥ, TA:) as used in the language of the law, *he said to him,* سَلَامٌ عَلَيْكَ. (Mṣb.) — [See also تَحِيَّةٌ, below.] = حَيَّا ٱلْخَمْسِينَ [in the CḲ (erroneously) حَيَّ ٱلْخَمْسِينَ] *He approached* [*the age of*] *fifty.* (IAạr, Ḳ, TA.) = حَيَّيْتُ حَآءً حَسَنَةً and حَيًّا حَسَنًا [*I wrote a beautiful* ح]. (TA in باب الالف اللّيّنة.)

3. حَايَيْتُ ٱلنَّارَ, (inf. n. مُحَايَاةٌ, TA,) *i. q.* ▼ أَحْيَيْتُهَا [i. e. † *I gave life to, enlivened,* or *revived, the fire*] by blowing. (Ḳ.) [See also 2 in art. وهب; where a similar meaning is assigned to the former verb; but perhaps it is there a mistranscription.] مُحَايَاةٌ also signifies *Nourishment* (غِذَآء [which is properly speaking a subst., but seems to be here used as an inf. n.,]) *for,* or *of, a child,* (Ḳ, TA,) *with that whereby is his life,* or, as in the M, because thereby is his life. (TA.)

4. احياهُ, (Ṣ, Mṣb, Ḳ, &c.,) inf. n. إِحْيَآءٌ, (TA,) said of God, (Ṣ, Mṣb,) *He made him alive, to live,* or *be a living being; quickened, endued with life, vivified,* [*revivified, revived,* or *resuscitated,*] *him.* (Ṣ,* Mṣb,* Ḳ.) Hence, in the Ḳur [lxxv. last verse], أَلَيْسَ ذٰلِكَ بِقَادِرٍ عَلَى أَنْ يُحْيِىَ ٱلْمَوْتَى, (Ṣ, TA,) i. e. *Is not That* Doer of these things (Jel) [*able to quicken the dead?*]. — See also 2, in two places. — † *He* (God) *endued him with the intellectual faculty:* as in the saying, in the Ḳur [vi. 122], أَوَمَنْ كَانَ مَيْتًا فَأَحْيَيْنَاهُ † [*And is he who was intellectually dead and whom we have endued with the intellectual faculty . . . ?*]. (Er-Rághib.) — See also 3. — احيا الأَرْضَ † *He* (God) *revived the earth,* or *land, by the rain, so as to produce in it plants,* or *herbage.* (TA.) And أُحْيِيَتِ ٱلْأَرْضُ † *The land was tilled, and made productive.* (AḤn.) And احيا ٱلْمَوَاتَ ‡ *He turned to use the unowned and unused and uncultivated land by walling it round or sowing it or building upon it and the like.* (TA.) — Also † *He* (a man) *found the land to be fruitful,* or *abundant in herbage,* (Ṣ, Ḳ,) *with fresh herbage.* (Ḳ.) — احيا ٱللَّيْلَ † *He passed the night in religious service, worship, adoration,* or *devotion, abstaining from sleep:* (TA:) or [simply] *he remained awake during the night.* (W p. 9.) And احيا لَيْلَتَهُ *He passed his night awake.* (MA.) = احيت ٱلنَّاقَةُ *The she-camel had living offspring;* (Ṣ, Ḳ;) *her offspring seldom,* or *never, died.* (Ṣ.) — احيا ٱلْقَوْمُ *The people,* or *company of men, had their cattle living:* (Ḳ:) or *had their cattle in good condition:* (AA, Ṣ, Ḳ:) or *had rain, so that their beasts obtained herbage to such a degree that they became fat:* (AZ:) or *they* (themselves) *became in a state of abundance of herbage,* or *plenty.* (Ṣ, Ḳ. [See also 1.])

5. تَحَيَّا مِنْهُ † *He shrank from it:* taken from ٱلْحَيَاةُ [or ٱلْحَيَآءُ]; because it is of the nature of the living to shrink: or it is originally تَحَوَّى; the و being changed into ى; or [in other words] from ٱلْحَىُّ "the act of collecting," [inf. n. of حَوَاهُ,] like تَحَيَّزَ from ٱلْحَوْزُ. (TA.) [See also 10.]

[6. تَحَايَا as quasi-pass. of 3, *He quickened, enlivened,* or *revived, himself.* See an ex. voce روح.]

10. استحياهُ *He spared him; let him live;* or *left him alive;* (Ṣ, Mṣb, Ḳ, TA;) *did not slay him;* (Mṣb;) in which sense the verb has but one form: (Mṣb, TA:) or *he left him; let him alone;* or *forbore from him.* (Mgh.) — Hence, as some say, (Ḳ,) إِنَّ ٱللّٰهَ لَا يَسْتَحْيِى أَنْ يَضْرِبَ مَثَلًا, (Ṣ, Ḳ,) in the Ḳur [ii. 24], i. e. لَا يَسْتَبْقِى [meaning *Verily God will not spare to propound,* or *refrain from propounding, a parable,* or *as a parable*]. (Ṣ.) [Hence, also,] إِنَّ ٱللّٰهَ يَسْتَحْيِى مِنْ ذِى ٱلشَّيْبَةِ ٱلْمُسْلِمِ أَنْ يُعَذِّبَهُ, i. e. [*Verily God*] *forbears from punishing* [*the hoary Muslim*]. (Er-Rághib.) — [اِسْتَحْيَا, or اِسْتَحَى, which latter is the more proper mode of writing it, also signifies *He felt,* or *had a sense of,* or *he was,* or *became, moved* or *affected with, shame, shyness,* or *bashfulness;* and particularly, but not always, *honest shame,* or *pudency,* or *modesty;* or *his soul shrank from foul things;* as also ▼ حَيِىَ, inf. n. حَيَآءٌ.] استحيى is of the dial. of El-Ḥijáz; and استحى, with a single ى, is of the dial. of Temeem; (Akh, Ṣ, Mṣb;) the former being the original, (Akh, Ṣ,) and that which is used in the Ḳur: (Mṣb:) in the latter, the first ى [of the original] is suppressed, [and its vowel is transferred to the ح,] to facilitate the pronunciation, because of the occurrence of the two ىs together: this is the opinion of Sb; and with it agrees that of Aboo-'Othmán [El-Mázinee]: the opinion ascribed to Sb in the Ṣ, namely, that اِسْتَحَيْتُ is changed from اِسْتَحْيَيْتُ in like manner as اِسْتَبَعْتُ is changed from اِسْتَبْيَعْتُ, is that of Kh, and is disallowed by El-Mázinee. (IB.) You say, استحيى مِنْهُ, (Ṣ, Ḳ, [in the CḲ, erroneously, عَنْهُ,]) or استحيا منه, (Mgh, Mṣb,) and استحى منه, and استحياهُ, (Ṣ, Mṣb, Ḳ, TA,) and استحاهُ; (TA, [and so in the CḲ in the place of استحياهُ;]) as also ▼ حَيِىَ منه, (Ṣ, Mgh, Mṣb, Ḳ,) aor. يَحْيَا, (Ṣ,) inf. n. حَيَآءٌ; (Ṣ,* Mgh, Mṣb, Ḳ;) *He was ashamed of it* or *on account of it,* or *ashamed to do it,* or *shy of doing it;* [generally meaning a foul thing;] *he was ashamed for himself,* or *of himself,* or *was bashful,* or *shy, with respect to it,* or *him; he was abashed at,* or *shy of, it,* or *him;* (Ḳ;) *he shrank from it,* or *him:* (Mṣb, Ḳ:*) and استحيى مِنْ كَذَا [or استحيا] *He disdained,* or *scorned, such a thing; abstained from it,* or *refused to do it, by reason of disdain and pride; he disliked,* or *hated, it, and his soul was above it; he shunned it, avoided it,* or *kept himself far from it;* syn. أَنِفَ مِنْهُ. (TA.)

حَيَّ a verbal noun, (Ṣ, M, Mgh, TA,) and therefore immediately followed by عَلَى, (M, TA,) used as an imperative [addressed to a single person, male or female, and to more than one]; (Ṣ;) used in calling, or summoning, or inviting, and urging; (Lth, T, TA;) and having no verb derived from it [alone, for حَيْعَلَ is derived from حَيَّ and عَلَى together]; (Lth, T, Mṣb, TA;) meaning *Come:* (Ṣ, M, Mgh, Mṣb, Ḳ, TA, &c.:) or *come quickly:* or *hasten.* (Mgh, TA.) Hence, حَيَّ عَلَى ٱلصَّلَاةِ (in the أَذَان, TA) *Come to prayer:* (IḲt, Ṣ, M, Mṣb, Ḳ:) or *come ye to prayer:* or *come ye quickly:* or *hasten ye.* (TA.) And حَيَّ عَلَى ٱلْفَلَاحِ [in the same: see art. فلح]. (Mgh, TA.) And حَيَّ عَلَى ٱلْغَدَآءِ [*Come to the morning-meal*]: and على ٱلْعَشَآءِ [*to the evening-meal*]: (Mṣb:) and على ٱلثَّرِيدِ [*to the crumbled bread moistened with broth*]: (Ṣ:) and على ٱلْخَيْرِ [*to good, good fortune, prosperity,* &c.]. (Lth, T, TA.) The saying of Ibn-Aḥmar,

حَيَّ ٱلْحُمُولَ فَإِنَّ ٱلرَّكْبَ قَدْ ذَهَبَا *

means *Keep thou to the loads* [*for the riders upon the camels have gone*]. (TA.) — In the phrase حَيَّ هَلَ, and حَيَّ هَلًا, followed by عَلَى and إِلَى كَذَا, and حَيَّ هَلَا كَذَا, (Ḳ,) which last is used in a case of pausation, but is bad in other cases, (Ṣ in art. هل,) and حَىَّ هَلْ, and حَيْهَلْ, [so in the copies of the Ḳ,] with the ى quiescent, (Ḳ,) and حَيَّ هَلَنْ [app. حَىّ هلن for حَيَّ هَلًا, or perhaps a mistranscription for حَيَّ هَلَكَ], (TA,) [the most common rendering of حَيَّ هَلَ &c. is like that of حَيَّ alone, namely, *Come:* or] حَيَّ signifies *hasten thou;* and [هل or] هلا, *come to it,* or *reach it;* [so that the meaning is *hasten thou: come to such a thing:*] or حَيَّ signifies *come;* and [هل or] هلا, *quickly* (حَثِيثًا), or *be thou quick;* [so that the meaning is *come quickly to such a thing:*] or [هل or] هلا signifies *be thou quiet;* and the meaning is أَسْرِعْ عِنْدَ ذِكْرِهِ وَٱسْكُنْ حَتَّى يَنْقَضِىَ [*Speed thou at the mention thereof, and be quiet until it is finished*]. (Ḳ.) Accord. to Abu-l-Khaṭṭáb, the Arabs used to say, حَيَّ هَلَ ٱلصَّلَاةَ, meaning *Come thou to prayer.* (TA.) And one says, حَيَّ هَلْ بِفُلَانٍ, (IAạr, and so in the CḲ,) and حَيَّ هَلًا بفلان, (IAạr, and so in MS. copies of the Ḳ,) and حَيَّ هَلَ بفلان, meaning *Hasten thou with such a one:* (IAạr:) or *keep thou to such a one,* and *call him.* (Ḳ.) It is said in a trad.,

إِذَا ذُكِرَ الصَّالِحُونَ فَحَيَّ هَلَ بِعُمَرَ, (Ṣ in art. هل,) or فَحَيَّهَلًا بِعُمَرَ, (TA,) i. e. [*When the good, or righteous, are mentioned, then*] *keep thou to 'Omar*, and *call 'Omar*, (Ṣ ubi suprà, and TA,) and *begin with him*, and *be quick in mentioning him*; (TA;) for he is of such. (Ṣ ubi suprà.) Accord. to some of the grammarians, (TA,) when you say حَيَّ هَلًا, with tenween, it is as though you said حَثًّا; and without tenween, it is as though you said الحَثَّ; the tenween being made a sign of indeterminateness; and the omission of it, a sign of determinateness: and so it is in all compounds of this kind. (Ḳ.) [See also art. هل.]

حَيٌّ *Living, having life, alive*, or *quick; contr. of* مَيِّتٌ [or مَيْتٌ]; (Ṣ, Mgh,* Mṣb,* Ḳ;) and حَيَوَانٌ ♦ is syn. with حَيٌّ [as meaning *having animal life*]: (IB:) dim. of the former ♦ حُيَىٌّ: (Mṣb:) and pl. أَحْيَآءٌ. (Mṣb, Ḳ.) When you say of a person, لَيْسَ بِحَيٍّ, you mean that he is dead: (Lḥ:) but ضُرِبَ ضَرْبَةً لَيْسَ بِحَاىٍ ♦ مِنْهَا, (Lḥ, Ḳ,) [in the CḲ ضَرَبَ, and] in [some of] the copies of the Ḳ, erroneously, بِحَآءٍ, (TA,) means [*He was struck a blow*] *in consequence of which he will not live*: (Lḥ, Ḳ:*) like as the saying لَا تَأْكُلْ هٰذَا فَإِنَّكَ مَارِضٌ means Thou wilt be sick if thou eat such a thing. (Lḥ, Ḳ.) Accord. to ISh, one says, أَتَانَا حَيُّ فُلَانٍ, meaning *Such a one came to us in his life* [*-time*]: and سَمِعْتُ حَيَّ فُلَانٍ يَقُولُ كَذَا, meaning *I heard such a one say thus in his life* [*-time*]: (TA:) [or the former may mean *Such a one himself came to us*: and the latter, *I heard such a one himself say thus*: for] حى [i. e. حَيّ as a prefixed noun] is sometimes redundant, like ذُو and آلُ: (Ḥam p. 308:) [and] IB says that حَيُّ فُلَانٍ means *Such a one himself*. (TA.) And they say, كَيْفَ أَنْتَ وَحَيَّةُ ♦ أَهْلِكَ, i. e. *How art thou, and those remaining alive of thy family?* (TA.) — Applied to God, *Deathless*. (Er-Rághib.) — *Possessing the faculty of growth, as an animal*, and *as a plant*: (Er-Rághib:) and, applied to a plant, *fresh, juicy*, or *succulent, and growing tall*. (TA.) — [*Sensitively alive;*] *possessing the faculty of sensation*. (Er-Rághib.) — †[*Intellectually alive;*] *possessing the faculty of intellect*. (Er-Rághib.) [Hence,] †*A Muslim*; like as مَيِّتٌ means an unbeliever. (TA.) — †*Lively*, as meaning *free from grief* or *sorrow*. (Er-Rághib.) — †*Whole, sound*, or *unbroken*. (L and TA in art. صاب.) — أَرْضٌ حَيَّةٌ ♦ †*Fruitful land*; or *land abounding with herbage*; (Ḳ, TA;) like as ارض مَيِّتَةٌ means unfruitful land. (TA.) — الشَّمْسُ حَيَّةٌ †*The sun is*, or *was, of a clear colour, unaltered by approaching the place of setting*; as though its setting were regarded as death: (TA:) or *still bright and white*: or *still hot and powerful*: but the former of these two meanings is the more probable. (Mgh.) — نَارٌ حَيَّةٌ † [*A live*, or *burning, fire*. (AḤn.) — طَرِيقٌ حَيٌّ †*An apparent*, or *a distinct, road* or *way*: (Ḳ:) pl. أَحْيَآءٌ. (TA.) = See also حَيَّةٌ, first sentence. = [*A tribe*] of the Arabs: (Ṣ:) the *children*, or *descendants, of one father* or *ancestor, whether many or few*: and *a* شَعْب *comprising* قَبَائِل: (Az, TA:) or *a* قَبِيلَة of the Arabs: (Mṣb:) or *a* بَطْن *of the* بُطُون of the Arabs: (Ḳ, TA:) pl. أَحْيَآءٌ. (Ṣ, Mṣb, Ḳ.) = The *vulva*, or *external portion of the organs of generation*, of a woman. (Az, Ḳ. See also حَيَآءٌ, of which حَيٌّ is also said to be a pl.) Hence, سَقَطُ الحَيِّ, applied by an Arab of the desert to *The paraphernalia of a bride*. (Az, TA.) = لَا حَيَّ عَنْهُ means لَا مَنْعَ مِنْهُ, (Ks, Ḳ,*) i. e. [*There is no forbiddance of him*; or] *nothing is forbidden him*. (Fr.) = لَا يَعْرِفُ الحَيَّ مِنَ اللَّيِّ *He knows not*, or *will not know, what is true from what is false*; (IAạr, Ḳ, Ḥar p. 236;) and so الحَوَّ من اللَّوِّ: (TA:) or *the* حَوِيَّة [or *winding gut*, &c.,] *from the twisting of the rope*: (Ḳ:) or *overt speech from covert*: or *the living from the dead*: or *the act of turning round*, or *winding*, [see حَوَاهُ, (in art. حوى,) of which حَيٌّ is an inf. n.,] *from the act of twisting*. (Ḥar ubi suprà.) = The act of *collecting*. (TA. [But in this and some other senses it is an inf. n. of حَوَى: see 1 in art. حوى.]) = أَحْيَآءٌ as pl. of حَىٌّ: see art. حوأ.

حَيٌّ: see حَيَآءٌ, of which it is said to be a syn.: and of which it is also said to be a pl.: = and see حَيَآءٌ, of which, also, it is said to be a pl.

حَيَّةٌ [*A serpent*;] *a certain thing well known*: (Ḳ, TA:) [improperly explained in the Mṣb as syn. with أَفْعَى:] applied to the *male* and the *female*; (Ṣ, Mṣb;) the ة being added to denote one of a kind, as in بَطَّةٌ and دَجَاجَةٌ; [although حَيٌّ is not used as a coll. gen. n.; and] although the saying رَأَيْتُ حَيًّا ♦ عَلَى حَيَّةٍ, as meaning [*I saw*] *a male* [*serpent*] *upon a female* [*serpent*], is related as having been heard from the Arabs: (Ṣ:) but حَيُّوتٌ ♦ is also applied to the *male*; (Az, Ṣ, Ḳ;) the ت being augmentative; for the word is originally حَيْوٌ [or حَيُّوٌ]: (Az, TA:) it is said that it does not die unless by an accident: and they say of a long-lived man, مَا هُوَ إِلَّا حَيَّةٌ [*He is none other than a serpent*]; and in like manner they say of a woman; as though it were called حيّة because of its long life: [for] some, including Sb, say that it is derived from حَيَاةٌ, as the rel. n. is حَيَوِىٌّ, not حَوَوِىٌّ: and to him who objects that one says رَجُلٌ حَوَّآءٌ [meaning "a man who collects serpents"], it is replied that حَيَّةٌ and حَوَّآءٌ are of different roots, like لُؤْلُؤٌ and لَآلٌ, &c.: but it may be from تَحَوَّى, because of its winding, or coiling, itself; and some say that it is originally حَوْيَةٌ; some, that it is originally حَيْوَةٌ: (TA:) [the dim. is حُيَيَّةٌ ♦:] the pl. is حَيَّاتٌ (Ḳ) and حَيَوَاتٌ, (Ḳ, TA,) or حَيْوَاتٌ. (So in some copies of the Ḳ [agreeably with the dial. of Hudheyl].) Hence the prov. هُوَ أَبْصَرُ مِنْ حَيَّةٍ [*He is more sharp-sighted than a serpent*]; because of the sharpness of its sight: and أَظْلَمُ مِنْ حَيَّةٍ [*more wrongful in conduct than a serpent*]; because it comes to the burrow of the [lizard called] ضَبّ, and eats its young one, and takes up its abode in its burrow. (TA.) And سَقَاهُ اللّٰهُ دَمَ الحَيَّاتِ [*May God give him to drink the blood of the serpents*]; i. e., †*destroy him*. (TA.) And لَا تَلِدُ الحَيَّةُ إِلَّا حُيَيَّةً ♦ [*The serpent does not bring forth* anything *save a little serpent*]: a prov. applied to the cunning and mischievous, or malignant. (TA.) And فُلَانٌ حَيَّةُ الوَادِى, or الأَرْضِ, or البَلَدِ, or الحَمَاطِ, †*Such a one is cunning and mischievous*, or *malignant*, (IAạr, Ḳ,) *and intelligent, in the utmost degree*: (IAạr:) [or] فلان حيّة الوادى means †*such a one is strong in resisting, a defender of his possessions*. (TA.) And حَيَّةُ الوَادِى is also an appellation applied to †*The lion*; (Ḳ, TA;) because of his cunning, or craftiness. (TA.) One says also, هُمْ حَيَّةُ الأَرْضِ, meaning †*They are cunning, guileful, malignant*, or *mischievous, and strong, not neglecting to take blood-revenge*: so in a saying of Dhu-l-Iṣba' El-'Adwánee cited voce عَذِيرٌ. (TA.) And فُلَانٌ حَيَّةٌ ذَكَرٌ (Ṣ, TA) †*Such a one is courageous and strong*. (TA.) And رَأْسُهُ رَأْسُ حَيَّةٍ †*He is clever*, or *ingenious; acute*, or *sharp; intelligent*. (TA.) And رَأَيْتُ فِى كِتَابِهِ حَيَّاتٍ وَعَقَارِبَ †*I saw in his letter slanders*, or *calumnies, addressed to the ruling power, in order to cause the object thereof to fall into embarrassment from which escape would be difficult*. (TA.) — الحَيَّةُ is also a name of †[*The constellation Draco*; commonly called التِّنِّينُ;] *certain stars* [*partly*] *between the* فَرْقَدَانِ [or *β and γ of Ursa Minor*] *and* بَنَاتُ نَعْشٍ [meaning *the stars in the tail of Ursa Major*]: (Ḳ:) so called by way of comparison. (TA.) — And حَيَّةٌ also signifies †*A certain mark made with a hot iron upon the neck*, and *upon the thigh, of a camel, twisting*, or *winding, like the* حيّة [*properly so called*]. (Ibn-Ḥabeeb, TA.) = See also حَيٌّ.

حَيًا *Rain*; (Ṣ, Mṣb, Ḳ;) as also حَيَآءٌ ♦: (Ḳ:) or *much rain*: (Ḥar p. 185:) as being the means of giving life to the earth: (TA:) and †*plenty*; or *abundance of herbage*, (Ṣ, Ḳ,) and the *means of giving life to the earth and to men*; as being caused by the rain; and so حَيَآءٌ ♦: (TA:) or [simply] *herbage*; because produced by the rain: and *fat*, and *fatness*; because produced by the herbage: (Ḥam p. 662:) dual. حَيَيَانِ: (Ṣ:) and pl. أَحْيَآءٌ. (TA.) حَيَا الرَّبِيعِ means *The rain* [*called* ربيع, or *of the season thus called*,] *that gives life to the earth*. (TA.) = See also the next paragraph.

حَيَآءٌ an inf. n. of حَيِىَ in the first of the senses explained in this art. (Ḳ.) — [Hence,] *syn. with* حَيًا, in two senses: see the next preceding paragraph, in two places. — Also inf. n. of حَيِىَ as syn. with إِسْتَحْيَى; (Ṣ,* Mgh, Mṣb, Ḳ;) *i. q.* إِسْتِحْيَآءٌ; (Ṣ;) *Shame*; *a sense of shame; shyness*, or *bashfulness*; [and particularly, but not always, *honest shame*, or *pudency*, or *modesty*;] syn. حِشْمَةٌ; (Ḳ;) *a shrinking of the soul from foul conduct*, (Bḍ in ii. 24, and Er-Rághib,) *through fear of blame*; (Bḍ ibid.;) *a languor that affects the animal faculty*, (Bḍ ibid, and Mgh,*) *and turns it back from its actions*: (Bḍ:) and *repentance*; syn. تَوْبَةٌ. (Ḳ.) — And hence, as being a thing that should be concealed, and of which one is ashamed to speak plainly, (TA,) The *vulva*, or *external portion of the female organs of generation*, (فَرْج, El-Fárábee, Mṣb, Ḳ, or رَحِم, [which here means the same,] Ṣ,) of a camel, (El-Fárábee, Ṣ, Mṣb, Ḳ,) or an animal having feet like those of the camel, and of a cloven-hoofed animal, and of a beast of prey: (Ḳ:) accord. to AZ, the دُبُر [here meaning the same as فَرْج] of any of these and of other animals: (Mṣb:) accord. to IAạr, it is of the ewe or she-goat, the cow, and the gazelle: (ISd, TA:) and [sometimes] the فَرْج of a girl, (El-Fárábee, Mṣb,) or of a woman: (Zj in his "Khalḳ el-Insán:" [see also حَىٌّ:]) and ✝ حَيًا signifies the same; (Ḳ;) but accord. to Az, this is not allowable except in poetry, in a case of necessity: (TA:) pl. أَحْيَآءٌ (AZ, IJ, Ḳ) and أَحْيِيَةٌ (Aṣ, Sb, Ṣ, Ḳ) and, by contraction, أَحِيَّةٌ, (Sb, IB, TA,) which is said to be preferable, (TA,) and [quasi-pl. n.] ✝ حَىٌّ and ✝ حِىٌّ [which two have been mistaken by Freytag for syns. of تَحِيَّةٌ, immediately following them in the Ḳ]. (Sb, Ḳ.)

حَيَاةٌ, or ✝ حَيَوٰةٌ, (as in different copies of the Ḳ, in the latter manner in copies of the Ṣ,) written with و in the Ḳur, to show that و follows ى in the pl. [حَيَوَاتٌ, like صَلَوَاتٌ], or because the sound of the ا is inclined towards that of و, (ISd, TA,) and ✝ حَيْوَةٌ, with sukoon to the و, (Ḳ,) which is substituted for the ا of حَيَاةٌ, as is done by the people of El-Yemen in the case of every ا that is changed from و, as in صَلَوٰةٌ and زَكَوٰةٌ, though the final radical letter of the verb of حَيَاةٌ is ى, (TA,) an inf. n. of حَيِىَ in the first of the senses explained in this art.; (IB, Mgh, Mṣb;) *Life*; *contr. of* مَوْتٌ; (Ṣ, Ḳ;) as also ✝ حَيَوَانٌ and ✝ حِىٌّ, (Ḳ,) or this last is asserted to be a pl. of حَيَوٰةٌ, (Ṣ,) and as also ✝ مَحْيًا, (Ṣ,* Ḥar pp. 25 and 350,) of which the pl. is مَحَايٍ: (Ṣ:) حَيَاةٌ signifies the *faculty of growth, as in an animal*, and *in a plant*: and the *faculty of sensation*: and ✝ the *faculty of intellect*: and ✝ *freedom from grief* or *sorrow*: and *everlasting life in the world to come*; to which one attains by that حياة which is intelligence and knowledge: and the حياة that is an attribute of God. (Er-Rághib.) يَا لَيْتَنِى قَدَّمْتُ لِحَيَاتِى, in the Ḳur [lxxxix. 25], means [*O, would that I had prepared*, or *laid up in store*,] *for my everlasting state of existence*. (Er-Rághib.) And ✝ وَإِنَّ الدَّارَ الآخِرَةَ لَهِىَ الحَيَوَانُ, in the Ḳur [xxix. 64], means [*And verily the last abode is*] the abode of *everlasting life*: (TA:) or الحيوان here means *the life that will not be followed by death*: or *much life*; like as مَوَتَانٌ signifies much death: (Mṣb:) and it is also the name of *a certain fountain in Paradise*, [*the water of*] *which touches nothing but it lives, by permission of God*. (TA.) الحَيَوٰةُ الطَّيِّبَةُ, accord. to I'Ab, explaining xvi. 99 of the Ḳur, (TA,) means *Lawful means of subsistence* (Ḳ, TA) *in the present world*: (TA:) or *Paradise*. (Ḳ, TA.) — Also ✝ *Advantage*, or *profit*; or *a cause*, or *means, thereof*: whence the saying, لَيْسَ لِفُلَانٍ حَيَاةٌ *There is not, in such a one, profit, nor good*: and so it is said to mean in the Ḳur [ii. 175], وَلَكُمْ فِى القِصَاصِ حَيٰوةٌ ✝ [*And there is to you, in retaliation, an advantage*]: (TA:) or this means that the knowledge of the law of retaliation restrains from slaughter, and so is a cause of life to two persons; and because they used to slay one who was not the slayer, retaliation upon the slayer is a cause of saving the lives of the rest: or the meaning is *life in the world to come*; for when the slayer has suffered retaliation in the present world, he is not punished for his act in the world to come. (Bḍ.) — حَيَاةُ الشَّمْسِ means ✝ *The remaining of the light and whiteness of the sun*: or *the remaining of its heat and power*: but the former of these two meanings is the more probable. (Mgh. [See حَىُّ الشَّمْسِ, voce حَيَّةٌ.])

حَيَوٰةٌ and حَيْوَةٌ: see the next preceding paragraph.

حَيِىٌّ *Having* حَيَآء [i. e. *shame*, *shyness*, *bashfulness*, *pudency*, or *modesty*]; (Ḳ;) part. n. of حَيِىَ as syn. with اِسْتَحْيَى; of the measure فَعِيلٌ: (Mṣb:) fem. حَيِيَّةٌ. (TA.) The saying of I'Ab, اَللّٰهُ حَيِىٌّ, means *God is one who acts with others in the manner of him who has* حَيَآء; for حَيَآء in its proper sense is not ascribable to Him: (Mgh:) or *one who leaves undone evil deeds, and does good deeds*. (Er-Rághib.)

حُيَىٌّ: see حَىٌّ, of which it is the dim.

حُيَيَّةٌ: see حَيَّةٌ, (of which it is the dim.,) in two places. ═ And dim. of حَآءٌ, q. v. in art. حوأ. (Lth, TA in باب الالف اللّيّنة.)

حَيَوِىٌّ [*Of, or relating to, the serpent*;] rel. n. of حَيَّةٌ. (Ṣ.) ═ [And rel. n. of حَا: see حَائِىٌّ in art. حوأ.]

حَيَوَانٌ an inf. n. of حَيِىَ, like حَيَاةٌ, (IB,) but having an intensive signification: (Mṣb:) see حَيَاةٌ, in two places. — See also حَىٌّ, first sentence. — Also *Any thing*, or *things, possessing animal life*, (Mṣb, Ḳ,*) whether *rational* or *irrational*; [*an animal*, and *animals*;] used alike as sing. and pl., because originally an inf. n.; (Mṣb;) *contr. of* مَوَتَانٌ [q. v.]. (Ṣ.) [حَيَوَانَاتٌ is used as its pl. of pauc. And hence,] الحَيَوَانَاتُ الخَمْسُ [*The five animals*] is applied to what may be killed by a person in the state of إِحْرَام, and by one engaged in prayer: (Mṣb in art. فسق:) these are *the rat*, or *mouse*, *and the biting dog*, *and* either *the serpent*, *the crow termed* أَبْقَع, *and the kite*, or *the serpent*, *the scorpion*, *and the kite*, or *the serpent*, *the scorpion*, *and the crow*, or *the scorpion*, *the crow*, *and the kite*. (Es-Suyooṭee, in "El-Jámi' eṣ-Ṣagheer," voce خمس.) It is originally حَيَيَانٌ; (Sb, Ḳ, TA;) the ى which is the final radical letter being changed into و because the occurrence of two ى together is disliked: (Sb, TA:) Aboo-'Othmán [El-Mázinee] holds the و to be a radical letter; but his opinion is said to be not admissible, because it is asserted that there is no instance in the language of a word of which the medial radical is ى, and the final و. (TA.)

حَيَوَانِىٌّ [*Of*, or *relating to, an animal* or *animals*]. — It is [also] particularly applied to *A seller of birds*. (TA.)

[حَيَوَانِيَّةٌ *Animality*; or *animal nature*.]

حُيُوتٌ: see حَيَّةٌ.

حَاىٍ, of the measure فَاعِلٌ, [said to be] originally حَايِوٌ, is *syn. with* حَاوٍ and حَوَّآءٌ, belonging to art. حوى [q. v.]. (Az, TA.)

أَحْىَ and أُحَىٌّ and أُحَيٌّ: see art. حو.

أُحَيْوٌ: see art. حو.

أَحْيَى in the saying أَحْيَى مِنْ ضَبٍّ [*More long-lived than a* ضَبّ, a kind of lizard, which is supposed to live seven hundred years,] is from الحَيَاةُ. (TA.) — In the sayings أَحْيَى مِنْ هَدِىٍّ [*More shy*, or *bashful, than the bride*] and أَحْيَى مِنْ مُخَدَّرَةٍ [*More shy*, or *bashful, than a girl kept behind the curtain*] it is from الحَيَآءُ. (TA.)

تَحْيَاةٌ: see the next paragraph.

التَّحَايِى *The two stars in the foot and before the foot of the foremost of Gemini*: (Ḳzw:) or *three stars over against* الهَنْعَة [which is *the Sixth Mansion of the Moon*]; (IḲt, Ḳ;) and sometimes the moon deviates from الهنعة, and makes its abode in التحايى: (IḲt:) they are *between the Milky Way and the stars that follow* العَيُّوق [or *Capella*]: Aboo-Ziyád El-Kilábee used to say that this name means الهنعة, and is also pronounced التَّحَائِى, with ء: but AḤn says that the moon makes its abode in these stars, and not in الهنعة itself: (TA:) its sing. is ✝ تَحْيَاةٌ; (IḲt, AḤn, TA;) if so, of the measure تَفْعَلَةٌ, like تَحْلَبَةٌ, not فَعْلَاةٌ, like عَزْهَاةٌ, because there is no such root as تحى; derived from الحَيَا, because its نَوْء [here meaning its auroral setting, in midwinter,] is attended with much rain: but التحائى, with ء, is irreg.; as though pl. of ✝ تَحِيَّةٌ likened to a word of the measure فَعِيلَةٌ. (IB.)

تَحِيَّةٌ inf. n. of 2. (Mgh, Mṣb, Ḳ.) — Also *A salutation*, or *greeting*, (A 'Obeyd, AHeyth, Mgh, Mṣb,) *pronounced by one person to another on their meeting*; (AHeyth;) particularly the *saying* سَلَامٌ عَلَيْكَ; (Mgh,* Mṣb;) and *the like*; (Mgh;) the most comprehensive form thereof, used by the believer [to his fellow-believer], being the *saying* السَّلَامُ عَلَيْكُمْ وَرَحْمَةُ ٱللّٰهِ وَبَرَكَاتُهُ: (AHeyth:) hence it is pluralized; the pl. being تَحِيَّاتٌ and تَحَايَا. (Mgh.) — Also *Continuance*, or *endurance*; or *endless*, or *everlasting, existence*:

(IAar, Msb, K, &c.:) and so it is said to mean in the following verse of Zuheyr Ibn-Jenáb El-Kelbee; who was a [kind of] king among his people:

* وَلَكُلُّ مَا نَالَ الفَتَى * قَدْ نِلْتُهُ إِلَّا التَّحِيَّهْ *

[*And indeed everything that the young man has attained, I have attained it, except endless existence*]: or, as some say, it here means *security from death and from evils.* (TA. [But more probably the meaning is that next following: for in the Mz, 49th نوع, where this verse is cited, but with مِنْ كُلِّ in the place of وَلَكُلُّ, he is said to have been (not a king, but,) "a nobleman." See also, respecting him, p. x. of my Preface.]) And ‡ *Dominion,* or *kingship:* (Fr, AA, S, Mgh, Msb, K, &c.:) because the people of the Time of Ignorance used to greet kings [or rather those of Himyer] by the saying أَبَيْتَ اللَّعْنَ, which they addressed to none other than a king; so that when any one of them became a king, it was said of him, فُلَانٌ نَالَ التَّحِيَّةَ [meaning ‡ *Such a one has attained the kingship*]. (Mgh: in which, and in the S, the foregoing verse is cited as an ex. of this last meaning.) التَّحِيَّاتُ لِلّٰهِ means *Endless existence belongs to God:* (Lth, Msb, TA:) or *dominion,* or *kingship:* (Lth, Yaakoob, S, Msb, TA:) or *freedom,* or *security, from all evils,* (Khálid Ibn-Yezeed, AHeyth,) and *from all causes of the cessation of existence:* (AHeyth:) or *endless existence, and security from evils, and dominion, and the like:* (Fr:) or *the expressions* [*of praise*] *that indicate and imply the ascription of dominion and endless existence:* (Kt:) or *salutations and benedictions are God's, and at his disposal.* (Mgh.) [التَّحِيَّاتُ is also a term applied to the following form of words repeated in the ordinary prayers: التَّحِيَّاتُ لِلّٰهِ وَالصَّلَوَاتُ وَالطَّيِّبَاتُ السَّلَامُ عَلَيْكَ يَا أَيُّهَا النَّبِيُّ وَرَحْمَةُ ٱللّٰهِ وَبَرَكَاتُهُ السَّلَامُ عَلَيْنَا وَعَلَى عِبَادِ ٱللّٰهِ الصَّالِحِينَ أَشْهَدُ أَنْ لَا إِلٰهَ إِلَّا ٱللّٰهُ وَأَشْهَدُ أَنَّ مُحَمَّدًا عَبْدُهُ وَرَسُولُهُ, or (instead of عبده ورسوله) رَسُولُ ٱللّٰهِ: see 5 in art. شهد.] — The assigning to this word, as used in the Kur iv. 88, the meaning of *A gift* is a mistake. (Mgh.) = See also التَّحَايِى.

مَحْيًا: see حَيَاةٌ. — Also *A time,* and *a place, of life.* (TA.)

مُحْيٍ and مُحْيِيَةٌ, applied to a she-camel, *Having living offspring; whose offspring seldom,* or *never, die.* (S.)

مُحَيَّا The *face* (S, K, Ham p. 23) of a man, because it is specified in salutation; [see 2;] (Ham ubi suprà;) a term used only in praise; (Ham p. 640;) i. e. the *face altogether:* or the حُرّ *of the face* [i. e. the *ball of the cheek;* or *what appears of the elevated part thereof;* or *what fronts one, of the face,* &c.]. (K.) — Of a horse, it is The *place where the flesh is separated* (حيث انفرق اللحم) *beneath the forelock.* (Ham p. 23.) And دَائِرَةُ المُحَيَّا, in a horse, [*The feather in*] *the place of separation* [*of the hair*] *beneath the forelock, in the upper part of the forehead.* (TA.)

مُحَيٍّ act. part. n. of 2; fem. مُحَيِّيَةٌ: (S, TA:) for in every noun in which three ى occur together, [the last of them being the final radical, and ending the word,] if it is not formed from a verb, the final radical letter is elided from it, as in عُطَيٌّ the dim. of عَطَآءٌ, and in أُحَيٌّ the dim. of أَحْوَى: but if it is formed from a verb, that letter remains, as in مُحَيِّى from حَيَّى. (S.)

أَرْضٌ مَحْيَاةٌ *i. q.* مَحْوَاةٌ, i. e. *A land containing serpents:* (Ibn-Es-Sarráj, S:) or *abounding with serpents.* (TA in art. حوى.)

أَبُو يَحْيَى *Death.* (TA, Har p. 218.)

حيث

حَيْثُ, (S, Msb, Mughnee, K,) indecl., (S, Msb,) with damm for its termination, (S, Msb, Mughnee,) as being likened to final words [such as قَبْلُ and بَعْدُ ending a proposition], (S, Mughnee,) because it does not [regularly] occur otherwise than prefixed to a proposition, (S,) for the being prefixed to a proposition is like the not being prefixed to anything, as the consequence of being prefixed, which is the sign of the gen. case, is not apparent: (Mughnee:) and حَيْثَ, (S, Mughnee, K,) also indecl., (S,) with fet-h, (S, Mughnee,) to render the pronunciation more easy, (Mughnee,) because damm with ى is deemed difficult to pronounce: (S:) and حَيْثِ, (Mughnee, K,) with kesr, accord. to the general rule observed to prevent the concurrence of two quiescent letters: (Mughnee:) and in like manner, حَوْثُ and حَوْثَ and حَوْثِ: (Mughnee, TA:) of which forms, حوث is asserted to be the original; (L;) though حَيْثُ is more chaste than حَوْثُ, and is the form used in the Kur-án: (Az and TA in art. حوث:) but some of the Arabs make حيث decl.: (Mughnee:) it is an adverbial noun of place, (S, Msb,) a vague adverbial noun of place, (L,) [signifying *Where,*] like حِينَ with respect to time: (S, K:) or it is a denotative of place, by general consent: but accord. to Akh it sometimes occurs as denoting time, [signifying *when,*] as in the following verse, (Mughnee, TA,) which is the strongest evidence of its use in this sense: (TA:)

* حَيْثُمَا تَسْتَقِمْ يُقَدِّرْ لَكَ ٱللّٰهُ *
* نَجَاحًا فِي غَابِرِ الأَزْمَانِ *

[*Whenever thou shalt pursue a right course, God will decree thee success in the time to come*]: (Mughnee, TA:) but in most instances it occupies the place of an accus., as an adverbial noun of place; or of a gen., governed by مِنْ, and sometimes by another prep., as in the saying (of Zuheyr, TA in art. قشعم),

* لَدَى حَيْثُ أَلْقَتْ رَحْلَهَا أُمُّ قَشْعَمِ *

[*At the place where Calamity,* or *Fate, has put down her saddle,* i. e., *made her abode*]: and sometimes it occurs as an objective complement, as it is said to do in ٱللّٰهُ أَعْلَمُ حَيْثُ يَجْعَلُ رِسَالَاتِهِ [in the Kur vi. 124], i. e. *God is knowing:* He knows *where to bestow his apostolic commissions;* يَعْلَمُ being suppressed, as implied by أَعْلَمُ; or أَعْلَمُ may be rendered by عَالِمٌ, and so may govern the accus. case. (Mughnee.) Accord. to rule, (Mughnee,) in every instance, (S, Mughnee,) it is prefixed to a proposition, (S, Msb, Mughnee,) nominal, or verbal, but in most cases the latter; (Mughnee;) as in أَقُومُ حَيْثُ يَقُومُ زَيْدٌ [*I will stand where Zeyd shall stand*]; and حَيْثُ تَكُونُ أَكُونُ [*Where thou shalt be, I will be*]; (S;) and جَلَسْتُ حَيْثُ زَيْدًا أَرَاهُ [*I sat where I saw Zeyd*], the accus. case being preferred in an instance like this; (Mughnee;) and اِذْهَبْ حَيْثُ شِئْتَ [*Go thou whither thou wilt.*] (Msb in art. حين.) You should not say حَيْثُ زَيْدٌ [alone]: (S:) or it occurs prefixed to a single word in poetry; (Msb, Mughnee;) as in the saying,

* وَنَطْعَنُهُمْ تَحْتَ الكُلَى بَعْدَ ضَرْبِهِمْ *
* بِبِيضِ المَوَاضِى حَيْثُ لَيِّ العَمَائِمِ *

[*And we pierce them beneath the kidneys, after smiting them, with the sharp swords, where the turbans are wound*]; (Mughnee;) but this is irregular; (Msb, Mughnee;) though Ks holds it to be regular. (Mughnee.) Lh relates, on the authority of Ks, that some make حيث to govern a noun in the gen. case, as in the saying,

* أَمَا تَرَى حَيْثُ سُهَيْلٍ طَالِعًا *

[*Seest thou not where Canopus is, rising?*]: but he says that this is not of respectable authority: (L:) some write حَيْثَ سُهَيْلٍ: and some, حَيْثُ سُهَيْلٌ, [which is the common reading, سهيل being an inchoative, and] the enunciative, مَوْجُودٌ, being suppressed. (Mughnee.) Abu-l-Fet-h says that he who prefixes حيث to a single word makes it declinable. (Mughnee.) [Accord. to Fei,] Benoo-Temeem say حَيْثَ when it occupies the place of an accus., as in the phrase, قُمْ حَيْثَ يَقُومُ زَيْدٌ [*Stand thou where Zeyd shall stand*]. (Msb.) Ks says, I have heard among Benoo-Temeem, of Benoo-Yarbooa and Tuheiyeh, those who say حَيْثُ in every case, when it occupies the place of a gen., and that of an accus., and that of a nom.; saying مِنْ حَيْثُ لَا يَعْلَمُونَ [*Whence they know not*], and حَيْثُ ٱلْتَقَيْنَا [*Where we met*]: and he says also, I have heard some of Benu-l-Hárith-Ibn-Asad-Ibn-El-Hárith-Ibn-Thaalabeh, and all Benoo-Fak'as, say حَيْثِ when it occupies the place of a gen., and حَيْثَ when it occupies the place of an accus.; saying مِنْ حَيْثِ لَا يَعْلَمُونَ, and حَيْثَ ٱلْتَقَيْنَا. (L.) Sometimes the proposition after حيث commences with إِنَّ, as in اِجْلِسْ حَيْثُ إِنَّ زَيْدًا جَالِسٌ [*Sit thou where Zeyd is sitting*]. (K in art. أن, and IAk p. 92.) — It sometimes comprises the meanings of two adverbial nouns of place, as when you say, حَيْثُ عَبْدُ ٱللّٰهِ قَاعِدٌ زَيْدٌ قَائِمٌ [*Where 'Abd-Allah is sitting, there Zeyd is standing*]. (AHeyth, L.) — The restrictive مَا (مَا كَافَّةٌ) is sometimes affixed to it, and in this case it implies a conditional meaning, [signifying *Wherever,* or *wheresoever,* and, accord. to Akh, *whenever,* or *whensoever,*] (Mughnee, TA,) and renders two verbs mejzoom, (Mughnee,) as in the saying, حَيْثُمَا تَجْلِسْ أَجْلِسْ [*Wherever thou shalt sit, I will sit*], (Ṣ,) and in the first of the verses cited above: (Mughnee, TA:) it is not [properly, though it is sometimes improperly,] used as a conditional without ما. (Ṣ.) — [It is also used, in scientific and other post-classical works, in senses different from those explained above. Thus, مِنْ حَيْثُ is used to signify *As to*, or *in respect of*: so in the phrase مِنْ حَيْثُ اللَّفْظُ وَالمَعْنَى *As to*, or *in respect of, the word and the meaning.* Also *As*, or *considered as, absolutely*, or *abstractedly*: so in the phrase مِنْ حَيْثُ هُوَ, or حَيْثُ هُوَ هُوَ, *As*, or *considered as, such, absolutely*, or *abstractedly*; and الإِنْسَانُ مِنْ حَيْثُ هُوَ إِنْسَانٌ *Man, as*, or *considered as, man, absolutely*, or *abstractedly*. And *As*, meaning *considered merely* or *only* or *simply as*: so in the saying, الإِنْسَانُ مِنْ حَيْثُ إِنَّهُ يَصِحُّ وَتَزُولُ عَنْهُ الصِّحَّةُ مَوْضُوعُ الطِّبِّ *Man, as*, or *considered merely* or *only* or *simply as, being healthy and ceasing to be healthy, is the object of therapeutics.* And *As*, meaning *since*, or *because*: so in the saying, النَّارُ مِنْ حَيْثُ إِنَّهَا حَارَّةٌ تُسَخِّنُ المَآءَ *Fire, as*, or *since*, or *because, it is hot, heats water.* بِحَيْثُ is also vulgarly used in this sense. And correctly as meaning *So that; so as that; in such a state*, or *condition, that*: often *syn.* with حَتَّى.]

حيج

1. حَاجَ, aor. يَحِيجُ, (Kr, Ḳ,) inf. n. حَيْجٌ, (TA,) *i. q.* حَاجَ, aor. يَحُوجُ: (Kr, Ḳ:) extr., because the ا of حَاجَةٌ [which is its logical root] is [originally] و. (TA.)

4. أَحْيَجَتِ الأَرْضُ, (Ḳ,) irregular, like أَحْوَجَ, (TA,) and أَحَاجَت, *The land produced the thorny plants*, or *trees, called* حَاج: (Ḳ:) or *abounded therewith.* (TA.)

حَاجٌ, a coll. gen. n., n. un. with ة, (TA,) *A certain kind of thorny plant*, or *tree*; (Ṣ, Ḳ;) *a plant of the sour kind* (مِنَ الحَمْضِ): accord. to ISd, *a kind of thorny plant* or *tree, i. q.* كَبَرٌ [or *caper*]: or *a certain plant different from that just named*: or *a certain kind of tree*: accord. to AḤn, *an evergreen, of which the roots extend far into the ground; which, cooked, is used as a medicine; having slender and long leaves, seeming as numerous as the thorns*: (TA:) [*asparago sylvestri similis*: (Golius, from Ibn-Beytár:) this name, and عَاقُول, are now applied by the Arabs to the *plant called by European botanists hedysarum alhagi*: see عَاقُول and تَرَنْجَبِين:] dim. ▼حُوَيْج; therefore its medial radical letter is ى; (Ḳ;) and ▼حُيَيْج is a chaste dial. var. of this dim., agreeably with a usage observed in similar cases [when the medial radical letter is ى]. (TA.)

حُوَيْج and حُيَيْج: see what next precedes.

حيد

1. حَادَ عَنْهُ, (Ṣ, A, Mṣb, Ḳ,) aor. يَحِيدُ, inf. n. حَيَدَانٌ and حَيْدَةٌ (Ṣ, Mṣb, Ḳ) and حَيْدٌ and حُيُودٌ and مَحِيدٌ (Ḳ) and حَيْدُودَةٌ, (Ṣ, Ḳ,) which last is originally حَيَدُودَةٌ, with the ى movent; this letter being afterwards made quiescent; for there is not in the language any word of the measure فَعْلُول, except صَعْفُوق; (Ṣ; [see the remarks on شَيْخُوخَةٌ, voce شَاخَ;]) *He declined*, or *turned aside* or *away, from it*; (Ṣ, A, Ḳ;) *removed, went away*, or *went far away, from it*; (Mṣb;) namely, a road, (Ṣ,) or a thing: (Mṣb:) *he shunned*, or *avoided, it*, from fear, or from disdain. (Az, L.) [See also 3.] You say, مَا لَكَ مَحِيدٌ عَنْ ذٰلِكَ *There is not, for thee, any avoiding that.* (L.) And حَادَتِ الدَّابَّةُ *The beast became scared*, or *shied, and quitted the middle of the road.* (L.) — حَادَ بِهِ, and ▼أَحَادَهُ, *He removed, took away*, or *took far away, him*, or *it*; [عَنْ شَىْءٍ *from a thing*;] similar to ذَهَبَ بِهِ and أَذْهَبَهُ. (Mṣb.)

2. قَدَّ السَّيْرَ فَحَيَّدَهُ *He cut*, or *cut lengthwise, the thong*, or *strip of skin* or *leather, and made it to have parts projecting beyond the rest.* (L, Ḳ.)

3. حَايَدَهُ, inf. n. مُحَايَدَةٌ and حِيَادٌ, *He went*, or *turned, aside from*, or *away from*, or *he avoided*, or *he went*, or *removed, to a distance from, him*, or *it*: (Ṣ, L, Ḳ:) [see also 1:] or, accord. to the A, *he inclined upon*, or *against, him*, or *it.* (TA.)

4: see 1.

حَيْدٌ *A rising*, or *protuberant*, or *prominent, part of a side* of a thing: (L, Ḳ:) so of the head; (Lth, L;) as also ▼حَيْدَةٌ: (A:) *a knot, knob*, or *protuberance*, of a stick or branch; [as also ▼حَيْدَةٌ: (AḤn, TA voce بَلْطٌ, q. v.:)] *a part* of a strap, or thong, *projecting beyond the rest*: (L:) *any rib*, (L, Ḳ,) or *other bone*, (L,) *that curves much* [*and is therefore prominent*]: (L, Ḳ:) [see an ex. voce حَابٍ, in art. حبو:] *a knot* in the horn of a mountain-goat; (A,* L, Ḳ;) or this is termed ▼حَيْدَةٌ: (Ṣ, L:) *a twisted part* of a horn: *a twisted internodal portion* of a horn: (L:) *any prominence* in a horn, and in a mountain, (Ṣ, L, Ḳ,) &c.: (Ṣ, L:) *a prominent and curved part* of a mountain: (T:) *a projecting portion*, or *ledge*, of a mountain, *resembling a wing*: (Ṣ, M, L, Ḳ:) pl. (of the former word, Ṣ) أَحْيَادٌ [a pl. of pauc.] and (of both words, Ṣ) حُيُودٌ and (of the latter, Ṣ) حِيَدٌ: (Ṣ, Ḳ:) the حُيُود of a camel are such parts as the *hips*, or *haunches*, and *thighs.* (L.) You say جَبَلٌ ذُو حُيُودٍ and أَحْيَادٍ, meaning *A mountain having projecting edges in its lower parts*, not in its upper parts. (Ṣ.) And قَعَدْتُ تَحْتَ حَيْدِ الجَبَلِ *I sat beneath the part of the mountain that projected like a wing.* (A.)

حَيْدَةٌ: see حَيْدٌ, in three places. — Also The *rugged part* of a road. (A.) — *An evil look*, (A, Ḳ,) *with a turning aside.* (A.) You say, مَا نَظَرَ إِلَىَّ إِلَّا الحَيْدَةَ, (A,) or إِلَّا نَظَرَ الحَيْدَةِ, (TA,) *He looked not towards me save with an evil look, with a turning aside.* (A, TA.)

حَيَدَى The *manner of walking of a proud and self-conceited person.* (Ḳ.) — حِمَارٌ حَيَدَى (Ṣ, Ḳ) and ▼حَيَدٌ, (Ḳ,) each occurring in a verse of [Umeiyeh the son of] Aboo-'Áïdh El-Hudhalee accord. to different relations thereof, (L, [see جَمَّازٌ,]) *An ass that turns aside from*, or *shies at, his shadow, by reason of his briskness, liveliness*, or *sprightliness*: (Ṣ, Ḳ:) or *that is wont often to turn aside from things*, or *to shy at them.* (Ṣ.) حَيَدَى is also applied as an epithet to a she-ass. (IAạr.) It is [said to be] the only masc. epithet of the measure فَعَلَى, (Ṣ, Ḳ,) except دَلَظَى a man "who thrusts vehemently," (IJ,) and وَقَرَى [but this is written in the Ḳ وَقَرَى] "a pastor of a وَقِير, or flock of sheep," and قَفَطَى vir "multum coiens," and جَمَزَى a "quick" ass. (MF.) But probably حَيَدٌ is the only correct word of the two above mentioned. (L.) [Or حِمَارٌ حَيَدَى is for حِمَارٌ ذُو حَيَدَى: see جَمَزَى, voce جَمَّازٌ.]

حَيَدَانٌ *Pebbles that become thrown aside from the legs of a beast as he goes along.* (Ṣ, Ḳ.)

حَيْدٌ: see حَيَدَى.

حَيَادِ, like قَطَامِ, (L,) indecl., with kesr for its termination, [and of the fem. gender,] occurs in the phrase (TA) حِيدِى حَيَادِ, similar to فِيحِى فَيَاحِ, (Ṣ, L, Ḳ,) meaning *Turn thou aside*, or *away*, [from me:] (A, L:) said by one when the time for fighting is come, (L,) and by one fleeing. (Ibn-Abi-l-Hadeed.)

حَيُودٌ [*That declines*, or *goes away, much*, or *frequently*]: an intensive epithet, applied by 'Alee to worldly prosperity (الدُّنْيَا). (L.)

مَحِيدٌ an inf. n. of حَادَ. (Ḳ.) — [It may also be used, agreeably with analogy, as a noun of place, signifying *A place to which one turns aside* or *away*; *to which one removes, goes away*, or *goes far away.*]

حير

1. حَارَ, [sec. pers. حِرْتَ,] aor. يَحَارُ, (Ṣ, A, Mgh, Mṣb, Ḳ,) and some say يَحِيرُ, but this is a mistake, (MF,) inf. n. حَيْرَةٌ (Ṣ, A, Mgh, Ḳ) and حَيْرٌ (Ṣ, Mṣb, Ḳ) and حَيَرٌ and حَيَرَانٌ, (Ḳ,) *He was*, or *became, dazzled by a thing at which he looked*, (T, Mṣb, Ḳ,) *so that he turned away his eyes from it*: this is the primary signification: (T, Mṣb:) and so ▼تحيّر (A,* Mgh,* Ḳ) and ▼استحار, (Ḳ,) and حَارَ بَصَرُهُ (A,* TA) and ▼تحيّر بصره. (Mgh, and Ṣ and A and Ḳ in art. قمر, &c.) — And hence, (T, Mṣb,) *He was*, or *became, confounded*, or *perplexed, and unable to see his right course*; (T, Mṣb, Ḳ,* TA;) as also ▼تحيّر (Mṣb, Ḳ) and ▼استحار. (Ḳ.) And حَارَ, (Ṣ, A,) or حَارَ فِى أَمْرِهِ, (Mṣb,) *i. q.* تحيّر فى امره [*He was*, or *became, confounded*, &c., *in his affair*, or *case*]. (Ṣ, A.) And [حَارَ (see its part. n. حَائِرٌ) and] ▼تحيّر [and ▼استحار] *He erred*, or *lost his way.* (TA.) — Also, said of water, (A, Mṣb, Ḳ,) and ▼تحيّر (Ṣ, A, Ḳ) and ▼استحار, (A, Ḳ,) ‡ *It became collected*, (Ṣ, A, Ḳ,) *and stayed*, (A,) or *went round*, (Ṣ, Ḳ,*) or *went to and fro*, or *fluctuated*, (Mṣb, Ḳ,) *in a place, as*

though it knew not which way to run. (A.) — See also 5.

2. حَيَّرَهُ *He,* or *it, caused him to become confounded,* or *perplexed, and unable to see his right course.* (Ṣ,* Mṣb, KL.) — [Accord. to Golius, as on the authority of the KL, حَيَّرَ, said of water, means † *It was whirled round in an eddy:* but to have this meaning, which I do not find in my copy of the KL, the verb should be تَحَيَّرَ.]

4. احار [*He,* or *it, caused* a thing *to descend easily down the throat:* or *it transmitted* food *to the stomach*: see 10: and see also 4 in art. حور]. (Ṣ and Ḳ voce مِشْفَرٌ, q. v.)

5. تَحَيَّرَ: see 1, in six places. — Also ‡ *It* (a cloud) *continued without motion, pouring forth its rain, and not being driven by the wind:* (IAạr:) or *went not in any direction:* (Ḳ:) [and so ▼ استحار: see مُتَحَيِّرٌ.] — Also † *It continued;* said of time; (TA;) and in like manner it is said of a man. (MF.) And ▼ حِيرُوا بِهِ [if not a mistranscription for تَحَيَّرُوا] occurs as meaning † *Remain ye therein;* referring to a place. (TA.) And ▼ استحار بِمَكَانٍ † *He alighted and abode some days in a place.* (TA.) — تحيّر بالمآءِ ‡ *It* (a place, Ṣ, Ḳ, and land, TA) *became full of water;* as also ▼ استحار. (Ṣ, Ḳ, TA.) — تَحَيَّرَتِ الجَفْنَةُ ‡ *The bowl became full of grease and food;* (Ḳ, TA;) like as a watering-trough or tank becomes full of water. (TA.) — See also what follows.

10: see 1, in four places: — and 5, in three places. — استحار الشَّبَابُ (Ṣ, IB, A, Ḳ) and ▼ تحيّر (Ḳ) ‡ *The sap* [or *vigour*] *of youth* (مَآءُ الشَّبَابِ) *flowed:* (IB:) or *became complete, and filled the body* of a woman: (A:) or *completely occupied the body:* (Ḳ:) or *filled it to the utmost:* (TA:) or *collected, and flowed to and fro, in the body* of a woman. (Aṣ, Ṣ.) = أُسْتُحِيرَ الشَّرَابُ *The beverage, or wine, was made to descend easily down the throat.* (Ṣ.)

حَيْرٌ [*An enclosure*] *like a* حَظِيرَة: or *a place of pasturage in which it is prohibited to the public to pasture their beasts.* (Ṣ, Ḳ.) — See also حَائِرٌ. = حَيْرَمًا [erroneously written by Golius حَارَمًا] *i. q.* دَهْرًا. (Ḳ.)

إِنَّهُ فِى حَيْرَ بَيْرَ and حِيرِ بِيرِ, like حُورِ بُورٍ; (Ḳ;) i. e. *Verily he is in a bad state,* and *a state of perdition:* or *in error.* (TA.) [See also art. حور.]

حَيَرٌ: see what next follows.

حِيَرٌ (IAạr, Ḳ) and ▼ حَيَرٌ (IB, Ḳ) *Much property,* or *many cattle;* and *a numerous family:* (Ḳ:) and أَنْعَامٌ حِيَرَاتٌ *many cattle.* (TA.) كَانَ حِيَرًا [app. for كَانَ ذَا حِيَرٍ] is expl. by Th as meaning *He was a possessor of much property, and of a numerous household and family.* (TA.)

حَيْرِىُّ الدَّهْرِ: see حِيرَ دَهْرٍ.

حَارَةٌ: see art. حور.

أَصْبَحَتِ الأَرْضُ حَيْرَةً *The land became green with plants* or *herbage,* (Ḳ,) *by reason of much collecting and continuance of water therein.* (TA.)

حَارِىٌّ *Made in the town of El-Ḥeereh:* applied to a sword, and a camel's saddle. (TA.) And *A kind of leathern housings, made in El-Ḥeereh, with which camels' saddles are ornamented.* (TA.) = حَارِىُّ الدَّهْرِ and حَارِىَ دَهْرٍ: see what next follows.

لَا آتِيهِ حَيْرِىَّ الدَّهْرِ (Ibn-'Omar,* Sh,* Ḳ) and حِيرِىَّ الدَّهْرِ (Sb, Akh, IAạr, Ḳ) and حِيرِىَ دَهْرٍ (Ṣ,) or حِيرِى دَهْرٍ, (CḲ,) or حَيْرِى دَهْرٍ, (Ḳ, TA,) with the last letter quiescent, (Ḳ,) and حِيرِىَ دَهْرٍ, or حَيْرِىَ دَهْرٍ, (accord. to different copies of the Ḳ,) and ▼ حَارِىَّ دَهْرٍ (ISh, Ḳ) and ▼ حَارِىَّ الدَّهْرِ (ISh) and ▼ حِيرَ دَهْرٍ, (IAạr, Ḳ,) ‡ [*I will not come to him,* or *it,* or *I will not do it,*] *while time lasts;* (A,* Ḳ,* TA;) or *ever:* (ISh, Ḳ:) or it may mean *while time returns;* from حَارَ of which the aor. is يَحُورُ. (A, TA.) Also حِيرِىَّ الدَّهْرِ, or حَيْرِىَّ الدَّهْرِ, ‡ *For an incalculable period of time.* (Ibn-'Omar, Sh, IAth.)

حَيْرَانُ (T, Ṣ, A, Ḳ) and ▼ حَائِرٌ (T, A, Ḳ) and ▼ مُتَحَيِّرٌ (TA) A man *in a state of confusion,* or *perplexity, and unable to see his right course:* (Ḳ,* TA:) *erring; having lost his way:* (T, TA:) fem. [of the first] حَيْرَى (Lḥ, T) and حَيْرَآءُ: (A, Ḳ:) and pl. [of the same] حَيَارَى (Ṣ, A, Ḳ) and حُيَارَى (Ḳ) and حَيْرَى, like the fem. sing. (Lḥ.) You say, لَا تَفْعَلْ ذٰلِكَ أُمُّكَ حَيْرَى [*Do not thou that: may thy mother become in a state of confusion, &c.*]: and لَا تَفْعَلُوا ذٰلِكَ أُمَّهَاتُكُمْ حَيْرَى [*Do not ye that: may your mothers become &c.*]. (Lḥ.) And رَجُلٌ حَائِرٌ ▼ بَائِرٌ *A man who does not apply himself rightly to an affair;* (Ṣ, TA;) *who knows not the right course to pursue in his affair;* as also ▼ مُتَحَيِّرٌ فِى أَمْرِهِ. (TA. [See also the same phrase in art. حور.]) — رَوْضَةٌ حَيْرَى ‡ *A meadow full of water.* (TA.) — [حَيْرَى is also applied as an epithet to the midday sun of summer: see a verse cited in the second paragraph of art. دوم.]

حَيِّرٌ: see مُتَحَيِّرٌ.

حَائِرٌ: see حَيْرَانُ, in two places. — Also ‡ *A place in which water collects* (Ṣ, Ḳ, TA) *and goes to and fro:* (TA:) *a watering-trough,* or *tank, to which a stream of rain-water flows:* (Ḳ:) or *what resembles a watering-trough,* or *tank, in which the rain-water collects and remains:* (A:) *a depressed place* (Ḳ, TA) *in which water collects and remains,* or *goes round,* or *goes to and fro, not passing forth from it:* (TA:) or *a place in the ground depressed in the middle and having elevated edges or borders,* (AḤn, TA,) *in which is water:* (TA voce يَعْبُوبٌ:) and hence, (TA,) *a garden;* as also ▼ حَيْرٌ; (Ḳ;) which is the form used by most persons, and by the vulgar; like as they say عَيْشَةٌ for عَائِشَةٌ: or this form is wrong: it is disallowed by AḤn, notwithstanding its being mentioned by A 'Obeyd; but he mentions it only in one place, and it is not found in every copy of his work: (ISd:) pl. حِيرَانٌ (Ṣ, A, Ḳ) and حُورَانٌ. (Ṣ, Ḳ.) Ḥassán Ibn-Thábit uses the phrase حَائِرُ البَحْرِ [in a verse which I have cited in the first paragraph of art. ربّ, app. as meaning † *The depth of the sea;* or *part of the sea in which is a confluence of the water, and where it goes round,* or *to and fro*]. (TA.) = Also *Grease; oily animal matter, that flows from flesh or fat.* (Ḳ.)

أَحْيَرُ مِنْ ضَبٍّ, and أَحْيَرُ مِنْ وَرَلٍ, [*More confounded,* or *perplexed, and unable to see his right course, than a ḍabb,* and *than a waral,*] are two proverbs; (Meyd;) accord. to Ḥamzeh El-Iṣfahánee, said because the ḍabb, [a kind of lizard, as is also the waral,] when it quits its hole, is confounded, and cannot find the right way to to it; and the like is said of the waral. (Ḥar p. 166.)

مُتَحَيِّرٌ: see حَيْرَانُ, in two places. — الكَوَاكِبُ المُتَحَيِّرَةُ † [*The erratic stars;* i. e. *the planets;*] *the stars that* [*at one time appear to*] *retrograde and* [*at another time to*] *pursue a direct* [*and forward*] *course;* also called الخُنَّسُ. (Ṣ in art. خنس.) — سَحَابٌ مُتَحَيِّرٌ † *Clouds continuing without motion, pouring forth rain, and not driven by the wind:* (IAạr:) and ▼ مُسْتَحِيرٌ † clouds (سحاب) *heavy, and moving to and fro,* (Ṣ, Ḳ) *not having any wind to drive them along:* (Ṣ:) and ▼ حَيِّرٌ ‡ *clouds,* or *clouds covering the sky,* syn. غَيْمٌ, (AZ, Ḳ, TA,) *rising with rain, and continuing without motion,* or *moving to and fro, but remaining, in the sky:* (AZ, TA:) or this last signifies ‡ *clouds* (سحاب) *raining, and continuing without motion,* or *moving to and fro, but remaining, in the sky.* (A, TA.) — See also what follows, in two places.

مُسْتَحِيرٌ *A way leading across a desert, of which the place of egress is not known.* (Ḳ.) — † Anything (TA) *continuing endlessly:* (IAạr, TA:) or *hardly,* or *never, ending;* as also ▼ مُتَحَيِّرٌ. (Sh, TA.) See also this latter word. = جَفْنَةٌ مُسْتَحِيرَةٌ ‡ *A full bowl:* (A:) or † *a bowl containing much grease.* (Ḳ.) And ▼ مَرَقَةٌ مُسْتَحِيرَةٌ † *Broth containing much grease.* (TA.)

حيز

1. حَازَهُ, aor. يَحِيزُهُ, inf. n. حَيْزٌ: see 1 in art. حوز, in three places.

Quasi 5. تَحَيَّزَ: see 5, and 7, and Q. Q. 2, in art. حوز; the first in six places.

حَيْزٌ and حَيِّزٌ: see art. حوز.

حيس

1. حَاسَ, aor. يَحِيسُ, inf. n. حَيْسٌ, *He mixed* [a thing or things]. (Ṣ, A, Ḳ; but in this sense, only the inf. n. is mentioned.) — *He made,* or *prepared,* what is called حَيْسٌ: (Ṣ, Mṣb, Ḳ:)

and ▼ حيّس, inf. n. تَحْيِيس, *he mixed and prepared* what is so called. (TA.)

2 : see above.

حَيْس, originally an inf. n., (Mṣb,) *Dates mixed with clarified butter and [the preparation of dried curd called]* اقط, (Ṣ, A, Mgh, Ḳ,) *and kneaded* (A, Ḳ) *vehemently,* (A,) or *rubbed and pressed with the hand until they mingle together,* (Mgh,) *whereupon their stones come forth;* (Ḳ;) *and sometimes* سَوِيق *[or meal of parched barley or wheat] is put into it;* (A, Ḳ;) *and a little crumbled bread instead of the* اقط: (TA:) or *dates, of the kind called* بَرْنِىّ, *and* اقط, *bruised together, and kneaded vehemently with clarified butter until the stones come forth from it one by one, and then made like* ثَرِيد: it is the same as وَطْبَة, except that حيس sometimes has سويق put in it, but وطبة has not: (L:) or *dates of which the stones have been taken out, bruised with* اقط, *and then kneaded, and rubbed and pressed with the hand until the whole becomes like* ثريد; *and sometimes* سويق *is put with it:* (Mṣb:) accord. to Ibn-Waddáh El-Andalusee, *dates of which the stones have been taken out, mixed with* سويق; but this is not known, (MF, TA,) because of the deficiency of the ingredients: (TA:) Hr is related to have described it as *a* ثَرِيدَة *composed of* أَخْلَاط *[or various mixtures].* (TA.) A rájiz says,

* اَلتَّمْرُ وَالسَّمْنُ مَعًا ثُمَّ الأَقِطْ *
* اَلْحَيْسُ إِلَّا أَنَّهُ لَمْ يَخْتَلِطْ *

[*Dates and clarified butter together, then* اقط, *are* حيس, *except that it is not yet mixed*]: (Ṣ, MF, TA:) from which it might be understood, that these components, when mixed, are not حيس: but this is the contrary of what is meant: (MF:) the meaning seems to be, that these three things, when brought, are virtually حيس, as being the materials thereof, though not mixed. (TA.)

حيص

1. حَاصَ عَنْهُ, aor. يَحِيصُ, inf. n. حَيْصٌ and حُيُوصٌ and مَحِيصٌ and مَحَاصٌ (Ṣ, Mṣb, Ḳ) and حَيَصَانٌ (Ṣ, Ḳ) and حَيْصُوصَةٌ (TA) and ▼ حَيْصَةٌ, (Ḳ,) [or the last is an inf. n. of un.,] *He turned away from him, or it:* (Ṣ, A, Mṣb, Ḳ:) *and he returned, or went back, and fled, from him, or it:* (TA:) and ▼ انحاص عنه signifies the same; (Ṣ, Ḳ, TA;) and so does ▼ تحايص عنه: (TA:) or (Ḳ) one says of friends, (Ṣ, Ḳ,) حَاصُوا, (Ḳ,) or حاصوا عَنِ الْعَدُوِّ *[they turned away from the enemy]*; (Ṣ;) [and in like manner, انْحَازُوا;] and of enemies, انْهَزَمُوا; (Ṣ, Ḳ;) [and in like manner, وَلَّوْا مُدْبِرِينَ.] You say also, حَاصَ عَنِ الْقِتَالِ *He turned away from the fight.* (A.) And حَاصَ عَنِ الْحَقِّ *He turned away from the truth; he deviated from it.* (Mṣb.) And حَاصَ عَنِ الشَّىْءِ *He turned away from the thing, and became safe from it.* (TA.) And it is said in a trad. respecting the battle of Ohod, فَحَاصَ الْمُسْلِمُونَ ▼ حَيْصَةً *And the Muslims made a wheel away, desiring flight:* or, accord. to one relation, فَجَاضَ جَيْضَةً, which means the same. (TA.) حَيْصٌ also signifies The *retiring,* or *going back,* from a thing. (Ṣ, TA.) [See also مَحِيصٌ, below.]

3. حَايَصَهُ, (Ḳ,) inf. n. مُحَايَصَةٌ, (TA,) *He acted towards him with artifice, or guile;* (Ḳ;) *vied with him;* (TA;) *and strove to overcome him.* (Ḳ.) Hence the saying of Mutarrif, related in a trad., when he was asked respecting his going forth from the pestilence, هُوَ الْمَوْتُ نُحَايِصُهُ وَلَا بُدَّ مِنْهُ, may be interpreted as meaning, [*It is death:*] *we are eager to flee from it* [*though there is no avoiding it*]. (AO, TA.)

6 : see 1.

7 : see 1.

وَقَعَ فِى حَاصِ بَاصِ : see what next follows.

وَقَعَ, (AA, Ṣ, A, and Ḳ in art. بيص,) and وَقَعُوا, (Ṣ,) فِى حَيْصَ بَيْصَ, (Ṣ, A, and Ḳ ubi suprà,) and حِيصَ بِيصَ, (Ṣ in this art. and in art. بيص; and so in the CḲ ubi suprà, and in a MS. copy of the Ḳ, [though app. contradicted by what follows in that work, as will be seen below,] or حِيصٍ بِيصٍ, (Ḳ ubi suprà, accord. to some copies, and so in the TA,) and حَيْصٍ بَيْصٍ, (Ḳ ubi suprà,) with fet-ḥ to the first [letter] of each and to the last of each, (Ḳ ubi suprà,) and with kesr to the first of each, (Ṣ in art. بيص,) or to both, and with fet-ḥ to the first of each and kesr to the last of each; and sometimes each of the two [vars., namely, حَيْصَ بَيْصَ and حِيصَ بِيصَ,] is made perfectly decl. in the second [word], (Ḳ ubi suprà,) [so that you say also حَيْصَ بَيْصٍ, and حِيصَ بِيصٍ, and حَيْصِ بَيْصٍ; (though the copies of the Ḳ differ in respect of these forms, two, for instance, giving one form, which is written حَيْصِ بَيْصٍ, and one adding حِيصِ بِيصٍ;) for it is said,] the whole make six dial. vars.; and, accord. to MF, each of the two [vars.] is sometimes made perfectly decl. in the first [word] also, [so that you say حَيْصٍ بَيْصٍ, and حِيصٍ بِيصٍ,] but this he may have inferred from what will be afterwards mentioned on the authority of ISk, (TA in art. بيص,) and فِى حَاصِ ▼ بَاصِ, (Ḳ ubi suprà,) indecl., with kesr for the termination, the ا being [originally] ى; (TA ubi suprà;) *He fell,* (Ṣ, Ḳ,) and *they fell,* (Ṣ,) *into confusion in respect of their case, or affair, from which there was no escape for them:* (Ṣ and Ḳ ubi suprà:) or *into straitness and difficulty:* (Ṣ:) حَيْصَ بَيْصَ and حِيصَ بِيصَ are each two nouns made into one, and made indecl. with fet-ḥ for their termination, as in the instance of جَارِى بَيْتَ بَيْتَ: or, as some assert, they are two nouns, from حيص meaning the "turning away," and "retiring," or "going back," and بَوْص meaning the "outstripping," and "fleeing;" and بوص is altered to assimilate it to حيص; and the meaning is, *an affair,* or *a case, of any kind, from which one retires, or goes back, and flees.* (Ṣ.) You say also, جَعَلْتُمُ الأَرْضَ عَلَيْهِ حَيْصَ بَيْصَ, (Ṣ and Ḳ, both in art. بيص,) or حِيصَ بِيصَ, (Ṣ ubi suprà,) and حَيْصًا بَيْصًا, (ISk, and Ḳ ubi suprà,) with fet-ḥ to each, and حِيصًا بِيصًا, with kesr to each, not compounded, (ISk, and ṬA ubi suprà,) *Ye have straitened* [*the earth, or land,*] *to him,* (Ṣ and Ḳ ubi suprà,) *so that he may not act as he pleases therein:* (Ḳ:) or *so that he may not travel therein in search of sustenance, nor employ himself as he would to make gain.* (Nh.) And إِنَّكَ لَتَحْسَبُ عَلَىَّ الأَرْضَ حَيْصًا بَيْصًا, or, as some say, حِيصَ بِيصَ, [i. e., حَيْصَ بَيْصَ or حِيصَ بِيصَ,] meaning, *Verily thou thinkest the earth to be straitened to me, so that I may not act as I please therein:* &c.] (Ṣ.) — حَيْصَ بَيْصَ also signifies *The hole of the rat or mouse.* (TA in art. بيص.)

حَيْصَة : see 1, in two places.

حَيُوصٌ A beast (دَابَّة) *that takes fright, and runs away at random;* (Ḳ;) *turning away from that which its master desires:* (TA:) a mule *evil in disposition.* (TA.)

مَحِيصٌ [an inf. n.: (see 1:) and also a n. of place, signifying *A place to which one turns away,* or *aside; to which one flees; a place of refuge:*] syn. مَحِيدٌ (Ṣ, Ḳ) and مَعْدِلٌ (Mṣb, Ḳ) and مَمِيلٌ (Ḳ) and مَهْرَبٌ. (Ṣ, Ḳ.) You say, مَا عَنْهُ مَحِيصٌ [*There is no turning away,* &c., or *no place to which to turn away,* &c., *from it*]. (Ṣ.)

حيض

1. حَاضَتْ, (Ṣ, A, Mṣb, Ḳ,) aor. تَحِيضُ, inf. n. حَيْضٌ and مَحِيضٌ (Ṣ, Mṣb, Ḳ) and مَحَاضٌ, (Aboo-Is-ḥáḳ, Ḳ,) said of a woman, (Ṣ, A, Mṣb, Ḳ,) *She menstruated;* i. e. *her blood flowed;* (A;) or [rather] *blood came forth from her womb;* (Mgh;) [*not in consequence of disease nor of childbirth, nor before she had attained to puberty;* as explained in the law-books of the Muslims;] as also ▼ تَحَيَّضَتْ: or this latter signifies *she likened herself to the* حَائِض. (TA.) — Also *She attained the age of menstruation.* (TA.) [See حَائِض.] — حَاضَتِ السَّمُرَةُ, (Ṣ, A, Mṣb,) aor. as above, (Mṣb,) inf. n. حَيْضٌ, (Ṣ, Mṣb,) ‡ *The* سمرة [*or gum-acacia-tree*] *flowed with,* (Ṣ,) or *emitted,* (A,) *a matter resembling blood;* (Ṣ, A;) *flowed with its gum:* (Mṣb:) the matter which it emits is called دُودِم; and is applied to the head of a new-born infant to scare away the jánn, or genii. (A, TA: [but in a copy of the A, and in one instance in the TA, الشَّجَرَةُ is put in the place of السَّمُرَةُ, app. by mistake.]) [I have marked this signification as tropical on the authority of the A and TA; but the author of the Mṣb has commenced the art. with it; and shows that he held the opinion, which some others have shared with him, that it is the primary signification.] — حَاضَ السَّيْلُ † *The torrent overflowed;* or *poured out,* or *forth, from fulness;* or *ran;* syn. فَاضَ. (TA.)

2. حَيَّضَ الْمَرْأَةَ *He attributed* حَيْض [*or menstruation*] *to the woman.* (Mṣb.) — حَيَّضَ جَارِيَتَهُ, (TḲ,) inf. n. تَحْيِيض, (Ṣgh, Ḳ,) *Tempore men-*

struorum inivit ancillam suam. (Ṣgh, Ḳ, TḲ.) — حَيَّضَ المَآءَ, (TḲ,) inf. n. as above, (Ḳ,) † *He made the water to flow.* (Ḳ, TA.)

5. تَحَيَّضَتْ : see 1. — Also *She abstained from prayer* (الصلاة) *during the days of her* حَيْض [or *menstruation*]; (Ṣ, Mṣb, Ḳ, TA;) *waiting for the stopping of the blood:* (TA:) or *she abstained, and did as the* حائض *does:* (A,* Mgh:) or *she reckoned herself* حائض, *and did as the* حائض *does.* (TA.)

10. اُسْتُحِيضَتْ, (Ṣ, Mgh, Mṣb,) in the pass. form, (Mṣb,) with ḍamm to the ت, (Mgh,) [as though originally signifying *She was reckoned to be menstruating,*] found in the handwriting of Aboo-Zekereeyà اِسْتَحَاضَتْ, (TA,) [which I hold to be a mistake, as being at variance with general usage,] *She continued to have a flow of blood* (Ṣ, Mgh) *after her days* [*of menstruation*]: (Ṣ:) or *she had an exuberance of blood* [*flowing from the vagina*]; *not what is termed* الحَيْض : (Mṣb:) or *her blood flowed without stopping, not on certain days, nor from the vein* [or *veins*] *of menstruation, but from a vein called* العَاذِل. (TA.)

حَيْضٌ [an inf. n. of 1: explained in the KT as applied to The *menstrual blood* itself; which seems to have been more properly called حِيضَةٌ and مَحِيضٌ and حِيَاضٌ : though what here follows may be considered as rendering it probable that حيض was also used in this sense in the classical times, for دَمُ حَيْضٍ]. — حَيْضُ السَّمُرِ † *A thing which flows from the* سَمُر [or *gum-acacia-trees*], *resembling* [*what is called*] دَمُ الغَزَالِ. (Mgh.) [See also 1.]

حَيْضَةٌ *A single time,* or *turn, of menstruation,* or *of the flow of the menstrual blood:* (Ṣ,* A,* Mgh, Mṣb:*) pl. حِيَضٌ; (A, Mṣb;) like as بِدَرٌ is pl. of بَدْرَةٌ, and ضِيَعٌ of ضَيْعَةٌ, and حِيَدٌ of حَيْدَةٌ, and خِيَمٌ of خَيْمَةٌ; though by rule it should be حَيْضَاتٌ. (Mṣb.) You say, حَاضَتْ حَيْضَةً وَاحِدَةً [*She menstruated one single time of menstruation*]: and حَيْضَةً طَوِيلَةً [*a long single time thereof*]: and ثَلَاثَ حِيَضٍ [*three single times thereof*]. (A.) — As used by the professors of practical law, The *accustomed days thereof.* (Mgh.) — Also † *A single flow* [of water &c.]: pl. حَيْضَاتٌ. (TA.)

حِيضَةٌ [*Menstruation;*] the subst. from حَاضَتِ المَرْأَةُ : (Ṣ, Ḳ, TA:) or *a mode,* or *manner,* or *state, of* حَيْض [or *menstruating*]: (Mṣb:) or the *state* (Mgh, TA) *of the* حَائِض, (TA,) *which is one of avoidance* (Mgh, TA) *of prayer and fasting and the like:* (Mgh:) pl. حِيَضٌ. (Ṣ, Mṣb.) — Also The *menstrual blood;* the *blood of menstruation;* and so ↑ مَحِيضٌ and ↑ حِيَاضٌ. (TA.) [See also حَيْضٌ.] — Also The *piece of rag which the* حَائِض *binds over her vulva;* (Ṣ, Mgh,* Mṣb,* Ḳ;) and so ↑ مَحِيضَةٌ : (Ṣ:) which latter also signifies † *a piece of rag thrown away:* (TA:) pl. of the latter, مَحَايِضُ. (Ṣ, TA.)

[حَيْضِيٌّ *Menstrual; of,* or *relating to, menstruation.*]

حِيَاضٌ : see حِيضَةٌ.

حَائِضٌ, applied to a woman, [*Menstruating;*] act. part. n. from حَاضَتْ; (Ṣ, Mgh, Mṣb, Ḳ;) thus, [without ة,] because it is an epithet of particular application [to a female]; (Mṣb;) and with ء, being like قَائِمٌ and ضَائِرٌ &c.; (TA;) [because the ى in its verb suffers alteration;] and in like manner حَائِضَةٌ also, (Ṣ, Mgh, Mṣb, Ḳ,) on the authority of Fr: (Ṣ:) pl. (of the former, Mṣb) حُيَّضٌ, (Ṣ, Mgh, Mṣb, Ḳ,) like as رُكَّعٌ is pl. of رَاكِعٌ, (Mṣb,) and حَاضَةٌ, like as حَاكَةٌ is pl. of حَائِكٌ, (TA,) and of the latter, حَائِضَاتٌ, (Mṣb,) or حَوَائِضُ. (Ṣ, Mgh, Ḳ.) — In a certain trad., in which it is said that God will not accept the prayer of a حائض unless she be [attired] with a خِمَار [or head-covering], this does not mean one who is menstruating while actually occupied by prayer, (Mṣb, TA,*) but † *One who has attained to the age of menstruation;* (TA;) or *one who has attained to womanhood:* (Mgh:) or it has not this meaning; for if it had, one would understand that a girl not arrived at puberty might pray with her head uncovered, which is not the case; but it means ‡ [*one of*] *the menstruating kind, whether she have attained to puberty or not;* as though the term female had been used in its place. (Mṣb.)

مَحِيضٌ is a simple subst. as well as an inf. n.: (Zj, Ḳ:) as the former, it is a n. of place; and as such it is [said to be] used in the Ḳur ii. 222; meaning *A woman's* مَأْتًى; (Zj;) *her* فَرْج; (Mgh;) because it is the *place of* الحَيْض. (Zj, Mgh.) Some say that حَوْضٌ is hence derived; because the water flows to the حوض: (Az, Ḳ:) for the Arabs put و in the place of ى, and ى in that of و. (Az, TA.) — It is also a n. of time [signifying The *time of menstruating*]. (TA.) — See also حِيضَةٌ. — When it is a simple subst., it has a pl., namely مَحَايِضُ. (TA.)

مَحِيضَةٌ : see حِيضَةٌ.

مُسْتَحَاضَةٌ A woman *continuing to have a flow of blood after her days* [*of menstruation*]: (Ṣ:) or *having an exuberance of blood* [*flowing from her vagina*]; *not what is termed* الحَيْض : (Mṣb:) or *having her blood flowing without stopping, not on certain days, nor from the vein* [or *veins*] *of menstruation, but from a vein called* العَاذِل. (Ḳ,* TA.)

حيط

8. اِحْتَاطَ : حَيْطَةٌ : تَحْيِيطٌ and تَحِيطٌ and التَّحِيطُ : تَحِيطٌ :	see art. حوط.

حيعل

Q. Q. 1. حَيْعَلَ, (Ṣ in art. هل,) inf. n. حَيْعَلَةٌ, (Ḳ,) *He* (a مُؤَذِّن) *said,* حَيَّ عَلَى الصَّلَاةِ حَيَّ عَلَى الفَلَاحِ : (Ḳ, TA:) composed of two words, like حَوْقَلَ. (Ṣ ubi suprà, and TA.)

حيف

1. حَافَ, aor. يَحِيفُ, (Ṣ, Mṣb,) inf. n. حَيْفٌ, (Ṣ, Mṣb, Ḳ,) *He* (a judge, or any other person, Mṣb) *acted wrongfully, unjustly, injuriously,* or *tyrannically,* (Ṣ, Mṣb, Ḳ,) عَلَيْهِ *against him:* (Ṣ:) or, as some say, *he inclined* [*to that which was wrong*], or *declined* [*from that which was right*], *in his judgment.* (TA.) حَيْفُ النَّاحِلِ means *A man's giving to certain of his children exclusively of others:* he should treat them equally, and not prefer some before others. (T, TA.)

2. تَحْيِيفٌ The *taking a thing,* or *something, from the side:* and *diminishing* [*from the side*]. (KL.)

5. تَحَيَّفْتُهُ *I took by little and little from its sides;* (Ṣ, Ḳ;) as also تَحَوَّفْتُهُ. (Ṣ.)

حِيفَةٌ *A side,* or *lateral part:* pl. حِيَفٌ; (Ḳ;) which is likewise pl. of حَافَةٌ. (TA.) [See also حَائِفٌ, below.]

حَائِفٌ *Acting wrongfully, unjustly, injuriously,* or *tyrannically:* (Lth, Mṣb, Ḳ: [in some copies of the Ḳ, حَائِرٌ is erroneously put for جَائِرٌ:] pl. حَافَةٌ (Mṣb, Ḳ) and حُيَّفٌ (Ḳ, Mṣb*) and حُيُفٌ. (TA.) — An arrow *declining from the right direction.* (TA.) — And, as being likened thereto, ‡ An *impotent man, who does not attain the object of his want.* (TA.) ═ The *side* of a mountain. (Ḳ, TA.) [See also حِيفَةٌ.]

أَرْضٌ حَيْفَآءُ, and بَلَدٌ أَحْيَفُ, *A district* or *the like,* and *a land, upon which rain has not fallen:* (Ibn-'Abbád, Ḳ:) as though the rain treated it wrongfully. (TA.)

حيق

1. حَاقَ بِهِ, aor. يَحِيقُ, (Ṣ, Ḳ,) inf. n. حَيْقٌ and حُيُوقٌ and حَيَقَانٌ, (Ḳ,) *It surrounded, encompassed, encircled,* or *beset, him,* or *it;* (Ṣ, Ḳ;) only used in relation to evil; (Bḍ in xvi. 36;) [as also حَاقَ بِهِ, aor. يَحُوقُ, inf. n. حَوْقٌ;] and so ↑ أَحَاقَ بِهِ. (Ibn-'Abbád, Ḳ.) So in the Ḳur [xxxv. 41], وَلَا يَحِيقُ المَكْرُ السَّيِّئُ إِلَّا بِأَهْلِهِ [*And evil artifice shall not beset any save the authors thereof*]: (Ṣ:) or this means *shall not befall:* (Mṣb:) [for] حَاقَ بِهِ, (Ibn-'Arafeh, Mṣb, Ḳ,*) aor. as above, (Mṣb, TA,) inf. n. حَيْقٌ and حَاقٌ, (TA,) signifies [also] *it clave to him,* and *became his due,* (Ibn-'Arafeh, Ḳ,) and *befell him:* (Ibn-'Arafeh, Mṣb, Ḳ:) and thus it is said to mean in the Ḳur [xi. 11, &c.], وَحَاقَ بِهِمْ مَا كَانُوا بِهِ يَسْتَهْزِئُونَ [*And that at which they used to mock shall cleave to them,* or *be their due,* or *befall them;* namely, the punishment at which they mocked.] (TA.) You say also, حَاقَ بِهِمُ العَذَابُ

Punishment beset them, and befell them. (Ṣ.) And حاق بِهِ البَلَآءُ *Trial, or trouble, beset him.* (TA in art. حق.) ═ حاق فِيهِ, (Ḳ,) inf. n. حَيْقٌ, (TA,) *i. q.* حالَ. (Ḳ.)

3. حايقهُ, (AA, Ḳ,) inf. n. مُحَايَقَةٌ, (AA, TA,) *He envied him, and hated him.* (AA, Ḳ.)

4: see 1. ═ احاق ٱللّٰهُ بِهِمْ مَكْرَهُمْ *God made their artifice to beset them:* (Lth, TA:) or *made it to befall them.* (Th, Ḳ,* TA.)

حَاقُ الجُوعِ *Vehemence of hunger.* (TA.) See also حَاقٌّ, in art. حق.

حَيْقٌ *What besets a man,* ('Eyn, Ḳ,) *and befalls him, of artifice,* ('Eyn, TA,) or *of an evil deed, that he has done.* ('Eyn, Ḳ.)

مَحْيُوقٌ: see art. حوق.

حيك

1. حَاكَ, aor. يَحِيكُ, inf. n. حَيْكٌ and حَيَكٌ and حِيَاكَةٌ, accord. to Lth, signifies *He wove* a piece of cloth: [and it is said in the Ḳ, in art. حوك, that the root of the verb in this sense is with و and with ى:] but Az says that this is a mistake; and that the verb is only حاك having for its aor. يَحُوكُ, inf. n. حَوْكٌ. (TA. [See, however, what follows.]) ═ حاكَ, (Ṣ, Ḳ,) or حاك فِى مِشْيَتِهِ, (TA,) aor. يَحِيكُ, inf. n. حَيَكَانٌ (Ṣ, Ḳ) and حَيْكٌ (Ḳ) and حَيْكَى (Mbr, TA) [and حِيَاكَةٌ, which see in what follows], *He* (a man, TA) *walked with an elegant and a proud and self-conceited gait, with an affected inclining of the body from side to side, or with a twisting of the back:* or *he moved about his shoulder-joints and his body in walking,* (Ḳ, TA,) *having much flesh;* which manner of walking in women is commended, but in men it is discommended; for the woman walks thus by reason of the largeness of her thighs; but the man, when his thighs, or legs, are wide apart: or, as some say, it signifies *he trod the ground vehemently:* (TA:) or *he moved about his shoulder-joints, and parted his legs widely, in walking,* [*as short persons do:* for] حَيَكَانٌ signifies the *manner of walking of him who is short:* (Ṣ:) or *a walking in which a man moves about his posteriors:* all which meanings are borrowed from the action of the حَائِكٌ [or weaver, who straddles when at work]: حِيَاكَةٌ, likewise, signifies *a walking with an elegant and a proud and self-conceited gait, with an affected inclining of the body from side to side,* or *with a twisting of the back, and in a lagging manner.* (TA.) And you say also, جَآءَ يَتَحَيَّكُ ▼, and ▼ يَتَحَايَكُ, meaning *He came walking with his legs parted as though there were something between them.* (TA.) ═ حاكَ, (Ḳ,) aor. يَحِيكُ, inf. n. حَيْكٌ, (TA,) said of a sword, (Ḳ,) and of an axe, (TA,) *It made an impression,* or *had effect;* as also ▼ احاكَ. (Ḳ, TA.) حاك فيه (Ṣ) and ▼ احاك فيه (Ṣ, Ḳ) and ▼ أَحَاكَهُ, (Ḳ,) said of a sword, signify the same: (Ṣ, Ḳ:) one says, ضَرَبْتُهُ فَمَا أَحَاكَ ▼ فِيهِ السَّيْفُ, i. e. [*He struck him, but the sword*] *made no impression,* or *had no effect, upon him.* (Ṣ, TA.) And حَاكَتِ الشَّفْرَةُ *The* [*knife called*] شفرة *cut;* as also ▼ احاكت. (Ḳ.) And مَا تَحِيكُ المُدْيَةُ اللَّحْمَ [*The butcher's knife does not cut the flesh-meat*], and ما تحيك فيه: both signify alike. (El-Ámidee, TA.) — [Hence,] حاك القَوْلُ فِى القَلْبِ, inf. n. حَيْكٌ, † *The saying took effect upon the heart;* (Sh, Ṣ, Ḳ, TA;) and *became fixed therein.* (Sh, TA.) And مَا يَحِيكُ فِيهِ المَلَامُ † *Blame does not make any impression upon him.* (Ṣ.) And مَا يَحِيكُ كَلَامُكَ فِى فُلَانٍ † *Thy speech does not make any impression upon such a one.* (TA.) And it is said [in a trad., as some read it], الإِثْمُ مَا حَاكَ فِى صَدْرِكَ وَكَرِهْتَ أَنْ يَطَّلِعَ عَلَيْهِ النَّاسُ † *Sin is that which makes an impression upon thy mind, and becomes fixed* [*therein, and with which thou dislikest that men should become acquainted*]. (Az, TA. [See also حَزَّ; and see حَاكَ.])

4: see 1, in five places.

5: } see 1.
6: }

8. احتاك, mentioned in this art. in the Ḳ: see 5 in art. حوك.

حَيْكَى and حِيكَى: see حَائِكٌ.

حَيَكَانَةٌ and حِيكَانَةٌ and حُيَكَانَةٌ: see حَائِكٌ. The first also signifies *A man who walks with his legs parted as though there were something between them.* (TA.) And *A bulky* [lizard such as is called] ضَبّ; *that moves about its shoulder-joints, and parts its legs widely, in going along;* (Ṣ;) as also the second and third. (Ibn-'Abbád, TA.)

حِيَاكَةٌ: see art. حوك.

حُيَيْكَةٌ *Short, and thick and compact in body;* applied to a woman. (Ibn-'Abbád, Ḳ.)

حَيَّاكٌ and حَيَّاكَةٌ: see حَائِكٌ; for the former, in two places. — The latter also signifies *A female ostrich;* as being likened to the حَائِك in her [manner of] walking. (TA.)

حَائِكٌ: see art. حوك. [In the present day, ▼ حَيَّاكٌ signifies the same; i. e. *A weaver.*] ═ Also, and ▼ حَيَّاكٌ, applied to a man; and ▼ حَيَّاكَةٌ and ▼ حَيَكَانَةٌ and ▼ حِيكَانَةٌ and ▼ حُيَكَانَةٌ, (Ḳ, TA,) and, accord. to the Ḳ, ▼ حَيْكَى, but this is an inf. n., and is here a mistake for ▼ حِيكَى, originally حُيْكَى, mentioned by Sb, (TA,) applied to a woman; *Walking, or who walks, in the manner denoted by the verb* حاكَ, i. e., *with an elegant and a proud and self-conceited gait,* &c. (Ḳ, TA.) ═ And the first, i. e. حَائِكٌ, *Becoming fixed in the heart, and disquieting one.* (Az, TA in art. حوك.) [See 1.]

حيل

1. حَالَ, aor. يَحِيلُ, inf. n. حُيُولٌ *It became altered,* or *changed:* (Ḳ:) a dial. var. of حالَ, aor. يَحُولُ, inf. n. حُؤُولٌ. (TA.) — حال المَآءُ, aor. as above, *The water remained,* or *stagnated, and collected;* or *remained long, and became altered;* or *became yellow and altered;* in the bottom of a valley. (TA.)

4. مَا أَحْيَلَهُ a dial. var. of مَا أَحْوَلَهُ. (Fr, Ṣ.) See 4 (last sentence) in art. حول.

5. تَحَيَّلَ: see 8 in art. حول.

حَيْلِ حَيْلِ *A cry with which goats are chidden.* (Ḳ.)

حَيْلٌ *Water that remains, or stagnates, and collects,* or *that remains long, and becomes altered,* or *that becomes yellow and altered, in the bottom of a valley:* pl. [of pauc.] أَحْيَالٌ and [of mult.] حُيُولٌ. (Ḳ.) ═ Also a subst. from الاِحْتِيَالُ; (Ṣ, Ḳ;) and so ▼ حِيلَةٌ, with kesr; (Ṣ;) or ▼ حَيْلَةٌ [perhaps a mistake for حِيلَةٌ]; (Ḳ;) and ▼ مَحَالَةٌ, and ▼ مَحَالٌ. (AZ, Ṣ.) [See حِيلَةٌ in art. حول.] — *Strength, power, might,* or *force;* syn. قُوَّةٌ; as also حَوْلٌ; (TA;) of which it is a dial. var. (Ṣ, Mṣb.) So in the saying, لَا حَيْلَ وَلَا قُوَّةَ إِلَّا بِٱللّٰهِ. (Ṣ,* Mṣb, TA. [See حَوْلٌ.]) So, too, in the phrase, in a form of prayer, اَللّٰهُمَّ ذَا الحَيْلِ الشَّدِيدِ [*O God, Possessor of great might*]: perverted by the relators of traditions into ذا الحَبْلِ, with ب. (TA.) If it be a contraction of حَيِّلٌ, originally حَيْوِلٌ, its proper place is art. حول: otherwise, this is its proper place. (TA.)

حَيْلَةٌ *A large number of goats:* (Ṣ:) or *a herd of goats:* and *a flock of sheep.* (Ḳ.) — *Stones rolled down from the side of a mountain to its bottom until they become many:* (Ḳ:) or *an overhanging mass of rock that falls down from the head of a mountain to its bottom.* (Abu-l-Mekárim, O.) ═ See also حَيْلٌ.

حِيلَةٌ: see حَيْلٌ, above; and see art. حول.

حِيلَى: see its syn. حُوَلٌ, in art. حول.

حَيْلُولَةٌ an inf. n. of حَالَ, [aor. يَحُولُ,] (Mgh and Mṣb in art. حول,) like كَيْنُونَةٌ [&c.]. (Mgh in that art. [See 1 in that art.])

حِيَالٌ; and حِيَالَهُ and بِحِيَالِهِ: and عَلَى حِيَالِهِ: see art. حول.

حَيِّلٌ: see 4 in art. حول, in the latter half of the paragraph.

حَيَّالٌ: see its syn. حُوَّلٌ, in art. حول.

أَحْيَلُ: see art. حول.

مَحَالٌ: } see حَيْلٌ, above; and see حِيلَةٌ in
مَحَالَةٌ: } art. حول.

حين

1. حَانَ, (Mṣb, Ḳ,) or حان حِينُهُ, (Ṣ,) aor. يَحِينُ, (Ṣ, Mṣb,) [inf. n. as in the exs. following,] *It,* (Mṣb, Ḳ,) or *its time,* or *season,* (Ṣ,) *was,* or *became,* or *drew, near;* or *was at hand:* (Ṣ, Mṣb, Ḳ:) and *its time came.* (Mṣb, Ḳ.*) You say, حَانَ لَهُ أَنْ يَفْعَلَ كَذَا, aor. as above, inf. n. حِينٌ [and as in the next ex.], *The time came,* or *drew near, for him to do,* or *that he should*

do, such a thing; syn. آنَ. (Ṣ.) And حَانَتِ الصَّلَاةُ, (Mṣb, TA,) inf. n. حَيْنٌ and حِينٌ and حَيْنُونَةٌ, *The time of prayer came:* (Mṣb:) or *the prayer was, or became, or drew, near.* (TA.) — حان السُّنْبُلُ *The ears of corn became dry,* (Ḳ, TA,) *so that the time of the reaping thereof came, or drew near.* (TA.) — حان حَيْنُ النَّفْسِ *The soul died, or perished.* (TA.) — And حان, inf. n. حَيْنٌ, *He* (a man) *died, or perished.* (Ṣ.) — *He experienced a trial, or trying affliction.* (Ḳ: a meaning indicated therein, but not expressed.) — Also, (Az, Ḳ,) aor. as above, inf. n. حَيْنٌ, (Az, TA,) *It* (anything) *was not accommodated, adapted, or disposed, to the right way or course or direction;* (Az, Ḳ, TA;) as also ▼ تحيّن. (Ḳ.) — And *i. q.* اِتَّفَقَ [*It happened,* &c.]. (Ḥar p. 382.)

2. حيّنهُ, [inf. n. تَحْيِينٌ,] *He assigned, or appointed, for him, or it, a time.* (Ḳ.) حيّنوا ضُيُوفَهُمْ and ▼ أَحَانُوهُمْ have the same meaning [app. *They assigned, or appointed, a time for their guests*]. (TA.) — حيّن النَّاقَةَ *He appointed for the she-camel a time in every day and night in which he should milk her;* (Ṣ, Ḳ;) as also ▼ تَحَيَّنَهَا; (Ḳ;) said when one milks her in the day and night once: Aṣ says that تَحْيِينٌ is like تَوْجِيبٌ [the *milking* a camel *but once in the course of each day and night*]; but is only after she has shown herself to be pregnant, and her milk has become little in quantity. (TA.) = *He* (God) *made him, or it, to be not accommodated, adapted, or disposed, to the right way or course or direction.* (Ḳ.) — See also 4.

3. مُحَايَنَةٌ and حِيَانٌ [are the inf. ns. of حَايَنَ]. You say, عَامَلَهُ مُحَايَنَةً (Ṣ, Ḳ) and حِيَانًا (Lḥ, TA) [*He bargained or contracted with him for work for a certain time*]; like مُسَاوَعَةً; (Ṣ, Ḳ;) from الحِينُ meaning الوَقْتُ. (Lḥ, TA.) And in like manner, اِسْتَأْجَرَهُ مُحَايَنَةً (TA) and حِيَانًا (Lḥ, TA) [*He hired him, or took him as a hired man or a hireling, for a certain time*].

4. احان *i. q.* أَزْمَنَ [i. e. *Time, or a long time, passed over him, or it; he, or it, endured, or continued, for a time, or for a long time*]. (TA.) — أَحْيَنَ *He remained, stayed, abode, or dwelt,* (Ṣ, Ḳ,) *for a time* in a place. (Ṣ.) — أَحْيَنَتِ الإِبِلُ *The time came, or drew near, for the camels to be milked: or for the camels to have their loads bound upon them.* (AA, Ḳ. [In the CḲ, يُعْلَمَ is erroneously put for يُعْكَمَ.]) — أَحْيَنَ القَوْمُ *The time of what they desired, or sought, came, or drew near, to the company of men:* (Ḳ:) *the time of their attainment of what they hoped for came, or drew near, to them.* (IAạr, TA.) = As a trans. verb: see 2. — احانهُ اللّٰهُ *God caused him to die, or destroyed him;* (Ṣ;) as also ▼ حيّنهُ, inf. n. تَحْيِينٌ. (KL: but only the inf. n. is there given.) — *God tried him, or afflicted him with a trial.* (Ḳ: a meaning indicated therein, but not expressed.)

5. تحيّن, said of a spunger (وَارِش), *He watched for the time of eating, in order that he might enter.* (Ṣ.) And تحيّن الطَّعَامَ [*He watched for the time of the food*]. (Ḳ voce حَضَرَ.) And تَحَيَّنْتُ رُؤْيَةَ فُلَانٍ *I watched for the time of seeing such a one.* (TA.) And تحيّن وَقْتَ الصَّلَاةِ *He sought* [*to know*] *the time of prayer.* (TA.) [And accord. to Freytag's Lex., ▼ استحان has a similar meaning; i. e. *He waited for the just time of a thing; delayed a thing till the fit time.*] — تحيّن النَّاقَةَ: see 2. = See also 1. = As meaning اِسْتَغْنَى, it is a vulgar word. (TA.)

10: see 5.

حَيْنٌ *Death; a state of destruction or perdition:* (Ṣ, Ḳ, TA:) or the *time of the appointed term* [*of life*]; or *time of death.* (Ḥar p. 322.) — *A trial, or trying affliction.* (Ḳ.)

حِينٌ *i. q.* دَهْرٌ [*Time; or a time; or a space, or period, of time;* &c.]: (Ḳ:) or, accord. to Esh-Sháfi'ee, *time, from the beginning of the world to its end;* as also دَهْرٌ: (Az voce دَهْرٌ:) or *a time,* (Az, Ṣ, Mgh, Mṣb, Ḳ,) *in a vague sense,* (Mgh, Ḳ,) applicable to *any time,* (Az, Ḳ,) *little or much,* (Mgh, Mṣb,) *long or short, that may be a year* and *more:* or [in some cases] particularly meaning *forty years:* or *seven years:* or *two years:* or *six months:* or *two months:* or *any morning and evening:* (Ḳ:) also *a space of time;* (Ṣ, Ḳ;) as in the Ḳur lxxvi. 1, (Ṣ,) and in the Ḳur xxxvii. 178: (Ḳ:) and *a continuous time:* (Ḥam p. 381:) and the *day of resurrection;* (Ḳ;) or the *coming to pass of the resurrection;* as in the Ḳur xxxviii. last verse: (Mgh, TA:) or it has two meanings; namely, *a time of unknown limit,* and also, as in the Ḳur xiv. 30, *six months:* (Fr, Mṣb:) accord. to Er-Rághib, the *time of* a thing's *arriving,* or *coming,* and *happening;* having a vague meaning, and rendered particular, or special, by that to which it is prefixed: some say that it occurs applied in different ways: to *an appointed term;* as that to which God makes one to live: and *a year;* as in the Ḳur xiv. 30: and to the *time* when an event takes place; as in the Ḳur xxx. 16: and to *time absolutely:* accord. to El-Munáwee, in the [genuine] language of the Arabs, it is applied to [the *time of*] *a glance of the eye,* and *more than that,* to *time without end:* (TA:) the pl. is أَحْيَانٌ, (Ṣ, Mṣb, Ḳ,) and pl. pl. أَحَايِينُ, (Ṣ, * Ḳ;) as in the saying, فُلَانٌ يَفْعَلُ كَذَا أَحْيَانًا and فِى الأَحَايِينِ [*Such a one does so at times, or sometimes*]. (Ṣ.) In the Ḳur [xiv. 30], تُؤْتِى أُكُلَهَا كُلَّ حِينٍ means [*Which yieldeth its fruit*] *every six months:* (Fr, Mṣb, TA:) or *every year:* or *every morning and evening:* or, accord. to Az, *in every season, uninterruptedly.* (TA.) ▼ مِحْيَانُ شَىْءٍ, also, means حِينُهُ [*The time, or season, of a thing*]. (Ḳ.) [You say, إِلَى حِينٍ *For a time, or season.*] And حِينًا *At one time; sometime; at some time; awhile.* (Mgh.) حِينَ in the phrase قُمْتُ حِينَ قُمْتَ [*I stood in the time when thou stoodest,* or *I stood when thou stoodest,*] is an adv. n. of time; (Mṣb;) [see also an ex. in a verse cited voce حَشَفَ, and the remarks there subjoined:] and one may well employ in its place لَمَّا and إِذَا (Mṣb, TA) and إِذْ and مَتَى and سَاعَةَ (TA) and وَقْتَ (Mṣb, TA) and the like; but not, as many have said, حَيْثُ; for this is an adv. n. of place. (Mṣb.) — When they make the two times to be distant, the one from the other, [i. e. the time of speaking and the time spoken of,] they do so by means of إِذْ, and thus, (Ḳ,) they say حِينَئِذٍ [meaning *At that time; then*]: (Ṣ, Ḳ:) and sometimes they suppress the ء, substituting for it ى. (TA.) — Sometimes, also, they prefix تَ to حِينَ; (Ṣ, TA;) and say لَا تَحِينَ, meaning *It is not,* or *was not, a time* [of such a thing; but this is generally written لَاتَ حِينَ]; as in the Ḳur xxxviii. 2 [respecting which see art. ليت]. (TA.) Aboo-Wejzeh Es-Saạdee says,

* اَلْعَاطِفُونَ تَحِينَ مَا مِنْ عَاطِفٍ *

* وَالمُطْعِمُونَ زَمَانَ أَيْنَ المُطْعِمُ *

[*The persons who return to the attack when there is none other that returns to the attack,* (as is said in the Ṣ and L in art. عطف,) or it may mean *who act affectionately in the time when there is none other that acts affectionately;* as is said in the L in that art.;) *and the feeders in the time when* it is said, *Where is the feeder?*]: (Ṣ:) ISd says that ت is thus prefixed to حين like as it is in تَلَانَ meaning الآنَ: but IB says that Ibn-Es-Seeráfee cites the former hemistich thus:

* اَلْعَاطِفُونَهْ حِينَ مَا مِنْ عَاطِفٍ *

[with the ه of pausation]: and some say that the ه of pausation is likened to the fem. ة, and is then made movent with fet-ḥ. (TA. [See more in art. ليت.]) — See also حِينَةٌ, in two places.

حَانَةٌ: see art. حون.

حَيْنَةٌ: see what next follows.

حِينَةٌ [The *time appointed for a she-camel to be milked in every day and night;*] a subst. from حَيَّنَ النَّاقَةَ; as also ▼ حَيْنٌ: you say, مَتَى حِينَةُ نَاقَتِكَ, meaning *When is the time of the milking of thy she-camel?* and كَمْ حِينَتُهَا, meaning *How many times is she milked?* (Ḳ.) One says also, of a man, (Ṣ,) يَأْكُلُ الحِينَةَ and ▼ الحَيْنَةَ, meaning *He eats once in the day and the night:* (Ṣ, Ḳ:) or, accord. to Aboo-'Amr Ez-Záhid, الوَجْبَةُ is used as meaning a man's *eating once in the day,* and الحينة as meaning a she-camel's *being milked once in the day.* (IB, TA.) And one says, مَا أَلْقَاهُ إِلَّا الحِينَةَ بَعْدَ الحِينَةِ, i. e. الحِينَ بَعْدَ ▼ الحِينِ [*I do not meet him save time after time;* meaning, *occasionally*]. (Ḳ.)

حَانَاةٌ: see art. حنو.

حَانُوتٌ: see arts. حون and حنو.

حَانِيَةٌ: } see art. حنو.
حَانِىٌّ: }

حَانِيَّةٌ: see arts. حون and حنو.

حَائِنٌ *Stupid; foolish;* or *having little,* or *no, intellect,* or *understanding.* (K.)

حَائِنَةٌ *A deadly,* or *destructive, calamity that befalls one:* (K, TA:) *a calamity in which is* الحَيْن: (TA:) pl. حَوَائِنُ. (K.) [In the CK it is, in one place, erroneously substituted for حَانِيَةٌ, as meaning "wine."]

مَحُونَةٌ [mentioned in the K in art. محن] is from الحَيْنُ or المِحْنَةُ. (TA.)

مِحْيَانٌ: see its syn. حِينٌ.

حيهل

حَيْهَلٌ (En-Nadr, AHn, K) and حَيَّهَلٌ and حَيْهِلٌ (K) *A certain shrub of the smaller kind of* حَمْض, *having no leaves:* (En-Nadr, AHn, K:) or the *species of* حمض *termed* هَرْم: (AA, TA:) a coll. gen. n.: (TA:) n. un. with ة: (K:) so called because, *when rain falls upon it, it grows quickly; and when the camels eat it, and do not speedily void their excrement, they die:* (AA, TA: [see the next paragraph:]) at the end of a verse of Homeyd Ibn-Thowr, the name is written الحَيْهَلْ, with the vowel of the ل transferred to the ه. (AHn, K.)

حَيَّهَلْ and حَيَّهَلَ and حَيَّهَلًا and حَيَّهَلَنْ and حَيَّهَلَا words used in inciting, or urging on: see more in art. حى; (K;) and in art. هل. (TA.)

حيو

For words that might perhaps be correctly mentioned under this head, see the art. headed حى: or حى and حيو.

خ

The seventh letter of the alphabet: called خَآءٌ [and خَا (respecting which latter see the letter ب)]. It is one of the letters termed مَهْمُوسَة [or non-vocal, i. e. pronounced with the breath only, without the voice]; and of those termed حَلْقِيَّة [i. e. faucial, or guttural]: its place of utterance is the same as that of غ [which differs from it in being vocal]. (L.) ═ [As a numeral, it denotes *Six hundred*.]

خا

خَا and خَآءٌ: see the letter خ, and arts. خوأ and خى.

خاتون

خَاتُونٌ: see art. ختن.

خانقاه

خَانْقَاهُ and خَانَقَاهِىٌّ: see art. خنق.

خب

1. خَبَّ, (A, L, Ḳ,) sec. pers. خَبِبْتَ, like عَلِمْتَ, aor. يَخَبُّ, inf. n. خِبٌّ; (Ṣ, L, Ḳ;*) or خَبَّ, [sec. pers. خَبَبْتَ,] aor. يَخُبُّ, like يَقْتُلُ, inf. n. خَبٌّ; (Mṣb; [in which خب is regarded as a simple subst.; but I doubt the correctness of this, and of the verb's being like قَتَلَ;]) *He* (a man) *was*, or *became, deceitful*, (Mṣb, Ḳ, TA,) *wicked, dishonest*, or *dissimulating*, (Ḳ, TA,) and *a mischief-maker*: (TA:) [or] *he was*, or *became, a great deceiver*, or *very deceitful*, (Ṣ, A, L, Ḳ, TA,) *wicked and deceitful*, and *a mischief-maker*. (Ṣ,* A,* L, TA.) [In the Ḳ and TA, neither the aor. nor the inf. n. of خَبَّ as signifying "he was, or became, a great deceiver, or very deceitful," is specified; nor the sec. pers. of the pret., which indicates the form of the aor.] — [Hence, app.,] خَبَّ signifies also *He alighted and abode in a depressed tract of ground, in order that his place might be unknown, from a motive of niggardliness* [*to avoid claims upon his hospitality, thus deceiving passers by*]. (Ḳ, TA.) — And *He denied*, or *refused, what he possessed*. (Ḳ.) ═ خَبَّ, (Ṣ, A, Ḳ,) aor. يَخُبُّ, with ḍamm, (Ṣ, A,) deviating from a general rule, accord. to which an intrans. v. of this class [of the measure فَعَلَ] should be with kesr, (MF,) inf. n. خَبٌّ (Ṣ, Ḳ) and خَبَبٌ and خَبِيبٌ, (Ṣ, A, Ḳ,) *He went the pace*, or *in the manner, denoted by* خَبَبٌ *as explained below*; said of a horse; (Ṣ, A, Ḳ;) as also ↓ اِخْتَبَّ: (Th, Ḳ:) and in like manner said of a man. (TA.) You say, جَاؤُوا تَخُبُّ بِهِمِ الدَّوَابُّ [*They came, the beasts going with them the pace*, or *in the manner, termed* خَبَب]. (A.) And خَبَّ فِى الأَمْرِ, aor. ـُ, inf. n. خَبَبٌ, *He hastened to begin the affair*. (Mṣb.) — [Hence,] خَبَّ, (aor. يَخُبُّ, TA, [inf. n. خَبٌّ,]) said of the sea, ‡ *It was*, or *became, agitated*, or *in a state of commotion*; (T, Ṣ, A, Ḳ, TA;) *the waves dashing together, and the winds whirling*; (T, A, TA;) such being the case at a certain period, when the ships make for the shore, for safety, or cast anchor. (T, TA.) You say, أَصَابَهُمُ الخَبُّ ‡ *Agitation*, or *commotion, of the sea, with a whirling of the winds, befell them*: (T, A, TA:) or اصابهم خَبٌّ, i. e. خَبَّ بِهِمُ البَحْرُ ‡ *The sea became agitated*, or *in a state of commotion, with them*. (Ṣ, TA.) — Also, said of the dust, ‡ *It rose high*: (JK, TA:) and ‡ *it ran along*. (TA.) — And said of a plant, or of herbage, ‡ *It became tall*. (JK, Ṣ, A, Ḳ.)

2. خَبَّبَ, (JK, Ṣ, A, Ḳ,) inf. n. تَخْبِيبٌ, (JK, TA,) *He deceived* another; (Ṣ, Ḳ;) namely, another's young man, or slave: (Ṣ:) or *deceived much* or *greatly*: (Ḥar p. 591:) and *he corrupted* another; (JK, A, TA;) namely, another's male or female slave. (TA.) You say, خَبَّبَ عَلَيْهِ صَدِيقَهُ, (Aboo-Bekr, TA,) and عَبْدَهُ, and حَلِيلَتَهُ, (A,) *He corrupted, and rendered disaffected to him*, (Aboo-Bekr, A,) *his friend*, (Aboo-Bekr,) and *his male slave*, and *his wife*. (A.) ═ *He bound his arm*, or *hand, with a* خِبَّة, i. e. *a piece of rag like a fillet*. (A, TA.) ═ *It* (one's flesh) *wasted so that there appeared streaks upon the skin*. (TA.) ═ See also R. Q. 1.

3. خَابَّ: see مُخَابٌّ. [It seems that خَابَّهُ, if used, signifies *He acted treacherously towards him, and took him unawares*.] — And [the inf. n.] مُخَابَّةٌ signifies The *being heavy*, or *sluggish, and holding back from a thing*. (JK.)

4. اخبّ *He made* a horse *to go the pace*, or *in the manner, denoted by* خَبَبٌ *as explained below*. (Ṣ, A, Ḳ.)

8. اختبّ: see 1. ═ اِخْتَبَّ مِنْ ثَوْبِهِ خِبَّةً, (Ṣ,) or, خُبَّةً, and اِخْتَبَّ ثَوْبَهُ, (JK,) *He took forth* [or, app., *tore*] *from his garment a piece of rag like a fillet*. (Ṣ.)

R. Q. 1. خَبْخَبَ *He* (a man, TA) *acted perfidiously, unfaithfully, faithlessly*, or *treacherously*. (Ḳ.) ═ *He was*, or *became, lax, flaccid*, or *flabby, in the belly*. (Ḳ.) [See also خَبْخَبَةٌ, below; and see R. Q. 2.] ═ خَبْخَبَ عَنِ الظَّهِيرَةِ *He stayed until the mid-day heat had become assuaged, and the air was cool*: (Ḳ:) or the phrase is خَبْخَبَ عَنْهُ مِنَ الظَّهِيرَةِ. (TA.) [Hence, in a trad. relating to the postponement of the noon prayers,] خَبْخِبُوا عَنْكُمْ مِنَ الظَّهِيرَةِ, (Ṣ,) or فِى الظَّهِيرَةِ, (JK, [but the former is more probably the correct phrase,]) *Stay ye until the mid-day heat shall have become assuaged, and the air be cool*: (JK, Ṣ:) it is originally ↓ خَبِّبُوا, and is altered therefrom for the purpose of distinction: (Ṣ in the present art.:) or originally بَخْبِخُوا. (Ṣ in art. بخ.)

R. Q. 2. تَخَبْخَبَ *It was*, or *became, lax, flaccid*, or *flabby*: said of a thing in a state of commotion, moving to and fro, quivering, or the like. (Ḳ.) [See also خَبْخَبَةٌ, below; and see R. Q. 1.] — *He was*, or *became, empty* [*in the belly*], *after repletion*. (JK.) And جَآءَ يَتَخَبْخَبُ *He came hungry*. (JK.) — تَخَبْخَبَ لَحْمُهُ, (JK,) or بَدَنُهُ, (Ḳ,) *He became lean after having been fat*, (JK, Ḳ, TA,) *so that his skin became lax, flaccid*, or *flabby*, (TA,) *and a sound was heard to proceed from him* [*when he moved*], (JK, TA,) *by reason of his leanness*. (TA.) — تَخَبْخَبَ الحَرُّ *The heat became allayed*, or *assuaged*, (Ḳ, TA,) *somewhat*, (TA,) *in its vehemence*. (Ḳ, TA.)

خَبٌّ (JK, Ṣ, A, L, Mṣb, Ḳ) [said in the Mṣb to be originally an inf. n.] and ↓ خِبٌّ (Ṣ, L, Ḳ) [originally an inf. n. accord. to most authorities] and ↓ خُبٌّ, (MF,) applied to a man, (Ṣ, A, L, Mṣb,) fem. خَبَّةٌ, [which casts doubt upon the assertion that خَبٌّ is originally an inf. n., for were it so the masc. and fem. accord. to a general rule would be the same, as well as the sing. and pl.,] applied to a woman, (JK, A,) *A great deceiver*, or *very deceitful*; (JK,* Ṣ, A, L, Mṣb,* Ḳ;) *wicked and deceitful*; *a mischief-maker*; (Ṣ,* A,* L, Ḳ,* TA;) *deceitful, guileful, artful, crafty*, or *cunning*; syn. مَكَّارٌ. (Ḥam p. 537, in explanation of the first and second.) ═ Also the first of these words, *A long, elevated tract* (حَبْل, in some copies of the Ḳ erroneously written جبل, TA) *of sand, cleaving to the ground*. (Ḳ, TA.) — And *A plain*, or *soft*, *tract, between two rugged tracts, in which* (i. e. in the former of which) *are truffles.* (AA, Ḳ.)

خُبّ: see خَبّ. ═ See also خُبَّة. Hence ثَوْبٌ أَخْبَابٌ: see, again, خُبَّة. — أَخْبَابُ الفَحِثِ *The* حَوَايَا [or *winding guts, or intestines into which the food passes from the stomach*]: (Ḳ:) thus used in the pl. form, as though pl. of خُبّ. (TA.) ═ Also The *bark* (لِحَآء) *of a tree.* (JK, Ḳ.) ═ And *Low, or depressed,* land: (JK, Ḳ:) pl. [of pauc.] أَخْبَابٌ and [of mult.] خُبُوبٌ. (TA.)

خِبّ *Deceit;* (JK, Mṣb, Ḳ;) *wickedness; dishonesty,* or *dissimulation;* (Ḳ, TA;) *mischief-making;* as also ▼خَبِيبٌ: (TA:) *guile, art, craft,* or *cunning.* (Ḥam p. 537.) ═ See also خَبّ. ═ ‡*A rising,* or *state of agitation and commotion, of the sea;* (JK, Ḳ, TA;) as also ▼خِبَابٌ. (IAạr, Ḳ.)

خِبَّة: see خُبَّة.

خَبَّة: see خُبَّة and خَبِيبَة. — Also *A place where water collects and remains* or *stagnates,* (AA, Ḳ, TA,) *and around which grow herbs,* or *leguminous plants:* (TA:) *a tract of land neither fruitful nor unfruitful, between two other tracts of land;* pl. خُبُبٌ: (AḤn:) *a tract of land between that which abounds with herbage and that which is unproductive:* (Ru-beh:) *a narrow tract of soft land abounding with herbage, not rugged nor plain, but inclining to be plain;* (ISh;) but ADḳ disapproves of this explanation: (TA:) or *a tract producing herbage between two long and elevated tracts of sand;* as also ▼خَبِيبَة: (Ibn-Nujeym:) and, accord. to AA, also *pasture,* or *herbage.* (TA.) Also, or ▼مَخَبَّة, (accord. to different copies of the Ḳ, or both, TA,) and ▼خَبِيبَة, The *bottom* (بَطْن) *of a valley.* (Ḳ.)

خُبَّة and ▼خِبَّة and ▼خَبَّة *A narrow tract,* or *streak,* of sand; [in one copy of the A, I find خُبَّة and خَبِيبَة thus explained; but in another, ▼مَخَبَّة is written in the place of the former of these two words;] or of clouds; (Ṣ, Ḳ;) as also ▼خَبِيبَة: (Aṣ:) or, of sand, *what resembles a* فَالِق [or *depressed tract between two hills*], *except in its being wider and more spreading, and not having abrupt sides;* so says AḤn in explaining خُبَّة [thus in the TA] and ▼خَبِيبَة: (TA:) or all three signify *a piece of rag like a fillet;* as also ▼خَبِيبَة (Ṣ, Ḳ) and ▼خُبّ: (Lḥ:) or the last two (خَبِيبَة and خُبّ) signify *a piece of rag from a garment, with which one binds his arm* or *hand.* (JK, TA.) [Hence,] ▼ثَوْبٌ أَخْبَابٌ, (Lḥ, Ḳ,) [like أَهْبَابٌ,] and ثوب خُبَبٌ, (Lḥ, JK, Ḳ,) like هُبَبٌ, (JK,) and ▼ثوب خَبَائِبُ, like هَبَائِبُ; (Ṣ;) [the latter word in the first of these phrases being pl. of خُبّ; that in the second, pl. of خُبَّة; and that in the third, pl. of خَبِيبَة;] *A garment, or piece of cloth, rent in pieces, ragged,* or *tattered.* (Lḥ, JK, Ṣ, Ḳ.) [See also خَبِيبَة, below.] It is also said that the خُبَّة of a garment, or piece of cloth, is [*A portion* thereof] *like the* طُرَّة [q. v.]: and accord. to Sh, the خُبَّة thereof is its طُرَّة. (TA.) And خَبَّة [so in the TA] signifies *A piece of rag which a woman wears, covering her head with it:* erroneously written by Lth جَنَّة. (Az, TA.) — Also, i. e. خُبَّة and its two vars., and ▼خَبِيبَة, of which the pl. is خَبَائِبُ, *A streak* of the flesh *appearing in the skin, occasioned by the loss of flesh.* (TA.)

خَبَبٌ: see خَبّ. ═ Also *A kind of run,* (Ṣ, Mgh, Mṣb, Ḳ,) *with wide steps, but falling short of that termed* عَنَق; (Mgh, Mṣb;) i. e. *a quick pace:* (TA:) or *a certain pace which is not quick:* (Ḥar p. 157:) or *i. q.* رَمَل [q. v.]: or *a pace of a horse,* (Ḳ,) and *of a camel,* (TA,) *in which he removes both his right legs together and both his left legs together;* i. e. *an amble:* (Ḳ, TA:) or *in which a horse rests on his right and left fore legs alternately,* (يُرَاوِحُ بَيْنَ يَدَيْهِ, Ṣ, Ḳ, TA,) *and in like manner on his hind legs:* (Ṣ, TA: [app., as thus explained in the Ṣ and TA, meaning the same as the explanation next before it:] and (accord. to some, TA) *quickness.* (Ḳ.)

خِبَابٌ: see خِبّ.

خَبِيبٌ *A trench,* or *furrow,* (خَدّ,) *in the ground.* (Ḳ.)

خَبِيبَة, and its pl. خَبَائِبُ: see خُبَّة, in five places. It is also said to signify *A fillet,* or *bandage.* (TA.) — Also ‡*A long strip,* or *slice,* of flesh, or flesh-meat; (JK, Ṣ, Ḳ;) and so ▼خُبَّة; (A, TA;) pl. of the former as above: (JK:) or *any compact and long portion of flesh:* any such portion is also termed خَصِيلَة: either *in the arm* or *elsewhere:* (AO, TA:) or a [*portion such as is termed*] خصيلة thereof, *intermixed with* [*sinews,* or *tendons, such as are termed*] عَقَب. (TA.) And خَبَائِبُ المَتْنَيْنِ *The flesh of the two corresponding portions extending along the two sides of the backbone.* (TA.) [Hence,] لَحْمُهُ خَبَائِبُ *His flesh is dissundered,* or *cut in pieces.* (TA.) — See also خُبَّة, in two places. — Also The *wool of a* ثَنِىّ [or *sheep in its third year*]; (Ṣ, L;) which is better than that termed عَقِيقَة, i. e. the wool of a جَذَع [or sheep in or before its second year], and cleaner, and more abundant: (ISk, Ṣ:) so accord. to most of the leading lexicologists; though said in the Ḳ to be a mistake of J, for جَنِيبَة. (TA.)

خَبْخَبَة [by rule an inf. n. of R. Q. 1:] *Laxness, flaccidity,* or *flabbiness;* and *a state of commotion, moving to and fro, quivering,* or *the like:* (Ṣ:) or *laxness, flaccidity,* or *flabbiness,* of a thing in a state of commotion, moving to and fro, quivering, or the like; (TA;) as also ▼خَبَخَابٌ. (JK, Ḳ, TA.) [See also R. Q. 2.]

خَبَخَابٌ: see what next precedes.

خِبَابٌ, (Ṣ, Ḳ, TA,) in one copy of the Ḳ خِبَابَة, [as in the CḲ,] but the former is the more correct, (TA,) *Relationship;* (Ṣ, Ḳ;) and *affinity,* syn. صِهْر: (Ṣ:) pl. خَوَابٌ. (Ṣ, Ḳ.) You say, لِى مِنْ فُلَانٍ خَوَابٌ [*I have ties of relationship,* or *affinity, to such a one*]. (Ṣ.)

مَخَبَّة: see خُبَّة: ═ and see also خَبَّة.

مُخَابٌ, as though from ▼خَابٌّ, One *who acts treacherously towards another, and takes him unawares.* (TA.)

خبأ

1. خَبَأَهُ, (Ṣ, Mgh, Mṣb, Ḳ,) aor. ـَ, (Mṣb, Ḳ,) inf. n. خَبْءٌ, (Ṣ, Mṣb,) *He hid,* or *concealed, it;* (Mgh, Mṣb, Ḳ;) as also ▼خَبَّأَهُ, [but app. in an intensive sense, or applying to a number of things,] (Ḳ,) inf. n. تَخْبِئَة; (TA;) and ▼اختبأهُ. (Ḳ.) — *He kept it, preserved it, guarded it,* or *took care of it:* and ▼خَبَّأَهُ *he did so much;* and *well,* or *carefully.* (Mṣb.) [*He laid it up; stored it,* or *reposited it, in a place of safety.*]

2: see 1, in two places. [Hence, خَبَّأَ جَارِيَةً *He kept a girl carefully concealed from view:* see the pass. part. n., below.]

3. خَابَأَهُ مَا كَذَا, (Ḳ,) inf. n. مُخَابَأَة, (TḲ,) *I proposed to him as an enigma, What is such a thing?* syn. حَاجَيْتُهُ. (Ḳ. [See also 8.])

8. اختبأ *It was,* or *became, hidden,* or *concealed:* (Mgh:) *he hid,* or *concealed, himself.* (Ṣ.) ═ It is also trans.: see 1. — [Hence,] ▼اختبأ لَهُ خَبِيئًا *He expressed a thing enigmatically to him, and then asked him respecting it.* (IDrd, Ḳ. [See also 3.])

خَبْءٌ: (Ṣ, Mṣb, Ḳ) and ▼خِبْءٌ: (TA) and ▼خَبَاءٌ, of the measure فَعْلَة from الخبأ [or rather الخَبْءُ], like غُرْفَة and قُبْضَة from الغَرْف and القَبْض, (Ḥar p. 426,) and ▼خَبِىءٌ: (Ṣ, Ḳ) and ▼خَبِيئَة, (Ḳ,) of which last the pl. is خَبَايَا, (TA,) *A thing that is hidden,* or *concealed,* (Ṣ,* Mṣb, Ḳ,) and *absent,* or *unseen.* (Ḳ.) [Hence,] خَبْءُ السَّمَآءِ *The rain.* (Th, Ṣ, Ḳ.) And خَبْءُ الأَرْضِ *The plants,* or *herbage.* (Ṣ, Ḳ.) And ▼خَبَايَا الأَرْضِ *The seed which the sower has hidden in the earth:* or *what God has hidden in the mines of the earth.* (TA, from a trad.) الَّذِى يُخْرِجُ الخَبْءَ فِى السَّمٰوَاتِ وَالأَرْضِ, in the Ḳur [xxvii. 25], is held by Az to mean *Who knoweth what is unseen in the heavens and the earth;* agreeably with an explanation of الخَبْء by Fr. (TA.)

خِبْءٌ: see the next preceding paragraph.

خُبَأَة *A daughter;* syn. بِنْت. (Ḳ, TA. [In the CḲ, خُبَأَة, البِنْتُ is put for البِنْت.]) Hence the prov., خُبَأَةٌ خَيْرٌ مِنْ يَفَعَةِ سَوْءٍ [*A daughter is better than a grown-up boy of evil deeds*]. (TA.) [In Freytag's Arab. Prov., i. 438, the first word in this prov. is written خُبَأَة, and followed by صِدْق.] Aboo-Zeyd Sa'eed Ibn-Ows El-Ansáree entitled one of his books كتاب خبأة because he commenced it by mentioning خبأة in the sense of بنت, quoting the foregoing prov. in confirmation thereof. (TA.)

خُبَآءٌ : see خَبْءٌ.

امْرَأَةٌ خُبَأَةٌ *A woman who shows herself and then hides herself:* (Ṣ, O, TA:) [like قُبَعَةٌ:] or *a woman who keeps to her house,* or *tent.* (Ḳ.)

خِبَآءٌ *A well-known kind of structure;* (Ḳ;) [i. e.] *a kind of tent,* (Mgh, TA,) *made of wool,* (Mgh, Mṣb,) or *of camels' fur,* or sometimes *of* [*goats'*] *hair,* sometimes *upon two poles,* or *three;* what is above this kind being termed بَيْتٌ: (Mṣb:) or *a tent having one pole;* that which has more than one pole being termed بيت: (AZ, TA in art. ربع:) [or] also applied to *a* بيت [or *tent*] *of any kind:* (Towsheeḥ, TA voce بَيْتٌ, q. v.:) pl. أَخْبِئَةٌ, (TA,) or أَخْبِيَةٌ: (Mṣb:) it is from خَبَأَهُ "he hid it," or "concealed it:" (Mgh:) or it belongs to art. خبى: (Ḳ:) most of the lexicologists hold that its radical letters are خبى: some, that they are خبو: IDrd asserts that they are خبأ. (TA:) [See also art. خبى.] ═ *A mark made with a hot iron upon some secret part of an excellent she-camel:* pl. أَخْبِئَةٌ. (Lth, Ḳ.)

خَبِىْءٌ: see خَبْءٌ: and see also 8.

خَبِيْئَةٌ, and its pl. خَبَايَا: see خَبْءٌ, in two places.

كَيْدٌ خَابِىٌّ *An artifice,* or *a stratagem, resulting in disappointment;* i. q. خَائِبٌ; (AḤei, Ḳ;) formed [from the latter] by transposition. (AḤei.)

خَابِئَةٌ, as sometimes pronounced, (Mṣb,) or خَابِيَةٌ, with the ء suppressed, (Ṣ, Mṣb, Ḳ,) because of frequent usage, (Mṣb,) *i. q.* حُبٌّ [q. v.]; (Ṣ, Ḳ;) i. e. *A large jar:* pl. خوابى [i. e. خَوَابِئُ or خَوَابٍ]: (TA:) from خَبَأَهُ "he hid it," or "concealed it." (Ṣ, Mṣb.) — [Hence,] بِنْتُ الخَابِيَةِ † *Wine.* (Ḥar p. 365.)

مَخْبَأٌ *A place,* or *chamber, for hiding* or *concealing* [*anything*]; *a secret place* or *chamber:* pl. مَخَابِئُ. (MA.)

جَارِيَةٌ مُخْبَأَةٌ; so in the [Ṣ and] O, and in some of the correct copies of the Ḳ; in other copies of the Ḳ مُخَبَّأَةٌ; (TA;) [and thus in the CḲ;] *A girl that is* [*kept in the house,* or *tent,*] *concealed from view;* or *that conceals herself;* (Ṣ;) *that is kept behind,* or *within, the curtain;* (Ḳ, TA;) *not going forth:* or (TA) *that is not yet married.* (Lth, Ḳ, TA.)

مُخْتَبِئٌ One *who conceals himself in order that he may see without the knowledge of him who is seen.* (Mgh.)

خبت

1. خَبَتَ ذِكْرُهُ *The mention of him,* or *it, was,* or *became, concealed:* (L:) [app. meaning *he,* or *it, was,* or *became, obscure;* or *of no reputation,* or *repute.*] ═ خَبُتَ, accord. to Z, *i. q.* خَبُثَ [q. v.]: occurring in a trad. (TA.) [See خَبِيتٌ.]

4. اخبت *He became in what is termed* خَبْتٌ [q. v.]. (A, TA.) — And, (Ṣ, Mṣb, Ḳ, TA,) [hence, or] from خَبْتٌ, (Ksh and Bḍ in xi. 25, and TA,) or from خَبَتَ ذِكْرُهُ, (L,) inf. n. إِخْبَاتٌ, (Ṣ, Mṣb,) ‡ *He* (a man, Mṣb, TA) *was,* or *became, lowly, humble,* or *submissive,* (Ṣ, Mṣb, Ḳ, TA,) *in heart,* (Mṣb,) *and obedient,* (TA,) لِلّٰهِ *to God.* (Ṣ, TA.) And in like manner, in the Ḳur [xi. 25], (TA,) وَأَخْبَتُوا إِلَى رَبِّهِمْ means ‡ *And who have become lowly, humble,* or *submissive,* [*and obedient,*] *to their Lord;* or *have lowered, humbled,* or *abased, themselves to their Lord;* or *have trusted to their Lord:* (A,* TA:) for the Arabs put إِلَى in the place of لِ. (TA.)

خَبْتٌ *A low,* or *depressed, tract of ground:* (TA:) or *a low,* or *depressed,* (Ṣ,) or *concealed and low,* (TA,) *tract of ground, in which is sand:* (Ṣ, TA:) or *a wide,* or *spacious, low tract of ground:* (IAṣr, A, Ḳ:) or *a plain,* or *soft, tract of ground in a* [*stony tract such as is termed*] حَرَّة: (TA:) and *a wide bottom,* or *bed,* or *interior, of a valley:* (A:) or *a deep valley, easy to be walked or ridden through, extended* [*to a great length*], *and in which grow varieties of the* عِضَاه: (TA:) pl. [of pauc.] أَخْبَاتٌ (Ḳ) and [of mult.] خُبُوتٌ: (A, Ḳ:) it is a genuine Arabic word. (TA.)

فِيهِ خَبْتَةٌ ‡ *In him is lowliness, humility,* or *submissiveness.* (Ṣ, TA.)

خَبِيتٌ A thing that is *contemptible,* or *despicable;* (Ḳ, TA;) *bad, corrupt, abominable, vile, base,* or *disapproved;* [&c.;] (TA;) and [thus] *i. q.* خَبِيثٌ. (Aṣ, Ḳ.) The Jew of Kheyber says,

* يَنْفَعُ الطَّيِّبُ القَلِيلُ مِنَ الرِّزْ *
* قِ وَلَا يَنْفَعُ الكَثِيرُ الخَبِيتُ *

[*The lawful, but small, supply of the means of subsistence is beneficial, but the large and unlawful is not beneficial*]. (TA.) Kh asked Aṣ respecting الخبيت in this verse; and the latter replied that the poet meant الخَبِيث; the former word being of the dial. of Kheyber: but Kh rejoined, "If so, the poet would have said الكثير: it behooves you only to say that the people of Kheyber change ث into ت in some words:" AM thinks that الخبيت in this verse is a mistranscription for الخَتِيت, which means the thing that is "contemptible and bad," and is syn. with الخَسِيس. (TA.) — It is also applied to a man; meaning as above; or *Bad, corrupt, vitious,* or *depraved.* (TA.)

مُخْبِتٌ † *Still; motionless:* as also مُخْبِطٌ. (TA in art. خبط.)

خبث

1. خَبُثَ, (Ṣ, Mgh, Mṣb, Ḳ, &c.,) aor. ـُ, (Mṣb, Ḳ,) inf. n. خَبَاثَةٌ, (Ṣ,) or خُبْثٌ, the former being a simple subst., (Mṣb,) or both, (Mgh, Ḳ, [the latter word erroneously written in the CḲ خَبْث,]) and خَبَاثِيَةٌ, (Ḳ,) said of a thing, (Ṣ, Mgh, Mṣb,) *It was,* or *became,* خَبِيث [q. v., meaning *bad,* &c.]; *contr. of* طَابَ. (Ṣ, Mgh, Mṣb, Ḳ.) [Hence,] خَبُثَتْ رَائِحَتُهُ ‡ [*Its,* or *his, odour was,* or *became, bad, foul,* or *abominable*]. (A.) And خَبُثَ طَعْمُهُ ‡ [*Its taste was,* or *became, bad, foul, abominable,* or *nauseous*]. (A.) And خَبُثَتْ نَفْسُهُ ‡ *His soul* [or *stomach*] *became heavy;* (TA;) *it heaved,* or *became agitated by a tendency to vomit;* syn. لَقِسَتْ: (A and TA in the present art., and Ṣ and Ḳ in art. لقس: [see also مَذِرَتْ نَفْسُهُ, in art. مذر:]) a phrase forbidden by Mohammad to be used; as though he disliked the word خَبُثَ. (TA.) One says of certain food, تَخْبُثُ عَنْهُ النَّفْسُ ‡ [*The soul,* or *stomach, becomes heavy,* or *heaves,* or *becomes agitated by a tendency to vomit, in consequence of it*]. (TA.) — خَبُثَ, (Ṣ, A, Ḳ,) inf. n. خُبْثٌ, (Ṣ, Ḳ,) said of a man, signifies [in like manner] *He was,* or *became,* خَبِيث, (Ṣ, A, Ḳ,) meaning *bad, corrupt, base,* or *abominable; wicked, deceitful, guileful, artful, crafty,* or *cunning.* (Ṣ, Ḳ, TA. [See also 4.]) [Hence,] خَبَثَ بِهَا ‡ *He committed adultery,* or *fornication, with her.* (A, Mgh, Mṣb, Ḳ.) — [It is also said of a venomous reptile and the like, meaning *It was,* or *became, malignant,* or *noxious; impure, unclean, foul,* or *filthy.*]

2. هٰذَا مِمَّا يُخَبِّثُ النَّفْسَ, (TA,) or ↓ يُخْبِثُ النفس, (so in a copy of the A, [but the former I believe to be the right,]) *This is of the things that cause the soul* [or *stomach*] *to become heavy,* or *to heave,* or *become agitated by a tendency to vomit.* (TA.)

4. اخبث *He* (a man) *was,* or *became, characterized by* خُبْث (Mṣb, TA) and شَرّ (Mṣb) [meaning *badness, wickedness, deceit, &c.*: see also خَبُثَ]. — *He had bad, wicked,* or *deceitful, companions* or *friends,* and *a bad, wicked,* or *deceitful, family:* (L:) or *his companions,* or *friends, became bad, wicked,* or *deceitful:* (Ṣ in art. فلس:) or *he took to himself bad, wicked,* or *deceitful, companions* or *friends* (Ṣ, L, Ḳ) or *connexions* or *assistants.* (TA.) ═ اخبثهُ *He taught him to be bad, wicked,* or *deceitful:* and *rendered him bad, corrupt, vitious,* or *depraved.* (Ṣ.) — See also 2.

5: see what next follows.

6. تخابث (A, TA) *He made a show of being,* or *pretended to be, bad, wicked,* or *deceitful.* (TA.) And you say also ↓ تخبّث [either in the same sense, or as meaning *He affected,* or *endeavoured, to be bad, wicked,* or *deceitful;* or *to do that which was* خَبِيث, or *bad, &c.*]. (A, TA.)

10. استخبث [*He deemed,* or *esteemed,* خَبِيث, i. e. *bad, &c.*]. كَانَتِ العَرَبُ تَسْتَخْبِثُ مِثْلَ الحَيَّةِ وَالعَقْرَبِ [*The Arabs used to deem impure, unclean, foul,* or *filthy, such as the serpent and the scorpion*]. (Mṣb.) — ‡ *He deemed bad,* or *corrupt,* a word, or a dialectic variant. (A, TA.)

خُبْثٌ an inf. n. of خَبُثَ: (Ṣ, Mgh, Mṣb, Ḳ:) [used as a simple subst., it means *Any of the qualities denoted by the epithet* خَبِيث, q. v., i. e. *badness, &c.*:] and ↓ خِبِّيثَى signifies the same: (Ḳ:) or this is a subst. from أَخْبَثَ meaning "he had a bad, wicked, or deceitful, family;" (TA;) and signifies the *state of having bad, wicked,* or *deceitful, companions* or *friends* or *connexions:* (L:) ↓ خَابِثَةٌ, also, is syn. with

[خُبْثٌ, and so is] † خَبَاثَةٌ, (Ḳ,) [for] this last is another inf. n. of خَبُثَ, like خُبْثٌ, (Ṣ, Mgh, Ḳ,) or it is a simple subst. (Mṣb.) [Hence,] the first particularly signifies ‡ *Adultery*, or *fornication*. (Ḳ, TA.) See also خَبِيثٌ, in three places.

خَبَثٌ The *dross* of iron, (Ṣ, TA,) and of silver, when they are molten. (TA.) [Hence the saying,] لَيْسَ الإِبْرِيزُ كَالخَبَثِ [lit. *Pure gold is not like dross*]; meaning ‡ *the good is not like the bad*. (A, TA.) — *Adulterating alloy in gold and iron &c.* (Ḥar p. 135.) — *A thing wherein is no good*. (TA.) — ‡ *Excrement*, or *ordure*: *impurity*, or *filth*. (Mgh in art. قل, and TA.) Hence the saying in a trad., إِذَا بَلَغَ المَآءُ قُلَّتَيْنِ لَمْ يَحْمِلْ خَبَثًا [explained in art. حمل]. (Mgh ubi suprà, and TA.)

يَا خَبَثُ : see خَبِيثٌ.

يَا خُبْثَةُ : see خَبِيثٌ. — خِبْثَةٌ with respect to a slave signifies † *Unlawful capture; capture from a people whom it is unlawful to make slaves*, (Mgh,* Ḳ, TA,) *by reason of a treaty*, or *league, made with them*, (Mgh, TA,) or *of some sacred*, or *inviolable, right, originally belonging to them*. (TA.) You say of a slave, لَا خِبْثَةَ فِيهِ مِنْ إِبَاقٍ وَلَا سَرِقَةٍ ‡ [*There is no unlawful capture in his case, from having run away, nor from having been stolen*]. (A.) — فُلَانٌ لِخِبْثَةٍ is like the saying لِزِنْيَةٍ † [*Such a one is the offspring of adultery, or fornication*]. (Ṣ.) And وُلِدَ فُلَانٌ لِخِبْثَةٍ means ‡ *Such a one was born spuriously*. (A,* L.)

خَبَاثِ : see the next paragraph, in two places.

خَبِيثٌ contr. of طَيِّبٌ ; (Ṣ, Mgh, Mṣb, Ḳ ;) applied to objects of the senses and to those of the intellect; (Kull p. 177 ;) to sustenance, or victuals, and to offspring, and men, and to other things : (TA :) *Bad; corrupt*: (Mṣb, TA :) *disapproved, hated*, or *abominable*; (Mṣb, TA ;) this, accord. to IAạr, being its primary signification : (TA :) or *so in respect of taste*, and *of odour*: (Mgh :) [*nasty, nauseous, loathsome*, or *disgusting* :] *impure, unclean, foul*, or *filthy*: (Mgh, Mṣb, TA :) *unlawful*; (Mgh, Mṣb ;) applied in this sense to certain food : and, applied to certain beverage, *injurious*: (TA :) applied to medicine such as is forbidden in a certain trad., it means either *impure and unlawful*, such as wine &c., or *nauseous to the taste*: (IAth, TA :) you say that a thing is خَبِيثٌ in taste, [and in odour,] and in colour : and you apply this epithet to adultery, or fornication ; and to property unlawfully acquired ; and to blood, and to the like things which God has forbidden : (TA :) also to such things as garlic and onions (Mṣb, TA) and leeks, (TA,) which are disagreeable in taste and odour : (TA :) and to such things as the serpent and the scorpion : (Mṣb :) applied to language, it means † *opprobrious*, or *of a reviling nature*; (TA ;) and ‡ *bad* or *corrupt* [*in respect of authority*; or *of a bad dialect*] : (A, TA :) applied to religion, † *infidel*, or *of the nature of infidelity*: (TA :) applied to a man, *bad, corrupt, base*, or *abominable*; *wicked, deceitful, guileful, artful, crafty*, or *cunning*; (Ṣ, Ḳ, TA ;) as also † خَابِثٌ : (Ḳ :) and *an adulterer*, or *a fornicator*: (Mṣb :) and *a blamer*, or *censurer*: or *a slanderer*, or *calumniator*: (Ḥar p. 611 :) [and, applied to a venomous reptile and the like, *malignant*, or *noxious*; as well as *impure, unclean, foul*, or *filthy* :] the fem. is خَبِيثَةٌ : (Mṣb :) the pl. masc. is خُبَثَآءُ (A, TA) and خُبُثٌ, for which it is allowable to say † خُبْثٌ, accord. to the dial. of Temeem, (Mṣb,) and خَبَثَةٌ, (Ṣ, A, Mṣb, TA,) like شَرَفَةٌ [pl. of شَرِيفٌ], (Mṣb,) and أَخْبَاثٌ, like أَشْرَافٌ [another pl. of شَرِيفٌ], (Mṣb, MF, TA,) and خَبَثَةٌ, (Kr, Mṣb, MF, TA,) like ضَعَفَةٌ pl. of ضَعِيفٌ, (Mṣb, MF, TA,) two instances of which the like can scarcely be found, (Mṣb,) or is not found among sound words, for سَرَاةٌ pl. of سَرِىٌّ is an unsound word, (MF, TA,) and خُبُوثٌ, (AZ, TA,) which is also extr., (TA,) [and خَبَاثَى, (like as حَزَانَى is a pl. of حَزِينٌ,) applied in the A, in art. خنث, to خَنَاثَى, pl. of خُنْثَى,] and خَبِيثُونَ [applied only to rational beings] : (Mgh :) and the pl. fem., i. e. of خَبِيثَةٌ, is خَبَائِثُ (Mṣb, TA) and خَبِيثَاتٌ. (Mgh.) الشَّجَرَةُ الخَبِيثَةُ, mentioned in the Ḳur [xiv. 31], (TA,) means *The colocynth*: or *the* كَشُوث, (Ḳ ; TA,) which is *a certain plant that clings to the branches of trees and has no root in the earth*; (Ṣ and Ḳ in art. كشث ;) [*a species of cuscuta*, or *dodder* ;] or *yellow* عُرُوق *that cling to trees*: (TA in the present art. :) also occurring in a trad., as meaning *the garlic-plant*; and *the onion*; and *the leek*; because of their disagreeable taste and odour. (IAth, TA.) It is said in a trad. respecting the slain at Bedr, أُلْقُوا فِى قَلِيبٍ خَبِيثٍ مُخْبِثٍ † *They were cast into a well corrupt, and corrupting what fell into it*. (TA.) خَبِيثٌ مُخْبِثٌ †, (Ṣ, L,) or خَبِيثٌ and † مُخْبِثٌ, (Ḳ,) and † خَابِثٌ (TA) and † مَخْبَثَانٌ, (Ṣ, L, Ḳ,) applied to a man, signify One *who takes to himself bad, wicked*, or *deceitful, companions* or *friends* (Ṣ, L, Ḳ, TA) or *connexions* or *assistants*: (TA :) or † مَخْبَثَانُ, [without the article ال,] as a determinate noun, is only used in calling to, or addressing, a person : (Ḳ :) you say, يَا مَخْبَثَانُ ; (Ṣ ;) fem. † مَخْبَثَانَةُ : and to a man and woman together, يَا مَخْبَثَانُ : (L, TA :) and in the phrase † خَبِيثٌ مُخْبِثٌ, the former word signifies *bad, wicked*, or *deceitful, in himself*; and the latter, *having bad, wicked*, or *deceitful, companions* or *friends* and *assistants*. (A 'Obeyd, TA.) One says also, † يَا خُبَثُ, meaning يَا خَبِيثُ [*O bad* or *wicked* or *deceitful* man!]; and to a woman, † يَا خَبَاثِ, (Ṣ, Ḳ,) indecl., with kesr for its termination, (Ṣ,) and يَا خَبِيثَةُ. (Ḳ [accord. to SM : so in all the copies in his hands; but not found by him in any other lexicon : not in the CḲ, nor in my MS. copy of the Ḳ.]) † خَبَاثِ also occurs, in a saying of El-Ḥasan, addressed to the present world, الدُّنْيَا. (L.) And † يَا خُبْثَةُ was said by El-Ḥajjáj to Anas, as meaning يَا خَبِيثُ : and is also used as meaning *O* [*thou of*] *bad, wicked*, or *deceitful, qualities* or *dispositions!* [app. addressed to a woman, as the context seems to show; and agreeably with an assertion in Ḥam p. 810, that خُبْثَةُ is sometimes used in speaking of an old woman]. (L, TA.) خَبِيثُ النَّفْسِ means ‡ *Having the soul* [or *stomach*] *heavy*, [or *heaving*, or *agitated by a tendency to vomit*,] *and in a disagreeable state*. (TA.) And † مَخْبَثَانٌ applied to a lie occurs in a trad. as meaning خَبِيثٌ app. in an intensive sense [i. e. *Very abominable*]. (TA.) In the saying, اَللّٰهُمَّ إِنِّى أَعُوذُ بِكَ, (Mgh,) or أَعُوذُ بِاللّٰهِ, (Mṣb,* Ḳ,* TA,) مِنَ الخُبُثِ وَالخَبَائِثِ, (Mgh, Mṣb, TA,) or مِنَ الخُبْثِ † وَالخَبَائِثِ, (Mṣb, Ḳ, TA,) a form of words which Mohammad directed his followers to pronounce on entering a privy, or place of retirement for the relief of nature, because devils are in such a place, (Mgh, TA,) الخُبُثُ is pl. of الخَبِيثُ, (Mgh, Mṣb, TA,) and so is الخُبْثُ accord. to the dial. of Temeem, (Mṣb, TA,) and الخَبَائِثُ is pl. of الخَبِيثَةُ ; (Mgh, TA ;) and the meaning is, *I seek protection by God*, or *O God, I seek protection by Thee, from the male devils and the female devils*, (IAth, Mgh, Mṣb, Ḳ, TA,) *of the genii and of mankind*: (Mgh :) or, reading † الخُبْثِ, [as a subst.,] *from infidelity and the devils*: (Aboo-Bekr, TA :) or, [so reading, and regarding الخبائث as pl. of † الخَبِيثَةُ used as a subst.,] *from infidelity and acts of disobedience*: (Mṣb, TA :) or, *from wicked*, or *unrighteous, conduct, such as adultery and the like, and culpable actions and evil qualities* or *dispositions*: El-Khaṭṭábee asserts that the reading الخُبْثِ, with the ب quiescent, is a mistake of the relaters of traditions; but En-Nawawee rejects this assertion. (TA.)

خَبَاثَةٌ : see خُبْثٌ.

خَبِيثَةٌ fem. of the epithet خَبِيثٌ. (Mṣb.) — Also, [used as a subst.,] *A bad, wicked*, or *deceitful, quality* or *disposition*; and *a culpable action*: pl. خَبَائِثُ. (L, TA.) [Hence,] † أُمُّ الخَبَائِثِ [*The mother of bad qualities* &c. ; meaning] *wine*. (T in art. ام.) See also خَبِيثٌ, last sentence. — الخَبَائِثُ also signifies *Those things which the Arabs deemed foul*, or *filthy*, or *unclean, and which they did not eat*; such as vipers, and scorpions, and the برص [i. e. either بُرْص or بِرَص], and the وَرَل, and beetles, and the rat, or mouse. (L.)

خِبِّيثٌ, applied to a man, (TA,) signifies كَثِيرُ الخُبْثِ [i. e. *Very bad* or *wicked* or *deceitful*; or *much addicted to adultery* or *fornication*] : pl. خِبِّيثُونَ. (Ḳ.)

خِبِّيثَى : see خُبْثٌ.

خَابِثٌ : see خَبِيثٌ, in two places.

خَابِثَةٌ : see خُبْثٌ.

أَخْبَثُ [compar. and superl. of خَبِيثٌ] : pl. أَخَابِثُ. (TA.) You say, هُمْ أَخَابِثُ النَّاسِ [*They are the worst*, or *the most wicked* or *deceitful, of men*]. (TA.) And هُوَ مِنَ الأَخَابِثِ [*He is of the worst*, &c., *of men*]. (A, TA.) And هِىَ

أَخْبَثُ اللُّغَتَيْنِ ‡ *It is the worse*, or *more corrupt*, [*in respect of authority*,] *of the two words*, or *dialectic variants*. (A, TA.) —— الأَخْبَثَانِ *Urine and dung* (Ṣ, A, Mṣb, Ḳ:) *of a human being*: (Ṣ, Mṣb, Ḳ:) or *vomit and human ordure* or *thin human ordure*: (Fr, TA:) or *fetor of the mouth, and sleeplessness*: or *sleeplessness, and disquietude of mind by reason of grief*. (Ḳ.) It is said in a trad., لَا يُصَلِّي الرَّجُلُ وَهُوَ يُدَافِعُ الأَخْبَثَيْنِ [*The man shall not pray while he is striving to suppress the urine and ordure*]. (TA.)

وَقَعَ فِي وَادِي تُخُبِّثَ, (Ḳ,* TA,) in which the last word, also pronounced تُخُبَّثَ, is imperfectly decl., (TA,) is similar to وقع فى وادى تُضُلِّلَ [and means *He fell into a state of things that was bad, corrupt, disapproved*, &c.]. (Ḳ, TA.)

مُخْبِثٌ One *who teaches others to be bad, wicked*, or *deceitful*: and some allow it to be applied to one *who attributes*, or *imputes, to others what is bad, wicked*, or *the like*. (TA.) —— See also خَبِيثٌ, in four places.

مَخْبَثَةٌ *A cause of evil* or *corruption*: (Ṣ, Ḳ:) pl. مَخَابِثُ. (TA.) So in the saying of 'Antarah,

نُبِّئْتُ عَمْرًا غَيْرَ شَاكِرِ نِعْمَتِى
وَالكُفْرُ مَخْبَثَةٌ لِنَفْسِ المُنْعِمِ

[*I have been told that 'Amr is not thankful for my beneficence: and ingratitude is a cause of evil to the soul of the benefactor*]. (Ṣ.) One says also, فِيهِ مَخَابِثُ جَمَّةٌ [*In him*, or *it, are many causes of evil* or *corruption*]. (A.) And طَعَامٌ مَخْبَثَةٌ ‡ *Food that is a cause of heaviness to the soul* [or *stomach*]; or *of heaving, or becoming agitated by a tendency to vomit*: or *that is unlawful*. (TA.)

مَخْبَثَانَةٌ and مَخْبَثَانُ and مَخْبَثَانُ: see خَبِيثٌ, in four places.

خبر

1. خَبُرَ, aor. ـُ, (Ḳ,) inf. n. خُبُورٌ; (TA;) and ▾اختبر, and ▾تخبّر; (Ḳ;) *He knew; or had, or possessed, knowledge*; بِشَىْءٍ [*of a thing*; generally meaning, *with respect to its internal*, or *real, state*]. (Ḳ, TA.) = خَبِرَهُ, (Ṣ, A, Mṣb,) aor. ـُ, (Mṣb, MṢ,) inf. n. خُبْرٌ; (Mṣb, MṢ;*) and خَبِرَهُ, [aor. ـَ,] (A,) inf. n. خَبَرٌ; (TA;) and ▾اختبره, and ▾تخبّره; (TA;) *He knew it*; syn. عَلِمَهُ; (Ṣ, A, Mṣb;) [generally meaning, *with respect to its internal*, or *real, state*; like خَبِرَ بِهِ: see خُبْرٌ, its simple subst., as distinguished from its inf. n.] You say, مِنْ أَيْنَ خَبَرْتَ هٰذَا الأَمْرَ, (so in a copy of the Ṣ,) or خَبِرْتَ, (so in another copy of the Ṣ, and so in the A, where it is expressly said to be with kesr,) *Whence knewest thou this thing?* (Ṣ, A.*) —— And خَبَرَهُ, (Ṣ, Ḳ,) aor. ـُ, (Ṣ,) inf. n. خُبْرٌ and خِبْرَةٌ, (Ṣ, Ḳ,) or the latter is a simple subst.; (Mṣb;) and ▾اختبره [which is the more common in this sense]; (Ṣ, Mṣb, Ḳ;) *He tried, made trial of, made experiment of, tested, proved, assayed, proved by trial* or *experiment* or *experience, him*, or *it*. (Ṣ, Mṣb, Ḳ.) Hence the phrase, (Ṣ,) لَأَخْبُرَنَّ خُبْرَكَ, (Ṣ, Ḳ,) in some good lexicons خَبَرَكَ, (TA, [and so in the CḲ, but this I think to be a mistake, suggested by the explanation, which is not literal,]) i. q. لَأَعْلَمَنَّ عِلْمَكَ [which properly signifies *I will assuredly know thy knowledge*, or *what thou knowest*, but here means, as is shown by the manner in which the phrase that it explains is mentioned in the Ṣ, *I will assuredly try, prove*, or *test, thy state, and so know what thou knowest*]. (Ṣ, Ḳ.) [Hence, also,] the saying of Abu-d-Dardà, وَجَدْتُ النَّاسَ اُخْبُرْ تَقْلِهِمْ, (Ṣ,) or تَقْلِهُ, (A, Ḳ,) *I found the people* to be persons of whom it is said thus: [*Try, prove*, or *test*, them, or him, and *thou wilt hate them*, or *him*:] i. e. there is not one [of them] but his conduct is hated when it is tried, or proved, or tested: (Ḳ:) or when thou triest, provest, or testest, them, thou wilt hate them: the imperative form being used, but the meaning being that of an enunciative: (Ṣ, A, L, B:) [وَجَدْتُ is a verb of the kind called أَفْعَالُ القُلُوبِ, which govern two objective complements; therefore اُخْبُرْ تَقْلِهِمْ and اُخْبُرْ تَقْلِهُ are for مَقْلِيِّينَ عِنْدَ الخِبْرَةِ and مَقْلِيًّا عند الخبرة.] = خَبَرَ الأَرْضَ, [and, as appears from a passage in the L, ▾خبرها, (see خَبْرٌ,)] *He furrowed*, or *ploughed, the land for sowing*. (Mṣb.) = خَبَرَ الطَّعَامَ, (Ḳ,) aor. ـُ, inf. n. خَبْرٌ, (TA,) *He made the food greasy; or put grease to it*. (Ḳ, TA.) = خَبِرَ *It* (a place) *was*, or *became, what is termed* خَبْرَآءُ: (Ṣ:) or *abounded with* سِدْر [or *lote-trees*]. (TA.) —— And خَبِرَتِ الأَرْضُ, aor. ـَ, (Ḳ,) inf. n. خَبَرٌ, (TA,) *The land*, or *ground, abounded with* خَبَار [app. meaning *soft soil*: see 3]. (Ḳ.) = خَبُرَت, [probably خَبَرَتْ, like غَزُرَتْ &c.,] inf. n. خُبُور, ‡ *She* (a camel) *abounded with milk*. (Lḥ, TA. [See خَبْرٌ.])

2: see 4, in two places: = and see 1.

3. خَابَرَهُ, (TA,) inf. n. مُخَابَرَةٌ, (Ṣ, A, Mgh, Mṣb, Ḳ, &c.,) [*He made a contract*, or *bargain, with him to till and sow and cultivate land for a share of its produce*:] the inf. n. signifies i. q. مُزَارَعَةٌ [i. e. *the making a contract*, or *bargain, with another to cultivate land for a share of its produce*], (AO, Lḥ, Ṣ, A, IAth, Mgh, Mṣb,) *for somewhat of its produce*, (Ṣ, Mṣb,) or *for a third or a quarter*, (AO, Mgh,) or *for a determined share, such as a third or a quarter or some other portion*, (IAth,) or *for half or the like*: (so in some copies of the Ḳ and in the TA:) or *the tilling the ground for half or the like*: (so in other copies of the Ḳ:) and i. q. مُؤَاكَرَةٌ: (Ḳ:) and ▾خَبْرٌ is syn. with مُخَابَرَةٌ: (Ṣ, Ḳ:) it is a forbidden practice: (A, Mgh, TA:) it is from خَبِيرٌ signifying "a tiller, or cultivator, of land:" (Ṣ, Mgh:) or from خَبَرَ "he furrowed, or ploughed (land) for sowing;" whence خَبِيرٌ also: (Mṣb:) or from خَبِرَتِ الأَرْضُ "the land abounded with خَبَار:" or from [the fortress of] خَيْبَر, because the Prophet made it to remain in the possession of its inhabitants for half of its revenue; and therefore it was said, خَابَرَهُمْ. (TA.)

4. اخبرهُ, [inf. n. إِخْبَارٌ;] (Ṣ, A, Mṣb, Ḳ;) and ▾خَبَّرَهُ, (Ṣ, A, Ḳ,) inf. n. تَخْبِيرٌ; (Ḳ;) are syn. [as signifying *He informed him, told him*, or *acquainted him*]. (Ṣ, A, Ḳ.) You say, أَخْبَرْتُهُ بِكَذَا, (Ṣ, Mṣb,) [and عَنْ كَذَا,] and ▾خَبَّرْتُهُ, (Ṣ,) [*I informed him*, or *told him, of such a thing*; or *acquainted him with such a thing*; or *made him to know the internal*, or *real, state of such a thing*.] And ▾اخبرهُ خُبُورَهُ, i. e. أَنْبَأَهُ مَا عِنْدَهُ [*He informed him*, or *told him, of what he had*, or *knew*]. (Ḳ. [Whether it be meant that اخبر is doubly trans. without a particle, in this instance, like أَعْلَمَ, or whether خبورة be a quasi-inf. n., is not explained.]) One says also, تُخْبِرُ عَنْ مَجْهُولِهِ مَرْآتُهُ ‡ [*His aspect acquaints* one *with his unknown state* or *qualities*]. (A.) [And اخبر عَنْهُ *He predicated of him*, or *it*.] = أَخْبَرْتُ اللِّقْحَةَ ‡ *I found the milch camel to be abounding with milk*. (Ḳ. [See 1, last sentence.])

5: see 1, in two places: = and see 10, in four places. = تَخَبَّرُوا خُبْرَةً, (Ḳ,) or تَخَبَّرُوا خَبْرَةً, (Ṣ,) *They bought a sheep or goat*, (Ṣ, Ḳ,) *for different sums*, (TA,) *and slaughtered it*, (Ṣ, Ḳ,) *and divided its flesh among themselves*, (Ṣ, TA,) *each of them receiving a share proportioned to the sum that he had paid*. (TA.)

8: see 1, in three places. = مَا اخْتَبَرْتَ لِأَهْلِكَ *What* خُبْرَة, or *flesh-meat, hast thou bought for thy family?* (TA.)

10. استخبرهُ (A, Ḳ) and ▾تخبّرهُ (Ḳ) *He asked*, or *sought*, or *desired, of him information*, or *news*, or *tidings*: (A,* Ḳ:) or *he asked him respecting news, or tidings, and desired that he should inform him thereof*. (TA.) And استخبر and ▾تخبّر, (Ṣ,) or استخبر الخَبَرَ and ▾تخبّرهُ, (TA,) *He asked*, or *inquired, after the news*, or *tidings*, (Ṣ, TA,) *that he might know the same*: (TA:) and ▾تخبّر الأَخْبَارَ *He searched after the news*, or *tidings, diligently*, or *time after time*. (A, TA.)

خَبْرٌ: see خَبْرٌ. = Also *Trees of the kind called* سِدْر [or *lote-trees*], (Lth, Ḳ,) and أَرَاك, *with abundant herbage around them*; (Lth;) as also ▾خَبْرٌ: (Lth, Ḳ:) [both coll. gen. ns.:] ns. un. خَبْرَةٌ and خِبْرَةٌ. (TA.) —— *Seed-produce*. (Ḳ.) —— *A place where water rests*, or *stagnates, in a mountain*: (Ḳ:) *a place where water has fallen, such as the water-course has furrowed* (خَبَرَ [perhaps a mistranscription for حَفَرَ]) *in the summits* (رُؤُوس) [*of mountains*], *and through which one wades*. (L.) = *A large* [*leathern water-bag of the kind called*] مَزَادَة [q. v.]; (Ṣ, Ḳ;) as also ▾خَبْرَاءُ (Kr, Ḳ) and ▾خَبِرٌ: (Ḳ:) but this last is disallowed, in the sense above-explained, by AHeyth; and others say that the first word is better: (TA:) pl. of the first خُبُورٌ. (Ṣ, Ḳ.) —— Hence, by way of comparison thereto, (Ṣ,) ‡ *A she-camel abounding with milk*; (Ṣ, Ḳ;) as also ▾خَبِرٌ, (Ḳ,) and ▾مَخْبُورَةٌ [نَاقَةٌ]. (TA.)

خُبْرٌ (Ṣ, A, Ḳ) and ▾خِبْرٌ (Ḳ) and ▾خُبُرٌ, an inf. n., (Mṣb,) and ▾خُبُورٌ, also an inf. n., (TA,) and ▾خِبْرَةٌ and ▾خُبْرَةٌ and ▾مَخْبُرَةٌ, (Ḳ,) *Knowledge*, syn. عِلْمٌ, (Ṣ, A, Mṣb, Ḳ,) بِشَىْءٍ [*of a thing*]:

(A, Ḳ:) or, accord. to some, خُبْرٌ signifies *knowledge of the secret internal state:* and ↓ خِبْرَةٌ and ↓ خُبْرَةٌ signify *knowledge of the external and internal state;* or, as some say, *of secret internal circumstances* or *properties;* but this necessarily involves acquaintance with external things. (TA.) You say, لِى بِهِ خُبْرٌ and ↓ خِبْرَةٌ [&c.] *I have knowledge of it.* (TA.) And مَا لِى بِهِ خُبْرٌ [&c.] *I have not knowledge of it.* (A.) — See also خِبْرَةٌ. ═ And see خَبِيرٌ: ═ and خُبْرَةٌ.

خُبْرٌ: see خَبِيرٌ: ═ and see also 3: ═ and see خَبَرٌ, in two places.

خَبَرٌ [originally] an inf. n. of خَبُرَ: see خُبْرٌ. (TA.) — Also *Information; a piece of information; a notification; intelligence; an announcement; news; tidings; a piece of news; an account; a narration,* or *narrative; a story;* syn. نَبَأٌ; (T, Ḳ;) *that comes to one from a person of whom he asks it:* (TA:) or خَبَرٌ and نَبَأٌ are not synonymous; for, accord. to Er-Rághib and others, the latter relates to a thing of great importance: and accord. to the leading authorities in lexicology and the science of conventional language, the former signifies properly, and in its common acceptation, *what is related from another* or *others:* to which authors on the Arabic language add, that it may be *true* or *false:* (MF:) or *what is related from another* or *others, and talked of:* (Mṣb:) pl. أَخْبَارٌ, (Ṣ, Mṣb, Ḳ,) and pl. pl. أَخَابِيرُ. (Ḳ.) — By the relaters of traditions, it is used as *syn.* with حَدِيثٌ [signifying *A tradition;* or *narrative relating,* or *describing, a saying* or *an action &c. of Mohammad*]: (TA:) or this latter term is applied to what comes from the Prophet; and خَبَرٌ, to *what comes from another than the Prophet;* or *from him* or *another;* and أَثَرٌ, to what comes from a Companion of the Prophet; but it may also be applied to a saying of the Prophet. (Kull p. 152.) — [In grammar, as correlative of مُبْتَدَأٌ, *An enunciative:* and as correlative of اِسْمٌ, the *predicate* of the non-attributive verb كَانَ and the like, and of مَا &c.] — Also A man's *state,* or *case;* الأَمْرُ الَّذِى هُوَ عَلَيْهِ. (Ḥar p. 20.)

خَبْرٌ: see خَبِيرٌ, in two places. ═ See also خَبْرَآءُ: — خَبِرَةٌ, or أَرْضٌ خَبِرَةٌ; and مَوْضِعٌ خَبِرٌ, and خَبِرٌ alone: see خَبْرَآءُ.

خُبْرَةٌ: see خُبْرٌ, in two places. ═ Also *A portion,* or *share,* (A'Obeyd, Ṣ, A, Mgh, Ḳ,) *which one takes, of flesh-meat* or *fish.* (A'Obeyd, Ṣ, Ḳ.) — *A sheep,* or *goat, which is bought by a number of persons,* (Ṣ, Ḳ,) *for different sums,* (TA,) *and slaughtered,* (Ṣ, Ḳ,) *and of which the flesh is then divided by them among themselves,* (Ṣ,) *each of them receiving a share proportioned to the sum that he has paid;* (TA; [see 5;]) as also ↓ خُبْرَةٌ: (Ḳ:) and ↓ شَاةٌ خُبْرَةٌ *a sheep,* or *goat, divided among several persons;* thought by ISd to be formed by rejection of the augmentative letter [in its verb تَخَبَّرَ]. (TA.) — *What one buys for his family;* as also ↓ خُبْرٌ: (Ḳ:) accord. to some, (TA,) *flesh-meat* (Ḳ, TA) *which one buys for his family.* (TA.) — *Food,* (Ḳ, TA,) *consisting of flesh-meat* and *other kinds.* (TA.) — *A thing brought forward* or *offered* [*for entertainment*]. (Lḥ, Ḳ.) So in the saying, اِجْتَمَعُوا عَلَى خُبْرَتِهِ [*They congregated over what he had brought forward,* or *offered, for their entertainment*]. (Lḥ.) — *A mess of crumbled,* or *broken, bread, moistened with broth, large,* (Ḳ, TA,) *and greasy.* (TA.) — *A bowl in which are bread and flesh-meat for four or five* [*persons*]. (Ḳ.) — *Food which the traveller carries in his journey,* (Ḳ,) *and provides for himself.* (TA.) — *Seasoning, condiment,* or *savoury food;* as also ↓ خَبِيرٌ: whence the saying, أَتَانَا بِخُبْزَةٍ وَلَمْ يَأْتِنَا بِخُبْرَةٍ [*He brought us a cake of bread, but he brought us not any seasoning*]. (TA.) — Hence, by the Karaj, whose land is adjacent to 'Irák el-'Ajam, applied to *A date;* and by some of them pronounced خُبْلَةٌ. (TA.)

خِبْرَةٌ *Trial, proof,* or *test;* (Ṣ, Mṣb, Ḳ;) and so ↓ خُبْرٌ, (Ṣ, Ḳ,) as in the saying, صَدَّقَ الخَبَرَ الخُبْرُ [*The trial, proof,* or *test, verified the information*]. (Ṣ.) — See also خُبْرٌ, in three places.

خَبْرَآءُ, (Lth, Ṣ, Ḳ,) and أَرْضٌ خَبْرَآءُ, (Ṣ,) and ↓ خَبِرَةٌ, (Lth, Ḳ, [in the CK خَبْرَةٌ,]) or أَرْضٌ خَبِرَةٌ, (Ṣ,) *A plain,* or *level, tract of land, that produces* سِدْر [or *lote-trees*]: (Ṣ, Ḳ:) or *a tract abounding with trees, in the lower part of a meadow, in which water remains until the hot season, and in which grow trees of the kinds called* سِدْر and أَرَاك, *with abundant herbage around them:* (Lth:) the pl. of خَبْرَآءُ is خَبَارَى and خَبَارٍ and خَبْرَاوَاتٌ (Ṣ, Ḳ) and خَبَارٌ; (Ḳ;) and the pl. of خَبِرَةٌ is ↓ خَبِرٌ; (TA;) [or this is neither a pl. nor a quasi-pl. n.: it may be a coll. gen. n.: but it is probably only an epithet, of which خَبِرَةٌ is the fem.; for] one says also ↓ مَوْضِعٌ خَبِرٌ, (Ṣ, TA,) meaning *a place abounding with* سِدْر. (TA.) — خَبْرَآءُ also signifies *A place where water collects and stagnates:* (TA:) or *where water collects and stagnates at the roots of trees of the kind called* سِدْر: (Ḳ, TA:) or *a round low tract of level ground in which water collects.* (T.) — See also خَبَارٌ. ═ And see خَبْرٌ.

خَبَارٌ *Soft land* or *soil,* (IAạr, Ṣ, A, Mgh, Ḳ,) *in which are burrows* (IAạr, Ṣ, A) *and hollows;* (IAạr;) as also ↓ خَبْرَآءُ: (A:) or *soft land* or *soil, in which beasts sink and are embarrassed:* or *crumbling ground, in which the feet of beasts sink.* (TA.) It is said in a prov., مَنْ تَجَنَّبَ الخَبَارَ أَمِنَ العِثَارَ [*He who avoids soft ground in which the feet sink will be secure from stumbling*]. (A, Ḳ.) — Also *Heaps of earth,* or *dust, collected at the roots of trees.* (Ḳ,* TA.) — And *Burrows of* جِرْذَان [or *large field-rats*]: (Ḳ:) [a coll. gen. n.:] n. un. with ة. (TA.)

الخَبُورُ *The lion.* (Ḳ.)

خَبِيرٌ *Knowing; having knowledge;* (Ṣ, A, Mṣb;) as also ↓ خَبْرٌ: (AḤn:) or *possessing much knowledge with respect to internal things;* like شَهِيدٌ with respect to external things: (L in art. شهد:) or *possessing knowledge of matters of information, news, tidings, accounts, narratives,* or *stories; of what is termed* خَبَرٌ; (Ḳ;) or *of what are termed* أَخْبَار; (TA;) as also ↓ خَابِرٌ and ↓ خَبِرٌ, (Ḳ,) which last is thought by ISd to be a possessive [as distinguished from a verbal] epithet, (TA,) [or it is from خَبِرَ, a form which ISd may not have known,] and ↓ خُبْرٌ, (Ḳ,) which is an intensive epithet: (TA:) also *informed; possessing information.* (TA.) You say, أَنَا بِهِ خَبِيرٌ *I have knowledge of it.* (A.) And [hence] الخَبِيرُ is a name of God, meaning *He who knoweth what hath been and what is or will be:* (TA:) or *He who well knoweth the internal qualities of things.* (Sharḥ Et-Tirmidhee.) — Also *Possessing knowledge of God,* (Ḳ, TA,) *by being acquainted with his names and his attributes.* (TA.) — *A lawyer; one skilled in the law,* or *practical religion.* (TA.) — *A head,* or *chief.* (TA.) ═ *A tiller,* or *cultivator, of land.* (Ṣ, Mgh, Mṣb, Ḳ.) ═ *Fur,* or *soft hair,* syn. وَبَر, (Ṣ, Ḳ,) of camels, and ‡ of the wild ass. (TA.) — *Hair that has fallen:* and with ة, *a portion thereof.* (Ḳ.) [See also خَبِيرَةٌ below.] — ‡ *Plants,* or *herbage;* (Ṣ, Ḳ, TA;) *fresh herbage:* (Ḳ, TA:) likened to the وَبَر of camels, because growing like the latter: and *seed-produce.* (TA.) It is said in a trad., نَسْتَخْلِبُ الخَبِيرَ ‡ *We cut* (Ṣ, TA) *with the reaping-hook,* (TA,) *and eat, the plants,* or *herbage.* (Ṣ, TA.) — *Froth,* or *foam:* (TA:) or the *froth,* or *foam, of the mouths of camels.* (Ṣ, Ḳ, TA.) ═ *Seasoned,* or *made savoury.* (TA.) — See also خُبْرَةٌ.

خُبُورَةٌ: see 4.

خَبِيرَةٌ: see خُبْرَةٌ, in two places. ═ Also *Good wool, of the first shearing.* (Ḳ.) [See also خَبِيرٌ.] ═ *An invitation to the* عَقِيقَة [q. v.] *of a boy.* (TA.)

خَابِرٌ: see خَبِيرٌ. — Also One *who tries, proves,* or *tests, things; having experience.* (TA.)

خَابُورٌ *A certain plant:* (Ḳ:) or *a kind of tree, having a blossom beautiful and bright, yellow, and of good odour, with which gardens are adorned:* MF says, I do not think it to be found in the East. (TA.)

الخَيْبَرَى, (Ḳ, TA,) in some copies of the Ḳ written الخَيْبُرَى, (TA,) *The black serpent.* (Ḳ.) So in the saying, بَلَاهُ اللّٰهُ بِالخَيْبَرَى [*May God afflict him,* or *it, with the black serpent*]: app. because a ruined place becomes the resort of deadly serpents. (TA.) ═ One says also, عَلَيْهِ الدَّبَرَى وَحُمَّى خَيْبَرَى [*May perdition befall him, and the fever of Kheyber:* الدبرى being app. an inf. n., syn. with الدَّبَار, which is used in a similar phrase (عَلَيْهِ الدَّبَارُ) mentioned in the TA in art. دبر, and خَيْبَر being altered to خَيْبَرَى, as is indicated in the Ṣ, in order to assimilate it in form to الدبرى]: (Ṣ, TA:) the fever of

Kheyber is مُتَنَاذَرَةٌ [i. e. a fever "against which people warn one another," because it is generally fatal]. (TA.) [See also خَابِرٌ.]

أَخْبَارِىٌّ *A historian:* a rel. n. formed from the pl., like أَنْصَارِىٌّ and أَنْمَاطِىٌّ. (TA.)

مَخْبَرٌ (Ṣ) and ↓ مَخْبَرَةٌ (Ṣ, Ḳ) and ↓ مَخْبُرَةٌ (Ṣ, M) *The internal state; an internal,* or *intrinsic, quality;* the *intrinsic,* or *real, as opposed to the apparent, state,* or *to the aspect,* of a thing; [whether *pleasing* or *displeasing;* but when used absolutely, meaning the former;] opposite of مَرْآةٌ (Ṣ, Ḳ) and of مَنْظَرٌ [q. v.]. (Ṣ.) See also مَخْبَرَانِىٌّ.

مَخْبُرَةٌ: see خُبْرٌ: — and see مَخْبَرٌ. — Also [*A privy;*] *a place where excrement,* or *ordure, is voided.* (Ḳ.)

مَخْبَرَةٌ: see مَخْبَرٌ.

رَجُلٌ مَخْبَرَانِىٌّ *A man of goodly internal,* or *intrinsic, qualities;* syn. ↓ ذُو مَخْبَرٍ; like مَنْظَرَانِىٌّ as meaning ذُو مَنْظَرٍ. (TA.)

مَخْبُورٌ *Well seasoned;* (Ḳ;) *having much grease.* (TA.) — نَاقَةٌ مَخْبُورَةٌ: see خَبْرٌ, last sentence.

مُخْتَبَرٌ † A camel *having much flesh.* (TA.)

خبز

1. خَبَزَ خُبْزًا, (Ṣ, Ḳ,) aor. ـِ, (Ḳ,) inf. n. خَبْزٌ, (Ṣ, Ḳ,) *He made* [or *kneaded and baked*] خُبْز [or *bread*]; (Ḳ, TA;) as also ↓ اِخْتَبَزَهُ: (Sb, Ṣ, TA:) or the ↓ the latter signifies *he made* [or *kneaded and baked*] *it for himself:* (Ḳ:) or ↓ اختبز signifies *he kneaded flour, and made dough of it, and then baked it in a* مَلَّة [see خُبْزَة below] or *in an oven:* (T, TA:) [and ↓ يُخْتَبَزُ signifies *it is made into bread:* see Ṣ and Ḳ voce فَتٌّ.] — خَبَزَ القَوْمَ, (Ṣ, A,) aor. ـِ, (TA,) inf. n. خَبْزٌ, (A, Ḳ,) *He fed the people,* or *company of men, with* خُبْز [or *bread*]: (Ṣ, A, Ḳ:*) like as تَمَرَهُمْ signifies "he fed them with تَمْر:" (A:) but Lḥ quotes the saying of certain of the Arabs, أَتَيْتُ بَنِى فُلَانٍ فَخَبَزُوا وَحَاسُوا وَأَقَطُوا, meaning [*I came to the sons of such a one, and*] *they fed me with* خُبْز *and* حَيْس *and* أَقِط: he does not say خَبَزُونِى وَحَاسُونِى وَأَقَطُونِى. (TA.) — خَبَزَهُ, aor. ـِ, (TḲ,) inf. n. خَبْزٌ, (Ḳ,) † *He beat him,* or *it:* (Ḳ,* TḲ:) accord. to some, *with the hand:* or *with the two hands:* (TA:) and some say that خُبْز [or bread] is thus called because they beat it with their hands: but this assertion is not valid: (TA:) and you say also, خَبَطَنِى بِرِجْلِهِ, and خَبَزَنِى, ‡ [*He beat me with his foot,*] and تَخَبَّطَنِى and ↓ تَخَبَّزَنِى. (A, TA.) And خَبَزَ البَعِيرُ, (TḲ,) inf. n. خَبْزٌ, (Ṣ, Ḳ,) ‡ *The camel beat the ground with his fore foot,* (Ṣ,* Ḳ,* TA,) or, as in some lexicons, *with his fore feet.* (TA.) And ↓ تَخَبَّزَتِ الإِبِلُ السَّعْدَانَ † *The camels beat the* [*herbage called*] سعدان *with their legs.* (TA.)

5: see 1, latter part, in two places.

8: see 1, first sentence, in four places.

خُبْزٌ a word of well-known meaning; (Ḳ;) [*Bread;*] *that which is eaten.* (Ṣ.) It is said in a prov., كُلُّ أَدَاةِ الخُبْزِ عِنْدِى غَيْرُهُ [*All the apparatus of bread is in my possession except it,* namely, the bread itself]: the origin of which was this: a company of men demanded hospitality of a certain man; and when they sat down, he threw down a [piece of leather such as is called] نِطْع, and put upon it a mill-stone, and adjusted its pivot, and covered it [with the upper stone]: and the presence of his apparatus made the company to wonder: then he took the handle of the mill, (هَادِى الرَّحَى,) and began to turn it: whereupon they said to him, What dost thou? and he answered in the words of this proverb. (Ḳ.) — [Hence,] الخُلَّةُ خُبْزُ الإِبِلِ ‡ [*Sweet herbage is the bread of camels:* and الحَمْضُ فَاكِهَتُهَا, or أَدَمُهَا, *sour herbage is their fruit,* or *fleshmeat*]. (A, TA.)

خُبْزَةٌ *i. q.* طُلْمَةٌ; (Ṣ, A, Ḳ;) meaning *Dough put in a* مَلَّة, *until it is thoroughly baked,* (Ṣ, TA,) i. e., *in ashes,* and *earth, in which fire is kindled;* (TA;) *a cake of bread,* (MA, KL,) [or *lump of dough,*] *baked in ashes* (KL) [or *in any way*]; *i. q.* قُرْصٌ and قُرْصَةٌ. (Ḳ in art. قرص.) — Also *A large* ثَرِيدَة [or *mess of crumbled or broken bread moistened with broth*]: or, as some say, *flesh-meat.* (TA.) [See also خَبِيزٌ.]

خَبِيزٌ Bread *made* [or *kneaded and baked*], (Ḳ, TA,) of whatever grain it be. (TA.) — Also *i. q.* ثَرِيد [Bread *crumbled or broken, and moistened with broth*]. (Ṣgh, Ḳ.) [See also خُبْزَة.] — Also a vulgar term for خَبِيصَة. (Esh-Shereeshee, in Ḥar p. 21.)

خِبَازَةٌ The *trade,* or *occupation, of the* خَبَّاز. (Ḳ.)

خُبَّازَى: see خُبَّازٌ.

خَبَّازٌ *A maker of bread; one whose office it is to make bread:* (TA:) *a baker;* syn. فَرَّانٌ. (Mṣb in art. فرن.)

خُبَّازٌ (IDrd, Ṣ, Ḳ) and ↓ خُبَّازَةٌ, (Ḳ,) [or the former is a coll. gen. n., and the latter the n. un.,] and ↓ خُبَّازَى (IDrd, Ṣ, Ḳ) [which last is the most common form] and ↓ خُبَازَى, (Ḳ,) or when with teshdeed the ى is elided, (IDrd,) and ↓ خُبَيْزٌ, (Ḳ,) [*Malva,* or *mallow;*] *a certain plant, well known,* (Ṣ, Ḳ, TA,) *of the leguminous kind, having broad leaves and a round fruit;* [whence perhaps its name;] accord. to the Minháj, *a species of the* مُلُوخِيَّة [*corchorus olitorious,* or *Jew's mallow*]: or, as some say, the مُلُوخِيَّة is the garden-kind, and the خُبَّازِى is *the wild kind:* some also say that *the* بَقْلَة يَهُودِيَّة [*sonchus,* or *sow-thistle,*] *is one of the species of* خُبَّازَى; and *there is a kind thereof that turns with the sun.* (TA.)

خُبَيْزٌ:
خُبَّازَةٌ: } see the next preceding paragraph.
خُبَّازَى:

خَابِزٌ A man *possessing* خُبْز [or *bread*]: (Ṣ, Ḳ:) like تَامِرٌ [possessing dates] and لَابِنٌ [possessing milk]. (Ṣ.)

مَخْبَزٌ *An oven;* syn. فُرْنٌ. (M and Ḳ in art. فرن.)

مَخْبَزَةٌ *A place where bread is made:* pl. مَخَابِزُ. (Meyd, in Golius.)

خبص

1. خَبَصَهُ, aor. ـِ, (A, Mṣb, Ḳ,) inf. n. خَبْصٌ, (Mṣb,) *He mixed it.* (A, Mṣb, Ḳ.) — خَبَصَ, aor. ـِ; and ↓ خَبَّصَ, inf. n. تَخْبِيصٌ; *He turned over and mixed and made* [خَبِيص, q. v.]: and ↓ تَخَبَّصَ (Ḳ) and ↓ اختبص (A, Ḳ) *he made,* or *prepared, for himself* خَبِيص. (Ḳ,* TA.)

2:
5: } see 1.
8:

10. استخبص ضَيْفُهُمْ *Their guest asked for,* or *demanded, a mess of* خَبِيص [q. v.]. (A, TA.)

خَبِيصٌ *Mixed;* syn. مَخْبُوصٌ. (TA.) — *A kind of food,* (Mṣb, TA,) *sweet,* (TA,) *well known,* (Ṣ, Mṣb, TA,) *made of dates and clarified butter,* (A, Ḳ, TA,) *mixed together:* (TA:) [Golius adds, on the authority of Ibn-Maaroof, "aut amylo et defruto;" app. meaning, or *of starch and of new wine of which half or a third part has been boiled away:* and one kind, called خَبِيصٌ مُرَمَّلٌ, was made *with coarse flour:* (see جَرِيش:)] of the measure فَعِيلٌ in the sense of the measure مَفْعُولٌ: (Mṣb:) خَبِيصَةٌ is a more particular term [signifying *a mess of* خَبِيص]. (Ṣ.)

مِخْبَصَةٌ *A spoon, or thing like a spoon, with which* خَبِيص *is made;* (Ṣ, as in two different copies;) *a spoon with which* خبيص *is stirred about, or turned over,* (A, L, Ḳ,) *in the* [*vessel called*] طِنْجِير: (Ḳ:) or the *thing in which* خبيص *is stirred about, or turned over.* (L, TA.)

خبط

1. خَبَطَهُ, aor. ـِ, (Mṣb, Ḳ, TA,) inf. n. خَبْطٌ, (Mṣb, TA, &c.,) *He struck,* or *beat,* (Mṣb, TA,) anything: (TA:) or *he struck,* or *beat,* it, or him, *vehemently:* (M, Ḳ, TA:) or خَبْطٌ signifies a camel's *striking,* or *beating,* a thing *with his fore foot:* (T, TA:) or in the cases of beasts, (دَوَابّ, [generally meaning horses and mules and asses,]) the *striking,* or *beating, with the fore feet;* not with the hind feet: and in the case of the camel, *with the fore foot and the hind foot:* or *vehement treading;* or *of the fore feet of beasts* (دوابّ): (TA:) or, accord. to the Keshsháf, the act of *striking,* or *beating, in a way*

that is not right: or, as some say, the *going*, or *journeying*, *upon what is not the middle*, or *main part*, *of the road*, or *what is not the main road*, or *upon a road not open to view*: or *continuous*, or *consecutive*, *striking*, or *beating*, *in different ways*: and afterwards tropically applied to any ‡*beating*, or *striking*, *that is not approved*: or originally, the *striking*, or *beating*, *with the fore foot* or *the hind foot*, and *the like*: (MF, TA:) with the fore feet or legs, it is like رَمْحٌ with the hind feet or legs. (TA.) You say, of a camel, خَبَطَ الأَرْضَ بِيَدِهِ, (Mṣb,) or خَبَطَ, (Ṣ, Ḳ,) inf. n. as above, (Ṣ,) *He struck*, or *beat*, *the ground with his fore foot*: (Ṣ, Mṣb:) or *he struck*, or *beat*, *vehemently the ground with his fore foot*; and ↓تخبّط and ↓اختبط signify the same: (Ḳ:) it is said in the O that خَبَطَهُ signifies *he struck him with his fore foot*, or *hand*, *and prostrated him*, as also ↓تَخَبَّطَهُ: and ↓اختبط, said of a camel, is *syn. with* خَبَطَ: and in the T, that ↓تَخَبَّطَنِى بِرِجْلِهِ is *syn. with* خَبَطَنِى. (TA.) Hence the trad., لَا تَخْبِطُوا خَبْطَ الجَمَلِ [lit. *Ye shall not beat* the ground *as the camel does with his fore foot* in rising]; meant to forbid a man's putting forward his foot in rising from prostration [in prayer]. (TA.) And خَبَطَهُ, (Ḳ, TA,) aor. and inf. n. as above, (TA,) signifies also *He trod him*, or *it*, *vehemently*, (Ḳ, TA,) *as the camel does with his fore foot*. (TA.) — Hence, (Ṣ,) فُلَانٌ يَخْبِطُ خَبْطَ عَشْوَآءَ (Ṣ,* TA) ‡[*Such a one goes at random*, *in a headstrong and reckless manner*,] *like the weak-sighted she-camel that beats the ground with her fore feet* (تَخْبِطُ) *as she goes along, not guarding herself from anything*. (Ṣ, TA.) It is a prov., applied to him who turns away from a thing as though he were not cognizant of it: or to him who is continually falling into a thing. (Ḥar p. 239.) Zuheyr says,

* رَأَيْتُ المَنَايَا خَبْطَ عَشْوَآءَ مَنْ تُصِبْ *
* تُمِتْهُ وَمَنْ تُخْطِئْ يُعَمَّرْ فَيَهْرَمِ *

I saw the fates [*treading mankind*] *like the treading of the weak-sighted she-camel*: *whom they smote, him they killed*: *and whom they missed, he was made to continue in life so that he lived to extreme old age*. (TA, and EM p. 132.) In like manner you say, فُلَانٌ يَخْبِطُ فِى عَمْيَآءَ ‡*Such a one undertakes what he undertakes with ignorance*. (TA.) And خَبَطَ أَمْرَهُ عَلَى غَيْرِ بَصِيرَةٍ †[*He prosecuted his affair without mental perception*, or *without certainty*]. (Ṣ in art. عشو, q. v.) And يَخْبِطُ فِى الظَّلَامِ ‡*He goes in the night without a lamp, and so becomes confounded and perplexed, and unable to see his right course, and errs from the way, and perchance may fall into a well*. (TA.) And خَبَطَ اللَّيْلَ, (Ḳ, TA,) aor. and inf. n. as above, (TA,) ‡*He went*, or *journeyed*, *in the night without direction*. (Ḳ, TA.) And بَاتَ يَخْبِطُ الظَّلْمَآءَ ‡[*He passed the night traversing the darkness without direction*]. (TA.) خَبْطٌ is said to signify †The act of *journeying*, or *going*, *without direction*: or *upon what is not the middle*, or *main part*, *of the road*, or *what is not the main road*. (TA.) — [And hence, perhaps,] خَبَطَهُ ‡*He asked of him a benefit*, or *favour*, *without any tie of relationship*; (Ḳ,* TA;) as also ↓اختبطهُ: (IB, Ḳ:) or this is from خَبَطَ وَرَقَ الشَّجَرِ [explained in what follows]: (Ḥar p. 425:) or the latter, [or both,] *he came to him seeking his beneficence without any such tie*: (Ṣ:) or *he came to him seeking a gift*; because he who does so must beat the ground with his feet: (IF:) and you also say, ↓اختبط مَعْرُوفَهُ. (Aboo-Málik, TA.) [The latter verb is the more common. See also 10.] — And ‡*He conferred a benefit upon him without there having been any acquaintance between them*, (Ṣ, Ḳ, TA,) *and without there being anything to draw them near, and without there being any relationship*: (TA:) and خَبَطَهُ بِخَيْرٍ signifies *the same*: (TA:) or *he bestowed on him a benefit*, (Ḳ, TA,) *being asked*: (TA:) and you say also, ↓اختبطه بِخَيْرٍ: (Aboo-Málik, TA:) and خَبَطَ قَوْمًا بِخَيْرٍ signifies †*He benefited them*. (TA.) 'Álḳameh Ibn-'Abadeh says, (Ṣ, TA,) praising El-Ḥárith Ibn-Abee-Shemir, (TA,)

* وَفِى كُلِّ حَىٍّ قَدْ خَبَطْتَ بِنِعْمَةٍ *

†[*And upon every tribe thou hast conferred benefit*, app. meaning *without being related to them*]: (Ṣ, TA:) but it is said in a marginal note to the Ṣ, that خَبَطْتَ would be better; and so it is accord. to one relation: in the L, however, it is said that خَبَتَّ would be more agreeable with analogy. (TA.) Accord. to AZ, خَبَطْتُ الرَّجُلَ, inf. n. خَبْطٌ, signifies †*I held loving communion, commerce*, or *intercourse*, *with the man*. (TA.) — [In respect of the places which I have given to the above-mentioned significations of asking and conferring a benefit, I have followed the opinion of IF; but it is said in the TA, and, I think, with greater probability, that they are from what here next follows.] — خَبَطَ الوَرَقَ مِنَ الشَّجَرِ, aor. ـِ, (Mṣb,) inf. n. خَبْطٌ, (Lth, T, Mṣb,) *He made the leaves to fall from the trees*: (Mṣb:) or *he beat the leaves of the trees*, (Lth, T,) meaning large trees of the kind called طَلْح, [acacia, or mimosa, gummifera,] *with a staff*, or *stick*, (Lth,) *so that they fell off*, or *became scattered*, (Lth, T,) after which he gave them as food to camels; (Lth;) *refraining from injuring thereby the trunks and branches of the trees*: (T:) and ↓اختبط لَهُ خَبَطًا signifies the same as خَبَطَ. (TA.) And خَبَطَ الشَّجَرَةَ, (Ṣ, Ḳ,) aor. as above, (TA,) and so the inf. n., (Ṣ,) *He beat the tree with a staff*, or *stick*, *in order that its leaves might fall off*: (Ṣ:) or *he bound the tree, and then made its leaves to fall*, (Ḳ, TA,) *by beating it* with a staff, or stick, to give them as food to camels and other beasts. (TA.) The leaves are stored up for the camels; and in winter-time are bruised, or broken up, for them, and moistened with water, and given to them as fodder. (Ḥar p. 218.) Mohammad was asked, Does الغَبْطُ [i. e. "the wishing for a blessing on the condition that it shall not become transferred from its possessor"] injure [its author]? and he answered, لَا إِلَّا كَمَا يَضُرُّ العِضَاهَ الخَبْطُ [*No, save as the beating off the leaves injures the* trees called *'iḍáh*]; i. e., it only diminishes, without annulling, its author's recompense, like the beating off the leaves of the 'iḍáh, without cutting them down and extirpating them; for the leaves will grow again. (TA.) [See also art. غبط.] — Hence, (A, TA,) خَبَطَ القَوْمَ بِسَيْفِهِ ‡*He struck the people with his sword*. (A, Ḳ, TA.) — خَبَطَهُ الشَّيْطَانُ ‡*The devil touched him with a hurt*, (Ḳ, TA,) *so as to corrupt him*, or *disorder him, and render him insane*; (TA;) as also ↓تَخَبَّطَهُ: (Ḳ, TA:) or the latter, [which is the more common,] *the devil corrupted him*, or *disordered him*: (Ṣ, Mgh, Mṣb:*) lit., *struck him*: (Mgh, Mṣb:) or *prostrated him, and sported with him*: or *trampled upon him, and prostrated him*. (TA.) It is said in the Ḳur [ii. 276], لَا يَقُومُونَ إِلَّا كَمَا يَقُومُ الَّذِى يَتَخَبَّطُهُ↓ الشَّيْطَانُ مِنَ المَسِّ ‡[*They shall not rise save as he riseth whom the devil prostrateth by reason of possession*, or *insanity*]; i. e., as he who is affected by diabolical possession rises, in his state of possession, when he is prostrated, and falls: or it means, *whom the devil corrupts*, or *disorders*, *by rendering him insane*. (Ḳ,* TA.) [You say also, of a drug, ↓خبّط العَقْلَ †*It disordered the intellect*: see the act. part. n., below.] — خَبَطَ also signifies †*He* (a man) *threw himself down* (Ṣ, L, Ḳ) *where he was*, (Ṣ, L,) *to sleep*, (Ṣ, Ḳ,) or *and slept*. (L.) And †*He* (a man) *slept*. (A'Obeyd, TA.) In the Ḳ, قَامَ is erroneously put for نَامَ. (TA.) — خَبَطَ عَلَى البَابِ *He knocked upon the door*, or *at the door*. (TA.) — خَبَطَ العِرْقُ *The vein beat*, or *pulsated*. (TA.)

2: see 1, near the end of the paragraph.

5. تخبّط *It was*, or *became*, *in a state of commotion*, *agitation*, *convulsion*, *tumult*, or *disturbance*; syn. اِضْطَرَبَ. (Az, TA in art. حبط.) — It is also trans.: see 1; second sentence, in three places; and again, near the end of the paragraph, in two places.

8: see 1, in six places. — You say also, النَّاقَةُ تَخْتَبِطُ الشَّوْكَ †*The she-camel eats the thorns*. (Th, TA.)

10. استخبطهُ †*He asked of him a means of access*, *nearness*, *intimacy*, or *ingratiation*. (TA.)

خَبَطٌ *What is beaten by beasts*, (Ḳ, TA,) *with their feet*, (TA,) *and broken*. (Ḳ, TA.) — *Leaves* (Mṣb, Ḳ) *of any kind* (Ḳ) *that have been made to fall from a tree*; (Mṣb, Ḳ;) *by its being beaten with a staff*, or *stick*; (Ḳ,* TA;) *used as food for camels*: (TA:) and *leaves that have been beaten off with staves*, or *sticks*, *then dried, and ground, and mixed with flour or other substance, and beaten with the hand, and moistened in a basin, with water, until they have become viscous*, or *cohesive*, *when they are put into the mouths of camels*. (AḤn, Ḳ.) The word is of the measure فَعَلٌ in the sense of the measure مَفْعُولٌ, like many other instances that have been heard, (Mṣb,) as نَفَضٌ and هَدَمٌ. (TA.)

خَبْطَةٌ ‡*A touch*, or *stroke*, *of diabolical possession*, or *insanity*. (TA.) You say also, بِفُلَانٍ خَبْطَةٌ بِالمَسِّ ‡[*In such a one is a touch of diabolical possession*, or *insanity*]. (TA.) — †*A single act of a stallion-camel's covering of the female.* (TA.)

خُبَاطٌ *A certain malady,* (Ḳ,) *like diabolical possession,* or *insanity,* (Ṣ, Ḳ,) *but not identical therewith:* (Ṣ, TA:) the word is also related with ح (TA.) [See also خُبَاطٌ.]

فَرَسٌ خَبُوطٌ and ▾خَبِيطٌ *A horse that strikes,* or *beats, with his hind feet:* (Ḳ:) or *with his fore feet.* (T, TA.)

خَبِيطٌ *A watering-trough beaten by the feet of the camels, and so demolished:* (Ḳ:) or *a watering-trough;* so called because its clay is beaten with the feet at its construction: (TA:) or *a small watering-trough:* (Aboo-Málik, TA:) pl. خُبُطٌ (Ḳ.) — See also خَبُوطٌ.

خَبَاطِ, determinate, [and imperfectly decl.,] †*The stupid:* like خَضَارِ applied to "the sea." (TA.)

خَبَّاطُ عَشَوَاتٍ ‡*One who* [*frequently*] *goes in the night without a lamp, and so becomes confounded and perplexed, and unable to see his right course, and errs from the way, and perchance may fall into a well:* occurring in a trad. of 'Alee. (TA.)

خَابِطٌ *Going,* or *journeying, without direction:* or one *who beats the ground with his foot, and knows not in what land he is going;* either because of the darkness or because he is blind. (Ḥar p. 55.) You say, مَا أَدْرِى أَىُّ خَابِطِ لَيْلٍ هُوَ, (Ṣ, TA,) and أَىُّ خَابِطِ اللَّيْلِ, (TA,) ‡*I know not what man he is.* (Ṣ, TA.) — مَا لَهُ خَابِطٌ وَلَا نَاطِحٌ, ‡*He has not a camel nor a bull;* meaning *he has not anything.* (TA.) — خَابِطٌ also signifies *A beating,* or *throbbing, in the head.* (TA.)

أَخْبَطُ *That strikes,* or *beats,* (Ḳ, TA,) *the ground,* (TA,) *with his feet:* (Ḳ, TA:) by poetic license written أَخْبَطُ: (TA:) pl. خُبْطٌ. (Ḳ.)

مُخْبِطٌ *Still; motionless;* like مُخْبِتٌ: (TA in art. خبت:) or i. q. مُطْرِقٌ [*silent; not speaking:* or *lowering his eyes, looking towards the ground*]. (JK, Ḳ, TA. [In the CḲ, مُخْبَطٌ and مُطْرَقٌ.]) — See also مُخْتَبِطٌ.

مِخْبَطٌ *A staff,* or *stick, with which the leaves of trees are beaten off:* (Ḳ:) and ▾مِخْبَطَةٌ, also, signifies *a staff,* or *stick;* and *a rod,* or *twig:* (TA:) pl. of the former, مَخَابِطُ. (Ḳ, TA.)

مِخْبَطَةٌ: see what next precedes.

مُخَبِّطٌ لِلْعَقْلِ [*Disordering the intellect;* said of a drug]. (Ḳ in art. بنج.)

مُخْتَبِطٌ ‡One *who asks* [*a benefit* or *favour*] *of another without there being anything to draw him near, and without acquaintance.* (JK, TA.* [In the latter, ▾مُخْبِطٌ, which is doubtless a mistake, is explained in one place as signifying ‡One *who seeks a gift without any previous acquaintance.*])

خبعثن

خَبَعْثَنٌ: see خَبَعْثَنَةٌ.

خُبَعْثِنٌ: see خَبَعْثَنَةٌ. — Also *Plump in body;* or *fat, soft, thin-skinned, and plump, in body;* applied to any thing [i. e. to any animal]. (Ḳ.)

خَبَعْثَنَةٌ, like قَذَعْمِلَةٌ, *Big and strong;* (AO, Ṣ, Ḳ;) applied to a man; and (in like manner, TA) a lion; (Ḳ;) and so ▾خُبَعْثِنٌ, (Ṣ, Ḳ,) as in the phrase خُبَعْثِنُ الخَلْقِ [*big and strong of make*]; (AA, Ṣ;) and ▾خَبَعْثَنٌ: (Ḳ:) the first is applied as an epithet to a lion by Aboo-Zubeyd Eṭ-Ṭá-ee: and [the pl.] خَبَعْثَنَاتٌ, by El-Farezdaḳ, to camels. (Ṣ.)

خبل

1. خَبِلَ, aor. ـَ, inf. n. خَبَالٌ (JK, Ḳ, TA) and خَبَلٌ, (TA,) *He was,* or *became, corrupted, unsound, vitiated,* or *disordered,* [*in an absolute sense;* and particularly] *in his reason,* or *intellect:* (TA:) [or *he was,* or *became, in a corrupt,* an *unsound, a vitiated,* or *a disordered, state, occasioning him agitation like that of possession* or *insanity, by disease affecting the reason and thought:* (see خَبَالٌ, below:) and hence,] *he was,* or *became, possessed,* or *insane.* (JK, Ḳ, TA.) — And خَبِلَ, [inf. n. خَبَلٌ, (q. v. voce خَبَالٌ,)] *He was,* or *became, affected with* [*the palsy termed*] فَالِج. (JK.) — And خَبِلَتْ يَدُهُ *His arm,* or *hand, became corrupted, unsound, vitiated,* or *disordered, so as to be rendered motionless;* or *dried up;* or *became lost;* (Ḳ, TA;) or *was cut off.* (TA.) = خَبَلَهُ, (JK, Ṣ, Mṣb, Ḳ,) aor. ـِ, (Mṣb,) inf. n. خَبْلٌ; (TA;) and ▾خبّلهُ, (Ṣ, Ḳ,) inf. n. تَخْبِيلٌ; (TA;) and ▾اختبلهُ; (Ṣ, Ḳ;) *It* (grief, JK, Ḳ, and love, JK, T, TA, and time, or fortune, and the ruling power, and disease, T, TA) *rendered him possessed,* or *insane:* (JK, Ḳ, TA:) and *it corrupted, rendered unsound, vitiated,* or *disordered, his reason,* or *intellect;* or *his limb,* or *member:* (Ṣ, Ḳ:) or *it* (grief) *deprived him of his heart:* and *he corrupted, rendered unsound, vitiated, or disordered, one of his limbs* or *members:* or *deprived him of his reason,* or *intellect.* (Mṣb.) And خَبَلَ قَلْبَهُ, aor. ـِ and ـُ, *It* (love) *corrupted,* or *rendered unsound, his heart.* (JM.) And خَبَلَ فُلَانٌ يَدَ فُلَانٍ *Such a one corrupted, rendered unsound, vitiated,* or *disordered, the arm,* or *hand, of such a one, so that it became motionless;* or *caused it to dry up;* or *deprived him of it.* (JK.) — Also خَبَلَهُ, (TA,) [aor. ـِ,] inf. n. خَبْلٌ, (Ḳ, TA,) *He restrained, withheld,* or *debarred, him:* (Ḳ,* TA:) and ▾اختبل likewise signifies *he restrained, withheld,* or *debarred.* (TA.) You say, مَا خَبَلَكَ عَنَّا *What withheld,* or *has withheld, thee from us?* (TA.) And خَبَلَهُ عَنْ كَذَا, aor. and inf n. as above, *He prevented,* or *hindered, him from doing such a thing.* (Ḳ,* TA.) = خَبَلَ عَنْ فِعْلِ أَبِيهِ *He fell short of the doing of his father.* (JK, Ḳ.)

2: see 1.

4. إِخْبَالٌ signifies The act of *lending;* (JK;) as also ▾اِخْتِبَالٌ. (TA.) You say, أَخْبَلْتُهُ المَالَ, (Ṣ,) i. e. النَّاقَةَ, or الفَرَسَ, (Ṣ, Ḳ,*) *I lent him the she-camel* (Ṣ, Ḳ*) *in order that he might ride her* (TA) or *in order that he might make use of her milk and her fur,* or *the horse in order that he might go on a hostile,* or *hostile and plundering, expedition upon him.* (Ṣ, Ḳ, TA.) And اخبلهُ إِبِلًا, and غَنَمًا, *He lent him camels,* and *sheep* or *goats.* (M, TA.) See also 10. — Also The *dividing one's camels into two halves, that one half might breed in each year;* like as one does in land and sowing. (Ibn-'Abbád, Ḳ.*)

8: see 1, in two places: — and see 4. = اختبلت الدَّابَّةُ *The beast remained not in its accustomed place.* (Lth, ISd, Ḳ.)

10. استخبل مَالَ فُلَانٍ *He sought to corrupt, render unsound, vitiate,* or *disorder, some of the camels of such a one.* (Er-Rághib.) — [And hence,] استخبلهُ مِنْ مَالِهِ *He asked of him the loan of some of his camels,* or *the like,* until the time of abundance of herbage. (JK, O,* TA.*) And استخبلنى نَاقَةً, or فَرَسًا, *He asked of me the loan of a she-camel* (Ḳ, TA) *in order that he might ride her* (TA) or *in order that he might make use of her milk and her fur,* or *a horse in order that he might go on a hostile,* or *hostile and plundering, expedition upon him.* (Ḳ, TA.) And استخبلهُ إِبِلًا, and غَنَمًا, *He asked of him the loan of camels,* and *sheep* or *goats.* (M, TA.) Zuheyr says,

* هُنَالِكَ إِنْ يُسْتَخْبَلُوا المَالَ يُخْبِلُوا ▾ *

[*There, if they be asked to lend cattle, they lend*]. (Ṣ, TA. [See also 10 in art. خول.])

خَبْلٌ: see خَبَالٌ, in four places. — Also *Corruptness, unsoundness,* or *a vitiated* or *disordered state, of the limbs* or *members,* (M, Ḳ,) of a man, *so that one knows not how to walk;* (Az, TA;) and so ▾خَبَلٌ; (Ḳ;) which likewise signifies *the same in the legs of a beast.* (JK, Ḳ.*) — — And The [*palsy termed*] فَالِج; as also ▾خَبَلٌ. (Ḳ.) — Also The *cutting off of arms* or *hands, and legs* or *feet:* (JK, Az, ISd, Ḳ:) pl. خُبُولٌ. (Ḳ.) So in the saying, قَوْمِى يُطَالِبُونَ بَنِى فُلَانٍ بِدِمَآءٍ وَخَبْلٍ [*My people,* or *party, prosecute the sons of such a one for blood* (lit. *bloods*) *and the cutting off of arms* or *hands, and legs* or *feet*]. (JK.) And so خُبُولٌ in the saying, لَنَا فِى بَنِى فُلَانٍ دِمَآءٌ وَخُبُولٌ [*We have a claim, upon the sons of such a one, to blood* (lit. *bloods*) *and the cutting off of arms* &c.]. (Ṣ.) — And *Wounds:* (JK:) and ▾خَبَلٌ signifies *a wound:* and is so explained as used in the saying, بَنُو فُلَانٍ يُطَالِبُونَنَا بِخَبَلٍ [*The sons of such a one prosecute us for a wound*]. (TA.) — And i. q. فِتْنَةٌ and هَرْج [i. e. *Trial, punishment, slaughter, civil war, conflict and faction, discord, dissension,* &c.]. (TA.) = Also *A loan:* and *a demand of a loan:* (Ḳ, TA:) relating to anything. (TA.) — And *An addition which one gives, beyond what the* جَمَّال [i. e. *owner,* or *attendant, of a camel* or *camels* (in the CḲ حَمَّال i. e. *porter*)] imposes on one by stipulation. (M, Ḳ, TA.) ═ See also the next paragraph.

خُبْلٌ: see خَبَالٌ ═ وَقَعَ فِى خُبْلِى It came into my mind; (JK, Ḳ;) a phrase like the saying, سُقِطَ فِى يَدِى: (JK, Ḳ:* [in the Ḳ, meaning the same as this saying:]) and sometimes (JK) one says also ↓فِى خُبْلِى. (JK [and so in the Ḳ accord. to the TA, but not in the CḲ, nor in my MS. copy of the Ḳ].)

خَبَلٌ: see خَبَالٌ, in four places: — and see خَبْلٌ, in three places. ═ Also The jinn, or genii; (IAạr, Fr, Ṣ, Ḳ;) and so ↓خَابِلٌ: (JK, Ḳ:) or the latter has this signification; and the former is a quasi-pl. n. of the latter, or, as some say, a pl., as is also خُبَّلٌ: (TA:) and ↓خَابِلٌ signifies also a devil, or the devil. (Ḳ.) One says, بِهِ خَبَلٌ, meaning In him is somewhat of [the jinn, or genii, called] أَهْلُ الأَرْضِ. (Ṣ. [See الأَرْضُ, near the end of the paragraph: and see other explanations of خَبَلٌ voce خَبَالٌ, which may apply in this case.]) — Accord. to IAạr and Fr, it is also applied to Mankind. (TA.) — Also A certain bird, that cries all the night, with one cry, resembling مَاتَتْ خَبَلْ. (M, Ḳ,* TA.) ═ Also A مَزَادَة [or leathern water-bag]. (Fr, Ḳ.) — And A full قِرْبَة [or water-skin]. (Fr, Ḳ.)

خَبِلٌ and ↓أَخْبَلُ (Ḳ, TA) Corrupted, unsound, vitiated, or disordered, [in an absolute sense; and particularly] in his reason, or intellect; as also ↓مَخْبُولٌ: (TA:) [or in a corrupt, an unsound, a vitiated, or a disordered, state, occasioning him agitation like that of possession or insanity, by disease affecting the reason and thought: (see خَبَالٌ, below:) and hence,] possessed, or insane; (Ḳ, TA;) as also ↓مُخَبَّلٌ and ↓مُخْتَبَلٌ: (TA:) or ↓مُخَبَّلٌ signifies a man having no heart; (JK;) as also ↓مَخْبُولٌ: (JK, Mṣb:) or this last, having one of his limbs, or members, corrupted, rendered unsound, vitiated, or disordered: (Mṣb:) and خَبِلٌ and ↓مُخَبَّلٌ signify also a beast corrupted, rendered unsound, vitiated, or disordered, in the legs, so as not to know how to walk: (JK:) or ↓مُخَبَّلٌ signifies a man who is as though his extremities were amputated. (Ṣ.) — دَهْرٌ خَبِلٌ ‡ A time difficult to the people thereof; (T, Ṣ, Ḳ, TA;) in which they see not happiness. (T, TA.)

خَبَلَةٌ Corruption from a wound. (TA.) ═ See also خُبْرَةٌ, last signification.

خَبَالٌ Corruptness, unsoundness, or a vitiated or disordered state, [in an absolute sense;] (Ṣ, Mṣb, TA;) said in the O and the Mufradát [of Er-Rághib] to be the primary signification; (TA;) as also ↓خَبْلٌ (Ḥam p. 542) and ↓خَبَلٌ, of which last the pl. is خُبُولٌ: (Ṣ:) [and particularly in the reason, or intellect: (see خَبِلَ, of which it is an inf. n.:)] and in actions, as well as in bodies and in minds: (TA:) or, primarily, such as is incident to an animal, occasioning him agitation like that of possession or insanity, by disease affecting the reason and thought; as also ↓خَبَلٌ and ↓خَبْلٌ: (Er-Rághib, TA:) or ↓خَبْلٌ signifies possession, or insanity; (Ḳ;) and so ↓خَبَلٌ (JK, Mṣb, Ḳ) and ↓خَبْلٌ (Ḳ) and خَبَالٌ; (Mṣb;) or ↓خَبْلٌ signifies an affection, in the heart, resembling possession or insanity; (Az, TA;) or egregious stupidity or foolishness, without possession or insanity; (TA;) and ↓خَبْلٌ also signifies a state, or quality, resembling possession or insanity, such as stupidity, or foolishness; and heedlessness, or weakness of intellect, and the like. (Mṣb.) مَا زَادُوكُمْ إِلَّا خَبَالًا, in the Ḳur [ix. 47], means They had not added to you aught save corruption and evil. (Bḍ, TA.) And لَا يَأْلُونَكُمْ خَبَالًا, in the same [iii. 114], They will not fall short, or flag, or be remiss, in corrupting, or vitiating, your affairs. (TA.) — Hence, (TA,) Loss, or a state of diminution; syn. نُقْصَانٌ: (O, Ḳ, Er-Rághib:) or this is the primary signification. (TA.) — And hence, (TA,) A state of perdition or destruction: (O, Ḳ, Er-Rághib:) or a thing's going, passing, or wasting, away; or being consumed or destroyed. (Zj, TA.) — Also The condition of a well when it is hollowed in the sides, and old, so that sometimes the bucket enters into its hollowed part and becomes lacerated. (Fr, Ḳ.) — And Fatigue, weariness, distress, embarrassment, affliction, trouble, or difficulty. (JK, Ṣ, O, Ḳ.) So in the saying, فُلَانٌ خَبَالٌ عَلَى أَهْلِهِ [Such a one is a cause of fatigue, &c., to his family]. (JK,* Ṣ, O.) — And A deadly poison. (IAạr, Ḳ.) — And The fluid squeezed, or wrung, (IAạr, TA,) or flowing, (Ṣ, Ḳ, TA,) from the inhabitants of Hell, or from their skins. (IAạr, Ṣ, Ḳ, TA.) [See also رَدْغَةٌ.]

خَابِلٌ Corrupting, rendering unsound, vitiating, or disordering, [in an absolute sense;] (M, Ḳ;) and particularly in the reason, or intellect. (TA.) — See also خَبَلٌ, in two places. — It is also added to خَبْلٌ to give intensiveness to the signification. (TA.)

أَخْبَلُ: see خَبِلٌ.

مُخَبَّلٌ: see خَبِلٌ, in four places.

مُخَبِّلٌ a [proper] name of Time. (Ṣ, Ḳ.)

مَخْبُولٌ: see خَبِلٌ, in two places.

مُخْتَبَلٌ: see خَبِلٌ ═ مُخْتَبَلُ دَابَّةٍ The legs of a beast. (JK. [But this I do not find in any other lexicon; and I doubt its correctness.])

خبن

1. خَبَنَهُ, aor. ـِ, inf. n. خَبْنٌ (Ṣ, Mṣb, Ḳ) and خِبَانٌ (Ṣ, Ḳ) and خَبَانٌ, (Ṣ, ISd,) He folded it, namely, a garment, (Ṣ, Mṣb, Ḳ,) &c., (Ṣ, Ḳ,) in its skirt, (Mṣb,) and sewed it, (Ṣ, Ḳ,) in order that it might become shorter; (Ṣ, Mṣb, Ḳ;) he contracted it [in its length], and sewed it; (M, TA;) he raised its (a garment's) skirt, or lower part, and sewed it higher up, in order that it might become contracted and shortened, as is done with the garment of a child; (Lth, TA;) [he made a tuck in it, to shorten it;] i. q. ثَبَنَهُ. (Ṣ in art. ثبن.) — [Hence, † He shortened it; namely, a period.] You say, [of a she-camel, or of camels,] خُبِنَ مِنْ طُولِ ظِمْئِهَا, meaning † The length of the interval between her, or their, two waterings was shortened. (TA.) — Also, (Mṣb, TA,) aor. ـِ, (Mṣb,) [or. ـُ,] He hid it, or concealed it; (Mṣb, TA;) kept it, or preserved it; or stored it; namely, a thing. (TA.) You say, خَبَنَ الطَّعَامَ He hid, or concealed, kept, or preserved, or stored, (Ṣ, Ḳ,) and prepared, (Ṣ,) wheat, or food, for [a time of] dearth, or adversity. (Ṣ, Ḳ.) — يَخْبِنُ الكَذِبَ † He prepares falsehood. (Ḳ,* TA.) — ↓خَبَنَتْهُ خَبُونُ [as though signifying † Death hid him, or perhaps death shortened his existence,] is a phrase like شَعَبَتْهُ شَعُوبُ, meaning he died. (Ḳ. [In copies of the Ḳ, خَبُونٌ and شَعُوبٌ: but both are imperfectly decl., as fem. proper names of more than three letters.])

4. اخبن He (a man, TA) hid, or concealed, a thing in the خُبْنَة [q. v.] of his trousers, (Ḳ, TA,) next the back: اثبن signifies "he hid, or concealed, [a thing] in his ثُبْنَة, next the belly." (TA.) [See also what next follows.]

8. اختبن الشَّىْءَ He took [and carried] the thing beneath the part extending from his armpit to his flank. (Ḥar p. 552.) [See also what next precedes.]

خُبْنٌ The part of a مَزَادَة [or leathern water-bag that is hung on either side of a camel] which is between its عُرْوَة [or loop at either of its upper corners, whereby it is suspended, (in the CḲ خُرْب, which may signify the same,)] and its mouth [which is in the middle of the upper part]: (JK, Ḳ:) [thus] there are two such parts, [on either side of the mouth,] together called خُبْنَانِ. (JK, TA.)

خُبْنَةٌ The doubled upper border of the trousers, (IAth, TA,) next the back, in which one hides, or conceals, [or carries,] a thing; (IAạr, TA;) the ثُبْنَة being [similar to it, but] in the waist-wrapper, (IAth, TA,) next the belly: (IAạr, TA:) or the raised skirt, or lower part, of the garment, in which one carries a thing: pl. خُبَنٌ. (Ḥar p. 427.) And What one carries in the حِضْن [or part between the armpit and the flank, &c.]: (Ṣ, Ḳ:) or what one carries beneath the armpit, (JK, Mṣb,) and in the sleeve: (JK:) or what is put, of food, and carried under the armpit or in the sleeve. (Ḥar p. 427.) It is said in a trad. of 'Omar, إِذَا مَرَّ أَحَدُكُمْ بِحَائِطٍ فَلْيَأْكُلْ مِنْهُ وَلَا يَتَّخِذْ خُبْنَةً [When any one of you passes by a garden of palm-trees, let him eat thereof, but not make, or take for himself, a خبنة]. (Ṣ,* TA. [See another reading voce ثِبَانٌ.])

خَبَنَتْهُ خَبُونُ: see 1.

خَابِنٌ [applied to a she-camel, or to a number of camels,] † Whose interval between two waterings has been shortened. (IAạr.) ═ †One who prepares falsehood. (JK,* Ḳ,* TA.) ═ I. q. شَدِيدٌ [Strong, &c.]. (JK, Ḳ.)

خبو

1. خَبَتِ النَّارُ, aor. تَخْبُو, inf. n. خُبُوٌّ (JK, Ṣ, Mṣb, Ḳ) and خُبُوٌّ, (Ḳ,) *The fire became allayed; subsided;* (Ḳ;) *became extinguished;* (Ṣ, Ḳ;) *the flaming,* or *blazing, of the fire ceased,* or *became allayed or extinguished.* (JK, Mṣb.) [The verb is used, in this sense, of Hell, in the Ḳur xvii. 99.] — [Hence,] خَبَتِ الحَرْبُ ‡ *The war became allayed, assuaged,* or *appeased;* (JK, Ḳ, TA;) *became extinguished.* (Ḳ, TA.) And خَبَتْ حِدَّةُ النَّاقَةِ ‡ *The sharpness,* or *irascibility, of the she-camel became allayed, assuaged,* or *appeased.* (JK, Ḳ,* TA.) And خَبَا لَهَبُهُ ‡ *The heat,* or *vehemence, of his anger became allayed, assuaged,* or *appeased.* (TA.)

4. اخبى النَّارَ *He extinguished the fire;* (Ṣ, Ḳ;) *he allayed,* or *extinguished, the flaming,* or *blazing, of the fire.* (JK, Mṣb.) — And [hence,] اخبى الحَرْبَ ‡ *He extinguished* [or *allayed*] *the fire of the war.* (Ḳ, TA.) And اخبى حِدَّةَ النَّاقَةِ ‡ *He extinguished* [or *allayed*] *the sharpness,* or *irascibility, of the she-camel.* (Ḳ,* TA.)

خِبَآءٌ accord. to some belongs to this art. (TA in art. خبأ.) See arts. خبأ and خبى.

خبى

2: see what next follows.

4. اخبى, inf. n. إِخْبَآءٌ; (Ks, TA;) or اخبى خِبَآءً; (Ṣ, Ḳ;) and ↓ تخبّى, (Ks, TA,) or تخبّى خِبَآءً; (Ṣ, Ḳ;) and ↓ خبّى خِبَآءً, (Ṣ,* Ḳ,) inf. n. تَخْبِيَةٌ; (Ṣ;) *He made a* [*tent such as is called*] خِبَآء: (Ks, Ṣ, Ḳ:) and *he set it up.* (Ḳ.) [See also 10.] And أَخْبَيْتُ كِسَائِى *I made my* [*garment called*] كساء *to be* [*as*] *a* خِبَآء. (TA.)

5: see 4.

10. استخبى خِبَآءً *He set up a* [*tent such as is called*] خِبَآء, *and entered into it.* (Ṣ, Ḳ.) [See also 4.]

خِبَآءٌ *A kind of structure;* (Ḳ;) [i. e.] *one of the* بُيُوت [or *kinds of tents*] *of the Arabs,* (IAth, TA,) peculiarly *of wool,* (Yaạḳoob, Th,) or *of camels' fur,* or *of wool,* (IAạr, Ṣ, IAth, Ḳ,) or *of* [*goats'*] *hair,* (Ḳ,) or *not of* [*goats'*] *hair,* (Ṣ,) [*except in cold countries and in fertile regions, where the goats have abundant hair, for the goats of the Arabs of the desert have short hair, not long enough to be spun,* (see بَيْتٌ,)] *smaller than the* مِظَلَّة, (IAạr,) *upon two poles,* or *three;* what is above this kind being termed بَيْتٌ; (Ṣ;) or *a tent having one pole;* that which has more than one pole being termed بَيْتٌ: (AZ, TA in art. ربع:) pl. أَخْبِيَةٌ, (Ṣ, TA,) without ء: (TA:) most of the lexicologists hold that its radical letters are خبى. (TA in art. خبأ, q. v.) — Sometimes, † *A dwelling, such as is in a city.* (TA, from a trad.) — Also ‡ *A receptacle for oil or ointment.* (Ḳ, TA.) — And ‡ The *calyx* of a flower. (TA.) — And ‡ The *husk of a grain of wheat,* and *of a grain of barley, in the ear.* (Ḳ, TA.) — And الخِبَآءُ signifies ‡ *Certain round stars,* [or *stars in a circle,*] (Ḳ, TA,) *forming one of the Mansions of the Moon,* [namely, *the Twenty-fifth,*] *also called* الأَخْبِيَةُ [or سَعْدُ الأَخْبِيَةِ: see سَعْدٌ]. (TA.)

خَابِيَةٌ, originally خَابِئَةٌ: see art. خبأ.

ختر

1. خَتَرَتْ نَفْسُهُ *His soul* [or *stomach*] *became heavy;* or *heaved,* or *became agitated by a tendency to vomit;* syn. خَبُثَتْ: (IAạr, Ḳ:) and *became in a corrupt,* or *disordered, state.* (Ḳ.) — خَتَرَهُ, (Ṣ, Ḳ,*) aor. ـِ and ـُ, (Ḳ,) inf. n. خَتْرٌ (Ṣ, A, Ḳ) and خُتُورٌ, (Ḳ,) *He acted,* or *behaved, towards him with perfidy, treachery,* or *unfaithfulness:* (Ṣ, Ḳ:*) or *with the foulest perfidy* or *treachery* or *unfaithfulness:* (A, Ḳ:) or *with deceit, guile,* or *circumvention:* (Ḳ:) or *in a bad,* or *corrupt, manner.* (Ibn-'Arafeh.) And خَتَرَ بِالعَهْدِ [*He was unfaithful,* &c., *to the compact,* or *covenant*]. (TA, from a trad.)

2. خَتَّرَهُ, inf. n. تَخْتِيرٌ, said of wine, or beverage, *It corrupted,* or *disordered, his soul* [or *stomach*], (Ibn-'Arafeh, Ḳ,) *and rendered him relaxed.* (Ibn-'Arafeh.)

5. تختّر *He* (a man, TA) *was,* or *became, languid,* (تَفَتَّرَ,) *and relaxed, and heavy,* or *sluggish, and fevered:* (Ḳ:) *he was,* or *became, languid in body, in consequence of disease* &c.: (TA:) and *his intellect became confused, from drinking milk and the like.* (Ḳ.) And تختّرت نَفْسُهُ *His soul* [or *stomach*] *was,* or *became, languid.* (TA.) — *He walked with the gait of him who is heavy,* or *sluggish.* (Ḳ.)

خَتَرٌ *Languor* (خَدَرٌ), (Ḳ,) or the *like thereof,* (TA,) *that betides on the occasion of drinking medicine or poison,* (Ḳ, TA,) *such as weakens and intoxicates.* (TA.)

خَتُورٌ : } see what follows.
خَتِيرٌ : }
خَتَّارٌ : }
خِتِّيرٌ : }

خَاتِرٌ (Ḳ) and [in an intensive sense] ↓ خَتَّارٌ (Ṣ, A, Ḳ) and ↓ خَتُورٌ and ↓ خِتِّيرٌ (Ḳ) One *who acts,* or *behaves, with perfidy, treachery,* or *unfaithfulness:* (Ṣ, Ḳ:) or *with the foulest perfidy* or *treachery* or *unfaithfulness:* (A, Ḳ:) or *with deceit, guile,* or *circumvention:* (Ḳ:) [or *in a bad,* or *corrupt, manner:* (see 1:) the second and following epithets signifying one *who does so much,* or *frequently,* or *habitually.*]

ختعر

Q. 1. خَتْعَرَ, inf. n. خَتْعَرَةٌ, *It passed away, and came to nought:* (Ḳ: but only the inf. n. is there mentioned:) said of the mirage. (Ḳr.)

خَيْتَعُورٌ *Anything that does not remain in one state; and that passes away, and comes to nought;* (Ṣ, Ḳ:) or *that has no real existence.* (IAth.) — *The mirage;* syn. سَرَابٌ: (Ṣ, Ḳ:) or *what remains of the latter part of the mirage, when it becomes dispersed, and delays not to pass away and come to nought.* (Kr, L.) — [*Gossamer;* i. e.] *what descends from the air in a time of intense heat, like spiders' webs;* (Ṣ;) *a thing like spiders' webs, which appears in a time of heat,* (Ḳ,) *descending from the sky,* (TA,) *resembling threads,* (Ḳ,) or *white threads,* (TA,) *in the air.* (Ḳ.) [See لُعَابُ الشَّمْسِ, in art. لعب.] — The *present world or life.* (Ḳ.) — The *devil:* (Fr, Ḳ:) or the *devil of the 'Aḳabeh, called* أَزَبُّ العَقَبَةِ. (IAth.) — The [*imaginary creature called*] غُول: (Ṣ, Ḳ:) because it changes its appearance. (TA.) — *A certain insect,* (Ḳ,) *of a black colour,* (TA,) *that is upon the surface of water, and that does not remain in one place* (Ḳ) *save as long as the time of the winking of an eye.* (TA.) — *A perfidious,* or *an unfaithful,* man. (TA.) — A woman *whose affection does not last:* (TA:) a woman *evil in disposition:* (Ḳ:) likened to a غُول inasmuch as her love does not last. (TA.) — The *wolf:* (Ṣ, Ḳ:) because of his unfaithfulness. (TA.) — The *lion:* (Ḳ:) because of his perfidiousness. (TA.) — *A calamity;* syn. دَاهِيَةٌ. (Ṣ, Ḳ.) — نَوًى خَيْتَعُورٌ [*A point towards which one journeys*] *that is not in a right direction;* syn. الَّتِى لَا تَسْتَقِيمُ: (L:) or *far distant.* (Ḳ.)

ختل

1. خَتَلَهُ, (Ṣ, M, Ḳ, &c.,) aor. ـِ (M, Ṣgh, Ḳ) and ـُ, (M, Ḳ,) inf. n. خَتْلٌ and خَتَلَانٌ, (Ḳ,) i. q. خَدَعَهُ, or خَادَعَهُ; (so accord. to different copies of the Ṣ;) and ↓ خاتله signifies the same: (Ṣ:) or the former, i. q. خَدَعَهُ (Mgh, Ḳ, TA) [i. e.] *He deceived, deluded, beguiled, circumvented,* or *outwitted, him,* (KL, PṢ,) *unawares:* (TA:) and ↓ خاتله, (Ḳ,) inf. n. مُخَاتَلَةٌ, (TA,) i. q. خَادَعَهُ [which means the same as خَدَعَهُ; or *he practised with him mutual deceit, delusion,* &c.; or *he strove, endeavoured,* or *desired, to deceive, delude, beguile, circumvent,* or *outwit, him*]; (Ḳ, TA;) and رَاوَغَهُ [which means *he endeavoured to turn him,* or *to entice him to turn,* to, or from, a thing, &c.]. (TA.) — خَتْلُ الصَّائِدِ means *The sportsman's going along by little and little, stealthily, lest he should make a sound to be heard.* (JM.) And خَتَلَ الصَّيْدَ, (Ḳ,) inf. n. خَتْلٌ, (TA,) said of a wolf, *He concealed himself to seize the prey.* (Ḳ.)

3: see 1, in two places.

6. تَخَاتَلَ i. q. تَخَادَعَ [as meaning *The deceiving one another;* thus explained in the KL: and app. also as meaning *the pretending deceit,* &c.; or *the pretending to be deceived,* &c.: accord. to the PṢ, the *being deceived;* but for this I know not any other authority]. (Ṣ.) You say, تخاتلوا, meaning تخادعوا [*They deceived, deluded, beguiled, circumvented,* or *outwitted, one another*]. (Ḳ.) And تخاتل عَنْ غَفْلَةٍ [app. meaning *He pretended deceit* &c., or *pretended to be deceived* &c., *unawares*]. (TA.)

8. اِخْتَتَلَ *He* (a man, TA) *listened* to the secret of a party of men. (T, K,* TA.) El-Aạshà says,

* وَلَا تَرَاهَا لِسِرِّ الجَارِ تَخْتَتِلُ *

[*Nor wilt thou see her listen to the secret of the neighbour*]. (TA.)

مَخْتَلٌ *A place of retreat,* or *concealment.* (K.) And *Any similar place in which one listens* to hear secrets [&c.]. (TA.) The *form of a hare,* or *burrow of a rabbit.* (K.)

خَتُولٌ : see خَاتِلٌ and خَتَّالٌ.

خَتَّالٌ *i. q.* خَدَّاعٌ [i. e. One *who deceives, deludes, beguiles, circumvents,* or *outwits, much,* or *often; very deceitful* &c.; as also † خَتُولٌ, like خَدُوعٌ]. (TA.)

خَاتِلٌ and [in an intensive sense] † خَتُولٌ [like خَتَّالٌ, q. v.,] *Deceiving, deluding, beguiling, circumventing,* or *outwitting.* (K, TA.)

خَوْتَلٌ *Elegant; polite; acute,* or *sharp,* or *quick, in intellect; clever, ingenious, skilful, knowing,* or *intelligent:* (K,* TA:) thought by ISd to be, perhaps, from الخَتْلُ signifying "the act of deceiving" &c. (TA.)

خَوْتَلَى *A stealthy walk* or *gait:* (O, K, TA:) or *a walking on one side.* (T, TA.) Hence the saying, هُوَ يَخْلِجُنِى بِعَيْنِهِ وَيَمْشِى لِى الخَوْتَلَى [*He makes a sign to me with his eye, and walks to me stealthily,* or *sideways*]. (TA.)

أَخْتَلُ [*More, and most, deceitful, deluding, guileful,* &c.]. You say أَخْتَلُ مِنْ ذِئْبٍ *More deceitful,* &c., *than the wolf.* (Mgh.)

ختم

1. خَتَمَهُ, (S, Mgh, Msb, K,) aor. ـِ, (Msb, K,) inf. n. خَتْمٌ (S, Mgh, Msb, K) and خِتَامٌ, (Lh, K, TA,) with kesr, (TA,) [in the CK, erroneously, خَتَام,] *i. q.* طَبَعَهُ [*He sealed, stamped, imprinted,* or *impressed, it*]: (Msb, K:) or *he put the* خَاتَم [or *signet*] *upon it:* (Mgh:) namely, a thing, (S, Mgh,) or a writing or book and the like: and خَتَمَ عَلَيْهِ signifies the same [or *he put a seal,* or *the impression of a signet, upon it*]. (Msb.) Accord. to Er-Rághib, خَتْمٌ and طَبْعٌ signify The *impressing* a thing *with the engraving of the signet and stamp:* and the former [as is indicated, but not plainly expressed, as distinguished from the latter,] is tropically used, sometimes, as meaning the *securing oneself* from a thing, and *protecting* [*oneself*] from it; in consideration of protection by means of sealing upon writings and doors: and sometimes as meaning the *producing an impression,* or *effect,* upon a thing from another thing; in consideration of the impress produced [by the signet]: and sometimes it is used as relating to † the *reaching the end* [of a thing]: (TA:) or the primary signification of خَتْمٌ is the act of *covering over* [a thing]: (Az, TA:) accord. to Zj, the proper meaning of خَتْمٌ and طَبْعٌ is the *covering over* a thing, *and securing oneself from a thing's entering* it: some say that the former signifies the *concealing a notification* of a thing, [*as*] *by putting one's fingers over it, by way of guarding oneself thereby.* (TA.) — Hence, خَتْمُ الشَّهَادَةِ [*The sealing of the testimony*]; which is thus described by El-Hulwánee: the witness, when he wrote his name upon a صَكّ [q. v.], caused his written name to be beneath a piece of lead, [i. e. covered it with a piece of lead,] and put upon it the impress of his signet, in order that there might be no falsification of it or substitution for it. (Mgh.) — As to خَتْمُ الأَعْنَاقِ [*The sealing of the necks*], the case is related, in the "Risáleh Yoosufeeyeh," to have been this: 'Omar sent Ibn-Honeyf to seal the عُلُوج [or unbelievers] of the Sawád; and he sealed five hundred thousand of them, in classes; that is, he marked them twelve dirhems, and twenty-four, and forty-eight; tying a thong upon the neck of each, and putting upon the knot a seal of lead. (Mgh.) — خَتْمٌ also signifies The *protecting* what is in a writing *by marking* [or *stamping*] *a piece of clay* [upon it, or *by means of a seal of any kind*]. (TA.) — And you say, of a man, خَتَمَ عَلَيْكَ بَابَهُ [*He sealed his door against thee*]; meaning ‡ *he turned away from thee, avoided thee,* or *shunned thee.* (TA.) — And خَتَمَ لَكَ بَابَهُ [*He sealed for thee his door*]; meaning ‡ *he preferred thee to others.* (TA.) — خَتَمَ عَلَى قَلْبِهِ [which may be rendered *He sealed his heart*] means ‡ *he made him to be such that he understood not, and such that nothing proceeded from him;* or *he made his heart,* or *mind, to be such that it understood not,* &c. (K, TA.) خَتَمَ ٱللّٰهُ عَلَى قُلُوبِهِمْ, in the Kur [ii. 6], is like the phrase in the same [xvi. 110 and xlvii. 18] طَبَعَ ٱللّٰهُ عَلَى قُلُوبِهِمْ: (TA:) it points to what God has made to be usually the case when a man has ended in believing what is false and in committing that which is forbidden, so that he turns not his face to the truth; this occasioning, as its result, his becoming inured to the approval of acts of disobedience, so that he is as though this habit were impressed upon his heart: the assertion of El-Jubbáee, that it means *God hath put a seal upon their hearts,* as a sign, to the angels, of their infidelity, is nought: (Er-Rághib, TA:) الخَتْمُ is explained by IAạr as meaning *the preventing* of the heart *from believing.* (L in art. خدع.) [See also طَبَعَ.] — خَتَمَ الشَّىْءَ, inf. n. خَتْمٌ, also signifies [as indicated above] † *He reached the end of the thing.* (K.) And † اِخْتَتَمْتُ الشَّىْءَ † [*I ended,* or *finished, the thing,*] *contr. of* اِفْتَتَحْتُهُ. (S, TA.) You say, خَتَمَ القُرْآنَ † *He reached the end of the Kur-án* [*in reciting it*]; (S, Er-Rághib;) [*he recited the whole of the Kur-án;*] *he completed* [*the recital of*] *the Kur-án:* (Mgh:) [and] *he retained in his memory the last portion of the Kur-án;* meaning *he retained the whole of it in his memory.* (Msb.) It is said of Suleymán El-Aạmash, كَانَ يَقْرَأُ خَتْمًا, meaning † *He used to recite the whole of the Kur-án;* at one time, according to the reading of Ibn-Mes'ood; at another time, from the edition of 'Othmán. (Mgh.) — You say also, خَتَمَ ٱللّٰهُ لَهُ بِخَيْرٍ † [*May God make his end to be good*]. (S.) — خَتَمَ البَذْرَ † *He covered over the sown seed:* (Az, TA:) or خَتَمُوا عَلَى البَذْرِ † *they turned up the earth over the sown seed, and then watered it:* (Et-Táifee, TA:) or خَتَمَ الزَّرْعَ, (JK, K, TA,) aor. ـِ, inf. n. خَتْمٌ, (TA,) ‡ *he watered the seed-produce,* or *sown field, the first time;* (JK, K, TA;) because, when it is watered, it is finished (خُتِمَ) with the وحا [app. وَحًا, which here seems to mean the "roller," as being likened to a mill-stone, though I find no authority for this meaning]; (TA;) as also خَتَمَ عَلَيْهِ: (K:) or خَتَمُوا عَلَى زُرُوعِهِمْ means † *they watered their sown fields while these were as yet* كِرَاب [app. meaning *clear of vegetation*]. (JK, TA.) — خَتَمَ النَّحْلُ ‡ *The bees filled with honey the place in which they deposited it:* (JK, A, TA:) or خَتْمٌ signifies bees' *collecting some thin wax, thinner than the wax of the comb, and smearing with it the orifices of their* خَلَايَا [or *hives*]. (M, K, TA.)

2. خَتَّمَهُ, inf. n. تَخْتِيمٌ, *He sealed it, stamped it,* &c., *much.* (TA.) — [In modern Arabic, *He put a* خَاتَم, or *signet-ring, upon his* (another's) *finger.*]

5. تَخَتَّمَ, or تَخَتَّمَ خَاتَمًا, (accord. to different copies of the S,) or تَخَتَّمَ بِخَاتَمٍ, (K, [agreeably with a trad. cited in the TA,]) *He put on* [i. e. *put on his own finger*] *a* خَاتَم [or *signet-ring*]. (S, K.) — And تَخَتَّمَ ‡ *He put on a turban:* (K, TA:) or تَخَتَّمَ بِعِمَامَتِهِ *he put on his turban in the manner of a* نِقَاب [q. v.]; syn. تَنَقَّبَ بِهَا. (Z, TA.) The subst. [signifying the act or mode, of doing so] is † تَخْتِيمَةٌ [q. v. infrà]. (K.) — تَخَتَّمَ بِأَمْرِهِ † *He concealed his affair,* or *case.* (Z, K.) — تَخَتَّمَ عَنْهُ ‡ *He feigned himself heedless of him, and was silent* [*to him*]. (K, TA.)

8: see 1, in the latter half of the paragraph.

خَتْمٌ inf. n. of 1 [q. v.]. (S, Mgh, Msb, K.) — Also The *impress produced by the engraving of a signet.* (Er-Rághib, TA.) — See also خَاتَمٌ. — أَعْطَانِى خَتْمِى means ‡ *He gave me my sufficiency,* or *what sufficed me:* because what suffices a man is the last [or utmost] of his desire, or demand. (TA.) — Also ‡ *Honey.* (K, TA.) — And ‡ The *orifices of the* خَلَايَا [or *hives*] *of bees.* (K, TA.)

خَتَمٌ : see خَاتَمٌ. — Also *A sealed piece of clay* [or *wax*]: like نَقَضٌ in the sense of مَنْقُوضٌ, and قَبَضٌ in the sense of مَقْبُوضٌ: so in the saying of El-Aạshà,

* وَصَهْبَآءَ طَافَ يَهُودِيُّهَا * وَأَبْرَزَهَا وَعَلَيْهَا خَتَمْ *

[*And a jar of reddish-coloured wine, the Jew-vender whereof came, and brought it out, with a sealed piece of clay upon it*]. (S.) [See also خِتَامٌ.]

خَتْمَةٌ [an inf. n. of un. of 1, † *A recitation of the whole of the Kur-án:* used in this sense in the present day: pl. خَتَمَاتٌ. — And also] vulgarly used as meaning † *A copy of the Kur-án:* and so † خِتْمَةٌ. (TA.)

خَتْمَة : see what next precedes.

خَتَامٌ : see the last sentence of the next paragraph.

خِتَامٌ an inf. n. of 1 [q. v.]. (Lḥ, Ḳ.) —— And a subst. signifying The *first watering of seed-produce*, or *of a sown field:* (JK, TA:) or the *turning up the earth over sown seed, and then watering it.* (Eṭ-Ṭáifee, TA.) ══ See also خَاتَمٌ. —— Also The *clay*, (JK, Ṣ, Ḳ,) and the *wax*, (TA,) *with which one seals*, or *stamps*, (JK, Ṣ, Ḳ,) *upon a writing*, (JK,) or *upon a thing:* (Ḳ:) or *which is sealed*, or *stamped, upon a writing.* (Mṣb.) [See also خَتْمٌ. And see an ex. in a verse of Lebeed cited in art. دكن.] —— ‡ [The *hymen;* as being the *seal of virginity;* as also † خَاتَمٌ.] You say, زُفَّتْ إِلَيْكَ بِخِتَامِهَا ‡ [*She was conducted as a bride to thee with her seal of virginity*], and بِخَاتَمِ رَبِّهَا [*with the seal of her Lord*]. (TA.) And [hence, app.,] سِيقَتْ هَدِيَّتُهُمْ إِلَيْهِ بِخِتَامِهَا ‡ [if it mean, as I suppose it to do, *Their present was sent to him with what rendered it perfect* or *complete*, or *with what appertained to it*]. (TA.) —— † The *furthest part* of a valley. (JK, TA.) † The *last* of a company of men; (Lḥ, TA;) as also † خَاتِمٌ and † خَاتَمٌ: (Ḳ:) whence خَاتِمُ النَّبِيِّينَ † [*The last of the prophets*], in the Ḳur [xxxiii. 40]; accord. to one reading, † خَاتَمُ, with ḍamm to the ت; (TA;) or خَاتِمُ الأَنْبِيَآءِ, i. e. Moḥammad; (Ṣ;) also called † الخَاتِمُ and † الخَاتَمُ. (TA.) And † The *last portion* of anything that is drunk [&c.]. (TA.) خِتَامُهُ مِسْكٌ, in the Ḳur [lxxxiii. 26], means † *The last* that they will perceive *thereof* will be *the odour of musk:* (Ṣ, TA:) or, accord. to 'Alḳameh and Mujáhid, *its admixture* shall be *musk:* accord. to Ibn-Mes'ood, *its result* shall be the *taste of musk:* Fr says, † خَاتَمٌ and † خَاتِمٌ and خِتَامٌ are nearly the same in meaning; whence the reading of 'Alee, خَاتَمُهُ † مِسْكٌ: and the explanation is this; that when any one shall drink thereof, he will find the last cup thereof to have the odour of musk: Er-Rághib says that the meaning is, *the end*, and *the last draught*, i. e. *what shall remain, thereof* shall be *in perfume* [like] *musk:* and that the assertion that it means it shall be sealed with musk is nought. (TA.) [See also خَاتِمٌ and خَاتِمَةٌ.] ══ Also, (IAạr, Ḳ,) and † خَاتَمٌ, (Ḳ,) or, accord. to IAạr, † خِتَامٌ, (TA,) sings. of خُتُمٌ, which signifies † The *places of separation* (فُصُوص [q. v.]) *of the joints* (مَفَاصِل) *of horses.* (IAạr, Ḳ.)

خَاتَمٌ (JK, Ṣ, Mṣb, Ḳ) and خَاتِمٌ (TA) and خَاتِمٌ, (Ṣ, Mṣb, Ḳ,) which is more commonly known than خَاتَمٌ, (Mṣb, [but see what follows,]) and † خَاتَامٌ and † خِيتَامٌ (JK, Ṣ, Ḳ, the last in the CḲ and TA † خَيْتَامٌ) and † خِتَامٌ (Ḳ and TA but omitted in the CḲ) and † خَتْمٌ (ISd, IHsh, Ḳ) and † خَتَمٌ (Ez-Zeyn El-'Irákee, TA) and † خَيْتَمٌ (Ibn-Málik, TA) and † خَيْتُومٌ (Ez-Zeyn El-'Irákee, TA) and † خَاتِيَامٌ (Ḳ) signify the same; (JK, Ṣ, Mṣb, Ḳ, TA;) [*A signet;* generally *a signet-ring;* i. e.] *a certain ornament* (حَلْىٌ, M, Ḳ) *for the finger*, (Ḳ,) app., at the first, *used for sealing*, or *stamping, therewith;* so that the word is of the same class as طَابَع: afterwards, in consequence of frequency of usage, applied to *one not used for that purpose:* (ISd, TA:) or *a ring having a* فَصّ *of a substance different therefrom* [*set in it;* i. e., *having a stone*, or *gem, set in it*]: if without a فَصّ, it is called فَتَخَةٌ: (Mṣb:) or † خَاتِمٌ signifies the *agent* [i. e. the person *sealing*, or *stamping*]: (JK, Az, Mṣb:) خَاتَمٌ, the *thing that is put upon the piece of clay* [or *wax, for the purpose of sealing*, or *stamping*]: (Az, Mṣb, Ḳ:) the pl. [of خَاتَمٌ and خَاتِمٌ] is خَوَاتِمُ (Ḳ) and [properly of خَاتَامٌ] خَوَاتِيمُ: (Ṣ, in which the former pl., though more common, is not mentioned, and Ḳ:) Sb says that those who use the latter pl. make it to be pl. of a sing. of the measure فَاعَال, though it be not in their language; which shows that he knew not خَاتَامٌ: the pl. of خَتْمٌ is خُتُومٌ. (TA.) —— خَاتَمٌ also signifies *A seal*, or *stamp*, and *a mark:* so in a trad., in which it is said, آمِينَ خَاتَمُ رَبِّ العَالَمِينَ عَلَى عِبَادِهِ المُؤْمِنِينَ, i. e. آمِينَ [(or *Amen*) *is*] *the seal*, or *stamp*, and *the mark*, [*of the Lord of the beings of the whole world upon his servants the believers*,] which removes from them accidents, and causes of mischief; for the seal of the writing protects it, and precludes those who look from [seeing] what is within it. (TA.) —— See also خِتَامٌ, in seven places. [It is nearly syn. with خِتَامٌ, as Fr says: and thus,] it signifies also, (JK, Ḳ,) and so does † خَاتِمَةٌ, (Ṣ, Ḳ,) † The *end*, or *last part* or *portion*, (JK, Ṣ, Ḳ,) and *result*, or *issue*, (Ḳ,) of a thing (JK, Ṣ, Ḳ) of any kind: (JK, Ḳ:) † the latter [particularly] signifies † the *last part* or *portion* (JK, Mṣb) of a chapter of the Ḳur-án, (JK,) and of the Ḳur-án itself: (Mṣb:) [and † *a concluding chapter* or *section: an epilogue:* and *an appendix:*] and † مُخْتَتَمٌ signifies [in like manner] the *contr. of* مُفْتَتَحٌ; as in the saying, التَّحْمِيدُ مُفْتَتَحُ القُرْآنِ وَالاِسْتِعَاذَةُ مُخْتَتَمُهُ † [*The declaration of the praises of God is the opening portion of the Ḳur-án, and the prayer for the protection of God is its closing portion*]; (A, TA;) and it is a chaste word, of frequent occurrence, though the contr. has been asserted. (TA.) One says also, الأَعْمَالُ بِخَوَاتِيمِهَا [*Actions are characterized, or to be judged, as good or evil, by their results*]. (TA.) —— Also, i. e. خَاتَمٌ, of a mare, ‡ The *lower ring* (الحَلْقَةُ الدُّنْيَا) [app. meaning the *extremity, in which is the orifice*, see حَلْقَتَا الرَّحِمِ and حَلْقَةُ الدُّبُرِ, in art. حلق,] *of the* طُبْيَة [evidently here used as a dial. var. of طُبْى, i. e. the *teat*, though I do not find it mentioned in its proper art. in any lexicon; unless مِنْ طُبْيَتِهَا be a mistranscription for [مِن طُبْيِهَا]: (Ḳ, TA:) so called by way of comparison [to a signet or seal]. (TA.) —— And ‡ The *hollow* (نُقْرَة) *of the back of the neck;* (JK, Ḳ, TA;) which is the *cupping-place.* (TA.) —— And † The *least* وَضَح [or *whiteness*] *of the legs* (JK, Ḳ, TA) *of horses;* (JK, TA;) i. e. *a slight whiteness in the parts next the hoof, less than what is termed* تَحْجِيل. (TA.)

خِتَامٌ : see خَاتَمٌ.

خَاتِمٌ : see خَاتَمٌ, in two places: —— and see also خِتَامٌ, in five places.

خَيْتَمٌ : see خَاتَمٌ, first sentence.

خَاتِمَةٌ : see خَاتَمٌ, in two places, in the latter half of the paragraph.

خَاتَامٌ : خَيْتَامٌ : خِيتَامٌ : خَيْتُومٌ : خَاتِيَامٌ :	see خَاتَمٌ, first sentence.

تَخَتُّمٌ : see 5. You say, مَا أَحْسَنَ تَخَتُّمَهُ [*How good*, or *beautiful, is his act*, or *mode, of putting on the turban!* or, *of putting it on in the manner of the* نِقَاب!]. (Ez-Zejjájee, TA.)

مُخَتَّمٌ *Sealed*, or *stamped*, &c., *much.* (Ṣ,* TA.) —— Applied to a horse, (TA,) † *Having the whiteness of the legs which is termed* خَاتَمٌ. (Ḳ, TA. [See the latter word, last sentence.])

مَخْتُومٌ *Sealed*, or *stamped*, &c. (Ṣ,* TA.) —— Also The [*measure commonly called*] صَاع: (A 'Obeyd, Mgh, Ḳ:) or the *sixth part of the* [*measure called*] قَفِيز. (Mgh in art. حكر. [It is there added that the قفيز is the tenth part of the جَرِيب: but it seems that this is the قفيز which is a measure of land; not what is here meant in the explanation of مختوم, which is a measure of corn and the like.]) [Pl. مَخَاتِيمُ.]

مُخْتَتَمٌ : see خَاتَمٌ, in the latter half of the paragraph.

ختن

1. خَتَنَ, (Ṣ, Mgh, Mṣb, Ḳ,) aor. ِ (Mṣb, Ḳ) and ُ, (Ḳ, TA, but omitted in the CḲ,) inf. n. خَتْنٌ, (Ṣ, Mgh, Mṣb, Ḳ,) *He circumcised* (Ḳ, TA) a boy, (Ṣ, Mgh, Mṣb, Ḳ,) and a girl also: or, as some say, خَتْنٌ relates to men [or boys], and خَفْضٌ to women [or girls]. (TA.) خَتَنَ: see 8. —— And † خِتَانٌ [which see below, app. as an inf. n. of which the verb is as above,] signifies The *making a feast*, or *banquet, to which people are invited, on account of a wedding*, and *of a circumcision* also. (KL.) —— [And accord. to Golius, as on the authority of a gloss. in the KL, خَتَنَ also signifies *He diminished;* h. *rendered imperfect:* and *he acted unjustly.*] ══ خَتَنَهُ is also *syn. with* خَتَلَهُ [*He deceived, deluded, beguiled, circumvented*, or *outwitted, him, unawares*]: and † مُخَاتَنَةٌ is *syn. with* مُخَاتَلَةٌ [which signifies in like manner the act of *deceiving, deluding*, &c.; or *practising mutual deceit*, &c.; or *striving, endeavouring*, or *desiring, to deceive*, &c.]. (TA.)

3. خاتنه He allied himself to him by marriage; syn. تَزَوَّجَ إِلَيْهِ. (Ḳ.) مُخَاتَنَةٌ is syn. with مُصَاهَرَةٌ [The becoming that kind of relation that is termed صِهْرٌ]: (ISh, Mgh:) as some say, مصاهرة on the side of the wife, and on the side of the husband: so that one says خَاتَنْتُهُمْ as meaning صَاهَرْتُهُمْ [I became a relation to them on the side of the wife, and on the side of the husband]. (Mṣb.) — See also 1, last sentence.

8. اختتن He (a boy) was circumcised; (TA;) syn. ▼ خُتِنَ: or he circumcised himself; syn. خَتَنَ نَفْسَهُ. (Mgh.)

خَتَنٌ i. q. صِهْرٌ, (Lth, Mgh, Ḳ, &c.,) as meaning A man married among a people: (Lth, Mgh:) [such a man is said to be that people's خَتَن:] or any relation on the side of the wife; (Ṣ, IAạr, Mgh, Mṣb, Ḳ;) such as a man's wife's father, (Lth, IAạr, Ṣ, Mgh, Mṣb, Ḳ,) and wife's mother, (Lth, Mgh,) and wife's brother, (IAạr, Ṣ, Mgh, Mṣb, Ḳ,) and the like; (Ḳ;) so it signifies with the Arabs: (Ṣ, Mgh, Mṣb:) thus Aboo-Bekr was the Prophet's ختن, and so was 'Omar: (Mgh, TA:) and [it is said that] with the vulgar it signifies a man's daughter's husband: (Ṣ, Mgh, Mṣb:) but it is used in this sense by a rájiz; and, in a trad., 'Alee is called the Prophet's ختن: (TA:) accord. to Az, it signifies a man's wife's father: (Mṣb:) and خَتَنَةٌ is applied to the female; and means a man's wife's mother: (Az, Mgh, Mṣb, Ḳ, TA:) the pl. is أَخْتَانٌ: (Az, Ṣ, Mgh, Mṣb, Ḳ:) accord. to Aṣ, (Mgh,) the أَخْتَان are [the relations] on the side of the wife; and the أَحْمَآء, on the side of the husband; and the أَصْهَار, on either side: (Mgh, Mṣb:) or a man's اختان are his wife's relations; and a woman's اختان are her husband's relations: and a man's اختان are also said to be his daughters' husbands and sisters' husbands and paternal aunts' husbands and maternal aunts' husbands, and the husbands of any women whom, by reason of relationship, it is unlawful for him to marry, and any relations on the side of these husbands to whom marriage is unlawful, of men and of women. (Mgh.)

خِتَانٌ Circumcision, of a boy, (Ṣ,* Mgh,* Mṣb,* Ḳ, TA,) and of a girl; (TA;) a subst. from 1 in the first of the senses explained above; (Ṣ, Mgh, Mṣb, Ḳ;) as also ▼ خِتَانَةٌ. (Ṣ, Mṣb, Ḳ.) You say, ▼ أُطْهِرَتْ خِتَانَتُهُ His circumcision was made to be extirpative. (Ṣ, TA.) — And A feast, or banquet, to which people are invited on account of a circumcision. (JK, Ṣ, TA.) You say, كُنْتُ فِى خِتَانِ فُلَانٍ I was at the feast, or banquet, &c., of such a one. (TA.) — See also 1, third sentence. — Also The part, of the male, which is the place of circumcision; (T, Ṣ, Mgh, Ḳ;) and of the female likewise; (T, Mgh, TA;) the part, of the فَرْج, which is the place of circumcision. (Mṣb.) Hence, in a trad., إِذَا ٱلْتَقَى ٱلْخِتَانَانِ [When the two places of circumcision meet together]: (Ṣ,* Mṣb,* TA:) اِلْتِقَآءُ الخِتَانَيْنِ is a euphemism, metonymically denoting the disappearing, or causing to disappear, of the part of the penis that is above, or beyond, the place of circumcision (Mgh,* Mṣb, TA) in the vulva of the woman. (Mgh,* TA.)

خُتُونٌ: see خُتُونَةٌ.

خَتِينٌ Circumcised, applied to a boy, (Mṣb, Ḳ, TA,) as also ▼ مَخْتُونٌ (JK, Mṣb, Ḳ) and ▼ مُخْتَتَنٌ; (TA;) and to a girl likewise, (Mṣb, TA,) as also ▼ مَخْتُونَةٌ. (Mṣb.)

خِتَانَةٌ: see خِتَانٌ, in two places. — Also The art, or business, of circumcising. (JK, Ḳ, TA. [In the CḲ, او الخِتَانَةُ is erroneously put for وَالخِتَانَة.])

خُتُونَةٌ The alliance by which one acquires the relationship of a خَتَن, (Az, Mgh,) or of a صِهْر; (Ḳ;) as also ▼ خُتُونٌ. (Az, Mgh, Ḳ.) And A man's marrying, or taking to wife, a woman. (Ḳ.)

خَاتِنٌ A circumciser. (JK,* Mṣb,* TA.)

خَاتُونٌ A lady, or noble woman; a foreign word, (Ḳ, TA,) used by the Persians and Turks: pl. خَوَاتِينُ. (TA.)

مَخْتُونٌ; and its fem., with ة: see خَتِينٌ. — عَامٌ مَخْتُونٌ ‡ A year of drought, or barrenness, or dearth. (A, TA.)

مُخْتَتَنٌ: see خَتِينٌ.

خثر

1. خَثَرَ, (Ṣ, Mṣb, Ḳ,) aor. ـُ; (Ṣ, Mṣb;) and خَثُرَ, aor. ـُ, (Ṣ, A, Mṣb, Ḳ,) a rare dial. var.; (Fr, Ṣ;) and خَثِرَ, (Ṣ, A, Mṣb, Ḳ,) aor. ـَ, (Mṣb,) a form heard by Kṣ; (Ṣ;) inf. n. (of the first, TA) خَثْرٌ and خُثُورٌ and خَثَرَانٌ, (Ḳ, TA,) which last is irregular, because this word does not imply motion, (TA,) [but this assertion requires consideration,] and (of the second [accord. to rule], TA, or of the first, Mṣb, [or used as inf. n. of the first because it is the most common form,]) خُثُورَةٌ (Ṣ, A, Mgh, Mṣb, Ḳ) and [of the second accord. to rule] خَثَارَةٌ (Ḳ) and of the third خَثَرٌ; (Mṣb, TA;) [and probably ▼ تَخَثَّرَ, (mentioned by Freytag, though without any indication of his authority,) as quasi-pass. of خَثَّرَهُ; but I have more than once found it erroneously written for تَخَتَّرَ, which has a different signification;] It (milk, Ṣ, A, Mgh, Mṣb, Ḳ, and honey, and the like, TA, and a liniment or the like, A, or some other thing, Mṣb) was, or became, thick, (Ṣ, A, Mgh, Mṣb, Ḳ,) and strong. (Mṣb.) — [Hence,] خَثُرَتْ نَفْسُهُ, (Ṣ, Ḳ,) or خَثِرَتْ, (A,) or خَثَرَتْ, (Mgh,) inf. n. خُثُورٌ and خَثَارَةٌ, but not, as some write it, خُثَارَةٌ, (TA,) ‡ His soul [or stomach] heaved, became agitated by a tendency to vomit, or became heavy; (A, Mgh, Ḳ, TA;) as also خَثُرَ alone; (IAạr, TA;) became disordered; syn. اِخْتَلَطَتْ. (Ṣ, Ḳ.) — And خَثَرَ; (Ṣ;) or خَثِرَ, aor. ـَ; (Ḳ;) or خَثِرَ فِى الحَىِّ; (A;) ‡ He remained among the tribe, (Ṣ, A, Ḳ,) not going forth with people to procure wheat or corn or other provisions, (Ṣ, Ḳ,) by reason of shame, or of heaviness of the soul [or stomach]. (TA.) — And خَثِرَ † He felt, or had a sense of, or he was, or became, moved or affected with, shame, shyness, bashfulness, or honest shame. (Ḳ.)

2: see 4.

4. اخثرهُ; (Mṣb, Ḳ;) and ▼ خَثَّرَهُ, (A, Mṣb, Ḳ,) inf. n. تَخْثِيرٌ; (TA;) He thickened it; made it thick, (Mṣb, A, Ḳ,) and strong; (Mṣb;) namely, milk, (A, Mṣb, Ḳ,) and honey, and the like, (TA,) and a liniment or the like, (A,) or some other thing. (Mṣb.) — And the former signifies also He left it in a thick state; namely, fresh butter; (Aṣ, Ṣ, Ḳ;) not melting it. (Aṣ, Ṣ.) It is said in a prov., مَا يَدْرِى أَيُخْثِرُ أَمْ يُذِيبُ [He knows not whether to leave in a thick state or to melt]: (Ṣ, Ḳ; in one copy of the former of which, the fem. forms of the verbs are used:) applied to him who is confounded, or perplexed, and unable to see his right course, and who wavers, or vacillates: its origin being this: a woman melts fresh butter, and what is thick thereof becomes mixed with what is thin, and she is vexed and wearied by her case, and knows not whether to raise the fire with fuel, in order that it may become clear; fearing that, if she do so, it will burn: thus she is perplexed. (Ḳ.)

5: see 1.

خَثْرَآءُ الانفس and خَثْرَى الأَنْفُسِ: see خَاثِرٌ.

خُثَارٌ What remains upon a table of food. (Ṣ.)

خُثَارَةٌ What remains (Ṣ, Ḳ) of a thing, (Ṣ,) or of milk: (Ḳ:) the dregs; lees; or thick, or turbid, portion that sinks to the bottom of a thing, beneath the clear portion. (TA in art. ثفل.) You say, ذَهَبَ صَفْوُهُ وَبَقِيَتْ خُثَارَتُهُ [The clear part of it went away, and the thick part of it remained]. (A.)

خَاثِرٌ, applied to milk, (Mgh, Mṣb,) and to a liniment or the like, (A,) &c., (Mṣb,) Thick, or thickening, (A, Mgh, Mṣb,) and strong. (Mṣb.) — [Hence,] خَاثِرُ النَّفْسِ ‡ Having the soul [or stomach] in a heaving state, agitated by a tendency to vomit, or heavy: (TA:) or not in a good state: (A, Mgh:) and languid: (A:) and ▼ قَوْمٌ خَثْرَآءُ الأَنْفُسِ (Ṣ, Ḳ) and ▼ خَثْرَى الأَنْفُسِ (Ḳ) a party disordered [in souls or stomachs]. (Ṣ, Ḳ.) And خَاثِرَةٌ † A woman feeling a little pain (Ḳ) and languor; as also ▼ مُخْثِرَةٌ [or, more probably مُخْثَرَةٌ]. (TA.)

خَاثِرَةٌ [fem. of خَاثِرٌ, q. v. — And also, as a subst.,] ‡ A party of men: (Ḳ, TA:) or ‡ a dense body of men. (A, TA.)

مُخْثِرَةٌ: see خَاثِرٌ.

خثى

1. خَثَى, aor. يَخْثِى, inf. n. خَثْىٌ, said of a beast of the ox-kind, (JK, Ṣ, Mṣb, Ḳ,) or of a bull, but not [خَثَتْ] of a cow, (A 'Obeyd, TA,) and of an elephant, (Ḳ,) He dunged. (JK, Ṣ, Mṣb, Ḳ.) [See also خِثْىٌ.]

4. اخثى He (a man, TA) kindled [dry dung such as is called] خِثْى. (Ḳ.)

خِثْىٌ *Dung* of a beast of the ox-kind; (JK, Ṣ, Mgh, Mṣb, Ḳ;) as also ▼ خَثْىٌ: (Mṣb:) or of a bull: (IAạr, TA:) and of the elephant: (Ḳ:) and metaphorically, of the camel; as used in a trad.: or, accord. to AZ, *compact dung* of a beast of the ox-kind, and of the sheep or goat, and of any cloven-hoofed animal, and of the camel: (TA:) pl. أَخْثَآءٌ [a pl. of pauc.] (JK, Ṣ, Mgh, Mṣb, Ḳ) and خُثِىٌّ, (CḲ, [a quasi-pl. n. like عُبَيْدٌ,]) or خِثِىٌّ, (Ḳ accord. to the TA, [like جِثِىٌّ, q. v., a pl. of جِثْوٌ,]) and خُثِىٌّ [originally خُثُوىٌ], (Ḳ,) these two from Fr. (TA.) — خِثْىٌ also signifies †*A number of people in a state of dispersion:* (Ṣgh, TA:) or so خِثْىٌ مِنَ النَّاسِ. (JK.)

خَثْىٌ: see the next preceding paragraph.

مِخْثَىٌ, (JK, TA,) so in the Tekmileh, (TA,) [or مخثًا accord. to the CḲ, there, with the article, written المخثا,] or ▼ مِخْثَاةٌ, (Ḳ accord. to the TA,) The [*kind of pouch called*] خُرَيْطَة, (JK, Ḳ,) and *small* [*bag such as is termed*] جِرَاب, (JK,) *of the gatherer of* [*wild*] *honey,* (JK, Ḳ,) *which he puts beneath the part between his armpit and his flank,* (TA,) *and in which he deposits the honey.* (JK.)

مِخْثَاةٌ: see what next precedes.

خجل

1. خَجِلَ, (Ṣ, Mṣb, Ḳ, &c.,) aor. ـَ, (Ḳ,) inf. n. خَجَلٌ, (Ṣ, Mṣb, &c.,) but not خَجَالَةٌ, [though authorized by the KL, in my copy of which I find it thus written (not خُجَالَةٌ as written by Golius),] for this is a vulgar mistake for خَجْلَةٌ or خَجَلٌ, (Mgh, [so in my copy, but correctly ▼ خَجْلَةٌ (which may be either a simple subst. or an inf. n. of un.) or خَجَلٌ,]) *He was, or became, confounded, or perplexed, and unable to see his right course, by reason of shame:* (Ṣ, O:) or *he was, or became, ashamed, and confounded,* or *perplexed, and unable to see his right course,* (T, M, Ḳ,) [or, simply, *ashamed,* (see خَجِلٌ,)] *in consequence of a deed that he had done:* (T, M, TA:) thus الخَجَلُ has a more particular signification than الحَيَآءُ: (TA:) or it is like الاِسْتِحْيَآءُ. (Mṣb.) — And *He remained silent,* (T, Ḳ,) or *still,* (M,) *not speaking nor moving.* (Ḳ.) — And *He was, or became, in a confused and dubious case,* (JK, M,* Ḳ,*) *so that he knew not how to extricate himself from it.* (M, Ḳ.) — Also, said of a camel, ‡*He went in mud, and became like him who is confounded,* or *perplexed, and unable to see his right course:* (JK,* M, Ḳ, TA:) or *he stuck fast in mire.* (T, TA.) — And in like manner, †*He became agitated,* or *convulsed,* or *he struggled,* or *floundered, with his load:* (JK:) or خَجِلَ بِالْحِمْلِ *he was oppressed by the load,* (Ḳ, TA,) *so that he was agitated,* or *convulsed,* or *he struggled,* or *floundered, beneath it.* (TA.) — And, said of a plant, or of herbage, ‡*It was,* or *became, tall, and tangled,* or *luxuriant,* or *abundant and dense;* (ISd, Ḳ, TA;) and so ▼ اخجل said of the kind of trees termed حَمْض. (JK, Ḳ.) — [And, as inf. n. of خَجِلَ,] خَجَلٌ also signifies The *bearing richness ill; as when, being rich, one exults,* or *exults greatly* or *excessively, and behaves insolently and unthankfully:* (Ṣ,* Ḳ:) or the *taking a wide,* or *an ample, range,* or *being profuse, when rich.* (TA.) It is related in a trad. that he [Mohammad] said to the women, إِذَا جُعْتُنَّ دَقِعْتُنَّ وَإِذَا شَبِعْتُنَّ خَجِلْتُنَّ, (Ṣ,* TA,) i. e. *When ye are hungry, ye become lowly, humble,* or *submissive, and cleave to the dust,* or *earth;* (Ṣ and TA in art. دقع;) or *ye bear poverty ill;* (TA in the present art.;) *and when ye are satiated,* [*ye bear richness ill;* or] *ye exult,* or *exult greatly* or *excessively, and behave insolently and unthankfully.* (Ṣ in the present art.) [See also a verse of El-Kumeyt cited in the first paragraph of art. دقع.] — And *i. q.* بَرَمٌ [The *being affected with disgust, loathing,* or *aversion;* the *being vexed, grieved, disquieted by grief,* &c.] (Ḳ, TA. [In the CḲ, البُرْمُ is erroneously put for البَرَمُ.]) — And The *being remiss in seeking subsistence.* (Ḳ.) — And The *being lazy,* or *indolent:* (Az, ISd, Ḳ:) from the verb in the sense explained in the second sentence of this paragraph. (TA.) — And *i. q.* فَسَادٌ [The *being bad, corrupt,* &c.]. (M, Ḳ.) — Also, in a shirt, †The *being much slit,* or *rent, in the lower parts,* or *skirts.* (Fr, Ḳ.)

2: see what next follows.

4. اخجلهُ (Ṣ, Mṣb, Ḳ) *i. q.* ▼ خجّلهُ, (Mṣb,* Ḳ, TA,) inf. n. تَخْجِيلٌ; (TA;) *He,* (Ṣ,) or *it,* namely, an affair, or event, (TA,) *caused him to become confounded,* or *perplexed, and unable to see his right course, by reason of shame:* (Ṣ in explanation of the former:) [or *caused him to become ashamed, and confounded,* or *perplexed, and unable to see his right course, in consequence of a deed that he had done:* (see 1:)] or *he said to him* خَجِلْتَ. (Mṣb. [But it is not clear whether this meaning be there assigned to both of these verbs, or only to the latter of them.]) — See also 1.

خَجِلٌ part. n. of خَجِلَ; (Mṣb;) [*Confounded,* or *perplexed, and unable to see his right course, by reason of shame:* or *ashamed, and confounded,* or *perplexed, and unable to see his right course, in consequence of a deed that he has done:* or, simply,] *ashamed.* (Ṣ, Mṣb.*) — [Other meanings are shown by explanations of the verb.] — Applied to herbage, ‡*Tall,* (Ḳ, TA,) *and tangled,* or *luxuriant,* or *abundant and dense, and goodly,* and ISd adds, *full-grown:* and ▼ مُخْجِلٌ [in like manner], applied to the kind of trees termed حَمْض, *dense,* or *tangled, and tall:* or, applied to herbage, or pasturage, *wide, abundant, full-grown, that detains one so that he stays among it, not passing beyond.* (TA.) — And, applied to a place, and a valley, †*Abounding with tangled,* or *luxuriant,* or *abundant and dense, herbage:* (Ṣ:) or, applied to a valley, (JK, Ḳ,) as also ▼ مُخْجِلٌ, (Ḳ,) ‡*exceedingly abundant in herbage:* (Ḳ, TA:) or *tangled,* or *luxuriant,* or *abundant and dense, therewith;* (JK, Ḳ, TA;) *resounding with the humming of flies.* (JK.) — Also, applied to a garment, †*Wide and long:* (ISh, Ḳ:) or *ample:* or *such that the wearer is impeded and clogged therein:* (TA:) and, so applied, †*old, and worn out:* (Ḳ:) or †*much slit,* or *rent, in the lower parts,* or *skirts.* (Fr, TA.) — And, applied to a جُلّ [or horse-cloth, or covering for a beast], (ISh, Ḳ,) [or] such as is put upon a camel, (ISh,) *That moves to and fro,* or *from side to side,* (ISh, Ḳ,) *upon the camel,* (ISh,) or *upon the horse,* (Ḳ,) *by reason of its width.* (ISh.)

خَجْلَةٌ: see 1: [it seems to be most probably a subst. signifying *Confusion,* or *perplexity, and inability to see one's right course, by reason of shame:* or *shame, and confusion,* or *perplexity, and inability to see one's right course, in consequence of a deed that one has done:* or simply,] *i. q.* حَيَآءٌ [*shame,* or *a sense of shame,* &c.]. (Ṣ.)

مُخْجِلٌ: see خَجِلٌ, in two places.

خد

1. خَدَّ الأَرْضَ, aor. ـُ, (Ṣ,) or خَدَّ فِى الأَرْضِ, (A,) [aor., if accord. to rule, ـُ,] inf. n. خَدٌّ, (T,) *He furrowed,* or *trenched,* or *clave, the ground;* (Ṣ, L;) *he made a furrow,* or *trench,* [or *furrows,* or *trenches,*] *in the ground.* (T, A.) The latter (خدّ فى الارض) is also said of a torrent, meaning *It furrowed,* or *clave, the ground by its course.* (L.) — خَدَّ, (L,) inf. n. ـُ, (L, Ḳ,) also signifies *He,* or *it, marked, scored,* or *impressed,* a thing: (L:) and *made a mark* or *marks,* or *an impression* or *impressions,* upon a thing. (L, Ḳ.*) You say, خَدَّ الفَرَسُ الأَرْضَ بِحَوَافِرِهِ *The horse marked,* or *scored,* [or *furrowed,*] *the ground with his hoofs.* (L.) And خَدَّ الدَّمْعُ فِى خَدِّهِ *The tears made marks upon his cheeks.* (L.) — Also *He* (a camel) *clave a thing with his* ناب [or *tush*]. (L.) — And *He cut* a thing. (IAạr.)

2. خَدَّدَ لَحْمُهُ, (as in the Ṣ and Ḳ,) or خَدَّدَ, (as in one place in the L,) [both of which may be correct, for the verb is said in the Ḳ to be both intrans. and trans.,] ‡*His flesh became contracted, shrunk,* or *wrinkled;* (Ṣ, TA;) as also ▼ تَخَدَّدَ: (Ṣ, A,* Ḳ:) or *his flesh wasted so that there appeared streaks upon his skin:* (TA in art. خب:) or *he* (a beast) *became lean,* or *lank,* or *light of flesh,* or *slender* or *lank in the belly, so that his flesh became furrowed,* or *wrinkled:* and ▼ تخدّد لَحْمُهُ *his flesh became flaccid and quivering, by reason of leanness.* (L.) And خَدَّدَهُ ‡*It* (travel) *rendered him lean and wasted:* (Ḳ:) and so evilness of state or condition. (A,* TA.)

3. خَادَّهُ ‡*He opposed him, being opposed by him:* (A:) or *he was,* or *became, angered,* or *enraged, against him, and opposed him in his deed,* or *work.* (Ḳ.)

5. تخدّد *It* (the ground) *became furrowed,* or *cleft,* by a torrent. (L in art. خد.) — See also 2, in two places. — تخدّد القَوْمُ ‡*The people became divided into distinct bodies, or parties.* (L in the present art.)

6. تخادّوا ‡ *They opposed each other.* (A, TA.)

خَدٌّ, (Ṣ, Mṣb, Ḳ, &c.,) of the masc. gender (Lḥ, Ḳ) only, (Lḥ,) and ▼ خُدَّةٌ, with ḍamm, (Ḳ,) but the latter is rare, (TA,) The *cheek;* the *part extending from the circuit of the eye* (المَحْجِر) [in the CḲ المِحْجَن] *to the part where the beard grows, on either side of the face:* (L, Mṣb, Ḳ:) or *from the outer angle of each eye to the extremity of the side of the mouth:* or the *part bordering upon the nose, on either side:* (L, Ḳ:) *each of the* خَدَّان, *in the face:* (Ṣ:) pl. of the former خُدُودٌ: (Mṣb, TA:) it has no other pl. (TA.) — The former is also sing. of خُدُودٌ meaning ‡ The *planks on the right and left of the* دَفَّتَانِ [or *two boards that lie against the camels' sides*] of the [vehicles termed] هَوَادِج (Aṣ, A, TA) and غُبُط: (Aṣ, TA:) or خَدٌّ signifies the *plank* (صَفِيحَة) *of the* هَوْدَج: and the pl. is أَخِدَّةٌ, (Ḳ,) [a pl. of pauc., but] contr. to rule, (TA,) and (pl. of mult., TA) خِدَادٌ and خِدَّانٌ. (Ḳ.) — Also ‡ *A side,* or *lateral portion,* of a [tract of high and rugged ground such as is termed] قُفّ. (A, TA.) — And ‡ *An assembly, a company,* or *a congregated body,* (Ḳ, TA,) of men: (TA:) *a rank,* or *class,* of men: (A, L:) and *a race,* or *generation,* of men. (L.) You say, رَأَيْتُ خَدًّا مِنَ النَّاسِ † *I saw a rank,* or *class, of men.* (L.) And قَتَلْنَا, (A,) or قَتَلَهُمْ, (L,) خَدًّا فَخَدًّا, (A, L,) ‡ *We slew,* (A,) or *he slew them,* (L,) *rank after rank,* (A, L,) and *class after class.* (A.) And مَضَى خَدٌّ مِنَ النَّاسِ † *A race, or generation, of men passed away.* (L.) ═ *A road.* (IAạr, Ḳ.) — See also أُخْدُودٌ, in four places.

خُدَّةٌ: see أُخْدُودٌ, in four places: ═ and see also خَدٌّ.

خُدَدٌ: see أُخْدُودٌ.

خِدَادٌ *A mark made with a hot iron upon the cheek* (Ṣ, A, Ḳ) *of a camel.* (A.)

أُخْدُودٌ *A furrow, trench,* or *channel, in the ground;* (Ṣ, A, Ḳ;) as also ▼ خَدٌّ (A, Ḳ) and ▼ خُدَّةٌ: (Ḳ:) *a deep trench in the ground;* as also ▼ خَدٌّ: applied in the Ḳur [lxxxv. 4] to a deep trench into which, it being filled with fire, some believers in the true God were cast by some idolaters among whom they dwelt: (TA:) *a hole, hollow, cavity, pit,* or *the like, dug,* or *excavated;* (Mṣb;) as also ▼ خُدَّةٌ: (Ṣ:) pl. أَخَادِيدُ: (A, Mṣb:) the pl. of ▼ خَدٌّ is خُدُودٌ; (A;) and the pl. of ▼ خُدَّةٌ is خُدَدٌ. (TA.) — *A rivulet,* or *streamlet;* syn. جَدْوَلٌ; (Mṣb;) as also ▼ خَدٌّ; (L, Ḳ;) of which the pl. [of pauc.] is أَخِدَّةٌ, contr. to rule; and of mult. خِدَادٌ and خِدَّانٌ. (L.) — أَخَادِيدُ and ▼ خُدَدٌ [the latter probably a mistranscription for خُدَدٌ, pl. of ▼ خُدَّةٌ,] The *main* or *middle parts,* or *open* or *obvious tracks,* (شَرَك,) of a road: (L:) [because furrowed by the feet of beasts and men.] — أَخَادِيدُ الأَرْشِيَةِ *The furrows,* or *grooves, of well-ropes, in a well; made by drawing them.* (L.) — أَخَادِيدُ السِّيَاطِ *The marks of whips;* (Ḳ;) *the furrows made by whips* upon the back. (L.) — ضَرْبَةٌ أُخْدُودٌ, (Ṣ,) or ضَرْبَةُ أُخْدُودٍ, (so in a copy of the A,) ‡ *A blow,* or *stroke,* or *beating, that furrows the skin.* (Ṣ, A,* TA.)

مِخَدٌّ *Each of the tushes,* or *canine teeth;* the two together being termed the مِخَدَّانِ. (L.)

مِخَدَّةٌ [*A pillow,* or *cushion*]: so called because it is put beneath the cheek: (Ṣ, A, Mṣb:) pl. مَخَادُّ. (A, Mṣb.) ═ Also *An iron implement with which the ground is furrowed, trenched,* or *cleft.* (Ṣ, L.)

مَخْدُودٌ A camel *having the mark called* خِدَاد *made upon his cheek.* (Ṣ, A.)

مُتَخَدِّدٌ ‡ A man (L) *lean,* or *emaciated;* (Ṣ, L;) *having little flesh.* (L.) And the fem., with ة, ‡ A woman *fat but wasted in body* [*so that she is furrowed,* or *wrinkled*]. (L.)

خدب

1. خَدَبَهُ, (Ṣ, Ḳ,) aor. ِ, inf. n. خَدْبٌ, (TA,) *He smote him* with a sword: (Ṣ, Ḳ:) or *he cut him:* (AZ:) or *he smote him* with a sword, *cutting the flesh, without cutting the bone:* (T, Ḳ:) or *he smote him on the head,* (JK, Ḳ,* TA,) and *the like,* (JK, TA,) with a sword. (JK.) — خَدْبٌ signifies also The act of *rending,* or *cutting open, the skin and the flesh together,* (JK, Ṣ, TA,) with the canine tooth. (JK, TA.) — And The act of *biting.* (Ḳ.) You say, خَدَبَتْهُ الحَيَّةُ *The serpent bit him.* (Ṣ.) — And The act of *lying,* or *saying what is untrue.* (Ḳ.) You say, خَدَبَ *He lied.* (JK, Ṣ.) — And The act of *milking much.* (Ḳ.)

5. تخدّب *He went a middling pace.* (Aṣ, Ḳ.)

خَدَبٌ *Stupidity;* (JK;) i. q. هَوَجٌ [which means the same; or *tallness combined with stupidity;* &c.]: (Ṣ, Ḳ:) and [simply] *tallness,* or *length;* (Ṣ,* Ḳ, TA;) as also ▼ خُدْبَةٌ. (TA.) One says, كَانَ بِنَعَامَةَ خَدَبٌ *There was, in Na'ámeh,* هَوَج [i. e. *stupidity,* &c.]: (Ṣ, TA:) Na'ámeh being the surname of Beyhes, (TA,) who became the subject of a prov. respecting the taking of blood-revenge: (Ḳ in art. بيهس:) Beyhes El-Fezáree, surnamed En-Na'ámeh, [or simply Na'ámeh,] was one of seven brothers, who were slain [in a case of blood-revenge], except himself; he being left because of his stupidity; for which he became proverbial. (TA in art. بيهس.) One says also, فِى لِسَانِهِ خَدَبٌ [*He is long-tongued;* lit.] *in his tongue is length.* (Ṣ.) — Also The act of *making haste.* (KL.)

خَدِبٌ *Cutting;* or *sharp:* (IAạr, Ḳ:) applied to a sword, and a spear-head, and a dog-tooth. (IAạr.) ═ See also أَخْدَبُ, in three places.

خُدْبَةٌ: see خَدَبٌ.

وَقَعُوا فِى وَادِى خَدِبَاتٍ [*They fell into the valley of Khadibát*]: a prov., applied to a man who falls into perdition; or to one who misses the object of his aim, or pursuit: (Aṣ, Ḳ:*) see, respecting it, [جَذَبَات, in] art. جذب. (TA.)

خِدَبٌّ An *old* man. (A, Ḳ.) *Great, big,* or *bulky;* (Ṣ, A, Ḳ;) applied to a man, (Ṣ,) or to an ostrich, &c.; (Ḳ;) fem. with ة, applied to a girl: (Ṣ:) *great, and thick* or *coarse* or *rude* [*in make*]: (TA:) *big in the sides:* (Ḥam p. 439:) *big and strong;* applied to an old man: (JK:) a man, and a camel, *perfect and strong in make:* (A:) a camel *strong and hardy* (A, Ḳ, TA) *and big.* (TA.)

خَادِبَةٌ, (JK, TA,) or شَجَّةٌ خَادِبَةٌ, (TA,) *A severe wound breaking the head.* (JK, TA.)

خَيْدَبٌ, (Ṣ, Ḳ,) or ▼ خَيْدَبَةٌ, (JK,) *A conspicuous road:* (JK, Ṣ, Ḳ:) so says Esh-Sheybánee. (Ṣ.) — And the latter, *A way of acting* or *conduct:* so in the saying, فُلَانٌ عَلَى خَيْدَبَةٍ صَالِحَةٍ [*Such a one is following a good, righteous,* or *just, way of acting* or *conduct*]. (TA.) — *Opinion, judgment,* or *counsel:* (Ḳ:) so in the saying, تَرَكْتُهُ وَخَيْدَبَتَهُ [*I forsook him and his counsel*]. (TA.) — One's *first,* or *original, affair, concern,* or *business:* (AZ, JK, Ṣ, Ḳ:) so in the saying, أَقْبِلْ عَلَى خَيْدَبَتِكَ [*Apply thyself to thy first,* or *thine original, affair,* &c.]. (AZ, JK, Ṣ.)

خَيْدَبَةٌ: see the next preceding paragraph.

أَخْدَبُ (Ṣ, Ḳ) and ▼ خَدِبٌ (JK) and ▼ مُتَخَدِّبٌ, (Ṣ, Ḳ,) applied to a man, (JK, Ṣ,) *Stupid;* (JK;) *characterized by* هَوَج [i. e. *stupidity,* or *tallness combined with stupidity,* &c.]; (Ṣ, Ḳ;) and [simply] *by tallness:* (Ḳ: [in which أَخْدَبُ is said to be *syn. with* طَوِيلٌ:]) fem. [of the first], applied to a woman, خَدْبَآءُ: (Ṣ:) and the first, one *who cannot restrain himself, by reason of stupidity:* (TA:) and the same, (Ḳ,) or ▼ خَدِبٌ, (JK,) one *who goes his own way, at random,* or *heedlessly, without any certain aim* or *object,* or *without consideration, not obeying a guide to the right course,* (JK, Ḳ, TA,) *by reason of ignorance,* (JK,) or *from boldness.* (TA.) — Also [the fem.] خَدْبَآءُ, applied to any [she-] animal, *That is wont to wound, and kill, and seize its prey and break its neck.* (IAạr.) And حَرْبَةٌ خَدْبَآءُ and ▼ خَدِبَةٌ [*A dart,* or *javelin,*] *that inflicts a wide wound.* (Ḳ.) And طَعْنَةٌ خَدْبَآءُ, (Ṣ,) and ضَرْبَةٌ خَدْبَآءُ, (Ḳ,) [*A thrust, and a blow,*] *with which the* جَوْف [or *belly,* &c.,] *is assaulted:* (Ṣ, Ḳ:) or the former, *a wide thrust* or *wound with a spear* or *the like:* and the latter, *a continuous and long act of beating.* (TA.) — And دِرْعٌ خَدْبَآءُ *A coat of mail that is smooth,* or *pliable,* (لَيِّنَة, JK, Ṣ, Ḳ,) *and* (so in the JK, but in the Ḳ "or") *wide.* (JK, Ḳ.)

مُتَخَدِّبٌ: see the next preceding paragraph.

خدج

1. خَدَجَتْ, aor. ِ (Ṣ, Mṣb, Ḳ) and ُ, (Ḳ,) inf. n. خِدَاجٌ, (Ṣ, Ḳ,*) or this is a simple subst.,

(Mṣb,) *She* (a camel, AZ, Ṣ, Mṣb, Ḳ, and any cloven-hoofed or solid-hoofed animal, AZ, Mṣb, TA, or sometimes another female, TA) *cast her young one before the completion of the days of gestation,* (AZ, IḲṭ, Ṣ, Mṣb, Ḳ,) to which IḲooṭ adds, (Mṣb,) *even if it were completely formed;* (Ṣ, Mṣb;) as also خَدَجَتْ وَلَدَهَا; (IḲṭṭ, Mṣb;) and ↓ خدّجت, inf. n. تَخْدِيجٌ: (TA:) or, as some say, *she cast her young one when her pregnancy had become apparent, between the commencement of its formation and the period a little before the completion;* as also ↓ اخدجت وَلَدَهَا; but see the latter verb below: (Mṣb:) or *she* (a camel) *cast her young one in a defective state, before the time:* or *without any hair upon it:* or *miscarried, bringing forth merely blood:* and Thábit mentions the last but one of these significations as relating to a human being: (TA:) or you say of a woman, خدجت وَلَدَهَا and ↓ أَخْدَجَتْهُ in one and the same sense, (Aboo-Kheyreh, TA,) meaning *she cast her fœtus in such a state that its form had become apparent:* and خدجت *she miscarried, bringing forth merely blood:* when she has brought it forth before its hair has grown, one says of her غَضَنَتْ: so says Az. (TA.) — And خدج ‡ *He* (a man) *was,* or *became, deficient in a limb,* or *member.* (A.) — See also 4, in two places.

2: see 1.

4. اخدجت *She* (a camel) *brought forth her young one imperfectly formed,* (Ṣ, Mgh, Ḳ,) *even if the period of gestation were complete:* (Ṣ, Ḳ:) or so اخدجت وَلَدَهَا: and the former, *she brought forth her young one imperfectly formed at the completion of the period:* (IḲt, Mṣb:) or *she brought forth her young one completely formed before the proper time of bringing forth.* (TA.) See also 1, in two places. — Hence, (TA,) اخدجت الشِّتْوَةُ, (IAạr, Ṣ,) or الصَّيْفَةُ, (Ḳ,) ‡[*The winter,* or *the spring* or *summer* (but more probably the former),] *had little rain.* (IAạr, Ṣ, Ḳ, TA.) And اخدجت الزَّنْدَةُ, (T, TA,) or ↓ خَدَجَت, (TA,) ‡[*The lower of the two wooden instruments for producing fire*] *failed to emit fire.* (T, TA.) And اخدج ‡ *He rendered* a man *defective in a limb, or member:* said of God. (A.) And ‡ *He performed incompletely* his prayer; (Es-Saraḳustee, A, Mṣb;) or so ↓ خدّج: (Mṣb:) and in like manner, (i. e. the former verb,) † a salutation: (TA, from a trad.:) and ‡ *he performed unsoundly* his affair: (A, TA:) and ‡ *he formed unsoundly* his opinion. (A.)

خِدْج: see مُخْدَج.

خِدَاجٌ inf. n. of 1 in the first of the senses explained above; (Ṣ, Ḳ;*) or a simple subst. therefrom; (Mṣb;) or a subst. from 4 in the first of the senses explained above. (Mgh.) See also مِخْدَاجٌ. — And hence, (Mṣb,) ‡ *Defect,* or *deficiency.* (Aṣ, A, Mgh, Mṣb, TA.) — صَلَاةٌ خِدَاجٌ, (Ṣ, A, Mgh, Ḳ,) in which the latter word is an inf. n. used as an epithet, (A,) or for ذَاتُ خِدَاجٍ, (Mgh, TA,) ‡ *Prayer incompletely performed;* (Ṣ, A, Mgh, Ḳ;) as also ↓ مُخْدَجَةٌ and ↓ خَادِجَةٌ: (A:) applied in a trad. to prayer in which the Fátiḥah is not recited. (Ṣ, Mgh, TA.) And ↓ حَجٌّ إِخْدَاجٌ, in which [likewise] the latter word is for ذُو إِخْدَاجٍ, or is an inf. n. used as an epithet, † *Pilgrimage incompletely performed.* (Ḥar p. 392.) — Also a pl. of خَدُوجٌ. (TA.)

خَدُوجٌ: see خَادِجٌ: — and see also مُخْدَجٌ.

خَدِيجٌ The young one of a camel *brought forth before the completion of the days* [*of gestation*], (Ṣ, Ḳ,) *even if it be completely formed;* (Ṣ;) [and so, accord. to rule, ↓ مَخْدُوجٌ.] — See also مُخْدَجٌ. — It is applied in a trad. respecting the poor-rate to *A calf one year old, or under, that still follows its mother;* resembling a خَدِيج [properly so called] in the smallness of its limbs, and in its having less strength than a ثَنِيّ or a رَبَاعِيّ. (TA.)

خَادِجٌ A she-camel (or a female of another kind, TA) *casting her young one before the completion of the days* [*of gestation*], (Ṣ, A, Ḳ,) *even if it be completely formed;* (Ṣ, A;) as also ↓ خَدُوجٌ, of which the pl. is خُدُجٌ and خِدَاجٌ and خَدَائِجُ: or a she-camel *casting her young one in a defective state, before the time.* (TA. [See 1, of which it is the act. part. n. And see also مُخْدِجٌ.]) — [Hence,] ‡ A man *deficient in a limb,* or *member.* (A. [See also مُخْدَجٌ.]) صَلَاةٌ خَادِجَةٌ: see خِدَاجٌ.

حَجٌّ إِخْدَاجٌ: see خِدَاجٌ.

مُخْدَجٌ The young one of a camel *brought forth imperfectly formed, even if the period of gestation have been completed;* (IḲt, Ṣ, A, Ḳ;) as also ↓ مَخْدُوجٌ (TA) and ↓ خَدِيجٌ [q. v.] (A, TA) and ↓ خَدُوجٌ and ↓ خِدْجٌ: (TA:) or *brought forth completely formed before the proper time of bringing forth.* (TA.) — ‡ A man *made deficient in a limb,* or *member:* (A:) or *defective in make.* (TA, from a trad. [See also خَادِجٌ.]) And ‡ مُخْدَجُ اليَدِ A man *deficient in the arm,* or *hand.* (Ṣ, A, Mgh, Ḳ.) صَلَاةٌ مُخْدَجَةٌ: see خِدَاجٌ.

مُخْدِجٌ (Ṣ, A, Ḳ,) and مُخْدِجَةٌ (TA) A she-camel *bringing forth her young one imperfectly formed, even if the period of gestation be complete:* (Ṣ, A, Ḳ, TA:) or *bringing forth her young one completely formed before the proper time of bringing forth.* (TA. [See also خَادِجٌ.])

مِخْدَاجٌ and ↓ ذَاتُ خِدَاجٍ [A she-camel *that usually casts her young before the completion of the days of gestation, even if completely formed:* (see 1:) or] a she-camel *that usually brings forth her young imperfectly formed,* (A, TA,) *even if at the proper time,* (A,) or *before the proper time:* (TA:) or *that usually brings forth her young completely formed before the proper time of bringing forth.* (TA.)

مُخْدَجٌ: see خَدِيجٌ: and see also مَخْدُوجٌ.

خدر

1. خَدَرَ and خَدِرَ, as intrans. vs.: see 4, in six places: ═ and for the former, as a trans. v.: see 2, in two places. ═ خَدِرَ, aor. ـَ, inf. n. خَدَرٌ, said of a limb, (Mṣb, Ḳ,) and of the body, (TA,) and خَدِرَتْ, inf. n. as above, said of the leg or foot, (Ṣ, A,) and of the arm or hand, (TA,) *It was,* or *became, benumbed,* or *torpid,* or *affected by a languidness,* or *laxity,* (Ṣ, Mṣb, Ḳ,) or *by a heaviness,* (IAạr,) *and an impotence of exercising motion,* (IAạr, Mṣb,) or *by a contraction of the sinews;* (TA;) said of the leg or foot [&c.], *it became asleep.* (TA in art. بسر.) — Also خَدِرَ, inf. n. as above, † *He became languid from drinking wine or medicine.* (TA: but only the inf. n. of the v. in this sense is there mentioned.) And † *He was,* or *became, lazy,* or *slothful, and languid.* (Ḳ,* TA: but in this instance, also, only the inf. n. is mentioned.) And خَدِرَتْ عِظَامُهُ (Ṣ, A) ‡ *His bones became feeble.* (A.) And خَدِرَتْ عَيْنُهُ, (A,) inf. n. as above, (Ḳ,) ‡ *His eye became languid:* (Ḳ, TA:) or *became heavy,* (A, Ḳ,) *by reason of rubbing,* (A,) or *from a mote in it.* (A, Ḳ.) — And خَدِرَ, (TA,) inf. n. as above, (Ḳ, TA,) said of the day, (TA,) † *It became intensely hot:* — and † *intensely cold:* (Ḳ, TA: [see also the part. n. خَدِرٌ:]) — and ‡ *it was,* or *became, calm; without wind,* and *without a breeze.* (A, TA.)

2. خدّر, (A, Mṣb,) inf. n. تَخْدِيرٌ; (Ḳ;) and ↓ اخدر, (A, Mṣb,) inf. n. إِخْدَارٌ; (Ḳ;) and ↓ خَدَرَ, (Mṣb,) inf. n. خَدْرٌ; (Ḳ;) *He,* (Mṣb,) or *they,* namely, her family, (A, Mṣb,) *made* a girl *to keep herself behind,* or *within, the curtain;* (A, Mṣb, Ḳ;) *and kept her from menial employments and from going out to accomplish her wants.* (Mṣb.) — [Hence,] ↓ خَدَّرَتْ † *She* (a gazelle) *concealed* her young one in a covert of trees or the like, or in a hollow. (TA.) And ↓ اخدر ‡ *It* (a lurking-place) *concealed* a lion; (Ḳ, TA;) [as also ↓ خَدَرَ: (see مَخْدُورٌ:)] and † *it* (anything) *prevented* a thing *from being seen.* (TA.) — [And hence,] خدّر † *It* (rain) *confined* people in their houses or tents. (TA.) And ↓ اخدر † *It* (night) *confined, detained,* or *withheld,* a person. (TA.) ═ See also 4, where it is app. a mistranscription for تخدّر. ═ خدّر (A) and ↓ اخدر (Ḳ) also signify *It made* a limb, (Ḳ,) and the body, (TA,) and a leg or foot, (A,) and an arm or a hand, (TA,) *to become* خَدِر, i. e. *benumbed,* &c. (A, Ḳ, TA. [See خَدِرَ.]) You say, خَدَّرَتْهُ المَقَاعِدُ, meaning *Long sitting* [lit. *the sitting-places*] *made his legs,* or *feet, to be in that state.* (A, TA.)

3. خَادَرَنِى [*He acted covertly with me*]. (A, TA. [In both, يُخَادِرُنِى is coupled with يُسَاتِرُنِى.])

4. أَخْدَرَتْ *She* (a girl) *kept herself behind,* or *within, the curtain;* (Es-Saraḳustee, Mṣb;) as also ↓ تَخَدَّرَت, (A, TA,) and ↓ اختدرت, and

تخدّر ▾ خَدِرَتْ ▾ فِى خِدْرِهَا. (TA.) — [Hence,] [in the CḲ ▾ خدر (app. a mistranscription)] and ▾ اختدر † He concealed, or hid, himself; (Ḳ, TA;) as also ▾ خَدِرَ, like فَرِحَ [in measure]: (TA:) whence the saying, اِخْتَدَرَتْ ▾ القَارَةُ بِالسَّرَابِ, i. e. [The small isolated mountain, or the like,] became concealed by the mirage. (TA.) [Hence also,] اخدر ‡ He (a lion) kept himself in his lurking-place; (Ṣ, A, Ḳ;) as also ▾ خَدِرَ and ▾ خَدَرَ, (TA,) or خَدَرَ فِى عَرِينِهِ. (A, TA.) And † It (a bird) remained in its nest. (Ṣ.) And † He (a man) remained, stayed, or abode; (Ṣ, Ḳ;) بِمَكَانٍ in a place; as also ▾ خَدِرَ, inf. n. خَدَرٌ; (Ḳ;) and فِى أَهْلِهِ among his family. (Ṣ.) And ▾ خَدَرَ, (Ṣ,) inf. n. خُدُورٌ (Ḳ,) † He (a gazelle) remained behind the herd; not going with it: (Ṣ, Ḳ:) and he (a beast) remained behind; not overtaking, or coming up with, the others. (TA.) And اخدروا † They entered upon night [and so became concealed from view]. (TA.) And † They entered upon a day of rain, and of clouds or mist, and of wind: (Ḳ:) or rain came upon them. (Ṣ.) = اخدر as a trans. v.: see 2, in four places.

5: see 4, in two places.

8: see 4, in three places.

خِدْرٌ A curtain (Ṣ, A, Mṣb, Ḳ) that is extended for a girl in a part of a house, or chamber, or tent; as also ▾ أُخْدُورٌ: (Ḳ:) and hence, (M,) any chamber, or house, or tent, or the like, that conceals a person: (M, Ḳ:) or a chamber, or house, or tent, in which is a woman; not otherwise: (Mṣb:) pl. [of mult.] خُدُورٌ (A, Mṣb, Ḳ) and [of pauc.] أَخْدَارٌ, and pl. pl. [i. e. pl. of the latter of these two, or pl. of أُخْدُورٌ,] أَخَادِيرُ. (Ḳ.) — [And hence, A vehicle composed of] pieces of wood set up over the saddle (قَتَب) of the camel, and curtained with a piece of cloth; (Ḳ;) i. e. a هَوْدَج. (TA.) — [Hence also,] ‡ The lurking-place of a lion. (Ṣ, Ḳ, TA.) — See also what next follows.

خَدَرٌ: inf. n. of خَدِرَ [q. v.]. (Mṣb, Ḳ.) = Also, and ▾ خَدْرٌ, † The darkness of night: (Ḳ:) or darkness absolutely; as also ▾ خُدْرَةٌ: (TA:) or this last signifies intense darkness: (Ḳ, TA:) or, accord. to some, the night consists of five divisions, سُدْفَة and سُتْفَة and جُهْمَة and يَعْفُور and خُدْرَة; so that this last signifies the last [of five divisions] of the night: or, accord. to Kr, the division next before this is called هَزِيع. (TA.) — † A dark place: (Ḳ:) or a dark, and low or depressed, place. (Ḥam p. 234.) — See also خُدَارِىٌّ. — † Rain: (Ṣ, Ḳ:) or clouds, or mist, and rain. (ISk.) = See also خُدْرَةٌ.

خَدْرٌ: see خُدَارِىٌّ.

خَدِرٌ, applied to a limb, Affected with خَدَر, or numbness, &c. (Ḳ.) — [Hence,] عَيْنٌ خَدِرَةٌ and ▾ خَدْرَآءُ ‡ An eye in a languid state: or heavy, by reason of rubbing, or from a mote in it. (TA.) And يَعْفُورٌ خَدِرٌ ‡ [A gazelle, or young gazelle, &c., with languid eyes,] as though drowsy, (Ṣ, A,) by reason of the motionless state of its eye, and its weakness. (A.) — يَوْمٌ ▾ خَدِرٌ † A day intensely hot: (Lth:) — and [intensely cold: (see خَدَرٌ:) or] cold and damp: (TA:) or damp: (Ṣ:) or rainy, and cloudy or misty: (Az:) and لَيْلَةٌ خَدِرَةٌ † A night cold and damp: (TA:) or damp. (Ṣ.) — See also خُدَارِىٌّ.

خَدْرَةٌ † A rain. (TA.)

خُدْرَةٌ [i. q. ▾ خَدَرٌ (inf. n. of خَدِرَ) as meaning Numbness, &c., or] heaviness of a leg, and inability thereof to walk. (IAạr.) — See also خَدَرٌ.

خُدْرِىٌّ † A black ass: (Ḳ:) as though a rel. n. from خُدْرَةُ اللَّيْلِ [The darkness, or intense darkness, of night]. (TA. [See also خُدَارِىٌّ.])

خَدُورٌ: see خَادِرٌ, in two places.

خُدَارِىٌّ ‡ A dark night; (Ṣ, A, Ḳ;) as also ▾ أَخْدَرُ (Ḳ) and ▾ مُخْدِرٌ (A) and ▾ خَدِرٌ and ▾ خَدَرٌ and ▾ خَدْرٌ. (Ḳ.) † A black cloud. (Ṣ.) † A camel intensely black: (Ṣ, Ḳ:) fem. with ة. (Ṣ. [See also خُدْرِىٌّ.]) ‡ Black hair. (A.) And خُدَارِيَّةُ الشَّعَرِ ‡ A black-haired girl. (A.) — خُدَارِيَّةٌ also signifies † An eagle; (Ṣ, Ḳ;) because of its colour; (Ṣ;) i. e. its intense blackness. (IB.) In the following verse,

كَأَنَّ عُقَابًا خُدَارِيَّةً
تُنَشِّرُ فِى الجَوِّ مِنْهَا جَنَاحَا

[which may be rendered, As though a black eagle spread in the sky its wing], Th says that the poet may mean, by عُقَابًا, the bird [so called], or a banner, or garments of the kind called أَبْرَاد, which they spread over them. (TA.)

خَادِرٌ (Ṣ, A, Ḳ) and ▾ مُخْدِرٌ (A, TA) [originally Keeping behind, or within, the خِدْر, or curtain. — And hence,] ‡ A lion keeping, or abiding, in his lurking-place: (A,* Ḳ,* TA:) or entering into it. (Ṣ, TA.) And the former, and ▾ خَدُورٌ, † A gazelle remaining behind the herd; not going with it: and † a beast that remains behind; not overtaking, or coming up with, the others: and ▾ خَدُورٌ likewise signifies † a camel that is in the rear of the other camels; that remains behind them, and when it sees them go on, goes on with them. (TA.) = خَادِرٌ signifies also † Languid, and lazy, or slothful. (Ṣ.) — And ‡ A gazelle having feeble bones. (TA.)

أَخْدَرُ: [fem. خَدْرَآءُ:] see خُدَارِىٌّ. = عَيْنٌ خَدْرَآءُ: see خَدِرٌ. = بَنَاتُ الأَخْدَرِ: see what next follows.

أَخْدَرِىٌّ A wild ass: (Ṣ, Ḳ:) so called from a certain stallion named الأَخْدَرُ: (TA:) some say, (TA,) this was a horse, (A, TA,) belonging to Ardasheer, that became wild: (A:) and some say that he was an ass: or so called in relation to El-'Irák, but ISd says, I know not how this is: (TA:) the pl. is أَخْدَرِيَّاتٌ; (A;) and بَنَاتُ ▾ الأَخْدَرِ is used as a pl.; (TA;) and [in like manner] بنات الأَخْدَرِىّ means the [wild] she-asses. (TA in art. بنى.) — الأَخْدَرِيَّةُ A certain race of horses: so called from a stallion named أَخْدَر. (Ḳ.)

أُخْدُورٌ: see خِدْرٌ.

مُخْدَرَةٌ and مُخْدِرَةٌ: see مَخْدُورَةٌ.

مُخْدِرٌ: see خَادِرٌ: — and مُخْدَرَةٌ: = and see also خُدَارِىٌّ.

مَخْدُورَةٌ ▾ (Ṣ, A, Ḳ) and ▾ مُخْدَرَةٌ and ▾ مُخَدَّرَةٌ (Ḳ) A girl kept behind, or within, the curtain. (Ṣ, A, Ḳ.) — And مُخَدَّرٌ (TA) and ▾ مَخْدُورٌ (A, TA) A curtained [vehicle of the kind called] هَوْدَج. (A, TA.) — [And hence,] ▾ مَخْدُورٌ and ▾ مُخْدِرٌ (in some copies of the Ḳ and in the TA ▾ مُخْدَرٌ) and ▾ مُخَدَّرٌ ‡ A lion concealed in his lurking-place. (Ḳ, TA.)

مَخْدُورٌ and مَخْدُورَةٌ: see what next precedes, in three places.

خدش

1. خَدَشَهُ, (Az, Ṣ, A, &c.,) aor. ـِ, (Ṣ, Mṣb, Ḳ,) inf. n. خَدْشٌ, (Mgh, Mṣb,) He scratched him, or it, (namely, the face, Az, Ṣ, Mgh,) with the nails, so as to cause bleeding or not; (Az, Ṣ,* Mgh, TA;) i. q. خَمَشَهُ: (Az, A, Ḳ, TA:) he wounded him in the outer skin, so as to make it bleed or not: (Mṣb:) he lacerated it, namely, the skin, (A, Ḳ,) little or much: or tore off its surface with a stick or the like. (Ḳ.) You say, خَدَشَتْ وَجْهَهَا عِنْدَ الْمُصِيبَةِ She scratched her face with her nails in the upper parts of what appeared thereof, so as to make it bleed or not, on the occasion of affliction. (Az. TA.)

2. خدّشه, (Ṣ, TA,) inf. n. تَخْدِيشٌ, (A, TA,) [meaning He scratched him, or it, (namely, the face,) with the nails, vehemently, or much,] is with teshdeed to denote intensiveness, or muchness. (Ṣ, TA.) — [Hence,] وَقَعَ فِى الأَرْضِ تَخْدِيشٌ ‡ A little rain [such as scratched the ground in many places] fell upon the land. (A, TA.)

3. خَادَشْتُ الرَّجُلَ, inf. n. مُخَادَشَةٌ and خِدَاشٌ, I scratched the man's face with my nails, he scratching my face in like manner. (TA.)

خَدْشٌ, an inf. n. used as a subst., (Mgh, Mṣb,) The mark made by scratching with the nails, (Mgh, Mṣb,* Ḳ,*) whether it cause bleeding or not: (Mgh:) pl. خُدُوشٌ, (Ṣ, A, Mgh, Mṣb, Ḳ,) which is syn. with كُدُوحٌ. (Ṣ, TA.)

بِقَلْبِهِ خَدْشَةٌ ‡ In his heart is somewhat of hurt. (A, TA.)

خدع

1. خَدَعَهُ, (TA,) [aor. ـَ,] inf. n. خَدْعٌ, (Bḍ in ii. 8,) He hid it, or concealed it; (TA;) as also ▾ اخدعه, (Mṣb, TA,) inf. n. إِخْدَاعٌ. (TA.) —

[And hence, app.,] (Lḥ, Ḳ,) inf. n. as above, (Lḥ,) ‡ *He doubled it*, or *folded it, one part upon another*; namely, a garment, or piece of cloth. (Lḥ, Ḳ, TA.) — [And hence, also, accord. to some,] خَدَعَهُ, aor. ـَ, inf. n. خَدْعٌ (Ṣ, Mgh, Mṣb, Ḳ) and خِدْعٌ, (AZ, Ṣ, Ḳ,) or the latter is a simple subst., (Mṣb, TA,) and خَدِيعَةٌ, (TA,) or this [also] is a simple subst., (Mṣb, TA,) like خِدَاعٌ, [which is also an inf. n. of 3,] and like خُدْعَةٌ, (TA,) *He deceived, deluded, beguiled, circumvented,* or *outwitted, him*; syn. خَتَلَهُ; (Ṣ, Mgh, Ḳ;) and *desired to do to him a foul, an abominable,* or *an evil, action, without his knowing whence it proceeded*: (Ṣ, Ḳ:) or *he dissembled* [or *acted deceitfully*] *with him; pretended to him the contrary of what he concealed*: (TA:) or *he made him to resign,* or *relinquish, the object that he had in view, by pretending to him something the contrary of what he concealed*: (Er-Rághib, B:) and ▼ خادعه, (Ṣ, TA,) inf. n. مُخَادَعَةٌ (Ṣ) [and خِدَاعٌ], signifies the same; (Ṣ,* TA;) as also ▼ اختدعه; and ▼ تخدّعه; and ▼ خدّعه, inf. n. تَخْدِيعٌ: (TA:) or this last signifies *he deceived him, deluded him, beguiled him, circumvented him,* or *outwitted him, much*: (KL:) [and of another of these verbs we find the following various explanations:] ▼ خادعه is *syn. with* كَايَدَهُ [which has the first of the meanings assigned in this sentence to خَدَعَهُ; or signifies *he practised with him mutual deceit, delusion, guile,* or *circumvention; he deceived him,* &c., *being deceived,* &c., *by him*; and this latter meaning, if not each meaning, may be intended here by كايده; for Bḍ says, (in ii. 8,) that مُخَادَعَةٌ is between two]: (TA:) or it signifies *he strove, endeavoured,* or *desired, to deceive, delude, beguile, circumvent,* or *outwit, him*; (AAF, L;) [agreeably with what is said by Kemál Páshá Zádeh, (as I find in a marginal note in a copy of the MṢ, and also in the Kull p. 178,) that one says of a man خادع when he has not attained his desire, and خَدَعَ when he has attained his desire;] for many a verb of the measure فَاعَلَ relates to one only; as in the instances of عَاقَبْتُ اللِّصَّ and طَارَقْتُ النَّعْلَ: (L:) or it signifies, [like خَدَعَهُ,] *he pretended to him something different from that which was in his mind.* (Ḳ.) It is said in the Ḳur [ii. 8], يُخَادِعُونَ ▼ ٱللّٰهَ وَٱلَّذِينَ آمَنُوا, (TA,) meaning *They pretend, to God and to those who have believed, something different from that which is in their minds,* by concealing unbelief and pretending belief; for when they do thus to the believers, they do thus to God: (Ḳ:) and again, [in iv. 141], يُخَادِعُونَ ▼ ٱللّٰهَ وَهُوَ خَادِعُهُمْ [*They strive, endeavour,* or *desire, to deceive God,* or] *they think that they deceive God, but He is* [*their deceiver*, i. e.,] *the requiter, to them, of their* خِدَاع [or *deceit,* &c.]: (TA:) or the meaning is [*they deceive*] *the friends of God*: (Ṣ:) and [agreeably with this last rendering, and that given in the Ḳ,] Aboo-Ḥayáh reads, in the former passage, يَخْدَعُونَ ٱللّٰهَ: (TA:) [which passage continues thus:] وَمَا يُخَادِعُونَ ▼ إِلَّا أَنْفُسَهُمْ [*but they do not deceive any save themselves*]; i. e., *the result of their* خِدَاع [or *deceit*] *does not befall any save themselves*: (Ḳ:) here, again, Aboo-Ḥayáh reads يَخْدَعُونَ: (TA:) Muärrik reads وَمَا يَخْتَدِعُونَ ▼, meaning يَخْتَدِعُونَ. (Ḳ.) Accord. to IAạr, الخَدْعُ signifies مَنْعُ الحَقِّ [meaning *The preventing from discovering,* or *accepting, the truth*]. (L.) ["He deceived him," or the like, seems to be generally regarded as the primary signification of خَدَعَهُ, for it occupies the first place in all the lexicons to which I have access: but Bḍ says (in ii. 8) that this meaning is from خَدَعَ said of the ضَبّ, and that the primary signification of خَدْعٌ is the act of "concealing:" the action of the ضبّ, however, as will appear in what follows, implies, and originates from, a desire of deceit; and so, often, does the act of concealing.] — [Hence, app.,] خَدَعْتُهُ *I gained the mastery over him.* (TA.) — خَدَعَ, (Lth, TA, &c.,) aor. ـَ, inf. n. خَدْعٌ, (TA,) said of a [lizard of the kind called] ضَبّ, [as though meaning either خَدَعَ المُحْتَرِشَ *It deceived the hunter*, or خَدَعَ نَفْسَهُ *it concealed itself,*] signifies *it entered into its hole*; (Lth;) as also خَدَعَ فِي جُحْرِهِ: (Ṣ, Ḳ:) or *it scented a man, and therefore entered its hole, in order that it might not be caught*; as also ▼ انخدع: (TA:) or *it entered into its hole in a tortuous manner*: and in like manner, a gazelle *into its covert*: but mostly said of a ضبّ: (Abu-l-'Omeythil:) also said of other things: (Lth:) of a fox, meaning *he took to going to the right and left, deceitfully,* or *guilefully*: and of a man, meaning *he hid himself* from another: and *he assumed a disposition not his own.* (TA.) [See also خَادِعٌ, below.] — Hence, i. e. from خَدَعَ said of the ضبّ, (A, TA,) خَدَعَتْ عَيْنُ الشَّمْسِ ‡ *The disc of the sun set*; (A, Ḳ, TA;) like خَضَعَتْ. (TA in art. خضع.) — [And] خَدَعَتْ عَيْنُهُ ‡ *His eye sank,* or *became depressed, in his head.* (Lḥ, Ḳ, TA.) [Also meaning † *His eye did not sleep*: for] خَدَعَتِ العَيْنُ signifies † *the eye did not sleep.* (TA.) — [Hence also, as indicated in the Ṣ,] مَا خَدَعَتْ فِي عَيْنِي نَعْسَةٌ † [*A slumber did not enter my eye*]: (Ṣ:) or مَا خَدَعَتْ بِعَيْنِهِ نَعْسَةٌ, (so in the L,) or خَدْعَةً, i. e. نَعْسَةً, (so in the TA,) meaning ‡ *a slumber did not pass by his eye.* (L, TA.) [And from the same source have originated several other tropical significations, of which exs. here follow.] — خَدَعَتِ الأُمُورُ ‡ *The affairs varied in their state*; or *were,* or *became, variable.* (Ibn-'Abbád, Ḳ.) — خَدَعَتِ السُّوقُ, (Ṣ, Ḳ,) inf. n. خَدْعٌ, (TA,) ‡ [*The market varied in its state; at one time being brisk, and at another time dull, in respect of traffic*: (see خَادِعٌ, below:) or] *the market became dull in respect of traffic*; (Ṣ, Ḳ;) as also ▼ انخدعت; (Lḥ, TA;) or انخدع: (Ḳ: [but سوق is generally fem.:]) and, as some say, *it became brisk in respect of traffic*: thus it appears to have two contr. significations: (TA:) and خَدَعَ السِّعْرُ ‡ *The price became high,* or *dear.* (TA.) — Said of a man, خَدَعَ also signifies ‡ *His wealth,* (Ḳ, TA,) and *the like,* (TA,) *became small in amount,* or *little.* (Ḳ, TA.) — Said of a time, inf. n. خَدْعٌ, ‡ *Its rain became little*: (TA:) and of rain, ‡ *It became little.* (Ḳ, TA.) — Said of spittle, or saliva, ‡ *It dried*: (Ṣ, Ḳ, TA:) or *it became little, and dried,* in the mouth: (A, TA.) or *it became deficient*; and when it becomes deficient, it becomes thick; and when it becomes thick, it becomes stinking: (TA:) or *it became corrupt*: (IAạr, TA:) and in like manner, said of a thing, *it became corrupt,* or *bad.* (TA.) [See also خَادِعٌ, below.] — Said of a generous man, (Ḳ,) ‡ *He refrained* [from giving], (Ṣ, L, Ḳ,) *and refused.* (L.) You say, كَانَ فُلَانٌ يُعْطِى ثُمَّ خَدَعَ ‡ [*Such a one used to give; then he refrained, and refused*]. (Ṣ.) — خَدَعَهُ, aor. ـَ, inf. n. خَدْعٌ, *He cut,* or *severed, his* [*vein called the*] أَخْدَع. (TA.)

2. خدّعهُ, inf. n. تَخْدِيعٌ: see 1, third sentence. — خُدِّعَ *He was deceived, deluded, beguiled, circumvented,* or *outwitted, repeatedly, so that he became experienced*: or *he was deceived,* &c., *in war, time after time, so that he became skilful*: or *he became experienced in affairs*: or *he became experienced in affairs, sound in judgment, cunning, and guileful.* (TA.)

3. خادعهُ, inf. n. مُخَادَعَةٌ and خِدَاعٌ: see 1, in five places. — مُخَادَعَةُ العَيْنِ means *The causing the eye to doubt respecting that which it sees.* (Ḥam p. 541.) — خَادَعَ المَجْدَ, (Aṣ, Ḳ,*) or الحَمْدَ, (AA,) a phrase used by Er-Rá'ee, (TA,) *He forsook,* or *relinquished,* (Aṣ, AA, Ḳ,) *glory*, (Aṣ,) or *praise,* not being worthy of it. (AA.)

4. اخدعهُ: see 1, first signification. — *He incited him to deceive, delude, beguile, circumvent,* or *outwit*; or *to desire to do to another a foul, an abominable,* or *an evil, action, without the latter's knowing whence it proceeded*; or *to pretend to another something different from that which was in his mind.* (Ḳ.) In the Ḳur ii. 8, quoted above, Yaḥyà Ibn-Yaạmar reads, وَمَا يُخْدِعُونَ. (TA.)

5. تخدّع *He constrained himself to deceive, delude, beguile, circumvent, outwit,* or *the like.* (Ḳ,* TA.) — تخدّعهُ: see 1, third sentence.

6. تخادعوا *They deceived, deluded, beguiled, circumvented,* or *outwitted, one another*; or *practised deceit, guile, circumvention,* or *the like, one to another.* (TA.) — تخادع *He pretended deceit, delusion, guile,* or *circumvention,* (Ṣ,* PṢ,) *on his part*: (Ṣ:) or *he pretended to be deceived, deluded, beguiled, circumvented,* or *outwitted, not being so*; (Ḳ, TA;) as also ▼ انخدع. (TA.)

7. انخدع quasi-pass. of خَدَعْتُهُ [i. e. *He became deceived, deluded, beguiled, circumvented,* or *outwitted*]: (Ṣ, Mṣb, TA:) or *he was content to be deceived, deluded, beguiled, circumvented,* or *outwitted.* (Lth, Ḳ.) — See also 6. — See also 1, latter half, in two places.

8. اختدعهُ; and يَخَدِّعُونَ, for يَخْتَدِعُونَ: see 1, in the former half of the paragraph, in two places.

خَدْعٌ : see خَدِيعَةٌ.

خَدِعٌ : see خَادِعٌ, in two places.

خَدْعَةٌ *A single act of deceit, delusion, guile, circumvention,* or *outwitting.* (Mgh.) It is said in a trad., (Mgh, TA,) الحَرْبُ خَدْعَةٌ, and ▼خُدْعَةٌ, (Ṣ, Mgh, Mṣb, Ḳ,) and ▼خُدَعَةٌ, (Ḳ,) and ▼خَدَعَةٌ, (AZ, Ks, Ṣ, Mgh, Ḳ,) accord. to different relaters; (Th, Mgh, Ḳ;) the first being the most chaste, (Ṣ, Mgh, Mṣb,) said to be the form used by Mohammad; (Th, Mgh, Mṣb;) the second ascribed by El-Khaṭṭábee to the vulgar; (TA;) the last the best in point of meaning: (Mgh:) i. e., accord. to the first reading, (Mgh,) *War is finished* by *a single act of deceit,* &c.; (Mgh, O, Ḳ;) accord. to the second, *war is a thing by which one is deceived;* (Mgh, Mṣb;) or *war is deceived;* for when one of the two parties deceives the other, it is as though the war itself were deceived; (IAth, TA;) [accord. to the third, *war is a mode,* or *manner, of deceiving;*] and accord. to the fourth, *war is a deceiver of those engaged in it* (IAth, Mgh, TA) *by the frequent deceits which occur therein.* (Mgh.) ═ ‡ *A slumber.* (TA.)

خُدْعَةٌ *A thing by which,* or *with which, one deceives, deludes, beguiles, circumvents,* or *outwits;* (Mgh, Mṣb;) like as لُعْبَةٌ signifies "a thing with which one plays." (Mṣb.) See خَدْعَةٌ. — One *whom people deceive, delude, beguile, circumvent,* or *outwit,* (Ṣ, Ḳ,) *much;* (Ḳ;) like as لُعْنَةٌ signifies one "who is much cursed." (TA.) [See, again, خَدْعَةٌ. The comparison of خُدْعَةٌ in one sense with لُعْبَةٌ, and in another sense with لُعْنَةٌ, suggests that one of the explanations above may perhaps be founded upon a mistranscription. On فُعْلَةٌ as the measure of a word having the sense of a pass. part. n., see a remark of IB voce لُقَطَةٌ.] — See also خَدِيعَةٌ.

خِدْعَةٌ [*A mode,* or *manner, of deceiving, deluding, beguiling, circumventing,* or *outwitting*]: see خَدْعَةٌ.

خُدَعَةٌ : see خَادِعٌ, in three places: — and see also خَدْعَةٌ.

خِدَاعٌ : see خَدِيعَةٌ; [and see also 3.] — خِدَاعُ الضَّبِّ signifies *The procedure of the* [*lizard called*] ضبّ *when it is attacked by a serpent, or hunted by a man feeling the head of its hole in order that it may imagine him to be a serpent: if the* ضبّ *be experienced, it puts forth its tail to half the length of the hole, and if it feel a serpent, it strikes it, and cuts it in halves; and if it be a hunter, it does not suffer him to lay hold upon its tail, and so it escapes, for the hunter does not dare to put his hand into its hole, because it may not be free from a scorpion, of which he fears the sting, as a strong friendship subsists between the* ضبّ *and the scorpion, and the former makes use of the latter to defend itself from the hunter:* or, as some say, it signifies *its concealing itself, and remaining long in its hole, and seldom appearing, and being very cautious.* (O, TA.)

خَدُوعٌ : see خَادِعٌ, in three places. — Also †A she-camel *that yields milk abundantly at one time, and withholds it at another.* (Ḳ.)

خَدِيعَةٌ *Deceit, delusion, guile, circumvention,* or *outwitting;* and *a desire to do to another a foul, an abominable,* or *an evil, action, without the latter's knowing whence it proceeds;* (Ṣ, Ḳ;) a subst. from خَدَعَهُ; (Ṣ, Mṣb, Ḳ;) as also ▼خَدْعٌ, (Mṣb, TA,) or this is an inf. n.; (AZ, Ṣ, Ḳ;) and ▼خُدْعَةٌ; (TA;) and ▼خِدَاعٌ; (TA;) which [is also an inf. n. of خَادَعَهُ, and] originally signifies *concealment:* (Ḥam p. 541:) [and hence as above: and] also signifies *prevention* (مَنْعٌ); and *art, artifice, cunning,* or *skill, in the management of affairs;* (IAạr, Ṣgh, Ḳ;) or *a making another to resign,* or *relinquish, the object that he has in view, by pretending to him something the contrary of what he conceals.* (Er-Rághib, B.)

خَدَّاعٌ; and its fem., with ة: see the next paragraph, in two places.

خَادِعٌ [*Deceiving, deluding, beguiling, circumventing, outwitting,* or *the like;*] act. part. n. of خَدَعَهُ; as also ▼خَدُوعٌ; (Mṣb;) or [rather] this latter is an intensive epithet, signifying one *who deceives,* &c., *much,* or *often;* or *very deceitful,* &c.; or *a great deceiver,* &c.; (Mgh, Ḳ;) and ▼خَدَّاعٌ, (Mṣb, TA,) and ▼خَدِعٌ, and ▼خَيْدَعٌ; (TA;) [but these three are also intensive epithets, like خَدُوعٌ;] and ▼خُدَعَةٌ signifies one *who deceives,* &c., *other men;* (Ṣ;) or [rather] this last is syn. with خَدُوعٌ as explained above, (Ḳ,) or خَدَّاعٌ: (Mgh:) [the pl. of خَادِعَةٌ, fem. of خَادِعٌ, is خَوَادِعُ:] and the pl. of ▼خَدُوعٌ is خُدُعٌ. (Mgh.) — [Hence,] ضَبٌّ خَادِعٌ *A* [*lizard of the kind called*] ضبّ *that deceives, beguiles,* or *circumvents;* (Z, TA;) as also ▼خَدِعٌ. (Ṣ, Ḳ.) — And دَهْرٌ خَادِعٌ and ▼خُدَعَةٌ † [*Deceiving,* or *varying,* and *very deceitful,* or *very variable, fortune,* or *time*]. (TA.) [Whence, or] because of its variableness, (TA,) ▼الخُدَعَةُ is ‡ a name for *Fortune,* or *time.* (Ḳ, TA.) — And فُلَانٌ خَادِعُ الرَّأْىِ ‡ *Such a one remains not steadily in one opinion.* (TA.) — And خُلُقٌ خَادِعٌ ‡ *A varying,* or *variable, disposition.* (Ṣ, Ḳ, TA.) — And سُوقٌ خَادِعَةٌ ‡ *A market varying,* or *variable, in its state;* (Ṣ, A, O, Ḳ;) *at one time being brisk, and at another time dull, in respect of traffic:* (A, TA:) or *a market dull in respect of traffic:* or *a market in which one cannot obtain a thing because of its dearness.* (TA.) خَادِعٌ also signifies ‡ Anything *unsaleable,* or *difficult of sale, and in little demand.* (TA.) And accord. to Fr, the Benoo-Asad use the epithet ▼مُخَادِعٌ [perhaps a mistake for خَادِعٌ] in the sense of ‡ *High,* or *dear,* applied to a price. (TA.) — And طَرِيقٌ خَادِعٌ ‡ *A road that appears at one time and disappears at another;* as also ▼خَدُوعٌ: (Ḳ:) *a road which one does not know:* (TA:) *a road deviating from the right course;* (TA;) as also ▼خَيْدَعٌ; (Ṣ, Ḳ, TA;) *which one does not know.* (Ṣ, TA.) And مَآءٌ خَادِعٌ ‡ *A water to which one does not know the way.* (TA.) — [Hence also,] سِنُونَ خَوَادِعُ ‡ *Years in which is little good; bad years:* (Sh, TA:) and سِنُونَ ▼خَدَّاعَةٌ ‡‡ *years in which is little increase:* (Ṣ, Ḳ, TA:) from خَدَعَ said of rain, or of spittle; and therefore doubly tropical: (TA:) or, as some say, *years in which is much rain, and in which the produce is little.* (Ṣgh.) خَادِعٌ also signifies † *Corrupt,* or *bad;* applied to food and other things. (TA.) And you say, دِينَارٌ خَادِعٌ † *A deficient,* or *defective, deenár.* (Ṣ.) And رَجُلٌ خَادِعٌ ‡ *A man who brings evil upon others.* (TA.)

خَيْدَعٌ : see خَادِعٌ, first sentence. [Hence,] ‡A wolf *that acts deceitfully,* or *mischievously;* or *that practises artifice.* (Z, Ṣgh, Ḳ. [In the CḲ, المُخْتَالُ is erroneously put for المُحْتَالُ.]) — Also A person *in whose love,* or *affection, no confidence is placed.* (Ḳ.) — And hence, (TA,) الخَيْدَعُ is also applied to ‡ *The mirage;* (Ṣ, Ḳ, TA;) accord. to some. (Ṣ.) You say, غَرَّهُمُ الخَيْدَعُ ‡ *The mirage deceived them.* (TA.) — [For the same reason,] it is also applied to † *The cat.* (IB.) — And from the former of the last two meanings is derived the phrase (TA) غُولٌ خَيْدَعٌ (Ṣ, Ḳ, TA) ‡‡ *A very deceitful,* or *guileful, ghool;* (Ḳ, TA;) so that it is doubly tropical. (TA.) — طَرِيقٌ خَيْدَعٌ : see خَادِعٌ; in the latter part of the paragraph.

خَادِعَةٌ fem. of خَادِعٌ [q. v.]. — Also *A small door in a large door.* (Ibn-'Abbád, Ḳ.) — See also مُخْدَعٌ.

أَخْدَعُ [*More,* and *most, deceitful, deluding, guileful, outwitting,* or *the like*]. [Hence,] أَخْدَعُ مِنْ ضَبٍّ [*More deceitful,* or *guileful, than a ḍabb*]; a prov.; (Ṣ, Ḳ;) applied to a person over whom one has not power, by reason of deceit, or guile. (IAạr.) They said also, إِنَّكَ لَأَخْدَعُ مِنْ ضَبٍّ حَرَشْتَهُ [*Verily thou art more deceitful,* or *guileful, than a ḍabb that I have hunted*]. (AZ, AAF, O.) [See خِدَاعٌ.] ═ الأَخْدَعُ [app. *Each of the two branches of the occipital artery which are distributed upon the occiput;*] *a certain vein,* (Ṣ, Ḳ,) *one of a pair of veins, called the* أَخْدَعَانِ, (Ṣ, Mgh, Mṣb,) *in the cupping-place* (Mgh, Mṣb) *of the neck,* (Mgh,) or *in the place* [*of the application*] *of the two cupping-instruments; being a branch from the* وريد [or *carotid artery*]; (Ṣ, Ḳ;) sometimes the scarification [for cupping] happens to be upon one of them, and the patient consequently is exhausted by loss of blood: (Ṣ:) the اخدعان are *two concealed veins in the place of the cupping of the neck:* Lḥ says, they are *two veins in the neck:* some say that they are *the* وَدَجَانِ, q. v.: (TA:) the pl. is أَخَادِعُ. (Ḳ.) — فُلَانٌ شَدِيدُ الأَخْدَعِ means *Such a one is strong in the place of the* اخدع. (Aṣ, Ṣ, O.) — It also means † [*Such a one is*] *a person who resists; unyielding; uncomplying.* (TA.) And لَيِّنُ الأَخْدَعِ † *One who does not resist; yielding; complying.* (TA.) — You say also, لَوَى فُلَانٌ أَخْدَعَهُ ‡ *Such a one turned away,* or *aside, and behaved proudly*, or *haughtily*. (TA.) And سَوَّى أَخْدَعَهُ ‡ *He relinquished pride*, or *haughtiness*. (TA.) And to him who is proud, one says, لَأُقِيمَنَّ أَخْدَعَيْكَ, meaning † *I will assuredly dispel thy pride*. (Ḥam p. 432.)

مَخْدَعٌ : see what next follows.

مُخْدَعٌ and ▼ مِخْدَعٌ (Fr, Yaạkoob, Ṣ, Mṣb, Ḳ) and ▼ مَخْدَعٌ ; (Mṣb, TA;) the first of which is the original form, the second being adopted because the first was found to be difficult of utterance; (Fr, Yaạkoob, Ṣ;) and the first is the only proper subst. of the measure مُفْعَل ; other words of that measure being epithets; (Sb;) *A closet, or small chamber, in which a thing is kept*, or *preserved*; (Mṣb;) *i. q.* خِزَانَةٌ ; (Fr, Yaạkoob, Ṣ, Ḳ;) by which is meant *a small chamber within a large chamber*: (TA:) from أَخْدَعَهُ meaning "he hid it," or "concealed it:" (Mṣb:) and [in like manner] ▼ خَادِعَةٌ signifies *a chamber within a chamber*: (Ḳ:) Er-Rághib says, as though its builder made it a deceiver of him who might seek, or desire, to take, or reach, a thing in it. (TA.)

مِخْدَعٌ : see the next preceding paragraph.

مُخَدَّعٌ : see مَخْدُوعٌ, in two places.

مَخْدُوعٌ and ▼ مُخَدَّعٌ are syn. [as signifying *Deceived, deluded, beguiled, circumvented, outwitted*, or *the like*: or rather, the latter signifies *much deceived*, &c.]. (TA.) — And [hence] ▼ the latter, A man (Ṣ, L) *Deceived, deluded, beguiled, circumvented*, or *outwitted*, (Ṣ, L, Ḳ,) *in war*, (Ṣ, L,) *time after time*, (Ṣ, L, Ḳ,) *so that he has become experienced*, (Ṣ, Ḳ,) or *so that he has become skilful*: (L:) or *experienced in affairs*: (TA:) or *experienced in affairs, sound in judgment, cunning and guileful*: (ISh:) or *characterized by deceit, delusion, guile*, or *circumvention, in war*. (AO.) ═ Also the former, One *whose* [*vein called the*] أَخْدَع *is cut*, or *severed*. (Ṣ, Ḳ.)

مُخَادِعٌ : see خَادِعٌ.

خدل

1. خَدُلَ, inf. n. خَدَالَةٌ [and app. خُدُولَةٌ also], *He was*, or *became, large, and full* [or *plump*], *in the shank and fore arm*. (TA.) [And in like manner, خَدُلَتْ ; or خَدِلَتْ, inf. n. خَدَلٌ [q. v.]; *She* (a woman) *was*, or *became, full*, or *plump, in the shanks and fore arms*.] And خَدُلَتْ, inf. n. خَدَالَةٌ [and خُدُولَةٌ]; (JK;) or خَدِلَتْ, [inf. n. خَدَلٌ ;] (Ḳ;) said of the shank (السَّاقُ, JK, Ḳ), *It was*, or *became, round*: (JK:) or *full* [or *plump*]. (Ḳ.)

خَدْلٌ *Full* [or *plump*]: and *large, big*, or *bulky*: (Ḳ:) or *large, and full* [or *plump*], *in the shank and fore arm*: or, as some say, *large, big*, or *bulky*. TA.) You say اِمْرَأَةٌ خَدْلَةٌ and ▼ خَدِلَةٌ (Ḳ, TA, [in the CḲ, by an omission, the latter is made to be خِدْلَةٌ,]) *A woman thick and round in the shank*: pl. خِدَالٌ : [in the CḲ, erroneously, أَخْدَالٌ :] or *full of flesh in the limbs, with slenderness of the bones*; as also ▼ خَدْلَاءُ and ▼ خِدْلِمٌ : (Ḳ:) or these two, in the latter of which the م is augmentative, a woman *full* [or *plump*] *in the shanks and fore arms*. (Ṣ.) And خَدْلَةُ السَّاقِ A woman *round in the shank*. (JK.) And مُخَلْخَلُهَا خَدْلٌ *Her place of the anklet* [i. e. *her ankle*] *is large*, or *big*. (Ṣ.) And سَاقٌ خَدْلَةٌ *A full* [or *plump*] *shank*: (Ḳ:) or *a round shank*: pl. خِدَالٌ. (JK.)

خَدَلٌ (Ṣ, Ḳ) and ▼ خَدَالَةٌ (JK, Ṣ, Ḳ) and ▼ خُدُولَةٌ (Ḳ) *Fulness* [or *plumpness*], (Ṣ, Ḳ,) or *roundness*, (JK,) of the shank, (JK, Ḳ,) or of the shanks and fore arms. (Ṣ.) [All are properly inf. ns.: see 1.]

خَدْلَةٌ fem. of خَدْلٌ [q. v.]. (JK, Ḳ.) — Also *A grape that is small and worthless by reason of blight, or the like, and want of moisture*. (AḤát, Ḳ,* TA.) — And The *stem of the tree called* صَاب, (M, Ḳ,) which is a sort of bitter tree; (TA;) as also ▼ خُدْلَةٌ. (M, Ḳ.)

خُدْلَةٌ : see what next precedes.

خَدِلَةٌ :
خَدْلَاءُ : } see خَدْلٌ.
خِدْلِمٌ :

خَدَالَةٌ :
خُدُولَةٌ : } see خَدَلٌ.

خدلج

خَدَلَّجٌ *Large, big*, or *bulky*; (Mgh, Mṣb;) applied to the shank, (Mgh,) or to a man: (Mṣb:) or *large*, or *big, in the shank, with a goodly fulness therein*; applied to a female: (Lth, L:) or, with ة, applied to a woman, (Ṣ, Ḳ, TA,) *juicy*, (TA,) *full in the fore arms and the shanks*. (Ṣ, Ḳ, TA.) You say خَدَلَّجُ السَّاقَيْنِ, meaning *Large*, or *big, in respect of the shanks*; like خَدْلٌ. (Mgh, TA.)

خدم

1. خَدَمَهُ, aor. ِ (Ṣ, Mṣb, Ḳ) and ُ, (Lḥ, Ḳ,) inf. n. خِدْمَةٌ (Ṣ, Mṣb, Ḳ) and خَدْمَةٌ, (Lḥ, Ḳ,) or, as some say, the latter is the inf. n., and the former [though generally used as the inf. n.] is a simple subst., (TA,) *He served him; did service for him; ministered to him*; (PṢ, TA;) syn. مَهَنَهُ. (TA.) And خَدَمَ بِطَعَامِ بَطْنِهِ [*He served for*, meaning *in return for, the food of his belly*]. (Ṣ and A &c. in art. وغد.) — One says also, هٰذَا القَمِيصُ يَخْدُمُ سَنَةً ‡ [*This shirt will serve*, or *last, a year*]: and ثَوْبٌ سَخِيفٌ لَا يَخْدُمُ ‡ [*A thin*, or *flimsy, garment, that will not serve*, or *last, long*; or *that will not be serviceable*]. (TA.)

2. خدّمهُ خَدَّمَهُ *He occupied*, or *busied, him with service*. (TA.) — خَدَّمَهَا *He gave her several*, or *many, female servants*. (Mṣb.) ═ [خدّم البَعِيرَ *He attached a* خَدَمَة, meaning *the thong thus called, upon the pastern of the camel*.

— Hence,] تَخْدِيمٌ, [as inf. n. of خَدَّمَ,] in a horse, † The *having a whiteness* (Ṣ, Mgh, TA) *such as is termed* تَحْجِيل (Ṣ, TA) *surrounding the pastern of each hind foot*, (Ṣ, Mgh, TA,) *above the* أَشَاعِر [or *extremities next the hoof*], *and stopping short of the shank*; (Ṣ, TA;) *but not in the fore foot*. (Ṣ, Mgh, TA.) [See also خَاتِمٌ, last sentence.] When it is in one hind leg, the horse is termed أَرْجَلُ. (Ṣ, TA.) — [Hence, also,] خَدَّمَهَا زَوْجُهَا † *Her husband attired her with the* خَدَمَة [or *anklet*]. (TA.)

4. اخدمهُ *He gave him a servant*. (Ṣ, Ḳ, TA. [كَاخْدَمَهُ in the CḲ is erroneously put for فَأَخْدَمَهُ.]) And أَخْدَمْتُهَا *I gave her a female servant*. (Mṣb.) — El-Ḳuṭb Er-Ráwendee asserts that one says, أَخْدَمْتُهُ لِنَفْسِى, peculiarly; [*I made him a servant to myself*;] but Ibn-Abi-l-Ḥadeed says, This is of the things that I know not. (TA.) [See اختدمهُ.]

8. اختدم *He served himself*; (Lḥ, Ḳ;) as one must do who has not a servant. (Lḥ.) ═ اختدمهُ *He made him a servant*. (TA.) — See also what next follows, in two places.

10. استخدمهُ *He asked him to give him a servant*; as also ▼ اختدمهُ. (Ḳ, TA.) — And اِسْتَخْدَمْتُهُ *I asked him to serve me*; (Mṣb, TA;) as also ▼ اِخْتَدَمْتُهُ : (TA:) [or *I took him as my servant*:] or *I made him to serve me*. (Mṣb.) Accord. to El-Ḳuṭb Er-Ráwendee, one says, اِسْتَخْدَمْتُهُ لِنَفْسِى and لِغَيْرِى [*I took him as a servant for myself* and *for another than myself*: or *I made him to be a servant to myself* and *to another than myself*]. (TA.)

خَدَمٌ : see خَادِمٌ : ═ and see also خَدَمَةٌ, in four places.

خَدْمَةٌ, (Ḳ,) or ▼ خِدْمَةٌ, (JK,) *A space*, or *period*, (سَاعَةٌ,) *of the night* (JK, Ḳ) *or of the day*: (Ḳ:) and خَذْمَةٌ [or خِذْمَةٌ, q. v.,] is a dial. var. thereof. (TA.)

خُدْمَةٌ a subst. signifying † The *characteristic denoted by the epithet* خَدْمَآءُ [fem. of أَخْدَمُ], in a sheep or goat; (JK, Ḳ, TA;) i. e. *whiteness in the lower end of the shank*, (JK, Ḳ,) *upon blackness*; (Ḳ;) such whiteness resembling the خَدَمَة, (JK, TA,) or being likened to خَدَم, or anklets: (TA:) or *blackness upon whiteness*: &c. (Ḳ.) [See أَخْدَمُ : and see also خَدَمَةٌ.]

خِدْمَةٌ, accord. to some, an inf. n. of 1: accord. to others, a simple subst. [signifying *Service*]. (TA.) — [Also *Pay for service*: but in this sense probably post-classical.] ═ See also خَدْمَةٌ.

خَدَمَةٌ *A thong*, (JK, Ṣ, Ḳ,) *plaited*, (TA,) *thick and strong, like a ring*, (JK, Ḳ,) *which is fastened upon the pastern of a camel*, (JK, Ṣ, Ḳ,) *and to which is attached the* سَرِيحَة [or *thong*] *of the* نَعْل [or *leathern shoe* with which the foot is sometimes protected], (Ṣ,) or *to which are attached the* سَرَائِح [or *thongs*] *of the* نعل : (Ḳ:) [it is the n. un. of ▼ خَدَمٌ : and its pl. is خِدَامٌ ; as below: in the TA, said to be tropical; but this is probably a mistake: the other significations here following are all tropical:] accord. to AA, [the pl.] خِدَامٌ signifies *shackles*, or *hobbles*; syn. قُيُود. (TA.) —— Hence, (Ṣ,) ‡ *i. q.* خَلْخَالٌ [meaning *An anklet*]; (JK, Ṣ, Mgh, Ḳ;) because sometimes made of thongs, with gold and silver affixed thereon: (Ṣ:) pl. خِدَامٌ, (Ṣ,) and [coll. gen. n.] ▼خَدَمٌ. (Ḥam p. 612.) كَالْمَمْهُورَةِ إِحْدَى خَدَمَتَيْهَا is a prov. [meaning † *Like her who has been dowered with one of her two anklets*]. (JK, TA. [See also مَمْهُورَةٌ.]) —— [And hence, † *A ring of white a little above the hoof* of a horse &c.] You say of a horse, لَهُ خَدَمَتَانِ مِنْ خِلَافٍ, meaning † *He has a whiteness* [or rather *a ring of white a little above the hoof*] *in his fore leg* [or *right fore leg*] *and another in his left fore leg* [probably a mistake of a copyist for his *left hind leg*]. (TA in art. خلف.) [The coll. gen. n. ▼خَدَمٌ is used in the Deewán of the Hudhalees, as stated by Freytag, in the sense of † *A place where the colour differs, like an anklet, on the foot*: and *a whiteness on the foot of a bull, surrounding it like a circle*: and خِدَامٌ as meaning *whiteness*: or, as some say, *streaks* ("striæ"). See also خَدَمَةٌ.] —— [Hence, also,] † The *place where each foot comes forth from the trousers*. (TA.) —— Also † The *shank*; (Ḳ;) because it is the place of the خَدَمَة, i. e. the anklet: (TA:) pl. خِدَامٌ and [coll. gen. n.] ▼خَدَمٌ. (Ḳ.) Hence, in a trad. of Selmán, كَانَ عَلَى حِمَارٍ وَعَلَيْهِ سَرَاوِيلُ وَخَدَمَتَاهُ تَذَبْذَبَانِ [*He was upon an ass, and upon him were trousers, and his shanks were dangling*]: or, as some say, the meaning here is, *the parts from which his two feet came forth, of the trousers*. (TA.) And one says, أَبْدَتِ الحَرْبُ عَنْ خِدَامِ المُخَدَّرَاتِ ‡ [*The war made apparent the shanks*, or *the anklets, of the girls that had been kept behind the curtains*]; meaning *the war became vehement*. (A, TA.) —— Also † *A ring of* people; (Ṣ, Ḳ;) *a compact ring* thereof: likened to the thong described in the first sentence of this paragraph. (TA.) Hence the saying of Khálid Ibn-El-Weleed, in a letter that he wrote to the Satraps of Persia, (TA,) الحَمْدُ لِلّٰهِ الَّذِى فَضَّ خَدَمَتَكُمْ, meaning [*Praise be to God,*] *who hath dispersed*, or *broken up, your congregation*: (Ṣ,* Ḳ,* TA:) for when the thong above mentioned is broken, or parted, the سَرَائِح [or thongs of the leathern shoe] become loosed, and the shoe falls off: so says IAth, and A'Obeyd says the like. (TA.)

خِدْمَةٌ *A thong*: (Ḳ:) or *a plaited thong*. (TA.)

خَدُومٌ: see what next follows.

خَدَّامٌ One *who does much service*; (TA in the present art.;) as also ▼خَدُومٌ. (TA in art. عقرب.) And also applied to *A* خَادِم [q. v.]. (TA in the present art. [It is commonly used in the latter sense in the present day: fem. with ة.])

خَادِمٌ *A servant*; (Ṣ,* Mgh,* Mṣb,* Ḳ,* TA;) applied to *a young man*, (Ṣ, Mgh, Mṣb,) or *a male*: (Ḳ:) and, (Ṣ, Mgh, Mṣb, Ḳ,) as also خَادِمَةٌ, (Mṣb, Ḳ,) each in chaste Arabic, (TA,) but the latter rare, (Mṣb,) to *a young woman*, (Ṣ, Mgh, Mṣb,) or *a female*: (Ḳ:) and خَدَّامٌ [q. v.] is also used in the same sense as خَادِمٌ: (TA:) the pl. of خَادِمٌ is ▼خَدَمٌ, (JK, Ṣ, Mgh, Mṣb, Ḳ,) or rather this is a quasi-pl. n., (TA,) and خُدَّامٌ (JK, Mṣb, Ḳ) and خَدَمَةٌ, and vulgarly خُدْمَانٌ. (TA.) It is said in a trad., of 'Abd-Er-Raḥmán, (TA,) طَلَّقَ ٱمْرَأَتَهُ فَمَتَّعَهَا بِخَادِمٍ سَوْدَآءَ, meaning [*He divorced his wife, and gave her on that occasion*] *a black young woman* [as a slave]. (Mgh,* TA.) In the saying فُلَانَةُ خَادِمَةٌ غَدًا, the word خادمة is not properly an epithet: the meaning is, *Such a woman will be a servant to-morrow*; like as one says حَائِضَةٌ غَدًا. (Mṣb.)

[خَادِمِيَّةٌ *Servitude*; or the *state*, or *condition, of a servant*: a term in common use; and mentioned by Freytag on the authority of Meyd.: opposed to مَخْدُومِيَّةٌ.]

أَخْدَمُ *i. q.* ▼مُخَدَّمٌ, (Ṣ, Mgh, Ḳ,) as meaning, applied to a horse, ‡ *Having a whiteness* (Ṣ, Mgh) *such as is termed* تَحْجِيل (Ṣ) *surrounding the pastern of each hind foot*, (Ṣ, Mgh,) *above the parts next the hoof, and stopping short of the shank*; (Ṣ;) *but not in the fore foot*: (Ṣ, Mgh: [see 2:]) or, so applied, *whose* تَحْجِيل *encircles* [*the pastern*] *above the* أَشَاعِر [or *extremities next the hoof*]: or the latter epithet, so applied, *whose whiteness passes beyond the pasterns or part thereof*. (Ḳ.) And خَدْمَآءُ, [fem. of أَخْدَمُ,] applied to a sheep or goat, (JK, Ṣ, Ḳ,) i. e. to a شَاة, (Ṣ, Ḳ,) † *Having in the lower end of her shank a whiteness* (JK, Ḳ) *like the* خَدَمَة [or *anklet*], (JK,) *upon blackness*; or *a blackness upon whiteness*; and in like manner applied to a mountain-goat: (Ḳ:) or *having white shanks*; (AZ, Ṣ, Ḳ;) like حَجْلَآءُ; [but see this latter word;] and so مُخَدَّمٌ applied to a mountain-goat: (Ṣ:) or *having one white shank; the rest of her being black*. (Ḳ.)

مُخَدَّمٌ [pass. part. n. of 2, q. v.]. You say قَوْمٌ مُخَدَّمُونَ *A people, or party, having many servants and other dependents*. (Ṣ, Ḳ.) And in like manner مُخَدَّمَةٌ is applied to a woman. (A, TA.) ═ And مُخَدَّمَةٌ ‡ A woman *attired*, or *adorned, with anklets*. (A, TA.) —— See also أَخْدَمُ, in two places. ═ And مُخَدَّمٌ and ▼مُخَدَّمَةٌ (Ṣ, Ḳ) The *place of the thong* [*called* خَدَمَة] (Ḳ, TA) *in the leg of the camel*: (TA: [there said to be above the كَعْب; but this is a mistake:]) and † the *place of the anklet* [*so called*] (Ḳ, TA) *in the shank of a woman*: (TA:) the *place of the* خِدَام *in the shank*: (Ṣ:) the *place of the* خَدَمَة; like as مُسَوَّرٌ is the place of the سِوَار. (M in explanation of the former, in art. سور.) —— And the former ‡ The *band of the trousers*, (JK, Ḳ, TA,) *at the lower part of the leg of a woman*: (Ḳ, TA:) or the *band at the lower part of the leg of the trousers*: (M, TA:) the woman seems to be specified in the Ḳ because women generally tie the legs of the trousers upon the middle [or upper part] of the shank, and then make them to fall down over it. (TA.)

مُخَدَّمَةٌ as a subst.: see مُخَدَّمٌ; of which it is also the fem.

مَخْدُومٌ [lit. *Served*: —— and hence, *A master*;] *a head*, or *chief*: pl. مَخَادِيمُ. (TA.) —— And A man *having a servant of the jinn*, or *genii*. (Ṣ, Ḳ.)

[مَخْدُومِيَّةٌ The *state*, or *condition, of a master*: opposed to خَادِمِيَّةٌ.]

خدن

3. خَادَنَهُ, (Ṣ, Ḳ,) inf. n. مُخَادَنَةٌ, (Mgh, TA,) *He was*, or *became, his* خِدْن [or *friend*; or his *secret*, or *private, friend*; &c.]: (JK, Ṣ, Ḳ, TA:) *he associated*, or *kept company, with him as a friend*: (Mgh:) or [simply] *he associated*, or *kept company, with him*. (TA.) —— مُخَادَنَةٌ also signifies The *contracting of the eyes* (Mgh, JM, TA) *in holding amatory and enticing talk*, or *conversation, with another, the latter doing the same*. (Mgh, JM.) —— The saying of certain of the lawyers, (JM,) لَا يَجُوزُ شَهَادَةُ صَاحِبِ الغِنَآءِ الَّذِى يُخَادِنُ عَلَيْهِ means *The testimony of the singer who has made singing to be his habitual occupation*, or *means of subsistence, and thereby associates as a* خِدْن *with people, and collects them to him, is not allowable*. (Mgh, JM.)

خِدْنٌ (Ṣ, Mgh, Mṣb, Ḳ, &c.) and ▼خَدِينٌ (Ṣ, Ḳ) *A friend*: (Ṣ, TA:) or *a secret*, or *private, friend*; *a friend in secrecy* or *privacy*: (Mgh, Mṣb, and Ksh and Bḍ in iv. 29:) or *a companion*, or *an associate, who converses, or talks, with one*: (M, TA:) or [simply] *a companion*, or *an associate*: (Ḳ:) but accord. to Er-Rághib, mostly used as meaning *an amorous companion* or *associate*; *a companion*, or *an associate, affected with sensual appetency*: (TA:) the former is applied alike to the *male* and the *female*: (Ksh and Bḍ in v. 7:) and ▼the latter signifies also *one who is thy friend*, &c., (مَنْ يُخَادِنُكَ, Ḳ,) *and who is with thee*, (TA,) *in every affair*, or *case, open and secret*: (Ḳ:) pl. (of the former, Ṣ, Mgh, Mṣb) أَخْدَانٌ (Ṣ, Mgh, Mṣb, TA) and [of the latter] خُدَنَآءُ. (TA.) Hence, خِدْنُ الجَارِيَةِ [*The friend, &c., of the girl*]: (Ṣ:) or *he who converses, or talks, with the girl*. (JK.) It is said in the Ḳur [iv. 29], وَلَا مُتَّخِذَاتِ أَخْدَانٍ, (Ṣ,) meaning [*Nor taking to themselves*] *friends* [or *associates*] *to commit fornication with them in secret*. (Jel.) خَدِينُ العُلَى is a metaphorical expression, used by a poet, [meaning ‡ *Lover of eminence*,] like عَشِيقُ العُلَى. (TA.)

خُدَنَةٌ One *who associates with men as their* خِدْن [or *friend*, or *secret friend*, &c.,] (يُخَادِنُ النَّاسَ) *much*. (Ṣ, Ḳ.)

خَدِينٌ: see خِدْنٌ, in three places.

أَخْدَنُ *Having* أَخْدَان [or *friends*, or *secret friends*, &c.]. (TA.)

خذرف

Q. 1. خَذْرَفَ, (Ḳ,) inf. n. خَذْرَفَةٌ, (TA,) *He hastened, sped*, or *went quickly*. (Ḳ.) And

خَذْرَفَتْ, said of a [wild] she-ass, *She hastened, sped, or went quickly, and threw out her legs.* (TA.) And, said of camels, *They threw the pebbles with their feet by reason of quickness.* (Ḳ.) — خَذْرَفَهُ *He cut off his* (a man's) *extremities with a sword.* (Ḳ.) — *He sharpened it;* namely, a sword. (Ḳ.) — *He filled it;* namely, a vessel. (Ibn-'Abbád, Ḳ.)

Q. 2. تَخَذْرَفَ *It* (a garment, or piece of cloth,) *became rent,* or *pierced with holes.* (TA.) — تَخَذْرَفَتْهُ النَّوَى [*The place which was the object of a journey*] *caused him to go forth,* or *depart;* (L, Ḳ, TA;) as also تَخَذْرَمَتْهُ. (TA.)

خَذْرَفَةٌ inf. n. of Q. 1. — Also *Roundness of the legs of an animal,* or *quadruped.* (TA.) — And *A piece of a garment* or *cloth.* (TA.)

خَذْرَافٌ *A certain plant of the spring, which dries up when it feels the* [*heat of*] *summer:* (Lth, Ḳ:) or *a species of the* [*kind called*] حَمْض, (AḤn, Ṣ, Ḳ,) *having a small leaf, and rising to the height of a cubit:* (AḤn:) the latter is said by Az to be the correct explanation: the former he disapproves: (TA:) n. un. with ة. (Ṣ.)

خُذْرُوفٌ [*A kind of whirling plaything;*] *a thing which a boy turns round by means of a thread, or string, in his hands, causing it to make a sound such as is termed* دَوِىّ: (Ṣ, Ḳ:) or *a small piece of wood, or a slit, or split, reed or cane, in the middle of which is cut a notch, and which is then tied with a thread, or string, which being pulled, it turns round, and is heard to make a sound such as is termed* حَفِيف: *boys play with it; and it is also called* خَرَّارَةٌ: (Lth, TA: [in the latter of which, in art. خر, the خرّارة is said to be *a piece of wood like the half of a sandal, tied with a thread or string, which, being put in motion, draws along the piece of wood and causes it to make a sound:*]) or *a small, round, piece of skin, attached to which are two connected threads or strings, which being pulled by a boy, with his fingers, it turns round, causing a sound to be heard such as is termed* دَوِىّ: (EM p. 43:) pl. خَذَارِيفُ. (Ṣ.) Imra-el-Ḳeys likens to it a swift horse: (Ṣ:) and it is applied as an epithet to a horse; (Lth;) meaning *Swift in his running,* (Lth, Ḳ, TA,) or *in going.* (TA.) And one says, تَرَكَتِ السُّيُوفُ رَأْسَهُ خَذَارِيفَ, meaning † *The swords made his head to be pieces, each piece like the* خذروف. (Ṣ, Ḳ.) — *Clay kneaded, and made like sugar,* (يُعْمَلُ مِنْهُ شَبِيهًا بِالسُّكَّرِ, in the CḲ يُعْمَلُ منهُ شَيْئًا بالسُّكَّرِ, [the latter reading evidently wrong, and the former I think doubtful,]) *with which boys play.* (Ḳ.) — The *piece of wood that is put* [app. *as a handle*] *in the hole of the upper millstone.* (TA.) — *A herd of camels:* and *one separate therefrom.* (Ḳ.) — *Lightning gleaming,* or *shining brightly, in the clouds, and separate therefrom.* (Ḳ.) — *Anything scattered from,* or *of, a thing.* (L, O, Ḳ.) — The خذاريف of the هَوْدَج are *The pieces of wood* (سَقَائِف) *with which the* هودج *is made of a square form.* (Ḳ.) [See also دَامِغَةٌ.]

مُتَخَذْرِفٌ رَجُلٌ *A man of good natural disposition.* (TA.)

خذف

1. خَذَفَ, (Mgh, Mṣb,) aor. ـِ, (Ṣ, Mgh, Mṣb,) inf. n. خَذْفٌ, (Ṣ, Mgh, Mṣb, Ḳ,) trans. without a particle, (Mṣb,) and by means of بِ, (Ṣ,) *He threw* a pebble, (Ṣ, Mgh, Mṣb, Ḳ,) or a datestone, (Mgh, Ḳ,) or the like, (Mgh, Mṣb, Ḳ,) *with the fingers;* (Ṣ;) or *by taking it between his two fore fingers;* (Mgh, Ḳ;) or *with the extremity of the thumb and that of the fore finger,* (Mgh, Mṣb,) *by putting the extremity of the thumb upon that of the fore finger* [*and then jerking the latter forward*]; (Mgh;) or *with a* مِخْذَفَة *of wood.* (Lth, Ḳ.) The doing this was forbidden by Moḥammad, because game cannot be taken thereby, nor an enemy defeated, but a tooth may be broken, and an eye put out. (TA.) حَصَى الخَذْفِ [properly] *Pebbles that are thrown* [*in the manner described above*], (Mṣb,) is used as meaning ‡ *small pebbles.* (Mṣb, TA.) — [Hence,] خَذَفَ بِبَوْلِهِ *He emitted his urine and then stopped it.* (TA.) — And خَذْفُ النُّطْفَةِ *The emitting of the sperma genitale into the midst of the womb.* (TA.) — And خَذَفَ بِهَا, aor. and inf. n. as above, † *He broke wind with a sound.* (TA.) — And الخَذْفُ *The hastening, speeding,* or *going quickly,* of camels. (TA. [See also خَذَفَانٌ.]) — And The act of *cutting,* or *cutting off.* (Kr, TA.)

6. [تَخَاذَفَا app. signifies *They vied with each other in throwing in the manner described in the first sentence above.* — And hence,] عَيْنَاهُ تَخَاذَفَتَا بِالدَّمْعِ ‡ *i. q.* اسرتا [app. a mistranscription for أَجْرَتَاهُ, i. e. *His eyes shed tears*]: so in the A. (TA.)

خَذَفَانٌ *A certain pace,* or *manner of going, of camels.* ('Eyn, T, Ḳ. [See also 1, last sentence but one.])

خَذُوفٌ A horse or the like (Lth) *quick in pace.* (Lth, Ḳ.) — A she-ass *that throws the pebbles by reason of her quickness:* (Ṣ, Ḳ:) or *whose navel is near to the ground by reason of her fatness:* (Ḳ:) or *so fat that, if a pebble is thrown at her with the fingers, or with two fore fingers, or with the extremity of the thumb and that of the fore finger, it sinks into her fat:* (Z:) and *that raises her legs to the side of her belly:* (TA:) pl. خُذُفٌ. (Aṣ, TA.)

الخَذَّافَةُ † *The anus;* syn. الاِسْتُ; (TA;) as also ▼ المِخْذَفَةُ. (Ḳ.)

مِخْذَفٌ The *loops of the* مِقْرَن [q. v.] *with which the* [*quiver termed*] كِنَانَة *is connected with the* [*case termed*] جَعْبَة: (Ibn-'Abbád, Ḳ:) pl. مَخَاذِفُ. (Ibn-'Abbád.)

مِخْذَفَةٌ *A kind of sling;* syn. مِقْلَاعٌ: (Ṣ, Ḳ:) or *a thing with which one throws:* (Ṣ:) or it signifies also *a wooden instrument with which one throws in the manner termed* خَذْفٌ: (Ḳ:) or *a thing into which stones are put, and with which they are thrown at birds &c., like a* مِقْلَاع. (ISd.) [In the TA in art. دث, it is said (in my opinion erroneously) to signify *a bullet.*] — See also الخَذَّافَةُ.

خذل

1. خَذَلَهُ, (Ṣ, Mṣb, Ḳ,) and خَذَلَ عَنْهُ, (Mṣb, Ḳ,) aor. ـُ, (JK, Mṣb,) inf. n. خِذْلَانٌ (JK, Ṣ, Ḳ) and خَذْلٌ, (JK, Mṣb,* Ḳ,) or ▼ خِذْلَانٌ is a simple subst.; (Mṣb;) and ▼ اخذله; (TA;) *He abstained from,* or *neglected, aiding him,* or *assisting him;* (JK,* Ṣ, Mṣb, Ḳ;) and *held back from him:* (Mṣb:) and خَذَلَهُ *he left, forsook,* or *deserted, him:* (MA:) and, accord. to AZ, خَذَلَ عَنْهُ, aor. ـُ, inf. n. خُذُولٌ, *he left him, and held back from going with him.* (T in art. تلو.) And ▼ خِذْلَانٌ signifies God's *failing to preserve* a man; accord. to Az, *from an evil action;* so that he falls into it. (TA.) It is said in the Ḳur [iii. 154], وَإِنْ يَخْذُلْكُمْ فَمَنْ ذَا الَّذِى يَنْصُرُكُمْ مِنْ بَعْدِهِ [*But if He fail to aid you,* or *to preserve you, then who is he that will aid you after Him?*]: here 'Obeyd Ibn-'Omeyr read ▼ يُخْذِلْكُمْ, with ḍamm to the ى and kesr to the ذ. (TA.) — And خَذَلَ *He* (a gazelle) *held back from going with the herd;* or *remained behind the herd.* (Aṣ, Ṣ.) And خَذَلَتْ *She* (a gazelle, &c., Ḳ, as, for instance, a [wild] cow, TA) *held back from going with her female companions, and remained alone:* or *held back,* or *remained behind, not overtaking* [*the others*]: (Ḳ:) or you say, of a female gazelle, and of a [wild] cow, تَخْذُلُ صَوَاحِبَهَا *She holds back from going with her female companions,* or *remains behind them, with her young one,* in the place of pasturage: or *remains alone with him;* as A 'Obeyd says, on the authority of Aṣ. (T, TA.) And خَذَلَتْ *She* (a wild animal, Ṣ, O, or a gazelle, Ḳ) *remained attending to her young one* [when the rest of the herd had gone]; (Ṣ, O, Ḳ;) said by some to be inverted [as to the meaning], because she is [not the one that leaves, but] the one that is left; (Ṣ;) as also ▼ تخاذلت, (Ṣ, Ḳ,) and ▼ اخذلت. (Ḳ.) — A poet says, (namely, 'Adee Ibn-Zeyd, TA,) describing a horse,

* فَهْوَ كَالدَّلْوِ بِكَفِّ المُسْتَقِى *
* خَذَلَتْ عَنْهُ العَرَاقِى فَانْجَذَمْ *

meaning [*And he is,* or *was, like the bucket in the hand of the drawer of water,*] *of which the cross pieces of wood have become separated from it* [*so that it has become severed* from the well-rope]. (Ṣ.)

2. خَذَّلَهُ, inf. n. تَخْذِيلٌ, *He incited him,* or *induced him, to abstain from,* or *to neglect, aiding him,* or *assisting him;* and *hindered him, withheld him,* or *prevented him, from aiding his companion,* or *assisting him.* (T, TA.) And خَذَّلَ عَنْهُ أَصْحَابَهُ, inf. n. as above, *He incited,* or *induced, his companions to abstain from,* or *to neglect, aiding him,* or *assisting him.* (Ṣ.) — *He induced him to be cowardly and weak-hearted, and to abstain from fighting.* (Mṣb.)

4: see 1, in three places. — Also اخذل, (Ḳ,) or اخذل امّه, (T, TA,) said of the young one of a wild animal, (Ḳ,) or of the young one of a gazelle or [wild] cow, (T,) means وَجَدَ أُمَّهُ تَخْذُلُهُ (Ḳ, TA) [which may be rendered *He found his mother to hold back from him*; or *to leave him, and hold back from going with him*: or it means] *he found his mother to hold back from going with her female companions*, or *to remain behind them, with him*; or *to remain alone with him*. (T.)

6. تخاذلوا *They abstained from*, or *neglected, aiding*, or *assisting, one another*: (Ṣ:) i. q. تدابروا [which has the foregoing signification accord. to the TA, but properly means *they turned their backs, one upon another*]: (Ḳ:) *they left, forsook*, or *deserted, one another* in trouble. (MA.) [And, accord. to Freytag's Lex., *They fled from fight*, or *battle*.] — تخاذلت رِجْلَاهُ (said of an old man, TA) *His legs became weak* (Ṣ, Ḳ, TA) *in consequence of some evil affection*. (TA.) — And تخاذل *He was*, or *became, behind*. (KL.) — See also 1.

7. [انخذلوا, accord. to Freytag's Lex., *They were destitute of aid*, or *assistance*: *they fled*.]

خُذَلَةٌ A man *constantly abstaining from*, or *neglecting, the rendering of aid*, or *assistance*. (IAạr, Ṣ, Ḳ.* [In the Ḳ, this is not distinguished from خاذل in the first of the senses assigned to this latter below. See also خذول.])

خِذْلَانٌ: see 1, in two places.

خَذُولٌ One *who often abstains from*, or *neglects, the rendering of aid*, or *assistance*: so in the Ḳur xxv. 31, applied to the Devil; (TA;) because he forsakes the unbeliever, and declares himself to be clear of him, on the occasion of trial, or affliction. (Jel.) — See also خاذل. — Also A mare *that, when suffering from the pain of parturition, moves not from her place*. (ISd, Ḳ.) — خَذُولُ الرِّجْلِ A man *whose leg aids him not, in consequence of weakness*, (Ṣ,* TA,) or *of some evil affection*, or *of intoxication*. (TA.)

خَاذِلٌ *Abstaining from*, or *neglecting, the rendering of aid*, or *assistance*. (Ṣ, Ḳ.) — And Any one *leaving, forsaking*, or *deserting*. (TA.) — Also, and ▾ خَذُولٌ, (T, Ḳ,) A female gazelle, and a [wild] cow, *holding back from going with her female companions*, or *remaining behind them, with her young one*, in the place of pasturage: or *remaining alone with him*: (T:) or a female gazelle, &c., (Ḳ,) as, for instance, a [wild] cow, (TA,) *holding back from going with her female companions, and remaining alone*: or *holding back*, or *remaining behind, not overtaking* [*the others*]: (Ḳ:) or the latter signifies a female gazelle *that has left, forsaken*, or *deserted, her young ones*. (EM p. 60.) — And the former, A female gazelle *remaining attending to her young one*; as also ▾ مُخْذِلٌ. (Ḳ. [See 1.]) — Also *Routed, defeated*, or *put to flight*. (IAạr, Ḳ.)

مُخْذِلٌ: see the next preceding paragraph.

مَخْذُولٌ *Having no aider*, or *assister*. (Jel in xvii. 23.)

خذم

1. خَذَمَهُ, (Ṣ, Ḳ,) aor. ـِ, (Ḳ,) inf. n. خَذْمٌ, (JK, Ṣ, TA,) *He cut it*, or *cut it off*: (Ṣ, Ḳ:) or he did so *quickly*: (JK, Z, TA:) and ▾ خذّمه and ▾ تخذّمهُ signify the same: (Ḳ:) or ▾ the second, inf. n. تَخْذِيمٌ, *he cut it in pieces*; (Ṣ, in which only the inf. n. is mentioned;) and so ▾ the third. (TA.) [It is like جَذَمَهُ.] Hence, in a trad., خَذَمُوا بِالسُّيُوفِ, said of men stopping the road, and robbing and slaying passengers, *They cut*, and *smote*, men *with the swords*. (TA.) And خَذَمَ لَهُ مِنْ مَالِهِ خَذْمَةً *He cut off for him a portion of his property*. (JK.) — Also, said of a hawk, *He struck him*, or *it, with his talon*, or *claw*. (IAạr, Ḳ.*) — خَذِمَ, aor. ـَ, (Ḳ,) inf. n. خَذَمٌ, (TḲ, [as indicated in the Ḳ by its being said that the verb is like سَمِعَ,]) *It became cut*, or *cut off*; as also ▾ تخذّم: (Ḳ:) [or] the latter is quasi-pass. of خَذَّمَهُ [and therefore signifies *it became cut in pieces*]. (TA.) And خَذِمَتِ النَّعْلُ *The* [thong called] شِسْع [q. v.] *of the sandal became cut*, or *severed*. (TA.) — خَذِمَ, aor. ـَ, (Ḳ,) inf. n. خَذَمٌ (Ṣ, TA) [and app. خَذَمَانٌ, q. v. infrà]; or خَذَمَ, [so in the JK, which may be also correct, agreeably with the primary signification of this verb, though the aor. is there said to be ـَ, which may perhaps be a mistranscription for ـِ,] inf. n. خَذْمٌ [there mentioned in two places] and خَذَمٌ; (JK;) ‡ *He was quick* (JK, Ṣ, Ḳ, TA) *in pace*, or *rate of going*; (JK, Ṣ, TA;) as also ▾ اخذم. (JK. [But this last may be erroneously taken from the phrase مَرَّ مُخْذِمًا, there immediately following; for this phrase may be mistranscribed for مَرَّ يَخْذِمُ, which I find in the TA.]) — [And hence,] خَذَمٌ signifies also †[The *being liberal*, or *bountiful*; (see the part. n. خَذِمٌ;) or] the *giving a gift*, or *making a present*. (KL.)

2: see 1, first sentence, in two places.

4: see 1, in the latter part. — اخذم النَّعْلَ *He repaired the* [thong called] شِسْع *of the sandal*. (AA, TA.) [Here, as in many other instances, the ا has a privative effect: see خَذِمَتِ النَّعْلُ, above.]

5, as an intrans. v.: see 1, near the middle of the paragraph: — as a trans. v.: see 1, first sentence, in two places.

8. اِخْتِذَامٌ, [accord. to some, seems to signify The *cutting off* a thing *for oneself*:] accord. to some, طِيبُ النَّفْسِ [as meaning the *being pleased, content*, or *willing*: see خَذِمٌ]. (Ḥam p. 703, q. v.)

خَذِمٌ A *cutting*, or *sharp*, sword; (Ḳ;) as also ▾ خَذُومٌ (JK, Ḳ) and ▾ مِخْذَمٌ; (JK, T, Ṣ, ISd, TA;) the last erroneously said in the Ḳ to be like مُعَظَّم. (TA.) One of the swords of Moḥammad was named ▾ المِخْذَمُ: it was formerly the sword of El-Ḥárith El-Ghassánee. (TA.) — A garment, or piece of cloth, *altogether*, or *wholly, old and worn out*. (TA.) [And in like manner, accord. to Freytag (who does not name his authority), ▾ خَذُومٌ, applied to a scabbard of a sword, *Lacerated and worn by use*.] — ‡ *Quick*, or *swift of pace*, applied to a horse; (JK, Ṣ;) and so ▾ خَذُومٌ, applied to an ostrich. (Ṣ.) — [And hence,] ‡ A man (JK, Ṣ) *liberal*, or *bountiful*, (Ṣ, Ḳ,) *in giving*; (Ṣ;) *pleased, content*, or *willing*, (JK, Ḳ, TA,) *to be liberal* or *bountiful*; *plenteous* in gifts: (TA:) pl. خَذِمُونَ: (JK, Ḳ:) it has no broken pl. (TA.)

خَذْمَةٌ *A snatching away*; *a snatch*: and *a stroke*, or *blow*. (TA.) — *A certain mark* (سِمَةٌ) *of camels*: (JK, Ḳ:) an Islámee term: (Ḳ:) *a certain mark of sheep* or *goats*, (JK, T, TA,) being *a slit across the ear*, (T, TA,) *which leaves the ear dried up*. (JK, T, TA. [See خَذْمَآءُ.]) — Also, [or خِذْمَةٌ, q. v.] *A space*, or *period*, syn. سَاعَةٌ, (Ḳ,) [of the night, or of the day:] and خَذْمَةٌ [or خِذْمَةٌ] is a dial. var. thereof. (TA.)

خِذْمَةٌ [like جِذْمَةٌ] *A piece cut off* of a garment. (JK. [See also خُذَامَةٌ.]) *A portion cut off* from property. (JK.) *A portion* of the night. (JK. [See also خَذْمَةٌ.]) — *A company* of men. (JK. [But perhaps this may be a mistake for خِذْمَةٌ.])

خَذْمَآءُ A sheep or goat, (شَاةٌ, JK, Ḳ,) or a she-goat, (Ṣ,) *having the mark in the ear termed* خَذْمَةٌ; (JK;) [i. e.] *having the ear slit across, without its being separated*: (Ṣ, Ḳ:) or a ewe *having the extremity of her ear cut off*. (T, TA.)

خَذَمَانٌ [app. an inf. n. of خَذِمَ] *Quickness of pace*, or *of rate of going*. (TA.)

خَذُومٌ: see خَذِمٌ, in three places.

أُذُنٌ خَذِيمٌ *An ear cut*, or *cut off*: (Ḳ:) pl. خُذُمٌ. (TA.)

خُذَامَةٌ *A piece cut off*. (JK, Ḳ. [See also خِذْمَةٌ.])

مِخْذَمٌ: see خَذِمٌ, in two places.

خر

1. خَرَّ, (Ṣ, A, Ḳ,) aor. ـِ, (Ṣ, M, Ḳ) and ـُ, (M, Ḳ,) [the latter of which is anomalous,] inf. n. خَرِيرٌ, (Ṣ, A, Mṣb, Ḳ,) *It* (water) *sounded*; [i. e., *murmured*; *rumbled*; or *gurgled*;] (Ṣ, A, Mṣb, Ḳ;) as also ▾ خَرْخَرَ, (A,) inf. n. خَرْخَرَةٌ: (TA:) or خَرَّ, aor. ـِ, (IAạr, T,) inf. n. خَرٌّ, (IAạr,) *it* (water) *ran vehemently* [*so as to make a noise*: as used in the present day, *it ran, flowed*, or *trickled down*; and *leaked*, or *oozed, out*: and also *it ran so as to make a murmuring*, or *similar, sound*]: (IAạr, T:) and خُرَّ *it was made to run*: (TA: but the subject of this verb is not mentioned.) And خَرَّتْ, (A, Ḳ,) aor. ـِ and ـُ, (Ḳ,) inf. n. خَرِيرٌ, *It* (the wind, الرِّيحُ) *sounded*; [i. e., *murmured*; *rumbled*; or *rustled*;] (A, Ḳ;) among reeds or canes; as also ▾ خَرْخَرَتْ, (A,) inf. n. as above: (TA:) or the latter signifies

it made a quick خرير [or *rustling*] among reeds or canes or the like. (Lth.) Also *It* (an eagle, عُقَاب) *made a rustling* (حَفِيف) *with its wings*, in flying. (Lth, Ḳ.) And خَرَّ, (Ṣ,) inf. n. خَرِيرٌ, (Ḳ,) *He* (a man sleeping) *snored*, or *made a sound in breathing*; (Ṣ, Ḳ,* TA;) as also ↑خَرْخَرَ, (Ṣ,) inf. n. خَرْخَرَةٌ: (Ṣ, Ḳ:) and the latter, (TA,) inf. n. as above, (Ḳ,) has the same signification when said of a leopard (Ḳ,* TA) in his sleep; (TA;) as also the former, inf. n. as above; (TA;) and when said of a cat; (Ḳ,* TA;) as also the former, (TA,) inf. n. خُرُورٌ (so in the CḲ and in a MS. copy of the Ḳ, but in the text of the Ḳ in the TA, خَرُورٌ,) and خَرِيرٌ; (TA;) [i. e. *he made a loud purring in his sleep*;] and it is likewise significant of the sound, or sounds, made by a person suffering strangulation. (Ṣ, TA: but of the verb applied in this last manner, only the inf. n., خَرْخَرَةٌ, is mentioned.) And خَرَّ, aor. ـِ, *It* (a stone) *made a sound in its descent.* (TA.) — [Hence,] خَرَّ, aor. ـَ (Ṣ, A, Mṣb, Ḳ) and ـُ, (Ḳ,) the latter of which is anomalous, (TA,) inf. n. خُرُورٌ (Ṣ, Ḳ) and خَرٌّ, (Ḳ,) *He*, or *it*, *fell*, or *fell down*: (Ṣ, A, Mṣb, Ḳ:) originally, *he*, or *it*, *fell, making a sound to be heard at the same time*: afterwards used in the sense of falling absolutely: you say خَرَّ البِنَآءُ *The building fell down*: (TA:) and خَرَّ لِلّٰهِ سَاجِدًا *He fell down prostrating himself to God*: (Ṣ, A,* TA:) or خَرَّ signifies *He*, or *it*, *fell from a high to a low place*: (Ḳ, TA:) so in the Ḳur [xxii. 32], فَكَأَنَّمَا خَرَّ مِنَ السَّمَآءِ [*He is as though he fell from the sky*]. (A,* TA.) And خَرُّوا لِأَذْقَانِهِمْ, inf. n. خُرُورٌ, [*They fell down prostrate, with their chins to the ground*: see the Ḳur xvii. 108 and 109:] (A:) [whence the saying,] عَصَفَتْ رِيحٌ فَخَرَّتِ الأَشْجَارُ لِلْأَذْقَانِ ‡ [*A wind blew violently, so that the trees fell, or bent themselves, down to the ground*]. (A.) — You say also, خَرَّ, (TA,) inf. n. خَرٌّ, (Ḳ,) meaning † *He died*: (Ḳ, TA:) because a man, when he dies, falls down. (TA.) In the Ḳur xxxiv. 13, it may mean † *He died*, or *he fell down.* (TA.) — Also (خَرَّ) *He stumbled after going right.* (TA.) — And خَرَّتْ خَطَايَاهُ † *His sins fell*; [or *fell from him*; q. d. *fell to the ground*;] *went away*; or *departed.* (TA, from a trad.) — And خَرَّ مِنَ الجَبَلِ, inf. n. خُرُورٌ, *He*, (a man,) and *it*, (a stone, &c.,) *rolled down from the mountain.* (TA.) — And الأَعْرَابُ يَخِرُّونَ مِنَ البَوَادِى إِلَى القُرَى ‡ *The Arabs of the desert come down*, or *descend, from the deserts to the towns or villages.* (A.) — And خَرُّوا عَلَيْنَا, (TA,) inf. n. خَرٌّ, (Ḳ,) † *They came upon us suddenly, or unawares, from a place unknown.* (Ḳ,* TA.) — And خَرُّوا † *They came from one town, or country, or the like, to another.* (TA.) — And † *They passed along, or away, or by.* (TA.) — مِنْ يَدَيْكَ, or خَرَرْتَ عَنْ يَدَيْكَ, is a metonymical phrase, well known, meaning ‡ *Mayest thou be confounded and stupified by shame*; or, *so as to be speechless and motionless*: خَرَرْتَ [properly] signifying سَقَطْتَ. (TA in the present art. and in art. ارب.)

4. اخرّ يَدَهُ *He made his arm, or hand, to fall*, by a stroke of the sword. (Yaạḳoob, Ṣ, Ḳ.)

7. انخرّ *He*, (a man, TḲ,) or *it*, (the belly, TḲ,) *became lax, or flabby.* (Ḳ, TḲ.) [See also R. Q. 2.]

R. Q. 1: see 1, in three places.

R. Q. 2. تَخَرْخَرَ *It* (a man's belly) *quivered, or shook about, with bigness*: (Ṣ, Ḳ:) or *by reason of leanness.* (TA.) [See also 7.]

خُرٌّ and ↑خُرِّىٌّ The لُهْوَة of a mill or mill-stone; i. e. the *place into which the wheat is thrown with the hand*; (Ṣ;) the *mouth* of a mill or mill-stone. (Ḳ.)

هِرَّةٌ خَرُورٌ *A she-cat that makes much loud breathing* or *purring* (خَرِير) *in her sleep.* (TA.)

خَرِيرٌ *A depressed tract between two hills*, (Ṣ, Ḳ,) *stretching along*: (Ṣ:) pl. أَخِرَّةٌ. (Ṣ, Ḳ.) The pl. occurs in a verse of Lebeed, commencing بِأَخِرَّةِ الثَّلَبُوتِ, as Khalaf El-Aḥmar heard the Arabs recite it: so says A'Obeyd: (Ṣ:) but the common reading is بِأَجِزَّةِ, with ج and ز. (TA.)

خُرِّىٌّ: see خُرٌّ.

عَيْنٌ خَرَّارَةٌ *A sounding* [i. e. *murmuring* or *gurgling*] *spring* or *source*: (Ṣ:) or *a running spring or source*: so called because of the sound of its water: (IAạr:) or *a spring*, or *source*, *welling forth abundantly.* (Mṣb.) — See also خَارٌّ.

خَرْخَرٌ The *sound* of water: and of wind: and of an eagle making a rustling with its wings in flying. (Ḳ.) [See 1.]

خَرْخَارٌ Water *flowing* (Ḳ) *vehemently*, (TA,) or *copiously.* (Ḥam p. 821.)

خَارٌّ part. n. of 1. (TA.) — خُرَّارٌ [is a pl. thereof: and] signifies † Men *coming from one town, or country, or the like, to another*; as also ↑خَرَّارَةٌ: which latter signifies also † Men *passing along, or away, or by.* (TA.)

خرأ

1. خَرِئَ, aor. ـَ, (Ṣ, Mgh, Mṣb, Ḳ,) inf. n. خَرَآءَةٌ (Ṣ, Mgh, Ḳ) and ↑خِرَآءَةٌ, (Ḳ,) which is the more chaste form, (O,) or this may be a quasi-inf. n., (El-Khaṭṭábee, Mṣb,) and, accord. to some, ↑خِرَآءٌ, (MF,) or this is a quasi-inf. n., (Lth,* Mṣb, Ḳ,* MF,) like صِيَامٌ for صَوْمٌ, or, as some say, a pl. of خَرْءٌ, (Mṣb, MF,) and خَرْءٌ also is an inf. n., (Ṣ, Ḳ,) and خُرُوءٌ, (Ḳ, TA,) like قُعُودٌ, (TA,) or خُرُوءَةٌ, (so in a MS. copy of the Ḳ and in the CḲ,) *He voided his excrement*, or *ordure*: (Ṣ, Mgh, Mṣb, Ḳ:) or خِرَآءَةٌ signifies the *retiring alone, and sitting down, to satisfy a want* [*of nature*]. (IAth.)

خَرْءٌ: see what next follows.

خُرْءٌ *Human excrement*, or *ordure*: (Ṣ, Mgh,* Mṣb, Ḳ;) as also ↑خَرْءٌ: (Mṣb, TA:) and sometimes applied to the *dung* of the dog, and of the field-rat, and of the ant, and of the fly: (TA:) pl. خُرُوءٌ, (Ṣ, Mgh, Mṣb, Ḳ,) which is also a pl. of خَرْءٌ, (Mṣb, TA,) and خُرْآنٌ, (Ḳ,) which is anomalous, and خُرُؤٌ; (TA;) and some say that خِرَآءٌ is a pl. of خَرْءٌ. (Mṣb, MF.)

خِرَآءٌ: } see 1.
خِرَآءَةٌ: }

خَارِئٌ act. part. n. of 1. (TA.)

مَخْرَأَةٌ *A place where excrement*, or *ordure*, *is voided*; [*a privy*;] (Ṣ, Ḳ;) as also مَخْرُأَةٌ, (Ḳ,) and ↑مَخْرُؤَةٌ; (T, Ṣ, Ḳ, TA;) the last written in some copies of the Ḳ مَخْرِئَةٌ; and in one, مِخْرَأَةٌ. (TA.)

مَخْرُؤَةٌ: see what next precedes.

خرب

1. خَرِبَ, (JK, Ṣ, A, Mṣb, &c.,) aor. ـَ, (JK, Ḳ,) inf. n. خَرَابٌ, (JK, Ṣ,* A, Mgh,* Mṣb, KL, TA,) said of a place, (Ṣ,) or a country, (A, Mgh,*) or a dwelling, or place of abode, (Mṣb,) or a house, (TḲ,) *It was*, or *became, in a state of ruin, waste, uninhabited, depopulated, deserted, desolate, uncultivated*, or *in a state the contrary of flourishing*: (JK, Ṣ, A, Mgh, KL, TA:) خَرَابٌ is the *contr.* of عِمَارَةٌ. (Ṣ.) — خَرِبَ, aor. ـَ, (Mṣb,) inf. n. خَرَبٌ, (Ṣ, Mṣb, Ḳ,) *It had in it a slit*, or *a round perforation*: (Ṣ:) or *he had his ear slit*, (Mṣb, Ḳ,) or *bored with a round perforation.* (Mṣb.) — خَرِبَ: see 4. — Also, (Mṣb, Ḳ,) aor. ـُ, inf. n. خِرَابَةٌ, (Mṣb,) *He stole*: (Mṣb:) or *he became a thief*, or *robber.* (Ḳ.) And خَرَبَ بِإِبِلِ فُلَانٍ, (Ṣ, A,*) or خَرَبَ إِبِلَ فُلَانٍ, (Ḳ,) both mentioned as on the authority of Lḥ, (TA,) aor. ـُ, (Ṣ, A,) inf. n. خِرَابَةٌ (Ṣ, A, Ḳ) and خَرَابَةٌ and خَرْبٌ and خُرُوبٌ, (Ḳ,) *He stole the camels of such a one.* (Ṣ, Ḳ.) — خَرَبَهُ, (Ḳ,) aor. ـُ, inf n. خَرْبٌ, (TA,) *He bored it, perforated it, or made a hole through it*: or *he slit it*: (Ḳ, TA:) namely, a thing. (TA.) — And *He struck his* خُرْبَة, (Ḳ,) meaning *the part where the head of his thigh-bone was inserted*; or خُرْبَة here has some other of the significations assigned to it in this article. (TA.)

2: see 4, in four places.

4. اخرب, (Ṣ, A, Mṣb, Ḳ,) inf. n. إِخْرَابٌ, (TA,) *He reduced to ruin*; or *rendered waste, uninhabited, depopulated, deserted, desolate, uncultivated*, or *in a state the contrary of flourishing*; (Ṣ, A, Ḳ;) a house, (Ṣ, Ḳ,) or a dwelling, or place of abode, (Mṣb,) or a country; (A;) as also ↑خرّب, (A, Mṣb, Ḳ,) inf. n. تَخْرِيبٌ; (TA;) and ↑خَرِبَ, (Ḳ,) [inf. n. خَرَابٌ, as in the Ḳur ii. 108:] or ↑خرّب signifies the same, but in a more extensive, or a superlative or an intensive sense: you say, خَرَّبُوا بُيُوتَهُمْ [*They ruined their houses*; the ر being doubled because the verb has many objects: or *they demolished their houses*]. (Ṣ, TA.) يُخَرِّبُونَ ↑ بُيُوتَهُمْ, in the Ḳur [lix. 2],

means *They demolishing their houses:* this is the reading of AA: all others read يُخْرِجُونَ بُيُوتَهُمْ, meaning *they going forth from their houses, and leaving them;* (TA;) or *evacuating their houses;* or *leaving them in a state of ruin.* (Bḍ.) —— [Hence the saying,] عِنْدَهُ تُخْرَبُ ▼ الأَمَانَاتُ ‡ [*Deposits entrusted to him become lost, or perish*]. (A.)

5. تخرّب *It* (a building) *became demolished.* (TA.)

10. استخرب *It* (a skin for water or milk) *became perforated with many holes; became full of holes.* (A, TA.) —— † *He became broken by misfortune.* (JK, Ḳ.) —— اِسْتَخْرَبْتُ لِهٰذَا الأَمْرِ † *I lamented for this event, or case.* (JK.) —— استخرب إِلَيْهِ † *He was angry with him;* or *was angry with him with the anger that proceeds from a friend;* (وَجَدَ عَلَيْهِ;) namely, one who had separated himself from him: (JK:) or *he yearned towards, longed for, or desired, him.* (Ḳ.)

Q. Q. 1. تَخْرَبَ [in the CḲ, erroneously, تَخَرَّبَ,] *It* (the canker-worm) *corroded* a tree: (Ḳ, TA:) but accord. to some, this verb is [radically] quadriliteral, and as such it occurs again in the Ḳ [in art. تخرب]. (TA.)

خُرْبٌ: see خُرْبَةٌ, in five places: — and see also خَرِبٌ.

خُرَبٌ: see خُرْبَةٌ, in seven places. — Also The *place where an elevated accumulation of sand terminates,* (JK, Ṣ,* TA,) *producing trees of the kind called* غَضًا. (TA.)

خَرَبٌ The *male of the* [*species of bustard called*] حُبَارَى: (Ṣ, Ḳ:) or *i. q.* حُبَارَى, absolutely: (TA:) pl. خِرْبَانٌ. (Ṣ.) —— And hence, (A,) or ▼خِرْبَانٌ, (JK,) or ▼خُرَبٌ, (TA,) and ▼خَرِبٌ, (Ḳ,) ‡ *Cowardly;* or *a coward.* (A, Ḳ, TA.) — See also خُرْبَةٌ, near the end of the paragraph.

خَرِبٌ (Ṣ, TA) and ▼خَرَابٌ (A, Mṣb) *In a state of ruin, waste, uninhabited, depopulated, deserted, desolate, uncultivated,* or *in a state the contrary of flourishing;* (Ṣ, A, TA;) applied to a place, (Ṣ,) or a country, (A,) or a dwelling, or place of abode. (Mṣb.) You say دَارٌ خَرِبَةٌ *A house which its owner has reduced to ruin, or rendered uninhabited,* &c. (Ṣ, TA.) [In the phrase, هٰذَا جُحْرُ ضَبٍّ خَرِبٍ, meaning *This is a deserted hole of a* lizard of the kind called *ḍabb*, the word خرب is put in the gen. case عَلَى الجِوَارِ, i. e. because of its proximity to a preceding word in that case, not being so properly.] —— [Hence,] هُوَ خَرِبُ العَظْمِ ‡ [*He is without marrow in the bone*]. (A, TA.) And خَرِبُ الأَمَانَةِ ‡ [One *in whom trust is not safely reposed*]. (A, TA.) —— See also خَرَابٌ. — And see خَرَبٌ.

خَرْبَةٌ: see the next paragraph, in two places. — See also خُرْبَةٌ.

خُرْبَةٌ *A hole, perforation,* or *bore;* (Mṣb, TA;) *whether round or not:* (TA:) or *any round hole* or *perforation* or *bore;* (Ṣ, Ḳ, TA;) *such as that of the ear;* (TA;) [and] so ▼خُرْبٌ: (A:) pl. [of mult.] of the former (in this and in other senses here following, Ḳ,* TA) خُرَبٌ (Mṣb, Ḳ) and خُرُوبٌ, which latter is extr. [with respect to rule], and [of pauc.] أَخْرَابٌ [which is irregular as pl. of the former, but regular as pl. of the latter]. (Ḳ. [See also خُرْتٌ and خُرْتَةٌ.]) [Hence,] خُرْبَةُ السِّنْدِيِّ *The bore of the lobe of the ear, when not slit:* [the Sindee being particularly noted by the Arabs for his pierced ears:] *when slit,* it is termed ▼خَرْبَةُ السِّنْدِيِّ. (TA.) —— *Width of the hole,* or *perforation, of the ear;* (JK;) as also ▼خُرْبٌ, (JK,) or ▼خُرْبٌ, (A,) and ▼أَخْرَبُ, (Ḳ,) this last being a subst. like أَفْكَلُ. (TA.) —— The *eye of a needle:* [like خُرْتٌ and خُرْتَةٌ:] and the *foramen* of the anus: as also ▼خُرْبٌ and ▼خُرَبٌ and ▼خُرَابَةٌ and ▼خُرَّابَةٌ and ▼خُرَّابَةٌ; (Ḳ, MF;) in both of these senses, though this is not clearly shown in the Ḳ: (MF:) and likewise, of the vagina; the dual of خُرْبَةٌ occurring in a trad., as some relate it, applied to the foramen of the anus and that of the vagina together: (TA:) and the last, ▼خُرَّابَةٌ, also signifies *any perforation like the eye of a needle.* (Ḳ.) —— الخُرْبَةُ and ▼الخُرْبُ and ▼الخُرَابَةُ and ▼الخُرَّابَةُ *The hole* [or *socket*] *of the hip,* (Ṣ, TA,) *where the head of the thigh-bone is inserted;* as also خُرْبَةُ الوَرِكِ and ▼خُرْبُ الوَرِكِ and ▼خُرَابَةُ الوَرِكِ [or, probably, ▼خُرَابَةُ] and ▼خُرَّابَةُ الوَرِكِ and خُرَّابَةُ الوَرِكِ: and the pl., أَخْرَابٌ, also signifies the *lower extremities of the shoulder-blades.* (TA.) —— And الخُرْبَةُ, (A'Obeyd, Ṣ, Mgh, Mṣb, Ḳ,) or خُرْبَةُ المَزَادَةِ, (A,) and ▼الخُرَّابَةُ, and sometimes ▼الخُرَابَةُ without teshdeed, (TA,) [and perhaps ▼الخُرْبُ also, (see خُبْنٌ,)] *The loop of the* [*leathern water-bag called*] مَزَادَة; (A'Obeyd, Ṣ, A, Mgh, Mṣb, Ḳ;) because of its round form: every مزادة having two loops [whereby it is suspended upon either side of the camel], each of which is thus called; and two kidney-shaped pieces of leather (كُلْيَتَانِ) [at the two upper corners]; and the two loops are sewed to these. (TA. [See also خُرْتَةٌ.]) — *A vice,* or *fault;* (IAth, TA;) as also ▼خَرْبَةٌ: (Ḳ:) and *corruption,* or *unsoundness, in religion;* (JK, Ḳ;) as also ▼خَرْبَةٌ (JK, TA) and ▼خِرْبَةٌ [like خِرْقَةٌ] (Mgh,* Ḳ) and ▼خُرْبٌ and ▼خَرَبٌ (Ḳ) and ▼خَرَبٌ: and *a quality inducing suspicion,* or *evil opinion:* [a meaning app. belonging to all of the foregoing words:] (TA:) pl., of the first, خُرُبَاتٌ; and of the second, خَرَبَاتٌ: (JK:) also, the first (i. e. خُرْبَةٌ), *a crime: a bad, an evil,* or *a foul, word* or *saying:* and *a trial,* or *an affliction.* (TA.) You say, مَا فِيهِ خُرْبَةٌ *There is not in him a vice,* or *fault.* (TA.) And مَا رَأَيْنَا مِنْ فُلَانٍ خُرْبَةً (JK, TA) and ▼خُرْبًا, (TA,) or ▼خَرْبَةً, (JK,) *We have not seen in such a one unsoundness of religion nor anything disgraceful.* (JK, TA.) —— فَارٌّ بِخُرْبَةٍ, occurring in a trad., means One *who flees with a thing desiring to appropriate it to himself and to take possession of it unlawfully.* (TA.)

خَرْبَةٌ: see خَرَابٌ, in three places. — Also The *state,* or *condition,* or *guise, of him who is termed* خَارِبٌ: (Ḳ:) also explained as signifying *a thing whereof one is ashamed:* or as derived from [خَرْبَةٌ, meaning] "contemptibleness, and disgrace, or ignominy:" or it may be ▼خُرْبَةٌ, meaning *a single act* [*of a shameful nature, or the like*]. (Et-Tirmidhee, TA.)

خِرْبَةٌ: see خُرْبَةٌ, in three places, near the end of the paragraph. —— Also *i. q.* ذِلَّةٌ [*Baseness, vileness,* &c.]: (Ḳ, TA:) in one copy of the Ḳ, زَلَّةٌ [*a slip, lapse, fault,* &c.]: (TA:) and *disgrace,* or *ignominy,* and *contemptibleness.* (TA.) —— And الخِرْبَةُ signifies العَوْرَةُ [*The part, or parts, of the person, which it is indecent to expose*]. (Ḳ.)

خَرِبَةٌ and its pls.: see خَرَابٌ, in five places.

خِرْبَانٌ: see خَرَبٌ.

خَرَابٌ inf. n. of خَرِبَ in the first of the senses explained above. (JK, Ṣ,* A, &c. [See 1, first sentence.]) —— [Then used as an epithet:] see خَرِبٌ. —— [And then used as an epithet in which the quality of a subst. predominates, as appears from what follows;] *contr. of* عُمْرَانٌ: (JK, A, Ḳ:) and ▼خَرِبَةٌ signifies [the *same;* or] مَوْضِعٌ خَرَابٌ; (A, Ḳ;) as also ▼خِرْبَةٌ: (Lth, Ḳ:) [all may be rendered *A ruin,* or *waste; a place, country, place of abode,* or *house, in a state of ruin, waste, uninhabited, depopulated, deserted, desolate, uncultivated,* or *in a state the contrary of flourishing:*] the pl. of خَرَابٌ is أَخْرِبَةٌ, (JK, Ḳ,) a pl. of pauc., (JK,) and خِرَبٌ, which latter is mentioned by El-Khaṭṭábee, (Ḳ,) as occurring in a trad. respecting the building of the mosque of El-Medeeneh: كَانَ فِيهِ نَخْلٌ وَقُبُورُ المُشْرِكِينَ وَخِرَبٌ فَأَمَرَ بِالخِرَبِ فَسُوِّيَتْ [*There were in it palm-trees, and the graves of the believers in a plurality of gods, and ruins; and he gave orders respecting the ruins, and they were levelled*]: but IAth says that خِرَبٌ may be pl. of ▼خِرْبَةٌ, or of ▼خَرِبَةٌ: or it may be ▼خَرِبٌ [coll. gen. n.] of ▼خَرِبَةٌ: and accord. to one reading of the trad., the word is حُرُثٌ, meaning "a place ploughed for sowing:" (TA:) [accord. to F,] the pl. of ▼خَرِبَةٌ, also, is خَرِبٌ: and the pl. of ▼خَرِبَةٌ is خَرَائِبُ [mentioned above] and خَرِبَاتٌ [which is anomalous] and خَرِبَاتٌ. (Ḳ.) [Hence,] وَقَعُوا فِي وَادِي خَرِبَاتٍ ▼ [*They fell into a valley of ruins,* or *waste places,* &c.]: (A, TA:) i. e., *into destruction:* (TA:) [a prov., of which there are various readings: see جَذَبَاتٌ, in art. جذب.] — [Also inf. n. of خَرِبَ as syn. with اخرب, q. v.]

خُرَابَةٌ: see خُرْبَةٌ.

خُرَّابَةٌ; see خُرْبَةٌ, in five places.

خَرُّوبٌ (Az, Ṣ, Mgh, Ḳ) and ▼خُرْنُوبٌ, (Az, Ṣ, Ḳ,) mentioned by Az as radically quadriliteral, (TA in art. خرنب,) and ▼خَرْنُوبٌ, (Mgh, Ḳ,) but this last is of weak authority, (TA,) or not allowable, (Ṣ,) a coll. gen. n.; n. un. with ة; (TA;) *A kind of tree, growing upon the mountains of Syria, having grains* (حَبّ) *like those of the* يَنْبُوت [q. v.], *called by the children of El-*

'Irák القَتَادُ الشَّامِىُّ, *dry*, or *tough, and black*: (Az, TA in art. خرنب:) *a certain plant*, (Ṣ, Mgh,) *well known*: (Ṣ:) said by some to be the *kind of tree* [or *plant*] *called* خَشْخَاش [i. e. *poppy*]: (Mgh:) *certain trees, of which there are two kinds, wild* (بَرِّىٌّ), and *Syrian* (شَامِىٌّ): (AḤn, Ḳ:) *the former kind is also called* يَنْبُوتَةٌ; (AḤn;) *and this is thorny*, (AḤn, Ḳ,) *used as fuel, rising to the height of a cubit, having branches*, (AḤn,) *with a fruit* (AḤn, Ḳ) *black* (أَحْمَر) *and light, like bubbles*, (AḤn, TA,) in the copies of the Ḳ كَالنُّفَّاخ, but correctly كَالتُّفَّاح, (TA,) *disagreeable in taste*, (AḤn, Ḳ,) *not eaten except in cases of difficulty*, or *distress; having grains* (حَبّ) *which are hard and lubricous*: (AḤn:) *the Syrian kind* [*is that to which the name of* خَرُّوب *is now commonly applied*, the *carob*, or *locust-tree; ceratonia siliqua; the fruit of which*] *is sweet, and is eaten; having grains* (حَبّ) *like those of the* يَنْبُوت, *but larger*; (AḤn;) *the fruit of this kind is like the* خِيَار [or *cassia fistula*], *but wide; and from it are prepared an inspissated juice and* [*a kind of*] سَوِيق [or *parched meal*]. (AḤn, Ḳ.) [Its grain is used as a weight: see قِيرَاطٌ and دِرْهَمٌ and دِينَارٌ.]

خُرَابَةٌ: see خُرْبَةٌ, in two places.

خُرَّابَةٌ: see خُرْبَةٌ, in four places.

خَرُوبٌ and خَرْنُوبٌ: see خَرُّوبٌ.

خَارِبٌ *A stealer of camels*: (Aṣ, Ṣ, A:) and (by extension of its original meaning, TA) *any thief*, or *robber*: (JK, Ṣ:) dim. ▼خُوَيْرِبٌ: (TA:) and pl. خُرَّابٌ, (Ṣ, A, TA,) or أَخْرَابٌ. (JK.) [See also خِرْبَةٌ.]

خُوَيْرِبٌ: see what next precedes.

أَخْرَبُ *Slit*: or *having a round hole* or *perforation*: (Ṣ:) [fem. خَرْبَآءُ; as in] أُذُنٌ خَرْبَآءُ *An ear having the lobe slit*. (Ḳ.) — A man, (Ṣ,) or a ram, (Mṣb,) *having his ear slit*; (Ṣ, Mṣb, Ḳ;) as also ▼مُخَرَّبٌ and مُخَرَّمٌ; (TA;) from ▼مَخْرُوبٌ signifying *slit*: (Ṣ:) and (so in the Ṣ and TA, but in the Mṣb "or") *having his ear pierced*, or *bored*: when it is slit (after the piercing, Ṣ, TA), he is said to be أَخْرَمُ: (Ṣ, Mṣb, TA: [but see this last in art. خرم:]) and أَخْرَبُ الأُذُنَيْنِ *having the ears pierced*, or *bored*: (AM, TA in art. خرت:) and خَرْبَآءُ a female slave *having the lobe of her ear slit* [or *pierced*, or *bored*]: and ▼مُخَرَّبَةٌ a female slave *having her ear* [*slit* or] *pierced*, or *bored*: (TA:) and خَرْبَآءُ a she-goat *having her ear slit, but so that the slit is not long nor wide*. (Ḳ.) — أَخْرَبُ: see خُرْبَةٌ.

خَلِيَّةٌ مُخْرِبَةٌ *An empty bee-hive*, (Ḳ,) *in which honey has not been collected*. (TA.)

مُخَرَّبٌ, and its fem. (with ة): see أَخْرَبُ.

مَخْرُوبٌ: see أَخْرَبُ.

نُخْرُوبٌ sing. of نَخَارِيبُ, (TA,) which latter signifies *Holes like those of hornets' nests*: and the *holes*, or *cells*, (*prepared with wax*, Ḳ in art. نخرب,) *in which the bees deposit their honey*. (Ḳ, TA. [In the CḲ, erroneously, تخاريب.]) Accord. to some, the ن is a radical letter. (TA.)

خرت

1. خَرَتَ, (Ḳ,) aor. ـُ, inf. n. خَرْتٌ, (TḲ,) *He perforated, bored*, or *pierced*, (Ḳ,) the ear, (TḲ,) or a thing. (TA.) And خَرَتَ أَنْفَ الجَمَلِ [*It perforated*, or *slit*, (see the pass. part. n., below,) *the nose of the camel*]: said of the خِشَاش [or wooden thing that is inserted in the bone of the camel's nose]. (A.) — خَرَتْنَا الأَرْضَ *We knew the land and its roads*. (Ks, Ṣ.) [Golius omits this; but mentions, as on the authority of Ibn-Maạroof, خَرِتَ, signifying *He was skilful*, or *expert, in showing the way*. What Ibn-Maạroof says, however, is that the inf. n. خَرْتٌ signifies the *being acquainted with a road*; and, *with a place*. See خِرِّيتٌ.]

خُرْتٌ: see what next follows, in two places.

خَرْتٌ The *perforation, bore*, or *hole*, (Ṣ, A, Ḳ,) of a needle; [i. e. its *eye*;] (Ṣ, A; [see also خُرْتَةٌ;] and of the ear, (Ṣ,) or in the ear, [but see خُرْتَةٌ,] &c.; (A, Ḳ;) and of the فَأْس, [i. e. hoe, or adz, or axe,] (Ṣ, A, TA,) meaning, of the handle thereof; (A, TA;) as also خُرَةٌ; (Fr, TA in art. خرو;) [see again خُرْتَةٌ;] and ▼خُرْتٌ signifies the same: (A, Ḳ:) pl. [of pauc.] أَخْرَاتٌ (Ṣ, A) and [of mult.] خُرُوتٌ. (Ṣ.) You say أَضْيَقُ مِنْ خَرْتِ الإِبْرَةِ [*Narrower than the eye of the needle*]. (A.) And مَضَايِقُ كَأَخْرَاتِ الإِبَرِ [*Narrow passes like the eyes of needles*]. (A.) The خَرْت of a sandal is The *hole*, or *perforation*, *of the* ذُؤَابَة [q. v.], *into which the thong* [*called the* شِرَاك] *enters*. (An anon. Arabic MS. in my possession.) — Also The *rings at the heads* [or *extremities*] *of* [*camels' plaited fore-girths of the kind called*] نُسُوع; and so [the pls.] خُرُوتٌ (Ḳ) and أَخْرَاتٌ: (Ṣ, Ḳ:) and ▼خُرْتَةٌ signifies *one of these*; (Ḳ;) i. e. the *ring in which is* [*inserted the end of*] *the* نِسْعَة. (TA.) [Hence the phrase,] قَلِقَ خُرُوتُ فُلَانٍ [lit. *The rings of the fore-girths of the camels of such a one became unsteady*; meaning] ‡ *the state of such a one became disordered*, or *perverted*. (A, TA.) And similar to this are the phrases, زَادَ خُرُوتُ القَوْمِ and زَادَتْ أَخْرَاتُهُمْ, [in the TA زاد and زادت, but the comparison evidently shows that the verbs should be راد and رادت,] said of a people when they do not receive or entertain hospitably him who alights at their place of abode: so says IAạr on the authority of Es-Saloolee. (TA.) — See also خُرْتَةٌ. — Also *A small rib, at*, or *near, the breast*; and so ▼خُرْتٌ (Ḳ:) pl. أَخْرَاتٌ, which Lth explains as meaning the *ribs at*, or *near, the breast, collectively*. (TA.) — And [the pl.] أَخْرَاتٌ signifies The *obscure roads* or *ways*, and the *narrow passes*, of a desert. (TA.)

خُرْتَةٌ: see خَرْتٌ. — Its pl., أَخْرَاتٌ, [also pl. of ▼خُرْتٌ, (see خُبْنٌ,)] in the formation of which the ة of the sing. seems to have been considered as elided, also signifies The *loops* of a [leathern water-bag such as is called] مَزَادَة: it is said in the T that in the مزادة are its اخرات, the *loops between which is the* قَصَبَة [commonly signifying *cane*, or *reed*, but here app. meaning the *mouth*, which has the form of a short cylinder, and is in the middle of the upper part of the مزادة, between the two loops, these being at the two upper corners], *whereby* [app. referring to the اخرات] *it is carried* [*and suspended on the side of a camel*, counterpoised by another مزادة on the other side of the camel]: and AM adds that one says [also] أَخْرَابُ المَزَادَةِ, sing. خُرْبَةٌ [q. v.]; and in like manner, خُرْبَةُ الأُذُنِ ["the bore of the ear"]; with ب: and غُلَامٌ أَخْرَبُ الأُذُنَيْنِ ["a boy having his ears pierced, or bored"]: he says, also, that the خُرْتَة, with ت, is [the *hole*] in the iron of the فَأْس, and [the *eye*] of the needle; and the خُرْبَة, with ب, is in the skin: and AA says that خُرْتَة signifies the *eye* of the [kind of needle called] شُغَيْزَة, i. e. the مِسَلَّة: (TA:) and Lth says that it signifies *a round hole*. (TA in art. حرت.)

الخَرَاتَانِ *Two stars*, (Ḳ,) *of the stars of the Lion, two whips' lengths apart*, [(see سَوْطٌ,) *in*] *the two shoulder-blades of the Lion*, (TA,) *also called* زُبْرَةُ الأَسَدِ, (Ḳ,) [*composing the Eleventh Mansion of the Moon*: (see زُبْرَةٌ: and see also مَنَازِلُ القَمَرِ, in art. نزل:)] the word is mentioned here in the Ḳ, as though it were of the measure فَعَالَانِ; but accord. to Kr and others, it is dual of خَرَاةٌ, belonging to art. خرو, in which it is again mentioned in the Ḳ: (TA:) accord. to ISd, however, only the dual form is known, and the radical ت and the augmentative ت [by which latter is meant ة] are in the dual alike: (TA in art. خرو:) Zj asked Th respecting the خراتان, and he answered, IAạr says that they are *two stars, of those of the Lion*; and Aboo-Naṣr, the companion of Aṣ, says that they are *two stars in the* زبرة *of the Lion*, i. e. *in the middle thereof*; but in my opinion they are *two stars after* [i. e. *to the eastward of*] *the* جَبْهَة *and the* قَلْب: Zj disapproved of this, and replied, I say that they are *two stars in that part of the breast which is the stabbing-place*, derived from خُرْتُ الإِبْرَةِ, "the eye of the needle:" but Th rejoined, that this was an error, because the word is the dual of خَرَاةٌ; and he cited some verses in which a poet speaks of certain stars in the Lion, and, among them, of الخَرَاة. (MF, TA.)

خِرِّيتٌ (Ṣ, A, Ḳ) and خِرِّيتٌ مِرِّيتٌ (Sh) *A skilful*, or *an expert, guide of the way*; (Sh, Ṣ, Ḳ;) one *who pursues the right course to the* أَخْرَات, i. e. the *obscure roads* or *ways*, and the *narrow passes, of the deserts*; or *who pursues the right course in a way that may be likened to the* خُرْت [or *eye*] *of the needle*: (TA:) or *skilful*; applied to a man, and [particularly] to a guide: (A:) pl. خَرَارِتُ, occurring in a verse [perhaps used by poetic licence for the regular pl. خَرَارِيتُ]. (Ṣ.)

مَخْرَتٌ *A strait, direct,* or *right, road* or *way.* (K.)

مَخْرُوتٌ originally *Perforated, bored,* or *pierced.* (TA.) — Then, (TA,) *Having the nose slit;* (K;) [and] so مَخْرُوتُ الأَنْفِ, applied to a camel: (A, TA:) or مخروت signifies *having a slit lip.* (S, K.)

خرث

خُرْثِىٌّ The *utensils and furniture of a house* or *tent:* (S, Mgh, K:) and the *refuse,* or the *worthless* or *mean* or *vile articles, thereof;* (S, A, Mgh;) thus accord. to the lawyers: (Mgh:) or *bad furniture and utensils:* (A:) or the *worst thereof,* and *of spoils:* (K:) or [*rubbish,*] *such as a fragment of a bowl, and ashes, and* [*the dung that is termed*] بَعْر, of a house: (AZ, TA in art. اسى:) [pl. خَرَاثِىُّ.] You say, نَقَلُوا خُرْثِىَّ مَتَاعِهِمْ *They removed the worthless articles of their furniture and utensils.* (A.) — [Hence,] فُلَانٌ يَسْمَعُ خُرْثِىَّ الكَلَامِ ‡ *Such a one hears language in which is no good.* (A, TA.) And أَلْقَى فُلَانٌ خَرَاثِىَّ صَدْرِهِ and خَرَاثِىَّ قَوْلِهِ ‡ [*Such a one cast forth the unprofitable things of his mind* and *the unprofitable things of his speech*]. (A, TA.)

خرج

1. خَرَجَ, (S, Msb, K, &c.,) aor. ـُ, (L,) inf. n. خُرُوجٌ and مَخْرَجٌ, (S, Msb, K,) *He,* or *it, went, came, passed,* or *got, out,* or *forth; issued, emanated, proceeded, went,* or *departed;* contr. of دَخَلَ; (TA;) مِنَ المَوْضِعِ [*from the place*]. (Msb.) One says, خَرَجَ مَخْرَجًا حَسَنًا [*He,* or *it, went, came, passed,* or *got, out,* or *forth,* &c., *well:* and *it turned out well*]. (S.) [And خَرَجَ مِنْ طَاعَتِهِ: see طَالِعٌ, in art. طوع. When خرج means *It was disbursed,* or *expended,* the inf. n. is خَرْجٌ.] خَرَجَ بِهِ [lit. *He went out,* &c., *with him,* or *it*]: see 4. (TA.) يَوْمُ الخُرُوجِ [*The day of going forth*] means *the day of the* عيد [or *festival*]. (A, TA, from a trad.) And [as used in the Kur l. 41] *The day when men shall come forth from their graves;* (TA;) a name of *the day of resurrection.* (AO, K.) — [† *It became excluded* by a definition or a rule or the like, or by a portion thereof.] مَنْصُوبٌ عَلَى الخُرُوجِ is a phrase of the Basree grammarians, said of the objective complement of a verb, meaning † *Put in the accus. case as being out of the predicament of the subject and that of the attribute.* (TA.) — خَرَجَ مِنْ أَمْرٍ † [*He got out of, escaped from, extricated himself from, evaded,* or *became quit of, a affair,* or *a state*]. (ISh, TA in art. نكس.) [And خَرَجَ مِنْ حَالٍ إِلَى حَالٍ † *He passed from one state to another state.* And خَرَجَ مِنْ دِينِهِ † *He quitted,* or *forsook, his religion.* And خرج مِنْ دَيْنِهِ, and مِنْ مَرَضِهِ, † *He became quit of his debt,* and *of his disease.*] And خَرَجَ إِلَى فُلَانٍ مِنْ دَيْنِهِ † *He paid such a one his debt:* a phrase used in law. (TA.) [And خَرَجَ عَلَى السُّلْطَانِ, and عَنْ أَمْرِ السُّلْطَانِ, † *He rebelled against the Sultán.*] And خَرَجَتْ عَلَى خِلْقَةِ الجَمَلِ ‡ [*She became formed like the he-camel*]; said of a she-camel that is termed ▼ مُخْتَرَجَةٌ. (S, A, K.) And خَرَجَ إِلَى البَذَآءِ † [*He became foul,* or *obscene, in his language*]. (L and K in art. خنذ.) And خَرَجَ فِى العِلْمِ وَالصِّنَاعَةِ, inf. n. خُرُوجٌ, ‡ *He was,* or *became, conspicuous in science and art.* (A, TA. [See also 5.]) — مَا أَحْسَنَ خُرُوجَهَا, said of a cloud (سَحَابَة), ‡ *How good is its first rising from the horizon!* (A.) [You say also, خَرَجَ السَّحَابُ, inf. n. خُرُوجٌ, meaning † *The clouds became extended,* or *expanded:* see خَرْجٌ.] And خَرَجَتِ السَّمَآءُ ‡ *The sky became clear, after having been cloudy.* (T, A.)

2. خرّج, inf. n. تَخْرِيجٌ, [sometimes resembles in signification أَخْرَجَ:] see the inf. n. voce خَرِيجٌ. — [† *He resolved, explained,* or *rendered,* a saying. عَلَى هٰذَا خَرَّجُوا قَوْلَ كَذَا † *According to this meaning* &c. *they have resolved, explained,* or *rendered, such a saying,* is a phrase of frequent occurrence in the larger lexicons &c.] — † *He educated, disciplined,* or *trained, well* a youth: and in like manner, a horse [and a camel; for مُخَرَّجٌ, as is indicated in the K voce مُدَرَّبٌ, applied to a camel, is syn. with مُؤَدَّبٌ]. (IAar.) You say, خَرَّجَهُ فِى الأَدَبِ, (S, A,* K,) inf. n. as above, ‡ *He educated, disciplined,* or *trained, him well in polite accomplishments;* i. e. a teacher, his pupil. (TA.) ═ [*He,* or *it, rendered* a thing أَخْرَج, i. e. *of two colours, white and black:* &c.] You say, النُّجُومُ تُخَرِّجُ اللَّوْنَ *The stars render the colour* [of a thing, such as an expanse of water,] *a mixture of black and white,* by reason of its blackness and their whiteness. (TA.) And خرّج اللَّوْحَ, (A, K,) inf. n. as above, (K,) ‡ *He* (a boy, A) *wrote upon part of the tablet and left part of it without writing.* (A,* K.) And خرّج كِتَابًا ‡ *He wrote a book leaving* [*blank*] *the places* [*of the titles*] *of the sections and chapters.* (A.) And خرّج العَمَلَ, (A, K,) inf. n. as above, (TA,) ‡ *He made the work to be of different kinds.* (A, K,* TA.) And خرّجت الرَّاعِيَةُ المَرْعَى, inf. n. as above, *The pasturing animals ate part of the pasture and left part.* (S,* A, K,* TA. [See also 4.]) And أَرْضٌ فِيهَا تَخْرِيجٌ: and عَامٌ فِيهِ تَخْرِيجٌ, and عام ذُو تَخْرِيجٍ: see أَخْرَجُ.

3. المُخَارَجَةُ i. q. المُنَاهَدَةُ بِالأَصَابِعِ, (S, TA,) i. e. (TA) *One person's putting forth as many of his fingers as he pleases, and the other's doing the like:* (K, TA:) [or *the playing at the game called morra; micare digitis:* see خَرِيجٌ. You say, خارجهُ *He played with him at the game of morra.* See also 6.] — خَارَجَهُمْ, [inf. n. as above,] *He contributed with them to the expenses of a journey or an expedition against an enemy, sharing equally with each of them;* like نَاهَدَهُمْ. (L in art. نهد.) — And خارجهُ *He made an agreement with him,* namely, his slave, *that he* (the latter) *should pay him a certain impost at the expiration of every month;* (Mgh, L, TA;) *the slave being left at liberty to work:* (L, TA:) in which case the slave is termed ▼ عَبْدٌ مُخَارَجٌ. (Mgh, L, TA.)

4. اخرجهُ, (S, Msb, K, &c.,) inf. n. [إِخْرَاجٌ and] مُخْرَجٌ, (S, K,) *He made,* or *caused, him,* or *it, to go, come, pass,* or *get, out,* or *forth; to issue, emanate, proceed,* or *depart:* [*he put, cast,* or *thrust, him,* or *it, out,* or *forth; expelled, ejected,* or *dislodged, him,* or *it: he took, led, drew,* or *pulled, him,* or *it, out,* or *forth: he gave it forth: he,* or *it, produced it:*] as also خَرَجَ بِهِ ▼: [but it should be observed that this latter properly and generally denotes accompaniment, like ذَهَبَ بِهِ; and may be literally rendered *he went, came, passed,* or *got, out,* or *forth, with him,* or *it:*] and ▼ اخترج, also, is syn. with أَخْرَجَ; as in the saying, in a trad., فَٱخْتَرَجَ تَمَرَاتٍ مِنْ قِرْبَةٍ [*And he took forth,* or *took forth for himself* (accord. to a property of many verbs of this form), *some dates from a water-skin*]: (TA:) [so, too, is ▼ استخرج; as meaning *he took, led, drew,* or *pulled, out,* or *forth:* but this generally implies some degree of effort, or labour; as does also ▼ اخترج; and likewise, desire: i. e. it means *he sought,* or *endeavoured, to make* a thing *come forth:* the former is also syn. with أَبْدَعَهُ (q. v.) and أَحْدَثَهُ: and both of them signify, and so does اخرج in many instances, *he drew out,* or *forth; extracted; educed; produced; elicited; fetched out by labour* or *art; got out;* or *extorted:* this is what is meant by its being said that] ▼ الاِسْتِخْرَاجُ is syn. with الاِسْتِنْبَاطُ, (S, K,) and so is ▼ الاِخْتِرَاجُ. (K.) أَخْرِجْنِى مُخْرَجَ صِدْقٍ, in the Kur xvii. 82, means *Cause Thou me to go forth* from Mekkeh *in a good,* or *an agreeable, manner,* so that I may not turn my heart [or affections] towards it: (Jel: [see also various similar explanations in Bd:]) or مخرج is here a n. of place, or, accord. to the more approved opinion, of time. (TA.) — اخرج مَا فِى صَدْرِهِ † [*He vented that which was in his bosom,* or *mind*]. (TA in art. سرح.) — [اخرج, said of a definition or a rule or the like, or of a portion thereof, means † *It excluded* something.] — اخرجهُ مِنَ الأَمْرِ † [*He excluded him from participation in the affair*]. (TA in art. حصن, &c.) ═ اخرج [intrans.] *He paid his* خَرَاج; (K;) i. e. his *land-tax,* and *poll-tax.* (TA.) ═ *He hunted ostriches such as are termed* خُرْج, (K, TA, [in the CK الخَرْج is erroneously put for الخُرْج,]) pl. of أَخْرَجُ. (TA.) — *He married to a woman of brown complexion, white intermixed with black, whose parents were, one, white, and the other, black.* (T, K.) — ‡ *He passed a year of fruitfulness and sterility,* (K, TA,) or *half fruitful and half sterile.* (TA.) — اخرجت الرَّاعِيَةُ ‡ *The pasturing animals ate part of the pasture and left part.* (K, TA. [See also 2.])

5. تخرّج [† *It* (a saying) *was resolved, explained,*

or rendered. عَلَى هٰذَا يَتَخَرَّجُ قَوْلُ كَذَا † *According to this* meaning &c. *is,* or *may be, resolved, explained,* or *rendered, such a saying,* is a phrase of frequent occurrence in the larger lexicons &c. —] ‡ *He was,* or *became, well educated* or *disciplined* or *trained,* (A,* TA,) in polite accomplishments, (Ṣ, Ḳ, TA,) or in science and art. (A. [See also 1: and see 2, of which it is quasi-pass.])

6. تَخَارَجَ i. q. تَنَاهَدَ; (Ṣ;) *similar to* مُخَارَجَةٌ *with the fingers, as explained above.* (TA.) You say, تخارجوا, meaning تناهدوا [i. e. *They played together, one putting forth as many of his fingers as he pleased, and another doing the like: or they played together at the game called morra:* see خَرِيجٌ]. (A.) — تخارجوا is also *syn.* with تناهدوا as meaning *They contributed equally to the expenses which they had to incur on the occasion of a journey, or an expedition against an enemy; or contributed equal shares of food and drink.* (L in art. نهد.) — And تخارجا *They* (two copartners, Ḳ, TA, or two coinheritors, TA) *became quit of claim to sharing property by one's taking the house and the other's taking the land;* (Ḳ,* TA;) or *by selling the property by mutual consent and then dividing it; or by one's taking ready money and the other's taking a debt.* (TA.)

8: see 4, in three places: and see also 10.

9. اخرجّ *He* (a ram, Ḳ, or an ostrich, Ṣ, Ḳ) *was,* or *became,* أَخْرَج, i. e., *of two colours, white and black;* as also ▼ اخراجّ. (Ṣ, Ḳ.)

10. استخرج: see 4, in two places. You say, اِسْتَخْرَجْتُ الشَّىْءَ مِنَ المَعْدِنِ *I extracted the thing from the mine, clearing it from its dust.* (Mṣb.) And اِسْتِخْرَاجُ المُعَمَّى مَتْعَبَةٌ لِلْخَوَاطِرِ †[*The eliciting of the meaning of that which is made enigmatical is a cause of fatigue to minds*]. (A in art. تعب.) — [Also †*He tilled* land, *and made it productive.* (See Ḳ voce غَامِرٌ.]) And اُسْتُخْرِجَتِ الأَرْضُ †*The land was put into a good state for sowing* or *planting.* (AḤn, TA.) — استخرجه and ▼ اخترجه *He asked him,* or *petitioned him, to go,* or *come, out,* or *forth;* or *he desired of him that he should go,* or *come, out,* or *forth.* (TA.)

11: see 9.

خَرْجٌ [originally an inf. n.] *Outgoings, disbursements, expenditure,* or *expenses; what goes out,* or *is expended, of a man's property; contr. of* دَخْلٌ. (Ṣ, Ḳ.) — See also خَرَاجٌ, throughout. — Also, (Ṣ, L, Ḳ,) and ▼ خُرُوجٌ, (L,) *Clouds when first rising and appearing:* (Ṣ, L, Ḳ:) or the *rain that comes forth from clouds:* (Akh:) or the خُرُوج of clouds is their *becoming extended,* or *expanded.* (TA. [See 1.])

خُرْجٌ: see خَرَاجٌ. = Also *A well-known kind of* وِعَآء; [*a pair of saddle-bags;* i. e. *a double bag,* or *double sack, for the saddle;*] (Ṣ, Mṣb, Ḳ;) *a* جُوَالِق *having two corresponding receptacles* [*the mouths whereof are generally closed by means of loops which are inserted one into another*]: (TA:) [also, app., *a single saddle-bag;* and خُرْجَانِ *a pair of saddle-bags:* (see بَدِيدٌ:)] an Arabic word, (Ṣ,) accord. to the more correct opinion; but said by some to be arabicized: (TA:) pl. [of mult.] خِرَجَةٌ (Ṣ, Mṣb, Ḳ) and [of pauc.] أَخْرَاجٌ. (TA.)

خَرَجٌ [The *quality of being of*] *two colours, white and black.* (Ṣ, Ḳ. [See أَخْرَجُ.])

خَرْجَةٌ [n. un. of 1: pl. خَرَجَاتٌ]. You say, مَا خَرَجَ إِلَّا خَرْجَةً وَاحِدَةً *He went not,* or *came not, out,* or *forth, save once:* and مَا أَكْثَرَ خَرَجَاتِكَ *How many are thy goings,* or *comings, out,* or *forth!* (A.)

رَجُلٌ خَرْجَةٌ وَلَجَةٌ (Ṣ, Ḳ*) and ▼ خَرَّاجٌ وَلَّاجٌ and ▼ خُرُوجٌ وَلُوجٌ (TA in art. ولج) *A man frequently going,* or *coming, out and in:* (Ṣ, Ḳ, TA:) and the second phrase [and app. the others likewise] ‡a man *of much cleverness, ingenuity,* or *acuteness, and artifice,* or *cunning;* (Ḳ, TA;) ‡a man *who uses art, artifice,* or *cunning, in the disposal,* or *management, of affairs:* (A:) or ‡one *who does not hasten in an affair from which he cannot easily escape when he desires to do so.* (TA.)

خَرَاجٌ (Ṣ, A, Mgh, Mṣb, Ḳ) and ▼ خَرْجٌ, (Ṣ, Mṣb, Ḳ,) both also written with ḍamm, [i. e. ▼ خُرَاجٌ and ▼ خُرْجٌ,] (Ḳ,) but the former mode of writing them is that which more commonly obtains, (TA,) i. q. إِتَاوَةٌ; (Ṣ, Ḳ;) *A tax,* or *tribute, which is taken from the property of people; an impost,* or *a certain amount of the property of people, which is given forth yearly; a tax upon lands &c.:* (TA:) or the *revenue,* or *gain, derived from land,* (A, Mgh, Mṣb,) or *from a slave,* (Mgh,) or also *from a slave:* (A:) and then applied to the *land-tax, which is taken by the Sulṭán:* (A, Mgh:) and the *poll-tax paid by the free non-Muslim subjects of a Muslim government:* (A, Mgh, Mṣb:) or خَرَاجٌ signifies especially *a land-tax:* and ▼ خَرْجٌ, *a poll-tax:* (IAṭr:) or the former also signifies the *poll-tax paid by the free non-Muslim subjects of a Muslim government:* it is a term which was applied to *a yearly land-tax which 'Omar imposed upon the people of the Sawád* [*of El-'Irák*]: then, to the *land-tax which the people of a land taken by convention agreed to pay;* and their lands were termed خَرَاجِيَّةٌ: accord. to Bḍ, it is a name for the *proceeds of land:* and has then been used to signify the *profits arising from possessions;* such as the *revenue derived from the increase of lands,* and *from slaves* and *animals:* accord. to Er-Ráfi'ee, its primary signification is *an impost which the master requires to be paid him by his slave:* accord. to Zj, ▼ خَرْجٌ is an [obsolete] inf. n.: and خَرَاجٌ, a name for *that which comes forth:* and he also explains the latter word by فَىْءٌ: and ▼ خَرْجٌ, by ضَرِيبَةٌ and جِزْيَةٌ: (TA:) the pl. (of خَرَاجٌ, L, TA) is أَخْرَاجٌ and أَخَارِيجُ [a pl. pl.] and أَخْرِجَةٌ. (Ṣ, Ḳ.) اَلْخَرَاجُ بِالضَّمَانِ, a saying ascribed to Moḥammad, (Ḳ, TA,) occurring in a trad. of 'Áïsheh, of disputed authority, but affirmed by several authors to be genuine, means, accord. to most of the lawyers, (TA,) *The revenue derived from the slave is the property of the purchaser because of the responsibility which he has borne for him:* (A,* Mgh,* Ḳ, TA:) for one purchases a slave, and imposes upon him the task of producing a revenue for a time, and then may discover in him a fault which the seller had concealed; wherefore he has a right to return him and to receive back the price; but the revenue which he had required the slave to produce is his lawful property, because he had been responsible for him; and if he had perished, part of his property had perished: (Ḳ,* TA:) in a similar manner IAth explains it, as relating to a male or female slave or to other property. (TA.) — ▼ خَرْجٌ and خَرَاجٌ as used in the Ḳur xxiii. 74 mean *A recompense,* or *reward.* (Fr.) Some, for ▼ خَرْجًا, in this instance, read خَرَاجًا. (TA.) — And خَرَاجٌ is also used as meaning ‡*The taste* of fruit; this being likened to the خراج of lands &c. (TA, from a trad.) — See also خَرِيجٌ, in five places.

خُرَاجٌ *Pimples,* or *small swellings* or *pustules:* [a coll. gen. n.:] n. un. with ة: (Mgh, Mṣb:) or [the *kind of pustule termed*] دُمَّلٌ, and *the like, that come forth upon the body:* (Mgh:) or *purulent pustules,* or *imposthumes,* (Ṣ, Ḳ,) *that come forth upon the body:* (Ṣ:) or *a spontaneous swelling that comes forth upon the body:* or *an ulcerous swelling that comes forth upon a beast of the equine kind and upon other animals:* pl. [of pauc.] أَخْرِجَةٌ and [of mult.] خِرْجَانٌ. (TA.) = See also خَرَاجٌ.

خَرُوجٌ: see خَارِجٌ, and خُرَجَةٌ. — Also A horse *that outstrips in the race.* (TA.) — And ‡A horse *having a neck so long that, by reason of its length, he plucks away at unawares* (يَخْتَذِلُ) *every bridle that is attached to his bit:* (A,* L, Ḳ:*) and in like manner, without ة, a mare. (TA.) — And A she-camel *that lies down apart from the* [*other*] *camels:* (Ḳ:) and one *excellent in the pace termed* عَنَقٌ, *that goes before others:* (TA:) pl. خُرُجٌ, (Ḳ, TA,) [in the CḲ خُرَّجٌ, but it is] with two ḍammehs. (TA.)

خُرُوجٌ an inf. n. of 1. (Ṣ, Mṣb, Ḳ.) — See also خَرْجٌ.

خَرِيجٌ (Ṣ, Ḳ) and ▼ خَرَاجِ and ▼ تَخْرِيجٌ (TA) *A certain game,* (Ṣ, Ḳ, TA,) *played by the Arab youths,* (TA,) *in which they say* ▼ خَرَاجِ خَرَاجِ: (Ṣ, Ḳ, TA:) accord. to ISk, you say, لَعِبَ الصِّبْيَانُ ▼ خَرَاجِ [*The boys played at* خراج], with kesr to the ج: Fr says, خراج is the name of *a well-known game of the Arabs, in which one of the players holds a thing in his hand and says to the others, Elicit ye* (أَخْرِجُوا) *what is in my hand:* in the T, ▼ خراج and خريج are explained by the word مُخَارَجَةٌ [meaning *micare digitis;* and hence it appears that the game thus termed, accord. to the T, is the *morra,* a game common in ancient and modern Italy, and in very remote times in Egypt, *in which one of the players puts forth some, or all, of his fingers, and another is required to name instantly the number put forth,* or *to do the same*]; and it is there added, that it is *A game of the Arab youths:* Aboo-Dhu-eyb El-Hudhalee says,

أَرِقْتُ لَهُ ذَاتَ العِشَآءِ كَأَنَّهُ
مَخَارِيقُ يُدْعَى تَحْتَهُنَّ خَرِيجُ

I was sleepless in consequence of it, (referring to lightning,) *at nightfall, as though it were kerchiefs twisted for the purpose of beating with them, under which was uttered the cry* خريج; likening the thunder to the cry of the players: but Aboo-'Alee says that خريج [thus used] is incorrect; that he should have said ▼ خَرَاجِ, but that the rhyme required him to say خريج. (TA.)

بِلَادٌ خَرَاجِيَّةٌ *Countries subject to a* [خَرَاج, or] *tax upon their lands.* (MF.)

خَرَّاجٌ: see خَارِجٌ, and خُرَجَةٌ.

خِرِّيجٌ has the meaning of a pass. part. n.: (Ṣ, Ḳ:) you say, هُوَ خِرِّيجُ فُلَانٍ ‡ *He is,* or *has been, well educated* or *disciplined* or *trained by such a one* (Ṣ, A,* Ḳ*) in polite accomplishments, (Ṣ, Ḳ,) or in science and art. (A.)

خَارِجٌ and [in an intensive sense] ▼ خُرُوجٌ and [in an intensive or a frequentative sense] ▼ خَرَّاجٌ *Going, coming, passing,* or *getting, out,* or *forth; issuing, emanating, proceeding,* or *departing:* [the second signifying *doing so much:* and the third, *doing so much* or *frequently.*] (TA.) — [*External; extrinsic; foreign:*] the *exterior,* or *outside,* of anything. (TA.) You say, كُنْتُ خَارِجَ الدَّارِ [*I was outside the house*]: (A:) [or,] accord. to Sb, خارج is not used adverbially unless with the particle [فِى]. (TA.) — [Hence, الخَارِجُ as meaning † *What is external,* or *extrinsic, to the mind; what is objective; reality.* (See also خَارِجِىٌّ.) And فِى الخَارِجِ † *In what is external,* or *extrinsic, to the mind;* &c.].

خَارِجَةٌ [fem. of خَارِجٌ: and sing. of خَوَارِجُ used as a subst.]. — الخَوَارِجُ in the phrase الدَّوَاخِلُ وَالخَوَارِجُ means *The arches,* or *vaults, and niches, in the inner side of a wall;* الدواخل meaning the figured forms, and inscriptions, upon a wall, executed with gypsum or otherwise: or الدواخل والخوارج means *the ornamental* [*depressed and*] *projecting forms of a building, differing from the forms adjacent thereto.* (Mṣb, from a saying of Esh-Sháfi'ee.) — خَوَارِجُ المَالِ † *The mare and the female slave and the she-ass.* (Ḳ.) — خَرَجَتْ خَوَارِجُهُ ‡ *His generosity became apparent, and he applied himself to the sound management of affairs,* (Ḳ,* TA,) *and became intelligent like others of his class, after his youth,* or *ignorant and youthful conduct.* (TA.)

خَارِجِىٌّ One *who makes himself a lord,* or *chief,* (Ṣ, Ḳ, TA,) *and goes forth* [*from his party,* or *fellows*], *and becomes elevated,* or *exalted,* (TA,) *without his having noble ancestry:* (Ṣ, Ḳ, TA:) and it is also said to signify anything *that surpasses, or excels its kind and fellows:* (TA:) accord. to Abu-l-'Alà, in ancient times, before El-Islám, it was applied to a *courageous,* or *generous,* man, *the son of a coward* or *niggard,* and *the like:* — and in like manner, to a *A fleet,* or *swift,* horse; or one *excellent in running;* or *that outstrips others; not the offspring of a sire and dam possessing the like qualities:* [and in the TA, the coll. gen. n. خَارِجِيَّةٌ is explained as applied to such horses:] — then, in the times of El-Islám, it was applied to *A rebel:* and *a heretic.* (Ḥam p. 188.) [The pl.] الخَوَارِجُ is the appellation of *A party* [*of heretics,* or *schismatics,*] *of those following erroneous opinions, having a singular,* or *particular, persuasion:* (Ḳ:) *they are* [*said by some to be*] *the* حَرُورِيَّة [q. v.]; *and the* خَارِجِيَّة *are* [*said to be*] *a sect of them; and they consist of seven sects:* (TA:) they were so called because they went forth from, (as in one copy of the Ḳ,) or against, (as in other copies,) the rest of the people; (Ḳ, TA;) or from the religion, or from the truth, or from 'Alee after [the battle of] Ṣiffeen. (TA.) — [Also † *Relating to what is external,* or *extrinsic, to the mind; objective; real.* Hence, الأُمُورُ الخَارِجِيَّةُ † *The things that are external,* or *extrinsic, to the mind; the things that are considered objectively; real things;* opposed to الأُمُورُ الذِّهْنِيَّةُ. (See also خَارِجٌ.)]

[خَارِجِيَّةٌ fem. of خَارِجِىٌّ: — and also a coll. gen. n., of which the n. un. is خَارِجِىٌّ.]

خَارُوجٌ *A certain sort of palm-trees,* (L, Ḳ,*) *well known.* (Ḳ.)

[خَوَارِجُ pl. of خَارِجَةٌ: — and also of خَارِجِىٌّ as an epithet applied to a man &c., not as a rel. n.]

أَخْرَجُ A ram, (Ṣ, Ḳ,) and (so in the Ṣ, but in the Ḳ "or") a male ostrich, (AA, Ṣ, A, Ḳ,) *of two colours, white and black:* (Ṣ, A,* Ḳ:) or a male ostrich *of a colour in which black predominates over white, like the colour of ashes:* and in this sense also applied to a mountain: (Lth, TA:) and a goat *half white and half black:* and a horse *of which the belly, and the sides as far as the back, but not the back itself, are white, and the rest of any colour:* (TA:) fem. خَرْجَآءُ: (A, TA:) which is applied to a female ostrich: (A:) and to a ewe or she-goat *having white hind legs and flanks:* (AZ, Ṣ:) or a ewe *that is black, with one hind leg, or both hind legs, and the flanks, white; the rest being black:* (TA:) or a ewe *white in the hinder part, half of her being white, and the other half of any colour:* (T, TA:) and a small isolated mountain (قَارَةٌ) *of two colours,* (A, TA,) *white and black:* (A:) pl. خُرْجٌ. (Ḳ.) Also ‡ *A garment white and red;* rendered so by being besmeared with blood. (TA.) El-'Ajjáj says,

إِنَّا إِذَا مُذْكِى الحُرُوبِ أَرَّجَا
وَلَبِسَتْ لِلْمَوْتِ ثَوْبًا أَخْرَجَا

(so in the TA: in the Ṣ, جُلًّا اخرجا:) meaning ‡ [*Verily we, when the inflamer of wars excites them, and*] *they* (the wars) *have put on, for death, a garment white and red, rendered so by being besmeared with blood:* i. e., have been rendered notable like a thing that is black and white. (Ṣ, TA.) — الأَخْرَجُ *The* [*bird called*] مُكَّآء; (Ḳ;) because of its colour. (TA.) — أَرْضٌ خَرْجَآءُ (TA) and ▼ مُخَرَّجَةٌ (Sh, Ṣ, Ḳ) and ▼ فِيهَا تَخْرِيجٌ (TA) ‡ *Land having plants,* or *herbage, in one place and not in another:* (Ṣ, Ḳ, TA:) *that has been rained upon, and has produced herbs, in some parts and not in others:* (Sh:) or the second means *land upon which rain has not fallen.* (L in art. صح.) — عَامٌ أَخْرَجُ (TA) and ▼ مُخَرَّجٌ (A, TA) and ▼ فِيهِ تَخْرِيجٌ (Ṣ, A, Ḳ) and ذُو تَخْرِيجٍ (Ḳ) ‡ *A year of fruitfulness,* or *of abundant herbage, and of sterility:* (Ṣ, A, Ḳ, TA:) or *half fruitful,* or *abundant in herbage, and half sterile.* (TA.)

مَخْرَجٌ an inf. n. of 1. (Ṣ, Mṣb, Ḳ.) — Also *A place of* خُرُوج [i. e. of *going, coming, passing,* or *getting, out,* or *forth; a place of egress,* or *exit; an outlet*]: (Ṣ, Ḳ, TA:) pl. مَخَارِجُ. (TA.) You say, وَجَدْتُ فِى الأَمْرِ مَخْرَجًا † *I found, in the affair,* or *case, a place* [or *way*] *of escape, evasion,* or *safety.* (Mṣb.) And فُلَانٌ يَعْرِفُ مَوَالِجَ الأُمُورِ وَمَخَارِجَهَا ‡ *Such a one knows the ways of entering into affairs and those of withdrawing himself out of them.* (A, TA.) — [Hence, *A privy:* used in this sense in the Ṣ and Ḳ in art. حش, &c. — And The *anus:* used in this sense in the Mṣb in art. حقن.] — Also *A time of* خُرُوج [i. e. of *going,* &c., *out,* or *forth; of egress,* or *exit*]. (TA.) — فُلَانٌ حَسَنُ المَدْخَلِ وَالمَخْرَجِ means † *Such a one is good, and laudable, in his way of acting,* or *conduct.* (TA in art. دخل.)

مُخْرَجٌ an inf. n. of the trans. v. أَخْرَجَ. (Ṣ, Ḳ.) [So accord. to some in a phrase in the Ḳur xvii. 82, respecting which see 4.] — Also pass. part. n. of the same. (Ṣ, Ḳ.) — And n. of place of the same. (Ṣ, Ḳ.) — And n. of time of the same. (Ṣ.)

مُخَرَّجٌ; and its fem., with ة: see أَخْرَجُ.

يَوْمٌ مَخْرُوجٌ occurs in poetry for يَوْمٌ مَخْرُوجٌ فِيهِ [*A day in which one goes forth;* or *in which people go forth*]. (TA.)

عَبْدٌ مُخَارِجٌ: see 3, last sentence.

نَاقَةٌ مُخْتَرِجَةٌ ‡ *A she-camel formed like the he-camel:* (Ṣ, A, Ḳ, TA:) or *like the male* بُخْتِىّ *camel.* (TA.) See 1.

خرد

1. خَرِدَتْ, aor. ـَ, (L, K,) inf. n. خَرَدٌ, (L,) *She was an untouched virgin;* (L, Ḳ;) as also ▼ اخرودت, inf. n. إِخْرِوَادٌ; (L;) and ▼ تخرّدت: (L, Ḳ:) or *she was bashful and grave* or *staid* or *sedate,* or *very bashful, long silent, low in voice,*

one who concealed herself from public view, (L, Ḳ,) and did not remain long in her parents' house, or tent, unmarried, after having attained the usual proper age for marriage. (L.) —— And خَرِدَ He was, or became, affected with shame, bashfulness, or pudency; (IAạr;) as also ↓ اخرد. (Ḳ.) —— He was, or became, abject. (IAạr.) —— Also, inf. n. as above; and ↓ اخرد; He kept long silence: (L, Ḳ:) and the latter signifies also he kept silence by reason of abjectness; not by reason of bashfulness: so accord. to the Ḳ: but accord. to the A, he kept silence by reason of bashfulness; and اقرد signifies "he kept silence by reason of abjectness:" (TA:) and so says IAạr: (TA in art. قرد:) or اخرد accord. to IAạr signifies he spoke little. (Ḥar p. 250.) [See also خَارِدٌ.]

4: see 1, in three places. —— اخرد إِلَى اللَّهْوِ He inclined to play, sport, or diversion. (Ḳ.)

5: see 1.

خَرُودٌ: see خَرِيدَةٌ, in two places.

خَرِيدٌ: see the next paragraph. —— صَوْتٌ خَرِيدٌ A gentle voice, characterized by bashfulness, or modesty. (IAạr, Ḳ.)

خَرِيدَةٌ A virgin: (IAạr, Ṣ, A:) and a bashful, or modest, woman: and sometimes they said ↓ جَارِيَةٌ خَرُودٌ, meaning a girl bashful and grave or staid or sedate; or very bashful: (Ṣ:) or خَرِيدَةٌ and ↓ خَرِيدٌ and ↓ خَرُودٌ signify an untouched virgin: or a female bashful and grave or staid or sedate, or very bashful, long silent, low in voice, who conceals herself from public view, (L, Ḳ,) and does not remain long in her parents' house, or tent, unmarried, after having attained the usual proper age for marriage: (L:) pl. خَرَائِدُ and خُرُدٌ (Ṣ, A, L, Ḳ) and خُرَّدٌ; (Ṣ, L, Ḳ;) the second of which is contr. to rule. (L.) —— Also ‡ An unbored pearl. (Lth, IAạr, Ṣ, A, Ḳ.)

خَارِدٌ Silent by reason of bashfulness; not by reason of abjectness: and ↓ مُخْرِدٌ silent by reason of abjectness; not by reason of bashfulness: so accord. to AA: and the latter, simply, silent. (L.) [See also 1.]

مُخْرِدٌ: see the next preceding paragraph.

خردل

خَرْدَلٌ [Mustard-seed;] the grain of a certain tree, (Ḳ,) well known; (Ṣ, Ḳ;) a species of حُرْف [q. v.]; (JK;) heating; emollient; drawing; a phlegmagogue; lenitive; digestive; used as a liniment, good for the نِقْرِس [or gout], and [especially] the نَسَا [or sciatica], and the [malignant species of leprosy termed] بَرَص, (Ḳ,) and the [mild species thereof termed] بَهَق; clearing to the face; good for the alopecia, especially the wild sort thereof; (TA;) its smoke drives away serpents, or, as in the Ḳánoon, venomous or noxious reptiles or the like; (TA;) its juice, dropped, allays earache, (Ḳ,) and in like manner its oil; (TA;) and its powder, upon the aching tooth, is extremely efficacious, (Ḳ,) especially when حِلْتِيت [or assa] has been cooked with it: (TA: [in which many other properties assigned to it are mentioned:]) n. un. with ة. (Ṣ.) —— الخَرْدَلُ الفَارِسِىُّ is A certain plant in Egypt known by the name of حَشِيشَةُ السُّلْطَانِ. (Ḳ.)

خرز

1. خَرَزَ, aor. ـِ and ـُ, inf. n. خَرْزٌ, He sewed (Mṣb, Ḳ, TA) a skin, or hide, (Mṣb, TA,) or a boot, &c. (Ṣ, A, Ḳ.) You say, كَلَامُ فُلَانٍ كَخَرْزِ الإِمَآءِ [The language of such a one is like the female slaves' sewing of skins]; i. e., [its ornaments, lit.] its pearls, and its cowries, are far apart. (A, TA.)

خَرَزٌ [a coll. gen. n.,] a word of well-known meaning, (Mṣb,) [i. e., Beads;] what are strung: (Ṣ:) a thing that is hung [or rather things that are hung] upon the neck, made of coloured stone, red and green: (Ḥar p. 431:) or gems, or similar stones, both good and bad: (JK:) [also factitious gems, and the like: (see فُسَيْفِسَآءُ as explained in the Ḳ &c.:)] n. un. خَرَزَةٌ: (Ṣ, Mṣb:) the latter signifying [a single bead;] what is strung: (Ḳ:) and also, (i. e. the latter,) a gem, or precious stone, (Ḳ, TA,) such [for instance] as is set in a ring, whether good or bad: (TA:) pl. of the latter, خَرَزَاتٌ. (TA.) Hence, خَرَزَاتُ المَلِكِ, (Ṣ, Ḳ,) and المُلْكِ, (Ṣ, A,) The gems of the king's crown: when the king had reigned a year, a خَرَزَة was added to his crown, in order that the number of the years of his reign might be known: (Ṣ, Ḳ:) such is said to have been the case. (Ṣ.) You say, أُوتِىَ خَرَزَاتِ المُلْكِ سِتِّينَ حِجَّةً ‡ [meaning He reigned sixty years: lit., he received the gems of the crown sixty years]. (A.) —— خَرَزَةُ العَيْنِ signifies The حَدَقَة [or lens] of the eye. (TA in art. حدق.) —— And خَرَزٌ is also applied to The small shells called وَدْع. (Ṣ* and Ḳ* and TA in art. ودع.) —— It also signifies ‡ The vertebræ of the back, (Ṣ, A, TA,) and of the neck: each one is called خَرَزَةٌ: which latter is also explained as meaning ‡ what is between two vertebræ. (TA.)

خَرْزَةٌ, with fet-ḥ, A single puncture [or stitch-hole, made in sewing a skin or a boot; and so خُرْزَةٌ]; syn. خُرْزَةٌ. (TA.)

خُرْزَةٌ i. q. كُتْبَةٌ; (Ṣ, Ḳ;) A seam, or suture, in a skin, or hide, (KL, PṢ, TḲ,*) or in a boot, &c.; (PṢ;) [app. made by sewing together two edges so that one laps over the other: and app. also a single stitch in such a seam;] what is between two punctures; i. e., every puncture with its thread: (TA:) also, a puncture, or stitch-hole, in a skin [&c.]: (TA voce وِذَابٌ: [its pl. being there said to be syn. with خُرُوبٌ, pl. of خُرْبَةٌ: and this last meaning, (for evidences of the correctness of which see أُثْرٌ and أُتُومٌ and خُصْفَةٌ &c.,) common to it and to خَرْزَةٌ, it perhaps bears in exs. here following:]) and any round perforation: (JM:) pl. خُرَزٌ (Ṣ, Ḳ) [and app. خُرُوزٌ, which see below]. اِجْمَعْ سَيْرَيْنِ فِى خُرْزَةٍ [lit., Conjoin thou two thongs in a single seam, or stitch, or puncture,] is a prov., meaning ‡ accomplish thou two wants at once. (TA.) And you say to him who seeks to attain two wants together, سَيْرَيْنِ فِى خُرْزَةٍ [lit., Wilt thou conjoin two thongs in a single seam, or stitch, or puncture? the first word being in the accus. case because أَتَجْمَعُ is understood]. (A, TA.) —— Also † The foramen podicis: and † the foramen vaginæ. (TA voce خُرْبَةٌ.)

خَرَزَةٌ; pl. خَرَزَاتٌ: n. un. of خَرَزٌ [q. v.]. (Ṣ, Mṣb, TA.)

خُرُوزُ السَّفِينَةِ [The seams of the ship: خُرُوز being app. a pl. of خُرْزَة, like as خُرُوبٌ is a pl. of خُرْبَةٌ]. (Ḳ voce جُمَّةٌ q. v.)

خِرَازَةٌ The art, or occupation, of sewing [skins, or hides, or] boots, &c. (A, Ḳ.)

خَرَّازٌ A sewer of [skins, or hides, or] boots, &c. (Ṣ, A.)

مِخْرَزٌ [and vulg. ↓ مِخْرَازٌ] The instrument [i. e. the needle, or awl,] with which one sews [skins, or hides, or] boots, &c. (Ṣ, Ḳ.)

مُخَرَّزٌ Any bird, (A, Ḳ,) as a pigeon &c., (TA,) having upon its wings marks resembling خَرَز [or beads]. (A, Ḳ, TA.)

مِخْرَازٌ: see مِخْرَزٌ.

خرس

1. خَرِسَ, (Mṣb, Ḳ,) aor. ـَ, (Ḳ,) inf. n. خَرَسٌ, (Ṣ, A, Mṣb, Ḳ,) He (a man) was dumb; was naturally, by conformation, prevented from speaking; (Mṣb;) [he was destitute of the faculty of speech, by natural conformation, like the beast: see أَخْرَسُ:] or he was, or became, tonguetied, or withheld from speech, (A, Ḳ,) either from inability to find words to express what he would say, or by natural conformation [of the organs of speech]. (TA.) You say also خَرِسَ المَجْلِسُ The assembly was, or became, mute, or speechless. (A.) ═ خَرَسَ المَرْأَةَ, (Lḥ, IAth,) aor. ـُ, (Lḥ,) [inf. n., app., خَرْسٌ,] He fed the woman with what is termed خُرْسَة; (Lḥ, IAth;) he fed her on the occasion of child-birth; (Lḥ;) as also ↓ خرّس عَلَيْهَا, inf. n. تَخْرِيسٌ (Ṣ, Ḳ) and تَخْرِسَةٌ. (TA.) In like manner you say, ↓ خرّسها, inf. n. تَخْرِيسٌ; and خرس عَنْهَا; [so in the TA, without any syll. signs to the verb;] He made for her what is termed خُرْسَة. (TA.) And ↓ خُرِّسَتْ, (Ṣ, L,) or خُرِسَتْ, (so in a copy of the A,) She was fed with what is so termed: (A:) or a feast on the occasion of her having given birth to a child was made for her. (Ṣ, L.) ═ خَرِسَ, aor. ـَ, He drank from the [kind of wine-jar called] خُرْس, (Ṣgh, Ḳ,) i. e. the دَنّ. (TA.)

2: see خَرِسَ, in three places.

4. اخرسهُ اللّٰهُ [God made him to be dumb: see خَرِسَ]: (Ṣ:) God made him to be tonguetied, or speechless, (A, Ḳ,) either from inability to find words to express what he would say, or by natural conformation [of the organs of speech]. (TA.)

5. تخرّست She made for herself the food for the occasion of child-birth, (A,* Ḳ,) i. e., what is

called خُرْسَة. (TA.) Hence the prov. تَخَرَّسِى يَا نَفْسُ لَا مُخَرِّسَةَ لَكِ (A,* TA) *Make thou the food for child-birth for thyself, O self: there is no maker of it for thee:* said by a woman who had given birth to a child and had not any one to care for her: alluding to a man's taking care for himself: (Ḳ, TA:) and also related [in the A] without the words يا نفس. (TA.)

6. تخارس [*He feigned himself dumb,* or *speechless,*] is from خَرِسَ المَجْلِسُ: hence إِذَا شَهِدْتَ مَنْ لَا يَفْهَمُ عَنْكَ فَتَخَارَسْ [*When thou art present with,* or *beholdest, him who will not understand what thou sayest, then feign thyself dumb,* or *speechless*]. (A.)

خُرْسٌ *A* [*wine-jar such as is called*] دَنّ; (JK, Ṣ, Ḳ;) as also ↓خِرْسٌ, (Kr, Ḳ,) and خِرْصٌ: (TA:) pl. [of pauc.] أَخْرَاسٌ (JK) and [of mult.] خُرُوسٌ (Ḳ.)

خُرْسٌ *Food that is prepared on the occasion of the birth of a child;* (Ṣ, A, Mṣb, Ḳ;) as also ↓خُرَاسٌ. (Lḥ, TA.) IJ seems to assign to it also the sense of خُرْسَة. (TA.) — Hence, as also ↓ the second word, *A feast,* or *banquet, that is prepared on that occasion, and to which people are invited:* so in a trad. of Ḥassán, كَانَ إِذَا دُعِىَ إِلَى طَعَامٍ قَالَ إِلَى عُرْسٍ أَمْ خُرْسٍ أَمْ إِعْذَارٍ [*He used, when he was invited to food, to say, To a wedding-feast, or a feast for child-birth, or a circumcision-feast?*]; and if it were for one of these, he consented: but if not, he did not consent. (TA.)

خِرْسٌ: see خُرْسٌ.

خُرْسَةٌ *Food for a woman who has given birth to a child;* (Ṣ, A, Ḳ;) *what is made for her, such as* فَرِيقَة *and the like;* as also [خُرْصَة,] with ص. (TA.) خُرْسَةُ مَرْيَمَ [*The child-birth-food of Maryam*], applied to dates, and occurring in a trad., alludes to verse 25 of chap. xix. of the Ḳur, وَهُزِّى إِلَيْكِ الخ: and Khálid Ibn-Ṣafwán uses in the same manner the phrase ↓ تَخْرِسَةُ مَرْيَمَ; in which تخرسة is an inf. n. used as a subst.; or it may be a subst., like تَوْدِيَة. (TA.)

خَرُوسٌ A she-camel *that does not utter the cry termed* رُغَاء. (Ibn-'Abbád, Ṣgh, Ḳ.)

خُرَاسٌ: see خُرْسٌ, in two places.

خَرُوسٌ *A* بِكْر [or *female that has not yet brought forth*] *in the first period of her pregnancy:* and (some say, Ṣ) one *for whom* خُرْسَة *is made:* (Ṣ, Ḳ:) and one *having a scanty flow of milk.* (Ṣgh, Ḳ.)

خَرَّاسٌ *A maker* (Ṣ, TA) and *seller* (Ḳ, TA) *of the* [*kind of wine-jar called*] خُرْس. (Ṣ, Ḳ, TA.) — And *A vintner.* (JK, TA.)

أَخْرَسُ (Ṣ, &c.) *Dumb; prevented from speaking by natural conformation;* (Mṣb;) *speechless, or destitute of the faculty of speech, by natural conformation,* (T and Mṣb in art. بكم,) *like the beast that lacks the faculty of articulation;* (T ibid;) differing from أَبْكَم, q. v.: (T and Mṣb ibid:) or *tonguetied,* or *speechless,* (Ḳ, TA,) either *from inability to find words to express what he would say, or by natural conformation* [*of the organs of speech*]: (TA:) fem. خَرْسَاءُ: (Mṣb:) pl. خُرْسٌ (Mṣb, Ḳ) and خُرْسَانٌ. (Ḳ.) — جَمَلٌ أَخْرَسُ *A he-camel that has no perforation to his* شِقْشِقَة, *for his bray to issue therefrom, so that he reiterates it therein:* such a one they like to send among the [she-camels in the state in which they are termed] شَوْل, because in most instances he begets females: and نَاقَةٌ خَرْسَاءُ *a she-camel that is not heard to utter the cry termed* رُغَاء. (TA.) — عَلَمٌ أَخْرَسُ ‡ *A sign of the way,* (Ḳ,) [or *a mountain,*] *from which,* (A,) or *at,* or *in, which,* (فيه, Ḳ,) *no echo is heard:* (A, Ḳ:) or *a sign of the way in the mountain whereof no echo is heard:* (Lth, T:) or [it seems to be applied to *a mountain where no echo is heard;* for] it is said when no sound of an echo is heard in the mountain. (Ṣ.) — سَحَابَةٌ خَرْسَاءُ ‡ *A cloud that does not thunder:* (A:) or *a cloud in which is no thunder nor lightning,* (Ṣ, Ḳ, TA,) *and of which no sound is heard;* which is mostly in winter. (TA.) — عَيْنٌ خَرْسَاءُ † *A spring of the running whereof no sound is heard.* (TA.) — صَخْرَةٌ خَرْسَاءُ † *A hard and solid rock:* (Akh, TA:) and عِظَامٌ خُرْسٌ †*hard and solid bones.* (Th, TA.) — كَتِيبَةٌ خَرْسَاءُ ‡ *An army,* or *a portion thereof, without any clamour or confused noise:* (A:) or *whereof no sound is heard, by reason of their staidness in war:* (Ṣ, Ḳ:) or *that is silent, by reason of the multitude of the coats of mail, without any clashing of arms.* (A 'Obeyd, Ṣ, Ḳ.) — لَبَنٌ أَخْرَسُ ‡ *Thick milk,* (Ṣ, A, Ḳ,) *that makes no sound in the vessel,* (Ṣ, Ḳ,) or *that does not shake about in its vessel:* (A:) or *thick milk, of which no sound is heard when it is poured out:* (Az, TA:) and شَرْبَةٌ خَرْسَاءُ † *a thick draught of milk.* (M, TA.) — وَلَّانِى عِرْضًا أَخْرَسَ [or عُرْضًا?] أَمْرَسَ † *He turned from me, and would not speak to me.* (Fr, TA.) — خَرْسَاءُ † *A viper:* (A:) pl. خُرْسٌ. (TA.) — Hence, (A,) رَمَاهُ بِخَرْسَاءَ ‡ *He smote him with a calamity.* (A, Ḳ.*)

تَخْرِسَةُ مَرْيَمَ: see خُرْسَة.

خرش

1. خَرَشَهُ, (Ṣ, A, Ḳ,) aor. ـِ, (Ṣ, Ḳ,) inf. n. خَرْشٌ, (Lth,) *He,* or *it,* (namely, a cat, A,) *scratched him,* or *it,* (namely, the skin, A,) *with the nails,* (Lth, Ṣ, A,* Ḳ,) in any part of the body, (Lth,) *so as to cause bleeding or not;* or *wounded him in the outer skin, so as to make it bleed or not;* or *lacerated it* (namely, the skin) *little* or *much;* or *tore off its surface with a stick or the like;* syn. خَدَشَهُ; (Ṣ, Ḳ;) [as also خَرَّشَهُ;] and so ↓ احترشهُ: (Ṣ, Ḳ:) and ↓ احترشهُ بِظُفْرِهِ *He scratched him, &c., with his nail.* (TA.) You say also, خَرَشَهُ بِحَدِيدَةٍ وَنَحْوِهَا [*He scratched it with an iron thing, and the like*]. (Ḳ.) — *It* (a fly) *bit him;* (A, TA;) as also خَرَّشَهُ. (TA.) — خَرَشَ البَعِيرَ, (Ṣ, Ḳ,) aor. and inf. n. as above, (TA,) *He drew the camel* (Ṣ, Ḳ) *towards him* (Ṣ) *with the* مِخْرَاش, i. e., the مِحْجَن: (Ṣ, Ḳ:) or *he struck the camel and then drew him towards him therewith; desiring thereby to excite him to quicken his pace:* and *he struck him with its extremity in the side of his neck,* or [*elsewhere*] *in his skin, so as to scrape off from him his fur.* (TA.) [See also خَرَّشَهُ.] — خَرَشَ لِعِيَالِهِ, aor. as above, (Ṣ, Ḳ,) and so the inf. n., (TA,) † *He earned for his family,* or *household;* (Ṣ, Ḳ;) *collected, and practised artifice, for them;* (TA;) and *sought sustenance for them;* (Ṣ, Ḳ;) as also ↓ احترش. (Ḳ.) — You say also, فُلَانٌ يَخْرِشُ مِنْ فُلَانٍ الشَّىْءَ بَعْدَ الشَّىْءِ ‡ *Such a one takes from such a one thing after thing;* as also ↓ يَحْتَرِشُهُ. (A.) And مَا خَرَشَ شَيْئًا ‡ *He has not taken anything.* (TA.) And فُلَانٌ يَتَخَرَّشُ ↓ مِنْ فُلَانٍ الشَّىْءَ ‡ *Such a one takes and acquires from such a one the thing.* (TA.) And ↓ مُخَارَشَة signifies † The *taking against one's will.* (TA.)

2. خرّشهُ, inf. n. تَخْرِيشٌ, [*He scratched him,* or *it, with the nails, vehemently,* or *much;* like خدّشهُ.] (TA.) — خرّش الزَّرْعُ, inf. n. as above, † *The seed-produce put forth the first extremity of the ear.* (Ibn-'Abbád, Ṣgh, Ḳ.)

3. خارشهُ, inf. n. مُخَارَشَةٌ and خِرَاشٌ, [*He scratched him with the nails, being scratched by him in like manner;* like خادشهُ.] (TA.) — [Hence,] كَلْبٌ خِرَاشٌ [*An irritable,* or *a quarrelsome, dog*]; like كَلْبٌ هِرَاشٌ: (Ṣ, Ḳ:) IF holds that the خ is a substitute for ه. (TA.) — See also 1, last signification.

5: see 1, last sentence but one.

6. تخارشت الكِلَابُ, (A, Ḳ,) and السَّنَانِيرُ, (A,) *The dogs,* and *the cats,* [*scratched,* or] *fought and assailed,* (تهارشت, Ḳ, TA,) *and lacerated,* or *tore, one another.* (TA.)

8: see 1, in four places.

خُرُوشٌ *Gain;* or *earning:* pl. خُرُوش. (TA.)

خِرْشَاءُ The *skin,* (Ṣ, Ḳ,) or *slough,* (A, TA,) *of a serpent.* (Ṣ, A, TA.) To this one likens a thin, transparent, shirt. (A, TA.) AḤn applies its pl., خَرَاشِىّ, metaphorically, to [the skins of] all [animals of the class termed] حَشَرَات. (TA.) — The *upper covering,* (Ṣ, Ḳ, TA,) *which is dry,* or *tough,* or *hard,* [app. meaning the *shell,*] (TA,) *of an egg,* (Ṣ, Ḳ, TA,) *after it has been broken, and what was in it,* (Ṣ, TA,) *of moisture,* (TA,) *has been extracted:* (Ṣ, TA:) or the *inner skin,* (T, TA,) or *inner covering,* (A,) *of an egg;* (T, A, TA;) also called غِرْقِئ: (TA:) pl. as above. (TA.) — To this is likened † *Anything in which are an inflation and a bursting and holes.* (AZ,* Ṣ.) — The *thin skin upon the surface of milk:* (Ḳ:) or the *bubbles that rise upon the surface of milk:* (A:) *froth,* or *foam.* (Ṣ, TA. [See an ex. in a verse cited voce قَصْر.]) — ‡ *Phlegm:* (Ṣ, A, Ḳ:) or *viscous phlegm in the chest:* (TA:) and *what is termed* نُخَامَة, q. v. (A, TA.) You say, أَلْقَى خَرَاشِىَّ صَدْرِهِ ‡ *He cast forth the phlegm* (lit. *phlegms*) *of his chest.* (Ṣ.) [See also a tropical meaning below.] And أَلْقَى مِنْ صَدْرِهِ خَرَاشِىَّ ‡ *He cast forth from his chest what is termed* نُخَامَة (Az, A, TA) *and phlegm:*

(A:) or *thick spittle.* (Ḳ.) — خُرْشَةُ العَسَلِ The *wax of honey, and the dead bees in it:* (TA:) or the *young bees,* or the *wings, that are upon honey.* (M and L in art. جحش.) — طَلَعَتِ الشَّمْسُ فِى خُرْشَاءَ ‡ *The sun rose in dust;* (Ṣ, A;) syn. غُبْرَة. (Ṣ, A, Ḳ.) — أَلْقَى فُلَانٌ خُرَاشَى صَدْرِهِ ‡ *Such a one cast forth what he concealed of rancours and enmities, and of grief.* (A, TA.)

خُرَاشَةٌ *What falls from a thing when it is scratched with an iron instrument or the like.* (Ḳ.)

مِخْرَشٌ: see what next follows.

مِخْرَاشٌ A مِحْجَن [or *stick of which the end is naturally curved,* or *crooked*]; (Ṣ, Ḳ;) [see 1;] sometimes written [مِحْرَش,] with ح: (Ṣ:) *a stick with a curved,* or *crooked, head, like the* صَوْلَجَان; as also ↓ مِخْرَشٌ. (TA.)

خرص

1. خَرَصَ النَّخْلَ, (Ṣ, A, Mgh, Mṣb,) aor. ـُ (A, Mṣb) [and ـِ, as appears from what follows,] inf. n. خَرْصٌ, (Ṣ, A, Mgh, Mṣb,) [and خِرْصٌ is also allowable, (see what follows,)] *He computed by conjecture the quantity of the fruit upon the palm-trees:* (A, Mgh, Mṣb:) or *he computed by conjecture the quantity that was upon the palm-trees of fresh ripe dates that would be dried dates.* (Ṣ.) خَرْصٌ signifies The *computing quantity by conjecture;* (A, Ḳ, TA;) *conjecturing; opining:* or *forming an opinion of a thing of which one is not certain:* you say, خَرَصَ العَدَدَ, aor. ـُ and ـِ, inf. n. خَرْصٌ and خِرْصٌ, the latter said by ISh to be like عِلْمٌ as inf. n. of عَلِمَ, and said by Az to be allowable because a simple subst. is put in the place of an inf. n., *He computed by conjecture the number:* and hence خَرْصُ النَّخْلِ, and التَّمْرِ, [*the computing by conjecture the quantity of fruit upon palm-trees,* and *of dates,*] because خَرْصٌ is the *computing quantity by opinion, not by knowledge.* (TA.) — خَرْصٌ also signifies Any *speaking by opinion,* or *conjecture.* (Ḳ,* TA.) You say, خَرَصَ فِيهِ *He spoke of him, or it, by opinion,* or *conjecture.* (TḲ.) — And hence, (TA,) خَرَصَ, (Ṣ, Mṣb, Ḳ,) aor. ـُ, (Ṣ,) inf. n. خَرْصٌ, (Ṣ, Mṣb, Ḳ,) ‡ *He lied; spoke falsely; said what was untrue;* (Ṣ, Mṣb, Ḳ;*) as also ↓ تخرّص. (Ṣ.) You say also, ↓ تخرّص عَلَيْهِ ‡ *He forged a lie against him.* (A, Ḳ, TA.) And قَالَ ذٰلِكَ ↓ تَخَرُّصًا ‡ [*He said that forging a lie*]. (A.) And ↓ اخترص القَوْلَ, (A, Ḳ,*) and ↓ تخرّصه, (A,) ‡ *He forged the saying.* (A, Ḳ.*) = خَرِصَ, (Ṣ, Ḳ,) aor. ـَ, (Ḳ,) inf. n. خَرَصٌ, (Ṣ, A,) *He* (a man) *was hungry and cold:* (Ṣ, A:*) or *hungry in cold.* (Ḳ.) Being hungry without being cold is not termed خَرَصٌ: but being cold without being hungry is termed خَصَرٌ. (Ṣ.)

5: see 1, in four places, near the end.

8: see 1, near the end.

خُرْصٌ (Ṣ, A, Mṣb, Ḳ) and ↓ خِرْصٌ (Ṣ, Ḳ) *A ring:* (Mṣb:) or *a ring of gold, and of silver:* (Ṣ, Ḳ:) or *an ear-ring with one bead,* (A, TA,) *of the kind called* قُرْط: (TA:) or the *ring of a* قُرْط: (Ḳ:) or *a small ring; one of the ornaments of women;* (Sh, Ḳ;) *in the form of the* قُرْط *or some other thing:* (Sh:) pl. خِرْصَانٌ. (Ṣ, Ḳ.) [Hence,] مَا تَمْلِكُ فُلَانَةُ خُرْصًا, meaning ‡ *Such a woman has not in her possession anything.* (A.)

خَرْصٌ a subst. from خَرَصَ as explained above; (Ṣ, A, Mṣb, Ḳ;) *Conjectural computation of quantity:* (ISh:) or *quantity computed by conjecture* of the fruit upon palm-trees. (A, Mgh.) You say, كَمْ خَرْصُ نَخْلِكَ [*What is the conjectural computation of the quantity,* or *the quantity computed by conjecture, of the produce of thy palm-trees?*]. (TA.) And كَمْ خَرْصُ أَرْضِكَ [*What is the conjectural computation,* or *the quantity computed by conjecture, of the produce of the palm-trees of thy land?*]. (Ṣ, A, Ḳ.) = See also خُرْصٌ. = And see خَرَصٌ.

خَرِصٌ A man *hungry and cold:* (Ṣ, TA:) or *hungry in cold:* (Ḳ:) and ↓ خَارِصٌ signifies the same. (TA.) [See also خَصِرٌ.]

خُرْصَةٌ *Food for a woman who has given birth to a child:* (Ḳ:) app. a dial. var. of خُرْسَةٌ. (TA.)

خَرَّاصٌ: see the next paragraph.

خَارِصٌ One *who computes by conjecture the quantity of the fruit upon palm-trees:* pl. خُرَّاصٌ. (A,* TA.) — ‡ *A liar;* (Mṣb, TA;) as also ↓ خَرَّاصٌ. (Ṣ, A, Mṣb, TA.) قُتِلَ الخَرَّاصُونَ, in the Ḳur [li. 10], (TA,) means, ‡ *Slain be the liars;* (Fr, Zj, A, Bḍ, TA;) i. e., *cursed be they;* (Bḍ;) who say that Mohammad is a poet, and the like thereof, conjecturing that which they know not: (Fr, TA:) or it may mean, *they who only opine, and do not ascertain;* and therefore act according to that which they know not. (Zj, TA.) = See also خَرِصٌ.

خرط

1. خَرَطَ الوَرَقَ, (Ṣ, Mṣb,) aor. ـِ and ـُ, inf. n. خَرْطٌ, (Mṣb,) *He rubbed off the leaves* (Ṣ, Mṣb) *from the branches,* (Mṣb,) *by grasping the upper part, and passing the hand along it to the lower part.* (Ṣ.) — خَرَطَ الشَّجَرَ, aor. as above, (Ḳ,) and so the inf. n., (TA,) *He pulled off the leaves,* (Ḳ, TA,) and *the bark,* or *peel,* (TA,) *from the trees* (Ḳ, TA) *with his hand* [*in the manner above described*]. (TA.) It is said in a prov., دُونَهُ خَرْطُ القَتَادِ [*Before one can attain it he has to strip the tragacanth of its leaves by grasping each branch and drawing his hand down it:* i. e. he has to perform what will be extremely difficult, if not impossible]. (Ṣ, TA. [In the Ṣ and L in art. قتد, we find مِنْ دُونِهِ.]) You say also, خَرَطَ العُنْقُودَ, *He pulled off the grapes,* or *the like, from the bunch with all his fingers:* (AHeyth:) or *he put the bunch in his mouth and drew forth its stalk bare;* as also ↓ اخترطه. (Ḳ.) It is said of Mohammad, كَانَ يَأْكُلُ العِنَبَ خَرْطًا [*He used to eat grapes by putting the bunch in his mouth and drawing forth its stalk bare:* or *by stripping them off with all his fingers*]. (TA.) — خَرَطَ العُودَ, aor. as above, (Ṣ, Ḳ,*) and so the inf. n., (Ṣ,) *He removed the bark,* or *peel, from the wood,* or *stick,* (Ṣ, Ḳ,) *and planed it,* or *made it even,* (Ḳ,) *with the* مِخْرَط, which is also called بَلْط and بُلْط, (TA in art. بلط,) or *with his hand.* (TA in the present art.) — [Hence, in modern Arabic, *He turned the wood,* or *stick;* i. e., *shaped it,* or *made it round, with a lathe.*] — خَرَطَ الحَدِيدَ, inf. n. as above, *He made the iron long, like a column,* or *pole,* or *rod.* (Ṣ.) = خَرَطَ الجَوَاهِرَ *He collected the jewels in a* خَرِيطَة [q. v.]. (MF.)

4. اخرط الخَرِيطَةَ *He bound,* or *made fast, the* خريطة [q. v.]; or *closed it by inserting its loops one into another;* syn. أَشْرَجَهَا. (Ṣ, Ḳ.*)

7. انخرط [*It* (a piece of wood, or a stick,) *had its bark,* or *peel, removed, and was planed,* or *made even, with the* مِخْرَط, (as appears from what here follows,) or *with the hand:* see 1]. — [And hence,] انخرط جِسْمُهُ ‡ *His body became slender;* (Ṣ, Ḳ, TA;) as though it were barked and planed (خُرِطَ) with the مِخْرَط. (TA.)

8: see 1. — [Hence,] اخترط سَيْفَهُ, (Ṣ,) or السَّيْفَ, (Mgh, Ḳ,) ‡ *He drew his sword,* or *the sword,* (Ṣ, Mgh, Ḳ,) *from its scabbard.* (Mgh, TA.)

خُرَاطَةٌ The *parings,* or *shavings, that fall from the work of the* خَرَّاط; like نُجَارَةٌ and نُحَاتَةٌ. (TA.) — *What falls from a bunch of grapes,* or *the like, when the fruit is pulled off with all the fingers.* (AHeyth.)

خِرَاطَةٌ The *art,* or *craft, of the* خَرَّاط. (Ḳ.)

خَرِيطَةٌ *A receptacle,* (Ṣ, Ḳ,) [*a pouch,*] or *thing like a* كِيس [or *purse*], (Lth, Mṣb,) *of leather,* (Lth, Ṣ, Mṣb, Ḳ,) or *of rag,* (Lth,) or *other material,* (Ṣ, Ḳ,) *which is bound,* or *made fast,* or *closed by the insertion of its loops one into another,* (يُشْرَج, Lth, Ṣ, Mṣb, Ḳ,) *upon its contents:* (Lth, Ṣ, Ḳ:) pl. خَرَائِط. (Mṣb.) — Also *A thing likened thereto, which is made for the letters of the sulṭán, and of prefects,* or *agents, to be sent therein.* (Lth, L.) — Also *A similar thing* [*which was formerly, in the time of paganism,*] *put upon the head of the she-camel* [*that was*] *confined* [*to perish*] *at the tomb of a dead person.* (Lth.) — [Also The *pod,* or *oblong capsule* or *pericarp,* of sesamum and the like: pl. as above. Used in this sense by writers on botany, and in the spoken language of the present day.] — See also بِدَادٌ.

خَرَائِطِىٌّ [*A maker,* or *seller, of* خَرَائِط, pl. of خَرِيطَة]; a rel. n. formed from a pl., like أَنْمَاطِىٌّ. (TA.)

خَرَّاطٌ *One whose occupation is to remove the bark,* or *peel, of wood,* or *sticks, and to plane it,* or *make it even,* (Ḳ,) *with the* مِخْرَط, which is also called بَلْط and بُلْط, (TA in art. بلط,) or *with the hand.* (TA in the present art.) — [Hence, in modern Arabic, *A turner* of wood &c.]

مِخْرَطٌ The *iron instrument with which the* خَرَّاط *performs his work; also called* بُلْطٌ and بَلْطٌ. (TA in art. بلط, q. v.)

مَخْرُوطٌ [pass. part. n. of 1]. — †A man (TA) *having a scanty beard:* (Ḳ, TA:) [or you say,] رَجُلٌ مَخْرُوطُ اللِّحْيَةِ †*A man having a beard in which is length without breadth.* (Ṣ.) And لِحْيَةٌ مَخْرُوطَةٌ †*A beard that is scanty in its side,* (Ḳ,) or, correctly, *in its two sides,* (TA,) *and lank and long in the part on and beneath the chin.* (Ḳ.) — †A face *in which is length* (Ḳ, TA) *without breadth.* (TA.) You say, رَجُلٌ مَخْرُوطُ الوَجْهِ †*A man whose face has length without breadth.* (Ṣ.) — بِئْرٌ مَخْرُوطَةٌ ‡*A narrow well.* (A, TA.) — [مَخْرُوطٌ and مَخْرُوطَةٌ ▼, in mathematics, signify *A cone.*]

مَخْرُوطَةٌ: see what next precedes.

[مَخْرُوطِىٌّ, in mathematics, *Conical.*]

خرطم

Q. 1. خَرْطَمَهُ, (Ḳ,) inf. n. خَرْطَمَةٌ, (TḲ,) *He hit,* or *struck, his* خُرْطُوم [or nose, &c.]: or *he twisted it.* (Ḳ.)

Q. 3. اِخْرَنْطَمَ *He* (a man, TA) *elevated his nose:* (Ḳ:) or *twisted it, and was silent, in his anger:* (TA:) and *was proud and angry,* (Ḳ, TA,) *raising his head.* (TA.)

خُرْطُمٌ and خُرْطُومٌ: see خُرْطُومٌ.

خُرْطُمَانٌ *Long,* or *tall:* (JK, Ḳ:) or *long-nosed.* (TA.) And A man *having a large nose.* (IKh, IB.)

خُرْطُومٌ The *nose,* (AZ, JK, Ṣ, Mṣb, Ḳ,) [properly,] accord. to Th, of a beast of prey: (TA:) or the *fore part thereof:* (Ḳ:) or *a large,* or *an elevated, nose:* (MA:) or the *part upon which a man contracts,* or *closes, the* خَتْكَان [or *upper and lower portions of the mouth*]: as also ▼خُرْطُمٌ; (Ḳ;) sometimes, by poetic license, written ▼خُرْطُمٌ: (TA:) or it signifies also the *part upon which contracts,* or *closes, the front of the* خَتْكَان: (JK:) [and *a snout:* often used in this sense; and so, in describing the fish termed كَوْسَج, in the Ṣ and Ḳ:] and the *proboscis* of an elephant; and, as being likened thereto, of a flea: (Th, TA:) pl. خَرَاطِيمُ. (Mṣb.) سَنَسِمُهُ عَلَى الخُرْطُومِ, in the Ḳur [lxviii. 16], means ‡[*We will brand him*] *upon the nose;* the nose of a man being thus termed metaphorically: (ISd, TA:) or it is like the phrase جَدَعْتُ أَنْفَهُ; and means †*we will stigmatize him with indelible disgrace;* the term خرطوم, which signifies the "proboscis" of an elephant, being applied to his nose because it is regarded as unseemly: (Er-Rághib, TA:) or it means †[*we will brand him*] *upon the face.* (Fr, Th, TA.) — [Hence, †*A spout.* You say] الإِبْرِيقُ إِنَاءٌ لَهُ خُرْطُومٌ †[*The* ابريق *is a vessel having a spout*]. (Mgh in art. برق, and Bḍ and Jel in lvi. 18.) — [Hence also, †The *pointed toe* of a boot and the like: pl. as above: see خِفَافٌ مُخَرْطَمَةٌ, below. — And app. †The *point* of a sword: whence,] ذُو الخُرْطُومِ the name of a certain sword. (Ḳ.)] — [Hence, also,] خَرَاطِيمُ القَوْمِ ‡*The chiefs of the people* or *party;* (JK, Ṣ, Ḳ, TA;) *those who are made the foremost of the people* or *party, in affairs,* (JK, TA,) and *in the military forces.* (JK.) = Also *Wine:* (JK, Ṣ:) or *wine that quickly intoxicates:* (Ḳ:) and the *juice that first flows from the grapes, before they are trodden.* (Ḳ.)

خُرَاطِمٌ A woman *advanced in age.* (M, Ḳ.)

[مُخَرْطَمٌ †*Elongated like a snout* or *nose;* and so, app., ▼مُخْرَنْطِمٌ: see a verse cited voce مُتَاوِحٌ. Hence,] خِفَافٌ مُخَرْطَمَةٌ [in the sing. خُفٌّ مُخَرْطَمٌ (occurring in the Ḳ in art. فقع)] *i. q.* ذَاتُ ▼خَرَاطِيمَ; i. e. †[*Boots*] *having their fore parts pointed.* (TA.)

مُخْرَنْطِمٌ: see the next preceding paragraph. — Also *Angry and proud, with his head raised:* (Ṣ:) or *angry when his nose is twisted.* (JK.) [See اِنْبَاعٌ, in art. بوع.]

خرع

1. خَرَعَهُ, (Ṣ,) aor. ـَ, (Ḳ,) inf. n. خَرْعٌ, (Ṣ, Ḳ,) *He cut it,* or *divided it, lengthwise; slit it; split it;* (Ṣ, Ḳ;) as also ▼اِخْتَرَعَهُ. (Ḳ.) — خَرَعَ أُذُنَ الشَّاةِ, aor. and inf. n. as above, *He slit the ear of the sheep,* or *goat:* or *he slit it in the middle, cutting the upper part thereof lengthwise, so that it became three pieces, and the middle piece hung down upon the cavity of the ear.* (TA.) = خَرِعَ, aor. ـَ, inf. n. خَرَاعَةٌ (IDrd, Ḳ) and خُرُوعَةٌ, (IDrd, TA,) in the Ḳ خُرُوعٌ, which is a mistake, (TA,) and خَرَعٌ, (Ibn-'Abbád, Ḳ,) [and quasi-inf. n. خَرَعٌ, which see below,] *He was,* or *became, supple in the joints:* (IDrd, Ḳ:) and *it* (a thing, TA) *was,* or *became, soft, yielding, flaccid, flabby, lax,* or *fragile.* (Ḳ.) — خَرِعَ, (Ṣ, Ḳ,) aor. ـَ, (Ḳ,) [inf. n. خَرَعٌ, app. signifies *the same as* خَرَعٌ: (see خَرَعٌ, below:) and also] *He* (a man, Ṣ, IAth) *was,* or *became, weak,* or *infirm;* (Ṣ, IAth, Ḳ;) as also ▼اِنْخَرَعَ: (Lth, Ḳ:) *he was, or became, weak,* or *infirm, in body, after firmness, strength,* or *hardiness:* (IAạr:) and *he* (a man, Lth) *was,* or *became, languid,* or *languishing,* or *broken in spirit;* syn. اِنْكَسَرَ; as also ▼اِنْخَرَعَ: (Lth, Ḳ:) and [in like manner] ▼تَخَرَّعَ *he* (a man) *was,* or *became, relaxed,* or *flaccid, and weak,* or *infirm, and soft,* or *tender.* (TA.) — †*He* (a man) *was,* or *became, weak in his opinion, after being strong.* (اِسْتَرْخَى) (IAạr.) — Also, (Sh, IAth,) inf. n. خَرَعٌ, (Sh, Ḳ,) †*He* (a man, Sh, IAth) *became confounded, or perplexed, and unable to see his right course;* or *bereft of his reason or intellect, in consequence of shame,* or *fear,* or *grief,* or *the like.* (Sh, IAth, Ḳ.) Hence the saying of Aboo-Tálib, when death overtook him, [and he was urged to make profession of El-Islám,] لَوْ لَا رَهْبَةُ أَنْ تَقُولَ قُرَيْشٌ دَهَرَهُ الخَرَعُ لَفَعَلْتُ †[*Were it not for fear that Kureysh would say, "Confusion,"* &c. "*put him into a state from which there was no escape for him," I would do what thou desirest*]. (TA.) — خَرَعَتِ النَّخْلَةُ *The palm-tree lost the stumps of its lopped branches* [*from its trunk, which thus became comparatively smooth*]. (Ṣ, Ḳ.)

5: see 1: — and 7.

7. انخرع *It became cut,* or *divided, lengthwise; it became slit,* or *split; it slit,* or *split.* (Ṣ.) And انخرعت القَنَاةُ *The spear-shaft split, and broke into pieces,* or *into small pieces.* (Ḳ.) — *I. q.* انخلع; (Ḳ;) a dial. var. of the latter; as in the phrase انخرعت كَتِفُهُ [*His shoulder-blade became dislocated*]. (Ṣ.) You say also, انخرعت أَعْضَاءُ البَعِيرِ *The limbs of the camel became displaced;* as also ▼تخرّعت. (TA.) — See also two significations above, voce خَرِعَ, in two places. — [Whence,] اِنْخَرَعْتُ لَهُ †*I became gentle,* or *mild, to him.* (TA.)

8. اخترعهُ: see 1, first signification. — *He broke it off;* namely a branch, or piece of wood, from a tree. (TA.) — *He cut it off for himself;* or *took it;* namely, a man's property, &c. (TA.) — *He acted treacherously towards him,* (Aboo-Sa'eed, Ḳ,) *and took of his property;* (Ḳ;) as also اخترعهُ, with ز. (TA.) — *He expended it; exhausted it; caused it to pass away and come to an end,* or *to cease;* or *made an end of it.* (ISh, Ḳ.) — اخترع الدَّابَّةَ *He made use of,* or *rode, the beast of another person for some days, and then restored it.* (Ibn-'Abbád, Ḳ.) = *He did it,* or *produced it, without premeditation;* syn. اِرْتَجَلَهُ: (TA:) or اِشْتَقَّهُ [app. here meaning *he constructed it,* or *founded it,* (بَنَاهُ) *without premeditation;* this meaning being one assigned to اشتقّه in its proper art. in the TA]: (Ṣ:) or *he originated it; invented it; devised it; excogitated it; innovated it; made it, did it, produced it, caused it to be or exist,* or *brought it into existence, newly, for the first time, it not having been or existed before, and not after the similitude of anything preexisting;* syn. أَنْشَأَهُ, (Ṣ, O, Ḳ,) and اِبْتَدَعَهُ, (Ṣ, O,) or اِبْتَدَأَهُ. (Ḳ.) You say, اخترع ٱللّٰهُ الأَشْيَآءَ *God originated, brought into existence,* or *created, the things* [*that exist*], *without any means,* or *second cause.* (TA.) And اخترع بَاطِلًا *He forged a falsehood.* (A, TA.)

خَرْعٌ *A mark in the ear of a sheep or goat; the upper part* [أَعْلَى, for which we find in the CḲ عَلَى,] *of the animal's ears being cut* (يُقْطَعُ, in the CḲ تُقْطَعُ,) *lengthwise, so that the ear becomes three pieces, and the middle piece hangs down upon the cavity of the ear.* (Ḳ, TA.) = *Suppleness of the joints:* (IDrd, Ḳ:) and *softness, yieldingness, flaccidity, flabbiness, laxness,* or *fragility,* (Ṣ, Ḳ,) in a thing. (Ṣ.) [See خَرَعٌ.] — ‡ *Cowardice;* and *weakness,* or *feebleness, and languor,* or *languidness;* in a man. (TA.) [See also خَرَعٌ.]

خَرِعٌ (Ṣ, Ḳ) and ▼خَرِيعٌ (Ḳ) *Weak,* or *infirm;* (Ṣ, Ḳ;) applied to a man: (Ṣ:) anything *weak;* and *soft, yielding, flaccid, flabby, lax,* or *fragile:* (O:) also the former, applied to a young camel, *weak;* or, as some say, *small, that is* [or *may be*] *raised,* or *lifted:* and, applied to a branch, *soft, tender,* or *supple.* (TA.)

خَرْعَةٌ a subst. from اِخْتَرَعَ الشَّىْءَ signifying اِرْتَجَلَهُ [&c.; app. meaning *A thing done,* or *produced, without premeditation;* &c.]; like بِدْعَةٌ from اِبْتَدَعَهُ, a syn. of اِخْتَرَعَهُ, q. v.]. (TA.)

خَرُوعٌ: see خَرِيعٌ.

خِرْوَعٌ Any *weak, bending* plant, of whatever kind it be: (Aṣ, Ṣ:) any plant *weak,* or *fragile,* (قَصِيف,) *and sappy,* whether it be a tree or a herb: (TA:) a plant *weak by reason of its softness,* or *tenderness, and sappiness.* (Ṣgh.) [See also خَرِعٌ.] — Hence, as some say, (TA,) [The *ricinus communis; common palma Christi;* or *castor-oil-plant;*] *a certain plant,* (Ṣ, Mṣb, Ḳ,) *well known,* (Ṣ,) *soft, tender,* or *pliant,* (Mṣb,) *not serving for pasturage,* (Ḳ,) *bearing a berry resembling sparrows' eggs, called* السِّمْسِم; accord. to Ibn-Jezleh, *the best thereof is that called* الهِنْدِىّ; *it has the property of loosening phlegm, and it is useful for counteracting the colic and palsy and the* [*disease in the face called*] لَقْوَة, *the dose extending to a* مِثْقَال. (TA.) The word is of the measure فِعْوَلٌ; (Mṣb;) and J says [in the Ṣ] that there is no other word of the same measure except عِتْوَدٌ, which is the name of a certain valley; but to this have been added ذِرْوَدٌ, the name of a certain mountain; and عِتْوَرٌ, the name of a certain valley, and not a mistranscription of عِتْوَدٌ; and جِدْوَلٌ, a dial. var. of جَدْوَلٌ. (TA.) — [Hence also] اِمْرَأَةٌ خِرْوَعَةٌ *A beautiful, and soft,* or *tender, woman:* and [the pl.] خَرَاوِيع, applied to women, signifies [*the same,* or merely] *beautiful.* (TA.) — And خِرْوَعٌ is likewise applied to youth, or youthfulness, and to life, meaning ‡ *Soft,* or *delicate.* (TA.)

خَرِيعٌ: see خَرِعٌ. — Anything *that quickly breaks.* (TA.) — *Soft;* applied to a lip (شَفَة): (TA:) and *pendulous;* applied to the lip of a camel. (Ṣ, Ḳ.*) — Applied to a woman, (Ṣ, Mṣb, Ḳ, TA,) *Youthful, and soft, tender,* or *delicate:* or *beautiful:* (TA:) or *that walks with an affected bending of the body, and with softness,* or *delicacy:* (Mṣb:) or *that affects a bending of the body by reason of softness,* or *delicacy;* (Aṣ, Ṣ, Ḳ;) as also خَرِيعَةٌ and ▼ خَرُوعٌ: (Ibn-'Abbád, Ḳ:) or † *vitious;* or *immoral;* or *an adulteress;* or *a fornicatress;* (Ṣ, Ḳ;) but this explanation is disallowed by Aṣ: (Ṣ:) or † *that does not repel the hand of a feeler,* or *toucher;* as though she were gentle, or mild, (تَتَخَرَّعُ,) to him; as also with ة: or *hard,* or † *impudent, not caring for what is said or done, and inordinately brisk, lively,* or *sprightly:* pl. خُرُوعٌ and خَرَائِعُ and خُرُعٌ. (TA.) Also † One *who induces,* or *is an object of, suspicion;* because such a person fears, and is therefore as though he were weak. (TA.) — Also, [as a subst.,] *A branch;* because of its softness, or tenderness, and its bending. (TA.)

خَرَاعَةٌ a dial. var. of خَلَاعَةٌ, which is syn. with دَعَارَةٌ [i. e. *Vice,* or *immorality;* or *vitious,* or *immoral, conduct;* &c.]. (Ṣ.)

شَاةٌ مَخْرُوعَةٌ *A sheep,* or *goat, having the mark termed* خَرْعٌ [q. v.] *in the ear.* (Ḳ.)

خرعب

خَرْعَبٌ (Ḳ) and ▼ خَرْعَبَةٌ (TA) and ▼ خُرْعُوبٌ and ▼ خُرْعُوبَةٌ (Ḳ) *A branch,* or *twig, until a year old;* or *of a year's growth:* or *fresh,* or *juicy, and tall:* (Ḳ, TA:) or (TA) *soft,* or *tender, and of recent growth,* (Ḳ, TA,) *that has not yet become hard:* (TA:) or ▼ خُرْعُوبٌ [is an epithet, and] signifies a *bending* branch or twig. (Ṣ.) [Compare خَرِعٌ, and خِرْوَعٌ, and خَرِيعٌ.] — Also, [i. e. all the words above,] (Ḳ,) or ▼ خَرْعَبَةٌ [only], (TA,) *A young woman of goodly make, soft,* or *tender:* (Ḳ:) or *a young woman large in body, and of goodly make:* or *soft,* or *tender, and pliant:* (TA:) or *fair, tender,* or *pliant, fat, large in body, fleshy, with small,* or *delicate, bones:* (Ḳ:) or *fair:* (TA:) or *soft,* or *pliant, in the waist, and tall:* (Aṣ, TA:) or *large in the body, and fleshy:* (TA:) or ▼ خُرْعُوبَةٌ and ▼ خَرْعَبَةٌ signify a girl *slender in the bones,* (Ṣ, TA,) *having much flesh,* (TA,) *and soft,* or *tender:* (Ṣ, TA:) or *a young woman of goodly stature, resembling a twig* (خُرْعُوبَة) *of a year's growth.* (Lth, TA.) And خَرْعَبٌ signifies A *soft,* or *tender,* body. (TA.) And A *tall and fleshy* man. (Ḳ.)

خَرْعَبَةٌ: see above, in three places.

خُرْعُوبٌ: see خَرْعَبٌ, in two places. — Also A *tall and well-made* camel: (Ṣ:) or a *tall and large* she-camel: and one *having much milk.* (Ḳ.)

خُرْعُوبَةٌ: see خَرْعَبٌ, in two places. — Also *A piece of a gourd,* and *of a cucumber,* and *of fat;* as in the L &c.: in the Ḳ written خُذْعُوبَة. (TA.)

خرف

1. خَرَفَ, (Ṣ, Mṣb, Ḳ,) aor. ِ, (Ṣ, Mṣb,) inf. n. خَرْفٌ (Mṣb, Ḳ) and مَخْرَفٌ and خَرَافٌ and خِرَافٌ; (Ḳ;) and ▼ اخترف; (Ṣ, Mṣb, Ḳ;) *He gathered,* or *plucked,* fruit: (Ṣ, Ḳ:) or *cut it off.* (Mṣb.) Accord. to the M, خَرَفَ النَّخْلَ signifies *He cut off the fruit of the palm-trees:* and accord. to AḤn, ▼ اِخْتِرَافٌ signifies the *picking up the fruit of the palm-trees, whether unripe or ripe.* (TA.) — خَرَفَ فُلَانًا, (Ḳ,) aor. ُ, inf. n. خَرْفٌ, (TA,) *He picked up, for such a one, dates* (تَمْرًا), or *fruit* (ثَمَرًا), accord. to different copies of the Ḳ: from Sh. (TA.) — يَخْرُفُ مِنْ هَاهُنَا وَمِنْ هَاهُنَا, said of a lamb, means *He depastures,* and *eats, from this place and from this.* (Mṣb, TA.*) — And خَرَفَ, said of a man, (JK, TA,) aor. ُ, (JK,) or ِ, (TA,) *He took of the* طُرَف [app. meaning the *choice part*] *of the fruits.* (JK, TA.) — خَرَفَ also signifies *He remained, stayed,* or *abode, in the* [*season called*] خَرِيف: (Ḥam p. 676:) and in like manner, ▼ اخرفوا *they remained, stayed,* or *abode,* in a place *during their* خَرِيف. (TA.) You say, خَرَفُوا فِى حَائِطِهِمْ *They remained, stayed,* or *abode, in their* حائط [or *garden,* or *walled garden of palm-trees,*] *in the time of the gathering of the fruits.* (TA, from a trad. of 'Omar.) — خُرِفْنَا *We were rained upon by the rain called* الخَرِيف. (Ṣ, Ḳ.) And خُرِفَتِ الأَرْضُ, (Ṣ,) inf. n. خَرْفٌ, (TA,) *The land was rained upon by the rain so called.* (Ṣ, TA.) And خُرِفَتِ البَهَائِمُ *The beasts were rained upon by the rain so called:* or *had that upon which they might pasture produced for them by that rain.* (TA.) — خَرِفَ, aor. َ, *He* (a man, TA) *was,* or *became, fond of,* or *addicted to, the eating of* خُرْفَة, (Ḳ,) i. e. *gathered,* or *plucked, fruit* (Ṣ, Ḳ, TA) *of the palm-tree.* (TA.) — خَرِفَ, (Ṣ, L, Mṣb, Ḳ,) aor. َ, (Mṣb, Ḳ,) inf. n. خَرَفٌ; (Ṣ,* Mṣb;) and خَرُفَ, aor. ُ; and خَرَفَ, aor. ُ; (Ḳ;) *He* (a man, Ṣ, Mṣb) *doted;* or *was,* or *became, corrupted, rendered unsound,* or *disordered, in his intellect;* (Ṣ, Mṣb, Ḳ;) *in consequence of old age.* (Ṣ, Mṣb.) [The first of these three verbs, in the present day, is used as meaning *He doted; told stories such as are termed* خُرَافَات, i. e. *fictions,* &c.; and *talked nonsense:* as also ▼ خرّف.] — ▼ خَرَّفَتْهُ أَخَارِيفُ [app. *Stories such as are termed* اخاريف, i. e. ▼ خُرَافَات, or *fictions,* &c., *caused him to dote,* or *talk nonsense*]. (JK, TA.* [Mentioned in the former immediately after خُرَافَةٌ explained as meaning "a fiction that is deemed pretty." See also 4.])

2. خرّفهُ, inf. n. تَخْرِيفٌ, *He attributed to him* خَرَف, (Ḳ, TA,) i. e. [*dotage;* or] *a corrupt,* an *unsound,* or *a disordered, state of intellect.* (TA.) — See also 1, near the end of the paragraph.

3. خارفهُ, (Ḳ,) inf. n. مُخَارَفَةٌ and خِرَافٌ, (TA,) *He bargained,* or *made an engagement, with him, for work, for the* خَرِيف [or *autumn*]; (Ḳ;) from الخَرِيف, like المُشَاهَرَةُ from الشَّهْرُ; (O, TA;) as also عَامَلَهُ مُخَارَفَةً (Ṣ, TA) and خِرَافًا: and so اِسْتَأْجَرَهُ مُخَارَفَةً and خِرَافًا [*He hired him,* or *took him as a hired man* or *a hireling, for the autumn*]. (Lḥ, TA.)

4. اخرف, said of the palm-tree, *It attained,* or *nearly attained, the time for its fruit to be cut off.* (JK, Ḳ.) — And, said of a people, or party, *They entered upon the* [*season called*] خريف. (Ṣ, Ḳ.) See also 1. — اخرفت, said of a ewe, or she-goat, *She brought forth in the* [*season called*] خريف. (Ṣ, Ḳ.) — Said of a she-camel, *She brought forth in the like of the time* [*of the year*] *in which she became pregnant* (Ṣ, Ḳ) *in the preceding year:* so says El-Umawee: (Ṣ:) [or, so applied, it means the same as when said of a ewe or she-goat; for] the epithet applied to her in this case is ▼ مُخْرِفٌ; (Ṣ, Ḳ;) but this is more correctly explained as signifying, applied to a she-camel and to a ewe or she-goat, *that brings forth in the* خريف. (TA.) — Also, said of ذُرَة [or millet], *It became very tall.* (JK, Ibn-'Abbád, Ḳ.) — اخرفهُ نَخْلَةً *He assigned to him a palm-tree of which he should cut,* or *gather, the fruit.* (Lth, Ḳ.) — Also, (said of anxiety, JK, or of time,

or fortune, TA,) *It corrupted him*, or *disordered him*; (Ḳ, TA;) [app., *in his intellect*; or *caused him to dote*; as is indicated in the JK;] namely, an old man. (JK.)

8: see 1, first and second sentences.

خَرَفٌ *A corrupt, an unsound*, or *a disordered, state of the intellect; dotage.* (Ṣ. [See خَرِفَ, of which it is the inf. n.]) —— The [*bad sort of*] *dates called* شِيص. (Ḳ,* TA.)

خَرِفٌ *Corrupted, unsound*, or *disordered, in his intellect*, (Ṣ, Mṣb, Ḳ,) *in consequence of old age; doting*: (Ṣ, Mṣb:) fem. with ة. (TA.)

خُرْفٌ *A time of going forth* of camels, (Nh,) or of men, (O, Ḳ,) *to the* [*herbage of the season called*] خَرِيف: so in the saying of El-Járood, يَا رَسُولَ ٱللّٰهِ قَدْ عَلِمْتَ مَا يَكْفِينَا مِنَ ٱلظَّهْرِ ذَوْدٌ نَأْتِى عَلَيْهِنَّ فِى خُرْفٍ [*O Apostle of God, verily thou knowest that a* number such as is termed ذود, *of camels for riding or carriage, whereon we come in a time of going forth* &c., *is not sufficient for us*]. (Nh, O, Ḳ.)

خُرْفَةٌ *Gathered*, or *plucked, fruits*; (Ṣ, Mgh, Ḳ;) and particularly of the palm-tree: (TA:) and ↓ خُرَافَةٌ signifies the same. (Mgh, Ḳ, TA. [See also خَرِيف.]) It is said in a trad., التَّمْرُ خُرْفَةُ الصَّائِمِ [*Dates are the gathered fruit of the faster*]; (Ṣ, TA;) because breaking the fast upon them is approved: and in another, النَّخْلَةُ خُرْفَةُ الصَّائِمِ, meaning *The palm-tree is that of which the fruit is eaten by the faster.* (TA.) See also مَخْرَفٌ, last sentence.

خُرْفِىٌّ The جلبان, (i. e. جُلْبَان, or جُلُبَّان, &c., accord. to different copies of the Ḳ, [see art. جلب,]) *a well-known grain* or *seed*, (AḤn, Ḳ,) *of the kind called* قِطْنِيّ [i. e. *pulse*]: (AḤn:) an arabicized word, from خُرْفَى, (AḤn, Ḳ,) which is Persian; also called خُلَّر. (AḤn.)

خَرْفِىٌّ and خِرْفِىٌّ: see what next follows.

خَرَفِىٌّ and ↓ خَرْفِىٌّ, (Ṣ, Mṣb, Ḳ,) the latter a contraction of the former, (Mṣb,) and ↓ خِرْفِىٌّ, (Ḳ,) *Of*, or *relating to, the season called* خَرِيف; (Ṣ, Mṣb, Ḳ;) and applied to the rain *of that season*; (JK;) rel. ns. from الخَرِيف; (Ṣ, Mṣb, Ḳ;) irregularly formed. (Ṣ, Mṣb.) —— The first also signifies The *increase* (نِتَاج) [*of sheep and goats*] *in the end of the* [*season called*] قَيْظ. (Aboo-Naṣr, TA voce صَفَرِىٌّ, q. v.)

خَرَافٌ and ↓ خِرَافٌ The *time of the gathering*, or *plucking, of fruits*: (Ks, Ḳ:) like حَصَادٌ and حِصَادٌ [&c.]. (TA.) —— Also inf. ns. of خَرَفَ in the first of the senses explained above. (Ḳ.)

خِرَافٌ: see the next preceding paragraph: and see مَخْرَفٌ, last sentence.

خَرُوفٌ *A lamb*; syn. حَمَلٌ [q. v.]: (Ṣ, Mṣb:) or the *male young one of the sheep-kind*: or *such as has pastured, and become strong*: (Lth, Ḳ:) *younger than the* جَذَع: (Lth, TA:) so called because it depastures from this place and this: (Mṣb, TA: [see 1:]) fem. with ة: (Ḳ:) pl. (of pauc., TA) أَخْرِفَةٌ and (of mult., TA) خِرْفَانٌ. (Mṣb, Ḳ, TA.) The latter pl. is sometimes used as meaning † *Young and ignorant persons*; like as كِبَاش is used as meaning aged and learned persons. (TA.) And hence the prov., كَٱلْخَرُوفِ أَيْنَمَا أَتَّكَأَ ٱتَّكَأَ عَلَى صُوفٍ [*Like the lamb: wherever he reclines, he reclines upon wool*]: (JK, TA: but in the latter, اتّكى:) applied to him who leads a soft and delicate life. (TA.) —— Also, (sometimes, Ṣ,) *A colt*; the *male offspring of a mare*; *when he has attained the age of six months*, or *seven months*; (Ṣ, Ḳ;) a meaning assigned to it by Aṣ, in the "Book of the Horse;" but unknown to Abu-l-Ghowth: (Ṣ:) or, *until a year old*: (ISk, Ḳ:) it is said by some to be applied *to a horse*: in the L it is said that the خروف of horses is *such as is brought forth in the* [*season called*] خَرِيف: but Khálid Ibn-Jebeleh says that it means *such as pastures upon the* [*herbage of the season called*] خَرِيف: and Suh thinks that it is an epithet applied to a horse, and any beast, as meaning *that depastures the trees and herbage.* (TA.)

خَرِيفٌ Fresh ripe dates, (Ḳ, TA,) or fruits [in general], (Ṣ, TA,) *gathered*, or *plucked*; (Ṣ, Ḳ, TA;) as also ↓ مَخْرُوفٌ. (Ṣ, TA. See also مَخْرَفٌ, last sentence. [And see خُرْفَةٌ.]) —— And hence, ‡ *Fresh* milk; milk *recently drawn from the udder.* (Hr, TA.) —— Palm-trees (نَخْلٌ) *whereof the quantity of the fruit that is upon them is computed by conjecture.* (Ḳ. [See also خَرَائِفُ, voce خَرِيفَةٌ.]) —— [*The autumn*;] *one of the divisions of the year*, (Ṣ, Mgh,) the *division* (Mṣb) [*consisting of*] *three months between the end of the* قَيْظ [or *summer*] *and the beginning of the* شِتَآء [or *winter*], (Lth, Ḳ,*) *in which the fruits are gathered.* (Lth, Ṣ, Mgh, Mṣb, Ḳ.) —— And hence, (Mgh, TA,) † *A year*: (Mgh, Ḳ, TA:) so in the saying, مَنْ صَامَ يَوْمًا فِى سَبِيلِ ٱللّٰهِ بَاعَدَهُ ٱللّٰهُ مِنَ ٱلنَّارِ أَرْبَعِينَ خَرِيفًا أَوْ سَبْعِينَ, i. e. [*Whoso fasteth a day in the way of God, God will remove him from the fire* of Hell] *to the distance of a journey of forty years, or seventy.* (Mgh: and similar exs. are given in the TA, from three trads.: see also an ex. voce إِنَّ.) —— Also The *rain of the season so called*: (Ṣ, Ḳ:) or the *rain*, (JK,) or the *first of the rain*, (Ḳ,) *in the beginning of the* شِتَآء [or *winter*], (JK, Ḳ,) *which comes at the time of the cutting off of the fruit of the palm-trees*: then follows the وَسْمِىّ, at the coming in of the winter; then, the رَبِيع; then, the صَيْف; and then, the حَمِيم: so says Aṣ: El-Ghanawee says that the خَرِيف is *between the* [*auroral*] *rising of* الشِّعْرَى [or *Sirius*, which commenced, in central Arabia, about the epoch of the Flight, on *the 13th of July, O.S.*,] *and the* [*auroral*] *setting of* العَرْقُوَتَان [or الفَرْغَان, *the 26th and 27th of the Mansions of the Moon*, commencing, in the same region and period, on *the 8th and 21st of Sept., O.S.*, and continuing thirteen days]: El-Ghowr and Rekeeyeh [? (imperfectly written)] and El-Ḥijáz are all rained upon by the خَرِيف; but Nejd is not: AZ says, the first rain is the وَسْمِىّ; then follows the شَتَوِىّ; then, the دَفَئِىّ; then, the صَيْف; then, the حَمِيم; then, the خَرِيف: and therefore the year is made to consist of six seasons: accord. to AḤn, [who seems in this matter to differ from most others,] الخريف is not originally the name of the division of the year; but the name of *the rain of the* قَيْظ [or *summer*]; and then the season was named thereby. (TA.) [See also نَوْءٌ.] —— [Also The *herbage of the season so called*, or *of the rain so called*; like as رَبِيع signifies the "herbage of the season, or of the rain, so called." So in the phrase used by Khálid Ibn-Jebeleh (in explaining the word خَرُوف), مَا رَعَى الخَرِيفَ *Such as pastures upon the* خَرِيف.] —— Also, accord. to AA, (TA,) *A rivulet, streamlet*, or *small channel for irrigation.* (JK, Ḳ, TA.)

خُرَافَةٌ *i. q.* خُرْفَةٌ, q. v. (Mgh, Ḳ.) —— Hence خُرَافَاتٌ meaning *Stories that are deemed pretty*: similar to فُكَاهَةٌ from فَاكِهَةٌ: (Mgh:) [or] خُرَافَةُ was the name of a man, (Ṣ, Mgh, Ḳ,) of [the tribe of] 'Odhrah, (Ṣ, Ḳ,) whom the Jinn (or Genii) fascinated, (Ṣ, Mgh, Ḳ,) as the Arabs assert, (Mgh,) and carried off, (TA,) and who related what he had seen, (Ṣ, Mgh, Ḳ,) of them, when he returned, (Mgh,) and they pronounced him a liar, and said, (Ṣ, Mgh, Ḳ,) of a thing that was impossible, (Mgh,) حَدِيثُ خُرَافَةَ [*a story of Khuráfeh*]: (Ṣ, Mgh, Ḳ:) but it is related of the Prophet, that he said, خُرَافَةُ حَقٌّ, (Ṣ, Mgh,) meaning *What Khuráfeh relates* [as heard] from the Jinn [*is true*]: (Mgh:) the ر is without teshdeed; and the article ال is not prefixed, because the word is determinate [by itself], unless one mean thereby خُرَافَاتٌ as signifying *fictitious night-stories*: (Ṣ:) or خُرَافَةٌ signifies *a fictitious story that is deemed pretty*: (Lth, Ḳ:) [and ↓ أَخَارِيفُ app. signifies the same as خُرَافَاتٌ, as though its sing. were أُخْرُوفَةٌ, like as أَسَاطِيرُ and أَحَادِيثُ, which have similar meanings, are pls. of which the sings. are said to be أُسْطُورَةٌ and أُحْدُوثَةٌ:] see 1, last signification.

خَرُوفَةٌ: see what next follows.

خَرِيفَةٌ and ↓ خَرُوفَةٌ *A palm-tree* (نَخْلَةٌ) *of which a man gathers*, or *plucks, the fruit for himself and his household*; as also ↓ مَخْرَفٌ: (AḤn:) or *a palm-tree which one takes for the picking up of its fresh ripe dates*: (Sh, O, Ḳ:) or the latter signifies *a palm-tree of which the fruit is cut off*; being of the measure فَعُولَةٌ in the sense of the measure مَفْعُولَةٌ: and the former is said to signify *one that is set apart for its fruit that is* [*to be*] *gathered*, or *plucked*: (TA:) or *a selected palm-tree*: (JK:) and its pl. is خَرَائِفُ: (JK, TA:) or خَرَائِفُ signifies *palm-trees whereof the quantity of the fruit that is upon them is computed by conjecture.* (AZ, Ṣ, Ḳ. [See also خَرِيف.]) Also, the former, [*A palm-tree set in the manner described in the following explanation*:] *one's digging, for a palm-tree, in a water-course*, or *channel of a torrent, in which are pebbles, until reaching hard ground, and then filling up the hollow with sand, and setting the palm-tree therein.* (O, Ḳ.)

خَارِفٌ *A keeper*, or *watcher, of palm-trees:* (Ḳ:) pl. خُرَّافٌ. (TA.)

أَخَارِيفُ: see خُرَافَةٌ; and see 1, last signification.

مَخْرَفٌ The *place of the gathering*, or *plucking*, or *cutting off, of fruit.* (Mṣb.) *A place of abode* of a people, or party, *during their* خَرِيف. (TA. [It is there added, "as though formed from أَخْرَفُوا, by the rejection of the augmentative letter:" but it is rather to be regarded as regularly formed, from خَرَفُوا: see 1.]) — Also *A garden;* (Mgh, TA;) and so ▼ مَخْرَفَةٌ: (Ṣ, Ḳ:) or *a garden of palm-trees;* as also ▼ مَخْرِفٌ and ▼ مَخْرَفَةٌ: (TA:) *a single palm-tree:* or *a few palm-trees, up to ten;* more than these being termed a بُسْتَان or a حَدِيقَة: (El-Ḥarbee, TA:) see also خَرِيفَةٌ: or *a small collection of palm-trees, six or seven, which a man purchases for the fruit that is* [*to be*] *gathered*, or *plucked:* or *any collection of palm-trees:* (L, TA:) or *a walled garden of palm-trees:* (IAth, TA:) or *palm-trees* [absolutely]: (Mgh:) and *an avenue between two rows of palm-trees, such that one may gather*, or *pluck, the fruit from whichever of them he will;* (Ḳ;) as also ▼ مَخْرَفَةٌ: (Sh, Ḳ:) and, (Ṣ, Mgh, Ḳ,) as also ▼ مَخْرَفَةٌ, (Ṣ, Ḳ,) *a road,* (Ṣ, Mgh, Ḳ,) *such as is conspicuous, clear,* or *open:* (Ḳ:) pl. مَخَارِفُ. (Mgh, TA.) It is said in a trad., عَائِدُ الْمَرِيضِ عَلَى مَخَارِفِ الْجَنَّةِ حَتَّى يَرْجِعَ, i. e. *The visitor of the sick is as though he were in the gardens of Paradise until he returns:* or *upon the palm-trees of Paradise;* gathering, or plucking, their fruits: or *upon the roads of Paradise:* (Mgh,* TA:) or, as some relate it, على ▼ مَخْرَفَةِ الجنّة. (TA. [See also another explanation, and other readings, in what follows.]) And it is said in a trad. of 'Omar, تَرَكْتُكُمْ عَلَى مَخْرَفَةِ ▼ النَّعَمِ, (Ṣ,) or تُرِكْتُمْ, (TA,) i. e. [*I have left you,* or *ye have been left,*] upon a conspicuous road, like *the road of the camels,* (Aṣ, Ṣ,* TA,) *which they have trodden with their feet so that it has become plainly apparent.* (Aṣ, TA.) — Also *Gathered*, or *plucked, fruit of palm-trees:* (Aṣ, A'Obeyd, IAmb, Ḳ:) a correct meaning, though IḲt says that the proper word in this sense is only مَخْرُوفٌ: it is like مَشْرَبٌ and مَطْعَمٌ and مَرْكَبٌ as meaning طَعَامٌ مَأْكُولٌ and مَرْكُوبٌ and مَشْرُوبٌ; and may signify *fresh ripe dates gathered* or *plucked:* (IAmb, TA:) pl. as above. (Aṣ, &c.) So in the former of the two trads. mentioned above accord. to Aṣ and A'Obeyd: (TA:) and this interpretation is corroborated by another reading, i. e., على ▼ خُرْفَةِ الجنّة: (Mgh:) another reading is, فِى ▼ خِرَافِ الجنّة: [see خِرَافٌ:] and another, لَهُ ▼ خَرِيفٌ فِى الْجَنَّةِ, i. e. [The visitor of the sick *shall have*] *gathered fruits in Paradise.* (TA.)

مَخْرِفٌ: see the next preceding paragraph.

مُخْرِفٌ: see 4.

مِخْرَفٌ The *thing in which fruits are gathered;* (Ṣ, Ḥar p. 374;) called by the Arabs خَافَةٌ: (Ḥar ib.:) *a* [*basket of the kind called*] مِكْتَل, (Mṣb,) or زِنْبِيل, *of small size, in which the best fresh ripe dates are gathered:* (O, Ḳ:) pl. مَخَارِفُ. (A, TA.) One says, خَرَجُوا إِلَى الْمَخَارِفِ بِالْمَخَارِفِ, i. e. *They went forth to the gardens with the baskets* (زُبُل) [*for gathering fruit*]. (A, TA.) — And hence, †The *basket* (زنبيل) *in which the importunate beggar puts his food.* (Ḥar ubi suprà.)

مَخْرَفَةٌ: see مَخْرَفٌ, in six places.

مَخْرُوفٌ: see خَرِيفٌ. = Also *Rained upon by the rain called* خَرِيف; pl., applied to men, مَخْرُوفُونَ: (TA:) [so, too, applied to a beast:] and so, with ة, applied to land (أَرْض). (Aṣ, Ṣ.)

مُخَارَفٌ *Denied,* or *refused, good,* or *prosperity; prevented,* or *withheld, from obtaining good, good fortune,* or *sustenance;* (Ḳ;) *i. q.* مُحَارَفٌ; (JK, TA;) as also مُجَارَفٌ. (TA.)

خرفج

Q. 1. خَرْفَجَهُ, (L, Ḳ,) inf. n. خَرْفَجَةٌ, (TA,) *He took it plentifully,* or *largely.* (L, Ḳ.) — *He made it to be of the best kind;* namely, food, or meat and drink. (Er-Riyáshee.)

خَرْفَجٌ: see مُخَرْفَجٌ: — and خِرْفِيجٌ.

خُرْفُجٌ and ▼ خُرَافِجٌ [app. خُرْفُجٌ and خُرَافِجٌ, but in the CḲ خَرْفَجٌ and خُرَافَجٌ,] and ▼ خِرْفَاجٌ and ▼ خِرْفِيجٌ *A plentiful and pleasant state of life.* (Ḳ.)

خُرْفُجٌ *Fat,* as an epithet, (L, Ḳ,) applied to a lamb; as also ▼ خُرَافِجٌ. (L.) — See also خِرْفِيجٌ.

خَرْفَجَةٌ *Goodness of food, aliment,* or *nutriment,* or *of meat and drink, with plenty.* (TA.) — Also inf. n. of the verb above. (TA.)

خِرْفَاجٌ: see خُرْفُجٌ: — and see also خِرْفِيجٌ.

خُرْفُنْجٌ *A flourishing and fresh condition,* or *softness,* or *tenderness,* of a plant, or of herbage. (L.) — See also what next follows.

خِرْفِيجٌ: see خُرْفُجٌ. — Also *Fresh,* or *juicy; flourishing and fresh,* or *soft,* or *tender;* (L, Ḳ;) applied to a plant, or to herbage; and so ▼ خِرْفَاجٌ and ▼ خُرَافِجٌ and ▼ خُرْفُجٌ and ▼ خُرْفُنْجٌ. (L.) [And ▼ خَرْفَجٌ seems to have a similar meaning.] The rájiz [El-'Ajjáj (so in a copy of the Ṣ)] says,

* جَارِيَةٌ شَبَّتْ شَبَابًا خَرْفَجَا *

[app. meaning *A girl that had attained to flourishing,* or *soft,* or *tender, youthfulness*]. (Ṣ.)

خُرَافِجٌ: see خُرْفُجٌ: — and خُرْفُجٌ: — and خِرْفِيجٌ: — and what here follows.

مُخَرْفَجٌ *Ample:* (Ḳ:) applied to anything. (TA.) It is said in a trad., كَرِهَ السَّرَاوِيلَ الْمُخَرْفَجَةَ, *He disliked,* or *disapproved of, ample trousers:* (A'Obeyd:) or they say it means *trousers reaching down to the upper part of the foot.* (Ṣ.) And you say, عَيْشٌ مُخَرْفَجٌ *A plentiful life.* (Ṣ.) — Also The best of food, or of meat and drink; and so ▼ خَرْفَجٌ and ▼ خُرَافِجٌ. (Er-Riyáshee.)

خرق

1. خَرَقَهُ, (JK, Ṣ, Mgh, Mṣb, Ḳ,) aor. ـُ (Mgh, Mṣb, Ḳ) and ـِ, (Ḳ,) the former of which is the more chaste, (TA,) inf. n. خَرْقٌ, (Ṣ, Mgh, Mṣb, KL,) *He made a hole in it, perforated it, pierced it,* or *bored it;* (Mṣb,* KL;) syn. جَابَهُ [in this sense, as well as in another to be explained below], (Ḳ, [in the CḲ, erroneously, جاءَ بِهِ,]) and ثَقَبَهُ: (TA:) and *he cut it* [*so as to make a hole* or *a slit in it*]: (Mṣb:) and *he rent it,* or *tore it.* (JK, Ḳ, KL.) You say, خَرَقَ الثَّوْبَ, (JK, Ṣ, Mgh, Ḳ,) aor. ـُ [and ـِ], inf. n. as above, (Mgh,) *He* [*made a hole in,* or] *rent,* or *tore, the garment,* or *piece of cloth;* (JK, Ḳ;) and الخُفَّ [*the boot*]; and the like. (Mgh.) And خَرَقَ الصَّخْرَةَ *He made a hole in the rock;* syn. جَابَهُ. (A in art. جوب.) [And خَرَقَ الحَائِطَ *He made a hole in, or through, the wall:* see خَرْقٌ, below.] And خَرَقَهُ بِالْمِثْقَبِ *He made a hole in it* or *through it, perforated it, pierced it,* or *bored it, with a drill* or *the like;* syn. ثَقَبَهُ. (Mṣb in art. ثقب.) خَرَقَ السَّفِينَةَ [*He made a hole in the ship*], in the Ḳur xviii. 70, means that he did so by taking out, from the ship, with an axe, (Ksh, Bḍ, Jel,) a plank, (Jel,) or two planks. (Ksh, Bḍ.) — [Hence,] خَرَقَ الأَرْضَ, (JK, Ṣ, Mṣb,) or المَفَازَةَ, (Mgh, Ḳ,*) ‡*He traversed, crossed,* or *cut through by journeying,* (JK, Ṣ, Mgh, Mṣb, Ḳ,) *the earth,* or *land,* (JK, Ṣ, Mṣb,) or *the desert;* (Mgh, Ḳ;) syn. قَطَعَهَا; (JK, Mgh, Ḳ;*) or جَابَهَا; (Ṣ, Mṣb;) *so as to reach the furthest part thereof.* (Mgh, TA.) [See also 8.] It is said in the Ḳur [xvii. 39], إِنَّكَ لَنْ تَخْرِقَ الأَرْضَ, meaning, *For thou shalt not reach the extremities of the earth:* or, accord. to Az, *thou shalt not traverse the earth in length and breadth:* (TA:) or it means *thou shalt not bore through the earth,* (Jel, TA,) *so as to reach the end thereof:* (Jel:) or *thou shalt not make a hole in the earth* by thy vehement treading: (Ksh, Bḍ:) accord. to one reading, لن تَخْرُقَ. (Ksh, TA.) — [And خَرَقَتِ الرِّيحُ †*The wind passed along:* and †*blew:* for] the inf. n. خَرْقٌ signifies †the *passing* of the wind: and †the *blowing* thereof. (KL.) [See also 7 and 8.] — خَرَقَ الكَذِبَ ‡*He forged,* or *feigned, the lie;* as also ▼ اخترقهُ. (Ḳ, TA.) It is said in the Ḳur vi. 100, وَخَرَقُوا لَهُ بَنِينَ وَبَنَاتٍ, i. e. ‡*And they have feigned Him to have,* or *falsely attributed to Him, sons and daughters.* (Ksh, Bḍ, Jel. [See also 2.]) And خَرَقَ [alone, the object being understood], (Ḳ,) inf. n. as above, (KL,) signifies ‡*He lied; told a lie:* (Ḳ, KL, TA:) and ▼ تَخَرَّقَ ‡*he forged,* or *feigned, a lie.* (Ṣ, Ḳ, TA.) = خَرِقَتِ الشَّاةُ, aor. ـَ, inf. n. خَرَقٌ, *The sheep had in its ear a* خَرْق, i. e. *a round hole* or *perforation.* (Mṣb.) = خَرِقَ فِى البَيْتِ, aor. ـَ, (JK, Ḳ,) inf. n. خُرُوقٌ, (JK,) or

خَرِقَ; (TK;) and خَرَقَ, inf. n. خُرُوقٌ; (K;) *He remained in the house,* or *tent, not quitting it.* (JK,* K.) —— And خَرِقَ, aor. َ, inf. n. خَرَقٌ, said of a gazelle, or young gazelle, (Msb, K, TA,) when hunted, (TA,) or when overtaken by the dog, (IAar,) *It was frightened,* (Msb, K, TA, [in the CK, أَنْ يَخْرَقَ is erroneously put for أَنْ يَخْرَقَ,]) *so as to be unable to go away,* (Msb,) or *so as to be unable to rise,* (K, TA,) *and clave to the ground:* (IAar, TA:) and in like manner said of a bird, (Msb, K,) *it became frightened,* (K,) or *impatient,* (TA,) *so as to be unable to fly away.* (K, TA.) —— And hence, (Msb,) the same verb, (S, Msb, K,) with the same aor., (Msb, K,) and the same inf. n., (S, Msb, K,) said of a man, (Msb,) *He became confounded,* or *perplexed, and unable to see his right course; or he became bereft of his reason* or *intellect, because of fear,* or *of shame:* (S, Msb, K:) or *he was confounded, perplexed,* or *amazed,* [for يَتَهَيَّبُ in the CK, I read يَبْهَتُ, as in other copies of the K and in the TA,] *opening his eyes, and looking:* (K, TA:) and *he remained confounded,* or *perplexed, and unable to see his right course, by reason of anxiety,* or *of hardship,* or *distress.* (TA.) وَقَعَ فَخَرِقَ [*He fell down and clave to the ground*], occurring in a trad., means *he fell down dead.* (TA.) ═ خَرِقَ, aor. َ, (JK, S, Msb, K,) inf. n. خَرَقٌ; (S, Msb, K;*) and خَرُقَ, aor. ُ, (JK, K,) [of which خُرْقٌ, said in the S and Msb to be a simple subst., may be the inf. n., like as حُسْنٌ is of حَسُنَ;] *He was rough, ungentle, clumsy,* or *awkward,* (S, Msb, K,) *in doing,* or *making, a thing:* (Msb:) and *he was unskilful in work,* and *in the management of affairs:* and *he was foolish; stupid;* or *unsound,* or *deficient, in intellect* or *understanding:* (K:) or *he was ignorant:* (JK:) or the latter verb signifies *he knew not his work with his hand,* or *his handicraft.* (Msb.) And خَرِقَ بِالشَّىْءِ *He was ignorant of the thing,* (K, TA,) *and did it not well.* (TA.)

2. خرّقه, (S, Msb,) inf. n. تَخْرِيقٌ, (Msb, K,) is similar to خَرَقَهُ, but has an intensive signification; [*He made holes in it; perforated it, pierced it,* or *bored it, in several,* or *many, places: he cut it so as to make holes* or *slits in it:*] (Msb:) *he rent it,* or *tore it, much,* or *in several,* or *many, places:* (K, TA:) namely, a garment, (S, TA,) &c. (TA.) —— And خرّق, (TA,) inf. n. as above, (K,) † *He lied much.* (K, TA.) Aboo-Jaafar and Náfi' read, [in the Kur vi. 100,] وَخَرَّقُوا لَهُ بَنِينَ وَبَنَاتٍ † [*And they have very falsely attributed to Him sons and daughters*]. (TA. [See also 1.])

4. اخرقهُ *He* (a man, S) *caused him to be confounded,* or *perplexed, so that he was unable to see his right course; or caused him to be bereft of his reason,* or *intellect.* (S, K.)

5. تخرّق quasi-pass. of خَرَّقَ; [thus signifying *It had holes made in it; became perforated, pierced,* or *bored, in several,* or *many, places: it became cut so as to have holes* or *slits made in it: it became rent,* or *torn, much,* or *in several,* or *many, places:*] (S,* K:) as also ▼انخرق; (K:) [or rather the latter, as is indicated in the S, is quasi-pass. of خَرَقَ, and thus signifies *it had a hole made in it; became perforated, pierced,* or *bored: it became cut so as to have a hole* or *slit made in it: it became rent,* or *torn:*] and ▼اخروّق signifies the same [as the former or as the latter]: all said of a garment [&c.]: (S:) and ▼انخرق signifies also *it became wide,* or *expanded.* (TA.) —— [Hence,] تخرّق فِى السَّخَآءِ ‡ *He took a wide,* or *an ample, range,* or *was profuse, in liberality, bounty,* or *munificence;* syn. توسّع. (S, K, TA.) —— See also 1, in the middle of the paragraph. —— And see what next follows, in two places.

7: see 5, in two places. —— انخرقت الرِّيحُ ‡ *The wind blew* [app. *in any manner,* (see مُنْخَرَقٌ,) or] *irregularly; not in one uniform manner:* (TA:) [and ▼تخرّقت app. signifies the same: for you say,] بَلَدٌ وَاسِعٌ تَتَخَرَّقُ بِهِ الرِّيَاحُ [† *A wide country in which the winds blow,* or *blow irregularly*]: (El-Muärrij, TA:) and أَرْضٌ وَاسِعَةٌ تَتَخَرَّقُ ▼ فِيهَا الرِّيَاحُ [app. meaning, in like manner, † *A wide land in which the winds blow,* &c.]. (S, K.)

8. اخترق ‡ *He,* or *it, passed through,* or *over,* or *across.* (Mgh, K,* TA.) [See also 1, in the former half of the paragraph.] ‡ *He traversed,* or *crossed,* (Mgh, TA,) a desert, (Mgh,) or a land, (TA,) *not following a road.* (Mgh, TA.) [† *He travelled* a road: see an ex. voce ثُغْرَةٌ.] ‡ *He made* a mosque, (Mgh, TA,) and a house, (TA,) *to be his way,* or *thoroughfare.* (Mgh, TA.) Hence, اخترق الحِجْرَ † *He entered into the midst of the* حِجْر [q. v.], *without going around the* حَطِيم. (Mgh.) And الخَيْلُ تَخْتَرِقُ مَا بَيْنَ القُرَى وَالأَرْضِ ‡ *The horses,* or *horsemen, pass through the midst of the intervening spaces of the towns,* or *villages, and the land.* (TA.) And اخْتَرَقْتُ القَوْمَ † *I stepped into the midst of the people,* or *party.* (TA.) And الرِّيحُ تَخْتَرِقُ الأَشْجَارَ † [*The wind passes,* or *blows, through the trees.*] (JK.) اِخْتِرَاقُ الرِّيَاحِ signifies † *The passing* [or *blowing*] *of the winds.* (S.) [See also 1, in the middle of the paragraph; and see 7.] —— اخترق الكَذِبَ: see 1, in the middle of the paragraph.

12: see 5.

خَرْقٌ, originally an inf. n., of 1: (S, Mgh, Msb, TA:) *A hole,* or *perforation,* (Mgh, Msb, KL,) in a garment, (S, TA,) and in a wall, (Msb, TA,) &c.: (Msb:) and *a round hole,* or *perforation,* in the ear of a sheep: (S, Msb:) pl. خُرُوقٌ. (S, Mgh, Msb.) Hence the saying, اِتَّسَعَ الخَرْقُ عَلَى الرَّاقِعِ [*The hole was,* or *became, wide to the patcher*]. (TA.) —— And *A part that has a hole made in it,* or *that is rent,* or *torn, of,* or *from, a thing.* (TA.) —— Also *A desert;* and so ▼مُنْخَرَقٌ: (K:) or the former, *a desert far extending,* (JK, TA,) whether *level* or *not level:* and ▼the latter, *a wide desert in which the winds* [*blow,* or] *blow irregularly:* (TA:) and the former, (El-Muärrij, K,) as also ▼خَرْقَآءُ, (K,) signifies likewise *a wide land,* (K,) or *a wide country,* (El-Muärrij,) *in which the winds* [*blow,* or] *blow irregularly:* (El-Muärrij, K: [see 7:]) ISh says, the space between El-Basrah and Hafr Abee-Moosà is a خَرْق, and that between En-Nibáj and Dareeyeh is a خَرْق: (TA:) pl. خُرُوقٌ. (K.) You say also مَفَازَةٌ خَرْقَآءُ ▼ خَوْقَآءُ *A far-extending desert.* (TA.) And قَطَعْنَا إِلَيْكُمْ أَرْضًا خَرْقَآءَ ▼ and خُرُوقًا ▼ [*We have traversed,* in journeying *to you, a wide land,* or *a wide land in which the winds blow,* &c.]. (TA.) ═ Also *A certain plant, resembling the* قُسْط [q. v.], (JK, Ibn-'Abbád, K,) *having leaves.* (JK, Ibn-'Abbád.)

خُرْقٌ a subst. from خَرِقَ, (S, Msb,) [or perhaps inf. n. of خَرُقَ as syn. with خَرِقَ, (see 1, last two sentences,)] and ▼خُرُقٌ, (TA,) *Roughness, ungentleness, clumsiness,* or *awkwardness;* contr. of رِفْقٌ; (JK, S,* Mgh, Msb,* K, TA;) *in doing,* or *making, a thing:* (Msb:) *unskilfulness in work,* and *in the management of affairs: foolishness; stupidity;* or *unsoundness,* or *deficiency, in intellect* or *understanding;* as also ▼خُرْقَةٌ: (K:) and *ignorance.* (TA.) [Hence,] نَوْمَةُ الخُرْقِ *The sleep of* [*the time of morning called*] *the* ضُحَى. (Har p. 223. [See also خُمْقٌ and خُلْقٌ.]) ═ The first of these words is also pl. of أَخْرَقُ and of [its fem.] خَرْقَآءُ. (K.) ═ Also *The she-camel's vulva.* (JK.)

خِرْقٌ and ▼خِرِّيقٌ ‡ *Liberal, bountiful, munificent,* or *generous;* (S, K;) as also ▼مِخْرَاقٌ: (IAar, K:) or ▼the second signifies *very liberal* or *bountiful* &c.: (K: [so in a later part of the art.:]) or this and the first signify one *who takes a wide,* or *an ample, range,* or *is profuse, in liberality* or *bounty* &c.: (TA:) or a youth, or young man, (JK,) *excellent,* or *elegant, in mind, manners, address, speech, person, and the like;* or *clever,* or *ingenious; with liberality, bounty, munificence,* or *generosity,* (Lth, JK, K,) *and courage:* (Lth, JK:) and a *goodly* youth or young man, [for الفَتَى in the CK, I read الفَتِىُّ, as in other copies of the K,] *of generous disposition:* (K:) the pl. (of خِرْق, TA) is أَخْرَاقٌ [a pl. of pauc.] (K) and خِرَاقٌ, or خُرَّاقٌ, (accord. to different copies of the K, [both anomalous, and perhaps it is خِرَاقٌ, agreeably with analogy,]) and خُرُوقٌ; (K;) and the pl. of ▼خِرِّيقٌ is خِرِّيقُونَ; no broken pl. of it having been heard. (T, TA.) One says also, هُوَ مُتَخَرِّقُ ▼ الكَفِّ بِالنَّوَالِ ‡ [*He has a liberal hand, largely beneficent*]. (TA.) And هُوَ مَخْرُوقُ ▼ الكَفِّ بِالنَّوَالِ ‡ *He is liberal, bountiful, munificent,* or *generous.* (TA. [But see مَخْرُوق below.]) —— خِرْقٌ is also applied to a spear, meaning † *Highly esteemed* or *prized; excellent;* or *rare.* (TA.)

خُرُقٌ: see خُرْقٌ.

خَرِقٌ [part. n. of خَرِقَ, q. v.:] A young gazelle *weak in the legs,* (K, TA,) *cleaving to the ground, and not rising:* (TA:) a gazelle, or young gazelle, (K, TA,) when hunted, (TA,) *frightened, so as to be unable to rise:* (K, TA:) and in like manner a bird (K, TA) *frightened,* (K,) or *impatient,* (TA,) *so as to be unable to fly away:* (Ḳ, TA:) fem. with ة. (Ḳ.) — And [hence,] A man (Mṣb) *confounded,* or *perplexed, and unable to see his right course;* or *bereft of his reason* or *intellect, by reason of fear,* or *of shame:* (Ṣ, Mṣb, Ḳ:) or *confounded, perplexed,* or *amazed, opening his eyes, and looking.* (Ḳ, TA.) See also أَخْرَقُ. — Also *Ashes:* because they remain [cleaving to the ground] while the people thereof go away. (Ḳ.)

خُرْقٌ : } see خَرْقٌ.
خُرْقَةٌ :

خِرْقَةٌ *A piece,* (Ṣ, Mṣb, Ḳ,) or *piece torn off,* (TA,) of a garment, or of cloth; [*a rag;*] pl. خِرَقٌ. (Ṣ, Mṣb, TA.) — [*A ragged, patched, garment:* and particularly *one worn by a devotee;* also called مُرَقَّعَةٌ, q. v.: but this is probably post-classical. Hence, أَصْحَابُ الخِرَقِ *The devotees.*] — ‡ *A portion* of a swarm of locusts, (Ḳ, TA,) less than a رِجْل; as also حِزْقَةٌ. (TA.)

خَرُوقٌ : see the next paragraph: — and see also خَرِقٌ.

خَرِيقٌ A womb *rent by the fœtus, and that consequently does not conceive* (Ḳ, TA) *afterwards;* (TA;) [of the measure فَعِيلٌ in the sense of the measure مَفْعُولٌ;] as also ▼ مُتَخَرِّقَةٌ. (Ḳ.) — And A she-camel *whose womb has been rent.* (JK.) Applied to a well (بِئْر), it signifies الَّتِى كُسِرَ جَبْلَتُهَا عَنِ المَآءِ: (JK, Ibn-'Abbád, Ḳ:) [in the CḲ جَبْلَتُها: neither of these readings affords an admissible meaning: the right reading I believe to be جِيلُهَا; and the meaning, *Of which the side,* or *lateral part, is broken, from the water upwards:*] pl. خَرَائِقُ (JK, Ibn-'Abbád, Ḳ, TA) and خُرُقٌ, (Ibn-'Abbád, Ḳ, TA, [the latter erroneously written in the CḲ خُرُوقٌ,]) like سَفَائِنُ and سُفُنٌ. (TA.) — *A channel of water that is not deep, and not without trees.* (JK, Ibn-'Abbád, Ḳ.) — The *place of expanding of a valley, where it ends.* (JK, Ḳ.) — *A low,* or *depressed, tract* of land, *containing herbage:* pl. خُرُقٌ. (Ṣ, Ḳ.) One says, مَرَرْتُ بِخَرِيقٍ مِنَ الأَرْضِ بَيْنَ مَسْحَاوَيْنِ [*I passed by a low tract of land, containing herbage, between two plain tracts containing small pebbles and without herbage*]. (Fr, Ṣ.) — *Hard ground.* (A, TA.) — ‡ *A violent wind;* (A, TA;) as also ▼ رِيحٌ خَرْقَآءُ: (Ṣ, Ḳ:) the latter signifies ‡ *a wind that blows violently:* or, *that does not continue to blow in the same direction:* (TA:) or the former signifies ‡ *a cold* wind *that blows violently;* (Ṣ, Ḳ;) as also ▼ خَرُوقٌ: (Ḳ:) [it is an epithet; for] one says رِيحٌ خَرِيقٌ which is anomalous, as by rule one should say خَرِيقَةٌ: (Ṣ:) it is [also] one of the names for ‡ *a cold wind that blows violently;* (JK, T, TA;) as though it perforated, or rent; the agent [رِيح] being unused: (T, TA:) and (as some say, TA) it signifies also † a *gentle, soft,* wind; thus bearing two contr. meanings: or *that returns, and* [*then*] *continues its course:* (Ḳ:) or, as in the L, *does not continue its course:* (TA:) or *that blows long.* (Ḳ.)

خُرَّقٌ *A certain bird,* (JK, IDrd, Ḳ,) *smaller than the* قُنْبُر [or *lark*], (JK,) *that cleaves to the ground:* (IDrd:) or *a kind of sparrow:* (Ḳ:) so says AḤát, in the "Book of Birds:" (TA:) pl. خَرَارِقُ. (JK, IDrd, Ḳ.)

خِرِّيقٌ : see خَرِقٌ, in three places.

خَارِقٌ [act. part. n. of خَرَقَ]. — [And hence,] سَيْفٌ خَارِقٌ *A sharp,* or *cutting, sword:* pl. خُرُقٌ. (TA.) — [Hence also,] أَمْرٌ خَارِقٌ لِلْعَادَةِ † [*An event breaking through,* or *infringing, the usual course* of nature]. (KT, in a definition of مُعْجِزَةٌ, q. v.) — [In the present day, خَارِقٌ signifies also † *Profound,* or *penetrating,* in learning or science.]

أَخْرَقُ and [its fem.] خَرْقَآءُ have for their pl. خُرْقٌ. (Ḳ.) — The fem., applied to a ewe, signifies *Having her ear perforated* (Ṣ, Mgh, Mṣb, Ḳ) *with a round hole.* (Ṣ, Mṣb.) And, applied to an ear, *Perforated,* or *bored.* (TA.) — And the masc., applied to a camel, *That puts his* مَنْسِم [or *toe*] *upon the ground before* [*the sole of*] *his* خُفّ [or *foot*]: the doing of which is a result of generous quality. (JK, Ibn-'Abbád, Ḳ.) And the fem., applied to a she-camel, *That does not retread* (لَا تَتَعَاهَدُ, in the L لا تتعهّد,) [*with her hind feet*] *the places of her* [*fore*] *feet* (JK, L, Ḳ) *upon the ground:* mentioned by Ibn-'Abbád and Z. (TA.) — Applied to a man, (Mgh, Mṣb,) *Rough, ungentle, clumsy,* or *awkward,* (JK, Ṣ, Mṣb,) *in doing,* or *making, a thing:* (Mṣb:) *unskilful in work* [*and in the management of affairs* (see خَرِقَ]; as also ▼ خَرِقٌ and ▼ خُرُقٌ: (Ḳ:) or *foolish; stupid;* or *unsound,* or *deficient, in intellect* or *understanding:* (Mgh, Ḳ:) *ignorant:* (TA:) *not knowing his work with his hand,* or *his handicraft:* (Mṣb:) fem. as above. (JK, Ṣ, Mgh, Mṣb, &c.) It is said in a prov., (JK, Ṣ,) لَا تَعْدَمُ الخَرْقَآءُ عِلَّةً [*The stupid woman is not in want of an excuse*]: (JK, Ṣ, Ḳ:) used in forbidding excuses: (Ḳ:) i. e., excuses are many: the stupid woman is skilled in making them: how then must be the clever? (Ṣ, Ḳ:) applied to every one who excuses himself being able. (Ḳ in art. عل.) — خَرْقَآءُ applied to a desert, and to a land: see خَرْقٌ, in three places. — And applied to a wind: see خَرِيقٌ. — Hence, رِحْلَةٌ خَرْقَآءُ † *A hard journey.* (Ḥar p. 177.)

مَخْرَقٌ : see خَرْقٌ, in two places. — مَخْرَقُ حَوْضٍ *A stone that is at the* عُقْر [or *hinder part*] *of a watering-trough, for the purpose of their* [*standing upon it, and*] *drawing forth the water from it,* [i. e. the trough,] *when they will.* (Ḳ.)

مَخْرِقٌ, though unheard by us, is the sing. of مَخَارِقُ signifying The *orifices of the body; such as the mouth and the nose and the ears and the anus and the like.* (Mgh.)

مُخْرَقٌ [pass. part. n. of 4; *Confounded,* &c.: and hence,] *silent.* (JK: but there written without the vowel of the ر.)

غَيْرُ مُخْرِقٍ, applied to a road, means [*That does not cause one to be confounded,* or *perplexed, and unable to see his right course;* or] *in which one is not confounded,* or *perplexed, so as to be unable to see his right course.* (IAạr in TA: but the latter word is there written without the vowel of the ر.)

مِخْرَاقٌ *A kerchief twisted for the purpose of beating therewith:* (JK, Ṣ:) a genuine Arabic word: (Ṣ:) or *a thing made of twisted rags, with which boys play:* (TA:) or *a twisted kerchief,* or *an inflated* [*skin such as is termed*] زِقّ, or *the like, with which boys play, beating one another therewith:* so called because it rends (يَخْرِقُ) the air when they make use of it: (Ḥam p. 702:) pl. مَخَارِيقُ. (Ṣ, TA.) 'Amr Ibn-Kulthoom says,

* كَأَنَّ سُيُوفَنَا مِنَّا وَمِنْهُمْ *
* مَخَارِيقٌ بِأَيْدِى لَاعِبِينَا *

[*As though our swords, ours and theirs, were kerchiefs twisted for beating therewith, in the hands of players*]: (Ṣ:) or مخاريق in this verse [written with tenween by poetic license] is the pl. of مِخْرَاقٌ signifying *a wooden sword with which boys play:* the poet means, we cared not for the smiting with the swords, like as the players care not for the smiting with the مخاريق. (EM p. 198.) [See also another ex., in a verse cited voce خَرِيعٌ.] 'Alee is related, in a trad., to have said that the lightning is the مَخَارِيقُ of the angels; (Ṣ, TA;) meaning thereby the instruments with which the angels chide and drive the clouds. (TA.) — Also *A garment,* or *piece of cloth.* (JK. [But this I find not elsewhere.]) — And ‡ *A sword* [in the ordinary sense of the word]: so in the A and O and L: in the Ḳ, السَّيِّدُ is erroneously put for السَّيْفُ. (TA.) — † A man *goodly in body,* or *person,* whether *tall* or *not tall.* (JK, Ḳ.) — † One *who falls not into a case without escaping,* or *extricating himself, therefrom.* (Sh, TA.) — † One *who exercises art in the management of affairs.* (Ḳ.) — ‡ *A wild bull:* (Aṣ, Ḳ:) so called because he traverses far-extending districts: (Aṣ, TA:) or because the dogs pursue him and he escapes from them: said in the A to be called مِخْرَاقُ المَفَازَةِ. (TA.) — † A man *who engages in wars,* or *fights, and is active therein.* (Ṣ, Ḳ.) — See also خِرْقٌ.

مَخْرُوقٌ ‡ One who is *denied good,* or *prosperity; into whose hand wealth falls not.* (Ḳ, TA.) And مَخْرُوقُ الكَفِّ † A man *who gains not,* or *gets not, anything.* (JK.) See also خَرِقٌ.

مُخَرْوِقٌ One *who goes round about camels,* [meaning *who has them within the compass of his rule and care,*] (JK, Ḳ, TA,) *and urges them against their will,* (TA,) *and is active, and exercises art in his management* [*of them*]: (JK, Ḳ, TA:) mentioned by Ṣgh on the authority of Ibn-'Abbád. (TA.)

مُخْتَرَقٌ † *A passage,* or *place of passing.* (Ṣ. [See خَوْخَةٌ, in two places.]) — [Hence,] بَلَدٌ بَعِيدُ المُخْتَرَقِ † [*A country,* or *district, wide to traverse;* lit., *far extending in respect of the place of passing*]. (TA.) —— مُنْخَرَقُ الرِّيَاحِ † *A place in which the winds blow:* (Ḳ:) and مُنْخَرِقُ ↑ الرِّيحِ † *a place in which the wind blows* [*in any manner,* or *irregularly:* see 7]. (Ṣ.)

مُتَخَرِّقٌ: see the last paragraph in this art.: and see also خَرِيقٌ: —— and خِرْقٌ.

مُخْتَرَقٌ: see مُنْخَرَقٌ.

مُنْخَرِقٌ [*Having a hole made in it,* &c.: see its verb]. رَجُلٌ مُنْخَرِقُ السِّرْبَالِ *A man having his clothing rent,* or *torn,* (JK, Ḳ,) *by long travel;* as also مُتَخَرِّقُ ↑ السِّرْبَالِ. (Ḳ.) —— Also † *Quick,* or *swift.* (Ḥam p. 42.)

خرم

1. خَرَمَهُ, aor. ـِ, inf. n. خَرْمٌ, *He perforated,* or *pierced, it;* namely, a thing. (Mṣb.) [And so خَزَمَهُ. (Mgh in art. خزم.)] —— And *He cut it,* or *cut it off.* (Mṣb.) You say, مَا خَرَمْتُ مِنْهُ شَيْئًا *I did not diminish,* and *did not cut off, from it,* or *him, anything.* (Ṣ.) And مَا خَرَمَ مِنَ الحَدِيثِ حَرْفًا *He did not diminish* [*from the narrative,* or *tradition, a letter,* or *a word*]. (TA.) And خَرَمَ فُلَانًا, (Ḳ,) aor. and inf. n. as above, (TA,) *He slit the partition between the nostrils of such a one:* (Ḳ:) or خَرْمٌ signifies the *mutilating* (قَطْع) *of the nose:* (JK:) or *mutilation* (قَطْع) *in the partition between the nostrils and in the* نَاشِرَتَانِ [or *two alæ*]; or *in the extremity of the* أَرْنَبَة [or *lobule of the nose*]; *not amounting to what is termed* جَدْع; (Lth, TA;) and the epithet is أَخْرَمُ↑, fem. خَرْمَآءُ: (Lth, JK, TA:) and *the like in the lip;* or *in the upper part of the* فُوق [app. meaning the *front edge of the lobe,* which at its termination above forms a crena,] *of the ear:* (Lth, TA: [see أَخْرَمُ, as relating to the ear:]) accord. to Sh, it is both *in the nose* and *in the ear;* but in the nose, it is the *mutilation* (قَطْع) *of the fore part of the nostril* of a man, *and the* أَرْنَبَة [or *lobule of the nose*], *after the mutilation of the upper part of this, so as to reach the interior of the nose;* and the epithet applied to the man is أَخْرَمُ↑. (TA.) And خَرَمَهُ, inf. n. as above, also signifies *He hit,* or *hurt, his* خُورَمَة [q. v.]. (TA.) You say also, خَرَمَ الخُرْزَةَ, (Ḳ, TA, in the CḲ [erroneously] الخَرَزَةَ,) aor. as above, (Ḳ,) and so the inf. n.; (TA;) and خَرَّمَهَا↑, (Ḳ,) inf. n. تَخْرِيمٌ; (TA;) *He cracked,* or *tore without separating, the suture,* or *seam,* of a skin; syn. فَصَمَهَا: (Ḳ, TA, in the CḲ قَصَمَهَا:) or خَرَمْتُ الخَرْزَ, aor. and inf. n. as above, *i. q.* أَثْأَيْتُهُ [meaning *I spoiled the sewing* of the skin, or hide; *as when one uses a thick instrument for sewing or perforating, and a thin thong;* or *as when one rends two stitch-holes into one*]. (Ṣ.) [And خَرَمَهُ سَيْلٌ app. *A torrent cut into it,* or *trenched it;* namely, the ground, or the side of a mountain: see خَرْمٌ.] And خَرَمَتْهُ خَوَارِمُ↑ [lit. *Cutting-off events cut him off*]; meaning †*he died:* like as one says, شَعَبَتْهُ شَعُوبُ. (TA. [See also 8.]) —— خَرَمَ القِرْطَاسَ *He hit the target with his arrow without perforating it.* (TA.) —— مَا خَرَمَ عَنِ الطَّرِيقِ *He* (a guide) *did not turn aside from the way.* (JK, Ṣ.) ═ خَرِمَ, (Ṣ, Ḳ,) aor. ـَ, (Ḳ,) inf. n. خَرَمٌ, (Ṣ,) said of a man, (Ṣ, Ḳ,) *He had the partition between his nostrils cut,* or *mutilated:* or *the extremity of his nose, but not to the extent denoted by the term* جَدْع: the epithet applied to the man in this case is أَخْرَمُ↑: (Ṣ:) or *he had the partition between his nostrils slit;* i. e. تَخَرَّمَتْ↑ وَتَرَتُهُ. (Ḳ.) ═ خَرِمَ, aor. ـَ, *He cared not for what he did nor for what was said to him.* (Ḳ.)

2: see 1. —— [Hence,] ضَرْعٌ فِيهِ تَخْرِيمٌ *An udder in which are incisions* [or *crackings of the skin*]; and so فِيهِ تَشْرِيمٌ. (TA.)

5. تَخَرَّمَتْ وَتَرَتُهُ: see 1, last sentence but one. تَخَرَّمَتِ الخُرْزَةُ (Ḳ, TA, in the CḲ [erroneously] الخَرَزَةُ,) *The suture,* or *seam,* of a skin *cracked,* or *became torn without separating;* quasi-pass. of خَرَّمَهَا [q. v.]. (Ḳ, TA.) انخرم↑ [in like manner] signifies *It became slit;* said of the bore of the ear. (Ṣ.) And you say also, تَخَرَّمَ الزَّنْدُ [*The wooden instrument for producing fire cracked,* or *split*]. (TA.) Hence the phrase, أَرَاكَ يَتَخَرَّمُ زَنْدُكَ, mentioned by IAạr, meaning †*I see thee to have no good in thee:* for when the زند cracks, or splits, (إِذَا تَخَرَّمَ,) one cannot produce fire by means of it, and there is no good in it. (TA.) [Hence likewise,] تَخَرَّمَ زَنْدُهُ means also ‡*His anger became appeased:* [or,] accord. to the Ṣ, تَخَرَّمَ زَبْدُهُ has this meaning: and accord. to the A, تَخَرَّمَ أَنْفُهُ has the same meaning. (TA.) Accord. to IAạr, (TA,) جَآءَنَا فُلَانٌ يَتَخَرَّمُ زَبْدُهُ means ‡*Such a one came to us doing to us that which was wrongful,* or *injurious, and foolish,* or *stupid.* (Ḳ,* TA.) ═ See also 8, in two places. ═ تخرّم also signifies *He followed,* or *adopted, the religion of the* خُرَّمِيَّة. (Ṣ, Ḳ, TA: in the CḲ, the مَخْرُمِيَّة.)

7. انخرم *It became cut,* or *cut off.* (Mṣb.) See also 5. Said of a writing, or book, it means *It became deficient; part of it went.* (TA.) And said of a generation, *It went away; came to an end.* (TA.) See also أُخْتُرِمَ.

8. اِخْتَرَمَهُمُ الدَّهْرُ *Time,* or *fortune, cut them off;* and *extirpated them;* as also تَخَرَّمَهُمْ↑: (Ṣ:) or *destroyed them by its calamities.* (Mṣb.) And تَخَرَّمَتْهُمُ↑ and اِخْتَرَمَتْهُمُ المَنِيَّةُ *Death,* or *the decree of death, cut them off;* and *extirpated them.* (Ḳ.) And اِخْتَرَمَتْهُ المَنِيَّةُ *Death,* or *the decree of death,* [*cut him off,* or] *took him away,* (JK, Ḳ, TA,) مِنْ بَيْنِ أَصْحَابِهِ [*from amidst his companions.* (TA. [A phrase similar to خَرَمَتْهُ خَوَارِمُ: see 1.]) And أُخْتُرِمَ عَنَّا (JK,* Ḳ) [*He was cut off from us by death;*] *he was taken away* [*from us by death*]; (JK;) *he died,* (Ḳ, TA,) and *went away* [*from us*]. (TA.) Accord. to some, اِخْتِرَامٌ [as an inf. n. of أُخْتُرِمَ] signifies The *dying suddenly.* (Ḥar p. 123.)

خَرْمٌ *A prominence,* or *projecting part, of a mountain:* (JK, Ṣ, Ḳ:) pl. خُرُومٌ. (JK.) And *A bed trenched by a torrent* (مَا خَرَمَ سَيْلٌ [see 1]): (JK:) so some say: (TA:) or *a road in a* [*tract of high ground such as is termed*] قُفّ; or *on the summit of a mountain.* (JK, TA.) [See also مَخْرِمٌ.]

خُرْمٌ The *place of the bore,* or *perforation,* of a thing. (Mṣb.) The *eye* of a needle. (TA.) —— See also مَخْرِمٌ.

خَرَمَةٌ The *place of perforation of the ear:* (Ṣ:) or the *place of slitting, of the nose, in the partition between the nostrils* [and *in either of the alæ,* as appears from what here follows]. (Ḳ.) It is said in a trad., فِى الخَرَمَاتِ الثَّلَاثِ مِنَ الأَنْفِ الدِّيَةُ, by الخرمات being app. meant المَخْرُومَاتُ↑, i. e. [*In the case of* the mutilation of] *the two alæ and the partition between the nostrils* [*the blood-wit,* or *fine for homicide,* shall be paid]. (TA.)

خُرْمَانٌ *A lie,* or *falsehood.* (Ṣ, Ḳ.) One says, جَآءَ فُلَانٌ بِالخُرْمَانِ (Ṣ, TA) i. e. [*Such a one uttered*] *that which was a lie.* (TA.)

خَرِيمٌ One *who cares not for what he does nor for what is said to him.* (Ḳ.)

الخُرَّمِيَّةُ *The sect who held the doctrine of the transmigration of the soul, and allowed general license:* (Ṣ, Ḳ, TA:) they were in the time of El-Moạtaṣim: their sheykh, Bábak [El-Khurramee, i. e. of Khurram, in Persia], was then slain, and they scattered themselves in the countries; and there remains of them a remnant in the mountains of Syria. (TA.)

خُرَّامٌ [a pl. of which the sing. is not mentioned,] Young men (TA) *such as follow the licentious ways of the* خُرَّمِيَّة [so I render مُتَخَرِّمُونَ↑] *in acts of disobedience.* (Ḳ, TA.) [See what next follows.]

خَارِمٌ [act. part. n. of خَرَمَ: fem. with ة; and pl. of the latter خَوَارِمُ]. One says, خَرَمَتْهُ خَوَارِمُ [explained above]: see 1. (TA.) —— *Corrupting; acting corruptly; doing evil,* or *mischief.* (Ḳ.) [See خُرَّامٌ, which is probably a pl. thereof.] —— *Neglecting;* or *leaving undone* [*what ought to be done*]. (Ḳ.) ═ *Cold,* as an epithet. (Ḳ.) —— *A cold wind:* (Ḳ:) so accord. to A 'Obeyd: but accord. to Kr, [خَازِمٌ,] with زاى. (TA.)

خُورَمٌ: see what next follows.

خُورَمَةٌ The *end,* or *tip, of the nose* (JK, Ṣ) of a man: (Ṣ:) or the *fore part of the nose:* or the *part between the nostrils.* (Ḳ.) —— Also, (JK, Ḳ,) as being likened thereto, (TA,) †*A rock in which are holes;* (JK, Ḳ;) n. un. of خُورَمٌ↑: (Ḳ:) [or] the latter has this signification. (Ṣ. [But this seems to be a mistake.])

أَخْرَمُ [*Having the nose mutilated in any of the manners explained in the first paragraph of this art.*]: fem. خَرْمَآءُ: see 1, in three places. —— And *Having the ear perforated, when it is not slit:*

(Ṣ:) or *having the ear slit after it has been pierced:* (Ṣ and Mṣb* and TA in art. خرب:) and مُخَرَّمٌ ↑, likewise, signifies *having the ear slit;* as also أَخْرَبُ and مُخَرَّبٌ. (TA in that art.) And the fem., applied to a she-goat, *Having her ear slit crosswise.* (Ḳ. [See also خَذْمَآءُ.]) Also, the fem., applied to an ear, *Slit,* or *perforated,* or *mutilated.* (Ḳ,* TA.) ═ Also *A pool of water left by a torrent;* because one part thereof passes away (يَنْخَرِمُ ↑) to another: pl. خُرْمٌ. (TA.) —— And the fem. also signifies *Any hill,* or *rising ground, sloping down into a [hollow such as is termed]* وَهْدَة; (Ḳ;) and so the masc.: (TA:) or *any [hill such as is termed]* أَكَمَة *having a side whereby it cannot be ascended.* (Ḳ.) ═ أَخْرَمُ الكَتِفِ *The extremity of the spine of the scapula:* (Ṣ:) or *a notch,* or *small hollowed place,* [app. *the glenoid cavity,*] *at the extremity of the spine of the scapula,* (JK, T, TA,) *next the socket:* (T, TA:) pl. أَخَارِمُ: (JK, T, TA:) or أَخْرَمَا الكَتِفَيْنِ, in the Ḳ miswritten آخِرُ مَا فِى الكَتِفَيْنِ, signifies *the heads of the two scapulæ, next the upper arms:* (Ḳ,* TA:) or *the two extremities,* or *edges, of the lower portion of the two scapulæ, which surround,* or *border, the* كُعْبُرَة [app. here meaning the *thick part next to the inferior angle*] *of the scapula:* and الأَخْرَمُ *the end of the spine [of the scapula].* (Ḳ,* TA. [In the Ḳ is here added, accord. to different copies, حَيْثُ يَنْخَدِعُ, as in the TA; or حيث يَنْخَدِمُ, as in the CḲ; or حيث يَنْخَذِمُ: the right reading seems to be حَيْثُ يَنْخَرِمُ, *where it forms a kind of cleft;* app. meaning *where it forms the glenoid cavity.* In the CḲ, for مُنْقَطَعُ العَيْرِ, is erroneously put مُنْقَطَعُ العَيْنِ; and و is erroneously prefixed to the former noun.]) —— الأَخْرَمَانِ *Two cleft bones at the extremity of the interior of the upper part of the inside of the mouth.* (Ḳ.) ═ أَخْرَمُ الرَّأْيِ ‡ *A man weak in judgment.* (JK, TA.)

مَخْرِمٌ *The end of a prominence,* or *projecting part, of a mountain:* pl. مَخَارِمُ: (Ṣ:) or مَخْرِمُ جَبَلٍ signifies *the prominence,* or *projecting part, of a mountain:* [like خُرْمٌ:] and مَخْرِمُ سَيْلٍ, *the extremity of a torrent:* (Ḳ: [accord. to the TḲ, *of a sword;* for مخرم السيف is there put in the place of مخرم السيل:]) pl. as above: (TA:) and مَخْرِمُ أَكَمَةٍ and خُرْمٌ ↑ اكمة signify *the place where a hill such as is termed* اكمة *ends.* (Ḳ.) Also *A [road such as is termed]* ثَنِيَّة, *between two mountains:* (TA:) [or the pl.] مَخَارِمُ signifies the *mouths of [mountain-roads such as are termed]* فِجَاج: (Ṣ:) or *roads in rugged tracts:* (Skr, Ḳ:) or *roads in mountains,* and *in sands.* (IAth, TA.) [Hence,] يَمِينٌ ذَاتُ مَخَارِمَ ‡ *An oath in which are ways of evasion.* (Ṣ, TA.) And لَا خَيْرَ فِى يَمِينٍ لَا مَخَارِمَ لَهَا ‡ *There is no good in an oath that has not ways of evasion:* from مَخْرِمٌ signifying "a ثَنِيَّة between two mountains." (TA.) And هٰذِهِ يَمِينٌ قَدْ طَلَعَتْ فِى المَخَارِمِ ‡ [*This is an oath that has come forth in expressions that admit of ways of evasion*]: said of an oath that affords a way [or rather ways] of evasion to the utterer thereof. (AZ, TA.) —— [The pl.] مَخَارِمُ [or مَخَارِمُ اللَّيْلِ] also signifies *The first portions of the night.* (Ḳ.) It occurs in an instance in which some read المَخَارِمُ [pl. of مَخْرَمٌ, q. v.]. (TA.)

مُخَرَّمٌ: see أَخْرَمُ. It is said in a trad., نَهَى أَنْ يُضَحَّى بِالمُخَرَّمَةِ الأُذُنِ *He forbade the sacrificing as an* أُضْحِيَّة [q. v.] *the animal having the ear cut,* or *cut off,* or *mutilated:* or *having many perforations,* and *slits, in its ear.* (TA.)

مَخْرُومَاتٌ: see خُرْمَةٌ.

مُتَخَرِّمُونَ: see خُرَّامٌ.

خرنب

خُرْنُوبٌ and خَرْنُوبٌ: see خَرُّوبٌ, in art. خرب.

خِرْنَابَتَانِ: see خِنَّابَتَانِ, in art. خنب.

خرو

خُرَةُ الفَأْسِ, as in the Tekmileh, on the authority of Fr; in the Ḳ and accord. to Ṣgh, خُرُوَةُ ↑ الفأس, which is a mistake; (TA;) *The* خُرْتٌ [q. v. in art. خرت] *of the* فأس: pl. خُرَاتٌ; (Fr, Ṣgh, Ḳ, TA;) like as ثُبَةٌ has for its pl. ثُبَاتٌ. (TA.)

الخَرَاتَانِ *Two stars,* (Ḳ,) *mentioned* [*and described*] *in art.* خرت, (TA,) *each of which is* [*said to be*] *called* خَرَاةٌ: (Ḳ:) accord. to ISd, only the dual form of the word is known; and the radical ت and the augmentative ت [by which latter is meant ة] are in the dual alike: (TA:) but Kr and others say that it is dual of خَرَاةٌ, and belongs to this art. (TA in art. خرت.)

خُرُوَةٌ: see the first paragraph in this art.

خز

خَزٌّ *A certain kind of cloth,* (Ṣ, A, Ḳ,) *well known,* (Ḳ, TA,) *woven of wool and silk:* (TA:) and also *a kind of cloth entirely of silk;* and this is the kind which one is forbidden to ride upon and to sit upon; not the former kind, which is allowable, and was sometimes worn by companions of the Prophet and by the next succeeding generation, as IAth has ascertained: (TA:) derived from خُزَزٌ, (Ḳ, TA,) accord. to some: (TA:) or it is the name of *a certain beast* [thought by Golius to be the *beaver*]: and afterwards applied to the *cloth made of its fur:* (Mgh, Mṣb:) pl. خُزُوزٌ. (Ṣ, A, Mṣb, Ḳ.) [Golius seems to derive it from the Persian قز, meaning *raw silk;* and assigns to it also the meaning of *a coarser kind of spun silk.*] خُزُوزٌ وَبُزُوزٌ signifies *Good cloths,* or *stuffs,* or *garments.* (A in art. بز.)

خُزَزٌ *The male of the* أَرْنَب [or *hare*]: (Ṣ, A, Mṣb, Ḳ:) or the *offspring of the* ارنب: (TA:) pl. [of pauc.] أَخِزَّةٌ (Ḳ) and [of mult.] خِزَّانٌ. (Ṣ, Mṣb, Ḳ.) Hence the saying, مَسُّهُ مَسُّ الخُزَزِ [*The feel of him,* or *it, is like the feel of the male,* or *young, hare*]. (A, TA.)

خَزَّازٌ *A seller of* خَزّ. (TA.)

أَرْضٌ مَخَزَّةٌ *A land containing,* (Ḳ,* TA,) or *abounding with,* (TA,) خِزَّان, pl. of خُزَزٌ. (Ḳ, TA.)

خزر

1. خَزِرَتِ العَيْنُ, aor. ـَ, (Mṣb,) inf. n. خَزَرٌ, (Ṣ, A, Mgh, Mṣb, Ḳ,) *The eye was,* or *became, narrow and small:* (Ṣ, A, Mgh, Mṣb, Ḳ:) or *it contracted its sight, naturally:* (Ḳ:) or خَزِرَ, aor. ـَ, (Ḳ,) inf. n. as above, (Ṣ, Ḳ,) signifies *he* (a man) *was as though he looked from the outer angle of the eye:* (Ṣ, A:*) or *he looked as though on one side:* or *he opened and closed his eyes;* (Ḳ;) or, *his eye:* (M:) or *he had a distortion* (حَوَل) *of one of his eyes:* (Ḳ:) [or *he had eyes looking towards his nose;* or, *looking sideways;* (see أَخْزَرُ;) or, *looking towards their outer angles;* (see خَزْرَةٌ;) see also 2, and 6, and Q. Q. 1.] ═ خَزَرَهُ, aor. ـِ, (TA,) inf. n. خَزْرٌ, (Ḳ,) *He looked at him from the outer angle of the eye;* (Ḳ,* TA;) as one does in pride, and in light estimation of the object at which he looks. (MF.) A poet says,

* لَا تَخْزُرِ القَوْمَ شَزْرًا عَنْ مُعَارَضَةٍ *

[*Look not thou at the people from the outer angle of the eye, askew, sideways*]. (TA.) ═ خَزَرَ [as an intrans. v.] *He affected,* or *pretended, to be cunning;* i. e. *intelligent,* or *sagacious;* or *intelligent with a mixture of craft and forecast;* syn. تَدَاهَى. (IAạr, Ḳ. [See also 2.]) ═ Also *He fled.* (Ḳ.)

2. خَزَّرَ, (TA,) inf. n. تَخْزِيرٌ, (Ḳ,) *He made narrow.* (Ḳ, TA.) You say, خَزَّرَ عَيْنَيْهِ *He* (an old man) *narrowed his eyes; contracted his eyelids as though they were sewed together; to collect the light:* when a young man does so, يَتَدَاهَى بِذٰلِكَ [i. e. he affects, or pretends, thereby, to be cunning; i. e. intelligent, or sagacious; or intelligent with a mixture of craft and forecast]. (IAạr. [See also خَزَرَ: and see 6.])

6. تخازر *He looked from the outer angle of his eye.* (TA. [See also Q. Q. 1.]) —— *He pretended,* or *made a show of, what is termed* خَزَرٌ: [see 1.] (TA, and Ḥar p. 62.) —— *He contracted his eyelids, to sharpen the sight:* (Ṣ, Mṣb, Ḳ:) a verb similar to تعامى and تجاهل. (Ṣ. [See also 2.])

Q. Q. 1. خَنْزَرَ *He looked from the outer angles of his eyes:* from the subst. خِنْزِيرٌ, because the animal so called is أَخْزَرُ. (A. [See also 6.]) —— Also *He acted like the swine.* (TA in art. خنزر.)

خَزْرٌ [commonly known only as inf. n. of خَزَرَ or خَزِرَتِ العَيْنُ]: see خَزِيرٌ.

أَخْزَرُ: see خَزَرُ العَيْنِ.

خَزْرَةٌ: see خَزَرَةٌ.

خُزْرَةٌ *A turning of the pupil towards the outer angle of the eye.* (TA. [See 1.])

خُزْرَةٌ (ISk, Ṣ, Ḳ) and ▼ خُزَرَةٌ (Ḳ) *A pain in the back:* (Ḳ:) *a pain in a vertebra of the back:* (Ṣ:) *a pain in the slender part of the back, in* [*the vertebra called*] فَقْرَةُ القَطَنِ: (TA:) the pl. of the former is خُزُرَاتٌ. (Ṣ, TA.)

خَزِيرٌ and ▼ خَزِيرَةٌ *A kind of food like* عَصِيدَةٌ *with flesh-meat;* (Ḳ;) *made of flesh-meat* (Ṣ, TA) *that has remained throughout a night,* (TA,) *cut into small pieces, and put into a cooking-pot with abundance of water,* (Ṣ, TA,) *and with salt;* (TA;) *and when it is thoroughly cooked, some flour is sprinkled upon it,* (Ṣ, TA,) *and it is stirred about with it, and seasoned with any seasoning that the maker pleases to add:* (TA:) when there is no flesh-meat, it is called عَصِيدَةٌ: (Ṣ, Ḳ, TA:) or *a broth made with the water in which bran has been soaked,* (Mgh, Ḳ, TA,) *which water is strained, and then cooked:* (Mgh, TA:) this is what is called by the Persians سُبُوسَبَا: (Mgh:) [see also حَرِيرَةٌ:] or خَزِيرَةٌ is *flour thrown upon water* or *upon milk, and cooked, and then eaten with dates,* or *supped:* it is also called سَخِينَةٌ and سَخُونَةٌ and نَفِيتَةٌ and حُذْرُقَةٌ: حَرِيرَةٌ is thinner: (AHeyth, on the authority of an Arab of the desert:) and *a soup made of grease* or *gravy* (Ḳ) *and flour;* (TA;) as also ▼ خَزِرٌ: (Ḳ:) but no one except the author of the Ḳ mentions this last form: in the other lexicons, soup of grease or gravy is said only to be called خَزِيرٌ and خَزِيرَةٌ. (TA.)

خَزِيرَةٌ: see the next preceding paragraph.

خَازِرٌ A man *possessing much cunning;* i. e. *intelligence,* or *sagacity;* or *intelligence with a mixture of craft and forecast.* (AA, Ḳ.)

خِنْزِيرٌ [The *swine;* the *hog;* the *pig;*] *a certain foul animal,* (Mṣb,) *well known;* (Ḳ;) *said to be forbidden* [*to be eaten*] *by every prophet:* (Mṣb:) [fem. with ة:] pl. خَنَازِيرُ: (Ṣ, Mṣb, Ḳ:) not, as some say, خُزْرٌ: [though this is an epithet applicable to swine:] (TA:) accord. to some, it is of the measure فِعْلِيلٌ; because ن is not [generally] added as a second letter: but accord. to others, of the measure فِنْعِيلٌ; because ن is sometimes added as a second letter, and because it is held to be derived from خَزَرٌ, since all خنازير are خُزْر; as it is said in the A, كُلُّ خِنْزِيرٍ أَخْزَرُ. (TA.) — خَنَازِيرُ also signifies *A well-known disease;* (Ṣ;) [*scrofula;* or *glandular swellings in the neck;*] *ulcers,* (Ḳ,) or *hard ulcers,* (Ṣ,) *which arise in the neck:* (Ṣ, Ḳ:) or *ganglions, or hard or nodous lumps beneath the skin, in the neck, and in soft parts, such as the armpits; but most frequently in the neck.* (Mgh.)

خَوْزَرَى: see what next follows, in two places.

خَيْزَرَى and ▼ خَوْزَرَى *A certain mode of walking, with a looseness of the joints,* (Ṣ, A, Ḳ,) *as though the limbs were dislocated;* (A;) as also خَيْزَلَى and خَوْزَلَى: (Ṣ in art. خزل, and TA:) or *a limping,* or *halting, manner of walking:* or *an elegant, and a proud and self-conceited, gait, with an affected inclining of the body from side to side.* (TA.) You say, هُوَ يَمْشِى الخَيْزَرَى and ▼ الخَوْزَرَى *He walks with a looseness of the joints,* &c. (A.)

خَيْزُرَانٌ, (Ṣ, Ḳ, &c.,) vulgarly pronounced خَيْزَرَان, (TA,) [a coll. gen. n., The *kind of cane called rattan;* so in the present day;] *a kind of Indian tree, which consists of roots extending upon the ground;* as also ▼ خَيْزُورٌ: (Ḳ:) or [*a kind of tree*] *not growing in the country of the Arabs, but only in that of the Greeks;* whence the saying of En-Nábighah El-Jaạdee,

* بِلَادُهُمْ بِلَادُ الخَيْزُرَانِ *

[*Their lands are the lands of the kheyzurán*]: it is *a kind of plant with pliable and smooth twigs:* (ISd:) or *a kind of tree,* (Ṣ,) *the roots of the* قَنَا [by which are app. meant the *canes of which spear-shafts are made*]: (Ṣ, Mṣb:) pl. خَيَازِرُ. (Ṣ.) — *Reed,* or *reeds; cane,* or *canes.* (Ṣ, Ḳ.) — And hence, *Musical reeds* or *pipes.* (TA.) — *Spears:* (IAạr, Ḳ:) because of their pliableness: (TA:) [or because commonly made of canes:] pl. as above. (TA.) — *Any pliable twig or rod;* (Mbr, Ḳ;) *any piece of wood that is pliable.* (AHeyth.) [Often applied in the present day to the *osier;* as well as to the rattan: n. un. with ة.] — *The rod which kings hold in their hands, and with which they amuse themselves* (يَتَعَبَّثُونَ) *and make signs.* (Ḥam p. 710.) — The *pole with which a ship,* or *boat, is pushed* or *propelled,* (Mbr, Ḳ,) *when pliable,* or *bending;* as also ▼ خَيْزَارَةٌ. (Mbr, TA.) — Also, (AO, Mṣb, Ḳ,) and with ة, (Ṣ, TA,) The سُكَّان (Ṣ, Mṣb, Ḳ) of a ship, (Ḳ,) i. e. its كَوْثَل [meaning the *rudder*]: (TA:) or, accord. to 'Amr Ibn-Baḥr, the لِجَام [lit. the bridle and bit, app. meaning the *tiller*] of a ship, *by means of which the* سُكَّان, *which is the* ذَنَب, *is directed.* (TA: [but instead of التى بها يقوم السُّكَّانُ وهو فى الذنب, I read الذى به يُقَوَّمُ السُّكَّانُ وهو الذَّنَبُ.]) En-Nábighah says, describing the Euphrates in the time of its increase, or fulness,

* يَظَلُّ مِنْ خَوْفِهِ المَلَّاحُ مُعْتَصِمًا *
* بِالخَيْزُرَانَةِ بَعْدَ الأَيْنِ وَالنَّجَدِ *

[*By reason of his fear, the sailor becomes in a state of cleaving,* or *laying fast hold, upon the* خيزرانة, (which may here mean the *pole* above mentioned, or the *rudder,* or the *tiller,*) *after fatigue and distress*]. (Ṣ, TA.) In a trad. it is said that the devil, when he had been commanded by Noah to go forth from the ark, mounted upon the خيزران of the ark, i. e. its سُكَّان. (TA.)

خَيْزُورٌ: } see the last paragraph above.
خَيْزَارَةٌ: }

أَخْزَرُ A man *having narrow and small eyes:* (Ṣ, A, Mṣb, Ḳ:) or *having eyes of which the sight is contracted, naturally:* (Ḳ:) or *who looks from the outer angle of his eye:* (A:) or *who is as though he so looked:* (Ṣ:) or *who looks as though on one side:* or *who opens and closes his eyes;* (Ḳ;) or, *his eye:* (M:) or *who has a distortion* (حَوَل) *of one of his eyes:* (Ḳ:) or *whose eyes look towards his nose:* (TA:) [or *whose eyes look towards their outer angles:* (see خُزْرَةٌ:)] and أَخْزَرُ العَيْنِ one *who looks askew,* or *sideways;* as also ▼ خَزِرُ العَيْنِ, an epithet applied to an enemy: (TA:) the fem. of أَخْزَرُ is خَزْرَآءُ: (A, Mṣb:) and the pl. is خُزْرٌ. (Ḳ.) You say also أَعْيُنٌ خُزْرٌ [meaning *Eyes that are narrow and small:* &c.]. (TA.)

خزعبل

خُزَعْبَلٌ *Stories that are held to be clever, ingenious,* or *elegant,* (IDrd, Ḳ,) *and at which one laughs.* (IDrd.)

خُزَعْبَلٌ i. q. بَاطِلٌ [app. as a subst., meaning *A false,* or *vain, saying* or *deed* or *affair* or *thing*]; as also ▼ خُزَعْبِيلٌ: (Ḳ:) or, accord. to El-Jarmee, (Ṣ, TA,) the latter, (Ṣ,) or each, (TA,) *false,* or *vain, sayings* or *deeds* or *affairs* or *things.* (Ṣ, TA.)

خُزَعْبَلَةٌ or خُزَعْبِلَةٌ (accord. to different copies of the Ḳ) *A wonderful thing.* (IAạr, Ḳ.) And [the pls.] خُزَعْبِلَاتٌ and خُزَعْبَلَاتٌ *False,* or *vain, stories.* (Ḥar p. 16.)

خُزَعْبِيلٌ: see خُزَعْبَلٌ.

خُزَعْبِيلَةٌ *A laughable thing; a thing that makes people laugh.* (Ṣ, Ḳ.) One says, هَاتِ بَعْضَ خُزَعْبِيلَاتِكَ [*Give me some of thy laughable things* or *stories*]. (Ṣ.)

خزف

خَزَفٌ *Pottery; jars;* or *earthern vessels;* syn. جَرٌّ; (Lth, Ṣ, Ḳ; [and so in the present day;]) and *anything made of clay, and baked, so that it becomes* فَخَّار: (IDrd, Ḳ:) or *clay made into vessels, before it is baked;* i. q. صَلْصَالٌ: when baked, it is called فَخَّار. (Mṣb.) [See an ex. in a verse cited voce إِنْ, p. 107.]

خَزَفِيٌّ [*Of,* or *relating to, pottery,* or *jars,* &c.;] rel. n. of خَزَفٌ. (TA.) — See also what follows.

خَزَّافٌ and ▼ خَزَفِيٌّ *A seller* [or *maker*] *of* خَزَف [or *pottery, jars,* &c.]. (TA.)

خزق

1. خَزَقَ, aor. ـِ, (Ṣ, Mgh, Ḳ,) inf. n. خَزْقٌ and خُزُوقٌ, (TA,) *It* (an arrow) *hit the target;* (Ṣ, Ḳ;) or *the object at which it was shot;* (ISd, TA;) as also ▼ خَسَقَ [q. v.]: (TA:) or *transpierced,* or *passed through,* or *its extremity passed through,* (Mgh, TA,) *making the blood to flow:* (TA:) [and app. also *it stuck fast* therein: (see خَازِقٌ, below:)] or خَزَقَ القِرْطَاسَ *it* (an arrow) *transpierced,* or *passed through, the target;* or *pierced it so that its extremity passed through.* (Mṣb.) — خَزَقَهُ, aor. ـِ, (Mṣb, Ḳ,) inf. n. خَزْقٌ, (Ṣ, Mṣb,) *He pierced him* [with a spear or the like]. (Ṣ, Mṣb, Ḳ.) And *He pierced him slightly* with a spear. (TA.) And *It* (an arrow) *hit him.*

(JK.) And خَزَقْتُهُمْ بِالنَّبْلِ *I hit them with the arrows.* (Ṣ.) [Whence,] خَزَقَهُ بِعَيْنِهِ ‡ *He looked sharply at him;* and *cast his eye at him,* or *smote him with his eye:* (Lḥ, TA:) or *he wounded him with his eye.* (Ibn-'Abbád, Z, TA.) — Also *He stuck it* (i. e. anything sharp) into the ground (Lth, JK, TA) &c. (Lth, TA.)

7. انخزق *He became pierced* [with a spear or the like]. (Ḳ.) — *It* (anything sharp, JK) *became stuck* into the ground [&c.]. (JK, TA.) — Also, or ↓ اختزق, (accord. to different copies of the Ḳ, the latter in the CḲ,) *It* (a sword) *became drawn from its scabbard.* (Ḳ, TA.)

8: see what next precedes.

خَزْقٌ inf. n. of 1. (Ṣ,* Mṣb, TA.) — [I also find in the TA, الخزق ما يثبت والخزق ما ينفذ, as though meaning that خَزْقٌ signifies A thing *that becomes fixed* or *fast:* and also a thing *that transpierces,* or *passes through:* but these may be loose explanations, intended to mean that خَزْقٌ is said of that which becomes fixed or fast (as an arrow in the body pierced by it), and also of that which transpierces, or passes through: see خَسَقَ, of which the inf. n., خَسْقٌ, is said to be used in these two senses.]

خَزُوقٌ A she-camel *that pierces the ground with her toes,* (Ḳ, TA,) *making marks upon it:* (TA:) or *that furrows the ground with her toe turned up, as she goes.* (Ḳ, TA.) [See also خَسُوقٌ.]

خَازِقٌ, applied to an arrow, *Hitting,* or *that hits, the target;* (Ṣ, Mgh, Ḳ;) and (Mgh, TA) *transpiercing,* or *passing through,* or *that transpierces* or *passes through,* or *piercing,* or *that pierces, so that its extremity passes through,* (Mgh, Mṣb, Ḳ,) the target [or the object shot at]; (Mṣb;) as also خَاسِقٌ: (TA:) pl. خَوَازِقُ. (Mṣb.) And [hence, used as a subst.,] *A spear-head:* (JK, Ṣ, Ḳ:) and the *iron head* or *blade* of an arrow &c. (TA.) One says, هُوَ أَمْضَى مِنْ خَازِقٍ [*He is more penetrating than a spear-head*]: (Ṣ:) and أَنْفَذُ مِنْ خَازِقٍ, which means the same; (JK, Meyd;) or *more penetrating than a transpiercing arrow:* (TA:) the latter is a prov.; (Meyd, TA;) applied to him who is penetrating in affairs. (Meyd.) And إِنَّهُ لَخَازِقُ وَرَقَةٍ [lit. *Verily he is a piercer,* or *transpiercer, of a leaf*]; meaning لَا يُطْمَعُ فِيهِ [i. e. † *he is a person whom one may not hope to overcome*]: (IAạr, Ḳ:) or † *he is bold and skilful.* (Ḳ.)

[خَازُوقٌ *A stake for impaling a man:* but this I believe to be post-classical: pl. خَوَازِيقُ.]

مِخْزَقٌ *A small stick,* (JK, Ḳ,) *with a pointed end,* (JK,) or *having at its end a sharp nail,* (Ḳ,) *that is with the seller of full-grown unripe dates for date-stones* [which are used as food for camels]: (JK, Ḳ:) he has many مَخَازِقُ [pl. of مِخْزَقٌ]; and a boy comes to him with date-stones, which he takes from him on the condition that he shall strike with the مخزق a certain number of times and have as many of the dates as become thus transfixed for him, whether many or few, but nothing if he miss. (Ḳ.)

مِخْزَقَةٌ *A dart,* or *javelin.* (TA.)

مُخْتَزَقٌ *Chase,* or *game;* i. e. *an object of the chase or the like.* (JK, TA.)

خزل

1. خَزَلَهُ, (Mṣb, Ḳ,) aor. ِ, (Mṣb,) or ُ, (Ḳ,) inf. n. خَزْلٌ, (Mṣb,) *He cut it off.* (Mṣb, Ḳ.) [See also 8.] — خَزَلَهُ عَنْ حَاجَتِهِ, aor. ُ, † *He impeded, hindered, prevented,* or *withheld, him from attaining the object of his want.* (Ḳ,* TA.) == خَزِلَ, aor. َ, (JK, Ḳ,) inf. n. خَزَلٌ, (JK,) *He had a fracture in his back.* (JK, Ḳ.) — Also, (TḲ,) inf. n. as above; (Ḳ;) and ↓ تخزّل and ↓ انخزل; *He walked in a heavy,* or *sluggish, manner:* (Ḳ:) or, as in the 'Eyn, *with a looseness of the joints.* (TA.)

5: see 1. — [Hence,] تَخَزَّلَ السَّحَابُ *The clouds appeared as though they were retreating, by reason of the heaviness, or sluggishness, of their motion.* (M, Ḳ,* TA.)

7. انخزل *It became cut off.* (Ṣ, Mṣb.) — [Hence,] انخزل فِى كَلَامِهِ † *He became cut short,* or *stopped, in his speech.* (IDrd, Ḳ.) — انخزل عَنْ جَوَابِى *He cared not for my answer,* or *reply.* (M, Ḳ.) — See also 1.

8. اختزلهُ *He cut it off for himself.* (Ṣ,* Mṣb, Ḳ.*) One says, اختزل المَالَ *He cut off for himself the property.* (TA.) And اختزلهُ عَنِ القَوْمِ *He cut him off from the people,* or *party.* (Ṣ.) — And hence, اِخْتَزَلْتُ الوَدِيعَةَ † *I was unfaithful in respect of the trust,* or *deposit;* even if meaning *I refused to restore it:* because the doing so is a cutting off, for one's self, of the property of the owner. (Mṣb.) — Accord. to Sb and the Ḳ, الاِخْتِزَالُ is also *syn. with* الحَذْفُ [*The cutting off* a thing; or *cutting it so as to lessen it at its extremity;* &c.]: but ISd says, I know not this on the authority of any other than Sb. (TA.) == الاختزال also signifies *The being alone* (Ḳ, TA) *in opinion; with none to share therein.* (TA.) == And اختزل, *He* (a man) *became lame;* or *walked lamely.* (TA.)

خَزَلٌ: } see what next follows.
خَزَلَةٌ: }

خَزْلَةٌ, (JK, Ḳ,) or, as in the M, ↓ خَزَلَةٌ and ↓ خَزَلٌ, (TA,) *A fracture in the back.* (JK, M, Ḳ.)

خُزَلَةٌ One *who impedes, hinders, prevents,* or *withholds, another from that which he desires to do.* (Az, Ḳ,* TA.)

خَوْزَلَى: see خَيْزَلٌ.

خَوْزَلَةٌ *i. q.* إِعْيَآءٌ [*Fatigue:* or *a fatiguing*]. (TA.)

خَوْزَلَى and ↓ خَيْزَلَى *A certain mode of walking, with a looseness of the joints;* like خَوْزَرَى and خَيْزَرَى: (Ṣ:) or *a heavy,* or *sluggish, manner of walking;* as also ↓ خَيْزَلٌ: (Ḳ:) the dual of خَوْزَلَى [to which that of خَيْزَلَى is similar] is خَوْزَلَانِ. (IAmb, TA in art. قهقر.) Accord. to the T, one says, ↓ الخَوْزَلَى and هُوَ يَمْشِى الخَيْزَلَى, meaning *He walks with an elegant, and a proud and self-conceited, gait, with an affected inclining of the body from side to side:* or *as though thorns pricked his feet.* (TA.)

خَيْزَلَى: see what next precedes, in two places.

أَخْزَلُ *Having a fracture in his back;* as also ↓ مَخْزُولٌ: (O, Ḳ:) or *having a fracture in the middle of his back;* as also ↓ مَخْزُولُ الظَّهْرِ. (Lth, TA.) — And A camel *whose hump has entirely gone away:* (Ḳ:) so says Lth; but Az thinks that this may be a mistranscription for أَجْزَلُ; or خ and ج may be interchangeable in this case. (TA.) — Also *Lame.* (AA, TA.)

مَخْزُولٌ: see what next precedes, in two places.

خزم

1. خَزَمَهُ, aor. ِ, (Mgh, Ḳ,) inf. n. خَزْمٌ, (JK, TA,) *He pierced it,* or *perforated it;* (JK, Mgh, Ḳ;) namely, a thing of any kind: for instance, the nose of a camel, for the [ring called] خِزَامَة [to which the rein is attached]: and the letter of a Ḳádee to another Ḳádee; for such a letter is pierced for the سَحَاءَة [or strip of paper with which it is bound], and is then sealed [upon this strip]; and when this is done, the letter is said to be ↓ مَخْزُومٌ. (Mgh.) You say, خَزَمَ البَعِيرَ, aor. as above, (Mgh, Mṣb,) and so the inf. n., (Mṣb,) *He pierced the nose of the camel* (Mgh, Mṣb) *for the* خِزَامَة: (Mgh:) or خَزَمَ البَعِيرَ بِالخِزَامَةِ (Ṣ, Ḳ), (Ṣ,) aor. and inf. n. as above, (TA,) *he put the* خزامة *in the side of the nostril,* (Ḳ,) or *in the partition between the nostrils,* (Ṣ,) [but the former is the more common practice,] *of the camel;* (Ṣ, Ḳ;) as also ↓ خزّمهُ; (Ḳ;) [or] this signifies the doing so much, or often. (TA.) [Hence,] خَزَمَ أَنْفَهُ † *He brought him under, or into, subjection; rendered him submissive, tractable,* or *manageable.* (TA.) And خَزَمْتُ الجَرَادَ فِى العُودِ *I spitted the locusts in a series upon the* [*skewer,* or] *piece of stick* or *wood.* (Ṣ.) And خَزَمَ شِرَاكَ نَعْلِهِ *He pierced and tied the* [*thong called*] شراك [q. v.,] *of his sandal* [app. *so as to attach to it the two branches* (عَضُدَا) *of the strap that encompasses the heel*] (الشِّرَاك). (TA. [See also خِزَامَةٌ.]) — [Also *He cut it.* for] الخَزْمُ is *syn. with* القَطْعُ. (Ḥam p. 166. [It is there also said to be syn. with الشدّ; i. e. الشَّدُّ; but this is app. a mistake for الشَّكُّ, a meaning assigned to it in the JK, agreeably with the Ḳ.])

2: see 1.

3. خَازَمَهُ, (Ṣ,) or خَازَمَهُ الطَّرِيقَ, (Ḳ,) inf. n. مُخَازَمَةٌ (Ṣ in art. حصر) [and خِزَامٌ], *He* (a man, Ṣ) *took a different way from his* (another's) *until they both met in one place:* (Ṣ, Ḳ:) the doing so is also termed مُخَاصَرَةٌ: (Ṣ in art. حصر, and TA:) it is as though it were a rivalling, or imitating, in travelling. (TA.) Ibn-Fesweh says,

* إِذَا هُوَ نَحَّاهَا عَنِ القَصْدِ خَازَمَتْ *
* بِهِ الجَوْرَ حَتَّى تَسْتَقِيمَ ضُحَى الغَدِ *

i. e. *When he turns her*, meaning his she-camel, *from the right way, she pursues with him a way different from the wrong, so that she overcomes* him, and *takes the right way, in the early day-time of the morrow.* (TA.) مُخَازَمَةٌ also signifies The act of *rivalling*, or *imitating.* (JK, TA.) And one says, لَقِيتُهُ خِزَامًا (JK, TA) and مُخَازَمَةً, (JK,) meaning *I met him face to face:* (TA:) or *suddenly*, or *unexpectedly, and face to face.* (JK.)

5. تَخَزَّمَ الشَّوْكُ فِى رِجْلِهِ *The thorns pierced his foot*, or *leg, and entered into it.* (Ḳ,* TA.)

6. تَخَازَمَ الجَيْشَانِ *The two armies rivalled*, or *imitated, each other;* or *opposed each other;* syn. تَعَارَضَا. (TA.)

خَزَمٌ *A certain kind of tree,* (JK, T, Ṣ, Mṣb, Ḳ,) *of the bark of which ropes are made:* (Ṣ, Mṣb:) *it is like the* دَوْم [or *Theban palm*]; (Ḳ;) *having branches with small dates, which become black when ripe, bitter, astringent,* or *disagreeable and choking; not eaten by men; but the crows are greedy of them, and come to them time after time:* so says AḤn.: (TA:) n. un. with ة. (Ṣ, Mṣb.) — See also خَزُومَةٌ.

خُزَامَى : see خَزَمٌ.

خَزَمٌ [a pl. of which the sing. is not mentioned] *Sewers of skins or hides or boots and the like;* syn. خَرَّازُونَ. (TA.)

خَزَمَةٌ n. un. of خَزَمٌ, explained above. (Ṣ, Mṣb.) — Also The *leaves* (خُوص) *of the* مُقْل [or *Theban palm*]; (JK, Ḳ;) *of which are made women's* أَخْفَاش [i. e. *receptacles for their perfumes and other similar things*, pl. of خِفْش]. (TA.)

مَخْزُومٌ : see إِبِلٌ خَزَمَى.

خِزَامٌ : see خِزَامَةٌ, in two places.

خَزُومٌ : see خَزُومَةٌ.

خِزَامَةٌ *A ring of* [goat's] *hair, which is put* [*in the side of the nostril* (see 1) or] *in the partition between the nostrils of the camel,* (Ṣ, Mṣb,* TA,) *and to which is tied the rein;* (Ṣ, TA;) as also ▼خِزَامٌ: (TA, and Ḥar p. 73:) or *a* بُرَة *in the nose of a she-camel:* (JK:) or, accord. to the Ḳ, *a* بُرَة *which is put in the side of the nostril of the camel:* but Lth says that when it is of brass it is termed بُرَة; and when of [goat's] hair, خِزَامَة: (TA:) pl. خَزَائِمُ (JK, Mṣb, TA) and خِزَامَاتٌ. (Mṣb.) [Hence,] أَعْطِى القُرْآنَ خِزَامَتَهُ : from a trad. of Abu-d-Dardà, in which it is said, مُرْهُمْ أَنْ يُعْطُوا القُرْآنَ بِخَزَائِمِهِمْ ‡ *Command ye them that they submit themselves to the judgment,* or *decision, of the Ḳur-án;* خزائم being here pl. of خِزَامَة: (IAth, TA:) or أَعْطُوا القُرْآنَ خِزَامَتَهُ † *Render ye to the Ḳur-án its due.* (JK.) [In the present day, ▼خِزَامٌ, vulgarly pronounced خُزَام, is applied to *A woman's nose-ring, of gold* or *other metal.*] — The خِزَامَة of the sandal is *A slender thong which is pierced and tied between* [*the two thongs called*] *the* شِرَاكَانِ [app. here meaning the عَضُدَانِ of the شِرَاك: see 1]. (Ḳ,* TA.)

خُزَامَى *A certain plant,* (JK, Ḳ, TA,) called also ▼خُزَامٌ, (JK,) *of sweet odour:* (TA:) or *i. q.* خِيرِىُّ البَرِّ [q. v.]; (Ṣ, Mṣb, Ḳ;) accord. to El-Fárábee: *one of the plants of the desert:* said by Az to be *a certain herb of sweet odour, having a flower like that of the violet:* (Mṣb:) [accord. to the book entitled مَا لَا يَسَعُ الطَّبِيبَ جَهْلُهُ, as stated by Golius, *a certain wild herb, having a long stalk, small leaves, red flower, and very sweet odour:*] *its flower is the sweetest of flowers in odour; the fumigation therewith dispels every fetid odour; the use thereof as a suppository in the vagina promotes pregnancy; and the taking it internally restores to a right state the liver and the spleen, and the brain affected with cold:* (Ḳ:) [in the present day, applied to the *common lavender; lavandula spica:*] n. un. خُزَامَاةٌ. (TA.)

خَزُومَةٌ *i. q.* بَقَرَةٌ [app. as meaning both *A bull* and *a cow*], (JK, Ṣ, Ḳ,) in the dial. of Hudheyl; (Ṣ;) or *such as is advanced in age, and short:* (M, Ḳ:) pl. ▼خَزُومٌ, [or rather this is a coll. gen. n.,] and [the pl. is] خَزَائِمُ (JK, Ḳ) and خَزُومَاتٌ (Ṣ) and [quasi-pl. n.] ▼خَزَمٌ. (TA.)

خَزَّامٌ *A maker of ropes of the bark of the tree called* خَزَم. (JK, Ṣ.*) [Accord. to the Ḳ, *A seller of the kind of tree called* خَزَم: but this is a mistake, app. caused by an omission in the Ḳ.]

رِيحٌ خَازِمٌ † *A cold wind;* as though it pierced through the extremities: so says Kr: A'Obeyd says خَارِم: accord. to the Ḳ, both signify the same. (TA.)

مُخَزَّمٌ; and its fem., with ة: see what follows, in four places.

مَخْزُومٌ Anything *pierced* or *perforated:* (Ṣ, Mgh:) applied in this sense, (JK,) or as meaning *pierced and tied,* (TA, [see 1,]) to the [thong, of a sandal, called] شِرَاك; (JK, TA;) or, thus applied, it means *cut.* (Ḥam p. 166.) See also 1, first sentence, for its meaning as applied to a letter. — Any animal *having the nose pierced.* (Mṣb.) And ▼إِبِلٌ خَزَمَى means *Camels having rings such as are termed* خَزَائِم (pl. of خِزَامَة) *in their nostrils;* (IAạr, Ḳ, TA;) as also ▼مُخَزَّمَةٌ. (IAạr, TA.) All birds, also, are said to be مَخْزُومَة, (Ṣ, Ḳ,) and ▼مُخَزَّمَة; (Ḳ;) because perforated in the partition between the nostrils: (Ṣ, Ḳ:) and particularly the ostrich is said to be مَخْزُوم (Ṣ, Ḳ*) and ▼مُخَزَّم. (JK, Ḳ.*) One says, مَا هُمْ إِلَّا كَالأَنْعَامِ المُخَزَّمَةِ▼, meaning ‡ *They are none others than stupid,* or *foolish, persons.* (TA. [But كالانعام is app., here, a mistranscription for كَالنَّعَامِ; for the ostrich is proverbial for stupidity: one says أَحْمَقُ مِنْ نَعَامَةٍ "More stupid than an ostrich:" because, as Meyd says, when an ostrich happens to see the eggs of another ostrich, it will sit upon them, and forget its own eggs.])

خزن

1. خَزَنَهُ, (JK, Ṣ, Mṣb, Ḳ,) aor. ـُ, (JK, Mṣb,) inf. n. خَزْنٌ, (Mṣb,) *He reposited it, stowed it,* or *put it,* (Ṣ, Mṣb,) *laid it up, kept it, preserved it,* or *guarded it,* (JK, Ḳ,) *in a* خِزَانَة, (JK, Ṣ,) or *in a* مَخْزَن [or مَخْزِن]; (Mṣb;) namely, a thing, (JK, Mṣb,) or property; (Ṣ, Ḳ;) as also ▼اِخْتَزَنَهُ, (Ṣ, Ḳ,) and ▼اِسْتَخْزَنَهُ: or ▼ the second of these signifies he did so *for himself.* (TA.) — And [hence] the first (Ṣ, Mṣb) and ▼ the second, (Ṣ,) † *He concealed it;* namely, a secret. (Ṣ, Mṣb.) — And خَزَنَ عَنْهُ عَطَاءَهُ † *He withheld from him his gift.* (TA.) = خَزِنَ, (JK, Ṣ, Mṣb, Ḳ,) formed by transposition from خَنِزَ, (Ṣ, Mṣb,) aor. ـَ, inf. n. خَزَنٌ; (Mṣb;) and خَزُنَ; and خَزَنَ, [aor. ـُ,] inf. n. خَزْنٌ and خُزُونٌ; (Ḳ;) said of flesh-meat, (JK, Ṣ, Mṣb, Ḳ,) *It became altered* [*for the worse*] (JK, Mṣb, Ḳ) *in odour,* (Mṣb,) or *stinking:* (Ṣ, TA:) or, accord. to Z, خَزِنَ, said of flesh-meat, means *it became stinking in consequence of its being laid up,* or *stored:* and Er-Rághib says that خَزِنَ, in relation to flesh-meat, means the *laying up,* or *storing:* and hence is metonymically used as meaning its *becoming stinking.* (TA.) Ṭarafeh says,

* ثُمَّ لَا يَخْزَنُ فِينَا لَحْمُهَا *
* إِنَّمَا يَخْزَنُ لَحْمُ المُدَّخِرْ *

[*Then the flesh thereof will not become stinking among us: only the flesh-meat of him who lays it up becomes stinking*]. (Ṣ, TA.)

4. اخزن *He became rich,* or *in a state of competence* or *sufficiency, after poverty.* (Ḳ.)

8: see 1, in three places. = اِخْتَزَنَ طَرِيقًا *He took the nearest road,* or *way.* (Ḳ, TA.)

10: see 1.

خِزَانَةٌ : see خَزِينَةٌ: — and see also خِزَانَةٌ.

خَزِينٌ A thing *reposited, stowed,* or *put,* [*laid up, kept, preserved,* or *guarded,*] *in a* مَخْزَن [or مَخْزِن or خِزَانَة]. (Mṣb.) — Flesh-meat *altered* [*for the worse in odour*]; (Ḳ;) *stinking.* (TA.)

خِزَانَةٌ *A small chamber within a large chamber;* (TA in art. خدع;) [*a closet;* also called in the present day ▼خَزْنَة: and *a cupboard:*] *a place in which things are reposited, stowed, laid up, kept, preserved,* or *guarded; a repository;* [*a magazine; a store-room;*] (JK, Mṣb, Ḳ, TA;) and so ▼مَخْزَنٌ, (Ṣ, Ḳ,) with fet-ḥ to the ز, (Ṣ,) like مَقْعَد, (Ḳ,) or ▼مَخْزِنٌ, like مَجْلِس, [which is irreg., as the aor. of خَزَنَ is يَخْزُنُ:] (Mṣb:) the former should not be pronounced with fet-ḥ [i. e. خَزَانَة], (Ḳ,) as the vulgar are given to pronounce it: (TA:) the pl. of the former is خَزَائِنُ; (Ṣ, Mṣb;) and that of مخزن is مَخَازِنُ. (Mṣb, TA.) [Hence, خِزَانَةُ كُتُبٍ *A library;* and *a bookcase.* And خِزَانَةُ سِلَاحٍ *An armoury.*] — And ‡ The *heart;* (Ḳ, TA;) because the secret is concealed in it. (TA.) See an ex. voce خَازِنٌ. — Also The *occupation*, (JK, TA,) and *act*, (Ḳ, TA,) *of the* خَازِن. (JK, Ḳ, TA.)

خَزِينَةٌ and ↓ خِزَانَةٌ *Wealth, or property, reposited, stored, laid up, kept, preserved,* or *guarded.* (TA.) [In the present day, both signify also *A treasury.* The pl. of the former is خَزَائِنُ.] — [Hence,] خَزَائِنُ ٱللّٰهِ [in the Ḳur. vi. 50 and xi. 33, accord. to some,] means †*The hidden things that are known of God:* (TA:) or †*the events decreed by God:* (Bḍ in vi. 50:) or †*the treasures of the means of subsistence that are supplied by God.* (Bḍ and Jel* ibid.)

خَزَّانٌ One *who stores up wheat,* or *food:* of the dial. of Egypt. (TA.) — See also خَازِنٌ. — Also, as a subst., like جَبَّانٌ, (AḤn,) *Ripe dates becoming black in the interior by reason of some bane:* (AḤn, Ḳ:) n. un. with ة. (AḤn.)

خَازِنٌ [One *who reposits, stows, lays up, keeps, preserves,* or *guards,* property, &c.; *a treasurer*]: (Ḳ, TA:) pl. خَازِنُونَ and خَزَنَةٌ. (TA.) [The latter of these pls. is applied in the Ḳur. xxxix. 73 to The *keepers,* or *guardians,* of Paradise: and in xxxix. 71 and xl. 52 and lxvii. 8 to those of Hell.] مَا أَنْتُمْ لَهُ بِخَازِنِينَ, in the Ḳur [xv. 22, accord. to some], means †*Ye are not bearing it in mind with thankfulness.* (TA.) — ‡The *tongue;* as also ↓ خَزَّانٌ. (Ḳ, TA.) Hence the saying of Luḳmán to his son, إِذَا كَانَ خَازِنُكَ حَفِيظًا وَخِزَانَتُكَ ↓ أَمِينَةً رَشَدْتَ فِى أَمْرَيْكَ دُنْيَاكَ وَآخِرَتِكَ i. e. ‡[*When*] *thy tongue* [*is such as keeps the secret confided to thee*], *and thy heart* [*is trustworthy, thou wilt follow the right way in respect of thy two states, thy state in the present world and thy state in the world to come*]. (TA.)

مَخَازِنُ ٱلطَّرِيقِ — خِزَانَةٌ: see مَخْزَنٌ and مَخْزِنٌ. *The nearest roads,* or *ways.* (Ḳ,* TA. [In the CḲ, مَخَاصِرُهُ is erroneously put for مَخَاصِرُهُ.])

خزو

1. خَزَاهُ, (Ṣ, Ḳ,) aor. يَخْزُوهُ, (Ṣ,) inf. n. خَزْوٌ, (Ṣ, Ḳ,) *He ruled,* or *governed, him; and subdued him.* (Ṣ, Ḳ.) And خَزَا ٱلدَّابَّةَ *He broke,* or *trained, the beast.* (Ḳ.) — Also *He withheld,* or *restrained, him from* [*indulging*] *his natural desire.* (Ḳ.) And خَزَا ٱلنَّفْسَ, inf. n. as above, *He withheld,* or *restrained, the soul from its purpose,* or *intention.* (JK.) One says, أُخْزُ فِى طَاعَةِ ٱللّٰهِ نَفْسَكَ *Withhold thou,* or *restrain thou, thy soul from its purpose,* or *intention, and make it to endure patiently the right course of events, in obedience to God.* (TA.) — And *He slit his tongue* [*to prevent his sucking*]; namely, a young camel's. (Ḳ.) — *He possessed him,* or *it.* (Ḳ.) — *He treated him,* or *regarded him, with enmity,* or *hostility.* (Ḳ.) — And خَزْوٌ (JK, TA) and خِزَايَةٌ (JK) signify The act of *piercing,* or *thrusting,* (JK, TA,) *with spears.* (JK.)

خزى

1. خَزِىَ, aor. يَخْزَى, inf. n. خِزْىٌ (Ṣ, Mgh,* Mṣb, Ḳ, &c.) and خَزًى (Sb, Ḳ) and خَزْيَةٌ and مَخْزَاةٌ, (MA, [or these two are simple substs.,]) *He was,* or *became, base, abased, abject, vile, despicable,* or *ignominious:* (Ṣ, Mgh, Mṣb:) or, accord. to ISk, *he fell into trial,* or *affliction:* (Ṣ:) or *he fell into trial,* or *affliction,* (Ḳ, TA,) *and evil,* (TA,) *and a thing that exposed him to disgrace, and thereby became base, abased, abject, vile, despicable,* or *ignominious;* as also ↓ اِخْزَوَى, (Ḳ, TA,) like اِرْعَوَى: (TA:) or خِزْىٌ signifies *disgrace,* or *ignominy:* so in the Ḳur v. 37 [&c.]: (Sh, TA:) or its primary signification is *a state of abasement, vileness,* or *ignominy, of which one is ashamed:* so accord. to Er-Rághib and Bḍ and the Ksh: (MF, TA:) or the *manifesting foul actions* or *qualities, for the manifesting of which one deserves punishment:* (El-Ḥarállee, TA:) or خِزْىٌ signifies *he became disgraced, and was confounded,* or *perplexed, by reason of disgrace.* (TA.) — And خَزِىَ, (Ṣ, Mṣb, Ḳ,) aor. as above, (Ṣ, Mgh,) inf. n. خَزَايَةٌ (Ṣ, Mgh, Mṣb, Ḳ) and خَزًى, (Ḳ,) *He was,* or *became, moved,* or *affected, with shame;* (Ṣ, Mgh, Mṣb, Ḳ;) [as also ↓ اِسْتَخْزَى: see the part. n. of this latter below:] or خَزَايَةٌ signifies the *being moved,* or *affected, with much,* or *intense, shame.* (JK.) — خَازَانِى فَخَزَيْتُهُ: see the next paragraph. — خِزَايَةٌ: see art. خزو.

3. خَازَانِى, inf. n. مُخَازَاةٌ, *He vied,* or *contended, with me in mutual abasing;* or *rendering base, abject, vile, despicable,* or *ignominious.* (TḲ.) You say, خَازَانِى فَخَزَيْتُهُ ↓, (Ks, JK, Ṣ, Ḳ,) aor. of the latter أَخْزِيهِ, (Ks, JK, Ṣ,) [inf. n. app. خَزْىٌ,] *I vied,* or *contended, with him* (JK, TḲ) *in mutual abasing,* &c., (TḲ,) *and I surpassed,* or *overcame, him* [*therein*]: (JK:) [or it may signify, *in abasement,* &c.; for] the meaning [of the latter verb with its pronoun] is كُنْتُ أَشَدَّ خِزْيًا مِنْهُ. (Ḳ: in the CḲ, خَزْيًا: but in a MS copy of the Ḳ, خِزْيًا.)

4. اخزاه *He* (God) *abased him;* or *rendered him base, abject, vile, despicable,* or *ignominious:* or *may He abase him;* &c.: (Ṣ, Mṣb:) or *He disgraced him,* or *put him to shame:* or *may He disgrace him,* &c. (Ḳ.) Hence, in the Ḳur [xi. 80], the saying of Lot to his people, وَلَا تُخْزُونِ فِى ضَيْفِى (TA) *And disgrace ye not me in respect of my guests:* (Bḍ, Jel, TA:) or *make not me ashamed* &c. (Bḍ.) One says also, of him who has done or said that which is approved, مَا لَهُ أَخْزَاهُ ٱللّٰهُ [*What aileth him? May God abase him,* or *disgrace him!*]; and so without ما له: (Ḳ:) it is like مَا لَهُ قَاتَلَهُ ٱللّٰهُ, said of a man whose action pleases: (Ṣ in art. نفر:) it is used in lieu of praise, to charm a person against the evil eye; and means a prayer for him, not an imprecation against him. (TA.) See also مُخْزٍ, below. — Also *He made him to be ashamed for himself* (Ḥam pp. 114 and 397, and TA) *in respect of him, for his shortcoming.* (TA.) [See the citation from the Ḳur above.] — *He compelled him,* or *constrained him, to admit an evidence,* or *a proof, whereby he abased him,* or *disgraced him.* (TA.) — And *He aided him in, and made him to keep to, a* مَخْزَاة [i. e. a *thing that was a cause of shame,* or *of abasement* or *disgrace*]. (TA.)

9. اِخْزَوَى: see 1.

10: see 1.

خَزٍ [part. n. of خَزِىَ] *Base, abased, abject, vile, despicable,* or *ignominious.* (MA.) [See also خَزْيَانُ and مُخْزًى.]

خَزْيَةٌ *A habit, a quality, a practice,* or *an action,* (JK,) or *a crime, a sin,* or *an offence,* or *act of disobedience,* (TA,) *of which one is,* or *should be, ashamed:* (JK, TA: [and ↓ مَخْزَاةٌ, which is coupled therewith in the JK, app. as syn. with it, properly signifies *a thing,* or *an action,* &c., *that is a cause of shame,* or *of abasement* or *disgrace;* being a noun of the class of مَجْبَنَةٌ and مَبْخَلَةٌ; originally مَخْزَيَةٌ: see 4, last sentence:]) and ↓ مُخْزِيَةٌ, of the form of an act. part. n., from أَخْزَى, signifies [the same; or] *a bad, an evil,* or *a foul, habit, quality, practice,* or *action:* the pl. of this last is مُخْزِيَاتٌ, and of the same [and of مَخْزَاةٌ also] مَخَازٍ. (Mṣb.) Or the first signifies *A habit, a practice,* or *an action, in which one becomes base, despicable,* or *ignominious.* (Mgh.) — Also *A trial,* or *an affliction,* (Ḳ, TA,) *into which one is made to fall;* (TA;) and so ↓ خِزْيَةٌ. (Ḳ.)

خِزْيَةٌ: see what next precedes.

خَزْيَانُ *Moved,* or *affected, with shame;* (Ṣ, Mṣb, Ḳ, and Ḥam p. 36;) as also ↓ مُسْتَخْزٍ: (Ḥar p. 482:) or, *with much shame,* (Lth, JK, TA,) *on account of a bad,* or *foul, deed that he has done:* (Lth, TA:) or it may signify [like خَزٍ] *base, abased, abject, vile, despicable,* or *ignominious:* (Ḥam ubi suprà:) fem. خَزْيَا (Lth, JK, Ṣ, Ḳ) and خَزْيَانَةٌ, which is irreg.: (TA:) pl. خَزَايَا. (Lth, JK, Ṣ, Ḳ.)

مُخْزًى *Rendered base, abject, vile, despicable,* or *ignominious, by a thing proved against him.* (TA.) [See also خَزٍ and خَزْيَانُ.]

كَلَامٌ مُخْزٍ *Speech,* or *language, that is approved, so that one says of its author,* أَخْزَاهُ ٱللّٰهُ. (TA.) They relate that El-Farezdaḳ gave utterance to an excellent verse, and said, هٰذَا بَيْتٌ مُخْزٍ, i. e. [*This is a verse such as that,*] *when it is recited, people will say,* [or rather, *such as will make it to be said of me,*] مَا أَشْعَرَهُ قَاتَلَهُ ٱللّٰهُ ↓ أَخْزَى [*May God abase,* or *disgrace, the sayer thereof! How good,* or *excellent, a poet is he!*]. (TA. [See 4.]) [Hence,] قَصِيدَةٌ مُخْزِيَةٌ [*An ode*] *that is extremely good.* (TA.)

مَخْزَاةٌ, said in the MA to be an inf. n. of خَزِىَ: see خَزْيَةٌ.

مُخْزِيَةٌ: see خَزْيَةٌ.

مُسْتَخْزٍ: see خَزْيَانُ.

خس

1. خَسَّ, (Mṣb, TA,) sec. pers. خَسِسْتَ, (Fr, Ṣ,

A, Mṣb, Ḳ,) aor. يَخَسُّ ; (A, Mṣb, TA;) and خَسَّ, sec. pers. خَسِسْتَ, aor. يَخَسُّ ; (Mṣb, TA;) inf. n. خِسَّةٌ (Fr, Ṣ, A, Ḳ) and خَسَاسَةٌ (Fr, Ṣ, A, Mṣb, Ḳ) and خُسُوسٌ, (TA,) *He* (a man) *was*, or *became, low* or *ignoble, base, vile, mean* or *sordid, weak;* (Fr, Ṣ, A, Ḳ;) [or, more commonly,] *contemptible:* (Fr,* Ṣ,* A, Ḳ:) and *it* (a thing) *was*, or *became, base, vile*, or *mean;* (TA;) or *contemptible, paltry*, or *inconsiderable;* (Mṣb,* TA;) and *bad, corrupt, abominable*, or *disapproved.* (TA.) — خَسَّ فِعْلُهُ, and قَوْلُهُ, and رَأْيُهُ, and حَظُّهُ, ‡ [*His action*, and *his saying*, and *his opinion*, and *his fortune, was*, or *became, low* or *ignoble*, &c.] (A.) — خَسَّ, aor. يَخِسُّ, *It was*, or *became, light of weight, not equal to what corresponded with it.* (Mṣb.) — خَسَّ, aor. يَخَسُّ ; (Mṣb;) and ↓ اخسّ, (ISk, Ṣ, Mṣb, Ḳ,) inf. n. إِخْسَاسٌ ; (ISk, Ṣ;) ‡ *He did what was low* or *ignoble, base, vile, mean* or *sordid, weak;* [or, more commonly,] *contemptible:* (ISk,* Ṣ,* Mṣb,* Ḳ:) or ↓ the latter signifies *he did something low* or *ignoble*, &c., *in actions.* (TA.) ═ خَسَّ نَصِيبَهُ, (Ṣ, A, Mgh,* Ḳ,) aor. يَخُسُّ ; (Ṣ, TA;) and ↓ اخسّهُ ; (Mgh;) *He made his lot, portion*, or *share, to be low* or *ignoble, base, vile, mean;* [or, more commonly,] *contemptible:* (Ṣ,* A, Mgh,* Ḳ:) and خَسَّ الحَظَّ, and ↓ اخسّهُ, *He made the lot, portion*, or *share, little*, and *incomplete.* (TA.) — Also خَسَّ حَظَّهُ, [and ↓ اخسّهُ,] ‡ [*He made his fortune to be low* or *ignoble*, &c.; or *contemptible;* differing from the signification immediately preceding, being said to be tropical.] (A.) The Arabs say of a man without any good fortune in the present life, أَخَسَّ ٱللّٰهُ حَظَّهُ †[*God made his fortune mean*, or *contemptible*]; as also أَحَتَّهُ. (AM, TA.)

4. اخسّ : see خَسَّ, in two places. ═ اخسّهُ : see خَسَّ نَصِيبَهُ and what follows it, in four places. — Also *He found him to be low* or *ignoble, base, vile, mean* or *sordid, weak;* [or, more commonly,] *contemptible.* (Ṣ,* Ḳ.)

6. تَخَاسُّوهُ *They did it by turns:* or *they hastened together*, or *vied in hastening, to do it.* (Ṣgh, Ḳ.)

10. استخسّهُ *He reckoned, accounted*, or *esteemed, him low* or *ignoble, base, vile, mean* or *sordid, weak;* [or, more commonly,] *contemptible.* (Ṣ,* Ḳ.) — استخسّ حَظَّهُ ‡[*He accounted his fortune low* or *ignoble*, &c.]. (A.)

خَسٌّ [*Lettuce; lactuca;*] *a certain plant*, (Ṣ, Mṣb, Ḳ,) *of the kind called* بَقْل, (Ṣ, Ḳ,) *well known*, (Mṣb, Ḳ,) *of the description termed* أَحْرَار, [i. e., *that are eaten without being cooked*, or *that are slender and succulent*, or *slender and soft*,] *with broad leaves: it increases the blood: the wild kind has the property of the black poppy: the best is the garden-kind*, [*lactuca sativa*,] *which is succulent, yellow, and broad* [*in the leaf*]: *it is cold and moist in temperament: the most nutritious is that which is cooked; and it is useful for counteracting contrariety of the fluids; but the eating it constantly weakens the sight, and is injurious to the venereal faculty:* (TA:) n. un. with ة. (Mṣb.)

خُسَاسٌ : see خَسِيسٌ.

هٰذِهِ الأُمُورُ خِسَاسٌ بَيْنَهُمْ *These things*, or *affairs, are done by them by turns.* (JF, Ḳ.)

خَسِيسٌ, applied to a man, and to a lot or portion or share, (Ṣ, A, Ḳ,) or a thing, (Mṣb, TA,) *Low* or *ignoble, base, vile, mean* or *sordid, weak;* (Ṣ, A, Ḳ;) [or, more commonly,] *contemptible;* (A, Mṣb, Ḳ;) as also ↓ مُسْتَخَسٌّ and ↓ مُسْتَخِسٌّ ; (Ḳ;) and, applied to a thing, also, *paltry*, or *inconsiderable;* and so ↓ خُسَاسٌ and ↓ مَخْسُوسٌ ; *bad, corrupt, abominable*, or *disapproved:* (TA:) fem. with ة: (Mgh, Mṣb:) pl. masc. أَخِسَّآءُ and خِسَاسٌ (Mṣb, TA) and أَخِسَّةٌ ; (A;) pl. fem. خَسَائِسُ. (Mgh, Mṣb.) — حَظٌّ خَسِيسٌ, and ↓ مَخْسُوسٌ, ‡ *A low*, or *mean, fortune, that is not held in any estimation.* (A.) — هُوَ لَا يَدْخُلُ فِى خَسَائِسِ الأُمُورِ ‡[*He will not enter into low, mean*, or *contemptible, affairs*]. (A.) — خَسِيسٌ is also applied to *A disbeliever, an unbeliever*, or *infidel.* (TA.)

خَسِيسَةٌ fem. of خَسِيسٌ. (Mgh, Mṣb.) — You say also, رَفَعَ ٱللّٰهُ خَسِيسَةَ فُلَانٍ *God raised the condition of such a one after it had been low:* (Az, TA:) or رَفَعْتُ مِنْ خَسِيسَتِهِ *I did to him a deed whereby he became raised to a high condition.* (Ṣ, Ḳ.) ═ The *teeth* of a she-camel *within the period of the shedding of the central incisors:* you say, جَاوَزَتِ النَّاقَةُ خَسِيسَتَهَا [*The she-camel passed beyond the period of her* خَسِيسَة]: this is in the sixth year, when she sheds her central incisor: she is then such as is allowable for sacrifice. (Ṣ, Ḳ.)

خَسَّآءُ *The state of him, or it, that is* خَسِيس [i. e. *low* or *ignoble*, &c.]. (TA.) ═ *A foul*, or an *ugly*, woman. (TA.) [See also مُسْتَخَسٌّ.]

أَخَسُّ [*More, and most*, خَسِيس, i. e., *low* or *ignoble*, &c.]. You say, مَا رَأَيْتُ أَخَسَّ مِنْهُ [*I have not seen any more low* or *ignoble*, or, more commonly, *contemptible, than he*, or *it*]. (A.)

مَخْسُوسٌ A man *made*, or *rendered, low* or *ignoble*, &c., *and weak.* (TA.) — See also خَسِيسٌ, in two places.

مُسْتَخَسٌّ and مُسْتَخِسٌّ : see خَسِيسٌ. — Also, both words, *A foul*, or an *ugly*, face: fem. with ة. (Ḳ, TA.) [See also خَسَّآءُ.]

خسأ

1. خَسَأَ, (Ṣ, Ḳ,) aor. ـَ, (Ḳ,) inf. n. خَسْءٌ (Ṣ, Ḳ) and خُسُوٌّ, (Ḳ,) *He drove away* a dog: (Ṣ, Ḳ:) *he chid* him. (Lth.) ═ It is also intrans., (Ṣ,) and signifies *He* (a dog) *went away, to a distance;* (Ṣ,* Ḳ;) [*being driven away*, or *chidden;*] as also خَسِئَ, (Ḳ,) and ↓ انخسأ. (Ṣ, Ḳ.) — [Hence,] tropically, said to a man, اِخْسَأْ إِلَيْكَ, meaning اِخْسَأْ عَنِّى ‡[*Go thou away:* or *go thou away from me*]. (TA.) اِخْسَئُوا فِيهَا وَلَا تُكَلِّمُونِ, in the Ḳur [xxiii. 110], is expressive of removal to a distance with anger; [meaning † *Go ye away into it*, (i. e. the fire of Hell,) *and speak not unto Me.*] (Zj.) — [And hence, † *He was*, or *became, vile and despised and hated:* so says Golius, as on the authority of the KL; but this meaning is not in my copy of that work: it agrees, however, with a signification of the part. n. خَاسِئٌ, q. v.] ═ Also, inf. n. خَسْءٌ and خُسُوٌّ [as above], said of the sight, (AZ, Ṣ, Ḳ,) ‡ *It was*, or *became, dazzled*, or *confused*, (AZ, Ṣ, TA,) and *dim.* (Ḳ, TA.)

3. خَاسَئُوا, (Ḳ,) inf. n. مُخَاسَأَةٌ, (Ṣ,) ‡ *They threw stones, one at another;* (Ṣ, Ḳ;) as also ↓ تَخَاسَئُوا بِالحِجَارَةِ, (Ḳ,) or تخاسؤا. (Ṣ.) And كَانَتْ بَيْنَهُمْ مُخَاسَأَةٌ ‡[*There was between them a contending in throwing*, or *throwing of stones*]. (Ṣ, TA.) [See also art. خسى.] — And هُوَ يُخَاسِئُ means يُقَامِرُ [*He contends in a game of hazard*]. (IB, TA in art. خسو.)

6: see 3.

7: see 1.

خَسِىْءٌ *Bad* wool. (O, Ḳ.)

خَاسِئٌ, applied to a dog, and to a swine, (Ḳ,) and to a devil, (TA,) *Driven away, repelled, and not suffered to come near to men.* (Ḳ, TA.) — And [hence,] † *Contemptible, despicable, vile*, or *abject.* (TA.) — Applied to the sight, ‡ *Dazzled*, or *confused*, (Ṣ, TA,) and *dim.* (TA.) So in the words of the Ḳur [lxvii. 4], يَنْقَلِبْ إِلَيْكَ البَصَرُ خَاسِئًا ‡[*The sight will recoil to thee dazzled*, or *confused*, or *dim*]: (Ṣ, TA:) or the meaning here is † *contemptible:* or *withdrawing far away:* or it is of the measure فَاعِل in the sense of the measure مَفْعُول, [meaning *repelled far away*,] like عِيشَةٌ رَاضِيَةٌ, in the Ḳur [lxix. 21 and ci. 5], for مَرْضِيَّة. (TA.)

خسر

1. خَسِرَ, (Ṣ, A, Mṣb, Ḳ, &c.,) aor. ـَ; (Ḳ;) and خَسَرَ, aor. ـِ; (Ḳ;) but the latter is an unusual form [except in the sense of أَخْسَرَ]; (B, TA;) inf. n. خُسْرَانٌ (Ṣ, A, Mṣb, Ḳ) and خُسْرٌ (Ṣ, Mṣb, Ḳ) and خَسَارَةٌ (Mṣb, Ḳ) [which are the only forms assigned in the TA to the verb when used with reference to traffic] and خُسُرٌ and خَسْرٌ and خَسَرٌ and خَسَارٌ ; (Ḳ;) *He lost*, or *suffered loss* or *diminution:* or *he was deceived, cheated, beguiled*, or *circumvented:* (Ḳ:) فِى البَيْعِ *in selling;* (Ṣ;) or فِى بَيْعِهِ *in his selling;* (A;) or فِى تِجَارَتِهِ *in his traffic:* (Mṣb, Ḳ: [see also 4:]) the former is the original signification: (TA:) *he suffered diminution of his capital; he lost part thereof:* (B, TA:) and *he lost his capital altogether.* (Bḍ in iv. 118; &c.) خُسْرَانٌ is also attributed to an action, as well as to a man: (B, TA:) you say, (but in this case the verb is used tropically, A,) خَسِرَتْ تِجَارَتُهُ ‡[*His traffic was losing;* or *an occasion of loss*]; (A, B;) opposed

to رَبِحَتْ. (A.) It is also used in relation to personal acquisitions; such as health, and safety, and intellect, and faith, and the recompense or reward of obedience [to God], which God has declared [Ḳur xxii. 11 and xxxix. 17] to be manifest خُسْرَان, (B,) since there is none like it. (Bḍ.) For instance, you say, خَسِرَ عَقْلَهُ, and مَالَهُ, *He lost his intellect*, and *his property*. (IAạr.) [In a phrase of this kind, the noun which immediately follows the verb may be considered as put in the accus. case on account of the rejection of a prep., namely فِى: for] it is said that خَسِرَ is never used otherwise than intransitively: though this has been contradicted, on the ground of the following phrase in the Ḳur [xxii. 11], خَسِرَ الدُّنْيَا وَالْآخِرَةَ [*He hath lost*, or *he loseth*, *the things of the present life and of the latter life*]; and the like; as الَّذِينَ خَسِرُوا أَنْفُسَهُمْ وَأَهْلِيهِمْ [*Who shall have lost themselves*, or *their own souls*, *and their families*, or *their wives*; Ḳur xxxix. 17 and xlii. 44]; (MF, TA;) i. e., themselves, or their own souls, by their having erred, and their families by their having caused them to err, or by being separated from them for ever; (Bḍ;) or by being themselves made to remain for ever in Hell, and by their not gaining access to the حُور prepared in Paradise [as wives] for the believers: (Jel:) or the meaning is, accord. to Fr, *who shall be deceived of their own souls, &c.*: or, accord. to others, *who shall have destroyed their own souls, &c.* (TA.) — Also [*He experienced*, or *saw*, *that he was a loser*; or] *his having lost became manifest to him*: so in the Ḳur xl. [78 and] last verse. (TA.) — Also (with all the forms of the inf. n. above mentioned, Ḳ,) *He erred; went astray; deviated from*, or *lost*, or *missed*, *the right way*: or *he became lost; he perished*; or *he died*: syn. ضَلَّ, (Ḳ,) and هَلَكَ. (Mṣb.) ══ خَسَرَهُ, (A 'Obeyd, IAạr, Zj, Ṣ, A, &c.,) aor. ـِ (Zj, Mṣb) and ـُ, (Bḍ in lv. 8,) inf. n. خَسْرٌ (Mṣb, Ḳ) and خُسْرَانٌ; (Ḳ;) and ▼اخسرهُ, (A 'Obeyd, Zj, Ṣ, A, Mṣb,) inf. n. إِخْسَارٌ; (Mṣb, Ḳ;) and ▼خسّرهُ; (A;) *He made it defective*, or *deficient*; (A 'Obeyd, IAạr, Zj, Ṣ, A, Mṣb, Ḳ;) namely, the weight, and the measure; (Zj, TA;) and the thing weighed; (TA;) and the balance, (A 'Obeyd, IAạr, Zj, A, Mṣb,) by diminishing the weight. (Mṣb.) ▼The second of these forms is more common, in this sense, than the first (Zj, TA) [and than the third]. For وَلَا تُخْسِرُوا ▼الْمِيزَانَ, in the Ḳur lv. 8, there are three other readings; namely تَخْسِرُوا and تَخْسُرُوا and تَخْسَرُوا; in the last of which, the prep. فِى is omitted after the verb. (Bḍ.) — [And *He*, or *it*, *made him to lose*, or *suffer loss*; *to err*, or *go astray*; *to become lost*, or *to perish*.]

2. خَسَّرَهُ, (A, Ḳ,) inf. n. تَخْسِيرٌ, (Ṣ, Ḳ,) *i. q.* خَسَرَهُ, q. v.: (A:) [and particularly] *He*, or *it*, *destroyed him; caused him to perish*. (Ṣ, Ḳ.) You say, خسّرهُ سُوءُ عَمَلِهِ ‡*The evilness of his conduct caused him to perish*. (A.) — *He put him away*, or *far away; removed, alienated*, or *estranged, him;* (IAạr, Mṣb;) *from good*, or *prosperity*. (IAạr.) — *He attributed*, or *imputed, to him* خُسْرَان [i. e. *loss;* or *error*, or *deviation from the right way*]: like كَذَّبَهُ meaning "he attributed, or imputed, to him lying," &c. (Mṣb.)

4. اخسرهُ *i. q.* خَسَرَهُ, which see in three places: (A 'Obeyd, Zj, Ṣ, A, Mṣb:) [and particularly] *He made him to lose*, or *suffer loss, in his traffic; contr. of* أَرْبَحَهُ. (A.) ══ And اخسر *He fell into loss;* (A;) *he met with loss in his traffic*. (TA. [See also 1.])

خُسْرٌ an inf. n. of خَسِرَ. (Ṣ, Mṣb, Ḳ.) In the Ḳur ciii. 2, accord. to some, it means *Punishment for sin*. (TA.)

خَسِرٌ : see خَاسِرٌ.

خُسْرَانٌ an inf. n. of خَسِرَ. (Ṣ, A, Mṣb, Ḳ.) [For particular usages thereof, see 1. As a simple subst., it generally signifies *Loss*, or the *state of suffering loss* or *diminution*: the *state of being deceived* or *cheated*: *error*, or *deviation from the right way*: (see also خَسَارٌ:) or the *state of becoming lost, of perishing*, or *of dying*.] — It is also an inf. n. of خَسَرَهُ. (Ḳ.)

خُسْرَوِىٌّ : see what next follows.

خُسْرُوَانِىٌّ, (A, Ḳ,) or خُسْرَوَانِىٌّ, (TA, [but the former is the better known,]) *A certain kind of garment* or *cloth*; (A, Ḳ;) so called in relation to Khusrow Sháh, one of the [kings of Persia called] أَكَاسِرَة [pl. of كِسْرَى or كَسْرَى]; as also ▼خُسْرَوِىٌّ. (A, TA.) — And *A certain wine* or *beverage*. (Ḳ.)

خَسَارٌ and ▼خَسَارَةٌ, [both inf. ns. of خَسِرَ, q. v.,] (Ṣ,) and ▼خَيْسَرَى, (Ṣ, M, Ḳ, in some copies of the Ḳ written خَنْسَرَى, with ن, TA,) *Error*; or *deviation from the right way*: [like خُسْرَانٌ:] (Ṣ:) and *perdition*; or *death*; (Ṣ, Ḳ;) as also ▼خَنَاسِيرُ, (Ṣ, and Ḳ in art. خنسر,) which last [is of a pl. form, but] has no sing. (Ṣ.) — And all the foregoing words, including ▼خناسير, *Baseness, ignobleness, ungenerousness*, or *meanness*; (Ḳ;) the last, in poetry, shortened to ▼خَنَاسِرُ: (TA:) and ▼خَيْسَرَى, (Ḳ,) and, as some say, ▼خَنَاسِرُ, (TA,) *perfidy, unfaithfulness*, or *treachery*. (Ḳ, TA.)

خَسِيرٌ : see خَاسِرٌ.

خَسَارَةٌ : see خَسَارٌ.

خَاسِرٌ *Losing*, or *suffering loss, in his traffic*. (Lth.) And [hence,] تِجَارَةٌ خَاسِرَةٌ ‡ [*Losing traffic; traffic which is an occasion of loss*]; opposed to رَابِحَةٌ. (A.) And صَفْقَةٌ خَاسِرَةٌ † *A bargain that does not bring gain* [*but on the contrary occasions loss*]. (TA.) And كَرَّةٌ خَاسِرَةٌ † *An unprofitable charge* or *assault*. (Ḳ.) — One *who has lost his property*, and *his intellect*. (IAạr.) — *Erring; going astray; deviating from*, or *losing*, or *missing, the right way*: or *becoming lost; perishing;* or *dying*: syn. ضَالٌّ: (Ḳ:) and so ▼خَسِرٌ (TA) and ▼خَسِيرٌ and ▼خَيْسَرَى, (Ḳ, TA, but the last written in the CḲ خَيْسَرِىّ,) or ▼خَيْسَرٌ, for it is said to occur [as an epithet] only in the following saying, in which خَيْسَرَى is said to be put for خَيْسَرٌ to assimilate it to preceding words: بِفِيهِ الْبَرَى وَحُمَّى خَيْبَرَى وَشَرُّ مَا يَرَى فَإِنَّهُ خَيْسَرَى [*In his mouth be dust, and may the fever of Kheyber befall him, and evil be that which he shall see, for he is one who goeth astray*: but in the TA, in art. ورى, is another reading; for بفيه البرى, substituting بِهِ الْوَرَى, meaning a certain disease]. (TA.) [Hence,] أَحْمَقُ خَاسِرٌ دَابِرٌ دَامِرٌ [*Foolish*, or *stupid, erring, and utterly perishing*]. (T in art. بت. [See بَاتٌّ: and see also دَامِرٌ.]) — Also One *who makes the measure, and the balance, defective*, or *deficient, when he gives, and demands excess when he receives*. (AA.)

خَاسُورَةٌ : see the next paragraph.

خَنْسَرٌ and ▼خَنْسَرِىٌّ A man *in a place* [or *condition*] *of* خُسْرَان [or *loss*, &c.]: (Ḳ in the present art. and in art. خنسر:) pl. خَنَاسِرَةٌ. (Ḳ in art. خنسر.) — And [the pl.] الخَنَاسِرَةُ, in several copies of the Ḳ, in other copies of the Ḳ ▼الخَاسُورَةُ, but correctly ▼الخَنَاسِرُ, (TA,) *The weak of mankind;* (Ḳ, TA;) and *the small*, or *little, of them;* (TA;) as also ▼الخَنَاسِيرُ, in the former sense, (Ḳ and TA in art. خنسر,) and in the latter sense also: (TA in that art.:) and أَهْلُ الخِيَانَةِ; (Ḳ and TA in this art.; and Ḳ in art. خنسر, accord. to several copies;) i. e. *The people of perfidy, unfaithfulness*, or *treachery;* and *of baseness, ignobleness, ungenerousness*, or *meanness*: (TA in the present art.:) or اهل الجبانة; because of their weakness; (TA in art. خنسر;) [as though meaning *the people of cowardice* (الجَبَانَة): or it may mean *the people of the burial-ground* (الجَبَّانَة)]; for, accord. to AḤát, ▼الخناسير signifies *those who conduct* [*to the burial-ground*] *the corpse* or *the bier with the corpse*; perhaps from خَنَاسِرُ meaning "small, or little, and weak men." (TA.)

خِنْسِرٌ, (Ḳ in art. خنسر, [in the CḲ, erroneously, خِنْسَر,]) or ▼خِنْسِيرٌ, (Ibn-'Oṣfoor, AḤei, and Ḳ in the present art.,) *Base, ignoble, ungenerous*, or *mean*: (Ḳ:) and *perfidious, unfaithful*, or *treacherous*. (TA in explanation of the latter.) ══ Also (the former accord. to the Ḳ in art. خنسر, and the latter likewise accord. to the TA in the present art.,) *A calamity*, or *misfortune*: (Ḳ, TA:) pl. [of the latter] in this sense خَنَاسِيرُ, like خَنَازِيرُ. (IAạr, TA.)

خَيْسَرٌ : see خَاسِرٌ.

خَيْسَرَى : see خَسَارٌ, in two places: ══ and see also خَاسِرٌ. — Also One *who will not accept an invitation to partake of food, lest he should be required to make a requital*: so in a trad. of 'Omar. (TA.)

خَنْسَرِىٌّ : see خَنْسَرٌ.

خِنْسِيرٌ : see خِنْسِرٌ.

خَنَاسِرُ : see خَسَارٌ: ══ and see also خَنْسَرٌ.

خَنَاسِيرُ a word [of a pl. form] having no sing.: (Ṣ:) see خَسَارٌ, in three places. ══ [Also pl. of خِنْسِيرٌ, q. v.] ══ See also خَنْسَرٌ, in two places. ══ Also The *urine of the mountain-goats upon the herbage and the trees* [or *shrubs*]: (Ḳ in this art. and in art. خنسر :) in which sense, also, it has no singular. (TA in the present art.)

أَخْسَرُ sing. of أَخْسَرُونَ, which occurs in the Kur [xi. 24 and] xviii. 103 [and xxi. 70 and xxvii. 5], (Akh, Ṣ,) and signifies The *greatest losers; those who suffer,* or *shall suffer, the greatest loss.* (Bḍ.)

[مَخْسَرَةٌ *An occasion,* or *a cause, of loss;* or *of error,* or *going astray;* or *of being lost,* or *perishing,* or *of dying:* a word of the same class as مَبْخَلَةٌ and مَجْبَنَةٌ &c.: pl. مَخَاسِرُ. Hence the saying,] المَسَاخِرُ مَخَاسِرُ ‡ [*Occasions,* or *causes, of mockery,* or *derision,* or *ridicule, are occasions,* or *causes, of loss,* &c.]. (A.)

خسف

1. خَسَفَ, aor. ـِ, (JK, Ṣ, Mṣb, Ḳ,) inf. n. خَسْفٌ, (JK,) or خُسُوفٌ, (Ṣ, Ḳ,) or both; (Mṣb;) [and ▼انخسف;] *It* (a place) *sank,* (JK, Mṣb,) or *went away, into the ground,* or *earth,* (Ṣ, Mṣb, Ḳ,) *with what was upon it.* (JK.) You say, خَسَفَتِ الأَرْضُ ▼انخسفت, [and خَسَفَت,] *The ground sank* [*into the earth*] *with what was upon it.* (TA.) And انخسفت ▼ بِهِ الأَرْضُ, (JK,) or خُسِفَ بِهِ الارض, and انخسف ▼ به الارض, (TA,) and خَسَفَتْ, (Mṣb in art. سوخ,) *The ground sank with him,* or *it:* (JK:) or *the ground,* or *earth,* [*swallowed up him,* or *it;* or] *took and enclosed him,* or *it.* (TA.) And انخسفت ▼ البِئْرُ *The well* [*sank and collapsed;* or] *went away into the earth with its casing of stones and wood.* (Mgh.) And خُسِفَ بِهِ and خَسَفَ فِى الأَرْضِ [*He,* or *it, sank into the ground,* or *earth, and became swallowed up,* or *enclosed,* or *concealed, therein*]. (Ṣ.) It is said in the Ḳur [xxviii. 82], accord. to one reading, لَخُسِفَ بِنَا [*We had been swallowed up by the earth*]: (Ṣ:) accord. to another reading, (that of 'Abd-Allah, Ṣ, i. e. Ibn-Mes'ood, TA,) لَانْخُسِفَ ▼ بِنَا, (Ṣ, Ḳ,) in the pass. form; (Ḳ;) [meaning the same;] like as one says, أُنْطُلِقَ بنا. (Ṣ.) You say also, خَسَفَتْ عَيْنُ المَآءِ *The spring of water sank,* or *went away, into the earth.* (Mṣb, Ḳ.*) And انخسفت ▼ العَيْنُ *The eye sank,* or *became depressed, in the head;* syn. غَارَتْ; (Mṣb in art. غور;) [and so خَسَفَتْ, inf. n. خُسُوفٌ; for] خُسُوفُ العَيْنِ signifies *The eye's going away into the head:* (Ṣ:) or ▼انخسفت signifies *its black,* or *part surrounded by the white, disappeared in the head:* (Mgh:) or this last, (Ḳ,) as quasi-pass. of the trans. v. خَسَفَ, (TA,) ‡ *it* (the eye) *became blind;* as also ▼أَخْسَفَتْ; (Ḳ, TA;) and [in like manner] خَسَفَتْ † *it* (the eye) *lost its light* [or *sight*]. (Mṣb.) — [Hence, app.,] خَسَفَ القَمَرُ, inf. n. خُسُوفٌ; (Ṣ, Mṣb, Ḳ;) and خُسِفَ; (TA;) † *The moon* [*suffered eclipse,* or *became eclipsed,* or] *lost its light,* or *part of its light;* (Mṣb;) i. q. كَسَفَ: (Ṣ,* Mṣb,* Ḳ:) and خَسَفَتِ الشَّمْسُ and كَسَفَتْ both signify the same [i. e. *the sun suffered eclipse,* &c.]: (Mgh:) or one says كَسَفَتْ of the sun, and خَسَفَ of the moon, (Th, Ṣ, Mṣb, Ḳ,) accord. to the more approved usage: (Th, Ṣ, Mṣb:) or, in the common conventional language, الكُسُوفُ is the partial loss of the light of the sun, and الخُسُوفُ is *the total loss of the light* thereof: (AḤát, Mṣb:) or الخسوف is *the partial loss of the light* of the sun, and الكسوف is the total loss thereof, (Ḳ, TA,) accord. to AḤát: (TA:) الخسوف often occurs in the trads., as said of the sun; though the term commonly known in the classical language is الكسوف [in this case]: and it is said in a trad., إِنَّ الشَّمْسَ وَالقَمَرَ لَايَخْسِفَانِ لِمَوْتِ أَحَدٍ أَوْ لِحَيَاتِهِ [*Verily the sun and the moon suffer not eclipse for the death of any one or for his life*]; predominance being in this instance attributed to the moon, as being masc., over the sun, which is fem. (IAth.) — Also, inf. n. خَسْفٌ, † *It* (a thing) *became defective* or *deficient; suffered loss* or *diminution.* (Ḳ.) — ‡ *It* (the body) *became lean,* or *emaciated.* (TA.) And خَسَفَتْ, said of camels and of sheep or goats, ‡ *They became lean,* or *emaciated.* (TA. [This meaning is there indicated, but not clearly expressed. See خَسْفَةٌ. Accord. to the KL, the inf. n. خَسْفٌ signifies The *being vile, abject,* or *contemptible:* and also the *being lean,* or *emaciated:* and hence Golius, on that authority, has rendered the verb as meaning *vilis et macer fuit.*]) — Also † *It* (the colour, or complexion, of a person) *became altered,* or *altered for the worse.* (TA.) — And ‡ *It* (a thing, Ḳ, as, for instance, a roof, TA) *became pierced with a hole,* or *rent;* (Ḳ, TA;) as also ▼انخسف. (TA.) — And, خَسَفَتْ, said of a she-camel, ‡ *She, after yielding abundant milk, soon stopped* [*its flow*] *in winter.* (Ḳ, TA.) — And, said of a well, *It was,* or *became, such as is termed* خَسِيفٌ [q. v.]. (TA.) — And خَسَفَ, said of a man, ‡ *He recovered from a disease.* (IDrd, Ḳ, TA.) = خَسَفَ, (JK, Mṣb, TA,) aor. ـِ, (Ḳur xvi. 47, &c.,) inf. n. خَسْفٌ, *He* (God) *made* a place, (JK, Mṣb,) or the ground, (TA,) *to sink,* (JK, Mṣb, TA,) or *go away, into the earth,* (Mṣb,) *with what was upon it.* (JK, TA.) And خَسَفَ بِهِ الأَرْضَ, (Ṣ, Ḳ,) inf. n. خَسْفٌ, (Ṣ,) *He* (God) *made him,* or *it, to disappear in the earth,* or *ground:* (Ṣ, Ḳ:) [or *made the earth,* or *ground, to sink with, and swallow up, him,* or *it:*] whence, in the Ḳur [xxviii. 81], فَخَسَفْنَا بِهِ وَبِدَارِهِ الأَرْضَ [*And we made the ground to sink with, and swallow up, him and his mansion*]. (Ṣ.) And خَسَفْتُ عَيْنَ المَآءِ *I made the spring of water to sink,* or *go away, into the earth.* (Mṣb.) — خَسَفَ عَيْنَ فُلَانٍ, (Ḳ, TA,) aor. ـِ, inf. n. خَسْفٌ, (TA,) ‡ *He put out,* or *blinded, the eye of such a one,* (Ḳ,* TA,) *so that the black,* or *part surrounded by the white, disappeared in the head.* (TA.) — خَسَفَ الشَّىْءَ, (Ḳ,) aor. and inf. n. as above, (TA,) ‡ *He made a hole in,* or *rent, the thing.* (Ḳ, TA.) — And † *He cut,* or *cut off, the thing.* (Ḳ.) — خَسَفَ البِئْرَ, (Ḳ,) inf. n. as above, (TA,) † *He dug the well in stones, so that it yielded an abundant and unceasing flow of water:* (Ḳ, TA:) or *he dug the well by piercing through its mountain* [or *rock*] *to the water beneath so that it would never become exhausted:* or *he dug the well so as to reach an unceasing,* or *a copious, source of water.* (TA.) Hence the saying of 'Omar, in reply to a question of El-'Abbás respecting the poets, امْرُؤُ القَيْسِ سَابِقُهُمْ خَسَفَ لَهُمْ عَيْنَ الشِّعْرِ, i. e. † [*Imra-el-Ḳeys is he who has the precedence of them:*] *he has made the source of poetry to well forth abundantly to them.* (TA.) — خَسَفَ النَّاقَةَ, inf. n. as above, ‡ *He* (God) *made the she-camel, after yielding abundant milk, soon to stop* [*its flow*] *in winter.* (Ḳ, TA.) — خَسْفٌ also signifies The *confining* a beast *without fodder:* (Ḳ, TA:) or *making* a beast *to pass the night without fodder:* (Ḥam p. 290:) and (hence, TA) ‡ the *constraining* a man *to do that which he dislikes,* or *hates;* (JK, Ḥam ibid., Ḳ, TA;) as also خُسْفٌ: (JK:) and (hence, Ḥam) ‡ the *lowering, humbling,* or *abasing,* another: (Ḥam,* Ḳ, TA:) whence, سُمْتُهُ الخَسْفَ, (Ḥam,) or سَامَهُ خَسْفًا, &c.: [explained below: see خَسْفٌ:] (TA:) and the verb of خَسْفٌ in these three senses is خَسَفَ. (T, Ḳ.)

4. أَخْسَفَتِ العَيْنُ: see 1. = اخسف, said of a well-sinker, † *He found his well to be such as is termed* خَسِيفٌ [q. v.]: (JK:) or *he produced an abundant flow of water.* (TA.)

7: see 1, in nine places.

خَسْفٌ [an inf. n. of 1: and hence several of the significations here following.] *Deep places in the ground* (عُمُوقُ ظَاهِرِ الأَرْضِ; in the CḲ عُمُوقُ ماءِ الارضِ;) as also ▼خُسْفٌ. (Ḳ, TA.) — *The place whence the water* of a well *issues.* (AZ, Ṣ, Ḳ.) In the following saying of Sá'ideh El-Hudhalee,

* أَلَا يَا فَتًى مَا عَبْدُ شَمْسٍ بِمِثْلِهِ *
* يُبَلُّ عَلَى العَادِى وَتُؤْبَى المَخَاسِفُ ▼ *

the last word is pl. of خَسْفٌ [app. as signifying *A source of water*], after the manner of مَشَابِهُ and مَلَامِحُ: (TA:) the meaning is, [*Truly, O young man, what is 'Abd-Shems?* i. e.] *how great a person is 'Abd-Shems! by the like of him the enemy is overcome* [*and the sources of water become difficult of access*]. (M in art. بل.) — *A cloud,* or *collection of clouds, that has risen and appeared from the direction of the extreme west,* [as North-western Africa is called by the Arabs,] *from* [*the quarter of*] *the right of the Ḳibleh* [*to one who is on the north-east of Mekkeh, towards El-'Irák*]: (Lth, Ḳ:) or it signifies, (JK, TA,) [and] so ▼خُسْفٌ and ▼خَسِيفٌ, (Ḳ,) *a cloud,* or *collection of clouds, that has risen and appeared* مِنْ قِبَلِ العَيْنِ, *bearing much water;* (JK, Ḳ, TA;) i. e., *from* [*the quarter of*] *the right of the Ḳibleh* [*as explained above*]. (TA.) = ‡ *Deficiency,* or *imperfection; a fault;* or *a low,* or *base, quality;* (Ṣ, Ḳ, TA;) as also ▼خَسِيفَةٌ. (TA.) One says, رَضِىَ فُلَانٌ بِالخَسْفِ ‡ *Such a one was content with deficiency,* or *imperfection;* &c. (Ṣ, TA.) — † *Leanness,* or *emaciation;* (TA;) as also ▼خَسِيفَةٌ. (JK.) — [See also 1, last sentence. — Hence,] بَاتَ القَوْمُ عَلَى الخَسْفِ ‡ *The party passed the night in a state of hunger, not having anything wherewith to feed themselves:* (TA:) and بَاتَ فُلَانٌ الخَسْفَ ‡ *Such a one passed the night hungry:* (Ṣ, Ḳ, TA:) and شَرِبْنَا عَلَى الخَسْفِ ‡ *We drank without eating.* (IAạr, IDrd, Ḳ, TA.) A poet says,

* بِتْنَا عَلَى الخَسْفِ لَا رِسْلَ نُقَاتُ بِهِ *
* حَتَّى جَعَلْنَا حِبَالَ الرَّحْلِ فُصْلَانَا *

[*We passed the night in a state of hunger: there was no milk wherewith we might be fed, until we made the ropes of the camel's saddle* to serve as *young camels*]: i. e. we had no food until we bound the she-camels with ropes in order that they might yield us milk [as though they had young ones to suckle], and we might feed ourselves with their milk. (O, TA.) [See also another ex., in a verse of Dhu-r-Rummeh, cited voce إِلَّا, p. 78.] — [Hence, also,] سَامَهُ خَسْفًا and ▼خُسْفًا, (Ṣ, Ḳ,) and سَامَهُ الخَسْفَ, (Ṣ, Mṣb,) ‡ *He brought upon him abasement,* or *ignominy:* (Ṣ, Mṣb, Ḳ:) or *he required,* or *constrained, him to do an affair of difficulty; and to become in a state of abasement,* or *ignominy.* (Ṣ, TA.) [See also two similar phrases voce خُطَّة.] — [And hence,] خَسْفٌ signifies also † *Wrong, wrongdoing, injustice, injuriousness,* or *tyranny.* (TA.) [And سَامَهُ خَسْفًا, or الخَسْفَ, sometimes means † *He brought upon him wrong,* &c.] == See also the next paragraph.

خُسْفٌ: [see 1, last sentence: and] see خَسْفٌ, in two places. — دَعِ الأَمْرَ بِخُسْفِهِ means † *Leave thou the thing,* or *affair, as it is.* (Ṣgh, Ḳ.) == The [*fruit called*] جَوْز, *which is eaten;* [i. e. the *walnut,* or *walnuts;*] (AA, AḤn, Ḳ;) of the dial. of the people of Esh-Shiḥr; (AA;) as also ▼خُسُفٌ: (AA, Ḳ:) accord. to ISd, the former is the correct word: (TA:) n. un. with ة. (JK.)

خُسُفٌ: see خُسْفٌ.

خَسْفَةٌ [app. *A leanness,* or *an emaciation:* see 1, and see also خَسْفٌ]: this befalls camels, and sheep or goats, in the heat and in the cold. (A, TA.) == Also sing. of ▼أَخَاسِيفُ, (JK,) which signifies *Soft tracts* of land: (Ṣ, Ḳ,* TA:) or *level lands:* (JK:) and one says also ▼أَخَاسِفُ [and thus the word is written in the CḲ]. (Fr, TA.) One says, وَقَعُوا فِي أَخَاسِيفَ ▼ مِنَ الأَرْضِ *They became in soft tracts of land.* (Ṣ.) [See also أَخَاشِفُ, in art. خشف.]

خُسُوفٌ: see the next paragraph.

خَسِيفٌ ‡ A spring, or source, (عَيْنٌ, [shown in the TA to have this meaning here,]) *sinking,* or *going away* [*into the earth*]; as also ▼خَاسِفٌ; (Ḳ, TA;) in like manner without ة. (TA.) — † A well (بِئْرٌ) *dug in stones, so that it yields an abundant and unceasing flow of water;* (Ṣ, Ḳ;) as also خَسِيفَةٌ and ▼خَسُوفٌ and ▼مَخْسُوفَةٌ; (Ḳ;) or, as some say, خَسِيفٌ only: (TA:) or this signifies a well *pierced through its mountain* [or *rock*] *to the water beneath so that it never becomes exhausted;* (JK, TA;) as also ▼مَخْسُوفَةٌ: (JK:) or a well *dug so as to reach an unceasing,* or *a copious, source of water:* (TA:) pl. [of pauc.] أَخْسِفَةٌ (JK, Ḳ) and [of mult.] خُسُفٌ. (Ṣ, Ḳ.) — ‡ A she-camel *that yields abundant milk, but soon stops* [*its flow*] *in winter.* (Ḳ, TA.) [And] with ة, † A she-camel *that yields abundant milk.* (JK.) — See also خَسْفٌ — عَيْنٌ خَسِيفَةٌ (Mgh, Ḳ, TA) and ▼خَاسِفَةٌ (JK, Mgh) ‡ *An eye put out,* or *blinded;* (JK, Ḳ, TA;) *of which the black,* or *part surrounded by the white, has disappeared in the head.* (JK, Mgh, TA.) == الخَسِيفَانِ, thus correctly written, as in the L, and so in the Nawádir of Aboo-'Amr Esh-Sheybánee, and in the Tedhkireh of Aboo-'Alee El-Hejeree, who asserts that the ن is the ن of the dual, and in one dial. with ḍamm, [so that the word is written الخَسِيفَانِ and الخَسِيفَانُ,] and on whose authority is mentioned the saying هُمَا خَلِيلَانُ, with ḍamm to the ن, [so that each is a dual in form, though not in signification,] but in the O and the Ḳ ▼الخَيْسَفَانُ, [in the CḲ الخَيسَفَانُ,] with fet-ḥ to the س, and [▼الخَيْسُفَانُ,] with ḍamm to that letter, (TA,) *Bad dates:* (O, Ḳ:) so in the Nawádir and Tedhkireh above mentioned: (TA:) or *a palm-tree that bears a small quantity of fruit, and of which the unripe dates turn bad.* (O, Ḳ.)

خَسِيفَةٌ [as an epithet, fem. of خَسِيفٌ, q. v.:] as a subst.: see خَسْفٌ, in two places.

خَاسِفٌ, and its fem., with ة: see خَسِيفٌ, in two places. — Also ‡ *Lean,* or *emaciated.* (Ṣ, Ḳ.) — † A body *altered,* or *altered for the worse.* (A, TA.) † A man (JK) *altered,* or *altered for the worse, in colour,* or *complexion,* (JK, Ibn-'Abbád, Ḳ,) *and in aspect.* (JK.) — ‡ *Hungry.* (AHeyth, TA.) — † A boy *light,* or *active,* (Ḳ, TA,) *and brisk, lively,* or *sprightly;* as also خَاشِفٌ. (TA.) — † A man *convalescent;* or *recovering from disease;* syn. نَاقِهٌ: (AA, Ḳ: [see 1:]) pl. خُسُفٌ. (Ḳ.)

الخَيْسَفَانُ and الخَيْسُفَانُ: see خَسِيفٌ.

أَخَاسِفُ and أَخَاسِيفُ: see خَسْفَةٌ, in three places.

المُخْسِفُ *The lion.* (TṢ, Ḳ.)

مَخْسُوفَةٌ, applied to a well: see خَسِيفٌ, in two places.

المَخَاسِفُ: see خَسْفٌ.

خسق

1. خَسَقَ, aor. ـِ, (Mṣb, Ḳ,) inf. n. خَسْقٌ and خُسُوقٌ, (Mṣb, TA,) *It* (an arrow) *hit the target:* (Ḳ:) or *passed through the object at which it was shot; or penetrated into its inside, and its extremity went forth from the other side, the rest remaining therein:* (IKṭṭ, Mṣb:) or *stuck fast therein:* (IF, Mṣb:) or خَسَقَ الهَدَفَ (Mṣb, TA) *it hit the object at which it was shot, and passed through, or its extremity passed through;* like خَزَقَ: (TA:) or *passed through,* or *pierced so that its extremity passed through, not with vehemence.* (Mṣb.) And رَمَى فَخَسَقَ *He shot,* or *cast, and clave the skin.* (Az, TA.) IF says that it is not a primitive; that the س is substituted for ز; and that the word is altered because of the alteration of the meaning. (TA.)

خَسُوقٌ, applied to a she-camel, i. q. خَزُوقٌ; (Ḳ;) or *Evil in disposition; that pierces the ground with her toes, as she goes;* (JK, TA;) *furrowing the ground with her toe turned up.* (TA.)

خَاسِقٌ i. q. خَازِقٌ [q. v.]. (Ṣ.)

خسو

2. خَسَّى, (so in some copies of the Ḳ, and in the TA,) in [some of] the copies of the Ḳ تَخَسَّى, but the former is the right, (TA,) inf. n. تَخْسِيَةٌ, *He played with walnuts at the game of odd or even;* (Ḳ, TA;) as also ▼اخسى: and ▼خَاسَاهُ, (inf. n. مُخَاسَاةٌ, TA,) *He played with him at that game:* (Ḳ:) or you say, هُوَ يُخَسِّي وَيُزَكِّي *He plays, and says, "Is it even or odd?"* (TA.)

3 and 4: see above.

6. تَخَاسَا *They* (two men) *played together at the game of odd or even.* (JK,* TA.)

خَسَا, (JK, Ḳ,* and TA in art. زكو,) without tenween, and accord. to some with tenween, and not having the article ال prefixed to it; and in like manner زَكَا, which is coupled with the former, is without tenween, and accord. to some with tenween, and not having the article ال prefixed to it; (TA in that art.;) [but each has ال prefixed to it in the Ḳ;] accord. to Fr, some make it quasi-coordinate to فَتًى [i. e. with tenween, and masc.]; some, to زفر [app. meaning the proper name زُفَرُ, i. e. without tenween, and masc.]; and some, to سَكْرَى [app. meaning without tenween, and fem.]: (TA:) a word that is said in playing with walnuts; (JK;) *An odd number:* (Lth, Fr, Ḳ, TA:) and زكا signifies an even number: (Lth, Fr, TA:) accord. to IB, its final letter is hemzeh, for one says, يُخَاسِئُ, meaning يُقَامِرُ [he contends in a game of hazard]; but it is pronounced without ء to assimilate it to زَكًا: (TA:) the pl. is أَخْسَاءٌ, (Ḳ, TA,) accord. to the M, مَخَاسٍ, like مَسَاوٍ, (TA, [but the former pl. occurs in a verse cited in the TA,]) which is anomalous. (Ḳ.) One says, خَسًا أَوْ زَكًا, [so in my copies of the Ṣ, with tenween,] i. e. *Odd or even?* (Ṣ:) [or خَسَا أَوْ زَكَا, as shown above:] and some say, خَسَا زَكَا, like خَمْسَةَ عَشَرَ. (IB, TA.) And it is said in a trad., مَا أَدْرِي كَمْ حَدَّثَنِي أَبِي عَنْ رَسُولِ ٱللّٰهِ صَلَّى ٱللّٰهُ عَلَيْهِ وَسَلَّمَ أَخَسًا أَمْ زَكًا, or أَخَسَا أَمْ زَكَا, [for the vowel-signs are not written in my original,] meaning فَرْدًا أَمْ زَوْجًا [i. e. *I know not how many times my father told me, from the Apostle of God, God bless and save him; whether an odd or an even number:* or, perhaps, *once or twice*]. (TA.)

خسى

6. التَّخَاسِي *The throwing of pebbles, one at another.* (Ḳ.) You say, تَخَاسَتْ قَوَائِمُ الدَّابَّةِ بِالحَصَا *The legs of the beast threw the pebbles, one at another.* (TA.) [See also 3 in art. خسأ.]

خَسِيٌّ The *like of a* [*garment of the kind called*]

حِسَآء, or of a [tent such as is called] خِبَآء, woven of wool. (JK, Ḳ.)

خش

1. خَشَّ فِيهِ, (Ṣ, Ḳ,) aor. ـُ, (JM,) [vulgarly, and irregularly, ـِ,] inf. n. خَشٌّ, (TA,) *He* (a man) *entered into it*; (Ṣ, Ḳ;) namely, a thing; (Ṣ;) as also ↓انخشّ; (Ḳ, A, TA;) and ↓خَشْخَشَ, (TA,) inf. n. خَشْخَشَةٌ; (Ḳ, TA;) and in like manner, into a collection of trees, and a company of people: (A,* TA:) or ↓خَشْخَشَ فِيهِ, (IDrd,) and ↓تَخَشْخَشَ فِيهِ, (IDrd, Ḳ,) *he entered into it,* (namely, a thing, IDrd, or a collection of trees, Ḳ, and in like manner a company of men, TA,) *so as to become hidden,* or *concealed*: (IDrd, Ḳ:) and خَشَّ, *he* (a man) *went,* or *went away,* or *advanced,* [into a thing,] *and penetrated.* (TA.) —— Hence, (TA,) خَشَّ البَعِيرَ, (Ṣ, Ḳ,) aor. ـُ, [agreeably with general rule in this case,] inf. n. خَشٌّ, (Ṣ,) *He put into the camel's nose the thing termed* خِشَاش; (Ṣ, Ḳ;) as also ↓أَخَشَّ. (Zj, Ḳ.) —— And hence the saying in a trad., خُشُّوا بَيْنَ كَلَامِكُمْ لَا إِلٰهَ إِلَّا ٱللّٰهُ, meaning, †*Introduce ye,* or *insert ye, in your speech* the words *There is no deity but God.* (TA.) And ↓خَشْخَشَهُ likewise signifies *He introduced,* or *inserted, him* or *it.* (TA.) —— Also خَشَّهُ, aor. and inf. n. as above, *He pierced him,* or *stabbed him.* (TA.)

4: see خَشَّ البَعِيرَ.

7: see خَشَّ فِيهِ.

8. اختشّ مِنَ الأَرْضِ *He ate of the* خِشَاش *of the earth.* (TA.)

R. Q. 1. خَشْخَشَ: see خَشَّ فِيهِ, in two places. —— خَشْخَشَهُ: see 1, last signification but one. —— Also *He caused it to make a sound such as is described below, voce* خَشْخَشَةٌ. (Ṣ, TA.*) See an ex. in the next paragraph.

R. Q. 2. تَخَشْخَشَ: see خَشَّ فِيهِ. —— Also *It made a sound* (Ṣ, Ḳ) *such as is described below, voce* خَشْخَشَةٌ. (Ṣ.) 'Alḳameh Ibn-'Abadeh says,

* تَخَشْخَشُ أَبْدَانُ الحَدِيدِ عَلَيْهِمُ *

* كَمَا خَشْخَشَتْ ↓ يَبْسَ الحَصَادِ جَنُوبُ *

[*The short coats of mail of iron rustled upon them, like* as when *a south wind has caused to rustle the dry reaped corn*]. (Ṣ.)

مَخْشُوشٌ: see خَشَّ.

خَشَاشٌ: } see خِشَاشٌ; for the former, in three
خُشَاشٌ: } places.

خِشَاشٌ The *wooden thing that is inserted in the bone of the nose of the camel,* (Ṣ, A, Ḳ,) *to which the nose-rein is tied, in order that he may be quickly submissive*: (TA:) the بُرَة is of brass, (Ṣ, TA,) or of silver; (TA;) and the خِزَامَة is of hair: (Ṣ:) or the *thing that is put in the nose*; and the بُرة is the thing that is put in the flesh: (Lḥ:) or *what is in the bone, when it is wood,* or *a stick*; and the عِرَان is what is in the flesh, above the nose: (Aṣ:) *a wooden thing,* or *stick, that is put in the bone of the nose of the camel*: (Mṣb:) n. un. with ة: (Ṣ, Mṣb:) pl. أَخِشَّةٌ. (A, Mṣb.) [Hence the saying,] جَعَلَ الخِشَاشَ فِى أَنْفِهِ وَقَادَهُ إِلَى الطَّاعَةِ بِعُنْفِهِ ‡[*He put the* خشاش *in his nose, and drew him to obedience by his violence*]. (A, TA.) [And hence, also,] it signifies غَضَب, (IAạr, Ḳ,) as used in the saying, حَرَّكَ خِشَاشَهُ, [lit., *He put in motion his* خشاش: meaning, †*he roused,* or *excited, his anger*; or] *he made him angry.* (IAạr.) —— الخِشَاشُ, and ↓الخَشَاشُ, (Ṣ, Ḳ,) the latter form being sometimes used, (Ṣ,) which indicates that the former is the more chaste, but, accord. to MF, several authorities say the contrary, (TA,) and ↓الخُشَاشُ, (Ḳ,) or خِشَاشُ ↓ الأَرْضِ, (A'Obeyd, Mṣb,) and خَشَاشُ الأَرْضِ, (Mṣb,) *The creeping things of the earth*: n. un. with ة, which is *syn. with* الحَشَرَةُ *and* الهَامَّةُ: (Mṣb:) *the* حَشَرَات (A'Obeyd, Ṣ, Ḳ) *of the earth,* (A'Obeyd, Ḳ,) *and its* هَوَامّ, *and* [*other*] *creeping things,* (A'Obeyd,) *such as sparrows and the like*: (A'Obeyd, Ḳ:*) or خشاش الأَرْضِ, and الطَّيْرِ, signify *the small ones of beasts* or *creeping things* [*of the earth*], and *of birds*: (A:) IAạr is related to have said that it is ↓خَشْخَاش, contr. to what is said by the lexicologists in general: and these things are said to have their appellation from their entering into the earth and concealing themselves; but this assertion is not valid: (ISd:) in a trad., for من خشاش الارض, one relation substitutes ↓مِنْ خَشِيشِهَا, which has the same meaning: and some say that it is ↓خُشَيْش, a contracted dim. of خشاش; or ↓خُشَيِّش, without contraction: (TA:) and ↓الخَشَاشُ signifies *the bad* [meaning *ignoble*] *kinds of birds*; this being with fet-ḥ only: (Aṣ:) or *birds that do not prey*: (IAạr, TA voce عُقَاب:) خِشَاشٌ, with kesr, also signifies the *serpent of the mountain*; which does not suffer one to survive; and the أَفْعَى is the serpent of the plain; (El-Faḳ'asee, Ḳ;) which likewise does not suffer one to survive: (K:) or *a great and abominable* ثُعْبَان: or *a serpent like the* أَرْقَم, *but smaller*: or *a small, tawny serpent, smaller than the* ارقم: (TA:) or *a white serpent, which seldom hurts, between the* حُفَّاث *and the* ارقم: (Aboo-Kheyreh:) or *such as is light,* or *active, and small in the head, of serpents*: explained also as signifying the *serpent,* without restriction: (TA:) and, (Ḳ,) or as some say, (TA,) *such as has no power of defence,* [مَا لَا دِفَاعَ لَهُ, as in the CḲ and a MS copy of the Ḳ, for which we find in some copies of the Ḳ, and in the TA, مَا لَا دِمَاغَ لَهُ *such as has no brains,* which is doubtless a mistake,] *of beasts* or *creeping things of the earth,* and *of birds,* (Ḳ,) *such as the ostrich, and the* [*bustard called*] حُبَارَى, *and the* كَرَوَان [*or stone-curlew*], *and* [*the bird called*] مُلَاعِبُ ظِلِّهِ, *and the* [*harmless kinds of*] *serpent*: (TA:) or *what is small in the head, and slender,* of beasts or creeping things: and the *kite*; and [the *bird called*] مُلَاعِبُ ظِلِّهِ: (Aboo-Muslim:) the pl. is خِشَآء [app. خُشَّآء, originally خُشَشَآء, unless a mistake for أَخِشَّآء, originally أَخْشِشَآء]. (TA.)

خَشِيشٌ, and خُشَيْشٌ, and خُشَيِّشٌ: see خِشَاشٌ.

خُشَّآءٌ *The bone which is protuberant behind the ear,* (Ṣ, Mṣb, Ḳ,) *and which is thin, and bare of hair*: (TA:) originally خُشَشَآءٌ, (Ṣ, Mṣb, Ḳ,) of the measure فُعَلَآءٌ; (Ṣ;) [but masc., and perfectly decl., as being quasi-coordinate to قُرْطَاس, whereas the original is fem., and imperfectly decl.; (see قُوبَآء;)] like قُوبَّآء, which is originally قُوَبَآء; (Ṣ, Mṣb;) and these two words are the only instances of their kind: (ISk, Mṣb:) dual خُشَّاوَانِ. (Ṣ, Ḳ.)

خَشْخَشَةٌ The [*clashing, clattering, chinking, jingling, rattling,* or *rustling,*] *sound of arms,* or *weapons,* (Ṣ, A,* Ḳ,) *and the like*; (Ṣ;) as also شَخْشَخَةٌ, but this latter is a dial. var. of weak authority: (TA:) and *of any dry* or *hard thing rubbing against another such thing*: (Ḳ:) accord. to IDrd, such [sound or thing (for his words are ambiguous)] is termed ↓خَشْخَاش: (TA:) and the [*rustling*] *sound of a new garment* or *piece of cloth, when it is put in motion*; as also نَشْنَشَةٌ: (IAạr:) and [*a confused sound*] *such as is heard to proceed from the inside of an animal on its being hit by an arrow*: (JK in art. خشف:) and *a motion having a sound like the sound of arms,* or *weapons*; (TA;) or *an audible motion.* (Mgh in art. خشف.)

خَشْخَاشٌ *A company*: (ISd, TA:) or *a numerous company of men*: (Az, TA:) or *a company* (Ṣ, Ḳ) *in,* (Ḳ,) or *having upon them,* (Ṣ,) *arms,* or *weapons, and coats of mail.* (Ṣ, Ḳ.) —— See also خَشْخَشَةٌ. —— Also *A certain plant,* (Ṣ, Mṣb,) *well known*; (Ṣ, Mṣb, Ḳ;) [namely, the *poppy*;] *which is of several species*; (Ḳ;) i. e., *four*; (TA;) بُسْتَانِىّ, [or *garden-poppy,*] (Ḳ,) *which is the white, and this is the most fit for eating, and the best thereof is the fresh and heavy*; (TA;) and مَنْثُور, (Ḳ,) *which is the wild Egyptian*; (TA;) and مُقَرَّن, [app. the *horned poppy,*] (Ḳ,) *the produce of which has an elongated extremity like the horn of the bull*; (TA;) and زَبَدِىّ, [app. the *spattling poppy,*] (Ḳ,) *which is known by the name of* بلبس [a word which I have not been able to find elsewhere]: (TA:) *every one of these is soporiferous, and produces torpidness, and cools*: (Ḳ:) *used as a suppository, it produces sleep*: *and the integument* [*of the capsule*] *has a stronger power of producing sleep than the seeds*: (TA:) [or rather *the seeds have no narcotic power*:] *from half a drachm of the integument, with cold water, as a draught, taken early in the morning, and the like at sleep, has a wonderful effect in stopping a looseness characterized by a mixture of humours and by blood, when accompanied by heat and inflammation*: (Ḳ:) *it is wonderful also that its solid part confines, and its juice relaxes*: *and when the root,* or *lower part, is taken with water,* [*and boiled*] *so that the water is reduced to half its quantity, it is beneficial as a remedy for diseases of the liver arising from thick humour*: so says the author of the Minháj: (TA:) the n. un. is with ẗ: (Mṣb:) and the pl. is خَشَاخِشُ [app. a mistake for خَشَاخِيشُ]. (TA.) [See also أُفْيُونٌ.]

خِشَاشٌ: see خِشْخَاشٌ.

مَخْشُوشٌ A camel *having a* خِشَاش *put in his nose*; as also ▼ خَشٌّ. (Ibn-'Abbád, Ḳ.)

خشب

1. خَشَبَ, (Ṣ, Ḳ,) aor. ـِ, (Ḳ,) inf. n. خَشْبٌ, (TA,) *He mixed* a thing (Ṣ, Ḳ) with (بِ) another thing. (Ṣ.) —— And *He picked out, chose out*, or *selected*, a thing: the verb thus having two contr. significations. (Ḳ, TA.) ═ Also, (Ṣ, Ḳ,) aor. and inf. n. as above, (Ṣ,) *He polished* a sword, (Ṣ, Ḳ,) *by laying on it a broad and smooth spear-head and rubbing it therewith*: so accord. to El-Aḥmar, who relates that an Arab of the desert said to him, I said to a sword-polisher, "Hast thou finished my sword?" and he answered, نَعَمْ إِلَّا أَنِّى لَمْ أَخْشِبْهُ [*Yes, except that I have not polished it*]. (Ṣ.) And [or, as in the TA, "or"] *He sharpened* it. (Ḳ, TA.) —— And *He forged* a sword: (Ḳ:) or *fashioned it with the file, without polishing* it: (TA:) or *he made it imperfectly, not thoroughly*, or *not well*: (A:) thus, again, the verb has two contr. significations: (Ḳ:) also *he thus made* an arrow: (A:) or *he shaped out* a bow, (AḤn, Ḳ,) and an arrow, (TA,) [*in a rough manner*, or] *by the first operation*, (AḤn, Ḳ, TA,) *without perfecting it*, or *making it smooth*, or *even*. (TA.) You say of a sword, before it has been filed, مَا أَحْسَنَ مَا خُشِبَ [*How well has it been forged!*]: and in like manner one says of an arrow, when it has been filed, before the سَفَن [with which it is smoothed] has been applied to it. (Skr, on a verse of Ṣakhr, cited below, voce خَشِيبَةٌ.) —— [Hence,] خَشَبَ الشِّعْرَ, (ISk, Ṣ, A, Ḳ,) aor. and inf. n. as above, (A,) ‡ *He said, spoke*, or *uttered, the poetry* (ISk, Ṣ, A, Ḳ) *as it came*, (ISk, Ṣ,) [*unpolished, and unstudied*,] *without affecting nicety*, or *refinement, therein*, (ISk, Ṣ, A, Ḳ,) and *without study*, or *labour*: (A, Ḳ:) Jereer did thus, and Farezdaḳ trimmed his verses; but the verses of Jereer thus produced are better than the trimmed verses of Farezdaḳ: (A, TA:) and ▼ اِخْتَشَبَهُ signifies the same. (A, Ḳ.) You say also, هُمْ يَخْشِبُونَ الكَلَامَ وَالعَمَلَ ‡ [*They say, speak, or utter, words, and do work, without affecting nicety*, or *refinement*, and *without study*, or *labour*]: (A:) or *imperfectly*, or *not thoroughly*; *inelegantly*, or *not well*. (TA.) And اِتَّخَذَ السَّيْفَ خَشْبًا: see 8.

5. تَخَشَّبَ: see 8. ═ تَخَشَّبَتِ الإِبِلُ *The camels ate thick branches*: (Ḳ:) or *ate dry herbage*. (Ṣ.) And تَتَخَشَّبُ عِيدَانَ الشَّجَرِ *They take with the mouth, and eat, the branches of the trees*. (TA.)

8. اختشب السَّيْفَ signifies ▼ اِتَّخَذَهُ خَشْبًا; *He took the sword without choosing the best by taking it from this place or that*; (L, TA;) as also ▼ تَخَشَّبَهُ. (TA.) —— See also 1, near the end.

12. اخشوشب *He* [a man or a camel (see خَشِب)] *was*, or *became, tall, and gross, rude*, or *coarse, with bones uncovered by flesh, and hard*, or *hardy*. (Ḳ.) *He* (an ostrich) *was*, or *became, rough*, or *coarse*. (Ṣ.) —— † *He* (a man) *became hard*, or *hardy, and rough*, or *coarse, in his religion, clothing, food*, and *in all respects*. (TA.) † *He employed himself in work, and in walking barefoot, in order that his body might become thick, gross*, or *coarse*. (Ṣ, TA.) And اخشوشب فِى عَيْشِهِ † *He endured with patience a life of hardship*, or *difficulty*: or *he subjected himself to a life of hardship*, or *difficulty, in order to render himself the more able to bear it*. (Ḳ, TA.) اِخْشَوْشِبُوا is thus used in a trad. of 'Omar: (Ṣ, TA:) or, as some relate it, the word is [اجشوشبوا,] with ج; or, accord. to some, اخشوشنوا, with خ and ن. (TA.)

رَجُلٌ قَشْبٌ خَشْبٌ *A man in whom is no good*: (Ṣ, Ḳ:) or *with whom is no good*: (TA:) [in some copies of the Ḳ, خَشْبٌ وَقَشْبٌ; but this, as is said in the TA, is incorrect:] خَشْبٌ being an imitative sequent to قَشْبٌ. (Ṣ, TA.)

خَشَبٌ [*Wood, such as is used in carpentry and the like; timber;*] *thick wood*: (A, Ḳ:) [a coll. gen. n.:] n. un. خَشَبَةٌ [signifying *a piece of wood* or *timber*]: (Mṣb:) the pl. of the latter, (Ṣ, Mṣb,*) or of the former, (Ḳ,) is خَشَبٌ, (Ṣ, Ḳ, [i. e., accord. to the Ḳ, the pl. is the same as the sing., but properly speaking, as said above, this is a coll. gen. n.,]) and خُشُبٌ and خُشْبٌ (Ṣ, Mṣb, Ḳ) and خُشْبَانٌ, (Ṣ, Ḳ,) [which last is agreeable with analogy as pl. of خَشَبٌ,] or خُشْبَانٌ is pl. of خُشُبٌ, and خُشُبٌ is pl. of خَشَبَةٌ. (JK.) The hypocrites are described in a trad. as خُشُبٌ بِاللَّيْلِ صُخُبٌ بِالنَّهَارِ Like *timbers*, or *pieces of wood, in the night*; [*clamorous in the day*;] meaning that they pass the night in sleep, without prayer. (TA.) —— مَالٌ خَشَبٌ † *Cattle that are lean*, or *emaciated*, syn. هَزْلَى, (Ḳ,) *in consequence of their feeding upon dry herbage*. (TA.) [And it seems that ▼ خَشِبٌ signifies the same: for I find in the TA, and in a copy of the A which I believe to have been used by the author of the TA, mentioned as tropical, مَالٌ خَشِبٌ وَحَطِبٌ جَزْلٌ, app. meaning that مَالٌ خَشِبٌ and حَطِبٌ signify جَزْلٌ; but جَزْلٌ, I think, is here evidently a mistranscription for هَزْلَى; as حَطِبٌ is explained in the Ṣ and Ḳ as signifying "very lean or meagre."]

خَشِبٌ *Rough*, or *coarse*; as also ▼ أَخْشَبُ: (Ḳ:) the former applied in this sense to a male ostrich: (Ṣ:) and both signify anything *gross*, or *big*, and *rough*, or *coarse*; (A'Obeyd, Ṣ;) as also ▼ خَشِيبٌ: (TA:) and the first, (Ḳ,) applied to a man and to a camel, (TA,) *tall, and gross, rude*, or *coarse, with bones uncovered by flesh, and hard*, or *hardy, and strong*; (Ḳ,* TA;) as also ▼ خَشِيبٌ and ▼ خَشَبِيٌّ: (Ḳ:) or these three signify, or signify also, *dry*, or *rigid*, or *tough*: (Kr, ISd:) and خَشْبٌ, a man *hard*, or *hardy, strong, and vigorous, in body*: (A, TA:) and the same, (JK,) or ▼ خَشِيبٌ, (TA,) a man *whose bones are uncovered by flesh, and whose sinews are apparent*; (JK, TA;) *hard*, or *hardy, and strong*: (JK:) and the last, a *gross, big*, or *coarse*, camel: (Ṣ, TA:) a camel *gross, coarse*, or *rude, in make, and ugly*: (TA:) and a horse *thick*, or *big, in the bones*. (Ḥam p. 207.) See also خَشَبٌ. And see أَخْشَبُ, in two places. —— Also † Life *in which one is not dainty, nice*, or *scrupulous*. (Ḳ.)

خُشْبَةٌ The *first filing* of a sword, *before the polishing*. (TA.)

خُشْبَانٌ: خُشَّابٌ: } see أَخْشَبُ.

خُشَابٌ, from the Persian خُوشْ آبْ, [The *beverage properly called in Arabic*] نَبِيذ. (TA.)

خَشِيبٌ and ▼ مَخْشُوبٌ *Mixed*. (TA.) —— And the former, (Ḳ,) or both, (TA,) *Picked out, chosen*, or *selected*: (Ḳ, TA:) both words thus having two contr. significations. (TA.) ═ Also the former (Ṣ, Ḳ) and latter, (Ḳ,) A sword *polished*: (Ṣ, Ḳ:*) this is [said to be] the prevailing signification: (TA:) or both signify a *sharpened* sword. (JK, TA.) —— And the former, (Aṣ, Ṣ, Ḳ,) or both, (JK, A,) A sword *of which the forging is commenced*; thus [again] having two contr. significations: (Ṣ:) or *forged*, (Ḳ,* TA,) or *fashioned with the file, but not yet polished*: (Aṣ, TA:) or *newly made*: (TA:) or *imperfectly, not thoroughly*, or *not well, wrought*; (JK, A;) and thus both words applied to an arrow: (A:) or the former, (Ṣ, Ḳ,) or both, (TA,) applied to an arrow, (Ṣ, Ḳ,) and to a bow, (Ḳ,) *shaped out* (Ṣ, Ḳ) [*in a rough manner*,] *by the first operation*, (Ṣ, TA,) *not yet perfected*, or *made smooth*, or *even*: (TA:) pl. of the former (accord. to the TA as applied to a bow [but I see no reason for this restriction]) خُشُبٌ and خَشَائِبُ. (Ḳ.) مَخْشُوبٌ ▼ لَمْ يُنَقَّحْ [*Rough hewn, not yet trimmed*,] is a prov., mentioned by Meyd and Z. (MF, TA.) —— [Hence,] شِعْرٌ خَشِيبٌ and ▼ مَخْشُوبٌ ‡ *Poetry said, spoken*, or *uttered, as it has come to the speaker*, [*unpolished, and unstudied*,] *without his affecting nicety*, or *refinement, therein*, and *without study*, or *labour*. (A,* TA.) And جَآءَ ▼ بِالمَخْشُوبِ ‡ [*He said*, or *uttered, that which came to him, as it came, unpolished, and unstudied*]. (A, TA.) —— See also خَشِيبٌ voce خَشِبٌ, in three places. —— It also signifies *Bad, corrupt*, or *vile*. (Ḳ.)

خَشِيبَةٌ The *natural quality* [*of the metal*] of a sword, (Skr on the verse here following, Ṣ, TA,) *before the making* thereof *is completed*: (Skr:) or its *blade*, or *iron*: (A:) or its *edge*: or its *polish*. (JK.) Ṣakhr says,

* وَصَارِمٍ أُخْلِصَتْ خَشِيبَتُهُ *
* أَبْيَضَ مَهْوٍ فِى مَتْنِهِ رُبَدُ *

And a sharp sword of which the natural quality [of the metal] before the completion of the making thereof has been refined, [white, or a sword,] thin in the two edges or sides, having [in its broad side] diversified marks. (Skr.)

خَشَبِيٌّ: see خَشِبٌ.

خَشَّابٌ: see what next follows.

خَشَّابَةٌ [a coll. gen. n., of which the n. un. is ▼ خَشَّابٌ,] *Sellers of* خَشَب [i. e. *wood*, or *timber*]. (TA.) —— *Fighters with staves.* ═ Accord. to El-Hejeree, خشابة [so in the TA, without any syll. sign,] signifies *A slender* [*implement of the kind called*] مطرق [i. e. مِطْرَق, q. v.,] *which the polisher, when he has finished the polishing of a sword, passes over it, in consequence of which the scabbard does not alter its state.* (TA.)

مُخْتَشِبٌ : see خَاشِبٌ.

أَخْشَبُ : see خَشِبٌ. Also *A great mountain:* (A:) or *a rugged,* or *rough, and great mountain;* (Ṣ, Ḳ;) and so ▼ جَبَلٌ خَشِبٌ : or *such as is not to be ascended:* (TA:) *an elevated place, rugged, with rough stones:* (JK:) *a tract of the kind termed* قُفّ, *rugged and stony:* (TA:) pl. أَخَاشِبُ, (A, TA,) because the quality of a subst. is predominant in it: and the fem. خَشْبَآءُ is also sometimes used in the same sense; or as syn. with غَيْضَةٌ [i. e. *a thicket*, &c.]; but the former meaning is better known: and this [likewise] is thought to be rather a subst. than an epithet, because of the pl., mentioned above: (TA:) and ▼ خُشْبَانٌ [also seems to be a pl. of أَخْشَبُ, or of خَشِبٌ; for it is said that it] signifies *rugged,* or *rough, mountains, neither great nor small:* (Ḳ:) and *rugged ground.* (TA in art. ذنب.) خَشْبَآءُ also signifies *Hard land* or *ground;* (Ḳ,* TA;) *land,* or *ground, in which are stones and pebbles and earth* or *clay.* (IAmb, TA.) And أَرْضٌ ▼ خَشَابٌ (Ḳ, TA) *Hard land* or *ground,* like خَشْبَآءُ, (TA,) *that flows with the least rain.* (Ḳ, TA.) And أَكَمَةٌ خَشْبَآءُ (Ṣ, TA) *A hill of which the stones are scattered, but near together.* (TA.) And جَبْهَةٌ خَشْبَآءُ *A displeasing forehead;* as also ▼ خَشِبَةٌ : (TA:) or *a displeasing, rigid forehead;* (JK, Ṣ, Ḳ;*) *not even.* (JK.) And أَخْشَبُ الجَبْهَةِ A man *having a displeasing and rigid forehead.* (TA.)

بَيْتٌ مُخَشَّبٌ [so in the present day, but written in the TA without any syll. sign,] *A house having* خَشَب [i. e. *wood,* or *timber, employed in its construction*]. (TA.)

مَخْشُوبٌ : see خَشِيبٌ, in four places. —— It is applied to a horse, by El-Aạshà; (Ṣ, TA;) meaning *Of mixed pedigree:* (A 'Obeyd, TA:) or *not broken; not well trained;* from what next follows; and thus used only by El-Aạshà. (IKh, TA.) —— جَفْنَةٌ مَخْشُوبَةٌ *A wooden bowl imperfectly made.* (IKh, TA.) —— طَعَامٌ مَخْشُوبٌ [*Food imperfectly prepared;* i. e.], if flesh-meat, *not thoroughly cooked;* and if not flesh-meat, (but grain, TA,) *without any seasoning,* or *condiment, to render it pleasant,* or *savoury.* (Ḳ,* TA.)

مُخْتَشِبٌ One *who eats what he can;* as also ▼ خَاشِبٌ. (JK.)

خشر

1. خَشَرَهُ, aor. ـِ, (Ṣ, Ḳ,) inf. n. خَشْرٌ, (Ṣ,) *He picked it,* (Lḥ, Ṣ, Ḳ,) namely, a collection of goods, or commodities, (Lḥ,) *removing from it what was bad.* (Lḥ, Ṣ, Ḳ.) —— And خَشَرَ, aor. (and inf. n., TA) as above, *He left upon the table refuse of food:* thus the verb bears two contr. significations. (Ḳ.) ═ Also خَشَّرَهُ *He rendered it* (a thing) *bad,* or *vile.* (TA.)

خُشَارٌ (Ḳ) and ▼ خُشَارَةٌ (Ṣ, A, Ḳ) *The bad part* or *parts* [or *the refuse*] (Lḥ, Ṣ, Ḳ) of goods, or commodities, (Lḥ,) or of anything. (Ṣ, Ḳ.) And the latter, The *worst kind,* syn. شِيص, of dates. (A.) —— *What contains no* لُبّ [or *heart*], of barley. (A, Ḳ.) —— And the latter, The *refuse of food remaining upon a table; that in which is no good.* (Ṣ, A, Ḳ.*) —— Also the latter, (Ṣ, A, Ḳ,) and the former, (Ḳ,) and ▼ خُشَّارٌ, (IAạr,) ‡ The *refuse,* or *lowest* or *basest* or *meanest sort,* of mankind, or of people; (IAạr, Ṣ, A, Ḳ;) as also ▼ خَاشِرٌ, accord. to the Ḳ, but correctly ▼ خَاشِرَةٌ, as related by AA from IAạr. (TA.) El-Ḥoṭei-ah says,

* وَبَاعَ بَنِيهِ بَعْضُهُمْ بِخُشَارَةٍ *
* وَبِعْتَ لِذُبْيَانَ العَلَآءَ بِمَالِكِ *

[*And some of them have sold their sons for the refuse of mankind; but*] *thou hast purchased eminence for Dhubyán with Málik:* in the Ṣ we find بِمَالِكَا [*with thy property*]: but it is correctly as above: Málik was a son of 'Oyeyneh Ibn-Ḥiṣn: the Benoo-'Ámir slew him: wherefore 'Oyeyneh made war upon them, and obtained his blood-revenge, and spoil: and to this event El-Ḥoṭei-ah refers in the verse above. (IB, TA.)

خُشَارَةٌ :
خُشَّارٌ :
خَاشِرٌ :
خَاشِرَةٌ : } see the paragraph next preceding.

خشع

1. خَشَعَ, aor. ـَ, inf. n. خُشُوعٌ, *He was,* or *became, lowly, humble,* or *submissive;* (Ṣ, Mṣb, Ḳ;) as also ▼ اختشع (Ṣ, Ḳ) and ▼ تخشّع; (Abu-l-Fet-ḥ, Ḥam pp. 24 and 127;) خُشُوعٌ being *syn.* with خُضُوعٌ: (Ṣ, Mṣb, Ḳ:) or خشوع is *nearly the same as* خضوع: (Lth, Ḳ:) or the former is mostly used as meaning *in the voice;* and the latter, in the necks: (Mṣb:) or the latter is in the body; and the former is *in the voice* and *in the eyes:* (Ḳ:) or, as we read in the 'Eyn, the former is nearly the same as the latter, except that the latter is in the body, and signifies the acknowledging of humility and submission, and the former is in the voice and in the eyes; and the like is said in the Nh [and in the Mṣb in art. خضع]. (TA.) You say, خَشَعَتِ الأَصْوَاتُ *The voices were* [or *shall be* (as in the Ḳur xx. 107)] *still and low:* (Mṣb:) or *low:* or, as some say, *still.* (TA.) And خَشَعَ بِبَصَرِهِ *He lowered his eye.* (Ṣ.) And ▼ اختشع and ▼ تخشّع *He cast his eye towards the ground, and lowered his voice.* (TA.) Lth says that you say, ▼ اختشع فُلَانٌ, but not اختشع بِبَصَرِهِ. (TA.) And خَشَعَ بَصَرُهُ *His eye became contracted.* (TA.) And خَشَعَتْ دُونَهُ الأَبْصَارُ ‡ [meaning *The eyes were cast down before him, or it*]. (TA.) خُشُوعٌ also signifies The *being,* or *becoming, still:* and the *abasing oneself;* or *lowering oneself.* (Ḳ, TA.) And ▼ اختشع, *He lowered,* or *stooped,* or *bent down, his breast.* (TA.) —— Also, inf. n. as above, *He feared;* for instance, in prayer: (TA:) or خَشَعَ فِى صَلَاتِهِ and فِى دُعَائِهِ signifies *He applied himself with his heart to* [or *in*] *his prayer, and his supplication.* (Mṣb.) —— خَشَعَتِ الكَوَاكِبُ, (Aboo-'Adnán,) inf. n. as above, (Ḳ,) ‡ *The stars approached to the place of setting;* (Aboo-'Adnán;) or *approached to setting:* (Ḳ:) or *sank, and nearly disappeared in their setting-place.* (Aboo-Ṣáliḥ El-Kilábee.) [The corresponding phrase in Hebrew, occurring in Gen. xxxvii. 9, probably has the same meaning.] —— خَشَعَتِ الشَّمْسُ ‡ *The sun became eclipsed.* (TA.) —— خَشَعَ السَّنَامُ ‡ *The hump for the most part went away;* (O, Ḳ;) i. e. the hump of the camel: (TA:) or *became lean; its fat going away, and its height becoming lowered.* (L.) —— فُلَانٌ جِذْلُ حِكَاكٍ خَشَعَتْ عَنْهُ الأُبَنُ is a saying of the Arabs, explained in art. حك. (TA in that art.) —— خَشَعَتِ الوَرَقُ ‡ *The leaves withered.* (TA.) —— خَشَعَتِ الأَرْضُ ‡ *The earth,* or *land, dried up, not being rained upon.* (TA.) ═ خَشَعَ فُلَانٌ خَرَاشِىَّ صَدْرِهِ *Such a one ejected the viscous saliva* [or *phlegm of his chest*]. (O, Ḳ.) —— And خَشَعَتْ خَرَاشِىُّ صَدْرِهِ *The viscous saliva* [or *phlegm of his chest*] *became ejected.* (O, Ḳ.) The verb is thus intrans., as well as trans. (O.)

5. تخشّع *He lowered, humbled,* or *abased, himself:* (Lth, Ḳ:) or *he constrained himself to be,* or *to become, lowly, humble,* or *submissive;* or *to be so,* or *to become so, in voice,* or *in the eyes.* (Ṣ.) See also 1, in two places.

6. تخاشع [*He feigned lowliness, humility,* or *submissiveness, in demeanour,* or *in voice,* or *in the eyes*]. (TA in art. موت; &c.)

8: see 1, in four places.

خُشْعَةٌ *A low hill:* (Ṣ:) or *a hill cleaving to the ground:* (IAạr, Ḳ:) and *a piece of rugged ground:* (IDrd, Ḳ:) or [*elevated ground such as is termed*] قُفّ *that is for the most part soft,* i. e. *neither stone nor clay:* (Lth:) and *a rock growing in the sea:* (TA:) pl. خُشَعٌ. (Ḳ.) It is said in a trad., كَانَتِ الأَرْضُ خُشْعَةً عَلَى المَآءِ ثُمَّ دُحِيَتْ [*The earth was a low hill, &c., upon the water: then it was spread out*]: (Ṣ:) but this trad. is variously related. (TA.)

خَاشِعٌ *Lowly, humble,* or *submissive,* (Ḳ, TA,) *and still:* (TA:) [or *so in the voice* and *in the eyes:* (see 1:)] pl. خَاشِعُونَ and خُشَّعٌ; the latter also signifying men *lowering, humbling,* or *abasing, themselves:* or *constraining themselves to be,* or *to become, lowly, humble,* or *submissive;* or *to be so,* or *to become so, in voice,* or *in the eyes:* or *casting their eyes towards the ground, and lowering their voices.* (TA.) Hence, in the Ḳur [lxviii. 43, and lxx. 44], accord. to different readings, خَاشِعًا أَبْصَارُهُمْ and خَاشِعَةً أَبْصَارُهُمْ [*Having their eyes cast down*]: the accus. case being used as denotative of state. (Zj, TA.) — *Bowing*; or *bending down the head and body.* (Ḳ.) — *Fearing.* (TA.) — ‡ A camel's foot (خُفّ) *cleaving to the ground.* (TA.) — ‡ A wall *that has cracked, and given notice of its falling, and* [*then*] *become even with the ground.* (TA.) — ‡ A herb *dried up, and falling down upon the ground.* (TA.) — Applied to a place, (Ṣ, Ḳ,) and, with ة, to a بَلْدَة [or portion of country], (Ṣ,) ‡ *Overspread with dust,* [in the CḲ المُعْتَبِرُ is erroneously put for المُغْبَرُّ,] *and having in it no place of alighting,* or *of abiding*: (Ṣ, Ḳ:) and to land (أَرْض), meaning *of which the wind raises the surface, by reason of its softness, so as to efface its traces,* or *tracks*: (L:) or in this case it is with ة, as in the Ḳur xli. 39, and means *altered* (مُتَغَيِّرَة [probably a mistranscription for مُتَغَبِّرَة *overspread with dust*]), *and having its herbage broken in pieces*: (Zj,* TA:) or *dried up, and containing no herbage*: (Jel:) or *containing no green herbage*: or *low,* or *depressed, and still*: (TA:) and, without ة, applied to a place, *to which one finds not his way*: (Sgh, Ḳ:) pl. خُشَّع. (TA.)

خشف

1. خَشَفَ, aor. ـِ (Ṣ, Sgh, L, Ḳ) and ـُ, (L, Ḳ,) inf. n. خَشْفٌ, (Ṣ,) *He,* or *it, made a sound,* (L, Ḳ,) or *what is termed* حِسّ [i. e. *a low, faint, gentle,* or *soft, sound*], (Ṣ,) and *an audible motion*: (Ṣ, L:) said of a man: and said also of snow, as meaning *it caused one to hear a* [*sound such as is termed*] خَشْفَة *in walking* [*upon it*]; as is the case in intense cold. (Ṣ.) And خَشَفَ, aor. ـِ, inf. n. خُشُوفٌ, said of snow, *It was rough, so that it caused one to hear a* خَشْفَة *in walking* [*upon it*]: and in like manner said of ice; i. e. *it was soft,* or *yielding* [*to the feet, crackling*], or *easily broken.* (TA.) A poet says, (Ṣ,) namely, El-Ḳatámee, (TA,)

* إِذَا كَبَّدَ النَّجْمُ السَّمَآءَ بِشَتْوَةٍ *
* عَلَى حِينَ هَرَّ الكَلْبُ وَالثَّلْجُ خَاشِفُ ▼ *

[*When the asterism* of the Pleiades *culminates in winter, at the time when the dog whines* by reason of the cold, *and the snow causes one to hear a slight sound in walking upon it*]: (Ṣ:) or, accord. to IB, the right reading is, بِسُحْرَةٍ [*a little before daybreak,* or *in the last third of the night*]: (TA:) حين is here manṣoob because على is made to be redundant, and because it is prefixed to a verbal proposition: (Ṣ:) this is the more approved way in a case of this kind, when the verb commencing the proposition is a pret.; but some say على حِينِ. (I'Aḳ p. 199.) — خَشَفَ said of water, *It froze.* (Ḳ.) — Said of cold, *It was,* or *became, intense.* (Ḳ.) — خَشَفَ فِى السَّيْرِ, (Ḳ,) inf. n. خَشَفَانٌ, (JK,) *He hastened, made haste,* or *sped,* [app. *so as to cause a slight sound to be heard,*] *in going, journeying,* or *pace.* (JK,* Ḳ.) And مَرَّ يَخْشِفُ *He passed along hastening.* (TA.) — خَشَفَ, (Ṣ, Ḳ,) aor. ـُ (Ṣ, TA) and ـِ, (TA,) inf. n. خُشُوفٌ (Ṣ, Ḳ) and خَشَفَانٌ, (Ḳ,) *He went away in,* or *into, the land,* or *country.* (Ṣ, Ḳ.) And خَشَفَ فُلَانٌ *Such a one journeyed away, went away,* or *departed,* or *became hidden* or *concealed,* syn. تَغَيَّبَ, (Ḳ, TA,) *in the land,* or *country.* (TA.) — خَشَفَ فِى الشَّىْءِ, (aor. ـِ, TA,) *He entered into the thing;* as also ▼انخشف. (Ḳ.) — خَشَفَ, inf. n. خَشَفَانٌ, *He* (a man) *went,* or *travelled, by night.* (Ḳ.) — And *He was bold,* or *daring, in night-journeying*: or *he went about,* or *round about, by night*; (L, Ḳ, TḲ; but in the first and second, only the inf. n. is mentioned in this case;) and *journeyed much by night.* (L.) And خَشَفَ بِالقَوْمِ, aor. ـِ, said of a guide of the way, *He went about,* or *round about, by night, and hastened,* or *sped, with the party*: (JK:) or خَشَفَ بِهِمْ, inf. n. خَشَافَةٌ, *he* (a guide of the way) *acted with a penetrative energy,* or *with sharpness, vigorousness, and effectiveness, with them* [*in conducting them*]; as also ▼خشّف, inf. n. تَخْشِيفٌ. (Ḳ,* TA.) — خَشَفَتْ بِوَلَدِهَا *She* (a woman) *cast forth her child* [from the womb]. (Ḳ.) And خُشِفَ بِهِ *He,* or *it, was cast,* or *thrown*; as also خُفِشَ بِهِ, and حُفِشَ بِهِ. (TA.) ═ خَشَفَ رَأْسَهُ بِالحَجَرِ *He* (a man, Ṣ) *broke his head with the stone.* (Ṣ, Ḳ.)

2: see 1.

3. خاشف, (Ḳ,) inf. n. مُخَاشَفَةٌ, (JK, TA,) *It* (an arrow) *caused a* [*sound such as is termed*] خَشْفَة *to be heard on its hitting the object aimed at*: (Ḳ,* TA:) or *it* (an arrow) *caused a* [*confused sound such as is termed*] خَشْخَشَة *to be heard from the inside of the animal hit thereby.* (JK.) — خاشف فِى الشَّرِّ, and إِلَى الشَّرِّ, *He hastened in doing,* and *to do, evil,* or *mischief.* (TA.) — خاشف فى ذِمَّتِهِ *He hastened in breaking,* or *violating, his compact, covenant,* or *promise of protection* or *safeguard,* or *of security* or *safety.* (Ḳ,* TA.) — خاشف الإِبِلَ لَيْلَتَهُ *He went,* or *kept pace, with the camels during his night*; syn. سَايَرَهَا. (Ḳ.)

7: see 1.

خَشْفٌ: see خَشْفَةٌ, in two places: ═ and خِشْفٌ: ═ and خَاشِفٌ.

خُشْفٌ: see what next follows.

خِشْفٌ (Aṣ, JK, IDrd, Mṣb, Ḳ) and ▼خُشْفٌ and ▼خَشْفٌ, (Ḳ,) the second of which is said by MF to be the most common, and then the first, (TA,) [but this is a mistake, for the first is the only form commonly occurring,] The *young one of the gazelle*; (JK, Mṣb;) applied to the *male* and the *female*; (Mṣb;) or the female is termed خشفة [i. e. خِشْفَةٌ and خُشْفَةٌ and خَشْفَةٌ]: (Ḳ:) or the *young one of the gazelle in the first stage after its birth*: (Ḳ:) or *after it is termed* طَلًا; for it is called by the latter appellation when just born: (Aṣ, TA: [see شَصَرٌ:]) or *after it is termed* جِدَايَة: (TA:) or *when it first walks*: or *she that flees,* or *goes away, from,* or *of,* (مِنْ,) *her young ones*: (Ḳ: [a strange (as well as an ambiguous) explanation, seeing that the fem. is said in the Ḳ to be with ة:] pl. خُشُوفٌ, (Mṣb,) or خِشَفَةٌ. (Ḳ.)

خَشَفٌ and ▼خَشِيفٌ *Rough snow,* (JK, Ḳ,) *that causes one to hear a* [*sound such as is termed*] خَشْفَة *in walking* [*upon it*]: (TA:) and (in like manner, TA) *ice that is soft,* or *yielding* [*to the feet, crackling*], or *easily broken*: (Ḳ:) or the latter signifies [simply] *snow.* (Ṣ.) One says, ▼أَصْبَحَ المَآءُ خَشِيفًا [*The water became ice such as was soft,* &c.]. (JK, TA.) [See also خَاشِفٌ.]

خَشْفَةٌ (JK, Ṣ, Ḳ) and ▼خَشَفَةٌ (Mgh, Ḳ) and ▼خَشْفٌ (Ḳ) *A sound*: (Ḳ:) or *such as is termed* حِسّ (Ṣ;) [i. e.] *a low, faint, gentle,* or *soft, sound*: (JK:) and *a motion*: (JK, Ṣ, Ḳ:) or *a low, faint, gentle,* or *soft,* حِسّ: or the first signifies the *sound of the creeping of serpents*; and the *sound of the hyena*: (Ḳ:) or *a sound that is not loud,* or *vehement*; (A'Obeyd;) and so ▼the second: (Mgh:) or *a single sound*; so accord. to Fr: (Az, TA:) and ▼the last signifies the *sound of a sword falling upon flesh,* and *upon a weapon* or *weapons*: and the *sound of feet, such as is not loud,* or *vehement.* (TA.) ═ Also the first, *A* [*tract of high ground such as is termed*] قُفّ *that is mostly soft.* (L, Ḳ.)

خَشَفَةٌ: see the next preceding paragraph, in two places.

خَشُوفٌ *A quick,* or *swift,* man. (Ṣ.) — One *going away,* or *who goes away, into the land,* or *country*; as also ▼خَشِيفٌ and ▼خَاشِفٌ. (Ḳ.) — One *entering,* or *who enters,* into a thing; as also ▼خَشِيفٌ and ▼خَاشِفٌ and ▼مِخْشَفٌ. (Ḳ.) One *who enters into affairs* (Ḳ, TA) *and fears not,* or *dreads not*; as also ▼مِخْشَفٌ. (TA.) — Also and ▼خَاشِفٌ and ▼خَاشِفَةٌ sings. of خُشَّفٌ, which signifies Camels *that journey by night*: (Ṣ:) or, accord. to IB, the sing. of this pl. is خَاشِفٌ only: and the pl. of خَشُوفٌ is خُشُفٌ [a mistranscription for خُشُفٌ]. (L.) — See also مِخْشَفٌ: — and the paragraph here next following.

خَشِيفٌ: see خَشَفٌ, in two places: ═ and خَشُوفٌ also, in two places. — Also A *sharp,* or *penetrating,* sword; and so ▼خَاشِفٌ and ▼خَشُوفٌ: (Ḳ:) or, applied to a sword, *i. q.* خَشِيبٌ [q. v.]. (JK.) — And Water *that runs in a* [*watercourse such as is termed*] بَطْحَآء, *beneath the pebbles, two or three days, and then goes away.* (TA.)

الخَشَّافُ: see مِخْشَفٌ. — أُمُّ خَشَّافٍ *Calamity,* or *misfortune*; (JK, Ḳ;) as also خَشَّافٌ, without أُمّ. (TA.)

خُشَّافٌ *A certain nocturnal flying thing*; (Mṣb;) the خُفَّاش [or *bat*], (Ṣ, Ḳ, Mṣb,) *that flies by night*: so says El-Fárábee, in section ش: (Mṣb:) formed by transposition from the latter word, which is the more chaste: (Sgh, Mṣb, TA:) or rather so called because of its خَشَفَان, i. e. its going about, or round about, by night: (Lth, TA:) or *a certain flying thing, having two small eyes:* (TA:) as some say, (Ṣ,) and thus also says El-Fárábee, (Mṣb,) the خُطَّاف [or *swallow*]: (Ṣ, Mṣb, TA:) he who says خُفَّاش derives its name from the smallness of its eyes. (Lth, TA.)

خَاشِفٌ [act. part. n. of 1; fem. with ة]: see 1. — Water *freezing*, or *in a state of congelation*, and so ▼خَشِفٌ. (TA.) [See خَشِفٌ.] — See also خَشُوفٌ, in four places: — and see خَشِيفٌ: — Also A boy *light*, or *active, and brisk, lively*, or *sprightly;* like خَاسِفٌ. (TA in art. خسف.)

أَخَاشِفُ *Hard tracts* of land: with س, it signifies such as are soft. (Fr, Ḳ.)

مَخْشَفٌ *A place of ice;* (Ṣgh, Ḳ;) [*an ice-house;*] this is the meaning of the term by which Lth explains it, namely, يَخْدَان, (Ṣgh, TA,) [or يَخْدَانْ,] which is Persian, and which the author of the L has mistranscribed نَجْران, adding thereto الذى يجرى عليه الباب. (TA.)

ظَبْيَةٌ مُخْشِفٌ *A she-gazelle having a* خِشْف. (Ṣgh, Ḳ.)

مِخْشَفٌ: see خَشُوفٌ, in two places. — Also A guide of the way (Lth, JK, Ḳ) *who travels*, or *goes about, with people by night:* (Lth, JK:) or *who acts with a penetrative energy, and with sharpness, vigorousness, and effectiveness.* (Ḳ.) A man (AA, Ṣ) *bold*, or *daring*, (AA, Ṣ, Ḳ,) *to encounter the night*, (Ṣ,) or *to encounter the terror of the night*, (AA,) or *in night-journeying:* (Ḳ:) or *who goes about, or round about, by night;* as also ▼خَشُوفٌ: (Ḳ:) or this last, one *who fears not by night:* (JK:) or *who goes away boldly in the night or in any case.* (AA, IB.) — المِخْشَفُ *The lion;* (Ḳ;) because of his boldness in going about: (TA:) and so ▼الخَشَّافُ. (JK.)

خشم

1. خَشَمَهُ, (Ṣ, Ḳ,) aor. ـُ, (Ḳ,) inf. n. خَشْمٌ, (JK, Ṣ,) *He broke his* خَيْشُوم [q. v.]. (JK, Ṣ, Ḳ.) ═ خَشِمَ, aor. ـَ, inf. n. خَشَمٌ, (Ḳ, TA,) agreeably with rule, (TA, [accord. to the CḲ خَشْمٌ,]) and خُشُومٌ, (Ḳ,) which is irreg., (TA,) *He* (a man, TA) *was*, or *became, wide in the nose.* (Ḳ.) — And *It* (the nose) *became altered for the worse in odour*, or *stinking, by reason of a disease therein;* (Ḳ, TA;) i. e., *by reason of a stoppage therein, affecting the passage of the breath, and preventing respiration:* or *had one of its three bones broken.* (TA.) — And خَشِمَ, (JK, Mgh, Mṣb, Ḳ,) aor. ـَ, (Mgh, Mṣb, Ḳ,) inf. n. خَشَمٌ, (JK, Mgh, and so in some copies of the Ḳ,) or خُشَمٌ, (Ṣ, Mṣb, and so in some copies of the Ḳ and in the TA,) and خُشَامٌ, (Ḳ, [but mentioned in the JK as though a simple subst.,]) said of a man, (Ṣ,* Mṣb, Ḳ,) *He became affected with a certain disease in the nose*, (JK, Ṣ, Mgh, Mṣb,) *which stopped the passage of the breath;* (JK;) or *which caused it to become altered for the worse in odour*, or *stinking;* (Zj, Mgh;) or *which rendered it corrupt*, or *unsound, so that the person could not smell:* (Mṣb:) or *his* [*cartilages of the nose called the*] خَيَاشِيم [pl. of خَيْشُوم q. v.] *delapsed*, (Ḳ, TA,) *and the passage of his breath became stopped.* (TA.) — And خَشِمَ, (JK, Mṣb, Ḳ,) aor. ـَ, (Ḳ,) inf. n. خَشَمٌ; (TA;) and ▼اخشم; (JK, Ḳ;) and ▼خشّم, (Ṣ, JM, TA,) inf. n. تَخْشِيمٌ; (JM;) for which last, the Ḳ erroneously substitutes ▼تخشّم; (TA;) *It* (flesh-meat) *became altered for the worse in odour*, or *stinking:* (Ṣ,* Mṣb, Ḳ:) or *became very stinking; stank much.* (JK.)

2. خَشَّمَهُ الشَّرَابُ, inf. n. تَخْشِيمٌ, *The odour of the wine rose into his* خَيْشُوم, *and intoxicated him:* (M, Ḳ:) or *the odour of the wine rose into his* خيشوم, *and became infused in his brain, and so dispelled his reason.* (T, TA.) ═ See also 1, last sentence.

4: see 1, last sentence.

5. تخشّم *His reason became dispelled by the rising of the odour of wine into his* خَيْشُوم *and its becoming infused in his brain.* (T, TA.) — See also 1, last sentence.

خَشْمٌ The *nose:* [see also خَيْشُومٌ:] and the *mucus that flows from it.* (TA, from a trad.: and the latter signification is mentioned in the TA voce سَلْتٌ; as well as in the present article.) — [In modern Arabic, it signifies The *mouth:* and hence, *a spout.*] ═ In Persian, it signifies *Anger:* and this meaning is with probability deducible from the literal root of this art.; for he who is angry raises his nose and makes it pointed. (TA.)

خَشْمَةٌ [*Intoxication produced by the odour of wine rising into the* خَيْشُوم;] a subst. from خَشَّمَهُ الشَّرَابُ. (Ḳ.)

خَشِمٌ, applied to flesh-meat, [*Stinking:* (see 1, last sentence:) or] *stinking much.* (JK.)

خُشَامٌ *A certain disease in the nose, and a stoppage of the passage of the breath* [therein]. (JK. [See also 1.]) ═ A man *having a large nose:* (Ṣ:) [or] *a large nose;* (Zj, JK, Ḳ;) and so though not elevated, or *prominent.* (Zj, TA.) — And † *A mountain having a thick prominence:* (Ṣ:) or *a long mountain*, (AA, JK, TA,) *having a prominence*, (AA, TA,) or *having a thick prominence:* (TA:) or *a great mountain.* (Ḳ.) — And الخُشَامُ *The lion:* (JK, Ḳ:) because of the greatness of his nose. (TA.)

خُشَامَةٌ *Refuse; anything remaining after the good has been picked out.* (JK.)

خَيْشُومٌ The *extreme*, or *most remote*, [meaning *innermost*,] *part of the nose:* (Ṣ, Mṣb:) or the *interior of the nose:* (MA:) or the *upper part of the interior of the nose:* and the *bone of the nose:* (KL:) or the *part that is above the* نُخْرَة [which here seems to mean the *end*, or *tip*, or *flexible part*,] *of the nose, of the bone thereof:* and what is beneath this [is] of [the thin cartilages called] the خَشَارِم of the head: (M, Ḳ:) and the *nose* [altogether] (Mṣb, KL) is so called by some: (Mṣb:) the word is of the measure فَيْعُولٌ: (Mṣb, TA:) and its pl. is خَيَاشِيمُ: (Mṣb:) which [also] signifies *certain cartilages in the extreme* [or *innermost*] *part of the nose, between it and the brain:* or *certain ducts*, (عُرُوق, [meaning, or including, the *air-passages*, see خَشَمٌ, and نَخَرٌ, &c.,]) in *the interior* (بَاطِن M, or بَطْن Ḳ) *of the nose.* (M, Ḳ.) — [Hence,] the pl. signifies also ‡ *Prominences*, or *projecting parts*, of mountains. (JK, Ṣ, TA.) — And the sing., [as a coll. gen. n.,] *Small, thin, black things, resembling flesh;* and *morbose nodes; upon a bone.* (TA.)

أَخْشَمُ *Wide in the nose:* (Ḳ:) applied to a man. (TA.) — And, so applied, *Having a certain disease in the nose*, (Ṣ, Mṣb,) *whereby it is rendered corrupt*, or *unsound, so that he cannot smell:* (Mṣb:) or *whose* خَيْشُوم *has a fetid odour;* (Mgh, Mṣb;) from خَشِمَ said of flesh-meat, explained above: (Mṣb:) or *that cannot smell anything*, (JK, Az, Mgh, Ḳ, TA,) *whether sweet or stinking*, (Az, Mgh, TA,) *by reason of a stoppage in his* خَيَاشِيم, *from having one of the three bones broken:* (TA:) and ▼مَخْشُومٌ [in like manner] signifies *having his nose altered for the worse in odour*, or *stinking, by reason of a stoppage therein, affecting the passage of the breath, and preventing respiration;* or *having one of its three bones broken:* (TA:) fem. of the former خَشْمَاءُ. (Mṣb.) — And, applied to the nose, *Altered for the worse in odour*, or *stinking, by reason of a disease therein*, (Ḳ, TA,) i. e., *by a stoppage therein, affecting the passage of the breath, and preventing respiration:* or *having one of its three bones broken.* (TA.)

مُخَشَّمٌ *Intoxicated;* as also ▼مَخْشُومٌ and ▼مُتَخَشِّمٌ: (Ḳ:) or *much intoxicated.* (Ṣ, TA.) — And *Broken in pieces.* (TA.)

مَخْشُومٌ: see أَخْشَمُ: — and see also مُخَشَّمٌ.

مُتَخَشِّمٌ: see مُخَشَّمٌ.

خشن

1. خَشُنَ, aor. ـُ, (Ṣ, Mṣb, Ḳ,) inf. n. خُشُونَةٌ and خُشْنَةٌ (Ṣ, Mṣb, Ḳ, KL) and خَشَانَةٌ (CḲ, TA, but omitted in some copies of the Ḳ) and خَشْنٌ and مَخْشَنَةٌ, (Ḳ,) *It was*, or *became, rough, harsh*, or *coarse;* (Ḳ,* KL, PṢ;) *contr. of* لَانَ, (Ṣ, Ḳ,) or *of* نَعُمَ; (Mṣb;) as also ▼اخشوشن. (Ḳ.) — [Hence,] هُوَ ذُو خُشُونَةٍ and مَخْشَنَةٍ (Ḳ) and خُشْنَةٍ (TA) ‡ *He is difficult, refractory*, or *stubborn; not to be coped with.* (Ḳ, TA. [See also خَشِنٌ.]) And خَشُنَ عَلَيْهِ صَدْرُهُ and ▼اخشوشن عليه صدره † *He was*, or *became, angry with him.* (Sh, TA.) See also 3.

2. [خَشَّنَهُ *He made it rough, harsh*, or *coarse.* — Hence,] خَشَّنَ صَدْرَهُ, inf. n. تَخْشِينٌ, ‡ *He exasperated him; made him to be affected with wrath*, or *rage.* (Ṣ, Ḳ, TA.) A poet says,

* وَخَشَّنْتَ صَدْرًا جَيْبُهُ لَكَ نَاصِحُ *

[explained in art. جيب]. (Ṣ.)

3. خَاشَنَهُ, (Ṣ, M, Ḳ,) inf. n. مُخَاشَنَةٌ, (TA,) † *He was rough, harsh*, or *coarse, to him;* syn. ▼خَشُنَ عَلَيْهِ; in speech, and in action; (M, TA;)

[*he acted towards him, or with him, roughly, harshly, or coarsely;*] *contr. of* لَايَنَهُ. (Ṣ, Ḳ.)

5: see 1: — and see also 12.

10. اِسْتَخْشَنَهُ *He found it* [or *esteemed it*] *rough, harsh, or coarse.* (Ḳ.) — Hence, in a trad. of 'Alee, making mention of pious men of learning, وَٱسْتَلَانُوا مَا ٱسْتَخْشَنَ المُتْرَفُونَ † [*And they esteemed, or esteem, soft, or smooth, what those leading a life of ease and plenty esteemed, or esteem, rough, harsh, or coarse*]. (TA.) And اِستخشن المَقَامَ فِى مَحَلِّ كَذَا † [*He esteemed unpleasant, or uncomfortable, the remaining in such a place of abode*]. (TA in art. بشع.)

12. اِخْشَوْشَنَ (JK, Ṣ, Ḳ) and ↓ تَخَشَّنَ (Ḳ) *It was, or became, very rough, harsh, or coarse:* (Ṣ, Ḳ:) or (Ḳ) *he wore rough, harsh, or coarse, clothes:* (JK, Ḳ:) or the former signifies also *he accustomed himself to the wearing of such clothes:* (Ṣ:) or each, *he ate rough, harsh, or coarse, food:* (TA:) or the former, (JK,) or each, (Ḳ,) † *he said what was rough, harsh, or coarse:* (JK, Ḳ, TA:) or † *he lived a rough, or coarse, life.* (Ḳ.) The former verb is more intensive in all its senses (Ḳ, TA) than خَشُنَ and تَخَشَّنَ, because of the repetition of the medial radical and the addition of the و; and the same is the case of every verb of this class, as اعشوشب and the like, as is indicated in the Ṣ. (TA.) — See also 1.

خَشِنٌ *Rough, harsh,* or *coarse;* (Ṣ,* Mṣb,* Ḳ, KL, PṢ;) applied to a thing (Ṣ, Mṣb, Ḳ) of any kind; (Ḳ;) as also ↓ أَخْشَنُ: (Ṣ, Mṣb, Ḳ:) in relation to a stone, they seldom or never say otherwise than أَخْشَنُ: (Mṣb:) the fem. of the former is with ة; (Mṣb, Ḳ;) and the pl. is خُشْنٌ, (Mṣb,) [also said to be a pl. by poetic license of أَخْشَنُ, for خُشْنٌ, as will be seen below,] or خِشَانٌ, (Ḳ,) which is applied in the sense explained above to land [or lands]: (TA:) the fem. of ↓ أَخْشَنُ is خَشْنَآءُ; (Ḳ;) and the pl. is خُشْنٌ. (Ṣ.) You say أَرْضٌ خَشِنَةٌ *Rough,* or *rugged, ground* or *land.* (Mṣb.) And أَرْضٌ ↓ خَشْنَآءُ *Rugged ground or land,* (JK, TA,) *in which are stones and sand.* (TA.) And مُلَآءَةٌ ↓ خَشْنَآءُ [*A wrapper for the body*] *in which is roughness, harshness, or coarseness,* either *from newness* or *from make.* (TA.) — [Hence,] † A *strong* man. (Mṣb.) And هُوَ خَشِنُ الجَانِبِ and ↓ أَخْشَنُهُ ‡ *He is difficult, refractory,* or *stubborn; not to be coped with.* (Ḳ, TA. [See also 1.]) And عَاشَ عَيْشًا خَشِنًا † *He lived a rough,* or *coarse, life.* (Ḳ.) See also أَخْشَنُ.

خُشَيْنَآءُ [dim. of خَشْنَآءُ] *A certain small herb,* or *leguminous plant, green, found in meadows and plains;* so called because of its roughness, or harshness, or coarseness. (TA.) See also خَشْنَآءُ at the end of the next paragraph.

أَخْشَنُ, and its fem. خَشْنَآءُ: see خَشِنٌ, in five places. You say also كَتِيبَةٌ خَشْنَآءُ ‡ [*An army,* or *a portion thereof, bristling with weapons:* or] *having many weapons:* (JK, Ṣ, Ḳ, TA:) [and in like manner, ↓ جَيْشٌ خَشِنٌ, occurring in the TA in art. خمس:] and مَعْشَرٌ خُشْنٌ and خُشُنٌ; the latter allowable in poetry: (Ṣ: [it is there implied that this has a similar meaning:]) or the last signifies † [*a company of men*] *who resist harm,* or *injury.* (Ḥam p. 5.) And أَلْفٌ خَشْنَآءُ † *A great number* [*of people*]. (Ṣ in art. ثقى.) — Also, (Ḳ, TA,) or ↓ أُخَيْشِنُ, (JK,) ‡ A man *whose state, or condition, is discommended.* (JK, Ḳ, TA. [See also مُخَشَّنَةٌ.]) And the fem., † A *she-camel lean,* or *emaciated.* (JK, Ḳ.) And سَنَةٌ خَشْنَآءُ † *A year of drought* or *distress.* (JK.) — A rájiz says,

* مِنْ يَثْرِبِيَّاتٍ قِذَاذٍ خُشْنِ *

meaning [*Of the fabric of Yethrib,*] *new* [*unfeathered arrows*]. (Ṣ, TA.) — خَشْنَآءُ also signifies *A certain green herb,* or *leguminous plant,* (AḤn, JK, Ḳ,) *having short leaves,* (JK,) *that spreads upon the ground,* (AḤn,) *rough to the feel, but soft in the mouth, viscous like purslane;* (AḤn, Ḳ;) *its blossom is yellow, and it is eaten* [*by men*], *and is likewise a pasture:* (AḤn:) also called ↓ خُشَيْنَآءُ. (TA.)

أُخَيْشِنُ dim. of أَخْشَنُ as syn. with خَشِنٌ. (TA.) أُخَيْشِنُ فِى ذَاتِ ٱللّٰهِ is a phrase occurring in a trad. [app. as meaning † *Somewhat rough* or *coarse* in clothing, or in mode of living, *for the sake, or to obtain the approbation, of God*]. (Ṣ, TA.) See also أَخْشَنُ.

مُخَشَّنَةٌ A she-camel *whose* طِرْق [or *condition in respect of fatness*] *is discommended.* (JK, Ḳ. [See also أَخْشَنُ.])

خشو

1. خَشَتِ النَّخْلَةُ, aor. تَخْشُو, (JK, Ṣ, Ḳ,) inf. n. خَشْوٌ, (JK, TA,) *The palm-tree bore dates such as are termed* خَشْو, i. e. حَشَف: (JK, Ṣ, Ḳ:) so says El-Umawee, (Ṣ,) or IAạr. (TA.)

خَشًا *Black wheat.* (IAạr, Ḳ,* TA.)

خَشْوٌ *Dates such as are termed* حَشَف: (JK, Ṣ, Ḳ:) so says El-Umawee, (Ṣ,) or IAạr, who adds, i. e., *of which the lower portion has become bad and rotten, while in its place:* he says that it is of the dial. of Belḥárith Ibn-Kaab. (TA.)

خشى

1. خَشِىَ, aor. يَخْشَى, inf. n. خَشْيَةٌ (JK, Ṣ, M, Mṣb, Ḳ) and خَشْىٌ (JK, M, Ḳ) and خِشْىٌ (Ṣgh, Ḳ) [the second and third erroneously written in the CḲ خَشًا and (by indication) خِشًا] and خَشَاةٌ (M, Ḳ) and خَشْيَانٌ, (JK, M, Ḳ,) though it has been said that the only instances of this kind are شَنْآنٌ and لَيَّانٌ, [see the former of these two,] but in one copy of the M found written خِشْيَانٌ, (TA,) and مَخْشَاةٌ (JK, M, Ḳ) and مَخْشِيَةٌ, (M, Ḳ,) *He feared;* syn. خَافَ: (JK, Ṣ, M, Mṣb, Ḳ:) or, accord. to Er-Rághib and others, *he dreaded;* or *feared with reverence, veneration, respect, honour,* or *awe.* (TA.) You say, خَشِيَهُ *He feared him,* or *it;* [or *he dreaded him,* or *it;* i. e. *feared him,* or *it, with reverence,* &c.;] as also ↓ تَخَشَّاهُ. (Ḳ.) [And خَشِىَ مِنْهُ, meaning the same: or *He feared,* or *dreaded, what might happen to him from him,* or *it.* And خَشِىَ عَلَيْهِ شَيْئًا *He feared,* or *dreaded, for him a thing.*] And فَعَلْتُ ذٰلِكَ خَشَاةَ أَنْ يَكُونَ كَذَا [*I did that in fear,* or *dread, that such a thing might happen*]. (IAạr, TA.) — خَشْيَةٌ also signifies *Hope.* (Er-Rághib, TA.) And the saying of Ibn-'Abbás to 'Omar, لَقَدْ أَكْثَرْتَ مِنَ الدُّعَآءِ بِالمَوْتِ حَتَّى خَشِيتُ أَنْ يَكُونَ ذٰلِكَ أَسْهَلَ لَكَ عِنْدَ نُزُولِهِ, is explained as meaning [*Verily thou hast prayed much for death, so that*] *I hope* [*that it may be easier to thee when it happens*]. (TA.) — And sometimes خَشِيتُ means عَلِمْتُ [*I knew,* or *know*]. (Mṣb.) So it is said to mean in the saying of the poet,

* وَلَقَدْ خَشِيتُ بِأَنَّ مَنْ تَبِعَ الهُدَى *
* سَكَنَ الجِنَانَ مَعَ النَّبِىِّ مُحَمَّدِ *

[*And I know assuredly that he who follows the right direction shall dwell in the gardens of* Paradise *with the Prophet Mohammad*]: (Ṣ, TA:) or the meaning may be, *I hope.* (TA.) — In the saying in the Ḳur [xviii. 79], فَخَشِينَا أَنْ يُرْهِقَهُمَا طُغْيَانًا وَكُفْرًا, the meaning is said, by Akh, to be *And we disapproved* [*that he should make excessive disobedience, and ingratitude, to come upon them twain*]; (Ṣ;) and so says Zj, explaining it as the saying of El-Khiḍr: or, accord. to Fr, the meaning is, *and we knew.* (TA. [See also أَرْهَقَ.]) — خَاشَانِى فَخَشِيتُهُ: see 3.

2. خَشَّاهُ, inf. n. تَخْشِيَةٌ, *He frightened him,* or *made him to fear;* (Ṣ, Ḳ;) [or *he made him to dread; or to fear with reverence,* &c.; (see 1;)] بِالأَمْرِ [*with the thing,* or *event*]. (TA.) One says, خَشِّ ذُؤَالَةَ بِالحِبَالَةِ, meaning [*Frighten thou*] *the wolf* [*with the snare*]. (Ṣ. [See art. ذأل.]) And لَقَدْ كُنْتُ وَمَا أُخَشَّى بِالذِّئْبِ [*Verily I used to be in a state when I was not frightened by the wolf*]: a prov. (JK, TA.)

3. خَاشَانِى ↓ فَخَشِيتُهُ, (A'Obeyd, Ṣ, Ḳ,) aor. of the latter أَخْشِيهِ, (A'Obeyd, Ṣ,) [*I vied with him in fear* or *dread, and*] *I was more fearful* [or *dreading*] *than he.* (A'Obeyd, Ṣ, Ḳ.) — خَاشَى فُلَانًا, (JK, TA,) inf. n. مُخَاشَاةٌ, (TA,) *He left, forsook, relinquished,* or *abandoned, such a one,* *being left,* &c., *by him.* (JK, TA.) — خاشى بهم *He guarded himself against them in an extraordinary degree, and was cautious,* or *wary,* (JK, TA,) *and therefore turned away,* or *withdrew.* (TA.)

5: see 1, second sentence.

خَشٍ: see what next follows.

خَشْيَانُ *Fearful,* or *fearing;* (Ṣ, Mṣb, TA;) [or *dreading;* i. e. *fearing with reverence,* &c.: (see 1:)] as also ↓ خَاشٍ and ↓ خَشٍ: (Ḳ:) fem. خَشْيَا, (Ṣ, Mṣb, Ḳ, [in the CḲ, erroneously, خَشْيَآء,]) like غَضْبَى fem. of غَضْبَانُ, (Mṣb,) accord. to rule, (TA,) and خَشْيَانَةٌ, mentioned by El-

Marzookee, and thought by MF to be of the dial. of Asad; (TA;) or this signifies a woman *who fears,* or *dreads,* (تَخْشَى,) *everything;* (JK, TA;) so in the Tekmileh: (TA:) pl. خَشَايَا, (Ḳ,) pluralized in a similar manner to epithets significant of diseases, like حَبَاطَى &c., because الخَشْيَةُ is like a disease. (TA.)

خَشَاةٌ Land *such as is termed* جِهَادٌ [i. e. *hard;* or *having no herbage;* or *hard, and having no herbage;* or *level;* or *rugged,* &c.]. (JK, Ṣgh, Ḳ.)

خَشِىٌّ *Dry;* (Aṣ, JK, Ṣ, Ḳ;) like خَشِىٌّ; (Aṣ, Ṣ;) applied to herbage; (Aṣ, Ṣ;) or to herbs and trees: (JK:) or *dry and rotten.* (IAạr, TA.) A rájiz says, (Ṣ,) namely, Ṣakhr, (TA,)

* سَمُّ ذَرَارِيحَ رِطَابٍ وَخَشِى *

[*Poison of moist cantharides, and dry*]; (Ṣ, TA;) meaning وَخَشِيٍّ, suppressing one of the two ى by poetic license. (IB, TA.)

خَشْيَانُ: see خَاشٍ.

أَخْشَى meaning *More* [and *most*] *fearful,* or *feared,* [or *dreaded,*] (Ṣ, Ḳ,) is anomalous, (Ḳ,) being from the pass. [verb, like its syn. أَخْوَفُ]. (TA.) You say, هٰذَا المَكَانُ أَخْشَى مِنْ ذَاكَ *This place is more fearful,* or *feared,* [or *dreaded,*] *than that.* (Ṣ, Ḳ.*)

مَخَاشٍ [pl. of مَخْشَاةٌ, originally مَخْشَيَةٌ,] *Causes of fear* [or *dread*]; syn. مَخَاوِفُ [pl. of مَخَافَةٌ: like which, مَخْشَاةٌ is also an inf. n.]. (Ḥar p. 138.)

خص

1. خَصَّهُ بِالشَّىْءِ, (Ṣ, Ḳ,) or بِكَذَا, (A, Mṣb,) aor. ـُ, (Mṣb, TA,) inf. n. خَصٌّ (Ḳ) and خُصُوصٌ (Ṣ, Mṣb, Ḳ) and خَصُوصٌ (TA) and خُصُوصِيَّةٌ and خَصُوصِيَّةٌ, (Ṣ, Mgh,* Mṣb, Ḳ,) of which last two the former is the more chaste, (Ṣ, TA,) and is the form mentioned in the Fṣ and its Expositions, and the ى in it is said to be the relative ى, and the ى which is characteristic of an inf. n., whereas in the latter it is said to be a characteristic of intensiveness, but MF thinks that this requires consideration, because the ى is also said to be, and by some more commonly, without teshdeed, as in كَرَاهِيَةٌ and عَلَانِيَةٌ, (TA,) and خِصِّيصَى, (Ṣ, Ḳ,) like مِكِّيثَى, which is said to be the only other instance of this measure, (TA,) [but some others might be added, as دِلِّيلَى and خِطِّيبَى and قِتِّيتَى,] and خِصِّيصَآءٌ, (IAạr, Kr, Ḳ,) [like مِكِّيثَآءٌ,] the former of which last two [each of which has an intensive signification] is the chaste, and commonly known, form, (TA,) and خَصِّيَّةٌ, or خِصِّيَّةٌ, or خُصِّيَّةٌ, (accord. to different copies of the Ḳ,) written by Ṣgh with damm, (TA,) and تَخِصَّةٌ, (Ibn-'Abbád, Ḳ,) or, as some say, خُصُوصِيَّةٌ and خِصِّيَّةٌ [or خُصِّيَّةٌ or خَصِّيَّةٌ] are each a quasi-inf. n., as also خَاصَّةٌ, (TA,) *He distinguished him particularly, peculiarly,* or *specially,* i. e., *above,* or *from,* or *exclusively of, others, by the thing,* or *by such a thing; he particularized him,* or *particularly* or *peculiarly* or *specially characterized him, thereby;* syn. فَضَّلَهُ (A, Ḳ, TA) دُونَ غَيْرِهِ, and مَيَّزَهُ; (TA;) *he appropriated,* or *assigned,* [*the thing* or] *such a thing,* or *made it to belong, to him alone,* or *in particular,* or *peculiarly,* or *specially, exclusively of others;* (Mṣb;) and ↓اخْتَصَّهُ بِهِ signifies the same; (Ṣ, A, Mṣb, Ḳ;) as also ↓اخَصَّهُ, (A, [but for this I know not any other authority,]) and ↓خَصَّصَهُ; (A;) or this last has an intensive signification. (Mṣb.) You say, خَصَّهُ بِالوُدِّ *He distinguished him* &c. *by love,* or *affection;* or *favoured him in preference to another,* or *others, thereby.* (Ḳ, TA.) As to the saying of AZ,

* إِنِ امْرَأً خَصَّنِى عَمْدًا مَوَدَّتَهُ *
* عَلَى التَّنَائِى لَعِنْدِى غَيْرُ مَكْفُورِ *

[*If a man distinguish me above,* or *from,* or *exclusively of, others, purposely, by his love,* or *because of his love of me, notwithstanding distance of each from the other, it will not be disacknowledged with me,*] the meaning is, خَصَّنِى بِمَوَدَّتِهِ; or it may be خَصَّنِى لِمَوَدَّتِهِ إِيَّاىَ [in the TA بمودّته, which is evidently a mistranscription]; for, says ISd, we have not heard خَصَّهُ [or rather خَصَّ] doubly transitive. (TA.) And [hence] خَصَّهُ بِكَذَا also signifies *He gave him such a thing in large quantity,* or *abundantly.* (TA.) [You say also, خَصَّهُ بِالذِّكْرِ *He distinguished,* &c., or *singled out, him,* or *it, by mention:* or *he particularized, peculiarized,* or *specified, him,* or *it, thereby; he particularly, peculiarly,* or *specially, mentioned him,* or *it.* And خَصَّهُ, alone, *He pointed particularly,* or *peculiarly, to him,* or *it,* in what he said; or *he meant particularly,* or *peculiarly, him,* or *it.* And خَصَّ مِنْهُ كَذَا *He distinguished, particularized, peculiarized,* or *specified, thereof such a thing:* and *he distinguished therefrom such a thing; he particularly, peculiarly,* or *specially, excepted therefrom such a thing.*] You also say, خَصَّهُ لِنَفْسِهِ (TA) and ↓اخْتَصَّهُ لِنَفْسِهِ (T, A, TA) [*He appropriated,* or *took,* or *chose, him,* or *it, particularly,* or *specially, to,* or *for, himself;* as also خَصَّ بِهِ نَفْسَهُ and ↓اخْتَصَّ بِهِ نَفْسَهُ]. And هُوَ يَسْتَخْلِصُهُ and ↓يَسْتَخِصُّ فُلَانًا (A, TA) [*He appropriates such a one purely to himself, exclusively of any partner;* (see the latter verb;)] *he chooses such a one for himself; he appropriates him to himself as his particular,* or *special, intimate;* (TA in art. خلص;) both signify the same. (Ṣ and Ḳ in art. خلص.) [And ↓اخْتَصَّهُ *He treated him,* or *behaved towards him, with partiality; was partial towards him:* a signification implied by the first explanation in this art.: and in this sense it is often used.] — خَصَّ, aor. ـَ, [contr. to general rule, by which it should be ـِ, for it is intrans., and of the measure فَعَلَ, accord. to the Mṣb,] inf. n. خُصُوصٌ, [and app. خُصُوصِيَّةٌ and خَصُوصِيَّةٌ, accord. to modern usage,] *It was,* or *became, particular, peculiar,* or *special; restricted,* or *confined, to one* or *more* of persons, places, or things; *distinct,* or *distinguished, from others; not common,* or *general; contr. of* عَمَّ; as also ↓اخْتَصَّ: (Mṣb:) [each, also, followed by لَهُ, signifies *He,* or *it, belonged, pertained,* or *appertained, to him,* or *it, particularly, peculiarly, specially,* or *exclusively; it so related to him,* or *it; it was,* or *became, peculiar to him,* or *it:* see also the latter verb below.] — خَصَّ, sec. pers. خَصِصْتَ, (in the CḲ خَصُصْتَ,) [inf. n., app., خَصَاصَةٌ and خَصَاصٌ and خَصَاصَآءٌ,] ‡ *He was,* or *became, poor; in a state of poverty;* (Fr, Ṣgh, Ḳ;) as also ↓اخْتَصَّ. (A, TA.)

2. خَصَّصَهُ, inf. n. تَخْصِيصٌ, *He made it,* or *rendered it, particular, peculiar,* or *special; distinct,* or *distinguished, from others; not common,* or *general; he individuated it; particularized it; distinguished it from the generality; singled it out;* تَخْصِيصٌ being the *contr. of* تَعْمِيمٌ. (Ḳ. [But only the inf. n. is there mentioned.]) — See also 1, first sentence.

4: see 1, first sentence.

5. تَخَصَّصَ quasi-pass. of 2; *It was,* or *became, made,* or *rendered, particular, peculiar,* or *special;* &c.; *not common,* or *general.* (TA.) — See also 8, in two places. — It is also said to mean † *He was,* or *became, in a peculiar, unparticipated state of pressing want and poverty.* (Ḥar p. 94.)

8. اخْتَصَّهُ: see 1, in four places. — اخْتَصَّ as an intrans. v.: see 1, last sentence but one. — اخْتَصَّ بِالشَّىْءِ, (Ḳ,) or بِكَذَا, (A, Mṣb,) quasi-pass. of خَصَّهُ بِهِ; (A, Mṣb, Ḳ;) *He was,* or *became, distinguished particularly, peculiarly,* or *specially,* i. e., *above,* or *from,* or *exclusively of, others, by the thing,* or *by such a thing; he was,* or *became, particularized,* or *particularly* or *peculiarly* or *specially characterized, thereby;* (A,* Ḳ,* TA;) *he had* [*the thing* or] *such a thing appropriated,* or *assigned,* or *made to belong, to him alone,* or *in particular,* or *peculiarly,* or *specially, exclusively of others;* (Mṣb;) and ↓تَخَصَّصَ signifies the same. (A, Mṣb, Ḳ.) You say, اخْتَصَّ فُلَانٌ بِالأَمْرِ, and ↓تَخَصَّصَ لَهُ, [or بِهِ, accord. to general usage,] *Such a one was,* or *became, alone, with none to share or participate with him, in the affair;* syn. اِنْفَرَدَ. (TA.) — See also 1, last sentence.

10: see 1, latter half.

خُصٌّ *A booth of reeds,* or *canes,* (Ṣ, Mgh, Mṣb, Ḳ, TA,) or *of* [*boughs of*] *trees:* (TA:) or *a house roofed with a piece of wood, in the form of the* [*oblong vaulted structure called*] أَزَج: (JK, Ḳ:) so called because of the خَصَاص, or "narrow interstices," which are in it; (T, TA;) or because one sees what is in it through its خَصَاص, or "interstices:" (TA:) pl. [of pauc.] أَخْصَاص (JK, Mṣb, TA) and [of mult.] خِصَاص (JK, L, Ḳ [in the CḲ خَصَاص, which is wrong,]) and خُصُوص (JK, Ḳ) and خُصُوصَة. (JK.) — Also *The shop of a vintner,* (Aṣ, Ḳ,) *although it be not of reeds,* or *canes.* (Ḳ.)

خَصَاصٌ: see خَصَاصَةٌ, from the beginning to the last sentence but two.

خُصُوصٌ an inf. n. of 1, trans. and intrans. — [Used as a simple subst., *Particularity; peculiarity; speciality,* or *specialty;* as also the inf. ns. ↓خُصُوصِيَّةٌ, and ↓خَصُوصِيَّةٌ.] You say, لَهُ خُصُوصٌ and ↓خُصُوصِيَّةٌ [*To him belongs a particularity,* &c.], and بِى [*in me is* &c.]. (A.) [Hence خُصُوصًا *Particularly; specially;* as also ↓خَاصَّةً.] Th was heard to say, ↓إِذَا ذُكِرَ الصَّالِحُونَ فَبِخَاصَّةٍ أَبُو بَكْرٍ وَإِذَا ذُكِرَ الأَشْرَافُ فَبِخَاصَّةٍ عَلِىٌّ [meaning فَخُصُوصًا], i. e., *When the righteous are mentioned, then in particular, or peculiarly, Aboo-Bekr is* virtually mentioned; *and when the shereefs are mentioned, then in particular, or peculiarly, 'Alee*]. (L, TA.)

خَصَاصَةٌ *An interstice, interval,* or *intervening space* or *opening;* (Ṣ, Ḳ;) as also ↓خَصَاصٌ [which is commonly used as a coll. gen. n.] and ↓خَصَاصَآءُ: (Ḳ:) or *an interstice,* &c., *in the* ثَغْر [app. meaning the *front teeth*]; as also ↓the second of these words: (TA:) or the first and ↓second, (TA,) or ↓all, (Ḳ,) *any interstice,* &c., or *hole* or *perforation, in a door, and sieve, and* [*veil of the kind called*] بُرْقُع, *and the like,* (Ḳ, TA,) *such as a cloud, and a strainer,* &c.: (TA:) [*a crevice, cranny, chink,* or *fissure:*] or the first, (TA,) or ↓all, (Ḳ,) *a small hole* or *perforation:* (Ḳ, TA:) or the first has this signification as well as the signification first mentioned: (Ṣ:) or the ↓second, the *like of a* كُوَّة [or *mural aperture*] *in a* [*structure of the kind called*] قُبَّة, *or the like, when as wide as the face;* or, accord. to some, *whether wide or narrow:* (TA:) and the same, *interstices, intervals,* or *intervening spaces* or *openings,* in a خُصّ; (TA;) or *narrow interstices,* &c., therein: (T, TA:) and the same, (TA,) or ↓all, (Ḳ,) the *intervening spaces between the three stones upon which a cooking-pot is placed;* (Ḳ, TA;) and *between the fingers:* (TA:) and the first, the *intervening spaces between the feathers of an arrow:* (IAạr, TA:) pl. [of the first,] خَصَاصَاتٌ. (TA.) You say of the moon, بَدَا مِنْ خَصَاصَةِ الغَيْمِ [*It appeared from the gap of the cloud,* or *clouds*]. (Ṣ, A.) — Also *A cloud* itself; or *clouds;* syn. غَيْمٌ. (TA.) ═ Also the first, (Ṣ, A, Mgh, Mṣb, Ḳ,) and ↓second, (Ṣ, Ḳ,) and ↓third, (IDrd, Ḳ,) ‡*Poverty;* (Ṣ, Mgh, Mṣb, Ḳ;) *need;* (A, Mṣb, TA;) *straitness,* or *difficulty;* (Mgh;) *an evil state* or *condition:* (TA:) from خَصَاصَات meaning the "holes" of a sieve: whence a saying cited voce تَجَمَّلَ: (Mgh:) or from the first of the senses explained in this paragraph; because a thing, when it opens so as to form an interstice, becomes weak and unsound. (TA.) You say also, ↓سَدَدْتُ خُصَاصَةَ فُلَانٍ, with ḍamm, meaning, ‡*I repaired the broken fortune of such a one.* (A, TA.) — Also the first, ‡*Thirst;* as in camels when they return from water without having satisfied themselves with drink: and *hunger;* as in a man when he has not satisfied himself with food. (TA.)

خُصَاصَةٌ: see خَصَاصَةٌ, last sentence but one.

خَصَاصَآءُ: see خَصَاصَةٌ, from the beginning to the last sentence but two.

خُصُوصِيَّةٌ and خَصُوصِيَّةٌ: see خُصُوصٌ, in three places.

خُصَّانٌ and خِصَّانٌ: see the next paragraph, in three places.

خَاصٌّ *Particular; peculiar; special; distinct,* or *distinguished, from others;* contr. *of* عَامٌّ. (Mṣb, TA.) — [And hence, *Choice; select.* — And *Pure; unmixed; unadulterated.*] — [Used as a subst.,] it is syn. with ↓خَاصَّةٌ; (Ks, Mṣb, Ḳ;) in which the ة is a corroborative; (Mṣb;) and which signifies *Distinguished people; persons of distinction;* the *distinguished sort;* contr. *of* عَامَّةٌ: (Ṣ, Mṣb, Ḳ:) or the former is *contr. of* عَامٌّ, and ↓the latter is *contr. of* عَامَّةٌ: (TA:) [the pl. of both is خَوَاصُّ and ↓خُصَّانٌ and ↓خِصَّانٌ: or, accord. to rule, the first of these is pl. only of خَاصَّةٌ; and judging from other instances, we should regard the second and third as more probably pls. of خَاصٌّ: but however the case may be,] خُصَّانٌ is *syn. with* خَوَاصُّ, (Ṣ, Ḳ,) and so is خِصَّانٌ. (Ḳ.) [You say, الخَاصُّ وَالعَامُّ, and ↓الخَاصَّةُ وَالعَامَّةُ, *The distinguished and the common people; the persons of distinction and the vulgar.*] You also say, ↓إِنَّمَا يَفْعَلُ هٰذَا خُصَّانٌ مِنَ النَّاسِ meaning خَوَاصٌّ مِنْهُمْ [*Only distinguished persons of mankind do this*]. (Ṣ.) — [It seems to be also, in some instances, *syn. with* ↓خَاصَّةٌ as signifying *A particular, peculiar,* or *special, friend, intimate, familiar, companion, associate, attendant, dependent,* or *servant:*] the latter is explained in the T [and JK] as meaning *a person whom thou hast appropriated, particularly distinguished, taken,* or *chosen,* (اِخْتَصَصْتَهُ,) [*as a friend,* &c.,] *to,* or *for, thyself:* (TA:) [and it is used as a sing. and as a pl.: for] you say, هٰذَا خَاصَّتِى [*This is my particular,* or *special,* or *choice,* or *choicest, friend,* &c.]: and هُمْ خَاصَّتِى [*They are my particular,* or *peculiar,* or *special,* or *choice,* or *choicest, friends,* &c.]. (A.) You say also, فُلَانٌ خَاصٌّ لِفُلَانٍ, (Kull p. 174,) or بِفُلَانٍ, (so in the L,) [app. meaning لِفُلَانٍ, unless it be mistranscribed, and the latter be the correct reading, which I think much the more probable;] i. e., *Such a one belongs exclusively* [*as a particular,* or *peculiar,* or *special, friend,* &c.,] *to such a one;* (Kull;) and ↓مُخِصٌّ signifies the same. (L.) — See the dim. of خَاصَّةٌ, (namely خُوَيْصَّةٌ,) below.

خَاصَّةٌ: see خَاصٌّ, in four places. — It also signifies *A property* of a thing, *not found,* or *not existing, either wholly or partly, in another thing:* and ↓خَاصِّيَّةٌ [thus correctly written, and thus I have always found it written except by Golius and those who have probably imitated him, who write it without the sheddeh to the ى,] is used as denoting [*a property,* or *particular* or *peculiar virtue, which is*] *an unknown cause of a known effect; as that by which a medicine operates:* the former differs from the latter in being conventionally applied to *an effect,* [or *effective property,*] *whether the cause of its existence be known or not:* [the pl. of the former is خَوَاصُّ, agreeably with analogy and usage, like as عَوَامُّ is pl. of عَامَّةٌ:] the pl. of the latter is خَاصِّيَّاتٌ [and خَصَائِصُ]; and خَوَاصُّ is a quasi-pl. n., not a pl., of the same. (Kull p. 174. [All the above-mentioned words here cited from that work are there without syll. signs, as being well known. Both خَاصَّةٌ and ↓خَاصِّيَّةٌ, as here explained, are perhaps post-classical; but of this I am not certain: and both are sometimes used as meaning The *peculiar nature* of a thing; also termed its *essence.*]) — خَاصَّةً and بِخَاصَّةٍ: see خُصُوصٌ.

خَاصِّيَّةٌ: see خَاصَّةٌ, in two places.

خُوَيْصَّةٌ dim. of خَاصَّةٌ; (A, Ḳ;) [like دُوَيْبَّةٌ, q. v., dim. of دَابَّةٌ;] originally خُوَيْصِصَةٌ; (TA;) the ى being quiescent because the ى of the dim. cannot be movent; (A, Ḳ;) [properly signifying *A little,* or *young, particular,* or *peculiar,* or *special, friend, companion, associate, attendant,* or *servant;* and used in other senses, like other diminutives; implying littleness of estimation; and also affection, and awe.] It is said in a trad., (TA,) عَلَيْكَ بِخُوَيْصَّةِ نَفْسِكَ [*Keep thou to the little,* or *dear, particular friend of thine own self:* so it seems to mean accord. to Z, being mentioned by him among the proper expressions belonging to this art.: but accord. to the TḲ, it appears to be tropical; for the meaning is there said to be, †*thine own particular state,* or *condition*]. (A, TA.) In another trad., خُوَيْصَّة is used as signifying *A little, young, particular,* or *peculiar,* or *special, servant.* (TA.) And in another trad. it is said, بَادِرُوا بِالأَعْمَالِ سِتًّا الدَّجَّالَ وَكَذَا وَكَذَا وَخُوَيْصَّةَ أَحَدِكُمْ, i. e. †*Strive ye to be before six things with* [*good*] *works; Antichrist, and such and such things, and the event of death which is specially,* or *peculiarly, appointed to any one of you:* [or, I would rather say, *the awful special awaiter of any one of you;* though it is asserted that] the diminutive form is here used to denote low estimation of that which it signifies in comparison with what follows it, namely, the resurrection, &c. (TA.)

مُخِصٌّ: see خَاصٌّ, last sentence but one.

خصب

1. خَصِبَ and خَصُبَ: see 4.

[2. خَصَّبَ, inf. n. تَخْصِيبٌ, *It rendered fruitful; it fecundated:* so in the present day: see an instance voce بَاقِلَّى.]

4. اخصب, (A, Mṣb, Ḳ,) inf. n. إِخْصَابٌ; (TA;) [and some add خِصْبٌ, as another inf. n.; but ISd holds this to be a simple subst.; (see 4 in art. ريف;)] and ↓خَصِبَ, (A, Mṣb, Ḳ,) aor. ـَ; (Mṣb, Ḳ;) and ↓خَصُبَ, aor. ـُ, inf. n. خِصْبٌ; (Ḳ;) *It* (a place) *abounded,* or *became abundant, with herbage* [or *with the produce of the earth*], and *with the goods, conveniences,* or *comforts, of life;* (A, Ḳ;) [*was,* or *became, fruitful;*] *had increase; had plenty,* or *abundance;* (Mṣb;) [*contr. of* جَدُبَ or جَدِبَ and أَجْدَبَ:]

and الأَرْضُ اخصبت [the land, or earth, abounded, or became abundant, with herbage &c.]. (JK, Ṣ.) إِخْصَابٌ and ↑ اِخْتِصَابٌ are both from الخِصْبُ [but the precise meaning of the latter is not explained]. (Lth, JK, TA.) In the saying of the rájiz,

* لَقَدْ خَشِيتُ أَنْ أَرَى جَدْبًا *
* فِى عَامِنَا ذَا بَعْدَ أَنْ أَخْصَبَّا *

[Verily I feared to see drought, or barrenness, or dearth, in this our year, after it had been abundant in herbage &c.], أَخْصَبَّا is put for أَخْصَبَا: but accord. to one reading, it is ↑اِخْصَبَّا, of the measure اِفْعَلَّ, though this is generally employed for colours; and the incipient ا is rendered disjunctive of necessity, for the sake of the metre. (L. [Respecting جَدْبًا, see جَدْبٌ.]) You say also, اخصب جَنَابُ القَوْمِ, meaning The tract surrounding the people [became abundant with herbage &c.]. (Ṣ, TA.) — اخصبوا They attained, obtained, had, or became in the condition of having, abundance of herbage [or of the produce of the earth], and of the goods, conveniences, or comforts, of life. (Ṣ,* Ḳ.) [They became in the condition of persons whose food and milk, and the pasture of whose land, were abundant. (See the part. n., مُخْصِبٌ, below.)] And اخصبت الشَّاةُ The ewe, or she-goat, obtained abundance of herbage. (TA.) ═ اخصب اللّٰهُ المَوْضِعَ God caused the place to produce herbs and pasture. (Mṣb.) ═ اخصبت العِضَاهُ, mentioned as on the authority of Lth, [and in the Ḳ,] is, accord. to Az, a gross mistranscription, for اخضبت [q. v.]. (TA.)

8: } see 1.
9: }

خَصْبٌ: see خَصْبَةٌ, in two places.

خِصْبٌ Abundance of herbage [or of the produce of the earth], and of the goods, conveniences, or comforts, of life; (A, Ḳ;) contr. of جَدْبٌ; (JK, Ṣ, Mṣb;) [fruitfulness;] increase; plenty, or abundance: (Mṣb:) abundance of good, or of good things: (Ḳ:) [abundant herbage, and the like:] truffles are included in the term خِصْبٌ; and also locusts, when they come after the herbage has dried up and the people are secure from being injured by them. (AḤn.) ═ بَلَدٌ خِصْبٌ and أَخْصَابٌ, (Ṣ, Ḳ,) like بَلَدٌ سَبْسَبٌ and سَبَاسِبُ &c., the sing. being used [in بلد اخصاب] as a pl., as though made to consist of parts, or portions, [each termed خِصْب,] (Ṣ, TA,) A country, or region, abounding with herbage [or with the produce of the earth], or with the goods, conveniences, or comforts, of life; [fruitful; or plentiful;] (Ṣ,* Ḳ;) as also ↑مُخْصِبٌ (Ṣ,* A, Mṣb,* Ḳ) and ↑خَصِيبٌ (Ṣ,* A, Ḳ) and ↑خَصِبٌ. (A, Mṣb.*) And أَرْضٌ خِصْبٌ and ↑خَصِيبَةٌ, (AḤn, TA,) and أَرَضُونَ خِصْبٌ [because خِصْب is originally an inf. n.] and خِصْبَةٌ and ↑خَصْبَةٌ, which last word is either an inf. n. used as an epithet, or a contraction of ↑خَصِبَةٌ, (K,) A land, and lands, abounding with herbage &c. (Ḳ, TA.) — And عَيْشٌ خِصْبٌ and ↑مُخْصِبٌ [A life of abundance or plenty]. (TA.)

خَصِبٌ; and its fem., with ة: see خِصْبٌ, in two places.

خَصْبَةٌ: see خِصْبٌ. — Also, [app. as an epithet in which the quality of a subst. is predominant,] A palm-tree having much fruit: pl. خِصَابٌ (Ṣ, Ḳ) and ↑خَصْبٌ: (Ḳ, TA:) or خَصْبٌ [is properly a coll. gen. n., and] signifies palm-trees [absolutely:] (Ḳ:) and خَصْبَةٌ signifies a palm-tree of the kind called نَخْلَةُ الدَّقَلِ, in the dial. of the people of El-Baḥreyn, (Az, TA,) or of Nejd; (TA;) and its pl. is خِصَابٌ. (Az, TA.) — It is said that ↑خَصْبٌ signifies also The spadix of the palm-tree: so in the Ḳ: and accord. to Lth, خَصْبَةٌ signifies a single spadix of a palm-tree: but [it is probably a mistranscription for خَضْبَةٌ, with the pointed ض:] Az says that he who assigns to it this meaning errs. (TA.)

خَصِيبٌ; and its fem., with ة: see خِصْبٌ, in two places. — رَجُلٌ خَصِيبٌ A man abounding with good, or with good things; (Ḳ;) i. e., whose abode abounds therewith; (TA;) as also خَصِيبُ الرَّحْلِ (A, TA) and خَصِيبُ الجَنَابِ: (TA:) or this last means one whose region, or quarter, is خَصِيب: (Ṣ:) or it is tropical, (A in art. جنب,) as is also the expression immediately preceding, (A in the present art.,) and means ‡ Generous or bountiful [or hospitable]. (A in art. جنب.)

أَخْصَبُ [More, and most, abundant with herbage &c.]

مُخْصِبٌ: see خِصْبٌ, in two places. — قَوْمٌ مُخْصِبُونَ A people, or party, whose food and milk, and the pasture of whose land, have become abundant. (TA.)

مخصبة [so in the TA, either مُخْصِبَةٌ (like مُبْقِلَةٌ &c.) or مَخْصَبَةٌ,] A land (أَرْضٌ) abounding with pasture or herbage. (TA.)

بَلَدٌ مِخْصَابٌ (Ḳ) A country, or region, scarcely ever, or never, sterile, barren, unfruitful, or afflicted with dearth or scarcity or drought. (TA.) — And قَوْمٌ مَخَاصِيبُ [A people, or party, scarcely ever, or never, without abundance of herbage &c.]. (TA in art. ربع.)

خصر

1. خَصِرَ, (Ṣ, A,) aor. َ, inf. n. خَصَرٌ, (TḲ,) It (a day) was, or became, intensely cold. (Ṣ, A.) He (a man) suffered pain from the cold in his extremities. (Ṣ.) And خَصِرَتْ يَدِى, (Ṣ, TA,) and أَنَامِلِى, (TA,) My arm, or hand, and my fingers' ends, were pained by the cold. (Ṣ,* TA.)

2. تَخْصِيرٌ [an inf. n. of which the verb, if it have one, is خَصَّرَ]: see مُخَصَّرٌ.

3. خاصر المَرْأَةَ, (A,) inf. n. مُخَاصَرَةٌ, (TA,) He laid hold upon the woman's خَاصِرَة [or flank], (A,) or put his hand to her خَصْر [or waist], (TA,) in compressing her. (A, TA.) — And خاصرهُ He took his hand in walking, or walked with him hand in hand, (Ṣ, A, IAth, Ḳ,) so that the hand of each was by the waist (خَصْر) of the other: (IAth:) and, (so in the Ṣ, but in the Ḳ "or,") inf. n. as above, (Ṣ,) he took a different way from his (another's) until he met him in a place: (Ṣ, Ḳ:) مخاصرة as the inf. n. of the verb in this sense is syn. with مُخَازَمَةٌ: (Ṣ:) or خاصرهُ signifies he walked with him, and then parted from him, and so continued until he met him at a time, or place, at which they had not appointed to meet: (IAṣr:) or he walked by his side. (Ḳ.)

4. اخصر It (cold) pained a man's arms, or hands, and his fingers' ends. (A,* TA.)

5: see 8, in the first sentence: ═ and again, in the last two sentences.

6. تخاصر: see 8. — تخاصروا They took one another by the hand in walking, or walked together hand in hand [so that the hand of each was by the waist (خَصْر) of another: see 3]. (Ṣ, Ḳ.*)

8. اختصر (A, Mgh, L, Mṣb, Ḳ) and ↑تخصّر, (Mgh, Mṣb, Ḳ,) or ↑تخاصر, (A, L,) He put his hand upon his خَصْر [or waist], (A, Mgh, L, Mṣb,) or upon his خَاصِرَة [or flank], (Mgh, Ḳ,) in prayer. (Mgh, L, Mṣb.) The doing this in prayer [except in the night, when tired, (see المُتَخَصِّرُونَ,)] is forbidden, or disapproved. (Mgh, TA.) ═ اختصر الطَّرِيقَ He went the nearest way. (Ṣ, A, Mṣb, Ḳ.) — And hence, (Mṣb, TA,) اختصر الكَلَامَ ‡ He abridged the language, or the discourse; syn. أَوْجَزَهُ: (Ṣ, A, Ḳ:) [and in like manner, الكِتَابَ the book, or writing:] or, accord. to some, the latter (اوجزه) signifies "he expressed its correct meaning concisely, without regard to the original words;" and the former, he curtailed its words, preserving the meaning: (MF:) or properly, he abridged the expressions, making the words fewer, but preserving the entire meaning: (Mṣb:) or he abridged the language by omitting superfluities, and choosing from it concise expressions which conveyed the meaning. (L.) [You say, اختصرهُ عَلَى الرُّبْعِ † He reduced it by abridgment to the fourth of its original bulk.] And اختصر السَّجْدَةَ † He recited the chapter in which a prostration should be performed, omitting the verse requiring prostration, in order that he might not prostrate himself: or he recited only the verse requiring a prostration, to prostrate himself in so doing: both which practices are forbidden. (T,* Mgh,* Mṣb,* Ḳ.) And the verb alone † He recited a verse, or two verses, of the last part of the chapter, in prayer; (Ḳ;) not the whole chapter. (TA.) — Also, the verb alone, He curtailed a thing of its superfluities, (Ḳ,) in a general sense. (TA.) — And اختصر فِى الجَزِّ, (JK, Ḳ, TA,) in some copies of the Ḳ فِى الحَزِّ, with ح, (TA,) or اختصر الجَزَّ, (A,) He did not extirpate in cutting; did not cut off entirely, or utterly: (A, Ḳ:) or he extirpated in cutting; cut off utterly. (JK.) ═ اختصر also signifies He took a مِخْصَرَة [in his hand]: (Ṣ,* Ḳ:) and ↑تخصّر بِهَا he took it in his hand; namely, a مخصرة: (Ḥar p. 122:)

or the former, *he leaned upon it in walking:* (TA:) or *he took* a مخصرة or a staff *in his hand, to lean upon it.* (Mgh.) You say also, اختصر العَنَزَةَ [*He took in his hand the* عنزة: *or he leaned upon the* عنزة *in walking*]: it is a thing [i. e. a kind of staff, or short spear,] like the عُكَّازَة: and in like manner, ▼ تخصّر; as in the L &c.: (TA:) and اختصر بِالعَصَا *He leaned upon the staff in walking.* (A.)

خَصْرٌ The *middle,* or *waist,* of a man or woman: (Ṣ, A, Mṣb, Ḳ;) i. e. the *slender part above the hips* or *haunches:* (Mṣb:) pl. خُصُورٌ. (A, Ḳ.) See also الخَاصِرَةُ, in two places. — ‡ The *hollow part of the sole* of the foot, *which does not touch the ground:* (A, Ḳ:) pl. as above. (Ḳ.) — ‡ The *narrow part* of a sandal, *before the* أُذُنَانِ [which are the *two loops whereto is attached the strap that passes behind the wearer's heel*]: (TA:) or خَصْرَانِ [the dual] signifies the *narrow part* of a sandal. (IAạr, TA.) — ‡ The *part which is between the base of the notch and the feathers* of an arrow: (AḤn, A,* Ḳ:) pl. as above. (Ḳ.) — ‡ *A way between the upper and lower parts* of a heap of sand: (Ḳ, TA:) or ‡ the *lower part* of a heap of sand; the *thin part* thereof; as also ▼ مُخْصَرٌ: (A, TA:) pl. as above. (Ḳ.) — † The *place of the* بُيُوت [or *tents*] *of the Arabs of the desert:* (Ḳ:) or, as some say, of such بيوت, *a clean place:* (TA:) pl. as above. (Ḳ.)

خَصَرٌ *Cold* (Ṣ, Ḳ) *which a man feels in his extremities.* (TA.)

خَصِرٌ, applied to a day, *Painfully cold.* (A, TA.) — *Cold,* as an epithet, (Ṣ, Ḳ,) applied to water, (Ṣ,) and to anything. (TA.) — A man *feeling cold* [especially *in his extremities:* see 1]: to signify cold and hungry, the epithet خَرِصٌ is used. (A'Obeyd.) — ثَغْرٌ خَصِرٌ [*A mouth,* or *front teeth,*] *cold,* or *cool, in the place that is kissed.* (A, TA. [See also مُخَصَّرٌ.])

خُصَّيْرَى, (Ḳ, TA,) in some copies of the Ḳ خُصَيْرَى, (TA,) [but the former is shown to be the right reading by a verse cited in the TA,] The *curtailment of the superfluities* of a thing; like اِخْتِصَارٌ. (Ḳ,* TA.)

الخَاصِرَةُ [*The flank;* i. e. *each of the ilia;*] *i. q.* الشَّاكِلَةُ; (Zj, in his "Khalḳ el-Insán;" Ṣ, Ḳ;) i. e. *the* طَفْطَفَة [or *quivering flesh*] *of the side, that reaches to the extremities of the ribs:* (Zj, ibid.:) and [so in the Ḳ, but more properly "or,"] الخَاصِرَةُ, (Ḳ,) or الخَاصِرَتَانِ (JK, TA) and ▼ الخَصْرَانِ, (TA,) *what is between the* حَرْقَفَة [or *crest of the hip*] and the lowest rib; (JK, Ḳ, TA;) i. e. *the part from which retires each of the lowest ribs, and in advance of which projects each of the* حَجَبَتَانِ: [explained by the words ما قلص عنه القُصَيْرَيَانِ وتقدّم من الحجبتين: but for من الحجبتين, I read مِنْهُ الحَجَبَتَانِ; referring, for corroboration, to explanations of this last word; and therefore I have rendered the passage as above: the meaning seems evidently to be *the part between the lowest rib and the crest of the hip, on each side:*] the thin skin which is above the خَصْر is called the طَفْطَفَة: so in the M, agreeably with the saying of Ibn-El-Ajdábee, that ▼ الخَصْرُ and الخَاصِرَةُ are syn.; i. e., in this sense: [this assertion, however, requires consideration; for all the explanations of الخاصرة are easily reconcileable:] pl. خَوَاصِرُ [which is also used in the sense of the sing. or dual]. (TA.) You say رَجُلٌ ضَخْمُ الخَوَاصِرِ [*A man large in the flank* or *flanks*]: and Lḥ mentions the phrase إِنَّهَا لَمُنْتَفِخَةُ الخَوَاصِرِ [*Verily she is inflated,* or *swollen, in the flank* or *flanks*]; as though the term خاصرة were applicable to every portion [of the flank]. (TA.) — Also *A pain in the* خَاصِرَة [or *flank*]: or *in the kidneys.* (TA.) — And it is also said to signify *A certain vein* (عِرْق) *in the kidney, which occasions pain to the person when it is in motion.* (TA.)

خِنْصَرٌ: see art. خنصر.

أَخْصَرُ [*Shorter:* and *shortest*]. You say, هٰذَا أَخْصَرُ مِنْ ذَاكَ *This* [road] *is shorter than that.* (A.) But this is irregular; أَخْصَرُ being formed from اُخْتُصِرَ, a verb of more than three letters. (I'Aḳ p. 237.)

مِخْصَرَةٌ *A thing like a whip:* and *anything that a man takes* (يَخْتَصِرُ) *with his hand, and holds, such as a staff* and *the like:* (Ṣ:) *a thing which a man takes in his hand, and upon which he leans, such as a staff* and *the like:* (Ḳ,* TA:) *a rod* [or *sceptre*] *which a king used to take in his hand, with which he made signs,* or *pointed, in holding a discourse,* or *addressing,* (A, Ḳ,*) *and accompanied what he said,* (A,) *and in like manner the* خَطِيب *in reciting a* خُطْبَة: (Ḳ,* TA:) it was *one of the insignia of kings:* (TA:) *a rod,* or *what is termed* عَنَزَة, or *the like, with which the* خطيب *makes signs,* or *points, in addressing the people:* (Mṣb:) *a thing which a man holds in his hand, such as any of the things termed* عَصًا *and* مِقْرَعَة *and* عَنَزَة *and* عُكَّازَة *and* قَضِيب, or *the like; and upon which he sometimes leans:* (A'Obeyd:) pl. مَخَاصِرُ. (Ṣ, TA.)

مُخْصَرٌ, applied to a man, (TA,) *Slender* (Ḳ, TA) *in the waist:* (TA:) *lean,* or *lank in the belly:* (Ḳ:) or, *in the* خَاصِرَة [or *flank*]: (TA:) and ▼ مَخْصُورُ البَطْنِ is also applied to a man [as meaning *lank in the belly*]. (A, TA.) — كَشْحٌ مُخْصَرٌ *A thin* [*flank* or rather *waist:* see a verse of Imra-el-Ḳeys cited voce مُذَلَّل]. (Ṣ, A, Ḳ.) — قَدَمٌ مُخَصَّرَةٌ (JK, A, TA) and ▼ مَخْصُورَةٌ (JK, TA) ‡ [*A foot that touches the ground with its fore part and heel; the middle of the sole being hollow and narrow:* this meaning, or a meaning similar to that of يَدٌ مُخَصَّرَةٌ explained below, seems to be indicated in the TA: the latter is the meaning accord. to the JK; but this I think doubtful, on account of what here follows]. مُخَصَّرُ القَدَمَيْنِ means ‡ A man *whose feet touch the ground with the fore part and the heel; the middle of the sole being hollow and narrow:* (Ṣ, Ḳ:) and you say also ▼ مَخْصُورُ القَدَمَيْنِ. (A, TA.) — يَدٌ مُخَصَّرَةٌ, or ▼ مَخْصُورَةٌ, (as in different copies of the Ḳ,) or both, (TA,) ‡ *An arm,* or *a hand, in the wrist of which is what is termed* ▼ تَخْصِير, *as though it were bound:* or *which has an encircling groove-like depression.* (Ḳ, TA.) — نَعْلٌ مُخَصَّرَةٌ ‡ *A sandal narrow in the middle.* (Ṣ,* A,* Ḳ, TA.) — See also خَصْرٌ. = ثَغْرٌ بَارِدُ المُخَصَّرِ [*A mouth,* or *front teeth,*] *cold,* or *cool, in the place that is kissed.* (TA. [See also خَصِرٌ.])

مَخْصُورٌ A man *having a complaint of,* or *a pain in, his* خَصْر [or *waist*], or *his* خَاصِرَة [or *flank*]. (TA.) — See also the next preceding paragraph, in four places.

مَخَاصِرُ pl. of مِخْصَرَةٌ. (Ṣ, TA.) = مَخَاصِرُ الطَّرِيقِ *The nearest roads or ways;* (Ḳ;) as also ▼ المُخْتَصَرَاتُ: (TA:) or مُخْتَصِرَاتُ الطُّرُقِ signifies *The roads,* or *ways, that are near, notwithstanding their ruggedness, but not so easy as those that are longer.* (L.)

مُخْتَصِرَاتُ الطُّرُقِ, or المُخْتَصَرَاتُ: see the paragraph next preceding.

المُتَخَصِّرُونَ, (Ḳ,) or المُتَخَصِّرُونَ فِى الصَّلَاةِ, (Mgh,) *Those who, in praying in the night, becoming tired thereby, put their hands upon their* خَوَاصِر [or *flanks*]: of such it is said (in a trad., IAth, Ḳ) that light shall be [seen] on their faces (IAth, Mgh, Ḳ) on the day of resurrection: (IAth, Ḳ:) [in other cases, this action is forbidden, or disapproved: see 8:] or, in the instance mentioned above, it may mean *those who shall rest upon their righteous works* on the day of resurrection: (IAth, Mgh, TA:) this latter is apparently the right meaning: otherwise, two trads. contradict each other. (MF.)

خصف

1. خَصْفٌ [inf. n. of خَصَفَ] signifies The act of *adjoining,* and *putting together.* (TA.) — Hence, (TA,) خَصَفَ, (Ṣ, Mṣb, Ḳ, TA,) aor. ـِ, (Mṣb, Ḳ,) inf. n. خَصْفٌ, (Mṣb,) *He sewed* a sole (Ṣ, Ḳ, TA) [*so as to make it double*], *covering,* or *facing, one piece with another:* (TA:) or *he patched* a sole; *mended it by sewing on another piece.* (Mṣb.) And *He made* anything *double, putting one piece upon another; he faced it.* (TA.) — And [hence,] خَصَفَ عَلَى نَفْسِهِ, (JK,) or خَصَفَ الوَرَقَ عَلَى بَدَنِهِ, (Ṣ,* Ḳ,) aor. as above, (Ṣ, TA,) and so the inf. n.; (TA;) and ▼ اختصف; (Ṣ, Ḳ;) and ▼ اخصف; (Ḳ;) and ▼ خصّف, inf. n. تَخْصِيف; (TA;) ‡ *He stuck* [or *sewed*] *the leaves together, one to another,* (Ṣ, Ḳ,* TA,) *and covered his person with them, leaf by leaf,* (Ḳ,) *to conceal therewith his pudenda:* (Ṣ, TA:) or the first phrase, (JK,) as also ▼ اختصف, (Lth, JK,) signifies *he* (a naked man) *put upon his pudenda wide leaves,* (Lth, JK,) *or the like:* (Lth:) you say, اختصف ▼ بِكَذَا [*he covered his pudenda with such a thing*]. (Lth, JK.) It is said in the Ḳur

[vii. 21 and xx. 119], وَطَفِقَا يَخْصِفَانِ عَلَيْهِمَا مِنْ وَرَقِ الجَنَّةِ; and ▼ يَخِصِّفَانِ, originally يَخْتَصِفَانِ, by some pronounced ▼ يَخَصِّفَانِ, (Ṣ, TA,) and by some, ▼ يَخْصِّفَانِ, with two quiescent letters together; (TA; [but this appears to be incorrect; see 8 in art. خصم;]) and ▼ يُخْصِفَانِ, from أَخْصَفَ; and ▼ يُخَصِّفَانِ, from خَصَّفَ; (Ksh and Bḍ in vii. 21, and TA;) thus accord. to different readings; i. e. ‡ *And they betook themselves to sticking* [or *sewing*] *together, one to another, of the leaves of Paradise, to conceal therewith their pudenda.* (Ṣ, TA.) And hence, also, the saying, in a trad., إِذَا دَخَلَ أَحَدُكُمُ الحَمَّامَ فَعَلَيْهِ بِالنَّشِيرِ وَلَا يَخْصِفْ ▼, i. e. ‡ [*When any one of you enters the bath,*] *let him take the waist-wrapper, and not put his hand upon his pudendum*: and like this in meaning is تخصفه [app. a mistranscription for ▼ يَتَخَصَّفُ, or ▼ يَخْصِفُ or the like, for يَخْتَصِفُ; if not, it must be ▼ تَخَصَّفَهُ, meaning *he put his hand upon it*]. (TA.) — [Hence also the saying,] فَمَا زَالُوا يَخْصِفُونَ أَخْفَافَ المَطِيِّ بِحَوَافِرِ الخَيْلِ حَتَّى لَحِقُوهُمْ ‡ *And they ceased not to make the prints of the feet of the camels to be covered by the prints of the hoofs of the horses* [*until they overtook them*]; as though they sewed these upon the others, like as one sews a sole by covering, or facing, one piece with another. (TA.) — And خُصِفَتِ الكَتِيبَةُ مِنْ وَرَائِهَا بِخَيْلٍ, i. e. † [*The body of troops*] *was followed* [*by horsemen*]. (Ṣ.) — And خَصَفَ, aor. ـِ, inf. n. خَصْفٌ, † *He lied.* (Munjid of Kr. [See خَصَّافٌ.]) — And خَصَفْتُ فُلَانًا ‡ *I exceeded such a one in reviling* [as though adding reviling upon reviling]. (TA.) ═ خَصَفَتْ, (AZ, Ṣ, Ḳ,) aor. ـِ, (AZ, Ṣ,) inf. n. خِصَافٌ, said of a she-camel, *She cast her young one in the ninth month*: (AZ, Ṣ, Ḳ:) the epithet applied to her in this case is ▼ خَصُوفٌ: (AZ, Ṣ:) or, as some say, (Ṣ,) this epithet signifies one *that brings forth a year and a month,* (Ṣ, and so in some copies of the Ḳ,) in [some of] the copies of the Ḳ *a year and two months,* which is wrong, (TA,) *after the time when she was covered*: (Ṣ, Ḳ:) جَزُورٌ is applied to one that brings forth a year and two months after that time: (Ṣ, TA:) or ▼ the former epithet signifies one *that brings forth on the completion of the year*: (IAạr, TA:) or one *of the camels termed* مَرَابِيع [pl. of مِرْبَاعٌ q. v.] *that brings forth at the completion of the year*; or one *of such camels that bring forth when she comes to the time of the year in which she was covered, completely*: (TA:) and ▼ اختصفت signifies *she* (a camel) *became such as is termed* خَصُوف. (JK, TA.)

2: see 1, in three places. — [From the primary signification of the verb is derived the phrase,] خَصَّفَهُ الشَّيْبُ, inf. n. تَخْصِيفٌ, ‡ *Hoariness rendered his hair white and black in equal proportions*; (IAạr,* Ḳ,* TA;) *syn. with* خَوَّصَهُ, inf. n. تَخْوِيصٌ; *and with* ثَقَّبَ فِيهِ, inf. n. تَثْقِيبٌ. (IAạr.) And خَصَّفَ الشَّيْبُ لِمَّتَهُ ‡ *Hoariness rendered* ▼ خَصِيف [i. e. *white and black*] *his* لِمَّة [or *hair hanging down below his ears*]. (A, TA.)

4: see 1, in two places.

5: see 1, in two places.

8. اختصف, and three variations of the aor.: see 1, in seven places: ═ and اختصفت said of a she-camel: see 1, last sentence.

خَصْفٌ *A sole having another sole sewed upon it*; (Ṣ, Ḳ;) and so ▼ نَعْلٌ خَصِيفٌ, (Ṣ,* TA,) *i. q.* ▼ مَخْصُوفَةٌ. (Ḳ.)

خَصَفٌ † *A mixed colour, black and white.* (Freytag, from the Deewán of the Hudhalees.) ═ See also خَصَفَةٌ, in two places. ═ Also a dial. var. of خَزَفٌ [q. v.]. (Lth, TA.)

خَصْفَةٌ *Any sole,* or *matching piece, that is sewed upon a sole* [*so as to make it double*]; (JK, Ṣ, Ḳ;) as also ▼ خَصِيفَةٌ. (Ṣ* and Ḳ voce طِرَاقٌ.)

خُصْفَةٌ *A puncture,* or *stitch-hole, in a skin*; syn. خُرْزَةٌ. (Ḳ.) — And [hence,] † The *anus,* or *orifice of the rectum*: and † the *orifice of the vagina.* (TA voce خُرْبَةٌ.)

خَصَفَةٌ *A receptacle for dates, such as is termed* جُلَّة, (Ṣ, Mgh, Mṣb, Ḳ,) *made of palm-leaves*; (Ṣ, Ḳ;) *wherein they are stored*: of the dial. of El-Baḥreyn: (TA:) and *a mat upon which* أَقِط *&c. are put to dry*: (TA in art. شر:) and [it is said to signify] *a very thick kind of cloth*: (Lth, Ḳ:) pl. ▼ خَصَفٌ, (Ṣ, Ḳ,) [or rather this is a coll. gen. n.,] and [the pl. properly speaking is] خِصَافٌ: (Ṣ, Mṣb, Ḳ:) Lth says that a certain Tubbaʼ [a king of El-Yemen] clothed the House [i. e the Kaạbeh] with ▼ خَصَف, meaning *very thick cloths*; so called as being likened to the خَصَف of woven palm-leaves: but Az says that this is wrong; and that it means *pieces of matting made of palm-leaves woven together, oblong pieces of which were used as coverings for the tents of the Arabs of the desert, and sometimes made into* جِلَال [pl. of جُلَّةٌ] *for dates*: (TA:) ▼ خَصَّافٌ, also, signifies *a piece of matting of palm-leaves*; and its pl. is خَصَاصِيفُ. (JK.)

خَصُوفٌ: see 1, last sentence, in two places. Applied to a woman, One *who brings forth in the ninth* [*month*], *not entering upon the tenth.* (TA.)

خَصِيفٌ: see خَصْفٌ. — † A thing *in which are united any two colours.* (Ṣ, TA.) See also 2. And see أَخْصَفُ in two places. [Hence,] † *Ashes*; (Ḳ;) because there are two colours therein, blackness and whiteness: but one says more commonly رَمَادٌ خَصِيفٌ, using the latter word as an epithet. (TA.) And كَتِيبَةٌ خَصِيفٌ, (Ṣ,) or كَتِيبَةٌ خَصِيفَةٌ, (Ḳ,) † [*A body of troops*] *having two colours,* (Ḳ,) *having the colour of iron* (Ṣ, Ḳ) *and another colour*: (Ḳ:) or so called because of the rust of the iron &c.: (L:) or the former phrase means, as some say, *followed by horsemen*; and therefore the epithet is without ة, because it has the signification of a pass. part. n.: for were it to denote the colour of the iron, they had said خَصِيفَةٌ, because it would in this latter case have the signification of an act. part. n. (Ṣ.) — Also † *Fresh milk upon which is poured* رَائِب [i. e. *curdled,* or *thick,* or *churned, milk*]: (Ṣ, Ḳ:) if dates and clarified butter are put into it, it is [termed] عَوْبَثَانِيٌّ. (Ṣ.)

خَصِيفَةٌ [fem. of خَصِيفٌ, q. v. — And also a simple subst.]: see خَصْفَةٌ.

خَصَّافٌ One *who sews soles* [*so as to make them double, covering, or facing, one piece with another*: see 1]: (Kr, Ḳ:) or one *who patches soles*; *who mends them by sewing on other pieces.* (Mṣb.) — ‡ One *who covers his pudendum with his hand*: on the authority of Seer. (TA. [See 1.]) — ‡ *A liar*: (Kr, Ḳ, TA:) as though he sewed one saying upon another, and [thus] embellished it. (TA.)

خَصَّافَةٌ: see خَصَفَةٌ.

أَخْصَفُ † *Of a colour like that of ashes, in which are blackness and whiteness*; (JḲ, Ṣ;) as also ▼ خَصِيفٌ. (JK.) In this sense, (TA,) applied to a mountain, (Ṣ, Ḳ,) as also ▼ خَصِيفٌ, (TA,) and to a male ostrich, meaning † *In which are blackness and whiteness*: (Ṣ, Ḳ:) fem. خَصْفَاءُ. (TA.) † A rope, or cord, *of two colours, having one strand black and another strand white.* (JK.) † A horse, and a sheep, *white in the flanks*; (Ṣ, Ḳ:) the rest being of any colour: and sometimes *in one side*: (TA:) or *whose* بَلَق [or *blackness and whiteness*] *extends from his belly to his sides*: (Ṣ, TA:) or a horse *white in the side.* (Mgh.)

مِخْصَفٌ The *awl*; or *instrument for boring,* or *perforating*; (JK, TA;) *used in the sewing of soles* [*and the like;*] (JK;) *i. q.* إِشْفَى [q. v.]: (Ṣ, Mṣb, TA:) [pl. مَخَاصِفُ.]

مَخْصُوفَةٌ, applied to a sole: see خَصْفٌ. — Applied to a ewe or she-goat, † *Smooth*: or *of two colours, black and white*: (Ḳ, TA:) so in the O. (TA.)

خصل

1. خَصَلَ, (Ḳ,) inf. n. خَصْلٌ, (TA,) *He cut,* or *cut off,* a thing; (Ḳ;) as also قَصَلَ. (TA.) [Accord. to the TA, this is the proper, or primary signification.] — خَصَلَهُمْ, [aor., accord. to rule, ـُ,] inf. n. خَصْلٌ and خِصَالٌ, *He overcame them, or surpassed them, in shooting.* (Ṣ, Ḳ. [In the CḲ, فَضَلَهُمْ is erroneously put for نَضَلَهُمْ.]) See also 3.

2. خَصَّلَهُ, inf. n. تَخْصِيلٌ, *He cut it,* or *divided it, in pieces.* (M, Ḳ.) — خَصَّلَ الشَّجَرَ, (Ḳ,) inf. n. as above, (TA,) *He lopped the branches of the trees*: (Ḳ, TA:) or تَخْصِيلٌ signifies the *cutting off slender extremities and branches from the* [*species of mimosa called*] عُرْفُط, *in the interior parts thereof.* (JK.) — خَصَّلَ البَعِيرَ *He cut off, for the camel, the* خَصْلَة, (Ḳ,) i. e. the *soft and tender branch of a tree.* (TA.)

3. خَاصَلْتُهُمْ فَخَصَلْتُهُمْ ▼, inf. n. of the former مُخَاصَلَةٌ, *I vied, competed,* or *contended for superiority, with them in shooting, and I overcame them,* or *surpassed them, therein.* (TA.)

4. اخصل *He* (a shooter) *hit the target:* (Ḳ, TA:) or *made his arrow to fall close by the side of the target.* (JK, Ḳ.)

6. تخاصلوا *They vied, competed,* or *contended for superiority, in shooting:* (Az, TA:) or *they contended together for stakes,* or *wagers, laid by them to be taken by the winner in shooting.* (Ṣ, Ḳ.)

خَصْلٌ *A stake,* or *wager, laid in a shooting-match.* (Ṣ, TA.) One says, أَحْرَزَ خَصْلَهُ and أَصَابَ خَصْلَهُ [*He won his stake,* or *wager;*] *he overcame* (Ṣ, Ḳ, TA) *in the case of laying stakes* or *wagers* [*in a shooting-match*]. (TA.) — And *A thing for which persons contend together in a game of hazard.* (Ḥar p. 640.) — See also خَصْلَةٌ, in two places.

خَصِلٌ [One *who overcomes much,* or *often, in shooting-matches:* occurring in the Deewán of the Hudhalees: expl. by Freytag as meaning *multum vincens in ludo alearum*].

خَصْلَةٌ *i. q.* خَلَّةٌ: (Ṣ, Ḳ:) i. e. *A property, quality, nature,* or *disposition:* and *a habit,* or *custom:* (KL, PṢ, TḲ:) [and *a practice,* or *an action:* it is used in these various senses in different trads.: in one trad., avarice is termed a خَصْلَة; and so is evilness of nature: in another, fasting, and praying: in another, the inflicting of castigation, and the executing of retaliation, in a mosque:] it signifies *an excellent quality* or *the like;* and *a low, base,* or *mean, quality* or *the like;* (Ḳ, TA;) in a man: (TA:) or its predominant application is to *an excellent quality* or *the like:* (Ḳ:) so in the M: (TA:) [it is said that] it is used only in commendation; whereas خَلَّةٌ is used in relation to good and evil: (Ḥam p. 525:) [but this is a mistake, as I have shown above:] accord. to Az, it signifies the *states,* or *conditions, of things* or *affairs:* (TA:) [or this is a signification of the pl.:] the pl. is خِصَالٌ (Ḳ) and خَصَلَاتٌ; (TA;) [and ▾خَصَائِلُ is a pl. pl., i. e. pl. of خِصَال, with which it is explained in the KL as syn.: see an ex. in a verse cited voce دَوْنٌ.] — *A hitting of the target;* (Ḳ;) in shooting: (TA:) or, (Ḳ,) as also ▾خَصْلٌ, (JK, Ḳ,) in a shooting-match, (JK,) it is [*a shot*] *in the case in which the arrow goes close by the target:* (JK, Ḳ:*) thus accord. to Lth, who says that the former explanation is erroneous; (TA;) [as appears also from the assertion that] what are termed خَصْلَتَانِ, in a shooting-match, are reckoned as equivalent to a shot that goes right to the target. (T, Ḳ, TA.) — And accord. to Ṣgh, *A single act of overcoming in a shooting-match.* (TA.) — Also, and ▾خُصْلَةٌ, *A raceme,* or *bunch,* of grapes or the like; syn. عُنْقُودٌ. (Ḳ.) — And (both words) *A stick, branch,* or *twig,* (عُودٌ,) *in which are thorns.* (Ḳ.) — And خَصْلَةٌ and ▾خُصْلَةٌ, or this latter only, The *extremity of a fresh, pliant, soft,* or *tender, twig,* or *rod:* (Ḳ,* TA:) and (some say, TA) *a soft and tender twig* or *rod, of the* [*species of mimosa called*] عُرْفُط: (Ḳ, TA:) and ▾خَصْلٌ [of which خَصْلَةٌ is the n. un.] signifies the *slender extremities and branches of the* عُرْفُط: (JK:) and ▾خُصْلَةٌ, *a soft and tender branch of any tree:* (T, TA:) and [its pl.] خُصَلٌ, the *pendent extremities of trees.* (Ṣ, TA.)

خُصْلَةٌ *A* لَفِيفَة, (Ṣ,) [i. e.] *a lock,* or *flock,* (PṢ,) or *a plexus,* (KL,) or *a quantity collected* [or *hanging*] *together,* (Ḳ,) *of hair,* (Ṣ, Ḳ, KL, PṢ,) and *of wool,* (PṢ, and Ṣ and Ḳ in art. جز,) *&c.:* (PṢ:) or *a small quantity of hair;* as also ▾خُصَيْلَةٌ, (Ḳ,) as in the M: pl. خُصَلٌ. (TA.) — See also خَصْلَةٌ, in two places. — Also *A portion of flesh forming a distinct limb* or *member* or *organ* (عُضْوٌ مِنَ ٱللَّحْمِ). (Ḳ.)

خَصَلَةٌ: see خَصْلَةٌ.

خَصِيلٌ: see خَصِيلَةٌ, in two places. — Also *Overcome* [*in a shooting-match,* or] *in a contest for stakes* or *wagers.* (JK, Ḳ.) — And *A tail;* (Ḳ, TA;) as, for instance, of a [wild] bull. (TA.)

خُصَالَةٌ a dial. var. of خُصَالَةٌ, (JK, Ḳ, TA,) meaning The *remains of wheat in the sieve, after the sifting, with what are mixed therewith:* but the latter word is the more known. (JK, TA.)

خَصِيلَةٌ *A piece,* or *portion, of flesh,* (M, Ḳ,) *small* or *large:* (M, TA:) or the *flesh of the thighs* and *of the upper arms* and *of the fore arms:* (Ḳ:) or *any portion of flesh, by itself,* of *the flesh of the thighs* and *of the upper arms* (JK, T, Ṣ, TA) and *of the shanks* and *of the fore arms:* (JK, T, TA:) or the *portion of flesh of the thigh:* (TA:) or *any compact and long portion of flesh, in the arm* or *elsewhere;* also called خَبِيبَةٌ: (AO, TA in art. خب:) or (Ḳ, TA, but in the CḲ "and") [*any muscle, of those that are termed voluntary muscles;* as also عَضَلَةٌ and عَضِيلَةٌ; i. e.] *any tendon,* or *sinew, upon which is thick flesh:* (Ḳ:) or *any portion of flesh that is oblong, and intermixed with tendons,* or *sinews:* (O, TA:) or, as some say, the طِفْطِفَة [or *flank,* &c.]: (TA:) pl. ▾خَصِيلٌ [or rather this is a coll. gen. n.] and [the pl. is] خَصَائِلُ. (Ḳ.) A certain person has described a horse as being سَبْطُ ▾الخَصِيلِ [app. meaning *Lank in the muscles;* or *long and even therein*]: and sometimes خَصَائِل is used in relation to a man. (TA.) — See also خَصْلَةٌ. — And for the pl. خَصَائِلُ see also خَصْلَةٌ.

مِخْصَلٌ *A very sharp* sword (JK, Ṣ, Ḳ) &c.: (M:) a dial. var. of مِقْصَلٌ. (Ṣ.) مِخْصَلٌ is said by A'Obeyd to be a mistranscription for مِخْضَلٌ; but AḤei and others authorise it.

مِخْصَالٌ *A* مِنْجَل [or *reaping-hook*]: (Ḳ:) or *an instrument with which the branches of trees are lopped,* (JK, Ibn-'Abbád, TA,) *like the* فَأْس. (Ibn-'Abbád, TA.)

خصم

1. خَصَمَ, aor. ـِ, [inf. n., app., خَصْمٌ,] *He contended in an altercation, disputed,* or *litigated, in a valid,* or *sound, manner.* (Mṣb.) — See also 3: — and 8.

3. خاصمهُ, inf. n. مُخَاصَمَةٌ (Ṣ, Mṣb, Ḳ) and خِصَامٌ (Ṣ, Mṣb) and [quasi-inf. n.] خُصُومَةٌ, (Ḳ,) the last said in the Ṣ to be a simple subst., (TA,) *He contended with him in an altercation, disputed with him,* or *litigated with him;* (Ḳ, TA;) *i. q.* نَازَعَهُ: (Mgh and Mṣb and Ḳ in art. نزع:) accord. to El-Ḥarállee, الخِصَامُ signifies *the saying which the listener is made to hear, and which is made to enter his ear-hole, such as may cause him to refrain,* or *desist, from his assertion, and his plea,* or *claim.* (TA.) You say, خَاصَمْتُهُ ▾فَخَصَمْتُهُ, (Ṣ, Mgh, Mṣb, Ḳ,*) aor. of the latter أَخْصِمُهُ, with kesr, (Ṣ, Ḳ,*) or أَخْصُمُهُ, with ḍamm, (Mgh, Mṣb,) or not with ḍamm, (Ṣ,) or both these forms of the aor. are used, accord. to AḤei; the latter agreeable with analogy; (MF;) the former anomalous; for the regular aor. of an unaugmented sound verb in a case of this kind is with ḍamm, (Ṣ, Ḳ,) as in the instance of عَالَمْتُهُ فَعَلَمْتُهُ, aor. أَعْلُمُهُ; (Ṣ;) if it has not a faucial letter (Ṣ, Ḳ) for its medial radical, (Ḳ,) in which case it is with fet-ḥ, as in the instance of فَاخَرْتُهُ فَفَخَرْتُهُ, aor. أَفْخَرُهُ, (Ṣ, Ḳ,) accord. to the opinion of Ks, but this is contr. to the opinion generally held: (MF:) the inf. n. of خَصَمْتُهُ is خَصْمٌ: (Ṣ,* TA:) and the meaning is, [*I contended with him in an altercation,* or *I disputed,* or *litigated, with him, and*] *I overcame him in the altercation,* &c. (Mgh, Mṣb.) — خاصمهُ also signifies *He put it in,* or *by, the* خُصْم, i. e. *edge,* or *side, of the bed.* (TA.)

4. اخصمهُ *He dictated to him his plea against his adversary in an altercation* or *a dispute* or *litigation,* (JK, TA,) *whereby he might overcome the latter.* (JK.)

6: see the next paragraph, in two places.

8. اختصموا *They contended in altercation, disputed,* or *litigated, one with another;* (Mṣb, TA;) *i. q.* ▾تخاصموا; (Ṣ, Ḳ, TA;) both signifying as above. (TA.) He who reads يَخِصِّمُونَ [in the Ḳur xxxvi. 49] means يَخْتَصِمُونَ; changing the ت into ص, and incorporating [it into the other ص], and transferring its vowel to the خ: some read يَخِصِّمُونَ, without transferring that vowel; (Ṣ, Ḳ;) because a quiescent letter, when it is made movent, is [regularly] made so with kesr: (Ṣ:) AA slurred the vowel of the خ: the pronunciation [يَخْصِّمُونَ] with two quiescent letters together is incorrect: (Ṣ, Ḳ:) Ḥamzeh read ▾يَخْصِمُونَ, (Ṣ,) i. e., with the خ quiescent and with kesr to the ص. (TA.) — اِخْتَصَمَا إِلَيْهِ *They two applied to him for the decision of a cause, each of them claiming the right.* (TA in art. قمط.) And تُخُوصِمَ إِلَيْهِ [*An application was made to him by litigants for the decision of a cause*]. (Mgh in art. دلو.) — السَّيْفُ يَخْتَصِمُ جَفْنَهُ, said by J to signify *The sword cuts* (lit. *eats*) *its scabbard, by reason of its sharpness,* is a mistake; the verb being correctly with ض, (Ḳ,* TA,) dotted. (TA.)

خَصْمٌ *An adversary in contention or altercation, in dispute, or in litigation; an antagonist; a litigant:* (JK, Ḳ, TA:) as also ▾خَصِيمٌ: (JK, Ṣ, Ḳ, TA:) the former is used alike as masc. and fem. (Ṣ, Mṣb, Ḳ) and sing. (JK, Ṣ, Mṣb, Ḳ) and dual (Mṣb, Ḳ) and pl.; (JK, Ṣ, Mṣb, Ḳ;) because it is originally an inf. n.: (Ṣ, TA:) [see an ex. of its use in a pl. sense in a verse cited voce جَنَفٌ:] but it also has the dual form, خَصْمَانِ; (Ṣ, Mṣb;) and the pl. خُصُومٌ (JK, Ṣ, Mṣb, Ḳ) and خِصَامٌ (Mṣb) and perhaps أَخْصَامٌ, [which is a pl. of pauc.,] or this may be pl. of خَصِمٌ: (TA:) the pl. of ▼خَصِيمٌ is خُصَمَآءُ (JK, Ṣ, Ḳ) and خُصْمَانٌ. (Ḳ.)

خُصْمٌ The *side* (Ṣ, Ḳ) of anything; (Ṣ, TA;) as, for instance, of a load such as is called عِدْل; (Ṣ;) and of a bed; and the *edge* thereof: (TA:) written by Aboo-Moosà with ض; but IAth says that it is correctly with ص: (TA in art. خضم:) *a lateral part* or *portion* (Ṣ, Ḳ) of anything: (Ṣ:) *a corner*, (Ṣ, Ḳ,) as well as *a side*, (Ṣ,) of an عِدْل, and of a receptacle, such as a خُرْج or a جُوَالِق or an عَيْبَة: (Ṣ:) and the [*anterior lower*] *extremity* of a [water-bag of the kind termed] رَاوِيَة, *that is opposite to the* عَزْلَآء; (JK, Ḳ, TA; [in the CḲ, الزّاوِيَة and العَزْلاء are erroneously put for الرَّاوِيَة and العَزْلَآء;]) the upper extremity [correctly extremities, at which are the loops whereby it is suspended upon the side of the camel,] being called the عصم [i. e. عُصْم, pl. of عِصَامٌ]: (TA:) pl. [of pauc.] أَخْصَامٌ and [of mult.] خُصُومٌ: (Ḳ:) but some say that the أخصام of the [water-bag termed] مَزَادَة, and its خصوم, are its *corners*: the خصوم of a cloud are its *sides*: (TA:) and أَخْصَامُ العَيْنِ signifies *the part*, or *parts*, *of the eye upon which the edges of the lids close together*. (Ṣ, Ḳ.) — [Also *A gap*, or *an intervening space*: it is said in the TA that] الأَخْصَامُ [pl. of الخُصْمُ] signifies الفرج [i. e. الفُرَجُ, pl. of الفُرْجَةُ: and it is added,] one says, of an unsound, a corrupt, or a disordered, affair, لَا يُسَدُّ مِنْهُ خُصْمٌ إِلَّا انْفَتَحَ خُصْمٌ آخَرُ ‡ [*A gap of it will not be stopped up but another gap will open*]; occurring in a trad., meaning, the state of affairs is disordered and distressing, and not to be rectified and repaired. (TA.) — [The pl.] خُصُومٌ also signifies The *mouths of valleys*. (JK, Ḳ.) — And The *lower parts*, or *stocks*, syn. أُصُولٌ, (JK, Ḳ,) of [trees of the kind called] سَرَحَات [pl. of سَرْحَة]; used in this sense by Eṭ-Ṭirimmáḥ. (JK.)

خَصِمٌ *Vehement in altercation* or *dispute* or *litigation*; (Ṣ, Ḳ,* TA;) as also ▼خَصُومٌ: (Ḥam p. 628:) [or each signifies *contentious*, *disputatious*, or *litigious*:] or the former, *knowing*, or *skilled, in altercation &c., though not practising it*: (IB, TA:) or *valid*, or *sound, therein*; as also ▼خَصِيمٌ: (Mṣb:) or this last signifies one *who contends with another in an altercation, disputes with him, or litigates with him*: (IB, TA:) the pl. of the first is خَصِمُونَ, (Ḳ,) occurring in the Ḳur xliii. 58; and perhaps أَخْصَامٌ, or this may be a pl. of خَصْمٌ. (TA.)

خُصْمَةٌ *A certain bead*, or *gem*, or *the like, used by men* [*as an amulet*], in the Ḳ, مِنْ خُرُوزِ الرِّجَالِ, but correctly, as in the M, مِنْ خَرَزِ الرِّجَالِ, (TA,) *worn on the occasion of contending in an altercation, or disputing, or litigating, or on going into the presence of the Sulṭán*; (Ḳ, TA;) *and sometimes it is beneath the gem of the man's signet-ring, when it is small; and it may be in his button; and sometimes they put it in the* ذُؤَابَة [*or cord by which the hilt is occasionally attached to the guard*] *of the sword*: (TA:) also called خُضْمَة. (Ḳ and TA in art. خضم.)

خُصْمَةٌ and خُصْمَانِيَّةٌ: } see خُصُومَةٌ.

خُصُومٌ: see خَصِمٌ.

خَصِيمٌ: see خَصْمٌ, in two places; and خَصِمٌ.

خُصُومَةٌ *Contention* or *altercation*; *disputation*; *litigation*; (Ḳ, TA;) a subst. from 3 (Ṣ, TA) or 8 (JK,* TA) and 6, as also ▼خُصْمَةٌ and ▼خُصْمَانِيَّةٌ. (TA.) فَصْلُ الخُصُومَةِ: see art. فصل. [See also an ex. voce حُكْمٌ.]

أُخْصُومٌ The *loop of the* [*sack called*] جُوَالِق, (JK, TA,) and *of the* [*load called*] عِدْل; (TA;) i. q. أُخْسُومٌ; (Ḳ;) but the latter is a dial. var. of weak authority, and disapproved. (TA in art. خسم.)

خصو

خُصْوَةٌ a dial. var. of خُصْيَةٌ, q. v. (Sh, TA.)

خصى

1. خَصَاهُ, (Ṣ, Mgh, Mṣb, Ḳ,) aor. يَخْصِيهِ, (JK, Mgh, Mṣb,) inf. n. خِصَآءٌ (JK, Ṣ, Mgh, Mṣb, Ḳ) and خَصَآءٌ, mentioned, by MF, from Expositions of the Fṣ, (TA,) and خَصْىٌ, agreeably with analogy, occurs in a trad. of Esh-Shaabee, though we have not heard it, (Mgh,) *He drew forth*, or *extracted, his testicles*; (Ṣ, Mgh, Mṣb, Ḳ;) [*he gelded*, or *castrated, him*;] namely, a stallion, (Ṣ,) a sheep or goat, or a horse or similar beast, (Lth, JK, TA,) and a man or boy, (TA,) or a slave. (Mṣb.) One says, بَرِئْتُ إِلَيْكَ مِنَ الخِصَآءِ [*I am irresponsible to thee for castration*]. (Ṣ.) — [Hence,] كَانَ جَوَادًا فَخُصِىَ [lit. *He was a fleet and excellent horse, and he was gelded*]; meaning ‡ *he was rich, and he became poor*. (TA.) — The poets term satire, and the act of overcoming, خِصَآء: one of them says,

> خَصَيْتُكَ يَا ابْنَ حَمْزَةَ بِالقَوَافِى
> كَمَا يُخْصَى مِنَ الحَلَقِ الحِمَارُ

[‡ *I have emasculated thee, O son of Hamzeh, with rhymes, like as the he-ass is emasculated in consequence of the disease termed* حَلَق; for which, it is asserted, (as is said in the TA, art. حلق,) there is no remedy but gelding]. (IB, TA.)

4. اخصى [as though meaning † *He did away with that which rendered him like one emasculated*; the ا, app., having a privative property;] ‡ *he learned one science*. (Ṣgh, Ḳ, TA.) — The use of إِخْصَآء [its inf. n.] in the sense of خِصَآء inf. n. of خَصَى is a mistake. (Mgh.)

8. اختصى *He castrated himself*; or *made himself a eunuch*. (KL.)

خَصٍ *Having a complaint of his* خُصْى [or *testicles*]. (Ḳ.)

خُصْىٌ; and the dual خُصْيَانِ: see خُصْيَةٌ, in five places.

خِصْىٌ: see what next follows.

خُصْيَةٌ *A testicle*; (El-Umawee, Ṣ;) sing. of خُصًى; (Ṣ, Mgh, Mṣb, Ḳ;) it is [*one*] *of the organs of generation*; (Ḳ;) *well known*; (Mṣb;) and ▼خِصْيَةٌ signifies the same, (Ṣ, Ḳ,) and so does ▼خُصْىٌ, (Mṣb, Ḳ,) and ▼خِصْىٌ; (Ḳ;) and خُصْوَةٌ is a dial. var., occurring in a trad., but is extr.: (Sh, TA:) accord. to some, (Mṣb,) the sing. is خُصْيَةٌ [alone], (T, Mṣb,) of the fem. gender; (T, TA;) and the dual is ▼خُصْيَانِ, (El-Umawee, T, Ṣ, Mgh, Mṣb, Ḳ,) of the masc. gender, (T, TA,) without ت, (El-Umawee, Ṣ, Mgh, Mṣb,) irreg., (El-Umawee, Ṣ, Mṣb,) like أَلْيَانِ dual of أَلْيَةٌ, (El-Umawee, Ṣ,) and خُصْيَتَانِ also, (T, Mgh, Ḳ,) this latter being sometimes used, (T, Mgh,) though rarely, (MF on the authority of the Expositions of the Fṣ,) both mentioned by ISh: (T, TA:) AO says, I have heard خُصْيَة, with ḍamm, but I have not heard خِصْيَة, with kesr; and I have heard ▼خُصْيَاهُ [as the dual], though they did not use ▼خُصْىٌ as the sing.; (Ṣ;) IB, however, cites exs. of this last as a sing.: (TA:) AA says that الخُصْيَتَانِ signifies *the two testicles*; and ▼الخُصْيَانِ, *the two skins* [*which compose the scrotum*, i. e.,] *in which are the two testicles*; citing an ex. in which the latter dual is used in this sense; (Ṣ;) and ISk says the like; whereas IḲooṭ makes الخُصْيَة to signify [*the scrotum*, i. e.] *the skin containing the testicle*. (Mṣb.) — Also ‡ *An earring* (قُرْط) *in the ear*: (JK, Ṣgh, Ḳ:) thus called by way of comparison: (TA:) pl. خُصًى. (JK.)

خِصْيَةٌ: see the next preceding paragraph, in two places.

خَصِىٌّ *Whose testicles have been drawn forth*, or *extracted*; (Ṣ, Mgh, Mṣb, Ḳ;) [*gelded*, or *castrated*; *a eunuch*;] applied to a sheep or goat, and a horse or similar beast, (TA,) and a man (Ṣ, TA) or boy, (TA,) or a slave; (Mṣb;) as also ▼مَخْصِىٌّ: (Ḳ:) pl. خِصْيَانٌ (Ṣ, Mgh, Ḳ) and خِصْيَةٌ: (Ṣ, Ḳ:) in giving it the former pl., they liken it to a subst., like ظَلِيمٌ, of which the pl. is ظِلْمَانٌ: so says Sb; meaning that فِعْلَانٌ is generally the pl. of فَعِيلٌ as a subst. (TA.) One says also خَصِىٌّ نَصِىٌّ; using the latter word as an imitative sequent. (Lḥ, TA.) — Also ‡ Poetry *in which is no amatory effusion*. (Ḳ, TA.)

خَاصٍ [act. part. n. of 1]. They say, جَآءَ كَخَاصِى العَيْرِ [lit. *He came like the gelder of the ass*], meaning *he came ashamed*: (JK, and TA in art. جوح:) and also, *disappointed*, or *unsuccessful*. (TA in that art.)

مَخْصًى The *place of cutting* [or *gelding* or *castration*]. (Ṣ.)

خَصِىٌّ : see مَخْصِىٌّ.

خض

R. Q. 1. خَضْخَضَهُ, (Ṣ, O,) inf. n. خَضْخَضَةٌ, (Ṣ, O, Ḳ,) *He agitated it; moved it about; stirred it; shook it;* namely, water, (Ṣ, O, Ḳ,) and the like, (Ṣ, O,) and سَوِيق, and the like. (O, Ḳ.) You say also, خَضْخَضْتُ دَلْوِى فِى المَآءِ [*I agitated,* or *dashed about, my bucket in the water*]. (O, TA.) Ṣakhr-el-Gheí El-Hudhalee says, describing a water to which he had come for drink,

* فَخَضْخَضْتُ صُفْنِى فِى جَمِّهِ *
* خِيَاضَ المُدَابِرِ قِدْحًا عَطُوفَا *

(O, TA,) i. e. [*And I dashed about my leathern bucket in the main body thereof,*] *as he who is overcome in the game of el-meysir introduces among the other arrows a borrowed arrow* [*that comes forth winning*], in the luck of which he has confidence. (TA in art. خوض.) This verb, though mentioned here, is [said to be] from خَاضَ, aor. يَخُوضُ; not from خَضَّ; therefore the poet here uses خِيَاض for its inf. n. (O, TA.) You also say, جَآءَهُ بِالخَنْجَرِ فَخَضْخَضَ بِهِ بَطْنَهُ [*He came to him with the dagger, and stirred about with it his belly*]. (TA.) — [Hence a meaning of the inf. n. explained in the first paragraph of art. جلد.]

R. Q. 2. تَخَضْخَضَ *It* (water, and the like, Ṣ) *became agitated, moved about, stirred,* or *shaken.* (Ṣ, Ḳ.)

خضب

1. خَضَبَ, (Ṣ, A, Mṣb, Ḳ,) aor. ـِ, (Mṣb, Ḳ,) inf. n. خَضْبٌ (Ṣ, Mṣb) [and accord. to MF خِضَابٌ and خُضُوبٌ, but respecting these two inf. ns. (the latter of which seems to be peculiar to the intrans. verb خَضَبَ see what follows], *He coloured,* or *tinged,* (A, Ḳ,) a thing; (Ṣ, A,* Ḳ;*) or *changed* it in colour to red, or yellow, &c.; (TA;) and ▼خَضَّبَ signifies the same, [but app. in an intensive sense, or as applying to a number of objects, (see its pass. part. n. voce خَضِيبٌ,)] (Ḳ,) inf. n. تَخْضِيبٌ: (TA:) and the former, particularly, *he tinged,* or *dyed,* his white hair, (TA,) or the hand, &c., (Mṣb,) with ḥinnà: (Mṣb, TA:) but when a man has dyed his hair with any other dye than ḥinnà, you say, صَبَغَ شَعْرَهُ: (Mṣb, TA:) or you say also, خَضَبَ بِالسَّوَادِ [*He dyed* his hair *with black*]. (Suh, TA.) When one does not mention the hair (Mṣb, TA) or the white hair [&c.], (Mṣb,) he says خَضَبَ, inf. n. خِضَابٌ; (IKṭṭ, Mṣb;) and ▼اختضب, (Ṣ, IKṭṭ, Mṣb, TA,) and ▼تخضّب; (A, TA;) [meaning *He dyed his hair,* &c.,] with ḥinnà, (Ṣ, IKṭṭ, Mṣb, TA,) and the like: (Ṣ, TA:) and in like manner one says of a woman, خَضَبَتْ, aor. ـِ; and ▼اختضبت: (TA:) which last also signifies [particularly] *She dyed her hands* with ḥinnà. (T, TṢ, TA, in art. غمس.) — Hence, in a trad., بَكَى حَتَّى خَضَبَ دَمْعُهُ الحَصَى ‡ *He wept so that his tears wetted the pebbles:* or, more probably, *so that his tears became red, and dyed the pebbles:* (IAth, TA:) [or most probably, *so that his tears caused the pebbles to appear of a reddish colour;* for such is commonly the case when pebbles are wetted.] — خَضَبَ, aor. ـِ; and خَضِبَ, aor. ـَ; and خُضِبَ; inf. n. of each خُضُوبٌ; and ▼اخضوضب; ‡ *It* (a tree) *became green.* (Ḳ, TA.) And خَضَبَ, inf. n. خُضُوبٌ, † *Its small leaves came forth in the spring, and its twigs lengthened;* said of the قَتَاد, at the commencement of its vegetation; and likewise of the عَوْسَج and عَرْفَج; but of no other tree of the kind called عِضَاه: or said also of the عُرْفُط and سَمُر; meaning † *it dropped its leaves, and became red and yellow:* (TA:) and you say also, خَضَبَتِ العِضَاهُ ‡ *the* عضاه *became green, and broke forth;* (A;) or *the sap of the* عضاه *flowed in their branches, and they became green;* as also ▼أَخْضَبَتْ, (TA,) for which اخصبت, with the unpointed ص, is said by Az to be a gross mistranscription; explained by Lth, on whose authority it is written with ص [as also in the Ḳ in art. خصب,] as meaning *the sap flowed in the branches of the* عضاه *so as to reach the roots.* (T and TA in art. خصب.) And خَضَبَ النَّخْلُ, (Ṣ, Ḳ,) inf. n. خَضْبٌ, (Ḳ,) *The palm-trees,* (Ṣ,) or *the spadices of the palm-trees,* (Ḳ,) *became green.* (Ṣ, Ḳ.) And خَضَبَتِ الأَرْضُ, (A, Ḳ,) inf. n. خَضْبٌ; (TA;) and ▼اخضبت, (Ḳ,) inf. n. إِخْضَابٌ; (TA;) or ▼اختضبت; and ▼تخضّبت; (A;) *The earth,* or *land, exposed to view,* (A,) or *produced,* (Ḳ,) *its herbage,* (A, Ḳ,) *and it* (the latter) *became green.* (TA.)

2: see 1, first sentence.

4: }
5: } see 1, each in two places.

8: see 1, in three places.

12: see 1.

خَضْبٌ The *colour of a tree,* or *of the spadix of a palm-tree, when it becomes green:* pl. خُضُوبٌ. (Ḳ.) — A plant *fresh,* or *new,* and *green in consequence of rain;* as also ▼خُضُوبٌ: (Ḳ:) or *watered by rain, and imparting a colour to the ordure:* (TA:) or the *green colour that appears in trees when they begin to put forth their leaves:* (Ḳ:) pl. خُضُوبٌ. (TA.)

خَضَبَةٌ *A spadix of a palm-tree:* خَصَبَةٌ, [q. v.,] with the unpointed ص, is erroneously said to have this signification. (TA.)

خُضَبَةٌ A woman *who uses* خِضَاب *for herself* [i. e. *for dyeing her hair or hands* &c.] *much,* or *often.* (Ṣ, A, Ḳ.)

خِضَابٌ *Ḥinnà* (حِنَّآء), and *the like:* (Mṣb:) or the *thing with which one dyes,* or *tinges, his,* or *her, hair* &c.; (Ṣ, Ḳ, TA;) *such as* حِنَّآء *and* كَتَم *and the like.* (TA.)

خُضُوبٌ : see خَضْبٌ.

خَضِيبٌ Anything *dyed, tinged,* or *changed in colour;* [generally, with ḥinnà;] as also ▼مَخْضُوبٌ: the former is both masc. and fem.: and its pl. is خُضُبٌ. (TA.) You say كَفٌّ خَضِيبٌ (Ṣ, A, Ḳ) and ▼خَاضِبٌ (TA voce ضَارِبٌ) [*A hand dyed* with ḥinnà]: and ▼مَخْضُوبٌ and بَنَانٌ خَضِيبٌ (Ḳ) and ▼مُخَضَّبٌ (Ṣ, A, Ḳ) [*fingers,* or *fingers' ends, dyed* with ḥinnà]; but the last of these has an intensive signification. (Ṣ.) — And hence, (TA,) الكَفُّ الخَضِيبُ † *A certain star;* (Ṣ, A, Ḳ;) *the star β of Cassiopeia;* (so in the Egyptian almanacs;) [i. e.] *the bright star of the constellation called* ذَاتُ الكُرْسِىِّ; which star is [termed] the extended right hand of الثُّرَيَّا [or the Pleiades; corresponding to the star called الكَفُّ الجَذْمَآءُ]. (Kzw. [See أَجْذَمُ.]) — And امْرَأَةٌ خَضِيبٌ [*A woman having her hands,* or *feet,* or *hair,* &c., *dyed* with ḥinnà or the like]. (Ḳ.)

خَاضِبٌ A man *dyeing,* or *who dyes, his hair* with ḥinnà. (Mṣb.) — See also خَضِيبٌ. — Also ‡ A male ostrich (Ṣ, A, Ḳ, &c.) *whose shanks* (A, Ḳ) *and legs* (A) *have become red,* (A, Ḳ,) or *green,* [app. meaning *of a dark,* or *an ashy, dust-colour,*] or *yellow,* (A,) *in consequence of his lusting after the female,* (A, Ḳ,) or *in consequence of his having eaten the* [*herbage termed*] رَبِيع: (A:) or *the front edges of whose shanks have become red,* (Ṣ, Ḳ,) or *green,* (Ḳ,) or *yellow, in consequence of his having eaten the* [*herbage termed*] ربيع: (Ṣ, Ḳ:) or *whose beak and shanks have become red from his having eaten the* [*herbage termed*] ربيع: in the summer (الصَّيْف) he becomes bald (يقرع), and his shanks become white: (L:) or *whose shanks have become green by reason of lust in the* [*season termed*] رَبِيع: (ADk:) accord. to some, (TA,) it is applied only to the male ostrich: (Ṣ, Ḳ:) but some explain it without this restriction; and Lth mentions [the fem.] خَاضِبَةٌ as applied to an ostrich: [it is said that] the skin of the neck, and that of the breast, and that of the thighs, of the male ostrich, but not his feathers, become intensely red when he lusts after the female: or, as some say, خاضب signifies an ostrich *that has eaten green food:* (TA:) or *the extremities of whose feathers are dyed by* [*the eating of*] *blossoms, and the slender parts of whose legs have become red by the same cause:* accord. to an Arab of the desert, supposed to be Aboo-Kheyreh, in the [season termed] ربيع, when it eats أَسَارِيع [app. meaning certain worms so called], its legs and beak assume the red hue of the عُصْفُر [or safflower]: (AḤn, L:) or خاضب is applied to a male ostrich *the slender parts of whose legs become red when the dates begin to become red, and cease to be so when the redness of the dates ceases:* (AḤn, Ḳ:) so that it is not from eating اساريع, which, it is said, no ostrich is known to eat: accord. to Aṣ, the cause [of the redness above mentioned] is only the dye of blossoms; but were it so, the bird would also become yellow, and green, &c., [and some assert that it partially does, as has been shown above,] accord. to the colours of the blossoms and herbs;

and the green colour would predominate: [but, as the Arabs say, this requires consideration:] whatever be the cause, the bird, it is said, is termed خاضب on account of the redness that affects its shanks: and this word is [said to be] an epithet used as a proper name of the bird: (AḤn, L:) but this is a mistake, unless it mean that, because of its prevailing application, it is used in the same manner as الحَرِثُ and العَبَّاسُ, not that it may be used [in a determinate sense] without the article ال: (L:) the pl. is خَوَاضِبُ. (TA.) It is also said to be applied as an epithet to Any animal *that eats* خَضْب [q. v.]: (TA:) and particularly to [the species of bovine antelope called] the wild bull (الثَّوْرُ الوَحْشِىُّ). (L.) — [See also a saying of Dukeyn cited voce رَاوُوقٌ.]

مِخْضَبٌ *I. q.* مِرْكَنٌ, (Ṣ, Ḳ,) or إِجَّانَةٌ: (A:) or *a vessel resembling that called* إِجَّانَةٌ, *in which clothes are washed.* (TA.) — مَخَاضِبُ [is its pl.; and also] signifies The *rags of the* خِضَاب [or *ḥinnà* or *the like*]: (A:) [or] *of the* حَيْض [or *catamenia*]. (TA.) [If these two significations be correct, the latter is app. tropical: but حيض may be a mistranscription for خضاب.]

مُخْتَضِبٌ: see خَضِيبٌ.

مَخْضُوبٌ: see خَضِيبٌ, in two places.

خضد

1. خَضَدَ, aor. ـِ, (L, Ḳ,) inf. n. خَضْدٌ, (L,) *He broke* wood, or a branch, or twig, whether moist or dry, (L, Ḳ,) or a soft thing, (L,) *so that its parts did not separate.* (L, Ḳ.) — *It* (carriage) *broke in pieces* fruit. (A.) — *He bent,* (Ṣ, A, L,) *without breaking,* (Ṣ, L,) wood, or a branch, or twig. (Ṣ, A, L.) You say also, خَضَدَ البَعِيرُ عُنُقَ آخَرَ (L, Ḳ) *The camel broke,* (L,) or *bent,* (Ḳ,) *the neck of another* camel: (L, Ḳ:) or ‡ *the camel fought another* camel. (Lth, A.) — Also, (Ṣ, Ḳ,) inf. n. as above, (Ṣ,) *He cut,* or *cut off,* (Ṣ, Ḳ,) anything moist, or soft, or supple; and so ▾ خضّد, inf. n. تَخْضِيدٌ. (Ṣ.) And خَضَدَ الشَّجَرَ *He cut off,* (Ṣ, A, Ḳ,) or *pulled off,* or *removed,* (Fr, Zj,) *the thorns of the trees.* (Fr, Zj, Ṣ, A, Ḳ.) — Also, (A, Ḳ,) aor. and inf. n. as above, (Ṣ,) ‡ *He* (a man, Ḳ) *ate vehemently:* (Ṣ, A, Ḳ:) or [*he craunched* a thing;] *he ate something moist, or soft, or supple;* as a cucumber, and a carrot, (Ḳ,) and the like: (TA:) and † *he* (a horse) *ate in the manner termed* خَضْم, (L,) or قَضْم. (TA.) An Arab of the desert, who liked the cucumber, being asked what pleased him therein, answered, خَضْدُهُ [app. meaning *The craunching thereof*]: (Ṣ, L:) or he was asked what pleased him thereof, and answered, خَضَدُهُ, meaning *what is broken in pieces, thereof.* (A.) ═ خَضِدَ, [aor. ـَ,] (TA,) inf. n. خَضَدٌ, (Ḳ,) *It* (fruit) *became shrunk and shrivelled.* (Ḳ, TA.)

2: see 1.

5: see 7, in two places. — [Also, app., † *He affected an inclining of his body, or a bending,* or *he inclined his body, or bent, from side to side, by reason of languor;* syn. تَثَنَّى: (see its part. n., voce أَخْضَدُ:) as said of a drunken man, in the Deewán of the Hudhalees, Freytag renders it *fractus fuit et quasi fractus corruit.*]

7. انخضد *It* (wood, or a branch, or twig, whether moist or dry,) *broke so that its parts did not separate;* as also ▾ تخضّد. (L, Ḳ.) — *It* (fruit, A, Ḳ, fresh, or moist, TA, being carried from one place to another, A, TA) *became broken in pieces,* (A, Ḳ,) or *crushed;* (Ḳ;) and so ▾ تخضّد. (A.) — *It* (wood, or a branch, or twig,) *bent without breaking* (AZ, Ṣ, L) *asunder.* (AZ, L.)

خَضَدٌ *Feebleness and weakness* in a plant. (Ḳ.) — † *A malady* (L, Ḳ) in a man, (L,) *affecting the limbs, not amounting to a fracture;* as also ▾ خُضَادٌ. (L, Ḳ.) — † *Languor and pain of the body, with laziness.* (L.) — خَضَدُ السَّفَرِ ‡ *The fatigue and weariness that are occasioned* to a man *by travel.* (TA.) ═ *Whatever is cut off* from wood, or a branch, or twig, (Ṣ, A, L, Ḳ,) that is moist, or soft, or supple: (Ṣ, L, Ḳ:) or *what breaks in pieces,* or *broken pieces,* of trees; as also ▾ مَخْضُودٌ: (Ḳ:) or *what is cut off, and removed,* of trees: and *broken pieces, heaped up,* of the papyrus, and of any pieces of wood, or of branches, or twigs, that are moist, or soft, or supple: (L:) and *broken pieces* of cucumbers [&c.]. (A.) ═ *A certain plant:* (Ḳ:) or *a certain soft,* or *flaccid, kind of tree, without thorns.* (L.) [See also خُضَادٌ.]

خَضِدٌ † *Lacking power to rise,* (Ḳ, TA,) *from languor of the body, and pain, with laziness;* (TA;) as also ▾ مَخْضُودٌ. (Ḳ.)

خُضَادٌ: see خَضَدٌ. ═ Also *A certain kind of tree,* (Ṣ, Ḳ,) *soft,* or *flaccid, and without thorns:* (Ṣ:) [see also خَضَدٌ:] or, *of the kind of trees called* جَنْبَة, *the leaves of which have edges like those of the* حَلَقَة, *which are pulled with the hand like* حلفاء. (L.)

خَضُودٌ A horse *that eats in the manner termed* قَضْم. (TA. [See 1.])

خَضِيدٌ, applied to wood, or a branch, or twig, whether moist or dry, *Broken so that its parts are not separated;* as also ▾ مَخْضُودٌ. (L.) — Also, and ▾ مَخْضُودٌ (Ṣ, A) and ▾ مُخَضَّدٌ, (A,) A tree, (Ṣ,) or lote-tree, (A,) *having its thorns cut off,* (Ṣ, A,) or *pulled off,* or *removed:* (Fr and Zj in explanation of the second word as occurring in the Ḳur lvi. 27:) or the second, in the Ḳur (ubi suprà), may mean *having the branches bent by reason of the abundance of the fruit.* (Bḍ.)

أَخْضَدُ and ▾ مُتَخَضِّدٌ *i. q.* مُتَثَنٍّ [app. as meaning † *Affecting an inclining of the body, or a bending, or inclining the body, or bending, from side to side, by reason of languor:* see 5; and see also خَضَدٌ and خَضِدٌ]. (Ḳ.)

مِخْضَدٌ ‡ *A vehement eater:* (A, L, Ḳ:) one *who eats with coarseness, or rudeness, and quickness.* (L.)

مُخَضَّدٌ: see خَضِيدٌ.

مَخْضُودٌ: see خَضِيدٌ, in two places: — and خَضَدٌ.

مُتَخَضِّدٌ: see أَخْضَدُ.

يَخْضُودٌ: see خَضَدٌ.

خضر

1. خَضِرَ: see 9, in two places. ═ خَضَرَ: see 8, in two places.

2. خَضَّرَهُ, [inf. n. تَخْضِيرٌ,] *He rendered it* أَخْضَر [i. e. *green,* &c.]. (Ṣ.) — [Hence,] it is said in a trad., إِذَا أَرَادَ ٱللّٰهُ بِعَبْدٍ شَرًّا خَضَّرَ لَهُ فِى اللَّبِنِ وَالطِّينِ حَتَّى يَبْنِىَ, (TA,) i. e. † [*When God desires evil to befall a man,*] *He makes him to have pleasure in unburnt bricks and clay, so that he may build,* and thus be diverted from the things of the world to come, if his building be beyond his need, or not such a structure as a mosque or the like. (Marginal note in a copy of the "Jámi' eṣ-Ṣagheer" of Es-Suyootee.) [Hence also,] خُضِّرَ لَهُ فِيهِ, inf. n. as above, ‡ *He was blessed in it.* (L, Ḳ.) You say, مَنْ خُضِّرَ لَهُ فِى شَىْءٍ فَلْيَلْزَمْهُ, (L,) or مَنْ خُضِرَ مِنْ شَىْءٍ فليلزمه, (so in a copy of the Mgh,) i. e. † *Whosoever is blessed in a thing,* (Mgh, L,) meaning an art or a trade or traffic, or a means of subsistence, *let him keep to it.* (L.)

3. خاضرهُ, (TḲ,) inf. n. مُخَاضَرَةٌ, (Ṣ, A, Mgh, Mṣb, Ḳ,) *He sold to him fruits before they were in a good,* or *sound, state:* (A:) or *before their goodness,* or *soundness, became apparent:* (Ṣ, Mgh, Mṣb, Ḳ, TḲ:) the doing of which is forbidden: (Ṣ:) accord. to some, (TA,) the prohibition includes the sale of fresh ripe dates, [app. if not fully ripe,] and herbs, or leguminous plants, and the like; and therefore some disapprove of selling a greater quantity of fresh ripe dates than is cut at once. (Ṣ.)

4. اخضر *It* (plenty of moisture) *rendered* seed-produce *soft,* or *tender.* (TA.)

8. اختضر *He cut* herbage, (Ṣ, Ḳ,) or a tree, (A,) *while it was green;* (Ṣ, A, Ḳ;) as also ▾ خَضَرَ, (A, TA,) aor. ـُ, inf. n. خَضْرٌ. (TA.) And أُخْتُضِرَ *It* (herbage, TA) *was taken,* (Ḳ,) *and pastured upon,* (TA,) *while fresh and juicy,* (Ḳ,) *and green, before it had attained its full height.* (TA.) See also 9, last sentence. — Hence, (Ṣ, TA,) the pass. form, † *He died in his youth;* (Ṣ, Ḳ;) *in his fresh and flourishing state.* (Ṣ.) Young men used to say to an old man, أَجْزَرْتَ يَا شَيْخُ † [Thou hast attained to the time for dying, (lit. for being cut,) O old man]: and he replied, أَىْ بَنِىَّ وَتُخْتَضَرُونَ † [*O my sons, and ye shall be cut off,* or *die, in your youth*]. (Ṣ. [See also أَجْزَرَ.]) — Also, the act. v., *He cut off* the green branches of a palm-tree with his مِخْلَب; (TA;) and so ▾ خَضَرَ, (Ḳ, * TA,) aor. ـُ, inf. n. خَضْرٌ: (TA:) and *he cut off* a thing, as a man's nose, *entirely:* (TA:) or, simply, *he cut off* a man's nose. (IAạr.) — And *He ate fruit* [*while it was green,* or] *before it was ripe.* (A.) — And hence, (TA,) ‡ *He deflowered* a girl: (Ḳ, TA:) or, *before she had attained to puberty;* (Mṣb in art. قض, and Ḳ;) as also ابتسر and ابتكر. (TA.) — Also † *He took* a camel *in a refractory state, not trained, and attached the nose-rein to him, and drove*

him. (TA.) — And † *He took up* a load, or burden. (Ḳ.)

9. اخضرّ, (Ṣ, A, Ḳ,) inf. n. اِخْضِرَارٌ; (Ṣ, A;) and ▼ اخضوضر, (Ṣ, Ḳ,) [inf. n. اِخْضِيضَارٌ, in the TA written by mistake اِخْضِيرَارٌ;] and ▼ خَضِرَ, aor. ـَ, (Mṣb, Ḳ,) inf. n. خَضَرٌ; (Mṣb;) *It* (a colour, Mṣb, or seed-produce, Ḳ) *was,* or *became, of the colour termed* خُضْرَة [i. e. *green:* and *he,* (a camel, and a horse, and an ass, and sometimes a bird,) and *it,* (a garment of the kind called كِسَآء, and the like, or any other thing,) *was,* or *became, of a dark,* or *an ashy, dust-colour;* or *dingy ash-colour;* or *dark dust-colour:* and *he,* (a man,) or *it,* (a thing,) *was,* or *became, of a tawny,* or *brownish, colour;* or *blackish;* or *of a blackish hue inclining to green;* or *black;* or *intensely black:* see خُضْرَة and أَخْضَرُ]. (Ṣ, A, Mṣb, Ḳ.) [Hence,] اخضرّ إِزَارِى (*The place of*) *my* ازار *became black:* or, rather, *became of a [blackish] hue inclining to green:* because the hair when it first grows is of that hue. (Ḥar p. 494.) And اخضرّ شَارِبُهُ [*His mustache grew so as to appear dark*]; said of a boy; a phrase similar to بَقَلَ وَجْهُهُ. (Mgh in art. بقل.) And اخضرّ اللَّيْلُ ‡ *The night became dark and black.* (Ḳ,* TA.) And اخضرّت الظُّلْمَةُ ‡ *The darkness became intensely black.* (A.) — اخضرّ جِلْدُهُ [properly *His skin became green* from carrying the produce of his land; meaning] ‡ *he became in a state of plenty.* (TA. [See هُمْ خُضْرُ المَنَاكِبِ, voce أَخْضَرُ.]) — اخضرّ said of seed-produce, *It was,* or *became, soft,* or *tender;* as also ▼ اخضوضر; and ▼ خَضِرَ, aor. ـَ, (TA,) inf. n. خَضَرٌ. (Ḳ,* TA.) == اخضرّ and ▼ اِخْتَضَرَ, (Ḳ,) or this may be of the pass. form, [أُخْتُضِرَ,] so as to agree with what occurs before, [see 8,] (TA,) *It* (herbage, TA) *was,* or *became, cut.* (Ḳ,* TA.)

12: see 9, first sentence: — and last sentence but one.

خَضْرٌ Trees (شَجَرٌ) *that are soft,* or *tender, when cut;* as also ▼ مَخْضُورٌ. (TA.)

خَضْرٌ: see خُضَارَةٌ.

خَضْرٌ [*i. q.* خَضِرٌ]. You say, أَخَذَهُ خَضْرًا مَضْرًا, and ▼ خَضِرًا مَضِرًا, *He took it without price:* or *in its fresh,* or *juicy, state:* (Ḳ:) مضرا being an imitative sequent. (TA.) Whence the saying, الدُّنْيَا خَضِرَةٌ ▼ مَضِرَةٌ [in the Ṣ حُلْوَةٌ خَضِرَةٌ] *The goods of this world are delicate, fresh, and pleasant:* or *pleasing.* (TA.) And ▼ الغَزْوُ حُلْوٌ خَضِرٌ [*Predatory warfare is sweet and*] *fresh* [or *refreshing*] *and loved;* because of the victory and spoil attending it. (TA, from a trad. of Ibn-'Omar [which see fully quoted voce الثُّمَامُ].) — You say also, هُوَ لَكَ خَضْرًا مَضْرًا *It is thine,* or *for thee: may it be attended with enjoyment and a wholesome result.* (Ḳ.) — And ذَهَبَ دَمُهُ خَضْرًا مَضْرًا, (Ṣ, Ḳ,) and ▼ خَضِرًا مَضِرًا, (Ḳ,) *His blood went unrevenged,* or *unretaliated,* or *unexpiated by a mulct:* (Ṣ, Ḳ:) مضرا being an imitative sequent [here as in the former instance]. (TA.)

خَضَرٌ inf. n. of خَضِرَ: [see 9, first sentence: — and last sentence but one; and] see also خُضْرَةٌ. == Also *Green palm-branches with the leaves upon them:* and *green palm-branches stripped of their leaves:* (Fr, Ḳ:) pl. أَخْضَارٌ. (AḤn.)

خَضِرٌ: see أَخْضَرُ. — Also A place *having much verdure;* and so ▼ يَخْضُورٌ and ▼ مَخْضَرَةٌ. (Ḳ.) And أَرْضٌ خَضِرَةٌ and ▼ يَخْضُورٌ *Land in which is much verdure:* and ▼ ارض مَخْضَرَة, as in the Ḳur xxii. 62, accord. to one reading, *verdant land.* (TA.) — See also خَضْرٌ, in four places. — Also, [as a subst.,] *What is green:* (Akh, Ṣ, and Bḍ in vi. 99:) *seed-produce;* (Lth, Bḍ, Ḳ;) and so ▼ خُضَارَى: (Ṣ:) so the former in the Ḳur ubi suprà: (Lth, Bḍ:) or *goodly green herbage:* (A:) and *a branch:* (Ḳ:) *any branch.* (TA.) — And الخَضِرُ *The plant called* ▼ البَقْلَةُ الخَضْرَآءُ; as also ▼ الخَضِرَةُ and ▼ الخَضِيرُ (Ḳ) and ▼ الخُضْرَةُ: (TA:) it is *a green and rough herb* or *leguminous plant, the leaves and fruit of which are like those of millet; it rises to the height of a cubit; and fills the mouth of the camel.* (TA.) Also *A species of plant of the kind called* جَنْبَة; (Ḳ;) which latter term is applied to herbage whereof the root is deep in the earth, like the نَصِىّ and صِلِّيَان: (TA:) [a coll. gen. n.:] n. un. with ة: (Ḳ:) it is *not of the slender and succulent herbs* or *leguminous plants, which dry up in summer.* (TA.) Hence آكِلَةُ الخَضِرِ, occurring in a trad., [properly signifying A she-camel *that eats the plant above mentioned,*] applied to a man who acts justly and moderately with respect to worldly enjoyments: for the خضر is not of the slender and succulent herbs, as above observed, nor of those excellent plants which the spring produces by its consecutive rains, and which therefore become goodly and soft or tender; but of those upon which beasts pasture after others have dried up, because they find no others, and which the Arabs call جَنْبَة; and the beasts do not eat much of it, nor do they find it wholesome. (IAth, TA.)

خَضِرَةٌ [if not a mistranscription for خُضْرَةٌ] *Fresh cut herbage, to be eaten quickly.* (TA.)

خُضْرَةٌ [*Greenness; a green colour; verdure;*] *a certain colour,* (Ṣ, A, Ḳ,) *well known;* (Ḳ;) [and] *a colour between black and white:* it is in plants and in animals &c., and, accord. to IAạr, in water also: (TA:) in camels, (Ṣ,) and horses, (Ṣ, Ḳ,) [and asses, and sometimes in birds, and in a garment of the kind called كِسَآء, and the like, and in other things, *a dark,* or *an ashy, dust-colour;* or *dingy ash-colour;* or *dark dust-colour;*] *a dust-colour intermixed with* دُهْمَة [i. e. *blackness* or *deep ash-colour*]: (Ṣ, Ḳ:) in men, [and in other things,] *a tawny,* or *brownish, colour;* syn. سُمْرَة: (Ṣ:) [and *a blackish hue:* and *a blackish hue inclining to green:*] and *blackness:* (TA:) [and *intense blackness:* see 9; and see also أَخْضَرُ:] pl. خُضَرٌ and خُضْرٌ. (Ḳ.) — And *A green plant:* pl. خُضَرٌ: (TA:) or the latter signifies *herbs,* or *leguminous plants;* as though pl. of the former. (Mṣb.) [See خُضَارٌ: and خَضِرَةٌ.] — See also خَضِرٌ. — Also *Softness,* or *tenderness,* (IAạr, Ḳ,) of seed-produce [and the like]; (TA;) and so ▼ خَضَرٌ, (Ḳ,) inf. n. of خَضِرَ. (TA.) — And *What is soft,* or *tender; fresh,* or *juicy; and pleasant to the eater.* (TA, from a خُطْبَة of 'Alee, delivered at El-Koofeh.)

الخَضِرَةُ: see خَضِرٌ.

خُضْرِيَّةٌ A palm-tree (نَخْلَةٌ) *that bears good green dates.* (Az, Ḳ.) — *A kind of dates, green, resembling glass, of a colour that is admired.* (AḤn.)

خُضَارٌ *Herbs,* or *leguminous plants, in the first state of their growth.* (Ṣ,* Ḳ,* TA.) [See also خُضْرَةٌ.] — Also *Milk mixed with much water:* (Ṣ, Ḳ:) AZ says that it is *like* سَمَارٌ, meaning as above, *diluted so as to be of a dark,* or *an ashy, dust-colour* (حَتَّى اخْضَرَّ): like as the rájiz says,

* جَاءُوا بِضَيْحٍ هَلْ رَأَيْتَ الذِّئْبَ قَطْ *

[They brought milk mixed with much water. Hast thou ever seen the wolf?]: meaning that the milk was of an ash-colour (أَوْرَق), like the colour of the wolf, by reason of the great quantity of the water: or, as some say, *milk and water in the proportion of one third of the former to two thirds of the latter:* it is *of any milk, that has been kept in a skin* or *that is fresh,* and *from any beast:* some say that the word is a pl., [or rather a coll. gen. n.,] and that the sing., or n. un., is with ة. (TA.)

خَضُورٌ: see أَخْضَرُ.

خَضِيرٌ: see أَخْضَرُ: — and see also خَضِرٌ.

خُضَيْرٌ: see what next follows.

خُضَارَةٌ: see أَخْضَرُ, in the latter half of the paragraph. — خُضَارَةُ, determinate, and imperfectly decl., (ISk, Ṣ, Ḳ,) because it has the quality of a proper name and the fem. gender with ة, like أُسَامَةُ &c., (TA,) ‡ *The sea;* (ISk, Ṣ, A, Ḳ;) as also ▼ الأُخَيْضِرُ, and ▼ خُضَيْرٌ, (A, TA,) or ▼ خَضْرٌ. (So in a copy of the A.) [But it is used as a masc. proper name; for] you say, هٰذَا خُضَارَةُ طَامِيًا [*This is the sea, in a state of rising,* or *becoming full,* or *becoming high and full*]. (Ṣ, TA. [In one copy of the Ṣ, I find هٰذِهِ; but in others, هٰذَا; and in all, طَامِيًا.])

خُضَيْرَةٌ A palm-tree (نَخْلَةٌ) *of which the dates fall while unripe and green;* (Ṣ, Ḳ;) as also ▼ مِخْضَارٌ. (TA.)

خُضَيْرَةٌ dim. of خُضْرَةٌ. == Also ‡ A woman *who scarcely ever,* or *never, completes the fruit of her womb, so that she casts it.* (TA.)

خُضَارِىٌّ *A certain bird;* (Ṣ, Ḳ;) *also called* the أَخْيَل; (Ṣ;) *regarded as of evil omen when it alighted upon the back of a camel:* it is أَخْضَرُ [i. e. *green,* or *of a dark* or *an ashy dust-colour*], *with redness in the* حَنَك [or *part beneath the beak*], and is *larger than the* قَطَا: or *certain green,* or *dark* or *ashy dust-coloured, birds,* (طَيْرٌ خُضْرٌ,) *also called* قَارِيَة: A'Obeyd asserts that the Arabs loved them, and likened to them a liberal, or bountiful, man: but ISd says, on the authority of the 'Eyn, that they regarded them as of evil omen: (TA:) [Golius states, on the authority of Meyd, that the خُضَارِىّ is *a bird of a blackish colour, called in Persian* كَرَايَة. See Bochart's Hieroz. p. ii. col. 61; referred to by Freytag.] — Also The [*tree*, or *shrub, called*] رِمْث, *when it has grown tall.* (TA.)

خُضَّارٌ *A certain bird,* (Ḳ,) *green* or *of a dark* or *an ashy dust-colour* (أَخْضَر). (TA.)

خُضَّارَى: see خَضِرٌ. — Also *A certain plant.* (Ḳ.)

أَخْضَرُ [*Green; verdant;*] *of the colour termed* خُضْرَة; (Ṣ, A, Mṣb, Ḳ;) as also ↓خَضِرٌ (Mṣb, Ḳ) and ↓خَضِيرٌ and ↓خُضُورٌ and ↓يَخْضُورٌ and ↓يَخْضِيرٌ: (Ḳ, TA: the last two written in the CḲ تَخْضُور and تَخْضِير:) applied to a horse, [and to a camel, (see خُضْرَة,) and to an ass, and sometimes to a bird, and to a garment of the kind called كِسَآء, and the like, and to various other things, *of a dark,* or *an ashy, dust-colour;* or *dingy ash-colour;* or *dark dust-colour;*] *of a dust-colour intermixed with* دُهْمَة [i. e. *blackness* or *deep ash-colour*]; which is the same as دَيْزَج; (Ṣ;) in horses being distinguished as أَخْضَرُ أَدْغَم and أَخْضَرُ أَطْحَل and أَخْضَرُ أَوْرَق: (TA: [see the latter epithet in each of these cases:]) applied to a man, [and to other things,] *tawny,* or *brownish:* (Ṣ:) [and *blackish:* and *of a blackish hue inclining to green:*] and *black;* (Ṣ, Ḳ;) *black-complexioned:* (TA:) [and *intensely black:* it is said in the Mṣb, art. حتر, that الأَخْضَر is, with the Arabs, أَسْوَد; which may mean either that green is, with the Arabs, termed اسود, or that الاخضر is, with the Arabs, black: but the truth is, that each of the epithets أَخْضَر and أَسْوَد is sometimes used for the other: see what here follows, and see أَسْوَدُ: in Ḥar p. 495, it is erroneously said, on the authority of Er-Rázee, that the اسود is not termed by the Arabs اخضر, although the اخضر is termed by them اسود because of its intense خُضْرَة and رِىّ:] the fem. is خَضْرَآءُ: and the pl. is خُضْرٌ. (Mṣb, TA.) You say شَجَرَةٌ خَضْرَآءُ *A green, and fresh,* or *juicy, tree.* (TA.) And مَآءٌ أَخْضَرُ *Water inclining to a green colour,* by reason of its clearness. (TA.) And أَخْضَرُ الجِلْدَةِ [lit. *Tawny of skin:*] meaning †*of pure race;* because the complexions of the Arabs are tawny; (Ṣ;) *of genuine Arab race:* (IB:) as in the saying of El-Lahabee, (Ṣ, TA,) El-Faḍl Ibn-'Abbás, (TA,)

• وَأَنَا الأَخْضَرُ مَنْ يَعْرِفُنِى •
• أَخْضَرُ الجِلْدَةِ فِى بَيْتِ العَرَبْ •

[*And I am the tawny: who knows me? the tawny of skin* (or *pure of race*), *of the family that comprises the nobility of the Arabs*]. (Ṣ, IB.) And فُلَانٌ أَخْضَرُ القَفَا [lit. *Such a one is blackish,* or *black, in the back of the neck:*] meaning ‡*such a one is the son of a black woman:* (Az, A:) or ‡*one who is slapped on the back of his neck:* (A:) or ‡*a freedman,* or *an emancipated slave.* (TA.) And أَخْضَرُ البَطْنِ ‡*A weaver:* (A, TA:) because his belly, being stuck close to his loom, becomes blackened by it. (TA.) And أَخْضَرُ النَّوَاجِذِ ‡*An eater of onions and leeks:* or *a tiller,* or *cultivator, of the ground;* because he eats herbs, or leguminous plants. (A.) And هُمْ خُضْرُ المَنَاكِبِ [lit. *They are green in the shoulders,* from carrying the produce of their land:] meaning ‡*they are in a state of great plenty.* (Ḳ, TA.) And [hence, perhaps,] فُلَانٌ أَخْضَرُ ‡*Such a one possesses abundant* خَيْر [or *wealth,* or *prosperity*]: (A, TA:) [or it may mean *goodness:* for] الأَخْضَرُ, applied to a man, is an epithet of praise, whereby he may be likened to the sea, because it is described as green, or to the [rain or herbage called] رَبِيع; in both cases meaning †*liberal,* or *bountiful;* and it is so applied because خُضْرَة is of the colours of the Arabs: and it is also an epithet of dispraise, as meaning †*black by reason of baseness, ignobleness,* or *meanness.* (Ḥam p. 282.) And شَابٌّ أَخْضَرُ ‡*A young man whose hair has begun to grow upon the sides of his face.* (TA.) And كَتِيبَةٌ خَضْرَآءُ ‡*An army,* or *a troop of horse, overspread with the blackness of iron:* (Ṣ, TA:) or *a great army* or *troop of horse* (Ḳ, TA) *of which most of the men are clad in iron;* like جَأْوَآءُ: (TA:) because of the خُضْرَة of the iron: (A:) [i. e.] because of the blackness thereof. (TA.) And اللَّيْلُ أَخْضَرُ ‡*Night is black.* (TA.) And [hence,] جَنَّ عَلَيْهِ أَخْضَرُ الجَنَاحَيْنِ ‡*Night* [lit. *the black-winged*] *veiled him, concealed him,* or *covered him with its darkness.* (A.) مُدْهَامَّتَانِ, in the Ḳur [lv. 64, relating to two gardens of Paradise], is explained by خَضْرَاوَانِ because it means *Inclining to blackness,* by reason of abundance of moisture, or irrigation. (Ṣ.) — الأَخْضَرُ used as a subst.: see خُضَارَةٌ. — The fem. خَضْرَآءُ [is also used as a subst., and] signifies *Green herbs* or *leguminous plants;* (Mṣb, Ḳ;) as also ↓خُضَارَةٌ: (Ḳ:) pl. خَضْرَاوَاتٌ: by rule it should be خُضْرٌ; but as the quality of a subst. predominates in it, it has a pl. like the pl. of a subst., like صَحْرَاوَاتٌ pl. of صَحْرَآءُ: (Mṣb:) this pl. occurs in the saying (in a trad., TA) لَيْسَ فِى الخَضْرَاوَاتِ صَدَقَةٌ *There is no poor-rate in the case of green herbs* or *leguminous plants;* (Mṣb;) or *fresh fruits and herbs* or *leguminous plants;* (TA;) or *fruits,* such as the apple and the pear &c.; or *herbs* or *leguminous plants,* such as leeks and smallage and rue and the like; and خُضَرٌ, pl. of خُضْرَةٌ, is sometimes substituted for it. (Mgh.) [Hence,] إِيَّاكُمْ وَخَضْرَآءَ الدِّمَنِ, meaning ‡*Avoid ye the beautiful woman that is of bad origin:* (Ṣ, A, Mṣb:) because what grows in a دِمْنَة [or place which men have blackened by their cooking, and where their camels or other beasts have staled and dunged], though it may be beautiful and bright, does not bear fruit [because it is neglected, and left unwatered], (Ṣ, Mṣb,) and soon becomes corrupt, or bad. (Mṣb. [See also دِمْنَةٌ: and see عُشْبَةُ الدَّارِ, in art. عشب.]) — And الخَضْرَآءُ, as an epithet in which the quality of a subst. predominates, (TA,) ‡*The sky,* or *heaven;* (Ṣ, A, Ḳ;) because of its greenness; like as the earth is called الغَبْرَآءُ. (TA.) You say, مَا تَحْتَ الخَضْرَآءِ أَبْغَضُ مِنْهُ ‡[*There is not under the sky one more hateful than he*]. (A.) — And خَضْرَآءُ ‡*A bucket* (A, Ḳ) *with which water has been drawn long, so that it has become green* or *blackish* &c. (حَتَّى ٱخْضَرَّتْ). (Ḳ.) — And †The *congregated* or *collective body,* and *mass,* or *bulk,* of a people. (Ṣ, Ḳ.) So in the saying, أَبَادَ ٱللّٰهُ خَضْرَآءَهُمْ †[*May God destroy the congregated* or *collective body, mass,* or *bulk, of them*]: (Ṣ:) or this means, ‡*their stock* (شَجَرَة) *from which they have branched off;* (A;) [for] خَضْرَآءُ signifies the *origin* of anything: (TA:) or, *their life in this present world:* (Fr, TA:) or, as some say, *their enjoyment and plenty;* (TA;) [for] خَضْرَآءُ signifies *prosperity,* and *plenty,* and *enjoyment:* (TA in a later part of this art.:) or the right reading is غَضْرَآءَهُمْ, meaning "their prosperity, and their pleasantness of life, or plenty and prosperity." (Ṣ. [See art. غضر.]) — البَقْلَةُ الخَضْرَآءُ: see خَضِرٌ. — الخُضْرُ, (T,) or الخَضْرَآءُ, (Ḳ,) *The domestic pigeons;* (T, Ḳ;) so called although of various colours, because their predominant colour is وُرْقَة [or ash-colour], or خُضْرَة [meaning a dark, or an ashy, dust-colour]: the خُضْر and the نُمْر [or spotted with white and black, &c.,] are especially characterized by the faculty of rightly directing their course. (T, TA.) — الأَخَاضِرُ [a pl. of الأَخْضَرُ used as a subst.] ‡*Gold and flesh-meat and wine;* as also الأَحَامِرَةُ [as some explain this latter]. (TA.) — أَخْضَرُ also signifies ‡*Fresh,* or *recent:* so in the saying, الأَمْرُ بَيْنَنَا أَخْضَرُ ‡[*The affair between us is fresh,* or *recent*]: and in like manner you say, المَوَدَّةُ بَيْنَنَا خَضْرَآءُ ‡*Love,* or *affection, between us is fresh.* (A.) And *Soft,* or *tender;* applied to herbage, or seed-produce. (TA.) — [Hence,] عِيشَةٌ خَضْرَآءُ †*A mode of life soft,* or *delicate, and plentiful and pleasant.* (Ḥar p. 639.) — الأَخْضَرُ is also the name of [*A certain star,* or *asterism;* most probably either *α of Piscis Australis* or *ε of Pegasus,* or *some star or asterism nearly in a line with those two;*] *one of the three* أَنْوَآء *of the rain called* الخَرِيف; namely, *the middle* نَوْء *of those three* أَنْوَآء; the first being the نَسْرَانِ; and the last, the foremost of the فَرْغَانِ: see نَوْءٌ. (AZ, T and TA in art. نوأ.)

الأُخَيْضِرُ dim. [of الأَخْضَرُ], (TA,) [*Cantharides;*] *a kind of fly,* (Ḳ,) *green, of a dark* or *an ashy dust-colour,* (أَخْضَر,) *of the size of the black fly, and called the Indian fly* [as cantharides are by the Arabs in the present day]; *having properties and uses mentioned in medical books.* (TA.) — Also *A certain disease in the eye.* (Ḳ.)

مَخْضَرَةٌ: see خَضِرٌ, in two places.

مِخْضَارٌ: see خَضِيرَةٌ.

مَخْضُورٌ: see خَضِرٌ.

يَخْضُورٌ: see خَضِرٌ, in two places: and see also أَخْضَرُ, first sentence.

يَخْضِيرٌ: see أَخْضَرُ, first sentence.

خضرم

Q. 1. خَضْرَمَ الأُذُنَ, inf. n. خَضْرَمَةٌ, *He cut somewhat of the extremity of the ear* of a camel, *and left it dangling*: or *he cut the ear in halves*: and you say also, خَضْرَمُوا نَعَمَهُمْ [*They so cut the ears of their camels*]: (TA:) or خَضْرَمَةٌ signifies the *cutting one of the ears* only. (JK.) —— And خَضْرَمَ *He mixed* [a thing or things]: (IKh, TA:) [and so حَضْرَمَ, as is indicated in the Ḳ in art. حضرم, by an explanation of the inf. n.] —— And خَضْرَمَةٌ signifies also The *making* a thing *to be of an intermediate*, or *a middling, kind* or *quality*. (TA.)

Q. 2. تَخَضْرَمَ, said of butter [in the process of formation], *It became dissundered*, or *separated* [*into clots*], *by reason of intense cold; and did not coalesce*; as also تحضرم. (TA in art. حضرم.)

خِضْرِمٌ A well *having much water*; (JK, Ḳ;) and so, [as some say,] a sea, or great river; but its application as an epithet to a sea, or great river, is disallowed by Aṣ: (Ṣ:) or a *great* sea: (Ḳ:) accord. to some, so called because of its greenness; and if so, the م is augmentative: (MF:) or water *copious and wide in the utmost degree*: (Mz 49th نوع:) and anything *much in quantity*, or *copious*, (Ṣ, Ḳ,) and *wide*, or *ample*: (Ṣ:) or it signifies also *wide*, or *ample*, (Ḳ,) applied to anything: (TA:) pl. خَضَارِمُ: (Ṣ:) and ↑مُخَضْرَمٌ and ↑خُضَارِمٌ, also, signify *much in quantity*, or *copious*, applied to water: (TA:) and خِضْرِمٌ is applied in this sense as an epithet to نَبِيذ. (Ṣ,* TA.) —— ‡ *Bountiful*, or *munificent*; (JK, Ḳ;) *who gives many gifts*: (Ṣ, Ḳ:) said to be likened to the sea, or great river, to which this epithet is applied; though Aṣ disallowed its application to a sea, or great river: (Ṣ:) or likened to the well to which the same epithet is applied: (JK:) and a *forbearing*, or *clement*, lord, or chief; as also ↑خُضَارِمٌ: pl. خَضَارِمُ and خَضَارِمَةٌ and خِضْرِمُونَ: all applied peculiarly to men: (Ḳ:) not to women. (TḲ.)

خُضْرُمٌ *Sweet* water: or water *between sweet and bitter*: (Ḳ:) on the authority of Yaḳoob. (TA.) ═ The *young of the* [*kind of lizard called*] ضَبّ: (Ṣ, Ḳ:) accord. to IDrd, in its first stage it is called حِسْل; [after which he should have said, then, غَيْدَاق;] then it is called مُطَبِّخ [q. v.]; then, خُضْرُمٌ; and then, ضَبٌّ: he does not mention the term غَيْدَاق, but AZ mentions it. (Ṣ.)

خِضْرِمِىٌّ: see الخَضَارِمَةُ.

خُضَارِمٌ: see خِضْرِمٌ, in two places.

الخَضَارِمَةُ *A certain people of the* عَجَم [i. e. *Persians*], (Ṣ, Ḳ,) *of the sons of Fáris*, (Ṣ,) *who went forth* [*from their country*] *in the beginning of El-Islám, and dwelt in Syria*: (Ṣ, Ḳ:) i. e., those people who went forth at that period dispersed themselves in the countries of the Arabs; some of them settling in El-Baṣrah, and these are the أَسَاوِرَة; and some of them, in El-Koofeh, and these are the أَحَامِرَة; and some of them, in Syria, and they are the خَضَارِمَة; and some of them, in El-Jezeereh, and they are the جَرَاجِمَة; and some of them, in El-Yemen, and they are the أَبْنَآء; and some of them, in El-Mowṣil, and they are the جَرَامِقَة: (Ṣ, TA:) the n. un. is ↑خِضْرِمِىٌّ. (Ḳ.)

مُخَضْرَمٌ [pass. part. n. of خَضْرَمَ]. You say نَاقَةٌ مُخَضْرَمَةٌ *A she-camel having the extremity of her ear cut*. (Ṣ, Ḳ.) And أُذُنٌ مُخَضْرَمَةٌ *An ear cut*. (Mz 49th نوع.) —— Hence, as some say, (Mz ubi suprà,) A man, (Ḳ, Mz,) or a poet, (Ṣ, Ḳ,) *who lived in the Time of Ignorance and in that of El-Islám*; (Ṣ, Ḳ, and Mz ubi suprà;) as though he were cut off from paganism to El-Islám; (Mz ubi suprà;) or from infidelity; (IB, TA;) as Lebeed, (Ṣ, Ḳ, and Mz 20th نوع,) and Ḥassán Ibn-Thábit, and Nábighah of the Benoo-Jaạdeh, and Aboo-Zubeyd, and 'Amr Ibn-Sha-s, and Ez-Zibriḳán Ibn-Bedr, and 'Amr Ibn-Maạdee-Kerib, and Kaạb Ibn-Zuheyr, and Maạn Ibn-Ows: (Mz 20th نوع: [see also إِسْلَامِىٌّ, and شَاعِرٌ:]) or a person *who passed half of his life in the Time of Ignorance, and half thereof in that of El-Islám*: (Ḳ:) [I have generally found the word thus written;] but IB says that, accord. to most of the lexicologists, it is ↑مُخَضْرِمٌ, with kesr to the ر; for the pagans, when they became Muslims, cut somewhat of the extremities of the ears of their camels (خَضْرَمُوا آذَانَ إِبِلِهِمْ) as a sign of their being Muslims in case of their being attacked and plundered, or their being made war with; (IB, TA;) and this they were ordered to do in a manner different from that of the pagans: (TA:) accord. to some, the epithet applied to a poet of the class above mentioned is محضرم, with the unpointed ح, [i. e. مُحَضْرَمٌ,] from الحَضْرَمَةُ signifying الخَلْطُ, [like الخَضْرَمَةُ, as shown above,] because of his mixing paganism with El-Islám: (Mz 49th نوع:) or مُحَضْرَمٌ: (Ḳ in art. حضرم:) and in like manner IKh explains the epithet ↑مُخَضْرِمٌ. (TA.) —— Also A *black* man *whose father is white*. (IKh, Ḳ.*) —— And *Deficient in respect of* الحَسَب; (Ḳ;) meaning *not of generous parentage*. (TA.) —— And One *whose origin is suspected*; or *who claims for his father one who is not*: (Ḳ:) and so مُخَضْرَمٌ فِى نَسَبِهِ or مُخَضْرَمُ النَّسَبِ: (Ṣ, TA:) means *of mixed parentage*. (TA.) One *whose father is unknown*: [or, app., accord. to the TA, *whose parents are unknown*:] or one *sprung from* [*a succession of*] *concubines*. (Ḳ.) —— *Uncircumcised*. (Ḳ.) And, with ة, applied to a woman, *Circumcised*: (Ṣ, Ḳ:) or, as some say, *cut, by a mistake of the woman operating, in a place not that of circumcision*. (TA.) —— Flesh-meat *such that one knows not whether it be from a male or a female*. (Ṣ, Ḳ.) —— Accord. to IAạr, who does not explain it, (ISd, TA,) it is applied also to food, (ISd, Ḳ,) ISd thinks (TA) as meaning *Insipid*; (ISd, Ḳ;) *neither sweet nor bitter*. (ISd.) —— And Water *between heavy and light*: (T, Ḳ, TA:) or *not sweet*. (TA.) —— See also خِضْرِمٌ.

مُخَضْرِمٌ: see مُخَضْرَمٌ, in two places.

مُتَخَضْرِمٌ, applied to butter, [as also مُتَحَضْرِمٌ and مُحَضْرَمٌ,] *Dissundered*, or *separated* [*into clots*]; *not coalescing; by reason of cold*. (Ḳ. [See Q. 2.])

خضع

1. خَضَعَ, aor. َ, inf. n. خُضُوعٌ, (Ṣ, Mṣb, Ḳ) and خَضَعٌ and خُضْعَانٌ, or خِضْعَانٌ, (TA,) *He was*, or *became, lowly, humble*, or *submissive*, (Ṣ, Mṣb, Ḳ,) لَهُ *to him*, (Mṣb, TA,) [for instance,] to his creditor, (Mṣb,) or to God; (TA;) as also ↑اختضع, (Ṣ, Ḳ,) [and ↑انخضع, (Ḳ in art. خذا,)] and ↑اخضوضع: (Ṣgh, Ḳ:) خُضُوعٌ is *nearly the same as* خُشُوعٌ, except that the latter is mostly used in relation to the voice [or the eyes]; but the former is used as meaning *in the necks*: (Mṣb:) or the former is *in the body*, ('Eyn and Ḳ in art. خشع,) and signifies the acknowledgment of humility and submission; ('Eyn;) and the latter is in the voice and in the eyes. ('Eyn and Ḳ ubi suprà.) It is said in a trad. respecting the [devils'] hearing [the words of the angels] by stealth, خُضْعَانًا لِقَوْلِهِ [*With submissiveness to his saying*, or *to what he said*]; or, accord. to one relation, خِضْعَانًا; but it may be a pl. of خَاضِع; and accord. to another relation, it is خُضَّعًا, which is a pl. of خَاضِع. (TA.) —— *He was*, or *became, still*, (Ḳ, TA,) *and tractable*, or *submissive*. (TA.) —— *He made his words soft* to a woman; as also ↑اخضع: (L:) or the latter signifies *his speech was soft* to a woman. (O, Ḳ.) It is said in the Ḳur [xxxiii. 32], فَلَا تَخْضَعْنَ بِالْقَوْلِ *Then be ye not soft in speech*. (TA.) And you say, خَضَعَ لَهَا بِكَلَامِهِ وَخَضَعَتْ لَهُ وَتَطَمَّعَ فِيهَا [*He was soft to her in speech, and she was soft to him, and he became excited to feel an eager desire for her, or to lust after her*]; (TA;) and in like manner, ↑خَاضَعَهَا, (Ḳ,* TA,) inf. n. مُخَاضَعَةٌ, (TA,) [*he was soft in his speech to her, she being soft in her speech to him*.] And خَضَعَا بَيْنَهُمَا حَدِيثًا *They two* (a man and a woman) *made soft discourse together, saying that which excited each to feel an eager desire for, or to lust after, the other*. (TA from a trad.) —— خَضِعَ, aor. َ, inf. n. خَضَعٌ, [or, as in two copies of the Ṣ, خَضَعَ, though it seems that the verb is correctly خَضِعَ, not خَضَعَ,] *He had a natural stooping of the neck*: (TA:) and *he bent himself*, or *became bent*; as also ↑اخضع. (Zj.) And ↑اختضع, said of a hawk, *He lowered his head to make a swoop*, or *to pounce down*. (Z, TA.) —— [Hence,] خَضَعَتِ الإِبِلُ ‡ *The camels strove*, or *exerted themselves*, or *hastened, in their pace*, or *going*; (Ḳ;) because, when they do so, they lower their necks. (TA.) And ↑اختضع, (Ḳ,) said of a horse, (IAạr,) [for the same reason,] † *He went quickly*, or *swiftly*. (IAạr, Ḳ.) —— خَضَعَ النَّجْمُ ‡ *The star*, or *asterism, inclined* (Ṣ, Ḳ, TA) *to the place of setting*, (Ṣ, TA,) or *to setting*: (Ḳ, TA:) and in like manner, خَضَعَتِ الشَّمْسُ ‡ *the sun inclined*

&c.; like خَدَعَتْ: (TA:) and خَضَعَتْ أَيْدِى الكَوَاكِبِ ‡ *the stars inclined to setting.* (Aboo-'Adnán, TA in art. خشع.) — خَضَعَهُ *He, or it, rendered him still* (Ḳ, TA) [*and submissive:* see 1]: the verb being both intrans. and trans. (TA.) [See also 4.] — Also, (Ḳ,) inf. n. خَضْعٌ and خُضُوعٌ, (TA,) *He, or it, caused him to have a stooping neck;* as also ▼ اخضعهُ; (Ḳ;) i. e., *bent him:* (TA:) said of old age. (TḲ.) Jereer says,

* أَعَدَّ ٱللّٰهُ لِلشُّعَرَآءِ مِنِّى *
* صَوَاعِقَ يَخْضَعُونَ لَهُ ٱلرِّقَابَا *

[*God hath prepared, for the poets, from me, thunderbolts which make the necks to stoop to Him*]. (TA.) — خَضَعَ فُلَانًا إِلَى السَّوْءَةِ; in the Ḳ الى السوء, but the former is the right; inf. n. خُضُوعٌ; (TA;) *He, or it, invited such a one to that which was foul, abominable, or evil.* (Ḳ, TA.)

2: see 4.

3: see 1, near the middle of the paragraph.

4. اخضع, intrans.: see 1, in two places. — اخضعهُ *It* (poverty) *lowered, humbled,* or *abased, him;* (Mṣb;) [as also ▼ خَضَّعهُ; for its inf. n.] تَخْضِيعٌ signifies the *rendering lowly, humble,* or *submissive;* in Persian, فُرُوتَنْ كَرْدَانِيدَنْ. (KL. [But Golius, from the same source, explains the verb as signifying "Submissum humilemque se commonstravit."]) [Hence,] أَخْضَعَتْنِى إِلَيْكَ الحَاجَةُ (Zj, Ṣ, TA) *Want,* or *need,* [*made me lowly, humble,* or *submissive, to thee;* or] *constrained me to have recourse to thee, and to require thine aid.* (TA.) — See also 1, near the end of the paragraph.

7: see 1, first sentence.

8: see 1, in three places.

12. اخضوضع: see 1, first sentence.

خَضِعٌ A plant *bending by reason of softness,* or *tenderness:* ISd holds it to be formed after the manner of a relative, or possessive, noun, because there is no verb [of the measure خَضِعَ] to which it may be referred. (TA.) [The regular form, if it were a part. n., would be خَاضِعٌ, q. v.]

خَضْعَةٌ, or ▼ خُضَعَةٌ: see بَضْعَةٌ and بَاضِعٌ.

خَضْعَةٌ: see what next precedes.

خُضَعَةٌ A man (Ṣ) *who is lowly, humble,* or *submissive, to everyone.* (Ṣ, Ṣgh, Ḳ.) — And One *who overcomes,* or *subdues, his adversaries,* or *opponents,* (Ḳ, TA,) *and humbles and abases them.* (TA.)

خَضُوعٌ: see the next paragraph, in two places.

خَاضِعٌ *Lowly, humble,* or *submissive;* (Mṣb;) and ▼ خَضُوعٌ signifies the same: (Ṣ, Ḳ:) [or rather the latter is an intensive epithet, signifying *very lowly,* &c.:] the pl. of the former is خَاضِعُونَ and خُضَّعٌ and خُضْعَانٌ, or خِضْعَانٌ: (TA:) [respecting the last two of which, see 1, second sentence:] and the pl. of ▼ خَضُوعٌ is خُضُعٌ; (Ṣ, Ḳ;) as in the phrase قَوْمٌ خُضُعُ الرِّقَابِ [*A people,* or *company of men, very submissive in the necks*]. (Ṣ.) It is said in the Ḳur [xxvi. 3], فَظَلَّتْ أَعْنَاقُهُمْ لَهَا خَاضِعِينَ *And their necks shall continue,* the pret. being used in the sense of the aor., meaning تَدُومُ, (Jel,) *submissive to it:* (Jel,* TA:) the original of the phrase is فَظَلُّوا لَهَا خَاضِعِينَ; and اعناق is redundantly inserted to show the place of خُضُوع, and the predicate is left in its original state: (Bḍ:) or as the خضوع is only that of the اعناق, it is allowable to make the predicate relate to [the pronoun هم, which is] the complement of the latter word: (Sb, Kh:) or since the خضوع is ascribed to the necks but really belongs to the persons, the epithet has that form of pl. which is proper to rational beings: (Jel: [and the like is said by Bḍ:]) or اعناقهم means *their chiefs:* or *their companies:* but there is another reading [which is literally grammatical], namely خَاضِعَةً. (Bḍ.) — The pl. خُضَّعٌ is also applied to Women *who have been* [or *who are*] *soft in speech, and still.* (IAạr.) [See 1.] — نَعَامٌ خَوَاضِعُ [pl. of خَاضِعَةٌ] *Ostriches inclining their heads towards the ground in their places of pasture;* and in like manner, ظِبَآءٌ [*gazelles*]. (TA.) — إِبِلٌ خَوَاضِعُ ‡ *Camels striving,* or *exerting themselves,* or *hastening, in their pace,* or *going;* because, when they do so, they lower their necks. (TA.) — مَنْكِبٌ خَاضِعٌ and ▼ أَخْضَعُ *A low,* or *depressed, shoulder-joint.* (TA.) — نُجُومٌ خَوَاضِعُ ‡ *Stars inclining to setting,* or *to their places of setting.* (A, TA.) — *Inviting* to that which is foul, abominable, or evil. (TA.)

أَخْضَعُ *Content with abasement;* fem. خَضْعَآءُ. (Lth, Ḳ.) — *Having a natural stooping of the neck;* (Ṣ, Ḳ;) applied to a man, (TA,) and to a horse, (Ṣ, TA,) and a camel, and an ostrich, and a gazelle. (TA.) — See also خَاضِعٌ, near the end of the paragraph.

خضل

1. خَضِلَ, aor. ـَ, inf. n. خَضَلٌ: see 9, in two places.

2: see the next paragraph.

4. اخضلهُ *He moistened it;* or *wetted it;* (Ṣ, Ḳ;) as also ▼ خضّلهُ, inf. n. تَخْضِيلٌ. (TA.) You say, أَخْضَلَتْ دُمُوعُهُ لِحْيَتَهُ *His tears moistened,* or *wetted, his beard.* (JK,* TA.) And أَخْضَلَتْهُ السَّمَآءُ *The rain wetted him:* (TA:) or *wetted him much.* (JK.) — See also 9.

8. اختضل بِصَاحِبِهِ *He* (a man) *became united with his companion.* (Fr, TA.)

9. اخضلّ, (Ṣ, Ḳ,) inf. n. اِخْضِلَالٌ, (Ṣ,) *It was,* or *became, moistened,* or *wetted;* (Ṣ, Ḳ;) as also ▼ أَخْضَلَ, (Ḳ, TA, [but not in the CḲ,]) inf. n. إِخْضَالٌ; (TA; [perhaps a mistranscription for اخضالّ, inf. n. اِخْضِيلَالٌ; but said by Freytag to occur in the Deewán of the Hudhalees;]) and ▼ اخضوضل, (Fr, Ṣ, Ḳ,) inf. n. اِخْضِيضَالٌ; (Ṣ;) and ▼ خَضِلَ: (Ḳ: [but see what follows:]) it is said of a garment, and of the beard: (TA:) and the first of these verbs signifies also *it was,* or *became, moist, so that its moisture became sprinkled,* or *scattered in drops;* (Ḳ,* TA; [accord. to the explanation of the part. n. خَضِلٌ (q. v.) in the JK and M;]) and so ▼ اخضالّ, (Ḳ,) inf. n. اِخْضِيلَالٌ; (TA;) and ▼ خَضِلَ, aor. ـَ, (Ḳ,) inf. n. خَضَلٌ; but Lth says, I have not heard them say خَضِلَ. (TA.) — اخضلّ اللَّيْلُ *The night became dark:* (JK, Ibn-'Abbád, Ḳ:) or *the pleasant coolness of the night came.* (T, TA.)

11: see 9, in two places. — Also اِخْضَالَّتِ الشَّجَرَةُ, inf. n. اِخْضِيلَالٌ; (Ṣ;) or اخضالّ الشَّجَرُ, and ▼ اِخْضَالَّ; (IDrd, Ḳ;) *The tree,* or *trees, had many branches and leaves:* (IDrd, Ṣ, Ḳ:) or *became green, and fresh,* or *sappy, in the branches thereof.* (TA.)

12. اِخْضَوْضَلَ: see 9.

Q. Q. 4. اِخْضَأَلَّ: see 11.

خَضْلٌ *Moisture.* (TA. [But perhaps this may be a mistranscription for خَضَلٌ, inf. n. of خَضِلَ.]) — Also, (JK, T, Ḳ,) and ▼ خَضَلٌ, (ISd, Ḳ,) *Pearls:* (JK, Ḳ:) or *clear large pearls:* (Ḳ:) or *good, clear, lustrous, large pearls:* of the dial. of Yethrib. (TA.) — And *A well-known kind of beads:* (ISk, Ḳ:) or *a red bead:* or *a bead of ivory:* (TA:) n. un. with ة. (Ḳ.) — And دُرَّةٌ خَضْلَةٌ, (TA,) or ▼ خَاضِلَةٌ, (JK,) *A clear large pearl.* (JK, TA.)

خَضَلٌ: see the next preceding paragraph.

خَضِلٌ A thing, (Ṣ,) or roasted meat, (JK, T, M, Ḳ,) *moist,* or *juicy,* (JK, T, Ṣ,) *and well cooked:* (T:) or *dripping with its gravy;* or *succulent, and dripping with its juice:* (M, Ḳ:) and anything *moist so that its moisture becomes sprinkled,* or *scattered in drops;* (JK, M, Ḳ;*) as also ▼ خَاضِلٌ. (Ḳ: in the copies of which we find يَتَرَشَّفُ in the place of يَتَرَشَّشُ; the latter being the reading in the [JK and] M, TA.) — Applied to a plant, or herbage, *Soft,* or *tender.* (Ṣ, TA.) — And hence metaphorically applied to life: you say عَيْشٌ خَضِلٌ, (Ḥar pp. 54-55,) and ▼ مُخْضِلٌ, and ▼ مُخْضَلٌ, (Ḳ,) ‡ *A soft and delicate life.* (Ḳ, and Ḥar ubi suprà.)

خُضْلَةٌ *A plentiful, and a pleasant* or *an easy, and a soft* or *delicate, state of life.* (Ḳ, TA. [In the CḲ, النَّعَمَةُ is erroneously put for النَّعْمَةُ.]) You say, هُمْ فِى خُضْلَةٍ مِنَ العَيْشِ *They are in a plentiful, and a pleasant* or *an easy, and a soft* or *delicate, state of life.* (TA.) And يَوْمٌ خُضْلَةٌ *A day of plenty, and pleasure* or *ease, and softness* or *delicacy;* (JK, Ḳ;) or *a day of plenty;* or *of abundance of herbage,* or *of the goods* or *conveniences* or *comforts of life;* and *of pleasantness and easiness of life.* (Ṣ, TA.) And نَزَلْنَا فِى خُضْلَةٍ مِنَ العُشْبِ *We alighted among green, soft* or *tender, fresh herbage.* (TA.) — Also *A wife:* (Ḳ:) or a name for *a woman:* (Ḳ,* TA:) and *a soft*, or *tender, woman*. (JK, Ḳ.) — And The *rainbow*. (JK, Ibn-'Abbád, Ḳ.) And *A halo round the moon*. (AA, TA.) — One says also, دَعْنِى مِنْ خُضَّلَاتِكَ, meaning *Let me alone, and cease from thy vain*, or *false, sayings* or *actions*. (TA.)

رَوْضَةٌ خَضِيلَةٌ *A* [or *meadow*] (IDrd, Ṣ, Ḳ) *that is luxuriant and moist*. (IDrd.)

خَضِلٌ : see خَضِلٌ — دُرَّةٌ خَاضِلَةٌ : see خَاضِلٌ.

مُخْضِلٌ : see خَضِلٌ.

مِخْضَلٌ : see مِخْضَلٌ.

مُخْضِلٌ : see خَضِلٌ.

خضم

1. خَضِمَ, aor. ـَ ; (Ṣ, Ḳ;) and خَضَمَ, aor. ـِ ; (JK, Ḳ;) inf. n. خَضْمٌ; (JK, Ṣ, Ḳ;) *He* (a man) *ate* a thing *with the whole of the mouth*: (Aṣ, Ṣ:) or *he ate*, (Ḳ,) in a general sense: (TA:) or *he ate with the more remote of the teeth*: (Ḳ:) قَضْمٌ signifies the "eating with the nearer of the teeth;" (TA;) [i. e., "with the teeth of the fore part of the mouth:" or the "eating with the extremities of the teeth:" see art. قضم :] or [*he ate so that*] *he filled his mouth with that which he ate*: or it relates peculiarly to the thing that is moist, or juicy, as the cucumber, (Ḳ,) and the like: (TA:) or *he ate in the enjoyment of a plentiful and pleasant life*: (JK, TA:) or خَضْمٌ referring to a man is like قَضْمٌ referring to a beast. (TA.) — And خَضَمَهُ, aor. ـِ , (JK, Ḳ,) inf. n. as above; (TA;) and ↑ اخْتَضَمَهُ ; (JK, Ḳ;) *He cut it; or cut it off*: (Ḳ:) or *he cut it in pieces*. (JK.) — خَضَمَ لَهُ مِنْ مَالِهِ, (Ḳ,) accord. to IAạr, (TA,) signifies *He gave him of his property*; (Ḳ;) [as though he cut off for him a portion thereof;] but Th rejects this, and says that it is خَضَمَ. (TA.)

8: see 1. — [Hence,] السَّيْفُ يَخْتَضِمُ جَفْنَهُ *The sword cuts*, and *eats, its scabbard*, (Ḳ,) *by reason of its sharpness*; mentioned by J as a meaning of يَخْتَضِمُ : [see 8 in art. خضم :] and يَخْتَضِمُ العَظْمَ *cuts the bone*: and الذِّرَاعَ [*the fore arm*]. (TA.) — And اختضم الطَّرِيقَ *He stopped the way, robbing and slaying passengers*. (Ḳ.)

خُضُمٌ : see خُضَمٌ.

خُضَمَةٌ *i. q.* خُضَمَةٌ, (Ḳ,) i. e. *A certain bead*, or *gem*, mentioned before. (TA.)

خُضَامٌ : see what next follows.

خُضَامَةٌ *A thing that is eaten in the manner termed* خَضْمٌ ; [see 1;] (Ḳ;) as also ↑ خُضَامٌ [expressly said to be like غُرَابٌ, otherwise it would seem to be خَضَامٌ, like قَضَامٌ, to which it is opposed,] (TA,) [and ↑ مَخْضَمٌ, as is indicated in the Ḳ in art. قضم, opposed to مَقْضَمٌ in that art. (q. v.) in the Ṣ and Ḳ.]

مَخْضَمٌ : see what next precedes.

خط

1. خَطَّ, aor. ـُ , inf. n. خَطٌّ, *He made* [*a line*, or *lines*, or] *a mark*, عَلَى الأَرْضِ, *upon the ground*. (Mṣb.) You say, خَطَّ الزَّاجِرُ فِى الأَرْضِ, aor. and inf. n. as above, *The diviner made a line*, or *a mark*, or *lines*, or *marks*, *upon the ground, and then divined*. (TA.) And الزَّاجِرُ يَخُطُّ بِإِصْبَعِهِ فِى الرَّمْلِ وَيَزْجُرُ [*The diviner makes lines*, or *marks*, *with his finger upon the sand, and divines*]. (Ṣ.) Th says, on the authority of IAạr, that عِلْمُ الخَطِّ is عِلْمُ الرَّمْلِ [or *geomancy*]: I 'Ab says that it is an ancient science, which men have relinquished: but Lth says that it is practised to the present time; [to which I may add, that it has not even now ceased; being still practised on sand and the like, and also on paper;] and they have conventional terms which they employ in it, and they elicit thereby the secret thoughts &c., and often hit upon the right therein: the diviner comes to a piece of soft ground, and he has a boy, with whom is a style; and the master makes many lines, or marks, in haste, that they may not be counted; then he returns, and obliterates leisurely lines, or marks, two by two; and if there remain two lines, or marks, they are a sign of success, and of the attainment of the thing wanted: while he obliterates, his boy says, for the sake of auguring well, اِبْنَىْ عِيَانْ أَسْرِعَا البَيَانْ [*O two sons of 'Iyán* (meaning two lines or marks), *hasten ye the manifestation*]: I 'Ab says that when he has obliterated the lines, or marks, and one remains, it is the sign of disappointment: and AZ and Lth relate the like of this. (TA.) It is said in a trad. of Mo'áwiyeh Ibn-El-Ḥakam Es-Sulamee, traced up by him to its author, كَانَ نَبِىٌّ مِنَ الأَنْبِيَآءِ يَخُطُّ فَمَنْ وَافَقَ خَطَّهُ عَلِمَ مِثْلَ عِلْمِهِ [*A prophet of the prophets used to practise geomancy; and he who matches his geomancy knows the like of his knowledge*]. (TA.) You say also, when a man is meditating upon his affair, and considering what may be its issue, or result, فُلَانٌ يَخُطُّ فِى الأَرْضِ ‡ [*Such a one makes lines*, or *marks*, *upon the ground*]. (TA.) [See also نَكَتَ : and see St. John's Gospel, ch. viii. verses 6 and 8.] And خَطَّ بِرِجْلِهِ الأَرْضَ means ‡ *He walked*, or *went along*. (TA.) — Also, (Ṣ, Mṣb,) aor. as above, (Mṣb,) and so the inf. n., (Mṣb, Ḳ,) *He wrote* (Ṣ, Mṣb, Ḳ) a writing, or book, (Mṣb,) or a thing, (TA,) with the reed prepared for that purpose, (Ṣ, Ḳ,) or with some other thing; (Ḳ, TA;) [and so ↑ خطّط, for] تَخْطِيطٌ is *syn. with* تَسْطِيرٌ, or, as in the T, *like* تسطير ; whence the saying, خُطِّطَتْ عَلَيْهِ ذُنُوبُهُ *His sins were written* [or *registered*] *against him*. (TA.) — خَطَّهَا لِنَفْسِهِ, and خَطَّ الخِطَّةَ : see 8; for the latter, in two places. — [Hence,] خَطَّ عَلَيْهِ † *He prohibited it*; or *took it for himself*; relating to anything. (Ḳ, TA.) — خَطَّ وَجْهُهُ : or خَطَّ : see 8. — خَطَّ الغُلَامُ : see 8. — خَطَطْتُ بِالسَّيْفِ وَجْهَهُ وَوَسَطَهُ ‡ [*I cut*, or *clave, with the sword his face and his waist*]. (TA.) And خَطَّهُ بِالسَّيْفِ نِصْفَيْنِ ‡ [*He clave him, or it, in halves with the sword*].

(TA.) And جَارَاهُ فَمَا خَطَّ غُبَارَهُ ‡ *He contended with him in running, and did not cleave his dust*. (JK, Ṣ,* A, L.) — خَطَّ اللّٰهُ نَوْءَهَا ‡ [*God made its* (a land's) *rain-giving star* or *asterism* (see نَوْءٌ) *to pass it over*: or *may God make* &c.]: from خَطِيطَةٌ signifying "a land not rained upon between two lands that have been rained upon:" (Ṣ, TA:*) it was said by I 'Ab [in a tropical sense, by way of imprecation, with reference to a woman], when he was asked respecting a man who had put the affair of his wife in her own hand and she had in consequence divorced him by a triple sentence: (Ṣ:) accord. to one relation, it is خَطَّأَ ; the meaning being "may he make its rain to miss it:" (Ṣ, TA:) and accord. to another, ↑ خَطَّى, originally خَطَّطَ, like تَقَضَّى البَازِى : the former, or first, is the weaker, or weakest, in authority, of these relations. (TA. [See also 2 in art. خطأ.])

2. خطّط, inf. n. تَخْطِيطٌ, [*He marked with lines, streaks*, or *stripes*. Also] *He wove* a piece of cloth *with lines, streaks*, or *stripes*. (KL.) And *He drew lines well and elegantly*. (KL.) — See also 1, in two places, in the latter half of the paragraph.

4: see 8, in three places.

8. اختطّ الخِطَّةَ, (Mṣb, Ḳ,) or اختطّها لِنَفْسِهِ, (Ṣ,) *He took the* خِطَّة [q. v.] *to himself, and* (Ḳ) *made a mark upon it*, (Ṣ, Ḳ,) *in order to its being known that he had chosen it to build there a house*; (Ṣ, TA;) as also ↑ خَطَّهَا ; (as in some copies of the Ḳ;) or ↑ اخطّها ; (as in other copies of the Ḳ, and as in the TA;) and ↑ خَطَّهَا لِنَفْسِهِ : (TA:) and *he alighted and took up his abode in the* خِطَّة, *none having done so before him*; as also ↑ خَطَّهَا لِنَفْسِهِ. (Ḳ.) [And hence, اختطّ signifies also *He founded* a town or the like.] — اختطّ وَجْهُهُ ‡ *His face became marked with lines* [app. *by the growth of his beard*]; (Ḳ, TA;) as also ↑ خَطَّ ; (Ḳ, L, TA;) or ↑ خُطَّ ; (JK;) or ↑ اخطّ : (CḲ:) or ‡ *the hair of his beard extended* [*so as to form lines*] *upon the two sides of his face*. (A, TA.) — اختطّ الغُلَامُ ‡ *The two sides of the boy's*, or *young man's*, *beard grew forth*; (Ṣ, L, Ḳ, TA;) as also ↑ خَطَّ ; or ↑ اخطّ. (Ḳ, accord. to different copies.)

خَطٌّ *A line, streak*, or *stripe*; in, or upon, a thing: (Ḳ:) pl. خُطُوطٌ (Ṣ, Ḳ) and أَخْطَاطٌ ; (Ḳ;) the latter, [a pl. of pauc.,] used by El-'Ajjáj: (TA:) and ↑ خُطَّةٌ is [syn. with خَطٌّ as explained above, being] a subst. from [the inf. n.] الخَطُّ, like as نُقْطَةٌ is from النَّقْطُ : (Ṣ, Ḳ:*) you say, عَلَى ظَهْرِ الحِمَارِ خُطَّتَانِ ↑ *Upon the back of the ass are two lines*, or *streaks, differing in colour from the rest of the body*. (TA.) — [In mathematics, *A line*. And hence, خَطُّ الاِسْتِوَآءِ *The equinoctial line*.] — *A slight track*, or *path*, or *road*, *in plain*, or *smooth*, or *soft, ground*: pl. as above. (Ḳ.) And *A road*, or *path*: (Th, Ḳ:) as in the saying, اِلْزَمْ ذٰلِكَ الخَطَّ وَلَا تَظْلِمْ مِنْهُ شَيْئًا [*Keep thou to that road, or path, and do not deviate* *from it at all*]: or in this sense the word is ↑ خُطَّة. (TA.) Also *A road*, or *way*, or *street, that is a common thoroughfare*; and so ↑ خُطّ. (IDrd, Ḳ.) — ‡ [*A streak*, or *stripe, of herbage.*] You say, الكَلَأُ خُطُوطٌ فِى الأَرْضِ ‡ *The herbage consists of streaks*, or *stripes, upon the ground;* the rain not having watered the country in common. (L, TA.) — *Writing*, and *the like.* (TA.) [*Handwriting. Character;* or *particular form of letters.* — See also 1.]

خُطٌّ : see خَطٌّ, in two places. — The *place of the tribe.* (AA, Ḳ.)

خِطٌّ : see خِطَّةٌ : — and see also خَطِيطٌ, in two places.

خُطَّةٌ : see خَطٌّ, in two places. — Also *An affair: a matter: a case: an event: a state*, or *condition:* syn. أَمْرٌ : (Ṣ, Ḳ:) and قِصَّةٌ : (Ṣ:) or the *like of* قِصَّة : (JK, Ḳ:) and خَطْبٌ : and حَالٌ : (TA:) or حَالَةٌ : (Mṣb:) or, as some say, *a dubious affair, of great magnitude* or *moment, to accomplish which*, or *to perform which, one finds not the way:* (Ḥar p. 436:) and *a quality*, or *property.* (Mṣb.) You say, سُمْتُهُ خُطَّةَ خَسْفٍ [*I required*, or *constrained, him to do an affair of difficulty;* or *to become in a state of abasement*, or *ignominy*]: and خُطَّةَ سُوْءٍ [*an evil affair*]. (L.) And هُوَ يُكَلِّفُنِى خُطَّةً مِنَ الخَسْفِ [*He requires*, or *constrains, me to do an affair of difficulty; &c.*]. (JK. [See also خَسْفٌ.]) And it is said in a trad., of Ḳeyleh, أَيُلَامُ ٱبْنُ هٰذِهِ أَنْ يَفْصِلَ الخُطَّةَ وَيَنْتَصِرَ مِنْ وَرَآءِ الحَجَزَةِ [*Is the son of this* woman *to be blamed for deciding the affair*, or *matter*, or *case, &c., and defending himself in the absence of the wrongdoers who would prevent his obtaining his right;* or, *of those who defend men, one from another, and decide between them justly?*]: i. e., when a dubious event, to the encountering of which he does not find the right way, befalls him, that he should not care for it, but decide it so as to settle it and extricate himself from it. (Ṣ, TA.) [See also حَاجِزٌ.] Also, in a trad. respecting El-Ḥodeybiyeh, لَا يَسْأَلُونِى خُطَّةً يُعَظِّمُونَ فِيهَا حُرُمَاتِ ٱللّٰهِ تَعَالَى إِلَّا أَعْطَيْتُهُمْ إِيَّاهَا [*They shall not ask of me a matter wherein they honour the sacred things of God, (exalted be He,) but I will grant it to them*]. (TA.) And in the same, قَدْ عَرَضَ عَلَيْكُمْ خُطَّةَ رُشْدٍ فَٱقْبَلُوهَا *He hath proposed to you a case of evident rectitude; therefore do ye accept it.* (TA.) And Taäbbata-sharrà says,

* هُمَا خُطَّتَا إِمَّا إِسَارٌ وَمِنَّةٌ *
* وَإِمَّا دَمٌ وَالقَتْلُ بِالحُرِّ أَجْدَرُ *

[*They are two cases; either bondage and reproach, or else blood; and slaughter is more befitting to the free*, or *ingenuous*]: he means خُطَّتَانِ. (Ṣ. [See Ḥam p. 34.]) — Also *A course:* as in the phrase خُطَّةٌ نَائِيَةٌ *A distant*, or *far-extending, course.* (Ṣ, TA.) You say also, خُذْ خُطَّةً, i. e. خُذْ خُطَّةَ الاِنْتِصَافِ [*Take thou the course of exacting thy right*, or *due, with equity*]; meaning اِنْتَصِفْ [*exact thou thy right*, or *due, with equity*]. (Ṣ.) — *A proof; an evidence; a testimony; an argument; a plea;* or *an allegation;* syn. حُجَّةٌ. (O, TA.) So in the phrase, أَقِمْ عَلَى هٰذَا الأَمْرِ بِخُطَّةٍ [app. *Establish thou a proof, &c., against this thing*, or *case*]; as is said in the "Nawádir." (TA.) — *An object of want which one has determined to accomplish:* as in the saying, جَآءَ وَفِى رَأْسِهِ خُطَّةٌ *He came having in his mind* [lit. *his head*] *an object of want &c.:* [but see the last sentence in this paragraph:] the vulgar say خِطَّةٌ : (Ṣ, L:) the former is the word used by the Arabs: (L:) the latter, however, occurs in the "Nawádir" of AZ; therefore the attribution of it to the vulgar demands consideration. (TA.) — *Boldness to undertake affairs.* (Ḳ.) — *Ignorance.* (Ḳ.) You say, فِى رَأْسِهِ خُطَّةٌ *In his head is ignorance:* or, as some say, *some affair:* and it has another meaning explained above. (TA.)

خِطَّةٌ *A piece of ground*, or *land, which a man takes to himself, and upon which he makes a mark, in order to its being known that he has chosen it to build there a house;* whence the خِطَط of El-Koofeh and of El-Baṣrah: (Ṣ:) or *a piece of ground, and a house, which a man takes to himself, and upon which he makes a mark, in land not possessed, that he may have it for himself exclusively, and build there;* this being done when the Sulṭán gives permission to a number of the Muslims to found houses in a particular place, and to make their abodes there, as they did in El-Koofeh and El-Baṣrah: (L:) or *a place which is taken and marked for building houses*, or *for habitation*, or *the like:* (Mgh, Mṣb:) or, as is said in the Bári', *a piece of ground*, or *land, which a man takes to himself, and upon which he makes a mark, it not having belonged to any one before him;* as also ↑ خِطٌّ ; (Mṣb;) which latter is explained by IDrd as signifying *a place which one takes to himself, and marks, from other places:* (IB, L:) or both signify *a piece of land in which one alights and takes up his abode, none having done so before him:* (Ḳ:) the pl. of the former is خِطَطٌ. (Ṣ, Mṣb.) — [Hence,] فُلَانٌ بَيِّنُ خِطَطِ المَكَارِمِ ‡ [*Such a one exhibits in himself the marks of generous*, or *honourable, qualities*]. (TA.)

خَطُوطٌ A wild bull, (Ṣ, L,) and any beast, (L,) *that marks the ground with the extremities of his hoofs.* (Ṣ, L.)

خَطِيطٌ وَادٍ خَطِيطٌ [*A valley not rained upon*]. (AO, TA voce خَطْوَةٌ, q. v.) And أَرْضٌ خَطِيطَةٌ [or خَطِيطَةٌ] *Land not rained upon;* (TA;) as also ↑ خِطٌّ : (Ḳ:) or *land not rained upon between two lands that have been rained upon:* (Ṣ, Ḳ:) or *land of which part has been rained upon*, (Ḳ, TA,) *and part has not:* (TA:) or *land not rained upon surrounded by land that has been rained upon;* (ISh;) as also ↑ the latter word: (AḤn:) pl. of the former, خَطَائِطُ. (Ṣ.) — Hence the saying of a certain Arab, to his son, اِلْزَمْ خَطِيطَةَ الذُّلِّ مَخَافَةَ مَا هُوَ أَشَدُّ مِنْهُ ‡ [*Keep thou to the condition of abasement in fear of what is more grievous than it*]. (IAạr, M.) — خَطِيطَةٌ also signifies *A strip of ground differing in roughness and smoothness from what is on either side of it:* pl. as above. (L.)

خَطَّاطٌ *A practiser of what is termed* عِلْمُ الخَطِّ [*or geomancy*]. (Lth.) — [Also *A practiser of the art of writing:*] *a caligraphist.* (KL.)

رِمَاحٌ خَطِّيَّةٌ *Spears of El-Khaṭṭ;* so called from الخَطّ, a place in El-Yemámeh, (Ṣ, Mṣb,) also called خَطُّ هَجَرٍ, (Ṣ,) because they are brought thither (Ṣ, Mṣb) from India, (Ṣ,) and straightened in that place, (Ṣ, Mṣb,*) which is a coast for ships; not that the canes grow there: (Mṣb:) or they are so called from الخَطّ which is the station for ships in El-Baḥreyn, because they are sold there; not that it is the place of their growth: this place is also called الخُطّ : (Ḳ:) but this demands consideration; for it is said [in the 'Eyn, i. e.] by Lth, (TA,) or by Kh, (Mṣb,) that when you convert the rel. n. into a subst., you say ↑ خِطِّيَّةٌ, (Mṣb, TA,) with kesr to the خ, (Mṣb,) without رماح, like as you say, ثِيَابٌ قِبْطِيَّةٌ, (Mṣb, TA,) with kesr, (Mṣb,) but when you convert the rel. n. into a subst., you say, قُبْطِيَّةٌ, (Mṣb, TA,) with ḍamm, to distinguish the subst. from the rel. n., without ثياب : (Mṣb:) a single spear of this kind is called رُمْحٌ خَطِّىٌّ : (TA:) AḤn says that الخَطِّىُّ signifies *the spears;* and that it is a rel. n. used in the manner of a proper name; being a rel. n. from الخَطّ, which is خَطُّ البَحْرَيْنِ, where ships moor when they come from India. (TA.)

خِطِّيَّةٌ : see the next preceding paragraph.

أَخَطُّ † *Delicate in beauties.* (IAạr.) [See also مُخْتَطٌّ.]

مَخَطٌّ [*A place marked with a line* or *lines, with a streak* or *streaks*, or *with a stripe* or *stripes*]. (TA in art. طر.)

مِخَطٌّ *A wooden instrument with which one makes lines* or *marks* or *the like:* (Ṣ:) or *the wooden instrument with which the weaver makes lines* or *marks* or *the like, in*, or *upon, a piece of cloth.* (L, Ḳ.)

مُخَطَّطٌ A [garment of the kind called] كِسَآء, (Ṣ, TA,) and a date, and a wild animal, (TA,) or anything, (Ḳ, TA,) *marked with lines, streaks*, or *stripes.* (Ṣ, Ḳ, TA.) — ‡ *Beautiful;* (Ḳ, TA;) applied to a boy [whose hair of his beard has appeared upon the sides of his face, forming lines]; as also ↑ مُخْتَطٌّ [originally مُخْتَطَطٌ : see 8]. (TA.)

مِخْطَاطٌ [*A wooden ruler;*] *an instrument of wood by means of which lines are made even.* (Ṣ, O.)

مَخْطُوطٌ A book or the like *written in*, or *upon*. (TA.)

مُخْتَطٌّ : see مُخْتَطَطٌ.

خطأ

1. خَطِئَ is *syn. with* ▼ اخطأ, inf. n. إِخْطَآءٌ and ▼ خَاطَئَةٌ, (Ḳ,) which latter, mentioned by AAF, on the authority of AZ, is extr. in the case of a triliteral [unaugmented] verb, and more so in the case of a quadriliteral [i. e. a triliteral augmented by one letter]; (TA;) and with ▼ تَخَطَّأَ; signifying *He did wrong*; or *committed a mistake*, or *an error*: (Ḳ:) [and if this and similar explanations be correct, خَطَأٌ may be an inf. n. of the first of these verbs, and a quasi-inf. n. of the second and third:] or ▼ اخطأ and ▼ تَخَطَّأَ have this signification: (Ṣ:) and خَطِئَ, aor. ـَ, inf. n. خِطْءٌ and خِطْأَةٌ, (Ṣ, Ḳ,) signifies *he committed a sin, a crime*, or *an act of disobedience for which he deserved punishment*: (Ṣ, Ḳ:*) or *he committed a fault* or *an offence* or *an act of disobedience* [in an absolute sense]: (Ḳ,* TA:) or, accord. to AO, (Mṣb,) or A 'Obeyd, (TA,) خَطِئَ, inf. n. خِطْءٌ, signifies *he committed a fault, an offence*, or *an act of disobedience, unintentionally*; as also ▼ اخطأ: (Mṣb, TA:) or, as others say, خَطِئَ means [*he committed a fault*, &c.,] *in religion*; and ▼ اخطأ, *in anything*; *intentionally* or *unintentionally*: (Mṣb:) خَطِئَ, *in religion*; and ▼ اخطأ, *in calculation* [&c.]: (Aṣ, M, TA:) or, accord. to Ibn-'Arafeh, (TA,) you say, خَطِئَ فِى دِينِهِ, (Ḳ, TA, [in a MS. copy of the Ḳ and in the CḲ,* فى ذَنْبِهِ,]) and ▼ اخطأ, meaning *he pursued a wrong way in his religion, intentionally* or *otherwise*: (Ḳ, TA:) or خَطِئَ signifies *he committed an act of disobedience intentionally*; (Mṣb, TA;) so accord. to the 'Ináyeh, and the like is said in the A; (TA;) and ▼ اخطأ, *he did wrong, meaning to do right*: (Aṣ, Mṣb, TA:) [and this distinction is agreeable with general usage:] accord. to AHeyth, you say, خَطِئْتَ بِمَا صَنَعْتَهُ [*Thou didst wrong, in that which thou didst*,] *intentionally*; and ▼ أَخْطَأْتَ مَا صَنَعْتَهُ [or فيما صنعته or بما صنعته *Thou didst wrong, in that which thou didst*,] *unintentionally*. (TA.) — See also 4, in two places. — خَطِئَتِ القِدْرُ بِزَبَدِهَا, aor. ـَ, ‡ *The cooking-pot threw up its froth*, or *foam*, or *scum*, (Ḳ, TA,) *in boiling*. (TA.)

2. خَطَّأَهُ, (Ṣ, Mṣb, Ḳ,) inf. n. تَخْطِئَةٌ and تَخْطِىءٌ, (Ṣ, Ḳ,) *He said* to him, أَخْطَأْتَ [meaning *Thou hast done wrong*, or *committed a mistake* or *an error*]: (Ṣ, Mṣb, Ḳ:) or *he pronounced him*, or *asserted him, to be doing wrong*, or *committing a mistake* or *an error*. (Mṣb.) You say, إِنْ أَخْطَأْتُ فَخَطِّئْنِى [*If I do wrong*, &c., *tell me that I have done so*]. (Ṣ.) — Also *He made it to miss*: so in the saying, خَطَّأَ ٱللّٰهُ نَوْءَهَا *God made*, or *may God make, its* [i. e. a land's] *star*, or *asterism, to miss*; so that the rain which the star or asterism should have brought did not, or shall not, fall upon it. (TA.) This was [also] said by I 'Ab [in a tropical sense] with reference to a woman, as an imprecation, in disapproval of her conduct. (Mgh.) As some relate this saying, the verb is خَطَّى, (Mgh, TA,) and the meaning, *God made*, or *may God make, its* [*rain-giving*] *star* or *asterism, to pass it over, and not send rain upon it*: and in this case it may be, (TA,) or it is, (Mgh,) from خَطِيطَةٌ, signifying "a land not rained upon (Mgh, TA) between two lands that have been rained upon;" (Mgh;) the verb being originally خَطَّطَ, and the final ط being changed into ى. (Mgh, TA. [See art. خط.]) نَوْء is [here] the sing. of أَنْوَآء meaning the "Mansions of the Moon," also called the "stars, or asterisms, of rain." (Mgh.) [See more in the first paragraph of art. خط: and see also 4 in the present art.] Accord. to Fr, خَطَّى السَّهْمَ and خَطَّأَهُ are syn. [as meaning *He made the arrow to pass over*, or *to miss*, the mark]. (TA.) One says also, خُطِّئَ عَنْكَ السُّوْءُ [*May evil be made to miss thee*;] i. e. *may evil be repelled from thee*. (ISk.) And خَطَأً عَنْكَ السُّوْءُ [app. for ▼ تَخَطَّأَ] *Evil missed thee*, or *may evil miss thee*. (AZ.)

4. اخطأ, inf. n. إِخْطَآءٌ and خَاطِئَةٌ: see 1, in eight places. أَخْطَيْتُ, for أَخْطَأْتُ, should not be said: (Ṣ:) it is a word of weak authority; or a mispronunciation: (Ḳ:) but some use it; (Ṣ, Ṣgh, TA;) because a change of this kind is generally allowed by some of the writers on inflection. (TA.) See also 5. — اخطأه, (Ṣ, Ḳ,) which signifies, *He* [or *it*] *missed*, or *failed of hitting, it* [or *him*], (TA,) and ▼ تَخَاطَأَهُ (Ṣ, Ḳ) and ▼ تَخَطَّأَهُ (Ḳ) and ▼ تَخَطَّأَ لَهُ (TA) [and ▼ خَطِئَهُ, as will be seen from what follows,] are syn. (Ṣ, Ḳ, TA.) [See also 2, last sentence.] You say, اخطأ الرَّامِى الغَرَضَ *The archer*, or *thrower, missed the mark*; or *failed of hitting it*. (TA.) And اخطأه السَّهْمُ *The arrow* [*missed it*, or *him*, or] *passed beyond it*, or *him*: and you may also say, أَخْطَاهُ, suppressing the ء. (Mṣb.) And اخطأ الطَّرِيقَ [*He missed the way*; or] *he deviated from the way*. (TA.) And اخطأ نَوْؤُهُ [†*His star*, or *asterism, missed*]; said of him who has sought an object of want and not succeeded in attaining it: (TA:) and to a person in this case one says, اخطأ نَوْؤُكَ [†*Thy star*, or *asterism, has missed*]. (Mgh. [See also 2.]) And اخطأه الحَقُّ *The right*, or *due, was*, or *became*, [*out of his reach*,] or *far from him*. (Mṣb.) Owfà Ibn-Maṭar El-Mázinee says,

* تَخَاطَأَتْ ▼ النَّبْلُ أَحْشَآءَهُ *

[meaning *The arrows missed his bowels*]. (Ṣ.) And AO, (Ṣ,) or A 'Obeyd, (TA,) says that ▼ خَطِئَ and اخطأ are syn.; citing, as an ex., the saying of Imra-el-Ḳeys,

* يَا لَهْفَ هِنْدٍ إِذْ خَطِئْنَ كَاهِلَا *

(Ṣ, TA,) meaning [*O the grief of Hind*,] *when they* (the troop of horse) *missed* the sons of *Káhil*; (TA;) خَطِئْنَ being here used in the sense of أَخْطَأْنَ, (Ṣ, TA,) which latter, accord. to Az, is the more proper in this case. (TA.) — مَا أَخْطَأَهُ is an expression of wonder [meaning *How sinful*, or *criminal*, or *intentionally-disobedient*, or *intentionally-wrongdoing, is he!*] from خَطِئَ, not from أَخْطَأَ. (Ṣ.)

5: see 1, in two places: — and see also 2, last sentence; and 4, in two places. — تخطّأ لَهُ فِى المَسْأَلَةِ *He addressed to him the question with the desire of causing him to make a mistake*: (TA:) or i. q. ▼ أَخْطَأَ. (Ṣ.) — تَخَطُّؤٌ also signifies The *feigning a wrong action, a mistake*, or *an error*. (KL. [See also 6.]) — And The *charging* another *with a wrong action, a mistake*, or *an error*. (KL. [See also 2.])

6. تخاطأ *He imputed to himself a wrong action, a mistake*, or *an error, not having committed any*. (KL. [See also 5.]) — See also 4, in two places.

10. استخطأت *She* (a camel) *did not conceive*, or *become pregnant*. (TA. [See also the part. n., below.])

خِطْءٌ : see خَطَأٌ.

خِطْأَةٌ : see خَطِيئَةٌ.

خَطَأٌ *A wrong action*; *a mistake*, or *an error*; *contr. of* صَوَابٌ; as also ▼ خَطَآءٌ (Ṣ, Mṣb, Ḳ) and ▼ خِطْءٌ: (Ḳ:) accord. to some, it is syn. with خَطِيئَةٌ and خِطْءٌ; and is an inf. n. used as a simple subst.; but accord. to others, (TA,) it signifies *an unintentional fault* or *offence* or *disobedience*; (Ḳ, TA;) a subst. from أَخْطَأَ: (M, Mṣb: [see 1, first sentence:]) and accord. to the M, ▼ خِطْءٌ is a subst. from خَطِئَ [and therefore *syn. with* خَطِيئَةٌ accord. to the general acceptation of خَطِئَ]. (TA.)

خِطْأَةٌ [so in the TA, app. خَطْأَةٌ,] *A land which the rain misses, while it falls upon another near it*. (TA. [See 2.])

خَطَآءٌ : see خَطَأٌ, in two places.

خَطِيئَةٌ (Ṣ, Ḳ) and خَطِيَّةٌ, a change of this kind being allowable in this and in similar cases, (Ṣ, TA,) *A fault, an offence*, or *an act of disobedience*; (Ṣ, Ḳ;) or *such as is intentional*; (Ḳ;) like ▼ خِطْءٌ, (Ṣ, Ḳ,) which is an inf. n., thus used as a subst.; (Mṣb;) meaning *a sin, a crime*, or *an act of disobedience for which one deserves punishment*: (Ṣ:) pl. خَطَايَا, (Lth, Ṣ, Ḳ,) originally خَطَائِئُ; (Lth, Ṣ;) and خَطَائِى also, (Ḳ, TA, [in a MS. copy of the Ḳ خَطَائِئُ,]) or this is [anomalous and] incorrect, unless with the art. ال, being otherwise خَطَاءٍ; (MF;) and خَطَائِى, [an anomalous pl.,] of which Th gives an ex. in the following verse, related to him by IAạr:

* لِكُلِّ آمْرِئٍ مَا قَدَّمَتْ نَفْسُهُ لَهُ *

* خَطَائِيهَا إِنْ أَخْطَأَتْ وَصَوَابُهَا *

[*For every man* is appointed, in the world to come, the recompense of *what his soul has prepared*, or *laid up in store, for him*: *its wrong actions, if it have done wrong; and its right action*]. (L.) —— خَطِيَّةُ لَيْلَةٍ and خَطِيَّةُ يَوْمٍ are expressions like طِيلُ لَيْلَةٍ and طِيلُ يَوْمٍ: you say, خَطِيَّةُ يَوْمٍ يَمُرُّ بِى إِلَّا أَرَى فِيهِ فُلَانًا [app. meaning *It were a crime that a day should pass with me without my seeing in it such a one*; or perhaps, *it is a rare event that a day passes with me* &c.: see what follows]. (TA.) —— *A little*, or *small quantity*; or *a few*, or *small number*; of anything. (Ḳ, TA.) You say, عَلَى النَّخْلَةِ خَطِيَّةٌ مِنْ رُطَبٍ [*Upon the palm-tree are a few fresh ripe dates*]: and بِأَرْضِ بَنِى فُلَانٍ خَطِيَّةٌ مِنْ وَحْشٍ [*In the land of the sons of such a one is*] *a small number of wild animals that have missed their* [*wonted*] *places and are in what are not their accustomed places.* (TA.)

خَطَّآءٌ A man *who constantly adheres to faults, offences, sins, crimes, or acts of disobedience for which he deserves punishment.* (TA.)

خَاطِئٌ *Intentionally doing that which is not right*; (El-Umawee, Ṣ;) *intentionally pursuing a wrong way* in his religion; (Ḳ;) *intentionally doing that which he is forbidden to do.* (Mṣb.) [See خَطِئَ, of which it is the part. n.] —— [Also *Missing the mark.* Hence the saying,] مَعَ الخَوَاطِئِ سَهْمٌ صَائِبٌ [*With those that miss is an arrow that goes right*, or *hits the mark*]; (Ṣ, Ḳ;) خَوَاطِئُ being pl. of خَاطِئَةٌ, meaning *that misses the butt*: (Ḥar p. 481:) a prov., (Ṣ,) applied to him who frequently errs, but sometimes does right; (Ṣ, Ḳ;) or to the niggard who sometimes gives notwithstanding his niggardliness. (A 'Obeyd.)

خَاطِئَةٌ: see 1, first sentence.

مُخْطِئٌ [act. part. n. of 4, q. v.;] One *who does wrong, meaning to do right.* (El-Umawee, Ṣ.)

[مُتَخَطِّئٌ signifies the same as مُتَخَطٍّ, or nearly so: and hence the saying,] نَاقَتُكَ مِنَ المُتَخَطِّئَاتِ الجِيفَ, (TA in the present art.,) or نَاقَتُكَ هٰذِهِ مِنَ المُتَخَطِّيَاتِ الجِيفَ, [the last word being app. الجِيَفَ, and the lit. meaning, *Thy she-camel*, or *this thy she-camel, is of those that step over the carcasses*;] i. e. she is *hardy and strong, such as will go on, and leave behind* [*others that have fallen down and died*] (تخلف [so in the TA, app. تُخَلِّفُ,]) until she [herself] has fallen down (الى مد سقطت). (AZ, TA in art. خطو.)

مُتَخَطِّئَةٌ, applied to a she-camel, ‡ *i. q.* حَائِلٌ [i. e. *Not conceiving*, or *not becoming pregnant during a year*, or *two years*, or *some years*; &c.: see its verb, 10]. (Ḳ, TA.)

خطب

1. خَطَبَ, (Ṣ, Ḳ,) or خَطَبَ خُطْبَةً, (A,) aor. ـُ, (TA,) inf. n. خُطْبَةٌ, (Ṣ, Ḳ,) or this, accord. to some, is a subst. used as an inf. n., (TA,) and خَطَابَةٌ, (Ḳ,) said of a خَطِيب (A) or خَاطِب, (Ḳ,) عَلَى المِنْبَرِ [*He recited a* خُطْبَة (q. v. infrà),] [*upon the pulpit*]; (Ṣ, Ḳ;) as also ▼اختطب. (Ṣ.) And خَطَبَ القَوْمَ, (Mṣb,) and عَلَى القَوْمِ, (Th, Mṣb,) aor. ـُ, (Mṣb,) inf. n. خُطْبَةٌ, (Th, Mṣb,) [*He recited a* خُطْبَة *to the people*, and *over the people*, i. e. on the pulpit, beneath which they sat: or] *he delivered an exhortation*, or *admonition, to the people.* (Mṣb.) —— خَطَبَ المَرْأَةَ, (Ṣ, A,* Mṣb, Ḳ,) aor. as above, (TA,) inf. n. خِطْبَةٌ, (Ṣ, A, Ḳ,) or this is a simple subst., (Mṣb,) and خَطْبٌ (Lḥ, Ḳ) and خِطِّيبَى; (T, Ṣ,* Ḳ;) and ▼اختطبها; (Ṣ, Ḳ;) *He asked*, or *demanded, the woman in marriage.* (Mṣb.) In the following verse of 'Adee Ibn-Zeyd, respecting Jedheemeh El-Abrash, and his asking in marriage Ez-Zebbà,

* لِخِطِّيبَى الَّتِى غَدَرَتْ وَخَانَتْ *
* وَهُنَّ ذَوَاتُ غَائِلَةٍ لُحِينَا *

[*For the asking in marriage of her who acted perfidiously and treacherously: for they* (i. e. women) *are possessed of secret malevolence: may they be disgraced and accursed*:] خِطِّيبَى is syn. with خِطْبَةٌ: (Ṣ:) accord. to Lth, it is a simple subst.; but AM says that he is in error, and that it is an inf. n. (TA.) You say also, خَطَبَ المَرْأَةَ إِلَى القَوْمِ *He asked*, or *demanded, the woman in marriage, of the people.* (Mṣb.) And خَطَبَ إِلَى فُلَانٍ [*He asked*, or *demanded, a woman in marriage, of such a one*]. (TA.) And خَطَبَ عَلَى خِطْبَةِ أَخِيهِ *He asked*, or *demanded, a woman in marriage, when another had done so, and she had inclined to the latter, and he and she had agreed to a certain dowry, and had approved each other, and nothing remained but to conclude the contract*; the doing of which is forbidden: but it is not forbidden to ask in marriage a woman when another has done so if she and the latter have not agreed, nor approved each other, nor has either of them inclined to the other. (TA.) —— فُلَانٌ يَخْطُبُ عَمَلَ كَذَا ‡ *Such a one seeks*, or *desires, to do such a thing.* (A, TA.) = خَطُبَ, [aor. ـُ,] inf. n. خَطَابَةٌ, *He became a* خَطِيب. (Ṣ.) = خَطِبَ, aor. ـَ, (JK, Ḳ,) inf. n. خَطَبٌ, (Ṣ, TA,) *He*, or *it, was*, or *became, of the colour termed* خُطْبَةٌ: (Ṣ,* Ḳ:) or *his*, or *its, colour was*, or *became, what is thus termed.* (JK.)

2. خَطَّبَهُ *He granted his request of a woman in marriage*; as also ▼اخطبه. (TA.)

3. خَاطَبَهُ, (A, Mṣb, TA,) or خَاطَبَهُ بِالكَلَامِ, (Ṣ, TA,) inf. n. مُخَاطَبَةٌ and خِطَابٌ, (Ṣ, Mṣb,) *He talked, spoke, conversed*, or *discoursed, with him*; *held a colloquy, dialogue, conversation*, or *discourse, with him*: (Mṣb, TA:) *he talked to him, spoke to him*, or *addressed him, face to face*; *accosted him with speech or words.* (A.) [Hence,] حَرْفُ خِطَابٍ [*A particle of allocution*]: such is the تَ in أَنْتَ and أَنْتِ, (Mughnee and Ḳ on the letter ت,) and such is the ك in ذَاكَ &c. (I 'Aḳ p. 36; &c.) —— *He consulted with him.* (TA.) —— فَصْلُ الخِطَابِ [as used in the Ḳur xxxviii. 19] means *The deciding a case*, or *passing sentence*, or *judging, with evident demonstration*, or *proof*; or *by testimony confirmed by oath*: (Ḳ, TA:) or *the deciding between truth and falsehood, and distinguishing between just judgment and the contrary thereof*: (TA:) or *understanding, intelligence, sagacity*, or *knowledge, in judging* or *passing sentence*: or *the pronouncing the phrase* أَمَّا بَعْدُ, (Ḳ, TA,) which David [it is said] was the first to utter, and which means, accord. to Abu-l-'Abbás, *Now, after these preliminary words*, [I proceed to say] thus and thus; (TA;) or this last phrase means *after my prayer for thee*; (Ḳ in art. بعد;) or *after praising God.* (TA in art. بعد. [See also art. فصل.])

4: see 2. —— [Accord. to the KL, إِخْطَابٌ signifies The *inviting* one *for the purpose of marriage*: but I think it is only اِخْتِطَابٌ that has this signification.] —— أَخْطَبَكَ الصَّيْدُ ‡ *The game*, or *object of the chase, has become within thy power*, or *reach*; (Ṣ, A;) and *has become near thee*; (Ṣ;) فَارْمِهِ [*therefore shoot it, or cast at it*]. (A.) And أَخْطَبَكَ الأَمْرُ ‡ *The thing*, or *affair, has become within thy power*, or *reach.* (JK, A.*) = اخطب, said of the colocynth, (JK, Ṣ, Ḳ,) *It became striped with green*: (JK, Ḳ:) or *it became yellow, with green stripes.* (Ṣ.) And اخطبت الحِنْطَةُ *The wheat became coloured.* (TA.)

6. تخاطبا *They two talked, spoke, conversed*, or *discoursed, each with the other*; *held a colloquy, dialogue, conversation*, or *discourse, each with the other.* (TA.) [Hence, عُرْفُ التَّخَاطُبِ *The generally-known, generally-received*, or *conventional, language of conversation.*]

8: see 1, in two places. —— اِخْتَطَبُوهُ *They invited him to marry a woman of their family*: (Ṣ, Mṣb, Ḳ:) or *they invited him to ask*, or *demand, in marriage, a woman of their family.* (AZ,* A. [See also 4.])

خَطْبٌ ‡ *A thing, an affair*, or *a business*, (A, Ḳ, MF,) *small* or *great*, (Ḳ,) *that one seeks*, or *desires, to do*, (A,) syn. أَمْرٌ مَخْطُوبٌ, (Ḥam p. 33,) or *that is*, or *may be, a subject of discourse*: this is the primary signification: (MF:) or *a great thing* or *affair*: or *a thing*, or *an affair, that is disliked*; not one that is liked: or *that is liked* also: (Ḥam p. 127:) or *the cause*, or *occasion, of a thing* or *an event*: (JK, Ṣ:) or *an affliction*; *a calamity*: (Mṣb:) [often used in this last sense in the present day:] and *a state*, or *condition*: (TA:) pl. خُطُوبٌ; (A, Mṣb, Ḳ;) for which خُطُب is used in a verse below. (TA.) You say, مَا خَطْبُكَ ‡ *What is the thing*, or *affair*, or *business, that thou seekest*, or *desirest, to do?* (A:) or *what is thy cause* [of coming &c.]? (Ṣ.) خَطْبٌ يَسِيرٌ and جَلِيلٌ ‡ [*A little*, or *an unimportant*, and *a great*, or *an important, thing* or *affair*]. (A.) And هُوَ يُقَاسِى خُطُوبَ الدَّهْرِ ‡ [*He endures*, or *he contends*, or *struggles, with*, or *against, the afflictions*, or *calamities, of fortune*]. (A.) El-Akhṭal says,

* كَلَمْعِ أَيْدِى مَثَاكِيلَ مُسَلِّبَةٍ *
* يَنْدُبْنَ ضَرْسَ بَنَاتِ الدَّهْرِ وَالخُطُبِ *

† [*Like the wavings of the hands of mothers bereft of many children, in mourning on account of*

them, bewailing the biting cruelty of the daughters of misfortune and afflictions]: using الخطب for الخطوب. (L.)

خَطْبٌ: see the next paragraph, in two places.

خِطْبٌ A man *who asks*, or *demands, a woman in marriage*; (Ṣ, A,* Ḳ;*) as also ▼ خُطْبٌ (MF) and ▼ خَاطِبٌ (A, Mṣb, Ḳ) and ▼ خِطِّيبٌ: pl. of the first أَخْطَابٌ, (Ḳ,) and of the second خُطَّابٌ, (A,) and of the last خِطِّيبُونَ. (Ḳ.) You say, هُوَ خِطْبُهَا [and ▼ خَاطِبُهَا] and ▼ خِطِّيبُهَا *He is her asker*, or *demander, in marriage*. (Ḳ,* TA.) It was a custom, in the Time of Ignorance, for a man to stand up and to say خِطْبٌ, (A, Ḳ,*) and ▼ خُطْبٌ, (Ḳ,) meaning I am *an asker*, or *demander, in marriage*; (MF;) and he who desired to give to him in marriage would reply نِكْحٌ, (A, Ḳ,*) and نُكْحٌ, (Ḳ,) [meaning I am "a giver in marriage,"] and thus marriage was effected: there was a woman among them, called Umm-Khárijeh, and the man who asked her in marriage used to stand at the door of her tent, and say, خِطْبٌ; and she used to reply, نِكْحٌ; (Ṣ,* TA;) and hence the prov., أَسْرَعُ مِنْ نِكَاحِ أُمِّ خَارِجَةَ [Quicker than the marriage of Umm-Khárijeh]. (TA.) — Also A woman *asked*, or *demanded, in marriage*; (Ṣ, A,* Ḳ;*) and so ▼ خِطْبَةٌ (Ṣ, Ḳ) and ▼ خِطْبَى (Kr, Ḳ) and ▼ خِطِّيبَةٌ (A, Ḳ) and ▼ خِطِّيبَى [which is also an inf. n.: see 1]: (Ḳ:) or this last signifies a woman *often asked*, or *demanded, in marriage*. (JK.) You say, هِيَ خِطْبُهُ and ▼ خِطْبَتُهُ (Ṣ, Ḳ) &c. (Ḳ) *She is the person asked*, or *demanded, in marriage by him*. (Ṣ, Ḳ.*)

خُطْبَةٌ, a word of the measure فُعْلَةٌ in the sense of the measure مَفْعُولَةٌ, like نُسْخَةٌ in the sense of مَنْسُوخَةٌ, and غُرْفَةٌ مِنَ المَآءِ in the sense of مَغْرُوفَةٌ; meaning *An exhortation* or *admonition* [*recited by a* خَطِيب]: (Mṣb:) *a form of words, a discourse, a sermon, a speech, an oration*, or *a harangue, which the* خطيب *recites on the pulpit*: (Ḳ,* TA:) [in the noon-service of the congregational mosque on Friday, the خطيب recites two forms of words, each of which is thus termed: the former chiefly consists of expressions of praise to God, blessings on Moḥammad and his family and companions, and exhortation to the congregation; and is termed خُطْبَةُ الوَعْظِ: the latter, of praise to God, exhortation, blessings on Moḥammad and his family and companions, and prayer for the Muslims in general, and especially for the Sovereign; and is termed خُطْبَةُ النَّعْتِ: (see my "Modern Egyptians," ch. iii.:)] or, [accord. to its original signification,] with the [Pagan] Arabs, *a discourse, a speech, an oration*, or *a harangue*, [generally applied to one *delivered in public*,] *in rhyming prose*; and *the like*: (Aboo-Is-ḥáḳ, Ḳ:) or the old Arabian خُطْبَة, in the Pagan and the early Muslim ages, was, *in most instances, not in rhyming prose*; and the term "prose," as here used, does not exclude what contains poetry introduced by way of testimony and the like: (MF:) or [*a tract*, or *small treatise or discourse*,] *like a* رِسَالَة, *which* [*is complete in itself, or*, in other words,] *has a beginning and an end*: (T, TA:) the pl. is خُطَبٌ: (Mṣb:) and ▼ مَخَاطِبُ, occurring in the following words of a trad., مِنْ أَهْلِ المَحَاشِدِ وَالمَخَاطِبِ, meaning *of those who congregate, and harangue people*, exciting them to go forth and assemble for seditious purposes, is said to be used in the same sense as خُطَب, and to be a pl. [of خُطْبَة], contr. to rule, like مَشَابِهُ [pl. of شِبْهٌ] and مَلَامِحُ [pl. of لَمْحَةٌ]: or it is pl. of ▼ مَخْطَبَةٌ, which is syn. with خُطْبَةٌ: (TA:) or it [is pl. of ▼ مَخْطَبٌ, and] signifies *places of haranguing*. (L in art. حشد.) You say, خَطَبَ الخَطِيبُ ▼ خُطْبَةً حَسَنَةً [*The* خطيب *recited a beautiful* خطبة]. (A.) — See also خُطْبٌ. — Also *A turbid*, or *dusky, colour*, (Ḳ,) or *a colour inclining to turbidness* or *duskiness*, (TA,) *mixed with yellowish red*; (Ḳ, TA;) *like the colour of wheat before it dries, and that of some wild asses*: (TA:) and *a green* [app. here meaning *a dark*, or *an ashy, dust-*] *colour*: (TA:) or *a dust-colour suffused with* خُضْرَة: [or *a dark*, or *an ashy, dust-colour*: see أَخْطَبُ:] (A, Ḳ:) or خُضْرَة *mixed with black*. (TA.) — The saying, أَنْتَ الأَخْطَبُ ▼ البَيِّنُ الخُطْبَةِ, which might be imagined to ascribe to the person addressed perspicuity, or eloquence, in his خُطْبَة, really means *Thou art* [*the asinine*;] *he who bears evidence of* الحِمَارِيَّة [i. e. *asinineness*]. (A.)

خِطْبَةٌ an inf. n. of خَطَبَ المَرْأَةَ: (Ṣ, A, Ḳ:) or a simple subst. (Mṣb.) — See also خِطْبٌ, in two places.

خُطْبَانٌ, and خُطْبَانَةٌ: see أَخْطَبُ, in four places. — The former is also the name of *A certain plant*, (Ḳ,) *of the most bitter of herbs*, (TA,) *resembling the* هِلْيَوْن [or *asparagus*], (Ḳ,) or *like the tails of serpents, with thin extremities resembling* [*in colour*] *the violet, or blacker; the part next below being green; and the part next below that, to the roots, white*: whence the saying, أَمَرُّ مِنَ الخُطْبَانِ [*More bitter than the* خطبان]; in which خطبان has been erroneously said to be pl. of أَخْطَبُ, like as سُودَانٌ is pl. of أَسْوَدُ. (TA.)

خِطْبَانٌ: see أَخْطَبُ.

أَوْرَقُ خُطْبَانِيٌّ [*Of a dusky colour, inclining to black, in a great degree*; or *very dusky*]: the latter word is added to give intensiveness to the signification. (Ḳ.)

خَطِيبٌ [*A speaker*; generally *a public speaker*; *an orator*; *a preacher*;] *a reciter of a* خُطْبَة, (A, Mṣb, TA,) [and particularly] *on the pulpit*; (TA;) *i. q.* ▼ خَاطِبٌ [in these senses; but the latter is generally used in another sense, explained above, voce خِطْبٌ]: (Ṣ, TA:) or *one who recites a* خُطْبَة *well*; (Ḳ, TA;) [*a good speaker* or *orator*:] pl. خُطَبَآءُ. (Mṣb, TA.) See خُطْبَة. You say also, هُوَ خَطِيبُ القَوْمِ, meaning *He is the speaker for the people* or *party*. (Mṣb.)

خِطَابَةٌ The *office of a* خطيب *of a mosque*. (TA.)

خِطَّابٌ A man *practised in*, or *accustomed to, the asking*, or *demanding, women in marriage*. (Ḳ,* Mṣb, TA.)

خِطِّيبٌ: see خِطْبٌ, in two places.

خِطِّيبَةٌ: }
خِطِّيبَى: } see خِطْبٌ.

خَاطِبٌ: see خِطْبٌ, in two places: — and see also خَطِيبٌ.

أَخْطَبُ *Of the colour termed* خُطْبَة. (Ḳ.) — An ass, (Ṣ, A, Ḳ,) i. e. a wild ass, (TA,) *of a colour tinged with* خُضْرَة [here meaning *a dark*, or *an ashy, dust-colour*]: (Ṣ, Ḳ:) or *of a dust-colour suffused with* خُضْرَة: (A:) or *having a black line*, or *stripe, along the middle of the back*: (Fr, Ṣ, Ḳ:) fem. خَطْبَآءُ, applied to a she-ass; (Fr, Ṣ;) and likewise to a she-camel. (Ṣ, A.) — See also خُطْبَة. — حَمَامَةٌ خَطْبَآءُ القَمِيصِ [*A pigeon of the colour termed* خُطْبَة]. (A.) — يَدٌ خَطْبَآءُ, (Ḳ,) and أَنَامِلُ خُطْبٌ, (TA, [خُطْبٌ being the pl.,]) [*A hand, and fingers' ends*,] *of which the darkness of the dye imparted by* حِنَّآء *has faded*: (Ḳ, TA:) and in like manner the epithet [أَخْطَبُ] is sometimes applied to the hair. (TA.) One says also اِمْرَأَةٌ خَطْبَآءُ الشَّفَتَيْنِ [*A woman pale in the lips; whose lips have lost their deep red hue*]. (A.) — حَنْظَلٌ أَخْطَبُ, (Ḳ,) or ▼ خُطْبَانٌ, (Ṣ,) *Colocynths that are yellow*, (Ṣ,) *with green stripes*: (Ṣ, Ḳ:) fem. (applied to a single colocynth, which is termed حَنْظَلَةٌ, TA) خَطْبَآءُ, with which ▼ خُطْبَانَةٌ is syn.: the pl. [or rather coll. gen. n.] of this last [or pl. of أَخْطَبُ] is ▼ خُطْبَانٌ, and ▼ خِطْبَانٌ, which is extr. [with respect to rule]. (Ḳ.) And ▼ خُطْبَانٌ (a pl. of أَخْطَبُ, JK) also signifies *Green leaves of the* سَمُر. (JK, Ḳ.) — الأَخْطَبُ *The* [*bird called*] شِقِرَّاق; (Ṣ, Mgh, Mṣb, Ḳ;) *called in Persian*, accord. to a marginal note in a copy of the Ṣ, كَاسْكِينَهْ: (TA:) or *the* [*bird called*] صُرَد; (Ṣ, Mgh, Mṣb, Ḳ;) because it has a mixture of black and white. (TA.) — And *The* صَقْر [or *hawk*]. (Ḳ.) — And *A certain creeping thing* (دُوَيْبَّة) *of a green colour, longer than the locust, having six legs; called in Persian* شش پايه, *and* سوشكتك. (Mgh.)

أُخْطُبَانٌ a [proper] name of *A certain bird*; (Ḳ, TA;) so called because of a خُطْبَة, i. e. خُضْرَة, in its wings. (TA.)

مَخْطَبٌ: }
مَخْطَبَةٌ: } see خُطْبَة.
مَخَاطِبُ: }

خطر

1. خَطَرَ بِذَنَبِهِ, (Ṣ, A, Mgh, Ḳ, TA,) aor. ـِ, [in the CḲ, erroneously, ـُ,] inf. n. خَطْرٌ and خَطَرَانٌ (Ṣ, Mgh, Ḳ) and خَطِيرٌ, (JK, Ḳ,) *He* (a camel, Ṣ, Mgh, or a stallion [camel], A, Ḳ) *raised his tail time after time, and struck his thighs with*

it: (Ṣ:) or *lashed with it to the right and left:* (Ḳ:) or *moved about his tail:* (A,* Mgh, TA:) the stallion does so in threatening, through pride; (T, TA;) or in fighting with others, as though threatening; (A;) or by reason of emaciation occasioned by severe drought; or by reason of sprightliness: but a she-camel, to inform the stallion that she has become pregnant. (TA.) You say also, خَطَرَ بِذَنبِهِ, aor. ـِ: the غ being a substitute for the خ: (TA:) or each form may be original; but the latter is the less used. (IJ, TA.) — [Hence,] خَطَرَ بِرُمْحِهِ, (A,* Ḳ,) and بِسَيْفِهِ, (Ḳ,) and بِقَضِيبِهِ, and بِسَوْطِهِ, (TA,) inf. n. خَطَرَانٌ, (Ḳ,) ‡ *He moved his spear up and down,* and *his sword,* (Ḳ, TA,) and *his rod,* and *his whip.* (TA.) A man does so with the spear when he walks between the two [opposing] ranks. (A.) — And خَطَرَ بِيَدِهِ فِى مَشْيِهِ ‡ [*He moved his arm up and down in his walking*]. (A.) And خَطَرَ فِى مِشْيَتِهِ, (Ḳ,) aor. ـِ, (TA,) inf. n. خَطَرَانٌ (Ḳ) and خَطِيرٌ, (TA,) † *He moved his arms up and down in his mode of walking,* (Ḳ, TA,) *inclining his body from side to side at the same time.* (TA.) And خَطَرَ, aor. ـِ, (TA,) inf. n. خَطَرَانٌ, (Ṣ,) † *He* (a man) *shook himself in walking;* (Ṣ;) and *walked with an elegant and a proud and self-conceited gait, with an affected inclining of the body from side to side.* (Ṣ, TA.) الجُنْدُ يَخْطِرُونَ حَوْلَ قَائِدِهِمْ † [*The troops strut around their leader*] is said when they show their energy to their leader; and in like manner, when they assemble and equip themselves in war. (TA.) — And خَطَرَ, aor. ـِ, inf. n. خَطْرٌ, † *He* (a man) *raised his arm,* or *hand, with a stone which he lifted for the purpose of trying his strength, to cast,* or *throw, and shook the stone in lifting it.* (TA.) — And خَطَرَ بِإِصْبَعِهِ إِلَى السَّمَآءِ ‡ *He moved his finger,* [or *raised it towards the sky,*] in supplication. (A.) [This one does in the ordinary prayers, in uttering the profession of belief in the unity of God; raising the first finger only (of the right hand, which is placed on the thigh, while sitting on the left foot), and not the hand itself.] — And خَطَرَ, (Ṣ, Ḳ,) aor. ـِ, inf. n. خَطَرَانٌ, (Ṣ,) † *It* (a spear) *quivered, vibrated,* or *shook:* (Ṣ, Ḳ:) or *moved up and down previously to a thrusting with it.* (Ṣ.) — خَطَرَ بِبَالِى, (Ṣ, A, Mgh, Mṣb, Ḳ,) and عَلَى بَالِى, (JK, Mgh, Mṣb, Ḳ,) aor. ـُ (Ṣ, ISd, IḲṭṭ, Mgh, Ḳ) and ـِ, (ISd, IḲṭṭ, Ḳ,) inf. n. خُطُورٌ, (JK, Ṣ, Mgh, Ḳ,) or خَطْرٌ, (Mṣb,) and خَطَرَانٌ, (JK,) or this last is a mistranscription, (Mgh,) ‡ *It bestirred itself in my mind:* (A: [see خَاطِرٌ:]) or *it moved my mind:* (Mṣb:) or *it occurred to my mind* [absolutely, or] *after I had forgotten it.* (Ḳ.) — خَطَرَ الشَّيْطَانُ بَيْنَهُ وَبَيْنَ قَلْبِهِ † *The devil put vain suggestions into his mind.* (TA.) خَطَرَ الدَّهْرُ خَطَرَانَهُ, (Ṣ,) or مِنْ خَطَرَانِهِ, (TA,) ‡ [*Fortune,* or *time, produced,* or *brought to pass, its events,* or *among its events* such and such things]: a phrase like ضَرَبَ الدَّهْرُ ضَرَبَانَهُ, (Ṣ,) or مِنْ ضَرَبَانِهِ. (T, A. [See art. ضرب.]) ═ خَطُرَ, aor. ـُ, (Ṣ, A, Mṣb, Ḳ,) inf. n. خُطُورَةٌ, (Ṣ, Ḳ,) or خَطَرٌ, (Mṣb,) ‡ *He* (a man, Ṣ &c.) *was,* or *became, eminent, noble,* or *of high rank,* (Mṣb, Ḳ,) or *characterized by rank* or *station.* (Ṣ, A.) And خَطَرَ, [or this is probably a mistranscription for خَطُرَ,] aor. ـُ, inf. n. خَطْرٌ and خُطُورٌ, [or, more probably, خُطُورَةٌ,] † *He was,* or *became, great in estimation, rank,* or *dignity, after having been little in respect thereof.* (TA.)

2. خَطَّرَ, inf. n. تَخْطِيرٌ, † *He took, got,* or *won, a bet, wager,* or *stake.* (L in art. ندب, and TA.)

3. خَاطَرَ بِنَفْسِهِ, (Ṣ, A, Mṣb, Ḳ,) and بِقَوْمِهِ, (A,) inf. n. مُخَاطَرَةٌ; (TḲ;) and ▼اخطر بقومه; (A;) *He placed himself at the point of,* or *near to, destruction; perilled, imperilled, endangered, jeoparded, hazarded,* or *risked, himself;* (Ṣ, A;) and *his people* or *party:* (A:) or خاطر بنفسه signifies *he did that in which fear predominated:* (Mṣb:) or *he caused himself to be on the brink of destruction or of attaining dominion.* (Ḳ.) And خاطر بِنَفْسِهِ وَمَالِهِ *He threw himself and his property into destruction.* (TA.) And ▼اخطر لَهُ كَذَا *He hazarded,* or *risked, to him such a thing.* (L.) [See also 4, below.] — خاطرهُ عَلَى كَذَا, (Ṣ, A,) or عَلَى مَالٍ, (Mṣb,) inf. n. as above, (TA,) ‡ *He laid a bet, wager,* or *stake, with him,* (Ṣ,* A, Mṣb,) *for such a thing,* (Ṣ, A,) or *for property.* (Mṣb.) [See, again, 4.]

4. اخطرهُ ٱللّٰهُ بِبَالِى, (Ṣ, Ḳ,) [and عَلَى بَالِى, (see 1,)] ‡ *God caused it* [*to bestir itself in my mind:* or *to move my mind:* or] *to occur to my mind after I had forgotten it.* (Ḳ.) ═ See also 3, in two places. — اخطر المَالَ, (Ṣ, Ḳ, &c.,) inf. n. إِخْطَارٌ, (Mṣb,) ‡ *He made the property a stake* (Ṣ, A, Mṣb, Ḳ) *between the parties betting.* (Ṣ, Mṣb, Ḳ.) And أَخْطَرَ لِى وَأَخْطَرْتُ لَهُ ‡ [*He laid me a bet and I laid him a bet;*] *we laid bets, wagers,* or *stakes, one to another.* (Ḳ. [See also 3.]) And اخطر المَوْتَ نَفْسَهُ ‡ *He made his soul a stake to death* [by exposing it to be taken by death, like as a stake is taken by one of two parties who have betted]. (TA.) And اخطر [alone] ‡ *He made himself,* or *his soul, a stake to his adversary, and sallied forth against him.* (Ḳ.) — اخطر لَهُمْ خَطَرًا and أَخْطَرَهُمْ خَطَرًا † *He gave them liberally,* or *freely, a lot, portion,* or *share,* or *a compensation, such as contented them.* (TA.) — اخطرهُ *He* (God) *made him to be characterized by rank,* or *station.* (A.) — اخطر فُلَانٌ فُلَانًا † *Such a one became like in rank,* or *station, to such a one.* (Ḳ.) And أُخْطِرْتُ لِفُلَانٍ † *I was made like to such a one in rank,* or *station.* (Lth, TA.) And أُخْطِرَ بِهِ *He was made equal.* (TA. [See أُنْظِرَ بِهِ.])

6. تَخَاطَرَتِ الفُحُولُ بِأَذْنَابِهَا [*The stallions* of the camels *lashed with their tails*] previously to their attacking one another. (A.) ═ تخاطروا ‡ *They laid bets, wagers,* or *stakes, one with another,* (Ḳ, TA,) عَلَى أَمْرٍ *for a thing.* (TA.) And تَخَاطَرَا عَلَيْهِ ‡ *They two laid bets, wagers,* or *stakes, for it.* (A.)

خَطْرٌ: see خِطْرٌ: ═ and خَطَرٌ, in two places: ═ and خَاطِرٌ.

خِطْرٌ *A large number of camels:* (Ṣ, Ḳ:) or *forty:* (Ḳ:) or *two hundred;* (AḤát, Ḳ;) and the like *of sheep or goats:* (TA:) or *a thousand thereof:* (Ḳ:) and *more:* (TA:) and ▼خَطْرٌ signifies the same: (Ḳ:) pl. أَخْطَارٌ. (Ṣ, K.) ═ *A certain plant, with which one dyes,* or *tinges, himself* or *his hair,* (Ṣ, Ḳ,) *its leaves being put into black dye:* (TA:) *it resembles the plant called* كَتَمٌ, *with which it often grows; and old men dye their hair with it:* (AḤn:) or [the *plant called*] وَسْمَةٌ: (Ḳ:) [a coll. gen. n.:] n. un. with ة. (AḤn, Ḳ.) — Hence, (Ṣ,) ‡ *Milk mixed with much water:* (Ṣ, Ḳ, TA:) as though it were tinged [with the plant so called]. (TA.) — And *A branch* (Ḳ) *of a tree:* pl. خِطَرَةٌ, which is extr.; or as though the ة were imagined to be elided. (TA.)

خَطَرٌ The *being at the point of,* or *near to, destruction;* (JK, Ṣ, A, Mṣb, Ḳ;) [*imminent danger; peril; jeopardy; risk; hazard;*] and *fear of perishing:* (Mṣb:) pl. أَخْطَارٌ. (A.) You say, هُوَ عَلَى خَطَرٍ عَظِيمٍ *He is* [*in a state of great peril,* or] *on the brink of destruction.* (A.) And رَكِبُوا الأَخْطَارَ [*They embarked in perilous undertakings; or braved perils*]. (A.) [And أَمْرٌ لَهُ خَطَرٌ and ذُو خَطَرٍ *A perilous affair* or *event* or *case:* and hence, *a momentous,* or *an important, affair* or *event* or *case; an affair,* or *event,* or *a case, of moment* or *importance* or *magnitude:* see also خَطِيرٌ.] And خَطَرٌ [alone] signifies *A thing,* or *an affair,* &c., *of great magnitude:* and *a trial,* or *an affliction:* pl. as above. (Ḥar p. 264.) — ‡ *A bet, wager, stake,* or *thing wagered; a thing staked at a shooting-match* or *a race, and taken by the winner:* (T, Ṣ,* A,* Mgh,* Mṣb,* Ḳ,* TA:) [accord. to the TA, this is the primary signification; but accord. to the A, it is tropical:] pl. as above; (Mṣb;) or خِطَارٌ; and pl. pl. خُطُرٌ: (Ḳ: [but in some copies of the Ḳ, the last is written أَخْطَارٌ; and so in the TA, where it is added that some say it is pl. of خَطَرٌ, like as أَسْبَابٌ is of سَبَبٌ, and أَنْدَابٌ of نَدَبٌ:]) خَطَرٌ and سَبَقٌ and نَدَبٌ all signify the same. (TA.) You say, وَضَعُوا خَطَرًا ‡ [*They laid a bet*]. (A.) And أَحْرَزَ فُلَانٌ الخَطَرَ ‡ [*Such a one won the bet*]. (A.) — Hence, [app. as being likened to a stake won,] (TA,) ‡ *Eminence; nobility;* as also ▼خَطْرٌ: (Ḳ, TA:) in which sense it has become so much used as to be, in this acceptation, conventionally regarded as proper: (TA:) also *excellence:* (TA:) and (as also ▼خَطْرٌ, TA) *rank; degree of dignity; station;* of a man: (Ṣ, A, Ḳ, TA:) and *highness of rank* or *account* or *estimation:* and *wealth:* (TA:) pl. أَخْطَارٌ: (A:) accord. to some, it is only used to signify *high rank:* but accord. to others, you say, إِنَّهُ لَعَظِيمُ الخَطَرِ ‡ [*Verily he is of great dignity*] with respect to his good actions and his nobility, and صَغِيرُ الخَطَرِ ‡ [*of little rank*] with respect to his evil actions and his ignobleness. (TA.) Also † *A lot,* or *portion,* or *share.* (TA.) — And † *A compensation.* (TA.) ═ † *A like,* or *fellow,* (Ṣ, Ḳ,) *in rank* or *station,* (Ṣ,) or *in eminence;* (Ḳ;) as also ▼خَطِيرٌ. (Ṣ, Ḳ.) You say, هٰذَا خَطَرٌ لِهٰذَا, and ↑خَطِيرٌ, † *This is like to that.* (Ṣ.) And الجَنَّةُ لَا خَطَرَ لَهَا † *Paradise has not its like.* (TA.) And ↑فُلَانٌ لَيْسَ لَهُ خَطِيرٌ † *Such a one has not his like* or *fellow.* (TA.) ═ Also [an inf. n. of خَطَرَ in the phrase خَطَرَ بِبَالِى and عَلَى بَالِى, accord. to the Mṣb. And hence,] † *A vain suggestion* of the devil. (JK.) [See خَاطِرٌ.]

خَطِرٌ: see خَاطِرٌ.

خَطْرَةٌ [inf. n. of un. of خَطَرَ: and hence,] † *A going away;* and *walking with an elegant and a proud and self-conceited gait, with an affected inclining of the body from side to side.* (Ḥar p. 35.) — See also خَاطِرٌ. — مَا لَقِيتُهُ إِلَّا خَطْرَةً ‡ [*I met him not save*] *sometime;* (A;) or *sometimes.* (Ḳ.) And مَا ذَكَرْتُهُ إِلَّا خَطْرَةً بَعْدَ خَطْرَةٍ ‡ [*I remembered not,* or *mentioned not, him,* or *it, save sometime after sometime;* i. e., *save*] *sometimes.* (A.) — أَصَابَتْهُ خَطْرَةٌ مِنَ الجِنِّ † *A touch,* or *stroke, from the jinn,* or *genii, befell him;* or *madness,* or *insanity,* [*proceeding*] *from the jinn;* syn. مَسٌّ. (Ḳ,* TA.) — بَيْنِى وَبَيْنَهُ خَطْرَةُ رَحِمٍ (IAạr, TA) app. means † *Between me and him is a tie of relationship.* (TA.) — رَعَيْنَا خَطَرَاتِ الوَسْمِىِّ † *We pastured* [our beasts] *upon the patches of herbage produced by the* [*rain called*] وسمىّ. (Ḳ,* TA.) — خَطْرَةٌ also signifies † *A small quantity* [or *shower*] *of rain:* pl. خِطَارٌ (JK) [and probably خَطَرَاتٌ also]. — And one says, لَا جَعَلَهَا اللّٰهُ خَطْرَتَهُ وَلَا جَعَلَهَا آخِرَ ↑مُخْطَرٍ مِنْهُ [app. referring to rain, and meaning † *May God not make it to be the only shower,* or *fall, thereof,* or *the only time thereof; nor make it to be the last time thereof*]; (TA;) آخِرَ مُخْطَرٍ meaning آخِرَ عَهْدٍ. (Ḳ,* TA.)

خِطَارٌ: see what next follows.

خَطِيرٌ The *falling of a camel's tail between the parts above his thighs, when he moves it about;* [see 1, first sentence;] as also ↑خطار. (TA: in which the latter is written without any syll. signs.) ═ *A camel's nose-rein;* (Ṣ, Ḳ;) *a nose-rein by which a she-camel is led:* (Kr:) *a rope:* (Sh, Ḳ:) these, says Meyd, are one and the same thing. (TA.) It is related in a trad. of 'Alee that he said to [a mistake for "respecting"] 'Ammár, جُرُّوا لَهُ الخَطِيرَ مَا ٱنْجَرَّ لَكُمْ [*Pull ye his nose-rein as long as it will be pulled by you*]: or, as some relate the saying, مَا جَرَّهُ لَكُمْ [*as long as he pulls it to you*]: meaning follow him as long as there is ground for doing so: or, accord. to some, as Sh says, act patiently towards 'Ammár as long as he acts patiently towards you: Meyd mentions it as a proverb. (TA.) ═ ‡ *Eminent; noble; of high rank:* (Mṣb, Ḳ, TA:) *characterized by rank* or *station:* (Ṣ, A:) pl. خُطَرٌ (Ḳ) and خَطِيرُونَ (A.) And † Anything *excellent.* (TA.) You say أَمْرٌ خَطِيرٌ † *A thing,* or *an affair, of high account* or *estimation.* (TA.) — Also † *Ignoble; of low rank;* (AZ, TA;) *contemptible.* (AZ, Mṣb.) — See also خَطَرٌ, in three places.

خَطَّارٌ [is probably applied to a he-camel in a sense like that of the fem., here following]. خَطَّارَةٌ, applied to a she-camel, *That lashes with the tail to the right and left:* (Ḳ:) or *that moves about her tail, when going, in a brisk,* or *sprightly, manner:* (A:) or *that raises her tail, in going along, by reason of briskness, and exceeding sprightliness.* (Ḥar p. 557.) [See 1, first sentence.] — [Hence,] ‡ A spear *that quivers, vibrates,* or *shakes:* (Ṣ, A, Ḳ:) or *that does so much:* and in like manner, a man. (TA.) And خَطَّارٌ بِالرُّمْحِ ‡ A man *who thrusts much with the spear.* (Ṣ, Ḳ, TA.) — † A man *who raises his arm,* or *hand,* (Ḳ, TA,) *with a stone which he lifts for the purpose of trying his strength,* (TA,) *to cast,* or *throw,* (Ḳ, TA,) *and who shakes the stone in lifting it.* (TA.) — † *A sling.* (Ḳ.) † The [*engine of war called*] مَنْجَنِيق; (Ḳ;) as also ↑خَطَّارَةٌ: its casting being likened to the action termed خَطَرَانٌ [inf. n. of 1, q. v.], of the stallion-camel. (TA.) — الخَطَّارُ † *The lion:* (Ḳ:) because of his proud walk, and self-admiration: or because of his shaking himself in his walk. (TA.) — مِسْكٌ خَطَّارٌ ‡ *Musk that diffuses much odour* or *fragrance.* (A.)

خَطَّارَةٌ: see the next preceding paragraph.

خَاطِرٌ [part. n. of 1, q. v.:] ‡ *Walking with an elegant and a proud and self-conceited gait, with an affected inclining of the body from side to side;* (Ḳ;) as also ↑خَطِرٌ, (Ḳ, TA,) or ↑خَطْرٌ. (So in the CḲ and in a MS. copy of the Ḳ.) ═ ‡ *An opinion,* or *an idea,* or *object of thought, bestirring itself in the mind;* (A and Kull p. 179;) i. q. هَاجِسٌ, (M, Ḳ,) i. e. *a thing coming at random into the mind:* (Ṣ in art. هجس:) or *a cogitation which bestirs itself,* or *occurs,* (يَخْطُرُ,) *in the mind, with a view to the end, issue,* or *result, of a thing:* (Mṣb:) pl. خَوَاطِرُ: (A, Ḳ:) [and ↑خَطْرَةٌ signifies the same; for] خَطَرَاتٌ [which is its pl.] is syn. with خَوَاطِرُ; (A;) [whence the phrase,] خَطَرَاتُ الشَّيَاطِينِ † *The vain suggestions of the devils.* (Ṣ and TA in art. همز, &c.) [See also خَطَرٌ, last sentence.] — Hence it is applied to † The *mind* itself. (Kull p. 179.)

مُخْطَرٌ: see خَطْرَةٌ.

بَادِيَةٌ مُخْطِرَةٌ † [*A perilous,* or *dangerous, desert;*] as though it made the traveller a stake between safety and perdition. (Mṣb.)

مُخَاطِرٌ [act. part. n. of 3, q. v.:] † One *who contends with another in shooting* or *casting* [app. *for a wager*]. (JK, TA.)

خطف

1. خَطِفَهُ, aor. ـَ, (Ṣ, Mgh, Mṣb, Ḳ,) inf. n. خَطْفٌ; (Ṣ, TA;) this is the approved form of the verb; (T, Ṣ;) and خَطَفَهُ, aor. ـِ, (Ṣ, Mṣb, Ḳ,) inf. n. as above; (Mṣb;) a form of the verb mentioned by Akh, (Ṣ,) but this is rare, (Ṣ, Ḳ,) or (Ḳ) bad, (Ṣ, Ḳ,) scarcely, or not at all, known; (Ṣ;) and ↑اختطفه, (Ṣ, Mgh, Mṣb, TA,) and ↑تخطّفه; (Ṣ, Mṣb, TA;) *He seized it;* or *took it,* or *carried it off, by force:* (Ṣ, Ḳ:) or *he did so quickly; snatched it away:* (Mgh, Mṣb, TA:) and ↑خطّف has been said to imply repetition of the action [unless it be a variation of اختطف as in a case mentioned below]; but this is strange, and not known on any other authority than that of the "Aḳáneem et-Taạleem" by El-Khuweiyee, a disciple of El-Fakhr Er-Rázee. (MF, TA.) Hence, in the Ḳur [xxix. 67], وَيُتَخَطَّفُ ↑النَّاسُ مِنْ حَوْلِهِمْ [*And men are carried off by force from around them*]. (TA.) — [And hence,] هٰذَا سَيْفٌ يَخْطَفُ الرَّأْسَ ‡ [*This is a sword that will strike off the head*]. (TA.) — And خَطِفَ البَصَرَ and خَطَفَهُ, said of lightning, (Ḳ,) and of a ray of light, and of a [glistening] sword, and of any polished body, (TA,) ‡ *It took away the sight:* (Ḳ, TA:) and ↑أُخْتُطِفَ بَصَرُهُ ‡ *His sight was suddenly taken away.* (M and Ḳ in art. ملس.) It is said in the Ḳur [ii. 19], يَكَادُ البَرْقُ يَخْطَفُ أَبْصَارَهُمْ ‡ [*The lightning almost taketh away their sight,* lit. *sights*]: (TA:) Yoo read يَخْطِفُ ابصارهم; (Ṣ, TA;) and so did Aboo-Rejà and Mujáhid: and some read ↑يِخِطِّفُ, and ↑يَخَطِّفُ, originally يَخْتَطِفُ, accord. to the opinion of the Baṣrees, disputed by Fr, but confirmed by Zj. (TA.) — And خَطِفَ السَّمْعَ, (Ḳ,) aor. ـَ, (Ṣ,) said of a devil, ‡ *He stole* [*an opportunity of*] *hearing* [the speech of the angels, from the confines of the lowest Heaven; or *snatched it*]; (Ṣ, Ḳ, TA;) as also ↑اختطفه: (Ḳ:) the two verbs being like نَزَعَهُ and ٱنْتَزَعَهُ. (Sb, TA.) Hence, in the Ḳur [xxxvii. 10], إِلَّا مَنْ خَطِفَ ↑الخَطْفَةَ ‡ *Except him who steals the* [*opportunity of*] *hearing:* (TA:) or *who snatches unawares and by stealth,* (Bḍ,) or *hears and snatches,* (Jel,) *the speech* of the angels: (Bḍ, Jel:) El-Ḥasan read الّا من ↑خَطِّفَ الخطفة, originally ٱخْتَطَفَ: (Ṣ, TA:) and another reading, ascribed to him and others, is ↑خِطِّفَ; but this is very weak. (TA.) — خَطِفَ, aor. ـَ; and خَطَفَ, aor. ـِ; inf. n. خَطَفَانٌ; (Ḳ;) thus in all the copies of the Ḳ, but correctly خَطْفٌ, as in the L; (TA;) said of a camel, † *He went along quickly.* (Ḳ, TA.) And مَرَّ يَخْطِفُ خَطْفًا مُنْكَرًا † *He went along at a quick rate* [*such as was deemed strange,* or *disapproved*]. (TA.) And خَطِفَتِ السَّفِينَةُ, and خَطَفَت, † *The ship sailed,* or *voyaged:* you say, خَطِفَتِ اليَوْمَ مِنْ عُمَانَ † *She sailed,* or *voyaged, to-day, from 'Omán.* (TA.)

2: see 1, first sentence.

4. اخطف بِالأَمْرِ *He said, Seize thou this* [*thing*], O man; or *take it,* or *carry it off, by force;* or *snatch it away.* (Sgh.) ═ أَخْطَفَ لِى مِنْ حَدِيثِهِ شَيْئًا ثُمَّ سَكَتَ, inf. n. إِخْطَافٌ, † *He cut short somewhat of his discourse, or narrative, which he had begun to me, on some other thing's occurring to his mind, and was silent.* (TA.) — أَخْطَفَتْهُ الحُمَّى, (Lḥ, O, TA,) or أَخْطَفَتْ عَنْهُ, (JK,) or ↑اختطفته, (Ḳ,) † *The fever left him,* or *quitted him.* (Lḥ, JK, O, Ḳ.) — اخطفه المَوْتُ ‡ [*Death missed him by a little;*] *he escaped death by a little.* (JK.) And اخطف الرَّمِيَّةَ † *He missed the*

animal at which he shot or *cast*, (JK, Ṣ, Ḳ,) *nearly hitting it*: (JK:) and in like manner, الشَّىْءَ *the thing*. (Ibn-Buzurj, TA.) And *He captured*, or *caught*, *the animal at which he shot* or *cast*; expl. by إِذَا كَانَ يُصِيدُهَا [perhaps a mistranscription for يُصِيبُهَا, and, if so, meaning *he hit*]. (JK.) And اخطف said of an arrow, † [*It missed*: or *it fell upon the ground, and then glided along upon the ground to the butt*, or *object of aim*: (see خَاطِفٌ:) and] *it went straight*. (TA.) — اخطف said of a man, † *He became affected with a slight sickness, and then speedily recovered*. (TA.) — إِخْطَافُ الحَشَا i. q. اِنْطِوَآؤُهُ [meaning † *The state of being lean*, or *lank*, *in the belly*: see مُخْطَفٌ]. (Ṣ, TA.) الإِخْطَافُ in horses is a fault: it is † *The contr. of* الإِنْتِفَاخُ: AHeyth says that it is, in horses, † *smallness of the* جَوْف [here meaning the *belly*, or *abdomen*]. (TA.)

5: see 1, in two places.

6. تخاطفوا الكُرَةَ بَيْنَهُمْ [*They contended together in snatching away the ball*] with the goff-sticks. (Ḳ* and TA in art. جحف.)

8. اختطف; and its variations خَطَّفَ and خِطِّفَ; and يَخِطِّفُ and يَخَطِّفُ, variations of its aor.: see 1, in seven places. — كَأَنَّهُ يَخْتَطِفُ فِى مَشْيِهِ عُنُقَهُ, said of a swift camel, means *As though he were straining*, or *stretching*, (يَجْتَذِبُ,) *in his going along, his neck*. (Ṣ.) ═ See also 4.

خَطْفٌ † *A slight disease*; as also ↓ خَطْفَةٌ. (JK.) — مَا مِنْ مَرَضٍ إِلَّا وَلَهُ خَطْفٌ † *There is no disease but there is for it a cure*. (JK, Ḳ.) — خَطَفٌ and ↓ خُطْفٌ † *Leanness*; or *lankness of the belly*: and † *lightness of the flesh of the side*. (TA.)

خُطْفٌ: see what next precedes. — بِهِ خُطْفٌ † *In him* (namely, a man, JK) *is madness*, or *diabolical possession*; (JK, TA;) as also ↓ خُطُفٌ: but this latter may be either a pl., like ضُرُبٌ [pl. of ضَارِبٌ], or a sing. (TA.)

خَطْفَةٌ *A single act of seizing*; or, *of taking*, or *carrying off, by force*: (TA:) or, *of doing so quickly*; *of snatching away*. (Mgh, Mṣb, TA.) Hence, [in a trad.,] accord. to one reading, نَهَى عَنْ كُلِّ ذِى خَطْفَةٍ, meaning *He prohibited the prey of whatever snatches away the prey, and goes away with it, not withholding it for its owner*: or, as some say, *what snatches away with its talon*, or *claw*: but the reading commonly known is, نَهَى عَنِ الخَطْفَةِ: (Mgh:) and الخَطْفَةُ signifies *what the wolf*, (Mgh, Mṣb, TA,) or *the like*, (Mṣb,) *snatches away*, (Mgh, Mṣb, TA,) *of the limbs*, or *members*, *of a living sheep* or *goat*, (Mgh, TA,) or *of a living animal*; (Mṣb;) or *what the dog snatches away from the limbs*, or *members*, *of the animal of the chase, of flesh &c., while the animal is alive*: (Mgh, TA:) or *the limb*, or *member*, *which the beast or bird of prey seizes*, or *carries off by force*, or *which a man cuts off, from the beast that is alive*: (Ḳ, TA:) for whatever is separated from the living animal, (Mgh, TA,) of flesh or fat, (TA,) is carrion, (Mgh, TA,) unlawful to be eaten: the prohibition originated from the Prophet's finding, when he came to El-Medeeneh, that the people loved and ate the humps of camels and the tails of sheep: (TA:) the reading الخَطَفَة, of the measure فَعَلَة, with fet-ḥ to the medial radical letter, as pl. of خَاطِفٌ, is a mistake. (Mgh.) — Also † *A single suck of a small quantity of milk quickly taken by a child from the breast*. (TA.) — For its meaning in the Ḳur xxxvii. 10, see 1. — See also خَطْفٌ.

خَطَفَى † *Quickness in pace* or *going*, (Ṣ, Ḳ,) of a camel, as though he were straining, or stretching, his neck, in going along; (Ṣ; [see 8;]) as also ↓ خَيْطَفَى, (Ḳ,) and ↓ خَيْطَفٌ. (JK, TA.) — See also the last of these words below.

خَيْطَفٌ: see خَطِيفٌ.

خَطِيفَةٌ The *act of seizing*, or *carrying off by force*; or, *of snatching away at unawares*. (TA.) ═ *Flour sprinkled upon milk*, (Ṣ,) or *flour upon which milk is sprinkled*, (JK, Ḳ,) *then cooked*, (JK, Ṣ, Ḳ,) *and licked*, or *eaten with the finger*, (Ṣ, Ḳ,) *and snatched up with spoons*: (Ḳ:) IAạr says that it is [*what is called*] جَبُولَآءُ [a word I have not found in any other instance]: (Ṣ:) or, with the Arabs, it is *a food made with milk* (لَبَنِيَّةٌ), *which is heated, then flour is sprinkled upon it, and then it is cooked, and people lick it*, or *eat it with the finger, snatching it up hastily*. (Az, TA.)

خُطُفٌ: see خُطْفٌ.

الخُطَّافُ [lit. *He that is wont to seize*, &c.: and particularly † *he that is wont to snatch*, or *steal*, *opportunities of hearing* the speech of the angels, from the confines of the lowest Heaven: and hence,] applied in a trad. to † *the Devil*, or *Satan*: (Ṣ, TA:) or, as some say, it is in this instance ↓ الخُطَّافُ, as pl. of خَاطِفٌ, [and therefore meaning † *the devils*,] or as being likened to the hooked iron called خُطَّافٌ. (TA.) — أَبُو الخُطَّافِ a surname of *The* حِدَأَة [or *kite*]. (TA in art. حدأ.)

خُطَّافٌ [The *swallow*; thus called in the present day;] *a certain bird*, (JK, Ṣ, Mgh,) *well known*; (JK, Mgh;) *a certain black bird*; (Ḳ;) the عُصْفُور [or *passerine bird*] *which the common people call* عُصْفُورُ الجَنَّةِ [*the* عصفور *of Paradise*]: pl. خَطَاطِيفُ. (ISd, TA.) [See also خُشَّافٌ.] — The *bent*, or *crooked*, *piece of iron which is on each side of the sheave of a pulley, and in which is the pin whereon the sheave turns*: (Aṣ,* JK, Ṣ, Ḳ:) *it confines the sheave on each side*: (TA:) that which is of wood is termed قَعْوٌ. (Aṣ, TA.) Also (Ṣ [in the Ḳ "or"]) *Any crooked*, or *hooked*, *iron*: (Ṣ, Ḳ, TA:) pl. as above. (TA.) [*An iron hook*: *a grapple*: *a grapnel*: and *the like*.] The خُطَّافَانِ of a bit are The *two bent pieces of iron in the* مِسْحَل *and the* شَكِيمَة, *on the right and left*. (IDrd in his "Book on the Saddle and Bridle.") And خَطَاطِيفُ signifies ‡ The *claws*, or *talons*, of a beast or bird of prey; (Ṣ, TA;) as being likened to a hooked iron. (TA.) — † *A wicked thief*: so in the saying of Abu-n-Nejm,

وَٱسْتَصْحَبُوا كُلَّ عَمٍ أُمِّىِّ
مِنْ كُلِّ خُطَّافٍ وَأَعْرَابِىِّ

† [*And they took as companions every blind illiterate man, of every wicked thief and Arab of the desert*]. (TA.) يَا ٱبْنَ خُطَّافٍ [app. meaning † *O son of a wicked thief*] was said by a woman to Jereer, in derision. (TA.) — See also the paragraph next preceding this. — † *A mark made with a hot iron upon a camel, like the* خُطَّاف *of the sheave of a pulley*. (JK, L, Ḳ.*) — † The *part*, of a horse, *which is the place of the heel of the rider*. (JK.) ═ Also pl. of خَاطِفٌ. (TA. See الخُطَّافُ.)

خَاطِفٌ [act. part. n. of 1, *Seizing*, &c.]: pl. خُطَّافٌ. (TA.) — الخَاطِفُ *The wolf*; (JK, Ṣ, Ḳ;) because he seizes, or carries off by force, his prey. (TA.) — خَاطِفُ ظِلِّهِ *A certain bird*, (JK, Ṣ, Ḳ,) said by Ibn-Selemeh to be *called* الرَّفْرَافُ; (Ṣ, [so in three copies, not رَقْرَاق as in Freytag's Lex.,] TA;) *that sees its shadow, and thinks it to be a bird*; (JK;) or *when it sees its shadow in the water, it advances to it to seize it*, (Ṣ, L, Ḳ,) *thinking it to be a prey*: (L, TA:) [see خَيَال:] it is *one of the birds of the deserts*, *and is* [*said to be*] *thus called because of the swiftness with which it pounces down*; it is *green*, or *of a dark*, or *an ashy, dust-colour*, (أَغْبَرُ,) *in the back*; *white in the belly*; *long in the wings, and short in the neck*: (Mṣb in art. لعب:) *also called* مُلَاعِبُ ظِلِّهِ. (Ṣ and Mṣb in that art.) — بَرْقٌ خَاطِفٌ ‡ *Lightning that takes away the sight*. (JK, Ṣ,* TA.*) — سَهْمٌ خَاطِفٌ † *An arrow that falls upon the ground, and then glides along upon the ground to the butt*, or *object of aim*; as though snatching something from the ground: pl. خَوَاطِفُ: (Ḥam p. 573:) or خَوَاطِفُ signifies arrows *that miss*; for مُخْطِفَاتٌ. (TA.)

خَيْطَفٌ, (Ḳ,) or ↓ خَطِيفٌ, (Ṣ, [so in my copies,]) † *A quick*, or *swift*, camel; (Ṣ, Ḳ, TA;) as though he strained, or stretched, his neck, in going along: (Ṣ: [see 8:]) and the former, † a camel *of the* [*excellent and swift*] *kind called* مَهَارِىّ: pl. خَيَاطِفُ. (TA.) — خَيْطَفٌ, (TA,) or ↓ خَطَفَى, (JK,) [as meaning † *Quick*,] is also applied to [the pace termed] عَنَقٌ; (JK, TA;) and so ↓ خَيْطَفَى. (JK.) — See also خَطَفَى.

خَيْطَفَى: see خَطَفَى: — and see also خَيْطَفٌ.

خَاطُوفٌ *A thing like a reaping-hook, which is tied to a snare, and by which the gazelle is caught*. (JK, O, L, Ḳ.)

أَخْطَفُ الحَشَا: see what next follows.

مُخْطَفُ الحَشَا, applied to a horse, † *Lean*, or *lank*, *in the part of the belly that is behind the place of the girth*: (Ṣ:) and مُخْطَفٌ [alone] is applied to a man [in a similar sense]; as also

▼ مَخْطُوفٌ : (TA:) and مُخْطَفُ البَطْنِ † lean, or lank, in the belly; syn. مُنْطَوِيهِ; (Lth, Ḳ;) applied to a camel, and to an ass: (Lth, TA:) and ▼ أَخْطَفُ الحَشَا and ▼ مَخْطُوفُهُ, applied to a man, [signify the same,] ‡ i. q. ضَامِرُهُ. (TA.)

مِخْطَفٌ ‡ A sword *that takes away the sight by its glistening.* (TA.)

مَخْطُوفٌ : see مُخْطَفُ الحَشَا, in two places. ═ Also † A camel *branded with a mark like the* خُطَّاف *of the sheave of a pulley.* (JK, L, Ḳ.)

خطل

1. خَطِلَتِ الأُذُنُ, aor. ـَ, (Mṣb,) inf. n. خَطَلٌ, (Ṣ, Mṣb,) *The ear was,* or *became, flaccid, flabby,* or *pendulous.* (Ṣ,* Mṣb.) [In like manner, also,] خَطَلٌ in flesh signifies The *being flaccid, flabby,* or *pendulous.* (KL.) And in a man, and a horse or a bow, (فرس or قوس, accord. to different copies of the Ḳ, the former being the reading in the TA,) and a spear, (Ḳ,) and the like, (TA,) The *being tall,* or *long, and shaky,* or *quivering.* (Ḳ.) — خَطَلٌ, as an inf. n. [of خَطِلَ], also signifies The *being unsteady,* or *restless.* (KL.) — And خَطِلَ, aor. ـَ, (Ḳ,) inf. n. خَطَلٌ, (JK, Ḳ,) *He was,* or *became, light, and quick.* (JK, Ḳ. [In the former it seems to be implied that it is said of a stupid man.]) — And خَطِلَ, (JK, Ḳ,) or خَطِلَ فِى كَلَامِهِ, (Ṣ,) aor. ـَ, (Ḳ,) inf. n. خَطَلٌ, (JK, Ṣ, Ḳ,) *He was foul, and obscene, in his speech;* (JK, Ṣ;) as also ▼ اخطل, (JK,) or اخطل فى كلامه: (Ṣ, TA:) or *he was corrupt in his speech, and loquacious:* (Ḳ:) or خَطِلَ, aor. and inf. n. as above, *he erred,* or *was wrong,* in his speech; as also ▼ اخطل; and in his opinion, or judgment. (Mṣb.) [See also خَطَلٌ below.]

4: see 1, in two places.

5. تَخَطَّلَ فِى مِشْيَتِهِ* *He twisted himself in his gait, and walked with an elegant and a proud and self-conceited gait, with an affected inclining of the body from side to side.* (Ḳ. [See also خَطَلٌ, an inf. n. of which the verb (خَطِلَ) is not mentioned as having this meaning.])

خَطَلٌ inf. n. of 1 [q. v.]. (Ṣ, Mṣb, &c.) — *A twisting of one's self, and walking with an elegant and a proud and self-conceited gait, with an affected inclining of the body from side to side.* (Ḳ. [See 5.]) — *Corrupt speech:* or *loquacity:* (TA:) or *corrupt and conflicting speech:* (Ṣ, O:) or *corrupt speech with loquacity.* (Ḳ.) In a woman, *Foul,* or *obscene, speech; and conduct that induces doubt,* or *suspicion,* or *evil opinion.* (Ḳ.)

خَطِلٌ *Long and quivering;* applied to a spear. (Ṣ, TA.) — And simply *Long:* so in the phrase, رَجُلٌ خَطِلُ القَوَائِمِ [*A man having long legs*]. (TA.) — And, applied to a garment, *That drags upon the ground by reason of its length.* (T, O, Ḳ.) — Also, applied to a garment, (JK, Ḳ,) and a بَدَن [app. here meaning a short coat of mail], (Ḳ,) *Rough and thick:* (JK, Ḳ:) pl. أَخْطَالٌ. (TA.) — Also *Stupid,* or *foolish,* (Ḳ, TA,) *and hasty.* (TA.) — *Light and quick;* (JK, Ḳ;) applied to a stupid, or foolish person; (JK;) and ▼ أَخْطَلُ signifies the same. (Ḳ.) — A fighting man (JK) *quick in thrusting* or *piercing.* (JK, Ḳ.) — ‡ One *who gives,* or *bestows, quickly;* (Ṣ;) or so خَطِلُ اليَدَيْنِ; (JK;) or so خَطِلُ اليَدَيْنِ بِالمَعْرُوفِ: (Ḳ, TA:) a tropical phrase. (TA.) And خَطِلُ اليَدَيْنِ A man *having rough arms* or *hands.* (Ḳ.) — Also *Corrupt in speech, and loquacious;* and so ▼ أَخْطَلُ: (Ḳ:) or *erring,* or *wrong,* in speech, and in opinion, or judgment. (Mṣb.) — An arrow (JK, Ḳ) *that goes to the right and left;* (JK;) *that does not take a direct course towards the butt.* (Ḳ.) ═ *The cord of the sportsman* [*who catches,* or *snares, game therewith*]. (Ḳ.) — And The *border,* or *extremity, of a* [*tent of the kind called*] فُسْطَاط: (Ḳ:) pl. as above: so in the O. (TA.)

خَطَّالَةٌ A woman *very foul,* or *obscene, in speech:* (JK,* Ḳ:) or *whose conduct induces doubt,* or *suspicion,* or *evil opinion:* (Ḳ:) so in the M and O. (TA.)

أَخْطَلُ *Having* [*flaccid,* or *flabby,* or *pendulous,* or] *long, and flaccid* or *flabby* or *pendulous, ears:* (Mgh:) خَطْلَآءُ [is its fem., and therefore signifies the same; or this], applied to a ewe or she-goat, (JK, Ḳ,) signifies *having broad ears;* (Ḳ;) or *having very broad ears;* (JK, TA;) or *long-eared:* (Ḥam p. 741:) pl. خُطْلٌ (Ḳ) and خُطُلٌ. (TA.) You say ثَلَّةٌ خُطْلٌ (Ṣ) or خُطُلٌ (TA) *A flock of sheep,* or *goats, having flaccid,* or *flabby,* or *pendulous, ears:* and the same epithet is applied to dogs: (Ṣ, TA:) all dogs of the chase have such ears. (TA.) And أُذُنٌ خَطْلَآءُ *A flaccid,* or *flabby,* or *pendulous, ear:* (Ṣ, Mṣb, Ḳ:) or *a long ear, that shakes about:* (TA:) or *a very broad ear.* (JK.) — Also, the fem., A woman *thick, coarse,* or *rude,* (T, Ḳ,) *of make:* (T, TA:) or, as some say, (TA,) *long in the breasts.* (Ḳ, TA.) — See also خَطِلٌ, in two places. — رَجُلٌ أَخْطَلُ اللِّسَانِ *A man having a wagging tongue; able in speech.* (TA.) — بَعِيرٌ أَخْطَلُ القَوَائِمِ *A camel that does not put his legs in their proper places.* (JK.) — دَهْرٌ أَخْطَلُ *A time,* or *fortune, that brings calamities.* (JK.)

خطم

1. خَطَمَهُ, aor. ـِ, (Ḳ,) inf. n. خَطْمٌ, (TA,) *He struck his* خَطْم, i. e. his *nose.* (Ḳ,* TA.) And *He struck the very middle of his nose* with a sword. (TA.) And خُطِمَ أَنْفُهُ *His nose was broken.* (Ḥam p. 528.) — † *He branded him* [i. e. a camel] *on his nose with the mark called* خَطْم [or خِطَام]. (TA.) [Hence,] خَطَمَ أَنْفَهُ ‡ [*He branded him with disgrace;*] *he made disgrace to cleave to him manifestly.* (TA.) And خَطَمَهُ بِاللَّوْمِ ‡ [*He branded him with blame*]; and عَذَرَهُ [i. e. باللوم signifies the same]. (TA.) — *He attached the* زِمَام [or خِطَام, q. v.,] *to him;* namely, a camel: (Ṣ:) or خَطَمَهُ بِالخِطَامِ, (Ḳ,) aor. and inf. n. as above, (TA,) *he put the* خِطَام *upon his nose;* as also ▼ خَطَّمَهُ بِهِ: (Ḳ:) [but the latter verb seems to be more properly used in relation to a number of camels:] or the former, (Ḳ,) or simply خَطَمَهُ, and ▼ خَطَّمَهُ, (TA,) *he made a cut,* or *notch, in his nose,* (حَزَّ أَنْفَهُ, so in the Ḳ accord. to the TA,) *not deep,* (TA,) or *he drew his nose* [*down*], (جَرَّ أَنْفَهُ, so in my MS. copy of the Ḳ and in the CḲ,) *in order to put upon it the* خِطَام. (Ḳ, TA.) — [Hence,] † *He withheld him,* or *prevented him,* from going forth [&c.]. (TA.) And خَطَمَهُ بِالكَلَامِ ‡ *He overcame him,* or *subdued him, by speech, and prevented him from speaking,* (Ḳ, TA,) and *from answering,* or *replying.* (TA.) — خَطَمَ الكَلِمَةَ, inf. n. as above, ‡ *He made the word,* or *saying, valid and strong;* alluding to prudence and precaution as to what one utters. (TA.) — خَطَمَ أُمُورًا ‡ *He conducted,* or *managed, affairs.* (TA.) — خَطَمَ القَوْسَ بِالوَتَرِ, inf. n. خَطْمٌ and خِطَامٌ, ‡ *He suspended the bow by the suspensory called* وَتَر and خِطَام. (AḤn, Ḳ, TA.) And خَطَمَ القَوْسَ بِخِطَامِهَا † *He strung the bow with its string.* (TA.) — خَطَمَ الأَدِيمَ, (Ḳ,) inf. n. خَطْمٌ, (TA,) ‡ *He sewed the edges of the skin,* or *hide.* (Kr, Ḳ, TA.) — خَطَمَتْهُ لِحْيَتُهُ, and خُطِمَ بِلِحْيَتِهِ, ‡ *His beard grew upon his two cheeks.* (TA.) — خَطَمَ أَنْفَ الرَّمْلِ ‡ *He passed over,* or *crossed, the extremity,* or *prominent portion, of the tract of sand:* (Aṣ, TA:) or *he faced it, crossing it.* (TA.)

2: see 1, in two places. — تَخْطِيمٌ, [as inf. n. of خَطَّمَ or خُطِّمَ, (see the part. ns. below,)] said of unripe dates, signifies † The *putting forth colours.* (KL.)

8. اخْتَطَمَ الثَّوْبَ *He bound the garment over the* مَخْطِم, i. e. the *nose;* or *over the* خَطْم, i. e. the *fore part of the nose:* and اختطم بلثام [*he so bound a* لثام, q. v.]. (Ḥar p. 433.)

خَطْمٌ The *muzzle,* i. e. the *fore part of the nose and mouth,* of a دَابَّة [i. e. beast], (JK, Ṣ, Mṣb, Ḳ, TA,) whatever it be, (Ṣ, Mṣb,) as a dog, and a camel, but originally of a beast of prey, and of a sheep or goat: (TA:) or, of a beast of prey, *i. q.* خُرْطُوم: (IAṭr, TA:) or, as some say, of a beast of prey, [the *lip,* i. e.] *what corresponds to the* جَحْفَلَة *of the horse:* (TA:) or of a camel, the *nose.* (Mgh.) And ‡ The *bill,* or *beak,* of a bird, (JK, Ṣ, Ḳ, TA,) whatever it be, (JK, Ṣ,) as a hawk, or falcon, (JK,) and a قَطَاة. (TA.) And of a man, ‡ The *nose;* (Ḳ;) as also ▼ مَخْطِمٌ (JK, Ṣ, Mṣb, Ḳ, TA) and ▼ مَخْطَمٌ; (Ḳ, TA;) pl. مَخَاطِمُ: (JK, Ṣ, Mṣb, Ḳ:) or the *fore part of the nose:* (Ḥar p. 433:) and the ▼ مخطم is also of a camel. (IAth, TA.) Also, of a man, ‡ The *fore part of the face.* (TA.) — خَطْمُ اللَّيْلِ ‡ *The first approach of night:* like as one says أَنْفُ اللَّيْلِ. (TA.) — See also خِطَامٌ. ═ *A thing, an affair,* or *a business, of magnitude.* (IAṭr, Th, Ḳ.) It is related in a trad. that Mohammad promised a certain man to go forth to him, and delayed to do so; and when he went forth, he said to him, شَغَلَنِى خَطْمٌ, meaning *A thing*, &c., *of magnitude* [*occupied me so as to divert me*]; as though the م were a substitute for ب: (IAạr, Th, TA:) but IAth says that it may mean †*a thing*, &c., *that withheld* me, or *prevented* me, [see 1,] from going forth. (TA.)

خَطْمَةٌ ‡*A prominent portion of a mountain.* (Ṣ, TA.)

خِطْمِىٌّ (JK, Ṣ, Mṣb, Ḳ) and خَطْمِىٌّ, (Mṣb, Ḳ,) or, accord. to Az, the latter only, the former being incorrect, (TA,) but the former is the more common, (Mṣb,) [*Althæa*; the *althæa officinalis* of Linn.; i. e. *marsh-mallow*;] *a certain plant* (JK, Ḳ) *with which*, (Ṣ, TA,) or *with a preparation of which*, (JK,) *the head is washed*; (JK, Ṣ, TA;) *a well-known preparation for washing the head*: (Mṣb:) *it is a dissolvent, suppurative, lenitive; good for dysury, and the stone, and sciatica, and ulcer of the bowels, and tremour, and for the suppuration of wounds, and the allaying of pain; and, with vinegar, for the* [*species of leprosy termed*] بَهَق; *and for toothache, used as a gargle; and for the sting or bite of venomous reptiles and the like, and for burns; the mixing of its seed with water, or its bruised stem or root, causes it to congeal; and its mucilage, extracted by hot water, is beneficial to the sterile woman.* (Ḳ.)

خِطَامٌ [*A kind of halter for a camel; a cord of which one end is fastened round the nose and jaws of a camel*; accord. to J,] *i. q.* زِمَامٌ: (Ṣ:) [but the following explanations are more correct:] *a certain thing well known*; so called because [*a portion of*] *it lies upon* [or *surrounds*] *the fore part of the nose and the mouth of the camel*: (Mṣb:) or *anything that is put upon the nose of the camel in order that he may be led thereby*: (M, Ḳ:) or *a cord*, or *rope, which is put upon the neck of the camel, and folded* [for يُثْنَى, in my copy of the work from which this is taken, I read يُثْنَى, as in another explanation, below,] *upon*, or *over, his nose*: (Mgh:) or *a cord*, or *rope, which is attached to an iron that surrounds the nose and jaws* [*of the camel*]: (JK:) or *any cord*, or *rope, that is suspended upon the throat of the camel and then tied upon*, or *over, his nose, whether of skin or of wool or of fibres of the palm-tree or of hemp*: (ISh, TA:) but if of plaited leather, it is said to be called جَرِيرٌ: (TA:) or the خطام of the camel is *a cord*, or *rope, of fibres of the palm-tree, or of* [*goats'*] *hair*, or *of flax, at one end of which is put a ring, then the other end is tied to it*, [i. e. to the rope, as the relative pronoun in the original shows, or to some part of it,] *so that it becomes like a ring* [or *loop*], *then it is put upon the neck of the camel, and then it is folded upon*, or *over, his nose*: what is put in the nose, [attached to a ring, or the like, therein,] and is slender, is termed زِمَامٌ: (IAth, TA:) pl. خُطُمٌ. (Mṣb, Ḳ.) مَنَعَ خِطَامَهُ, said of a camel, means *He refused to have his* خطام *put upon him.* (TA.) And تَزَوَّجَ عَلَى خِطَامٍ means †*He married two wives, so that they became like a* خطام *to him.* (TA.) — †*A brand*, or *mark made with a hot iron, upon the nose of a camel*; (Ḳ;) as also ↓خِطَامٌ: *it* (the خطام) *spreads upon the camel's two cheeks*: so says Aboo-'Alee, in the "Tedhkireh:" (TA:) or *such a mark upon the side* (عُرْض, in the CḲ عَرْض,) *of his face, extending to the cheek*, (En-Naḍr, Ḳ, TA,) *in the form of a line*: (En-Naḍr, TA:) sometimes the camel is branded with one such mark, and sometimes with two; and one says جَمَلٌ مَخْطُومٌ ↓خطام or خِطَامَيْنِ, making مخطوم to govern the gen. case as a prefixed noun; (En-Naḍr, Ḳ, TA;) and بِهِ خِطَامٌ and خِطَامَانِ. (En-Naḍr, TA.) — †The *rope* of a bucket. (TA.) — ‡The *suspensory* of a bow. (AḤn, Ḳ, TA.) And †The *string* of a bow. (Ḳ, TA.)

خَطِيمٌ *Struck upon the nose.* (Ḳ.) *Having the nose broken.* (Ḥam p. 528.)

مِسْكٌ خَطَّامٌ (like شَدَّادٌ, TA, in the CḲ [erroneously] without teshdeed,) ‡*Musk that fills with its odour the innermost parts of the nose*: (Aṣ, Ḳ:) or *musk sharp*, or *pungent, in odour*; as though striking the nose (كَأَنَّهُ يَخْطِمُ الأَنْفَ). (Z, TA.)

فُلَانٌ خَاطِمُ أَمْرِ بَنِى فُلَانٍ ‡*Such a one is the leader, and the conductor*, or *manager, of the affairs, of the sons of such a one.* (TA.)

أَخْطَمُ A man (Ṣ) *having a long nose.* (Ṣ, Ḳ.) — And *Black.* (JK, Ḳ.)

مَخْطَمٌ *A woman.* (Ḳ.)

مَخْطِمٌ and مِخْطَمٌ: see خَطْمٌ, in three places.

مُخَطَّمٌ: see مَخْطُومٌ. — †A horse *having a whiteness extending from the fore part of his nose and his mouth to the part beneath his lower jaw*, (ISd, Ḳ, TA,) *so as to resemble the* خطام: in which sense it has no verb. (ISd, TA.) — Full-grown unripe dates (بُسْر) *upon which are lines* (Ṣ, Ḳ) *and streaks* [*of colour*]; (Ṣ;) as also ↓مُخْطَمٌ. (Kr, Ḳ.) [See 2: and see also بُسْرٌ.] — The *part of the nose* of the camel *which is the place of the* خطام. (TA.)

مُخْطَمٌ: see the next preceding paragraph.

مَخْطُومٌ [pass. part. n. of 1]. You say نَاقَةٌ مَخْطُومَةٌ *A she-camel having a* خطام *put upon her*: and ↓نُوقٌ مُخَطَّمَةٌ *she-camels having* خُطُم *put upon them.* (Ṣ, TA.) — See also خِطَامٌ.

خطو

1. خَطَا, (Ṣ, Mṣb, Ḳ, &c.,) aor. يَخْطُو, (Mṣb,) inf. n. خَطْوٌ; (Mṣb, Ḳ;) and ↓اختطى; (Ṣ, Ḳ;) said of a man (Ṣ, Mṣb, TA) [and of a beast]; both signify the same; (Ṣ, Ḳ;) *He stepped, paced*, or *walked*; (MA, KL;) *i. q.* مَشَى; (M, Mṣb, Ḳ;) as also اِخْتَاطَ, [which see in art. خيط,] formed by transposition. (Ḳ.) You say, خَطَوْتُ خَطْوَةً وَاحِدَةً [*I stepped one step*]. (JK.) [And وَسَّعَ الخَطْوَ *He stepped wide.*] See also the last sentence of the next paragraph.

2. خطّى *He made to pass over*: so in the saying, خَطَّى ٱللّٰهُ نَوْءَهَا *God made*, or *may God make, its* (a land's) [*rain-giving*] *star* or *asterism to pass it over, and not send rain upon it*: (TA in art. خطأ:) but in this case the verb is, (Mgh in that art.,) or may be, (TA ibid.,) originally خَطَّأَ, the final ء being changed into ى. (Mgh and TA ibid. [See 2 in art. خطأ.]) Accord. to Fr, خَطَّى السَّهْمَ and خَطَّأَهُ are syn. [as meaning *He made the arrow to pass over*, or *to miss*, the mark]. (TA in art. خطأ.) One says also, in praying for a man, خُطِّىَ عَنْهُ السُّوْءُ [*May evil be made to pass him*; or] *may evil be repelled from him*: and one says also خُطِّىَ عَنْكَ *May it be removed*, or *put away, from thee*: (Ṣ, TA:) or خُطِّىَ عَنْكَ السُّوْءُ. (ISk, TA in art. خطأ.) — The vulgar say [to a she-ass and to a she-camel or other beast in a slippery or difficult place] خَطِّى, meaning اِمْشِى [for اِمْشِى رُوَيْدًا *Step thou leisurely*]: but the correct word is ↓أَخْطِى [imperative fem. of أَخْطَا]. (TA.)

4. اخطاهُ *He* (a man) *made him* (another man) *to step, pace*, or *walk.* (Ṣ, TA.) — أَخْطَيْتُ for أَخْطَأْتُ: see the latter.

5. تَخَطَّيْتُهُ *I stepped*, or *walked, over him*, or *it*: (Mṣb:) or *I passed over and beyond him*, or *it*: (Ṣ:) or تخطّى النَّاسَ *He went over the people*, (رَكِبَهُمْ,) *and passed beyond them*; and so ↓اِخْتَطَاهُمْ. (Ḳ.) One says, تَخَطَّيْتُ رِقَابَ النَّاسِ [*I stepped over, walked over, passed over and beyond*, or *went over and passed beyond, the necks of the people*]. (Ṣ, TA.) It is said in a trad. respecting Friday, [of one who came too late to the Friday-prayers, as is shown in the TA in art. انى,] رَأَى رَجُلًا يَتَخَطَّى رِقَابَ النَّاسِ *He saw a man passing step by step* [*over the necks of the people* who were already in their ranks in the mosque]. (TA. [See also Ḥar p. 83.]) One says also, فُلَانٌ لَا يَتَخَطَّى عَنِ الطُّنُبِ [*Such a one will not step over*, or *beyond*, or *from, the tent-rope*], meaning, *will not go far from the tent for the purpose of voiding his excrement*, by reason of his foulness and vileness and uncleanness. (TA.) And تَخَطَّيْتُ إِلَى كَذَا (Ṣ, TA) *I passed over* [*to such a thing* or *place* or *person*]: (TA:) one should not say تَخَطَّأْتُ [in this sense], with ء. (Ṣ, TA.) [Hence the following tropical phrases.] تخطّاهُ المَكْرُوهُ ‡[*What was disliked* or *hated*, or *evil, passed over him*; not alighting upon him]. (TA.) And تَخَطَّيْتُ إِلَيْهِ بِالمَكْرُوهِ ‡[*I passed over* others *to him with that which was disliked* or *hated*, or *evil*; *i. q.* تَجَاوَزْتُ]. (TA.) And تخطّى عَنِّى بَصَرُكَ †[*Thine eye*, or *thy sight, passed me over*]. (Aboo-Turáb, TA in art. تيه.) — [Also †*I overstepped it*, or *transgressed it*; namely, a limit prescribed to me, &c.]

8: see 1: — and see also 5.

خَطْوَةٌ *A step*, or *pace*, as meaning a *single act of stepping* or *pacing* or *walking*: (JK, Ṣ, Ḳ, Mṣb:) pl. [of pauc.] خَطَوَاتٌ (Ṣ, Mṣb, Ḳ) and [of mult.] خِطَاءٌ. (Ṣ.) Imra-el-Ḳeys says,

* لَهَا وَثَبَاتٌ كَوَثْبِ الظِّبَآءِ *

* فَوَادٍ خِطَآءٌ وَوَادٍ مَطِرْ *

[*She has bounds like the bounding of gazelles; and a valley is* stepped over by her with leisurely *steps, and a valley is* trampled over by her rapidly as though it were *rained upon*]: (Ṣ:) i. e., one time she steps, and refrains from running; and one time she runs with a running resembling rain: but AO relates it otherwise, saying, فَوَادٍ خَطِيطٌ [lit. *and a valley is not rained upon*]: and some substitute كَصَوْبِ الخَرِيفِ [*like the pouring rain of the autumn*]. (IB, TA.) [See also what next follows.]

خُطْوَةٌ *A step*, or *pace*, as meaning the *space between the two feet* [*in walking* or *running*]: pl. (of pauc., Ṣ) خُطُوَاتٌ (Ṣ, Mṣb, Ḳ) and خُطْوَاتٌ and خُطُوَاتٌ (Ṣ, Mṣb) and (of mult., Ṣ) خُطًى. (Ṣ, Mṣb, Ḳ.) One says, قَرَّبَ ٱللّٰهُ عَلَيْكَ الخُطْوَةَ, meaning *May God make short to thee the space*, or *distance*. (TA.) And بَيْنَ القَوْلَيْنِ خُطًى يَسِيرَةٌ †*Between the two sayings is little difference*. (TA.) لَا تَتَّبِعُوا خُطُوَاتِ الشَّيْطَانِ, in the Ḳur [ii. 163 &c.], means [*Follow not ye*] *the ways of the Devil*: (TA:) or *the footsteps of the Devil*: (JK:) here some read خُطْوَات; and some, accord. to Lth, خُطُؤَات, which Az pronounces to have no meaning. (TA.)

خَطِيَّةٌ for خَطِيئَةٌ: see the latter, in art. خطأ. [Freytag, evidently from his having found it incorrectly written for خَظِيَّةٌ, has assigned to it the meaning of "amica," and "amata."]

نَاقَتُكَ هٰذِهِ مِنَ المُتَخَطِّيَاتِ الجِيفَ [the last word being app. الجِيَفَ] is a saying mentioned by AZ: (TA in the present art.:) or مِنَ المُتَخَطِّئَاتِ الجِيفَ. (TA in art. خطأ, where see the explanation.)

خظو

1. خَظَا لَحْمُهُ, (Ṣ, Ḳ,) aor. يَخْظُو, (Ṣ,) inf. n. خُظُوٌّ, (Ḳ,) *His flesh was*, or *became, compact*; (Ṣ, Ḳ;) as also خَظِىَ, (Ḳ in art. خظى,) mentioned, as well as the former, by IF, and also by Ḳz, who does not mention the former, (TA,) aor. يَخْظَى, inf. n. خَظًى; (Ḳ;) or خَظِىَ should not be said; (Ṣ;) or خَظَا is more common. (IF.) See also خَظَا and خَظَتَا voce خَاظٍ, below. ═ خَظَاهُ ٱللّٰهُ *God made him, or it*, (namely, flesh,) *big*, or *large*; (Ḳ accord. to the TA;) or *big and thick*; (CḲ;) as also ↓ اخظاهُ. (Ḳ.)

4. اخظاهُ: see what immediately precedes. [Freytag also mentions اخظى, on the authority of the "Deewán el-Hudhaleeyeen," as signifying *He, or it, caused* the flesh in the arm *to become prominent, so that the muscles appeared*.] And اخظى also signifies *He*, or *it, fattened*, or *rendered fat*, (Ḳ and TA in art. خظى,) the body. (TA.) ═ اخظى [as an intrans. v.] (said of a man, IAạr) *He became fat*. (IAạr, Ḳ in art. خظى.)

خَظَا, followed by بَظَا; and the fem. خَظَاةٌ; and خَظَاتَا: see خَاظٍ.

خَظٍ, followed by بَظٍ; and the fem. خَظِيَةٌ, followed by بَظِيَةٌ: see خَاظٍ.

خَظَوَانٌ One *whose flesh is such* [*in thickness*, or *abundance*, or *brawniness*,] *that one part overlies another*. (Ṣ, Ḳ.)

خَاظٍ *Compact*, applied to flesh; (TA;) and so ↓ خَظَا بَظَا, (Ṣ, TA,) likewise applied to flesh, (TA,) and to a horse, (AHeyth, TA,) [each] originally a verb, (Ṣ, TA,) and the latter an imitative sequent; and the fem. ↓ خَظَاةٌ, applied to anything; (TA;) and خَظَاةٌ بَظَاةٌ, applied to a woman, the [radical] ى being changed into ا accord. to the dial. of Ṭeiyi; (AHeyth, TA;) and ↓ خَظٍ بَظٍ, applied to a horse; and ↓ خَظِيَةٌ بَظِيَةٌ, applied to a woman. (AHeyth, Ḳ in art. خظى.) The Saadee says, (accord. to the TA, 'Ámir Ibn-Eṭ-Ṭufeyl,)

* رِقَابٌ كَٱلْمَوَاجِنِ خَاظِيَاتٌ *

* وَأَسْتَاهٌ عَلَى الأَكْوَارِ كُومُ *

[*Compact necks like the bleachers' beating-implements, and rumps elevated upon the camels' saddles*]. (Ṣ.) In the saying of Imra-el-Ḳeys,

* لَهَا مَتْنَتَانِ خَظَاتَا ↓ كَمَا *

* أَكَبَّ عَلَى سَاعِدَيْهِ النَّمِرْ *

[*She has two compact portions of flesh and sinew confining her back-bone, like as appear when the leopard falls prostrate upon his fore shanks*], he means خَظَاتَانِ, suppressing the ن for the purpose of alleviation of the utterance: (Fr, Ṣ:) or, as some say, he means ↓ خَظَتَا, restoring the ا that fell out on account of the concurrence of two quiescent letters in the sing. [i. e. خَظَتْ, for خَظَاتْ is formed from خَظَوَتْ, which is from خَظَاتْ,] when the ت has become movent. (Ṣ.) — خَاظٍ also signifies *Thick, and hard, firm*, or *rigid*: (TA:) and [so, or simply] *thick*, applied to an arrow. (AḤn, TA.) And one says سَاعِدٌ خَاظِى البَضِيعِ [*A fore arm, or an upper arm*,] *full of flesh*. (IB, TA in art. بضع.) And رَجُلٌ خَاظِى البَضِيعِ (Aṣ, Ṣ in art. بضع) *A fat man*. (TA in that art.)

خظى

1. خَظِىَ لَحْمُهُ: see 1 in art. خظو.

4: see 4 in art. خظو.

خَظٍ بَظٍ, and خَظِيَةٌ بَظِيَةٌ: see خَاظٍ, in art. خظو.

خف

1. خَفَّ, aor. ـِ, inf. n. خِفَّةٌ (JK, Ṣ, Mṣb, Ḳ, &c.) and خَفٌّ (Mṣb, Ḳ) and خَفَّةٌ and تَخْفُوفٌ, but this last belongs to art. خوف, (Ḳ,) *contr. of* ثَقُلَ [both properly and tropically]; (Mṣb;) properly, (TA,) *It* (a thing, Ṣ, Mṣb) *was*, or *became*, خَفِيف [i. e. *light*] (JK, Ṣ, Mṣb, Ḳ, TA) *of weight*, (JK,) *in body*, or *material substance*. (TA.) Hence the saying of 'Aṭà, in a trad., خِفُّوا عَلَى الأَرْضِ, meaning [*Be ye, or bear ye, lightly upon the ground*] *in prostration*: (A'Obeyd, TA:) or, as some relate it, ↓ خَفِّفُوا: i. e. *prostrate yourselves* [*lightly upon the ground*;] *not heavily, so as to make marks, or impressions, upon your foreheads*: and in another trad. it is said, إِذَا سَجَدْتَ فَتَخَافَّ ↓ *When thou prostratest thyself, put thy forehead upon the ground lightly*: but A'Obeyd says that some say فَتَجَافَ, [i. e. فَتَجَافِ عَضُدَيْكَ عَنْ جَنْبَيْكَ, with ج] (TA) [See 3 in art. جفو.]) [Hence also,] خَفَّ المِيزَانُ *The balance had one of its two scales light, so that it rose*. (TA.) — [Used tropically, it means †*It*, or *he, was*, or *became, light in estimation, lightly esteemed*, or *of little account*.] — And ‡ *He was*, or *became*, خَفِيف [i. e. *light* as meaning *active, agile*, &c.,] in work: (TA:) *he was*, or *became, brisk, lively, sprightly, active, agile, prompt*, and *quick*; syn. نَشِطَ. (Mṣb and TA in art. نشط.) You say, خَفَّ فِي عَمَلِهِ وَخِدْمَتِهِ ‡ *He was*, or *became*, [*brisk*, &c., or] *obedient and submissive, in his work and his service*: (TA:) and خَفَّ لَهُ فِي الخِدْمَةِ [‡ *He was*, or *became, brisk*, &c., *to him in service*], aor. يَخِفُّ, inf. n. خِفَّةٌ: (Ṣ:) and in like manner, خَفَّ لِفُلَانٍ ‡ *He was*, or *became*, [*promptly*] *obedient and submissive, to such a one*. (TA.) [Hence,] خَفَفْتُ إِلَى فُلَانٍ †[*I was*, or *became, brisk, lively*, or *sprightly, in behaviour to such a one*]. (Ṣ in art. هش.) And خَفَّتِ الأُنْثَى لِلْفَحْلِ ‡ *The female was*, or *became, submissive to the male*. (A, TA.) And خَفَّتِ الأُتُنُ لِعَيْرِهَا ‡ *The she-asses obeyed their he-ass*. (Ḳ, TA.) And خَفَّ إِلَى العَدُوِّ, inf. n. خُفُوفٌ, †*He hastened to the enemy*. (Mṣb.) And خَفَّ القَوْمُ عَنْ وَطَنِهِمْ, (Ḳ,* TA,) inf. n. خُفُوفٌ, (TA,) ‡*The people, or party, removed*, or *departed*, or *journeyed, quickly from their home*: or, as some say, simply *removed*, or *departed*, or *journeyed, from it*. (TA.) — ‡ *He was*, or *became*, خَفِيف [or *light*] in intellect, or understanding: (TA:) [and in conduct, or behaviour: generally meaning] †*he was*, or *became, light, inconstant, unsteady, irresolute*, or *fickle*; or *light of intellect*; *light-witted*; syn. طَاشَ: (Mṣb:) the inf. n. of the verb in this sense is خِفَّةٌ. (Mṣb and Ḳ &c. in art. طيش, and TA in the present art.) [But sometimes, when relating to the intellect, or understanding, it means, †*He was*, or *became, quick, acute*, or *sharp*; and *clever*, or *ingenious*: see خَفِيف.] And you say of him whose hearing is good, فِى أُذُنِهِ خِفَّةٌ ‡ [*In his ear is quickness, acuteness*, or *sharpness, of hearing*]. (TA in art. ثقل.) — [† *He was*, or *became, flurried, agitated*, or *excited*, by reason of fear, and by anger, or the like: see 10. — †*He was*, or *became, light-hearted*, or *cheerful*; *one whose company*, or *converse, was acceptable and cheering*.] You say, خَفَّ فُلَانٌ عَلَى المَلِكِ †*Such a one was*, or *became, acceptable and cheering to the king*. (TA.) —

[† *It* (an action, or an affair, and a case, or the like,) *was*, or *became, light*, or *easy*: and *it became alleviated*.] You say, خَفَّتْ عَلَيْهِ الحَرَكَةُ †[*Motion*, or *moving, was*, or *became, easy to him*]; opposed to ثَقُلَتْ. (TA.) And خَفَّتْ حَالُهُ: see 4. — [† *It* (a word) *was light*, or *easy, of utterance*: and in like manner said of a sound, † *it was*, or *became, light to the ear*; or *slight*. — † *It* (food) *was*, or *became, light to the stomach; easy of digestion*. — Said of the hair of the head, and of the beard, † *It was*, or *became, light, thin*, or *scanty*.] — Said of a people, or company of men, (قَوْم,) inf. n. خُفُوفٌ, it means قَلُّوا وَقَدْ خَفَّتْ زَحْمَتُهُمْ [i. e. † *They became few in number, their crowding having diminished*]. (Ṣ.) — Said of rain, [&c.,] † *It diminished; decreased*; or *was*, or *became*, [*light*, or] *deficient*. (TA.) — [Said of a blow, a disease, an affection of the mind, &c., *It was*, or *became, light, slight*, or *inconsiderable*.] — خَفَّتْ مَنَازِلُهُمْ مِنْهُمْ وَمَضَوْا †[means *Their abodes became clear of them, and they went away*]. (Ḳ* and TA in art. شول.)

2. خَفَّفَهُ *He made it*, or *rendered it*, خَفِيف [i. e. *light*, both properly and tropically: the tropical significations are shown by the preceding paragraph, and by explanations of خَفِيف; and some by what here follows]: (Mṣb:) تَخْفِيفٌ is the *contr. of* تَثْقِيلٌ. (Ṣ, Ḳ.) — Hence, in the Ḳur [ii. 174], ذٰلِكَ تَخْفِيفٌ مِنْ رَبِّكُمْ † [*That is an alleviation from your Lord*]. (TA.) Hence also, in a trad. [respecting the estimates to be made by the collectors of the poor-rate], خَفِّفُوا الخَرْصَ † [*Make ye the conjectural computation* of the quantity of the fruit upon palm-trees &c. *light* to the owners, or *moderate*;] *go not to the utmost length in the* خرص. (TA.) [And خَفَّفَ عَنْهُ † *He made light*, or *alleviated, his burden, suffering, distress, uneasiness*, or *the like, by removing from him somewhat thereof; he alleviated him; he relieved him*: see Ḳur iv. 32 and viii. 67 &c.] And خَفِّفُوا عَلَى الأَرْضِ: see 1; second sentence. [And خَفَّفَ فِي عَمَلِهِ † *He relaxed*, or *remitted, in his work*.] — [تَخْفِيفٌ also signifies † The *making* a word *light*, or *easy, of utterance, by the suppression of hemzeh*, or *by its conversion into* ا or و or ى; opposed to تَثْقِيلٌ and تَحْقِيقٌ: and *by making a double consonant single*; opposed to تَثْقِيلٌ and تَشْدِيدٌ: and *by making a movent consonant quiescent*; opposed to تَثْقِيلٌ and تَحْرِيكٌ: each of these changes in a word is said to be لِلتَّخْفِيفِ *for the purpose of alleviating the utterance*. Also, in like manner, † The *making* a sound *light to the ear*, or *slight*; opposed to تَثْقِيلٌ. And † The *suppressing* of hemzeh; opposed to تَحْقِيقٌ.]

4. اخفّ *He made* an arrow *light, by scraping* or *paring it*. (L in art. حوذ.) — See also 10. — *He was*, or *became, unburdened*, or *unencumbered*, or *without anything that burdened him heavily*: (Mṣb:) or *he was*, or *became, little burdened* or *encumbered*, in journeying, (JK, TA,) or in his residence at home. (TA.) — And i. q. خَفَّتْ حَالُهُ ♦ [i. e. ‡ *His state*, or *condition, was*, or *became, light, little encumbered, easy*, or *alleviated*: or *it was*, or *became, that of one having a small family to maintain*: or *that of having little property*: or *that of having little property and a small family to maintain*]: (JK, Ṣ, Ḳ, TA: [see حَالٌ:]) and, as some add, رَقَّتْ [i. e., † *it was*, or *became, narrow in its circumstances*, or *evil*: it is used in contr. senses: though رَقَّتْ seems to be here intended as explanatory of خَفَّتْ]. (TA.) — اخفّ القَوْمُ † *The people's beasts were*, or *became*, خِفَاف [i. e. *light* as meaning *active, agile*, or *brisk*]: (AZ, Ṣ:) or *the people had such beasts*. (Ḳ.)

5. تخفّف [*He lightened his clothing*; or *clad himself lightly*: but for this I know no other authority than modern usage]. — تخفّف مِنْهُ: see 10. — Also *He put on*, or *wore, a* خُفّ [i. e. *boot*], or خِفَاف [i. e. *boots*]: (Ḳ, accord. to different copies:) or تخفّف بِالخُفِّ, (JK,) or تخفّف الخُفَّ, (TA,) *he put on*, or *wore, the* خُفّ (JK, TA) on the foot. (TA.)

6. تخافّ *He pressed*, or *bore, lightly* [upon a thing]; *contr. of* تثاقل. (Ḳ,* TA.) Hence the saying, in a trad., إِذَا سَجَدْتَ فَتَخَافَّ explained above: see 1, second sentence. (TA.)

10. استخفّهُ *contr. of* اِسْتَثْقَلَهُ; (Ṣ, Ḳ, TA;) *He deemed it*, or *him*, خَفِيف [i. e. *light*, properly and tropically]. (TA.) *He found it light*, or *easy, to carry*, (Bḍ in xvi. 82, and TA,) *and to remove*. (Bḍ ibid.) — استخفّ بِهِ † *He held him*, or *it*, (namely, a man's right, or due, or just claim, Mṣb, TA,) *in light*, or *little, estimation* or *account*, or *in contempt*; *he contemned*, or *despised, him*, or *it*. (Ṣ, Mṣb, TA.) — استخفّ الهَمْزَةَ † [*He deemed the hemzeh light*, or *easy, of utterance*]. (TA.) — استخفّهُ also signifies † *He demanded*, or *desired, his* خِفَّة [i. e. *briskness*, or *promptness*]; as also ♦ تَخَفَّفَ مِنْهُ: (TA:) and † *it* (a thing) *incited him*, or *excited him, to briskness, liveliness*, or *sprightliness*; syn. اِسْتَنْشَطَهُ: (JK and Ḳ and TA in art. هش;) and أَطْرَبَهُ: (Ḥar p. 139:) and † *he incited him*, or *excited him, to lightness, levity*, or *unsteadiness, so as to induce him to follow him in his error*: (TA:) [and simply] ‡ *he excited him to lightness, levity*, or *unsteadiness*; (Ksh and Bḍ and Jel in the Ḳur xxx. last verse;) *flurried him*, or *disquieted him*: (Ksh and Bḍ ibid.:) and ‡ *it flurried him, so that he became unsteady*; said of impatience; and of a lively emotion of the heart or mind; (TA;) and of fear; (MA;) and of anger: (T in art. حمل:) and ♦ اخفّهُ † *he angered him*, (TA,) *and deprived him of his forbearance, moderation, patience, staidness*, or *calmness, and incited him*, or *excited him, to levity*, or *unsteadiness*. (Ḳ,* TA.) فَاسْتَخَفَّ قَوْمَهُ, in the Ḳur xliii. 54, means † *And he demanded*, or *desired, of his people, briskness*, or *promptness, in obeying him*: or فاستخفّ أَحْلَامَهُمْ [*and he held in light estimation their qualities of forbearance, moderation, patience*, or *staidness*]: (Bḍ:) or *he incited*, or *excited, his people to be promptly obedient and submissive* (أَنْ يَخِفُّوا) *to him and to that which he desired of them*; like اسْتَفَزَّ: (Ksh:) or *he incited*, or *excited, his people to levity*, or *unsteadiness*, (الخِفَّة,) *and ignorance, foolishness*, or *wrong conduct*. (Mṣb.) And you say, استخفّ فُلَانًا عَنْ رَأْيِهِ † *He incited*, or *excited, such a one to ignorance, foolishness*, or *wrong conduct, and levity*, or *unsteadiness, so as to make him swerve from his right sentiment, opinion*, or *judgment*; (Az, Ḳ, TA;) as also استفزّهُ عن رأيه. (Az, TA.)

خُفٌّ *A boot*; (KL, PṢ, &c.;) *a certain thing that is worn* (JK, Ṣ, Mṣb, Ḳ, TA) *upon the foot*: (TA:) pl. خِفَافٌ (Ṣ, L, Mṣb, Ḳ) and أَخْفَافٌ [which is a pl. of pauc.]. (L, TA.) Hence, رَجَعَ بِخُفَّيْ حُنَيْنٍ [*He returned with the two boots of Ḥoneyn*]; a saying which originated thus: (A'Obeyd, TA:) an Arab of the desert bargained with Ḥoneyn the إِسْكَاف [or maker of shoes and boots], (Ḳ,) who was of the people of El-Ḥeereh, (TA,) for a pair of boots, until he angered him, (Ḳ,) and Ḥoneyn desired to anger the Arab: (TA:) so when the Arab of the desert departed, Ḥoneyn took one of his two boots and threw it down in the way, and then he threw down the other in another place; and when the Arab passed by one of them, he said, "How like is this to the boot of Ḥoneyn! and if the other were with it, I would take it:" and he went on: and when he came to the other, he repented of having left the former: and Ḥoneyn had lain in wait for him: therefore when the Arab went away in search of the former [boot], Ḥoneyn betook himself to the camel that he rode, and what was upon it, and went away therewith; and the Arab came, having with him nothing but a pair of boots; and it was said, (Ḳ,) i. e. his people said to him, (TA,) "What hast thou brought from thy journey?" and he answered, جِئْتُكُمْ بِخُفَّيْ حُنَيْنٍ ["*I have brought you the two boots of Ḥoneyn*"]: and this became a prov., applied on the occasion of one's despairing of an object of want, and returning disappointed: (Ḳ:) thus the case is related by A'Obeyd, and by most others after him. (TA.) Accord. to ISk, Ḥoneyn was a strong man, who asserted his relationship to Asad Ibn-Háshim Ibn-'Abd-Menáf, and came to 'Abd-El-Muṭṭalib, wearing a pair of red boots, [formerly distinctive of kings and men of high rank,] and said, "O my paternal uncle, I am the son of Asad the son of Háshim the son of 'Abd-Menáf:" but 'Abd-El-Muṭṭalib said, "No, by the garments of my father Háshim, I know not in thee the natural qualities of Háshim; therefore return thou:" so he returned: and it was said, رَجَعَ حُنَيْنٌ بِخُفَّيْهِ [*Ḥoneyn returned with his pair of boots*]. (O, Ḳ, &c.) As to the saying of the rájiz,

* يَحْمِلُ فِي سَحْقٍ مِنَ الخِفَافِ *
* تَوَادِيًا سُوِّينَ مِنْ خِلَافِ *

he means thereby [*He carries, in*] *a pastor's bag* (كِنْف) made *of the leg of a* خُفّ [or *boot, wooden implements to be tied upon the dugs of she-camels, made of different trees*]. (Ṣ. See خِلَاف.) — The *foot* (KL, PṢ) of the camel; (Ṣ, Mṣb, KL,

P Ṣ;) the *whole* (مُجْتَمَع) *of the* فِرْسِن of the camel; (JK, Ḳ, TA;) of the male and of the female; *corresponding to the* حَافِر [or *hoof*] *of the horse:* (TA:) and sometimes of the ostrich, (Ḳ,) because resembling that of the camel: (TA:) but of no other than these two: (Ḳ:) of the masc. gender; whereas فِرْسِن [its syn.] is fem.: (TA:) pl. أَخْفَافٌ. (Ṣ, Mṣb, Ḳ.) — [And hence, by a synecdoche, for ذَوَاتُ خُفٍّ,] ‡ *Camels;* coupled with حَافِر as meaning horses, [and sometimes asses or mules], (Mgh, TA,*) and ظِلْف [as meaning sheep or goats or other cloven-hoofed beasts]. (TA.) You say, مَا لَهُ خُفٌّ وَلَا حَافِرٌ وَلَا ظِلْفٌ ‡ [meaning *He possesses not camels, nor horses* or *asses* or *mules, nor sheep* or *goats* or *other cloven-hoofed beasts*]. (TA.) You say also, جَآءَتِ الإِبِلُ عَلَى خُفٍّ وَاحِدٍ, meaning ‡ *The camels came following one another, the head of each* [*except the first*] *being at the tail of the next* [*before it*], *whether tied together in a file or not.* (L.) — *An aged camel:* (Ḳ:) [and *a weak camel:*] or, as some say, *a bulky camel:* pl. أَخْفَافٌ. (TA.) It is said in a trad., يُحْمَى مِنَ الأَرَاكِ مَا لَمْ تَنَلْهُ أَخْفَافُ الإِبِلِ [*Of the trees called* اراك, *what the aged and weak of camels cannot reach may be prohibited*]: i. e. what is near, thereof, to the place of pasturage is not to be prohibited, but is to be left for the aged and weak camels, that cannot go far in search of pasture: (Aṣ, O, Mṣb:) or what camels cannot reach (Mṣb, TA) by means of their اخفاف, (Mṣb,) by walking thereto, (TA,) may be prohibited: (Mṣb, TA:) or it means, what camels cannot reach with their heads may be prohibited [to be shaken or beaten off for them]. (Mgh.) — ‡ The *sole,* or *part that touches the ground, of the foot* of a man. (M, Ḳ, TA.) — ‡ *A tract* of ground (Ṣ, A, O, L) *more rugged,* (Ṣ, O, L,) or *longer,* (A,) *than such as is termed* نَعْل: (Ṣ, A, O, L:) or *a rugged piece of* ground. (Ḳ.)

خُفٌّ: see خَفِيفٌ, in four places. — Also *A company consisting of few persons.* (Ṣ, Ḳ.) You say, خَرَجَ فُلَانٌ فِى خُفٍّ مِنْ أَصْحَابِهِ *Such a one went forth among a company consisting of few persons of his companions.* (Ṣ.)

خِفَّةٌ an inf. n. of 1 [in almost all of its senses, proper and tropical; and much used as a simple subst., signifying *Lightness:* † *levity:* &c.]. (JK, Ṣ, Mṣb, Ḳ, &c.)

خُفَافٌ: see what next follows, in four places.

خَفِيفٌ a part. n. of 1 [in all its senses, proper and tropical, signifying *Light:* &c.]: (JK, Ṣ,* Mṣb, Ḳ,* TA:) as also ▼ خُفٌّ [in the proper sense] (JK,* Ṣ,* Mṣb, Ḳ,* TA) and ▼ خُفَافٌ: (Ṣ,* Ḳ,* TA:) the first is applied to a thing; as also ▼ the second, (Mṣb,) which signifies anything *light to carry,* (TA,) [as also the first;] and *light in weight but heavy in price, not incommoding the bearer:* (Ḥar p. 139:) and the first and ▼ third are also applied to a man: (Ṣ, TA:) but, as some say, the first means [*light*] *in body* [as well as in tropical senses]; and ▼ the third, † [*light*] *in* [*the sense of possessing*] *quickness* or *acuteness* or *sharpness, and cleverness* or *ingeniousness:* and [in like manner] خَفِيفُ القَلْبِ signifies † *quick, acute,* or *sharp, in intellect;* and خَفِيفُ الرُّوحِ, *the same;* or *clever,* or *ingenious:* the pl. of the first is خِفَافٌ and أَخْفَافٌ and أَخِفَّآءُ; the first of which three pls. is also pl. of ▼ خُفَافٌ: and hence, in the Ḳur [ix. 41], انْفِرُوا خِفَافًا وَثِقَالًا [explained in art. ثقل]. (TA.) ▼ خُفٌّ is also applied to a boy, (Ṣ, TA,) meaning *Light to carry;* (TA;) as in the saying of Imra-el-Ḳeys,

يُزِلُّ الغُلَامَ الخُفَّ عَنْ صَهَوَاتِهِ

[*The boy that is light to carry slips from the parts of his* (the horse's) *back whereon the rider sits*]: (Ṣ: so in my copies:) or يُزِلُّ الغُلَامَ الخُفَّ [*he makes the boy that is light to carry to slip*]: and [it is said that] it means also † *the hardy, strong,* or *sturdy, boy.* (TA.) And خَفِيفٌ signifies also *Little burdened* or *encumbered* in journeying, or in residence at home; like ▼ خُفٌّ and ▼ مُخِفٌّ. (TA.) [Hence,] رَجُلٌ خَفِيفُ ذَاتِ اليَدِ † *A poor man.* (TA.) — [† *Brisk, lively, sprightly, active, agile, prompt,* and *quick.* Hence,] خَفِيفٌ إِلَى الخَيْرِ [† *Prompt,* or *quick, to do good*]. (TA in art. هش.) — [† *Light,* or *easy, of utterance:* and † *light to the ear; light in sound.* Hence,] النُّونُ الخَفِيفَةُ [† *The light-sounding* ن; as in يَفْعَلَنْ &c.]; *contr. of* الثَّقِيلَةُ: and also applied to *the tenween.* (TA.) — [† *Light, thin,* or *scanty;* applied to the hair of the head &c. Hence,] هُوَ خَفِيفُ العَارِضَيْنِ [† *He is light, thin,* or *scanty,*] *in the hair of the two sides of the cheeks,* (Ṣ and O and Mṣb in art. عرض,) and *of the beard.* (O in that art.) — الخَفِيفُ *A certain kind of metre of verse;* [namely, *the eleventh;*] *the measure of which consists of* فَاعِلَاتُنْ مُسْتَفْعِلُنْ فَاعِلَاتُنْ [*in each hemistich*]. (Ḳ [in which is added "six times," a mistake for "six feet"].)

خَفَّافٌ [*A maker, or seller, of boots* (خِفَاف, pl. of خُفّ).] (TA.)

مُخِفٌّ: see خَفِيفٌ.

[العَوْرَةُ المُخَفَّفَةُ † *The part,* or *parts, of the person which it is improper, but not grossly indecent, to expose:* so in the law-books: see art. عور.]

خفت

1. خَفَتَ, (Ṣ, A, Mṣb, Ḳ,) aor. ـِ, (Mṣb,) inf. n. خُفُوتٌ (Ṣ, A, Ḳ) and خَفْتٌ (Mṣb, TA) and خُفَاتٌ, (TA,) *It* (the voice, Ṣ, A, Mṣb) *was,* or *became, still,* (Ṣ, Ḳ,) or *silent;* (A;) *was,* or *became, low,* (Mṣb,) or *soft,* or *gentle,* or *slender;* and *became weak, by reason of vehement hunger.* (TA.) — Hence, said of a dying man, *He ceased speaking;* (Ṣ;) *he was,* or *became, silent;* (Ṣ, A, Ḳ;) *he spoke not.* (A.) — And [hence,] خَفَتَ, (A, TA,) inf. n. خُفُوتٌ, † *He died:* (A, TA:) and خَفَتَ, inf. n. خُفَاتٌ, ‡ *he died suddenly;* (AA, Ṣ, Ḳ, TA;) as also مَاتَ خُفَاتًا. (A.) And the latter, i. e. خَفَتَ inf. n. خُفَاتٌ, accord. to AM, † *He was,* or *became, weak, and abject,* or *abased.* (TA.) — خَفْتٌ also signifies The *speaking with a suppressed voice;* and so ▼ تَخَافُتٌ, (Ṣ, Ḳ, TA,) and ▼ مُخَافَتَةٌ. (Ḳ, TA.) And you say, خَفَتَ بِصَوْتِهِ, (Mṣb,) and ▼ خَافَتَ بِهِ, (TA,) *He lowered his voice; spoke with a low voice.* (Mṣb, TA.) And خَفَتَ بِقِرَآءَتِهِ; (TA;) and ▼ خَافَتَ بِهَا, (A, Mṣb,) inf. n. مُخَافَتَةٌ; (Mṣb;) and ▼ قَرَأَ مُخَافَتَةً; (TA;) *He lowered his voice in his reading* or *reciting; read,* or *recited, with a low voice:* (A, Mṣb, TA:) or the second of these signifies *he read,* or *recited, indistinctly, not with raised voice.* (Lth, TA.) — خَفَتَ, said of seed-produce, † *It was,* or *became, such as is termed* خَافِتٌ [explained below]. (Mṣb.)

2. خفّت [app. *It silenced,* or *killed:* said of a smiting with a sword or the like: see فَوَاق]. (TA.)

3. خافت, inf. n. مُخَافَتَةٌ: see 1, in four places. — [Hence,] الإِبِلُ تُخَافِتُ المَضْغَ *The camels ruminate.* (TA.)

4. اخفتت *She* (a camel) *brought forth on the day* [*of the year*] *in which she was impregnated* [or *just a year after she was covered*]. (Ḳ.)

6. تخافتوا *They consulted together secretly.* (TA.) See also 1. — And تخافت † *He feigned,* or *made a false show of, weakness and stillness.* (TA.)

خَفْتٌ (Ṣ) and ▼ خُفَاتٌ, (A,) applied to speech, (Ṣ, A,) *Uttered with a low,* or *suppressed, voice.* (Ṣ,* A.) [See also خَافِتٌ.] — The former is also syn. with خَبْتٌ [*A low,* or *depressed, tract of ground:* &c.]. (Ḳ. [So accord. to my MS. copy of the Ḳ, and accord. to the TA: but in the CḲ this signification is omitted; for instead of وَالخَفْتُ بِالفَتْحِ والخَبْتُ وَبِالضَّمِّ السَّذَابُ, we find وَالخَفْتُ وَبِالضَّمِّ السَّذَابُ, meaning that خَفْتٌ and خُفْتٌ signify the same as سَذَابٌ.])

خُفْتٌ [and accord. to the CḲ خَفْتٌ, but this is app. a mistake, (see what next precedes,)] *i. q.* سَذَابٌ [i. e. *Rue*]; (Ḳ;) as also خُفْتُفٌ. (T, TA.)

خُفَاتٌ [an inf. n. of 1, q. v. — And also used as an epithet; for ذُو خُفَاتٍ]: see خَفْتٌ. — Also *Weak* hearing. (TA, from a trad.) — لَسْتُ خُفَاتًا [*I am not weak, and abject,* or *abased*]. (T, from a verse of El-Jaadee.)

خَفُوتٌ ‡ *A lean,* or *an emaciated,* woman: (Lḥ, Ḳ:) or a woman *who is scarcely seen distinctly, by reason of leanness,* or *emaciation:* (TA:) or a woman *who is deemed goodly,* or *beautiful,* (Ḳ,) *whom the eye regards as worthy of notice,* (A, TA,) *as long as she is alone, not when she is among other women.* (A, Ḳ, TA.) You say امْرَأَةٌ خَفُوتٌ لَفُوتٌ: (Lth, A, TA:) لفوت meaning *wont to calumniate,* or *slander.* (A.) But AM says, I have not heard خفوت as an epithet applied to a woman on any authority but that of Lth. (TA.)

خَفِيتٌ: see what follows, in two places.

خَافِتٌ A voice *becoming still,* or *silent;* or *low,* or *suppressed;* as also ▼ خَفِيتٌ. (A.) You say ▼ صَوْتٌ خَفِيضٌ خَفِيتٌ. (TA.) —— One *whose voice is still, or silent, by reason of his weakness.* (Ḥar p. 76.) Applied to a dying man, *Ceasing to speak; silent.* (Ṣ.) —— سَحَابٌ خَافِتٌ *Clouds in which is no water.* (Aboo-Sa'eed, Ḳ.) A cloud like these does not move from its place: only that which contains water travels along: that which sends forth a slightly-flashing lightning scarcely ever, or never, does so. (Aboo-Sa'eed.) —— زَرْعٌ خَافِتٌ ‡ *Dying,* or *dead, seed-produce:* (A:) or *seed-produce that has not grown tall:* (Mṣb,* Ḳ, TA:) or *that has not attained the full height.* (TA.) The weak believer is likened, in a trad., to خَافِتُ الزَّرْعِ, (A, TA,) which at one time inclines, and at another time stands straight; accord. to A 'Obeyd, خافت meaning what is *fresh,* or *juicy, and soft,* or *tender:* or, accord. to one reading, to خَافِتَةُ الزَّرْعِ, meaning *fresh,* or *juicy, seed-produce, that is soft,* or *tender, and weak;* the ة being added in خافتة as though by زرع were meant سُنْبُلَة. (TA.)

خفر

1. خَفَرَهُ, (Ṣ, A, Mṣb, Ḳ,) and خَفَرَ بِهِ, and خَفَرَ عَلَيْهِ, (A, Ḳ,) aor. ـِ (Ṣ, Ḳ) and ـُ, (Ḳ,) inf. n. خَفْرٌ; (Ṣ, Ḳ;) and ▼خفّرهُ, (Aṣ, Ṣ, Ḳ,) inf. n. تَخْفِيرٌ; (Aṣ, Ṣ;) and ▼تخفّر بِهِ; (Ḳ;) *He protected him; granted him refuge; preserved, saved, rescued,* or *liberated, him;* (Ṣ, A, Mṣb, Ḳ;) *from him who sought* or *pursued him.* (Mṣb.) And خَفَرَ القَوْمَ, inf. n. خُفَارَةٌ, *He protected the people and became responsible for their safety.* (TA.) —— And خَفَرَهُ *He received from him hire for protecting him* (Ḳ) *and being responsible for his safety.* (TA.) —— And خَفَرَ, (TḲ,) inf. n. خِفَارَةٌ, (Ḳ,) *He guarded* palm-trees *from injury:* and seed-produce *from the birds:* syn. of the inf. n. of the verb in the latter sense, شِرَاحَةٌ. (Ḳ, TA: in the CḲ شَرَاحَةٌ, with fet-ḥ to the ش, and with ج in the place of ح.) —— See also 5. —— خَفَرَ بِالعَهْدِ, (Mgh, Mṣb,) and بِعَهْدِى, (A,) aor. ـِ (Mgh, Mṣb) and ـُ, (Mṣb,) inf. n. خِفَارَةٌ, (Mgh,) *He fulfilled the covenant,* or *engagement,* (Mgh, Mṣb,) and *my covenant,* or *engagement.* (A.) —— See also 4, in two places. == وِرِيحُهُ تَخْفِرُ شَهْوَةَ النِّسَآءِ *Its odour puts a stop to the carnal desire of women.* (R, TA.) == خَفِرَ, (Ṣ, Mṣb,) or خَفِرَتْ, (Ḳ,) for most assert that this verb is only used in relation to a woman, and it seems to be seldom, if ever, otherwise used, (MF,) aor. ـَ, (Ḳ,) inf. n. خَفَرٌ (Ṣ, A, Mṣb, Ḳ) and خَفَارَةٌ, (Ḳ, TA,) or the latter is a simple subst.; (Mṣb;) and ▼تخفّر, (Ḳ,) or تخفّرت; (M, L;) *He,* or *she, was bashful;* or *felt,* or *had a sense of,* or *was affected with, shame, shyness,* or *bashfulness;* (Mṣb, TA;) and *was grave, staid,* or *sedate:* (Mṣb:) or *was very bashful;* &c. (Ṣ, M, A, Ḳ.)

2: see 1. == تَخْفِيرٌ [its inf. n.] is also syn. with تَشْوِيرٌ [The *doing an action of which one should be ashamed*]: (Ṣ, and some copies of the Ḳ, and so in the CḲ:) or تَسْوِيرٌ [the *act of walling* a town]: (so in other copies of the Ḳ, and in the TA:) and تَحْصِينٌ [the *act of fortifying*]. (TA.)

4. اخفرهُ *He sent,* (Ṣ, Ḳ,) or *appointed,* (A,) *with him a* خَفِير [or *protector*], (Ṣ, A, Ḳ,) *to defend and guard him.* (Abu-l-Jarráḥ El-'Oḳeylee, TA.) == *He broke,* or *violated, his covenant,* or *engagement, with him;* (Ṣ, A, Mṣb, Ḳ;) the ا having a privative effect; (TA;) *he behaved perfidiously, treacherously,* or *unfaithfully, to him;* as also ▼خَفَرَ بِهِ, (Mṣb, Ḳ,) aor. ـِ, (Mṣb,) inf. n. خَفْرٌ (Ḳ, TA, in the CḲ خَفَرٌ) and خُفُورٌ: (Ḳ, TA:) or خُفُورٌ is an inf. n. syn. with إِخْفَارٌ as inf. n. of اخفر in the sense above explained, but having no verb, such as خَفَرَ, belonging to it: (TA:) or you say, ▼خَفَرَتْ ذِمَّةُ فُلَانٍ, inf. n. خُفُورٌ, *such a one's compact, covenant,* or *obligation, by which he had become responsible for the safety, or safe-keeping, of a person or thing, or the like, was unfulfilled:* and اخفرها الرَّجُلُ *the man broke, or violated, or failed of performing, it:* (Sh, TA:) and اخفر العَبْدَ *he broke, or violated, the covenant,* or *engagement:* (Mgh:) and اخفر alone signifies the same. (IAth.)

5: see 1, in two places. —— تخفّر بِهِ also signifies *He had recourse to him for refuge, protection,* or *preservation; sought his protection; asked him to be his* خَفِير [or *protector*]; (Ṣ, A,* Ḳ;) and so ▼خَفَرَهُ; (TA;) [and ▼استخفرهُ, occurring in the Ṣ, in art. قوب:] *he protected,* or *defended, himself by means of him;* syn. اِحْتَمَى بِهِ. (Mṣb.)

10: see 5.

خُفْرٌ: see خِفَارَةٌ.

خَفِرٌ, applied to a man; (Mṣb;) and the same, (Ḳ,) without ة, (TA,) and خَفِرَةٌ and ▼مُتَخَفِّرَةٌ, applied to a female, (Ṣ, Ḳ,) as also ▼مِخْفَارٌ, (Ḳ,) as a possessive or an intensive epithet; (TA;) *Bashful;* or *feeling, having a sense of,* or *affected with, shame, shyness,* or *bashfulness;* (Mṣb, TA;) *and grave, staid,* or *sedate:* (Mṣb:) or *very bashful,* &c.; (Ṣ, Ḳ;) as also ▼خَفِيرٌ applied to a man: (TA, from a trad.:) pl. [of the first, applied to a female, and of the second,] خَفَائِرُ. (Ḳ.)

خُفْرَةٌ (Ṣ, A, Ḳ) and ▼خُفَارَةٌ (Ṣ, A, Mṣb, Ḳ) [both, in Ḥam p. 677, said to be inf. ns., but they are rather to be regarded as quasi-inf. ns.,] and ▼خِفَارَةٌ (Ṣ, Mṣb, Ḳ) and ▼خَفَارَةٌ (Ḳ) *Protection, refuge, preservation, rescue,* or *liberation,* (Ṣ,* A, Mṣb, Ḳ,) *from one seeking* or *pursuing:* (Mṣb:) *a compact, a covenant,* or *an obligation, by which one becomes responsible for the safety,* or *safe-keeping, of a person* or *thing;* or *the like;* or simply *responsibility,* or *suretiship;* syn. ذِمَّةٌ: (Ṣ, TA:) pl. of the first, خُفَرٌ. (TA.) It is said in a trad., مَنْ صَلَّى الصُّبْحَ فَهُوَ فِى خُفْرَةِ اللّٰهِ *Whoso performeth the prayer of daybreak, he is in the protection of God.* (TA.) And in another trad., الدُّمُوعُ خُفَرُ العُيُونِ *Tears are the protections of the eyes from Hell-fire when they weep from the fear of God.* (TA.) And وَفَتْ خُفْرَتُكَ, (Ṣ, A,) and ▼خُفَارَتُكَ, (A,) *May thy compact, covenant,* or *obligation, which hath made thee responsible for my safety, be fulfilled,* (Ṣ,) is said by the object of protection to his protector when he has not as yet preserved him in safety. (A.)

خَفَرَةٌ: see the next paragraph, in two places.

خَفِيرٌ *A protector; one who protects, grants refuge, preserves, saves, rescues,* or *liberates;* (Ṣ, A, Mṣb, Ḳ;) *from one who seeks* or *pursues;* (Mṣb;) as also ▼خُفَرَةٌ: (A, Ḳ, TA:) *a protector* of a people, *in whose safeguard they are as long as they remain in his district:* (Lth:) pl. خُفَرَآءُ. (A.) *One who guards seed-produce from the birds.* (TA.) —— *One who is protected, to whom refuge is granted, who is preserved, saved, rescued,* or *liberated.* (Ḳ.) The Ḳ might be thought to imply that ▼خُفَرَةٌ is also used in this sense; but it is not. (TA.) —— See also خَفِرٌ.

خِفَارَةٌ (A, Mṣb, Ḳ) and خُفَارَةٌ and خَفَارَةٌ (Mṣb, Ḳ) The *hire,* or *pay, of a* خَفِير [or *protector*]: (A, Mṣb, Ḳ:) the vulgar say ▼خَفَرٌ: and some erroneously change the خ into غ. (TA.) —— See also خُفْرَةٌ, first and last sentences.

خَافُورٌ *A certain plant,* (Aṣ, Ṣ, Ḳ,) *which ants collect in their habitations,* (TA,) *resembling tares,* or *darnel,* (زُوَان,) (Ḳ,) i. e., *in form;* said to be so called because its odour puts a stop to the carnal desire of women; also called مَرْوٌ and زَغْبَرٌ: so says Suh in the R. (TA.)

مِخْفَارٌ:
مُتَخَفِّرَةٌ: } see خَفِرٌ

خفش

1. خَفِشَ, aor. ـَ, inf. n. خَفَشٌ, *He had that quality of the eyes, or sight, which is termed* خَفَش *as this word is explained below.* (Mṣb.) And خَفِشَتْ عَيْنُهُ *His eye had that quality.* (A.)

خَفَشٌ *Smallness of the eye,* (Ṣ, A, Ḳ,) or *of the eyes,* (Mṣb,) *and weakness in the sight, by nature:* (Ṣ, A, Mṣb, Ḳ:) or *a natural narrowness in the eye:* (TA:) and *sometimes it is a disease:* (Ṣ, Mṣb:*) or *a corrupt state in the eyelids,* (Kh, A, Ḳ,) *and redness, which causes the eyes to become narrow,* (Kh,) *without pain,* (Kh, A, Ḳ,) *and without ulceration:* (Kh:) and [so in the Ṣ and A and Mṣb, but in the Ḳ "or"] *nyctalopia;* or the *seeing by night,* (Ṣ, A, Ḳ,) *but not by day:* (Ṣ, Ḳ:) or the *seeing by night more than by day:* (Mṣb:) and *in a cloudy day, but not in a clear one:* (Ṣ, A, Mṣb, Ḳ:) and sometimes, ‡the *being affected with ophthalmia,* or *inflammation of the eye with pain and swelling.* (Mṣb.) كَأَنَّهُمْ مِعْزَى حَظِيرَةٍ فِى خَفَشٍ [*As though they were the goats of a pen, in respect of weakness of sight,*] is a prov., applied to him who falls into blindness or perplexity or the darkness of night; because goats are the weakest of the غَنَم in rain and cold: originally said by 'Áïsheh. (TA.)

خُفَّاشٌ The bat; syn. وَطْوَاطٌ; (Ḳ;) a certain flying thing; (Mṣb;) that flies by night: (Ṣ:) so called because it can scarcely see by day; (Mṣb;) or because of the smallness of its eyes and the weakness of its sight (Ḳ, TA) by day: (TA:) its brain, if the hollows of the soles of the feet be anointed with it, excites the venereal passion: and if burnt, and used as a collyrium, it removes, or stops, (according to different copies of the Ḳ,) whiteness of the eye, (Ḳ, TA,) and sharpens the sight: (TA:) its blood, if smeared upon the pubes of one who has nearly attained the age of puberty, prevents the growth of hair; (El-Minháj, Ḳ;) as some say; but this is not true: (El-Minháj:) and if the pudendum of her who has difficulty in bringing forth, be rubbed gently with its gall-bladder, she brings forth immediately: (Ḳ,* TA:) the pl. is خَفَافِيشُ. (Ṣ, Ḳ.) [See also خُشَّافٌ.]

أَخْفَشُ One who has that quality of the eyes, or sight, which is termed خَفَشٌ as this word is explained above: (Ṣ, A, Mṣb:) and one who contracts his eyes when he looks: (TA:) and one who has in his eyes white fluid matter, or motes, or the like: (AZ:) fem. خَفْشَآءُ. (Mṣb.)

خفض

1. خَفَضَهُ, [aor. ـِ,] inf. n. خَفْضٌ, *He lowered it; depressed it;* namely, a thing; *contr. of* رَفَعَهُ. (A.) — *He* (God) *abased him;* (Ṣ, Mṣb;) namely, an unbeliever. (Mṣb.) You say, اللّٰهُ يَخْفِضُ مَنْ يَشَآءُ وَيَرْفَعُ † *God abaseth whom He will, and exalteth.* (Ṣ.) — خَفَضَ جَنَاحَهُ *He* (a bird) [*lowered or*] *relaxed his wing, and contracted it to his side, in order that he might rest,* or *cease, from his flying.* (TA.) — And the same phrase, ‡ *He made himself gentle, easy to deal with, compliant,* or *obsequious.* (TA.) It is said in the Ḳur [xv. 88], وَٱخْفِضْ جَنَاحَكَ لِلْمُؤْمِنِينَ ‡ *And make thyself gentle, &c., to the believers:* (Jel, TA:) or *be thou condescending to the believers, and treat them with gentleness.* (Bḍ.) And again, in the same [xvii. 25], (TA,) وَٱخْفِضْ لَهُمَا جَنَاحَ الذُّلِّ مِنَ الرَّحْمَةِ ‡ *And humble,* or *abase, thou thyself to them both, from compassion:* (Bḍ, Ḳ, TA:) or *make thyself submissively gentle to them both, from compassion:* (Bḍ,* Jel:) or there is a transposition in the sentence, and the meaning is واخفض لهما جناح الرحمة من الذلّ [*and make thyself compassionately gentle to them both, from submissiveness*]. (O, Ḳ.) — إِنَّ ٱللّٰهَ يَخْفِضُ القِسْطَ وَيَرْفَعُهُ, in a trad., means *Verily God,* at one time, *bringeth down to the ground the just,* or *equitable; and,* at another time, *exalteth him:* (ISh:) or *maketh ample* [*the means of subsistence &c.*] *to whom He will, and maketh scanty to whom He will:* (Ṣgh, Ḳ:) or *maketh little the portion of the means of subsistence which is the share of any created being, and maketh it much.* (TA in art. قسط, q. v.). خَفْضُ العَدْلِ وَرَفْعُهُ is also explained as signifying *The just's being overcome by the unjust,* when men act corruptly, *and the just's overcoming the unjust,* when they repent, and act righteously. (TA.) [See also art. رفع.]

مَا زَالَتْ تَخْفِضُنِى أَرْضٌ وَتَرْفَعُنِى أُخْرَى حَتَّى وَصَلْتُ إِلَيْكُمْ ‡ [app. means *One land ceased not to make me go a gentle pace, and another to make me go a vehement pace, until I came unto you:* for خفض as relating to pace is probably not only intrans., as it will be seen to be below, but also trans., like its contr. رفع: or it may mean *one land ceased not to make me go down, and another to make me go up, &c.*; though its being tropical if having this meaning may be doubted]. (A, TA.) — خَفَضَ صَوْتَهُ, (A, Mṣb,) aor. ـِ, (Mṣb,) inf. n. خَفْضٌ, (Ṣ, Mṣb, Ḳ,) ‡ *He* (a man, Mṣb) *lowered his voice;* (Ṣ, Ḳ;) *did not raise his voice;* (Mṣb;) [*contr. of* رَفَعَهُ, as is indicated in the A.] — [In most of the above-mentioned senses, ↓ خَفَّضَ is nearly, if not exactly, syn. with خَفَضَ.] — خَفَضَ الحَرْفَ فِى الإِعْرَابِ † *He made the* [final] *letter to have kesreh, in inflection.* (Mṣb.) خَفْضٌ is *syn. with* جَرٌّ [q. v.] (Ṣ, Ḳ) in the inflection of words: (Ḳ:) these two terms, in the inflection of words, are like كَسْرٌ in the non-inflection, in the conventional language of the grammarians. (Ṣ.) — خَفَضَ عَيْشُهُ, aor. ـُ, [inf. n., app., خَفْضٌ, q. v. infrà,] † *His life was,* or *became, easy; free from trouble or inconvenience, and toil or fatigue; tranquil; and plentiful.* (JK, Ḳ.*) — خَفَضَ صَوْتُهَا † *Her* (a woman's) *voice was,* or *became,* [*low, soft,*] *gentle and easy.* (TA.) — خَفُضَتْ † *She* (a woman) *was,* or *became, low, soft,* or *gentle, in voice.* (TA.) — خَفَضَتِ الإِبِلُ, [inf. n. خَفْضٌ and مَخْفُوضٌ, like the contr. رَفْعٌ and مَرْفُوعٌ, (see خَفْضٌ below,)] ‡ *The camels went a gentle pace;* (A, TA;) *contr. of* رَفَعَتْ. (A.) — خَفَضَ بِالمَكَانِ, aor. ـِ, † *He remained, stayed,* or *abode, in the place.* (Ḳ.) [See also خَافِضٌ.] A poet says, [app. using the verb in this sense,]

إِنَّ شَكْلِى وَإِنَّ شَكْلَكِ شَتَّى
فَٱلْزَمِى الخُصَّ وَٱخْفِضِى تَبْيَضِضِى

[*Verily the like of me, and verily the like of thee, are different: therefore keep thou to the booth* which is our home, *and remain at rest: thou wilt become fair*]: the last word is for تَبْيَضِّى; a ض being added. (Ṣ.) — خَفَضَ, inf. n. خُفُوضٌ, also signifies † *He died;* said of a man. (TA.) — خَفَضَتِ الجَارِيَةَ, [aor. and inf. n. as below,] *She circumcised the girl:* [see بَظْرٌ:] (Mṣb:) خَفَضْتُ الجَارِيَةَ, (Ṣ,) or خَفَضَتِ الجَارِيَةَ, (A, Ḳ,) is like خَتَنْتُ الغُلَامَ, (Ṣ,) or, خَتَنَ الغُلَامَ: (A, Ḳ:) the former verb applies only to a girl: (Mṣb, Ḳ:) or you say sometimes, خَفَضَ الصَّبِىَّ, aor. ـِ, inf. n. خَفْضٌ, meaning *he circumcised the boy.* (TA.)

2: see 1, in the latter half of the paragraph. — خَفِّضْ رَأْسَ البَعِيرِ *Draw thou the camel's head towards the ground,* that thou mayest mount him. (Lth, Ḳ.) — خَفَّضَهُ † *He weakened,* and *lowered,* or *abased, his state,* and *his rank.* (TA.) — † *He quieted him,* or *tranquillized him, and rendered the affair,* or *case,* or *state, easy to him.* (TA, from a trad.) — خَفِّضْ عَلَيْكَ جَأْشَكَ † *Quiet,* or *calm, thy heart.* (TA.) — خَفِّضِ القَوْلَ يَا فُلَانُ ‡ *Make thou thy words* (lit. *the saying*) *gentle,* or *soft, O such a one:* (Ḳ, TA:) and خَفِّضْ عَلَيْكَ القَوْلَ [signifies the same]. (Ṣ.) — خَفِّضْ عَلَيْكَ الأَمْرَ, (Ṣ, Ḳ,*) or [simply] خَفِّضْ عَلَيْكَ, (A,) ‡ *Make thou the case,* or *affair, light,* or *easy,* (Ṣ, A, Ḳ,) *to thyself:* (A:) [i. e. *regard it lightly:* for] خَفِّضِى عَلَيْكِ, occurring in a trad., as said by Aboo-Bekr to 'Áisheh, means *make thou the case,* or *affair, light,* or *easy; and do not grieve for it.* (TA.) — أُصِيبَ بِمَصَائِبَ تُخَفِّضُ المَوْتَ † *He was smitten by afflictions which brought near to him death, and from which he could not escape.* (IAạr, L.)

3. رَافَعَنِى وَخَافَضَنِى : see art. رفع.

5: see what next follows.

7. انخفض, (JK, Ṣ, Ṣgh,) or ↓ اختفض, (Ḳ,) or both, (TA,) [but the latter seems to be very rare, whereas the former is of very frequent occurrence,] and ↓ تخفّض, (A,) *It was,* or *became, lowered,* or *low,* or *depressed.* (JK, Ṣ, A, Ṣgh, Ḳ.)

8. اختفض: see 7. — اختفضت *She* (a girl) *was,* or *became, circumcised.* (Ṣ, Ḳ.) [See 1, last signification.]

خَفْضٌ : [see خَفَضَ, (of which it is the inf. n.,) throughout. —] *A state of abatement,* or *remissness,* or *the like:* (A, TA:) † *ease; repose; freedom from trouble or inconvenience, and toil or fatigue; tranquillity; quietness; quietude; stillness;* syn. دَعَةٌ; (Ṣ, A, Ḳ;) and رَاحَةٌ; (Mṣb;) and سُكُونٌ; (TA;) of life: (Mṣb:) or *ampleness of the circumstances* of life; (El-Marzookee, Mṣb;) *plentifulness and pleasantness* thereof: (El-Marzookee:) *softness, delicateness,* or *easiness:* (A, TA:) *pleasant life:* (L:) and [in like manner] ↓ خَفِيضَةٌ, *softness, delicateness,* or *easiness, of life; and ampleness of the circumstances thereof:* (TA:) and the former, † *gentleness and easiness* of voice. (TA.) You say, هُمْ فِى خَفْضٍ مِنَ العَيْشِ † *They are in an easy,* or *a tranquil,* [*or a plentiful and pleasant,* or *a soft or delicate,*]* *state of life.* (Ṣ.) [This phrase is said in the A to be tropical; but why, I do not see; since خفض in the sense of دعة is proper accord. to the same authority.] And هُوَ فِى خَفْضِ العَيْشِ † *He is in an ample, and an easy,* or *a tranquil, state of life.* (Mṣb.) And a poet says,

لَا يَمْنَعَنَّكَ خَفْضَ العَيْشِ فِى دَعَةٍ
نُزُوعُ نَفْسٍ إِلَى أَهْلٍ وَأَوْطَانِ
تَلْقَى بِكُلِّ بِلَادٍ إِنْ حَلَلْتَ بِهَا
أَهْلًا بِأَهْلٍ وَجِيرَانًا بِجِيرَانِ

(Ḥam p. 137, and Ṣgh;) i. e. † [*Let not yearning of soul for family and homes prevent thee from enjoying*] *ampleness of the circumstances of life,* or *plentifulness and pleasantness thereof, in ease and tranquillity:* [*thou wilt find in every country, if thou take up thine abode in it, a family for a family, and neighbours for neighbours:*] (El-Marzookee, MF:) another reading, which is preferable, though each is allowable, is نُزَاعٌ in the place of نُزُوعٌ. (Ham ubi suprà.) — [It is also used as an epithet; app. for ذُو خَفْضٍ.] You say, عَيْشٌ خَفْضٌ, (JK, TA,) and ↓ خَافِضٌ, (Ş, A, Ķ,) and ↓ خَفِيضٌ, and ↓ مَخْفُوضٌ, (TA,) † *An easy, or a tranquil,* (JK, Ş, Ķ, TA,) *and plentiful,* (JK, TA,) *and soft,* or *delicate,* (TA,) *life:* (JK, Ş, &c.:) and ↓ مُخْفِضٌ signifies the same as خَفْضٌ. (TA: there mentioned in the same place as here.) [It is said in the A, that عيش ↓ خافض is like عِيشَةٌ رَاضِيَةٌ, (meaning that it is for عَيْشٌ مَخْفُوضٌ,) and that it is tropical.] — Also ‡ *A gentle pace; contr. of* رَفْعٌ; (Ş, A,* Ķ;) and so ↓ مَخْفُوضٌ; (Ş,* A;) *contr. of* مَرْفُوعٌ. (A, TA.) [See خَفَضَتِ الإِبِلُ.] — Also *Low,* or *depressed,* land: (TA:) and [in like manner] ↓ خَافِضَةٌ *a low,* or *depressed, tract* (تَلْعَةٌ مُطْمَئِنَّةٌ) *of land:* (ISh, Ķ:) رَافِعَةٌ signifying [the contr., i. e.] a hard and elevated tract of land. (ISh.)

هُوَ فِى حَالِ خَفْضَةٍ and حَالِ رِفْعَةٍ [*He is in a state of abasement* and *in a state of elevation:* or perhaps the word خفضة should be written خِفْضَة, to agree in form with رِفْعَة, and because in itself denoting a state]. (A.)

صَوْتٌ خَفِيضٌ ‡ *A low, soft,* or *gentle, voice.* (TA.) And كَلَامٌ خَفِيضٌ and ↓ مَخْفُوضٌ ‡ [*Low, soft,* or *gentle, speech*]. (A, TA.) See also خَافِضٌ. — عَيْشٌ خَفِيضٌ: see خَفْضٌ.

خَفِيضَةٌ, as a subst.: see خَفْضٌ, near the beginning of the paragraph.

الخَافِضُ, one of the names of God called الأَسْمَآءُ الحُسْنَى, *The Abaser of the proud, haughty,* or *insolent:* (Ķ:) *the Abaser of everything which He desireth to abase.* (TA.) — خَافِضَةٌ رَافِعَةٌ in the Ķur [lvi. 3, applied to the resurrection, (القِيَامَة,)] means *Abasing certain persons to Hell: exalting certain persons to Paradise:* (O, Ķ:) or *abasing the disobedient: exalting the obedient.* (Zj.) — A rájiz [of the tribe of Asad (Ş in arts. شول and صن)] says, censuring a collector of the poor-rate,

* أَإِبِلِى تَأْكُلُهَا مُصِنَّا * خَافِضَ سِنٍّ وَمُشِيلًا سِنَّا *

[*Dost thou devour my camels, elevating the nose with pride, lowering age* in one case *and raising age* in another?]: or, accord. to IAạr, this was a man addressing his wife, and censuring her father, who had required as her dowry twenty camels, all to be بَنَاتُ لَبُونٍ, and demanded them of him; and when he saw among his camels a fat حِقَّة, he said "This is a بِنْتُ لَبُونٍ," that he might take her; and when he saw a lean بنت لبون, he said "This is a بِنْتُ مَخَاضٍ," that he might leave her. (Ş.) — هُوَ خَافِضُ الجَنَاحِ ‡ *He is gentle, easy to deal with, compliant,* or *obsequious:* (A, TA:) ‡ *he is grave, staid, sedate,* or *calm;* (TA;) and so هُوَ خَافِضُ الطَّيْرِ. (Ķ, TA.) — اِمْرَأَةٌ خَافِضَةٌ الصَّوْتِ, and ↓ خَفِيضَةُ الصَّوْتِ, † *A woman low, soft,* or *gentle, in voice:* (TA:) *not clamorous and foul-tongued.* (T, TA.) — عَيْشٌ خَافِضٌ: see خَفْضٌ, in two places. — أَرْضٌ خَافِضَةُ السُّقْيَا † *Land easy of irrigation.* (Ķ.) The contr. is termed رَافِعَةُ السقيا. (TA.) — بَيْنِى وَبَيْنَكَ لَيْلَةٌ خَافِضَةٌ ‡ *Between me and thee is a night of easy journeying.* (Ş, TA.) — قَوْمٌ خَافِضُونَ † *A people,* or *company of men, remaining at a water:* when going in search of pasture and of the places where rain has fallen, they are not so called. (IAạr.) — خَافِضَةٌ *A woman who circumcises girls.* (Ş, A, Mṣb, Ķ.*) And خَافِضٌ is sometimes applied to *A man who circumcises boys.* (TA.)

خَافِضَةٌ, as a subst., or an epithet in which the quality of a subst. predominates: see خَفْضٌ, last signification.

مَخْفِضُ قَوْمٍ *A place where a people are in a state of ease, or tranquillity;* or *in a plentiful and pleasant state of life.* (TA.) — See also خَفْضٌ, in the latter part of the paragraph. — مَخْفِضٌ also signifies The *place of a girl where the operation of circumcision is performed.* (Lḥ and Az, in TA, voce عُذْرَة.)

مَخْفُوضٌ: see خَفْضٌ, in two places, in the latter part of the paragraph: and see خَفِيضٌ. — مَخْفُوضَةٌ A girl *circumcised.* (Mgh, Mṣb.)

الحُرُوفُ المُنْخَفِضَةُ *All the letters of the alphabet except* خ, ص, ض, ط, ظ, غ, *and* ق; (Ķ;) which latter are called المُسْتَعْلِيَةُ. (TA.)

خفق

1. خَفْقٌ signifies The *striking,* or *slapping,* (JK, Ş, Ķ,) a thing, [*so as to make a slight sound,*] with a دِرَّة [q. v.], (JK, Ķ,) or with something broad. (JK, Ş, Ķ.) You say, خَفَقَهُ, (Mgh, Mṣb,) aor. ـِ [and ـُ], inf. n. خَفْقٌ, (Mṣb,) *He struck,* or *slapped, him,* or *it,* [*so as to make a slight sound,*] with something broad, (Mgh, Mṣb,) such as a دِرَّة. (Mṣb.) And خَفَقَهُ, aor. ـِ and ـُ, *He struck him slightly,* [or *so as to make a slight sound,*] with a sword, (Ş, Ķ,) and with a whip, and with a دِرَّة. (TA.) And خَفَقَ الأَرْضَ بِنَعْلِهِ *He struck the ground* [*so as to make a sound*] *with his sandal.* (Ş, TA.) — And hence, i. e. from خَفَقَهُ as first explained above, (Mgh,) The *sounding* [*of the patting,* or *pattering,*] (JK, Mgh, Ķ) of the sandal, (JK, Ķ,) or of the sandals, (Mgh,) and the like: (TA:) and خَفْقُ الأَقْدَامِ عَلَى الأَرْضِ [*the sounding of the patting,* or *pattering, of the feet upon the ground*]. (Az, in TA, art. همس.) You say, خَفَقَ النَّعْلُ *The sandal made a sound,* or *sounds.* (Mṣb.) — And خَفَقَتِ الرَّايَةُ, aor. ـِ and ـُ, inf. n. خَفْقٌ and خَفَقَانٌ (Ş, Ķ) and خُفُوقٌ, (TA,) *The banner,* or *standard, was,* or *became, in a state of commotion; moved,* or *went, to and fro; trembled; fluttered;* or *quivered;* (Ş, Ķ;) as also ↓ اختفقت: (JK, Ķ:) and in the same sense the former verb is used in speaking of the heart; (Ş, Mṣb;) خَفَقَانُ القَلْبِ signifying *the fluttering,* or *palpitating, of the heart;* (JK, T, Ķ;) and in like manner خَفَقَانُ الجَنَاحِ *the fluttering,* or *flapping, of the wing:* (JK:) so, too, the former verb is used in speaking of the mirage; (Ş, Ķ;) and ↓ the latter verb likewise; (Lth, Ķ;) and Ru-beh, by poetic license, makes the ف of [the inf. n.] الخَفْق to be with fet-ḥ, in his saying,

* مُشْتَبِهِ الأَعْلَامِ لَمَّاعِ الخَفَقْ *

[*Indistinct in respect of the signs of the way, glistening much in the quivering,* or *fluttering*]: (Ş, Ķ:) in like manner, also, the former verb is used in speaking of lightning, (Ş,* TA,) inf. n. خَفْقٌ; (Ş;) and of a sword, and of the wind, and the like: and ↓ اخفق, said of the heart, and of lightning, and of a sword, and [اخفقت said] of a banner, or standard, and of the wind, signifies the same: (TA:) or خَفَقَتِ الرِّيحُ, (Ş,) inf. n. خَفَقَانٌ, (Ş, TA,) signifies *The wind made a rustling,* or *murmuring,* or *confused and continued, sound.* (Ş, TA.*) And خَفَقَتِ النَّاقَةُ *The she-camel broke wind, with a sound.* (Ķ.) — خَفَقَ said of a bird, [because of the flapping, or sound, of its wings,] *It flew.* (Ş, Ķ.) See also 4, first sentence. And said of an arrow, [because of its whizzing,] *It went swiftly.* (TA.) And خَفَقَ فِى البِلَادِ, inf. n. خُفُوقٌ, *He went away into,* or *in, the countries,* or *lands,* &c. (TA.) — Also, said of a man, [in the CĶ, فُلَانًا is erroneously put for فُلَانٌ,] *He moved,* or *shook, his head,* (Ş, Ķ,) or *bent* [*down*] *his head,* (TA,) [or *nodded,*] *being drowsy,* or *dozing;* (Ş, Ķ, TA;) as also ↓ اخفق: (Ṣgh, Ķ:) or *he drowsed,* or *dozed:* (Mgh:) or *he had a fit of drowsiness,* or *dozing, and then awoke:* (TA:) or *he slept;* (JK, TA;) so says Ibn-Hání; (TA;) aor. ـِ and ـُ, (JK,) inf. n. خُفُوقٌ. (TA.) And خَفَقَ بِرَأْسِهِ خَفْقَةً أَوْ خَفْقَتَيْنِ, (Mgh, Mṣb,) occurring in a trad., (Mgh,) *He bent* [*down*] *his head, without the rest of his body,* [or *nodded,*] *once,* or *twice, being taken by a fit of drowsiness,* or *dozing.* (Mṣb.) It is said in another trad., كَانَتْ رُؤُوسُهُمْ تَخْفِقُ خَفْقَةً ↓ أَوْ خَفْقَتَيْنِ [*Their heads used to nod by reason of drowsiness,* or *dozing, once* or *twice*]. (Ş.) And in another, كَانُوا يَنْتَظِرُونَ العِشَآءَ حَتَّى تَخْفِقَ رُؤُوسُهُمْ, i. e. [*They used to wait for nightfall until*] *they slept so that their chins dropped upon their breasts.* (TA.) — خَفَقَتِ النُّجُومُ, inf. n. خُفُوقٌ, *The stars set,* or *disappeared.* (Ş.) And خَفَقَ النَّجْمُ, (JK, Mgh, Ķ,) aor. ـِ, inf. n. خُفُوقٌ, (Ķ,) *The star,* or *the asterism,* [or *the Pleiades,*] *set,* or *disappeared;* (JK, Mgh, Ķ;) as also ↓ اخفق: (JK:) or the former signifies *the star,* &c., *went down in the place of setting;* and in like manner the verb is used in speaking of the moon; (TA;) and of the sun: (IAạr, TA:) and ↓ اخفقت النُّجُومُ signifies *the stars retired to the place of setting:* (Ş, Ķ:) or, as some say, *shone with a flickering light,* or *glistened,* or *shone brightly:* [because of their twinkling, or apparent quivering: or] as though the ا in the verb had a privative effect. (TA.) One says, وَرَدْتُ خُفُوقَ النَّجْمِ, meaning [*I came*] *at the time of the setting of the Pleiades;* making the inf. n. an adv. n. [of time]. (Ş, TA.) — Hence, (Mgh, TA,) or, as some say, from the same word as signifying "the act of striking [or slapping]," (TA,) الخَفْقُ signifies *The act of inserting;* (Mgh;) [i. e.] *the causing the penis to become concealed in the vulva;* (Ḳ;) or *the act of copulation:* (JK:) or [rather] *the penis' becoming concealed in the vulva.* (Az, TA.) — خَفَقَ اللَّيْلُ *The night for the most part passed away:* (JK, Ḳ:) [and in like manner the verb is said of the day:] see خَافِقٌ. — خَفَقَ المَكَانُ *The place was,* or *became, void,* or *unoccupied.* (TA.) — خُفُوقٌ in a horse is The *being slender,* or *lean, in the belly.* (AO, Ḳ. [See خَفِقٌ.])

4. اخفق, said of a bird, *It beat with* [or *flapped*] *its wings:* (Ṣ, Ḳ:) and ▼ خَفَقَ بِجَنَاحَيْهِ [signifies the same]. (Ṣ and Ḳ in art. رنق.) And اخفق بِثَوْبِهِ *He* (a man) *made a sign with his garment, by raising it, and waving it.* (Ṣ, Z, Ṣgh, Ḳ.) — Said of the heart, and of lightning, and of a sword, &c.: see 1. — And said of a drowsy, or dozing, man: see 1. — Also *He* (a man who had gone on a warring and plundering expedition) *failed of obtaining any spoil:* (A'Obeyd, Ṣ, Mgh, Ḳ, and Ḥam p. 157, and Ḥar p. 26:) because he becomes in a shaky, or unsteady, condition, at that time: or because his travelling-bags become unsteady, or shake about, by reason of their lightness and emptiness: so that the verb is of the same category as أَعْطَشَ [meaning "his camels thirsted"] and أَجْرَبَ [meaning "he had his camels affected with the mange, or scab]: (Ḥar ubi suprà:) or the proper signification is, *he found the spoil to be not stationary:* (TA:) or it means *he returned disappointed of spoil,* or *of predatory warfare:* (JK:) or *he was disappointed of that for which he hoped.* (Ḥam p. 157.) And *He* (a hunter or fowler) *returned without having taken any game.* (Ṣ, Ḳ.) And *His property became little.* (TA.) You say also, اخفق فِى زَادِهِ *He* (a man) *had his travelling-provisions all spent,* or *consumed,* [so that his provision-bags, being empty, shook about.] (JK.) And طَلَبَ حَاجَةً فَأَخْفَقَ (Ṣ, Ḳ) *He sought an object of want, and failed of obtaining it.* (Ḳ.) — اخفق النَّجْمُ, and اخفقت النُّجُومُ: see 1. ═ اخفق فُلَانًا *He threw down,* or *prostrated, such a one on the ground.* (AA, Ḳ.)

8: see 1, in two places.

مِشْفَرٌ خَفِقٌ *A camel's lip flaccid,* or *pendulous.* (JK.) — فَرَسٌ خَفِقٌ and خَفِقَةٌ and ▼ خُفُقٌ and خَفِقَةٌ *A horse,* or *mare, slender,* or *lean, in the belly:* sometimes the animal is so by nature; sometimes, by reason of loss of flesh; and sometimes, by being jaded: (AO, Ḳ:) sometimes, also, they said خَفِقُ الأَحْشَآءِ; at other times using the epithet by itself: (AO, TA:) pl. [masc.] خِفَاقٌ and [fem.] خَفِقَاتٌ and خُفُقَاتٌ. (AO, Ḳ.)

خُفُقٌ and [its fem.] with ة: see خَفِقٌ.

خَفْقَةٌ [*A single nodding of the head, by reason of drowsiness,* or *dozing*]: see 1, in two places: (Mgh, Mṣb:) *a slight,* or *light, sleep.* (TA.) It is said in a trad. respecting Ed-Dejjál [or Antichrist], يَخْرُجُ فِى خَفْقَةٍ مِنَ الدِّينِ, explained as meaning [*He will come forth*] *in a time when religion will be drowsy,* or *dozing, by reason of weakness.* (TA.) — مَضَى خَفْقَةٌ مِنَ اللَّيْلِ means *A period* (سَاعَة) *of the night passed.* (JK.) — And one says, سَيْرُ اللَّيْلِ الخَفْقَتَانِ وَسَيْرُ النَّهَارِ البَرْدَانِ [The time of] *the journeying of the night is the first part thereof and the last part thereof, and* [that of] *the journeying of the day is the morning, between daybreak and sunrise, and the evening, between sunset and nightfall.* (TA.) ═ See also the next paragraph, in two places.

خَفْقَةٌ, (Ḳ,) or, as in the Tekmileh, ▼ خِفْقَةٌ, (TA,) *A thing with which one strikes,* or *beats, such as a thong,* or *strap, or a* دِرَّة [q. v.]. (Ḳ, TA.) [See also مِخْفَقَةٌ.] ═ And the former, (Ḳ,) or ▼ the latter, (JK,) *A smooth desert in which is* [*the kind of mirage termed*] آل: (JK, Ḳ:) so says Lth. (TA.) [See also مَخْفَقٌ.]

خِفَاقٌ *A garment with which one makes a sign, by raising it, and waving it.* (JK.)

خَفُوقٌ: see خَافِقٌ. — Also A she-camel *that breaks wind* [*often*], *with a sound.* (Ḳ.)

خَفَّاقٌ [*Flapping,* or *flapping much;*] applied to a wing. (TA.) And applied to a bird, [because of the sound of its wings,] meaning *Flying.* (TA.) — أَرْضٌ خَفَّاقَةٌ *A land in which the* سَرَاب [or *mirage*] *quivers.* (TA.) — خَفَّاقُ القَدَمِ A man *broad in the fore part of the foot:* (Ṣ, Ḳ:) or *broad in the under part,* or *sole, of the foot:* (JK, TA:) or *having the foot light upon the ground; not heavy, nor slow:* (IAạr, TA:) or *quick in step, beating the ground much with the foot so that it makes a sound of flapping to be heard by reason of the vehemence of his tread.* (Ḥam p. 173.) — خَفَّاقَةُ الحَشَى A woman *lank in the belly.* (Ṣ, L, Ḳ, TA.)

خَفَّاقَةٌ fem. of خَفَّاقٌ [q. v.]. — [Hence,] الخَفَّاقَةُ *The anus.* (IDrd, Ḳ.)

خَافِقٌ [act. part. n. of خَفَقَ in all its senses]. It is applied as an epithet to the سَرَاب [or mirage, as meaning *Quivering*]: and so ▼ خَفُوقٌ [but with an intensive signification]. (JK.) And [the fem. pls.] خَوَافِقُ and خَافِقَاتٌ are used as [substs.] signifying *Banners,* or *standards,* [because of their *fluttering.*] (TA.) — Applied to a man, *Moving,* or *shaking, his head,* or *bending it* [*down,* or *nodding*], *when drowsy,* or *dozing.* (TA.) — [Hence, app.,] رَأَيْتُ فُلَانًا خَافِقَ العَيْنِ ‡ *I saw such a one with the eye cast down, and depressed in the head* [as though drowsy]. (TA.) — أَيَّامُ الخَافِقَاتِ *Certain days in which the stars* [*in great number*] *became scattered* (تَنَاثَرَتْ) [in the CḲ, erroneously, تَنَاصَرَتْ]), [causing a belief that the day of judgment was at hand, (see Ḳur lxxxii. 2,)] *in the time of Abu-l-'Abbás and Aboo-Jaafar,* (Ḳ, TA,) *the 'Abbásees.* (TA.) — الخَافِقَانِ is a term applied to *The place of sunrise and the place of sunset,* (AHeyth, JK, Mgh, Ḳ,) by the attribution of predominance to the latter; for الخَافِقُ, meaning *the disappearing,* is applied to *the place of sunset:* (AHeyth, TA:) or *the horizon* (أُفُق) *of the place of sunrise and that of the place of sunset;* (Ṣ, Ḳ;) accord. to Lth (TA) and ISk, (Ṣ, TA,) because the night and the day for the most part pass away (▼ يَخْفِقَانِ, so in the T and Ṣ, but in the Ḳ, erroneously, يَخْتَلِفَانِ, TA) between them, (T, TA,) or in them: (Ṣ, TA:) or *the two* [*opposite*] *extremities of the heaven and the earth:* (Aṣ, Sh, Ḳ:) or *the end of the heaven and earth:* (Khálid Ibn-Jembeh, Ḳ:) or *two vacant spaces* (هَوَاآنِ) *next to the two* [*opposite*] *sides of the earth:* (Khálid Ibn-Jembeh, TA:) and خَوَافِقُ السَّمَآءِ signifies *The regions of the heaven from which issue the four* [*cardinal*] *winds.* (Khálid Ibn-Jembeh, Ḳ.) One says, مَا بَيْنَ الخَافِقَيْنِ مِثْلُهُ *There is not between the place of sunrise and the place of sunset the like of him.* (TA.) And أَلْحَقَهُ ٱللّٰهُ بِالخَافِقِ and بِالخَوَافِقِ [*May God remove him to the place of sunset* and *to the four cardinal regions* of the heaven or earth]. (TA.) — خَافِقٌ also signifies A place *void of,* or *unoccupied by, any one to cheer by his presence.* (TA.)

خَيْفَقٌ, applied to a desert (فَلَاة), *Wide,* (Ṣ, Ḳ,) *in which the* سَرَاب [or *mirage*] *quivers.* (Ṣ.) — Applied to a horse or mare, (JK, Ṣ, Ḳ, TA,) mostly to a female, (IDrd, TA,) and a she-camel, (IDrd, JK, Ḳ,) and a male ostrich, (IDrd, Ṣ, Ḳ,) *Quick,* or *swift:* (Ḳ:) or *very quick* or *swift:* (JK, Ṣ:) and ▼ خَنْفَقِيقٌ, (so in some copies of the Ḳ,) or ▼ خَيْفَقِيقٌ, (so in other copies of the Ḳ and in the JK and O, and so written by A'Obeyd,) each correct, the ن in the former and the ى in the latter augmentative, (MF, TA,) is applied to a she-camel and a male ostrich, (JK, Ḳ,) in the former sense, as is also خَيْفَقٌ, (JK,) or in the latter sense. (A'Obeyd, Ḳ.) Accord. to some, applied to a she-camel, it signifies *Lean,* or *lank, in the belly; having little flesh.* (TA.) And, applied to a woman, *Long in the* رُفْغَانِ [app. here meaning the *two inguinal creases*], *slender in the bones, and wide in step.* (El-Kilábee, Ḳ.) Also, applied to a woman, *Quick and bold;* and so ▼ خَنْفَقِيقٌ: (TA:) or the latter, so applied, signifies *light, active,* or *agile, and bold:* and Sb says that the ن in it is augmentative; deriving it from خَفْقُ الرِّيحِ [explained above: see 1]. (Ṣ.) — Also *i. q.* دَاهِيَةٌ [meaning either *A calamity,* or, as an epithet, *very cunning*]; (AA, Ḳ;) and so ▼ خَنْفَقِيقٌ; which latter occurs in a verse, variously related, applied to a child brought forth by a woman who had been in labour a whole night; (Ṣ, Ḳ;) meaning داهية; or, as some explain it, in this instance, meaning *imperfectly formed;* (TA;) [and is also used as a corroborative of داهية; for] one says ▼ دَاهِيَةٌ خَنْفَقِيقٌ [*a great,* or *severe, calamity;* or *extremely cunning*]. (Ṣ.)

خَنْفَقِيقٌ: see the next preceding paragraph, in four places. ═ Also, (as in some copies of the Ḳ,) or ▼ خَيْفَقِيقٌ, (as in other copies of the Ḳ and in the JK, and thus written by Lth,) a word imitative of The *sound of the running of horses* (JK, Ḳ) *in which is a quivering,* or *convulsive, motion.* (Ḳ.)

خَنْفَقِيقٌ : see خَيْفَقٌ : —— and see also خَيْفَقِيقٌ.

مَخْفَقٌ *A place,* (TA,) or *a level land,* (Aṣ, TA,) *in which the* سَرَاب [or *mirage*] *quivers.* (Aṣ, TA.) [See also خَفْقَةٌ.] —— And [the pl.] مَخَافِقُ signifies The *places of setting* [of stars]: and is used as [a sing.,] meaning the *place of setting* of a star. (Ḥam p. 152.) [See also خَافِقٌ.]

مِخْفَقٌ *A broad sword:* (JK, Ṣ, Ḳ:) or *anything broad with which one strikes.* (Mgh.)

مِخْفَقَةٌ *A* دِرَّة [q. v.] (JK, Ṣ, Ḳ) *with which one strikes* [or *flogs*]: (Ṣ:) or (so in the Ḳ, but in the JK "and") *a whip of wood:* (JK, Ḳ:) so says Lth. (TA.)

مَخْفُوقٌ A man (T) *having a fluttering,* or *palpitation, of the heart.* (IDrd,* T, Ḳ.*) —— And *Possessed, bereft of reason,* or *insane;* syn. مَجْنُونٌ: (AA, Ḳ:) fem. with ة. (AA.)

خفو

1. خَفَا, (Ṣ, Ḳ,) aor. يَخْفُو, (Ṣ,) inf. n. خَفْوٌ (Ṣ, Ḳ) and خُفُوٌّ; (ISd, Ḳ;) and خَفَى, aor. يَخْفِى, inf. n. خَفْىٌ; (Ṣ, TA;) and خَفِىَ, aor. يَخْفَى, inf. n. خَفًى; (Kr, TA;) said of lightning, *It flashed, gleamed,* or *shone,* (Ṣ, Ḳ, TA,) *faintly, extending sideways in the adjacent tracts of cloud:* when it flashes, gleams, or shines, a little, and then ceases, not extending sideways, it is termed وَمِيضٌ; and when it cleaves the clouds, and extends high, into the midst of the sky, without going to the right and left, it is termed عَقِيقَةٌ: (Ṣ, TA:) accord. to A 'Obeyd, خَفْوٌ signifies the *extending* of lightning *sideways in the tracts of the sky.* (TA.) —— And خَفَا, (Ḳ,) inf. n. خَفْوٌ, (TA,) *It* (a thing) *appeared.* (Ḳ, TA. [See also خَفَى, in art. خفى.])

خَفْوَةٌ i. q. خَفْيَةٌ, (Ḳ, TA, [in the CḲ, erroneously, خَفِيَّة,]) an inf. n. of خَفِيتُ لَهُ as syn. with اِخْتَفَيْتُ, (Ḳ* and TA in art. خفى, q. v.,) the و and ى being interchangeable. (TA.)

خفى

1. خَفِىَ, (Mgh, Mṣb, Ḳ,) aor. يَخْفَى, (Mṣb, Ḳ,) inf. n. خَفَآءٌ, (Mgh, Mṣb, Ḳ,) has two contr. significations: (Mgh, Mṣb:) *It was,* or *became, unperceived* or *imperceptible,* [or *hardly perceived* or *perceptible, by any of the senses,* or only *by the eye* or *ear,* or *by the mind;* mostly] *unapparent,* or *not apparent;* (Ḳ;) [*latent; obscure;*] *hidden, concealed,* or *covered;* (Mgh, Mṣb;) [or *unconspicuous;* but also *faint,* or *dim, to the sight; suppressed,* or *stifled,* said of the voice; or *low, faint, gentle,* or *soft, to the ear;* and *obscure to the mind, abstruse, recondite, occult,* or *covert;* and *secret, private,* or *clandestine:*] and *the contr.,* i. e. *it appeared; it was,* or *became, apparent, open, manifest, plain,* or *evident;* (Mgh, Mṣb;) [as also ↑ استخفى: (see مُسْتَخْفٍ, below:)] or, accord. to some, the particle that connects it with its subject distinguishes one meaning from the other: (Mṣb:) you say, خَفِىَ عَلَيْهِ, aor. and inf. n. as above, (Ṣ, Mgh, Mṣb, TA,) *it* (a thing, or an affair, Mgh) *was,* or *became, unperceived* or *imperceptible,* [&c., *by him;*] *unapparent,* or *not apparent,* [or *obscure,* &c., *to him;*] (TA;) or *hidden,* or *concealed, from him:* (Mgh, Mṣb:) and خَفِىَ لَهُ *it appeared to him; it was,* or *became, perceptible, apparent, open,* &c., *to him:* [but see what follows:] (Mgh, Mṣb:) whence the sayings of Mohammad, [app. the Ḥanafee Imám,] referring to spoils, فَخَفِىَ لَهُمْ أَنْ يَذْهَبُوا بِهَا وَيَكْتُمُوهَا أَهْلَ الشِّرْكِ, i. e. *It appeared* [*to them that they should go away with them,* or *take them away, and conceal them from the believers in a plurality of gods*], and خَفِىَ لَهُمْ أَنْ يُخْرِجُوهَا إِلَى دَارِ الإِسْلَامِ [*It appeared to them that they should take them forth to the territory of El-Islám*]: but this is said only in relation to that which appears from a state of concealment or from a hidden quarter. (Mgh.) [Hence,] بَرِحَ الخَفَآءُ *The affair,* or *case, became manifest:* (Ṣ, Ḳ:) or *the state of concealment departed,* or *ceased;* but the former explanation is better: or, as some say, ↑ الخَفَآءُ here signifies *the secret;* and the meaning is, *the secret became apparent:* (TA:) or, lit., *the low ground became high and apparent;* meaning † *what was concealed became revealed.* (Ḥar pp. 133-4. [See also art. برح.]) [And عَلَى خَفَآءٍ means *Covertly, secretly, privately, stealthily,* or *clandestinely.* (See also what follows.)] —— خَفِيتُ لَهُ, aor. ـَ, (Ḳ,) inf. n. خَفْيَةٌ and خُفْيَةٌ (Mṣb, Ḳ) and خَفْوَةٌ, (Ḳ,) the ى and و being interchangeable, (TA,) signifies ↑ اِخْتَفَيْتُ [i. e. *I made myself unapparent to him, lurked,* or *lay hid* or *in ambush, for him; cloaked,* or *disguised, myself to him; hid,* or *concealed, myself from him*]: (Ḳ:) [for] اختفى signifies *he hid,* or *concealed, himself,* (Fr,* El-Fárábee, JK,* Mṣb, Ḳ,) مِنْهُ *from him;* (TA;) as also ↑ استخفى, (Fr,* JK,* Mṣb, Ḳ,) and ↑ اخفى, (IAạr, Ḳ,) and ↑ تخفّى also is syn. with اختفى [in this sense]: (Z, TA:) or you say, ↑ اِسْتَخْفَيْتُ مِنْكَ, meaning *I hid,* or *concealed, myself from thee;* but not ↑ اِخْتَفَيْتُ: (IḲṭ, Th, Ṣ, Mṣb:) or ↑ اختفى in the sense of خَفِىَ is not of high authority, nor is it disallowed, (Az, Mṣb, TA,) but ↑ استخفى is more usual. (Az, TA.) You say, فَعَلْتُهُ خِفْيَةً and خُفْيَةً [*I did it covertly, secretly, privately, stealthily,* or *clandestinely*]. (Mṣb.) And قُتِلَ خِفْيَةً and خِفْوَةً [*He was slain covertly, secretly,* &c.]. (JK.) And يَأْكُلُهُ خِفْوَةً [lit. *He eats it covertly,* &c.,] means *he steals it.* (Ḳ.) In the saying in the Ḳur [vii. 53], اُدْعُوا رَبَّكُمْ تَضَرُّعًا وَخُفْيَةً, the meaning [of the last word] is, *Submissively, devoting yourselves to his service:* or, accord. to Zj, *adhering to his service in your minds:* or, accord. to Th, *celebrating Him in your minds:* or, accord. to Lḥ, *in quietness, and stillness:* (TA:) or *secretly;* and so in the similar passage in the Ḳur vi. 63. (Jel, and so Bḍ on this latter passage.) خُفْيَةٌ is from أَخْفَيْتُ الصَّوْتَ [explained below in this paragraph]: (JK:) the intrans. v. whereof is ↑ اختفى [signifying *It* (the voice) *was,* or *became, suppressed,* or *stifled;* or *low, faint, gentle,* or *soft;* like خَفِىَ, which is more common]. (Lth, TA.) —— خَفَاهُ, aor. يَخْفِيهِ, (JK, Ṣ, Mṣb, Ḳ,) inf. n. خَفْىٌ (JK, Mṣb,* Ḳ) and خُفِىٌّ, (Ḳ,) also has two contr. significations: (Ṣ, Mṣb, TA:) *He made it perceptible, apparent, open, manifest, plain,* or *evident:* (JK, Ṣ, Mṣb, Ḳ:) and *he hid,* or *concealed,* or *covered, it;* (Ṣ, Mṣb, Ḳ; [in this latter sense, erroneously written in the CḲ خَفَّاهُ;]) as also ↑ اخفاه: (Ṣ, Ḳ:) or, accord. to some, this latter has the latter meaning; and the former verb has [only] the former meaning: but accord. to others, the reverse is the case: (Mṣb:) or, accord. to Aboo-'Alee El-Ḳálee, the former verb has the former meaning only; and ↑ the latter verb has both meanings: (IB, TA:) ↑ the latter is also explained as signifying *he removed its* خِفَآء, i. e. *its covering:* (TA:) and the former, as meaning *he made it to come forth from a state of concealment:* (JK:) and *he drew it forth;* (Ḳ;) as also ↑ اختفاه. (Ṣ, Mṣb, Ḳ.) One says, خَفَى المَطَرُ الفِأَرَ *The rain made the rats,* or *mice, to come forth from their holes.* (Ṣ.) It is said in the Ḳur [xx. 15], إِنَّ السَّاعَةَ آتِيَةٌ أَكَادُ أَخْفِيهَا, (JK, TA,) or ↑ أُخْفِيهَا, (JK, Ṣ, TA,) accord. to different readers: (TA:) the former means [*Verily the hour of the resurrection is coming:*] *I am almost making it to appear:* (JK, TA:) and the latter, *I am almost removing that which conceals it:* (Ṣ, IJ, TA:) or *I almost conceal it:* (JK, TA:) or, as Ubei reads it, أَكَادُ أُخْفِيهَا مِنْ نَفْسِى [*I almost conceal it from Myself*]: and Fr says [that the meaning is], *I almost conceal it from Myself,* and how then should I acquaint you therewith? (TA.) And it is said in a trad. respecting the flight [from Mekkeh], اخفى عَنَّا خَبَرَكَ [written without the vowel-signs, so that it may be اِخْفِى or ↑ أَخْفِى,] i. e. *Conceal thou thine information from such as may ask thee respecting us.* (TA.) And in another trad., كَانَ يَخْفِى صَوْتَهُ بِآمِينَ, thus with fet-ḥ to the ى, meaning *He used to make his voice perceptible* [or *audible, with moaning*]. (TA.) And you say, ↑ أَخْفَيْتُ الصَّوْتَ [meaning *I suppressed,* or *stifled, the voice;* or *made it low, faint, gentle,* or *soft*]. (Lth, JK, TA.) [And ↑ أَخْفَى الكَلَامَ *He uttered speech,* or *the speech, in a low, faint, gentle,* or *soft, tone; he spoke in a low, faint, gentle,* or *soft, manner;* lit. *he made speech,* or *the speech, to be low,* &c.] —— خَفَى, aor. يَخْفِى; and خَفِىَ, aor. يَخْفَى; inf. n. of each خَفْىٌ; said of lightning: see خَفَا, in art. خفو.

4, as an intrans. v.: see 1, in the former half of the paragraph, near the middle. —— As a trans. v.: see 1, in seven places, in the latter half of the paragraph.

5: see 1, in the former half of the paragraph, near the middle.

8, as an intrans. v.: see 1, in four places. —— اِخْتَفَى, said of a man, [if it be not a mistranscription for أُخْتُفِىَ, like أُخْتُفِىَ دَمُهُ,] signifies also *He was slain covertly, secretly,* or *clandestinely.* (JK.) —— As a trans. v.: see 1. —— You say also اختفى مَيْتًا *He drew forth a dead body from the grave,* to steal the grave-clothes. (TA.) —— And اختفى البِئْرَ *He dug,* or *cleared out, the well.* (Mṣb.) —— And اختفى دَمَهُ *He slew him without its being known.* (Ḳ.)

10: see 1, in four places, in the former half of the paragraph.

خَفًا [more properly written خَفًى] *A thing that is unperceived or imperceptible,* [or *hardly perceived or perceptible,*] *unapparent,* or *not apparent;* [*latent; obscure;* &c.; (see 1, first sentence;)] (JK, Ḳ;) as also ▼ خَافِيَةٌ and ▼ خَافٍ [for شَىْءٌ خَافٍ, the explanation in the JK]. (Ḳ.) [See also خَفَآءٌ.]

خَفِىُّ البَطْنِ A man *lank in the belly.* (IAạr, TA.)

خَفَآءٌ inf. n. of خَفِىَ. (Mgh, Mṣb, Ḳ.) —— Also *A thing that is unperceived* or *imperceptible* [&c. (see خَفًا)] *by one; unapparent,* or *not apparent,* [*latent,* or *obscure,*] *to one;* or *hidden,* or *concealed, from one.* (TA.) *A secret:* so, accord. to some, in the phrase بَرِحَ الخَفَآءُ mentioned above: see 1, in the former half of the paragraph. (TA.) And [in the same phrase, accord. to some,] *Low,* or *depressed, ground.* (TA.)

خِفَآءٌ *A* [*garment of the kind called*] رِدَآء, *which a woman wears over her other clothes:* (Lth, JK:) or *a* [*garment of the kind called*] كِسَآء: (Ṣ, Ḳ:) and *any covering* of a thing, (Lth, JK,*) *whatever it be with which one covers a thing, such as a* كساء *and the like:* (Lth:) pl. أَخْفِيَةٌ. (Lth, JK, Ṣ, Ḳ.) —— [Hence,] أَخْفِيَةُ النَّوْرِ *The calyxes of flowers:* (Ḳ:) sing. as above. (TA.) —— And أَخْفِيَةُ الكَرَا [*The coverings of drowsiness;* meaning] *the eyes.* (Ḳ.)

خَفِىٌّ *i. q.* ▼ خَافٍ; (Ṣ, Ḳ;) applied to a thing; (Ṣ;) i. e. *Unperceived* or *imperceptible,* [or *hardly perceived* or *perceptible, by any of the senses,* or only *by the eye* or *ear,* or *by the mind;* mostly] *unapparent,* or *not apparent;* (Ḳ;) [*latent; obscure; hidden,* or *concealed;* or *unconspicuous;* but also *faint,* or *dim, to the sight; suppressed,* or *stifled,* applied to the voice; or *low, faint, gentle,* or *soft, to the ear;* and *obscure to the mind, abstruse, recondite, occult,* or *covert;* and *secret, private,* or *clandestine:* see 1, first sentence:] pl. خَفَايَا. (Ṣ.) [You say نَجْمٌ خَفِىٌّ *A dim star* or *asterism.* And مَكَانٌ خَفِىٌّ *An obscure,* or *a concealed, place.* And صَوْتٌ خَفِىٌّ *A low, faint, gentle,* or *soft, voice* or *sound.*] And اِمْرَأَةٌ خَفِيَّةُ الصَّوْتِ *A woman having a low, faint, gentle,* or *soft, voice.* (TA in art. خفض.) And النُّونُ الخَفِيَّةُ *i. q.* الخَفِيفَةُ [q. v.]. (Ḳ.) And some of the Arabs say, (Yaạḳoob, Ṣ,) إِذَا حَسُنَ مِنَ المَرْأَةِ خَفِيَّاهَا حَسُنَ سَائِرُهَا, meaning [*When*] *the voice and the foot-mark of the woman* [*are good,* or *pleasing, the rest,* or *the whole, of what pertains to her is good,* or *pleasing*]: (Yaạḳoob, JK, Ṣ, Ḳ:) for when her voice is soft, or gentle, this indicates her being bashful, or shy; and when her foot-marks are near together, and firmly impressed, they indicate that she has [large] buttocks and haunches. (Yaạḳoob, Ṣ.) One says also, لَقِيتُهُ خَفِيًّا *I met him covertly, secretly, privately,* or *clandestinely.* (TA.) [And مَشَى مِشْيَةً خَفِيَّةً *He walked with a soft,* or *stealthy, gait.*] —— Also One *who secludes himself from* [*other*] *men; whose place is concealed from them.* (TA.)

خَفِيَّةٌ *A well:* (Ṣ, Ḳ:) or *a deep well;* because its water is not perceived, or not apparent: (TA:) or *a well of ancient times, that has become filled up and then dug again:* (JK, TA:) or *any well that has been dug and then left until it has become filled up, then dug again, and cleared out:* (ISk, Ṣ:) [opposed to بَدِىٌّ:] accord. to A 'Obeyd, it is so called because it is made to appear: (Ṣ:) pl. خَفَايَا and خَفِيَّاتٌ. (JK, TA.) —— And *A tangled,* or *luxuriant,* or *dense, thicket,* (JK, Ḳ, TA,) *which the lion takes as his covert:* (JK, TA:) or خَفِيَّةٌ is the name of a certain place frequented by lions; (Ṣ, IB;) and is properly imperfectly decl., so that you say أُسُودُ خَفِيَّةَ; but it may be perfectly decl. in poetry. (IB.) == Also *A slight taint,* or *infection,* or *a touch,* or *stroke, of insanity:* so in the phrase بِهِ خَفِيَّةٌ *In him is a slight taint,* &c., *of insanity.* (Ibn-Menádhir, Ṣ, Ḳ.*)

خَافٍ: see خَفِىٌّ: —— and see also خَفًا. —— الخَافِى *The jinn,* or *genii;* (Aṣ, Lḥ, JK, Ṣ, Ḳ;) because they conceal themselves from the eyes [of men]; (TA;) as also ▼ الخَافِيَةُ (JK, Ḳ) and ▼ الخَافِيَآءُ: (Ḳ:) or this last signifies *what conceals itself in the body, of the jinn,* or *genii:* (Ibn-Menádhir, Ṣ:) the pl. (of the first, Lḥ, JK, [and of the second and third also accord. to analogy,]) is خَوَافٍ; (Lḥ, JK, Ḳ;) [and of the first, خَافُونَ also, like قَاضُونَ; for] the bare piece of ground amid herbage is said, in a trad., to be مُصَلَّى الخَافِينَ, i. e. [*The praying-place*] *of the jinn,* or *genii.* (TA.) The first (الخَافِى) also signifies *Mankind;* thus bearing two contr. [or rather opposite] meanings. (TA.) And one says, مَا أَدْرِى أَىُّ خَافِيَةٍ ▼ هُوَ, (Ḳ and TA voce خَالِفَةٌ, q. v.,) or ▼ خَافِيَةٍ, (CḲ ibid.,) *I know not what one of mankind he is.* (Ḳ ibid.) —— أَرْضٌ خَافِيَةٌ [and ▼ أَرْضٌ خَافِيَةٌ, the latter word in the former case being an epithet, fem. of خَافٍ, and in the latter case a subst., or an epithet in which the quality of a subst. is predominant,] *A land in which are jinn,* or *genii.* (Ḳ.) —— خَافِى الغُرَابِ: see the last sentence but one in the next paragraph.

خَافِيَةٌ contr. of عَلَانِيَةٌ [app. meaning that it signifies *A state of being unapparent* or *not apparent, covert, secret, private,* or *clandestine:* though explained in the TḲ (followed by Freytag) as an epithet applied to a man, meaning *whose actions are always covert*]. (Ḳ.) —— See also خَفًا. —— And see خَافٍ, in four places. —— Also *One,* i. e. *a single feather, of the feathers called* الخَوَافِى: (TA:) الخَوَافِى signifies *the feathers below the ten that are in the fore part of the wing:* (Aṣ, Ṣ:) or *certain feathers that are concealed when the bird contracts its wing:* (Ḳ:) or *the four feathers that are* [*next*] *after those called* المَنَاكِب, (Lḥ, Ḳ,*) *and next before those called* الأَبَاهِر: (Ṣ in art. بهر, and L in art. نكب:) or *seven feathers in the wing, after the seven foremost:* (Ḳ,* TA:) but the people [generally] mention them as *four:* or they are *the small feathers in the wing of a bird.* (TA.) خَنْجَرٌ مِثْلُ خَافِيَةِ النَّسْرِ [*A dagger like the* خافية *of the vulture*], occurring in a trad., means *a small* خنجر. (TA.) One says also خَافِيَةُ الغُرَابِ [*The* خافية *of the crow*]: and the pl. is [sometimes expressed by using the coll. gen. n., saying] خَافِى ▼ الغُرَابِ. (JK.) —— الخَوَافِى also signifies *The palm-branches* [*next*] *below the* قِلَبَة [which latter are the *branches that grow forth from the heart of the tree*]: (Ṣ, TA:) thus called in the dial. of Nejd: (TA:) in the dial. of El-Ḥijáz called العَوَاهِنُ. (Ṣ, TA.)

الخَافِيَآءُ: see خَافٍ.

مُخْتَفٍ *A rifler of graves:* (JK, Ṣ, Mṣb, Ḳ:) because he extracts the grave-clothes; (Ṣ, Mṣb, TA;) or because he steals covertly: a word of the dial. of the people of El-Medeeneh: fem. مُخْتَفِيَةٌ. (TA.)

مُسْتَخْفٍ *Hiding,* or *concealing, himself:* and accord. to Akh, *appearing:* in both of which senses it is said to be used in the words of the Ḳur [xiii. 11], مُسْتَخْفٍ بِاللَّيْلِ وَسَارِبٌ بِالنَّهَارِ [*Hiding himself by night, and appearing by day:* or *appearing by night, and hiding himself by day:* see art. سرب]. (TA.) —— اليَدُ المُسْتَخْفِيَةُ *The hand of the thief,* and *of the rifler of graves:* opposed to اليَدُ المُسْتَعْلِيَةُ, which is the hand of him who takes by force, and of the plunderer, and the like: the Sunneh ordains that the former shall be cut off [except in certain cases], but not the latter. (TA.)

خل

1. خَلَّ لَحْمُهُ, aor. ـِ, (Ks, Ṣ, Ḳ, TA, in the CḲ ـُ,) [irreg. in the case of an intrans. v. of this class, unless the verb be of the measure فَعُلَ,] and ـُ, (Ḳ,) [agreeably with general rule,] inf. n. خَلٌّ and خُلُولٌ; (Ks, Ṣ, Ḳ;) and ▼ اختلّ; (Ṣgh, Ḳ;) *His flesh became little,* or *scanty;* (Ks, Ṣ;) or *his flesh decreased, diminished,* or *wasted:* (Ḳ:) *he became lean,* or *spare.* (Ks, Ṣ, Ḳ.) [But it seems, from what follows, that the verb may be of the measure فَعِلَ, aor. ـَ; as well as of the measure فَعَلَ, aor. ـِ or ـُ; or perhaps of the measures فَعِلَ and فَعَلَ and فَعُلَ, so that the aor. may be regularly ـَ and ـِ and ـُ.] —— You say also خَلِلْتُ مِنْ كَذَا *I missed such a thing.* (JK.) And خَلَّ البَعِيرُ مِنَ الرَّبِيعِ *The camel missed the* [*herbage called*] ربيع, *and became lean in consequence thereof.* (JK, Ibn-'Abbád, TA.) —— And خَلَّ, (JK, Ṣ, Ḳ,) inf. n. خَلٌّ; (TA;) and ▼ أَخَلَّ, (JK, Mṣb, TA,) or ▼ أُخِلَّ, (Ḳ,) and ▼ أُخِلَّ بِهِ; (Ṣ, TA;) and ▼ اختلّ; (MA, KL;) said of a man, (JK, Ṣ, Mṣb,) *He was,* or *became, poor,* or *in want* or *need.* (JK, Ṣ, MA, KL, Mṣb, Ḳ,

TA.) — خَلَّ الشَّىْءَ, (K,) aor. ُ , inf. n. خَلٌّ, (TA,) *He, or it, perforated the thing; transpierced it, or pierced it through;* as also ▼ تخلّلهُ: (K:) so in the M. (TA.) You say, خَلَلْتُ الشَّىْءَ بِالخِلَالِ, aor. ُ , *I transfixed,* or *transpierced, the thing with the* [*pin called*] خلال. (JK.) [And خَلَّ اللَّحْمَ *He skewered the flesh-meat.*] And خَلَلْتُهُ بِالرُّمْحِ *I pierced him with the spear.* (JK.) And ▼ اختلّهُ بِالرُّمْحِ *He transpierced him,* or *transfixed him, with the spear;* (T, M, K, TA;) and so بِالسَّهْمِ *with the arrow:* (S:) or the former signifies *he pierced him with the spear and transfixed his heart:* (TA:) accord. to AZ, الاِخْتِلَالُ relates to the heart and the liver. (M in art. نظم.) And ▼ يَخْتَلُّ الثَّوْرُ الكَلْبَ بِقَرْنِهِ [*The bull pierces the dog with his horn*]. (JK. [It is there vaguely indicated that ▼ خِلَّة signifies The *act,* or perhaps the *effect, of a bull's piercing a dog with his horn.*]) And ▼ تخلّلهُ بِالرُّمْحِ *He pierced him time after time with the spear.* (M, K.) — And خَلَّ الفَصِيلَ, (K,) inf. n. خَلٌّ, (TA,) *He slit the tongue of the young camel, and inserted into it a wooden pin called* خِلَال*, in order that he might not suck:* (K:) or [simply] *he slit the tongue of the young camel, in order that he might not be able to suck* [*any longer*]*, so that he became lean;* as also خَلَّ لِسَانَ الفَصِيلِ: (S:) or الخَلُّ signifies *the fixing a* خِلَال *above the nose of the young camel, to prevent his sucking.* (TA in art. لهج.) — And خَلَّهُ, (T, Mgh, Msb, K, TA,) aor. ُ , inf. n. خَلٌّ, (S, Msb, TA,) namely, a thing, (TA,) a garment, (T, TA,) a [garment such as is called] كِسَآء (S, K, TA) or رِدَآء (Mgh, Msb) &c., (TA,) and a [tent such as is called] خِبَآء, (S, TA,) *He pinned it with the* [*pin called*] خِلَال; (T, TA;) *he conjoined* (Mgh, Msb, TA) *its two edges,* (Mgh, Msb,) or *its edges,* (TA,) or *he fastened it,* (K,) *with a* خِلَال: (S, Mgh, Msb, K, TA:) and ▼ خلّلهُ has a similar, but intensive, signification. (Msb.) A poet says,

* سَمِعْنَ بِمَوْتِهِ فَظَهَرْنَ نَوْحًا *
* قِيَامًا مَا يُخَلُّ لَهُنَّ عُودُ *

meaning, لَا يُخَلُّ لَهُنَّ ثَوْبٌ بِعُودٍ [i. e. *They* (the women) *heard of his death, and appeared, wailing, standing; no garment of theirs having its edges fastened together with a pointed piece of wood*]. (TA.) — خَلَّ الإِبِلَ, (K,) aor. ُ , inf. n. خَلٌّ, (TA,) *He removed, transferred,* or *shifted, the camels to what is termed* خُلَّة [after they had been pasturing upon حَمْض]; as also ▼ أَخَلَّهَا: (K:) or the latter signifies *he pastured them upon* خُلَّة. (S.) — خَلَّ, (Lh, S, K,) [aor. ُ ,] inf. n. خَلٌّ, (TA,) is also *syn. with* خَصَّ [*He particularized,* or *specified*]; (Lh, S, K;) contr. of عَمَّ; (K;) and so ▼ خلّل: (JK, S, TA:) thus in the phrase, عَمَّ فِي دُعَائِهِ وَخَلَّ (S, TA) and وَخَلَّلَ (JK, S, TA) [*He included,* or *comprehended,* persons or things *in common,* or *in general,* in his prayer or supplication &c., *and particularized,* or *specified,* some person or thing, or some persons or things].

2. خلّل أَسْنَانَهُ, inf. n. تَخْلِيلٌ, [*He picked his teeth;*] *he extracted the remains of food between his teeth* with a خِلَال [or toothpick]; (Msb, K,*) in which latter the pass. form of the verb is mentioned;) and so ▼ تخلّل, alone; (T, S,* O, TA;) but accord. to the K, you say, تخلّلهُ [*he extracted it*], meaning the remains of food between the teeth. (TA.) — خلّل الشَّعَرَ بِالمُشْطِ [*He separated the hair with the comb; he combed the hair*]. (Mgh voce تَسْرِيح.) — خلّل لِحْيَتَهُ, (S,* Msb, K,) and أَصَابِعَهُ, (S,* K,) inf. n. as above, (S,) *He made the water to flow into the interstices of his beard,* (Msb, K,) and *of his fingers* or *toes,* (K,) in the ablution termed وُضُوء; (S, TA;) and ▼ تخلّل, alone, signifies the same. (S.) It (the former) is as though it were taken from تَخَلَّلْتُ القَوْمَ meaning "I entered amid the breaks, or interspaces, of the people." (Msb.) Hence the trad., خَلِّلُوا أَصَابِعَكُمْ لَا تُخَلِّلُهَا نَارٌ قَلِيلٌ بُقْيَاهَا [*Make ye the water to flow into the interstices of your fingers or toes, lest fire that shall spare little be made to flow into their interstices*]. (TA.) — خلّلهُ كِلْسًا *He put* صَارُوج [or كِلْس, i. e. *quick lime,* &c.,] *into the interstices of its* (a building's) *stones.* (TA in art. كلس.) — خلّل القِثَّآءَ, and البِطِّيخَ, inf. n. as above, *He investigated the state of the cucumbers, and the melons,* or *water-melons, so as to see every one that had not grown, and put another in its place.* (AA, TA.) — See also 1, in the latter half of the paragraph. — And see 1 again, last sentence. — خلّل, inf. n. تَخْلِيلٌ, said of wine and of other beverages, *It became acid,* or *sour; and spoiled:* (K:) or, said of شَرَاب [i. e. wine and the like], (Mgh,) or of نَبِيذ [i. e. must and the like], (Msb,) or of expressed juice, (K,) *it became vinegar;* (Mgh, Msb, K;) as also ▼ اختلّ; (Lth, K;) but this is disallowed by Az; (TA;) and ▼ تخلّل; but this is of the language of the lawyers; (Mgh;) or, said of نبيذ, this last signifies *it was made into vinegar:* (Msb:) or خلّل, said of شراب, signifies *it spoiled,* (JK, T,) *and became vinegar.* (T.) — تَخْلِيلٌ also signifies The *making vinegar;* (S;) and so ▼ اِخْتِلَالٌ; (K;) i. e. of the expressed juice of grapes and of dates. (TA.) You say, خلّل الخَمْرَ, (K,) or الشَّرَابَ, (Mgh,) or النَّبِيذَ, inf. n. as above, (Msb,) the verb being trans. as well as intrans., (Mgh, Msb, K,) and ▼ تخلّل النَّبِيذَ, (TA,) *He made the wine,* or *beverage,* or *must* or *the like, into vinegar.* (Mgh, Msb, K, TA.) — And خلّل البُسْرَ *He put the full-grown unripe dates in the sun, and then sprinkled them* (نَضَحَهُ, in some copies of the K نضجه,) *with vinegar, and placed them in a jar:* (K:) so in the M: and in like manner, other things than بُسْر; as cucumbers, and cabbage, and بَاذَنْجَان [q. v.], and onions. (TA.) [Accord. to modern usage, the verb signifies *He pickled.*]

3. خَالَّهُ, (JK, Mgh, K,) inf. n. مُخَالَّةٌ and خِلَالٌ (JK, S, K) and [quasi-inf. n.] ▼ خُلَّةٌ, (JK,) *He acted,* or *associated, with him as a friend,* or *as a true,* or *sincere, friend.* (JK, S,* Mgh, K.) لَا بَيْعٌ فِيهِ وَلَا خِلَالٌ, in the Kur [xiv. 36], is said to mean [*Wherein shall be no buying or selling*] *nor mutual befriending:* or [*and no friends,* or *true friends,* for], as some say, خِلَال is here pl. of ▼ خُلَّة, like as جِلَال is pl. of جُلَّة. (TA.)

4. أَخَلَّ and أَخَلَّ بِهِ and أُخِلَّ بِهِ: see 1, near the beginning. — أَخَلَّ بِهِ *He* (a man) *fell,* or *stopped, short in it; fell short of accomplishing it; fell short of doing what was requisite,* or *due,* or *what he ought to have done, in it,* or *with respect to it;* or *flagged,* or *was remiss, in it;* namely, a thing; syn. قَصَّرَ فِيهِ; (Msb;) as, for instance, in belief, and in confession thereof, and in works: (Ksh and Bd in ii. 2:) *he left it, neglected it, omitted it;* or *left it undone:* (Har p. 402:) or *i. q.* أَجْحَفَ بِهِ [app. as meaning *he was near to falling short of accomplishing it,* or *of doing what was requisite in it;* or *was near to being remiss in it*]; namely, a thing. (K.) — *He failed of fulfilling his compact with him,* or *his promise to him.* (K.) — *He became absent,* or *he absented himself, from it; he left, abandoned,* or *quitted, it;* namely, a place &c. (K.) You say, اخلّ بِمَرْكَزِهِ *He* (a man, S, or a horseman, Mgh) *left, abandoned,* or *quitted, his station* (S, Mgh) *which the commander had appointed him.* (Mgh.) And اخلّ بِهِمْ *He became absent,* or *he absented himself, from them.* (JK.) — اخلّ الوَالِى بِالثُّغُورِ *The prefect made the frontiers to be kept by a small body of troops.* (K.) — أَخَلَّ إِلَيْهِ: see 8. — اخلّهُ *He made him,* or *caused him, to want,* or *be in need.* (JK, S, K.) You say, مَا أَخَلَّكَ إِلَى هٰذَا *What has made thee,* or *caused thee, to want, or be in need of, this?* (S.) And مَا أَخَلَّكَ اللّٰهُ إِلَيْهِ *What has God made thee,* or *caused thee, to want,* or *be in need of?* (Lh, K.) — اخلّ الإِبِلَ: see 1, near the end of the paragraph. — اخلّوا, (K,) inf. n. إِخْلَالٌ, (TA,) *Their camels pastured upon what is termed* خُلَّة. (K.) — Hence, اخلّ said of a man signifies † أَخَذَ مِنْ قُبُلٍ [i. e. *He took frontways*]: opposed to أَحْمَضَ [and حَمَّضَ, q. v.], meaning أَخَذَ مِنْ دُبُرٍ. (TA.) — اخلّت النَّخْلَةُ *The palm-tree produced bad fruit.* (A 'Obeyd, JK, S, K.) — And *The palm-tree produced dates such as are termed* خَلَال: [like أَبْلَحَت from بَلَح:] thus it bears two contr. significations. (K.)

5. تخلّل [primarily signifies *It entered,* or *penetrated,* or *passed through, the* خِلَال, i. e. *interstices,* &c., *of a thing*]. You say, تَخَلَّلْتُ القَوْمَ *I entered amid the breaks,* or *interspaces, of the people.* (S, M, Msb, K.*) And تَخَلَّلُوا الدِّيَارَ [*They went through the midst of the houses*]. (S in art. جوس.) And تخلّل الرَّمْلَ *He passed through the sands.* (Az, TA.) And تخلّل القَلْبَ

† [*It penetrated the heart*]; said of admonition. (TA in art. بهر.) And تخلّل الشَّىْءُ *The thing* [i. e. anything] *went*, or *passed, through.* (JK,* Ṣ, Ḳ.) — [Hence, *It intervened;* said of a time &c. And hence the phrase مِنْ غَيْرِ تَخَلُّلٍ *Without interruption.*] — And تخلّل المَطَرُ *The rain was confined to a particular place, or to particular places; was not general.* (Ṣ, Ḳ.) — See also 1, in two places, in the former half of the paragraph. — تخلّل الرُّطَبَ *He sought out the fresh ripe dates in the interstices of the roots of the branches* (M, Ḳ) *after the cutting off of the racemes of fruit.* (M.) And تخلّل النَّخْلَةَ *He picked the dates that were among the roots of the branches of the palm-tree;* as also تَكَرَّبَهَا. (AḤn, TA.) — For other significations, see 2, in four places.

6. تَخَالٌّ [said of several persons] The *being friendly, one with another.* (KL.) [You say, تَخَالُّوا *They acted together*, or *associated, as friends*, or *as true friends.*]

8. اختلّ [primarily signifies] *It had interstices, breaks, chinks*, or *the like.* (MA. [See خَلَلٌ.]) — [And hence,] *It was*, or *became, shaky, loose, lax, uncompact, disordered, unsound, corrupt,* (Mṣb,) *faulty*, or *defective,* (KL, Mṣb,) [and *weak*, or *infirm,* (see خَلَلٌ and مُخْتَلٌّ,)] said of a thing or an affair; (KL;) *it became altered for the worse.* (Mṣb.) [You say, اختلّ مِزَاجُهُ *His constitution*, or *temperament, became in a corrupt* or *disordered state.* And اختلّ alone *He was*, or *became, disordered in temper;* (see تَحَمَّضَ;) but this seems to be from the same verb said of a camel; (see اختلّت الإِبِلُ, below;) for the camel becomes disordered in his stomach by pasturing long upon خُلَّة, without shifting to حَمْض. And اختلّ عَقْلُهُ *His mind*, or *intellect, was*, or *became, unsound*, or *disordered.*] And اختلّ أَمْرُهُ [*His affair*, or *state, was*, or *became, unsound, corrupt*, or *disordered*]; (Ṣ, voce اِضْطَرَبَ;) i. e. وَقَعَ فِيهِ الخَلَلُ. (JM.) — *He was*, or *became, lean, meagre*, or *emaciated;* (KL;) and so اختلّ جِسْمُهُ. (Ṣ.) See 1, first sentence. — See also خَلَّ as syn. with أَخَلَّ or أُخِلَّ &c., near the beginning of the first paragraph. [Hence,] اختلّ إِلَيْهِ *He wanted it*, or *needed it;* (Ṣ, Mṣb, Ḳ;) namely, a thing; (Ṣ, Mṣb;) as also ▼ أَخَلَّ اليه: (TA:) whence the saying of Ibn-Mes'ood, عَلَيْكُمْ بِالعِلْمِ فَإِنَّ أَحَدَكُمْ لَا يَدْرِى مَتَى يُخْتَلُّ إِلَيْهِ [*Keep ye to* the pursuit *of knowledge*, or *science; for any one of you knows not*, or *will not know, when it will be wanted*, or *needed*]; i. e., when men will want, or need, that [knowledge] which he possesses. (Ṣ.) You say also, أُخْتُلَّ إِلَى فُلَانٍ *Such a one was wanted*, or *needed.* (JK.) ═ See also 2, in two places. ═ اختلّه بِالرُّمْحِ, and بِالسَّهْمِ: and يَخْتَلُّ الثَّوْرُ الكَلْبَ بِقَرْنِهِ: see 1, in the former half of the paragraph. — اختلّ also signifies *He sewed together.* (KL.) — أُخْتُلَّ said of herbage: see خُلَّةٌ, near the end of the paragraph. ═ اختلّ المَكَانُ *The place had in it* خُلَّة [q. v.]. (MA.) — And اختلّت الإِبِلُ *The camels were confined in* [*pasturage such as is termed*] خُلَّة. (Ḳ.)

R. Q. 1. خَلْخَلَهَا *He attired her with the* خَلْخَال [or *anklet*, or *pair of anklets*]. (TA.) ═ خلخل العَظْمَ *He took the flesh that was upon the bone.* (Ḳ.)

R. Q. 2. تَخَلْخَلَتْ *She attired herself with the* خَلْخَال [or *anklet*, or *pair of anklets*]. (Ḳ.) ═ تخلخل *It* (a garment, or piece of cloth,) *was*, or *became, old, and worn out.* (JK.)

خَلٌّ a word of well-known meaning, (Ṣ, Mṣb.) *Vinegar;* i. e. *expressed juice of grapes* (JK, Mgh, Ḳ) and *of dates* (JK) *&c.* (Ḳ) *that has become acid*, or *sour:* (JK,* Mgh, Ḳ:) so called because its sweet flavour has become altered for the worse (اختلّ): (Mṣb:) a genuine Arabic word: (IDrd, Ḳ:) *the best is that of wine: it is composed of two constituents* (Ḳ) *of subtile natures,* (TA,) *hot and cold,* (Ḳ,) *the cold being predominant:* (TA:) *and is good for the stomach; and for the gums,* (Ḳ,) *which it strengthens, when one rinses the mouth with it;* (TA;) *and for foul ulcers* or *sores; and for the itch; and for the bite*, or *sting, of venomous reptiles; and as an antidote for the eating of opium; and for burns; and for toothache; and its hot vapour is good for the dropsy, and for difficulty of hearing, and for ringing in the ears:* (Ḳ: [various other properties &c. are assigned to it in the TA:]) ▼ خَلَّةٌ signifies *somewhat* (lit. *a portion*) *thereof;* [being the n. un.:] (Aboo-Ziyád, Ḳ;) or it may be a dial. var. thereof, like as خَمْرَةٌ is [said by some to be] of خَمْرٌ: (Aboo-Ziyád, TA:) see also خَلَّةٌ: the pl. is خُلُولٌ [meaning *sorts*, or *kinds, of vinegar*]. (Mṣb.) It is said in a trad., نِعْمَ الإِدَامُ الخَلُّ [*Excellent*, or *most excellent, is the seasoning, vinegar!*]. (TA.) — [Hence,] أُمُّ الخَلِّ [*The mother of vinegar;* meaning] *wine.* (JK, TA.) — [Hence also the saying,] مَا فُلَانٌ بِخَلٍّ وَلَا خَمْرٍ, (A'Obeyd, JK, Ṣ,) or مَا لَهُ خَلٌّ وَلَا خَمْرٌ, (Ḳ,) or مَا عِنْدَ فُلَانٍ خَلٌّ وَلَا خَمْرٌ, (Ṣ in art. خمر,) *Such a one*, or *he, possesses neither good nor evil:* (A'Obeyd, JK, Ṣ, Ḳ:) [or *neither evil nor good:* for] AA says that some of the Arabs make الخَمْرُ to be *good*, and الخَلُّ to be *evil;* [and thus the latter is explained in one place, in this art., in the Ḳ;] and some of them make الخمر to be *evil*, and الخلّ to be *good.* (Ḥar p. 153.) ═ *I. q.* حَمْضٌ [i. e. *A kind of plants in which is saltness:* or *salt and bitter plants:* or *salt*, or *sour, plants* or *trees:* &c.: opposed to خُلَّةٌ]. (Ḳ.) A poet says,

* لَيْسَتْ مِنَ الخَلِّ وَلَا الخِمَاطِ *

[*She is not*, or *they are not, of the plants* or *trees called* خلّ, *nor of the kind called* خماط (pl. of خَمْطٌ)]. (TA.) ═ *A road in sands:* (Ṣ:) or *a road passing through sands:* or *a road between two tracts of sand:* (Ḳ:) or *a road passing through heaped-up sands:* (JK, Ḳ:) masc. and fem. [like طَرِيقٌ]: (Ṣ, Ḳ:) pl. [of pauc.] أَخُلٌّ and [of mult.] خِلَالٌ. (Ḳ.) One says حَيَّةُ خَلٍّ [*A serpent of a road in sands, &c.*]; like as one says أَفْعَى صَرِيمَةٍ. (Ṣ.) — *An oblong tract of sand.* (Ḥam p. 709.) — *A vein in the neck* (JK, Ḳ) *and in the back,* (Ḳ,) *communicating with the head.* (JK, TA.) — *A slit*, or *rent, in a garment*, or *piece of cloth.* (K.) ═ *An old and worn-out garment*, or *piece of cloth,* (JK, Ṣ, Ḳ, TA,) *in which are streaks:* (TA:) [or so ثَوْبٌ خَلٌّ:] and ▼ خَلْخَلٌ and ▼ خَلْخَالٌ, applied to a garment, or piece of cloth, (JK, Ḳ,) signify *old and worn out,* (JK,) or *thin,* (Ḳ,) like هَلْهَلٌ and هَلْهَالٌ. (TA.) — *A bird having no feathers:* (JK:) or *having few feathers.* (Ḳ.) — A man (JK, Ṣ) *lean, meagre*, or *emaciated;* (JK, Ṣ, Ḳ;) as also ▼ خَلِيلٌ (Ḳ) [a meaning said in the TA to be tropical] and ▼ مَخْلُولٌ and ▼ مُخْتَلٌّ: (TA:) or *light in body:* (IDrd, TA:) and [the fem.] خَلَّةٌ, applied to a woman, *light* (Ḳ, TA) *in body, lean*, or *spare:* (TA:) the pl. of خَلٌّ is خُلُولٌ. (JK.) Also *Fat:* thus bearing two contr. significations: (Ḳ:) and so ▼ مَخْلُولٌ. (TA.) It is applied to a man and a camel. (TA.) Accord. to the Ḳ, it also signifies *A* [*young camel such as is termed*] فَصِيلٌ: (TA:) but it means *such as is lean*, or *emaciated;* (TA;) and so ▼ مَخْلُولٌ, applied to a فصيل as an epithet, for a reason mentioned above, in an explanation of the phrase خَلَّ الفَصِيلَ. (Ṣ, TA.) — Also *i. q.* اِبْنُ مَخَاضٍ [i. e. *A male camel in his second year*]; (JK, Ḳ;) and so ▼ خَلَّةٌ; which is also applied to the *female:* (Aṣ, Ṣ, Ḳ:) and *i. q.* اِبْنُ لَبُونٍ [i. e. *a male camel in*, or *entering upon, his third year*]; and in like manner ▼ خَلَّةٌ is applied to the *female;* (JK;) or, as in the M, *to a she-camel;* (TA;) and, as some say, (JK,) *a large she-camel:* (JK, TA:) and ▼ اِبْنُ الخَلَّةِ signifies *the same as* اِبْنُ اللَّبُونِ (T in art. بنى) or اِبْنُ مَخَاضٍ [or ابن المخاض]. (TA in that art.) You say, أَتَاهُمْ بِقُرْصٍ كَأَنَّهُ فِرْسِنُ ▼ خَلَّةٍ, (Ṣ, TA,) or كَأَنَّهُ خُفُّ خَلَّةٍ, (JK,) [*They brought them a round cake of bread as though it were the foot of a camel in its second*, or *third, year,*] meaning *small.* (JK. [In the TA, meaning سَمِينَة (i. e. *fat*); but this seems to be a mistranscription.]) ═ *A cautery.* (TA.)

خُلٌّ: see خَلِيلٌ, in two places.

خِلٌّ: see خُلَّةٌ, in two places: — and see خَلِيلٌ, in four places.

خَلَّةٌ *A road between two roads.* (TA.) — *A hole, perforation*, or *bore, that penetrates*, or *passes through, a thing, and is small:* or, *in a general sense:* (Ḳ:) or *a gap*, or *breach, in a booth of reeds or canes.* (T, TA.) [See also خَلَلٌ.] — [And hence,] The *gap that is left by a person who has died:* (Aṣ, T, Ṣ, TA:) or the *place, of a man, that is left vacant after his death.* (Ḳ.) One says, of him who has lost a person by death, اَللَّهُمَّ اخْلُفْ عَلَى أَهْلِهِ بِخَيْرٍ وَاسْدُدْ خَلَّتَهُ, i. e. [*O God, supply to his family, with that which is good, the place of him whom they have lost,*] *and fill up the gap which he has left by his death.* (Aṣ, T, Ṣ,* TA.) — And The *interval*, or *inter-*

vening space, between the piercer, or *thruster, and the pierced,* or *thrust:* whence the saying, رَقِّعْ خَلَّةَ القَارِىءِ, explained in art. رقع. (O and Ḳ and TA in that art.) — [Hence also,] *Want,* or *a want: poverty;* (Ṣ, Mṣb, Ḳ;) *need, straitness,* or *difficulty.* (Lḥ, Ḳ.) One says, بِهِ خَلَّةٌ شَدِيدَةٌ *He has pressing,* or *severe, need* or *straitness* or *difficulty.* (Lḥ, TA.) And سَدَّ ٱللّٰهُ خَلَّتَهُ *May God supply his want.* (TA.) And it is said in a prov., الخَلَّةُ تَدْعُو إِلَى السَّلَّةِ *Want invites to theft.* (Ḳ,* TA.) ═ *I. q.* خَصْلَةٌ; (JK, Ṣ, Mgh, Mṣb, Ḳ;) both signify *A property, quality, nature,* or *disposition:* and *a habit,* or *custom:* (KL, PṢ, TḲ:) [and app. also *a practice,* or *an action:*] in a man: (TA: [see the latter word:]) pl. خِلَالٌ. (JK, Mgh, Mṣb, Ḳ.) One says, فُلَانٌ خَلَّتُهُ حَسَنَةٌ [*Such a one, his nature,* or *disposition, is good*]. (IDrd, TA.) And hence, خَيْرُ خِلَالِ الصَّائِمِ السِّوَاكُ [*The best of the habits,* or *customs, of the faster is* the use of *the tooth-stick*]. (Mgh.) — See also خُلَّةٌ. ═ *An isolated tract of sand,* (Fr, Ḳ,) *separate from other sands.* (Fr, TA.) — And *i. q.* هَضْبَةٌ [which signifies *An elevated tract of sand:* but more commonly *a hill;* or *a spreading mountain;* &c.]. (JK, TA.) ═ *Wine,* (Ḳ,) in a general sense: (TA:) or *acid,* or *sour, wine:* (Ṣ, Ḳ:) or *wine altered for the worse,* (Ḳ, TA,) *in flavour,* (TA,) *without acidity,* or *sourness:* (Ḳ, TA:) pl. [or coll. gen. n.] ▼خَلٌّ. (Ḳ.) — See also خَلٌّ, first sentence. ═ And see this last word near the end of the paragraph, in four places.

خُلَّةٌ an inf. n. [or rather quasi-inf. n.] of خَالَّهُ, q. v.: (JK:) *True,* or *sincere, friendship, love,* or *affection;* as also ▼خُلُولَةٌ and ▼خَلَالَةٌ and ▼خُلَالَةٌ and ▼خِلَالَةٌ: (Ṣ:) or all these signify *a particular true* or *sincere friendship,* or *love,* or *affection, in which is no unsoundness,* or *defect, and which may be chaste and may be vitious:* (Ḳ: [in which all are said to be substs., except خُلَّةٌ, as though this were properly speaking an inf. n., though having a pl., as shown below:]) [and sometimes simply *friendship:* see an ex. in a verse cited voce مُرَحِّبٌ, in art. رحب:] or خُلَّةٌ and ▼خِلَّةٌ, (Mṣb,) or ▼خِلٌّ and ▼خِلَّةٌ, each with kesr, (Ḳ,) signify *true,* or *sincere, friendship,* or *love,* or *affection,* (Mṣb, Ḳ,) *and brotherly conduct:* the last two as used in the phrases, إِنَّهُ لَكَرِيمُ الخِلِّ ▼ and ▼الخِلَّةِ [*Verily he is generous in respect of true,* or *sincere, friendship,* &c.]: (Ḳ:) the pl. of خُلَّةٌ in the sense explained above is خِلَالٌ. (Ṣ, Ḳ.) — See also خَلِيلٌ, in three places. ═ *A kind of plants* or *herbage* [or *trees*]; (JK, Ṣ, Mṣb, Ḳ;) namely, the *sweet kind thereof;* (Ṣ, Ḳ;) *not* حَمْض: (JK:) or *any pasture,* or *herbage, that is not* حَمْض; all pasture, or herbage, consisting of حَمْض and خُلَّة, and حَمْض being such as has in it saltness [or sourness]: (TA:) the [*kind of plant, or tree, called*] عَرْفَج; and *every tree that remains in winter:* (JK:) accord. to Lḥ, it is [applied to *certain kinds*] *of trees &c.:* accord. to IAạr, *peculiarly of trees:* but accord. to A'Obeyd, [*shrubs,* i. e.] *not including any great trees:* (TA:) and *a certain thorny tree:* also *a place of growth,* and *a place in which is a collection,* of [the plants, or trees, called] عَرْفَج: (Ḳ:) and *any land not containing* [*the kind of plants,* or *herbage,* or *trees, called*] حَمْض; (AḤn, Ḳ;) *even though containing no plants,* or *herbage:* (AḤn, TA:) the pl. is خُلَلٌ: (Ḳ:) one says أَرْضٌ خُلَّةٌ and أَرْضُونَ خُلَلٌ: ISh says that خُلَّةٌ and خُلَلُ الأَرْضِ mean *land,* and *lands, in which is no* حَمْض, *sometimes containing* [*thorny trees such as are called*] عِضَاه, *and sometimes not containing such;* and that خُلَّةٌ is also applied to *land in which are no trees nor any herbage:* (TA:) some say that خُلَّةٌ, as meaning the *pasture,* or *herbage, which is the contrary of* حَمْض, has for a pl. خِلَالٌ, and then, from خِلَالٌ is formed the pl. أَخِلَّةٌ: and some say that this last means *herbage that is cut* (أُخْتُلَ ▼ وَٱجْتُزَّ) [in which the latter verb seems to be an explicative adjunct to the former]) *while green.* (Ḥam p. 602, q. v.) They say that the خُلَّة is the bread of camels, and the حَمْض is their fruit, (JK, T, Ṣ, TA,) or their flesh-meat, (Ṣ, TA,) or their خَبِيص. (TA.) — Hence, by way of comparison, it is applied to ‡*Ease,* or *repose; freedom from trouble* or *inconvenience, and toil* or *fatigue;* or *tranquillity;* and *ampleness of circumstances:* and حَمْض, to evil, and war: (T, TA:) and the former, to *life:* and the latter, to death. (Ḥam p. 315.) — Also *Acid,* or *sour, leaven* or *ferment.* (IAạr, TA.)

خِلَّةٌ: see 1, near the middle of the paragraph: ═ and see also خُلَالَةٌ, in four places: ═ and خُلَّةٌ, first sentence, in two places: ═ and خَلِيلٌ, in two places. ═ Also The جَفْن [i. e. the *scabbard,* or the *case,*] *of a sword, covered with leather:* (Ḳ:) or *a lining with which the* جفن *of a sword is covered,* (Ṣ, Ḳ, and Ḥam pp. 330 et seq.,) *variegated,* or *embellished, with gold &c.;* (Ṣ;) but the pl. is also used as meaning *scabbards:* (Ḥam p. 331:) and *a thong that is fixed upon the outer side of the curved extremity of a bow:* (Ṣ, Ḳ:) in the T it is explained as meaning the *inner side of the thong of the* جَفْن, *which is seen from without, and is an ornament,* or *a decoration:* (TA:) and *any piece of skin that is variegated,* or *embellished:* (M, Ḳ:) the pl. is خِلَلٌ (Ṣ, Ḳ, and Ḥam p. 330) and خِلَالٌ, and pl. pl. أَخِلَّةٌ, (Ḳ,) i. e. pl. of خِلَالٌ. (TA.)

خَلَلٌ *An interstice, an interspace* or *intervening space, a break, a breach, a chink,* or *a gap, between two things;* (JK, Ṣ, Mṣb, Ḳ;) pl. خِلَالٌ: (JK, Ṣ, Mṣb:) and particularly the *places,* (Ḳ,) or *interstices,* (Ṣ,) of the clouds, *from which the rain issues;* as also ▼خِلَالٌ; (Ṣ, Ḳ;) both occurring in this sense, accord. to different readings, in the Ḳur xxiv. 43 and xxx. 47: (Ṣ, TA:) the latter may be [grammatically] a sing. [syn. with the former], or it may be pl. of the former: (MF, TA:) and ▼خِلَالُ الدَّارِ signifies *what is around the limits of the house;* (JK, Ḳ;) or *around the walls thereof;* thus in the M; (TA;) *and what is between the chambers thereof.* (Ḳ.) You say, دَخَلْتُ بَيْنَ خَلَلِ القَوْمِ and ▼خِلَالِهِمْ [*I entered amid the breaks,* or *interspaces, of the people*]. (Ṣ, Mṣb.) And هُوَ خَلَلَهُمْ and ▼خِلَالَهُمْ (M, Ḳ) and ▼خُلَالَهُمْ (Ḳ [but in the CḲ these words are with ḍamm to the second ل]) *He is amid them.* (M, Ḳ.) And جُسْنَا خِلَالَ ▼ بُيُوتِ الحَيِّ, and خِلَالَ ▼ دُورِ القَوْمِ, i. e. [*We went,* or *went to and fro,* or *went round about,* &c.,] *amid the tents of the tribe,* and *in the midst of the houses of the people;* like a phrase in the Ḳur xvii. 5. (TA.) — And [hence] *Shakiness, looseness, laxness,* or *want of compactness,* and *disorder,* or *want of order,* of a thing; (Mṣb;) *unsoundness,* or *corruptness,* (Ṣ, Mṣb,*) in an affair or a thing, (Ṣ,) or of a thing; (Mṣb;) [*a flaw* in a thing;] *defect, imperfection,* or *deficiency;* (Ḥam p. 300;) *weakness,* or *infirmity,* in an affair, (JK, Ḳ, TA,) as though some place thereof were left uncompact, or unsound, (TA,) and in war, (JK,) and in men: (JK, Ḳ:*) and ‡*unsettledness* in an opinion. (Ḳ,* TA.) — الخَلَلُ *The night.* (JK, Ibn-'Abbád.)

خُلَلٌ: see خُلَالَةٌ, in two places.

خِلَلٌ: see خُلَالَةٌ, in three places.

خَلَلَةٌ: see خُلَالَةٌ, in two places.

خَلَالٌ [*Dates in the state in which they are termed*] بَلَح, (JK, T, Ṣ, Ḳ,) in the dial. of the people of El-Baṣrah; (T, TA;) i. e. *green dates:* (JK:) [but see بَلَحٌ and بُسْرٌ:] n. un. with ة. (JK, TA.) ═ هُوَ خَلَالَهُمْ: see خَلَلٌ.

خُلَالٌ: see خُلَالَةٌ. ═ Also *An accident that happens in anything sweet so as to change its flavour to acidity,* or *sourness.* (Ḳ.)

خِلَالٌ *A thing with which one perforates,* or *transpierces, a thing,* (JK, Ḳ,) either *of iron* or *of wood:* (JK:) pl. أَخِلَّةٌ. (Ḳ.) — *A wooden thing* [or *pin*] (Ṣ, Mṣb) *with which one pins a garment,* (T, Ṣ, Mgh, Mṣb, Ḳ,) *conjoining its two edges:* (Mgh, Mṣb:) pl. as above: (Ṣ, Mṣb:) which also signifies the *small pieces of wood with which one pins together the edges of the oblong pieces of cloth of a tent.* (TA.) — [*A skewer for flesh-meat.*] — *A wooden pin which is inserted into the tongue of a young camel, in order that he may not suck:* (Ḳ:) or *which is fixed above the nose of a young camel, for that purpose.* (TA in art. لهج.) — [*A toothpick;*] *a thing* (*of wood,* Ṣ, Mṣb) *with which one extracts the remains of food between his teeth;* (Ṣ, Mṣb, Ḳ;) as also ▼خِلَالَةٌ. (Ḥar p. 101.) — [*A long thorn* or *prickle:* such being often used as a pin and as a toothpick.] ═ See also خُلَالَةٌ. ═ And see خَلَلٌ, in six places.

خَلِيلٌ *Perforated,* or *transpierced;* like ▼مَخْلُولٌ. (Ḳ.) — See also خَلٌّ, in the latter half of the paragraph. — *Poor; needy; in want;* (JK, Ṣ, Mṣb, Ḳ;) as also ▼مُخِلٌّ, (so in some copies of the Ḳ and in the M,) or ▼مُخَلٌّ, (so in other copies of the Ḳ,) and ▼ مُخْتَلٌّ and ▼ أَخَلُّ: (Ḳ:) and أَخِلَّةٌ may be a pl. of خَلِيلٌ in this sense. (Ḥam p. 662.) — *A friend; or a true, or sincere, friend;* (Ṣ, Mgh, Mṣb;) as also ▼ خِلٌّ, and ▼ خُلَّةٌ, which is used alike as masc. and fem., because originally an inf. n., [or a quasi-inf. n., i. e. of 3, q. v.,] (Ṣ,) or ▼ خِلَّةٌ, [thus in the copies of the Ḳ, but what precedes it, though not immediately, seems to show that the author perhaps meant خُلَّةٌ,] used alike as masc. and fem. and sing. and pl.: (Ḳ:) or *a special,* or *particular, friend* or *true* or *sincere friend;* as also ▼ خِلٌّ and ▼ خُلٌّ; or this latter is only used in conjunction with وُدٌّ, as when you say, كَانَ لِى وُدًّا وَخُلًّا [*He was to me an an object of love and a friend* &c.]; (Ḳ;) or, as ISd says, ▼ خِلٌّ is the more common, and is applied also to a female; (TA;) as is also ▼ خُلَّةٌ, (Ḳ,) and ▼ خِلَّةٌ: (TA:) خَلِيلٌ also signifies *veracious;* (Ḳ;) thus accord. to IAạr: (TA:) or *a friend in whose friendship is no* خَلَلٌ [i. e. *unsoundness,* or *defect,* or *imperfection*]: (Zj, TA:) or *one who is pure and sound in friendship,* or *love:* (IDrd, Ḳ:) the pl. is أَخِلَّآءُ (Mṣb, Ḳ) and خُلَّانٌ (JK, Ḳ) and أَخِلَّةٌ: (Ḥam p. 662, and MA:) the fem. is خَلِيلَةٌ; (Ṣ, M, Ḳ;) of which the pl. is خَلِيلَاتٌ and خَلَائِلُ: (M, Ḳ:) the pl. of ▼ خِلٌّ or ▼ خُلٌّ is أَخْلَالٌ: (Ḳ:) and the pl. of ▼ خُلَّةٌ is خِلَالٌ, (Ṣ,) mentioned before, see 3, second sentence. It is applied in the Ḳur iv. 124 to Abraham; who is called خَلِيلُ ٱللّٰهِ, (TA,) and ٱلْخَلِيلُ. (Ḳ.) And it is said that the pl. أَخِلَّةٌ means also *Pastors;* because they act to their beasts like أَخِلَّآء [or friends, &c.], in labouring to do good to them. (Ḥam p. 662.) — Also *One who advises,* or *counsels,* or *acts, sincerely, honestly,* or *faithfully.* (IAạr, TA.) — And ٱلْخَلِيلُ also signifies *The heart.* (IAạr, JK, Ḳ.) — And *The liver.* (JK, TA.) — And *The nose.* (JK, Ḳ.) — And *The sword.* (IAạr, TA.) [And] *A sword of Sa'eed Ibn-Zeyd Ibn-'Amr Ibn-Nufeyl.* (Ḳ.) — And *The spear.* (IAạr, TA.)

خَلَالَةٌ: see خُلَّةٌ, first sentence.

خُلَالَةٌ *i. q.* كُرَابَةٌ; (AḤn, JK;) i. e. The *scattered dates that remain at the roots of the branches* [*after the racemes of fruit have been cut off*]; (AḤn, TA;) the *fresh ripe dates that are sought out in the interstices of the roots of the branches;* as also ▼ خُلَالٌ. (Ḳ.) — Also *What comes forth from the teeth when they are picked;* (JK, Ṣ,* Mṣb;) as also ▼ خِلَلٌ (JK, Ṣ) and ▼ خُلَلٌ (Ṣ) and ▼ خِلَّةٌ: (JK:) or خِلَلٌ and خِلَالٌ and خُلَالَةٌ (Ḳ) and خِلَّةٌ (Ṣ) and ▼ خَالٌّ (TA) signify the *remains of food between the teeth;* (Ṣ, Ḳ;) and the sing. [of خِلَلٌ] is ▼ خِلَّةٌ and [the n. un. of the same] ▼ خِلَلَةٌ. (Ḳ, TA. [In the CḲ, for خِلَّةٌ is erroneously put خُلَّةٌ.]) You say, فُلَانٌ يَأْكُلُ خُلَالَتَهُ and ▼ خِلَلَهُ (JK, Ṣ) and ▼ خُلَلَهُ (Ṣ) and ▼ خِلَّتَهُ (JK) and ▼ خِلَلَتَهُ (TA) *Such a one eats what comes forth from his teeth when they are picked.* (JK, Ṣ,* TA.) ═ See also خُلَّةٌ, first sentence.

خِلَالَةٌ: see خُلَّةٌ, first sentence: ═ and see also خِلَالٌ.

خُلُولَةٌ: see خُلَّةٌ, first sentence.

خَلَّالٌ *A seller of vinegar.* (Ḳ,* TA.)

خُلِّىٌّ a rel. n. from خُلَّةٌ as meaning the "sweet kind of plants or herbage." (Ṣ.) You say بَعِيرٌ خُلِّىٌّ, (Yaḳoob, Ṣ,) and إِبِلٌ خُلِّيَّةٌ (Yaḳoob, Ṣ, Ḳ) and ▼ مُخِلَّةٌ and ▼ مُخْتَلَّةٌ, (Ḳ,) meaning [*A camel, and camels,*] *pasturing upon* خُلَّة. (Ḳ.) And hence the prov., إِنَّكَ مُخْتَلٌّ ▼ فَتَحَمَّضْ † [meaning *Verily thou art disordered in temper, therefore sooth thyself;* or] *shift from one state,* or *condition, to another:* accord. to IDrd, said to him who is threatening: (TA. [See also 5 in art. حمض:]) [or it may mean *verily thou art weary of life, therefore submit to death:* see Ḥam p. 315.] And the saying of El-'Ajjáj,

* كَانُوا مُخِلِّينَ ▼ فَلَاقَوْا حَمْضَا *

[lit. *They were pasturing upon* خُلَّة, *and they found* حَمْض; meaning † *they were seeking to do mischief, and found him who did them worse mischief*]: applied to him who threatens, and finds one stronger than he. (TA. [See also حَمْض.])

خَلْخَلٌ: see خَلٌّ, in the latter half of the paragraph: ═ and see also خَلْخَالٌ.

خُلْخُلٌ: see the next paragraph.

خَلْخَالٌ: see خَلٌّ, in the latter half of the paragraph. — رَمْلٌ خَلْخَالٌ *Rough sand.* (TA.) ═ Also, and ▼ خَلْخَلٌ, (JK, Ṣ, Ḳ,) which is a dial. var. of the former, or a contraction thereof, (Ṣ,) and ▼ خُلْخُلٌ, (JK, Ḳ,) *A well-known ornament* (Ḳ) *of women;* (Ṣ, Ḳ;*) i. e. *an anklet:* (KL:) [or *a pair of anklets;* for you say,] فِى سَاقَيْهَا خَلْخَالٌ [*Upon her legs is a pair of anklets*]: (TA in art. حجل:) pl. (of the first, Ṣ) خَلَاخِيلُ (Ṣ, TA) and [of the second and third] خَلَاخِلُ. (TA.)

خَالٌّ (Ḳ) and ▼ مُتَخَلْخِلٌ (Mgh, Ḳ) [and ▼ مُخْتَلٌّ] all signify *Having interstices, breaks, chinks,* or *the like:*] *uncompact,* or *incoherent:* (Mgh, Ḳ:) the first and second applied in this sense to an army. (Ḳ.) — For the first, see also خَالٌ, in art. خيل. ═ And see خُلَالَةٌ.

أَخَلُّ *More,* and *most, poor,* or *needy:* (Ḳ, TA:) from أَخَلَّ إِلَيْهِ signifying "he wanted it," or "needed it." (TA.) Hence the phrase أَخَلُّ إِلَيْهِ [meaning *More,* or *most, in need of him,* or *it*]. (TA.) — See also خَلِيلٌ.

مُخِلٌّ: see خَلِيلٌ.

مُخَلٌّ: see خَلِيلٌ: ═ and see also خُلِّىٌّ, in two places: — and what here follows.

أَرْضٌ مُخَلَّةٌ, or ▼ مُخِلَّةٌ, (accord. to different copies of the Ṣ,) *A land abounding with* خُلَّة, *not containing any* حَمْض. (Ṣ.)

مَخْلُولٌ: see خَلِيلٌ, first sentence: ═ and see also خَلٌّ, in the latter half of the paragraph, in three places.

مُخَلْخَلٌ The *part, of the leg, which is the place of the* خَلْخَال [or *anklet*]; (JK, Ḳ;) i. e., *of the leg of a woman.* (TA.)

مُخْتَلٌّ: see خَالٌّ: — and see خَلٌّ, in the latter half of the paragraph: — and خَلِيلٌ. — Also *Vehemently thirsty.* (ISd, Ḳ.) — أَمْرٌ مُخْتَلٌّ *An affair in a weak,* or *an unsound, state.* (Ḳ.) ═ See also خُلِّىٌّ, in two places.

مُتَخَلْخِلٌ: see خَالٌّ.

خلأ

1. خَلَأَتْ, (Ṣ, Ḳ,) aor. َ, (Ḳ,) inf. n. خِلَآءٌ, (Ṣ, Ḳ,) in an expos. of the Mo'allaḳát written خِلَا, (TA,) and خِلَآءٌ, (Ṣ, Ḳ,) with kesr and medd, (Ṣ,) so accord. to IḲooṭ and IḲṭṭ and 'Iyáḍ and IAth and Z and Hr, but in some copies of the Ḳ خِلَآءٌ, and so many assert it to be, (TA,) and خُلُوْءٌ, (Ḳ,) said of a she-camel, (Ṣ, Ḳ) *She lay down,* or *kneeled and lay down, upon her breast:* (Lḥ, Ḳ:) or *she was,* or *became, refractory,* or *stopped and was refractory,* (Ṣ, Ḳ,) *and lay down,* or *kneeled and lay down, upon her breast, without disease or other like cause,* (Ṣ,) *and would not move from her place:* (Lḥ, Ḳ:) like أَلَحَّ said of a he-camel, and حَرَنَ said of a horse: (Ṣ:) the epithet applied to her that does thus is ▼ خَالِئٌ, (Lḥ, Ḳ,) without ة; (Lḥ;) and [app. to her that does so much, or often,] ▼ خَلُوْءٌ. (Ḳ.) And in like manner خَلَأَ is said of a he-camel; (Ḳ;) accord. to ISh, only of a he-camel: (TA:) or the verb is used only in speaking of a female [when relating to a camel]: (Ḳ:) one should not say of a he-camel خَلَأَ: (Az, Ṣ, Z, Ṣgh:) but it is also said of a man, (Ḳ,) tropically, (TA,) inf. n. خُلُوْءٌ, meaning ‡ *He moved not from his place.* (Ḳ, TA.) — [See also what next follows.]

3. خَالَأَ ٱلْقَوْمُ [in the CḲ خَلَأَ] *The people,* or *party,* or *company of men, left one thing, and betook themselves to another.* (Th, Ḳ, TA.) [خَالَى (see 3 in art. خلو) has a similar meaning.]

خَلُوْءٌ: } see 1.
خَالِئٌ: }

خلب

1. خَلَبَهُ, (A, Mgh, Mṣb, Ḳ,) aor. ِ, (A, Mṣb,) or ُ, (Mgh,) or ُ and ِ, (Mgh, Ḳ,) inf. n. خَلْبٌ, (Lth, Mgh, TA,) *He wounded him,* or *scratched him,* or *cut him, with his nail;* (A, Ḳ;) as also ▼ اِسْتَخْلَبَهُ: (Ḳ:) *he* (a beast or bird of prey, TA) *seized him,* i. e. the prey, *with his claw* or *talon:* (Ḳ:) or *he* (a beast of prey) *rent his skin with his dog-tooth:* (TA:) or *he rent it* (the skin) *with his dog-tooth:* (Lth, Mgh, TA:) or

he (a bird) *cut and rent it* (i. e. the skin) بِمِخْلَبِهِ *with his talon:* (Mṣb:) *he rent it,* or *slit it.* (Ḳ.) One says of a woman, قَلَبَتْ قَلْبِى وَخَلَبَتْ ↑خِلْبِى [*She smote,* or *overturned, my heart, and rent my midriff,* or, more probably, *liver,* which is regarded as a seat of passion]. (A, TA.) And خَلَبَتْ فُلَانًا ↑خِلْبَ *She* (a woman) *smote the* [app. here, also, meaning *liver*] *of such a one.* (Ḥam p. 343.) —— Also *It* (a venomous or noxious reptile or the like, TA) *bit him.* (Ḳ.) —— And خَلَبَ النَّبَاتَ, aor. ـُ, inf. n. خَلْبٌ, *He cut the plants,* or *herbage;* (Ṣ, Mṣb;) as also ↑اِسْتَخْلَبَهُ. (Ṣ.) —— And خَلَبَ بِالْمِخْلَبِ *He worked,* and *cut, with the reaping-hook.* (TA.) —— The root denotes the making a thing to incline: for الطَّائِرُ يَخْلِبُ بِمِخْلَبِهِ الشَّىْءَ إِلَى نَفْسِهِ [*The bird makes to incline, with its talon, the thing towards himself*]. (IF, Mgh.) —— [Hence,] خَلَبَ فُلَانًا عَقْلَهُ, aor. ـُ, and ـِ, *He despoiled,* or *deprived, such a one of his reason:* (Ḳ:) or خَلَبَ الْمَرْأَةَ عَقْلَهَا, inf. n. خَلْبٌ, *he despoiled,* or *deprived, the woman of her reason:* and خَلَبَتْ عَقْلَهُ, inf. n. as above, *she took away his reason;* as also ↑اِخْتَلَبَتْهُ. (L.) —— And [hence,] ↑خِلَابَةٌ signifies The *endeavouring to deceive* or *beguile* (IF, IAth, Mgh) *with blandishing speech:* (IAth:) or *deceiving with the tongue:* (Ṣ:) or a woman's *captivating the heart of a man by the most blandishing and deceiving speech.* (Lth.) You say, خَلَبَهُ, (Ṣ, A, Mgh, Mṣb, Ḳ,) aor. ـُ (Ṣ, Mgh, Mṣb, Ḳ,) or ـِ and ـَ, (Mgh,) inf. n. خَلْبٌ (Mṣb,* Ḳ) and خِلَابَةٌ, (A, Ḳ,) or this is a simple subst., (Mṣb,) and خِلَابٌ; (Ḳ;) and ↑اِخْتَلَبَهُ; (Ṣ, A, Ḳ;) and ↑خَالَبَهُ; (Ḳ;) *He deceived him* (Ṣ, Mṣb, Ḳ) *with his tongue:* (Ṣ:) or *he despoiled,* or *deprived, him of his reason,* بِمَنْطِقِهِ [*by his speech*]: (A:) or, followed by بِمَنْطِقِهِ, *he made his heart to incline* [*to him*] *by the most blandishing speech.* (Mgh.) It is said in a prov., إِذَا لَمْ تَغْلِبْ فَاخْلُبْ, (Ṣ, TA,) or فَاخْلِبْ; accord. to the former reading, which is that of Aṣ, (TA,) *When thou dost not overcome, use deceit:* (Ṣ, IAth, TA:) accord. to the latter reading, [it is said to mean *when thou dost not overcome,*] *grasp little after little;* as though it were taken from مِخْلَبٌ signifying "a claw" or "talon." (TA.)

3: see 1.

8: see 1, in two places.

10: see 1, in two places. اِسْتَخْلَبَ also signifies *He cut,* (Ṣ, TA,) *with the reaping-hook,* (TA in art. خبر,) *and craunched* (خَضَدَ, TA) *and ate,* plants, or herbage. (Ṣ, TA.)

خِلْبٌ *i. q.* ظُفُرٌ, (Ḳ,) used in a general sense [as meaning The *nail* of a man, and the *talon* of a bird, and the *claw* of a beast: see also مِخْلَبٌ]: pl. أَخْلَابٌ only. (TA.) —— The *diaphragm,* or *midriff;* syn. حِجَابُ الْقَلْبِ; (JK, L;) or حِجَابُ الْكَبِدِ; (A, Ḳ;) the *partition intervening between the heart and the liver;* (IAạr, Ṣ;) the *partition between the heart and the belly;* (Zj in his "Khalḳ el-Insán;") or *a small and thin piece of flesh forming a connection between the ribs* [app. *of one side and those of the other*]: or the *liver* [itself]: (Ḳ:) or *its* زِيَادَة [or زَائِدَة]: (A, Ḳ:) or *a white thin thing adhering to the liver:* (Ḳ:) or *a certain thing in,* or *upon, the liver, like a* فُدَّة: (JK:) or *a small bone, resembling a man's nail, adhering to one side of the midriff, next the liver.* (TA.) See 1, in two places. —— *A friend;* [app. because he cleaves to another;] as also خِلْمٌ. (JK.) —— [And hence, app.,] خِلْبُ نِسَآءٍ, (Ṣ, A, Ḳ,) a phrase like حِدْثُ نِسَآءٍ and زِيرُ نِسَآءٍ, (TA,) A man *whom women love:* (Ṣ:) or one *who loves women for the sake of discourse,* or *for the sake of vitious* or *immoral conduct,* or *adultery,* or *fornication,* (A, Ḳ,) *and whom they love* (Ḳ) *in like manner:* (TA:) and one *who endeavours to deceive,* or *beguile, women* [*with blandishing speech:* see 1]: (TA:) pl. أَخْلَابٌ and ↑خِلَبَةُ نِسَآءٍ: (Ḳ, TA:) the latter [in the CḲ خِلْبَةٌ] extr. [with respect to rule]. (TA.)

—— *I. q.* وَشْىٌ [app. as meaning *A kind of variegated,* or *figured, cloth* or *garment*]. (TA.) [See also مُخْلَبٌ.] —— The *radish.* (Ḳ, TA.) In a copy of the Ḳ, الفَحْل is erroneously put for الفُجْل. (TA.) —— The *leaves,* (Ḳ,) or *broad leaves,* (Lth,) *of the grape-vine.* (Lth, Ḳ.)

خَلِبَةٌ: } see خَالِبٌ.
خَلَّابَةٌ: }

خِلَبَةُ نِسَآءٍ: see خِلْبٌ.

خَلَبُوتٌ: see خَالِبٌ.

خَلَبُوتٌ: } see خَالِبٌ, for each in two places.
خَلُوبٌ: }

خِلَابَةٌ: see 1. [And see also خِلِّيبَى.]

خُلَّبٌ †*Clouds* (سَحَابٌ, Ṣ, Ḳ, TA) *that thunder and lighten,* (TA,) *containing no rain:* (Ṣ, Ḳ, TA:) or *whereof the lightning flashes slightly, so that one hopes for their raining, but which deceive the expectation, and become dispersed:* as though derived from خِلَابَةٌ, the "deceiving with blandishing speech." (IAth.) And الْبَرْقُ الخُلَّبُ and بَرْقُ الخُلَّبِ (Ṣ, Ḳ) and بَرْقٌ خُلَّبٌ (Ḳ) and بَرْقٌ خُلَّبٌ (A) ‡*Lightning with which is no rain;* (Ṣ, A;) as though deceiving: (Ṣ:) *that excites hope* [*of rain*] *and breaks its promise.* (Ḳ.) Hence the saying, to him who promises and does not fulfil his promise, إِنَّمَا أَنْتَ كَبَرْقٍ خُلَّبٍ ‡[*Thou art only like lightning with which is no rain*]. (Ṣ.). And فُلَانٌ خُلَّبٌ قُلَّبٌ †*Such a one is sharp in intellect, clever, ingenious, skilful, knowing,* or *intelligent.* (JK.)

خَلَّابٌ and خَلَّابَةٌ: see خَالِبٌ.

خِلِّيبَى *Deceit,* or *guile.* (Ḳ.) [See also خِلَابَةٌ, in the first paragraph.]

خَالِبٌ, applied to a man, *Deceiving:* (Ḳ:) and in like manner, [but in an intensive sense,] ↑خَلَّابٌ (ISk, Ṣ, Ḳ) and ↑خَلُوبٌ (Kr, Mṣb, TA) and ↑خَلَبُوتٌ (ISk, Ṣ, Ḳ) and ↑خَلَبُوبٌ (Ḳ) *Very deceitful* (ISk, Ṣ, Kr, Mṣb, Ḳ,* TA) *and lying:* (ISk, Ṣ:) and so, applied to a woman, ↑خَلَّابَةٌ and ↑خَلُوبٌ (A, Ḳ) and ↑خَلِبَةٌ (Ṣ, Ḳ) and ↑خَلَبُوتٌ (Ḳ) and ↑خَلَبَةٌ (TA) *very deceitful:* (Ṣ, A,* Ḳ,* TA:) خَلَبَةٌ is a pl. [of خَالِبٌ], and means men *who deceive women.* (Ṣ.) You say also اِمْرَأَةٌ خَالِبَةٌ لِلْفُؤَادِ [meaning *A woman who captivates the heart by the most blandishing and deceitful speech*]. (TA.)

أَخْلَبُ [*More,* and *most, deceiving* or *deceitful*]. You say of a woman, تَخْلُبُ قَلْبَ الرَّجُلِ بِأَلْطَفِ الْقَوْلِ وَأَخْلَبِهِ [*She captivates the heart of the man by the most blandishing and deceiving speech*]. (Lth.)

مِخْلَبٌ [The *talon,* or *claw,* of a bird or beast of prey; *a tearing talon* or *claw;*] the *same to the bird* (Ṣ, Mgh, Mṣb) *and to the beast of prey* (Ṣ, Mṣb) *as the* ظُفُر *to man;* (Ṣ, Mgh, Mṣb;) because the bird [or beast] cuts and rends with it the skin: (Mṣb:) the ظُفُر [or *nail*] (A, Ḳ) of any beast or bird of prey: or it is of a bird of prey; and the ظُفُر is of a bird that does not prey: (Ḳ:) pl. مَخَالِبُ. (A.) [See also خِلْبٌ.] You say, أَنْشَبَ فِيهِ مَخَالِبَهُ, meaning ‡*He clung,* or *caught, to him,* or *it.* (A.) —— Also *A* مِنْجَل [or *reaping-hook*] (Ṣ, Mṣb, Ḳ) in a general sense: or (TA) *that has no teeth.* (Ṣ, Mṣb, TA.)

عُقَابٌ مُخْلِبَةٌ *An eagle with sharp talons.* (JK.)

مُخَلَّبٌ, applied to a garment, or piece of cloth, (Ṣ, TA,) means كَثِيرُ الْوَشْىِ, (Ṣ, Ḳ, TA,) i. e. [*Much variegated* or *figured;* or] *of many colours.* (TA.) [See also خِلْبٌ.]

خلج

1. خَلَجَ, (Ṣ, A, L, Mṣb, Ḳ,) aor. ـِ, (Ṣ, Ḳ,) or ـُ, (Mṣb,) inf. n. خَلْجٌ; and ↑اِخْتَلَجَ; (Ṣ, L, Mṣb, TA;) and ↑تَخَلَّجَ; (L, TA;) *He drew, dragged, pulled, strained, stretched, extended, lengthened,* or *protracted,* (Ṣ, L, Ḳ,) a thing: (Ṣ,* L, TA:) and *he pulled out* or *up, displaced, removed,* or *took away,* (Ṣ, A, Mṣb, Ḳ,) a thing, (Ṣ,* A, Mṣb, TA,) and a person. (A.) Thus in the saying, أَخَذَ بِيَدِهِ فَخَلَجَهُ مِنْ بَيْنِ صَحْبِهِ [*He took his hand, and pulled him out from amid his companions*]: and خَلَجَ رُمْحَهُ مِنَ الْمَطْعُونِ [*He pulled out his spear from the person pierced*]: and اِخْتَلَجَ ↑رُمْحًا مَرْكُوزًا [*He pulled out a spear stuck in the ground*]. (A, TA.) [See also an ex. in a verse cited voce مُطَرَّبٌ.] El-'Ajjáj says,

* فَإِنْ يَكُنْ هٰذَا الزَّمَانُ خَلَجَا *
* فَقَدْ لَبِسْنَا عَيْشَهُ الْمُخَرْفَجَا *

meaning †*And if this time has taken away,* and exchanged for another, a state [in which we were, *we have long enjoyed its plentiful life*]. (Ṣ.) —— [Hence,] خُلِجَ, said of a stallion-camel, *He was taken away from the females that had passed seven or eight months since the period when they last brought forth, before he had become too languid to cover any longer.* (Lth, A, L.) And خَلَجَ, aor. ـُ, †*He weaned* his offspring, or the offspring of his she-camel: (Ḳ:) ‡*he separated* a young camel from the mother. (A.) And خَلَجَتْ وَلَدَهَا ‡*She* (a mother) *weaned her offspring:* (M, A:) so accord. to Lḥ, who does not particularize any kind [of animal]. (M.) And خَلَجَ نَاقَةً †*He weaned the offspring of a she-camel.* (Ṣ.) And أُخْتُلِجَ ▼ مِنْ بَيْنِهِمْ ‡[*He was taken away from among them*]: said of the dead. (A, TA.) — خَلَجَنِى كَذَا, (Ṣ, Ḳ,*) aor. ـِ, (Ḳ,) †*Such a thing occupied me; busied me; or diverted me, by employing my attention,* from other things. (Ṣ, Ḳ,* TA.) You say, خَلَجَتْهُ أُمُورُ الدُّنْيَا †[*The affairs of the world occupied him, &c.*]. (Ṣ, TA.) And خَلَجَتْهُ الخَوَالِجُ ▼ †*Busying* [or *distracting*] *affairs busied* [or *distracted*] *him.* (Lth.) And a poet says,

* وَأَبِيتُ تَخْلِجُنِى الهُمُومُ كَأَنَّنِى *
* دَلْوُ السُّقَاةِ تُمَدُّ بِالأَشْطَانِ *

[*And I pass the night,*] *anxieties busying me* [*as though I were the bucket of the waterers, drawn from the well by the ropes*]. (IAạr.) — تَخْلِجُ السَّيْرَ, said of a fleet she-camel, (L, Ḳ,) †*She goes, journeys,* or *travels, quickly.* (L.) And خَلَجَ فِى مِشْيَتِهِ: see 5. — خَلَجَ, aor. ـِ, also signifies †*He put* (a thing, TA) *in motion,* or *into a state of commotion.* (A, Ḳ, TA.) You say, خَلَجَ حَاجِبَيْهِ, and عَيْنَيْهِ, ‡*He put in motion,* or *into a state of commotion, his eyebrows,* and *his eyes.* (A.) — And خَلَجَ, aor. ـِ (L, Ḳ) and ـُ, inf. n. خَلْجٌ, (L, TA,) †*He made a sign* [*by a motion*] (L, Ḳ, TA) بِعَيْنِهِ *with his eye,* and بِحَاجِبَيْهِ *with his eyebrows.* (L, TA.) And خَلَجَهُ بِحَاجِبِهِ, aor. and inf. n. as above, †*He made a sign to him with his eyebrow.* (L.) And خَلَجَهُ بِعَيْنِهِ †*He made a sign to him with his eye; winked to him.* (Ṣ, L.) And خَلَجَتْنِى بِعَيْنِهَا ‡*She made a sign to me with her eye,* or *winked to me, to indicate a time or place of appointment,* or *something that she desired.* (A, TA.) — See also 8, in two places.

3. خالجهُ, (A, Mṣb, TA,) inf. n. مُخَالَجَةٌ, (Mgh,) *He contended with him,* (A, Mgh,* Mṣb, TA,) [*as though drawing,* or *pulling, him,* (see 6,)] namely, a man. (TA.) You say, خالجهُ الشَّىْءَ *He contended with him for the thing.* (A.) And خَالَجَنِى القِرَآءَةَ †*He vied with me in reciting* the words of prayer, (Mgh,* TA,) *uttering aloud what I uttered aloud, so that he took from my tongue what I was reciting, and I did not* [or *could not*] *continue to do so.* (TA, from a trad.) And خالج قَلْبِى أَمْرٌ ‡*A thing,* or *an affair, troubled my heart with contending thoughts.* (Ḳ, TA.) And مَا يُخَالِجُنِى فِى ذٰلِكَ الأَمْرِ شَكٌّ ‡[*Doubt does not contend with me respecting that affair*], meaning *I doubt not respecting that affair.* (Sh, TA.)

4. اخلج حَاجِبَيْهِ عَنْ عَيْنَيْهِ †[*He drew up his eyebrows from his eyes*]. (Lth.) = اخلج is also quasi-pass. of خَلَجَ, though this is extr. with respect to analogy, like ابشر [q. v.] &c.; (TA;) signifying *It was,* or *became, drawn, dragged, pulled,* &c. (L, TA.)

5. تخلّج: see 1, first sentence. — [Hence,] تخلّج فِى مِشْيَتِهِ *He* (a paralytic, Ṣ, Ḳ, or an insane, or a possessed, man, A) *walked in a loose manner, as though disjointed, and inclined from side to side,* (Ṣ, A, Ḳ, TA,) *as one dragging a thing:* (A, TA:) it is similar to تخلّع: (TA:) and signifies also *he* (an insane, or a possessed, man) *inclined from side to side in his gait,* (Mgh,* TA,) *as though he were drawing along, now to the right and now to the left;* and so خَلَجَ ▼ فى مشيته, aor. ـِ, inf. n. خَلَجَانٌ. (TA.) — See also 8, in two places. — And see 6. = [*It branched off,* like a خَلِيج, from a large river: occurring in this sense in art. دجل of the T and TA; where دُجَيْل is described as نَهْرٌ صَغِيرٌ يَتَخَلَّجُ مِنْ دِجْلَةَ.]

6. تَخَالَجَتْهُ الهُمُومُ ‡*Anxieties contended with him, one on one side and another on another side, as though each were drawing him to it.* (A, L.) And تخالج فِى صَدْرِى شَىْءٌ ‡ (Ṣ, A, Ḳ) and اختلج ▼ (TA) ‡*A thing was,* or *became, unsettled in my bosom,* or *mind;* (TA;) meaning *I was in doubt* [*respecting a thing*]; (Ṣ, A, Ḳ;) as also تخلّج ▼ and تحلّج, (Lth,* Aṣ, TA in art. حلج,) or these two mean nearly the same. (Sh, TA in that art.; in which see 5, in three places.) [See also 8.]

8. اختلج, as a trans. v.: see 1, in three places. = Also ‡*It* (a thing) *was,* or *became, in a state of commotion,* or *agitation; it quivered, quaked,* or *throbbed;* (Sh, TA;) and so تخلّج ▼ (Sh, Ḳ) [and خَلَجَ ▼, as will be seen from what follows]. You say اختلج حَاجِبَاهُ †*His eyebrows quivered,* or *were in a state of commotion.* (Lth.) And اختلجت عَيْنُهُ; (Ṣ, Ḳ;) and تخلّجت ▼; (TA;) and خَلَجَتْ ▼, aor. ـِ and ـُ, inf. n. خُلُوجٌ (Ṣ, Ḳ) and خَلَجَانٌ; (Sh;) †*His eye quivered, throbbed,* or *was in a state of commotion;* (Sh, L;) i. q. طَارَتْ, (Ṣ, Ḳ,) i. e., *throbbed.* (PṢ, TḲ.) And اختلج العُضْوُ †*The member* (i. e. any member, L) *quivered,* &c. (Mgh, L, Mṣb.) — †*He trembled, quivered,* or *quaked.* (TA.) And اختلج بِوَجْهِهِ †*He moved about his lips and his chin, mocking and imitating a person talking.* (TA, from a trad.) — اختلج فِى صَدْرِى هَمٌّ ‡[*Anxious thought fluttered in my bosom*]. (TA.) See also 6.

خَلُوجٌ: see خِلْجٌ.

خُلُجٌ †Persons *trembling in the bodies.* (Ḳ.) — †Persons *tired,* or *fatigued.* (IAạr.) — †A people *whose lineage,* or *origin, is doubted,* (T, Ḳ,) *so that different persons dispute, one with another, respecting it.* (T.) See also مُخْتَلَجٌ.

خَلَنْجٌ: see art. خلنج.

خَلُوجٌ Clouds (سَحَاب) *separated,* or *scattered,* (Ḳ, TA,) *as though drawn away from the mass;* of the dial. of Hudheyl: (TA:) or clouds, (سحاب, Ḳ,) and a cloud, (سَحَابَةٌ, TA,) *abounding with water,* (Ḳ, TA,) *and lightening vehemently.* (TA.) — And hence, †A she-camel *abounding with milk, and yearning towards her young one.* (T, TA.) — Also †A she-camel, (Ṣ, Ḳ,) or other female, (TA,) *whose young one has been taken from her* (Ṣ, Ḳ) *by slaughter or death, and that yearns towards it,* (TA,) *and whose milk in consequence has become little in quantity.* (Ṣ, Ḳ.) Accord. to some, (L,) †A she-camel *that goes, journeys,* or *travels, quickly, by reason of her* [*natural, not forced,*] *fleetness.* (L, Ḳ.*) Pl. خُلُجٌ ▼ [or, rather, this is a quasi-pl. n., like لَبُونٌ as is of لَبْنٌ,] and خِلَاجٌ. (L.)

خَلِيجٌ *A canal,* or *cut, from a large river;* syn. شَرْمٌ مِنْ بَحْرٍ: (Ṣ, A, Ḳ:) *what is cut off from the main mass of water;* so called because it is drawn from it: (ISd, TA:) *a river cut off from a larger river, extending to a place where use is made of it: a river on one side of a larger river:* (TA:) and [simply] *a river:* (Ṣ, A, Ḳ:) and خَلِيجَا نَهْرٍ is said to signify *the two sides of a river:* (Ṣ:) or *the two wings thereof:* and some explain the sing. (خليج) as meaning *a branch from a valley, conveying its water to another place:* (TA:) pl. خُلْجَانٌ (A, TA) and خُلُجٌ. (TA.)

خَالِجٌ [act. part. n. of 1]. — It is said in a trad. of 'Alee, respecting life (الحَيَاة), إِنَّ ٱللّٰهَ جَعَلَ المَوْتَ خَالِجًا لِأَشْطَانِهَا, meaning †*Verily God has made death to be quick in seizing its cords;* i. e. the cords of life. (L.) — [Hence,] الخَالِجُ is applied to †*Death;* because it draws away mankind. (TA.)

[خَالِجَةٌ †*A busying,* or *distracting, affair:* pl. خَوَالِجُ. Hence,] خَلَجَتْهُ الخَوَالِجُ: see 1.

مُخَلَّجٌ †*Fat, so that his flesh quivers.* (TA.)

مُخْتَلَجٌ ‡A man *whose name has been transferred from the register of his own people to that of another people, to whom his lineage,* or *origin, is consequently ascribed,* (A, TA,) *and respecting whose lineage,* or *origin, people differ and dispute:* (TA:) accord. to some, i. q. خُلُجٌ ▼ as meaning a people *whose reputed origin is transferred so as to be ascribed to another people:* and the former signifies also a man *whose lineage,* or *origin, is disputed;* as though he were drawn, and pulled away, from his people. (TA.) — †One *whose flesh and strength are taken away.* (TA.) — †A face (Lth, ISd, Ḳ) *lean,* (Lth, ISd,) *having little flesh.* (Ḳ.)

خلد

1. خَلَدَ, aor. ـُ, inf. n. خُلُودٌ (Ṣ, A, L, Mṣb, Ḳ) and خُلْدٌ, (Ṣ,* A, L, Ḳ,* [but the latter is not said to be an inf. n. in the first nor in the last of these lexicons, and is perhaps a simple subst.,]) *He remained, stayed, dwelt,* or *abode;* syn. أَقَامَ: (L, Mṣb, Ḳ:) or *he remained, stayed, dwelt,* or

abode, long; syn. أَطَالَ الإِقَامَةَ: (A:) بِمَكَانٍ (Ṣ, A, Mṣb, Ḳ) and إِلَى مَكَانٍ (Ḳ) [in a place]; as also ↓ اخلد (Ṣ, A, L, Mṣb, Ḳ) and ↓ خلّد: (Ḳ:) and *he remained,* or *continued, incessantly, always, endlessly,* or *for ever;* (Ṣ, A, L, Ḳ;*) syn. بَقِىَ, (A, L, Ḳ,) and دَامَ, (Ḳ,) or دَامَ بَقَاؤُهُ; (Ṣ, L;) فِى دَارٍ *in a house,* or *an abode, not going forth from it:* (L:) *he remained, stayed, dwelt,* or *abode, for ever,* or *perpetually,* in Paradise, (A, L,) or in Hell. (A.) — [Hence,] خَلَدَ, (L, Ḳ,) aor. ـِ and ـُ, (Ḥam p. 70, and L,) inf. n. خَلْدٌ, (Ḳ,) or خُلْدٌ, (thus in the L,) and خُلُودٌ; (L, Ḳ;) and ↓ اخلد; (Ḥam ubi suprà;) ‡ *He was slow in becoming hoary,* (Ḥam, L, Ḳ,) *when advanced in years;* (Ḳ;) as though he were created to continue for ever. (L.) — See also 4, in two places.

2. خلّد, as a trans. v.: see 4. = Also *He adorned* a girl [*with bracelets,* or *other ornaments* (see the pass. part. n.), or] *with earrings.* (AA.) = As an intrans. v.: see 1: — and see also 4.

4. اخلدهُ, (Ṣ, A, L,) inf. n. إِخْلَادٌ; (Ṣ, L;) and ↓ خلّدهُ, (Ṣ, A, L,) inf. n. تَخْلِيدٌ; (Ṣ, L;) *He* (God, Ṣ, L) *caused him to remain, stay, dwell,* or *abide:* (L:) or *caused him to remain, stay, dwell,* or *abide, long,* in a place: (A:) or *caused him to remain,* or *continue, incessantly, always, endlessly,* or *for ever,* (Ṣ, L,) in a house, or an abode, *not going forth* from it: (L:) or *caused him to remain, stay, dwell,* or *abide, for ever,* or *perpetually,* in Paradise, (A, L,) or in Hell. (A.) يَحْسَبُ أَنَّ مَالَهُ أَخْلَدَهُ, in the Ḳur civ. 3, means *He thinketh that his wealth hath made him to be one that shall continue for ever;* that he shall not die: (Jel:) i. e. he acteth as one that thinketh, with his opulence, he shall not die. (L.) = As an intrans. v.: see 1, in two places. — You say also, اخلد بهِ, (inf. n. as above, AA,) *He kept,* or *clave, to him;* (AA, AZ, Ṣ, Ḳ;) i. e., to his companion. (AZ, Ṣ, Ḳ.) — And اخلد إِلَيْهِ ‡ *He inclined,* or *propended, to him,* (L, Ḳ, TA,) *and liked him:* (L, TA:) *he inclined to, and relied upon,* (Ṣ, A, Mṣb,) *him,* (Ṣ, A,) or *it;* (Mṣb;) as also ↓ خَلَدَ. (Mṣb.) اخلد إِلَى الأَرْضِ, in the Ḳur [vii. 175], (Ks, Ṣ, A, L,) as also ↓ خَلَدَ, and ↓ خلّد, but this last is rare, (Ks, L,) and so is the second, (L,) means ‡ *He inclined to, and relied upon, the earth:* (Ṣ, A:) or *he inclined,* or *propended, to the world;* (Bḍ, Jel;) *and relied thereon:* (Jel:) or *he inclined,* or *propended, to lowness, baseness,* or *meanness.* (Bḍ.)

خَلَدٌ: see the next paragraph.

خُلْدٌ an inf. n. of خَلَدَ, (A, L,) [or a simple subst.] syn. with [the inf. n.] خُلُودٌ. (Ḳ.) — [Hence,] الخُلْدُ, (T, Ḳ,) or دَارُ الخُلْدِ, (L,) [the latter signifying *The abode of the state of perpetual existence;*] *Paradise:* (Ḳ:) or *the Paradises:* (T:) or *the world to come.* (L.) = *A bracelet:* and *an earring;* as also ↓ خَلَدَةٌ: pl. خِلَدَةٌ: (Ḳ:) which last signifies [also] *ornaments for the person,* collectively; (TA;) and so ↓ خُلْدَةٌ [if this be not a mistranscription for خِلَدَةٌ]. (L.) = [*The mole;*] *the blind rat;* (L, Ḳ;) as also ↓ خَلَدٌ, (Ḳ,) and جُلْدٌ [q. v.]: (Ḳ in art. جلد:) or *a species of rat;* as also ↓ خِلْدٌ: (L:) or one of the names of the فَأْر [or *rat*]: (IAạr:) or *a species of the* [*kind of rats called*] جُرْذَان, *blind* (Lth, Ṣ, L, Mṣb) *by nature,* (Lth, L, Mṣb,) *having no eyes,* (Lth, L,) *inhabiting the deserts:* (Mṣb:) Lth says that the sing. is ↓ خَلَدٌ, and the pl. خِلْدَانٌ: in the T it is said that the sing. is ↓ خُلْدَةٌ, and the pl. خِلْدَانٌ; which is very strange: (L:) or *a blind beast* [*that lives*] *beneath the ground,* (Ḳ,) *having no eyes,* (TA,) *that likes the smell of onions and leeks; so that if either of these be put over its hole, it comes forth and is caught: if its upper lip be hung upon a person affected with a quartan fever, it cures him; and its brain, mixed,* or *moistened, with oil of roses, and used as an ointment, dispels the maladies termed* البَرَص *and* البَهَق *and* القَوَابِى *and* الجَرَب *and* الكَلَف *and* الخَنَازِير, *and every eruption upon the body:* (Ḳ:) مَنَاجِذُ, (L, Ḳ,) or, as in some copies of the Ḳ, مَنَاجِدُ, with the unpointed د, (TA,) is used as its pl., like as مَخَاضٌ is used as pl. of خَلِفَةٌ. (L, Ḳ.) — Also *A species of the* قُبَّرَة [or *lark*]. (Ḳ.)

خِلْدٌ: see خُلْدٌ, in two places.

خَلَدٌ *The mind:* (Ṣ, A, Ḳ:) *the heart:* (Ṣ, Ḳ:) pl. أَخْلَادٌ. (TA.) You say, وَقَعَ ذٰلِكَ فِى خَلَدِى *That came into my mind,* or *heart.* (Ṣ.)

خَلَدَةٌ: }
خِلْدَةٌ: } see خُلْدٌ.
خُلْدَةٌ: }

خَالِدٌ [*Remaining, staying,* &c.]. — [Hence,] الخَوَالِدُ [as though pl. of الخَالِدَةُ] † *The three pieces,* or *portions, of stone,* or *rock, called* الأَثَافِى, *upon which the cooking-pot is placed,* (Ṣ, A, L, Ḳ,) *remaining in their places:* (L:) so called because of their remaining (Ṣ, L) a long time (L) after the standing relics of a house have become effaced. (Ṣ, L.) [See an ex., from a poem of El-Mukhabbal Es-Saadee, voce إِلَّا, p. 78; where خوالدٍ is with tenween for the sake of the metre.] — Also ‡ *The mountains:* and *the stones:* (L, Ḳ:) and *the rocks:* so called for the same reason. (L.) = [As a proper name, خَالِدٌ is often written خلد.]

مُخْلَدٌ: see the next paragraph, in two places.

مُخْلِدٌ ‡ *Slow in becoming hoary;* (A;) as also ↓ مُخْلَدٌ and ↓ مُخَلَّدٌ: (Ḥar p. 588:) *whose teeth do not fall out* (T, A) *by reason of extreme old age:* (T:) or, as some say, it is ↓ مُخْلَدٌ; as though [meaning] made by God to continue for ever in such a state: (A:) a man *who is not hoary when advanced in age:* (ISk, Ṣ:) *whose hair of his head and beard remains black in old age.* (T.) — Also † *Still; motionless.* (TA in art. خمد.)

مُخَلَّدٌ: see the paragraph next preceding. — مُخَلَّدُونَ in the Ḳur [lvi. 17 and lxxvi. 19] means *Always of the same age; never altering in age:* (Fr:) or [*endowed with perpetual vigour;*] *that never become decrepit:* (Ḳ:) or *that never exceed the fit age for service:* (L, Ḳ:) = or it means *adorned with earrings:* (L, Ḳ:) or, *with bracelets;* (AO, L, Ḳ;) accord. to the dial. of El-Yemen: (L:) or, *with ornaments.* (Zj.)

خلس

1. خَلَسَهُ, (Ṣ, A, Mṣb,) aor. ـِ, (Mṣb, MṢ,) inf. n. خَلْسٌ (Mgh, Mṣb, Ḳ) and خِلِّيسَى; (Ḳ;) and ↓ اختلسهُ; (Ṣ, A, Mṣb, Ḳ;) and ↓ تخلّسهُ; (Ṣ, TA;) *He seized it,* or *carried it off, by force;* (Ṣ, A, Ḳ;) مِنْ يَدِهِ *from his hand:* (A:) *he took it at an opportunity, with deceit, guile,* or *circumvention:* (TA:) or *he took it hastily and openly:* (Mgh:) or *he snatched it at unawares; seized it hastily when its owner was unaware:* (Mṣb:) or ↓ اِخْتِلَاسٌ denotes a quicker action than خَلْسٌ. (Lth, Ḳ.) You say also, خَلَسَهُ إِيَّاهُ [*He seized it,* or *carried it off, from him by force;* &c.]: (TA:) and خَالَسْتُهُ ↓ الشَّىْءَ *I seized the thing,* or *snatched it away, from him.* (Ḥar p. 139.) And أَخَذَهُ خِلِّيسَى, i. e. ↓ اِخْتِلَاسًا [*He took it by forcible seizure;* &c.]. (TA.) خَلْسٌ is also in fighting and prostrating: and you say, ↓ اختلس طَعْنَةً [*He seized an opportunity to inflict a thrust,* or *wound, with a spear* or *the like*]. (TA.) And ↓ أُخْتُلِسَ بَصَرُهُ † *His sight was suddenly taken away.* (A* and TA in art. ملس.)

3. خالسهُ الشَّىْءَ, (A, TA,*) inf. n. مُخَالَسَةٌ and خِلَاسٌ, (TA,) [*He contended with him in a mutual endeavouring to seize,* or *carry off, the thing by force;* or *to take it at an opportunity, with deceit, guile,* or *circumvention;* or *to take it hastily and openly;* or *to snatch it at unawares:* see 1 and 6.] A poet says, [app. using the verb tropically,]

* نَظَرْتُ إِلَى مَىٍّ خِلَاسًا عَشِيَّةً *
* عَلَى عَجَلٍ وَٱلْكَاشِحُونَ حُضُورُ *

[*I looked at Mei, vying with her in snatching glances, in the evening, in haste, while the secret enemies were present*]. (Th, TA.) — See also 1.

4. اخلس *It* (the hair, A, TA, and the head, [meaning the hair of the head,] AZ, Ṣ, TA) *became a mixture of black and white,* (Ṣ, A, TA,) *in equal proportions:* or *more black than white:* (TA:) [or *mostly white:* (see خَلِيسٌ:)] or *partly white.* (AZ.) — [Hence,] ‡ *It* (herbage) *became a mixture of fresh and dry:* (Ṣ, Ḳ, TA:) or *partly green and partly white, in drying up.* (Az, TA.) † *It* (حَلِىّ [q. v.]) *put forth fresh verdure.* (IAạr, TA.) And اخلست الأَرْضُ †*The land put forth somewhat of herbage.* (TA.)

5: see 1.

6. تَخَالَسَا الشَّىْءَ [*They both contended together, each endeavouring to seize, or carry off, the thing by force; or to take it at an opportunity, with deceit, guile, or circumvention; or to take it hastily and openly; or to snatch it at unawares:*

see 1 and 3]: (A:) the inf. n., تَخَالُسٌ, signifies *i. q.* تَسَالُبٌ. (Ṣ, Ḳ.) — تخالس القِرْنَانِ, and تخالسا نَفْسَيْهِمَا, *The two opponents sought to seize each other by force*; &c. (T, TA.)

8: see 1, in five places. — [اختلس also signifies † *He slurred* a vowel; i. e., *pronounced* it *slightly*: and *he suppressed* it *altogether*.]

خَلْسٌ : } see خَلِيسٌ.
خُلْسٌ : }

خَلْسَةٌ inf. n. of un. of خَلَسَهُ [*A single act of seizing*, or *carrying off, by force*; &c.]. (Mṣb.) = It also occurs in a trad., where, if correctly related, it is syn. with خُلْسَةٌ. (Mgh.)

خُلْسَةٌ a subst. from خَلَسَهُ: (Ṣ, Ḳ:) [which may perhaps mean that it has the abstract sense of the inf. n.: and] *A thing that is* [*seized*, or *carried off, by force*; or *taken at an opportunity, with deceit, guile*, or *circumvention*; or] *taken hastily and openly*; (Mgh;) or *snatched at unawares*: (Mṣb:) or *spoil*; *plunder*; *booty*; *a thing taken by spoliation and force*; as also ▼خَلِيسَةٌ; which last also signifies *an animal that is snatched from a beast or bird of prey and dies before it has been legally slaughtered*; in consequence of which it is forbidden [to be eaten]. (TA.) Hence, لَا قَطْعَ فِى الخُلْسَةِ [*There shall be no amputation* (of the right hand) *in the case of a thing seized*, or *carried off, by force*; &c.]. (Mgh, Mṣb.) [See also an ex. voce حُذَيًّا, in art. حذى.] — *An opportunity*. (Ṣ,* A, TA.) You say, هٰذِهِ خُلْسَةٌ فَانْتَهِزْهَا *This is an opportunity, therefore do thou take it*, or *seize it*. (A, TA.) = *A mixture of whiteness with blackness* in the hair, (Ṣ, Ḥam p. 387, TA,) *in equal proportions*: or *of more blackness than whiteness*: (TA:) [or *a predominance of whiteness*: see خَلِيسٌ.] — [Hence,] ‡ *A mixture of fresh and dry portions* [or *green and white* (see 4)] in herbage. (Ṣ, Ḳ, TA.)

خَلَسَاتٌ: see the next paragraph.

خَلِيسٌ [an epithet having the sense of the pass. part. n. of خَلَسَهُ; i. e., *Seized*, or *carried off, by force*; &c.]. [Hence,] طَعْنَةٌ خَلِيسٌ *A thrust*, or *wound, with a spear or the like, which one has seized an opportunity to inflict, by means of his skill*. (TA.) = A *courageous* man; as also ▼مُخَالِسٌ and ▼خَلَّاسٌ. (TA.) = Also, and ▼مُخْلِسٌ, Hair *having whiteness mixed with its blackness*, (A, TA,) *in equal proportions*: or *with more blackness than whiteness*: (TA:) or *mostly white*: (Mgh:) or *partly white*. (AZ, TA.) And the former, *Having a mixture of whiteness with the blackness of his hair*. (Ṣ, Ḳ,) — [Hence,] the former also signifies ‡ Herbage *drying up*, or *dried up*, (Ṣ, Ḳ, TA,) *part yellow and part green*; as also ▼مُخْلِسٌ: (TA:) or both signify *having its dry and green portions intermixed*: (A:) and the former signifies *dry* herbage, *upon the lower part of which fresh has grown and mixed with the former*; as also ▼خَلْسٌ. (Ḳ.) — Also † *White* (أَحْمَرُ [q. v.]) *whose whiteness is mixed with blackness*: and so, applied to women, ▼خُلْسٌ, (syn. سُمْرٌ, TA,) of which the sing. may be ▼خَلْسَاءُ; or خَلِيسٌ; or ▼خِلَاسِيَّةٌ, supposing the two augments (ىة TA) to be elided. (Ḳ.) — Also † *i. q.* خَلِيطٌ [q. v.]. (TA.) — And † The young one of a she-camel *begotten by a stallion not prepared for her*. (Ṣgh, TA.)

خَلِيسَةٌ: see خُلْسَةٌ.

خِلَاسِىٌّ ‡ A boy *whose mother is black, and his father a white*, or *tawny, Arab, and who is born of a colour between those of his two parents*; fem. with ة: (Az, TA:) or ‡ a child *whose parents are* (*one*) *white and* (*the other*) *black*, (A, Ḳ, TA,) *a white man and a black woman*, or *a black man and a white woman*. (TA.) See also خَلِيسٌ. — And ‡ A domestic fowl, (A,) or cock, (Ḳ,) *begotten between an Indian and a Persian fowl*. (A, Ḳ.)

خَلَّاسٌ: see خَالِسٌ: — and see خَلِيسٌ.

خَالِسٌ One *who seizes*, or *carries off*, a thing *by force*: *who takes* it *at an opportunity, with deceit, guile*, or *circumvention*: [or *who takes it hastily and openly*: or *who snatches at unawares*:] as also ▼خَلَّاسٌ: [or this latter has an intensive signification:] and [in like manner] ▼مُخْتَلِسٌ one *who seizes*, or *carries off*, a thing *at a time of inadvertence*. (TA.) — [Hence,] الخَالِسُ *Death*: because it seizes people unawares. (TA.)

مُخْلِسٌ: see خَلِيسٌ, in two places.

مُخَالِسٌ: see خَلِيسٌ.

مُخْتَلِسٌ: see خَالِسٌ.

خلص

1. خَلَصَ, (Ṣ, A, Ḳ, &c.,) aor. ـُ, (Ṣ, TA,) inf. n. خُلُوصٌ (Ṣ, A, Ḳ) and خَلَاصٌ (TA) and خَالِصَةٌ, (Ḳ,) or the second and third of these are simple substs. [used as inf. ns., i. e., quasi-inf. ns.]; (TA;) and خَلْصٌ also; (Et-Towsheeḥ, TA;) but the former is that which is commonly known; (TA;) *It* (a thing, Ṣ, TA) *was*, or *became*, خَالِص, (Ṣ, A, Ḳ,) which signifies [here] *clear, pure, sheer, free from admixture, unmingled, unmixed*, or *genuine*; (B, TA;) and *white*. (Ḳ.) You say, خَلَصَ المَآءُ مِنَ الكَدَرِ *The water became clear from turbidness*. (Mṣb.) And خَلَصَ الزُّبْدُ مِنَ الثُّفْلِ [*The butter became clear from the dregs*, or *sediment*,] in being cooked. (Ṣ.) — خَلَصَ مِنَ الوَرْطَةِ, (A,) or التَّلَفِ, aor. ـُ, (Mṣb,) inf. n. خَلَاصٌ (A, Mṣb) and خُلُوصٌ and مَخْلَصٌ, (Mṣb,) ‡ *He became safe*, or *secure*, or *free, from embarrassment* or *difficulty*, or *from destruction*, (A, Mṣb,) *like as a thing becomes clear from its turbidness*. (A.) [See also 5.] — خَلَصَ مِنَ القَوْمِ ‡ *He withdrew, retired*, or *went away* or *apart, from the people*, or *company of men*. (A, TA.) It is said in the Ḳur [xii. 80], خَلَصُوا نَجِيًّا ‡ *They retired, conferring privately together*. (Bḍ, Jel, TA.) — خَلَصَ إِلَيْهِ, (Ṣ, A, Ḳ,) and بِهِ, (TA,) inf. n. خُلُوصٌ, (Ḳ,) ‡ *He*, or *it*, (a thing, Ṣ, and grief, and happiness, A, TA,) *came to*, or *reached, him*: (Ṣ, A, Ḳ, TA:) *he came to, reached*, or *arrived at, it*; namely, a place. (TA.) — Also خَلَصُوا إِلَيْهِ *They came to him* (namely a judge or governor) *and referred to him their cause*, or *suit, for judgment*. (T and L in art. نفذ.) = خَلَصَهُ, inf. n. خَلَاصٌ and خُلُوصٌ; (TA;) or ▼خَلَّصَهُ, (Ḳ,) inf. n. تَخْلِيصٌ; (TA;) but the former is that which is found in the correct lexicons; (TA;) *He took the* خُلَاصَة [q. v.] (Ḳ, TA) of, or from, clarified butter; (TA;) and ▼اخلصهُ, inf. n. إِخْلَاصٌ, signifies the same. (TA.) [See also this last below.]

2. خَلَّصَهُ, (A,) inf. n. تَخْلِيصٌ, (TA,) *He made*, or *rendered, it clear* or *pure* [&c. (see 1, first signification)]; *he cleared, clarified, purified*, or *refined, it*; (A, Mgh, TA;) [as also ▼اخلصهُ, q. v.] — † *He separated it from another thing* or *other things*. (Mṣb.) You say also خلّص بَيْنَهُمَا [*He separated them, each from the other*]. (M in art. قلص.) — ‡ *He* (God, A, TA, or a man, Ṣ) *saved, secured*, or *freed, him*, (Ṣ, A, Ḳ,) مِنْ كَذَا *from such a thing*, (Ṣ,) [as, for instance, a snare, and embarrassment or difficulty, or destruction, *like as one renders a thing clear from its turbidness*, (see 1,)] *after he had become caught*, or *entangled*; (TA;) as also ▼اخلصهُ. (TA.) Also † [*He disentangled it*; *unravelled it*:] said of spun thread that has become entangled. (Lth and Az and Ṣgh, in TA, art. عسر.) — † *He made it clear*; or *explained, expounded*, or *interpreted, it*; as also لَخَّصَهُ. (A in art. لخص.) — خلّص, inf. n. as above, also signifies † *He gave* [a man (for the verb in this case, as in others, is trans., accord. to the TḲ,)] *the* خَلَاص, (Ḳ, TA,) i. e., the *equivalent* of a thing, or *requital*, or *hire for work*. (TA.) = See also 1, last signification.

3. خالصهُ, (Ṣ, Ḳ,) inf. n. مُخَالَصَةٌ, (TḲ,) † [*He regarded him*, or *acted towards him, with reciprocal purity of mind*, or *sincerity*: and particularly, as also خالصهُ الوُدَّ, mentioned in this art. in the A, but not explained,] ‡ *he regarded him*, or *acted towards him, with reciprocal purity*, or *sincerity, of love*, or *affection*; syn. صَافَاهُ (Ṣ, Ḳ, TA) and وَادَّهُ; (TA;) فِى العِشْرَةِ [*in social intercourse*]. (Ṣ, TA.) You say also, خالص ٱللّٰهَ دِينَهُ ‡ [*He acted with reciprocal purity*, or *sincerity, towards God, in his religion*]. (A.) And one says, خَالِصِ المُؤْمِنَ وَخَالِفِ الكَافِرَ ‡ [*Act thou with reciprocal purity*, or *sincerity, towards the believer, and act thou with contrariety to the unbeliever*]. (A. [See 3 in art. خلق, where a similar saying is mentioned.]) [See also the next paragraph.]

4. اخلصهُ: see 2, first signification. You say, اخلص السَّمْنَ, inf. n. إِخْلَاصٌ, *He clarified the cooked butter by throwing into it somewhat of the meal of parched barley or wheat* (سَوِيق), *or dates*, or *globules of gazelles' dung*: (Ṣ,* L:) or *he took the* خُلَاصَة [q. v.] *of the cooked*, or *clarified, butter*. (Fr, Ḳ.) See also 1, last signification.

And أَخْلَصَتْهُ النَّارُ [*The fire clarified it*, or *purified it*,] namely, butter, and gold, and silver. (Ḳ.) — You say also, أَخْلَصُوا النَّصِيحَةَ and الحُبَّ ‡ [lit. *They made good advice* or *counsel*, and *love, pure*, or *sincere*; meaning, *they were pure*, or *sincere, in giving good advice*, and *in love*]. (TA.) And أَخْلَصَ لَهُ المَوَدَّةَ ‡ [*He was pure*, or *sincere, to him in love*, or *affection*]. (A.) And أَخْلَصَ لِلّٰهِ العَمَلَ † [*He was pure*, or *sincere, towards God in works*]. (Mṣb.) And أَخْلَصَ لِلّٰهِ الدِّينَ, (Ṣ, TA,) or دِينَهُ, (A,) ‡ *He was pure*, or *sincere, towards God in religion*, [or *in his religion*;] *without hypocrisy*. (Ṣ,* TA.) And أَخْلَصَ لِلّٰهِ, [elliptically,] † *He was without hypocrisy* [*towards God*]. (Ḳ.) or إِخْلَاصٌ properly signifies † *The asserting oneself to be clear*, or *quit*, *of* [*believing in*] *any beside God*. (B, TA.) [Hence,] سُورَةُ الإِخْلَاصِ is † a title of *The* [112*th*] *chapter of the Kur-án commencing with the words* قُلْ هُوَ ٱللّٰهُ أَحَدٌ: (IAth, Mṣb:) and سُورَتَا الإِخْلَاصِ † *the same together with the* [109*th*] *chapter commencing with the words* يَا أَيُّهَا ٱلْكَافِرُونَ. (Mṣb.) And كَلِمَةُ الإِخْلَاصِ is applied to ‡ *The sentence which declares belief in the unity of God*. (A,* TA.) أَخْلَصْنَاهُمْ بِخَالِصَةٍ: see below, voce خَالِصَةٌ. — See also 2, third signification. — And see 10.

5. تَخَلَّصَ ‡ *He became saved, secured*, or *freed; he escaped*, or *freed himself*; or *became safe, secure, free*, or *in a state of freedom* or *immunity*; (Ṣ, Ḳ;) from a thing; (Ṣ;) as, for instance, a gazelle, and a bird, from a snare, (A,) [and a man from embarrassment or difficulty, or destruction, like as a thing becomes cleared from its turbidness, (see 1,) or] like spun thread when it has been entangled. (TA.) — [See also نَشِبَ بِالمَرْأَةِ.]

6. تَخَالَصُوا ‡ *They regarded one another*, or *acted reciprocally*, [*with purity*, or *sincerity*: and particularly,] *with purity*, or *sincerity, of love*, or *affection*. (A,* TA.)

10. اِسْتَخْلَصَ الزُّبْدَ مِنَ اللَّبَنِ *He extracted the butter from the milk*. (ADḳ, A, L.) — اِسْتَخْلَصَهُ لِنَفْسِهِ *He appropriated him* [or *it*] *purely to himself*, (Bḍ and Jel in xii. 54,) *exclusively of any partner*: (Jel:) *he chose him* [or *it*] *for himself; took him* [or *it*] *in preference for himself*; (IAạr, in L, art. قرع; and TA in the present art.;) *he appropriated him to himself as his particular*, or *special, intimate*; (TA;) syn. اِسْتَخَصَّهُ; (Ṣ, Ḳ, TA;) and ▼ أَخْلَصَهُ signifies the same. (TA.)

خِلْصٌ (Ṣ, A, Ḳ) and ▼ خُلْصَانٌ (Ṣ, A, TA) and ▼ خَالِصَةٌ (Ṣ, TA) ‡ *A man's friend*; [or his *sincere*, or *true*,] or his *secret*, or *private, friend*; or his *companion*, or *associate, who converses*, or *talks*, with him; syn. خِدْنٌ; (Ṣ, Ḳ, TA;) his *particular*, or *special, friend*: (TA:) ▼ the second is also used in a pl. sense: (Ṣ, TA:) pl. of the first, خُلَصَاءُ. (Ḳ.)

خَلَصٌ *A kind of tree like the grape-vine* (Ḳ) *in its manner of growth*, (TA,) *that clings to other trees, and rises high*; (Ḳ;) *having leaves of a dust-colour, thin, round, and wide; and a blossom like that of the* مرو [?]; *and tinged in the lower parts of its stems*; (TA;) *sweet in odour; and having berries* (Ḳ) *like those of* [*the plant called*] عِنَبُ الثَّعْلَبِ, [see art. ثعلب,] *three and four together, red*, (TA,) *like the beads of* عَقِيق [q. v.]; (Ḳ;) *not eaten* [*by men*], *but depastured*: (TA:) n. un. with ة: (Ḳ:) thus described by [AḤn] Ed-Deenawaree, on the authority of an Arab of the desert. (TA.) See the end of the next paragraph.

ذُو الخَلَصَةِ, (Ṣ, Ḳ,) and ذُو الخُلَصَةِ, (Hishám, Ḳ,) and ذُو الخَلْصَةِ, accord. to IDrd, and some write it ذُو الخِلْصَةِ, but the first is the form commonly obtaining with the relaters of trads., (TA,) *A certain temple*, (Ṣ, Ḳ,) *called* كَعْبَةُ اليَمَامَةِ, (Ṣ,) or الكَعْبَةُ اليَمَانِيَّةُ, (El-Háfidh Ibn-Hajar, Ḳ,) *and also* الكَعْبَةُ الشَّامِيَّةُ, because its door faced Syria, (TA,) *belonging to the tribe of Khath'am*, (Ṣ, Ḳ,) *and Dows and Bejeeleh and others*, (TA,) *in which was an idol called* الخَلَصَةُ, (Ṣ, Ḳ,) *which was demolished* (Ṣ, TA) *by command of Mohammad*: (TA:) or ذُو الخَلَصَةِ was *the idol itself*, as some say; but, says IAth, this requires consideration, because [it is asserted that] ذو is not prefixed to any but generic names: (TA: [but see ذُو:]) or the temple was so called because it was the place of growth of a tree of a kind called ▼ خَلَصٌ. (Ḳ,* TA.)

خُلْصَانٌ: see خِلْصٌ, in two places.

خَلَاصٌ an inf. n. of 1. — يَوْمُ الخَلَاصِ is *The day of the coming forth of* الدَّجَّال [or *Antichrist*]; because then the believers will be distinguished. (TA.) ═ Also † *An equivalent*; *a requital*, or *compensation*; *hire, pay*, or *wages, for work*: pl. أَخْلَاصٌ. (TA.) — See also خُلَاصَةٌ.

خُلَاصٌ: see خُلَاصَةٌ, in two places.

خِلَاصٌ: see خُلَاصَةٌ, throughout.

خُلُوصٌ: see خُلَاصَةٌ, in three places. — Also an inf. n. of 1.

خُلَيْصٌ: see خَالِصٌ.

خُلَاصَةُ السَّمْنِ (Ṣ, A, L, Mṣb, Ḳ) and خِلَاصَتُهُ (Fr, Ṣgh, Ḳ) *What has become clear, of cooked butter*; (Ṣ, A, L, Ḳ;) or *cooked butter into which some dates have been thrown, or into which some* سَوِيق [i. e. *meal of parched barley or wheat*] *has been thrown, in order that thereby it may become clear from the remains of the milk*: (Mṣb:) *for when they cook fresh butter, to make it* سَمْن, *they throw into it somewhat of* سَوِيق, *or dates*, or *globules of gazelles' dung; and when it becomes good, and clear from the dregs*, or *sediment, that* سمن *is called* الخُلَاصَةُ, *and* ▼ الخِلَاصُ *also*, (Ṣ, L,) mentioned by A'Obeyd, (Ṣ,) and this, namely the خلاص, is the إِثْر: (Ṣ, L, Ḳ:) and the terms ▼ خُلُوصٌ (Ṣ, L, Ḳ) and قِلْدَةٌ (Ṣ, L) and قِشْدَةٌ (Ṣ, L, Ḳ) and كُدَادَةٌ (Ṣ, L) are applied to the *dregs*, or *sediment, remaining at the bottom*; (Ṣ, L, Ḳ;) as also خُلَاصَةٌ: (AHeyth, L in art. قشد:) the inf. n. is إِخْلَاصٌ; and you say, أَخْلَصْتُ السَّمْنَ: (Ṣ, L:) or خُلَاصَةٌ and ▼ خِلَاصٌ signify *dates and* سَوِيق *that are thrown into* سمن; and أَخْلَصَ السَّمْنَ signifies "he threw dates and سويق into the سمن [and so clarified it]:" and ▼ خُلَاصٌ [thus I find it written] signifies *what has become clear, of* سمن, *when it is cooked*: and خِلَاصٌ also signifies, and so ▼ إِخْلَاصٌ, and ▼ إِخْلَاصَةٌ, *butter when clear from the dregs*, or *sediment*: and ▼ خُلُوصٌ, the *dregs*, or *sediment, at the bottom of the milk*: (L:) ▼ إِخْلَاصٌ and ▼ إِخْلَاصَةٌ are *syn. with* إِذْوَابٌ and إِذْوَابَةٌ: (TA:) or, accord. to AZ, the latter two terms are applied to butter when it is put into the cooking-pot to be cooked into سمن; and *when it has become good, and the milk has become clear from the dregs*, or *sediment, that milk is called* إِثْر *and* ▼ إِخْلَاصٌ: Az says, I have heard the Arabs apply the term ▼ خِلَاصٌ to *that with which* سمن *is cleared, in the cooking-pot, from the water and milk and dregs; for when it is not clear, and the milk is mixed with the butter, they take dates*, or *flour*, or سويق, *which they throw therein, that the* سمن *may become clear from the remains of the milk mixed with it*: this is the خلاص: but the خلاصة [i. e. خُلَاصَة] *is what remains, of the* خلاص *and dregs* or *milk &c., in the bottom of the cooking-pot*: (L, TA:) [or] ▼ خِلَاصٌ (Ḳ) [accord. to some, ▼ خُلَاصٌ, but this is app. wrong, (see Ḥar p. 311,)] and خُلَاصَةٌ (Hr, TA) also signify *what fire has clarified*, or *purified*, (مَا أَخْلَصَتْهُ النَّارُ,) of butter, and of gold, and of silver: (Hr, Ḳ, TA:) or خِلَاصُ اللَّبَنِ, means *what is extracted from milk*; i. e. *butter*; (ADḳ, L, TA;) and so does خُلَاصَةُ اللَّبَنِ: (A: [but there mentioned among tropical expressions:]) خُلَاصَةٌ being applied in the manner first mentioned in this paragraph, by a secondary application is made to signify *what is clear*, or *pure*, of other things; (Mṣb;) [as also ▼ خَالِصٌ: and hence both of them often signify † the *choice, best*, or *most excellent, part* of anything; and so, probably, does ▼ خِلَاصٌ:] and خُلَاصَةٌ and ▼ خُلَاصٌ also signify *Inspissated juice* (رُبّ) *made from dates*; (JK;) or this is called ▼ خُلُوصٌ. (TA.)

خَالِصٌ *Clear; pure; sheer; free from admixture; unmingled; unmixed; genuine*: (B, TA:) *clear*, or *pure*, applied to any colour: (Lḥ, TA:) ‡ *white*; as also ▼ خُلَيْصٌ; [which latter appears to me doubtful, though I know not why Freytag has substituted for this, or for the former word, خَلِصٌ;] both applied to anything. (Ḳ, TA.) You say, ثَوْبٌ خَالِصٌ ‡ *A garment*, or *piece of cloth, of a clear*, or *pure, white*: and قَبَآءٌ أَزْرَقُ خَالِصُ البِطَانَةِ ‡ *A garment of the kind called* قباء *blue with a white lining*. (A.) — [Also † *Pure*, or *sincere*, love, religion, &c.] — See also خُلَاصَةٌ, near the end of the paragraph.

خَالِصَةٌ [fem. of خَالِصٌ: used as a subst.,] † *A pure property*, or *quality*. (Bḍ in xxxviii. 46; and Ḳ. [In the CḲ, خُلَّة is erroneously put for خَلَّة: the corresponding word in Bḍ is خَصْلَة.])

So in the Ḳur [xxxviii. 46], أَخْلَصْنَاهُمْ بِخَالِصَةٍ ذِكْرَى الدَّارِ † *We have rendered them pure by a pure quality,* (Bḍ, Ḳ,* TA,) *the keeping in memory the final abode:* (Bḍ, TA:) ذكرى الدار being a substitute for خالصة: or it may mean [*by*] their *keeping in memory much the final abode* and the return to God: (TA:) some also, (TA,) namely Náfi' and Hishám, (Bḍ,) read بخالصة, making it a prefix to ذكرى (Bḍ, TA) as an explicative; or an inf. n., in the sense of خُلُوص, prefixed to its agent. (Bḍ.) — You say also, هٰذَا الشَّىْءُ خَالِصَةٌ لَكَ † *This thing is a property of thine:* (so in a copy of the Ṣ, and so the phrase is written in the TA:) or *is a thing purely thine, exclusively of others:* (TA:) or هذا الشىء خالصة لك *this thing is particularly,* or *specially, thine,* or *for thee.* (So accord. to other copies of the Ṣ, and a copy of the JK.) — See also خلص. — خَالِصَةٌ is also *syn. with* إِخْلَاصٌ [in some sense not pointed out: see the latter below; and see also 4]. (TA.)

إِخْلَاصٌ [inf. n. of 4, used as a subst.]: see خُلَاصَة, in three places.

إِخْلَاصَةٌ: see خُلَاصَة, in two places.

[مَخْلَصٌ ‡ *A place of safety,* or *security,* or *escape* from an event.]

مُخْلَصٌ *Chosen:* (JK:) *chosen* by God, *and pure from pollution;* applied to a man. (Zj, TA.) [It is implied in the A and TA that it is also syn. with مُخْلِصٌ in the sense explained below.]

مُخْلِصٌ ‡ *Pure,* or *sincere,* towards God *in religion; without hypocrisy:* (TA:) or *purely believing in the unity of God.* (Zj, TA.)

يَاقُوتٌ مُتَخَلِّصٌ *Picked* [*sapphires*]. (A, TA.)

خلط

1. خَلَطَهُ, (Ṣ, Mṣb, Ḳ,) aor. ـِ, (Mṣb, Ḳ,) inf. n. خَلْطٌ, (Ṣ, Mṣb,) *He mixed it; mingled it; incorporated,* or *blended, it;* (Mṣb, Ḳ;) or *put it together;* (Mṣb;) بِغَيْرِهِ *with another thing;* (Ṣ, Mṣb;) *inseparably,* as in the case of fluids; and *separably,* as in the case of animals, (Mṣb, TA,) and grains; (TA;) as also ↓خَلَّطَهُ, (Ḳ,) inf. n. تَخْلِيطٌ: (TA:) [or the latter relates to many, or several, objects; or signifies *he mixed it much:*] El-Marzookee says that the primary signification of خَلْطٌ is the *intermingling of the particles of a thing, one with another.* (Mṣb, TA.) [And hence, † *He confused, confounded,* or *disordered, it.*] — خَلَطَ القَوْمَ; and خَلَطَ: see 3, near the end of the paragraph.

2: see 1. — [Its inf. n. is pluralized: you say,] جَمَعَ مَالَهُ مِنْ تَخَالِيطَ [*He collected together his property,* or *camels, &c., from states of confusion*]. (TA.) — التَّخْلِيطُ فِى الأَمْرِ signifies *The creating confusion,* or *disorder,* (الإِفْسَادُ,) *in the affair,* or *case.* (Ṣ.) And you say, هُوَ فِى تَخْلِيطٍ فِى أَمْرِهِ [and مِنْ امره, *He is in a state of confusion,* or *disorder, in,* or *with respect to, his affair,* or *case*]. (TA.) [And خَلَّطَ عَلَيْهِ الأَمْرَ *He rendered the affair, or state,* or *case, confused,* or *disordered,* or *perplexed, to him.* And خَلَّطَ بَيْنَ القَوْمِ *He created confusion,* or *disorder,* or *disturbance, among the people,* or *company of men.*]

3. خَالَطَهُ, inf. n. مُخَالَطَةٌ (Ṣ, Mgh, Ḳ) and خِلَاطٌ, (Ṣ, Ḳ,) *It mixed, mingled, commingled, intermixed,* or *intermingled, with it; it became incorporated,* or *blended, with it;* syn. مَازَجَهُ; (Mgh, Ḳ;) and خَامَرَهُ; (Ṣ, A, Ḳ, all in art. خمر;) [as, for instance,] water with milk. (A in art. خمر, and Mgh in the present art.) خِلَاطٌ in relation to camels, and men, and beasts, also signifies Their *being mixed together.* (Ḳ.) A poet says,

* يَخْرُجْنَ مِنْ بُعْكُوكَةِ الخِلَاطِ *

[*They come forth from the crowding and dust* (of the beasts) *occasioned by the being mixed together*]. (Th, TA.) And it is said in a trad., لَا خِلَاطَ وَلَا وِرَاطَ (Ṣ, Mgh,) *There shall be no putting together what is separate, nor separating what is put together,* from fear of the poor-rate: (Ṣ:) for the Prophet made it incumbent on a person having possessed forty sheep or goats a whole year to give one sheep or goat; and so on one having possessed more thereof to the number of a hundred and twenty, to give one sheep or goat; but if they exceeded a hundred and twenty by one, two sheep or goats were to be given of them: (Az, TA:) i. e. *there shall be no putting together what is separate; as, for instance, when three persons possess a hundred and twenty sheep or goats, every one of them having forty, they not having been partners for a whole year, and it being incumbent on every one of them to give a sheep or goat; and when the collector of the poor-rate comes to them, they put them together, assigning them to one pastor, in order that they may not be obliged to give for them more than one sheep or goat:* (Ḳ,* TA:) accord. to IAth, this is termed إِخْلَاطٌ [app. a mistake for خِلَاطٌ]: *nor shall there be any separating of what is put together; i. e., when there are two partners, each of them having a hundred and one sheep or goats, for which together they are bound to give three sheep or goats; and when the collector of the poor-rate comes to them, they separate their sheep or goats, so that each of them shall not have to give more than one sheep or goat:* [see also art. ورط:] (TA:) or خلاط signifies *a man's mixing his sheep or goats when they are eighty in number with those of another which are forty in number, both together being bound to give two sheep or goats while they are separate, in order that one* [*only*] *may be taken:* and وراط, *a man's giving to another the half of his sheep or goats when they are forty in number, in order that the collector of the poor-rate may not take anything:* (Mgh:) or خلاط is, *when there are, between two partners, a hundred and twenty sheep or goats, one of them having eighty and the other forty, and the collector of the poor-rate has taken two of these sheep or goats, the former partner's restoring to the latter the third of a sheep or goat; so that the former has had to give a sheep or goat and a third; and the latter, two thirds of one: and if the collector have taken, from the hundred and twenty, one sheep or goat, the former partner's restoring to the latter one third* [in some copies of the Ḳ, erroneously, two thirds] *of a sheep or goat; so that the former has had to give two thirds of a sheep or goat; and the latter, one third of one:* (ISd, Ḳ,* TA:) and وراط is *deceiving,* and *acting dishonestly:* (ISd, L, TA:) in the place of وراط, we find, accord. to one relation, شِنَاق, followed by فِى الصَّدَقَةِ. (TA.) — El-'Ajjáj contended with Ḥomeyd El-Arḳaṭ in two poems of the metre termed رَجَز ending with ط, and Ḥomeyd said, الخِلَاطَ يَا أَبَا الشَّعْثَآءِ, i. e. [*Beware thou of mixing;* or] *do not thou mix* my أُرْجُوزَة with thine [*O father of her with the shaggy hair*]; to which El-'Ajjáj replied, الفِجَاجُ أَوْسَعُ مِنْ ذٰلِكَ يَا ٱبْنَ أَخِى [*The roads are wider than* to require my doing *that, O son of my brother*]. (AO, Ṣ.) — خَالَطَ الذِّئْبُ الغَنَمَ [lit. signifying *The wolf mixed with the sheep,* or *goats,*] means ‡ *the wolf fell upon the sheep,* or *goats:* (Ḳ, TA:) the inf. n. is خِلَاطٌ. (TA.) — خَالَطَهَا, (Az, Mṣb, Ḳ,) inf. n. خِلَاطٌ and مُخَالَطَةٌ, (Az, Mṣb,) ‡ *He had carnal intercourse with her;* (Az, Mgh,* Mṣb, Ḳ;) i. e., a man with his wife, (Az, Mṣb,) or with a woman: (Ḳ:) the lawyers say, خالطها مُخَالَطَةَ الأَزْوَاجِ: (Mṣb:) Th explains the inf. n. خِلَاطٌ by رَفَثٌ, q. v. (TA.) Also, in like manner, with the same inf. ns., ‡ a stallion-camel with the female. (Lth, Ḳ, TA.) [See also 4.] IAạr explains خِلَاط in relation to camels as signifying † *A man's coming to the nightly resting-place of another's camels, and taking thence a male camel, and making him to cover his she-camel without his owner's knowledge.* (TA.) — خَالَطَهُ السَّهْمُ † [*The arrow penetrated into him*]. (TA.) — خَالَطَهُ الشَّيْبُ [*Hoariness,* or *whiteness, became intermixed in his hair*]. (Ṣ and Ḳ in art. وخط; &c.) — خَالَطَهُ الدَّآءُ ‡ *The disease infected,* or *pervaded, him;* [*as though commingling with him;*] syn. خَامَرَهُ: (Sh, Ḳ:) or *infected,* or *pervaded, his inside.* (Lth, Ṣ.) — خَالَطَ قَلْبَهُ هَمٌّ عَظِيمٌ ‡ [*Great anxiety,* or *disquietude of mind, infected,* or *pervaded, his heart*]. (TA.) It is said in a trad., وَرَجَعَ الشَّيْطَانُ يَلْتَمِسُ الخِلَاطَ ‡ *And the devil returned seeking to infect* (يُخَالِط) the heart of the man praying by suggesting what was vain. (TA.) — الخَمْرُ تُخَالِطُ العَقْلَ ‡ [*Wine infects the intellect*]. (Ṣ and Ḳ in art. خمر.) And خُولِطَ فِى عَقْلِهِ, inf. n. خِلَاطٌ, ‡ [*He became infected, corrupted, disordered,* or *confused, in his intellect.*] (Ṣ, Ḳ.) And خُولِطَ عَقْلُهُ, and ↓اخْتَلَطَ عَقْلُهُ, ‡ *His intellect became corrupted,* or *disordered;* (TA; [in which only the latter phrase is thus explained, though both are mentioned;]) and so ↓اخْتَلَطَ alone: (Ṣ, Ḳ:) and ↓اخْتَلَطَتْ نَفْسُهُ † [*His soul,* or *stomach, became disordered*]: (Ṣ and Ḳ in art. خثر:) and ↓أَخْلَطَ, said of a man, signifies the same as اختلط. (TA.) — خَالَطَ القَوْمَ † *He mixed with the people,* or *company of men, in familiar,* or *social, inter-*

course; conversed with them; or *became intimate with them;* or *mixed with them in,* or *entered with them into, their affairs;* syn. دَاخَلَهُمْ; as also ↑خَالَطَهُمْ, inf. n. خِلَاطٌ; (TA;) and ↑خَلِطَ, like فَرِحَ, is used in a similar manner, in the sense of خَالَطَ: (IAar, TA:) and you say also ↑اختلط بالنَّاسِ † [*he mixed,* or *associated,* or *conversed, with men*]. (TA.) And خَالَطْتُ فُلَانًا † *I mixed with such a one in familiar,* or *social, intercourse; conversed with him;* or *became intimate with him;* syn. خَامَرْتُهُ, (A in art. خمر,) and عَاشَرْتُهُ. (Ṣ, Mṣb, Ḳ, all in art. عشر.) And خالطه فِى أَمْرٍ † [*He mixed,* or *joined, with him in an affair*]. (Mgh.) And hence خالطهُ signifies † *He was,* or *became, copartner with him; he shared with him.* (Mgh.) خَالَقَهُمْ also signifies خَالَفَهُمْ [evidently a mistranscription, for حَالَفَهُمْ † *He entered into a confederacy, league, compact,* or *covenant, with them*]. (TA.) And you say also خالط الأُمُورَ † [*He mixed in, engaged in,* or *entered into, affairs*]. (Ṣ, Ḳ.)

4. اخلطهُ, (AZ, Ṣ, Ḳ,) and اخلط لَهُ, (IAar, Ḳ,) *He put,* (Ṣ,) or *inserted,* (AZ,) or *directed* (Ḳ, TA) *and inserted,* (TA,) *his* (a camel's) قَضِيب *into the* حَيَآء, (AZ, Ṣ, Ḳ,) *he having missed it;* (AZ, Ḳ;) as also أَلْطَفَهُ: (AZ:) IF makes إِخْلَاطٌ and ↑اسْتِخْلَاطٌ to be the same. (TA.) — اخلط [intrans.] ‡ *He* (a stallion) *covered the female.* (Ḳ.) [This seems to be taken from IF, who, as shown above, makes it syn. with استخلط. See also 3.] — As syn. with اختلط, see 8, near the end of the paragraph. — Said of a horse, *He fell short,* or *flagged, in his running;* as also ↑اختلط. (IDrd, Ḳ.)

6. تخالطوا فِى الحَرْبِ ‡ *They commingled;* or *became mixed,* or *confounded, together, in war,* or *battle;* as also ↑اختلطوا فى الحرب. (TA.) — تخالطوا also signifies † *They commingled,* or *mixed together, in familiar,* or *social, intercourse;* [*conversed together;* or *became intimate, one with another;* or *they mixed, one in another's affairs:* see 3, near the end;] syn. تعاشروا. (Ṣ, Mṣb, Ḳ, all in art. عشر.)

8. اختلط *It was,* or *became, mixed, mingled, commingled, incorporated* or *blended together,* (Ṣ,* Mṣb, Ḳ,) or *put together.* (Mṣb.) [And hence, † *It was,* or *became, confused, confounded, indiscriminate, promiscuous, without order, disordered,* or *perplexed.*] — اختلط اللَّيْلُ بِالتُّرَابِ † [*The night became confused,* or *confounded, with the dust,* or *earth*]: (AZ, Ḳ:) and الحَابِلُ بِالنَّابِلِ (Ḳ) † *the setter of the snare with the shooter of arrows;* or *the warp with the woof:* (TA:) and المَرْعَى بِالهَمَلِ † [*the place of pasturage with the camels left to pasture by themselves*]: (AZ, Ḳ:) and الخَاثِرُ بِالزُّبَّادِ (as in some copies of the Ḳ and in the TA) † *the thick milk with the butter that had become bad,* or *spoiled, in the churning;* or, as some say, *with the thin milk;* (TA;) or بِالزُّبَادِ (as in other copies of the Ḳ and in the TA) *with the herb* [*so called*], which, when it falls into the رَائِب [or milk that is thick, and fit for churning, &c.], is with difficulty separated from it: (TA:) [but see art. زبد:] proverbs, alluding to the dubiousness and confusedness of an affair or a case: (Ḳ:) or the first, to the dubiousness of an affair or case; and the second, to its confusedness; and the third is applied when a people's affair or case is confused or perplexed to them; and the last relates to the confusedness of truth with falsity; and to a people whose affair or case is dubious to them, so that they do not decide upon anything. (TA.) — [اختلط الظَّلَامُ † *The darkness,* or *the beginning of night, became confused,* is a phrase of frequent occurrence. And so اِخْتِلَاطُ الظَّلَامِ † *The confusedness of the darkness,* &c.] — اختلط عَلَيْهِمْ أَمْرُهُمْ † [*Their affair, or case, became confused,* or *perplexed, to them*]. (Ṣ.) — See also 3, in four places, near the end of the paragraph: and see 6. — Said of a camel, ‡ *He became fat;* (ISh, Ḳ;) *his fat and flesh becoming mixed together.* (ISh.) — Said of a horse: see 4, last sentence.

10. استخلط *He* (a camel) *inserted,* (AZ,) or *directed* (Ḳ, TA) *and inserted,* (TA,) *his* قَضِيب *into the* حَيَآء, *by himself:* (AZ, Ḳ, TA:) or *he leaped the female;* syn. قَعَا. (Ṣ.) See also 4.

خَلْطٌ: see the next paragraph.

خِلْطٌ *Anything that mixes, mingles, commingles,* or *becomes incorporated* or *blended, with a thing; an admixture;* (Ḳ, TA;) *any kind of mixture;* as *a medicinal mixture;* and *the like:* pl. أَخْلَاطٌ. (TA.) — *A kind of* [*mixed*] *perfume,* (Ṣ,* Mṣb,) *well known:* (Mṣb:) pl. as above. (Ṣ, Mṣb.) — [Sing. of اخلاط in the term] أَخْلَاطُ الإِنْسَانِ *The four humours of man,* (Ḳ, TA,) *which are the constituents of his composition;* (TA;) namely, المِرَّتَانِ [*the black bile and the yellow bile*] and البَلْغَمُ [*the phlegm*] and الدَّمُ [*the blood*]. (TA in art. مزج.) — *Mixed dates of various sorts:* pl. as above. (Ḳ.) — ‡ A man *who mixes with others, and manifests love to them;* (TA;) and خِلْطَةٌ a woman *who does so:* (Ḳ,* TA:) and the former, (IAar, TA,) or ↑خَلْطٌ, (Ḳ,) or this signifies [simply] مُخَالِطٌ, [see 3,] and is an inf. n. used in that sense, (TA,) and ↑خِلَطٌ, (Lth, Ḳ,) and ↑خِلَّطٌ, (Ḳ,) which is mentioned by Sb and explained by Seer, (TA,) ‡ a man *who mixes with others,* (Ḳ, TA,) *and manifests love to them,* (TA,) *and behaves in a blandishing manner to them;* and one *who casts his women and goods among men;* (Ḳ, TA;) and ↑خِلَطَةٌ in like manner, applied to a female. (TA.) — † A man *of mixed race:* or *a bastard.* (Aṣ.) You say رَجُلٌ خِلْطٌ مِلْطٌ † *A man of mixed race:* (Ḳ,* TA:) or *of faulty race:* (O, TA:) or مِلْطٌ signifies one *whose race and father are unknown.* (Aṣ, TA.) And أَخْلَاطٌ مِنَ النَّاسِ † *A medley,* or *mixed* or *promiscuous multitude* or *collection, of men,* or *people;* or *of the lowest* or *basest* or *meanest sort,* or *refuse,* or *riffraff, thereof;* (Ḳ,* TA;) as also ↑خَلِيطٌ, (Ibn-'Abbád, Ḳ,) and ↑خُلَّيْطَى, (Ḳ,) and ↑خُلَيْطَى: (Ibn-'Abbád, Ḳ:) to these (لَهُنّ) [in the CḲ لَهُمْ] there is no sing.: (Ḳ, TA:) but خليط is also a sing., and has pls., as will be seen below. (TA.) — ‡ *Stupid; foolish; having little sense;* (IAar, Ḳ;) as also ↑خَلِطٌ: (IAar, Ṣgh, Ḳ:) pl. of the former أَخْلَاطٌ; (IAar, TA;) with which ↑خِلَطٌ is syn. (TA.) — *A crooked* bow, and arrow; (Ḳ;) an arrow *of which the wood has grown crookedly, and which ceases not to be crooked even if it have been straightened;* (Ṣ;) as also ↑خَلِيطٌ, applied to either of these. (Ḳ.) And in like manner, † a man; he being likened to such an arrow: and † a people, or company of men. (TA.)

خَلِطٌ; fem. with ة: see خِلْطٌ, in three places. — Also † *Good in disposition; good-natured.* (TA.)

خُلُطٌ: see خِلْطٌ, in two places: — [and see خَلِيطٌ, of which it is a pl.]

خِلَطٌ: see خِلْطٌ, last sentence but one.

خِلْطَةٌ [*A state of mixing,* or *mingling, together;*] a subst. from اختلط. (Mṣb.) — [And hence,] † *Copartnership.* (Ṣ, Mgh, TA.) You say بَيْنَهُمَا خِلْطَةٌ † *Between them two is a copartnership.* (Mgh.) [See also what next follows.]

خُلْطَةٌ † *Social,* or *familiar, intercourse.* (Ṣ, Mṣb, TA.) [See also what next precedes.]

خَلِيطٌ [*Mixed; mingled; incorporated,* or *blended:* of the measure فَعِيلٌ in the sense of the measure مَفْعُولٌ; like قَتِيلٌ &c. And hence,] (Ḳ,) or عَلَفٌ خَلِيطٌ, (Ṣ, TA,) [*The kind of trefoil called*] قَتّ *and cut straw* (Ṣ, TA) *mixed together:* (TA:) or *clay mixed with cut straw:* (Ḳ, TA:) or *with* قَتّ. (Ḳ.) Also, (Ḳ,) or لَبَنٌ خَلِيطٌ, (TA,) *Sweet milk mixed with sour* or *such as bites the tongue.* (Ḳ, TA.) Also, (Ḳ,) or سَمْنٌ خَلِيطٌ, (TA,) *Clarified butter in which are fat and flesh-meat.* (Ḳ, TA.) [Hence also,] it is said in a trad. respecting [the beverage called] نَبِيذ, (TA,) نُهِىَ عَنِ الخَلِيطَيْنِ (Ṣ, Ḳ) أَنْ يُنْبَذَا (Ṣ) or فِى الأَنْبِذَةِ (Ḳ) [*Two sorts of things mixed together are forbidden in the beverages of the kind called* نبيذ, or *that* نبيذ *should be made of them*]; i. e. it is forbidden to mix together [for making نبيذ] two sorts of things; (Ṣ, TA;) meaning *dried dates and raisins;* (Ṣ, Mgh, Ḳ;) or *fresh grapes and fresh ripe dates;* (Ṣ;) or *dried dates and full-grown unripe dates,* (T, Mgh, Ḳ,) *thoroughly cooked by fire;* (Mgh;) or *fresh grapes and raisins;* (T, Ḳ;) and *the like;* because such نبيذ quickly alters, and becomes intoxicating: (Ḳ:) and some hold that نبيذ so made is forbidden even if it do not intoxicate. (TA.) — See also أَخْلَاطٌ مِنَ النَّاسِ, voce خِلْطٌ. — † *One who mixes much with men:* (Mṣb, TA:) [see also مِخْلَاطٌ:] † *one who mixes with others in familiar,* or *social, intercourse;* or *becomes intimate with them;* or *mixes with them in,* or *enters with them into, their affairs;* syn. with ↑مُخَالِطٌ; (Ṣ, Ḳ;) like as نَدِيمٌ is syn. with مُنَادِمٌ, and جَلِيسٌ with مُجَالِسٌ: (Ṣ:) pl. خُلَطَآءُ (Ṣ, Mṣb, Ḳ) and خُلُطٌ: (Ṣ, Ḳ:) it sometimes has these pls., but is itself both sing. and pl.: (Ṣ, TA:) and as a pl. signifies †*a people*, or *company of men, whose affair*, or *case*, or *state, is one:* (Ḳ, TA:) it occurs frequently in the poems of the Arabs because they used to assemble in the days of the fresh herbage, sundry tribes of them congregating in one place, and familiar intercourse took place between them, and when they separated and returned to their homes, it grieved them: (Ṣ, TA:) or, accord. to some, it relates only to partnership: (TA:) it signifies †*a partner, copartner*, or *sharer;* (Mgh, Mṣb, Ḳ, TA;) as, for instance, in merchandise, and sheep or goats: (Mgh:) or †*one who has mixed his property with that of his copartner:* (Bḍ in xxxviii. 23:) or †*one who shares in merchandise*, or *in a debt*, or *in commerce*, or *in neighbourship:* (Ibn-'Arafeh, TA:) and †*a sharer in the rights of possession*, or *property; such as water, and a road:* (Ḳ:) the pl. is خُلَطَآءُ; (Mgh, TA;) occurring in the Ḳur xxxviii. 23: (TA:) and the sing. also signifies †*a neighbour;* syn. جَارٌ [which has also other significations here assigned to خَلِيطٌ]; (TA;) and مُجَاوِرٌ: (Mṣb:) and †*a husband:* and †*the son of a paternal uncle:* (Ḳ:) and [the pl.] خُلُطٌ is also explained by IAạr as †*i. q.* مَوَالٍ [pl. of مَوْلًى, which has several of the significations here assigned to خليط]: and as signifying also †*neighbours of sincere friendly conduct.* (TA.) It is said in a trad. (Ḳ, TA) respecting [the right termed] الشُّفْعَة, (TA,) الشَّرِيكُ أَوْلَى مِنَ الخَلِيطِ وَالخَلِيطُ أَوْلَى مِنَ الجَارِ †*The sharer in what is not divided is more deserving than the sharer in the rights of possession*, or *property;* [*and the sharer in the rights of possession*, or *property, is more deserving than the neighbour:*] (Ḳ, TA:) [or the trad. is as follows:] الخَلِيطُ أَحَقُّ مِنَ الشَّرِيكِ وَالشَّرِيكُ أَحَقُّ مِنَ الجَارِ وَالجَارُ أَحَقُّ مِنْ غَيْرِهِ †*the sharer in the thing itself that is sold has more right than the sharer in the rights thereof; and the sharer in the rights thereof has more right than the adjoining neighbour; and the adjoining neighbour has more right than another:* or the meaning here is, *he between whom and thyself are acts of receiving and giving, and affairs of debt and credit;* not the sharer, or partner. (Mgh.) And in another trad., مَا كَانَ مِنْ خَلِيطَيْنِ فَإِنَّهُمَا يَتَرَاجَعَانِ بَيْنَهُمَا بِالسَّوِيَّةِ †*Whatever two copartners there be that have not divided the beasts* [*belonging to them*], *they shall make claim for restitution, one of the other, with equality;* i. e., *if they be copartners in camels for which it is incumbent to give sheep or goats, and the camels be found in the possession of one of them, and the poor-rate for them be taken from him, he shall make a claim for restitution* [*of what he has given above his own share*] *upon his copartner, with equality:* (Esh-Sháfi'ee, Ḳ, TA:) the two persons are not خليطان unless they be *such as drive back their beasts to the nightly resting-place, and drive them forth in the morning to the pasturage, and water them, together, and have their stallions mixed together, and have been copartners for a year;* and if so, they give the poor-rate as one: otherwise, they are not خليطان; and they give the poor-rate as two: (Esh-Sháfi'ee, TA:) the trad. applies, for instance, to the case of two copartners who have mixed their property together; one of them having forty bulls or cows or of both kinds; and the other, thirty; and the collector of the poor-rates takes from the forty a مُسِنَّة [q. v.], and from the thirty a تَبِيع [q. v.]; then the giver of the مسنّة makes a claim for restitution of three sevenths thereof upon his copartner; and the giver of the تبيع, of four sevenths thereof upon his copartner; for it is incumbent to give the beasts of these two ages [the مسنّة and the تبيع] when the property is not divided, as though it were the property of one: and the saying بالسويّة shows that if the collector of the poor-rate wrong one of them, and take from him more than the law imposes upon him, he cannot make a claim for restitution thereof upon his copartner, who is only responsible to him for the value of what falls upon him in particular, of what is incumbent by the law: and the making claim for [just] restitution, by one upon the other, shows that the partnership holds good notwithstanding the distinction of the things which compose the possessions, with such as hold this to be the case. (IAth, TA.)

خَلَاطَةٌ ‡*Stupidity; foolishness; paucity of sense.* (IAạr, Ḳ.)

خَلِيطَةٌ *Camel's milk milked upon that of sheep or goats:* or *sheep's milk upon that of goats:* and *the reverse.* (Ḳ.)

خُلَّيْطَى: see خِلْطٌ:—and see what next follows, in two places.

خُلَيْطَى: see خِلْطٌ. — وَقَعُوا فِى خُلَيْطَى, (Ṣ, Ḳ,) and ▼خُلَّيْطَى, (Ḳ,) †*They fell into a state of confusion:* (Ḳ:) *their affair*, or *case, became confused*, or *perplexed*, (اِخْتَلَطَ,) *to them.* (Ṣ.) And ▼كُنَّا خُلَّيْطَى †[*We were in a state of confusion*]: cited by Az, from an Arab of the desert. (TA.) [▼خُلَّيْطَآءُ, which probably signifies the same, is mentioned in the TA, voce لُغَّيْزٌ, on the authority of Sb.]

خِلِّيطَى The *creating confusion*, or *disorder*, (إِفْسَادٌ,) in an affair, or a case. (TA.) [See also 2.] — مَالُهُمْ خِلِّيطَى [in the CḲ مالُهُمْ] *Their possessions*, or *camels &c.*, *are mixed together.* (Ḳ,* TA.)

خُلَّيْطَآءُ: see خُلَيْطَى.

أَخْلَطُ مِنَ الحُمَّى ‡[*More insinuating than fever*]; a saying of the Arabs; meaning that it manifests an affection for a person by its access to him, like the lover and blandisher. (TA.)

مِخْلَطٌ †*One who renders things confused*, or *dubious, to the hearers and beholders.* (TA.) — †*One who mixes in*, or *enters into*, (يُخَالِطُ,) *affairs*, (Ṣ, Ḳ, TA,) *and relinquishes them;* (TA; [but this addition seems rather to apply to مِزْيَل in what follows;]) as also ▼مِخْلَاطٌ: (Ḳ:) or this latter signifies †*one who mixes much with men.* (Ṣgh, TA.) [See also خَلِيطٌ.] You say, هُوَ مِخْلَطٌ مِزْيَلٌ †[*He is one who mixes in*, or *enters into, affairs;* (and, accord. to an explanation of مِزْيَل in the TA, in art. زيل, on the authority of IAth,) *one who is vehement in altercation*, or *litigation, relinquishing one plea*, or *argument, and taking to another*]; like as you say, هُوَ رَاتِقٌ فَاتِقٌ. (Ṣ, Ḳ.)

مِخْلَاطٌ: see مِخْلَطٌ.

مُخَالَطٌ ‡*Infected, corrupted, disordered*, or *confused, in his intellect;* as also ▼مُخْتَلِطٌ: (TA:) or *mad; insane;* or *affected by diabolical possession.* (TA in art. لبس.)

مُخَالِطٌ: see خَلِيطٌ.

مُخْتَلِطٌ: see مُخَالَطٌ. — Also ‡A camel *that has become fat, so that the fat is mixed with the flesh:* fem. with ة, applied to a she-camel. (ISh, Ḳ.)

خلع

1. خَلَعَهُ, (Ṣ, Mgh, Mṣb,) aor. َ, (TA,) inf. n. خَلْعٌ, (Ṣ, Mgh, Mṣb, Ḳ,) *He pulled it off;* syn. نَزَعَهُ; (Mgh, Mṣb;) or *stripped it off;* or *took it off;* (TA;) or *put it*, or *threw it*, or *cast it, off from him;* (IAth;) namely, his garment, (Ṣ, IAth, Mgh, Mṣb,) عَنْ بَدَنِهِ *from his body;* (Mgh;) and his sandal, (Ṣ, Mgh, Mṣb,) عَنْ رِجْلِهِ *from his foot;* (Mgh;) &c.; (Mṣb, TA;) [as also ▼اختلعه, as appears from its being said that] اِخْتِلَاعٌ is *syn. with* خَلْعٌ: (TA:) accord. to some, خَلْعٌ is *syn. with* نَزْعٌ; but accord. to Lth, (TA,) *the former is like the latter, except that the former is a somewhat leisurely action.* (Ḳ, TA.) The phrase in the Ḳur [xx. 12], فَاخْلَعْ نَعْلَيْكَ is said to be used in its proper sense, [*And do thou pull off*, or *put off, thy sandals*,] because his sandals were of the skin of a dead ass: or, as the Ṣoofees say, it is a command to stay; like as you say to him whom you desire to stay, "Pull off thy garment and thy boots," and the like; and is tropical: (TA:) or, accord. to some, †*make thy heart vacant from* [*care for*] *family and property.* (Bḍ.) — خَلَعَ عَلَيْهِ, (B, TA,) and ▼خَلَعَ عَلَيْهِ خِلْعَةً, (Ṣ, TA,) [*He took off from himself, and bestowed upon him, a garment:* and hence,] *he bestowed upon him*, or *gave him, a garment;* [generally meaning, *a robe of honour;*] the meaning of giving being inferred from the connective على, not from the verb alone. (B, TA.) — It is said in a trad. respecting 'Othmán, إِنَّ اللّٰهَ سَيُقَمِّصُكَ قَمِيصًا وَإِنَّكَ تُلَاصُ عَلَى خَلْعِهِ (L,) meaning ‡*Verily God will invest thee with the apparel of the office of Khaleefeh*, (Ḳ and TA in art. قمص,) *and thou wilt be urged with enticement, and solicited, to divest thyself of it.* (TA in art. لوص.) — خَلَعَ الفَرَسُ عِذَارَهُ †*The horse threw off his head-stall, or halter, and wandered about at random.* (Mgh.) — [And hence,] خَلَعَ عِذَارَهُ [said of a man,] ‡‡ *He threw off from himself his* عذار, [meaning *restraint*,] *and acted in a wrongful and evil manner towards others, with none to repress him.* (TA.) —

خَلَعَ أَوْصَالَهُ He removed its اوصال [meaning the bones so called, as is indicated by the context]. (TA.) — خَلَعَ مَالَ صَاحِبِهِ ‡ [He took away the property of his companion]; said of a person gambling with another. (A, TA.) — خَلَعَ قَلْبَ النَّاظِرِ إِلَيْهِ † [It drew away the heart of the beholder towards it]; said of the best of property. (Aboo-Sa'eed.) — خَلَعَ قَيْدَهُ ‡ [He took off his shackles; or] he released him from his shackles: and in like manner, خَلَعَ دَابَّتَهُ, and ↓ خَلَّعَهَا, he released his beast from its shackles. (TA.) — خَلَعَ الرِّبْقَةَ عَنْ عُنُقِهِ ‡ He annulled his compact, or covenant. (TA.) — خَلَعَ يَدًا مِنْ طَاعَةٍ ‡ He [threw off his allegiance, or] forsook obedience to his Sultán, and acted in a wrongful and evil manner towards him: (TA:) obedience being likened to a garment which a man puts off, or throws off, from him. (IAth, TA.) — يُخْلَعُ المَيِّتُ [for يُخْلَعُ الكَفَنُ عَنِ المَيِّتِ, like خَلَعَ الدَّابَّةَ (mentioned above) for خَلَعَ قَيْدَ الدَّابَّةِ,] The corpse shall have its grave-clothes pulled off from it. (Mgh.) — [In like manner you say,] خَلَعْتُ الوَالِيَ عَنْ عَمَلِهِ ‡ I removed the ruler, or governor, or the like, from his office; or deposed him. (Mṣb.) And خُلِعَ الوَالِي ‡ [The ruler, or governor, or the like, was divested of his authority; or] was removed from his office; or was deposed; (Ṣ, TA;) and so العَامِلُ [the agent, or the exactor of the poor-rates]; and الخَلِيفَةُ [the Khaleefeh]. (TA.) And خَلَعَ قَائِدَهُ ‡ [He divested his leader of his authority; or removed him from his office; or dismissed him]. (Ṣ, TA.) But IF says, This is scarcely, or never, said, except of an inferior who forsakes, or relinquishes, his superior; so that [خَلَعَهُ signifies, in a case of this kind, ‡ He threw off his allegiance to him; or forsook obedience to him; like another phrase, mentioned above; and] one does not say, [or seldom says,] خَلَعَ الأَمِيرُ وَالِيَهُ عَلَى بَلَدِ كَذَا [as meaning † The prince deposed his ruler over such a province, or the like]; but only, [or rather,] عَزَلَهُ. (TA.) — خَلَعَ آمْرَأَتَهُ, (Az, Ṣ, Mgh, Mṣb,) inf. n. خُلْعٌ, with damm, (Ṣ, Ḳ,*) or this is a simple subst., (Az, Mgh, Mṣb,) and the inf. n. is خَلْعٌ, (Az, Mṣb, TA,) and some add خِلَاعٌ; (TA;) and ↓ خَالَعَهَا, (Az, TA,) inf. n. مُخَالَعَةٌ; (Ḳ;) [and خِلَاعٌ seems to be another inf. n. of this latter verb, rather than of the former;] ‡ He divorced his wife (Az, Mgh, Mṣb, Ḳ) for a ransom given by her, (Mṣb,) or for her property given by her as a ransom to release herself from him, (Az, Mgh, TA,) or for a gift, or a compensation, from her, (Ḳ, accord. to different copies; some having بِبَذْلٍ; and others, بِبَدَلٍ;) or from another: (Ḳ:) because the wife is [as] a garment to the husband, and the husband to the wife, (Az, Mgh, Mṣb, TA,) as is said in the Ḳur ii. 183: (Az, TA:) [it is also said that] ↓ تَخَالَعَ is syn. with خَلَعَ: (Ḳ:) [but see 6, below:] and اِخْلَعْهَا, occurring in a trad., is explained as signifying Divorce thou her, and quit her. (TA.) — خَلَعَهُ أَهْلُهُ ‡ [His family cast him off, repudiated him, or renounced him;] so that if he committed a crime, or an offence rendering him liable to punishment, they should not be prosecuted for it. (Ṣ, TA.) In the Time of Ignorance, when one said, (Ḳ, TA,) proclaiming in the fair, or festival, (TA,) يَا أَيُّهَا النَّاسُ هٰذَا ٱبْنِى قَدْ خَلَعْتُهُ, (Ḳ,* TA,) meaning [O men, this, my son, I cast off, repudiate, or renounce, him, or] I declare myself to be clear of him; so that if he commit a crime, or an offence rendering him liable to punishment, I am not responsible; and if a crime, or an offence rendering liable to punishment, be committed against him, I will not pursue [for redress, or retaliation]; (TA;) he was not punished afterwards for any such act committed by him: (Ḳ, TA:) this was when the person doing so feared some foul action or treachery from his son: and in like manner, they said, إِنَّا قَدْ خَلَعْنَا فُلَانًا [Verily we cast off, &c., such a one]. (TA.) In like manner, also, خَلَعُوهُ, inf. n. خَلْعٌ, signifies ‡ [They cast him off, repudiated him, or renounced him, or] they declared themselves to be clear of him; meaning a confederate; so that they should not be punished for a crime, or an offence rendering liable to punishment, committed by him, nor should he be punished for such an act committed by them. (IAth, L.) In the same sense the verb is used in the saying, نَخْلَعُ وَنَتْرُكُ مَنْ يَفْجُرُكَ [We repudiate, or renounce, or] we declare ourselves clear of, and forsake, him who disobeys, or opposes, Thee: (Mgh, TA:) or نَخْلَعُ وَنَهْجُرُ مَنْ يَكْفُرُكَ we hate, and [repudiate, or renounce, or] declare ourselves clear of, [and forsake,] him who denies, or disacknowledges, thy favour, or who is ungrateful, or unthankful, for it. (Mṣb.) ═ خَلُعَ, aor. ـُ, (Ḳ,) inf. n. خَلَاعَةٌ, (Ṣ,* TA,) ‡ He became cast off, repudiated, or renounced, by his family; (صَارَ خَلِيعًا; TA; i. e. خَلَعَهُ أَهْلُهُ; Ṣ, TA;) so that if he committed a crime, or an offence rendering him liable to punishment, they were not prosecuted for it: (Ṣ, Ḳ, TA:) he became alienated or estranged [from his family]; syn. تَبَاعَدَ: (TA:) [he became vitious, or immoral; notorious for drinking and play; a gambler; or the like: see خَلَاعَةٌ, below; and see خَلِيعٌ.] ═ خُلِعَ He became affected with what is termed خَالِعٌ, i. e., a twisting of the عُرْقُوب [or hock-tendon]. (Ḳ.)

2. خلّع دَابَّتَهُ: see 1. — تَخْلِيعٌ as signifying a certain manner of walking: see 5.

3. خَالَعَتْ بَعْلَهَا, (Ṣ,) or زَوْجَهَا, (Mgh, Mṣb,) inf. n. مُخَالَعَةٌ, (Mṣb,) † She incited, urged, or induced, her husband to divorce her for a gift, or a compensation, (بِبَذْلٍ, or بِبَدَلٍ, accord. to different copies of the Ṣ,) from her to him: (Ṣ, Mgh:*) or † she ransomed herself from him, and he divorced her for the ransom. (Mṣb.) — خالع آمْرَأَتَهُ: see 1. — خَالَعَهُ ‡ He contended with him in a game of hazard: because he who does so takes away the property of his companion. (TA.)

5. تخلّع It (a bond, or chain,) came off, or fell off, from the hand or foot. (KL.) [See also 7.] — تَخَلَّعَتِ السَّفِينَةُ The ship parted asunder; became disjointed; became separated in its places of joining. (Mgh.) — تخلّع, in walking, i. q. تَفَكَّكَ; (Ṣ, Ḳ, TA;) i. e. ‡ [He was, or became, loose in the joints; or] he shook his shoulder-joints and his arms, and made signs with them: (TA:) and ↓ تَخْلِيعٌ also signifies a certain manner of walking, (so in some copies of the Ḳ, and in the TA,) in which one shakes his shoulder-joints and his arms, and makes signs with them: (TA:) or the walking of him whose buttocks are apart, or parted. (CḲ, and so in a MS. copy of the Ḳ.) [See also تَخَلَّجَ.] — تخلّع فِى الشَّرَابِ † He persisted in the drinking of intoxicating beverage, (Ḳ,* TA,) or became intoxicated, so that his joints became lax, or loose. (TA.) — تَخَلَّعَ القَوْمُ † The people, or company of men, stole away, slipped away, or went away secretly. (IAạr.)

6. تخالعوا ‡ They annulled, dissolved, or broke, the confederacy, league, compact, or covenant, that was between them. (Ṣ,* Ḳ,* TA.) — تَخَالَعَا ‡ They divorced themselves, namely, a husband and his wife, for a gift, or a compensation, (بِبَذْلٍ, or بِبَدَلٍ, accord. to different copies of the Ṣ,) from the wife to the husband. (Ṣ.) See also خَلَعَ آمْرَأَتَهُ.

7. انخلع It became pulled off, stripped off, taken off, or removed, from its place; it became displaced. (Mgh.) [See also 5.] — Hence, آنْخَلَعَ قِنَاعُ قَلْبِهِ مِنْ شِدَّةِ الفَزَعِ ‡ [He became as though] the integument of his heart became pulled off, in consequence of violence of fear, or fright. (Mgh.) And hence also, آنْخَلَعَ فُؤَادُ الرَّجُلِ ‡ [The heart of the man became removed from its place; meaning] the man became frightened. (Mgh.) — انخلع العُضْوُ, or العَظْمُ, The limb, or the bone, became dislocated. See اِنْخَزَعَ. — انخلع عَنْ عَمَلِهِ ‡ He (a ruler, or governor, or the like,) became removed from his office; became deposed. See 1.] — انخلع مِنْ مَالِهِ ‡ He became stripped of his property, like as a man is stripped of his garment. (TA.)

8. اِخْتَلَعَتْ (Ṣ, Mgh, Ḳ) مِنْ زَوْجِهَا, (Mgh,) ‡ She became divorced from her husband (Ṣ, Mgh, Ḳ) for a gift, or a compensation, from her, (Ṣ, Ḳ, accord. to different copies; some having بِبَذْلٍ; and others, بِبَدَلٍ;) or from another, (Ḳ,) or for her property given by her as a ransom to release her from him. (Mgh.) ═ اختلعهُ: see 1, first sentence. — اختلعوهُ ‡ They took his property: (Ḳ, TA:) from the "Nawádir el-Aạráb." (TA.)

خَلْعٌ Flesh-meat cooked with seeds that are used for seasoning, then put into a receptacle of skin, (Ṣ, Ḳ,*) which is called قُرْفٌ: (Ṣ:) or flesh-meat cut into strips or oblong pieces, and dried, or salted, and dried in the sun, roasted, (Ḳ, TA,) and, as Lth says, (TA,) put into a receptacle with its melted grease: (Ḳ,* TA:) or flesh-meat having its bones pulled out, then cooked, and seasoned with seeds, and put into a skin, and used as provision for travelling: (Z, TA:) and ▼ خَلِيعٌ, also, signifies *flesh-meat of which the bones have been pulled out, and which is seasoned with seeds, and laid up* (يُرْفَع) [*for future use*]: (TA:) and ▼ خَوْلَعٌ, *flesh-meat which is boiled in vinegar, and then carried in journeys.* (TA.) ═ *A state of dislocation of the joint, of the arm or hand, or of the leg or foot; its becoming displaced, without separation;* as also ▼ خَلْعٌ. (TA.)

خُلْعٌ [accord. to the Ṣ, and app. accord. to the Ḳ, an inf. n., (see خَلَعَ ٱمْرَأَتَهُ,) or] a simple subst., signifying ‡ The *act of divorcing a wife* (Az, Mgh, Mṣb) *for a ransom given by her,* (Mṣb,) or *for her property given by her as a ransom to release her from her husband,* (Az, Mgh, TA,) or *for a gift,* or *a compensation, from her;* or *from another:* (Ḳ: see 1:) IAth says that it annuls the return to the wife unless by means of a new contract: accord. to Esh-Sháfi'ee, there is a difference of opinion respecting it; whether it be *an annulment of the marriage,* or *a divorcement:* [if the latter, it is not irrevocable unless preceded by two divorcements:] sometimes it is called by the latter term. (TA.) [See also خُلْعَةٌ.]

خَلَعٌ: see خَلْعٌ.

خُلْعَةٌ *A state of divorcement* [*for a ransom given by the wife,* or *for her property given by her as a ransom to release her from her husband,* or] *for a gift,* or *a compensation, from the wife,* (Ṣ,* Ḳ,) or *from another.* (Ḳ.) [See 8: and see also خُلْعٌ.] You say, وَقَعَتْ بَيْنَهُمَا الخُلْعَةُ [*Divorcement,* or] *separation,* [*for a ransom,* &c., or] *for* [*a gift,* or] *a compensation, took place between them two.* (TḲ.) ═ † The *best,* or *choice part,* of property, or of camels or the like; (Aboo-Sa'eed, Ṣ, Ṣgh, Ḳ;) so called because it takes away the heart of him who looks at it; (Aboo-Sa'eed;) as also ▼ خِلْعَةٌ. (Aboo-Sa'eed, Ṣgh, Ḳ.) ═ † *Weakness* in a man. (TA.)

خِلْعَةٌ *Any garment which one pulls off,* or *takes off, from himself:* (TA:) and particularly, (TA,) *a garment which is bestowed upon a man,* [generally meaning *a robe of honour,*] (Ḳ,* TA,) *whether it be put upon him or not:* (TA:) or *a gift,* or *free gift,* [*of any kind,*] *which a man bestows upon another:* (Mṣb:) or *a sewed garment:* (KL:) pl. خِلَعٌ. (Mṣb, TA.) You say, خَلَعَ عَلَيْهِ خِلْعَةً [explained above]: see 1, near the beginning. (Ṣ, TA.) — See also خُلْعَةٌ.

خُلْعِىٌّ, with ḍamm, *One who sells* [*cast-off* or] *old garments.* (Ibn-Nuḳṭah, TA.)

خِلْعِىٌّ, with kesr to the خ, and with the ل quiescent, *One who sells the garments bestowed by kings.* (TA.)

خُلَاعٌ † *An affection resembling what is termed* خَبَل [q. v.], (Ḳ, TA,) *and insanity,* or *diabolical possession,* (TA,) *which befalls a man:* (Ḳ, TA:) or *weakness, and fear* or *fright:* (TA:) and ▼ خَوْلَعٌ and ▼ خَيْلَعٌ [in like manner] signify ‡ *fear,* or *fright, affecting the heart,* (Ṣ, Ḳ, TA,) *occasioning evil imagination, and weakness,* (TA,) *as though it were a touch of insanity,* or *of diabolical possession,* (Ṣ, Ḳ, TA,) in a man, and in the heart. (Ṣ.)

خَلِيعٌ *Pulled off; stripped,* or *taken, off; put,* or *thrown,* or *cast, off; i. q.* ▼ مَخْلُوعٌ; applied [to a garment, and a sandal, or the like, or], accord. to some, to anything. (TA.) — [Hence, used as a subst., *A cast-off,* or] *an old, and worn-out, garment.* (Ḳ, TA.) You say, هُوَ يَكْسُوهُ مِنْ خَلِيعِهِ [*He clothes him with some of his cast-off, or old, and worn-out, apparel*]. (TA.) — See also خَلْعٌ. — † A person *whose property is won from him in a game of hazard;* as also ▼ مَخْلُوعٌ. (L.) — خَلِيعُ العِذَارِ †† A man *who does and says what he pleases; not caring, nor fearing God nor the blame of men; like the beast that has no halter on its head.* (Ḥar p. 676.) Also applied to a woman in a state of estrangement [from her husband; lit., *Having her headstall,* or *halter, pulled off,* or *thrown off; she being likened to a mare;* meaning, †† *without restraint*]; *having none to command or forbid her:* [see 1:] incorrectly written خَلِيعَةُ العِذَارِ; for خليع is here of the measure فَعِيلٌ in the sense of the measure مَفْعُولٌ: or you say خَلِيعَةٌ without mentioning the عذار, from خَلَاعَةٌ, like ظَرِيفَةٌ and لَطِيفَةٌ [from ظَرَافَةٌ and لَطَافَةٌ]. (Mgh.) [See also خَالِعٌ.] — خَلِيعٌ is also applied to a Khaleefeh, and a prince or the like, meaning ‡ *Divested of his authority; removed from his office; deposed;* (L;) as also ▼ مَخْلُوعٌ: and it is a strange thing, noticed by Dmr and others, that every sixth is مخلوع. (TA.) — Also ‡ A young man, (Ṣ,) or a son, (Ḳ,) and a confederate, (IAth, L,) *cast off, repudiated,* or *renounced,* (Ṣ, IAth, L, Ḳ,) *by his family,* (Ṣ,) or *father,* (Ḳ,) or *confederates,* (IAth, Ḳ,) *so that if he commit a crime,* or *an offence rendering liable to punishment, they,* i. e. *his family,* or *he,* i. e. *his father,* or *they,* i. e. *his confederates, shall not be prosecuted,* or *punished, for it;* (Ṣ, IAth, L, Ḳ;) as also ▼ مَخْلُوعٌ: (Ḳ:) pl. of the former, خُلَعَآءُ: (Ḳ:) and ‡ a young man (Ḳ, TA) *thus cast off by his family,* (TA,) *who commits,* or *has committed, many crimes,* or *offences rendering him liable to punishment;* as also ▼ خَوْلَعٌ: (Ḳ, TA:) † one *alienated* or *estranged* [*from his family*]: (TA:) ‡ one *who has broken off from his family, and disagreed with them, and wearied them by his wickedness and baseness and guile;* (Mgh, Ḳ,* TA;*) as though he had thrown off his headstall or halter, [i. e., restraint,] *and who does what he will;* or because his family have cast him off, and declared themselves clear of him; (Mgh;) or because he has cast off his kinsfolk, and they have declared themselves clear of him; or because he is divested of religion and shame; (TA;) fem. with ة: (Ḳ: [indicating that it is a part. n. of خَلُعَ; not of the measure فَعِيلٌ in the sense of the measure مَفْعُولٌ, as is implied by some portions of the explanations here given:]) † *bad, evil, wicked,* or *mischievous:* (TA:) ‡ [*vitious,* or *immoral:* (see خَلَاعَةٌ, whence it is derived:)] † *notorious for drinking and play:* (TA:) [in the present day commonly used in this sense; and as signifying † *waggish;* or *a wag:*] † *a player, with another, at a game of hazard,* or *for stakes laid by both of them to be taken by the winner;* (IDrd, Ḳ;) as also ▼ مُخَالِعٌ; because the best, or choice part, of his property (خُلْعَتُهُ) is [often] won from him: (Ṣ:) † one *who applies himself constantly to games of that kind:* (TA:) and ▼ خَوْلَعٌ signifies ‡ *a player at games of that kind, who has had the punishment termed* حَدّ *inflicted upon him, and is always overcome in such games,* or *who is fortunate, and always overcomes in such games.* (Ḳ, accord. to different copies; in some of which we read المُقَامِرُ المَحْدُودُ الَّذِى يُقْمَرُ أَبَدًا; and in others, المقامر المَجْدُودُ الذى يَقْمُرُ ابدا.) — † *A hunter, fowler,* or *fisherman;* (Ṣ, Ṣgh, Ḳ;) so called because he is alone. (Ṣgh.) — ‡ *A* [*demon,* or *devil,* &c., *of the kind called*] غُول; (Ṣ, Ḳ, TA;) because of its evil nature; (TA;) as also ▼ خَوْلَعٌ (Ḳ) and ▼ خَيْلَعٌ. (TA.) — † *A wolf;* (Ṣ, Ḳ;) as also ▼ خَوْلَعٌ (Ḳ) and ▼ خَيْلَعٌ. (Ṣgh, Ḳ.) — † The *gaming-arrow that does not win* (Ṣ, Kr, Ḳ) *at first:* (Ṣ, Kr:) or, accord. to some, the *gaming-arrow that wins at first;* as is said by Ṣgh and in the L: (TA:) pl. خِلْعَةٌ. (Kr.)

خَلَاعَةٌ: [see خَلُعَ:] it is *syn. with* دَعَارَةٌ [i. e. ‡ *Vice,* or *immorality;* or *vitious,* or *immoral, conduct;* &c.]; as also خَرَاعَةٌ; a dial. var. thereof; (Ṣ in art. خرع;) and ▼ خَلِيعَةٌ signifies the same. (TA.)

خَلِيعَةٌ: see what next precedes.

خَالِعُ العِذَارِ [† A horse *throwing off his headstall,* or *halter, and wandering about at random.* — And hence, †† A man *throwing off from himself restraint, and acting in a wrongful and an evil manner towards others, with none to repress him.* See also خَلِيعٌ. — And hence,] ††† A *beardless* youth, or young man; or one *whose mustache has grown forth, but not his beard.* (TA.) — خَالِعٌ † *A kid.* (TA.) [App. because of its playful disposition.] — † A woman *who incites, urges,* or *induces, her husband to divorce her for a gift,* or *a compensation, from her to him:* [see 3:] (Ṣ:) or *who causes herself to be divorced for a gift,* or *a compensation, from her to her husband:* [see 6:] and in like manner, a husband *who divorces his wife for a gift,* or *a compensation, from her.* (Ḳ.) — رُطَبٌ خَالِعٌ *Dates that are all ripe,* or *ripe throughout,* or *soft;* syn. مُنْسَبِتٌ; (Ṣ, Ḳ;) because their skins strip off by reason of their succulency: (TA:) and بُسْرَةٌ خَالِعٌ, (Ḳ, TA,) and خَالِعَةٌ, (TA,) *a date that has become wholly fit to be eaten.* (Ḳ,* TA.) [See بُسْرٌ.] — خَالِعٌ also signifies *A twisting of the* عُرْقُوب [or *hock-tendon*]: (Ḳ:) or *a certain disease that attacks the* عُرْقُوب *of a she-camel.* (TA.) And you say, بَعِيرٌ خَالِعٌ, (Ṣ,) or بَعِيرٌ بِهِ خَالِعٌ, (Ḳ,) *A camel that is unable to rise* (Ṣ, Ḳ) *when a man sits upon the part called* غُرَاب [q. v.] *of its haunch,* (Ṣ,) *in* consequence, as some say, *of a dislocation of the tendon of the hock.* (TA.) — جُبْنٌ خَالِعٌ ‡ *Vehement cowardice; as though the vehemence of the man's fear removed his heart from its place;* accord. to IAth, *an affection arising from yearning thoughts, and weakness of the heart, on an occasion of fear.* (TA.)

خَوْلَعٌ : see خُلَعٌ. — It also signifies هَبِيد [i. e. *Colocynth, or its pulp, or seed,*] *when it is cooked until its* سَمْن [*or decocted juice*] *comes forth, whereupon it is cleared, and put aside; and bruised dates of which the stones have been taken out are put upon it, and flour, and it is stirred about and beaten until it becomes mixed; then it is left, and put down; and when it becomes cold, its* سمن *is restored to it:* or, as some say, *colocynth* (حَنْظَل) *bruised, moistened with something to sweeten it, and then eaten; also called* مُبَسَّل. (TA.) [See هَبِيدٌ.] ═ See also خُلَاعٌ: ═ and خَلِيعٌ, in four places. ═ Also *Stupid;* (Ḳ;) applied to a man. (TA.) ═ And *A skilful guide.* (Ṣgh, Ḳ.)

خَيْلَعٌ : see خُلَاعٌ: ═ and see خَلِيعٌ, in two places, near the end. — Also † *A weak* man. (TA.) [See also مُخَلَّعٌ.]

مُخْلَعُ الأَلْيَتَيْنِ A man (Ṣ) *having the buttocks apart, or parted.* (Ṣ, Ḳ.) — And مُخْلَعٌ *A weak, and soft, or flabby,* man. (Lth, Ḳ.) [See also خَيْلَعٌ.] — ‡ A man (TA) *in whom is what resembles a loss of reason, or a touch of insanity or of diabolical possession:* (Ḳ,* TA:) and ‡ a man *insane,* or *possessed by a jinnee.* (TA.)

مَخْلُوعٌ : see خَلِيعٌ, in four places. — رَجُلٌ مَخْلُوعُ الفُؤَادِ ‡ *A man frightened,* or *terrified; as though his heart were removed from its place.* (TA.)

مُخَالِعٌ : see خَلِيعٌ, in the latter part of the paragraph.

مُخْتَلِعَةٌ ‡ A woman *divorced from her husband for a gift, or a compensation, from him,* (Ṣ, Ḳ,) or *from another:* (Ḳ:) [see 8:] and [the pl.] مُخْتَلِعَاتٌ [is explained as signifying] ‡ *women who incite, urge, or induce, their husbands to divorce them for a gift, or a compensation, without any injurious conduct from the latter.* (TA.) — † A woman *affected with lust.* (Ṣgh, Ḳ.)

خلف

1. خَلَفَ, aor. ـُ, inf. n. خَلْفٌ, *He came after, followed, succeeded,* or *remained after, another,* or *another that had perished or died.* (TA.) Hence, in the Ḳur [vii. 168 and xix. 60], فَخَلَفَ مِنْ بَعْدِهِمْ خَلْفٌ (TA) *And there hath succeeded them, or come after them,* [*a posterity, or*] *an evil posterity.* (Bḍ in xix. 60.) And خَلَفَهُ *He came after him,* (Ṣ, A in art. دبر, Mgh, Mṣb, TA,) or *behind him,* (A ubi suprà, Mgh,) or *following him nearly;* (A ubi suprà;) inf. n. as above, (Mgh, TA,) and خِلَافَةٌ also: (Mgh:) or *he remained after him:* (Ḳ:) and ↓ جَآءَ خِلَافَهُ likewise signifies [the same as جَآءَ خَلْفَهُ; an inf. n. of خَالَفَ being thus used as an adv. n.; i. e.] *he came after him.* (TA.) You say also, خَلَفَ اللَّيْلُ النَّهَارَ, inf. n. خَلْفٌ and خِلْفَةٌ, *The night followed,* or *came after, the day.* (MA.) — [Hence,] خَلَفْتُهُ, [aor. as above,] inf. n. خَلْفٌ, [perhaps a mistranscription for خَلَفٌ,] *I was, after him, a substitute for him:* (TA:) [*I supplied his place: and I superseded him.*] And خَلَفَهُ, (aor. as above, TA,) inf. n. خِلَافَةٌ (Ṣ, Mgh, Mṣb, Ḳ, TA) and خَلْفٌ (TA) and خِلِّيفَى, (Ṣ,* Ḳ,* TA,) which last is an inf. n. of the intensive kind, (Ṣgh, MF, TA,) *He was,* or *became, his* خَلِيفَة [i. e. *successor,* or *vice-agent,* &c.], (Ṣ, Mgh, Mṣb, Ḳ,) or *his substitute;* (TA;) فِى قَوْمِهِ [*among,* or *in respect of, his people*], (Ṣ, TA,) and أَهْلِهِ [*his family*]; relating to good and to evil; wherefore one says, أَوْصَى لَهُ بِالخِلَافَةِ [*he charged him by his will with the being his successor, or vice-agent,* &c.]; (TA;) or عَلَى أَهْلِهِ وَمَالِهِ [*over his family and his property*]: (Mṣb:) and ↓ اختلفهُ signifies the same; (Lḥ, Ibn-'Abbád, Ḳ;) *he was,* or *became, his* خَلِيفَة (Ibn-'Abbád, Ḳ, TA) *after him.* (Ibn-'Abbád, TA.) And خَلَفَ فُلَانًا [alone] *He was,* or *became, the* خَلِيفَة *of such a one among,* or *in respect of, his family* (Ḳ, TA) *and his children.* (TA.) And خَلَفَهُ رَبُّهُ فِى أَهْلِهِ (Ḳ, TA) and وَلَدِهِ, (TA,) inf. n. خِلَافَةٌ, (Ḳ,) *His Lord was* [*for him*] *a* خَلِيفَة [*or supplier of his place*] *to his family* (Ḳ, TA) *and his children.* (TA. [In the CḲ, اخلف is made to signify the same; but this is in consequence of an omission.]) And one says, خَلَفَ اللّٰهُ عَلَيْكَ *May God be to to thee a* خَلِيفَة [*or supplier of the place*] (Ṣ, Mṣb, Ḳ) *of thy father,* (Ṣ, Mṣb,) or *of the one whom thou hast lost:* (Ṣ, Mṣb, Ḳ:) thus one says to one who has lost by death his father (Ṣ, Mṣb, Ḳ) or mother (Ḳ) or paternal uncle (Ṣ, Mṣb) or any other who cannot be replaced: (Mṣb, Ḳ:) and خَلَفَ اللّٰهُ عَلَيْكَ خَيْرًا, (Ḳ,) or بِخَيْرٍ, (AZ, Mṣb, Ḳ,) or both, (L,) and خَلَفَ اللّٰهُ لَكَ بِخَيْرٍ, (AZ, Mṣb,) and لَكَ خَيْرًا and أَخْلَفَ ↓ اللّٰهُ عَلَيْكَ خَيْرًا: (Ḳ: [in which it is implied that these phrases mean *May God supply to thee well the place of him whom thou hast lost:* but it is implied in the Mṣb that the two of them there mentioned mean *May God restore to thee good in the place of that which has gone from thee:* and it appears from what here follows that all of these phrases have the latter meaning, whether or not they have the former meaning also:]) to him who has lost property or a child or a thing [of any other kind] of which the replacement may be asked, (Ṣ,) or to him of whom that which may be replaced has perished, or died, (Ḳ,) one says, أَخْلَفَ ↓ اللّٰهُ عَلَيْكَ (Ṣ, Mṣb, Ḳ) *May God restore to thee the like of that which has gone from thee,* (Ṣ, Mṣb,) or *may God restore to thee what has gone from thee;* (Ḳ in a later part of the art.;) and اخلف ↓ الله لك; and خَلَفَ اللّٰهُ عَلَيْكَ : or خلف الله لك is allowable in relation to property and the like; and يَخْلُفُ, like يَمْنَعُ, is allowable as its aor., though extr., (Ḳ,) as it has no faucial letter to occasion the fet-ḥ: (TA:) and one says also, خَلَفَ اللّٰهُ لَكَ بِخَيْرٍ, meaning *May God give thee good in the place of that which has gone from thee;* (TA;) and اخلف ↓ عَلَيْكَ خَيْرًا, (Mṣb, TA,) meaning the same; (TA;) and [so] اخلف ↓ لَكَ خَيْرًا and بِخَيْرٍ: and لَكَ مَالَكَ and اخلف ↓ اللّٰهُ عَلَيْكَ مَالَكَ [*May God restore,* or *replace, to thee thy property*]. (Mṣb.) خَلَفَ أَبَاهُ, (Ḳ,) aor. ـُ, (TA,) means *He became behind his father;* (Ḳ;) and if so, its inf. n. is خَلْفٌ: (TA:) or it means *he became in the place of his father;* (Ḳ;) and if so, its inf. n. is خَلَفٌ: (TA:) and خَلَفَ مَكَانَ أَبِيهِ, inf. n. خِلَافَةٌ (Ḳ) and خَلَفٌ, (TA,) *he became in the place of his father, exclusively of every other.* (Ḳ.) You say also, خَلَفَتِ الفَاكِهَةُ بَعْضُهَا بَعْضًا, (JK, Ḳ,) inf. n. خَلْفٌ, (JK, TA,) or خَلَفٌ, (TA, [the former being there altered to the latter (which is the more probably correct), or the latter to the former,]) and خِلْفَةٌ, (JK, TA,) *The fruit replaced other fruit; or became substituted for other fruit.* (JK,* Ḳ,* TA. [In the CḲ, صَارَ خَلْفًا is erroneously put for صَارَتْ خَلَفًا.]) And خَلَفَ فُلَانٌ عَلَى فُلَانَةَ, inf. n. خِلَافَةٌ, [like عَقَبَ عَلَيْهَا,] *Such a man took as his wife such a woman after another husband* [*and thus supplied his place*]. (Z, TA.) — خَلَفَ فُلَانًا, (aor. ـُ, TA,) *He took,* or *seized, such a one from behind him;* (JK,* Ḳ;) as also ↓ اختلفهُ. (TA.) And hence, (TA,) خَلَفَ لَهُ بِالسَّيْفِ (JK, TA) *He came to him from behind him, and smote his neck,* or *struck off his head, with the sword.* (TA.) — خَلَفَ فُلَانٌ بِعَقِبِى [is explained as meaning] *Such a one stayed,* or *abode, after me.* (Mṣb in art. عقب.) [But] — خَلَفَ فُلَانٌ بِعَقِبِ فُلَانٍ is said by some to mean خَالَفَهُ ↓ إِلَى أَهْلِهِ [q. v.]: accord. to Aṣ, however, it means *He parted with such a one on the condition of doing a certain thing, and then came behind him* [or *behind his back*] *and did another thing after parting with him:* and Az says that this is a more correct explanation than the former one. (TA.) [Hence, app.,] one says also, إِنَّ ٱمْرَأَةَ فُلَانٍ تَخْلُفُ زَوْجَهَا بِالنِّزَاعِ إِلَى غَيْرِهِ إِذَا غَابَ عَنْهَا [*Verily the wife of such a one is unfaithful to her husband by yearning towards another when he is absent from her:* or *deceives her husband behind his back by yearning towards another;* for it is implied, by an ex. given, that اذا غاب عنها is added by way of explanation]. (TA.) خَلَفَهُ also signifies *He spoke of him,* or *mentioned him,* [*behind his back,* or] *when he was not present:* so in the phrase, خَلَفَهُ بِخَيْرٍ or بِشَرٍّ [*He spoke of him behind his back well* or *ill*]. (TA.) And one says, يَخْلُفُ النَّاسَ مِنْ وَرَائِهِمْ [meaning *He defames men behind their backs*]: the action signified hereby is like غِيبَة, and may be [*by making signs*] *with the side of the mouth,* and *with the eye,* and *with the head.* (TA in art. همز.) — خَلَفَ عَنْ أَصْحَابِهِ, (Ḳ, TA,) aor. ـُ, *He remained behind,* or *after, his companions; did not go forth with them;* as also قَعَدَ خِلَافَ ↓ أَصْحَابِهِ [similar to a phrase mentioned near the beginning of this

art.]; (TA;) i. q. ▼ تخلّف; (Ḳ in explanation of the former phrase;) which is syn. with تَأَخَّرَ (Ṣ, Ḳ;) as in the phrase تخلّف عَنِّى [which means *He remained behind me*, or *after me*]; (Ṣ;) [for] تخلّف عَنْهُ means بَقِىَ خَلْفَهُ; (Mgh;) and [in like manner] you say, تخلّف عَنِ القَوْمِ *He remained behind*, or *after*, *the people*, or *party*, *not going with them*; [*he held back*, or *hung back*, *from them*;] and ▼ قَعَدْتُ خِلَافَهُ [i. e. *I remained*] *behind him*, or *after him*; (Mṣb;) and ▼ خالف عَنَّا *He remained behind us*, or *after us*; syn. ▼ تخلّف. (TA, from a trad.) It is said in the Ḳur [xvii. 78], وَإِذًا لَا يَلْبَثُونَ خِلَافَكَ ▼ إِلَّا قَلِيلًا, i. e. [*But in that case they should not have remained*] *after thee* [*save a little while*]: (JK, TA:*) so accord. to one reading [instead of خَلْفَكَ, which means the same]. (TA.) And in like manner, فَرِحَ المُخَلَّفُونَ بِمَقْعَدِهِمْ خِلَافَ ▼ رَسُولِ ٱللّٰهِ, in the Ḳur [ix. 82], means [*Those who were left behind rejoiced in their remaining*] *behind the Apostle of God*: (Ṣ, TA:) or the meaning here is, مُخَالَفَةَ رسول الله [i. e. *in disagreement with the Apostle of God*]: (JK, Ṣ:) thus says Lḥ; but IB disagrees with him; saying that خلاف here means بَعْدَ; and cites six exs. in which it has this meaning, from poets. (TA.) — [Hence,] خَلَفَ فُلَانٌ عَنْ كُلِّ خَيْرٍ † [*Such a one was*, or *became*, *kept back from all good*; i. e.,] *did not prosper*, or *was not successful*. (TA. [It is there added, that it is explained in the A as tropical, and as meaning تَغَيَّرَ وَفَسَدَ: but this is perhaps a mistake, occasioned by the accidental omission of وَخَلَفَ اللَّبَنُ or the like, of which this is a correct explanation: or the phrase thus explained in the TA, as from the A, may correctly mean *He became altered for the worse, and corrupt*; agreeably with other explanations of the verb below.]) — خَلَفَ, aor. ـُ, also signifies *He* (a man) *retired*, *withdrew*, or *went away* or *apart*. (JK.) And خَلَفَتْ نَفْسُهُ عَنِ الطَّعَامِ, aor. ـُ, inf. n. خُلُوفٌ, † *His soul turned away from*, *avoided*, or *shunned*, *the food*, *in consequence of disease*. (JK, TA.) — And *He fled*. (Ḥam p. 411.) — And *He* (a man, Ṣgh) *ascended a mountain*. (Ṣgh, Ḳ.) = See also 2, first sentence. = خَلَفَ, (Ṣ, Mṣb, Ḳ, &c.,) aor. ـُ, (Lḥ, Mṣb, TA, and Ḥam p. 670,) [inf. n. خُلُوفٌ,] said of the taste of water, *It was*, or *became*, *different from*, or *contrary to*, *what it was thought to be*: and [hence,] *it was*, or *became*, *altered* [*for the worse*]: (Ḥam ubi suprà:) [and] said of milk, (Ṣ, Ḳ,) and of food, (Lḥ, Ṣ, Mṣb, Ḳ,) and the like, (Lḥ, TA,) and some say خَلُفَ, (TA,) aor. ـُ, inf. n. خُلُوفٌ, (Lḥ, TA,) of both verbs, (TA,) *it was*, or *became*, *altered* [*for the worse*] (Lḥ, Ṣ, Mṣb, Ḳ) *in taste*, or *in odour*; (Ṣ, Mṣb, Ḳ;) as also ▼ اخلف: or, said of milk, the first signifies *it became bad from being long kept*; or, as in the A, ‡ *what was good thereof became mixed* (خَلَفَ, i. e. خَلَطَ,) *with other milk*: and ▼ اخلف, said of milk, signifies also *it was*, or *became*, *sour*: (TA:) and the first, said of [the beverage called] نَبِيذ, *it became bad*. (Ḳ.) Also, inf. n. خُلُوفٌ (Ṣ, Mgh, Mṣb, Ḳ) and خُلُوفَةٌ (Ḳ) and خِلْفَةٌ, (L, TA,) said of the mouth (Ṣ, Mgh, Mṣb, Ḳ) of a person fasting, (Ṣ, Mṣb, K,) *It was*, or *became*, *altered* [*for the worse*] *in odour*; (Ṣ, Mgh, Mṣb, Ḳ;) as also ▼ اخلف. (Ṣ, Mṣb, Ḳ.) It is said in a trad. that the خُلُوف of the mouth of one who is fasting is sweeter in the estimation of God than the odour of musk: or, accord. to some of the lawyers and of the relaters of traditions, خَلُوف; but [SM says,] I think this to be a mistake, as several affirm it to be, while others say that it is of a bad dial.: accord. to one reading, it is خِلْفَة. (TA.) — [Hence,] خَلَفَ عَنْ خُلُقِ أَبِيهِ, (Ḳ,) aor. ـُ, inf. n. خُلُوفٌ, (TA,) † *He was*, or *became*, *altered* [*for the worse*] *from the natural disposition of his father*. (Ḳ, TA.) — And خَلَفَ, (ISk, Ṣ, Ḳ,) inf. n. خَلْفٌ [or خُلُوفٌ]; or خَلُفَ, aor. ـُ, inf. n. خَلَافَةٌ and خُلُوفٌ; (TA;) † *He* (a man) *was*, or *became*, *bad*, or *corrupt*. (ISk, Ṣ, Ḳ, TA.) — And خَلُفَ, (Ḳ,) inf. n. خَلَافَةٌ (IAth, Ḳ) and خُلُوفٌ, (Ḳ,) † *He* (a man) *was*, or *became*, *stupid*, or *foolish*; or *one who had little*, or *no*, *intellect* or *understanding*: (Ḳ,* TA:) or *unprofitable*: or *a frequent promise-breaker*: (IAth:) or خَلَفَ and ▼ اخلف, said of a slave, *he was*, or *became*, *idiotic*, *deficient in intellect*, or *bereft of his intellect*. (JK.) = خَلَفَ, (Ḳ,) inf. n. خَلْفٌ; (Ṣ,* Ḳ,* TA;) and ▼ اخلف, and ▼ استخلف; (Ṣ, Ḳ;) *He drew water*, (Ṣ, Ḳ, TA,) لِأَهْلِهِ *for his family*: [app. because he who does so leaves his family behind him: see 2, first sentence:] (Ḳ, TA:) [or] ▼ استخلف, said of a man, signifies ٱسْتَعْذَبَ المَآءَ [app. as meaning *he sought*, or *drew*, or *brought*, *sweet water*: see art. عذب]: and, accord. to IAạr, you say, أَخْلَفْتُ ▼ القَوْمَ, meaning properly *I carried sweet water to the people*, or *party*, *when they were in the* [season, or herbage, called] رَبِيع *and without sweet water*, or *when they were by salt water*: إِخْلَافٌ [as meaning the *carrying*, or *drawing*, *of water*,] being [properly] only in the رَبِيع: in other cases, metaphorically applied. (TA.) El-Ḥoṭeiah says,

* لِزُغْبٍ كَأَوْلَادِ القَطَا رَاثَ خَلْفُهَا ▼ *
* عَلَى عَاجِزَاتِ النَّهْضِ حُمْرٍ حَوَاصِلُهْ *

† [*To*, or *for*, *downy ones*, *like the young ones of the ḳaṭà*, *whose procurer of water has been slow in coming to those lacking the power of spreading their wings for flight*, *red in their crops*]: he means ▼ مُخْلِفُهَا [or خَالِفُهَا], and has put in the place of this the inf. n.: and by حواصله, accord. to Ks, he means حَوَاصِلُ مَا ذَكَرْنَا [*the crops of what we have mentioned*]: but Fr says that the ه relates to the زغب, exclusively of the عاجزات, which [latter] has the sign of the pl.; for every pl. that has the form of a sing. may be imagined to be a sing., as in the saying of the poet,

* مِثْلُ الفِرَاخِ نُتِفَتْ حَوَاصِلُهْ *

[meaning "like the young birds of which the crops have been plucked"]; for الفراخ has not the sign of the pl., but has the form of a sing., like الكِتَاب and الحِجَاب: another says, [but this is very far-fetched,] that the ه relates to النهض, which [sometimes] means a place in the shoulder-blade of the camel; and that the poet has used it metaphorically as belonging to the قطا. (Ṣ.) = خَلَفَ الثَّوْبَ, (Ṣ,) or القَمِيصَ, aor. ـُ, (Mṣb, Ḳ,) inf. n. خَلْفٌ (Kr, TA) and خِلْفَةٌ, in some copies of the Ḳ خَلْفَةٌ, [so in my MS. copy of the Ḳ, and so in the TḲ,] and [in some] خُلْفٌ also, but these require consideration, (TA,) *He took out from the garment*, or *shirt*, *the part that was worn out*, (Ṣ, Mṣb, Ḳ,) that is, *the middle part*, *which was worn out*, (Ṣ, Mṣb,) *and then sewed the* [cut] *edges together*. (Ṣ, Mṣb, Ḳ.) And ▼ اخلف الثَّوْبَ signifies the same as خَلَفَهُ, i. e. *He repaired the garment* [app. in any manner, or, as is implied in the Ṣ and TA, *by substituting one piece for another*]. (Ṣ, Ḳ, TA.) — The saying, in a trad. of Ḥamneh, فَإِذَا خَلَفَتْ ذٰلِكَ فَلْتَغْتَسِلْ, meaning † *And when she has discriminated that* period of days and nights during which she has been حَائِض, [*she shall perform a complete ablution of herself*,] is from خَلَفَ القَمِيصَ signifying as explained above. (Mṣb.) = خَلَفَ signifies also *He mixed* [a thing with another thing; as, for instance, (see خَلَفَ in a passage above,) milk with other milk]: and *he mixed* saffron, and medicine, *with water*. (TA.) = خَلَفَ بَيْتَهُ *He put to*, or *made for*, *his tent*, *a pole*, (Ḳ, TA,) *termed a* خَالِفَة, (TA,) *in the hinder part thereof*. (Ḳ, TA.) = خَلِفَ, aor. ـَ, (Ḳ,) inf. n. خَلَفٌ, (Ṣ, Ḳ,) *He* (a camel) *inclined towards one side*. (Ṣ, Ḳ.) — خَلَفٌ is also an inf. n. (of خَلِفَ, said of a man, TḲ) signifying The *being* أَخْلَف as meaning *contrarious*, *hard in disposition*, *as though going with a leaning towards one side*: — and the *being left-handed*: — and the *being* أَحْوَل [or *squint-eyed*]. (Ḳ.) = خَلِفَتْ, aor. ـَ, (Mṣb, Ḳ,) inf. n. خَلَفٌ, (Mṣb, TA,) *She* (a camel) *was*, or *became*, *pregnant*. (Mṣb, Ḳ.)

2. خلّفهُ, (Mṣb,) and خلّفهُ وَرَآءَهُ, (Ṣ, TA,) inf. n. تَخْلِيفٌ, (TA,) *He left him behind him*; (Mṣb;) namely, a man: (Ṣ, Mṣb, TA:) and ▼ خَلَفَهُ [signifies the same: or] *he made him to be behind him*; as also ▼ اخلفهُ [q. v.], and ▼ اختلفهُ: (TA:) [whence the saying,] أَلْحَحْتُ عَلَى فُلَانٍ فِى الإِتْبَاعِ حَتَّى ٱخْتَلَفْتُهُ ▼, i. e. [*I pressed upon such a one in following*] *until I made him to be behind me*. (ISk, TA.) You say also, خلّفهُمْ, inf. n. as above, meaning *He was*, or *became*, or *went*, *before them*; *and left them behind him*. (TA.) And خلّفوا أَثْقَالَهُمْ, inf. n. as above, *They left their loads*, or *baggage &c.*, *behind their backs*; (O, Ḳ;) when they went away to draw water. (TA.) — [Hence,] خلّفهُ *He made him*, or *appointed him*, *his* خَلِيفَة [i. e. *successor*, or *vice-agent*, &c.]; (Ḳ;) and so ▼ استخلفهُ. (Ṣ, Mṣb, Ḳ.) So in the Ḳur [xxiv. 54], ▼ لَيَسْتَخْلِفَنَّهُمْ فِى الأَرْضِ كَمَا ٱسْتَخْلَفَ ٱلَّذِينَ مِنْ قَبْلِهِمْ [*That He will assuredly make them to be successors in the earth*, *like as He made to be successors those who were before them*]. (TA.) = خلّف بِنَاقَتِهِ, (Ṣ, Ḳ,) inf. n. as above, (Ṣ,) *He bound one of the teats of his she-camel with the thing termed* صِرَار [in order that her young one might not suck it]: (Ṣ, Ḳ:) from Yaạḳoob. (Ṣ.)

3. خَالَفَهُ, (Mgh, Mṣb, TA,) inf. n. خِلَافٌ (Ṣ, Mgh, Mṣb, Ḳ, TA) and مُخَالَفَةٌ, (Ṣ, Mṣb, Ḳ, TA,) *He* [or *it*] *disagreed with*, or *differed from, him* [or *it*]; or *he dissented from him;* (Mgh, Mṣb;) *contr. of* وَافَقَهُ فِى كَذَا [*in*, or *in respect of, such a thing*]: (Mgh:) and *he*, or *it, was contrary, opposed*, or *repugnant, to him*, or *it:* (TA:) [and *he acted contrarily, contrariously, adversely*, or *in opposition, to him*, or *it; he*, or *it, contravened*, or *opposed, him*, or *it:*] and *he* [or *it*] *contradicted him* [or *it*]. (M in art. نقض.) It is said in a prov., إِنَّمَا أَنْتَ خِلَافُ الضَّبُعِ الرَّاكِبَ, i. e. تُخَالِفُ خِلَافَ الضَّبُعِ [*Verily thou art one who acts with the contrariousness of the hyena towards the rider*]: for the hyena [attacks a man on foot, but], when it sees the rider, flees from him. (IAạr, TA.) You say also, خَالَفَنِى عَنْ كَذَا *He turned away from such a thing* [*in opposition to me*, or] *when I betook myself to it:* [see also the last sentence but one of the first paragraph of art. بهت:] and خالفنى إِلَى كَذَا *He betook himself to such a thing* [*in opposition to me*, or] *when I turned away from it:* (Mgh:) or خالفهُ إِلَى الشَّىْءِ means *he disobeyed him by betaking himself to the thing; or betook himself to the thing after he had forbidden him it.* (TA.) And hence, (Mgh,) هُوَ يُخَالِفُ إِلَى آمْرَأَةِ فُلَانٍ, (Ṣ, Mgh,*) or إِلَى فُلَانَةَ, (O, L, TA,) in the Ḳ, erroneously, هو يخالف فلانة, (TA,) *He comes to the wife of such a one when he* [the latter] *is absent from her*, (Ṣ,) or *to such a woman when her husband is absent from her:* (Mgh,* O, L, Ḳ, TA:) and خالفهُ إِلَى أَهْلِهِ [*he came to his* (another's) *wife in his* (the husband's) *absence*]: see 1, in the former half of the paragraph. (Az, TA.) And خالفهُ إِلَيْهِمْ *He watched to see him, and, when he was absent from them*, namely, his family, *he went in to them:* (Jm, O, TA:) and, accord. to AZ, اخلف ♦ فُلَانٌ صَاحِبَهُ *Such a one watched to see his companion, and, when he was absent, he came, and went in to him* [or rather *to his wife* or *to his family*]: (TA:) [or] اختلف ♦ صَاحِبَهُ *he watched to see his companion, and, when he was absent, went in to his wife:* (Ḳ, and the like is said in the JK:) thus says IDrd, on the authority of AZ. (TA.) And خالف إِلَى قَوْمٍ *He came to a party, or company of men, from behind them* [or *behind their backs*]: or *he feigned to them the contrary of that which he conceived in his mind, and took them unawares.* (TA.) Aboo-Dhu-eyb says, [describing a collector of wild honey,]

* إِذَا لَسَعَتْهُ النَّحْلُ لَمْ يَرْجُ لَسْعَهَا *

* وَخَالَفَهَا فِى بَيْتِ نُوبٍ عَوَاسِلِ *

(Ṣ in the present art., in which only the former hemistich is cited, and in art. رجو,) i. e. [*When the bees sting him,*] *he fears not nor minds* [*their stinging*], (Ṣ in art. رجو,) [*but comes, during their absence, to the hiring-place of bees occupied in gathering honey:*] meaning, he comes to their honey, (Ṣ, TA, [in the latter of which, in the place of النحل, is put الدَّبْرُ "the swarm of bees,"]) and takes it, (TA,) while they are feeding; (Ṣ, TA;) or, as AA says, he comes behind them to the honey while they are absent: AO explains it by خَالَفَهَا إِلَى مَوْضِعٍ آخَرَ which [he says] means *he keeps with them* [*to another place*]; syn. لَازَمَهَا; [and thus this phrase (which is strangely misinterpreted in the TḲ and in Freytag's Lexicon) is explained in the Ḳ, but without any reference to the verse;] as also حَالَفَهَا, with the unpointed ح: (TA:) and some read the verse thus; but this reading is said to be a mistake. (TA in art. حلف.) — جَآءَ خِلَافَهُ: see 1, near the beginning of the paragraph. And see also five other exs. in the middle portion of the same paragraph. — خالف بَيْنَ رِجْلَيْهِ *He put one of his legs forward and the other backward:* and [hence,] الْمُخَالَفَةُ بَيْنَ الرِّجْلَيْنِ [as signifying *the alternate shifting of the legs to and fro*] is metonymically used as meaning *the act of dancing.* (Ḥar p. 108.) [And خالف بَيْنَ الشَّيْئَيْنِ *He put*, or *placed, the two things contrariwise; or on contrary sides;* or *in contrary directions.* Hence,] أَوْ تُقَطَّعَ أَيْدِيهِمْ وَأَرْجُلُهُمْ مِنْ خِلَافٍ, in the Ḳur v. 37, [*Or that their hands and their feet shall be cut off on contrary sides,*] means that their right hands and left feet shall be cut off. (Bḍ, Jel. [See also similar exs. in the Ḳur vii. 121 and xx. 74 and xxvi. 49.]) [Hence also,] فَرَسٌ بِهِ شِكَالٌ مِنْ خِلَافٍ, (JK,) or ذُو شِكَالٍ مِنْ خِلَافٍ, (TA,) *A horse having a whiteness in his right fore leg and his left hind leg* [or *the reverse*]: (JK, TA:) and some say, لَهُ خَدَمَتَانِ مِنْ خِلَافٍ when *he has a whiteness* [or rather *a ring of white a little above the hoof*] *in his fore leg* [or *right fore leg*] *and another in his left fore leg* [probably a mistake of a copyist for his *left hind leg*]. (TA.)

4. اخلفهُ: see 2, first sentence. Also *He put him, turned him*, or *made him to go back* or *stand back, behind him.* (Ḳ, TA.) And اخلف يَدَهُ *He put his hand behind him.* (Az, TA.) And also, (Fr, TA,) or اخلف يَدَهُ إِلَى السَّيْفِ, (JK,) or simply اخلف [used elliptically], (Ṣ, Ḳ,) *He put* [*back*] *his hand to his sword,* (Fr, Ṣ, Ḳ, TA,) in order to draw it, (JK, Ṣ, Ḳ, TA,) *it being hung behind him.* (Fr,* TA.) And اخلف السَّيْفَ [*He hung the sword behind him;* or *kept it hung behind him*]: said, in a trad., of a man on the day of Bedr. (TA.) And اخلف عَنِ البَعِيرِ [for اخلف عَنْهُ الحَقَبَ] *He shifted* [*backwards*] *the hind girth of the camel, putting it next to his testicles,* on account of its hurting the sheath of his penis, and causing a suppression of his urine; (Aṣ, Ṣ, Ḳ;) as also اخلف البَعِيرَ: (TA:) or you say only, أَخْلِفِ الحَقَبَ, meaning *remove thou the hind girth from the sheath of the penis.* (Lḥ, TA.) And اخلف الدَّابَّةَ بِالسَّوْطِ *He struck the beast on the hinder part with the whip.* (JK.) — اخلف البَازِلَ [*He* (a camel) *exceeded in age the* بازل, which is generally *one that has entered his ninth year:* as though he made the بازل to be behind him: and so, app., اخلف alone; البازل being understood: see مُخْلِف]. El-Jaadee says,

* أَيِّدِ الكَاهِلِ جَلْدٍ بَازِلٍ * أَخْلَفَ البَازِلَ عَامًا أَوْ بَزَلْ *

[*Strong in the withers, hardy, a* بازل; *that has exceeded in age him who has just become a* بازل *by a year, or that has* himself *just become a* بازل]. (Ṣ, TA.) Some say that الإِخْلَافُ is [a term denoting] the last of the ages [that have words to signify them] with respect to all beasts. (TA.) — اخلف فُلَانٌ صَاحِبَهُ: see 3, near the middle of the paragraph. — اخلفهُ مَا وَعَدَهُ, (Ṣ,) or وَعْدَهُ, (Mṣb,) or مَوْعِدَهُ, (Mgh,) or الوَعْدَ, (Ḳ,) inf. n. إِخْلَافٌ, (Mgh,) *He broke,* (Mgh,) or *failed to perform,* (Ṣ, Ḳ,) *his promise,* or *the promise, to him:* (Ṣ, Mgh, Ḳ:) restricted to future time: (Mṣb:) الإِخْلَافُ is, *in respect of the future, like* الكَذِبُ *in respect of the past:* (Ṣ, Ḳ:) or *the making a promise and not fulfilling it:* (Lḥ, Ḳ:) and some say that it signifies one's *seeking an object of want,* or *water, and not finding it.* (TA.) It is said in a trad., إِذَا وَعَدَ أَخْلَفَ, i. e. *When he promises, he does not fulfil his promise, and is not true* [*to it*]. (TA.) [Hence,] أَخْلَفَتِ النُّجُومُ, i. e. ‡[*The stars broke their promise;* meaning,] *were attended with drought, not attended with rain:* (Ṣ, Ḳ, TA:) a saying of the people in the Time of Ignorance: (Ṣ, TA:) and so اختلفت ♦ عَنْ أَنْوَائِهَا: for they used to believe and say that they were rained upon by such and such a نَوْء. (TA. [See نَوْء.]) Hence also, أَخْلَفَتِ الحُمَّى † *The fever,* being tertian or quartan, *came not in its time,* or *turn.* (Mgh.) And أَخْلَفَتْ said of a she-camel, † *She, having been covered by the stallion, did not become pregnant:* (JK, TA:) and † *she proved to be not pregnant when thought to be pregnant.* (JK.) And in like manner said of a palm-tree; (JK;) ‡ *It bore not one year:* and ‡ *it* (a tree) *bore no fruit:* or *lost the fruit that it had.* (L, TA. [The verb, said of trees, has also another meaning, which see below.]) — اخلفهُ is also said, by El-Fárábee, to occur as meaning *He acted according to his promise* [or *fulfilled his promise*] *to him;* thus bearing two contr. significations: but this is strange. (MF.) — Also *He found him to be a breaker of his promise;* (JK;) or *he found his promise to be broken,* or *unfulfilled.* (Ṣ, Ḳ.) — اخلف عَلَيْكَ and لَكَ, each with an objective complement (مَالَكَ or خَيْرًا) expressed or understood: see 1, in six places, in the former half of the paragraph. You say also, اخلف فُلَانٌ لِنَفْسِهِ, (Ṣ, Ḳ,) or لِغَيْرِهِ, (TA,) *Such a one replaced to himself,* (Ṣ, Ḳ,) or *to another,* (TA,) *a thing that had gone from him, with another thing.* (Ṣ, Ḳ.) Ibn-Muḳbil says,

* فَأَخْلِفْ وَأَتْلِفْ إِنَّمَا المَالُ عَارَةٌ *

* وَكُلْهُ مَعَ الدَّهْرِ الَّذِى هُوَ آكِلُهْ *

[*Then replace thou, and consume: wealth is but a loan: and devour it with time, which is a devourer thereof*]: he means, gain a substitute for what thou hast consumed. (Ṣ, TA.) And the Arabs say to him who has put on a new garment, أَبْلِ وَأَخْلِفْ وَآحْمَدِ الكَاسِى [*Wear out thy garment, and replace it with another, and praise the Clother,* meaning God]. (TA.) And

أَبْلِ وَيُخْلِفُ ٱللّٰهُ [*Wear out* thy garment, *and God will replace* it *with another; or, may God replace &c.*]. (Ṣ in art. بلو.) — See also اخلف الثَّوْبَ near the end of the first paragraph. — اخلف said of a plant, or of herbage, *It put forth the* خِلْفَة, (Ṣ, Mṣb, Ḳ,) meaning *leaves that come forth after the first leaves, in the* [*season called*] صَيْف; (TA;) and in like manner said of trees: (Mṣb, TA:) or اخلف الشَّجَرُ means *the trees put forth fruit after other fruit.* (JK.) And, said of fruit, *It came forth, some thereof after other thereof.* (TA.) And اخلفت الأَرْضُ *The land became affected by the cold of the latter part of the* [*season called*] صَيْف, *and some of its trees consequently become green.* (TA.) — Also, said of a bird, ‡ *It put forth feathers after the first feathers:* (Ḳ, TA:) from the same verb said of a plant, or of herbage. (TA.) — And, said of a boy, † *He nearly attained to puberty.* (JK, Az, Ḳ, TA.) — And, said of a solid-hoofed beast, † *He completed a year after the* قُرُوح [or *finishing teething,* or *shedding the corner-nipper*]. (JK.) = اخلفه said of medicine, *It weakened him* (Ḳ, TA) *by causing him to go frequently to the privy.* (TA.) — And الإِخْلَافُ also signifies *The bringing the stallion again to the she-camel when she has not conceived at once.* (Ḳ.) = See also 1, in six places, in the latter half of the paragraph.

5: see 1, in two places, in the middle of the paragraph. [Hence, تخلّف عَنِ الأَمْرِ *He held back from,* or *fell short of, doing the thing.*]

6: see the next paragraph, in three places.

8. اِخْتِلَافٌ signifies The *following reciprocally;* or *alternating.* (Mgh.) So in the phrase in the Ḳur [ii. 159 and iii. 187 and xlv. 4], وَٱخْتِلَافِ ٱللَّيْلِ وَٱلنَّهَارِ *And the alternating of the night and the day.* (Mgh.) [And in a verse of El-'Ajjáj cited voce أَبْلَى, in art. بلو.] And hence the phrase, اِخْتَلَفَا ضَرْبَةً *Each of them beat,* or *struck, the other in turn.* (Mgh.) And the saying, in a trad. of 'Alee, فَٱخْتَلَفَتْ بَيْنَ عُبَيْدَةَ بْنِ ٱلْحٰرِثِ وَٱلْوَلِيدِ بْنِ عُقْبَةَ ضَرْبَتَانِ [*And two blows were interchanged between 'Obeydeh Ibn-El-Ḥárith and El-Weleed Ibn-'Okbeh*]. (Mgh.) And the saying, in a trad. of Umm-Ṣabeeyeh, اِخْتَلَفَتْ يَدِى وَيَدُ رَسُولِ ٱللّٰهِ فِى إِنَآءٍ وَاحِدٍ, meaning *My hand and the hand of the Apostle of God were both put* [*by turns*] *into one vessel.* (Mgh.) And اِخْتَلَفُوا signifies *They followed,* or *succeeded, one another; whenever one went, another coming after him.* (TA in art. عور.) — Also The *going,* or *moving, repeatedly, to and fro; so coming and going;* or *reciprocating;* syn. تَرَدُّدٌ [in this sense, as is shown in this art. in the Ḳ and TA, and in the Ṣ and Ḳ in art. ردد, &c.: and also as meaning the *returning,* or *repairing, time after time,* or *repeatedly,* or *frequently,* to a person or place; because it implies coming and going: and sometimes it means simply the *returning;* because this cannot be without a previous going]. (Ḳ.) You say, هُوَ يَخْتَلِفُ إِلَى فُلَانٍ, i. e. يَتَرَدَّدُ [*He returns,* or *repairs, time after time, repeatedly,* or *frequently, to such a one*]: and اِخْتَلَفَ إِلَيْهِ ٱخْتِلَافَةً وَاحِدَةً [*He returned to him once*]. (TA.) And هُوَ يَخْتَلِفُ إِلَى مَجَالِسِ ٱلْعِلْمِ *He repairs frequently to,* or *frequents, the assemblies of science;* syn. يَتَرَدَّدُ. (A in art. ردد.) And اِخْتَلَفَ إِلَى ٱلْمُتَوَضَّا [*He returned,* or *repaired, time after time,* &c., *to the privy*]. (Ṣ.) And اِخْتَلَفَ إِلَى ٱلْخَلَآءِ [properly signifies the same: and hence, † *He had a looseness of the bowels,* or *a diarrhœa*]. (Ḳ.) And [perhaps as implying coming and going,] اختلف also signifies *He supplied,* or *gave,* or *offered, water.* (TA.) — [Also The *disagreeing, differing,* or *varying, in state* or *condition* or *quality* &c.; *being dissimilar, different, diverse, various, incongruous, discordant,* or *dissentient:*] اختلاف is the contr. of اِتِّفَاقٌ; (Ḳ, TA;) and is said of anything that is dissimilar [in the parts or members &c. of which it is composed]; as also ▼ تخالف. (TA.) You say, ▼ تخالف الأَمْرَانِ [and اختلف الامران], i. e. لَمْ يَتَّفِقَا [*The two things,* or *affairs,* or *cases, were,* or *became, dissimilar,* &c.]. (TA.) And اختلفوا and ▼ تخالفوا (Mgh, Mṣb) [*They disagreed,* &c., فِى أَمْرٍ *in a thing* or *an affair* or *a case;*] *every one of them took to,* or *held, a way,* or *an opinion, different from,* or *contrary to, that of another:* (Mṣb:) both signify the same. (Mgh.) It is said in a trad., سَوُّوا صُفُوفَكُمْ وَلَا تَخْتَلِفُوا فَتَخْتَلِفَ قُلُوبُكُمْ [*Make ye your ranks even* when ye place yourselves to pray together, *and be not dissimilar in your positions, for in that case your hearts would disagree*]; meaning, when one of you advances, or stands, before another in the ranks, your hearts will be affected, and disagreement in respect of friendship and amity will arise among you: or, as some say, it means, *your hearts will be made to recoil:* or *the* صُورَة [or *specific character*] *of your hearts will become changed into another* صورة. (TA.) [Hence,] اِخْتَلَفَتْ عَنْ أَنْوَائِهَا, said of stars: see 4, near the middle of the paragraph. — Also The *being complicated, intricate,* or *confused.* (KL.) [You say, اختلف الأَمْرُ بَيْنَهُمْ *The affair,* or *case, was,* or *became, complicated, intricate,* or *confused, so as to be a subject of disagreement,* or *difference, between them:* a phrase of frequent occurrence.] = اختلفهُ: see 1, in two places, in the former half of the paragraph. — See also 2, in two places. — اختلف صَاحِبَهُ: see 3, near the middle of the paragraph.

10. استخلفهُ: see 2, in two places. — Also *He took it* (a thing) *as a substitute,* or *in exchange, for another thing;* or *in the place of another thing;* syn. اِسْتَعْوَضَهُ and اِسْتَبْدَلَهُ. (TA.) — استخلفت الأَرْضُ *The land produced the herbage of the* [*season called*] صَيْف. (TA.) = See also 1, in the middle of the latter half of the paragraph, in two places.

خَلْفٌ [meaning The *location,* or *quarter, that is behind;* and the *time past;*] (Ḳ; [so in my MS. copy, and thus it should be written as a simple noun; but in the CK خَلْفَ;]) or ٱلْخَلْفُ; (Lth, Ḳ;) *contr. of* قُدَّامٌ [or ٱلْقُدَّامُ]: (Lth, Ḳ:) [and] خَلْفَ [*Behind;* and *after;*] contr. of قُدَّامَ: (Ṣ: [thus in my two copies; and said in the margin of one of them to be thus in the copy of IB, and in that of El-Jawáleekee:]) a simple noun: and an adv. n.: of the fem. gender [as meaning the جِهَة; but otherwise it seems to be masc.]. (TA.) You say, جَآءَ خَلْفَهُ [and مِنْ خَلْفِهِ, both meaning *He came behind him,* and *after him*]. (Mgh.) And جَلَسْتُ خَلْفَ فُلَانٍ *I sat after,* or *behind, such a one;* syn. بَعْدَهُ. (Ṣ.) And بَقِيتُ خَلْفَهُ *He remained after him.* (Ḳ.) Some read, in the Ḳur [xvii. 78], وَإِذًا لَا يَلْبَثُونَ خَلْفَكَ: others read خِلَافَكَ [which means the same, as mentioned above: see the middle of the first paragraph of this art.]. (TA.) — خَلْفٌ signifies also The *back* (Ḳ, TA) itself: so says IAạr: and particularly, of a house; the *side corresponding to,* or *over against, that in which is the door;* and as a house may have two doors, [in two different sides,] it may be said to have two backs, each of which may be thus termed; and the dual of this word seems to be used as meaning two backs in a trad. [respecting the building of the Kaạbeh]. (TA.) — And *One who comes after another;* (Ṣ, TA;) as also ▼ خَلَفٌ, or, accord. to some, there is a difference between these two, as will be shown in what follows; (Ṣ;) and ▼ خَالِفٌ and ▼ خَالِفَةٌ: it is originally an inf. n.: (TA:) and signifies *one who remains after another, whether this other be dead or living:* and *one remaining after another who is dead; his follower,* or *successor;* the *follower,* or *successor, of one who has gone:* used in praise and in dispraise: pl. خُلُوفٌ: and the sing. also signifies [like the pl.] *persons remaining after others;* accord. to some: (IB, TA:) *a remnant of people:* (Lḥ, TA:) and *a generation after a generation;* (Lth, Ṣ, Ḳ;) as also ▼ خَلَفٌ: (Lth, TA:) but Lth says that the former is applied to the evil, and ▼ the latter to the good, (Ḳ, TA,) whether meaning a generation or a son: (TA:) the latter means *a good son* (Ḳ, TA) *remaining after his father:* (TA:) and the former, *a bad son:* (Ḳ, TA:) [therefore] one says, هُوَ خَلْفُ سَوْءٍ مِنْ أَبِيهِ [*He is a bad son*] *who has taken the place of his father,* and ▼ خَلَفُ صِدْقٍ من ابيه [*a good son*] &c.: (Ṣ:) but sometimes each is used in the place of the other; so that one says, هو خَلْفُ صِدْقٍ من ابيه: (Ḳ:) or both signify the same: (Ṣ, Ḳ:) so says Akh: some, he says, use the former; and some, the latter: but some say ▼ خَلَفُ صِدْقٍ and خَلْفُ سَوْءٍ, meaning thus to distinguish between them: (Ṣ:) accord. to IB, ▼ خَلَفٌ correctly signifies a man's *successor who is a substitute* for him, good and bad: and is originally an inf. n.: (TA:) and the pl. of this is أَخْلَافٌ: (AZ, IB, TA:) accord. to IAth, خَلْفُ صِدْقٍ means *a good generation:* and خَلْفُ سَوْءٍ, *an evil generation:* (TA:) and خَلْفٌ likewise signifies *progeny* [without restriction]. (Ḳ.) One says also, (Ṣ, Ḳ,) of a people following people more in number than they, (Ṣ,) هٰؤُلَآءِ خَلْفُ سَوْءٍ [*These are a bad generation.* (Ṣ, Ḳ.) And بَقِينَا فِى خَلْفِ سَوْءٍ *We remained among an evil remnant.* (Lḥ, TA.) And فَخَلَفَ مِنْ بَعْدِهِمْ خَلْفٌ, in the Ḳur [vii. 168

and xix. 60], is explained as meaning *And there remained after them a remnant.* (TA.) —— [Hence,] ‡ *One in whom is no good.* (IB, Ḳ.) [And app. also *Persons in whom is no good.*] —— And ‡ *A thing in which is no good:* (IB, TA:) [and particularly] † *a bad saying;* (ISk, Ṣ, Mṣb, Ḳ;) *a wrong, bad saying, like the* خَلْف *of mankind.* (A 'Obeyd, Mṣb.) See also خُلْف. It is said in a prov., سَكَتَ أَلْفًا وَنَطَقَ خَلْفًا † *He held his tongue from a thousand* words (سَكَتَ عَنْ أَلْفِ كَلِمَةٍ), *and then uttered what was wrong.* (ISk, Ṣ, Mṣb.) An Arab of the desert, who had been guilty of a breach of manners (حَبْقَة), pointed with his thumb towards his اِسْت, and said, إِنَّهَا خَلْفٌ نَطَقَتْ خَلْفًا [which may be rendered, *Verily it is a thing in which is no good: it uttered a thing in which was no good:* but it obviously admits of being rendered otherwise]. (IAạr, Ṣ.) —— Also *People who have gone away from the tribe* (T, Ḳ) *to draw water, and have left their baggage &c. behind them:* (T, TA:) and *such as are present,* [*remaining behind,*] *of the tribe:* thus bearing two contr. significations: pl. خُلُوف. (Ḳ.) You say حَىٌّ خُلُوفٌ *A tribe who are absent; none of them remaining behind:* (Ṣ, TA:) or *a tribe of which the men are absent and the women remaining:* (TA:) and خُلُوف also signifies the contr., i. e. *such as are present,* (Ṣ, TA,) *remaining behind.* (Ṣ.) It is said of Moḥammad, in a trad., لَمْ يَتْرُكْ أَهْلَهُ خُلُوفًا, i. e. *He did not leave his family neglected, with no pastor nor protector.* (TA.) See also a verse of El-Ḥoṭeäh in the latter part of the first paragraph of this article. ═ *Old and worn out;* applied to a وَطْب [or skin for milk, or for clarified butter and milk: as though it were a remnant thereof]. (Ibn-'Abbád, Ḳ.) ═ *A* مِرْبَد; (Ṣ, Ḳ;) i. e. *a place of confinement for camels:* (TA:) or *such as is behind the tent* or *house.* (JK, Ṣ,* Ḳ.) ═ *A large* فَأْس [i. e. *hoe* or *adze* or *axe*]: or *such as has one head:* and the *edge of a* فأس: or the *head thereof:* (Ḳ:) you say فَأْسٌ ذَاتُ خَلْفَيْنِ *a two-headed* فأس: (Ṣ, TA:) or ذَاتُ خَلْفَيْنِ and ↓ ذَاتُ خِلْفَيْنِ are names of *the* فأس (Ḳ, TA) *when two-headed:* (TA:) and the pl. is ذَوَاتُ الخَلْفَيْنِ: (Ḳ:) pl. خُلُوف. (JK.) —— And The *head of a razor.* (Ḳ.) —— And The [*pointed*] *head of a* مِنْقَار, [*an iron instrument like the* فأس, (A and Ḳ in art. نقر,) *with which mill-stones &c. are pecked,* or *wrought into shape,* (see مِنْقَار, and] *with which wood is cut.* (TA.) ═ See also خِلْف.

خُلْف a subst. from إِخْلَاف, (Ṣ, Mṣb, Ḳ,) relating to a promise, and restricted to future time; (Mṣb;) i. e. a subst. used in the place of إِخْلَاف; (Lḥ, TA;) meaning The *breach,* or *non-fulfilment, of a promise;* as also ↓ خَلْف, which is said to be the original form of the word, and ↓ خُلُوف: (TA:) it is, *in respect of the future, like* كَذِب *in respect of the past:* (Ṣ, Ḳ:) some say that it signifies *a false,* or *wrong, saying;* which is a meaning of ↓ خَلْف, with fet-ḥ, before mentioned: but perhaps these two words may be syn. dial. vars. (MF, TA.) —— Also, (Mṣb,) or ↓ خُلْفَة, and ↓ خِلْفَة, (Ḳ, TA,) *Disagreement, difference, dissension, contrariety, contrariousness,* or *opposition,* (Mṣb, Ḳ,* TA,) *in opinions* or *the like,* (Mṣb,) or *in respect of friendship and amity,* (TA in explanation of the second of these words,) or in natural disposition; (Ḳ;) as also ↓ خِلِفَّة (Ṣ, Ḳ) and ↓ خِلِفَّاة and ↓ خَالِف and ↓ خَالِفَة. (Ḳ.) ═ It is also pl. of خَلِيف, in its various senses.

خِلْف : see خِلْفَة, in the latter half of the paragraph, in seven places. —— Also, applied to a man, (Ṣgh,) *i. q.* لَجُوج [app. as meaning One *who perseveres much in opposition* or *contention* or *the like*]; (Ṣgh, Ḳ;) as also ↓ خَالِفَة. (TA.) ═ Also a subst. from إِخْلَاف meaning The *act of drawing water;* and so ↓ خِلْفَة: (A 'Obeyd, Ḳ:*) [whence the saying,] مِنْ أَيْنَ خِلْفَتُكُمْ *Whence do ye draw water?* (Ṣ, Ḳ.) ═ The *teat* (حَلَمَة) *of the udder of the she-camel:* (Ṣ, Ḳ:) and the *two fore ones,* and the *two hinder ones:* (Ṣ:) or the *part of the udder upon which the milker lays hold:* (TA:) or the *extremity of the udder of the she-camel:* (Mṣb, Ḳ:) or the *hinder of the* أَطْبَاء [or *teats*]: (Ḳ:) or the *udder itself;* (Lth, TA;) [i. e.] it is, *to the she-camel,* (Mṣb,* Ḳ,) *like the* ثَدْى *to the human being,* (Mṣb,) or *like the* ضَرْع *to the ewe* or *she-goat:* (Ḳ:) or the خلف is *of the camel* and *of the cloven-hoofed animal;* and the طُبْى, of the solid-hoofed animal and of the animal that has a claw: (Lḥ, TA:) the pl. [properly of pauc.] is أَخْلَاف (Mṣb, TA) and [of mult.] خُلُوف. (TA.) One says, دَرَّتْ لَهُ أَخْلَافُ الدُّنْيَا ‡ [*The world yielded him abundance of its good things*]. (TA.) ═ The *shortest of the ribs of the side;* (Ṣ;) [and] so ↓ خَلْف; (Ḳ;) likewise called ضِلَعُ الخَلْف and الخِلْف; it is the *furthest and thinnest of the ribs;* (TA;) [i. e.] the خلف is *that next to the belly, of the small ribs; their* قُصَيْرَى: (Ḳ: [see القُصْرَى:]) pl. of the former (Ṣ) [and] of the latter (Ḳ) خُلُوف. (Ṣ, Ḳ.) ═ ذَاتُ خِلْفَيْنِ: see خَلْف, near the end of the paragraph.

خَلَف *A substitute; a thing given,* or *received,* or *put,* or *done, instead of, in place of,* or *in exchange* for, another thing. (A 'Obeyd, Th, Ṣ, Mṣb, Ḳ, TA.) You say, اِجْعَلْ هٰذَا خَلَفًا مِنْ هٰذَا *Make thou this to be a substitute for this.* (Mṣb.) And هٰذَا خَلَفٌ مِمَّا أُخِذَ لَكَ *This is a substitute for what has been taken to thee.* (IB.) And فِى هٰؤُلَآءِ القَوْمِ خَلَفٌ مِمَّنْ مَضَى *In these people are such as supply the place of those who have gone.* (TA.) And فِى فُلَانٍ خَلَفٌ مِنْ فُلَانٍ [*In such a one is a substitute for such a one*]. (TA.) And هُوَ مِنْ أَبِيهِ خَلَفٌ *He is a substitute for his father.* (IB.) See also خَلْف, in six places, in the former half of the paragraph.

خَلِف, applied to she-camels, *i. q.* مَخَاض, i. e. *Pregnant:* n. un. with ة: (Ṣ, Ḳ:) accord. to some, (TA,) the pl. of خَلِفَة, which signifies a *pregnant* camel, (Mgh, Mṣb, TA,) or, as some say, one *that has completed a year after bringing forth and has then been covered and has conceived, until she enters upon the term called* التَّعْشِير, (TA, [*from the time when her pregnancy has become manifest,* (see قَارِح and لَاقِح,)]) is مَخَاض, (Mgh, Mṣb, TA,) like as the pl. of اِمْرَأَةٌ is نِسَآءٌ; (Mṣb, TA;) and sometimes خَلِفَات (Mgh, Mṣb, TA) and خِلَاف: (TA:) but خَلِف occurs in the saying of the rájiz,

مَا لَكِ تَرْغِينَ وَلَا تَرْغُو الخَلِفْ *

[*What aileth thee that thou utterest a grumbling cry, when the pregnant camels utter not that cry?*]. (IB.)

خَلُف : see خَلْفَة.

خُلُف : see خُلْف.

خَلْفَة : see the next paragraph, in two places.

خُلْفَة : see خُلْف. —— Also *A vice, a fault,* or *an imperfection:* (Ḳ:) and *badness, corruptness, vitiousness,* or *dishonesty:* (TA:) and *foolishness,* or *stupidity;* or *paucity,* or *want, of intellect* or *understanding;* as also ↓ خَلَافَة [properly an inf. n., of خَلُفَ, and before mentioned as such; (see 1, in the latter half of the paragraph;)] and *idiocy.* (Ḳ.) All of these meanings have been assigned to it in explanations of the saying, أَبِيعُكَ هٰذَا العَبْدَ وَأَبْرَأُ إِلَيْكَ مِنْ خُلْفَتِهِ [*I sell to thee this slave, but I am irresponsible to thee for his vice, &c.*]: or, accord. to IAạr, the meaning is, خِلَافِه [*his contrariousness*]. (TA.) —— Also The *last taste* of food; (Ḳ;) as in the saying, إِنَّهُ لَطَيِّبُ الخُلْفَةِ [*Verily it is good,* or *sweet, in respect of the last taste*]; (TA;) and so ↓ خَلْفَة: pl. خُلَف: and it (خُلْفَة) signifies also *loss of appetite for food, in consequence of disease:* (so accord. to the CḲ:) [or,] accord. to some copies of the Ḳ, ↓ خَلْفَة has this latter signification; and so ↓ خُلُف: accord. to other copies, خُلَف is pl. of خُلْفَة in this sense: but both these readings require consideration: what is found in the Lexicons is, خَلَفَتْ نَفْسُهُ عَنِ الطَّعَامِ, aor. ـُ, inf. n. خُلُوف; meaning as explained above, in the latter half of the first paragraph. (TA.)

خِلْفَة a subst. signifying *A mode,* or *manner, of coming after* [or *behind*]; like قِعْدَة signifying "a mode, or manner, of sitting." (Mṣb.) —— See also خُلْف. —— It signifies also *Difference* [*of any kind*]: (Ḳ,* TA:) or the *coming and going* of the night and the day; (Ṣ, Ḳ,* TA;) and likewise of wild animals. (Ḳ.) Hence the saying in the Ḳur [xxv. 63], وَهُوَ ٱلَّذِى جَعَلَ ٱللَّيْلَ وَٱلنَّهَارَ خِلْفَةً, (Ṣ, Ḳ,*) meaning ذَوَىْ خِلْفَةٍ, (Bḍ,) i. e. [*And He it is who hath made the night and the day*] *so that each replaces the other:* or *each follows the other:* (Ḳ,* TA:) or *so that he who is unable to accomplish a thing in the night may do it in the day, and the reverse.* (Fr, L, Ḳ.) Zuheyr says, of wild animals, يَمْشِينَ خِلْفَةً, meaning *They go to and fro.* (Ṣ, TA. [See EM p. 109.]) And one says, أَخَذَتْهُ خِلْفَةٌ, meaning *He was* *taken with* [*an affection causing*] *a frequent going to and from the privy.* (Ṣ, Ḳ.) [And hence,] خِلْفَةٌ signifies also, *A discharging of the bowels; or a purging and vomiting together;* (Ḳ;) or *a disordered state of the stomach arising from* [*unwholesome*] *food;* (TA;) *a looseness,* or *diarrhœa.* (JK, TA.) — See also خِلْفٌ. — Also The *bringing* of camels *to the watering-place in the evening, after the people have gone away.* (L, Ḳ.) — And A man's *watching to see another,* (أَنْ يُنَاظِرَ in some copies of the Ḳ, and ان يُنَاصِرَ in other copies, being put for ان يُبَاصِرَ, which is the right reading, agreeably with an explanation of اِخْتَلَفَ صَاحِبَهُ, [for which see 8, near the middle of the paragraph,] TA,) *and when he is absent from his family, going in to them,* (Ḳ, TA,) or [rather], *when he is absent from his wife, going in to her.* (TA, after the explanation of the phrase above mentioned.) = *A thing that is suspended behind the rider;* (JK, Ḳ;) *such as is suspended behind the* [*kind of vehicle called*] مَحْمِل. (TA.) — *Remains* of water in a trough or tank. (TA.) — *What remains, of food, between the teeth.* (Lḥ, Ḳ.) — *A plant,* or *herbage, that comes forth after another plant,* or *other herbage,* (Ṣ, Mṣb, Ḳ,) *which has become dry, and broken in pieces:* (Ṣ, TA:) or *that comes forth not from rain, but by reason of the cold of the latter part of the night.* (Aboo-Ziyád El-Kilábee, Ḳ. [See also رَبْلٌ.]) — *What the trees disclose in the beginning of the cold,* (Ḳ, TA,) *by reason of the* [*rain called*] صَفَرِيَّة [q. v.]: (TA:) or *fruit that comes forth after other fruit:* (Ḳ:) or *fruit that comes forth after abundant fruit;* (Ṣ, Mgh, TA;) this being termed the خِلْفَة of trees: (Ṣ, Mgh:) or *a growth of leaves after the falling away of other leaves:* (Ḳ,* TA: دُونَ in the Ḳ is a mistake for بَعْدَ: TA:) or *leaves that come forth after the first leaves, in the* [*season called*] صَيْف. (Nh, TA.) — *What grows in the* صَيْف [or *summer*]; so says A'Obeyd: (Ṣ, Ḳ:) or, (JK, Mgh, Ḳ,) as also ▼خِلْفٌ, (Ḳ,) the *herbage produced by the* صيف, (JK, Ḳ,) or *in the* صيف, (Mgh,) *after the spring-herbage has dried up.* (JK, Mgh.) — *A produce of grape-vines after the grapes have turned black; the grapes being gathered while it is fresh and green, it then ripens:* and *so other fruits:* or *a new produce, by the vine, of fresh sour grapes.* (Ḳ.) — *Grain that is sown* (JK,* Mgh, Ḳ*) *after the former has come to maturity:* (Mgh, TA:) because taken as a substitute for wheat and barley: (Ḳ:) pl. خِلَفٌ. (Mgh.) — *A piece with which a garment is patched* (Ḳ) *when it is old and worn out.* (TA.) — *A time after a time.* (IAạr, Ḳ.) = *Differing* [*one from another* or *others*]; as also ▼خِلْفٌ: (Ḳ:) it is applied in this sense to a people, or company of men: (AZ, Ṣ, Ḳ:) and to beasts, or horses or the like, as meaning *differing* (Ḳ, TA) *in their colours and appearances:* (TA:) and خِلْفَتَانِ is applied to any two things that are *different;* (Ks, TA;) as also ▼خِلْفَانِ: (Ks, Mṣb, TA:) and خِلْفَةٌ, (Ḳ,) or خِلْفَتَانِ, (Ks, TA,) to any two colours that are *combined* [because *different*]. (Ks, Ḳ, TA.) AZ cites, as an ex., the saying [of a rájiz],

* دَلْوَايَ خِلْفَانِ ▼ وَسَاقِيَاهُمَا *

[*My two buckets are different, and their two suppliers with water*]; (Ṣ, TA;) meaning that one of them [i. e. of the buckets] is ascending and full, and the other is descending and empty; or that one of them is new, and the other is old and worn out. (TA, in two places.) And one says of two children, or two male slaves; or two female slaves, that they are خِلْفَتَانِ, (Ks, Ḳ,) and ▼خِلْفَانِ, (Ḳ,) applying to the male and the female, (TA,) meaning *One tall and the other short:* or *one white and the other black.* (Ks, Ḳ.) One says also, بَنُو فُلَانٍ خِلْفَةٌ, meaning *The children of such a one are half males and half females.* (Ṣ.) And نِتَاجُ فُلَانٍ خِلْفَةٌ *The offspring of the beasts of such a one are one year male and another year female.* (JK, TA.) And وَلَدَتْ خِلْفَيْنِ ▼, said of a ewe or goat, (Ḳ,) or of a camel, (L,) *She brought forth one year a male and another year a female.* (L, Ḳ.) The pl. [of ▼خِلْفٌ] (Ḳ, TA) in all its senses (TA) is أَخْلَافٌ and خِلَفَةٌ; (Ḳ,* TA;) the latter, [in the CḲ خِلْفَةٌ,] like قِرَدَةٌ as pl. of قِرْدٌ. (TA.)

خُلُفٌ and خُلُفٌ and خُلَفَةٌ and خُلَفَةٌ: see خَالِفٌ, in the latter half of the paragraph. — أُمُّ خُلَفَفٍ (Sgh, Ḳ) and خُلَفَفٌ *Calamity,* or *misfortune:* or *the greatest calamity* or *misfortune.* (Ḳ.)

خِلْفَنَةٌ and خِلْفَنَاةٌ: see خُلُفٌ: — and see also خَالِفٌ, in the middle of the paragraph.

خِلَافٌ an inf. n. of 3 [q. v. passim]. (Ṣ, &c.) — Also The *contrary,* or *opposite,* of a thing; syn. ضِدٌّ.* (Mṣb in art. ضد. [Very often used in this sense.]) You say, الاِخْتِلَافُ خِلَافُ الاِتِّفَاقِ [i. e. الاختلاف *is the contrary of* الاتّفاق]. (TA.) = Also, (Ṣ, Mṣb, Ḳ,) by the vulgar (O, Mṣb, TA) incorrectly pronounced with teshdeed (O, Mṣb, Ḳ, TA) and fet-ḥ [to the خ, i. e. خَلَّافٌ], (TA,) *A well-known kind of tree;* (Ṣ;) the [*kind of tree called*] صَفْصَاف: (Mṣb:) or *a species of the* صفصاف, *but not the* صفصاف *itself:* (Ḳ:) [the *salix Aegyptia* of Linnæus; called by this name in the present day; and by some, improperly, بَانٌ, q. v.:] *it abounds in the land of the Arabs; and is* [*also*] *called* سُوجَرٌ [or سَوْجَرٌ]; *and there are many varieties thereof; all of them soft and weak;* (TA;) *but it is seldom, or never, found in the desert:* (Mṣb:) they assert that it is thus called because the torrent brings it from one locality to another, so that it grows in a place different from that of its origin; (AḤn, Mṣb, Ḳ,* TA;) but this is not a valid assertion: (TA:) [it is a coll. gen. n.:] n. un. with ة. (Mṣb, TA.) سُوَيْنٌ مِنْ خِلَافٍ, in the saying of the rájiz cited voce خُفٌّ, means *Made of different trees:* it does not mean *of the tree called* خِلَاف; because this is seldom, or never, found in the desert. (Ṣ, TA.) — Also The *sleeve of a shirt.* (IAạr, Ḳ.)

خُلُوفٌ: see خُلْفٌ. = It is also, as stated above, pl. of خَلْفٌ: (IB, Ḳ, TA:) — and a pl. of خِلْفٌ. (TA.)

خَلِيفٌ: see خَلِيفَةٌ, in three places. — Also One *who holds back from the place,* or *time, of promise:* and one *who breaks a promise.* (TA.) — And A woman *that has let down her hair behind her.* (JK, O, Ḳ.) — And A woman *that has attained to the period of one day,* or *two days, after her having brought forth.* (IAạr.) [Perhaps from the signification next following.] — A she-camel *in the second day after her having brought forth:* pl. خُلُفٌ and خُلُفٌ: (Ḳ, TA:) these two pls. are mentioned in the Ḳ in different places in this art., but both are correct, like رُسُلٌ and رُسْلٌ. (TA.) Hence, (TA,) one says, رَكِبَهَا يَوْمَ خَلِيفِهَا [*He rode her on the second day after she had brought forth*]. (Ḳ.) — And The *milk that is after the biestings:* (AA, Ḳ:) pl. as above. (Ḳ.) One says also, حَلَبَهَا خَلِيفَ لِبَئِهَا *He drew from her the milk that came after the biestings had passed away.* (JK.) And ايتنا بِلَبَنِ نَاقَتِكَ يَوْمَ خَلِيفِهَا, i. e. [*Bring thou to us the milk of thy she-camel of the day*] *after the cessation of her biestings;* i. e., *of the milking that is after her bringing forth by a day* or *two days.* (AA, TA.) = Applied to a garment, (Ṣ, Ḳ,) or a shirt, (Mṣb,) *Having the middle, worn-out part taken out, and the* [*cut*] *edges then sewed together:* (Ṣ, Mṣb, Ḳ*:) and ▼مَخْلُوفٌ signifies the same; (JK;) or a garment *composed of two pieces sewed together:* or, as some say, this signifies a garment *pledged.* (TA.) = Also, accord. to A'Obeyd, The *part beneath the armpit:* and the خَلِيفَانِ of the camel are *like the* إِبْطَانِ *of man:* accord. to the Ṣ and the O, خَلِيفَا النَّاقَةِ signifies *the two armpits of the she-camel* (إِبْطَاهَا): but the author of the Ḳ, following the [first] explanation given by A'Obeyd, says that this is wrong, and that the meaning is *the parts beneath the two armpits of the she-camel.* (TA.) = And *A gap between two mountains,* (JK,) or *between two mountain-tops,* (TA,) *of little breadth and length:* (JK, TA:) or *a road between two mountains:* (Ṣ, Ḳ:) or *a valley between two mountains:* (Ḳ:) or *a place where water pours forth* (Ḳ, TA) *between two mountains,* or *between two valleys, passing thence into a wide tract:* (TA:) and *any road in a mountain,* (Skr, Ḳ,) or *behind a mountain,* or *behind a valley:* (TA:) or simply *a road;* as also ▼مُخْتَلَفَةٌ; (JK, Ḳ;) this being either in a plain or in a mountain: (TA:) pl. of the former as above. (Ḳ.) One says ذِيخُ الخَلِيفِ i. e. [*The hairy male hyena*] *of the road between two mountains,* (Ṣ, Ḳ,) or *of the valley between two mountains;* (Ḳ;) like as one says ذِئْبُ غَضًا. (Ṣ.) = And A *sharp* arrow: (AḤn, Ḳ:) or, accord. to Skr, the word in this sense is حَلِيفٌ, with the unpointed ح; and this is more probably correct. (TA.)

خِلَافَةٌ: see خِلْفَةٌ.

خِلَافَةٌ inf. n. of خَلَفَهُ as meaning "he was, or became, his خَلِيفَة" [q. v.]. (Ṣ, Mgh, Mṣb, Ḳ.) — [And hence, as a simple subst., The *office of* خَلِيفَة.]

خَلِيفَةٌ *A successor:* and *a vice-agent, vice-*

gerent, lieutenant, substitute, proxy, or deputy: (KL:) one who has been made, or appointed, to take the place of him who has been before him: (JK:) an act. part. n. of خَلَفَهُ, inf. n. خَلْفٌ and خِلَافَةٌ; as also ▾ خَلِيفٌ: (TA:) or it may have the meaning of an act. part. n. or that of a pass. part. n.: and so in the sense next following: (Mṣb:) the *supreme*, or *greatest*, *ruler* or *sovereign*, (Ṣ, Mṣb, Ḳ, TA,) *who supplies the place of him who has been before him;* (TA;) [particularly the *successor of the Prophet;* whence "Caliph," commonly used by English writers for "Khaleefeh;"] as also ▾ خَلِيفٌ, (Ḳ,) which is the original form, (Mṣb,) without ة; (Mṣb, TA;) disapproved by some, but mentioned by AḤát and Ibn-'Abbád and IB, and occurring in a verse of Ows Ibn-Ḥajar cited by IB: (TA:) the ة in the former is to denote intensiveness of signification, (Nh, Mṣb, TA,) as in عَلَّامَةٌ and نَسَّابَةٌ: (Mṣb:) or, as some say, لِلنَّقْلِ [i. e. for the transference of the word from the category of epithets to that of substs.]: (TA:) it is also said that the word may be an epithet of which the subst. qualified thereby is suppressed, for نَفْسٌ خَلِيفَةٌ; but this requires consideration: (MF, TA:) it is an epithet applied to a man peculiarly: (Mṣb:) some make it fem.; (Fr, Ṣ, Mṣb, Ḳ;) saying هٰذَا خَلِيفَةٌ أُخْرَى [*This is another Khaleefeh*]; though the proper way is to make it masc.: (Mṣb:) a poet says,

* أَبُوكَ خَلِيفَةٌ وَلَدَتْهُ أُخْرَى *
* وَأَنْتَ خَلِيفَةٌ ذَاكَ الْكَمَالُ *

[*Thy father was a Khaleefeh, whom another* Khaleefeh *begat; and thou art a Khaleefeh: that is perfection*]: (Fr, Ṣ:) the pl. is خَلَائِفُ [generally applied to any people that have succeeded others, and supplied their places, as in the Ḳur x. 15], (Ṣ, Ḳ,) like as كَرَائِمُ is pl. of كَرِيمَةٌ; (Ṣ;) and خُلَفَآءُ [generally applied to successors of the Prophet], (Ṣ, Ḳ,) because, as it applies only to the male, and has ة added, the ة is dropped in forming the pl., which is thus like ظُرَفَآءُ as pl. of ظَرِيفٌ: (Ṣ:) thus says ISk, and the like is said in the O: but what AḤát and Ibn-'Abbád say requires not this straining: (TA:) [i. e.] خَلَائِفُ is pl. of خَلِيفَةٌ; and خُلَفَآءُ, of ▾ خَلِيفٌ: (JK:) or some, having regard to the original, make the pl. to be خُلَفَآءُ, like as شُرَفَآءُ is pl. of شَرِيفٌ; (Mṣb;) and this pl. is masc. only, so that you say ثَلَاثَةُ خُلَفَآءَ: (ISk, Mṣb, TA:) and some, having regard to the word itself [in its altered and used state], make the pl. to be خَلَائِفُ; (Mṣb;) and this pl. may have prefixed to it either a masc. or a fem. n. of number, so that you say ثَلَاثُ خَلَائِفَ and ثَلَاثَةُ خَلَائِفَ; (ISk, Mṣb, TA;) both of which are chaste. (Mṣb.) You say, كَانَ ٱللّٰهُ خَلِيفَةَ وَالِدِكَ عَلَيْكَ [*May God be to thee a supplier of the place of thy father*]: (Ṣ, Mṣb:*) and in like manner you say, to a person, of any one whom he has lost by death, (Ṣ, Mṣb,) and who cannot be replaced; as the paternal uncle; (Mṣb;) or the mother. (Ḳ.) Some say that the application of the title خَلِيفَةُ ٱللّٰهِ [*The Vicegerent of God*] is not allowable, except to Adam and David because there is express authority in these instances [in the Ḳur ii. 28 and xxxviii. 25]; but others allow it in other cases, like سُلْطَانُ ٱللّٰهِ and جُنُودُ ٱللّٰهِ and حِزْبُ ٱللّٰهِ and خَيْلُ ٱللّٰهِ; all of which have been heard: (Mṣb:) and Zj says that it is allowable to say of the Imáms that they are خُلَفَآءُ ٱللّٰهِ فِى أَرْضِهِ [*The Vicegerents of God in his earth*]. (TA.)

خِلِّيفَى: see the middle of the next paragraph.

خَالِفٌ: see خَلْفٌ, in the former half of the paragraph. — Also One *who remains behind*, or *after, another*, (Yz, Ḳ, TA,) or *others, in the case of a war*, or *a warring and plundering expedition*, and *in other cases:* (TA:) pl. خَالِفُونَ (Yz, Ḳ, TA) and خَوَالِفُ, which latter is extr. [in this case], but is also said to be a [reg.] pl. of ▾ خَالِفَةٌ, and as such to signify persons *who do not go forth on a warring*, or *warring and plundering, expedition:* and ▾ خَالِفَةُ الغَازِى signifies *he who remains behind*, or *after, him who goes forth on such an expedition, being of his family.* (TA.) فَٱقْعُدُوا مَعَ ٱلْخَالِفِينَ, in the Ḳur [ix. 84], means *Then stay ye with those who remain behind.* (Yz, Ḳ.*) خَوَالِفُ is also pl. of ▾ خَالِفَةٌ [as fem. of خَالِفٌ], (TA,) and signifies *Women* (Ḳ, TA) *remaining behind in the houses or tents:* but some assign to it the first of the meanings explained above: and some say that it means the *children remaining behind.* (TA.) It is said in the Ḳur [ix. 88 and 94], رَضُوا بِأَنْ يَكُونُوا مَعَ ▾ الخَوَالِفِ, i. e. [*They chose to be*] *with the women:* (Ṣ, Ḳ:*) thus it is explained by Ibn-'Arafeh: but some say that the meaning is, *with the bad*, or *corrupt*, persons; and that خوالف is here a pl. [of خَالِفٌ,] like فَوَارِسُ. (TA.) For — خَالِفٌ is applied to a man [as meaning *Bad*, or *corrupt*]: and ▾ خَالِفَةٌ to a woman as meaning *bad*, or *corrupt, and remaining behind in her abode:* and the former to a slave as meaning *bad*, or *corrupt:* and also *contrarious:* and in this last sense it is likewise applied to a companion: and some of the grammarians say that there is no word of the measure فَاعِلٌ having its pl. of the measure فَوَاعِلُ, except خَالِفٌ and هَالِكٌ and فَارِسٌ: but see this last: (TA:) and ▾ خَلِيفَةٌ, also, has this last signification; (JK, TA;) or [rather] signifies *very contrarious;* (Ḳ;) as also ▾ خَالِفَةٌ; (JK;) and so ▾ خِلَفْنَةٌ, and ▾ خِلَفْنَاةٌ, (Lḥ, JK, Ḳ,) in each of which the ن is augmentative, and each of which is applied to a man and to a woman and to a pl. number; (Lḥ, Ḳ;) but خِلَفْنَيَاتٌ has been mentioned as pl. [of خلفناة], and as applied to males and females: (TA:) and خَالِفُونَ is likewise used in this sense applied to a number of men. (JK.) — Also, applied to a slave, [and app. to any man, but in this latter case I find it written خلف, which I believe to be a mistranscription,] One *who has withdrawn from the people of his house:* so says Lḥ. (TA.) — Also *Stupid; foolish;* or *having little*, or *no, intellect* or *understanding;* as also ▾ خَالِفَةٌ, (Ḳ, TA,) but in an intensive sense, and also applied to a woman; (TA;) and ▾ أَخْلَفُ, (JK, Ḳ,) of which the fem. is خَلْفَآءُ; (JK, TA;) and ▾ خُلْفُفٌ, (Ḳ,) or ▾ خُلَفُفٌ, (L,) or both, (JK,) likewise applied to a woman, as also ▾ خُلْفُفَةٌ, (Ḳ,) or ▾ خُلَفُفَةٌ: (JK:) or, as some say, خَالِفٌ signifies one *in whom is no good:* and, as also ▾ خَالِفَةٌ, one *who has not what suffices:* or *who often breaks his promises:* (TA:) or both of these mean one *who has not what suffices, and in whom is no good:* or *very contrarious.* (JM.) One says that a man is خَالِفُ and ▾ خَالِفَةُ أَهْلِ بَيْتِهِ, and خَالِفُ اهل بيته, meaning *The one in whom is no good, of the people of his house:* (Ṣ, Ḳ:) and *the ungenerous:* (Ḳ:) or *the stupid*, or *foolish:* or *the bad*, or *corrupt*, and *the evil:* and it is [said to be] tropical. (TA.) And ▾ قَوْمٌ خَوَالِفُ *Persons in whom is no good.* (JK.) — And, [app. because he leaves his family behind him,] *A drawer of water;* (Ṣ, TA;) as also ▾ مُسْتَخْلِفٌ [q. v.]: both mentioned in the Ḳ; but السَّقَّآءُ is there erroneously put for المُسْتَقِى. (TA.) — And *Weak, without appetite for food.* (TA.) — And Flesh-meat *from which a slight smell is perceived, but in the chewing of which is no harm.* (Lth, TA.) = See also خَلْفٌ.

خَالِفَةٌ: see خَلْفٌ, in the former half of the paragraph. Also, particularly, [or perhaps خَالِفَةٌ أُمَّةٌ only in this sense,] *A nation, people*, or *race, remaining after another that has gone before.* (I'Abbád, Ḳ.) And *One who comes to the water after him who has returned* [*from it*]: whence Aboo-Bekr applied this appellation to himself, from a motive of humility, when asked if he were the Khaleefeh of the Apostle of God. (IAth, TA.) See also خَالِفٌ, in eight places: and see its pl., خَوَالِفُ, in the same paragraph, in two places. — Also, applied to a man, [like خَلِيفَةٌ as explained in the Ḳ,] *Very contrarious*, or *adverse, and inimical.* (Ṣ,* Ḳ,* TA.) See also خَلْفٌ. — مَا أَدْرِى أَىُّ خَالِفَةَ هُوَ, the word خالفة being here imperfectly decl., (Ṣ, Ḳ,) because of the fem. gender and determinate, being explained by النَّاسِ, (Ṣ,) or because determinate and occupying the place of a pl., like as one says أَىُّ تَمِيمَ and أَىُّ أَسَدَ, [or rather, I think, because used as a proper name, as MF, says, (though SM disputes this,) and with the sign of the fem. gender,] means *I know not what one of mankind he is;* (Ṣ, Ḳ;) as also أَىُّ خَالِفَةٍ, perfectly decl.; and أَىُّ الخَوَالِفِ; and أَىُّ الخَالِفَةِ; (Ḳ;) and so أَىُّ خَافِيَةٍ, (Ḳ, TA, [in the CḲ اىّ خالفة again,]) imperfectly decl. (TA.) Lḥ says that الخَالِفَةُ, writing it thus with ال, signifies النَّاسُ. (TA.) = Also *One of the poles of a* [*tent of the kind called*] خِبَآء: or *one of the poles of a* بَيْت [or *tent*] *in the hinder part thereof:* (Ḳ:) Lḥ says that the خَالِفَة is the *hinder part*, or *in the hinder part*, (آخِر, [i. e. آخِرُ or آخِرَ, app. the latter,]) *of a* بَيْت; and one says بَيْتٌ ذُو خَالِفَتَيْنِ [app.

meaning *a tent having two poles in its hinder part*]: (TA:) the pl. is خَوَالِفُ: (Ṣ, TA:) which is hence applied to the *angles*, or *corners, of a* بَيْت: AZ says that the خَالِفَة of a بيت is [app. the *skirt thereof*,] *beneath the* [ropes called] أَطْنَاب, *in the* [part called] كِسْر [q. v.]; and it is also called the خِمَاصَة, and the فرجة: [thus I find these two words written, without any syll. signs:] and he cites, as an ex.,

* مَا خِفْتُ حَتَّى هَتَّكُوا الخَوَالِفَ *

[app. meaning *And I feared not until they rent open the skirts of the tent*, or *tents*]: (TA:) or, as some say, the خَالِفَتَانِ are the *two sides* of a tent, and its رِوَاق is its fore part, and its كِفَآء is its hinder part. (TA in art. روق.) — خَوَالِفُ, (Yz, Ḳ,) or خَوَالِفُ مِنَ الأَرْضِ, (TA,) *Lands that produce not plants*, or *herbage, save among the last of lands*. (Yz, Ḳ,* TA.) — See also خُلْفٌ.

أَخْلَفُ: see خَالِفٌ, in the latter half of the paragraph. — Also *Contrarious, hard in disposition, as though going with a leaning towards one side*: (Ḳ:) and [simply] *leaning towards one side*; applied to a camel: (Ṣ, Ḳ:) so says A'Obeyd; (Ṣ, TA;) and so Aṣ. (TA.) — Also A camel *that has the sheath of his penis slit, and that will not remain stationary, by reason of pain*: (TA:) and ▼ مَخْلُوفٌ signifies a camel *having the sheath of his penis slit in the hinder part*, (JK, TA,) *when suffering suppression of his urine in consequence of the pressure of his hind girth upon his sheath*: so says El-Fezáree. (TA.) — And *Left-handed*. (JK, Ḳ.) — And *Squint-eyed*; syn. أَحْوَلُ. (Ḳ.) — Accord. to some, (TA,) *A torrent*: (Ḳ, TA:) or, as some say, *a river*. (Skr, TA.) — And *A male serpent*. (Ibn-'Abbád, Ḳ.) [All these meanings seem to have been assigned to the word as occurring in a verse of Aboo-Kebeer El-Hudhalee, in which he likens the course of a wolf in a narrow road to the course of the أَخْلَف.] — [Also *More*, and *most, wont to break promises*. Hence the prov., mentioned by Meyd, أَخْلَفُ مِنْ عُرْقُوبٍ *More wont to break promises than 'Orḳoob*: a certain man who rendered himself notorious for breaking his promises. See Freytag's Arab. Prov. i. 454. — And *More*, and *most, disagreeing, differing, dissentient, contrary, contrarious*, or *opposing*. See an ex. in a prov. cited voce لَيْل. — And app. *More*, and *most, offensive in the odour of the mouth*. See Freytag's Arab. Prov. ubi suprà.]

تَخَالِيفُ *Different colours*. (TA.)

مَخْلَفَةٌ: see مَخْلَفٌ.

مُخْلِفٌ A camel *that has exceeded in age the* بَازِل; [which latter is generally *one that has entered the ninth year*;] (Ṣ, M, Ḳ;) beyond which there is no age [having an epithet to denote it]; therefore, (TA,) one says مُخْلِفُ عَامٍ and مُخْلِفُ عَامَيْنِ [*that has exceeded in age the* بازل *by a year* and *by two years*]; (Ṣ, TA; [see 4;]) applied alike to the male and the female; (Ṣ, Ḳ;) and the female is also termed مُخْلِفَةٌ: (Ḳ:) or this latter signifies ‡ a she-camel *that appears*, (Ṣ, Ḳ,) or *is thought*, (A,) *to be pregnant, and is not pregnant*: (Ṣ, A, Ḳ:) and the pl. is مَخَالِيفُ. (TA.) — See also مِخْلَافٌ. — Also A man *whose cattle have not obtained the* [*herbage termed*] رَبِيع. (JK.) — رَجُلٌ مُخْلِفٌ مُتْلِفٌ, or ▼ مِخْلَفٌ مِتْلَفٌ, and ▼ مِخْلَافٌ مِتْلَافٌ: see art. تلف. — نَوْمَةُ الضُّحَى مُخْلِفَةٌ لِلْفَمِ, (Ḳ, TA,) also written ▼ مَخْلَفَةٌ, and in some copies نَوْمُ الضُّحَى, [which requires the reading مَخْلَفَةٌ,] (TA,) i. e. [*The sleep*, or *sleeping, in the period of the morning when the sun is yet low is*] *a cause of the mouth's becoming altered* [*for the worse*] *in odour*. (Ḳ, TA.) — مُخْلِفُ جَنْبٍ *Having one half of his face and of his mouth turning sideways*. (JK.) — See also the explanation of the verse of El-Ḥoṭeiäh cited in the last quarter of the first paragraph. The قَطَا are termed مُخْلِفَاتٌ because they draw water for their young ones. (JK.)

مِخْلَفٌ: see the next preceding paragraph.

مَخْلَفَةٌ: see مُخْلِفٌ. — See also خَلِيفٌ, near the end of the paragraph. ▼ المَخْلَفُ [as a coll. gen. n.] signifies *The roads along which the people pass in Minè*; (Ḳ;) which are three: one says, أُطْلُبْهُ بِالمَخْلَفَةِ الوُسْطَى مِنْ مِنًى [*Seek thou him in the middle road of Minè*]. (TA.) And مَخْلَفَةُ بَنِى فُلَانٍ *The place of alighting*, or *descending and stopping* or *sojourning* or *abiding* or *lodging* or *settling, of the sons of such a one*. (Ḳ,* TA.) And مَخْلَفَةُ مِنًى *The place of alighting*, or *descending and stopping* &c., *of the people in Minè*. (Ḳ.) — *A place in which are trees of the kind called* خِلَاف. (Ṣ, Ḳ.)

مِخْلَفَانُ البَلَدِ *The ruler*, or *sovereign*, (سُلْطَان,) *of the country*; as also ▼ مِخْلَافُهُ. (TA.)

مِخْلَافٌ A man *who often breaks his promises*; (Ṣ, Ḳ;) as also ▼ مُخْلِفٌ: (TA:) [whence the latter (which properly signifies simply *breaking a promise*) is applied to a star, or an asterism, as meaning ‡ *Unattended with rain*: (see 4:) and in the same sense to clouds (سَحَاب): or, accord. to Freytag's Lex., in this or in the contr. sense.] — See also مُخْلِفٌ. — And see مِخْلَفَان. — Also A كُورَة [i. e. *province, district*, or *region*] (Ṣ, Mgh, Mṣb) *pertaining to the people of El-Yemen*, (Ṣ,) or in the dial. of El-Yemen; (Mgh, Mṣb;) pl. مَخَالِيفُ; (Ṣ, Mṣb;) every مخلاف thereof having a [distinctive] name whereby it is known; (Ṣ;) the مخاليف of the people of El-Yemen being like the أَجْنَاد of the people of Syria and the كُوَر of the people of El-'Iráḳ and the رَسَاتِيق of the people of El-Jibál and the طَسَاسِيج of the people of El-Ahwáz: (IB:) or مِخْلَافٌ signifies *a* كُورَة (JK, M, Ḳ) *to which a man comes*; (M;) [*in any country*;] and hence the مخاليف of El-Yemen, (Ḳ,) i. e. its كُوَر: (TA:) some say that there is a مخلاف in every country; (Mṣb;) so says Khálid Ibn-Jembeh; (TA;) i. e. *a* نَاحِيَة [as meaning a *district* &c.]; (Mṣb;) and thus one says the مخلاف of El-Medeeneh, and of El-Yemámeh, (Khálid Ibn-Jembeh, TA,) and the مخاليف of Eṭ-Ṭáïf: (AA, Mṣb, TA:) but properly it is peculiar to the dial. of El-Yemen. (TA.) — Also *i. q.* بنكرد [a foreign word, and perhaps mistranscribed], i. e. The *poor-rate* of any particular people or party, which is given by them to [the poor of] their own community: so says Aboo-Mo'ádh: (L:) and ▼ مَخَالِفُ [is its pl., as also, app., مَخَالِيفُ, agreeably with rule, and] signifies the *poor-rates* of the Arabs; (JK, TA;) [as in the saying,] أُسْتُعْمِلَ فُلَانٌ عَلَى مَخَالِفِ بَنِى فُلَانٍ [*Such a one was employed as collector of the poor-rates of the sons of such a one*]. (JK.)

مَخْلُوفٌ: see خَلِيفٌ: — and أَخْلَفُ. — Also A man *affected with a looseness*, or *diarrhœa*. (JK, TA.)

مَخَالِفُ: see مِخْلَافٌ, last sentence.

مَخَالِيفُ: pl. of مِخْلَافٌ. (Ṣ, Mṣb, Ḳ, &c.) — Also Camels *that have pastured upon fresh herbs*, or *leguminous plants, and have not fed upon dry herbage, and to which their pasturing upon the former has been of no avail*. (IAạr, TA.)

قَوْلٌ مُخْتَلِفٌ [*Discordant speech*;] *speech expressing different opinions*. (Bḍ and Jel in li. 8.) — [طُرُقٌ مُخْتَلِفَةٌ *Roads leading in different directions*.]

مُسْتَخْلِفٌ: see خَالِفٌ, near the end of the paragraph. — ذَهَبَ المُسْتَخْلِفُونَ يَسْتَقُونَ a saying mentioned by Lḥ as meaning *Those going before* [or *leaving others in their places*] *went away to draw water*. (TA.)

خلق

1. خَلْقٌ signifies The act of *measuring*; or *determining the measure, proportion*, or *the like*, of a thing; and the *making* a thing *by measure*, or *according to the measure of another thing*; or *proportioning* a thing *to another thing*; syn. تَقْدِيرٌ: (Ṣ, Mṣb, Ḳ, TA, and Bḍ in ii. 19:) this is the primary meaning. (Mṣb, TA, and Bḍ ubi suprà.) You say, خَلَقَ الأَدِيمَ, (Ṣ, Mṣb, Ḳ,) aor. ـُ, (Ṣ, TA,) inf. n. خَلْقٌ (JK, Ṣ, Mṣb, Ḳ) and خِلْقَةٌ, (Ḳ,) *He measured*, or *proportioned*, (قَدَّرَ,) *the hide, and sewed it*: (Ḳ:) or *he measured*, or *proportioned*, (قدّر,) *the hide*, (JK, Ṣ, Mṣb, Ḳ,) لِمَا يُرِيدُ [*for*, or *to, that which he desired* to make of it], (JK,* TA,) or لِلسِّقَآءِ [*for*, or *to, the skin for water or milk* that he desired to make], (Mṣb,) *before cutting it*; (Ṣ, Ḳ, TA;) *he measured it* (قَاسَهُ) *to cut from it a water-bag*, or *a water-skin*, or *a boot*: (TA:) and in like manner, خَلَقَ النِّطْعَ *he measured*, &c., *the* نطع [q. v.]: when one cuts it, one says, فَرَاهُ. (Ḳ.) And خَلَقَ النَّعْلَ *He determined the measure of the sandal*, or *proportioned it*; (قَدَّرَهَا;) and *made it by measure*. (Ksh and Bḍ in ii. 19.) Hence the saying of Zuheyr, (Ṣ,) praising Herim Ibn-Sinán, (TA,)

* وَلَأَنْتَ تَفْرِى مَا خَلَقْتَ وَبَعْ *
* ضُ القَوْمِ يَخْلُقُ ثُمَّ لَا يَفْرِى *

[† *And thou indeed cuttest what thou hast measured; but some of the people measure, then will not cut*]: (Ṣ, TA:) i. e., when thou determinest upon a thing thou executest it; but others determine upon that which they do not execute. (TA.) And El-Ḥajjáj said, مَا خَلَقْتُ إِلَّا فَرَيْتُ وَلَا وَعَدْتُ إِلَّا وَفَيْتُ [† *I have not measured unless I have* afterwards *cut, and I have not promised unless I have* afterwards *performed*]. (Ṣ.) أَخْلُقُ لَكُمْ مِنَ الطِّينِ كَهَيْئَةِ الطَّيْرِ, in the Ḳur iii. 43, means *I will form for you,* (Jel,) or *I will make according to its proper measure* (أُقَدِّرُ) *for you,* (Ksh, Bḍ,) and *will form,* (Bḍ,) *of clay, a thing like the form of the bird,* or *of birds.* (Ksh, Bḍ, Jel.) — [Hence,] it signifies also The *bringing* a thing *into existence according to a certain measure,* or *proportion, and so as to make it equal* [*to another thing*], or *uniform* [*therewith*]: (Ksh and Bḍ in ii. 19:) or the *originating,* or *producing,* [a thing] *after a pattern,* or *model, which one has devised, not after the similitude of anything preexisting:* this is another meaning which it has in the [classical] language of the Arabs. (TA.) As the act of God, it signifies The *originating,* or *bringing into being* or *existence,* anything, *not after the similitude of anything pre-existing:* (TA:) [and the *creating* a thing; and thus it is generally best rendered; as meaning the *bringing into existence from a state of non-existence:* for] خَلَقَ ٱللّٰهُ الشَّىْءَ, inf. n. خَلْقٌ, means *God brought the thing into existence* (Mgh,* TA) *after it had not been:* (TA:) [or خَلْقٌ, as the act of God, signifies the *creating out of nothing:* for it is said that] اُعْبُدُوا رَبَّكُمُ ٱلَّذِى خَلَقَكُمْ, in the Ḳur ii. 19, means [*Serve ye your Lord*] *who brought you into existence when ye were nothing.* (Jel. [But in other passages of the Ḳur (vi. 2 &c.) it is said that God created (خَلَقَ) mankind of clay.]) Accord. to the A, خَلَقَ ٱللّٰهُ الخَلْقَ is a tropical phrase, meaning ‡ *God brought into existence the creation,* or *created beings,* or *mankind, according to a predetermination* (تَقْدِير) *required by wisdom.* (TA.) You say, هٰذِهِ خَلِيقَتُهُ ٱلَّتِى خُلِقَ عَلَيْهَا and ٱلَّتِى خُلِقَهَا: see خُلُقٌ. (Lḥ.) — [Hence, also,] خَلَقَ, (Ṣ, Mṣb, Ḳ, TA,) inf. n. خَلْقٌ, (TA,) ‡ *He fabricated* speech, or a saying or sentence, &c.: (Ḳ,* TA:) ‡ *he forged* (Ṣ, Mṣb, Ḳ, TA) a saying, (Mṣb,) or a lie, or a falsehood; (Ṣ, Ḳ, TA;) as also ▼ اختلق (Ṣ, Mṣb, Ḳ) and ▼ تخلّق. (Ṣ, Ḳ.) The Arabs say, حَدَّثَنَا فُلَانٌ بِأَحَادِيثِ الخَلْقِ ‡ *Such a one related to us fictitious tales* or *stories, such as are deemed pretty,* or *such as are told by night* [*for entertainment*]. (TA.) And it is said in the Ḳur [xxvi. 137], accord. to one reading, إِنْ هٰذَا إِلَّا خَلْقُ الأَوَّلِينَ, meaning ‡ *This is nought but the lying, and forging, of the ancients.* (TA.) And in the same [xxxviii. 6], إِنْ هٰذَا إِلَّا ▼ ٱخْتِلَاقٌ ‡ *This is nought but forging, and lying.* (TA.) — خَلَقَهُ, (Ḳ,) inf. n. خَلْقٌ, (TA,) also signifies *He made it smooth;* (Ḳ;) and so ▼ خلّقهُ; namely, an arrow, (Ṣ,) [and any other thing; for] of anything that has been made smooth one says, خُلِّقَ: (TA:) *he made it equable,* or *even;* namely, wood, or a stick; and so ▼ خلّقهُ, (Ḳ,) inf. n. تَخْلِيقٌ. (TA.) — خَلُقَتْ, inf. n. خَلَاقَةٌ, said of a woman, (JK, Ḳ,) *She had* [*a goodly*] *body and make:* (JK:) or *she was,* or *became, goodly in make,* or *well made.* (Ḳ. [In the CḲ, instead of حَسُنَ خَلْقُهَا, is put حَسُنَ خُلُقُهَا, meaning *She was,* or *became, good in nature,* &c.]) — And خَلُقَ, aor. ـُ, (JK, Ḳ,) inf. n. خَلْقٌ; (JK, Ṣ;*) and خَلِقَ, aor. ـَ, (Ḳ,) inf. n. خُلُوقَةٌ (TA) [and خَلَاقَةٌ, and perhaps خَلَقَةٌ q. v. infrà]; *It* (a thing) *was,* or *became, smooth,* (JK, Ḳ, TA,) and *equable,* or *even.* (TA.) [See also 12. And it seems that one says, خَلِقَتِ الصَّخْرَةُ, inf. n. خَلَقٌ, q. v. infrà, meaning *The rock was free from crack or fracture.*] — And خَلُقَ, (JK, Ṣ, Mṣb, Ḳ,) aor. ـُ; (Ḳ;) and خَلِقَ, aor. ـَ; and خَلَقَ, aor. ـُ; (Ḳ;) inf. n. (of the first, JK, Ṣ) خُلُوقَةٌ (JK, Ṣ, Ḳ) and خَلَاقَةٌ (JK, TA) and [of the second] خَلَقٌ (Ḳ) and [of the third] خُلُوقٌ; (JK, TA;) *It* (a garment) *was,* or *became, old, and worn out;* as also ▼ اخلق, (JK, Ṣ, Mṣb,) inf. n. إِخْلَاقٌ; (JK, TA;) and ▼ اخلولق. (TA.) [Hence,] اخلق ▼ دِيبَاجُهُ [lit. *His face became worn out;* meaning ‡ *it became used for mean service* [*so that it lost its grace, or was disgraced,*] by his begging. (Ḥar p. 476. [See also 4 below.]) [Hence also,] اخلق ▼ شَبَابُهُ † *His youth declined,* or *departed.* (TA.) — And خَلُقَ, (Ṣ, Ḳ,) aor. ـُ, (Ḳ,) inf. n. خَلَاقَةٌ, (Ḥam p. 522,) *He was,* or *became,* خَلِيق, i. e. جَدِير [meaning *adapted* or *disposed by nature, apt, meet,* &c.: see خَلِيقٌ, below]. (Ṣ, Ḳ.) You say, خَلُقَ لِذٰلِكَ [and بِذٰلِكَ (see خَلِيقٌ) *He was,* or *became, adapted, disposed,* &c., *for that*]; as though he were one of those in whom that was reckoned to be, and in whom the symptoms, signs, or tokens, thereof were seen. (Ṣ.) [And خَلُقَ أَنْ يَفْعَلَ ذٰلِكَ and مِنْ أَنْ and لِأَنْ يفعل ذلك and بِأَنْ يفعل ذلك *He was, or became, adapted, &c., to do that:* see خَلِيقٌ. And خَلُقَ may signify also *It was,* or *became, probable;* or *likely to happen* or *be,* or *to have happened* or *been:* see, again, خَلِيقٌ.]

2. خلّقهُ: see 1, latter half, in two places. — Also, (Ṣ, Ḳ,) inf. n. تَخْلِيقٌ, (Ḳ,) *He rubbed him over with* خَلُوق [q. v.]: (Ṣ:) or *he perfumed him:* (Ḳ:) or خلّقهُ بِخَلُوقٍ *he perfumed him with* خلوق. (TA.) And خَلَّقْتُ المَرْأَةَ بِالخَلُوقِ [*I perfumed the woman,* or *rubbed her over, with the* خلوق]. (Mṣb.) And خَلَّقَتْ جِسْمَهَا *She* (a woman) *rubbed her body and limbs over with* خلوق. (TA.)

3. خَالَقَهُمْ, (Ḳ,) inf. n. مُخَالَقَةٌ, (TA,) *He consorted* [or *comported himself*] *with them* (Ḳ, TA) *according to their natures,* or *moral characters* or *qualities;* (TA;) or *with good nature,* or *moral character* or *qualities:* (Ḳ:) or خالقهم بِخُلُقٍ حَسَنٍ has this latter meaning. (TA.) One says, خَالِصِ المُؤْمِنَ وَخَالِقِ الفَاجِرَ, (Ṣ,) or وخالق الكَافِرَ, (TA,) [*Act thou with reciprocal sincerity towards the believer, and comport thyself with the vitious,* or *the unbeliever, according to his nature,* &c. See also 3 in art. خلص, where a similar saying is mentioned.]

4. اخلق: see 1, latter part, in three places. — Also *He had old and worn-out garments.* (TA.) — اخلقهُ *He wore it out;* namely, a garment; the verb being trans. as well as intrans. (Ṣ, Mṣb, Ḳ.) [Hence,] اخلق الدَّهْرُ الشَّىْءَ † *Time wore out,* or *wasted, the thing.* (TA.) [Hence also,] one says to the beggar, أَخْلَقْتَ وَجْهَكَ ‡ (TA) [lit. *Thou hast worn out thy face;*] meaning ‡ *thou hast used thy face for mean service* [*so that it has lost its grace,* or *has become disgraced*]: and in like manner one says, أَخْلَقَ لَهُ دِيبَاجَتِى, i. e. وَجْهِى: and يُخْلِقُ دِيبَاجَتَهُ ‡ *He uses his face for mean service by begging.* (Ḥar pp. 15 and 476.) — Also, (Ḳ,) اخلقهُ ثَوْبًا, (Ṣ,) *He clad him with an old and worn-out garment.* (Ṣ, Ḳ.) And اخلقنى ثَوْبَهُ *He gave me his old and worn-out garment.* (JK.) And some say, اخلقهُ خَلَقًا *He gave him an old and worn-out garment.* (TA.) — And إِخْلَاقُ الثَّوْبِ also signifies *The cutting out of the garment:* whence the saying, to Umm-Khálid, أَبْلِى وَأَخْلِقِى [*Wear out, and cut out new*]; or, as some relate it, وَأَخْلِفِى, i. e., "and replace," which is the more likely. (TA.) — مَا أَخْلَقَهُ and أَخْلِقْ بِهِ [have both of the following significations; though it is said that] the former signifies *How likely is he, or it!* (JK, TA;) and the latter, *How well adapted* or *disposed,* or *how apt, meet, suited, suitable, fitted, fit, competent,* or *proper,* or *how worthy, is he, or it!* i. q. أَجْدِرْ بِهِ and أَحْرِ بِهِ. (TA. [See 4 in arts. جدر and حرى.])

5: see 1, a little after the middle of the paragraph. — تخلّق بِغَيْرِ خُلُقِهِ means *He affected a* خُلُق [or *nature, &c.,*] *that was not his own.* (Ṣ, Ḳ.) And تخلّق بِكَذَا *He feigned such a thing, it not being in his nature,* or *not being created in him.* (TA.) And تخلّق لِلنَّاسِ بِمَا لَيْسَ مِنْ نَفْسِهِ, occurring in a trad., [*He affected, to men, a nature, &c., that did not belong to him;* or] *he pretended* [*to men*] *that there was in his nature* (فِى خُلُقه) *that which was contrary to his real intention;* (Mbr, TA;) or *that which was contrary to what he had in his heart:* the verb is similar to تَصَنَّعَ and تَجَمَّلَ. (TA.) — تخلّق بِهِ; (Ṣ, Ḳ;) and تخلّقت بِهِ; (Mṣb;) *He was,* or *became, rubbed over,* (Ṣ,) or *perfumed;* (Ḳ;) and *she was,* or *became, so;* (Mṣb;) [or *he rubbed himself over,* or *perfumed himself;* and *she did so;*] *with it;* (Ṣ, Mṣb, Ḳ;) namely, with خَلُوق. (Ṣ, Mṣb.)

8: see 1, latter half, in two places.

12. اخلولق, said of the back (مَتْن) of a horse, *It was,* or *became, smooth;* (Ḳ;) [like خَلُقَ and خَلِقَ; or *very smooth;* for] the verb is of a form intensive in signification. (TA. [See its part. n., مُخْلَوْلِقٌ, below.]) — Said of a رَسْم, [i. e. a trace,

or a remain or relic marking the place of a house or the like and cleaving to the ground,] *It was*, or *became, even with the ground.* (Ṣ, Ḳ.) —— اخلولق السَّحَابُ *The clouds became equable*, or *uniform*, (JK, Ṣ, Ḳ, TA,) *their sides becoming conjoined*; or, as some say, *they became smooth*; (TA;) and, (Ḳ,) or as some say, (Ṣ, TA,) *they became adapted*, or *disposed, to rain*; (Ṣ, Ḳ, TA;) as though they were rendered smooth: or *they became collected together* after separation, *and prepared to rain.* (TA.) And اخلولقت السَّمَآءُ أَنْ تَمْطُرَ *The sky was near, and likely, to rain.* (TA.) —— See also 1, latter part.

خَلْقٌ inf. n. of خَلَقَ. (JK, Ṣ, Mṣb, Ḳ, &c.) You say رَجُلٌ تَامُّ الخَلْقِ [*A man complete*, or *perfect, in respect of make*, or *proportion*, &c.]. (Ṣ, Ḳ.*) [See also خِلْقَةٌ.]) [In this and similar instances,] الخَلْقُ signifies *The fashion of the outer man, and its* [*peculiar*] *qualities and attributes*; like as الخُلُقُ signifies "the fashion of the inner man," &c. (TA.) —— الخَلْقُ is also used in the sense of ▼ المَخْلُوقُ [meaning *What is created*; *the creature*]: (TA, and Bḍ in xxiii. 17, &c.:) [and, collectively, *the creation*; as meaning *the beings*, or *things, that are created*;] *all created things*: (Bḍ ubi suprà, &c.:) and [particularly] *mankind*; as also ▼ الخَلِيقَةُ: (Ṣ,* Ḳ:) and *mankind and the jinn*, or *genii, and others*: (Jel in lv. 9, &c.:) and ▼ الخَلِيقَةُ and [its pl.] خَلَائِقُ signify the same: you say, هُمْ خَلِيقَةُ ٱللّٰهِ and also هُمْ خَلْقُ ٱللّٰهِ [*They are the creatures of God*]: الخَلْقُ being originally an inf. n.: (Ṣ, TA:) and Lḥ mentions [an instance of its having a pl., in] the saying, لَا وَٱلَّذِى خَلَقَ الخُلُوقَ مَا فَعَلْتُ كَذَا meaning [*No, by Him who created*] *all creatures*, [*I did not such a thing.*] (TA.) In the saying, فَلَيُغَيِّرُنَّ خَلْقَ ٱللّٰهِ, in the Ḳur [iv. 118, lit. *And they shall alter the creature of God*], some say that castration is meant: (TA: [and Bḍ includes, with this, other unnatural actions:]) or the meaning is, *the religion of God*; (Bḍ, Jel, TA;) accord. to El-Ḥasan and Mujáhid. (TA.) And لَا تَبْدِيلَ لِخَلْقِ ٱللّٰهِ, in the Ḳur [xxx. 29], means, accord. to Ḳatádeh, [*There shall be no changing*, or *altering*,] *of the religion of God.* (TA.) —— خَلْقٌ also signifies Anything *made smooth.* (TA.) [See also مُخْلَقٌ.]

خُلْقٌ: see خُلُقٌ, in four places.

خَلَقٌ inf. n. of خَلِقَ: as such, signifying The *being smooth* [&c.]. (JK, Ṣ.*) [As such also,] in a rock, *Freedom from crack or fracture.* (Ṣ, Ḳ.) —— [And, as such,] The *being old, and worn out.* (Ḳ.) —— [Hence, used as an epithet,] *Old, and worn out*: (Ṣ, Mṣb, Ḳ:) [and as an epithet in which the quality of a subst. is predominant; meaning *an old and worn-out garment* or *piece of cloth*:] pl. خُلْقَانٌ (Ṣ, Ḳ) and أَخْلَاقٌ. (Ṣ,* Ḳ,* TA.) And [as an epithet] it is masc. and fem.; (Ṣ, Ḳ;) because it is originally an inf. n., the inf. n. of أَخْلَقَ meaning "smooth," (Ṣ,) [or rather of خَلِقَ meaning "it was, or became, old, and worn out;" although it has pls.; and] IB mentions an instance of its dual, خَلَقَانِ: (TA:) Ks says, We have not heard them say, خَلَقَةٌ in any instance: (Lḥ, TA:) Fr says that it is without ة [as a fem. epithet] because it was originally used as a prefixed noun; for one said, أَعْطِنِى خَلَقَ عِمَامَتِكَ and خَلَقَ جُبَّتِكَ [lit. meaning *Give thou to me what is old, and worn out, of thy* جُبَّة and *of thy turban*]; but Ez-Zejjájee says that this is nought. (TA.) You say ثَوْبٌ خَلَقٌ [*An old and worn-out garment* or *piece of cloth*], and مِلْحَفَةٌ خَلَقٌ [*an old and worn-out outer wrapping garment*]: (Ṣ:) also رُمَّةٌ خَلَقٌ [*an old and worn-out piece of rope*]: and دَارٌ خَلَقٌ [*an old and decayed house*]: and جِسْمٌ خَلَقٌ [*an old and wasted body*]. (TA.) One says also ثَوْبٌ أَخْلَاقٌ, meaning *A garment*, or *piece of cloth, altogether*, or *wholly, old and worn out*; (Fr, Ṣ, Ḳ;) every portion of it being خَلَقٌ; (Fr;) like as they said بُرْمَةٌ أَعْشَارٌ &c.: (Ṣ:) and in like manner, مُلَآءَةٌ أَخْلَاقٌ. (IAạr.) And Ks mentions the saying, أَصْبَحَتْ ثِيَابُهُمْ خُلْقَانًا وَخَلَقُهُمْ جُدُدًا [*Their garments became old, and worn out*; *and their old and worn-out garments* became replaced by *new*]; with the sing. [in the latter clause] in the place of the pl. خُلْقَان: (TA:) or جُدُدًا may be here put for جَدِيدًا. (L in art. جد.) In the phrase ▼ مُلَيْحِفَةٌ خُلَيْقٌ [*An outer wrapping garment that is a little*, or *somewhat, old, and worn out*], the dim. is without ة because it is [the dim. of] an epithet [applied without ة to a fem. n.], and ة is not affixed to the dims. of epithets [of this kind]: it is like نُصَيْفٌ dim. of نَصَفٌ an epithet applied to a woman. (Ṣ, Ḳ.*) [See Lumsden's Arab. Gram. p. 623: but some of the grammarians consider these instances as anomalous.]) —— بَيْعَ ذِى الخَلَقِ, and بَاعَهُ بِيعَةَ الخَلَقِ, the latter as used by a poet, [lit. *He bought it*, or *sold it*, (app. the former,) *as one buys*, or *sells, the old and worn-out garment*, like as we say "dog-cheap," and "cheap as dirt"], are phrases mentioned, but not explained, by IAạr, who cites the following saying:

* أَبْلِغْ فَزَارَةَ أَنِّى قَدْ شَرَيْتُ لَهَا *
* مَجْدَ الحَيَاةِ بِسَيْفِى بَيْعَ ذِى الخَلَقِ *

[app. meaning *Tell thou Fezárah that I have purchased for them life-long glory* (lit. *the glory of life*), *with my sword*, as cheaply, i. e. as easily, *as one purchases the old and worn-out garment*]. (TA.) —— سَحَابَةٌ خَلَقَةٌ: see the next paragraph.

خَلِقٌ [part. n. of خَلِقَ]. —— [Hence,] سَحَابَةٌ خَلِقَةٌ *A cloud in which is a sign*, or *trace, of rain*; as also ▼ خَلِيقَةٌ: (Ṣ, Ḳ:) or *a cloud giving hope of rain*; as also ▼ خَلَقَةٌ; (JK;) both are said by IAạr to signify the same: (TA:) and سَحَابَةٌ ▼ خَلِقَةٌ [alone, as a subst., or probably خَلَقَةٌ,] *a cloud that is equable*, or *uniform, giving hope of rain.* (Aboo-Sa'eed, Ḳ.)

خُلُقٌ (Ṣ, Mṣb, Ḳ) and ▼ خُلْقٌ (Ṣ, Ḳ) *A nature*; or *a natural, a native*, or *an innate, disposition* or *temper* or *the like*; syn. سَجِيَّةٌ, (Ṣ, Mṣb, Ḳ, TA,) and طَبْعٌ; (Ḳ, TA;) *of which one is created*: (TA:) and ▼ خَلِيقَةٌ signifies [the same; i. e.] the فِطْرَة [or *nature*, &c.,] (Ṣ, Mṣb, Ḳ, TA) *of which a man is created*; (TA;) like [خُلُقٌ and] ▼ خُلْقٌ: (Ḳ, TA: [in the CḲ, erroneously, خَلْق:]) and ▼ خَلِيقَةٌ [also] signifies [*the same*; i. e.] the طَبِيعَة [or *nature*, &c.] (Ṣ, Ḳ, TA) *with which a man is created*: (TA:) the proper signification of خُلُقٌ is [the *moral character*; or] the *fashion of the inner man*; i. e. his *mind*, or *soul, and its peculiar qualities and attributes*; like as خَلْقٌ signifies the "fashion of the outer man, and its [peculiar] qualities and attributes:" it signifies also *custom* or *habit* [as being *a second nature*]: (TA:) and, as also ▼ خُلْقٌ, [which is merely a contraction thereof, and therefore identical with it in all its senses,] *manliness*; syn. مُرُوَّةٌ: and *religion*: (IAạr, Ḳ:) the pl. is أَخْلَاقٌ only: (TA:) [this is often used as signifying *morals*: and *ethics*:] and the pl. of ▼ خَلِيقَةٌ in the sense explained above [said in Ḥar p. 193 to be that of خُلُقٌ] is خَلَائِقُ. (Ṣ.) It is said in a trad., لَيْسَ شَىْءٌ فِى المِيزَانِ أَثْقَلَ مِنْ حُسْنِ الخُلُقِ [*Nothing is heavier in the balance* in which good and evil will be weighed *than goodness of the moral character*, &c.] (TA.) And one says, هٰذِهِ خَلِيقَتُهُ ▼ الَّتِى خُلِقَ ▼ عَلَيْهَا and ▼ خَلِيقَتُهَا and ▼ الَّتِى خُلِقَ *This is his nature*, &c., *of which he was created.* (Lḥ.) And ▼ إِنَّهُ لَكَرِيمُ الخَلِيقَةِ *Verily he is generous in respect of nature*, &c. (AZ.) And صَارَ ذٰلِكَ لَهُ خُلُقًا *That became to him* [*a second nature, a habit*, or] *a thing to which he was habituated.* (TA.) It is said in the Ḳur [xxvi. 137], إِنْ هٰذَا إِلَّا خُلُقُ الأَوَّلِينَ *This is nought but a custom of the ancients.* (TA.) And in the same [lxviii. 4], وَإِنَّكَ لَعَلَى خُلُقٍ عَظِيمٍ *And verily thou art of a great religion.* (Jel, TA.) And in a trad. of 'Áïsheh, كَانَ خُلُقُهُ القُرْآنَ, meaning *That whereto he clung was the Ḳur-án*, with its rules of discipline and its commands and its prohibitions, and the excellences and beauties and gracious things comprised in it. (TA.) —— نَوْمَةُ الخُلُقِ [i. e. الخُلُقِ or الخُلْقِ] *The sleep of midday*, which was prescribed by the Prophet. (Ḥar p. 223. [See also حُمْقٌ and خُرْقٌ.])

خُلُوقَةٌ *Smoothness*; (Ḳ, TA;) as also ▼ خُلُوقَة and ▼ خَلَاقَةٌ: (Ḳ:) but the second of these three, correctly speaking, [as also the third, accord. to analogy, and perhaps the first also,] is an inf. n. of خَلِقَ. (TA.)

خِلْقَةٌ [primarily signifies *A mode*, or *manner*, of خَلْق, generally as meaning *creation*; *a particular make*: and hence,] *constitution*; syn. تَرْكِيبٌ: (Mgh:) [and particularly the *natural constitution* of an animated being, *as created in the womb of the mother*; also termed فِطْرَة:] see also خَلْقٌ. You say رَجُلٌ حَسَنُ الخِلْقَةِ [*A man goodly*, or *beautiful, in respect of make*]. (A, TA.) فِى طَرِيقٍ فِى مَسْلَكٍ هُوَ خِلْقَةٌ means ▼ خِلْقِىٍّ أَصْلِىٍّ [*In a way*, or *road, that is natural*, and *original*]. (Mgh.)

خَلَقَةٌ: see خَلِقٌ.

خِلْقِىٌّ *Natural; not accidental: [constitutional: of, or relating to, or belonging to, the natural constitution of an animated being, as created in the womb of the mother:]* rel. n. of خِلْقَةٌ. (Mṣb.) You say عَيْبٌ خِلْقِىٌّ *A natural fault* or *imperfection* &c. (Mṣb.) And صِفَةٌ خِلْقِيَّةٌ [*A natural quality*]; opposed to اِخْتِيَارِيَّةٌ. (Mṣb in art. مدح.) See also خِلْقَةٌ.

خَلَقِىٌّ One *who wears old and worn-out clothes.* (TA.)

خُلْقَانِىٌّ *A seller of old and worn-out clothes.* (TA.)

خَلَاقٌ *A share,* or *portion:* (JK, Ṣ, Mṣb:) and *a good, just,* or *righteous, share* or *portion:* (JK:) or *a full, a complete,* or *an abundant, share* or *portion of good,* (Ḳ, TA,) and *of goodness,* or *righteousness:* (TA:) and *religion:* or *a share,* or *portion, thereof.* (TA.) One says, لَا خَلَاقَ لَهُ فِى الآخِرَةِ *There is no share,* or *portion,* [*of good*] *for him in the final state of existence.* (Ṣ. [See the Ḳur iii. 71, &c.]) And لَا خَلَاقَ لَهُ *He has no desire for good, nor righteousness in religion.* (TA.)

خِلَاقٌ: see the next paragraph.

خَلُوقٌ *A certain species of perfume;* (JK, Ṣ, Mgh, Mṣb, Ḳ;) also termed ▼ خِلَاقٌ; (Lḥ, Mṣb, Ḳ;) accord. to some of the lawyers, (Mṣb,) *fluid,* (Mgh, Mṣb,) *but of thick consistence;* (L, voce نَضْخٌ;) *and in which is a yellowness:* (Mgh, Mṣb:) *it is composed of saffron and other things; and redness and yellowness are predominant in it:* it is forbidden [to men], because it is of the perfumes of women, who use it more than do men. (TA.)

خَلِيقٌ, applied to a man, (Ṣ, TA,) *Perfect,* or *complete, in make;* (TA;) as also ▼ مُخْتَلَقٌ: (Ḥam p. 561:) or *perfect,* or *complete, in make, and just in proportion;* (Ṣ, TA;) and so ▼ the latter; (Ṣ, Ḳ, TA; [in the CḲ, erroneously, مُخْتَلِق; in the TA expressly said to be of the pass. form;]) fem. of the former with ة: (TA:) or ▼ both signify *goodly,* or *beautiful, in make:* or the former is not applied to a man; but ▼each, with ة, signifies a woman *having* [*a goodly*] *body and make:* (TA, in which this signification is said to be tropical:) and خَلِيقٌ and خَلِيقَةٌ are alike, (JK, TA,) accord. to Lḥ, (TA,) in this last sense: (JK:) or the former of these two may be pl. [or coll. gen. n.] of the latter, like as شَعِيرٌ is of شَعِيرَةٌ: (TA:) and ▼ مُخْتَلَقٌ signifies anything *just in proportion:* (IF, TA:) ▼ مُخَلَّقٌ, also, signifies *perfect,* or *complete, in make;* applied to a camel (جمل): (TA:) [or جمل, here may be a mistranscription for حَمْل; for] ▼ مُضْغَةٌ مُخَلَّقَةٌ signifies [*a fœtus when it has become like a lump of flesh*] *perfect,* or *complete, in make;* (Fr, Ṣ, Ḳ;) so in the Ḳur xxii. 5; (Fr, TA;) or *of which the make has become apparent.* (IAạr, TA.) — Also *Adapted* or *disposed* [*by nature*], *apt, meet, suited, suitable, fitted, fit, proper, competent,* or *worthy;* (KL, PṢ;) syn. جَدِيرٌ (Ṣ, Ḳ) and حَرِىٌّ (TA) [and حَقِيقٌ &c.: pl. خُلَقَآءُ, and Freytag adds خُلُقٌ]. You say, فُلَانٌ خَلِيقٌ لِكَذَا, i. e. جَدِيرٌ بِهِ [*Such a one is adapted* or *disposed by nature,* &c., *for such a thing*]; as though he were one of those in whom that was reckoned to be, and in whom the symptoms, signs, or tokens, thereof were seen. (Ṣ.) [And هُوَ خَلِيقٌ لِلْخَيْرِ *He is adapted* or *disposed by nature to good;* i. e., *to be,* or *to do,* or *to effect,* or *to produce, what is good.*] And إِنَّهُ لَخَلِيقٌ أَنْ يَفْعَلَ ذٰلِكَ and بِأَنْ يفعل ذلك and مِنْ أَنْ يفعل ذلك and لِأَنْ يفعل ذلك [*Verily he is adapted* or *disposed* &c. *for doing that;* or *worthy to do it*]: so says Lḥ: and he adds that the Arabs say, يَا خَلِيقُ بِذٰلِكَ, using the nom. case; and يَا خَلِيقًا بِذٰلِكَ, using the accus. case; [the latter being the usual form; both meaning *O thou who art adapted* or *disposed* &c. *for that;*] but ISd says, I know not the reason of this. (TA.) And هٰذَا ▼ مَخْلَقَةٌ لِذَاكَ, i. e. مَجْدَرَةٌ لَهُ [*This is one that is adapted* or *disposed* &c. *for that*]: (Ṣ, Ḳ:*) and هٰذَا الأَمْرُ مَخْلَقَةٌ ▼ لَكَ [*This affair,* or *thing, is one that is adapted* &c. *for thee*]: and إِنَّهُ مَخْلَقَةٌ ▼ مِنْ ذٰلِكَ [*Verily it is adapted* &c. *for that*]: like مَجْدَرَةٌ and مَحْرَاةٌ: and مَقْمَنَةٌ: and in like manner one says of two, and of more than two, and of a feminine: so says Lḥ. (TA.) [▼ مَخْلَقَةٌ properly signifies *A place,* and hence *a thing, an affair,* and *a person, adapted* or *disposed* &c.: it is of the same class as مَعْسَاةٌ and مَظِنَّةٌ and مَئِنَّةٌ.] خَلِيقٌ also signifies *Habituated,* or *accustomed.* (PṢ, TA.*) And one says, إِنَّهُ لَخَلِيقٌ, i. e. لَحَرِىٌّ, meaning *Verily it is probable;* or *likely to happen* or *be,* or *to have happened* or *been.* (TA.) And هُوَ خَلِيقٌ لَهُ *He,* or *it, is like to him,* or *it.* (JK, TA.) — سَحَابَةٌ خَلِيقَةٌ: see خَلْقٌ. — [See also خَلِيقَةٌ, which, in several senses, is a fem. epithet used as a subst.]

خُلَيْقٌ: see خَلْقٌ (of which it is the dim.), in the latter half of the paragraph.

خَلَاقَةٌ: } see خُلُوقَةٌ.
خُلُوقَةٌ: }

خَلِيقَةٌ: see خَلْقٌ, in two places. — Also The *beasts,* or *brutes.* (En-Naḍr, Ḳ.) The saying, respecting the خَوَارِج [a sect of heretics, or schismatics], هُمْ شَرُّ الخَلْقِ وَالخَلِيقَةِ is explained by En-Naḍr as meaning [*They are the worst of mankind and*] *of the beasts,* or *brutes.* (TA.) — And *A well* (بِئْرٌ) *just dug:* (AA, Ḳ:) or *a well in which is no water:* or *a hollow, cavity, pit,* or *hole, formed by nature in the ground:* or *a small hollow* or *cavity, in a mountain, in which water remains and stagnates:* accord. to IAạr, خلق [app. خُلُقٌ, pl. of خَلِيقَةٌ, like as مُدُنٌ and صُحُفٌ are pls. of مَدِينَةٌ and صَحِيفَةٌ,] signifies *wells recently dug.* (TA.) — And *Land* (أَرْضٌ) *that is dug.* (TA.) — See also خُلُقٌ, in four places.

خُلَيْقَآءُ [dim. of خَلْقَآءُ fem. of أَخْلَقُ]: see أَخْلَقُ, in three places.

خَمَائِرُ q. v. i. e. الخَلَائِقُ [pl. of خَلِيقَةٌ]. — الخَلَائِقُ, i. e. الْمَآءِ, i. e. *Four large and smooth masses of stone at the head of the well, upon which the drawer of the water stands.* (TA.) Accord. to Ibn-'Abbád, حَوْضٌ بَادِى الخَلَائِقِ means [*A watering-trough of which*] *the* [*stones termed*] نَصَائِب [*appear*]. (JK, TA. [See نَصِيبَةٌ.])

أُخْلُقٌ: see أَخْلَقُ.

خَلَّاقٌ: see the next paragraph.

خَالِقٌ [act. part. n. of خَلَقَ:] *A worker in leather* and *the like;* (Ḳ, TA;) because he measures first, and then cuts. (TA.) To خَالِقَات, meaning Women *working in leather,* as engaged in dividing a hide (أَدِيم), El-Kumeyt likens genealogists. (TA.) — الخَالِقُ, as an epithet applied to God, (Ḳ, Mṣb, TA,) properly, *He who brings into existence according to the proper measure,* or *proportion,* or *adaptation;* (TA;) [and hence, *the Creator;* or] *the Originator, not after the similitude of anything pre-existing:* (Ḳ:) or *He who hath brought into existence all things after they had not been in existence:* (Az, TA:) and ▼ الخَلَّاقُ signifies the same; (Mṣb,* TA;) [i. e. *the Creator of all things;* or, as an intensive epithet, *the Great Creator;*] or *the Creator of many creatures:* (Ksh and Bḍ and Jel, in xxxvi. 81:) Az says that this epithet, with the article ال, may not be applied to any but God. (Mṣb.) Accord. to IAmb, تَبَارَكَ اللّٰهُ أَحْسَنُ الخَالِقِينَ means احسن المُقَدِّرِينَ [i. e. *Blessed be God, the Best of those who make things according to their proper measures,* or *proportions,* or *adaptations*]. (TA.)

خَوَالِقُ [a pl. of which the sing. is not mentioned] *Smooth mountains:* so in the saying of Lebeed,

* وَالأَرْضُ تَحْتَهُمُ مِهَادًا رَاسِيًا *
* ثَبَتَتْ خَوَالِقُهَا بِصُمِّ الجَنْدَلِ *

[*And the earth beneath them a firm expanse; its smooth mountains being rendered fast by hard and solid stones*]. (Ḳ, TA. [In the CḲ, بِصُمِّ is erroneously put for بِصُمِّ.])

أَخْلَقُ *Smooth:* (JK, Ḳ:) *smooth and solid;* (Ṣ, Ḳ, TA;) applied in this sense to anything: (TA:) *smooth and firm:* (JK:) fem. خَلْقَآءُ. (JK, Ṣ, Ḳ.) You say حَجَرٌ أَخْلَقُ *Stone that is smooth* (Ḳ, TA) *and solid, upon which nothing makes an impression.* (TA.) And صَخْرَةٌ خَلْقَآءُ *A rock,* or *great mass of stone, smooth* (Ḳ, TA) *and solid:* (TA:) or *free from crack and fracture.* (Ṣ, Ḳ, TA.) And فِرْسِنٌ خَلْقَآءُ *A camel's foot in which is no crack.* (Ibn-'Abbád, Ḳ.) And هَضْبَةٌ خَلْقَآءُ [*A hill, or the like,*] *destitute of herbage or vegetation.* (TA.) — [Hence,] † *Poor;* syn. فَقِيرٌ. (Ḳ.) You say رَجُلٌ أَخْلَقُ مِنَ المَالِ † *A man destitute of property.* (TA.) And it is said in a trad., لَيْسَ الفَقِيرُ فَقِيرَ المَالِ إِنَّمَا الفَقِيرُ الأَخْلَقُ الكَسْبِ, i. e. † [*The poor in respect of property is not the poor: the poor is only*] *he who has no good deeds for which he will be rewarded* in the world to come. (TA, in two places.) — الأَخْلَقُ also signifies *The exterior of a horse's hoof.* (JK.) — And خَلْقَآءُ, (JK, Ṣ, Ḳ,) applied to a woman, (JK, Ṣ,) *Impervia coëunti;* (Ṣ, Ḳ, TA;) as also ▼ خُلَّقٌ. (Ibn-'Abbád, Ḳ.) — See also خَلَقٌ. — And الخَلْقَآءُ [used as a subst.] *The sky;* because of its smoothness and evenness. (TA.) — And *The side* of a camel &c. (Ḳ.) One says also, ضَرَبْتُ خَلْقَآءَ جَنْبِهِ (Ḳ, TA [in the CḲ (على خَلْقاءِ جَنْبِهِ)]) *I struck the outer part of his side.* (TA.) — And *The interior* (Lth, Ḳ, TA) *and smooth part* (Lth, TA,) of the غَار, (Ḳ,) i. e., of [the upper part of the interior of the mouth, or] what is termed الغَارُ الأَعْلَى; (Lth, TA;) as also ▼ الخُلَيْقَآءُ [the dim. of الخَلْقَآءُ]: (Lth, Ḳ, TA:) or both signify *what appears of* the غار: and the dim. form is that which is predominant in this case. (TA.) — And *The part* of the forehead *that is even* (JK, Ḳ, TA) *and smooth;* (TA;) as also ▼ الخُلَيْقَآءُ. (JK, Ḳ, TA.) One says, سُحِبُوا عَلَى خَلْقَاوَاتِ جِبَاهِهِمْ [*They were dragged along upon the even and smooth parts of their foreheads*]. (TA [in which this is said to be tropical].) — ▼ خُلَيْقَآءُ الفَرَسِ *That* [*part*] *of the horse which is like the* عِرْنِين [or *upper part of the nose*] *of man;* (Ṣ, Ḳ;) *the part where the forehead of the horse meets the narrow portion of the bone of the nose:* AO says that the خُلَيْقَاوَانِ in the face of the horse are [the *two parts*] *where his forehead meets the bone of his nose, on the right and left of the* خُلَيْقَآء, *sloping towards the eye;* and the خُلَيْقَآء is [the *part*] *between the eyes;* and some call it the خَلْقَآء. (TA.) — إِنَّ أَخْلَقَ بِكَ أَنْ تَفْعَلَ كَذَا is a phrase mentioned by Ks, as meaning *Verily the most apt, meet, suitable, fit,* or *proper, thing for thee to do is such a thing.* (TA.)

مُخْلَقَةٌ: see خَلِيقٌ, in four places, in the latter half of the paragraph.

مُخَلَّقٌ: see خَلِيقٌ, in two places, in the former half of the paragraph. — Also, applied to an arrow, *Made smooth* (Ṣ, Ḳ, TA) *and even.* (TA.) [See also خَلَقٌ, last signification; and مُخْتَلَقٌ.]

مَخْلُوقٌ [pass. part. n. of خَلَقَ. When used as a subst., signifying *A creature,* or *created thing,* its pl. is مَخْلُوقَاتٌ]. See خَلْقٌ. — قَصِيدَةٌ مَخْلُوقَةٌ ‡ [*An ode that is forged;* or] *ascribed to a person not its author.* (Ṣ, Ḳ,* TA.)

مُخْتَلَقٌ: see خَلِيقٌ, first sentence, in five places. — Also *Made smooth.* (TA.) [See also مُخَلَّقٌ.] — And *Generous in* [*nature,* or] *natural dispositions.* (Ḥam p. 561.) — مُخْتَلَقٌ لِلْمُلْكِ, in a verse of Dhu-r-Rummeh, means *Created of a nature fitting for dominion:* (Ṣ, TA:) and so لِلْأَصْحَابِ [*for companions*]; as in a verse of Ibn-Aḥmar. (TA.)

مُخْلَوْلِقٌ *Very smooth;* its measure being one of those that denote intensiveness. (Ḥam p. 358.)

خلنج

خَلَنْجٌ a Persian word, arabicized, (Ṣ,) *A kind of tree,* (Ṣ, Ḳ,) *of the wood of which vessels are made:* (TA:) or *any* [*bowl of the kinds called*] جَفْنَة *and* صَحْفَة, or *other vessel, made of wood having variegated streaks:* (L:) pl. خَلَانِجُ. (Ṣ, Ḳ.) The word is mentioned [in the Ṣ and Ḳ in art. خلج; but] in the L and some other lexicons in a separate art., because all the letters of a word which is not Arabic are regarded as radical. (MF.)

خلو

1. خَلَا, (Ṣ, Mṣb, Ḳ,) aor. يَخْلُو, (Ṣ, Mṣb,) inf. n. خُلُوٌّ, (Ṣ, Mṣb, Ḳ,) or خَلَآءٌ, (Mṣb,) or both, (Ḳ,) said of a place, (Ḳ,) of a place of alighting or abode, (Mṣb,) and of a thing, (Ṣ, TA,) *It was,* or *became, empty, vacant, void, devoid, destitute,* or *unoccupied;* (Ḳ, TA;) *had none,* and *nothing, in it;* (TA;) as also ▼ اخلى, (Mṣb, Ḳ,) and ▼ استخلى. (Ḳ.) خَلَا المَكَانُ مِنَ النَّاسِ وَالمَآءِ وَالكَلَإِ means *The place was,* or *became, devoid,* or *destitute, of human beings and water and herbage* or *pasturage; without human beings* &c.] Of a place of alighting or abode, you say, خَلَا مِنْ أَهْلِهِ and ▼ اخلى [*It was,* or *became, devoid,* or *destitute, of its occupants*]. (Mṣb.) And of a vessel, خَلَا مِمَّا فِيهِ *It was,* or *became, empty of what was in it.* (Mgh.) And خَلَوْتُ عَنِ الطَّعَامِ (Ṣ) *I became empty, in the belly, of food;* (PṢ;) and ▼ أَخْلَيْتُ عَنْهُ signifies the same. (Ṣ.) And خَلَا مِنَ العَيْبِ, (Mṣb,) or عَنِ الأَمْرِ, and مِنْهُ, (Ḳ,) inf. n. خُلُوٌّ, *He was,* or *became, free* (Mṣb, Ḳ) *from fault,* (Mṣb,) or *from the thing,* or *affair:* (Ḳ:) and, accord. to IAạr, خلا alone signifies *he was,* or *became, free from a fault,* or *the like, of which he was accused,* or *suspected.* (TA.) And خَلَتْ عَنْ مَانِعِ النِّكَاحِ, inf. n. خُلُوٌّ, is said of a woman [as meaning *She was,* or *became, free from any obstacle to marriage*]. (Mṣb.) Accord. to the Ḳ, خَلَا مَكَانُهُ [lit. *His place became vacant*] means ‡ *he died:* but accord. to IAạr, خَلَا alone has this signification [from the same verb signifying مَضَى, explained below]: and if you add مكانه, you say خَلَّى, with teshdeed; which see below. (TA.) You say also, خَلَا لَكَ الشَّىْءُ and ▼ اخلى, both signifying the same, (AA, Ṣ, TA,) *i. q.* فَرَغَ [i. e. *The thing was,* or *became, vacant,* or *unoccupied, for thee:* (see an ex. of the former verb in a saying of Ṭarafeh cited voce جَوٌّ:) and hence, *the thing was,* or *became, exclusively for thee*]. (TA.) AA cites as an ex. the saying of Maạn Ibn-Ows,

أَعَاذِلُ هَلْ يَأْتِى القَبَائِلَ حَظُّهَا
مِنَ المَوْتِ أَمْ أَخْلَى ▼ لَنَا المَوْتُ وَحْدَنَا

[*O censurer, does their share of death come to the tribes* in common, *or is death exclusively for us alone?*]. (Ṣ, TA.) See also the paragraph, below, commencing with خَلَا as a word denoting exception. — [Hence,] خَلَا and ▼ اخلى, (Ṣ, Ḳ,) said of a man, (TA,) or the same two verbs followed by بِنَفْسِهِ, said of a man, (Mṣb,) both signify the same; (Ṣ;) *He was,* or *became,* [*without any companion,* i. e.] *alone, by himself;* (Mṣb;) or *he became* (وَقَعَ [q. v.]) *in a vacant place, in which he was not pressed against,* or *straitened.* (Ḳ.) And خَلَا بِهِ, (Ṣ, Mṣb, Ḳ,) and إِلَيْهِ, (Ṣ, Ḳ,) and مَعَهُ, (Ḳ,) inf. n. خَلْوَةٌ (Ṣ, Mṣb, Ḳ) and خَلَآءٌ (Ṣ, Ḳ) and خُلُوٌّ, (Ḳ, TA,) or خَلْوٌ, (CḲ,) or the first of these, i. e. خَلْوَةٌ, is a simple subst., and the second and third are the inf. ns.; (TA;) and ▼ اخلى بِهِ, (Lḥ, Ḳ,) and ▼ اخلاهُ, (Ṣ, Ḳ,) and ▼ استخلى بِهِ; (Ḳ; [the last omitted in the CḲ;]) *He was,* or *became, alone with him;* (Mṣb;) *he was,* or *became, in company with him,* or *he met him,* or *had a meeting* or *an interview with him, in a vacant place,* or *a place unoccupied* [*by others,* i. e., *in a private place*]. (Ṣ, Ḳ.) In the saying in the Ḳur [ii. 13], وَإِذَا خَلَوْا إِلَى شَيَاطِينِهِمْ, it is said that إِلَى is used in the sense of مَعَ, [so that the meaning is *And when they are alone with their devils,*] as in that other saying in the Ḳur [iii. 45 and lxi. 14], مَنْ أَنْصَارِى إِلَى ٱللّٰهِ. (Ṣ.) A man says to another man, أَخْلُ مَعِى حَتَّى أُكَلِّمَكَ, i. e. *Be* [or *come*] *thou alone with me* [*that I may speak to thee* in private]. (TA.) And one says, خَلَا بِزَوْجَتِهِ, inf. n. خَلْوَةٌ, [but see what is said of this noun above,] *He was,* or *became, alone with his wife:* but [properly speaking, according to the law,] the term خَلْوَةٌ [or خَلْوَةٌ صَحِيحَةٌ, in this case,] is not used unless it be *with the enjoyment of* المُفَاخَذَة, [see 3 in art. فخذ,] and then it has an effect upon the circumstances of the marriage [by its rendering obligatory the payment of the dowry, though consummation has not taken place]: if with consummation, the act is termed دُخُولٌ. (Mṣb.) You say also, ▼ أَخْلِ بِأَمْرِكَ and *Be thou alone in thine affair, with none to take part with thee in it; confine thyself to it exclusively of other things.* (TA. [See also 5.]) And ▼ أَخْلِ إِلَيْكَ *Keep thou to thine affair, and be alone in it, with none to take part with thee therein.* (JK.) And ▼ استخلى البُكَآءَ [app. for بِالبُكَآءِ] *He was,* or *became, alone in weeping, with none to participate with him in it.* (TA.) [And خَلَا لِلأَمْرِ: see 5.] And خَلَا عَلَى بَعْضِ الطَّعَامِ *He restricted himself to a portion of the food.* (Ḳ.) Temeem say, خَلَا فُلَانٌ عَلَى اللَّبَنِ وَاللَّحْمِ (JK,* TA) i. e. *Such a one fed upon milk and flesh-meat alone;* (JK;) or *such a one ate not, nor mixed, anything with milk and flesh-meat:* and Kináneh and Ḳeys say ▼ أَخْلَى. (Lḥ, JK,* TA.) [And it seems to be indicated in the T that خَلَوْا signifies *They selected a she-camel for a* خَلِيَّة, q. v.: or *i. q.* تَخَلَّوْا بِخَلِيَّةٍ: see 5.] — خَلَا also signifies *He devoted himself to religious services* or *exercises* [app. *in solitude,* or *seclusion,* or *in a* خَلْوَة; or because one generally does so in solitude; or because the doing so involves abstraction from other affairs]. (TA. [See also 5; and see مُسْتَخْلٍ.]) — And خَلَا بِهِ

[sometimes] signifies ‡ *He mocked at, scoffed at, laughed at, derided,* or *ridiculed, him:* (Lḥ, Ṣ, Z, Ḳ, TA:) said by Az to be strange, and not known by him or any other authority than that of Lḥ: (TA:) from the saying, خَلَا فُلَانٌ بِعِرْضِ فُلَانٍ يَعْبَثُ بِهِ [*Such a one occupied himself alone with the honour,* or *reputation, of such a one, making sport with it*]. (Ksh in ii. 13.) And *i. q.* خَادَعَهُ ‡ [*He deceived, deluded, beguiled, circumvented,* or *outwitted, him;* &c.: or *he strove to do so*]: (TA:) as also ↓ خالاهُ, (JK, and Ḳ in art. خلى,) inf. n. مُخَالَاةٌ. (JK.) — And خَلَا عَلَيْهِ *He relied upon him;* [as though he betook himself to him alone;] syn. اِعْتَمَدَ. (TA.)

— And خَلَا, (JK, Ḳ,) inf. n. خُلُوٌّ, (TA,) or خَلَآءٌ, (JK,) said of a man (JK) and of a thing, (JK, TA,) *He,* or *it, went, went away,* or *passed away.* (JK, Ḳ.) Hence, (TA,) وَإِنْ مِنْ أُمَّةٍ إِلَّا خَلَا فِيهَا نَذِيرٌ, in the Ḳur [xxxv. 22], means [*And there is not any people but a warner*] *hath gone,* and *hath been sent, among them.* (Ṣ, TA.) [Hence also خَلَا explained above as meaning *He died.*] And خَلَا مِنْهَا [an elliptical phrase] *She became old; the greater part of her life passed.* (TA from a trad.) And خَلَاكَ ذَمٌّ [for خَلَا عَنْكَ ذَمٌّ] *Blame passed away from thee;* or *may blame pass away from thee.* (Ksh and Bḍ in ii. 13.) You say, اِفْعَلْ كَذَا وَخَلَاكَ ذَمٌّ *Do thou such a thing, and thou wilt have an excuse;* [i. e.] *blame will fall from thee.* (Ṣ. [See art. ذم.]) And خَلَاهُ الحُزْنُ *Grief passed away from him,* and *quitted him.* (Ḥar p. 590, from the Tekmileh.) — خَلَا عَنِ الشَّىْءِ: see 2. ═ خَلَاهُ [or خَلَى, probably belonging to art. خلى, though mentioned in the present art.,] *He ate what was good, sweet,* or *pleasant.* (TA.)

2. خَلَّى, inf. n. تَخْلِيَةٌ, [*He left* a place, &c., *empty, vacant, void, devoid, destitute,* or *unoccupied.*] Hence, خَلَّى مَكَانَهُ [*He left his place vacant;*] meaning ‡ *he died:* (TA, and so in Ḥam p. 478:) a meaning assigned in the Ḳ to ↓ خَلَا مَكَانَهُ, and by IAạr to خَلَا alone, without teshdeed; but when مكانه is added, it is with teshdeed. (TA.) And † *He went his way.* (Ḥam p. 379.) And خَلَّى سَبِيلَهُ [*He left his way free,* or *open, to him*]. (Ṣ, TA.) And خَلَّى بَيْنَهُمَا [*He left the way,* or *space, free between them two;* meaning *he left them two free, each to do to the other as he pleased*]. (TA.) [And خَلَّى بَيْنَهُ وَبَيْنَ كَذَا *He left him free access to such a thing.*] And خَلَّى بَيْنَهُ وَبَيْنَ نَفْسِهِ *He left him,* or *it, alone;* syn. أَهْمَلَهُ. (Ṣ and O and Ḳ in art. همل.) [And خَلَّاهُ وَفُلَانًا *He left him to do as he pleased with such a one.*] And خَلَّى الأَمْرَ *He left, left alone,* or *let alone, the thing,* or *affair;* as also ↓ تَخَلَّى عَنْهُ and مِنْهُ; and ↓ خالاهُ, (Ḳ, TA,) inf. n. خِلَآءٌ. (TA.) For تَخْلِيَةٌ signifies The *leaving,* and *making* a thing *to be alone.* (Ḥar p. 123.) [خَلَّاهُ and خَلَّى عَنْهُ both signify *He left,* or *left alone, it,* or *him.*] It is said in a trad., خَلَّى عَنْهُمْ أَرْبَعِينَ عَامًا *He* (God) *left them,* or *left them alone, and turned from them, forty years.* (TA.) [And خَلَّاهُ لِكَذَا *He made him,* or *left him, vacant, unoccupied, unemployed,* or *at leisure, for such a thing.*] — تَخْلِيَةٌ also signifies The act of *loosing; contr. of* شَدٌّ. (IAạr, Ḳ in art. ابض.) [Hence,] خَلَّى عَنِ الشَّىْءِ, (JK, Ṣ,* TA,) in the Ḳ ↓ خَلَا, without teshdeed, but this requires consideration, (TA,) *He dismissed, loosed, let loose,* or *let go, the thing.* (JK, Ḳ, TA.) — [And hence خَلَّاهُ meaning *He left it, permitted it,* or *allowed it:* see the pass. part. n., below.] ═ خَلَّيْتُ, said of a she-camel such as is termed خَلِيَّة; and hence, of a cooking-pot: see 1 in art. خلى.

3. خالاهُ *He left, forsook, relinquished, abandoned, deserted,* or *quitted, him, being left,* &c., *by him;* namely, another man; syn. تَارَكَهُ; (Ṣ;) inf. n. مُخَالَاةٌ, *syn. with* مُوَادَعَةٌ, (JK,) [and خِلَآءٌ also: and *he was,* or *became, distant, remote, far off, aloof,* or *apart, from him;* for] خِلَآءٌ is *syn. with* مُبَاعَدَةٌ and مُجَانَبَةٌ (TA in art. خلا) and فُرْقَةٌ. (TA in the present art.) And خالى الأَمْرَ, inf. n. خِلَآءٌ: see 2. — [Also *He went,* or *came, out,* or *forth, to him, in the field;* for] مُخَالَاةٌ is also *syn. with* مُبَارَزَةٌ. (Sh, TA.) — Also, (Lth, JK, Ḳ,) inf. n. مُخَالَاةٌ, (Lth, JK,) *He wrestled with him, each endeavouring to throw down the other; contended with him in wrestling:* (Lth, JK, Ḳ: mentioned in the Ḳ in art. خلى:) because, when one does so, he is alone with the other, so that neither of them seeks aid from any other. (Az, TA.) And in like manner the word مُخَالَاةٌ is used [app. as meaning The act of *contending with another, by oneself,*] in relation to any affair, or case. (Lth, JK, TA. [See its act. part. n., below.]) — See also 1, in the latter part of the paragraph.

4: see 1, in eleven places. ═ اخلى المَكَانَ, (Ṣ, Ḳ,) or المَنْزِلَ, (Mṣb,) *He made the place,* (Ḳ,) or *the place of alighting or abode,* (Mṣb,) *empty, vacant, void, devoid, destitute,* or *unoccupied:* (Mṣb, Ḳ:) or it signifies, (Ṣ, Ḳ,) or signifies also, (Mṣb,) *he found it empty,* &c. (Ṣ, Mṣb, Ḳ.) One says in praying for another that he may have a long life, لَا أَخْلَى ٱللّٰهُ مَكَانَكَ [*May God not make thy place vacant*]. (TA.) — اخلاهُ مَعَهُ [*He made him,* or *found him, to be alone with him*]. (Ḳ.)

5. تخلّى *He went forth into the field,* or *open country, to satisfy a want of nature.* (TA.) And تخلّى فِي الخَلَآءِ *He went forth into the vacant tract,* or *into the privy, to satisfy a want of nature:* or *he satisfied a want of nature therein.* (TA.) — Also *He was,* or *became,* or *made himself, vacant from occupation,* or *business;* [*unoccupied; unemployed;* or *at leisure;*] syn. تَفَرَّغَ: (Ṣ:) or so تخلّى مِنَ الشُّغْلِ. (Ḳ in art. فرغ.) You say, تخلّى لِلْعِبَادَةِ *He was,* or *became,* or *made himself, vacant for,* or *he confined himself exclusively to, the service of God.* (TA.) [See also 1, in the latter part of the paragraph. In like manner, one says also, ↓ خَلَا لِلْأَمْرِ *He was,* or *became,* or *made himself, vacant for,* or *he confined himself exclusively to, the affair.*] And تَخَلَّوْا بِخَلِيَّةٍ (Ṣ, Ḳ, TA) *They confined themselves exclusively to a she-camel,* or *to she-camels, such as they termed* خَلِيَّة, (Ḳ, TA,) يَحْلُبُونَهَا [*milking* only *her,* or *them*]. (Ṣ, TA.) And تخلّى خَلِيَّةً *He took for himself a* خَلِيَّة. (TA.) — And تخلّى مِنَ الأَمْرِ and عَنْهُ: see 2. — And تَخَلَّتِ الإِبِلُ بِلَا رَاعٍ [*The camels were left to themselves without a pastor*]. (Ḳ in art. سوع.)

10: see 1, in three places. [And see also مُسْتَخْلٍ.] ═ استخلاهُ مَجْلِسَهُ *He asked him to leave his sitting-place vacant,* or *unoccupied, for him.* (Ṣ. [But found by me in only one copy of that work.]) — استخلى المَلِكَ *He asked the king to have a meeting,* or *an interview, with him in a vacant place,* or *a place unoccupied* [*by others,* i. e., *in a private place; he asked the king to grant him a private meeting* or *interview*]. (Ḳ.)

خَلَا as a word denoting exception, (Ṣ, Mughnee, Ḳ,) when it governs a gen. case, (Ṣ, Mughnee,) as when you say, جَاؤُونِى خَلَا زَيْدٍ [*They came to me, except Zeyd*], is a particle, (Ṣ, Mughnee, Ḳ,) accord. to some of the grammarians, like حَاشَى; but accord. to some, a prefixed inf. n. (Ṣ.) ↓ It also governs an accus. case, as a verb: (Ṣ, Mughnee:) so that you say, جَاؤُونِى خَلَا زَيْدًا [meaning as above]; the agent of خلا being implied, (Ṣ, Mughnee,*) like that of حَاشَى [used as a verb]: it is as though you said, خَلَا مَنْ جَآءَنِى مِنْ زَيْدٍ [i. e. *those who came to me were without Zeyd*]: (Ṣ:) or correctly, accord. to IB, خَلَا بَعْضُهُمْ زَيْدًا [for خَلَا بَعْضُهُمْ مِنْ زَيْدٍ, like as you say, خَلَاكَ ذَمٌّ, for خَلَا عَنْكَ ذَمٌّ]. (TA.) When you say مَا خَلَا, it is followed only by an accus., because ما خلا is equivalent to an inf. n.; (Ṣ, Mughnee;) so that when you say, جَاؤُونِى مَا خَلَا زَيْدًا [meaning as above], it is as if you said, جَاؤُونِى خُلُوَّ زَيْدٍ [or خُلُوًّا زَيْدًا], i. e. خُلُوَّهُمْ مِنْ زَيْدٍ, (Ṣ,) which two phrases mean جَاؤُونِى خَالِينَ مِنْ زَيْدٍ [*They came to me, they being without Zeyd*]: (Ṣ, Ḳ:) [for] accord. to Seer, ما خلا occupies the place of a noun in the accus. as a denotative of state: but some say, as an adv. n. of time; so that, accord. to these, مَا خَلَا زَيْدًا means وَقْتَ خُلُوِّهِمْ عَنْ زَيْدٍ [*in the time of their being without Zeyd*]. (Mughnee.) You say also, مَا أَرَدْتُ مَسَآءَتَكَ خَلَا أَنِّى وَعَظْتُكَ, meaning [*I desired not to displease thee,*] but *I admonished thee* (إِلَّا أَنِّى وَعَظْتُكَ). (JK, TA.)

خِلْوٌ, and its fem. (with ة), and dual: see خَالٍ, in seven places.

خَلْوَةٌ said by some to be an inf. n.: [see خَلَا بِهِ &c. in the first paragraph of this art.:] by others said to be a simple subst.; (TA;) meaning *Loneliness; solitude; lonesomeness; solitariness; desolateness;* syn. وَحْشَةٌ. (Ṣ and Ḳ in art. وحش.) [Hence, app.,] رَجُلٌ سَهْلُ الخَلْوَةِ [*A man easy in private conference*]. (Mṣb in art. سلس. [See also a phrase in the latter part of the next paragraph.]) — Also *An empty, a vacant, a void,* or *an unoccupied, place.* (KL. [See also خَلَآءٌ.]) [In the present day, it is often applied to *A closet to which one retires for privacy;* and particularly to *a cell for religious retirement:* and is vulgarly pronounced خِلْوَة.] You say, اِجْتَمَعَ مَعَهُ فِى خَلْوَةٍ (Ṣ) or اجتمع بِهِ فى خلوة (Ḳ) [*He had a meeting,* or *an interview, with him in a vacant place,* or *a place unoccupied by others,* i. e., *in a private place*]. — Also *Each of the two sharp sides* or *edges* of an arrow-head (AḤn, JK, TA) or of a spear-head: (AḤn, TA:) both together are called the خَلْوَتَانِ. (AḤn, JK, TA.)

خَلَآءٌ is primarily an inf. n. (MF, TA. [See 1, first sentence.]) — [Then it is used as an epithet, syn. with خَالٍ:] see خَالٍ, in five places. — Then it is used [as a subst.] in the sense of *A vacant place* [*in a general sense*]: (MF, TA:) or *a place in which is nothing:* (Ṣ, Ḳ:) [often applied in the present day to *any open tract of country* or *desert:*] and then, *particularly, such as one takes for the purpose of satisfying a want of nature;* (MF, TA;) *i. q.* مُتَوَضَّأٌ, (Ṣ, Mṣb, Ḳ,) but *not as meaning only a place for the performance of* الوُضُوء, as might be imagined from this explanation: pl. أَخْلِيَةٌ. (MF, TA.) It is said in a prov., (Ṣ, Meyd,) خَلَاؤُكَ أَقْنَى لِحَيَائِكَ, (Ṣ, Meyd, Ḳ,) [in Freytag's Arab. Prov., (i. 436,) بِحَيَائِكَ,] i. e. [*Thy place of retirement is*] *most preservative* (أَلْزَمُ) [*of thy sense of shame,* or *modesty*]; meaning it is most fit for thee to be alone in thine abode; (Ṣ,* Meyd;) for he who is so needs not to be careful for his shame, or modesty: it is used in blaming the mixing with others. (Meyd.) حُصِرَ عَلَيْهِ خَلَاؤُهُ [*His place of retirement for satisfying a want of nature was straitened to him*] is used as meaning *he suffered suppression of the feces,* or *constipation of the bowels.* (Ibn-Buzurj, TA in art. حصر.) — إِنَّهُ لَحُلْوُ الخَلَآءِ, (TA,) or [as written in a verse in which it occurs in the TA in the present art., and in art. خلى,] الخَلَا, (JK, TA,) [without ء, but whether this be the right reading, or only required by poetic license, seems to be doubtful,] is a phrase mentioned by Th, (TA,) meaning *Verily he is good in speech.* (JK, TA. [If the former reading be right, the meaning may be similar to that of سَهْلُ الخَلْوَةِ, mentioned above: if the latter only, or rather انّه لحلو الخَلَى, be right, it probably belongs to art. خلى, and is tropical, from the herbage termed خَلًى; and this may also be the case if the former reading be right.])

خَلِىٌّ; and its fem. خَلِيَّةٌ: see خَالٍ, in twelve places. — The fem. also signifies, applied to a she-camel, (Ṣ, Mṣb,) *Loosed from the cord,* or *rope, with which her fore shank and her arm have been bound together,* (Ṣ, Mṣb, Ḳ,) *and left alone,* or *free,* (Ṣ,) *so that she pastures where she will.* (Mṣb.) Hence, (Mṣb,) it is used by way of metonymy as meaning *Divorced:* (Lḥ, Ṣ, Mṣb, Ḳ:) one says to a woman, أَنْتِ خَلِيَّةٌ *Thou art divorced;* (Lḥ, Ṣ;) and thus a man used to say in the Time of Ignorance: (TA:) and one says, هِىَ خَلِيَّةٌ *She is divorced:* (Mṣb:) and a woman is divorced thereby when divorce is meant. (Lḥ, TA.) Applied to a woman, it signifies also *Free from any obstacle to marriage:* pl. خَلِيَّاتٌ. (Mṣb.) — Also A she-camel *that is made to affect, with another she-camel, one young one, so that both yield their milk to it, and to which the people of a tent, or house, confine themselves exclusively of the other for the purpose of milking her:* (Ṣ:) or a she-camel *that is chosen as the one more abundant in milk, when one has brought forth and her young one is drawn away* (يُجَرُّ) *as soon as born, before she smells it, and the young one of another, that has brought forth before her, is brought near to her, and she affects it;* the other is left to suckle the young one, and is termed بَسُوطٌ, pl. بسط [app. بُسْطٌ or بُسُطٌ]: (Az, TA:) or a she-camel *that is left,* or *left alone, to be milked:* (Ḳ:) or *that affects a young one* [*not her own*], or *is destitute of her young one,* (JK, M, Ḳ,) *whether she incline to another's young one or do not,* or *that is destitute of her young one by death or slaughter,* (M, TA,) *and whose milk one causes to flow by means of the young one of another; but only by her affecting a young one, and not suckling it:* (M, Ḳ:*) or *that brings forth, when abundant in milk, and has her young one drawn* (يُجَرُّ) *from beneath her, and another put beneath her, and is then left,* or *left alone, to be milked;* (Lḥ, Ḳ;) this being done because of her generous quality: (Lḥ:) or a she-camel, or two she-camels, *to which the people of a tent, or house, confine themselves exclusively, for milking, when two or three she-camels are made to affect one young one, and to yield their milk to it; the young one* [*afterwards*] *sucking from one of them only:* (Ḳ,* TA:) or a she-camel *that brings forth, and whose young one is drawn away* (يُجَرُّ) *in order that her milk may continue for their use, she being made to yield her milk by means of the young one of another, which is then withdrawn from her, and she is milked:* sometimes, also, they bring together three and four خَلَايَا [pl. of خَلِيَّةٌ] to one young camel: and the doing so is termed تَلْشِنٌ: (IAạr, TA:) in this case they take as a خَلِيَّة whichever of them they will. (ISh, TA.) [Applied to a she-camel in any of these senses, it seems to be an epithet in which the quality of a subst. is predominant; i. e., used without its having نَاقَةٌ prefixed to it.] — See also the paragraph next following, in two places.

خَلِيَّةٌ [as fem. of the epithet خَلِىٌّ: see the next preceding paragraph, and the places there referred to in its first sentence. — As a subst. it signifies] *A great ship:* (T, Ṣ, Ḳ:) or *a ship that goes of itself, without its being made to do so by the sailor:* (JK, Ḳ:) or *one that is followed by a small boat:* (Ḳ:) the first held by Az to be the right meaning: (TA:) pl. خَلَايَا. (JK, Ṣ.) — Also, (Ṣ, Mgh, Mṣb, Ḳ, &c.,) and ▼خَلِىٌّ, (JK, Mṣb, Ḳ,) The *habitation* (بَيْت) *of bees,* [whether it be *a manufactured hive* or *a hollow in the trunk of a tree* or *in a rock,*] *in which they deposit their honey;* (Ṣ;) the *places in which bees deposit their honey:* (Mgh:) or the *thing in which bees deposit their honey,* (Ḳ, TA,) *not manufactured for them:* (TA:) or *a thing like the* [*kind of jar called*] رَاقُود, *of clay,* (Ḳ, TA,) *made for bees:* (TA:) or *a certain thing for bees, well known, of clay* or *of wood:* (Mṣb:) or, accord. to Lth, if made of clay, it is called كُوَارة, (Mṣb, TA,) i. e. [كُوَارَةٌ and كُوَّارَةٌ and كِوَارَةٌ] with kesr: (Mṣb:) or *a piece of wood hollowed out for honey to be deposited therein* [by bees]: or the *lower part of a tree that is called* خَزَمَة, [n. un. of خَزَم, q. v., *hollowed out for that purpose,*] *resembling the* [*kind of jar called*] رَاقُود: (Ḳ:) or ▼خَلِىٌّ signifies the *part of the* كُوَارة *which is the place of the honey:* (JK:) pl. as above. (Mṣb, TA.)

خَلَاوَةٌ: see the next paragraph.

خَالٍ *Empty, vacant, void, devoid, destitute,* or *unoccupied;* (Mgh, TA;) *having none, and nothing, in it:* (TA:) applied to a place, (Mṣb, TA,) as also ▼خَلِىٌّ (TA) and ▼مُخْلٍ; (Mṣb;) and to a thing, as also ▼خَلِىٌّ; (TA;) or a vessel. (Mgh.) You say also ▼مَكَانٌ خَلَآءٌ, [as well as خَلَآءٌ alone,] meaning *A place in which is none* (Ḳ, TA) *and nothing.* (TA.) And وَجَدْتُ الدَّارَ ▼مُخْلِيَةً, meaning خَالِيَةً [i. e. *I found the house empty,* &c.]. (TA.) — *Vacant,* or *free;* from a thing or an affair; or *devoid,* or *destitute,* of a thing; (TA;) and so ▼خَلِىٌّ and ▼خِلْوٌ; which last is the same as masc. and fem., though it has خِلْوَةٌ also for fem., and أَخْلَآءٌ for pl.; (Ḳ;) but properly, accord. to Lḥ, it has no dual form, nor pl., nor fem., though some give it such forms: (TA:) or ▼خَلِىٌّ, which has a dual, [i. e. خَلِيَّانِ,] and pl., (Ṣ, Mṣb,) i. e. خَلِيُّونَ and أَخْلِيَآءُ, (Ḳ,) signifies *free* [from a thing]; or *clear* or *quit* [of a thing or person]; as also ▼خَلَآءٌ, (Ṣ, Mṣb,) which, being [originally] an inf. n., has no dual nor pl. [nor fem.]; (Ṣ;) and ▼خِلْوٌ. (Mṣb.) You say, أَنْتَ خَلِىٌّ ▼ مِنْ هٰذَا الأَمْرِ and خَالٍ, i. e. *Thou art free from this thing,* or *affair.* (TA.) And أَنَا خَلِىٌّ ▼ مِنَ الهَمِّ, meaning خَالٍ [i. e. *I am free from anxiety*]. (Mgh.) And أَنَا خِلْوٌ ▼ مِنْ كَذَا, meaning خَالٍ [i. e. *I am free from such a thing*]: (Ṣ:) and هُمَا خِلْوٌ, and هُمْ خِلْوٌ; and some say, هُمَا خِلْوَانِ, and هُمْ أَخْلَآءٌ, which is not proper. (T, TA.) And أَنْتَ خِلْوٌ ▼ مِنْ مُصِيبَتِى *Thou art free in mind from my affliction,* or *misfortune.* (TA from a trad.) And أَنَا خَلِىٌّ ▼ مِنْكَ *I am clear,* or *quit, of thee.* (Ṣ.) And أَنَا مِنْكَ خَلَآءٌ ▼ signifies the same. (Ṣ.) And نَحْنُ مِنْكَ الخَلَآءُ ▼ and البَرَآءُ [q. v.] *We are clear,* or *quit, of you.* (Fr, T in art. برأ.) And أَنْتَ خَلَآءٌ ▼ مِنْ هٰذَا الأَمْرِ *Thou art clear,* or *quit, of this affair.* (TA.) And أَنَا مِنْ هٰذَا الأَمْرِ كَفَالِجِ بْنِ خَلَاوَةَ ▼ [lit. *I am, with respect to this affair, like Fálij Ibn-Kha-*

láweh], (Ṣ,) or فَالِجُ بْنُ خَلَاوَةَ, (so in the JK and Ḳ in this art., and in the Ṣ and Ḳ in art. فلج,) meaning بَرِئْتُ [i. e. *I am clear*, or *quit*, *of this affair*]: (JK, Ṣ, Ḳ:) a saying originating from its being asked of Fálij Ibn-Khaláweh, on the day of Er-Raḳam, when Uneys killed the captives, "Dost thou," or "wilt thou," "aid Uneys?" and his answering, "I am clear," or "quit," "of him." (Ṣ and Ḳ in art. فلج.) And ▼خَلِىٌّ [alone] signifies خَالٍ مِنَ الهَمِّ [*Free from anxiety*]; *contr. of* شَجِىٌّ. (Ṣ.) It is said in a prov., وَيْلٌ لِلشَّجِىِّ مِنَ الخَلِىِّ ▼, i. e. *Woe to him who is occupied by anxiety from him who is free therefrom:* (TA:) and in another, مَا يَلْقَى الشَّجِىُّ مِنَ الخَلِىِّ ▼, i. e. *What will he who is occupied by anxiety experience from him who is free therefrom?* meaning, accord. to AO, that the latter will not aid the former against his anxieties, but will censure him: it is said in the Tekmileh that الخَلِىّ [in these provs.] is from خَلَاهُ الحُزْنُ meaning "Grief passed away from him," and "quitted him." (Ḥar p. 500.) And أَنْتِ خَلِيَّةٌ ▼ means خَالِيَةٌ مِنَ الخَيْرِ [i. e. *Thou, O woman, art devoid*, or *destitute, of good*]. (Mgh.) — Also *A man having no wife;* (Ṣ, Ḳ;) [for خَالٍ مِنَ الزَّوْجَاتِ, a phrase occurring in the TA:] and a woman *having no husband;* (Ḳ;) thus without ة: (TA:) pl. أَخْلَآءُ: (Ḳ:) and ▼خِلْوٌ, also, has the latter meaning; dual خِلْوَانِ, and pl. خِلْوَاتٌ: and so has ▼مُخْلِيَةٌ: and ▼خَلِيَّةٌ means a woman *having no husband nor children;* pl. خَلِيَّاتٌ. (TA.) — [And *Alone;* as also ▼مُخْلٍ, and ▼خِلْوٌ.] It is said in a prov., الذِّئْبُ مُخْلِيًا ▼ أَشَدُّ *The wolf when* [*alone* or] *in a vacant place* [*is most courageous*, or *violent*]; (TA;) or خَالِيًا [which means the same]. (JK. [And another reading is أَسَدُ. See Freytag's Arab. Prov., i. 500.]) And one says, وَجَدْتُ فُلَانَةَ مُخْلِيَةً ▼, meaning خَالِيَةً [i. e. *I found such a woman alone*]. (TA.) And وَجَدَهُمَا خِلْوَيْنِ ▼ i. e. خَالِيَيْنِ [*He found them two alone*]. (Ḳ.) — [Also *Past*, or *past away:* as well as *going, going away*, or *passing away*.] القُرُونُ الخَالِيَةُ means [*The generations*] *that have passed*. (JK, Ṣ, TA.)

مُخْلٍ, and its fem. مُخْلِيَةٌ: see خَالٍ, in six places. = لَسْتُ لَكَ بِمُخْلِيَةٍ, occurring in a trad., means *I did not find thee destitute of wives beside me:* it is not from اِمْرَأَةٌ مُخْلِيَةٌ signifying "a woman having no husband." (TA.)

مُخْلَاةٌ A she-camel *left alone, away from her young one*. (IDrd, JK.)

مُخَلًّى pass. part. n. of 2. (Ṣ, TA.) — *Left, permitted*, or *allowed*. (M in art. بسل.)

مُخَالٍ [act. part. n. of 3, q. v.]. Accord. to IAạr, it signifies *Contending with another in war*. (TA in art. خلأ.)

مُتَخَلٍّ *Devoting himself to religious services* or *exercises* [app. *in solitude* or *seclusion*, or in a خَلْوَة; or because one generally does so in solitude; or because the doing so involves abstraction from other affairs: see also 1 and 5]. (TA.)

خلى

1. خَلَى, (Ṣ, Mgh, Mṣb, Ḳ,) aor. يَخْلِى, (Mṣb,) inf. n. خَلْىٌ; (Mṣb, Ḳ;) and ▼اختلى; (Ṣ, Mgh, Mṣb, Ḳ;) *He cut* the herbage called خَلًى: (Ṣ, Mgh, Mṣb, Ḳ:) or *he plucked* it *up*. (Lḥ, Ḳ.) Hence, in a trad., (Mgh, Mṣb, TA,) respecting the declaration of the sacredness of Mekkeh, (TA,) لَا يُخْتَلَى ▼ خَلَاهَا, (Mgh, Mṣb, TA,) i. e. [*Its fresh herbage*] *shall not be cut*. (Mṣb.) — خَلَى الدَّابَّةَ, (Ṣ,) or المَاشِيَةَ, (Ḳ,) aor. as above; (Ṣ, Ḳ;) or ▼اخلى الماشية, inf. n. إِخْلَآءٌ; (TA, as from the Ḳ;) *He cut the herbage called* خَلًى (Ṣ, Ḳ) *for the beast*, (Ṣ,) or *for the cattle:* (Ḳ:) and *he fed the beast*, or *the cattle, with* خَلًى. (TA.) — خَلَى الشَّعِيرَ *He collected the barley* in a مِخْلَاة. (Ḳ.) — خَلَى القِدْرَ, (inf. n. as above, TA,) ‡ *He put firewood beneath the cooking-pot:* or *he put flesh-meat into the cooking-pot*. (IAạr, Ḳ, TA.) And ▼اخلى القِدْرَ † *He kindled a fire for the cooking-pot with camels'*, or *similar, dung;* as though he put خَلًى to it. (TA.) And ▼أُخْلِيَتْ, said of a cooking-pot, †*It had* firewood *put to it, like as a she-camel has* خَلًى *put to her, and kindled beneath it:* or, as some relate a verse in which it occurs, ▼خُلِّيَتْ, [belonging to art. خلو,] having a similar meaning, from this verb said of a she-camel such as is termed خَلِيَّة, meaning "she had" a young one "put to her." (Ḥam p. 663.) — خَلَى الفَرَسَ, (Ḳ,) or خَلَاهُ اللِّجَامَ, (JK, TA,) ‡ *He put the bit in the mouth of the horse*, (JK, Ḳ, TA,) like fresh خَلًى. (JK.) — And خَلَى اللِّجَامَ عَنِ الفَرَسِ, (Ḳ, TA,) aor. and inf. n. as above, (TA,) † *He pulled out the bit* [*from the mouth of the horse*]. (Ḳ, TA.) = See also 1, last sentence, in art. خلو.

2: see خَلًى, below: = and see also 1.

3, mentioned in this art. in the Ḳ: see art. خلو.

4. اخلى المَاشِيَةَ, (inf. n. إِخْلَآءٌ, TA,) said of God, *He made* خَلًى *to grow for the cattle*. (Lḥ, Ḳ.) — See also 1, in three places. = اخلت الأَرْضُ *The land became abundant in* خَلًى. (JK, Ṣ, Ḳ.) — [And hence,] اخلى † *He uttered words*, or *expressions, without any great meaning*. (Ḥam p. 391.)

7. انخلى, said of خَلًى, *It was cut*. (Ṣ.)

8: see 1, in two places. [Hence,] السَّيْفُ يَخْتَلِى الأَيْدِىَ وَالأَرْجُلَ ‡ *The sword cuts off the arms and the legs*. (JK, Ṣ,* TA.)

12. اخلولى *He constantly drank milk*. (IAạr, Ḳ.)

خَلًى *Fresh, green*, or *juicy, herbage:* (Ṣ, IB, Mgh, Mṣb, Ḳ:) that which is dry is termed حَشِيش [but see this word]: (Mṣb, from the Kf:) or *dry* حَشِيش: (so in one place in the Ṣ: [app. a mistake occasioned by an omission:]) or i. q. رُطْب, (IB, Mṣb,) with ḍamm, (IB,) [i. e. *fresh*, or *green, pasture;* or *such as consists of the herbs*, or *leguminous plants, of the* رَبِيع; or *of these and of trees* or *shrubs:*] or *herbage that is cut, of the herbs*, or *leguminous plants, of the* رَبِيع: (Lth, JK:) or, accord. to [the Imám] Mohammad, *anything that is eaten as pasture, not* [*growing*] *upon a stem:* (Mgh:) or *slender herbage as long as it remains fresh, green*, or *juicy:* (IAth, TA:) it is also written ▼خَلَآء, with medd, like قَضَآء: (Mṣb:) n. un. خَلَاةٌ: (Ṣ, Mṣb, Ḳ:) or this signifies *any herb* (بَقْلَة) *that one pulls up:* (Ḳ,* TA:) pl. أَخْلَآءٌ, (Ḳ,) a pl. sometimes used [app. as meaning *sorts of* خَلًى]. (TA.) [Hence,] عَبْدٌ وَخَلًى فِى يَدَيْهِ [*A slave with fresh herbage in his hands*, or *arms*]; meaning, though a slave, yet rich, or possessing sufficiency: (Ṣ, TA:) a prov., (Ṣ, Meyd,) applied to the case of property possessed by him who does not deserve it: or, as some relate it, وَخُلِّىَ ▼ فى يديه [*having fresh herbage put in his hands*, or *arms*]: (Meyd:) but this latter reading is disallowed by Yaḳoob: (Ṣ:) [see other readings, not belonging to this art., in Freytag's Arab. Prov. ii. 75:] هٰذَا, or هُوَ, is understood before عَبْدٌ. (Meyd.) And ‡ مَا كُنْتُ خَلَاةً لِمَوْعِدَةٍ *I was not a breaker of a promise*. (TA.) And [hence also, app., if this be the right reading,] إِنَّهُ لَحُلْوُ الخَلَا [or rather الخَلَى]; or ▼الخَلَآء: see خَلَآء, in art. خلو.

خَلَآء: see the next preceding paragraph, in two places.

خَالٍ and ▼مُخْتَلٍ *A cutter of* خَلًى. (Mṣb. [The pls. خَالُونَ and مُخْتَلُونَ are mentioned in the Ṣ and TA.])

مِخْلًى *The thing* [or *instrument*] *with which* خَلًى *is cut*. (Ṣ, TA.)

مِخْلَاةٌ *A thing* [or *bag*] *into which* خَلًى *is put:* (Ṣ, Ḳ:) [and hence *a nose-bag for a horse or the like;* so in the present day;] *a small sack that is hung to the head of a horse* [*or the like*], *in which he eats barley* [*&c.*]: (Ḥar p. 76:) so called because they used to cut [and put] خَلًى therein for their beasts: (JK:) pl. مَخَالٍ. (TA.) — مَخَالِى القَتَبِ: see جِذْعٌ.

مُخْتَلٍ: see خَالٍ. — المُخْتَلِى ‡ *The lion:* (Ḳ, TA:) because of his courage. (TA.)

خمر

1. خَمِرَ, aor. َ (JK, Ṣ, Ḳ) and ُ, (Ḳ,) [the latter irreg.,] inf. n. خُمُورٌ (JK, Ḳ) and خَمَرٌ, (Ḳ,) *It* (flesh-meat) *was*, or *became, stinking;* (Ṣ, Ḳ;) said of what is roasted, or cooked; (Ṣ;) or mostly said of what is cooked, and what is roasted: (IDrd, Ḳ:) or *became altered for the worse in odour;* said of roasted meat, and of meat cut into strips and dried: (A'Obeyd, TA:) or

became stinking after having been thoroughly cooked: (TA:) and said also of milk, (JK, Ḳ,) in like manner, (JK,) it became altered by the bad odour of the skin, (Ḳ, TA,) and corrupt: (TA:) and ↓اخمّ signifies the same, (JK, Ṣ, Ḳ,) in both cases: (TA:) and خَمَّ also, said of a cake of bread not thoroughly baked, signifies the becoming altered in odour. (TA.) [Hence,] خَمَّ, inf. n. خُمُومٌ, is likewise said of a man. (TA. [See also 10.]) And one says, هُوَ لَا يَخِمُّ, meaning †He will not become altered (JK, TA) from his state, or condition, (JK,) or from his liberality, and generosity. (TA.) And هُوَ السَّمْنُ لَا يَخِمُّ, (Ṣ, TA,) [lit.] meaning [It is the clarified butter] that will not become altered [for the worse]: (TA:) a prov., relating to a man when one speaks well of him, and praises him. (Ṣ, TA.) And هُوَ السُّمُّ لَا يَخِمُّ, i. e. †It is unmixed poison. (TA.) — خَمٌّ signifies also The act of sweeping violently. (Ḳ.) You say, هُوَ يَخُمُّ He sweeps violently. (TḲ.) ═ خَمَّ, (JK, Ṣ, Ḳ,) aor. ـُ, (Ṣ,) inf. n. خَمٌّ, (TḲ,) He cleaned out a well: (Ṣ, Ḳ:*) and he swept a tent, or house, or chamber: (JK, Ṣ, Ḳ:) and ↓اختمّ signifies the same, (Ṣ, Ḳ,) in both cases. (TA, and so in some copies of the Ḳ.) — [Hence,] هُوَ يَخُمُّ ثِيَابَهُ ‡He eulogizes him, commends him, or speaks well of him: (Ḳ, TA:) and خَمَّهُ بِثَنَآءٍ حَسَنٍ, aor. ـُ, inf. n. خَمٌّ, ‡He eulogized him: (TA:) [and so, app., خَمَّهُ alone; for] خَمٌّ signifies The act of eulogizing. (Ḳ, TA.) — خَمٌّ also signifies The act of cutting; and so ↓اِخْتِمَامٌ. (Ḳ.) — And خَمَّ النَّاقَةَ, (Ḳ,) aor. ـُ, inf. n. خَمٌّ, (TA,) He milked the she-camel: (Ḳ:) or خَمَّ, aor. ـُ, signifies he turned in his thumb upon his palm when milking. (JK.) ═ خُمِّمَ, said of a domestic fowl, It was confined in a خُمّ, i. e. cage, or coop. (Ḳ.)

4: see 1.

5. تخمّم مَا عَلَى الخِوَانِ †He ate what remained, of fragments, and scattered particles, upon the table, (Ḳ, TA,) by reason of his greediness. (TA.) [See also R. Q. 1.]

8: see 1, in two places. — اختمّ بِهِ He took it away. (JK.) — And He threw it down prostrate; and, from the foundation; or uprooted it. (JK.)

10. It is said in a trad. of Mo'áwiyeh, مَنْ أَرَادَ أَنْ يَسْتَخِمَّ لَهُ النَّاسُ قِيَامًا: thus, accord. to Eṭ-Ṭaḥáwee, with the pointed خ, meaning [Whoso desireth that men] should become altered in their odour to him by reason of their long standing in his presence: but it is also related otherwise, يستجمّ [q. v.: see also 1 in the present art.]. (TA.) [See also 2 in art. جمّ.]

R. Q. 1. خَمْخَمَةٌ [inf. n. of خَمْخَمَ] i. q. خَنْخَنَةٌ, (Ṣ, Ḳ,) i. e. The [snuffling, or] speaking [indistinctly, through the nose,] as though one were مَخْنُونٌ [app. here meaning affected with the disease termed خُنَان], (so in a copy of the Ṣ and in the TA,) or مَجْنُونٌ [i. e. bereft of reason, or mad, insane, &c.; and this is another meaning of مَخْنُونٌ], (so in another copy of the Ṣ,) by reason of pride. (Ṣ.) [See خَمْخَمٌ.] — Also †The eating in a certain foul manner; (JK, Ṣ, TA;) and so ↓تَخَمْخُمٌ [inf. n. of تَخَمْخَمَ]. (TA.) Hence, ↓خَمْخَامٌ [app. meaning †One who so eats], used as a proper name. (JK, TA.) [See also 5.]

R. Q. 2: see the next preceding paragraph.

خَمٌّ, (Ḳ,) or ↓خَامٌّ, (AA, Ṣ,) applied to flesh-meat (AA, Ṣ, Ḳ) that is roasted or cooked, (AA, Ṣ,) or mostly to what is cooked and what is roasted, (Ḳ,) Stinking; (AA, Ṣ, Ḳ;) as also ↓مُخِمٌّ: (AA, Ṣ:) or this last signifies altered in odour, but not yet corrupt (Lth, JK, TA) like a stinking dead body. (Lth, TA.)

خُمٌّ A cage, or coop, for domestic fowls: (ISd, Ḳ:) [and so, in modern Arabic, خُنّ:] thought by ISd to be so called because of its foul smell. (TA.) — A [receptacle made of matting or of reeds, such as is called] قَوْصَرَة, in which straw is put, for the domestic hen to lay her eggs therein, (Ḳ,) or to hatch therein. (TA.) — A hollow dug in the ground, in the bottom of which are put ashes, and then new-born lambs or kids are put therein: pl. خِمَمَةٌ. (Ḳ.)

خُمَامٌ The refuse of anything. (JK.) [See also خُمَّانٌ.]

خَمِيمٌ Heavy, or sluggish, in spirit: (Ḳ:) from خُمَامَةٌ signifying "sweepings." (TA.) — †Praised: (Ḳ:) from خَمٌّ signifying the act of "eulogizing." (TA.) — Milk just milked. (Ḳ.)

خُمَامَةٌ Sweepings; (Ḳ;) like قُمَامَةٌ: (JK, Ṣ:) and the earth that is cleared out from a well: (Ṣ:) the dust, or earth, of a tent or house or chamber, and of a well, that is swept, or cleared, out, and thrown in a heap. (Lth, TA.) — Also, (Ḳ,) or خُمَامَةُ مَائِدَةٍ, (TA,) Scattered fragments of food, which are [gathered up, or swept together, and] eaten, and on account of which a recompense is hoped for [from God]. (Ḳ,* TA.)

خُمَامَةٌ A corrupt, bad, feather, beneath the other feathers. (Ḳ,* TA.)

خِمَّانٌ: see the next paragraph. — Also A weak spear. (Ṣ, Ḳ.)

خُمَّانٌ (JK, Ḳ) and ↓خُمَانٌ, (JK, IDrd, TA,) or ↓خِمَّانٌ, (Ḳ,) What is bad of household goods, or furniture, or utensils; (JK, IDrd, Ḳ, TA;) and of trees. (Ḳ.) Also the first and second, (JK, Ṣ,) or the first and third, (Ḳ,) †The refuse, or the low, ignoble, or mean, (Ṣ, Ḳ,) or the bad, (JK,) of mankind: (JK, Ṣ, Ḳ:) the lowest, basest, or meanest, sort, and the mass, thereof: or the weak thereof. (TA.) You say, ذَاكَ رَجُلٌ مِنْ خُمَّانِ النَّاسِ and خُمَانِ النَّاسِ †That is a man of the refuse, &c., of mankind. (Ṣ.) [See also خُمَامٌ.]

خُمَانٌ: see the next preceding paragraph.

خَمْخَمٌ One who speaks with [or through] his nose. (TA.) [See R. Q. 1.]

خَمْخَامٌ: see R. Q. 1.

خَامٌّ: see خَمٌّ.

[خُمُومَةٌ, mentioned in this art. by Golius and Freytag, belongs to art. خيم.]

مُخِمٌّ: see خَمٌّ.

مِخَمَّةٌ A broom; a thing with which one sweeps. (Ḳ.) — [Hence,] هُوَ مِخَمَّةٌ وَمِئَمَّةٌ †He is a vehement eater [and one who sweeps together the good and the bad]. (JK.) [See also art. ثم.]

قَلْبٌ مَخْمُومٌ ‡A heart clear from malevolence, malice, or spite, and envy. (Ṣ, TA.) And مَخْمُومُ القَلْبِ ‡Having the heart clear from malevolence, malice, or spite, and envy: (Ḳ, TA:) or from dishonesty, or dissimulation, and envy; as explained by Mohammad himself, when used by him: or from dishonesty, or dissimulation, and corruption: or from pollution: all these explanations being from خَمَّ signifying "he cleaned out" a well. (TA.)

خمد

1. خَمَدَتِ النَّارُ, aor. ـُ; (Ṣ, Mṣb, Ḳ;) and خَمِدَت, (A, Ḳ,) aor. ـَ; (Ḳ;) inf. n. خُمُودٌ (Ṣ, A, Mṣb, Ḳ) and خَمْدٌ; (IḲṭṭ, Ḳ;) The fire subsided; its flaming, or blazing, ceasing; (Ṣ, A, Mṣb, Ḳ;) but its embers remaining unextinguished: (Ṣ, Mṣb, Ḳ:) when its embers have become extinguished, you say of it, هَمَدَت: (Ṣ:) or it died away, and became utterly extinguished. (Mṣb.) — [Hence,] خَمَدَتِ الحُمَّى, (Ṣ, Mṣb, Ḳ,) or خَمِدَت, (A,) ‡The fever became allayed: (A, Mṣb:) or the vehemence of the fever became allayed. (Ṣ, Ḳ.) — And خَمَدَ ‡He (a sick man, Ṣ) fainted, or swooned: (Ṣ, A, Mṣb, Ḳ:) or he died. (Ṣ, A, Mṣb.)

4. اخمد النَّارَ He allayed the flaming, or blazing, of the fire; leaving its embers unextinguished: (Ṣ, Mṣb, Ḳ:) or he extinguished the fire utterly. (Mṣb.) And اخمدتها الرِّيحُ The wind allayed its flaming, or blazing. (A.) ═ ‡He was, or became, still, or motionless, and silent. (Ḳ, TA.)

خَمُودٌ A place in which fire is buried in order that its flaming, or blazing, may cease; its embers remaining unextinguished: (Ṣ, Ḳ:) [or in order that it may become utterly extinguished: see 1.]

خَامِدٌ ‡Silent; from whom no voice is heard: and in like manner, ↓مُخْمِدٌ signifies still, or motionless, and silent: still, or motionless; having disposed and submitted himself to an affair, or event. (L.) خَامِدُونَ in the Ḳur xxxvi. 28 means †Silent and dead: (Jel:) or silent; having died, and become like extinguished ashes. (Zj, Bḍ.*)

مُخْمِدٌ: see the paragraph next preceding.

خمر

1. خَمَرَ, aor. ـُ, (TA,) inf. n. خَمْرٌ, (Ḳ,) He

veiled, covered, or *concealed,* a thing; (Ḳ,* TA;) as also ▼ خمّر, inf. n. تَخْمِيرٌ, (Mgh, Mṣb,) which also signifies *he covered over* a thing; (Ṣ, Mṣb, Ḳ;) and ▼ اخمر, (TA,) inf. n. إِخْمَارٌ. (Ḳ.) [Hence,] خَمَرَهَا [and app. ▼ خمّرها also, for the quasi-pass. is تخمّرت as well as اختمرت, *He veiled her with a muffler;*] *he put on her a* خِمَار. (A.) And ▼ خمّر إِنَآءَهُ, and وَجْهَهُ, *He covered over his vessel,* and *his face.* (Ṣ.) And ▼ خمّر بَيْتَهُ *He concealed his house,* or *chamber,* or *tent,* [meaning its interior,] *and ordered it aright.* (TA, from a trad.) And ▼ أَخْمَرَتْهُ الأَرْضُ عَنِّى and مِنِّى and عَلَىَّ *The land,* or *ground, concealed him,* or *it, from me.* (Ḳ.) And ▼ اخمره † *He concealed it,* or *conceived it, in his mind.* (Ṣ, Ḳ.) And ▼ اخمر فُلَانٌ عَلَىَّ ظِنَّةً † *Such a one concealed,* or *conceived, in his mind a suspicion,* or *an evil opinion, of me.* (T, TA.) And خَمَرَ شَهَادَتَهُ, (Ṣ, Mṣb,) and ▼ خمّرها, (A, Mgh,) and ▼ اخمرها, (TA,) ‡ *He concealed his testimony.* (Ṣ, A, Mgh, Mṣb, TA.) And الخَمْرُ تَخْمُرُ العَقْلَ † *Wine veils* [or *obscures*] *the intellect;* (Ḳ;) and so ▼ تُخَامِرُهُ, lit. *covers it:* (Mṣb:) or the latter signifies † *infects it;* [as though acting like leaven; and if so, from خَمَرَ العَجِينَ, which see in what follows; nearly the same as "intoxicates," which properly signifies "empoisons," or "infects with poison;"] syn. تُخَالِطُهُ. (Ṣ,* Ḳ. [See خَمْرٌ.]) —— خَمِرَ, aor. خَمَرَ, (Ṣ, Ḳ,) inf. n. خَمَرٌ, (Ṣ,) *He became concealed,* or *hidden;* or *he concealed,* or *hid, himself;* (Ṣ, Ḳ;) عَنِّى *from me;* (Ṣ;) as also ▼ خامر, (Ṣ, Ḳ,) inf. n. مُخَامَرَةٌ; (Ḳ;) and ▼ اخمر: (Ḳ:) or this last signifies *he concealed,* or *hid, himself in a* خَمَر [or *covert of trees* or *the like*]. (TA.) One says also, خَمِرَ عَنِّى الخَبَرُ † *The news,* or *story, became concealed from me.* (Ṣ.) And one says to the hyena, ▼ خَامِرِى أُمَّ عَامِرٍ *Hide thyself, O Umm-'Ámir:* (Ṣ, Ḳ:) which is a prov.: (TA:) and is said to be also a phrase used as a surname of the hyena, in the manner of تَأَبَّطَ شَرًّا. (Ḥam p. 242.) And ▼ خَامِرِى حَضَاجِرُ أَتَاكِ مَا تُحَاذِرُ [*Hide thyself, O hyena: what thou fearest has come to thee*]: thus we have found it: (Ḳ:) and this is the reading commonly obtaining accord. to the authors on proverbs: (TA:) but it should properly be خَامِرْ [and أَتَاكَ] or تُحَاذِرِينَ. (Ḳ.) —— خَمَرٌ also signifies The *becoming changed,* or *altered, from a former state* or *condition.* (Ḳ.) You say, خَمِرَ الشَّىْءُ *The thing became changed,* &c. (TḲ.) —— خَمَرَ العَجِينَ, (Ks, Ṣ, A, Mṣb, Ḳ,) aor. خَمُرَ (Ṣ, Mṣb, Ḳ) and خَمِرَ, (Ṣ, Ḳ,) inf. n. خَمْرٌ, (Ṣ, Mṣb, Ḳ,) [*He leavened the dough;*] *he put* خَمِيرَة, (Ks, A,) or خَمِير, (Ṣ, A, Mṣb,) *into the dough;* (Ks, Ṣ, A, Mṣb, TA;) as also ▼ خمّره: (TA:) or *he left the dough until it became good* [or *mature*]; (Ḳ;) and in like manner, accord. to the Ḳ, الطِّينَ [*the clay,* or *mud:* see فَطَرَ]: or, as in other lexicons, الطِّيبَ [*the perfume*]; (TA;) and the like; as also ▼ خمّره, inf. n. تَخْمِيرٌ, in relation to any of these things; and ▼ اخمره in relation to the first [and probably to the others also]: (Ḳ:) and خَمَرَ النَّبِيذَ [*he fermented the beverage called* نبيذ;] *he put* خُمْرَة *into the* نبيذ. (A.) [Mṭr says, in the Mgh, ▼ خَمَّرَ العَصِيرَ I have not found, nor ▼ تخمّر as its quasi-pass.] —— خَمَرَهُ, aor. خَمُرَ, (TA,) inf. n. خَمْرٌ; (Ḳ;) and ▼ اخمره; (Mgh;) *He gave him* (namely, a man, and a beast, such as a horse and the like, TA) *wine* (خَمْر) *to drink.* (Ḳ,* Mgh, TA.) —— خُمِرَ, (Mgh, TA,) inf. n. خُمَارٌ, (TA,) *He suffered,* or *was affected with,* خُمَار [i. e. *the remains of intoxication*]. (Mgh, TA.) [See also 5.] —— خَمِرَهُ, aor. خَمَرَ, (AA, Ṣ,) inf. n. خَمَرٌ, (Ḳ,) *He was ashamed for himself,* or *of himself,* or *was bashful,* or *shy, with respect to him; was abashed at him,* or *shy of him.* (AA, Ṣ, Ḳ.*)

2: see 1, in eight places: —— and see also 3.

3. خامر as an intrans. v.: see 1, in three places. —— خامرهُ, inf. n. مُخَامَرَةٌ, *It mixed, mingled, commingled, intermixed,* or *intermingled, with it; became incorporated,* or *blended, with it; infected,* or *pervaded, it;* syn. خَالَطَهُ. (Ṣ, A, Mgh,* Ḳ.) You say, خامر المَآءُ اللَّبَنَ *The water mixed with the milk.* (A.) And خَامَرْتُ فُلَانًا ‡ *I mixed with such a one in familiar,* or *social, intercourse; conversed with him;* or *became intimate with him;* syn. خَالَطْتُهُ. (A.) And الخَمْرُ تُخَامِرُ العَقْلَ: see 1. And خامرهُ الدَّآءُ † *The disease infected,* or *pervaded, him;* syn. خَالَطَهُ: (Sh:) or *infected,* or *pervaded,* (خالط,) *his inside.* (Lth.) —— Also, (TA,) inf. n. as above, (Ḳ,) † *He approached it;* or *was,* or *became, near to it;* (Ḳ,* TA;) namely, a thing. (TA.) —— And خامر المَكَانَ, (Ṣ, A,) inf. n. as above, (Ḳ,) ‡ *He kept,* or *clave, to the place;* (Ṣ, A, Ḳ;) *did not quit it;* (A;) *remained, stayed, dwelt,* or *abode, in it;* (Ḳ;) and in like manner, بَيْتَهُ *his house,* or *tent;* and so ▼ خمّرهُ. (TA.) —— خامر, (TḲ,) inf. n. as above, (IAạr, Ḳ,) [app. in the dial. of El-Yemen, (see 10,)] also signifies *He sold a free person as being a slave.* (IAạr, Ḳ, TḲ.)

4. اخمر: see 1 in the former half of the paragraph, in six places. —— أَخْمَرَتِ الأَرْضُ *The land abounded with* خَمَر, (Ṣ, Ḳ,) meaning *tangled trees.* (TA.) —— See also 1, latter part, in two places. —— اخمرهُ الشَّىْءَ *He gave him the thing,* or *put him in possession of it,* (Ḳ,) is a phrase common in El-Yemen: (Moḥammad Ibn-Ketheer, TA:) a man says, أَخْمِرْنِى كَذَا, meaning *Give thou me such a thing as a free gift: put me in possession of it:* and the like. (Moḥammad Ibn-Ketheer, Ṣ.)

5. تَخَمَّرَتْ: see 8. —— Also *She* (a woman) *applied* خُمْرَة *as a liniment to her face, to beautify her complexion.* (TA.) —— تخمّر *He was affected with languor by wine.* (TA.) [See خُمِرَ.] —— See also 1, near the end of the paragraph.

8. اِخْتَمَرَتْ *She wore,* or *put on* [*her head*], *a* خِمَار; (Ṣ, A, Mgh, Mṣb, Ḳ;) as also ▼ تخمّرت. (A, Mgh, Mṣb, Ḳ.) —— اختمر, said of dough, [*It became fermented;*] *it had* خُمْرَة *put into it:* and in like manner one says of the beverage called نَبِيذ [*it became fermented*]: (A:) or, said of dough, and of clay, or mud, (طِين, as in the Ḳ, but accord. to other lexicons perfume, طِيب, TA,) and the like, *it was left until it became good* [or *mature*]: (Ḳ:) and اختمرت الخَمْرُ *the wine became mature* [*and fermented*]; (Mgh, Mṣb, Ḳ;) as it does when it becomes changed in odour: (TA:) or *became changed in odour.* (Ṣ.)

10. استخمرهُ *He made him,* or *took him as, a slave:* (Ṣ, Mgh, Ḳ:) of the dial. of El-Yemen. (Mgh, TA.) [See 3.] So in the trad. of Mo'ádh, مَنِ ٱسْتَخْمَرَ قَوْمًا أَوَّلُهُمْ أَحْرَارٌ وَجِيرَانٌ مُسْتَضْعَفُونَ فَلَهُ مَا قَصَرَ فِى بَيْتِهِ [*Whosoever hath made slaves,* or *taken as slaves, persons the first state of whom hath been that of freemen and neighbours, regarded as weak, to him shall belong what he hath held in possession in his house* or *tent*]: (Ṣ,* L:) i. e., *hath taken them by force, and obtained possession of them:* (Ṣ:) meaning, whosoever hath made slaves, or taken as slaves, persons in the Time of Ignorance, and then El-Islám hath come, to him shall belong those whom he hath held in possession in his house or tent: they shall not go from his hand. (Az, TA.) Moḥammad Ibn-Ketheer says, This is a phrase known to us in El-Yemen, where any other is scarcely ever used [in its stead]. (Ṣ.)

خَمْرٌ [*Wine:* or *grape-wine:*] *what intoxicates, of the expressed juice of grapes:* (ISd, Ḳ:) or the *juice of grapes when it has effervesced, and thrown up froth, and become freed therefrom, and still:* (Mgh:) or it has a common application to *intoxicating expressed juice of anything:* (Ḳ, TA:) or *any intoxicating thing, that clouds,* or *obscures,* (lit. *covers,*) *the intellect;* as some say: (Mgh,* Mṣb: [but see what follows:]) and the general application is the more correct, because خمر was forbidden when there was not in El-Medeeneh any خمر of grapes; the beverage of its inhabitants being prepared only from dates in their green and small state, or full-grown but unripe, or fresh and ripe, or dried: (Ḳ,* TA:) or the arguing thus, from this fact alone, requires consideration: (MF:) AḤn says, it is † *sometimes prepared from grains:* but ISd holds this to be an improper signification: (TA:) it is also sometimes applied to the † *beverage called* نَبِيذ, like as نبيذ is sometimes applied to wine expressed from grapes: (L in art. نبذ:) applied to ‡ *expressed juice from which* خمر [*properly so called*] *is made,* [i. e., to *must,* or *unfermented* نَبِيذ,] it is tropical: it is so used in a trad. in which خمر is said to have been sold by [a companion of Moḥammad named] Samurah: خمر [in its proper acceptation] is so called because it veils (تَسْتُرُ, i. e. تَخْمُرُ,) the intellect: (Ḳ:) or because it infects (تُخَامِرُ, i. e. تُخَالِطُ,) the intellect: (Ṣ, Ḳ:) [as though acting like leaven: (see 1:)] so said 'Omar: (TA:) or because it is left until it has become mature [and fermented]; (Ḳ;) or until its odour has changed: (IAạr, Ṣ:) [see 8:] the proper application of the root is to denote "covering," and "commingling in a hidden manner:" (Ṣgh, Er-Rághib, TA:) it is of the fem. gender, and sometimes masc.: (Mṣb, Ḳ:) you say هُوَ الخَمْرُ as well as هِىَ الخَمْرُ:

but Aṣ does not allow it to be masc.: (Mṣb:) and ↑ خَمْرَةٌ signifies the same: (Ḳ:) [or *a kind of wine:*] or خَمْر and خَمْرَةٌ are like تَمْر and تَمْرَةٌ; [the former a coll. gen. n., and the latter its n. un.;] (Ṣ;) and خَمْرَةٌ [thus] signifies *some wine;* lit., *a portion of* خَمْر: (Mṣb:) the pl. of خَمْرٌ is خُمُورٌ. (Ṣ, Mṣb.) You say [also] خَمْرَةٌ ↑ صِرْفٌ [*Some pure,* or *unmixed, wine;* using a masc. epithet, contr. to rule]. (Ṣ.) — [Hence the saying,] مَا عِنْدَ فُلَانٍ خَلٌّ وَلَا خَمْرٌ, (Ṣ,) or مَا هُوَ بِخَلٍّ وَلَا خَمْرٍ, (Ḳ,) ‡ *Such a one,* (Ṣ,) or *he,* (Ḳ,) *possesses neither good nor evil:* (Ṣ, Ḳ:) [or *neither evil nor good:* for] AA says that some of the Arabs make الخَمْر to be *good,* and الخَلّ to be *evil;* and some of them make الخمر to be *evil,* and الخلّ to be *good.* (Ḥar p. 153.) — خَمْرٌ also signifies † *Grapes;* (AḤn, M, Ḳ;) in the dial. of El-Yemen:) (M:) like as عِنَبٌ signifies "wine" in that dial. (AḤn, TA in art. عنب.) It is said in the Ḳur [xii. 36], إِنِّى أَرَانِى أَعْصِرُ خَمْرًا *Verily I thought myself pressing grapes:* (ISd:) or the meaning is, *pressing out wine* from grapes. (Ibn-'Arafeh.)

خَمَرٌ *A covert of trees &c.:* (ISk, Ṣ, Mgh, Ḳ:) *or a place where the ground is eaten away by a torrent, or an oblong tract of sand collected together and elevated, forming a place for concealment:* (ISk, Ṣ:) and *a hollow,* or *cavity, in which a wolf conceals himself:* and *tangled trees.* (TA.) You say, تَوَارَى الصَّيْدُ مِنِّى فِى خَمَرِ الوَادِى [*The game,* or *wild animal* or *animals, concealed itself,* or *themselves, from me in the covert,* &c., *of the valley*]. (Ṣ.) And هُوَ يَدِبُّ لَهُ الضَّرَآءَ وَيَمْشِى لَهُ الخَمَرَ † [*He creeps to him in the thicket,* or *place overgrown with trees; and he walks to him in the covert of trees,* &c.: see Freytag's Arab. Prov. 918]: speaking of a man when he deceives, or circumvents, his companion. (Ṣ.) And جَآءَنَا عَلَى خَمَرٍ and ↑ على خَمْرَةٍ † *He came to us secretly; unexpectedly; clandestinely.* (Ḳ.) — Hence, (Ṣ,) خَمَرٌ and ↑ خَمَارٌ and ↑ خُمَارٌ (Ṣ, Ḳ) and ↑ خَمْرَةٌ (Ḳ) † *A crowding,* (Ṣ,) or *congregation,* (Ḳ,) *and multitude,* of men or people. (Ṣ, Ḳ.) You say, دَخَلَ فِى خَمَارِ↑ النَّاسِ and ↑ خُمَارِهِمْ, dial. vars. of غَمَار and غُمَار, i. e. ‡ *He entered among the crowding and multitude of the men* or *people;* (Ṣ;) and in like manner, فِى خَمَرِهِمْ and فى خُمَرَتِهِمْ↑; (TA;) as also فى غَمَرِهِمْ and غُمَرِهِمْ: (TA in art. غمر:) or *among such* [*a crowd*] *of the people as hid him.* (ISk, Ṣ.)

خَمِرٌ A place *abounding with coverts of the description termed* خَمَر; (IAạr, Ṣ, Ḳ;) a place *concealing by dense trees.* (TA.) = † A man *infected,* syn. مُخَامَرٌ, (Sh, IAạr, Ṣ,) *by a disease:* (TA:) thought by ISd to be a possessive epithet: (TA:) or *in the last stage of the remains of intoxication.* (Ṣ.) [See also مَخْمُورٌ.])

خَمْرَةٌ: see خَمْرٌ, in two places. — Also, (Ṣ, A, Ḳ,) and ↑ خَمَرَةٌ (Kr, Ḳ) and ↑ خُمْرَةٌ, (Ḳ,) The *odour* of perfume: (Ṣ, A:) or *a sweet odour:* (Ḳ:) and the last signifies also *an odour which has infected* (خَامَرَ, i. e. خَالَطَ,) *a person;* (Ḳ;) as also ↑ خَمَرَةٌ. (AZ, Ḳ.) You say, وَجَدْتُ خَمَرَةَ الطِّيبِ *I experienced,* or *smelt, the odour of the perfume.* (Ṣ, A.) = See also خَمْرٌ, in two places.

خُمْرَةٌ: see خَمِيرٌ, in two places. — Also a dial. var. of غُمْرَةٌ [q. v.], *A thing* [or *composition*] *which is used as a liniment for beautifying the complexion;* (Ṣ;) [the *plant called*] وَرْس *and certain perfumes which a woman uses as a liniment* (so in the Ḳ, or *applies as a liniment to her face,* as in other lexicons, TA) *to beautify her face.* (Ḳ.) = *Pain, and headache, and annoyance, occasioned by wine* (خَمْر, for which in some copies of the Ḳ we find خَمْسٌ erroneously put, TA); as also ↑ خُمَارٌ: or the *intoxication thereof, which has infected* (خَالَطَ) [*a person*]; (Ḳ;) and so ↑ خُمَارٌ: (TA:) or this latter signifies the *remains of intoxication:* (Ṣ:) pl. of the former خُمَرٌ. (TA.) — See also خَمْرَةٌ. = *A small pot or jar:* and *a vessel for leaven.* (KL.) = *A small mat,* (Ṣ, A,* Mgh, Mṣb, Ḳ,) [*of an oblong shape,*] *large enough for a man to prostrate himself upon it,* (Mgh, Mṣb,) *used for that purpose* [*in prayer*], (Ṣ, A,) *made of palm-leaves* (Ṣ, Ḳ) *woven* (تُرْمَلُ) *with threads or strings:* (Ṣ:) so called because it veils the ground from the face of the person praying [upon it]: (Zj,* Mgh:) or because its threads or strings are hidden by its palm-leaves. (TA.)

خِمْرَةٌ *A hiding,* or *concealing, oneself:* (IAạr, TA:) [or, accord. to analogy, *a mode,* or *manner, of doing so.*] — See also خَمَرٌ. — *A mode, manner,* or *way, of wearing the* خِمَار. (Ḳ,* TA.) You say, إِنَّهَا لَحَسَنَةُ الخِمْرَةِ [*Verily she has a beautiful mode of wearing the* خمار]. (Ṣ.) And hence the saying of 'Omar to Mo'áwiyeh, مَا أَشْبَهَ عَيْنَكَ بِخِمْرَةِ هِنْدَ [*How like is thine eye to Hind's* (when she practises her) *mode of wearing the* خمار!]. (TA.) Hence also, (TA,) إِنَّ العَوَانَ لَا تُعَلَّمُ الخِمْرَةَ [*Verily she who has had a husband will not require to be taught the mode of wearing the* خمار]: (Ṣ, Ḳ,* TA:) a prov., (Ṣ, TA,) applied to him who is experienced and knowing: (Ḳ:) i. e. the experienced woman is not to be taught how she should act. (TA.) = See also خُمْرَةٌ.

خَمَرَةٌ: see خَمْرَةٌ.

خَمْرِىٌّ Grapes (عِنَبٌ) *fit for wine.* (TA.) — A colour *resembling the colour of wine.* (TA.)

خُمُرٌ: see خِمَارٌ.

خَمَارٌ: see خَمَرٌ, in two places.

خُمَارٌ: see خَمَرٌ, in two places: = and see also خُمْرَةٌ, in two places.

خِمَارٌ [*A woman's muffler,* or *veil, with which she covers her head and the lower part of her face, leaving exposed only the eyes and part or the whole of the nose:* such is the خمار worn in the present day: *a kind of veil which is called in Turkish* يَشْمَقْ; as in the TḲ:] *a woman's head-covering;* (Mgh, TA;) *a piece of cloth with which a woman covers her head;* (Mṣb;) *i. q.* نَصِيفٌ, (Ḳ,) *pertaining to a woman;* (Ṣ;) as also ↑ خِمِرٌّ: (Th, Ḳ:) and *any covering of a thing; anything by which a thing is veiled,* or *covered:* (Ḳ:) pl. [of pauc.] أَخْمِرَةٌ (Ḳ) and [of mult.] خُمُرٌ (Mṣb, Ḳ) and خُمْرٌ. (Ḳ.) — Also A man's *turban;* because a man covers his head with it in like manner as a woman covers her head with her خمار: when he disposes it in the Arab manner, he turns [a part of] it under the jaws [nearly in the same manner in which a woman disposes her خمار]. (TA.) [Hence,] مَا شَرُّ خِمَارِكَ, a prov., (TA,) [meaning] † *What hath changed thee from the state in which thou wast? What hath befallen thee?* (Ḳ.)

خَمِيرٌ (Ḳ) and ↑ مَخْمُورٌ and ↑ مُخْمَرٌ, (TA,) applied to dough, [*Leavened;*] *having had* خَمِير [as meaning *leaven*] *put into it:* (TA:) or, applied to dough, and to clay or mud (طِين, as in the Ḳ, but accord. to other lexicons perfume, طِيب, TA), and the like, *left until it has become good* [or *mature*]: (Ḳ:) pl. [of the first] خَمْرَى. (TA.) You say also خُبْزٌ خَمِيرٌ *Bread* [*leavened,* or] *into which leaven* (خَمِير) *has been put:* (Lḥ, TA:) or *yesterday's bread; bread that has been kept over a night:* (Ṣ:) and خُبْزَةٌ خَمِيرٌ, without ة [in the epithet]. (Lḥ, TA.) And خَمِيرٌ is also applied to *Bread itself:* or *leavened bread.* (Sh, TA.) — خَمِيرٌ [used as a subst.] (Ṣ, A, Mṣb, Ḳ) and ↑ خَمِيرَةٌ and ↑ خُمْرَةٌ (Ṣ, A, Ḳ) signify *Leaven,* or *ferment,* expl. by مَا خُمِرَ بِهِ, (Ḳ,) of dough, and of perfume; (TA;) *what is put into dough,* (Ṣ, A, Mṣb,) *and into the beverage called* نَبِيذ; (A;) and ↑ خُمْرَةٌ also signifies *what is put into perfume,* as well as what is put into dough and into نبيذ: (Ks:) the خُمْرَة of نبيذ is its *dregs,* (Ḳ,) and its [*ferment which is called*] دُرْدِىّ; (TA;) or *what is put into it, of wine* (خَمْر) *and of* دُرْدِىّ; and so too of perfume; (Ṣ;) and the خُمْرَة of milk is its *ferment* (رَوْبَة) *which is poured upon it in order that it may quickly curdle, or coagulate, or thicken, or become thick and fit for churning.* (TA.) — [Hence,] اِجْعَلْهُ فِى سِرِّ خَمِيرِكَ ‡ *Conceal thou it* (i. e. a secret, A) *in thy mind.* (A, TA.) And أَخْرَجَ مِنْ سِرِّ خَمِيرِهِ سِرًّا ‡ *He revealed,* or *disclosed, a secret.* (TA.) — See also مَخْمُورٌ.

خَمِيرَةٌ: see the next preceding paragraph.

خَمَّارٌ *A vintner; a seller of* خَمْر [or *wine*]. (Ḳ.)

خِمِّيرٌ (Ṣ) and ↑ مُسْتَخْمِرٌ (Ḳ) One who *constantly drinks wine;* (Ṣ, Ḳ;) *a great drinker; devoted to drink.* (Ḳ.)

مُخَمَّرٌ † A horse *having a white head, whatever be the rest of his colour;* but not ↑ مُخْتَمِرٌ: (Lth:) and مُخَمَّرَةٌ, applied to a ewe or she-goat, (AZ, T, Ṣ, A,) accord. to Lth and the Ḳ ↑ مُخْتَمِرَةٌ, but the former is the right term, (TA,) [in the CḲ مُخْتَمَرٌ,] † *whose head is white, and the rest of her black;* like رَخْمَآءُ: (Ṣ:) or *having a white head;*

(AZ, T, A;) and in like manner, a mare: (Ḳ:) or a *black ewe with a white head*: from the خِمَار of a woman. (TA.) ═ See also خَمِيرٌ. — And see مَخْمُورٌ.

مُخَمِّرٌ *A maker of* خَمْر [or *wine*]. (Ḳ.)

مَخْمُورٌ: see خَمِيرٌ — Also, (Ṣ,) and ▼مُخْمَرٌ and ▼خَمِيرٌ, (TA,) A man *affected with* خُمَار, (Ṣ, TA,) i. e. *the remains of intoxication*. (Ṣ. [Like مَخْمُورٌ. See also خَمِيرٌ.])

مُخْتَمِرٌ, and with ة: see مُخَمَّرٌ.

مُسْتَخْمِرٌ: see خَمِيرٌ.

خمس

1. خَمَسَ القَوْمَ, (Ṣ, A, Mgh, Ḳ,) aor. ـُ, (Ṣ, Mgh, Ḳ,) [inf. n. خَمْسٌ,] *He took the fifth part of the possessions of the people.* (Ṣ, A, Mgh, Ḳ.) And خَمَسَ المَالَ, (A, Mṣb,) aor. ـُ, inf. n. خَمْسٌ, (Mṣb,) *He took the fifth part of the property.* (A, Mṣb.) خَمْسٌ signifies The *taking one from five*: and hence the saying of 'Adee Ibn-Ḥátim, رَبَعْتُ فِى الجَاهِلِيَّةِ وَخَمَسْتُ فِى الإِسْلَامِ [*I took the fourth part* of the spoil *in* the Time of *Ignorance, and I took the fifth part* thereof *in* the time of *El-Islám*]; meaning, I headed the army in both those states; for the commander, in the Time of Ignorance, used to take the fourth part of the spoil; and in El-Islám, the fifth part was assigned to him. (TA.) — خَمَسَ القَوْمَ, (Ṣ, A, Mṣb, Ḳ,) aor. ـِ, (Ṣ, Mṣb, Ḳ,) inf. n. خَمْسٌ, (Mṣb,) *He was, or became, the fifth of the people*: (Ṣ, A, Mṣb, Ḳ:) or *he made them five by* [*adding to their number*] *himself*. (Ṣ, Ḳ.) — خمس also signifies *He made* fourteen *to be fifteen*. (T in art. ثلث.) — And *He made* forty-nine *to be fifty with himself*. (A'Obeyd, Ṣ in that art.) — خَمَسَ الحَبْلَ, aor. ـُ, inf. n. خَمْسٌ, *He made the rope of five strands twisted together.* (TA.) ═ خَمَسَتِ الإِبِلُ *The camels drank on the fifth day, counting the day of the next preceding drinking as the first.* (TA.) [See خِمْسٌ.] — خَمَسَ, said of a horse, *He came fifth in the race.* (T, M, L; all in art. ثلث.)

2. خَمَّسَهُ, inf. n. تَخْمِيسٌ, *He made it five.* (Esh-Sheybánee and Ḳ, voce وَحَّدَهُ.) — *He made it to be five-cornered; five-angled; pentagonal.* (Ḳ.) — خَمَّسَتْ *She brought forth her fifth offspring.* (TA in art. بكر.) — And خَمَّسَهُ *He made it five-fifths.* (Mṣb.) — خَمَّسَ لِامْرَأَتِهِ, or عِنْدَهَا, *He remained five nights with his wife*: and in like manner the verb is used in relation to any saying or action. (TA voce سَبَّعَ.) — تَخْمِيسٌ also signifies [The *watering* of land or seed-produce *on the fifth day, counting the day of the next preceding watering as the first*;] the *watering* of land *that is* [*next*] *after the* تَرْبِيع. (TA.)

4. اخمس القَوْمُ *The party of men became five*: (Ṣ, Ḳ:) — also, *The party of men became fifty.* (M and L in art. ثلث.) — اخمس الرَّجُلُ *The man was, or became, one whose camels came to water on the fifth day, counting the day of the next preceding drinking as the first.* (Ṣ,* Ḳ,* TA.) [See خِمْسٌ.]

خُمْسٌ fem. of خَمْسَةٌ [q. v.].

خُمُسٌ: see خُمْسٌ.

خِمْسٌ The *drinking of camels on the fifth day, counting the day of the next preceding drinking as the first; their drinking one day, then pasturing three days, then coming to the water on the fifth day, the first and last days, on which they drink, being thus reckoned*: this is the correct explanation, accord. to Aboo-Sahl El-Khowlee; and Aboo-Zekereeyà says the like; (TA;) or *their pasturing three days, and coming to the water on the fourth day* [*not counting the day of the next preceding watering*; for it is evident that this explanation is virtually the same as that preceding]: (Ṣ, Ḳ:) accord. to Lth, the *drinking of camels on the fourth day, counting the day on which they returned from* [*the next preceding*] *watering*; but Az says, that this is a mistake; the day of returning from watering not being counted [when it is explained as meaning the drinking on the fourth day]: (TA:) pl. أَخْمَاسٌ, the only pl. form. (Sb, TA.) [See ظِمْءٌ.] Hence, فَلَاةٌ خِمْسٌ [as in copies of the Ḳ, or it may be فَلَاةُ خِمْسٍ,] *A desert in which the water is far distant, so that the camels come to the water on the fourth day, exclusive of the* [*next preceding*] *day on which they drank.* (Az, Ḳ, TA.) Hence also the saying, فُلَانٌ يَضْرِبُ أَخْمَاسًا لِأَسْدَاسٍ (Ṣ, Ḳ*) ‡*Such a one makes a pretence of* اخماس [or *fifth-day waterings*] *for the purpose of* اسداس [or *sixth-day waterings*]: i. e., he advances his camels from the خِمْس to the سِدْس: (Ḳ:) a prov.: (TA:) meaning, *such a one strives to deceive, or circumvent*: (Ṣ, Ḳ:) applied to him who acts towards another with artifice, pretending that he obeys him, or complies with his desire: (TA:) or to him who pretends one thing while he means another: (Ḳ:) and taken from the saying, related by AO and IAạr, ضَرَبَ أَخْمَاسًا لِأَسْدَاسٍ [*He made a pretence of* اخماس *for the purpose of* اسداس]; said of him who proposes a thing whereby he means another thing, which he commences and by slow degrees accomplishes: (TA:) for a man, when he desires to make a long journey, accustoms his camels to drink خِمْسًا سِدْسًا [i. e. *on the fifth day* and then *on the sixth, in each case counting the day of the next preceding drinking as the first*]: (Ḳ, TA:) the origin of the saying, accord. to IAạr, being this: an old man was among his camels, accompanied by his sons, men, who pastured them, and who had been long far distant from their families; and he told them one day to pasture their camels رِبْعًا [i. e. watering on the fourth day, counting the day of the next preceding watering as the first], which they did, proceeding in the way towards their families: then they proposed to do so خِمْسًا; and then, سِدْسًا: whereupon the old man, understanding what they meant, said, ye are doing nothing but making a pretence of اخماس for the purpose of اسداس: the object of your desire is not the pasturing of them, but it is only your families. (TA.) [See below, voce خُمْسٌ, a saying similar in words but different in meaning.] — It is also used for سَيْرٌ خِمْسٌ [*A journey in which the camels are watered only on the first and fifth days; a journey in which the second and third and fourth days are without water*]. (L in art. جلذ.) You say خِمْسٌ بَصْبَاصٌ, [and صَبْصَابٌ,] and قَعْقَاعٌ, and حَثْحَاثٌ, [and حَصْحَاصٌ, &c.,] i. e. *A journey* [*in which the camels are watered only on the first and fifth days,*] *in the course of which, to the water, there is no flagging, by reason of its remoteness.* (TA.) El-'Ajjáj uses the expression

خِمْسٌ كَحَبْلِ الشَّعَرِ المُنْحَتِّ

meaning, *A* [*journey of the kind termed*] خمس without any deviation, *like a rope* made *of hair that has fallen off* and that is free from any unevenness. (L, TA.) — خِمْسٌ also signifies The *fifth young one*, or *offspring*. (A in art. ثلث.) ═ *A* [*garment of the kind called*] بُرْد, (Ṣ, Ḳ,) *of the fabric of El-Yemen*; (Ṣ;) so called because first made for a king of El-Yemen named خِمْس, (AA, Ṣ,) or الخِمْس; (Ḳ, TA;) as also ▼خَمِيسٌ. (TA.) For the latter word, we find in the work of Bkh, خَمِيص, with ص; which, if correct, is masc. of خَمِيصَة, which is a small kind of كِسَآء. (IAth, and L.) [The pl. of خِمْس applied to a بُرْدَة is أَخْمَاس.] See also مَخْمُوس, in four places.

خُمْسٌ and ▼خُمُسٌ *A fifth part*; (Mṣb, Ḳ;) as also ▼خَمِيسٌ, (Ṣ, in art. ثلث, and IAmb and Mṣb,) agreeably with a rule applicable in the case of every one of the units, except ثَلِيث: (TA:) some allow this last; but AZ disallows it, and خميس also: (Ṣ in art. ثلث:) pl. أَخْمَاس. (Mṣb, TA.) — [Hence, app.,] ضَرَبَ أَخْمَاسَهُ فِى أَسْدَاسِهِ *He turned his five senses towards his six relative points*; [namely, above, below, before, behind, right, and left:] an allusion to the collecting all the thoughts to examine a thing, and turning the attention in all directions. (MF.)

خَمْسَةٌ, (Ṣ, Ḳ,) masc.; and خَمْسٌ, fem.; (Ṣ;) [*Five*;] *a certain number.* (Ṣ, Ḳ.) You say خَمْسَةُ رِجَالٍ [*Five men*], and خَمْسُ نِسْوَةٍ [*Five women*]. (Ṣ.) You say also, عِنْدِى خَمْسَةُ دَرَاهِمَ [*I have five dirhems*], with refạ: and if you please, you incorporate the ة into the د [and say, خَمْسَدَّرَاهِمَ]: but when you prefix ال to دراهم, you say, عِنْدِى خَمْسَةُ الدَّرَاهِمِ [*I have the five dirhems*], with ḍamm; and may not incorporate, because you have incorporated the ل into the د: and in the case of a fem. n. you say, عِنْدِى خَمْسُ القُدُورِ [*I have the five cooking-pots*]: also, هٰذِهِ الخَمْسَةُ الدَّرَاهِمُ [*These five dirhems*]; and, if you please, الدَّرَاهِمِ, using it in the manner of an epithet: and in like manner [you use the other nouns of number] to عَشَرَة [inclusive]. (Ṣ.) You say also, صُمْنَا خَمْسًا مِنَ الشَّهْرِ [*We fasted* during a period of *five nights of the month* with their days]; making لَيَالٍ to predominate over أَيَّام, when you do not mention the word أَيَّام, though the fasting is in the day; because the night of each day precedes the day: but when you mention the word أَيَّام, you say, صُمْنَا خَمْسَةَ أَيَّامٍ [*We fasted five days*]. (ISk, TA.) يَعَضُّ بِالخَمْسِ means *He bites the fingers:* these being [five in number and] of the fem. gender: (Ham p. 790:) [i. e.] خَمْس means the *five fingers.* (Har p. 76.) [Respecting a peculiar pronunciation of the people of El-Hijáz, and a case in which خَمْسَة is imperfectly decl., see ثَلَاثَة.] — خَمْسَةَ عَشَرَ, masc.; and خَمْسَ عَشْرَةَ, fem.; *Fifteen.* For variations thereof, see art. عشر.]

خَمْسُونَ [*Fifty,* and *fiftieth,*] is also written and pronounced خِمِسُونَ, with kesr to the م, by poetic license, as related by Ks; or خَمَسُونَ, with fet-h, as related by others, after the manner of خَمَسَة and خَمَسَاتٌ: (Fr, TA:) accord. to the T, the variation خِمِسُونَ, with kesr to the م, is [dialectic, being] similar to خَمْسَ عَشِرَة, with kesr to the ش [in the dial. of Nejd]. (TA.)

جَاؤُوا خُمَاسَ, and ▼مَخْمَسَ, *They came five and five;* [or *five and five together;* or *five at a time and five at a time;*] (K, TA;) like as they say, ثُنَاءَ and مَثْنَى, and رُبَاعَ and مَرْبَعَ: (TA:) or, accord. to A'Obeyd, not more than أُحَادَ and ثُنَاءَ and ثُلَاثَ and رُبَاعَ has been heard, except عُشَارَ occurring in a verse of El-Kumeyt. (TA in art. عشر.)

خَمِيسٌ: see خُمْسٌ: — and مَخْمُوسٌ, in two places. — *An army;* because consisting of five parts, namely, the van, the body, the right wing, the left wing, and the rear; (S, A, K;) or because the spoils are divided into fifths among it; but this latter assertion requires consideration; (ISd, MF;) for this division of the spoils is an affair of the Muslim law, whereas خميس [thus applied] is an old term: (MF:) or *an army having numerous weapons;* syn. جَيْشٌ خَشِنٌ. (TA.) — يَوْمُ الخَمِيسِ, (S, Msb, K,) and simply الخَمِيسُ, *Thursday; the fifth day of the week;* thus used for الخَامِسُ, in like manner as الدَّبَرَانُ is applied to the star [that follows the Pleiades, for الدَّابِرُ]: (TA:) pl. [of pauc.] أَخْمِسَةٌ and [of mult.] أَخْمِسَاءُ (S, Msb, K) and أَخَامِسُ. (Fr, TA.) AZ used to say, مَضَى الخَمِيسُ بِمَا فِيهِ [*Thursday passed with what happened in it*], making it sing. and masc.: but Abu-l-Jarráh used to say, مَضَى الخَمِيسُ بِمَا فِيهِنَّ, making it pl. and fem., and using it as a n. of number. (Lh, TA.) It has no dim. (Sb, S in art. امس.) = See also خُمْسٌ, last signification. = مَا أَدْرِي أَيُّ خَمِيسِ النَّاسِ هُوَ means *I know not what company of men it is.* (Ibn-'Abbád, Sgh, K.)

خُمَاسِيٌّ A boy *five spans* (أَشْبَار) *in height:* (S, Mgh, Msb,* K:) said of him who is increasing in height [but has not attained his full stature]: (Msb:) fem. with ة: (Lth, TA:) and in like manner you say رُبَاعِيٌّ: (S, Msb:) but you do not say سُبَاعِيٌّ, (Lth, S, K,) nor سُدَاسِيٌّ; (Lth, K;) [i. e., in speaking of a boy;] for when he has attained seven spans, (S,) or six spans, (Lth, K,) he is a man: (Lth, S, K:) or, to a slave you apply the epithet سداسيّ also; and to a garment, or piece of cloth, سباعيّ. (Msb.) — See also مَخْمُوسٌ. — [Also A word *composed of five letters, radical only,* or *radical and augmentative.*]

خَمِيسِيٌّ One *who fasts alone on Thursday.* (IAar, Th.)

خَامِسٌ [*Fifth*]: for this you also say خَامٍ; (ISk, S, K;) whence the phrase, جَآءَ فُلَانٌ خَامِيًا [*Such a one came fifth*], for خَامِسًا: (ISk, S:) [fem. with ة.] — خَامِسَ عَشَرَ and خَامِسَةَ عَشْرَةَ, the former masc. and the latter fem., meaning *Fifteenth,* are subject to the same rules as ثَالِثَ عَشَرَ and its fem., explained in art. ثلث, q. v.] = إِبِلٌ خَامِسَةٌ (TA) and خَوَامِسُ (S, K) *Camels that drink on the fifth day, counting the day of the next preceding drinking as the first:* [see خِمْسٌ:] (TA:) or *that pasture three days, coming to the water on the fourth day* [*not counting the day of the next preceding watering*]. (S, K.)

مَخْمَسَ: see خُمَاسَ. جَاؤُوا مَخْمَسَ.

مُخَمَّسٌ A thing *five-cornered; five-angled; pentagonal.* (S.) [See also مُثَلَّثٌ.]

مَخْمُوسٌ, *Five cubits in length;* applied to a spear, (S, A, K,) as also ▼خَمِيسٌ; (K;) and to a garment, or piece of cloth, (S, A, K,) as also ▼خَمِيسٌ, (S, A, Mgh, K,) which occurs in a trad. as meaning a *small* garment or piece of cloth, (Mgh,) and ▼خُمَاسِيٌّ [q. v. suprà]; (TA;) and in like manner, ▼بُرْدَةٌ أُخْمَاسٌ a [*garment of the kind called*] بردة *five cubits long.* (ISk, TA.) Hence the saying, ▼هُمَا فِي بُرْدَةٍ أَخْمَاسٍ † *They two have become near together, and in a state of agreement.* (K.) A poet says,

* صَيَّرَنِي جُودُ يَدَيْهِ وَمَنْ *
* أَهْوَاهُ فِي بُرْدَةِ أَخْمَاسٍ ▼*

i. e., † *The bounty of his hands has made me and the person whom I love to be near together, as though we were in a* بردة *five cubits long:* (Th, TA:) app. meaning that the person thus spoken of had purchased for him a female slave, or had given for him the dowry of his wife. (Az, Sgh, TA.) You also say, ▼لَيْتَنَا فِي بُرْدَةٍ أَخْمَاسٍ, a prov., meaning † *Would that we were near together.* (ISk, TA.) [See also بُرْدٌ.] — Also A rope *made of five strands twisted together.* (S, A, K.)

خمش

1. خَمَشَهُ, (S, A, K,) aor. ـِ (S, Msb, K) and ـُ, (S, K,) inf. n. خَمْشٌ, (Msb,) *He scratched it,* namely, the face, *with the nails, so as to cause bleeding* or *not;* syn. خَدَشَهُ: (S,* A, K:) only used in relation to the face: (A:) or also used in relation to the rest of the person: (TA:) and ▼خَمَّشَهُ, inf. n. تَخْمِيشٌ, signifies the same: (TA:) [or denotes intensiveness, or muchness, like خَدَّشَهُ.] And خَمَشَتِ المَرْأَةُ وَجْهَهَا بِظُفُرِهَا *The woman wounded the exterior of the scarf-skin of her face with her nail.* (Msb.) One says also, by way of imprecation, خَمْشًا [*May thy,* or *his,* or *her, face be scratched*]; like as one says جَدْعًا and قَطْعًا. (TA.) — *He slapped it;* namely, the face. (A, K.) — *He beat him,* or *it,* (K, TA,) with a staff, or stick. (TA.) — *He cut off from him a limb,* or *member.* (K.)

2: see 1.

خَمْشٌ The *mark made by scratching with the nails upon the face:* (Msb, TA:) pl. خُمُوشٌ. (S, A, Msb.)

لَا تَفْعَلْ ذٰلِكَ أُمُّكَ خَمْشَى (Lh) *Do not thou that: may thy mother,* being bereft of thee by death, *scratch her face* for thee. (ISd.) One says also, on the occasion of a thing at which one wonders, خَمْشَى عَقْرَى حَلْقَى. (S and TA in art. حلق: see 1 in that art.)

خُمُوشٌ *Gnats:* (S, A, K:) in the dial. of Hudheyl: (S:) n. un. with ة: or it has no n. un.; (TA;) one thereof being called بَقَّةٌ. (S.)

خُمَاشَةٌ *A wound,* (S, A, K,) or *mutilation,* (S,) *for which there is no fine,* or *mulct,* (A,) or *for which there is no certain fine,* or *mulct;* (S, K:) or *what is below the bloodwit; as the cutting off of an arm or a hand,* or *of an ear, and the like:* (K:) or *a wound,* or *mutilation, of any kind below slaughter and the bloodwit; such as amputation,* or *a wound;* or *a blow,* or *plunder,* or *a similar injury.* (L.) It is related in a trad., that Keys Ibn-'Ásim collected his sons at his death, and said, كَانَ بَيْنِي وَبَيْنَ فُلَانٍ خُمَاشَاتٌ فِي الجَاهِلِيَّةِ [*There were, between me and such a one, wounds, &c., in the Time of Ignorance*]. (L.) And you say, قَدْ أَخَذْتُ خُمَاشَتِي مِنْ فُلَانٍ *I have retaliated upon such a one* [*my wound, &c.*]. (TA.) — خُمَاشَاتٌ, also, (S, TA,) or خُمَاشَاتُ ذَحْلٍ, (A, TA,) signifies ‡ *Remains of* ذَحْل [or *desire of retaliation, or the like*]. (S, A, TA.)

خمص

1. خَمِصَتِ القَدَمُ, aor. ـَ, inf. n. خَمَصٌ, *The man's foot rose from the ground,* [or *was hollow in the middle of the sole,*] *so that it did not touch it.* (Msb.) — خَمَصَ البَطْنُ, (A, K,) aor. ـُ; (TK;) and خَمِصَ, aor. ـَ; and خَمُصَ, aor. ـُ; (A, K, TK;) inf. n. خمص [i. e. خَمْصٌ or خَمَصٌ, or probably both] and خُمُوصٌ and مَخْمَصَةٌ; (TK;) *The belly was,* or *became, empty;* (A, K, TK;) i. e., *hungry:* (TK:) [and *lank:* see خَمِيصٌ.] And خَمَصَ الشَّيْءُ, aor. ـُ, (Msb,) inf. n. خَمْصٌ (Msb, TA) and خَمَصٌ (A, TA) and مَخْمَصَةٌ, (S, A, Msb,* K,) the last an inf. n. like مَغْضَبَةٌ and مَعْتَبَةٌ, (S,) [but in art. عتب in the S, مَعْتِبَةٌ is said to be a subst.,] *The thing was,* or *became, hungry.* (S,* A,* Msb, K.*) = خَمَصَهُ الجُوعُ, (S, K,) aor. ـُ, (TK,) inf. n. خَمْصٌ and

مُخَمَّصَةٌ (Ṣ, Ḳ) and خُمُوصٌ, (TḲ, [but this last I think doubtful,]) *Hunger rendered him lank in the belly.* (TḲ.)

6. تخامص عَنْهُ ‡ *He shrank,* or *drew away, from it;* (A, Ḳ;*) i. e., from anything of which he disliked the nearness. (A.) You say, مَسِسْتُهُ بِيَدِى وَهِىَ بَارِدَةٌ فَتَخَامَصَ مِنْ بَرْدِ يَدِى ‡ [*I touched him with my hand, it being cold, and he shrank from the coldness of my hand*]. (A, TA.) تَخَامَصْ لِفُلَانٍ عَنْ حَقِّهِ ‡ [*Relinquish thou,* i. e.,] *give thou, to such a one, his right,* or *due.* (A, Ḳ.*) — تخامص اللَّيْلُ ‡ [*The night retreated;*] *the darkness of the night became thin a little before daybreak.* (A, Ḳ.)

خَمْصَةٌ *A hungering.* (Ṣ, Ḳ.) You say, لَيْسَ لِلْبِطْنَةِ خَيْرٌ مِنْ خَمْصَةٍ تَتْبَعُهَا [*There is not anything better for repletion of the belly than a hungering which follows it*]. (Ṣ, A.)

خَمْصَى : see خَمِيصٌ.

خُمْصَانٌ : see أَخْمَصُ : — and see also خَمِيصٌ, in two places.

خَمْصَانٌ : see خَمِيصٌ.

خَمِيصٌ *Empty;* applied to the belly: (TA:) *hungry.* (Mṣb.) — خَمِيصُ البَطْنِ, (A,) or خَمِيصُ الحَشَا, (Ṣ, Ḳ,) and ▼خُمْصَانٌ, (Ṣ, A, Ḳ,) and ▼خَمْصَانٌ, (A, Ḳ,) A man *empty in the belly,* (A,) or *lank in the belly;* (Ṣ, K;) as also ▼خَامِصُ البَطْنِ: (Ḳ* and TA in art. رهف:) and *slender in make:* (TA:) fem. of the first with ة, (Ṣ, A, Ḳ,) and so of the second, (Yaạḳoob, Ṣ, A, Ḳ,) and so of the third; (TA;) and IAạr mentions ▼خَمْصَى as a fem., occurring prefixed to الحَشَا in a verse of El-Aṣamm Ed-Dubeyree: (TA:) pl., (Ṣ, A, Ḳ,) masc., (A, Ḳ,) خِمَاصٌ; (Ṣ, A, Ḳ;) and fem., [i. e., of خميصة,] خَمَائِصُ: (A, Ḳ:) ▼خُمْصَانٌ has no pl. formed by the addition of و and ن, though its fem. is formed by the addition of ة; being made to accord with the measure فُعْلَانٌ, of which the fem. is فُعْلَى. (TA.) خِمَاصٌ [also] signifies *Hungry,* in a pl. sense, (Ḳ,) *and lank in the bellies:* (TA:) ▼مَخَامِصُ also signifies the same as خَمِيصٌ; and [its pl.] مَخَامِيصُ, *lank in the bellies* (خُمْصُ البُطُونِ [whence it appears that ▼أَخْمَصُ, sing. of خُمْصٌ, is also syn. with خَمِيصٌ]). (TA.) You say also, هُوَ خَمِيصُ البَطْنِ مِنْ أَمْوَالِ النَّاسِ, meaning ‡ *He is one who abstains from* [*devouring*] *the possessions of men.* (A.) And خِمَاصُ البُطُونِ مِنْ أَمْوَالِ النَّاسِ خِفَافُ الظُّهُورِ مِنْ دِمَائِهِمْ, (A, TA,) meaning ‡ *Persons who abstain from* [*devouring*] *the possessions of men, whose backs are light with respect to* [*the*] *burden* [*of their blood*]. (TA, from a trad.) — زَمَنٌ خَمِيصٌ ‡ *A time of hunger.* (A, TA.)

خَمِيصَةٌ *A* [*garment of the kind called*] كِسَآء, *black, square, and having* عَلَمَانِ [i. e. *two ornamental* or *coloured* or *figured borders*]: (Ṣ, A, Mgh, Ḳ:) or *a black* كساء, *having a border such as is above described* (مُعْلَم) *at each end, and which is of* خَزّ, [q. v.], or *of wool:* (Mṣb:) if not bordered, it is not so called: (Ṣ, Mṣb:) or, accord. to Aṣ, *a* مُلَآءَة *of wool,* or *of* خَزّ, *bordered* (مُعْلَمَة); not unless bordered: so called because of its softness and thinness, and smallness of bulk when it is folded: Aḥmad Ibn-Fáris says that it is the *black* كِسَآء: and he says that it may be thus called because a man wraps himself with it, so that it is against his أَخْمَص, meaning by this his waist: (Ḥar p. 21:) pl. خَمَائِصُ: or خمائص are *garments of* خَزّ, *thick, black,* and *red, and having thick* أَعْلَام [*or borders such as above described*]; *worn by people of old.* (TA.) El-Aạshà says,

* إِذَا جُرِّدَتْ يَوْمًا حَسِبْتَ خَمِيصَةً *
* عَلَيْهَا وَجِرْيَالَ النَّضِيرِ الدُّلَامِصَا *

[*When she is stripped of her clothing, any day, thou wouldst think there was upon her a khameeṣah, and the glistening redness of gold*]: Aṣ says, he likens her [long and spreading] hair to a خميصة, which is black. (Ṣ.) [See also خَمِيسٌ, voce خَمْسٌ, near the end of the paragraph.]

خَامِصُ البَطْنِ : see خَمِيصٌ.

أَخْمَصُ القَدَمِ A man *whose foot rises from the ground,* [or *is hollow in the middle of the sole,*] *so that it does not touch it:* fem. خَمْصَآءُ: and pl. خُمْصٌ: (Mṣb:) and ▼خُمْصَانٌ signifies *having the middle of the sole of the foot moderately rising from the ground;* which is a goodly quality; but when it is flat, or rises much, it is dispraised: so explained by IAạr when he was asked by Th respecting 'Alee's saying of Moḥammad, [cited, but not explained, in the K,] كَانَ خُمْصَانَ الأَخْمَصَيْنِ: or, accord. to Az, خُمْصَانٌ signifies *having the part* [*of the sole*] *of the foot which does not cleave to the ground in treading very much retiring from the ground.* (TA.) — الأَخْمَصُ [when without the article ال also written without tenween accord. to the best authorities, because the quality of an epithet is original to it, and that of a subst. is accidental,] also signifies *The part* [*of the sole*] *of the human foot which does not cleave to the ground in treading;* (Az, TA;) *the part of the sole of the human foot which is hollow, so that it does not touch the ground;* (Ṣ, Ḳ;*) *the part of the bottom of the human foot which is thin, and retires from the ground;* or, as some explain it, [meaning the same,] *the* خَصْر *of the human foot:* (TA:) pl. أَخَامِصُ. (Mṣb.) — See also خَمِيصٌ. — Also *The waist* of a man. (Ḥar p. 21.)

مَخْمَاصٌ : see خَمِيصٌ.

خمط

خَمْطٌ The [*kind of tree called*] أَرَاك: (Bḍ in xxxiv. 15:) or *a species of the* اراك, *having a fruit which is eaten:* (Lth, Ṣ:) or the *fruit of the* اراك: (IB, Ḳ:) or *any trees having no thorns:* (IDrd, Bḍ, Ḳ:) or *trees having thorns;* cited from Fr; and by Z, in the Ksh, on the authority of A'Obeyd: (TA:) or *certain trees like the* سِدْر, (Ḳ, TA,) *the fruit of which is like the mulberry:* (TA:) or *certain deadly trees:* (Ḳ:) or *deadly poison:* (TA:) or *any plant that has acquired a taste of bitterness,* (Zj, Bḍ, Ḳ,) *so that it cannot be eaten:* (Zj, TA:) or *scanty fruit of any trees:* (AḤn, Ḳ:) or the *fruit of what is called* فَسْوَةُ الضَّبُعِ: (Ḳ:) or *a certain fruit called* فَسْوَةُ الضَّبُعِ, *having the form of the poppy, friable, and of no use:* (IAạr:) or it signifies, in the Ḳur xxxiv. 15, *fruit that is disagreeable in taste, and choking:* (Bḍ:) or, [as an epithet,] *bitter, and disagreeable in taste, and choking:* (Jel:) or *bitter;* applied to anything: or *acid.* (Ḳ.) In the Ḳur, ubi suprà, some read, ذَوَاتَىْ أُكُلِ خَمْطٍ: (Ṣ, IB, Jel:) this is the right reading accord. to him who makes خمط to mean the اراك: but accord. to him who makes it to mean the fruit of the اراك, the right reading of اكل is with tenween, and خمط is a substitute for that word. (IB.) [The pl. is خُمُوطٌ: see an ex. voce خَلٌّ.]

خمع

1. خَمَعَ, (Lth, Ṣ, Ḳ,) aor. ـَ, inf. n. خَمْعٌ (Lth, IDrd, Ḳ) and خُمُوعٌ (Lth, Ḳ) and خَمَعَانٌ, (Az, Ḳ,) said of a hyena, (Ḳ,) &c., (TA,) *He limped,* or *had a slight lameness,* (IDrd, Ṣ,) in his gait, or manner of going; (Ṣ;) *he went as though he had a lameness.* (Ḳ, TA.)

خِمْعٌ *A wolf:* (Ṣ, Ḳ:) pl. أَخْمَاعٌ. (TA.) — And hence, (TA,) † *A thief.* (Ṣ, Ḳ.)

خُمَاعٌ *A limping,* or *slight lameness;* (IDrd, S;) *a manner of going as though with a lameness;* a subst. from the above-mentioned verb. (Ḳ.) You say, بِهِ خُمَاعٌ [*He has a limping,* &c.]. (Ṣ.)

خَمُوعٌ, applied to a woman, *Vitious,* or *immoral; an adulteress,* or *a fornicatress;* as also ▼خَيْمَعٌ. (Ibn-'Abbád, Ḳ.)

خَامِعٌ [*Limping,* or *having a slight lameness;*] *going as though having a lameness.* (TA.) — And hence, (TA,) خَامِعَةٌ *A hyena,* or *a female hyena:* (Ṣ, Ḳ:) pl. خَوَامِعُ. (Ḳ.)

خَيْمَعٌ : see خَمُوعٌ.

خمل

1. خَمَلَ, aor. ـُ, inf. n. خُمُولٌ, *It* (a place of alighting or abode, Mṣb, TA, and a tattooing, TA) *was,* or *became, effaced,* or *obliterated;* (Mṣb, TA;) *and imperceptible,* or *unapparent.* (TA.) — And hence, (Mṣb,) said of a man, aor. and inf. n. as above, *He was,* or *became, obscure, unnoted, reputeless,* or *of no reputation:* (Ṣ, Mṣb:) [and] said of a man's reputation (ذِكْرُهُ, JK, Ḳ, and صَوْتُهُ, Ḳ), aor. as above, (JK,) and so the inf. n., (JK, Ḳ,) *it was,* or *became, obscure.* (JK, Ḳ.) Some mention also خَمُلَ,

inf. n. خَمَالَةٌ; and this inf. n. occurs in a description of the Prophet; but it is only for the purpose of assimilation to its contr. نَبَاهَةٌ. (TA.) —— [It is app. also said of speech, meaning *It was*, or *became*, *low*, *soft*, or *gentle*: see خَامِلٌ.] ═ خَمِلَ *He* (a man, and a beast, Ḳ, a horse, a sheep or goat, and a camel, TA) *had*, or *was affected with*, *the malady termed* خُمَالٌ. (Ḳ.)

4. اخمل *He* (a man, Ṣ, or God, Ḳ) *rendered* a person *obscure, unnoted, reputeless*, or *of no reputation*; (Ṣ, Ḳ;*) *contr. of* نَبَّهَ. (TA.) ═ *He made* a [garment such as is termed] قَطِيفَة, and the like, *to have what is termed* خَمْل [i. e. *a nap*, or *pile*, or *villous substance on its surface*]. (Ḳ.)

8. اختمل *He pastured*, or *depastured*, خَمَائِل, (Ḳ,) i. e. *meadows* [&c., pl. of خَمِيلَة]. (TA.)

خَمْلٌ [The *nap*, or *pile*, or *villous substance on the surface, of cloth*;] *i. q.* هُدْبٌ; (Ṣ, Mṣb;) or the هُدْب *of the* [*kind of garment called*] قَطِيفَة [q. v.] and *the like*, (Ḳ, TA,) *of woven cloths whereof portions* [*of the substance*] *are redundant*; (TA;) or [rather] *what resembles* هُدْب *on the surface of a* كِسَآء [*or the like*]; (Mgh;) the خَمْل *of a* [*carpet such as is called*] طُنْفُسَة [or طِنْفِسَة &c.] and *of a garment*; as also ▼خَمِيلَةٌ, of which the pl. [or rather coll. gen. n.] is ▼خَمِيلٌ. (JK.) —— Also *A* طُنْفُسَة [*itself*]: (Ṣ, Ḳ:) or so ▼خَمِيلَةٌ: and خَمْل signifies as first explained above, and also *a* قَطِيفَة [*itself*]: (Mṣb:) or ▼خَمِيلَةٌ has this last meaning; as also ▼خَمْلَةٌ and ▼خِمْلَةٌ; (Ḳ,* TA; [in the CḲ, كالخَمِيلَةِ والخِمْلَةِ is erroneously put for كالخَمْلَةِ والخِمْلَةِ;]) or signifies *a* قطيفة *having* خَمْل [*or nap*]: (TA:) and its pl. [or coll. gen. n.] is as above. (Mṣb, TA.) —— And The *feathers*, or *plumage*, *of the ostrich*; (JK, T, M, Ḳ;) as also ▼خَمَالَةٌ and ▼خَمِيلَةٌ; (T, M, Ḳ;) of which last the pl. [or coll. gen. n.] is as above. (TA.)

خَمْلَةٌ: see the next preceding paragraph. —— Also, and ▼خِمْلَةٌ, *A garment* (Lth, Ḳ) *of wool*, (Lth,) *having* خَمْل [*or nap*], *such as the* كِسَآء *and the like*: (Lth, Ḳ:) or *an* عَبَآء *of the fabric of Ḳaṭawán, white, and with short* خَمْل [*or nap*]. (Az, TA.)

خِمْلَةٌ: see خَمْلٌ: and خَمْلَةٌ. ═ Also A man's *secret, which he conceals*: and his *secret disposition of the mind*. (Ḳ.) One says, اِسْأَلْ عَنْ خِمْلَاتِهِ [in the CḲ خَمْلَاتِه] *Ask thou concerning his secrets*, and *his bad, evil*, or *foul, qualities, dispositions, habits, practices*, or *actions*. (Ḳ, TA.) And هُوَ لَئِيمُ الخِمْلَةِ [*He is base, ignoble*, or *mean, in respect of the secret disposition of the mind*], and كَرِيمُهَا [*generous in respect thereof*]: (Fr, Ḳ:) or it is applied peculiarly to baseness, ignobleness, or meanness: (AZ, Ḳ:) حَسَنُ الخِمْلَةِ has not been heard. (AZ.)

خُمَالٌ *Lameness*: or, accord. to A'Obeyd, *a limping*, or *slight lameness, in the legs of camels, which is cured by cutting the vein*: (Ṣ:) or *a malady in the joints of a man*, (Ḳ,) *resembling lameness*, (TA,) and *in the legs of a beast*, (Ḳ,) *a horse, a sheep or goat, and a camel*, (TA,) *occasioning a limping*, or *slight lameness*: (Ḳ:) or *a malady that affects the horse*, (T, TA,) or *the camel*, (JK,) *in consequence of which he will not move until he has a vein cut*; *otherwise he dies*: (JK, T, TA:) and also *a malady that affects a leg of the sheep or goat, and then shifts to the other legs, going the round of them*. (T, TA.)

خَمِيلٌ pl. [or rather coll. gen. n.] of خَمِيلَةٌ in three senses explained above: see خَمْلٌ. —— Also *garments having* خَمْل [*or nap*]. (Ḳ.) —— *A black garment*. (JK.) —— ‡ *Dense clouds*. (IDrd, Ḳ, TA.) —— ‡ *Soft food*; (Ḳ, TA;) meaning *such as is termed* ثَرِيد: mentioned by ISd. (TA.) ═ See also مَخْمُولٌ.

خَمَالَةٌ: see خَمْلٌ.

خَمِيلَةٌ: see خَمْلٌ, in four places. —— Also *A dense collection of trees*; (JK, Ṣ;) so says Aboo-Ṣá'id: (Ṣ:) or *numerous tangled*, or *luxuriant*, or *dense, trees*, (Ḳ, TA,) *among which one sees not a thing when it falls in the midst thereof*: (TA:) and *a place abounding in trees, wherever it be*, (Ḳ,) or, accord. to Az, *only in plain, level*, or *soft, ground*: (TA:) and *a low*, or *depressed, tract of ground*, (Ḳ,* TA,) *or of sand*, (M, TA,) or *an intervening tract between low*, or *depressed, and hard, ground*, (T, TA,) or *an intervening tract amid sands, in low*, or *depressed, and hard, ground*, (JK,) *and producing good herbage* or *plants*: (JK, T, Ḳ, TA:) or *plain*, or *soft, land, producing herbage* or *plants, which are likened to the* خَمْل [*or nap*] *of the* قَطِيفَة: or *a place where water remains and stagnates, and which produces trees*; *but only in plain, level*, or *soft, ground*: (TA:) or *a meadow* (رَوْضَة) *in which are trees*; that in which are no trees being termed جَلْحَآء: (Ḥar p. 118:) or *a tract of sand producing trees*: (Aṣ, Ṣ, Ḳ:) or *a place where a tract of sand becomes thin*, or *shallow*; *where the main portion of it passes away, and somewhat of the soft part of it remains*: pl. خَمَائِلُ: which is also explained as signifying meadows (رِيَاض). (TA.)

خَامِلٌ A man *obscure, unnoted, reputeless*, or *of no reputation*; (Ṣ, Mṣb, Ḳ;) *unknown*, (JK, T,) *and unmentioned*; (T;) *destitute of good fortune*: (Mṣb:) and one says also خَامِنٌ, by substitution [of ن for ل]: (TA:) pl. خُمَّلٌ (Ḳ) and خَمَلَةٌ, explained as signifying the *lower* or *lowest*, or *meaner* or *meanest, sort of mankind*. (TA.) You say also قَوْلٌ خَامِلٌ *A low, soft*, or *gentle, saying* or *speech*. (Az, TA.) And it is said in a trad., اُذْكُرُوا ٱللّٰهَ ذِكْرًا خَامِلًا *Celebrate ye God with a low, soft*, or *gentle, voice*, in reverence of his greatness, or majesty. (TA.)

مُخْمَلٌ A garment, (JK, TA,) or a كِسَآء, (Mgh, Mṣb,) *having* خَمْل [*or nap*], (JK,* Mgh, Mṣb, TA,) i. e. *what resembles* هُدْب *on its surface*. (Mgh.)

مَخْمُولٌ (JK, TA) and مَخْمُولَةٌ, (JK,) applied to a young camel, (JK,) or to a camel, and a horse, (TA,) and a sheep or goat, (شَاة, JK, TA,) *Having*, or *affected with*, *the disease termed* خُمَال: (JK, TA:) and so ▼خَمِيلٌ, applied to a young camel; pl. خَمْلَى. (JK.)

خمن

1. خَمَنَ, [aor. ـُ,] inf. n. خُمُونٌ, said of [a man's] reputation (الذِّكْر), *It was*, or *became, obscure*; *i. q.* خَمَلَ, inf. n. خُمُولٌ: and, said of a thing, *it was*, or *became, obscure, unapparent, hidden*, or *concealed*: whence خَمَّنَهُ as syn. with خَمَّنَهُ, q. v. (Mṣb.)

2. خَمَّنَهُ, (Mṣb, Ḳ,) inf. n. تَخْمِينٌ; (Ṣ, Mṣb;) and ▼خَمَنَهُ, (Mṣb, Ḳ,) aor. ـُ, inf. n. خَمْنٌ; (Mṣb;) *He spoke of it conjecturally*, (Ṣ, Mṣb, Ḳ,) and *opining*; (TA;) or *surmising*: (Ḳ:) or *he formed a surmise respecting it*; or *an opinion*. (Mṣb.) IDrd says, I think it to be post-classical: (TA:) AḤát says that it is of Persian origin, (Mṣb, TA,) arabicized, (TA,) from خمانا, [app. a mistranscription for گُمَانَا, or گُمَانَهْ, or گُمَانْ,] applied to "an opinion," and "a conjecture," or "conjectural saying." (Mṣb, TA.) —— تَخْمِينٌ is also *syn. with* تَحْرِير. (So in the TA. [But this, I doubt not, is a mistranscription for تَحْزِير, which, though perhaps post-classical, signifies The act of *conjecturing*.])

خَمَنٌ *Stink*. (Ḳ.)

[خَمَانٌ The *elder-tree*: and خَمَانٌ صَغِيرٌ *Dwarf elder-tree*: so in the present day.]

خُمَانٌ A *weak* spear: and قَنَاةٌ خُمَانَةٌ [*a weak spear* or *spear-shaft*]. (A'Obeyd, Ṣ, Ḳ.) —— *What is bad* of household-goods, or furniture, or utensils: (TA:) and The *refuse*, or *lowest* or *basest* or *meanest sort*, of mankind; (Ṣ, Ḳ;) the *bad* thereof. (Ḳ.) [Mentioned also in art. خمر.]

خَامِنُ الذِّكْرِ A man (TA) *obscure, unnoted, reputeless, of no reputation*; *i. q.* خَامِلُ الذِّكْرِ. (Ḳ, TA.)

Quasi خمى

خَامٍ (accus. خَامِيًا) for خَامِسٌ: see the latter word.

خن

1. خَنَّ, (Sh, Ṣ, Ḳ,) aor. يَخِنُّ, (Ṣ, Ḳ,) inf. n. خَنِينٌ, (Sh, Ṣ,* Ḳ,* TA,) *He made a sound from the nose, like* حَنِين *from the mouth*: (TA:) *he made a sound like weeping*, (Ṣ,* Ḳ,* TA,) and (so in the Ṣ, but in the Ḳ "or") *like laughing, in the nose*: (Ṣ,* Ḳ,* TA:) *he reiterated a sound of weeping in the air-passages of the nose*; and sometimes خَنِين is [the *reiterating a sound in the nose*] *from faint laughing*: (Sh, TA:) or *he laughed faintly*. (JK.) [See also خَنِين below.] ═ خُنَّ *He* (a camel) *was affected with the disease termed* خُنَان: (JK, TA:) [and in like manner, a bird: see مَخْنُون.]

4. اخنّهُ اللّٰهُ i. q. أَجَنَّهُ [*God caused him to be bereft of reason; or mad, insane, &c.*]. (Lḥ, Ḳ.)

R. Q. 1. خَنْخَنَ, (TA,) inf. n. خَنْخَنَةٌ, (JK, Ṣ, Ḳ, TA,) [like خَشْخَمَ,] *He snuffled*; i. e., *spoke through his nose*: (TA:) *he spoke indistinctly, making a sort of twang* (يُخَنْخِنُ) *in his* خَيَاشِيم [or *air-passages of the nose*]. (JK, Ṣ, Ḳ.) A poet says,

* خَنْخَنَ لِى فِى قَوْلِهِ سَاعَةً *
* فَقَالَ لِى شَيْئًا وَلَمْ أَسْمَعِ *

[*He snuffled to me in his speech awhile, and said to me something, but I heard not*]. (TA.) —— خَنْخَنَةٌ also signifies The *crying of the ape*. (IAạr, TA.)

خُنَّةٌ i. q. غُنَّةٌ [i. e. *A nasal sound* or *twang*; or *a snuffling sound*]; (JK, Ḳ;) the latter word explained by Mbr as meaning *a mixture of the sound of the* خَيْشُوم [or *air-passage of the nose*] *in the pronunciation of a letter* or *word*; (TA;) as also ▼ مَخَنَّةٌ: (Ḳ:) or the first is *like* غُنَّةٌ; (Ṣ, Ḳ;) as also ▼ خُنَنٌ: (ISd, TA:) or, (Mbr, Ḳ,) as also ▼ مَخَنَّةٌ, (TA,) *louder than* غُنَّةٌ: (Mbr, Ḳ, TA:) or *more open than* غُنَّةٌ: (Ḳ, TA. [In the CḲ, أَقْبَح is put in the place of أَفْتَح.])

خُنَنٌ: see what next precedes.

خُنَانٌ *A certain disease that attacks in the nose*: (Ṣ, TA:) *a disease that attacks camels in their nostrils, and from which they die*; (Aṣ, TA;) *a rheum that affects camels*; (Ḳ;) *in camels, like the* زُكَام *in human beings*. (JK.) زَمَنُ الخُنَانِ [*The time of the* خنان] was in the age of El-Mundhir Ibn-Má-es-Semà; in consequence thereof the camels died: (Ḳ:) it is well known with the Arabs, is mentioned in their verses, (TA,) and became an era to them. (Aṣ, TA.) —— Also *A certain disease that attacks birds in their throats*. (Ṣ, M, Ḳ.) —— And *A certain disease in the eye*. (M, Ḳ.)

خَنِينٌ The *issuing of a sound from the nose, like* حَنِين *from the mouth*: [see حَنِينٌ, in two places:] this is the primary signification: (TA:) and it is [the *making a sound*] *like weeping*, and (so in the Ṣ, but in the Ḳ "or") *like laughing, in the nose*: (Ṣ, Ḳ:) IB says that there is a kind of خنين like weeping in the nose: (TA:) or *a weeping of women*, (JK,) or *a kind of weeping*, (IAth, TA,) *less than what is termed* اِنْتِحَاب: (JK, IAth, TA:) and *a faint laughing*. (JK.) [See also 1.] —— And *Stoppages in the* خَيَاشِيم [or *air-passages of the nose*]. (TA.)

أَخَنُّ i. q. أَغَنُّ [as meaning *Having a nasal twang*]; (Ṣ, Ḳ, TA;) *who snuffles*; i. e., *speaks from* [i. e. *through*] *his nose*: (TA voce أَدْغَم:) [or] as meaning *having the* خَيَاشِيم [or *air-passages of the nose*] *stopped up*: or, as some say, *having the* خياشيم [here app. meaning certain *cartilages in the upper*, or *inmost, part of the nose*] *delapsed*: [see 1 in art. خشم:] fem. خَنَّآءُ: (TA:) and pl. خُنٌّ. (Ṣ, Ḳ.)

مَخَنَّةٌ: see خُنَّةٌ, in two places. —— Also The *nose*: (Ṣ, Ḳ:) written by J [accord. to some of the copies of the Ṣ, but not accord. to all,] with kesr to the م: (TA:) or the *extremity thereof*. (Ḳ.) —— And i. q. مَأْكَلَةٌ: so in the phrase, فُلَانٌ مَخَنَّةٌ لِفُلَانٍ [*Such a one is to such a one a person from whom to obtain what to eat*]. (Ṣ, Ḳ.) —— You say also, البِطِّيخُ لِى مَخَنَّةٌ i. e. [*The melon*, or *water-melon, is to me*] *a usual food*. (JM.)

مَخْنُونٌ A camel, and a bird, *affected with the disease termed* خُنَان. (TA.) —— And i. q. مَجْنُونٌ [*Bereft of reason; or mad, insane, &c.*]. (Lḥ, Ḳ.) [See R. Q. 1 in art. خم.]

خنث

1. خَنَثَ, (Lth, L,) aor. ـِ, inf. n. خَنْثٌ, (L,) *He folded*, or *doubled*, a skin for water or milk, and a sack. (Lth, L.) And خَنَثَ السِّقَآءَ, (Ṣ, A, Mgh, Ḳ, TA,) and قَبَعَ السِّقَآءَ, (A, TA,) and ▼ اختنث السِّقَآءَ, (Ṣ, A,* Mgh, Ḳ,) *He doubled the skin*, (Ṣ, Ḳ,) or *the mouth of the skin*, (A, Mgh, TA,) *outwards*, (Ṣ, A, Mgh, Ḳ,) or *inside-out*, (TA,) *and drank from it*; (Ṣ, Mgh, Ḳ;) the doing of which is forbidden (Mgh, TA) by Moḥammad: (TA:) when you double it inwards, you say, قَبَعْتُهُ: (Ṣ, A, Mgh:) or خَنَثَ فَمَ السِّقَآءِ signifies *he turned the mouth of the skin outside-in* or *inside-out*: and خَنْثٌ signifies *any kind of inverting*, or *turning upside-down* or *inside-out* or *the like*. (TA.) —— [Hence, app.,] خَنَثَ لَهُ بِأَنْفِهِ [*He contracted his nose at him*]; as though he mocked at, scoffed at, derided, or ridiculed, him: so in the A: but in the Ḳ, خَنَثَهُ, aor. ـِ, *he mocked at, scoffed at, derided, or ridiculed, him*. (TA.) —— خَنِثَ, aor. ـَ, (L, Mṣb, Ḳ,) inf. n. خَنَثٌ; (A, Mṣb, TA;) and ▼ انخنث, (Ṣ, A, L, Mṣb, Ḳ,) and ▼ تخنّث; (A, L, Ḳ;) *He* (a man, L) *affected a bending, or an inclining of his body, from side to side, and languor*, or *languidness*; or *he became bent and languid*; syn. تَثَنَّى وَتَكَسَّرَ: (Ṣ, A, L, Ḳ:) [or *he was*, or *became, flaccid*, or *flabby, and affected a bending, or an inclining of his body, from side to side*: (see خُنْثَى:) or *he was*, or *became, effeminate*: (see خَنِثٌ:)] or *he was*, or *became, soft, delicate, tender, flabby, lax*, or *limber, and affected languor*, or *languidness*; expl. by كَانَ فِيهِ لِينٌ وَتَكَسُّرٌ. (Mṣb.) 'Áisheh, describing the death of Moḥammad, says, ▼ فَٱنْخَنَثَ فِى حَجْرِى, meaning *He became bent and languid* (اِنْثَنَى وَتَكَسَّرَ), *by reason of the flaccidness of his limbs, in my bosom*. (TA.)

2. خنّثهُ, (Ṣ, Ḳ,) inf. n. تَخْنِيثٌ, (Ḳ,) *He bent it*; (Ṣ, Ḳ;) namely, a thing. (Ṣ.) Hence the epithet مُخَنَّثٌ. (Ṣ, Ḳ.) —— *He made him to be*, or *become, such as is termed* خُنْثَى. (Mṣb.) —— خنّث كَلَامَهُ *He made his speech like that of women, in softness and gentleness*: so some say. (Mṣb.) —— تَخْنِيثٌ also signifies The *doing what is excessively foul*, or *obscene*; [i. e. the *acting the part of a catamite*;] but this meaning was unknown to the Arabs [of the classical ages]. (MF.)

5. تخنّث *It* (a thing, Ṣ) *bent*, or *became bent*. (Ṣ, Ḳ.) —— Also i. q. خَنِثَ, q. v. (A,* L, Ḳ.) And *He* (a man) *acted in the manner of the* مُخَنَّث [or *effeminate*, &c.]. (TA.) [*He became a* مُخَنَّث: used in this sense in the Ṣ and Ḳ in art. طوس.] And تَخَنَّثَ فِى كَلَامِهِ [*He was soft*, or *effeminate, in his speech*]. (Ṣ, Mgh.) —— *He* (a man &c.) *fell down by reason of weakness*. (TA.)

7. انخنثت القِرْبَةُ *The water-skin became folded*, or *doubled*. (L.) —— انخنثت عُنُقُهُ *His neck inclined*, or *bent*. (TA.) —— See also 1, in two places.

8: see 1, second sentence.

خُنْثٌ a subst. from انخنث [*An affectation of a bending, or of an inclining of the body, from side to side, and of languor*, or *languidness*; or *a bending and languidness*: or *flaccidity* or *flabbiness, and an affectation of a bending, or of an inclining of the body, from side to side*: or *effeminacy*: or *softness, delicacy, tenderness, flabbiness, laxness*, or *limberness, and an affectation of languor*, or *languidness*]: (Ṣ, L:) as also ▼ خُنَاثَةٌ. (Mṣb.) Jereer says,

* أَتُوعِدُنِى وَأَنْتَ مُجَاشِعِىٌّ *
* أَرَى فِى خُنْثِ لِحْيَتِكَ ٱضْطِرَابَا *

[*Dost thou threaten me, thou being a Mujáshi'ee? I see, in the softness and weakness of thy beard*, or *in the bending and languidness*, or *the effeminacy, of thy person*, (for the beard is sometimes, by a synecdoche, put for the whole person,) an evidence of *unsoundness, uncompactness*, or *weakness*]. (Ṣ.)

خِنْثٌ, with kesr, sing. of أَخْنَاثٌ and خِنَاثٌ, (TA,) which signify The *creases*, or *places of folding*, of a garment, or piece of cloth. (Ḳ, TA.) You say, طَوَى الثَّوْبَ عَلَى أَخْنَاثِهِ and خِنَاثِهِ *He folded the garment*, or *piece of cloth, at its creases*. (TA.) And [hence,] أَلْقَى اللَّيْلُ أَخْنَاثَهُ عَلَى الأَرْضِ † *The night cast the folds of its darkness upon the earth*. (TA.) —— Also the former pl., (TA,) and the latter also, (Ḳ,) The *parts* of the دَلْو [or bucket] *whence the water pours forth, between the* عَرَاقِى. (Ḳ, TA.) —— The sing. also signifies The *interior of the part of the cheek by the side of the mouth, next the molar teeth*, (Ḳ, TA,) *above and below*. (TA.) —— And *A company in a state of dispersion*. (Ḳ.)

خَنِثٌ One *in whom is an affectation of a bending, or of an inclining of the body, from side to side, and of languor*, or *languidness*; or *in whom is a bending and languidness*; expl. by مَنْ فِيهِ تَثَنٍّ وَتَكَسُّرٌ: (A, L, Ḳ:) or *flaccid*, or *flabby, and affecting a bending, or an inclining of the body, from side to side*: (Ṣ:) [or *effeminate*; like مُخَنَّثٌ]: or one *in whom is softness, delicacy, tenderness, flabbiness, laxness*, or *limberness, and an affectation of languor*, or *languidness*: (Mṣb:) fem. with ة. (TA.) And ▼ خُنُثٌ (TA) and ▼ مِخْنَاثٌ, (Ḳ,) applied to a woman, (Ḳ, TA,) *Soft, delicate, tender, flabby, lax*, or *limber*, (TA,) *and affecting languor*, or *languid-*

ness: (Ḳ, TA:) pl. of the latter مَخَانِيثُ. (TA.) One says to such a woman, ▼ يَا خَنَاثِ, (Ḳ,) [indecl.,] like قَطَامِ; (TA;) and to a man [of the same description], ▼ يَا خُنَثُ. (Ḳ.)

يَا خُنَثُ : } see the next preceding paragraph.
خُنَثُ :

خُنْثَى [*A hermaphrodite;*] *one who has what is proper to the male and what is proper to the female:* Kr makes it an epithet, and says رَجُلٌ خُنْثَى; (TA;) *one who has what is proper to men and what is proper to women,* (Ṣ, Mgh, Ḳ,) *together;* (Ṣ, Ḳ;) *one who has, by creation, the anterior pudendum of a man and that of a woman:* (Mṣb:) in the language of the lawyers, *one who has what are proper to both sexes;* or *who has neither that of a man nor that of a woman:* but some of them say that the former meaning is the proper one; and that he who has no external organ of generation is adjoined to the class of the خنثى as being subject to the same special laws: (MF, TA:) the pl. is خَنَاثَى (Ṣ, Mgh, Mṣb, Ḳ) and خِنَاثٌ. (Mṣb, Ḳ.) — Also The *plant called* بَرْوَاق [i. e. the *asphodel*]. (Ḳ in art. برق.)

خَنَثِ : see يَا خَنَاثِ.

خَنِيثٌ A skin of the kind called قِرْبَة *folded,* or *doubled.* (L.)

خُنَاثَةٌ : see مُخَنَّثٌ.

خِنَاثَةٌ : see خُنَثٌ.

خُنَيْثَةٌ : see مُخَنَّثٌ.

أَخْنَثُ مِنْ دَلَالٍ [*More effeminate,* or *more incapable of venery, than Delál*]: a prov. (Ṣ, TA.) Delál was a certain man of El-Medeeneh, (TA,) who was made a eunuch, together with several other مُخَنَّثُون. (TA in art. دل. [See Freytag's Arab. Prov. i. 451; where the name is erroneously written دَلَّال.])

مُخَنَّثٌ, from خَنَثَ "he bent," (Ṣ, Ḳ,) because of his softness, delicacy, tenderness, flabbiness, laxness, or limberness, and affectation of languor, or languidness; (TA;) or from خُنْثَى; (Kh, JK, MṢ;) An *effeminate* man; (T in art. انث, and TA;) one *who resembles a woman in gentleness, and in softness of speech, and in an affectation of languor of the limbs:* (TA voce مُؤَنَّثٌ, q. v.: [see also خَنِثٌ:]) it is written thus and ▼ مُخَنِّثٌ: (TA:) this latter is explained by some as meaning one *who makes his speech like that of women, in softness and gentleness:* (Mṣb, TA:) it is also said that both these epithets are used to signify one *who affects languor,* or *languidness, of the limbs;* one *who makes himself like women in the bending of himself, and in affecting languor,* or *languidness, and in speech:* but that one uses the latter epithet only when he means one *who does what is excessively foul,* or *obscene;* [i. e. *a catamite;* though this is a meaning often borne by the former also;] notwithstanding that تَخْنِيثٌ, as signifying the "committing such an action," was unknown to the Arabs [of the classical ages], and is not found in their language: (MF, TA:) [often, also,] the former epithet signifies a man *incapable of venery:* (MA:) it is said in a trad. that they used to reckon the مخنّث as one of those having no need of نِكَاح. (TA in art. ارب.) The مُخَنَّث is also called ▼ خُنَاثَةٌ and ▼ خُنَيْثَةٌ [each imperfectly decl.]. (Ḳ, TA.)

مُخَنِّثٌ : see what next precedes.

مِخْنَاثٌ : see خَنِثٌ.

خنجر

خَنْجَرٌ (Ṣ, Mgh, Mṣb, Ḳ) and خِنْجَرٌ (Mṣb) and خُنْجَرٌ, (Ḳ,) the last of a rare form, like دُرْهَمٌ, (TA,) *A knife:* or *a great knife:* (Ḳ:) or *a kind of large knife* [or *dagger, generally curved, and double-edged*], (Ṣ, Mgh, Mṣb,) *called in Persian* دَشْنَهْ: (Mgh:) pl. خَنَاجِرُ. (Mṣb.) Some say, [as, for instance, the author of the Mṣb,] that the ن is augmentative, the measure being فنعل. (TA.) — Also the first, (Ḳ,) and ▼ خُنْجُورٌ, (Aṣ, Ṣ, TA,) or ▼ خُنْجُورَةٌ, and ▼ خَنْجَرَةٌ, (Ḳ,) A she-camel *abounding with milk:* (Aṣ, Ṣ, Ḳ:) pl. خَنَاجِرُ. (Ṣ.) And ▼ خُنْجُورَةٌ A *bulky* she-camel. (Ḳ.)

خَنْجَرَةٌ : see above.

خُنْجُورٌ and خُنْجُورَةٌ : see خَنْجَرٌ, in three places.

خندرس

خَنْدَرِيسٌ *Wine;* (Ṣ, Ḳ;) so called because of its oldness: (Ṣ:) or *old wine:* (TA:) [accord. to some,] derived from خَدَرْسَةٌ, a word not explained: (Ḳ:) by some said to be of the measure فَنْعَلِيسٌ, so that its radical letters are خدر; because wine is مُخَدِّر [i. e. a cause of torpor or languor]: by some said to be from خَرْسٌ; but to this it is objected that د is not augmentative: the truth is, that it is of the measure فَعْلَلِيلٌ, as Sb says: (MF:) by the author of the L and others, it is mentioned after art. خنس: (TA:) or it is a Greek word, arabicized: (Ḳ:) [but I know not how this is, unless, as an epithet applied to wheat, (see what follows,) it be supposed to be from χόνδρος:] IDrd thought it to be an arabicized word: it may be an arabicized word from the Persian خَنْدَهْ رِيش, meaning "having a laughing beard," [or rather "one whose beard is laughed at," i. e., "a laughing-stock,"] because he who makes use of it [namely of wine] has his beard laughed at. (TA.) — You also say, حِنْطَةٌ خَنْدَرِيسٌ *Old wheat:* (IDrd, Ṣ, Ḳ:) and تَمْرٌ خَنْدَرِيسٌ *old dates.* (TA.)

خندق

Q. 1. خَنْدَقَهُ (Ḳ) and خَنْدَقَ حَوْلَهُ (TA) *He dug a* خَنْدَق, i. e. *fosse,* or *moat, around it.* (Ḳ,* TA.) [In the CḲ the words of this art. are with ذ in the place of د.]

خَنْدَقٌ *A fosse,* or *moat,* [*such as is*] *dug around the walls of cities:* arabicized, from كَنْدَهْ, (IDrd, Ḳ,) which is Persian: (IDrd:) pl. خَنَادِقُ. (TA.) — And *A valley.* (TA.)

مُخَنْدِقٌ One *who makes a* خَنْدَق [i. e. *fosse,* or *moat*]. (JK.)

خَنْدَقُوقٌ *Tall.* (TA. [But perhaps this is a mistranscription for حَنْدَقُوقٌ, q. v.])

خنر

خَنَوَّرٌ : and الخَنَوَّرُ : and أُمُّ خَنَوَّرٍ : see what follows, in four places.

خُنُورٌ and ▼ خَنَوَّرٌ *Any soft and weak tree:* (Ḳ:) or such is called خنورة [i. e. خُنُورَةٌ or خَنَوَّرَةٌ: each, without ة, being app. a coll. gen. n.; and with ة, a n. un.]. (TA.) — And hence, accord. to AḤn, (TA,) The *reeds of* [*which are made*] *arrows.* (Ḳ.) — أُمُّ خَنُّورٍ (Ṣ, Ḳ) and أُمُّ خِنَّوْرٍ (Ḳ) *The female hyena:* (Ṣ, Ḳ:) or she is called by the latter appellation, accord. to Aboo-Riyásh: or, as some say, these are surnames of *the hyena.* (TA.) And *The cow* [probably *the wild cow*]. (Aboo-Riyásh, Ḳ.) — Also *Calamity,* or *misfortune.* (Ṣ, Ḳ.) You say, وَقَعَ القَوْمُ فِى أُمِّ خَنُّورٍ *The people,* or *party, fell into calamity,* or *misfortune.* (L, TA. But see two other explanations in what follows.) — And *Plentifulness, and pleasantness* or *easiness, and softness* or *delicacy, of life;* or *a life of softness* or *delicacy, and ease, comfort,* or *affluence;* syn. النَّعْمَةُ: [in the CḲ النِّعْمَةُ; which is in many instances in the CḲ, as I hold it to be in the present instance, erroneously substituted for النَّعْمَةُ:] thus bearing two contrary significations: and الخُنُورُ and ▼ الخَنَوَّرُ signify the same, *such as is apparent;* (Ḳ;) or, as some say, *abundant.* (TA.) Some explain the saying above-mentioned as meaning *The people,* or *party, fell into a state of plenty,* or *abundance, and softness* or *delicacy* or *easiness of life.* (TA.) — And *The present world;* or *the present life,* or *state of existence;* as also ▼ أُمُّ خَنَوَّرٍ: (L:) or so الخَنُّورُ and ▼ الخَنَوَّرُ. (Ḳ.) — And *The deserts;* syn. الصَّحَارَى: and so accord. to some in the saying mentioned above. (TA.) — And *The podex,* or *the anus;* syn. الإِسْتُ: (Ḳ:) but AḤát doubts respecting the teshdeed of the ن; [app. whether this letter be doubled, or the و;] Aboo-Sahl says that it is أُمُّ خَنُّورٍ [only]: and IKh says that it means *the* إِسْت *of the bitch.* (TA.)

خنز

1. خَنِزَ, aor. ـَ; (Ṣ, A,* Mṣb, Ḳ;) and خَنَزَ, aor. ـُ; (Mṣb;) inf. n. of the former, خَنَزٌ, (Ṣ, Mṣb, Ḳ,) and of the former also, (Ḳ,) or of the latter, (Mṣb,) خُنُوزٌ; (Mṣb, Ḳ;) *It* (flesh-meat, Ṣ, A, Mṣb, Ḳ, and a date, and a walnut, TA) *became stinking:* (Ṣ, A, Ḳ:) or *altered* [*in odour*]: (Mṣb:) or *maggotty and stinking:* (TA:) like خَزِنَ. (Ṣ.)

خَنَزٌ : see what next follows.

خَنِزٌ, applied to flesh-meat, (Mṣb, Ḳ,) and to a date, and a walnut, (TA,) *Stinking:* (Ḳ:) or *altered* [*in odour*]: (Mṣb:) or *maggotty and stinking:* (TA:) as also ▼خَنْزٌ. (Yaạḳoob, Ḳ.)

خُنْزُوَةٌ: } see what next follows.
خُنْزُوَانٌ: }

خُنْزُوَانَةٌ *Pride; self-magnification;* (Ṣ, A, Ḳ;) as also ▼خُنْزُوَانٌ, and ▼خُنْزُوَانِيَّةٌ, and ▼خُنْزُوَةٌ: (Ḳ:) so called because it changes one from the right state: (TA:) pl. of the first, خُنْزُوَانَاتٌ. (Ṣ.) You say, هُوَ ذُو خُنْزُوَانَاتٍ [*He possesses proud feelings*]. (Ṣ.) And فِيهِ خُنْزُوَانَةٌ *In him is pride.* (A.) And لَأَنْزِعَنَّ خُنْزُوَانَتَكَ [*I will assuredly pluck out thy pride*]. (TA.)

خُنْزُوَانِيَّةٌ: see the next preceding paragraph.

خَنَازِ *Stinking:* (Ḳ:) used as a proper name, (TA,) applied to a woman: (Ḳ, TA:) from خَنِزَ said of flesh-meat. (TA.)

خنزر

خَنْزَرٌ: and خِنْزِيرٌ: see art. خزر.

خنس

1. خَنَسَ, (Ṣ, A, Mgh, Mṣb, Ḳ,) aor. ـُ, (Ṣ,) or ـِ, (Mgh, Mṣb,) or both, (Ḳ,) inf. n. خُنُوسٌ, (A,) or خَنْسٌ, (Mṣb,) or both, (Ḳ, TA,) and خِنَاسٌ, (TA,) *He went,* or *drew, back* or *backwards; receded; retreated; retired;* or *retrograded:* or *he remained behind; held back; hung back;* or *lagged behind:* syn. تَأَخَّرَ: (Ṣ, A, Mgh, Mṣb, Ḳ:) مِنْ عَنْهُ *from him* or *it:* (Ṣ, Ḳ:) or بَيْنِ القَوْمِ *from among the company of people: and hid himself:* (A:) or *and shrank,* or *drew himself together:* (TA:) and ▼انخنس signifies the same; (Mṣb, Ḳ;) and so ▼اختنس; and خَنِسَ, aor. ـَ, is mentioned by Ṣgh: (TA:) or خَنَسَ signifies *he went back,* &c., syn. تَأَخَّرَ: and also, *he shrank,* or *drew himself together:* (Mgh, Mṣb:*) and *he hid himself; became hidden* or *absent.* (TA.) You say, خَنَسَ الكَوْكَبُ ‡ *The star returned,* or *went back,* or *retrograded:* syn. رَجَعَ; a tropical signification: (A: [and in the TA it is said that خُنُوسٌ is syn. with رُجُوعٌ, and is tropical in this sense:]) or *became hidden,* (Ḳ, TA,) like a gazelle in its covert, (TA,) or like the devil when he hears the mention of God: (Ḳ, TA:) or *became concealed in the day-time:* (TA:) and خُنُوسٌ signifies also the *being,* or *becoming, depressed.* (Ḥam p. 332.) And خَنَسَ عَنِ القَوْمِ *He held back,* or *hung back, from the company of people; remained behind them, not going with them;* syn. تَخَلَّفَ; (Aṣ, on the authority of an Arab of the desert, of the Benoo-'Okeyl;) as also ▼انخنس. (Ḳ.) [This is said in the TA to be tropical; but why, I see not.] And خَنَسَتِ النَّخْلُ † *The palm-trees were backward to receive fecundation,* (تَأَخَّرَتْ عَنْ قَبُولِ التَّلْقِيحِ, lit., *held back from receiving fecundation,*) *so that it had not any effect upon them, and they did not bear fruit that year.* (TA.) And يَخْنِسُ الشَّيْطَانُ إِذَا سَمِعَ ذِكْرَ اللّٰهِ *The devil shrinks when he hears the mention of God.* (Mṣb.) And خَنَسَ مِنْ بَيْنِ أَصْحَابِهِ *He hid himself from among his companions.* (TA.) And خَنَسَ عَنِّى ‡ *He,* or *it,* [app. the latter,] *became hid from me.* (A.) And خَنَسَ بِهِ *He went away with him; took him away; so that he was not seen;* (ISh, Ḳ;) as also ▼تخنّس بِهِ: (Ḳ:) and *he hid him,* or *it.* (TA.) ― See also 4, in four places. ― خَنِسَ الأَنْفُ, aor. ـَ, (Mṣb,) inf. n. خَنَسٌ, (Ṣ, A, Mṣb, Ḳ,) *The nose was,* or *became,* [*camous,* or *camoys,* i. e.,] *depressed in its bone:* (Mṣb:) or *depressed in its bone,* (A,) or *contracted therein,* (TA,) *and wide in the end:* (A, TA:) or *retiring from the face, with a slight elevation in the end;* (Ṣ, Ḳ, TA;) خَنَسٌ being nearly the same as فَطَسٌ: (TA:) or *retiring towards the head, and rising from the lip, not being long nor prominent:* or *its bone lay close upon the elevated part of the cheek, and it was large in the end.* (TA.) See also the inf. n. voce أَخْنَسُ. ― خَنِسَتِ القَدَمُ, inf. n. خَنَسٌ, *The foot was,* or *became, flat in the hollow part of the sole, and fleshy.* (TA: but only the inf. n. of the verb in this sense is there mentioned.)

4. اخنسهُ *He made him to go back* or *backwards; to recede, retreat, retire,* or *retrograde:* or *he put him,* or *placed him,* or *made him to be, behind,* or *after:* or *he made him to remain behind, hold back, hang back,* or *lag behind:* or *he kept him back:* or *he delayed,* or *retarded, him:* syn. أَخَّرَهُ: (T, A, Mṣb, Ḳ:) as also ▼خَنَسَهُ, (Fr, T, A, Mgh, Mṣb, Ḳ,) aor. ـِ, (Mṣb,) [and app., accord. to the Ḳ, ـُ also,] inf. n. خَنْسٌ; (Mṣb;) but the former is the more common: (TA:) *and* ‡*hid him,* or *it:* or *made him,* or *it, to hide himself* or *itself;* (A;) or *he left behind,* (Aṣ, Ṣ,) *and went away from,* (Ṣ,) *him,* or *it:* (Aṣ, Ṣ:) or both signify *he contracted,* or *drew together,* or *made to contract or draw together, him* [or *it*]: (Mṣb:) or the latter verb has this signification as well as that of أَخَّرَهُ: (Mgh:) [and so has the former also, as will be seen below:] and the former also signifies ‡*he hid,* or *concealed, him,* or *it;* (A;) as also ▼خَنَسَ بِهِ, as mentioned above. (TA.) You say, أَخْنَسْتُ عَنْهُ بَعْضَ حَقِّهِ *I kept back* (أَخَّرْتُ) *from him part of his right,* or *due.* (Fr, TA.) And أَخْنَسُوا الطَّرِيقَ ‡ *They passed beyond the road:* (AA, TA:) or *left it behind them:* (TA:) or *passed beyond it and left it behind them.* (A.) And أَشَارَ بِأَرْبَعٍ وَأَخْنَسَ إِبْهَامَهُ, (A,) and ▼خَنَسَهَا, (Mgh, Mṣb, Ḳ,) *He* [*made a sign with four fingers and*] *contracted his thumb.* (Mgh, Mṣb, Ḳ.) It is related of Mohammad, that he said, "The month is thus and thus," [twice extending the fingers and thumb of each hand,] and that, the third time, ▼خَنَسَ إِصْبَعَهُ, i. e., *he contracted his finger,* [meaning, one of his fingers,] to inform them that the month is nine and twenty [nights with their days]. (TA.)

5. تخنّس بِهِ: see 1.

7: see 1, in two places.

8: see 1.

خُنَّسٌ: see أَخْنَسُ. ― *A place of gazelles:* (Ḳ:) or *a place to which gazelles betake themselves for covert.* (L.)

أَخْنَسُ: see خَنَّاسٌ.

الخَنَّاسُ *The devil:* (Ṣ, Ḳ:) an epithet applied to him, (Mṣb,) because he retires, or shrinks, or hides himself, (يَخْنِسُ, Ṣ, Mṣb, Ḳ, i. e., يَتَأَخَّرُ, as is implied in the Ṣ, or يَتَقَبَّضُ, Mṣb, or يَغِيبُ, Ḳ,) at the mention of God; (Ṣ, Mṣb, Ḳ;*) being an intensive act. part. n. from خَنَسَ. (Mṣb.)

خَنُوسٌ: see أَخْنَسُ, in two places.

خَانِسٌ *Going back* or *backwards; receding; retreating; retiring;* or *retrograding:* or *remaining behind; holding back; hanging back;* or *lagging behind:* syn. مُتَأَخِّرٌ: pl. خُنَّسٌ. (TA.) ― [Hence,] الخُنَّسُ, (in the Ḳur lxxxi. 15, Ṣ,) ‡*The stars;* (Ṣ, Ḳ;) i. e., all of them; because they retire, or hide themselves, (تَخْنِسُ,) at setting; or because they become concealed in the day-time: (Ṣ:) or *the planets:* (Ṣ, Ḳ:) or *the five stars, Saturn, Jupiter, Mars, Venus, and Mercury:* (Fr, Ṣ, Ḳ, Jel:) because they return, (تَخْنِسُ, i. e., تَرْجِعُ, Jel,) in their course: (Fr, Ṣ, Jel:) when you see a star [thereof] in the end of a sign of the zodiac, it returns to the beginning of it: (Jel:) or because of their retrogression; for they are the erratic stars (الكَوَاكِبُ المُتَحَيِّرَةُ), which [at one time appear to] retrograde, and [at another time to] pursue a direct [and forward] course: (Ṣ:) or because they sometimes return (تخنس) in their course until they become concealed in the light of the sun: (TA:) or because they hide themselves, as the devil does at the mention of God. (Ḳ, TA.) ― And hence, i. e., from خَانِسٌ in the sense of مُتَأَخِّرٌ, the saying in a trad. of El-Ḥajjáj, الإِبِلُ ضُمْزٌ خُنَّسٌ, meaning, †[*Camels are lean, and lank in the belly,* and] *patient of thirst.* (TA.) ― And اللَّيَالِي الخُنَّسُ *The three nights of the lunar month during which the moon retires* [*from view*]. (TA.)

أَخْنَسُ [*Having a camous,* or *camoys, nose;*] *having the configuration termed* خَنَسٌ *in the nose:* (Ṣ, Mṣb, Ḳ:) [see خَنِسَ الأَنْفُ:] accord. to some, *having a nose of which the bone is short, and the end turning back towards its bone:* (TA:) fem. خَنْسَاءُ: (Ṣ, Mṣb:) pl. خُنْسٌ. (Ṣ, A.) ▼خُنُسٌ in its original application is in gazelles and bulls and cows: (TA:) all bulls and cows are خُنْسٌ, (Ṣ, A, TA,) and so are all gazelles: (TA:) or ▼خُنُسٌ, with two ḍammehs, (Ḳ,) but written by Sh خُنْسٌ, (TA,) is used to signify *gazelles:* and *bulls or cows:* (Ḳ:) and خَنْسَاءُ is an epithet applied to the *wild cow:* (Ḳ:) also أَخْنَسُ, to the *tick:* (Ṣgh, Ḳ:) and the *lion;* and so ▼خَنُوسٌ; (Ḳ;) which last is an epithet so applied as relating to his face and his nose: (Fr, TA:) and the last, ▼خَنُوسٌ, is also applied to *a young pig:* (Aṣ, TA:) or in this sense it is with ص: (Fr, TA:) and ▼خَنَّاسٌ is syn. with خَنُوسٌ. (TA.) ― [Hence,] خُنْسٌ is metaphorically applied [as an epithet] to arrows, in the following verse, describing a coat of mail:

* لَهَا عُكَنٌ تَرُدُّ النَّبْلَ خُنْسًا *
* وَتَهْزَأُ بِٱلْمَعَابِلِ وَٱلْقِطَاعِ *

[*It has folds which repel the arrows turned up at the points, and mock at the broad and long, and the small and broad, arrow-heads*]. (TA.) — قَدَمٌ خَنْسَآءُ *A foot flat in the hollow part of the sole, and fleshy.* (TA.)

خنسر

خَنْسَرٌ : and خِنْسِرٌ : and خَنْسَرِىٌّ : and خِنْسِيرٌ : and خَنَاسِرُ : and خَنَاسِيرُ : see art. خسر.

خنص

خِنَّوْصَةٌ : see خِنْصِيصٌ.

خِنَّوْصٌ The *young of the swine*: (Ibn-'Abbád, Ṣ, Ḳ:) and the *young*, or *little*, of anything: (Ḳ:) pl. خَنَانِيصُ. (Ṣ, Ḳ.) [See also خِنَّوْسٌ.]

خِنَّوْصَةٌ The *young of the* بَبْر, q. v.; (Ḳ;) as also ▼خِنْصِيصٌ. (Ṣgh, Ḳ.) — Also *A palm-tree which does not rise beyond reach of the hand.* (Ibn-'Abbád, Ḳ.)

خنصر

خِنْصِرٌ (Ṣ, Mṣb, Ḳ) and خِنْصَرٌ, (A, Ḳ,) the latter [of a rare form,] like دِرْهَمٌ, (TA,) The *little finger*: (Ṣ, A, Ḳ:) or the *middle finger*: (Ḳ:) the latter signification said by MF to be unknown; but it is mentioned in the L, as from the Book of Sb: (TA:) [and the *little toe*:] of the fem. gender: (Mṣb, Ḳ:) pl. خَنَاصِرُ: (Sb, Ṣ, Ḳ:) like فِرْسِنٌ, it has no pl. formed by the addition of ات: (Sb:) its pl. is also used as a sing., as though every part were termed خنصر; as in the phrase إِنَّهُ لَعَظِيمُ الخَنَاصِرِ [*Verily he has a large little finger*]. (Lḥ.) You say, فُلَانٌ تُثْنَى بِهِ الخَنَاصِرُ [*The little fingers are bent in mentioning such a one* with others of his class]: i. e., one begins with him in mentioning persons of his class. (Mṣb.) [See 1 in art. ثنى.] And in like manner you say, عَدُّوهُ بِالخِنْصِرِ [*They counted him with the little finger*]: i. e., they commenced with him in counting. (MF.)

خنع

1. خَنَعَ إِلَيْهِ, and لَهُ, (ISd,) [aor. ـَ ,] inf. n. خُنُوعٌ (Ṣ, ISd, Ḳ) and خَنْعٌ; (ISd;) [and app. خَنِعَ, aor. ـَ , inf. n. خَنَعٌ; (see خَنَعٌ, below; and خَنِعٌ;)] *He was, or became, lowly, humble*, or *submissive*, (Ṣ, ISd, Ḳ,) *to him, and petitioned him, or solicited him, he, the latter, not being a fit person to be petitioned*, or *solicited*: (ISd:) or خُنُوعٌ signifies the *being low, vile, base, abject*, or *submissive*; almost always *in an improper case*. (Ḥam p. 44.) — Accord. to Lth, (TA,) خَنْعٌ signifies The act of *playing, toying*, or *dallying, and conversing* with one of the other sex, *enticing*, or *striving to induce*, the latter *to yield to one's desire, and behaving in a soft, tender*, or *blandishing, manner.* (Ḳ, TA.) You say, خَنَعَ النِّسَآءَ, [or more probably, لِلنِّسَآءِ,] *He played, &c., with women, and behaved in a soft, tender*, or *blandishing, manner to them.* (TḲ.) — خَنَعَ, aor. ـَ , (Lth, Ḳ,) inf. n. خَنْعٌ and خُنُوعٌ, (Lth,) also signifies *He acted in a suspicious manner*, or *so as to induce suspicion* or *evil opinion*; (Ḳ;) *he acted vitiously*, or *immorally*; or *committed adultery*, or *fornication.* (Lth, Ḳ.) You say, خَنَعَ إِلَيْهَا *He came to her for a vitious*, or *an immoral, purpose*; or *for the purpose of adultery*, or *fornication*: (Lth, TA:) or, as some say, *he listened to her.* (TA.) — And خَنَعَ بِهِ, aor. ـَ , inf. n. خُنُوعٌ, *He acted perfidiously, unfaithfully*, or *treacherously, to him*; or *broke his compact, contract, covenant*, or *the like, with him.* (TA.) — خَنَعَ فُلَانًا إِلَى السَّوْءَةِ *He*, or *it, invited such a one to that which was foul, abominable*, or *evil*; as also خَضَعَ. (TA in art. خضع.)

4. أَخْنَعَتْنِى إِلَيْكَ الحَاجَةُ (Ṣ, Ḳ*) *Want*, or *need, made me lowly, humble*, or *submissive, to thee*; or *constrained me to have recourse to thee, and to require thine aid.* (Ṣ,* Ḳ.)

خَنَعٌ [app. inf. n. of خَنِعَ,] *Lowness, vileness, baseness, abjectness*, or *submissiveness*; almost always, *in an improper case.* (Ḥam p. 44.)

خَنِعٌ [app. part. n. of خَنِعَ,] *Low, vile, base, abject*, or *submissive.* (KL.) [See what next precedes.]

خُنْعَةٌ *A thing that induces suspicion* or *evil opinion*; (Ṣ, Ḳ;) *a vitious*, or *an immoral, act*; or *adultery*; or *fornication.* (Ḳ.) [See a remark on one of the pls. of خَانِعٌ.] You say, اِطَّلَعْتُ مِنْ فُلَانٍ عَلَى خُنْعَةٍ *I became acquainted with*, or *got knowledge of, a vitious*, or *an immoral, act of such a one.* (TA.) And وَقَعَ فِى خُنْعَةٍ *He fell into a thing of which one is ashamed.* (TA.) — *A vacant place.* (O, L, Ḳ.) You say, لَقِيتُهُ بِخُنْعَةٍ فَقَهَرْتُهُ *I found him*, or *met him, in a vacant place, and I overcame him.* (Ḳ,* TA.)

خُنْعَةٌ *Necessity*, or *constraint*: and *excuse.* (TA.) — رَجُلٌ ذُو خُنْعَاتٍ *A man in whom is corruptness*, or *vitiousness*, or *corrupt* or *vitious conduct.* (TA.)

خَنُوعٌ *Perfidious, unfaithful*, or *treacherous*; one *who breaks his compact, contract, covenant, or the like.* (Ibn-'Abbád, Ḳ.) — One *who turns away from*, or *shuns*, or *avoids, another.* (Ibn-'Abbád, Ḳ.)

خَنَاعَةٌ The *state of being bad, evil, abominable, foul, unseemly, ugly*, or *hideous*; or *excessively bad*, &c. (TA.)

خَانِعٌ One *who acts in a suspicious manner*, or *so as to induce suspicion* or *evil opinion*; *who acts vitiously*, or *immorally*; or *commits adultery*, or *fornication*: (Ṣ, Ḳ:) pl. خُنَّعٌ (Ḳ, TA) and خَنَعَةٌ. (TA.) El-Aạshà says,

* هُمُ الخَضَارِمُ إِنْ غَابُوا وَإِنْ شَهِدُوا *
* وَلَا يَرَوْنَ إِلَى جَارَاتِهِمْ خُنُعَا *

[*They are the bountiful, if they be absent and if they be present; and they do not see persons acting in a suspicious manner, &c., towards their wives*]. (TA.) [The latter hemistich of this verse is cited in the Ṣ; in one copy of which I find يُرَوْنَ in the place of يَرَوْنَ: and it seems to be there implied that خُنُعٌ is pl. of خُنْعَةٌ; but I do not know any instance of فُعُلٌ as the measure of a pl. of a word of the measure فُعْلَةٌ.] — One *who commits a foul action whereof the disgrace returns upon him, and is ashamed of it, and hangs down his head towards the ground.* (Aṣ, on the authority of an Arab of the desert.)

أَخْنَعُ الأَسْمَآءِ عِنْدَ ٱللّٰهِ مَلِكُ الأَمْلَاكِ, (Ḳ,) or إِلَى ٱللّٰهِ, (TA,) *The vilest and most abasing* (أَذَلُّ and أَقْهَرُ) *of names*, (Ḳ,) for a man, *and the most effectual to bring into a state of humility and humiliation, in the estimation of God, is "king of kings;"* like [the Persian] شَاهَنْشَاهْ; because this name belongs to God himself: a trad., which is variously related: (TA:) accord. to different relations, thus, and أَنْخَعُ, (Ḳ,) meaning "most effectual to kill, and destroy," its owner, (TA,) and أَبْخَعُ, [which means the same,] (Ḳ, TA, [in the CḲ أَنْجَعُ,]) and أَخْنَى, (Ḳ,) meaning "most foul, abominable, or the like." (TA in art. خنى.)

مُخَنَّعٌ applied to a camel, *Broke*; *trained*; *rendered submissive*, or *manageable.* (Ḳ, TA.) And in like manner applied to a place [app. as meaning *Rendered easy to sit*, or *lie, upon*; or, *to travel*]. (TA.)

خنفس

خُنْفُسٌ and its variations: see what here follows.

خُنْفُسَآءُ (Ṣ, Mṣb, Ḳ) [in two copies of the Ṣ, in which it is without the article, written without tenween, but in the Mṣb and Ḳ it has the article prefixed, and is therefore necessarily without tenween,] and خُنْفَسَآءُ, (Mṣb, TA,) which is more common, (Mṣb,) [but this I doubt, for I have found it nowhere else,] and ▼خُنْفَسٌ (Ṣ, Ḳ) and خُنْفُسٌ (TA) and خِنْفِسٌ, (Ḳ,) which last is of the dial. of the people of El-Baṣrah, (TA,) and خُنْفُسَةٌ and خُنْفَسَةٌ, (Ḳ,) [The *black beetle*: or *a certain species thereof*:] *a well-known creeping thing*; (Mṣb;) *a certain insect*, (Ṣ, Ḳ,) *black*, (Ḳ,) *of fetid odour, smaller than the* جُعَل, *found in the bottoms of walls*: (TA:) the first and second of these words are both applied to the *male* and the *female*: (Mṣb:) or خُنْفُسٌ is applied to the *male*, (AA, Mṣb,) by some of the Arabs, (Mṣb,) and is syn. with عُنْظُبٌ and خُنْظُبٌ, (AA, TA,) and خُنْفَسٌ is not disallowable, being agreeable with analogy: (Mṣb:) and خُنْفُسَآءَةٌ is applied to the *female*, [which, if correct, shows خنفسآء to be with tenween,] and so is خُنْفُسَةٌ: (Ṣ:) or خنفساءة, with ة, is not allowable; [and if so, خنفسآء is without tenween;] (Aṣ, TA;) and خنفسة is [not applied to the female, but] used by the Benoo-Asad for خنفسآء, as though they made the ة a substitute for the ا: (Mṣb:) [and this seems to indicate that the ا is a fem. ا, grammatically speaking, and that the word is therefore without tenween:] the pl. is خَنَافِسُ. (Mṣb.) One says, هُوَ أَلَحُّ مِنَ الخُنْفُسَآءِ [*He is more persevering than the black beetle*]: because it returns to thee as often as thou throwest it away. (TA.)

[Accord. to the Ḳ, the ن in the words of this art. is radical; but accord. to the Ṣ and Mṣb, augmentative.]

خنق

1. خَنَقَهُ, (Ṣ, Mgh, Mṣb, Ḳ,) aor. ـُ, (Ṣ, Mṣb,) inf. n. خَنِقٌ (Ṣ, Mgh, Mṣb, Ḳ) and خَنْقٌ, (Mṣb, TA,) the latter a contraction of the former, (Mṣb,) or, accord. to El-Fárábee, the latter is not allowable, (Mgh,) *He throttled him*, or *strangled him*, i. e. *squeezed his throat* (Mgh, Mṣb) *that he might die*; (Mṣb;) [but it does not always mean *he squeezed his throat so that he died*; often meaning, simply, *he*, or *it, throttled him, strangled him*, or *choked him*; and frequently said of a disease in the throat, and of food;] and ▼ خنّقهُ signifies the same, (Ṣ, Ḳ,) [or has an intensive meaning,] and its inf. n. is تَخْنِيقٌ. (TA.) — [Hence,] خَنَقَتْهُ ▼ العَبْرَةُ [and خَنَّقَتْهُ] † *Weeping* [or *sobbing*] *choked him*; as though the tears throttled him. (Mgh.) — And خُنِقَ *He* (a horse) *was affected with the disease*, or *wind in the throat, termed* خُنَاقِيَّةٌ. (TA.) — And خَنَقَ الوَقْتَ, aor. as above, † *He postponed*, or *deferred, and* [*so*] *straitened, the time*: and خَنَقَ الصَّلَاةَ † *He straitened the time of prayer by postponing it*, or *deferring it*. (TA.)

2: see 1, in two places. — You say also, خنّق السَّرَابُ الجِبَالَ, inf. n. تَخْنِيقٌ, ‡ *The mirage nearly covered the heads of the mountains*. (Ḳ, TA.) — And خنّق الإِنَآءَ ‡ *He filled the vessel*: (Ḳ, TA:) or *filled it up* (سَدَّدَ مَلْأَهُ): and in like manner, الحَوْضَ [*the watering-trough*]. (Aboo-Sa'eed, TA.) — And خنّق الأَرْبَعِينَ ‡ *He* (a man) *nearly attained to* [the age of] *forty* [years]. (Ḳ, TA.)

7: see what next follows, in three places.

8. اختنق *He was*, or *became, throttled*, or *strangled*; i. e. *he had his throat squeezed that he might die*; (JK,* Ṣ,* Mṣb, Ḳ;*) [and simply *he was, or became, throttled, strangled*, or *choked*;] as also ▼ انخنق: (JK,* Mṣb:) or you say, ▼ انخنقت الشَّاةُ بِنَفْسِهَا [*the sheep, or goat, became throttled*, or *strangled*, or *choked, by itself*]: (Ṣ, Ḳ:) or ▼ الإِنْخِنَاقُ signifies *the having the* خِنَاق [q. v.] *compressed upon one's throat*: and الإِخْتِنَاقُ, *the compressing it upon one's own throat*. (TA.)

خَنْقٌ: see مُخَنَّقٌ.

خَنِقٌ inf. n. of خَنَقَهُ: (Ṣ, Mgh, Mṣb, Ḳ:) — and *i. q.* مَخْنُوقٌ, q. v. (JK, Ḳ.)

خُنَقٌ: see خِنَاقٌ.

خُنَاقٌ [*Quinsy*;] *a certain disease which prevents the passage of the breath to the lungs and heart*; (Ḳ;) as also [▼ خُنَاقَةٌ, thus in modern Arabic, and] ▼ خَنَّاقٌ: pl. خَوَانِيقُ (TA) [and خَنَانِيقُ, thus in modern Arabic]. — See also مُخَنَّقٌ, in three places.

خِنَاقٌ *A cord*, (JK, Ṣ, Mgh, Ḳ,) or *bow-string*, or *the like*, (Mgh,) *with which one is strangled*; (JK,* Ṣ,* Mgh, Ḳ;*) also, metaphorically, termed ▼ مِخْنَقَةٌ ▼ خُنَاقٌ. (Mgh.) — See also مُخَنَّقٌ, in two places. — فَلْهَمٌ خِنَاقٌ † *A narrow vulva of a woman*: (Abu-l-'Abbás, TA:) and ▼ خُنُقٌ [or فُرُوجٌ خُنُقٌ, for خُنُقٌ is app. pl. of خِنَاقٌ, like as كُنُزٌ is pl. of كِنَازٌ,] † *narrow vulvas* (IAạr, Ḳ) *of women*. (IAạr.) — هُمْ فِي خِنَاقٍ مِنَ المَوْتِ † *They are in straitness by reason of death*. (TA.)

خَنِيقٌ: see مَخْنُوقٌ, in three places.

خُنَاقَةٌ: see خُنَاقٌ.

خُنَاقِيَّةٌ *A certain disease in the throats of birds and horses*: (Ḳ:) or *a certain disease that attacks the bird in its head, and the horse in its throat, and chokes it*: (JK:) or *a certain disease*, or *wind, that attacks men and horses or similar beasts in the throat, and sometimes attacks birds in the head and throat, mostly appearing in pigeons*. (TA.)

خَنَّاقٌ: see خَانِقٌ. — Also One *who sells fish* [*taken*] *with the* خُنَّاقَة [q. v.]. (TA.)

خَنَّاقٌ: see خُنَاقٌ.

خُنَّاقَةٌ *A snare with which beasts of prey are taken* (JK, TA) *by the throat*: and *a snare with which fish are taken in El-Andalus*. (TA.)

خَانِقٌ One *who strangles*; (Mṣb, TA;*) as also ▼ خَنَّاقٌ: (Mgh, Mṣb:) or the latter signifies one *whose office it is to strangle*. (TA.) — [Hence,] خَانِقُ النَّمِرِ and خَانِقُ الذِّئْبِ and خَانِقُ الكِرْسِنَّةِ and الكَلْبِ [in the CḲ الكَرْسَنَّةِ] *Four herbs*: (Ḳ:) [the first and second, in the present day, *wolfsbane*, or *aconite*: or, as Golius says, referring for the former and latter respectively to Diosc. iv. 78 and 77, the former is the *aconitum lycoctonon*; and the latter, the *aconitum pardalianches*: the third, *dogsbane*, or *colchicum*; or, as Golius says, referring to Diosc. iv. 81, *apocynon*: and the fourth, *strangle-weed*, (because it strangles the كِرْسِنَّة, or bitter vetch,) or *broomrape*, i. e., as Golius says, referring to Diosc. ii. 172 and Ibn-Beyṭár, *orobanche*:] the first is *high* (مشرف [but perhaps this should be مُشْرِق i. e. *shining*]) *in the leaves, downy, and resembling the* دلب [?]: the second is *like the tail of the scorpion, glistening, about a span* [*in height*], *and has not more than five leaves*: *each of these is of the* [*season called*] رَبِيع; *and they are poisonous*; *they kill all animals*; the ذِئْب and the نَمِر being particularized only because of the quickness of its acting in them: Ibn-Seenà says, in the "Ḳánoon," the leaves of خانق النمر, when mixed with fat, and kneaded and baked with bread, and given as food to wolves and dogs and foxes and leopards (نمر), kills them: whence it seems that this may be two herbs or one herb. (TA.) — خَانِقٌ signifies also ‡ *A narrow road* or *ravine, in a mountain*: (Ṣ,* Ḳ,* TA:) or *a narrow place* or *pass, between two mountains*, and *between two tracts of sand*. (JK.) [See also مُخْنَقٌ.] — And ‡ *A narrow street*; syn. زُقَاقٌ; (Ṣ, Ḳ, TA;) in the dial. of the people of El-Yemen. (Ṣ, TA.) — See also مَخْنُوقٌ, in two places.

خَانَقَاهُ *A convent inhabited by righteous and good men, and Ṣoofees*; an arabicized word, from [the Persian] خَانَهْ گَاهْ; [and post-classical, for] El-Maḳreezee says that the خانقاه was instituted in the fifth century of the Flight, for Ṣoofee recluses to employ themselves therein in the service of God: (TA:) [pl. خَوَانِقُ.]

خَانَقَاهِيٌّ A man *of*, or *belonging to, a* خانقاه. (TA.)

مَخْنَقٌ: see مُخَنَّقٌ, in two places.

مِخْنَقَةٌ *A necklace*, syn. قِلَادَةٌ, (Ṣ, Mgh, Mṣb, Ḳ,) *that surrounds the neck*; (Mgh, Mṣb;) wherefore it is thus called; (Mṣb;) [i. e., because] *it lies against the* مُخَنَّق: pl. مَخَانِقُ. (TA.) — See also خِنَاقٌ. [And see زِرَادٌ.]

مُخَنَّقٌ: see مَخْنُوقٌ. — [Hence,] غُلَامٌ مُخَنَّقُ الخَصْرِ † *A boy slender in the waist*. (Ḳ.) — Also The *part, of the neck, which is the place of the cord* [or *the like*] *called* خِنَاق [*wherewith one is strangled*]; (Ṣ, Ḳ;*) i. e., (TA,) the *throat*; (Ḳ, accord. to the TA; in the CḲ and in my MS. copy of the Ḳ ▼ مَخْنَقٌ;) and so ▼ خِنَاقٌ, (Ṣ, Ḳ, in the former said to be syn. with مُخَنَّق,) and ▼ خُنَاقٌ (Ḳ) and ▼ خَنْقٌ. (TA; and so, accord. to the TA, in the Ḳ.) You say, أَخَذْتُ بِمُخَنَّقِهِ [*I seized his throat*]. (Ṣ.) And أَخَذَهُ بِمُخَنَّقِهِ (Ḳ, accord. to the TA, but accord. to the CḲ and my MS. copy of the Ḳ ▼ بِمَخْنَقِهِ,) and ▼ بِخِنَاقِهِ and ▼ بِخُنَاقِهِ, i. e. [*He took him*, or *seized him*,] *by his throat*. (Ḳ.) And أَخَذَ بِمُخَنَّقِهِ (A in art. زرد) and ▼ بِخُنَاقِهِ (Ṣ) [properly *He*, or *it, seized his throat*, or *throttled him*, or *choked him*; meaning] ‡ *he*, or *it, straitened him*; as also أَخَذَ بِمُزَرَّدِهِ. (A in art. زرد.) And أَخَذَ مِنْهُ بِالمُخَنَّقِ ‡ *He*, or *it, constrained him, and straitened him*. (TA.) And بَلَغَ مِنْهُ المُخَنَّقَ [properly *It reached his throat*; meaning † *it straitened him*, or *distressed him*]. (Ṣ.) بَلَغَ الأَمْرُ المُخَنَّقَ signifies the same as بَلَغَ المُذَمَّرَ, (A in art. ذمر,) which means † *The affair*, or *case*, or *event, reached a distressing pitch*. (Ḳ in art. ذمر.)

مَخْنُوقٌ and ▼ خَنِقٌ and ▼ خَنِيقٌ (JK, Ḳ) and ▼ خَانِقٌ, applied to a man, [and to any animal, as also ▼ مُنْخَنِقٌ, *Throttled*, or *strangled*, i. e. *having his throat squeezed that he may die*; but not always meaning, *so as to be killed thereby*; often meaning, simply, *throttled, strangled*, or *choked*;] (JK;) all signify the same; from خَنَقَهُ: (JK, Ḳ:) or ▼ خَانِقٌ, in the place of ▼ خَنِيقٌ, signifies ذو خناق [app. meaning having a خِنَاق, or *cord*, &c., *by which he is throttled*, or *strangled*, *round his neck*; or perhaps *having a* خُنَاق, or *quinsy*]: (TA:) and ▼ شَاةٌ خَنِيقَةٌ and ▼ مُنْخَنِقَةٌ signify *a sheep*, or *goat, throttled*, or *strangled*, i. e. *having its throat squeezed that it may die*: (Mṣb:) or the latter of these two means *a sheep*, or *goat, throttled*, or *strangled*, or *choked, by itself* (اِنْخَنَقَتْ بِنَفْسِهَا). (Ṣ, TA. [See 8.]) It is said in a prov., (Meyd,) اِفْتَدِ مَخْنُوقُ, (Meyd, Ḳ,) i. e. يَا مَخْنُوقُ [*Ransom thyself, O* thou who art *throttled*, or *strangled*, or *choked*]; applied to any one distressed and constrained; (Meyd;) meaning free thyself from difficulty (Ḳ, TA) and harm: (TA:) or, as some relate it, اِفْتَدَى مَخْنُوقٌ [*One throttled*, or *strangled*, or *choked, ransomed himself*]. (Meyd.)

مُخْتَنَقٌ *A narrow place* or *pass*. (Ṣ, TA.) [See also خَانِقٌ, near the end of the paragraph.]

مُخْتَنِقٌ ‡ A horse *whose blaze occupies his jaws*, (Ḳ, TA,) *extending to the roots of his ears*. (TA.)

شَاةٌ مُنْخَنِقَةٌ: see مَخْنُوقٌ.

خنو

1. خَنَا, aor. يَخْنُو, inf. n. خَنْوٌ and خَنًا: see خَنِىَ, in the art. here following.

خنى

1. خَنِىَ (Ṣ, Ḳ) عَلَيْهِ (Ṣ) فِى مَنْطِقِهِ, (TA,) aor. يَخْنَى, inf. n. خَنًى; (TA;) as also خَنَا, (JK, Ḳ,) aor. يَخْنُو, (JK, TA,) inf. n. خَنْوٌ (Ḳ, TA) and خَنًا; (JK, TA;) and ▼ اخنى (JK, Ṣ, Ḳ) عَلَيْهِ فِى مَنْطِقِهِ, (Ṣ, TA,) or فِى كَلَامِهِ, (JK,) *He uttered foul, abominable, unseemly*, or *obscene, speech* (JK,* Ṣ, Ḳ) *against him*. (Ṣ, TA.) — خَنَى, (JK, Ḳ,) [aor. يَخْنِى,] inf. n. خَنْىٌ, (TA,) *He cut*, or *severed*, the trunk of a palm-tree (JK, Ḳ) with an axe; (JK;) as also خَنَأَ. (JK, TA.)

4: see 1. — اخنى عَلَيْهِ also signifies *He*, (a man, Ṣ,) or *it*, (misfortune, Ḥam p. 430,) *corrupted*, or *marred, his state*. (Ṣ, Ḥam, TA.) And the same, (Ṣ,) or اخنى عَلَيْهِمْ, (JK, Ḳ,) said of time, or fortune, (JK, Ṣ,) *It destroyed him, or them*. (JK, Ṣ, Ḳ.) And the former, said of time, *It became long to him*. (Ḳ.) — اخنى بِهِ signifies أَسْلَمَهُ وَخَفَرَ ذِمَّتَهُ [or, app., بِذِمَّتِهِ]: a phrase which admits of two contr. meanings; *He betrayed him, and broke his covenant*, or *the like*; and *he preserved him in safety, and fulfilled his covenant*, &c.]. (TA.) — اخنى الجَرَادُ *The locusts had many eggs*. (AḤn, Ḳ.) — اخنى المَرْعَى *The pasturage had much and luxuriant herbage*. (AḤn, Ḳ.*)

خَنًا [more properly written خَنًى, originally an inf. n.,] *Foul, abominable, unseemly*, or *obscene, speech*: (Ṣ, TA, and Ḥam p. 489:) or the *most foul, abominable, unseemly*, or *obscene*, of speech. (JK, T, TA.) And ▼ خناية [written without any syll. signs, app. خَنَايَةٌ, like فَحَاشَةٌ,] is of the measure فعالة from خَنًا [and seems to be syn. therewith as an inf. n. or as a subst.]: it occurs in the saying of El-Ḳaṭámee,

* دَعُوا التَّمْرَ لَا تُثْنُوا عَلَيْهَا خِنَايَةً *
* فَقَدْ أَحْسَنَتْ فِى جُلِّ مَا بَيْنَنَا التَّمْرُ *

[*Leave ye dates: speak not of them in a foul manner*, or *speak not of them foul speech; for dates have benefited in most of what* has occurred *between us*, or *among us*]. (TA.) — خَنَا الدَّهْرِ *The calamities of time* or *fortune*. (JK, Ḳ.)

كَلَامٌ خَنٍ *Foul, abominable, unseemly*, or *obscene, speech*; and كَلِمَةٌ خَنِيَةٌ *a foul, an abominable, an unseemly*, or *an obscene, word* or *expression* or *sentence*: (Ṣ:) or *most foul*, &c.: (JK:) [or *having a foul, an abominable, an unseemly*, or *an obscene, meaning*; for] خَنٍ is not a verbal epithet, since we know not خَنِيَتِ الكَلِمَةُ, but a possessive epithet; like the instance, mentioned by Sb, in رَجُلٌ طَعِمٌ, meaning ذُو طَعَامٍ; and نَهِرٌ, meaning صَاحِبُ نَهَارٍ [or سَيْرٌ بِالنَّهَارِ]; &c. (TA.)

خناية: see خَنًا.

اخنى الأَسْمَآءِ *The most foul*, or *abominable*, or *unseemly, of names*. (TA.) [See أَخْنَعُ.]

خو

خَوٌّ *Hunger*. (IAạr, Ḳ.) [And so خَوًى and خَوَآءٌ, belonging to art. خوى; with which the present art. is intimately connected.] — *A wide valley*: (Ḳ:) *any wide valley in a soft*, or *plain*, [*low ground such as is termed*] جَوّ; (Az, TA;) as also خَوًى: (Az, TA in art. خوى:) and *low*, or *depressed, ground*; (TA;) as also ▼ خَوَّةٌ: (JK, TA:) or the former, *a soft place*: (JK:) or *soft ground*; as in the saying, وَقَعَ غَرْسُكَ بِخَوٍّ *What thou hast planted has chanced to be in soft ground, in which it will take root and not fail to be productive*. (TA.)

[خَوٌّ, by a mistake in the CḲ, in art. خوى, is made to signify *Honey*: see خَوَآءٌ in that art.]

خوّة [app. خَوَّةٌ] *Languor*: occurring in a trad., in which it is said, وَأَخَذَ أَبَا جَهْلٍ خوّة فَلَا يَنْطِقُ [*And a languor seized Aboo-Jahl, so that he spoke not*]. (IAth, TA.)

خَوَّةٌ *A vacant land*. (Ḳ.) — See also خَوٌّ. — Also a dial. var. of أُخُوَّةٌ. (IAth, TA in art. اخو: see 1 in that art.)

خوأ

خَآءٌ *The name of the letter* خ, q. v.; as also خَا: [it is called خَا in a case of pause, and خَآءٌ when made a noun:] it is fem. and masc.: [its dim. is خُيَيَّةٌ, meaning *a* خ *written small, or indistinct*: (see خَآءٌ in art. خوأ:)] and its pl. is أَخْوَآءٌ and أَخْيَآءٌ and خَآءَاتٌ. (TA in باب الالف اللينة.)

خَآءِ بِكَ عَلَيْنَا, (AZ* and Ṣ* and TA in باب الالف اللينة, and Ḳ and TA in the present art.,) in which خَآءِ is a صَوْتٌ [or noun significant of a sound], indecl., with kesr for its termination; (Ṣ, TA;) and خَائِى بِكَ, as in the Book of the Nawádir by Ibn-Háni, (Az, TA,) in which latter, ISd says, the ى is not for a sign of the fem. gender, because the word is a صوت; and, as Sh says, on the authority of A 'Obeyd, خَائِبِكَ; but correctly written as in the Book of Ibn-Háni; (Az, TA;) *Hasten thou* [*to us*]: (AZ, Az, Ṣ, Ḳ, TA:) it is also used in addressing a female, and two persons, and a pl. number: (Ṣ, TA:) you say خَآءِ بِكِ, and] خَائِى بِكِ; and خَآءِ بِكُمَا [and خَائِى بِكُمَا; and خَآءِ بِكُمْ, and] خَائِى بِكُمْ. (TA.) In the saying of El-Kumeyt,

* بِخَائِى بِكَ ٱلْحَقْ يَهْتِفُونَ وَحَىِّ هَلْ *

[which app. means *Calling out "Hasten thou, come up with us," and "Come,"* or *"come quickly,"* &c.,] it is held by Ibn-Selemeh to be used as an imprecation, meaning خِبْتَ [*mayest thou be disappointed of attaining that which thou desirest*]; the poet saying بِخَائِبِكَ for بِأَمْرِكَ الَّذِى خَابَ وَخَسِرَ [meaning *By thine affair that shall result in disappointment, and be an occasion of loss*]: which, as you see, is at variance with the explanation of AZ. (Ṣ, TA.)

خَائِىٌّ and خَاوِىٌّ and خَوِىٌّ [a mistranscription for خَيَوِىٌّ, like بَيَوِىٌّ and تَيَوِىٌّ] rel. ns. of خَآءٌ and خَا the names of the letter خ. (TA in باب الالف اللينة.)

خوب

1. خَابَ, aor. يَخُوبُ, inf. n. خَوْبٌ, *He was*, or *became, poor, needy*, or *indigent*. (IAạr, Ḳ.) — See also خَابَ in art. خيب.

خَوْبَةٌ *A state of utter destitution, in which nothing remains in possession*: so in the saying, أَصَابَتْهُمْ خَوْبَةٌ [*A state of utter destitution befell them*]. (A 'Obeyd, T.) — *Hunger*: (AA, T, Ṣ, A, Ḳ:) pl. خَوْبَاتٌ. (TA.) So in the phrase, أَصَابَتْهُ خَوْبَةٌ [*Hunger befell him*]. (AA, Ṣ.) Sh knew not this word, and thought it to be a mistake for جَوْبَةٌ; (T;) which latter signifies "want." (Ṣ.) [See the latter word.] — *A tract of land upon which rain has not fallen, between two tracts of land watered by rain*. (AA, Ṣ, Ḳ.) — *Land that is bad*, (Ṣ,) *in which is no pasture* (Ṣ, Ḳ) *nor water*. (TA.) So in the saying, نَزَلْنَا بِخَوْبَةٍ مِنَ الأَرْضِ [*We alighted in bad land, without pasture or water*]. (Ṣ.) [See also جَوْبَةٌ.]

خوت

1. خَاتَ, (Ṣ, A, Ḳ,) aor. يَخُوتُ, inf. n. خَوْتٌ (Ṣ, TA) and خَوَاتَةٌ; (TA;) and ▼اختات, (Ṣ, Ḳ,) and ▼انخات; (Ḳ;) *He* (a hawk, or falcon, Ṣ, Ḳ, and an eagle, A, TA) *pounced down,* or *made a stoop, upon the prey,* or *quarry,* (Ṣ, A, Ḳ,) *to take it,* or *seize it,* (Ṣ,) *making a sound to be heard with his wings.* (TA.) And خَاتَتِ العُقَابُ, aor. تَخُوتُ, inf. n. خَوَاتٌ, *The eagle made a sound by the motion of its wings* [*in pouncing down*]. (Ṣ.) — Also خات, (Ḳ,) inf. n. خَوْتٌ, (TA,) *He* (for instance an eagle, TA) *seized,* or *snatched away,* (Ḳ, TA,) a thing; (TA;) and so ▼تخوّت, (IAạr, Ḳ,) and ▼اختات. (TA in art. خيت [where this last is said to belong to the present art.].) You say of the hawk, ▼اختات الطَّيْرَ *He seized,* or *snatched away, the birds:* (TA in art. خيت:) and of the wolf, ▼اختات الشَّاةَ (Fr, Ṣ, TA) *He seized,* or *snatched away, the sheep* or *goat;* (TA in art. خيت;) or *he stole away the sheep* or *goat by stratagem.* (Fr, Ṣ, Ḳ.) — خات مَالَهُ *He* (a man) *lessened,* or *diminished, his property;* or *took from it by little and little;* syn. تَنَقَّصَهُ; (Ḳ;) as also ▼تخوّت مَالَهُ, (Ṣ, Ḳ,) and تَخَوَّفَهُ: (Ṣ:) and خَاتَهُ مَالَهُ, aor. يَخُوتُهُ and يَخِيتُهُ, [*he lessened,* or *diminished, to him his property;* or *did so by taking from it by little and little;*] both signify the same. (Ḳ in art. خيت.) — And خات *He* (a man, TA) *lessened, diminished,* or *impaired, his provision of corn* or *food* (مِيرَتَهُ); or *made it defective,* or *deficient.* (Ḳ.) — خات, (aor. as above, Mṣb,) also signifies *He* (a man) *broke his promise:* (IAạr, Ṣ, Mṣb, Ḳ:) and *broke,* or *dissolved, his compact, contract,* or *covenant.* (Ḳ.)

3. خاوت طَرْفَهُ دُونِى, (Ḳ,) inf. n. مُخَاوَتَةٌ, (TA,) i. q. سَارَقَهُ [i. e. *He cast his eye furtively towards a spot between me and him;* meaning, *towards me:* see similar phrases in art. دون]. (Ḳ.)

5: see 1, in two places: — and see also 8, in two places.

7: see 1, first sentence.

8: see 1, in four places. — One says also, اختات بَلَدًا, and ▼تخوّته, *He cut off for himself a land,* or *district.* (IAạr, TA in art. خوى.) — And فُلَانٌ يَخْتَاتُ حَدِيثَ القَوْمِ, (Ṣ, Ḳ,*) and ▼يَتَخَوَّتُ, (Ṣ,) *Such a one takes of the discourse of the people, or party, and learns it by heart* [*one part after another*]. (Ṣ, Ḳ.*) In [some of] the copies of the Ḳ, in the explanation of اختات الحَدِيثَ, for فَتَحَفَّظَهُ is erroneously put فَتَخَطَّفَهُ. (TA.) — And إِنَّهُمْ يَخْتَاتُونَ اللَّيْلَ, meaning يَسْرُونَ وَيَقْطَعُونَ الطَّرِيقَ [*Verily they journey by night, and make the road to be feared, relying upon their strength, and robbing and slaying men*]. (Ṣ: [in two copies of which I find يَسِيرُونَ in the place of يَسْرُونَ.])

خَوَاتٌ *The sound made by the motion of the wings of an eagle.* (Ṣ, Ḳ. [In a copy of the Ṣ, I find it said that this word is fem., though its meaning is masc.; and the same is said in the TA, doubtless from another copy of the Ṣ; but this observation, which is omitted in one copy of the Ṣ in my possession, applies to خَوَاةٌ, mentioned in art. خوى in the Ṣ, as meaning صَوْتٌ.]) And *A sound* [absolutely]; (Ḳ;) as also ▼خَوَاتَةٌ: (TA:) or the *sound,* or *noise, of thunder,* and *of a torrent,* or *flow of water.* (AḤn, Ḳ.) A poet says,

* فَلَا حِسَّ إِلَّا خَوَاتُ السُّيُولِ *

[*And there was no sound but the noise of the torrents*]; (AḤn, TA;) which shows that we should read in the Ḳ وَالسُّيُولِ; not والسيل, as in some copies. (TA.)

خَوَاتَةٌ: see the next preceding paragraph.

خَوَّاتٌ A *bold, daring, brave,* or *courageous, man.* (Ṣ, Ḳ.) — One *who is always eating, but does not eat much.* (Ḳ.) — *A great promise-breaker.* (Mṣb.)

خَائِتٌ [act. part. n. of 1]. خَائِتَةٌ signifies An eagle (عُقَابٌ) *pouncing down,* or *making a stoop,* (Ṣ, Ḳ,) *and thereby causing a sound to be heard.* (Ṣ.) — *Breaking his promise;* or one *who breaks a promise.* (Mṣb.)

خوخ

خَوْخٌ [The *peach,* or *peaches;*] *a well-known kind of fruit,* (Ḳ, TA,) *which is eaten;* (TA;) i. q. دُرَّاقِنٌ: (Ḳ in art. درقن:) sing., (Ḳ,) or n. un., (Ṣ,) ▼خَوْخَةٌ. (Ṣ, Ḳ.) — See also what follows.

خَوْخَةٌ: see above. — Also *An aperture* (Ṣ, A, Mgh, Ḳ) *in a wall,* (Ṣ, A, Mgh,) *admitting the light* (Ṣ, A, Ḳ) *to a house,* or *chamber.* (Ḳ.) — *A passage* (مُخْتَرَقٌ) *between any two houses, not having a door,* or *gate:* (Ḳ:) of the dial. of El-Ḥijáz; (TA;) [and of Egypt, where it is applied to *a lane leading from one street or quarter to another:* coll. gen. n. ▼خَوْخٌ:] accord. to some, *a passage* (مُخْتَرَقٌ) *between any two things:* and *a small door-way between two houses,* or *chambers, with a door affixed to it.* (L.) — *A wicket,* or *small door,* (A, Mgh,) *in a large door.* (A.) [Golius, as on the authority of Meyd, explains it as meaning *Fenestella in medio januæ.*] — [Also applied to *A sluice* in a rivulet: see دَرْقَةٌ.] — ‡The *anus;* syn. دُبُرٌ. (Ḳ, TA.) — *A kind of green garment:* (Az, Ḳ:) of the dial. of Mekkeh. (TA.)

خود

2. خَوَّدَ, (L,) inf. n. تَخْوِيدٌ, (Ṣ, L, Ḳ,) *He* (a camel, L) *went quickly;* or *was quick in his pace* or *going:* (Ṣ, L, Ḳ:) *he went quickly, and threw out his legs:* or, as some say, *he shook, as though he were convulsed:* and in like manner, a male ostrich: and sometimes it is said of a man, signifying as first explained above. (L.) — خَوَّدَ رَأْلَهُ فِى الإِبِلِ: see art. رأل. — *He sent* the stallion *among the camels.* (L, Ḳ.) — *He obtained* somewhat of food. (Ḳ.)

5. تخوّد *He* (a camel) *shook* in going, *by reason of briskness, liveliness,* or *sprightliness.* (A.) — *It* (a branch) *inclined,* (A, TA,) or *bent.* (Ḳ.)

خَوْدٌ, applied to a girl, (Ṣ,) or young woman, (L,) or a female, (A, Ḳ,) *Soft,* or *tender,* (Ṣ, A, L, Ḳ,) *and youthful:* (A:) or *goodly,* or *beautiful, in make, and youthful,* (L, Ḳ,) *not yet of middle age:* (L:) pl. خَوْدَاتٌ (L, Ḳ) and خُودٌ; (Ṣ, L, Ḳ;) the latter like لُدْنٌ as pl. of لَدْنٌ, an epithet applied to a spear. (Ṣ, L.)

خور

1. خَارَ, aor. يَخُورُ, (Ṣ, A, Mgh,) inf. n. خُوَارٌ (JK, Ṣ, A, Mgh, Ḳ*) and خَوْرٌ, (JK,) *He* (a bull) *uttered his cry;* [i. e. *lowed,* or *bellowed;*] (Lth, JK, Ṣ, A, Mgh, Ḳ;) this being its primary signification: (Er-Rághib:) the inf. n. خوار, used agreeably with this explanation, occurs in the Ḳur xx. 90 [and vii. 146]: (Ṣ:) it signifies the *loud crying* [i. e. the *lowing* or *bellowing*] of a cow and of a calf: (Lth:) and the *crying* [i. e. *bleating*] of sheep, or that of goats, and of gazelles, (Ḳ,) and of any beast: (Er-Rághib:) and the *sounding* [i. e. *whizzing*] of arrows: (Ḳ:) of any of these, you say, خَارَ, aor. and inf. n. as above. (TA.) [Hence,] لَهُ صَوْتٌ كَخُوَارِ الثَّوْرِ *He has a voice like the bellowing of the bull.* (A.) — [And hence, (see 10,)] خار عَلَيْهِ ‡ *He bent, turned,* or *inclined, towards him.* (A.) — خار, aor. as above, (Ṣ, Mṣb,) inf. n. خُؤُورٌ, (Ṣ, Ḳ, [for which Golius, as on these authorities, substitutes خُؤُورَةٌ,]) said of a man, (Ṣ,) and of anything, (TA,) *He,* or *it, was,* or *became, weak,* or *feeble,* (Ṣ, Mṣb, Ḳ,) *and languid;* (Ṣ, TA;) as also خَوِرَ, (TA,) aor. يَخْوَرُ, (JK,) inf. n. خَوَرٌ; (Ṣ,* Ḳ,* TA;) and ▼خوّر, (JK, TA,) inf. n. تَخْوِيرٌ. (Ḳ.) خار and خَوِرَ both signify *It was soft,* or *fragile;* said of anything, like a reed. (JK.) It is said in a trad. of 'Omar, لَنْ يَخُورَ قَوِىٌّ مَا دَامَ صَاحِبُهَا يَنْزِعُ وَيَنْزُو, meaning *A possessor of strength* (صَاحِبُ قُوَّةٍ) *will not be weak as long as he can pull* his *bow and leap* to his beast. (TA.) In a camel that is drinking, خَوَرٌ denotes, or implies, a quality that is praised; i. e. *Patient enduring of thirst and fatigue:* and a quality dispraised; i. e. the *lacking patience to endure thirst and fatigue.* (TA.) — Also, said of heat, (Ṣ, TA,) and of cold, inf. n. خُؤُورٌ and خُؤُورَةٌ, (JK,) ‡ *It became faint;* it *remitted,* or *abated;* (JK, Ṣ, TA;) and so خَوِرَ, inf. n. خَوَرٌ; and ▼خوّر. (TA.) And خار عَنَّا, said of cold, *It ceased from us; quitted us.* (A.) — خَارَهُ, (JK, Ṣ,) inf. n. خَوْرٌ, (Ṣ, Ḳ,) *He hit,* or *hurt, his* خَوْرَان, (JK, Ṣ, Ḳ,*) in thrusting or piercing him with a spear or the like. (JK, Ṣ.)

2. خوّر: see 1, in two places. — خوّرهُ *He attributed to him weakness,* or *feebleness, and languor.* (TA.)

4. اخارهُ, (Ṣ, Ḳ,) inf. n. إِخَارَةٌ, (Ṣ,) [app., in its primary acceptation, *He caused him to utter a cry.* (See 10.) — And hence,] † *He bent, turned,* or *inclined, him,* or *it.* (Ṣ, Ḳ.) You say, أَخَرْنَا المَطَايَا إِلَى مَوْضِعِ كَذَا † *We bent,* or *turned, the riding-camels to such a place.* (Ṣ.)

6. تخاورت الثّيران *The bulls lowed,* or *bellowed, one to another.* (A.)

10. استخارها [*He endeavoured to make her* (namely, a gazelle, or a wild cow,) *to utter her cry;* or] *he uttered a cry in order that she should do the same.* (TA.) The sportsman, coming to a place in which he thinks the young one of a gazelle or [wild] cow to be, utters a cry like that of her young one; and the mother, hearing it, if she have a young one, thinks the cry to be that of her young one, and follows the cry. (Ṣ,* TA.) — Hence, (Ṣ, TA,) استخارهُ ‡ *He endeavoured to make him bend, turn,* or *incline:* (JK, Ṣ, A, Ḳ, TA:) and *he called him to him:* and *he interrogated him;* or *desired him to speak;* syn. اِسْتَنْطَقَهُ: namely, a man. (JK.) [استخار المَنْزِلَ is explained in the L and Ḳ as meaning اِسْتَنْظَفَهُ: to which is added in the TA, كأنّه طلب خيره, with the remark that it should therefore properly be mentioned in art. خير: but an explanation in the sentence immediately preceding, and a verse cited below, evidently show that استنظفه is a mistranscription for اِسْتَنْطَقَهُ, and that خيره should be خَبَرَهُ: so that the meaning is *He interrogated the place of abode.*] The author of the L cites, as an ex., the saying of El-Kumeyt,

* وَلَنْ يَسْتَخِيرَ رُسُومَ الدِّيَارِ *
* لِعَوْلَتِهِ ذُو الصِّبَى المُعْوِلُ *

[*And he who is affected with youthful amorousness, wailing, will not ask the remains,* or *traces, of the dwellings to reply to his wailing:* but for لعولته I would rather read بِعَوْلَتِهِ; i. e., *will not interrogate* them *with his wailing*]. (TA.) — استخار الضَّبَّ, (Ḳ, TA, [in some copies of the Ḳ, erroneously, الضَّبُعَ,]) and اليَرْبُوعَ, (TA,) *He placed a piece of wood in the hole of the burrow of the* [*lizard called*] ضَبّ, (Ḳ, TA,) and *of the jerboa,* i. e. *in the* قاصعآء, (TA,) *in order that it should come forth from another place,* (Ḳ, TA,) i. e. *the* نافقآء, *so that he might catch it.* (TA.) Lth falsely assigns the act of الاِسْتِخَارَة to the ضبّ and the jerboa. (Az, TA.)

خُورٌ *Low,* or *depressed, ground* or *land,* (JK, Ṣ, Ḳ,) *between two elevated parts;* (JK, Ṣ;) like خور: (TA:) *an inlet* (lit. *a neck*) *from a sea* or *large river, entering into the land:* (Sh:) *a place,* or *channel, where water pours into a sea* or *large river:* (JK, Ḳ:) or *a wide place* or *channel, where waters pour, running into a sea* or *large river;* (TA:) or (as in the TA, but in the Ḳ "and,") *a canal,* or *cut, from a sea* or *large river:* (Ḳ, TA:) and *i. q.* رَحَبَة [app. as meaning the *part in which the water flows from the two sides* of a valley]: (JK:) pl. خُؤُورٌ. (TA.)

خُورٌ a pl. of خَوَّارَةٌ, (Ṣ, Ḳ,) contr. to rule; (MF, TA;) and of خَوَّار in the phrase خَوَّارُ العِنَانِ. (JK, TA.) See خَوَّار, in five places.

خُورَةُ الإِبِلِ, with damm, [app. originally خُيْرَةٌ,] *The best of camels,* or *of the camels;* (IAạr, Ḳ;) [see خُيُورٌ, (in art. خير,) near the end of the paragraph;] and so ↓ خُوَارُهَا, and ↓ مِنْهَا الخُورَى. (Fr, TA.)

خُورَى fem. of أَخْيَرُ, and properly belonging to art. خير: see what next precedes.

خَوْرَانٌ *The* مَبْعَر [or *rectum*], *which comprises the* حِتَار [or *anus, with the extremities of its skin,*] *of the* صُلْب [or *back*], (Ḳ,) of a man &c.: (TA:) or the *passage of the* رَوْث [or *dung,* properly of a horse or the like, but here app. meaning of a man also]: (Ṣ:) or the *head* [or *extremity*] *of the* مَبْعَرَة [or *rectum*]: or the *part in which is the* دُبُر [or *anus*]: (Ḳ:) or the دُبُر [or *anus*] itself; (TA;) or it has this meaning also; (JK;) and so ↓ خَوَّارَةٌ, syn. اِسْتٌ; (Ḳ;) the دُبُر being so called because it is like a depressed place between two hills: (TA: [see خُورٌ:]) or the *gap in which is the* دُبُر [or *anus*] of a man; and that *in which is the* قُبُل [or *anterior pudendum*] of a woman: (TA:) or the *gap in which is the* دُبُر *and the place of the* ذَكَر *and that of the* قُبُل *of the woman:* (Zj in his "Khalḳ el-Insán:") pl. خَوْرَانَاتٌ and خَوَارِينُ: (Ḳ:) the former pl. of a form which any sing. subst. not significant of a human being may receive. (TA.)

خُوَارٌ an inf. n. of خَارَ as explained in the first sentence in this art. (Ṣ, A, &c.) — خُوَارُ الإِبِلِ: see خُورَة.

خُؤُورٌ: see the next paragraph, in two places.

خَوَّارٌ *Weak,* or *feeble;* (JK, Ṣ, Mṣb, Ḳ;) applied to a man; (Ṣ;) as also ↓ خَائِرٌ, (Ḳ,) and ↓ خُؤُورٌ: (AHeyth:) a *weak* man, *who cannot endure difficulty* or *distress:* (Lth:) and ‡ *cowardly,* or *a coward:* (A:) pl. of the first خَوَّارُونَ, and of the third خُوَرَةٌ. (AHeyth.) Applied to a camel, *Slender* (رَقِيق) *and beautiful:* (Ḳ, TA: [for الحسّ in the CḲ, I read الحَسَنُ, as in other copies of the Ḳ and in the TA:]) and the fem., with ة, applied to a she-camel, *having soft flesh and fragile bones:* (TA:) pl. of the former [and of the latter] خَوَّارَاتٌ. (Ḳ.) Applied to a spear, *Weak:* (Ṣ:) *not hard:* (Mṣb:) or *weak and soft;* (TA;) and in the same sense applied to an arrow, (A, TA,) as also ↓ خُؤُورٌ; (TA;) and so the fem. of the former, with ة, applied to a reed or cane (قَصَبَة); (A, TA;) and to land or ground (أَرْض) as meaning *weak,* (Ṣ,) or *soft:* (A, Mṣb:) pl. ↓ خُورٌ. (Ṣ.) And خَوَّارُ العِنَانِ ‡ A horse (A) *that turns easily,* (JK, A, Ḳ,) *and runs much:* (Ḳ:) pl. ↓ خُورٌ. (JK, TA.) And بَكْرَةٌ خَوَّارَةٌ † *A sheave of a pulley of which the pin runs* [or *turns*] *easily in the cheeks.* (TA.) And ↓ خُورُ الحَشَايَا *Beds, or the like, stuffed with soft substances.* (TA, from a trad.) And خَوَّارُ الصَّفَا *Smooth stones that sound* [*when struck*] *by reason of their hardness.* (IAạr.) And زَنْدٌ خَوَّارٌ *A* زند [q. v.] *that emits much fire;* syn. قَدَّاحٌ. (AHeyth, Ḳ.) [Hence,] هُوَ خَوَّارُ العُودِ [meaning † *He is lavish when asked*]: an expression of dispraise. (TA in art. عكس.) [Hence also,] خَوَّارَةٌ ‡ A she-camel *abounding with milk;* pl. ↓ خُورٌ; (Ṣ, Ḳ, TA;) which is contr. to rule, and said by MF to be without a parallel: (TA:) and so a ewe or she-goat: (TA:) or a she-camel *whose milk flows easily;* and so a ewe or she-goat: (A:) or a she-camel *thin-skinned, and abounding with milk:* (AHeyth:) or one *that is of a hue between dust-colour and red, with a thin skin;* and such is the most abundant in milk: (Kf:) or *of a red colour inclining to dust-colour, thin-skinned, and having long fur with* [*coarse*] *hair protruding through it, longer than the rest:* such a she-camel is less hardy than others, but abounds with milk. (ISk.) Also ‡ A palm-tree (نَخْلَة) *that bears much fruit.* (JK, A, Ḳ.) — ↓ خُورٌ as meaning † Women *much suspected, on account of their corruptness,* (Ḳ, TA,) *and the weakness of their forbearance,* (TA,) is [a pl.] without a sing. (Ḳ.)

خَوَّارَةٌ fem. of خَوَّار [q. v.]. — As a subst.: see خَوْرَانٌ.

خَائِرٌ: see خَوَّار, first sentence.

خوص

1. خَوِصَ, aor. ـَ, inf. n. خَوَصٌ, *He* (a man, Ṣ) *had the eye sunk,* or *depressed:* (Ṣ, Ḳ:) or the inf. n. signifies the eye's *being narrow,* or *contracted, and sunk,* or *depressed:* (Mṣb:) or its *being small, and sunk,* or *depressed:* (A:) or its *being sunk,* or *depressed, and narrow,* or *contracted, and small:* or one eye's *being smaller than the other:* or the eye's *being narrow in the slit, naturally,* or *by reason of disease:* or, accord. to AM, all that they have related respecting this word is correct except narrowness of the eyes; for the Arabs, when they mean this, use the term حَوَص, with [the unpointed] ح; but when they mean the eye's being sunk, or depressed, this they term خَوَص, with [the pointed] خ: (TA:) and accord. to A 'Obeyd's relation on the authority of his companions, (TA,) [and accord. to Mṭr also,] خَوِصَتْ عَيْنُهُ signifies *his eye became sunk,* or *depressed;* (Mgh, TA;) but حَوِصَتْ, "it became narrow, or contracted." (Mgh.) — Also خَوِصَتْ, inf. n. as above; and ↓ اِخْوَاصَّتْ, inf. n. اِخْوِيصَاصٌ; *She* (a ewe) *had one of her eyes black, and the other white.* (AZ, TA.)

2. خوّص الشَّجَرُ, inf. n. تَخْوِيصٌ, said of palm-trees, [and some others, see خُوص,] *The trees put forth leaves,* [or only *leaves of the kinds called* خُوص,] *little by little.* (L, TA.) See also 4. — خوّص التَّاجَ, inf. n. as above, *He ornamented the crown with plates of gold* (Ḳ, TA) *of the width of palm-leaves.* (TA.)

3: see 6, in two places.

4. أَخْوَصَتِ النَّخْلُ *The palm-trees put forth their* خُوص [or *leaves*]: (Ṣ, Ḳ:) or, accord. to the A, you say, ↓ خَوَّصَتِ النَّخْلُ, meaning *the palm-trees put forth their leaves.* (TA.) [See also 2, above.] اخوص is also said of the عَرْفَج, (Ṣ, Ḳ,) and of the رِمْث, (TA,) [and of other trees, (see خُوص,)] or of trees in general, (TA,) or of trees (الشَّجَر) you say أَخَاصَ, inf. n. إِخْوَاص,

(AḤn, ISd,) the verb being thus made infirm, and the inf. n. sound, (ISd,) or of all trees except thorny trees and herbs or leguminous plants, (TA,) meaning, *It broke out with leaves*: (Ṣ, Ḳ:) or, when said of the عرفج, *its* خوص *became perfect*. (AA, TA voce عَرْفَج; and Ṣ voce ثَقَّبَ.) And you say also, أَخْوَصَتِ الخُوصَةُ *The* خوصة [see خُوص] *appeared*. (TA.)

6. تخاوص, (A, Ḳ,) or تخاوص فِى نَظَرِهِ, (TA,) *He blinked*, or *contracted his eyelids*, (A, Ḳ,) *somewhat*, (Ḳ,) *looking intently, as though he were aiming an arrow*; and so in *looking at the sun*; (A, Ḳ;) as also ▼خاوص. (Ḳ.) [But the latter is trans.] You say, إِنَّهُ يُخَاوِصُ ▼ فُلَانًا, and يَتَخَاوَصُ لَهُ, *Verily he blinks*, or *contracts his eyelids, looking intently, at such a one, as though he were aiming an arrow*. (A.) [See also هُوَ يُخَاوِصُ فُلَانًا.] — تَخَاوَصَ إِلَى الشَّمْسِ; and [Hence,] تَخَاوَصَتِ النُّجُومُ, (A,) or تخاوصت النجوم لِلْغُرُوبِ, (TA,) ‡ *The stars inclined to setting*. (A, TA.)

11: see 1, last signification.

خُوصٌ The *leaves of the date-palm*, (T, Ṣ, A, Mṣb, Ḳ,) and *of the* مُقْل [or *Thebaïc palm*], (T, TA,) and *of the* نَارَجِيل [or *cocoa-nut-tree*], and *the like*, (TA,) and *of the* عَرْفَج, (T, Ḳ,) and *of the* ثُمَام, (T, TA,) and *of the* نَصِىّ, (Ṣ voce أُمْصُوخَة, q. v.,) and *of the* أَرْطَى, and *of the* أَلَآء, and *of the* سَبَط: (Ibn-'Eiyásh Ed-Ḍabbee, Ḳ:) n. un. with ة: (T, Ṣ, Ḳ, &c.:) *the* خوصة *of the* عرفج *is the green* [*part*] *thereof when it appears upon the white thereof*; (TA;) [or] *it resembles the leaves of the* حِنَّآء: *that of the* ارطى *is like the* هَدَب [or *evergreen leaves*] *of the* أَثْل: *that of the* الآء *has the form of the ears of sheep*, or *goats*: *and that of the* سبط *has the form of the* حَلْفَآء: (Ibn-'Eiyásh, TA:) *there is also the* خوصة *of the* [*class of trees or plants called*] جَنْبَة, *which is of the plants*, or *herbage, of the* [*season called*] صَيْف, or, *as some say, it is what grows upon a root-stock* or *rhizoma* (عَلَى أُرُومَة): (TA:) but to herbs, or leguminous plants, of which the leaves fall and become scattered when they dry up, there is no خوصة. (T, TA.)

خِيَاصَةٌ The *trade*, or *art*, *of the* خَوَّاص. (A, TA.)

خَوَّاصٌ *A seller of* خُوص: (Ṣ, Ḳ:) or *a weaver thereof* [*into baskets and mats and the like*]: (A:) or *both*. (TA.)

أَخْوَصُ A man (Ṣ, Mgh) *having the eye sunk*, or *depressed*; (Ṣ, Mgh, Ḳ;) *having the quality of the eye termed* خَوَص: [see 1:] fem. خَوْصَآءُ: (TA:) which is [also] applied to the eye, meaning *sunk*, or *depressed*: (Mgh:) or *small, and sunk*, or *depressed*: (A:) and to a ewe, meaning *having one of her eyes black, and the other white*: (AZ, Ḳ:) or *having one eye black, and the other, with the rest of the body, white*: (TA:) pl. خُوصٌ, which, prefixed to العُيُون, is applied to camels. (A.) — [Hence,] بِئْرٌ خَوْصَآءُ ‡ *A deep well*; *a well of which the bottom is deep*; (A, Ḳ, TA;) *of which the beasts see not the water*: (TA:) because one contracts his eyelids (يَتَخَاوَصُ) in looking into it: (A, TA:) or خوصاء applied to a well (رَكِيَّة), signifies *of which the water has sunk into the earth*. (TA.) And the same epithet applied to a [mountain of the kind called] هَضْبَة, (A,) or قَارَة, (Ḳ,) ‡ *High*; *lofty*: (A, Ḳ:) because one contracts his eyelids in looking at it. (A, TA.) And رِيحٌ خَوْصَآءُ ‡ *A hot wind*: (Ḳ:) or *a vehemently-hot wind*: (A:) *that makes the eye to blink, or contract the eyelids*, (تُكَسِّرُهَا,) *by reason of heat*: (Ḳ,* TA:) *in which one does not see without blinking*, or *contracting the eyelids*. (A.) And ظَهِيرَةٌ خَوْصَآءُ ‡ *A summer mid-day vehemently hot*: (A:) or *most vehemently hot*; (Ḳ, TA;) *in which one cannot look without blinking*, or *contracting the eyelids*. (TA.)

مُخَوَّصٌ applied to a crown, *Ornamented with plates of gold like* خوص *in width*: (A,* TA:) and applied to a vessel, *having in it what resemble* خوص *in shape*. (TA.) مُخَوَّصٌ بِالذَّهَبِ, applied to دِيبَاج [or silk brocade], *Woven with gold in the form of* خوص. (TA.)

أَرْضٌ مُخْوِصَةٌ *Land in which are* خُوص *of the* أَرْطَى *and* أَلَآء *and* عَرْفَج *and* سَبَط. (Ibn-'Eiyásh Ed-Ḍabbee, Ḳ.)

خوض

1. خَاضَ المَآءَ, aor. يَخُوضُ, inf. n. خَوْضٌ (Ṣ, A, Mṣb, Ḳ) and خِيَاضٌ, (Ṣ, A, Ḳ,) [*He waded*, or *forded, through the water*;] *he passed through the water walking* or *riding*: (Ṣ:) or *he entered into the water*; (A, Ḳ;) as also ▼خَوَّضَهُ, (Ḳ,) inf. n. تَخْوِيضٌ; (TA;) [or this latter has an intensive signification, as it is said to have in a phrase below;] and ▼اختاضهُ: (Ḳ:) or *he walked in*, or *through, the water*; (Mṣb;) as also ▼تخوّضهُ: (TA:) or *he entered into the water and walked in it*, or *through it*. (TA.) You say also, خاض بِالفَرَسِ, meaning *He brought the horse to the water*; as also ▼اخاضهُ, (Ḳ,) inf. n. إِخَاضَةٌ; (AZ;) and ▼خاوضهُ, (Ḳ,) or خاوضهُ فِى المَآءِ, inf. n. مُخَاوَضَةٌ, as in the A: (TA:) or ▼اخاضوا دَوَابَّهُمْ signifies خَاضُوهُ بِدَوَابِّهِمُ المَآءَ [*They waded* or *forded through the water, or entered into it*, &c., *with their beasts*]: and you say also, ▼خَاوَضْتُهُمْ فِى المَآءِ [*I waded* or *forded with them through the water*; &c.; meaning with men, not with beasts]: (so I find in a copy of the A:) and خَاضَتْ خَيْلُهُمُ المَآءَ signifies ▼اخاض القَوْمُ [*The people's horses waded* or *forded through the water*]. (Ṣ.) — خَاضَتِ الإِبِلُ لُجَّ السَّرَابِ ‡ [*The camels passed through the great expanse of mirage*]. (A.) — خَاضَ البَرْقُ الظَّلَامَ ‡ [*The lightning penetrated through the darkness*]. (A, TA.) — خَاضَ إِلَيْهِ الرِّمَاحَ حَتَّى أَخَذَهُ ‡ [*He forced his way to him through the spears until he took him*, or *it*]. (A, TA.*) — خَاضَ القَوْمُ فِى الحَدِيثِ, (Ṣ, A,) and ▼تخاوضوا فيه, (Ṣ, A, Ḳ,) ‡ *The people*, or *company of men, entered* [or *waded*] *together into discourse*. (Ṣ, A, Ḳ.) — خَاضَ الغَمَرَاتِ, (Ṣ, Ḳ,) aor. as above, inf. n. خَوْضٌ, (TA,) ‡ *He plunged into the submerging floods of ignorance*, or *the like*; syn. اِقْتَحَمَهَا. (Ṣ, Ḳ, TA.) — خَاضَ فِى الأَمْرِ † *He entered* [or *plunged*] *into the affair*. (Mṣb.) — In like manner you say, [خَاضَ فِى البَاطِلِ and] ▼اخاض البَاطِلَ † *He entered* [or *plunged*] *into false*, or *vain, discourse* or *speech*: (Mṣb:) and خَاضَ, alone, signifies ‡ *He said*, or *spoke, what was false*. (A.) It is said in the Ḳur [lxxiv. 46], (TA,) وَكُنَّا نَخُوضُ مَعَ الخَائِضِينَ, i. e. [فِى البَاطِلِ ‡] *And we used to enter into false, or vain, discourse* or *speech, with those who entered thereinto*]; (Bḍ, Jel, Ḳ;) syn. نَشْرَعُ: (Bḍ:) or *and we used to follow the erring*, &c. (O, K.) And again, [lii. 12,] الَّذِينَ هُمْ فِى خَوْضٍ يَلْعَبُونَ ‡ [*Who amuse themselves in entering into false, or vain, discourse or speech*]; (TA;) فى الباطل being here, again, understood. (Bḍ.) And again, [ix. 70,] وَخُضْتُمْ كَالَّذِى خَاضُوا, i. e. كَخَوْضِهِمْ ‡ [*And ye have entered into false, or vain, discourse or speech, like their entering thereinto*]. (Ḳ.) And again, [vi. 67,] الَّذِينَ يَخُوضُونَ فِى آيَاتِنَا ‡ [*Who enter into false, or vain, discourse or speech respecting our signs*; meaning the Ḳur-án]. (TA.) خَاضَ فِيهِ is also explained as signifying † *He said what was false respecting it*. (TA.) And خَوْض signifies † The *confusing*, or *confounding, in an affair*. (TA.) — خَاضَ, (Ṣ, A, Mgh, Ḳ,) and ▼خوّض, (A, TA,) also signify *He mixed*, (Ṣ,* Ḳ, TA,) *and stirred about*, (TA,) the beverage, or wine: (Ṣ, Ḳ, TA:) or *he stirred about* the سَوِيق with the مِخْوَض. (A, Mgh.*) — خَاضَهُ بِالسَّيْفِ ‡ *He moved about the sword in him, having smitten him*: (Ṣ, Ḳ,* TA:) or *he put* [or *thrust*] *the sword into the lower part of his belly, and then raised it upwards*. (A,* TA.) — خُضْتُ بِقِدْحٍ فِى القِدَاحِ, (A, TA,) inf. n. خِيَاضٌ; and ▼خَاوَضْتُ القِدَاحَ, inf. n. خِوَاضٌ; (TA;) ‡ *I put an arrow*, (A, TA,) *which I had borrowed, and by which I expected to have good luck*, (TA,) *among the* [*other*] *arrows* (A, TA) *used in the game called el-meysir*: (TA:) see an ex. (a verse of Ṣakhr-el-Gheí) in art. خضض.

2: see 1, first signification: — and again in the latter part of the paragraph. — خَوَّضَ فِى نَجِيعِهِ [app. meaning *He wallowed in his effused blood*] is with teshdeed to render the signification intensive. (Ṣ.)

3: see 1, second sentence, in two places: and again in the last sentence.

4. أَخَضْتُ فِى المَآءِ دَابَّتِى [*I made my beast to wade, or ford, through the water*]. (Ṣ, A.*) اخاض القَوْمُ خَيْلَهُمُ المَآءَ [*The people*, or *company of men, made their horses to wade*, or *ford, through the water*] is said when they wade, or ford, with their horses through the water. (TA.) اخاض القَوْمُ: and اخاضوا المَآءَ: [which are evidently elliptical phrases:] and اخاض الفَرَسَ: see 1, second sentence. — اخاض البَاطِلَ: see 1.

═ اخاض المَآءُ *The water admitted of being walked* [or *waded* or *forded*] *in* or *through*: contr. to general rule; being intrans. while the unaugmented verb is trans. (Mṣb.)

5: see 1, first sentence. — تخوّض also signifies *He constrained himself to wade*, or *ford, in*, or *through, water.* (Ḳ,* TA.) This is the primary signification: and hence, تخوّض فى الأَمْرِ ‡ *He employed*, or *occupied, himself in the affair*: and *he used art* or *artifice* or *cunning*, or *his own judgment* or *discretion, in the affair*, or *in the disposal* or *management thereof*: and so in the phrase تخوّض فى المَالِ: or, accord. to some, this means *he acted wrongly in acquiring the property in an improper manner, in whatsoever way it was possible.* (TA.)

6: see 1, near the middle of the paragraph.

8: see 1, first sentence.

مَخَاضَةٌ: see مَخَاضٌ.

مِخْوَضٌ [The *instrument with which beverage*, or *wine, is mixed and stirred about*]; it is, *for beverage*, or *wine, like the* مِجْدَح *for* سَوِيق: (Ṣ, Ḳ:) or the *instrument with which* سويق *is stirred about.* (A, Mgh.)

مَخَاضَةٌ [*A ford*; i. e.] *a place where people pass through water, walking or riding*: (Ṣ, A, Ḳ:) or *a place where one walks through water*: (Mṣb:) pl. ▼ مَخَاضٌ, (Ṣ, Ḳ,) [or this is rather a coll. gen. n.,] or مَخَائِضُ, (as in one copy of the Ṣ,) and مَخَاوِضُ (AZ, Ṣ, Ḳ) and مَخَاضَاتٌ. (Mṣb, TA.)

خوف

1. خَافَ, (Ṣ, Mṣb, Ḳ, &c.,) originally خَوِفَ, (Lth, L, &c.,) first pers. خِفْتُ, (TA,) aor. يَخَافُ, (Ṣ, Ḳ, &c.,) originally يَخْوَفُ, (L,) imperative خَفْ, (Ṣ,) inf. n. خَوْفٌ (Ṣ, Mṣb, Ḳ, &c.) and ▼ خِيفٌ, [originally خِوْفٌ,] (Lḥ, TA,) erroneously written in the Ḳ with fet-ḥ [to the خ], but some say that this is a simple subst., not an inf. n., (TA,) and ▼ خِيفَةٌ, (Lḥ, Ṣ, Mṣb, Ḳ, &c.,) originally خِوْفَةٌ, (Ḳ,) but some say that this also is a simple subst., not an inf. n., (TA,) and [therefore] its pl. is خِيَفٌ, (Lḥ, JK, Ṣ, and so in the CḲ,) in [some of] the copies of the Ḳ erroneously written خِيْفٌ, (TA,) or this [as well as the next preceding] may be an inf. n., for some few inf. ns. have pls., (IJd, TA,) and مَخَافَةٌ, (Ṣ, Mṣb, Ḳ, &c.,) originally مَخْوَفَةٌ, for which last, the first of these inf. ns. is used by a poet, and therefore made fem., (TA,) *He feared*; *he was afraid* or *frightened* or *terrified*; syn. فَزِعَ. (Ḳ.) It is also trans.: (Mṣb:) you say, خَافَهُ and ▼ تخوّفهُ [*He feared*, or *was afraid of, him*, or *it*]; (Mṣb, TA;) both signifying the same: (TA:) [and so خَافَ مِنْهُ; or this may mean *he feared what might happen to him from him*, or *it*:] and ▼ تخوّف عَلَيْهِ شَيْئًا, meaning خَافَهُ [i. e. خَافَ عَلَيْهِ شَيْئًا *He feared for him a thing*]: (Ṣ, Ḳ:) and ▼ تخوّفهُ عَلَيْهِ and خَافَهُ عَلَى مَالِهِ [*He feared him*, or *it, for his property*]. (Mgh.) — [Hence,] it is also used in the sense of ظَنَّ [*He thought*, or *opined*]: and in this case, the Arabs sometimes use it in the same manner as a verb signifying an oath, and give it the same kind of complement; as in an ex. cited voce وَرَدَ [q. v.]. (Ṣ in art. ورد.) And *He knew.* (Lḥ, Kr, Ḳ.) Hence, وَإِنِ ٱمْرَأَةٌ خَافَتْ مِنْ بَعْلِهَا نُشُوزًا [*And if a woman know* that there is, *on the part of her husband, injurious treatment*, or *unkindness*, or *estrangement*], (Ḳ,) in the Ḳur [iv. 127]. (TA.) And hence also, فَمَنْ خَافَ مِنْ مُوصٍ جَنَفًا [*And he who knoweth* that there is, *on the part of the testator, an inclining to a wrong course*, or *a declining from the right course*, &c.], (Ḳ,) in the Ḳur [ii. 178]; thus explained by Lḥ. (TA.) ═ خَافَهُ, (Ṣ,) first pers. خُفْتُهُ, (Ḳ,) aor. يَخُوفُهُ, (Ṣ,) *He exceeded him in fear.* (Ṣ, Ḳ.*) You say, ▼ خَاوَفَهُ فَخَافَهُ, (Ṣ,) inf. n. of the former مُخَاوَفَةٌ, (TA,) i. e. [*He vied with him to see which of them would exceed the other in fear, and*] *he exceeded him in fear.* (Ṣ.)

2. خوّفهُ, (Mṣb, Ḳ,) inf. n. تَخْوِيفٌ, (TA,) i. q. أَخَافَهُ. (Mṣb, Ḳ.) See the latter, in two places. *He put fear into him.* (JK, TA.) خَوِّفْنَا [app. addressed to God] is mentioned by Lḥ as meaning *Render the Ḳur-án and the Traditions beautiful to us in order that we may* [*give heed thereto and*] *fear.* (TA.) — *He made him to be in such a state*, or *condition, that men feared him*; (JK, Ḳ;) *he made him to be feared by men.* (M.) Hence, in the Ḳur [iii. 169], إِنَّمَا ذٰلِكُمُ ٱلشَّيْطَانُ يُخَوِّفُ أَوْلِيَآءَهُ, i. e. [*Verily that is the devil*:] *he causeth his friends to be feared* by you: [or *that devil causeth* &c.:] or, as Th says, *causeth* you *to fear by his friends.* (TA.) ═ *He diminished it, lessened it*, or *took from it*; and so خوّف مِنْهُ. (TA.) [See also 5.] — خوّف غَنَمَهُ *He sent away his sheep*, or *goats, flock by flock.* (TA.)

3: see 1, last sentence.

4. اخافهُ, (Mṣb, Ḳ,) inf. n. إِخَافَةٌ (Ṣ) and إِخَافٌ, like كِتَابٌ, (Lḥ, TA,) [but the latter is irreg. and rare,] *He*, or *it*, (an affair, a case, or an event, Mṣb,) *caused him*, or *made him, to fear*, or *be afraid*; *put him in fear*; *frightened*, or *terrified, him*; (TA;) and ▼ خوّفهُ, (Mṣb, Ḳ,) inf. n. تَخْوِيفٌ, (Ṣ, TA,) signifies the same. (Ṣ, Mṣb, Ḳ.) So in the phrase اخاف الثَّغْرُ [*The enemies' frontier caused to fear*, &c.; *was insecure*:] or *fear entered from it.* (TA.) You say also, مَالَ الحَائِطُ فَأَخَافَ النَّاسَ [*The wall leaned, and caused the people to fear*]. (Mṣb.) And أَخَافَ اللُّصُوصُ الطَّرِيقَ [for أَخَافَ اللُّصُوصُ أَهْلَ الطَّرِيقِ *The robbers caused the people of the road*, or *the passengers thereof, to fear*, &c.; or it may be rendered *the robbers caused the road to be insecure*]. (Mṣb.) And أَخَفْتُهُ الأَمْرَ فَخَافَهُ [*I caused him to fear the thing*, or *affair*, &c., *and he feared it*; making the verb doubly trans.]; as also ▼ خَوَّفْتُهُ إِيَّاهُ فَتَخَوَّفَهُ. (Mṣb.) It is said in a trad., أَخِيفُوا الهَوَامَّ قَبْلَ أَنْ تُخِيفَكُمْ, *Make ye the venomous reptiles and the like to fear before they make you to fear*; (TA;) i. e. *kill ye them before they kill you.* (JM, TA.) — مَا أَخْوَفَنِى عَلَيْكَ [*How greatly do I fear for thee!*]. (TA.)

5. تخوّفهُ: see 1, in three places. ═ Also *He took by little and little* (Ṣ, L, Ḳ) *from it*, (Ṣ, Ḳ,) or *from its sides*; (L;) as also تحوّفهُ: (Ṣ and Ḳ* in arts. حوف and حيف:) or *he took from its extremities*; so in the A; in which it is said to be tropical: accord. to IF, it is originally [تَخَوَّنَ,] with ن [in the place of the ف]. (TA.) Dhu-r-Rummeh says, (Ṣ,) or not he, but some other poet, for it is ascribed to several different authors, (L,)

* تَخَوَّفَ الرَّحْلُ مِنْهَا تَامِكًا قَرِدًا *
* كَمَا تَخَوَّفَ ظَهْرَ النَّبْعَةِ السَّفَنُ *

[*Her saddle abraded from a long and high, compact hump, like as when the piece of skin used for smoothing arrows has abraded from the back of a rod of the tree called* نبعة]. (Ṣ. [See also 5 in art. حوف, where another reading of this verse is given. In the TA, in the present art., in the places of الرحل and ظهر النبعة, I find السَّيْرُ and عُودَ.]) Hence, (Ṣ, Ḳ,) accord. to Fr, (TA,) أَوْ يَأْخُذَهُمْ عَلَى تَخَوُّفٍ, (Ṣ, Ḳ,) in the Ḳur [xvi. 49], (Ṣ,) which Az explains as meaning [*Or are they secure from his destroying them*] *by causing them to suffer loss* [*by little and little*] *in their bodies and their possessions*, or *cattle, and their fruits*: or, accord. to Zj, it may mean, *after causing them to fear*, by destroying a town, so that the one next to it shall fear. (TA.) You say also, تخوّف مِنْ مَالِى *He took by little and little from my property.* (JK.) And تَخَوَّفَنَا السَّنَةُ [*The year of drought*, or *sterility, took from us by little and little*]. (JK.) And تَخَوَّفَنِى حَقِّى [*He diminished to me by little and little my right*, or *due*]. (JK.) And تَخَوَّفَهُ حُمْقُهُ ‡ *i. q.* اهْضَمَهُ [an evident mistranscription for اِهْتَضَمَهُ or هَضَمَهُ, meaning *His stupidity deprived him of his right*, or *due*]. (TA.)

خَافٌ A man *very fearful* or *timorous*; (Ṣ, Ḳ;) [and so, in the present day, ▼ خَوَّافٌ; the former originally] of the measure فَعِلٌ, like فَرِقٌ and فَزِعٌ; and similar to صَاتٌ, meaning a man "having a strong, or loud, voice:" (Ṣ:) or *i. q.* ▼ خَائِفٌ: (TA:) accord. to Kh, it may be [originally خَاوِفٌ,] of the measure فَاعِلٌ, having the medial radical rejected; or [خَوِفٌ,] of the measure فَعِلٌ; and in either case, the dim. is [▼ خُوَيْفٌ,] with و: so says Sb. (TA.)

خَوْفٌ inf. n. of 1. (Ṣ, Mṣb, Ḳ, &c.) — Also *Slaughter*: whence, وَلَنَبْلُوَنَّكُمْ بِشَىْءٍ مِنَ ٱلْخَوْفِ [*And we will assuredly try you with somewhat of slaughter*]; (Lḥ, Ḳ;) in the Ḳur [ii. 150]. (TA.) [See also 4.] — And *Fighting*: whence, فَإِذَا جَآءَ الخَوْفُ [*But when fighting cometh*; in the Ḳur xxxiii. 19]. (Ḳ.) ═ See also خَائِفٌ. ═ Also *A red hide from which are cut strips like thongs*, (Kr, Ḳ, TA,) *and then upon these are put* [*ornaments of the kind termed*] شَذْرٌ; *worn by a girl*:

(TA:) a dial. var. of حَوْفٌ [q. v.]: (K:) but this latter is preferable. (L, TA.)

خِيفٌ: see 1, first sentence.

خَافَةٌ A [*coat of the kind called*] جُبَّةٌ, *of hide, or leather, which the collector of honey wears;* (Akh, JK, K;) and *also worn by the water-carrier:* (JK:) or *a fur-garment,* or *hide with the fur or wool on it, worn by him who enters into the places occupied by bees, in order that they may not sting him:* (TA:) or a [*pouch of the kind termed*] خَرِيطَةٌ, (S, K,) *of hide,* or *leather,* (S,) *narrow in the upper part and wide in the lower part,* (TA,) *in which honey is collected:* (S, K:) or a [*round piece of leather with a running string by means of which it may be converted into a bag, such as is termed*] سُفْرَةٌ, *like the* خَرِيطَة, *made,* or *sewed, small,* [for مُصْعَدَةٌ or مُصَعَّدَةٌ, which I find in different copies of the K, and to which no appropriate meaning is assignable, I read مُصْغَرَةٌ or مُصَغَّرَةٌ, (see 2 in art. صغر, and particularly أَصْغَرَ القِرْبَةَ,)] *having its head* [or *border*] *raised, for honey;* (K;) so says Skr, in explaining the following verse: or, as IB says, accord. to Aboo-'Alee, it is from the phrase النَّاسُ أَخْيَافٌ, meaning "men," or "the people," "are different, one from another;" for it is *a* خَرِيطَة *of hide,* or *leather, embellished with different kinds of embellishment;* and if so it should be mentioned in art. خيف: (TA:) [but] the dim. is ▼ خُوَيْفَةٌ. (JK.) Aboo-Dhu-eyb says, [describing a collector of wild honey,]

* تَأَبَّطَ خَافَةً فِيهَا مِسَابٌ *
* فَأَصْبَحَ يَقْتَرِى مَسَدًا بِشِيقِ *

(S,) [*He put beneath his armpit a* خافة *in which was a receptacle for honey, and betook himself to making successive endeavours to reach the most difficult part of a mountain by means of a rope,* or *rope of palm-fibres;* for] he means شِيقًا بِمَسَدٍ; the phrase being inverted: (S and TA in art. شيق:) or he means, [*betook himself to*] *taking successive holds of a rope* (يَتَتَبَّعُ حَبْلًا) *tied to a* شيق [here best rendered *mountain-top*] in his descent to the place of the honey; so that there is no inversion. (TA in that art.) — Also *i. q.* عَيْبَةٌ [*A kind of basket,* or *receptacle, of hide,* or *leather*]; (TA;) the *thing in which fruits are gathered;* also called مِخْرَفٌ. (Har p. 374.) — And خَافَةُ الزَّرْعِ is said to mean *The envelope of the grain of seed-produce;* so called because it protects it: to this the believer is likened in a trad. [as some relate it]; but the reading [commonly known] is [خَامَةٌ,] with م. (TA.) [See خامة, in art. خمر.]

خِيفَةٌ; pl. خِيفٌ: see 1, first sentence. — [Sometimes it may mean, agreeably with analogy, *A kind of fear.*] == See also art. خيف.

خُوَافٌ *Vociferation, clamour,* or *a confused noise,* of a company of men. (JK, Sgh, K.)

خُوَيْفٌ: see خَافٌ.

خُوَيْفَةٌ: see خَافَةٌ.

خُوَّافٌ: see خَافٌ. — [Hence, perhaps,] *A certain black bird:* ISd says, I know not why it is thus called. (TA.)

خَائِفٌ *Fearing; being afraid* or *frightened* or *terrified:* (S,* TA:) pl. خُوَّفٌ (S, K) and خُيَّفٌ, (S,) or خِيَفٌ, (K,) or, accord. to Ks, خُيَّفٌ and خَيْفٌ and خُوفٌ, (L,) [but the second and third of these three should be خُيَّفٌ and خُوَّفٌ, for all are said to be of the measure فُعَّلٌ,] and ▼ خَوْفٌ; or this last is a quasi-pl. n.; (K;) whence, in the Kur [vii. 54], خَوْفًا وَطَمَعًا, meaning Worship ye Him *fearing* his punishment *and eagerly desiring* his recompense. (TA.) See also خَافٌ. — And see مَخُوفٌ.

طَرِيقٌ مُخَافٌ [for مُخَافٌ أَهْلُهُ, *A road of which the people,* or *passengers, are caused to fear,* by robbers]. (Msb.) [See also what next follows.])

طَرِيقٌ مَخُوفٌ *A road in which people fear:* (S,* Msb, K:) or *a road that is feared;* (JK, TA;) as also ▼ مَخِيفٌ, and ▼ خَائِفٌ; which last is tropical, of the measure فَاعِلٌ in the sense of the measure مَفْعُولٌ; (TA;) or, thus applied, this last [is a possessive epithet, and thus] means *having fear:* (JK: [see also مُخَافٌ:]) you should not say ▼ طَرِيقٌ مُخِيفٌ, because the road does not cause fear, but only he who robs and slays therein. (S,* K,* TA.) One says also ثَغْرٌ ▼ مُخِيفٌ and ▼ مُتَخَوِّفٌ *An enemies' frontier* [*that is feared,* or] *from which one fears,* or *from the direction of which fear comes.* (TA.) مَخُوفٌ signifies A thing [of any kind] that is *feared;* as a lion, and a serpent, and fire, and the like. (Har p. 360.) [Hence,] حَائِطٌ مَخُوفٌ *A wall of which the falling is feared.* (Lh, Msb, TA. [See also مُخِيف.]) And وَجَعٌ مَخُوفٌ [*A pain that is feared*]. (TA. [See, again, مُخِيف.]) And أَمْرٌ مَخُوفٌ [*An affair,* or *event, that is feared*]. (Mgh, Msb. [See, again, مُخِيف.]) And فَاسِقٌ مَخُوفٌ عَلَى مَالِهِ *A transgressor who is feared for his property, that he will consume it, and expend it in that which is not right.* (Mgh.)

مَخِيفٌ: see the next preceding paragraph, in two places: and see also what next follows.

حَائِطٌ مُخِيفٌ (Msb, K, in the CK ▼ مَخِيفٌ,) *A wall that causes one to fear that it will fall.* (Msb, K.* [See also مَخُوف.]) And وَجَعٌ مُخِيفٌ (S, K) *A pain that causes him who sees it to fear.* (S. [See, again, مَخُوف.]) And أَمْرٌ مُخِيفٌ *An affair,* or *event, that is formidable; that causes him who sees it to fear.* (Msb. [See, again, مَخُوف.]) And المُخِيفُ means *The lion,* (K, TA,) *that frightens him who sees him.* (TA.) See also مَخُوفٌ, first sentence.

أَخْوَفُ [*More,* and *most, formidable, fearful,* or *feared:* anomalous, like its syn. أَخْشَى, being from the pass. verb. Hence,] أَخْوَفُ مَا أَخَافُ عَلَيْكُمْ كَذَا [*The most formidable,* or *fearful, of what I fear for you is such a thing*]. (Mgh,* TA.)

مَخَافَةٌ an inf. n. of 1, (S, Msb, K, &c.,) originally مَخْوَفَةٌ. (TA.) — [Also *A cause of fear:* a word of the same category as مَجْبَنَةٌ and مَبْخَلَةٌ &c.: pl. مَخَاوِفُ. Hence,] أَوَّلُ كُتُبِهِ المَخَاوِفُ [*The first of his letters,* or *epistles, consisted of the causes of fear*]. (TA.) — And مَخَاوِفُ also signifies *Places of fear.* (KL.)

مُتَخَوَّفٌ: see مَخُوفٌ.

خوق

1. خُقْ خُقْ [imperative of خَاقَ] *Ornament thy young woman,* or *female slave, with the earring.* (K. [See خُوق.]) == خَاقَهَا *He* (a man, TA) *made with her* (a woman, TA) *the sound termed* خَاقِ بَاقِ [*during the act of* نِكَاح]. (K.) == خاق الشَّىْءَ *He took away the thing,* and *extirpated it,* or *removed it utterly.* (TA.)

2. خَوَّقَهُ, (K,) inf. n. تَخْوِيقٌ, (TA,) *He made it wide;* (K;) namely, an earring. (TA.)

4. اخاق *He* (a man, TA) *went away into,* or *in, the country,* or *land.* (Sgh, K.)

5. تخوّق *It* (an earring, TA) *was,* or *became, wide,* or *widened.* (K.) [See also 7.] — تخوّق عَنْهُ *He went, retired,* or *withdrew himself, far away from him,* or *it.* (K,* TA.) And *He left, quitted,* or *forsook, it;* namely, a course that he desired, or meant, to pursue. (TA.)

7. انخاقت المَفَازَةُ *The desert was wide within.* (JK, K,* TA.) [See also 5.]

خَاقٌ The *length* of a desert. (JK, TA.) == خَاقِ بَاقِ, (S, K,) or الخَاقِ بَاقِ, (IB, TA,) or the latter also, without ال, (K, accord. to the TA, [but not in the CK nor in my MS. copy of the K,) *The sound of the* ذَكَر *in the flesh of the interior of the* فَرْج; (IAar, K,* TA;) or *the sound of the* فرج *on the occasion of the act of* نِكَاح. (IB, TA.) — And hence, (IB, TA,) or because of its خَوْق, i. e. width, (S, K,*) *The* فرج [or *vulva, itself*]. (S, IB, K.) الخَاقِ بَاقِ is indecl., with kesr for its termination, (S,) like الخَازِ بَازِ. (S, K.)

خُوقٌ [in the L خُوق, which is evidently a mistranscription; and in one copy of the S, in one place, written خُوقَةٌ;] *A ring* (S, L, TA) *of gold* and *of silver:* (L, TA:) or, accord. to Th, *a ring* [*that is worn*] *in the ear:* he does not say of gold nor of silver: (TA:) or the *ring of the* [*kind of earring called*] قُرْط and *of the* [*kind called*] شَنْف. (Lth, K.)

خَوْقٌ *Width* (S, K) of a desert, and of a well, and of a vulva: (S:) or, of a desert, *width of the interior:* (JK:) or *length,* and *breadth of expanse,* and *width of the interior:* and of a well, *depth and width.* (TA.) == And The *mange,* or *scab,* in camels: (El-Umawee, S, K:) or [*a disease*] *like the mange* or *scab.* (TA.)

أَخْوَقُ [*Wide;* or *wide in the interior;* or *far-extending:* fem. خَوْقَاءُ: pl. خُوقٌ]. You say

خَرْقٌ أَخْوَقُ (JK, Ḳ, TA, [in the CḲ جَوْفٌ, and in my MS. copy of the Ḳ خَوْقٌ,]) *A wide* [*desert such as is termed*] خَرْق: (Ḳ:) or *a* خَرق *wide in the interior.* (JK.) And مَفَازَةٌ خَوْقَآءُ (JK, Ṣ, Ḳ) and ▼ مُنْخَاقَةٌ (JK, Ḳ) *A wide desert:* (Ṣ, Ḳ:) or *a desert wide in the interior:* (JK:) and the former, also, *a desert in which is no water.* (TA.) And بَلَدٌ أَخْوَقُ *A wide, far-extending region* or *country.* (TA.) And بِئْرٌ خَوْقَآءُ (Ṣ, Ḳ) and ▼ مُنْخَاقَةٌ (Ḳ) *A wide well:* (Ṣ, Ḳ:) or رَكِيَّةٌ خَوْقَآءُ *a deep and wide well.* (JK, TA.) — And the fem., خَوْقَآءُ, applied to a woman, *Having no partition between her vulva and her anus:* or *having her vagina and rectum united:* or *wide in the vulva:* (TA:) or *tall and slender.* (JK, TA.) — And, so applied, *Foolish,* or *stupid:* (ISh, JK, Ḳ:) pl. خُوقٌ. (ISh, Ḳ.) — And the masc., (applied to a man, JK,) *Blind of one eye;* or *one-eyed:* (JK, Ḳ:) pl. as above. (JK.) ═ Also, applied to a camel, *Mangy,* or *scabby:* (Ṣ, Ḳ:) or *having what resembles the mange* or *scab:* (TA:) fem. as above. (Ṣ, Ḳ.)

مُخَوَّقٌ An earring *having a large* خَوْق [or *ring*]. (IAạr.)

بِئْرٌ مُنْخَاقَةٌ and مَفَازَةٌ مُنْخَاقَةٌ: see أَخْوَقُ.

خول

1. خَالَ, aor. يَخُولُ, inf. n. خَوْلٌ, *He became possessed of* خَوَل [so I read, meaning *slaves,* or *servants, and other dependents,* in the place of خوال, an evident mistranscription, in the TA,] *after having been alone.* (TA.) — فُلَانٌ يَخُولُ عَلَى أَهْلِهِ *Such a one pastures for his family:* (Ṣ:) or يَخُولُ عَلَيْهِمْ signifies *he milks and waters and pastures for them.* (T, TA.) And خَالَ عَلَيْهِمْ *He ruled,* or *governed, them.* (JK.) And خَالَ مَالَهُ, (Ḳ,* TA,) aor. as above, (TA,) inf. n. خَوْلٌ and خِيَالٌ, (Ḳ,) *He pastured his cattle,* or *camels &c., and managed them,* or *tended them, and sustained them,* (Ḳ,* TA,) *well:* (Ḳ:) or خُلْتُ المَالَ, aor. أَخُولُ, *I managed the cattle,* &c., *well:* (Ṣ:) and خَالَ عَلَى المَالِ, aor. يَخُولُ, *he pastured the cattle, &c., and managed them well;* as also خَالَ, aor. يَخِيلُ. (TA in art. خيل.) ═ خَالَ, aor. يَخُولُ and يَخَالُ or يَخِيلُ, see اختال (with which it is syn.) in art. خيل.

2. خَوَّلَهُ ٱللّٰهُ ٱلشَّىْءَ, (JK, Ṣ,) or مَالًا, (Mṣb,) or المَالَ, (Ḳ,) inf. n. تَخْوِيلٌ, (Ṣ,) *God made him to possess,* (JK, Ṣ,) or *gave him,* (Mṣb, Ḳ,) or *conferred upon him, as a favour,* (Ḳ,) *the thing,* (JK, Ṣ,) or *property,* (Mṣb,) or *the property.* (Ḳ.) So in the Ḳur vi. 94 and xxxix. 11 [and 50]. (TA.)

4. أَخْوَلَ (JK, Mṣb, Ḳ) and أُخْوِلَ (Ḳ) *He* (a man, JK, Mṣb) *had maternal uncles:* (JK, Ḳ:) or *he had many maternal uncles:* (Mṣb:) [both signify the same accord. to the Ḳ: but the latter properly signifies *he was made to have maternal uncles,* or *many maternal uncles:* see مُخْوَلٌ.] اخال فِيهِ خَالًا ▼ مِنَ الخَيْرِ *He perceived,* or *discovered, in him an indication,* or *a symptom, sign, mark,* or *token, of good;* as also ▼ تَخَوَّلَ (JK, Ṣ, Ḳ) and تَخَيَّلَ. (Ḳ.) [See also 2 in art. خيل.] ═ See also 10, in two places.

5: see 4: ═ and see also 10, in three places. ═ تَخَوَّلَهُ also signifies *He paid frequent attention,* or *returned time after time,* (JK, Ṣ, Ḳ,) *to it,* (JK,) or *to him;* syn. تَعَهَّدَهُ. (JK, Ṣ, Ḳ.) You say, تَخَوَّلْتُهُمْ بِالمَوْعِظَةِ *I paid frequent attention to them with exhorting,* or *admonishing;* syn. تَعَهَّدْتُهُمْ. (Mṣb.) It is said in a trad., of the Prophet, كَانَ يَتَخَوَّلُنَا بِالمَوْعِظَةِ مَخَافَةَ السَّآمَةِ [*He used to pay frequent attention to us with exhorting,* or *admonishing, for fear of loathing* on our part, or *disgust*]; (Ṣ;) or يَتَخَوَّلُهُمْ, i. e. يَتَعَهَّدُهُمْ: (TA:) Aṣ used to say يَتَخَوَّنُنَا, i. e. يَتَعَهَّدُنَا; (Ṣ;) or يَتَخَوَّنُهُمْ: and some read يَتَحَوَّلُهُمْ, with the unpointed ح, explained in art. حول. (TA.) And sometimes they said, تَخَوَّلَتِ الرِّيحُ الأَرْضَ, i. e. تَعَهَّدَتْهَا [app. meaning *The wind returned to the land time after time*]. (Ṣ.)

10. اِسْتَخْوَلَهُمْ *He took them as* خَوَل, (Ḳ, TA,) i. e. *slaves,* or *servants, and other dependents.* (TA.) ═ استخول فِيهِمْ and استخال *He took,* or *adopted, them as maternal uncles:* and تَخَوَّلَ ▼ خَالًا *he took,* or *adopted, a maternal uncle;* (Ḳ;) like as one says, تَعَمَّمَ عَمًّا: and ▼ تَخَوَّلَتْهُ *She called him her maternal uncle.* (TA.) You say, اِسْتَخْوِلْ خَالًا غَيْرَ خَالِكَ and ▼ تَخَوَّلْ (JK, Ṣ) and اِسْتَخِلْ (JK) *Adopt thou a maternal uncle other than thy* [proper] *maternal uncle.* (JK,* Ṣ.) ═ الاِسْتِخْوَالُ is also like الاِسْتِخْبَالُ [as meaning *The asking* one *to lend* cattle, or camels &c.: and ▼ الإِخْوَالُ is like الإِخْبَالُ as meaning *The lending* cattle, or camels &c.]: and AO used to recite thus the saying of Zuheyr:

* هُنَالِكَ إِنْ يُسْتَخْوَلُوا المَالَ يُخْوِلُوا ▼ *

[*There, if they be asked to lend cattle, they lend*]. (Ṣ, TA. [See also 10 in art. خبل.])

خَالٌ *A maternal uncle;* one's *mother's brother:* (JK, Ṣ, Ḳ:) pl. أَخْوَالٌ (Ṣ, Mṣb, Ḳ) and أَخْوِلَةٌ, (Ḳ,) [both pls. of pauc.,] the latter anomalous, (TA,) and (of mult., TA) خُؤُولٌ and خُؤُولٌ (Ḳ) and خُؤُولَةٌ: (Mṣb, Ḳ:) the fem. is خَالَةٌ, (JK, Ṣ, Ḳ,) *a maternal aunt;* one's *mother's sister:* (JK, Ṣ:) and the pl. of this is خَالَاتٌ. (Mṣb.) One says, هُمَا ٱبْنَا خَالَةٍ [meaning *Each of them two is a son of a maternal aunt of the other*]; but one cannot say, ٱبْنَا عَمَّةٍ: (Ḳ:) and in like manner one says, ٱبْنَا عَمٍّ; but one cannot say, ٱبْنَا خَالٍ. (TA.) ═ *An owner* of a horse: you say, أَنَا خَالُ هٰذَا الفَرَسِ *I am the owner of this horse.* (Ḳ.) [See also خَالٌ in art. خيل.] — خَالُ مَالٍ and ▼ خَائِلُ مَالٍ *He is a manager,* or *tender, of cattle,* or *camels &c.;* (Ḳ;) or *a good manager* or *tender thereof;* (Ṣ, Ḳ;*) and so ▼ خَوَلِيُّ مَالٍ: (Ṣ:) ▼ خَائِلٌ signifies also *a keeper,* or *guardian,* of a thing; (T, Ṣ;) or *a pastor;* (Fr, TA;) *a people's pastor, who milks and waters and pastures for them;* and *one who pays frequent attention to a thing, puts it into a good* or *right state,* or *restores it to such a state, and undertakes the management of it:* (T, TA:) خُوَّلٌ [is a pl. of خَائِلٌ, like as نُوَّمٌ is of نَائِمٌ, &c., and] signifies *pastors who take care of cattle,* or *camels &c.:* (TA:) and ▼ خَوَلِيٌّ, (Ḳ,) or, accord. to the M, ▼ خُوَلِيٌّ, (TA,) signifies *a pastor who is a good manager of cattle,* or *camels, and sheep* or *goats;* (M, Ḳ,* TA;) or *a good manager and orderer of the affairs of men;* (TA;) and its pl. [or quasi-pl. n. or n. un.] is ▼ خَوَلٌ; (M, Ḳ;) accord. to the M, like as عَرَبٌ is of عَرَبِيٌّ. (TA.) [See also خَالٌ in art. خيل.] ═ *An indication,* or *a symptom, sign, mark,* or *token,* of good (Ṣ,* Ḳ, TA) in a person. (Ṣ, TA.) See 4. ═ *A mole;* i. e. [*a thing resembling*] *a pimple in the face, inclining to blackness:* dim. ▼ خُوَيْلٌ and خُيَيْلٌ: and pl. خِيلَانٌ. (JK. [See also art. خيل.]) ═ The [*kind of banner called*] لِوَآء, *of an army* or *a military force.* (Ṣ, Ḳ. [See also art. خيل.]) — *A kind of soft garment,* or *cloth, of the fabric of El-Yemen:* (JK:) *a kind of* بُرْد, (Ṣ, Ḳ,) *well known,* (Ḳ,) *having a red* [or *brown*] *ground, with black lines* or *stripes.* (TA. [Mentioned also in art. خيل.]) ═ *A black stallion-camel.* (IAạr, Ḳ. [See also art. خيل.])

خَوَلٌ A man's *slaves,* or *servants, and other dependents:* (Ṣ, Mṣb, TA:) or *slaves, and cattle,* or *camels &c.:* (JK:) or the *cattle, camels &c.,* [in the CḲ, النَّعَم is erroneously put for النِّعَم,] *and male and female slaves, and other dependents, given to one by God:* (Ḳ:) said to be (Ṣ) from 2 [q. v.]: (JK, Ṣ, TA:) it is said to be a quasi-pl. n.; (TA;) and the sing. is ▼ خَائِلٌ; (Ṣ, Ḳ, TA;) though used as sing. and pl., and masc. and fem.: (Ḳ:) sometimes used as a sing. applied to *a male slave* and *a female slave:* but Fr says that it is pl. [or quasi-pl. n.] of ▼ خَائِلٌ meaning *a pastor.* (Ṣ.) You say هٰؤُلَآءِ خَوَلُ فُلَانٍ, meaning *These are persons who have been subjected, and taken as slaves, by such a one.* (TA.) — See also خَالٌ. — Also *A gift,* or *gifts:* [and this seems to be the primary signification; whence "a slave" &c., and "slaves" &c., as being given by God:] so in the phrase, هُوَ كَثِيرُ الخَوَلِ [*He is a person of many gifts*]. (TA.) ═ Accord. to Lth, (TA,) it signifies also The *lower part* (أَصْل) *of the* فَأْس [q. v.] *of a bit:* (JK, Ḳ, TA:) but Az says, "I know not the خول of the bit nor what it is. (TA.) [See خَالٌ, last sentence but one, in art. خيل.]

خَوْلَةٌ *A female gazelle.* (IAạr, Ḳ.)

خَوَلِيٌّ: see خَالٌ, in two places. — Also *A measurer of land with the measuring-cane.* (TA.)

خُوَلِيٌّ: see خَالٌ.

خُوَيْلٌ: see خَالٌ, of which it is the dim.

خُؤُولَةٌ The *relationship of a maternal uncle* [and *of a maternal aunt*]: (JK, Ṣ, Ḳ, TA:) an inf. n. (JK, TA) having no verb. (TA.) You say, بَيْنِى وَبَيْنَهُ خُؤُولَةٌ [*Between me and him is a relationship of maternal uncle*]. (Ṣ, Ḳ.) —— Also a pl. of خَالٌ in the first of the senses assigned to the latter above. (Mṣb, Ḳ.)

خَوَّالٌ *A giver of many gifts.* (TA.)

خَائِلٌ : see خَالٌ, in two places: — and خَوَلٌ, also in two places.

تَطَايَرَ الشَّرَرُ أَخْوَلَ أَخْوَلَ *The sparks flew about scattered;* meaning the sparks that fly about from hot iron when it is beaten; as in a verse of Ḍábi [El-Burjumee] cited in art. سقط : see 3 in that art. (Ṣ.) And ذَهَبُوا أَخْوَلَ أَخْوَلَ *They went away scattered,* (JK, Ṣ, Ḳ,) *one after another, like as sparks are scattered from iron:* or, as some say, الأَخْوَلُ itself means *sparks:* (JḲ:) [but here,] اخول اخول are two nouns made into one, and indecl., with fet-ḥ for the termination: (Ṣ:) Sb says that they may be like شَغَرَ بَغَرَ, or like يَوْمَ يَوْمَ. (TA.) —— هُوَ أَخْوَلُ مِنْ فُلَانٍ *He is prouder than such a one.* (Suh, TA.) [See also أَخْيَلُ, in art. خيل.]

مُخْوَلٌ : see مُخَالٌ.

مَخُولٌ : see أَخْيَلُ, in art. خيل.

مُخْوَلٌ and ♦ مُخْوِلٌ A man *having maternal uncles:* (TA:) or the former signifies a man *made to have many maternal uncles;* and ♦ the latter, *having many maternal uncles:* (Mṣb:) and رَجُلٌ مُعِمٌّ مُخْوَلٌ (Mṣb, Ḳ) and ♦ مُخَالٌ, (JK, Ḳ,) and ♦ مُعِمٌّ مُخْوِلٌ, *A man who has generous paternal and maternal uncles:* (Mṣb, Ḳ:) but Aṣ disallows مُعِمٌّ and ♦ مُخْوِلٌ: (Mṣb:) and the latter word in each case is not used, (Ḳ,) or is scarcely ever used, (TA,) without the former. (Ḳ, TA.)

مُخْوِلٌ : see what next precedes, in four places.

مَخِيلٌ, إِنَّهُ لَمَخِيلٌ لِلْخَيْرِ, (Ḳ in this art.,) or مَخِيلٌ, (Ṣ in art. خيل,) *Verily he is adapted* or *disposed by nature to good* [i. e. *to be,* or *to do,* or *to effect, or to produce, what is good*]. (Ṣ, Ḳ.) [See also مَخِيلٌ in art. خيل.]

خوم

4. اخام: }
خَامٍ: } see art. خيم.
خَامَةٌ: }

خون

1. خَانَهُ, (Ṣ, Ḳ,) aor. يَخُونُهُ, (Ṣ,) inf. n. خِيَانَةٌ and خَوْنٌ and مَخَانَةٌ (Ṣ, Ḳ) and خَانَةٌ (Ḳ) and خَائِنَةٌ, of the measure فَاعِلَةٌ, like لَاغِيَةٌ &c.; (TA;) and ♦ اختانهُ; (Ṣ, Ḳ;) *He was unfaithful,* or *he acted unfaithfully, to the confidence,* or *trust, that he reposed in him;* (Ḳ;) [*he was treacherous, perfidious,* or *unfaithful, to him;* or *he acted treacherously, perfidiously,* or *unfaithfully, towards him;*] فِى كَذَا [*in such a thing*]: (Ṣ:) خِيَانَةٌ is the *contr. of* أَمَانَةٌ; and does not relate only to property, but also to other things: (Mgh:) or the *neglecting,* or *failing in,* أَمَانَة [which is *trustiness,* or *faithfulness*]: (El-Ḥarállee, TA:) or *i. q.* نِفَاقٌ, except that خيانة regards a compact or covenant or the like, and trustiness, or faithfulness, and نفاق regards religion; so that the former is the *acting contrary to what is right, by breaking a compact* or *covenant* or *the like:* (Er-Rághib, TA:) but [it is said that] the primary signification of خَوْنٌ is the *making to suffer loss,* or *diminution;* because the خَائِن makes the مَخُون to suffer loss, or diminution, of something. (TA.) Hence, in the Ḳur [ii. 183], كُنْتُمْ تَخْتَانُونَ ♦ أَنْفُسَكُمْ [lit. *Ye used to act unfaithfully to yourselves*] means *ye used to act unfaithfully, one to another:* (Ṣ,* TA:) or *ye used to act wrongfully to yourselves:* اِخْتِيَانٌ has a more intensive signification than خِيَانَةٌ. (Bḍ.) One says also, خَانَ العَهْدَ *He broke the compact* or *covenant* or *the like:* whence, تَقُولُ النِّعْمَةُ كُفِرْتُ وَلَمْ أُشْكَرْ وَتَقُولُ الأَمَانَةُ خُنْتُ وَلَمْ أُحْفَظْ [*The benefit says, I have been disacknowledged, and have not been requited with thankfulness; and the trust says, I have been betrayed, and have not been faithfully kept*]: the verb [خُنْتُ] being here of the measure فُعِلْتُ, a verb of which the agent is not named. (Mgh.) And خَانَهُ العَهْدَ, (Mṣb, Ḳ,) and فِى العَهْدِ, (Mṣb,) and خَانَهُ الأَمَانَةَ, (Mṣb, Ḳ,) aor. as above, inf. n. خَوْنٌ and خِيَانَةٌ and مَخَانَةٌ, (Mṣb,) [*He was unfaithful to him in respect of the compact* or *covenant* or *the like, and the trust.*] — [Hence,] خَانَ سَيْفُهُ †[*His sword was unfaithful;*] i. e., *failed of taking effect upon the thing struck with it.* (TA.) A certain person, being asked respecting the sword, said, أَخُوكَ وَرُبَّمَا خَانَكَ †[It is *thy brother, but sometimes it is unfaithful to thee*]. (TA.) — And خَانَتْهُ رِجْلَاهُ †[*His two legs were unfaithful to him;*] *he was unable to walk.* (TA.) — And خَانَ الدَّلْوَ الرِّشَآءُ †*The well-rope broke off,* or *became severed, from the bucket.* (TA.) — And خَانَهُ الدَّهْرُ, inf. n. خَوْنٌ; (T, TA;) and ♦ تَخَوَّنَهُ; (TA;) †*Time altered his state,* or *condition,* (T, TA,) *from softness,* or *easiness, to hardness,* or *difficulty,* (TA,) or *to evil;* (T, TA;) and in like manner, النَّعِيمُ [*enjoyment,* &c.]: and of everything that has altered thy state, or condition, [for the worse,] one says, ♦ تَخَوَّنَكَ. (T, TA.)

2. خَوَّنَهُ, (Ṣ, Ḳ,) inf. n. تَخْوِينٌ, (Ḳ,) *He attributed to him* خِيَانَة [i. e. *treachery, perfidy,* or *unfaithfulness*]. (Ṣ, Ḳ.) — See also 5, in two places.

5. تخوّنهُ : see 1, last sentence, in two places. You say also, تَخَوَّنَهُمْ meaning *He sought* [*to discover,* or *show,*] *their* خِيَانَة [i. e. *treachery, perfidy,* or *unfaithfulness*], and *their slip, lapse,* or *wrong action;* and *suspected them,* or *accused them.* (TA.) — Also *He,* or *it, diminished it, wasted it, impaired it,* or *took from it;* and so ♦ خَوَّنَهُ, and خَوَّنَ مِنْهُ: (Ḳ:) or *diminished it, wasted it, impaired it,* or *took from it, by little and little;* syn. تَنَقَّصَهُ. (JK,* Ṣ, Mṣb.) You say, تَخَوَّنَنِى فُلَانٌ حَقِّى *Such a one took from me by little and little of my right,* or *due.* (Ṣ, TA.) And Dhu-r-Rummeh says,

لَا بَلْ هُوَ الشَّوْقُ مِنْ دَارٍ تَخَوَّنَهَا
مَرًّا سَحَابٌ وَمَرًّا بَارِحٌ تَرِبُ

[*No, but it is,* or *was, yearning of the soul arising from a place of abode from which sometimes raining clouds, and sometimes a hot wind carrying with it dust, took away by little and little,* so as gradually to efface the traces thereof]. (Ṣ, TA.) And Lebeed says, (Ṣ, TA,) describing a she-camel, (TA,)

تَخَوَّنَهَا نُزُولِى وَٱرْتِحَالِى

[*Which my alighting and my journeying had wasted by little and little;*] i. e. *whose flesh and fat my alighting and my journeying had diminished by little and little.* (Ṣ, TA.) —— Also *He paid frequent attention to him,* or *it;* or *he,* or *it, returned to him,* or *it, time after time;* syn. تَعَهَّدَهُ; (JK, Ṣ, Ḳ;) and so ♦ خَوَّنَهُ: (Ḳ:) in this sense, the former verb is [said to be] from تخوّلهُ, by the substitution of ن for ل. (TA.) Dhu-r-Rummeh says, [describing a young gazelle,]

لَا يَنْعَشُ الطَّرْفَ إِلَّا مَا تَخَوَّنَهُ
دَاعٍ يُنَادِيهِ بِٱسْمِ المَآءِ مَبْغُومُ

(Ṣ,) [*He raises not his eye,* or *eyes, except when a caller calling him by the sound of* مَآءِ *returns to him time after time, addressed by the cry termed* بُغَام:] i. e. except when he hears the مَآءِ مَآءِ of his mother calling him by the cry بُغَام: (TA in art. بغم: [it is there added, that the pass. part. n. مَبْغُوم is used in this instance for the act. part. n.; but for this I see no sufficient reason:]) he says that the young gazelle is slumbering, not raising his eye, or eyes, unless his mother comes to him time after time: or, as some say, unless his mother's call to him takes by little and little from his sleep. (Ṣ in the present art.) One says also الحُمَّى تَخَوَّنُهُ [for تَتَخَوَّنُهُ] *The fever returns to him time after time:* (Ṣ:) or *in its time.* (TA.)

8: see 1, in two places.

خَانٌ *A place in which travellers lodge:* (Mṣb:) *a place in which travellers pass the night:* and the دَيْر [i. e. *monastery,* or *convent,*] is the خان of the Christians: (Kull pp. 96 and 97:) or the خان is *for merchants;* (Ṣ, Ḳ;) *i. q.* فُنْدُقٌ; (Ḥar p. 325;) [*a building for the reception of merchants and travellers and their goods, generally surrounding a square or an oblong court, having, on the ground-floor, vaulted magazines for merchandise, which face the court, and lodgings,* or *other magazines, above:* a Persian word, arabicized:] pl. خَانَاتٌ. (Mṣb.) — Also *A shop:* or *a shop-keeper:* (Ḳ:) a Persian word, arabicized. (TA.) —— [It is also a title of honour, used by the Tartars (who apply it to their Emperor), the

Turks (who apply it to the reigning and to a deceased Sultán), and the Persians (who apply it to the governor of a province, and to a man of rank).]

خَوْنٌ an inf. n. of 1. (Ṣ, Mṣb, Ḳ.) — And [hence,] ‡ *Weakness.* (JK, Ḳ, TA.) One says فِى ظَهْرِهِ خَوْنٌ ‡ *In his back is weakness.* (JK, TA.) — And † *Languidness in the sight.* (Ḳ.)

خَانِىٌّ *Of, or belonging to, a* خَان *of the merchants.* (TA.)

خِوَانٌ (JK, Ṣ, Mgh, Mṣb, Ḳ) and خُوَانٌ (ISk, Mṣb, Ḳ) and ▾ إِخْوَانٌ, (IF, Mṣb, Ḳ,) the first of which is the most common, (Mṣb,) *A table;* (JK;) *a thing upon which one eats;* (Ṣ, Mgh, Mṣb;) *a thing upon which food is eaten:* (Ḳ:) but said to be not so called except *when food is upon it:* (Ḥar p. 360:) arabicized [from the Persian]: (Ṣ, Mṣb:) the pl. (of pauc., of the first, Ṣ, Mṣb) is أَخْوِنَةٌ and (of mult., Ṣ, Mṣb) خُونٌ, (Ṣ, Mgh, Mṣb, Ḳ,) said by IB to be the only instance of its kind except بُونٌ pl. of بِوَانٌ, (TA,) originally خُوُنٌ, like كُتُبٌ pl. of كِتَابٌ, (Mṣb,) but خُوُنٌ is not used: (Ṣ:) the pl. of ▾ اخوان is أَخَاوِنُ, (Mṣb,) or أَخَاوِينُ. (TA, from a trad.)

خَؤُونٌ [for خَوُونٌ]: see خَائِنٌ.

خَوَّانٌ: see خَائِنٌ. — [Hence,] الخَوَّانُ *The lion:* (JK, Ṣ:) because he is [very] treacherous. (JK.) And † *Time, or fortune.* (TA.) — أَعُوذُ بِٱللّٰهِ مِنَ الخُوَّانِ means † [*I seek protection by God*] *from the day of the exhaustion of provisions.* (A, TA.) ═ Also, and ▾ خُوَّانٌ, [accord. to the CḲ, each is with ال, but this seems to be a mistake, (see شَهْرٌ,)] *The month* [*latterly called*] رَبِيعُ الأَوَّلُ: pl. أَخْوِنَةٌ: (Ḳ:) but ISd says, "I know not how this is." (TA.)

خُوَّانٌ: see what next precedes.

الخَوَّانَةُ *i. q.* الاِسْتُ [meaning † *The anus*]. (TA.)

خَائِنٌ and ▾ خَائِنَةٌ, (Ṣ, Mṣb, Ḳ,) the latter an intensive epithet, (Ṣ, Mṣb,) like عَلَّامَةٌ and نَسَّابَةٌ, (Ṣ,) [and also fem. of خَائِنٌ,] and ▾ خَؤُونٌ and ▾ خَوَّانٌ, (Ḳ,) [which are likewise intensive epithets,] *Unfaithful, or acting unfaithfully, to the confidence, or trust, reposed in him;* (Ḳ;) [*treacherous, perfidious, or unfaithful; or acting treacherously, perfidiously, or unfaithfully:* thus the first signifies: the others signifying *very unfaithful,* &c.:] pl. [of the first] خَوَنَةٌ, (Ṣ, M, Ḳ,) which is anomalous, (M,) like حَوَكَةٌ [pl. of حَائِكٌ], (Ṣ,) and خُوَّانٌ. (Ḳ.) [Hence,] خَائِنُ النَّظَرِ *Looking treacherously, and clandestinely, at a thing at which it is not allowable to look.* (TA.) — † خَائِنُ العَيْنِ † [*The languid in respect of the eye*] is an appellation applied to *the lion;* (Ḳ, TA;) because of a languidness in his eye when he looks. (TA.)

خَائِنَةٌ: see خَائِنٌ. ═ It is also an inf. n. of خَانَ. (TA.) [Hence,] خَائِنَةُ الأَعْيُنِ (as used in the Ḳur xl. 20, TA) *A surreptitious look* (JK, Mgh, Ḳ) *at a thing at which it is not allowable to look:* (JK, Ḳ:) or *the looking with a look that induces suspicion or evil opinion:* (Th, Ḳ:) or *the making a sign with the eye to indicate a thing that one conceals in the mind:* (TA:) or, as some say, *the contracting of the eye, or eyes, by way of making an obscure indication:* or *the looking intentionally* [*at a thing at which it is not allowable to look*]. (Mṣb.)

إِخْوَانٌ: see خِوَانٌ, in two places. ═ [It is also a pl. of أَخٌ: see art. اخو.]

مُتَخَوَّنٌ One *to whom* خِيَانَة [i. e. *treachery, perfidy,* or *unfaithfulness,*] *is attributed.* (TA.)

خوى

1. خَوَتِ الدَّارُ, (JK, Ṣ, Mṣb, Ḳ,) aor. تَخْوِى, (Mṣb,) inf. n. خَوَآءٌ (Ṣ, Mṣb, TA) and خُوِىٌّ, (Mṣb,) *The house became empty, vacant,* or *unoccupied:* (Ṣ:) in [some of the copies of] the Ṣ with teshdeed, [i. e. خَوَّت,] which is incorrect: (TA:) or *became devoid, or destitute, of its occupants;* as also خَوِيَت, aor. تَخْوَى, inf. n. خَوَآءٌ: (Mṣb:) or *its occupants perished:* (JK:) and *it fell down:* (Ṣ:) or *it became demolished:* and خَوَت and خَوِيَت, inf. n. خَىٌّ and خُوِىٌّ and خَوَآءٌ and خَوَايَةٌ, *it became devoid, or destitute, of its occupants,* (Ḳ, TA,) *standing, without inhabitant.* (TA.) And خَوَى المَكَانُ, aor. يَخْوِى, inf. n. خَىٌّ, *The place became empty, vacant, or unoccupied.* (Mgh.) And خَوِىَ البَطْنُ, aor. يَخْوَى, inf. n. خَوًى, *The belly became empty of food.* (Mgh.) — [Hence,] خَوِىَ, aor. يَخْوَى, inf. n. خَوًى [and خَوَآءٌ], *He was, or became, hungry;* (JK;) as also ▾ اخوى: (Ḳ:) or *his belly became empty of food:* (Ḥar p. 167:) or خَوِىَ, like رَمِىَ, [aor. يَخْوَى,] inf. n. خَوًى and خَوَآءٌ, *he was affected with uninterrupted hunger.* (Ḳ.) And خَوَتْ, said of a woman, *She became empty in her belly on the occasion of childbirth;* as also خَوِيَتْ: (Ṣ, Ḳ:) in [some of] the copies of the Ḳ, خَوَّتْ is here erroneously put for خَوِيَتْ. (TA.) And *She abstained from food on the occasion of childbirth;* (JK, Ḳ;) as also خَوِيَتْ. (JK, TA.) — خَوَتِ النُّجُومُ *The stars inclined to setting:* (so in two copies of the Ṣ:) or so ▾ خَوَّتْ, inf. n. تَخْوِيَةٌ. (JK, Mṣb, and so in some copies of the Ṣ.) And the former, (JK, Ṣ, Mṣb, Ḳ,) aor. تَخْوِى, (JK, Ṣ,) inf. n. خَىٌّ, (JK, Ṣ, Ḳ,) *The stars set;* as also ▾ اخوت: and *the stars brought no rain:* (JK:) or *the stars set aurorally and brought no rain;* (Ṣ, Mṣb,* Ḳ;*) as also ▾ اخوت (A'Obeyd, Ṣ, Mṣb, Ḳ) and ▾ خَوَّت. (Ḳ.) — خَوَى الزَّنْدُ, (Ḳ,) inf. n. خَوًى (TA) [or خُوِىٌّ ?], *The* زند [q. v.] *failed to produce fire;* as also ▾ اخوى. (Ḳ.) ═ خَوَى, aor. يَخْوِى, *He called,* or *cried, out.* (JK.) ═ خَوَى, (TA,) inf. n. خَىٌّ, (Ḳ, TA,) *i. q.* قَصَدَ [*He, or it, pursued a right,* or *direct, course;* &c.: if trans., it may mean *he aimed at, intended,* or *purposed,* a thing]. (Ḳ,* TA.) ═ خَوَاهُ, inf. n. خَوًى and خَوَايَةٌ, *He seized it; took it,* or *carried it off, by force;* or *snatched it away;* (Ḳ;) and so ▾ اختواهُ. (IAạr, TA.)

2. خَوَّتِ الإِبِلُ, inf. n. تَخْوِيَةٌ, *The camels became empty* (JK, M, Mṣb) *and drawn up* (M) *in their bellies.* (JK, M, Mṣb.) — See also 4. — And خَوَّى, inf. n. as above, said of a camel, (JK, Ṣ, TA,) *He lay down upon his breast, and then set firmly upon the ground his* [*callous protuberances called the*] ثَفِنَات: (JK:) or *he made his belly to be separated by some interval from the ground, in lying upon his breast,* (Ṣ, TA,*) *and set firmly* [*upon the ground*] *his* ثَفِنَات. (TA.) And in like manner, (Ṣ,) خَوَّى فِى سُجُودِهِ, (Ṣ, Mgh, Mṣb, Ḳ,) said of a man, (Ṣ, Mgh, Mṣb,) *He raised his belly from the ground in his prostration:* (Ṣ,* Mṣb:) or *he put, or set, his upper arms apart,* or *remote, from his sides therein:* (Mgh, Mṣb:) or *he drew up his body, and made a space between his upper arms and his sides, in his prostration:* (Ḳ:) thus a man is directed to do in prostrating himself in prayer. (Mgh, TA.) — Said of a man, it signifies also *He lowered his eyes,* or *looked towards the ground, desiring to be silent.* (JK.) — Said of a bird, *It hung down its wings:* (Ṣ:) or *it spread its wings,* (JK, TA,) *and stretched out its legs,* (TA,) *desiring to alight.* (JK, TA.) — خَوَّتِ النُّجُومُ: see 1, in two places. ═ خَوَّيْتُهَا, inf. n. as above, *I dug* (Ḳ, TA) *for her,* namely, a woman, (TA,) *a hole, or hollow, in the ground, and kindled* [*fire*] *in it, and then seated her in it, or upon it,* (فيها [i. e. فِى الحَفِيرَةِ],) *on account of a disease that she had.* (Ḳ, TA.) One says of a woman for whom this is done, خُوِّيَتْ. (Aṣ, TA.) ═ And خَوَّى لَهَا, (JK, Kr, Ṣ, Ḳ,) and خَوَّاهَا, (Ḳ,) inf. n. as above, *He made for her* (namely, a woman,) *the food called* خُوَيَّةٌ, (JK, Kr, Ṣ, Ḳ,) *that she might eat it.* (Ṣ.)

4: see 1, in four places. — اخوى المَالُ *The cattle, or camels &c., attained the utmost degree of fatness;* as also ▾ خَوَّى, inf. n. تَخْوِيَةٌ. (Fr, Ḳ.) ═ See also 8.

8. اختوى *He abstained from food:* and it may mean *he became void of everything but anger.* (Ḥam p. 219.) — *He lost his reason, or intellect.* (Ḳ.) ═ See also 1, last sentence. — Also *He took away* a thing. (JK.) *He took altogether, or entirely,* what another possessed; as also ▾ اخوى. (JK, Ḳ.) — *He* (a beast of prey) *stole and ate* the young one of a cow. (IAạr, Ḳ.) — *He cut off for himself* (اِقْتَطَعَ) a land, or district; (IAạr, Ḳ;) as also اختات and تخوّت. (IAạr, TA.) — *He thrust at a horse in his* خَوَآء; i. e., *the space between his fore legs and his hind legs.* (JK, Ḳ.)

خَوًى *Emptiness of the belly;* (JK, Ḳ;) i. e. *its emptiness of food;* as also ▾ خَوَآءٌ; (Ḳ;) [both inf. ns.;] the former of higher authority than the latter. (TA.) And *Hunger;* (JK, Mṣb;) as also ▾ خَوَآءٌ, (TA,) *i. q.* خَوٌّ. (Ḳ, TA. [In the CḲ, والخُوُّ بالضَّبِّرِ العَسَلُ is erroneously put for والخَوُّ وَبِالضَّمِّ العَسَلُ. The word خَوٌّ belongs to art. خو, q. v.]) — See also خَوَآءٌ. — Also, [الخَوِىُّ in the CḲ being a mistake for الخَوَى,]

A low, or depressed, tract between two mountains: and *a soft tract of land:* (Ḳ, TA:) or *a low,* or *depressed, tract, in plain,* or *soft,* and *in rugged and hard, ground, sinking into the earth, larger than the* [*tract termed*] شِبْث, *producing much herbage:* (AḤn, TA:) or *any wide valley in a soft,* or *plain,* [*low ground such as is termed*] جَوّ; (Az, TA;) as also خَوٌّ: (Az, TA in art. خو:) or *a soft, far-extending, valley.* (Aṣ, TA.) — Also *i. q.* ثَابِتٌ [*Continuing, subsisting, lasting,* &c.]: (Ḳ:) of the dial. of Ṭeiyi. (TA.) — And *A flow of blood from the nose;* or *blood flowing from the nose.* (Ḳ. [In this instance the word is correctly given in the CḲ.])

خَوَاةٌ *The space between the udder and the vulva* in the she-camel and other cattle; (Ḳ,* TA; [accord. to the CḲ and JK, † خُوَيَّةٌ; but this is app. a mistake;]) also with medd [i. e. † خَوَايَةٌ, for خَوَاةٌ is originally خَوَيَةٌ]. (Ḳ.) — Also *A sound:* (A 'Obeyd, Ṣ, TA:) and *the confused and continued sound* (حَفِيف) *of pouring* of rain: (IAạr, TA:) and † خَوَايَةٌ signifies the *confused and continued sound* (حَفِيف [in the CḲ, erroneously, خَفِيف,]) *of the running* of horses: (Ḳ, TA:) and *a sound like what one fancies.* (Aboo-Málik, TA.)

خَوَآءٌ: see خَوًى, in two places. — Also *An intervening space between two things.* (JK, Mgh,* TA.) *The space between the fore legs and the hind legs* of a horse; (JK, Ḳ;) as also † خَوًى. (JK.) *A vacant space between two things;* (Ḳ;) *such between the heaven and the earth;* (TA;) like هَوَآءٌ. (Ḳ, TA.) — *A wide,* or *spacious, open tract* of the earth, *containing no herbage nor trees nor habitations.* (TA.)

خُوَآءٌ (like غُرَابٌ [in measure], TA) *Honey.* (Ez-Zejjájee, Ḳ, TA. [By a mistake in the CḲ, mentioned above, voce خَوًى, the word thus explained is there made to be خَوٌّ.])

خَوِيٌّ, of the measure فَعِيلٌ, *A low,* or *depressed, soft,* or *plain, tract of land.* (Ṣ, TA.)

خَوَايَةٌ: see خَوَاةٌ, first sentence. — Also The *part that a horse closes with his tail, of the space between his hind legs.* (TA.) — The *part of a* spear-head *into which the shaft enters.* (Ḳ,* TA.) — And The *wide part of the interior* of a رَحْل [or camel's saddle]. (Ḳ, TA. [In the CḲ, مِنَ الرَّجُلِ is erroneously put for مِنَ الرَّحْلِ.]) — See also خَوَاةٌ, second sentence.

خُوَيَّةٌ *Food prepared for a woman on the occasion of childbirth.* (Ṣ, Ḳ.) — See also خَوَاةٌ.

[خَاوٍ part. n. of 1.] فَتِلْكَ بُيُوتُهُمْ خَاوِيَةً, in the Ḳur [xxvii. 53], means [*And those are their houses,*] *empty;* or, as some say, *fallen down:* like the phrase in the same [ii. 261 and xxii. 44], خَاوِيَةٌ عَلَى عُرُوشِهَا *having fallen down upon its roofs:* (Ṣ:) or this means empty; its walls having fallen upon its roofs. (Bd in ii. 261. [See also عَرْشٌ.]) You say also أَرْضٌ خَاوِيَةٌ *A land devoid of its inhabitants:* (Ḳ:) and sometimes it means, *of rain.* (TA.) كَأَنَّهُمْ أَعْجَازُ نَخْلٍ خَاوِيَةٍ, in the Ḳur [lxix. 7], means [*As though they were trunks of palm-trees*] *torn up:* (TA:) or *eaten within:* (Bd:) or *fallen down and empty.* (Jel.)

خَاوِيَةٌ [fem. of خَاوٍ: and hence, as a subst.,] *A calamity,* or *misfortune.* (Kr, TA.)

خَاوِىٌّ: see خَائِيٌّ, in art. خوأ.

مُخَوًّى The *place of* a camel's *lying down in the manner described above voce* خَوَّى: [and so app. مُخَوَّاةٌ; for] the pl. is مُخَوَّيَاتٌ. (JK.)

خى

2. خَيَّيْتُ خَاءً *I wrote a* خ. (JK, TA.)

[خُيَيَّةٌ dim. of خَاءٌ, q. v. in art. خوأ.]

خيب

1. خَابَ, (Ṣ, A, Mṣb, Ḳ, &c.,) aor. يَخِيبُ (Mṣb, Ḳ) and يَخُوبُ, (TA,) inf. n. خَيْبَةٌ, (Ṣ, Mṣb, Ḳ,) *He* (a man, Ṣ) *was disappointed of attaining what he desired or sought; was balked; was unsuccessful; failed of attaining his desire:* (Ṣ, Mṣb, Ḳ:) *he was denied, refused, prohibited from attaining,* or *debarred from, what he desired* or *sought.* (A, Ḳ.) You say, مَنْ هَابَ خَابَ [*He who fears will be disappointed*]. (A, TA.) And خَابَ سَعْيُهُ وَأَمَلُهُ [*His labour, and his hope,* or *expectation, resulted in disappointment; were disappointed, balked,* or *frustrated;*] *he attained not what he sought* or *desired.* (A, TA.) [See also خَيْبَةٌ, below.] — Also *He suffered loss.* (Ḳ.) — And *i. q.* كَفَرَ [*He disbelieved;* or *became an unbeliever,* or *infidel;* &c.]. (Ḳ.)

2. خَيَّبَهُ, (Ṣ, A, Mṣb, Ḳ,) inf. n. تَخْيِيبٌ, (Ṣ,) *He* (God, A, Mṣb, Ḳ, or a man, Ṣ) *disappointed him;* or *caused him to be disappointed of attaining what he desired or sought, to be balked, to be unsuccessful,* or *to fail of attaining his desire:* (Ṣ, Mṣb:) *he denied him, refused him, prohibited him from attaining,* or *debarred him from, that which he desired* or *sought.* (A, Ḳ.)

خَيْبَةٌ inf. n. of 1 [q. v.]. (Ṣ, Mṣb, Ḳ.) It is said in a prov., (Ṣ, Mṣb, Ḳ,) الهَيْبَةُ خَيْبَةٌ [*Fear is a cause of disappointment*]. (Ṣ, A, Mṣb, Ḳ.) And one says, خَيْبَةً لِزَيْدٍ [May God send *disappointment to Zeyd*], and خَيْبَةٌ لِزَيْدٍ [*Disappointment* be *to,* or befall, *Zeyd*]: (Ṣ, Ḳ:) خيبة in the former instance being in the accus. case as governed by a verb understood; and in the latter, in the nom. case as an inchoative: (Ṣ:) each being a form of imprecation. (Ḳ.)

قِدْحٌ خَيَّابٌ ‡ [*An endeavour to produce fire* with a زَنْد] *that does not produce fire* (لَا يُورِى). (A, Ḳ.* [In some copies of the Ḳ, for قِدْح, we find قَدْح. Golius explains خَيَّابٌ as meaning *ignitabulum fallens, quod non excudit semina ignis:* but I cannot anywhere find قِدْح or قَدْح in the sense which he assigns to one of these words, which is that of مِقْدَحَةٌ &c.]) — In the following verse, quoted by Th,

* أَسْكُتْ وَلَا تَنْطِقْ فَأَنْتَ خَيَّابْ *
* كُلُّكَ ذُو عَيْبٍ وَأَنْتَ عَيَّابْ *

خَيَّاب may be [an epithet] of the measure فَعَّال from الخَيْبَةُ; [so that the meaning may be † *Be thou silent, and speak not, for thou art habitually unsuccessful; thou art altogether vitious,* or *faulty, and thou art a great imputer of vices,* or *faults, to others;*] or the person there mentioned may be meant to be likened to the قدح above mentioned. (TA.) — One says also, سَعْيُهُ فِى خَيَّابِ بْنِ هَيَّابٍ ‡ *His labour* [has ended, or ends, or will end,] *in loss.* (A, Ḳ.)

خَائِبٌ part. n. of 1. (Mṣb, TA.) — خَائِبَكَ and بِخَائِبِكَ: see خَاءَ بِكَ in art. خوأ.

قِدْحٌ أَخْيَبُ † *An arrow of those employed in the game called* المَيْسِر, *to which no lot,* or *portion, pertains:* for there are three such arrows; namely, the مَنِيح, the سَفِيح, and the وَغْد: occurring in a trad. of 'Alee. (TA.)

وَقَعُوا فِى وَادِى تُخُيِّبَ, (Ks, Ṣ, A, Ḳ, but in the last وَقَعَ,) and تُخَيِّبَ, (A, Ḳ,) the last word being imperfectly decl. [in each of these instances], (Ks, Ṣ, A, Ḳ,) meaning فِى البَاطِلِ [i. e. *They fell into that which was vain, unreal, nought, futile,* or *the like,* and consequently, *into disappointment*], (Ks, Ṣ, Ḳ,) or فِى بَاطِلٍ [*into a state of things that was vain,* &c.]. (A.)

خيت

1. خَاتَ, aor. يَخِيتُ, (IAạr,) inf. n. خَيْتٌ and خُيُوتٌ, (IAạr, Ḳ,) and inf. n. of un. خَيْتَةٌ, (IAạr,) *He made a sound,* syn. صَوَّتَ; (IAạr, Ḳ;*) [app. *with his wings, in pouncing down,* or *making a stoop;* see 1 in art. خوت;] said of a bird. (IAạr.) — خَاتَهُ مَالَهُ, aor. يَخِيتُهُ: see 1 in art. خوت.

خير

1. خَارَ, aor. يَخِيرُ, (Ḳ,) inf. n. خَيْرٌ, (TA,) *He* (a man, TA) *was,* or *became, possessed of* خَيْر [or *good,* &c.]. (Ḳ, TA.) — [*He was,* or *became, good:* and *he did good:* contr. of شَرَّ.] You say, خِرْتَ يَا رَجُلُ [*Thou hast been good;* or *thou hast done good,* or *well; O man*]. (Ṣ.) And خَارَ ٱللّٰهُ لَكَ فِى هٰذَا الأَمْرِ [*May God do good to thee, bless thee, prosper thee,* or *favour thee, in this affair:* or] *may God cause thee to have,* or *appoint to thee, good in this affair:* (Ḳ:) or *may God choose for thee the better thing* [*in this affair*]. (A.) اَللّٰهُمَّ خِرْ لِى occurs in a trad., meaning *O God, choose for me the better of the two things.* (TA.) — See also 3. — خَارَهُ عَلَى صَاحِبِهِ, aor. as above, inf. n. خِيَرَةٌ and خَيْرٌ (Mṣb, Ḳ*) and خِيْرَةٌ (Ḳ) and خِيَرٌ; (Mṣb, TA;) and † خَيَّرَهُ, (Ḳ,) inf. n. تَخْيِيرٌ; (TA;) *He preferred him before his companion.* (Mṣb, Ḳ.*) — خَايَرَهُ, فَخَارَهُ: see 3.

2. خَيَّرَهُ *He gave him the choice,* or *option,* (Ṣ, A,* Mgh,* Mṣb,* Ḳ,) بَيْنَ الشَّيْئَيْنِ [*between the two things*], (Ṣ, Mgh, Mṣb,) or بَيْنَ الأَمْرَيْنِ [*between the two affairs*]: فَتَخَيَّرَ ♦ [*so he had the choice,* or *option, given him*]. (A.) — See also 1. It is said in a trad., خَيَّرَ بَيْنَ دُورِ الأَنْصَارِ, meaning *He preferred some among the houses of the Assistants before others of them.* (TA.) And in another trad., خُيِّرَ, meaning *He was preferred, and pronounced to have surpassed,* or *overcome,* or *won, in a contest,* or *dispute.* (IAth.)

3. خَايَرَهُ فَخَارَهُ ♦, (A, Ḳ,) inf. n. مُخَايَرَةٌ, (A,) *He vied with him,* or *strove to surpass him,* or *contended with him for superiority, in goodness,* or *excellence,* (A, Ḳ,) in, or with respect to, (فِى,) a thing, (A,) *and he surpassed him therein.* (A, Ḳ.)

4. مَا أَخْيَرَ فُلَانًا, (A,) and مَا خَيْرَهُ ♦, which latter is extr. [with respect to form, though more commonly used than the former], (TA,) [*How good is such a one!*] phrases similar to مَا أَشَرَّهُ and مَا شَرَّهُ [which have the contr. meaning]. (TA.) مَا خَيْرَ اللَّبَنَ لِلْمَرِيضِ [*How good is milk for the diseased!*], (Ḳ,* TA,) with naṣb to the ر and ن, is an expression of wonder: (Ḳ:) it was said to Khalaf El-Aḥmar, by an Arab of the desert, in the presence of Aboo-Zeyd; whereupon Khalaf said to him, "What a good word, if thou hadst not defiled it by mentioning it to the [common] people!" and Aboo-Zeyd returned to his companions, and desired them, when Khalaf El-Aḥmar should come, to say, all together, these words (ما خير اللبن للمريض), [in order to vex him], and they did so. (TA.)

5. تخيّر, as an intrans. v.: see 2. — As a trans. v.: see 8.

6. تخايروا فيه إِلَى حَكَمٍ *They contended together for superior goodness,* or *for excellence, in it,* or *with respect to it, appealing to a judge,* or *an arbiter.* (A.)

8. اختارهُ; and تخيّرهُ ♦, (Ṣ,* A, Mgh, Mṣb, Ḳ,) inf. n. [or rather quasi-inf. n.] خِيَرَةٌ ♦, said by IAth to be the only instance of the kind except طِيَرَةٌ; (TA voce تَطَيَّرَ;) and استخارهُ ♦; (A;) and خَارَهُ ♦; (Ḳ;) *He chose, made choice of, selected, elected,* or *preferred, him,* or *it.* (Ṣ, Mṣb,* Ḳ.) You say also, اِخْتَرْتُهُ الرِّجَالَ, and مِنَ الرِّجَالِ, [*I chose him from the men,*] and عَلَيْهِمْ, (Ḳ,) which last signifies *in preference to them.* (TA.) It is said in the Ḳur [vii. 154], وَاخْتَارَ مُوسَى قَوْمَهُ سَبْعِينَ رَجُلًا [*And Moses chose from his people seventy men*]. (TA.) وَلَقَدِ اخْتَرْنَاهُمْ عَلَى عِلْمٍ, in the Ḳur [xliv. 31, *Verily we have chosen them with knowledge*], may be indicative of God's producing good, or of his preferring them before others. (TA.)

10. استخار *He sought, desired,* or *asked for,* خِيرَة (Ṣ, Mṣb, Ḳ) or خِيَرَة (as in some copies of the Ḳ) [i. e. *the blessing, prospering,* or *favour, of God;* &c.]. [And it is trans.; for] one says, اِسْتَخِرِ اللّٰهَ يَخِرْ لَكَ [*Desire thou,* or *ask thou for, the blessing, prospering,* or *favour, of God;* &c.; and *He will bless, prosper,* or *favour, thee;* &c.]. (Ṣ.) And اِسْتَخَرْتُ اللّٰهَ فِيهِ فَخَارَ لِى *I desired,* or *asked, of God, the better of the two things,* [or rather *the better in it,* meaning a case, or an affair,] *and He chose it for me.* (A.) — See also 8.

خَيْرٌ [*Good, moral or physical; anything that is good, real or ideal,* and *actual or potential;* and, being originally an inf. n., used as sing. and pl.;] *a thing that all desire; such as intelligence,* for instance, *and equity;* (Er-Rághib, and so in some copies of the Ḳ;) [or *goodness;*] *and excellence; and what is profitable* or *useful; benefit;* (Er-Rághib;) *contr. of* شَرٌّ: (Ṣ, A, Mṣb:) pl. خُيُورٌ, (Mṣb, Ḳ,) and also, accord. to the Mṣb, خِيَارٌ ♦: (TA:) [but this latter seems to be properly pl. only of خَيْرٌ used as an epithet (see below) and as a noun denoting the comparative and superlative degrees: it may however be used as an epithet in which the quality of a subst. is predominant:] خير is of two kinds: namely, absolute خير, which is *what is desired in all circumstances and by every person:* and *what is* خير [or *good*] *to one and* شرّ [or *evil*] *to another;* as, for instance, (Er-Rághib,) *wealth,* or *property:* (Zj, L in art. شد, Er-Rághib, Ḳ:) it has this last signification, namely *wealth,* or *property,* in the Ḳur, ii. 176 (Ṣ, TA) and ii. 274 and xxiv. 33 and xli. 49: or in the first and second of these instances it is thus called to imply the meaning of *wealth,* or *property, that has been collected in a praiseworthy manner,* or it means *much wealth* or *property;* and this is its meaning in the first of the instances mentioned above, agreeably with a trad. of 'Alee; and also in the Ḳur, c. 8: (TA:) [being used as a pl. (as well as a sing.), it may be also rendered *good things:*] and it is also used by the Arabs to signify *horses;* (Ḳ,* TA;) and has this meaning in the Ḳur, xxxviii. 31: (TA:) [it is often best rendered *good fortune; prosperity; welfare; wellbeing; weal; happiness;* or *a good state* or *condition:* and sometimes *bounty,* or *beneficence.*] رَجُلٌ قَلِيلُ الخَيْرِ means [*A man possessing little,* or *no, good; possessing few,* or *no, good things;* or *poor:* and *in whom is little,* or *no, good* or *goodness;* or *niggardly:* and also] *a man who does little good:* (TA in art. عص:) or [*who does no good;*] *who is not near to doing good;* denoting the nonexistence of good in him. (Mṣb in art. قل.) [Thus it sometimes means the same as رَجُلٌ لَا خَيْرَ فِيهِ *A man in whom is no good* or *goodness; devoid of goodness; worthless.*] And قِلَّةُ خَيْرٍ means *Poverty:* and also *niggardliness.* (A and TA in art. جحد.) هُوَ مِنْ أَهْلِ الخَيْرِ وَالخِيرِ is explained voce خِيرٌ. عَلَى يَدَيِ الخَيْرِ وَاليُمْنِ [May it be *with the aid of good fortune and prosperity*] is a prayer used with respect to a marriage. (A 'Obeyd, TA.) And إِنَّكَ مَا وَخَيْرًا means مَعَ خَيْرٍ, i. e., *Mayest thou meet with, or attain, good.* (Ḳ.) — خَيْرٌ in the phrase فُلَانٌ خَيْرٌ resembles an epithet [like خَيِّرٌ ♦, and signifies *Good;* or *possessing good*]; (Akh, Ṣ;) therefore the fem. is خَيْرَةٌ, of which the pl. is خَيْرَاتٌ, (Akh, Ṣ, Mṣb,*) as occurring in the Ḳur, lv. 70; and they do not [there] mean by it [the comparative or superlative signification of the measure] أَفْعَلُ: (Akh, Ṣ:) you say رَجُلٌ خَيِّرٌ ♦, (Ṣ, A, Mṣb,) meaning [*A good man;* or] *a man possessing* خير [or *good*]; (Mṣb;) and رَجُلٌ خَيْرٌ: (Ṣ:) and in like manner, امْرَأَةٌ خَيِّرَةٌ ♦ and خَيْرَةٌ, (Ṣ, Mṣb,) meaning [*A good woman;* or] *a woman excellent in beauty and disposition:* (Mṣb:) or خير and خيّر ♦ signify *possessing much* خير [or *good*], (Ḳ,) applied to a man; (TA;) and in the same sense you say رَجُلٌ خَيْرِىٌّ ♦, and خُيَرِىٌّ ♦, and خِيرِىٌّ ♦: and the fem. of the first is خَيْرَةٌ; and of the second, خَيِّرَةٌ ♦: (Ḳ:) and the pl. [of pauc.] (of the first, TA) is أَخْيَارٌ, and [of mult.] خِيَارٌ: (A, Mṣb, Ḳ:) you say also خِيَارُ المَالِ, meaning *The excellent of the camels* or *the like:* (Mṣb, Ḳ:) and in like manner you say of men &c.: (TA:) [see also below:] and the fem. is خَيْرَةٌ, of which the pl. is خَيْرَاتٌ: (Mṣb:) خيار is *contr. of* أَشْرَارٌ, (Ṣ, Mgh,) [thus] used as an epithet: (Mgh:) and خَيْرَةٌ ♦ [used as a subst.] signifies *anything excellent;* and the pl. thereof in this sense, خَيْرَاتٌ, occurs in the Ḳur, ix. 89: (Ṣ:) or خَيْرٌ, (Ḳ,) or the fem. خَيْرَةٌ, (Lth,) or each, (Ḳ,) signifies *excellent in beauty:* (Lth, Ḳ:) and خيّر ♦ and خيّرة signify *excellent in righteousness* (Lth, Ḳ) *and religion:* (Ḳ:) or there is no difference in the opinion of the lexicologists [in general] between خَيْرَة and خَيِّرَة ♦: (Az:) accord. to Zj, خَيْرَات and خَيِّرَات ♦, both occurring in different readings of the Ḳur, lv. 70, signify *good in dispositions:* accord. to Khálid Ibn-Jembeh, خَيْرَةٌ, applied to a woman, signifies *generous in race, exalted in rank* or *quality* or *reputation, goodly in face, good in disposition, possessing much wealth, who, if she bring forth, brings forth a generous child:* (TA:) [خِيَارٌ ♦ is also applied as an epithet to a sing. subst., either masc. or fem.:] you say جَمَلٌ خِيَارٌ and نَاقَةٌ خِيَارٌ, meaning *A he-camel* [*that is excellent* or] *excellent and brisk* and so *a she-camel.* (TA.) See also مُخْتَارٌ, in three places. In the saying لَعَمْرُ أَبِيكَ الخَيْرِ, the word خير is in the nom. case as an epithet of عمر; [so that the phrase lit. means *By the good life of thy father;*] but properly it should be لَعَمْرُ أَبِيكَ الخَيْرِ [*By the life of thy good father*]: and the like is said with شَرّ. (TA.) [See also art. عمر.] — خَيْرٌ is also used to denote superiority: one says, هٰذَا خَيْرٌ مِنْ هٰذَا *This is better than this:* and in the dial. of the Benoo-'Ámir, هٰذَا أَخْيَرُ ♦ and أَشَرُّ مِنْ هٰذَا, with أ, and in like manner, أَشَرُّ; but the rest of the Arabs drop the أ in each case: (Mṣb:) you say, هُوَ أَخْيَرُ ♦ مِنْكَ [*He is better than thou*], and in like manner, أَشَرُّ مِنْكَ; and هُوَ خَيْرٌ مِنْكَ, and in like manner, شَرٌّ مِنْكَ; and, [using the dim. form of خَيْرٌ,] خُيَيْرٌ ♦ مِنْكَ, and in like manner, شُرَيْرٌ مِنْكَ. (Ibn-Buzurj, TA.) You also say, when you mean to express the signification of superiority, فُلَانَةُ خَيْرُ النَّاسِ [*Such a* *woman is the best of mankind*]; but not خَيْرَةٌ [see, however, what will be found cited hereafter from the Ḳ,] and فُلَانٌ خَيْرُ ٱلنَّاسِ [*Such a man is the best of mankind*]; but not ▼ أَخْيَرُ [unless in the dial. of the Benoo-'Ámir]: and [it is said that] خَيْرٌ when thus used does not assume the dual form nor the pl., because it has the signification of [the measure] أَفْعَلُ: for though a poet uses the dual form, he uses it as a contraction of the dual of خَيِّرٌ, like مَيْتٌ and مَيِّتٌ, and هَيْنٌ and هَيِّنٌ: (Ṣ:) [but this remark in the Ṣ is incorrect: for both خَيْرٌ and ▼ أَخْيَرُ, when used in such phrases as those to which J here refers, have pl. forms of frequent occurrence, and of which examples will be found below; and, as is said by I 'Aḳ (p. 239), and by many other grammarians, you may say, الزَّيْدَانِ أَفْضَلَا ٱلْقَوْمِ, and الزَّيْدُونَ أَفْضَلُو ٱلْقَوْمِ and أَفَاضِلُ ٱلْقَوْمِ, and also هِنْدٌ فُضْلَى ٱلنِّسَآءِ, &c.; and such concordance is found in the Ḳur, vi. 123; and is even said by many to be more chaste than the mode prescribed by J:] it is said in the Ḳ, that you say, هُوَ ▼ أَخْيَرُ مِنْكَ, like خَيْرٌ; and when you mean the signification of superiority, you say فُلَانَةُ خَيْرَةُ ٱلنَّاسِ, with ة, and فُلَانَةٌ خَيْرُهُمْ, without ة: but [SM says,] I know not how this is; for in the Ṣ is said what is different from this, and in like manner by Z in several places in the Ksh; and what is most strange is, that the author of the Ḳ quotes in the B the passage of J [from the Ṣ], and adopts the opinion of the leading authorities [as given in the Ṣ]: (TA:) or you say, فُلَانَةُ ٱلْخَيْرَةُ مِنَ ٱلْمَرْأَتَيْنِ [*Such a woman is the better of the two women*]: and هِىَ ٱلْخَيْرَةُ, and ▼ ٱلْخِيَرَةُ, [so in the TA, but in the CK ٱلْخِيْرَةُ,] and ▼ ٱلْخُيْرَى, and ▼ ٱلْخُورَى, [the last being fem. of أَخْيَرُ, originally خُيْرَى, and so, app., the last but one, *She is the better*, or *best*:] (Ḳ:) and [using the dim. form of أَخْيَرُ] you say, هُوَ ▼ أُخَيِّرُ أَهْلِهِ [*He is the best of his family*]: (Ibn-Buzurj, TA:) one says also, to one coming from a journey, خَيْرَ مَا رُدَّ فِى أَهْلٍ وَمَالٍ, meaning May God make that with which thou comest [back] to be *the best of what is brought back* by the absent *with family and property*; (Aṣ, Meyd, TA;) or, as some relate it, خَيْرَ, i. e. رَدَّكَ خَيْرَ رَدٍّ [may thy bringing back be *the best bringing back*]; and فِى is used in the sense of مَعَ: (Meyd:) [أَخْيَارٌ is pl. of pauc., and خِيَارٌ pl. of mult., and so app. is خِيرَانٌ, of خَيْرٌ thus used; and ▼ أَخَايِرُ is pl. of أَخْيَرُ, and so is أَخْيَرُونَ applied to rational beings: in the TA, أَخَايِرُ is said to be a pl. pl. of خِيرَانٌ, and so أَخْيَرُ; but this is app. a mistake, probably of transcription:] you say رَجُلٌ مِنْ خِيَارِ ٱلنَّاسِ and أَخْيَارِهِمْ and ▼ أَخَايِرِهِمْ [*A man of the best of mankind*]: (A, TA:) and لَكَ خِيَارُ هٰذِهِ ٱلْإِبِلِ, and ▼ خِيَرَتُهَا, [*Thine are*, or *is*, or *shall be*, *the best of these camels*,] alike with respect to a sing. and a pl.: (TA:) and نَحَرَ خِيرَةَ ▼ إِبِلِهِ and ▼ خُورَةَ إِبِلِهِ [*He slaughtered the best of his camels*]: (IAạr, TA:) and ▼ هُمُ ٱلْأَخْيَرُونَ [*They* (meaning men) *are the better*, or *best*]. (Ibn-Buzurj, TA.) —— مَا خَيْرَ for مَا أَخْيَرَ: see 4, in two places. —— خَيْرُ بُوَا [from the Persian خير بوا *Lesser cardamom*;] *a kind of small grain, resembling the* قَاقُلَّة [or *common cardamom*], (Ḳ,) *of sweet odour*. (TA.)

خِيرٌ *Generousness; generosity*; (Ṣ, A, Mṣb, Ḳ;) *liberality; munificence*. (Mṣb.) You say, فُلَانٌ ذُو خِيرٍ *Such a one is a possessor of generousness, or generosity*, &c. (Mṣb.) And هُوَ مِنْ أَهْلِ ٱلْخَيْرِ وَٱلْخِيرِ [*He is of the people of good*, or *of wealth*, &c., *and of generosity*]. (A.) —— *Eminence; elevated state* or *condition; nobility*. (IAạr, Ḳ.) —— *Origin*. (Lḥ, Ḳ.) —— *Nature*, or *disposition*. (A, Ḳ.) You say, هُوَ كَرِيمُ ٱلْخِيرِ *He is generous in nature*, or *disposition*. (A.) —— *Form, aspect*, or *appearance; figure, person, mien, feature*, or *lineaments; guise*, or *external state* or *condition*; or *the like*; syn. هَيْئَةٌ. (Lḥ, Ḳ.)

خُورَةٌ [app. originally خُيْرَةٌ]: see خَيْرٌ, near the end of the paragraph; and see also art. خور.

خَيْرَةٌ fem. of خَيْرٌ [q. v.] used as an epithet: pl. خَيْرَاتٌ. (Akh, Ṣ, Mṣb.) —— [Also, used as a subst., or as an epithet in which the quality of a subst. is predominant, *A good thing, of any kind: a good quality; an excellency*: and *a good act* or *action*: &c.: pl. as above:] see خَيْرٌ, in the former half of the paragraph.

خِيرَةٌ: see خَيْرٌ, in three places, towards the end of the paragraph: —— and see خِيَرَةٌ, in four places: —— and خِيَارٌ. —— It is also a subst. from خَارَ ٱللّٰهُ لَكَ فِى هٰذَا ٱلْأَمْرِ, (Ṣ,) and so ▼ خِيَرَةٌ; both signifying [The *blessing, prospering*, or *favour*, of God; his *causing one to have*, or *appointing to one, good in an affair*: or his *choosing for one the better thing in an affair*: or] the *state that results to him who begs God to cause him to have good*, or *to choose for him the better thing, in an affair*. (TA.) You say, كَانَ ذٰلِكَ خِيرَةً مِنَ ٱللّٰهِ [*That was through God's blessing, prospering*, or *favour*; &c.: or *through God's choosing the better thing in the affair*]. (A.)

خِيَرَةٌ and ▼ خِيرَةٌ (of which the former is the better known, TA) are substs. from ٱخْتَارَهُ, (Ḳ,) or from ٱخْتَارَهُ ٱللّٰهُ, (Ṣ,) both signifying *A thing, man*, or *beast*, and *things*, &c., *that one chooses*: (TA:) or [*a thing*, &c.,] *chosen, selected*, or *elected*: (Mgh:) as in the saying, مُحَمَّدٌ خِيَرَةُ ٱللّٰهِ مِنْ خَلْقِهِ and ▼ خِيرَتُهُ [*Mohammad is the chosen*, or *elect, of God, from his creatures*]: (Ṣ, Mgh:*) or ▼ خِيَرَةٌ is a subst. from ٱلْإِخْتِيَارُ, like فِدْيَةٌ from ٱلْإِفْتِدَآءُ; and خِيرَةٌ is syn. with خِيَارٌ and ٱخْتِيَارٌ; or is from تَخَيَّرْتُ ٱلشَّىْءَ: or, as some say, خِيرَةٌ and خِيَرَةٌ are syn.: (Mṣb:) see 8; and see also خِيَارٌ: and ▼ هٰذِهِ خِيرَتِى (Mṣb, TA) or خِيَرَتِى (TA) means *This is what I choose*; (Mṣb, TA;) and so هٰؤُلَآءِ خِيرَتِى: and هٰذَا خِيرَتِى: *These are what I choose*. (TA.) [See مُخْتَارٌ.] —— See also خِيرَةٌ.

خُورَى: see خَيْرٌ, in two places.

خُيْرَى: see خَيْرٌ.

خِيرَى: see خَيْرٌ, in two places.

[خَيْرِىٌّ *Of*, or *relating to*, خَيْرٌ, or *good*, &c.]

خِيرِىٌّ *Of*, or *relating to*, or *possessing, generousness, generosity, liberality*, or *munificence*. (Mṣb.) —— And hence, (Mṣb,) or [thus applied] it is an arabicized word, (Ṣ,) [from the Persian خيرى,] The مَنْثُور [or *gilliflower*:] but generally applied to the *yellow species thereof*; [so in the present day;] for it is this from which is extracted its oil, which is an ingredient in medicines. (Mṣb.) [Accord. to Golius, "*Viola alba, ejusque genera*: Diosc. iii. 138:" and he adds, as on the authority of Ibn-Beyṭár, "*spec. luteum*."] —— And خِيرِىُّ ٱلْبَرِّ *The* خُزَامَى [q. v.]; because it is the most pungent in odour of the plants of the desert. (Mṣb.)

[خَيْرِيَّةٌ The *quality of* خَيْرٌ; i. e. *goodness*.]

خِيَارٌ a subst. from ٱلْإِخْتِيَارُ; (Ṣ, Mgh, Ḳ;) meaning *Choice*, or *option*; (Mṣb;) and so ▼ خِيَرَةٌ in the Ḳur [xxviii. 68], مَا كَانَ لَهُمُ ٱلْخِيَرَةُ *They have not choice*, or *option*; (Mgh;) or the meaning of these words is, *it is not for them to choose* in preference to God; (Fr, Zj;) and so, accord. to Lth, ▼ خِيرَةٌ, as being an inf. n. [or rather a quasi-inf. n., though this seems doubtful,] of ٱخْتَارَ. (TA.) You say, إِنَّ فِى ٱلشَّرِّ خِيَارًا [*Verily in evil there is a choice*, or *an option*]; i. e. *what may be chosen*: a prov. (TA.) And أَنْتَ بِٱلْخِيَارِ and ▼ بِٱلْمُخْتَارِ [in some copies of the Ḳ بالمخيار, which, as is said in the TA, is a mistranscription, *Thou hast the choice*, or *option*]; i. e. *choose thou what thou wilt*. (Ḳ.) And ٱلْبَيْعُ صَفْقَةٌ أَوْ خِيَارٌ *Selling is decisive or with the option of returning*. (Mgh in art. صفق.) Hence, خِيَارُ ٱلرُّؤْيَةِ *The choice of returning* [on seeing it] *a thing which one has purchased without seeing it*. (Mgh,* Mṣb,* KT.) And خِيَارُ ٱلْمَجْلِسِ [*The choice of returning a thing purchased while sitting with the seller*]. (TA.) And خِيَارُ ٱلْعَيْبِ [and ٱلنَّقِيصَةِ] *The choice of returning a thing to the seller when it has a fault, a defect*, or *an imperfection*. (KT.) And خِيَارُ ٱلشَّرْطِ *The choice of returning a thing purchased when one of the two contracting parties has made it a condition that he may do so within three days or less*. (KT.) And خِيَارُ ٱلتَّعْيِينِ *The choice of specifying* [*for instance*] *one of two garments*, or *pieces of cloth, which one has purchased for ten pieces* [*of money*, or *some other sum*,] *on the condition of so doing*. (KT.) —— See also مُخْتَارٌ, in three places. And see خَيْرٌ, in the middle of the paragraph, where it is explained as an epithet applied to a sing. subst., either masc. or fem. See also the first sentence of that paragraph. —— It is also a pl. of خَيْرٌ [q. v.] as an epithet, (A, Mṣb, Ḳ,) [and as a noun denoting the comparative and superlative degrees.] —— Also [*A species of cucumber*; *cucumis sativus* Linn. *α fructu minore*: (Delile, Flor. Aeg.

Illustr., no. 927:)] *i. q.* قِثَّآءٌ: (Ṣ:) or *resembling the* قِثَّآء; (Ḳ, &c.;) which is the more suitable explanation: (TA:) or *i. q.* قَثَدٌ [q. v.]: an arabicized word: (Mgh:) [from the Persian خِيَارْ:] not Arabic. (Ṣ.) —— خِيَارُ شَنْبَرٍ [*The cassia fistula* of Linn.;] *a well-known kind of tree;* (Ḳ;) *a species of the* خُرُّوب, *resembling a large peach-tree;* (TA;) *abounding in Alexandria and Miṣr;* (Ḳ;) *and having an admirable yellow flower:* (TA:) the latter division [or rather the whole] of the name is arabicized [from the Persian خِيَارْ چَنْبَر]. (TA.)

خُيَيْرٌ: see خَيْرٌ, [of which it is the dim.,] in two places, in the latter half of the paragraph.

خَيِّرٌ, and its fem. خَيِّرَةٌ, and pl. fem. خَيِّرَاتٌ: see خَيْرٌ, (used as an epithet,) in eight places, in the former half of the paragraph.

خَائِرٌ [*Doing good, or well:* &c.:] act. part. n. of خَارَ. (Ṣ, TA.)

أَخْيَرُ, and its pls. أَخَايِرُ and أَخْيَرُونَ: see خَيْرٌ, in eight places, in the latter half of the paragraph.

اِخْتِيَارِيٌّ [*Of, or relating to, the will, or choice*]. صِفَةٌ اخْتِيَارِيَّةٌ [meaning *A quality which originates from, or depends upon, the will, or choice,* i. e. *an acquired quality,*] is opposed to خِلْقِيَّةٌ. (Mṣb in art. مدح, &c.)

مَخْيَرَةٌ [*A cause of good:* and hence,] *excellence,* and *eminence, or nobility:* so in the phrase, فُلَانٌ ذُو مَخْيَرَةٍ [*Such a one is a possessor of eminence,* &c.]. (A, TA.)

مُخَيْيِرٌ: see what follows.

مُخْتَارٌ act. part. n. [of 8, signifying *Choosing, selecting,* or *electing*]. (TA.) —— And pass. part. n. [of the same, signifying *Chosen, selected, elected,* or *preferred:* and *choice, select,* or *elect;* as also †خِيَارٌ, which signifies likewise the *best* of anything; often used in this sense, as a sing. and as a pl.; and *excellent,* or *excellent and brisk,* applied to a he-camel and to a she-camel; as mentioned above, voce خَيْرٌ]. (TA.) You say also جَمَلٌ خِيَارٌ† in the sense of مُخْتَارٌ [*A choice he-camel*], and نَاقَةٌ خِيَارٌ† in the sense of مُخْتَارَةٌ [*A choice she-camel*]. (TA.) [See also خِيَرَةٌ.] The dim. of مُخْتَارٌ is مُخَيْيِرٌ†: the ت is thrown out because it is augmentative; and the ا is changed into ى because it was changed from ى in مختار: (Ṣ:) one should not say مُخَيْتِيرٌ. (El-Ḥareeree's Durrat el-Ghowwáṣ, in De Sacy's Anthol. Gr. Ar. p. 49 of the Arabic text.) —— See also خِيَارٌ.

خيش

خَيْشٌ *Garments, or pieces of cloth, of the worst of flax:* (Ṣ:) or *garments, or pieces of cloth, of thin texture, and of coarse threads, made of the hards, or hurds, of flax,* (Ḳ,* TA,) *and of the worst thereof:* (TA:) or *of the coarsest of* [*the stuff called*] عصب [i. e. عَصْب, q. v., in the copies of the Ḳ in my hands incorrectly written عَضْب]: (Lth, Ḳ:) or *coarse flax:* (Mgh:) or *a cloth of coarse flax.* (Ḥar p. 544.) —— [Hence,] †*A low, vile, or mean, man.* (Ḳ.)

خَيْشِيٌّ and †خَيَّاشٌ [*A weaver, or seller, of* خَيْش. The former mentioned in the Ḳ, and the latter in the TA, as surnames of men.]

خَيَّاشٌ: see what next precedes.

خيط

1. خَاطَ, (Mṣb, TA,) first pers. خِطْتُ, (Ṣ,) aor. يَخِيطُ, (Mṣb,) inf. n. خِيَاطَةٌ, (Ṣ, TA,) or this is a simple subst., (Mṣb, TḲ,) and the inf. n. is خَيْطٌ, (TḲ,) which is said in the Ḳ to be syn. with خِيَاطَةٌ, but this last is a mistake for خِيَاطٌ as signifying "thread," (TA,) or "a thread," (AZ, TA,) though خِيَاطٌ is also syn. with خِيَاطَةٌ, (TA,) *He sewed, sewed together,* or *sewed up,* a garment, or piece of cloth; (Ṣ, Mṣb, TA;) as also †خَيَّطَهُ, inf. n. تَخْيِيطٌ. (TA.) —— [Hence,] خَاطَ بَعِيرًا بِبَعِيرٍ ‡*He coupled a camel with a camel* [*by tying the end of the halter of one to the tail of the other*]. (TA.) —— خَاطَتِ الحَيَّةُ, (TA,) inf. n. خَيْطٌ, (Ḳ, TA,) ‡*The serpent ran along upon the ground.* (Ḳ,* TA.) —— خَاطَ إِلَيْهِ خَيْطَةً† ‡*He passed by him, or it,* [*or to, or towards, him or it,*] *once:* or خاط خَيْطَةً† *he passed along quickly:* (Ḳ,* TA:) and so اختاط† and اِخْتَطَى. (Ḳ.) It is said by Kr to be formed by transposition from الخَطْوُ: but this is a mistake; for, were it so, they would have said, خاط خَوْطَةً, not خَيْطَةً. (ISd.) Accord. to Lth, خاط خَيْطَةً† وَاحِدَةً and means †*He made his journey* [*or a journey*] *without interruption.* (TA.) In the A it is said that خاط فُلَانٌ خَيْطَةً† means ‡*Such a one journeyed on, not pausing for anything:* and in like manner, خاط إِلَى مَقْصِدِهِ ‡[*He journeyed on, not pausing for anything, to his place, or object, of aim*]. (TA.)

2: see 1. —— خَيَّطَ الشَّيْبُ فِى رَأْسِهِ, (Ṣ, Ḳ,) inf. n. تَخْيِيطٌ, (Ḳ,) means ‡*Whiteness of the hair, or hoariness, appeared upon his head* (Ḳ, TA) *in streaks, or lines:* (TA:) it is like وَخَطَ: (Ṣ, TA:) or *became like threads:* (Ḳ:) and in like manner, فِى لِحْيَتِهِ *in his beard.* (TA.) Bedr Ibn-'Ámir El-Hudhalee says,

أَقْسَمْتُ لَا أَنْسَى مَنِيحَةَ وَاحِدٍ
حَتَّى تُخَيِّطَ بِالبَيَاضِ قُرُونِى

(Ṣ, TA) [*I swear that I will not forget the loan* (here meaning the قَصِيدَة, Skr) *of one* (meaning Abu-l-'Iyál [with whom he was carrying on a controversy], Skr)] *until the sides of my head become streaked with whiteness:* (TA:) but some read تُخَيَّطَ; and Ibn-Ḥabeeb says that خَيَّطَ الشَّيْبُ الرَّأْسَ signifies †*Whiteness of the hair, or hoariness, became conjoined and continuous upon the head,* as though one part thereof were sewed to another: (IB, TA:) some read تَخَيَّطَ†; and accord. to the Ḳ, you say, تَخَيَّطَ رَأْسَهُ بِالشَّيْبِ, meaning †*His head became streaked, or marked as with threads, by whiteness of the hair, or hoariness:* [the best reading seems to be تَخَيَّطَ, for تَتَخَيَّطُ:] and some read تَوَخَّطَ [for تَتَوَخَّطَ, from تَوَخَّطَ as having the meaning here assigned to تَخَيَّطَ]. (TA.)

5: see 2.

8: see 1.

خَائِطٌ: see خَيَّاطٌ.

خَيْطٌ *Thread, or string;* or *a thread* or *string;* syn. سِلْكٌ; (Ṣ, Ḳ;) the *thing with which one sews;* (Mṣb;) [often used as a coll. gen. n.; n. un. with ة;] and †خِيَاطٌ [likewise] signifies the *thing with which a garment, or piece of cloth, is sewed;* as also †مِخْيَطٌ; besides having another signification, common to it with the last, namely "a needle;" (Ḳ;) the pl. of خَيْطٌ is أَخْيَاطٌ [a pl. of pauc.] (IB, Ḳ) and خُيُوطٌ (Ṣ, Mṣb, Ḳ) and خُيُوطَةٌ [both pls. of mult.]. (Ṣ, Ḳ.) It is said in a trad., أَدُّوا الخِيَاطَ† وَالمِخْيَطَ, meaning [*Bring ye*] *the* خَيْط *and the needle.* (TA.) And you say, أَعْطِنِى خِيَاطًا† and نِصَاحًا, i. e. [*Give thou to me*] *a single* خَيْط. (AZ, TA.) [أَعْطِنِى خِيَاطًا وَنِصَاحًا may, however, mean *Give thou to me a needle and thread.*] —— خَيْطُ الرَّقَبَةِ †*The* نُخَاع [or *spinal cord*] *of the neck.* (Ṣ, Ḳ.) You say, جَاحَشَ فُلَانٌ عَنْ خَيْطِ رَقَبَتِهِ, meaning ‡*Such a one defended his blood.* (Ṣ, O, L.) —— الخَيْطُ الأَبْيَضُ and الخَيْطُ الأَسْوَدُ, mentioned in the Ḳur ii. 183, mean †*The true dawn,* and *the false dawn:* (Mṣb:) or *the whiteness of the dawn,* and *the blackness of night;* (Ḳ, TA;) likened to a thread because of its thinness: (TA:) or *the whiteness of day,* and *the blackness of night:* (A'Obeyd, Nh:) or *the dawn that extends sideways,* and *the dawn that rises high,* or, as some say, *the blackness of night:* (Ṣ:) or *what appears of the true dawn, which is the* مُسْتَطِير, and *what extends with it of the darkness of night, which is the dawn termed the* مُسْتَطِيل: (Mgh:) or *what first appears of the dawn spreading sideways in the horizon, and what extends with it of the darkness of the last part of the night:* (Bd:) or *the dawn that rises high, filling the horizon,* and *the dawn that appears black, extending sideways:* (Aboo-Is-ḥáḳ:) or the real meaning is *the day* and *the night.* (TA.) الخَيْطَانِ also signifies †*The night and the day.* (L in art. وسد.) تَبَيَّنَ الخَيْطُ مِنَ الخَيْطِ means ‡[*The night became distinct from the day:* or] *what is termed* الخَيْطُ الأَبْيَضُ *became distinct from what is termed* الخَيْطُ الأَسْوَدُ. (TA.) And خَيْطٌ مِنَ الصُّبْحِ is also said to signify †*A tint of the dawn.* (TA.) [See بَرِيمُ الصُّبْحِ in art. برم.] —— خَيْطُ بَاطِلٍ ‡*What is called* مُخَاطُ الشَّيْطَانِ *and* لُعَابُ الشَّمْسِ, (Ṣ, TA,) which last is explained by Z and IB as meaning *what comes forth from the mouth of the spider:* (TA: [the author of which says that, accord. to this explanation, this term differs from لعاب الشمس: but in so saying he seems to be in error: both evidently signify *gossamer*:]) it was applied as a surname, or nickname, to Marwán Ibn-El-Ḥakam; because he was tall, and loose, or uncompact, in frame: (Ṣ:) or it signifies *the air*; syn. الهَوَآءُ [perhaps a mistranscription for الهَبَآءُ, occurring in another explanation hereafter]: (Ḳ:) or *light entering from an aperture in a wall* [*into a dark place*]: (Th, Ḳ:) or خَيْطُ البَاطِلِ signifies *the scattered* هَبَآء [or *atoms that are seen in the rays of the sun*] *entering from an aperture in a wall* [*into a dark place*] *when the sun is hot*: and one says, فُلَانٌ أَدَقُّ مِنْ خَيْطِ البَاطِلِ ‡[*Such a one is less in estimation than the scattered atoms that are seen in the rays of the sun*]; a prov., applied to him who is in an abject state; thus related, on the authority of Aḥmad Ibn-Yaḥyà, by Az and others; but by Ṣgh, erroneously, أَدَقُّ مِنْ خَيْطٍ بَاطِلٍ. (TA.) — See also what next follows, in two places.

خِيطٌ (Aṣ, IDrd, Ṣ, Ḳ) and ↓خَيْطٌ (IDrd, Mṣb, Ḳ) and ↓خَيْطَى (Ṣ, Ḳ) †*A collection*, or *flock*, of ostriches, (Ṣ, Mṣb, Ḳ, &c.,) and *a swarm* of locusts, (Ḳ,) and a ↓خِيط is sometimes of [wild] bulls or cows: (L, TA:) pl. [of pauc.] أَخْيَاطٌ (IB) and [of mult.] خِيطَانٌ: (Ḳ:) ↓which last, as also ↓خَيْطَانٌ, signifies likewise *a company* of men. (TA.) [خِيطٌ may perhaps be originally خِيَطٌ, pl. of خَيْطَةٌ, q. v.]

خَيْطٌ ‡ *Length of the neck* of an ostrich, (Ṣ, TA,) and *of the* [*bones, such as are termed*] قَصَب thereof: or, as some say, *a constant mixture of blackness with whiteness* therein: or their *being in an uninterrupted line, like an extended* خَيْط [or *thread*]. (TA.)

خَيْطَةٌ [n. un. of خَيْطٌ, q. v. — Also,] in the dial. of Hudheyl, (Ṣ,) *A wooden peg* or *stake*, (Skr, Ṣ, Ḳ,) *which is fixed in a mountain, in order that one may let himself down* [*by means of a rope attached thereto*] *over against the place where* [*wild*] *honey is deposited* [*to gather it*]. (Skr.) Aboo-Dhu-eyb says, (Ṣ, TA,) describing the gatherer of honey, (TA,)

تَدَلَّى عَلَيْهَا بَيْنَ سِبٍّ وَخَيْطَةٍ
بِجَرْدَآءَ مِثْلِ الوَكْفِ يَكْبُو غُرَابُهَا

(Ṣ, TA,) i. e. *He let himself down* [*over against it*, meaning the place of the honey, *partly*] *by means of a rope* (for so سِبّ signifies) *and* [*partly by means of*] *a wooden peg* or *stake* [to which the rope was attached, fixed] *in a rock smooth like the* [*leather termed*] وَكْف, i. q. نِطَع, [*the crow of which rock would fall prone upon its face* for want of something therein to which to cling:] (TA:) or (in the Ḳ "and") خيطة signifies *a rope*; (Aṣ, Az, Ḳ, TA;) [and if so, سِبّ here means "a wooden peg," which is a signification assigned to it in the Ḳ in art. سب:] or, accord. to AA, *a slender rope* (Ṣ, L, TA) *made* [*of the bark*] *of the tree called* سَلَب: (L, TA:) and (accord. to some, TA) *a string which is with the gatherer of honey*, (Ḳ, TA,) *and with which he pulls the rope* [app. *when he has detached himself from the latter to gather the honey*], *it being tied to him*: (TA:) or *a* [*tunic of the kind called*] دُرَّاعَة, [*of leather*,] *which he wears*. (Ibn-Ḥabeeb, Ḳ, TA. [In the CḲ, دَرّاعَة is erroneously put for دُرّاعَة.]) = See also 1, in four places. One says also, مَا آتِيكَ إِلَّا الخَيْطَةَ †*I do not come to thee save sometime*. (TA.)

خَيْطَى: see خِيطٌ.

خَيْطَآءُ ‡A she-ostrich *long in the neck*. (Ṣ, Ḳ, TA.)

خَيْطَانٌ and خِيطَانٌ: see خِيطٌ.

خِيَاطٌ *A needle*; as also ↓مِخْيَطٌ. (Ṣ, Mṣb,* Ḳ.) Hence the saying in the Ḳur [vii. 38], حَتَّى يَلِجَ الجَمَلُ فِي سَمِّ الخِيَاطِ [*Until the camel enter into the eye of the needle*]. (Ṣ.) — See also خَيْطٌ, in three places. — And see مَخِيط.

خِيَاطَةٌ The *art of sewing*. (Mṣb, TA.) [See also 1.]

خَيَّاطٌ *A seamster; one whose occupation is that of sewing*; (Mṣb, Ḳ;) as also ↓خَائِطٌ (Ḳ) and ↓خَاطٌ. (Ṣgh, K. [in the CḲ خَاطٌ.]) [In the present day, its predominant application is to *A tailor*.] — Also †One *who passes along quickly*. (TA.)

خَائِطٌ: see خَيَّاطٌ.

مَخِيطٌ and ↓مَخْيُوطٌ A garment, or piece of cloth, *sewed*: (Ṣ, Mṣb, Ḳ:) the ى in the former is the و of the measure مَفْعُول, changed into ى because of its being quiescent and the preceding letter's being with kesr; the letter preceding it being made movent because it and the و are quiescent after the ى has fallen out; [for by dropping the ى it becomes changed from مَخْيُوط to مَخُوط;] and it is made movent with kesr [and thus changed from مَخُوط to مَخِوط, which necessarily becomes مَخِيط,] in order to its being known that the letter which has dropped out is ى: some say that the ى in مَخِيط is the radical, and that the letter thrown out is the و of the measure مَفْعُول, in order that the word with و [for its medial radical] may be known from that with ى; [so that it is changed from مَخْيُوط to مَخْيُط, and then to مَخِيْط, and then to مَخِيط;] but the former saying is the right, because the و is a formative augment, and it is not proper that such should be thrown out. (Ṣ.) — Also, the former, †The *whole of the exterior* of the belly. (ISh.) — And †*A place of passage*; (O, L, TA;) a meaning erroneously assigned in the Ḳ to ↓خِيَاط and ↓مِخْيَط: (TA:) and particularly, ‡of a serpent; (TA;) the *place of creeping along* of a serpent. (Ḳ, TA.)

مِخْيَطٌ: see خِيَاطٌ: — and خَيْطٌ. — See also مَخِيطٌ.

مَخْيُوطٌ: see مَخِيطٌ.

خيف

1. خَيِفَ The *having one of the eyes blue and the other black*: (JK, Ṣ, Mgh, Mṣb, Ḳ:) inf. n. of خَيِفَ, aor. يَخْيَفُ: (JK, Mṣb:*) said of a horse, (Ṣ, Mgh,* Mṣb, Ḳ,) &c., (Ṣ, Ḳ,) i. e. of any animal. (Ṣ, TA.) — Also The *being wide in the sheath of the penis*: (Ṣ, Ḳ:) in this sense [likewise] inf. n. of خَيِفَ: (Ṣ:) said of a camel. (Ṣ, Ḳ.) — And [app. in like manner having for its verb خَيِفَتْ] A she-camel's *being such as is termed* خَيْفَآء [i. e. *wide in the udder*, or *in the skin thereof*, or only *when it is empty of milk, and flaccid*]. (Ṣ)

2. خَيَّفَتْ, (JK,) or خَيَّفَتْ أَوْلَادَهَا, (TA,) ‡*She* (a woman) *brought forth her children different, one from another*. (JK, TA.) — خُيِّفَ بَيْنَهُمْ, (JK, A, Ḳ,) inf. n. تَخْيِيفٌ, (Ḳ,) ‡*It* (a thing, JK, Ḳ, or property, A) *was divided*, or *distributed, among them*. (JK, A, Ḳ.) — خَيَّفَتْ بَيْنَ الأَسْنَانِ, عُمُورُ اللِّثَةِ مِنَ الأَسْنَانِ, (JK,) or (Ḳ,) ‡*The portions of the flesh of the gums between the teeth became separated* (JK, Ḳ) *from the teeth*. (JK.) = خَيَّفَ *He* (a man, JK) *alighted*, or *descended and stopped* or *sojourned* or *abode*, in a place; (JK, Ḳ;) as also خَيَّمَ. (JK.) — خَيَّفَ عَنْدَ القِتَالِ, (JK,) or عَنِ القِتَالِ, (Ḳ,) *He receded, drew back*, or *desisted*, (JK, Ḳ,) *on the occasion of fight*, (JK,) or *from fight*. (Ḳ.)

4. اخاف, (JK, Ṣ, Ḳ,) inf. n. إِخَافَةٌ; (TA;) and أَخْيَفَ; (JK, Ḳ;) said of a man, (JK, TA,) or of a party of men, (JK, Ṣ,) *He*, or *they, alighted*, or *descended and stopped* or *sojourned* or *abode, in a* [*tract such as is termed*] خَيْف: (JK:) and [particularly] *came to the* خَيْف *of Minè, and there alighted*, or *descended and stopped* &c.; (JK,* Ṣ, Ḳ;) as also ↓اختاف. (Yoo, Ḳ.) = اخاف السَّيْلُ القَوْمَ *The torrent made the party*, or *company of men, to alight*, or *descend and stop* or *sojourn* or *abide, in a* [*tract such as is termed*] خَيْف. (JK, Ibn-'Abbád, Ḳ.)

5. تخيّف أَلْوَانًا *He* (a man, TA) *altered so as to become of different colours*. (Ḳ, TA.) — تخيّفت الإِبِلُ *The camels took different directions* in the place of pasturage (Lḥ, JK) &c. (Lḥ.) = تخيّفهُ *He took by little and little from it*; (IAạr, JK;) as also تخوّفهُ [q. v.]. (JK.)

8: see 4.

خَيَفٌ [sing. of أَخْيَافٌ]. You say, النَّاسُ أَخْيَافٌ ‡*Men, or the people, are different, one from another*, (JK, Ṣ, A, Ṣgh, Mṣb, Ḳ,*) *in their states*, or *conditions*, (JK,) or *in their forms, shapes*, or *semblances*; (Ṣgh;) or *of various sorts in natural dispositions, and in forms, shapes, or semblances*: (L:) from خَيَف signifying the "having one of the eyes blue and the other black." (Ṣ. [See 1.]) And أَخْيَافٌ, (Mgh, Mṣb,) or إِخْوَةٌ أَخْيَافٌ, (Ṣ, Ḳ,) †*Brothers who are sons of one mother but of different fathers*: (Ṣ, Mgh, Mṣb,* Ḳ:) and in like manner, بَنُو الأَخْيَافِ, if of good authority. (Mgh.) — And hence, أَبْيَاتٌ أَخْيَافٌ ‡ *Verses* *diversified by having one word thereof composed of dotted letters and another composed of letters not dotted.* (Har p. 611 and 612.) — Also, the sing., *Land,* (ISd, TA,) or *a place,* (Mgh,) *of which the stones are of different colours.* (ISd, Mgh, TA.) — *A side, region, quarter,* or *tract;* syn. نَاحِيَةٌ. (Ḳ.) — The *part that slopes down from the rugged portion of a mountain and rises from the channel in which the water flows;* (Ṣ, Ḳ;) whence مَسْجِدُ الخَيْفِ [the mosque of the خيف] in Minè: (Ṣ:) or *an elevated place, like the* خيف *of Minè:* (Mgh:) or the *part, of a valley, that rises a little from the channel in which the water flows,* and only *between two mountains;* and hence مَسْجِدُ خَيْفٍ, originally مسجد الخيف, مِنًى: (Mṣb:) and *any declivity and acclivity at the foot of a mountain:* and *a white place in the black mountain that is behind Aboo-Kubeys;* and hence the name of مسجد الخيف; or this is so called because it is [in] a نَاحِيَة [or side &c.] of Minè; or because it is at the foot of a mountain: (Ḳ:) pl. [of pauc.] أَخْيَافٌ (TA) and [of mult.] خُيُوفٌ. (Mgh, TA.) — Also The *skin of the udder:* (Ṣ, Ḳ:) or the *side of the udder:* or the *skin of the she-camel's udder:* (Ḳ:) or *a she-camel's udder:* or the *anterior part of her neck:* and the *skin of her podex.* (JK.)

خِيفٌ: see 1 in art. خوف, first sentence.

خَافَةٌ, accord. to Aboo-'Alee belonging to this art.: see art. خوف. (TA.)

خَيْفَةٌ *A knife,* (AA, Ḳ,) *such as is termed* رَمِيضٌ [q. v.]. (AA, TA.) — Also, (thus in the Ḳ,) or ▼ خِيفَةٌ, (so in the JK, [and app. accord. to Ṣgh,]) The *place of resort of a lion:* (JK, Ḳ:) mentioned in this art. by Ibn-'Abbád; but accord. to Ṣgh, it may be from الخَوْفُ. (TA.)

خِيفَةٌ: see what next precedes: — and see also art. خوف.

خَيْفَانٌ *Locusts before their wings are full-grown:* (Lth,* Ḳ, TA:) [see جَرَادٌ:] or *when they have upon them streaks of different colours, white and yellow:* (Ṣ, Ḳ:) or *when they have changed from their first black or yellow colour to red:* (Aṣ, Ḳ:) or *when yellowness has appeared in their red colour, but some of the redness remains:* (AHát, TA:) or [in the CḲ "and"] *emaciated red locusts of the brood of the next preceding year:* (Ḳ:) accord. to Lḥ, you say جَرَادٌ خَيْفَانٌ, meaning *locusts of different colours:* (TA:) [but خيفان is generally used as a subst.:] the n. un. is with ة. (Ṣ.) — Hence the n. un. is applied to a mare, as meaning † *Brisk, sprightly, active,* or *agile, and leaping.* (Ṣ, TA.) — [Hence also, app.,] خَيْفَانٌ مِنَ النَّاسِ † *A multitude of men.* (Ibn-'Abbád, Ḳ.*) — Also *A certain plant of the mountains;* (Ibn-'Abbád, Ḳ;) *a certain herb growing in the mountain, having no leaves, rising more than a cubit in height, having a* سَنْمَة [or *head resembling an ear of corn*], *which is green in the upper part and white below, with a white awn,* or *beard.* (L.)

أَخْيَفُ, applied to a horse, (Ṣ, Mgh, Mṣb,) and a camel, (TA,) and any animal, (Ṣ, TA,) *Having one of the eyes blue and the other black:* (Ṣ, Mgh, Mṣb, TA:) fem. خَيْفَآءُ. (Ḳ, TA.) — And, applied to a camel, *Wide in the sheath of the penis.* (Ṣ, Ḳ.) — And the fem., applied to a she-camel, *Wide in the udder,* (Ḳ,) or *in the skin thereof,* (Ṣ,* Ḳ,) or only *when it is empty of milk, and flaccid:* pl. خَيْفَاوَاتٌ; (Ḳ;) which is extr., for a pl. like this belongs [regularly] only to a subst., and to an epithet in which the quality of a subst. predominates. (TA.) — The pl. of أَخْيَفُ is خِيفٌ and خُوفٌ, (Ḳ, TA, [the latter erroneously written in the CḲ خَوْفٌ,]) with kesr and ḍamm. (TA.)

مَخِيفٌ: see art. خوف.

مُخَيَّفٌ [*Diversified in colour*]; applied by El-Kumeyt to a horse of which one part was of the colour termed وَرْدٌ, and the rest جَوْنٌ. (L and TA voce مُخَضَّبٌ.)

مِخْيَافٌ A woman *who brings forth one year a boy and another year a girl.* (JK.)

خيل

1. خَالَ is syn. with ظَنَّ and تَوَهَّمَ: (TA:) you say, خَالَ الشَّىْءَ, (Mṣb, Ḳ,) first pers. خِلْتُ, (JK, Ṣ,) aor. يَخَالُ, (Mṣb, Ḳ,) first pers. إِخَالُ and أَخَالُ, (JK, Ṣ, Mṣb, Ḳ, &c.,) the former irregular, (Mṣb,) but the more chaste of the two, (Ṣ,) and the more used, (Mṣb,) of the dial. of Ṭeiyi, but commonly used by others also, (El-Marzookee, TA,) the latter of the dial. of Benoo-Asad, accord. to rule, (Ṣ, Mṣb,) but of weak authority, (Ḳ,) though some assert it to be the more chaste, (TA,) inf. n. خَيْلٌ (Ṣ, Mṣb, Ḳ) and خَيْلَةٌ and خِيلٌ (Ḳ) and خِيلَةٌ (Ṣ, Ḳ) and خَالٌ and خَيَلَانٌ, (Ḳ, TA, [the last accord. to the CḲ خَيْلَانٌ,]) or, as in the T [and JK], خِيلَانٌ, (TA,) and خَيْلُولَةٌ and مَخِيلَةٌ (Ṣ, Ḳ) and مَخَالَةٌ; (Ḳ;) and خَالَ, aor. يَخِيلُ, is a dial. var. thereof; (Mṣb;) meaning ظَنَّهُ [*He thought,* or *opined, the thing:* and sometimes (see I 'Aḳ p. 109) *he knew the thing:* but it seems to have originally signified تَوَهَّمَ الشَّىْءَ, i. e. *he surmised,* or *fancied, the thing:* see خَالٌ, below]. (Ṣ, Mṣb, Ḳ.) This verb, being of the class of ظَنَّ, occurs with an inchoative and an enunciative; if commencing the phrase, governing them; but if in the middle or at the end, it may be made to govern or to have no government. (Ṣ.) You say, إِخَالُ زَيْدًا أَخَاكَ [and, if you will, زَيْدٌ إِخَالُ أَخُوكَ and زَيْدٌ أَخُوكَ إِخَالُ, *I think Zeyd is thy brother* and *Zeyd I think is thy brother* and *Zeyd is thy brother I think*]. (JK.) Hence the prov., مَنْ يَسْمَعْ يَخَلْ, (Ṣ, TA,) i. e. *He who hears* the things related of men and of their vices, or faults, *will think* evil of them: meaning that it is most safe to keep aloof from other men: or, accord. to some, it is said on the occasion of verifying an opinion. (TA.) — See also 8. — خَالَ عَلَى المَالِ, aor. يَخِيلُ: see خَالَ in art. خول. — خَالَ said of a horse, (JK, Ḳ, TA,) aor. يَخَالُ, (Ḳ,) inf. n. خَالٌ, (JK, Ḳ,) *He limped,* or *halted,* or *was slightly lame.* (JK, Ḳ.*)

2. تَخْيِيلٌ signifies The *imaging* a thing *in the mind,* or *fancying it;* the *forming an image,* or *a fancied image,* thereof *in the mind:* (TA:) [and ▼ تَخَيُّلٌ has the same, as well as a quasi-pass., signification.] You say, [▼ خَيَّلْتُهُ فَتَخَيَّلَ and] تَخَيَّلْتُهُ ▼ فَتَخَيَّلَ لِى [*I imaged it in the mind,* or *fancied it, and it became imaged in the mind to me,* or *an object of fancy to me*]; like as you say, [تَصَوَّرْتُهُ فَتَصَوَّرَ لِى and] صَوَّرْتُهُ فَتَصَوَّرَ لِى: (Ṣ:) for ▼ تَخَيُّلٌ [as inf. n. of a quasi-pass. verb] signifies a thing's *being imaged in the mind,* or *fancied:* (Er-Rághib, TA:) and تَخَيَّلَ ▼ الشَّىْءُ لَهُ means تَشَبَّهَ. (Ḳ. [And the same is indicated in the Mṣb.]) You say also, خُيِّلَ لَهُ كَذَا [*Such a thing was imaged to him in the mind;* i. e. *such a thing seemed to him*]; from الوَهْمُ and الظَّنُّ: (Mṣb:) and خُيِّلَ إِلَيْهِ أَنَّهُ كَذَا (Ṣ) *It was imaged to him* [*in the mind,* i. e. *it seemed to him,*] *that it was so;* syn. شُبِّهَ; (PṢ;) from التَّخْيِيلُ and الوَهْمُ: (Ṣ, TA:) and تَخَيَّلَ ▼ لَهُ أَنَّهُ كَذَا signifies [in like manner *it became imaged* &c.; i. e.] تَشَبَّهَ; as also ▼ تَخَايَلَ: (Ṣ:) and so the first of these three verbs is used in the Ḳur xx. 69. (TA.) And فُلَانٌ يَمْضِى عَلَى مَا خَيَّلَتْ, (JK and Ṣ in explanation of the phrase فُلَانٌ يَمْضِى عَلَى المُخَيَّلِ ▼,) i. e. شَبَّهَتْ [*Such a one goes on, notwithstanding what* (*the mind,* or *the case,*) *may image* to him, or *what is fancied* by him, of danger or difficulty; النَّفْسُ, or الحَالُ, accord. to Z, (see Freytag's Arab. Prov. ii. 94,) being understood]; meaning, *notwithstanding peril,* or *risk; without any certain knowledge.* (Ṣ.) Whence the prov.,

* عَلَى مَا خُيِّلَتْ وَعْثُ القَصِيمِ *

i. e. I will go on, *notwithstanding what the soft tracts abounding in sand in which the feet sink may be imagined to be:* [or the right reading is probably خَيَّلَتْ, i. e. *notwithstanding what the soft tracts* &c. *may image to the mind,* of danger or difficulty:] the ت in خيّلت relates to the word وعث, which is [regarded as] pl. of وَعْثَةٌ; and على is a connective of a suppressed verb, namely, أَمْضِى, with what follows it: the meaning is, I will assuredly venture upon the affair, notwithstanding its terribleness. (Meyd.) And اِفْعَلْ ذٰلِكَ عَلَى مَا خَيَّلَتْ, i. e. عَلَى مَا شَبَّهَتْ [*Do thou that, notwithstanding what* (*the mind,* or *the case,* as explained above,) *may image* to thee, of danger or difficulty]; (JK;) meaning, *in any case.* (TA.) — [Hence,] خَيَّلَ لِلنَّاقَةِ, and أَخْيَلَ, *He put a* خَيَال [q. v.] *near the she-camel's young one, in order that the wolf might be scared away from him,* (JK,* Ṣ, Ḳ,*) *and not approach him.* (JK, Ṣ.) — And خَيَّلَ فِيهِ الخَيْرَ *He perceived,* or *discovered, in him an indication,* or *external sign, of good;* as also ▼ تَخَيَّلَهُ (Ḳ, TA) and تَخَوَّلَهُ: (TA: [see also 4 in art. خول:]) or you say, تَخَيَّلْتُ ▼ عَلَيْهِ, (T, Ṣ, TA,) meaning *I knew him;* or *knew his internal,* or *real, state;* (تَخَبَّرْتُهُ, T,

TA;) or *I chose him*; (اِخْتَرْتُهُ, Ṣ, TA;) *and perceived, or discovered, in him an indication, or external sign, of good.* (T, Ṣ, TA.) — And خَيَّلَ عَلَيْهِ, (Ṣ, Mṣb, Ḳ,) inf. n. تَخْيِيلٌ (Mṣb, Ḳ) and ▼ تَخَيُّلٌ, (Ḳ,) [the latter anomalous, being properly inf. n. of تَخَيَّلَ,] *He conveyed doubt, or suspicion,* (الشُّبْهَةَ, Ṣ, Ḳ, or الوَهْمَ, Mṣb,) *to him;* so in the M, on the authority of AZ; (TA;) i. q. لَبَّسَ عَلَيْهِ [*he made (a thing, or case) dubious to him*]. (Mṣb.) — And خَيَّلَتْ عَلَيْنَا السَّمَآءُ *The sky thundered and lightened* [*over us*], *and prepared to rain:* but when the rain has fallen, the term ▼ تَخَيُّلٌ [so in my two copies of the Ṣ, app. used as an inf. n. of the verb in this phrase, as in a case above, or perhaps a mistranscription for تَخْيِيلٌ, though it will be seen from what follows that خيّلت and تخيّلت are both said of the sky in the same sense,] is not used: (Ṣ:) or خَيَّلَت السَّمَآءُ signifies *the sky became clouded, but did not rain;* (JK, and Ḥar p. 36;) as also ▼ اخالت and ▼ تخيّلت and ▼ خايلت: (Ḥar ibid.:) or, as also ▼ تخيّلت (Mṣb, Ḳ) and ▼ اخالت, (Mṣb,) or ▼ أَخْيَلَتْ, (Ḳ,) *the sky prepared to rain,* (Mṣb, Ḳ, TA,) *and thundered and lightened, but did not yet rain:* (TA:) or, accord. to Az, ▼ اخالت السَّمَآءُ signifies *the sky became clouded:* (Mṣb, TA:) and ▼ تخيّلت السَّمَآءُ *the sky became clouded, and prepared to rain.* (Ṣ.) [In like manner,] one says also, ▼ أَخْيَلَتْ and ▼ اخالت السَّحَابُ and ▼ خايلت *The clouds gave hope of rain:* (Ṣ:) or ▼ اخالت السَّحَابَةُ *the cloud showed signs of rain, so that it was thought* [or *expected*] *to rain.* (Mṣb.) ═ خَيَّلَ also signifies, (JK, TA,) or ▼ تخيّل, (Ḥam p. 39,) [or each of these,] *He* (a man) *was cowardly, or weak-hearted, on the occasion of fight,* (JK, TA, and Ḥam,) *and did not act, or proceed, firmly, or steadily.* (Ḥam.) And خيّل عَنِ القَوْمِ and ▼ أَخْيَلَ, [but the former only is explained in this sense in the TA,] *He held back from the people, or party, through cowardice:* (Ḳ, TA:) so says Az, on the authority of 'Arrám. (TA.)

3. خايلهُ, (JK, TA,) inf. n. مُخَايَلَةٌ, (Ṣ, Ḳ,) *He vied with him, rivalled him, or imitated him,* (JK, Ṣ,* Ḳ,* TA,) *in pride and self-conceit;* (JK;) *did as he did.* (TA.) — خايلت السَّمَآءُ, and السَّحَابُ: see 2, in the latter part of the paragraph.

4. اخال *It* (a thing) *was, or became, dubious, or confused, or vague,* (JK, Ṣ, Mgh, Mṣb, TA,) عَلَيْهِ *to him.* (JK, Mgh.) One says, هٰذَا أَمْرٌ لَا يُخِيلُ [*This is a thing, or an affair, or a case, that will not be dubious, &c.*]. (Ṣ.) And لَا يُخِيلُ ذٰلِكَ عَلَى أَحَدٍ *That will not be dubious, &c., to any one.* (JK.) — اخال الشَّىْءُ إِلَى الخَيْرِ, and المَكْرُوهِ, *The thing exhibited an indication, or indications, of good,* and *of evil,* or *what was disliked or hated.* (Mṣb.) [Hence,] اخالت السَّمَآءُ and أَخْيَلَتْ: see 2, in the latter part of the paragraph, in four places. And اخالت السَّحَابُ and أَخْيَلَتْ, or اخالت السَّحَابَةُ: see, again, 2, in the latter part of the paragraph, in three places. — And hence, in the opinion of ISd, the she-camel in this case being likened to clouds [giving hope, or showing signs, of rain], (TA,) اخالت النَّاقَةُ ‡ *The she-camel had milk in her udder,* (JK, Ḳ, TA,) *and was in good condition of body.* (JK, TA.) — اخالت الأَرْضُ بِالنَّبَاتِ, (Ḳ,) or, as in the M, ▼ اختالت, (TA,) ‡ *The land became adorned, or embellished, with plants, or herbage.* (Ḳ, TA. [See also 5.]) ═ اخال فِيهِ خَالًا مِنَ الخَيْرِ: see 4 in art. خول; and see خَالٌ, below. — أَخْيَلْنَا and أَخَلْنَا *We watched, or observed, or looked at, a cloud which it was thought would rain, to see where it would rain.* (Ḳ,* TA.) And أَخَلْتُ السَّحَابَةَ and أَخْيَلْتُهَا *I saw the cloud to be such as gave hope of rain.* (Ṣ. [See also 10.]) ═ أَخْيَلَ لِلنَّاقَةِ: see 2, in the middle of the paragraph. ═ أَخْيَلَ عَنِ القَوْمِ: see 2, last sentence.

5. تخيّل, as a trans. v., syn. with خَيَّلَ; and its inf. n., syn. with تَخْيِيلٌ: see 2, first two sentences, in two places. — تخيّل فِيهِ الخَيْرَ, as syn. with خَيَّلَ: and تَخَيَّلْتُ عَلَيْهِ: see 2, in the latter half of the paragraph. ═ Also, as a quasi-pass. v., similar in signification to خُيِّلَ; and its inf. n.: see 2, first three sentences, in five places. — And تَخَيُّلٌ used as an inf. n. of خَيَّلَ عَلَيْهِ: and app. as an inf. n. of خَيَّلَتْ عَلَيْنَا السَّمَآءُ: see 2, latter half, in two places. — تخيّلت السَّمَآءُ: see 2, latter half, in three places. — تخيّل as syn. with اختال: see the latter verb. — [Hence, app.,] تخيّلت الأَرْضُ † *The land became abundant in its plants, or herbage:* (JK:) [and, (as is shown by an explanation of the part. n. of the verb, below,) † *the land had its plants, or herbage, in a state of full maturity, and in blossom;* and so ▼ تخايلت; whence,] a poet says,

* تَأَزَّرَ فِيهِ النَّبْتُ حَتَّى تَخَايَلَتْ ▼ *

* رُبَاهُ وَحَتَّى مَا تَرَى الشَّآءَ نُوَّمَا *

[*The herbage in it became, or had become, tangled, or luxuriant, and strong, so that its hills were clad with plants in full maturity, and in blossom, and so that the sheep, or goats, were seen sleeping*]. (Ṣ, TA. [In both, the meaning of the verb in this ex. is indicated by the context. See also 4, where a similar meaning is assigned to اختالت or اخالت.]) — تَخَيُّلٌ also signifies The *being, or becoming, of various colours.* (JK, Ḥam p. 39.*) [Hence the saying,] تَخَيَّلَ الخَرْقُ بِالسَّفْرِ, i. e. [*The desert, or far-extending desert*] *became of various colours with the travellers,* by reason of the آل [or mirage]. (JK.) — Also The *going on, or away; or acting with a penetrative energy;* and *being quick.* (JK, Ḥam p. 39.) — See also 2, last sentence but one.

6: see 2, third sentence: — and 8, in two places: — and see also 5, in two places.

8. اختال *He was proud, or haughty;* or *he behaved proudly, or haughtily;* (Ṣ;) as also ▼ خَالَ, (JK, Ṣ,) aor. يَخَالُ, (JK,) or يَخِيلُ, (Ḥam p. 122,) and يَخُولُ, (JK, Ḥam,) inf. n. خَالٌ and خَوْلٌ; (Ḥam;) and ▼ تخيّل and ▼ تخايل: (Ḳ, TA:) or *he was proud, or haughty, and self-conceited:* (Mṣb:) and *he walked with a proud, or haughty, and self-conceited, gait:* (MA, KL:) said of a man, and of a horse: (Mṣb:) and ▼ تَخَايُلٌ signifies the *behaving, or carrying oneself, with pride, or haughtiness, combined with slowness.* (JK.) You say of a horse, يَخْتَالُ فِي مِشْيَتِهِ [*He is proud and self-conceited in his gait*]. (TA.) — اختالت الأَرْضُ: see 4.

10. استخال السَّحَابَةَ *He looked at the cloud and thought it to be raining.* (TA. [See also 4, last sentence but two.])

خَالٌ i. q. ظَنٌّ and تَوَهُّمٌ [meaning *Thought,* or *opinion:* and *surmise,* or *fancy:* though تَوَهُّمٌ is often explained as syn. with ظَنٌّ]: (Ḳ:) an inf. n. of 1 [q. v.]. (TA.) So in the saying, أَصَابَ فِيهِ خَالِى [*My thought or opinion, or surmise or fancy, was right respecting him, or it*]. (TḲ.) — *I. q.* ▼ مَخِيلَةٌ, q. v., (Ḳ,) [accord. to the TA, which is followed in this instance, as usual, by the author of the TḲ, as meaning فِرَاسَةٌ: but this is a mistake: for وهى الفراسة, the explanation in the TA, we should read وَهِىَ مِنَ الفِرَاسَةِ; as is shown by its being there immediately added that one says, ▼ أَخَالَ فِيهِ خَالًا, explained in art. خول; (see 4, and خَالٌ, in that art.; and see also مَخِيلَةٌ in the present art.;) and by what here follows:] الخَالُ is *syn. with* المَخِيلَةُ and الشُّبْهَةُ. (JK.) — For another sense in which it is syn. with مَخِيلَةٌ, see the latter word, below. — *A nature; or a natural, a native, or an innate, disposition or temper* or *the like;* syn. خُلُقٌ. (TA.) — *I. q.* خُيَلَآءُ, q. v. (Ṣ, Ḳ.*) ═ *A limping, or halting, or slight lameness,* in a horse or similar beast: in this sense an inf. n. of خَالَ. (JK, Ḳ.*) — *Gout;* or *gout in the foot* or *feet;* syn. نِقْرِسٌ. (TA.) ═ *Lightning:* (Ḳ:) [app. as being a sign, or token, of coming rain.] — *Clouds;* syn. غَيْمٌ: (Ṣ:) or clouds (غيم) *lightening:* (JK, M, TA:) and also *rising, and seeming to one to be raining;* and the single cloud (سَحَابَةٌ) is termed ▼ مَخِيلَةٌ: (JK:) or *rising, and seeming to one to be raining, and then passing beyond one;* but when *having thunder, or lightning, therein,* termed ▼ مَخِيلَةٌ, though not when the rain has gone therefrom: (Ḥar p. 36, from the 'Eyn:) or clouds (سَحَابٌ) *raining:* (T, TA:) or clouds (سحاب) *that fail not to fulfil their promise of rain;* (Ḳ,* TA;) and a cloud of this description is termed ▼ مَخِيلَةٌ: (JK:) or *in which is no rain,* (Ḳ, TA,) *though thought, when seen, to be raining.* (TA.) — † *A liberal, bountiful, or generous, man:* (JK, T, M, Ḳ:) as being likened to the raining clouds, (T, TA,) or to the lightening clouds, (JK, M, TA,) which are so termed. (JK, T, M, TA.) — *A man in whom one sees an indication, or a sign, or token, of goodness.* (Ḳ, TA.) — *Free from* التُّهَمَةِ [as meaning *what occasions suspicion*]. (Ḳ.) — *A man who manages cattle, or camels &c.,* (Ḳ, TA,) *and pastures them,* (TA,) *well:* (Ḳ, TA:) or خَالُ مَالٍ

one who manages cattle, &c., and watches them, well. (JK.) And One *who keeps to a thing,* (Ḳ, TA,) *and manages, orders,* or *regulates, it.* (TA.) *A king who manages, orders,* or *regulates, the affairs of his subjects.* (JK.) [See also خَالٌ in art. خول.] — *An owner* of a thing: (Ḳ:) from خَالَهُ, aor. يَخُولُ, meaning "he managed it," &c. (TA.) You say, مَنْ خَالُ هٰذَا الفَرَسِ *Who is the owner of this horse?* (TA.) [See خَالٌ in art. خول.] — See also مُخْتَالٌ, in three places. — A man *free from an attachment of love.* (Ḳ.) — A man *having no wife.* (Ḳ.) — A man *weak in heart and body:* (Ḳ:) but this is most probably [خَالٌّ,] with teshdeed, from خَلَّ لَحْمُهُ, meaning "he became lean."* (TA.) ═ As meaning *A maternal uncle,* it is mentioned in art. خول. (TA.) ═ *A mole,* syn. شَامَةٌ, (Ḳ,) *a black* شامة, (TA,) *upon the person;* (Ṣ, Ḳ, Mṣb, TA;) [*a thing resembling*] *a pimple in the face, inclining to blackness;* (JK, T, Mgh, TA;) or *a small black spot upon the person:* (TA:) dim. ▼خُيَيْلٌ (JK, Ṣ) accord. to him who says مَخِيلٌ and مَخْيُولٌ [as meaning "marked with many moles upon the person"], (Ṣ,) and خُوَيْلٌ (JK, Ṣ, Mṣb) accord. to him who says مَخُولٌ, (Ṣ,) which shows it to be, in one dial., of the art. خول [in which it is also mentioned]: (Mṣb:) pl. [of mult.] خِيلَانٌ (JK, Ṣ, Mgh, Mṣb, Ḳ) and [of pauc.] أَخْيِلَةٌ (Mṣb.) ═ *A garment,* or *cloth, of the garments,* or *cloths, of the* جُهَّال [here meaning *people of the Time of Ignorance*]: (Ṣ:) *a soft garment* or *cloth* (JK, Ḳ, TA) *of the garments* or *cloths of El-Yemen:* (JK, TA:) and *a* [*garment of the kind called*] بُرْد, *of the fabric of El-Yemen,* (Ḳ, TA,) *red* [or *brown*], *with black lines* or *stripes, which used to be made in the first ages:* but Az makes these two to be one: it has been mentioned before, in art. خول, to which also it may belong. (TA.) — *A garment,* or *piece of cloth, with which a corpse is shrouded.* (Ḳ.) — The [*kind of banner called*] لِوَاء (JK, T, Ḳ) *that is tied* [to its spear-shaft] *for a commander,* (Ḳ,) or *to denote one's having the authority of a prefect, commander, ruler,* or *the like:* (T, TA:) [SM adds,] I do not think it to be so called for any other reason than that it was of the بُرُود of the kind termed خَال. (TA.) [See also خَالٌ in art. خول.] — The *office of Khaleefeh;* (Ḳ;) because belonging to one for whom a banner is tied [on the occasion of his appointment]. (TA.) ═ *A big mountain.* (Ḳ.) — And (as being likened thereto, TA) †*A big camel:* (JK, Ḳ:) pl. خِيلَانٌ: to such, a poet likens certain men, as resembling camels in their bodies and in their being devoid of intellect. (TA.) — And *A black stallion-camel.* (IAạr, Ḳ,* TA.) Mentioned also in art. خول. (TA.) ═ *A place in which is no one,* or *no one by whose company one may be cheered.* (Ḳ.) [Probably from خَالٍ, part. n. of خَلَا, aor. يَخْلُو.] — *A small* [*hill such as is termed*] أَكَمَة. (Ḳ.) ═ The لِجَام [i. e. *bit,* or *bit with its appertenances,*] *of a horse:* (Ḳ:) app. a dial. var. of خَوْلٌ, q. v. (TA.) ═ *A certain plant, having a blossom, well known in Nejd.* (Ḳ.)

خَالٌ, formed by transposition from خَائِلٌ: see مُخْتَالٌ.

خَيْلٌ *Horses,* (JK, Ṣ, Ḳ,) *collectively;* (JK, Ḳ;) as some say, (Mṣb,) applied to *Arabian horses and* [*such as are of inferior breed, termed*] بَرَاذِين; (Mgh, Mṣb;) the *males thereof* and the *females:* (Mgh, TA:*) but of the fem. gender: (Mṣb, TA:) a quasi-pl. n., (Mgh,) having no sing. (Mṣb, Ḳ) formed of the same radical letters: (Mṣb:) or the sing. is ▼خَائِلٌ: (Ḳ:) so called because of their اِخْتِيَال, (Mṣb, Ḳ,* TA,*) i. e. pride and self-conceit, (Mṣb,) in their gait: so says AO; but ISd says that this is not well known: (TA:) or because no one rides a horse without experiencing a feeling of pride: (Er-Rághib, TA:) pl., (Mṣb, CḲ,) or pl. pl., (so in copies of the Ḳ and in the TA,) [of mult.,] خُيُولٌ (Ṣ,* Mṣb, Ḳ) and خِيُولٌ and [pl. of pauc.] أَخْيَالٌ. (Ḳ.) And the dual form is used, [although خَيْلٌ has a pl. signification,] like as are [the duals] جِمَالَانِ and لِقَاحَانِ and] غَنَمَانِ and إِبِلَانِ. (ISd, TA.) One says, فُلَانٌ لَا تُسَايَرُ خَيْلَاهُ, or لَا تُوَاقَفُ, (Ḳ, TA,) and لَا تُسَايَرُ خَيْلَاهُ وَلَا تُوَاقَفُ, (TA, and so in the CḲ,) [*Such a one, his two troops of horses will not be competed with in going,* or *running, nor in standing still,*] meaning †*he is not to be endured in respect of calumny and lying:* (Ḳ, TA:) it is said of a great, or frequent, liar. (TA in art. سير.) And الخَيْلُ أَعْلَمُ مِنْ فُرْسَانِهَا [*The horses are more knowing than their riders*]; (Meyd, Ḳ;) a prov., (Meyd,) applied in relation to him of whom thou formest an opinion (Meyd, Ḳ, TA) that he possesses, or possesses not, what suffices, (TA,) and whom thou findest to be as thou thoughtest, (Meyd, Ḳ, TA,) or the contrary. (Meyd.) And الخَيْلُ أَعْلَمُ بِفُرْسَانِهَا [*The horses are possessed of most knowledge of their riders*]; a prov., meaning †seek thou aid of him who knows the case, or affair. (Meyd.) And الخَيْلُ تَجْرِى عَلَى مَسَاوِيهَا, another prov. [explained in art. سوأ]. (Meyd.) — Also *Horsemen,* or *riders on horses.* (Ṣ, Mṣb, Ḳ.) Thus in the Ḳur [xvii. 66], وَأَجْلِبْ عَلَيْهِمْ بِخَيْلِكَ وَرَجِلِكَ. (Ṣ. [See 1 in art. جلب.]) ═ See also خُيَلَاءُ.

خَيَلٌ:
خَيْلَةٌ:
خِيلَةٌ:
خَيَلَةٌ:
} see خُيَلَاءُ.

خَيْلِىٌّ *An equerry; one who has the superintendence of horses.* (TA.)

خُيَلَاءُ (Ṣ, Mṣb, Ḳ, &c.) and ▼خِيَلَاءُ (Ṣ, Ṣgh, TA) and ▼خَالٌ (Ṣ, Ḳ) and ▼خَيَلٌ, (Ḳ, TA,) or ▼خَيْلٌ, (CḲ,) and ▼خَيْلَةٌ, (Ḳ, TA,) or ▼خَيَلَةٌ, (CḲ,) or ▼خِيلَةٌ, (JK,) and ▼مَخِيلَةٌ (Ṣ, Ḳ) and ▼أَخْيَلُ, (Lth, JK, Ḳ,) [of all which the first is the most common,] *Pride* (Ṣ, Mṣb, Ḳ) *and self-conceit;* (Mṣb;) [or *vanity;* i. e.] *pride arising from some fancied,* or *imaginary, excellence in oneself.* (TA.) One says, هُوَ ذُو خُيَلَاءَ &c. *He is possessed of pride* [*and self-conceit,* or *vanity*]. (Ṣ.)

خِيَلَاءُ: see what next precedes.

خَيَالٌ primarily signifies *An incorporeal form* or *image; such as that which is imaged in sleep,* and *in the mirror,* and *in the heart* or *mind:* then applied to the *form of anything imaged;* and to *any subtile thing of a similar kind:* (Er-Rághib, TA:) *anything that one sees like a shadow:* and the *image* of a man *in a mirror,* (T, Mṣb, TA,) and *in water,* (Mṣb,) and *in sleep:* (T, TA:) and *a thing that sometimes passes by one, resembling a shadow:* (T, Mṣb, TA:) خَيَالٌ and ▼خَيَالَةٌ both signify the same; (JK, Ṣ, Ḳ;) i. e. *i. q.* طَيْفٌ [meaning *an apparition; a phantom; a spectre; a fancied image; an imaginary form;* and particularly *a form that is seen in sleep*]; (Ṣ, TA;) *anything that one sees like a shadow* [as the former word is explained above]; and *a thing that is seen in sleep;* (JK;) *a form that is imaged to one in the mind when awake,* and *when dreaming:* (Ḳ:) the former word is both masc. and fem.: (Ḥam p. 316:) pl. أَخْيِلَةٌ (Ḳ) [a pl. of pauc.; and probably خِيلَانٌ also, as a pl. of mult., mentioned as one of the pls. of خَيَالٌ in another sense, below]. You say, تَخَيَّلَ لِى خَيَالُهُ [*His apparition,* or *phantom,* &c., *became imaged to me in my mind*]. (Mṣb, TA.) And a poet says, (Ṣ,) namely, El-Boḥturee, (TA,)

فَلَسْتُ بِنَازِلٍ إِلَّا أَلَمَّتْ
بِرَحْلِى أَوْ خَيَالَتُهَا ▼الكَذُوبُ

[*And I do not alight but she visits my abode, or her false apparition*]. (Ṣ.) — [In philosophy it signifies] *A faculty that retains what the fancy perceives of the forms of objects of sense after the substance has become absent, so that the fancy beholds them whenever it turns towards them: thus it is the store-house of the fancy: its place is the hinder part of the first venter of* [*the three which are comprised by*] *the brain.* (KT. [In this sense, it is incorrectly written in Freytag's Lex. (in which only the Arabic words of the explanation are given, preceded by the rendering "phantasia,") خِيَال.]) — The خَيَال of a bird is *The shadow of himself which a bird sees when rising into the sky;* whereupon he pounces down upon it, thinking it to be a prey, and finds it to be nothing: he is [the bird] called خَاطِفُ ظِلِّهِ. (TA.) — خَيَالٌ and ▼خَيَالَةٌ signify also *The person,* or *body,* or *corporeal form* or *figure which one sees from a distance,* syn. شَخْصٌ, (Ṣ, Ḳ,) of a man; and his *aspect.* (Ḳ.) — And the former also signifies *A piece of wood with black garments upon it,* (Ṣ,) or *with a black* [*garment of the kind called*] كِسَاء *upon it,* (Ḳ,) *which is set up to make the beasts and birds fancy it to be a man:* (Ṣ, Ḳ:) or *a piece of wood with a garment thrown upon it, which is set up for the sake of the sheep or goats, in order that the wolf, seeing it, may think it to be a man:* (T, TA:) pl. [of pauc.] أَخْيِلَةٌ (Ks, TA) and [of mult.] خِيلَانٌ. (TA.) A poet says,

أَخِى لَا أَخَا لِى غَيْرُهُ غَيْرَ أَنَّنِى
كَرَاعِى خَيَالٍ يَسْتَطِيفُ بِلَا فِكْرِ

(Ṣ, TA) [cited by J as an ex. of خَيَال in the former of the senses explained in the sentence immediately preceding: but the meaning seems to be, *My brother: I have no brother but he: but I am like one watching an image dressed up to decoy; going round about without reflection:* for,] as some say, (TA,) رَاعِى الخَيَالِ means *the young ostrich for which the sportsman sets up a* خَيَال [i. e. *an image dressed up to decoy*], (JK, TA,) *in order that it may become familiar therewith, and the sportsman may then take it, and the young ostrich may follow him.* (TA.) — Also *A thing that is set up in land in order that it may be known to be prohibited to the public, and may not be approached.* (T, Mṣb.) = And *A certain plant.* (Ḳ.)

خُيَيْل : see خَيَال, of which it is a diminutive.

خَيَالَة : see خَيَال, in three places.

[خَيَالِىّ *Of, or relating to, the fancy:* a rel. n. from خَيَال.]

خَيَّالَة *Owners, or attendants, of horses.* (JK, Ṣ.) [In modern Arabic, *Horsemen*; and *a troop of horsemen.*]

خَائِل [act. part. n. of 1]: see مُخْتَال, in two places: — and see خَيْل, first sentence. — Applied to a horse, *Limping, halting,* or *slightly lame.* (JK, TA.)

[أَخْيَلُ *More,* and *most, proud and self-conceited.* (See also أَخْوَلُ, in art. خول.) Occurring in several provs.; as, for ex.,] أَخْيَلُ مِنْ غُرَابٍ [*More proud and self-conceited than a crow*]: because the غراب is proud and self-conceited in its gait. (Meyd.) = See also خَيْلَآءُ. = Also *Having a* خَال, meaning [*a thing resembling*] *a pimple, inclining to blackness,* [i. e. *a mole,*] *upon his face:* (Mgh:) or, as also ↓ مَخِيلٌ and ↓ مَخْيُولٌ (Ṣ, Mṣb, Ḳ) and مَخُولٌ, like مَقُولٌ, (Ṣ, Mṣb,) this last belonging to art. خول, as خَال, whence it is derived, does in one dial., (Mṣb,) *A man* (Ṣ, Mṣb) *having* [or *marked with*] *many* خِيلَان [or *moles* upon his person]: (Ṣ, Mṣb, K:*) fem. [of the first] خَيْلَآءُ. (Ḳ.) = الأَخْيَلُ, (Ṣ, Mṣb, Ḳ,) when indeterminate, [أَخْيَلٌ or أَخْيَلُ,] perfectly decl., [thus] used as a subst., but some make it imperfectly decl. both when determinate and when indeterminate, and assert it to be originally an epithet, from التَّخَيُّل, (Ṣ, O,) [though accord. to others it seems to be from أَخْيَلُ as meaning "having many moles,"] *A certain bird,* (JK, Ṣ, Mṣb, Ḳ,) *regarded as of evil omen,* (JK, Ṣ, Ḳ,) *that alights upon the rump of the camel, and is app. for that reason held to be of evil omen;* (TA; [see مَخْيُول;]) [applied in the present day to *the green wood-pecker, picus viridis;*] *the* صُرَد [q. v.]: (Ḳ:) or *the green* صُرَد: (TA:) or *the* شِقِرَّاق [a name likewise now applied to *the green wood-pecker,* and to *the common roller, coracias garrula*]: (Fr, Ṣ, Mṣb, Ḳ:) so called because upon its wings are colours differing from its general colour: (Skr, TA:) or so called because diversified with black and white: (Ḳ:) or *the* شَاهِين [q. v., *a species of falcon*]: (JK, TA, and Ḥam p. 705:) pl. أَخَايِلُ, (JK, T, Ṣ, Mṣb, TA, and Ḥam ubi suprà,) or خِيلٌ. (Ḳ.)

أَخَايِلُ : see مُخْتَال, in two places.

مَخِيلٌ : see its fem., with ة, in the next following paragraph, in three places: = and see also أَخْيَلُ.

مُخِيلٌ A thing *dubious, confused,* or *vague.* (TA.) — *Exhibiting a* خَيَال [or *fancied image,* or rather a خَال or مَخِيلَة, i. e. *an indication,* &c., (see 4,) of anything, as, for instance,] of good [and of evil]. (TA.) You say شَىْءٌ مُخِيلٌ إِلَى الخَيْرِ, and المَكْرُوهِ, *A thing exhibiting an indication,* or *indications, of good,* and *of evil,* or *what is disliked* or *hated.* (Mṣb.) Hence, (TA,) هُوَ مُخِيلٌ لِلْخَيْرِ, (Ṣ, TA,) said of a man, (Ṣ,) *He is adapted* or *disposed by nature to good* [i. e. *to be,* or *to do,* or *to effect,* or *to produce, what is good*]. (Ṣ, TA. [See also مُخِيل in art. خول.]) And سَحَابَةٌ مُخِيلَةٌ (JK, Mṣb, Ḳ) and ↓ مَخِيلَةٌ (Ḳ, TA, in the CḲ مُخَيَّلَة) and ↓ مَخِيلٌ and ↓ مُخْتَالَةٌ (Ḳ) and مخايلة [i. e. ↓ مُخَايِلَةٌ] (Ḥar p. 36) and ↓ مَخِيلَةٌ (Mṣb [but see what follows]) *A cloud thought* [or *expected*] *to rain,* (JK, Mṣb, Ḳ, TA, and Ḥar ubi suprà,) *when seen,* (TA, and Ḥar,) *because showing signs of rain:* مُخِيلَةٌ, with ḍamm, being an act. part. n., as meaning *causing to think;* and ↓ مَخِيلَةٌ, with fet-ḥ, being a pass. part. n., as meaning *thought:* and in like manner, accord. to Az, سَمَآءٌ مُخِيلَةٌ and ↓ مَخِيلَةٌ, meaning *a clouded sky:* (Mṣb:) or you say ↓ مَخِيلَةٌ, with fet-ḥ, when [you use the subst.] meaning *a cloud itself* [*showing signs of rain*]; and its pl. is مَخَايِلُ: (T, TA: see خَال, in the former half of the paragraph:) and سَحَابَةٌ مُخِيلَةٌ لِلْمَطَرِ *a cloud giving hope of rain.* (Ṣ.) See also خَال, in two places, in the former half of the paragraph. You say also, السَّمَآءُ مُخِيلَةٌ لِلْمَطَرِ, meaning *The sky is ready to rain.* (Ḥar p. 36.) — اِمْرَأَةٌ مُخِيلَةٌ *A woman having no husband.* (JK.)

مَخِيلَةٌ as fem. of the pass. part. n. مَخِيلٌ : see مُخِيل, in three places. — As a subst.: see, again, مُخِيل. And see خَال, mentioned a second time in the former half of the paragraph. — Hence, *A great banner* or *ensign;* as likened to a cloud that fails not to fulfil its promise of rain. (JK.) — Also *An indication; a symptom; a sign, mark,* or *token, by which the existence of a thing is known* or *inferred;* syn. شِيَةٌ (JK) and مَظِنَّةٌ; (TA;) and so ↓ خَال, q. v.: (JK:) pl. of the former مَخَايِلُ: originally used in relation to a cloud in which rain is thought to be. (TA.) You say, ظَهَرَتْ فِيهِ مَخَايِلُ النَّجَابَةِ, i. e. *The indications* &c. [*of generosity,* or *nobility, appeared in him*]. (TA.) — You say also, of a cloud (سَحَابَة), مَا أَحْسَنَ مَخِيلَتَهَا and ↓ خَالَهَا *How good is its* [*apparent*] *disposition to rain!* (Ṣ, TA.) — See also خَيْلَآءُ.

مُخَيَّلٌ [A thing *imaged* to one *by the mind* or *by a case;* or *fancied*]. You say, فُلَانٌ يَمْضِى عَلَى المُخَيَّلِ; explained above: see 2. (JK, Ṣ.) And وَقَعَ فِى مُخَيَّلِى كَذَا [*Such a thing occurred in what was imagined,* or *fancied, by me*], and فِى مُخَيَّلَاتِى [*among the things imagined,* or *fancied, by me*]. (TA.)

مُخَيِّلٌ; and its fem., with ة: see مُخِيل.

مَخْيُولٌ : see أَخْيَلُ. = Also A camel *lacerated in his rump by a bird of the kind called* أَخْيَل *that has alighted upon it.* (TA.) — And hence, † A man *whose reason has fled in consequence of fright:* a sense in which it is used by the vulgar; but correct. (TA.)

سَحَابَةٌ مُخَايِلَةٌ : see مُخِيل.

مُخْتَالٌ and ↓ خَائِلٌ (Ṣ, Ḳ) and ↓ خَالٌ and ↓ خَالٍ, which is formed by transposition, (Ḳ,) and ↓ أَخَايِلُ, (Ṣ, Ḳ,) like أُدَابِرُ and أُبَاتِرُ, (Ṣ,) which are [said to be] the only other epithets of this measure, (TA,) [i. e. of the measure أُفَاعِل, though there are many of the measure أَفَاعِل,] applied to a man, *Proud* (Ṣ, Ḳ, TA) *and self-conceited:* [or *vain:*] (TA in explanation of all, and Ḳ in explanation of خَال:) or ↓ خَال signifies *having much* خُيَلَآء [or *pride and self-conceit,* or *vanity*]: and ↓ أُخَايِل one *who walks with a proud and self-conceited gait, with an affected inclining of his body from side to side:* (JK:) ↓ خَائِل has for its pl. خَالَةٌ; (Ṣ, TA;) which is also fem. of ↓ خَال. (TA.) — See also مُخِيل.

أَرْضٌ مُتَخَيِّلَةٌ † *Land having its plants,* or *herbage, in a state of full maturity, and in blossom;* (JK, Ṣ;) as also ↓ مُتَخَايِلَةٌ. (Ṣ.)

أَرْضٌ مُتَخَايِلَةٌ : see what next precedes.

خيم

1. خَامَ as syn. with خَيَّمَ: see the latter. — خَامَ عَنْهُ, aor. يَخِيمُ, (JK, Ṣ, Ḳ,) inf. n. خَيْمُومَةٌ (Ṣ) and خَيْمٌ and خَيَمَانٌ and خُيُومٌ and خَيْمُومَةٌ and خِيَامٌ, (Ḳ, the last but one omitted in the CḲ,) *He held back,* or *refrained, from him,* or *it, through cowardice,* (JK, Ṣ, Ḳ,) and *fear:* (JK, Ḳ:) i. e., from war, (JK,) or from encounter, or conflict. (Ḥam p. 164.) And one says also, خَامُوا فِى القِتَالِ *They held back,* or *refrained, through cowardice, in fight, and attained no good:* and the poet Junádeh Ibn-'Ámir El-Hudhalee says, خَامَ القِتَالَ for خَامَ فِى القِتَالِ. (TA.) — And خَامَ also signifies *He drew back,* (JK, Ḳ,*) *through cowardice and fear,* (JK,) *from a stratagem,* or *plot, after he had devised it,* (JK, Ḳ,*) *not seeing in it what he liked,* or *approved.* (TA.) = خَامَ رِجْلَهُ, (Ṣ, Ḳ,) aor. as above, inf. n. خَيْمٌ, (Ṣ,) *He raised his leg,* or *foot.* (Ṣ, Ḳ.)

2. خَيَّمَ, (Ṣ, Mṣb, Ḳ,) inf. n. تَخْيِيمٌ, (TA,) [originally] *He pitched his tent*; (TA in art. رفض;) [like تخيّم: and hence,—] *He remained, stayed, dwelt,* or *abode,* in a place; (Ṣ, Mṣb, Ḳ, TA;) as also ↓ خَامَ, aor. يَخِيمُ: whence, in a trad., مَنْ أَحَبَّ أَنْ يَسْتَخِيمَ ↓ لَهُ الرِّجَالُ قِيَامًا [*Whoso loveth that men should remain before him standing*]; as some relate it; but others relate it differently, saying يَسْتَخِمَّ, and يَسْتَجِمَّ, as mentioned before [in arts. خم and جم]. (TA.) *He alighted,* or *descended and stopped* or *sojourned* or *abode,* in a place; as also خَيَّفَ. (JK in art. خيف.) And خَيَّمَ الوَحْشِىُّ فِى كِنَاسِهِ *The wild animal remained in his covert, not quitting it.* (TA.) And خَيَّمَتِ الرَّائِحَةُ, (TA,) or الرِّيحُ الطَّيِّبَةُ, (JK,) ‡ *The odour,* or *sweet odour, remained;* (JK, TA;) as also ↓ تَخَيَّمَتْ; (Ḳ, TA;) in a garment, (JK, Ḳ, TA,) and in a place. (TA.) — خَيَّمُوا *They entered into a* خَيْمَة [q. v.]. (JK, Ḳ.) = خَيَّمَهُ *He made it like a* خَيْمَة. (Ṣ, TA.) — See also 4. — Also † *He covered it with a thing in order that its odour might cling to it.* (Ḳ, TA.)

4. إِخَامَةٌ, [inf. n. of أَخَامَ,] in the phrase إِخَامَةُ الفَرَسِ, (Ḳ, TA,) which signifies *The horse's standing upon three legs and the extremity of the hoof of the fourth,* (TA,) or *raising one of his fore legs* or *one of his hind legs,* (JK,) belongs to the present art. and to art. خوم. (Ḳ, TA.) Accord. to Fr and IAạr, يُخِيمُ فِى إِحْدَى رِجْلَيْهِ, inf. n. as above, signifies *He* (a man, or a beast,) *is unable to place one of his legs,* or *feet, firmly upon the ground, and to rest upon it, by reason of a fault therein.* (L, TA.) = أَخَامَ خَيْمَةً, and أَخْيَمَهَا, *He constructed a* خَيْمَة; (IAạr, Ḳ;) as also ↓ خَيَّمَهَا. (TA.)

5. تَخَيَّمَ بِمَكَانِ كَذَا, (Ṣ,) or تَخَيَّمَ هُنَا, (Ḳ,) [like خَيَّمَ,] *He pitched,* or *fixed, his* خَيْمَة [*in such a place,* or *here*]. (Ṣ, Ḳ.) — See also 2.

10: see 2.

خَامٌ A skin, or hide, *untanned:* or *not tanned much,* or *thoroughly:* and a كِرْبَاس [or coarse garment or piece of cloth, or garment or piece of cloth of white cotton,] *unwashed:* (Ḳ:) a Persian word, (TA,) arabicized. (Ḳ.) *Unbleached* cloth. (Mṣb, in art. خوم.) Paper that is *polished,* [app. a mistake for *not polished,*] or *to be polished.* (TA.) [*Crude,* or *raw;* applied in this sense to silk, *contr. of* مَطْبُوخ; and to sugar, &c.: anything *unprepared for use.*] Honey of dates (دِبْس) *untouched by fire:* (AḤn, T:) which is the best thereof. (TA.) = See also خَامَةٌ.

خِيَمٌ: see خَيْمَةٌ, in three places.

خِيمٌ *Natural,* or *innate, dispositions* or *tempers* or *the like:* (A 'Obeyd, Ṣ, M, Ḳ:) a Persian word, arabicized: (TA:) having no sing. (Ṣ, Ḳ) of its own radical letters: (Ṣ:) or *largeness,* or *liberality, of disposition:* (JK, TA:) and the *original,* or *primary, state,* or *condition,* syn. أَصْل, of the soul, or mind. (TA.) You say, هُوَ كَرِيمُ الخِيمِ [*He is generous in respect of natural dispositions* &c.]. (TA.) — The *diversified wavy marks, streaks,* or *grain,* syn. فِرِنْد, of a sword. (Ḳ.) — And *I. q.* حَمْض [app. meaning the *kind of plants called* حَمْض]. (TA.)

خَامَةٌ belongs to the present art., and J has erred, (Ḳ,) in mentioning it in art. خوم: (TA:) [but why this is said, I know not; nor do the commentators on this passage, as is observed in the TA, give any explanation of it worthy of notice:] it signifies *A fresh,* or *juicy, plant:* (Ṣ, Mṣb: both in art. خوم:) or *an ear* of corn: (IAạr, TA:) or *a shoot of seed-produce when it first grows forth upon a single stalk:* (JK:) or خَامَةُ الزَّرْعِ signifies *what grows forth, of seed-produce, upon a single stalk:* (M, Ḳ:*) or *the fresh,* or *juicy, bunch thereof:* or *the fresh,* or *juicy, plant thereof:* (M, Ḳ:) pl. خَامَاتٌ (Mṣb) and [coll. gen. n.] ↓ خَامٌ. (Mṣb, TA.) It is said in a trad., مَثَلُ المُؤْمِنِ مَثَلُ الخَامَةِ مِنَ الزَّرْعِ تُمِيلُهَا الرِّيحُ مَرَّةً هٰكَذَا وَمَرَّةً هٰكَذَا [*The similitude of the believer is that of the fresh,* or *juicy, plant of seed-produce,* &c., *which the wind bends, at one time thus, and at one time thus*]: (Ṣ, TA:*) but Fr related it differently, saying الحَافَةُ, [app. a mistranscription for الخَافَةُ, which see in art. خوف,] and explained this as meaning "the bunch" of seed-produce. (TA.)

خَيْمَةٌ, (Ṣ, Mgh, Mṣb, Ḳ, &c.,) with which ↓ خِيَمٌ is syn., (Ṣ, Mṣb,) [though said by some to be a pl., as will be seen below,] *A* بَيْت [here meaning *booth,* or *the like,*] (Ṣ, Mṣb, Ḳ,) *of any kind* (Ḳ) *such as is built,* or *constructed,* (Ṣ, Mṣb, Ḳ,) *by the Arabs,* (Ṣ, Mṣb,) *of the branches of trees:* (Ṣ, Mṣb, Ḳ:) so says Aṣ, holding that the خيمة is only of trees, and that otherwise it is called بَيْت [q. v.]: but others hold that it is [*a tent;* i. e.] *made with pieces of cloth and tent-ropes;* because تَخْيِيم signifies the "remaining, staying, dwelling, or abiding;" wherefore it is thus called, as being used on the occasion of alighting: (IB, TA:) this latter is the meaning commonly known; but accord. to the saying of Aṣ, it is tropical: (TA:) or, accord. to IAạr, it is applied by the Arabs only to *a construction of four poles roofed over with* ثُمَام [or *panic grass*]; and is *not of cloths;* (Mgh, Mṣb, TA;) the مِظَلَّة, he says, being of cloths and of other things: or *i. q.* مِظَلَّة: (TA:) accord. to AḤát, the same as the Persian خَرْبُشْتَه [lit. "ass's back;" like the French "dos d'âne;" meaning *a high-pitched span-roof*]: (Mgh:) or *any round* بَيْت [which may here mean either *booth* or *tent*]: or *three poles,* or *four, over which is laid* ثُمَام; *by means of which one is shaded in the heat:* (Ḳ:) or *poles set up, with rafters laid across, covered with trees;* so that it is cooler than are أَخْبِيَة [pl. of خِبَاء]: or *poles upon which* خِيَام [pl. of خَيْمَة] *are constructed:* or *a construction of trees and palm-branches with their leaves upon them, which a man uses for shade when he brings his camels to water:* and applied by the Arabs to *a* بَيْت [or *tent* &c.], and *a place of abode* [*in an absolute sense*]: (TA:) the pl. of خَيْمَة is خَيْمَاتٌ and خِيَمٌ (Ṣ, Mṣb, Ḳ) and ↓ خَيْمٌ, (Ḳ,) or this last is [a coll. gen. n., or] syn. with خَيْمَة, (Ṣ, Mṣb,) and خِيَامٌ, (Ḳ,) or this is pl. of ↓ خِيَمٌ, (Ṣ, Mṣb,) and is applied also to ‡ [*women's vehicles of the kind called*] هَوَادِج; these being likened to خِيَام [properly so termed]. (TA.) It is said in a trad., الشَّهِيدُ فِى خَيْمَةِ اللّٰهِ تَحْتَ العَرْشِ † [*The martyr is in the tabernacle of God, beneath the empyrean:* this signification of خيمة being perhaps taken from the phrase ἡ σκηνὴ τοῦ θεοῦ in Rev. xxi. 3]. (TA.)

خَامِىٌّ (Ḳ, TA) *A preparer of skins,* or *hides, of the kind termed* خَام. (TA.)

خَيْمِىٌّ: see what next follows.

خَيَّامٌ and ↓ خَيْمِىٌّ *One who applies himself to the fabrication of the* [*kind of tent,* or *booth, called*] خَيْمَة. (TA.)

مَخْيَمٌ, like مَكْيَل, (in [some of] the copies of the Ḳ, erroneously, like مُكْتَل, TA, [in the CḲ like مِكْيَل,]) *A collection of bundles,* or *handfuls,* (جُزَز, [in the CḲ جُزَز,]) *of reaped seed-produce.* (Ḳ, TA.)

خيو

[خَيَوِىٌّ rel. n. of خَا: see خَائِىٌّ, in art. خوأ.]

END OF THE SECOND PART OF BOOK I.